ORGANIC

Chemistry  Now™

Turn the page for more details

ORGANIC
Chemistry ⬡ Now™

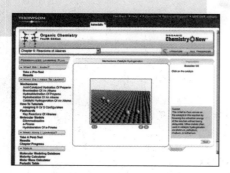
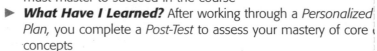

ORGANIC
Chemistry◆Now™
http://now.brookscole.com/bfi4

Welcome to your *Organic ChemistryNow*™ Media Integration Guide!

The **Media Integration Guide** on the next several pages provides you with a grid that links each chapter to the wealth of interactive media resources you will find at *Organic ChemistryNow,* a unique web-based, assessment-centered personalized learning system for Organic Chemistry students.

Chapter	Text Section	Organic ChemistryNow™ Resources: http://now.brookscole.com/bfi4
1 Covalent Bonding and Shapes of Molecules	**1.2** Lewis Model of Bonding	**Simulations** Electron Configuration and the Periodic Table (p. 5) Electron Configuration of Ions (p. 7) **Tutorials** Assigning Formal Change (p. 12) **"How to" Tutorials** Drawing Lewis Structures (p. 14)
	1.4 Bond Angles and Shapes of Molecules	**Tutorials** Recognition of Functional Groups (p. 23) VSEPR Model to Determine Molecular Shape (p. 27)
	1.5 Polar and Nonpolar Molecules	**Tutorials** How to Determine Molecular Polarity (p. 28)
	1.6 Resonance	**"How to" Tutorials** Drawing Curved Arrows and Pushing Electrons (p. 30)
2 Alkanes and Cycloakanes	**2.1** The Structure of Alkanes	**Simulations** Structure and Nomenclature of Alkanes (p. 58) **Active Figure** 2.1 Four Classes of Hydrocarbons (p. 57)
	2.2 Constitutional Isomerism in Alkanes	**Molecular Models** Alkane Model (p. 59)
	2.6 Conformations of Alkanes and Cycloalkanes	**"How to" Tutorials** Drawing Chair Conformations of Cyclohexanes and Identifying Axial and Equatorial Groups (p. 80) **Active Figures** 2.9 Energy of Butane (p. 73) 2.14 Cyclopentane (p. 77) 2.18 Interconversion (p. 79) 2.21 Chair Conformations of Methylcyclohexane (p. 82)
	2.7 *Cis,Trans* Isomerism in Cycloalkanes and Bicycloalkanes	**Tutorials** *Cis-* and *Trans*-Isomers in Cyclohexanes (p. 86)
3 Stereoisomerism and Chirality	**3.2** Chirality – The Handedness of Molecules	**"How to" Tutorials** Drawing Chiral Molecules (p. 114) **Tutorials** Recognizing Chiral Molecules (p. 116) **Active Figures** 3.2 Stereorepresentations of 1-bromo-1-chloroethane (p. 113)
	3.3 Naming Chiral Centers – The *R,S* System	**"How to" Tutorials** Assigning *R* or *S* Configuration to a Chiral Center (p. 118)
	3.5 Cyclic Molecules with Two or More Chiral Centers	**Active Figures** 3.5 Relationships Among Isomers (p. 129)
	Problems	**Simulations** 3-D Models (p. 147)

MEDIA INTEGRATION GUIDE

MEDIA INTEGRATION GUIDE

MEDIA INTEGRATION GUIDE

MEDIA INTEGRATION GUIDE

MEDIA INTEGRATION GUIDE

MEDIA INTEGRATION GUIDE

www.brookscole.com

www.brookscole.com is the World Wide Web site for Brooks/Cole and is your direct source to dozens of online resources.

At *www.brookscole.com* you can find out about supplements, demonstration software, and student resources. You can also send email to many of our authors and preview new publications and exciting new technologies.

www.brookscole.com
Changing the way the world learns®

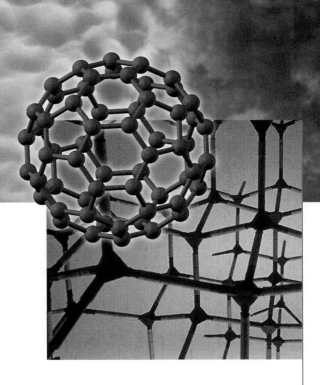

Organic Chemistry

Fourth Edition

William H. Brown
Beloit College

Christopher S. Foote
University of California, Los Angeles

Brent L. Iverson
University of Texas, Austin

THOMSON ™

BROOKS/COLE

Australia • Canada • Mexico • Singapore • Spain
United Kingdom • United States

THOMSON

BROOKS/COLE

Publisher, Physical Sciences: David Harris

Development Editor: Sandra Kiselica

Assistant Editors: Alyssa White and Annie Mac

Editorial Assistant: Candace Lum

Technology Project Manager: Donna Kelley

Marketing Manager: Amee Mosley

Marketing Assistant: Michele Colella

Advertising Project Manager: Nathaniel Bergeson-Michelson

Project Manager, Editorial Production: Lisa Weber

Creative Director: Rob Hugel

Print Buyer: Barbara Britton

Permissions Editor: Kiely Sexton

Production Service: Progressive Publishing Alternatives

Text Designer: Tani Hasegawa

Photo Researcher: Dena Digilio-Betz

Copy Editor: Sara Black

Illustrators: J/B Woolsey Associates and Progressive Information Technologies

Cover Designer: Didona Design

Cover Images: Molecules: Brent Iverson; blue images: Siede Preis/Getty Images and Joe Ginsberg/Getty Images

Cover Printer: Phoenix Color Corp

Compositor: Progressive Information Technologies

Printer: Quebecor World/Versailles

Printed in the United States of America

1 2 3 4 5 6 7 08 07 06 05 04

For more information about our products, contact us at:
Thomson Learning Academic Resource Center
1-800-423-0563
For permission to use material from this text or product, submit a request online at **http://www.thomsonrights.com**
Any additional questions about permissions can be submitted by email to **thomsonrights@thomson.com**

Library of Congress Control Number: 2004110543

Student Edition: ISBN 0-534-46773-3

Instructor's Edition: ISBN 0-534-46774-1

International Student Edition: ISBN 0-534-39597-X (Not for sale in the United States)

Thomson Brooks/Cole
10 Davis Drive
Belmont, CA 94002-3098
USA

Asia
Thomson Learning
5 Shenton Way #01-01
UIC Building
Singapore 068808

Australia/New Zealand
Thomson Learning
102 Dodds Street
Southbank, Victoria 3006
Australia

Canada
Nelson
1120 Birchmount Road
Toronto, Ontario M1K 5G4
Canada

Europe/Middle East/Africa
Thomson Learning
High Holborn House
50/51 Bedford Row
London WC1R 4LR
United Kingdom

Latin America
Thomson Learning
Seneca, 53
Colonia Polanco
11560 Mexico D.F.
Mexico

About the Authors

Chris Foote, Bill Brown, and Brent Iverson
(clockwise from upper left)

William H. Brown is Professor at Beloit College, where he was twice named Teacher of the Year. He is also the author of two other college textbooks; *Introduction to Organic Chemistry, Second Edition,* published in 2000, and *General, Organic, and Biochemistry, Seventh Edition,* coauthored with Fred Bettelheim and Jerry March and published in 2004. He received his PhD from Columbia University under the direction of Gilbert Stork and did postdoctoral work at California Institute of Technology and the University of Arizona. Twice he was Director of a Beloit College World Affairs Center seminar at the University of Glasgow, Scotland. Although officially retired from Beloit College, he writes and develops educational materials and continues to teach special topics in organic synthesis. Bill and his wife Carolyn enjoy hiking in the canyon country of the Southwest. In addition, they both enjoy quilting and quilts.

Christopher S. Foote is a Professor of Chemistry at the University of California, Los Angeles. He received his BS degree from Yale University and his PhD in Organic Chemistry from Harvard University. In 1995, he received the Tolman Award of the ACS Southern California Section for his contributions to the field of chemistry. Foote's research has focused on the chemistry of oxygen in organic and biological systems and on the chemistry of fullerenes. Other awards he has received include the Yale Science and Engineering Award for the Advancement of Basic and Applied Science, the ACS North Jersey Sections' Leo Hendrick Baekeland Award, and the 2000 American Society of Photo Biology Distinguished Research Award. He was also a Cope Scholar in 1994 and is the author of more than 250 research papers. In his spare time, Chris enjoys skiing.

Brent L. Iverson is a University Distinguished Teaching Professor at the University of Texas, Austin. He was raised in Palo Alto and San Jose, California. He received his BS degree from Stanford University and his PhD from California Institute of Technology. He planned to be a business major at Stanford until he enrolled in an organic chemistry course and became fascinated with the synthesis and design of new molecules. This attraction to science has lasted a lifetime. After postdoctoral training at the Scripps Research Institute, he began his academic career at the University of Texas at Austin. He teaches the large sophomore organic classes at UT and runs an interdisciplinary research group operating at the interface of organic chemistry and molecular biology. Brent has won several research awards including the National Science Foundation Presidential Young Investigator Award in 1991, the Chicago Community Trust Searle Scholar award in 1991, the Camille and Henry Dreyfus Foundation New Faculty and Teacher-Scholar awards in 1990 and 1995, respectively, an Alfred P. Sloan Foundation Research Fellowship in 1996, and the ACS Cope Scholar award for 2005. Brent and his wife Sheila have four active daughters, and he enjoys scuba diving and running marathons.

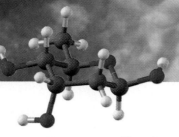

Brief Contents

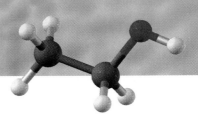

Contents

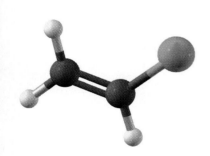

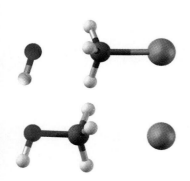

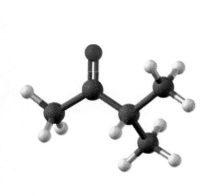

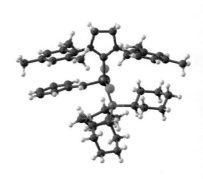

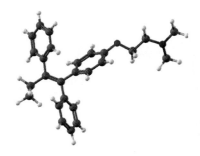

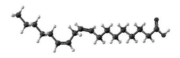

List of Mechanisms

Most mechanisms are set off in a box with a gold background. A few mechanisms are discussed in the text but not boxed.

Chapter 6 Reactions of Alkenes

- Electrophilic addition of HCl to 2-butene (Section 6.3A)
- Electrophilic addition of HCl to propene (Section 6.3A)
- Electrophilic addition of HCl to 2-methylpropene (Section 6.3A)
- Acid-catalyzed hydration of propene (Section 6.3B)
- Acid-catalyzed hydration of methylenecyclohexane (Section 6.3B)
- Carbocation rearrangement in the addition of HCl to an alkene (Section 6.3C)
- Acid-catalyzed addition of H_2O to 3-methyl-1-butene (Section 6.3C)
- Addition of bromine with anti stereoselectivity (Section 6.3D)
- Halohydrin formation and its anti stereoselectivity (Section 6.3E)
- Oxymercuration-reduction of an alkene (Section 6.3F)
- Hydroboration—a concerted regioselective and stereoselective reaction (Section 6.4)
- Oxidation of a trialkenylborane by alkaline hydrogen peroxide (Section 6.4)
- Addition of osmium tetroxide to an alkene (Section 6.5A)
- Formation of an ozonide (Section 6.5B)
- Catalytic hydrogenation of an alkene (Section 6.6A)
- Anti stereoselective addition of Br_2 to *cis*-2-butene (Section 6.7A)
- Anti stereoselective addition of Br_2 to *trans*-2-butene (Section 6.7A)

Chapter 7 Alkynes

- Addition of Br_2 to an alkyne (Section 7.6A)
- Addition of HBr to an alkyne (Section 7.6B)
- $HgSO_4/H_2SO_4$ catalyzed hydration of an alkyne (Section 7.7B)
- Reduction of an alkyne by sodium in liquid ammonia (Section 7.8C)

Chapter 8 Haloalkanes, Halogenation, and Radical Reactions

- Radical chlorination of ethane (Section 8.5B)
- Allylic bromination of propene using NBS (Section 8.6A)
- Radical autoxidation (Section 8.7)
- Vitamin E as an antioxidant: A scavenger of peroxy radicals (Section 8.7)
- Radical-initiated non-Markovinkov addition of HBr to alkenes (Section 8.8)

Chapter 16 Aldehydes and Ketones

- Addition of a Grignard reagent to formaldehyde (Section 16.5A)
- Addition of a Grignard reagent to an aldehyde other than formaldehyde (Section 16.5A)
- Addition of an alkyne anion to aldehydes and ketones (Section 16.5C)
- Formation of a cyanohydrin (Section 16.5D)
- Formation of a phosphonium ylide (Section 16.6)
- The Wittig reaction (Section 16.6)
- Base-catalyzed formation of a hemiacetal (Section 16.7B)
- Acid-catalyzed formation of a hemiacetal (Section 16.7B)
- Acid-catalyzed formation of an acetal (Section 16.7B)
- Acid-catalyzed formation of a tetrahydropyranyl ether (Section 16.7D)
- Formation of an imine from an aldehyde or ketone (Section 16.8A)
- Acid-catalyzed equilibration of keto and enol tautomers (Section 16.9A)
- $NaBH_4$ reduction of an aldehyde and ketone (Section 16.11A)
- NADH reduction of an aldehyde or ketone (Section 16.11A)
- Wolff-Kishner reduction (Section 16.11C)
- Acid-catalyzed α-halogenation of a ketone (Section 16.14C)
- Base-promoted α-halogenation of a ketone (Section 16.14C)

Chapter 17 Carboxylic Acids

- Addition of CO_2 to a Grignard reagent (Section 17.5A)
- Rhodium-catalyzed carbonation of methanol (Section 17.5B)
- Fischer esterification (Section 17.7B)
- Formation of a methyl ester using diazomethane (Section 17.7C)
- Reaction of a carboxylic acid with $SOCl_2$ (Section 17.8)
- Decarboxylation of a β-ketocarboxylic acid (Section 17.9A)
- Decarboxylation of a β-dicarboxylic acid (Section 17.9B)

Chapter 18 Functional Derivatives of Carboxylic Acids

- Nucleophilic acyl substitution (Section 18.3)
- Hydrolysis of an acid anhydride (Section 18.4B)
- Hydrolysis of a *tert*-butyl ester in aqueous acid (Section 18.4C)
- Hydrolysis of an ester in aqueous base (saponification) (Section 18.4C)
- Hydrolysis of an amide in aqueous acid (Section 18.4D)
- Hydrolysis of an amide in aqueous base (Section 18.4D)
- Hydrolysis of a cyano group to an amide in aqueous base (Section 18.4E)
- Reaction of acetyl chloride and ammonia (Section 18.6A)
- Reaction of an ester with a Grignard reagent (Section 18.8A)
- Reduction of an ester by $LiAlH_4$ (Section 18.10A)
- Reduction of an amide by $LiAlH_4$ (Section 18.10B)

Chapter 19 Enolate Anions and Enamines

- Base-catalyzed aldol reaction (Section 19.1)
- Acid-catalyzed aldol reaction (Section 19.1)
- Acid-catalyzed dehydration of an aldol product (Section 19.1)
- Claisen condensation (Section 19.2A)

- Alkylation of an enamine (Section 19.4A)
- Michael reaction: Conjugate addition of enolate anions (Section 19.7A)

Chapter 20 Conjugated Systems

- 1,2- and 1,4-Addition to a conjugated diene (Section 20.2A)

Chapter 21 Benzene and the Concept of Aromaticity

- Kolbe carboxylation of phenol (Section 21.4E)

Chapter 22 Reactions of Benzene and Its Derivatives

- Electrophilic aromatic substitution—chlorination (Section 22.1A)
- Formation of the nitronium ion (Section 22.1B)
- Nitration of benzene (Section 22.1B)
- Friedel-Crafts alkylation (Section 22.1C)
- Friedel-Crafts acylation—Generation of an acylium ion (Section 22.1C)
- Acid-catalyzed reaction of propene and benzene to give cumene (Section 22.1D)
- The methoxyl group as an ortho-para director (Section 22.2B)
- The carboxyl group as a meta director (Section 22.2B)
- Nucleophilic aromatic substitution via a benzyne intermediate (Section 22.3A)
- Nucleophilic aromatic substitution by addition-elimination (Section 22.3B)

Chapter 23 Amines

- Formation of the nitrosyl cation (Section 23.8)
- Reaction of a 2° amine with nitrous acid (Section 23.8C)
- Reaction of a 1° amine with nitrous acid (Section 23.8D)
- The Tiffeneau-Demjanov reaction (Section 23.8D)
- The Hofmann elimination (Section 23.9)
- The Cope elimination (Section 23.10)

Chapter 24 Carbon-Carbon Bond Formation and Synthesis

- The Heck reaction (Section 24.3B)
- The Suzuki coupling (Section 24.4B)
- Alkene metathesis (Section 24.5C)
- The Diels-Alder reaction (Section 24.6F)
- The Claisen rearrangement (Section 24.7A)
- Cope rearrangement (Section 24.7B)

Chapter 26 Lipids

- Activation of vitamin K by O_2 (Section 26.6D)

Chapter 27 Amino Acids and Proteins

- Cleavage of the peptide bond at methionine by BrCN (Section 27.4B)
- Edman degradation—Cleavage of an *N*-terminal amino acid (Section 27.4B)
- Acid-catalyzed removal of a benzyloxycarbonyl protecting group (Section 27.5C)
- Dicyclohexylcarbodiimide (DCC) and formation of a peptide bond (Section 27.5E)

Chapter 29 Organic Polymer Chemistry

- Reaction of an isocyanate with an alcohol (Section 29.5D)
- Radical polymerization of a substituted ethylene (Section 29.6A)
- Ziegler-Natta catalysis of ethylene polymerization (Section 29.6B)
- Homogeneous catalysis for Ziegler-Natta coordination polymerization (Section 29.6B)
- Initiation of anionic polymerization of alkenes (Section 29.6D)
- Initiation of anionic polymerization of butadiene (Section 29.6D)
- Polymerization of Superglue (Section 29.6D)
- Initiation of cationic polymerization of an alkene by $HF \cdot BF_3$ (Section 29.6D)
- Initiation of cationic polymerization of an alkene by a Lewis acid (Section 29.6D)
- Ring-opening alkene metathesis polymerization (Section 29.6D)

Preface

Introduction

The fourth edition of *Organic Chemistry* by Brown, Foote, and Iverson, takes the book in a bold new direction. Students taking an organic chemistry course have two objectives; the first is to learn organic chemistry, and the second is to establish the intellectual foundation for other molecular science courses. Most often, these other courses involve biochemistry or specialized topics such as materials science. This textbook addresses these two objectives head-on by first presenting mechanistic and synthetic organic chemistry geared toward giving students a fundamental understanding of organic principles, molecules, and reactions. The text then adds a new emphasis on bridging concepts that will prepare students for subsequent science courses.

Making Connections

All the important reaction mechanistic and synthetic details are still found throughout the text, but now important connections between different reaction mechanisms are emphasized in a clearer fashion. The intent is to make the study of mechanisms a process involving the learning and application of fundamental principles, and not an exercise in memorization. Mechanisms are presented in a clear, stepwise fashion and similarities between related mechanisms are emphasized. Written commentary accompanies each mechanistic step, so the more verbal learners can benefit alongside the more visually oriented learners. The uniting concept of nucleophiles reacting with electrophiles is highlighted throughout this edition. Especially helpful is the increased use of electrostatic potential maps of reacting molecules. These maps emphasize, in an easily interpreted, color-coded fashion, how the majority of reactions involve areas of higher electron density on one reactant (a nucleophile) interacting with areas of lower electron density on the other reactant (an electrophile).

Mastering Skills

The true mastery of organic chemistry requires the development of certain intellectual skills. To this end, nine *How To* boxes have been added to the first part of the book. These describe several "survival skills" for the organic chemistry students. Topics include *How To Draw Lewis Structures From Condensed Structural Formulas* (Section 1.2D), *How To Draw Alternative Chair Conformations of Cyclohexane* (Section 2.6B), and *How To Draw Curved Arrows and Push Electrons* (Section 1.6A). The importance of these skills will be emphasized on the new **Organic ChemistryNow** website. For example,

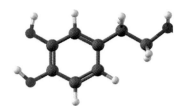

students will be able to test their understanding of how to draw curved arrows and push electrons. The students will answer some diagnostic questions and, based on their answers, will be directly linked to the corresponding *How To* tutorial.

Applying Organic Chemistry to Other Disciplines

Organic chemistry as a scientific discipline is most often applied to the synthesis of useful molecules. Synthetic applications of the reactions covered are emphasized throughout in this edition, partly through the introduction of many new challenging synthesis problems, the goal of which is to demonstrate clearly to students how synthetic organic chemistry is used in pharmaceutical research and in the production of useful pharmaceuticals. Even the latest organometallic transformations are included. Application of the reactions to the synthesis of important and well-known molecules, such as diazepam (Valium), fluoxetine (Prozac), meperidine (Demerol), albuterol (Proventil), tamoxifene, and sildefanil (Viagra) has been enhanced to provide a bridge to the practical applications of synthetic organic chemistry. These medicinal chemistry problems that were previously in a separate chapter have been integrated in chapters that emphasize the relevant chemistry.

The application of organic chemistry principles to important biological molecules is now integrated where appropriate in order to establish a bridge with subsequent biochemistry courses. In particular, an entirely new feature, called *Connections to Biological Chemistry*, has been added to give special attention to those aspects of organic chemistry that are essential to an understanding of the chemistry of living systems. For example, the organic chemistry of amino acids is now highlighted beginning in Section 3.8, along with the importance of alkene geometry to both membrane fluidity and nutrition (Section 5.4). How hydrogen bonding is involved with drug-receptor interactions is discussed (Section 10.2) as well as a description of how mustard gas derivatives are being used to treat cancer (Section 9.11). The fact that amide bonds are planar and have restricted bond rotation is newly emphasized as it relates to protein structure, and the biological oxidation of alcohols is discussed (Section 18.1). Importantly, these *Connections to Biological Chemistry* features have been added throughout the book, not just at the end, in recognition of the fact that not all instructors make it through the biological chemistry chapters at the end of the text.

Relevance to practical application is also emphasized in an expanded array of essays titled *Chemical Connections*. Topics include: medicines like penicillins and cephalosporins (Section 18.1), food supplements like antioxidants (Section 8.7) and vitamin C (Section 25.2), and materials science concepts such as spider silk (Section 27.7) and stitches that dissolve (Section 29.5). These sections provide a bridge between the theory of organic chemistry and well-known, current, practical applications. A list of the *Chemical Connections* as well as *Connections to Biological Chemistry* essays can be found on the inside back cover of this text. In effect, this fourth edition challenges students to apply the organic chemistry they are learning to the other aspects of their overall education.

Introducing Organic ChemistryNow

The completely new website is designed to engage students by helping them prepare for examination. **Organic ChemistryNow** is an assessment-centered learning tool developed in concert with the approach and pedagogy in the text. Students are given a

ORGANIC
Chemistry•ⵜ•Now™

personalized learning plan based on their pre-test results. The pre-test questions have been authored to reflect the level and approaches discussed in the text. The unique personalized learning plan will directly link students to Mechanism Tutorials, *How To* Tutorials, Molecular Models, animated Active Figures, and Reaction Flash Cards. Organic ChemistryNow is free with each new copy of the fourth edition.

Organic ChemistryNow is completely integrated within the text. When appropriate, Organic ChemistryNow icons are placed in the text and with figures to denote an activity or tutorial that can be explored on the website.

Summary

We hope students using our text will develop a mechanistic understanding of organic chemistry, will learn the important skills necessary to understand the language of organic chemistry, and will appreciate the application of organic chemistry to both biological systems and other fields such as materials science. Organic chemistry is a dynamic and ever-expanding area of science waiting for those who are prepared, both by training and inquisitive nature, to ask questions and to explore.

New to the Fourth Edition

In this edition, we have made major changes to unify the approach, to add new material that is at the forefront of organic synthesis, and to make the treatment of energy and mechanisms more complete.

- For almost every reaction for which a mechanism can be written, we present it. Each mechanism is clearly labeled and easily identified by a gold background. Steps are annotated, reaction conditions are clear, and arrow pushing is emphasized. A complete list of all mechanisms appears following the Contents.

- Chapter 6 has been expanded to include a more thorough discussion of the concepts of syn and anti stereoselectivity, and of stereospecificity. The chapter concludes with a discussion of the reactions of both achiral and chiral starting materials in achiral environments and, new to this edition, the reaction of achiral starting materials in a chiral environment, and enantioselective reactions. To emphasize the importance of stereochemistry among organic compounds, we have added a new *Summary of Stereochemical Terms* immediately following Chapter 6.

- To continue the chemistry of carbon-carbon pi bonds, we have moved alkynes from Chapter 10 to Chapter 7, immediately following the chemistry of alkenes (Chapters 5 and 6). In Chapter 7 we also introduce the concept of retrosynthesis and, as an application, introduce the alkylation of acetylide anions by methyl and primary haloalkanes. In our experience, alkylation of acetylide anions presents an easily assimilated example of nucleophilic substitution, which is then used as the conceptual foundation for the more detailed study of nucleophilic substitution in Chapter 9.

- The spectroscopy chapters (Chapters 12–14) have been expanded somewhat and have been made modular in that material for all functional groups are in this set of three chapters. Because these chapters are now complete and self-contained, IR and NMR could be used as early as after Chapter 3, but could also be introduced at any convenient point in a course, including much later.

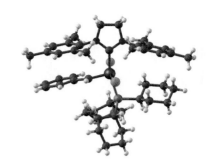

- The discussion of conjugated systems has been moved from Chapter 23 to Chapter 20 to provide a conceptual bridge to aromaticity (Chapter 21).

- New Chapter 24, *Carbon-Carbon Bonding Formation and Synthesis,* combines students knowledge from previous chapters with cutting edge organometallic reactions and challenges students to devise syntheses.

- The more than 40 synthesis problems, formerly in the Interchapter *Medicinal Chemistry—Problems in Organic Synthesis* in the third edition, have been integrated into the appropriate functional group chapters. In recognition of the fact that the majority of students taking introductory organic chemistry are interested in careers in the health and biological sciences, we have chosen problems primarily from the area of medicinal chemistry. In addition, the number of challenging, real-world application problems has been expanded to nearly 60 and are designated by a red problem number.

- Nine *How To* features discuss the most important skills for students to learn in this course, such as "How To Draw Curved Arrows and Push Electrons" and "How To Draw Chiral Molecules." Students will be able to test their mastery of these *How To* skills at the Organic ChemistryNow website. Students will be directed to *How To* tutorials based on individual learning needs.

- New *Connections to Biological Chemistry* boxes discuss the applications of organic chemistry to biology, such as "The Importance of *cis* Double Bonds in Fats Versus Oils."

- Many new electrostatic potential maps have been added to give students a greater appreciation for the charge distribution in molecules.

- A comprehensive *Summary of Stereochemical Terms* has been added following Chapter 6 to serve as a handy reference for the subtle concepts and terminology used to describe all aspects of stereochemistry in organic chemistry.

ORGANIC
Chemistry∙ᴥ∙Now™

- Organic ChemistryNow at **http://now.brookscole.com/bfi4** is a new book-specific website that correlates directly with the content of this book and contains additional problems, tutorials, mechanistic simulations, and a set of reaction flash cards for students to practice for quizzes and tests.

Special Features

- **Full-Color Art Program** One of the most distinctive features of this text is its visual impact. The text's extensive full-color art program includes over 250 pieces of art by professional artists John and Bette Woolsey. A large number of molecular models have been generated and energy minimized in CambridgeSoft's Chem3D, and then rendered by these artists to provide easily visualized pictures of three-dimensional molecular structures.

- **Electrostatic Potential Maps** are provided at appropriate places throughout the text to illustrate the important concepts of resonance, electrophilicity, and nucleophilicity.

- **Chemical Connections** These essays illustrate applications of organic chemistry to everyday settings. Topics range from "Chiral Drugs," "Penicillins and the Cephalosporins," and "Antioxidants" to "Drugs That Lower Plasma Levels of

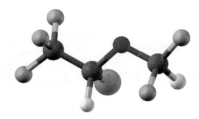

Cholesterol" and "The Chemistry of Superglue." A complete list can be found on the inside of the back cover.

- **Connections to Biological Chemistry** Application of organic chemistry to biology is emphasized throughout the text, in the *Connections to Biological Chemistry* essays and in end-of-chapter problems. See the inside of the back cover for a complete list of the essays.

- **How To** Nine *How To* features have been added to the first part of the book. These describe several "survival skills" for the organic chemistry students.

- **In-Chapter Examples** There is an abundance of in-chapter examples, each with a detailed solution, so students can *immediately* see how the concepts just discussed relate to specific questions and their answers. Following each in-chapter example is a comparable in-chapter problem designed to give students the opportunity to solve a related problem on their own.

- **End-of-Chapter Summaries and Summaries of Key Reactions** End-of-chapter summaries highlight all the important new reactions found in a chapter. In addition, each reaction is annotated and keyed to the section where it is discussed.

- **End-of-Chapter Problems** There are plentiful end-of-chapter problems, with the majority categorized by topic. A red problem number indicates an applied, real-world problem. New to this edition, particularly in the later functional group chapters, are numerous problems dealing with the synthesis of important pharmaceuticals.

- **Glossary of Key Terms** Throughout the book we place definitions for new terms in the margin. In addition, all definitions are collected in a glossary at the end of the text. Each glossary listing is keyed to the section of the text where the term is introduced.

- **Photos** Conceived and developed for this text, full-color photos show organic chemistry as it occurs in the laboratory and in everyday life and depict the natural sources of many organic compounds.

- **Color** Special colors are used to highlight parts of molecules and to follow the course of reactions.

- The **Organic ChemistryNow** icon alerts students to go to the Organic ChemistryNow website at http://now.brookscole.com/bfi4 for additional problems and tutorials associated with this text.

Support Package

For the Student

- **Student Study Guide** By Brent and Sheila Iverson of the University of Texas, Austin, this manual contains detailed solutions to all text problems. ISBN 0-534-46777-6

- **Pushing Electrons: A Guide for Students of Organic Chemistry, third edition** By Daniel P. Weeks, Northwestern University, this paperback workbook is designed to help students learn techniques of electron pushing. Its programmed approach emphasizes repetition and active participation. ISBN 0-030-20693-6

- **Organic ChemistryNow at** http://now.brookscole.com/bfi4 This first Web-based assessment-centered learning tool for the *Organic Chemistry* course was developed in concert with the text. Throughout each chapter, icons with captions alert students to media resources that enhance problem-solving skills and improve conceptual understanding. In Organic ChemistryNow, students take a diagnostic Pre-Test and are provided with a Personalized Learning Plan that targets their study needs and helps them to visualize, organize, practice, and master the material in the text. Passcode access to Organic ChemistryNow is packaged FREE with every new copy of the text.

- **OWL (Online Web-based Learning System) for *Organic Chemistry*** Developed over the past several years at the University of Massachusetts, Amherst, and class-tested by thousands of students, Organic OWL is a customizable and flexible Web-based homework system and assessment tool. This fully integrated testing, tutorial, and course-management system features over 3000 practice and homework problems. With built-in numerical and chemical parameterization, Organic OWL provides students with instant analysis and feedback to homework problems, modeling questions, molecular-structure-building exercises, and animations created specifically for *Organic Chemistry,* Fourth Edition. This powerful system maximizes the students learning experience and, at the same time, reduces faculty workload and facilitates instruction. A fee-based access code is required for access to Organic Owl. To learn more, visit http://owl.thomsonlearning.com or contact your Thomson Brooks/Cole representative for details.

For the Instructor

Supporting materials are available to qualified adopters. Please consult your local Thomson Brooks/Cole sales representative for details. Visit the *Organic Chemistry* website at http://now.brookscole.com/bfi4 to see samples of these materials, request a desk copy, locate your sales representative, or purchase a copy online.

- **iLrn Testing: Electronic Testing System** contains approximately 1000 multiple-choice problems and questions representing every chapter of the text. Available online and on a dual-platform CD-ROM. ISBN 0-534-46778-4

- **Test Bank** by David M. Collard, Georgia Institute of Technology. A multiple-choice bank of over 1000 problems for instructors to use for tests, quizzes, or homework assignments. ISBN 0-534-46776-8

- **Multimedia Manager CD-ROM** A dual-platform digital library and presentation tool that provides art and tables from the text in a variety of electronic formats that are easily exported into other software packages. This enhanced CD-ROM also contains simulations, molecular models, and QuickTime™ movies to supplement your lectures; PowerPoint™ lecture slides by William H. Brown; and a lecture outline with integrated media. You can customize your presentations by importing your personal lecture slides or other material you choose. ISBN 0-534-46779-2

- **Overhead Transparency Acetates** A selection of 125 full-color figures from the text. ISBN 0-534-46780-6

- **WebTutor™ ToolBox on WebCT and Blackboard** Preloaded with content and available free via PIN code when packaged with this text, WebTutor ToolBox pairs all the content of this text's rich Book Companion website with all the sophisticated course management functionality of WebCT or Blackboard. WebTutor ToolBox is ready to use as soon as you log on—or, you can customize its preloaded content by uploading images and other resources, adding Web links, or creating your own practice materials. Blackboard: ISBN 0-534-59125-6; Toolbox ISBN 0-534-59134-5

Acknowledgments

While one or a few persons are listed as "author" of any textbook, the book is in fact the product of collaboration of many individuals, some obvious, others not so obvious. It is with gratitude that we herein acknowledge the contributions of the many.

David Harris as publisher has masterfully guided the revision of the text. Sandi Kiselica has been a rock of support as senior developmental editor. We so appreciate her ability to set challenging but manageable schedules for us and then her constant encouragement as we worked to meet those deadlines. She was also an invaluable resource person with whom we could discuss everything from pedagogy to details of art. Others at the Brooks/Cole organization have helped to shape our words into this text, including Alyssa White, associate editor; Lisa Weber, production project manager; Rob Hugel, creative director; and Donna Kelley, technology project manager. Colleen Franciscus and Lisa McClanahan of Progressive Publishing Alternatives have helped us to keep on track through all phases of production.

We gratefully acknowledge the help of Eric Kantorowski, of the California Polytechnic State University, who read all of the page proofs for accuracy; Eric has a keen eye for detail. William Vining, at the University of Massachusetts, is the master of Organic ChemistryNow, the website that accompanies this book.

We are also indebted to the many reviewers of our manuscript who helped shape its contents. With their guidance we have revised this text to better meet the needs of their students.

Reviewers of the Fourth Edition

John Anthony, *University of Kentucky*
Julia Baker, *Winthrop University*
Robert Carlson, *University of Minnesota*
Dana S. Chatellier, *University of Delaware*
Michelle Chatellier, *University of Delaware*
Leland S. Endres, *California Polytechnic and State University*
John M. Ferguson, *University of Central Oklahoma*
Malcolm Forbes, *University of North Carolina*
Gamini U. Gunawardina, *Utah Valley State College*
Paul J. Kropp, *University of North Carolina*
Dominic McGrath, *University of Arizona*
Jennifer Muzyka, *Centre College*
Dallas G. New, *University of Central Oklahoma*
James Nowick, *University of California, Irvine*
Timothy P. O'Dea, *Indiana University*
Andrea Pace, *University of Wyoming*

Anne B. Padias, *University of Arizona*
James A. Pincock, *Dalhousie University*
Thomas Poon, *Claremont McKenna, Pitzer and Scripps Colleges*
Owen Priest, *Northwestern University*
Francisco M. Raymo, *University of Miami*
Brian Salvatore, *University of South Carolina*
Karl A. Scheidt, *Northwestern University*
Chad Stessman, *California State, Stanislaus*
Jennifer Swift, *Georgetown University*
Devvie Tahmassebi, *University of San Diego*
Emmanuel Theodorakis, *University of California, San Diego*
Philip Warner, *Northeastern University*
Peter Wepplo, *Monmouth University*
Jane Wissinger, *University of Minnesota*
Viktor V. Zhdankin, *University of Minnesota, Duluth*

Reviewers of Previous Editions

Neil T. Allison, *University of Arkansas;* Eric Anslyn, *University of Texas;* Rodney Badger, *Southern Oregon State College;* William Bailey, *University of Connecticut;* Shelton Bank, *State University of New York, Albany;* Nancy Barta, graduate student, *Michigan State University;* John Belletiri, *University of Cincinnati;* John Benbow, *Lehigh University;* Edwin Bryant, graduate student, *Michigan State University;* Thomas Bryson, *University of South Carolina;* Edward M. Burgess, *Georgia Institute of Technology;* Mary Campbell, *Mount Holyoke College;* James Canary, *New York University;* Robert G. Carlson, *University of Kansas;* Lyle Castle, *Idaho State University;* Claire Castro, *University of San Francisco;* Dana S. Chatellier, *University of Delaware;* Clair Cheer, *San Jose State University;* William D. Closson, *State University of New York, Albany;* Barry Coddens, *Northwestern University;* David Crich, *University of Illinois, Chicago;* Dennis D. Davis, *New Mexico State University;* Mark DeCamp, *University of Michigan, Dearborn;* James A. Deyrup, *University of Florida;* Thomas A. Dix, *University of California, Irvine;* Paul Dowd, *University of Pittsburgh;* Dale Drueckhammer, *Stanford University;* Michael B. East, *Florida Institute of Technology;* William Epstein, *University of Utah;* Morris Fishman, *New York University;* Raymond C. Fort, Jr., *University of Maine;* Warren Giering, *Boston University;* Jack Gilbert, *University of Texas, Austin;* Stanley I. Goldberg, *University of New Orleans;* Dorothy Goldish, *California State University, Long Beach;* Scott Gronert, *San Francisco University;* Steven Hardinger, *University of California, Los Angeles;* Leland Harris, *University of Arizona;* Dan Harvey, *University of California, San Diego;* John Helling, *University of Florida;* Gene Hiegel, *California State University, Fullerton;* John L. Hogg, *Texas A&M University;* John W. Huffman, *Clemson University;* Ian Hunt, *University of Calgary;* Brent Iverson, *University of Texas, Austin;* Francis Klein, *Creighton University;* Robert Kluger, *University of Toronto;* Joseph B. Lambert, *Northwestern University;* John Landgrebe, *University of Kansas;* Allan K. Lazarus, *Trenton State University;* Norman Lebel, *Wayne State University;* Robert Loeschen, *California State University, Long Beach;* Marco Lopez, *California State University, Long Beach;*

Richard Luibrand, *California State University, Hayward;* Gary Lyon, *Louisiana State University;* Andrew MacMillan, *University of Toronto;* Jerry March, *Adelphi University;* Kenneth L. Marsi, *California State University, Long Beach;* Eugene Mash, *University of Arizona;* Dominic McGrath, *University of Connecticut;* David M. McKinnin, *University of Manitoba;* Kirk McMichael, *Washington State University;* Richard Morrison, *West Virginia University;* James Mulvaney, *University of Arizona;* Kathy Nabona, *Austin Community College;* Gary Newton, *University of Georgia;* Bruce Norcross, *State University of New York, Binghamton;* Aaron Odom, *Michigan State University;* Walter Ott, *Emory University;* Anne Padias, *University of Arizona;* Brian Pagenkopf, *University of Texas;* E. Paul Papadopoulos, *University of New Mexico, Albuquerque;* Steven Pederson, *University of California, Berkeley;* Michael Rathke, *Michigan State University;* Russell C. Petter, *Sandoz Research Institute;* Joseph M. Prokipcak, *University of Guelph;* William A. Pryor, *Louisiana State University;* Carmelo Rizzo, *Vanderbilt University;* Michael Rathke, *Michigan State University;* Alan Rosan, *Drew University;* Charles B. Rose, *University of Nevada, Reno;* K. C. Russell, *University of Miami;* John Scheffer, *University of British Columbia;* James Schreck, *University of Northern Colorado;* Jonathan Sessler, *University of Texas;* William Shay, *University of North Dakota;* Valerie Sheares, *Iowa State University;* Daniel Singleton, *Texas A&M University;* Martin Sobczak, graduate student, *Michigan State University;* Steve Steffke, graduate student, *Michigan State University;* Robert Stern, *Oakland University;* John Stille, *Michigan State University;* J. William Suggs, *Brown University;* Michelle Sulikowski, *Texas A&M University;* Peter Trumper, *Bowdoin College;* Ken Turnbull, *Write State University;* Edward Waali, *University of Montana;* George Wahl, *North Carolina State University;* Michael Waldo, graduate student, *Michigan State University;* Daniel Weeks, *Northwestern University;* David F. Weimer, *University of Iowa;* Desmond Wheeler, *University of Nebraska;* Theodore Widlanski, *Indiana University;* Jeffrey Winkler, *University of Pennsylvania;* Darrell J. Woodman, *University of Washington;* Ali Zand, graduate student, *Michigan State University*

We have enjoyed writing this text, and hope that instructors and students alike find in it a measure of the excitement we feel for organic chemistry.

William H. Brown
Beloit College
brownwh@beloit.edu

Christopher Foote
University of California, Los Angeles
foote@chem.ucla.edu

Brent Iverson
University of Texas, Austin
biverson@mail.utexas.edu

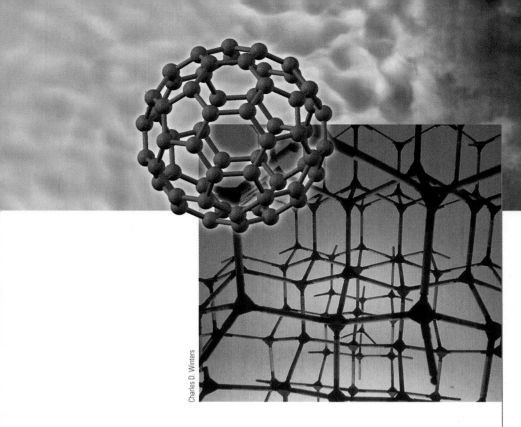

Charles D. Winters

■ A model of the structure of diamond, one form of pure carbon. Each carbon is bonded to four other carbons at the corners of a tetrahedron. Inset: a model of fullerene (C_{60}). See the box "Fullerene—A New Form of Carbon."

1

Covalent Bonding and Shapes of Molecules

A ccording to the simplest definition, **organic chemistry** is the study of the compounds of carbon. Perhaps its most remarkable feature is that most organic compounds consist of carbon and only a few other elements—chiefly, hydrogen, oxygen, and nitrogen. Chemists have discovered or made well over ten million compounds composed of carbon and these three other elements. Organic compounds are everywhere around us—in our foods, flavors, and fragrances; in our medicines, toiletries, and cosmetics; in our plastics, films, fibers, and resins; in our paints and varnishes; in our glues and adhesives; in our fuels and lubricants; and, of course, in our bodies and those of all living things.

But organic chemistry is more than the chemistry of carbon and the few other elements listed here. It is also the chemistry of compounds containing carbon-metal bonds. The chemistry of organometallic compounds is an extremely large area of study today and one of the major interfaces between organic chemistry, inorganic chemistry, and biochemistry. We introduce organometallics in Chapter 15 and continue their study in several later chapters.

Let us begin our study of organic chemistry with a review of how the elements of C, H, O, and N combine by sharing electron pairs to form

ORGANIC
Chemistry ⚛ Now™

Look for this logo in the text and go to Organic ChemistryNow at **http://now.brookscole.com/bfi4** to view tutorials and simulations, develop problem-solving skills, and test your conceptual understanding with unique interactive resources.

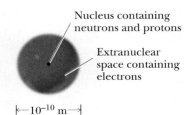

Nucleus containing neutrons and protons

Extranuclear space containing electrons

$\leftarrow\!10^{-10}\,\text{m}\!\rightarrow$

Figure 1.1
A schematic view of an atom. Most of the mass of an atom is concentrated in its small, dense nucleus.

Shell   A region of space around a nucleus that can be occupied by electrons, corresponding to a principal quantum number.

Orbital   A region of space that can hold two electrons.

molecules. There is a great deal of material in this chapter, but much of it should be familiar from your previous chemistry courses. However, because all subsequent chapters in this book use this material, it is essential that you understand it and can use it fluently.

1.1 Electronic Structure of Atoms

You should already be familiar with the fundamentals of the electronic structure of atoms. Briefly, an atom contains a small, dense nucleus made of neutrons and positively charged protons. Most of the mass of an atom is contained in its nucleus. The nucleus is surrounded by an extranuclear space containing negatively charged electrons. The nucleus of an atom has a diameter of 10^{-14} to 10^{-15} meters (m). The extranuclear space where its electrons are found is a much larger volume with a diameter of approximately 10^{-10} m (Figure 1.1).

Electrons do not move freely in the space around the nucleus but are confined to regions of space called **principal energy levels** or, more simply, **shells.** Electron shells are identified by the principal quantum numbers 1, 2, 3, and so on. Each shell can contain up to **$2n^2$ electrons,** where n is the number of the shell. Thus, the first shell can contain 2 electrons, the second 8 electrons, the third 18 electrons, the fourth 32 electrons, and so on (Table 1.1). Electrons in the first shell are nearest to the positively charged nucleus and are held most strongly by it; these electrons are said to be lowest in energy. To say that they are lowest in energy also means that it takes a greater amount of energy to remove an electron from the first shell of an atom than from any other shell. Electrons in higher numbered shells are farther from the positively charged nucleus. They are held less strongly, it takes less energy to remove them from the atom, and, accordingly, they are said to be higher in energy.

Shells are divided into subshells designated by the letters s, p, d, and f, and, within these subshells, electrons are grouped in orbitals (Table 1.2). An **orbital** is a region of space that can hold two electrons. The first shell contains a single orbital called a $1s$ orbital. The second shell contains one s orbital and three p orbitals. The three $2p$ orbitals are directed along the x-, y-, and z-axes and are designated $2p_x$, $2p_y$, and $2p_z$. The third shell contains one $3s$ orbital, three $3p$ orbitals, and five $3d$ orbitals. The shapes of s and p orbitals are shown in Figures 1.10 and 1.11, and are described in more detail in Section 1.7B.

A. Electron Configuration of Atoms

The electron configuration of an atom is a description of the orbitals its electrons occupy. Every atom has an infinite number of possible electron configurations. At this stage, we are concerned primarily with the **ground-state electron configuration**—the

Table 1.1	Distribution of Electrons in Shells	
Shell	**Number of Electrons Shell Can Hold**	**Relative Energies of Electrons in These Shells**
4	32	higher
3	18	
2	8	
1	2	lower

Table 1.2	Distribution of Orbitals Within Shells
Shell	**Orbitals Contained in That Shell**
3	$3s$, $3p_x$, $3p_y$, $3p_z$, plus five $3d$ orbitals
2	$2s$, $2p_x$, $2p_y$, $2p_z$
1	$1s$

electron configuration of lowest energy. We determine the ground-state electron configuration of an atom by using the following three rules.

Rule 1: The Aufbau ("Build-Up") Principle.   Orbitals fill in order of increasing energy, from lowest to highest. In this course, we are concerned primarily with the elements of the first, second, and third periods of the Periodic Table. Orbitals fill in the order $1s$, $2s$, $2p$, $3s$, $3p$, and so on.

Rule 2: The Pauli Exclusion Principle.   The Pauli exclusion principle requires that only two electrons can occupy an orbital and that their spins must be paired. To understand what it means to have paired spins, recall from general chemistry that just as the earth has a spin, so do electrons. And, just as the earth has magnetic north (N) and south (S) poles, so do electrons. As described by quantum mechanics, a given electron can exist in only two different spin states. Two electrons with opposite spins are said to have **paired spins.**

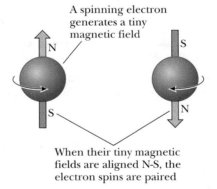

A spinning electron generates a tiny magnetic field

When their tiny magnetic fields are aligned N-S, the electron spins are paired

When filling orbitals with electrons, place no more than two in an orbital. For example, with four electrons, the $1s$ and $2s$ orbitals are filled and are written $1s^2\,2s^2$. With an additional six electrons, the set of three $2p$ orbitals is filled and is written $2p_x^2\,2p_y^2\,2p_z^2$. Alternatively, a filled set of three $2p$ orbitals may be written $2p^6$.

Rule 3: Hund's Rule.   When orbitals of equal energy are available but there are not enough electrons to fill all of them completely, then one electron is added to each orbital before a second electron is added to any one of them. Recall that electrons have negative charge; partially filling orbitals as much as possible minimizes electrostatic repulsion between electrons. After the $1s$ and $2s$ orbitals are filled with four electrons, a fifth electron is added to the $2p_x$ orbital, a sixth to the $2p_y$ orbital, and a seventh to

Ground-state electron configuration   The lowest-energy electron configuration for an atom or molecule.

Aufbau principle   Orbitals fill in order of increasing energy, from lowest to highest.

Pauli exclusion principle   No more than two electrons may be present in an orbital. If two electrons are present, their spins must be paired.

▄▄ The pairing of electron spins.

Hund's rule   When orbitals of equal energy are available but there are not enough electrons to fill all of them completely, one electron is put in each before a second electron is added to any.

Table 1.3 Ground-State Electron Configurations for Elements 1–18

First Period*			Second Period			Third Period		
H	1	$1s^1$	Li	3	[He] $2s^1$	Na	11	[Ne] $3s^1$
He	2	$1s^2$	Be	4	[He] $2s^2$	Mg	12	[Ne] $3s^2$
			B	5	[He] $2s^2 2p^1$	Al	13	[Ne] $3s^2 3p^1$
			C	6	[He] $2s^2 2p^2$	Si	14	[Ne] $3s^2 3p^2$
			N	7	[He] $2s^2 2p^3$	P	15	[Ne] $3s^2 3p^3$
			O	8	[He] $2s^2 2p^4$	S	16	[Ne] $3s^2 3p^4$
			F	9	[He] $2s^2 2p^5$	Cl	17	[Ne] $3s^2 3p^5$
			Ne	10	[He] $2s^2 2p^6$ octet	Ar	18	[Ne] $3s^2 3p^6$ octet

*Elements are listed by symbol, atomic number, and simplified ground-state electron configuration.

Gilbert N. Lewis (1875–1946) introduced the theory of the electron pair that extended our understanding of covalent bonding and of the concept of acids and bases. It is in his honor that we often refer to an "electron dot" structure as a Lewis structure.

the $2p_z$ orbital. Only after each $2p$ orbital contains one electron is a second electron added to the $2p_x$ orbital. Carbon, for example, has six electrons, and its ground-state electron configuration is $1s^2\ 2s^2\ 2p_x^1\ 2p_y^1\ 2p_z^0$. Alternatively, it may be simplified to $1s^2\ 2s^2\ 2p^2$. Table 1.3 shows ground-state electron configurations of the first 18 elements of the Periodic Table.

Example 1.1

Write the ground-state electron configuration for each element showing the occupancy of each p orbital.

(a) Lithium   **(b)** Oxygen   **(c)** Chlorine

Solution

(a) Lithium (atomic number 3): $1s^2\ 2s^1$
(b) Oxygen (atomic number 8): $1s^2\ 2s^2\ 2p_x^2\ 2p_y^1\ 2p_z^1$
(c) Chlorine (atomic number 17): $1s^2\ 2s^2 2p_x^2\ 2p_y^2\ 2p_z^2 3s^2\ 3p_x^2\ 3p_y^2\ 3p_z^1$

Problem 1.1

Write and compare the ground-state electron configurations for each pair of elements.

(a) Carbon and silicon   **(b)** Oxygen and sulfur   **(c)** Nitrogen and phosphorus

B. Lewis Dot Structures

Valence electrons Electrons in the valence (outermost) shell of an atom.

Valence shell The outermost occupied electron shell of an atom.

When discussing the physical and chemical properties of an element, chemists often focus on the electrons in the outermost shell of the atom because these electrons are involved in the formation of chemical bonds and in chemical reactions. Carbon, for example, with the ground-state electron configuration $1s^2\ 2s^2\ 2p^2$, has four outer-shell electrons. Outer-shell electrons are called **valence electrons,** and the energy level in which they are found is called the **valence shell.** To show the outermost elec-

Table 1.4	Lewis Dot Structures for Elements 1–18*						
1A	**2A**	**3A**	**4A**	**5A**	**6A**	**7A**	**8A**
H·							He :
Li·	Be :	B :	·C :	·N :	:O :	:F :	:Ne :
Na·	Mg :	Al :	·Si :	·P :	:S :	:Cl :	:Ar :

*Electron dots are arranged in this table in accordance with Hund's rule.

trons of an atom, we commonly use a representation called a **Lewis dot structure,** after the American chemist Gilbert N. Lewis (1875–1946) who devised this notation. A Lewis dot structure shows the symbol of the element surrounded by a number of dots equal to the number of electrons in the outer shell of an atom of that element. In Lewis dot structures, the atomic symbol represents the "core;" that is, the nucleus and all inner shells. Table 1.4 shows Lewis dot structures for the first 18 elements of the Periodic Table.

The noble gases helium and neon have filled valence shells. The valence shell of helium is filled with two electrons; that of neon is filled with eight electrons. Neon and argon have in common an electron configuration in which the s and p orbitals of their valence shells are filled with eight electrons. The valence shells of all other elements shown in Table 1.4 contain fewer than eight electrons.

For C, N, O, and F in period 2 of the Periodic Table, the valence electrons belong to the second shell. With eight electrons, this shell is completely filled. For Si, P, S, and Cl in period 3 of the Periodic Table, the valence electrons belong to the third shell. This shell is only partially filled with eight electrons; the $3s$ and $3p$ orbitals are fully occupied, but the five $3d$ orbitals can accommodate an additional ten electrons. Because of the differences in number and kind of valence-shell orbitals available to elements of the second and third periods, significant differences exist in the bonding patterns of oxygen and sulfur and of nitrogen and phosphorus (Section 1.2E). For example, although oxygen and nitrogen can accommodate no more than 8 electrons in their valence shells, many phosphorus-containing compounds have 10 electrons in the valence shell of phosphorus, and many sulfur-containing compounds have 10 or even 12 electrons in the valence shell of sulfur.

1.2 Lewis Model of Bonding

In 1916, Lewis devised a beautifully simple model that unified many of the observations about chemical bonding and reactions of the elements. He pointed out that the chemical inertness of the noble gases indicates a high degree of stability of the electron configurations of these elements: helium with a valence shell of two electrons ($1s^2$), neon with a valence shell of eight electrons ($2s^2\,2p^6$), and argon with a valence shell of eight electrons ($3s^2\,3p^6$). The tendency of atoms to react in ways that achieve an outer shell of eight valence electrons is particularly common among second-row elements of Groups 1A–7A (the main-group elements) and is given the special name **octet rule.**

Lewis dot structure The symbol of an element surrounded by a number of dots equal to the number of electrons in the valence shell of the atom.

ORGANIC
Chemistry·_>·Now™
Click Simulations to explore
Electron Configuration and the Periodic Table

Octet rule Group 1A–7A elements react to achieve an outer shell of eight valence electrons.

Example 1.2

Show how the loss of an electron from a sodium atom leads to a stable octet.

Solution

The ground-state electron configurations for Na and Na^+ are:

$$Na \text{ (11 electrons): } 1s^2 \, 2s^2 \, 2p^6 \, 3s^1$$

$$Na^+ \text{ (10 electrons): } 1s^2 \, 2s^2 \, 2p^6$$

Thus, Na^+ has a complete octet of electrons in its outermost (valence) shell and has the same electron configuration as neon, the noble gas nearest it in atomic number.

Problem 1.2

Show how each chemical change leads to a stable octet.

(a) Sulfur forms S^{2-}.   **(b)** Magnesium forms Mg^{2+}.

Anion   An atom or group of atoms bearing a negative charge.

Cation   An atom or group of atoms bearing a positive charge.

Covalent bond   A chemical bond formed between two atoms by sharing one or more pairs of electrons.

Electronegativity   A measure of the force of an atom's attraction for electrons.

A. Formation of Chemical Bonds

According to Lewis' model, atoms interact in such a way that each participating atom acquires a completed outer-shell electron configuration resembling that of the noble gas nearest to it in atomic number. Atoms acquire completed valence shells in two ways.

1. An atom may become ionic, that is lose or gain enough electrons to acquire a completely filled valence shell. An atom that gains electrons becomes an **anion** (a negatively charged ion), and an atom that loses electrons becomes a **cation** (a positively charged ion). A positively charged ion and a negatively charged ion attract each other. This attraction can lead to the formation of ionic crystals such as sodium chloride, in which each positive ion is surrounded by negative ions and vice versa, in a definite geometric arrangement that depends on the crystal.
2. An atom may share electrons with one or more other atoms to complete its valence shell. A chemical bond formed by sharing electrons is called a **covalent bond.**
3. Bonds may be partially ionic and partially covalent; these bonds are called polar covalent bonds.

■ Linus Pauling (1901–1994) was the first person ever to receive two unshared Nobel prizes. He received the 1954 Nobel prize in chemistry for his contributions to our understanding of chemical bonding. He received the 1962 Nobel peace prize for his efforts on behalf of international control of nuclear weapons testing.

B. Electronegativity and Chemical Bonds

How do we estimate the degree of ionic or covalent character in a chemical bond? One way is to compare the electronegativities of the atoms involved. **Electronegativity** is a measure of an atom's attraction for electrons that it shares in a chemical bond with another atom. The most widely used scale of electronegativities (Table 1.5) was devised by Linus Pauling in the 1930s and is based on bond dissociation enthalpies.

On the Pauling scale, fluorine, the most electronegative element, is assigned an electronegativity of 4.0, and all other elements are assigned values in relation to fluorine. As you study the electronegativity values in this table, note that they generally increase from left to right within a period of the Periodic Table and generally decrease from top to bottom within a group. Values increase from left to right because the

Table 1.5 Electronegativity Values for Some Atoms (Pauling Scale)

										H 2.1								

1A	2A	3B	4B	5B	6B	7B	8B			1B	2B	3A	4A	5A	6A	7A
Li 1.0	Be 1.5											B 2.0	C 2.5	N 3.0	O 3.5	F 4.0
Na 0.9	Mg 1.2											Al 1.5	Si 1.8	P 2.1	S 2.5	Cl 3.0
K 0.8	Ca 1.0	Sc 1.3	Ti 1.5	V 1.6	Cr 1.6	Mn 1.5	Fe 1.8	Co 1.8	Ni 1.8	Cu 1.9	Zn 1.6	Ga 1.6	Ge 1.8	As 2.0	Se 2.4	Br 2.8
Rb 0.8	Sr 1.0	Y 1.2	Zr 1.4	Nb 1.6	Mo 1.8	Tc 1.9	Ru 2.2	Rh 2.2	Pd 2.2	Ag 1.9	Cd 1.7	In 1.7	Sn 1.8	Sb 1.9	Te 2.1	I 2.5
Cs 0.7	Ba 0.9	La 1.1	Hf 1.3	Ta 1.5	W 1.7	Re 1.9	Os 2.2	Ir 2.2	Pt 2.2	Au 2.4	Hg 1.9	Tl 1.8	Pb 1.8	Bi 1.9	Po 2.0	At 2.2

- ▢ <1.0
- ▢ 1.0 – 1.4
- ▢ 1.5 – 1.9
- ▢ 2.0 – 2.4
- ▢ 2.5 – 2.9
- ▢ 3.0 – 4.0

increasing positive charge on the nucleus results in a greater force of attraction for the atom's valence electrons. Electronegativity decreases from top to bottom because the increasing distance of the valence electrons from the nucleus results in a lower attraction of the nucleus for them.

Example 1.3

Judging from their relative positions in the Periodic Table, which element in each set is the more electronegative?

(a) Lithium or carbon   **(b)** Nitrogen or oxygen   **(c)** Carbon or oxygen

Solution

The elements in these sets are all in the second period of the Periodic Table. Electronegativity in this period increases from left to right.

(a) $C > Li$ **(b)** $O > N$ **(c)** $O > C$

Problem 1.3

Judging from their relative positions in the Periodic Table, which element in each set is the more electronegative?

(a) Lithium or potassium   **(b)** Nitrogen or phosphorus   **(c)** Carbon or silicon

Formation of Ions

Ions are formed by the transfer of electrons from the valence shell of an atom of lower electronegativity to the valence shell of an atom of higher electronegativity. As a rough guideline, we say that ions will form if the difference in electronegativity between interacting atoms is 1.9 or greater. As an example, ions are formed from sodium (electronegativity 0.9) and fluorine (electronegativity 4.0). In the following

ORGANIC
Chemistry⚛Now™
Click Simulations to explore
Electron Configurations of Ions

equation, we use a single-headed (barbed) curved arrow to show the transfer of one electron from sodium to fluorine.

$$\text{Na} \cdot \overset{\frown}{+} \cdot \ddot{\underset{\cdot\cdot}{F}} \colon \longrightarrow \text{Na}^+ \colon \ddot{\underset{\cdot\cdot}{F}} \colon^-$$

In forming Na^+F^-, the single $3s$ valence electron of sodium is transferred to the partially filled valence shell of fluorine:

$$\text{Na}(1s^22s^22p^63s^1) + \text{F}(1s^22s^22p^5) \longrightarrow \text{Na}^+(1s^22s^22p^6) + \text{F}^-(1s^22s^22p^6)$$

As a result of this transfer of one electron, both sodium and fluorine form ions that have the same electron configuration as neon, the noble gas nearest each in atomic number. The attraction between ions leads to a strong crystal lattice with high melting points in ionic salts such as sodium fluoride.

Covalent Bonds

A covalent bond is a chemical bond formed between atoms by the sharing of one or more pairs of electrons to give a noble gas configuration at each atom. The simplest example of a covalent bond occurs in the hydrogen molecule. When two hydrogen atoms bond, the single electrons from each combine to form an electron pair. This shared pair completes the valence shell of each hydrogen. According to the Lewis model, a pair of electrons in a covalent bond functions in two ways simultaneously: It is shared by two atoms and at the same time fills the outer (valence) shell of each. We use a line between the two hydrogens to symbolize the covalent bond formed by the sharing of a pair of electrons.

$$\text{H}\cdot + \cdot\text{H} \longrightarrow \text{H}\colon\text{H}\quad \text{Symbolized}\ \ \text{H}\!\!-\!\!\text{H}$$

$$\Delta H^0 = -435\ \text{kJ}\ (-104\ \text{kcal})/\text{mol}$$

In this pairing, a large amount of energy is released. The same amount of energy, called the bond dissociation enthalpy (BDE) would have to be absorbed to break the bond. The Lewis model accounts for the stability of two covalently bonded atoms in the following way. In forming a covalent bond, an electron pair occupies the region between two nuclei and serves to shield one positively charged nucleus from the repulsive force of the other. At the same time, the electron pair attracts both nuclei. In other words, an electron pair in the space between two nuclei bonds them together and fixes the internuclear distance to within very narrow limits. The distance between nuclei participating in a chemical bond is called the **bond length.** Every covalent bond has a characteristic bond length. In H—H, it is 74 pm (picometer; 1 pm = 10^{-12} m). We use SI units of picometers in this book; many chemists still use Å (Ångstroms); 1 pm = 0.01 Å.

Since each bond requires two electrons, that means a maximum of four bonds can form with second row atoms. For each unshared pair of electrons on an atom (called a **lone pair**), one fewer bond is possible. For third row elements (e.g., phosphorus or chlorine) five or six bonds can form.

In many situations, filled valence shells can only be satisfied when bonded atoms share more than two electrons. In these cases, multiple covalent bonds form between the same two atoms. For example, four electrons shared between two atoms form a double bond. Six shared electrons form a triple bond.

To summarize, a good way to distinguish covalent bonds from ionic attraction is that covalent bonds have defined geometries and connectivities resulting from the sharing of electrons. In other words, the number and positions of atoms taking part in covalent bonds are defined. In crystals, the position of the ions depends on the particular crystal lattice.

Bond length   The distance between nuclei in a covalent bond in picometers (pm; 1 pm = 10^{-12} m) or Å (1Å = 10^{-10} m).

Table 1.6   Classification of Chemical Bonds

Difference in Electronegativity Between Bonded Atoms	Type of Bond
Less than 0.5	Nonpolar covalent
0.5 to 1.9	Polar covalent
Greater than 1.9	Ions

Polar Covalent Bonds

Although all covalent bonds involve the sharing of electrons, they differ widely in the degree of sharing. We divide covalent bonds into two categories, polar and nonpolar, depending on the difference in electronegativity between bonded atoms (Table 1.6).

A covalent bond between carbon and hydrogen, for example, is classified as **nonpolar covalent** because the difference in electronegativity between these two atoms is $2.5 - 2.1 = 0.4$.

An example of a **polar covalent bond** is that of H—Cl. The difference in electronegativity between chlorine and hydrogen is $3.0 - 2.1 = 0.9$. An important consequence of the unequal sharing of electrons in a polar covalent bond is that the more electronegative atom gains a greater fraction of the shared electrons and acquires a partial negative charge, indicated by the symbol $\delta-$. The less electronegative atom has a smaller fraction of the shared electrons and acquires a partial positive charge, indicated by the symbol $\delta+$. Alternatively, we show the direction of bond polarity by an arrow with the arrowhead pointing toward the negative end and a plus sign on the tail of the arrow at the positive end.

$$\overset{\delta+}{H}—\overset{\delta-}{Cl} \qquad \overset{+\longrightarrow}{H—Cl}$$

Nonpolar covalent bond   A covalent bond between atoms whose difference in electronegativity is less than approximately 0.5.

Polar covalent bond   A covalent bond between atoms whose difference in electronegativity is between approximately 0.5 and 1.9.

Example 1.4

Classify each bond as nonpolar covalent, polar covalent, or state that ions are formed.

(a) O—H   **(b)** N—H   **(c)** K—Br   **(d)** C—Mg

Solution

Based on differences in electronegativity between the bonded atoms, three of these bonds are polar covalent and one involves the formation of ions.

Bond	Difference in Electronegativity	Type of Interaction
(a) O—H	$3.5 - 2.1 = 1.4$	Polar covalent
(b) N—H	$3.0 - 2.1 = 0.9$	Polar covalent
(c) K—Br	$2.8 - 0.8 = 2.0$	Ions formed
(d) C—Mg	$2.5 - 1.2 = 1.3$	Polar covalent

Problem 1.4

Classify each bond as nonpolar covalent, polar covalent, or state that ions are formed.

(a) S—H   **(b)** P—H   **(c)** C—F   **(d)** C—Cl

Electronegativity varies somewhat depending on the chemical environment and oxidation state of an atom; therefore, these rules are only guidelines and must be used with caution. Lithium iodide, for example, has a high melting point (449°C) and boiling point (1180°C) characteristic of ionic compounds. Yet, based on the difference in electronegativity between these two elements of $2.5 - 1.0 = 1.5$ units, we would classify LiI as a polar covalent compound. In particular, carbon in compounds is not atomic carbon, and it often behaves as if its electronegativity is less than shown in Table 1.6. In fact, carbon often behaves as if it were less electronegative than hydrogen.

Example 1.5

Using the symbols $\delta-$ and $\delta+$, indicate the direction of polarity in each polar covalent bond.

(a) C—O   **(b)** N—H   **(c)** C—Mg

Solution

Electronegativity is given beneath each atom. The atom with the greater electronegativity has the partial negative charge; the atom with the lesser electronegativity has the partial positive charge.

$$\overset{\delta+}{\text{(a) }}\overset{}{\text{C}}\overset{\delta-}{\text{—O}} \qquad \text{(b) } \overset{\delta-}{\text{N}}\overset{\delta+}{\text{—H}} \qquad \text{(c) } \overset{\delta-}{\text{C}}\overset{\delta+}{\text{—Mg}}$$

$$\quad 2.5 \quad 3.5 \qquad\qquad 3.0 \quad 2.1 \qquad\qquad 2.5 \quad 1.2$$

Problem 1.5

Using the symbols $\delta-$ and $\delta+$, indicate the direction of polarity in each polar covalent bond.

(a) C—N   **(b)** N—O   **(c)** C—Cl

Bond dipole moment (μ)   A measure of the polarity of a covalent bond. The product of the charge on either atom of a polar bond times the distance between the nuclei.

The polarity of a covalent bond is measured by a vector quantity called a **bond dipole moment** and is given the symbol μ (Greek mu). Bond dipole moment is defined as the product of the charge, e (either the $\delta+$ or $\delta-$ because each is the same in

Table 1.7   Average Bond Dipole Moments of Selected Covalent Bonds

Bond	Bond Dipole (D)	Bond	Bond Dipole (D)	Bond	Bond Dipole (D)
⊢→		⊢→		⊢→	
H—C	0.3	C—F	1.4	C—O	0.7
H—N	1.3	C—Cl	1.5	C=O	2.3
H—O	1.5	C—Br	1.4	C—N	0.2
H—S	0.7	C—I	1.2	C≡N	3.5

absolute magnitude), on one of its atoms times the distance, d, separating the two atoms. The SI unit for a dipole moment is the coulomb·meter, but they are commonly reported instead in a derived unit called the debye (D: 1 D = 3.34×10^{-30} C·m). Table 1.7 lists bond dipole moments for the types of covalent bonds we deal with most frequently in this course.

C. Lewis Structures for Molecules and Polyatomic Ions

The ability to write Lewis structures for molecules and polyatomic ions is a fundamental skill for the study of organic chemistry. The following guidelines will help you do this. As you study these guidelines, look at the examples in Table 1.8.

1. Determine the number of valence electrons in the molecule or ion. To do this, add the number of valence electrons contributed by each atom. For ions, add one electron for each negative charge on the ion, and subtract one electron for each positive charge on the ion. For example, the Lewis structure for a water molecule, H_2O, must show eight valence electrons: one from each hydrogen and six from oxygen. The Lewis structure for the hydroxide ion, OH^-, must also show eight valence electrons: one from hydrogen, six from oxygen, plus one for the negative charge on the ion.

2. Determine the connectivity (arrangement) of atoms in the molecule or ion. Except for the simplest molecules and ions, this connectivity must be determined experimentally because alternative possibilities can lead to the possibility of **isomers.** Isomers are different compounds with the same molecular formula. For example, there are two different compounds with the molecular formula C_4H_{10}: one with four atoms connected in a row, and one in which there are three in a row and the fourth branches off. We discuss isomers extensively in Section 2.2 and Chapter 3.

Isomers are different compounds with the same molecular formula.

$$CH_3 - CH_2 - CH_2 - CH_3 \qquad CH_3 - \overset{\displaystyle CH_3}{\underset{\displaystyle |}{CH}} - CH_3$$

The two isomers of C_4H_{10}

For some molecules and ions given as examples in the text, you are asked to propose an arrangement of atoms. For most, however, you are given the experimentally determined arrangement.

Table 1.8   Lewis Structures for Several Compounds*

H_2O (8) Water	NH_3 (8) Ammonia	CH_4 (8) Methane	HCl (8) Hydrogen chloride
C_2H_4 (12) Ethylene	C_2H_2 (10) Acetylene	CH_2O (12) Formaldehyde	H_2CO_3 (24) Carbonic acid

*The number of valence electrons is shown in parentheses.

3. Connect the atoms with single bonds. Then arrange the remaining electrons in pairs so that each atom in the molecule or ion has a complete outer shell. Each hydrogen atom must be surrounded by two electrons. Each atom of carbon, oxygen, nitrogen, and halogen must be surrounded by eight electrons (per the octet rule).

4. Each pair of electrons (**bonding electrons**) shared between two atoms is shown as a single line between the atoms. Each unshared pair of electrons (often called a **lone pair** or **nonbonding electrons**) is shown as a pair of dots.

5. If two atoms only share a single pair of electrons, they form a single bond and a single line is drawn between them. If two pairs of electrons are shared between two atoms, they form a double bond (two lines). If three pairs of electrons are shared between two atoms, they form a triple bond (three lines).

Bonding electrons Valence electrons involved in forming a covalent bond (i.e., shared electrons).

Nonbonding electrons Valence electrons not involved in forming covalent bonds. Also called unshared pairs or lone pairs.

Table 1.8 shows Lewis structures, molecular formulas, and names for several compounds. The number of valence electrons each molecule contains is shown in parentheses. Notice that, in these molecules, each hydrogen is surrounded by two valence electrons, and each carbon, nitrogen, oxygen, and chlorine is surrounded by eight valence electrons. Furthermore, each carbon has four bonds, nitrogen has three bonds and one unshared pair of electrons, oxygen has two bonds and two unshared pairs of electrons (lone pairs), and chlorine (and other halogens as well) has one bond and three unshared pairs of electrons.

Example 1.6

Draw Lewis structures, showing all valence electrons, for these molecules.

(a) CO_2 **(b)** CH_3OH **(c)** CH_3Cl

Solution

The number of valence electrons each molecule contains appears under the Lewis structure.

(a) $\ddot{O}=C=\ddot{O}$ **(b)** $H-\underset{\underset{\displaystyle H}{|}}{\overset{\overset{\displaystyle H}{|}}{C}}-\ddot{O}-H$ **(c)** $H-\underset{\underset{\displaystyle H}{|}}{\overset{\overset{\displaystyle H}{|}}{C}}-\ddot{\underset{..}{Cl}}:$

Carbon dioxide
(16 valence electrons)

Methanol
(14 valence electrons)

Chloromethane
(14 valence electrons)

Problem 1.6

Draw Lewis structures, showing all valence electrons, for these molecules.

(a) C_2H_6 **(b)** CS_2 **(c)** HCN

ORGANIC
Chemistry·⚛·Now™
Click Tutorials to practice
Assigning Formal Charge

D. Formal Charge

Throughout this course, we deal not only with molecules but also with polyatomic cations and anions. Examples of polyatomic cations are the hydronium ion, H_3O^+, and the ammonium ion, NH_4^+. An example of a polyatomic anion is the bicarbonate ion, HCO_3^-. It is important that you are able to determine which atom or atoms in a neutral molecule or polyatomic ion bear positive or negative charge. The charge on an atom in a molecule or polyatomic ion is called its **formal charge.** To derive a formal charge:

1. Write a correct Lewis structure for the molecule or ion.
2. Assign to each atom all its unshared (nonbonding) electrons and half its shared (bonding) electrons.
3. Compare this number with the number of valence electrons in the neutral, unbonded atom. If the number of electrons assigned to a bonded atom is less than that assigned to the unbonded atom, then there are more positive charges in the nucleus than counterbalancing negative charges outside the nucleus, and the atom has a positive formal charge. Conversely, if the number of electrons assigned to a bonded atom is greater than that assigned to the unbonded atom, the atom has a negative formal charge.

$$
\text{Formal charge} = \begin{pmatrix}\text{Number of valence} \\ \text{electrons in the neutral,} \\ \text{unbonded atom}\end{pmatrix} - \begin{pmatrix}\text{All unshared} \\ \text{electrons} \end{pmatrix} + \begin{pmatrix} \text{One half of all} \\ \text{shared electrons}\end{pmatrix}
$$

4. The sum of all the formal charges is equal to the total charge on the molecule.

Example 1.7

Draw Lewis structures for these ions, and show which atom in each bears the formal charge.

(a) H_3O^+ **(b)** HCO_3^-

Solution

(a) The Lewis structure for the hydronium ion must show 8 valence electrons: 3 from the three hydrogens, 6 from oxygen, minus 1 for the single positive charge. An oxygen atom has 6 valence electrons. The oxygen atom in H_3O^+ is assigned 2 unshared electrons and 1 from each shared pair of electrons, giving it a formal charge of $6 - (2 + 3) = +1$.

assigned 5 valence electrons: formal charge of +1 on O

(b) The Lewis structure for the bicarbonate ion must show 24 valence electrons: 4 from carbon, 18 from the three oxygens, 1 from hydrogen, plus 1 for the single negative charge. Loss of a hydrogen ion from carbonic acid (Table 1.8) gives the bicarbonate ion. Carbon is assigned 1 electron from each shared pair and has no formal charge $(4 - 4 = 0)$. Two oxygens are assigned 6 valence electrons each and have no formal charges $(6 - 6 = 0)$. The third oxygen is assigned 7 valence electrons and has a formal charge of $6 - (6 + 1) = -1$.

assigned 7 valence electrons: formal charge of −1

Carbonic acid, H_2CO_3 Bicarbonate ion, HCO_3^-

| **How To** | *Draw Lewis Structures from Condensed Structural Formulas* |

ORGANIC
Chemistry⚛Now™
Click How To Tutorials to review
Drawing Lewis Structures

Drawing Lewis structures from condensed structural formulas is a survival skill for organic chemistry students. There are three steps you should follow to draw a correct structure.

1. From a structural formula, obtain information about which atoms are connected to each other in a molecule. Connect all of the appropriate atoms with single bonds (single lines) first.

Example: $CH_3CH_2CH_2COOCH_3$

Comment: The difficult part of this structure is deciding how to arrange the two oxygen atoms. Using the arrangement shown will produce a stable structure with filled valences for all of the atoms after you have completed Step 3. With practice you will begin to recognize common functional groups (Section 1.3) such as the carboxylic ester group (—COOCH₃ in this example). If you are unsure, you must draw the different possibilities you are considering, and upon completing the structure, determine which one produces the stable structure with the maximum number of filled valence shells around the atoms.

2. Determine how many electrons have been used for the bonds and how many remain. Add all of the additional valence electrons for each atom that does not already have a filled valence shell due to the single bonds. Remember to assign one electron to each atom taking part in a single bond for the purpose of counting valence electrons around atoms. Make sure to keep track of any formal charges that may be present in the condensed structural formula (the present example has none).

Comment: Recall that each neutral carbon atom has four valence electrons, and each neutral oxygen atom has six valence electrons. After taking into account all of the single bonds in the molecule, the carbon atom connected to both oxygen atoms has a single electron left over (4 total electrons −3 single bonds = 1 electron left over), the oxygen atom attached only to one carbon atom has five electrons left over (6 total electrons − 1 single bond = 5 electrons left over), and the other oxygen atom has four electrons left over (6 total electrons −2 single bonds = 4 electrons left over).

3. Add multiple bonds to eliminate unpaired electrons. Draw the remaining nonbonding electrons as lone pairs.

Comment: The only unpaired electrons were on carbon and oxygen, leading to one new bond being formed.

The Lewis structure is now complete. The good news is that drawing Lewis structures will get easier. After enough practice counting valence electrons, you will begin to recognize functional groups based on the numbers of bonds and lone pairs on the atoms. For example, hydrogen has one bond and no lone pairs, neutral carbon has four bonds and no lone pairs, neutral nitrogen has three bonds and one lone pair, neutral oxygen has two bonds and two lone pairs, and neutral halogens have one bond and three lone pairs. When counting bonds for this analysis, double bonds count as two bonds, and triple bonds count as three bonds. For atoms with a formal charge, the number of bonds and lone pairs is altered.

For example, positively charged nitrogen has four bonds and no lone pairs, positively charged oxygen has three bonds and one lone pair, and positively charged carbon has three bonds and no lone pairs (carbon has an unfilled valence shell). Negatively charged carbon has three bonds and one lone pair, negatively charged nitrogen has two bonds and two lone pairs, and negatively charged oxygen has one bond and three lone pairs.

Problem 1.7

Draw Lewis structures for these ions, and show which atom in each bears the formal charge.

(a) $CH_3NH_3^+$ (b) CO_3^{2-} (c) OH^-

When writing Lewis structures for molecules and ions, you must remember that elements of the second period, including carbon, nitrogen, oxygen, and fluorine, can accommodate no more than 8 electrons in the four orbitals ($2s$, $2p_x$, $2p_y$, and $2p_z$) of their valence shells. Following are two Lewis structures for nitric acid, HNO_3, each with the correct number of valence electrons (24):

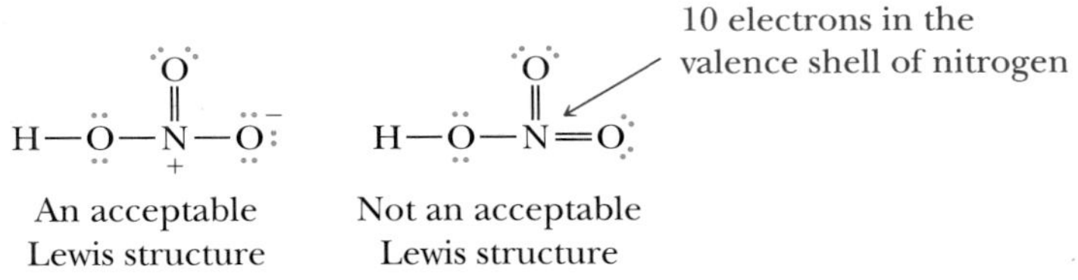

10 electrons in the valence shell of nitrogen

An acceptable Lewis structure

Not an acceptable Lewis structure

The structure on the left is an acceptable Lewis structure. It shows the required 24 valence electrons, and each oxygen and nitrogen has a completed valence shell of eight electrons. Further, it shows a positive formal charge on nitrogen and a negative formal charge on one of the oxygens. Note that the sum of the formal charges on the acceptable Lewis structure for HNO_3 is zero. The structure on the right is not an acceptable Lewis structure. Although it shows the correct number of valence electrons, it places 10 electrons in the valence shell of nitrogen.

E. Exceptions to the Octet Rule

The Lewis model of covalent bonding focuses on valence electrons and the necessity for each atom other than H participating in a covalent bond to have a completed valence shell of eight electrons. Although most molecules formed by main-group elements (Groups 1A–7A) have structures that satisfy the octet rule, there are two important exceptions to this rule.

The first group of exceptions consists of molecules containing atoms of Group 3A elements. Following is a Lewis structure for BF_3. In this uncharged covalent compound, boron is surrounded by only six valence electrons. Aluminum chloride is an example of a compound in which aluminum, the element immediately below boron in Group 3A, has an incomplete valence shell. Because their valence shells are only partially filled, trivalent compounds of boron and aluminum exhibit a

CHEMICAL CONNECTIONS

The Octet Rule

Because the octet rule of G. N. Lewis gives us a powerful and simple model for understanding bonding in organic compounds, experiments designed to prepare molecules with ten electrons in the valence shell of carbon or nitrogen atoms might seem pointless. However, because chemistry is an experimental science, the concepts such as the octet rule can be tested only by experiment. No matter how many molecules obey the octet rule, a single exception would result in major modifications to the rule, or even in its replacement.

In 1949, the German chemist Georg Wittig attempted to prepare pentamethylnitrogen, a compound with ten electrons in nitrogen's valence shell, by the following reaction:

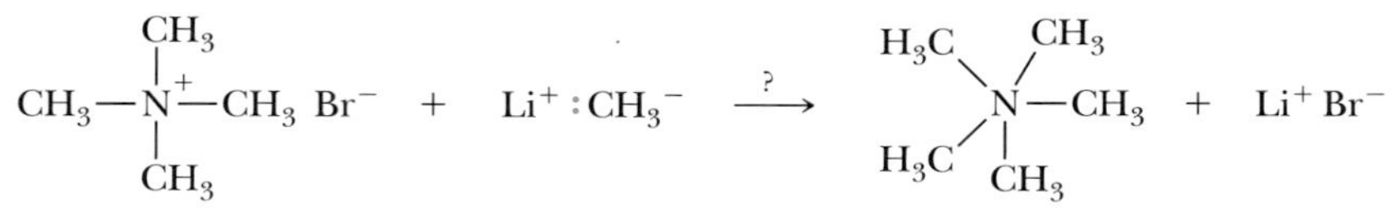

$$
\begin{array}{ccccc}
\underset{\substack{\text{Tetramethylammonium}\\\text{bromide}}}{\text{CH}_3\overset{\overset{\displaystyle\text{CH}_3}{|}}{\underset{\underset{\displaystyle\text{CH}_3}{|}}{\overset{+}{\text{N}}}}\text{CH}_3\ \text{Br}^-} & + & \underset{\text{Methyllithium}}{\text{Li}^+:\text{CH}_3{}^-} & \overset{?}{\longrightarrow} & \underset{\text{Pentamethylnitrogen}}{\text{(pentamethylnitrogen)}} & + & \text{Li}^+\ \text{Br}^-
\end{array}
$$

Instead, an acid-base reaction took place with one of the C—H bonds in the tetramethylammonium ion giving an unstable compound that has a positive charge on nitrogen and a negative charge on carbon. The curved arrows track the electron flow (this convention is explained in Section 1.6A).

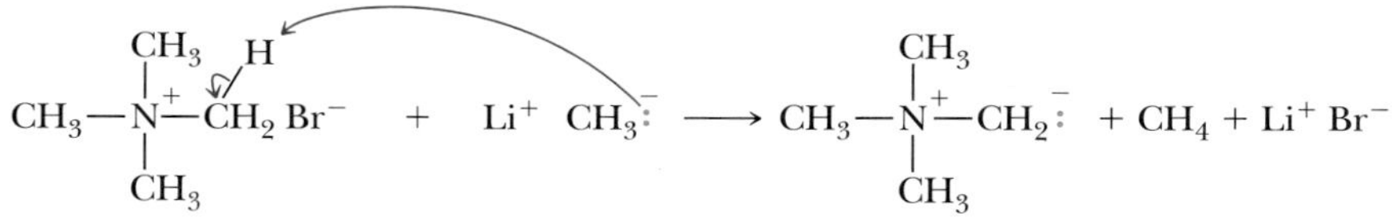

$$
\underset{\substack{\text{Tetramethylammonium}\\\text{bromide}}}{\text{CH}_3\overset{\overset{\displaystyle\text{CH}_3}{|}}{\underset{\underset{\displaystyle\text{CH}_3}{|}}{\overset{+}{\text{N}}}}\text{CH}_2\ \text{Br}^-}\ +\ \underset{\text{Methyllithium}}{\text{Li}^+\ \text{CH}_3{:}^-}\ \longrightarrow\ \underset{\text{A nitrogen ylide}}{\text{CH}_3\overset{\overset{\displaystyle\text{CH}_3}{|}}{\underset{\underset{\displaystyle\text{CH}_3}{|}}{\overset{+}{\text{N}}}}\text{CH}_2{:}^-}\ +\ \text{CH}_4 + \text{Li}^+\ \text{Br}^-
$$

high reactivity with compounds that have extra electrons, enabling them to fill their octets (Section 4.6).

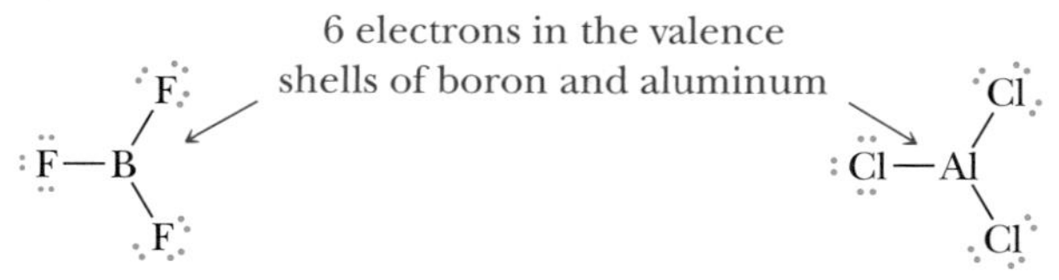

6 electrons in the valence shells of boron and aluminum

Boron trifluoride   Aluminum chloride

A second group of exceptions to the octet rule consists of molecules and ions that contain an atom with more than eight electrons in its valence shell. Atoms of second-period elements use $2s$ and $2p$ orbitals for bonding, and these orbitals can

Wittig gave this class of molecules the name ylide. This novel type of molecule had not been made before. Nitrogen ylides cannot be isolated as stable compounds, but they can be used as intermediates in other reactions.

Reasoning that phosphorus (just below nitrogen in the Periodic Table) is capable of expanding its octet and might form stable ylides, Wittig carried out an analogous reaction with phosphorus. Once again, an acid-base reaction took place, this time to form a phosphorus ylide. Phosphorus ylides, Wittig discovered, can be isolated as stable compounds, which illustrates a difference between the chemistry of nitrogen and that of phosphorus.

$$CH_3-\overset{\overset{\displaystyle CH_3}{|}}{\underset{\underset{\displaystyle CH_3}{|}}{\overset{+}{P}}}-CH_3 \; Br^- \; + \; Li^+:CH_3^- \longrightarrow CH_3-\overset{\overset{\displaystyle CH_3}{|}}{\underset{\underset{\displaystyle CH_3}{|}}{\overset{+}{P}}}-\overset{..}{C}H_2^- \; + \; Li^+ Br^- \; + \; CH_4$$

Tetramethylphosphonium   Methyllithium   A phosphorus ylide
bromide

This ylide is shown with eight electrons in the valence shell of phosphorus. It can also be written with the valence shell of phosphorus expanded to accommodate ten electrons.

$$CH_3-\overset{\overset{\displaystyle CH_3}{|}}{\underset{\underset{\displaystyle CH_3}{|}}{P}}=CH_2$$

An alternative prepresentation
for a phosphorus ylide

Wittig soon abandoned his attempts to make compounds with five bonds to nitrogen and instead studied the newly discovered phosphorus ylides (Section 16.6). He found that these ylides are extraordinarily useful reagents for preparing complex organic molecules, including such important compounds as vitamin A. Wittig shared the 1979 Nobel prize for chemistry for his discovery and work with phosphorus ylides.

contain only eight valence electrons, hence the octet rule. Atoms of third-period elements have $3d$ orbitals and may expand their valence shells to accommodate more than eight electrons. Following are Lewis structures for trimethylphosphine, phosphorus pentachloride, and phosphoric acid. The first compound has eight electrons in the valence shell of phosphorus; the second and third compounds have ten electrons in the valence shell of phosphorus.

$$CH_3-\overset{\overset{\displaystyle ..}{P}}{\underset{\underset{\displaystyle CH_3}{|}}{|}}-CH_3$$

$$\overset{\displaystyle :\overset{..}{Cl}:}{\underset{\displaystyle :\overset{..}{Cl}.\quad :\overset{..}{Cl}:}{:\overset{..}{Cl}\diagdown \underset{|}{P}\diagup\overset{..}{Cl}:}}$$

$$\overset{\displaystyle :\overset{..}{O}:}{H-\overset{..}{O}-\underset{\underset{\displaystyle :\overset{..}{O}-H}{|}}{\overset{||}{P}}-\overset{..}{O}-H}$$

Trimethylphosphine   Phosphorus pentachloride   Phosphoric acid

Sulfur, another third-period element, forms compounds in which its valence shell contains 8, 10, or 12 electrons.

$$H-\overset{..}{\underset{..}{S}}-H \qquad CH_3-\overset{\overset{..}{\overset{\displaystyle O}{\parallel}}}{\underset{..}{S}}-CH_3 \qquad H-\overset{..}{\underset{..}{O}}-\overset{\overset{..}{\overset{\displaystyle O}{\parallel}}}{\underset{\underset{..}{\overset{..}{O}}}{S}}-\overset{..}{\underset{..}{O}}-H$$

Hydrogen sulfide   Dimethyl sulfoxide   Sulfuric acid

1.3 Functional Groups

Functional group An atom or group of atoms within a molecule that shows a characteristic set of physical and chemical properties.

Carbon combines with other atoms (e.g., H, N, O, S, halogens) to form structural units called **functional groups.** Functional groups are important for three reasons. First, they are the units by which we divide organic compounds into classes. Second, they are sites of characteristic chemical reactions; a particular functional group, in whatever compound it is found, undergoes the same types of chemical reactions. Third, functional groups serve as a basis for naming organic compounds.

We introduce here several of the functional groups we encounter early in this course. At this point, our concern is only with pattern recognition. We shall have more to say about the structure and properties of these functional groups in following chapters. A complete list of the major functional groups we study in this text is presented on the inside front cover of the text.

A. Alcohols

Alcohol A compound containing an —OH (hydroxyl) group bonded to a tetrahedral carbon atom.

Hydroxyl group An —OH group.

The functional group of an **alcohol** is an —OH (**hydroxyl**) group bonded to a tetrahedral carbon atom (a carbon having single bonds to four other atoms). Here is the Lewis structure of ethanol.

$$-\overset{|}{\underset{|}{C}}-\overset{..}{O}-H \qquad\qquad H-\overset{\overset{\displaystyle H}{|}}{\underset{\underset{\displaystyle H}{|}}{C}}-\overset{\overset{\displaystyle H}{|}}{\underset{\underset{\displaystyle H}{|}}{C}}-\overset{..}{O}-H$$

Functional   An alcohol
group   (Ethanol)

We can also represent this alcohol in a more abbreviated form called a condensed structural formula. In a **condensed structural formula,** CH_3 indicates a carbon with three attached hydrogens, CH_2 indicates a carbon with two attached hydrogens, and CH indicates a carbon with one attached hydrogen. Unshared pairs of electrons are generally not shown in a condensed structural formula. Thus the condensed structural formula for the alcohol with molecular formula C_2H_6O is $CH_3—CH_2—OH$. It is also common to write these formulas in an even more condensed manner, by omitting all single bonds: CH_3CH_2OH.

$$H-\overset{\overset{\displaystyle H}{|}}{\underset{\underset{\displaystyle H}{|}}{C}}-\overset{\overset{\displaystyle H}{|}}{\underset{\underset{\displaystyle H}{|}}{C}}-\overset{..}{O}H \qquad CH_3-CH_2-OH \qquad CH_3CH_2OH$$

Lewis structure   Condensed   Fully
                     structural formula   condensed
                                         structural formula

Alcohols are classified as **primary** (1°), **secondary** (2°), or **tertiary** (3°) depending on the number of carbon atoms bonded to the carbon bearing the —OH group.

$$CH_3-\underset{\underset{H}{|}}{\overset{\overset{H}{|}}{C}}-OH \qquad CH_3-\underset{\underset{CH_3}{|}}{\overset{\overset{H}{|}}{C}}-OH \qquad CH_3-\underset{\underset{CH_3}{|}}{\overset{\overset{CH_3}{|}}{C}}-OH$$

A 1° alcohol   A 2° alcohol   A 3° alcohol

Primary (1°) A compound containing a functional group bonded to a carbon atom bonded to only one other carbon atom and two hydrogens.

Secondary (2°) A compound containing a functional group bonded to a carbon atom bonded to two other carbon atoms.

Tertiary (3°) A compound containing a functional group bonded to a carbon atom bonded to three other carbon atoms.

Example 1.8

Draw Lewis structures and condensed structural formulas for the two alcohols with molecular formula C_3H_8O. Classify each as primary, secondary, or tertiary.

Solution

Begin by drawing the three carbon atoms in a chain. The oxygen atom of the hydroxyl group may be bonded to the carbon chain in two ways: either to an end carbon or to the middle carbon.

C—C—C   C—C—C—OH   C—$\overset{\overset{OH}{|}}{C}$—C

The carbon chain   The two locations for the OH group

Finally, add seven more hydrogens for a total of eight shown in the molecular formula. Show unshared electron pairs on the Lewis structures but not on the condensed structural formulas.

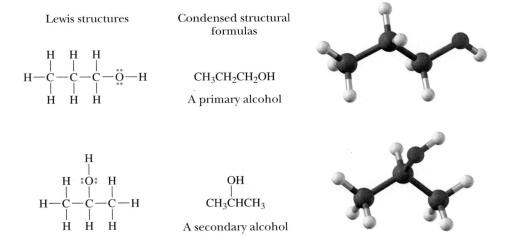

Lewis structures   Condensed structural formulas

$CH_3CH_2CH_2OH$
A primary alcohol

$CH_3\overset{\overset{OH}{|}}{C}HCH_3$
A secondary alcohol

The secondary alcohol, whose common name is isopropyl alcohol, is the major component in rubbing alcohol. We also describe other functional groups such as halides as primary, secondary, and tertiary.

$$H_3C-\underset{\underset{CH_3}{|}}{\overset{\overset{CH_3}{|}}{C}}-\ddot{C}l: \quad \text{A tertiary halide}$$

Problem 1.8

Draw Lewis structures and condensed structural formulas for the four alcohols with molecular formula $C_4H_{10}O$. Classify each alcohol as primary, secondary, or tertiary.

B. Amines

Amino group A compound containing an sp^3 hybridized nitrogen atom bonded to one, two, or three carbon atoms by single bonds.

Primary (1°) amine An amine in which nitrogen is bonded to one carbon and two hydrogens.

Secondary (2°) amine An amine in which nitrogen is bonded to two carbons and one hydrogen.

Tertiary (3°) amine An amine in which nitrogen is bonded to three carbons.

The functional group of an amine is an **amino group,** a nitrogen atom bonded to one, two, or three carbon atoms by single bonds. In a **primary (1°) amine,** nitrogen is bonded to one carbon atom. In a **secondary (2°) amine,** it is bonded to two carbon atoms, and in a **tertiary (3°) amine,** it is bonded to three carbon atoms. Notice that this classification scheme is different from that used with alcohols and halides!

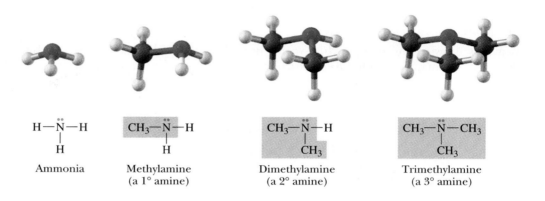

$H-\overset{..}{\underset{\underset{H}{\mid}}{N}}-H$	$CH_3-\overset{..}{\underset{\underset{H}{\mid}}{N}}-H$	$CH_3-\overset{..}{\underset{\underset{CH_3}{\mid}}{N}}-H$	$CH_3-\overset{..}{\underset{\underset{CH_3}{\mid}}{N}}-CH_3$
Ammonia	Methylamine (a 1° amine)	Dimethylamine (a 2° amine)	Trimethylamine (a 3° amine)

Example 1.9

Draw condensed structural formulas for the two primary amines with molecular formula C_3H_9N.

Solution

For a primary amine, draw a nitrogen atom bonded to two hydrogens and one carbon.

$C-C-C-\overset{..}{\underset{\underset{H}{\mid}}{N}}-H$ $C-\overset{\overset{\underset{\mid}{C}}{}}{\underset{\underset{H}{\mid}}{C}}-\overset{..}{\underset{\underset{H}{\mid}}{N}}-H$

The three carbons may be bonded to nitrogen in two ways.

$CH_3CH_2CH_2-\overset{..}{\underset{\underset{H}{\mid}}{N}}-H$ $CH_3\overset{\overset{\underset{\mid}{CH_3}}{}}{\underset{}{CH}}-\overset{..}{\underset{\underset{H}{\mid}}{N}}-H$

Add seven hydrogens to give each carbon four bonds and give the correct molecular formula.

Problem 1.9

Draw structural formulas for the three secondary amines with molecular formula $C_4H_{11}N$.

C. Aldehydes and Ketones

The functional group of both aldehydes and ketones is the **C=O (carbonyl) group.** In formaldehyde, CH_2O, the simplest **aldehyde,** the carbonyl carbon is bonded to two hydrogens. In all other aldehydes, it is bonded to one hydrogen and one carbon. In a condensed structural formula, the aldehyde group may be written showing the carbon-oxygen double bond as —CH=O; alternatively, it may be written —CHO. In a **ketone,** the carbonyl carbon is bonded to two carbon atoms.

Carbonyl group A C=O group.

Aldehyde A compound containing a —CHO group.

Ketone A compound containing a carbonyl group bonded to two carbons.

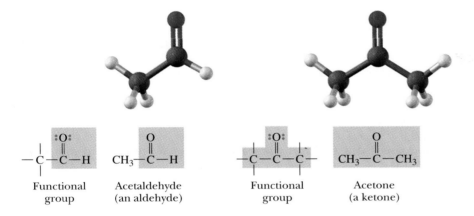

| Functional group | Acetaldehyde (an aldehyde) | Functional group | Acetone (a ketone) |

Example 1.10

Draw condensed structural formulas for the two aldehydes with molecular formula C_4H_8O.

Solution

First, draw the functional group of an aldehyde and then add the remaining carbons. These may be attached in two ways. Then, add seven hydrogens to complete the four bonds to each carbon.

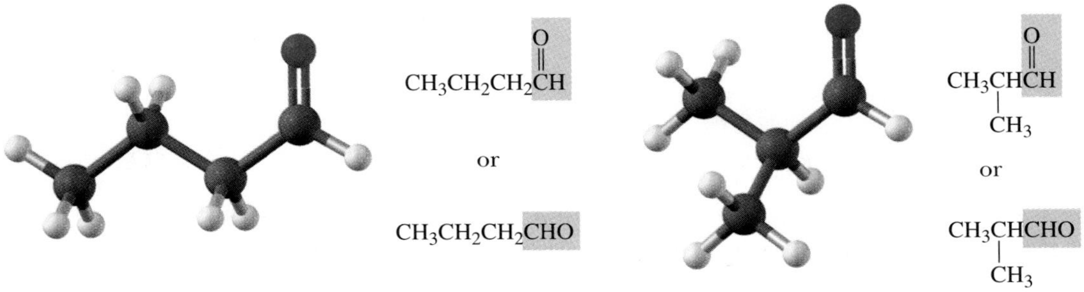

$$CH_3CH_2CH_2\overset{\displaystyle O}{\overset{\displaystyle \|}{C}}H$$

or

$$CH_3CH_2CH_2CHO$$

$$CH_3\overset{\displaystyle O}{\overset{\displaystyle \|}{C}}HCH$$
$$\quad\ |$$
$$\quad CH_3$$

or

$$CH_3CHCHO$$
$$\quad\ |$$
$$\quad CH_3$$

Problem 1.10

Draw condensed structural formulas for the three ketones with molecular formula $C_5H_{10}O$.

D. Carboxylic Acids

Carboxylic acid   A compound containing a carboxyl, —COOH, group.

Carboxyl group   A —COOH group.

The functional group of a **carboxylic acid** is a **—COOH** (**carboxyl:** *carb*onyl + hydr*oxyl*) **group.** In a condensed structural formula, a carboxyl group may also be written —CO_2H.

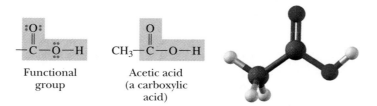

Functional
group

Acetic acid
(a carboxylic
acid)

Example 1.11

Draw a condensed structural formula for the single carboxylic acid with molecular formula $C_3H_6O_2$.

Solution

The only way the carbon atoms can be written is three in a chain, and the —COOH group must be on an end carbon of the chain.

Problem 1.11

Draw condensed structural formulas for the two carboxylic acids with molecular formula $C_4H_8O_2$.

E. Carboxylic Esters

Carboxylic ester   A derivative of a carboxylic acid in which H of the carboxyl group is replaced by a carbon.

A **carboxylic ester,** commonly referred to as an **ester,** is a derivative of a carboxylic acid in which the hydrogen of the carboxyl group is replaced by a carbon group. This group is written —COO— in this text.

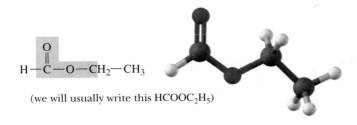

Functional group   Methyl acetate (an ester)

Example 1.12

The molecular formula of methyl acetate is $C_3H_6O_2$. Draw the structural formula of another ester of this same molecular formula.

Solution

There is only one other ester of this molecular formula. Its structural formula is

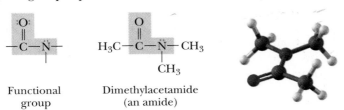

$$H-\overset{\overset{\displaystyle O}{\|}}{C}-O-CH_2-CH_3$$

(we will usually write this $HCOOC_2H_5$)

Problem 1.12

Draw structural formulas for the four esters with molecular formula $C_4H_8O_2$.

F. Carboxylic Amides

A **carboxylic amide,** commonly referred to as an **amide,** is a derivative of a carboxylic acid in which the hydrogen of the carboxyl group is replaced by an amine. As the model shows, the group is planar.

$$-\overset{\overset{\displaystyle :O:}{\|}}{C}-\overset{|}{\underset{|}{N}}-$$

$$H_3C-\overset{\overset{\displaystyle O}{\|}}{C}-\overset{}{\underset{\underset{\displaystyle CH_3}{|}}{N}}-CH_3$$

Functional group   Dimethylacetamide (an amide)

1.4 Bond Angles and Shapes of Molecules

In Section 1.2, we used a shared pair of electrons as the fundamental unit of a covalent bond and drew Lewis structures for several molecules and ions containing various combinations of single, double, and triple bonds. We can predict bond angles in

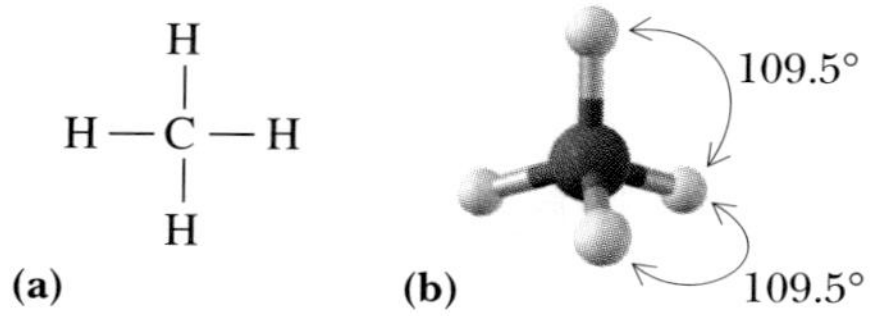

Figure 1.2
A methane molecule, CH_4.
(a) Lewis structure and
(b) shape.

H—C—H (a)

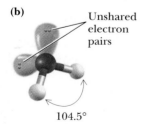

(b)   109.5°   109.5°

(a)

H—N—H

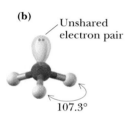

(b)   Unshared electron pair

107.3°

Figure 1.3
An ammonia molecule, NH_3.
(a) Lewis structure and
(b) shape.

(a)

H—O—H

(b)   Unshared electron pairs

104.5°

Figure 1.4
A water molecule, H_2O.
(a) Lewis structure and
(b) shape.

these and other molecules and ions in a very straightforward way using a concept referred to as **valence-shell electron-pair repulsion (VSEPR).** VSEPR is based on the electrons in an atom's valence shell. These valence electrons may be involved in the formation of single, double, or triple bonds, or they may be unshared (lone pair). Each combination creates a negatively charged region of space, and, because "like" charges repel each other, the various regions of electron density around an atom will spread out so that each is as far away from the others as possible.

We use VSEPR in the following way to predict the shape of a methane molecule, CH_4. The Lewis structure for CH_4 shows a carbon atom surrounded by four regions of electron density, each of which contains a pair of electrons forming a bond to a hydrogen atom. According to VSEPR, the four regions radiate from carbon so that they are as far away from each other as possible. This occurs when the angle between any two pairs of electrons is 109.5°. Therefore, we predict all H—C—H bond angles to be 109.5°, and the shape of the molecule to be **tetrahedral** (Figure 1.2). The H—C—H bond angles in methane have been measured experimentally and found to be 109.5°, identical to those predicted.

We predict the shape of an ammonia molecule, NH_3, in exactly the same manner. The Lewis structure of NH_3 shows nitrogen surrounded by four regions of electron density. Three regions contain single pairs of electrons forming covalent bonds with hydrogen atoms. The fourth region contains an unshared pair of electrons (Figure 1.3). Using VSEPR, we predict that the four regions of electron density around nitrogen are arranged in a tetrahedral manner, that all H—N—H bond angles are 109.5°, and that the shape of the molecule is **pyramidal** (like a triangular pyramid). The observed bond angles are 107.3°. This small difference between the predicted and observed angles can be explained by proposing that the unshared pair of electrons on nitrogen repels adjacent electron pairs more strongly than do bonding pairs.

Figure 1.4 shows a Lewis structure and a ball-and-stick model of a water molecule. In H_2O, oxygen is surrounded by four regions of electron density. Two of these regions contain pairs of electrons used to form single covalent bonds to the two hydrogens; the remaining two contain unshared electron pairs. Using VSEPR, we predict that the four regions of electron density around oxygen repel each other and are arranged in a tetrahedral manner. The predicted H—O—H bond angle is 109.5°.

Experimental measurements show that the actual bond angle is 104.5°, a value smaller than that predicted. This difference between the predicted and observed bond angles can be explained by proposing, as we did for NH_3, that unshared pairs of electrons repel adjacent pairs more strongly than do bonding pairs. Note that the distortion from 109.5° is greater in H_2O, which has two unshared pairs of electrons, than it is in NH_3, which has only one unshared pair.

A general prediction emerges from this discussion of the shapes of CH_4, NH_3, and H_2O molecules. If a Lewis structure shows four regions of electron density

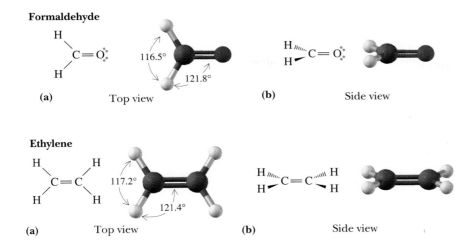

Formaldehyde

(a) Top view   (b) Side view

Ethylene

(a) Top view   (b) Side view

Figure 1.5
Shapes of formaldehyde,
CH_2O, and ethylene, C_2H_4.
Molecules shown from
(a) top view and (b) side view.

around a central atom, VSEPR predicts a tetrahedral distribution of electron density and bond angles of approximately 109.5°.

In many of the molecules we shall encounter, an atom is surrounded by three regions of electron density. Figure 1.5 shows Lewis structures and ball-and-stick models for formaldehyde, CH_2O, and ethylene, C_2H_4.

According to VSEPR, a double bond is treated as a single region of electron density. In formaldehyde, carbon is surrounded by three regions of electron density: two regions contain single pairs of electrons forming single bonds to hydrogen atoms; the third region contains two pairs of electrons forming a double bond to oxygen. In ethylene, each carbon atom is also surrounded by three regions of electron density: two contain single pairs of electrons, and the third contains two pairs of electrons.

Three regions of electron density about an atom are farthest apart when they are coplanar (in the same plane) and make angles of 120° with each other. Thus, the predicted H—C—H and H—C—O bond angles in formaldehyde and the predicted H—C—H and H—C—C bond angles in ethylene are all 120° and the atoms are coplanar.

In still other types of molecules, a central atom is surrounded by only two regions of electron density. Figure 1.6 shows Lewis structures and ball-and-stick models of carbon dioxide, CO_2, and acetylene, C_2H_2.

In carbon dioxide, carbon is surrounded by two regions of electron density: each contains two pairs of electrons and forms a double bond to an oxygen atom. In acetylene, each carbon is also surrounded by two regions of electron density. One contains a single pair of electrons and forms a single bond to a hydrogen atom, and the other contains three pairs of electrons and forms a triple bond to a carbon atom. In each case, the two regions of electron density are farthest apart if they form a straight line through the central atom and create an angle of 180°. Both carbon dioxide and acetylene are **linear** molecules.

Predictions of VSEPR are summarized in Table 1.9. In these three-dimensional drawings, a solid line indicates a bond in the plane of the paper. A solid wedge indicates a bond projecting out of the plane toward the reader, and a broken wedge indicates a bond projecting behind the plane away from the reader.

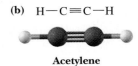

(a)

Carbon dioxide

(b)   H—C≡C—H

Acetylene

Figure 1.6
Shapes of (a) carbon dioxide,
CO_2, and (b) acetylene, C_2H_2,
molecules.

Table 1.9 Predicted Molecular Shapes (VSEPR)

Regions of Electron Density Around Central Atom	Predicted Distribution of Electron Density	Predicted Bond Angles	Examples
4	Tetrahedral	109.5°	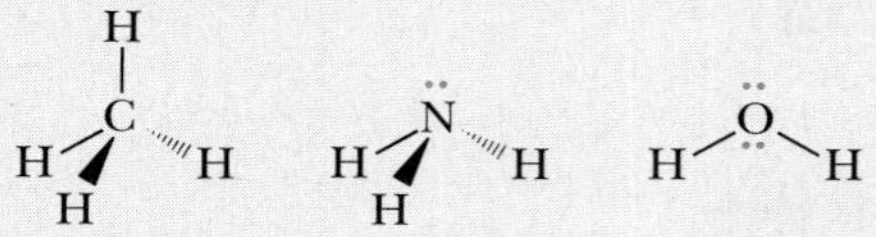
3	Trigonal planar	120°	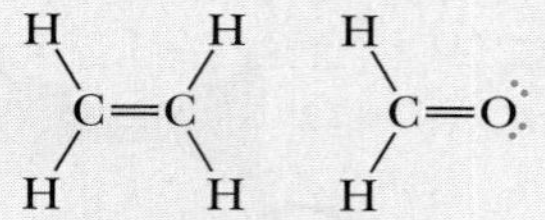
2	Linear	180°	

Example 1.13

Predict all bond angles in these molecules.

(a) CH_3Cl (b) $CH_2{=}CHCl$

Solution

(a) The Lewis structure for CH_3Cl shows carbon surrounded by four regions of electron density. Therefore, we predict the distribution of electron pairs about carbon to be tetrahedral, all bond angles to be 109.5°, and the shape of CH_3Cl to be tetrahedral. The actual H—C—Cl bond angle is 108°.

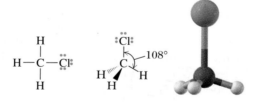

(b) The Lewis structure for $CH_2{=}CHCl$ shows each carbon surrounded by three regions of electron density. Therefore, we predict all bond angles to be 120°. The actual C—C—Cl bond angle is 122.4°.

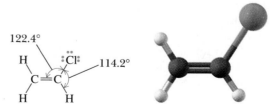

Problem 1.13

Predict all bond angles for these molecules.

(a) CH_3OH (b) PF_3 (c) H_2CO_3

Fullerene—A New Form of Carbon

A favorite chemistry examination question is: What are the elemental forms of carbon? For a long time, the answer was that pure carbon is found in two forms: graphite and diamond. These forms have been known for centuries, and it was generally believed that they are the only forms of carbon having extended networks of C atoms in well-defined structures. But that is not so! The scientific world was startled in 1985 when Richard Smalley of Rice University and Harry W. Kroto of the University of Sussex, UK, and their coworkers announced that they had detected a new form of carbon with the molecular formula C_{60}. They suggested that the molecule has a structure that resembles a soccer ball; it has 12 five-membered rings and 20 six-membered rings arranged such that each five-membered ring is surrounded by 5 six-membered rings. This structure reminded its discoverers of a geodesic dome, a structure invented by the innovative American engineer and philosopher R. Buckminster Fuller. Therefore, the official name of this allotrope of carbon has become fullerene. Most chemists refer to it as C_{60} or fullerene; some call it "buckyball." Kroto, Smalley, and Robert F. Curl were awarded the 1996 Nobel prize in chemistry for this work. Many higher fullerenes, such as C_{70} and C_{84}, have also been isolated and studied.

Fullerenes have a rich chemistry. They behave as if they have electron-deficient double bonds. Many different fullerene adducts with a variety of structures have been prepared. Cationic derivatives of these adducts, for example, bind tightly to DNA and can be used to visualize it by electron microscopy.

An astonishing recent development in this field has been the preparation of single-wall nanotubes, which are based on C_{60} or higher fullerenes and are extended for a very long distance to make long molecules that are hundreds of times stronger than steel and can act as molecular wires. Nanotubes can be used as the probe (a very sharp tip) in atomic force microscopes, which can be used to image single molecules. The nanotubes make the sharpest possible tips because they are of molecular dimensions. Nanotubes are also playing an increasingly important role in the new field of molecular electronics. They can be superconducting or semiconducting, depending on the exact structure of the nanotube.

■ A fullerene (C_{60}) molecule.

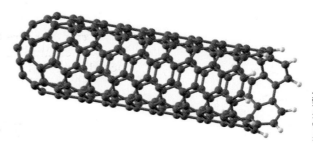

■ A nanotube.

Yves Rubin, UCLA

1.5 Polar and Nonpolar Molecules

We can now combine our understanding of bond polarity (Section 1.2B) and molecular geometry (Section 1.4) to predict the polarity of polyatomic molecules. As we shall see, to be polar, a molecule must have one or more polar bonds. But, as we shall also see, not every molecule with polar bonds is polar.

To predict whether a molecule is polar, we need to determine (1) if the molecule has polar bonds and (2) the arrangement of its atoms in space. The **molecular dipole moment (μ)** of a molecule is the vector sum of its individual bond dipoles. In carbon dioxide, for example, each C—O bond is polar with oxygen, the more electronegative atom, bearing a partial negative charge and with carbon bearing a partial positive charge. Because carbon dioxide is a linear molecule, the vector sum of its two bond dipoles is zero; therefore, the dipole moment of a CO_2 molecule is zero. Boron trifluoride is planar with bond angles of 120°. Although each B—F bond is polar, the vector sum of its bond dipoles is zero, and BF_3 has no dipole moment. Carbon tetrachloride is tetrahedral with bond angles of 109.5°. Although it has four polar C—Cl bonds, the vector sum of its bond dipoles is zero, and CCl_4 also has no dipole moment.

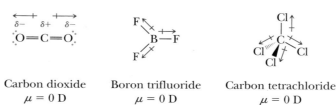

Carbon dioxide	Boron trifluoride	Carbon tetrachloride
$\mu = 0\ \text{D}$	$\mu = 0\ \text{D}$	$\mu = 0\ \text{D}$

Other molecules, such as water and ammonia, have polar bonds and dipole moments greater than zero; they are polar molecules. Each O—H bond in a water molecule and each N—H bond in ammonia are polar, with oxygen and nitrogen, the more electronegative atoms, bearing a partial negative charge and each hydrogen bearing a partial positive charge.

The charge densities are easily computed by modern desktop computer programs such as Spartan. Here are electrostatic potential maps that display the computed electronic charge density in water and ammonia. In these models, red represents negative charge and blue represents positive charge. In agreement with the dipole moment diagram and our expectations, the more electronegative atom has substantial negative charge in both molecules. For more information on how to interpret these plots, see Appendix 7.

Molecular dipole moment (μ)
The vector sum of individual bond dipoles.

ORGANIC
Chemistry Now™
Click Tutorials to examine **How to Determine Molecular Polarity**

direction
of dipole
moment
in water

Water
$\mu = 1.85\ \text{D}$

■ An electrostatic potential map of a water molecule.

direction
of dipole
moment
in ammonia

Ammonia
$\mu = 1.47\ \text{D}$

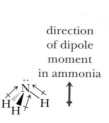

■ An electrostatic potential map of an ammonia molecule.

Example 1.14

Which of these molecules are polar? For each that is polar, specify the direction of its dipole moment.

(a) CH_3Cl **(b)** CH_2O **(c)** C_2H_2

Solution

Both chloromethane, CH_3Cl, and formaldehyde, CH_2O, have polar bonds and, because of their geometries, are polar molecules. Because of its linear geometry, acetylene, C_2H_2, has no dipole moment. The experimentally measured dipole moments are shown. The electrostatic potential map (elpot) of formaldehyde clearly shows this charge distribution.

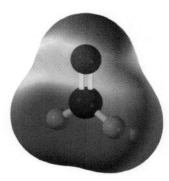

■ An electrostatic potential map of a formaldehyde molecule.

(a)   **(b)**   **(c)** $H—C\equiv C—H$

Chloromethane	Formaldehyde	Acetylene
$\mu = 1.87\ D$	$\mu = 2.33\ D$	$\mu = 0\ D$

Problem 1.14

Which molecules are polar? For each that is polar, specify the direction of its dipole moment.

(a) CH_2Cl_2 **(b)** HCN **(c)** H_2O_2

1.6 Resonance

As chemists developed a deeper understanding of covalent bonding in organic compounds, it became obvious that, for a great many molecules and ions, no *single* Lewis structure provides a truly accurate representation. For example, Figure 1.7 shows three Lewis structures for the carbonate ion, $CO_3{}^{2-}$, each of which shows carbon bonded to three oxygen atoms by a combination of one double bond and two single bonds. Each Lewis structure implies that one carbon-oxygen bond is different from the other two. However, this is not the case. All three carbon-oxygen bonds are identical.

The problem for chemists, then, was how to describe the structure of molecules and ions for which no single Lewis structure is adequate and yet still retain Lewis structures. As an answer to this problem, Linus Pauling proposed the theory of resonance.

A. Theory of Resonance

The **theory of resonance** was developed primarily by Pauling in the 1930s. According to this theory, many molecules and ions are best described by writing two or more Lewis structures and considering the real molecule or ion to be a composite of these structures. Individual Lewis structures are called **contributing structures.**

(a)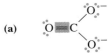

(b)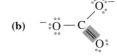

(c)

Figure 1.7
(a–c) Three Lewis structures for the carbonate ion.

Resonance   A theory that many molecules are best described as a hybrid of several Lewis structures.

Contributing structures   Representations of a molecule or ion that differ only in the distribution of valence electrons.

How To   *Draw Curved Arrows and Push Electrons*

ORGANIC
Chemistry ·Now™
Click How To Tutorials to review
Drawing Curved Arrows and Pushing Electrons

Curved arrow   A symbol used to show the redistribution of valence electrons in resonance contributing structures or reaction mechanisms, symbolizing movement of two electrons.

Resonance hybrid   A molecule, ion, or radical described as a composite of a number of contributing structures.

Double-headed arrow   A symbol used to show that structures on either side are resonance contributing structures.

Notice in Figure 1.8 that the only difference among contributing structures (a), (b), and (c) is the position of valence electrons. To generate one resonance structure from another, chemists use a symbol called a **curved arrow.** The arrow indicates where a pair of electrons originates (the tail of the arrow) and where it is positioned in the next structure (the head of the arrow).

A curved arrow is nothing more than a bookkeeping symbol for keeping track of electron pairs, or, as some call it, **electron pushing.** Do not be misled by its simplicity. Electron pushing will help you see the relationship among contributing structures. Later in the course, it will help you follow bond-breaking and bond-forming steps in organic reactions. Stated directly, electron pushing is a survival skill in organic chemistry.

Following are contributing structures for the nitrite and acetate ions. Curved arrows show how the contributing structures are interconverted. For each ion, the contributing structures are equivalent.

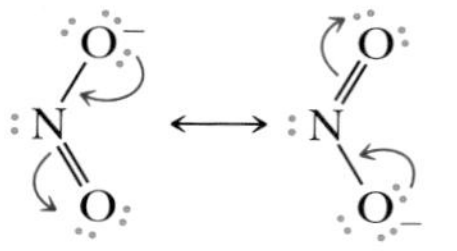

Nitrite ion
(equivalent contributing structures)

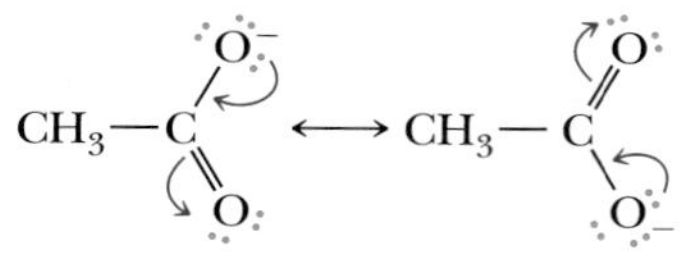

Acetate ion
(equivalent contributing structures)

A common mistake is to use curved arrows to indicate the movement of atoms or positive charges. This is never correct. Curved arrows must be used only to show the repositioning of electron pairs, when a new resonance hybrid is generated.

They are also sometimes referred to as **resonance structures** or **resonance contributors.** We show that the real molecule or ion is a **resonance hybrid** of the various contributing structures by interconnecting them with **double-headed arrows.** Do not confuse the double-headed arrow with the double arrow used to show chemical equilibrium. As we explain shortly, resonance structures are not in equilibrium with each other.

Three contributing structures for the carbonate ion are shown in Figure 1.8. These three contributing structures are said to be equivalent; they have identical patterns of covalent bonding.

The use of the term "resonance" for this theory of covalent bonding might suggest to you that bonds and electron pairs are constantly changing back and forth from one position to another over time. This notion is not at all correct. The carbonate ion, for example, has one and only one real structure. The problem is ours—how do we draw that one real structure? The resonance method is a way to describe the real structure and at the same time retain Lewis structures with electron-pair bonds. Thus, although we realize that the carbonate ion is not accurately represented by any one contributing structure shown in Figure 1.8, we continue to represent it by one of these for convenience. We understand, of course, that what is intended is the resonance hybrid.

(a) **(b)** **(c)**

Figure 1.8
(a–c) The carbonate ion represented as a resonance hybrid of three equivalent contributing structures. Curved arrows show the redistribution of valence electrons between one contributing structure and the next.

■ An electrostatic potential map of a carbonate ion shows that the negative charge is distributed equally among the three oxygens.

Example 1.15

Draw the contributing structure indicated by the curved arrow(s). Be certain to show all valence electrons and all formal charges.

(a) **(b)** **(c)**

Solution

(a) **(b)** **(c)**

Problem 1.15

Draw the contributing structure indicated by the curved arrows. Be certain to show all valence electrons and all formal charges.

(a) $H-C-O$ ⟷ **(b)** $H-C-O$ ⟷ **(c)** $CH_3-C-O-CH_3$ ⟷

B. Rules for Writing Acceptable Contributing Structures

Certain rules must be followed in writing acceptable contributing structures:

1. All contributing structures must have the same number of valence electrons.
2. All contributing structures must obey the rules of covalent bonding; no contributing structure may have more than two electrons in the valence shell of hydrogen or more than eight electrons in the valence shell of a second-period element. Third-period elements, such as phosphorus and sulfur, may have up to 12 electrons in their valence shells.
3. The positions of all nuclei must be the same in all resonance structures; that is, contributing structures differ only in the distribution of valence electrons.
4. All contributing structures must have the same number of paired and unpaired electrons.

Example 1.16

Which sets are valid pairs of contributing structures?

(a) $CH_3-\overset{\overset{\ddot{O}\cdot}{\|}}{C}-CH_3$ and $CH_3-\overset{\overset{:\ddot{O}:^-}{|}}{\underset{+}{C}}-CH_3$ (b) $CH_3-\overset{\overset{\cdot\ddot{O}\cdot}{\|}}{C}-CH_3$ and $CH_2=\overset{\overset{:\ddot{O}-H}{|}}{C}-CH_3$

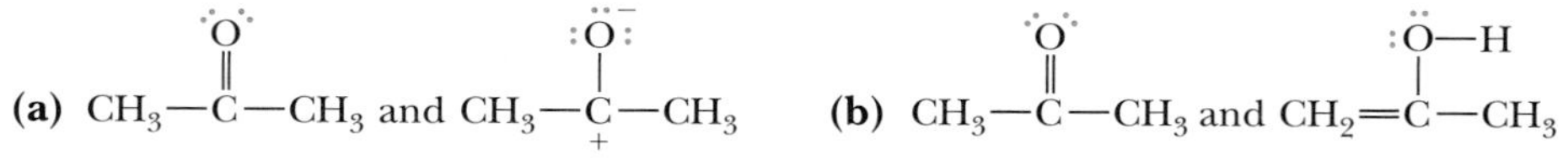

Solution

(a) These are valid contributing structures. They differ only in the distribution of valence electrons.
(b) These are not valid contributing structures. They differ in the connectivity of their atoms.

Problem 1.16

Which sets are valid pairs of contributing structures?

(a) $CH_3-\overset{\overset{\ddot{O}:}{\diagup}}{\underset{\diagdown\ddot{O}:^-}{C}}$ and $CH_3-\overset{\overset{\cdot\ddot{O}:^-}{\diagup}}{\underset{\diagdown\ddot{O}:^-}{\overset{+}{C}}}$ (b) $CH_3-\overset{\overset{\ddot{O}:}{\diagup}}{\underset{\diagdown\ddot{O}:^-}{C}}$ and $CH_3-\overset{\overset{\ddot{O}:}{\diagup}}{\underset{\diagdown\ddot{O}:}{C}}$

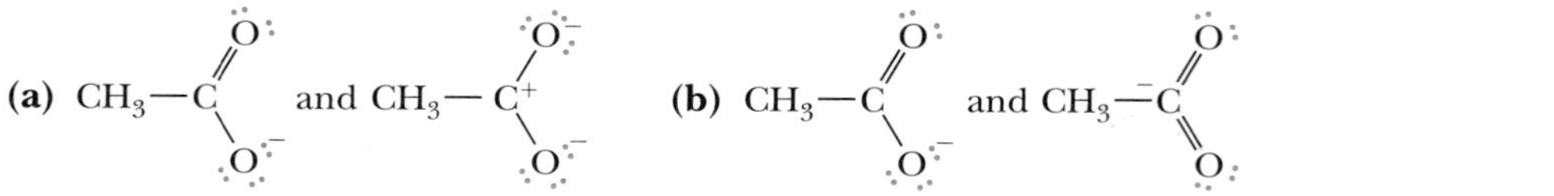

C. Estimating the Relative Importance of Contributing Structures

Not all structures contribute equally to a hybrid. The following preferences will help you to estimate the relative importance of various contributing structures. In fact, structures can be ranked by the number of preferences they follow. Those that follow the most preferences contribute most to the hybrid. Any structure that violates all four of these preferences can be ignored and never written.

Preference 1: Filled Valence Shells

Structures in which all atoms have filled valence shells (completed octets) contribute more than those in which one or more valence shells are unfilled. For example, the following are the contributing structures for Example 1.15(c) and its solution.

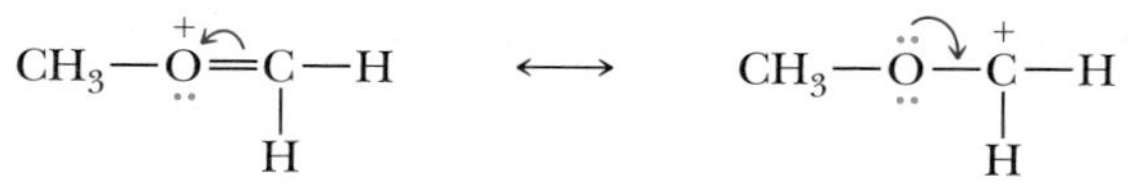

$$CH_3-\overset{+}{\ddot{O}}=\overset{|}{\underset{H}{C}}-H \longleftrightarrow CH_3-\overset{..}{\underset{..}{O}}-\overset{+}{\underset{H}{\overset{|}{C}}}-H$$

Greater contribution:
both carbon and oxygen have
complete valence shells

Lesser contribution:
carbon has only six electrons
in its valence shell

Preference 2: Maximum Number of Covalent Bonds

Structures with a greater number of covalent bonds contribute more than those with fewer covalent bonds. In the illustration for preference 1, the structure on the left has eight covalent bonds and makes the greater contribution to the hybrid. The structure on the right has only seven covalent bonds.

$$CH_3 - \overset{+}{\underset{..}{O}} = \overset{}{\underset{H}{C}} - H \longleftrightarrow CH_3 - \overset{}{\underset{..}{O}} - \overset{+}{\underset{H}{C}} - H$$

Greater contribution: Lesser contribution:
eight covalent bonds   seven covalent bonds

Preference 3: Least Separation of Unlike Charges

Structures involving separation of unlike charges contribute less than those that do not involve charge separation because separation of charges costs energy.

$$CH_3 - \overset{\overset{\displaystyle O}{\|}}{C} - CH_3 \longleftrightarrow CH_3 - \overset{\overset{\displaystyle :O:^-}{|}}{\underset{+}{C}} - CH_3$$

Greater contribution: Lesser contribution:
no separation of   separation of unlike
unlike charges   charges

Preference 4: Negative Charge on a More Electronegative Atom

Structures that carry a negative charge on a more electronegative atom contribute more than those with the negative charge on a less electronegative atom. Conversely, structures that carry a positive charge on a less electronegative atom contribute more than those that carry the positive charge on a more electronegative atom. Following are three contributing structures for acetone:

An electrostatic potential map of an acetone molecule.

$$\underset{\substack{\text{(a)} \\ \text{Lesser} \\ \text{contribution}}}{\overset{\displaystyle :\overset{..}{O}:^-}{\underset{CH_3}{\overset{|}{\underset{}{\overset{+}{C}}}} CH_3}} \quad\overset{1}{\longleftrightarrow}\quad \underset{\substack{\text{(b)} \\ \text{Greater} \\ \text{contribution}}}{\overset{\displaystyle \overset{1}{(}\overset{..}{O}\overset{..}{)}\,2}{\underset{CH_3}{\overset{\|}{\underset{}{C}}} CH_3}} \quad\overset{2}{\longleftrightarrow}\quad \underset{\substack{\text{(c)} \\ \text{Should not} \\ \text{be drawn}}}{\overset{\displaystyle :\overset{+}{O}:}{\underset{CH_3}{\overset{|}{\underset{}{\overset{-}{C}}}} CH_3}}$$

Structure (b) makes the largest contribution to the hybrid. Structure (a) contributes less because it involves separation of unlike charge and because carbon has an incomplete octet; thus, arrow 1 on structure (b) is preferred. Structure (c) violates all four preference rules and should not be drawn, and arrow 2 on structure (b) can be ignored.

It is important to realize that if resonance structures contribute unequally, the actual structure of the hybrid most resembles the structure that contributes most! The electrostatic potential map of acetone shows the negative charge (red) on oxygen and the positive charge (blue) on carbon in agreement with the results we derive from the resonance treatment.

Example 1.17

Estimate the relative contribution of the members in each set of contributing structures.

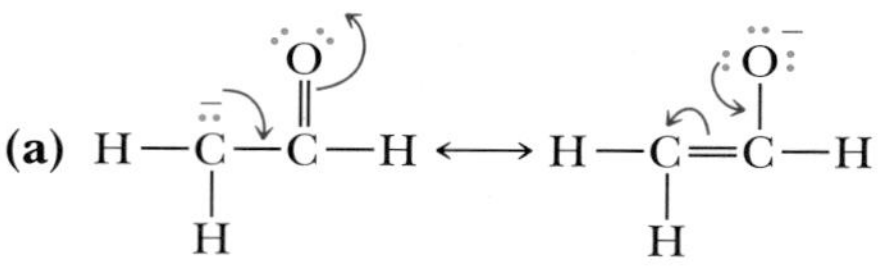

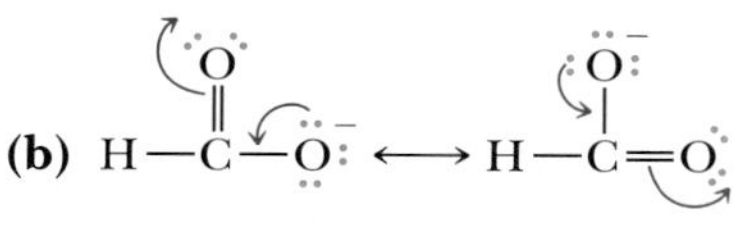

(a) H—C—C—H ⟷ H—C=C—H   **(b)** H—C—O: ⟷ H—C=O:

Solution

(a) The structure on the right makes a greater contribution to the hybrid because it places the negative charge on oxygen, the more electronegative atom.

(b) The structures are equivalent and make equal contributions to the hybrid.

Problem 1.17

Estimate the relative contribution of the members in each set of contributing structures.

(a)

H—C=C—H ⟷ H—C⁺—C:—H

(b) H—C—O—H ⟷ H—C=O—H

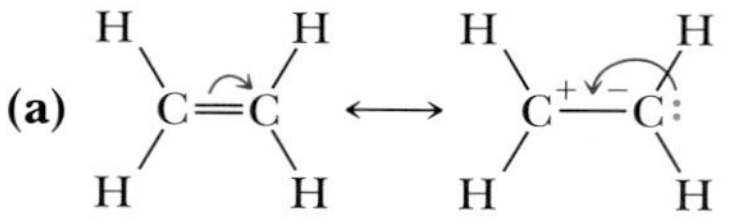

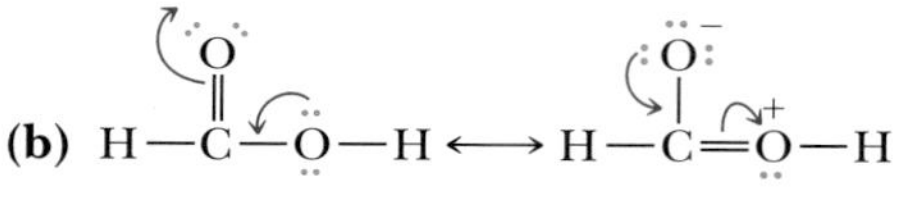

A final note: Do not confuse resonance contributing structures with equilibration among different species. A molecule described as a resonance hybrid is not equilibrating among individual electron configurations of the contributing structures. Rather, the molecule has only one structure, which is best described as a hybrid of its various contributing structures. The colors on the color wheel provide a good analogy. Purple is not a primary color; the primary colors of blue and red are mixed to make purple. You can think of molecules represented by resonance hybrids as being purple. Purple is not sometimes blue and sometimes red. Purple is purple. In an analogous way, a molecule described as a resonance hybrid is not sometimes one contributing structure and sometimes another; it is a single structure all the time.

1.7 Quantum or Wave Mechanics

Thus far in this chapter, we have concentrated on the Lewis model of bonding and on VSEPR. The Lewis model deals primarily with the coordination numbers of atoms (the number of bonds a given atom can form), and VSEPR deals primarily with bond angles and molecular geometries. Although each is useful in its own way, neither gives us any means of accounting in a quantitative or even semiquantitative way for the reasons atoms combine in the first place to form covalent bonds with the liberation of energy. At this point, we need to study an entirely new approach to the theory of covalent bonding, one that provides a means of understanding not only the coordination numbers of atoms and molecular geometries but also the energetics of chemical bonding.

A. Moving Particles Exhibit the Properties of a Wave

The beginning of this new approach to the theory of chemical bonding was provided by Albert Einstein (1879–1955), a German-born American physicist. In 1905, Einstein postulated that light consists of photons of electromagnetic radiation. The energy, E, of a photon is proportional to the frequency, ν (Greek nu), of the light. The proportionality constant in this equation is Planck's constant, h.

$$E = h\nu$$

In 1923, the French physicist Louis de Broglie followed Einstein's lead and advanced the revolutionary idea that if light exhibits properties of particles in motion, then a particle in motion should exhibit the properties of a wave. He proposed that a particle of mass m and speed v has an associated wavelength λ (Greek lambda), given by the equation

$$\lambda = \frac{h}{mv} \qquad \text{(the de Broglie relationship)}$$

Illustrated in Figure 1.9 is a wave such as might result from plucking a guitar string. The mathematical equation that describes this wave is called a **wave equation.** The numerical value(s) of the solution(s) of a wave equation may be positive (corresponding to a wave crest), negative (corresponding to a wave trough), or zero. A **node** is any point where the value of a solution of a wave equation is zero. A **nodal plane** is any plane perpendicular to the direction of propagation that runs through a node. Shown in Figure 1.9 are three nodal planes.

Erwin Schrödinger built on the idea of de Broglie and in 1926 proposed an equation that could be used to describe the wave properties associated with an electron in an atom or a molecule. **Quantum mechanics** (**wave mechanics**) is the branch of science that studies particles and their associated waves. Solving the Schrödinger equation gives a set of solutions called **wave functions.** Each wave function ψ (Greek psi) is associated with a unique set of quantum numbers and with a particular atomic or molecular orbital. This wave function occupies three-dimensional space and is called an **orbital.** Each orbital can contain no more than two electrons. The value of ψ^2 is proportional to the probability of finding an electron at a given point in space. Looked at in another way, the value of ψ^2 at any point in space is proportional to the electron density at that point. A plot of electron density (ψ^2) in a given orbital theoretically reaches to infinity but becomes vanishingly small at long distances from the nucleus. Notice that although the value of ψ at any point can be positive or negative, the value of ψ^2 will always be positive in an orbital. In other words, the electron density in two regions of an orbital will be equal if those regions have the same absolute value of ψ, regardless of whether that value is negative or positive. Of course, ψ^2 will be zero at a node. We often represent orbitals as a solid with the surface representing the volume within which some amount (such as 95%) of the electron density is contained.

In this course, we concentrate on wave functions and shapes associated with s and p atomic orbitals because they are the orbitals most often involved in covalent bond-

Node A point in space where the value of a wave function is zero.

Quantum mechanics The branch of science that studies the interaction of matter and radiation.

Wave function A solution to a set of equations that defines the energy of an electron in an atom and the region of space it may occupy.

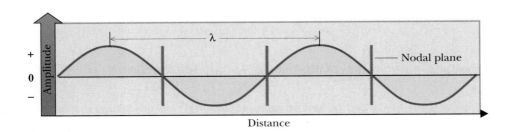

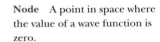
— Nodal plane

Figure 1.9
Characteristics of a wave associated with a moving particle. Wavelength is designated by the symbol λ.

ing in organic compounds. It is important to remember that when we talk about electrons and orbitals, we are really talking about wave equations. When we describe electron location, we are referring to electron density (not individual particles) in the context of wave equations. When we describe orbital interactions, remember that orbitals interact as waves interact; that is, they can interact constructively or destructively (adding or subtracting). For waves on the ocean or on a plucked guitar string, this characteristic is sometimes referred to as waves adding "in phase" and "out of phase."

It is convenient to add up the electron density in all of the orbitals in a molecule and then to determine which areas of a molecule have larger and smaller amounts of electron density. In general, the greater electron density is on the more electronegative atoms, especially those with lone pairs. Relative electron density distribution in molecules is important because it allows us to predict reactivity. Many reactions involve an area of relatively high electron density on one molecule reacting with an area of relatively low electron density on another molecule. It is convenient to keep track of overall molecular electron density distributions using computer graphics. This text presents electrostatic potential maps (elpots) in which areas of relatively high calculated electron density are shown in red and areas of relatively low calculated electron density are shown in blue, with intermediate electron densities represented by intermediate colors. The acetone molecule in Section 1.6C is an example of an electrostatic potential map.

B. Shapes of Atomic *s* and *p* Orbitals

All *s* orbitals have the shape of a sphere, with the center of the sphere at the nucleus. Shown in Figure 1.10 are probability distributions (ψ^2) for 1*s* and 2*s* orbitals. These orbitals are completely symmetric along all axes.

Shown in Figure 1.11 are the three-dimensional shapes (plots of ψ) of the three 2*p* orbitals, combined in one diagram to illustrate their relative orientations in space. Each 2*p* orbital consists of two lobes arranged in a straight line with the nucleus in the middle. The three 2*p* orbitals are mutually perpendicular and are designated 2*p_x*, 2*p_y*, and 2*p_z*. The sign of the wave function of a 2*p* orbital is positive in one lobe, zero at the nucleus, and negative in the other lobe. The plus or minus is

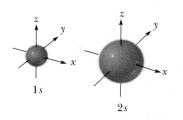

Figure 1.10
Probability distribution (ψ^2) for 1*s* and 2*s* atomic orbitals showing an arbitrary boundary containing about 95% of the electron density.

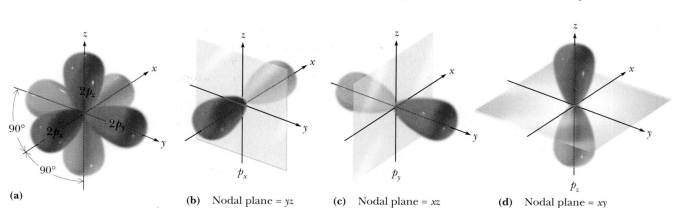

(b) Nodal plane = *yz*   **(c)** Nodal plane = *xz*   **(d)** Nodal plane = *xy*

Figure 1.11
Three-dimensional shapes of 2*p_x*, 2*p_y*, and 2*p_z* atomic orbitals and their orientation in space relative to one another.

simply the sign of the mathematical function ψ_{2p} and has no relationship to energy or electron distribution. These signs are shown by blue or red colors; however, these colors are not to be confused with the colors used to represent charge density. Recall that the value of ψ^2 is always positive, so the probability of finding electron density in the ($+$) lobe of a $2p$ orbital (i.e., value of ψ is positive) is the same as that of finding it in the ($-$) lobe (i.e., value of ψ is negative). Again, electron density is zero at the node.

Besides providing a way to determine the shapes of atomic orbitals, the Schrödinger equation also provides a way, at least in principle, to quantify the energetics of covalent bond formation. These approximations have taken two forms: (1) valence bond (VB) theory and (2) molecular orbital (MO) theory. Both theories of chemical bonding use the methods of quantum mechanics, but each makes slightly different simplifying assumptions. At sufficiently high levels of theory, both models converge. The VB approach provides the most easily visualized description of single bonds, while the MO method is most convenient for describing multiple bonds and for carrying out detailed calculations on computers.

1.8 The Molecular Orbital and Valence Bond Theories of Covalent Bonding

A. Molecular Orbital Theory; Formation of Molecular Orbitals

Molecular orbital (MO) theory begins with the hypothesis that electrons in atoms exist in atomic orbitals and assumes that electrons in molecules exist in molecular orbitals. Just as the Schrödinger equation can be used to calculate the energies and shapes of atomic orbitals, molecular orbital theory assumes that the Schrödinger equation can also be used to calculate the energies and shapes of molecular orbitals. Following is a summary of the rules used in applying molecular orbital theory to the formation of covalent bonds.

Molecular orbital theory   A theory of chemical bonding in which electrons in molecules occupy molecular orbitals that extend over the entire molecule and are formed by the combination of the atomic orbitals that make up the molecule.

1. Combination of n atomic orbitals (mathematically adding and subtracting wave functions) forms a set of n molecular orbitals (new wave functions); that is, the number of molecular orbitals formed is equal to the number of atomic orbitals combined.

2. Just like atomic orbitals, molecular orbitals are arranged in order of increasing energy. It is possible to calculate reasonably accurate relative energies of a set of molecular orbitals. Experimental measurements such as those derived from molecular spectroscopy can also be used to provide very detailed information about the relative energies of molecular orbitals.

3. Filling of molecular orbitals with electrons is governed by the same principles as the filling of atomic orbitals. Molecular orbitals are filled beginning with the lowest energy unoccupied molecular orbital (the Aufbau principle). A molecular orbital can accommodate no more than two electrons, and their spins must be paired (the Pauli exclusion principle). When two or more molecular orbitals of equal energy are available, one electron is added to each before any equivalent orbital is filled with two electrons.

To illustrate the formation of molecular orbitals, consider the shapes and relative energies of the molecular orbitals arising from combination of two $1s$ atomic orbitals. Combination by addition of their wave functions gives the molecular orbital shown in

Figure 1.12

Molecular orbitals (plots of ψ) derived from combination of two $1s$ atomic orbitals: (a) combination by addition and (b) combination by subtraction. Electrons in the bonding MO spend most of their time in the region between the two nuclei and bond the atoms together. Electrons in the antibonding MO are repulsive and decrease bonding.

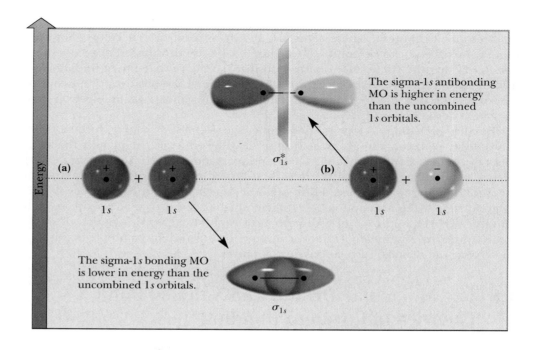

The sigma-$1s$ antibonding MO is higher in energy than the uncombined $1s$ orbitals.

σ^*_{1s}

The sigma-$1s$ bonding MO is lower in energy than the uncombined $1s$ orbitals.

σ_{1s}

Bonding molecular orbital A molecular orbital in which electrons have a lower energy than they would in isolated atomic orbitals.

Sigma (σ) molecular orbital A molecular orbital in which electron density is concentrated between two nuclei, along the axis joining them, and is cylindrically symmetrical.

Antibonding molecular orbital A molecular orbital in which electrons have a higher energy than they would in isolated atomic orbitals.

Figure 1.12(a). As with atomic orbitals, a molecular orbital is visualized as a plot of its wave function (ψ) in three-dimensional space. When electrons occupy this molecular orbital, electron density is concentrated in the region between the two positively charged nuclei and serves to offset the repulsive interaction between them. The molecular orbital we have just described is called a sigma bonding molecular orbital and is given the symbol σ_{1s} (pronounced sigma one ess). A **bonding molecular orbital** is one in which electrons have a lower energy than they would in the isolated atomic orbitals. A **sigma (σ) bonding molecular orbital** is one in which electron density lies between the two nuclei, along the axis joining them, and is *cylindrically symmetric* about the axis.

Combination of two $1s$ atomic orbitals by subtraction of their wave functions gives the molecular orbital shown in Figure 1.12(b). If electrons occupy this orbital, electron density is concentrated outside the region between the two nuclei. There is a node, or point of zero electron density, between the atoms. This molecular orbital is called a sigma antibonding molecular orbital and is given the symbol σ^*_{1s} (pronounced sigma star one ess). An **antibonding molecular orbital** is one in which the electrons in it have a higher energy (are more easily removed) than they would have in the isolated atomic orbitals. Thus this orbital is actually repulsive. An asterisk (*) is used to indicate that a molecular orbital is antibonding.

As with atomic orbitals, we use blue to indicate the area of a molecular orbital in which the sign of the wave function is negative and red where it is positive.

The **ground state** of an atom or molecule is its state of lowest energy. In the ground state of a hydrogen molecule, the two electrons occupy the σ_{1s} MO with paired spins. An **excited state** is any electronic state other than the ground state. In the lowest excited state of the hydrogen molecule, one electron occupies the σ_{1s} MO, and the other occupies the σ^*_{1s} MO. There is no net bonding in this excited state, and dissociation will result from the electrostatic repulsion of the two hydrogen

(a)   **(b)**

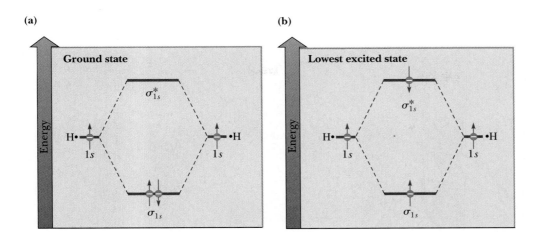

Figure 1.13
A molecular orbital energy diagram for the hydrogen molecule, H_2. (a) Ground state and (b) lowest excited state.

nuclei. Energy-level diagrams of the ground state and the lowest excited state of the hydrogen molecule are shown in Figure 1.13. Under normal circumstances, we do not have to consider antibonding orbitals because they are unoccupied. Only bonding molecular orbitals are filled and therefore important for molecular structure.

Combination of two $2s$ atomic orbitals produces two sigma molecular orbitals, designated σ_{2s} and σ^*_{2s}, that are larger in shape and higher in energy than the σ_{1s} and σ^*_{1s} molecular orbitals illustrated in Figure 1.13. Next let us consider the molecular orbitals formed by the combination of $2p$ orbitals. According to MO theory, two $2p$ atomic orbitals can overlap end-on to form a sigma bonding (σ_{2p}) and a sigma antibonding (σ^*_{2p}) molecular orbital. We will not encounter these kinds of MOs in this course. What we do encounter is combination of parallel $2p$ atomic orbitals by addition of their wave functions to give a **pi (π) bonding molecular orbital** (π_{2p}) shown in Figure 1.14(a). A pi bonding molecular orbital has a nodal plane that cuts through both atomic nuclei, with electron density above and below the plane concentrated between the nuclei. Combination of the parallel $2p$ atomic orbital wave functions by subtraction gives the pi antibonding (π^*_{2p}) molecular orbital shown in Figure 1.14(b).

We will have little need to refer to antibonding molecular orbitals throughout the remainder of the course, except for when we treat ultraviolet-visible spectroscopy in Section 20.3. Our concentration will be on bonding molecular orbitals and their participation in chemical reactions because only bonding molecular orbitals are occupied.

The key feature of MO theory is that molecular orbitals extend over entire molecules because all of the orbitals of all of the atoms take part in constructing molecular orbitals. This feature is extremely powerful when generating quantitative computational models of molecules. However, the full MO description of molecules is not particularly useful for students trying to understand and visualize covalent bonding and structures with sigma bonds. To understand and visualize sigma bonds in molecules, the valence bond approach is more useful. However, for reasons that will become clear later in the course, the MO description of pi orbitals is the best way to conceptualize the electron density in pi bonds, especially in molecules with multiple adjacent pi bonds, and we will return to MO theory in those sections dealing with these types of molecules.

Pi (π) molecular orbital   A molecular orbital formed by overlapping parallel p orbitals on adjacent atoms; its electron density lies above and below the line connecting the atoms.

Figure 1.14
Molecular orbitals (plots of ψ) formed by combination of parallel $2p$ orbitals; (a) combination by addition gives a pi bonding molecular orbital and (b) combination by subtraction gives a pi antibonding molecular orbital.

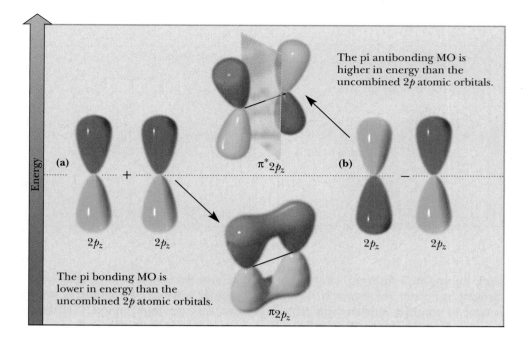

The pi antibonding MO is higher in energy than the uncombined $2p$ atomic orbitals.

(a)   + $\pi^{*}2p_z$ **(b)**   −

$2p_z$ $2p_z$ $2p_z$ $2p_z$

The pi bonding MO is lower in energy than the uncombined $2p$ atomic orbitals.

$\pi 2p_z$

B. Valence Bond Theory; Hybridization of Atomic Orbitals

Formation of a covalent bond between two hydrogen atoms is straightforward; two $1s$ orbitals overlap to give a sigma orbital, which holds two electrons. Formation of covalent bonds between atoms of second-period elements, however, presents a problem. In forming covalent bonds, atoms of carbon, nitrogen, and oxygen (all second-period elements) use $2s$ and $2p$ atomic orbitals. The three $2p$ atomic orbitals are at angles of 90° to each other (Figure 1.11), and, if atoms of second-period elements used these orbitals to form covalent bonds, we would expect bond angles around each would be approximately 90°. However, we rarely observe bond angles of 90° in organic molecules. What we find instead are bond angles of approximately 109.5° in molecules with only single bonds, 120° in molecules with double bonds, and 180° in molecules with triple bonds, as shown in Table 1.9.

To account for the observed bond angles in a way that is intuitive for chemists, Linus Pauling proposed that atomic orbitals for each atom should be thought of as first combining to form new orbitals, called **hybrid orbitals,** which then interact to form bonds by overlapping with hybrid orbitals from other atoms. The hybrid orbitals have the bond angles we observe around each atom, so molecular structure and bonding based on the overlap of hybrid orbitals provides an intuitive understanding. Being able to construct organic molecules from the overlap of hybrid orbitals is an essential organic chemistry survival skill.

Hybrid orbitals are formed by combinations of atomic orbitals, a process called **hybridization.** Mathematically, this is accomplished by combining the wave functions of the $2s$ (ψ_{2s}) and three $2p$ (ψ_{2p_x}, ψ_{2p_y}, ψ_{2p_z}) orbital wave functions. The number of hybrid orbitals formed is equal to the number of atomic orbitals combined. Elements of the second period form three types of hybrid orbitals, designated sp^3, sp^2, and sp, each of which can contain up to two electrons.

Hybrid orbital   An orbital formed by the combination of two or more atomic orbitals.

Hybridization   The combination of atomic orbitals of different types.

C. sp^3 Hybrid Orbitals—Bond Angles of Approximately 109.5°

The mathematical combination of the 2s atomic orbital and three 2p atomic orbitals forms four equivalent sp^3 **hybrid orbitals** described by four new wave functions. Plotting ψ for the four new wave functions gives a three-dimensional visualization of the four sp^3 hybrid orbitals. Each sp^3 hybrid orbital consists of a larger lobe pointing in one direction and a smaller lobe of opposite sign pointing in the opposite direction. The axes of the four sp^3 hybrid orbitals are directed toward the corners of a regular tetrahedron, and sp^3 hybridization results in bond angles of approximately 109.5° (Figure 1.15). Note that each sp^3 orbital has 25% s-character and 75% p-character because those are the percentages of the orbitals combined when constructing them (one 2s orbital, three 2p orbitals). We will refer back to these percentages several times in the text.

> sp^3 **Hybrid orbital**   A hybrid atomic orbital formed by the combination of one s atomic orbital and three p atomic orbitals.

You must remember that superscripts in the designation of hybrid orbitals tell you how many atomic orbitals have been combined to form the hybrid orbitals. You know that the designation sp^3 represents a hybrid orbital because it shows a combination of s and p orbitals. The superscripts in this case tell you that *one s* atomic orbital and *three p* atomic orbitals are combined in forming the hybrid orbital. Do not confuse this use of superscripts with that used in writing a ground-state electron configuration, as for example $1s^2 2s^2 2p^5$ for fluorine. In the case of a ground-state electron configuration, superscripts tell you the number of electrons in each orbital or set of orbitals.

In Section 1.2, we described the covalent bonding in CH_4, NH_3, and H_2O in terms of the Lewis model, and in Section 1.4 we used VSEPR to predict bond angles of approximately 109.5° in each molecule. Now let us consider the bonding in these molecules in terms of the overlap of hybrid atomic orbitals. To bond with four other atoms with bond angles of 109.5°, carbon uses sp^3 hybrid orbitals. Carbon has four valence electrons, and one electron is placed in each sp^3 hybrid orbital. Each partially filled sp^3 hybrid orbital then overlaps with a partially filled 1s atomic orbital of hydrogen to form the four sigma (σ) bonds of methane, and hydrogen atoms occupy the corners of a regular tetrahedron (Figure 1.16).

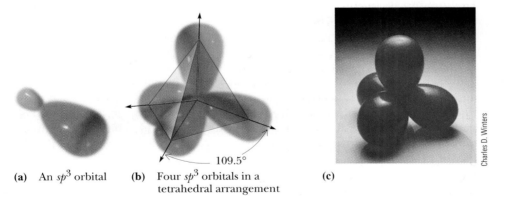

(a) An sp^3 orbital   **(b)** Four sp^3 orbitals in a   **(c)**
                                        tetrahedral arrangement

109.5°

Charles D. Winters

Figure 1.15

sp^3 Hybrid orbitals. (a) Representation of a single sp^3 hybrid orbital (plot of ψ) showing two lobes of unequal size. The sign of the wave function is positive in one lobe and negative in the other. (b) Three-dimensional representation of four sp^3 hybrid orbitals directed toward the corners of a regular tetrahedron. The smaller lobe of each sp^3 hybrid orbital is hidden behind the larger lobe. (c) If four balloons of similar size and shape are tied together, they will naturally assume a tetrahedral geometry.

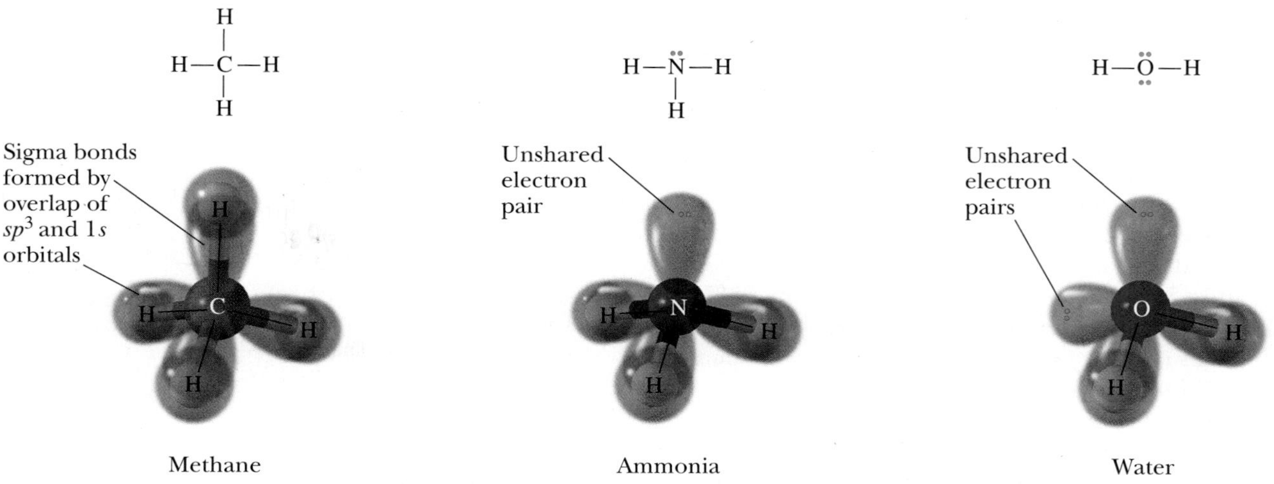

$$\begin{array}{c} H \\ | \\ H-C-H \\ | \\ H \end{array} \qquad H-\overset{..}{N}-H \qquad H-\overset{..}{\underset{..}{O}}-H \\ \qquad\qquad\qquad | \\ \qquad\qquad\qquad H$$

Sigma bonds formed by overlap of sp^3 and $1s$ orbitals

Unshared electron pair

Unshared electron pairs

Methane

Ammonia

Water

Figure 1.16
Orbital overlap pictures of methane, ammonia, and water.

In bonding with three other atoms, the five valence electrons of nitrogen are distributed so that one sp^3 hybrid orbital is filled with a pair of electrons (the lone pair) and the other three sp^3 hybrid orbitals have one electron each. Overlapping of these partially filled sp^3 hybrid orbitals with $1s$ atomic orbitals of three hydrogen atoms produces an NH_3 molecule (Figure 1.16).

In bonding with two other atoms, the six valence electrons of oxygen are distributed so that two sp^3 hybrid orbitals are filled, and the remaining two have one electron each. Each partially filled sp^3 hybrid orbital overlaps with a $1s$ atomic orbital of hydrogen, and hydrogen atoms occupy two corners of a regular tetrahedron. The remaining two sp^3 hybrid orbitals, each occupied by an unshared pair of electrons, are directed toward the other two corners of the regular tetrahedron (Figure 1.16).

D. sp^2 Hybrid Orbitals—Bond Angles of Approximately 120°

sp² **Hybrid orbital**   A hybrid atomic orbital formed by the combination of one *s* atomic orbital and two *p* atomic orbitals.

The mathematical combination of one $2s$ atomic orbital wave function and two $2p$ atomic orbital wave functions forms three equivalent sp^2 **hybrid orbital** wave functions. Because they are derived from three atomic orbitals, sp^2 hybrid orbitals always occur in sets of three. As with sp^3 orbitals, each sp^2 hybrid orbital (three-dimensional plot of ψ) consists of two lobes, one larger than the other. The axes of the three sp^2 hybrid orbitals lie in a plane and are directed toward the corners of an equilateral triangle; the angle between sp^2 hybrid orbitals is 120°. The third $2p$ atomic orbital (remember $2p_x$, $2p_y$, $2p_z$) is not involved in hybridization (its wave function is not mathematically combined with the other three) and remains as two lobes lying perpendicular to the plane of the sp^2 hybrid orbitals. Figure 1.17 shows three equivalent sp^2 orbitals along with the remaining unhybridized $2p$ atomic orbital. Each sp^2 orbital has 33% s-character and 67% p-character (one $2s$ orbital, two $2p$ orbitals).

Second-period elements use a combination of an sp^2 hybrid orbital and the unhybridized $2p$ atomic orbital to form <u>double bonds</u>. Consider ethylene, C_2H_4, a Lewis structure for which is shown in Figure 1.18(a). A sigma bond between the carbons in

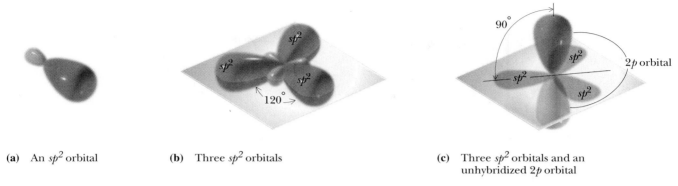

(a) An sp^2 orbital

(b) Three sp^2 orbitals

(c) Three sp^2 orbitals and an unhybridized $2p$ orbital

Figure 1.17

sp^2 Hybrid orbitals. (a) A single sp^2 hybrid orbital (plot of ψ) showing two lobes of unequal size. (b) The three sp^2 hybrid orbitals with their axes in a plane at angles of 120°. (c) The unhybridized $2p$ orbital that is perpendicular to the plane created by the three sp^2 hybrid orbitals.

ethylene is formed by overlapping sp^2 hybrid orbitals along a common axis as seen in Figure 1.18(b). Each carbon also forms sigma bonds to two hydrogens. The remaining $2p$ orbitals on adjacent carbons lie parallel to each other and overlap to form a pi bond [Figure 1.18(c)]. Recall that a **pi (π) bond** is a covalent bond formed by overlap of parallel p orbitals in a bonding arrangement (Figure 1.14). Because of the lesser degree of overlap of orbitals forming pi bonds compared with those forming sigma bonds, pi bonds are generally weaker than sigma bonds.

Pi (π) bond A covalent bond formed by the overlap of parallel p orbitals.

Just as with s orbitals, p orbitals can also combine repulsively to form antibonding pi* (π*) orbitals (Figure 1.14). Although they are not occupied in the ground state of molecules, these orbitals are important in photochemistry, where light can excite electrons from the pi orbitals to pi* orbitals, leading to chemical reactions. We will deal mainly with bonding orbitals in this text.

Valence bond theory describes all double bonds in the same manner we have used to describe carbon-carbon double bonds. In formaldehyde, $CH_2{=}O$, the simplest organic molecule containing a carbon-oxygen double bond, carbon forms sigma bonds to

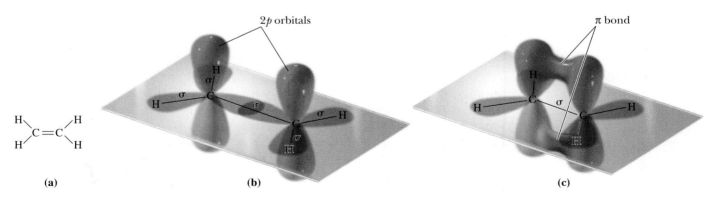

(a)

(b)

(c)

Figure 1.18

Covalent bond formation in ethylene. (a) Lewis structure, (b) overlap of sp^2 hybrid orbitals forms a sigma (σ) bond between the carbon atoms, and (c) overlap of parallel $2p$ orbitals forms a pi (π) bond.

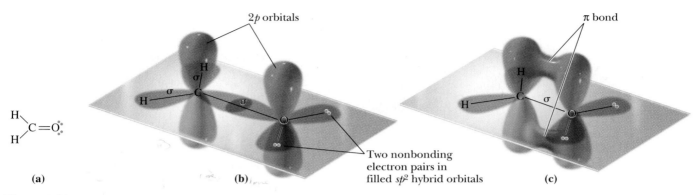

Figure 1.19
A carbon-oxygen double bond. (a) Lewis structure of H₂C=O, (b) the sigma (σ) bond framework and nonoverlapping parallel 2p atomic orbitals, and (c) overlap of parallel 2p atomic orbitals to form a pi (π) bond.

sp **Hybrid orbital**   A hybrid atomic orbital formed by the combination of one *s* atomic orbital and one *p* atomic orbital.

two hydrogens by overlapping sp^2 orbitals of carbon and 1s atomic orbitals of hydrogen. Carbon and oxygen are joined by a sigma bond formed by overlapping sp^2 hybrid orbitals and a pi bond formed by overlapping parallel 2p atomic orbitals (Figure 1.19).

E. *sp* Hybrid Orbitals—Bond Angles of Approximately 180°

The mathematical combination of one 2s atomic orbital and one 2p atomic orbital produces two equivalent *sp* **hybrid orbital** wave functions. Because they are derived from two atomic orbitals, *sp* hybrid orbitals always occur in sets of two. The three-dimensional plot of ψ shows that the two *sp* hybrid orbitals lie at an angle of 180°. The axes of the unhybridized 2p atomic orbitals are perpendicular to each other and to the axis of the two *sp* hybrid orbitals. In Figure 1.20, *sp* hybrid orbitals are shown on the x-axis, and unhybridized 2p orbitals are on the y- and z-axes. Each *sp* orbital has 50% s-character and 50% p-character because those are the percentages of the orbitals combined when constructing them (one 2s orbital, one 2p orbital).

Figure 1.21 shows an orbital overlap diagram for acetylene, C₂H₂. A carbon-carbon triple bond consists of one sigma bond formed by overlapping *sp* hybrid orbitals and two pi bonds. One pi bond is formed by overlapping a pair of parallel 2p atomic orbitals. The second pi bond is formed by overlapping another pair of parallel 2p atomic orbitals.

Figure 1.20
sp Hybrid orbitals. (a) A single *sp* hybrid orbital (plot of ψ) consisting of two lobes of unequal size. (b) Two *sp* hybrid orbitals in a linear arrangement. (c) Unhybridized 2p atomic orbitals are perpendicular to the line created by the axes of the two *sp* hybrid orbitals.

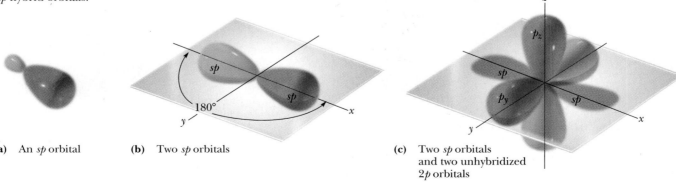

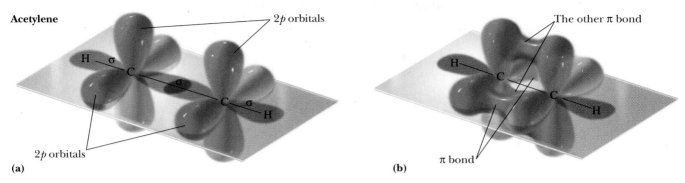

Figure 1.21
Covalent bonding in acetylene, C_2H_2. (a) The sigma bond framework shown along with nonoverlapping $2p$ atomic orbitals. (b) Formation of two pi bonds by overlapping two sets of parallel $2p$ atomic orbitals.

Table 1.10 Covalent Bonding of Carbon

Groups Bonded to Carbon	Orbital Hybridization	Predicted Bond Angles	Types of Bonds to Carbon	Example	Name
4	sp^3	109.5°	Four sigma bonds	H—C—C—H (ethane structure)	Ethane
3	sp^2	120°	Three sigma bonds and one pi bond	C=C (ethylene structure)	Ethylene
2	sp	180°	Two sigma bonds and two pi bonds	H—C≡C—H	Acetylene

The relationship among the number of atoms bonded to carbon, orbital hybridization, and types of bonds involved is summarized in Table 1.10.

Example 1.18

Describe the bonding in acetic acid, CH_3COOH, in terms of the atomic orbitals involved, and predict all bond angles.

Solution

Following are three identical Lewis structures. Labels on the first Lewis structure point to atoms and show the hybridization of each atom. Labels on the second Lewis structure point to bonds and show the type of bond, either sigma or pi. Labels on the third point to atoms and show bond angles about each atom as predicted by VSEPR.

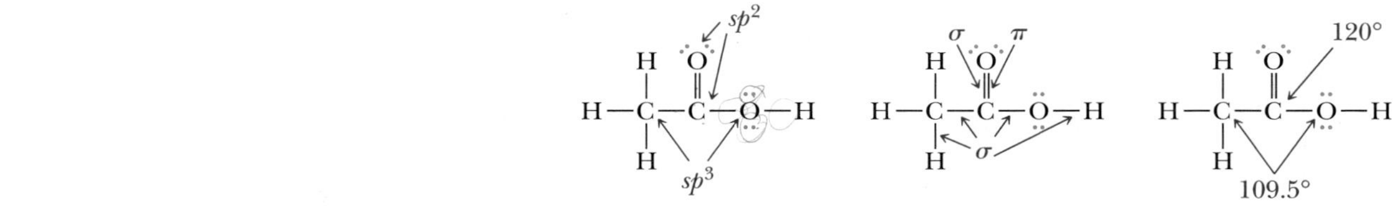

Problem 1.18

Describe the bonding in these molecules in terms of the atomic orbitals involved, and predict all bond angles.

(a) CH_3CH═CH_2 (b) CH_3NH_2

F. Bond Lengths and Bond Strengths in Alkanes, Alkenes, and Alkynes

Values for bond lengths and bond strengths (bond dissociation enthalpies) for ethane, ethylene, and acetylene are given in Table 1.11.

As you study Table 1.11, note the following points:

1. Carbon-carbon triple bonds are shorter than carbon-carbon double bonds, which in turn are shorter than carbon-carbon single bonds. This order of bond lengths exists because there are three versus two versus one bond holding the carbon atoms together.

2. The C—H bond in acetylene is shorter than that in ethylene, which in turn is shorter than that in ethane. The relative lengths of these C—H bonds are determined by the percent s-character in the hybrid orbital of carbon forming the sigma bond with hydrogen. The greater the percent s-character of a hybrid orbital, the closer electrons in it are held to the nucleus and the shorter the bond because s electrons are on average closer to the nucleus than p electrons. The relative lengths of C—H single bonds correlate with the fact that the percent s-character in an sp orbital is 50%, in an sp^2 orbital is 33.3%, and in an sp^3 orbital is 25%. Also, because electrons in s orbitals are bound more tightly than those in p

Table 1.11 Bond Lengths and Bond Strengths for Ethane, Ethylene, and Acetylene

Name	Formula	Bond	Bond Orbital Overlap	Bond Length (pm)	Bond Strength [kJ (kcal)/mol]
Ethane	H—C(H)(H)—C(H)(H)—H	C—C	sp^3–sp^3	153.2	376 (90)
		C—H	sp^3–$1s$	111.4	422 (101)
Ethylene	H(H)C═C(H)H	C—C	sp^2–sp^2, $2p$–$2p$	133.9	727 (174)
		C—H	sp^2–$1s$	110.0	464 (111)
Acetylene	H—C≡C—H	C—C	sp–sp, two $2p$–$2p$	121.2	966 (231)
		C—H	sp–$1s$	109.0	556 (133)

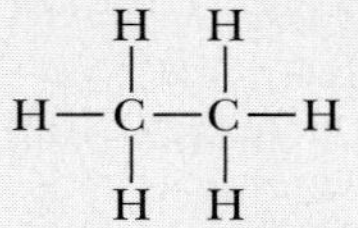

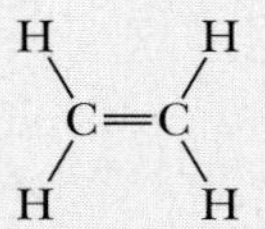

orbitals, the more *s*-character in a bond, the stronger it is. This is the first of several times we will refer to the percent *s*-character in hybrid orbitals. It has a large influence on a variety of molecular properties.

3. There is a correlation between bond length and bond strength; the shorter the bond, the stronger it is. A carbon-carbon triple bond is the shortest C—C bond; it is also the strongest. The carbon-hydrogen bond in acetylene is the shortest; it is also the strongest.

4. Although a C—C double bond is stronger than a C—C single bond, it is not twice as strong. By the same token, a C—C triple bond is stronger than a C—C single bond, but it is not three times as strong. These differences arise because the overlap of orbitals lying on the same axis and forming sigma bonds is more efficient (gives a greater bond strength) than overlap of orbitals lying parallel to each other and forming pi bonds.

Summary

Atoms consist of a small, dense nucleus and electrons distributed about the nucleus in regions of space called **principal energy levels** or **shells** (Section 1.1). Each shell can contain as many as $2n^2$ electrons, where n is the number of the shell. Each principal energy level is subdivided into regions of space called **orbitals.** The **Lewis dot structure** (Section 1.1B) of an element shows the symbol of the element surrounded by a number of dots equal to the number of electrons in the **valence shell** of the atom. According to the Lewis model of covalent bonding (Section 1.2), atoms bond together in such a way that each atom participating in a chemical bond acquires a completed valence-shell electron configuration resembling that of the noble gas nearest it in atomic number. Anions and cations attract but do not form bonds. A **covalent bond** is a chemical bond formed by the sharing of electron pairs between atoms. The tendency of main-group elements (those of Groups 1A–7A) to achieve an outer shell of eight valence electrons is called the **octet rule.**

Electronegativity (Section 1.2B) is a measure of the force of attraction by an atom for electrons it shares in a chemical bond with another atom. A **nonpolar covalent bond** (Section 1.2B) is a covalent bond in which the difference in electronegativity of the bonded atoms is less than 0.5. A **polar covalent bond** is a covalent bond in which the difference in electronegativity of the bonded atoms is between 0.5 and 1.9. In a polar covalent bond, the more electronegative atom bears a partial negative charge ($\delta-$) and the less electronegative atom bears a partial positive charge ($\delta+$). A polar bond has a bond dipole equal to the product of the absolute value of the partial charge times the distance between the dipolar charges (the bond length).

An acceptable **Lewis structure** (Section 1.2C) for a molecule or an ion must show (1) the correct connectivity of atoms, (2) the correct number of valence electrons, (3) no more than two electrons in the outer shell of hydrogen and no more than eight electrons in the outer shell of any second-period element, and (4) all formal charges. **Formal charge** is the charge on an atom in a molecule or polyatomic ion (Section 1.2D).

Functional groups (Section 1.3) are characteristic structural units by which we divide organic compounds into classes and that serve as a basis for nomenclature. They are also sites of chemical reactivity. A particular functional group undergoes the same types of chemical reactions in whatever compound it occurs.

Bond angles of molecules and polyatomic ions can be predicted using Lewis structures and **valence-shell electron-pair repulsion (VSEPR)** (Section 1.4). For atoms surrounded by four regions of electron density, VSEPR predicts bond angles of 109.5°; for atoms surrounded by three regions of electron density, it predicts bond angles of 120°; and for two regions of electron density, it predicts bond angles of 180°. The dipole moment of a polyatomic molecule is the vector sum of its bond moments (Section 1.5).

According to the **theory of resonance** (Section 1.6A), molecules and ions for which no single Lewis structure is adequate are best described by writing two or more **contributing structures** and considering the real molecule or ion to be a **resonance hybrid** of the various contributing structures. Contributing structures to the hybrid are interconnected by **double-headed arrows.** The manner in which valence electrons are redistributed from one contributing structure to the next is shown by **curved arrows** (Section 1.6A). Use of curved arrows in this way is commonly referred to as **electron pushing.** The most important contributing structures have (1) filled valence shells, (2) a maximum number of covalent bonds, (3) the least separation of unlike charges, and (4) any negative charge on the more electronegative atom and/or any positive charge on the less electronegative atom.

Quantum mechanics (Section 1.7) is the branch of science that studies particles and their associated waves. It provides a way to determine the shapes of atomic orbitals and to quantify the energetics of covalent bond formation. According to **molecular**

orbital theory (Section 1.8A), combination of n atomic orbitals gives n molecular orbitals. Molecular orbitals are divided into sigma and pi bonding and antibonding molecular orbitals. These orbitals are arranged in order of increasing energy, and their order of filling with electrons is governed by the same rules as for filling atomic orbitals. Although useful for quantitative calculations on computers, molecular orbital theory does not provide an intuitive understanding of sigma bonds in complex molecules, however it is of value when describing certain types of pi bonds.

For an intuitive understanding of sigma bonds in molecules, we use **valence bond theory** (Section 1.8B). Valence bond theory involves the combination of atomic orbitals on each atom in a process called **hybridization,** and the resulting atomic orbitals are called **hybrid orbitals.** Combination of one $2s$ atomic orbital and three $2p$ atomic orbitals produces four

equivalent sp^3 **hybrid orbitals,** each directed toward a corner of a regular tetrahedron at angles of 109.5° (Section 1.8C).

The combination of one $2s$ atomic orbital and two $2p$ atomic orbitals produces three equivalent sp^2 **hybrid orbitals,** the axes of which lie in a plane at angles of 120° (Section 1.8D). All C=C, C=O, C=N, N=N, and N=O double bonds are a combination of one sigma (σ) bond formed by the overlapping sp^2 hybrid orbitals and one pi (π) bond formed by overlapping parallel $2p$ orbitals.

The combination of one $2s$ atomic orbital and one $2p$ atomic orbital produces two equivalent sp **hybrid orbitals,** the axes of which lie at an angle of 180° (Section 1.8E). All C≡C and C≡N triple bonds are a combination of one sigma bond formed by the overlap of sp hybrid orbitals and two pi bonds formed by the overlap of two sets of parallel $2p$ orbitals.

Problems

Electronic Structure of Atoms

1.19 Write the ground-state electron configuration for each atom. After each atom is its atomic number in parentheses.

   (a) Sodium (11)   **(b)** Magnesium (12)   **(c)** Oxygen (8)   **(d)** Nitrogen (7)

1.20 Identify the atom that has each ground-state electron configuration.

   (a) $1s^2 2s^2 2p^6 3s^2 3p^4$ **(b)** $1s^2 2s^2 2p^4$

1.21 Define valence shell and valence electron.

1.22 How many electrons are in the valence shell of each atom?

   (a) Carbon   **(b)** Nitrogen   **(c)** Chlorine   **(d)** Aluminum

Lewis Structures and Formal Charge

1.23 Judging from their relative positions in the Periodic Table, which atom in each set is more electronegative?

   (a) Carbon or nitrogen   **(b)** Chlorine or bromine   **(c)** Oxygen or sulfur

1.24 Which compounds have nonpolar covalent bonds, which have polar covalent bonds, and which are ions?

   (a) LiF   **(b)** CH_3F **(c)** $MgCl_2$ **(d)** HCl

1.25 Using the symbols $\delta-$ and $\delta+$, indicate the direction of polarity, if any, in each covalent bond.

   (a) C—Cl   **(b)** S—H   **(c)** C—S   **(d)** P—H

1.26 Write Lewis structures for these compounds. Show all valence electrons. None of them contains a ring of atoms.

   (a) Hydrogen peroxide, H_2O_2 **(b)** Hydrazine, N_2H_4 **(c)** Methanol, CH_3OH

1.27 Write Lewis structures for these ions. Show all valence electrons and all formal charges.

   (a) Amide ion, NH_2^- **(b)** Bicarbonate ion, HCO_3^- **(c)** Carbonate ion, CO_3^{2-}
   (d) Nitrate ion, NO_3^- **(e)** Formate ion, $HCOO^-$ **(f)** Acetate ion, CH_3COO^-

ORGANIC
Chemistry•ᯤ•Now™
Assess your understanding of this chapter's topics with additional quizzing and conceptual-based problems at **http://now.brookscole.com/ bfi4**

1.28 Complete these structural formulas by adding enough hydrogens to complete the tetravalence of each carbon. Then write the molecular formula of each compound.

$$\text{(a) } C-C=C-\overset{\overset{\displaystyle C}{|}}{C}-C$$

$$\text{(b) } C-C-C-\overset{\overset{\displaystyle O}{\|}}{C}-OH$$

$$\text{(c) } C-C-\overset{\overset{\displaystyle O}{\|}}{C}-C$$

$$\text{(d) } C-\overset{\overset{\displaystyle O}{\|}}{\underset{\underset{\displaystyle C}{|}}{C}}-C-H$$

$$\text{(e) } C-\overset{\overset{\displaystyle C}{|}}{\underset{\underset{\displaystyle C}{|}}{C}}-C-C-NH_2$$

$$\text{(f) } C-\overset{\overset{\displaystyle O}{\|}}{C}-\underset{\underset{\displaystyle NH_2}{|}}{C}-OH$$

$$\text{(g) } C-\overset{\overset{\displaystyle OH}{|}}{C}-C-C-C$$

$$\text{(h) } C-\overset{\overset{\displaystyle OH}{|}}{C}-C-\overset{\overset{\displaystyle O}{\|}}{C}-OH$$

$$\text{(i) } C=C-C-OH$$

1.29 Some of these structural formulas are incorrect (i.e., they do not represent a real compound) because they have atoms with an incorrect number of bonds. Which structural formulas are incorrect, and which atoms in them have an incorrect number of bonds?

$$\text{(a) } H-\overset{\overset{\displaystyle H}{|}}{\underset{\underset{\displaystyle H}{|}}{C}}-\overset{\overset{\displaystyle H}{|}}{\underset{\underset{\displaystyle H}{|}}{C}}=O-H$$

$$\text{(b) } H-\overset{\overset{\displaystyle Cl}{|}}{C}=\overset{\overset{\displaystyle }{}}{\underset{\underset{\displaystyle H}{|}}{C}}-H \quad (\text{with H below left C})$$

$$\text{(c) } H-\overset{\overset{\displaystyle }{}}{\underset{\underset{\displaystyle H}{|}}{N}}-\overset{\overset{\displaystyle H}{|}}{\underset{\underset{\displaystyle H}{|}}{C}}-\overset{\overset{\displaystyle H}{|}}{\underset{\underset{\displaystyle H}{|}}{C}}-O-H$$

$$\text{(d) } H-C\equiv C-\overset{\overset{\displaystyle H}{|}}{\underset{\underset{\displaystyle H}{|}}{C}}-H$$

$$\text{(e) } H-O-\overset{\overset{\displaystyle H}{|}}{\underset{\underset{\displaystyle H}{|}}{C}}-\overset{\overset{\displaystyle H}{|}}{\underset{\underset{\displaystyle H}{|}}{C}}-\overset{\overset{\displaystyle O}{\|}}{C}-O-H$$

$$\text{(f) } H-\overset{\overset{\displaystyle H}{|}}{\underset{\underset{\displaystyle H}{|}}{C}}-\overset{\overset{\displaystyle H}{|}}{\underset{\underset{\displaystyle H}{|}}{C}}-\overset{\overset{\displaystyle O}{\|}}{C}-H$$

$$\text{(g) } H-\overset{\overset{\displaystyle H}{|}}{\underset{\underset{\displaystyle H}{|}}{C}}-C=C=C-\overset{\overset{\displaystyle H}{|}}{\underset{\underset{\displaystyle H}{|}}{C}}-H$$

$$\text{(h) } H-C\equiv C-\overset{\overset{\displaystyle H}{|}}{\underset{\underset{\displaystyle H}{|}}{C}}-H$$

1.30 Following the rule that each atom of carbon, oxygen, and nitrogen reacts to achieve a complete outer shell of eight valence electrons, add unshared pairs of electrons as necessary to complete the valence shell of each atom in these ions. Then assign formal charges as appropriate.

$$\text{(a) } H-\overset{\overset{\displaystyle H}{|}}{\underset{\underset{\displaystyle H}{|}}{C}}-\overset{\overset{\displaystyle H}{|}}{\underset{\underset{\displaystyle H}{|}}{C}}-O$$

$$\text{(b) } H-\overset{\overset{\displaystyle H}{|}}{\underset{\underset{\displaystyle H}{|}}{C}}-\overset{\overset{\displaystyle H}{|}}{\underset{\underset{\displaystyle H}{|}}{C}}$$

$$\text{(c) } H-\overset{\overset{\displaystyle H}{|}}{\underset{\underset{\displaystyle H}{|}}{N}}-\overset{\overset{\displaystyle H}{|}}{\underset{\underset{\displaystyle H}{|}}{C}}-C\underset{\diagdown O}{\overset{\diagup O}{}}$$

1.31 Following are several Lewis structures showing all valence electrons. Assign formal charges in each structure as appropriate.

$$\text{(a) } H-\overset{\overset{\displaystyle H}{|}}{\underset{\underset{\displaystyle H}{|}}{C}}-\overset{\overset{\displaystyle \ddot{O}}{\|}}{C}-\overset{\overset{\displaystyle }{}}{\underset{\underset{\displaystyle H}{|}}{C}}-H$$

$$\text{(b) } H-\overset{\overset{\displaystyle :\ddot{O}:}{|}}{\underset{\underset{\displaystyle H}{|}}{N}}-C=C-H$$

$$\text{(c) } H-\overset{\overset{\displaystyle \cdot\ddot{O}\cdot}{\|}}{C}-\overset{\overset{\displaystyle }{}}{\underset{\underset{\displaystyle H}{|}}{C}}-H$$

$$\text{(d) } H-\overset{\overset{\displaystyle H}{|}}{\underset{\underset{\displaystyle H}{|}}{C}}-\overset{\overset{\displaystyle :\ddot{O}:}{|}}{C}=C-H$$

$$\text{(e) } H-\overset{\overset{\displaystyle H}{|}}{\underset{\underset{\displaystyle H}{|}}{C}}-\overset{\overset{\displaystyle H}{|}}{\underset{\underset{\displaystyle H}{|}}{C}}-\overset{\overset{\displaystyle H}{|}}{\underset{\underset{\displaystyle H}{|}}{C}}-H$$

$$\text{(f) } H-\overset{\overset{\displaystyle H}{|}}{\underset{\underset{\displaystyle H}{|}}{C}}-\overset{\overset{\displaystyle \ddot{O}}{|}}{O}-H$$

Polarity of Covalent Bonds

1.32 Which statements are true about electronegativity?

(a) Electronegativity increases from left to right in a period of the Periodic Table.
(b) Electronegativity increases from top to bottom in a column of the Periodic Table.
(c) Hydrogen, the element with the lowest atomic number, has the smallest electronegativity.
(d) The higher the atomic number of an element, the greater its electronegativity.

1.33 Why does fluorine, the element in the upper right corner of the Periodic Table, have the largest electronegativity of any element?

1.34 Arrange the single covalent bonds within each set in order of increasing polarity.

(a) C—H, O—H, N—H   (b) C—H, B—H, O—H   (c) C—H, C—Cl, C—I
(d) C—S, C—O, C—N   (e) C—Li, C—B, C—Mg

1.35 Using the values of electronegativity given in Table 1.5, predict which indicated bond in each set is the more polar and, using the symbols δ+ and δ−, show the direction of its polarity.

(a) CH_3—OH or CH_3O—H   (b) CH_3—NH_2 or CH_3—PH_2
(c) CH_3—SH or CH_3S—H   (d) CH_3—F or H—F

1.36 Identify the most polar bond in each molecule.

(a) $HSCH_2CH_2OH$ (b) $CHCl_2F$ (c) $HOCH_2CH_2NH_2$

Bond Angles and Shapes of Molecules

1.37 Use VSEPR to predict bond angles about each highlighted atom.

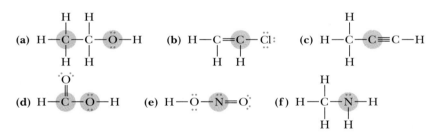

1.38 Use VSEPR to predict bond angles about each atom of carbon, nitrogen, and oxygen in these molecules.

(a) CH_3—CH=CH_2 (b) CH_3—N—CH_3 with CH_3 above N   (c) CH_3—CH_2—C—OH with O double-bonded above C

(d) H_2C=C=CH_2 (e) H_2C=C=O   (f) CH_3—CH=N—OH

1.39 Use VSEPR to predict the geometry of these ions.

(a) NH_2^- (b) NO_2^- (c) NO_2^+ (d) NO_3^-

Functional Groups

1.40 Draw Lewis structures for these functional groups. Be certain to show all valence electrons on each.

(a) Carbonyl group   (b) Carboxyl group   (c) Hydroxyl group
(d) Ester group   (e) Amide group

1.41 Draw condensed structural formulas for all compounds with molecular formula C_4H_8O that contain

(a) A carbonyl group (there are two aldehydes and one ketone).
(b) A carbon-carbon double bond and a hydroxyl group (there are eight).

1.42 What is the meaning of the term tertiary (3°) when it is used to classify alcohols? Draw a structural formula for the one tertiary (3°) alcohol with molecular formula $C_4H_{10}O$.

1.43 What is the meaning of the term tertiary (3°) when it is used to classify amines? Draw a structural formula for the one tertiary (3°) amine with molecular formula $C_4H_{11}N$.

1.44 Draw structural formulas for

(a) The four primary (1°) amines with molecular formula $C_4H_{11}N$.
(b) The three secondary (2°) amines with molecular formula $C_4H_{11}N$.
(c) The one tertiary (3°) amine with molecular formula $C_4H_{11}N$.

1.45 Draw structural formulas for the three tertiary (3°) amines with molecular formula $C_5H_{13}N$.

1.46 Draw structural formulas for

(a) The eight alcohols with molecular formula $C_5H_{12}O$.
(b) The eight aldehydes with molecular formula $C_6H_{12}O$.
(c) The six ketones with molecular formula $C_6H_{12}O$.
(d) The eight carboxylic acids with molecular formula $C_6H_{12}O_2$.
(e) The nine carboxylic esters with molecular formula $C_5H_{10}O_2$.

1.47 Identify the functional groups in each compound.

(a) $CH_3-\overset{\overset{\displaystyle OH}{|}}{CH}-\overset{\overset{\displaystyle O}{||}}{C}-OH$
Lactic acid

(b) $HO-CH_2-CH_2-OH$
Ethylene glycol

(c) $CH_3-\underset{\underset{\displaystyle NH_2}{|}}{\overset{\overset{\displaystyle O}{||}}{CH}}-\overset{\overset{\displaystyle O}{||}}{C}-OH$
Alanine

(d) $HO-CH_2-\overset{\overset{\displaystyle OH}{|}}{CH}-\overset{\overset{\displaystyle O}{||}}{C}-H$
Glyceraldehyde

(e) $CH_3-\overset{\overset{\displaystyle O}{||}}{C}-CH_2-\overset{\overset{\displaystyle O}{||}}{C}-OH$
Acetoacetic acid

(f) $H_2NCH_2CH_2CH_2CH_2CH_2CH_2NH_2$
1,6-Hexanediamine

Polar and Nonpolar Molecules

1.48 Draw a three-dimensional representation for each molecule. Indicate which ones have a dipole moment and in what direction it is pointing.

(a) CH_3F　　　(b) CH_2Cl_2　　　(c) CH_2ClBr　　　(d) $CFCl_3$　　　(e) CCl_4
(f) $CH_2=CCl_2$　(g) $CH_2=CHCl$　(h) $HC\equiv C-C\equiv CH$　(i) $CH_3C\equiv N$
(j) $(CH_3)_2C=O$　(k) $BrCH=CHBr$ (two answers)

1.49 Tetrafluoroethylene, C_2F_4, is the starting material for the synthesis of the polymer polytetrafluoroethylene (PTFE), one form of which is known as Teflon. Tetrafluoroethylene has a dipole moment of zero. Propose a structural formula for this molecule.

Resonance and Contributing Structures

1.50 Which statements are true about resonance contributing structures?

(a) All contributing structures must have the same number of valence electrons.
(b) All contributing structures must have the same arrangement of atoms.
(c) All atoms in a contributing structure must have complete valence shells.
(d) All bond angles in sets of contributing structures must be the same.

1.51 Draw the contributing structure indicated by the curved arrow(s). Assign formal charges as appropriate.

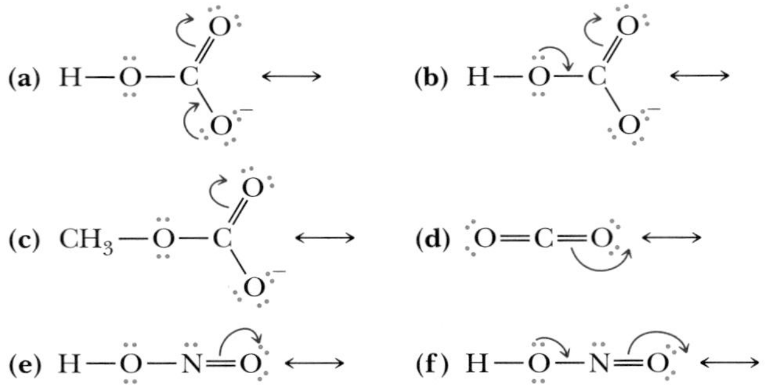

(a)

(b)

(c)

(d)

(e)

(f)

1.52 Using VSEPR, predict the bond angles about the carbon and nitrogen atoms in each pair of contributing structures in Problem 1.51. In what way do these bond angles change from one contributing structure to the other?

1.53 In Problem 1.51 you were given one contributing structure and asked to draw another. Label pairs of contributing structures that are equivalent. For those sets in which the contributing structures are not equivalent, label the more important contributing structure.

1.54 Are the structures in each set valid contributing structures?

(a)

(b)

(c)

(d)

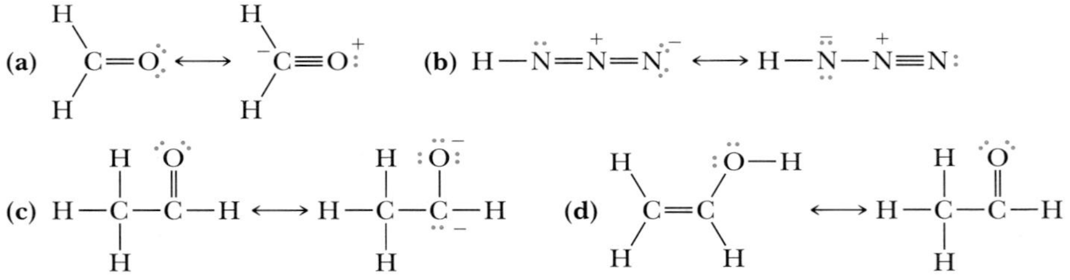

Valence Bond Theory

1.55 State the orbital hybridization of each highlighted atom.

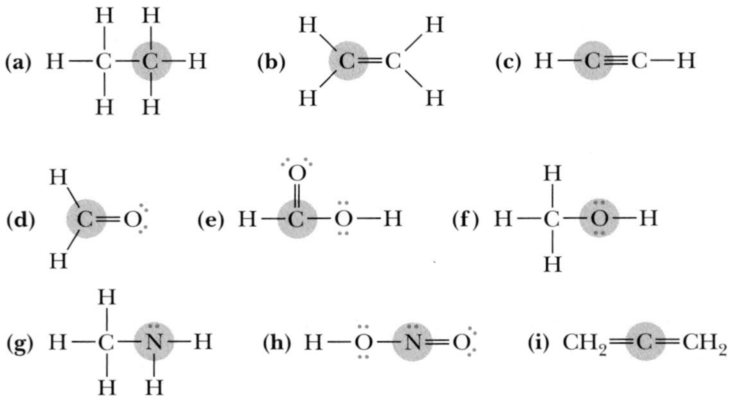

(a)

(b)

(c)

(d)

(e)

(f)

(g)

(h)

(i) $CH_2{=}C{=}CH_2$

1.56 Describe each highlighted bond in terms of the overlap of atomic orbitals.

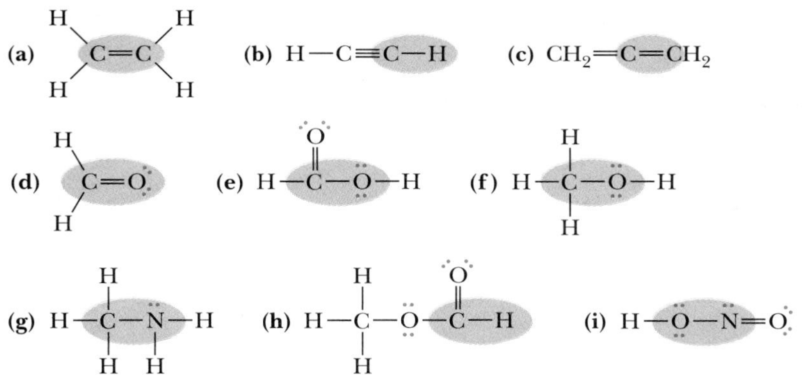

(a) C=C with H, H, H, H

(b) H—C≡C—H

(c) CH_2=C=CH_2

(d) C=O with H, H

(e) H—C—O—H with O double bond and H

(f) H—C—O—H with two H

(g) H—C—N—H with H, H, H

(h) H—C—O—C—H with H, H, H and O double bond

(i) H—O—N=O

1.57 Following is a structural formula of the prescription drug famotidine, manufactured by Merck Sharpe & Dohme under the name Pepcid. The primary clinical use of Pepcid is for the treatment of active duodenal ulcers and benign gastric ulcers. Pepcid is a competitive inhibitor of histamine H_2 receptors and reduces both gastric acid concentration and the volume of gastric secretions.

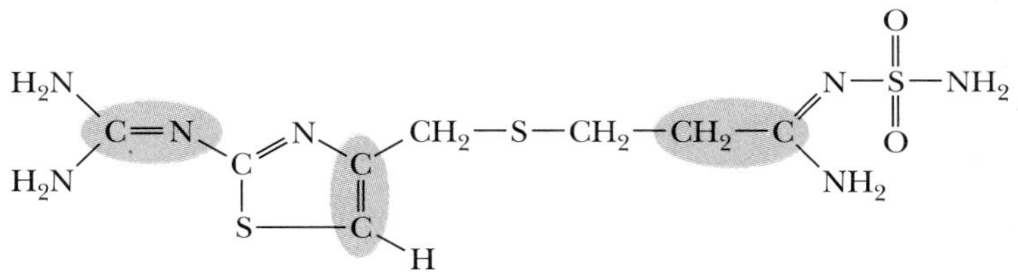

H_2N, H_2N C=N ... C=N ... CH_2—S—CH_2—CH_2—C ... N—S—NH_2 (O, O), NH_2; S, C—H

(a) Complete the Lewis structure of famotidine showing all valence electrons and any formal positive or negative charges.

(b) Describe each circled bond in terms of the overlap of atomic orbitals.

1.58 Draw a Lewis structure for methyl isocyanate, CH_3NCO, showing all valence electrons. Predict all bond angles in this molecule and the hybridization of each C, N, and O.

Additional Problems

1.59 Why are the following molecular formulas impossible?

(a) CH_5 **(b)** C_2H_7

1.60 Each compound contains both ions and covalent bonds. Draw the Lewis structure for each compound, and show by dashes which are covalent bonds and show by charges which are ions.

(a) Sodium methoxide, CH_3ONa **(b)** Ammonium chloride, NH_4Cl
(c) Sodium bicarbonate, $NaHCO_3$ **(d)** Sodium borohydride, $NaBH_4$
(e) Lithium aluminum hydride, $LiAlH_4$

1.61 Predict whether the carbon-metal bond in these organometallic compounds is nonpolar covalent, polar covalent, or ionic. For each polar covalent bond, show the direction of its polarity by the symbols $\delta+$ and $\delta-$.

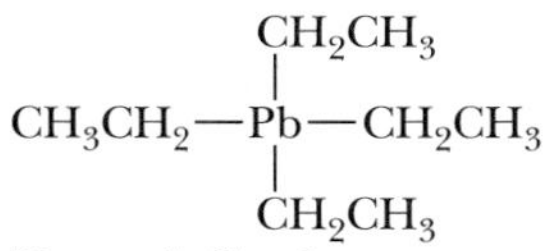

(a) CH_3CH_2—Pb—CH_2CH_3 with CH_2CH_3 above and CH_2CH_3 below
Tetraethyllead

(b) CH_3—Mg—Cl
Methylmagnesium chloride

(c) CH_3—Hg—CH_3
Dimethylmercury

1.62 Silicon is immediately under carbon in the Periodic Table. Predict the geometry of silane, SiH_4.

1.63 Phosphorus is immediately under nitrogen in the Periodic Table. Predict the molecular formula for phosphine, the compound formed by phosphorus and hydrogen. Predict the H—P—H bond angle in phosphine.

1.64 Draw a Lewis structure for the azide ion, N_3^-. (The order of atom attachment is N—N—N, and they do not form a ring.) How does the resonance model account for the fact that the lengths of the N—N bonds in this ion are identical?

1.65 Cyanic acid, HOCN, and isocyanic acid, HNCO, dissolve in water to yield the same anion on loss of H^+.

(a) Write a Lewis structure for cyanic acid.
(b) Write a Lewis structure for isocyanic acid.
(c) Account for the fact that each acid gives the same anion on loss of H^+.

Looking Ahead

1.66 In Chapter 6, we study a group of organic cations called carbocations. Following is the structure of one such carbocation, the *tert*-butyl cation.

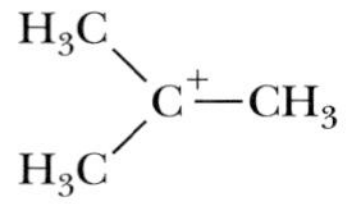

tert-Butyl cation

(a) How many electrons are in the valence shell of the carbon bearing the positive charge?
(b) Using VSEPR, predict the bond angles about this carbon.
(c) Given the bond angle you predicted in (b), what hybridization do you predict for this carbon?

1.67 Many reactions involve a change in hybridization of one or more atoms in the starting material. In each reaction, identify the atoms in the organic starting material that change hybridization, and indicate what the change is. We examine these reactions in more detail later in the course.

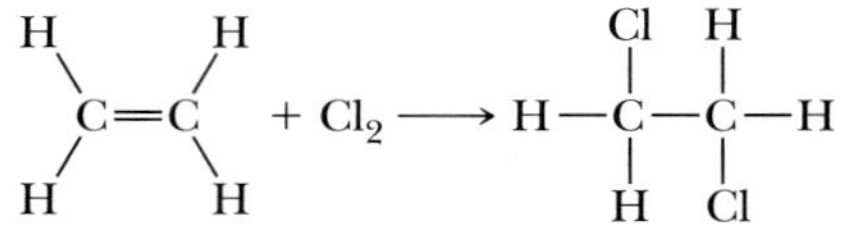

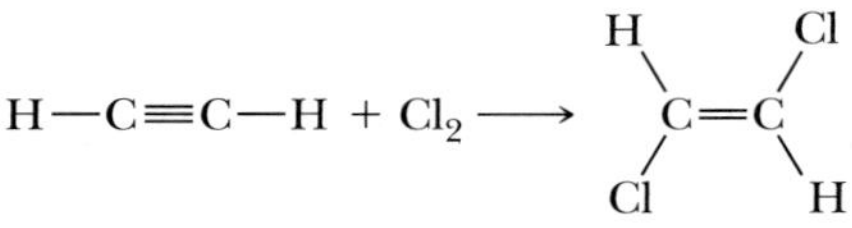

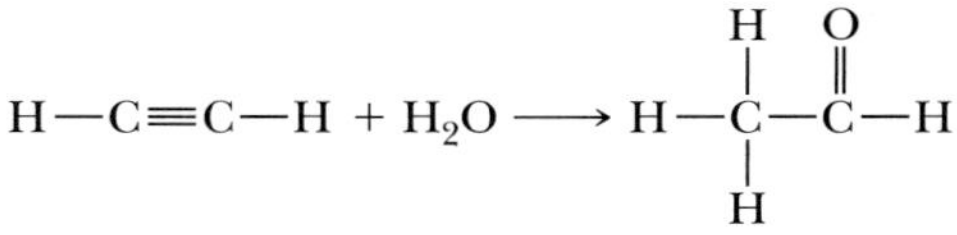

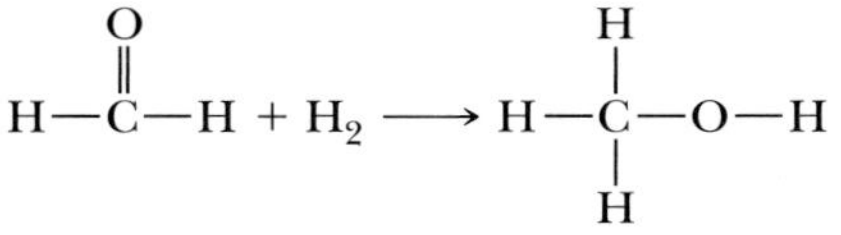

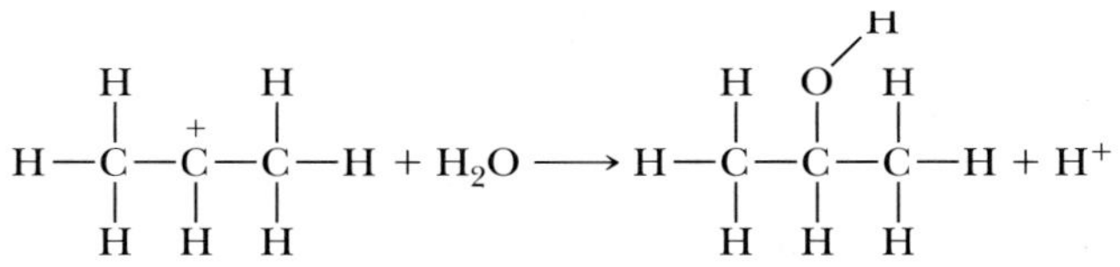

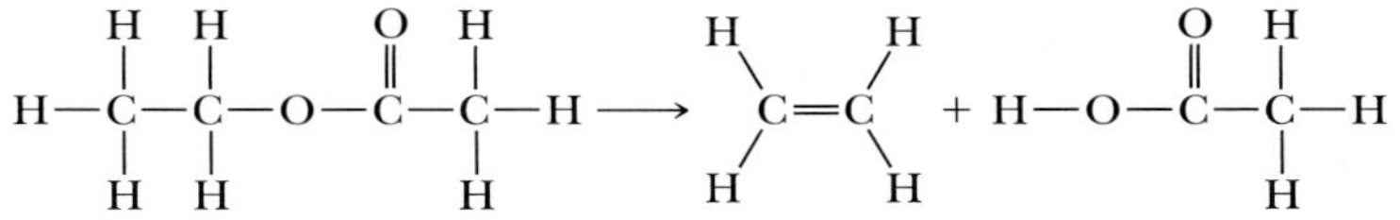

1.68 Following is a structural formula of benzene, C_6H_6, which we study in Chapter 21.

(a) Using VSEPR, predict each H—C—C and C—C—C bond angle in benzene.
(b) State the hybridization of each carbon in benzene.
(c) Predict the shape of a benzene molecule.
(d) Draw important resonance contributing structures.

1.69 Following are three contributing structures for diazomethane, CH_2N_2. This molecule is used to make methyl esters from carboxylic acids (Section 17.7C).

(a) Using curved arrows, show how each contributing structure is converted to the one on its right.
(b) Which contributing structure makes the largest contribution to the hybrid?

1.70 **(a)** Draw a Lewis structure for the ozone molecule, O_3. (The order of atom attachment is O—O—O, and they do not form a ring.) Chemists use ozone to cleave carbon-carbon double bonds (Section 6.5C).
(b) Draw four contributing resonance structures; include formal charges.
(c) How does the resonance model account for the fact that the length of each O—O bond in ozone (128 pm) is shorter than the O—O single bond in hydrogen peroxide (HOOH, 147 pm) but longer than the O—O double bond in the oxygen molecule (123 pm)?

1.71 Molecular Orbitals

The following two compounds are isomers, that is, they are different compounds with the same molecular formula. We discuss this type of isomerism in Chapter 5.

(a) Why are these different molecules that do not interconvert?
(b) Absorption of light by a double bond in a molecule excites one electron from a pi molecular orbital to a pi* molecular orbital. Explain how this absorption can lead to interconversion of the two isomers.

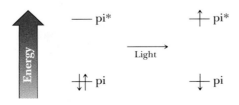

2

■ Bunsen burners burn natural gas, which is primarily methane with small amounts of ethane, propane, and butane (Section 2.10A). Inset: a model of methane, the major component of natural gas.

Outline

ORGANIC
Chemistry ⚛ Now™

Look for this logo in the chapter and go to Organic ChemistryNow at **http://now.brookscole.com/bfi4** for tutorials, simulations, and problems.

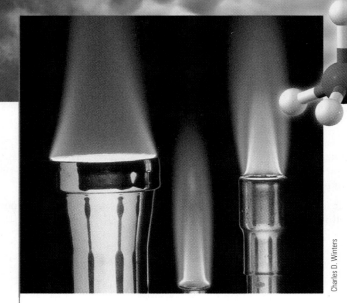

Charles D. Winters

Alkanes and Cycloalkanes

In this chapter, we begin our study of organic compounds with the physical and chemical properties of alkanes, the simplest types of organic compounds. Actually, alkanes are members of a larger group of organic compounds called hydrocarbons. A **hydrocarbon** is a compound composed of only carbon and hydrogen. Figure 2.1 shows the four classes of hydrocarbons, along with the characteristic pattern of bonding between the carbon atoms in each.

Alkanes are **saturated hydrocarbons;** that is, they contain only carbon-carbon single bonds. In this context, *saturated* means that each carbon has the maximum number of hydrogens bonded to it. We often refer to alkanes as **aliphatic hydrocarbons** because the physical properties of the higher members of this class resemble those of the long carbon-chain molecules we find in animal fats and plant oils (Greek: *aleiphar,* fat or oil).

A hydrocarbon that contains one or more carbon-carbon double bonds, triple bonds, or benzene rings is classified as an **unsaturated hydrocarbon.** We study alkanes (saturated hydrocarbons) in this chapter. We study alkenes and alkynes (both unsaturated hydrocarbons) in Chapters 5, 6, and 7, and we study arenes (also unsaturated hydrocarbons) in Chapters 21 and 22.

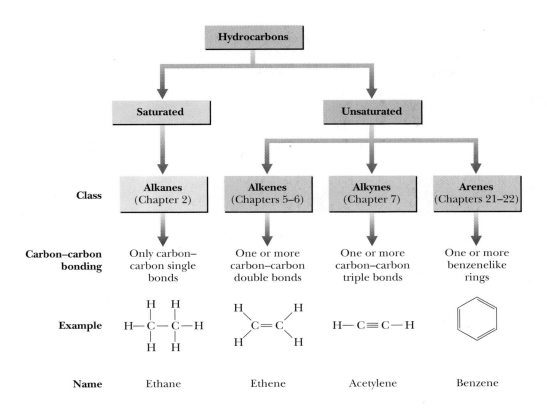

ORGANIC
Chemistry ⚛ Now™
Active Figure 2.1
The four classes of hydrocarbons. **See a simulation based on this figure, and take a short quiz on the concepts.**

2.1 The Structure of Alkanes

Methane (CH_4) and ethane (C_2H_6) are the first two members of the alkane family. Figure 2.2 shows Lewis structures and molecular models for these molecules. The shape of methane is tetrahedral and all H—C—H bond angles are 109.5°. Each carbon atom in ethane is also tetrahedral, and all bond angles are approximately 109.5°.

Although the three-dimensional shapes of larger alkanes are more complex than those of methane and ethane, the four bonds about each carbon are still arranged in a tetrahedral manner, and all bond angles are approximately 109.5°.

The next three alkanes are propane, butane, and pentane. In the following representations, these hydrocarbons are drawn first as condensed structural formulas that show all carbons and hydrogens. They are also drawn in an even more abbreviated form called a **line-angle formula.** In a line-angle formula, each vertex and line ending represents a carbon atom. Although we do not show hydrogen atoms in

Line-angle formula An abbreviated way to draw structural formulas in which vertices and line endings represent carbons.

Figure 2.2
Methane and ethane. Lewis structures and ball-and-stick models.

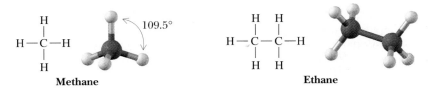

Table 2.1 Names, Molecular Formulas, and Condensed Structural Formulas for the First 20 Alkanes with Unbranched Chains

Name	Molecular Formula	Condensed Structural Formula	Name	Molecular Formula	Condensed Structural Formula
Methane	CH_4	CH_4	Undecane	$C_{11}H_{24}$	$CH_3(CH_2)_9CH_3$
Ethane	C_2H_6	CH_3CH_3	Dodecane	$C_{12}H_{26}$	$CH_3(CH_2)_{10}CH_3$
Propane	C_3H_8	$CH_3CH_2CH_3$	Tridecane	$C_{13}H_{28}$	$CH_3(CH_2)_{11}CH_3$
Butane	C_4H_{10}	$CH_3(CH_2)_2CH_3$	Tetradecane	$C_{14}H_{30}$	$CH_3(CH_2)_{12}CH_3$
Pentane	C_5H_{12}	$CH_3(CH_2)_3CH_3$	Pentadecane	$C_{15}H_{32}$	$CH_3(CH_2)_{13}CH_3$
Hexane	C_6H_{14}	$CH_3(CH_2)_4CH_3$	Hexadecane	$C_{16}H_{34}$	$CH_3(CH_2)_{14}CH_3$
Heptane	C_7H_{16}	$CH_3(CH_2)_5CH_3$	Heptadecane	$C_{17}H_{36}$	$CH_3(CH_2)_{15}CH_3$
Octane	C_8H_{18}	$CH_3(CH_2)_6CH_3$	Octadecane	$C_{18}H_{38}$	$CH_3(CH_2)_{16}CH_3$
Nonane	C_9H_{20}	$CH_3(CH_2)_7CH_3$	Nonadecane	$C_{19}H_{40}$	$CH_3(CH_2)_{17}CH_3$
Decane	$C_{10}H_{22}$	$CH_3(CH_2)_8CH_3$	Eicosane	$C_{20}H_{42}$	$CH_3(CH_2)_{18}CH_3$

line-angle formulas, we assume that they are there in sufficient numbers to give each carbon four bonds.

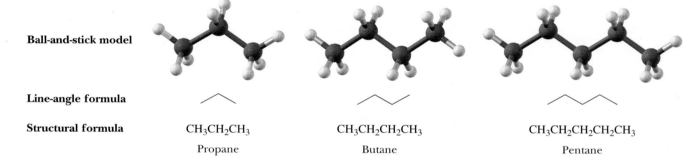

Ball-and-stick model

Line-angle formula

Structural formula

$CH_3CH_2CH_3$	$CH_3CH_2CH_2CH_3$	$CH_3CH_2CH_2CH_2CH_3$
Propane	Butane	Pentane

We can write structural formulas for alkanes in still another abbreviated form. The structural formula of pentane, for example, contains three CH_2 (methylene) groups in the middle of the chain. We can collect them and write the structural formula of pentane as $CH_3(CH_2)_3CH_3$. Table 2.1 gives the names and molecular formulas of the first 20 alkanes. Note that the names of all these alkanes end in *-ane*. We will have more to say about naming alkanes in Section 2.3.

Alkanes have the general molecular formula C_nH_{2n+2}. Thus, given the number of carbon atoms in an alkane, we can determine the number of hydrogens in the molecule and its molecular formula. For example, decane, with 10 carbon atoms, must have $(2 \times 10) + 2 = 22$ hydrogen atoms and a molecular formula of $C_{10}H_{22}$.

2.2 Constitutional Isomerism in Alkanes

Constitutional isomers
Compounds with the same molecular formula but a different connectivity of their atoms.

Constitutional isomers are compounds that have the same molecular formula but different structural formulas. By "different structural formulas," we mean that these compounds differ in the kinds of bonds they have (single, double, or triple) and/or in the connectivity of their atoms.

For the molecular formulas CH_4, C_2H_6, and C_3H_8, only one connectivity is possible. For the molecular formula C_4H_{10}, two connectivities are possible. In one of these, named butane, the four carbons are bonded in a chain; in the other, named 2-methylpropane, three carbons are bonded in a chain with the fourth carbon as a branch on the chain.

ORGANIC
Chemistry⋅⚛⋅Now™

Click Molecular Models to view a number of **Alkane Models**

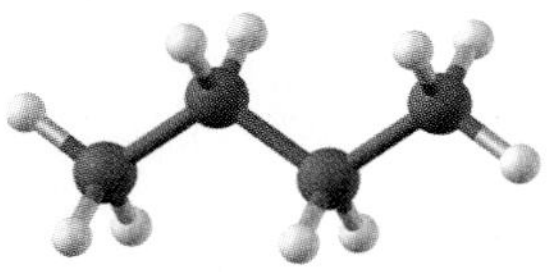

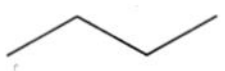

$$CH_3CH_2CH_2CH_3$$

Butane
(bp -0.5°C)

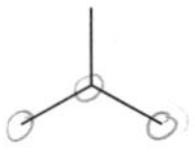

$$CH_3\overset{\overset{\displaystyle CH_3}{|}}{C}HCH_3$$

2-Methylpropane
(bp -11.6°C)

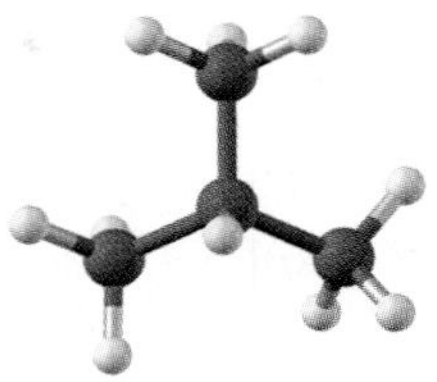

Butane and 2-methylpropane are constitutional isomers; they are different compounds and have different physical and chemical properties. Their boiling points, for example, differ by approximately 11°C.

In Section 1.3, we encountered several examples of constitutional isomers. We saw, for example, that there are two alcohols with molecular formula C_3H_8O, two aldehydes with molecular formula C_4H_8O, and two carboxylic acids with molecular formula $C_4H_8O_2$.

To determine whether two or more structural formulas represent constitutional isomers, write the molecular formula of each and then compare them. All compounds that have the same molecular formula, but different structural formulas, are constitutional isomers.

Example 2.1

Do the condensed formulas in each pair represent the same compound or constitutional isomers?

(a) $CH_3CH_2CH_2CH_2CH_2CH_3$ and $CH_3CH_2CH_2$ (each is C_6H_{14})
$$\qquad\qquad\qquad\qquad\qquad\qquad\qquad\qquad\overset{\displaystyle |}{CH_2CH_2CH_3}$$

(b) $CH_3\overset{\overset{\displaystyle CH_3}{|}}{C}HCH_2\overset{\overset{\displaystyle CH_3}{|}}{C}H$ and $CH_3CH_2\overset{\overset{\displaystyle CH_3}{|}}{C}HCHCH_3$ (each is C_7H_{16})
$$\qquad\quad\;\overset{\displaystyle |}{CH_3}\qquad\qquad\qquad\qquad\quad\overset{\displaystyle |}{CH_3}$$

Solution

(a) The molecules are drawn here as both condensed structural formulas and line-angle formulas. Each formula has an unbranched chain of six carbons; the two are identical and represent the same compound.

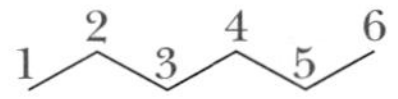

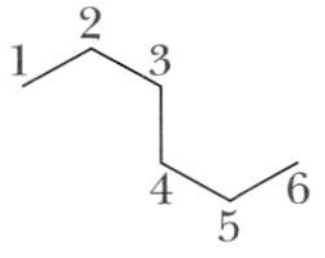

$$\underset{CH_3CH_2CH_2CH_2CH_2CH_3}{\overset{1\quad 2\quad 3\quad 4\quad 5\quad 6}{}} \quad \text{and} \quad \underset{\underset{CH_2CH_2CH_3}{\overset{|}{}}}{\overset{1\quad 2\quad 3}{CH_3CH_2CH_2}}$$

(b) Each formula has a chain of five carbons with two CH_3 branches. Although the branches are identical, they are at different locations on the chains; these formulas represent constitutional isomers.

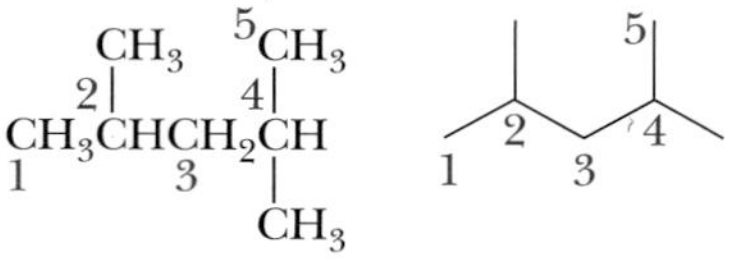

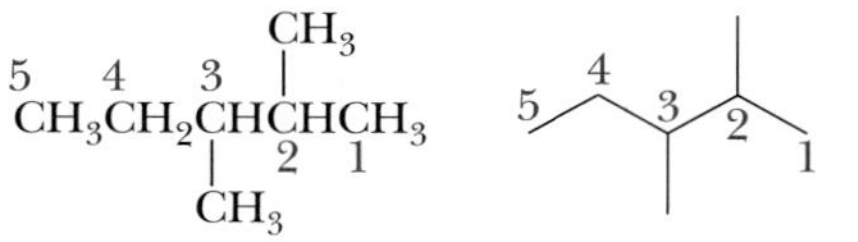

Problem 2.1

Do the line-angle formulas in each pair represent the same compound or constitutional isomers?

(a) and 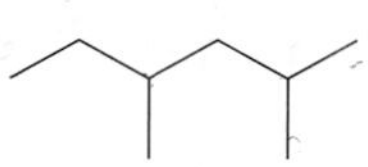   **(b)** 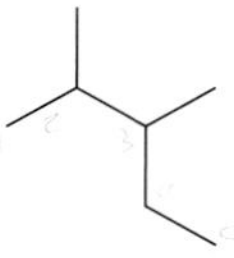 and

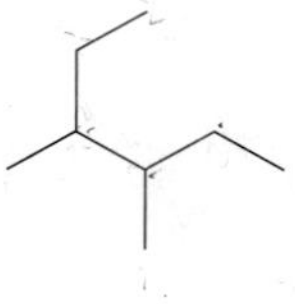

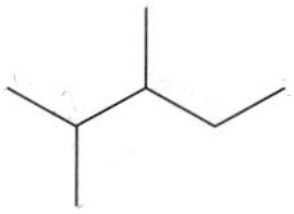

Example 2.2

Write line-angle formulas for the five constitutional isomers with molecular formula C_6H_{14}.

Solution

In solving problems of this type, you should devise a strategy and then follow it. Here is one such strategy. First, draw a line-angle formula for the constitutional isomer with all six carbons in an unbranched chain. Then, draw line-angle formulas for all constitutional isomers with five carbons in a chain and one carbon as a branch on the chain. Finally, draw line-angle formulas for all constitutional isomers with four carbons in a chain and two carbons as branches.

| Six carbons in an unbranched chain | Five carbons in a chain; with one carbon as a branch | Four carbons in a chain; two carbons as branches |

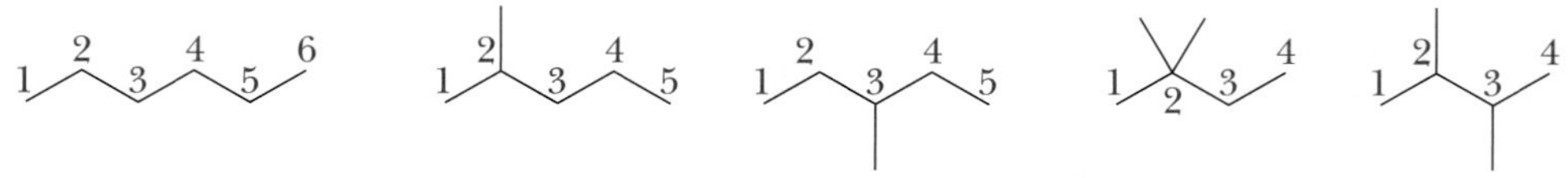

No constitutional isomers with only three carbons in the longest chain are possible for C_6H_{14}.

Problem 2.2

Draw line-angle formulas for the three constitutional isomers with molecular formula C_5H_{12}.

Carbon Atoms	Constitutional Isomers
1	0
5	3
10	75
15	4,347
25	36,797,588

   The ability of carbon atoms to form strong bonds with other carbon atoms results in a staggering number of constitutional isomers. As the table shows, there are 3 constitutional isomers with molecular formula C_5H_{12}, 75 constitutional isomers with molecular formula $C_{10}H_{22}$, and almost 37 million with molecular formula $C_{25}H_{52}$.
   Thus, for even a small number of carbon and hydrogen atoms, a very large number of constitutional isomers is possible. Because constitutional isomers have different chemical properties, a rich diversity of chemistry is possible within these sets.

2.3 Nomenclature of Alkanes

ORGANIC
Chemistry⋅ᵢ⋅Now™
Click Simulations to explore the
relationship between **Structure
and Nomenclature of Alkanes**

A. The IUPAC System

Ideally, every organic compound should have a name from which its structural formula can be drawn. For this purpose, chemists have adopted a set of rules established by the International Union of Pure and Applied Chemistry (IUPAC).

The IUPAC name of an alkane with an unbranched chain of carbon atoms consists of two parts: (1) a prefix that indicates the number of carbon atoms in the chain and (2) the suffix *-ane* to show that the compound is a saturated hydrocarbon. Table 2.2 gives the prefixes used to show the presence of 1 to 20 carbon atoms.

The first four prefixes listed in Table 2.2 were chosen by the IUPAC because they were well established in the language of organic chemistry. In fact, they were well established even before there were hints of the structural theory underlying the discipline. For example, the prefix *but-* appears in the name *butyric acid,* a compound of four carbon atoms formed by air oxidation of butter (Latin: *butyrum,* butter). Prefixes to show five or more carbons are derived from Greek or Latin numbers. See Table 2.1 for the names, molecular formulas, and condensed structural formulas for the first 20 unbranched alkanes.

The IUPAC name of an alkane with a branched chain consists of a parent name that indicates the longest chain of carbon atoms in the compound and substituent names that indicate the groups bonded to the parent chain.

4-Methyloctane

A substituent group derived from an alkane by the removal of a hydrogen atom is called an **alkyl group,** and is commonly represented by the symbol R—. We name

Alkyl group A group derived by removing a hydrogen from an alkane; given the symbol R—.

Table 2.2 Prefixes Used in the IUPAC System to Show the Presence of 1 to 20 Carbon Atoms in an Unbranched Chain

Prefix	Number of Carbon Atoms	Prefix	Number of Carbon Atoms
meth-	1	undec-	11
eth-	2	dodec-	12
prop-	3	tridec-	13
but-	4	tetradec-	14
pent-	5	pentadec-	15
hex-	6	hexadec-	16
hept-	7	heptadec-	17
oct-	8	octadec-	18
non-	9	nonadec-	19
dec-	10	eicos-	20

alkyl groups by dropping the -*ane* from the name of the parent alkane and adding the suffix -*yl*. The substituent derived from methane, for example, is methyl, CH_3— and that derived from ethane is ethyl, CH_3CH_2—.

The rules of the IUPAC system for naming alkanes follow:

1. The name for an alkane with an unbranched chain of carbon atoms consists of a prefix showing the number of carbon atoms in the chain and the ending -*ane*.
2. For branched-chain alkanes, take the longest chain of carbon atoms as the parent chain; its name becomes the root name.
3. Give each substituent on the parent chain a name and a number. The number shows the carbon atom of the parent chain to which the substituent is bonded. Use a hyphen to connect the number to the name.

$$CH_3$$
$$\overset{|}{CH_3CHCH_3}$$

2-Methylpropane

4. If there is one substituent, number the parent chain from the end that gives it the lower number.

$$CH_3$$
$$\overset{|}{CH_3CH_2CH_2CHCH_3}$$

2-Methylpentane   (not 4-methylpentane)

5. If there are two or more identical substituents, number the parent chain from the end that gives the lower number to the substituent encountered first. The number of times the substituent occurs is indicated by a prefix *di-, tri-, tetra-, penta-, hexa-,* and so on. A comma is used to separate position numbers.

2,4-Dimethylhexane   (not 3,5-Dimethylhexane)

6. If there are two or more different substituents, list them in alphabetical order, and number the chain from the end that gives the lower number to the substituent encountered first. If there are different substituents in equivalent positions on opposite ends of the parent chain, give the substituent of lower alphabetical order the lower number.

3-Ethyl-5-Methylheptane   (not 3-methyl-5-ethylheptane)

7. The prefixes *di-, tri-, tetra-,* and so on are not included in alphabetizing. Alphabetize the names of the substituents first, and then insert these prefixes. In the following example, the alphabetizing parts are *ethyl* and *methyl,* not ethyl and dimethyl.

4-Ethyl-2,2-dimethylhexane
(not 2,2-dimethyl-4-ethylhexane)

Table 2.3   Names for Alkyl Groups with One to Five Carbons. Common Names and Their Abbreviations are Given in Parentheses

Name	Condensed Structural Formula	Name	Condensed Structural Formula
Methyl (Me)	—CH₃	1,1-Dimethylethyl (*tert*-butyl, *t*-Bu)	$\begin{array}{c} CH_3 \\ | \\ -CCH_3 \\ | \\ CH_3 \end{array}$
Ethyl (Et)	—CH₂CH₃		
Propyl (Pr)	—CH₂CH₂CH₃		
1-Methylethyl (isopropyl, iPr)	$-CHCH_3$ $\quad|$ $\quad CH_3$	Pentyl	—CH₂CH₂CH₂CH₂CH₃
		3-Methylbutyl (isopentyl)	$-CH_2CH_2CHCH_3$ $\quad\quad\quad\quad|$ $\quad\quad\quad\quad CH_3$
Butyl (Bu)	—CH₂CH₂CH₂CH₃		
2-Methylpropyl (isobutyl, iBu)	$-CH_2CHCH_3$ $\quad\quad\;|$ $\quad\quad\; CH_3$	2-Methylbutyl	$-CH_2CHCH_2CH_3$ $\quad\quad|$ $\quad\quad CH_3$
1-Methylpropyl (*sec*-butyl, *s*-Bu)	$-CHCH_2CH_3$ $\quad|$ $\quad CH_3$	2,2-Dimethylpropyl (neopentyl)	$\begin{array}{c} CH_3 \\ | \\ -CH_2CCH_3 \\ | \\ CH_3 \end{array}$

8. Where there are two or more parent chains of identical length, choose the parent chain with the greater number of substituents.

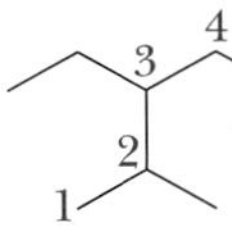

3-Ethyl-2-methylhexane 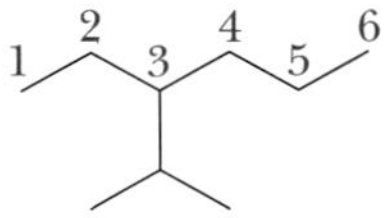   (not 3-Isopropylhexane)

Substituents are named following this same set of rules. Those with unbranched chains are named by dropping -*ane* from the name of the parent alkane and replacing it with -*yl*. Thus, unbranched alkyl substituents are named *methyl, ethyl, propyl, butyl, pentyl,* and so forth. Substituents with branched chains are named according to rules 2 and 3. The IUPAC names and structural formulas for unbranched and branched alkyl groups containing one to five carbon atoms are given in Table 2.3. Also given in parentheses are common names for these substituents.

Example 2.3

Write the IUPAC and common names for these alkanes.

(a) 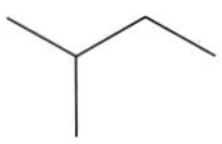   **(b)**

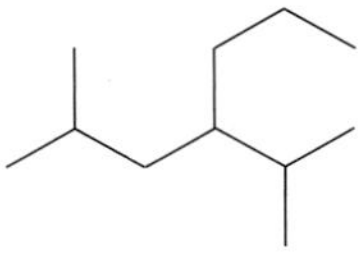

Solution

Number the longest chain in each compound from the end of the chain toward the substituent that is encountered first. For (a), the longest chain is four carbons (a butane) with a methyl group on carbon 2. For (b), the longest chain is seven carbons (a heptane), with substituents on carbons 2 and 4.

(a)

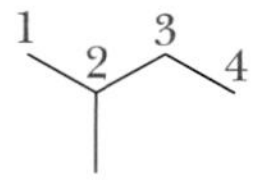

2-Methylbutane

(b)

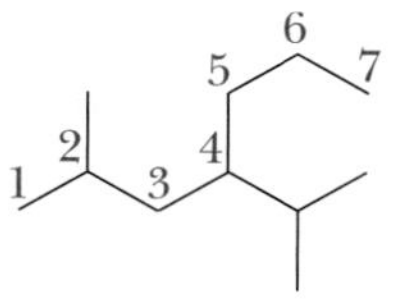

2-Methyl-4-(1-methylethyl)heptane
(4-Isopropyl-2-methylheptane)

Problem 2.3

Write IUPAC names for these alkanes.

(a) 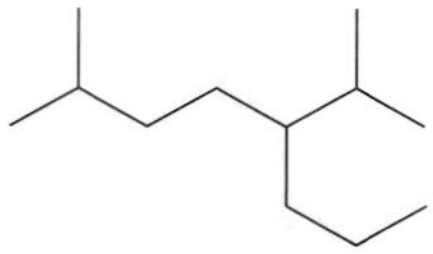   **(b)**

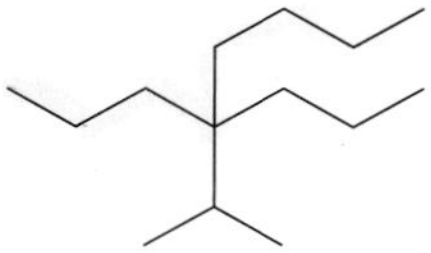

B. Common Names

In an alternative system known as common nomenclature, the total number of carbon atoms in an alkane, regardless of their arrangement, determines the name. The first three alkanes are methane, ethane, and propane. All alkanes of formula C_4H_{10} are called butanes, all those of formula C_5H_{12} are called pentanes, all those of formula C_6H_{14} are called hexanes, and so forth. The fact that an alkane chain is unbranched is sometimes indicated by the prefix *n-* (normal); an example is *n*-pentane for $CH_3CH_2CH_2CH_2CH_3$. For branched-chain alkanes beyond propane, *iso-* indicates that one end of an otherwise unbranched chain terminates in a $(CH_3)_2CH-$ group, and *neo-* indicates that it terminates in $-C(CH_3)_3$. Following are examples of common names.

$CH_3CH_2CH_2CH_3$ $CH_3\underset{|}{\overset{CH_3}{CH}}CH_3$ $CH_3CH_2CH_2CH_2CH_3$ $CH_3CH_2\underset{|}{\overset{CH_3}{CH}}CH_3$ $CH_3\underset{\underset{CH_3}{|}}{\overset{\overset{CH_3}{|}}{C}}CH_3$

Butane   Isobutane   Pentane   Isopentane   Neopentane

This system of common names has no good way of naming other branching patterns; for more complex alkanes, it is necessary to use the more flexible IUPAC system of nomenclature.

In this text, we concentrate on IUPAC names. However, we also use common names, especially when the common name is used almost exclusively in the everyday discussions among chemists. When both IUPAC and common names are given in the text, we give the IUPAC name first, followed by the common name in parentheses. In this way, you should have no doubt about which name is which.

C. Classification of Carbon and Hydrogen Atoms

We classify a carbon atom as primary (1°), secondary (2°), tertiary (3°), or quaternary (4°), depending on the number of carbon atoms bonded to it. A carbon bonded to one carbon atom is a primary carbon; a carbon bonded to two carbon atoms is a secondary carbon, and so forth. For example, propane contains two primary carbons and one secondary carbon; 2-methylpropane contains three primary carbons and one tertiary carbon; and 2,2,4-trimethylpentane contains five primary carbons, one secondary carbon, one tertiary carbon, and one quaternary carbon.

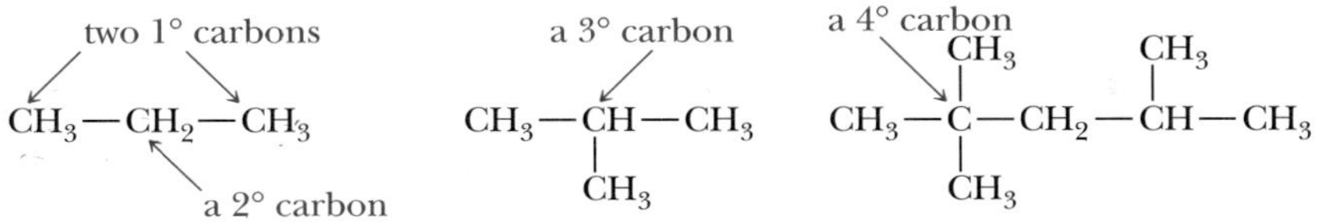

$$CH_3 - CH_2 - CH_3 \qquad CH_3 - \underset{\underset{CH_3}{|}}{CH} - CH_3 \qquad CH_3 - \underset{\underset{CH_3}{|}}{\overset{\overset{CH_3}{|}}{C}} - CH_2 - \underset{}{CH} - CH_3$$

two 1° carbons   a 3° carbon   a 4° carbon

a 2° carbon

Propane   2-Methylpropane   2,2,4-Trimethylpentane

Hydrogens are also classified as primary, secondary, or tertiary depending on the type of carbon to which each is bonded. Those bonded to a primary carbon are classified as primary hydrogens, those on a secondary carbon are secondary hydrogens, and those on a tertiary carbon are tertiary hydrogens.

2.4 Cycloalkanes

A hydrocarbon that contains carbon atoms joined to form a ring is called a **cyclic hydrocarbon.** When all carbons of the ring are saturated, the hydrocarbon is called a **cycloalkane.**

Cycloalkane A saturated hydrocarbon that contains carbons joined to form a ring.

A. Structure and Nomenclature

Cycloalkanes of ring sizes from 3 to over 30 are found in nature, and, in principle, there is no limit to ring size. Five-membered rings (cyclopentanes) and six-membered rings (cyclohexanes) are especially common and will receive special attention. Figure 2.3 shows structural formulas of cyclopropane, cyclobutane, cyclopentane, and cyclohexane. When writing structural formulas for cycloalkanes, chemists rarely show all carbons and hydrogens. Rather, they use line-angle formulas to represent cycloalkane rings. Each ring is represented by a regular polygon that has the same number of sides as there are carbon atoms in the ring. For example, chemists represent cyclobutane by a square, cyclopentane by a pentagon, and cyclohexane by a hexagon.

Cycloalkanes contain two fewer hydrogen atoms than an alkane with the same number of carbon atoms. For example, compare the molecular formulas of cyclohexane, C_6H_{12}, and hexane, C_6H_{14}. The general formula of a cycloalkane is $\mathbf{C_nH_{2n}}$.

To name a cycloalkane, prefix the name of the corresponding open-chain alkane with *cyclo-*, and name each substituent on the ring. If there is only one substituent on

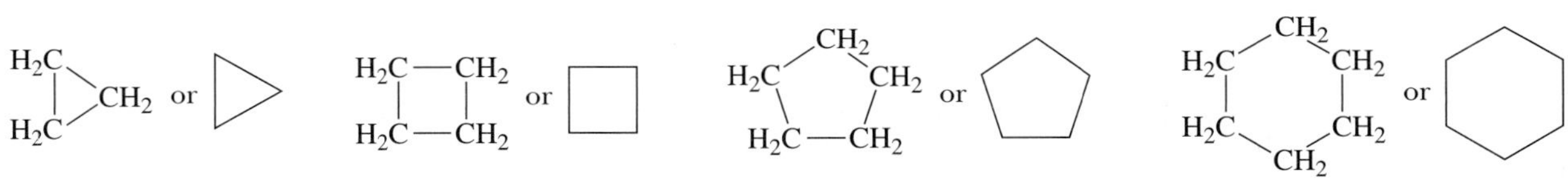

Figure 2.3
Examples of cycloalkanes.

Cyclopropane
C_3H_6

Cyclobutane
C_4H_8

Cyclopentane
C_5H_{10}

Cyclohexane
C_6H_{12}

the cycloalkane ring, there is no need to give it a number. If there are two sub-stituents, number the ring by beginning with the substituent of lower alphabetical order. If there are three or more substituents, number the ring so as to give the substituents the lowest set of numbers, and list them in alphabetical order.

Example 2.4

C_nH_{2n}

Write the molecular formula and the IUPAC name for each cycloalkane.

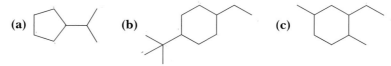

(a)   **(b)**   **(c)**

Solution

(a) First replace each vertex and line terminus with a carbon and then add hydro-gens as necessary to give each carbon four bonds. The molecular formula of this compound is C_8H_{16}. Because there is only one substituent on the ring, there is no need to number the atoms of the ring. This compound's name is (1-methylethyl)cyclopentane. Alternatively, the substituent can be named isopropyl, giving the name isopropylcyclopentane.

$$C_8H_{16}$$

(b) The two substituents are ethyl and 1,1-dimethylethyl, and the IUPAC name of the cycloalkane is 1-ethyl-4-(1,1-dimethylethyl)cyclohexane. Alternatively, the sub-stituents are named ethyl and *tert*-butyl, giving the cycloalkane the name 1-*tert*-butyl-4-ethylcyclohexane. Its molecular formula is $C_{12}H_{24}$.

(c) Number the ring to give the three substituents the lowest set of numbers and then list them in alphabetical order. The name of this compound is 2-ethyl-1,4-dimethylcyclohexane, and its molecular formula is $C_{10}H_{20}$.

Problem 2.4

Write the molecular formula, IUPAC name, and common name for each cycloalkane.

(a)   **(b)**   **(c)**

B. Bicycloalkanes

Bicycloalkane An alkane containing two rings that share two carbons.

An alkane that contains two rings that share two carbon atoms in common is classified as a **bicycloalkane.** The shared carbon atoms are called **bridgehead carbons,** and the carbon chains connecting them are called **bridges.** The general formula of a

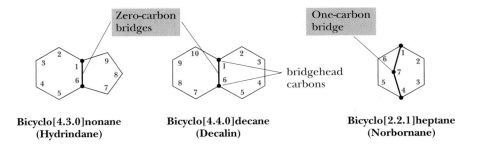

Figure 2.4
Examples of bicycloalkanes.

Bicyclo[4.3.0]nonane
(Hydrindane)

Bicyclo[4.4.0]decane
(Decalin)

Bicyclo[2.2.1]heptane
(Norbornane)

bicycloalkane is C_nH_{2n-2}. Figure 2.4 shows three examples of bicycloalkanes along with the IUPAC and common name of each.

Example 2.5

Write the general formula for an alkane, a cycloalkane, and a bicycloalkane. How do these general formulas differ?

Solution

General formulas are C_nH_{2n+2} for an alkane, C_nH_{2n} for a cycloalkane, and C_nH_{2n-2} for a bicycloalkane. Each general formula in this series has two fewer hydrogens than the previous member of the series.

Problem 2.5

Write molecular formulas for each bicycloalkane, given its number of carbon atoms.

(a) Hydrindane (9 carbons)   **(b)** Decalin (10 carbons)   **(c)** Norbornane (7 carbons)

IUPAC names of bicycloalkanes are derived in the following way:

1. The parent name of a bicycloalkane is that of the hydrocarbon with the same number of carbon atoms as are in the bicyclic ring system. For example, the first compound in Figure 2.4 contains nine carbons and is, therefore, a bicyclononane. The second compound contains ten carbons and is a bicyclodecane.
2. Begin numbering at one bridgehead carbon and proceed along the longest bridge to the second bridgehead carbon, then along the next longest bridge back to the original bridgehead carbon, and so on until all ring carbons are numbered. If there are two bridges of the same length, proceed along the one that gives the lower number to the first encountered substituent. The name and location of substituents are shown by the rules given in Section 2.3A. If there is a choice of numbering patterns, choose the one that gives substituents the lowest possible numbers.
3. Show the length of the bridges by counting the number of carbons linking the bridgeheads and placing them in decreasing order in brackets between the prefix *bicyclo-* and the parent name and with periods separating each number. For example, the first compound in Figure 2.4 has two bridgehead carbons. There are four carbons in the first bridge, three in the second, and none in the third; its name is bicyclo[4.3.0]nonane. Here are additional examples:

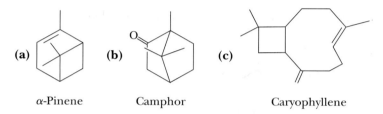

2-Methylbicyclo[4.4.0]decane   1,7,7-Trimethylbicyclo[2.2.1]heptane

---

Example 2.6

Following are structural formulas and common names for three bicyclic compounds. Write the molecular formula of each compound, and name the bicycloalkane from which it is derived.

(a) α-Pinene   **(b)** Camphor   **(c)** Caryophyllene

Solution

(a) The molecular formula of α-pinene is $C_{10}H_{16}$, and the bicycloalkane from which it is derived is bicyclo[3.1.1]heptane. α-Pinene is a major component, often as high as 65% by volume, of pine oil and turpentine.

(b) The molecular formula of camphor is $C_{10}H_{16}O$, and the bicycloalkane from which it is derived is bicyclo[2.2.1] heptane. Camphor, obtained from the camphor tree, *Cinnamonium camphora*, is used in the manufacture of certain plastics, lacquers, and varnishes.

(c) The molecular formula of caryophyllene is $C_{15}H_{24}$, and the bicycloalkane from which it is derived is bicyclo[7.2.0]undecane. Caryophyllene is one of the fragrant components of oil of cloves.

Problem 2.6

Draw structural formulas for the following bicycloalkanes:

(a) Bicyclo[3.1.0]hexane   **(b)** Bicyclo[2.2.2]octane
(c) Bicyclo[4.2.0]octane   **(d)** 2,6,6-Trimethylbicyclo[3.1.1]heptane

---

2.5 The IUPAC System—A General System of Nomenclature

The naming of alkanes and cycloalkanes in Sections 2.3 and 2.4 illustrates the application of the IUPAC system of nomenclature to two specific classes of organic compounds. Now let us describe the general approach of the IUPAC system. The name we give to any compound with a chain of carbon atoms consists of three parts: a **prefix**, an **infix** (a modifying element inserted into a word), and a **suffix.** Each part provides specific information about the structure of the compound.

1. The prefix indicates the number of carbon atoms in the parent chain. Prefixes that show the presence of 1 to 20 carbon atoms in an unbranched chain are given in Table 2.2.
2. The infix indicates the nature of the carbon-carbon bonds in the parent chain.

Infix	Nature of Carbon-Carbon Bonds in the Parent Chain
-an-	all single bonds
-en-	one or more double bonds
-yn-	one or more triple bonds

3. The suffix indicates the class of compound to which the substance belongs.

Suffix	Class of Compound
-e	hydrocarbon
-ol	alcohol
-al	aldehyde
-amine	amine
-one	ketone
-oic acid	carboxylic acid

Example 2.7

Following are IUPAC names and structural formulas for four compounds:

(a) $CH_2{=}CHCH_3$ **(b)** CH_3CH_2OH **(c)**    **(d)** $HC{\equiv}CH$

　　Propene　　　　　　　Ethanol　　　　　　　　Pentanoic acid　　　　　　　Ethyne

Divide each name into a prefix, an infix, and a suffix, and specify the information about the structural formula that is contained in each part of the name.

Solution

　　　　　　　 ⌐a carbon-carbon double bond　　　　　　　　 ⌐only carbon-carbon single bonds
(a) prop-en-e ←—a hydrocarbon　　　　**(b)** eth-an-ol ←—an alcohol
　　　 ⌐three carbon atoms　　　　　　　　　　　 ⌐two carbon atoms

　　　　　　　 ⌐only carbon-carbon single bonds　　　　　　 ⌐a carbon-carbon triple bond
(c) pent-an-oic acid ⌐a carboxylic acid　　**(d)** eth-yn-e ←—a hydrocarbon
　　　 ⌐five carbon atoms　　　　　　　　　　　 ⌐two carbon atoms

Problem 2.7

Combine the proper prefix, infix, and suffix, and write the IUPAC name for each compound:

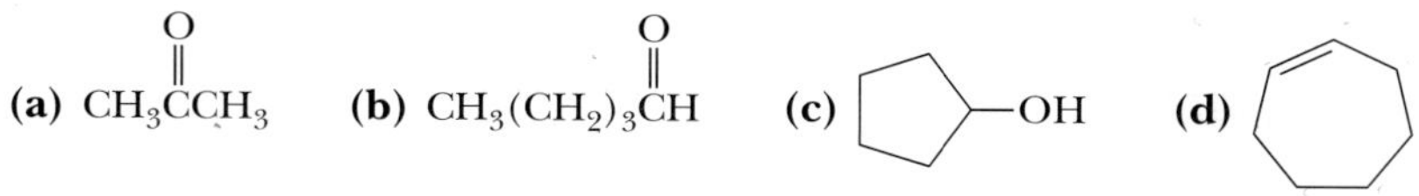

(a) $CH_3\overset{\displaystyle O}{\overset{\|}{C}}CH_3$ (b) $CH_3(CH_2)_3\overset{\displaystyle O}{\overset{\|}{C}}H$ (c) ⬠—OH   (d) ⬡

2.6 Conformations of Alkanes and Cycloalkanes

Conformation Any three-dimensional arrangement of atoms in a molecule that results by rotation about a single bond.

Staggered conformation A conformation about a carbon-carbon single bond in which the atoms or groups on one carbon are as far apart as possible from atoms or groups on an adjacent carbon.

Newman projection A way to view a molecule by looking along a carbon-carbon single bond.

Eclipsed conformation A conformation about a carbon-carbon single bond in which the atoms or groups on one carbon are as close as possible to the atoms or groups on an adjacent carbon.

Dihedral angle The angle created by two intersecting planes.

Torsional strain Strain that arises when nonbonded atoms separated by three bonds are forced from a staggered conformation to an eclipsed conformation. Torsional strain is also called eclipsed-interaction strain.

Structural formulas are useful for showing the connectivity of atoms in a molecule. However, they usually do not show three-dimensional shapes. As chemists try to understand more and more about the relationships between structure and the chemical and physical properties of compounds, it becomes increasingly important to know more about the three-dimensional shapes of molecules.

In this section, we ask you to look at molecules as three-dimensional objects and to visualize not only bond angles, but also distances between various atoms and groups of atoms within the molecules. We also describe intramolecular strain, which we divide into three types: torsional strain, steric strain, and angle strain. We urge you to build models (either physically or with desktop modeling programs such as Chem3D or Spartan) of the molecules discussed in this section so that you become comfortable in dealing with them as three-dimensional objects, and understand fully the origins of intramolecular strain.

A. Alkanes

Alkanes of two or more carbons can be twisted into a number of different three-dimensional arrangements of their atoms by rotating about one or more carbon-carbon bonds. Any three-dimensional arrangement of atoms that results from rotation about single bonds is called a **conformation.** Figure 2.5(a) shows a ball-and-stick model of a **staggered conformation** of ethane. In this conformation, the three C—H bonds on one carbon are as far apart as possible from those bonds on the adjacent carbon. Figure 2.5(b), called a **Newman projection,** (named for Melvin Newman, of Ohio State University, who developed these projections) is a shorthand way to represent this conformation of ethane. In a Newman projection, a molecule is viewed down the axis of a C—C bond. The three atoms or groups of atoms on the carbon nearer your eye are shown on lines extending from the center of the circle at angles of 120°. The three atoms or groups of atoms on the carbon farther from your eye are shown on lines extending from the circumference of the circle, also at angles

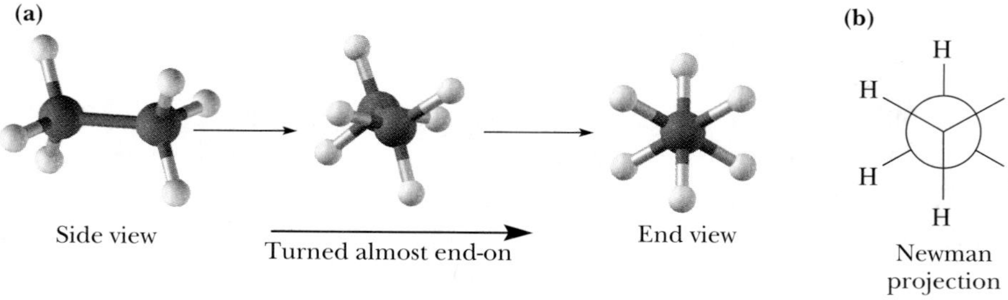

Figure 2.5

A staggered conformation of ethane. (a) Ball-and-stick models and (b) Newman projection.

(a) Side view → Turned almost end-on → End view

(b) Newman projection

of 120°. Remember that bond angles about each carbon in ethane are approximately 109.5° and not 120°, as this Newman projection might suggest. The three lines in front represent bonds directed toward you, whereas the three lines in back point away from you.

Figure 2.6 shows a ball-and-stick model and a Newman projection for an **eclipsed conformation** of ethane. In this conformation, the three C—H bonds on one carbon are as close as possible to the three C—H bonds on the adjacent carbon. In other words, hydrogen atoms on the back carbon are eclipsed by the hydrogen atoms on the front carbon (just as the sun is eclipsed when the moon passes in front of it). Note that different conformations that have long enough lifetimes to be characterized are often called **conformational isomers.**

If we are to discuss energy relationships among conformations, it is convenient to define the term *dihedral angle*. A **dihedral angle**, θ (Greek theta), is the angle created by two intersecting planes, each defined by three atoms. In the Newman projection of the eclipsed conformation of ethane in Figure 2.7(a), two H—C—C planes are shown. The angle at which these planes intersect (the dihedral angle) is 0°. A staggered conformation in which the dihedral angle of the two H—C—C planes is 60° is illustrated in Figure 2.7(b).

In principle, there are an infinite number of conformations of ethane that differ only in the degree of rotation about the carbon-carbon single bond. There is a small energy barrier between conformations, however, and rotation is not completely free. As we shall see, the lowest energy (the most stable) conformation of ethane is a staggered conformation. The highest energy (the least stable) conformation is an eclipsed conformation. At room temperature, ethane molecules undergo collisions with sufficient energy so that the barrier can be crossed and rotation about the carbon-carbon single bond from one conformation to another occurs rapidly.

The difference in energy between an eclipsed conformation and a staggered conformation of ethane is approximately 12.6 kJ (3.0 kcal)/mol, and is referred to as torsional strain. **Torsional strain** (also called eclipsed-interaction strain) arises when nonbonded atoms separated by three bonds are forced from a staggered conformation to an eclipsed conformation. In ethane, for example, torsional strain occurs when pairs of hydrogens H(4)-H(6), H(5)-H(8), and H(3)-H(7) on adjacent carbons are forced into eclipsed positions. The models shown here represent but one of the three different but equivalent eclipsed conformations of ethane.

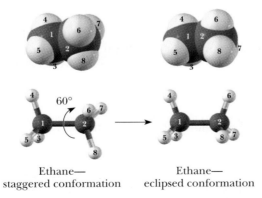

Ethane— staggered conformation

Ethane— eclipsed conformation

Figure 2.8 shows the relationship between energy and dihedral angle for the conformations of ethane. All energy diagrams in this book use *energy* as a vertical axis. Several types of energy—potential energy, free energy, and enthalpy—are important in various contexts. The diagrams for all these types of energy are often nearly

(a)

Side view

(b)

Turned almost end-on

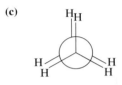

(c)

Newman projection

Figure 2.6
An eclipsed conformation of ethane. (a, b) Ball-and-stick models and (c) Newman projections.

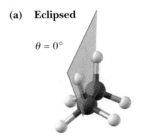

(a) Eclipsed

$\theta = 0°$

(b) Staggered

$\theta = 60°$

Figure 2.7
Dihedral angles in ethane. (a) An eclipsed conformation and (b) a staggered conformation.

Figure 2.8
The energy of ethane as a function of dihedral angle. The eclipsed conformations are approximately 12.6 kJ (3.0 kcal)/mol higher in energy than the staggered conformations.

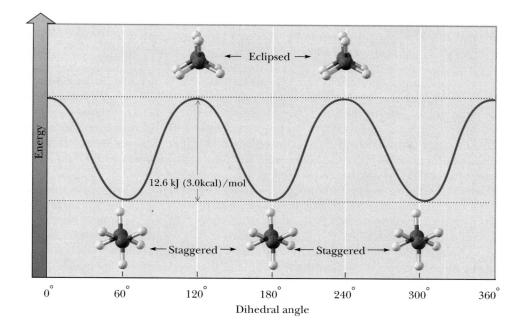

indistinguishable, and in most cases "energy" will do for the concepts we are introducing. When it is necessary to be more precise about the type of energy, we will do so and explain why.

There has been controversy over the years as to the origin of the torsional strain in the eclipsed conformations of ethane. It was originally thought that this strain is the result of the repulsion between eclipsed hydrogen nuclei; they are separated, it was reasoned, by 255 pm in a staggered conformation, but by only 235 pm in an eclipsed conformation. Alternatively, it has been held that the torsional strain is the result of repulsion between the electron clouds of the adjacent C—H bonds. Theoretical molecular orbital calculations, however, suggest that the energy difference between the conformational extremes arises not from destabilization of the eclipsed conformation, but rather from stabilization of the staggered conformation. This stabilization of the staggered conformation arises because of a small donor-acceptor interaction (donation of electron density from a filled orbital into an empty acceptor orbital) between a C—H bonding MO of one carbon and the C—H antibonding MO on the adjacent carbon with which it is aligned. This donor-acceptor stabilization is lost when a staggered conformation is converted to an eclipsed conformation. Thus, although the fact remains that eclipsed ethane is higher in energy than staggered ethane, the reason for this energy difference is still a matter of controversy among chemists.

Next let us look at the conformations of butane viewed along the bond between carbons 2 and 3. For butane, there are two types of staggered conformations and two types of eclipsed conformations. The staggered conformation in which the methyl groups are the maximum distance apart ($\theta = 180°$) is called the **anti conformation;** the staggered conformation in which they are closer together ($\theta = 60°$) is called the **gauche conformation.** In one eclipsed conformation ($\theta = 0°$), methyl is eclipsed by methyl. In the other ($\theta = 120°$), methyl is eclipsed by hydrogen. Figure 2.9 shows the energy relationships for rotation from $-180°$ to $180°$. Note that both the gauche and anti conformations of

Anti conformation   A conformation about a single bond in which two groups on adjacent carbons lie at a dihedral angle of 180°.

Gauche conformation   A conformation about a single bond of an alkane in which two groups on adjacent carbons lie at a dihedral angle of 60°.

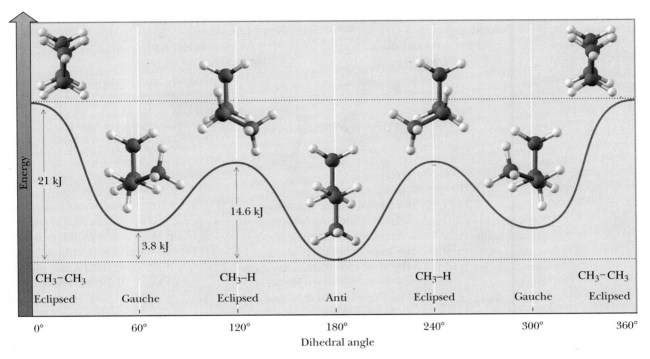

ORGANIC
Chemistry Now™
Active Figure 2.9
The energy of butane as a function of the dihedral angle about the bond between carbons 2
and 3. The lowest energy conformation occurs when the two methyl groups are the maximum
distance apart ($\theta = 180°$). The highest energy conformation occurs when the two methyl
groups are eclipsed ($\theta = 0°$). **See a simulation based on this figure, and take a short quiz on
the concepts.**

butane are staggered conformations, yet the gauche conformations are approximately
3.8 kJ (0.9 kcal)/mol higher in energy than the anti conformation.

In dealing with the relative stabilities of the various conformations of butane, we
again encounter torsional strain. In addition, we encounter two other types of strain:
angle strain and steric strain. **Angle strain** results when a bond angle in a molecule is
either expanded or compressed compared to its optimal values. Strain can also occur
when bond lengths are forced to become shorter or longer than normal. In general,
bond stretching is not as easy as bond bending. **Steric strain** (also called nonbonded
interaction or van der Waals strain) results when nonbonded atoms separated by four
or more bonds are forced closer to each other than their atomic (contact) radii
allow—that is, when they are forced to smash into each other. The total of all types of
strain can be calculated by molecular mechanics programs; angles and bond lengths
are chosen in an iterative procedure so that an optimum geometry with the lowest
total energy is calculated. Such routines are included with popular desktop programs
such as Chem3D and Spartan. Molecular mechanics calculations carried out with
these or other programs can determine the lowest energy arrangement of atoms in a
given conformation of a molecule, a process referred to as energy minimization.

Let us illustrate the origins of both angle strain and steric strain by comparing the
anti ($\theta = 180°$) and eclipsed ($\theta = 0°$) conformations of butane. In an energy-minimized

Angle strain   The strain that
arises when a bond angle is either
compressed or expanded compared to its optimal value.

Steric strain   The strain that
arises when nonbonded atoms
separated by four or more bonds
are forced closer to each other
than their atomic (contact) radii
would allow. Steric strain is also
called non-bonded interaction
strain, or van der Waals strain.

conformation of anti butane, the C—C—C bond angle is 111.9° and all H—C—H bond angles are between 107.4° and 107.9°. The calculated strain in this, the most stable conformation of butane, is approximately 9.2 kJ (2.2 kcal)/mol.

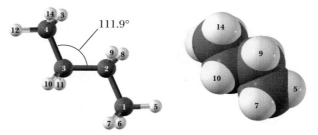

Now consider the eclipsed conformation. Figure 2.10(a) shows models of an eclipsed butane conformation in which C—C—C bond angles are set at 111.9°, the same value as in the energy-minimized anti conformation. Notice that in this eclipsed conformation, hydrogens H(5)-H(13) and H(5)-H(14) are forced to within 178 pm of each other, a distance closer than their contact radii would allow. When the energy of eclipsed butane is minimized [Figure 2.10(b)], the distance between these same pairs of hydrogens increases to 231 pm (which reduces steric strain). At the same time, the C(4)-C(3)-C(2) and C(3)-C(2)-C(1) bond angles increase from 111.9° to 119.0° (which increases angle strain).

Thus, the energy-minimized eclipsed conformation represents a balance between a decrease in steric strain and an increase in angle strain. The calculated strain in the energy-minimized eclipsed conformation of butane is 21 kJ (5.0 kcal)/mol, which means that the staggered conformation of butane is more stable than the eclipsed conformation by 21 kJ/mol.

An energy-minimized gauche conformation of butane (Figure 2.11) is approximately 3.8 kJ (0.90 kcal)/mol higher in energy than the anti conformation. This difference in energy is caused almost entirely by the steric strain (nonbonded interaction strain) between the two methyl groups. At any given instant, there is a larger number of butane molecules in the anti conformation than in the gauche conforma-

Figure 2.10

Eclipsed conformations of butane. (a) Non-energy-minimized and (b) energy-minimized eclipsed conformations. The calculated difference in energy between the non-energy-minimized and energy-minimized eclipsed conformations is 3.6 kJ (0.86 kcal)/mol.

Space-filling model, top view

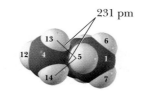

Space-filling model, top view

(a)   Ball-and-stick model, side view

(b)   Ball-and-stick model, side view

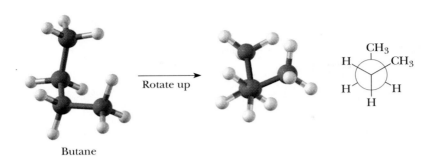

Figure 2.11
One of two equivalent
energy-minimized gauche
conformations of butane.

Butane

tion, and the number of molecules in the eclipsed conformation is vanishingly small. The percentage of the anti conformation present at 20°C is about 70%.

Note that, although the two gauche conformations ($\theta = 60°$ and 300°) have equal energies, they are not identical. They are related by reflection; that is, one gauche conformation is the reflection of the other, just as your right hand is the reflection of your left hand. Notice that the conformations with eclipsed —CH$_3$ and —H groups ($\theta = 120°$ and 240°) are also related by reflection. We shall have more to say about objects and their mirror reflections in Chapter 3.

Example 2.8

Following is the structural formula of 1,2-dichloroethane:

$$\begin{array}{c} \quad\text{H}\;\;\text{H} \\ \quad\;|\;\;\;| \\ \text{Cl}-\text{C}-\text{C}-\text{Cl} \\ \quad\;|\;\;\;| \\ \quad\text{H}\;\;\text{H} \end{array}$$

1,2-Dichloroethane

(a) Draw Newman projections for all staggered conformations formed by rotation from 0° to 360° about the carbon-carbon single bond.
(b) Which staggered conformation(s) has the lowest energy? Which has the highest energy?
(c) Which, if any, of these staggered conformations are related by reflection?

Solution

(a) If we take the dihedral angle when the chlorines are eclipsed as a reference point, staggered conformations occur at dihedral angles 60°, 180°, and 300°.

Gauche
$\theta = 60°$

Anti
$\theta = 180°$

Gauche
$\theta = 300°$

(b) We predict that the anti conformation (dihedral angle = 180°) has the lowest energy. The two gauche conformations (dihedral angle = 60° and 300°) are of higher but equal energy. We are not given data in the problem to calculate the actual energy differences.

(c) The two gauche conformations are related by reflection.

Problem 2.8

For 1,2-dichloroethane:

(a) Draw Newman projections for all eclipsed conformations formed by rotation from 0° to 360° about the carbon-carbon single bond.

(b) Which eclipsed conformation(s) has the lowest energy? Which has the highest energy?

(c) Which, if any, of these eclipsed conformations are related by reflection?

A word of caution. Although we talk about eclipsed along with staggered conformations, this can be misleading. Of all the conformations of butane, the eclipsed conformations are at the highest points on the energy profile. The gauche and anti conformations are in energy troughs. As a result, butane molecules spend their time in the anti and gauche conformations, and only fleetingly pass through the eclipsed conformations when interconverting between anti and gauche. A further nuance is that C—H and C—C bonds have vibrational motions at room temperature, and these vibrations occur simultaneously with bond rotations. The reality is that hydrocarbons are exceedingly complex and dynamic structures in solution, undergoing constant interconversions between low-energy conformations.

B. Cycloalkanes

i. Cyclopropane

The observed C—C—C bond angles in cyclopropane are 60° (Figure 2.12), a value considerably smaller than the bond angle of 109.5° predicted for sp^3 hybridized carbon atoms. This compression from the optimal bond angle introduces a considerable angle strain. Furthermore, there are six pairs of fully eclipsed C—H bonds, which introduce considerable torsional strain. The strain energy in cyclopropane is approximately 116 kJ (27.7 kcal)/mol. Because of its extreme degree of intramolecular strain, cyclopropane and its derivatives undergo several ring-opening reactions not shown by larger cycloalkanes.

ii. Cyclobutane

Nonplanar or puckered conformations are favored in all cycloalkanes larger than cyclopropane. If cyclobutane were planar [Figure 2.13(a)], all C—C—C bond angles would be 90° (which would introduce angle strain), and there would be eight pairs of eclipsed hydrogen interactions (which would introduce torsional strain). Puckering

(a)

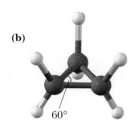

pairs of eclipsed
C-H interactions

(b)

Figure 2.12
Cyclopropane. (a) Structural formula and (b) ball-and-stick model.

Figure 2.13
Cyclobutane. (a) In the planar conformation, there are eight pairs of eclipsed C—H interactions. (b) The energy is a minimum in the puckered (butterfly) conformation.

 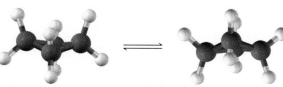

(a) Planar conformation   **(b)** Puckered or butterfly conformations

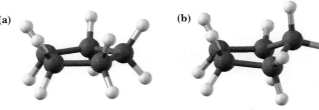

(a) Planar conformation

(b) Puckered envelope conformation

ORGANIC
Chemistry⫶Now™

Active Figure 2.14
Cyclopentane. (a) In the planar conformation, there are 10 pairs of eclipsed C—H interactions. (b) The most stable conformation is a puckered or envelope conformation. **See a simulation based on this figure, and take a short quiz on the concepts.**

of the ring [Figure 2.13(b)] alters the strain energy in two ways: (1) it decreases the torsional strain associated with eclipsed interactions, and (2) it increases further the angle strain caused by the compression of C—C—C bond angles. Because the decrease in torsional strain is greater than the increase in angle strain, puckered cyclobutane is more stable than planar cyclobutane. In the conformation of lowest energy, the measured C—C—C bond angles are 88°, and the strain energy in cyclobutane is approximately 110 kJ (26.3 kcal)/mol. Just like butane, cyclobutane is not static but undergoes interconversion between the puckered conformations.

iii. Cyclopentane

If cyclopentane were to adopt a planar conformation, all C—C—C bond angles would be 108° [Figure 2.14(a)]. This angle differs only slightly from the tetrahedral angle of 109.5°; consequently, there would be little angle strain in this conformation. In a planar conformation, however, there are 10 pairs of fully eclipsed C—H bonds creating a torsional strain of approximately 42 kJ (10 kcal)/mol. To relieve at least a part of this torsional strain, the ring twists into the **"envelope" conformation** shown in Figure 2.14(b). In this conformation, four carbon atoms are in a plane, and the fifth bends out of the plane, rather like an envelope with its flap bent upward. Cyclopentane exists as a dynamic equilibrium of five equivalent envelope conformations in which the average C—C—C bond angle is reduced to 105° (increasing angle strain). In the envelope conformation, the number of eclipsed C—H interactions is also reduced (decreasing torsional strain). Thus, the molecule is more stable in the envelope conformation than in a planar conformation. The strain energy in puckered cyclopentane is approximately 27 kJ (6.5 kcal)/mol.

iv. Cyclohexane

Cyclohexane adopts a number of puckered conformations, the most stable of which is the **chair conformation** (Figure 2.15). In this conformation, all C—C—C bond angles are 110.9° (minimizing angle strain), and all hydrogens on adjacent

Chair conformation The most stable nonplanar conformation of a cyclohexane ring; all bond angles are approximately 109.5°, and all bonds on adjacent carbons are staggered.

Figure 2.15
A chair conformation of cyclohexane.

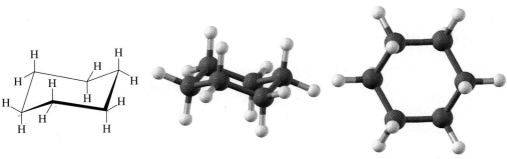

View from the side

View from above

Figure 2.16

A chair conformation of cyclohexane, showing axial and equatorial C—H bonds. The plane of the ring is defined by four carbons; the fifth carbon is above the plane, and the sixth carbon is below it.

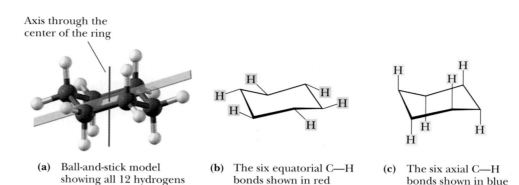

Axis through the center of the ring

(a) Ball-and-stick model showing all 12 hydrogens

(b) The six equatorial C—H bonds shown in red

(c) The six axial C—H bonds shown in blue

Axial bond   A bond to a chair conformation of cyclohexane that extends from the ring parallel to the imaginary axis through the center of the ring; a bond that lies roughly perpendicular to the equator of the ring.

Equatorial bond   A bond to a chair conformation of cyclohexane that extends from the ring roughly perpendicular to the imaginary axis through the center of the ring; a bond that lies roughly along the equator of the ring.

Boat conformation   A nonplanar conformation of a cyclohexane ring in which carbons 1 and 4 of the ring are bent toward each other.

Boat conformation

Twist-boat conformation

Figure 2.17

Boat and twist-boat conformations of cyclohexane.

carbons are staggered with respect to one another (minimizing torsional strain). In addition, no two atoms are close enough to each other for nonbonded interaction strain to exist. Thus, there is little strain in a chair conformation of cyclohexane.

The C—H bonds in a chair conformation of cyclohexane are arranged in two different orientations. Six C—H bonds are called **axial bonds,** and the other six are called **equatorial bonds.** One way to visualize the difference between these two types of bonds is to imagine an axis through the center of the chair, perpendicular to the floor (Figure 2.16).

Equatorial bonds are approximately perpendicular to the imaginary axis and form an equator about the ring. Equatorial bonds alternate, first slightly up and then slightly down as you move from one carbon of the ring to the next. Axial bonds are parallel to the imaginary axis. Three axial bonds point straight up; the other three axial bonds point straight down. Axial bonds also alternate, first up and then down as you move from one carbon of the ring to the next. Notice further that if the axial bond on a carbon points upward, then the equatorial bond on that carbon points slightly downward. Conversely, if the axial bond on a particular carbon points downward, then the equatorial bond on that carbon points slightly upward.

There are many other nonplanar conformations of cyclohexane, two of which, a **boat** and a **twist-boat,** are shown in Figure 2.17.

You can visualize interconversion of chair and boat conformations by twisting about two carbon-carbon bonds as illustrated in Figure 2.18. A boat conformation is considerably less stable than a chair conformation because of the torsional strain associated with four pairs of eclipsed C—H interactions and the steric strain between the two "flagpole" hydrogens. The difference in energy between chair and boat conformations is approximately 27 kJ (6.5 kcal)/mol.

Some of the strain in the boat conformation can be relieved by a slight twisting of the ring to a twist-boat conformation. It is estimated by computer modeling that a twist-boat conformation is favored over a boat conformation by approximately 6.3 kJ (1.5 kcal)/mol. Figure 2.19 shows an energy diagram for the interconversion between chair, twist-boat, and boat conformations. The large difference in energy between chair and boat or twist-boat conformations means that, at room temperature, molecules in the chair conformation make up more than 99.99% of the equilibrium mixture.

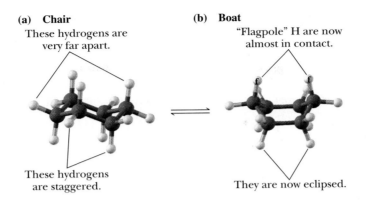

(a) Chair

These hydrogens are very far apart.

These hydrogens are staggered.

(b) Boat

"Flagpole" H are now almost in contact.

They are now eclipsed.

ORGANIC
Chemistry⚛Now™

Active Figure 2.18
Interconversion of (a) a chair conformation to (b) a boat conformation produces one set of flagpole steric interactions and four sets of eclipsed hydrogen interactions. **See a simulation based on this figure, and take a short quiz on the concepts.**

For cyclohexane, the two equivalent chair conformations can be interconverted by twisting one chair first into a boat and then into the other chair.

Twist-boat conformation   A non-planar conformation of a cyclohexane ring that is twisted from and slightly more stable than a boat conformation.

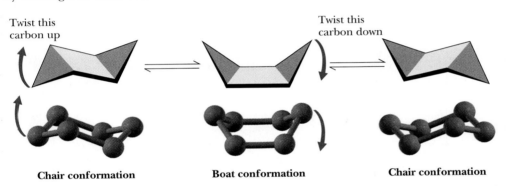

Twist this carbon up

Twist this carbon down

Chair conformation

Boat conformation

Chair conformation

When one chair is converted to the other, a change occurs in the relative orientations in space of the hydrogen atoms bonded to each carbon. All hydrogen atoms axial in

Figure 2.19
Energy diagram for interconversion of chair, twist-boat, and boat conformations of cyclohexane. The chair conformation is the most stable because angle, torsional, and steric strain are at a minimum.

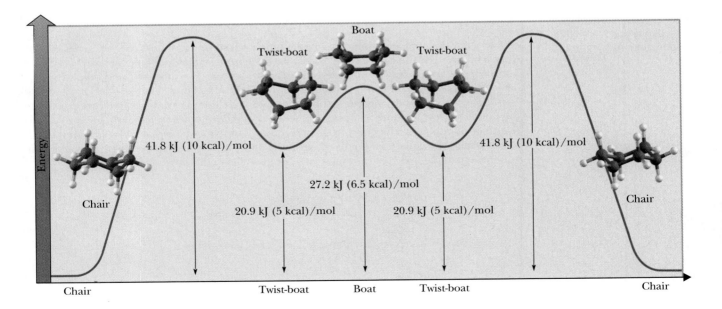

Energy

Chair

Twist-boat

Boat

Twist-boat

Chair

41.8 kJ (10 kcal)/mol

27.2 kJ (6.5 kcal)/mol

20.9 kJ (5 kcal)/mol

20.9 kJ (5 kcal)/mol

41.8 kJ (10 kcal)/mol

Chair

Twist-boat

Boat

Twist-boat

Chair

ORGANIC
Chemistry·⚛·Now™
Click How To Tutorials to review
Drawing Chair Conformations of Cyclohexane and Identifying Axial and Equatorial Groups

---

How To   *Draw Alternative Chair Conformations of Cyclohexane*

You will be asked to draw chair conformation of cyclohexane often because this conformation allows you to identify which substituents are axial and which are equatorial. Although drawing chair conformations takes practice, following a few simple guidelines will make you an expert at drawing even complicated substitution patterns.

Step 1: Start by drawing two sets of parallel lines, each set at a slight angle.

Step 2: Complete each chair by drawing the ends connected to the parallel lines, in each case making one end tip up, the other down.

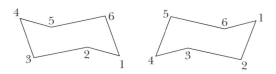

Step 3: Draw the axial bonds as vertical lines that are in the direction of the larger angle at each ring atom.

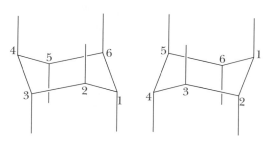

Axial bonds

Step 4: Draw the equatorial bonds using the bonds of the ring as guides for the angles. This is the tricky part. For the chair conformation on the left, the equatorial bonds on carbons 2 and 5 are parallel to the ring bonds between carbons 3–4 and 1–6 (the two ring bonds in red). The equatorial bonds of carbons 1 and 4 are parallel to the bonds between carbons 2–3 and 5–6 (the two ring bonds in green), and the equatorial bonds of carbons 3 and 6 are parallel to the bonds between carbons 1–2 and 4–5 (the two bonds in purple). Similarly, for the alternative chair on the right, the equatorial bonds on carbons 3 and 6 are parallel to the ring bonds in red bonds, the equatorial bonds of carbons 2 and 5 are parallel to the ring bonds in green, and the equatorial bonds of carbons 1 and 4 are parallel to the ring bonds in purple.

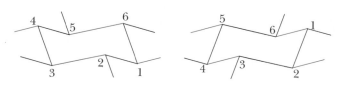

Equatorial bonds

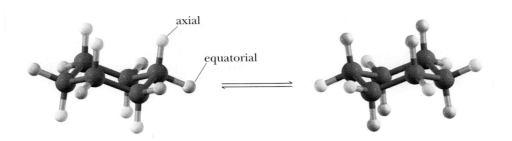

Figure 2.20
Interconversion of chair cyclohexanes. All C—H bonds equatorial in one chair are axial in the alternative chair and vice versa.

one chair become equatorial in the other and vice versa (Figure 2.20). The conversion of one chair conformation of cyclohexane to the other occurs rapidly at room temperature.

Example 2.9

Following is a chair conformation of cyclohexane showing one methyl group and one hydrogen.

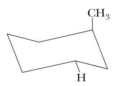

(a) Indicate by a label whether each group is equatorial or axial.
(b) Draw the alternative chair conformation and again label each group as axial or equatorial.

Solution

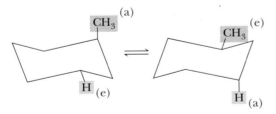

One chair   The alternative chair

Problem 2.9

Following is a chair conformation of cyclohexane with the carbon atoms numbered 1 through 6.

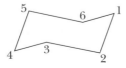

(a) Draw hydrogen atoms that are above the plane of the ring on carbons 1 and 2 and below the plane of the ring on carbon 4.

(b) Which of these hydrogens are equatorial? Which are axial?

(c) Draw the alternative chair conformation. Which hydrogens are equatorial? Which are axial? Which are above the plane of the ring? Which are below it?

Diaxial interactions Refers to the steric strain arising from interaction between an axial substituent and an axial hydrogen (or other group) on the same side of a chair conformation of a cyclohexane ring.

If we replace a hydrogen atom of cyclohexane with an alkyl group, the group occupies an equatorial position in one chair and an axial position in the other chair. This means that the two chairs are no longer equivalent and no longer of equal stability.

A convenient way to describe the relative stabilities of chair conformations with equatorial and axial substituents is in terms of a type of steric strain called **diaxial (axial-axial) interaction.** Diaxial interaction refers to the steric strain between an axial substituent and an axial hydrogen (or another group) on the same side of a cyclohexane ring. The axial positions on the same side of the ring are extremely close to each other, and any groups larger than hydrogen atoms will introduce steric strain between the larger group and the other two axial hydrogen atoms. Because this type of steric strain originates between groups on carbons 1 and 3 of a cyclohexane ring, it is often called a 1,3-diaxial interaction.

Consider methylcyclohexane (Figure 2.21). When —CH_3 is axial, it is parallel to the axial C—H bonds on carbons 3 and 5. Thus, for axial methylcyclohexane, there are two unfavorable methyl-hydrogen diaxial interactions. No such unfavorable interactions exist when the methyl group is in an equatorial position. For methylcyclohexane, the equatorial methyl conformation is favored over the axial methyl conformation by approximately 7.28 kJ (1.74 kcal)/mol.

Given the difference in strain energy between the axial and equatorial conformations of methylcyclohexane, we can calculate the ratio of the two conformations at equilibrium using the following equation, which relates the change in Gibbs free energy (ΔG°) for an equilibrium, the equilibrium constant (K_{eq}), and the temperature (T) in kelvins. R, the universal gas constant, has the value 8.314 J (1.987 cal) $\cdot K^{-1} \cdot mol^{-1}$.

$$\Delta G^0 = -RT \ln K_{eq}$$

Substituting the value of -7.28 kJ/mol (axial methyl $\rightarrow$ equatorial methyl) for ΔG° and solving the equation gives a value of 18.9 for the equilibrium constant at room temperature ($25°C = 298$ K).

$$\ln K_{eq} = \frac{-(-7280 \text{J} \cdot \text{mol}^{-1})}{8.314 \text{J} \cdot \text{K}^{-1} \cdot \text{mol}^{-1} \times 298 \text{K}} = 2.939$$

$$K_{eq} = \frac{18.9}{1} = \frac{\text{equatorial}}{\text{axial}}$$

ORGANIC
Chemistry◦♦◦Now™

Active Figure 2.21

Two chair conformations of methylcyclohexane. The steric strain introduced by two diaxial interactions makes the axial methyl conformation less stable by approximately 7.28 kJ (1.74 kcal)/mol. **See a simulation based on this figure, and take a short quiz on the concepts.**

Diaxial interactions

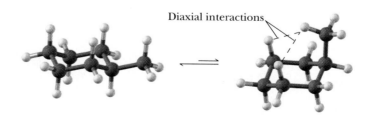

Table 2.4 ΔG^0 (Axial-Equatorial) for Monosubstituted Cyclohexanes at 25°C

| | | axial $\longrightarrow$ equatorial | | | |
| | $-\Delta G^0$ | | | $-\Delta G^0$ | |
Group	kJ/mol	kcal/mol	Group	kJ/mol	kcal/mol
C≡N	0.8	0.19	NH_2	5.9	1.41
F	1.0	0.24	COOH	5.9	1.41
C≡CH	1.7	0.41	$CH=CH_2$	7.1	1.70
I	1.9	0.45	CH_3	7.28	1.74
Cl	2.2	0.53	CH_2CH_3	7.3	1.75
Br	2.4	0.57	$CH(CH_3)_2$	9.0	2.15
OH	3.9	0.93	$C(CH_3)_3$	21.0	5.00

Thus, at any given instant at room temperature, there is a much larger number of methylcyclohexane molecules in the equatorial conformation than in the axial conformation. The percentage of equatorial is $100 \times$ (equatorial/(equatorial + axial)), or about 95%.

Table 2.4 shows the difference in free energy between axial and equatorial substituents for several monosubstituted cyclohexanes. Notice that as the size of the alkyl substituent increases, the preference for conformations with the group equatorial increases. With a group as large as *tert*-butyl, the energy of the axial conformer becomes so large that the equatorial conformation is approximately 4000 times more abundant at room temperature than the axial conformation. In fact, a chair with an axial *tert*-butyl group is so unstable that, if a *tert*-butyl group is forced into an axial position, the ring adopts a twist-boat conformation.

Note in Table 2.4 that the preference for the equatorial position among the halogens increases on the order F < I < Cl < Br. Yet the size of the halogen atoms increases in the order F < Cl < Br < I. This anomaly occurs because the C—I bond is so long that the center of the iodine atom is too far from the axial hydrogen to interact with it.

Example 2.10

Label all methyl-hydrogen diaxial interactions in the following chair conformation.

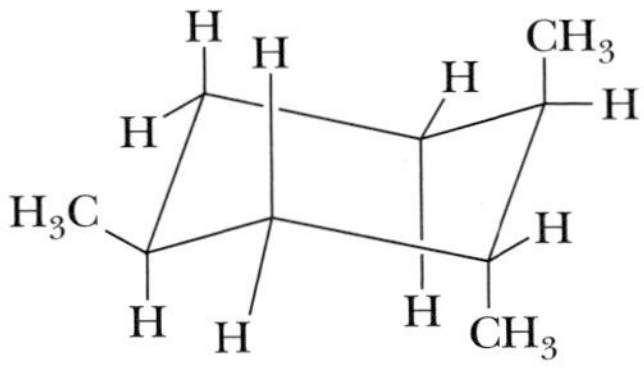

Solution

There are four methyl-hydrogen 1,3-diaxial interactions in this example; each axial methyl group has two sets of 1,3-diaxial interactions with parallel hydrogen atoms on the same side of the ring. The equatorial methyl group has no diaxial interactions.

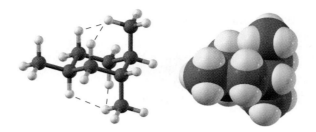

Problem 2.10

Draw the alternative chair conformation for the trisubstituted cyclohexane given in Example 2.10. Label all methyl-hydrogen 1,3-diaxial interactions in this chair conformation.

Example 2.11

Calculate the ratio of the diequatorial to diaxial conformation of this disubstituted cyclohexane at 25°C.

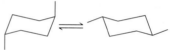

Solution

For these two chair conformations, $\Delta G°$ (2 axial $CH_3 \rightarrow$ 2 equatorial CH_3) = 2 × (−7.28 kJ/mol) = −14.56 kJ (3.5 kcal)/mol. Substituting this value in the equation $\Delta G° = -RT \ln K_{eq}$ gives a ratio of 357:1

$$\ln K_{eq} = \frac{-(-14{,}560\,\text{J/mol})}{8.314\,\text{J}\cdot\text{K}^{-1}\cdot\text{mol}^{-1} \times 298\,\text{K}} = 5.877$$

$$K_{eq} = \frac{357}{1} = \frac{\text{diequatorial chair conformation}}{\text{diaxial chair conformation}}$$

Thus, at any given instant at room temperature, more than 99.7% of the molecules of this compound are in the diequatorial chair conformation.

Problem 2.11

Draw a chair conformation of 1,4-dimethylcyclohexane in which one methyl group is equatorial and the other is axial. Draw the alternative chair conformation, and calculate the ratio of the two conformations at 25°C.

2.7 *Cis, Trans* Isomerism in Cycloalkanes and Bicycloalkanes

Stereoisomers Compounds that have the same molecular formula, the same connectivity of their atoms, but a different orientation of their atoms in space.

In this section we introduce the concept of stereoisomerism. **Stereoisomers** are compounds that have (1) the same molecular formula, (2) the same connectivity of their atoms, (3) but a different orientation of their atoms in space. Recall that constitutional

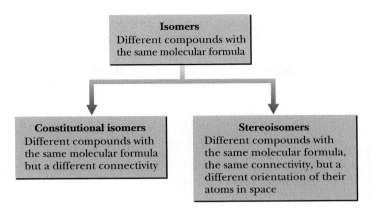

Figure 2.22
Relationship between
stereoisomers and
constitutional isomers.

isomers, the only other type of isomers we have studied so far, also have the same molecular formula but a different connectivity. The difference between constitutional isomers and stereoisomers are summarized in Figure 2.22.

We begin our study of stereoisomers with the study of *cis,trans* isomers in cycloalkanes.

A. *Cis,Trans* Isomerism in Cycloalkanes

Cycloalkanes with substituents on two or more carbons of the ring show a type of stereoisomerism called *cis,trans* isomerism, which we can illustrate by considering 1,2-dimethylcyclopentane. In the following structural formulas, the cyclopentane ring is drawn as a regular pentagon viewed through the plane of the ring. Carbon-carbon bonds of the ring projecting toward you are shown as heavy lines.

Cis,trans isomers   Stereoisomers that have the same connectivity but a different arrangement of their atoms in space as a result of the presence of either a ring or a carbon-carbon double bond.

cis-1,2-Dimethyl-
cyclopentane   *trans*-1,2-Dimethyl-
cyclopentane

In one isomer of 1,2-dimethylcyclopentane, the methyl groups are on the same side of the ring; in the other, they are on opposite sides of the ring. The prefix *cis* (Latin: on the same side) indicates that the substituents are on the same side of the ring; the prefix *trans* (Latin: across) indicates that they are on opposite sides of the ring. The *cis* isomer cannot be converted to the *trans* isomer and vice versa without breaking and reforming one or more bonds, a process that does not occur at or near room temperature. The *cis* isomer is approximately 7.1 kJ (1.7 kcal)/mol higher in energy (less stable) than the *trans* isomer because of the steric strain of the methyl groups on adjacent carbons in the *cis* isomer.

Alternatively, the cyclopentane ring can be viewed as a regular pentagon seen from above, with the ring in the plane of the page. Substituents on the ring then either project toward you (they project up, above the plane of the page) and are shown by solid wedges, or they project away from you (they project down, below the plane of

Cis   A prefix meaning on the same side.

Trans   A prefix meaning across from.

the page) and are shown by broken wedges. In the following structural formulas, only the two methyl groups are shown; hydrogen atoms of the ring are not shown.

cis-1,2-Dimethyl-
cyclopentane

trans-1,2-Dimethyl-
cyclopentane

Stereocenter A tetrahedral atom, most commonly carbon, about which exchange of two groups produces a stereoisomer.

Configuration Refers to the arrangement of atoms about a stereocenter.

We say that 1,2-dimethylcyclopentane has two stereocenters. A **stereocenter** is a tetrahedral atom, most commonly carbon, about which exchange of two groups produces a stereoisomer. Both carbons 1 and 2 of 1,2-dimethylcyclopentane, for example, are stereocenters; in this molecule, exchange of H and CH_3 groups at either stereocenter converts a *trans* isomer to a *cis* isomer, or vice versa. Alternatively, we refer to the stereoisomers of 1,2-dimethylcyclobutane as having either a *cis* or a *trans* configuration. **Configuration** refers to the arrangement of atoms about a stereocenter. We say, for example, that exchange of groups at either stereocenter in the *cis* configuration gives the isomer with the *trans* configuration.

Example 2.12

Which cycloalkanes show *cis,trans* isomerism? For each that does, draw the *cis* and *trans* isomers.

(a) Methylcyclopentane **(b)** 1,1-Dimethylcyclopentane **(c)** 1,3-Dimethylcyclobutane

Solution

(a) Methylcyclopentane does not show *cis,trans* isomerism. It has only one substituent on the ring.
(b) 1,1-Dimethylcyclobutane does not show *cis,trans* isomerism. Because both methyl groups are bonded to the same carbon, only one arrangement is possible for them; they must be *trans* to each other.
(c) 1,3-Dimethylcyclobutane shows *cis,trans* isomerism. In the following structural formulas, cyclobutane is drawn as a planar ring viewed first from the side, and then from above.

cis-1,3-Dimethylcyclobutane

trans-1,3-Dimethylcyclobutane

Problem 2.12

Which cycloalkanes show *cis,trans* isomerism? For each that does, draw both isomers.

(a)   **(b)**   **(c)**

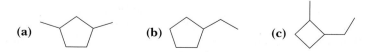

Two *cis,trans* isomers are possible for 1,4-dimethylcyclohexane. For the purposes of determining the number of *cis,trans* isomers in substituted cycloalkanes, it is adequate to draw the cycloalkane ring as a planar polygon as is done here.

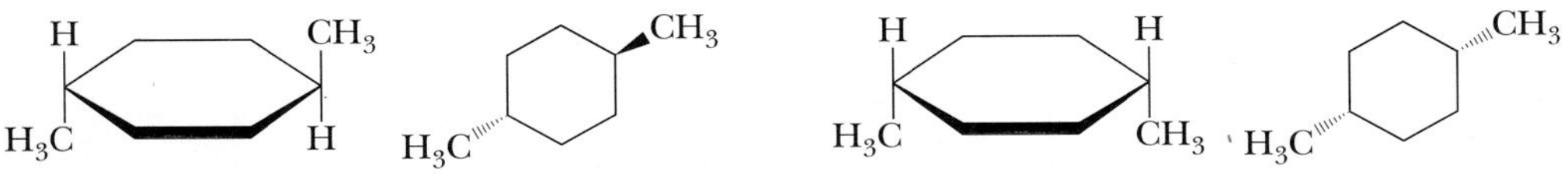

trans-1,4-Dimethylcyclohexane   *cis*-1,4-Dimethylcyclohexane

We can also draw the *cis* and *trans* isomers of 1,4-dimethylcyclohexane as nonplanar chair conformations. In working with alternative chair conformations, it is helpful to remember that all groups axial in one chair become equatorial in the alternative chair, and vice versa. In one chair conformation of *trans*-1,4-dimethylcyclohexane, the two methyl groups are axial; in the alternative chair conformation, they are equatorial. Of these chair conformations, the one with both methyl groups equatorial is considerably more stable.

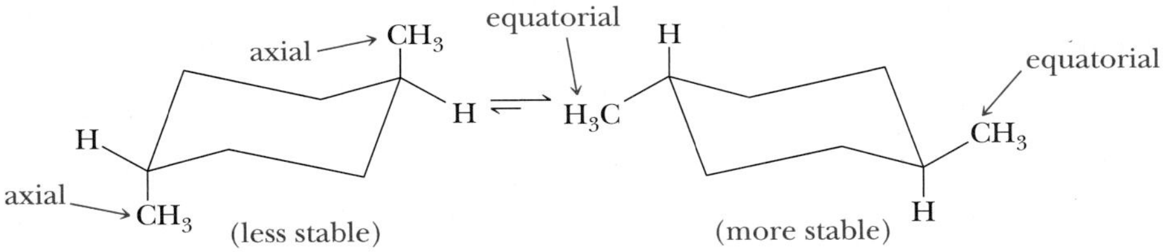

trans-1,4-Dimethylcyclohexane

The alternative chair conformations of *cis*-1,4-dimethylcyclohexane are of equal energy. In one chair, one methyl group is equatorial and the other is axial. In the alternative chair, the orientations in space of the methyl groups are reversed. The result is that a collection of *cis*-1,4-dimethylcyclohexane molecules is composed of rapidly equilibrating alternative chairs in equal proportions.

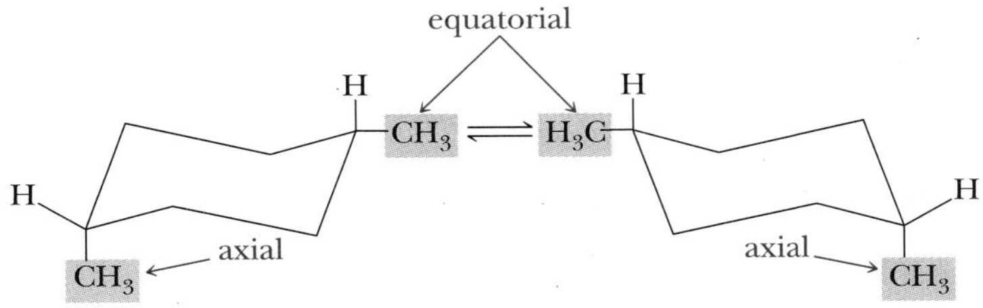

cis-1,4-Dimethylcyclohexane
(these conformations are of equal stability)

How To   Convert Planar Cyclohexanes to Chair Cyclohexanes

Following are three different stereorepresentations of *cis*-1, 2-dimethylcyclohexane, each with the ring drawn as a planar hexagon.

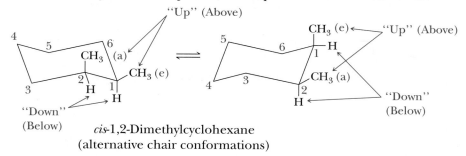

cis-1,2-Dimethylcyclohexane
(planar hexagon representations)

Students often find it difficult to convert substituted cyclohexanes from a planar hexagon representation such as these to a chair conformation. A good rule of thumb is that "up is up and down is down." If a substituent is *up* in a planar hexagon representation, place it *up* on the same carbon of the chair conformation. If a substituent is *down* on a planar hexagon representation, place it *down* on the same carbon of the chair conformation. Note that up or down on a chair conformation may be axial or equatorial, depending on which ring carbon you are considering. For *cis*-1,2-dimethylcyclohexane on which both methyl groups are up in the planar hexagon representation, the two methyl groups are also up in a chair conformation. Each of the alternative chair conformations has one methyl group axial and one equatorial. It is generally helpful to draw the hydrogen atoms bonded to the ring carbons bearing substituents, to make it absolutely clear which positions are equatorial and which are axial.

cis-1,2-Dimethylcyclohexane
(alternative chair conformations)

Example 2.13

Following is a chair conformation of 1,3-dimethylcyclohexane.

(a) Is this a chair conformation of *cis*-1,3-dimethylcyclohexane or of *trans*-1,3-dimethylcyclohexane?
(b) Draw the alternative chair conformation of this compound. Of the two chair conformations, which is the more stable?
(c) Draw a planar hexagon representation of the isomer shown in this example.

Solution

(a) The isomer shown is *cis*-1,3-dimethylcyclohexane; the two methyl groups are on the same side of the ring.

(b)

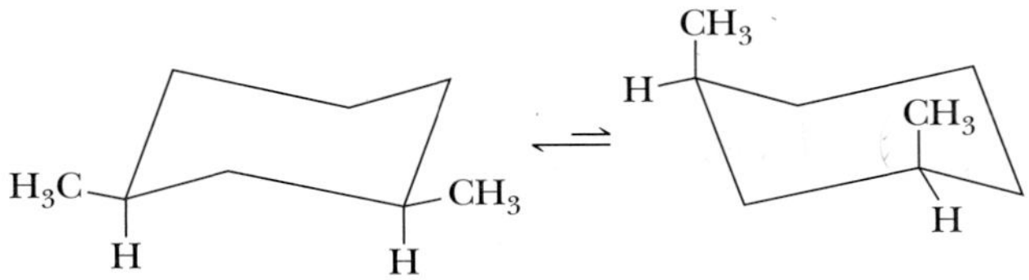

Diequatorial conformation
(more stable)

Diaxial conformation
(less stable)

(c)

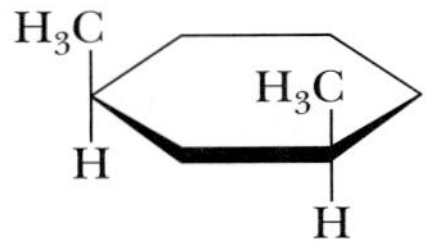

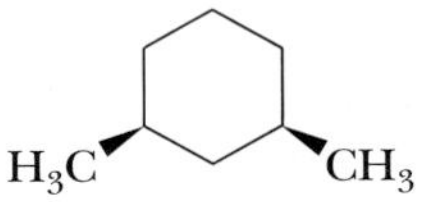

Problem 2.13

Following is a planar hexagon representation for one isomer of 1,2,4-trimethylcyclohexane. Draw the alternative chair conformations of this compound, and state which of the two is the more stable.

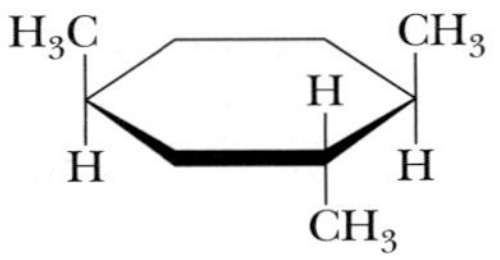

Example 2.14

Here is one *cis, trans* isomer of 2,4-dimethylcyclohexanol. Complete the alternative chair conformations on the right.

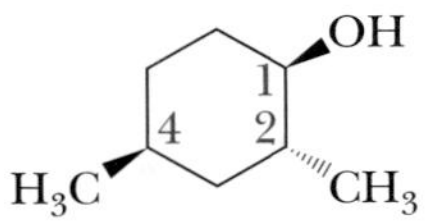

(a)

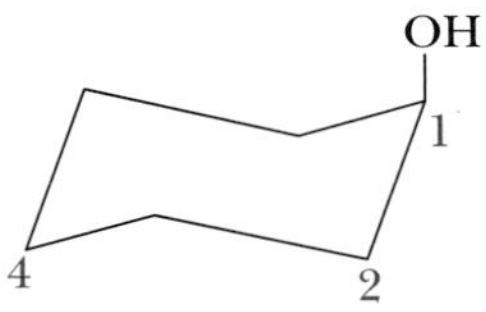

(b)

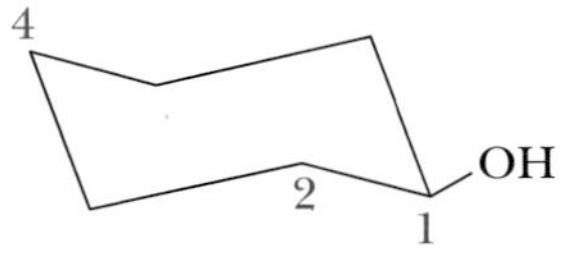

Solution

For (a), the CH_3 group on carbon 2 must be below the plane of the ring, which on this carbon is axial. The CH_3 group on carbon 4 must be above the plane of the ring, which on this carbon is equatorial. (b) The methyl group on carbon 2 is equatorial; the methyl group on carbon 4 is axial.

(a)

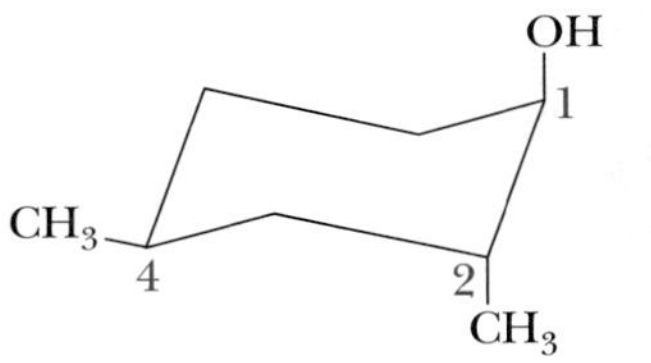

(b)

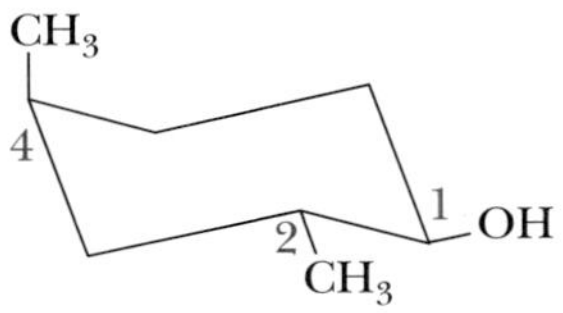

Problem 2.14

Here is one *cis,trans* isomer of 3,5-dimethylcyclohexanol. Complete the alternative chair conformations.

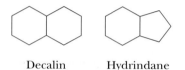

B. *Cis,Trans* Isomerism in Bicycloalkanes

By far the most common bicycloalkanes, and the ones we concentrate on in this section, are decalin and hydrindane (Section 2.4B).

Decalin   Hydrindane

Two stereoisomers of decalin and hydrindane are possible depending on whether the two hydrogen atoms at the ring junction are *trans* or *cis* to each other. If we draw conformations for the six-membered rings in the two decalins, we see that each ring can exist in its more stable chair conformation. In *trans*-decalin, the hydrogens at the ring junction are axial to both rings; that is, the ring-junction hydrogen above the plane of the rings is axial to ring A and to ring B. Likewise, the ring-junction hydrogen below the plane of the ring is axial to both rings. The situation is different in *cis*-decalin. Each ring-junction hydrogen is axial to one ring but equatorial to the other ring.

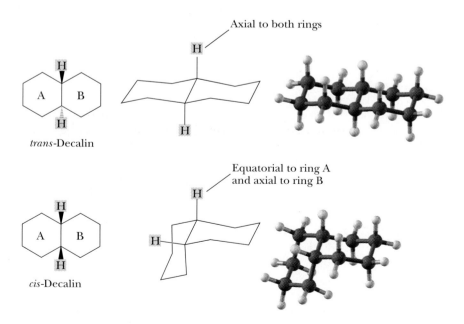

Let us look more closely at *trans*-decalin, by far the more common stereoisomer of decalin. An important feature of this bicycloalkane is that each ring is locked into one chair conformation; neither ring can invert to its alternative chair. This means, for example, that if an —OH group is equatorial in a decalinol (a decalin alcohol), it remains equatorial; it cannot become axial because the cyclohexane ring is locked into this one conformation. Likewise, if an —OH group is axial, it remains axial.

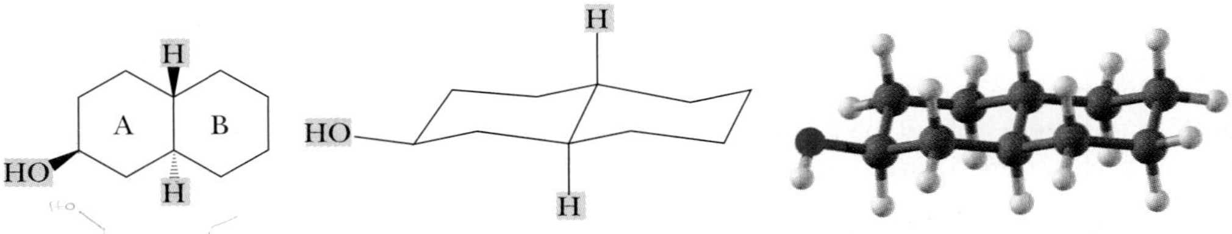

Suppose you are given the structural formula on the left for the decalinol. Can you tell from looking at this structure whether the —OH group is axial or equatorial? You can't tell directly, but you can figure it out. Remember that, in *trans*-decalin, the atoms at the ring junctions are axial to each ring. Remember also that in a chair cyclohexane, axial is up on one carbon, down on the next, up on the next, and so on. Therefore, if you start with the axial group at either ring junction and work your way from one carbon to the next until you come to the carbon bearing the —OH group, you come to the conclusion that the —OH on the structural formula is equatorial to ring A.

A good example of the occurrence of these types of ring systems is in the steroids, all of which contain a carbon skeleton consisting of three six-membered rings and one five-membered ring connected as show here. This ring system is present in both animal and plant steroids. Steroids are present in human metabolism as cholesterol, steroid hormones, and bile acids (Section 26.4).

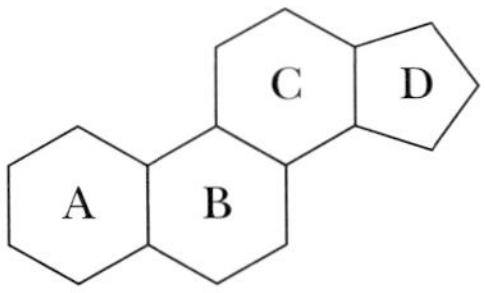

The steroid nucleus

Following are two stereorepresentations for cholestanol. In the conformational representation on the right, notice that all ring junctions are *trans,* all groups at each ring-junction atom are axial to the ring, and the —OH group on ring A is equatorial.

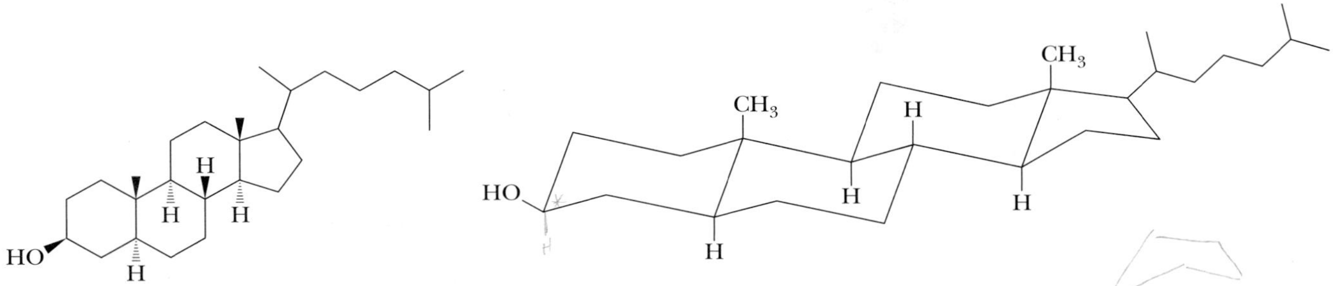

Cholestanol

CHEMICAL CONNECTIONS

The Poisonous Puffer Fish

Nature is by no means limited to carbon atoms in six-membered rings. Tetrodotoxin, one of the most potent toxins known, is composed of a set of interconnected six-membered rings, each in a chair conformation. All but one of these rings have atoms other than carbon in them. Tetrodotoxin is produced in the liver and ovaries of many species of *Tetraodontidae*, especially the puffer fish, so called because it inflates itself to an almost spherical spiny ball when it is alarmed. The puffer fish

is evidently a species that is highly preoccupied with defense, but the Japanese are not put off. They regard the puffer, called "fugu" in Japanese, as a delicacy. To serve it in a public restaurant, a chef must be registered as sufficiently skilled in removing the toxic organs so as to make the flesh safe to eat.

Symptoms of tetrodotoxin poisoning begin with attacks of severe weakness, progressing to complete paralysis and eventual death. Tetrodotoxin blocks sodium ion channels, which are essential for neurotransmission. This prevents communication between neurons and muscle cells and results in the fatal symptoms described.

A puffer fish with its body inflated.

Tetrodotoxin

Another type of bicycloalkane is a six-membered ring in which an added CH_2 group forms a bridge between carbons 1 and 4. You can view and draw this molecule from any number of perspectives. What becomes obvious if you view it from the side, as in (c), is that the one-carbon bridge locks the six-membered ring into a boat conformation. Notice that, even though (a) and (b) show the carbon skeleton of the molecule, it is not obvious from them that a locked boat conformation is embedded in the molecule. The lesson here is that it is essential to draw a molecule to best reveal what you want to show.

one carbon bridge between carbons 1 and 4 of the six-membered ring

(a)　　**(b)**　　**(c)**　　Camphor

An example of a natural product containing this bicyclic skeleton is camphor.

Another example of a carbon skeleton that contains several six-membered rings, all of which are locked into chair conformations, is adamantane (c). To understand how the carbon skeleton of adamantane can be constructed, imagine that you (a) start with a chair cyclohexane, (b) add the three axial bonds on the top side of the ring, and (c) then connect each of the axial bonds to a CH group. You now have adamantane, a compound first isolated from petroleum. Amantadine, a 1° amino derivative of adamantane, is an antiviral agent used to treat influenza A.

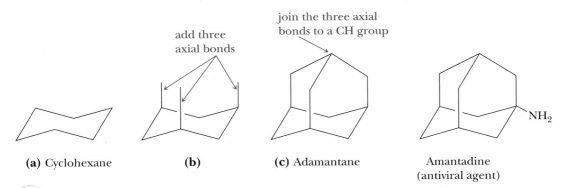

add three axial bonds

join the three axial bonds to a CH group

(a) Cyclohexane   **(b)**   **(c)** Adamantane   Amantadine (antiviral agent)

2.8 Physical Properties of Alkanes and Cycloalkanes

The most important property of alkanes and cycloalkanes is their almost complete lack of polarity. As we saw in Section 1.2B, the difference in electronegativity between carbon and hydrogen is $2.5 - 2.1 = 0.4$ on the Pauling scale, and, given this small difference, we classify a C—H bond as nonpolar covalent. Therefore, alkanes are nonpolar compounds, and there is only weak interaction between their molecules.

A. Boiling Points

The boiling points of alkanes are lower than those of almost any other type of compound of the same molecular weight. In general, both boiling and melting points of alkanes increase with increasing molecular weight (Table 2.5).

Alkanes containing 1 to 4 carbons are gases at room temperature, and those containing 5 to 17 carbons are colorless liquids. High-molecular-weight alkanes (those with 18 or more carbons) are white, waxy solids. Several plant waxes are high-molecular-weight alkanes. The wax found in apple skins, for example, is an unbranched alkane with molecular formula $C_{27}H_{56}$. Paraffin wax, a mixture of high-molecular-weight alkanes, is used for wax candles, in lubricants, and to seal home canned jams, jellies, and other preserves. Petrolatum, so named because it is derived from petroleum refining, is a liquid mixture of high-molecular-weight alkanes. Sold as mineral oil and Vaseline, petrolatum is used as an ointment base in pharmaceuticals and cosmetics and as a lubricant and rust preventative.

B. Dispersion Forces and Interactions Among Alkane Molecules

Methane is a gas at room temperature and atmospheric pressure. It can be converted to a liquid if cooled to $-164°C$, and to a solid if further cooled to $-182°C$. The fact that methane (or any other compound, for that matter) can exist as a liquid or a

Table 2.5   Physical Properties of Some Unbranched Alkanes

Name	Condensed Structural Formula	Melting Point (°C)	Boiling Point (°C)	Density of Liquid* (g/mL at 0°C)
Methane	CH_4	−182	−164	(a gas)
Ethane	CH_3CH_3	−183	−88	(a gas)
Propane	$CH_3CH_2CH_3$	−190	−42	(a gas)
Butane	$CH_3(CH_2)_2CH_3$	−138	0	(a gas)
Pentane	$CH_3(CH_2)_3CH_3$	−130	36	0.626
Hexane	$CH_3(CH_2)_4CH_3$	−95	69	0.659
Heptane	$CH_3(CH_2)_5CH_3$	−90	98	0.684
Octane	$CH_3(CH_2)_6CH_3$	−57	126	0.703
Nonane	$CH_3(CH_2)_7CH_3$	−51	151	0.718
Decane	$CH_3(CH_2)_8CH_3$	−30	174	0.730

*For comparison, the density of H_2O is 1 g/mL at 4°C.

solid depends on the existence of intermolecular forces of attraction between the particles of each pure compound. Although the forces of attraction between particles are all electrostatic in nature, they vary widely in their relative strengths. The strongest attractive forces are between ions, for example between Na^+ and Cl^- in NaCl [787 kJ (188 kcal)/mol]. Dipole-dipole interactions and hydrogen bonding [8–42 kJ (2–10 kcal)/mol] are weaker attractive forces. We shall have more to say about these intermolecular attractive forces in Chapter 10 when we discuss the physical properties of alcohols, compounds containing polar O—H groups.

Dispersion forces [0.08 − 8 kJ (0.02 − 2 kcal) mol] are the weakest intermolecular attractive forces. The existence of dispersion forces accounts for the fact that low-molecular-weight, nonpolar substances, such as hydrogen, neon, and methane, can be liquefied. To visualize the origin of dispersion forces, it is necessary to think in terms of instantaneous distributions of electron density rather than average distributions. Consider neon, for example. Neon is a gas at room temperature and 1 atm pressure. It can be liquefied when cooled to −246°C. From the heat of vaporization, it can be calculated that the neon-neon attractive interaction in the liquid state is approximately 0.3 kJ (0.07 kcal)/mol. We account for this intermolecular attractive force in the following way. Over time, the distribution of electron density in a neon atom is symmetrical, and there is no dipole moment [Figure 2.23(a)]. However, at any instant, there is a nonzero probability that its electron density is polarized (shifted) more toward one part of the atom than toward another. This temporary

Dispersion forces Very weak intermolecular forces of attraction resulting from the interaction between temporary induced dipoles.

Figure 2.23
Dispersion forces. (a) The average distribution of electron density in a neon atom is symmetrical, and there is no polarity. (b) Temporary polarization of one neon atom induces temporary polarization in adjacent atoms. Electrostatic attractions between temporary dipoles are called dispersion forces.

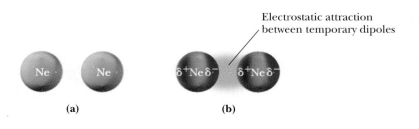

Electrostatic attraction between temporary dipoles

(a)   (b)

polarization creates a temporary dipole moment, which in turn induces temporary dipole moments in adjacent atoms [Figure 2.23(b)].

The strength of dispersion forces depends on how easily an electron cloud can be polarized. Electrons in smaller atoms and molecules are held closer to their nuclei and, therefore, are not easily polarized. Electrons in larger atoms and molecules are more easily polarized. For this reason, the strength of dispersion forces tends to increase with increasing molecular mass and size. Intermolecular attractive forces between Cl_2 molecules and between Br_2 molecules are estimated to be 2.9 kJ (0.7 kcal)/mol and 4.2 kJ (1.0 kcal)/mol, respectively.

Dispersion forces are inversely proportional to the sixth power of the distance between interacting particles. For them to be important, the interacting particles must be in virtual contact with one another.

Because interactions between alkane molecules consist only of these very weak dispersion forces, boiling points of alkanes are lower than those of almost any other type of compound of the same molecular weight. As the number of atoms and the molecular weight of alkanes increase, there is more opportunity for dispersion forces between their molecules and boiling points increase.

C. Melting Points and Density

Melting points of alkanes also increase with increasing molecular weight. The increase, however, is not as regular as that observed for boiling points because the packing of molecules into ordered patterns of solids changes as molecular size and shape change.

The average density of the alkanes listed in Table 2.5 is about 0.7 g/mL; that of higher-molecular-weight alkanes is about 0.8 g/mL. All liquid and solid alkanes are less dense than water (1.0 g/mL) and, therefore, float on it.

D. Constitutional Isomers Have Different Physical Properties

Alkanes that are constitutional isomers are different compounds and have different physical and chemical properties. Listed in Table 2.6 are boiling points, melting points, and densities of the five constitutional isomers with molecular formula C_6H_{14}. The boiling point of each of the branched-chain isomers of C_6H_{14} is lower than that of hexane itself, and the more branching there is, the lower the boiling point. These differences in boiling point are related to molecular shape in the following way. The

Table 2.6 Physical Properties of the Isomeric Alkanes of Molecular Formula C_6H_{14}

Name	Boiling Point (°C)	Melting Point (°C)	Density (g/mL)
Hexane	68.7	−95	0.659
2-Methylpentane	60.3	−154	0.653
3-Methylpentane	63.3	−118	0.664
2,3-Dimethylbutane	58.0	−129	0.661
2,2-Dimethylbutane	49.7	−98	0.649

Hexane

2,2-Dimethylbutane

only forces of attraction between alkane molecules are dispersion forces. As branching increases, the shape of an alkane molecule becomes more compact, and its surface area decreases. As surface area decreases, contact among adjacent molecules decreases, the strength of dispersion forces decreases, and boiling points also decrease. Thus, for any group of alkane constitutional isomers, the least branched isomer usually has the highest boiling point, and the most branched isomer usually has the lowest boiling point.

Example 2.15

Arrange the alkanes in each set in order of increasing boiling point.

(a) Butane, decane, and hexane
(b) 2-Methylheptane, octane, and 2,2,4-trimethylpentane

Solution

(a) All of these compounds are unbranched alkanes. As the number of carbon atoms in the chain increases, dispersion forces between molecules increase and so do boiling points. Decane has the highest boiling point, and butane has the lowest.

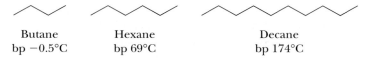

| Butane | Hexane | Decane |
| bp −0.5°C | bp 69°C | bp 174°C |

(b) These three alkanes are constitutional isomers with molecular formula C_8H_{18}. Their relative boiling points depend on the degree of branching. 2,2,4-Trimethylpentane, the most highly branched isomer, has the smallest surface area and the lowest boiling point. Octane, the unbranched isomer, has the largest surface area and the highest boiling point.

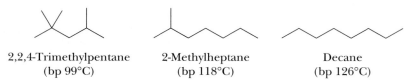

| 2,2,4-Trimethylpentane | 2-Methylheptane | Decane |
| (bp 99°C) | (bp 118°C) | (bp 126°C) |

Problem 2.15

Arrange the alkanes in each set in order of increasing boiling point.
(a) 2-Methylbutane, 2,2-dimethylpropane, and pentane
(b) 3,3-Dimethylheptane, 2,2,4-trimethylhexane, and nonane

2.9 Reactions of Alkanes

Alkanes and cycloalkanes are quite unreactive toward most reagents, a behavior consistent with the fact that they are nonpolar compounds and contain only strong sigma bonds. Under certain conditions, however, alkanes and cycloalkanes do react with O_2 and with the halogens Cl_2 and Br_2. At this point, we present only their combustion with oxygen. We discuss their reaction with halogens in Chapter 8.

A. Oxidation

The oxidation of alkanes by O_2 to give carbon dioxide and water is by far their most economically important reaction. Oxidation of saturated hydrocarbons is the basis for their use as energy sources for heat [natural gas, liquefied petroleum gas (LPG), and fuel oil] and power (gasoline, diesel fuel, and aviation fuel). Following are balanced equations for the complete oxidation of methane (the major component of natural gas) and propane (the major component of LPG).

$$CH_4 + 2O_2 \longrightarrow CO_2 + 2H_2O \qquad \Delta H^0 = -890.4 \text{ kJ } (-212.8 \text{ kcal})/\text{mol}$$

Methane

$$CH_3CH_2CH_3 + 5O_2 \longrightarrow 3CO_2 + 4H_2O \qquad \Delta H^0 = -2220 \text{ kJ } (-530.6 \text{ kcal})/\text{mol}$$

Propane

In this and all other hydrocarbon oxidations, the energy of the products is less than that of the reactants, with the difference in energy being given off as the **heat of combustion.** The heat of combustion is the energy of the products minus that of the reactants, and for the combustion of methane, for example, it is $(-890.4 - 0) = -890.4 \text{ kJ } (-212.8 \text{ kcal})/\text{mol}.$

Heat of combustion   Standard heat of combustion is the heat released when one mole of a substance in its standard state (gas, liquid, solid) is oxidized completely to carbon dioxide and water, and is given the symbol ΔH^0.

B. Heats of Combustion and Relative Stability of Alkanes and Cycloalkanes

One important use of heats of combustion is to give us information about the relative stabilities of isomeric hydrocarbons. To illustrate, consider the heats of combustion of the four constitutional isomers given in Table 2.7. All four compounds undergo combustion according to this equation.

$$C_8H_{18} + \frac{25}{2}O_2 \longrightarrow 8CO_2 + 9H_2O$$

We see that octane has the largest (most negative) heat of combustion. As branching increases, the ΔH^0 decreases (becomes less negative). Of these four

Table 2.7   Heats of Combustion of Four Constitutional Isomers of C_8H_{18}

Hydrocarbon	Structural Formula	ΔH^0 [kJ/mol (kcal/mol)]
Octane		−5470.6 (−1307.5)
2-Methylheptane		−5465.6 (−1306.3)
2,2-Dimethylhexane		−5458.4 (−1304.6)
2,2,3,3-Tetramethylbutane		−5451.8 (−1303.0)

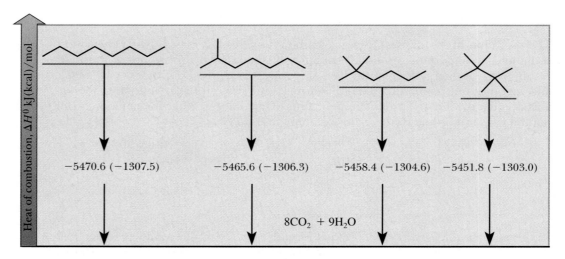

Heat of combustion, ΔH^0 kJ (kcal)/mol

−5470.6 (−1307.5)   −5465.6 (−1306.3)   −5458.4 (−1304.6)   −5451.8 (−1303.0)

$8CO_2 + 9H_2O$

Figure 2.24
Heats of combustion in kJ (kcal)/mol of four isomeric alkanes with molecular formula C_8H_{18}.

isomers, the isomer with four methyl branches has the lowest (least negative) heat of combustion. Therefore, we conclude that branching increases the stability of an alkane.

Figure 2.24 is a graphical analysis of the data in Table 2.7. Because all four compounds give the same products on oxidation, the only difference between them is their relative energies.

As we saw in Section 2.6B, there is considerable strain in small-ring cycloalkanes. We can measure this strain by measuring the heat of combustion versus the ring size. It has been determined by measurement of the heats of combustion of a series of unbranched alkanes that the average heat of combustion per methylene (CH_2) group is 658.7 kJ (157.4 kcal)/mol. Using this value, we can calculate a predicted heat of combustion for each cycloalkane. These results are displayed graphically in Figure 2.25. Strain energy is the difference between the predicted and actual heats of combustion.

Figure 2.25

Strain energy of cycloalkanes as a function of ring size.

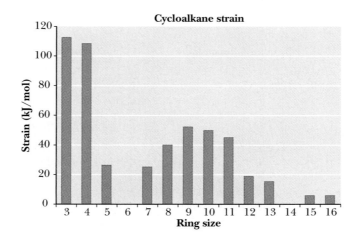

We see that cyclopropane has the largest strain energy of any cycloalkane, which is consistent with the extreme compression of its C—C—C bond angles to 60°. Cyclobutane and cyclopentane each have less strain, and cyclohexane, as expected, has zero strain. What is perhaps surprising is the presence of strain in rings of from 7 to 13 carbon atoms. This strain is primarily the result of torsional and steric strain caused by the fact that these rings are constrained to conformations that cannot achieve ideal bond and torsional angles.

2.10 Sources and Importance of Alkanes

The three major sources of alkanes throughout the world are the fossil fuels, namely natural gas, petroleum, and coal. These fossil fuels account for approximately 90% of the total energy consumed in the United States. Nuclear electric power and hydroelectric power make up most of the remaining 10%. In addition, these fossil fuels provide the bulk of the raw materials for the organic chemicals consumed worldwide.

A. Natural Gas

Natural gas consists of approximately 90–95% methane, 5–10% ethane, and a mixture of other relatively low-boiling alkanes, chiefly propane, butane, and 2-methylpropane. The current widespread use of ethylene as the organic chemical industry's most important building block is largely the result of the ease with which ethane can be separated from natural gas and cracked into ethylene. **Cracking** is a process whereby a saturated hydrocarbon is converted into an unsaturated hydrocarbon plus H_2. Ethane is cracked by heating it in a furnace at 800 to 900°C for a fraction of a second. The production of ethylene in the United States in 1997 was 51.1 billion pounds, making it the number one organic compound produced by the U.S. chemical industry, on a weight basis. The bulk of ethylene produced in this manner is used to create organic polymers, as described in Chapter 29.

$$CH_3CH_3 \xrightarrow{\text{800–900°C}} CH_2{=}CH_2 + H_2$$

Ethane   Ethene
(ethylene)

B. Petroleum

Petroleum is a thick, viscous, liquid mixture of thousands of compounds, most of them hydrocarbons, formed from the decomposition of marine plants and animals. Petroleum and petroleum-derived products fuel automobiles, aircraft, and trains. They provide most of the greases and lubricants required for the machinery of our highly industrialized society. Furthermore, petroleum, along with natural gas, provides nearly 90% of the organic raw materials for the synthesis and manufacture of synthetic fibers, plastics, detergents, drugs, dyes, and a multitude of other products.

It is the task of the petroleum refining industry to produce usable products with a minimum of waste, from the thousands of different hydrocarbons in the liquid mixture. The various physical and chemical processes for this purpose fall into two broad

K Straiton/Photo Researchers, Inc.

A petroleum refinery.

Figure 2.26 Fractional distillation of petroleum. The lighter, more volatile fractions are removed from higher up the column; the heavier, less volatile fractions are removed from lower down.

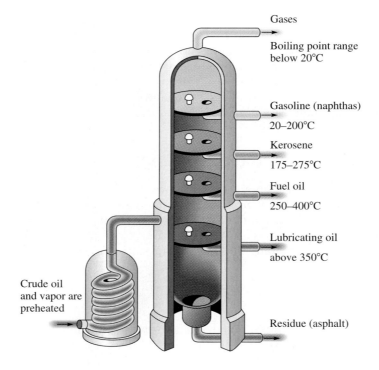

categories: separation processes, which separate the complex mixture into various fractions, and reforming processes, which alter the molecular structure of the hydrocarbon components themselves.

The fundamental separation process in refining petroleum is fractional distillation (Figure 2.26). Practically all crude petroleum that enters a refinery goes to distillation units, where it is heated to temperatures as high as 370 to 425°C and separated into fractions. Each fraction contains a mixture of hydrocarbons that boils within a particular range.

The two most common reforming processes are cracking, as illustrated by the thermal conversion of ethane to ethylene (Section 2.10A), and catalytic reforming. Catalytic reforming is illustrated by the conversion of hexane first to cyclohexane and then to benzene.

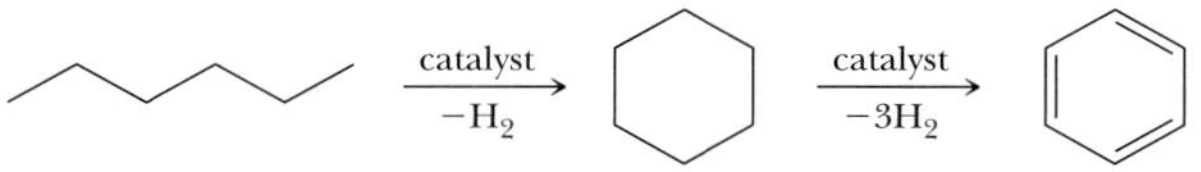

Hexane Cyclohexane Benzene

C. Coal

To understand how coal can be used as a raw material for the production of organic compounds, it is necessary to discuss synthesis gas. **Synthesis gas** is a mixture of carbon monoxide and hydrogen in varying proportions, depending on the means by which it is manufactured. Synthesis gas is prepared by passing steam over hot coal; it is also prepared by partial oxidation of methane with oxygen.

CHEMICAL CONNECTIONS

Octane Rating: What Those Numbers at the Pump Mean

Gasoline is a complex mixture of C_6 to C_{12} hydrocarbons. The quality of gasoline as a fuel for internal combustion engines is expressed in terms of an octane rating. Engine knocking occurs when a portion of the air-fuel mixture explodes prematurely (usually as a result of heat developed during compression) and independently of ignition by the spark plug. Two compounds were selected as reference fuels. One of these, 2,2,4-trimethylpentane (isooctane), has very good antiknock properties (the air-fuel mixture burns smoothly in the combustion chamber) and was assigned an octane rating of 100. (The name *isooctane* is a trivial name, its only relation to 2,2,4-trimethylpentane is that both compounds have eight carbons.) Heptane, the other reference compound, has poor antiknock properties and was assigned an octane rating of 0.

Heptane
(octane rating 0)

2,2,4-Trimethylpentane
(octane rating 100)

The **octane rating** of a particular gasoline is the percent of isooctane in a mixture of isooctane and heptane that has antiknock properties equivalent to that of the test gasoline. For example, the antiknock properties of 2-methylhexane are the same as those of a mixture of 42% isooctane and 58% heptane; therefore, the octane rating of 2-methylhexane is 42. Octane itself has an octane rating of -20, which means that it produces even more engine knocking than heptane.

$$C + H_2O \xrightarrow{\text{heat}} CO + H_2$$

Coal

$$CH_4 + \frac{1}{2}O_2 \xrightarrow{\text{catalyst}} CO + 2H_2$$

Methane

Two important organic compounds produced today almost exclusively from carbon monoxide and hydrogen are methanol and acetic acid. In the production of methanol, the ratio of carbon monoxide to hydrogen is adjusted to 2:1 and the mixture is passed over a catalyst at elevated temperature and pressure.

$$CO + 2H_2 \xrightarrow{\text{catalyst}} CH_3OH$$

Methanol

Treatment of methanol, in turn, with carbon monoxide over a different catalyst gives acetic acid.

$$CH_3OH + CO \xrightarrow{\text{catalyst}} CH_3\overset{\overset{\displaystyle O}{\displaystyle \|}}{C}OH$$

Methanol   Acetic acid

Because the processes for making methanol and acetic acid directly from carbon monoxide are commercially proven, it is likely that the decades ahead will see the development of routes to other organic chemicals from coal via methanol.

Summary

A **hydrocarbon** is a compound composed only of carbon and hydrogen. **Saturated hydrocarbons** (**alkanes** and **cycloalkanes**) contain only single bonds (Introduction). Alkanes have the general formula C_nH_{2n+2} (Section 2.1). Constitutional isomers (Section 2.2) have the same molecular formula but a different connectivity of their atoms. Alkanes are named according to a set of rules developed by the International Union of Pure and Applied Chemistry (IUPAC) (Section 2.3A). Substituents derived from alkanes are known as **alkyl groups.** A carbon atom is classified as **primary (1°), secondary (2°), tertiary (3°),** or **quaternary (4°)** depending on the number of alkyl groups bonded to it (Section 2.3C).

A saturated hydrocarbon that contains carbon atoms bonded to form a ring is called a **cycloalkane** (Section 2.4). To name a cycloalkane, name and locate each substituent on the ring and prefix the name of the open-chain alkane by *cyclo-*. Five-membered rings (cyclopentanes) and six-membered rings (cyclohexanes) are especially abundant in the biological world.

The IUPAC system is a general system of nomenclature (Section 2.5). The IUPAC name of a compound consists of three parts: (1) a **prefix** that tells the number of carbon atoms in the parent chain, (2) an **infix** that tells the nature of the carbon-carbon bonds in the parent chain, and (3) a **suffix** that tells the class to which the compound belongs.

A **conformation** is any three-dimensional arrangement of the atoms of a molecule resulting from rotations about one or more single bonds (Section 2.6). One convention for showing conformations is the **Newman projection.** A **dihedral angle** is the angle created by two intersecting planes. For ethane, staggered conformations occur at dihedral angles of 60°, 180°, and 300°. Eclipsed conformations occur at dihedral angles of 0°, 120°, and 240°. For butane viewed along the C_2—C_3 bond, the staggered conformation of dihedral angle 180° is called an **anti conformation;** the staggered conformations of dihedral angle 60° and 300° are called **gauche conformations.** The anti conformation of butane is lower in energy than the gauche conformations by approximately 3.8 kJ (0.9 kcal)/mol.

Intramolecular strain (Section 2.6) is of three types: (1) **torsional strain** (also called eclipsed interaction strain) arises when nonbonded atoms separated by three bonds are forced from a staggered conformation to an eclipsed conformation; (2) **angle strain** arises from creation of either abnormally large or abnormally small bond angles; and (3) **steric strain** (also called nonbonded interaction or van der Waals strain) arises when nonbonded atoms separated by four or more bonds are forced abnormally close to each other. The relationship between the change in Gibbs free energy, an equilibrium constant, and temperature in kelvins is given by the equation $\Delta G° = -RT\ln K_{eq}$.

In all cycloalkanes larger than cyclopropane, nonplanar conformations are favored. The lowest-energy conformation of cyclopentane is an envelope conformation (Section 2.6B). The lowest-energy conformations of cyclohexane are two interconvertible **chair conformations.** In a chair conformation, six bonds are **axial,** and six bonds are **equatorial.** Bonds axial in one chair are equatorial in the alternative chair. **Boat** and **twist-boat conformations** are higher in energy than chair conformations. The more stable conformation of a substituted cyclohexane is the one that minimizes diaxial interactions.

Stereoisomers are compounds that have the same connectivity but a different orientation of their atoms in space (Section 2.7). A **Stereocenter** is an atom about which exchange of two groups produces a stereoisomer. **Configuration** refers to the arrangement of atoms or groups of atoms bonded to a stereocenter. *Cis,trans isomers* have the same molecular formula and the same connectivity of their atoms, but the arrangement of their atoms in space cannot be interconverted by rotation about single bonds. *Cis* substituents are on the same side of the ring; *trans* substituents are on opposite sides of the ring. Most cycloalkanes with substituents on two or more carbons show *cis,trans* isomerism.

Alkanes are nonpolar compounds, and the only forces of attraction between their molecules are **dispersion forces** (Section 2.8B), weak electrostatic interactions between temporary induced dipoles of adjacent atoms or molecules. Low-molecular-weight alkanes are gases at room temperature and atmospheric pressure. Higher-molecular-weight alkanes are liquids. Very-high-molecular-weight alkanes are solids. Among a set of alkane constitutional isomers, the least branched isomer generally has the highest boiling point; the most branched isomer generally has the lowest boiling point.

As determined by **heats of combustion,** strain in cycloalkanes varies with ring size (Section 2.9). Cyclohexane, which has the most common ring size among organic compounds, is strain free.

Natural gas (Section 2.10A) consists of 90–95% methane with lesser amounts of ethane and other low-molecular-weight hydrocarbons. **Petroleum** (Section 2.10B) is a liquid mixture of literally thousands of different hydrocarbons. The most important processes in petroleum refining are fractional distillation, catalytic cracking, and catalytic reforming. **Synthesis gas** (Section 2.10C), a mixture of carbon monoxide and hydrogen, can be derived from natural gas, coal, or petroleum.

Key Reactions

1. Oxidation of Alkanes (Section 2.9A)

Oxidation of alkanes to carbon dioxide and water is the basis for their use as energy sources of heat and power.

$$CH_3CH_2CH_3 + 5O_2 \longrightarrow 3CO_2 + 4H_2O \qquad \Delta H^0 = -2220 \text{ kJ } (-530.6 \text{ kcal})/\text{mol}$$

Problems

2.16 Write a line-angle formula for each condensed structural formula.

(a) 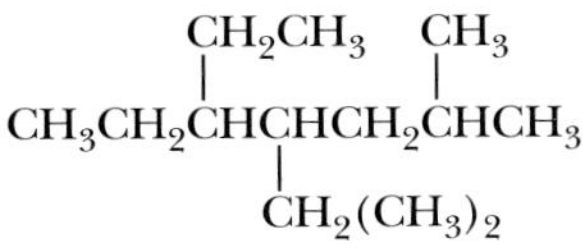 $CH_3CH_2\overset{\overset{\displaystyle CH_2CH_3}{|}}{C}H\overset{\overset{\displaystyle CH_3}{|}}{C}HCH_2\underset{\underset{\displaystyle CH_2(CH_3)_2}{|}}{C}HCH_3$

(b) 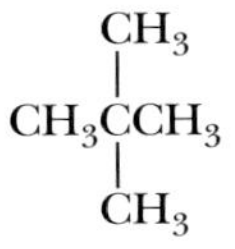 $CH_3\overset{\overset{\displaystyle CH_3}{|}}{\underset{\underset{\displaystyle CH_3}{|}}{C}}CH_3$

(c) $(CH_3)_2CHCH(CH_3)_2$

(d) 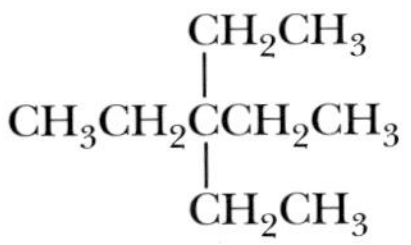 $CH_3CH_2\overset{\overset{\displaystyle CH_2CH_3}{|}}{\underset{\underset{\displaystyle CH_2CH_3}{|}}{C}}CH_2CH_3$

(e) $(CH_3)_3CH$

(f) $CH_3(CH_2)_3CH(CH_3)_2$

ORGANIC
Chemistry∙ᎧᎧ∙Now™
Assess your understanding of this chapter's topics with additional quizzing and conceptual-based problems at
http://now.brookscole.com/bfi4

2.17 Write the molecular formula of each alkane.

(a) 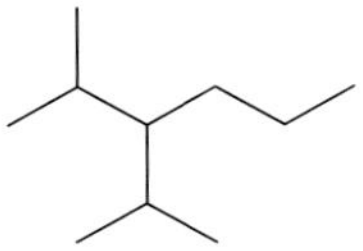   (b) 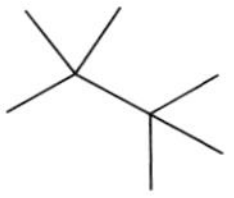   (c)

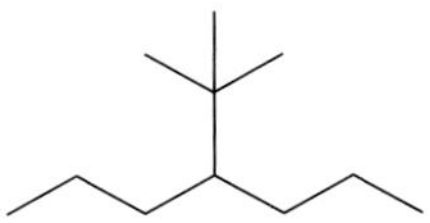

2.18 Provide an even more abbreviated formula for each structural formula, using parentheses and subscripts.

(a) 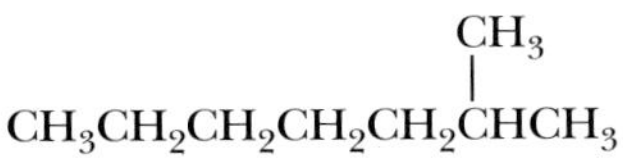 $CH_3CH_2CH_2CH_2CH_2\overset{\overset{\displaystyle CH_3}{|}}{C}HCH_3$

(b) $H\overset{\overset{\displaystyle CH_2CH_2CH_3}{|}}{\underset{\underset{\displaystyle CH_2CH_2CH_3}{|}}{C}}CH_2CH_2CH_3$

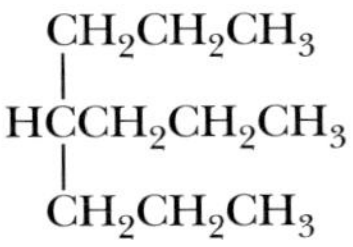

(c) $CH_3\overset{\overset{\displaystyle CH_2CH_2CH_3}{|}}{\underset{\underset{\displaystyle CH_2CH_2CH_3}{|}}{C}}CH_2CH_2CH_2CH_2CH_3$

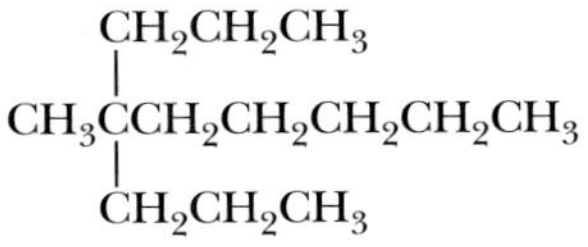

Constitutional Isomerism

2.19 Which statements are true about constitutional isomers?

(a) They have the same molecular formula.
(b) They have the same molecular weight.
(c) They have the same order of attachment of atoms.
(d) They have the same physical properties.

2.20 Indicate whether the compounds in each set are constitutional isomers.

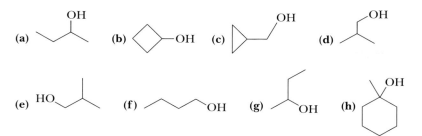

(a) CH₃CH₂OH and CH₃OCH₃

(b) CH₃CCH₃ and CH₃CH₂CH

(c) CH₃COCH₃ and CH₃CH₂COH

(d) CH₃CHCH₂CH₃ and CH₃CCH₂CH₃

(e) ⬠ and CH₃CH₂CH₂CH₂CH₃

(f) ⬠ and CH₂=CHCH₂CH₂CH₃

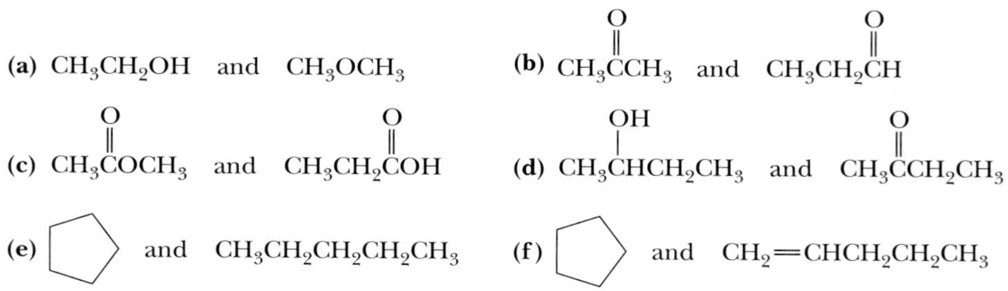

2.21 Each member of the following set of compounds is an alcohol; that is, each contains an —OH (hydroxyl group, Section 1.3A). Which structural formulas represent the same compound, and which represent constitutional isomers?

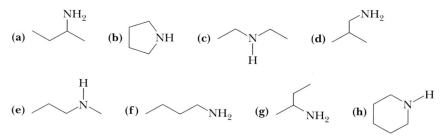

2.22 Each of the following compounds is an amine (Section 1.3B). Which structural formulas represent the same compound, and which represent constitutional isomers?

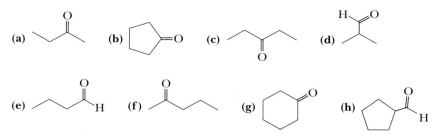

2.23 Each of the following compounds is either an aldehyde or a ketone (Section 1.3C). Which structural formulas represent the same compound and which represent constitutional isomers?

2.24 Draw structural formulas, and write IUPAC names for the nine constitutional isomers with molecular formula C₇H₁₆.

2.25 Draw structural formulas for all the following.

    (a) Alcohols with molecular formula $C_4H_{10}O$
    (b) Aldehydes with molecular formula C_4H_8O
    (c) Ketones with molecular formula $C_5H_{10}O$
    (d) Carboxylic acids with molecular formula $C_5H_{10}O_2$

Nomenclature of Alkanes and Cycloalkanes

2.26 Write IUPAC names for these alkanes and cycloalkanes:

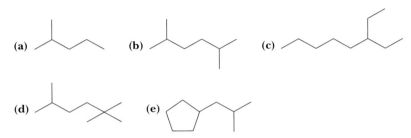

2.27 Write structural formulas for the following alkanes and cycloalkanes.

    (a) 2,2,4-Trimethylhexane   **(b)** 2,2-Dimethylpropane
    (c) 3-Ethyl-2,4,5-trimethyloctane   **(d)** 5-Butyl-2,2-dimethylnonane
    (e) 4-(1-Methylethyl)octane   **(f)** 3,3-Dimethylpentane
    (g) *trans*-1,3-Dimethylcyclopentane   **(h)** *cis*-1,2-Diethylcyclobutane

2.28 Explain why each is an incorrect IUPAC name, and write the correct IUPAC name for the intended compound.

    (a) 1,3-Dimethylbutane   **(b)** 4-Methylpentane
    (c) 2,2-Diethylbutane   **(d)** 2-Ethyl-3-methylpentane
    (e) 2-Propylpentane   **(f)** 2,2-Diethylheptane
    (g) 2,2-Dimethylcyclopropane   **(h)** 1-Ethyl-5-methylcyclohexane

The IUPAC System of Nomenclature

2.29 For each IUPAC name, draw the corresponding structural formula.

    (a) Ethanol   **(b)** Butanal   **(c)** Butanoic acid
    (d) Ethanoic acid   **(e)** Heptanoic acid   **(f)** Propanoic acid
    (g) Octanal   **(h)** Cyclopentene   **(i)** Cyclopentanol
    (j) Cyclopentanone   **(k)** Cyclohexanol   **(l)** Propanone

2.30 Write the IUPAC name for each compound.

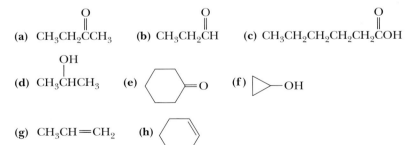

2.31 Assume for the purposes of this problem that, to be alcohol (-ol) or amine (-amine), the hydroxyl or amino group must be bonded to a tetrahedral (sp^3 hybridized) carbon atom.

Write the structural formula of a compound with an unbranched chain of four carbon atoms that is an:

(a) Alkane	**(b)** Alkene	**(c)** Alkyne
(d) Alkanol	**(e)** Alkenol	**(f)** Alkynol
(g) Alkanamine	**(h)** Alkenamine	**(i)** Alkynamine
(j) Alkanal	**(k)** Alkenal	**(l)** Alkynal
(m) Alkanone	**(n)** Alkenone	**(o)** Alkynone
(p) Alkanoic acid	**(q)** Alkenoic acid	**(r)** Alkynoic acid

(*Note:* There is only one structural formula possible for some parts of this problem. For other parts, two or more structural formulas are possible. Where two are more are possible, we will deal with how the IUPAC system distinguishes between them when we come to the chapters on those particular functional groups.)

Conformations of Alkanes and Cycloalkanes

2.32 Torsional strain resulting from eclipsed C—H bonds is approximately 4.2 kJ (1.0 kcal)/mol, and that for eclipsed C—H and C—CH$_3$ bonds is approximately 6.3 kJ (1.5 kcal)/mol. Given this information, sketch a graph of energy versus dihedral angle for propane.

2.33 How many different staggered conformations are there for 2-methylpropane? How many different eclipsed conformations are there?

2.34 Consider 1-bromopropane, CH$_3$CH$_2$CH$_2$Br.

(a) Draw a Newman projection for the conformation in which —CH$_3$ and —Br are anti (dihedral angle 180°).
(b) Draw Newman projections for the conformations in which —CH$_3$ and —Br are gauche (dihedral angles 60° and 300°).
(c) Which of these is the lowest energy conformation?
(d) Which of these conformations, if any, are related by reflection?

2.35 Consider 1-bromo-2-methylpropane, and draw the following.

(a) The staggered conformation(s) of lowest energy
(b) The staggered conformation(s) of highest energy

2.36 *Trans*-1,4-di-*tert*-butylcyclohexane exists in a normal chair conformation. *Cis*-1,4-di-*tert*-butylcyclohexane, however, adopts a twist-boat conformation. Draw both isomers, and explain why the *cis* isomer is more stable in a twist-boat conformation.

2.37 From studies of the dipole moment of 1,2-dichloroethane in the gas phase at room temperature (25°C), it is estimated that the ratio of molecules in the anti conformation to gauche conformation is 7.6 to 1. Calculate the difference in Gibbs free energy between these two conformations.

Cis,Trans Isomerism in Cycloalkanes

2.38 What structural feature of cycloalkanes makes *cis,trans* isomerism in them possible?

2.39 Is *cis,trans* isomerism possible in alkanes?

2.40 Draw structural formulas for the *cis* and *trans* isomers of 1,2-dimethylcyclopropane.

2.41 Name and draw structural formulas for all cycloalkanes with molecular formula C$_5$H$_{10}$. Be certain to include *cis* and *trans* isomers as well as constitutional isomers.

2.42 Using a planar pentagon representation for the cyclopentane ring, draw structural formulas for the *cis* and *trans* isomers of the following.

(a) 1,2-Dimethylcyclopentane　　(b) 1,3-Dimethylcyclopentane

2.43 Gibbs free energy differences between axial-substituted and equatorial-substituted chair conformations of cyclohexane were given in Table 2.4.

    (a) Calculate the ratio of equatorial to axial *tert*-butylcyclohexane at 25°C.

    (b) Explain why the conformational equilibria for methyl, ethyl, and isopropyl substituents are comparable but the conformational equilibrium for *tert*-butylcyclohexane lies considerably farther toward the equatorial conformation.

2.44 When cyclohexane is substituted by an ethynyl group, —C≡CH, the energy difference between axial and equatorial conformations is only 1.7 kJ (0.41 kcal)/mol. Compare the conformational equilibrium for methylcyclohexane with that for ethynylcyclohexane, and account for the difference between the two.

2.45 Calculate the difference in Gibbs free energy in kilojoules per mole between the alternative chair conformations of:

    (a) *trans*-4-Methylcyclohexanol   **(b)** *cis*-4-Methylcyclohexanol

    (c) *trans*-1,4-Dicyanocyclohexane

2.46 Draw the alternative chair conformations for the *cis* and *trans* isomers of 1,2-dimethylcyclohexane, 1,3-dimethylcyclohexane, and 1,4-dimethylcyclohexane.

    (a) Indicate by a label whether each methyl group is axial or equatorial.

    (b) For which isomer(s) are the alternative chair conformations of equal stability?

    (c) For which isomer(s) is one chair conformation more stable than the other?

2.47 Use your answers from Problem 2.46 to complete the table showing correlations between *cis,trans* and axial,equatorial for disubstituted derivatives of cyclohexane.

Position of Substitution	*cis*	*trans*
1,4-	a,e or e,a	e,e or a,a
1,3-	—— or ——	—— or ——
1,2-	—— or ——	—— or ——

2.48 There are four *cis,trans* isomers of 2-isopropyl-5-methylcyclohexanol:

2-Isopropyl-5-methylcyclohexanol

    (a) Using a planar hexagon representation for the cyclohexane ring, draw structural formulas for the four *cis,trans* isomers.

    (b) Draw the more stable chair conformation for each of your answers in part (a).

    (c) Of the four *cis,trans* isomers, which is the most stable? (*Hint:* If you answered this part correctly, you picked the isomer found in nature and given the name menthol.)

2.49 Draw alternative chair conformations for each substituted cyclohexane, and state which chair is more stable.

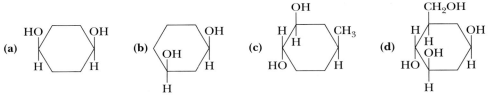

2.50 1,2,3,4,5,6-Hexachlorocyclohexane shows *cis, trans* isomerism. At one time a crude mixture of these isomers was sold as an insecticide. The insecticidal properties of the mixture arise from one isomer, known as lindane, which is *cis*-1,2,4,5-*trans*-3,6-hexachlorocyclohexane.

(a) Draw a structural formula for 1,2,3,4,5,6-hexachlorocyclohexane disregarding, for the moment, the existence of *cis,trans* isomerism. What is the molecular formula of this compound?

(b) Using a planar hexagon representation for the cyclohexane ring, draw a structural formula for lindane.

(c) Draw a chair conformation for lindane, and label which chlorine atoms are axial and which are equatorial.

(d) Draw the alternative chair conformation of lindane, and again label which chlorine atoms are axial and which are equatorial.

(e) Which of the alternative chair conformations of lindane is more stable? Explain.

Physical Properties

2.51 In Problem 2.24, you drew structural formulas for all isomeric alkanes with molecular formula C_7H_{16}. Predict which isomer has the lowest boiling point and which has the highest boiling point.

2.52 What generalization can you make about the densities of alkanes relative to the density of water?

2.53 What unbranched alkane has about the same boiling point as water? (Refer to Table 2.5 on the physical properties of alkanes.) Calculate the molecular weight of this alkane, and compare it with that of water.

Reactions of Alkanes

2.54 Complete and balance the following combustion reactions. Assume that each hydrocarbon is converted completely to carbon dioxide and water.

(a) Propane + O_2 →   (b) Octane + O_2 →
(c) Cyclohexane + O_2 →   (d) 2-Methylpentane + O_2 →

2.55 Following are heats of combustion per mole for methane, propane, and 2,2,4-trimethylpentane. Each is a major source of energy. On a gram-for-gram basis, which of these hydrocarbons is the best source of heat energy?

Hydrocarbon	Component of	ΔH^0 [KJ (kcal)/mol]
CH_4	Natural gas	−891 (−213)
$CH_3CH_2CH_3$	LPG	−2220 (−531)
$CH_3CCH_2CHCH_3$ (with CH_3, CH_3 and CH_3 substituents)	Gasoline	−5452 (−1304)

2.56 Following are structural formulas and heats of combustion of acetaldehyde and ethylene oxide. Which of these compounds is the more stable? Explain.

Acetaldehyde
−1164 kJ (−278.8 kcal)/mol

Ethylene oxide
−1264 kJ (−302.1 kcal)/mol

2.57 Without consulting tables, arrange these compounds in order of decreasing (less negative) heat of combustion: hexane, 2-methylpentane, and 2,2-dimethylbutane.

2.58 Which would you predict to have the larger (more negative) heat of combustion, *cis*-1,4-dimethylcyclohexane or *trans*-1,4-dimethylcyclohexane?

Looking Ahead

2.59 Following are structural formulas for 1,4-dioxane and piperidine. 1,4-Dioxane is a widely used solvent for organic compounds. Piperidine is found in small amounts in black pepper (*Piper nigrum*).

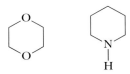

1,4-Dioxane   Piperidine

(a) Complete the Lewis structure of each compound by showing all unshared electron pairs.
(b) Predict bond angles about each carbon, oxygen, and nitrogen atom.
(c) Describe the most stable conformation of each ring, and compare these conformations with the chair conformation of cyclohexane.

2.60 Following is a planar hexagon representation of L-fucose, a sugar component of the determinants of the A, B, O blood group typing. For more on this system of blood typing, see Connections to Biological Chemistry "A, B, AB, and O Blood Types" in Chapter 25.

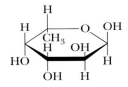

L-Fucose

(a) Draw the alternative chair conformations of L-fucose.
(b) Which of them is the more stable? Explain.

2.61 On the left is a stereorepresentation of glucose (we discuss the structure and chemistry of glucose in Chapter 25):

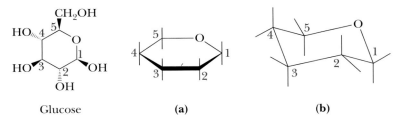

Glucose   (a)   (b)

(a) Convert the stereorepresentation on the left to a planar hexagon representation.
(b) Convert the stereorepresentation on the left to a chair conformation. Which substituent groups in the chair conformation are equatorial? Which are axial?

2.62 Following is the structural formula and a ball-and-stick model of cholestanol. The only difference between this compound and cholesterol (Section 26.4) is that cholesterol has a carbon-carbon double bond in ring B.

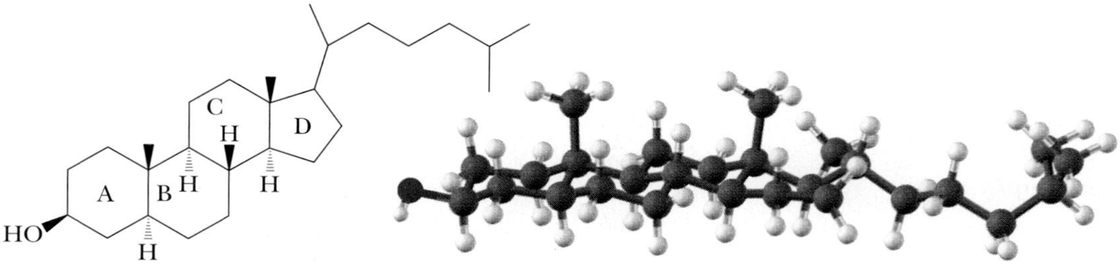

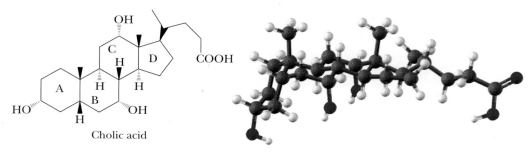

Cholestanol

(a) Describe the conformation of rings A, B, C, and D in cholestanol.
(b) Is the hydroxyl group on ring A axial or equatorial?
(c) Consider the methyl group at the junction of rings A and B. Is it axial or equatorial to ring A? Is it axial or equatorial to ring B?
(d) Is the methyl group at the junction of rings C and D axial or equatorial to ring C?

2.63 Following is the structural formula and a ball-and-stick model of cholic acid (Chapter 26), a component of human bile whose function is to aid in the absorption and digestion of dietary fats.

Cholic acid

(a) What is the conformation of ring A? of ring B? of ring C? of ring D?
(b) Are the hydroxyl groups on rings A, B, and C axial or equatorial to their respective rings?
(c) Is the methyl group at the junction of rings A and B axial or equatorial to ring A? Is it axial or equatorial to ring B?
(d) Is the hydrogen at the junction of rings A and B axial or equatorial to ring A? Is it axial or equatorial to ring B?
(e) Is the methyl group at the junction of rings C and D axial or equatorial to ring C?

Photo credit: Frank Wing/Photodisc Green/Getty Images

3

■ Tartaric acid (Section 3.4B) is found in grapes and other fruits, both free and as its salts. Inset: a model of the *R,R* enantiomer of tartaric acid.

Stereochemistry The study of three-dimensional arrangements of atoms in molecules.

Stereoisomerism and Chirality

The study of molecules as three-dimensional objects is called **stereochemistry.** The ability to visualize stereochemical relationships is a survival skill in organic chemistry and biochemistry. Our goal in this chapter is to expand your awareness of molecules as three-dimensional objects. We suggest you purchase a set of models (or, if you prefer, a computer modeling program such as Chem3D or Spartan). Alternatively, you may have access to a computer lab with a modeling program. Use your models heavily to aid you in visualizing the spatial concepts in this and later chapters.

3.1 Stereoisomerism

Isomers are different compounds with the same molecular formula. Thus far, we have encountered three types of isomers. Constitutional isomers (Section 2.2) have the same molecular formula but a different connectivity of atoms in their molecules. Examples of pairs of constitutional isomers are pentane and 2-methylbutane, and 1-pentene and cyclopentane.

Constitutional isomers

Pentane
(C_5H_{12})

and

2-Methylbutane
(C_5H_{12})

1-Pentene
(C_5H_{10})

and

Cyclopentane
(C_5H_{10})

Stereoisomers Isomers that have the same molecular formula and the same connectivity of their atoms but a different orientation of their atoms in space.

Configurational isomers Isomers that differ by the configuration of substituents on an atom.

A second type of isomerism is stereoisomerism. **Stereoisomers** have the same molecular formula and the same connectivity but different orientations of their atoms in space. One example of stereoisomerism we have seen thus far is that of *cis,trans* isomers in cycloalkanes (Section 2.7), which arise because substituents on a ring are locked into one of two orientations in space with respect to one another by the ring. Isomers of this type are called **configurational isomers** because they differ by the configuration of substituents on an atom.

Configurational isomers (*cis,trans* isomers)

H_3C━CH$_3$

and

H_3C━CH$_3$

cis-1,4-Dimethylcyclohexane

trans-1,4-Dimethylcyclohexane

The third type of isomerism is presented by conformational isomers, discussed in Section 2.6.

3.2 Chirality—The Handedness of Molecules

A **mirror image** is the reflection of an object in a mirror. When you look in a mirror, you see a reflection, or mirror image, of yourself. Now suppose your mirror image became a three-dimensional object. We could then ask, "What is the relationship between you and your mirror image?" To clarify what we mean by *relationship* we might instead ask, "Can your reflection be superposed on (placed on top of) the original 'you' in such a way that every detail of the reflection corresponds exactly to the original?" The answer is that you and your mirror image are not superposable. If you have a ring on the little finger of your right hand, for example, your mirror image has the ring on the little finger of its left hand. If you part your hair on your right side, it will be parted on the left side in your reflection. You and your reflection are different objects. You cannot exactly superpose one on the other.

■ The horns of this African gazelle show chirality and are mirror images of each other.

Molecules that are not superposable on their mirror images are said to be **chiral** (pronounced ki-ral, to rhyme with spiral; from the Greek: *cheir*, hand). That is, they show handedness. Chirality is encountered in three-dimensional objects of all sorts. Your left hand is chiral and so is your right hand (they are approximately mirror images of each other). A spiral binding on a notebook is chiral. A machine screw with a right-handed thread is chiral. A ship's propeller is chiral. As you examine objects around you, you will undoubtedly conclude that the vast majority of them are chiral as well.

Chiral From the Greek, *cheir*, hand; an object that is not superposable on its mirror image; an object that has handedness.

Achiral An object that lacks chirality; an object that has no handedness.

The contrasting situation to chirality occurs when an object and its mirror image are superposable. An object and its mirror image are superposable if one of them can be oriented in space so that all its features (corners, edges, points, designs, etc.) correspond exactly to those in the other member of the pair. If this can be done, the object and its mirror image are identical; the original object is **achiral** (that is, lacks chirality). Examples of objects lacking chirality are an undecorated cup, a regular tetrahedron, a cube, and a perfect sphere.

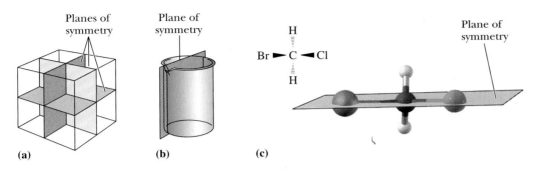

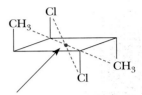

(d) Center of symmetry

Figure 3.1
Symmetry in objects. A cube has several planes of symmetry and a center of symmetry. The beaker and CH_2BrCl each have a single plane of symmetry. The cyclobutane has a center of symmetry.

An object will be achiral if it has one or more of certain elements of symmetry. The most common such elements in organic compounds are a plane and a center of symmetry. As we shall see, any molecule with either of these symmetry elements (and one other rarely encountered) is achiral and can be superposed on its mirror image. A **plane of symmetry** is an imaginary plane passing through an object dividing it such that one half is the reflection of the other half. The cube shown in Figure 3.1 has several planes of symmetry. Both the beaker and the compound bromochloromethane have a single plane of symmetry. A **center of symmetry** is a point so situated that identical components of the object are located equidistant and on opposite sides from the point along any axis passing through that point. The cube shown in Figure 3.1 has a center of symmetry as does the cyclobutane. Because it has a center of symmetry, the cyclobutane is identical to its mirror image, and is achiral.

Objects that lack both of these symmetry elements are chiral. We can illustrate the chirality of an organic molecule by considering 1-bromo-1-chloroethane. Figure 3.2 shows three-dimensional representations and ball-and-stick models for 1-bromo-1-chloroethane and its mirror image. This molecule has neither a plane nor a center of symmetry.

A model of 1-bromo-1-chloroethane can be turned and rotated in any direction in space, but as long as bonds are not broken and rearranged, only two of the four groups bonded to the central carbon of one molecule can be made to coincide with those of its mirror image. Because 1-bromo-1-chloroethane and its mirror image are nonsuperposable, they are enantiomers. **Enantiomers** are nonsuperposable mirror images. Note that the terms "chiral" and "achiral" refer to objects; the term "enantiomers" refers to the relationship between a pair of objects.

The most common cause of chirality in organic molecules is a tetrahedral atom, most commonly carbon, bonded to four different groups. A carbon atom with four

Plane of symmetry   An imaginary plane passing through an object dividing it so that one half is the mirror image of the other half.

Center of symmetry   A point so situated that identical components of an object are located on opposite sides and equidistant from that point along any axis passing through it.

Enantiomers   Stereoisomers that are nonsuperposable mirror images of each other; refers to a relationship between pairs of objects.

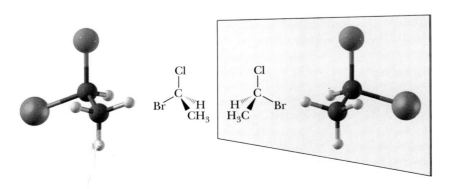

ORGANIC
Chemistry · Now™
Active Figure 3.2
Stereorepresentations of 1-bromo-1-chloroethane and its mirror image. **See a simulation based on this figure, and take a short quiz on the concepts.**

How To *Draw Chiral Molecules*

ORGANIC
Chemistry Now™
Click How to Tutorials to review
Drawing Chiral Molecules

It is worthwhile to notice that there are several different ways to represent the three-dimensional structure of chiral molecules on a two-dimensional page. For example, following are four different representations of one enantiomer of 2-butanol.

(1)   (2)   (3)   (4)

Representation (1) shows the tetrahedral geometry of the chiral center. We can turn (1) slightly in space and tip it a bit to place the carbon framework in the plane of the paper and give representation (2). In (2), we still have two groups in the plane of the paper, one coming toward us, and one going away from us. As discussed in Chapter 2, these are represented by wedged and dashed bonds, respectively. For an even more abbreviated representation, we can turn (2) into the line-angle formula (3). Although we don't normally show hydrogens in a line-angle formula, we do in (3) just to remind ourselves that the fourth group on this chiral center is really there and that it is H. Finally, we can carry the abbreviation a step further and write this enantiomer of 2-butanol as (4). Here we omit the H on the chiral center. We know it must be there because carbon needs four bonds, and we know it must be behind the plane of the paper because the OH is in front. Clearly the abbreviated formulas (3) and (4) are the easiest to write, and we will rely on these representations throughout the remainder of the text. When you need to write three-dimensional representations of chiral centers, try to keep the carbon framework in the plane of the paper and the other two atoms or groups of atoms on the chiral center toward and away from you. Using representation (4) as a model, here are two different representations for its enantiomers.

4   Enantiomer of 4   Enantiomer of 4

Chiral center A tetrahedral atom, most commonly carbon, that is bonded to four different groups; also called a **chirality center.**

Stereocenter An atom about which exchange of two groups produces a stereoisomer. Chiral centers are one type of stereocenter.

different groups bonded to it lacks the two key symmetry elements and is called a **chiral center,** or alternatively, a **chirality center.** The carbon atom of 1-bromo-1-chloroethane bearing the —Cl, —H, —CH₃, and —Br groups is a chiral center. The term **stereocenter** is also used but is broader; stereocenters (e.g., those involved in *cis,trans* isomerism, Section 2.7) need not be chiral centers, but chiral centers are a type of stereocenter.

Example 3.1

Each molecule has one chiral center. Draw stereorepresentations for the enantiomers of each.

(a)   (b)

Solution

You will find it helpful to view models of enantiomer pairs from different perspectives as is done in these representations. As you work with these pairs of enantiomers, notice that each has a tetrahedral carbon atom bonded to four different groups, which makes the molecule chiral.

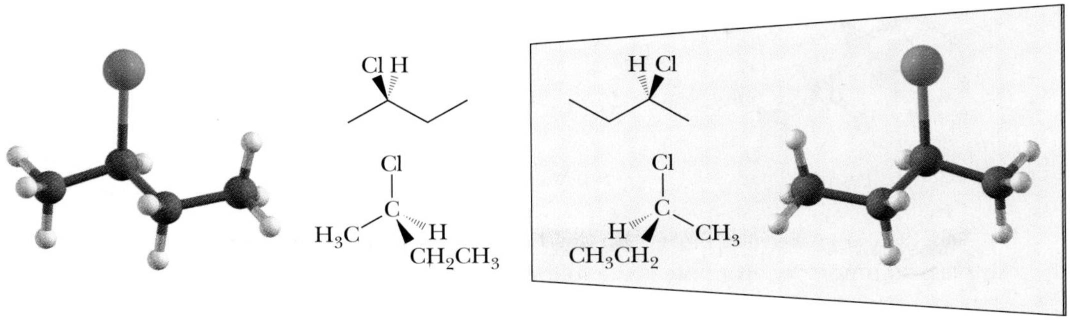

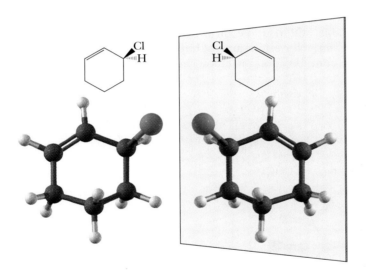

Problem 3.1

Each molecule has one chiral center. Draw stereorepresentations for the enantiomers of each.

(a) 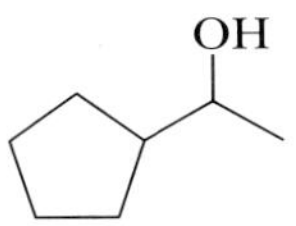   (b)

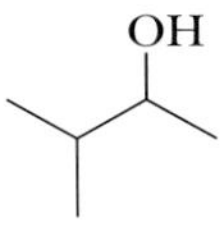

In all the molecules studied so far, chirality arises because of the presence of a tetrahedral carbon chiral center. Chiral centers are not limited to carbon. Following are stereorepresentations of a chiral cation in which the chiral center is nitrogen. We discuss the chirality of nitrogen centers in more detail in Section 23.3.

Enantiomers of tetrahedral silicon, phosphorus, and germanium compounds have also been isolated.

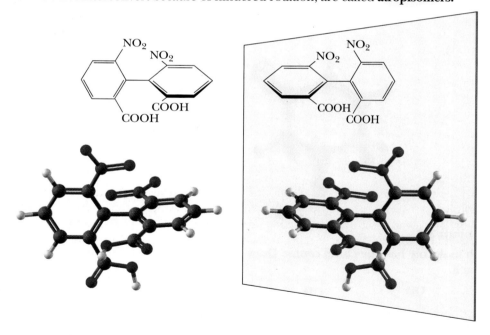

A pair of enantiomers

Chirality can also be present in molecules without chiral centers, although such cases are less common. An example is the following substituted biphenyl. Because of the large groups on the rings, there are huge nonbonded interactions in the planar conformer, and the twisted forms shown have far lower energies. These interactions result in a very high barrier to rotation around the carbon-carbon single bond connecting the rings, and rotation is very slow. Although this molecule has no chiral center, the mirror images are not superposable, and the molecule is chiral. Isomers of this sort, which lack a chiral center but do not interconvert because of hindered rotation, are called **atropisomers.**

Atropisomers   Enantiomers that lack a chiral center and differ because of hindered rotation.

3.3 Naming Chiral Centers — The *R,S* System

So far, we have discussed the fact that enantiomers exist. We have not considered the question of which isomer is which, that is, the **absolute configuration,** or which is the right-handed enantiomer and which is the left. For a given sample of a pure enantiomer, the correct arrangement must be determined by experiment. Experimental determination of absolute configuration can be accomplished by using x-ray analysis

Absolute configuration   Which of the two possible isomers an enantiomer is (i.e., whether it is the right- or left-handed isomer).

or making a derivative that has a chiral center with a known absolute configuration. In biological molecules, many absolute configurations were determined by comparison to absolute configurations of the chiral center in glyceraldehyde.

A system for designating the absolute configuration of a chiral center was devised in the late 1950s by R. S. Cahn and C. K. Ingold in England and V. Prelog in Switzerland and is named after them. The system, also called the **R,S system,** has been incorporated into the IUPAC rules of nomenclature. The orientation of groups about a chiral center is specified using a set of priority rules.

R,S System A set of rules for specifying absolute configuration about a stereocenter; also called the Cahn-Ingold-Prelog system.

Priority Rules

1. Each atom bonded to the chiral center is assigned a priority. Priority is based on atomic number; the higher the atomic number, the higher the priority. Following are several substituents arranged in order of increasing priority. The atomic number of the atom determining priority is shown in parentheses.

$$\overset{(1)}{-H}, \quad \overset{(6)}{-CH_3}, \quad \overset{(7)}{-NH_2}, \quad \overset{(8)}{-OH}, \quad \overset{(16)}{-SH}, \quad \overset{(17)}{-Cl}, \quad \overset{(35)}{-Br}, \quad \overset{(53)}{-I}$$

Increasing priority ⟶

2. If priority cannot be assigned on the basis of the atoms bonded directly to the chiral center (because of a "tie," that is, the same first atom on more than one substituent), look at the next set of atoms and continue until a priority can be assigned. Priority is assigned at the first point of difference. Following is a series of groups arranged in order of increasing priority. The atomic number of the atom on which the assignment of priority is based is shown above it.

$$\overset{(1)}{-CH_2-H} \quad \overset{(6)}{-CH_2-CH_3} \quad \overset{(7)}{-CH_2-NH_2} \quad \overset{(8)}{-CH_2-OH} \quad \overset{(17)}{-CH_2-Cl}$$

Increasing priority ⟶

If two carbons have substituents of the same priority, priority is assigned to the carbon that has more of these substituents. Thus, $-CHCl_2 > -CH_2Cl$.

3. Atoms participating in a double or triple bond are considered to be bonded to an equivalent number of similar "phantom" atoms (shown here in color) by single bonds; that is, atoms of the double bond are duplicated, and atoms of a triple bond are triplicated. The phantom atoms are bonded to no other atoms.

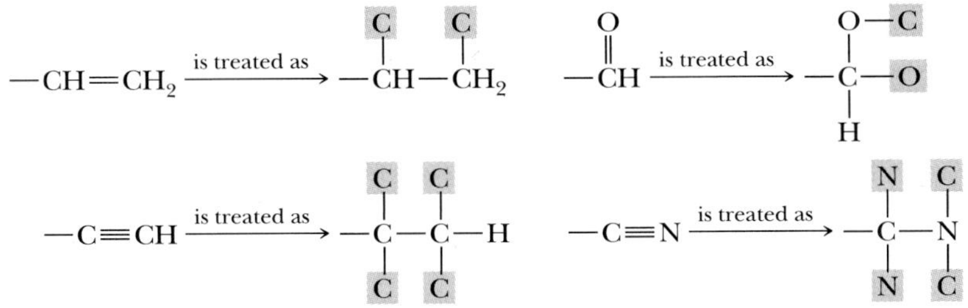

4. *Note:* priority assignment is made at the *first point of difference* between groups. A common mistake is to assume that larger groups must always have higher priority, but this might not necessarily be the case. For example, a $-CH_2Cl$ group has priority over a $-CH_2CH_2CH_2CH_3$ group because the Cl atom is the first point of difference.

ORGANIC
Chemistry⋅Now™

Click How To Tutorials to review
Assigning *R* or *S* Configuration to a Chiral Center

R From the Latin, *rectus,* straight, correct; used in the *R,S* convention to show that the order of priority of groups on a stereocenter is clockwise.

S From the Latin, *sinister,* left; used in the *R,S* convention to show that the order of priority of groups on a stereocenter is counterclockwise.

How To *Assign R or S Configuration to a Chiral Center*

1. Locate the chiral center, identify its four substituents, and assign a priority from 1 (highest) to 4 (lowest) to each substituent.
2. Orient the molecule in space so that the group of lowest priority (4) is directed away from you as would, for instance, the steering column of a car. The three groups of higher priority (1–3) then project toward you, as would the spokes of the steering wheel.
3. Read the three groups projecting toward you in order from highest priority (1) to lowest priority (3).
4. If the groups are read in a clockwise direction, the configuration is designated as ***R*** (Latin: *rectus,* straight, correct); if they are read in a counterclockwise direction, the configuration is ***S*** (Latin: *sinister,* left). You can also visualize this as follows: turning the steering wheel to the right (down the order of priority) equals *R*; turning it to the left equals *S*.

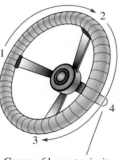

Group of lowest priority
points away from you

Example 3.2

Assign priorities to the groups in each set.

(a) $-\overset{\overset{\text{O}}{\|}}{\text{C}}\text{OH}$ and $-\overset{\overset{\text{O}}{\|}}{\text{C}}\text{H}$ **(b)** $-\text{CH}=\text{CH}_2$ and $-\text{CH(CH}_3)_2$

Solution

(a) The first point of difference is O of the —OH in the carboxyl group compared to —H in the aldehyde group. The carboxyl group has a higher priority.

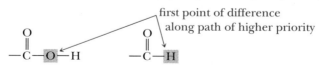

first point of difference
along path of higher priority

Carboxyl group
(higher priority)   Aldehyde group
(lower priority)

(b) Carbon 1 in each group has the same pattern of atoms; namely C(C,C,H) (i.e., carbon bonded to two carbons and a hydrogen). For the vinyl group, bonding at carbon 2 is C(C,H,H). For the isopropyl group, at carbon 2 it is C(H,H,H). The vinyl group is higher in priority than is the isopropyl group.

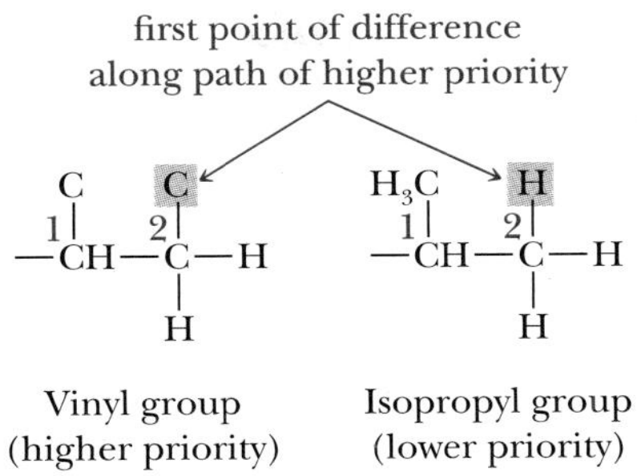

first point of difference
along path of higher priority

<pre>
   C   C   H₃C   H
   1│   2│   1│   2│
  ─CH─C─H   ─CH─C─H
       │   │
       H   H
</pre>

Vinyl group   Isopropyl group
(higher priority)   (lower priority)

Problem 3.2

Assign priorities to the groups in each set.

(a) —CH_2OH and —CH_2CH_2OH **(b)** —CH_2OH and —$CH=CH_2$
(c) —CH_2OH and —$C(CH_3)_3$

Example 3.3

Assign an *R* or *S* configuration to the chiral center in each molecule.

(a) 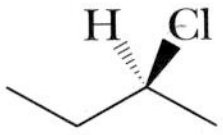   **(b)** 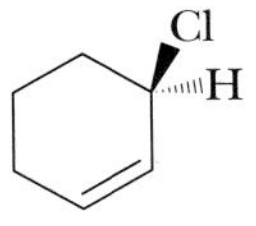   **(c)**

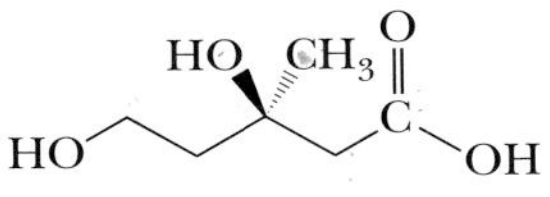

Solution

View each molecule through the chiral center along the bond from the chiral center toward the group of lowest priority. In (a), the order of priority is Cl > CH_2CH_3 > CH_3 > H; the configuration is *S*. In (b), the order of priority is Cl > CH=CH > CH_2 > H; the configuration is *R*. In (c), the order of priority is OH > CH_2COOH > CH_2CH_2OH > CH_3; the configuration is *R*.

(a)   **(b)**

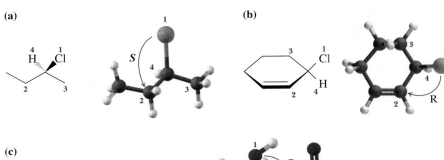

(c)

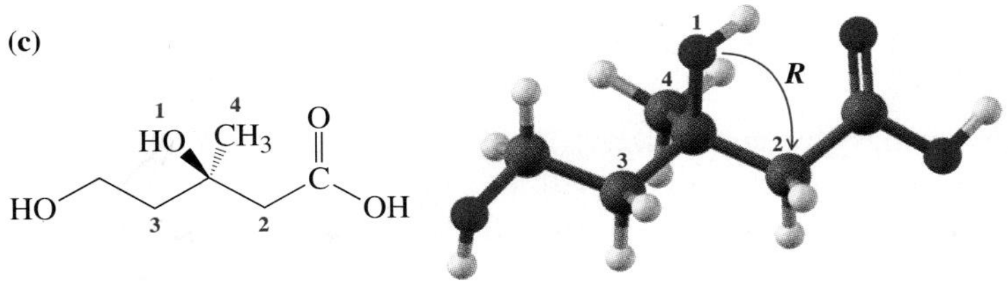

Problem 3.3

Assign an *R* or *S* configuration to the chiral center in each molecule.

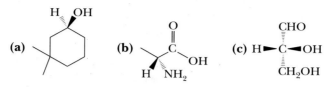

(a)   (b)   (c)

3.4 Acyclic Molecules with Two or More Chiral Centers

We have now seen several examples of molecules with one chiral center and verified that, for each, two stereoisomers (one pair of enantiomers) are possible. Now let us consider molecules with two or more chiral centers. To generalize, for a molecule with n chiral centers, the maximum number of stereoisomers possible is 2^n. We have already seen that, for a molecule with one chiral center, $2^1 = 2$ stereoisomers are possible. For a molecule with two chiral centers, $2^2 = 4$ stereoisomers are possible; for a molecule with three chiral centers, $2^3 = 8$ stereoisomers are possible, and so forth.

A. Enantiomers and Diastereomers

Let us begin our study of molecules with multiple chiral centers by considering 2,3,4-trihydroxybutanal, a molecule with two chiral centers, shown here in color.

$$HOCH_2-\boxed{CH}-\boxed{CH}-CHO$$
$$\qquad\qquad\ \ | \qquad\ |$$
$$\qquad\qquad\ OH \quad OH$$

2,3,4-Trihydroxybutanal

The maximum number of stereoisomers possible for this molecule is $2^2 = 4$, each of which is drawn in Figure 3.3. One of these pairs is called erythrose, and the other, threose.

Figure 3.3

The four stereoisomers of 2,3,4-trihydroxybutanal, a compound with two chiral centers.

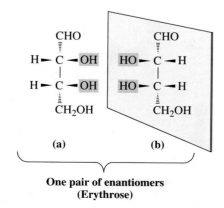

Stereoisomers (a) and (b) are nonsuperposable mirror images and are, therefore, a pair of enantiomers. Stereoisomers (c) and (d) are also nonsuperposable mirror images and are a second pair of enantiomers. One way to describe the four stereoisomers of 2,3,4-trihydroxybutanal is to say that they consist of two pairs of enantiomers. Enantiomers (a) and (b) of 2,3,4-trihydroxybutanal are given the names (2R,3R)-erythrose and (2S,3S)-erythrose; enantiomers (c) and (d) are given the names (2R,3S)-threose and (2S,3R)-threose. Note that all of the chiral centers in a molecule are reversed in its enantiomer. The molecule with the 2R,3S configuration is the enantiomer of the molecule with 2S,3R, and the molecule with 2S,3S is the enantiomer of the molecule with 2R,3R. Erythrose and threose belong to the class of compounds called carbohydrates, which we discuss in Chapter 25. Erythrose is found in erythrocytes (red blood cells), hence the derivation of its name.

We have specified the relationship between (a) and (b) and between (c) and (d); each represents a pair of enantiomers. What is the relationship between (a) and (c), between (a) and (d), between (b) and (c), and between (b) and (d)? The answer is that they are called **diasteromers.** Diastereomers are stereoisomers that are not mirror images (enantiomers). As we see in this example, molecules with at least two chiral centers can have diastereomers.

Diastereomers Stereoisomers that are not mirror images of each other; refers to relationships among two or more objects.

Example 3.4

Following are stereorepresentations for the four stereoisomers of 1,2,3-butanetriol. *R* and *S* configurations are given for the chiral centers in (1) and (4).

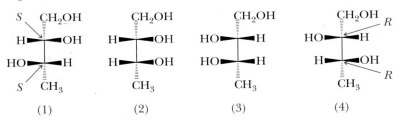

(a) Write the IUPAC names for each compound showing the *R* or *S* configuration of each chiral center.
(b) Which molecules are enantiomers?
(c) Which molecules are diastereomers?

Solution

(a) (1) (2S,3S)-1,2,3-Butanetriol   (2) (2S,3R)-1,2,3-Butanetriol
     (3) (2R,3S)-1,2,3-Butanetriol   (4) (2R,3R)-1,2,3-Butanetriol
(b) Enantiomers are stereoisomers that are nonsuperposable mirror images. As you see from their configurations, compounds (1) and (4) are one pair of enantiomers, and compounds (2) and (3) are a second pair of enantiomers.
(c) Diastereomers are stereoisomers that are not mirror images. Compounds (1) and (2), (1) and (3), (2) and (4), and (3) and (4) are pairs of diastereomers. Here is a diagram that shows the relationships between these isomers.

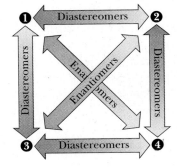

Problem 3.4

Following are stereorepresentations for the four stereoisomers of 3-chloro-2-butanol.

$$\underset{(1)}{\overset{CH_3}{\underset{CH_3}{\overset{|}{\underset{|}{H\text{—}C\text{—}OH}}\atop{Cl\text{—}C\text{—}H}}}}
\qquad
\underset{(2)}{\overset{CH_3}{\underset{CH_3}{\overset{|}{\underset{|}{H\text{—}C\text{—}OH}}\atop{H\text{—}C\text{—}Cl}}}}
\qquad
\underset{(3)}{\overset{CH_3}{\underset{CH_3}{\overset{|}{\underset{|}{HO\text{—}C\text{—}H}}\atop{H\text{—}C\text{—}Cl}}}}
\qquad
\underset{(4)}{\overset{CH_3}{\underset{CH_3}{\overset{|}{\underset{|}{HO\text{—}C\text{—}H}}\atop{Cl\text{—}C\text{—}H}}}}$$

(a) Assign an *R* or *S* configuration of each chiral center.
(b) Which compounds are enantiomers?
(c) Which compounds are diastereomers?

B. Meso Compounds

Certain molecules containing two or more chiral centers have special symmetry properties that reduce the number of stereoisomers to fewer than the maximum number predicted by the 2^n rule. One such molecule is 2,3-dihydroxybutanedioic acid, more commonly named tartaric acid.

$$\overset{O}{\overset{\|}{HOC}}\text{—}CH\text{—}CH\text{—}\overset{O}{\overset{\|}{COH}}$$
$$\qquad\quad\underset{OH}{|}\ \ \underset{OH}{|}$$

2,3-Dihydroxybutanedioic acid
(Tartaric acid)

Tartaric acid is a colorless, crystalline compound. During the fermentation of grape juice, potassium bitartrate (one carboxyl group is present as a potassium salt, —COO⁻K⁺) deposits as a crust on the sides of wine casks. When collected and purified, it is sold commercially as cream of tartar.

In tartaric acid, carbons 2 and 3 are chiral centers, and the maximum number of stereoisomers possible is $2^2 = 4$; these stereorepresentations are drawn in Figure 3.4. Structures (a) and (b) are nonsuperposable mirror images and are, therefore, a pair of enantiomers. Structures (c) and (d) are also mirror images, but they are superposable. To see this, imagine that you first rotate (d) by 180° in the plane of the paper, lift it out of the plane of the paper, and place it on top of (c). If you do this mental manipulation correctly, you find that (d) is superposable on (c). Therefore, (c) and

Figure 3.4
Stereoisomers of tartaric acid.
One pair of enantiomers and
one meso compound.

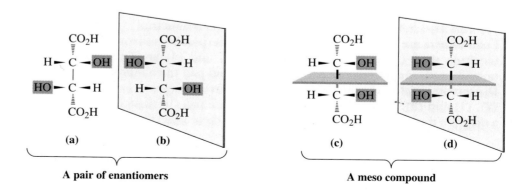

(a)   (b)

A pair of enantiomers

(c)   (d)

A meso compound

(d) are not different molecules; they are the same molecule, just oriented differently. Because (c) and its mirror image are superposable, (c) is achiral.

Another way to determine that (c) is achiral is to see that it has a plane of symmetry that bisects the molecule in such a way that the top half is the reflection of the bottom half. Thus, even though (c) has two chiral centers, it is achiral (Section 3.2).

The stereoisomer of tartaric acid represented by (c) or (d) is called a **meso compound.** A meso compound contains two or more chiral centers but is achiral. We can now answer the question: How many stereoisomers are there of tartaric acid? The answer is three: one meso compound and one pair of enantiomers. Note that the meso compound is a diastereomer of each member of the pair of enantiomers.

From this example, we can make this generalization about meso compounds: They have an internal mirror plane (or center of symmetry). If there are two chiral centers, one is *R* and the other is *S.*

Notice that the stereoisomers in Figure 3.4 are shown in only one conformation. Because conformational isomers in acyclic systems interconvert extremely rapidly, we need consider only the most symmetric conformers. Other conformers may be rotated to give the most symmetric one for determination of symmetry properties (see Problem 3.5).

It is sometimes helpful to identify stereochemical relationships among molecules based on assignment of configuration to all the chiral centers. In some cases, this can be easier than searching for and comparing the most symmetric conformations. Two stereoisomers will be enantiomers if all of the chiral centers present are reversed, and the stereoisomers will be diastereomers if only some of the chiral centers are reversed.

Meso compound An achiral compound possessing two or more chiral centers that also has chiral isomers.

Example 3.5

Following are stereorepresentations for the three stereoisomers of 2,3-butanediol. The carbons are numbered beginning from the left, as shown in (1).

(1)   (2)   (3)

(a) Assign an *R* or *S* configuration to each chiral center.
(b) Which are enantiomers?
(c) Which is the meso compound?
(d) Which are diastereomers?

Solution

(a) (1) (2*R*,3*R*)-2,3-Butanediol   (2) (2*R*,3*S*)-2,3-Butanediol
  (3) (2*S*,3*S*)-2,3-Butanediol
(b) Compounds (1) and (3) are enantiomers.
(c) Compound (2) is a meso compound. Note that compound (2) can be drawn in two symmetric conformations, one of which has a plane of symmetry and the other of which has a center of symmetry.

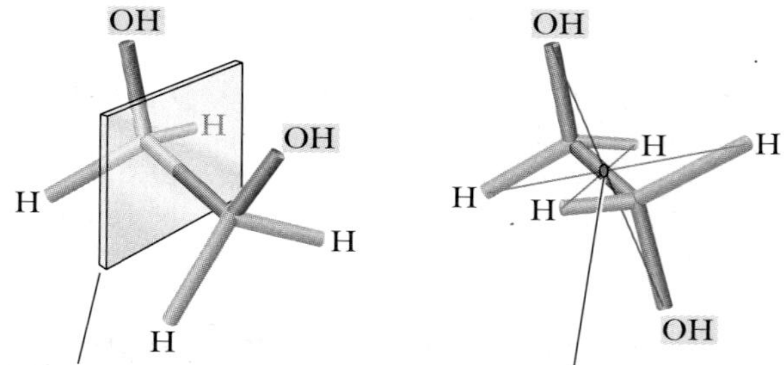

OH

H

OH

H

H

H

Plane of symmetry

OH

H   H

H

OH

Center of symmetry

(d) (1) and (2) are diastereomers; (2) and (3) are also diastereomers.

Problem 3.5

Following are four Newman projection formulas for tartaric acid.

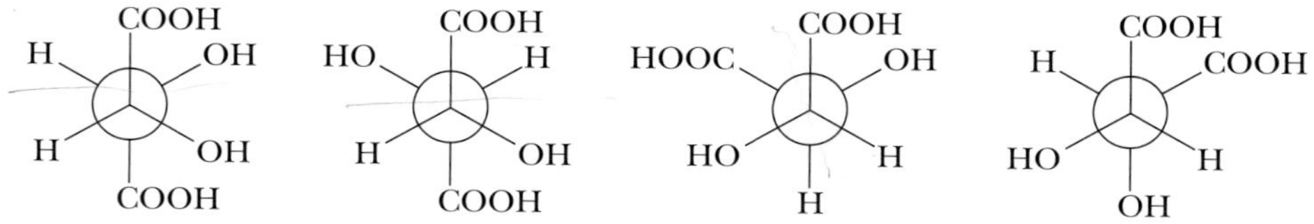

COOH	COOH	COOH	COOH
H ⟍ ⟋ OH	HO ⟍ ⟋ H	HOOC ⟍ ⟋ OH	H ⟍ ⟋ COOH
H ⟍ ⟋ OH	H ⟍ ⟋ OH	HO ⟍ ⟋ H	HO ⟍ ⟋ H
COOH	COOH	H	OH
(1)	(2)	(3)	(4)

(a) Which represent the same compound?
(b) Which represent enantiomers?
(c) Which represent a meso compound?
(d) Which are diastereomers?

C. Fischer Projection Formulas

Glyceraldehyde contains a chiral center and therefore exists as a pair of enantiomers.

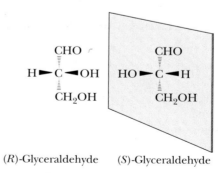

(R)-Glyceraldehyde   (S)-Glyceraldehyde

Fischer projection A two-dimensional projection of a molecule; in these projections, groups on the right and left are by convention in front, while those at the top and bottom are to the rear.

Chemists commonly use two-dimensional representations called **Fischer projections** to show the configuration of molecules with multiple chiral centers, especially carbohydrates. To write a Fischer projection, draw a three-dimensional representation of the molecule oriented so that the vertical bonds from the chiral center are directed away from you and the horizontal bonds from it are directed toward you. Then write the molecule as a two-dimensional figure with the chiral center indicated by the point at which the bonds cross.

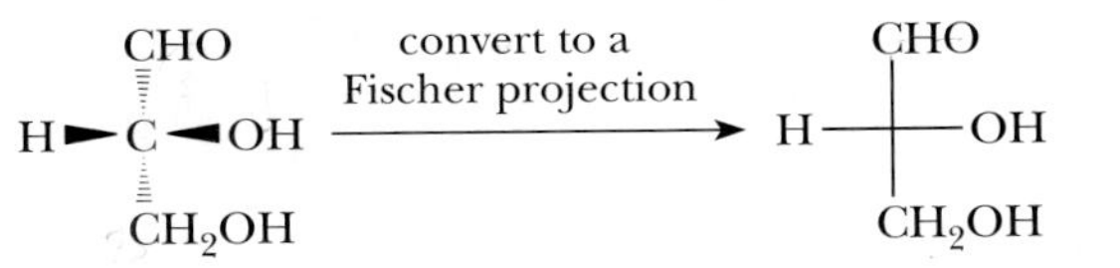

$$\begin{array}{c} \text{CHO} \\ | \\ \text{H} \blacktriangleright \text{C} \blacktriangleleft \text{OH} \\ | \\ \text{CH}_2\text{OH} \end{array} \quad \xrightarrow[\text{Fischer projection}]{\text{convert to a}} \quad \begin{array}{c} \text{CHO} \\ | \\ \text{H} \!\!-\!\!\!\!-\!\!\!\!-\!\! \text{OH} \\ | \\ \text{CH}_2\text{OH} \end{array}$$

(*R*)-Glyceraldehyde
(three-dimensional
representation)

(*R*)-Glyceraldehyde
(Fischer projection)

The horizontal segments of this Fischer projection represent bonds directed toward you, and the vertical segments represent bonds directed away from you. The only atom in the plane of the paper is the chiral center. Because a Fischer projection implies that the groups to each side are in front and those at top and bottom are behind the plane of the paper, rotations of these drawings by 90° are not permissible.

Example 3.6

Draw a Fischer Projection of (2*R*,3*R*) erythrose (Figure 3.3).

Solution

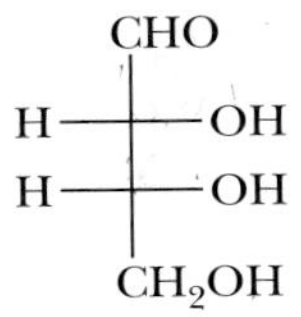

$$\begin{array}{c} \text{CHO} \\ | \\ \text{H} \!-\!\!\!\!|\!\!\!\!- \text{OH} \\ \text{H} \!-\!\!\!\!|\!\!\!\!- \text{OH} \\ | \\ \text{CH}_2\text{OH} \end{array}$$

(2*R*,3*R*)-Erythrose
(2,3,4-Trihydroxybutanal)

Problem 3.6

Give a complete stereochemical name for the following compound, which is a 1,2,3-butanetriol.

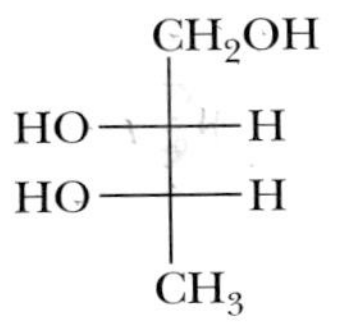

$$\begin{array}{c} \text{CH}_2\text{OH} \\ | \\ \text{HO} \!-\!\!\!\!|\!\!\!\!- \text{H} \\ \text{HO} \!-\!\!\!\!|\!\!\!\!- \text{H} \\ | \\ \text{CH}_3 \end{array}$$

3.5 Cyclic Molecules with Two or More Chiral Centers

In this section, we concentrate on derivatives of cyclopentane and cyclohexane containing two chiral centers. We can analyze stereoisomerism in cyclic compounds in the same way as in acyclic compounds.

A. Disubstituted Derivatives of Cyclopentane

Let us start with 2-methylcyclopentanol, a compound with two chiral centers. We predict a maximum of $2^2 = 4$ stereoisomers. Both the *cis* isomer and the *trans* isomer are chiral: The *cis* isomer exists as one pair of enantiomers, and the *trans* isomer exists as a second pair of enantiomers. The *cis* and *trans* isomers are stereoisomers that are not mirror images of each other, that is, they are diastereomers.

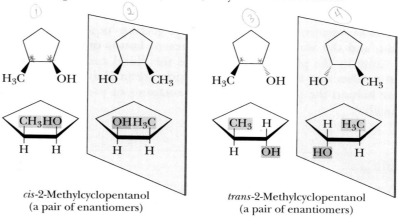

cis-2-Methylcyclopentanol
(a pair of enantiomers)

trans-2-Methylcyclopentanol
(a pair of enantiomers)

1,2-Cyclopentanediol also has two chiral centers; therefore, the 2^n rule predicts a maximum of $2^2 = 4$ stereoisomers. As seen in the following stereodrawings, only three stereoisomers exist for this compound. The *cis* isomer is achiral (meso) because it and its mirror image are superposable. An alternative way to identify the *cis* isomer as achiral is to notice that it possesses a plane of symmetry that bisects the molecule into two mirror halves. The *trans* isomer is chiral and exists as a pair of enantiomers.

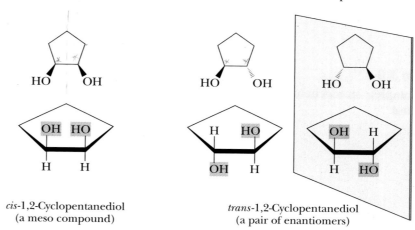

cis-1,2-Cyclopentanediol
(a meso compound)

trans-1,2-Cyclopentanediol
(a pair of enantiomers)

Example 3.7

How many stereoisomers exist for 3-methylcyclopentanol?

Solution

There are four stereoisomers of 3-methylcyclopentanol. The *cis* isomer exists as one pair of enantiomers; the *trans* isomer, as a second pair of enantiomers.

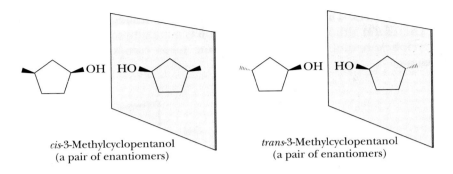

cis-3-Methylcyclopentanol
(a pair of enantiomers)

trans-3-Methylcyclopentanol
(a pair of enantiomers)

Problem 3.7

How many stereoisomers exist for 1,3-cyclopentanediol?

B. Disubstituted Derivatives of Cyclohexane

As an example of a disubstituted cyclohexane, let us consider the methylcyclohexa-nols. 4-Methylcyclohexanol can exist as two stereoisomers: a pair of *cis,trans* isomers. Both the *cis* and *trans* isomers are achiral. In each, a plane of symmetry runs through the —CH_3 and HO— groups and the carbons attached to them.

3-Methylcyclohexanol has two chiral centers and exists as $2^2 = 4$ stereoisomers. The *cis* isomer exists as one pair of enantiomers; the *trans* isomer, as a second pair of enantiomers.

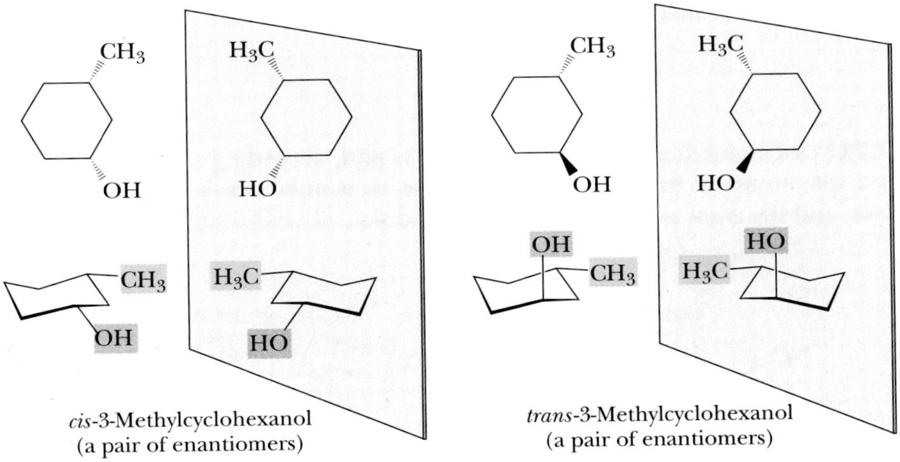

cis-3-Methylcyclohexanol
(a pair of enantiomers)

trans-3-Methylcyclohexanol
(a pair of enantiomers)

Similarly, 2-methylcyclohexanol has two chiral centers and exists as $2^2 = 4$ stereoiso-mers. The *cis* isomer exists as one pair of enantiomers; the *trans* isomer, as a second pair of enantiomers.

Example 3.8

How many stereoisomers exist for 1,3-cyclohexanediol?

Solution

1,3-Cyclohexanediol has two chiral centers and, according to the 2^n rule, has a maximum of $2^2 = 4$ stereoisomers. The *trans* isomer of this compound exists as a pair of enantiomers. The *cis* isomer has a plane of symmetry and is a meso compound. Therefore, although the 2^n rule predicts a maximum of four stereoisomers for 1,3-cyclohexanediol, only three exist: one meso compound and one pair of enantiomers.

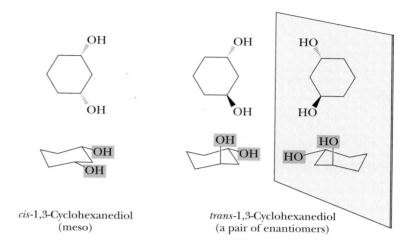

cis-1,3-Cyclohexanediol
(meso)

trans-1,3-Cyclohexanediol
(a pair of enantiomers)

Problem 3.8

How many stereoisomers exist for 1,4-cyclohexanediol?

1,2-Cyclohexanediol has two chiral centers and, according to the 2^n rule, can exist as a maximum of four stereoisomers. The *cis* isomer exists as one pair of enantiomers, and the *trans* isomer exists as a second pair of enantiomers.

(1)   (2)   (3)   (4)

cis-1,2-Cyclohexanediol
(a pair of enantiomers)

trans-1,2-Cyclohexanediol
(a pair of enantiomers)

The enantiomers of the *cis* isomer cannot be separated, however, because each is converted to the other by a rapid chair-to-alternative-chair conversion at room temperature. As shown in the following structural formulas, the alternative chair of (1) is, in fact, the mirror image of (1). Thus, each enantiomer of the *cis* isomer interconverts to its mirror image rapidly at room temperature. Therefore, the enantiomers cannot be separated and are always present as a mixture of equal parts of the enantiomers. Because these conformations interchange rapidly, we can treat them as if they were planar. This compound is effectively meso.

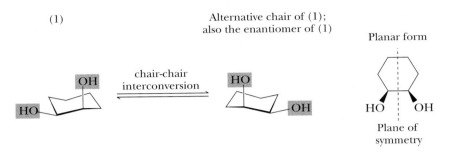

(1)

Alternative chair of (1);
also the enantiomer of (1)

chair-chair
interconversion

Planar form

Plane of
symmetry

Figure 3.5 summarizes the different types of isomers we have discussed so far and the relationships among them. In addition to configurational and conformational isomers (Chapters 1 and 2), we have examined in this chapter stereoisomers with chiral centers or other chiral elements and have distinguished enantiomers, diastereomers, and meso compounds.

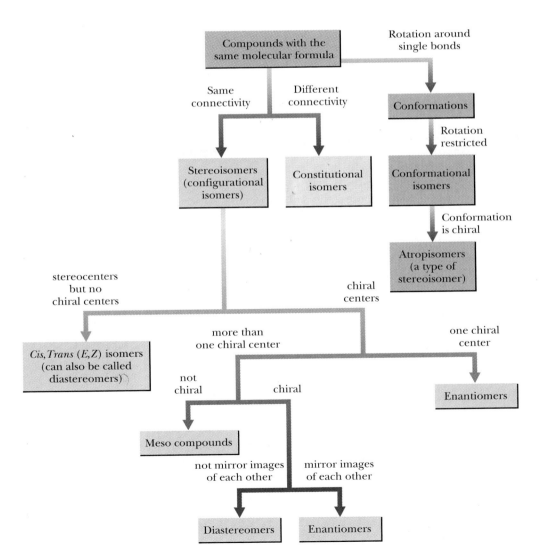

Table 3.1 Some Physical Properties of the Stereoisomers of Tartaric Acid

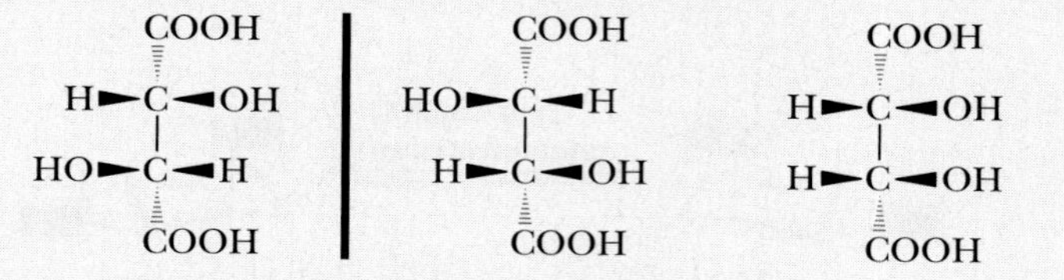

	(R,R)-Tartaric Acid	(S,S)-Tartaric Acid	Meso Tartaric Acid
Specific rotation*	+12.7	−12.7	0
Melting point (°C)	171–174	171–174	146–148
Density at 20°C (g/cm³)	1.7598	1.7598	1.660
Solubility in water at 20°C (g/100 mL)	139	139	125
pK_1 (25°C)	2.98	2.98	3.23
pK_2 (25°C)	4.34	4.34	4.82

*Specific rotation is discussed in Section 3.7B.

3.6 Properties of Stereoisomers

Enantiomers have identical physical and chemical properties in an achiral environment. Examples of achiral environments are solvents that have no chiral centers such as H_2O, CH_3CH_2OH, or CH_2Cl_2.

The enantiomers of tartaric acid (Table 3.1), for example, have the same melting point, the same boiling point, the same solubility in water and other common solvents, the same value of pK_a, and they undergo the same acid-base reactions. The enantiomers of tartaric acid do, however, differ in optical activity (the ability to rotate the plane of plane-polarized light), which we discuss in Section 3.7A. Diastereomers have different physical and chemical properties, even in achiral environments. Meso tartaric acid has different physical properties from those of the enantiomers and can be separated from them by methods such as crystallization.

3.7 Optical Activity—How Chirality Is Detected in the Laboratory

As we have already established, enantiomers are different compounds, and thus we must expect that they differ in some properties. One property that differs between enantiomers is their effect on the rotation of the plane of polarized light. Each member of a pair of enantiomers rotates the plane of polarized light in opposite directions, and for this reason, enantiomers are said to be **optically active.**

Optically active Refers to a compound that rotates the plane of polarized light.

The phenomenon of optical activity was discovered in 1815 by the French physicist Jean Baptiste Biot. To understand how it is detected in the laboratory, we must first understand something about plane-polarized light and a polarimeter, the device used to detect optical activity.

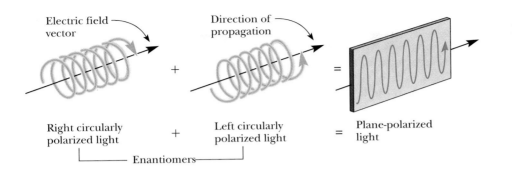

Figure 3.6
Plane-polarized light is a mixture of left and right circularly polarized light.

A. Plane-Polarized Light

Ordinary light consists of waves vibrating in all planes perpendicular to its direction of propagation (Figure 3.6). Certain materials such as calcite or a Polaroid sheet (a plastic film containing properly oriented crystals of an organic substance embedded in it) selectively transmit light waves vibrating only in parallel planes. Such radiation is said to be **plane polarized.**

    Plane-polarized light is the vector sum of left and right circularly polarized light that propagates through space as left- and right-Handed helices. These two forms of light are enantiomers, and because of their opposite handedness, each component interacts in an opposite way with chiral molecules. The result of this interaction is that each member of a pair of enantiomers rotates the plane of polarized light in an opposite direction.

Plane-polarized light Light oscillating in only a single plane.

B. Polarimeters

A **polarimeter** consists of a monochromatic light source, a polarizing filter and an analyzing filter (each made of calcite or Polaroid film), and a sample tube (Figure 3.7). If the sample tube is empty, the intensity of the light reaching the eye is at its minimum when the polarizing axes of the two filters are at right angles. If the analyzing filter is turned either clockwise or counterclockwise, more light is transmitted.

Polarimeter An instrument for measuring the ability of a compound to rotate the plane of polarized light.

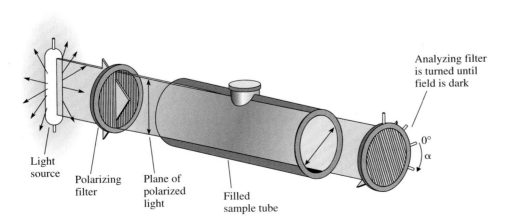

Figure 3.7
Schematic diagram of a polarimeter with its sample tube containing a solution of an optically active compound. The analyzing filter has been turned clockwise by α degrees to restore the dark field to the observer.

Observed rotation   The number of degrees through which a compound rotates the plane of polarized light.

Dextrorotatory   Refers to a substance that rotates the plane of polarized light to the right.

Levorotatory   Refers to a substance that rotates the plane of polarized light to the left.

Specific rotation   Observed rotation of the plane of polarized light when a sample is placed in a tube 1.0 dm in length and at a concentration of 1 g/100 mL. For a pure liquid, concentration is expressed in g/mL (density).

The ability of molecules to rotate the plane of polarized light can be observed using a polarimeter in the following way. First, a sample tube filled with solvent is placed in the polarimeter. The analyzing filter is adjusted so that the field is dark (the easiest position for the human eye to observe). This position of the analyzing filter is taken as 0°. When a solution of an optically active compound is placed in the sample tube, some light now passes through the analyzing filter; the optically active compound has rotated the plane of polarized light from the polarizing filter so that it is now no longer at an angle of 90° to the analyzing filter. The analyzing filter is then rotated to restore darkness in the field of view. The number of degrees, α, through which the analyzing filter must be rotated to restore darkness to the field of view is called the **observed rotation.** If the analyzing filter must be turned to the right (clockwise) to restore darkness, we say that the compound is **dextrorotatory** (Latin: *dexter,* on the right side). If the analyzing filter must be turned to the left (counterclockwise), we say that the compound is **levorotatory** (Latin: *laevus,* on the left side).

The magnitude of the observed rotation for a particular compound depends on its concentration, the length of the sample tube, the temperature, the solvent, and the wavelength of the light used. **Specific rotation, [α],** is defined as the observed rotation at a specific cell length and sample concentration.

$$\text{Specific rotation} = [\alpha]_\lambda^T = \frac{\text{Observed rotation (degrees)}}{\text{Length (dm)} \times \text{Concentration}}$$

The standard cell length is 1 decimeter (1 dm or 10 cm). For a pure liquid sample, the concentration is expressed in grams per milliliter (g/mL; density). The concentration of a sample dissolved in a solvent is expressed as grams per 100 milliliters of solution. Because specific rotation depends on temperature (T, in degrees Celsius) and wavelength λ of light, these variables are designated, respectively, as superscript and subscript. The light source most commonly used in polarimetry is the sodium D line (λ = 589 nm), the line responsible for the yellow color of sodium-vapor lamps.

In reporting either observed or specific rotation, it is common to indicate a dextrorotatory compound with a plus sign in parentheses, (+), and a levorotatory compound with a minus sign in parentheses, (−). For any pair of enantiomers, one enantiomer is dextrorotatory, and the other is levorotatory. For each member of the pair, the absolute value of the specific rotation is exactly the same, but the sign is opposite. Following are the specific rotations of the enantiomers of 2-butanol at 25°C using the D line of sodium. Note that for molecules with single chiral centers there is no absolute relationship between R and S and (+) and (−) rotation. For some molecules, the R enantiomer is (+), and for others, the S enantiomer is (+). This is why it is not redundant to report both the R or S and the rotation direction when naming a specific enantiomer.

Richard Megna, 1992, Fundamental Photographs

■ A polarimeter is used to measure the rotation of plane-polarized light as it passes through a sample.

(*S*)-(+)-2-Butanol
$[\alpha]_D^{25}$ +13.52

(*R*)-(−)-2-Butanol
$[\alpha]_D^{25}$ −13.52

Example 3.9

A solution is prepared by dissolving 400 mg of testosterone, a male sex hormone, in 10.0 mL of ethanol and placing it in a sample tube 10.0 cm in length. The observed rotation of this sample at 25°C using the D line of sodium is +436°. Calculate the specific rotation of testosterone.

Solution

The concentration of testosterone is 400 mg/10.0 mL = 4 g/100 mL. The length of the sample tube is 1.00 dm. Inserting these values in the equation for calculating specific rotation gives

$$\text{Specific rotation} = \frac{\text{Observed rotation}}{\text{Length (dm)} \times \text{Concentration (g/100 mL)}} = \frac{+436°}{1.00 \times 4.0} = +109$$

Problem 3.9

The specific rotation of progesterone, a female sex hormone, is +172. Calculate the observed rotation for a solution prepared by dissolving 300 mg of progesterone in 15.0 mL of dioxane and placing it in a sample tube 10.0 cm long.

C. Racemic Mixtures

An equimolar mixture of two enantiomers is called a **racemic mixture,** a term derived from the name "racemic acid" (Latin: *racemus,* a cluster of grapes). Racemic acid is the name originally given to an equimolar mixture of the enantiomers of tartaric acid (Table 3.1). Because a racemic mixture contains equal numbers of dextrorotatory and levorotatory molecules, its specific rotation is zero. Alternatively, we say that a racemic mixture is optically inactive. A racemic mixture is indicated by adding the prefix (±) to the name of the compound [or sometimes the prefix (d,l)].

Racemic mixture   A mixture of equal amounts of two enantiomers.

D. Optical Purity (Enantiomeric Excess)

When dealing with a pair of enantiomers, it is essential to have a means of describing the composition of that mixture and the degree to which one enantiomer is in excess relative to its mirror image. One way of describing the composition of a mixture of enantiomers is by its percent **optical purity,** a property that can be observed directly. Optical purity is the specific rotation of a mixture of enantiomers divided by the specific rotation of the enantiomerically pure substance.

$$\text{Percent optical purity} = \frac{[\alpha]_{\text{sample}}}{[\alpha]_{\text{pure enantiomer}}} \times 100$$

An alternative way to describe the composition of a mixture of enantiomers is by its **enantiomeric excess (ee),** which is the difference in the number of moles of each enantiomer in a mixture compared to the total number of moles of both. Enantiomeric excess in per cent is calculated by taking the difference in the percentage of each enantiomer.

$$\text{Enantiomeric excess (ee)} = \%R - \%S$$

Optical purity   The specific rotation of a mixture of enantiomers divided by the specific rotation of the enantiomerically pure substance (expressed as a percent). Optical purity is numerically equal to enantiomeric excess, but experimentally determined.

Enantiomeric excess (ee)   The difference between the percentage of two enantiomers in a mixture.

For example, if a mixture consists of 75% of the R enantiomer and 25% of the S enantiomer, then the enantiomeric excess of the R enantiomer is 50%. Enantiomeric excess and optical purity are numerically identical.

Example 3.10

Figure 3.8 presents a scheme for separation of the enantiomers of mandelic acid. The specific rotation of optically pure (S)-$(-)$-mandelic acid is -158. Suppose that, instead of isolating pure (S)-$(-)$-mandelic acid from this scheme, the sample is a mixture of enantiomers with a specific rotation of -134. For this sample, calculate the following.

(a) The enantiomeric excess of this sample of (S)-$(-)$-mandelic acid.

(b) The percentage of (S)-$(-)$-mandelic acid and of (R)-$(+)$-mandelic acid in the sample.

Solution

(a) The enantiomeric excess of (S)-$(-)$-mandelic acid is 84.8%.

$$\text{Enantiomeric excess} = \frac{-134}{-158} \times 100 = 84.8\%$$

(b) This sample is 84.8% (S)-$(-)$-mandelic acid and 15.2% (R,S)-mandelic acid. The (R,S)-mandelic acid is 7.6% S enantiomer and 7.6% R enantiomer. The sample, therefore, contains 92.4% of the S enantiomer and 7.6% of the R enantiomer. We can check these values by calculating the observed rotation of a mixture containing 92.4% (S)-$(-)$-mandelic acid and 7.6% (R)-$(+)$-mandelic acid as follows:

$$\text{Specific rotation} = 0.924 \times (-158) + 0.076 \times (+158) = -146 + 12 = -134$$

which agrees with the experimental specific rotation.

Problem 3.10

One commercial synthesis of naproxen (the active ingredient in Aleve and a score of other over-the-counter and prescription nonsteroidal anti-inflammatory drug preparations) gives the enantiomer shown in 97% enantiomeric excess.

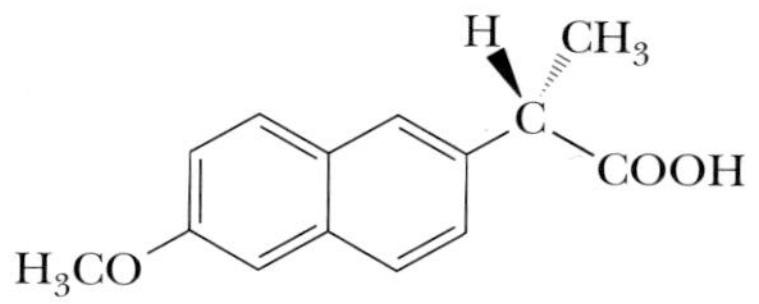

Naproxen
(a nonsteroidal anti-inflammatory drug)

(a) Assign an R or S configuration to this enantiomer of naproxen.

(b) What are the percentages of R and S enantiomers in the mixture?

3.8 Separation of Enantiomers—Resolution

The separation of a racemic mixture into its enantiomers is called **resolution.**

A. Resolution by Means of Diastereomeric Salts

One general scheme for separating enantiomers requires chemical conversion of a pair of enantiomers into two diastereomers with the aid of an enantiomerically pure chiral resolving agent. This chemical resolution is successful because the diastereomers thus formed are different compounds, have different physical properties, and often can be separated by physical means (most commonly fractional crystallization or column chromatography) and purified. The final step in this scheme for resolution is chemical conversion of the separated diastereomers back to the individual enantiomers and recovery of the chiral resolving agent.

A reaction that lends itself to chemical resolution is salt formation because it is readily reversible.

$$RCOOH \quad + \quad :B \rightleftharpoons RCOO^-HB^+$$

Carboxylic acid   Base   Salt

Chiral bases available from plants are often used as chiral resolving agents for racemic acids. More commonly, chemists now use commercially available chiral amines such as 1-phenylethylamine.

1-Phenylethanamine

The resolution of mandelic acid by way of its diastereomeric salts with the natural chiral base cinchonine is illustrated in Figure 3.8. Racemic mandelic acid and optically pure (+)-cinchonine (Cin) are dissolved in boiling water, giving a solution of a pair of diastereomeric salts. Diastereomers have different solubilities, and when the solution cools, the less-soluble diastereomeric salt crystallizes. This salt is collected and purified by further recrystallization. The filtrates, richer in the more soluble diastereomeric salt, are concentrated to give this salt, which is also purified by further recrystallization. The purified diastereomeric salts are treated with aqueous HCl to precipitate the nearly pure enantiomers of mandelic acid. Cinchonine remains in the aqueous solution as its water-soluble hydrochloride salt.

Optical rotations and melting points of racemic mandelic acid, cinchonine, the purified diastereomeric salts, and the pure enantiomers of mandelic acid are given in Figure 3.8. Note the following two points: (1) The diastereomeric salts have different specific rotations and different melting points. (2) The enantiomers of mandelic acid have identical melting points and have specific rotations that are identical in magnitude but opposite in sign.

Resolution of a racemic base with a chiral acid is carried out in a similar way. Acids that are commonly used as chiral resolving agents are (+)-tartaric acid, (−)-malic acid, and (+)-camphoric acid (Figure 3.9). These and other naturally occurring chiral resolving agents are produced in plant and animal systems as single enantiomers.

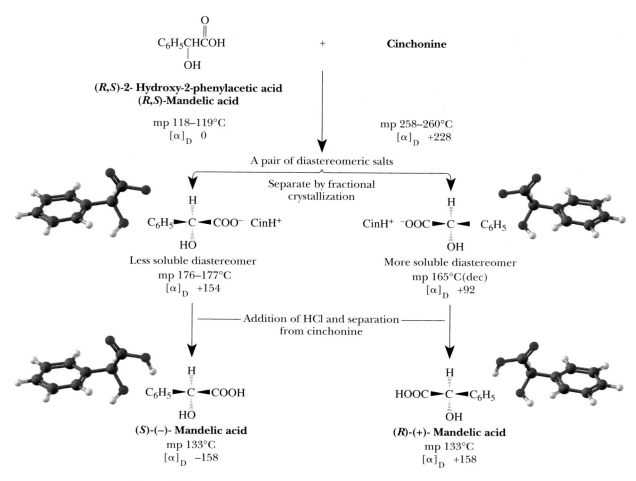

Figure 3.8
Resolution of mandelic acid.

B. Enzymes As Resolving Agents

In their quest for enantiomerically pure compounds, organic chemists have developed several new techniques for preparation of enantiomerically pure materials. One approach is to use enzymes. Enzymes are themselves chiral so they can produce

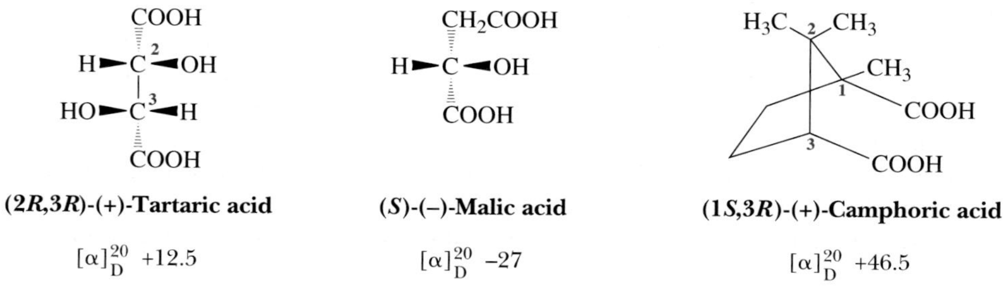

Figure 3.9
Some carboxylic acids used as chiral resolving agents.

(2R,3R)-(+)-Tartaric acid

$[\alpha]_D^{20}$ +12.5

(S)-(–)-Malic acid

$[\alpha]_D^{20}$ −27

(1S,3R)-(+)-Camphoric acid

$[\alpha]_D^{20}$ +46.5

single enantiomer products. A class of enzymes under study is the esterases, which catalyze the hydrolysis of esters to give an alcohol and a carboxylic acid.

The ethyl esters of naproxen crystallize in two enantiomeric crystal forms; one containing the *R* ester and the other containing the *S* ester. Each is insoluble in water. Chemists then use an esterase in alkaline solution to hydrolyze selectively the *S* ester to the (*S*)-carboxylic acid, which goes into solution as the sodium salt. The *R* ester is unaffected by these conditions. Filtering the alkaline solution recovers the crystals of the *R* ester. After crystals of the *R* ester are removed, the alkaline solution is acidified to give enantiomerically pure *S* naproxen.

The recovered *R* ester is racemized (converted to an *R,S* mixture) and treated again with the esterase. Thus, by recycling the *R* ester, all the racemic ester is converted to (*S*)-naproxen.

Ethyl ester of (*S*)-naproxen
1. Esterase   NaOH, H$_2$O
2. HCl, H$_2$O

Ethyl ester of (*R*)-naproxen
(Not affected by the esterase)

(*S*)-Naproxen

C. Resolution by Means of Chromatography on a Chiral Substrate

Chromatography is a term used to describe the purification of molecules in which a sample to be purified interacts with a solid material, and different components of the sample separate based on their different relative interactions with the solid material. Separation can be accomplished using the sample in either the gas phase (usually for analytical purposes) or the liquid phase (analytical or preparative separations are possible). The solid material is packed into a column and a solvent (or, in gas chromatography, a gas) passes down the column, carrying the more weakly bound components of the mixture with it more rapidly than the more tightly bound ones.

A common method of resolving enantiomers today is chromatography using a chiral column packing material. Each enantiomer interacts differently with the chiral molecules of the packing material, and the elution time will (in principle at least) be different for the two enantiomers. A wide variety of chiral column packings have been developed for this purpose.

Chromatography A separation method involving passing a vapor or solution mixture through a column packed with a material with different affinities for different components of the mixture.

CONNECTIONS TO BIOLOGICAL CHEMISTRY

Amino Acids

In organisms ranging from bacteria to humans, the 20 most common amino acids share the familiar structural motif of a central carbon atom bonded to a hydrogen atom, an amino group, and a carboxyl group. In addition, 19 of the 20 amino acids have an additional group of atoms, referred to as the "side chain," bonded to the central carbon. The central carbon atom of these 19 is a chiral center, having the configuration shown here. At neutral pH, amino acids exist as an internal salt (commonly called a zwitterion) as a result of the relative acid strength of the carboxyl group and base strength of the amino group, respectively (see Section 4.4).

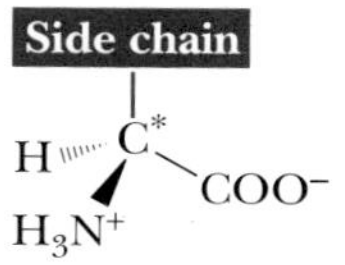

Side chain

H······$\overset{*}{C}$······COO⁻

H_3N^+

Ionized or zwitterion form

The amino acid side chains encompass a variety of functional groups including benzene rings, alkyl groups, amino groups, thiol (—SH) groups, and carboxyl groups. One amino acid, namely glycine, has no side chain; only another hydrogen is bonded to the central carbon.

According to the Cahn-Ingold-Prelog priority rules, 18 of the 19 side chain groups are lower priority than the amino and carboxyl groups, so these 18 have the *S* configuration. The exception is cysteine in which the side chain is a —CH_2—SH group. Because of the sulfur atom, the —CH_2—SH group is of higher priority than the carboxyl group, and the configuration of cysteine is *R*. Two of the amino acid side chains themselves have a chiral center (isoleucine and threonine). All 19 chiral amino acids are found in nature as single stereoisomers the vast majority of the time.

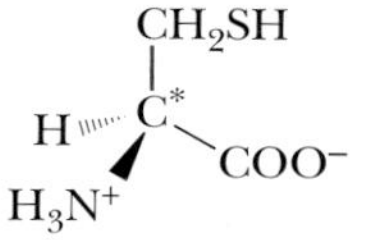

CH_2SH

H······$\overset{*}{C}$······COO⁻

H_3N^+

Cysteine

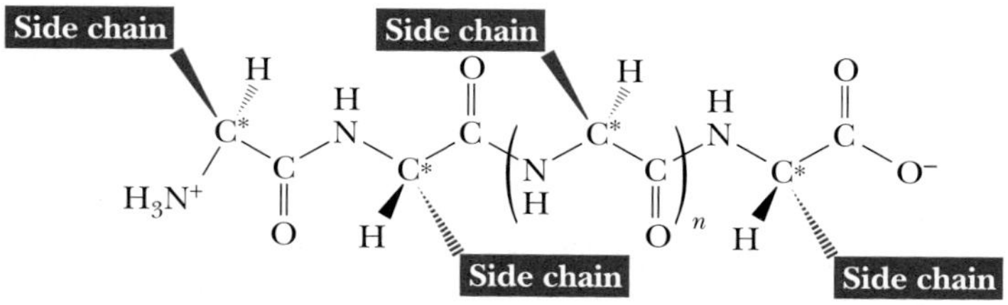

For most proteins, $n = 150$–750

Proteins are long chains of amino acids covalently bonded together by amide bonds (Section 1.3F) formed between the carboxyl group of one amino acid and the amino group of another amino acid. Because they are made of pure amino acid stereoisomers, proteins themselves are single stereoisomers despite having several hundred or more chiral centers. These chains fold into complex three-dimensional structures capable of a remarkable array of chemical functions.

Proteins are one of the key functional components of all living cells. This means that the chemistry of living systems is chiral, and that the presence of single stereoisomers is the rule rather than the exception for biological molecules. Another of the key biochemical building blocks, the carbohydrates, have multiple chiral centers, adding to the chiral nature of living systems.

You will encounter a different type of stereochemical designation when it comes to amino acids. In this designation, stereochemistry is related to the structure of glyceraldehyde, a molecule with one chiral center (Section 3.3, 3.4C). The chiral center in the 19 common chiral amino acids is structurally related to the glyceraldehyde stereoisomer that rotates plane-polarized light in the levorotatory or (−) direction, so this glyceraldehyde *and all the amino acids structurally related to it* are designated "L." The common 19 chiral amino acids are often referred to as L amino acids in biochemistry texts and literature. Note that this does *not* mean that each of the 19 common amino acids rotates plane polarized light in the levorotatory or (−) direction; they do not. Many students can be confused by this. The "L" designation refers *only* to the *structural* relationship of the common 19 chiral amino acids to the levorotatory (−) isomer of glyceraldehyde. For example, the specific rotation of L-serine is +14.45 in

1 M HCl, but −6.83 in water. The change in the ionization of the carboxyl group causes a change in specific rotation, but not a change in configuration of the chiral center.

Although relatively rare, amino acids with the opposite configuration are found in some organisms. These amino acids are structurally related to the dextrorotatory or (+) isomer of glyceraldehyde, so they are referred to as D amino acids. The D amino acids are the enantiomers of corresponding L amino acids.

L-Amino acids
(common)

D-Amino acids
(rare)

3.9 The Significance of Chirality in the Biological World

Except for inorganic salts and a relatively few low-molecular-weight organic substances, the molecules in living systems, both plant and animal, are chiral. Although these molecules can exist as a number of stereoisomers, almost invariably only one stereoisomer is found in nature. This occurrence is a consequence of the fact that their natural syntheses are catalyzed by enzymes, which are also chiral. Of course, instances do occur in which more than one stereoisomer is found, but these rarely exist together in the same biological system.

A. Chirality in Enzymes

Let us look more closely at the chirality of enzymes. An illustration is chymotrypsin, an enzyme in the intestines of animals, which catalyzes the hydrolysis of proteins during digestion. Chymotrypsin, like all proteins, is composed of a long molecular chain of amino acids that folds up into the active enzyme. Human chymotrypsin has 268 chiral centers that result from the amino acids, so the maximum number of stereoisomers possible is 2^{268}, a staggeringly large number, almost beyond comprehension.

Fortunately, because each chiral amino acid is only present as a single stereoisomer, only one of the possible stereoisomers of chymotrypsin is produced. Because enzymes are chiral substances and are present as single stereoisomers, most either produce or react only with substances that are single stereoisomers (if chiral).

As an interesting illustration of this enzyme chirality principle, Stephen B. Kent, then at the Scripps Research Institute in La Jolla, California, synthesized the enantiomer of a natural enzyme by using the enantiomers of the amino acids normally found in proteins as building blocks. As expected, this synthetic enzyme enantiomer only catalyzed a reaction with a substrate that was the enantiomer of the chiral substrate utilized by the natural enzyme.

B. How an Enzyme Distinguishes Between a Molecule and Its Enantiomer

Enzymes are chiral catalysts. Some are completely specific for the catalysis of the reaction of only one particular compound, whereas others are less specific and catalyze similar reactions of a family of compounds. An enzyme catalyzes a biological reaction of molecules by first positioning them at a binding site on its surface. These molecules may be held at the binding site by a combination of hydrogen bonds, electrostatic attractions, dispersion forces, or even temporary covalent bonds.

An enzyme with a specific binding site for a molecule with a chiral center can distinguish between a molecule and its enantiomer or one of its diastereomers. Assume, for example, that an enzyme involved in catalyzing a reaction of glyceraldehyde has a binding site with groups that interact with —H, —OH, and —CHO. Assume further that the binding sites are arranged in the enzyme binding site as shown in Figure 3.10. The enzyme can distinguish (*R*)-(+)-glyceraldehyde (the natural or biologically active form) from its enantiomer (*S*)-(−)-glyceraldehyde because the natural enantiomer can be bound to the binding site with three groups interacting with their appropriate binding sites; the other enantiomer can, at best, bind to only two of these sites.

Because interactions between molecules in living systems take place in a chiral environment, it should be no surprise that a molecule and its enantiomer or diastereomers have different physiological properties. The tricarboxylic acid (TCA) cycle, for example, produces and then metabolizes only (*S*)-(+)-malic acid. Because only one enantiomer is produced, both the production and metabolism of (*S*)-(+)-malic acid are said to be enantioselective.

$$HOOC \diagup \diagdown COOH$$
$$H \quad OH$$

(*S*)-(+)-Malic acid

That interactions between molecules in the biological world are highly enantioselective is not surprising, but just how these interactions are accomplished at the molecular level with such precision and efficiency is one of the great puzzles that modern science has only recently begun to unravel.

Figure 3.10

A schematic diagram of an enzyme surface capable of interacting with (*R*)-(+)-glyceraldehyde at three binding sites, but with (*S*)-(−)-glyceraldehyde at only two of these sites.

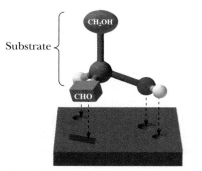

Enzyme surface

This enantiomer of glyceraldehyde fits the three specific binding sites on the enzyme surface.

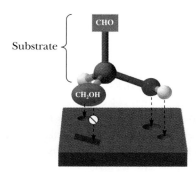

Enzyme surface

This enantiomer of glyceraldehyde does not fit the same binding sites.

CONNECTIONS TO BIOLOGICAL CHEMISTRY

Chiral Drugs

Some of the common drugs used in human medicine, for example aspirin (Section 18.5B), are achiral. Others are chiral and sold as single enantiomers. The penicillin and erythromycin classes of antibiotics and the drug captopril are all chiral drugs. Captopril, which is very effective for the treatment of high blood pressure and congestive heart failure, was developed in a research program designed to discover effective inhibitors for angiotensin-converting enzyme (ACE). It is manufactured and sold as the *S,S* stereoisomer. A large number of chiral drugs, however, are sold as racemic mixtures. The popular analgesic ibuprofen (the active ingredient in Motrin and many other non-aspirin analgesics) is an example.

Captopril

(*S*)-Ibuprofen

For racemic drugs, most often only one enantiomer exerts the beneficial effect, whereas the other enantiomer either has no effect or may even exert a detrimental effect. Thus, enantiomerically pure drugs are usually more effective than their racemic counterparts. A case in point is the drug dihydroxyphenylalanine used in the treatment of Parkinson's disease. The active drug is dopamine. Unfortunately, this compound does not cross the blood-brain barrier to the required site of action in the brain. What is administered, instead, is the prodrug, a compound that is not active by itself but is converted in the body to an active drug. 3,4-Dihydroxyphenylalanine is such a prodrug. It crosses the blood-brain barrier and then undergoes decarboxylation catalyzed by the enzyme dopamine decarboxylase. This enzyme is specific for the *S* enantiomer, which is commonly known as L-DOPA. It is essential, therefore, to administer the enantiomerically pure prodrug. Were the prodrug to be administered in a racemic form, there could be a dangerous buildup of the *R* enantiomer, which cannot be metabolized by the enzymes present in the brain.

(*S*)-(−)-3,4-Dihydroxyphenylalanine
(L-DOPA)
$[\alpha]_D^{13}$ −13.1

enzyme-catalyzed
decarboxylation

Dopamine

Recently, the U.S. Food and Drug Administration established new guidelines for the testing and marketing of chiral drugs. After reviewing these guidelines, many drug companies have decided to develop only single enantiomers of new chiral drugs. For this reason, there has been a tremendous recent interest in developing stereoselective synthetic methods that yield a single enantiomer or diastereomer. In addition to regulatory pressure, there are patent considerations. If a company has patents on a racemic drug, a new patent can often be taken out on one of its enantiomers. Only the *S* enantiomer of the pain reliever ibuprofen is biologically active. In the case of ibuprofen, however, the body converts the inactive *R* enantiomer to the active *S* enantiomer.

Summary

Configurational isomers are **stereoisomers** (Section 3.1) that have the same order of attachment of atoms in their molecules but a different three-dimensional orientation of their atoms in space. **Configurational isomers** are divided into enantiomers, diastereomers, and *cis,trans* isomers. **Enantiomers** are stereoisomers that are mirror images of each other. **Diastereomers** are stereoisomers that are not mirror images.

A **mirror image** is the reflection of an object in a mirror. Molecules that are not superposable on their mirror images are said to be **chiral** (Section 3.2). Chirality is a property of an object as a whole, not of a particular atom. An achiral object is one that lacks chirality; that is, it is an object that has a superposable mirror image. Almost all achiral objects possess at least one plane or center of symmetry. A **plane of symmetry** is an imaginary plane passing through an object dividing it such that one half is the reflection of the other half. A **center of symmetry** is a point so situated that identical components of the object are located on opposite sides and equidistant from the point along any axis passing through that point.

A **chiral center** (Section 3.2) is a tetrahedral atom, most commonly carbon, with four different groups bonded to it. Chiral centers are not limited to carbon. Chiral compounds of tetrahedral nitrogen, silicon, phosphorus, and germanium atoms have also been prepared. In some cases, molecules can be chiral, even if they have no chiral atom. Such molecules are called **atropisomers**.

The **absolute configuration** at any chiral center can be specified by the ***R,S* system** (Section 3.3). To apply this convention, each atom or group of atoms bonded to the chiral center is (1) assigned a priority and numbered from highest priority to lowest priority. (2) The molecule is oriented in space so that the group of lowest priority is directed away from the observer, and (3) the remaining three groups are read in order from highest priority to lowest priority. If the order of groups is clockwise, the configuration is ***R*** (Latin: *rectus*, right, correct). If the order is counterclockwise, the configuration is ***S*** (Latin: *sinister*, left).

For a molecule with n chiral centers, the maximum number of stereoisomers possible is 2^n (Section 3.4). Certain molecules have special symmetry properties that reduce the number of stereoisomers to fewer than that predicted by the 2^n rule. A compound is **meso** (Section 3.4B) if it contains two or more chiral centers assembled in such a way that its molecules are achiral.

Fischer projections (Section 3.4C) provide a way of displaying chiral centers in two dimensions. In these projections, groups at the right and left are considered to be in front, and those at the top and bottom, behind.

Enantiomers have identical physical and chemical properties in achiral environments (Section 3.6). They have different properties, however, in chiral environments, as for example, in the presence of plane-polarized light (Section 3.7A). They also have different properties in the presence of chiral reagents (Section 3.8A) and enzymes as chiral catalysts (Section 3.9B). **Diastereomers** have different physical and chemical properties even in achiral environments. **Meso compounds** have an internal mirror plane and are optically inactive.

Light that vibrates in parallel planes is said to be **plane-polarized** (Section 3.7A). Plane-polarized light contains equal components of left and right circularly polarized light. A **polarimeter** (Section 3.7B) is an instrument used to detect and measure the magnitude of optical activity. A compound is said to be **optically active** if it rotates the plane of polarized light. **Observed rotation** is the number of degrees the plane of polarized light is rotated. **Specific rotation** is the observed rotation of a solution measured in a cell 1 dm long and at a sample concentration of 1 g/100 mL (1 g/mL for a neat liquid). If the analyzing prism must be turned clockwise to restore the zero point, the compound is **dextrorotatory.** If the analyzing prism must be turned counterclockwise to restore the zero point, the compound is **levorotatory.** Each member of a pair of enantiomers rotates the plane of polarized light an equal number of degrees but opposite in direction (Section 3.7B). A **racemic mixture** (Section 3.7C) is a mixture of equal amounts of two enantiomers and has a specific rotation of zero. Percent **optical purity** (identical to **enantiomeric excess**) is defined as the specific rotation of a mixture of enantiomers divided by the specific rotation of the pure enantiomer times 100 (Section 3.7D).

Resolution (Section 3.8) is the experimental process of separating a mixture of enantiomers into the two pure enantiomers. A common chemical means of resolving organic compounds is to treat the racemic mixture with a chiral resolving agent that converts the mixture of enantiomers into a pair of diastereomers. The diastereomers are separated based on differences in their physical properties; each diastereomer is then converted to a pure stereoisomer, uncontaminated by its enantiomer. Enzymes are also used as resolving agents because of their ability to catalyze a reaction of one enantiomer but not that of its mirror image. Chromatography on a chiral substrate is also an effective separation method.

Enzymes catalyze biological reactions by first positioning the molecule or molecules at binding sites and holding them there by a combination of hydrogen bonds, electrostatic attractions, dispersion forces, and covalent bonds (Section 3.9). An enzyme with specific binding sites for three of the four groups on a chiral center can distinguish between a molecule and its enantiomer (Section 3.9B). Almost all enzyme-catalyzed reactions are stereoselective and give a single enantiomer or diastereomer as product.

Problems

Chirality

3.11 Think about the helical coil of a telephone cord or a spiral binding and suppose that you view the spiral from one end and find that it is a left-handed twist. If you view the same spiral from the other end, is it a right-handed or left-handed twist?

3.12 Next time you have the opportunity to view a collection of sea shells that have a helical twist, study the chirality of their twists. Do you find an equal number of left-handed and right-handed spiral shells or mostly all of the same chirality? What about the handedness of different species of spiral shells?

3.13 One reason we can be sure that sp^3-hybridized carbon atoms are tetrahedral is the number of stereoisomers that can exist for different organic compounds.

    (a) How many stereoisomers are possible for $CHCl_3$, CH_2Cl_2, and $CHClBrF$ if the four bonds to carbon have a tetrahedral arrangement?

    (b) How many stereoisomers would be possible for each of these compounds if the four bonds to the carbon had a square planar geometry?

Enantiomers

3.14 Which compounds contain chiral centers?

    (a) 2-Chloropentane   **(b)** 3-Chloropentane
    (c) 3-Chloro-1-pentene   **(d)** 1,2-Dichloropropane

3.15 Using only C, H, and O, write structural formulas for the lowest-molecular-weight chiral:

    (a) Alkane   **(b)** Alcohol   **(c)** Aldehyde
    (d) Ketone   **(e)** Carboxylic acid   **(f)** Carboxylic ester

3.16 Draw mirror images for these molecules. Are they different from the original molecule?

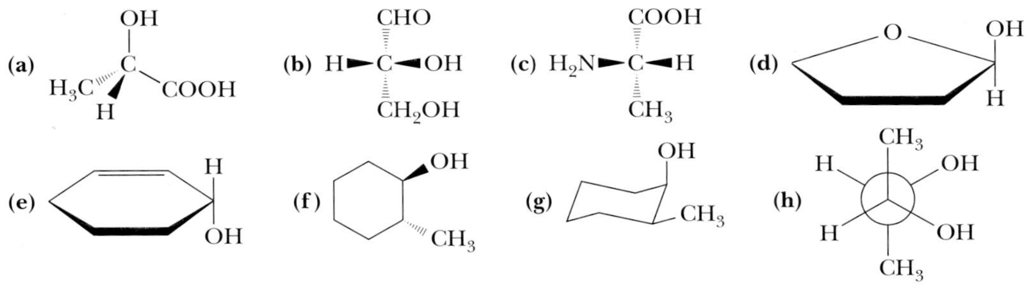

■ This Atlantic auger shell has a right-handed helical twist.

(a)
$$\underset{H}{\overset{OH}{\underset{H_3C}{\mid}}}\; C \;{}^{COOH}$$

(b) $H{-}C{-}OH$ with CHO above and CH_2OH below

(c) $H_2N{-}C{-}H$ with COOH above and CH_3 below

(d) ring with O, OH, H

(e) ring with H, OH

(f) ring with OH, CH_3

(g) ring with OH, CH_3

(h) Newman projection with CH_3, H, OH, CH_3

3.17 Following are several stereorepresentations for lactic acid. Take (a) as a reference structure. Which stereorepresentations are identical with (a), and which are mirror images of (a)?

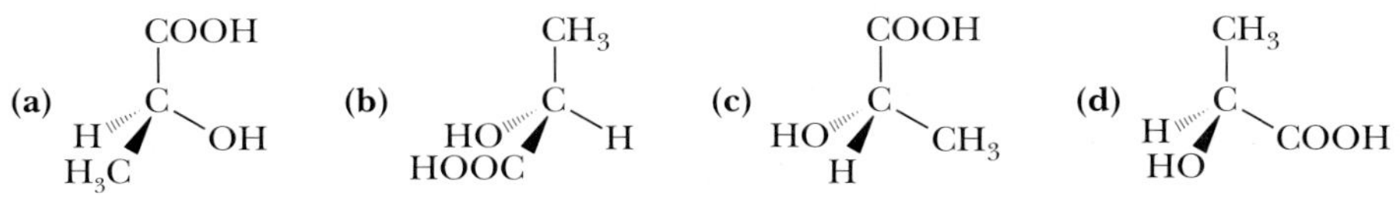

(a) $\underset{H_3C}{\overset{COOH}{H}}\; C \; OH$

(b) $\underset{HOOC}{\overset{CH_3}{HO}}\; C \; H$

(c) $\overset{COOH}{HO}\; C \; CH_3$ with H below

(d) $\underset{HO}{\overset{CH_3}{H}}\; C \; COOH$

3.18 Mark each chiral center in the following molecules with an asterisk. How many stereoisomers are possible for each molecule?

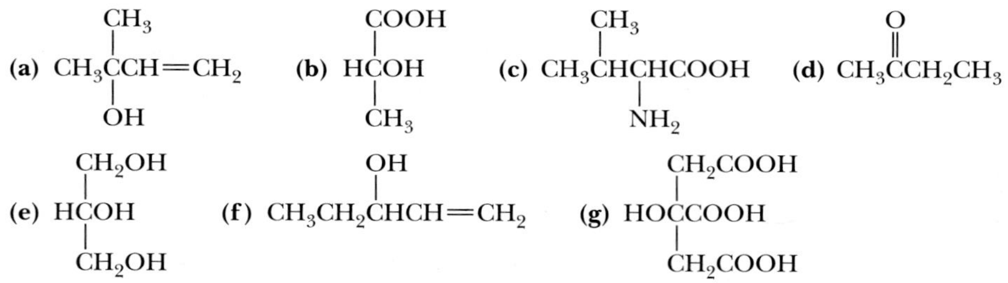

(a) $CH_3CCH{=}CH_2$
with CH_3 above and OH below

(b) $HCOH$
with $COOH$ above and CH_3 below

(c) $CH_3CHCHCOOH$
with CH_3 above and NH_2 below

(d) $CH_3CCH_2CH_3$
with O above (double bond)

(e) $HCOH$
with CH_2OH above and CH_2OH below

(f) $CH_3CH_2CHCH{=}CH_2$
with OH above

(g) $HOCCOOH$
with CH_2COOH above and CH_2COOH below

3.19 Show that butane in a gauche conformation is chiral. Do you expect that resolution of butane at room temperature is possible?

Designation of Configuration—The *R,S* System

3.20 Assign priorities to the groups in each set.
(a) —H —CH$_3$ —OH —CH$_2$OH
(b) —CH$_2$CH=CH$_2$ —CH=CH$_2$ —CH$_3$ —CH$_2$COOH
(c) —CH$_3$ —H —COO$^-$ —NH$_3{}^+$
(d) —CH$_3$ —CH$_2$SH —NH$_3{}^+$ —CHO

3.21 Following are structural formulas for the enantiomers of carvone. Each has a distinctive odor characteristic of the source from which it is isolated. Assign an *R* or *S* configuration to the single chiral center in each enantiomer. Why do they smell different when they are so similar in structure?

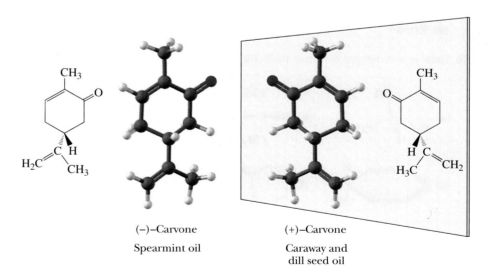

(−)–Carvone

Spearmint oil

(+)–Carvone

Caraway and
dill seed oil

3.22 Following is a staggered conformation for one of the enantiomers of 2-butanol.

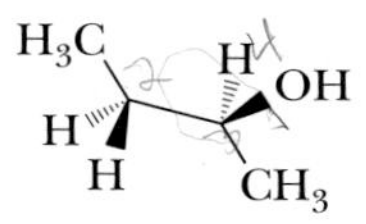

(a) Is this (*R*)-2-butanol or (*S*)-2-butanol?

(b) Draw a Newman projection for this staggered conformation, viewed along the bond between carbons 2 and 3.

(c) Draw a Newman projection for two more staggered conformations of this molecule. Which of your conformations is the most stable? Assume that —OH and —CH$_3$ are comparable in size.

3.23 For centuries, Chinese herbal medicine has used extracts of *Ephedra sinica* to treat asthma. *Ephedra* as an "herbal supplement" has been implicated in the deaths of several athletes, and has recently been banned as a dietary supplement. Phytochemical investigation of this plant resulted in isolation of ephedrine, a very potent dilator of the air passages of the lungs. Ephedrine also has profound effects on the cardiovascular system. The naturally occurring stereoisomer is levorotatory and has the following structure. Assign an *R* or *S* configuration to each chiral center.

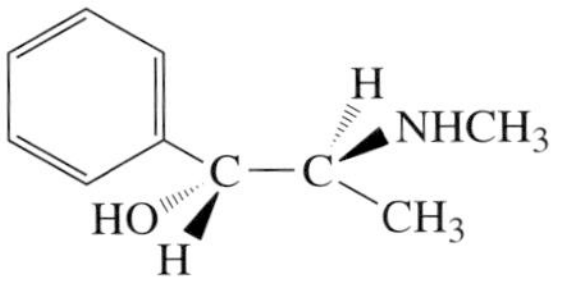

Ephedra sinica, the source of ephedrine, a potent bronchodilator.

Ephedrine $[\alpha]_D^{21}$ −41

3.24 When oxaloacetic acid and acetyl-coenzyme A (acetyl-CoA) labeled with radioactive carbon-14 in position 2 are incubated with citrate synthase, an enzyme of the tricarboxylic acid cycle, only the following enantiomer of [2-^{14}C]citric acid is formed stereoselectively. Note that citric acid containing only ^{12}C is achiral. Assign an *R* or *S* configuration to this enantiomer of [2-^{14}C]citric acid. *Note:* Carbon-14 has a higher priority than carbon-12.

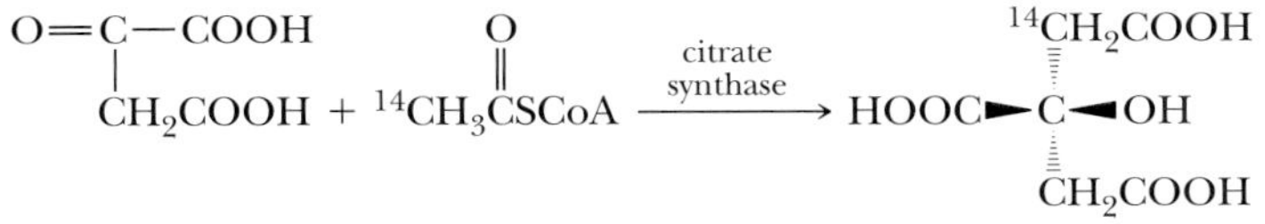

$$O{=}C{-}COOH$$
$$|$$
$$CH_2COOH \; + \; ^{14}CH_3\overset{O}{\overset{\|}{C}}SCoA \xrightarrow[\text{synthase}]{\text{citrate}} HOOC{\blacktriangleright}\overset{^{14}CH_2COOH}{\underset{CH_2COOH}{C}}{\blacktriangleleft}OH$$

Oxaloacetic acid   Acetyl-CoA   [2-^{14}C]Citric acid

Molecules with Two or More Chiral Centers

3.25 Draw stereorepresentations for all stereoisomers of this compound. Label those that are meso compounds and those that are pairs of enantiomers.

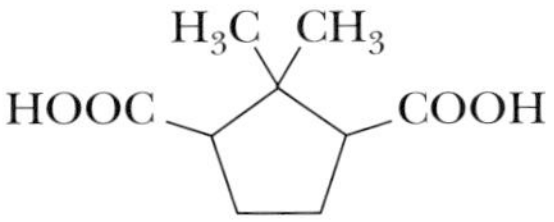

$$\underset{HOOC}{\overset{H_3C\;\;CH_3}{\diagup\diagdown}}COOH$$

3.26 Mark each chiral center in the following molecules with an asterisk. How many stereoisomers are possible for each molecule?

(a) 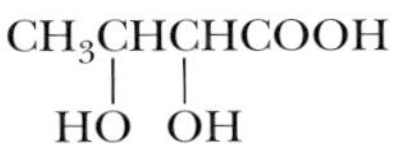 CH$_3$CHCHCOOH
         | |
         HO OH

(b) 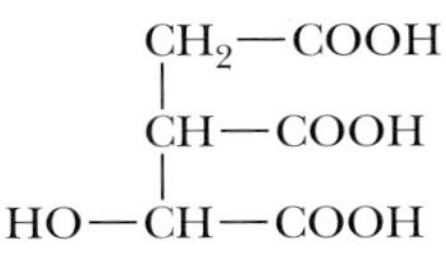
$$CH_2{-}COOH$$
$$|$$
$$CH{-}COOH$$
$$|$$
$$HO{-}CH{-}COOH$$

(c)

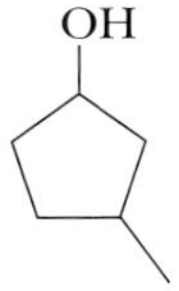

(d)

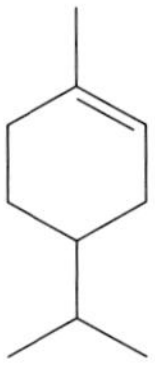

(e) 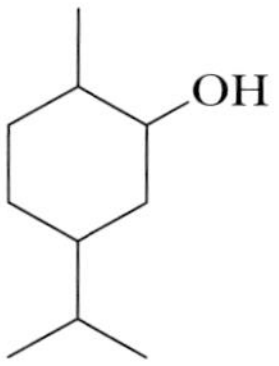   OH

(f) 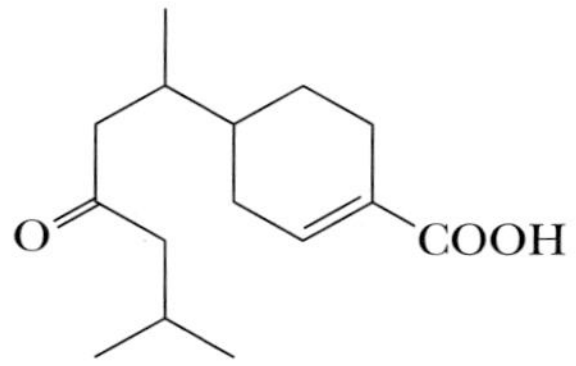   O   COOH

(g) 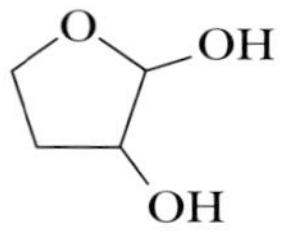   O   OH   OH

(h) 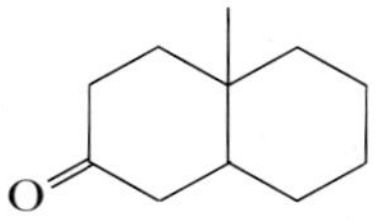   O

3.27 Label the eight chiral centers in cholesterol. How many stereoisomers are possible for a molecule with this many chiral centers?

HO

Cholesterol

3.28 Label the four chiral centers in amoxicillin, which belongs to the family of semisynthetic penicillins.

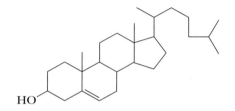

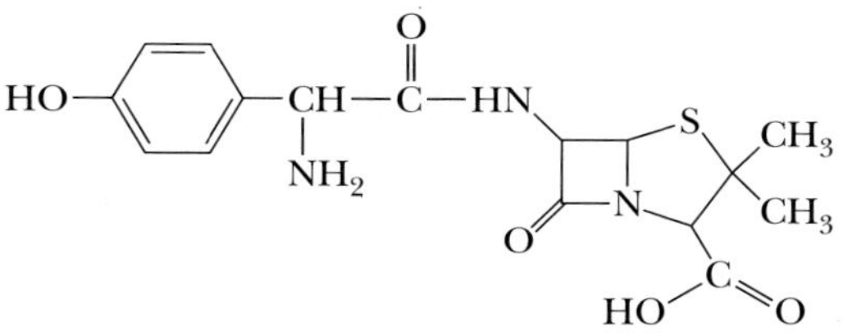

HO—CH—C—HN   S   CH₃

NH₂   N   CH₃

O   HO   C=O

Amoxicillin

3.29 If the optical rotation of a new compound is measured and found to have a specific rotation of +40, how can you tell if the actual rotation is not really +40 plus some multiple of +360? In other words, how can you tell if the rotation is not actually a value such as +400 or +760?

3.30 Are the formulas within each set identical, enantiomers, or diastereomers?

(a) 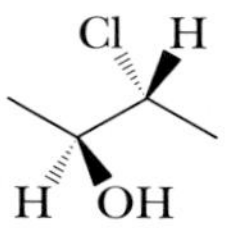   Cl H / H OH   and   H Cl / HO H

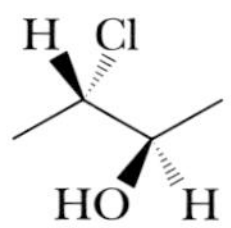

(b) 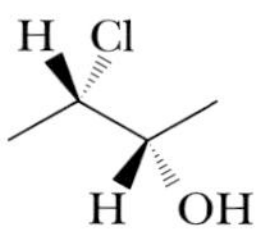   H Cl / H OH   and   HO H / H Cl

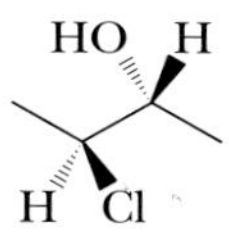

(c) 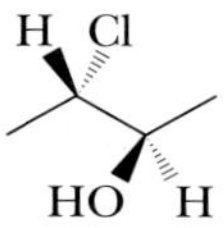   H Cl / HO H   and   HO H / Cl H

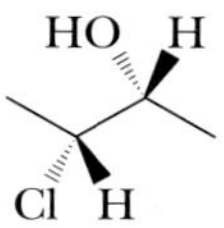

3.31 Which of the following are meso compounds?

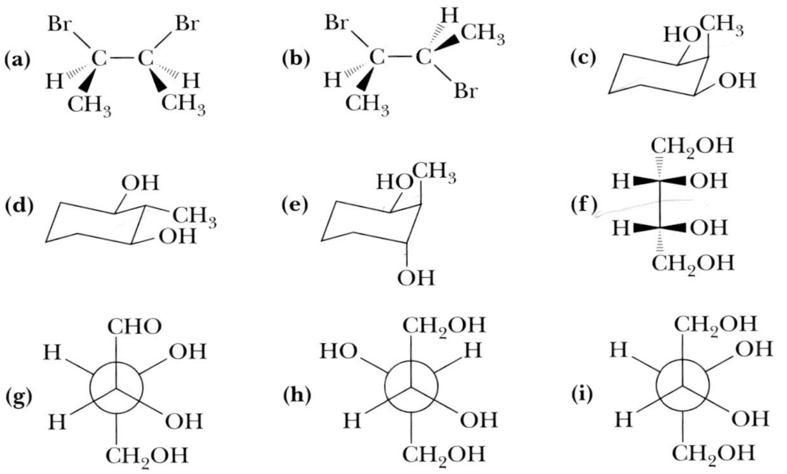

(a) Br, Br; C—C; H, H; CH₃, CH₃

(b) Br, H, CH₃; C—C; H, CH₃, Br

(c) HO, CH₃; OH

(d) OH; CH₃; OH

(e) HO, CH₃; OH

(f) CH₂OH; H—OH; H—OH; CH₂OH

(g) CHO; H, OH; H, OH; CH₂OH

(h) CH₂OH; HO, H; H, OH; CH₂OH

(i) CH₂OH; H, OH; H, OH; CH₂OH

3.32 Vigorous oxidation of the following bicycloalkene breaks the carbon-carbon double bond and converts each carbon of the double bond to a COOH group. Assume that the conditions of oxidation have no effect on the configuration of either the starting bicycloalkene or the resulting dicarboxylic acid. Is the dicarboxylic acid produced from this oxidation one enantiomer, a racemic mixture, or a meso compound?

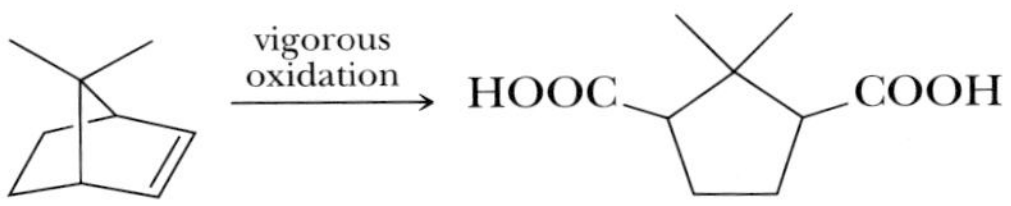

$$\xrightarrow{\text{vigorous oxidation}} \text{HOOC} \quad \text{COOH}$$

3.33 A long polymer chain, such as polyethylene ($-CH_2CH_2-$)$_n$, can potentially exist in solution as a chiral object. Give two examples of chiral structures that a polyethylene chain could adopt.

Molecular Modeling

These problems require molecular modeling programs such as ChemDraw, Chem3D, or Spartan to solve them. The prebuilt models can be found at **http://now.brookscole.com/bfi4.** Click **Molecular Models.**

3.34 ChemDraw provides a very easy way to make mirror images. If you have access to Chem-Draw, create a molecule with a chiral center, make a copy of the original, and place the copy adjacent to your original. (2) Select the copy and, (3) from the Object menu, select Flip Horizontal. As shown here, this procedure converts an enantiomer to its mirror image. These ChemDraw drawings can be pasted into Chem3D to view the molecules in three dimensions.

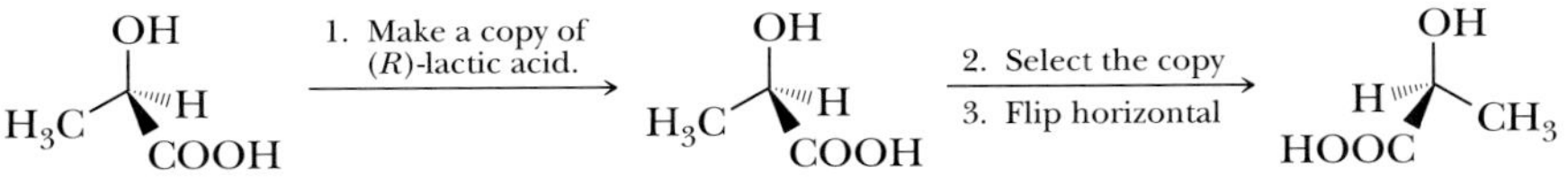

OH; H₃C, H, COOH — (*R*)-Lactic acid

$\xrightarrow{\text{1. Make a copy of (R)-lactic acid.}}$

OH; H₃C, H, COOH — Copy of (*R*)-lactic acid

$\xrightarrow{\text{2. Select the copy}}_{\text{3. Flip horizontal}}$

OH; H, CH₃, HOOC — (*S*)-Lactic acid

Alternatively, the models can be built directly in a modeling program such as Chem3D or Spartan; these programs will also allow inversion of configuration of models. Now try this procedure with these molecules chosen from the text.

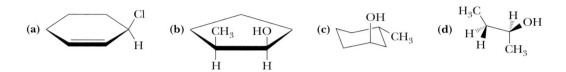

3.35 ChemDraw is able to assign an *R* or *S* configuration to a chiral center. To do this, use ChemDraw to build a model of a chiral compound. Then, from the Tools pulldown menu, select Show Stereochemistry. As practice, build structures 3.34(a–d) in Chem-Draw, and show the configuration of each chiral center.

3.36 The following molecule is an attractant pheromone for the olive fly. Go to http://now. brookscole.com/bfi4, and click Molecular Models to find this molecule.

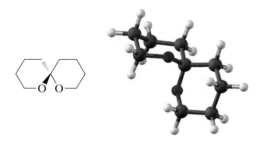

(a) Rotate the three-dimensional model using a molecular modeling program such as Chem3D or Spartan, and convince yourself that each six-membered ring has a strain-free chair conformation.
(b) This molecule has no chiral center and yet it is chiral. Examine the three-dimensional model, and convince yourself that it has no plane or center of symmetry and that it is, in fact, chiral.
(c) "The presence of a chiral center in an organic molecule is a sufficient condition for chirality, but it is not a necessary condition." Explain.

3.37 The following molecule belongs to the class of compounds called allenes. The functional group of an allene is two adjacent carbon-carbon double bonds. Disubstituted allenes of this type are chiral. The specific rotation of the enantiomer shown is −314. Go to http://now.brookscole.com/bfi4, and click Molecular Models to find this molecule.

$$[\alpha]_D^{25} \; -314$$

(a) Examine the three-dimensional model, and convince yourself that it has no plane or center of symmetry.

(b) Make its mirror image, and convince yourself that the original and the mirror image are nonsuperposable, that is, that they are a pair of enantiomers.

3.38 The following compound has a very high barrier to rotation around the bond between the benzene rings. How many stereoisomers can be prepared? What is their relationship? Go to http://now.brookscole.com/bfi4, and click Molecular Models to find this molecule.

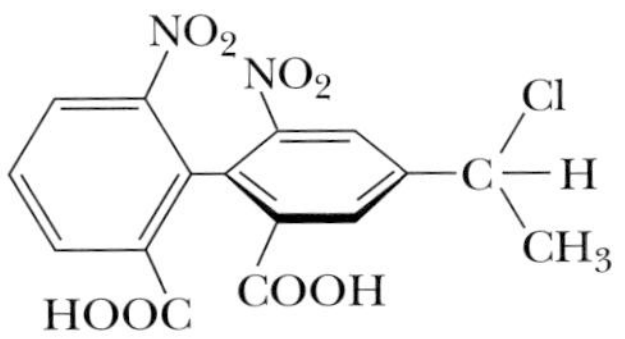

Looking Ahead

3.39 The chiral catalyst (*R*)-BINAP-Ru is used to hydrogenate alkenes to give alkanes (Section 6.7C). The products are produced with high enantiomeric excess. An example is the formation of (*S*)-naproxen, a pain reliever.

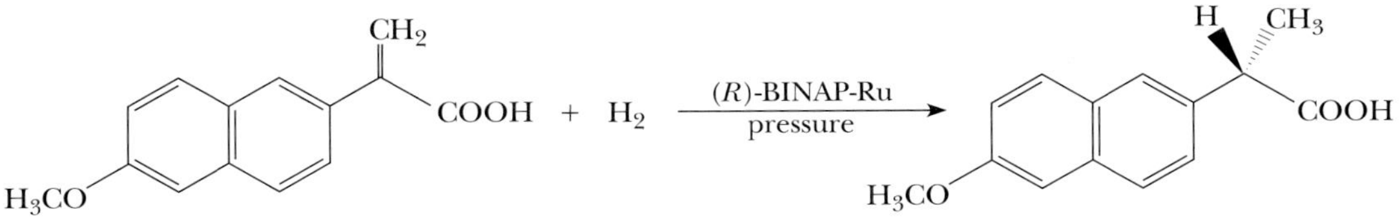

(*S*)-Naproxen
(ee > 98%)

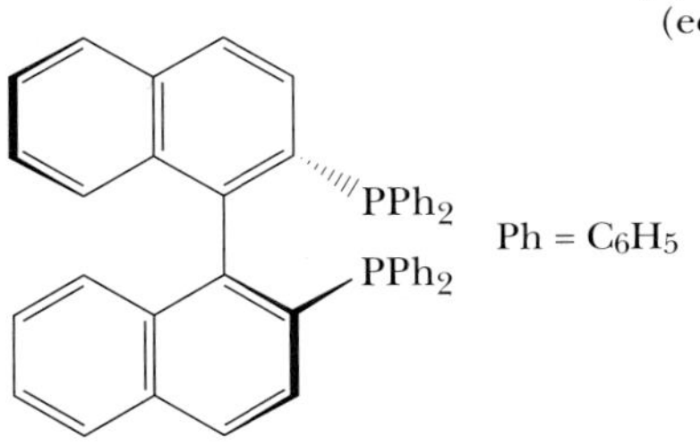

BINAP

(a) What kind of isomers are the enantiomers of BINAP?

(b) How can one enantiomer of naproxen be formed in such high yield?

3.40 In Section 10.5D, the following reactions are discussed. Ts is the toluenesulfonate group.

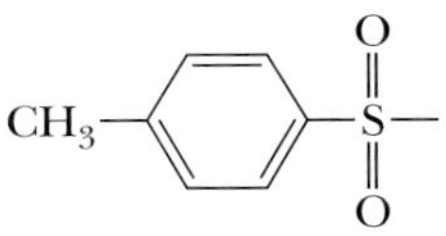

Toluenesulfonate group (Ts)

(a)

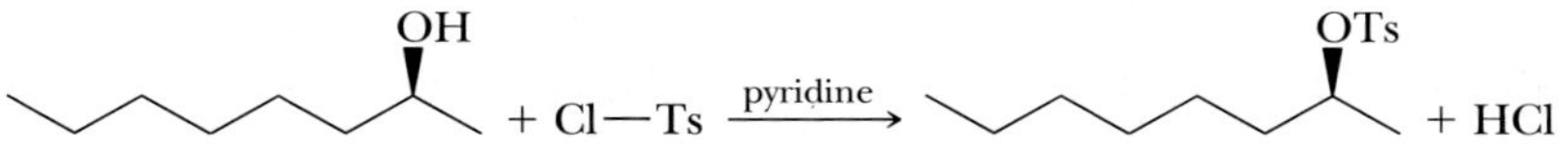

(S)-2-Octanol   (S)-2-Octyl tosylate

(b)

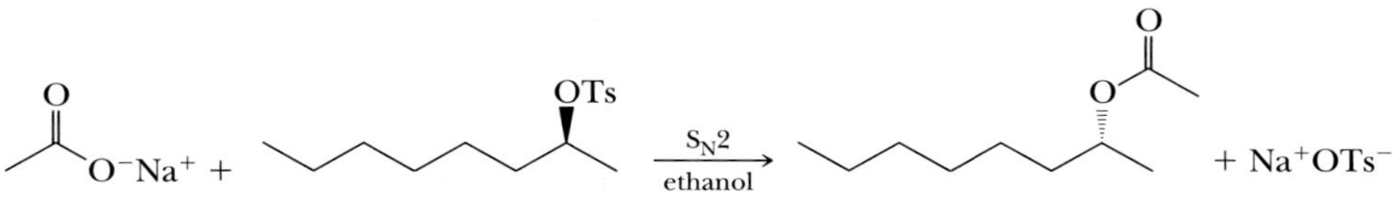

Sodium acetate   (S)-2-Octyl tosylate   (R)-2-Octyl acetate

In reaction (a), an *S* compound gives an *S* product. In reaction (b), an *S* compound gives an *R* product. Explain what is probably going on. (*Hint:* The oxygen atom in the starting material and product is the same in one reaction but not in the other.) What might this say about the second reaction?

Photodisc Blue/Getty Images

4

Acids and Bases

4 in the top-right corner

Citrus fruits are sources of citric acid. Lemon juice, for example, contains 5–8% citric acid. Inset: a model of citric acid.

Outline

4.1 Arrhenius Acids and Bases
4.2 Brønsted-Lowry Acids and Bases
4.3 Acid Dissociation Constants, pK_a, and Strengths of Acids and Bases
4.4 The Position of Equilibrium in Acid-Base Reactions
4.5 Molecular Structure and Acidity
4.6 Lewis Acids and Bases

A great many organic reactions either are acid-base reactions or involve catalysis by an acid or base at some stage. Of the reactions involving acid catalysis, some use proton-donating acids, such as H_2SO_4, H_3O^+, and $CH_3CH_2OH_2^+$. Others use Lewis acids, such as BF_3 and $AlCl_3$. It is essential, therefore, that you have a good grasp of the fundamentals of acid-base chemistry. In this and following chapters, we study the acid-base properties of the major classes of organic compounds.

4.1 Arrhenius Acids and Bases

The first useful definition of acids and bases was put forward by Svante Arrhenius in 1884. According to the original Arrhenius definition, an acid is a substance that dissolves in water to produce H^+ ions. A base is a substance that dissolves in water to produce OH^- ions. Today we know that H^+ ions do not exist in water. Therefore, we must modify the original Arrhenius definition to show that an H^+ ion in water immediately combines with a water molecule to give a hydronium ion, H_3O^+. Hydration of the hydronium ion itself gives the ion $H_5O_2^+$. Modern research suggests that the monohydrated and dihydrated forms (H_3O^+ and $H_5O_2^+$) are the major hydrated forms present in aqueous solution, and they are present in approximately equal concentrations. Throughout this text, we will represent a proton dissolved in aqueous solution as the hydronium ion, H_3O^+.

4 tutorials, simulations, and problems.

$$H^+(aq) + H_2O(l) \longrightarrow H_3O^+(aq)$$

Hydronium ion

$$H_3O^+(aq) + H_2O(l) \rightleftharpoons H_5O_2^+(aq)$$

Hydronium ion

Apart from these modifications, the Arrhenius definitions of acid and base are still valid and useful today, as long as we are talking about aqueous solutions.

4.2 Brønsted-Lowry Acids and Bases

Brønsted-Lowry acid   A proton donor.

Brønsted-Lowry base   A proton acceptor.

In 1923 the Danish chemist Johannes Brønsted and the English chemist Thomas Lowry independently proposed the following definitions: an **acid** is a proton donor, a **base** is a proton acceptor, and an acid-base reaction is a proton-transfer reaction.

A. Conjugate Acid-Base Pairs Differ by a Proton

Conjugate base   The species formed when an acid transfers a proton to a base.

Conjugate acid   The species formed when a base accepts a proton from an acid.

According to the Brønsted-Lowry definitions, any pair of molecules or ions that can be interconverted by transfer of a proton is called a **conjugate acid-base pair.** When an acid transfers a proton to a base, the acid is converted to its **conjugate base.** When a base accepts a proton, it is converted to its **conjugate acid.** Fundamental to these definitions is the fact that the members of a conjugate acid-base pair differ by a proton.

We can illustrate the relationships among conjugate acid-base pairs by examining the reaction of hydrogen chloride with water to form chloride ion and hydronium ion.

$$\underbrace{HCl(aq)}_{\substack{\text{Hydrogen}\\\text{chloride}\\\text{(acid)}}} + \underbrace{H_2O(l)}_{\substack{\text{Water}\\\text{(base)}}} \longrightarrow \underbrace{Cl^-(aq)}_{\substack{\text{Chloride}\\\text{ion}\\\text{(conjugate}\\\text{base of HCl)}}} + \underbrace{H_3O^+(aq)}_{\substack{\text{Hydronium}\\\text{ion}\\\text{(conjugate}\\\text{acid of }H_2O)}}$$

conjugate acid-base pair / conjugate acid-base pair

The acid HCl donates a proton and is converted to its conjugate base Cl^-. The base H_2O accepts a proton and is converted to its conjugate acid H_3O^+. Note that the members of each conjugate acid-base pair differ only by a proton.

We can show the transfer of a proton from an acid to a base by using a **curved arrow,** the same curved arrow symbol that we used in Section 1.6 to show the relocation of electron pairs among resonance contributing structures. For acid-base reactions, we write the Lewis structure of each reactant and product, showing all valence electrons on reacting atoms. We then use curved arrows to show the change in position of electron pairs during the reaction. The tail of the curved arrow is located at an electron pair, either a lone pair or a bonding pair, as the case may be. The head of the curved arrow shows the new location of the electron pair. Whenever we use curved arrows to show a change in position of an electron pair, an electron pair originating from an atom will form a new bond, while relocating an electron from a bond will break that bond.

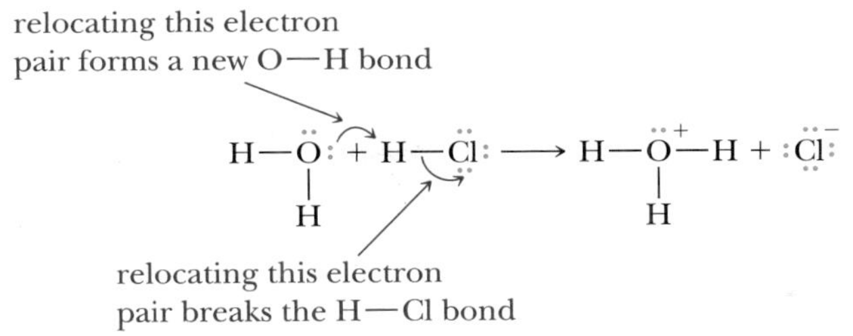

relocating this electron
pair forms a new O—H bond

$$H\!-\!\overset{..}{\underset{\overset{|}{H}}{O}}: + \,H\!-\!\overset{..}{\underset{..}{Cl}}: \longrightarrow H\!-\!\overset{..}{\underset{\overset{|}{H}}{O}}\!\overset{+}{-}\!H \,+\, :\overset{..}{\underset{..}{Cl}}:^{-}$$

relocating this electron
pair breaks the H—Cl bond

In this equation, the curved arrow on the left shows that an unshared pair of electrons on oxygen changes position to form a new covalent bond with hydrogen. The curved arrow on the right shows that the H—Cl bond breaks and that its electron pair is given entirely to chlorine to form a chloride ion. Thus, in the reaction of HCl with H_2O, a proton is transferred from HCl to H_2O. In the process, an O—H bond forms, and an H—Cl bond breaks.

We have illustrated the application of the Brønsted-Lowry definitions using water as a reactant. The Brønsted-Lowry definitions, however, do not require water as a reactant. Consider the following reaction between acetic acid and ammonia.

```
              ┌──── conjugate acid-base pair ────┐
    ┌─── conjugate acid-base pair ───┐
    ↓              ↓                  ↓           ↓
CH3COOH   +    NH3     ⇌     CH3COO⁻    +     NH4⁺
Acetic acid   Ammonia        Acetate           Ammonium
                               ion                ion
  (acid)       (base)     (conjugate base   (conjugate acid
                           acetic acid)      of ammonia)
```

We can use curved arrows to show how this reaction takes place. The curved arrow on the right shows that the unshared pair of electrons on nitrogen becomes shared between N and H to form a new H—N bond. At the same time that the H—N bond forms, the O—H bond breaks, and the electron pair of the O—H bond moves entirely to oxygen to form —O^- of the acetate ion, as depicted by the curved arrow on the right. The result of these two electron-pair shifts is the transfer of a proton from an acetic acid molecule to an ammonia molecule.

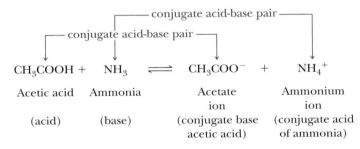

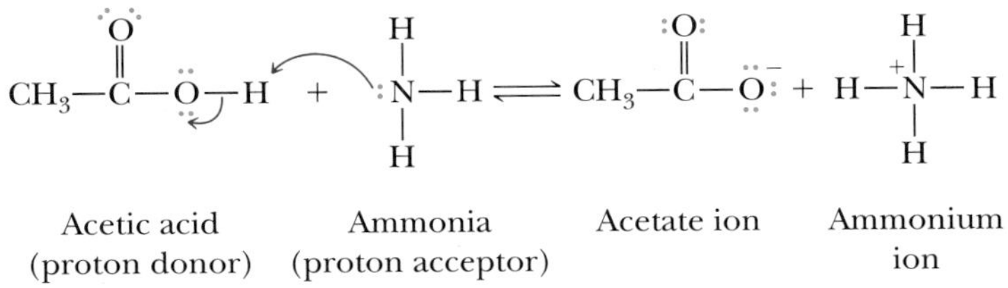

| Acetic acid | Ammonia | Acetate ion | Ammonium |
| (proton donor) | (proton acceptor) | | ion |

Example 4.1

For each conjugate acid-base pair, identify the first species as an acid or base and the second species as its conjugate base or conjugate acid. In addition, draw Lewis structures for each species, showing all valence electrons and any formal charges.

(a) CH_3OH, $CH_3OH_2^+$ **(b)** H_2CO_3, HCO_3^- **(c)** CH_3NH_2, $CH_3NH_3^+$

  base   conj. acid   acid   conj. base   base   conj. acid

Solution

(a) $CH_3 \overset{\cdot\cdot}{\underset{\cdot\cdot}{O}} - H$ $CH_3 \overset{\cdot\cdot}{\underset{\underset{H}{|}}{\overset{+}{O}}} - H$ (b) $H - \overset{\cdot\cdot}{\underset{\cdot\cdot}{O}} - \overset{\overset{\displaystyle\overset{\cdot\cdot}{\underset{\cdot\cdot}{O}}}{\|}}{C} - \overset{\cdot\cdot}{\underset{\cdot\cdot}{O}} - H$ $H - \overset{\cdot\cdot}{\underset{\cdot\cdot}{O}} - \overset{\overset{\displaystyle\overset{\cdot\cdot}{\underset{\cdot\cdot}{O}}}{\|}}{C} - \overset{\cdot\cdot}{\underset{\cdot\cdot}{O}} {\overset{-}{\vphantom{.}}}$

   Base   Conjugate   Acid   Conjugate
                        acid   base

(c) $CH_3 - \overset{\cdot\cdot}{\underset{\underset{H}{|}}{N}} - H$ $CH_3 - \overset{\overset{H}{|}}{\underset{\underset{H}{|}}{\overset{+}{N}}} - H$

   Base   Conjugate
                        acid

Problem 4.1

For each conjugate acid-base pair, identify the first species as an acid or base and the second species as its conjugate acid or conjugate base. In addition, draw Lewis structures for each species, showing all valence electrons and any formal charges.

(a) H_2SO_4, HSO_4^- (b) NH_3, NH_2^- (c) CH_3OH, CH_3O^-

Example 4.2

Write these reactions as proton-transfer reactions. Label which reactant is the acid and which is the base, which product is the conjugate base of the original acid, and which is the conjugate acid of the original base. In addition, write Lewis structures for each reactant and product, and use curved arrows to show the flow of electrons in each reaction.

(a) $H_2O + NH_4^+ \longrightarrow H_3O^+ + NH_3$
(b) $CH_3CH_2OH + NH_2^- \longrightarrow CH_3CH_2O^- + NH_3$

Solution

(a) Water is the base (proton acceptor), and ammonium ion is the acid (proton donor).

$$H - \overset{\cdot\cdot}{\underset{\underset{H}{|}}{O}} \quad + \quad H - \overset{\overset{H}{|}}{\underset{\underset{H}{|}}{\overset{+}{N}}} - H \quad \longrightarrow \quad H - \overset{\cdot\cdot}{\underset{\underset{H}{|}}{\overset{+}{O}}} - H \quad + \quad \overset{\cdot\cdot}{:}\overset{\overset{H}{|}}{\underset{\underset{H}{|}}{N}} - H$$

   Base   Acid   Conjugate acid   Conjugate base
                                                          of H_2O of NH_4^+

(b) Ethanol is the acid (proton donor), and amide ion (NH_2^-) is the base (proton acceptor).

$$CH_3CH_2 \overset{\cdot\cdot}{\underset{\cdot\cdot}{O}} - H \quad + \quad \overset{-}{\overset{\cdot\cdot}{:}}N - H \quad \longrightarrow \quad CH_3CH_2 \overset{\cdot\cdot}{\underset{\cdot\cdot}{O}} {\overset{-}{\vphantom{.}}} \quad + \quad H - \overset{\cdot\cdot}{\underset{\underset{H}{|}}{N}} - H$$

   Acid   Base   Conjugate base   Conjugate acid
                                               of CH_3CH_2OH of NH_2^-

Problem 4.2

Write these reactions as proton-transfer reactions. Label which reactant is the acid and which is the base, which product is the conjugate base of the original acid, and which is the conjugate acid of the original base. In addition, write Lewis structures for each reactant and product, and use curved arrows to show the flow of electrons in each reaction.

(a) $CH_3SH + OH^- \longrightarrow CH_3S^- + H_2O$

(b) $CH_2{=}O + HCl \longrightarrow CH_2{=}OH^+ + Cl^-$

B. Brønsted-Lowry Bases with Two or More Receptor Sites

Thus far, we have dealt with Brønsted-Lowry bases that have only one site that can act as a proton acceptor in an acid-base reaction. Many organic compounds have two or more such sites. In the following discussion, we restrict our discussion to compounds containing a carbonyl group in which the carbonyl carbon is bonded to either an oxygen or a nitrogen. The principle we develop here is an extremely important one and applies to other types of molecules as well, namely that the more stable protonated form is the one in which the charge is more delocalized, in this case by resonance (Section 1.6). Let us consider first the potential sites for proton transfer to an oxygen atom of a carboxylic acid such as acetic acid. Proton transfer to the carbonyl oxygen gives cation A, and proton transfer to the hydroxyl oxygen gives cation B.

A
(protonation on the carbonyl oxygen)

B
(protonation on the hydroxyl oxygen)

We now examine each cation and determine which is the more stable (lower in energy). For cation A, we can write three contributing structures. Two of these place the positive charge on oxygen, and one places it on carbon.

A-1
(C and O have complete octets)

A-2
(C has incomplete octet)

A-3
(C and O have complete octets)

Of these three structures, A-1 and A-3 make the greater contributions to the hybrid because all atoms in each have complete octets; A-2 makes a lesser contribution because its carbonyl carbon has an incomplete octet. Thus, on protonation of the carbonyl oxygen, the positive charge is delocalized over three atoms with the greater share of it being on the two equivalent oxygen atoms. (The two oxygens were not equivalent before protonation, but they are now.)

Protonation on the hydroxyl oxygen gives cation B for which we can write two resonance contributing structures.

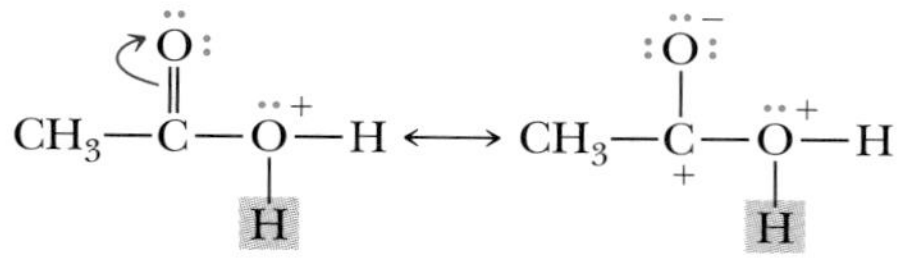

B-1   B-2
(charge separation and
adjacent positive charges)

Of these, B-2 makes at best only a minor contribution to the hybrid because of the adjacent positive charges; therefore, the charge on this cation is, in effect, localized on the hydroxyl oxygen.

From this analysis of cations A and B, we see that protonation of a carboxyl group occurs preferentially on the carbonyl oxygen because this allows greater resonance delocalization of the positive charge.

Example 4.3

The functional group created when the —OH of a carboxyl group is replaced by an NH_2 group is called an amide (Section 1.3F). Draw the structural formula of acetamide, which is derived from acetic acid, and determine if proton transfer to the amide group from HCl occurs preferentially on the amide oxygen or the amide nitrogen.

Solution

Following is a Lewis structure for acetamide and its two possible protonated forms.

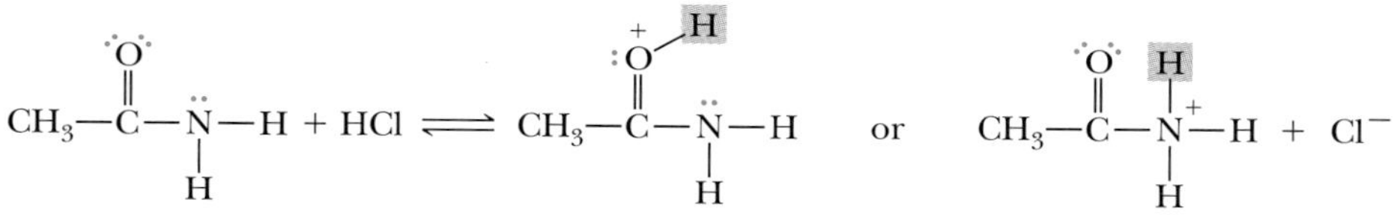

Acetamide   A   B
(an amide)   (protonation on the   (protonation on the
                  amide oxygen)   amide nitrogen)

Of the three contributing structures that can be drawn for cation A, structures A-1 and A-3 make the greater contributions to the hybrid because all atoms in each have complete octets; of these two contributors, A-3 has the positive charge on the less electronegative atom and, therefore, makes a greater contribution than A-1. The result is that the positive charge in cation A is delocalized over three atoms, the greater share of it being on nitrogen and oxygen.

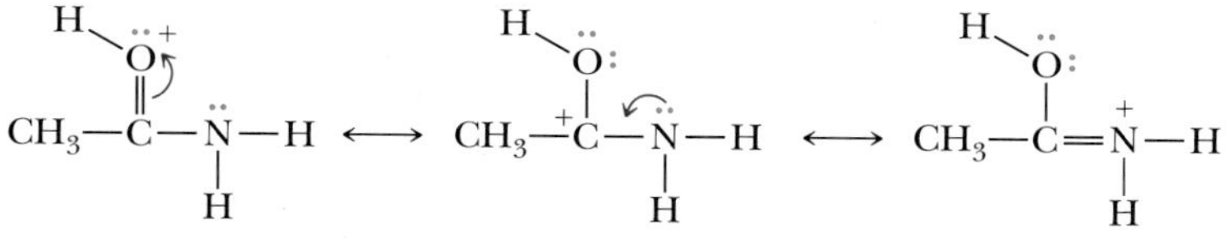

A-1   A-2   A-3
(complete octets;   (incomplete   (complete octets;
positive charge on more octet for C)   positive charge on less
electronegative atom)   electronegative atom)

Only two contributing structures can be drawn for cation B. Of these, B-2 requires creation and separation of unlike charges and places positive charges on adjacent atoms; it makes little contribution to the hybrid. Thus, the positive charge in cation B is essentially localized on the amide nitrogen.

$$CH_3-\overset{O}{\underset{}{C}}-\overset{H}{\underset{H}{N^+}}-H \longleftrightarrow CH_3-\overset{O^-}{\underset{+}{C}}-\overset{H}{\underset{H}{N^+}}-H$$

B-1
(protonation on the amide nitrogen)

B-2
(adjacent plus charges; charge separation)

From this analysis, we conclude that proton transfer to the carbonyl oxygen of the amide group gives the more stable cation; therefore, cation A is the predominant cation present at equilibrium.

Problem 4.3

Following is a structural formula for guanidine, the compound by which migratory birds excrete excess metabolic nitrogen. The hydrochloride salt of this compound is a white crystalline powder, freely soluble in water and ethanol.

(a) Write a Lewis structure for guanidine showing all valence electrons.
(b) Does proton transfer to guanidine occur preferentially to one of its —NH$_2$ groups (cation A) or to its =NH group (cation B)? Explain.

$$H_2N-\overset{NH}{\underset{}{C}}-NH_2 + HCl \longrightarrow H_2N-\overset{NH}{\underset{}{C}}-NH_3^+ + H_2N-\overset{NH_2^+}{\underset{}{C}}-NH_2 + Cl^-$$

Guanidine   A   B

C. Pi Electrons as Brønsted-Lowry Bases

Thus far we have considered proton transfer to atoms having a nonbonding pair of electrons. Proton-transfer reactions also occur with compounds having pi electrons (e.g., the pi electrons of carbon-carbon double and triple bonds). The pi electrons of the carbon-carbon double bond of 2-butene, for example, react with strong acids such as H_2SO_4, H_3PO_4, HCl, HBr, and HI by proton transfer to form a new carbon-hydrogen bond.

$$CH_3-CH=CH-CH_3 + H-Br: \longrightarrow CH_3-\overset{+}{C}-\overset{H}{\underset{H}{C}}-CH_3 + :Br:^-$$

2-Butene

sec-Butyl cation
(a 2° carbocation)

The result of this proton-transfer reaction is formation of a **carbocation,** a species in which one of its carbons has only six electrons in its valence shell and carries a charge of +1. Because the carbon bearing the positive charge in the sec-butyl cation has only two other carbons bonded to it, it is classified as a secondary (2°) carbocation. We will study the formation, structure, and reactions of carbocations in detail in Chapter 6.

Example 4.4

The acid-base reaction between 2-methyl-2-butene and HI can in principle form two carbocations. Write chemical equations for the formation of each carbocation and use curved arrows to show the proton transfer in each reaction.

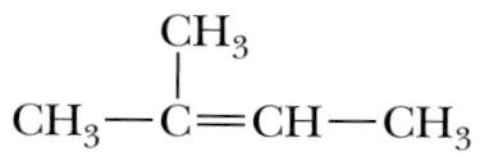

$$CH_3-\overset{\displaystyle CH_3}{\underset{}{C}}=CH-CH_3$$

2-Methyl-2-butene

Solution

Proton transfer to carbon 3 of this alkene gives a tertiary (3°) carbocation. Proton transfer to carbon 2 gives a secondary (2°) carbocation.

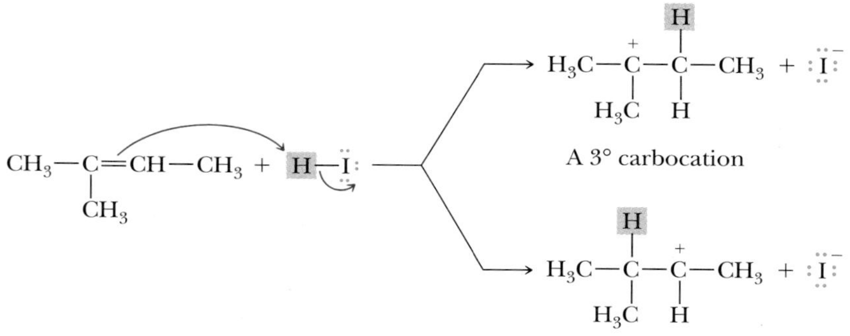

A 3° carbocation

A 2° carbocation

Problem 4.4

Write an equation to show the proton transfer between each alkene or cycloalkene and HCl. Where two carbocations are possible, show each.

(a) $CH_3CH_2CH{=}CHCH_3$

2-Pentene

(b)

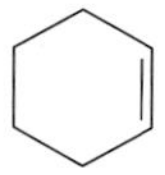

Cyclohexene

4.3 Acid Dissociation Constants, pK_a, and the Relative Strengths of Acids and Bases

Any quantitative measure of the acidity of organic acids or bases involves measuring the equilibrium concentrations of the various components in an acid-base equilibrium. The strength of an acid is then expressed by an equilibrium constant. The dissociation (ionization) of acetic acid in water is given by the following equation.

$$CH_3\overset{\overset{\displaystyle O}{\|}}{C}OH + H_2O \rightleftharpoons CH_3\overset{\overset{\displaystyle O}{\|}}{C}O^- + H_3O^+$$

Acetic acid   Water   Acetate ion   Hydronium
                                          ion

We can write an equilibrium expression for the dissociation of this or any other uncharged acid in a more general form; dissociation of the acid, HA, in water gives an anion, A$^-$, and the hydronium ion, H$_3$O$^+$. The equilibrium constant for this ionization is

$$HA + H_2O \rightleftharpoons A^- + H_3O^+$$

$$K_{eq} = \frac{[H_3O^+][A^-]}{[HA][H_2O]}$$

Because water is the solvent for this reaction and its concentration changes very little when HA is added to it, we can treat the concentration of water as a constant equal to 1000 g/L or approximately 55.6 mol/L. We can then combine these two constants (K_{eq} and the concentration of water) to define a new constant called an **acid dissociation constant,** given the symbol K_a.

$$K_a = K_{eq}[H_2O] = \frac{[H_3O^+][A^-]}{[HA]}$$

Because dissociation constants for most acids, including organic acids, are numbers with negative exponents, acid strengths are often expressed as **pK_a** where pK_a = $-\log_{10} K_a$. The pK_a for acetic acid is 4.76, which means that acetic acid is a weak acid because the major species present in aqueous solution is un-ionized CH$_3$COOH. Table 4.1 gives names, molecular formulas, and values of pK_a for some organic and inorganic acids. As you study the information in this table, note the following relationships:

- The larger the value of pK_a, the weaker the acid.
- The smaller the value of pK_a, the stronger the acid.
- The weaker the acid, the stronger its conjugate base.
- The stronger the acid, the weaker its conjugate base.

Example 4.5

For each value of pK_a, calculate the corresponding value of K_a. Which compound is the stronger acid?

(a) Ethanol, pK_a = 15.9   **(b)** Carbonic acid, pK_a = 6.36

Solution

(a) For ethanol, $K_a = 1.3 \times 10^{-16}$ **(b)** For carbonic acid, $K_a = 4.4 \times 10^{-7}$

Because the value of pK_a for carbonic acid is smaller than that for ethanol, carbonic acid is the stronger acid, and ethanol is the weaker acid.

Table 4.1   pK_a Values for Some Organic and Inorganic Acids

	Acid	Formula	pK_a	Conjugate Base	
Weaker acid	Ethane	CH_3CH_3	51	$CH_3CH_2^-$	Stronger conjugate base
	Ethylene	$CH_2{=}CH_2$	44	$CH_2{=}CH^-$	
	Ammonia	NH_3	38	NH_2^-	
	Hydrogen	H_2	35	H^-	
	Acetylene	$HC{\equiv}CH$	25	$HC{\equiv}C^-$	
	Ethanol	CH_3CH_2OH	15.9	$CH_3CH_2O^-$	
	Water	H_2O	15.7	HO^-	
	Methylammonium ion	$CH_3NH_3^+$	10.64	CH_3NH_2	
	Bicarbonate ion	HCO_3^-	10.33	CO_3^{2-}	
	Phenol	C_6H_5OH	9.95	$C_6H_5O^-$	
	Ammonium ion	NH_4^+	9.24	NH_3	
	Hydrogen sulfide	H_2S	7.04	HS^-	
	Carbonic acid	H_2CO_3	6.36	HCO_3^-	
	Acetic acid	CH_3COOH	4.76	CH_3COO^-	
	Benzoic acid	C_6H_5COOH	4.19	$C_6H_5COO^-$	
	Phosphoric acid	H_3PO_4	2.1	$H_2PO_4^-$	
	Hydronium ion	H_3O^+	−1.74	H_2O	
	Sulfuric acid	H_2SO_4	−5.2	HSO_4^-	
	Hydrogen chloride	HCl	−7	Cl^-	
Stronger acid	Hydrogen bromide	HBr	−8	Br^-	Weaker conjugate base
	Hydrogen iodide	HI	−9	I^-	

Problem 4.5

For each value of K_a, calculate the corresponding value of pK_a. Which compound is the stronger acid?

(a) Acetic acid, $K_a = 1.74 \times 10^{-5}$ (b) Chloroacetic acid, $K_a = 1.38 \times 10^{-3}$

Values of pK_a in aqueous solution in the range 2 to 12 can be measured quite accurately. Values of pK_a smaller than 2 are less accurate because very strong acids, such as HCl, HBr, and HI, are completely ionized in water, and the only acid present in solutions of these acids is H_3O^+. For acids too strong to be measured accurately in water, less basic solvents such as acetic acid or mixtures of water and sulfuric acid are used. Although none of the halogen acids, for example, is completely ionized in acetic acid, HI shows a greater degree of ionization in this solvent than either HBr or HCl; therefore, HI is the strongest acid of the three. Values of pK_a greater than 12 are also less precise. For bases too strong to be measured in aqueous solution, more basic solvents such as liquid ammonia and dimethyl sulfoxide are used. Because different solvent systems are used to measure relative strengths at either end of the acidity scale, pK_a values smaller than 2 and greater than 12 should be used only in a qualitative way when comparing them with values in the middle of the scale.

4.4 The Position of Equilibrium in Acid-Base Reactions

ORGANIC
Chemistry-⟨-Now™

Click Tutorials to review the **Position of Acid-Base Equilibria**

We know from the value of pK_a for an acid whether an aqueous solution of the acid contains more molecules of the undissociated acid or its ions. A negative pK_a value indicates that the majority of molecules of the acid are dissociated in water, while a positive value indicates that most acid molecules remain un-ionized in water. HCl, for example, a strong acid with a pK_a of -7, is almost completely dissociated at equilibrium in aqueous solution, and the major species present are H_3O^+ and Cl^-.

$$HCl + H_2O \longrightarrow H_3O^+ + Cl^- \qquad pK_a = -7$$

For acetic acid on the other hand, which is a weak acid with a pK_a of 4.76, the major species present at equilibrium in aqueous solution are CH_3COOH molecules.

$$CH_3\overset{\overset{\displaystyle O}{\|}}{C}OH + H_2O \rightleftharpoons CH_3\overset{\overset{\displaystyle O}{\|}}{C}O^- + H_3O^+ \qquad pK_a = 4.76$$

In these acid-base reactions, water is the base. But what if we have a base other than water as the proton acceptor, or what if we have an acid other than hydrogen chloride or acetic acid as the proton donor? How do we determine quantitatively or even qualitatively which species are present at equilibrium; that is, how do we determine where the position of equilibrium lies?

Let us take as an example the acid-base reaction of acetic acid and ammonia to form acetate ion and ammonium ion.

$$CH_3\overset{\overset{\displaystyle O}{\|}}{C}OH \; + \; NH_3 \; \rightleftharpoons \; CH_3\overset{\overset{\displaystyle O}{\|}}{C}O^- \; + \; NH_4^+$$

| Acetic acid | Ammonia | | Acetate ion | Ammonium ion |

There are two acids present in this equilibrium, acetic acid and ammonium ion. There are also two bases present, ammonia and acetate ion. A way to analyze this equilibrium is to view it as a competition between the two bases, ammonia and acetate ion, for a proton. The question we then ask is "Which of these is the stronger base?" The information we need to answer this question is in Table 4.1. We first determine which conjugate acid is the stronger acid and couple this with the fact that the stronger the acid, the weaker its conjugate base. From Table 4.1, we see that acetic acid, pK_a 4.76, is the stronger acid, which means that CH_3COO^- is the weaker base. Conversely, ammonium ion, pK_a 9.24, is the weaker acid, which means than NH_3 is the stronger base. We can now label the relative strengths of each acid and base.

$$CH_3\overset{\overset{\displaystyle O}{\|}}{C}OH \; + \; NH_3 \; \rightleftharpoons \; CH_3\overset{\overset{\displaystyle O}{\|}}{C}O^- \; + \; NH_4^+$$

| Acetic acid | Ammonia | | Acetate ion | Ammonium ion |

pK_a 4.76 (stronger acid) (stronger base) (weaker base) pK_a 9.24 (weaker acid)

In an acid-base reaction, the position of equilibrium always favors reaction of the stronger acid and stronger base to form the weaker acid and weaker base. Thus at equilibrium, the major species present are the weaker acid and weaker base.

Therefore, in the reaction between acetic acid and ammonia, the equilibrium lies to the right, and the major species present are acetate ion and ammonium ion.

Not only can we use values of pK_a for each acid to estimate the position of equilibrium, but we can also use them to calculate an equilibrium constant for the equilibrium. Consider an acid-base reaction represented by the following general equation.

$$HA \; + \; B \; \rightleftharpoons \; A^- \; + \; HB^+$$

Acid	Base	Conjugate base of HA	Conjugate acid of B

The equilibrium constant for this reaction is

$$K_{eq} = \frac{[A^-][BH^+]}{[HA][B]}$$

Multiplying the right-hand side of this equation by $[H_3O^+]/[H_3O^+]$ gives a new expression, which, on rearrangement, becomes the K_a of acid HA divided by the K_a of acid BH^+.

$$K_{eq} = \frac{[A^-][BH^+]}{[HA][B]} \times \frac{[H_3O^+]}{[H_3O^+]} = \frac{[A^-][H_3O^+]}{[HA]} \times \frac{[BH^+]}{[B][H_3O^+]} = \frac{K_{HA}}{K_{BH^+}}$$

Alternatively, we can take the logarithm of each side of this equation and arrive at this expression.

$$pK_{eq} = pK_{HA} - pK_{BH^+}$$

Thus, if we know the acid dissociation constants of each acid in the equilibrium, we can calculate the position of the acid-base equilibrium.

$$CH_3COOH \; + \; NH_3 \; \rightleftharpoons \; CH_3COO^- \; + \; NH_4^+ \qquad pK_{eq} = 4.76 - 9.24 = -4.48$$

Acetic acid	Ammonia	Acetate ion	Ammonium ion	$K_{eq} = 3.0 \times 10^4$
pK_a 4.76 (stronger acid)	(stronger base)	(weaker base)	pK_a 9.24 (weaker acid)	

From the fact that the stronger acid reacts with the stronger base to give the weaker acid and the weaker base, we conclude that the equilibrium for the reaction between acetic acid and ammonia lies to the right. Using values of pK_a, we calculate that the equilibrium constant for the reaction between acetic acid and ammonia is 3.0×10^4.

Consider now the reaction between aqueous solutions of acetic acid and sodium bicarbonate to give sodium acetate and carbonic acid. In the equation for this equilibrium, we omit the sodium ion, Na^+, because it does not undergo a chemical change in this reaction. Instead, we write the equilibrium as a net ionic equation, which shows only the species undergoing chemical change.

$$\overset{O}{\overset{\|}{CH_3C}}OH \; + \; HCO_3^- \; \rightleftharpoons \; \overset{O}{\overset{\|}{CH_3C}}O^- \; + \; H_2CO_3$$

Acetic acid	Bicarbonate ion	Acetate ion	Carbonic acid
pK_a 4.76 (stronger acid)			pK_a 6.36 (weaker acid)

Acetic acid is the stronger acid; therefore, the position of this equilibrium lies to the right. Carbonic acid is formed, which then decomposes to carbon dioxide and water.

Example 4.6

Predict the position of equilibrium, and calculate the equilibrium constant, K_{eq}, for each acid-base reaction.

(a) C_6H_5OH + HCO_3^- $\rightleftharpoons$ $C_6H_5O^-$ + H_2CO_3

    Phenol   Bicarbonate ion   Phenoxide ion   Carbonic acid

(b) $HC{\equiv}CH$ + NH_2^- $\rightleftharpoons$ $HC{\equiv}C^-$ + NH_3

    Acetylene   Amide ion   Acetylide ion   Ammonia

Solution

(a) Carbonic acid is the stronger acid; the position of this equilibrium lies to the left. Phenol does not transfer a proton to bicarbonate ion to form carbonic acid.

$$C_6H_5OH + HCO_3^- \rightleftharpoons C_6H_5O^- + H_2CO_3 \qquad pK_{eq} = 9.95 - 6.36 = 3.59$$

    pK_a 9.95 pK_a 6.36 $K_{eq} = 10^{-3.59} = 2.57 \times 10^{-4}$

    Weaker acid   Stronger acid

(b) Acetylene is the stronger acid; the position of this equilibrium lies to the right.

$$HC{\equiv}CH + NH_2^- \rightleftharpoons HC{\equiv}C^- + NH_3 \qquad pK_{eq} = 25 - 38 = -13$$

    pK_a 25 pK_a 38 $K_{eq} = 10^{13}$

    Stronger acid   Weaker acid

Problem 4.6

Predict the position of equilibrium, and calculate the equilibrium constant, K_{eq}, for each acid-base reaction.

(a) CH_3NH_2 + CH_3COOH $\rightleftharpoons$ $CH_3NH_3^+$ + CH_3COO^-

    Methylamine   Acetic acid   Methylammonium   Acetate
                                        ion   ion

(b) $CH_3CH_2O^-$ + NH_3 $\rightleftharpoons$ CH_3CH_2OH + NH_2^-

    Ethoxide ion   Ammonia   Ethanol   Amide ion

CONNECTIONS TO
BIOLOGICAL CHEMISTRY

The Ionization of Functional Groups at Physiological pH

The pH of living cells is generally between 7.0 and 8.5, a range often referred to as physiological pH. At physiological pH, several of the common functional groups found in biological molecules are ionized because they are either acids that are deprotonated or bases that are protonated. A good rule of thumb is that an acid will be substantially deprotonated if its pK_a is two or more units lower than the pH of the solution. The carboxyl group (the functional group of carboxylic acids) is present in all amino acids, as well as in the side chains of glutamic acid and aspartic acid. The pK_a values for carboxylic acids are typically between 4 and 5. At pH values of 7 or above, carboxylic acids are essentially fully deprotonated and therefore anionic.

$$R-\overset{\overset{\text{O}}{\|}}{C}-OH \qquad R-\overset{\overset{\text{O}}{\|}}{C}-O^-$$

Carboxylic acid   Form present at
pK_a 4–5   physiological pH

Another functional group found in biomolecules is the phosphodiester group. This group is found as part of the backbone of nucleic acids such as DNA and RNA. The pK_a values for phosphodiesters are between 1 and 3.

$$R-O-\overset{\overset{\text{OH}}{|}}{\underset{\underset{\text{O}}{\|}}{P}}-O-R' \qquad R-O-\overset{\overset{\text{O}^-}{|}}{\underset{\underset{\text{O}}{\|}}{P}}-O-R'$$

Phosphodiester   Form present
pK_a 1–3   at physiological pH

At physiological pH values, the phosphodiester group will be present in its anion form. Therefore, the backbones of nucleic acids (in which there is a repeating pattern of phosphodiester groups) will be polyanionic, a factor that has a major influence on their overall properties.

There are also basic functional groups in biological molecules. A good rule of thumb here is that a base will be protonated if the pK_a of its conjugate acid is two or more units higher than the pH of the solution. Two important examples include amino groups and guanidino groups. The pK_a values for the conjugate acids of

4.5 Molecular Structure and Acidity

Now let us examine in some detail the relationships between molecular structure and acidity. The overriding principle in determining the relative acidities of uncharged organic acids is the stability of the anion, A^-, resulting from loss of a proton; the more stable the anion, the greater the acidity of the acid, HA. As we discuss in this section, ways to stabilize A^- include

- Having the negative charge on a more electronegative atom
- Having the negative charge on a larger atom
- Delocalizing the negative charge through resonance
- Spreading the negative charge onto electron-withdrawing groups by the inductive effect (polarization of sigma bonds)
- Having the negative charge in an orbital with more *s* character

amines and guanidines are about 9 to 11 and 13 to 14, respectively. As a result, these groups are protonated and therefore positively charged at physiological pH.

R—NH$_2$

An amine group

R—NH$_3^+$

pK$_a$ 9–11
Form present at
physiological pH

$$\underset{\text{A guanidine group}}{\overset{\displaystyle \overset{NH}{\|}}{R-NH-C-NH_2}}$$

$$\underset{\substack{\text{pK}_a\ 13-14 \\ \text{Form present at} \\ \text{physiological pH}}}{\overset{\displaystyle \overset{NH_2^+}{\|}}{R-NH-C-NH_2}}$$

An interesting case is the imidazole group, which comprises the side chain of the amino acid histidine. An imidazole group, whose conjugate acid has a pK$_a$ between 6 and 7, will be present at physiological pH as a mixture of protonated and deprotonated forms. This ability to exist in both forms can be significant in situations in which proton transfer reactions are important for the function of the protein containing an imidazole group.

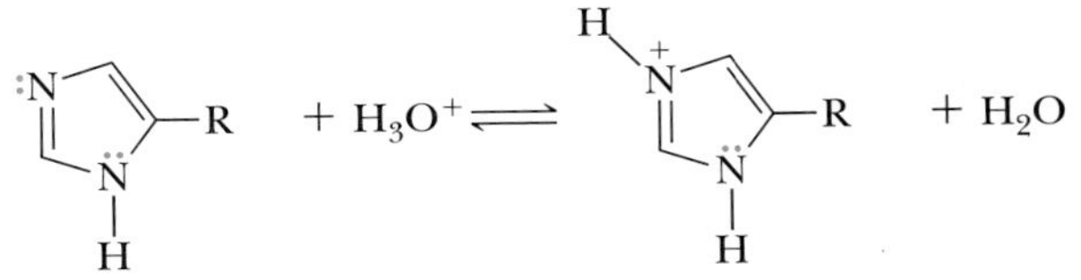

Imidazole group

pK$_a$ 6–7
Conjugate acid of
imidazole

Finally, no highlight section on the acid-base properties of biological molecules would be complete without a discussion of amino acids. At physiological pH, both the amino and carboxyl groups are ionized. Free amino acids are found in a number of situations in organisms, for example as neurotransmitters in mammals.

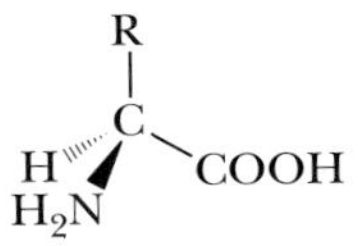

An amino acid contains
an amino group and a
carboxyl group

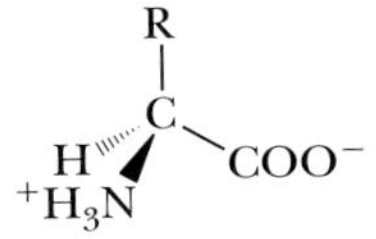

Ionized or zwitterion
form present at
physiological pH

A. Electronegativity of the Atom Bearing the Negative Charge

Let us consider the relative acidities of the following series of hydrogen acids, all of which are in the same period of the Periodic Table.

Acid				Conjugate base
Methanol pK$_a$ 16	CH$_3$—Ö—H		CH$_3$—Ö$^-$	Methoxide ion
Methylamine pK$_a$ 38	CH$_3$—N̈—H with H below	Increasing acidity / Increasing basicity	CH$_3$—N̈$^-$ with H below	Methylamide ion
Ethane pK$_a$ 51	CH$_3$—C—H with H above and below		CH$_3$—C̈$^-$ with H above and below	Ethyl anion

The pK_a value for ethane is given in Table 4.1, but values for methylamine and methanol are not. We can, however, make good guesses about the pK_a values of these acids by reasoning that the nature of the alkyl group bonded to nitrogen or oxygen has only a relatively small effect on the acidity of the hydrogen bonded to the heteroatom (in organic chemistry, an atom other than carbon). Therefore, we estimate that the pK_a of methylamine is approximately the same as that of ammonia (pK_a 38), and that the pK_a of methanol is approximately the same as ethanol (pK_a 15.9). As we see, ethane is the weakest acid in this series, and ethyl anion is the strongest conjugate base. Conversely, methanol is the strongest acid, and methoxide ion is the weakest conjugate base.

The relative acidity within a period of the Periodic Table is related to the electronegativity of the atom in the anion that bears the negative charge. The greater the electronegativity of this atom, the more strongly its electrons are held, and the more stable the anion is. Conversely, the smaller the electronegativity of this atom, the less tightly its electrons are held, and the less stable the anion is. Oxygen, the most electronegative of the three atoms compared, has the largest electronegativity (3.5 on the Pauling scale), and methanol forms the most stable anion. Carbon, the least electronegative of the three (2.5 on the Pauling scale), forms the least stable anion. Because methanol forms the most stable anion, it is the strongest acid in this series. Ethane is the weakest acid in the series.

It is essential to understand that this argument based on electronegativity applies only to acids within the same period (row) of the Periodic Table. Anions of atoms within the same period have approximately the same size, and their energies of solvation are approximately the same.

B. Size of the Atom Bearing the Negative Charge

To illustrate how the acidity of hydrogen acids varies within a group (column) of the Periodic Table, let us compare the acidities of methanol and methanethiol, CH_3SH. We estimated in the previous section that the pK_a of methanol is 16. We can estimate the pK_a of methanethiol in the following way. The pK_a of hydrogen sulfide, H_2S, is given in Table 4.1 as 7.04. If we assume that substitution of a methyl group for a hydrogen makes only a slight change in acidity, we then estimate that the pK_a of methanethiol is approximately 7.0. Thus, methanethiol is the stronger acid, and methanethiolate ion is the weaker conjugate base.

$$CH_3-\overset{..}{\underset{..}{S}}-H \ + \ CH_3-\overset{..}{\underset{..}{O}}:^- \ \rightleftharpoons \ CH_3-\overset{..}{\underset{..}{S}}:^- \ + \ CH_3-\overset{..}{\underset{..}{O}}-H$$

Methanethiol	Methoxide	Methanethiolate	Methanol
pK_a 7.0	ion	ion	pK_a 16
(stronger acid)	(stronger base)	(weaker base)	(weaker acid)

The relative acidity of these two hydrogen acids, and in fact any set of hydrogen acids within a group of the Periodic Table, is related to the size of the atom bearing the negative charge. Size increases from top to bottom within a group because the valence electrons are in increasingly higher principal energy levels. This means that (1) they are farther from the nucleus, and (2) they occupy a larger volume of space. Because sulfur is below oxygen in the Periodic Table, it is larger than oxygen. Accordingly, the negative charge on sulfur in methanethiolate ion is spread over a larger volume of space; therefore, the CH_3S^- anion is more stable. The negative charge on oxygen in methoxide ion is confined to a smaller volume of space; therefore, the CH_3O^- anion is less stable.

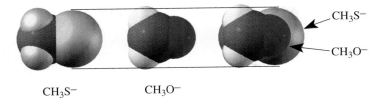

CH_3S^- CH_3O^-

We see this same trend in the strength of the halogen acids, HF, HCl, HBr, and HI, which increase in strength from HF (the weakest) to HI (the strongest). Of their anions, iodide ion is the largest; its charge is spread over the largest volume of space and, therefore, is the most stable. HI is the strongest of the halogen acids. Conversely, fluoride ion is the smallest anion; its charge is the most concentrated, and fluoride ion is the least stable. HF is, therefore, the weakest acid of the halogen acids.

We will return to the relative sizes of these ions in Chapter 9 at which point we will discuss the nature of their solvation in polar solvents. We will see there that, for the ions of the same charge, the smaller the ion, the greater its solvation, and that the degree of solvation has a profound effect on their relative reactivities.

C. Resonance Delocalization of Charge in the Anion

Carboxylic acids are weak acids. Values of pK_a for most unsubstituted carboxylic acids fall within the range 4 to 5. The pK_a for acetic acid, for example, is 4.76. Values of pK_a for most alcohols, compounds that also contain an —OH group, fall within the range 15 to 18; the pK_a for ethanol, for example, is 15.9. Thus, most alcohols are slightly weaker acids than water ($pK_a = 15.7$) but much weaker acids than carboxylic acids.

We account for the greater acidity of carboxylic acids compared with alcohols using the resonance model and looking at the relative stabilities of the alkoxide and carboxylate ions. Our guideline is this: The more stable the anion, the farther the position of equilibrium is shifted toward the right, and the more acidic the compound is.

Here we take the acid ionization of an alcohol as a reference equilibrium.

$$CH_3CH_2\ddot{O}-H + H_2O \rightleftharpoons CH_3CH_2\ddot{O}\dot{:}^- + H_3O^+$$

An alcohol   An alkoxide ion

Elpot of ethoxide ion

In the alkoxide anion, the negative charge is localized on oxygen. In contrast, ionization of a carboxylic acid gives an anion for which we can write two equivalent contributing structures that result in delocalization of the negative charge of the anion. Because of this delocalization of negative charge, a carboxylate anion is more stable than an alkoxide anion. Conversely, a carboxylic acid is a stronger acid than an alcohol.

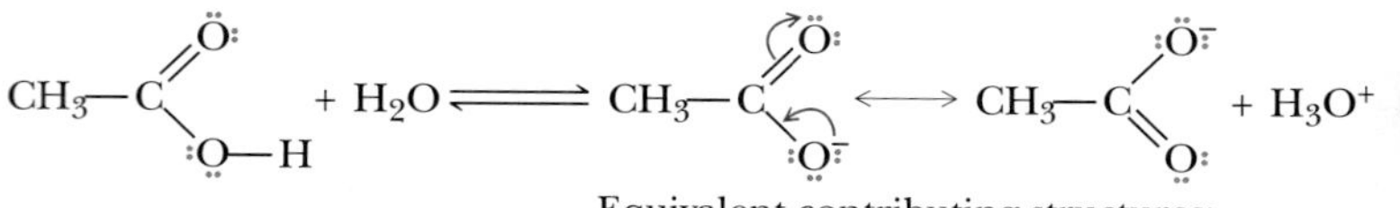

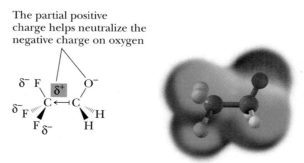

Equivalent contributing structures;
the carboxylate anion is stabilized by
delocalization of the negative charge

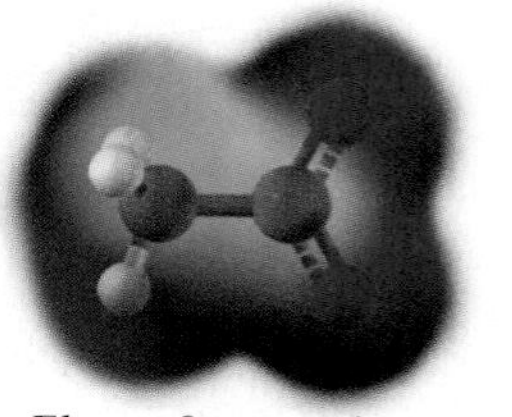

Elpot of acetate ion

D. Inductive Effect and Electrostatic Stabilization of the Anion

Inductive effect The polarization of the electron density of a covalent bond caused by the electronegativity of a nearby atom.

We see an example of the **inductive effect** in alcohols in the fact that an electronegative substituent adjacent to the carbon bearing the —OH group increases the acidity of the alcohol. Compare, for example, the acidities of ethanol and 2,2,2-trifluoroethanol. The acid dissociation constant for 2,2,2-trifluoroethanol is larger than that of ethanol by more than three orders of magnitude, which means that the 2,2,2-trifluoroethoxide ion is considerably more stable than the ethoxide ion.

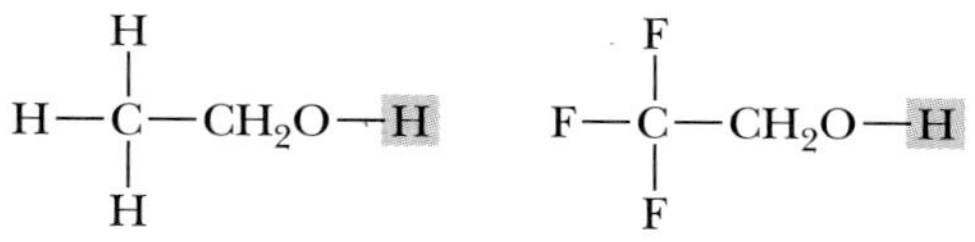

Ethanol
pK_a 15.9

2,2,2-Trifluoroethanol
pK_a 12.4

We account for the increased stability of the 2,2,2-trifluoroethoxide ion in the following way. Fluorine is more electronegative than carbon (4.0 versus 2.5); therefore, the C—F bond has a significant dipole, indicated in the following figure by directional arrows on the polar bonds as well as symbols to show the partial charges. There is an attractive stabilization by the interaction of the negatively charged oxygen and the partial positive charge on the carbon bearing the fluorines, which results in stabilization of the trifluoroethoxide ion.

The partial positive
charge helps neutralize the
negative charge on oxygen

Elpot of trifluoroethoxide ion

Stabilization by the inductive effect falls off rapidly with increasing distance of the electronegative atoms(s) from the site of the negative charge. Compare, for example, the pK_a values of alcohols substituted with fluorine on carbons 2 versus 3 versus 4. When fluorine atoms are more than two carbons away from the carbon bearing the —OH group, they have almost no effect on acidity.

$$CF_3—CH_2—OH \qquad CF_3—CH_2—CH_2—OH \qquad CF_3—CH_2—CH_2—CH_2—OH$$

2,2,2-Trifluoroethanol   3,3,3-Trifluoro-1-propanol   4,4,4-Trifluoro-1-butanol
    (pK_a 12.4)   (pK_a 14.6)   (pK_a 15.4)

We also see the operation of the inductive effect in the acidity of halogen-substituted carboxylic acids.

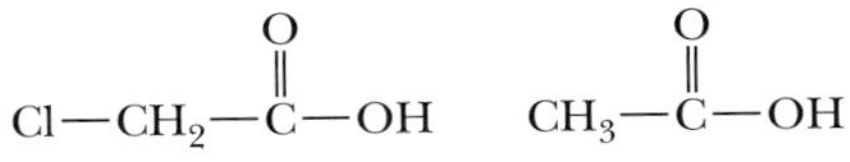

$$Cl—CH_2—\overset{\overset{\displaystyle O}{\|}}{C}—OH \qquad CH_3—\overset{\overset{\displaystyle O}{\|}}{C}—OH$$

Chloroacetic acid   Acetic acid
   pK_a 2.86 pK_a 4.76

The pK_a values of these two acids indicate that chloroacetic acid is approximately two orders of magnitude more acidic than acetic acid. In the case of chloroacetate anion, the negative charge is stabilized by electrostatic interaction between the partial negative charges on the oxygens and the partial positive charge on the carbon bearing the chlorine atom.

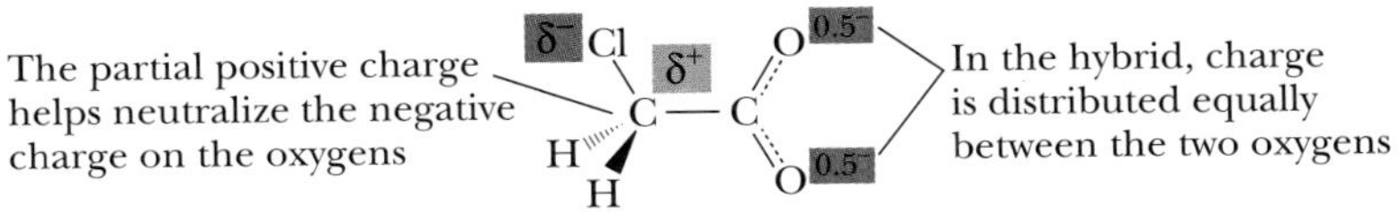

The partial positive charge helps neutralize the negative charge on the oxygens

In the hybrid, charge is distributed equally between the two oxygens

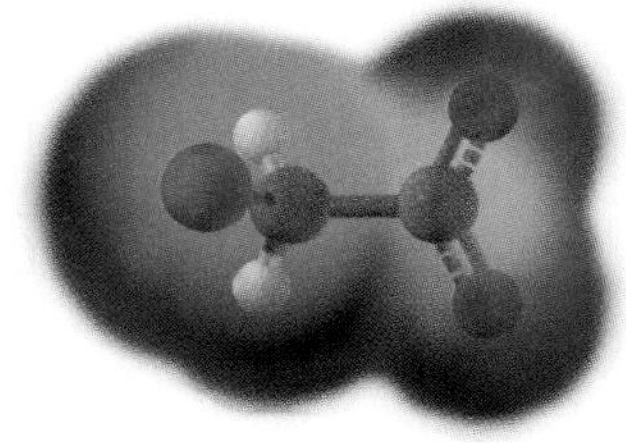

Elpot of chloroacetate ion

As was the case with halogen substitution and the acidity of alcohols, the acid-enhancing effect of halogen substitution in carboxylic acids falls off rapidly with increasing distance between the point of substitution and the carboxyl group.

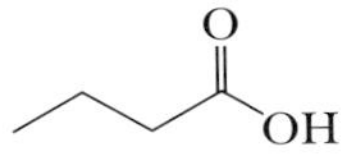

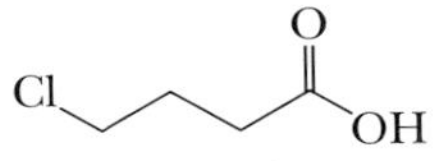

Butanoic   4-Chlorobutanoic   3-Chlorobutanoic   2-Chlorobutanoic
acid   acid   acid   acid
pK_a 4.82 pK_a 4.52 pK_a 3.98 pK_a 2.83

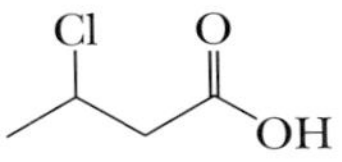

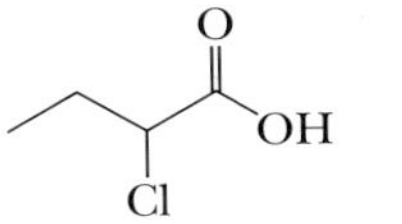

E. Hybridization and the Percent *s* Character of the Atom Bearing the Negative Charge

To see the effect of hybridization, we consider the case of two or more anions, each with the same charge and same element bearing the charge. The only difference is the hybridization of the atom bearing the negative charge. Of special importance for

Table 4.2   Acidity of Alkanes, Alkenes, and Alkynes

Weak Acid		Conjugate Base	pKa
Water	HO—H	HO$^-$	15.7
Alkyne	HC≡C—H	HC≡C$^-$	25
Ammonia	H_2N—H	H_2N$^-$	38
Alkene	CH_2=CH—H	CH_2=CH$^-$	44
Alkane	CH_3CH_2—H	$CH_3CH_2$$^-$	51

Increasing acidity →

us is the acidity of a hydrogen bound to a carbon of an alkane, an alkene, and an alkyne.

One of the major differences between the chemistry of alkynes and that of alkenes and alkanes is that a hydrogen attached to a triply bonded carbon atom is sufficiently acidic that it can be removed by a strong base, such as sodium amide or sodium hydride. Table 4.2 gives pK_a values for an alkyne, alkene, and alkane. Also given for comparison are values for ammonia and water.

We account for the greater acidity of alkynes in the following way. The lone pair of electrons on a carbon anion lies in a hybrid orbital: an sp orbital for an alkyne anion, an sp^2 orbital for an alkene anion, and an sp^3 orbital for an alkane anion. An sp orbital has 50% s character; an sp^2 orbital, 33%, and an sp^3 orbital, 25% (Section 1.8). Electrons in an s orbital are lower in energy than those in a p orbital; that is, they are held more tightly to the nucleus. Therefore, the more s character in a hybrid orbital, the more electronegative the atom will be, and the more acidic a hydrogen bonded to it will be (and the more stable the anion will be). Of the three types of compounds, the carbon in an alkyne (sp hybridized with 50% s character) is the most electronegative. Therefore, an alkyne anion is the most stable of the series, and an alkyne is the strongest acid of the series. By similar reasoning, the alkane carbon (sp^3 hybridized with 25% s character) is the least electronegative, and an alkane is the weakest acid of the series. An alkene anion, with 33% s character, is intermediate.

4.6 Lewis Acids and Bases

Lewis acid   Any molecule or ion that can form a new covalent bond by accepting a pair of electrons.

Lewis base   Any molecule or ion that can form a new covalent bond by donating a pair of electrons.

Gilbert N. Lewis, who proposed that covalent bonds are formed by sharing one or more pairs of electrons (Section 1.2A), further generalized the theory of acids and bases to include a group of substances not included in the Brønsted-Lowry concept. According to the Lewis definition, an acid is a species that can form a new covalent bond by accepting a pair of electrons; a base is a species that can form a new covalent bond by donating a pair of electrons. In the following general equation, the **Lewis acid,** A, accepts a pair of electrons in forming the new covalent bond and acquires a negative formal charge. The **Lewis base,** :B, donates the pair of electrons and acquires a positive formal charge.

$$A^{\curvearrowleft} \quad + \quad :B \quad \rightleftharpoons \quad \overset{-}{A}{-}\overset{+}{B}$$

new covalent bond
formed in this Lewis
acid-base reaction

Lewis acid
(electron pair
acceptor)

Lewis base
(electron pair
donor)

Note that, although we speak of a Lewis base as "donating" a pair of electrons, the term is not fully accurate. Donating in this case does not mean that the electron pair under consideration is removed completely from the valence shell of the base. Rather, donating means that the electron pair becomes shared with another atom to form a covalent bond.

An example of a Lewis acid-base reaction is that of a carbocation (a Lewis acid) with bromide ion, a Lewis base. The *sec*-butyl cation, for example, reacts with bromide ion to form 2-bromobutane.

$$CH_3{-}\overset{\overset{\displaystyle H}{|}}{\underset{\underset{\displaystyle H}{|}}{C}}{-}\overset{+}{\overset{\displaystyle |}{\underset{\underset{\displaystyle H}{|}}{C}}}{-}CH_3 \quad + \quad :\overset{..}{\underset{..}{Br}}:^{-} \quad \longrightarrow \quad CH_3{-}\overset{\overset{\displaystyle H}{|}}{\underset{\underset{\displaystyle H}{|}}{C}}{-}\overset{\overset{\displaystyle :\overset{..}{Br}:}{|}}{\underset{\underset{\displaystyle H}{|}}{C}}{-}CH_3$$

sec-Butyl cation
(a carbocation)

Bromide
ion

2-Bromobutane

The Lewis concept of acids and bases includes proton-transfer reactions; all Brønsted-Lowry bases (proton acceptors) are also Lewis bases, and all Brønsted-Lowry acids (proton donors) are also Lewis acids. The Lewis model, however, is more general in that it is not restricted to proton-transfer reactions.

Consider the reaction that occurs when boron trifluoride gas is dissolved in diethyl ether.

Diethyl ether
(a Lewis base)

Boron trifluoride
(a Lewis acid)

A BF_3-ether complex

Elpot of
diethyl ether

Elpot of BF_3

Boron, a Group 3A element, has three electrons in its valence shell, and, after forming single bonds with three fluorine atoms to give BF_3, boron still has only six electrons in its valence shell. Because it has an empty orbital in its valence shell and can accept two electrons into it, boron trifluoride is electron deficient and, therefore, a Lewis acid. In forming the O—B bond, the oxygen atom of diethyl ether (a Lewis base) donates an electron pair, and boron accepts the electron pair. The reaction between diethyl ether and boron trifluoride is classified as an acid-base reaction

CHEMICAL CONNECTIONS

The Strongest Acid?

What is the strongest acid? In recent years, organic chemists have prepared mixtures of protic and Lewis acids that have remarkable proton-donating power. Two of the most reactive of these mixtures, termed **superacids,** are HF with SbF_5 and HSO_3F (fluorosulfonic acid) with SbF_5. Either SO_2 or SO_2F_2 is used as a solvent for superacids. SbF_5 is a Lewis acid that reacts with the fluorine of the acid; the net effect is to pull electrons away from hydrogen.

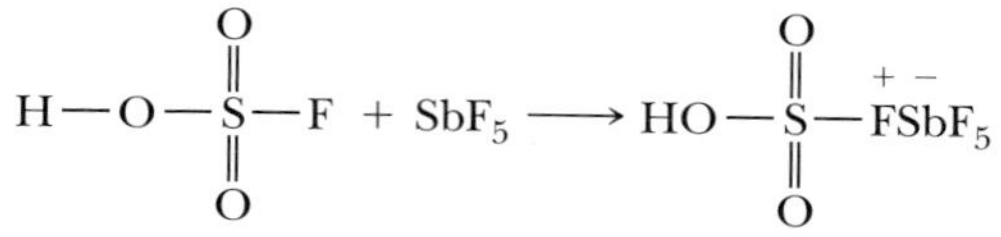

Both theoretical and experimental evidence exists that, in superacids, the electron pairs that make up the carbon-carbon and carbon-hydrogen bonds of hydrocarbons act as Lewis bases and become protonated. The resulting cations have unusual structures. In the case of ethane, for example, the ion $C_2H_7^+$ is produced.

The dashed lines in the structure shown for $C_2H_7^+$ indicate the formation of a three-center, two-electron bond. Notice that because the incoming

Fluorosulfonic acid

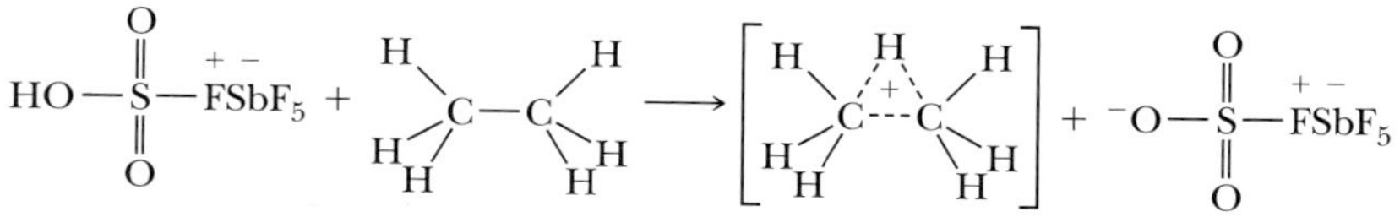

(Unstable; reacts further)

proton has no electrons of its own, the octet rule is not violated.

Even at low temperatures ($-78°C$), the $C_2H_7^+$ ion is not very stable. One of its decomposition products is methane, CH_4. The observation of methane as a product supports protonation of the carbon–carbon bond electron pair over one of the six carbon–hydrogen electron pairs.

Among those studying the reactions of superacids with alkanes was George Olah, at the University of Southern California, who received the 1994 Nobel prize for chemistry. Olah's discoveries completely transformed

our understanding of the chemistry of hydrocarbon cations.

Although the chemistry of alkanes in superacids may seem esoteric, in fact, these reactions provide a model for one of the most important reactions in industrial organic chemistry, namely the catalytic cracking of petroleum (Section 2.10B). Highly acidic sites on the solid catalysts used in petroleum refining promote protonation of C—C and C—H bonds. Protonation of C—C bonds leads to fragmentation and isomerization of larger hydrocarbons; protonation of C—H bonds leads to the production of hydrogen and alkenes.

Protic acid   An acid that is a proton donor in an acid-base reaction.

according to the Lewis model, but because there is no proton transfer involved, it is not classified as an acid-base reaction by the Brønsted-Lowry model. Said another way, all Brønsted-Lowry acids are **protic acids;** Lewis acids may be protic acids or they may be **aprotic acids.**

Aprotic acid   An acid that is not a proton donor; an acid that is an electron pair acceptor in a Lewis acid-base reaction.

Example 4.7

Write an equation for the reaction between each Lewis acid-base pair, showing electron flow by means of curved arrows.

(a) $BF_3 + NH_3 \longrightarrow$ **(b)** $(CH_3)_2 CH^+ + Cl^- \longrightarrow$

Solution

(a) BF_3 has an empty orbital in the valence shell of boron and is the Lewis acid. NH_3 has an unshared pair of electrons in the valence shell of nitrogen and is the Lewis base. In this example, each of these atoms takes on a formal charge; the resulting structure, however, has no net charge.

Lewis acid   Lewis base

(b) The trivalent carbon atom in the isopropyl cation has an empty orbital in its valence shell and is, therefore, the Lewis acid. Chloride ion is the Lewis base.

Lewis acid   Lewis base

Problem 4.7

Write an equation for the reaction between each Lewis acid-base pair, showing electron flow by means of curved arrows.

(a) $(CH_3CH_2)_3B + OH^- \longrightarrow$ **(b)** $CH_3Cl + AlCl_3 \longrightarrow$

Summary

By the Arrhenius definitions, an acid is a substance that dissolves in water to produce H_3O^+ ions (Section 4.1). A base is a substance that dissolves in water to produce OH^- ions. A **Brønsted-Lowry acid** is a proton donor, and a **Brønsted-Lowry base** is a proton acceptor (Section 4.2). Neutralization of an acid by a base is a **proton-transfer reaction** in which the acid is transformed into its **conjugate base,** and the base is transformed into its **conjugate acid.**

A **strong acid** or **strong base** is one that is completely ionized in water. A weak acid or weak base is one that is only partially ionized in water (Section 4.3). Among the most common weak organic acids are carboxylic acids, compounds that contain the —COOH (carboxyl) group. The value of K_a (the acid ionization constant) for acetic acid, a representative carboxylic acid, is 1.74×10^{-5}; the value of pK_a (the negative logarithm of K_a) for acetic acid is 4.76. The equilibrium position in an acid-base reaction favors reaction of the stronger acid with the stronger base to form the weaker acid and the weaker base (Section 4.4).

The acidity of hydrogen acids is determined by the stability of the anion formed on deprotonation (Section 4.5). Factors that influence the stability of an anion are (1) the electronegativity of the atom bearing the negative charge, (2) the size of the atom bearing the negative charge, (3) the delocalization of charge in the anion, (4) the **inductive effect,** and (5) the hybridization of the atom bearing the negative charge.

A **Lewis acid** (Section 4.6) is a species that can form a new covalent bond by accepting a pair of electrons; a **Lewis base** is a species that can form a new covalent bond by donating a pair of electrons.

Key Reactions

1. Proton-Transfer Reaction (Section 4.2)

A proton-transfer reaction involves transfer of a proton from a proton donor (a Brønsted-Lowry acid) to a proton acceptor (a Brønsted-Lowry base).

$$
\underset{\substack{\text{Proton} \\ \text{donor}}}{H-\overset{+}{\underset{H}{\overset{H}{O}}}-H} + \underset{\substack{\text{Proton} \\ \text{acceptor}}}{:\underset{H}{\overset{H}{N}}-H} \longrightarrow H-\underset{H}{\overset{\cdot\cdot}{O}}: + H-\overset{H}{\underset{H}{\overset{+}{N}}}-H
$$

2. Position of Equilibrium in an Acid-Base Reaction (Section 4.4)

The stronger acid reacts with the stronger base to give a weaker acid and a weaker base. K_{eq} for this equilibrium can be calculated from pK_a values for the two acids.

$$
\underset{\substack{pK_a\ 4.76 \\ (\text{stronger acid})}}{CH_3\overset{O}{\overset{\|}{C}}OH} + CN^- \rightleftharpoons \underset{}{CH_3\overset{O}{\overset{\|}{C}}O^-} + \underset{\substack{pK_a\ 9.31 \\ (\text{weaker acid})}}{HCN}
$$

$$pK_{eq} = 4.76 - 9.31 = -4.55$$
$$K_{eq} = 3.55 \times 10^4$$

3. Lewis Acid-Base Reaction (Section 4.6)

A Lewis acid-base reaction involves sharing an electron pair between an electron pair donor (a Lewis base) and an electron pair acceptor (a Lewis acid).

Problems

4.8 For each conjugate acid-base pair, identify the first species as an acid or base and the second species as its conjugate acid or base. In addition, draw Lewis structures for each species, showing all valence electrons and any formal charge.

(a) $HCOOH$ $HCOO^-$ (b) NH_4^+ NH_3 (c) $CH_3CH_2O^-$ CH_3CH_2OH
(d) HCO_3^- CO_3^{2-} (e) $H_2PO_4^-$ HPO_4^{2-} (f) CH_3CH_3 $CH_3CH_2^-$
(g) CH_3S^- CH_3SH

4.9 Complete a net ionic equation for each proton-transfer reaction using curved arrows to show the flow of electron pairs in each reaction. In addition, write Lewis structures for all starting materials and products. Label the original acid and its conjugate base; label the original base and its conjugate acid. If you are uncertain about which substance in each equation is the proton donor, refer to Table 4.1 for the relative strengths of proton acids.

(a) $NH_3 + HCl \longrightarrow$ (b) $CH_3CH_2O^- + HCl \longrightarrow$
(c) $HCO_3^- + OH^- \longrightarrow$ (d) $CH_3COO^- + NH_4^+ \longrightarrow$

4.10 Complete a net ionic equation for each proton-transfer reaction using curved arrows to show the flow of electron pairs in each reaction. Label the original acid and its conjugate base; then label the original base and its conjugate acid.

(a) $NH_4^+ + OH^- \longrightarrow$

(b) $CH_3COO^- + CH_3NH_3^+ \longrightarrow$

(c) $CH_3CH_2O^- + NH_4^+ \longrightarrow$

(d) $CH_3NH_3^+ + OH^- \longrightarrow$

4.11 Each molecule or ion can function as a base. Write a structural formula of the conjugate acid formed by reaction of each with HCl.

(a) CH_3CH_2OH

(b)

(c) $(CH_3)_2NH$

(d) HCO_3^-

4.12 In acetic acid, CH_3COOH, the OH hydrogen is more acidic than the CH_3 hydrogens. Explain.

Quantitative Measure of Acid and Base Strength

4.13 Which has the larger numerical value?

(a) The pK_a of a strong acid or the pK_a of a weak acid

(b) The K_a of a strong acid or the K_a of a weak acid

4.14 In each pair, select the stronger acid.

(a) Pyruvic acid (pK_a 2.49) and lactic acid (pK_a 3.08)

(b) Citric acid (pK_{a1} 3.08) and phosphoric acid (pK_{a1} 2.10)

4.15 Arrange the compounds in each set in order of increasing acid strength. Consult Table 4.1 for pK_a values of each acid.

(a) CH_3CH_2OH $HO\overset{\displaystyle O}{\overset{\|}{C}}O^-$ $C_6H_5\overset{\displaystyle O}{\overset{\|}{C}}OH$

   Ethanol   Bicarbonate ion   Benzoic acid

(b) $HO\overset{\displaystyle O}{\overset{\|}{C}}OH$ $CH_3\overset{\displaystyle O}{\overset{\|}{C}}OH$ HCl

   Carbonic acid   Acetic acid   Hydrogen chloride

4.16 Arrange the compounds in each set in order of increasing base strength. Consult Table 4.1 for pK_a values of the conjugate acid of each base.

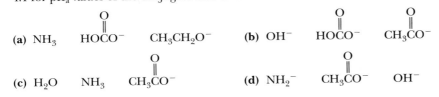

(a) NH_3 $HO\overset{\displaystyle O}{\overset{\|}{C}}O^-$ $CH_3CH_2O^-$

(b) OH^- $HO\overset{\displaystyle O}{\overset{\|}{C}}O^-$ $CH_3\overset{\displaystyle O}{\overset{\|}{C}}O^-$

(c) H_2O NH_3 $CH_3\overset{\displaystyle O}{\overset{\|}{C}}O^-$

(d) NH_2^- $CH_3\overset{\displaystyle O}{\overset{\|}{C}}O^-$ OH^-

Position of Equilibrium in Acid-Base Reactions

4.17 Unless under pressure, carbonic acid (H_2CO_3) in aqueous solution breaks down into carbon dioxide and water, and carbon dioxide is evolved as bubbles of gas. Write an equation for the conversion of carbonic acid to carbon dioxide and water.

4.18 Will carbon dioxide be evolved when sodium bicarbonate is added to an aqueous solution of these compounds? Explain.

(a) Sulfuric acid   (b) Ethanol   (c) Ammonium chloride

4.19   Acetic acid, CH_3COOH, is a weak organic acid, pK_a 4.76. Write an equation for the equilibrium reaction of acetic acid with each base. Which equilibria lie considerably toward the left? Which lie considerably toward the right?

(a) $NaHCO_3$ (b) NH_3 (c) H_2O (d) NaOH

4.20   Benzoic acid, C_6H_5COOH (pK_a 4.19), is only slightly soluble in water, but its sodium salt, $C_6H_5COO^- Na^+$, is quite soluble in water. In which solution(s) will benzoic acid dissolve?

(a) Aqueous NaOH   (b) Aqueous $NaHCO_3$ (c) Aqueous Na_2CO_3

4.21   4-Methylphenol, $CH_3C_6H_4OH$ (pK_a 10.26), is only slightly soluble in water, but its sodium salt, $CH_3C_6H_4O^- Na^+$, is quite soluble in water. In which solution(s) will 4-methyphenol dissolve?

(a) Aqueous NaOH   (b) Aqueous $NaHCO_3$ (c) Aqueous Na_2CO_3

4.22   One way to determine the predominant species at equilibrium for an acid-base reaction is to say that the reaction arrow points to the acid with the higher value of pK_a. For example,

$$NH_4^+ + H_2O \longleftarrow NH_3 + H_3O^+$$

$$pK_a \text{ 9.24} \qquad\qquad\qquad pK_a \text{ }-1.74$$

$$NH_4^+ + OH^- \longrightarrow NH_3 + H_2O$$

$$pK_a \text{ 9.24} \qquad\qquad\qquad pK_a \text{ 15.7}$$

Explain why this rule works.

4.23   Will acetylene react with sodium hydride according to the following equation to form a salt and hydrogen, H_2? Using pK_a values given in Table 4.1, calculate K_{eq} for this equilibrium.

$$HC{\equiv}CH \quad + \quad Na^+H^- \quad \longrightarrow \quad HC{\equiv}C^-Na^+ \quad + \quad H_2$$

$\quad$ Acetylene $\qquad\qquad$ Sodium $\qquad\qquad\qquad$ Sodium $\qquad$ Hydrogen
$\qquad\qquad\qquad\qquad$ hydride $\qquad\qquad\qquad$ acetylide

4.24   Using pK_a values given in Table 4.1, predict the position of equilibrium in this acid-base reaction, and calculate its K_{eq}.

$$H_3PO_4 + CH_3CH_2OH \rightleftharpoons H_2PO_4^- + CH_3CH_2OH_2^+$$

Lewis Acids and Bases

4.25   For each equation, label the Lewis acid and the Lewis base. In addition, show all unshared pairs of electrons on the reacting atoms, and use curved arrows to show the flow of electrons in each reaction.

(a) $F^- + BF_3 \longrightarrow BF_4^-$

(b) $\begin{array}{c} H \\ \diagdown \\ \diagup \\ H \end{array} C{=}O + H{-}Cl \longrightarrow \begin{array}{c} H \\ \diagdown \\ \diagup \\ H \end{array} C{=}O^+ \begin{array}{c} \\ \diagdown \\ H \end{array} + Cl^-$

4.26   Complete the equation for the reaction between each Lewis acid-base pair. In each equation, label which starting material is the Lewis acid and which the Lewis base; use curved arrows to show the flow of electrons in each reaction. In doing this problem, it is essential that you show valence electrons for all atoms participating in each reaction.

(a) $\begin{array}{c} CH_3 \\ | \\ CH_3{-}C{-}Cl \\ | \\ CH_3 \end{array} + \begin{array}{c} Cl \\ | \\ Al{-}Cl \\ | \\ Cl \end{array} \longrightarrow$

(b) $\begin{array}{c} CH_3 \\ | \\ CH_3{-}C^+ \\ | \\ CH_3 \end{array} + H{-}O{-}H \longrightarrow$

(c) $CH_3{-}\overset{+}{C}H{-}CH_3 + Br^- \longrightarrow$

(d) $CH_3{-}\overset{+}{C}H{-}CH_3 + CH_3{-}O{-}H \longrightarrow$

4.27 Each of these reactions can be written as a Lewis acid/Lewis base reaction. Label the Lewis acid and the Lewis base; use curved arrows to show the flow of electrons in each reaction. In doing this problem, it is essential that you show valence electrons for all atoms participating in each reaction.

(a) $CH_3-CH=CH_2 + H-Cl \longrightarrow CH_3-\overset{+}{C}H-\overset{\overset{\displaystyle H}{\displaystyle |}}{C}H_2 + Cl^-$

(b) $CH_3-\underset{\underset{\displaystyle CH_3}{\displaystyle |}}{C}=CH_2 + Br-Br \longrightarrow CH_3-\underset{\underset{\displaystyle CH_3}{\displaystyle |}}{\overset{+}{C}}-CH_2-Br + Br^-$

Additional Problems

4.28 The *sec*-butyl cation can react as both a Brønsted-Lowry acid (a proton donor) and a Lewis acid (an electron pair acceptor) in the presence of a water–sulfuric acid mixture. In each case, however, the product is different. The two reactions are

(1) $CH_3-\overset{+}{C}H-CH_2-CH_3 + H_2O \longrightarrow CH_3-\overset{\overset{\displaystyle H\diagdown\overset{+}{O}\diagup H}{\displaystyle |}}{C}H-CH_2-CH_3$

 sec-Butyl cation

(2) $CH_3-\overset{+}{C}H-CH_2-CH_3 + H_2O \longrightarrow CH_3-CH=CH-CH_3 + H_3O^+$

 sec-Butyl cation

(a) In which reaction(s) does this cation react as a Lewis acid? In which does it react as a Brønsted-Lowry acid?

(b) Write Lewis structures for reactants and products, and show by the use of curved arrows how each reaction occurs.

4.29 Write equations for the reaction of each compound with H_2SO_4, a strong protic acid.

(a) CH_3OCH_3 (b) $CH_3CH_2SCH_2CH_3$ (c) $CH_3CH_2NHCH_2CH_3$

(d) $CH_3\overset{\overset{\displaystyle CH_3}{\displaystyle |}}{N}CH_3$ (e) $CH_3\overset{\overset{\displaystyle O}{\displaystyle \|}}{C}CH_3$ (f) $CH_3\overset{\overset{\displaystyle O}{\displaystyle \|}}{C}OCH_3$

4.30 Write equations for the reaction of each compound in Problem 4.29 with BF_3, a Lewis acid.

4.31 Label the most acidic hydrogen in each molecule, and justify your choice by using appropriate pK_a values.

(a) $HOCH_2CH_2NH_2$ (b) $HSCH_2CH_2NH_2$ (c) $HOCH_2CH_2C\equiv CH$

(d) $HO\overset{\overset{\displaystyle O}{\displaystyle \|}}{C}CH_2CH_2SH$ (e) $CH_3\overset{\overset{\displaystyle HO\ \ O}{\displaystyle |\ \ \ \|}}{C}H\overset{}{C}OH$ (f) $H_3\overset{+}{N}CH_2CH_2\overset{\overset{\displaystyle O}{\displaystyle \|}}{C}OH$

(g) $H_3\overset{+}{N}CH_2CH_2\overset{\overset{\displaystyle O}{\displaystyle \|}}{C}O^-$ (h) $HSCH_2CH_2OH$

4.32 Explain why the hydronium ion, H_3O^+, is the strongest acid that can exist in aqueous solution. What is the strongest base that can exist in aqueous solution?

4.33 What is the strongest base that can exist in liquid ammonia as a solvent?

4.34 For each pair of molecules or ions, select the stronger base, and write its Lewis structure.

(a) CH_3S^- or CH_3O^- (b) CH_3NH^- or CH_3O^-
(c) CH_3COO^- or OH^- (d) $CH_3CH_2O^-$ or H^-
(e) NH_3 or OH^- (f) NH_3 or H_2O
(g) CH_3COO^- or HCO_3^- (h) HSO_4^- or OH^-
(i) OH^- or Br^-

4.35 Account for the fact that nitroacetic acid, O_2NCH_2COOH (pK_a 1.68), is a considerably stronger acid than acetic acid, CH_3COOH (pK_a 4.76).

4.36 Sodium hydride, NaH, is available commercially as a gray-white powder. It melts at 800°C with decomposition. It reacts explosively with water and ignites spontaneously on standing in moist air.

(a) Write a Lewis structure for the hydride ion and for sodium hydride. Is your Lewis structure consistent with the fact that this compound is a high-melting solid? Explain.
(b) When sodium hydride is added very slowly to water, it dissolves with the evolution of a gas. The resulting solution is basic to litmus. What is the gas evolved? Why has the solution become basic?
(c) Write an equation for the reaction between sodium hydride and 1-butyne, $CH_3CH_2C\equiv CH$. Use curved arrows to show the flow of electrons in this reaction.

4.37 Methyl isocyanate, $CH_3-N=C=O$, is used in the industrial synthesis of a type of pesticide and herbicide known as a carbamate. As a historical note, an industrial accident in Bhopal, India, in 1984 resulted in leakage of an unknown quantity of this chemical into the air. An estimated 200,000 persons were exposed to its vapors, and over 2000 of these people died.

(a) Write a Lewis structure for methyl isocyanate, and predict its bond angles. What is the hybridization of its carbonyl carbon? Of its nitrogen atom?
(b) Methyl isocyanate reacts with strong acids, such as sulfuric acid, to form a cation. Will this molecule undergo protonation more readily on its oxygen or nitrogen atom? In considering contributing structures to each hybrid, do not consider structures in which more than one atom has an incomplete octet.

4.38 Offer an explanation for the following observations.

(a) H_3O^+ is a stronger acid than NH_4^+.
(b) Nitric acid, HNO_3, is a stronger acid than nitrous acid, HNO_2.
(c) Ethanol and water have approximately the same acidity.
(d) Trifluoroacetic acid, CF_3COOH, is a stronger acid than trichloroacetic acid, CCl_3COOH.

Looking Ahead

4.39 Following is a structural formula for the *tert*-butyl cation. (We discuss the formation, stability, and reactions of cations such as this one in Chapter 6.)

$$CH_3-\overset{+}{\underset{\underset{CH_3}{|}}{C}}-CH_3$$

tert-Butyl cation
(a carbocation)

(a) Predict all C—C—C bond angles in this cation.
(b) What is the hybridization of the carbon bearing the positive charge?
(c) Write a balanced equation to show its reaction as a Lewis acid with water.
(d) Write a balanced equation to show its reaction as a Brønsted-Lowry acid with water.

4.40 Alcohols (Chapter 10) are weak organic acids, pK_a 15–18. The pK_a of ethanol, CH_3CH_2OH, is 15.9. Write equations for the equilibrium reactions of ethanol with each base. Which equilibria lie considerably toward the right? Which lie considerably toward the left?

(a) $NaHCO_3$ (b) $NaOH$ (c) $NaNH_2$ (d) NH_3

4.41 As we shall see in Chapter 19, hydrogens on a carbon adjacent to a carbonyl group are far more acidic than those not adjacent to a carbonyl group. The anion derived from acetone, for example, is more stable than is the anion derived from ethane. Account for the greater stability of the anion from acetone.

$$\overset{\overset{\displaystyle O}{\|}}{CH_3CCH_2}-H \qquad CH_3CH_2-H$$

Acetone   Ethane
pK_a 22 pK_a 51

4.42 2,4-Pentanedione is a considerably stronger acid than acetone (Chapter 19). Write a structural formula for the conjugate base of each acid, and account for the greater stability of the conjugate base from 2,4-pentanedione.

$$\overset{\overset{\displaystyle O}{\|}}{CH_3CCH_2}-\boxed{H} \qquad \overset{\overset{\displaystyle O \quad O}{\| \quad \|}}{CH_3CCHCCH_3}$$
$$\boxed{H}$$

Acetone   2,4-Pentanedione
pK_a 22 pK_a 9

4.43 Write an equation for the acid-base reaction between 2,4-pentanedione and sodium ethoxide, and calculate its equilibrium constant, K_{eq}. The pK_a of 2,4-pentanedione is 9; that of ethanol is 15.9.

$$\overset{\overset{\displaystyle O \quad O}{\| \quad \|}}{CH_3CCHCCH_3} + CH_3CH_2O^-Na^+ \rightleftharpoons$$
$$\boxed{H}$$

2,4-Pentanedione   Sodium ethoxide

4.44 An ester is a derivative of a carboxylic acid in which the hydrogen of the carboxyl group is replaced by an alkyl group (Section 1.3E). Draw a structural formula of methyl acetate, which is derived from acetic acid by replacement of the H of its —OH group by a methyl group. Determine if proton transfer to this compound from HCl occurs preferentially on the oxygen of the C=O group or the oxygen of the OCH_3 group.

4.45 Alanine is one of the 20 amino acids (it contains both an amino and a carboxyl group) found in proteins (Chapter 27). Is alanine better represented by the structural formula A or B? Explain.

$$CH_3-\overset{\overset{\displaystyle }{\underset{\underset{\displaystyle NH_2}{|}}{CH}}}{}-\overset{\overset{\displaystyle O}{\|}}{C}-OH \qquad CH_3-\overset{\overset{\displaystyle }{\underset{\underset{\displaystyle NH_3^+}{|}}{CH}}}{}-\overset{\overset{\displaystyle O}{\|}}{C}-O^-$$

(A)   (B)

4.46 Glutamic acid is another of the amino acids found in proteins (Chapter 27). Glutamic acid has two carboxyl groups, one with pK_a 2.10 and the other with pK_a 4.07.

$$\text{Glutamic acid} \quad \underset{}{\text{HO}}-\overset{\overset{\text{O}}{\|}}{\text{C}}-\text{CH}_2-\text{CH}_2-\underset{\underset{\text{NH}_3{}^+}{|}}{\text{CH}}-\overset{\overset{\text{O}}{\|}}{\text{C}}-\text{OH}$$

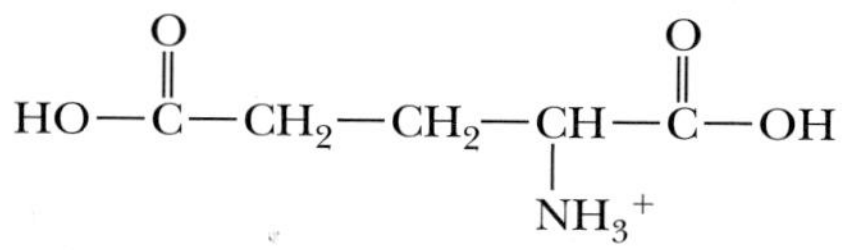

(a) Which carboxyl group has which pK_a?

(b) Account for the fact that one carboxyl group is a considerably stronger acid than the other.

4.47 Following is a structural formula for imidazole, a building block of the essential amino acid histidine (Chapter 27). It is also a building block of histamine, a compound all too familiar to persons with allergies and takers of antihistamines. When imidazole is dissolved in water, proton transfer to it gives a cation. Is this cation better represented by structure A or B? Explain.

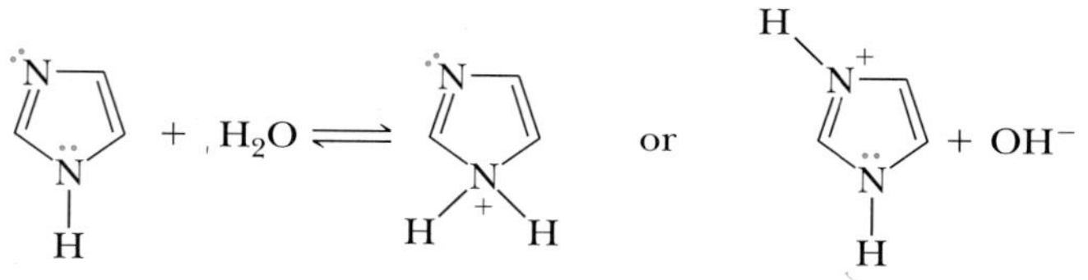

Imidazole   + H$_2$O ⇌   A   or   B   + OH$^-$

© Mark Muench/Stone/Getty Images

5

■ Haze in the Blue Ridge Mountains. This haze is caused by light-scattering from the aerosol caused by the photo-oxidation of isoprene and other hydrocarbons emitted by trees and other plants. Many naturally occurring hydrocarbons are formed from isoprene units (see Section 5.4). Inset: A model of isoprene.

Outline

Alkenes: Bonding, Nomenclature, and Properties

An **unsaturated hydrocarbon** contains one or more carbon-carbon double or triple bonds. The term "unsaturation" indicates that there are fewer hydrogens bonded to carbon than in an alkane, C_nH_{2n+2}. The three classes of unsaturated hydrocarbons are alkenes, alkynes, and arenes. Alkenes contain a carbon-carbon double bond and, with one double bond and no rings, have the general formula C_nH_{2n}. Alkynes contain a carbon-carbon triple bond and, with one triple bond and no rings, have the general formula C_nH_{2n-2}. The simplest alkene is ethylene, and the simplest alkyne is acetylene.

Unsaturated hydrocarbon
A hydrocarbon containing one or more carbon-carbon double or triple bonds. The three classes of unsaturated hydrocarbons are alkenes, alkynes, and arenes.

ORGANIC **Chemistry•Now™**
Click Molecular Models to view and manipulate **Models of Alkenes.**

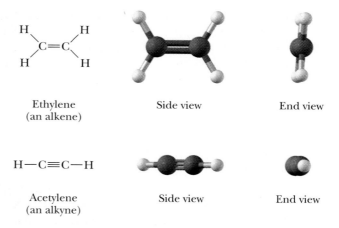

Ethylene
(an alkene)

Side view

End view

Acetylene
(an alkyne)

Side view

End view

In this chapter, we study the structure, nomenclature, and physical properties of alkenes. Alkynes are discussed separately in Chapter 7.

Arene A term used to classify benzene and its derivatives.

Arenes are the third class of unsaturated hydrocarbons. The Lewis structure of benzene, the simplest arene, is

ORGANIC **Chemistry•Now™**
Click Molecular Models to view and manipulate **Models of Arenes.**

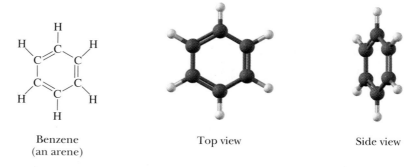

Benzene
(an arene)

Top view

Side view

Aryl group (Ar—) A group derived from an arene by removal of an H.

Phenyl group A group derived by removing an H from benzene; abbreviated C_6H_5— or Ph—.

Just as a group derived by removal of an H from an alkane is called an alkyl group and given the symbol R—(Section 2.3A), a group derived by removal of an H from an arene is called an **aryl group** and given the symbol **Ar—.**

When a benzene ring occurs as a substituent on a parent chain, it is named a **phenyl group.** You might think that, when present as a substituent, benzene would become benzyl, just as ethane becomes ethyl. This is not so! "Phene" is a now-obsolete name for benzene, and, although this name is no longer used, a derivative has persisted in the name "phenyl." Following is a structural formula for the phenyl group and two alternative representations for it. Throughout this text, we will represent benzene by a hexagon with three inscribed double bonds. It is also common to represent it by a hexagon with an inscribed circle. We explain the reasons for the alternative representations in Chapter 21.

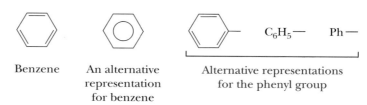

Benzene

An alternative representation for benzene

C_6H_5—   Ph—

Alternative representations for the phenyl group

The chemistry of benzene and its derivatives is quite different from that of alkenes and alkynes, but, even though we do not study the chemistry of arenes until Chapters 21 and 22, we will show structural formulas of compounds containing aryl groups before that time. What you need to remember at this point is that an aryl group is not chemically reactive under any of the conditions we describe in Chapters 6 through 20.

5.1 Structure of Alkenes

A. Shapes of Alkenes

Using valence-shell electron-pair repulsion (Section 1.4) for a carbon-carbon double bond, we predict a value of 120° for the bond angles about each carbon in a double bond. The observed H—C—C bond angle in ethylene is 121.1°, close to that predicted. In other alkenes, deviations from the predicted angle of 120° may be somewhat larger because of the strain introduced by nonbonded interactions created by groups attached to the carbons of the double bond. The C—C—C bond angle in propene, for example, is 123.9°.

Ethylene   Propene

B. Carbon-Carbon Double Bond Orbitals

In Section 1.8D, we described the formation of a carbon-carbon double bond in terms of the overlap of atomic orbitals. A carbon-carbon double bond consists of one sigma bond and one pi bond (Figure 5.1). Each carbon of the double bond uses its three sp^2 hybrid orbitals to form sigma bonds to three atoms. The unhybridized $2p$ atomic orbitals, which lie perpendicular to the plane created by the axes of the three sp^2 hybrid orbitals, combine to form two pi molecular orbitals: one bonding and the other antibonding. For the unhybridized $2p$ orbitals to be parallel, thus giving maximum overlap, the two carbon atoms of the double bond and the four attached atoms must lie in a plane.

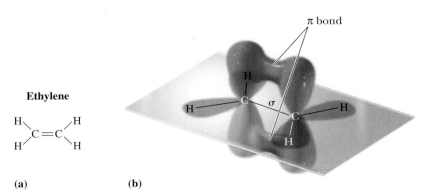

Ethylene

(a)   (b)

Figure 5.1
Covalent bonding in ethylene. (a) Lewis structure and (b) orbital overlap model showing the sigma and pi bonds.

How To *Calculate the Index of Hydrogen Deficiency*

ORGANIC
Chemistry Now™

Click How To Tutorials to calculate the **Index of Hydrogen Deficiency** for some compounds

Index of hydrogen deficiency
The sum of the number of rings and pi bonds in a molecule.

Valuable information about the structural formula of an unknown compound can be obtained by inspecting its molecular formula. In addition to learning the number of atoms of carbon, hydrogen, oxygen, nitrogen, and so forth in a molecule of the compound, we can also determine what is called its index of hydrogen deficiency. For each ring and pi bond, the molecular formula has two fewer hydrogens. The **index of hydrogen deficiency** is the sum of the number of rings and pi bonds in a molecule. It is determined by comparing the number of hydrogens in the molecular formula of a compound whose structure is to be determined ($H_{molecule}$) with the number of hydrogens in a reference alkane of the same number of carbon atoms ($H_{reference}$). The molecular formula of a reference acyclic alkane is C_nH_{2n+2} (Section 2.1).

$$\text{Index of hydrogen deficiency} = \frac{(H_{reference} - H_{molecule})}{2}$$

To compare the molecular formula for a compound containing elements besides carbon and hydrogen, write the formula of the reference hydrocarbon with the same number of carbon atoms, and make the following adjustments to the number of hydrogen atoms in the unknown.

1. Replace each monovalent atom of a Group 7 element (F, Cl, Br, I) with one hydrogen; halogen substitutes for hydrogen and reduces the number of hydrogens by one per halogen. The general formula of an acyclic monochloroalkane, for example, is $C_nH_{2n+1}Cl$; the general formula of the corresponding acyclic alkane is C_nH_{2n+2}.
2. No correction is necessary for the addition of divalent atoms of Group 6 elements (O, S, Se). Insertion of a divalent Group 6 element into a hydrocarbon does not change the number of hydrogens.
3. For each atom of a trivalent Group 5 element (N, P, As) present, add one hydrogen, because insertion of a trivalent Group 5 element adds one hydrogen to the molecular formula. The general molecular formula for an acyclic alkylamine, for example, is $C_nH_{2n+3}N$.

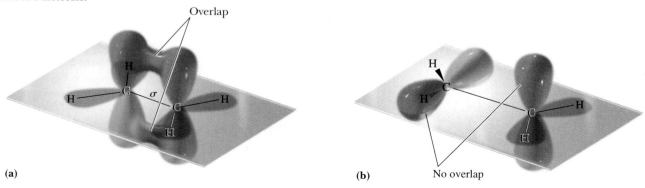

(a) **(b)**

Figure 5.2
Restricted rotation about a carbon-carbon double bond. (a) Orbital overlap model showing the pi bond. (b) The pi bond is broken by rotating the plane of one H—C—H group by 90° with respect to the plane of the other H—C—H group.

It takes approximately 264 kJ (63 kcal)/mol to break the pi bond in ethylene, that is, to rotate one carbon by 90° with respect to the other where zero overlap occurs between $2p$ orbitals on adjacent carbons (Figure 5.2). This energy is considerably greater than the thermal energy available at room temperature; consequently, rotation about a carbon-carbon double bond does not occur under normal conditions. You might compare rotation about a carbon-carbon double bond, such as in ethylene, with that about a carbon-carbon single bond, such as in ethane (Section 2.6A). Whereas rotation about the carbon-carbon single bond in ethane is relatively free [rotation barrier approximately 12.5 kJ (3.0 kcal)/mol], rotation about the carbon-carbon double bond in ethylene is severely restricted.

Example 5.1

Calculate the index of hydrogen deficiency for 1-hexene, C_6H_{12}, and account for this deficiency by reference to its structural formula.

Solution

The molecular formula of the reference acyclic alkane of six carbon atoms is C_6H_{14}. The index of hydrogen deficiency of 1-hexene $(14 - 12)/2 = 1$ and is accounted for by the one pi bond in 1-hexene.

Problem 5.1

Calculate the index of hydrogen deficiency of cyclohexene, C_6H_{10}, and account for this deficiency by reference to its structural formula.

Example 5.2

Isopentyl acetate, a compound with a banana-like odor, is a component of the alarm pheromone of honeybees. The molecular formula of isopentyl acetate is $C_7H_{14}O_2$. Calculate the index of hydrogen deficiency of this compound.

Solution

The molecular formula of the reference hydrocarbon is C_7H_{16}. Adding oxygens does not require any correction in the number of hydrogens. The index of hydrogen deficiency is $(16 - 14)/2 = 1$, indicating either one ring or one pi bond. Following is the structural formula of isopentyl acetate. It contains one pi bond, in this case, a carbon-oxygen pi bond.

Isopentyl acetate

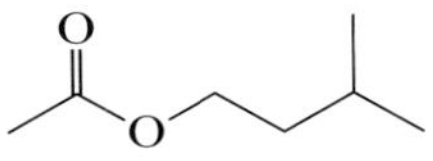

Problem 5.2

The index of hydrogen deficiency of niacin is 5. Account for this index of hydrogen deficiency by reference to the structural formula of niacin.

Nicotinamide
(Niacin)

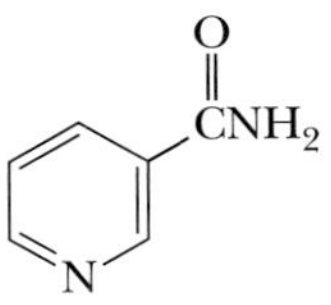

C. *Cis,Trans* Isomerism in Alkenes

Because of restricted rotation about a carbon-carbon double bond, any alkene in which each carbon of the double bond has two different groups attached to it shows ***cis,trans* isomerism.** For example, 2-butene has two stereoisomers. In *cis-*2-butene, the two methyl groups are on one side of the double bond, and the two hydrogens are on the other side. In *trans-*2-butene, the two methyl groups are on opposite sides of the double bond. These two compounds cannot be converted into one another at room temperature because of the restricted rotation about the double bond; they are different compounds, with different physical and chemical properties.

Cis,trans **isomers** Isomers that have the same order of attachment of their atoms but a different arrangement of their atoms in space owing to the presence of either a ring (Section 2.7) or a carbon-carbon double bond (Section 5.1C).

*cis-*2-Butene
mp −139°C, bp 4°C

*trans-*2-Butene
mp −106°C, bp 1°C

steric strain

Cis alkenes with double bonds in open chains are less stable than their *trans* isomers because of steric strain between alkyl substituents on the same side of the double bond, as can be seen in space-filling models of the *cis* and *trans* isomers of 2-butene. This is the same type of strain that results in the preference for equatorial methylcyclohexane over axial methylcyclohexane (Section 2.6B). *Trans-*2-butene is more stable than the *cis* isomer by about 5.8 kJ (1.4 Kcal)/mol because of the sum of steric strain and angle strain that results from the two methyls moving apart. At the energy minimum, the $C{=}C{-}CH_3$ angle is about 127°.

5.2 Nomenclature of Alkenes

Alkenes are named using the IUPAC system, but, as we shall see, some are usually referred to by their common names.

A. IUPAC Names

To form IUPAC names for alkenes, change the *-an-* infix of the parent alkane to *-en-* (Section 2.5). Hence, $CH_2{=}CH_2$ is named ethene, and $CH_3CH{=}CH_2$ is named propene. In higher alkenes, where isomers exist that differ in location of the double bond, a numbering system must be used. According to the IUPAC system,

1. Number the longest carbon chain that contains the double bond in the direction that gives the carbon atoms of the double bond the lowest possible numbers.
2. Indicate the location of the double bond by the number of its first carbon.

3. Name branched or substituted alkenes in a manner similar to alkanes.
4. Number the carbon atoms, locate and name substituent groups, locate the double bond, and name the main chain.

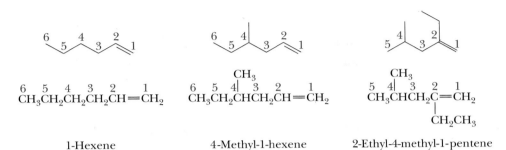

$\overset{6}{CH_3}\overset{5}{CH_2}\overset{4}{CH_2}\overset{3}{CH_2}\overset{2}{CH}=\overset{1}{CH_2}$	$\overset{6}{CH_3}\overset{5}{CH_2}\overset{4}{CH}\overset{3}{CH_2}\overset{2}{CH}=\overset{1}{CH_2}$ with CH_3 on C4	$\overset{5}{CH_3}\overset{4}{CH}\overset{3}{CH_2}\overset{2}{C}=\overset{1}{CH_2}$ with CH_3 on C4 and CH_2CH_3 on C2
1-Hexene	4-Methyl-1-hexene	2-Ethyl-4-methyl-1-pentene

Note that there is a chain of six carbon atoms in 2-ethyl-4-methyl-1-pentene. However, because the longest chain that contains the double bond has only five carbons, the parent hydrocarbon is pentane, and the molecule is named as a disubstituted 1-pentene.

Example 5.3

Write the IUPAC name of each alkene.

(a) [structure]   (b) [structure]

Solution

(a) 4-Ethyl-3,3-dimethyl-1-octene   (b) 2-Methyl-2-butene

Problem 5.3

Write the IUPAC name of each alkene.

(a) 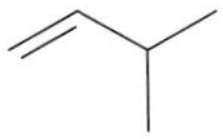 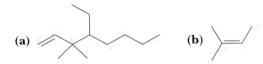   (b) 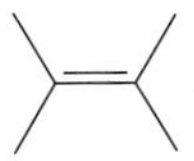 [structure]

B. Common Names

Some alkenes, particularly those of low molecular weight, are known almost exclusively by their common names, as illustrated by the common names of these alkenes.

	$CH_2=CH_2$	$CH_3CH=CH_2$	$CH_3\overset{\overset{\displaystyle CH_3}{|}}{C}=CH_2$
IUPAC name:	Ethene	Propene	2-Methylpropene
Common name:	Ethylene	Propylene	Isobutylene

Methylene   A CH_2 group.

Vinyl   A $CH=CH_2$ group.

Allyl   A $CH_2CH=CH_2$ group.

Furthermore, the common names **methylene** (a CH_2 group), **vinyl,** and **allyl** are often used to show the presence of the following alkenyl groups:

Alkenyl Group	IUPAC Name	Common Name	Example	IUPAC Name (Common Name)
$CH_2=$	Methylidene	Methylene	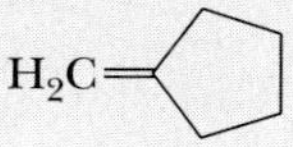 $H_2C=$⬠	Methylidenecyclopentane (Methylenecyclopentane)
$CH_2=CH—$	Ethenyl	Vinyl	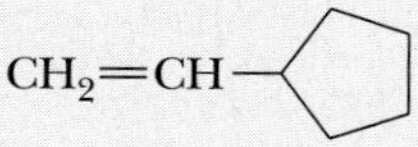 $CH_2=CH—$⬠	Ethenylcyclopentane (Vinylcyclopentane)
$CH_2=CHCH_2—$	2-Propenyl	Allyl	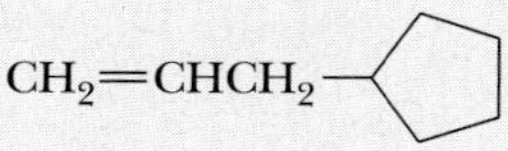 $CH_2=CHCH_2—$⬠	2-Propenylcyclopentane (Allylcyclopentane)

C. Systems for Designating Configuration in Alkenes

The *Cis, Trans* System

The most common method for specifying the configuration in alkenes uses the prefixes *cis* and *trans*. There is no doubt whatsoever which isomer is intended by the name *trans*-3-hexene. For more complex alkenes, the orientation of the atoms of the parent chain determines whether the alkene is *cis* or *trans*. On the right is a structural formula for the *cis* isomer of 3,4-dimethyl-2-pentene. In this example, carbon atoms of the main chain (carbons 1 and 4) are on the same side of the double bond and, therefore, this alkene is *cis*.

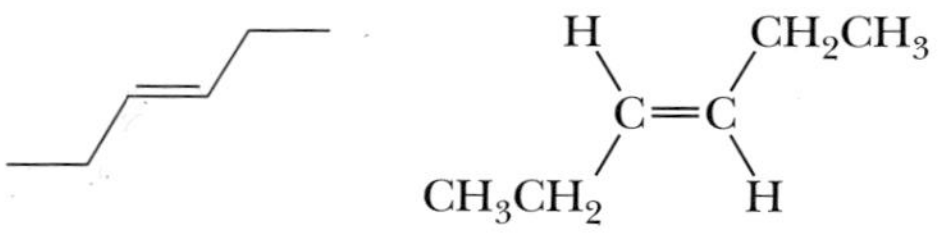

trans-3-Hexene

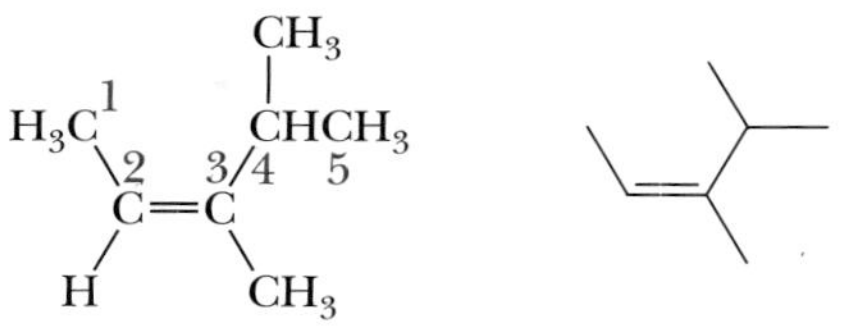

cis-3,4-Dimethyl-2-pentene

Example 5.4

Name each alkene and show the configuration about each double bond using the *cis, trans* system.

(a)

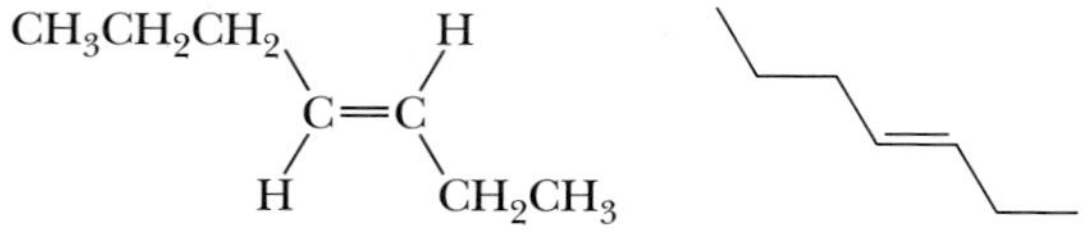

(b)

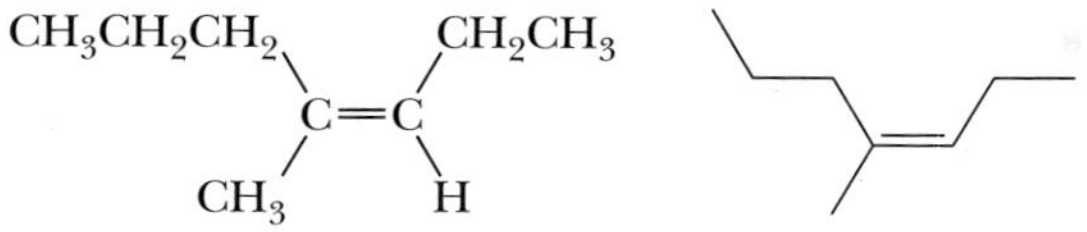

Solution

(a) The chain contains seven carbon atoms and is numbered from the end that gives the lower number to the first carbon of the double bond. Its name is *trans*-3-heptene.

(b) The longest chain contains seven carbon atoms and is numbered from the right so that the first carbon of the double bond is carbon 3 of the chain. Its name is *cis*-4-methyl-3-heptene.

Problem 5.4

Which alkenes show *cis,trans* isomerism? For each alkene that does, draw the *trans* isomer.

(a) 2-Pentene **(b)** 2-Methyl-2-pentene **(c)** 3-Methyl-2-pentene

The *E,Z* System

Because the *cis,trans* system becomes confusing with tri- and tetrasubstituted alkenes [see Problem 5.4(c) and Example 5.4(b)], and is not detailed enough to name all alkenes, chemists developed the **E,Z system**. This system uses the priority rules of the *R,S* system (Section 3.3) to assign priority to the substituents on each carbon of a double bond. Using these rules, we decide which group on each carbon has the higher priority. If the groups of higher priority are on the same side of the double bond, the configuration of the alkene is **Z** (German: *zusammen*, together). If they are on opposite sides of the double bond, the alkene is **E** (German: *entgegen*, opposite).

E,Z system A system to specify the configuration of groups about a carbon-carbon double bond.

Z From the German, *zusammen*, together. Specifies that groups of higher priority on the carbons of a double bond are on the same side.

E From the German, *entgegen*, opposite. Specifies that groups of higher priority on the carbons of a double bond are on opposite sides.

Z (*zusammen*)   E (*entgegen*)

Throughout this text, we use the *cis,trans* system for alkenes in which it is clear which is the main carbon chain. We use the *E,Z* system in all other cases. It should always be used if confusion is possible.

Example 5.5

Name each alkene and specify its configuration by the *E,Z* system.

(a)   **(b)**   **(c)**

Solution

(a) The group of higher priority on carbon 2 is methyl; that of higher priority on carbon 3 is isopropyl. Because the groups of higher priority are on the same side of the double bond, the alkene has the *Z* configuration. Its name is (*Z*)-3,4-dimethyl-2-pentene. Using the *cis,trans* system, its name is *cis*-3,4-dimethyl-2-pentene.
(b) Groups of higher priority on carbons 2 and 3 are —Cl and —CH_2CH_3. Because these groups are on opposite sides of the double bond, the configuration of this alkene is *E*, and its name is (*E*)-2-chloro-2-pentene. Using the *cis,trans* system, it is *cis*-2-chloro-2-pentene.

(c) The groups of higher priority are on opposite sides of the double bond; the configuration is *E*. The name of this bromoalkene is (*E*)-1-bromo-4-isopropyl-5-methyl-4-octene. Using the *cis,trans* system, it is *cis*-1-bromo-4-isopropyl-5-methyl-4-octene.

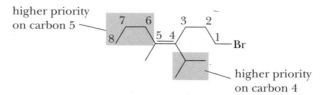

higher priority on carbon 5

higher priority on carbon 4

Problem 5.5

Name each alkene and specify its configuration by the *E,Z* system.

(a) 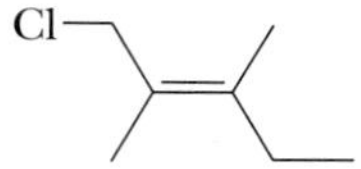 **(b)** 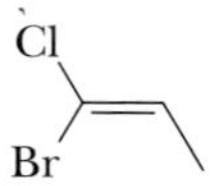 **(c)**

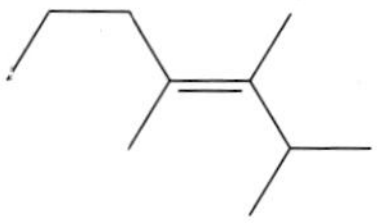

D. Cycloalkenes

In naming **cycloalkenes,** the carbon atoms of the ring double bond are numbered 1 and 2 in the direction that gives the substituent encountered first the smaller number.

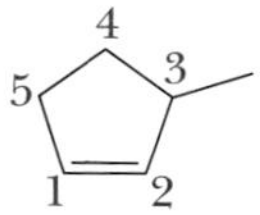

3-Methylcyclopentene 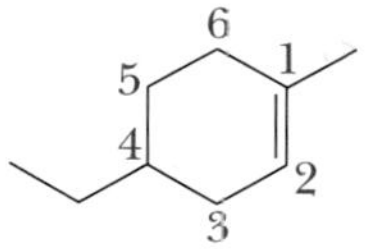   4-Ethyl-1-methylcyclohexene 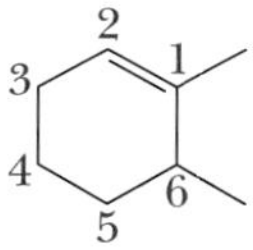   1,6-Dimethylcyclohexene

Example 5.6

Write the IUPAC and common name of each cycloalkene.

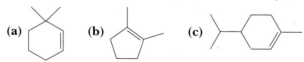

(a)   **(b)**   **(c)**

Solution

(a) 3,3-Dimethylcyclohexene   **(b)** 1,2-Dimethylcyclopentene
(c) 4-(1-methylethyl)-1-methylcyclohexene, (4-Isopropyl-1-methylcyclohexene)

Problem 5.6

Write the IUPAC and common name of each cycloalkene.

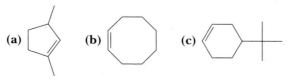

(a)   **(b)**   **(c)**

E. *Cis,Trans* Isomerism in Cycloalkenes

Following are structural formulas for four cycloalkenes:

Cyclopentene   Cyclohexene   Cycloheptene   Cyclooctene

In these representations, the configuration about each double bond is *cis*. Is it possible to have a *trans* configuration in these and larger cycloalkenes? To date, *trans*-cyclooctene is the smallest *trans* cycloalkene that has been prepared in pure form and is stable at room temperature. Yet, even in this *trans* cycloalkene, there is considerable angle strain; the double bond's *p* orbitals make an angle of 44° to each other. *Cis*-cyclooctene is more stable than its *trans* isomer by 38 kJ (9.1 kcal)/mol. Note that the *trans* isomer is chiral even though it has no chiral center.

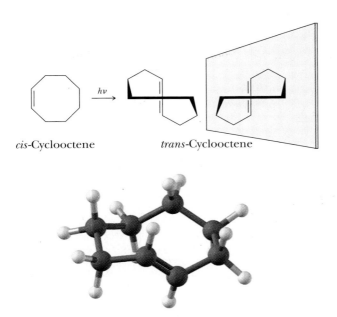

cis-Cyclooctene   *trans*-Cyclooctene

F. Dienes, Trienes, and Polyenes

For alkenes containing two or more double bonds, the infix *-en-* is changed to *-adien-*, *-atrien-*, and so on. Those that contain several double bonds are also referred to more generally as polyenes (Greek: *poly,* many). Following are examples of three dienes.

1,4-Pentadiene   2-Methyl-1,3-butadiene   1,3-Cyclopentadiene
                         (Isoprene)

G. *Cis,Trans* Isomerism in Dienes, Trienes, and Polyenes

Thus far we have considered *cis,trans* isomerism in alkenes containing only one carbon-carbon double bond. For an alkene with one carbon-carbon double bond that can show *cis,trans* isomerism, two stereoisomers are possible. For an alkene with *n* carbon-carbon double bonds, each of which can show *cis,trans* isomerism, 2^n stereoisomers are possible.

Example 5.7

How many stereoisomers are possible for 2,4-heptadiene?

Solution

This molecule has two carbon-carbon double bonds, each of which shows *cis,trans* isomerism. As shown in this table, there are $2^2 = 4$ stereoisomers. Two of these are drawn on the right.

Double Bond	
C_2-C_3	C_4-C_5
trans	*trans*
trans	*cis*
cis	*trans*
cis	*cis*

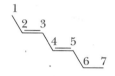

(2*E*,4*E*)-2,4-Heptadiene
trans,trans-2,4-Heptadiene

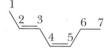

(2*E*,4*Z*)-2,4-Heptadiene
trans,cis-2,4-Heptadiene

Problem 5.7

Draw structural formulas for the other two stereoisomers of 2,4-heptadiene.

Example 5.8

How many stereoisomers are possible for 10,12-hexadecadien-1-ol?

$$CH_3(CH_2)_2CH=CHCH=CH(CH_2)_8CH_2OH$$

10,12-Hexadecadien-1-ol

Solution

Cis,trans isomerism is possible about both double bonds. Four stereoisomers are possible.

Problem 5.8

(10*E*,12*Z*)-10,12-hexadecadien-1-ol is a sex pheromone of the silkworm. Draw a structural formula for this compound.

■ Silkworms spinning cocoons on a loom, silk farm, Japan.

© Paul Chesley/Stone/Getty Images

An example of a biologically important compound for which a number of *cis,trans* isomers are possible is vitamin A. There are four carbon-carbon double

CHEMICAL CONNECTIONS

The Case of the Iowa and New York Strains of the European Corn Borer

Although humans communicate largely by sight and sound, chemicals are the primary means of communication for the vast majority of other species in the animal world. Often, communication within a species is specific for one of two or more configurational isomers. For example, a member of a given species may respond to a *cis* isomer of a chemical but not to the *trans* isomer. Or, alternatively, it might respond to a quite precise blend of *cis* and *trans* isomers but not to other blends of these same isomers.

Several groups of scientists have studied the components of the sex pheromones of both the Iowa and the New York strains of the European corn borer. Females of these closely related species secrete the sex attractant 11-tetradecenyl acetate. Males of the Iowa strain show maximum response to a mixture containing 96% of the *cis* isomer and 4% of the *trans* isomer. When the pure *cis* isomer is used alone, males are only weakly attracted. Males of the New York strain show an entirely different response. They respond maximally to a

■ The European corn borer, *Pyrausta nubilalis*.

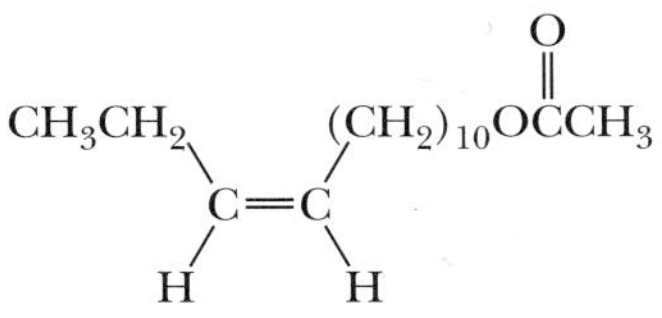

mixture containing 3% of the *cis* isomer and 97% of the *trans* isomer.

cis-11-Tetradecenyl acetate

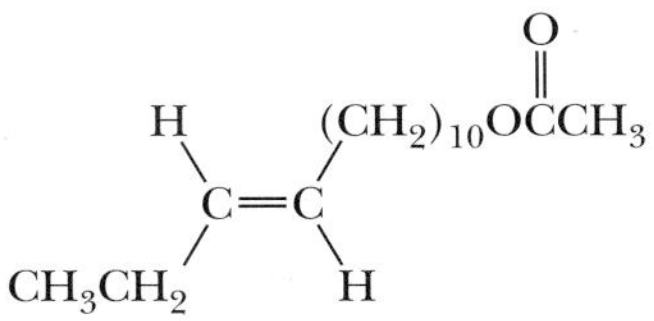

trans-11-Tetradecenyl acetate

There is evidence that optimum response to a narrow range of stereoisomers as we see here is widespread in nature and that a great many insects maintain species isolation for mating and reproduction by the stereochemistry of their pheromones.

bonds in the chain of carbon atoms bonded to the substituted cyclohexene ring, and each has the potential for *cis,trans* isomerism. There are $2^4 = 16$ stereoisomers possible for this structural formula. Vitamin A is the all-*E* (all-*trans*) isomer.

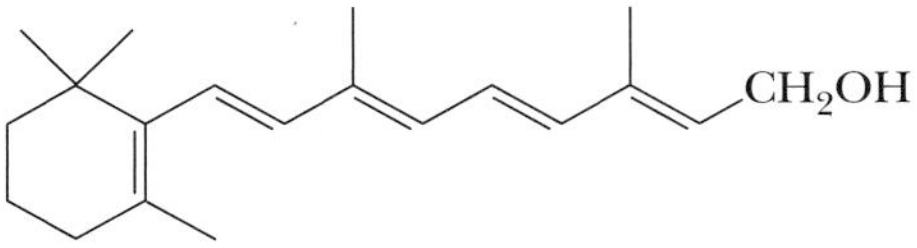

Vitamin A (retinol)

Table 5.1   Physical Properties of Some Alkenes

Name	Structural Formula	mp (°C)	bp (°C)
Ethylene	$CH_2{=}CH_2$	−169	−104
Propene	$CH_3CH{=}CH_2$	−185	−47
1-Butene	$CH_3CH_2CH{=}CH_2$	−185	−6
1-Pentene	$CH_3CH_2CH_2CH{=}CH_2$	−138	30
cis-2-Pentene		−151	37
trans-2-Pentene		−156	36
2-Methyl-2-butene		−134	39

5.3 Physical Properties of Alkenes

Alkenes are nonpolar compounds, and the only attractive forces between their molecules are dispersion forces (Section 2.8B). Their physical properties, therefore, are similar to those of alkanes. Alkenes of two, three, and four carbon atoms are gases at room temperature. Those of five or more carbons are colorless liquids that are less dense than water. Alkenes are insoluble in water but soluble in one another, in other nonpolar organic liquids, and in ethanol. Table 5.1 lists physical properties of some alkenes.

5.4 Naturally Occurring Alkenes— Terpene Hydrocarbons

Terpene A compound whose carbon skeleton can be divided into two or more units identical with the carbon skeleton of isoprene.

A **terpene** is a compound whose carbon skeleton can be divided into two or more units that are identical with the carbon skeleton of isoprene. Carbon 1 of an isoprene unit is called the head; carbon 4 is called the tail. Terpenes are formed by bonding the tail of one isoprene unit to the head of another. This is called the **isoprene rule.**

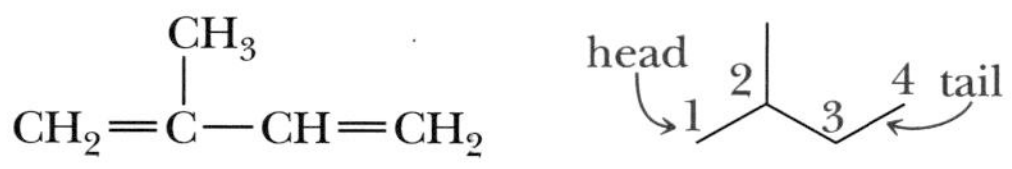

$$CH_2{=}\underset{\underset{CH_3}{|}}{C}{-}CH{=}CH_2$$

2-Methyl-1,3-butadiene
(Isoprene)

A study of terpenes provides a glimpse of the wondrous diversity that nature can generate from a simple carbon skeleton. Terpenes also illustrate an important principle of the molecular logic of living systems, namely, that in building large molecules, small subunits are bonded together enzymatically by an iterative process and then modified by subsequent precise enzyme-catalyzed reactions. Chemists use the same principles in the laboratory, but our methods do not have the precision and selectivity of the enzyme-catalyzed reactions of cellular systems.

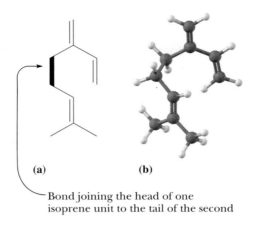

Figure 5.3
Myrcene. (a) Structural formula and (b) ball-and-stick model.

(a)   **(b)**

Bond joining the head of one
isoprene unit to the tail of the second

Probably the terpenes most familiar to you, at least by odor, are components of the so-called essential oils obtained by steam distillation or ether extraction of various parts of plants. Essential oils contain the relatively low-molecular-weight substances that are in large part responsible for characteristic plant fragrances. Many essential oils, particularly those from flowers, are used in perfumes.

One example of a terpene obtained from an essential oil is myrcene, $C_{10}H_{16}$, a component of bayberry wax and oils of bay and verbena. Myrcene is a triene with a parent chain of eight carbon atoms and two one-carbon branches [Figure 5.3(a)].

Head-to-tail bonds between isoprene units are vastly more common in nature than are the alternative head-to-head or tail-to-tail patterns. Figure 5.4 shows structural formulas of five more terpenes, all derived from two isoprene units. Geraniol has the same carbon skeleton as myrcene. In the last four terpenes of Figure 5.4, the carbon atoms

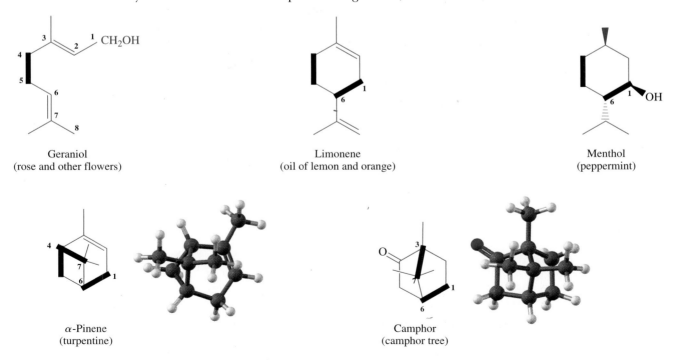

Geraniol
(rose and other flowers)

Limonene
(oil of lemon and orange)

Menthol
(peppermint)

α-Pinene
(turpentine)

Camphor
(camphor tree)

Figure 5.4
Five terpenes, each divisible into two isoprene units.

CONNECTIONS TO
BIOLOGICAL CHEMISTRY

The Importance of Cis Double Bonds in Fats Versus Oils

Fats and oils are very similar in that both are triesters of glycerol, hence the name *triglyceride*. Hydrolysis of a triglyceride in aqueous base followed by acidification gives glycerol and three carboxylic acids. Because these carboxylic acids can be derived from fats, they are called *fatty acids*:

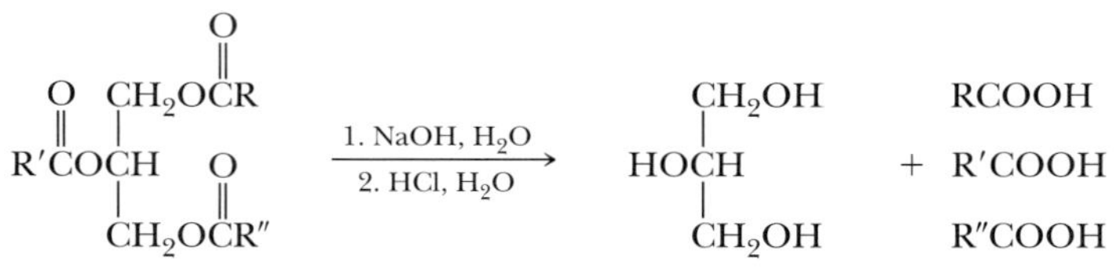

A triglyceride
(a triester of glycerol)

1,2,3-Propanetriol
(Glycerol)

Fatty acids

The most common fatty acids have between 12 and 24 carbon atoms in an unbranched chain.

The main difference between fats and oils is the temperature at which they melt. Fats are solids or semisolids at or near room temperature, while oils are liquids. The different physical properties of fats and oils result from the presence of different fatty acids.

Fatty acids with no double bonds are referred to as saturated fatty acids, those with a single double bond are called monounsaturated fatty acids, and those with more than one double bond are called polyunsaturated fatty acids. The double bonds in almost all naturally occurring fatty acids have *cis* configurations. The triglycerides of animal fats are richer in saturated fatty acids, whereas the triglycerides of plant oils (for example, corn, soybean, canola, olive, and palm oils) are richer in unsaturated fatty acids.

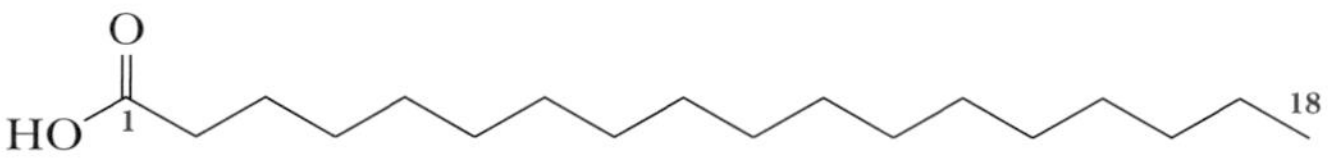

Stearic acid
mp 70°C
(a saturated C_{18} fatty acid)

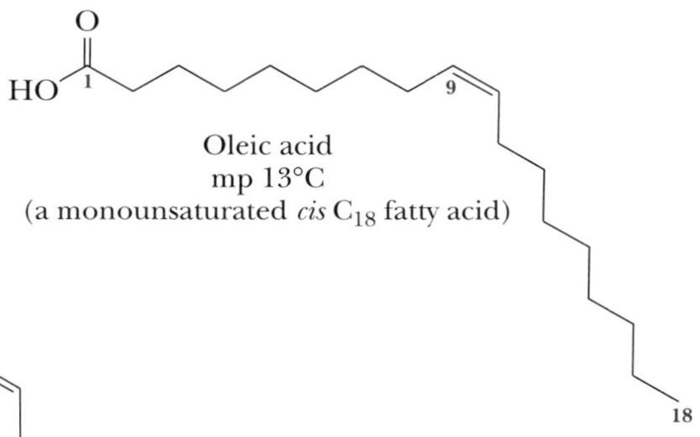

Oleic acid
mp 13°C
(a monounsaturated *cis* C_{18} fatty acid)

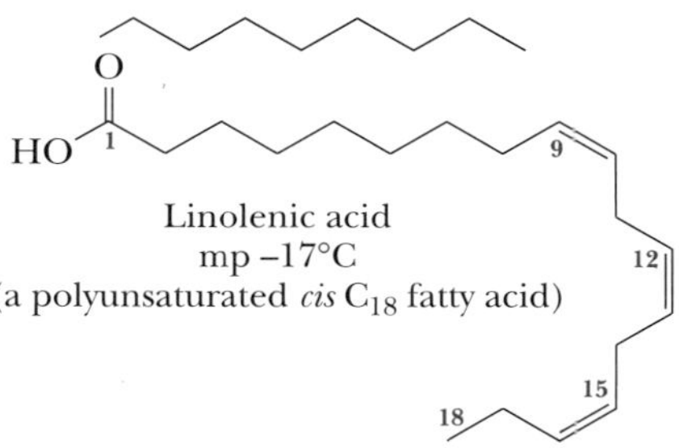

Linolenic acid
mp −17°C
(a polyunsaturated *cis* C_{18} fatty acid)

present in myrcene and geraniol are cross-linked to give cyclic structures. To help you identify the points of cross linkage and ring formation, the carbon atoms of the geraniol skeleton are numbered 1 through 8. This numbering pattern is used in the remaining terpenes to show points of cross-linking. In both limonene and menthol, a carbon-carbon bond is present between carbons 1 and 6. In α-pinene, carbon-carbon

The carbon-carbon single bonds of saturated fatty acid alkyl chains exist largely in the staggered, anti-conformation, so they can pack together relatively well, and are held together by dispersion forces (Section 2.8B). As a result, both saturated fatty acids and the triglycerides derived from them are solids at room temperature. However, the *cis* double bonds place a considerable "kink" in the chains of monounsaturated and polyunsaturated fatty acid chains, and they form weaker crystal lattices.

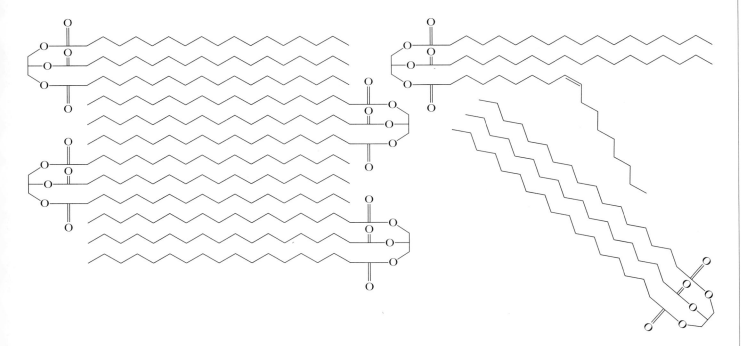

As a result, monounsaturated and polyunsaturated fatty acids and the triglycerides composed of them are liquid at room temperature. Overall, oils have fatty acids with more *cis* double bonds. Fats have fewer *cis* double bonds and more saturated fatty acids. For example, butter has a high content of saturated fats and is a solid at room temperature, whereas salad oil (from plant oils) has a high content of polyunsaturated fatty acids and is liquid, even at freezing temperatures. Olive oil, which has a high content of the monounsaturated fatty acid oleic acid (hence the name of oleic acid), will solidify in the refrigerator.

bonds are present between carbons 1 and 6 and between carbons 4 and 7. In camphor, they are between carbons 1 and 6 and between 3 and 7.

Vitamin A (Section 5.2G), a terpene of molecular formula $C_{20}H_{30}O$, consists of four isoprene units linked head-to-tail and cross-linked at one point to form a six-membered ring.

The synthesis of substances in living systems is a fascinating area of research and one of the links between organic chemistry and biochemistry. However tempting it might be to propose that nature synthesizes terpenes by joining together molecules of isoprene, this is not quite the way it is done. We will discuss the creation of new carbon-carbon bonds during terpene biosynthesis in Section 26.4B.

Summary

An **alkene** is an unsaturated hydrocarbon that contains a carbon-carbon double bond. The general formula of an alkene is C_nH_{2n}. A carbon-carbon double bond consists of one sigma bond formed by the overlap of sp^2 hybrid orbitals and one pi bond formed by the overlap of parallel $2p$ orbitals (Section 5.1B). The strength of the pi bond in ethylene is approximately 264 kJ (63 kcal)/mol, which is considerably weaker than the carbon-carbon sigma bond. The structural feature that makes *cis,trans* **isomerism** possible in alkenes is lack of rotation about the two carbons of the double bond (Section 5.1C).

According to the IUPAC system (Section 5.2A), the presence of a carbon-carbon double bond is shown by changing the infix of the parent hydrocarbon from -*an*- to -*en*. For compounds containing two or more double bonds, the infix is changed to -*adien*-, -*atrien*-, and so on. The names **methylene, vinyl,** and **allyl** are commonly used to show the presence of $=CH_2$, $-CH=CH_2$, and $-CH_2CH=CH_2$ groups, respectively.

Whether an alkene is *cis* or *trans* is determined by the orientation of the main carbon chain about the double bond

(Section 5.2C). The configuration of a carbon-carbon double bond is specified more precisely by the **E,Z system** using the same set of priority rules used for the R,S system (Section 3.3). If the two groups of higher priority are on the same side of the double bond, the alkene is designated **Z** (German: *zusammen,* together); if they are on opposite sides, the alkene is designated **E** (German: *entgegen,* opposite). To date, *trans*-cyclooctene is the smallest *trans* cycloalkene that has been prepared in pure form and is stable at room temperature.

Because alkenes are essentially nonpolar compounds and the only attractive forces between their molecules are **dispersion forces,** their physical properties are similar to those of alkanes (Section 2.8).

The characteristic structural feature of a **terpene** (Section 5.4) is a carbon skeleton that can be divided into two or more isoprene units. Terpenes illustrate an important principle of the molecular logic of living systems, namely that, in building large molecules, small subunits are strung together by an iterative process and then chemically modified by precise enzyme-catalyzed reactions.

Problems

Structure of Alkenes

5.9 Predict all approximate bond angles about each highlighted carbon atom. To make these predictions, use valence-shell electron-pair repulsion (VSEPR) (Section 1.4).

5.10 For each highlighted carbon atom in Problem 5.9, identify which atomic orbitals are used to form each sigma bond and which are used to form each pi bond.

5.11 Following is the structure of 1,2-propadiene (allene).

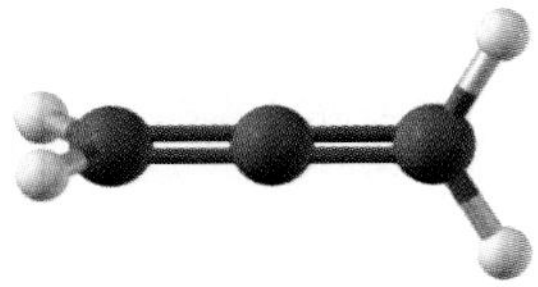

    (a) Predict all approximate bond angles in this molecule.

    (b) State the orbital hybridization of each carbon.

    (c) Show the three-dimensional geometry of allene, and explain it in terms of the orbitals used.

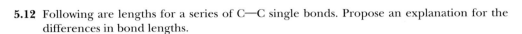

$$CH_2{=}C{=}CH_2$$

1,2-Propadiene
(Allene)

5.12 Following are lengths for a series of C—C single bonds. Propose an explanation for the differences in bond lengths.

Structure	Length of C—C Single Bond (pm)
$CH_3{-}CH_3$	153.7
$CH_2{=}CH{-}CH_3$	151.0
$CH_2{=}CH{-}CH{=}CH_2$	146.5
$HC{\equiv}C{-}CH_3$	145.9

Nomenclature of Alkenes

5.13 Draw structural formulas for these alkenes.

    (a) *trans*-2-Methyl-3-hexene

    (c) 2-Methyl-1-butene

    (e) 2,3-Dimethyl-2-butene

    (g) (*Z*)-1-Chloropropene

    (i) 1-Isopropyl-4-methylcyclohexene

    (k) 3-Cyclopropyl-1-propene

    (m) 2-Chloropropene

    (o) 1-Chlorocyclohexene

    (b) 2-Methyl-2-hexene

    (d) 3-Ethyl-3-methyl-1-pentene

    (f) *cis*-2-Pentene

    (h) 3-Methylcyclohexene

    (j) (*E*)-2,6-Dimethyl-2,6-octadiene

    (l) Cyclopropylethene

    (n) Tetrachloroethylene

5.14 Name these alkenes and cycloalkenes.

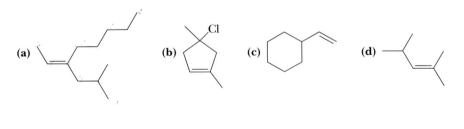

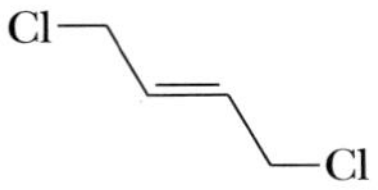

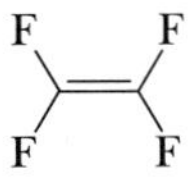

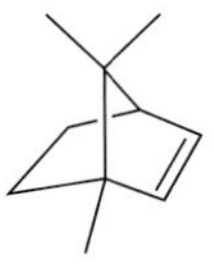

5.15 Arrange the following groups in order of increasing priority.

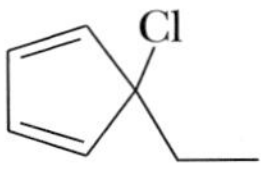

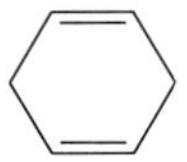

    (a) —CH_3 —H   —Br   —CH_2CH_3

    (b) —OCH_3 —$CH(CH_3)_2$ —$B(CH_2CH_3)_2$ —H

    (c) —CH_3 —CH_2OH —CH_2NH_2 —CH_2Br

5.16 Assign an *E* or *Z* and a *cis* or *trans* configuration to these dicarboxylic acids, each of which is an intermediate in the tricarboxylic acid cycle. Under each is its common name.

(a) 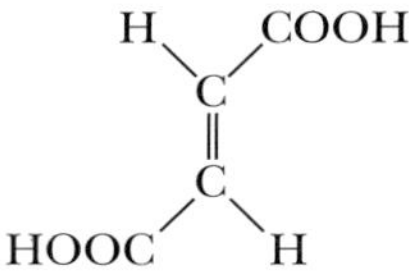 Fumaric acid

(b) 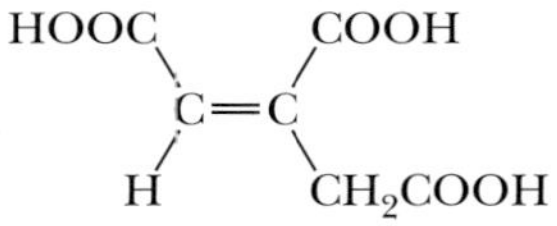 Aconitic acid

5.17 Name and draw structural formulas for all alkenes of molecular formula C_5H_{10}. As you draw these alkenes, remember that *cis* and *trans* isomers are different compounds and must be counted separately.

5.18 For each molecule that shows *cis,trans* isomerism, draw the *cis* isomer.

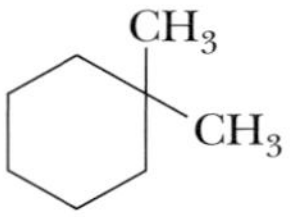

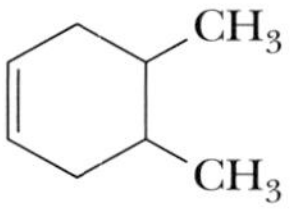

5.19 β-Ocimene, a triene found in the fragrance of cotton blossoms and several essential oils, has the IUPAC name (Z)-3,7-dimethyl-1,3,6-octatriene. Draw a structural formula for β-ocimene.

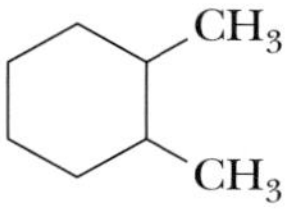

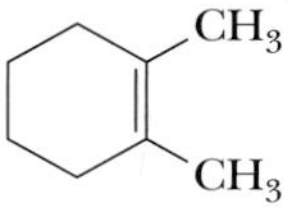

5.20 Draw the structural formula for at least one bromoalkene of molecular formula C_5H_9Br that shows:

(a) Neither E,Z isomerism nor chirality.
(b) E,Z isomerism but not chirality.
(c) Chirality but not E,Z isomerism.
(d) Both chirality and E,Z isomerism.

5.21 Following are structural formulas and common names for four molecules that contain both a carbon-carbon double bond and another functional group. Give each an IUPAC name.

(a) CH_2=CHCOH　Acrylic acid

(b) CH_2=CHCH　Acrolein

(c) 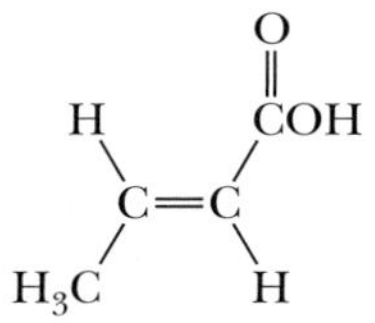 Crotonic acid

(d) CH_3CCH=CH_2　Methyl vinyl ketone

5.22 *Trans*-cyclooctene has been resolved, and its enantiomers are stable at room temperature. *Trans*-cyclononene has also been resolved, but it racemizes with a half-life of 4 min at 0°C. How can racemization of this cycloalkene take place without breaking any bonds? Why does *trans*-cyclononene racemize under these conditions but *trans*-cyclooctene does not? You will find it especially helpful to examine the molecular models of these cycloalkenes.

5.23 Which alkenes exist as pairs of *cis,trans* isomers? For each that does, draw the *trans* isomer.

(a) CH_2=CHBr
(b) CH_3CH=CHBr
(c) BrCH=CHBr
(d) $(CH_3)_2C$=$CHCH_3$
(e) $(CH_3)_2CHCH$=$CHCH_3$

5.24 Four stereoisomers exist for 3-penten-2-ol.

$$\underset{\text{3-Penten-2-ol}}{CH_3-CH=CH-\overset{\overset{\displaystyle OH}{|}}{CH}-CH_3}$$

(a) Explain how these four stereoisomers arise.
(b) Draw the stereoisomer having the *E* configuration about the carbon-carbon double bond and the *R* configuration at the stereocenter.

Molecular Modeling

These problems require molecular modeling programs such as Chem 3D, or Spartan to solve. Pre-built models can be found at **http://now.brookscole.com/bfi4.**

5.25 Measure the CH_3, CH_3 distance in the energy-minimized model of *cis*-2-butene, and the CH_3, H distance in the energy-minimized model of *trans*-2-butene. In which isomer is the nonbonded interaction strain greater?

5.26 Measure the C=C—C bond angles in the energy-minimized models of the *cis* and *trans* isomers of 2,2,5,5-tetramethyl-3-hexene. In which case is the deviation from VSEPR predictions greater?

5.27 Measure the C—C—C and C—C—H bond angles in the energy-minimized model of cyclohexene and compare them with those predicted by VSEPR. Explain any differences.

5.28 Measure the C—C—C and C—C—H bond angles in the energy-minimized models of *cis* and *trans* isomers of cyclooctene. Compare these values with those predicted by VSEPR. In which isomer are deviations from VSEPR predictions greater?

Terpenes

5.29 Show that the structural formula of vitamin A (Section 5.3G) can be divided into four isoprene units bonded head-to-tail and cross-linked at one point to form the six-membered ring.

5.30 Following is the structural formula of lycopene, $C_{40}H_{56}$, a deep-red compound that is partially responsible for the red color of ripe fruits, especially tomatoes. Approximately 20 mg of lycopene can be isolated from 1 kg of ripe tomatoes. Lycopene is an important antioxidant that may help prevent oxidative damage in atherosclerosis.

Lycopene

(a) Show that lycopene is a terpene, that is, its carbon skeleton can be divided into two sets of four isoprene units with the units in each set joined head-to-tail.
(b) How many of the carbon-carbon double bonds in lycopene have the possibility for *cis,trans* isomerism? Of these, which are *trans* and which are *cis*?

Carotene and lycopene are polyenes occurring in tomatoes and carrots. Carotene is a natural source of vitamin A.

5.31 As you might suspect, β-carotene, $C_{40}H_{56}$, precursor to vitamin A, was first isolated from carrots. Dilute solutions of β-carotene are yellow, hence its use as a food coloring. In plants, it is almost always present in combination with chlorophyll to assist in the harvesting of the energy of sunlight and to protect the plant against reactive species produced in photosynthesis. As tree leaves die in the fall, the green of their chlorophyll molecules is replaced by the yellows and reds of carotene and carotene-related molecules. Compare the carbon skeletons of β-carotene and lycopene. What are the similarities? What are the differences?

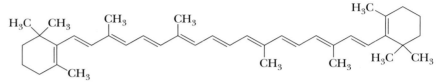

β-Carotene

5.32 Calculate the index of hydrogen deficiency for β-carotene and lycopene.

5.33 α-Santonin, isolated from the flower heads of certain species of Artemisia, is an anthelmintic (meaning against intestinal worms). This terpene is used in oral doses of 60 mg to rid the body of roundworms such as *Ascaris lumbricoides*. It has been estimated that over one third of the world's population is infested with these slender, thread-like parasites.

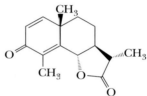

■ Santonin can be isolated form the flower heads of wormwood, *Artemisia absinthium*. This plant has also been used to make the drink absinthe, popular in nineteenth-century France, but now banned for its neurotoxicity.

(a) Locate the three isoprene units in santonin, and show how the carbon skeleton of farnesol might be coiled and then cross-linked to give santonin. Two different coiling patterns of the carbon skeleton of farnesol can lead to santonin. Try to find them both.

(b) Label all stereocenters in santonin. How many stereoisomers are possible for this molecule?

(c) Calculate the index of hydrogen deficiency for santonin.

5.34 Pyrethrin II and pyrethrosin are two natural products isolated from plants of the chrysanthemum family. Pyrethrin II is a natural insecticide and is marketed as such.

(a) Label all stereocenters in each molecule and all carbon-carbon double bonds about which there is the possibility for *cis,trans* isomerism.

(b) State the number of stereoisomers possible for each molecule.

(c) Show that the bicyclic ring system of pyrethrosin is composed of three isoprene units.

(d) Calculate the index of hydrogen deficiency for each of these natural products.

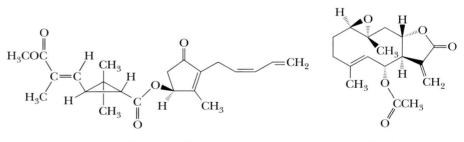

Pyrethrin II   Pyrethrosin

5.35 Limonene is one of the most common inexpensive fragrances. Two isomers of limonene can be isolated from natural sources. They are shown below. The one on the left has the odor of lemons, and the one on the right has the odor of oranges.

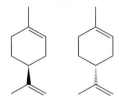

(a) What kind of isomers are they?
(b) Are E,Z isomers possible in limonene?
(c) Why do these two isomers smell different?

Looking Ahead

5.36 Bromine adds to *cis* and *trans*-2-butene to give different diastereomers of 2,3-dibromobutane. What does this say about the mode of addition of bromine to this alkene?

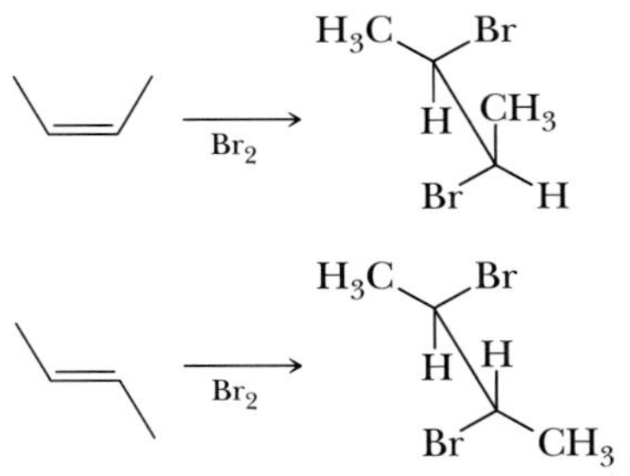

We discuss the addition of bromine to alkene in Chapter 6.

6

These laboratory squeeze bottles are fabricated from polyethylene. Inset: A model of ethylene, the monomer from which polyethylene is derived.

Outline

ORGANIC
Chemistry⚛Now™

Look for this logo in the chapter and go to Organic ChemistryNow at **http://now.brookscole.com/bfi4** for tutorials, simulations, and problems.

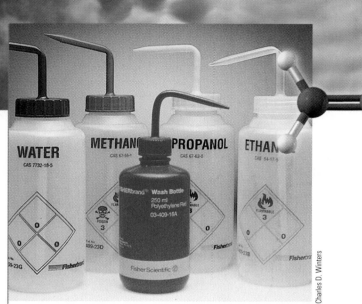

Charles D. Winters

Reactions of Alkenes

O ver 10 million organic compounds have been discovered or made by organic chemists! Surely it would seem to be an almost impossible task to learn the chemical properties of this many compounds. Fortunately, the study of organic compounds is not as formidable a task as you might think. Although organic compounds can undergo a wide variety of chemical reactions, only certain portions of their structures are changed in any particular reaction. As we will see in this chapter, the same functional group, in whatever organic molecule we find it, undergoes the same types of chemical reactions. Therefore, you do not have to study the chemical reactions of even a fraction of the 10 million known organic compounds. Instead, you need only to identify a few characteristic types of functional groups and then study the chemical reactions that each undergoes.

In this chapter, we begin our study of the chemical properties of organic compounds with the study of the chemistry of alkenes. To this end, we introduce one of the most important unifying concepts in organic chemistry: the concept of a reaction mechanism. We use the reactions of alkenes as the vehicle by which to introduce this concept.

6.1 Reactions of Alkenes—An Overview

The most characteristic reaction of alkenes is **addition** to the carbon-carbon double bond in such a way that the pi bond is broken and, in its place, sigma bonds form to two new atoms or groups of atoms. Table 6.1 gives several examples of reactions at a carbon-carbon double bond along with the descriptive name(s) associated with each. Some of these reactions are treated separately under oxidations (Section 6.5) and reductions (Section 6.6), but are included in this table because they are formally additions.

In the following sections, we study these alkene reactions in considerable detail, with particular attention to the mechanism by which each occurs.

A second characteristic reaction of alkenes is the formation of chain-growth polymers (Greek: *poly,* many, and *meros,* part). In the presence of certain catalysts called initiators, many alkenes form polymers made by the addition of monomers (Greek: *mono,* one, and *meros,* part) to a growing polymer chain as illustrated by the formation of polyethylene from ethylene.

Addition reaction A reaction in which two atoms or ions react with a double bond, forming a compound with the two new groups bonded to the carbons of the original double bond.

Table 6.1 Characteristic Alkene Addition Reactions

Reaction	Descriptive Name(s)
$\text{C=C} + \text{HCl (HX)} \longrightarrow \overset{\text{H}}{-}\text{C}-\text{C}-\underset{\text{Cl (X)}}{}$	Hydrochlorination (hydrohalogenation)
$\text{C=C} + \text{H}_2\text{O} \longrightarrow \overset{\text{H}}{-}\text{C}-\text{C}-\underset{\text{OH}}{}$	Hydration
$\text{C=C} + \text{Br}_2 (\text{X}_2) \longrightarrow \overset{(\text{X}) \text{ Br}}{-}\text{C}-\text{C}-\underset{\text{Br (X)}}{}$	Bromination (halogenation)
$\text{C=C} + \text{Br}_2 (\text{X}_2) \xrightarrow{\text{H}_2\text{O}} \overset{\text{HO}}{-}\text{C}-\text{C}-\underset{\text{Br (X)}}{}$	Bromo(halo)hydrin formation
$\text{C=C} + \text{Hg(OAc)}_2 \xrightarrow{\text{H}_2\text{O}} \overset{\text{HgOAc}}{-}\text{C}-\text{C}-\underset{\text{HO}}{}$	Oxymercuration
$\text{C=C} + \text{BH}_3 \longrightarrow -\text{C}-\text{C}-\underset{\text{H} \quad \text{BH}_2}{}$	Hydroboration
$\text{C=C} + \text{OsO}_4 \longrightarrow -\text{C}-\text{C}-\underset{\text{HO} \quad \text{OH}}{}$	Diol formation (oxidation)
$\text{C=C} + \text{H}_2 \longrightarrow -\text{C}-\text{C}-\underset{\text{H} \quad \text{H}}{}$	Hydrogenation (reduction)

$$n\text{CH}_2\text{=}\text{CH}_2 \xrightarrow{\text{initiator}} \left(\text{CH}_2\text{CH}_2\right)_n$$

In alkene polymers of industrial and commercial importance, n is a large number, typically several thousand. We discuss this alkene reaction in Chapter 29.

6.2 Reaction Mechanisms

Reaction mechanism A step-by-step description of how a chemical reaction occurs.

A **reaction mechanism** describes in detail how a reaction occurs. It describes which bonds are broken and which new ones are formed, as well as the order and relative rates of the various bond-breaking and bond-forming steps. If the reaction takes place in solution, the mechanism describes the role of the solvent. If the reaction involves a catalyst, it describes the role of the catalyst. A complete reaction mechanism describes the positions of all atoms and the energy of the entire system during each moment of the reaction. This ideal, however, is rarely approached in practice.

A. Energy Diagrams, Transition States, and Reaction Intermediates

Energy diagram A graph showing the changes in energy that occur during a chemical reaction; energy is plotted on the vertical axis, and reaction progress is plotted on the horizontal axis.

Reaction coordinate A measure of the change in the positions of atoms during a reaction; plotted on the horizontal axis in a reaction energy diagram.

Gibbs free energy change, ΔG A thermodynamic function relating enthalpy, entropy, and temperature, given by the equation $\Delta G = \Delta H - T\Delta S$. If $\Delta G < 0$, the position of equilibrium for the reaction favors the product(s). If $\Delta G > 0$, the position of equilibrium favors the reactant(s)

Think of a chemical bond as a spring. As a spring is stretched from its resting position, its energy increases. As it returns to its resting position, its energy decreases. Similarly, during a chemical reaction, bond breaking corresponds to an increase in energy, and bond forming corresponds to a decrease in energy. We use an **energy diagram** to show the changes in energy that occur in going from reactants to products. Energy is measured on the vertical axis, and the change in position of the atoms during the reaction is represented on the horizontal axis, called the **reaction coordinate.** The reaction coordinate corresponds to how far the reaction has progressed (it is not a time axis). Figure 6.1 shows an energy diagram for the reaction of compounds C + A—B to form C—A + B. This reaction occurs in one step, meaning that bond breaking in starting materials and bond forming to give products occur simultaneously.

Several types of changes in energy may be important to consider in reactions. In this text, we are most concerned with changes in the **Gibbs free energy** (ΔG^0) and the enthalpy (ΔH^0). A change in Gibbs free energy is directly related to chemical equilibria.

$$\Delta G^0 = -RT \ln K_{\text{eq}}$$

Figure 6.1
An energy diagram for a one-step reaction between C and A—B. The dashed lines in the transition state indicate that the A—B bond is partially broken and the new C—A bond is partially formed. On completion of the reaction, the A—B bond is fully broken, and the C—A bond is fully formed. The energy of the reactants is higher than that of the products.

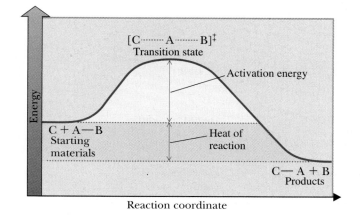

An **enthalpy change** is the energy contained in chemical bonds of the products compared to that in the reactants. Changes in enthalpy and Gibbs free energy are related by the term $T\Delta S^0$, where T is the temperature in kelvins and ΔS^0 is the change in entropy for a reaction:

$$\Delta G^0 = \Delta H^0 - T\Delta S^0$$

The change in entropy is determined by the change in the number of degrees of freedom (for example by how many molecules or how many bond rotations are on each side of a balanced chemical equation or by how many equivalent isomers can be formed in a reaction). If the number of molecules doesn't change in a reaction, ΔS^0 is often very small, in which case $\Delta G^0 = \Delta H^0$.

The $T\Delta S^0$ term is proportional to temperature, which means that entropy becomes more important at high temperatures. Because $T\Delta S^0$ is negative for reactions in which the number of molecules increases, high-temperature reactions often lead to bond breaking, with an increase in degrees of freedom or number of molecules and a resultant increase in entropy. Cracking of petroleum (Section 2.10B) is an example of a reaction that occurs with an increase in entropy.

The difference in enthalpy between the reactants and the products, each in their standard states, is called the **heat of reaction** (ΔH^0). The magnitude of the ΔH^0 for a reaction is a measure of the relative strengths of the bonds formed compared with the strengths of the bonds broken. If the bonds formed are stronger than the bonds broken, the enthalpy of the products is lower than that of the reactants, heat is released, and the reaction is **exothermic.** Conversely, if the bonds broken are stronger than the bonds formed, the enthalpy of the products is higher than that of the reactants, heat is absorbed, and the reaction is **endothermic.** The one-step reaction shown in Figure 6.1 is exothermic.

Similar relationships hold for Gibbs free energy. If the ΔG^0 for a reaction is negative, the reaction is exergonic, and, at equilibrium, products are favored over reactants. (It is important to remember that there is no necessary relationship between Gibbs free energy and rate; not all **exergonic** reactions proceed at measurable rates.) If the ΔG^0 for a reaction is positive, the reaction is **endergonic,** and, at equilibrium, reactants are favored over products. The relationships between ΔG^0, ΔH^0, ΔS^0, and the position of equilibrium in chemical reactions are summarized in the following table.

Enthalpy change, ΔH^0 The difference in total bond energy between reactants and products; a measure of bond making (exothermic) and bond breaking (endothermic).

Heat of reaction, ΔH^0 The difference in enthalpy between reactants and products. If the enthalpy of products is lower than that of reactants, heat is released and the reaction is exothermic. If the enthalpy of the products is higher than that of the reactants, energy is absorbed, and the reaction is endothermic.

Exothermic reaction A reaction in which the enthalpy of the products is lower than that of the reactants; a reaction in which heat is released.

Endothermic reaction A reaction in which the enthalpy of the products is higher than the enthalpy of the reactants; a reaction in which heat is absorbed.

Exergonic reaction A reaction in which the Gibbs free energy of the products is lower than that of the reactants. The position of equilibrium for an exergonic reaction favors products.

Endergonic reaction A reaction in which the Gibbs free energy of the products is higher than that of the reactants. The position of equilibrium for an endergonic reaction favors starting materials.

	$\Delta S^0 < 0$	$\Delta S^0 > 0$
$\Delta H^0 > 0$	$\Delta G^0 > 0$; the position of equilibrium favors reactants	At higher temperatures when $T\Delta S^0 < \Delta H^0$ and $\Delta G^0 < 0$, the position of equilibrium favors products
$\Delta H^0 < 0$	At lower temperatures when $T\Delta S^0 < \Delta H^0$ and $\Delta G^0 < 0$; the position of equilibrium favors products	$\Delta G^0 < 0$; the position of equilibrium favors products

Transition state An unstable species of maximum energy formed during the course of a reaction; a maximum on an energy diagram.

A **transition state** is a point on an energy diagram at which the energy is a maximum for a given step. At the transition state, sufficient energy has become concentrated in the proper bonds so that bonds in reactants break. As they break, energy is redistributed, and new bonds form, giving products. After a transition state is reached, the reaction proceeds to give products with the release of energy. A transition state has a definite geometry, a definite arrangement of bonding and nonbonding electrons, and a definite distribution of electron density and charge. Because a transition state is at an energy maximum, it cannot be isolated, and its structure cannot be determined experimentally. Its lifetime is fleeting, on the order of less than one picosecond (the duration of a single bond vibration). As we shall soon see, even though we usually cannot observe a transition state directly by any experimental means, we can often infer a great deal about its probable structure from other experimental observations.

Activation energy The difference in Gibbs free energy between reactants and a transition state.

The difference in energy between reactants and the transition state is called the **activation energy.** If we are discussing Gibbs free energy, activation energy is given the symbol $\Delta G^{\ddagger}$. (Often activation energy is discussed in terms of the closely related potential energy, in which case the activation energy is called E_a.) The activation energy is the minimum energy required for a reaction to occur and can be considered an energy barrier for the reaction. $\Delta G^{\ddagger}$ determines the rate of a reaction; that is, how fast the reaction occurs. If $\Delta G^{\ddagger}$ is large, only a very few molecular collisions occur with sufficient energy to reach the transition state, and the reaction is slow. If $\Delta G^{\ddagger}$ is small, many collisions generate sufficient energy to reach the transition state, and the reaction is fast.

Most of the organic reactions we deal with in this text have activation energies in the range of 42–146 kJ (10–35 kcal)/mol. Those with activation energies below 84 kJ (20 kcal)/mol proceed rapidly at room temperature. Those with higher activation energies may require heating or input of some other form of energy, such as light, to provide a sufficient number of molecules with enough energy to overcome the activation energy barrier.

Reaction intermediate A species, formed between two successive reaction steps, that lies in an energy minimum between the two transition states.

In a reaction that occurs in two or more steps, each step has its own transition state and activation energy. Figure 6.2 shows an energy diagram for the conversion of reactants to products in two steps. A **reaction intermediate** corresponds to an energy minimum between two transition states, in this case between transition states 1 and 2. Because the energies of the reaction intermediates we describe in this chapter are higher than that of either reactants or products, they are highly reactive and usually

Figure 6.2

An energy diagram for a two-step reaction involving formation of a reaction intermediate. The energy of the reactants is higher than that of the products, and energy is released in the conversion of A + B to C + D.

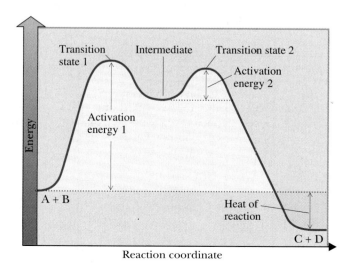

cannot be isolated. However, some have significant lifetimes and can be observed directly using special experimental strategies.

The slowest step in a multistep reaction, called the **rate-determining step,** is the step that crosses the highest energy barrier. In the two-step sequence shown in Figure 6.2, Step 1 crosses the higher energy barrier and is, therefore, the rate-determining step.

Rate-determining step The step in a multistep reaction sequence that crosses the highest energy barrier.

B. Developing a Reaction Mechanism

To develop a reaction mechanism, chemists begin by designing experiments that will reveal details of a particular chemical reaction. Next, through a combination of experience and intuition, they propose several sets of steps or mechanisms, each of which might account for the overall chemical transformation. Finally, chemists test each proposed mechanism against the experimental observations to exclude those mechanisms that are not consistent with the facts. Modern computational methods now allow detailed descriptions of mechanistic pathways, but the results must still be compared with experiment.

A mechanism becomes generally established by excluding reasonable alternatives and by showing that it is consistent with every test that can be devised. This, of course, does not mean that a generally accepted mechanism is a completely accurate description of the chemical events, only that it is the best mechanism chemists have been able to devise. It is important to keep in mind that, as new experimental evidence is obtained, it may be necessary to modify a generally accepted mechanism or possibly even discard it in favor of an alternative.

Before we go on to consider reactions and reaction mechanisms, we might ask why chemists go to the trouble of establishing them and you must spend time learning about them. One reason is very practical: Mechanisms provide a framework within which to organize a great deal of descriptive chemistry. For example, with insight into how reagents add to particular alkenes, it is possible to generalize and then to predict how the same reagents might add to other alkenes. A second reason lies in the intellectual satisfaction derived from constructing models that accurately reflect the behavior of chemical systems. Finally, to a creative scientist, a mechanism is a tool to be used in the search for new information and new understanding. A mechanism consistent with all that is known about a reaction can be used to make predictions about chemical interactions as yet unexplored, and experiments can be designed to test these predictions. Thus, reaction mechanisms provide a way not only to organize knowledge but also to extend it.

6.3 Electrophilic Additions

We begin our introduction to the chemistry of alkenes by examining five common addition reactions. We first study some of the experimental observations about each addition reaction and then look at its mechanism. Through the study of these particular reactions and their mechanisms, you will develop a general understanding of how alkenes undergo addition reactions.

A. Addition of Hydrogen Halides

The hydrogen halides HCl, HBr, and HI add to alkenes to give haloalkanes (alkyl halides). These additions may be carried out either with the pure reagents (neat) or in the presence of a polar solvent such as acetic acid. HCl reacts sluggishly compared to the other two acids. Addition of HBr to ethylene gives bromoethane (ethyl bromide):

$$CH_2{=}CH_2 + \boxed{HBr} \longrightarrow \overset{\overset{\text{H}}{|}}{CH_2}{-}\overset{\overset{\text{Br}}{|}}{CH_2}$$

Ethylene   Bromoethane
(Ethyl bromide)

Regioselective reaction An addition or substitution reaction in which one of two or more possible products is formed in preference to all others that might be formed.

Addition of HBr to propene gives 2-bromopropane (isopropyl bromide); hydrogen adds to carbon 1 of propene, and bromine adds to carbon 2. If the orientation of addition were reversed, 1-bromopropane (propyl bromide) would be formed. The observed result is that 2-bromopropane is formed to the virtual exclusion of 1-bromopropane. We say that addition of HBr to propene is highly regioselective. A **regioselective reaction** is a reaction in which one direction of bond forming or breaking occurs in preference to all other directions of bond forming or breaking.

$$CH_3CH{=}CH_2 + \boxed{HBr} \longrightarrow CH_3\overset{\overset{\text{Br}}{|}}{CH}{-}\overset{\overset{\text{H}}{|}}{CH_2} \; + \; CH_3\overset{\overset{\text{H}}{|}}{CH}{-}\overset{\overset{\text{Br}}{|}}{CH_2}$$

Propene   2-Bromopropane   1-Bromopropane
(not observed)

Markovnikov's rule In the addition of HX, H_2O, or ROH to an alkene, hydrogen adds to the carbon of the double bond having the greater number of hydrogens.

This regioselectivity was noted by Vladimir Markovnikov who made the generalization known as **Markovnikov's rule:** In the addition of H—X to an alkene, hydrogen adds to the double-bonded carbon that has the greater number of hydrogens already bonded to it. Although Markovnikov's rule provides a way to predict the products of many alkene addition reactions, it does not explain why one product predominates over other possible products.

Example 6.1

Name and draw a structural formula for the product of each alkene addition reaction:

(a) $CH_3\overset{\overset{\text{CH}_3}{|}}{C}{=}CH_2 + HI \longrightarrow$ (b) [cyclopentene with CH₃] $+ HCl \longrightarrow$

Solution

Using Markovnikov's rule, we predict that 2-iodo-2-methylpropane is the product in (a) and that 1-chloro-1-methylcyclopentane is the product in (b).

(a) $CH_3\overset{\overset{\text{CH}_3}{|}}{\underset{\underset{\text{I}}{|}}{C}}CH_3$ (b)   [cyclopentane with Cl and CH₃]

2-Iodo-2-methylpropane   1-Chloro-1-methylcyclopentane

Problem 6.1

Name and draw a structural formula for the product of each alkene addition reaction:

(a) ⬡—CH₃ + HBr ⟶   **(b)** ⬡=CH₂ + HI ⟶

Chemists account for the addition of HX to an alkene by a two-step mechanism, which we illustrate by the reaction of 2-butene with hydrogen bromide to give 2-bromobutane. We will first look at this two-step mechanism in overview and then go back and study each step in detail.

Mechanism *Electrophilic Addition of HCl to 2-Butene*

Step 1: Addition begins with the transfer of a proton from H—Br to 2-butene, as shown by the two curved arrows on the left side of the equation. The first curved arrow shows that the pi bond of the alkene breaks and its electron pair forms a new covalent bond with the hydrogen atom of H—Br. The second curved arrow shows that the polar covalent bond in H—Br breaks and its electron pair moves to bromine to form a bromide ion. The result of this step is the formation of an organic cation and bromide ion.

$$CH_3CH{=}CHCH_3 + H{-}\overset{\delta-}{\underset{}{Br}}{:} \xrightarrow[\text{determining}]{\text{slow, rate}} CH_3CH{-}\overset{+}{\underset{|}{C}}HCH_3 + {:}\overset{..}{\underset{..}{Br}}{:}^-$$

sec-Butyl cation
(a 2° carbocation intermediate)

Step 2: Reaction of the *sec*-butyl cation (an electrophile) with bromide ion (a nucleophile) completes the valence shell of carbon and gives 2-bromobutane.

$${:}\overset{..}{\underset{..}{Br}}{:}^- + CH_3\overset{+}{C}HCH_2CH_3 \xrightarrow{\text{fast}} CH_3\overset{\overset{..}{\underset{|}{Br}}{:}}{C}HCH_2CH_3$$

Bromide ion   *sec*-Butyl cation   2-Bromobutane
(a nucleophile)   (an electrophile)

Now that we have looked at this two-step mechanism in overview, let us go back and look at the individual steps in more detail. There is a great deal of important organic chemistry embedded in them.

The reaction commences when the electron-rich pi bond of the alkene interacts with the relatively electropositive H atom of H—Br. The pi bond is relatively electron-rich because the pi-bonding electron density is above and below the bond axis, not between the positively charged atomic nuclei as is the case with sigma-bonding electron density. The pattern of an electron-rich area of one molecule interacting with an electron-deficient area of another molecule is an extremely common pattern in organic reactions, analogous to reactions between Lewis bases and Lewis acids (Section 4.6). An electron-rich reactant is referred to as a **nucleophile** (meaning nucleus loving), and an electron-deficient reactant is referred to as an **electrophile** (meaning electron loving). In fact, most of the reactions of alkenes described in this chapter belong to this category, with the alkene pi bond acting as a weak nucleophile and interacting with a variety of electrophiles. Reactions proceed beyond the initial nucleophile–electrophile interaction, hence the name **electrophilic addition.**

Nucleophile From the Greek meaning nucleus loving. Any species that can donate a pair of electrons to form a new covalent bond; alternatively, a Lewis base.

Electrophile From the Greek meaning electron loving. Any species that can accept a pair of electrons to form a new covalent bond; alternatively, a Lewis acid.

Figure 6.3

The structure of the *tert*-butyl cation. (a) Lewis structure and (b) an orbital overlap picture.

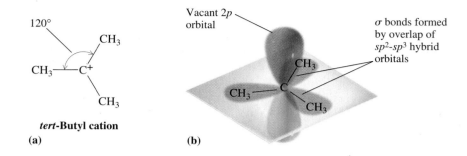

tert-**Butyl cation**

(a)

(b)

Carbocation   A species in which a carbon atom has only six electrons in its valence shell and bears a positive charge.

Step 1 results in formation of a cationic intermediate. One carbon atom in this intermediate has only six electrons in its valence shell and carries a charge of +1. A species containing a positively charged carbon atom is called a **carbocation** (*carbon* + *cation*). Such carbon-containing cations have also been called *carbonium ions* and *carbenium ions*. Even though the last two terms are obsolete, you may still encounter them. Carbocations are classified as primary (1°), secondary (2°), or tertiary (3°), depending on the number of carbon atoms bonded to the carbon bearing the positive charge. All carbocations are electrophiles as well as Lewis acids (Section 4.6).

In a carbocation, the carbon bearing the positive charge is bonded to three other atoms and, as predicted by VSEPR, the three bonds about the cationic carbon are coplanar and form bond angles of approximately 120^0. According to the orbital hybridization model, the electron-deficient carbon of a carbocation uses sp^2 hybrid orbitals to form sigma bonds to the three attached groups. The unhybridized $2p$ orbital lies perpendicular to the sigma bond framework and contains no electrons. Figure 6.3 shows a Lewis structure and an orbital overlap diagram for the *tert*-butyl cation.

Figure 6.4 shows an energy diagram for the two-step reaction of 2-butene with HBr. The slower, rate-determining step (the one that crosses the higher energy barrier) is Step 1, which leads to the formation of the 2° carbocation intermediate. This carbocation intermediate lies in an energy minimum between the transition states for Steps 1 and 2. As soon as the carbocation intermediate (an electrophile) forms, it reacts with bromide ion (a nucleophile) in a nucleophile-electrophile reaction to give 2-bromobutane. Note that the energy level for 2-bromobutane (the product) is lower than the energy level for 2-butene and HBr (the reactants). Thus, in this alkene addition reaction, energy is released; the reaction is exergonic.

Figure 6.4

An energy diagram for the two-step addition of HBr to 2-butene. The reaction is exergonic.

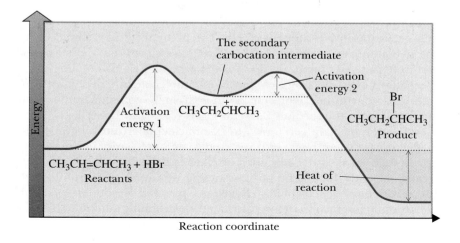

Regioselectivity and the Relative Stabilities of Carbocations

Reaction of HX and an alkene can, at least in principle, give two different carbocation intermediates depending on which of the doubly bonded carbon atoms the proton is transferred to, as illustrated by the reaction of HBr with propene.

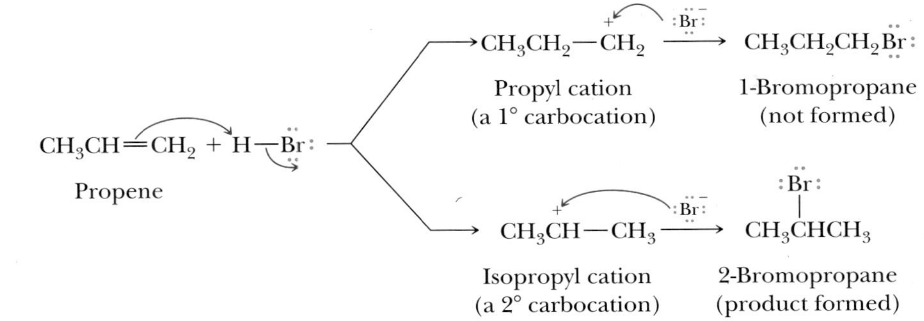

$CH_3CH_2—CH_2^+$:Br:⁻   ⟶ $CH_3CH_2CH_2Br:$

Propyl cation   1-Bromopropane
(a 1° carbocation)   (not formed)

$CH_3CH=CH_2 + H—Br:$

Propene

$CH_3CH^+—CH_3$:Br:⁻   ⟶ CH_3CHCH_3 (with Br)

Isopropyl cation   2-Bromopropane
(a 2° carbocation)   (product formed)

The observed product is 2-bromopropane, indicating that the 2° carbocation intermediate is formed in preference to the 1° carbocation intermediate.

Similarly, in the reaction of HBr with 2-methylpropene, proton transfer to the carbon-carbon double bond might form either the isobutyl cation (a 1° carbocation) or the *tert*-butyl cation (a 3° carbocation).

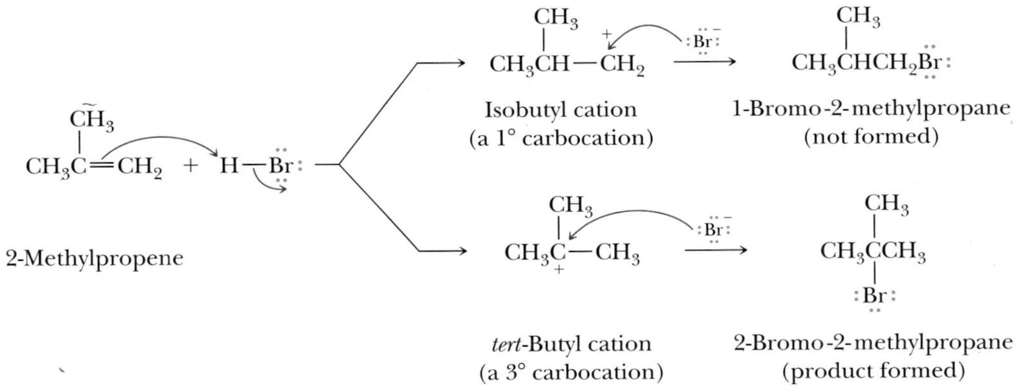

CH_3
|
$CH_3CH—CH_2^+$:Br:⁻   ⟶ $CH_3CHCH_2Br:$ (CH₃)

Isobutyl cation   1-Bromo-2-methylpropane
(a 1° carbocation)   (not formed)

CH_3
|
$CH_3C=CH_2 + H—Br:$

2-Methylpropene

CH_3
|
$CH_3C^+—CH_3$:Br:⁻   ⟶ CH_3CCH_3 (with Br and CH₃)

tert-Butyl cation   2-Bromo-2-methylpropane
(a 3° carbocation)   (product formed)

The observed product of this reaction is 2-bromo-2-methylpropane, indicating that the 3° carbocation is formed in preference to the 1° carbocation.

From experiments such as these and a great amount of other experimental evidence, we know that a 3° carbocation both is more stable and requires a lower activation energy for its formation than a 2° carbocation. A 2° carbocation, in turn, is more stable and requires a lower activation energy for its formation than a 1° carbocation. Methyl and 1° carbocations are so unstable they have never been observed in solution. It follows, then, that a more stable carbocation intermediate forms faster than a less stable carbocation intermediate. Following is the order of stability of four types of alkyl carbocations.

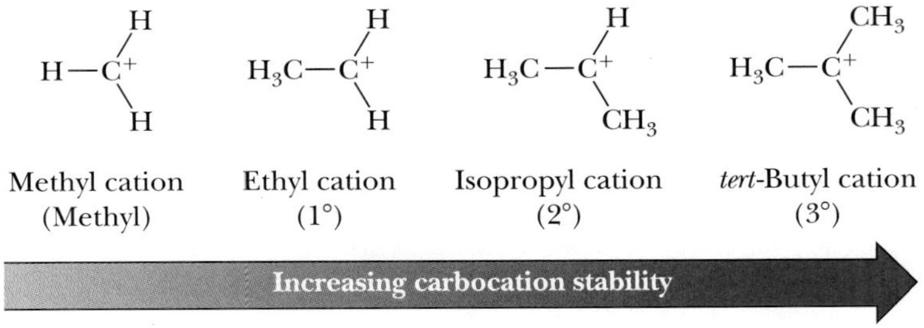

| Methyl cation (Methyl) | Ethyl cation (1°) | Isopropyl cation (2°) | *tert*-Butyl cation (3°) |

Increasing carbocation stability →

Now that we know the order of stability of carbocations, how do we account for this order? As we saw during the discussion of anion stability in Section 4.5, a system bearing a charge (either positive or negative) is more stable if the charge is delocalized. Using this principle, we can explain the order of stability of carbocations if we assume that alkyl groups bonded to a positively charged carbon release electron density toward that carbon and thereby help delocalize the positive charge on the cation. We account for the electron-releasing ability of alkyl groups bonded to a cationic carbon by two effects: the inductive effect and hyperconjugation.

The Inductive Effect

Inductive effect The polarization of the electron density of a covalent bond resulting from the electronegativity of a nearby atom.

The inductive effect operates in the following way. The electron deficiency of the cationic carbon exerts an electron-withdrawing **inductive effect** that polarizes electrons from adjacent sigma bonds toward it. In this way, the positive charge of the cation is not localized on the trivalent carbon but rather is delocalized over nearby atoms. The larger the volume over which the positive charge is delocalized, the greater the stability of the cation. Thus, as the number of alkyl groups bonded to the cationic carbon increases, the stability of the cation increases. Figure 6.5 illustrates the electron-withdrawing inductive effect of the positively charged carbon and the resulting delocalization of charge. According to quantum mechanical calculations, the charge on carbon in the methyl cation is approximately +0.645, and the charge on each of the hydrogen atoms is +0.118. Thus, even in the methyl cation, the positive charge is not localized entirely on carbon. Rather, it is delocalized over the volume of space occupied by the entire ion. Delocalization of charge is even much more extensive in the *tert*-butyl cation.

Hyperconjugation

Hyperconjugation Interaction of electrons in a sigma-bonding orbital with the vacant $2p$ orbital of an adjacent positively charged carbon.

The second effect by which alkyl groups stabilize carbocations is hyperconjugation. **Hyperconjugation** involves partial overlapping of the sigma-bonding orbital of an adjacent C—H or C—C bond of the alkyl group with the vacant $2p$ orbital of the cationic carbon (Figure 6.6). In other words, some electron density of the alkyl group C—H or C—C bond is mixed into the $2p$ orbital. The net result of hyperconjugation is a decrease of electron density on the cationic carbon, thereby delocalizing the positive charge. As more alkyl groups are bonded to a cationic carbon, the hyperconjugation effect becomes stronger and the carbocation becomes more stable.

Figure 6.5
Electrostatic potential plots showing the distribution of the positive charge in (a) the methyl cation and (b) the *tert*-butyl cation. Electron donation by the methyl groups decreases the positive charge (blue area) on the central carbon of the *tert*-butyl cation.

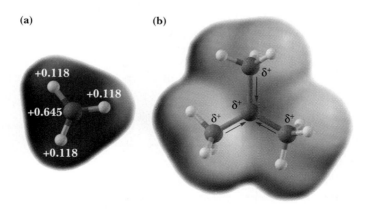

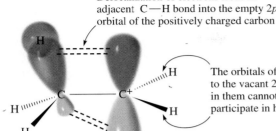

Delocalization of electrons from an
adjacent C—H bond into the empty $2p\square$
orbital of the positively charged carbon

The orbitals of these C—H bonds are perpendicular
to the vacant $2p$ orbital of the cationic carbon. Electrons
in them cannot flow into the vacant $2p$ orbital and cannot
participate in hyperconjugation.

Figure 6.6
Hyperconjugation.

Example 6.2

Arrange these carbocations in order of increasing stability.

(a)   (b)   (c)

Solution

Carbocation (a) is 2°, (b) is 3°, and (c) is 1°. In order of increasing stability they are
c < a < b.

Problem 6.2

Arrange these carbocations in order of increasing stability.

(a) $-CH_3$ (b) $-CH_3$ (c) $-CH_2$

Example 6.3

Propose a mechanism for the addition of HI to methylidenecyclohexane to give
1-iodo-1-methylcyclohexane. Which step in your mechanism is rate determining?

Solution

Propose a two-step mechanism similar to that proposed for the addition of HBr to
propene.

Step 1: A rate-determining proton transfer from HI to the carbon-carbon double bond gives a
3° carbocation intermediate.

$=CH_2$ + $H—I:$ $\xrightarrow{\text{slow, rate determining}}$ $-CH_3$ + $:I:^-$

Methylidenecyclohexane
(Methylenecyclohexane)

A 3° carbocation
intermediate

Step 2: Reaction of the 3° carbocation intermediate (an electrophile) with iodide ion (a nucleophile) completes the valence shell of carbon and gives the product.

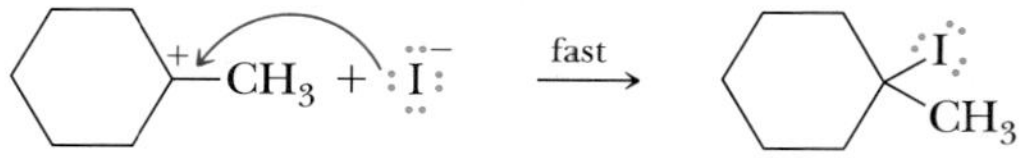

1-Iodo-1-methylcyclohexane

Problem 6.3

Propose a mechanism for addition of HI to 1-methylcyclohexene to give 1-iodo-1-methylcyclohexane. Which step in your mechanism is rate determining?

B. Addition of Water: Acid-Catalyzed Hydration

Hydration   The addition of water.

In the presence of an acid catalyst, most commonly concentrated sulfuric acid, water adds to an alkene to give an alcohol. The addition of water is called **hydration.**

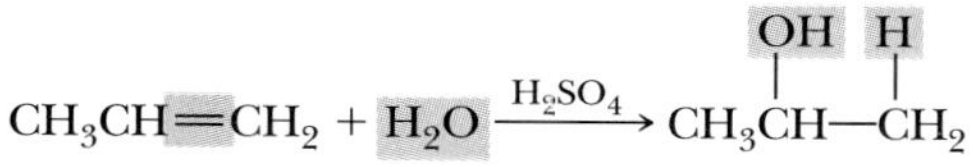

$$CH_3CH{=}CH_2 + H_2O \xrightarrow{H_2SO_4} CH_3CH{-}CH_2$$

Propene   2-Propanol

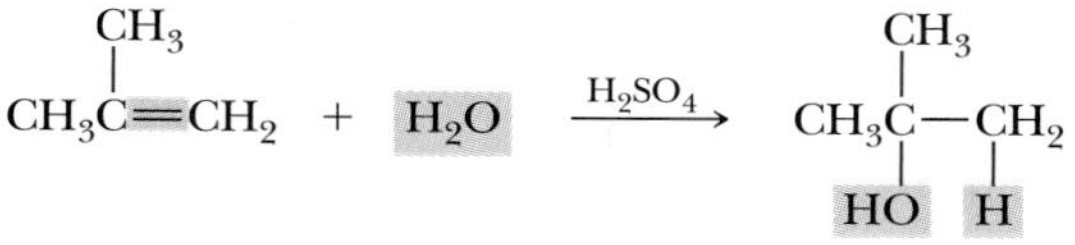

2-Methylpropene   2-Methyl-2-propanol

In the case of simple alkenes, H adds to the carbon of the double bond with the greater number of hydrogens, and OH adds to the carbon with the fewer hydrogens. Thus, H—OH adds to alkenes in accordance with Markovnikov's rule.

Example 6.4

Draw a structural formula for the product of the acid-catalyzed hydration of 1-methylcyclohexene.

Solution

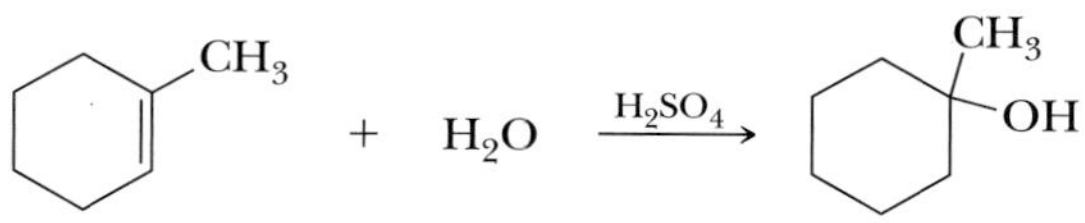

1-Methylcyclohexene   1-Methylcyclohexanol

Problem 6.4

Draw a structural formula for the product of each alkene hydration reaction:

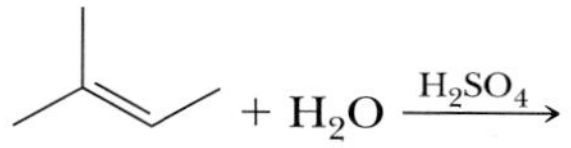

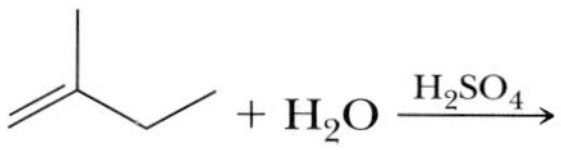

The mechanism for acid-catalyzed hydration of alkenes is quite similar to what we already proposed for addition of HCl, HBr, and HI to alkenes and is illustrated by conversion of propene to 2-propanol. As you study this mechanism, note that it is consistent with the fact that acid is a catalyst; an H_3O^+ is consumed in Step 1 but another is generated in Step 3.

Mechanism   *Acid-Catalyzed Hydration of Propene*

Step 1: Proton transfer from H_3O^+ to propene gives a 2° carbocation intermediate

$$CH_3CH{=}CH_2 + H{-}\overset{\cdot\cdot}{\underset{H}{O}}{-}H \xrightarrow[\text{determining}]{\text{slow, rate}} CH_3\overset{+}{C}HCH_3 + :\overset{\cdot\cdot}{\underset{H}{O}}{-}H$$

A 2° carbocation
intermediate

Step 2: The 2° carbocation intermediate (an electrophile) completes its valence shell by forming a new covalent bond with an unshared pair of electrons of the oxygen atom of water (a nucleophile) and gives an **oxonium ion.**

$$CH_3\overset{+}{C}HCH_3 + :\overset{\cdot\cdot}{\underset{H}{O}}{-}H \xrightarrow{\text{fast}} CH_3\overset{\overset{\displaystyle H\underset{+}{\overset{\cdot\cdot}{O}}H}{|}}{C}HCH_3$$

An oxonium ion

Oxonium ion An ion in which oxygen bears a positive charge.

Step 3: Proton transfer from the oxonium ion to water gives the alcohol and generates a new acid catalyst.

$$CH_3\overset{\overset{\displaystyle H\underset{+}{\overset{\cdot\cdot}{O}}H}{|}}{C}HCH_3 + :\overset{\cdot\cdot}{\underset{H}{O}}{-}H \xrightarrow{\text{fast}} CH_3\overset{\overset{\displaystyle H\overset{\cdot\cdot}{O}:}{|}}{C}HCH_3 + H{-}\overset{+}{\underset{H}{\overset{\cdot\cdot}{O}}}{-}H$$

ORGANIC
Chemistry Now™
Click Mechanisms to view an animation of the **Acid-Catalyzed Hydration of Propene**

Example 6.5

Propose a mechanism for the acid-catalyzed hydration of methylidenecyclohexane to give 1-methylcyclohexanol. Which step in your mechanism is rate determining?

Solution

Propose a three-step mechanism similar to that for the acid-catalyzed hydration of propene.

Step 1: Proton transfer from the acid catalyst to the alkene gives a 3° carbocation intermediate. Formation of the 3° carbocation intermediate is rate determining.

$$\text{(cyclohexane)}{=}CH_2 + H{-}\overset{\cdot\cdot}{\underset{H}{\overset{+}{O}}}{-}H \xrightarrow[\text{determining}]{\text{slow, rate}} \text{(cyclohexane)}\overset{+}{-}CH_3 + :\overset{\cdot\cdot}{\underset{H}{O}}{-}H$$

A 3° carbocation
intermediate

Step 2: Reaction of the 3° carbocation intermediate (an electrophile) with water (a nucleophile) completes the valence shell of carbon and gives an oxonium ion.

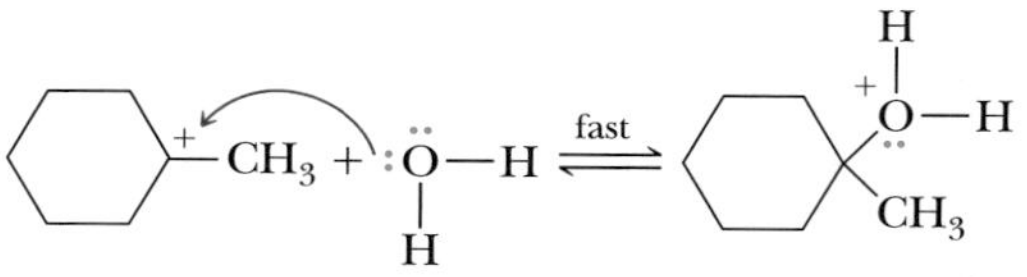

An oxonium ion

Step 3: Proton transfer from the oxonium ion to water gives the alcohol and generates a new acid catalyst.

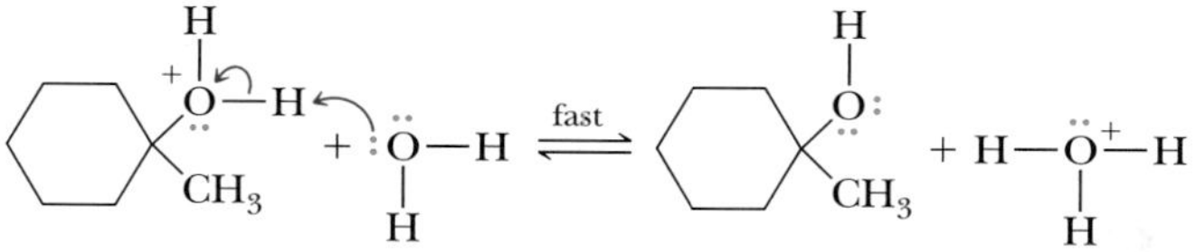

Problem 6.5

Propose a mechanism for the acid-catalyzed hydration of 1-methylcyclohexene to give 1-methylcyclohexanol. Which step in your mechanism is rate determining?

C. Carbocation Rearrangements

As we have seen in the preceding discussions, the expected product of electrophilic addition to a carbon-carbon double bond involves rupture of the pi bond and formation of two new sigma bonds in its place. In addition of HCl to 3,3-dimethyl-1-butene, however, only 17% of 2-chloro-3,3-dimethylbutane, the expected product, is formed. The major product is 2-chloro-2,3-dimethylbutane, a compound with a different connectivity of its atoms compared with that in the starting alkene. We say that formation of 2-chloro-2,3-dimethylbutane involves a **rearrangement.** Typically, either an alkyl group or a hydrogen migrates, each with its bonding electrons, from an adjacent atom to the electron-deficient atom. In the rearrangements we examine in this chapter, migration is to an adjacent electron-deficient carbon atom bearing a positive charge.

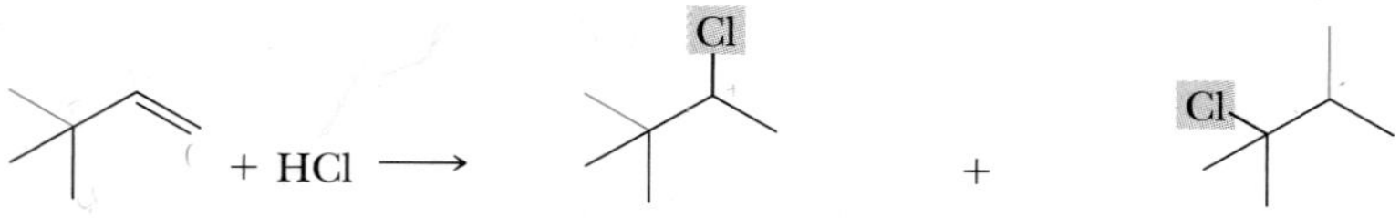

Rearrangement A change in connectivity of the atoms in a product compared with the connectivity of the same atoms in the starting material.

3,3-Dimethyl-1-butene   2-Chloro-3,3-dimethylbutane (the expected product; 17%)   2-Chloro-2,3-dimethylbutane (the major product; 83%)

1,2-Shift A type of rearrangement in which an atom or group of atoms with its bonding electrons moves from one atom to an adjacent electron-deficient atom.

Formation of the rearranged product in this reaction can be accounted for by the following mechanism, the key step of which is a type of rearrangement called a **1,2-shift.** In the rearrangement shown in Step 2, the migrating group is a methyl group with its bonding electrons.

Mechanism *Carbocation Rearrangement in the Addition of HCl to an Alkene*

Step 1: Proton transfer to the alkene gives a 2° carbocation intermediate.

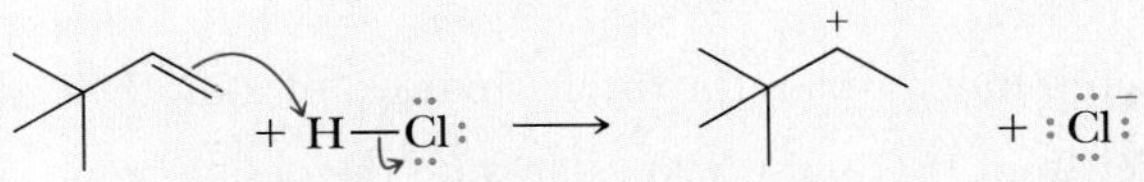

A 2° carbocation
intermediate

Step 2: Migration of a methyl group with its bonding electrons from an adjacent carbon to the positively charged carbon of the 2° carbocation gives a more stable 3° carbocation. In this rearrangement, the major movement is that of the bonding electron pair with the methyl group following.

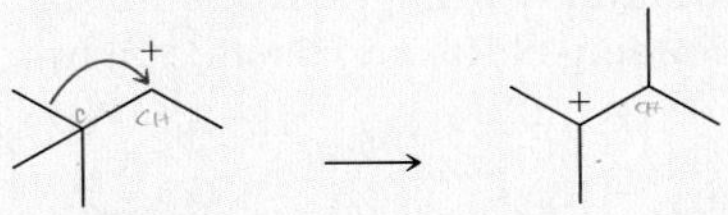

A 2° carbocation   A 3° carbocation
intermediate   intermediate

Step 3: Reaction of the 3° carbocation intermediate (an electrophile) with chloride ion (a nucleophile) gives the rearranged product.

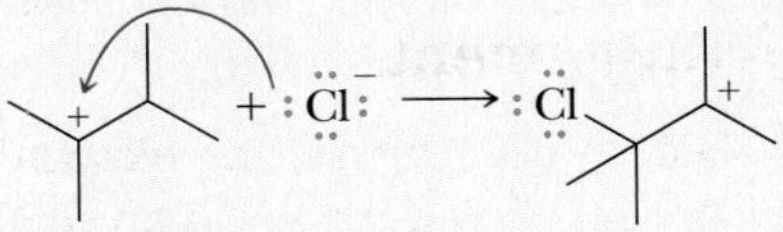

The driving force for this rearrangement is the fact that the less stable 2° carbocation is converted to a more stable 3° carbocation. From the study of this and other carbocation rearrangements, we find that 2° carbocations rearrange to more stable 2° or 3° carbocations. They rearrange in the opposite direction only under special circumstances such as where the relief of ring strain provides added driving force. 1° Carbocations have never been observed for reactions taking place in solution.

Rearrangements also occur in the acid-catalyzed hydration of alkenes, especially when the carbocation formed in the first step can rearrange to a more stable carbocation. For example, acid-catalyzed hydration of 3-methyl-1-butene gives 2-methyl-2-butanol. In this example, the group that migrates is a hydrogen with its bonding pair of electrons, in effect a hydride ion $H:^-$.

ORGANIC
Chemistry ·*·Now ™
Click Tutorials to review **Stability of Carbocations and Rearrangement of Carbocations**

$$CH_3CHCH{=}CH_2 + H_2O \xrightarrow{H_2SO_4} CH_3CCH_2CH_3$$

with CH_3 above the second carbon of the reactant and CH_3 above, OH below the second carbon of the product

3-Methyl-1-butene   2-Methyl-2-butanol

Example 6.6

Propose a mechanism for the acid-catalyzed hydration of 3-methyl-1-butene to give 2-methyl-2-butanol.

Solution

Following is a four-step mechanism for the formation of 2-methyl-2-butanol.

Step 1: Proton transfer from H_3O^+, the acid catalyst, to the double bond of the alkene gives a 2° carbocation intermediate.

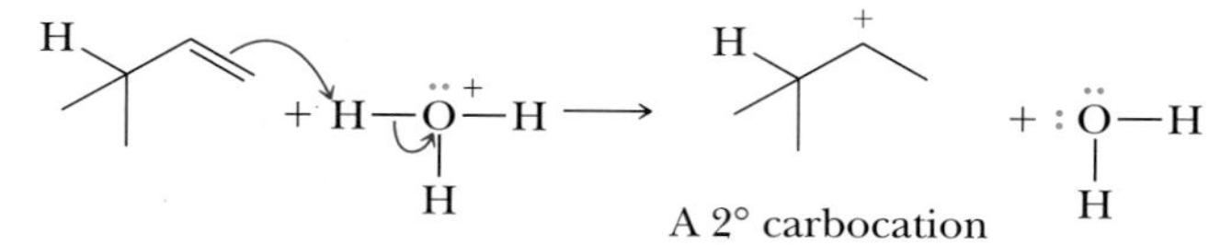

A 2° carbocation
intermediate

Step 2: The less stable 2° carbocation rearranges to a more stable 3° carbocation by migration of a hydrogen with its pair of bonding electrons (in effect, a hydride ion).

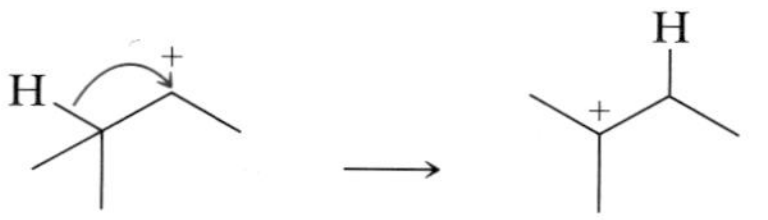

A 2° carbocation
intermediate

A 3° carbocation
intermediate

Step 3: Reaction of the 3° carbocation intermediate (an electrophile) with water (a nucleophile) completes the valence shell of carbon and gives an oxonium ion.

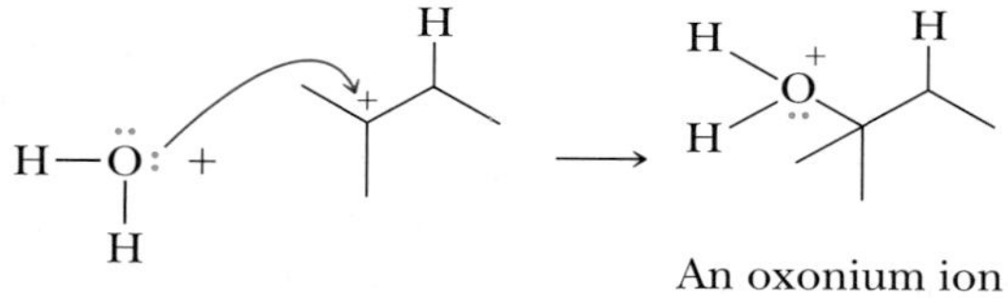

An oxonium ion

Step 4: Proton transfer from the oxonium ion to water gives the product and generates a new H_3O^+ to continue the hydration reaction.

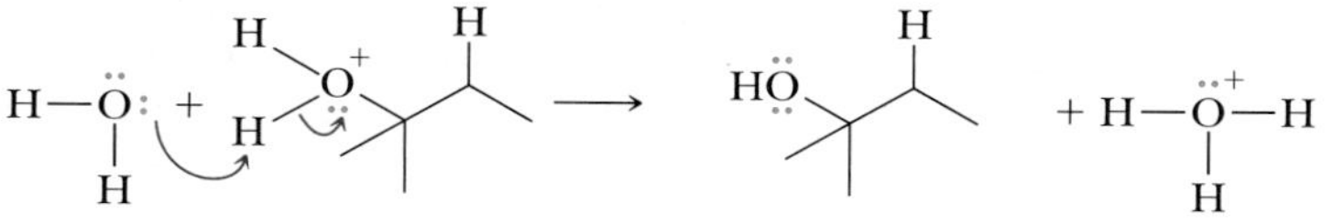

2-Methyl-2-butanol

Problem 6.6

The acid-catalyzed hydration of 3,3-dimethyl-1-butene gives 2,3-dimethyl-2-butanol as the major product. Propose a mechanism for the formation of this alcohol.

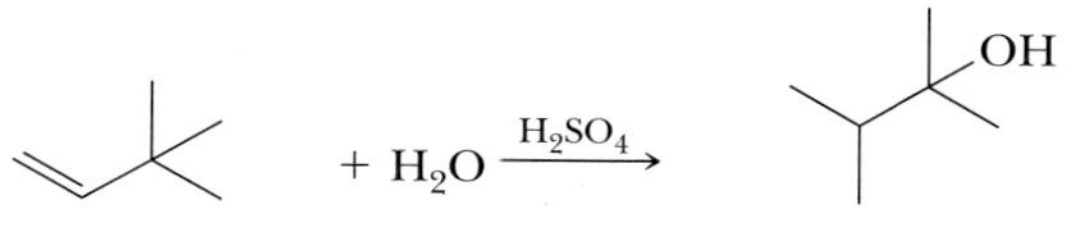

$$+ H_2O \xrightarrow{H_2SO_4}$$

3,3-Dimethyl-1-butene

2,3-Dimethyl-2-butanol

D. Addition of Bromine and Chlorine

Chlorine, Cl_2, and bromine, Br_2, react with alkenes at room temperature by adding halogen atoms to the two carbon atoms of the double bond with formation of two new carbon-halogen bonds. Fluorine, F_2, adds to alkenes, but, because its reactions are very fast and difficult to control, this reaction is not a useful laboratory procedure. Iodine, I_2, also adds, but the reaction is not preparatively useful. Halogenation with bromine or chlorine is generally carried out either with the pure reagents or by mixing them in an inert solvent such as CH_2Cl_2.

$$CH_3CH{=}CHCH_3 + Br_2 \xrightarrow{CH_2Cl_2} CH_3\overset{\overset{Br}{|}}{CH}-\overset{\overset{Br}{|}}{CH}CH_3$$

2-Butene   2,3-Dibromobutane

The addition of bromine or chlorine to a cycloalkene gives a *trans*-dihalocycloalkane formed as a racemic mixture. The addition of bromine to cyclohexene, for example, gives *trans*-1,2-dibromocyclohexane:

Cyclohexene   *trans*-1,2-Dibromocyclohexane
                                  (a racemic mixture)

At first glance, the two *trans* enantiomers may appear to be the same structure. However, there is no plane of symmetry or center of symmetry so they are both chiral. You should make molecular models of both enantiomers and convince yourself that they are indeed nonsuperposable mirror images, and not just the same molecule viewed from a different perspective.

We discuss the stereochemistry of addition on Cl_2 and Br_2 to alkenes in more detail in Section 6.7. For now, it is sufficient to point out that these reactions proceed with anti (from the opposite side or face) addition of halogen atoms; that is, they occur with **anti stereoselectivity.**

Reaction of bromine with an alkene is a particularly useful qualitative test for the presence of a carbon-carbon double bond. If we dissolve bromine in dichloromethane, the solution is red. Both alkenes and dibromoalkanes are colorless. If we now mix a few drops of the bromine solution with an alkene, a dibromoalkane is formed, and the red solution becomes colorless.

Stereoselective reaction   A reaction in which one stereoisomer is formed in preference to all others. A stereoselective reaction may be enantioselective or diastereoselective, as the case may be.

Anti stereoselectivity   The addition of atoms or groups of atoms to opposite faces of a carbon-carbon double bond.

Example 6.7

Complete these reactions, showing the stereochemistry of the product.

(a) $+ Br_2 \xrightarrow{CH_2Cl_2}$ **(b)** $+ Cl_2 \xrightarrow{CH_2Cl_2}$

Solution

Addition of both Br_2 and Cl_2 occurs with anti stereoselectivity, which means that the halogen atoms are *trans* to each other in each product.

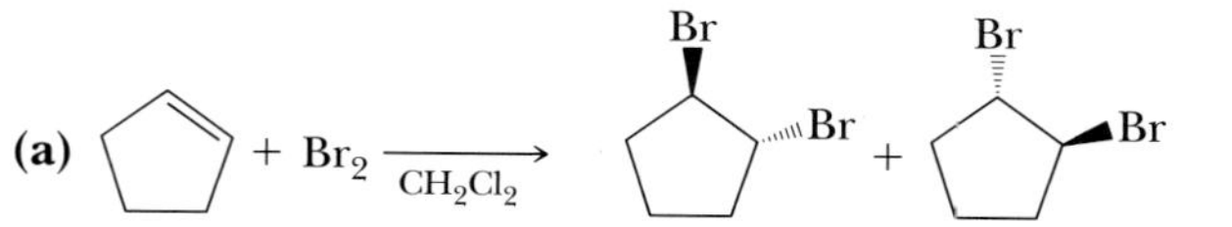

(a) + Br$_2$ $\xrightarrow{\text{CH}_2\text{Cl}_2}$ +

trans-1,2-Dibromocyclopentane
(a racemic mixture)

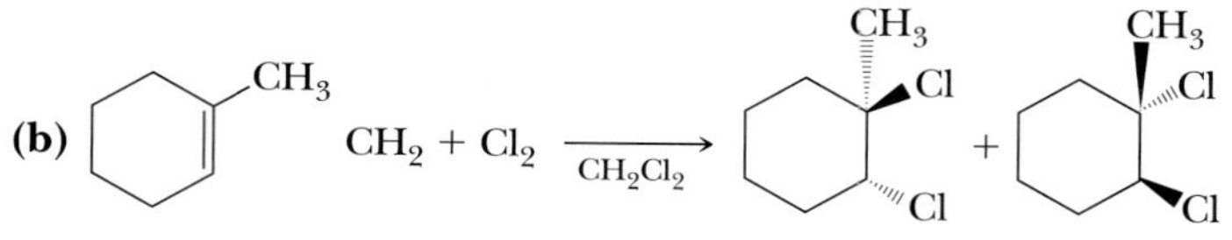

(b) CH$_2$ + Cl$_2$ $\xrightarrow{\text{CH}_2\text{Cl}_2}$ +

(a racemic mixture)

Problem 6.7

Complete these reactions.

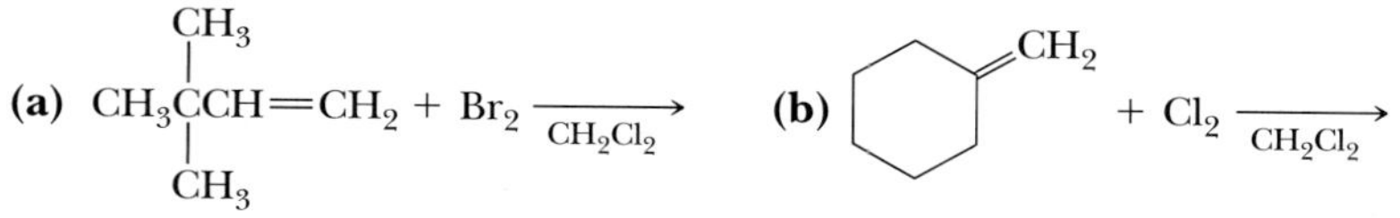

(a)
$$\underset{\underset{CH_3}{|}}{\overset{\overset{CH_3}{|}}{CH_3CCH}}=CH_2 + Br_2 \xrightarrow{\text{CH}_2\text{Cl}_2}$$

(b) CH$_2$ + Cl$_2$ $\xrightarrow{\text{CH}_2\text{Cl}_2}$

Anti Stereoselectivity and Bridged Halonium Ion Intermediates

We explain the addition of bromine and chlorine to alkenes and its anti stereoselectivity by the following two-step mechanism.

Mechanism   *Addition of Bromine with Anti Stereoselectivity*

Step 1: Reaction is initiated by interaction of the pi electrons of the alkene with bromine (or chlorine as the case may be) to form an intermediate in which bromine bears a positive charge. A bromine atom bearing a positive charge is called a **bromonium ion,** and the cyclic structure of which it is a part is called a **bridged bromonium ion.**

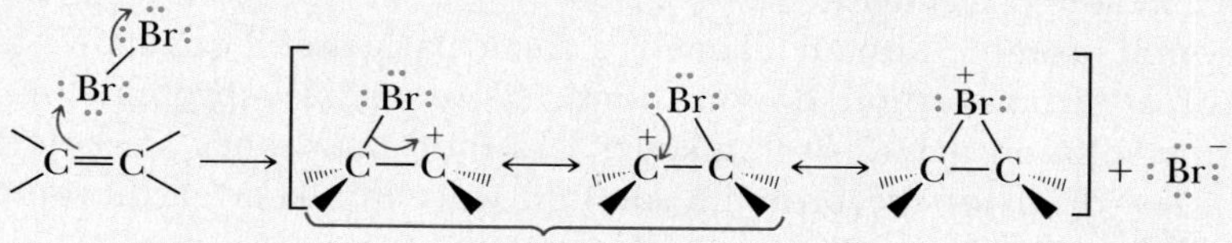

These carbocations
are major
contributing structures

The bridged
bromonium ion
retains the geometry

Although a bridged bromonium ion may look odd because it has two bonds to bromine, it is nevertheless an acceptable Lewis structure. Calculation of formal charge places a positive charge on bromine. This intermediate is a hybrid of three resonance-contributing structures. The open carbocation structures play a significant role. Because of the planarity of the atoms forming the pi bond, the bridged bromonium ion can form with equal probability on the top or bottom face of the alkene.

Step 2: Attack of bromide ion (a nucleophile) on carbon from the side opposite the bromonium ion (an electrophile) opens the three-membered ring to give the anti product.

The bromines are not only anti but also in the same plane (coplanar). Thus, we call this an anti-coplanar attack.

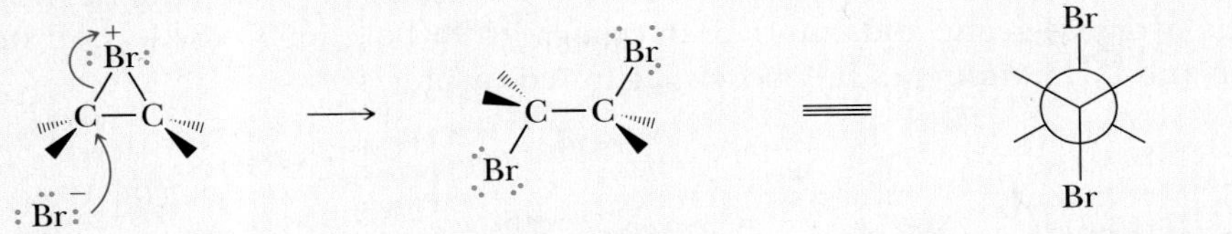

Anti (coplanar) orientation of added bromine atoms

Newman projection of the product

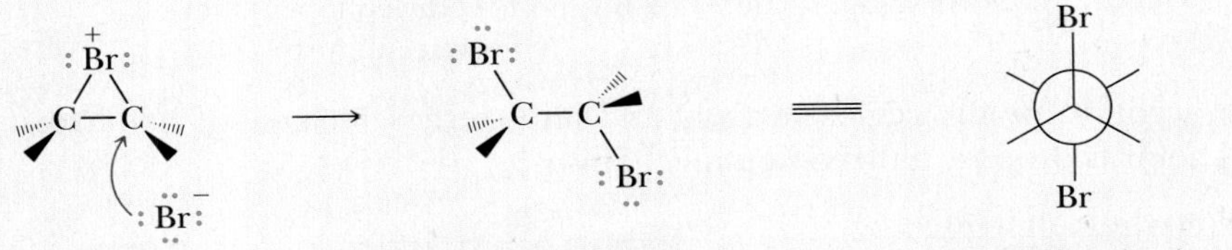

Anti (coplanar) orientation of added bromine atoms

Newman projection of the product

Addition of chlorine or bromine to cyclohexene and its derivatives gives a *trans* diaxial product because only axial positions on adjacent atoms of a cyclohexane ring are anti and coplanar. The initial *trans* diaxial conformation of the product is in equilibrium with the *trans* diequatorial conformation, and, in simple derivatives of cyclohexane, the *trans* diequatorial conformation is more stable and predominates. Because the original bromonium ion can form on either face of the double bond with equal probability, both *trans* enantiomers are formed as a racemic mixture.

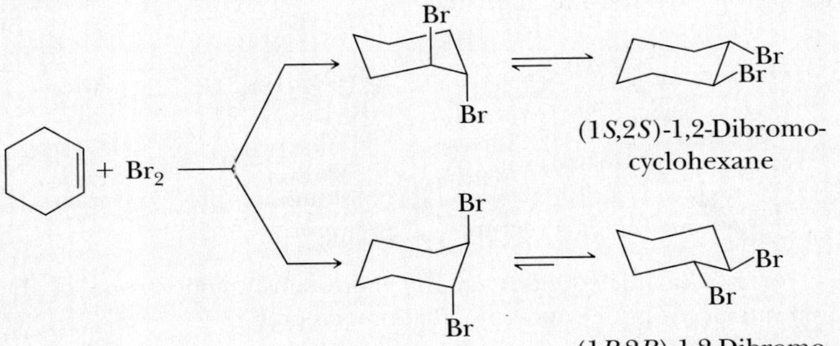

(1S,2S)-1,2-Dibromo-cyclohexane

(1R,2R)-1,2-Dibromo-cyclohexane

Trans diaxial conformations (less stable)

Trans diequatorial conformations (more stable)

ORGANIC
Chemistry ⚛ Now ™
Click Mechanisms to view an animation of the **Bromination of an Alkene**

E. Addition of HOCl and HOBr

Treating an alkene with Br_2 or Cl_2 in the presence of water results in addition of OH and Br, or OH and Cl, to the carbon-carbon double bond to give a **halohydrin.**

$$CH_3CH=CH_2 + Cl_2 + H_2O \longrightarrow CH_3\overset{\overset{\displaystyle HO}{|}}{C}H-\overset{\overset{\displaystyle Cl}{|}}{C}H_2 + HCl$$

Propene

1-Chloro-2-propanol
(a chlorohydrin)

Halohydrin A compound containing a halogen atom and a hydroxyl group on adjacent carbons; those containing Br and OH are bromohydrins, and those containing Cl and OH are chlorohydrins.

Addition of HOCl and HOBr is regioselective (halogen adds to the less substituted carbon atom) and anti stereoselective. Both the regioselectivity and anti stereoselectivity are illustrated by the addition of HOBr to 1-methylcyclopentene. Bromine and the hydroxyl group add anti to each other with Br bonding to the less substituted carbon and OH bonding to the more substituted carbon.

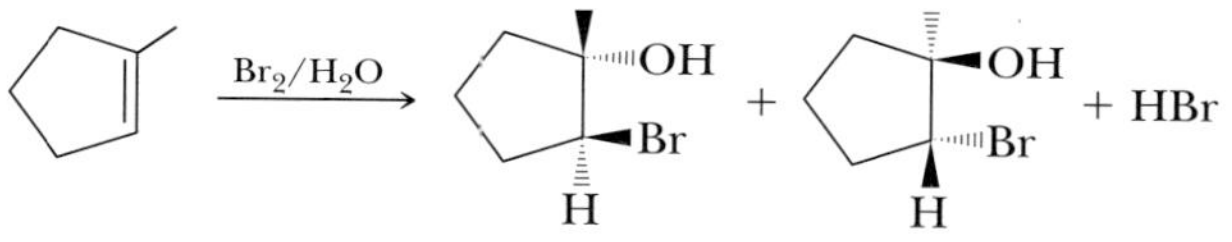

1-Methylcyclopentene   2-Bromo-1-methylcyclopentanol
                                    (a racemic mixture)

To account for the regioselectivity and anti stereoselectivity of halohydrin reactions, chemists propose a three-step mechanism.

Mechanism *Halohydrin Formation and Its Anti Stereoselectivity*

Step 1: Reaction of the pi electrons of the carbon-carbon double bond with bromine gives a bridged bromonium ion intermediate. This intermediate has some of the character of a carbocation (to account for the regioselectivity) and some of the character of a halonium ion (to account for the anti stereoselectivity). The secondary carbocation makes a substantial contribution to the structure of the hybrid; the primary carbocation is higher in energy and makes little contribution.

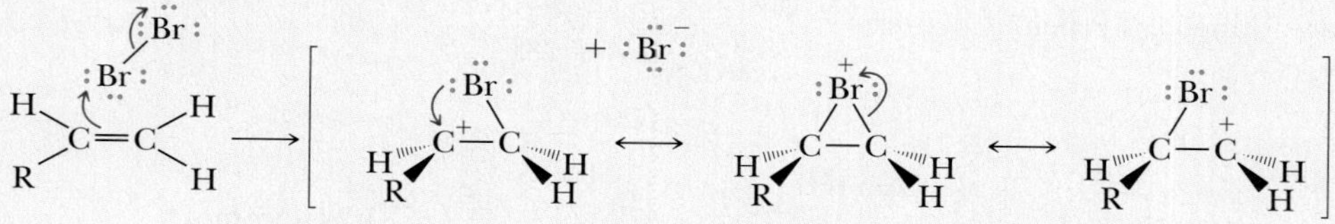

This carbocation   The bridged   This structure
is a major   bromonium ion plays   plays little role
contributing structure   an important role

Step 2: Attack of H_2O (a nucleophile) on the more substituted carbon of the bridged bromonium ion (an electrophile) opens the three-membered ring.

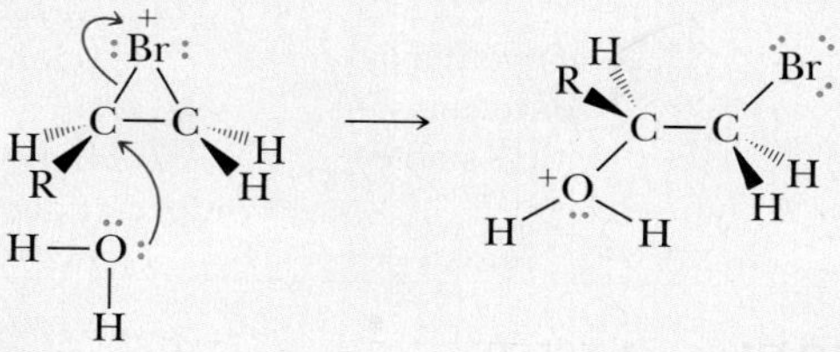

In the case of a bromonium ion derived from a symmetrical alkene, both carbons are attacked by H_2O with equal probability. In the case of unsymmetrical alkenes, as for example that derived from 2-methylpropene, there is preferential opening of the cyclic bromonium ion intermediate by attack of H_2O on the more substituted carbon of the alkene. At first glance, this may seem counterintuitive because the more substituted carbon might be considered less accessible to a nucleophile. However, the experimentally observed preferential attack at the more substituted carbon atom can be explained by a combination of two factors working together.

Carbocation Character

As the accompanying electrostatic potential maps show, there is more carbocation character on the more substituted carbon, which directs attack of the nucleophile preferentially to this carbon. Recall that alkyl groups stabilize carbocations, explaining the greater carbocation character at the more substituted carbon.

Activation Energy To Reach the Ring-Opening Transition State

As the accompanying electrostatic potential maps also show, the carbon-halogen bond to the more substituted carbon of the halonium ion is longer than the one to the less substituted carbon. This difference in bond lengths in the transition state means that the ring-opening transition state can be reached more easily by attack at the more substituted carbon.

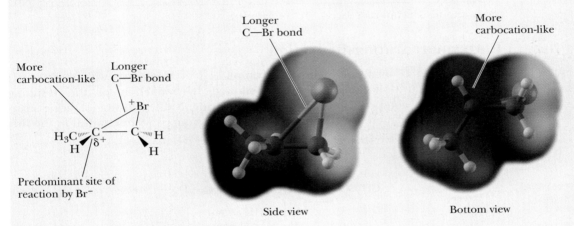

Bridged bromonim ion from propene.

Step 3: Proton transfer to water completes the reaction.

Example 6.8

Draw the structure of the bromohydrin formed by treating 2-methylpropene with Br_2/H_2O.

Solution

Addition is regioselective, with —OH adding to the more substituted carbon and —Br adding to the less substituted carbon.

$$CH_3-\underset{\underset{\displaystyle CH_3}{|}}{C}=CH_2 \ + \ Br_2 + H_2O \ \longrightarrow \ CH_3-\underset{\underset{\displaystyle HO \quad Br}{|\quad\;\;|}}{\overset{\overset{\displaystyle CH_3}{|}}{C}}-CH_2 \ + \ HBr$$

2-Methylpropene   1-Bromo-2-methyl-2-propanol
                                              (achiral)

Problem 6.8

Draw the structure of the chlorohydrin formed by treating 1-methylcyclohexene with Cl_2/H_2O.

F. Oxymercuration-Reduction

The hydration of an alkene can be accomplished by treating it with mercury(II) acetate (mercuric acetate) in water followed by reduction of the resulting organomercury compound with sodium borohydride, $NaBH_4$. In the following structural formulas for mercury(II) acetate, the acetate group is written in full in the first formula and abbreviated as AcO in second formula.

$$\underset{\text{Mercury(II) acetate}}{\underset{\text{(Mercuric acetate)}}{CH_3\overset{\overset{\displaystyle O}{\|}}{C}O-Hg-O\overset{\overset{\displaystyle O}{\|}}{C}CH_3 \qquad AcO-Hg-OAc}}$$

The result of oxymercuration followed by sodium borohydride reduction is Markovnikov addition of H—OH to an alkene.

$$\underset{\text{1-Hexene}}{\diagdown\diagup\diagdown\diagup} \quad \xrightarrow[\text{2. } NaBH_4]{\text{1. } Hg(OAc)_2, H_2O} \quad \underset{\text{2-Hexanol}}{\overset{\displaystyle OH}{\diagdown\diagup\diagdown\diagup}}$$

We discuss this reaction in two stages, first oxymercuration and then reduction.

Oxymercuration, the addition of mercury(II) to one carbon of the double bond and oxygen to the other, is illustrated by the first step in the two-step conversion of 1-hexene to 2-hexanol. Oxymercuration is regioselective: HgOAc becomes bonded to the less substituted carbon of the alkene, and OH of water becomes bonded to the more substituted carbon.

Oxymercuration-reduction A method for converting an alkene to an alcohol. The alkene is treated with mercury(II) acetate followed by reduction with sodium borohydride.

$$\underset{\text{1-Hexene}}{\diagdown\diagup\diagdown\diagup} + \underset{\substack{\text{Mercury(II)}\\\text{acetate}}}{Hg(OAc)_2} + H_2O \longrightarrow \underset{\substack{\text{An organomercury}\\\text{compound}}}{\overset{\displaystyle OH}{\underset{\displaystyle HgOAc}{\diagdown\diagup\diagdown\diagup}}} + \underset{\text{Acetic acid}}{CH_3\overset{\overset{\displaystyle O}{\|}}{C}OH}$$

Reduction of the organomercury compound by sodium borohydride, $NaBH_4$, replaces HgOAc by H.

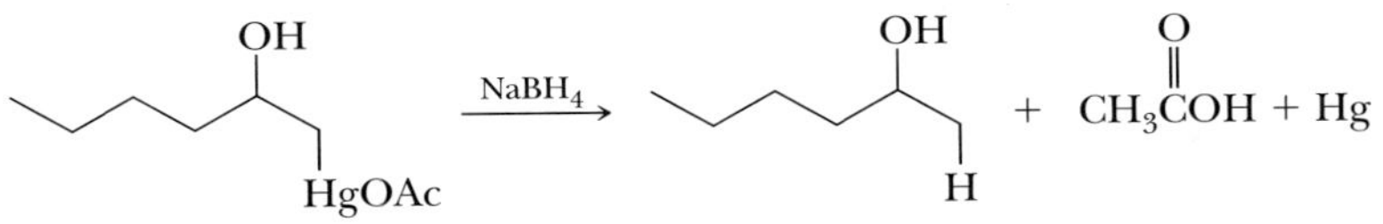

2-Hexanol   Acetic acid

Oxymercuration of 3,3-dimethyl-1-butene followed by $NaBH_4$ reduction gives 3,3-dimethyl-2-butanol exclusively and illustrates a very important feature of this reaction sequence: It occurs without rearrangement.

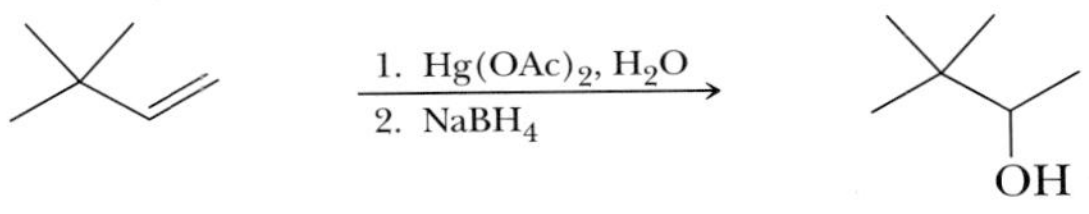

3,3-Dimethyl-1-butene   3,3-Dimethyl-2-butanol

You might compare the product of oxymercuration-reduction of 3,3-methyl-1-butene with the product formed by acid-catalyzed hydration of the same alkene (Section 6.3C). In the former, no rearrangement occurs. In the latter, the major product is 2,3-dimethyl-2-butanol, a compound formed by rearrangement. The fact that no rearrangement occurs during oxymercuration-reduction indicates that at no time is a free carbocation intermediate formed.

The stereoselectivity of oxymercuration is illustrated by the reaction of mercury(II) acetate with cyclopentene.

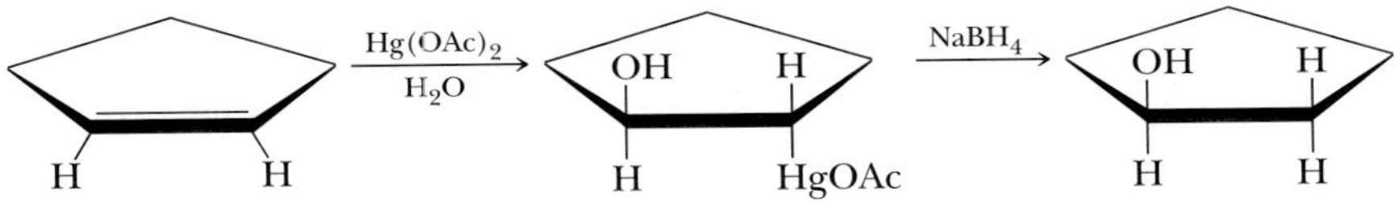

Cyclopentene   (Anti addition of   Cyclopentanol
                       OH and HgOAc)

The fact that oxymercuration is both regioselective and anti stereoselective has led chemists to propose the following mechanism for this reaction, which is closely analogous to that for the addition of Br_2 and Cl_2 to an alkene (Section 6.3D).

Mechanism   *Oxymercuration-Reduction of an Alkene*

Step 1: Dissociation of mercury(II) acetate gives $AcOHg^+$ (an electrophile) and acetate ion.

$$AcO-Hg-OAc \longrightarrow AcO-Hg^+ + AcO^-$$

(an electrophile)

Step 2: Attack of $AcOHg^+$ on the carbon-carbon double bond of the alkene forms a bridged mercurinium ion intermediate.

An open carbocation
intermediate
(a minor contributor)

A bridged mercurinium
ion intermediate
(the major contributor)

An open carbocation
intermediate
(a negligible contributor)

This intermediate closely resembles a bridged bromonium ion intermediate (Section 6.3D) with the difference that, unlike bromine, mercury has no electron pair to donate to form fully covalent bonds in the intermediate. Rather, in the bridged mercurinium ion intermediate, the two pi electrons of the carbon-carbon double bond form a ring containing three atoms bonded by two electrons. The open cation structure with the positive charge on the carbon 2° carbon is a minor contributing structure to the resonance hybrid. The open cation contributor with the positive charge on the 1° carbon is a negligible contributor.

Step 3: Anti attack of water (a nucleophile) on the bridged mercurinium ion intermediate (an electrophile) occurs at the more substituted carbon to open the three-membered ring.

Anti addition of
HgOAc and HOH

Proton transfer from this product to water completes oxymercuration of the alkene.

Step 4: Reduction of the C—HgOAc bond to a C—H bond gives the final product and metallic mercury.

The fact that oxymercuration occurs without rearrangement indicates that the intermediate formed in Step 2 is not a true carbocation but rather a resonance hybrid with largely the character of a bridged intermediate. Furthermore, the bridged structure allows us to account for the fact that the stereochemistry of this electrophilic addition is predominantly anti; the nucleophile attacks the bridged intermediate from the face opposite that occupied by mercury, as shown in Step 3.

We account for both the regioselectivity and the anti stereoselectivity of oxymercuration just as we did for the regioselectivity and anti stereoselectivity of halohydrin formation (Section 6.3E). Of the two carbons of the mercurinium ion intermediate, the more substituted carbon has a greater degree of partial positive charge and is attacked by the nucleophile, H_2O. In addition, computer modeling indicates that the carbon-mercury bond to the more substituted carbon of the bridged mercurinium ion intermediate is longer than the one to the less substituted carbon, which means that the ring-opening transition state is reached more easily by attack at the more substituted carbon.

6.4 Hydroboration-Oxidation

The net reaction from hydroboration and subsequent oxidation of an alkene is hydration of a carbon-carbon double bond. Because hydrogen is added to the more substituted carbon of the double bond and —OH to the less substituted carbon, we refer to the regiochemistry of hydroboration and subsequent oxidation as non-Markovnikov hydration:

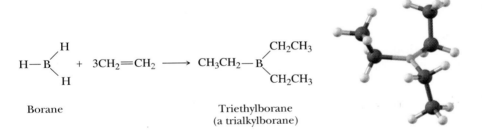

1-Hexene   1-Hexanol

Hydroboration-oxidation of alkenes is a valuable laboratory method for the regioselective and stereoselective hydration of alkenes. Furthermore, this sequence of reactions occurs without rearrangement.

Hydroboration is the addition of borane, BH_3, to an alkene to form a trialkylborane. The overall reaction occurs in three successive steps. BH_3 reacts with one molecule of alkene to form an alkylborane, then with a second molecule of alkene to form a dialkylborane, and finally with a third molecule of alkene to form a trialkylborane.

> **Hydroboration-oxidation** A method for converting an alkene to an alcohol. The alkene is treated with borane (BH_3) to give a trialkylborane, which is then oxidized with alkaline hydrogen peroxide to give the alcohol.

$$H-B\begin{matrix}H\\\\H\end{matrix} \quad + \quad 3CH_2{=}CH_2 \quad \longrightarrow \quad CH_3CH_2-B\begin{matrix}CH_2CH_3\\\\CH_2CH_3\end{matrix}$$

Borane   Triethylborane
(a trialkylborane)

Borane cannot be prepared as a pure compound because it dimerizes to diborane, B_2H_6, a toxic gas that ignites spontaneously in air.

$$2BH_3 \rightleftharpoons B_2H_6$$

Borane   Diborane

However, BH_3 forms stable Lewis acid-base complexes with ethers. Borane is most often used as a commercially available solution of BH_3 in THF.

$$2\ \overset{\frown}{\underset{\smile}{\bigcirc}}{:}\ddot{O}{:} \quad + \quad B_2H_6 \rightleftharpoons 2\ \overset{\frown}{\underset{\smile}{\bigcirc}}{:}\overset{+}{O}{-}\overset{-}{B}H_3$$

Tetrahydrofuran $BH_3 \cdot THF$
(THF)

Boron, atomic number 5, has three electrons in its valence shell. To bond with three other atoms, boron uses sp^2 hybrid orbitals. The unoccupied $2p$ orbital of boron is perpendicular to the plane created by boron and the three other atoms to which it is bonded. An example of a stable, trivalent boron compound is boron trifluoride, BF_3, a planar molecule with F—B—F bond angles of 120° (Section 1.2E). Because of the vacant $2p$ orbital in the valence shell of boron, BH_3, BF_3, and all other trivalent compounds of boron are electrophiles. These compounds of boron closely resemble carbocations, except that they are electrically neutral. BH_3 is a planar molecule with H—B—H bond angles of 120°.

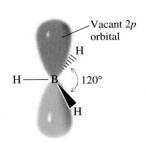

Addition of borane to alkenes is regioselective and stereoselective.

- Regioselective: In addition of borane to an unsymmetrical alkene, boron becomes bonded predominantly to the less substituted carbon of the double bond.
- Stereoselective: Hydrogen and boron add from the same face of the double bond; that is, the reaction is **syn** (from the same side) **stereoselective.**

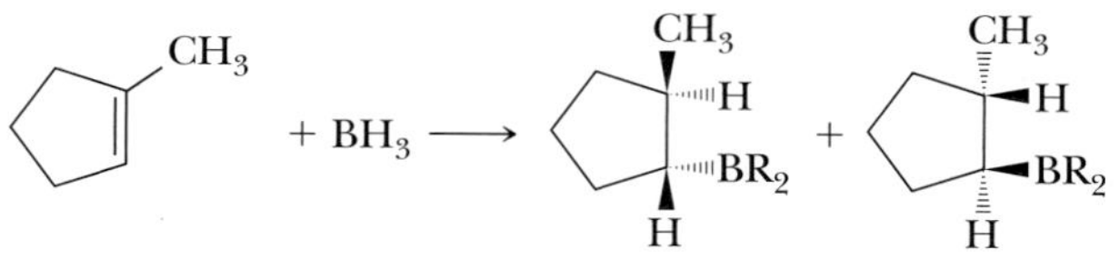

Syn stereoselective The addition of atoms or groups of atoms to the same face of a carbon–carbon double bond.

Both the regioselectivity and syn stereoselectivity are illustrated by hydroboration of 1-methylcyclopentene.

1-Methylcyclopentene

(Syn addition of BH_3)
(R = 2-methylcyclopentyl)
(a racemic mixture)

ORGANIC **Chemistry•Now**™
Click Mechanisms to view an animation of the **Hydroboration of an Alkene**

Mechanism *Hydroboration*

The addition of borane to an alkene is initiated by coordination of the vacant 2p orbital of boron (an electrophile) with the electron pair of the pi bond (a nucleophile). We account for the stereoselectivity of hydroboration by proposing the formation of a cyclic, four-center transition state. Boron and hydrogen add simultaneously and from the same face of the double bond, with boron adding to the less substituted carbon atom of the double bond.

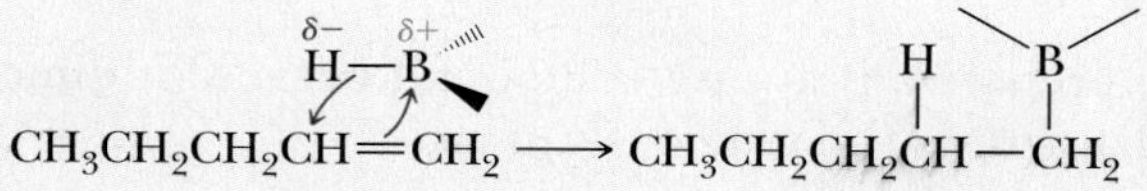

We account for the regioselectivity by a combination of steric and electronic factors. In terms of steric effects, boron, the larger part of the reagent, adds selectively to the less hindered carbon of the double bond, and hydrogen, the smaller part of the reagent, adds to the more hindered carbon. It is believed that the observed regioselectivity is due largely to steric effects.

Electronic effects probably also influence the regioselectivity. The electronegativity of hydrogen (2.1) is slightly greater than that of boron (2.0); hence, there is a small degree of polarity (approximately 5%) to each B—H bond, with boron bearing a partial positive charge and hydrogen a partial negative charge. It is proposed that there is some degree of carbocation character in the transition state and that the partial positive charge is on the more substituted carbon.

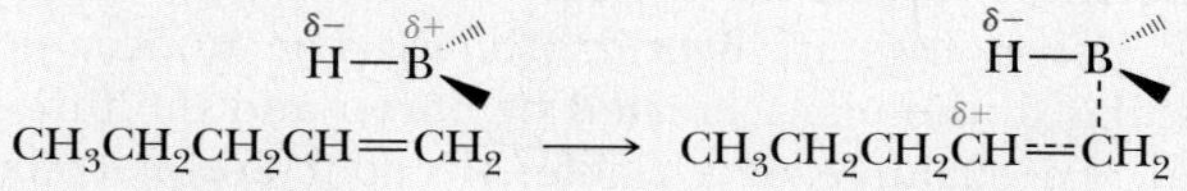

Starting reagents

Transition state
(some carbocation character exists in the transition state)

Trialkylboranes are rarely isolated. Rather, they are converted directly to other products formed by substitution of another atom (H, O, N, C, or halogen) for boron. One of the most important reactions of trialkylboranes is with hydrogen peroxide in aqueous sodium hydroxide. Hydrogen peroxide is an oxidizing agent and, under these conditions, oxidizes a trialkylborane to an alcohol and sodium borate, Na_3BO_3.

Mechanism   *Oxidation of a Trialkylborane by Alkaline Hydrogen Peroxide*

Step 1: Donation of a pair of electrons from a hydroperoxide ion (a nucleophile) to the boron atom of the trialkylborane (an electrophile) gives an intermediate in which boron has a filled valence shell and bears a negative formal charge.

$$HOOH \; + \; OH^- \; \rightleftharpoons \; HOO^- \; + \; HOH$$

Hydrogen   Hydroperoxide
peroxide   ion

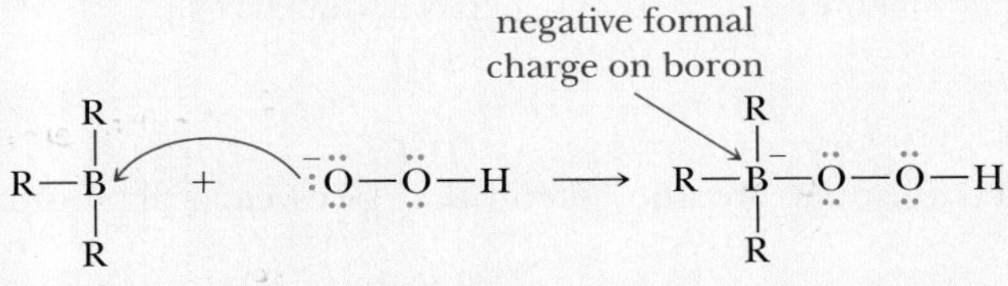

A trialkylborane   Hydroperoxide ion
(an electrophile)   (a nucleophile)

Step 2: Rearrangement of an R group with its pair of bonding electrons to an adjacent oxygen (a 1,2-shift) results in ejection of hydroxide ion.

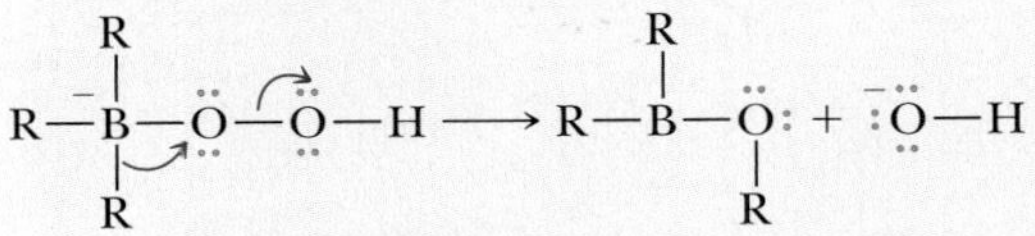

Two more reactions with hydroperoxide ion followed by rearrangements give a trialkylborate.

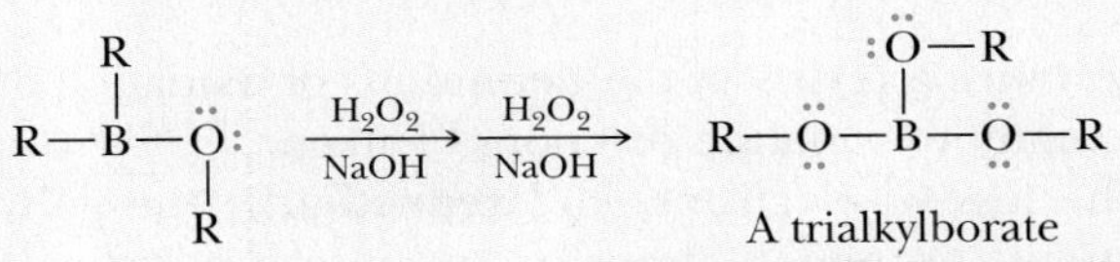

A trialkylborate
(a triester of boric acid)

Step 3: Reaction of the trialkylborate with aqueous NaOH gives the alcohol and sodium borate.

$$(RO)_3B \; + \; 3NaOH \longrightarrow 3ROH \; + \; Na_3BO_3$$

A trialkylborate   Sodium borate

Hydrogen peroxide oxidation of a trialkylborane is stereoselective in that the configuration of the alkyl group is retained; whatever the position of boron in relation to other groups in the trialkylborane, the OH group by which it is replaced occupies the same position. Thus, the net result of hydroboration-oxidation of an alkene is syn stereoselective addition of H and OH to a carbon-carbon double bond.

Example 6.9

Draw structural formulas for the alcohol formed by hydroboration-oxidation of each alkene.

(a) $CH_3\overset{\overset{\displaystyle CH_3}{|}}{C}{=}CHCH_3$ (b)

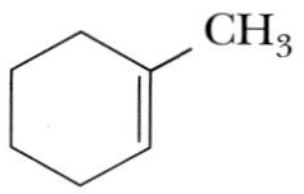

Solution

(a) 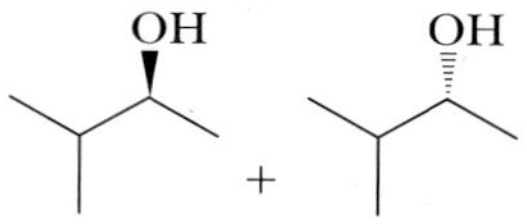   (a racemic mixture)

(b) 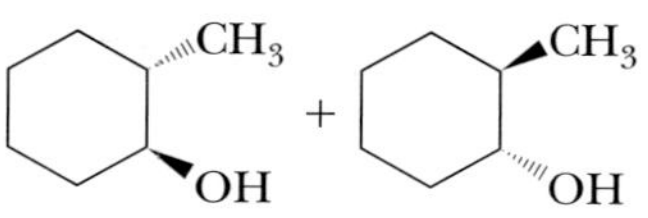   (a racemic mixture)

Problem 6.9

Draw structural formulas for the alkene that gives each alcohol on hydroboration-oxidation.

(a) 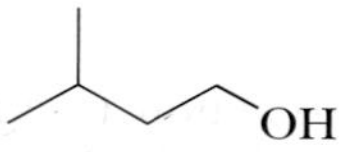   (b)

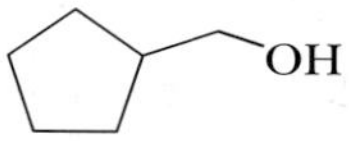

6.5 Oxidation

We begin this section with a review of the definitions of oxidation and reduction. We then consider two common oxidation reactions of alkene.

Oxidation   The loss of electrons. Alternatively, either the loss of hydrogens, the gain of oxygens, or both.

Reduction   The gain of electrons. Alternatively, either the gain of hydrogen, loss of oxygens, or both.

    Oxidation is the loss of electrons, and **reduction** is the gain of electrons. In the following reactions, propene is transformed into two different compounds by reactions we study in this and the following section. The first reaction involves reduction and the second involves oxidation. These equations, however, do not specify what reagents are necessary to bring about the particular transformation. Each does specify, however, that the carbon atoms of the products are derived from those of propene.

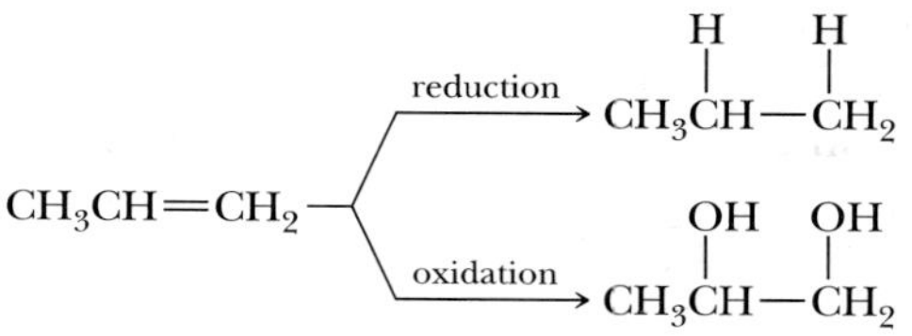

$$CH_3CH{=}CH_2 \begin{cases} \xrightarrow{\text{reduction}} CH_3\overset{\overset{\displaystyle H}{|}}{C}H{-}\overset{\overset{\displaystyle H}{|}}{C}H_2 \quad \text{Propane} \\[2em] \xrightarrow{\text{oxidation}} CH_3\overset{\overset{\displaystyle OH}{|}}{C}H{-}\overset{\overset{\displaystyle OH}{|}}{C}H_2 \quad \text{1,2-Propanediol} \end{cases}$$

How To   *Write a Balanced Half-Reaction*

One way to tell if these or other transformations involve oxidation, reduction, or neither is to use the method of balanced half-reactions.

To write a balanced half-reaction:

1. Write a half-reaction showing the organic reactant(s) and product(s).
2. Complete a material balance; that is, balance the number of atoms on each side of the half-reaction. To balance the number of oxygens and hydrogens for a reaction that takes place in acid solution, use H_2O for oxygens and then H^+ for hydrogens. For a reaction that takes place in basic solution, use H_2O and OH^-.
3. Complete a charge balance; that is, balance the charge on both sides of the half-reaction. To balance the charge, add electrons, e^-, to one side or the other. The equation completed in this step is a balanced half-reaction.

If electrons appear on the right side of the balanced half reaction, the reactant gives up electrons and is oxidized. If electrons appear on the left side of a balanced half reaction, the reactant has gained electrons and is reduced. If no electrons appear in the balanced half reaction, then the transformation involves neither oxidation nor reduction. Let us apply these steps to the transformation of propene to propane.

Step 1: Half-reaction $CH_3CH{=}CH_2 \longrightarrow CH_3CH_2CH_3$

Step 2: Material balance $CH_3CH{=}CH_2 + 2H^+ \longrightarrow CH_3CH_2CH_3$

Step 3: Balanced half-reaction $CH_3CH{=}CH_2 + 2H^+ + 2e^- \longrightarrow CH_3CH_2CH_3$

Because two electrons appear on the left side of the balanced half-reaction (Step 3), conversion of propene to propane is a two-electron reduction. To bring it about requires use of a reducing agent.

A balanced half-reaction for the transformation of propene to 1,2-propanediol requires two electrons on the right side of the equation for a charge balance; this transformation is a two-electron oxidation.

Balanced half-reaction: $CH_3CH{=}CH_2 + 2H_2O \longrightarrow \underset{\text{1,2-Propanediol}}{CH_3\overset{OH}{\underset{|}{C}}H{-}\overset{OH}{\underset{|}{C}}H_2} + 2H^+ + 2e^-$

Propene

Following is a balanced half-reaction for the transformation of propene to 2-propanol.

Balanced half-reaction: $CH_3CH{=}CH_2 + H_2O \longrightarrow \underset{\text{2-Propanol}}{CH_3\overset{OH}{\underset{|}{C}}HCH_3}$

Propene

Because no electrons are required to achieve an electrical balance in the half-reaction, conversion of propene to 2-propanol is neither oxidation nor reduction; this reaction can be brought about by acid-catalyzed hydration of propene.

It is important to realize that this strategy for recognizing oxidation and reduction is only that, a strategy. In no way does it give any indication of how a particular oxidation or reduction might be carried out in the laboratory. For example, the balanced half-reaction for the transformation of propene to propane requires $2H^+$ and $2e^-$. Yet by far the most common laboratory procedure for reducing propene to propane does not involve H^+ at all; rather it involves molecular hydrogen, H_2, and a transition metal catalyst (Section 6.6).

Example 6.10

Use a balanced half-reaction to show that each transformation involves an oxidation.

(a) $CH_3CH_2CH_2OH \longrightarrow CH_3CH_2\overset{\overset{\displaystyle O}{\|}}{C}H$ (b) $CH_3CH=CH_2 \longrightarrow CH_3\overset{\overset{\displaystyle O}{\|}}{C}H + H\overset{\overset{\displaystyle O}{\|}}{C}H$

Solution

First complete a material balance and then a charge balance.

(a) $CH_3CH_2CH_2OH \longrightarrow CH_3CH_2\overset{\overset{\displaystyle O}{\|}}{C}H + 2H^+ + 2e^-$

(b) $CH_3CH=CH_2 + 2H_2O \longrightarrow CH_3\overset{\overset{\displaystyle O}{\|}}{C}H + H\overset{\overset{\displaystyle O}{\|}}{C}H + 4H^+ + 4e^-$

The first transformation is a two-electron oxidation; the second is a four-electron oxidation. To bring each about requires an oxidizing agent.

Problem 6.10

Use a balanced half-reaction to show that each transformation involves a reduction.

(a)

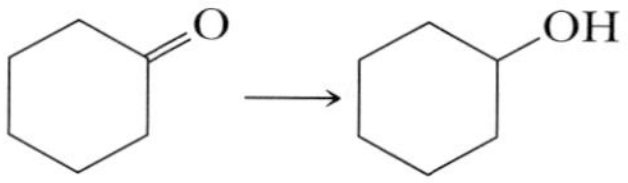

(b) $CH_3CH_2\overset{\overset{\displaystyle O}{\|}}{C}OH \longrightarrow CH_3CH_2CH_2OH$

As an alternative way to recognize oxidation/reduction, recall from your course in general chemistry that oxidation and reduction can be defined in terms of the loss or gain or oxygens or hydrogens. For organic compounds:

oxidation: The addition of O to and/or removal of H from a carbon atom

reduction: The removal of O from and/or addition of H to a carbon atom

Example 6.11

Tell which of these transformations are oxidations and which are reductions based on whether there is addition or removal of O or H.

(a) $CH_3CH_2OH \longrightarrow CH_3\overset{\overset{\displaystyle O}{\|}}{C}H$

(b) $CH_3\overset{\overset{\displaystyle O}{\|}}{C}CH_2\overset{\overset{\displaystyle O}{\|}}{C}OH \longrightarrow CH_3\overset{\overset{\displaystyle OH}{|}}{C}HCH_2\overset{\overset{\displaystyle O}{\|}}{C}OH$

(c) $CH_3CH=CH_2 \longrightarrow CH_3CH_2CH_3$

Solution

(a) Oxidation; there is a loss of two hydrogens.
(b) Reduction; there is a gain of two hydrogens.
(c) Reduction; there is a gain of two hydrogens.

Problem 6.11

Tell which of these transformations are oxidations and which are reductions based on whether there is addition or removal of O or H.

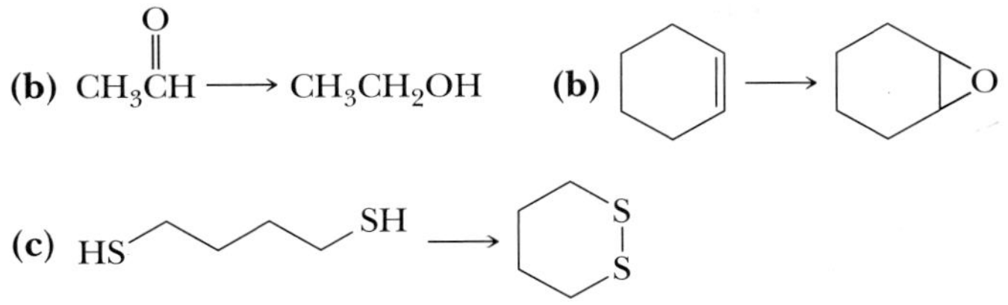

(b) CH$_3$CH $\longrightarrow$ CH$_3$CH$_2$OH (with C=O on first)   **(b)**

(c) HS $\diagup\diagdown\diagup$ SH $\longrightarrow$

A. OsO$_4$—Oxidation of an Alkene to a Glycol

Osmium tetroxide, OsO$_4$, and certain other transition metal oxides are effective oxidizing agents for the conversion of an alkene to a 1,2-diol (a **glycol**). Oxidation of an alkene by OsO$_4$ is syn stereoselective; it involves syn addition of an OH group to each carbon of the double bond. For example, oxidation of cyclopentene gives *cis*-1,2-cyclopentanediol. Note that both *cis* and *trans* isomers are possible for this glycol but that only the *cis* glycol forms in this oxidation.

Glycol A compound with hydroxyl (—OH) groups on adjacent carbons.

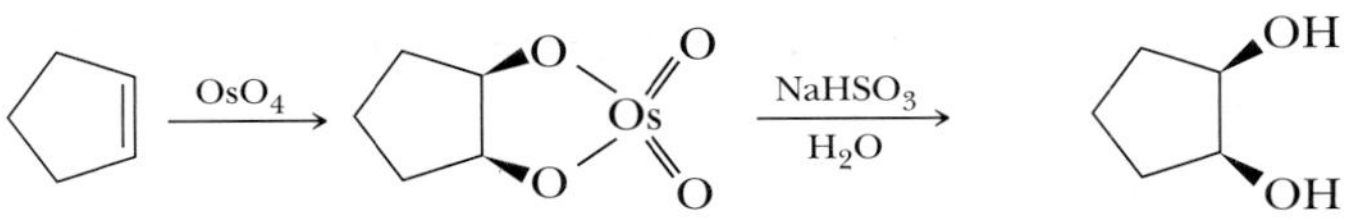

A cyclic osmate   *cis*-1,2-Cyclopentanediol (a *cis* glycol)

The syn stereoselectivity of the osmium tetroxide oxidation of an alkene is accounted for by the formation of a cyclic osmate in such a way that the five-membered osmium-containing ring is bonded in a *cis* configuration to the original alkene. Osmates can be isolated and characterized. Usually, the osmate is treated directly with a reducing agent, such as NaHSO$_3$, which cleaves the osmium-oxygen bonds to give a *cis* glycol and reduced forms of osmium.

| **Mechanism** | *Addition of Osmium Tetroxide to an Alkene* |

Step 1: The addition of OsO$_4$ to an alkene is probably a one-step reaction involving no intermediates (we study this type of reaction in more detail in Chapter 24.

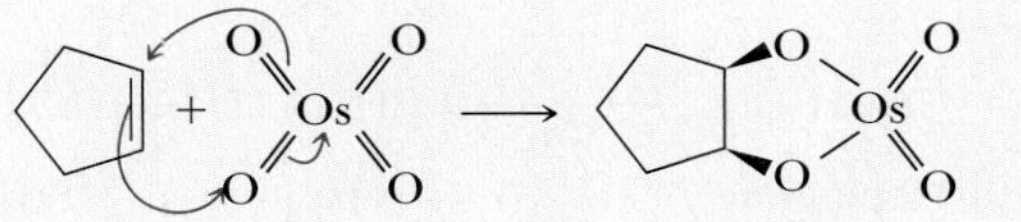

A cyclic osmate

Step 2: The mechanism of the reduction step is also uncertain, but it is thought to include both a reduction and a hydrolysis. We show this schematically as a pure reduction as if hydrogen is added across both Os—O bonds.

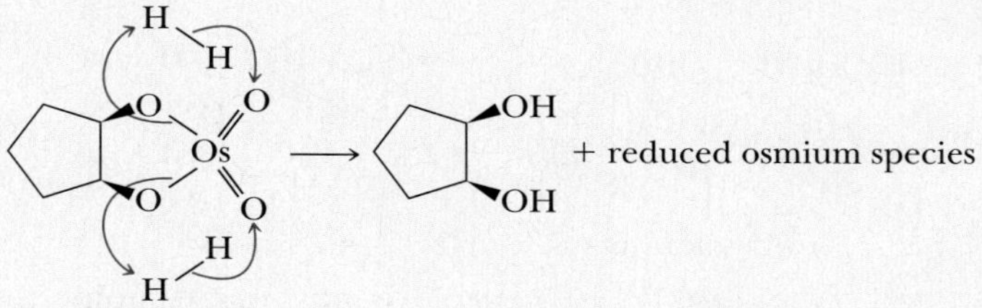

The drawbacks of OsO_4 are that it is both expensive and highly toxic. One strategy to circumvent the high cost is to use it in catalytic amounts along with a stoichiometric amount of another oxidizing agent whose purpose is to reoxidize the reduced forms of osmium and, thus, recycle the osmium reagent. Secondary oxidizing agents commonly used for this purpose are hydrogen peroxide and *tert*-butyl hydroperoxide. When this procedure is used, there is no need for a reducing step using $NaHSO_3$.

Hydrogen   *tert*-Butyl hydroperoxide
peroxide   (*t*-BuOOH)

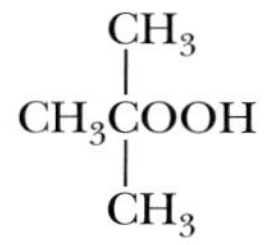

B. Ozone—Cleavage of a Carbon-Carbon Double Bond (Ozonolysis)

Treating an alkene with ozone, O_3, followed by a suitable work-up cleaves the carbon-carbon double bond and forms two carbonyl (C=O) groups in its place. This reaction is noteworthy because it is one of the very few organic reactions that breaks C—C bonds. The alkene is dissolved in an inert solvent, such as CH_2Cl_2, and a stream of ozone is bubbled through the solution. The products isolated from ozonolysis depend on the reaction conditions. Hydrolysis of the reaction mixture with water yields hydrogen peroxide, an oxidizing agent that can bring about further oxidations. To prevent side reactions caused by reactive peroxide intermediates, a weak reducing agent, most commonly dimethyl sulfide, $(CH_3)_2S$, is added during the work-up to reduce peroxides to water.

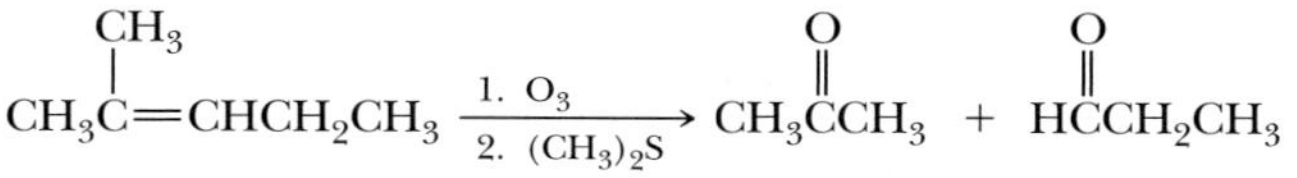

2-Methyl-2-pentene   Propanone   Propanal
                                      (a ketone)   (an aldehyde)

The initial product of reaction of an alkene with ozone is an adduct called a molozonide, which rearranges under the conditions of the reaction to an isomeric compound called an ozonide. Low-molecular-weight ozonides are explosive and are rarely isolated. They are treated directly with a weak reducing agent to give the carbonyl-containing products.

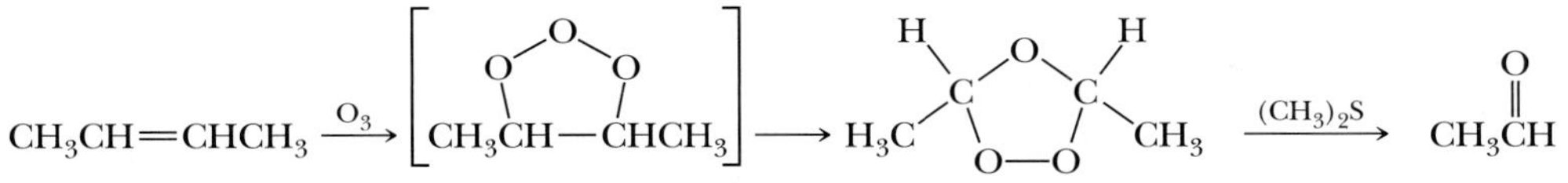

$$
CH_3CH{=}CHCH_3 \xrightarrow{O_3} \left[\overset{O-O}{\underset{CH_3CH-CHCH_3}{}} \right] \longrightarrow \underset{\text{An ozonide}}{} \xrightarrow{(CH_3)_2S} CH_3CH
$$

2-Butene · · · · · · A molozonide · · · · · · An ozonide · · · · · · Acetaldehyde

To understand how an ozonide is formed, we must first examine the structure of ozone. We can write this molecule as a hybrid of four contributing structures, all of which show separation of unlike charge.

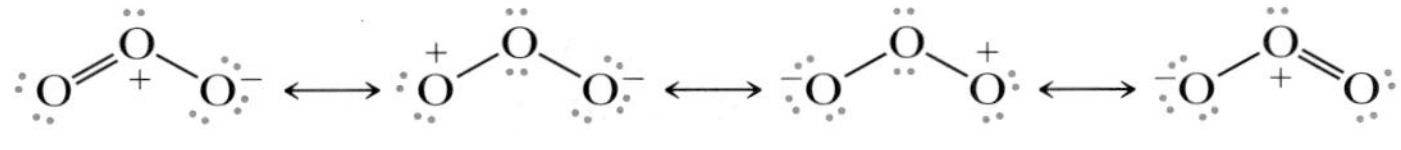

It is not possible to write a Lewis structure for ozone without separation of charges.

Mechanism   *Formation of an Ozonide*

Step 1: Ozone reacts with the alkene first as an electrophile and then as a nucleophile to give a molozonide. Even though this interaction is shown here as occurring in two steps, both steps are probably simultaneous (concerted).

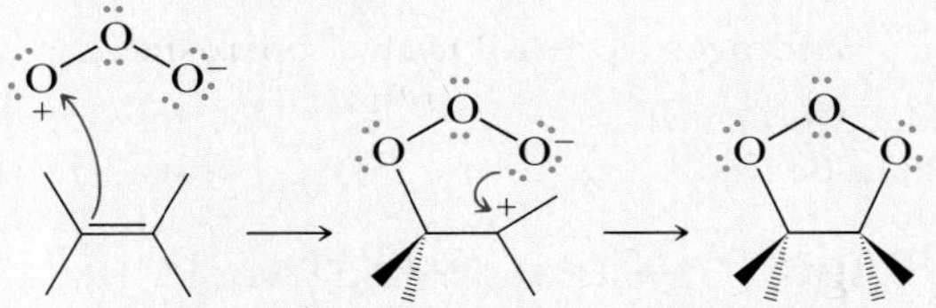

A molozonide

Step 2: Relocating valence electrons in the molozonide results in cleavage of one carbon-carbon and one oxygen-oxygen bond. The resulting fragments then recombine to form an ozonide.

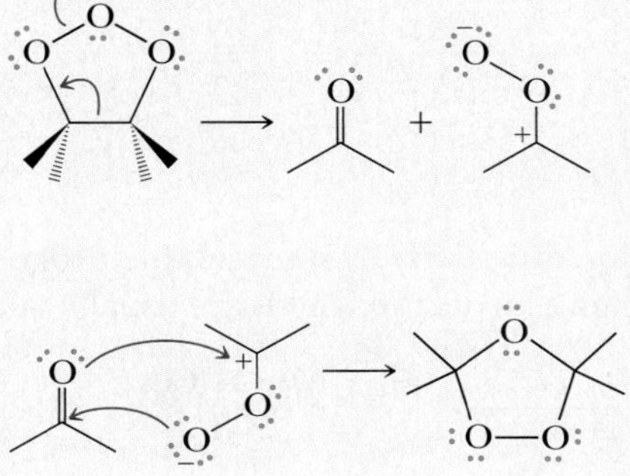

The zwitterionic fragment may also react with water or alcohols, but this need not concern us as the final outcome is the same.

Step 3: Reduction of the ozonide and cleavage results in the final carbonyl fragments.

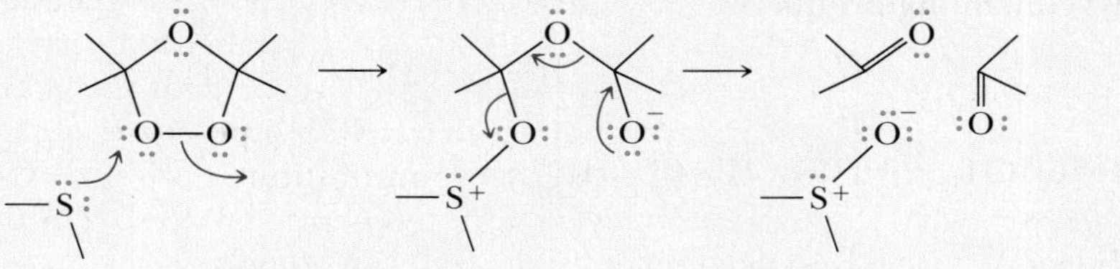

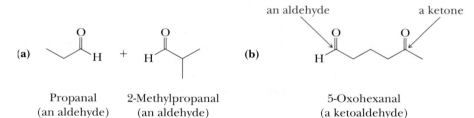

Example 6.12

Draw structural formulas for the products of the following ozonolysis reactions, and name the new functional groups formed in each oxidation.

(a) 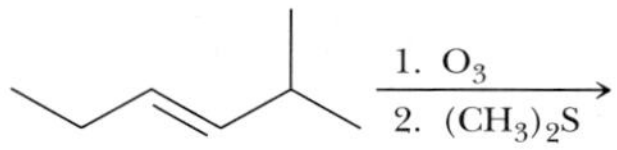 $\xrightarrow{\text{1. O}_3}$ $\xrightarrow{\text{2. (CH}_3)_2\text{S}}$ (b) 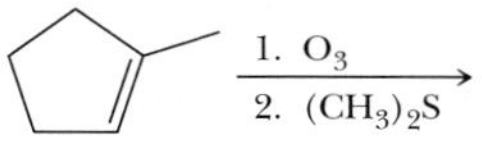 $\xrightarrow{\text{1. O}_3}$ $\xrightarrow{\text{2. (CH}_3)_2\text{S}}$

Solution

(a)

Propanal   2-Methylpropanal
(an aldehyde)   (an aldehyde)

(b)   an aldehyde   a ketone

5-Oxohexanal
(a ketoaldehyde)

Problem 6.12

What alkene of molecular formula C_6H_{12}, when treated with ozone and then dimethyl sulfide, gives the following product(s)?

(a)

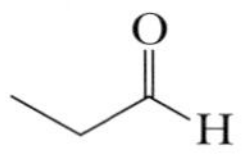

(only product)

(b)

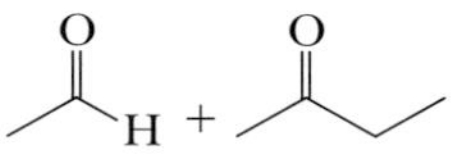

(equal moles of each)

(c)

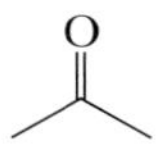

(only product)

6.6 Reduction

Most alkenes are reduced quantitatively by molecular hydrogen, H_2, in the presence of a transition metal catalyst to give an alkane. Yields are usually quantitative or nearly so.

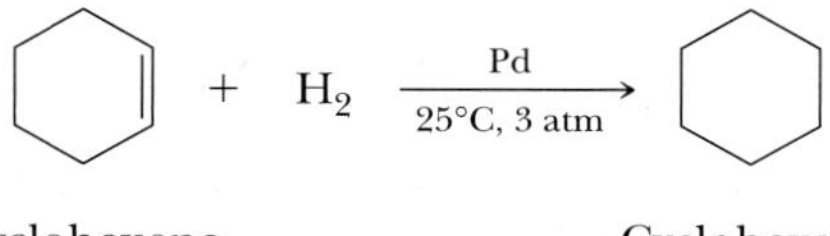

$\bigcirc$ + H_2 $\xrightarrow[\text{25°C, 3 atm}]{\text{Pd}}$ $\bigcirc$ $\Delta H^0 = -119.5$ kJ $(-28.6$ kcal$)/$mol

Cyclohexene   Cyclohexane

Although the addition of hydrogen to an alkene is exothermic, reduction is immeasurably slow in the absence of a catalyst. Commonly used transition metal catalysts include platinum, palladium, ruthenium, and nickel. Because the conversion of an alkene to an alkane involves reduction by hydrogen in the presence of a catalyst, the process is called **catalytic reduction** or, alternatively, **catalytic hydrogenation.**

Monosubstituted and disubstituted carbon-carbon double bonds react readily at room temperature under a few atmospheres pressure of hydrogen. Trisubstituted carbon-carbon double bonds require slightly elevated temperatures and pressures of up to 100 psi (pounds per square inch). Tetrasubstituted carbon-carbon double bonds are difficult to reduce and may require temperatures up to 275°C and hydrogen pressures of 1000 psi.

The metal catalyst is used as a finely powdered solid or may be supported on some inert material, such as finely powdered charcoal or alumina. The reaction is usually carried out by dissolving the alkene in ethanol or another nonreacting organic solvent, adding the solid catalyst, and then shaking the mixture under hydrogen gas at pressures of from 1 to 50 atm. Alternatively, the metal may be complexed with certain organic molecules and used in the form of a soluble complex (Section 6.7C).

Catalytic reduction is stereoselective, with the vast majority proceeding by syn addition of hydrogens to the carbon-carbon double bond.

■ Parr shaker-type hydrogenation apparatus.

A. Mechanism of Catalytic Reduction

The transition metals used in catalytic hydrogenation are able to adsorb large quantities of hydrogen onto their surfaces, probably by forming metal-hydrogen sigma bonds. Similarly, alkenes are also adsorbed on metal surfaces with formation of carbon-metal bonds. Addition of hydrogen atoms to the alkene occurs in two steps (Figure 6.7).

Under some experimental conditions, particularly with tetrasubstituted double bonds, some percentage of the product may appear to be formed by anti addition of hydrogen. Catalytic reduction of 1,2-dimethylcyclohexene, for example, yields predominantly *cis*-1,2-dimethylcyclohexane. Along with the *cis* isomer are formed lesser amounts of *trans*-1,2-dimethylcyclohexane as a racemic mixture.

(a)   **(b)**   **(c)**

Metal surface

Figure 6.7
The addition of hydrogen to an alkene involving a transition metal catalyst. (a) Hydrogen and the alkene are adsorbed on the metal surface and (b) one hydrogen atom is transferred to the alkene forming a new C—H bond. The other carbon remains adsorbed on the metal surface. (c) A second C—H bond is formed, and the alkane is desorbed.

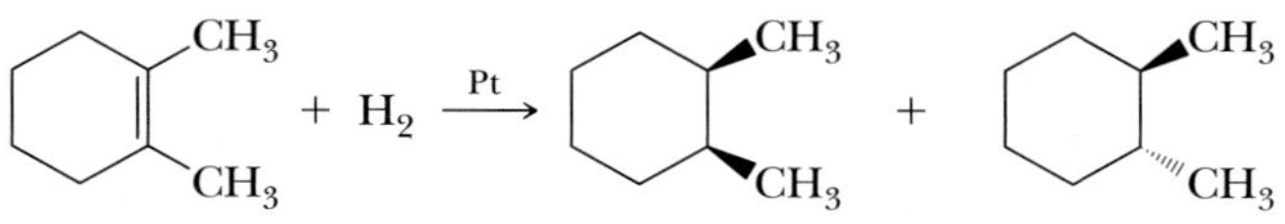

| 1,2-Dimethyl-cyclohexene | 70% to 85% cis-1,2-Dimethyl-cyclohexane | 30% to 15% trans-1,2-Dimethyl-cyclohexane |

(formed as a racemic mixture)

If addition of hydrogens is syn stereoselective, how then do we account for the formation of a *trans* product? It is proposed that before a second hydrogen can be delivered from the metal surface to complete the reduction, there is transfer of a hydrogen from a carbon atom adjacent to the original double bond to the metal surface. This hydrogen transfer, in effect, reverses the first step and forms a new alkene that is isomeric with the original alkene. As shown in the following equation, 1,2-dimethylcyclohexene undergoes isomerization on the metal surface to 1,6-dimethylcyclohexene. This alkene then leaves the metal surface. When it is later readsorbed and reduced, hydrogens are still added to it with syn stereoselectivity, but not necessarily from the same side as the original hydrogen.

ORGANIC
Chemistry Now™
Click Mechanisms to view an animation of **Catalytic Hydrogenation of an Alkene**

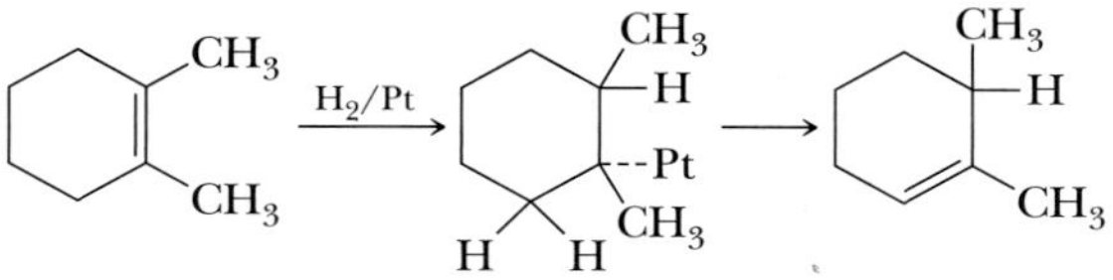

| 1,2-Dimethyl-cyclohexene | | 1,6-Dimethyl-cyclohexene |

B. Heats of Hydrogenation and the Relative Stabilities of Alkenes

The heat of hydrogenation of an alkene is its heat of reaction, ΔH^0, with hydrogen to form an alkane. Table 6.2 lists heats of reaction for the catalytic hydrogenation of several alkenes.

Three important points are derived from the information given in Table 6.2.

1. The reduction of an alkene to an alkane is an exothermic process. This observation is consistent with the fact that, during hydrogenation, there is net conversion of a weaker pi bond to a stronger sigma bond; that is, one sigma bond (H—H) and one pi bond (C=C) are broken, and two new sigma bonds (C—H) form. For a comparison of the relative strengths of sigma and pi bonds, refer to Section 1.8F.

2. Heats of hydrogenation depend on the degree of substitution of the carbon-carbon double bond; the greater the substitution, the lower the heat of hydrogenation. Compare, for example, heats of hydrogenation of ethylene (no substituents), propene (one substituent), 1-butene (one substituent), and the *cis* and *trans* isomers of 2-butene (two substituents).

Table 6.2 Heats of Hydrogenation of Several Alkenes

Name	Structural Formula	ΔH^0 [kJ (kcal)/mol]
Ethylene	$CH_2{=}CH_2$	−137 (−32.8)
Propene	$CH_3CH{=}CH_2$	−126 (−30.1)
1-Butene	$CH_3CH_2CH{=}CH_2$	−127 (−30.3)
cis-2-Butene	$\begin{array}{cc} H_3C & CH_3 \\ & C{=}C \\ H & H \end{array}$	−119.7 (−28.6)
trans-2-Butene	$\begin{array}{cc} H_3C & H \\ & C{=}C \\ H & CH_3 \end{array}$	−115.5 (−27.6)
2-Methyl-2-butene	$\begin{array}{cc} H_3C & CH_3 \\ & C{=}C \\ H_3C & H \end{array}$	−113 (−26.9)
2,3-Dimethyl-2-butene	$\begin{array}{cc} H_3C & CH_3 \\ & C{=}C \\ H_3C & CH_3 \end{array}$	−111 (−26.6)

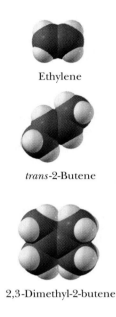

Ethylene

trans-2-Butene

2,3-Dimethyl-2-butene

3. The heat of hydrogenation of a *trans* alkene is lower than that of the isomeric *cis* alkene. Compare, for example, the heats of hydrogenation of *cis*-2-butene and *trans*-2-butene. Because reduction of each alkene gives butane, any difference in heats of hydrogenation must be caused by a difference in relative energy between the two alkenes (Figure 6.8). The alkene with the lower (less negative) value of ΔH^0 is the more stable.

We explain the greater stability of *trans* alkenes relative to *cis* alkenes in terms of steric strain (Section 2.6A), which can be visualized using space-filling models

Figure 6.8

Heats of hydrogenation of *cis*-2-butene and *trans*-2-butene. *Trans*-2-Butene is more stable than *cis*-2-butene by 4.2 kJ (1.0 kcal)/mol.

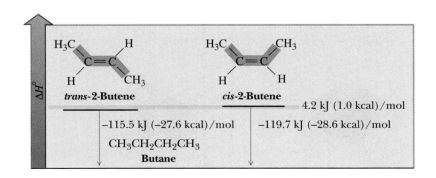

CONNECTIONS TO BIOLOGICAL CHEMISTRY

Trans *Fatty Acids: What They Are and How To Avoid Them*

Fats and oils were introduced in Chapter 5 (Connections to Biological Chemistry). Fats are added to processed foods to provide a desirable firmness along with a moist texture and pleasant taste. To supply the demand for dietary fats of the appropriate consistency, the *cis* double bonds of vegetable oils are partially hydrogenated using hydrogen in the presence of a Ni or other transition metal catalyst. The greater the extent of hydrogenation, the higher the melting point of the triglyceride. By controlling the degree of hydrogenation, a triglyceride with a melting point below room temperature can be converted to a semisolid or even a solid product.

Unfortunately, because of the reversible interaction of a carbon-carbon double bond with the Ni catalyst, some of the double bonds remaining in the triglyceride may be isomerized from the less stable *cis* configuration to the more stable *trans* configuration. Recall that a key step in catalytic hydrogenation involves the cleavage of the alkene pi bond and the bonding of its two carbons to the surface of the transition metal catalyst (Figure 6.7). This process is reversible, thus allowing equilibration between the *cis* and *trans* configurations.

The oils used for frying in fast-food restaurants are usually partially hydrogenated plant oils and, thus, contain substantial amounts of *trans* fatty acids that are transferred to the foods fried in them. Other major

sources of *trans* fatty acids in the diet include stick margarine, certain commercial bakery products, creme-filled cookies, chips, frozen breakfast foods, and cake mixes.

Recent studies have shown that consuming a significant amount of *trans* fatty acids can lead to serious health problems related to serum cholesterol levels. Low overall serum cholesterol and a decreased ratio of low-density lipoprotein (LDL) cholesterol to high-density lipoprotein (HDL) cholesterol are associated with good overall cardiovascular health. High serum cholesterol and an elevated ratio of LDL cholesterol to HDL cholesterol are linked to a high incidence of cardiovascular disease, especially atherosclerosis. Research has indicated that diets high in either saturated fatty acids or *trans* fatty acids raise the ratio of serum LDL cholesterol to HDL cholesterol and substantially increase the risk of cardiovascular disease.

The Food and Drug Administration announced recently that processed foods will soon be required to list the content of *trans* fatty acids. A diet low in saturated and *trans* fatty acids is recommended, along with more fish, whole grains, fruits, and vegetables. The recommendation is also for daily exercise, which is tremendously beneficial, regardless of diet.

(Figure 6.9). In *cis*-2-butene, the two —CH$_3$ groups are sufficiently close to each other that there is steric strain caused by repulsion between the two methyl groups. This repulsion is reflected in the larger heat of hydrogenation (decreased stability) of *cis*-2-butene compared with *trans*-2-butene [approximately 4.2 kJ (1.0 kcal)/mol]. Thus hydrogenation allows measurement of the strain energy of *cis*-2-butene directly.

Figure 6.9

Space-filling models of (a) *cis*-2-butene and (b) *trans*-2-butene.

Nonbonded interaction strain in *cis*-2-butene

No nonbonded interaction strain in *trans*-2-butene

6.7 Molecules Containing Chiral Centers as Reactants or Products

As the structure of an organic compound is altered in the course of a reaction, one or more chiral centers, usually at carbon, may be created, inverted, or destroyed. In Section 6.7A, we consider two alkene addition reactions in which a chiral molecule is created in an achiral environment. In doing so, we will illustrate the point that an optically active compound (that is, an enantiomerically pure compound or even an enantiomerically enriched compound) can never be produced from achiral starting materials reacting in an achiral environment. Then, in Section 6.7B, we consider the reaction of achiral starting materials reacting in a chiral environment, in this case in the presence of a chiral catalyst. We shall see that an enantiomerically pure product may be produced from achiral reagents if the reaction takes place in a chiral environment.

A. Reaction of Achiral Starting Materials in an Achiral Environment

The addition of bromine to 2-butene (Section 6.3D) gives 2,3-dibromobutane, a molecule with two chiral centers. Three stereoisomers are possible for this compound: a meso compound and a pair of enantiomers (Section 3.4). We now ask the following questions.

- What is the stereochemistry of the product; is it one enantiomer, a pair of enantiomers, the meso compound, or a mixture of all three stereoisomers?
- Is it optically active or optically inactive?

A partial answer is that the stereochemistry of the product formed depends on the configuration of the alkene. Let us first examine the addition of bromine to *cis*-2-butene.

The attack of bromine on *cis*-2-butene from either face of the planar part of the molecule gives the same bridged bromonium ion intermediate (Figure 6.10). Although this intermediate has two chiral centers, it has a plane of symmetry and is, therefore, meso. Attack of Br $^-$ on this meso intermediate from the side opposite that of the bromonium ion bridge gives a pair of enantiomers. Attack of bromide ion on carbon 2 gives the (2S,3S) enantiomer; attack on carbon 3 gives the (2R,3R) enantiomer. Attack of bromide ion occurs at equal rates at each carbon; therefore, the enantiomers are formed in equal amounts, and 2,3-dibromobutane is obtained as a racemic mixture (Figure 6.10). Thus, the product is chiral, but because it is a racemic mixture, it is optically inactive. We have described attack of Br—Br from one side of the carbon-carbon double bond. Attack from the opposite side followed by opening of the resulting bromonium ion intermediate produces the same two stereoisomers as a racemic mixture.

The addition of Br$_2$ to *trans*-2-butene leads to two enantiomeric bridged bromonium ion intermediates. Attack by Br$^-$ at either carbon atom of either bromonium ion intermediate gives the meso product, which is achiral and therefore optically inactive (Figure 6.11).

As we have seen in this discussion, the stereochemistry of the product formed by the addition of bromine to 2-butene depends on the stereochemistry of the starting alkene; the addition of halogen to *cis*-2-butene gives a racemic mixture, whereas the

Figure 6.10

Anti stereoselective addition of bromine to *cis*-2-butene gives 2,3-dibromobutane as a racemic mixture. The product is chiral, but because it is formed as a racemic mixture, it is optically inactive.

Step 1: Reaction of *cis*-2-butene with bromine forms bridged bromonium ions which are meso and identical.

cis-2-Butene
(achiral)

(meso bridged bromonium ion intermediate)

Step 2: Attack of bromide at carbons 2 and 3 occurs with equal probability to give enantiomeric products in a racemic mixture.

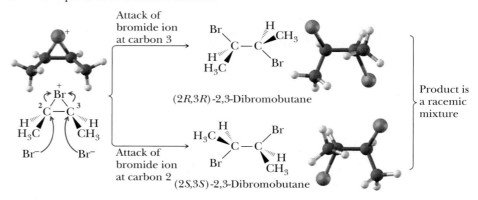

Attack of bromide ion at carbon 3

(2*R*,3*R*)-2,3-Dibromobutane

Attack of bromide ion at carbon 2

(2*S*,3*S*)-2,3-Dibromobutane

Product is a racemic mixture

Stereospecific reaction   (Section 6.7A) A special type of stereoselective reaction in which the stereochemistry of the product is dependent on the stereochemistry of the starting material.

addition of halogen to *trans*-2-butene gives meso product. Accordingly, we say that the addition of Br_2 or Cl_2 to an alkene is stereospecific. A **stereospecific reaction** is a special type of stereoselective reaction in which the stereochemistry of the product depends on the stereochemistry of the starting material.

In Section 6.5A, we studied oxidation of alkenes by osmium tetroxide in the presence of hydrogen peroxide. This oxidation results in syn stereoselective hydroxylation of the alkene to form a glycol. In the case of cycloalkenes, the product is a *cis* glycol. The first step in each oxidation involves formation of a cyclic osmate and is followed immediately by reaction with water to give a glycol.

cis-2-Butene
(achiral)

OsO_4
ROOH

(2*S*,3*R*)-2,3-Butanediol

(2*R*,3*S*)-2,3-Butanediol

identical; a meso compound

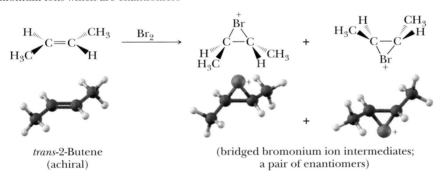

trans-2-Butene
(achiral)

OsO₄
ROOH

(2S,3S)-2,3-Butanediol

(2R,3R)-2,3-Butanediol

a pair of enantiomers; a racemic mixture

Syn hydroxylation of *cis*-2-butene gives meso-2,3-butanediol; because the meso com-pound is achiral, the product is optically inactive. Syn hydroxylation of *trans*-2-butene gives racemic 2,3-butanediol. Because the diol is formed as a racemic mixture, the product of the oxidation of the *trans* alkene is also optically inactive. Thus, the osmium tetroxide oxidation of an alkene is stereospecific; the stereochemistry of the product depends on the stereochemistry of the starting alkene.

Step 1: Reaction of *trans*-2-butene with bromine forms bridged bromonium ions which are enantiomers

trans-2-Butene
(achiral)

Br₂

(bridged bromonium ion intermediates; a pair of enantiomers)

Figure 6.11

Anti stereoselective addition of bromine to *trans*-2-butene gives meso-2,3-dibromobutane.

Step 2: Attack of bromide on either carbon of either enantiomer gives meso-2,3-dibromobutane

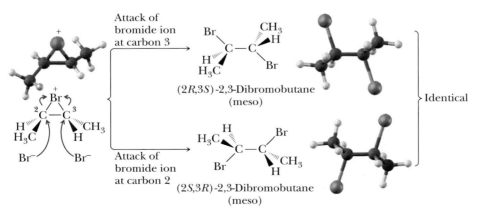

Attack of bromide ion at carbon 3

(2R,3S)-2,3-Dibromobutane
(meso)

Attack of bromide ion at carbon 2

(2S,3R)-2,3-Dibromobutane
(meso)

Identical

Notice that the stereochemistry of the product of the osmium tetroxide oxidation of *trans*-2-butene is opposite to that formed on the addition of bromine to *trans*-2-butene. Osmium tetroxide oxidation gives the glycol as a pair of enantiomers forming a racemic mixture. Addition of bromine to *trans*-2-butene gives the dibromoalkane as a meso compound. A similar difference is observed between the stereochemical outcomes of these reactions with *cis*-2-butene. The difference in outcomes occurs because bromination of an alkene involves anti addition, whereas oxidation by osmium tetroxide involves syn addition.

In this section, we have seen two examples of reactions in which achiral starting materials give chiral products. In each case, the chiral product is formed as a racemic mixture, (which is optically inactive) or as a meso compound (which is also optically inactive). These results illustrate a very important point about the creation of chiral molecules: an optically active product (that is, an enantiomerically pure compound or even an enantiomerically enriched compound) can never be produced from achiral starting materials and achiral reagents reacting under achiral conditions. Although the molecules of the product may be chiral, the product is always optically inactive (either meso or a pair of enantiomers).

We will encounter many reactions throughout the remainder of this course where achiral starting materials are converted into chiral products under achiral reaction conditions. For convenience, we often draw just one of the enantiomeric products, but you must always keep in mind that, under these experimental conditions, the product will always be optically inactive.

B. Reaction of a Chiral Starting Material in an Achiral Environment

Let us consider the bromination of (*R*)-4-*tert*-butylcyclohexene. Recall that, in derivatives of cyclohexane in which interconversion between one chair conformation and the other is not possible or is severely restricted, the *trans* diaxial product is isolated. If a cyclohexane ring contains a bulky alkyl group, such as *tert*-butyl (Section 2.6B), then the molecule exists overwhelmingly in a conformation in which the *tert*-butyl group is equatorial. Attack of bromine on enantiomerically pure (*R*)-4-*tert*-butylcyclohexene occurs at both faces of the six-membered ring. Because bromine atoms must add in an axial manner, each bromonium ion intermediate reacts with bromide ion to give the same product. In the favored chair conformation of this product, *tert*-butyl is equatorial, the bromine atoms remain axial, and only a single diastereomer is formed.

(*R*)-4-*tert*-Butyl-   (1*S*,2*S*,4*R*)-1,2-Dibromo-4-*tert*-butylcyclohexane
cyclohexene

In effect, the presence of the bulky *tert*-butyl group controls the orientation of the two bromine atoms added to the ring.

C. Reaction of Achiral Starting Materials in a Chiral Environment

The reduction of a carbon-carbon double bond can be carried out using hydrogen in the presence of a transition metal catalyst. Because hydrogen atoms are delivered to either side of the double bond with equal probability, if a new chiral center is created, equal amounts for both the *R* and *S* configurations will be produced.

Within the last three decades, chemists have discovered ways to embed transition metal hydrogenation catalysts in chiral molecules with the result that hydrogen can be delivered to only one face of the alkene. In catalytic reductions where a new chiral center is formed, a large enantiomeric excess of one enantiomer may be formed, and the reaction is said to be **enantioselective.** The most widely used of these chiral hydrogenation catalysts involve the chiral ligand 2,2-bis-(diphenylphosphanyl)-1,1'-binaphthyl, more commonly known as BINAP. BINAP has been resolved into its *R* and *S* enantiomers. The fact that BINAP can be resolved depends on restricted rotation about the single bond joining the two naphthalene rings. The two enantiomers are atropisomers (Section 3.2).

Enantioselective reaction A reaction that produces one enantiomer in preference to the other.

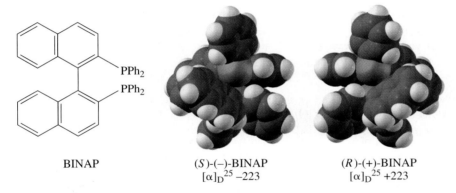

BINAP   (*S*)-(−)-BINAP   (*R*)-(+)-BINAP
                         $[\alpha]_D^{25}$ −223 $[\alpha]_D^{25}$ +223

Treating either enantiomer of BINAP with ruthenium chloride forms a complex in which ruthenium is bound as a complex ion in the chiral environment of the larger BINAP molecule. This complex is soluble in dichloromethane, CH_2Cl_2, and can be used as a homogeneous hydrogenation catalyst.

$$(R)\text{-BINAP} + RuCl_3 \longrightarrow (R)\text{-BINAP-Ru}$$

Using (*R*)-BINAP-Ru as a hydrogenation catalyst, (*S*)-naproxen is formed in greater than 98% enantiomeric excess. (*S*)-Naproxen is the antiinflammatory and pain reliever in Aleve and several other over-the-counter medications. Note that, in this reduction, neither the benzene rings of the naphthyl group nor the carboxyl group are reduced. We will have more to say about the reduction of these groups in later chapters.

(*S*)-Naproxen
(ee > 98%)

BINAP-Ru complexes are somewhat specific for the types of carbon-carbon double bonds they reduce. To be reduced, the double bond must have some kind of neighboring functional group that serves as a directing group during the reduction. The most common of these directing groups are the carboxyl group of carboxylic acids and esters and the hydroxyl group of alcohols. As seen in the following example, only the carbon-carbon double bond nearer the —OH group is reduced. Geraniol, as the name might suggest, is a natural product isolated from the rose and geranium oils. It is also present in citronella and lemon grass oils. With this chiral catalyst, only the *R* enantiomer is formed from the (*S*)-BINAP-Ru complex, and only the *S* enantiomer is formed from the (*R*)-BINAP-Ru complex.

(*E*)-3,7-Dimethyl-2,6-octadien-1-ol
(Geraniol)

(*R*)-3,7-Dimethyl-6-octen-1-ol

(*S*)-3,7-Dimethyl-6-octen-1-ol

Summary

A **reaction mechanism** (Section 6.2) is a detailed description of how and why a chemical reaction occurs as it does, which bonds are broken and which new ones are formed, the order in which the various bond-breaking and bond-forming steps take place and their relative rates, the role of the solvent if the reaction takes place in solution, and the role of the catalyst if one is present. Transition state theory provides a model for understanding the relationships among molecular structure, reaction rates, and energetics. An **energy diagram** shows the changes in energy that occur in going from reactants to products. Energy is measured on the vertical axis, and the change in position of the atoms during reaction, called the **reaction coordinate,** is measured on the horizontal axis. A **transition state** is a point on the reaction coordinate at which the energy is a maximum. The difference in energy between reactants and a transition state is called the **activation energy, $\Delta G^{\ddagger}$.** A **reaction intermediate** corresponds to an energy minimum between two transition states. In a multistep reaction, the step that crosses the highest energy barrier is called the **rate-determining step.**

A characteristic reaction of alkenes is addition, during which a pi bond is broken and sigma bonds to two new atoms are formed. A **regioselective reaction** is a reaction in which one direction of bond forming or bond breaking occurs in preference to all other directions (Section 6.3A). According to **Markovnikov's rule,** in addition of HX or H_2O to an alkene, hydrogen adds to the carbon of the double bond having the greater number of hydrogens. An **electrophile** (literally, electron loving) is any atom, molecule, or ion that can accept a pair of electrons to form a new covalent bond. A **nucleophile** (literally, nucleus loving) is any atom, molecule, or ion that can donate a pair of electrons to form a new covalent bond. The rate-determining step in electrophilic addition to an alkene is formation of a **carbocation** intermediate. A carbocation is a positively charged ion that contains a carbon atom with only six electrons in its valence shell. Carbocations are planar with bond angles of approximately 120° about the positive carbon. The order of stability of carbocations is 3° > 2° > 1° > methyl. Carbocations are stabilized by the electron-releasing **inductive effect** of alkyl groups bonded to the cationic carbon and by **hyperconjugation.**

The driving force for **carbocation rearrangement** (Section 6.3C) is conversion of a 2° carbocation to a more stable 2° or 3° carbocation. Rearrangement is by a 1,2-shift in which an atom or group of atoms with its bonding electrons moves from an adjacent atom to an electron-deficient atom.

A **stereoselective reaction** is a reaction in which one stereoisomer is formed in preference to all others that might be formed (Section 6.3D). A stereoselective reaction may be enantioselective or diastereoselective, as the case may be. Addition of new atoms or groups of atoms from opposite sides or faces of a double bond is called anti addition; alternatively, we say that the reaction occurs with **anti stereoselectivity.** In cyclic systems, anti addition is equivalent to *trans* coplanar addition. Syn addition is the addition of atoms or groups of atoms to the same side or face of a double bond.

Oxidation is the loss of electrons; **reduction** is the gain of electrons. Alternatively, oxidation is the gain of oxygen atoms and/or loss of hydrogen atoms. Reduction is the loss of oxygen atoms and/or gain of hydrogen atoms (Sections 6.5 and 6.6).

From **heats of hydrogenation** of a series of alkenes (Section 6.6B), we conclude that in general (1) the greater the degree of substitution of a carbon-carbon double bond, the more stable the alkene and (2) a *trans* alkene is more stable than a *cis* alkene.

In reactions in which chiral centers are made, only racemic mixtures are formed from achiral reagents in achiral media (Section 6.7). A **stereospecific reaction** is a reaction in which the stereochemistry of the product depends on the stereochemistry of the starting material. Optically active products can never be formed by the reaction of achiral starting materials in an achiral environment. Optically active products may be formed, however, by the reaction of achiral starting materials in a chiral environment.

Key Reactions

ORGANIC Chemistry Now™
Click Flashcards to review the **Key Reactions of Alkenes**

1. Addition of HX (Section 6.3A)

Addition is regioselective and follows Markovnikov's rule. Reaction involves a carbocation intermediate (an electrophile), which may rearrange before it combines with a halide ion (a nucleophile) to complete the addition:

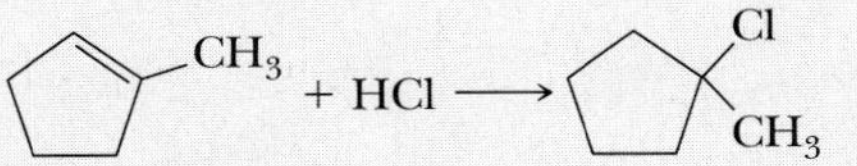

2. Acid-Catalyzed Hydration (Section 6.3B)

Addition is regioselective and follows Markovnikov's rule. Reaction involves a carbocation intermediate, which may rearrange before it combines with water (a nucleophile) to complete the hydration.

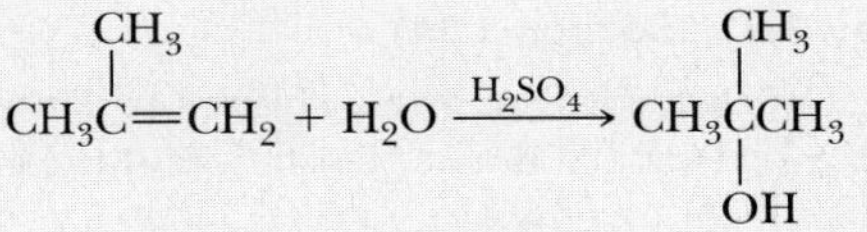

3. Addition of Bromine and Chlorine (Section 6.3D)

Addition is anti stereoselective; it involves anti addition of halogen atoms by way of a bridged halonium ion intermediate with no rearrangement. Addition is also stereospecific; for alkenes that exist as *cis* and *trans* isomers, the configuration of the product depends on the configuration of the alkene.

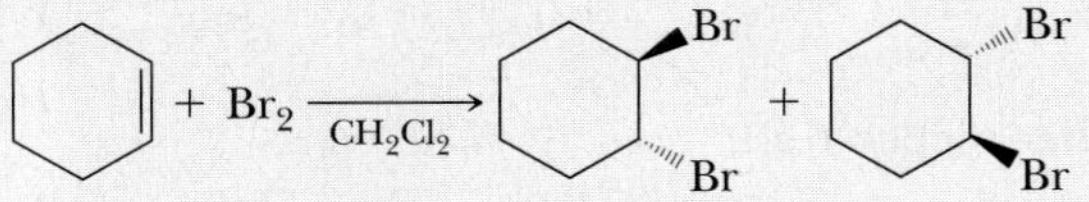

(a racemic mixture)

4. Addition of HOCl and HOBr (Section 6.3E)

Addition is regioselective (—X adds to the less substituted carbon via a bridged halonium ion intermediate and —OH adds to the more substituted carbon), anti stereoselective, and occurs without rearrangement. Addition is also stereospecific; for alkenes that exist as *cis* and *trans* isomers, the configuration of the product depends on the configuration of the starting alkene.

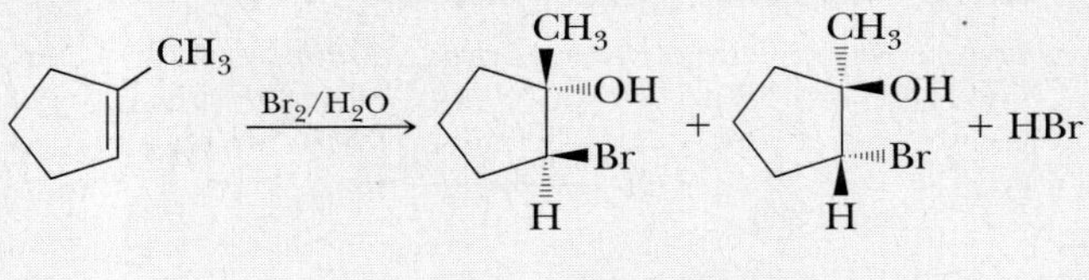

(a racemic mixture)

5. Oxymercuration-Reduction (Section 6.3F)

Oxymercuration is regioselective; HgOAc adds to the less substituted carbon, and OH adds to the more substituted carbon. It is also anti stereoselective; HgOAc and OH add from opposite faces of the alkene. Anti addition occurs via a bridged mercurinium ion intermediate with no rearrangement. The result of oxymercuration-reduction is Markovnikov hydration of an alkene.

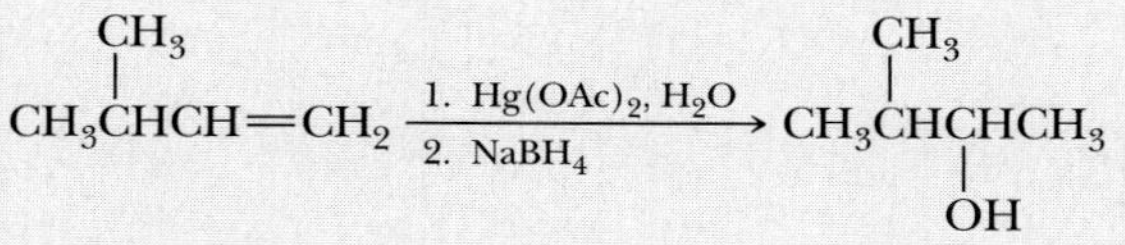
CH3
CH3CHCH=CH2
1. Hg(OAc)2, H2O
2. NaBH4
CH3
CH3CHCHCH3
OH

6. Hydroboration-Oxidation (Section 6.4)

Addition of BH_3 is syn stereoselective and regioselective; boron adds to the less substituted carbon, and hydrogen adds to the more substituted carbon. Hydroboration-oxidation results in non-Markovnikov hydration of the alkene without rearrangement.

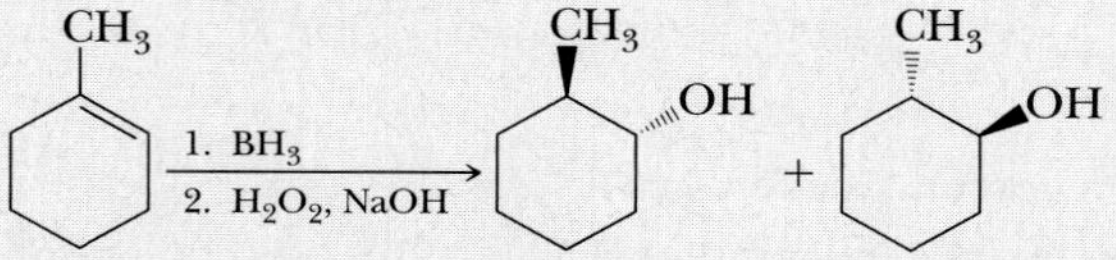
CH3
1. BH3
2. H2O2, NaOH
CH3
OH
+
CH3
OH

(a racemic mixture)

7. Oxidation to a Glycol by OsO_4 (Section 6.5A)

Oxidation occurs with syn stereoselectivity to give a glycol. Oxidation is also stereospecific; for alkenes that exist as *cis* and *trans* isomers, the configuration of the glycol depends on the configuration of the starting alkene.

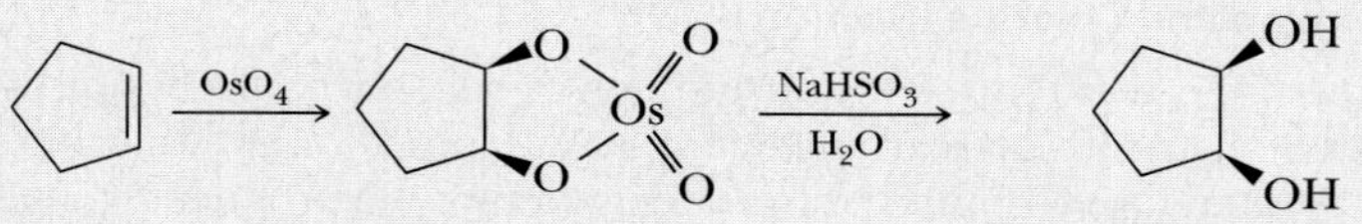
OsO4
O
O
Os
O
O
NaHSO3
H2O
OH
OH

A cyclic osmate

cis-1,2-Cyclopentanediol (a *cis* glycol)

8. Oxidation by Ozone (Section 6.5B)

Treatment with ozone followed by dimethyl sulfide cleaves a carbon-carbon double bond and gives two carbonyl groups in its place:

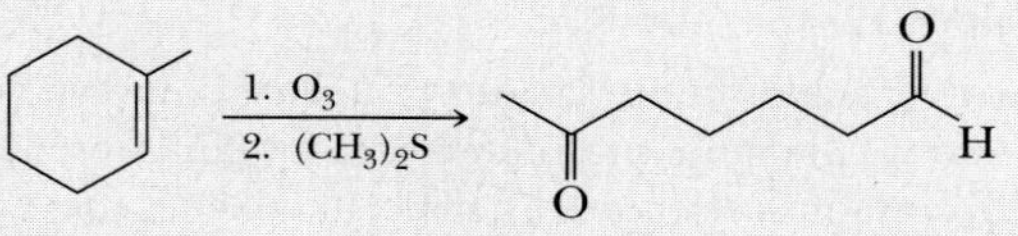
1. O3
2. (CH3)2S
O
O
H

9. Addition of H_2; Catalytic Reduction (Section 6.6)

Catalytic reduction is stereoselective and involves predominantly syn addition of hydrogens.

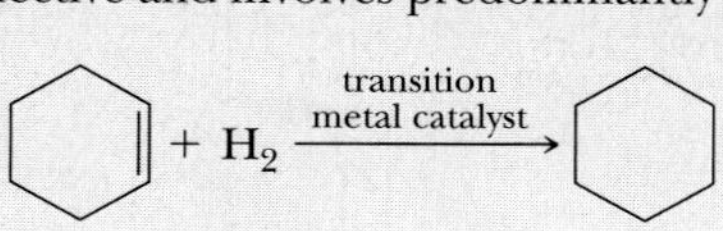
+ H2
transition
metal catalyst

10. Enantioselective Reduction (Section 6.7)

The most useful chiral hydrogenation catalysts involve a chiral phosphorus-containing ligand complexed with either ruthenium or rhodium.

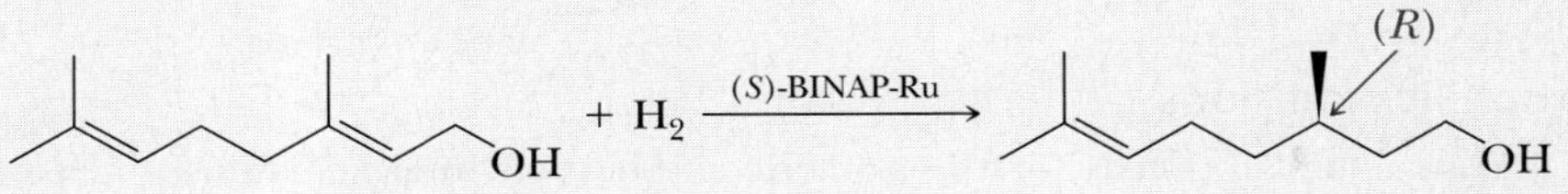

Problems

Energetics of Chemical Reactions

ORGANIC Chemistry•Now™ Assess your understanding of this chapter's topics with additional quizzing and conceptual-based problems at **http://now.brookscole.com/bfi4**

6.13 Using the table of average bond dissociation enthalpies at 25°C, determine which of the following reactions are energetically favorable at room temperature. Assume that $\Delta S = 0$.

Bond	Bond Dissociation Enthalpy kJ (kcal)/mol	Bond	Bond Dissociation Enthalpy kJ (kcal)/mol
H—H	435 (104)	C—I	238 (57)
O—H	439 (105)	C—Si	301 (72)
C—H (—CH_3)	422 (101)	C=C	727 (174)
C—H (=CH_2)	464 (111)	C=O (aldehyde)	728 (174)
C—H (≡CH)	556 (133)	C=O (CO_2)	803 (192)
N—H	391 (93)	C≡O	1075 (257)
Si—H	318 (76)	N≡N	950 (227)
C—C	376 (90)	C≡C	966 (231)
C—N	355 (85)	O=O	498 (119)
C—O	385 (92)		

(a) $CH_2{=}CH_2 + 2H_2 + N_2 \longrightarrow H_2NCH_2CH_2NH_2$

(b) $CH_2{=}CH_2 + CH_4 \longrightarrow CH_3CH_2CH_3$

(c) $CH_2{=}CH_2 + (CH_3)_3SiH \longrightarrow CH_3CH_2Si(CH_3)_3$

(d) $CH_2{=}CH_2 + CHI_3 \longrightarrow CH_3CH_2CI_3$

(e) $CH_2{=}CH_2 + CO + H_2 \longrightarrow CH_3CH_2\overset{\overset{\displaystyle O}{\|}}{C}H$

(f) 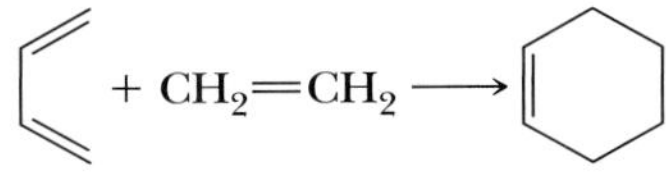

(g) 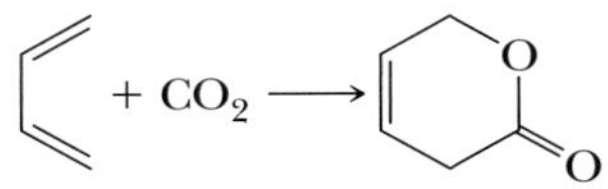

$$\textbf{(h)} \ HC\!\equiv\!CH + O_2 \longrightarrow \overset{\displaystyle O \quad O}{\underset{\displaystyle \| \quad \|}{HC\!-\!CH}}$$

(i) $2CH_4 + O_2 \longrightarrow 2CH_3OH$

Electrophilic Additions

6.14 Draw structural formulas for the isomeric carbocation intermediates formed on treatment of each alkene with HCl. Label each carbocation 1°, 2°, or 3°, and state which of the isomeric carbocations forms more readily.

(a) $CH_3CH_2\overset{\displaystyle CH_3}{\underset{\displaystyle |}{C}}\!\!=\!\!CHCH_3$ **(b)** $CH_3CH_2CH\!=\!CHCH_3$ **(c)** [structure: cyclopentene with CH₃]   **(d)** [structure: cyclohexane ring =CH₂]

6.15 Arrange the alkenes in each set in order of increasing rate of reaction with HI, and explain the basis for your ranking. Draw the structural formula of the major product formed in each case.

(a) $CH_3CH\!=\!CHCH_3$ and $CH_3\overset{\displaystyle CH_3}{\underset{\displaystyle |}{C}}\!\!=\!\!CHCH_3$ **(b)** [cyclohexene with methyl]   and   [cyclohexene]

6.16 Predict the organic product(s) of the reaction of 2-butene with each reagent.

(a) H_2O (H_2SO_4)   **(b)** Br_2 **(c)** Cl_2
(d) Br_2 in H_2O **(e)** HI   **(f)** Cl_2 in H_2O
(g) $Hg(OAc)_2$, H_2O **(h)** product (g) + $NaBH_4$

6.17 Draw a structural formula of an alkene that undergoes acid-catalyzed hydration to give each alcohol as the major product (more than one alkene may give each alcohol as the major product).

(a) 3-Hexanol   **(b)** 1-Methylcyclobutanol
(c) 2-Methyl-2-butanol   **(d)** 2-Propanol

6.18 Reaction of 2-methyl-2-pentene with each reagent is regioselective. Draw a structural formula for the product of each reaction, and account for the observed regioselectivity.

(a) HI   **(b)** HBr
(c) H_2O in the presence of H_2SO_4 **(d)** Br_2 in H_2O
(e) $Hg(OAc)_2$ in H_2O

6.19 Account for the regioselectivity and stereoselectivity observed when 1-methylcyclopentene is treated with each reagent.

(a) BH_3 **(b)** Br_2 in H_2O **(c)** $Hg(OAc)_2$ in H_2O

6.20 Draw a structural formula for an alkene with the indicated molecular formula that gives the compound shown as the major product (more than one alkene may give the same compound as the major product).

(a) $C_5H_{10} + H_2O \xrightarrow{\ H_2SO_4\ }$ [structure: tert-alcohol with OH]   **(b)** $C_5H_{10} + Br_2 \longrightarrow$ [structure with two Br]

(c) $C_7H_{12} + HCl \longrightarrow$ [structure: cyclohexane with CH₃ and Cl]

6.21 Account for the fact that addition of HCl to 1-bromopropene gives exclusively 1-bromo-1-chloropropane.

$$CH_3CH{=}CHBr + HCl \longrightarrow CH_3CH_2CHBrCl$$

1-Bromopropene   1-Bromo-1-chloropropane

6.22 Account for the fact that treating propenoic acid (acrylic acid) with HCl gives only 3-chloropropanoic acid.

$$CH_2{=}CHCOH + HCl \longrightarrow ClCH_2CH_2COH$$

Propenoic acid   3-Chloropropanoic acid   2-Chloropropanoic acid
(Acrylic acid)   (this product is not formed)

CH₃CHCOH

6.23 Draw a structural formula for the alkene with the molecular formula C_5H_{10} that reacts with Br_2 to give each product.

(a) Br   Br   **(b)** Br   Br   **(c)** Br   Br

6.24 Draw the alternative chair conformations for the product formed by the addition of bromine to 4-*tert*-butylcyclohexene. The Gibbs free energy differences between equatorial and axial substituents on a cyclohexane ring are 21 kJ (4.9 kcal)/mol for *tert*-butyl and 2.0–2.6 kJ (0.48–0.62 kcal)/mol for bromine. Estimate the relative percentages of the alternative chair conformations you drew in the first part of this problem.

6.25 Draw a structural formula for the cycloalkene with the molecular formula C_6H_{10} that reacts with Cl_2 to give each compound.

(a) Cl Cl   **(b)** Cl Cl CH₃   **(c)** H₃C Cl Cl   **(d)** Cl CH₂Cl

6.26 Reaction of this bicycloalkene with bromine in carbon tetrachloride gives a *trans* dibromide. In both (a) and (b), the bromine atoms are *trans* to each other. However, only one of these products is formed.

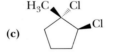

CH₃ + Br₂ $\xrightarrow{CH_2Cl_2}$ Br CH₃ Br H   or   Br CH₃ Br H

(a)   (b)

ORGANIC
Chemistry Now™
Click Molecular Models to view
two possible **Dibromodecalins**

Which *trans* dibromide is formed? How do you account for the fact that it is formed to the exclusion of the other *trans* dibromide?

6.27 Terpin, prepared commercially by the acid-catalyzed hydration of limonene, is used medicinally as an expectorant for coughs.

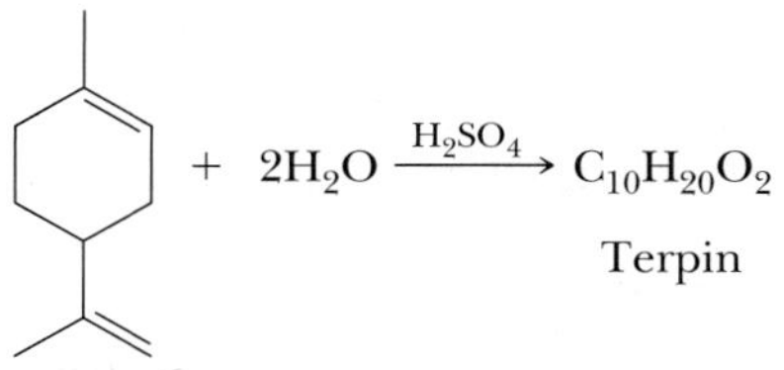

$$\text{Limonene} + 2H_2O \xrightarrow{H_2SO_4} C_{10}H_{20}O_2$$

Terpin

Limonene

(a) Propose a structural formula for terpin and a mechanism for its formation.
(b) How many *cis,trans* isomers are possible for the structural formula you propose?

6.28 Propose a mechanism for this reaction, and account for its regioselectivity.

$$CH_3-\underset{\underset{}{\overset{CH_3}{|}}}{C}=CH_2 + ICl \longrightarrow CH_3\underset{\underset{Cl}{|}}{\overset{\overset{CH_3}{|}}{C}}-CH_2I$$

6.29 Treating 2-methylpropene with methanol in the presence of sulfuric acid gives *tert*-butyl methyl ether.

$$CH_3\underset{\underset{}{\overset{CH_3}{|}}}{C}=CH_2 + CH_3OH \xrightarrow{H_2SO_4} CH_3\underset{\underset{CH_3}{|}}{\overset{\overset{CH_3}{|}}{C}}-OCH_3$$

Propose a mechanism for the formation of this ether.

6.30 When 2-pentene is treated with Cl_2 in methanol, three products are formed. Account for the formation of each product (you need not explain their relative percentages).

$$CH_3CH=CHCH_2CH_3 \xrightarrow[CH_3OH]{Cl_2} CH_3\underset{\underset{}{\overset{Cl}{|}}}{CH}\underset{\underset{}{\overset{OCH_3}{|}}}{CH}CH_2CH_3 + CH_3\underset{\underset{}{\overset{H_3CO}{|}}}{CH}\underset{\underset{}{\overset{Cl}{|}}}{CH}CH_2CH_3 + CH_3\underset{\underset{}{\overset{Cl}{|}}}{CH}\underset{\underset{}{\overset{Cl}{|}}}{CH}CH_2CH_3$$

50%   35%   15%

6.31 Treating cyclohexene with HBr in the presence of acetic acid gives bromocyclohexane (85%) and cyclohexyl acetate (15%).

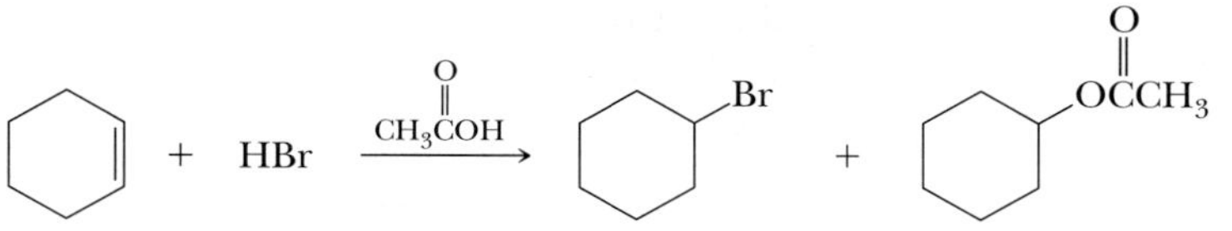

Cyclohexene + HBr $\xrightarrow{CH_3\overset{O}{\overset{||}{C}}OH}$ Bromocyclohexane (85%) + Cyclohexyl acetate (15%)

Propose a mechanism for the formation of the latter product.

6.32 Propose a mechanism for this reaction.

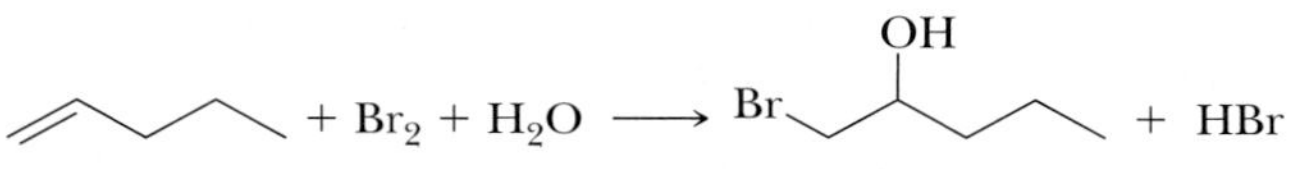

1-Pentene + Br_2 + H_2O $\longrightarrow$ 1-Bromo-2-pentanol + HBr

6.33 Treating 4-penten-1-ol with bromine in water forms a cyclic bromoether.

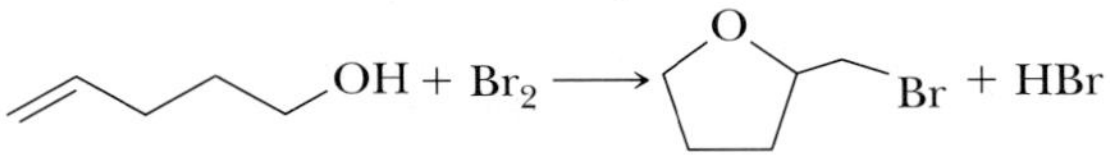

4-Penten-1-ol

Account for the formation of this product rather than a bromohydrin as was formed in Problem 6.32.

6.34 Provide a mechanism for each reaction.

(a) 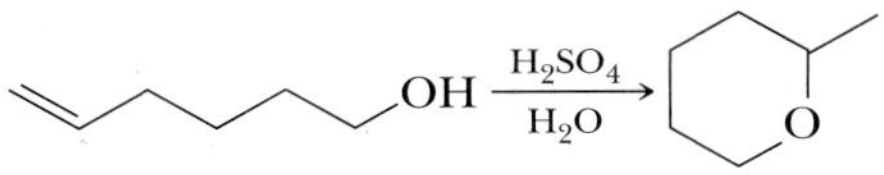

(b) 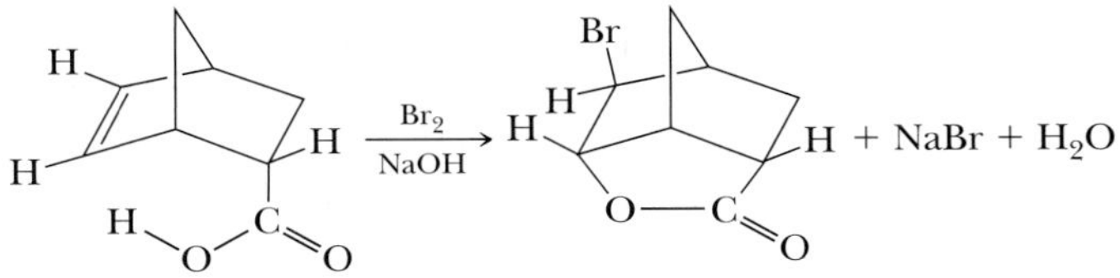

6.35 Treating 1-methyl-1-vinylcyclopentane with HCl gives mainly 1-chloro-1,2-dimethylcyclohexane.

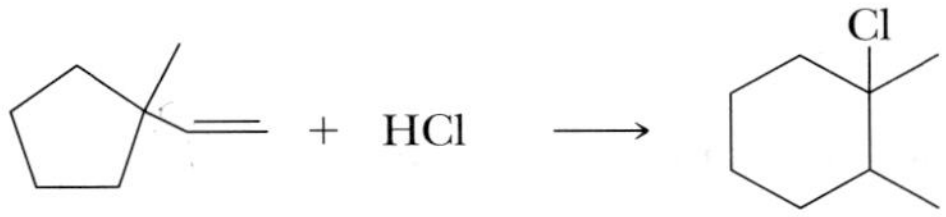

1-Methyl-1-vinylcyclopentane   1-Chloro-1,2-dimethylcyclohexane

Propose a mechanism for the formation of this product.

Hydroboration

6.36 Draw a structural formula for the alcohol formed by treating each alkene with borane in tetrahydrofuran (THF) followed by hydrogen peroxide in aqueous sodium hydroxide, and specify stereochemistry where appropriate.

(a) 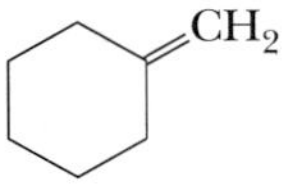

(b) 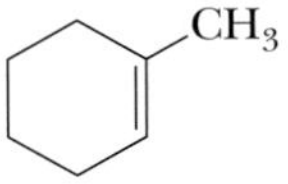

(c) 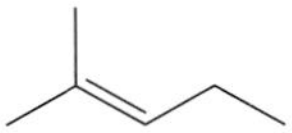

(d) 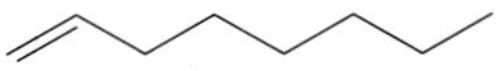

(e) 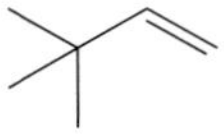

6.37 Reaction of α-pinene with borane followed by treatment of the resulting trialkylborane with alkaline hydrogen peroxide gives the following alcohol.

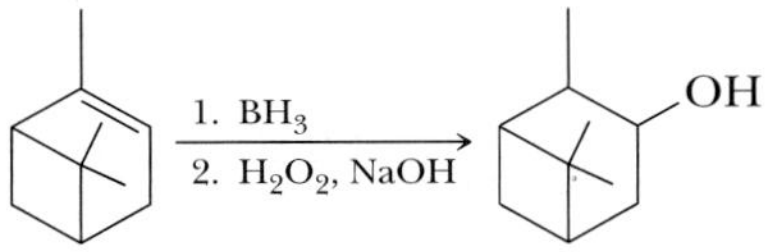

α-Pinene

ORGANIC Chemistry•Now™

Click Models to view a model of **α-Pinene**

Of the four possible *cis,trans* isomers, one is formed in over 85% yield.

(a) Draw structural formulas for the four possible *cis,trans* isomers of the bicyclic alcohol.
(b) Which is the structure of the isomer formed in 85% yield? How do you account for its formation? Make a model to help you make this prediction.

Oxidation

6.38 Write structural formulas for the major organic product(s) formed by reaction of 1-methylcyclohexene with each oxidizing agent.

(a) OsO_4/H_2O_2 **(b)** O_3 followed by $(CH_3)_2S$

6.39 Draw the structural formula of the alkene that reacts with ozone followed by dimethyl sulfide to give each product or set of products.

(a) 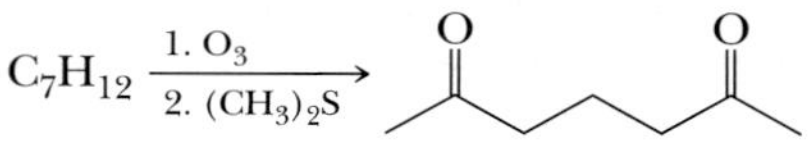 C_7H_{12} $\xrightarrow[\text{2. }(CH_3)_2S]{\text{1. }O_3}$

(b) $C_{10}H_{18}$ $\xrightarrow[\text{2. }(CH_3)_2S]{\text{1. }O_3}$

(c) $C_{10}H_{18}$ $\xrightarrow[\text{2. }(CH_3)_2S]{\text{1. }O_3}$

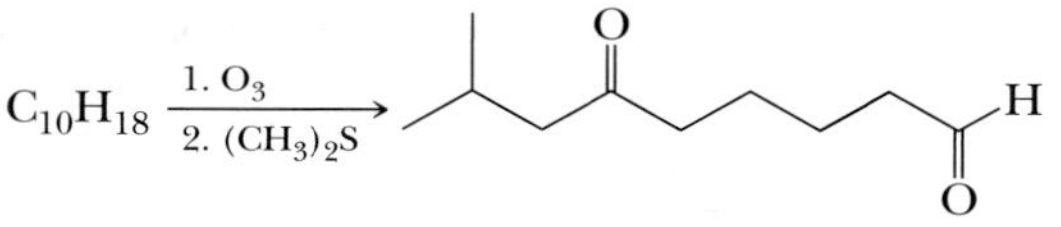

6.40 Consider the following reaction.

$$C_8H_{12} \xrightarrow[\text{2. }(CH_3)_2S]{\text{1. }O_3} \text{HC}-\text{CH}$$

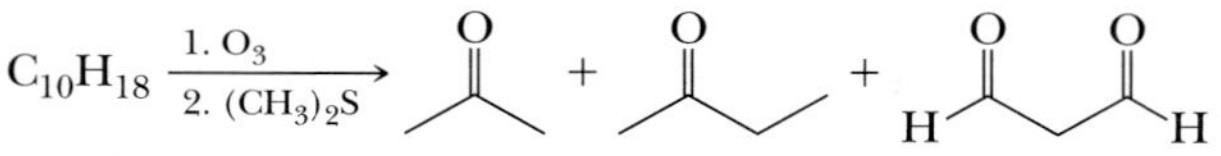

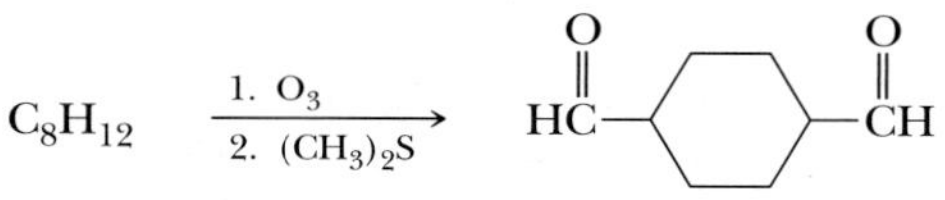

Cyclohexane-1,4-dicarbaldehyde

(a) Draw a structural formula for the compound with the molecular formula C_8H_{12}.
(b) Do you predict the product to be the *cis* isomer, the *trans* isomer, or a mixture of *cis* and *trans* isomers? Explain.
(c) Draw a suitable stereorepresentation for the more stable chair conformation of the dicarbaldehyde formed in this oxidation.

Reduction

6.41 Predict the major organic product(s) of the following reactions, and show stereochemistry where appropriate.

(a) 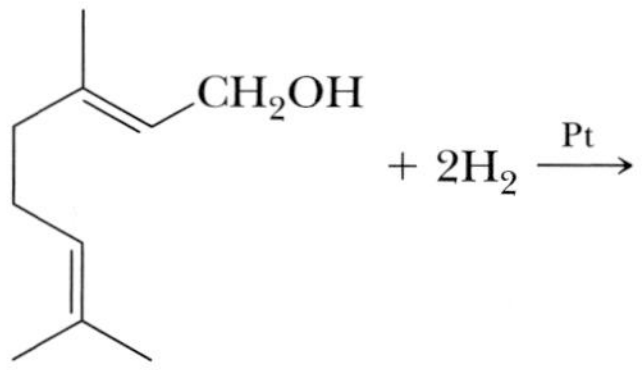 Geraniol CH_2OH $+ 2H_2 \xrightarrow{\text{Pt}}$

(b) 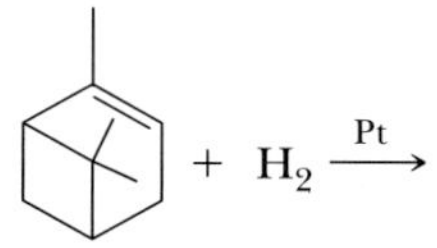 α-Pinene $+ H_2 \xrightarrow{\text{Pt}}$

6.42 The heat of hydrogenation of *cis*-2,2,5,5-tetramethyl-3-hexene is -154 kJ (-36.7 kcal)/mol, while that of the *trans* isomer is only -113 kJ (-26.9 kcal)/mol.

(a) Why is the heat of hydrogenation of the *cis* isomer so much larger (more negative) than that of the *trans* isomer?

(b) If a catalyst could be found that allowed equilibration of the *cis* and *trans* isomers at room temperature (such catalysts do exist), what would be the ratio of *trans* to *cis* isomers?

Synthesis

6.43 Show how to convert ethylene to these compounds.

(a) Ethane	**(b)** Ethanol	**(c)** Bromoethane
(d) 2-Chloroethanol	**(e)** 1,2-Dibromoethane	**(f)** 1,2-Ethanediol
(g) Chloroethane		

6.44 Show how to convert cyclopentene into these compounds.

(a) *trans*-1,2-Dibromocyclopentane	**(b)** *cis*-1,2-Cyclopentanediol
(c) Cyclopentanol	**(d)** Iodocyclopentane
(e) Cyclopentane	**(f)** Pentanedial

Reactions That Produce Chiral Compounds

6.45 State the number and kind of stereoisomers formed when (*R*)-3-methyl-1-pentene is treated with these reagents. Assume that the starting alkene is enantiomerically pure and optically active. Will each product be optically active or inactive?

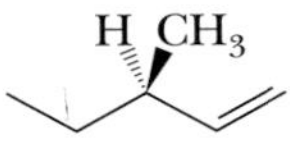

(*R*)-3-Methyl-1-pentene

(a) $Hg(OAc)_2$, H_2O followed by $NaBH_4$	**(b)** H_2/Pt
(c) BH_3 followed by H_2O_2 in NaOH	**(d)** Br_2 in CCl_4

6.46 Describe the stereochemistry of the bromohydrin formed in each reaction (each reaction is stereospecific).

(a) *cis*-3-Hexene + Br_2/H_2O **(b)** *trans*-3-Hexene + Br_2/H_2O

6.47 In each of these reactions, the organic starting material is achiral. The structural formula of the product is given. For each reaction, determine the following.

(1) How many stereoisomers are possible for the product?
(2) Which of the possible stereoisomers is/are formed in the reaction shown?
(3) Will the product be optically active or optically inactive?

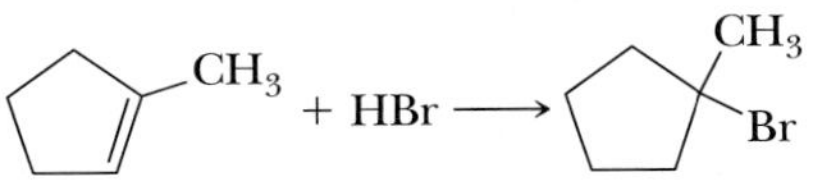

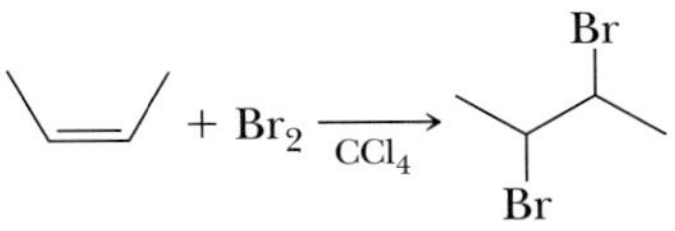

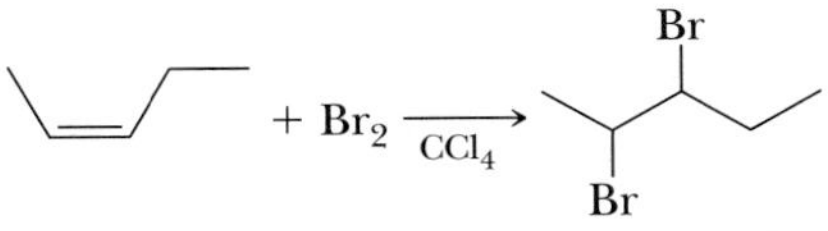

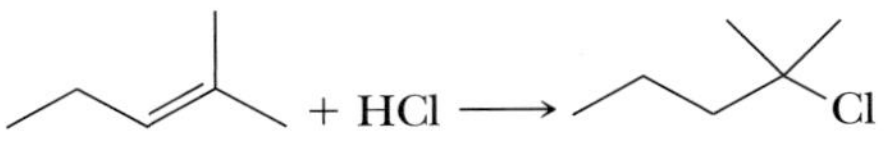

(e) 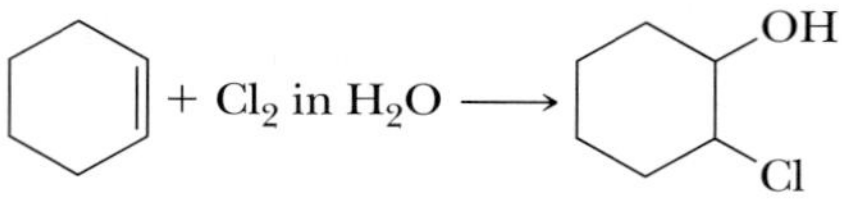 + Cl$_2$ in H$_2$O $\longrightarrow$

(f) 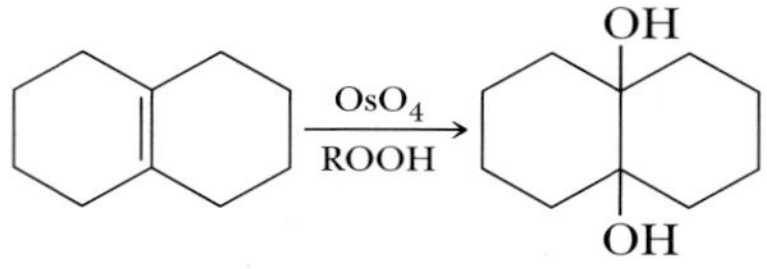 $\xrightarrow[\text{ROOH}]{\text{OsO}_4}$

(g) 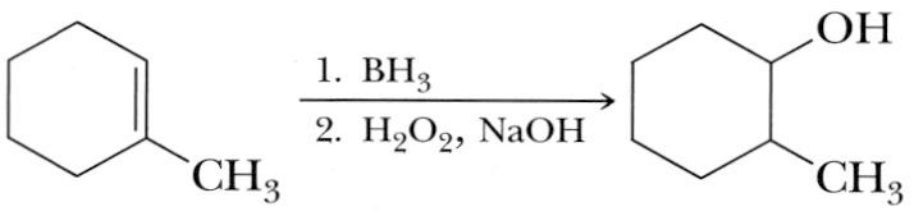 $\xrightarrow[\text{2. H}_2\text{O}_2,\ \text{NaOH}]{\text{1. BH}_3}$

Looking Ahead

6.48 The 2-propenyl cation appears to be a primary carbocation, and yet it is considerably more stable than a 1° carbocation such as the 1-propyl cation.

$$CH_2{=}CH{-}CH_2{}^+ \qquad CH_3{-}CH_2{-}CH_2{}^+$$

2-Propenyl cation   1-Propyl cation

How would you account for the differences in the stability of the two carbocation?

6.49 Treating 1,3-butadiene with one mole of HBr gives a mixture of two isomeric products.

$$CH_2{=}CH{-}CH{=}CH_2 + H{-}Br \longrightarrow CH_2{=}CH{-}\overset{\displaystyle Br}{\overset{\displaystyle |}{CH}}{-}CH_3 + CH_3{-}CH{=}CH{-}CH_2{-}Br$$

1,3-Butadiene   3-Bromo-1-butene   1-Bromo-2-butene

Propose a mechanism that accounts for the formation of these two products.

6.50 In this chapter, we studied the mechanism of the acid-catalyzed hydration of an alkene. The reverse of this reaction is the acid-catalyzed dehydration of an alcohol.

$$CH_3{-}\overset{\displaystyle OH}{\overset{\displaystyle |}{CH}}{-}CH_3 \xrightarrow{\text{H}_2\text{SO}_4} CH_3{-}CH{=}CH_2 + H_2O$$

2-Propanol   Propene
(Isopropyl alcohol)

Propose a mechanism for the acid-catalyzed dehydration of 2-propanol to propene.

6.51 As we have seen in this chapter, carbon-carbon double bonds are electron-rich regions and are attacked by electrophiles (for example, HBr); they are not attacked by nucleophiles (for example, diethylamine).

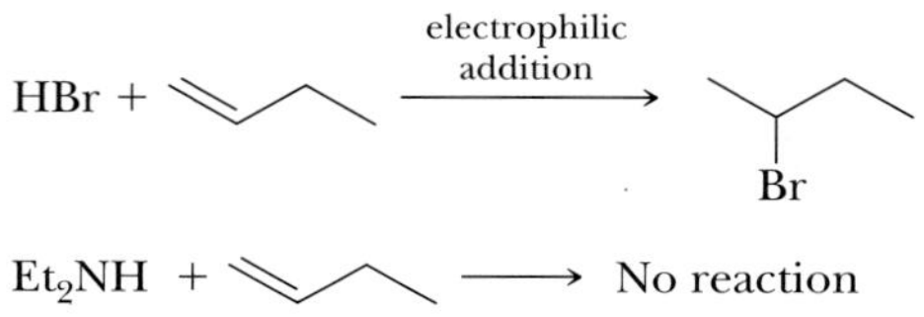

HBr + $\xrightarrow{\substack{\text{electrophilic} \\ \text{addition}}}$

Et$_2$NH + $\longrightarrow$ No reaction

Diethylamine
(a nucleophile)

However, when the carbon-carbon double bond has a carbonyl group adjacent to it, the double bond reacts readily with nucleophiles by nucleophilic addition (Section 19.8).

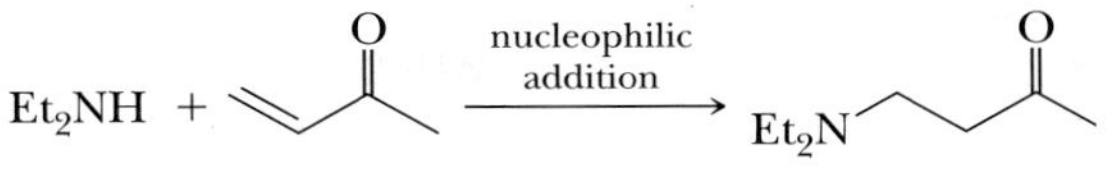

Diethylamine
(a nucleophile)

Account for the fact that nucleophiles add to a carbon-carbon double bond adjacent to a carbonyl group, and account for the regiochemistry of the reaction.

6.52 Following is an example of a type of reaction known as a Diels-Alder reaction (Chapter 24).

1,3-Pentadiene   Ethylene   3-Methylcyclohexene

The Diels-Alder reaction between a diene and an alkene is quite remarkable in that it is one of the few ways that chemists have to form two new carbon-carbon bonds in a single reaction. Given what you know about the relative strengths of carbon-carbon sigma and pi bonds, would you predict the Diels-Alder reaction to be exothermic or endothermic? Explain your reasoning.

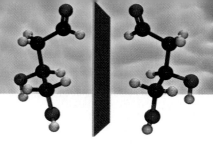

Summary of Stereochemical Terms

Absolute Configuration The actual configuration of groups about a tetrahedral chiral center; absolute configuration is specified by the R,S system.

Atropisomers Enantiomers that lack a chiral center and differ because of hindered rotation about a carbon-carbon single bond.

Center of symmetry A point so situated that identical components of an object are located on opposite sides and equidistant from that point along any axis passing through it.

Chemoselective reaction A reaction in which one functional group in a molecule containing two or more functional groups reacts selectively with a reagent.

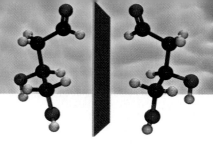

Chiral From the Greek, *cheir,* hand; an object that is not superposable on its mirror image; an object that has handedness.

Chiral center A tetrahedral atom, most commonly carbon, that is bonded to four different groups. In molecules containing one chiral center, the exchange of two groups makes an enantiomer. In molecules containing two or more chiral centers, the exchange of two groups on at least one (but not all) of the chiral centers gives a diastereomer.

***Cis,trans* isomers** Stereoisomers that have the same connectivity of their atoms but a different configurational arrangement of their atoms in space because of the presence of either a ring or a double bond. *Cis,trans* isomers are diastereomers; that is, they are stereoisomers that are not mirror images.

Configuration The arrangement of atoms or groups of atoms bonded to a stereocenter. Configuration in alkenes is designated by the E,Z system or the *cis,trans* system; configuration in molecules containing chiral centers is designated by the R,S system.

Diastereomers Stereoisomers that are not mirror images of each other; refers to relationships among two or more objects.

Diastereoselective reaction A reaction that produces one diastereomer in preference to all others.

Enantiomeric excess (ee)   The difference between the percentages of two enantiomers in a mixture. For example, if a sample contains 98% of one enantiomer and 2% of the other, the enantiomeric excess (ee) is $98\% - 2\% = 96\%$.

Enantiomers   Stereoisomers that are nonsuperposable mirror images; refers to a relationship between pairs of objects.

Enantioselective reaction   A reaction that produces one enantiomer in preference to the other. Catalytic reduction of the following alkene in the presence of an (*R*)-BINAP-Ru catalyst gives (*S*)-naproxen in greater than 98% enantiomeric excess ($> 99\% : < 1\%$).

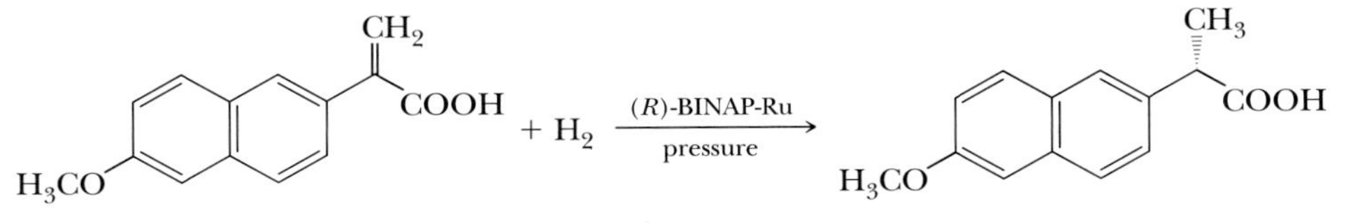

(*S*)-Naproxen
(ee > 98%)

Fischer projection   A two-dimensional projection of a chiral center in a molecule; groups on the right and left of the chiral center project toward the reader, whereas those above and below the chiral center project away from the reader. The only atom in the plane on which the projection is drawn is the chiral center itself.

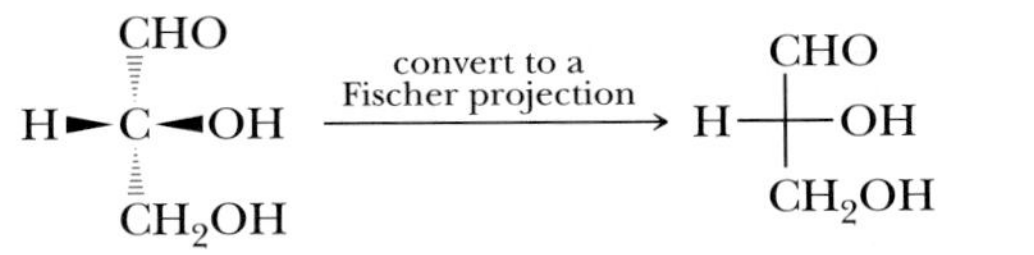

(*R*)-Glyceraldehyde
(three-dimensional
representation)

(*R*)-Glyceraldehyde
(Fischer projection)

Meso compound   An achiral compound possessing two or more chiral centers that also has chiral isomers. Examples of meso compounds are *cis*-1,2-cyclopentanediol and meso-2,3-butanediol. A meso compound has either a plane or a center of symmetry. Both of these examples as drawn have an internal plane of symmetry.

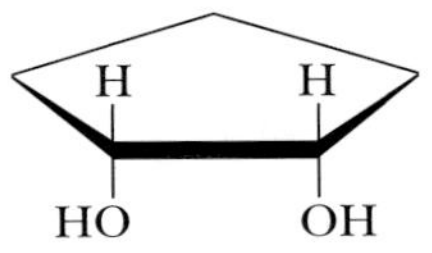

cis-1,2-Cyclopentanediol
(a meso compound)

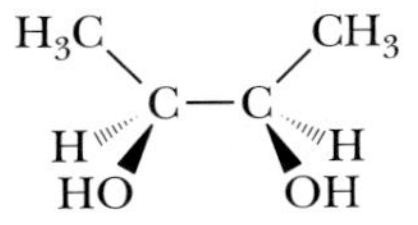

(2*R*,3*S*)-2,3-Butanediol
(a meso compound)

Optical activity   The ability of a compound to rotate the plane of polarized light.

Optical purity   The specific rotation of a mixture of enantiomers divided by the specific rotation of the enantiomerically pure substance (expressed as a percent). Optical purity is numerically equal to enantiomeric excess, but experimentally determined.

Plane of symmetry   An imaginary (mirror) plane passing through an object dividing it so that one half is the mirror image of the other half.

Racemic mixture   A mixture of equal amounts of two enantiomers.

Regioselective reaction   An addition or substitution reaction in which one of two or more possible products is formed in preference to all constitutional isomers that might be formed. Addition of HBr to 1-methylcyclohexene gives 1-bromo-1-methylcyclohexane to the virtual exclusion of 1-bromo-2-methylcyclohexane.

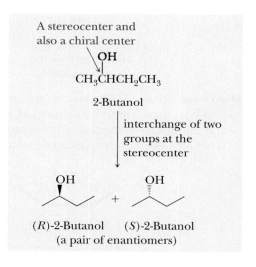

Hydroboration-oxidation of a cycloalkene is both regioselective and stereoselective, but it is not enantioselective (both enantiomers are produced in equal amounts).

Specific rotation   The observed rotation of the plane of polarized light when a sample is placed in a tube 1.0 dm in length and at a concentration expressed in g/mL (density) for a pure liquid and at a concentration of 1 g/100 mL for a solution. Specific rotation is in deg·mL/dm/g and is usually given without units.

Stereocenter   An atom, most commonly carbon, about which exchange of two groups produces a stereoisomer.

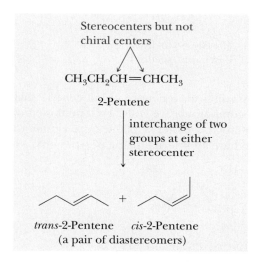
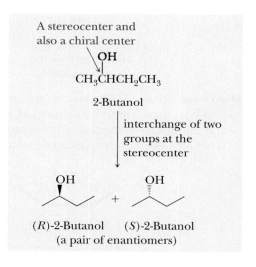

Stereoisomers   Compounds that have the same molecular formula, the same connectivity of their atoms, but a different orientation of their atoms in space. The term "stereoisomer" includes *cis,trans* isomers in cycloalkanes and alkenes as well as enantiomers, diastereomers, meso compounds, and atropisomers. Conformational isomers are also stereoisomers, whether they are isolable or not.

Stereoselective reaction   A reaction in which one stereoisomer is formed in preference to all others. A stereoselective reaction may be enantioselective or diastereoselective. For example, catalytic reduction of the following alkene in the presence of an (*R*)-BINAP-Ru catalyst is enantioselective; it gives (*S*)-naproxen in greater than 98% enantiomeric excess.

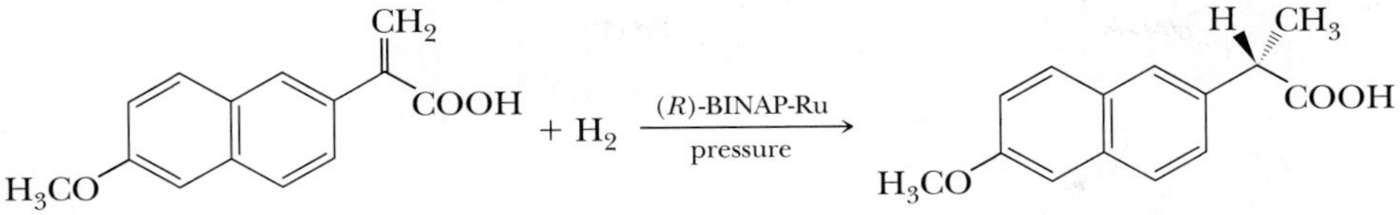

(*S*)-Naproxen
(ee > 98%)

Stereospecific reaction   A reaction in which the stereochemistry of the product is dependent on the stereochemistry of the starting material. For example, oxidation of 2-butene by osmium tetroxide is stereospecific: Oxidation of *cis*-2-butene gives meso-2,3-butanediol, whereas oxidation of *trans*-2-butene gives a racemic mixture of the enantiomers of 2,3-butanediol. (The term "regiospecific" is used analogously.)

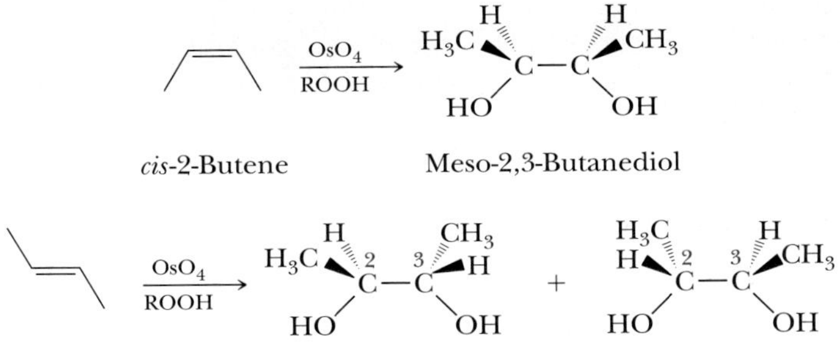

cis-2-Butene   Meso-2,3-Butanediol

trans-2-Butene   (2*S*,3*S*)-2,3-Butanediol   (2*R*,3*R*)-2,3-Butanediol

7

Adam Crowley/Photodisc Green/Getty Images

Cutting with an oxyacetylene torch. Inset: A model of acetylene.

Outline

Alkyne An unsaturated hydrocarbon that contains one or more carbon-carbon triple bonds.

ORGANIC
Chemistry Now™

Look for this logo in the chapter and go to Organic ChemistryNow at **http://now.brookscole.com/bfi4** for tutorials, simulations, and problems.

Alkynes

In this chapter, we continue our discussion of the chemistry of carbon-carbon pi bonds as we now consider the chemistry of alkynes. Because alkenes and alkynes are similar in that the multiple bond in each is a combination of sigma and pi bonds, the types of chemical reactions both functional groups undergo are similar. Alkynes undergo electrophilic additions of X_2, HX, and H_2O. They undergo hydroboration-oxidation, and the carbon-carbon triple bond can be reduced first to a double bond and then to a single bond. An important reaction of alkynes that is not typical of alkenes is the conversion of terminal alkynes to their alkali metal salts, which are good nucleophiles and, therefore, valuable building blocks for the construction of larger molecules through the formation of carbon-carbon bonds.

7.1 Structure of Alkynes

The functional group of an **alkyne** is a carbon-carbon triple bond. The simplest alkyne is ethyne, C_2H_2, more commonly named acetylene (Figure 7.1). Acetylene is a linear molecule; all bond angles are 180°. The carbon-carbon bond length in acetylene is 121 pm (Figure 7.1).

By comparison, the length of the carbon-carbon double bond in ethylene is 134 pm, and that of the carbon-carbon single bond in ethane

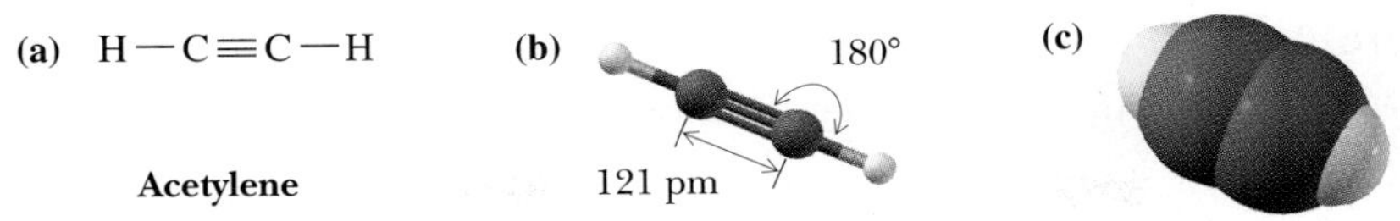

(a) $H-C\equiv C-H$ **(b)** [ball-and-stick model] 180°   **(c)** [space-filling model]

Acetylene   121 pm

Figure 7.1
Acetylene. (a) Lewis structure, (b) ball-and-stick model, and (c) space-filling model.

is 153 pm. Thus, triple bonds are shorter than double bonds, which, in turn, are shorter than single bonds. The bond dissociation enthalpy of the carbon-carbon triple bond in acetylene [966 kJ (231 kcal)/mol] is considerably larger than that for the carbon-carbon double bond in ethylene [727 kJ (174 kcal)/mol] and the carbon-carbon single bond in ethane [376 kJ (90 kcal)/mol].

A triple bond is described in terms of the overlap of sp hybrid orbitals of adjacent carbon atoms to form a sigma bond, the overlap of parallel $2p_y$ orbitals to form one pi bond, and the overlap of parallel $2p_z$ orbitals to form a second pi bond (Figure 1.21). In acetylene, each carbon-hydrogen bond is formed by the overlap of a $1s$ orbital of hydrogen with an sp orbital of carbon. Because of the 50% s-character of the acetylenic C—H bond, it is unusually strong (see Table 1.11 and related text).

ORGANIC
Chemistry ⚛ Now™
Click Molecular Models to view and manipulate **Models of Alkynes**

7.2 Nomenclature of Alkynes

A. IUPAC Names

According to the rules of the IUPAC system, the infix *-yn-* is used to show the presence of a carbon-carbon triple bond (Section 2.5). Thus, $HC\equiv CH$ is named ethyne, and $CH_3C\equiv CH$ is named propyne. The IUPAC system retains the name acetylene; therefore, there are two acceptable names for $HC\equiv CH$, ethyne and acetylene. Of these names, acetylene is used much more frequently.

There is no need to use a number to locate the position of the triple bond in ethyne and propyne; there is only one possible location for it in each compound. For larger molecules, number the longest carbon chain that contains the triple bond from the end that gives the triply bonded carbons the lower numbers. Show the location of the triple bond by the number of its first carbon. If a hydrocarbon chain contains more than one triple bond, we use the infixes *-adiyn-*, *-atriyn-*, and so forth.

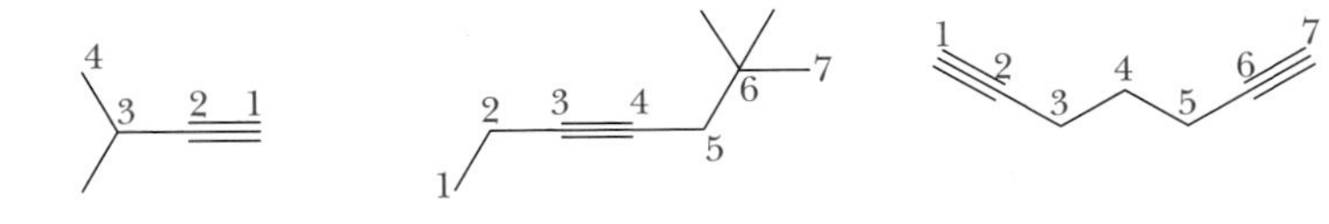

3-Methyl-1-butyne   6,6-Dimethyl-3-heptyne   1,6-Heptadiyne

Example 7.1

Write the IUPAC name of each compound.

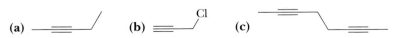

(a)   **(b)** Cl   **(c)**

Solution

(a) 2-Pentyne   **(b)** 3-Chloropropyne   **(c)** 2,6-Octadiyne

Problem 7.1

Write the IUPAC name of each compound.

(a) 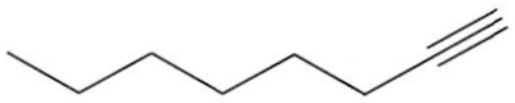 [structure]   (b) 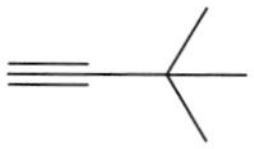 [structure]   (c) 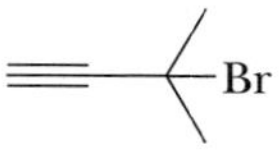 [structure with Br]

B. Common Names

Common names for alkynes are derived by prefixing the names of substituents on the carbon-carbon triple bond to the word *acetylene*. Note in the third example that, when a carbon-carbon double bond (indicated by *-en-*) and a carbon-carbon triple bond (indicated by *-yn-*) are both present in the same molecule, the IUPAC rules specify that the location of the double bond takes precedence in numbering the compound.

$$CH_3C{\equiv}CH \qquad CH_3C{\equiv}CCH_3 \qquad CH_2{=}CHC{\equiv}CH$$

| IUPAC name: | Propyne | 2-Butyne | 1-Buten-3-yne |
| Common name: | Methylacetylene | Dimethylacetylene | Vinylacetylene |

Alkynes in which the triple bond is between carbons 1 and 2 are commonly referred to as **terminal alkynes.** Examples of terminal alkynes are propyne and 1-butyne.

Example 7.2

Write the common name of each alkyne.

(a) 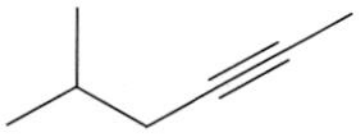 [structure]   (b) 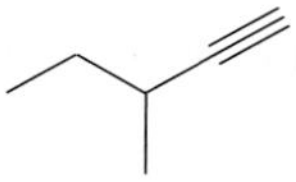 [structure]   (c) 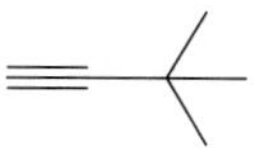 [structure]

Solution

(a) Isobutylmethylacetylene   (b) *sec*-Butylacetylene   (c) *tert*-Butylacetylene

Problem 7.2

Write the common name of each alkyne.

(a) 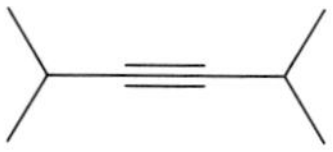 [structure]   (b) 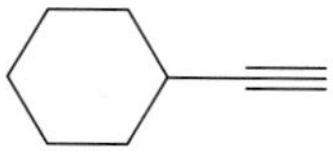 [structure]   (c)  [structure]

The smallest cycloalkyne that has been isolated is cyclooctyne. This molecule is quite unstable and polymerizes rapidly at room temperature. The C—C≡C bond angle in cyclooctyne calculated by molecular dynamics is approximately 155°, indicating a high degree of angle strain. Cyclononyne has also been prepared and is stable at room temperature. The calculated C—C≡C bond angles in this cycloalkyne are approximately 160°, which still represents a considerable distortion from the optimal

180°. You can see the distortion of the $C—C\equiv C$ bond angles in the accompanying optimized molecular model. You can also see the degree to which $C—C$ and $C—H$ bonds on adjacent carbons are staggered, thus minimizing torsional strain.

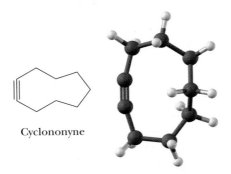

Cyclononyne

7.3 Physical Properties of Alkynes

The physical properties of alkynes are quite similar to those of alkanes and alkenes with analogous carbon skeletons. The lower-molecular-weight alkynes are gases at room temperature. Those that are liquids at room temperature have densities less than 1.0 g/mL (less dense than water). Listed in Table 7.1 are melting points, boiling points, and densities of several low-molecular-weight alkynes. Because, like alkanes and alkenes, alkynes are nonpolar compounds, they are insoluble in water and other polar solvents. They are soluble in each other and in other nonpolar organic solvents.

7.4 Acidity of 1-Alkynes

One of the major differences between the chemistry of alkynes and that of alkenes or alkanes is that a hydrogen attached to a triply bonded carbon atom of a terminal alkyne is sufficiently acidic that it can be removed by a strong base, such as sodium amide $NaNH_2$ (Table 4.1).

Table 7.1 Physical Properties of Some Low-Molecular-Weight Alkynes

Name	Formula	Melting Point (°C)	Boiling Point (°C)	Density at 20°C (g/mL)
Ethyne	$HC\equiv CH$	−81	−84	(a gas)
Propyne	$CH_3C\equiv CH$	−102	−23	(a gas)
1-Butyne	$CH_3CH_2C\equiv CH$	−126	8	(a gas)
2-Butyne	$CH_3C\equiv CCH_3$	−32	27	0.691
1-Pentyne	$CH_3(CH_2)_2C\equiv CH$	−90	40	0.690
1-Hexyne	$CH_3(CH_2)_3C\equiv CH$	−132	71	0.716
1-Octyne	$CH_3(CH_2)_5C\equiv CH$	−79	125	0.746
1-Decyne	$CH_3(CH_2)_7C\equiv CH$	−36	174	0.766

$$H-C\equiv\overset{\frown}{C}-H + :\overset{..}{N}H_2 \rightleftharpoons H-C\equiv C\overset{..}{:}{}^- + \overset{..}{N}H_3 \qquad K_{eq} = 10^{13}$$

pK$_a$ 25
(Stronger
acid)

pK$_a$ 38
(Weaker
acid)

Other strong bases commonly used to form acetylide anions are sodium hydride and lithium diisopropylamide (LDA).

$$Na^+H^-: \qquad [(CH_3)_2CH]_2\overset{..}{N}:{}^-Li^+$$

Sodium hydride   Lithium diisopropylamide
(LDA)

Because water is a stronger acid than acetylene, the hydroxide ion is not a strong enough base to convert a terminal alkyne to an alkyne anion.

$$H-C\equiv C-H + :\overset{..}{O}H^- \rightleftharpoons H-C\equiv C\overset{..}{:}{}^- + H-\overset{..}{O}H \qquad K_{eq} = 10^{-9.3}$$

pK$_a$ 25
(Weaker
acid)

pK$_a$ 15.7
(Stronger
acid)

The pK$_a$ values for alkene and alkane hydrogens are so large (they are so weakly acidic) that neither the commonly used alkali metal hydroxides nor sodium hydride, sodium amide, or lithium diisopropylamide are strong enough bases to remove a proton from alkanes or alkenes.

7.5 Preparation of Alkynes

A. Alkylation of Acetylide Anions with Methyl and 1° Haloalkanes

As we have already seen, an acetylide anion is a strong base. An acetylide anion is also a nucleophile; it has an unshared pair of electrons that it can donate to another atom to form a new covalent bond. In this instance, an acetylide anion donates its unshared pair of electrons to the carbon of a methyl or primary haloalkane, and, in so doing, the acetylide nucleophile substitutes for a halide ion. This type of reaction is called a nucleophilic substitution. For example, treating sodium acetylide with 1-bromobutane gives 1-hexyne.

HC≡C: Na⁺ + $\xrightarrow[\text{substitution}]{\text{nucleophilic}}$ + Na⁺Br⁻

:Br:

Sodium   1-Bromobutane   1-Hexyne
acetylide

Alkylation reaction Any reaction in which a new carbon-carbon bond to an alkyl group is formed.

Because an alkyl group (in this case a butyl group) is added to a molecule, this type of reaction is also called an **alkylation reaction.** We limit our discussion of nucleophilic substitution in this chapter to the reactions of acetylide anions with methyl and 1° haloalkanes. We discuss the scope and limitations of nucleophilic substitution in more general terms in Chapter 9. For reasons we discuss in Chapter 9, alkylation of acetylide anions is practical only with methyl and primary halides.

Because of the ready availability of acetylene and the ease with which it is converted to a good nucleophile, alkylation of an acetylide anion is the most convenient laboratory method for the synthesis of terminal alkynes. The process of alkylation can be repeated, and a terminal alkyne can, in turn, be converted to an internal alkyne. An important feature of this reaction is that new carbon-carbon bonds are made, allowing the construction of larger carbon backbones from smaller ones.

$$CH_3CH_2C{\equiv}C^-Na^+ + CH_3CH_2{-}Br \xrightarrow[\text{substitution}]{\text{nucleophilic}} CH_3CH_2C{\equiv}CCH_2CH_3 + Na^+Br^-$$

Sodium butynide   Bromoethane   3-Hexyne

B. Alkynes from Alkenes

To prepare an alkyne from an alkene, the alkene is first treated with 1 mole of either bromine (Br_2) or chlorine (Cl_2) to give a dihaloalkane (Section 6.3D). Treating the dihaloalkane with 2 moles of a strong base such as sodium amide ($NaNH_2$) in liquid ammonia [$NH_3(l)$] brings about two successive **dehydrohalogenations.** Recall that addition of HX to an alkene is called hydrohalogenation; removal of HX from a haloalkane is called dehydrohalogenation. The removal of atoms from adjacent carbons to form an alkene is also called an elimination reaction and is discussed fully in Chapter 9. The following example shows the conversion of 2-butene to 2-butyne.

Dehydrohalogenation The removal of HX from a molecule.

$$CH_3CH{=}CHCH_3 + Br_2 \xrightarrow{CH_2Cl_2} \overset{\overset{\displaystyle Br \quad\;\; Br}{|\quad\;\;\; |}}{CH_3CH{-}CHCH_3}$$

2-Butene   2,3-Dibromobutane

$$\overset{\overset{\displaystyle Br \quad\;\; Br}{|\quad\;\;\; |}}{CH_3CH{-}CHCH_3} + 2NaNH_2 \xrightarrow[-33°C]{NH_3(l)} CH_3C{\equiv}CCH_3 + 2NaBr + 2NH_3$$

                    Sodium   2-Butyne
                    amide

Given the ease of converting alkenes to dihaloalkanes and then to alkynes, alkenes are versatile starting materials for the preparation of alkynes.

With a strong base such as sodium amide, both dehydrohalogenations occur readily. However, with weaker bases, such as sodium hydroxide or potassium hydroxide in ethanol, it is often possible to stop the reaction after the first dehydrohalogenation and isolate the haloalkene.

$$\overset{\overset{\displaystyle Br \quad\;\; Br}{|\quad\;\;\; |}}{CH_3CH{-}CHCH_3} + KOH \xrightarrow{ethanol} \overset{\overset{\displaystyle Br}{|}}{CH_3CH{=}CCH_3} + KBr + H_2O$$

                                  2-Bromo-2-butene

In practice, it is much more common to use a stronger base and go directly to the alkyne.

The following equations show the conversion of 1-hexene to 1-hexyne. Note that 3 moles of sodium amide are used in this sequence. Two moles are required for the double dehydrohalogenation reaction, which gives 1-hexyne. As soon as any 1-hexyne (a weak acid, pK_a 25) forms, it reacts with sodium amide (a strong base) to give an alkyne salt. Thus, a third mole of sodium amide is required to complete the dehydro-halogenation of the remaining bromoalkene. Addition of water (a weak acid) or aqueous acid completes the sequence and gives 1-hexyne.

$$CH_3(CH_2)_3CH{=}CH_2 \xrightarrow{Br_2} \underset{\substack{Br \quad Br \\ | \quad | \\ \text{1,2-Dibromohexane}}}{CH_3(CH_2)_3CH{-}CH_2} \xrightarrow[-2HBr]{2NaNH_2} \underset{\text{1-Hexyne}}{CH_3(CH_2)_3C{\equiv}CH} \xrightarrow{NaNH_2}$$

$$\underset{\text{Sodium salt of 1-hexyne}}{CH_3(CH_2)_3C{\equiv}C^-Na^+} \xrightarrow{H_2O} \underset{\text{1-Hexyne}}{CH_3(CH_2)_3C{\equiv}CH}$$

In dehydrohalogenation of a haloalkene with at least one hydrogen on each adjacent carbon, a side reaction occurs, namely the formation of an allene.

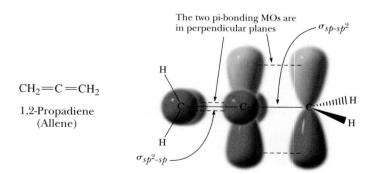

A haloalkene
(a vinylic halide)

An allene

An alkyne

Allene The compound $CH_2{=}C{=}CH_2$. Any compound that contains adjacent carbon-carbon double bonds; that is, any molecule that contains a $C{=}C{=}C$ functional group.

An **allene** has two adjacent carbon-carbon double bonds; that is, it contains a $C{=}C{=}C$ functional group. The simplest allene is 1,2-propadiene, commonly named allene. In it, each end carbon is sp^2 hybridized, and the middle carbon is sp hybridized. Each carbon-carbon sigma bond is formed by the overlap of sp and sp^2 hybrid orbitals. One pi bond is formed by the overlap of parallel $2p_y$ orbitals; the other, by the overlap of parallel $2p_z$ orbitals. The two pi-bonding molecular orbitals are in planes perpendicular to each other, as are the two H—C—H groups.

$$CH_2{=}C{=}CH_2$$

1,2-Propadiene
(Allene)

The two pi-bonding MOs are in perpendicular planes

$\sigma_{sp\text{-}sp^2}$

$\sigma_{sp^2\text{-}sp}$

Most allenes are less stable than their isomeric alkynes. For example, allene itself is less stable by 6.7 kJ (1.6 kcal)/mol than its constitutional isomer propyne, and 1,2-butadiene is less stable than 2-butyne by 16.7 kJ (4.0 kcal)/mol.

$$CH_2{=}C{=}CH_2 \longrightarrow CH_3C{\equiv}CH \qquad \Delta H^0 = -6.7 \text{ kJ } (-1.6 \text{ kcal})/\text{mol}$$

$$CH_2{=}C{=}CHCH_3 \longrightarrow CH_3C{\equiv}CCH_3 \qquad \Delta H^0 = -16.7 \text{ kJ } (-4.0 \text{ kcal})/\text{mol}$$

Because of their lower stability relative to isomeric alkynes, allenes are generally only minor products of alkyne-forming dehydrohalogenation reactions.

Example 7.3

Show how you might convert 1-pentene to 1-pentyne.

Solution

This synthesis can be done in two steps: Treating 1-pentene with 1 mole of bromine gives 1,2-dibromopentane. Treating this dibromoalkane with 3 moles of sodium amide followed by H_2O gives 1-pentyne.

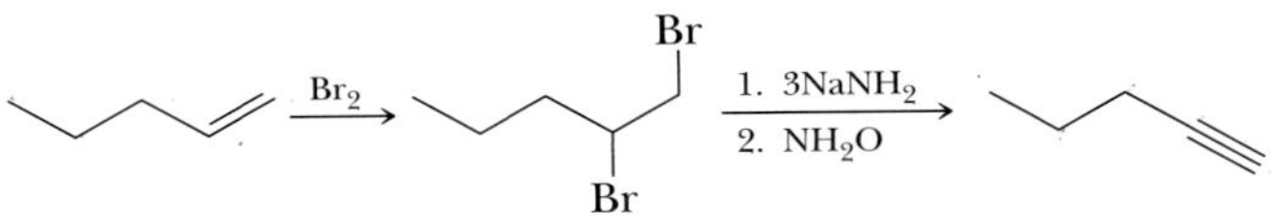

1-Pentene   1,2-Dibromopentane   1-Pentyne

Problem 7.3

Draw a structural formula for an alkene and dichloroalkane with the given molecular formula that yields the indicated alkyne by each reaction sequence.

(a) $C_6H_{12} \xrightarrow{Cl_2} C_6H_{12}Cl_2 \xrightarrow{2NaNH_2}$

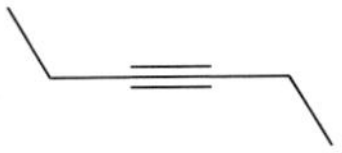

(b) $C_7H_{14} \xrightarrow{Cl_2} C_7H_{14}Cl_2 \xrightarrow{2NaNH_2}$

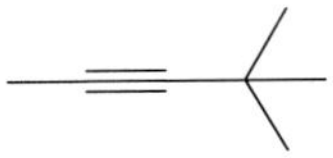

7.6 Electrophilic Addition to Alkynes

Alkynes undergo many of the same electrophilic additions as alkenes. In this section, we study both the addition of bromine and chlorine and the addition of hydrogen halides.

A. Addition of Bromine and Chlorine

Addition of 1 mole of Br_2 to an alkyne gives a dibromoalkene. Addition of bromine to a triple bond is stereoselective, as illustrated by the reaction of 2-butyne with 1 mole of Br_2. The major product corresponds to anti addition of the two bromine atoms. Carrying out the bromination in acetic acid with added bromide ion, for example LiBr, increases the preference for anti addition significantly.

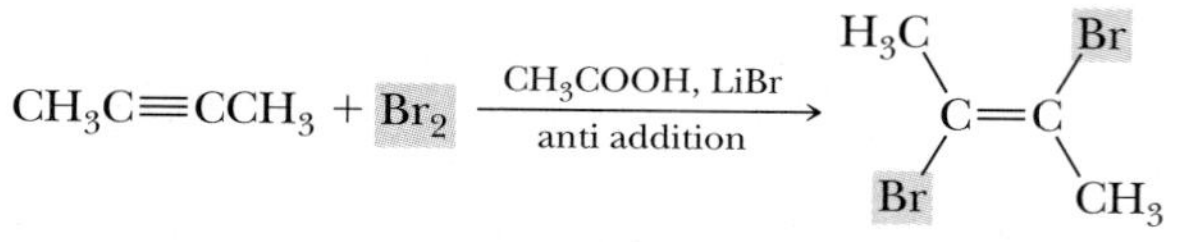

$$CH_3C{\equiv}CCH_3 + Br_2 \xrightarrow[\text{anti addition}]{CH_3COOH, LiBr}$$

H$_3$C, Br, C=C, Br, CH$_3$

2-Butyne   (E)-2,3-Dibromo-2-butene

Alkynes similarly undergo addition of Cl_2, although less stereoselectively than with Br_2.

Addition of bromine to alkynes follows much the same type of mechanism as it does for addition to alkenes (Section 6.3D), namely, formation of a bridged bromonium ion intermediate, which is then attacked by bromide ion from the face opposite that occupied by the positively charged bromine atom.

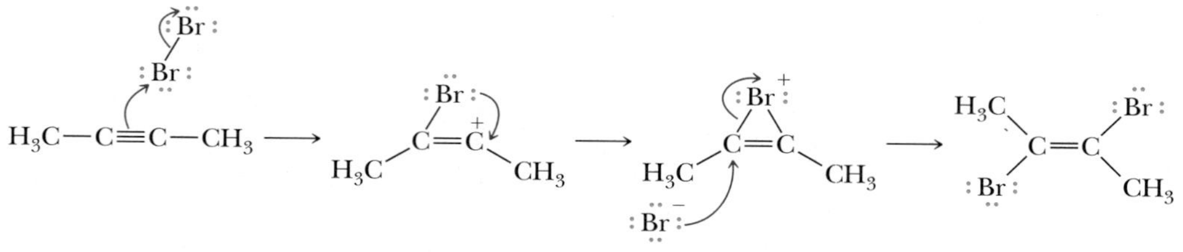

$$H_3C-C{\equiv}C-CH_3 \longrightarrow \quad \longrightarrow \quad \longrightarrow$$

Addition of a second mole of Br_2 gives a tetrabromoalkane.

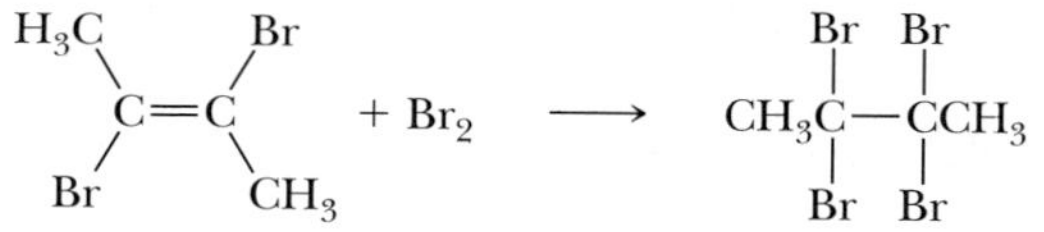

H$_3$C, Br, C=C, Br, CH$_3$ $\;+ Br_2 \longrightarrow\;$ CH_3C-CCH_3 (Br Br / Br Br)

2,2,3,3-Tetrabromobutane

B. Addition of Hydrogen Halides

Alkynes add either 1 or 2 moles of HBr and HCl, depending on the ratios in which the alkyne and halogen acid are mixed.

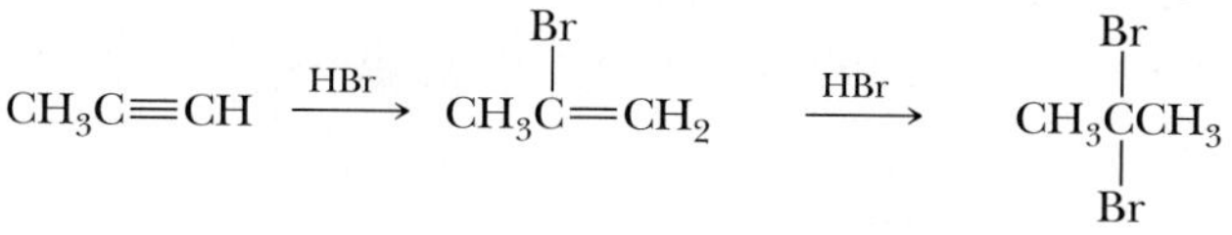

$$CH_3C{\equiv}CH \xrightarrow{HBr} CH_3C{=}CH_2 \xrightarrow{HBr} CH_3CCH_3$$

(Br on middle carbon)   (Br, Br on central carbon)

Propyne   2-Bromopropene   2,2-Dibromopropane

As shown in this equation, addition of both the first and second moles of HBr is regioselective. Addition of hydrogen halides follows Markovnikov's rule (Section 6.3A); hydrogen adds to the carbon that has the greater number of hydrogens. We can account for this regioselectivity of addition of HX by a two-step mechanism for each addition.

Mechanism   *Addition of HBr to an Alkyne*

Step 1: Proton transfer from HBr to the alkyne gives a **vinylic carbocation;** the more stable 2° vinylic carbocation is formed in preference to the less stable 1° vinylic carbocation.

$$CH_3C\equiv CH + H-\overset{\cdot\cdot}{\underset{\cdot\cdot}{Br}}: \longrightarrow CH_3\overset{+}{C}=CH_2 + :\overset{\cdot\cdot}{\underset{\cdot\cdot}{Br}}:^-$$

A 2° vinylic
carbocation

Step 2: Reaction of the vinylic carbocation (an electrophile) with bromide ion (a nucleophile) gives the vinylic bromoalkene.

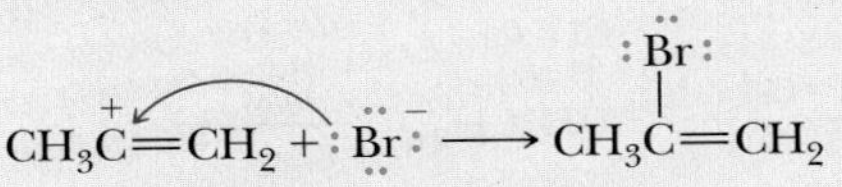

$$CH_3\overset{+}{C}=CH_2 + :\overset{\cdot\cdot}{\underset{\cdot\cdot}{Br}}:^- \longrightarrow CH_3\overset{\overset{\displaystyle :\overset{\cdot\cdot}{\underset{\cdot\cdot}{Br}}:}{|}}{C}=CH_2$$

2-Bromopropene

Vinylic carbocation A carbocation in which the positive charge is on one of the carbons of a carbon-carbon double bond.

Alkynes are considerably less reactive toward most electrophilic additions than are alkenes. The major reason for this difference is the instability of the *sp*-hybridized vinylic carbocation intermediate formed from an alkyne compared with the *sp²*-hybridized alkyl carbocation formed from an alkene.

In the addition of the second mole of HX, Step 1 is the reaction of the electron pair of the remaining pi bond with HBr to form a carbocation. Of the two possible carbocations, the one with the positive charge on the carbon bearing the halogen is favored because of the possibility for resonance stabilization by the adjacent halogen atom.

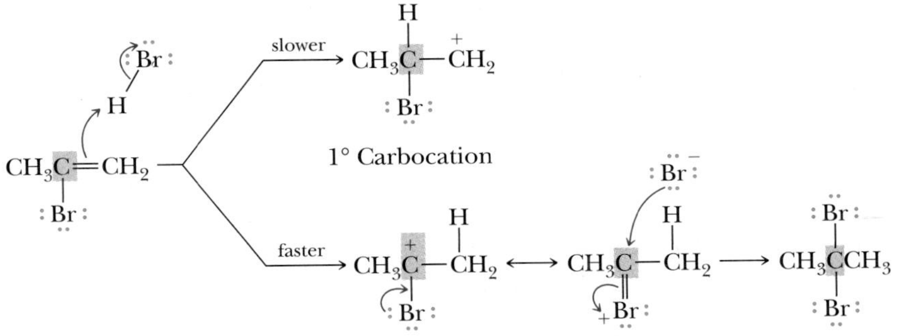

Resonance-stabilized 2° carbocation

Addition of 1 mole of HCl to acetylene gives chloroethene (vinyl chloride), a compound of considerable industrial importance.

$$HC\equiv CH + HCl \longrightarrow CH_2=CHCl$$

Ethyne   Chloroethene
(Acetylene)   (Vinyl chloride)

Vinyl chloride is the monomer in the production of the polymer poly(vinyl chloride), abbreviated PVC.

$$nCH_2=CHCl \xrightarrow{\text{catalyst}} \begin{array}{c} Cl \\ | \\ -(CH_2CH)_n \end{array}$$

Vinyl chloride   Poly(vinyl chloride)
                        (PVC)

PVC dominates much of the plumbing and construction market for plastics. Approximately 67% of all pipe, fittings, and conduit, along with 42% of all plastics used in construction at the present time, are fabricated from PVC. We will describe the synthesis of this polymer and its properties in Chapter 29. Our purpose here is to describe the synthesis of vinyl chloride.

At one time, hydrochlorination of acetylene was the major source of vinyl chloride. As the cost of production of acetylene increased, however, manufacturers of vinyl chloride sought other routes to this material. The starting material chosen was ethylene, which can be converted to vinyl chloride in two steps: Treating ethylene with chlorine gives 1,2-dichloroethane, which, when heated in the presence of charcoal or another catalyst, loses a molecule of HCl to form vinyl chloride.

$$CH_2{=}CH_2 \xrightarrow{Cl_2} \underset{\substack{| \quad | \\ CH_2CH_2}}{\overset{Cl \quad Cl}{}} \xrightarrow[\text{catalyst}]{\text{heat}} CH_2{=}CHCl \ + \ HCl$$

Ethylene   1,2-Dichloroethane   Vinyl chloride

Chlorine atoms of the byproduct are recycled by passing the HCl mixed with air over a copper(II) catalyst, which results in the oxidation of HCl to Cl_2.

$$4HCl \ + \ O_2 \xrightarrow{CuCl_2} 2H_2O \ + \ 2Cl_2$$

We described the production of vinyl chloride first from acetylene and then from ethylene to illustrate an important point about industrial organic chemistry. The aim is to produce a desired chemical from readily available and inexpensive starting materials by reactions in which byproducts can be recycled. All chemical companies now support this objective to minimize both costs and production of materials that require disposal or can harm the environment.

7.7 Hydration of Alkynes to Aldehydes and Ketones

The elements of H_2O can be added to the carbon-carbon triple bond of an alkyne by the same two reactions used for the hydration of alkenes, namely hydroboration-oxidation and acid-catalyzed hydration. Whereas the reagents are similar, the products from hydration of alkenes and alkynes are quite different.

A. Hydroboration-Oxidation

Borane adds readily to an internal alkyne as illustrated by its reaction with 3-hexyne.

3-Hexyne

A trialkenylborane
(R = *cis*-3-hexenyl group)

Notice that the hydroboration of an internal alkyne stops after the addition of 1 mole of borane. The product is a trialkenylborane (the infix -*enyl*- shows the presence of a carbon-carbon double bond on the carbon bonded to boron). As with hydroboration of alkenes (Section 6.4), hydroboration of alkynes is stereoselective; it involves syn addition of hydrogen and boron.

Terminal alkynes also react regioselectively with borane to form trialkenylboranes. In practice, however, the reaction is difficult to stop at this stage because the alkenyl group reacts further with borane to undergo a second hydroboration.

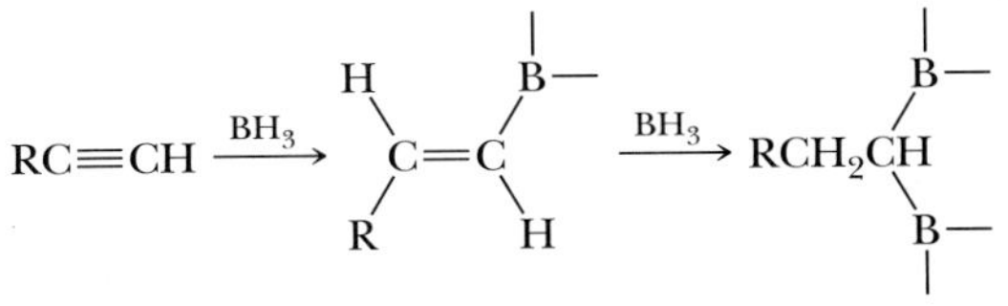

$$RC{\equiv}CH \xrightarrow{BH_3} \underset{\substack{| \\ R \qquad H}}{\overset{\substack{H \qquad B- \\ | }}{C=C}} \xrightarrow{BH_3} RCH_2\underset{\substack{| \\ B-}}{\overset{\substack{B- \\ |}}{CH}}$$

It is possible to prevent the second hydroboration step and, in effect, stop the reaction at the alkenylborane stage by using a sterically hindered disubstituted borane. One of the most widely used of these is disiamylborane, (sia)$_2$BH, prepared by treating borane with two equivalents of 2-methyl-2-butene (amyl is an older common name for pentyl).

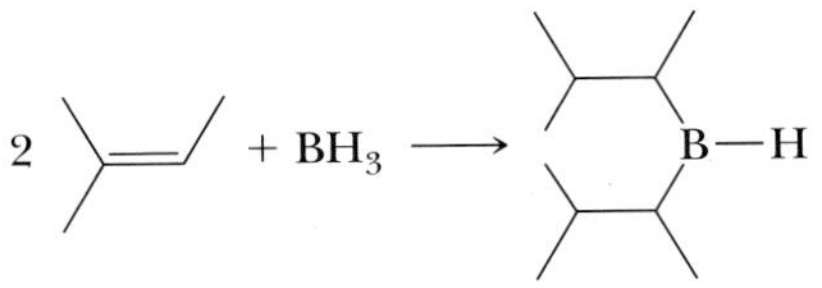

2-Methyl-2-
butane

Di-*sec*-isoamylborane
[(**sia**)$_2$BH]

Reaction of this sterically hindered dialkylborane with a terminal alkyne results in a single hydroboration and formation of an alkenylborane.

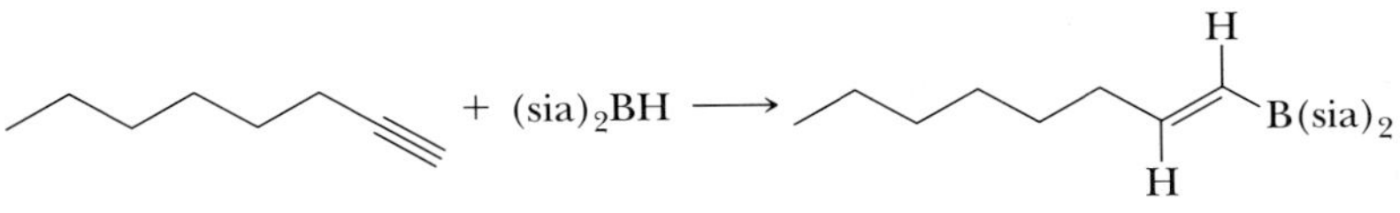

1-Octyne

An alkenylborane

Just as with hydroboration of unsymmetrical alkenes, the addition of (sia)$_2$BH to a carbon-carbon triple bond of a terminal alkene is regioselective; boron adds to the less substituted carbon.

Treatment of an alkenylborane with hydrogen peroxide in aqueous sodium hydroxide gives a product that corresponds to hydration of an alkyne; that is, it corresponds to addition of H to one carbon of the triple bond and OH to the other as illustrated by the hydroboration-oxidation of 2-butyne.

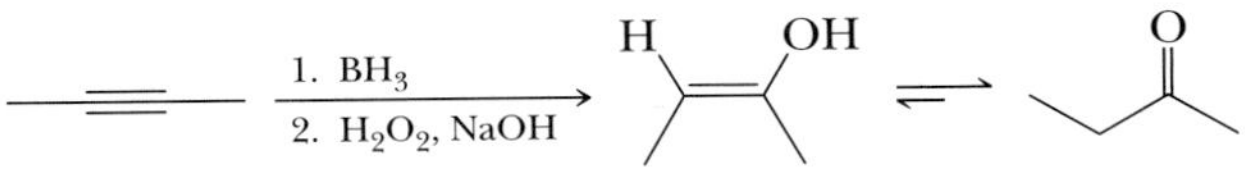

2-Butyne

2-Buten-2-ol
(an enol)

2-Butanone
(a ketone)

$K_{eq} = 6.7 \times 10^6$

(for keto-enol
tautomerism)

The initial product of hydroboration-oxidation of an alkyne is an **enol**, a compound containing a hydroxyl group bonded to a carbon of a carbon-carbon double bond. The name enol is derived from the fact that it is both an alkene (-*en*-) and an alcohol (-*ol*). To this point, hydroboration-oxidation of alkynes is identical to that of alkenes (Section 6.4).

Enol A compound containing a hydroxyl group bonded to a doubly bonded carbon atom.

Enols are in equilibrium with a constitutional isomer formed by migration of a hydrogen atom from oxygen to carbon and rearrangement of the carbon-carbon double bond to form a carbon-oxygen double bond. As can be seen from the value of K_{eq}, 2-butanone (the keto form) is much more stable than its enol. Keto forms in general are more stable than enol forms because (1) a C=O pi bond is generally stronger than a C=C pi bond, whereas (2) a C—H and O—H sigma bonds generally have similar bond strengths.

The keto and enol forms of 2-butanone are said to be tautomers. **Tautomers** are constitutional isomers that are in equilibrium with each other and differ only in the location of a hydrogen atom or other atom and a double bond relative to a heteroatom, most commonly O, N, and S. This type of isomerism is called tautomerism. Because the type of tautomerism we are dealing with in this section involves keto (from ketone) and enol forms, it is commonly called **keto-enol tautomerism.** We discuss keto-enol tautomerism in more detail in Section 16.9.

Hydroboration of a terminal alkyne using disiamylborane followed by oxidation in alkaline hydrogen peroxide also gives an enol that, in this case, is in equilibrium with the more stable aldehyde. Thus, hydroboration-oxidation of a terminal alkyne gives an aldehyde.

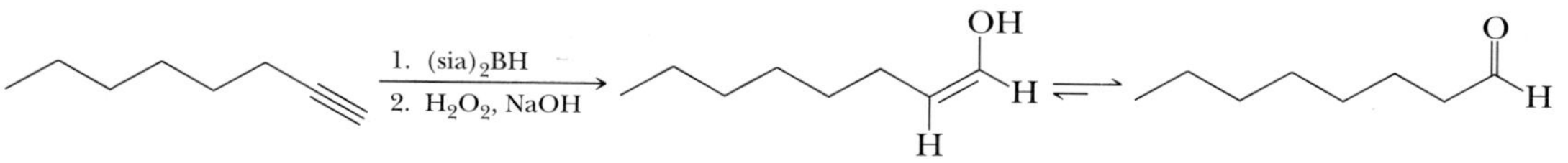

Tautomers Constitutional isomers in equilibrium with each other that differ in the location of a hydrogen atom and a double bond relative to a heteroatom, most commonly O, N, or S.

Keto-enol tautomerism A type of isomerism involving keto (from ketone) and enol tautomers.

1-Octyne   An enol   Octanal

Example 7.4

Hydroboration-oxidation of 2-pentyne gives a mixture of two ketones, each with molecular formula $C_5H_{10}O$. Propose structural formulas for these two ketones and for the enol from which each is derived.

Solution

Because each carbon of the triple bond in 2-pentyne has the same degree of substitution, very little regioselectivity occurs during hydroboration. Two enols are formed, and the isomeric ketones are formed from them.

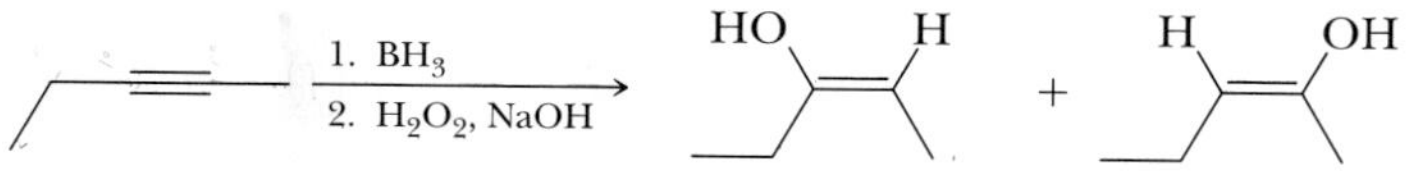

2-Pentyne   2-Pentene-3-ol   2-Pentene-2-ol
             (an enol)   (an enol)

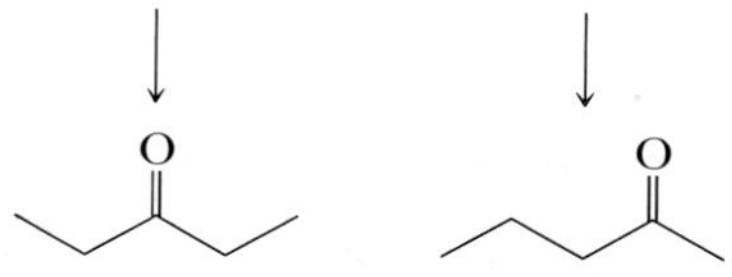

3-Pentanone   2-Pentanone

Problem 7.4

Draw a structural formula for a hydrocarbon of the given molecular formula that undergoes hydroboration-oxidation to give the indicated product.

(a) C_7H_{10} $\xrightarrow[\text{2. } H_2O_2, \text{ NaOH}]{\text{1. (sia)}_2\text{BH}}$

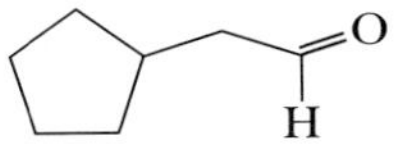

(b) C_7H_{12} $\xrightarrow[\text{2. } H_2O_2, \text{ NaOH}]{\text{1. BH}_3}$

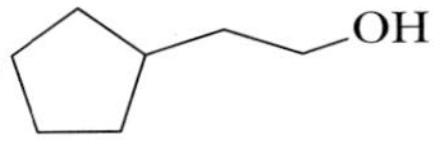

B. Acid-Catalyzed Hydration

In the presence of concentrated sulfuric acid and Hg(II) salts as catalysts, alkynes undergo the addition of water in a reaction analogous to the oxymercuration of alkenes (Section 6.3F). The Hg(II) salts most often used for this purpose are the sulfate or acetate. For terminal alkynes, addition of water follows Markovnikov's rule; hydrogen adds to the carbon atom of the triple bond bearing the hydrogen. The resulting enol is in equilibrium with the more stable keto form, so the product isolated is a ketone (an aldehyde in the case of acetylene itself).

$$CH_3C\equiv CH + H_2O \xrightarrow[\text{HgSO}_4]{\text{H}_2\text{SO}_4} CH_3\overset{\overset{\displaystyle OH}{|}}{C}=CH_2 \rightleftharpoons CH_3\overset{\overset{\displaystyle O}{||}}{C}CH_3$$

Propyne	Propen-2-ol	Propanone
	(an enol)	(Acetone)

The mechanism of this reaction is illustrated by the hydration of propyne to give propanone (acetone).

Mechanism   *HgSO₄/H₂SO₄ Catalyzed Hydration of an Alkyne*

Step 1: Attack of Hg^{2+} (an electrophile) on the carbon-carbon triple bond (a nucleophile) gives a bridged mercurinium ion intermediate, which contains a three-center two-electron bond. The mercurinium ion intermediate is best represented as a hybrid of contributing structures.

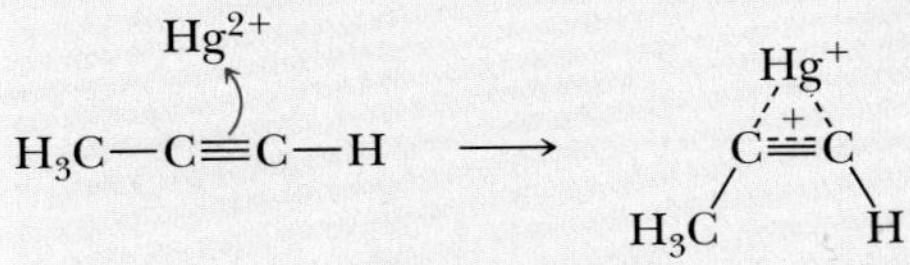

A bridged mercurinium
ion intermediate

Step 2: Attack of water (a nucleophile) on the bridged mercurinium ion intermediate (an electrophile) from the side opposite the bridge opens the three-membered ring.

Because the 2° vinylic cation structure makes a greater contribution to the hybrid than the 1° vinylic cation structure, attack of water occurs preferentially at the more substituted carbon, which accounts for the observed regioselectivity of the reaction.

Step 3: Proton transfer to solvent gives an organomercury enol.

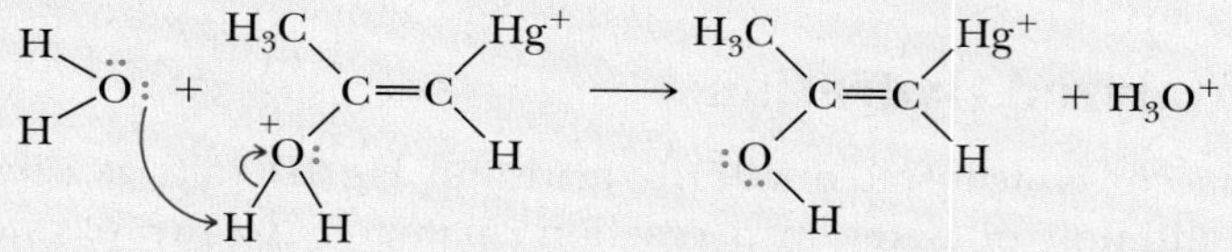

Step 4: Tautomerism of the enol gives the keto form.

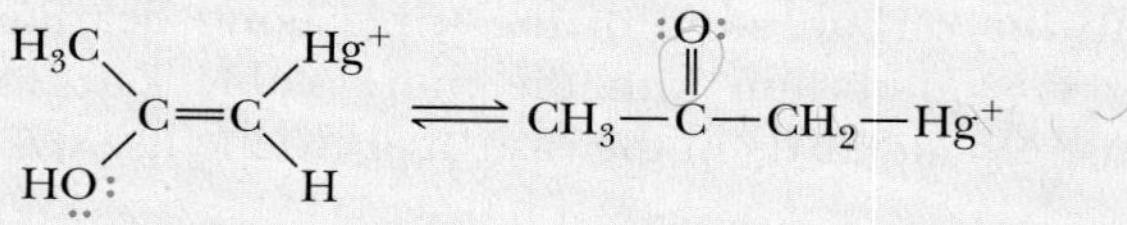

Enol form   Keto form

Step 5: Proton transfer to the carbonyl group of the ketone gives an oxonium ion.

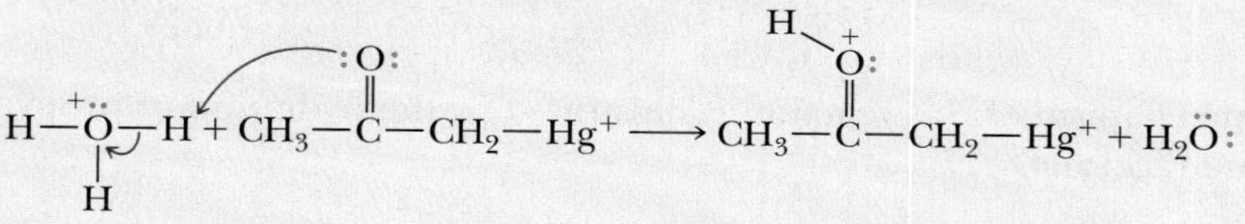

Steps 6 and 7: Loss of Hg^{2+} from the oxonium ion gives the enol form of the final product. Tautomerism of the enol gives the ketone.

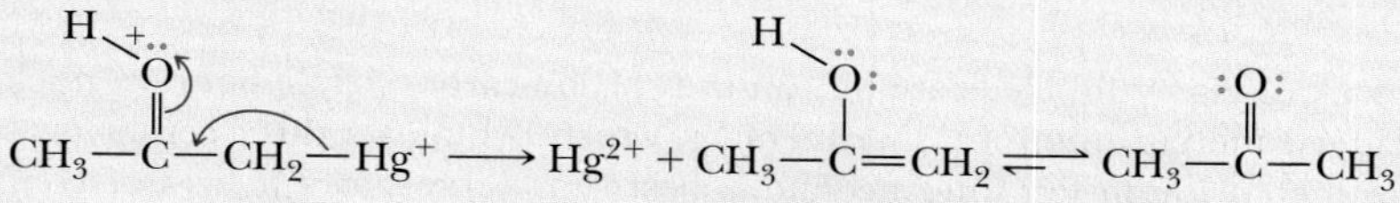

(Enol form)   (Keto form)

Example 7.5

Show reagents to bring about the following conversions.

(a)

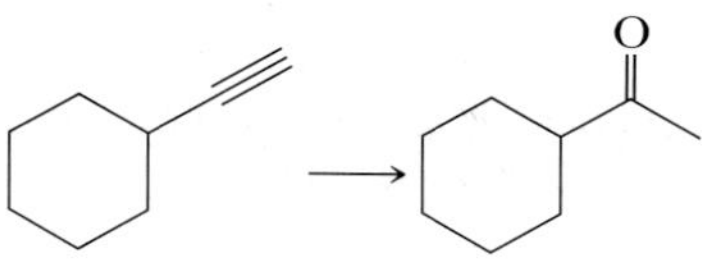

(b)

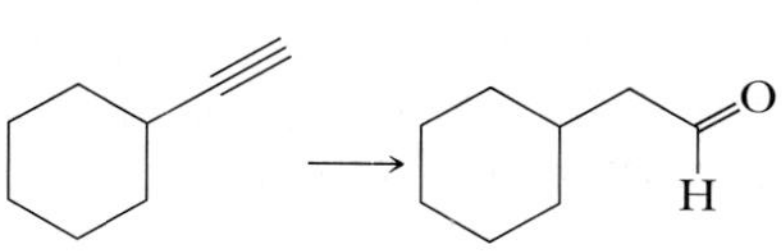

Solution

(a) Hydration of this monosubstituted alkyne using a mercuric ion catalyst gives an enol that is in equilibrium with the more stable keto form.

(b) Hydroboration using disiamylborane followed by treatment with alkaline hydrogen peroxide gives an enol that is in equilibrium with the more stable aldehyde.

Problem 7.5

Hydration of 2-pentyne gives a mixture of two ketones, each with molecular formula $C_5H_{10}O$. Propose structural formulas for these two ketones and for the enol from which each is derived.

7.8   Reduction of Alkynes

Three types of reactions are used to convert alkynes to alkenes and alkanes: catalytic reduction, hydroboration-protonolysis, and dissolving-metal reduction.

A.   Catalytic Reduction

Treatment of an alkyne with H_2 in the presence of a transition metal catalyst, most commonly palladium, platinum, or nickel, results in the addition of 2 moles of H_2 to the alkyne and its conversion to an alkane. Catalytic reduction of an alkyne can be brought about at or slightly above room temperature and with moderate pressures of hydrogen gas.

$$CH_3C{\equiv}CCH_3 + 2H_2 \xrightarrow[\text{3 atm}]{\text{Pd, Pt, or Ni}} CH_3CH_2CH_2CH_3$$

2-Butyne   Butane

Reduction of an alkyne occurs in two stages: first addition of 1 mole of H_2 to form an alkene and then addition of the second mole to the alkene to form the alkane. In most cases, it is not possible to stop the reaction at the alkene stage.

However, by careful choice of catalyst, it is possible to stop the reduction after the addition of 1 mole of hydrogen. The catalyst most commonly used for this purpose consists of finely powdered palladium metal deposited on solid calcium carbonate that has been specially modified with lead salts. This combination is known as the **Lindlar catalyst.** Reduction (hydrogenation) of alkynes over a Lindlar catalyst is stereoselective; **syn addition** of two hydrogen atoms to the carbon-carbon triple bond gives a *cis* alkene.

Lindlar catalyst   Finely powdered palladium metal deposited on solid calcium carbonate that has been specially modified with lead salts. Its particular use is as a catalyst for the reduction of an alkyne to a *cis* alkene.

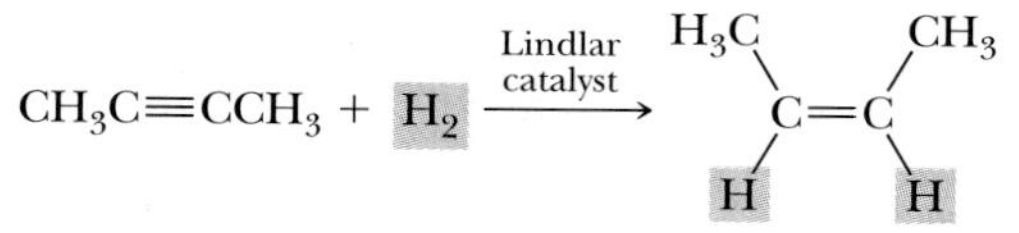

$$CH_3C \equiv CCH_3 + H_2 \xrightarrow[\text{catalyst}]{\text{Lindlar}} \begin{array}{c} H_3C \quad\quad CH_3 \\ C = C \\ H \quad\quad H \end{array}$$

2-Butyne   cis-2-Butene

Because addition of hydrogen in the presence of the Lindlar catalyst is stereose-lective for syn addition, it has been proposed that reduction proceeds by simultane-ous or nearly simultaneous transfer of two hydrogen atoms from the surface of the metal catalyst to the alkyne. We presented a similar mechanism in Section 6.6A for the catalytic reduction of an alkene to an alkane.

B. Hydroboration-Protonolysis

As we have just seen in Section 7.7A, internal alkynes react with borane to give a tri-alkenylborane. Treating a trialkenylborane with a carboxylic acid, such as acetic acid, results in stereoselective replacement of boron by hydrogen: a *cis* alkenyl group bonded to a boron is converted to a *cis* alkene.

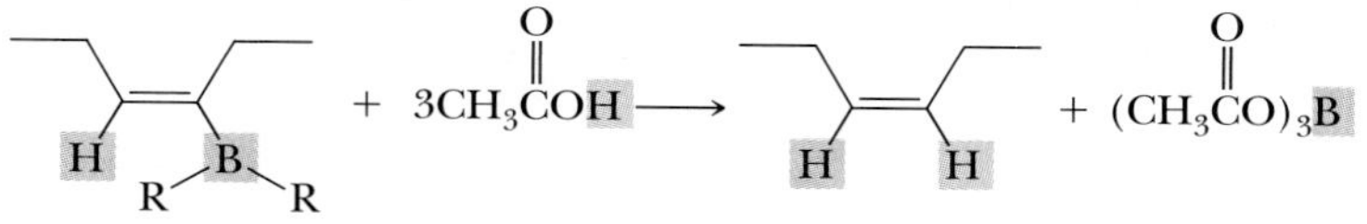

A trialkenylborane   cis-3-Hexene

The net effect of hydroboration of an internal alkyne followed by treatment with acetic acid is reduction of the alkyne to a *cis* alkene. Thus, hydroboration-protonoly-sis and catalytic reduction over a Lindlar catalyst provide alternative schemes for con-version of an alkyne to a *cis* alkene.

C. Dissolving-Metal Reduction

Alkynes can also be reduced to alkenes by using either sodium or lithium metal in liquid ammonia or in low-molecular-weight primary or secondary amines. The alkali metal is the reducing agent and, in the process, is oxidized to M^+, which dissolves as a metal salt in the solvent for the reaction. Reduction of an alkyne to an alkene by lithium or sodium in liquid ammonia, $NH_3(l)$, is stereoselective; it involves mainly **anti addition** of two hydrogen atoms to the triple bond.

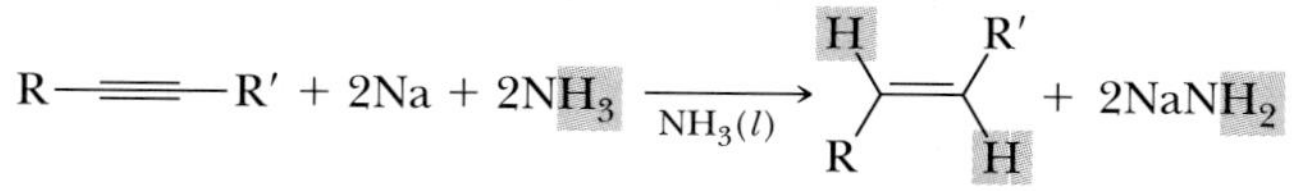

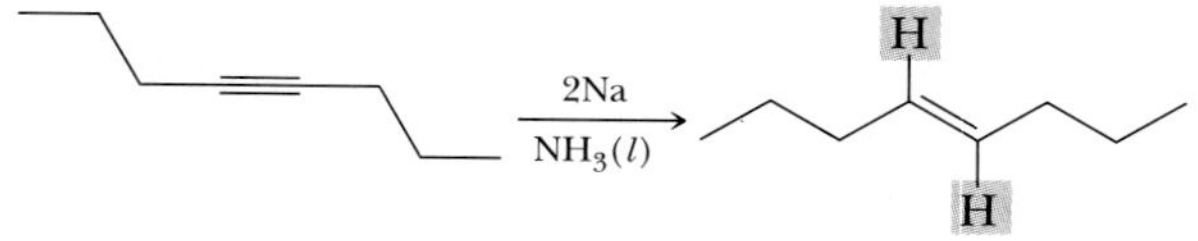

4-Octyne   trans-4-Octene

Thus, by the proper choice of reagents and reaction conditions, it is possible to reduce an alkyne to either a *cis* alkene (by catalytic reduction or hydroboration-protonolysis) or to a *trans* alkene (by dissolving-metal reduction).

The stereoselectivity of alkali metal reduction of alkynes to alkenes can be accounted for by the following mechanism. As you study this mechanism, note that it involves two one-electron reductions and two proton-transfer reactions. The stereochemistry of the alkene is determined in Step 3. Adding the four steps and canceling species that appear on both sides of the equation gives the overall equation for the reaction.

Mechanism   *Reduction of an Alkyne by Sodium in Liquid Ammonia*

Step 1: A one-electron reduction of the alkyne gives an alkenyl radical anion; that is, an ion containing an unpaired electron on one carbon and a negative charge on an adjacent carbon (note that we use a single-headed arrow to show the repositioning of single electrons).

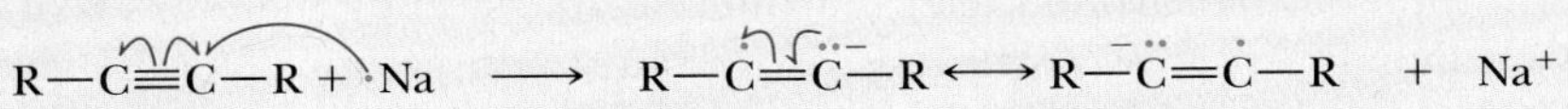

$$R-C\equiv C-R + \cdot Na \longrightarrow R-C{=}\overset{-}{\overset{\cdot\cdot}{C}}-R \longleftrightarrow R-\overset{\cdot\cdot}{\overset{-}{C}}{=}\overset{\cdot}{C}-R + Na^+$$

A resonance-stabilized alkenyl radical anion

Step 2: The alkenyl radical anion (a very strong base) abstracts a proton from a molecule of ammonia (under these conditions, a weak acid) to give an alkenyl radical.

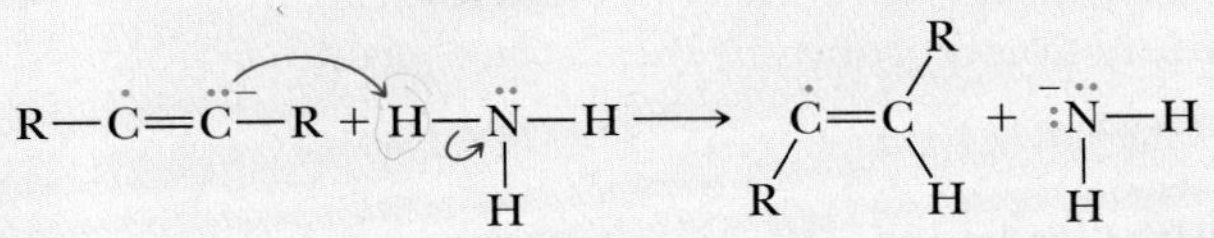

An alkenyl   Amide
radical   ion

Step 3: A one-electron reduction of the alkenyl radical gives an alkenyl anion. The *trans* alkenyl anion is more stable than its *cis* isomer, and the stereochemistry of the final product is determined in this step.

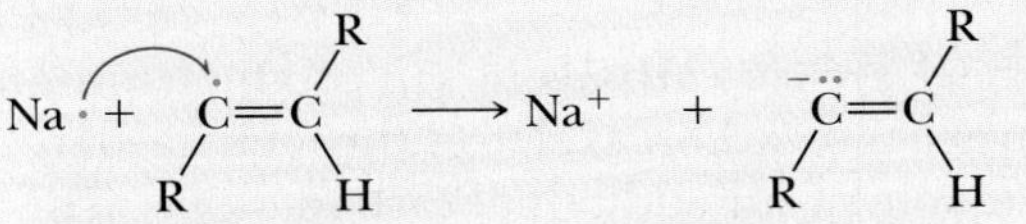

An alkenyl anion

Step 4: A second proton-transfer reaction completes the reduction and gives the *trans* alkene.

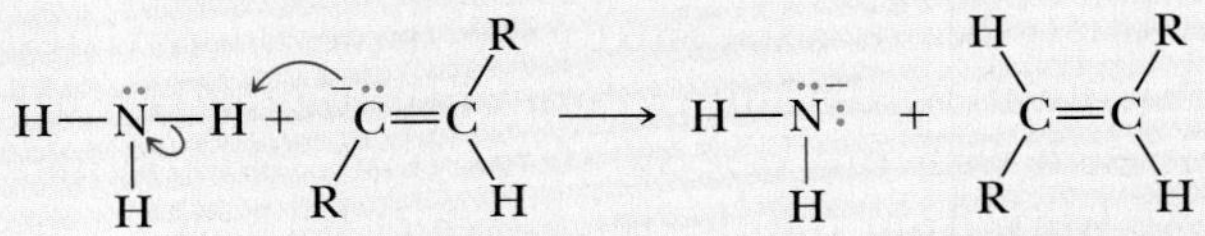

Amide ion   A *trans* alkene

ORGANIC
Chemistry·ᵢ·Now™
Click Mechanisms to view an animation of the **Reduction of an Alkyne by Sodium in Liquid Ammonia**

7.9 Organic Synthesis: Retrosynthetic Analysis

We have now seen how to prepare both terminal and internal alkynes from acetylene and substituted acetylenes, and we have seen several common reactions of alkynes, including addition (HX, X_2, and H_2O), hydroboration-oxidation, and

Organic synthesis   A series of reactions by which a set of organic starting materials is converted to a more complicated structure.

reduction. Now let us move a step farther to consider what might be called the art of **organic synthesis.** Synthesis is an important objective of organic chemists for the preparation of compounds for use as pharmaceuticals, agrochemicals, plastics, elastomers, or textile fibers. A successful synthesis must provide the desired product in a maximum yield and with a maximum control of stereochemistry at all stages of the synthesis. Furthermore, there is an increasing desire to develop "green" syntheses, that is, syntheses that do not produce or release byproducts harmful to the environment.

Our goal in this section is to develop an ability to plan a successful synthesis. The best strategy is to work backwards from the desired product. First, we analyze the target molecule in the following way.

1. Count the carbon atoms of the carbon skeleton of the target molecule. Determining how to build the carbon skeleton from available starting materials is often the most challenging part of a synthesis. If you must add carbons, you need to consider what carbon-carbon bond-forming reactions are available to you. At this stage in the course, you have only one such reaction, namely alkylation of acetylide anions (Section 7.5).
2. Determine the functional groups. What are they and how can they be changed to facilitate formation of the carbon skeleton? How can they then be changed to give the final set of functional groups in the desired product?

Target Molecule: *cis*-3-Hexene

As readily available starting materials, we use acetylene and haloalkanes.

Target molecule:

cis-3-Hexene

Analysis.   We note there are six carbons in the product and only two in acetylene. We will need to construct the carbon skeleton through carbon-carbon bond formation with haloalkanes totaling four additional carbon atoms. The functional group in the product is a *cis* carbon-carbon double bond, which can be prepared by catalytic reduction of a carbon-carbon triple bond using the Lindlar catalyst (Section 7.8A). We then disconnect the carbon skeleton into possible starting materials, which we can later reconnect by known reactions. In the example here, we disconnect at the two carbon-carbon single bonds adjacent to the triple bond. These bonds can be formed during the synthesis by alkylation of the acetylide dianion using two haloalkanes (Section 7.5A), each with two carbon atoms (e.g., bromoethane). This type of scheme, in which we work from the desired product back to a set of starting materials, is called a **retrosynthesis.** We use an open arrow to symbolize a step in a retrosynthesis.

Retrosynthesis   A process of reasoning backwards from a target molecule to a suitable set of starting materials.

disconnect
here

cis-3-Hexene   3-Hexyne   Acetylide   Bromoethane
                                           dianion

Synthesis. Our starting materials for this synthesis of *cis*-3-hexene are acetylene and bromoethane, both readily available compounds. This synthesis is carried out in five steps as follows.

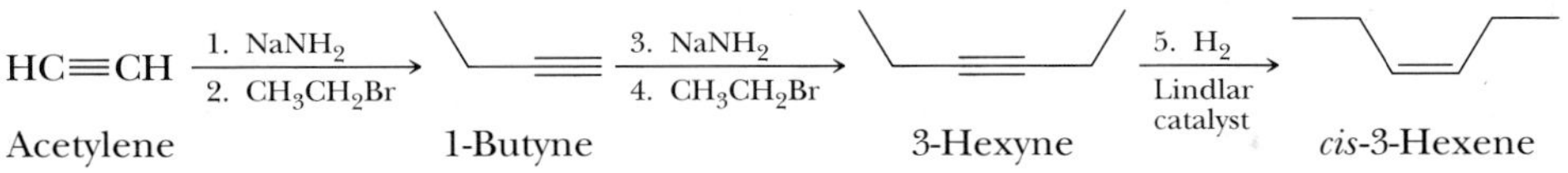

$$HC\equiv CH \xrightarrow[\text{2. } CH_3CH_2Br]{\text{1. } NaNH_2} \text{1-Butyne} \xrightarrow[\text{4. } CH_3CH_2Br]{\text{3. } NaNH_2} \text{3-Hexyne} \xrightarrow[\substack{\text{Lindlar} \\ \text{catalyst}}]{\text{5. } H_2} \textit{cis}\text{-3-Hexene}$$

Acetylene   1-Butyne   3-Hexyne   *cis*-3-Hexene

Target Molecule: 2-Heptanone

2-Heptanone is responsible for the "peppery" odor of cheeses of the Roquefort type. As readily available starting materials, we again use acetylene and haloalkanes.

Target molecule:

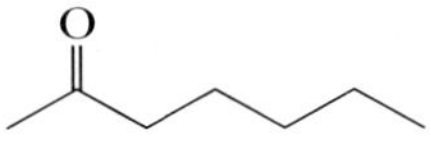

2-Heptanone

Analysis. We note that there are seven carbons in the product and only two in acetylene. We will need to construct the carbon skeleton through carbon-carbon bond formation with haloalkanes totaling five carbon atoms. The functional group in the target molecule is a ketone, which we can prepare by hydration of a carbon-carbon triple bond. Hydration of 1-heptyne gives only 2-heptanone, whereas hydration of 2-heptyne gives a mixture of 2-heptanone and 3-heptanone. Therefore, we choose a functional group interconversion via 1-heptyne.

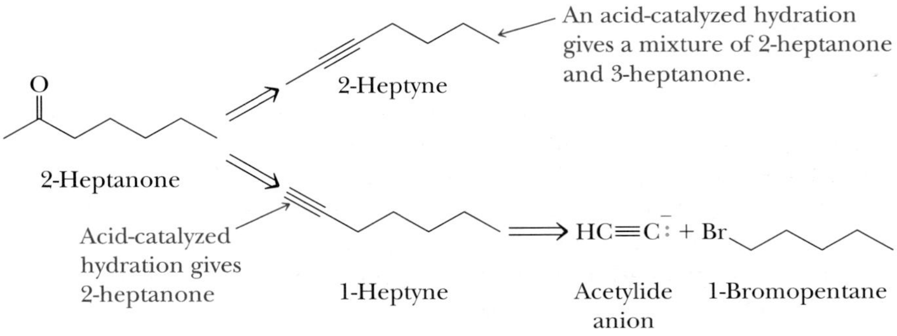

2-Heptyne — An acid-catalyzed hydration gives a mixture of 2-heptanone and 3-heptanone.

2-Heptanone

Acid-catalyzed hydration gives 2-heptanone   1-Heptyne $HC\equiv C:^- + Br\diagdown$ Acetylide anion   1-Bromopentane

Synthesis. This synthesis can be carried out in three steps as follows.

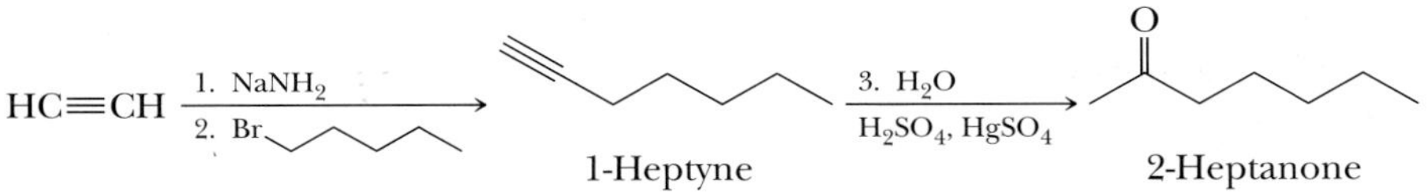

$$HC\equiv CH \xrightarrow[\text{2. } Br\diagdown]{\text{1. } NaNH_2} \text{1-Heptyne} \xrightarrow[H_2SO_4, HgSO_4]{\text{3. } H_2O} \text{2-Heptanone}$$

1-Heptyne   2-Heptanone

Example 7.6

How might the scheme for the synthesis of 2-heptanone be modified so that the product is heptanal?

Solution

Steps 1 and 2 are the same and give 1-heptyne. Instead of acid-catalyzed hydration of 1-heptyne, treat the alkyne with $(sia)_2BH$ followed by alkaline hydrogen peroxide (Section 7.7).

$$HC\equiv CH \xrightarrow[\text{2. Br}]{\text{1. NaNH}_2} \text{1-Heptyne} \xrightarrow[\text{H}_2\text{O}_2,\ \text{NaOH}]{\text{3. (sia)}_2\text{BH}} \text{Heptanal}$$

Problem 7.6

Show how the synthetic scheme in Example 7.6 might be modified to give the following.

(a) 1-Heptanol   **(b)** 2-Heptanol

Summary

Alkynes contain one or more carbon-carbon triple bonds (Section 7.1). The triple bond is a combination of one sigma bond formed by the overlap of sp hybrid orbitals and two pi bonds formed by the overlap of two sets of parallel $2p$ orbitals. The physical properties of alkynes are similar to those of alkanes and alkenes of comparable carbon skeleton.

The pK_a values of terminal alkynes are approximately 25 (Section 7.4); they are less acidic than water and alcohols but more acidic than alkanes, alkenes, and ammonia. Hydrogen bonded to a carbon-carbon triple bond is sufficiently acidic that it can be removed by a strong base, most commonly

sodium amide ($NaNH_2$), sodium hydride (NaH), or lithium diisopropylamide (LDA).

Tautomers (Section 7.7A) are constitutional isomers that are in equilibrium with each other but differ in the location of a hydrogen and a double bond relative to a heteroatom, most commonly O, N, and S. **Keto-enol tautomerism** is the most common type of tautomerism we encounter in this course. The functional group of an enol is an —OH group on a double-bonded carbon atom. The enol form is in equilibrium with the keto form, and equilibrium almost always lies far on the side of the keto form.

Key Reactions

ORGANIC
Chemistry ·ᐤ·Now ™

Click Reaction Flash Cards to review the **Key Reactions of Alkynes**

1. Acidity of Terminal Alkynes (Section 7.4)

Treatment of terminal alkynes (pK_a 25) with a strong base, most commonly $NaNH_2$, NaH, or lithium diisopropylamine (LDA), gives an acetylide salt.

$$HC\equiv CH + Na^+NH_2^- \longrightarrow CH_3C\equiv C^-Na^+ + NH_3$$

2. Alkylation of Acetylide Anions (Section 7.5A)

Acetylide anions are nucleophiles and will displace halide ion from methyl and 1° haloalkanes.

$$HC\equiv C^-Na^+ + \text{(alkyl bromide)} \longrightarrow \text{(alkyne)} + Na^+Br^-$$

3. Synthesis of an Alkyne from an Alkene (Section 7.5B)

Treating an alkene with Br_2 or Cl_2 gives a dihaloalkane. Treating the dihaloalkane with $NaNH_2$ or other strong base results in two successive dehydrohalogenations to give an alkyne.

$$RCH{=}CHR \xrightarrow{Br_2} \underset{\underset{Br}{|}}{RCH}{-}\underset{\underset{Br}{|}}{CHR} \xrightarrow[NH_3(l)]{2NaNH_2} RC{\equiv}CR$$

4. Addition of Br₂ and Cl₂ (Section 7.6A)

Addition of 1 mole of Br_2 or Cl_2 is anti stereoselective; anti addition of halogen to an alkyne gives an (*E*)-dihaloalkene. Addition of a second mole of halogen gives a tetrahaloalkane.

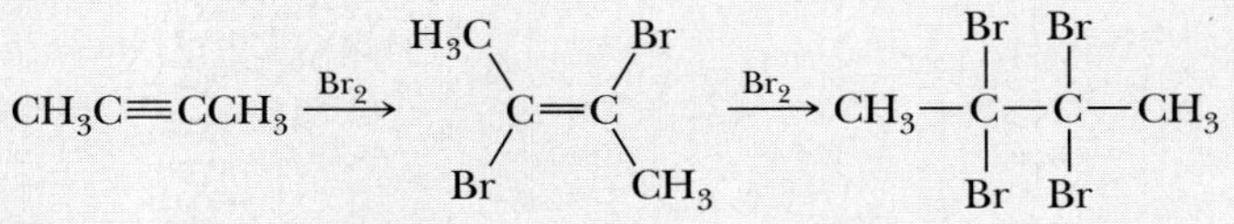

$$CH_3C{\equiv}CCH_3 \xrightarrow{Br_2} \underset{\underset{Br}{|}}{\overset{\overset{H_3C}{\diagdown}}{C}}{=}\underset{\underset{CH_3}{|}}{\overset{\overset{Br}{\diagup}}{C}} \xrightarrow{Br_2} CH_3{-}\underset{\underset{Br}{|}}{\overset{\overset{Br}{|}}{C}}{-}\underset{\underset{Br}{|}}{\overset{\overset{Br}{|}}{C}}{-}CH_3$$

5. Addition of HX (Section 7.6B)

Addition of HX is regioselective. Reaction by way of a vinylic carbocation intermediate follows Markovnikov's rule. Addition of 2HX gives a geminal dihaloalkane.

$$CH_3C{\equiv}CH \xrightarrow{HBr} \underset{\underset{Br}{|}}{CH_3C}{=}CH_2 \xrightarrow{HBr} CH_3\underset{\underset{Br}{|}}{\overset{\overset{Br}{|}}{C}}CH_3$$

6. Keto-Enol Tautomerism (Section 7.7A)

In an equilibrium between a keto form and an enol form, the keto form generally predominates.

$$\underset{\text{Enol form}}{\underset{\underset{OH}{|}}{CH_3CH}{=}CCH_3} \rightleftharpoons \underset{\text{Keto form}}{CH_3CH_2\overset{\overset{O}{\|}}{C}CH_3}$$

7. Hydroboration-Oxidation (Section 7.7A)

Hydroboration of an internal alkyne is syn stereoselective. Oxidation of the resulting trialkenylborane by $H_2O_2/NaOH$ gives an enol that is in equilibrium, through keto-enol tautomerism, with a ketone.

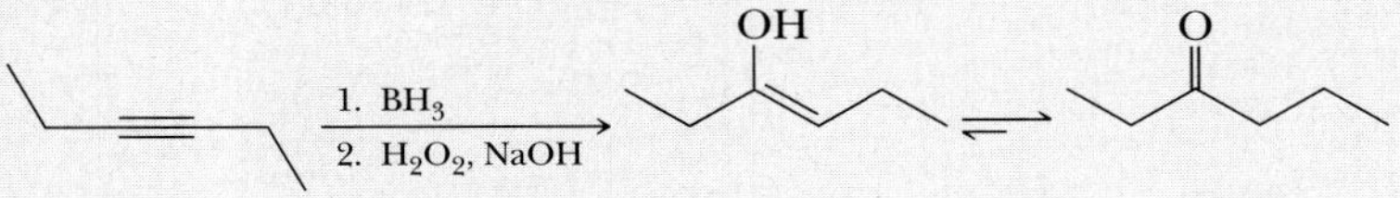

Hydroboration of a terminal alkyne using a hindered dialkylborane followed by oxidation of the resulting trialkenylborane with $H_2O_2/NaOH$ and then keto-enol tautomerism gives an aldehyde.

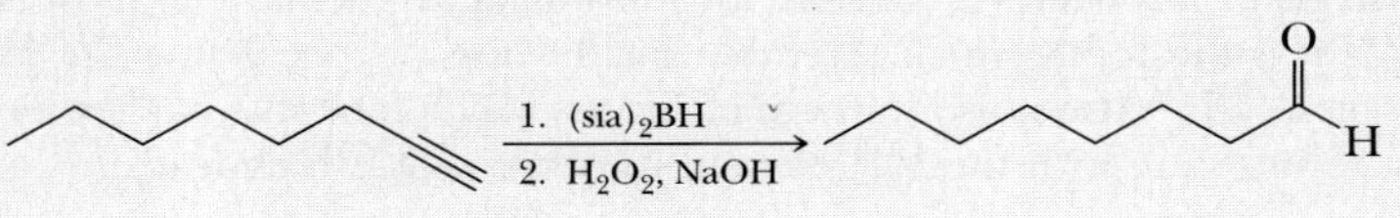

8. Acid-Catalyzed Hydration (Section 7.7B)

Acid-catalyzed addition of water in the presence of Hg(II) salts is regioselective. Keto-enol tautomerism of the resulting enol gives a ketone.

$$CH_3C\equiv CH + H_2O \xrightarrow[\text{HgSO}_4]{\text{H}_2\text{SO}_4} CH_3\overset{\overset{\displaystyle OH}{|}}{C}=CH_2 \rightleftharpoons CH_3\overset{\overset{\displaystyle O}{\|}}{C}CH_3$$

9. Catalytic Reduction (Section 7.8A)

Reaction of an alkyne with 2 moles of H_2 under moderate pressure in the presence of a transition metal catalyst at room temperature gives an alkane.

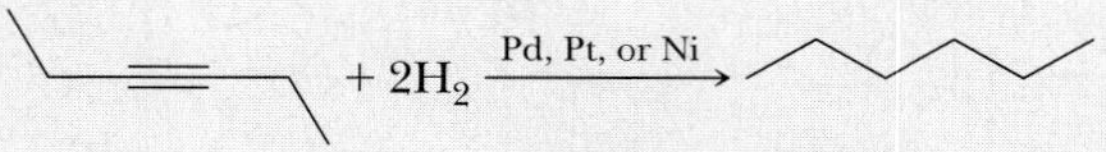

Catalytic reduction of an alkyne in the presence of the Lindlar catalyst is syn stereoselective; Lindlar reduction of an internal alkyne gives a *cis* alkene.

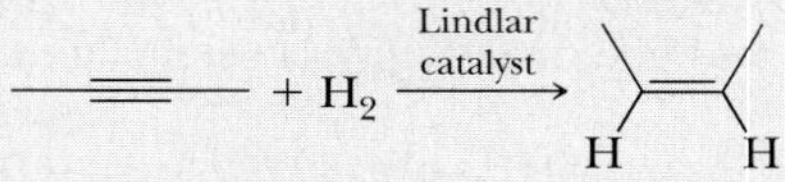

10. Hydroboration-Protonolysis (Section 7.8B)

Hydroboration of an alkyne followed by protonolysis converts an alkyne to a *cis* alkene.

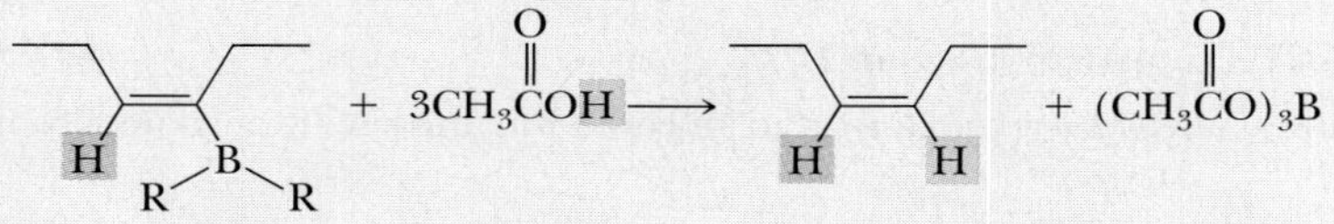

11. Reduction Using Na or Li Metal in NH₃(*l*) (Section 7.8C)

Alkali metal reduction is stereoselective: Anti addition of hydrogens to an internal alkyne gives a *trans* alkene.

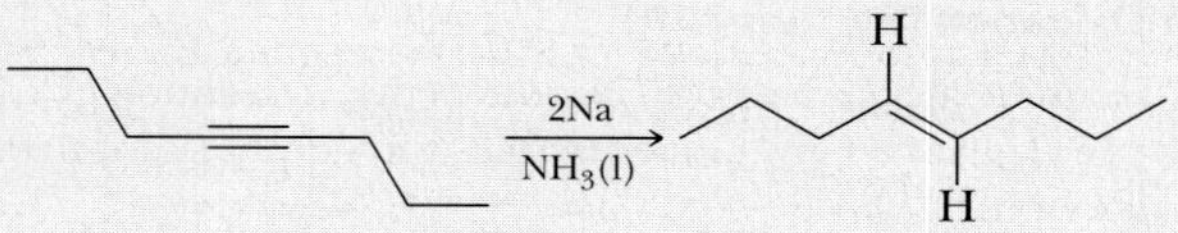

Problems

ORGANIC
Chemistry ⚛ Now™

Assess your understanding of this chapter's topics with additional quizzing and conceptual-based problems at
http://now.brookscole.com/bfi4

7.7 Enanthotoxin is an extremely poisonous organic compound found in hemlock water dropwort, which is reputed to be the most poisonous plant in England. It is believed that no British plant has been responsible for more fatal accidents. The most poisonous part of the plant is the roots, which resemble small white carrots, giving the plant the name "five finger death." Also poisonous are its leaves, which look like parsley. Enanthotoxin is thought to interfere with the Na^+ current in nerve cells, which leads to convulsions and death.

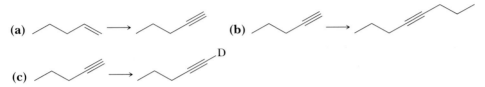

How many stereoisomers are possible for enanthotoxin?

Preparation of Alkynes

7.8 Show how to prepare each alkyne from the given starting material. In part (c), D indicates deuterium. Deuterium-containing reagents such as BD_3, D_2O, and CH_3COOD are available commercially.

(a) $\quad$ $\longrightarrow$ $\quad$ **(b)** $\quad$ $\longrightarrow$

(c) $\quad$ $\longrightarrow$

7.9 If a catalyst could be found that would establish an equilibrium between 1,2-butadiene and 2-butyne, what would be the ratio of the more stable isomer to the less stable isomer at 25°C?

$$CH_2{=}C{=}CHCH_3 \rightleftharpoons CH_3C{\equiv}CCH_3 \quad \Delta G^0 = -16.7 \text{ kJ } (-4.0 \text{ kcal})/\text{mol}$$

Reactions of Alkynes

7.10 Complete each acid-base reaction, and predict whether the position of equilibrium lies toward the left or toward the right.

(a) $CH_3C{\equiv}CH + CH_3CH_2O^-Na^+ \underset{CH_3CH_2OH}{\rightleftharpoons}$

(b) $CH_3C{\equiv}CCH_2CH_2OH + Na^+NH_2^- \underset{NH_3(l)}{\rightleftharpoons}$

(c) $CH_3C{\equiv}C^-Na^+ + CH_3\overset{\overset{\displaystyle O}{\|}}{C}OH \rightleftharpoons$

7.11 Draw structural formulas for the major product(s) formed by reaction of 3-hexyne with each of these reagents. (Where you predict no reaction, write NR.)

(a) H_2 (excess)/Pt	**(b)** H_2/Lindlar catalyst
(c) Na in $NH_3(l)$	**(d)** BH_3 followed by H_2O_2/NaOH
(e) BH_3 followed by CH_3COOH	**(f)** BH_3 followed by CH_3COOD
(g) Cl_2 (1 mol)	**(h)** $NaNH_2$ in $NH_3(l)$
(i) HBr (1 mol)	**(j)** HBr (2 mol)
(k) H_2O in H_2SO_4/$HgSO_4$	

7.12 Draw the structural formula of the enol formed in each alkyne hydration reaction, and then draw the structural formula of the carbonyl compound with which each enol is in equilibrium.

(a) $CH_3(CH_2)_5C{\equiv}CH + H_2O \xrightarrow[\text{H}_2\text{SO}_4]{\text{HgSO}_4}$ (an enol) $\longrightarrow$

(b) $CH_3(CH_2)_5C{\equiv}CH \xrightarrow[\text{2. NaOH/H}_2\text{O}_2]{\text{1. (sia)}_2\text{BH}}$ (an enol) $\longrightarrow$

7.13 Propose a mechanism for this reaction.

$$HC\equiv CH + CH_3\overset{\displaystyle O}{\overset{\|}{C}}OH \xrightarrow[HgSO_4]{H_2SO_4} CH_3\overset{\displaystyle O}{\overset{\|}{C}}OCH=CH_2$$

Acetylene   Acetic acid   Vinyl acetate

Vinyl acetate is the monomer for the production of poly(vinyl acetate), the major use of which is as an adhesive in the construction and packaging industry, but it is also used in the paint and coatings industry.

Syntheses

7.14 Show how to convert 9-octadecynoic acid to the following.

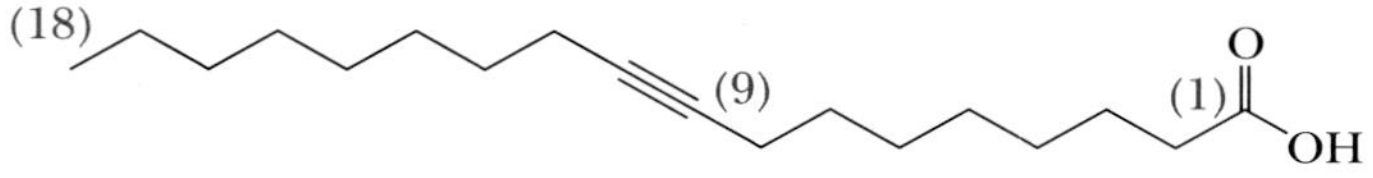

9-Octadecynoic acid

(a) (*E*)-9-Octadecenoic acid (eliadic acid)   **(b)** (*Z*)-9-Octadecenoic acid (oleic acid)
(c) 9,10-Dihydroxyoctadecanoic acid   **(d)** Octadecanoic acid (stearic acid)

7.15 For small-scale and consumer welding applications, many hardware stores sell cylinders of MAAP gas, which is a mixture of propyne (methylacetylene) and 1,2-propadiene (allene), with other hydrocarbons. How would you prepare the methylacetylene/allene mixture from propene in the laboratory?

7.16 Show reagents and experimental conditions you might use to convert propyne into each product. (Some of these syntheses can be done in one step, whereas others require two or more steps.)

(a) $CH_3\overset{\displaystyle Br}{\underset{\displaystyle Br}{\overset{|}{\underset{|}{C}}}}\!\!-\!\!\overset{\displaystyle Br}{\underset{\displaystyle Br}{\overset{|}{\underset{|}{CH}}}}$ **(b)** $CH_3\overset{\displaystyle Br}{\underset{\displaystyle Br}{\overset{|}{\underset{|}{C}}}}CH_3$ **(c)** $CH_3\overset{\displaystyle O}{\overset{\|}{C}}CH_3$ **(d)** $CH_3CH_2\overset{\displaystyle O}{\overset{\|}{C}}H$

7.17 Show reagents and experimental conditions you might use to convert each starting material into the desired product. (Some of these syntheses can be done in one step; others require two or more steps.)

(a)

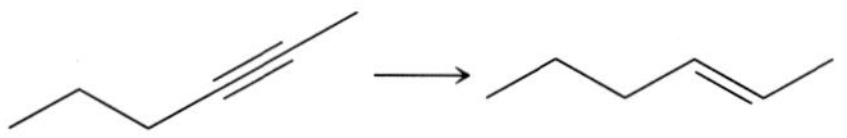

(b)

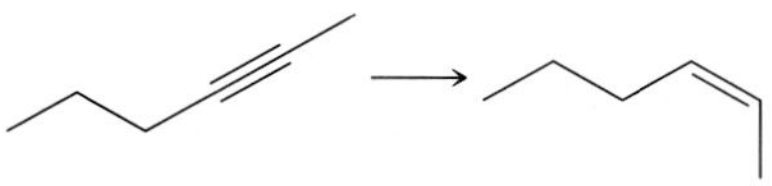

(c)

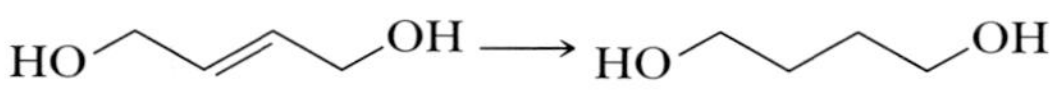

(d)

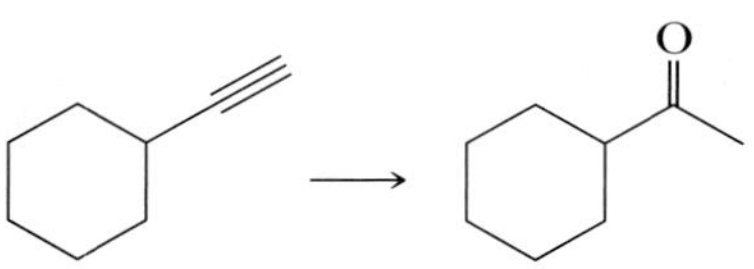

7.18 Show how to convert 1-butyne to each of these compounds.

(a) $CH_3CH_2C\equiv C^-Na^+$ **(b)** $CH_3CH_2C\equiv CD$

(c) 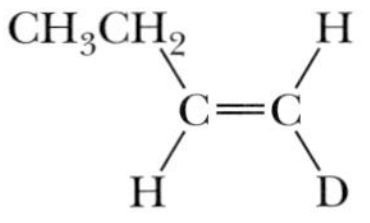 CH₃CH₂ and H on C=C with H and D

(d) CH₃CH₂ and H on C=C with D and H

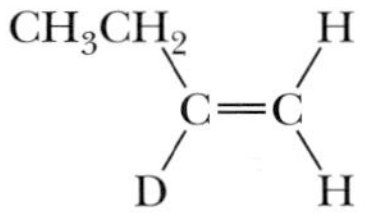

7.19 Rimantadine was among the first antiviral drugs to be licensed in the United States for use against the influenza A virus and in treating established illnesses. It is synthesized from adamantane by the following sequence. (We discuss the chemistry of Step 1 in Chapter 8 and the Chemistry of Step 5 in Section 16.8A)

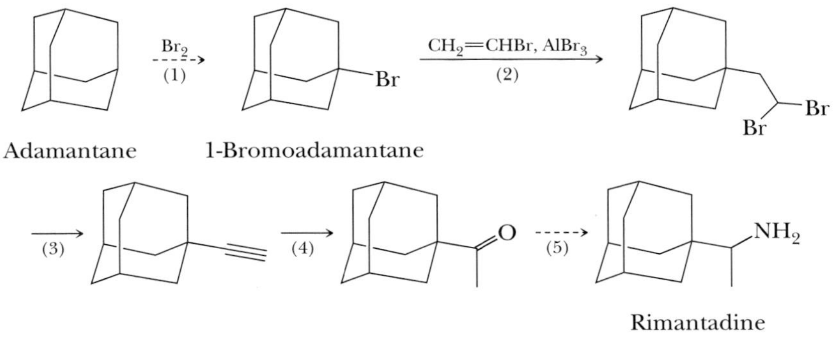

Adamantane $\xrightarrow[(1)]{Br_2}$ 1-Bromoadamantane $\xrightarrow[(2)]{CH_2=CHBr, AlBr_3}$ (adamantyl–CH₂–CHBr₂)

$\xrightarrow{(3)}$ (adamantyl–C≡CH) $\xrightarrow{(4)}$ (adamantyl–C(=O)CH₃) $\xrightarrow{(5)}$ (adamantyl–CH(CH₃)NH₂)

Rimantadine

Rimantidine is thought to exert its antiviral effect by blocking a late stage in the assembly of the virus.

(a) Propose a mechanism for Step 2. *Hint:* As we shall see in Section 21.1A, reaction of a bromoalkane such as 1-bromoadamantane with aluminum bromide (a Lewis acid, Section 4.6) results in the formation of a carbocation and $AlBr_4^-$. Assume that adamantyl cation is formed in Step 2, and proceed from there to describe a mechanism.

(b) Account for the regioselectivity of carbon-carbon bond formation in Step 2.

(c) Describe experimental conditions to bring about Step 3.

(d) Describe experimental conditions to bring about Step 4.

7.20 Show reagents and experimental conditions to bring about the following transformations.

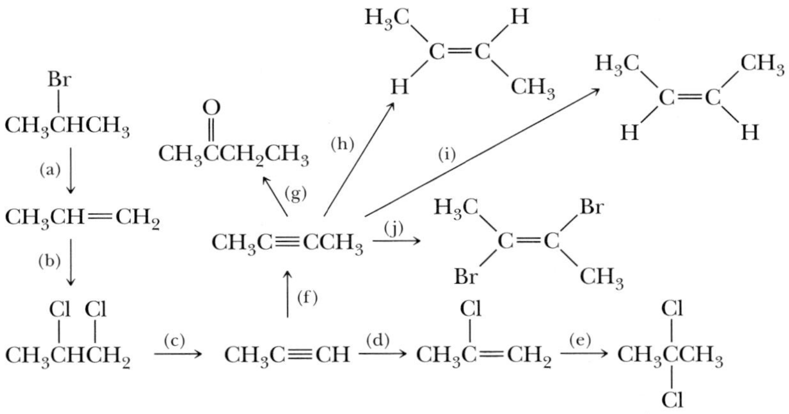

CH₃CHBrCH₃ $\xrightarrow{(a)}$ CH₃CH=CH₂ $\xrightarrow{(b)}$ CH₃CHClCH₂Cl $\xrightarrow{(c)}$ CH₃C≡CH $\xrightarrow{(d)}$ CH₃CCl=CH₂ $\xrightarrow{(e)}$ CH₃CCl₂CH₃

CH₃C≡CCH₃ via (f), (g), (h), (i), (j)

CH₃C(=O)CH₂CH₃ (h) → (E)-CH₃CH=CHCH₃ alkene; (i) → (Z) alkene

(j) → CH₃C(Br)=C(Br)CH₃ type dibromo alkene

7.21 Show reagents to bring about each conversion.

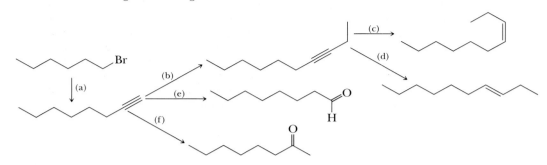

7.22 Propose a synthesis for (*Z*)-9-tricosene (muscalure), the sex pheromone for the common housefly (*Musca domestica*), starting with acetylene and haloalkanes as sources of carbon atoms.

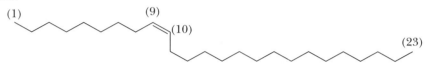

7.23 Propose a synthesis of each compound starting from acetylene and any necessary organic and inorganic reagents.

(a) 4-Octyne   (b) 4-Octanone   (c) *cis*-4-Octene
(d) *trans*-4-Octene   (e) 4-Octanol   (f) meso-4,5-Octanediol

7.24 Show how to prepare each compound from 1-heptene:

(a) 1,2-Dichloroheptane   (b) 1-Heptyne   (c) 1-Heptanol
(d) 2-Octyne   (e) *cis*-2-Octene   (d) *trans*-2-Octene

7.25 Show how to bring about this conversion.

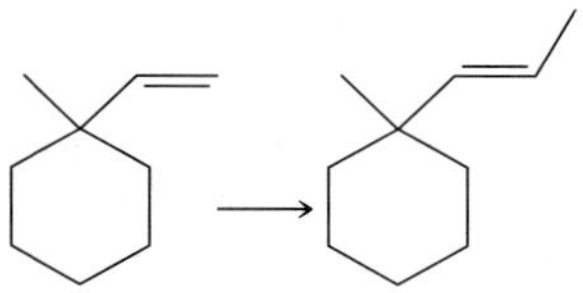

Looking Ahead

7.26 Alkyne anions react with the carbonyl groups of aldehydes and ketones to form alkynyl alcohols, as illustrated by the following sequence.

$$CH_3C{\equiv}C{:}^-Na^+ + H-\overset{\overset{\textstyle O}{\|}}{C}-H \longrightarrow [CH_3C{\equiv}C-CH_2O^-Na^+] \xrightarrow[\text{H}_2\text{O}]{\text{HCl}} CH_3C{\equiv}C-CH_2OH$$

An alkynyl alcohol

Propose a mechanism for the formation of the bracketed compound, using curved arrows to show the flow of electron pairs in the course of the reaction.

7.27 Following is the structural formula of the tranquilizer meparfynol (Oblivon).

Oblivon

Propose a synthesis for this compound starting with acetylene and a ketone. (Notice the- *yn-* and *-ol* in the chemical name of this compound, indicating that it contains alkyne and hydroxyl functional groups.)

7.28 The standard procedure for synthesizing a compound is the stepwise progress toward a target molecule by forming individual bonds through single reactions. Typically, the product of each reaction is isolated and purified before the next reaction in the sequence is carried out. One of the ways Nature avoids this tedious practice of isolation and purification is by the use of a domino sequence in which each new product is built on a preexisting one in stepwise fashion. The first laboratory equivalent of a domino reaction is William S. Johnson's elegant synthesis of the female hormone, progesterone. Johnson first constructed the polyunsaturated monocyclic 3° alcohol (A) and then, in an acid-induced domino reaction, formed Compound B, which he then converted to progesterone.

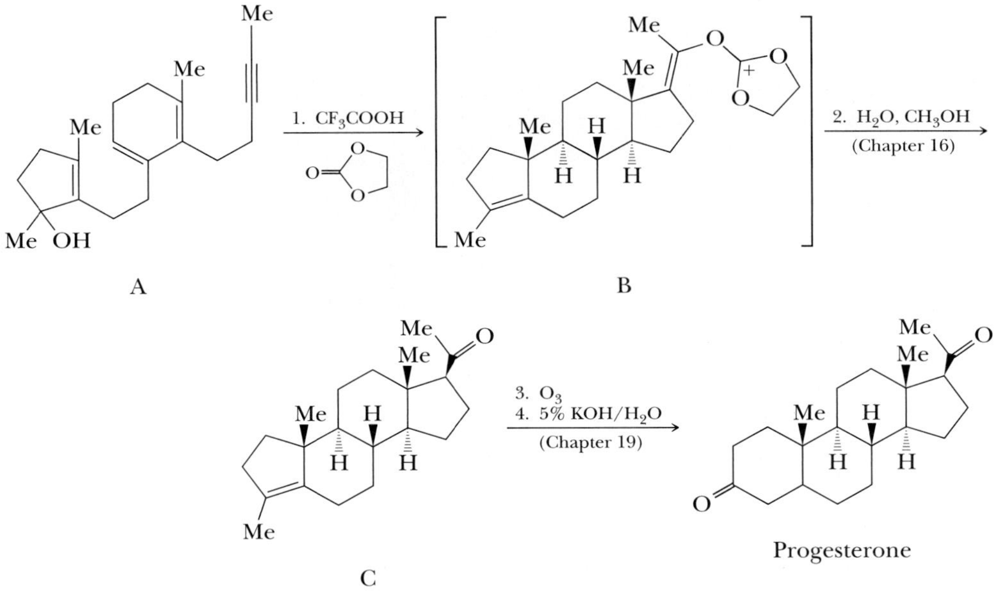

A remarkable feature of this synthesis is that compound A, which has only one stereocenter, gives compound B, which has five stereocenters, each with the same configuration as those in progesterone. We will return to the chemistry of Step 2 in Section 16.7, and to the chemistry of Steps 3 and 4 in Chapter 19. In this problem, we focus on Step 1.

(a) Assume that the domino reaction in Step 1 is initiated by protonation of the 3° alcohol in compound A followed by loss of H_2O to give a 3° carbocation. Show how the series of reactions initiated by the formation of this cation gives compound B.

(b) If you have access to a large enough set of molecular models or to a computer modeling program, build a model of progesterone and describe the conformation of each ring. There are two methyl groups and three hydrogen atoms at the set of ring junctions in progesterone. Which of these five groups occupies an equatorial position? Which occupies an axial position?

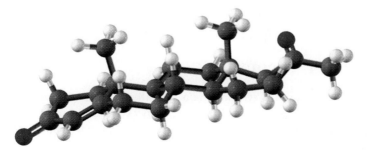

8

■ Many common objects such as this raft are made of poly(vinyl chloride). Inset: a model of vinyl chloride.

Haloalkanes, Halogenation, and Radical Reactions

Compounds containing a halogen atom covalently bonded to an sp^3 hybridized carbon atom are named haloalkanes, or, in the common system of nomenclature, alkyl halides. Several haloalkanes are important laboratory and industrial solvents. In addition, haloalkanes are invaluable building blocks for organic synthesis.

In this chapter, we begin with the structure and physical properties of haloalkanes. We then study radical halogenation of alkanes as a vehicle to introduce an important type of reaction mechanism, namely the mechanism of radical chain reactions. Reactions of oxygen with alkenes and a radical mechanism for HBr addition to alkenes complete the chapter.

8.1 Structure

The general symbol for a **haloalkane** is R—X, where —X may be —F, —Cl, —Br, or —I. If a halogen is bonded to a doubly bonded carbon of an alkene, the compound belongs to a class called **haloalkenes.** If it is bonded to a benzene ring, the compound belongs to a class called **haloarenes,** which have the general symbol Ar—X.

A haloalkane
(an alkyl halide)

A haloalkene
(an alkenyl
or vinylic halide)

A haloarene
(an aryl halide)

Haloalkane (alkyl halide) A compound containing a halogen atom covalently bonded to an sp^3 hybridized carbon atom. Given the symbol R—X.

Haloalkene (vinylic halide) A compound containing a halogen bonded to one of the carbons of a carbon-carbon double bond.

Haloarene (aryl halide) A compound containing a halogen atom bonded to a benzene ring. Given the symbol Ar—X.

ORGANIC
Chemistry·Now™
Click Tutorials to review the
Nomenclature of Haloalkanes

8.2 Nomenclature

A. IUPAC System

IUPAC names for haloalkanes are derived by naming the parent alkane according to the rules given in Section 2.3A.

- The parent chain is numbered from the direction that gives the first substituent encountered the lowest number, whether it is a halogen or an alkyl group. If two groups could have the same lowest number from the end of the chain, give the group of lower alphabetical order the lower number. An example is 2-bromo-4-methylpentane.
- Halogen substituents are indicated by the prefixes *fluoro-, chloro-, bromo-,* and *iodo-* and are listed in alphabetical order with other substituents.
- The location of each halogen atom on the parent chain is given by a number preceding the name of the halogen.
- In haloalkenes, numbering the parent hydrocarbon is determined by the location of the carbon-carbon double bond. Numbering is done in the direction that gives the carbon atoms of the double bond and substituents the lowest set of numbers.

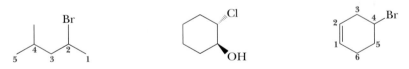

2-Bromo-4-methylpentane

trans-2-Chlorocyclohexanol

4-Bromocyclohexene

B. Common Names

Common names of haloalkanes and haloalkenes consist of the common name of the alkyl group followed by the name of the halide as a separate word. Hence, the name **alkyl halide** is a common name for this class of compounds. In the following examples, the IUPAC name of the compound is given first, and then its common name is given in parentheses:

2-Bromobutane
(*sec*-Butyl bromide)

Chloroethene
(Vinyl chloride)

3-Chloropropene
(Allyl chloride)

Haloform A compound of the type CHX₃ where X is a halogen.

Several polyhaloalkanes are important solvents and are generally referred to by their common names. Dichloromethane (methylene chloride) is the most widely used haloalkane solvent. Compounds of the type CHX_3 are called **haloforms.** The common name for $CHCl_3$, for example, is chloroform. The common name methyl chloroform for the compound CH_3CCl_3 derives from this name. Methyl chloroform and trichloroethylene (trichlor) are common solvents for industrial cleaning. Because they are somewhat toxic and cause environmental problems, they are being phased out.

CH_2Cl_2	$CHCl_3$	CH_3CCl_3	$CCl_2=CHCl$
Dichloromethane (Methylene chloride)	Trichloromethane (Chloroform)	1,1,1-Trichloroethane (Methyl chloroform)	Trichloroethylene (Trichlor)

Hydrocarbons in which all hydrogens are replaced by halogens are commonly called perhaloalkanes or perhaloalkenes. Perchloroethylene, commonly known as perc, is the most common dry cleaning solvent in use today. It is also being phased out.

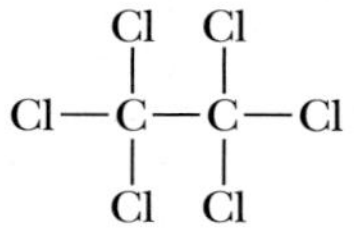

Perchloroethane 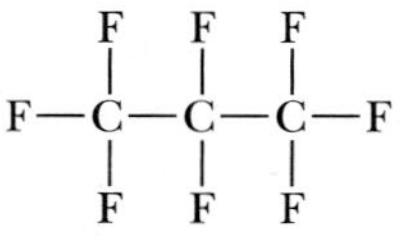   Perfluoropropane 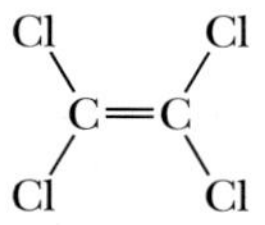   Perchloroethylene

---

Example 8.1

Write the IUPAC name and, where possible, the common name of each compound. Show stereochemistry where relevant.

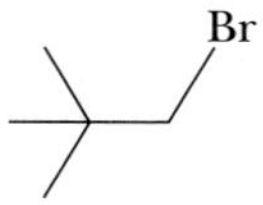

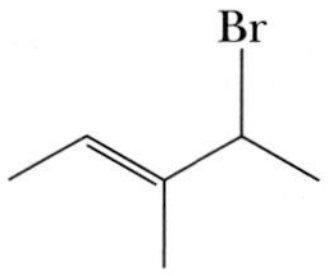

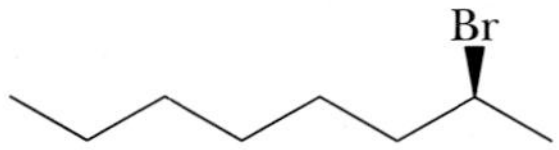

Solution

(a) 1-Bromo-2,2-dimethylpropane. Its common name is neopentyl bromide.
(b) (*E*)-4-Bromo-3-methyl-2-pentene.
(c) (*S*)-2-Bromooctane.

Problem 8.1

Write the IUPAC name, and where possible, the common name of each compound. Show stereochemistry where relevant.

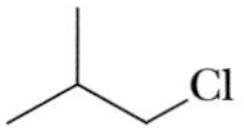

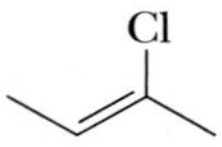

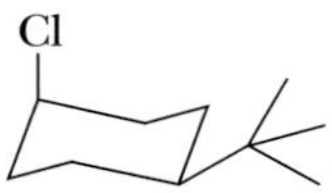

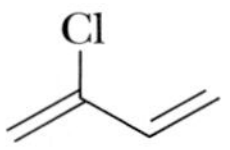

Table 8.1 Dipole Moments (Gas Phase) of Halomethanes

Halomethane	Electronegativity of Halogen	Carbon-Halogen Bond Length (pm)	Dipole Moment (debyes, D)
CH_3F	4.0	139	1.85
CH_3Cl	3.0	178	1.87
CH_3Br	2.8	193	1.81
CH_3I	2.5	214	1.62

8.3 Physical Properties of Haloalkanes

A. Polarity

Fluorine, chlorine, and bromine are all more electronegative than carbon (Table 1.5), and, as a result, C—X bonds with these atoms are polarized with a partial negative charge on halogen and a partial positive charge on carbon. Table 8.1 shows that each of the halomethanes has a substantial dipole moment. The electrostatic potential map of fluoromethane shows the large charge separation in this compound caused by the dipole.

The magnitude of a dipole moment depends on the size of the partial charges, the distance between them, and the polarizability of the three pairs of unshared electrons on each halogen. For the halomethanes, dipole moment increases as the electronegativity of the halogen and the bond length increase. These two trends run counter to each other, with the net effect that chloromethane has the largest dipole moment of the series. The experimental dipole moments also show clearly that the electronegativity of carbon is less than the standard Pauling value of 2.5 in many compounds (as we mentioned in Section 1.2B), or CH_3I would have no dipole moment.

Electrostatic potential map of fluoromethane.

B. Boiling Point

Haloalkanes are associated in the liquid state by a combination of dipole-dipole, dipole-induced dipole, and induced dipole-induced dipole (dispersion) forces. These forces are grouped together under the term **van der Waals forces,** in honor of J. D. van der Waals, the 19th century Dutch physicist. Van der Waals attractive forces pull molecules together. As atoms or molecules are brought closer and closer, van der Waals attractive forces are overcome by repulsive forces between electron clouds of adjacent atoms. The energy minimum is where the net attractive forces are the strongest. Nonbonded interatomic and intermolecular distances at these minima can be measured by X-ray crystallography of solid compounds, and each atom and group of atoms can be assigned an atomic or molecular radius called a **van der Waals radius.** Nonbonded atoms in a molecule cannot approach each other closer than the van der Waals radius without causing nonbonded interaction strain. Van der Waals radii for selected atoms and groups of atoms are given in Table 8.2.

Van der Waals forces A group of intermolecular attractive forces including dipole-dipole, dipole-induced dipole, and induced dipole-induced dipole (dispersion) forces.

van der Waals radius The minimum distance of approach to an atom that does not cause nonbonded interaction strain.

Table 8.2 Van der Waals Radii (pm) for Selected Atoms and Groups of Atoms

H	F	Cl	Br	CH_2	CH_3	I
120	135	180	195	200	200	215

Increasing van der Waals radius →

Table 8.3 Boiling Points of Some Low-Molecular-Weight Alkanes and Haloalkanes

Alkyl Group	Name	Boiling Point (°C)				
		H	**F**	**Cl**	**Br**	**I**
CH_3-	Methyl	−161	−78	−24	4	43
CH_3CH_2-	Ethyl	−89	−37	13	38	72
$CH_3(CH_2)_2-$	Propyl	−45	3	46	71	102
$(CH_3)_2CH-$	Isopropyl	−45	−11	35	60	89
$CH_3(CH_2)_3-$	Butyl	0	32	77	100	130
$CH_3CH_2(CH_3)CH-$	*sec*-Butyl	0	25	67	90	119
$(CH_3)_2CHCH_2-$	Isobutyl	−1	16	68	91	120
$(CH_3)_3C-$	*tert*-Butyl	−1	12	51	72	98
$CH_3(CH_2)_4-$	Pentyl	36	63	108	129	157
$CH_3(CH_2)_5-$	Hexyl	69	92	134	155	181

Notice from Table 8.2 that the van der Waals radius of fluorine is only slightly greater than that of hydrogen and that, among the halogens, only iodine has a larger van der Waals radius than methyl.

Boiling points of several low-molecular-weight haloalkanes and the alkanes from which they are derived are given in Table 8.3. There are several trends to be noticed from these data.

1. As with hydrocarbons, constitutional isomers with branched chains have lower boiling points than their unbranched-chain isomers (Section 2.8D). Compare, for example, the boiling points of unbranched-chain 1-bromobutane (butyl bromide, bp 100°C) with the more branched and compact 2-bromo-2-methylpropane (*tert*-butyl bromide, bp 72°C). Branched-chain constitutional isomers have lower boiling points because they have a more spherical shape and, therefore, decreased surface area, leading to smaller van der Waals forces between their molecules.

2. For an alkane and haloalkane of comparable size and shape, the haloalkane has a higher boiling point. Compare, for example, the boiling points of ethane (bp −89°C) and bromomethane (bp 4°C). Although both molecules are roughly the same size (the van der Waals radii of —CH_3 and —Br are almost identical) and have roughly the same effective contact area, the boiling point of bromomethane is considerably higher than that of ethane. This difference in boiling points is almost entirely a result of the greater **polarizability** of the three unshared pairs of electrons on the halogen compared with the shared electron pairs in the hydrocarbon of comparable size and shape. Recall from Section 2.8B that the strength of dispersion forces, the weakest of all intermolecular forces, depends on the polarizability of electrons, which, in turn, depends on how tightly they are held to the nucleus. The further electrons are from the nucleus, the less tightly they are held and the greater their polarizability. In addition, unshared electron pairs have a higher polarizability than electrons shared in a covalent bond.

3. The boiling points of fluoroalkanes are comparable to those of hydrocarbons of comparable molecular weight. Compare, for example, the boiling points of hexane (MW 86.2, bp 69°C) and 1-fluoropentane (MW 90.1, bp 63°C) and the

Polarizability A measure of the ease of distortion of the distribution of electron density about an atom or group in response to interaction with other molecules or ions. Fluorine, which has a high electronegativity, holds its electrons tightly and has a very low polarizability. Iodine, which has a lower electronegativity and holds its electrons less tightly, has a very high polarizability.

Table 8.4   Densities of Some Low-Molecular-Weight Haloalkanes

Alkyl Group	Name	Density of Liquid (g/mL) at 25°C		
		Cl	**Br**	**I**
CH_3-	Methyl	—	—	2.279
CH_3CH_2-	Ethyl	—	1.460	1.936
$CH_3(CH_2)_2-$	Propyl	0.891	1.354	1.749
$(CH_3)_2CH-$	Isopropyl	0.862	1.314	1.703
$CH_3(CH_2)_3-$	Butyl	0.886	1.276	1.615
$(CH_3)_3C-$	*tert*-Butyl	0.842	1.221	1.545
$CH_3(CH_2)_5-$	Hexyl	0.879	1.174	1.440

boiling points of 2-methylpropane (MW 58.1, bp −1°C) and 2-fluoropropane (MW 62.1, bp −11°C). This low boiling point is attributable to the small size of fluorine, the tightness with which its electrons are held, and their particularly low polarizability. The distinctive properties of fluorocarbons, for example, the non-stick properties of polytetrafluoroethylene (PTFE, one consumer end-product use of which is Teflon) are also a consequence of the uniquely low polarizability of the three unshared electron pairs on fluorine. Besides being useful for nonstick surfaces, fluoroalkanes have unique solvent properties that are being actively investigated, and their ability to dissolve oxygen has even been investigated for use as a synthetic blood alternative.

C. Density

The densities of liquid haloalkanes are greater than those of hydrocarbons of comparable molecular weight because of the halogens' large mass-to-volume ratio. A bromine atom and a methyl group have almost identical van der Waals radii, but bromine has a mass of 79.9 atomic mass units (amu) compared with 15 amu for methyl. Table 8.4 gives densities for some low-molecular-weight haloalkanes that are liquid at 25°C. The densities of all liquid bromoalkanes and iodoalkanes are greater than that of water.

Although the densities of liquid chloroalkanes are less than that of water, further substitution of chlorine for hydrogen increases the density to the point where di- and polychloroalkanes have a greater density than water (Table 8.5). These compounds sink in water and form the lower layer when mixed because they are immiscible with water.

Table 8.5   Density of Polyhalomethanes

Haloalkane	X =	Density of Liquid (g/mL) at 25°C		
		Cl	**Br**	**I**
CH_2X_2		1.327	2.497	3.325
CHX_3		1.483	2.890	4.008
CX_4		1.594	3.273	4.23

CHEMICAL CONNECTIONS

Freons

Of all the fluoroalkanes, **chlorofluorocarbons (CFCs)** manufactured under the trade name **Freons** are the most widely known. CFCs are nontoxic, nonflammable, odorless, and noncorrosive and seemed to be ideal replacements for the hazardous compounds such as ammonia and sulfur dioxide formerly used as heat-transfer agents in refrigeration systems. Among the CFCs most widely used for this purpose were trichlorofluoromethane (CCl_3F, Freon-11) and dichlorodifluoromethane (CCl_2F_2, Freon-12). They are particularly desirable refrigerants because of their low boiling points.

The CFCs found wide use as industrial cleaning solvents to prepare surfaces for coatings, to remove cutting oils and waxes from millings, and to remove protective coatings. CFCs were also used as propellants for aerosol sprays.

Concern about the environmental impact of CFCs arose in the 1970s when it was shown that more than 4.5×10^5 kg/yr of these compounds were being emitted into the atmosphere. Then, in 1974 Sherwood Rowland of the University of California, Irvine, and Mario Molina, now at the University of California, San Diego, announced their theory, which has since been amply confirmed, of ozone destruction by these compounds. When released into the air, CFCs escape to the lower atmosphere, but because of their inertness, they do not decompose there. Slowly, they find their way to the stratosphere where they absorb ultraviolet radiation from the sun and then decompose. As they do so, they set up a chemical reaction that leads to the destruction of the stratospheric ozone layer, which acts as a shield for the earth against short-wavelength ultraviolet radiation from the sun. Scientists believe that an increase in short-wavelength ultraviolet radiation reaching the earth will lead to the destruction of certain crops and agricultural species, and even to an increased incidence of skin cancer in light-skinned individuals.

The results of this concern were that in 1987 most countries subscribed to the so-called Montreal Protocol, which set limits on the production and use of ozone-depleting CFCs and urged a complete phaseout of their production by the year 1996. This phaseout has resulted in enormous costs and is not yet complete in developing countries. The fact that an international agreement on the environment that set limits on the production of any substance could be reached is indeed amazing and bodes well for the health of the planet. Rowland, Molina, and Paul Crutzen, a Dutch chemist at the Max Planck Institute for Chemistry in Germany, were awarded the 1995 Nobel prize in chemistry for their work on this topic.

The chemical industry has responded by developing less-ozone-depleting alternatives to CFCs, among which are the hydrofluorocarbons (HFCs) and hydrochlorofluorocarbons (HCFCs). These compounds are much more chemically reactive in the atmosphere than the Freons and are destroyed before reaching the stratosphere. However, they tend to act as "greenhouse gases" and may contribute to global warming. For this reason, they are likely to be replaced in turn.

HFC-134a   HCFC-123

We must not assume, however, that haloalkanes are introduced into the environment only by human action. It is estimated, for example, that annual production of bromomethane from natural sources is 2.7×10^8 kg, largely from marine algae, giant kelp, and volcanoes. Furthermore, global emission of chloromethane is estimated to be 4.5×10^9 kg/yr, most of it from terrestrial and marine biomass. These haloalkanes, however, have only short atmospheric lifetimes, and only a tiny fraction of them reach the stratosphere. The CFCs are the problem; they have longer atmospheric lifetimes, reach the stratosphere, and do their damage there.

Table 8.6 Average Bond Dissociation Enthalpies for C—H and C—X Bonds

Bond	Bond Length (pm)	Bond Dissociation Enthalpy [kJ (kcal)/mol]
C—H	109	414 (99)
C—F	142	464 (111)
C—Cl	178	355 (85)
C—Br	193	309 (78)
C—I	214	228 (57)

D. Bond Lengths and Bond Strengths

With the exception of C—F bonds, C—X bonds are weaker than C—H bonds as measured by bond dissociation enthalpies (BDE), which are one of the best measures of bond strength. A table of BDE values for many bonds is given in Appendix 3. Bond dissociation enthalpy is defined as the amount of energy required to break a bond homolytically into two radicals in the gas phase at 25°C.

$$A \overset{\frown}{\quad} B \longrightarrow A\cdot + \cdot B$$

A **radical,** sometimes called a free radical, is any chemical species that contains one or more unpaired electrons.

Radical Any chemical species that contains one or more unpaired electrons.

This reaction is a "virtual" one because it can't actually be carried out in most cases. Instead, the extensive tables of BDEs are collected from thermochemical data on heats of combustion, hydrogenation, and other reactions. The useful thing about these data is that by adding and subtracting them, heats of reaction can be calculated with confidence for reactions that have never been measured.

C—X BDEs are tabulated in Table 8.6. As the size of the halogen atom increases, the C—X bond length increases, and its strength decreases. These relationships between bond strength and bond length help us to understand the difference in the ease with which haloalkanes undergo reactions that involve carbon-halogen bond breaking. Fluoroalkanes, for example, with the strongest and shortest C—X bonds, are highly resistant to bond breaking under most conditions. This characteristic inertness is one of the factors that make perfluoroalkanes such as Teflon such useful materials.

8.4 Preparation of Haloalkanes by Halogenation of Alkanes

As we saw in Sections 6.3A and 6.3D, haloalkanes can be prepared by addition of HX and X_2 to alkenes. They are also prepared by replacement of the —OH group of alcohols by halogen (Section 10.5). Many of the simpler, low-molecular-weight haloalkanes are prepared by the halogenation of alkanes, illustrated here by treating 2-methylpropane with bromine at an elevated temperature.

$$
\underset{\substack{\text{2-Methylpropane} \\ \text{(Isobutane)}}}{\overset{\displaystyle CH_3}{\underset{\displaystyle CH_3}{CH_3CH}}} \quad + \quad Br_2 \quad \xrightarrow{\text{heat}} \quad \underset{\substack{\text{2-Bromo-2-methylpropane} \\ (\textit{tert}\text{-Butyl bromide})}}{\overset{\displaystyle CH_3}{\underset{\displaystyle CH_3}{CH_3CBr}}} \quad + \quad HBr
$$

Halogenation of alkanes is common with Br_2 and Cl_2. Fluorine, F_2, is seldom used because its reactions with alkanes are so exothermic that they are difficult to control and can actually cause C—C bond cleavage and even explosions. Iodine, I_2, is seldom used because the reaction is endothermic and the position of equilibrium favors alkane and I_2 rather than iodoalkane and HI.

If a mixture of methane and chlorine gas is kept in the dark at room temperature, no detectable change occurs. If, however, the mixture is heated or exposed to light, a reaction begins almost at once with the evolution of heat. The products are chloromethane and hydrogen chloride. What occurs is a **substitution** reaction, in this case substitution of a chlorine atom for a hydrogen atom in methane and the production of an equivalent amount of hydrogen chloride.

Substitution A reaction in which an atom or group of atoms in a compound is replaced by another atom or group of atoms.

$$
\underset{\text{Methane}}{CH_4} \quad + \quad Cl_2 \quad \xrightarrow{\text{heat}} \quad \underset{\substack{\text{Chloromethane} \\ \text{(Methyl chloride)}}}{CH_3Cl} \quad + \quad HCl
$$

If chloromethane is allowed to react with more chlorine, further chlorination produces a mixture of dichloromethane (methylene chloride), trichloromethane (chloroform), and tetrachloromethane (carbon tetrachloride). Notice that in the last equation, the reagent Cl_2 is placed over the reaction arrow and the equivalent amount of HCl formed is not shown. Placing reagents over reaction arrows and omitting byproducts is commonly done to save space.

$$
\underset{\substack{\text{Chloromethane} \\ \text{(Methyl chloride)}}}{CH_3Cl} \quad + \quad Cl_2 \quad \xrightarrow{\text{heat}} \quad \underset{\substack{\text{Dichloromethane} \\ \text{(Methylene chloride)}}}{CH_2Cl_2} \quad + \quad HCl
$$

$$
\underset{\substack{\text{Dichloromethane} \\ \text{(Methylene chloride)}}}{CH_2Cl_2} \quad \xrightarrow[\text{heat}]{Cl_2} \quad \underset{\substack{\text{Trichloromethane} \\ \text{(Chloroform)}}}{CHCl_3} \quad \xrightarrow[\text{heat}]{Cl_2} \quad \underset{\substack{\text{Tetrachloromethane} \\ \text{(Carbon tetrachloride)}}}{CCl_4}
$$

It is possible to prepare chloromethane or tetrachloromethane in relatively pure form by this reaction. In the case of chloromethane, a large excess of methane is used; for tetrachloromethane, a large excess of chlorine drives the reaction to complete halogenation. The other chlorinated methanes can be separated by distillation of partially chlorinated mixtures.

Treating ethane with bromine gives bromoethane (ethyl bromide).

$$
\underset{\text{Ethane}}{CH_3CH_3} \quad + \quad Br_2 \quad \xrightarrow{\text{heat}} \quad \underset{\substack{\text{Bromoethane} \\ \text{(Ethyl bromide)}}}{CH_3CH_2Br} \quad + \quad HBr
$$

In all cases, monosubstituted products are only obtained using an excess of ethane.

A. Regioselectivity

Treating propane with bromine gives a mixture consisting of approximately 8% of 1-bromopropane and 92% of 2-bromopropane:

$$CH_3CH_2CH_3 + Br_2 \xrightarrow[\text{or light}]{\text{heat}} CH_3\overset{\overset{\displaystyle Br}{|}}{C}HCH_3 + CH_3CH_2CH_2Br + HBr$$

Propane   2-Bromopropane 1-Bromopropane
            (92%)   (8%)

Propane contains eight hydrogens; one set of six equivalent primary hydrogens and one set of two secondary hydrogens (Section 2.3C). The hydrogens in each set are equivalent because of rapid bond rotation about C—C single bonds. Substitution of bromine for a primary hydrogen gives 1-bromopropane; substitution of bromine for a secondary hydrogen gives 2-bromopropane. If there were random substitution of any one of the eight hydrogens in propane, we would predict that the isomeric bromopropanes would be formed in the ratio of 6:2, or 75% 1-bromopropane and 25% 2-bromopropane. In fact, in the bromination of propane, substitution of a secondary hydrogen rather than a primary hydrogen is strongly favored. 2-Bromopropane is the major product and the reaction is highly regioselective.

Product Distribution	$CH_3CH_2CH_2Br$	$CH_3\overset{\overset{\displaystyle Br}{|}}{C}HCH_3$
Prediction based on ratio of six 1° H to two 2° H	75%	25%
Experimental observation	8%	92%

Other experiments have shown that substitution at a tertiary hydrogen is favored over both secondary and primary hydrogens. For example, monobromination of 2-methylpentane is very regioselective and gives almost exclusively 2-bromo-2-methylpentane.

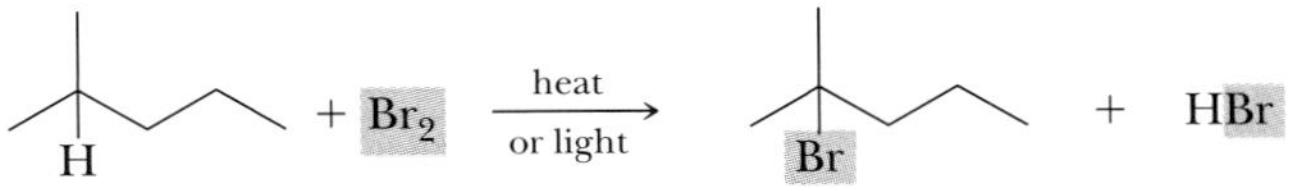

2-Methylpentane   2-Bromo-2-methylpentane

The reaction of bromine with an alkane occurs in the order $3° > 2° > 1°$ hydrogen. Chlorination of alkanes is also regioselective, but much less so than bromination. For example, treatment of propane with chlorine gives a mixture of approximately 57% 2-chloropropane and 43% 1-chloropropane.

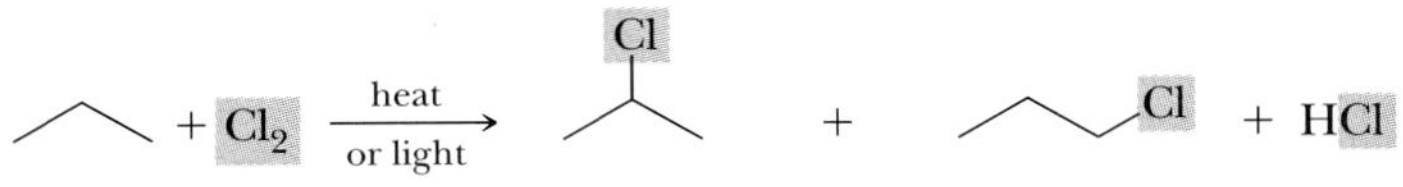

Propane   2-Chloropropane 1-Chloropropane
       (57%)   (43%)

Thus, we can conclude that, although both bromine and chlorine are regioselective in hydrogen replacement in the order $3° > 2° > 1°$, regioselectivity is far greater for bromination than for chlorination. From data on product distribution, it has been determined that regioselectivity per hydrogen for bromination is approximately 1600:80:1, whereas it is only about 5:4:1 for chlorination. We will discuss reasons for this difference in Section 8.5.

Example 8.2

Name and draw structural formulas for all monobromination products formed by treating 2-methylpropane with Br_2. Predict the major product based on the regioselectivity of the reaction of Br_2 with alkanes.

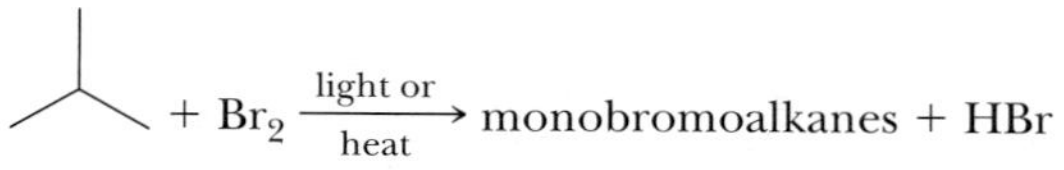

$$\text{2-Methylpropane} + Br_2 \xrightarrow[\text{heat}]{\text{light or}} \text{monobromoalkanes} + HBr$$

Solution

2-Methylpropane has nine equivalent primary hydrogens and one tertiary hydrogen. Substitution of bromine for a primary hydrogen gives 1-bromo-2-methylpropane; substitution for the tertiary hydrogen gives 2-bromo-2-methylpropane. Given that the regioselectivity per hydrogen of bromination for $3° > 2° > 1°$ hydrogens is approximately 1600:80:1, it is necessary to correct for the number of hydrogens: nine primary and one tertiary. The result is that 99.4% of the product is 2-bromo-2-methylpropane and 0.6% is 1-bromo-2-methylpropane.

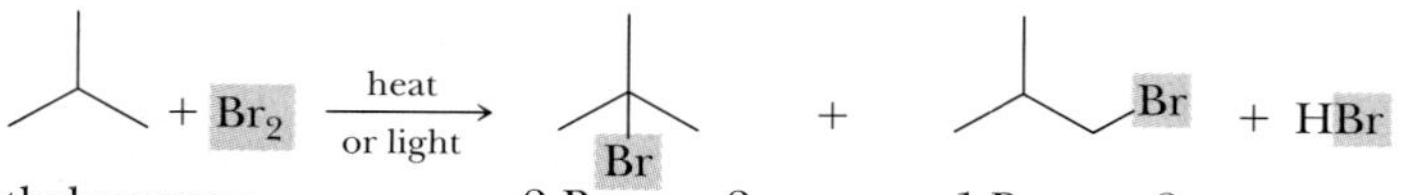

2-Methylpropane + Br_2 $\xrightarrow[\text{or light}]{\text{heat}}$ 2-Bromo-2-methylpropane (major product) + 1-Bromo-2-methylpropane + HBr

$$\text{Predicted \% 2-bromo-2-methylpropane} = \frac{1 \times 1600}{(1 \times 1600) + (9 \times 1)} \times 100 = 99.4\%$$

Problem 8.2

Name and draw structural formulas for all monochlorination products formed by treatment of 2-methylpropane with Cl_2. Predict the major product based on the regioselectivity of the reaction of Cl_2 with alkanes.

B. Energetics

We can learn a lot about these reactions by careful consideration of the energetics of each step. A selection of C—H BDE values is given in Table 8.7.

Table 8.7   Bond Dissociation Enthalpies for Selected C—H Bonds

Hydrocarbon	Radical	Name of Radical	Type of Radical	ΔH^0 [kJ (kcal)/mol]
CH_2=$CHCH_2$—H	CH_2=$CHCH_2 \cdot$	Allyl	Allylic	372 (89)
$C_6H_5CH_2$—H	$C_6H_5CH_2 \cdot$	Benzyl	Benzylic	376 (90)
$(CH_3)_3C$—H	$(CH_3)_3C \cdot$	*tert*-Butyl	3°	405 (97)
$(CH_3)_2CH$—H	$(CH_3)_2CH \cdot$	Isopropyl	2°	414 (99)
CH_3CH_2—H	$CH_3CH_2 \cdot$	Ethyl	1°	422 (101)
CH_3—H	$CH_3 \cdot$	Methyl	Methyl	439 (105)
CH_2=CH—H	CH_2=$CH \cdot$	Vinyl	Vinylic	464 (111)

Radical stability ↑

Note that the BDE values for saturated hydrocarbons depend on the type of hydrogen being abstracted and are in the order methane $> 1° > 2° > 3°$. This order follows because the resulting radicals are electron deficient (they have only seven electrons in the valence shell of the carbon bearing the radical); like carbocations, the more alkyl groups they have, the more stable they are. Note also that an sp^2 C—H bond is particularly strong, as mentioned in Section 1.8F.

Using data from Table 8.7 and Appendix 3, we can calculate the heat of reaction, ΔH^0, for the halogenation of an alkane. In this calculation for the chlorination of methane, energy is required to break CH_3—H and Cl—Cl bonds [439 and 247 kJ (105 and 59 kcal)/mol, respectively]. Energy is released in making the CH_3—Cl and H—Cl bonds [−351 and −431 kJ (−84 and −103 kcal)/mol respectively]. Summing these enthalpies, we calculate that chlorination of methane to form chloromethane and hydrogen chloride liberates 96 kJ (23 kcal)/mol.

$$CH_4 \; + \; Cl_2 \longrightarrow CH_3Cl \; + \; HCl \qquad \Delta H^0 = -96 \text{ kJ/mol}$$

BDE, kJ/mol	+439	+247	−351	−431	(−23 kcal/mol)
(kcal/mol)	(105)	(59)	(−84)	(−103)	

Example 8.3

Using the table of bond dissociation enthalpies in Appendix 3, calculate ΔH^0 for bromination of propane to give 2-bromopropane and hydrogen bromide.

Solution

Under each molecule is given the enthalpy for breaking or forming each corresponding bond. The calculated heat of reaction is −71 kJ (−17 kcal)/mol.

$$\underset{H}{CH_3CHCH_3} \; + \; Br—Br \longrightarrow \underset{Br}{CH_3CHCH_3} \; + \; H—Br \qquad \Delta H^0 = -71 \text{ kJ/mol}$$

BDE, kJ/mol	+414	+192	−309	−368	(−17 kcal/mol)
(kcal/mol)	(+99)	(+46)	(−74)	(−88)	

Problem 8.3

Using the table of bond dissociation enthalpies in Appendix 3, calculate ΔH^0 for bromination of propane to give 1-bromopropane and hydrogen bromide.

8.5 Mechanism of Halogenation of Alkanes

From detailed studies of the conditions and products for halogenation of alkanes, chemists have concluded that these reactions occur by a type of mechanism called a radical chain mechanism.

A. Formation of Radicals

Radicals are produced from a molecule by cleavage of a bond in such a way that each atom or fragment participating in the bond retains one electron, a process referred to as **homolytic bond cleavage.** This type of cleavage should be contrasted to the much more common **heterolytic bond cleavage,** where a bond breaks in such a way that one of the species retains both electrons. In the following equations, we use single-headed or **fishhook arrows** to show the change in position of single electrons and indicate a homolytic mechanism. BDEs of these reactions from Appendix 3 are shown on the right. Following are three reactions that result in homolytic cleavage.

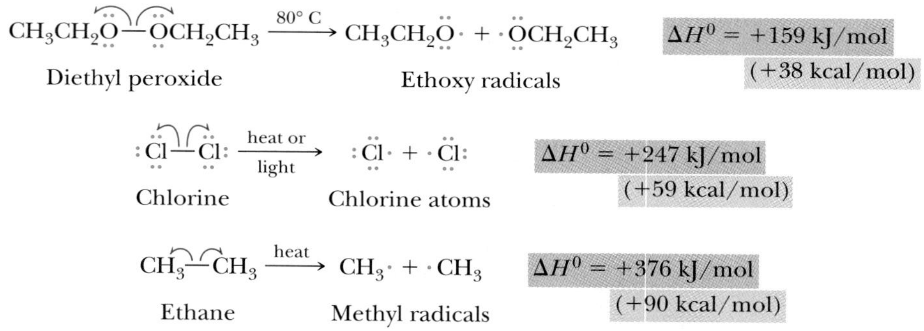

Homolytic cleavage Cleavage of a bond so that each fragment retains one electron; formation of radicals.

Heterolytic cleavage Cleavage of a bond so that one fragment retains both electrons and the other has none.

Fishhook arrow A barbed curved arrow used to show the change in position of a single electron.

$$CH_3CH_2O\!-\!OCH_2CH_3 \xrightarrow{80°\,C} CH_3CH_2O\cdot \; + \; \cdot OCH_2CH_3 \qquad \Delta H^0 = +159 \text{ kJ/mol}$$
Diethyl peroxide $\qquad\qquad$ Ethoxy radicals $\qquad\qquad\qquad$ (+38 kcal/mol)

$$:\!\ddot{C}l\!-\!\ddot{C}l\!: \xrightarrow[\text{light}]{\text{heat or}} :\!\ddot{C}l\cdot \; + \; \cdot\ddot{C}l\!: \qquad \Delta H^0 = +247 \text{ kJ/mol}$$
Chlorine $\qquad\qquad$ Chlorine atoms $\qquad\qquad$ (+59 kcal/mol)

$$CH_3\!-\!CH_3 \xrightarrow{\text{heat}} CH_3\cdot \; + \; \cdot CH_3 \qquad \Delta H^0 = +376 \text{ kJ/mol}$$
Ethane $\qquad\qquad$ Methyl radicals $\qquad\qquad$ (+90 kcal/mol)

Energy to cause bond cleavage and generation of radicals can be supplied either by light or heat. The energy of visible and ultraviolet radiation (wavelength from 200 to 700 nm) falls in the range of 585 to 167 kJ (140 to 40 kcal)/mol and is of the same order of magnitude as the bond dissociation enthalpies of halogen-halogen covalent bonds. The bond dissociation enthalpy of Br_2 is 192 kJ (46 kcal)/mol; that for Cl_2 is 247 kJ (59 kcal)/mol. Dissociation of these halogens can also be brought about by heating to temperatures above 350°C.

Oxygen-oxygen single bonds in peroxides (ROOR) and hydroperoxides (ROOH) have dissociation enthalpies in the range of 146 to 209 kJ (35 to 50 kcal)/mol, and compounds containing these bonds are cleaved to radicals at considerably lower temperatures than those required for rupture of carbon-carbon bonds. Diethyl peroxide, for example, begins to dissociate to ethoxy radicals at 80°C. Dissociation of ethane into two methyl radicals occurs only at very high temperature.

B. A Radical Chain Mechanism

To account for the products formed from halogenation of alkanes, chemists propose a radical chain mechanism involving three types of steps: (1) **chain initiation,** (2) **chain propagation,** and (3) **chain termination.** We illustrate radical halogenation of alkanes by the reaction of chlorine with ethane.

Mechanism   *Radical Chlorination of Ethane*

Chain initiation involves formation of radicals from nonradical species. Chlorine is homolytically dissociated by heat or light:

Step 1:
$$: \overset{..}{\underset{..}{Cl}} \!-\! \overset{..}{\underset{..}{Cl}} : \xrightarrow[\text{or light}]{\text{heat}} : \overset{..}{\underset{..}{Cl}} \cdot \; + \; \cdot \overset{..}{\underset{..}{Cl}} :$$

Chain propagation involves reaction of a radical and a molecule to form a new radical. The chlorine atom formed in Step 1 attacks the alkane, removing a hydrogen atom in another homolytic reaction:

Step 2:
$$CH_3 \overset{\frown}{CH_2} \!-\! H + \cdot \overset{..}{\underset{..}{Cl}} : \longrightarrow CH_3CH_2 \cdot + H \!-\! \overset{..}{\underset{..}{Cl}} :$$

Step 3:
$$CH_3CH_2 \cdot + : \overset{..}{\underset{..}{Cl}} \!-\! \overset{..}{\underset{..}{Cl}} : \longrightarrow CH_3CH_2 \!-\! \overset{..}{\underset{..}{Cl}} : + \cdot \overset{..}{\underset{..}{Cl}} :$$

Chain termination involves destruction of radicals. The first three possible chain termination steps involve coupling of radicals to form a new covalent bond. The fourth chain termination step, called disproportionation, involves transfer of a hydrogen atom from the beta position of one radical to another radical and formation of an alkane and an alkene.

Step 4:
$$CH_3CH_2 \cdot + \cdot CH_2CH_3 \longrightarrow CH_3CH_2 \!-\! CH_2CH_3$$

Step 5:
$$CH_3CH_2 \cdot + \cdot \overset{..}{\underset{..}{Cl}} : \longrightarrow CH_3CH_2 \!-\! \overset{..}{\underset{..}{Cl}} :$$

Step 6:
$$: \overset{..}{\underset{..}{Cl}} \cdot + \cdot \overset{..}{\underset{..}{Cl}} : \longrightarrow : \overset{..}{\underset{..}{Cl}} \!-\! \overset{..}{\underset{..}{Cl}} :$$

Step 7:
$$CH_3CH_2 \cdot + \overset{\overset{\displaystyle H}{|}}{CH_2} \!-\! CH_2 \cdot \longrightarrow CH_3CH_3 + CH_2 \!=\! CH_2$$

ORGANIC
Chemistry Now ™

Click Mechanisms to view an animation of the **Radical Halogenation of Ethane**

Chain initiation   A step in a chain reaction characterized by the formation of reactive intermediates (radicals, anions, or cations) from nonradical or noncharged molecules.

Chain propagation   A step in a chain reaction characterized by the reaction of a reactive intermediate and a molecule to give a new reactive intermediate and a new molecule.

Chain termination   A step in a chain reaction that involves destruction of reactive intermediates.

Initiation

The characteristic feature of a **chain initiation** step is formation of radicals from nonradical compounds. In the case of chlorination of ethane, chain initiation is by thermal or light-induced homolysis of the Cl—Cl bond to give two chlorine atoms (radicals).

Chain Propagation

The characteristic feature of a **chain propagation** step is reaction of a radical and a molecule to give a new radical. A chlorine atom is consumed in Step 2, but an ethyl radical is produced. Similarly, an ethyl radical is consumed in Step 3, but a chlorine radical is produced. Steps 2 and 3 can repeat thousands of times as long as neither radical is removed by a different reaction.

A second characteristic feature of chain propagation steps is that, when added together, they give the observed stoichiometry of the reaction. Adding Steps 2 and 3 and canceling structures that appear on both sides of the equation gives the balanced equation for the radical chlorination of ethane.

$$\text{Steps 2 + 3} \quad CH_3CH_3 + Cl_2 \longrightarrow CH_3CH_2Cl + HCl$$

The number of times a cycle of chain propagation steps repeats is called **chain length.** Chain lengths can range from a few to many thousand, depending on the relative rates of reactions and the concentrations of various species.

Chain length   The number of times the cycle of chain propagation steps repeats in a chain reaction.

Chain Termination

A characteristic feature of a **chain termination** step is destruction of radicals. Among the most important chain termination reactions during halogenation of alkanes are radical couplings, illustrated by Steps 4, 5, 6, and disproportionation illustrated by Step 7 in the mechanism for the halogenation of ethane.

Note that chain termination steps are usually relatively rare compared to chain propagation steps in radical chain reactions. This is because at any one time, the concentration of radical species is very low, making a collision between two radicals a relatively rare event. The relatively rare occurrence of chain termination steps explains why there can be chain lengths of many thousands of steps in a radical chain reaction.

A major concern with chlorofluorocarbon destruction of the ozone layer (mentioned in "Chemical Connections: Freons" in this chapter) is that the important reactions are thought to occur through a radical chain mechanism. The production of one radical species derived from a CFC can destroy a large number of ozone molecules before any termination steps occur.

The structures, geometries, and relative stabilities of simple alkyl radicals are similar to those of alkyl carbocations. They are planar, or almost so, with bond angles of 120° about the carbon with the unpaired electron. This geometry indicates that carbon is sp^2 hybridized and that the unpaired electron occupies the unhybridized $2p$ orbital. As mentioned, the order of stability of alkyl radicals, like alkyl carbocations, is 3° > 2° > 1° > methyl.

C. Energetics of Chain Propagation Steps

After the radical chain is initiated, the heat of reaction is derived entirely from the heat of reaction of the individual chain propagation steps. In Step 2 of radical chlorination of ethane, for example, energy is required to break the CH_3CH_2—H bond [422 kJ (101 kcal)/mol], but energy is released on formation of the H—Cl bond [−431 kJ (−103 kcal)/mol]. Similarly, energy is required in Step 3 to break the Cl—Cl bond [247 kJ (59 kcal)/mol], but energy is released on formation of the CH_3CH_2—Cl bond [−355 kJ (−85 kcal)/mol]. We see that, just as the sum of the chain propagation steps for radical halogenation gives the observed stoichiometry, the sum of the heats of reaction for each propagation step is equal to the observed heat of reaction:

	Reaction Step							ΔH^0, kJ/mol (kcal/mol)
Step 2:	CH_3CH_2—H +422 (+101)	+	·Cl	$\longrightarrow$	CH_3CH_2·	+	H—Cl −431 (−103)	−9 (−2)
Step 3:	CH_3CH_2·	+	Cl—Cl +247 (+59)	$\longrightarrow$	CH_3CH_2—Cl −355 (−85)	+	·Cl	−108 (−26)
Sum:	CH_3CH_2—H	+	Cl—Cl	$\longrightarrow$	CH_3CH_2—Cl	+	H—Cl	−117 (−28)

Example 8.4

Using the table of bond dissociation enthalpies in Appendix 3, calculate ΔH^0 for each propagation step in the radical bromination of propane to give 2-bromo-propane and HBr.

Solution

Here are the two chain propagation steps along with bond dissociation enthalpies for the bonds broken and the bonds formed. The first chain propagation step is endothermic, the second is exothermic, and the overall reaction is exothermic by 71 kJ (17 kcal)/mol.

$$\Delta H^0$$

$$
\underset{\substack{+414 \\ (+99)}}{\overset{\text{H}}{\text{Step 2:}\ \ \text{CH}_3\overset{|}{\text{C}}\text{HCH}_3}} + \text{Br}\cdot \longrightarrow \text{CH}_3\overset{\cdot}{\text{C}}\text{HCH}_3 + \underset{\substack{-368 \\ (-88)}}{\text{H---Br}}
$$

Step 2: $\text{CH}_3\overset{\text{H}}{\underset{+414\ (+99)}{\text{CHCH}_3}} + \text{Br}\cdot \longrightarrow \text{CH}_3\overset{\cdot}{\text{CHCH}_3} + \underset{-368\ (-88)}{\text{H---Br}}$ **+46 kJ/mol (+11 kcal/mol)**

Step 3: $\text{CH}_3\overset{\cdot}{\text{CHCH}_3} + \underset{+192\ (+46)}{\text{Br---Br}} \longrightarrow \text{CH}_3\overset{\text{Br}}{\underset{-309\ (-74)}{\text{CHCH}_3}} + \text{Br}\cdot$ **−117 kJ/mol (−28 kcal/mol)**

Sum: $\text{CH}_3\overset{\text{H}}{\text{CHCH}_3} + \text{Br---Br} \longrightarrow \text{CH}_3\overset{\text{Br}}{\text{CHCH}_3} + \text{H---Br}$ **−71 kJ/mol (−17 kcal/mol)**

Problem 8.4

Write a pair of chain propagation steps for the radical bromination of propane to give 1-bromopropane, and calculate ΔH^0 for each propagation step and for the overall reaction.

D. Regioselectivity of Bromination Versus Chlorination: Hammond's Postulate

The regioselectivity in halogenation of alkanes can be accounted for in terms of the relative stabilities of radicals (3° > 2° > 1° > methyl). As we will see, the energy of the transition state reflects the energy of the radicals; more stable radical products are formed with a lower activation energy, thus faster. But how do we account for the greater regioselectivity in bromination of alkanes compared with chlorination of alkanes? To do so, we need to consider **Hammond's postulate,** a refinement of transition state theory proposed in 1955 by George Hammond, then at Iowa State University. According to this postulate:

> The structure of the transition state for an exothermic reaction step is reached relatively early in the reaction, so it resembles the reactants of that step more than the products. Conversely, the structure of the transition state for an

Hammond's postulate The structure of the transition state for an exothermic step looks more like the reactants of that step than the products. Conversely, the structure of the transition state for an endothermic step looks more like the products of that step than the reactants.

endothermic reaction step is reached relatively late, so it resembles the products of that step more than the reactants.

It is important to realize that we cannot observe a transition state directly. Until the advent of modern computational theory, we could only infer its existence, structure, and stability from experiment. Hammond's postulate gives us a reasonable way of deducing something about the structure of a transition state by examining things we can observe: the structure of reactants and products and heats of reaction. Hammond's postulate applies equally well to multistep reactions. The transition state of any exothermic step in a multistep sequence looks like the starting material(s) of that step; the transition state of any endothermic step in the sequence looks like the product(s) of that step. Thus changes in starting material energy affect the transition state of an exothermic reaction more than changes in product energy. The converse is true for an endothermic reaction. Today, we can carry out high-level computations that reveal details of complex transition states and strongly support the validity of Hammond's postulate.

Shown in Figure 8.1 are energy diagrams for a highly exothermic reaction and a highly endothermic reaction, each occurring in one step.

Now let us apply Hammond's postulate to explain the relative regioselectivities of chlorination versus bromination of alkanes. In applying this postulate, we deal with the rate-determining step of the reaction, which, in radical halogenation of alkanes, is the abstraction of a hydrogen atom by a halogen radical. Given in Table 8.8 are heats of reaction, ΔH^0, for the hydrogen abstraction step in chlorination and bromination of the different hydrogens of 2-methylpropane (isobutane). Also given under the formulas of isobutane, HCl, and HBr are bond dissociation enthalpies for the bonds broken (1° and 3° C—H) and formed (H—Cl and H—Br) in each step. Because the 3° radical is more stable than the 1° radical, the BDE for the 3° H is lower than that of a 1° H by about 17 kJ (4 kcal)/mol and the difference in ΔH^0 between 1° and 3° for the two reactions is just this amount.

Abstraction of hydrogen by chlorine is exothermic, which, according to Hammond's postulate, means that the transition state for H abstraction by Cl· is reached early in the course of the reaction [Figure 8.2(a)]. Therefore, the structure of the transition state for this step resembles the reactants, namely the alkane and a chlorine atom, not the product radicals. As a result, there is relatively little radical character on carbon in this transition state, and regioselectively in radical chlorination is only slightly influenced by the relative stabilities of radical intermediates.

Figure 8.1

Hammond's postulate. Energy diagrams for two one-step reactions. In the exothermic reaction, the transition state occurs early, and its structure resembles that of the reactants. In the endothermic reaction, the transition state occurs late, and its structure resembles that of the products.

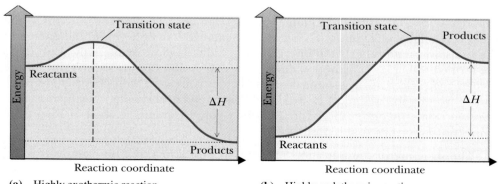

(a)   Highly exothermic reaction

(b)   Highly endothermic reaction

Table 8.8   Enthalpies of Reaction for Hydrogen Abstraction in 2-Methylpropane

Reaction Step	ΔH^0 [kJ (kcal)/mol]

Chlorination

+422(101)   Primary radical

$-9(-2)$

$-431(-103)$

+405(97)   Tertiary radical

$-26(-6)$

$\Big\}$ 17(4)

Bromination

+422(101)

$+54(+13)$

$-368(-88)$

+405(97)

$+37(+9)$

$\Big\}$ 17(4)

Products are determined more by whether a chlorine atom happens to collide with a 1°, 2°, or 3° H.

In Section 8.4A, we were given the fact that the selectivity for abstraction of a 3° H compared to a 1° H in chlorination is 5:1; this ratio directly reflects the relative reaction rates of these hydrogens with chlorine atoms. Using this ratio of reaction rates and the relationship between $\Delta G^{\ddagger}$ and rate constants (Section 6.2A), we can calculate that the difference in activation energies, $\Delta\Delta G^{\ddagger}$, for the abstraction of a 3° H versus a 1° H is about 4 kJ (1 kcal)/mol. However, we can calculate from the bond dissociation enthalpies [Figure 8.2(a)] that $\Delta\Delta H^0$ for the two reactions is about 17 kJ (4 kcal)/mol. Thus the difference in product stabilities is only slightly reflected in the transition states and the resulting reaction rates. Contrast this reaction with bromination [Figure 8.2(b)]. For bromination, the selectivity of 3° H to 1° H is 1600:1, which corresponds to $\Delta\Delta G^{\ddagger}$ of approximately 18 kJ (4.2 kcal)/mol. The $\Delta\Delta H^0$ for the formation of the primary and tertiary radicals is the same in bromination and in chlorination (it is just the difference in BDE of the two radicals). But the rate-determining step for bromination, because it is endothermic, has a transition state much more like the product radical, and the transition state reflects nearly all the energy difference of the primary and tertiary radicals. The later transition state, and the correspondingly larger $\Delta\Delta G^{\ddagger}$ (which causes a large difference in reaction rates), is the reason for the much larger regioselectivity in radical bromination than in radical chlorination.

BDE values are ΔH^0 and not ΔG^0. Recall that $\Delta G = \Delta H - T\Delta S$. Because we are dealing with similar reactions, we can assume that entropy differences between them are nearly zero and, therefore, $\Delta\Delta G^0 \sim \Delta\Delta H^0$ and $\Delta\Delta G^{\ddagger} \sim \Delta\Delta H^{\ddagger}$, which allows us to make these comparisons.

Figure 8.2
Transition states and energetics for hydrogen abstraction in the radical chlorination and bromination of 2-methylpropane (isobutane). The product is the intermediate radical, R·.

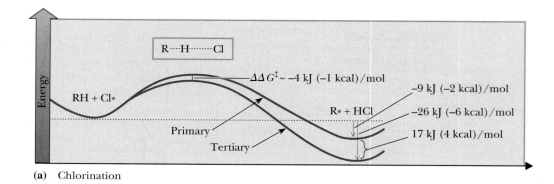

(a) Chlorination

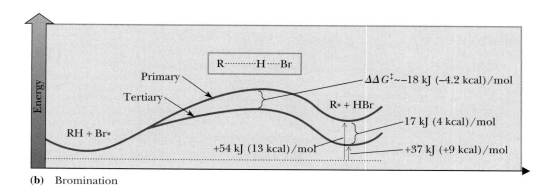

(b) Bromination

E. Stereochemistry of Radical Halogenation

When radical halogenation produces a stereocenter or takes place at a hydrogen on an existing stereocenter, the product is an equal mixture of R and S enantiomers. Consider, for example, radical bromination of butane, which produces 2-bromobutane.

$$CH_3CH_2CH_2CH_3 + Br_2 \xrightarrow[\text{or light}]{\text{heat}} CH_3CH_2\overset{\overset{\displaystyle Br}{\displaystyle |}}{C}HCH_3 + HBr$$

Butane (R,S)-2-Bromobutane

In this example, both of the starting materials are achiral, and, as is true for any reaction of achiral starting materials taking place in an achiral environment that gives a chiral product (Section 6.7A), the product is a racemic mixture.

In the case of the *sec*-butyl radical, the carbon bearing the unpaired electron is sp^2 hybridized, and the unpaired electron lies in the unhybridized $2p$ orbital. Reaction of the alkyl radical intermediate with halogen in the second chain propagation step occurs with equal probability from either face to give an equal mixture of the R and S configurations at the newly created stereocenter.

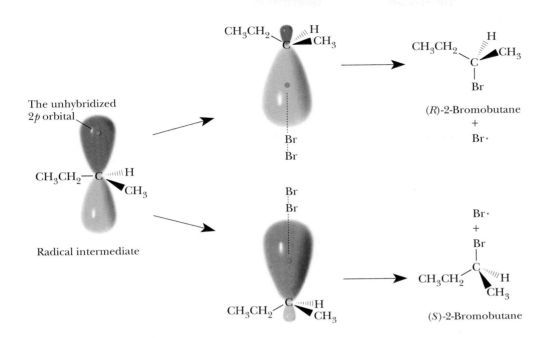

The unhybridized 2p orbital

Radical intermediate

(R)-2-Bromobutane
+
Br·

Br·
+
(S)-2-Bromobutane

8.6 Allylic Halogenation

We saw in Section 6.3D that propene and other alkenes react with Br_2 and Cl_2 at room temperature by addition to the carbon-carbon double bond. If, however, propene and one of these halogens are allowed to react at a high temperature, an entirely different reaction takes place, namely substitution of a halogen occurs at the **allylic carbon** (the carbon next to a carbon-carbon double bond). We illustrate **allylic substitution** by the reaction of propene with chlorine at high temperature:

$$CH_2{=}CHCH_3 + Cl_2 \xrightarrow{350°C} CH_2{=}CHCH_2Cl + HCl$$

Propene   3-Chloropropene
                       (Allyl chloride)

Allylic carbon A carbon adjacent to a carbon-carbon double bond.

Allylic substitution Any reaction in which an atom or group of atoms is substituted for another atom or group of atoms at an allylic carbon.

A comparable reaction takes place when propene is treated with bromine at an elevated temperature.

To predict which of the various C—H bonds in propene is most likely to break when a mixture of propene and bromine or chlorine is heated, we need to look at bond dissociation enthalpies. We find that the bond dissociation enthalpy of an allylic C—H bond in propene (Table 8.7) is approximately 92 kJ (22 kcal)/mol less than that of a vinylic C—H bond and 50 kJ (12 kcal)/mol less than a C—H bond of ethane. The allyl radical is even more stable than a 3° radical; as we shall see in Section 9.4E, this unusual stability also applies to carbocations. The reason that the allylic C—H bond is so weak is discussed in Section 8.6B. Note from Table 8.7 that the benzyl radical $C_6H_5CH_2$· is stabilized in exactly the same way as the allyl radical and for the same reason; benzylic compounds undergo many of the same reactions as allylic compounds (Section 21.5).

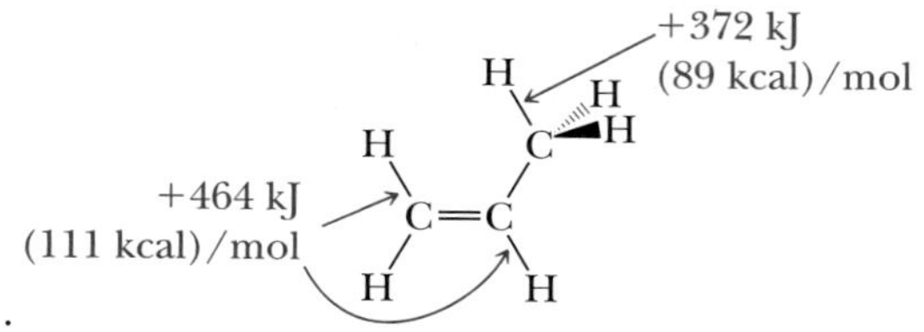

Treating propene with bromine or chlorine at elevated temperatures illustrates a very important point about organic reactions: It is often possible to change the product(s) by changing the mechanism through a change in reaction conditions. Under the high temperatures used in this reaction, the concentration of bromine radicals becomes much higher than at room temperature; this greatly accelerates the substitution reaction, which goes by the radical halogenation mechanism. At room temperature, there are far fewer radicals, and electrophilic addition is observed.

$$CH_2{=}CHCH_3 + Br_2 \xrightarrow{\text{high temp.}} CH_2{=}CHCH_2Br + HBr \qquad \text{(allylic substitution)}$$

Propene   3-Bromopropene

$$CH_2{=}CHCH_3 + Br_2 \xrightarrow[CH_2Cl_2]{\text{room temp.}} \underset{\underset{\text{Br Br}}{|\ \ |}}{CH_2CHCH_3} \qquad \text{(electrophilic addition)}$$

Propene   1,2-Dibromopropane

A very useful way to carry out allylic bromination in the laboratory at or slightly above room temperature is to use the reagent *N*-bromosuccinimide (NBS) in dichloromethane (CH_2Cl_2). Reaction between an alkene and NBS is most commonly initiated by light. This reaction involves a net double substitution: a bromine in NBS and a hydrogen in the alkene exchange places.

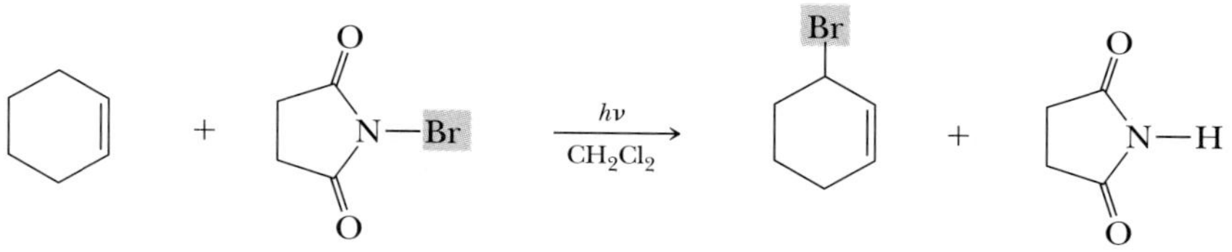

Cyclohexene   *N*-Bromosuccinimide   3-Bromocyclohexene   Succinimide
                (NBS)

A. Mechanism of Allylic Halogenation

Allylic bromination and chlorination proceed by a radical chain reaction involving the same type of chain initiation, chain propagation, and chain termination steps involved in the radical halogenation of alkanes.

Mechanism *Allylic Bromination of Propene Using NBS*

Chain Initiation Chain initiation involves formation of radicals from NBS by light-induced homolytic cleavage of the N—Br bond in NBS. This step is analogous to the homolytic dissociation of chlorine to chlorine radicals.

Step 1:

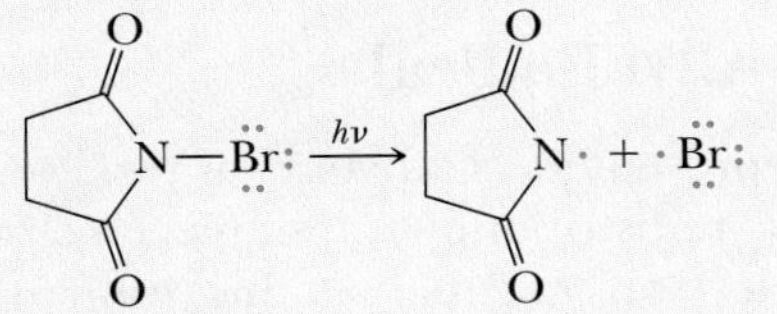

Chain Propagation Chain propagation involves the formation of products. Reaction of a radical and a nonradical gives a new radical. (Both radicals formed in the initiation can abstract hydrogen atoms. We show only the Br· reaction.) In the first propagation step, a bromine atom abstracts an allylic hydrogen (the weakest C—H bond in propene) to produce an allyl radical. The allyl radical, in turn, reacts with a bromine molecule to form allyl bromide and a new bromine atom.

Step 2: $CH_2{=}CHCH_2{-}H + \cdot Br: \longrightarrow CH_2{=}CHCH_2\cdot + H{-}Br:$

Allylic hydrogen ↗   Allyl Radical ↗

Step 3: $CH_2{=}CHCH_2\cdot + :Br{-}Br: \longrightarrow CH_2{=}CHCH_2{-}Br: + \cdot Br:$

Note that, as always, this combination of chain propagation steps adds up to the observed stoichiometry. This reaction is exactly like halogenation of alkanes, but is strongly regioselective for the allylic hydrogen because of its weak bond.

Chain Termination—The Destruction of Radicals Propagation of the chain reaction continues until termination steps produce nonradical products and thus stop further reaction.

Step 4: $:Br\cdot + \cdot Br: \longrightarrow :Br{-}Br:$

Step 5: $CH_2{=}CHCH_2\cdot + \cdot Br: \longrightarrow CH_2{=}CHCH_2{-}Br:$

Step 6: $CH_2{=}CHCH_2\cdot + \cdot CH_2CH{=}CH_2 \longrightarrow CH_2{=}CHCH_2{-}CH_2CH{=}CH_2$

The Br_2 necessary for Step 2 is formed by reaction of product HBr with NBS.

Step 7:

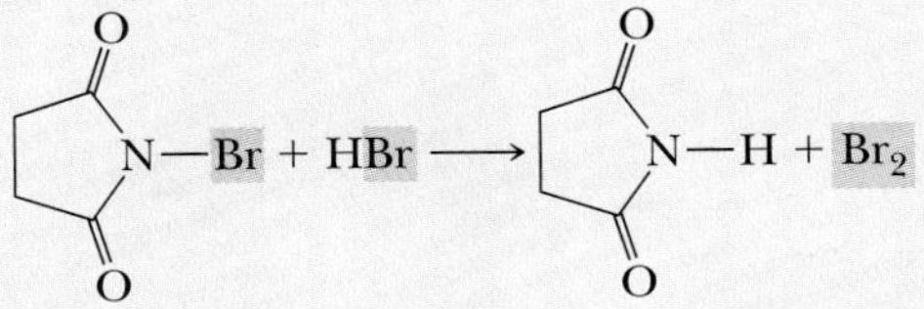

Bromine formed in this step then reacts with an allyl radical to continue the chain propagation reactions (Step 3).

The mechanism we described for allylic bromination by NBS poses the following problem. NBS is the indirect source of Br_2, which then takes part in chain propagation. But if Br_2 is present in the reaction mixture, why does it not react instead with the carbon-carbon double bond by electrophilic addition? In other words, why is the observed reaction allylic substitution rather than addition to the double bond? The answer is that the rates of the chain propagation steps are much faster than the rate of electrophilic addition of bromine to the alkene when radicals are present. Furthermore, the concentration of Br_2 is very low throughout the course of the reaction, which slows the rate of electrophilic addition.

B. Structure of the Allyl Radical

The allyl radical can be represented as a hybrid of two contributing structures. Here fishhook arrows show the redistribution of single electrons between contributing structures. Note three pi electrons take part in this resonance.

$$CH_2{=}CH{-}\overset{\cdot}{C}H_2 \longleftrightarrow \overset{\cdot}{C}H_2{-}CH{=}CH_2$$

(Equivalent contributing structures)

The eight atoms of the allyl radical lie in a plane, and all bond angles are approximately 120°. Each carbon atom is sp^2 hybridized, and the three $2p$ orbitals participating in resonance stabilization of the radical are parallel to one another as shown in Figure 8.3. Like charged systems, in which a delocalized charge is more stable than a localized one, delocalized unpaired electron density leads to more stable structures than localized unpaired electron density.

Based on bond dissociation enthalpies, we conclude that an allyl radical is even more stable than a 3° alkyl radical. Note that because of the larger amount of s character in its carbon sp^2 hybrid orbital, a vinylic C—H bond is stronger (has a larger bond dissociation enthalpy) than any sp^3 C—H bond and is never abstracted in homolytic reactions.

Figure 8.3
Molecular orbital model of covalent bonding in the allyl radical. Combination of three $2p$ atomic orbitals gives three pi molecular MOs. The lowest, a pi-bonding MO, has zero nodes. The next in energy, a pi-nonbonding MO, has one node, and the highest in energy, a pi-antibonding MO, has two nodes.

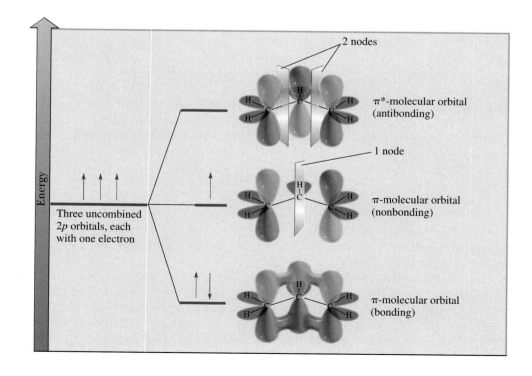

According to the molecular orbital description, the conjugated system of the allyl radical involves the formation of three molecular orbitals by overlap of three $2p$ atomic orbitals (Figure 8.3). The lowest energy MO has zero nodes, the next MO has one node, and the highest energy MO has two nodes. The molecular orbital of intermediate energy in this case leads to neither net stabilization nor destabilization and is, therefore, called a nonbonding MO. The lowest pi MO is at a lower energy than the isolated $2p$ orbitals.

In the lowest energy (ground) state of the allyl radical, two electrons of the pi system lie in the pi-bonding MO, and the third lies in the pi-nonbonding MO; the pi-antibonding MO is unoccupied. Because the lowest pi MO is at a lower energy than the isolated $2p$ atomic orbitals, putting two electrons in this MO releases considerable energy, which accounts for the stability of the allyl radical.

The lone electron of the allyl radical is associated with the pi-nonbonding MO, which places electron density on carbons 1 and 3 only. This localization is shown clearly in the unpaired electron density map in Figure 8.4. Thus both the resonance model and molecular orbital theory are consistent in predicting radical character on carbons 1 and 3 of the allyl radical but no radical character on carbon 2.

Figure 8.4
Unpaired electron spin density map for the allyl radical. Unpaired electron density (green cones) appears only on carbons 1 and 3.

Example 8.5

Account for the fact that allylic bromination of 1-octene by NBS gives these products.

1-Octene → NBS / CH_2Cl_2 → 3-Bromo-1-octene (racemic, 17%) + (E)-1-Bromo-2-octene (83%)

Solution

The rate-determining step of this radical chain mechanism is hydrogen abstraction from the allylic position on 1-octene to give a 2° allylic radical. This radical is stabilized by delocalization of the two pi electrons and the unpaired electron. Reaction of the radical at carbon 1 gives the major product. Reaction at carbon 3 gives the minor product. The more substituted (and more stable) alkene isomer predominates.

1-Octene + ·Br → [allylic radical] ↔ [allylic radical] + HBr

↓ Br_2 → 3-Bromo-1-octene (racemic) + ↓ Br_2 → 1-Bromo-2-octene (major product) + ·Br

■ Unpaired electron spin density map of the radical formed from 1-butene. Spin (blue) is on carbons 1 and 3, but more is on carbon 1.

Autoxidation Air oxidation of materials such as unsaturated fatty acids.

The reason for this regioselectivity seems to be that the resonance hybrid of the allylic radical with the more substituted double bond dominates. This hypothesis is borne out by the unpaired electron spin density, here calculated for the radical formed from 1-butene. The terminal carbon has the highest spin density (blue).

Problem 8.5

Given the solution to Example 8.5, predict the structure of the product(s) formed when 3-hexene is treated with NBS.

8.7 Radical Autoxidation

One of the most important reactions for materials, foods, and also for living systems is called **autoxidation;** that is, oxidation requiring oxygen and no other reactant. This reaction takes place by a radical chain mechanism very similar to that for allylic bromination. If you open a bottle of cooking oil that has stood for a long time, you will notice the hiss of air entering the bottle because there is a negative pressure caused by the consumption of oxygen by autoxidation of the oil.

Cooking oil contains **polyunsaturated fatty acid esters.** (See Section 5.4 "Connections to Biological Chemistry: The Importance of *cis* Double Bonds in Fats Versus Oils," as well as Section 26.1). The most common of these compounds have chains of 16 or 18 carbons containing 1,4-diene functional groups. (Both double bonds are *cis*; the nature of R_1 and R_2 need not concern us at this stage.) The hydrogens on the CH_2 group between the double bonds are doubly allylic; that is, they are allylic with respect to both double bonds. As you might expect, the radical formed by abstraction of one of these hydrogens is unusually stable because it is even more delocalized (stabilized by resonance) than an allylic radical. An allylic C—H bond is much weaker than a corresponding alkane C—H bond, and the doubly allylic C—H is even weaker.

Autoxidation begins when a radical initiator, X·, formed either by light activation of an impurity in the oil or by thermal decomposition of peroxide impurities, abstracts a doubly allylic hydrogen to form a radical. This radical is stabilized by resonance with both double bonds (1 and 2 in the following structure).

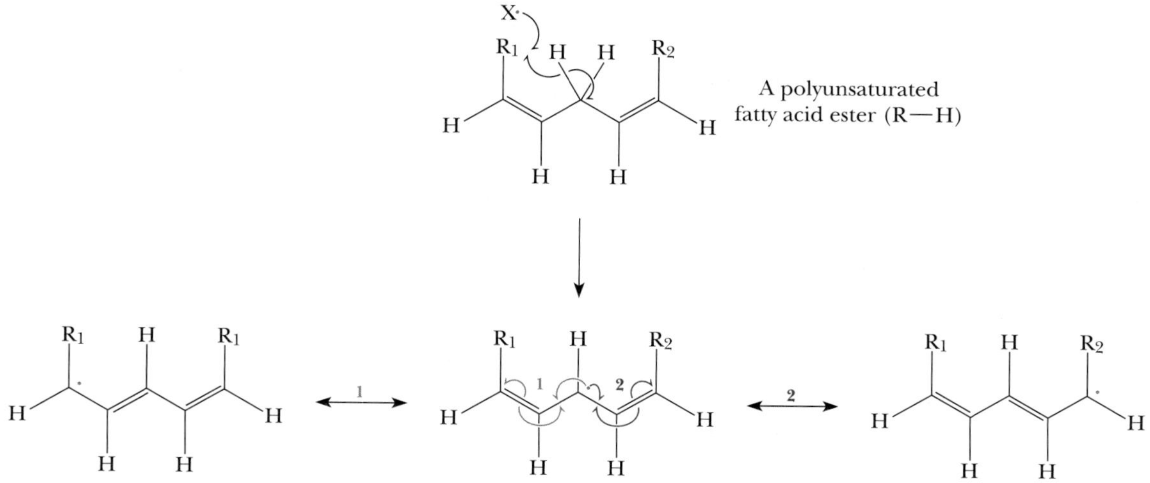

A polyunsaturated
fatty acid ester (R—H)

This radical reacts with oxygen, itself a (very unreactive) diradical, to form a peroxy radical, which then reacts with the CH_2 of another 1,4-diene fatty acid ester (R—H) to give a new radical (R·) and a hydroperoxide. Hydroperoxides are formed on both sides by reactions with the resonance hybrid; only one is shown. The new radical reacts again with oxygen, causing a radical chain reaction in which hundreds of molecules of fatty acid ester are oxidized for each initiator radical.

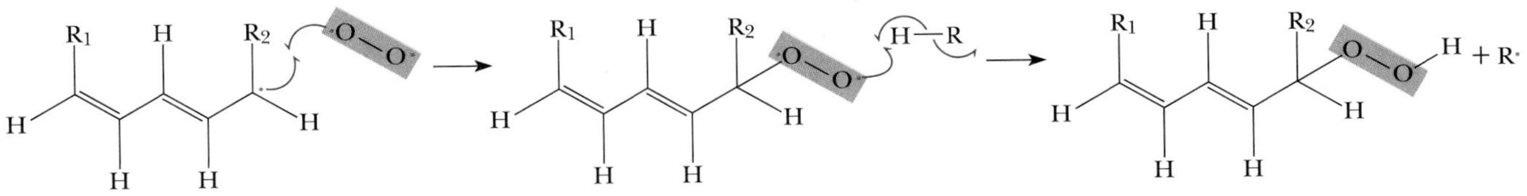

Peroxy radical   A hydroperoxide

The ultimate fate of the peroxide, and some of the peroxy radical as well, is complex. Some autoxidation products degrade to short-chain aldehydes and carboxylic acids with unpleasant "rancid" smells familiar to anyone who has smelled old cooking oil or aged foods that contain polyunsaturated oils. It has been suggested that some products of autoxidation of oils are toxic and/or carcinogenic. Oils lacking the 1,4-diene structure are much less easily oxidized.

Example 8.6

What products would you expect from the following reaction? Indicate the major one and specify stereochemistry if relevant.

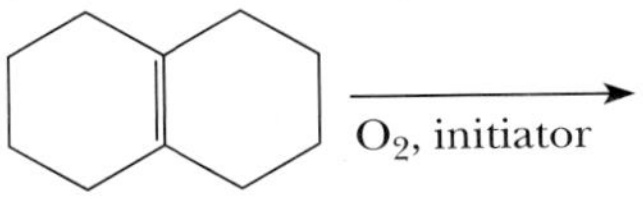

Solution

The major product has the more substituted double bond. (Both are racemic.)

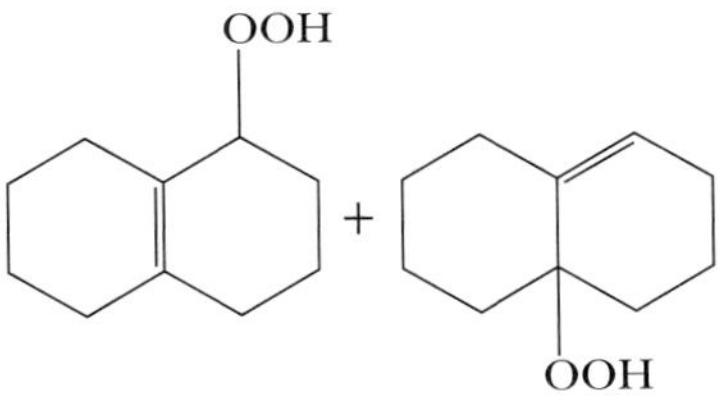

Major   Minor

Problem 8.6

Show the products of the following reaction and indicate the major one.

$$CH_3(CH_2)_4CH_2CH{=}CH_2 \xrightarrow{O_2,\ \text{initiator}}$$

CONNECTIONS TO BIOLOGICAL CHEMISTRY

Antioxidants

Many plants contain polyunsaturated fatty acid esters in their leaves or seeds. However, in the natural state, they are protected against autoxidation by a variety of agents, one of the most important of which is α-toco-

pherol (vitamin E, Section 26.6C). This compound is a phenol (Section 21.4). The characteristic of phenols that makes them effective as protective agents against autoxidation is their O—H bond, which is even weaker

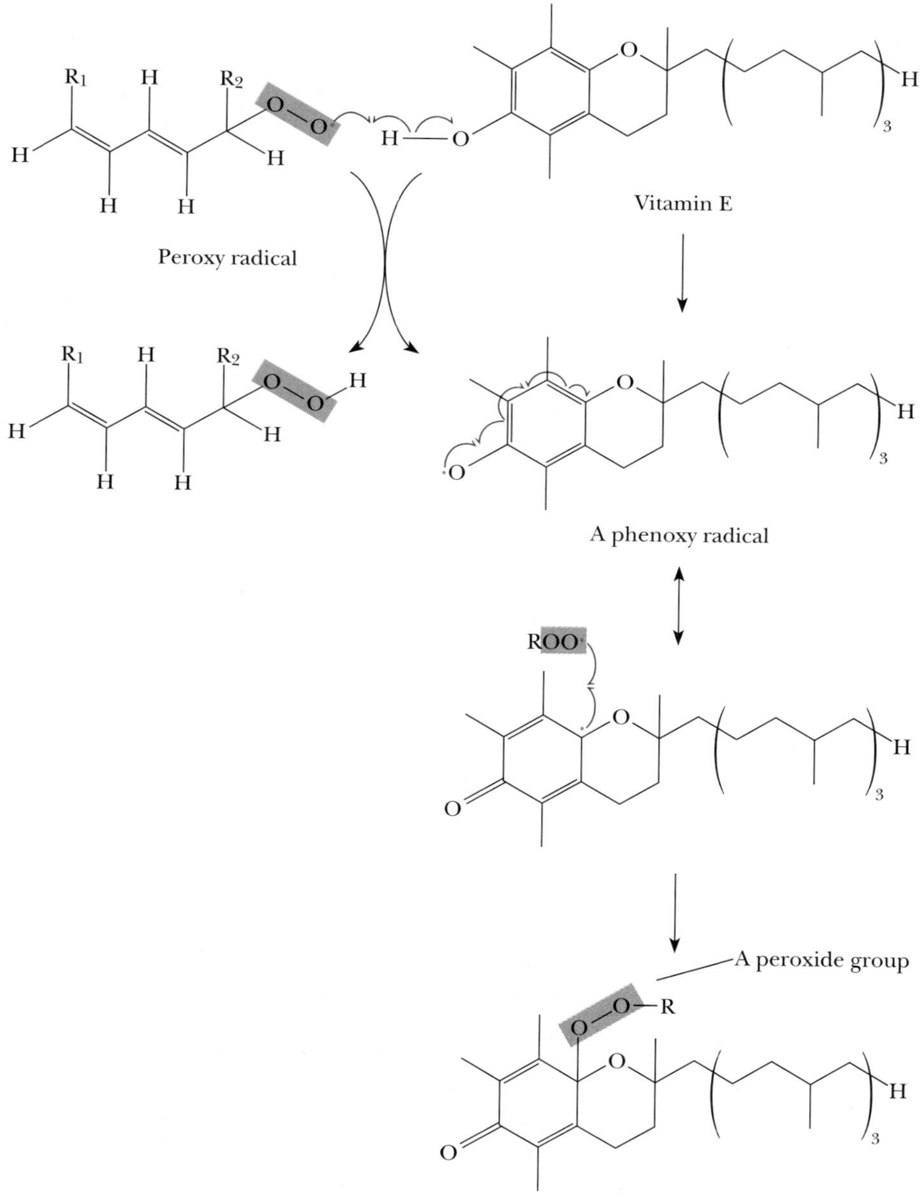

Vitamin E

Peroxy radical

A phenoxy radical

A peroxide group

A peroxide derived from vitamin E

than the doubly allylic C—H bond (see Appendix 3). Vitamin E reacts preferentially with the initial peroxy radical to give a resonance-stabilized phenoxy radical, which is very unreactive and survives to scavenge another peroxy radical, ROO·. The resulting peroxide is stable under the conditions. The antioxidant thus removes two molecules of peroxy radical and stops the radical chain oxidation dead in its tracks.

Because vitamin E is removed in the processing of many food products, similar phenols such as BHT (Problem 22.23) are often added to foods to prevent spoilage by autoxidation. These compounds are all called **radical inhibitors** because of their ability to terminate radical chains before significant damage to the food occurs. Radical inhibitors are often referred to as preservatives. Similar compounds are also added to many materials, such as plastics and rubber, to protect them against autoxidation.

BHT

Similar oxidative degradation of materials in low-density lipoproteins (LDL, Section 26.4A) deposited in arteriosclerotic plaques on the walls of arteries leads to cardiovascular disease in humans. Many effects of aging and possibly cancer in humans and damage to materials such as rubber and plastic occur by similar mechanisms.

Radical inhibitor A compound such as a phenol that selectively reacts with radicals to remove them from a chain reaction and terminate the chain.

8.8 Radical Addition of HBr to Alkenes

When the addition of hydrogen halides to alkenes was first studied systematically in the 1930s, chemists observed that the addition of HBr sometimes gives Markovnikov addition, and sometimes gives non-Markovnikov addition. These two modes of addition of HBr are illustrated for 2-methylpropene (isobutylene).

The puzzle was solved in 1933 when it was discovered that non-Markovnikov products are observed only in the presence of peroxides or other sources of radicals. In the absence of radicals, addition of HBr gives only the expected Markovnikov product.

Another puzzling observation was that these variable additions occur only with HBr. Both HCl and HI always add to alkenes according to Markovnikov's rule.

To account for the products of HBr addition to alkenes in the presence of peroxides, chemists proposed a radical chain mechanism like the one for halogenation (Section 8.6A). In the following mechanism, the source of initiating radicals is a dialkyl peroxide, which is frequently present as an impurity in the solvent or alkene.

Mechanism   *Radical-Initiated Non-Markovnikov Addition of HBr to Alkenes*

Chain Initiation

Heterolytic cleavage of a dialkyl peroxide is induced by light or heat to two alkoxy radicals. An alkoxy radical then reacts with HBr by hydrogen abstraction to give an alcohol and a bromine radical.

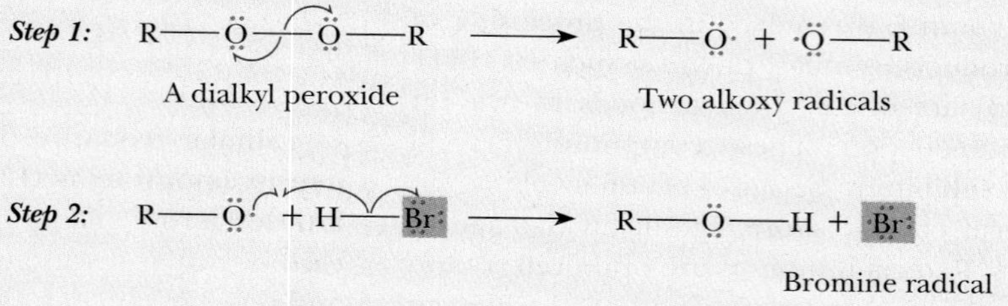

Step 1:   R—Ö—Ö—R ⟶ R—Ö· + ·Ö—R

A dialkyl peroxide   Two alkoxy radicals

Step 2:   R—Ö· + H—Br ⟶ R—Ö—H + ·Br·

Bromine radical

Chain Propagation

A bromine radical adds to the carbon-carbon double bond regioselectively to give the more substituted (and more stable) carbon radical. The carbon radical, in turn, reacts with a molecule of HBr to give the bromoalkane and generate a new bromine radical. Note that in each propagation step, one radical is consumed, but another is formed.

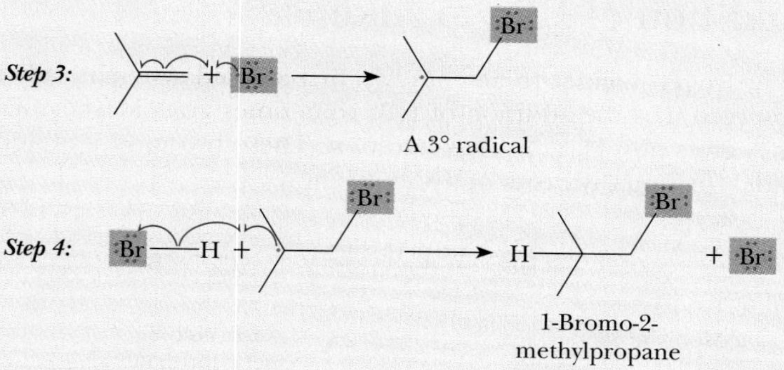

Step 3:   ⟩=⟨ + ·Br· ⟶

A 3° radical

Step 4:   ·Br—H + ⟶ H— + ·Br·

1-Bromo-2-
methylpropane

Chain Termination

The most important chain termination steps are the combination of a carbon radical with a bromine radical, and the combination of two bromine radicals. Each of these steps destroys one or both of the radical intermediates in the chain:

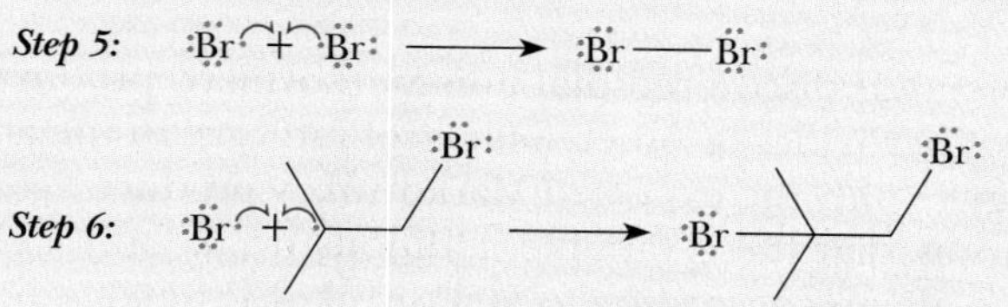

Step 5:   ·Br· + ·Br· ⟶ ·Br—Br·

Step 6:   ·Br· + ⟶ ·Br—

The observed non-Markovnikov regioselectivity of radical addition of HBr to an alkene is a combination of a steric factor and an electronic factor. First, a bromine radical attacks the less hindered carbon of the double bond (the steric factor). Second, as mentioned in Section 8.4C, the relative stabilities of radicals parallel those of carbocations (Section 6.3A).

Because the intermediate in Step 3 is a radical, it does not preserve stereochemistry in Step 4. Where more than one stereoisomer is possible, both may be expected, although steric factors can favor one over another.

Thus we have seen in Section 6.3A that polar addition of HBr to an alkene is regioselective, with bromine adding to the more substituted carbon. Radical addition of HBr to an alkene is also regioselective, with bromine adding to the less substituted carbon. This pair of alkene additions illustrates how the products of a reaction often can be changed by a change in experimental conditions and a change in mechanism; in this case, a change from a polar mechanism to a radical mechanism.

Example 8.7

Predict the product of the following reaction:

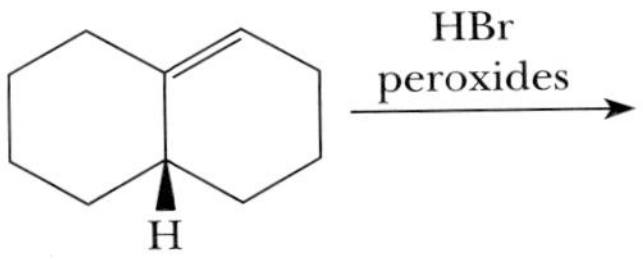

Solution

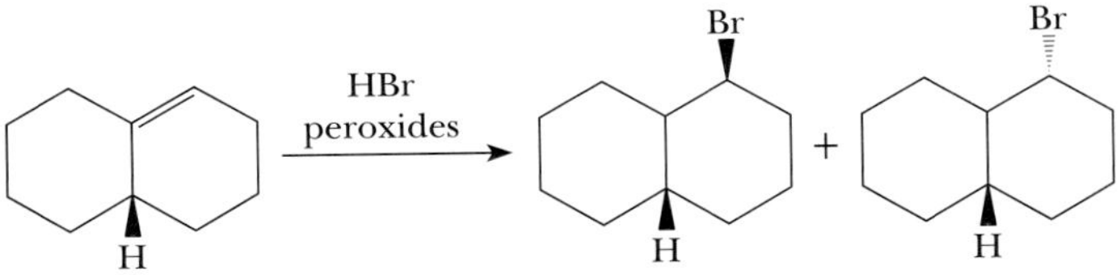

Mixture of *cis* and *trans*; the radical intermediate
does not preserve stereochemistry

Problem 8.7

Predict the major product of the following reaction:

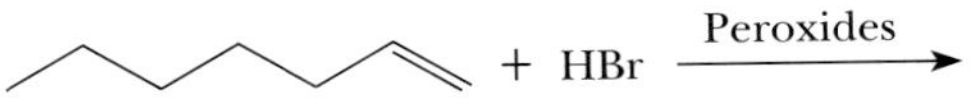

Summary

Haloalkanes contain a halogen covalently bonded to an sp^3 hybridized carbon (Section 8.1). In the IUPAC system, halogen atoms are named *fluoro-*, *chloro-*, *bromo-*, and *iodo-* substituents and are listed in alphabetical order with other substituents (Section 8.2A). In the common system, they are named alkyl halides. **Haloalkenes** contain a halogen covalently bonded to an sp^2 hybridized carbon of an alkene. In the common system, they are named alkenyl or vinylic halides. **Haloarenes** contain a halogen atom bonded to a benzene ring.

The van der Waals radius of fluorine is only slightly greater than that of hydrogen, and, among the other halogens, only iodine has a larger van der Waals radius than methyl.

Among alkanes and chloro-, bromo-, and iodoalkanes of comparable size and shape, the haloalkanes have the higher boiling points because of the greater polarizability of the unshared electrons of the halogen atom (Section 8.3B). Boiling points of fluoroalkanes are generally comparable to those of alkanes of comparable size and shape because of the uniquely low polarizability of the valence electrons of fluorine. The density of liquid haloalkanes is greater than that of hydrocarbons of comparable molecular weight because of the halogen's larger mass-to-volume ratio (Section 8.3C). Haloalkanes are prepared by halogenation of alkanes, among other reactions (Section 8.4). This reaction can give monohalogenation if the reaction is carried out with excess alkane. The regioselectivity is greater with bromination than with chlorination (Section 8.4A).

Halogenation proceeds by a **radical chain reaction** (Section 8.5A). A radical chain reaction consists of three types of steps: chain initiation, chain propagation, and chain termination (Section 8.5B). In **chain initiation,** radicals are formed from nonradical compounds. In a **chain propagation** step, a radical and a molecule react to give a new radical. When summed, chain propagation steps give the observed stoichiometry of the reaction. **Chain length** is the number of times a cycle of chain propagation steps repeats. In a **chain termination** step, radicals are destroyed. Simple alkyl radicals are planar or almost so with bond angles of 120° about the carbon with the unpaired electron. Heats of reaction for a radical reaction and for individual chain initiation, propaga-

tion, and termination steps can be calculated from bond dissociation enthalpies.

According to **Hammond's postulate** (Section 8.5D), the structure of the transition state of an exothermic reaction step occurs earlier and looks more like the reactant of that step than like the products, thus changes in the reactants have a large effect on the rate. Conversely, the structure of the transition state of an endothermic reaction step occurs later and looks more like the products of that step than like the reactants, and changes in products have a large effect on the rate. Hammond's postulate accounts for the fact that bromination of an alkane is more regioselective than chlorination. For both bromination and chlorination of alkanes, the rate-determining step is hydrogen abstraction to form an alkyl radical. Hydrogen abstraction is endothermic for bromination and exothermic for chlorination.

Allylic substitution is any reaction in which an atom or group of atoms is substituted for another atom or group of atoms at a carbon adjacent to a carbon-carbon double bond (Section 8.6). **Allylic halogenation** proceeds by a radical chain mechanism. Because of delocalization of electrons, the allyl radical is more stable than the *tert*-butyl radical.

Autoxidation of unsaturated compounds is an important process in aging and degradation of materials (Section 8.7).

Addition of HBr to alkenes in the presence of peroxides can lead to non-Markovnikov addition because of a change in mechanism to a radical chain (Section 8.8).

Key Reactions

1. Chlorination and Bromination of Alkanes (Section 8.4)

Chlorination and bromination of alkanes are regioselective in the order $3° \text{ H} > 2° \text{ H} > 1° \text{ H}$. Bromination has a higher regioselectivity than chlorination.

$$CH_3CH_2CH_3 + Br_2 \xrightarrow[\text{or light}]{\text{heat}} CH_3\overset{\overset{\displaystyle Br}{|}}{C}HCH_3 + CH_3CH_2CH_2-Br$$

$$92\% \qquad\qquad 8\%$$

2. Allylic Bromination (Section 8.6)

These reactions occur at high temperatures (heat is the radical initiator) using the halogens themselves. Bromination using *N*-bromosuccinimide (NBS) is initiated by light,

(racemic)

3. Autoxidation (Section 8.7)

Autoxidation involves reaction of a CH bond, especially an allylic one, with oxygen under radical initiation conditions. The primary product is a hydroperoxide.

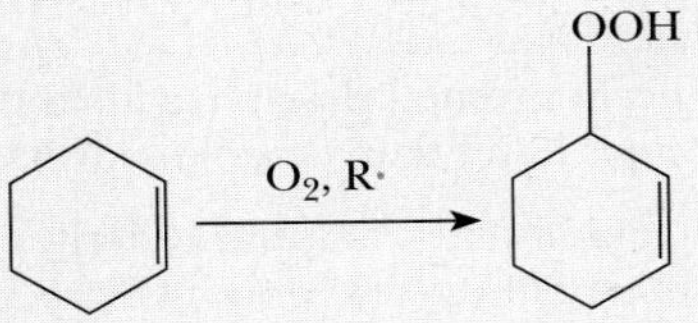

(racemic)

4. HBr Addition to Alkenes Under Radical Conditions

Non-Markovnikov addition of HBr to alkenes occurs by a radical mechanism in the presence of peroxides. The products are the opposite of the ordinary Markovnikov addition products that form under polar conditions.

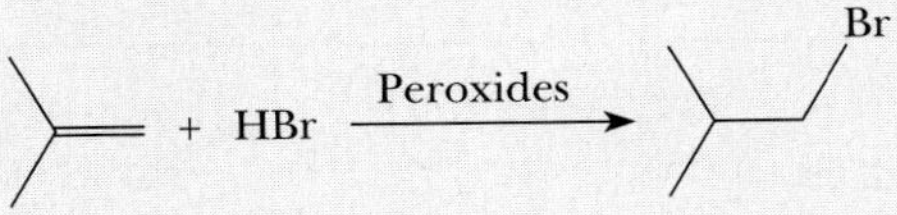

Problems

Nomenclature

8.8 Give IUPAC names for the following compounds. Where stereochemistry is shown, include a designation of configuration in your answer.

(a) 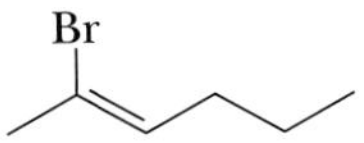

(b) 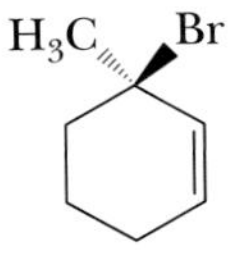

(c) 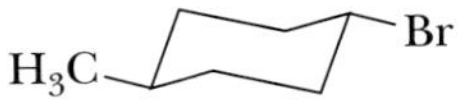

(d) 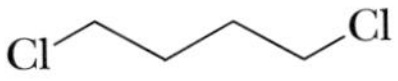

(e) 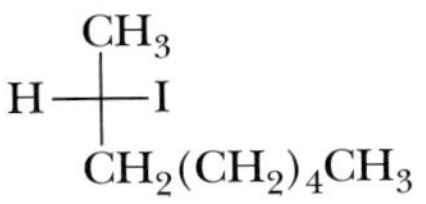

Fischer projection

(f) 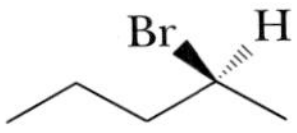

(g) 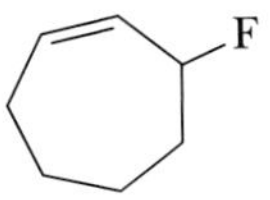

(h) 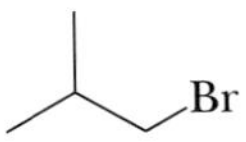

(i) 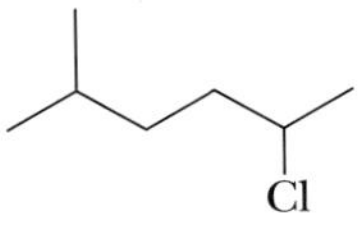

8.9 Draw structural formulas for the following compounds.

(a) 3-Iodo-1-propene
(b) (*R*)-2-Chlorobutane
(c) meso-2,3-Dibromobutane
(d) *trans*-1-Bromo-3-isopropylcyclohexane
(e) 1-Iodo-2,2-dimethylpropane
(f) Bromocyclobutane

Physical Properties

8.10 Water and dichloromethane are insoluble in each other. When each is added to a test tube, two layers form. Which layer is water and which is dichloromethane?

8.11 The boiling point of methylcyclohexane (C_7H_{14}, MW 98.2) is 101°C. The boiling point of perfluoromethylcyclohexane (C_7F_{14}, MW 350) is 76°C. Account for the fact that although the molecular weight of perfluoromethylcyclohexane is over three times that of methylcyclohexane, its boiling point is lower than that of methylcyclohexane.

8.12 Account for the fact that, among the chlorinated derivatives of methane, chloromethane has the largest dipole moment and tetrachloromethane has the smallest dipole moment.

Name	Molecular Formula	Dipole Moment (debyes, D)
Chloromethane	CH_3Cl	1.87
Dichloromethane	CH_2Cl_2	1.60
Trichloromethane	$CHCl_3$	1.01
Tetrachloromethane	CCl_4	0

Halogenation of Alkanes

8.13 Name and draw structural formulas for all possible monohalogenation products that might be formed in the following reactions.

(a) 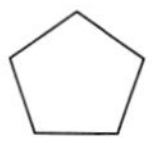 + Cl$_2$ $\xrightarrow{\text{light}}$

(b) 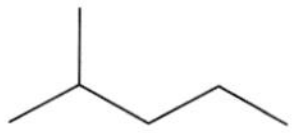 + Cl$_2$ $\xrightarrow{\text{light}}$

(c) 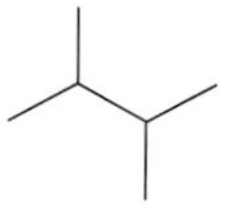 + Br$_2$ $\xrightarrow{\text{light}}$

8.14 Which compounds can be prepared in high yield by halogenation of an alkane?

(a) 2-Chloropentane
(b) Chlorocyclopentane
(c) 2-Bromo-2-methylheptane
(d) (R)-2-Bromo-3-methylbutane
(e) 2-Bromo-2,4,4-trimethylpentane
(f) Iodoethane

8.15 There are three constitutional isomers of molecular formula C_5H_{12}. When treated with chlorine at 300°C, isomer A gives a mixture of four monochlorination products. Under the same conditions, isomer B gives a mixture of three monochlorination products, and isomer C gives only one monochlorination product. From this information, assign structural formulas to isomers A, B, and C.

8.16 Following is a balanced equation for bromination of toluene.

$$C_6H_5CH_3 + Br_2 \longrightarrow C_6H_5CH_2Br + HBr$$

Toluene   Benzyl bromide

(a) Using the values for bond dissociation enthalpies given in Appendix 3, calculate ΔH^0 for this reaction.
(b) Propose a pair of chain propagation steps, and show that they add up to the observed reaction.
(c) Calculate ΔH^0 for each chain propagation step.
(d) Which propagation step is rate-determining?

8.17 Write a balanced equation and calculate ΔH^0 for reaction of CH_4 and I_2 to give CH_3I and HI. Explain why this reaction cannot be used as a method of preparation of iodomethane.

8.18 Following are balanced equations for fluorination of propane to produce a mixture of 1-fluoropropane and 2-fluoropropane.

$$CH_3CH_2CH_3 + F_2 \longrightarrow CH_3CH_2CH_2F + HF$$

  Propane   1-Fluoropropane

$$CH_3CH_2CH_3 + F_2 \longrightarrow CH_3\overset{\overset{\textstyle F}{|}}{C}HCH_3 + HF$$

  Propane   2-Fluoropropane

Assume that each product is formed by a radical chain mechanism.

(a) Calculate ΔH^0 for each reaction.
(b) Propose a pair of chain propagation steps for each reaction, and calculate ΔH^0 for each step.
(c) Reasoning from Hammond's postulate, predict the regioselectivity of radical fluorination relative to that of radical chlorination and bromination.

8.19 As you demonstrated in Problem 8.18, fluorination of alkanes is highly exothermic. As per Hammond's postulate, assume that the transition state for radical fluorination is almost identical to the starting material. With this assumption, estimate the fraction of each monofluoro product formed in the fluorination of 2-methylbutane.

8.20 Cyclobutane reacts with bromine to give bromocyclobutane, but bicyclo[1.1.0]butane reacts with bromine to give 1,3-dibromocyclobutane. Account for the differences between the reactions of these two compounds.

  Cyclobutane   Bromocyclobutane

  Bicyclo[1.1.0]butane   1,3-Dibromocyclobutane

8.21 The first chain propagation step of all radical halogenation reactions we considered in Section 8.5B was abstraction of hydrogen by the halogen atom to give an alkyl radical and HX, as for example

$$CH_3CH_3 + \cdot \ddot{B}\ddot{r}: \longrightarrow CH_3CH_2\cdot + H\ddot{B}\ddot{r}:$$

Suppose, instead, that radical halogenation occurs by an alternative pair of chain propagation steps, beginning with this step.

$$CH_3CH_3 + \cdot \ddot{B}\ddot{r}: \longrightarrow CH_3CH_2\ddot{B}\ddot{r}: + H\cdot$$

(a) Propose a second chain propagation step. Remember that a characteristic of chain propagation steps is that they add to the observed reaction.
(b) Calculate the heat of reaction, ΔH^0, for each propagation step.
(c) Compare the energetics and relative rates of the set of chain propagation steps in Section 8.5B with the set proposed here.

Allylic Halogenation

8.22 Following is a balanced equation for the allylic bromination of propene.

$$CH_2=CHCH_3 \ + \ Br_2 \longrightarrow CH_2=CHCH_2Br \ + \ HBr$$

(a) Calculate the heat of reaction, ΔH^0, for this conversion.
(b) Propose a pair of chain propagation steps, and show that they add up to the observed stoichiometry.
(c) Calculate the ΔH^0 for each chain propagation step, and show that they add up to the observed ΔH^0 for the overall reaction.

8.23 Using the table of bond dissociation enthalpies (Appendix 3), estimate the BDE of each indicated bond in cyclohexene.

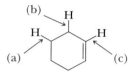

8.24 Propose a series of chain initiation, propagation, and termination steps for this reaction, and estimate its heat of reaction.

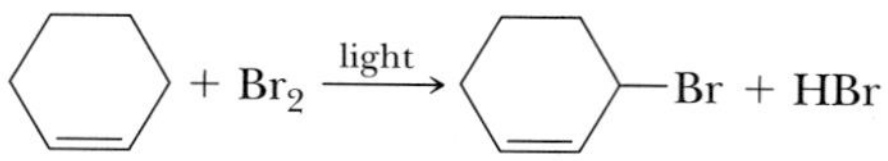

$$\bigcirc + Br_2 \xrightarrow{\text{light}} \bigcirc{-}Br + HBr$$

(racemic)

8.25 The major product formed when methylenecyclohexane is treated with NBS in dichloromethane is 1-(bromomethyl)-cyclohexene. Account for the formation of this product.

8.26 Draw the structural formula of the products formed when each alkene is treated with one equivalent of NBS in CH_2Cl_2 in the presence of light.

(a) $CH_3CH=CHCH_2CH_3$ **(b)**    **(c)**

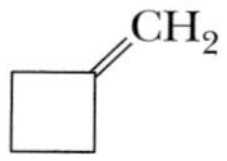

8.27 Calculate the ΔH^0 for the following reaction step. What can you say regarding the possibility of bromination at a vinylic hydrogen?

$$CH_2=CH_2 \ + \ Br \cdot \longrightarrow CH_2=CH \cdot \ + \ HBr$$

Synthesis

8.28 Show reagents and conditions to bring about these conversions, which may require more than one step.

(a)

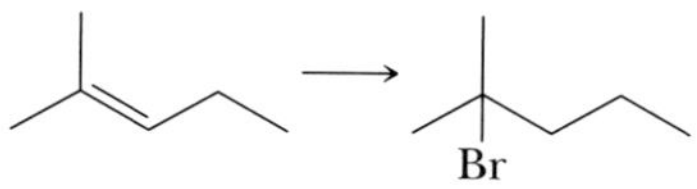

(b) $CH_3CH=CHCH_3 \longrightarrow CH_3CH=CHCH_2Br$

(c) $CH_3CH=CHCH_3 \longrightarrow CH_3CH-CHCH_3$
$$\qquad\qquad\qquad\qquad\qquad\quad \underset{Br}{|} \quad \underset{Br}{|}$$

(racemic)

(d) 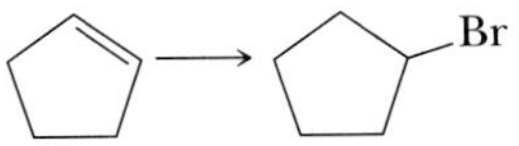
Br

(e) 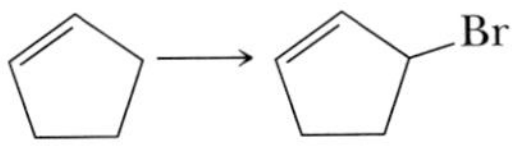
Br

(racemic)

(f) CH_4 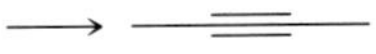 (Review section 7.5.)

(g) 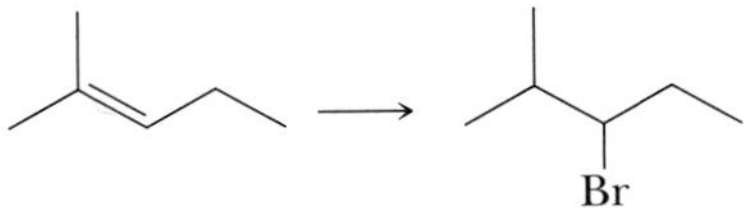
Br

(racemic)

Autoxidation

8.29 Predict the products of the following reactions. Where isomeric products are formed, label the major product.

(a) 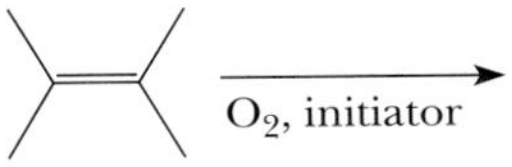
O₂, initiator

(b) 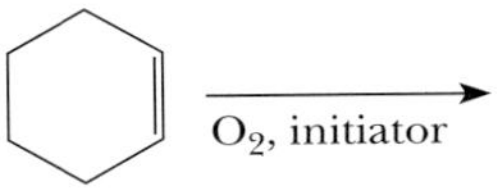
O₂, initiator

8.30 Give the major product of the following reactions.

(a) 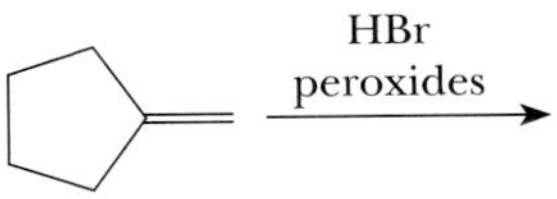
HBr
peroxides

(b) 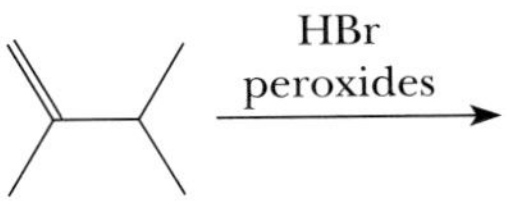
HBr
peroxides

Looking Ahead

8.31 A major use of the compound cumene is in the industrial preparation of phenol and acetone in the two-step synthesis, shown below.

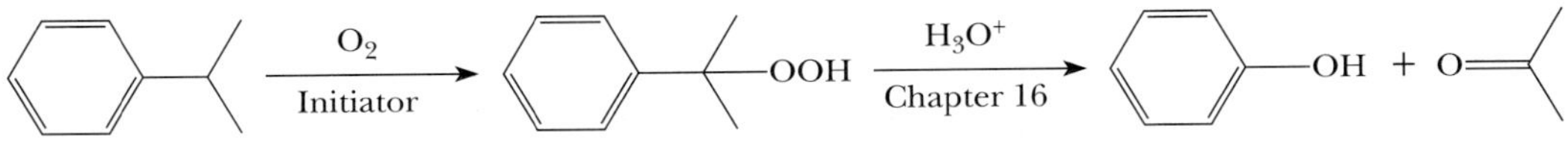
O₂
Initiator
OOH
H₃O⁺
Chapter 16
OH
+ O

Cumene Cumene hydroperoxide Phenol Acetone

Write a mechanism for the first step. We will see in Problem 16.63 how to complete the synthesis.

8.32 An important use of radical-chain reactions is in the polymerization of ethylene and sub-stituted ethylene monomers such as propene, vinyl chloride (the synthesis of which was discussed in Section 7.6 along with its use in the synthesis of poly(vinyl chloride), PVC), and styrene. The reaction for the formation of PVC, where n is the number of repeating units and is very large, follows.

$$R\cdot \ + \ CH_2{=}CHCl \ \xrightarrow{\quad} \ \text{Many steps} \ \xrightarrow[\substack{\text{Chain} \\ \text{termination} \\ \text{with } R\cdot'}]{\quad} \ \underset{\substack{\text{Poly(vinyl chloride)} \\ \text{(PVC)}}}{R{-}(CH_2{-}\overset{\overset{\textstyle Cl}{|}}{CH})_n{-}R'}$$

$$\underset{\text{Vinyl chloride}}{}$$

(a) Give a mechanism for this reaction (see Chapter 29).
(b) Give a similar mechanism for the formation of poly(styrene) from styrene. Which end of the styrene double bond would you expect $R\cdot$ to attack? Why?

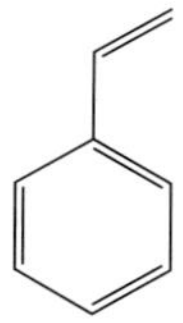

Styrene

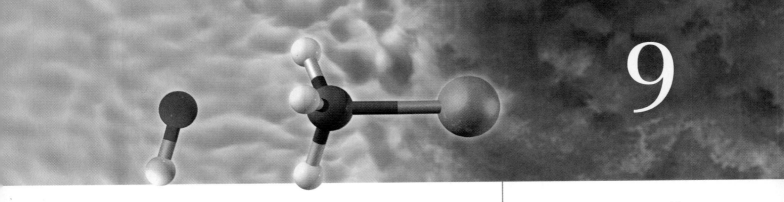

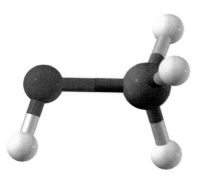

Hydroxide ion reacts with bromomethane (upper models) to give methanol and bromide ion (lower models) by an S_N2 mechanism (Section 9.3).

Outline

Nucleophilic Substitution and β-Elimination

Nucleophilic substitution refers to any reaction in which an electron-rich **nucleophile** (meaning nucleus loving) (Nu) replaces a **leaving group** (Lv). All nucleophiles are also Lewis bases (Section 4.6). With the exception of radical reactions (Chapter 8) and certain cyclic reactions (Chapter 24), essentially every reaction you will study involves a reaction of a Lewis acid with a Lewis base. In these reactions, the Lewis base, which is electron rich, reacts with the Lewis acid, which is electron poor. The Lewis acid is called an **electrophile** (meaning electron loving). The leaving group (Lv) can be a halide (X) or other electronegative group. Following is a general equation for nucleophilic substitution. Not all nucleophiles react with all electrophiles; the ability to recognize which do and which don't is part of what you should gain from this chapter.

Nucleophilic substitution Any reaction in which one nucleophile is substituted for another at a tetravalent carbon atom.

ORGANIC
Chemistry⚛Now™

Look for this logo in the chapter and go to Organic ChemistryNow at **http://now.brookscole.com/bfi4** for tutorials, simulations, and problems.

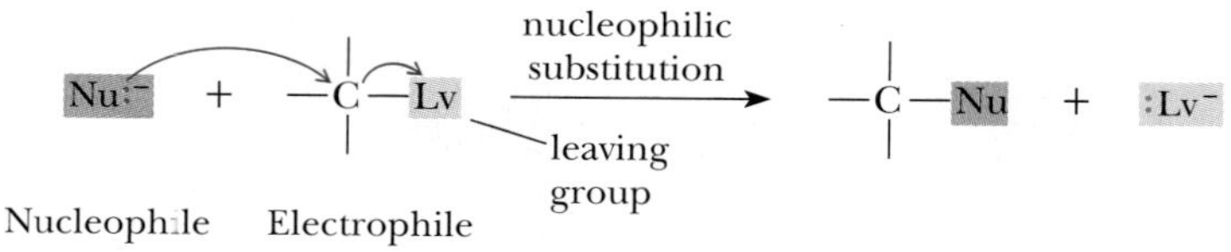

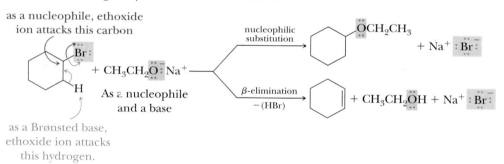

Nucleophile   Electrophile

An example of this reaction that you have already studied is the alkylation of terminal alkynes (Section 7.5A). Another is the reaction of hydroxide ion with chloromethane. In this reaction, chloromethane is the electrophile: Because of the electronegativity of chlorine, there is a partial positive charge on the carbon. An electrostatic potential map shows the negative electron density on HO⁻ (the nucleophile) interacting with the partial positive charge on the methyl group. Note that because the hydrogen and the electron pairs on HO⁻ are tetrahedrally arranged, the C, O, and H on the O are not in a straight line.

$$HO:^- + \delta^+ CH_3 - Cl \, \delta^- \longrightarrow CH_3 - OH + :Cl:^-$$

Nucleophile   Electrophile
(Lewis base)   (Lewis acid)

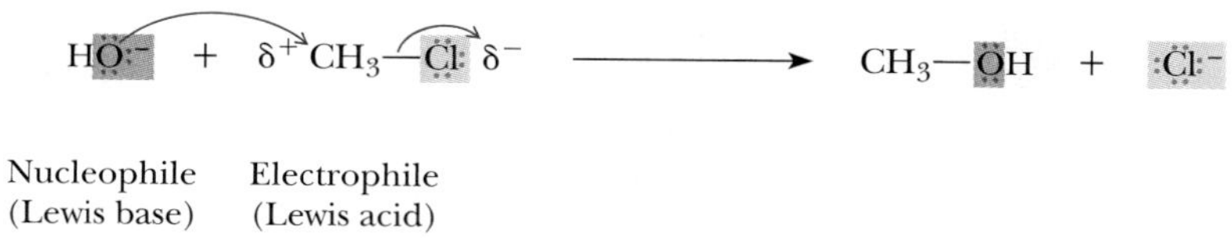

Nucleophiles are also Brønsted bases (Section 4.2), although some are very weak. The stronger ones can remove protons as well as attack at carbon centers. A reaction in which a halide and a hydrogen on the neighboring (β) carbon are removed is called **β-elimination.** Nucleophilic substitution and base-promoted β-elimination are therefore competing reactions. For example, ethoxide ion reacts with bromocyclohexane as a nucleophile to give ethoxycyclohexane (cyclohexyl ethyl ether), and as a Brønsted base to give cyclohexene and ethanol.

Electrostatic potential map showing the nucleophile (OH⁻) reacting at its negative (red) end with the electrophilic carbon (blue) in the reaction of hydroxide with chloromethane.

as a nucleophile, ethoxide ion attacks this carbon

As a nucleophile and a base

as a Brønsted base, ethoxide ion attacks this hydrogen.

nucleophilic substitution

β-elimination −(HBr)

+ Na⁺ :Br:

+ CH₃CH₂OH + Na⁺ :Br:

Nucleophile From the Greek, meaning nucleus loving. A molecule or ion that donates a pair of electrons to another atom or ion to form a new covalent bond; a Lewis base.

Leaving Group (Lv) The group that is displaced in a substitution or that is lost in an elimination.

Electrophile From the Greek, meaning nucleus loving. A molecule or ion that accepts a pair of electrons from another atom or molecule in a reaction; a Lewis acid.

β-Elimination A reaction in which a molecule, such as HCl, HBr, HI, or HOH, is split out or eliminated from adjacent carbons.

In this chapter, we study substitution and elimination. By using these reactions, we can convert haloalkanes to compounds with other functional groups including alcohols, ethers, thiols, sulfides, amines, nitriles, alkenes, and alkynes. Nucleophilic substitution and β-elimination open entirely new areas of organic chemistry and are a major method of interconverting functional groups. One of the most challenging aspects of the study of these reactions is deciding which of the two competing reactions is likely to prevail, and this will be the major focus of the last part of the chapter.

9.1 Nucleophilic Substitution in Haloalkanes

Nucleophilic substitution is one of the most important reactions of haloalkanes and can lead to a wide variety of new functional groups, many of which are illustrated in Table 9.1. Some of these reactions are efficient; others go only at elevated

Table 9.1 Some Nucleophilic Substitution Reactions

Reaction: $Nu:^- + CH_3Br: \longrightarrow CH_3Nu + :Br:^-$

Nucleophile	Product	Class of Compound Formed
$HO:^- \longrightarrow$	$CH_3\ddot{O}H$	An alcohol
$RO:^- \longrightarrow$	$CH_3\ddot{O}R$	An ether
$HS:^- \longrightarrow$	$CH_3\ddot{S}H$	A thiol (a mercaptan)
$RS:^- \longrightarrow$	$CH_3\ddot{S}R$	A sulfide (a thioether)
$HC{\equiv}C:^- \longrightarrow$	$CH_3C{\equiv}CH$	An alkyne
$:N{\equiv}C:^- \longrightarrow$	$CH_3C{\equiv}N:$	A nitrile
$:I:^- \longrightarrow$	$CH_3\ddot{I}:$	An alkyl iodide
$:\overset{-}{N}{=}\overset{+}{N}{=}\overset{\cdot\cdot}{N}:^- \longrightarrow$	$CH_3-\overset{+}{N}{=}N{=}\overset{\cdot\cdot}{N}:^-$	An alkyl azide
$:NH_3 \longrightarrow$	$CH_3NH_3^+$	An alkylammonium ion
$:\overset{\cdot\cdot}{O}-H$, $H \longrightarrow$	$CH_3\overset{\cdot\cdot+}{O}-H$, H	An alcohol (after proton transfer)
$:\overset{\cdot\cdot}{O}-CH_3$, $H \longrightarrow$	$CH_3\overset{\cdot\cdot+}{O}-CH_3$, H	An ether (after proton transfer)

temperatures, as we will see in later sections. As you study the entries in this table, note these points:

1. If the nucleophile is negatively charged, as for example OH^- and $HC{\equiv}C^-$, then the atom donating the pair of electrons in a substitution reaction becomes neutral in the product.
2. If the nucleophile is uncharged, as for example NH_3 and CH_3OH, then the atom donating the pair of electrons in the substitution reaction becomes positively charged in the product.
3. In the middle of the table are reactions involving $N{\equiv}C^-$ and $HC{\equiv}C^-$ nucleophiles. The products of these nucleophilic substitution reactions have new carbon-carbon bonds, as we have seen for alkynes in Section 7.5A. The formation of new carbon-carbon bonds is important in organic chemistry because it provides a means of extending a molecular carbon skeleton.

Example 9.1

Complete these nucleophilic substitution reactions. In each reaction, show all electron pairs on both the nucleophile and the leaving group.

(a) $\diagdown\diagup\diagdown$Br + $CH_3O^-Na^+ \xrightarrow[\text{methanol}]{}$ (b) $\diagdown\diagup\diagdown$Cl + $NH_3 \xrightarrow[\text{ethanol}]{}$

Solution

(a) Methoxide ion is the nucleophile, and bromide is the leaving group.

$$\overset{..}{\underset{..}{Br}}: \;+\; CH_3\overset{..}{\underset{..}{O}}:Na^+ \xrightarrow[\text{methanol}]{} \qquad \overset{..}{\underset{..}{O}}CH_3 \;+\; Na^+:\overset{..}{\underset{..}{Br}}:$$

| 1-Bromobutane | Sodium methoxide | 1-Methoxybutane (Butyl methyl ether) | Sodium bromide |

(b) Ammonia is the nucleophile, and chloride is the leaving group.

$$\overset{..}{\underset{..}{Cl}}: \;+\; :NH_3 \xrightarrow[\text{ethanol}]{} \qquad NH_3{}^+ \; :\overset{..}{\underset{..}{Cl}}:{}^-$$

| 1-Chlorobutane | Ammonia | Butylammonium chloride |

Problem 9.1

Complete the following nucleophilic substitution reactions. In each reaction, show all electron pairs on both the nucleophile and the leaving group.

(a)

$$\text{(cyclohexyl-Cl)} + CH_3\overset{O}{\overset{\|}{C}}O^-Na^+ \xrightarrow[\text{ethanol}]{}$$

(b)

$$\text{(2-iodobutane)} + \text{\ \ }S^-Na^+ \xrightarrow[\text{acetone}]{}$$

(c)

$$\text{(1-bromo-3-methylbutane)} + H_3C\!-\!C\!\equiv\!C^-Li^+ \xrightarrow[\text{dimethyl sulfoxide}]{}$$

As you will see, many factors affect the relative rates of nucleophilic substitutions and whether substitution or elimination reactions occur under a given set of conditions. These factors include solvent, structure of the substrate, the nucleophile, the leaving group, and temperature.

9.2 Solvents for Nucleophilic Substitution Reactions

Protic solvent A solvent that is a hydrogen-bond donor; the most common protic solvents contain —OH groups. Common protic solvents are water, low-molecular-weight alcohols such as ethanol, and low-molecular-weight carboxylic acids.

Aprotic solvent A solvent that cannot serve as a hydrogen-bond donor; nowhere in the molecule is there a hydrogen bonded to an atom of high electronegativity. Common aprotic solvents are dichloromethane, diethyl ether, and dimethyl sulfoxide.

Dielectric constant A measure of a solvent's ability to insulate opposite charges from one another.

Solvents dissolve reactants and provide the medium in which reactions take place. The properties of the solvent can have a dramatic influence over chemical reactions, especially nucleophilic substitution reactions.

Common solvents can be divided into two groups: **protic** and **aprotic.** Protic solvents are those with a relatively acidic proton such as an OH; aprotic solvents are those that lack a relatively acidic proton. Furthermore, solvents are classified as polar and nonpolar based on their **dielectric constant.** The greater the value of the dielectric constant of a solvent is, the smaller the interaction becomes between ions of opposite charge dissolved in it. As an arbitrary guideline, we say that a solvent is polar if it has a dielectric constant of 15 or greater. A solvent is nonpolar if it has a dielectric constant of less than 5. Solvents with a dielectric constant between 5 and 15 are borderline, and we will return to those later.

The common protic solvents for nucleophilic substitution reactions are water, low-molecular-weight alcohols, and low-molecular-weight carboxylic acids (Table 9.2). Each of these has a partially negatively charged oxygen bonded to a partially positively

Table 9.2 Common Protic Solvents

Solvent	Structure	Dielectric Constant (25°C)
Water	H_2O	79
Formic acid	HCOOH	59
Methanol	CH_3OH	33
Ethanol	CH_3CH_2OH	24
Acetic acid	CH_3COOH	6

charged hydrogen atom. Protic solvents solvate ionic substances by electrostatic interactions between anions and the partially positively charged hydrogens of the solvent and also between cations and partially negatively charged atoms of the solvent. By our guideline, water, formic acid, methanol, and ethanol are classified as polar protic solvents. Because of its smaller dielectric constant, acetic acid is classified as a moderately polar protic solvent.

The aprotic solvents most commonly used for nucleophilic substitution reactions are given in Table 9.3. Of these, dimethyl sulfoxide (DMSO), acetonitrile, *N,N*-dimethylformamide (DMF), and acetone are classified as polar aprotic solvents. Dichloromethane and tetrahydrofuran (THF) are moderately polar. Diethyl ether, toluene, and hexane are classified as nonpolar aprotic solvents. We discuss the role of aprotic solvents in Section 9.4B.

Table 9.3 Common Aprotic Solvents

Solvent	Structure	Dielectric Constant
Polar		
Dimethyl sulfoxide (DMSO)	$\overset{\displaystyle O}{\overset{\|}{CH_3SCH_3}}$	48.9
Acetonitrile	$CH_3C{\equiv}N$	37.5
N,N-Dimethylformamide (DMF)	$\overset{\displaystyle O}{\overset{\|}{HCN(CH_3)_2}}$	36.7
Acetone	$\overset{\displaystyle O}{\overset{\|}{CH_3CCH_3}}$	20.7
Moderately Polar		
Dichloromethane	CH_2Cl_2	9.1
Tetrahydrofuran (THF)	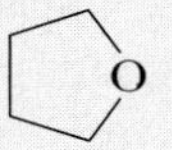 (cyclopentane ring with O)	7.6
Nonpolar		
Diethyl ether *ether = OR*	$CH_3CH_2OCH_2CH_3$	4.3
Toluene	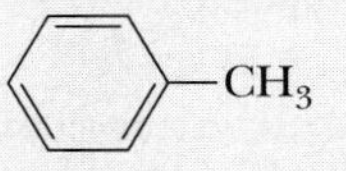 (benzene ring)—CH_3	2.3
Hexane	$CH_3(CH_2)_4CH_3$	1.9

Increasing solvent polarity ↑

ORGANIC **Chemistry Now™**
Click Tutorials for a summary of **Nucleophilic Reactions and S_N1 and S_N2 Mechanisms**

9.3 Mechanisms of Nucleophilic Aliphatic Substitution

On the basis of a wealth of experimental observations developed over a 70-year period, two limiting mechanisms for nucleophilic substitutions have been proposed. A fundamental difference between them is the timing of bond breaking between carbon and the leaving group and of bond forming between carbon and the nucleophile. At one extreme, the two processes are concerted, meaning that bond breaking and bond forming occur simultaneously. Thus, departure of the leaving group is assisted by the incoming nucleophile. This mechanism is designated **S_N2**. Here S stands for Substitution, N for Nucleophilic, and 2 for a **bimolecular reaction.** This type of substitution reaction is classified as bimolecular because both the haloalkane and the nucleophile are involved in the rate-determining step.

S_N2 reaction A bimolecular nucleophilic substitution reaction.

Bimolecular reaction A reaction in which two species are involved in the rate-determining step.

Following is an S_N2 mechanism for the reaction of hydroxide ion and bromomethane to form methanol and bromide ion.

Mechanism *An S_N2 Reaction*

The nucleophile attacks the reactive center from the side opposite the leaving group; that is, an S_N2 reaction involves backside attack of the nucleophile. In the following diagram, the dashed lines in the transition state represent partially formed or broken bonds.

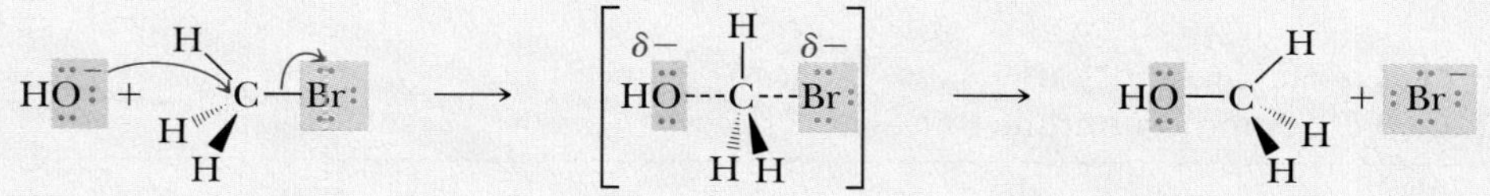

Transition state with simultaneous bond breaking and bond forming

Backside attack by the nucleophile is facilitated in two ways. First, because of the polarization of the C—Br bond, the backside of the carbon atom has a partial positive charge and therefore attracts the electron-rich nucleophile (as shown for methyl chloride at the beginning of the chapter). Second, the electron density of the nucleophile entering from the backside assists in breaking the carbon-bromine bond, thereby helping the bromide leave. In theoretical terms, the electron density of the nucleophile attacking from the backside can be thought of as populating the antibonding orbital of the carbon-bromine bond, weakening it as the new oxygen-carbon sigma bond becomes stronger. Other reaction geometries do not involve orbital overlap that leads to weakening of the C–leaving group bond, so they are higher in energy and not observed. Backside attack has important stereochemical consequences, as we shall see in Section 9.4C.

Figure 9.1 shows an energy diagram for an S_N2 reaction. Note that because there is just a single step in the S_N2 mechanism, the energy diagram has one energy barrier that corresponds to the transition state.

S_N1 reaction A unimolecular nucleophilic substitution reaction.

Unimolecular reaction A reaction in which only one species is involved in the rate-determining step.

In the other limiting mechanism, called S_N1, bond breaking between carbon and the leaving group is entirely completed before bond forming with the nucleophile begins. In the designation **S_N1**, 1 stands for **unimolecular** reaction. This type of substitution

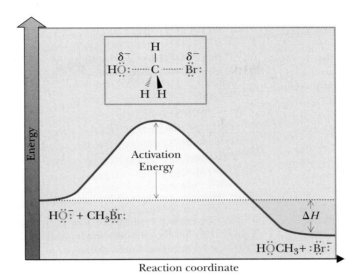

ORGANIC
Chemistry·⚛·**Now**™
Active Figure 9.1
An energy diagram for an S_N2 reaction. There is one transition state and no reactive intermediate. **See a simulation based on this figure, and take a short quiz on the concepts.**

is classified as unimolecular because only the haloalkane is involved in the rate-determining step. An S_N1 mechanism is illustrated by the **solvolysis** of 2-bromo-2-methylpropane (*tert*-butyl bromide) in methanol to form 2-methoxy-2-methylpropane (*tert*-butyl methyl ether) and HBr. In this reaction, the nucleophile (methanol) is the solvent, hence the name, **solvolysis.** The last step in this three-step mechanism is a proton-transfer reaction following the S_N1 reaction. The stereochemical consequences of the S_N1 mechanism are addressed in Section 9.8C.

Solvolysis A nucleophilic substitution in which the solvent is also the nucleophile.

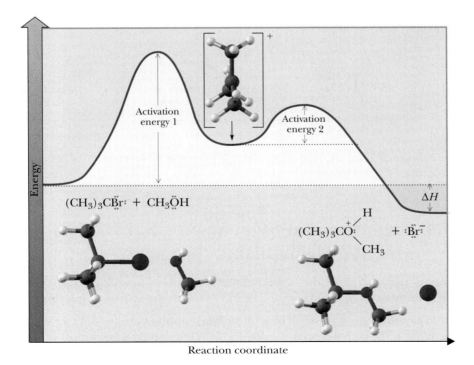

ORGANIC
Chemistry·⚛·**Now**™
Active Figure 9.2
An energy diagram for an S_N1 reaction. There is one transition state leading to formation of the carbocation intermediate in Step 1 and a second transition state for the reaction of the carbocation intermediate with methanol in Step 2 to give an oxonium ion. Step 1 crosses the higher energy barrier and, therefore, is rate-determining. **See a simulation based on this figure, and take a short quiz on the concepts.**

Mechanism *An S_N1 Reaction*

Step 1: Ionization of the C—Lv (Lv = Br) bond forms a carbocation intermediate. Because no nucleophile is assisting the departure of the halide anion, this is the relatively slow, rate-determining step of the reaction.

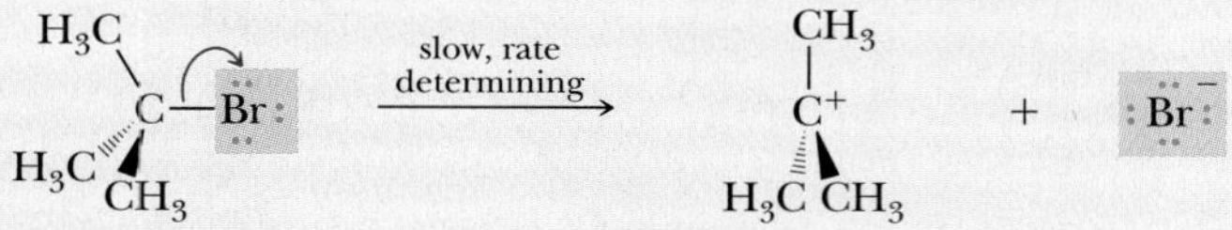

A carbocation intermediate;
its shape is trigonal planar

Step 2: Reaction of the carbocation intermediate (an electrophile) with methanol (a nucleophile) gives an oxonium ion. Attack of the nucleophile occurs with equal probability from either face of the planar carbocation intermediate.

Nucleophile   Electrophile   An oxonium ion

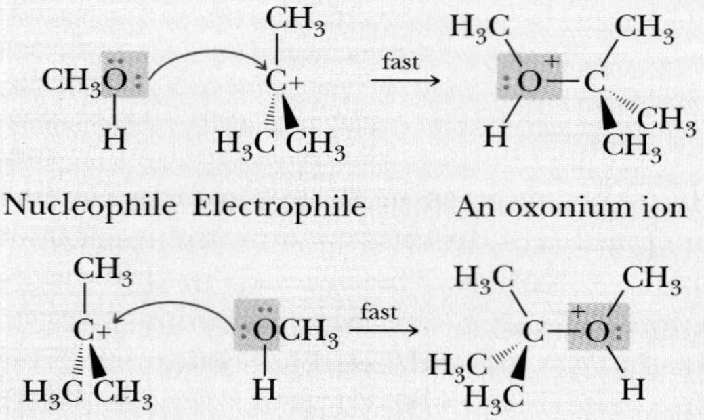

Step 3: Proton transfer from the oxonium ion to methanol completes the reaction and gives *tert*-butyl methyl ether.

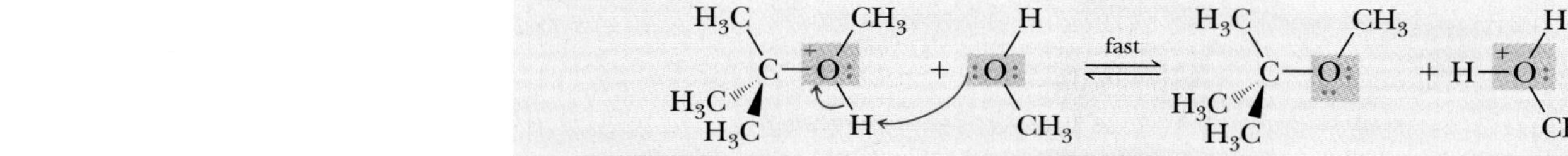

Figure 9.2 shows an energy diagram for the S_N1 reaction of 2-bromo-2-methyl-propane with methanol.

9.4 Experimental Evidence for S_N1 and S_N2 Mechanisms

Let us now examine the experimental evidence on which these two contrasting mechanisms are based. As we do, we will consider the following questions.

1. What effect does the structure of the nucleophile have on the rate of reaction?
2. What is the stereochemical outcome of nucleophilic substitution when the leaving group is displaced from a chiral center?

3. What effect does the structure of the haloalkane have on the rate of reaction?
4. What effect does the structure of the leaving group have on the rate of reaction?
5. What is the role of the solvent?
6. Under what conditions are skeletal rearrangements observed?

A. Kinetics

The kinetic order of nucleophilic substitutions can be studied by measuring the effect on rate of varying the concentrations of haloalkane and nucleophile. Those reactions whose rate is dependent only on the concentration of haloalkane are classified as S_N1; those reactions whose rate is dependent on the concentration of both haloalkane and nucleophile are classified as S_N2.

Because the transition state for formation of the carbocation intermediate in an S_N1 mechanism involves only the haloalkane and not the nucleophile, it is a unimolecular process. The result is a first-order reaction. In this instance, the rate of reaction is expressed as the rate of disappearance of the starting material, 2-bromo-2-methylpropane.

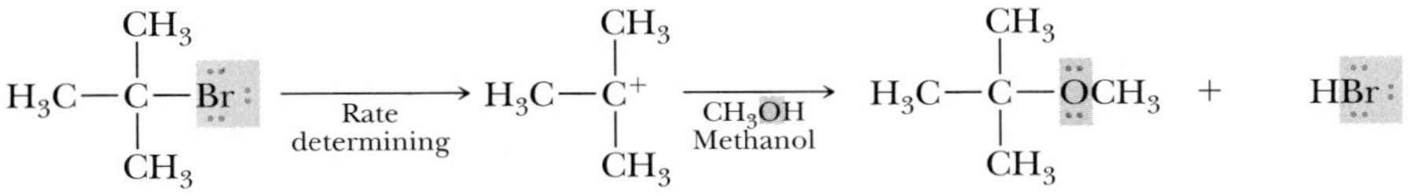

2-Bromo-2-methylpropane
(*tert*-Butyl bromide)

2-Methoxy-2-methylpropane

$$\text{Rate} = -\frac{d[(CH_3)_3CBr]}{dt} = k[(CH_3)_3CBr]$$

By contrast, there is only one step in the S_N2 mechanism. For the reaction of OH^- and CH_3Br, for example, both species must collide and are present in the transition state; that is, the reaction is bimolecular. The reaction between CH_3Br and NaOH to give CH_3OH and NaBr is second order: it is first order in CH_3Br and first order in OH^- so doubling the concentration of either increases the rate by a factor of two.

$$CH_3Br + Na^+OH^- \longrightarrow CH_3OH + Na^+Br^-$$

Bromomethane   Methanol

$$\text{Rate} = -\frac{d[CH_3Br]}{dt} = k[CH_3Br][OH^-]$$

Example 9.2

The reaction of *tert*-butyl bromide with azide ion (N_3^-) in methanol is a typical S_N1 reaction. What happens to the rate of the reaction if $[N_3^-]$ is doubled?

Solution

The rate remains the same because the nucleophile concentration does not appear in the rate equation in an S_N1 reaction.

Problem 9.2

The reaction of methyl bromide with azide ion (N_3^-) in methanol is a typical S_N2 reaction. What happens to the rate of the reaction if $[N_3^-]$ is doubled?

B. Structure of the Nucleophile

Nucleophilicity   A kinetic property measured by the rate at which a nucleophile causes nucleophilic substitution on a reference compound under a standardized set of experimental conditions.

Nucleophilicity is a kinetic property and is measured by relative rates of reaction. Relative nucleophilicities for a series of nucleophiles are established by measuring the rate at which each displaces a leaving group from a haloalkane, for example, the rate at which each displaces bromide ion from bromoethane in ethanol at 25°C.

$$CH_3CH_2\overset{..}{\underset{..}{Br}}: \; + :NH_3 \; \xrightarrow[\text{Ethanol}]{25°C} \; CH_3CH_2\overset{+}{N}H_3 \; + \; :\overset{..}{\underset{..}{Br}}:^-$$

From these studies, we can then make correlations between the structure of a nucleophile and its relative nucleophilicity. Listed in Table 9.4 are the types of nucleophiles we deal with most commonly in this text and their nucleophilicities in alcohol or water. The more rapidly a nucleophile reacts with a substrate in an S_N2 reaction, the more nucleophilic it is (the "better a nucleophile"), by definition.

Basicity   An equilibrium property measured by the position of equilibrium in an acid-base reaction, as for example the acid-base reaction between ammonia and water.

Because all nucleophiles are Brønsted bases as well (see Chapter 4), we also study correlations between nucleophilicity and **basicity.** Basicity and nucleophilicity are related because they both involve reaction at an electron pair, but they are not the same and should not be used interchangeably. Basicity also has a strong (inverse) correlation with leaving group ability, as we shall see in Section 9.4F.

The solvent in which nucleophilic substitutions are carried out has a marked effect on relative nucleophilicities. For a fuller understanding of the role of the solvent,

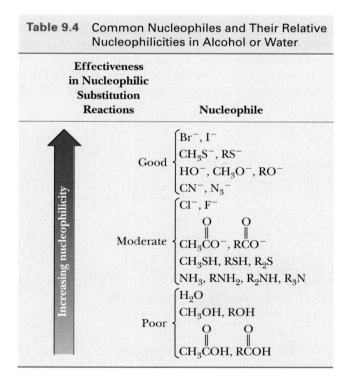

Table 9.4   Common Nucleophiles and Their Relative Nucleophilicities in Alcohol or Water

let us consider nucleophilic substitution reactions carried out in polar aprotic solvents and in polar protic solvents. An organizing principle for substitution reactions is the following: All other factors being equal, the stronger the interaction of the nucleophile with the solvent is, the lower its nucleophilicity will be. Conversely, the less the nucleophile interacts with solvent, the greater its nucleophilicity will be.

Nucleophilicity in Polar Aprotic Solvents

The most commonly used polar aprotic solvents (DMSO, acetone, acetonitrile, and DMF) are very effective in solvating cations (in addition to the attraction to the negative end of the dipole, the lone pairs on oxygen and nitrogen act as Lewis bases) but are not nearly as effective in solvating anions. Consider, for example, acetone. Because the negative end of its dipole and the lone pairs on oxygen can come close to the center of positive charge in a cation, acetone is effective in solvating cations. The positive end of its dipole, however, is shielded by the two methyl groups and is, therefore, less effective in solvating anions. The sodium ion of sodium iodide, for example, is effectively solvated by acetone and DMSO, but the iodide ion is only poorly solvated. Because anions are only poorly solvated in polar aprotic solvents, they are more free to participate in nucleophilic substitution reactions. Their relative nucleophilicities parallel their relative basicities. The relative nucleophilicities of halide ions in polar aprotic solvents, for example, are $F^- > Cl^- > Br^- > I^-$.

Solvation of NaI in acetone.

Nucleophilicity in Polar Protic Solvents

The relative nucleophilicities of halide ions in polar protic solvents are quite different from those in polar aprotic solvents (Table 9.5).

In polar protic solvents, iodide ion, the least basic of the halide ions, has the greatest nucleophilicity. Conversely, fluoride ion, the most basic of the halide ions, has the smallest nucleophilicity. The reason for this reversal of correlation between nucleophilicity and basicity lies in the degree of solvation of anions in protic solvents compared with aprotic solvents.

- In polar aprotic solvents, anions are only weakly solvated and, therefore, relatively free to participate in nucleophilic substitution reactions.

Table 9.5	Relative Nucleophilicities of Halide Ions in Aprotic and Protic Solvents
Solvent	**Increasing nucleophilicity** →
Polar aprotic	$I^- < Br^- < Cl^- < F^-$
Polar protic	$F^- < Cl^- < Br^- < I^-$

Table 9.6	Relative Nucleophilicities of Atoms Within a Row	
Period	**Increasing nucleophilicity** →	
Period 2	$F^- < OH^- < NH_2^- < CH_3^-$	
Period 3	$Cl^- < SH^- < PH_2^-$	

- In polar protic solvents, anions are highly solvated by hydrogen bonding with solvent molecules and, therefore, are less able to participate in nucleophilic substitution reactions.

The negative charge on the fluoride ion, the smallest of the halide ions, is concentrated in a small volume, and the very tightly held solvent shell constitutes a barrier between fluoride ion and substrate. The fluoride ion must be at least partially removed from its tightly held solvation shell before it can participate in nucleophilic substitution. The following figure shows a fluoride ion hydrogen-bonded by methanol (several solvent molecules are probably involved). The negative charge on the iodide ion, the largest of the halide ions, is far less concentrated, the solvent shell is less tightly held, and iodide is considerably freer to participate in nucleophilic substitution reactions.

We can make the following additional generalizations about nucleophilicity:

1. Within a row of the Periodic Table, nucleophilicity increases from left to right (Table 9.6); that is, it increases with basicity.
2. In a series of reagents with the same attacking atom, anionic reagents are stronger nucleophiles than neutral reagents (Table 9.7). This trend also parallels the basicity of the nucleophile.
3. When comparing groups of reagents in which the nucleophilic atom is the same, the stronger the base is, the greater the nucleophilicity will be. The oxygen nucleophiles in Table 9.8 are listed in order of increasing nucleophilicity. Below each, for comparison, is given the formula and pK_a of its conjugate acid. In this series,

Table 9.7	The Effect of Charge on Nucleophilicity
Increasing nucleophilicity →	
$H_2O < OH^-$	
$ROH < RO^-$	
$NH_3 < NH_2^-$	
$RSH < RS^-$	

Table 9.8 Correlation of Nucleophilicity and Basicity for Reagents with the Same Attacking Atom

Nucleophile	RCOO⁻ Carboxylate ion	HO⁻ Hydroxide ion	RO⁻ Ethoxide ion
		Increasing nucleophilicity →	
Conjugate acid	RCOOH	HOH	ROH
pK_a	4–5	15.7	16–18
	← Increasing acidity		

the carboxylic acid is the strongest acid; consequently, its anion is the weakest base and the poorest nucleophile.

4. In addition, the more polarizable a nucleophile is, the stronger its nucleophilicity. Polarizability is a measure of the ability of an electric field to distort the electrons around an atom. Thus, anions with weakly bound electrons such as azide ion or iodide ion are particularly nucleophilic.

C. Stereochemistry

S_N1 Reactions

Experiments in which nucleophilic substitution takes place at a chiral center provide us with information about the stereochemical outcome of the reaction. One of the compounds studied to determine the stereochemistry of an S_N1 reaction was the following chloroalkane. When either enantiomer of this molecule undergoes nucleophilic substitution by an S_N1 pathway, the product is racemic. The reason is that ionization of this secondary chloride forms an achiral carbocation. Attack of the nucleophile can occur from either side. Attack from the right gives the R enantiomer; attack from the left gives the S enantiomer. The R and S enantiomers are formed in equal amounts, and the product is a racemic mixture.

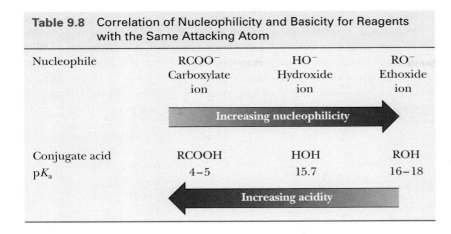

| R enantiomer | Planar carbocation (achiral) | S enantiomer | R enantiomer |

A racemic mixture

The S_N1 mechanism as initially described requires complete racemization of any product where the carbon at which substitution takes place is a chiral center. Although examples of complete racemization have been observed, it is common to find only partial racemization, with the predominant product being the one with inversion of configuration at the chiral center. Although bond breaking between carbon and the leaving group is complete, the leaving group (chloride ion in this example) remains associated for a short time with the carbocation in an ion pair.

Approach of the nucleophile from this side is less hindered.

Approach of the nucleophile from this side is partially blocked by chloride ion, which remains associated with the carbocation as an ion pair.

To the extent that the leaving group remains associated with the carbocation as an ion pair, it hinders approach of the nucleophile from that side of the carbocation. The result is that somewhat more than 50% of the product is formed by attack of the nucleophile from the side of the carbocation opposite that of the leaving group.

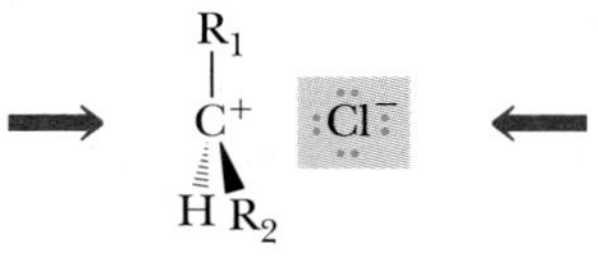

S_N2 Reactions

Every S_N2 reaction proceeds with backside attack by the nucleophile and therefore inversion of configuration. This was shown in an ingenious experiment designed by the English chemists E. D. Hughes and C. K. Ingold. They studied the exchange reaction between enantiomerically pure 2-iodooctane and iodine-131, a radioactive isotope of iodine. Iodine-127, the naturally occurring isotope of iodine, is stable and does not undergo radioactive decay.

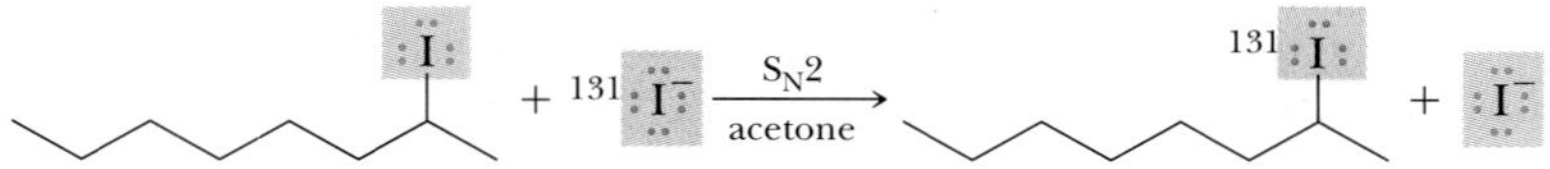

2-Iodooctane

Hughes and Ingold first demonstrated that the reaction is second order: first order in 2-iodooctane and first order in iodide ion. Therefore, the reaction proceeds by an S_N2 mechanism. They observed further that the rate of racemization of enantiomerically pure 2-iodooctane is exactly twice the rate of incorporation of iodine-131. This observation must mean, they reasoned, that each displacement of iodine-127 by iodine-131 proceeds with inversion of configuration, as illustrated in the following equation.

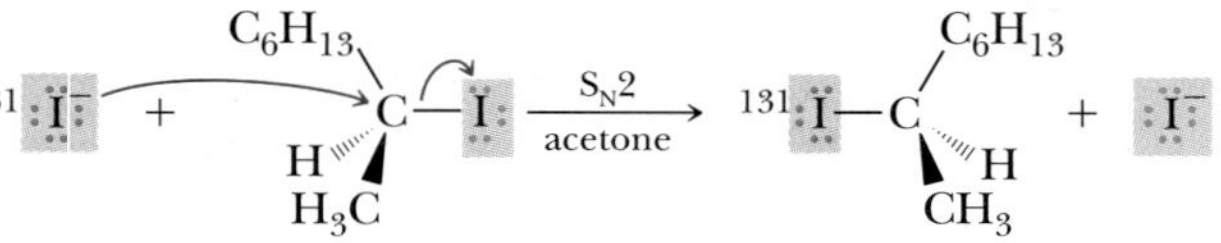

(*S*)-2-Iodooctane (*R*)-2-Iodooctane

Substitution with inversion of configuration in one molecule cancels the rotation of one molecule that has not reacted so that, for each molecule undergoing inversion, one racemic pair is formed. Inversion of configuration in 50% of the molecules results in 100% racemization.

Example 9.3

Complete these S_N2 reactions, showing the configuration of each product.

(a) [structure: cyclopentane ring with CH₃ (R) and Br (R)] + $CH_3COO^- Na^+$ ⟶   **(b)** [structure: chain with Br at R center] + CH_3NH_2 ⟶

Solution

S_N2 reactions occur with inversion of configuration at the chiral center. In (a), the starting material is the R,R diastereomer; the product is the R,S diastereomer. In (b), the starting material is the R enantiomer; the product is the S enantiomer.

(a) [structure: cyclopentane ring with CH₃ (R) and OCCH₃ ester group (S)]   **(b)** $CH_3{-}\overset{H}{\underset{}{N}}{-}H$ Br^- [chain, S configuration]

Problem 9.3

Complete these S_N2 reactions, showing the configuration of each product.

(a) [structure: H₃C substituted cyclohexane with Br] + $Na^+ N_3^-$ ⟶   **(b)** [structure: benzene ring with CHBr CH₃] + $CH_3S^- Na^+$ ⟶

D. Structure of the Haloalkane

The rates of S_N1 reactions are governed mainly by electronic factors, namely the relative stabilities of carbocation intermediates. The rates of S_N2 reactions, on the other hand, are governed mainly by steric factors and their transition states are particularly sensitive to **steric hindrance** about the site of reaction.

Steric hindrance The ability of groups, because of their size, to hinder access to a reaction site within a molecule.

Relative Stabilities of Carbocations

As we learned in Section 6.3A, 3° carbocations are the most stable (lowest activation energy for their formation) as a result of hyperconjugation, whereas 1° carbocations are the least stable (highest activation energy for their formation). In fact, 1° carbocations are so unstable that they rarely if ever are observed in solution. Because carbocations are high-energy intermediates, the transition states for their formation are very similar to the carbocation in energy (Hammond postulate, Section 8.5D). Therefore, 3° haloalkanes are most likely to react by carbocation formation; 2° haloalkanes are less likely to react by carbocation formation, and methyl and 1° haloalkanes react in this manner only when they are specially stabilized.

Steric Hindrance

To complete a substitution reaction, the nucleophile must approach the substitution center and begin to form a new covalent bond to it. If we compare the ease of approach to the substitution center of a 1° haloalkane with that of a 3° haloalkane, we see that the approach is considerably easier for bromoethane than for *tert*-butyl bromide. Two hydrogen atoms and one alkyl group screen the backside of the substitution center of a 1° haloalkane. In contrast, three alkyl groups screen the backside of the 3° haloalkane.

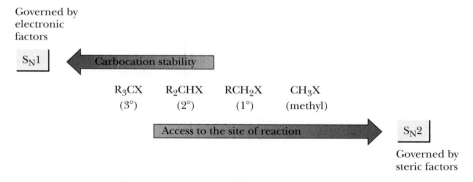

Bromoethane
(Ethyl bromide)

2-Bromo-2-methylpropane
(*tert*-Butyl bromide)

Given the competition between electronic and steric factors, we find that tertiary haloalkanes react by an S_N1 mechanism because tertiary carbocation intermediates are particularly stable and tertiary haloalkanes are very hindered toward backside attack; tertiary haloalkanes are never observed to react by an S_N2 mechanism. Halomethanes and primary haloalkanes have little crowding around the reaction site and react by an S_N2 mechanism; they are never observed to react by an S_N1 mechanism because methyl and primary carbocations are so unstable. Secondary haloalkanes may react by either S_N1 or S_N2 mechanisms, depending on the nucleophile and solvent. The competition between electronic and steric factors and their effects on relative rates of nucleophilic substitution reactions of haloalkanes are summarized in Figure 9.3.

We see a similar effect of steric hindrance on S_N2 reactions in molecules with branching at the β-carbon. The carbon bearing the halogen in a haloalkane is called the α-carbon, and the next carbon is called the β-carbon. Table 9.9 shows relative rates of S_N2 reactions on a series of primary bromoalkanes. In these data, the rate of nucleophilic substitution of bromoethane is taken as a reference and is

Figure 9.3
Effect of steric and electronic factors in competition between S_N1 and S_N2 reactions of haloalkanes. Methyl and primary haloalkanes react only by the S_N2 mechanism; they do not react by S_N1. Tertiary haloalkanes do not react by S_N2; they react only by S_N1. Secondary haloalkanes may be made to react by either S_N1 or S_N2 mechanisms depending on the solvent and the choice of nucleophile.

Table 9.9   Effect of β-Branches on the Rate of S_N2 Reactions

Alkyl bromide	⤵Br$_\beta$	⤵Br$_\beta$	⤵Br$_\beta$	⤵Br$_\beta$
β-Branches	0	1	2	3
Relative rate	1.0	4.1×10^{-1}	1.2×10^{-3}	1.2×10^{-5}

given the value 1.0. As CH_3 branches are added to the β-carbon, the relative rate of reaction decreases. Compare the relative rates of bromoethane (no β-branch) with that of 1-bromo-2,2-dimethylpropane (neopentyl bromide), a compound with three β-branches. The rate of S_N2 substitution of this compound is only 10^{-5} that of bromoethane. For all practical purposes, primary halides with three beta branches do not undergo S_N2 reactions.

As shown in Figure 9.4, the carbon of the C—Br bond in bromoethane is unhindered and open to attack by a nucleophile in an S_N2 reaction. The corresponding carbon in neopentyl bromide, however, is screened by the three β-methyl groups. Thus, although the carbon bearing the leaving group is primary, the approach of the nucleophile is so hindered that the rate of S_N2 reaction of neopentyl bromide is greatly reduced compared to bromoethane.

Steric hindrance can also be important for the nucleophile. The less steric hindrance there is around the nucleophilic atom, the better a nucleophile will be. Lack of steric hindrance is one reason why linear nucleophiles such as azide (N_3^-) and acetylide ions are unusually good nucleophiles in S_N2 reactions.

In looking at nucleophilic substitution reactions, it is important to realize that the relative energies of both the S_N1 and S_N2 transition states depend on the structures of both nucleophile and substrate. For methyl and primary halides, the S_N2 transition state is much lower in energy, while for tertiary halides, the S_N1 transition state leading to the carbocation intermediate is the lower. For secondary halides, the outcome depends on the details of the solvent and nucleophile.

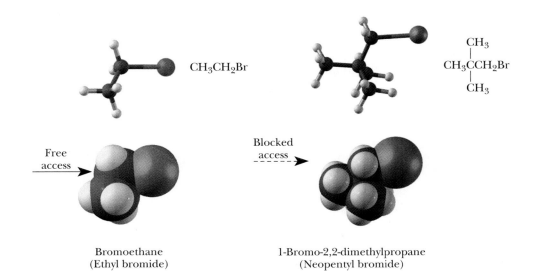

CH$_3$CH$_2$Br

$$CH_3CCH_2Br$$
with CH$_3$ above and CH$_3$ below the central C

Free access

Blocked access

Bromoethane
(Ethyl bromide)

1-Bromo-2,2-dimethylpropane
(Neopentyl bromide)

Figure 9.4

The effect of β-branching in S_N2 reactions on a primary haloalkane. With bromoethane, attack of the nucleophile is unhindered. With 1-bromo-2,2-dimethylpropane, the three β-branches block approach of the nucleophile to the backside of the C—Br bond, thus drastically reducing the rate of S_N2 reaction of this compound.

E. Allylic Halides

Allylic Next to a carbon-carbon double bond.

Allylic carbocation A carbocation in which an allylic carbon bears the positive charge.

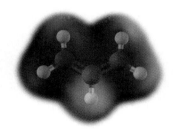

■ Electrostatic potential map for the allyl cation. The positive charge (blue) is on carbons 1 and 3.

At this point, we need to introduce a new type of carbocation, namely **allylic carbocations.** Allylic carbocations, like allylic radicals (Section 8.6) have a double bond next to the electron-deficient carbon. The **allyl cation** is the simplest allylic carbocation. Because the allyl cation has only one substituent on the carbon bearing the positive charge, it is a primary allylic carbocation.

We discussed the special stabilization of allylic radicals in Section 8.6. Allylic carbocations are also considerably more stable than comparably substituted alkyl carbocations because of the resonance interaction between the positively charged carbon and the adjacent vinyl group. The allyl cation, for example, can be represented as a hybrid of two equivalent resonance structures. The result is that the positive charge appears only on carbons 1 and 3, as shown in the accompanying electrostatic potential map.

$$CH_2{=\!\!=}CH{-}\overset{+}{C}H_2 \longleftrightarrow \overset{+}{C}H_2{-}CH{=\!\!=}CH_2$$

Allyl cation
(a hybrid of two equivalent contributing structures)

Because of this delocalization of the pi electrons of the double bond and the positive charge in the hybrid, the allyl cation is considerably more stable than a primary carbocation. Recall that, in general, a distributed charge in a molecule is stabilizing compared to the situation in which a charge is more localized. It has been determined experimentally that the double bond of one adjacent vinyl group provides approximately as much stabilization as two alkyl groups. Thus, the allyl cation and isopropyl cation are of comparable stability.

These cations are of comparable stability $$CH_2{=\!\!=}CH{-}\overset{+}{C}H_2 \qquad CH_3{-}\overset{+}{C}H{-}CH_3$$

1° Allylic cation   2° Alkyl cation

The classification of allylic cations as 1°, 2°, and 3° is determined by the location of the positive charge in the more important contributing structure. Following are examples of 2° and 3° allylic carbocations.

$$CH_2{=\!\!=}CH{-}\overset{+}{C}H{-}CH_3 \qquad CH_2{=\!\!=}CH{-}\overset{+}{\underset{\underset{\displaystyle CH_3}{|}}{C}}{-}CH_3$$

A 2° allylic carbocation   A 3° allylic carbocation

Benzylic carbocations show approximately the same stability as allylic carbocations. Both are stabilized by resonance delocalization of the positive charge owing to adjacent pi bonds.

$$\text{⟨benzene ring⟩}{-}CH_2{}^+ \qquad\qquad C_6H_5{-}CH_2{}^+$$

Benzyl cation   The benzyl cation is also written
(a benzylic carbocation)   in this abbreviated form

In Section 6.3A, we presented the order of stability of methyl, 1°, 2°, and 3° carbocations. We can now expand this order to include 1°, 2°, and 3° allylic, and benzylic carbocations.

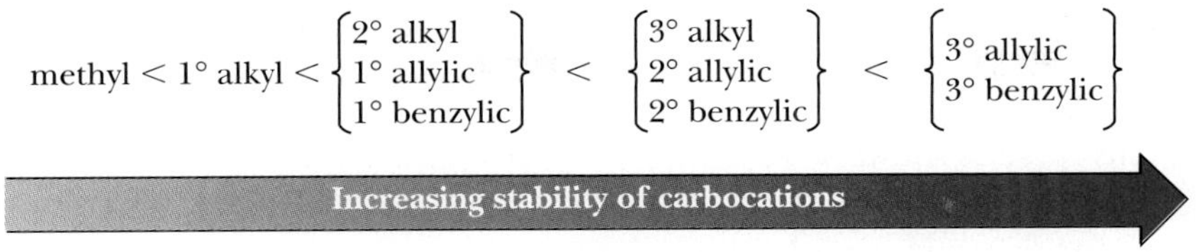

$$\text{methyl} < 1° \text{ alkyl} < \begin{Bmatrix} 2° \text{ alkyl} \\ 1° \text{ allylic} \\ 1° \text{ benzylic} \end{Bmatrix} < \begin{Bmatrix} 3° \text{ alkyl} \\ 2° \text{ allylic} \\ 2° \text{ benzylic} \end{Bmatrix} < \begin{Bmatrix} 3° \text{ allylic} \\ 3° \text{ benzylic} \end{Bmatrix}$$

Increasing stability of carbocations ⟶

Example 9.4

Write an additional resonance contributing structure for each carbocation, and state which of the two makes the greater contribution to the resonance hybrid. Classify each as a 1°, 2°, or 3° allylic cation.

(a) (b)

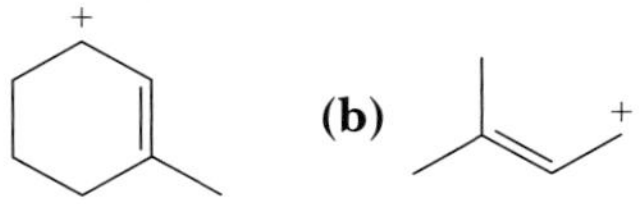

Solution

The additional resonance contributing structure in both cases is a 3° allylic cation. The contributing structure having the greater degree of substitution on the positively charged carbon makes the greater contribution to the hybrid.

(a) (b)

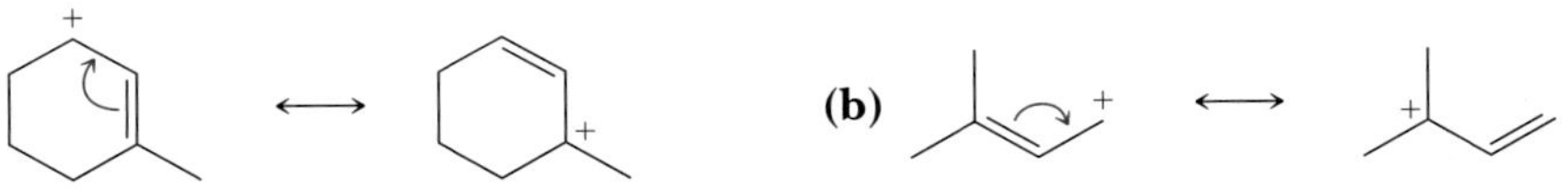

Greater contribution   Greater contribution

Problem 9.4

Write an additional resonance contributing structure for each carbocation, and state which of the two makes the greater contribution to the resonance hybrid.

(a) $=CH_2$ (b)

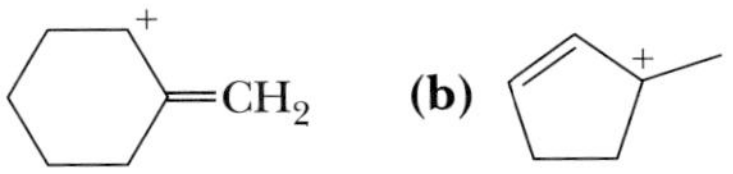

Primary allylic and benzylic halides can be made to react by either S_N1 or S_N2 mechanisms depending on the solvent and the nucleophile. They are primary (the steric factor favoring S_N2), and, at the same time, they can lose a halide ion to form a stable allylic or benzylic carbocation (the electronic factor favoring S_N1). In polar protic solvents, they undergo solvolysis by an S_N1 mechanism. They can be made to undergo S_N2 reactions in aprotic solvents by treatment with good nucleophiles. Secondary and tertiary allylic and benzylic halides react almost exclusively by an S_N1 mechanism.

F. The Leaving Group

In the transition state for nucleophilic substitution on a haloalkane, the leaving group develops a partial negative charge in both S_N1 and S_N2 reactions; therefore, the ability of a group to function as a leaving group is related to how stable it is as an anion. The most stable anions and the best leaving groups are the conjugate bases of strong acids. Thus, we can use the information on the relative strengths of organic and inorganic acids in Table 4.1 to determine which anions are the best leaving groups. This order is shown here.

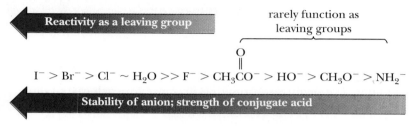

The best leaving groups in this series are the halides, I^-, Br^-, and Cl^-. Hydroxide ion, OH^-, methoxide ion, CH_3O^-, and amide ion, NH_2^-, are such poor leaving groups that they rarely, if ever, are displaced in nucleophilic aliphatic substitution. However, we shall see in Chapters 10 and 11 that protonation of —OH and —OR converts them to better leaving groups. Conversion to a sulfonate ester RSO_2OR' (the ester of a sulfonic acid, RSO_2OH, a strong acid) is another way to convert an alcohol to a good leaving group (see Section 10.5D).

G. The Solvent

To appreciate the important role of the solvent in nucleophilic substitution reactions, we need to be specific about whether the substitution is S_N2 or S_N1. Let us first take up the effect of solvent on S_N2 reactions.

The Effect of Solvent on S_N2 Reactions

The most common type of S_N2 reaction involves a negatively charged nucleophile and a negatively charged leaving group. The central carbon atom has a partial positive charge in the starting material; in the transition state, it may have either a smaller or larger charge depending on the conditions.

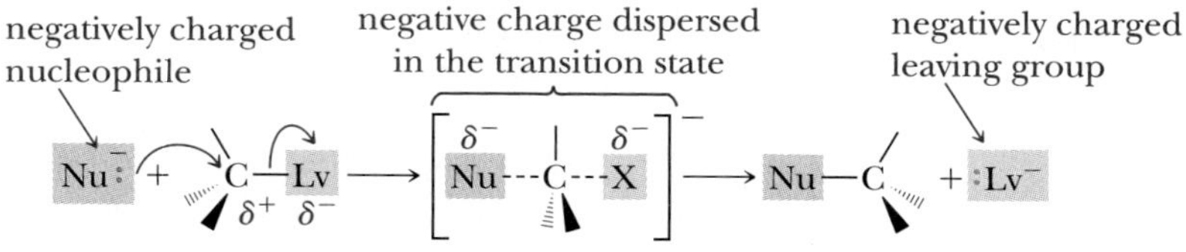

negatively charged nucleophile   negative charge dispersed in the transition state   negatively charged leaving group

$$Nu{:}^- \;+\; \overset{\delta^+}{C}\!-\!\overset{\delta^-}{Lv} \longrightarrow \left[\; \overset{\delta^-}{Nu}\text{---}C\text{---}\overset{\delta^-}{X} \;\right]^- \longrightarrow Nu\!-\!C \;+\; {:}Lv^-$$

Transition state

The stronger the solvation of the nucleophile is, the greater the energy required to remove the nucleophile from its solvation shell to reach the transition state will be, and hence the lower the rate of the S_N2 reaction will be.

As mentioned before, polar aprotic solvents solvate cations very well but solvate anions (nucleophiles) relatively poorly. For this reason, nucleophiles are freer and more reactive in polar aprotic solvents than in protic solvents, so the rates of S_N2 reactions are dramatically accelerated, often by several orders of magnitude compared to the same reaction in protic solvents. Table 9.10 shows ratios of specific rate constants for the S_N2 reaction of 1-bromobutane with sodium azide as a function of solvent. The rate of reaction in methanol is taken as a reference and assigned a relative rate of 1.

Table 9.10 Rates of an S_N2 Reaction as a Function of Solvent

$$\text{Br} + :N_3^- \xrightarrow[\text{solvent}]{S_N2} N_3 + :Br:^-$$

Solvent Type	Solvent		$\dfrac{k_{(solvent)}}{k_{(methanol)}}$
Polar aprotic	Acetonitrile	$CH_3C\equiv N$	5000
	DME	$(CH_3)_2NCHO$	2800
	DMSO	$(CH_3)_2S=O$	1300
Polar protic	Water	H_2O	7
	Methanol	CH_3OH	1

The Effect of Solvent on S_N1 Reactions

Nucleophilic substitution by an S_N1 pathway involves creation and separation of opposite charges in the transition state of the rate-determining step. For this reason, the rate of S_N1 reactions depends on both the ability of the solvent to keep opposite charges separated and its ability to stabilize both positive and negative sites by solvation. The solvents that meet these requirements best are polar protic solvents such as H_2O, low-molecular-weight alcohols such as methanol and ethanol, and, to a lesser degree, low-molecular-weight carboxylic acids such as formic acid and acetic acid. As seen in Table 9.11, the rate of solvolysis of 2-chloro-2-methylpropane (*tert*-butyl chloride) increases by a factor of 10^5 when the solvent is changed from ethanol to water.

Table 9.11 Rates of an S_N1 Reaction as a Function of Solvent

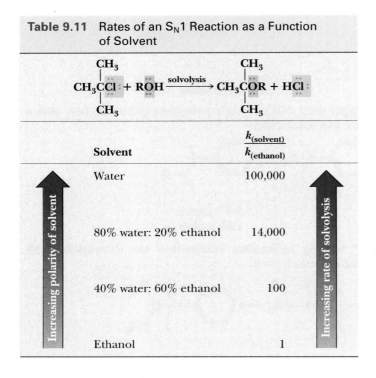

	Solvent	$\dfrac{k_{(solvent)}}{k_{(ethanol)}}$
	Water	100,000
	80% water: 20% ethanol	14,000
	40% water: 60% ethanol	100
	Ethanol	1

Increasing polarity of solvent

Increasing rate of solvolysis

H. Skeletal Rearrangement

As we saw in Section 6.3C, skeletal rearrangement is typical of reactions involving a carbo-cation intermediate that can rearrange to a more stable one. Because there is little or no carbocation character at the substitution center, S_N2 reactions are free of rearrangement. In contrast, S_N1 reactions often proceed with rearrangement. These and related carbo-cation rearrangements are often referred to as **Wagner-Meerwein** rearrangements. An ex-ample of an S_N1 reaction involving rearrangement is solvolysis of 2-chloro-3-phenylbutane in methanol, a polar protic solvent and a weak nucleophile. The major substitution prod-uct is the ether with rearranged structure. The chlorine atom in the starting material is on a 2° carbon, but the methoxyl group in the product is on the adjacent 3° carbon.

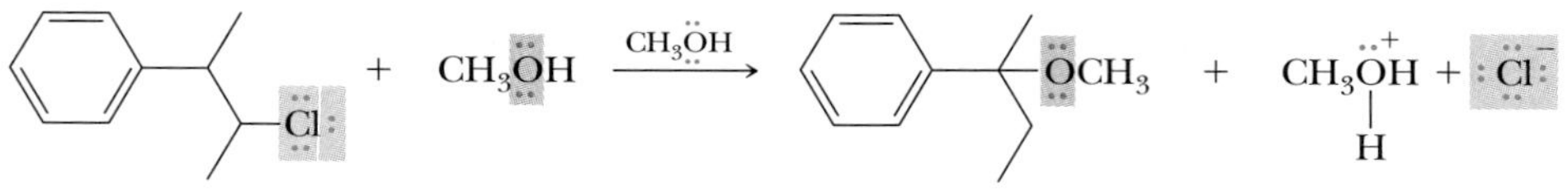

2-Chloro-3-phenylbutane   2-Methoxy-2-phenylbutane

As seen in the following mechanism, reaction is initiated by heterolytic cleavage of the carbon-chlorine bond to form a 2° carbocation, which rearranges to a considerably more stable 3° carbocation by shift of a hydrogen with its pair of electrons (a hydride ion) from the adjacent benzylic carbon. Note that not only is the rearranged carbocation tertiary (hyperconjugation stabilization), but it is also benzylic (resonance stabilization).

Mechanism	*Rearrangement During Solvolysis of 2-Chloro-3-phenylbutane*

Step 1: Ionization of the C—Cl bond gives a 2° carbocation intermediate.

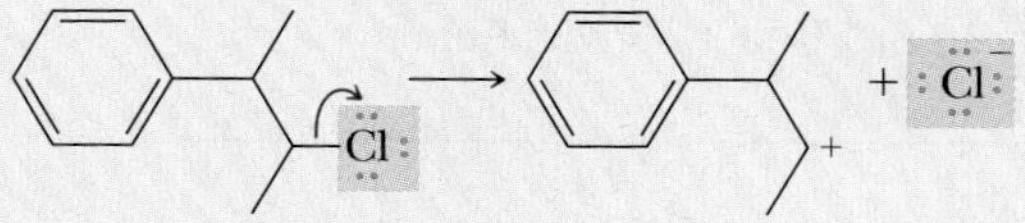

A 2° carbocation

Step 2: Migration of a hydrogen atom with its bonding electrons (a hydride ion) gives a more stable 3° benzylic carbocation intermediate.

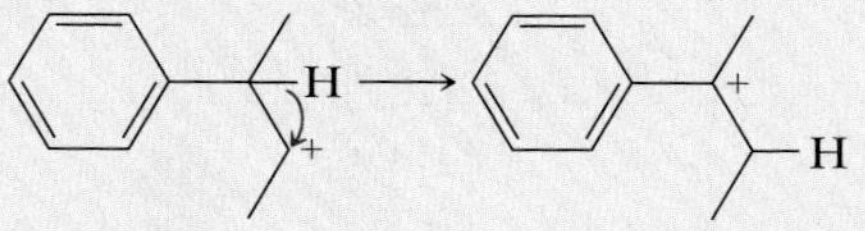

A 3° benzylic
carbocation

Step 3: Reaction of the 3° benzylic carbocation intermediate (an electrophile) with methanol (a nucleophile) forms an oxonium ion.

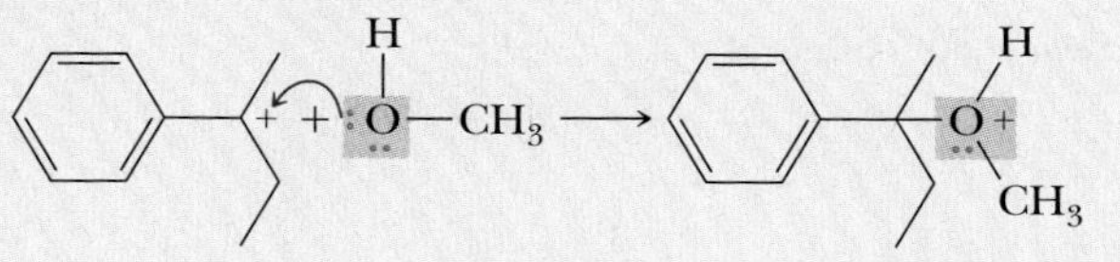

An oxonium ion

Step 4: Proton transfer to solvent (in this case, methanol) gives the final product.

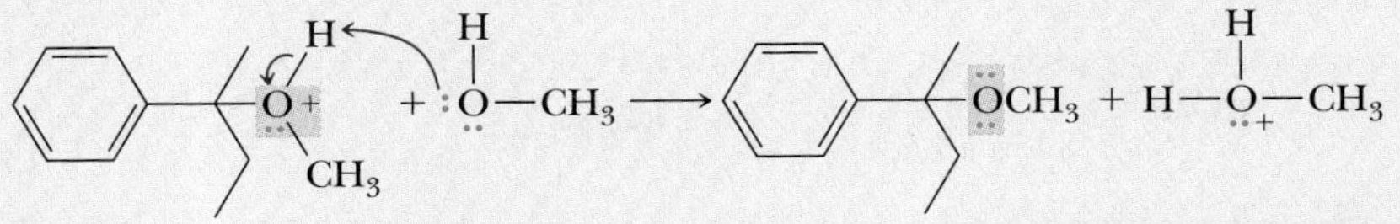

In general, migration of a hydrogen atom or alkyl group with its bonding electrons occurs when a more stable carbocation can be formed.

The factors favoring S_N1 or S_N2 reactions are summarized in Table 9.12. Also shown is the expected configurational result when substitution takes place at a chiral center.

9.5 Analysis of Several Nucleophilic Substitution Reactions

Predictions about the mechanism for a particular nucleophilic substitution reaction must be based on considerations of the structure of the haloalkane, the nucleophile, the leaving group, and the solvent. Following are five nucleophilic substitution reactions and an analysis of the factors that favor an S_N1 or S_N2 mechanism for each and the products that result from the mechanism used.

Table 9.12 Summary of S_N1 Versus S_N2 Reactions of Haloalkanes

Type of Alkyl Halide	S_N2	S_N1
Methyl CH_3X	**S_N2 is favored.**	**S_N1 does not occur.** The methyl cation is so unstable, it is never observed in solution.
Primary RCH_2X	**S_N2 is favored.**	**S_N1 rarely occurs.** Primary cations are so unstable, they are never rarely observed in solution.
Secondary R_2CHX	**S_N2 is favored** in aprotic solvents with good nucleophiles.	**S_N1 is favored** in protic solvents with poor nucleophiles. Carbocation rearrangements may occur.
Tertiary R_3CX	**S_N2 does not occur** because of steric hindrance around the reaction center.	**S_N1 is favored** because of the ease of formation of tertiary carbocations.
Substitution at a chiral center	**Inversion of configuration.** The nucleophile attacks the chiral center from the side opposite the leaving group.	**Racemization is favored.** The carbocation intermediate is planar, and attack of the nucleophile occurs with equal probability from either side. There is often some net inversion of configuration.

Nucleophilic Substitution 1

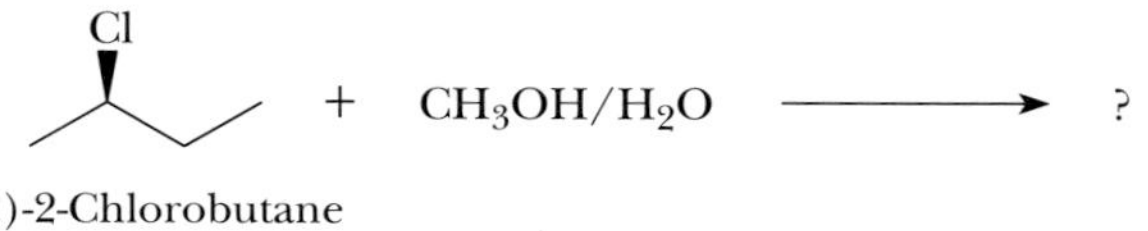

(R)-2-Chlorobutane

The mixture of methanol and water is a polar protic solvent and a good ionizing solvent in which to form carbocations. 2-Chlorobutane ionizes in this solvent to form a fairly stable 2° carbocation intermediate. Both water and methanol are poor nucleophiles. From this analysis, we predict that the reaction occurs by an S_N1 mechanism. Ionization of the 2° chloroalkane gives a carbocation intermediate, which then reacts with either water or methanol as the nucleophile to give the observed products. Each product is formed as an approximately 50 : 50 mixture of R and S enantiomers.

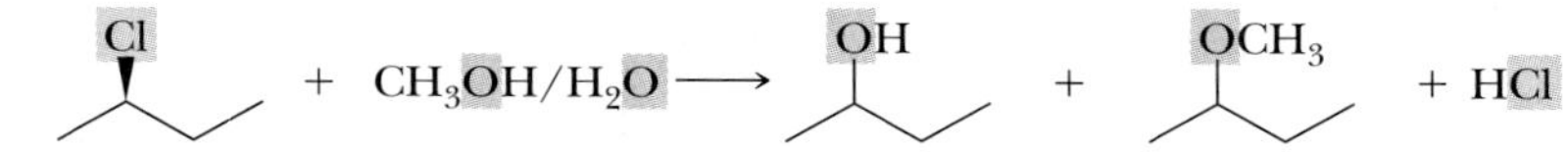

(R)-2-Chlorobutane   (R,S)-2-Butanol   (R,S)-2-Methoxy-
                                          50 : 50   butane

Nucleophilic Substitution 2

This is a primary bromoalkane with two beta branches in the presence of cyanide ion, a good nucleophile. Dimethyl sulfoxide (DMSO), a polar aprotic solvent, is a particularly good solvent in which to carry out nucleophile-assisted substitution reactions because of its good ability to solvate cations (in this case, Na^+) but its poor ability to solvate anions (in this case, CN^-). From this analysis, we predict that this reaction occurs by an S_N2 mechanism.

Nucleophilic Substitution 3

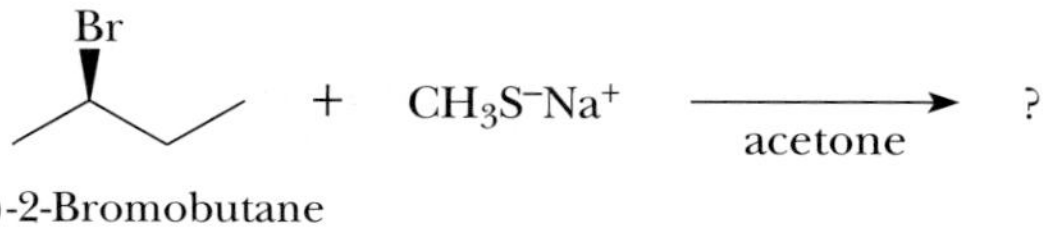

(R)-2-Bromobutane

Bromine is a good leaving group, and it is on a 2° carbon. The methylsulfide ion is a good nucleophile. Acetone, a polar aprotic solvent, is a good medium in which to carry out S_N2 reactions but a poor medium in which to carry out S_N1 reactions. From this analysis, we predict that this reaction occurs by an S_N2 mechanism and that the product is the S enantiomer.

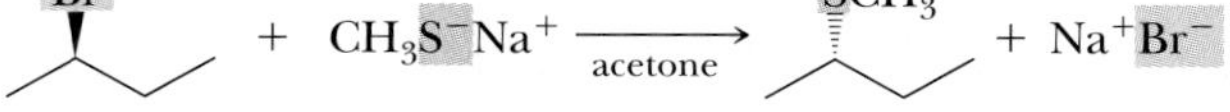

(R)-2-Bromobutane   (S)-2-Methylsulfanylbutane

Nucleophilic Substitution 4

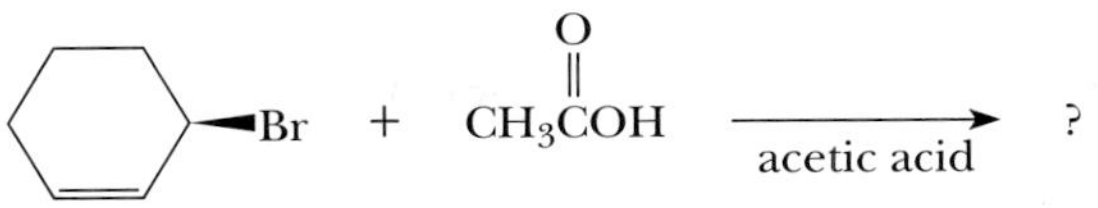

Ionization of the carbon-bromine bond forms a resonance-stabilized 2° allylic carbo-cation. Acetic acid is a poor nucleophile, which reduces the likelihood of an S_N2 reaction. Further, acetic acid is a moderately polar protic (hydroxylic) solvent that favors S_N1 reaction. From this analysis, we predict that this reaction occurs by an S_N1 mechanism and both enantiomers of the product are observed.

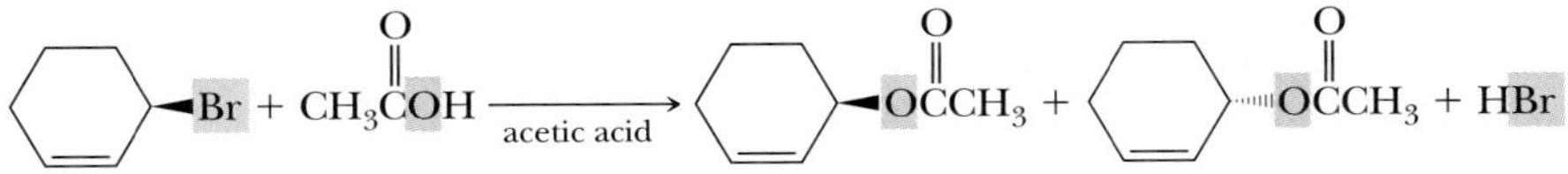

| (R)-3-Bromocyclo-hexene | (R)-3-Acetoxy-cyclohexene | (S)-3-Acetoxy-cyclohexene |

Nucleophilic Substitution 5

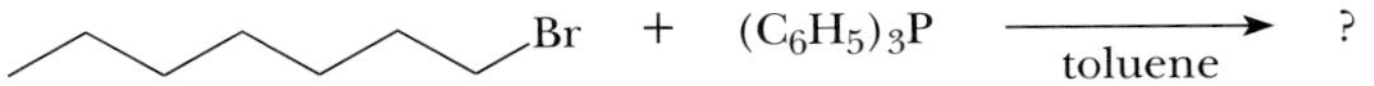

The bromoalkane is primary, and bromine is a good leaving group. Trivalent compounds of phosphorus, a third-row element, are good nucleophiles. Toluene is a nonpolar aprotic solvent. Given the combination of a primary halide, a good leaving group, a good nucleophile, and a nonpolar aprotic solvent, we predict that the reaction occurs by an S_N2 pathway.

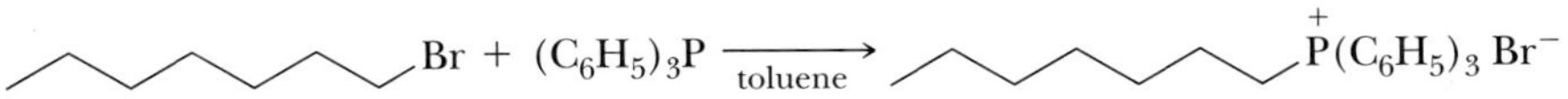

Example 9.5

Write the expected substitution product(s) for each reaction and predict the mechanism by which each product is formed.

(a)

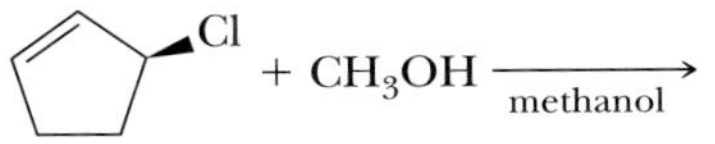

S enantiomer

(b)

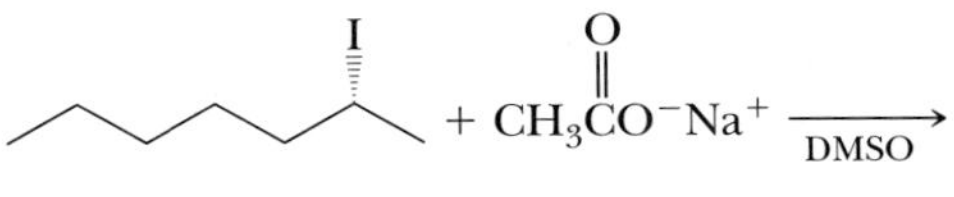

R enantiomer

Solution

(a) This 2° allylic chloride is treated with methanol, a poor nucleophile and a polar protic solvent. Ionization of the carbon-chlorine bond forms a resonance-stabilized secondary allylic cation. Therefore, we predict reaction by an S_N1 mechanism and formation of the product as a racemic mixture.

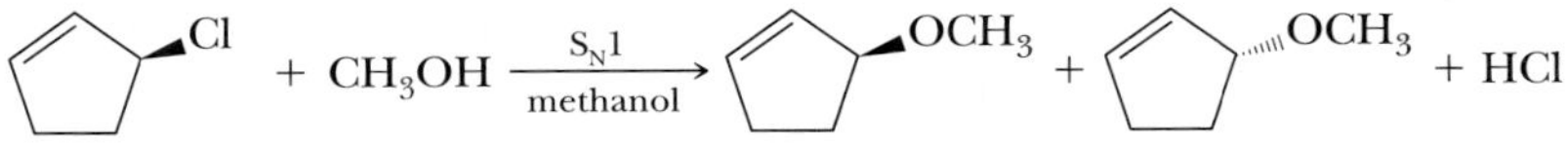

| *S* enantiomer | *S* enantiomer (50%) | *R* enantiomer (50%) |

(b) Iodide is a good leaving group on a moderately accessible secondary carbon. Acetate ion dissolved in a polar aprotic solvent is a moderate nucleophile. We predict substitution by an S_N2 pathway with inversion of configuration at the chiral center.

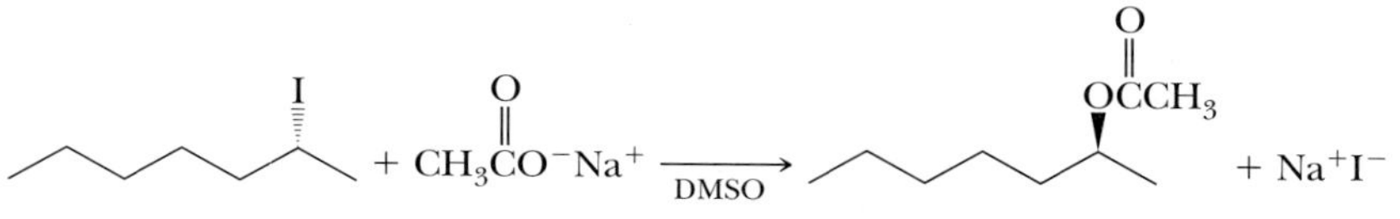

R enantiomer   *S* enantiomer

Problem 9.5

Write the expected substitution product(s) for each reaction and predict the mechanism by which each product is formed.

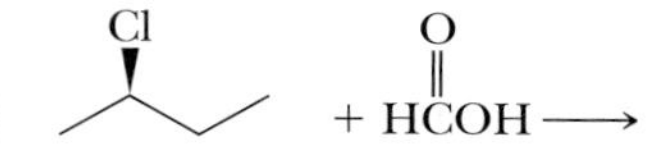

(*R*)-2-Chlorobutane

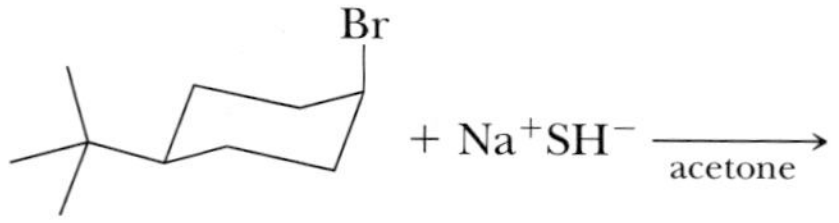

ORGANIC
Chemistry Now™
Click Tutorials for a summary of
Elimination Reactions

9.6 β-Elimination

As mentioned in the introduction to this chapter, β-elimination competes with nucleophilic substitution.

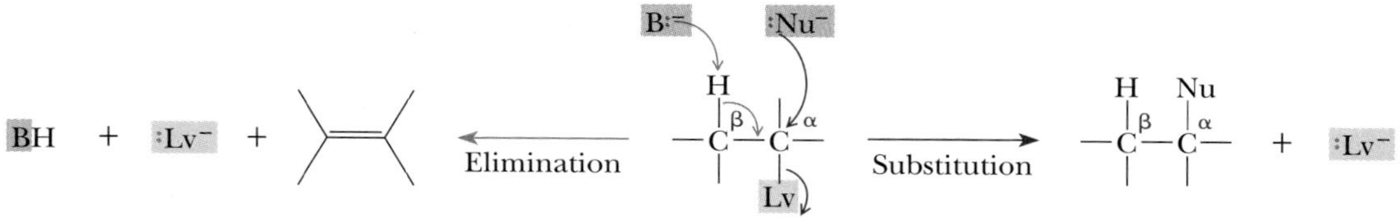

Dehydrohalogenation Removal of —H and —X from adjacent carbons; a type of β-elimination.

Here, we study a type of β-elimination called **dehydrohalogenation.** In the presence of a base, halogen is removed from one carbon of a haloalkane, and hydrogen is removed from an adjacent carbon to form an alkene.

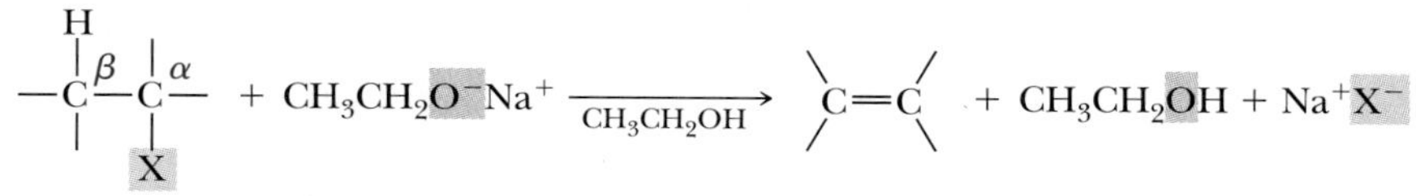

A haloalkane   Base   An alkene

It is important keep in mind that β-elimination and nucleophilic substitution are competing reactions that can occur when an electron-rich species reacts with a molecule containing a leaving group. Up until now, we have focused entirely on substitution. In this and the following two sections, we concentrate on β-elimination. In Section 9.9, we put all the concepts together and show how to analyze reactions with respect to competition between nucleophilic substitution and β-elimination.

Strong bases promote β-elimination reactions. Strong bases that serve effectively in β-eliminations of haloalkanes are OH^-, OR^-, and NH_2^-. Following are three examples of base-promoted β-elimination reactions. In the first example, the base is shown as a reactant. In the second and third examples, the base is a reactant but is shown over the reaction arrow.

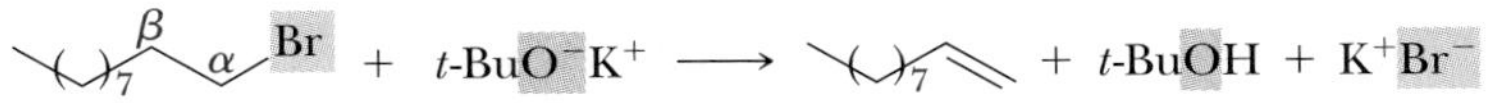

1-Bromodecane   Potassium   1-Decene
             tert-butoxide

2-Bromo-2-   2-Methyl-2-butene   2-Methyl-1-butene
methylbutane   (major product)   disubstituted
                    disubstituted

1-Bromo-1-methylcyclopentane   1-Methylcyclopentene   Methylenecyclopentane
                                 (major product)

In the second and third illustrations, there are two nonequivalent β-carbons, each bearing a hydrogen; therefore, two alkenes are possible. In each case, the major product of this and most other β-elimination reactions is the more substituted (and, therefore, the more stable) alkene (Section 6.6B). Formation of the more substituted alkene in an elimination is common, and when this is the outcome, the reaction is said to follow **Zaitsev's rule,** or to undergo Zaitsev elimination. Not all alkenes, however, undergo β-elimination to give the more stable alkene, and these exceptions give us important clues to the mechanism of β-elimination, as we shall see when we discuss the regio- and stereochemistry of β-elimination reactions.

Zaitsev's rule A rule stating that the major product of a β-elimination reaction is the most stable alkene; that is, it is the alkene with the greatest number of substituents on the carbon-carbon double bond.

Example 9.6

Predict the β-elimination product(s) formed when each bromoalkane is treated with sodium ethoxide in ethanol. If two or more products might be formed, predict which is the major product.

(a) 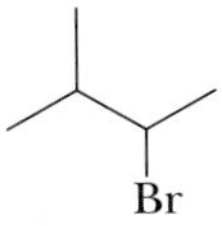   (b)

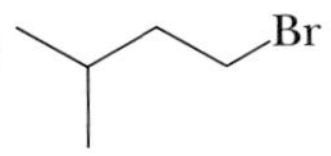

Solution

(a) There are two nonequivalent β-carbons in this bromoalkane, and two alkenes are possible. 2-Methyl-2-butene, the more substituted alkene, is the major product.

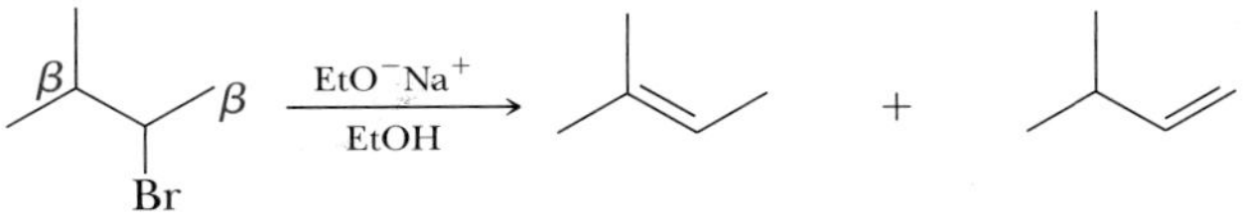

2-Methyl-2-butene   3-Methyl-1-butene
(major product)

(b) There is only one β-carbon in this bromoalkane, and only one alkene is possible.

$$\text{(structure)} \quad \xrightarrow[\text{EtOH}]{\text{EtO}^-\text{Na}^+} \quad \text{(structure)}$$

3-Methyl-1-butene

Problem 9.6

Predict the β-elimination product(s) formed when each chloroalkane is treated with sodium ethoxide in ethanol. If two or more products might be formed, predict which is the major product.

(a)   **(b)**   **(c)**

9.7 Mechanisms of β-Elimination

There are two limiting mechanisms for β-eliminations. A fundamental difference between them is the timing of the bond-breaking and bond-forming steps. Recall that we made the same statement about the two limiting mechanisms for nucleophilic substitution reactions (Section 9.3).

A. E1 Mechanism

E1 A unimolecular
β-elimination reaction.

At one extreme, breaking of the C—Lv bond to give a carbocation is complete before any reaction occurs with the base to lose a hydrogen and form the carbon-carbon double bond. This mechanism is designated **E1**, where E stands for *e*limination and 1

Figure 9.5

An energy diagram for an E1 reaction showing two transition states and one carbocation intermediate.

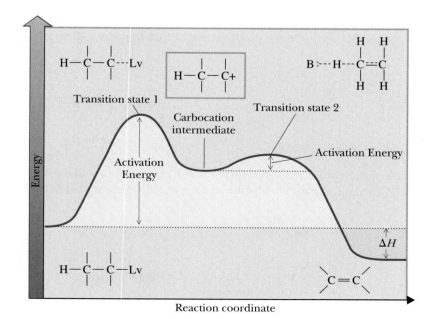

Reaction coordinate

stands for *uni*molecular. One species, in this case the haloalkane, is involved in the rate-determining step. The mechanism of an E1 reaction is illustrated here by the reaction of 2-bromo-2-methylpropane to form 2-methylpropene.

Mechanism   *E1 Reaction of 2-Bromo-2-methylpropane*

Step 1: Rate-determining ionization of the C—Br bond gives a carbocation intermediate.

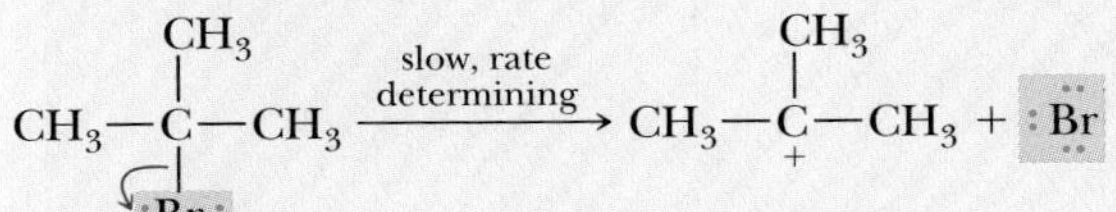

(A carbocation intermediate)

Step 2: Proton transfer from the carbocation intermediate to solvent (in this case methanol) gives the alkene. (This reaction competes with S_N1 substitution. E1 and S_N1 almost always occur together.)

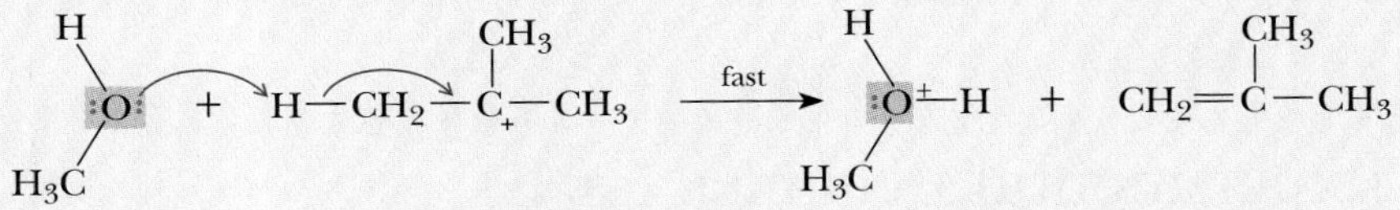

In an E1 mechanism, one transition state exists for the formation of the carbocation in Step 1, and a second exists for the loss of a hydrogen in Step 2 (Figure 9.5). Formation of the carbocation intermediate in Step 1 crosses the higher energy barrier and is the rate-determining step.

B. E2 Mechanism

At the other extreme of elimination mechanisms is a concerted process. In an E2 reaction, here illustrated by the reaction of 2-bromobutane with sodium ethoxide, proton transfer to the base, formation of the carbon-carbon double bond, and ejection of bromide ion occur simultaneously; all bond-breaking and bond-forming steps are concerted. Because base removes a β-hydrogen at the same time the C—Br bond is broken to form a halide ion, the transition state has considerable double bond character (Figure 9.6).

Mechanism   *E2 Reaction of 2-Bromopropane*

Bond breaking and bond forming are concerted; that is, they occur simultaneously.

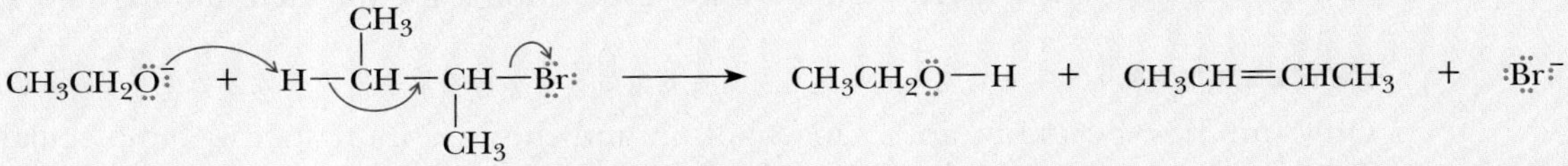

This mechanism is designated **E2**, where E stands for *e*limination and 2 stands for *bi*molecular; both the haloalkane and the base are involved in the transition state for the rate-determining step.

E2 A bimolecular β-elimination reaction.

Figure 9.6

An energy diagram for an E2 reaction. There is considerable double bond character in the transition state.

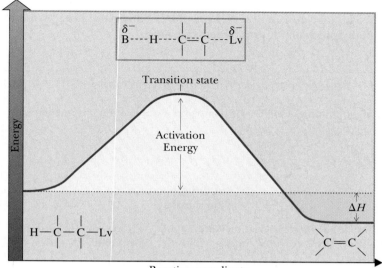

9.8 Experimental Evidence for E1 and E2 Mechanisms

As we examine some of the experimental evidence on which these two contrasting mechanisms are based, we consider the following questions.

1. What are the kinetics of base-promoted β-eliminations?
2. Where two or more alkenes are possible, what factors determine the ratio of the possible products?
3. What is the stereochemistry of β-elimination?

A. Kinetics

E1 Reactions

The rate-determining step in an E1 reaction is ionization of the halide to form a carbocation. Because this step involves only the haloalkane, the reaction is unimolecular and follows first-order kinetics.

$$\text{Rate} = -\frac{d[\text{RX}]}{dt} = k[\text{RX}]$$

Recall that the first step in an S_N1 reaction is also formation of a carbocation. Thus, for both S_N1 and E1 reactions, formation of the carbocation is the rate-determining step.

E2 Reactions

Only one step occurs in an E2 mechanism, and the transition state is bimolecular. The reaction is second order: first order in haloalkane and first order in base.

$$\text{Rate} = -\frac{d[\text{RX}]}{dt} = k[\text{RX}][\text{Base}]$$

B. Regioselectivity

E1 Reactions

The major product in E1 reactions is the more stable alkene; that is, the alkene with the more highly substituted carbon-carbon double bond (Zaitsev's rule). After the carbocation is formed in the rate-determining step of an E1 reaction, it may lose a hydrogen to complete β-elimination, or it may rearrange and then lose a hydrogen.

E2 Reactions

For E2 reactions using strong bases and in which the leaving group is a halide ion, the major product is also that formed following Zaitsev's rule unless special steric relations apply (Section 9.8C). Double bond character is so highly developed in the transition state that the relative stability of possible alkenes determines which alkene is the major product. Thus, the transition state of lowest energy is that leading to the most stable, most highly substituted alkene. (Note that E2 elimination predominates over S_N2 reaction with the strongly basic alkoxide ions.)

2-Bromohexane	2-Hexene	1-Hexene
	(74%)	(26%)

With larger, sterically hindered bases such as *tert*-butoxide, however, where isomeric alkenes are possible, the major product is often the less substituted alkene because reaction occurs primarily at the most accessible H atom. Sterically hindered bases like *tert*-butoxide are also noteworthy because the steric hindrance prevents them from reacting as nucleophiles, even with primary alkyl halides.

C. Stereoselectivity

The lowest-energy transition state of an E2 reaction is the one in which the —Lv and —H are oriented anti and coplanar (at a dihedral angle of 180°) to one another. The reason for this geometrical requirement is that it allows for proper orbital overlap between the base, the proton being removed, and the departing leaving group. As in the case of backside attack in an S_N2 reaction, this reaction geometry weakens the carbon-leaving group bond, facilitating its departure as the new pi bond is being made. Remembering the anti, coplanar geometry requirement is important because it allows prediction of alkene stereochemistry in E2 reactions, namely whether E or Z products are produced.

—H and —Lv are anti and coplanar
(dihedral angle 180°)

This reaction geometry is shown more clearly in a Newman projection with a bromide as the leaving group.

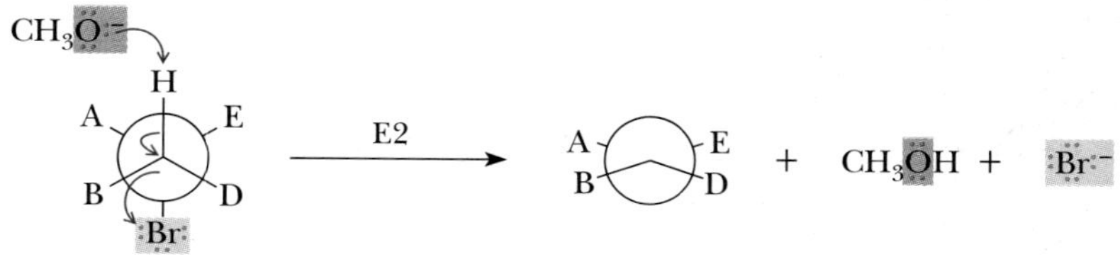

For example, treatment of 1,2-dibromo-1,2-diphenylethane with sodium methoxide in methanol gives 1-bromo-1,2-diphenylethylene. The meso isomer of 1,2-dibromo-1,2-diphenylethane gives (E)-1-bromo-1,2-diphenylethylene, whereas the racemic mixture of 1,2-dibromo-1,2-diphenylethane gives (Z)-1-bromo-1,2-diphenylethylene.

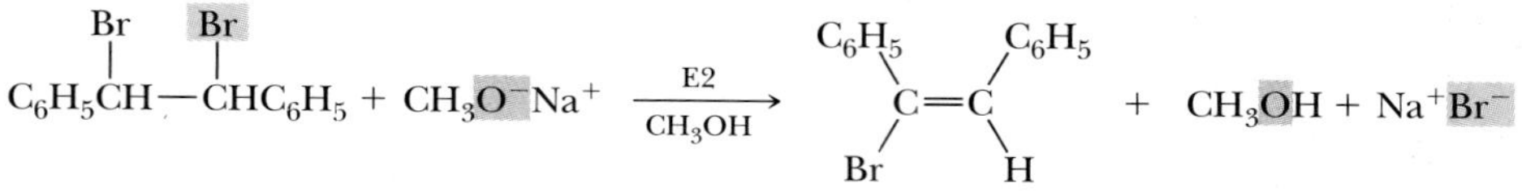

meso-1,2-Dibromo-
1,2-diphenylethane

(E)-1-Bromo-
1,2-diphenylethylene

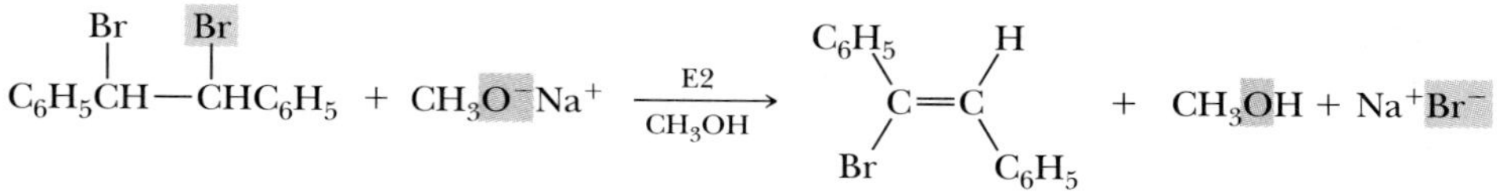

racemic-1,2-Dibromo-
1,2-diphenylethane

(Z)-1-Bromo-
1,2-diphenylethylene

We use the anti coplanar requirement of the transition state to account for the stereospecificity of these E2 β-eliminations. Because it is a requirement for an E2 reaction that —H and —Lv be anti and coplanar, it is important to identify the reactive conformation of a haloalkane starting material. Following is a stereorepresentation of the meso isomer of 1,2-dibromo-1,2-diphenylethane, drawn to show the plane of symmetry.

Mechanism E2 Reaction of meso-1,2-Dibromo-1,2-diphenylethane

Rotation of the left carbon by 60° brings —H and —Br into the required anti and coplanar relationship.

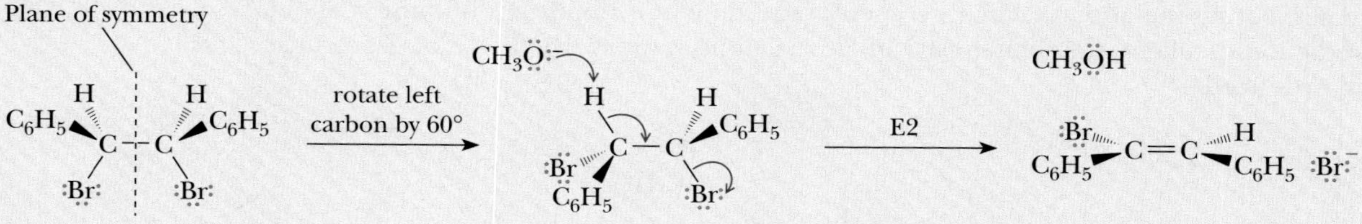

Plane of symmetry

Meso isomer

—H and —Br and now
anti and coplanar
as required for E2

(E)-1-Bromo-1,2-
diphenylethylene

E2 reaction on this conformation gives only the (E)-alkene, as shown here in a Newman projection. The —H and —Br are anti and coplanar. All bond-breaking and bond-forming steps are concerted.

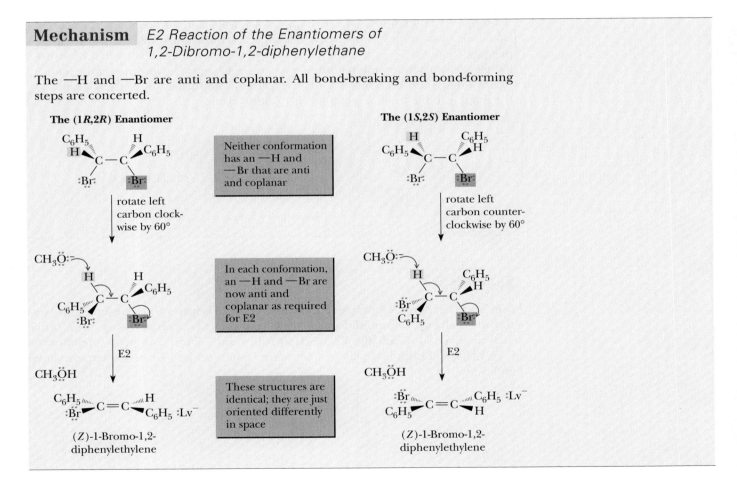

E2 reaction of either enantiomer of the racemic mixture of 1,2-dibromo-1,2-diphenyl-methane gives only the *(Z)*-alkene as predicted by analysis of the proper anti and coplanar conformations.

Mechanism   *E2 Reaction of the Enantiomers of 1,2-Dibromo-1,2-diphenylethane*

The —H and —Br are anti and coplanar. All bond-breaking and bond-forming steps are concerted.

The required *trans* and coplanar transition state geometry can also be used to predict products of E2 elimination in halocyclohexanes such as chlorocyclohexanes. In these molecules, anti and coplanar correspond to *trans*, diaxial. Consider the E2 reaction of the *cis* isomer of 1-chloro-2-isopropylcyclohexane. The major product is 1-isopropylcyclohexene, the more substituted cycloalkene.

Achiral

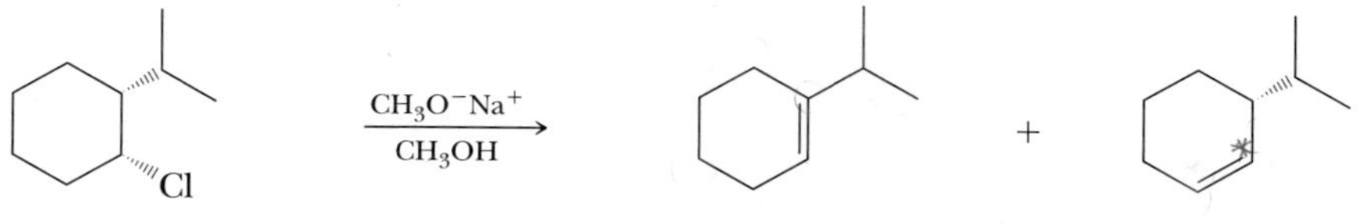

cis-1-Chloro-2-isopropylcyclohexane   1-Isopropylcyclohexene (*R*)-3-Isopropylcyclohexene
                                                        (major product)
                                                         trisubstituted   disubstituted

---

Mechanism   *E2 Reaction of cis-1-Chloro-2-isopropylcyclohexane*

In the more stable chair conformation of the *cis* isomer, the considerably larger isopropyl group is equatorial and the smaller chlorine is axial. In this chair conformation, —H on carbon 2 and —Cl on carbon 1 are anti and coplanar. Concerted E2 elimination gives 1-isopropylcyclohexene, a trisubstituted alkene, as the major product. Note that —H on carbon 6 and —Cl are also anti and coplanar. Dehydrohalogenation of this combination of —H and —Cl gives 3-isopropylcyclohexene, a disubstituted, and therefore less stable, alkene. The formation of the 1-isomer as the major product is in agreement with Zaitsev's rule.

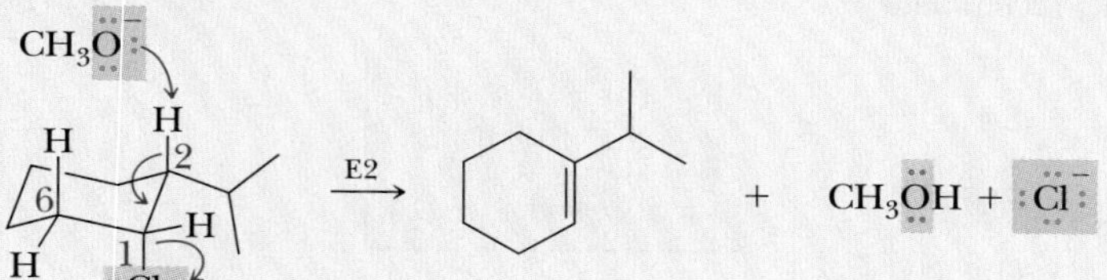

1-Isopropylcyclohexene

Example 9.7

From *trans*-1-chloro-2-isopropylcyclohexane, only 3-isopropylcyclohexene, the less substituted alkene, is formed. Explain why this product is observed using a conformational analysis. Also, will the E2 reaction with *trans*-1-chloro-2-isopropylcyclohexane or *cis*-1-chloro-2-isopropylcyclohexane occur faster under the same experimental conditions?

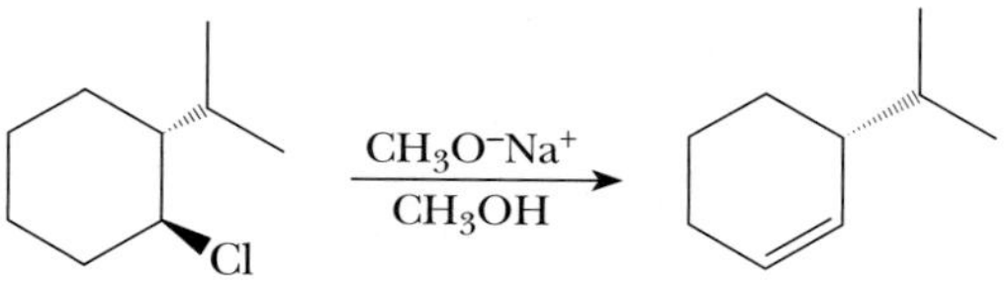

trans-1-Chloro-2   (*R*)-3-Isopropylcyclohexene
isopropylcyclohexane

Solution

Step 1: In the more stable chair conformation of the *trans* isomer, both isopropyl and chlorine are equatorial. In this conformation, the hydrogen atom on carbon 2 is *cis* to the chlorine atom. One of the hydrogen atoms on carbon 6 is *trans* to —Cl, but it is not anti and coplanar. Therefore, the reaction is not favored from this conformation. In the alternative, less stable chair conformation of the *trans* isomer, both isopropyl and chlorine are axial. In this conformation, the axial hydrogen in carbon 6 is *anti* and coplanar to chlorine, and E2 *β*-elimination can occur to give 3-isopropylcyclohexene. Thus, even though the diaxial conformation is less stable, the reaction occurs through this conformation because it is the only one with an anti-coplanar arrangement of the Cl and a *β*—H; consequently, the non-Zaitsev product is formed.

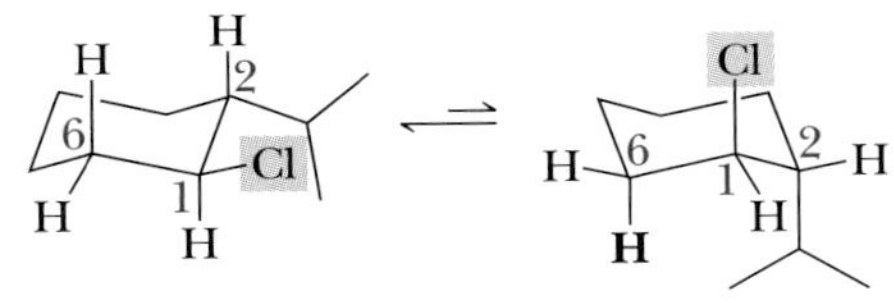

More stable chair
(no H is anti and
coplanar to Cl)

Less stable chair
(H on carbon 6 is
anti and coplanar to Cl)

Step 2: E2 reaction can take place now that an —H and —Cl are anti and coplanar. This reaction doesn't follow the Zaitsev rule because of the requirement in the mechanism of the reaction for the anti arrangement.

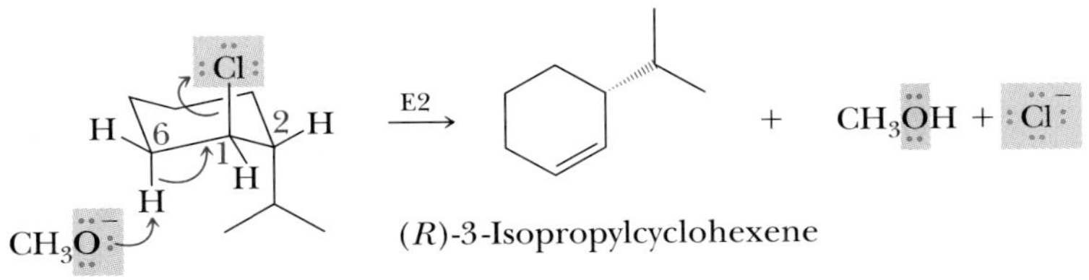

(*R*)-3-Isopropylcyclohexene

The rate at which the *cis* isomer undergoes E2 reaction is considerably greater than the rate for the *trans* isomer. We can account for this observation in the following manner. The more stable chair conformation of the *cis* isomer has —H and —Cl anti and coplanar, and the activation energy for the reaction is that required to reach the E2 transition state. The more stable chair conformation of the *trans* isomer, however, cannot undergo anti elimination. To react, it must first be converted to the less stable chair, and the transition state for elimination is correspondingly higher in energy because of the axial isopropyl group.

Problem 9.7

1-Chloro-4-isopropylcyclohexane exists as two stereoisomers: one *cis* and one *trans*. Treatment of either isomer with sodium ethoxide in ethanol gives 4-isopropylcyclohexene by an E2 reaction.

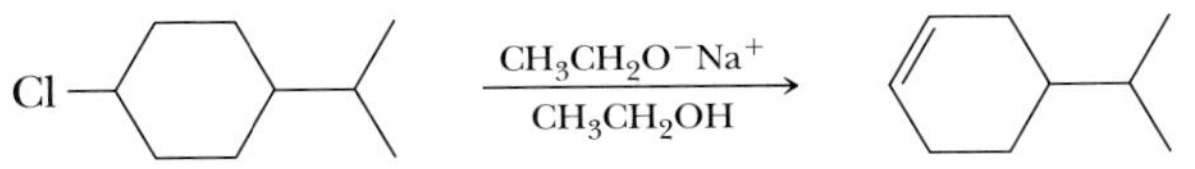

1-Chloro-4-
isopropylcyclohexane

4-Isopropylcyclohexene

The *cis* isomer undergoes E2 reaction several orders of magnitude faster than the *trans* isomer. How do you account for this experimental observation?

Table 9.13 Summary of E1 Versus E2 Reactions for Haloalkanes

Alkyl Halide	E1	E2
Primary RCH_2X	E1 not observed. Primary carbocations are so unstable that they are never observed in solution.	E2 is favored if elimination is observed. Usually requires sterically hindered base.
Secondary R_2CHX	Main reaction with weak bases such as H_2O, ROH.	Main reaction with strong bases such as OH^- and OR^-.
Tertiary R_3CX	Main reaction with weak bases such as H_2O, ROH.	Main reaction with strong bases such as OH^- and OR^-.

The factors favoring E1 or E2 elimination are summarized in Table 9.13.

ORGANIC **Chemistry•Now™**
Click Tutorials to use a flow diagram to predict **Substitution vs Elimination Pathways**

9.9 Substitution Versus Elimination

Nucleophilic substitution and β-elimination often compete with each other, and the ratio of products formed by these reactions depends on the relative rates of the two reactions. In this section, we consider factors that influence this competition.

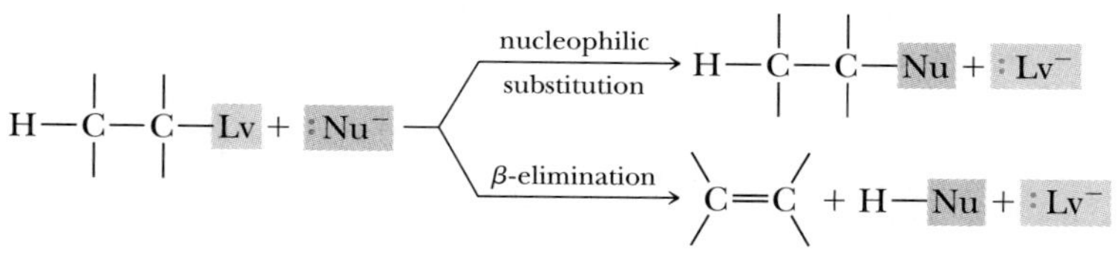

A. S_N1 Versus E1 Reactions

Reactions of secondary and tertiary haloalkanes in polar protic solvents give mixtures of substitution and elimination products. In both reactions, Step 1 is the formation of a carbocation intermediate. This step is then followed by one or more characteristic carbocation reactions: (1) loss of a hydrogen (E1) to give an alkene, (2) reaction with solvent (S_N1) to give a substitution product, or (3) rearrangement followed by reaction (1) or (2). In polar protic solvents, the products formed depend only on the structure of the particular carbocation. For example, *tert*-butyl chloride and *tert*-butyl iodide in 80% aqueous ethanol both react with solvent giving the same mixture of substitution and elimination products. Because iodide ion is a better leaving group than chloride ion, *tert*-butyl iodide reacts over 100 times faster than *tert*-butyl chloride. Yet the ratio of products is the same because the intermediate *tert*-butyl cation is the same.

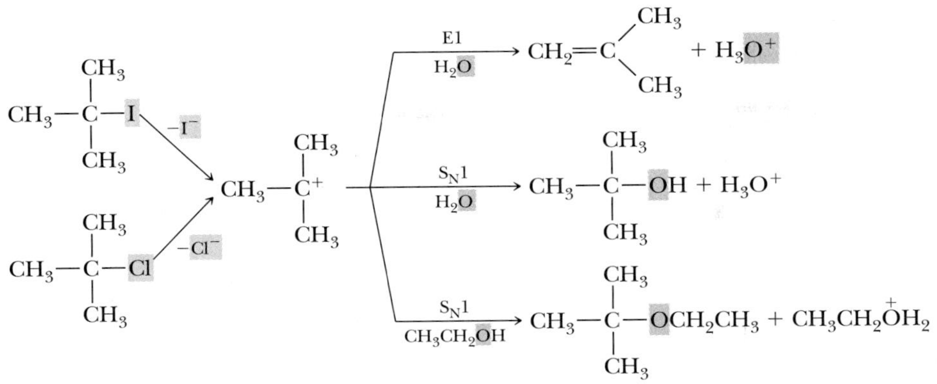

It is difficult to predict the ratio of substitution to elimination products for first-order reactions of haloalkanes. For the majority of cases, however, S_N1 predominates over E1 when weak bases are used.

B. S_N2 Versus E2 Reactions

It is considerably easier to predict the ratio of substitution to elimination products for second-order reactions of haloalkanes with reagents that act both as nucleophiles and bases. The guiding principles follow.

1. Branching at the α-carbon or β-carbon(s) increases steric hindrance about the α-carbon and significantly retards S_N2 reactions. Conversely, branching at the α-carbon or β-carbon(s) increases the rate of E2 reaction because of the increased stability of the alkene product.
2. The greater the nucleophilicity of the attacking reagent, the greater the S_N2-to-E2 ratio. Conversely, the greater the basicity of the attacking reagent is, the greater the E2-to-S_N2 ratio will be.

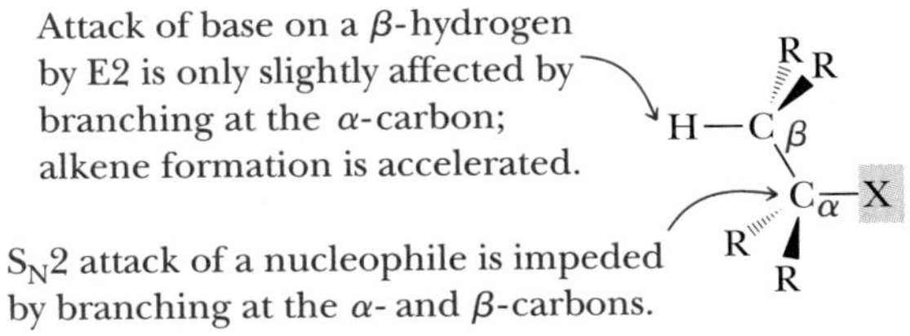

Attack of base on a β-hydrogen by E2 is only slightly affected by branching at the α-carbon; alkene formation is accelerated.

S_N2 attack of a nucleophile is impeded by branching at the α- and β-carbons.

Primary haloalkanes react with bases/nucleophiles to give predominantly substitution products. With strong bases, such as hydroxide ion and ethoxide ion, a percentage of the product is formed by an E2 reaction, but it is generally small compared with that formed by an S_N2 reaction. With strong bulky bases, such as *tert*-butoxide ion, there is too much steric hindrance for substitution, and the E2 product becomes the major product. Tertiary haloalkanes react with all strong bases to give only elimination products.

Secondary haloalkanes are borderline, and substitution or elimination may be favored depending on the particular base/nucleophile, solvent, and temperature at which the reaction is carried out. Elimination is favored with strong bases (those for which the pK_a of the conjugate acid is 11 or above), as for example hydroxide ion and ethoxide ion. Substitution is favored with weak bases/nucleophiles (the pK_a of the conjugate acid is 11 or below, as for example acetate ion). The reason for this

Table 9.14   Summary of Substitution Versus Elimination Reactions of Haloalkanes

Halide	Reaction	Comments
Methyl CH$_3$X	S$_N$2	S$_N$1 reactions of methyl halides are never observed. The methyl cation is so unstable that it is not observed in common solvents.
Primary RCH$_2$X	S$_N$2	The main reaction with good nucleophiles/weak bases such as I$^-$ and (CH$_3$)$_3$CO$^-$.
	E2	The main reaction with strong, bulky bases such as I$^-$ and CH$_3$COO$^-$.
	S$_N$1/E1	Primary cations are never observed in solution, and, therefore, S$_N$1 and E1 reactions of primary halides are never observed.
Secondary R$_2$CHX	S$_N$2	The main reaction with bases/nucleophiles where pK$_a$ of the conjugate acid is 11 or less, as for example I$^-$ and CH$_3$COO$^-$.
	E2	The main reaction with bases/nucleophiles where the pK$_a$ of the conjugate acid is 11 or greater, as for example OH$^-$ and CH$_3$CH$_2$O$^-$.
	S$_N$1/E1	Common in reactions with weak nucleophiles in polar protic solvents, such as water, methanol, and ethanol.
Tertiary R$_3$CX	E2	Main reaction with strong bases such as HO$^-$ and RO$^-$.
	S$_N$1/E1	Main reactions with poor nucleophiles/weak bases.
	S$_N$2	S$_N$2 reactions of tertiary halides are never observed because of the extreme crowding around the 3° carbon.

change in the ratio of S$_N$2/E2 is that as the pK$_a$ of the conjugate acid increases, basicity increases faster than nucleophilicity. In addition, raising the temperature favors elimination because the entropy change of a reaction becomes more important at high temperatures (remember, $\Delta G = \Delta H - T\Delta S$, and the more particles formed in a reaction, the larger the entropy). Elimination reactions give more particles than substitution reactions and have a larger entropy of reaction, and this is reflected in the transition state. These generalizations about substitution versus elimination reactions of methyl, primary, secondary, and tertiary haloalkanes are summarized in Table 9.14.

Example 9.8

Predict whether each reaction proceeds predominantly by substitution (S$_N$1 or S$_N$2) or elimination (E1 or E2) or whether the two compete. Write structural formulas for the major organic product(s).

(a) [structure with Cl] $+ NaOH \xrightarrow[\text{H}_2\text{O}]{80°C}$ **(b)** [structure with Br] $+ (C_2H_5)_3N \xrightarrow[\text{CH}_2\text{Cl}_2]{30°C}$

Solution

(a) A 3° halide is heated with a strong base/good nucleophile. Elimination by an E2 reaction predominates to give 2-methyl-2-butene as the major product.

$$\text{(structure: tert-chloride)} \quad + \text{NaOH} \xrightarrow[\text{H}_2\text{O}]{80°\text{C}} \quad \text{(structure: 2-methyl-2-butene)} \quad + \text{NaCl} + \text{H}_2\text{O}$$

(b) Reaction of a 1° halide with this moderate nucleophile/weak base gives substitution by an S_N2 reaction.

$$\text{(structure: bromide)} \quad + (C_2H_5)_3N \xrightarrow[\text{CH}_2\text{Cl}_2]{30°\text{C}} \quad \text{(structure: ammonium product)} \quad \overset{+}{N}(C_2H_5)_3 \ Br^-$$

Problem 9.8

Predict whether each reaction proceeds predominantly by substitution (S_N1 or S_N2) or elimination (E1 or E2) or whether the two compete. Write structural formulas for the major organic product(s).

(a) $\quad$ (structure: 2-bromobutane derivative) $\quad + \ CH_3O^-Na^+ \xrightarrow[\text{methanol}]{}$

(b) $\quad$ (structure: chlorocyclohexane) $\quad + \ Na^+I^- \xrightarrow[\text{acetone}]{}$

(c) $\quad C_6H_5$ (structure: bromide) $\ + \ Na^+CN^- \xrightarrow[\text{methanol}]{}$

An important lesson to be learned from this chapter is that understanding the key transition state or reactive intermediate geometries as well as the relative transition state energies allows the prediction of product stereochemistry and regiochemistry. Backside attack in S_N2 reactions, the anti and coplanar geometry of the H atom and leaving group in E2 reactions, and the presence of carbocation intermediates in S_N1 and E1 reactions are important examples of reaction geometries that dictate stereochemistry. Understanding the relative energies of alternative possible transition states is also important. In the case of β-elimination reactions, relative transition state energies provide a rationale for Zaitsev's rule of regiochemistry. As you go through the rest of this book, try to learn key features of reaction mechanisms that dictate the stereochemistry and regiochemistry of reaction products. You should think of mechanisms as more than just arrow pushing and electron flow. They involve three-dimensional molecular interactions with associated relative energies and completely control the formation of products.

9.10 Phase-Transfer Catalysis

Very often, nucleophilic substitution involves reactions between a covalent organic compound and an ionic compound, as for example between 1-bromooctane and sodium cyanide.

$$CH_3(CH_2)_6CH_2Br + Na^+CN^- \longrightarrow CH_3(CH_2)_6CH_2CN + Na^+Br^-$$

1-Bromooctane   Nonanenitrile

This reaction is simple to write, but, for it to occur, both compounds must be brought together so that they can react in the same solvent. The solubility characteristics of these reactants are quite different. Sodium cyanide is an ionic solid, soluble in water and a few other polar solvents, but insoluble in nonpolar organic solvents such as dichloromethane. 1-Bromooctane, on the other hand, is insoluble in water but quite soluble in dichloromethane and other nonpolar organic solvents. One way to bring these reactants together is to dissolve them in DMSO, DMF, or other polar aprotic solvents. The advantages of DMSO and DMF are that each dissolves both organic and ionic compounds. When 1-bromooctane and sodium cyanide are dissolved in DMSO, reaction between them occurs very readily.

Although DMSO and DMF are excellent solvents in which to carry out organic reactions, they have certain disadvantages. Both are several times more expensive than solvents such as dichloromethane and ethanol, and, on an industrial scale, solvent cost is an important consideration. Furthermore, because DMSO and DMF are so soluble in water, it is often difficult to recover them from mixtures with water. Finally, because they have higher boiling points than other common solvents (189°C for DMSO and 153°C for DMF compared with 78°C for ethanol and 40°C for dichloromethane), it is often difficult to remove them entirely from organic reaction products by distillation.

Another way to bring about reaction between 1-bromooctane and cyanide ion is by using a **phase-transfer catalyst.** The characteristics necessary for an effective phase-transfer catalyst for anions are a balanced combination of (1) **hydrophilic** character to dissolve in water and form an ion pair with the anion to be transported and (2) **hydrophobic** character to dissolve in the organic phase and transport the anion into it. One commonly used phase-transfer catalyst is the water-soluble salt tetrabutylammonium chloride.

$$(CH_3CH_2CH_2CH_2)_4N^+Cl^-$$

Tetrabutylammonium chloride
$(Bu_4N^+Cl^-)$

Phase-transfer catalyst A substance that transfers ions from an aqueous phase into an organic phase and vice versa.

Hydrophilic From the Greek, meaning water loving.

Hydrophobic From the Greek, meaning water fearing.

The tetrabutylammonium ion, Bu_4N^+, has both hydrophilic and hydrophobic regions. The positively charged nitrogen atom of this ion is a hydrophilic site that interacts with water, with other polar molecules, and with anions. Its four butyl groups are hydrophobic sites and do not interact with water. Thus, because of its hydrophilic positively charged nitrogen atom, the Bu_4N^+ ion is soluble in water, and, because of its four hydrophobic butyl groups, it is also soluble in nonpolar organic solvents.

The way a phase-transfer catalyst works is illustrated using tetrabutylammonium chloride. Suppose sodium cyanide is dissolved in water, 1-bromooctane is dissolved in dichloromethane, and the solutions are mixed. Because water and dichloromethane are immiscible, a two-phase system results [Figure 9.7(a)]. Dichloromethane is denser than water and forms the lower layer. No reaction takes place between 1-bromooctane and cyanide ion because they are in different phases.

When added to this two-phase system, $Bu_4N^+Cl^-$ dissolves in the upper aqueous phase [Figure 9.7(b)]. Bu_4N^+ and CN^- then form an ion pair that is transferred into the organic phase [Figure 9.7(c)]. CN^- displaces the bromine atom [Figure 9.7(d)], and Br^- is transferred as an ion pair with Bu_4N^+ to the aqueous phase [Figure 9.7(e)]. This process is repeated until all cyanide ions have been transferred to the organic phase and have reacted with 1-bromooctane. Chloride is also transferred, but it is a weak nucleophile and doesn't react appreciably.

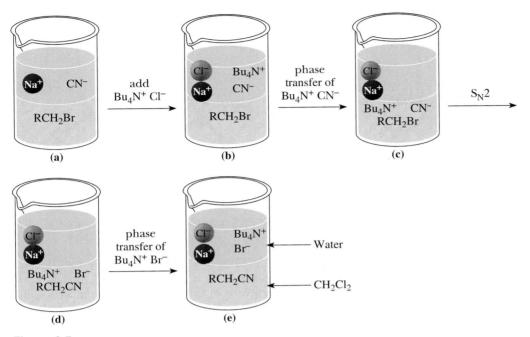

Figure 9.7

Phase-transfer catalysis. (a) NaCN is dissolved in water (the upper phase), and 1-bromooctane is dissolved in dichloromethane (the lower phase), the (b) phase-transfer catalyst $Bu_4N^+Cl^-$ is added, and the mixture is shaken. (c) CN^- is transferred from the aqueous phase to the organic phase as an ion pair with Bu_4N^+ and (d) displaces Br^- from 1-bromooctane. (e) Br^- forms an ion pair with Bu_4N^+ and is transferred from the organic phase to the aqueous phase.

9.11 Neighboring Group Participation

Thus far, we have considered two limiting mechanisms for nucleophilic substitutions that focus on the degree of covalent bonding between the nucleophile and the substitution center during departure of the leaving group. In an S_N2 mechanism, the nucleophile assists the leaving group in its departure. In an S_N1 mechanism, the leaving group is not assisted in this way. An essential criterion for distinguishing between these two pathways is the order of reaction. Nucleophile-assisted substitutions are second order: first order in RX and first order in nucleophile. Nucleophile-unassisted substitutions are first order: first order in RX and zero order in nucleophile.

Chemists recognize that certain nucleophilic substitutions with the kinetic characteristics of first-order (S_N1) substitution involve two successive displacement reactions. A characteristic feature of a great many of these reactions is the presence of an internal nucleophile, most commonly sulfur, nitrogen, or oxygen, on the carbon atom beta to the leaving group. This neighboring nucleophile participates in the departure of the leaving group to give an intermediate, which then reacts with an external nucleophile to complete the reaction.

The mustard gases are one group of compounds that react by participation of a neighboring group. The characteristic structural feature of a mustard gas is a two-carbon chain, with a halogen on one carbon and a divalent sulfur or trivalent nitrogen on the other carbon (S—C—C—Lv or N—C—C—Lv). An example of a mustard gas is *bis*(2-chloroethyl)sulfide, a poison gas used extensively in World War I and at one time, at least, manufactured by Iraq. This compound is a deadly vesicant (blistering agent) and quickly causes conjunctivitis and blindness.

ClCH₂CH₂SCH₂CH₂Cl

Bis(2-chloroethyl)sulfide
(a sulfur mustard gas)

$$CH_3$$
$$|$$
$$ClCH_2CH_2-N-CH_2CH_2Cl$$

Bis(2-chloroethyl)methylamine
(a nitrogen mustard gas)

Bis(2-chloroethyl)sulfide and *bis*(2-chloroethyl)methylamine are not gases at all. They are oily liquids with a high vapor pressure, hence the designation "gas." Nitrogen and sulfur mustards react very rapidly with moisture in the air and in the mucous membranes of the eye, nose, and throat to produce HCl, which then burns and blisters these sensitive tissues. What is unusual about the reactivity of the mustard gases is that they react very rapidly with water, a very poor nucleophile.

Mustard gases also react rapidly with other nucleophiles, such as those in biological molecules, which makes them particularly dangerous chemicals. Of the two steps in the mechanism of the hydrolysis of a sulfur mustard, the first is the slower and is rate determining. As a result, the rate of reaction is proportional to the concentration of the sulfur mustard but independent of the concentration of the external nucleophile. Thus, although this reaction has the kinetic characteristics of an S_N1 reaction, it actually involves two successive S_N2 displacement reactions.

Mechanism *Hydrolysis of a Sulfur Mustard—Participation by a Neighboring Group*

Step 1: The reason for the extremely rapid hydrolysis of the sulfur mustards is neighboring group participation by sulfur in the ionization of the carbon-chlorine bond to form a cyclic sulfonium ion. This is the rate-determining step of the reaction; although it is the slowest step, it is much faster than reaction of a typical primary chloride with water. At this point, you should review halogenation of alkenes (Sections 6.3D and 6.3F) and compare the cyclic halonium ion intermediates formed there with the cyclic sulfonium ion intermediate formed here.

A cyclic
sulfonium ion

Step 2: The cyclic sulfonium ion contains a highly strained three-membered ring and reacts rapidly with an external nucleophile to open the ring followed by proton transfer to H_2O to give H_3O^+. In this S_N2 reaction, H_2O is the nucleophile, and sulfur is the leaving group.

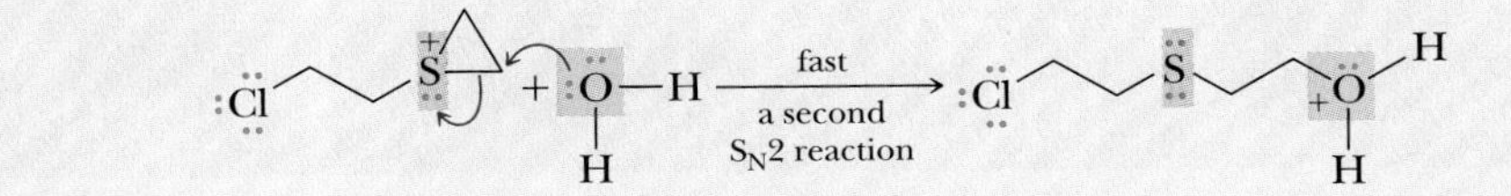

Step 3: Proton transfer to water completes the reaction.

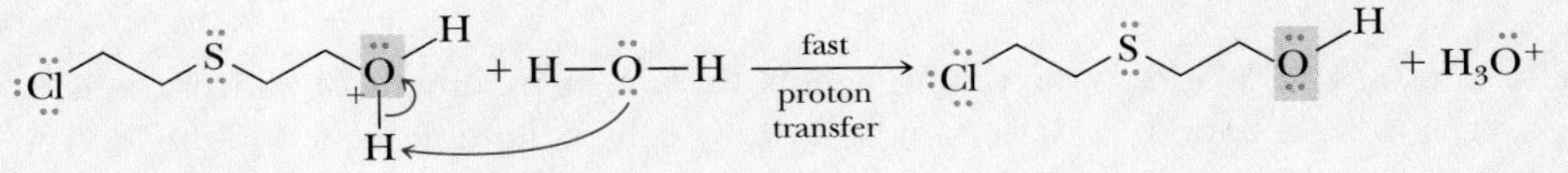

The net effect of these reactions is nucleophilic substitution of Cl by OH.

We continue to use the terms S_N2 and S_N1 to describe nucleophilic substitution reactions. You should realize, however, that these designations do not adequately describe all nucleophilic substitution reactions.

Example 9.9

Write a mechanism for the hydrolysis of the nitrogen mustard *bis*(2-chloroethyl)-methylamine.

Solution

Following is a three-step mechanism.

Step 1: An internal S_N2 reaction in which ionization of the C—Cl bond is assisted by the neighboring nitrogen atom to form a highly strained three-membered ring.

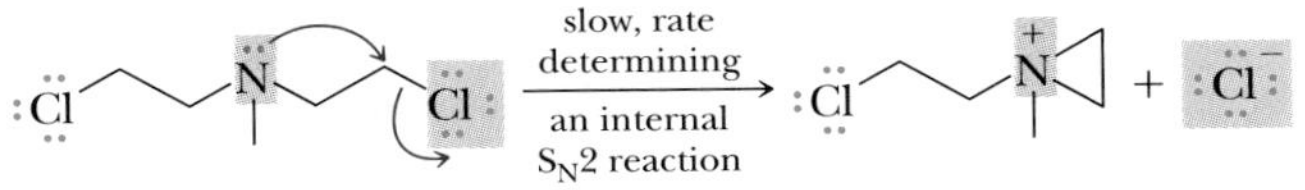

Step 2: Reaction of the cyclic ammonium ion with water opens the three-membered-ring. In this S_N2 reaction, H_2O is the nucleophile, and nitrogen is the leaving group.

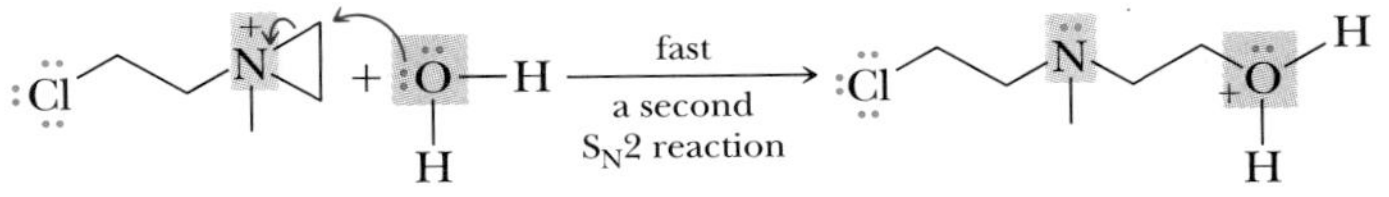

Step 3: Proton transfer to the basic nitrogen completes the reaction.

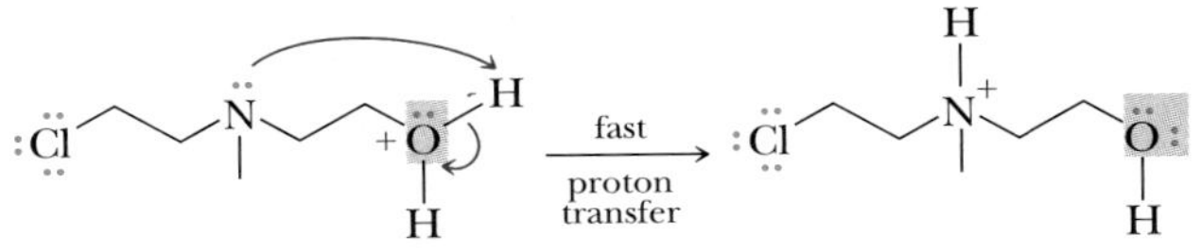

Problem 9.9

Knowing what you do about the regioselectivity of S_N2 reactions, predict the product of hydrolysis of this compound.

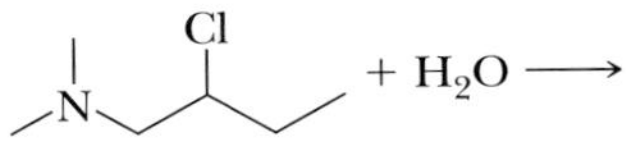

CONNECTIONS TO
BIOLOGICAL CHEMISTRY

Mustard Gases and the Treatment of Neoplastic Diseases

Autopsies of soldiers killed by sulfur mustard in World War I revealed, among other things, very low white blood cell counts and defects in bone marrow development. From these observations, it was realized that sulfur mustards have profound effects on rapidly dividing cells. This became a lead observation in the search for less toxic alkylating agents for use in treatment of cancers, which have rapidly dividing cells. Attention turned to the less reactive nitrogen mustards. One of the first compounds tested was mechlorethamine. As with other mustards, the reaction of mechlorethamine with nucleophiles is rapid because of the formation of an aziridinium ion.

$$H_3C-N \xrightarrow[\text{(internal } S_N2)]{} H_3C-N \xrightarrow[(S_N2)]{-H^+} H_3C-N + HCl$$

Mechlorethamine   (An aziridinium ion intermediate)

Mechlorethamine undergoes very rapid reaction with water (hydrolysis) and with other nucleophiles, so much so that within minutes after injection into the body, it has completely reacted. The problem for the chemist, then, was to find a way to decrease the nucleophilicity of nitrogen while maintaining reasonable water solubility. Substitution of phenyl for methyl reduced the nucleophilicity, but the resulting compound was not sufficiently soluble in water for intravenous injection. The solubility problem was solved by adding a carboxyl group. When the carboxyl group was added directly to the aromatic ring, however, the resulting compound was too stable and, therefore, not biologically active.

Adding a propyl bridge (chlorambucil) or an aminoethyl bridge (melphalan) between the aromatic ring and the carboxyl group solved both the solubility problem and the reactivity problem. Note that melphalan is chiral. It has been demonstrated that the R and S enantiomers have approximately equal therapeutic potency.

Chlorambucil   Melphalan

The clinical value of the nitrogen mustards lies in the fact that they undergo reaction with certain nucleophilic sites on the heterocyclic aromatic amine bases in DNA (see Chapter 28). For DNA, the most reactive nucleophilic site is N-7 of guanine. Next in reactivity is N-3 of adenine, followed by N-3 of cytosine.

The nitrogen mustards are bifunctional alkylating agents; one molecule of nitrogen mustard undergoes reaction with two molecules of nucleophile. Guanine alkylation leaves one free reactive alkylating group,

Nucleophilicity of nitrogen acceptable, but the compound is too insoluble in water for intravenous injection

Solubility in water is acceptable, but nucleophilicity of nitrogen reduced and compound is unreactive

which can react with another base, giving cross links that lead to miscoding during DNA replication. The therapeutic value of the nitrogen mustards lies in their ability to disrupt normal base pairing. This prevents replication of the cells, and the rapidly dividing cancer cells are more sensitive than normal cells.

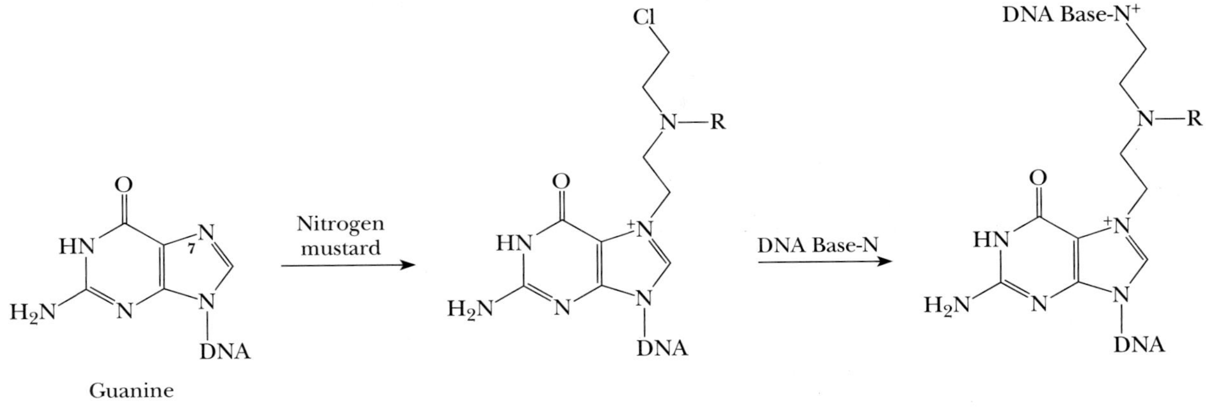

Summary

A **nucleophile (Nu:)** is an electron-rich molecule or ion that donates a pair of electrons to another atom or ion to form a new covalent bond. **Nucleophilic substitution** is any reaction in which a nucleophile replaces another electron-rich group called a **leaving group**. There are two limiting mechanisms for nucleophilic substitution, namely S_N2 and S_N1. In the S_N2 reaction mechanism, bond forming and bond breaking occur simultaneously. In the S_N1 mechanism, the leaving group departs first, leaving a carbocation that reacts with the nucleophile in a second step.

Protic solvents are hydrogen-bond donors (Section 9.2). The most common protic solvents are those containing —OH groups. **Aprotic solvents** cannot serve as hydrogen-bond donors. Common aprotic solvents are acetone, diethyl ether, dimethyl sulfoxide, and N,N-dimethylformamide. Polar solvents (Section 9.2) interact strongly with ions and polar molecules. Nonpolar solvents (Section 9.2) do not interact strongly with ions and polar molecules. The **dielectric constant** is the most commonly used measure of solvent polarity.

Solvolysis is a nucleophilic substitution reaction in which the solvent is the nucleophile (Section 9.3).

The **nucleophilicity** of a reagent is measured by the rate of its reaction in a reference nucleophilic substitution (Section 9.4B). Relative nucleophilicities depend on whether a reaction is carried out in a polar protic solvent or a polar aprotic solvent. A general principle is that the freer a nucleophile is from a surrounding solvation shell, the greater its nucleophilicity will be.

The ability of a group to function as a leaving group is related to its stability as an anion (Section 9.4F). The most stable anions and the best leaving groups are the conjugate bases of strong acids.

A **β-elimination** reaction (Section 9.6) involves removal of atoms or groups of atoms from adjacent carbon atoms. β-Elimination to give the more highly substituted alkene is called **Zaitsev elimination.** This reaction can proceed via unimolecular (**E1**) or bimolecular (**E2**) mechanisms (Section 9.7).

Nucleophilic substitution and β-elimination are competing processes. Prediction of which process will dominate in a given reaction can be made by evaluating the strength of the nucleophile compared to its basicity, the structure of the alkyl halide, and the solvent used (Section 9.9).

A thorough knowledge of reaction mechanisms is the key to predicting product stereochemistry and regiochemistry. In particular, known requirements for key transition state or intermediate geometries and their relative energies can be used to decide which product stereoisomers or regioisomers will be seen in the products.

A **phase-transfer catalyst** is a substance that transports ions from an aqueous phase into an organic phase and vice versa (Section 9.10). Effective phase-transfer catalysts for anions have a balanced combination of (1) hydrophilic character to dissolve in water and form an ion pair with the anion to be transferred and (2) hydrophobic character to dissolve in the organic phase and transfer the anion into it.

Certain nucleophilic displacements that have the kinetic characteristic of S_N1 reactions (first order in haloalkane and zero order in nucleophile) involve two successive S_N2 reactions. Many such reactions involve participation of a neighboring nucleophile. The mustard gases are one group of compounds whose nucleophilic substitution reactions involve neighboring group participation (Section 9.11).

Key Reactions

ORGANIC **Chemistry Now™**
Click Flashcards to review **Nucleophilic Substitution and β-Elimination Reactions**

1. Nucleophilic Aliphatic Substitution: S_N2 (Section 9.3)

S_N2 reactions occur in one step; departure of the leaving group is assisted by the incoming nucleophile, and both nucleophile and leaving group are involved in the transition state. The nucleophile may be negatively charged as in the first example or neutral as in the second example.

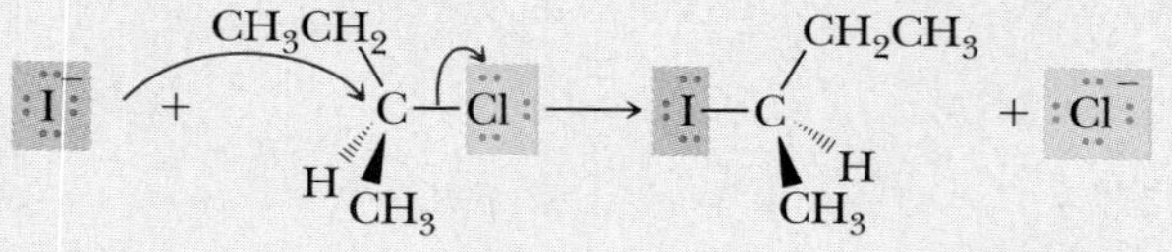

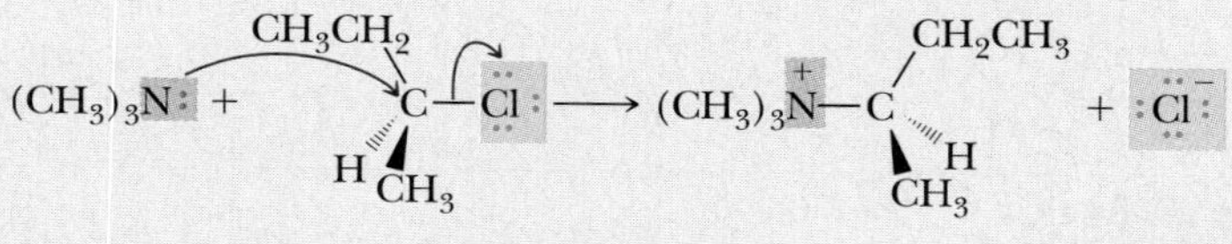

S_N2 reactions result in inversion of configuration at the reaction center. They are accelerated in polar aprotic solvents compared with polar protic solvents. The relative rates of S_N2 reactions are governed by steric factors, namely the degree of crowding around the site of reaction.

2. Nucleophilic Aliphatic Substitution: S_N1 (Section 9.3)

An S_N1 reaction occurs in two steps. Step 1 is a slow, rate-determining ionization of the C—Lv bond to form a carbocation intermediate followed in Step 2 by rapid reaction of the carbocation intermediate with a nucleophile to complete the substitution. Reaction at a stereocenter gives largely racemization, often accompanied with a slight excess of inversion of configuration. Reactions often involve carbocation rearrangements and are accelerated by polar protic solvents. S_N1 reactions are governed by electronic factors, namely the relative stabilities of carbocation intermediates. The following reaction involves an S_N1 reaction, with a hydride rearrangement.

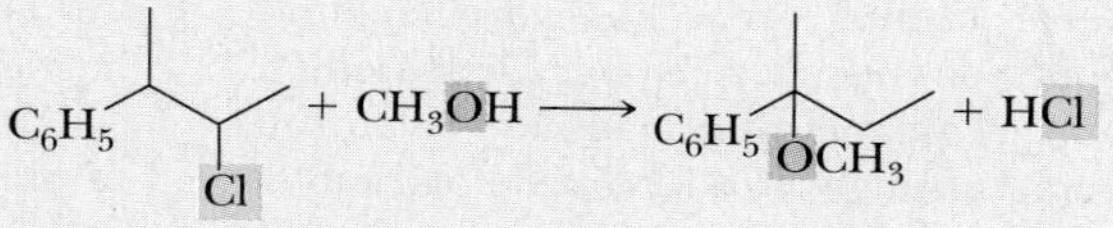

3. β-Elimination: E1 (Section 9.6, 9.7)

An E1 reaction occurs in two steps: slow, rate-determining breaking of the C—Lv bond to form a carbocation intermediate followed by rapid proton transfer to solvent to form an alkene. An E1 reaction is first order in haloalkane and zero order in base. Skeletal rearrangements are common.

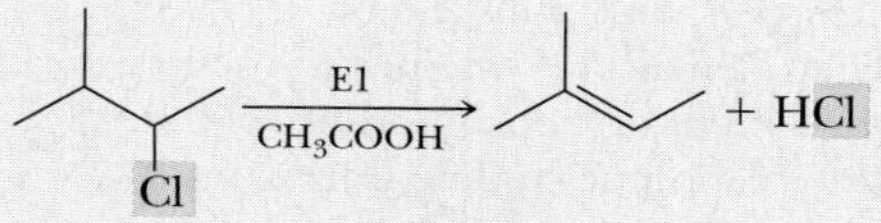

4. β-Elimination: E2 (Section 9.6, 9.7)

An E2 reaction occurs in one step: simultaneous reaction with the base to remove a hydrogen, formation of the alkene, and departure of the leaving group. Elimination is stereoselective, requiring an anti and coplanar arrangement of the groups being eliminated. It is also stereospecific, since different isomers give different products.

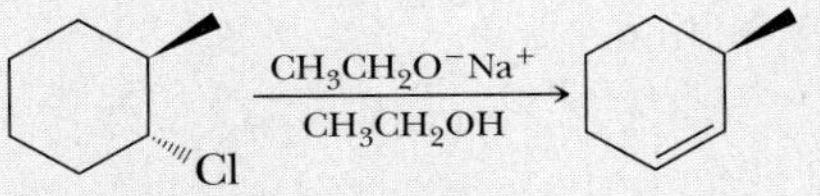

5. Neighboring Group Participation (Section 9.11)

Neighboring-group participation is characterized by first-order kinetics and participation of an internal nucleophile in departure of the leaving group, as in hydrolysis of a sulfur or nitrogen mustard gas. The mechanism for their solvolysis involves two successive nucleophilic displacements.

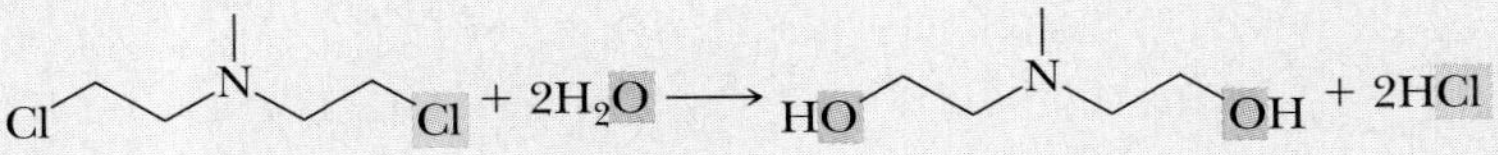

Problems

Nucleophilic Aliphatic Substitution

9.10 Draw a structural formula for the most stable carbocation of each molecular formula.

(a) $C_4H_9^+$ (b) $C_3H_7^+$ (c) $C_5H_{11}^+$ (d) $C_3H_7O^+$

9.11 The reaction of 1-bromopropane and sodium hydroxide in ethanol follows an S_N2 mechanism. What happens to the rate of this reaction under the following conditions?

(a) The concentration of NaOH is doubled.

(b) The concentrations of both NaOH and 1-bromopropane are doubled.

(c) The volume of the solution in which the reaction is carried out is doubled.

9.12 From each pair, select the stronger nucleophile.

(a) H_2O or OH^- (b) CH_3COO^- or OH^-
(c) CH_3SH or CH_3S^- (d) Cl^- or I^- in DMSO
(e) Cl^- or I^- in methanol   (f) CH_3OCH_3 or CH_3SCH_3

9.13 Draw a structural formula for the product of each S_N2 reaction. Where configuration of the starting material is given, show the configuration of the product.

(a) $CH_3CH_2CH_2Cl + CH_3CH_2O^-Na^+ \xrightarrow[ethanol]{}$ (b) $(CH_3)_3N + CH_3I \xrightarrow[acetone]{}$

(c) $-CH_2Br + Na^+CN^- \xrightarrow[acetone]{}$ (d) $H_3C$$Cl + CH_3S^-Na^+ \xrightarrow[ethanol]{}$

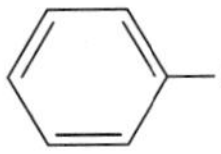

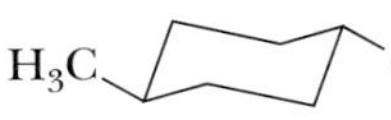

(e) $CH_3CH_2CH_2Cl + CH_3C{\equiv}C^-Li^+ \xrightarrow[diethyl\ ether]{}$ (f) $-CH_2Cl + NH_3 \xrightarrow[ethanol]{}$

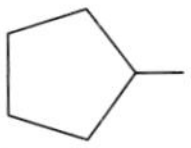

(g) O$NH + CH_3(CH_2)_6CH_2Cl \xrightarrow[ethanol]{}$ (h) $CH_3CH_2CH_2Br + Na^+CN^- \xrightarrow[acetone]{}$

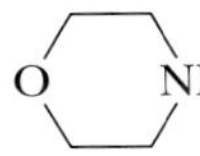

9.14 You were told that each reaction in Problem 9.13 proceeds by an S_N2 mechanism. Suppose that you were not told the mechanism. Describe how you could conclude from the structure of the haloalkane, the nucleophile, and the solvent that each reaction is in fact an S_N2 reaction.

9.15 Treatment of 1,3-dichloropropane with potassium cyanide results in the formation of pentanedinitrile. The rate of this reaction is about 1000 times greater in DMSO than it is in ethanol. Account for this difference in rate.

1,3-Dichloropropane   Pentanedinitrile

9.16 Treatment of 1-aminoadamantane, $C_{10}H_{17}N$, with methyl 2,4-dibromobutanoate in the presence of a nonnucleophilic base, R_3N, involves two successive S_N2 reactions and gives compound A. Propose a structural formula for compound A.

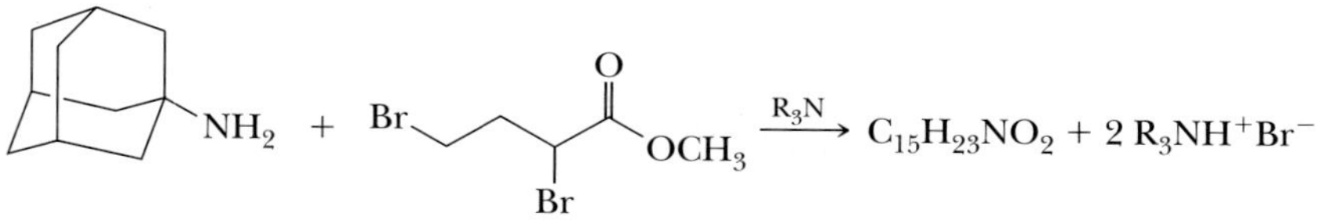

1-Aminoadamantane   Methyl   A
                         2,4-dibromobutanoate

9.17 Select the member of each pair that shows the greater rate of S_N2 reaction with KI in acetone.

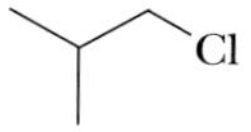

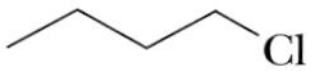

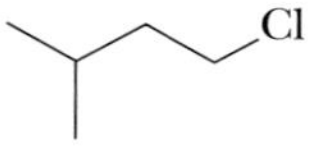

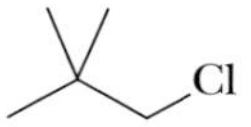

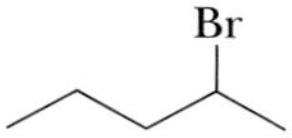

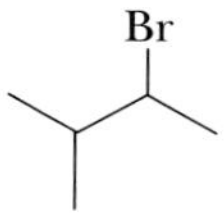

9.18 Select the member of each pair that shows the greater rate of S_N2 reaction with KN_3 in acetone.

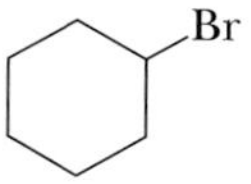

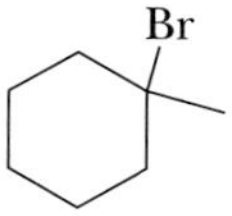

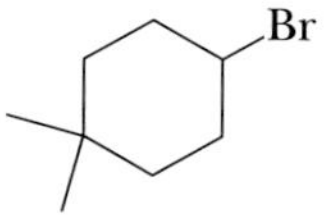

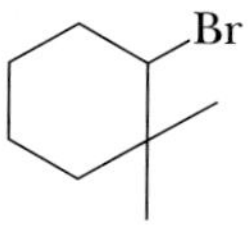

9.19 What hybridization best describes the reacting carbon in the S_N2 transition state?

9.20 Each carbocation is capable of rearranging to a more stable carbocation. Limiting yourself to a single 1,2-shift, suggest a structure for the rearranged carbocation.

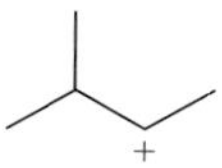

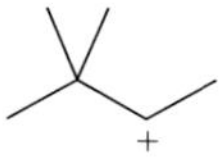

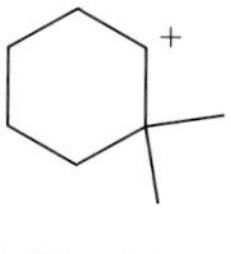

9.21 Attempts to prepare optically active iodides by nucleophilic displacement on optically active bromides using I^- normally produce racemic iodoalkanes. Why are the product iodoalkanes racemic?

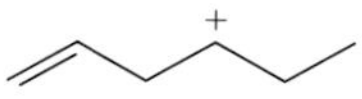

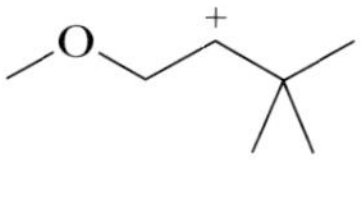

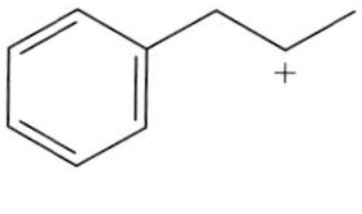

9.22 Draw a structural formula for the product of each S_N1 reaction. Where configuration of the starting material is given, show the configuration of the product.

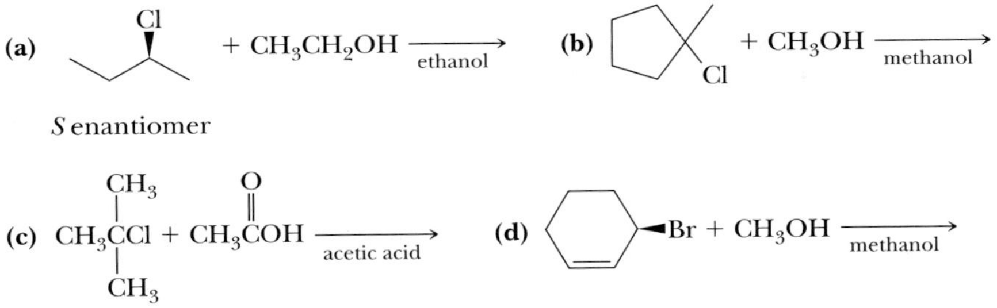

(a) [structure: Cl on secondary carbon] + CH_3CH_2OH $\xrightarrow{\text{ethanol}}$ (b) [cyclopentane structure with Cl] + CH_3OH $\xrightarrow{\text{methanol}}$

S enantiomer

(c) $CH_3\overset{\displaystyle CH_3}{\underset{\displaystyle CH_3}{\overset{|}{\underset{|}{C}}}}Cl$ + $CH_3\overset{\displaystyle O}{\overset{\|}{C}}OH$ $\xrightarrow{\text{acetic acid}}$ (d) [cyclohexene structure]◀Br + CH_3OH $\xrightarrow{\text{methanol}}$

9.23 Suppose that you were told that each reaction in Problem 9.22 is a substitution reaction but were not told the mechanism. Describe how you could conclude from the structure of the haloalkane or cycloalkene, the nucleophile, and the solvent that each reaction is in fact an S_N1 reaction.

9.24 Alkenyl halides such as vinyl bromide, $CH_2{=}CHBr$, undergo neither S_N1 nor S_N2 reactions. What factors account for this lack of reactivity?

9.25 Select the member of each pair that undergoes S_N1 solvolysis in aqueous ethanol more rapidly.

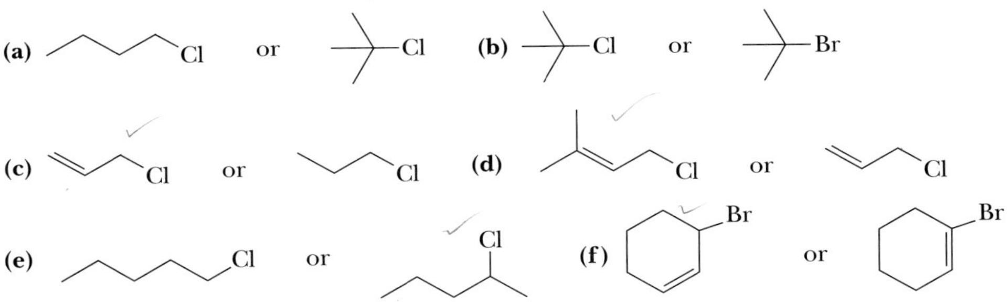

(a) [structure]Cl   or   [structure]Cl   (b) [structure]Cl   or   [structure]Br

(c) [structure]Cl   or   [structure]Cl   (d) [structure]Cl   or   [structure]Cl

(e) [structure]Cl   or   [structure with Cl]   (f) [cyclohexene with Br]   or   [cyclohexene with Br]

9.26 Account for the following relative rates of solvolysis under experimental conditions favoring S_N1 reaction.

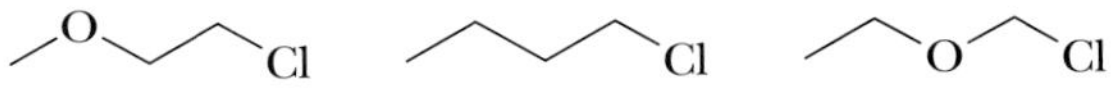

[structure: O–CH₂CH₂–Cl]   [structure: CH₂CH₂CH₂–Cl]   [structure: O–CH₂–Cl]

Relative rate of   0.2   1 10^9
solvolysis (S_N1)

9.27 Not all tertiary haloalkanes undergo S_N1 reactions readily. For example, the bicyclic compound shown here is very unreactive under S_N1 conditions. What feature of this molecule is responsible for such lack of reactivity? You will find it helpful to examine a model of this compound.

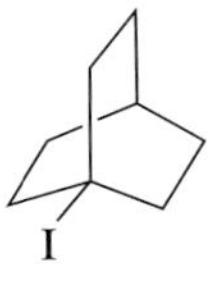

[structure with I]   1-Iodobicyclo[2.2.2]octane

9.28 Show how you might synthesize the following compounds from a haloalkane and a nucleophile.

(a) 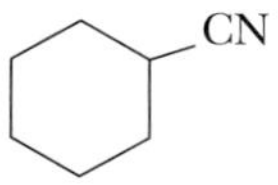 [structure: cyclohexyl-CN]   (b) 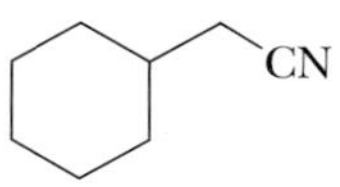 [structure: cyclohexyl-CH2-CN]   (c) 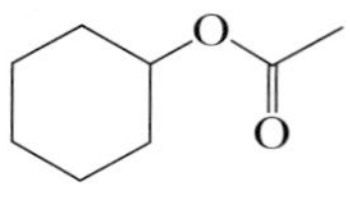 [structure: cyclohexyl-O-C(=O)-CH3]   (d) 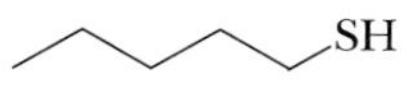 [structure: CH3CH2CH2CH2-SH]

(e) 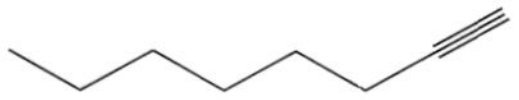 [structure: alkyne chain]   (f) 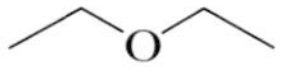 [structure: CH3CH2-O-CH2CH3 ether]   (g) 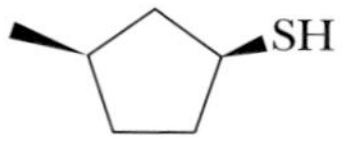 [structure: cyclopentane with SH]

9.29 3-Chloro-1-butene reacts with sodium ethoxide in ethanol to produce 3-ethoxy-1-butene. The reaction is second order; first order in 3-chloro-1-butene and first order in sodium ethoxide. In the absence of sodium ethoxide, 3-chloro-1-butene reacts with ethanol to produce both 3-ethoxy-1-butene and 1-ethoxy-2-butene. Explain these results.

9.30 1-Chloro-2-butene undergoes hydrolysis in warm water to give a mixture of these allylic alcohols. Propose a mechanism for their formation.

$$CH_3CH=CHCH_2Cl \xrightarrow{H_2O} CH_3CH=CHCH_2OH + CH_3\overset{\overset{\displaystyle OH}{|}}{C}HCH=CH_2$$

1-Chloro-2-butene   2-Buten-1-ol   3-Buten-2-ol
                                                          (racemic)

9.31 The following nucleophilic substitution occurs with rearrangement. Suggest a mechanism for formation of the observed product. If the starting material has the *S* configuration, what is the configuration of the stereocenter in the product?

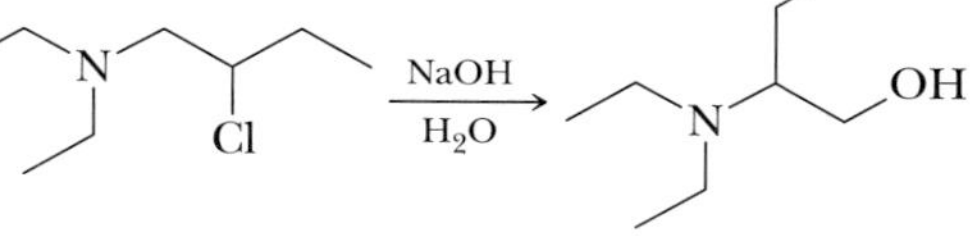

[structure: amine with Cl] $\xrightarrow[H_2O]{NaOH}$ [structure: amine with OH]

9.32 Propose a mechanism for the formation of these products in the solvolysis of this bromoalkane.

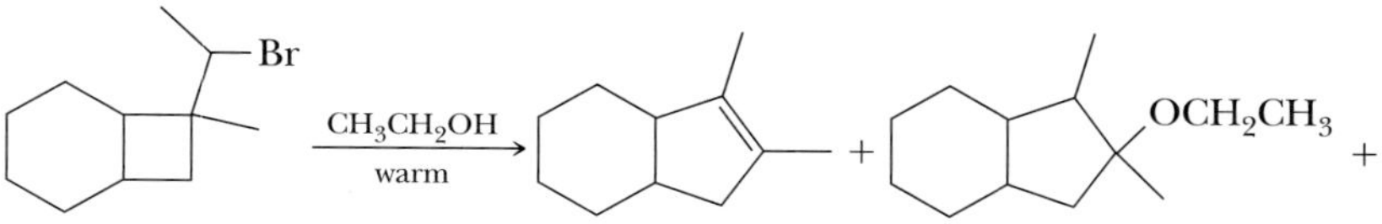

[structure: bicyclic with Br] $\xrightarrow[warm]{CH_3CH_2OH}$ [structure alkene] + [structure with OCH₂CH₃] + HBr

9.33 Solvolysis of the following bicyclic compound in acetic acid gives a mixture of products, two of which are shown. The leaving group is the anion of a sulfonic acid, ArSO₃H. A sulfonic acid is a strong acid, and its anion, $ArSO_3^-$, is a weak base and a good leaving group. Propose a mechanism for this reaction.

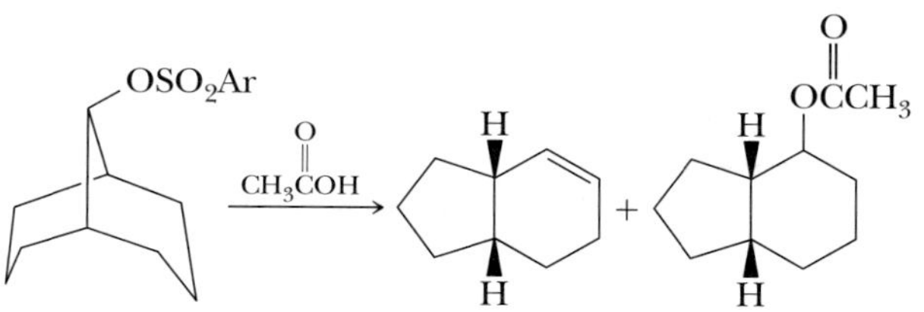

[structure: OSO₂Ar bicyclic] $\xrightarrow{CH_3COH}$ [structure with H] + [structure with OCCH₃]

9.34 Which compound in each set undergoes more rapid solvolysis when refluxed in ethanol? Show the major product formed from the more reactive compound.

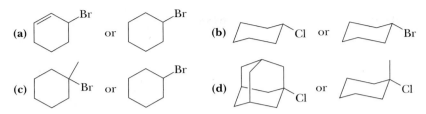

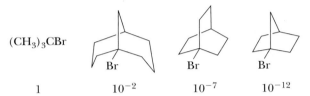

9.35 Account for the relative rates of solvolysis of these compounds in aqueous acetic acid.

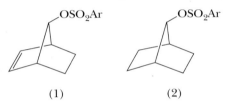

$(CH_3)_3CBr$

1   10^{-2} 10^{-7} 10^{-12}

9.36 A comparison of the rates of S_N1 solvolysis of the bicyclic compounds (1) and (2) in acetic acid shows that compound (1) reacts 10^{11} times faster than compound (2). Furthermore, solvolysis of (1) occurs with complete retention of configuration: The nucleophile occupies the same position on the one-carbon bridge as did the leaving OSO_2Ar group.

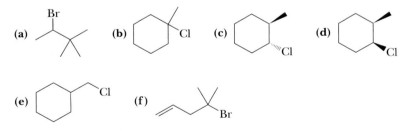

(1)   (2)

(a) Draw structural formulas for the products of solvolysis of each compound.
(b) Account for the difference in rate of solvolysis of (1) and (2).
(c) Account for complete retention of configuration in the solvolysis of (1).

β-Eliminations

9.37 Draw structural formulas for the alkene(s) formed by treatment of each haloalkane or halocycloalkane with sodium ethoxide in ethanol. Assume that elimination occurs by an E2 mechanism.

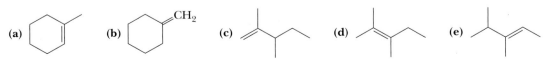

9.38 Draw structural formulas of all chloroalkanes that undergo dehydrohalogenation when treated with KOH to give each alkene as the major product. For some parts, only one chloroalkane gives the desired alkene as the major product. For other parts, two chloroalkanes may work.

9.39 Following are diastereomers (A) and (B) of 3-bromo-3,4-dimethylhexane. On treatment with sodium ethoxide in ethanol, each gives 3,4-dimethyl-3-hexene as the major product. One diastereomer gives the *E* alkene, and the other gives the *Z* alkene. Which diastereomer gives which alkene? Account for the stereoselectivity and stereospecificity of each β-elimination.

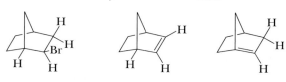

(A)   (B)

9.40 Treatment of the following stereoisomer of 1-bromo-1,2-diphenylpropane with sodium ethoxide in ethanol gives a single stereoisomer of 1,2-diphenylpropene. Predict whether the product has the *E* configuration or the *Z* configuration.

$$\underset{\text{1-Bromo-1,2-diphenylpropane}}{\text{H}_3\text{C}\overset{\text{HC}_6\text{H}_5}{\underset{\text{HC}_6\text{H}_5}{\text{C}}\text{---}\text{C}\text{---Br}}} \quad \xrightarrow[\text{CH}_3\text{CH}_2\text{OH}]{\text{CH}_3\text{CH}_2\text{O}^-\text{Na}^+} \quad \underset{\text{1,2-Diphenylpropene}}{\text{C}_6\text{H}_5\text{CH}=\overset{\text{CH}_3}{\underset{}{\text{C}}}\text{C}_6\text{H}_5}$$

9.41 Elimination of HBr from 2-bromonorbornane gives only 2-norbornene and no 1-norbornene. How do you account for the regioselectivity of this dehydrohalogenation? In answering this question, you will find it helpful to look at molecular models of both 1-norbornene and 2-norbornene and to analyze the strain in each.

2-Bromonorbornane   2-Norbornene   1-Norbornene

9.42 Which isomer of 1-bromo-3-isopropylcyclohexane reacts faster when refluxed with potassium *tert*-butoxide, the *cis* isomer or the *trans* isomer? Draw the structure of the expected product from the faster reacting isomer.

Substitution Versus Elimination

9.43 Consider the following statements in reference to S_N1, S_N2, E1, and E2 reactions of haloalkanes. To which mechanism(s), if any, does each statement apply?

   (a) Involves a carbocation intermediate.
   (b) Is first-order in haloalkane and first-order in nucleophile.
   (c) Involves inversion of configuration at the site of substitution.
   (d) Involves retention of configuration at the site of substitution.
   (e) Substitution at a stereocenter gives predominantly a racemic product.
   (f) Is first order in haloalkane and zero order in base.
   (g) Is first order in haloalkane and first order in base.
   (h) Is greatly accelerated in protic solvents of increasing polarity.
   (i) Rearrangements are common.
   (j) Order of reactivity of haloalkanes is 3° > 2° > 1°.
   (k) Order of reactivity of haloalkanes is methyl > 1° > 2° > 3°.

9.44 Arrange these haloalkanes in order of increasing ratio of E2 to S_N2 products observed on reaction of each with sodium ethoxide in ethanol.

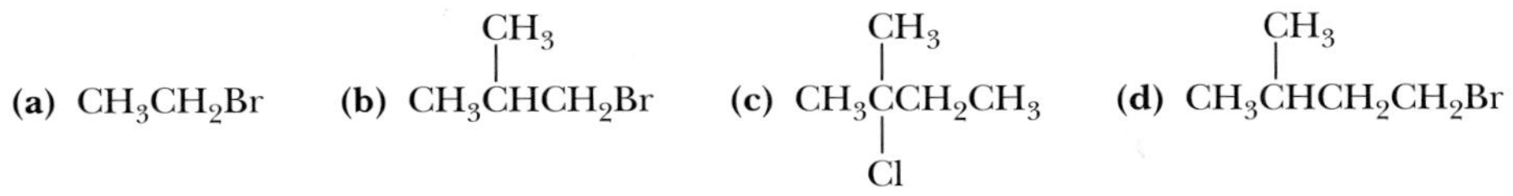

$$
\text{(a)} \ CH_3CH_2Br \qquad \text{(b)} \ CH_3\overset{\overset{\displaystyle CH_3}{|}}{C}HCH_2Br \qquad \text{(c)} \ CH_3\overset{\overset{\displaystyle CH_3}{|}}{\underset{\underset{\displaystyle Cl}{|}}{C}}CH_2CH_3 \qquad \text{(d)} \ CH_3\overset{\overset{\displaystyle CH_3}{|}}{C}HCH_2CH_2Br
$$

9.45 Draw a structural formula for the major organic product of each reaction and specify the most likely mechanism by which each is formed.

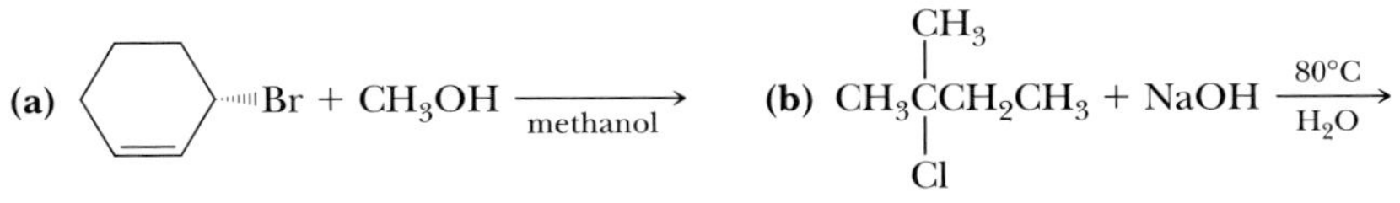

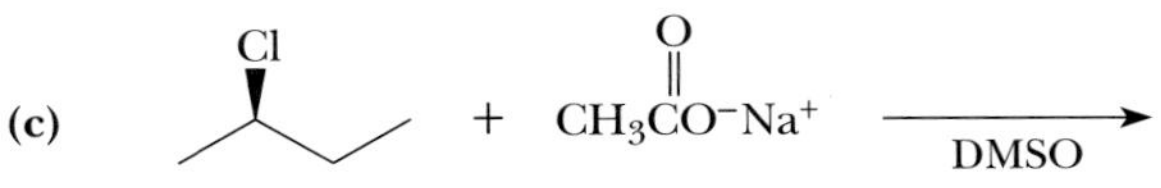

(a) $\overset{\dots}{}Br + CH_3OH \xrightarrow[\text{methanol}]{}$

(b) $CH_3\overset{\overset{\displaystyle CH_3}{|}}{\underset{\underset{\displaystyle Cl}{|}}{C}}CH_2CH_3 + NaOH \xrightarrow[H_2O]{80°C}$

(c) $+ \ CH_3\overset{\overset{\displaystyle O}{\|}}{C}O^-Na^+ \xrightarrow[\text{DMSO}]{}$

(*R*)-2-Chlorobutane

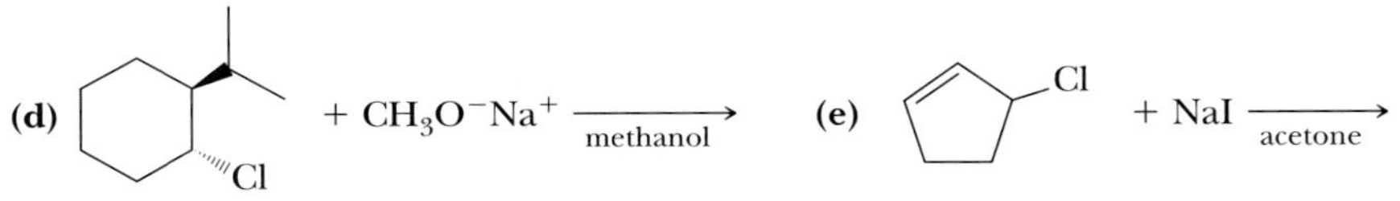

(d) $+ CH_3O^-Na^+ \xrightarrow[\text{methanol}]{}$

(e) $+ \ NaI \xrightarrow[\text{acetone}]{}$

R enantiomer

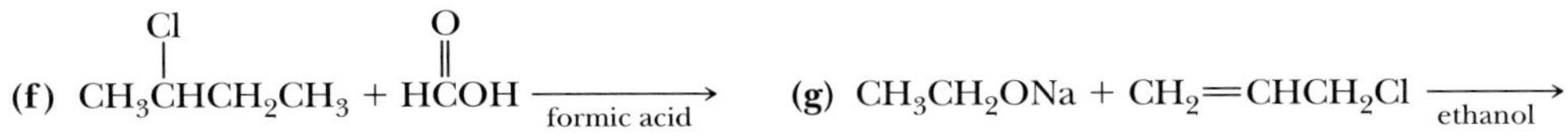

(f) $CH_3\overset{\overset{\displaystyle Cl}{|}}{C}HCH_2CH_3 + H\overset{\overset{\displaystyle O}{\|}}{C}OH \xrightarrow[\text{formic acid}]{}$

R enantiomer

(g) $CH_3CH_2ONa + CH_2{=}CHCH_2Cl \xrightarrow[\text{ethanol}]{}$

9.46 When *cis*-4-chlorocyclohexanol is treated with sodium hydroxide in ethanol, it gives mainly the substitution product *trans*-1,4-cyclohexanediol (1). Under the same reaction conditions, *trans*-4-chlorocyclohexanol gives 3-cyclohexenol (2) and the bicyclic ether (3).

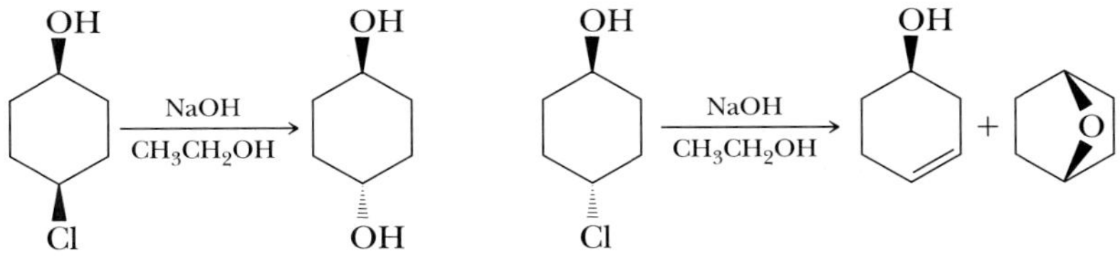

| *cis*-4-Chloro-cyclohexanol | (1) | *trans*-4-Chloro-cyclohexanol | (2) | (3) |

(a) Propose a mechanism for formation of product (1), and account for its configuration.
(b) Propose a mechanism for formation of product (2).
(c) Account for the fact that the bicyclic ether (3) is formed from the *trans* isomer but not from the *cis* isomer.

Synthesis

9.47 Show how to convert the given starting material into the desired product. Note that some syntheses require only one step, whereas others require two or more.

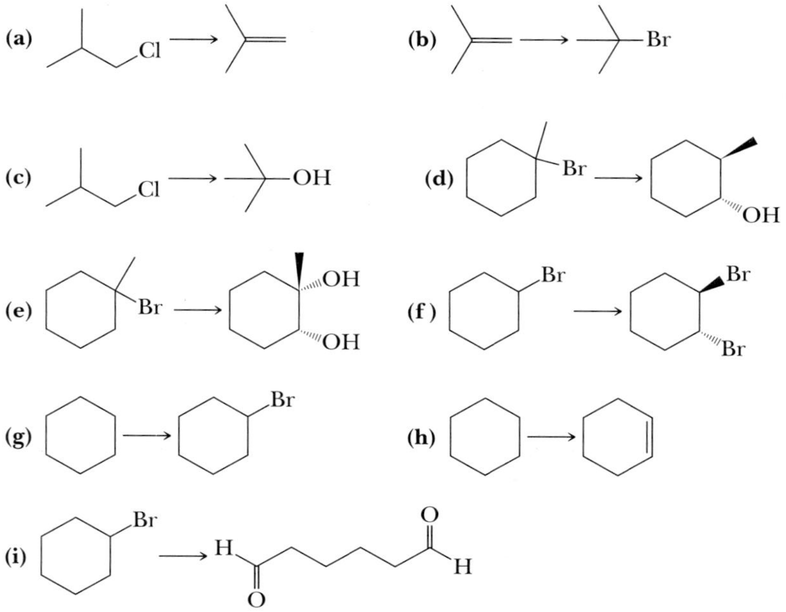

(a)

(b)

(c)

(d)

(e)

(f)

(g)

(h)

(i)

9.48 The Williamson ether synthesis involves treatment of a haloalkane with a metal alkoxide. Following are two reactions intended to give benzyl *tert*-butyl ether. One reaction gives the ether in good yield, the other reaction does not. Which reaction gives the ether? What is the product of the other reaction? How do you account for its formation?

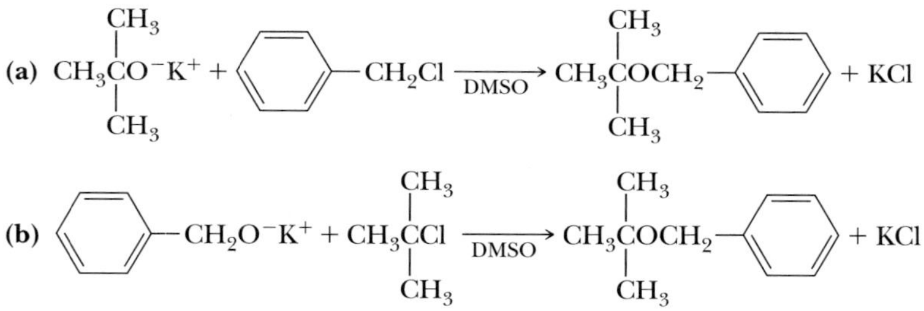

(a)

(b)

9.49 The following ethers can, in principle, be synthesized by two different combinations of haloalkane or halocycloalkane and metal alkoxide. Show one combination that forms ether bond (1) and another that forms ether bond (2). Which combination gives the higher yield of ether?

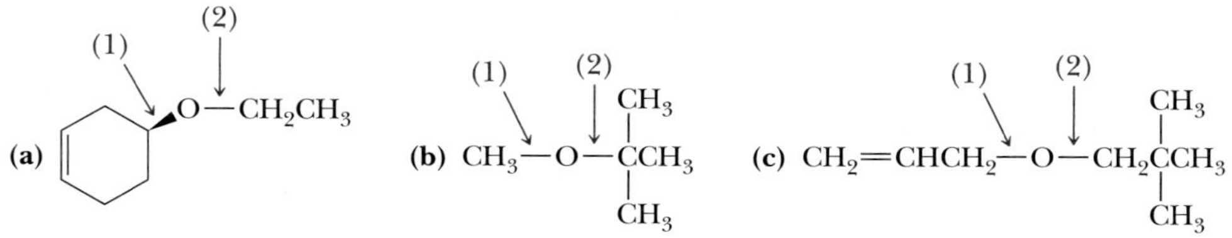

(a)

(b)

(c)

9.50 Propose a mechanism for this reaction.

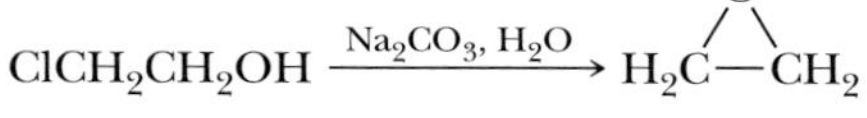

9.51 Each of these compounds can be synthesized by an S_N2 reaction. Suggest a combination of haloalkane and nucleophile that will give each product.

(a) CH_3OCH_3

(b) CH_3SH

(c) $CH_3CH_2CH_2PH_2$

(d) CH_3CH_2CN

(e) $CH_3SCH_2C(CH_3)_3$

(f) $(CH_3)_3NH^+Cl^-$

(g) $C_6H_5\overset{\displaystyle O}{\overset{\|}{C}}OCH_2C_6H_5$

(h) $(R)\text{-}CH_3\overset{\displaystyle N_3}{\overset{|}{C}}HCH_2CH_2CH_3$

(i) $CH_2{=}CHCH_2OCH(CH_3)_2$

(j) $CH_2{=}CHCH_2OCH_2CH{=}CH_2$

(k)

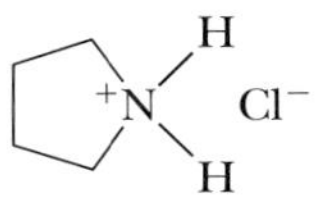

(l)

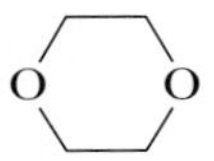

Looking Ahead

9.52 OH^- is a very poor leaving group; however, many alcohols react with alkyl or aryl sulfonyl chlorides to give sulfonate esters.

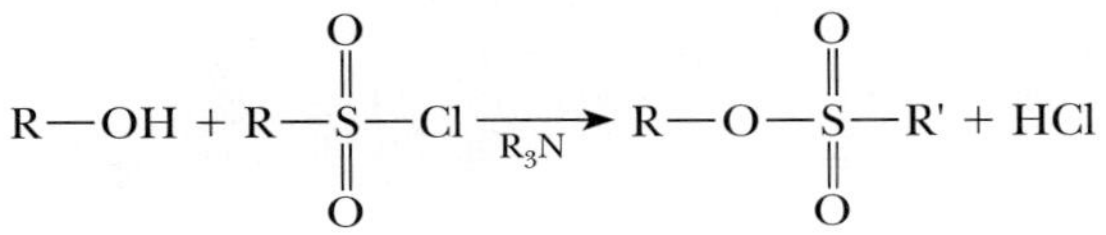

$$R\text{—}OH + R\text{—}\underset{\underset{\displaystyle O}{\|}}{\overset{\overset{\displaystyle O}{\|}}{S}}\text{—}Cl \xrightarrow{R_3N} R\text{—}O\text{—}\underset{\underset{\displaystyle O}{\|}}{\overset{\overset{\displaystyle O}{\|}}{S}}\text{—}R' + HCl$$

(a) Explain what this change does to the leaving group ability of the substituent.

(b) Suggest the product of the following reaction.

$$CH_3CH_2\text{—}O\text{—}\underset{\underset{\displaystyle O}{\|}}{\overset{\overset{\displaystyle O}{\|}}{S}}\text{—}C_6H_5 + CH_3S^-Na^+ \xrightarrow{DMSO}$$

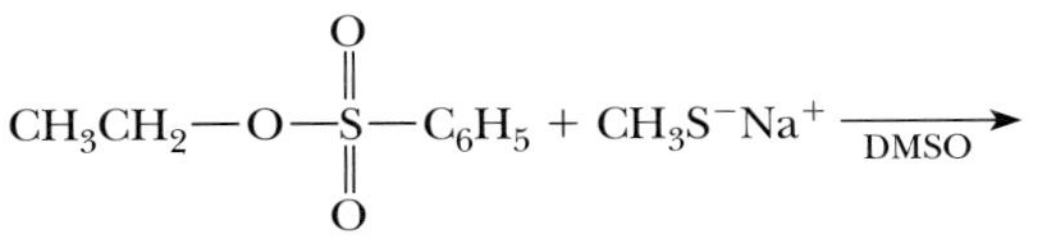

9.53 Suggest a product of the following reaction. HI is a very strong acid.

$$CH_3CH_2OCH_2CH_3 + 2HI \longrightarrow$$

10

Earl Roberge/Photo Researchers, Inc.

■ Fermentation vats of wine grapes at the Beaulieu Vineyards, California. Inset: A model of ethanol.

Outline

ORGANIC
Chemistry⚛Now™

Look for this logo in the chapter and go to Organic ChemistryNow at **http://now.brookscole.com/bfi4** for tutorials, simulations, and problems.

Alcohols

In this chapter, we study the physical and chemical properties of alcohols, a class of compounds containing the —OH (hydroxyl) group. We also study thiols, a class of compounds containing the —SH (sulfhydryl) group.

$$CH_3CH_2OH \qquad CH_3CH_2SH$$

Ethanol   Ethanethiol
(an alcohol)   (a thiol)

Ethanol is the fuel additive in gasohol, the alcohol in alcoholic beverages, and an important industrial solvent. Ethanethiol, like all other low-molecular-weight thiols, has a stench; such smells from skunks, rotten eggs, and sewage are caused by thiols or H_2S.

Alcohols are important because they can be converted into many other types of compounds, including alkenes, haloalkanes, aldehydes, ketones, carboxylic acids, and esters. Not only can alcohols be converted to these compounds, but these compounds can also be converted to alcohols. Thus, alcohols play a central role in the interconversion of organic functional groups.

Hydroxyl groups are found in carbohydrates and certain amino acids. Following are two representations for glucose, the most abundant organic compound in nature. On the left is a Fischer projection showing the configuration of all stereocenters. On the right is a cyclic structure, the predominant form in which this molecule exists in both the solid form and in solution. The amino acid L-serine is one of the 20 amino acid building blocks of proteins.

D-Glucose
(Fischer projection)

β-D-Glucopyranose

L-Serine

Because sulfur and oxygen are both Group 6 elements, thiols and alcohols undergo many of the same types of reactions. Sulfur, a third-row element, however, can expand its valence shell to include more than eight electrons; therefore, thiols undergo some reactions that are not possible for alcohols. In addition, sulfur's electronegativity and basicity are less than those of oxygen.

10.1   Structure and Nomenclature of Alcohols

A. Structure

The functional group of an alcohol is an —**OH** (**hydroxyl**) **group** (Section 1.3A) bonded to an sp^3-hybridized carbon. The oxygen atom of an alcohol is also sp^3 hybridized. Two sp^3 hybrid orbitals of oxygen form sigma bonds to atoms of carbon and hydrogen, and the remaining two sp^3 hybrid orbitals each contain an unshared pair of electrons. Figure 10.1 shows a Lewis structure and a ball-and-stick model of methanol, CH_3OH, the simplest alcohol. The measured C—O—H bond angle in methanol is 108.9°, very close to the tetrahedral angle of 109.5°.

B. Nomenclature

In the IUPAC system, the longest chain of carbon atoms containing the —OH group is selected as the parent alkane and numbered from the end closer to —OH. To show that the compound is an alcohol, change the suffix -*e* of the parent alkane to -*ol* (Section 2.5), and use a number to show the location of the —OH group. The location of the —OH group takes precedence over alkyl groups and halogen atoms in numbering the parent chain. For cyclic alcohols, numbering begins with the carbon bearing the —OH group. Because the —OH group is understood to be on carbon 1 of the ring, there is no need give its location a number. In complex alcohols, the number for the hydroxyl group is often placed between the infix and the suffix. Thus, for example, 2-methyl-1-propanol and 2-methylpropan-1-ol are both acceptable names.

Common names for alcohols are derived by naming the alkyl group bonded to —OH and then adding the word *alcohol*. Here are IUPAC names and, in parentheses, common names for several low-molecular-weight alcohols.

(a)

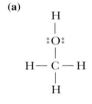

(b)

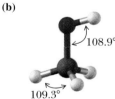

Figure 10.1
Methanol, CH_3OH. (a) Lewis structure and (b) ball-and-stick model.

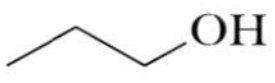

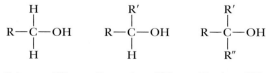

1-Propanol
(Propyl alcohol)

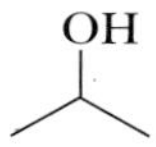

2-Propanol
(Isopropyl alcohol)

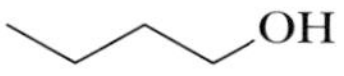

1-Butanol
(Butyl alcohol)

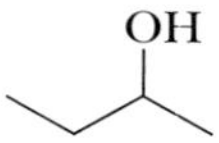

2-Butanol
(*sec*-Butyl alcohol)

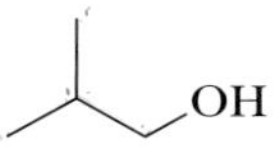

2-Methyl-1-propanol
(Isobutyl alcohol)

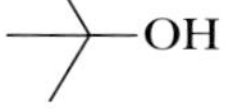

2-Methyl-2-propanol
(*tert*-Butyl alcohol)

Example 10.1

Write IUPAC names for these alcohols.

(a) 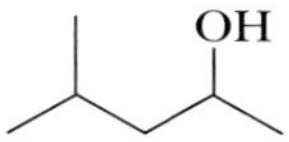   (b) 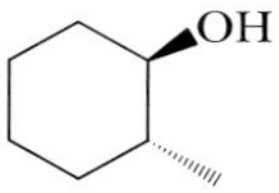   (c)

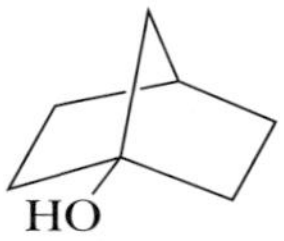

Solution

(a) 4-Methyl-2-pentanol.
(b) (1*R*,2*R*)-2-Methylcyclohexanol. Note that the designation of the configuration as *R,R* specifies not only the absolute configuration of each stereocenter but also that the —CH₃ and —OH groups are *trans* to each other on the ring. The alcohol can also be named *trans*-2-methylcyclohexanol and, while this name specifies that the hydroxyl and methyl groups are *trans* to each other, it does not specify the absolute configuration of either group.
(c) Bicyclo[2.2.1]-1-heptanol.

Problem 10.1

Write IUPAC names for these alcohols, and include the configuration for (a).

(a) 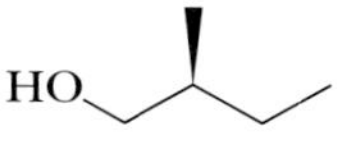   (b) 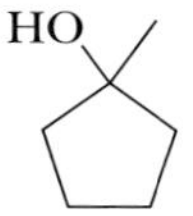   (c)

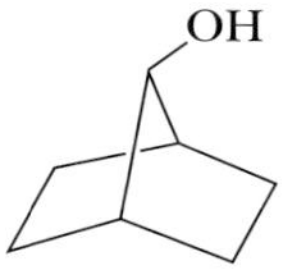

We classify alcohols as **primary** (**1°**), **secondary** (**2°**), or **tertiary** (**3°**), depending on whether the —OH group is on a primary, secondary, or tertiary carbon.

$$R-\underset{\underset{\displaystyle H}{|}}{\overset{\overset{\displaystyle H}{|}}{C}}-OH \qquad R-\underset{\underset{\displaystyle H}{|}}{\overset{\overset{\displaystyle R'}{|}}{C}}-OH \qquad R-\underset{\underset{\displaystyle R''}{|}}{\overset{\overset{\displaystyle R'}{|}}{C}}-OH$$

Primary (1°)   Secondary (2°)   Tertiary (3°)

Example 10.2

Classify each alcohol as primary, secondary, or tertiary.

(a) [structure: cyclohexyl group attached to a carbon bearing OH, CH₃, and H]

(b) $CH_3\overset{\displaystyle CH_3}{\underset{\displaystyle CH_3}{C}}OH$

(c) [structure: cyclopentane with CH₂OH substituent]

Solution

(a) Secondary (2°)   (b) Tertiary (3°)   (c) Primary (1°)

Problem 10.2

Classify each alcohol as primary, secondary, or tertiary.

(a) [branched structure with OH]   (b) [cyclopropyl—OH]   (c) [cyclohexene ring with OH]   (d) [cyclopentane with CH₃ and OH]

   In the IUPAC system, a compound containing two hydroxyl groups is named as a **diol,** one containing three hydroxyl groups as a **triol,** and so on. In IUPAC names for diols, triols, and so on, the final -*e* (the suffix) of the parent alkane name is retained, as for example in the name 1,2-ethanediol. As with many organic compounds, common names for certain diols and triols have persisted. Compounds containing two hydroxyl groups on adjacent carbons are often referred to as **glycols** (Section 6.5). Ethylene glycol and propylene glycol are synthesized from ethylene and propylene, respectively, hence their common names.

Diol   A compound containing two —OH groups.

Triol   A compound containing three hydroxyl groups.

1,2-Ethanediol	1,2-Propanediol	1,3-Propanediol	1,2,3-Propanetriol
(Ethylene glycol)	(Propylene glycol)		(Glycerol, glycerine)

   Compounds containing —OH and C=C groups are often referred to as **unsaturated alcohols** because of the presence of the carbon-carbon double bond. In the IUPAC system, the double bond is shown by changing the infix of the parent alkane from -*an*- to -*en*- (Section 2.5), and the hydroxyl group is shown by changing the suffix of the parent alkane from -*e* to -*ol.* Numbers must be used to show the location of both the carbon-carbon double bond and the hydroxyl group. The parent alkane is numbered to give the —OH group the lowest possible number; that is, the group shown by a suffix (in this case, -*ol*) takes precedence over the group shown by an infix (in this case, -*en*-).

Example 10.3

Write IUPAC names for these unsaturated alcohols.

(a) $CH_2{=}CHCH_2OH$ (b) HO [structure: HO attached to a chain with a double bond]

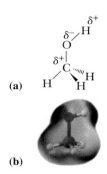

(a)

(b)

Figure 10.2
Polarity of the C—O—H bond in an alcohol.

Dipole-dipole interaction The attraction between the positive end of one dipole and the negative end of another.

Hydrogen bonding The attractive interaction between a hydrogen atom bonded to an atom of high electronegativity (most commonly O or N) and a lone pair of electrons on another atom of high electronegativity (again, most commonly O or N).

Solution

(a) 2-Propen-1-ol. Its common name is allyl alcohol.
(b) (*E*)-2-Hexen-1-ol (*trans*-2-Hexen-1-ol)

Problem 10.3

Write IUPAC names for these unsaturated alcohols.

(a) ∕∕∕∕OH   **(b)** (structure with OH)

10.2 Physical Properties of Alcohols

Because of the presence of the polar —OH group, alcohols are polar compounds, with partial positive charges on carbon and hydrogen and a partial negative charge on oxygen (Figure 10.2).

The attraction between the positive end of one dipole and the negative end of another is called **dipole-dipole interaction.** When the positive end of one of the dipoles is a hydrogen atom bonded to O or N (atoms of high electronegativity) and the negative end of the other dipole is an O or N atom, the attractive interaction between dipoles is particularly strong and is given the special name of **hydrogen bonding.** The length of a hydrogen bond in water is 177 pm, about 80% longer than an O—H covalent bond. The strength of a hydrogen bond in water is approximately 21 kJ (5 kcal)/mol. For comparison, the strength of the O—H covalent bond in water is approximately 498 kJ (118 kcal)/mol. As can be seen by comparing these numbers, an O---H hydrogen bond is considerably weaker than an O—H covalent bond. The presence of a large number of hydrogen bonds in liquid water, however, has an important cumulative effect on the physical properties of water. Because of hydrogen bonding, extra energy is required to separate each water molecule from its neighbors, hence the relatively high boiling point of water.

Similarly, there is extensive hydrogen bonding between alcohol molecules in the pure liquid. Figure 10.3 shows the association of ethanol molecules by hydrogen bonding between the partially negative oxygen atom of one ethanol molecule and the partially positive hydrogen atom of another ethanol molecule.

Table 10.1 lists the boiling points and solubilities in water for several groups of alcohols and hydrocarbons of similar molecular weight. Of the compounds compared in each group, the alcohols have the higher boiling points because more energy is needed to overcome the attractive forces of hydrogen bonding between their polar —OH groups. The presence of additional hydroxyl groups in a molecule further increases the extent of hydrogen bonding, as can be seen by comparing the boiling points of hexane (bp 69°C), 1-pentanol (bp 138°C), and 1,4-butanediol (bp 230°C), all of which have approximately the same molecular weight. Because of increased dispersion forces between larger molecules, boiling points of all types of compounds, including alcohols, increase with increasing molecular weight. Compare, for example, the boiling points of ethanol (bp 78°C), 1-propanol (bp 97°C), 1-butanol (bp 117°C), and 1-pentanol (bp 138°C).

The effect of hydrogen bonding in alcohols is illustrated dramatically by comparing the boiling points of ethanol (bp 78°C) and its constitutional isomer dimethyl

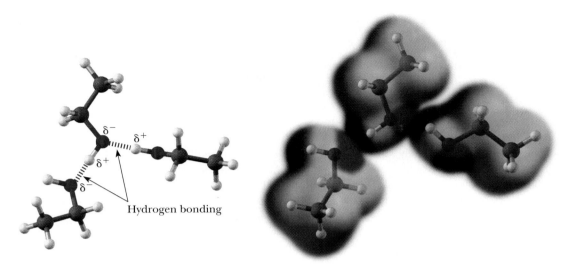

Figure 10.3
The association of ethanol molecules in the liquid state by hydrogen bonding. Each O—H can participate in up to three hydrogen bonds (one through hydrogen and two through oxygen). Only two of the three possible hydrogen bonds per molecule are shown.

ether (bp 24°C). The difference in boiling point between these two compounds is caused by the presence of a polar O—H group in the alcohol, which is capable of forming intermolecular hydrogen bonds. This hydrogen bonding increases the attractive forces between molecules of ethanol, and, thus, ethanol has a higher boiling point than dimethyl ether.

Table 10.1   Boiling Points and Solubilities in Water of Five Groups of Alcohols and Hydrocarbons of Similar Molecular Weight

Structural Formula	Name	Molecular Weight (g/mol)	Boiling Point (°C)	Solubility in Water
CH_3OH	Methanol	32	65	Infinite
CH_3CH_3	Ethane	30	−89	Insoluble
CH_3CH_2OH	Ethanol	46	78	Infinite
$CH_3CH_2CH_3$	Propane	44	−42	Insoluble
$CH_3CH_2CH_2OH$	1-Propanol	60	97	Infinite
$CH_3CH_2CH_2CH_3$	Butane	58	0	Insoluble
$CH_3CH_2CH_2CH_2OH$	1-Butanol	74	117	8 g/100 g
$CH_3CH_2CH_2CH_2CH_3$	Pentane	72	36	Insoluble
$HOCH_2CH_2CH_2CH_2OH$	1,4-Butanediol	90	230	Infinite
$CH_3CH_2CH_2CH_2CH_2OH$	1-Pentanol	88	138	2.3 g/100 g
$CH_3CH_2CH_2CH_2CH_2CH_3$	Hexane	86	69	Insoluble

CONNECTIONS TO BIOLOGICAL CHEMISTRY

The Importance of Hydrogen Bonding in Drug-Receptor Interactions

Hydrogen bonds have directionality in that the donor and acceptor groups must be oriented appropriately with respect to each other for a hydrogen bond to form. Important hydrogen bond donors in biological molecules include —OH groups (proteins, carbohydrates) and —NH groups (proteins, nucleic acids). Important hydrogen bond acceptors are any N or O with a lone pair of electrons, such as C=O groups (proteins, carbohydrates, nucleic acids), —OH groups (proteins, carbohydrates), and COO⁻ groups (proteins).

With directionality comes the potential for hydrogen bonds to organize molecules at many levels ranging from the folding of biological molecules to the specific binding and recognition between a pharmaceutical and its receptor. The drug atorvastatin (Lipitor) is used to treat high cholesterol. Cholesterol is synthesized in the liver from the two-carbon acetyl group of acetyl coenzyme A (acetyl-CoA). A key intermediate in the sequence of reactions leading to the synthesis of cholesterol is a six-carbon molecule named mevalonate (Section 26.4B). Atorvastatin specifically binds to, and blocks the action of, HMG-CoA reductase, a key enzyme in the biosynthesis of mevalonate. Atorvastatin binds to this enzyme in preference to the large number of other potential enzyme targets because (1) the drug has a shape complementary to the catalytic cavity (the active site) of HMG-CoA reductase (Figure 1), and (2) it can form at least nine specific hydrogen bonds with functional groups at the active site on the enzyme (Figure 2).

The complementary shape and pattern of hydrogen bonding ensure that atorvastatin binds to HMG-CoA reductase and inhibits its ability to catalyze the formation of mevalonate. The hallmark of this and other effective drugs is their ability to bind strongly with their intended target molecules, while at the same time not interacting with other molecules that could lead to unwanted side effects.

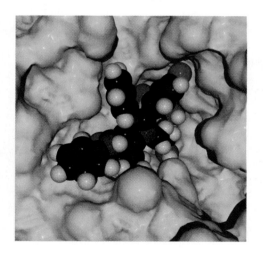

Figure 1

A space-filling model of the cholesterol-lowering drug atorvastatin (Lipitor) bound to the active site of its enzyme target HMG-CoA reductase (shown as a yellow surface). The shape of the drug is complementary to the active site of the enzyme.

Atorvastatin

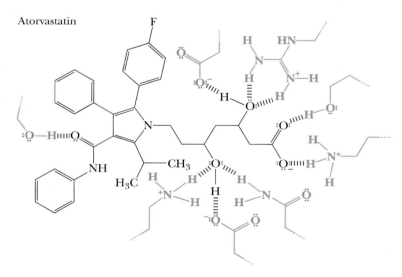

Figure 2

Hydrogen bonding between atorvastatin and the functional groups at the active site of the enzyme HMG-CoA reductase (shown in red). The nine hydrogen bonds, many of which involve hydroxyl groups on atorvastatin or the enzyme surface, help to provide the specificity that directs the binding of the drug to its target enzyme.

$$CH_3CH_2OH \qquad CH_3OCH_3$$

Ethanol   Dimethyl ether

bp 78°C   bp −24°C

Because alcohols can interact by hydrogen bonding with water, they are more soluble in water than alkanes and alkenes of comparable molecular weight. Methanol, ethanol, and 1-propanol are soluble in water in all proportions. As molecular weight increases, the physical properties of alcohols become more like those of hydrocarbons of comparable molecular weight. Higher molecular-weight alcohols are much less soluble in water because of the increase in size of the hydrocarbon portion of their molecules.

Example 10.4

Following are three alcohols with molecular formula $C_4H_{10}O$. Their boiling points, from lowest to highest, are 82.3°C, 99.5°C, and 117°C. Which alcohol has which boiling point?

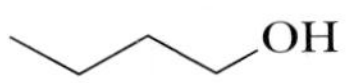

1-Butanol 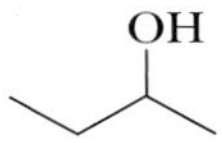   2-Butanol 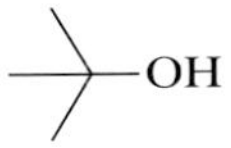   2-Methyl-2-propanol

Solution

Boiling points of these constitutional isomers depend on the strength of intermolecular hydrogen bonding. The primary —OH group of 1-butanol is the most accessible for intermolecular hydrogen bonding; this alcohol has the highest boiling point, 117°C. The tertiary —OH group of 2-methyl-2-propanol is the least accessible for intermolecular hydrogen bonding; this alcohol has the lowest boiling point, 82.3°C.

Problem 10.4

Arrange these compounds in order of increasing boiling point.

Heptane    1,5-Pentanediol    1-Hexanol

Example 10.5

Arrange these compounds in order of increasing solubility in water.

Hexane 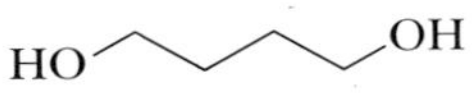   1,4-Butanediol 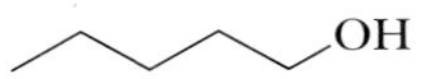   1-Pentanol

Solution

Hexane, C_6H_{14}, a nonpolar hydrocarbon, has the lowest solubility in water. Both 1-pentanol and 1,4-butanediol are polar compounds due to the presence of —OH groups, and each interacts with water molecules by hydrogen bonding. Because 1,4-butanediol has more sites within its molecules for hydrogen bonding than 1-pentanol, the diol is more soluble in water than 1-pentanol. The water solubilities of these compounds are given in Table 10.1.

Insoluble	2.3 g/100 mL	Infinitely soluble

Problem 10.5

Arrange these compounds in order of increasing solubility in water.

1-Butanol	1-Propanol	1,2-Dichloroethane

10.3 Acidity and Basicity of Alcohols

Alcohols can function as both weak acids (proton donors) and weak bases (proton acceptors). Table 10.2 lists the acid ionization constants for several low-molecular-weight alcohols.

In dilute aqueous solution, only methanol (pK_a 15.5) is more acidic than water.

$$CH_3\ddot{O}-H + \ddot{O}-H \rightleftharpoons CH_3\ddot{O}:^- + H-\overset{+}{O}-H \qquad K_a = \frac{[CH_3O^-][H_3O^+]}{[CH_3OH]} = 10^{-15.5}$$
$$\qquad\qquad H \qquad\qquad\qquad\qquad H \qquad\qquad pK_a = 15.5$$

Ethanol has about the same acidity as water. Higher molecular weight, water-soluble alcohols are slightly weaker acids than water. Thus, although alcohols have some acidity, they are not strong enough acids to react with weak bases such as sodium bicarbonate or sodium carbonate. (At this point, it would be wise to review Section 4.4, which discusses the position of equilibrium in acid-base reactions.)

For simple alcohols, such as methanol and ethanol, acidity depends primarily on the degree of solvation and stabilization of the alkoxide ion by water molecules. The negatively charged oxygen atoms of the methoxide and ethoxide ions are almost as

Table 10.2   pK_a Values for Selected Alcohols in Dilute Aqueous Solution*

Compound	Structural Formula	pK_a	
Hydrogen chloride	HCl	−7	Stronger acid
Acetic acid	CH_3COOH	4.8	
Methanol	CH_3OH	15.5	
Water	H_2O	15.7	
Ethanol	CH_3CH_2OH	15.9	
2-Propanol	$(CH_3)_2CHOH$	17	
2-Methyl-2-propanol	$(CH_3)_3COH$	18	Weaker acid

*Also given for comparison are pK_a values for water, acetic acid, and hydrogen chloride.

accessible for solvation as the hydroxide ion is; therefore, these alcohols are about as acidic as water. As the bulk of the alkyl group bonded to oxygen increases, the ability of water molecules to solvate the alkoxide ion decreases. 2-Methyl-2-propanol (*tert*-butyl alcohol) is a weaker acid than either methanol or ethanol, primarily because of the bulk of the *tert*-butyl group, which reduces solvation of the *tert*-butoxide anion by surrounding water molecules.

In the presence of strong acids, the oxygen atom of an alcohol is a base and reacts with an acid by proton transfer to form an oxonium ion.

$$CH_3CH_2-\ddot{O}-H \; + \; H-\overset{+}{\underset{H}{\ddot{O}}}-H \; \xrightarrow{H_2SO_4} \; CH_3CH_2-\overset{+}{\underset{H}{\ddot{O}}}-H \; + \; \ddot{O}-H$$

Ethanol	Hydronium ion	Ethyloxonium ion
	(pK_a −1.7)	pK_a −2.4
	(weaker acid)	(stronger acid)

An important mechanistic theme in many of the reactions of alcohols is that the HO⁻ group, a poor leaving group, reacts with protons or a variety of strong electrophiles to create the H_2O or analogous group, a much better leaving group, enabling subsequent substitution or elimination reactions to take place.

10.4 Reaction of Alcohols with Active Metals

Alcohols react with Li, Na, K, and other active metals to liberate hydrogen and form metal alkoxides. In this oxidation/reduction reaction, Na is oxidized to Na⁺, and H⁺ is reduced to H_2.

$$2CH_3OH + 2Na \longrightarrow 2CH_3O^-Na^+ + H_2$$

Sodium methoxide
(MeO⁻Na⁺)

To name a metal alkoxide, name the cation first, followed by the name of the anion. The name of the anion is derived from the prefix showing the number of carbon atoms and their arrangement (*meth*-, *eth*-, *isoprop*-, tert-*but*-, and so on) followed by the suffix -*oxide*.

Alkoxide ions are nearly the same or somewhat stronger bases than the hydroxide ion. In addition to sodium methoxide, the following metal salts of alcohols are commonly used in organic reactions requiring a strong base in a nonaqueous solvent, as for example sodium ethoxide in ethanol and potassium *tert*-butoxide in 2-methyl-2-propanol (*tert*-butyl alcohol).

$$CH_3CH_2O^-Na^+ \qquad\qquad CH_3\overset{\overset{CH_3}{|}}{\underset{\underset{CH_3}{|}}{C}}O^-K^+$$

Sodium ethoxide	Potassium *tert*-butoxide
(EtO⁻Na⁺)	(*t*-BuO⁻K⁺)

Alcohols can also be converted to salts by reaction with bases stronger than alkoxide ions. One such base is sodium hydride, NaH. Hydride ion, H:⁻, the conjugate base of H_2, is an extremely strong base.

$$CH_3CH_2OH + Na^+H^- \longrightarrow CH_3CH_2O^-Na^+ + H_2$$

Ethanol Sodium hydride Sodium ethoxide

Reactions of sodium hydride with compounds containing acidic hydrogens are irreversible and driven to completion by the formation of H_2, which is given off as a gas.

Example 10.6

Write a balanced equation for the reaction of cyclohexanol with sodium metal.

Solution

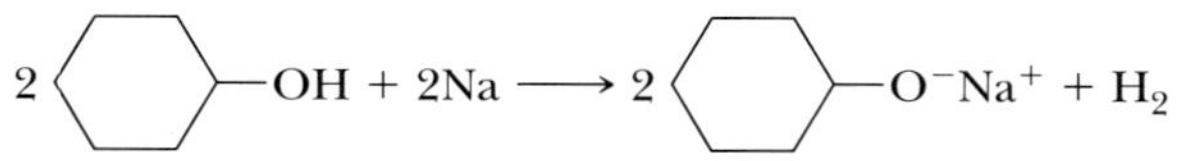

Problem 10.6

Predict the position of equilibrium for this acid-base reaction.

$$CH_3CH_2O^-Na^+ + CH_3\overset{\displaystyle O}{\overset{\|}{C}}OH \rightleftharpoons CH_3CH_2OH + CH_3\overset{\displaystyle O}{\overset{\|}{C}}O^-Na^+$$

10.5 Conversion of Alcohols to Haloalkanes and Sulfonates

Conversion of an alcohol to a haloalkane involves substitution of halogen for —OH at a saturated carbon. The most common reagents for this conversion are the halogen acids (HCl, HBr, and HI) and certain inorganic halides (PBr_3, $SOCl_2$, and $SOBr_2$).

A. Reaction with HCl, HBr, and HI

Tertiary alcohols react rapidly with HCl, HBr, and HI. Mixing a low-molecular-weight, water-soluble tertiary alcohol with concentrated hydrochloric acid for a few minutes at room temperature results in conversion of the alcohol to a chloroalkane.

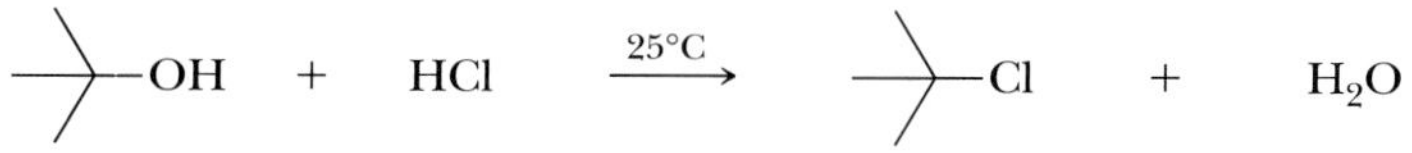

2-Methyl-2-propanol 2-Chloro-2-methylpropane

Reaction is evident by formation of a water-insoluble chloroalkane that separates from the aqueous layer. Low-molecular-weight, water-soluble primary and secondary alcohols are unreactive under these conditions.

Water-insoluble tertiary alcohols are converted to tertiary halides by bubbling gaseous HX through a solution of the alcohol dissolved in diethyl ether or tetrahydrofuran (THF).

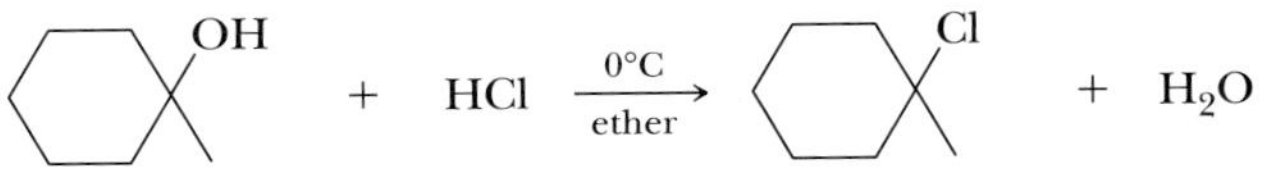

1-Methylcyclohexanol 1-Chloro-1-methyl-cyclohexane

Water-insoluble primary and secondary alcohols react only slowly under these conditions.

Primary and secondary alcohols are converted to bromoalkanes and iodoalkanes by treatment with hydrobromic and hydroiodic acids. For example, when heated to reflux with concentrated HBr, 1-butanol is converted smoothly to 1-bromobutane.

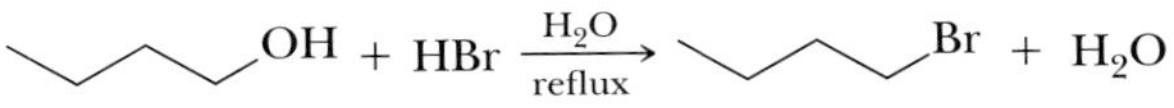

1-Butanol 1-Bromobutane

Many secondary alcohols give at least some rearranged product, evidence for the formation of carbocation intermediates during their reaction. For example, treating 3-pentanol with HBr gives 3-bromopentane as the major product, along with some 2-bromopentane.

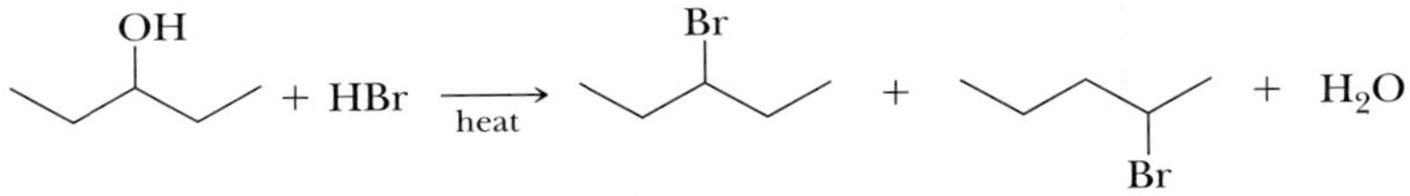

3-Pentanol 3-Bromopentane (major product) 2-Bromopentane (a product of rearrangement)

Primary alcohols with extensive β–branching give large amounts of a product derived from rearrangement. For example, treatment of 2,2-dimethyl-1-propanol (neopentyl alcohol) with HBr gives a rearranged product almost exclusively.

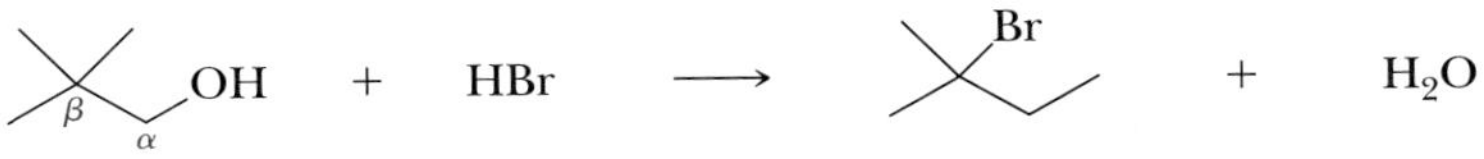

2,2-Dimethyl-1-propanol 2-Bromo-2-methylbutane (a product of rearrangement)

Based on observations of the relative ease of reaction of alcohols with HX (3° > 2° > 1°) and the occurrence of rearrangements, chemists propose an S_N1 mechanism for the conversion of tertiary and secondary alcohols to haloalkanes by concentrated HX, with the formation of a carbocation intermediate.

Mechanism *Reaction of a 3° Alcohol with HBr—An S_N1 Reaction*

Step 1: While we often show HBr as the acid present in solution, the actual acid involved in this reaction is H_3O^+ formed by dissociation of HBr in aqueous solution.

ORGANIC Chemistry•Now™

Click Mechanisms to view an animation of the **Acid-Catalyzed Substitution of a Tertiary Alcohol**

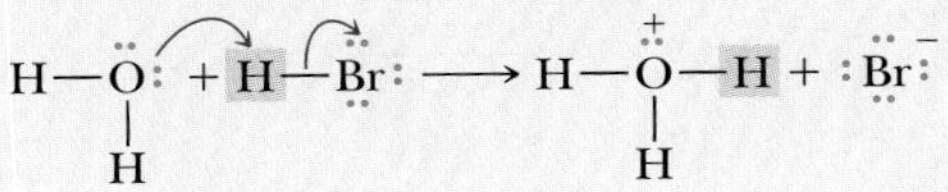

Rapid and reversible proton transfer from H_3O^+ to the —OH group of the alcohol gives an oxonium ion, which converts OH^-, a poor leaving group, into H_2O, a better leaving group.

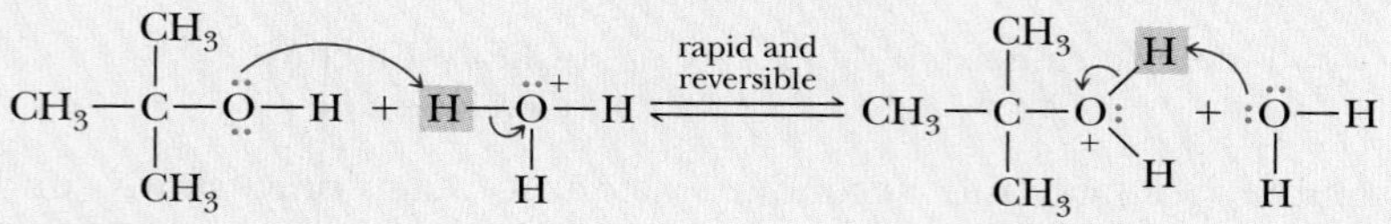

2-Methyl-2-propanol
(*tert*-Butyl alcohol)

An oxonium ion

Step 2: Loss of water gives a 3° carbocation intermediate.

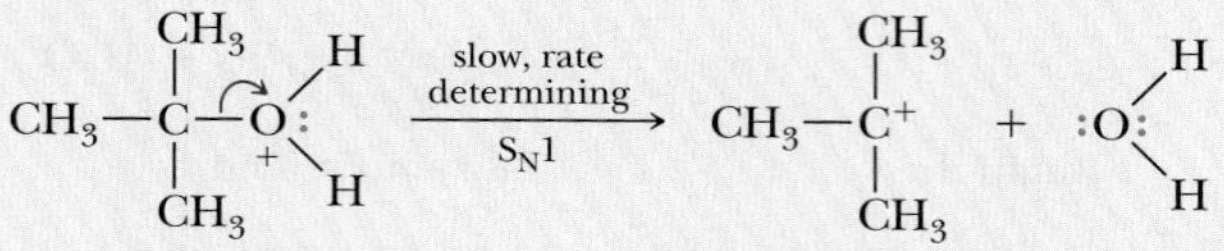

A 3° carbocation
intermediate

Step 3: Reaction of the 3° carbocation (an electrophile) with chloride ion (a nucleophile) gives the alkyl halide.

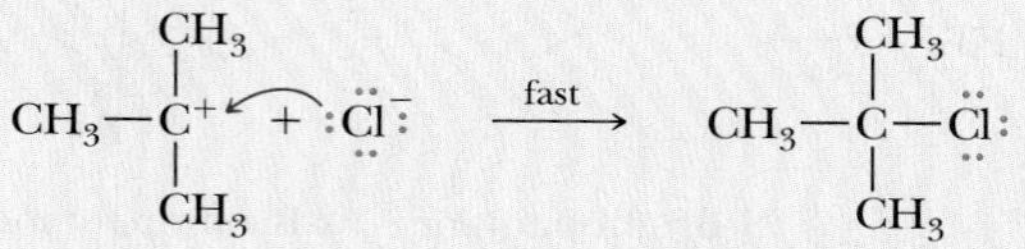

2-Chloro-2-methylpropane
(*tert*-Butyl chloride)

Primary alcohols react with HX by an S_N2 mechanism. In the rate-determining step, halide ion reacts at the carbon bearing the oxonium ion to displace H_2O and form the C—X bond.

ORGANIC
Chemistry Now™
Click Mechanisms to view an animation of the **Reaction of a Primary Alcohol with HCl**

Mechanism *Reaction of a 1° Alcohol with HBr—An S_N2 Reaction*

Step 1: Rapid and reversible proton transfer gives an oxonium ion, which transforms OH^-, a poor leaving group, into H_2O, a better leaving group.

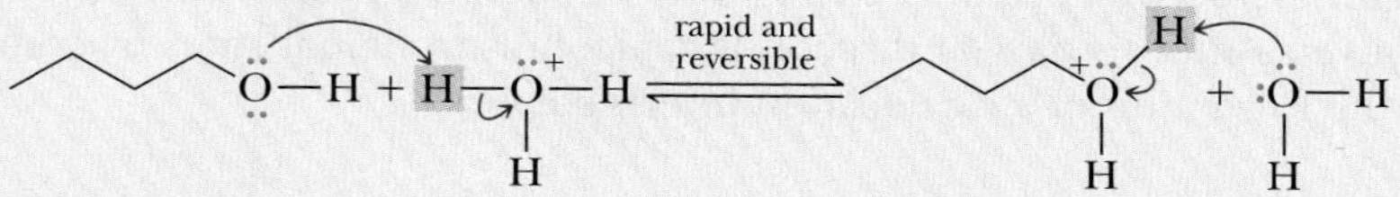

An oxonium ion

Step 2: Nucleophilic displacement of H_2O by Br^- gives the bromoalkane.

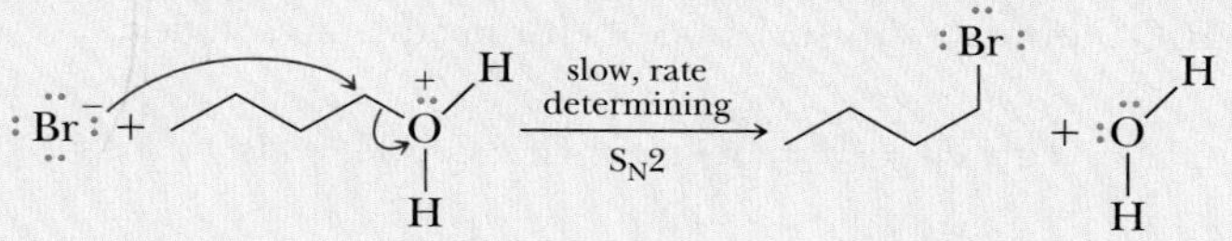

For primary alcohols with extensive β-branching, such as 2,2-dimethyl-1-propanol (neopentyl alcohol), it is difficult if not impossible for reaction to occur by direct displacement of H_2O from the primary carbon. Furthermore, formation of a 1° carbocation is also difficult, if not impossible. Instead, primary alcohols with extensive β-branching react by a mechanism involving formation of a 3° carbocation intermediate by simultaneous loss of H_2O and migration of an alkyl group, as illustrated by the conversion of 2,2-dimethyl-1-propanol to 2-chloro-2-methylbutane. Because the rate-determining step of this transformation involves only one reactant, namely the protonated alcohol, it is classified as an S_N1 reaction.

Mechanism   *Rearrangement upon Treatment of Neopentyl Alcohol with HX*

Step 1: Rapid and reversible proton transfer gives an oxonium ion. This step converts OH^-, a poor leaving group, into H_2O, a better leaving group.

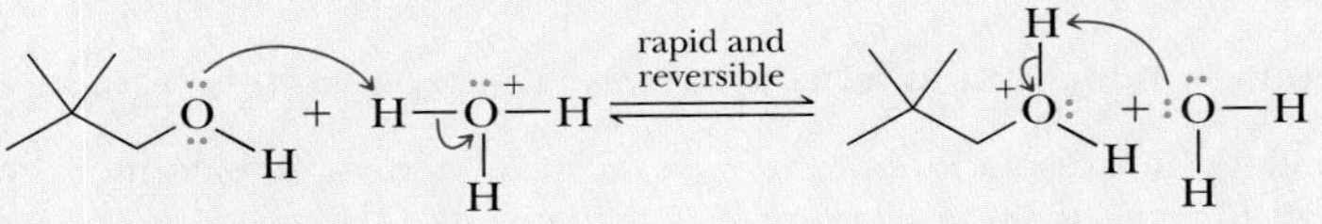

2,2-Dimethyl-1-propanol   An oxonium ion

Step 2: Two changes take place simultaneously in this step; the C—O bond breaks and a methyl group with its pair of bonding electrons migrates to the site occupied by the departing H_2O group. The result of these changes is loss of H_2O and the formation of a 3° carbocation.

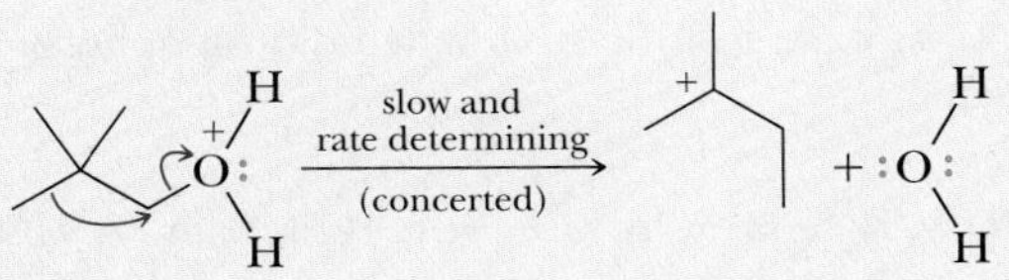

A 3° carbocation
intermediate

Step 3: Reaction of the 3° carbocation (an electrophile) with chloride ion (a nucleophile) gives the 3° alkyl halide.

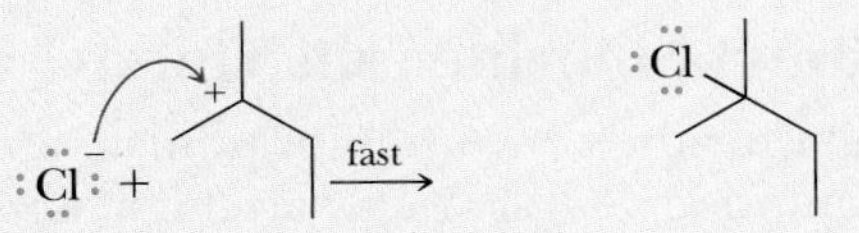

2-Chloro-2-methylbutane

In summary, preparation of haloalkanes by treatment of ROH with HX is most useful for primary and tertiary alcohols. The central theme in all these reactions is that protonation of HO^-, a very poor leaving group, transforms it into H_2O, a better leaving group so that an S_N1 or S_N2 reaction can take place with a halide nucleophile. Because of the possibility of rearrangement, this process is less useful for secondary alcohols (except for simple cycloalkanols) and for primary alcohols with extensive branching on the β-carbon.

B. Reaction with Phosphorous Tribromide

An alternative method for the synthesis of bromoalkanes from primary and secondary alcohols is through the use of phosphorous tribromide, PBr_3.

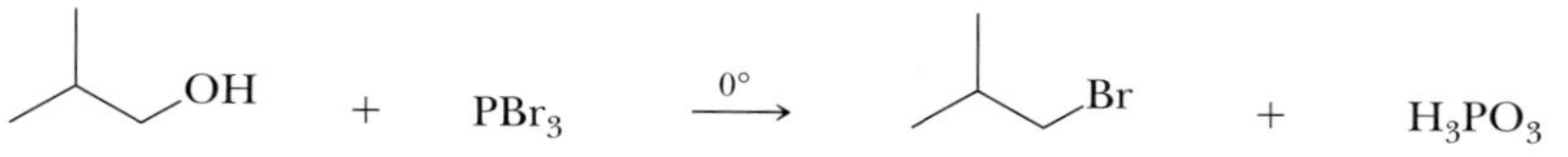

2-Methyl-1-propanol (Isobutyl alcohol) — Phosphorous tribromide — 1-Bromo-2-methylpropane (Isobutyl bromide) — Phosphorous acid

This method of preparation of bromoalkanes takes place under milder conditions than treatment with HBr. Although rearrangement sometimes occurs with PBr_3, the extent is considerably less than that with HBr, especially when the reaction mixture is kept at or below 0°C.

Mechanism *Reaction of a Primary Alcohol with PBr_3*

Conversion of an alcohol to a bromoalkane takes place in two steps.

Step 1: Nucleophilic displacement on phosphorus by the oxygen atom of the alcohol gives a protonated dibromophosphite group, which converts OH^-, a poor leaving group, into a good leaving group.

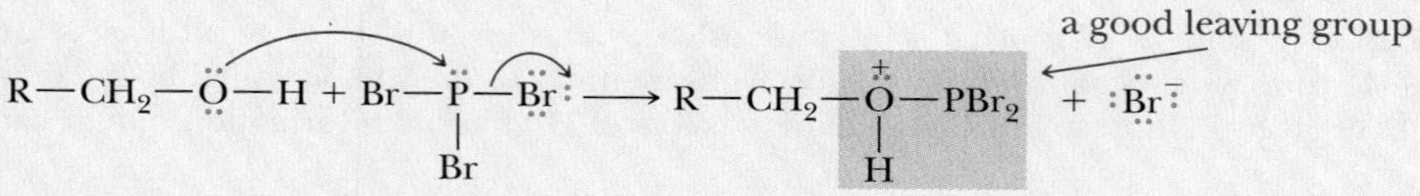

Step 2: Nucleophilic displacement of the protonated dibromophosphite group by bromide ion gives the bromoalkane.

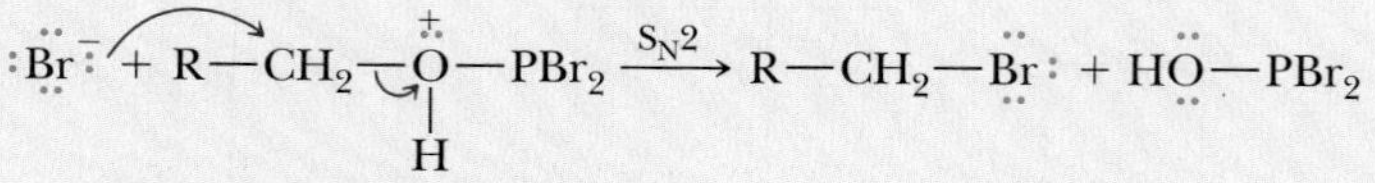

The other two bromine atoms on phosphorus are replaced in similar reactions, giving three moles of RBr and one mole of phosphorous acid, H_3PO_3.

ORGANIC Chemistry Now™
Click Mechanisms to view an animation of the **Bromination of a Primary Alcohol with Phosphorus Tribromide**

C. Reaction with Thionyl Chloride and Thionyl Bromide

The most widely used reagent for the conversion of primary and secondary alcohols to chloroalkanes is thionyl chloride, $SOCl_2$. Yields are high, and rearrangements are seldom observed. The byproducts of this conversion are HCl and SO_2.

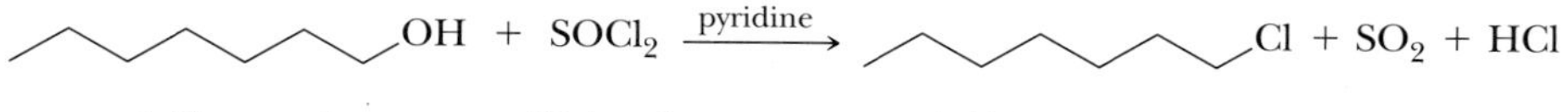

1-Heptanol — Thionyl chloride — 1-Chloroheptane

Similarly, thionyl bromide, $SOBr_2$ can be used to convert an alcohol to a bromoalkane.

Reactions with these reagents are most commonly carried out in the presence of pyridine (Section 23.1) or a tertiary amine such as triethylamine, Et_3N. The function of the amine (a weak base) is twofold. First, it catalyzes the reaction by forming a small amount of the alkoxide in equilibrium. The alkoxide is more reactive than the alcohol as a nucleophile. In addition, the amine neutralizes the HCl or HBr generated during the reaction and, in this way, prevents unwanted side reactions.

The 3° amine promotes
formation of an alkoxide:

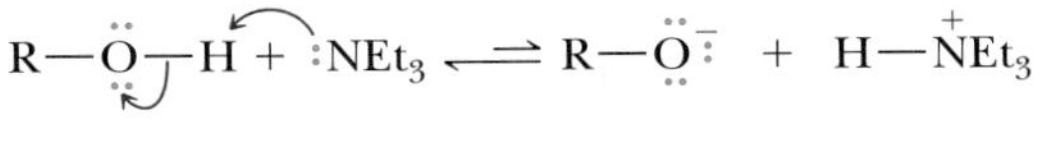

$$R-\overset{..}{\underset{..}{O}}-H + :NEt_3 \rightleftharpoons R-\overset{..}{\underset{..}{O}}:^- + H-\overset{+}{N}Et_3$$

It also neutralizes the HCl
formed during the reaction:

$$Et_3N: + H-\overset{..}{\underset{..}{Cl}}: \longrightarrow Et_3\overset{+}{N}-H \quad :\overset{..}{\underset{..}{Cl}}:^-$$

Triethylammonium
chloride

A particular value of thionyl halides is that their reaction with alcohols is stereoselective; it occurs with inversion of configuration. Reaction of thionyl chloride with (S)-2-octanol, for example, in the presence of a tertiary amine occurs with inversion of configuration and gives (R)-2-chlorooctane.

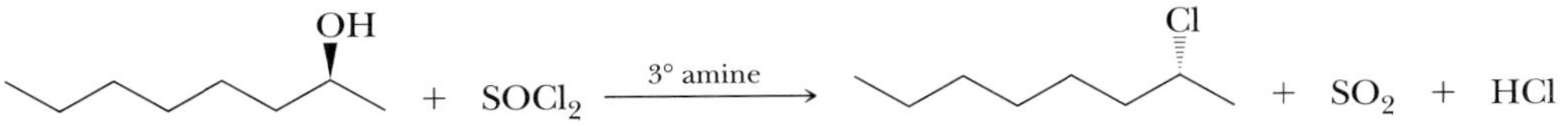

(S)-2-Octanol Thionyl
chloride
+ SOCl₂ $\xrightarrow{3° \text{ amine}}$ (R)-2-Chlorooctane + SO₂ + HCl

A key feature of the reaction of an alcohol with thionyl chloride is the formation of an alkyl chlorosulfite, which converts OH⁻, a poor leaving group, into a chlorosulfite, a good leaving group. If the reaction between the alcohol and thionyl chloride is carried out at 0°C or below, the alkyl chlorosulfite can be isolated.

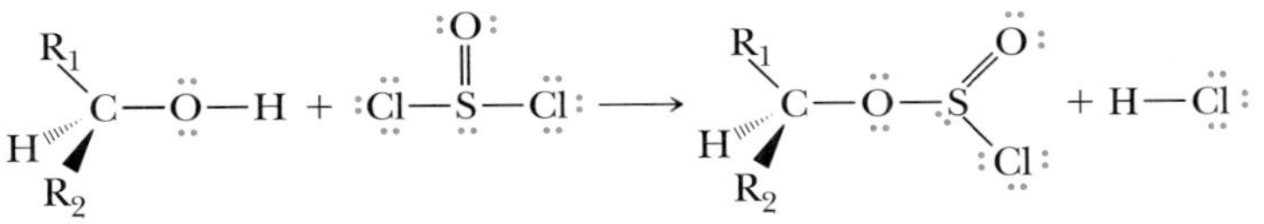

An alkyl chlorosulfite

Nucleophilic displacement of this leaving group by chloride ion gives the product.

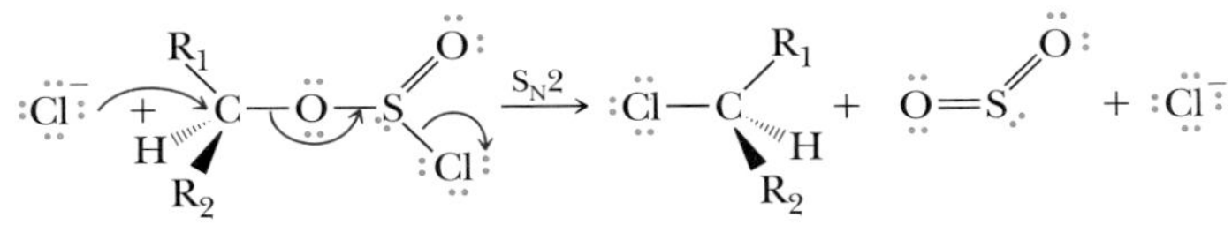

ORGANIC
Chemistry⋅Now™

Click Mechanisms to view an animation of the **Reaction of a Secondary Alcohol with Thionyl Chloride**

D. Formation of Aryl and Alkyl Sulfonates

As we have just seen, alcohols react with thionyl chloride to form alkyl chlorosulfites. Alcohols also react with compounds called sulfonyl chlorides to form alkylsulfonates. Sulfonyl chlorides are derived from sulfonic acids, compounds that are very strong acids, comparable in strength to sulfuric acid.

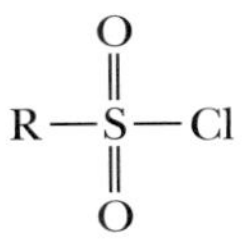

A sulfonyl chloride

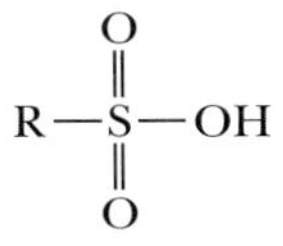

A sulfonic acid
(a very strong acid)

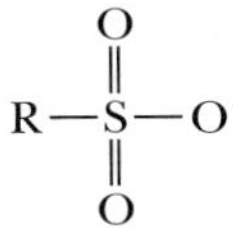

A sulfonate anion
(a very weak base and stable anion;
a very good leaving group)

What is important for us at this point is that a sulfonate anion is a very weak base and stable anion, and therefore a very good leaving group in nucleophilic substitution reactions.

Two of the most commonly used sulfonyl chlorides are *p*-toluenesulfonyl chloride (abbreviated tosyl chloride, TsCl) and methanesulfonyl chloride (abbreviated mesyl chloride, MsCl). Treating ethanol with *p*-toluenesulfonyl chloride in the presence of pyridine gives ethyl *p*-toluenesulfonate (ethyl tosylate). Pyridine is added to catalyze the reaction and to neutralize the HCl formed as a byproduct. Cyclohexanol is converted to cyclohexyl methanesulfonate (cyclohexyl mesylate) by a similar reaction of cyclohexanol with methanesulfonyl chloride.

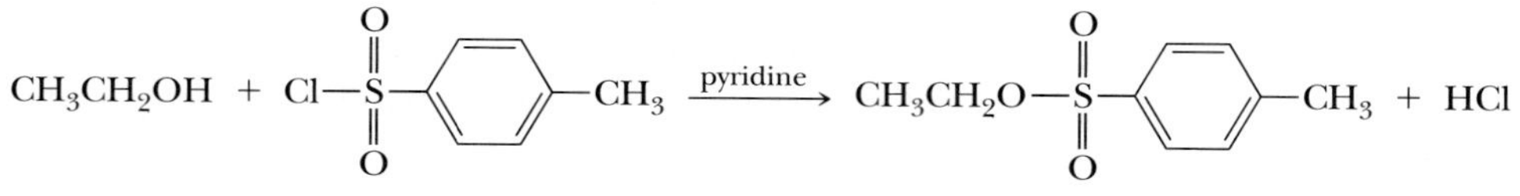

Ethanol   *p*-Toluenesulfonyl chloride   Ethyl *p*-toluenesulfonate (Ethyl tosylate)

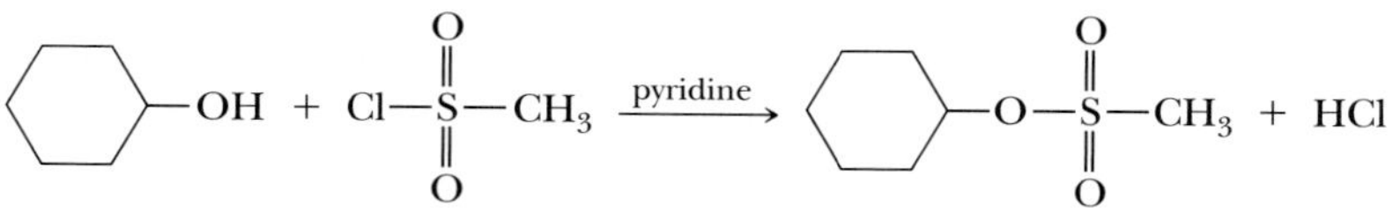

Cyclohexanol   Methanesulfonyl chloride   Cyclohexyl methanesulfonate (Cyclohexyl mesylate)

In formation of either a tosylate or a mesylate, the reaction involves breaking the O—H bond of the alcohol; it does not affect the C—O bond in any way. If the carbon bearing the —OH group is a stereocenter, sulfonate ester formation takes place with retention of configuration.

A particular advantage of sulfonate esters is that, through their use, a hydroxyl group, a very poor leaving group, can be converted to a tosylate or mesylate group, often shown as OTs and OMs, respectively, both very good leaving groups that are readily displaced by nucleophilic substitution.

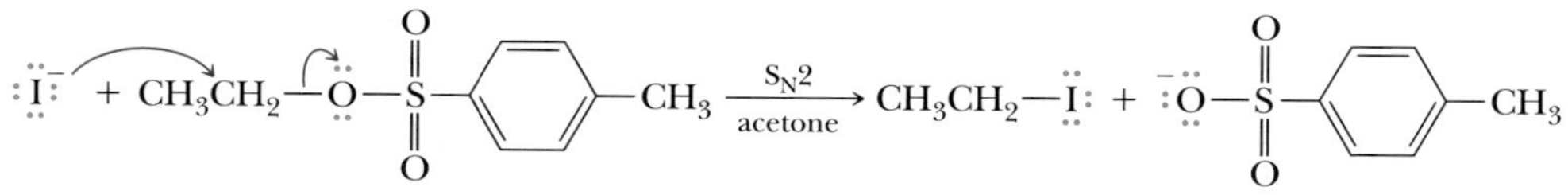

Ethyl *p*-toluenesulfonate   *p*-Toluenesulfonate anion (a good leaving group)

Following is a two-step sequence for conversion of (*S*)-2-octanol to (*R*)-2-octyl acetate via a tosylate. The first step involves cleavage of the O—H bond and proceeds with retention of configuration at the stereocenter. The second step involves S$_N$2 nucleophilic displacement of tosylate by acetate ion and proceeds with inversion of configuration at the stereocenter.

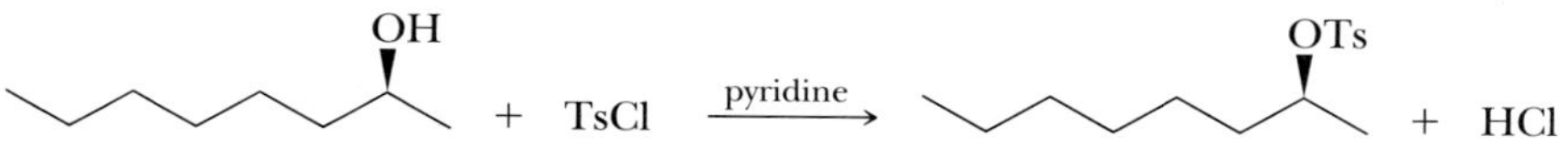

(*S*)-2-Octanol   Tosyl chloride   (*S*)-2-Octyl tosylate

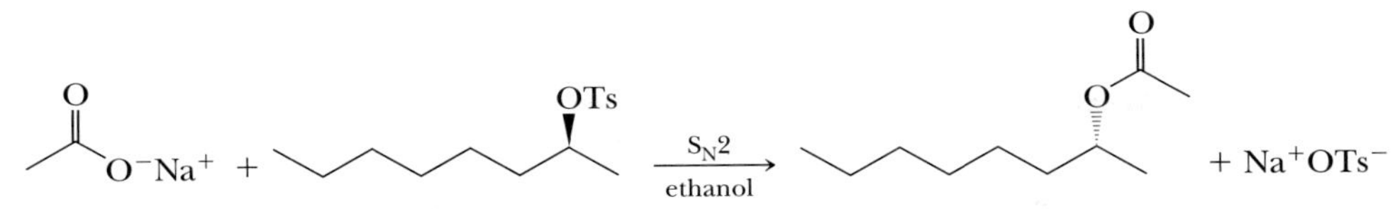

Sodium acetate   (S)-2-Octyl tosylate   (R)-2-Octyl acetate

Example 10.7

Show how to convert *trans*-4-methylcyclohexanol to *cis*-1-iodo-4-methylcyclohexane via a tosylate.

Solution

Treat the alcohol with *p*-toluenesulfonyl chloride in pyridine to form a tosylate with retention of configuration. Then treat the tosylate with sodium iodide in acetone. The S_N2 reaction with inversion of configuration gives the product. Because of the requirement for backside attack by the I^- nucleophile, the tosylate group must be in the axial position to react. Backside attack is not possible for an equatorial leaving group on a cyclohexane ring. The molecule must undergo a ring flip before the I^- displacement reaction can occur.

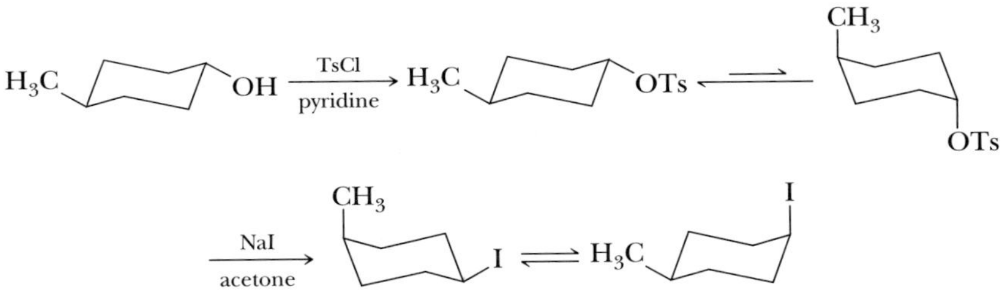

Problem 10.7

Show how to convert (R)-2-pentanol to (S)-2-pentanethiol via a tosylate:

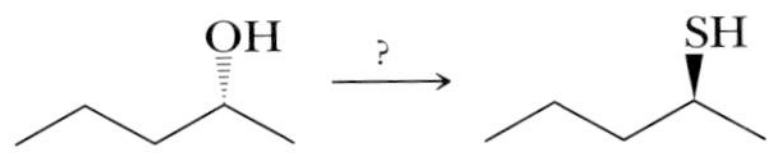

(R)-2-Pentanol   (S)-2-Pentanethiol

10.6 Acid-Catalyzed Dehydration of Alcohols

An alcohol can be converted to an alkene by **dehydration;** that is, by the elimination of a molecule of water from adjacent carbon atoms. Dehydration is most often brought about by heating the alcohol with either 85% phosphoric acid or concentrated sulfuric acid. Primary alcohols are the most difficult to dehydrate and generally require heating in concentrated sulfuric acid at temperatures as high as 180°C. Secondary alcohols undergo acid-catalyzed dehydration at somewhat lower temperatures. Acid-catalyzed dehydration of tertiary alcohols often requires temperatures only slightly above room temperature.

Dehydration   Elimination of water.

$$CH_3CH_2OH \xrightarrow[180°C]{H_2SO_4} CH_2=CH_2 + H_2O$$

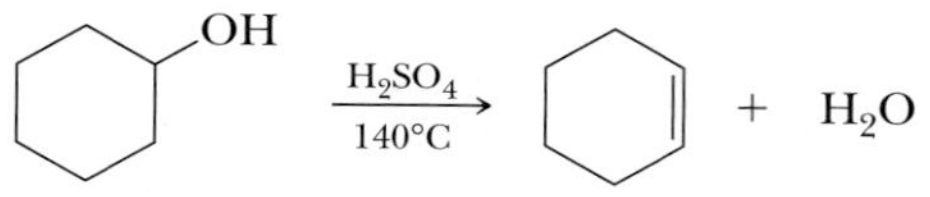

Cyclohexanol   Cyclohexene

$$\underset{\underset{CH_3}{|}}{\overset{\overset{CH_3}{|}}{CH_3COH}} \xrightarrow[50°C]{H_2SO_4} \overset{\overset{CH_3}{|}}{CH_3C}=CH_2 + H_2O$$

2-Methyl-2-propanol   2-Methylpropene
(*tert*-Butyl alcohol)   (Isobutylene)

Thus, the ease of acid-catalyzed dehydration of alcohols is in this order:

1° alcohol < 2° alcohol < 3° alcohol

Ease of dehydration of alcohols

When isomeric alkenes are obtained in acid-catalyzed dehydration of an alcohol, the alkene having the greater number of substituents on the double bond (the more stable alkene) generally predominates (Zaitsev's rule, Section 9.8).

$$\overset{\overset{OH}{|}}{CH_3CH_2CHCH_3} \xrightarrow[heat]{85\% \ H_3PO_4} CH_3CH=CHCH_3 + CH_3CH_2CH=CH_2 + H_2O$$

2-Butanol   (*E, Z*)-2-Butene   1-Butene
                       (80%)   (20%)

Example 10.8

Draw structural formulas for the alkenes formed on acid-catalyzed dehydration of each alcohol. Where isomeric alkenes are possible, predict which alkene is the major product.

(a) 3-Methyl-2-butanol   **(b)** 2-Methylcyclopentanol

Solution

(a) Elimination of H_2O from carbons 2-3 gives 2-methyl-2-butene; elimination from carbons 1-2 gives 3-methyl-1-butene. 2-Methyl-2-butene, with three alkyl groups (three methyl groups) on the double bond, is the major product (Zaitsev rule). 3-Methyl-1-butene, with only one alkyl group (an isopropyl group) on the double bond, is the minor product. A small amount of 2-methyl-1-butene is formed by rearrangement.

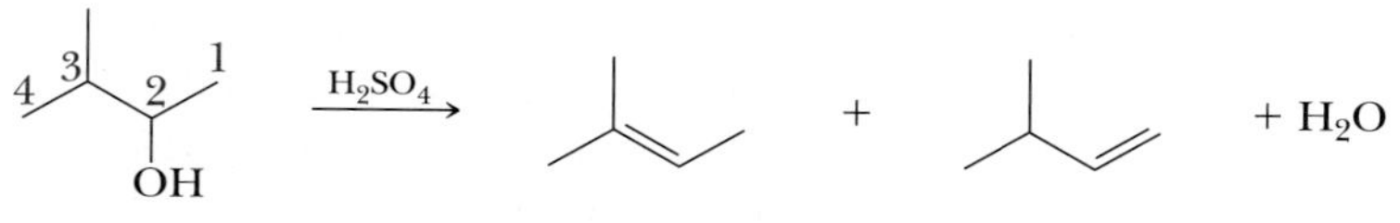

3-Methyl-2-butanol   2-Methyl-2-butene   3-Methyl-1-butene
                         (major product)

(b) The major product, 1-methylcyclopentene, has three alkyl substituents on the double bond. 3-Methylcyclopentene has only two substituents on the double bond.

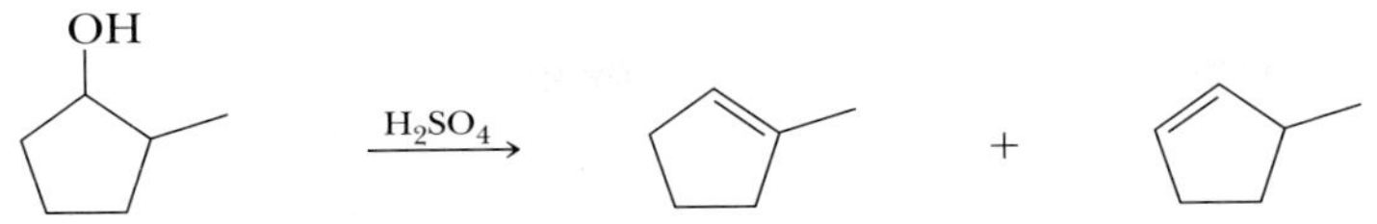

2-Methylcyclopentanol   1-Methylcyclopentene   3-Methylcyclopentene
                                    (major product)

Problem 10.8

Draw structural formulas for the alkenes formed by acid-catalyzed dehydration of each alcohol. Where isomeric alkenes are possible, predict which is the major product.
(a) 2-Methyl-2-butanol   **(b)** 1-Methylcyclopentanol

Dehydration of primary and secondary alcohols is often accompanied by rearrangement. Acid-catalyzed dehydration of 3,3-dimethyl-2-butanol, for example, gives a mixture of two alkenes, each of which is the result of a rearrangement.

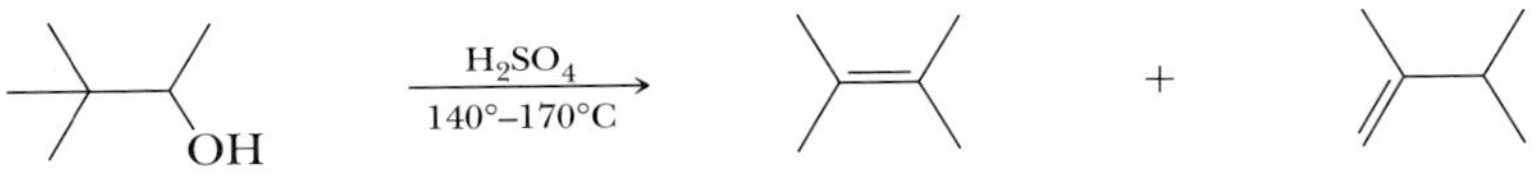

3,3-Dimethyl-2-butanol   2,3-Dimethyl-2-butene   2,3-Dimethyl-1-butene
                                         (80%)   (20%)

Based on the relative rates of dehydration of alcohols ($3° > 2° > 1°$) and the prevalence of rearrangement, particularly among primary and secondary alcohols, chemists propose a three-step mechanism for acid-catalyzed dehydration of secondary and tertiary alcohols. This mechanism involves formation of a carbocation in the rate-determining step and, therefore, is classified as an E1 mechanism.

Mechanism   *Acid-Catalyzed Dehydration of 2-Butanol—*
                 An E1 Reaction

Step 1: Proton transfer from H_3O^+ to the OH group of the alcohol gives an oxonium ion; OH^-, a poor leaving group, is converted to H_2O, a better leaving group.

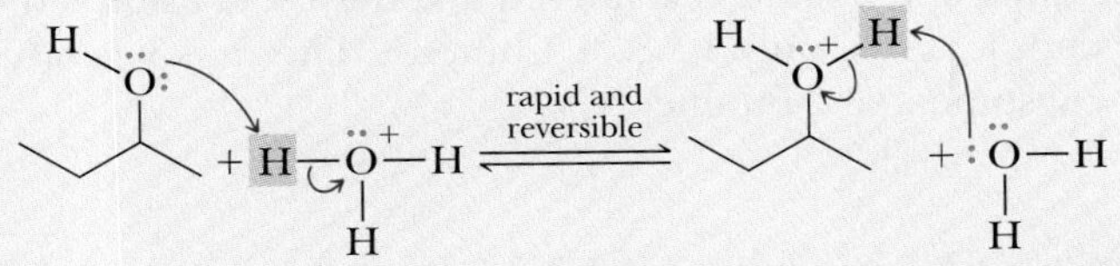

Step 2: Breaking of the C—O bond and loss of H_2O gives a 2° carbocation intermediate.

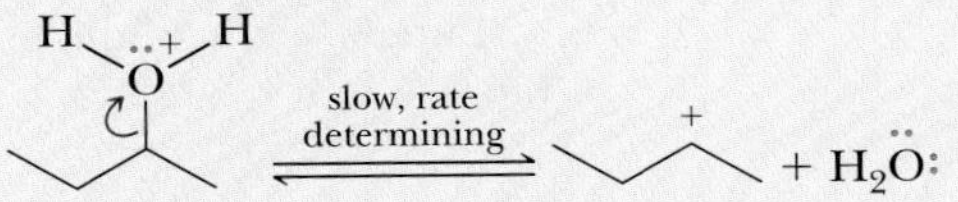

A 2° carbocation
intermediate

Step 3: Proton transfer from a carbon adjacent to the positively charged carbon to H_2O gives the alkene. In this step, the sigma electrons of the C—H bond become the pi electrons of the carbon-carbon double bond.

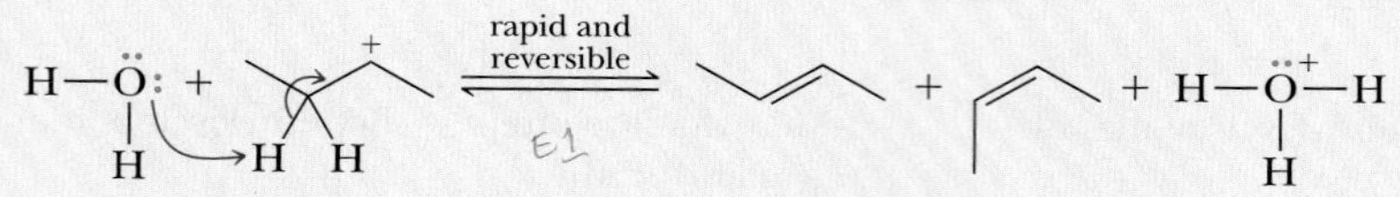

We account for acid-catalyzed dehydration accompanied by rearrangement in the same manner we did for the rearrangements that accompany S_N1 and E1 reactions (Section 9.4H). Rearrangement occurs through formation of a carbocation intermediate followed by migration of an atom or group, with its bonding pair of electrons, from the β-carbon to the carbon bearing the positive charge.

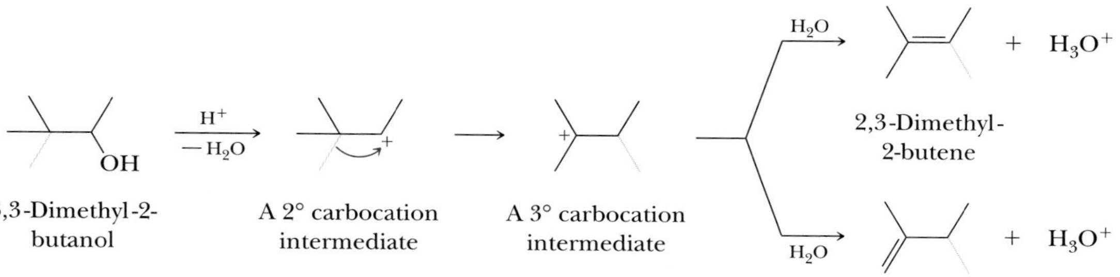

3,3-Dimethyl-2-butanol

A 2° carbocation intermediate

A 3° carbocation intermediate

2,3-Dimethyl-2-butene

2,3-Dimethyl-1-butene

The driving force for rearrangements of this type is conversion of a less stable carbocation to a more stable one. Proton transfer to H_2O then gives the alkenes. As in other cases of acid-catalyzed dehydration of alkenes, the Zaitsev rule applies, and the more substituted alkene predominates.

Primary alcohols with little or no β-branching undergo acid-catalyzed dehydration to give a terminal alkene and rearranged alkenes. Acid-catalyzed dehydration of 1-butanol, for example, gives only 12% of 1-butene. The major product is a mixture of the *trans* and *cis* isomers of 2-butene.

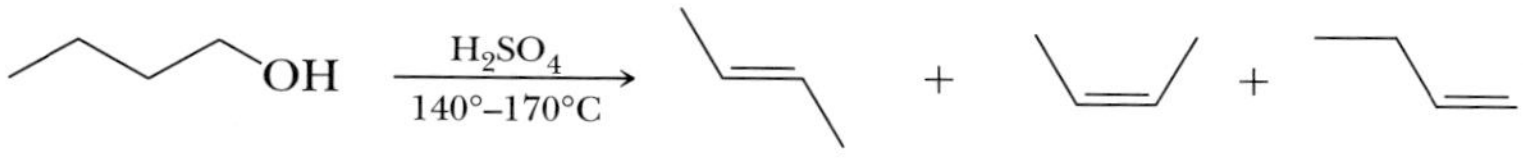

1-Butanol

trans-2-Butene (56%)

cis-2-Butene (32%)

1-Butene (12%)

We account for the formation of these products by a combination of E1 and E2 mechanisms.

Mechanism   *Acid-Catalyzed Dehydration of an Unbranched Primary Alcohol*

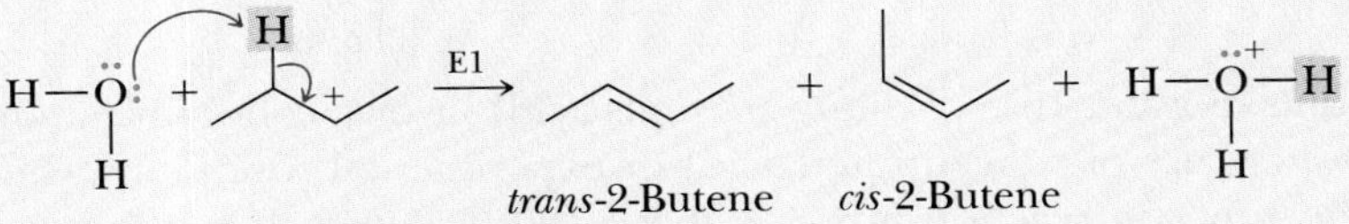

ORGANIC
Chemistry··Now™
Click Mechanisms to view an animation of the **Dehydration of an Unbranched Primary Alcohol to an Alkene**

Step 1: Proton transfer from H_3O^+ to the OH group gives an oxonium ion.

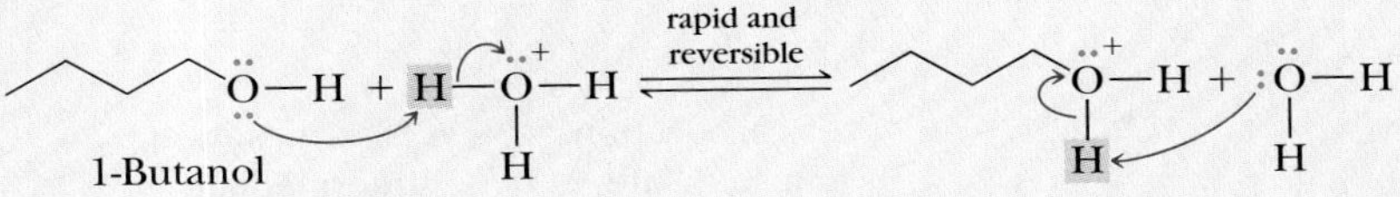

1-Butanol

Step 2: Simultaneous proton transfer to solvent and loss of H_2O gives the carbon-carbon double bond of the terminal alkene.

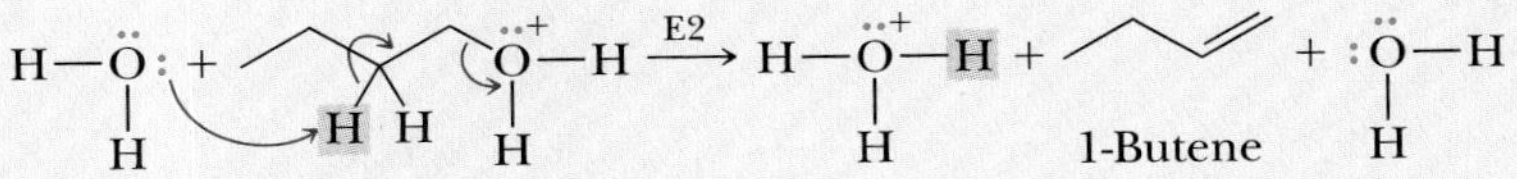

1-Butene

Step 3: Simultaneous shift of a hydride ion from the β-carbon to the α-carbon and loss of H_2O gives a carbocation intermediate.

A 2° carbocation

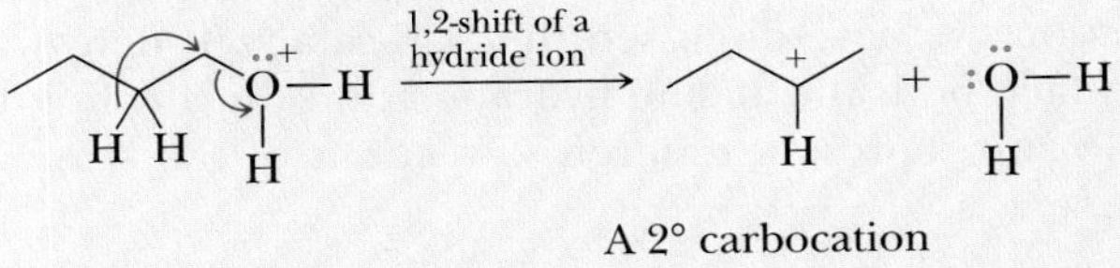

Step 4: Transfer of a proton from a carbon adjacent to the carbocation to solvent gives the rearranged alkenes.

trans-2-Butene   *cis*-2-Butene

Example 10.9

Propose a mechanism to account for this acid-catalyzed dehydration.

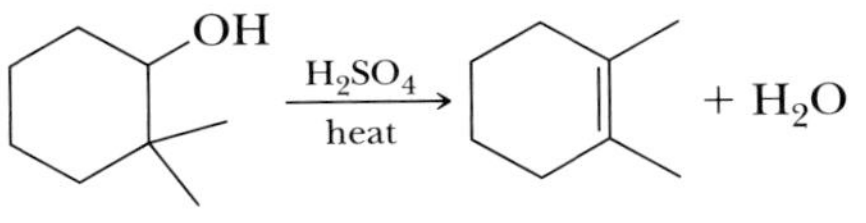

Solution

Proton transfer to the OH group to form an oxonium ion followed by loss of H_2O gives a 2° carbocation intermediate. Migration of a methyl group with its bonding pair of electrons from the adjacent carbon to the positively charged carbon gives a more stable 3° carbocation. Proton transfer from this intermediate to a base, here shown as H_2O, gives the observed product.

A 2° carbocation intermediate

A 3° carbocation intermediate

Problem 10.9

Propose a mechanism to account for this acid-catalyzed dehydration.

In Section 6.3B we discussed the acid-catalyzed hydration of alkenes to give alcohols. In the present section, we have discussed the acid-catalyzed dehydration of alcohols to give alkenes. In fact, both the alkene hydration and the alcohol dehydration reactions are reversible and represent different directions of the same process. The following equilibrium exists.

An alkene   An alcohol

Large amounts of water (use of dilute aqueous acid) favor alcohol formation, whereas scarcity of water (use of concentrated acid) or experimental conditions where water is removed (heating the reaction mixture above 100°C) favor alkene formation. Thus, depending on experimental conditions, it is possible to use the hydration-dehydration equilibrium to prepare either alcohols or alkenes in high yields.

This hydration-dehydration equilibrium illustrates a very important principle in the study of reaction mechanisms—the **principle of microscopic reversibility.** According to this principle, the sequence of transition states and reactive intermediates (that is, the mechanism) for any reversible reaction must be the same, but in reverse order, for the backward reaction as for the forward reaction.

As an illustration of the principle of microscopic reversibility, notice that the mechanism we presented in this section for the acid-catalyzed dehydration of 2-butanol to give 2-butene is exactly the reverse of that presented in Section 6.3B for the acid-catalyzed hydration of 2-butene to give 2-butanol.

Principle of microscopic reversibility This principle states that the sequence of transition states and reactive intermediates in the mechanism of any reversible reaction must be the same, but in reverse order, for the reverse reaction as for the forward reaction.

10.7 The Pinacol Rearrangement

Compounds containing hydroxyl groups on two adjacent carbon atoms are called *1,2-diols,* or alternatively, *glycols.* Such compounds can be synthesized by a variety of methods, including oxidation of alkenes by OsO_4 (Section 6.5A). The products of acid-catalyzed dehydration of glycols are quite different from those of acid-catalyzed dehydration of alcohols. For example, treating 2,3-dimethyl-2,3-butanediol (commonly called pinacol) with concentrated sulfuric acid gives 3,3-dimethyl-2-butanone (commonly called pinacolone):

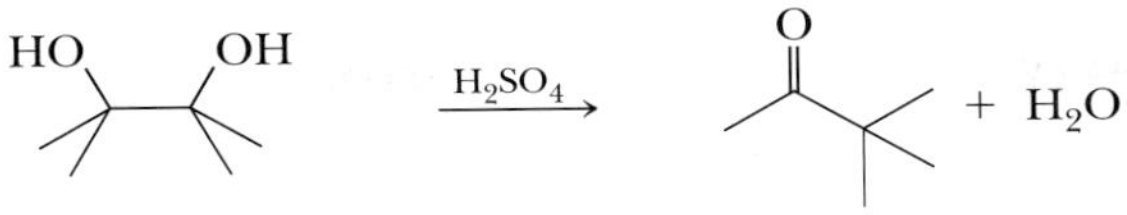

2,3-Dimethyl-2,3-butanediol   3,3-Dimethyl-2-butanone
(Pinacol)   (Pinacolone)

Note two features of this reaction: (1) It involves dehydration of a glycol to form a ketone, and (2) it involves migration of a methyl group from one carbon to an adjacent carbon. Acid-catalyzed conversion of pinacol to pinacolone is an example of a type of reaction called the **pinacol rearrangement.** We account for the conversion of pinacol to pinacolone in a four-step mechanism.

Mechanism   *The Pinacol Rearrangement of 2,3-Dimethyl-2,3-Butanediol (Pinacol)*

Step 1: Proton transfer from the acid catalyst to one of the —OH groups gives an oxonium ion, which converts OH⁻, a poor leaving group, into H_2O, a better leaving group.

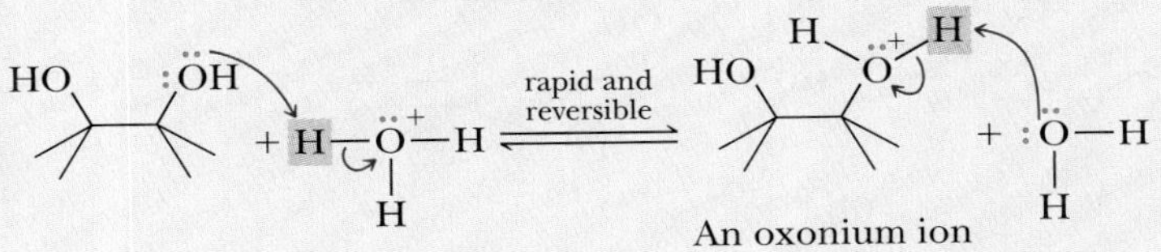

An oxonium ion

Step 2: Loss of H_2O from the oxonium ion gives a 3° carbocation intermediate.

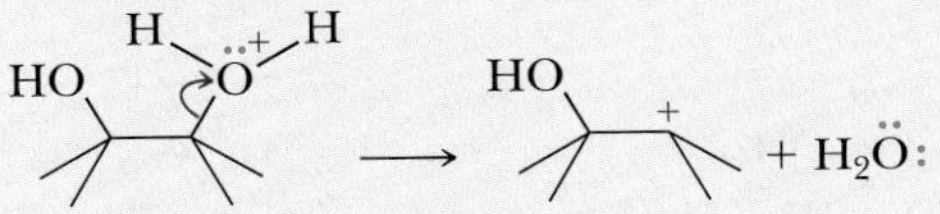

A 3° carbocation
intermediate

Step 3: Migration of a methyl group from the adjacent carbon with its bonding electrons gives a new, more stable resonance-stabilized cation intermediate. Of the two contributing structures we can draw for it, the one on the right makes the greater contribution because, in it, both carbon and oxygen have complete octets of valence electrons (Section 1.6C).

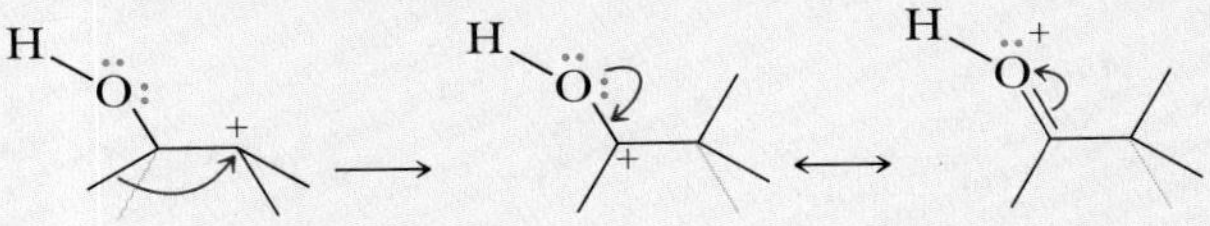

A resonance-stabilized cation intermediate

Step 4: Proton transfer to solvent gives pinacolone.

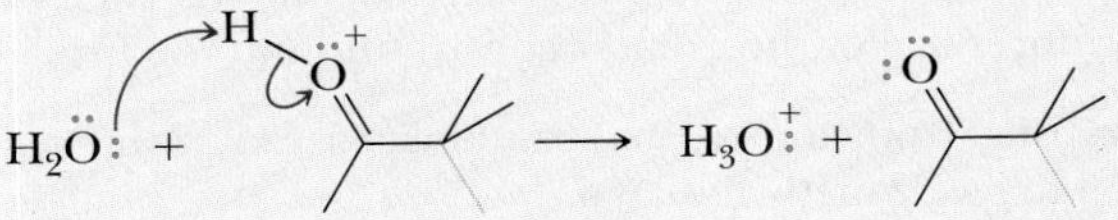

The pinacol rearrangement is general for all 1,2-diols. In the rearrangement of pinacol, a symmetrical diol, equivalent carbocations are formed no matter which —OH becomes protonated and leaves. Studies of unsymmetrical 1,2-diols reveal that the —OH group that becomes protonated and leaves is the one that gives rise to the more stable carbocation. For example, treatment of 2-methyl-1,2-propanediol with cold concentrated sulfuric acid gives a 3° carbocation. Subsequent migration of hydride ion (H: ⁻) followed by transfer of a proton from the new cation to solvent gives 2-methylpropanal.

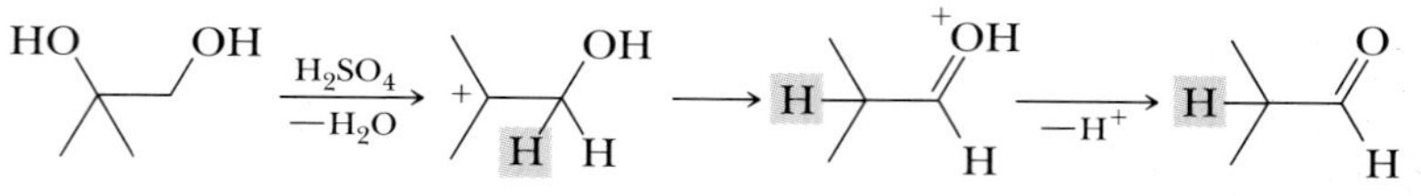

2-Methyl-1,2-   A 3° carbocation   2-Methylpropanal
propanediol   intermediate

Example 10.10

Predict the product formed by treating each glycol with H_2SO_4.

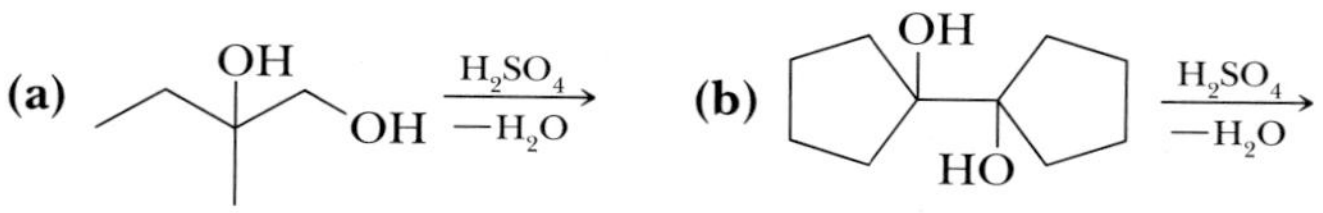

Solution

(a) **Step 1:** Protonation of the 3° hydroxyl group followed by loss of H_2O gives a 3° carbocation intermediate.

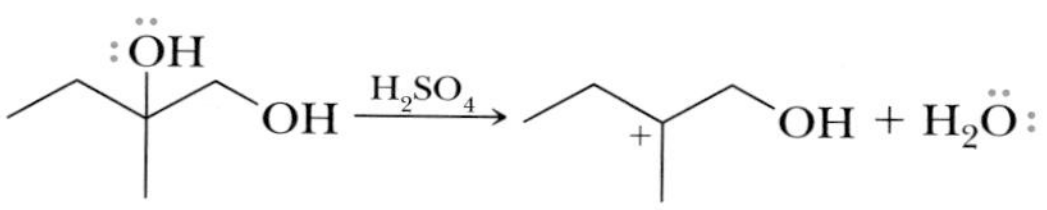

A 3° carbocation
intermediate

Step 2: Migration of a hydride ion from the adjacent carbon gives a resonance-stabilized cation intermediate.

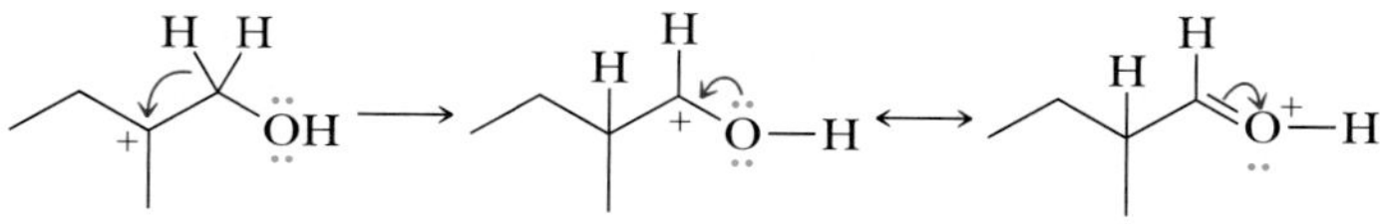

A resonance-stabilized cation intermediate

The contributing structure on the right has filled valence shells on both carbon and oxygen and, therefore, makes the greater contribution to the hybrid.

Step 3: Proton transfer from the resonance-stabilized cation intermediate to water completes the reaction to give 2-methylbutanal.

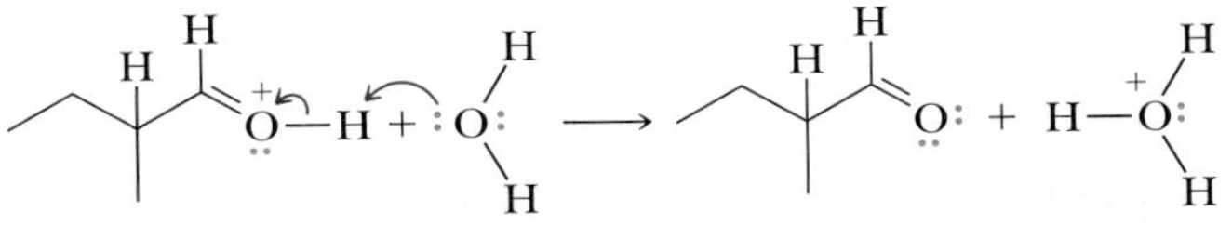

2-Methylbutanal

(b) Protonation of either hydroxyl group followed by loss of water gives a 3° carbo-cation. The group that then migrates is a CH_2 group of the five-membered ring, and the product is a bicyclic ketone.

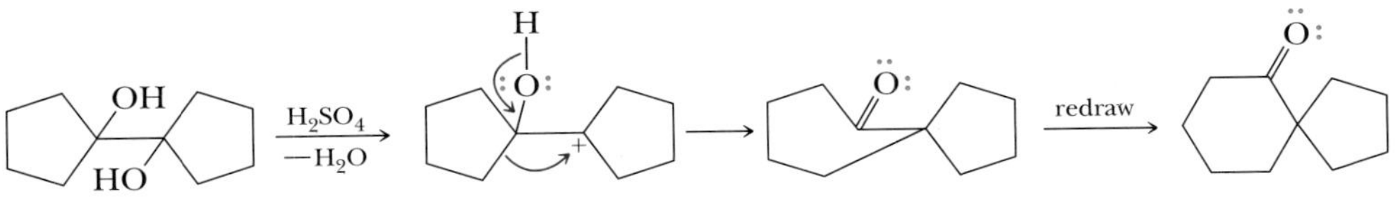

A 3° carbocation
intermediate

Spiro[4.5]decan-6-one

The product belongs to the class of compounds called spiro compounds, in which two rings share only one carbon atom.

Problem 10.10

Propose a mechanism to account for the following transformation.

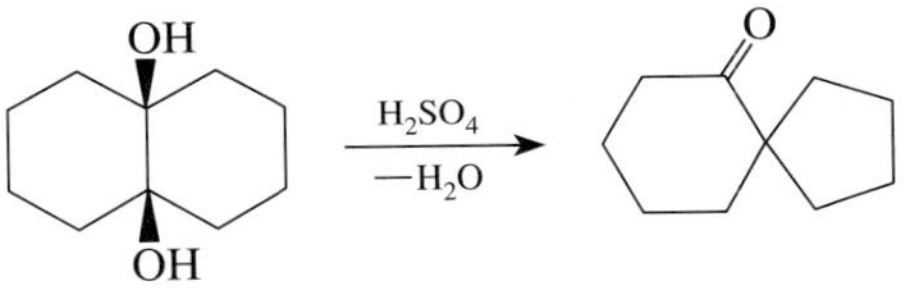

10.8 Oxidation of Alcohols

Oxidation of a primary alcohol gives an aldehyde or a carboxylic acid, depending on experimental conditions. Secondary alcohols are oxidized to ketones. Tertiary alcohols are not oxidized. Following is a series of transformations in which a primary alcohol is oxidized first to an aldehyde and then to a carboxylic acid. The fact that each transformation involves oxidation is indicated by the symbol O in brackets over the reaction arrow:

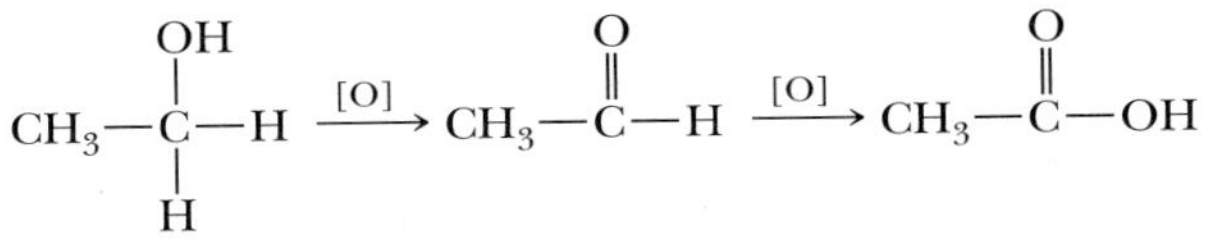

A primary
alcohol

An aldehyde

A carboxylic
acid

A. Chromic Acid

The reagent most commonly used in the laboratory for the oxidation of a primary alcohol to a carboxylic acid is chromic acid, H_2CrO_4. Chromic acid is prepared by dissolving either chromium(VI) oxide or potassium dichromate in aqueous sulfuric acid:

$$CrO_3 \; + \; H_2O \; \xrightarrow{H_2SO_4} \; H_2CrO_4$$

Chromium(VI)   Chromic acid
oxide

$$K_2Cr_2O_7 \; \xrightarrow{H_2SO_4} \; H_2Cr_2O_7 \; \xrightarrow{H_2O} \; 2H_2CrO_4$$

Potassium   Chromic acid
dichromate

A solution of chromic acid in aqueous sulfuric acid is known as the **Jones reagent.**

     Because of the low solubility of most organic compounds in water, their oxidation by chromic acid is commonly carried out by dissolving them in acetone and then adding a stoichiometric amount of Jones reagent to complete the oxidation.

     Oxidation of 1-hexanol, for example, using chromic acid in the mixed solvent of aqueous sulfuric acid and acetone gives hexanoic acid in high yield.

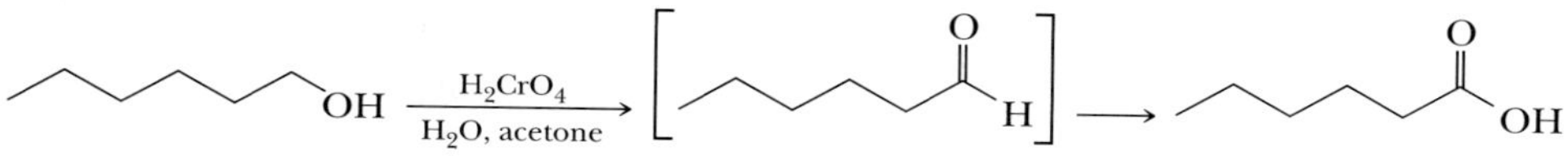

1-Hexanol   Hexanal   Hexanoic acid
                              (not isolated)

These experimental conditions are more than sufficient to oxidize the intermediate aldehyde to a carboxylic acid.

     Secondary alcohols are oxidized to ketones by chromic acid:

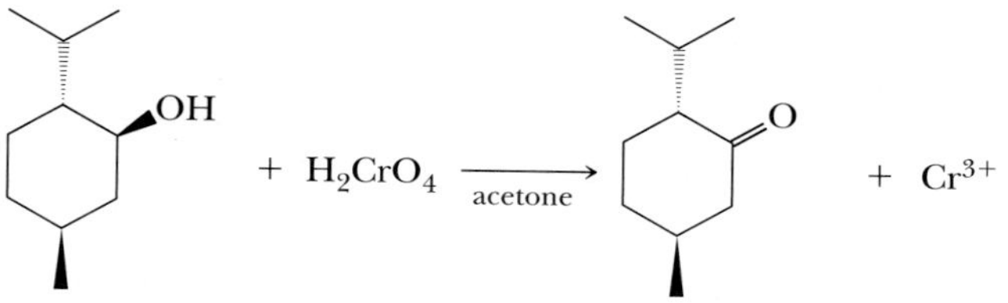

2-Isopropyl-5-methyl-   2-Isopropyl-5-methyl-
cyclohexanol   cyclohexanone
(Menthol)   (Menthone)

     Tertiary alcohols are resistant to oxidation because the carbon bearing the —OH is already bonded to three carbon atoms and, therefore, cannot form an additional carbon-oxygen bond.

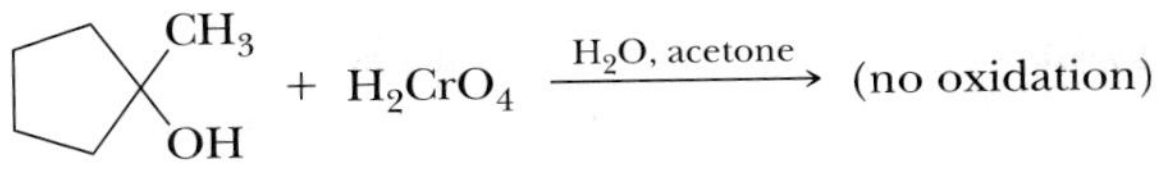

1-Methylcyclopentanol

The prerequisite for the oxidation of an alcohol to an aldehyde or ketone is at least one H on the carbon bearing the OH group.

Mechanism   *Chromic Acid Oxidation of an Alcohol*

Step 1: Reaction of the alcohol and chromic acid gives an alkyl chromate by a mechanism similar to that for the formation of a carboxylic ester (Section 17.7). There is no change in oxidation state of either carbon or chromium as a result of this step.

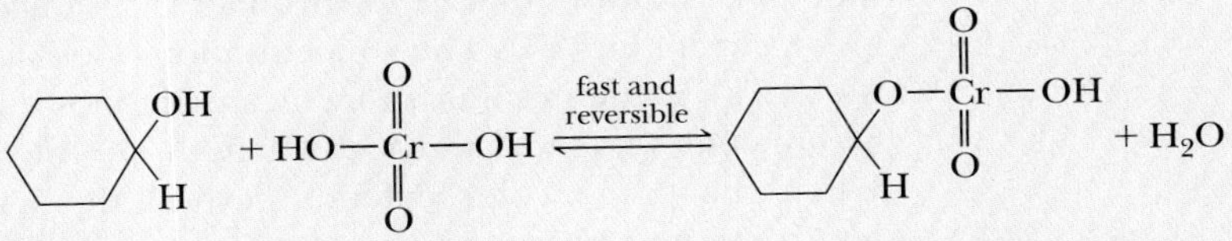

Cyclohexanol   An alkyl chromate

Step 2: Reaction of the alkyl chromate with a base, here shown as a water molecule, results in cleavage of a C—H bond, formation of the carbonyl group, and reduction of chromium(VI) to chromium(IV).

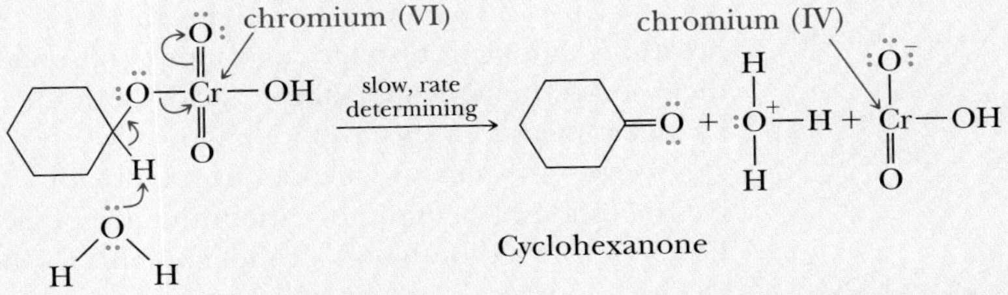

Cyclohexanone

This step is the oxidation-reduction step; carbon undergoes a two-electron oxidation and chromium(VI) undergoes a two-electron reduction to chromium(IV). Chromium(IV) then participates in further oxidations by a similar mechanism and eventually is transformed to Cr(III).

ORGANIC
Chemistry Now™

Click Mechanisms to view an animation of the **Chromic Acid Oxidation of a Secondary Alcohol**

We have shown that in aqueous chromic acid, a primary alcohol is oxidized first to an aldehyde and then to a carboxylic acid. In the second step, it is not the aldehyde that is oxidized, but rather the aldehyde hydrate formed by addition of a molecule of water to the aldehyde carbonyl group (hydration). We will study the hydration of aldehydes and ketones in more detail in Section 16.7. An —OH of the aldehyde hydrate now reacts with chromic acid to complete the oxidation of the aldehyde to a carboxylic acid.

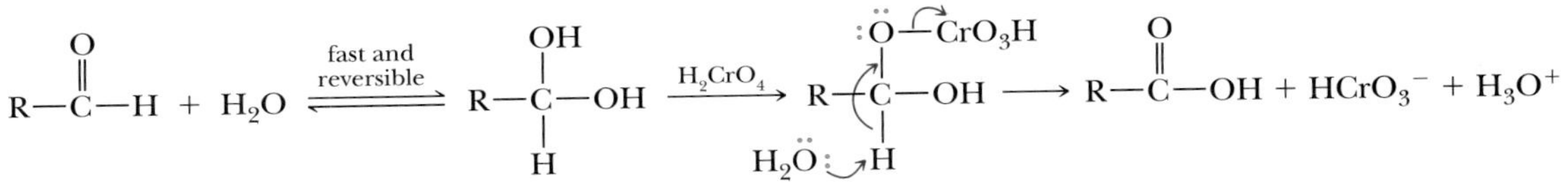

An aldehyde   An aldehyde   An alkyl   A carboxylic
                           hydrate   chromate ester   acid

CHEMICAL CONNECTIONS

Blood Alcohol Screening

Potassium dichromate oxidation of ethanol to acetic acid is the basis for the original breath alcohol screening test used by law enforcement agencies to determine a person's blood alcohol content. The test is based on the difference in color between the dichromate ion (reddish orange) in the reagent and the chromium(III) ion (green) in the product.

$$CH_3CH_2OH \quad + \quad Cr_2O_7^{2-}$$

Ethanol   Dichromate ion
(reddish orange)

$$\xrightarrow[\text{H}_2\text{O}]{\text{H}_2\text{SO}_4} \quad CH_3\overset{\displaystyle O}{\overset{\|}{C}}OH \quad + \quad Cr^{3+}$$

Acetic acid   Chromium(III) ion
(green)

Thus, color change from reddish orange to green can be used as a measure of the quantity of ethanol present in a sample of a person's breath.

In its simplest form, a breath alcohol screening test consists of a sealed glass tube containing a potassium

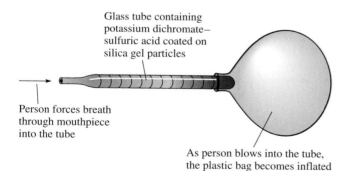

Glass tube containing potassium dichromate–sulfuric acid coated on silica gel particles

Person forces breath through mouthpiece into the tube

As person blows into the tube, the plastic bag becomes inflated

dichromate-sulfuric acid reagent impregnated on silica gel. To administer the test, the ends of the tube are broken off, a mouthpiece is fitted to one end, and the other end is inserted into the neck of a plastic bag. The person being tested then blows into the mouthpiece until the plastic bag is inflated.

As breath containing ethanol vapor passes through the tube, reddish orange dichromate is reduced to green chromium(III). The concentration of ethanol in the breath is then estimated by measuring how far the green color extends along the length of the tube. When the green color extends beyond the halfway point, the person is judged to have a sufficiently high blood alcohol content to warrant further, more precise testing.

The Breathalyzer, a more accurate testing device, operates on the same principle as the simplified screening test. In a Breathalyzer test, a measured volume of breath is bubbled through a solution of potassium dichromate in aqueous sulfuric acid, and the color change is measured spectrophotometrically.

These tests measure alcohol in the breath. The legal definition of being under the influence of alcohol, however, is based on blood alcohol content, not breath alcohol content. The chemical correlation between these two measurements is that air deep within the lungs is in equilibrium with blood passing through the pulmonary arteries, and an equilibrium is established between blood alcohol and breath alcohol. It has been determined by tests in a person drinking alcohol that 2100 mL of breath contains the same amount of ethanol as 1.00 mL of blood.

B. Pyridinium Chlorochromate

The form of Cr(VI) most commonly used for oxidation of a primary alcohol to an aldehyde is prepared by dissolving CrO_3 in aqueous HCl and adding pyridine to precipitate **pyridinium chlorochromate (PCC)** as a solid.

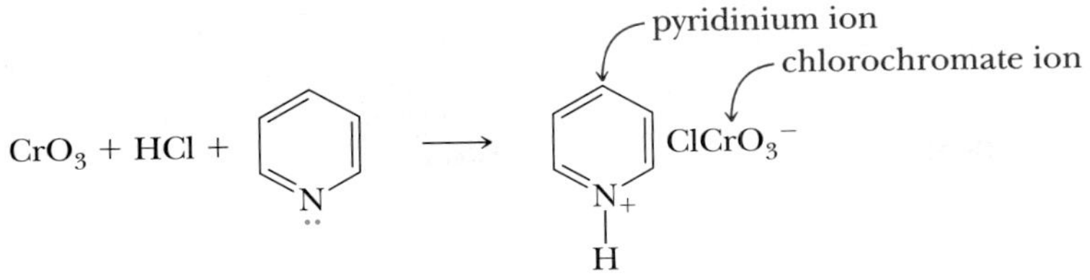

Pyridine   Pyridinium chlorochromate
(PCC)

PCC oxidations are carried out in aprotic solvents, most commonly dichloromethane, CH_2Cl_2.

PCC not only is selective for the oxidation of primary alcohols to aldehydes but also has little effect on carbon-carbon double bonds or other easily oxidized functional groups. In the following example, geraniol, a primary terpene alcohol, is oxidized to geranial without affecting either carbon-carbon double bond.

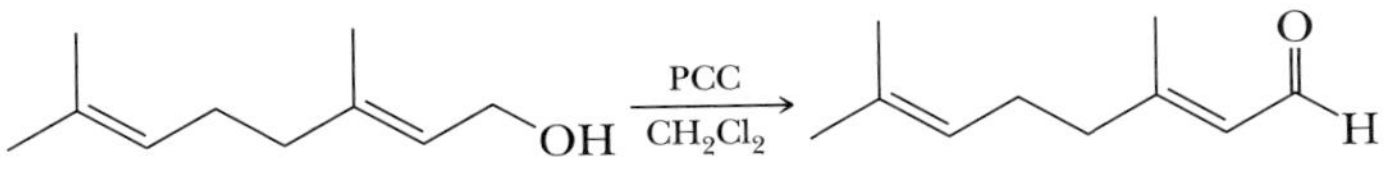

Geraniol   Geranial

Thus, we see why PCC is specific for the oxidation of primary alcohols to aldehydes; it does not oxidize aldehydes further because the PCC reagent is not used in water but rather in an organic solvent, usually CH_2Cl_2. Without added water, the aldehyde hydrate cannot form. Only the aldehyde hydrate is susceptible to further oxidation by Cr(VI). Both PCC and H_2CrO_4 can be used for the oxidation of a 2° alcohol to a ketone.

Example 10.11

Draw the product of treating each alcohol with PCC.
(a) 1-Hexanol   (b) 2-Hexanol   (c) Cyclohexanol

Solution

1-Hexanol, a primary alcohol, is oxidized to hexanal. 2-Hexanol, a secondary alcohol, is oxidized to 2-hexanone. Cyclohexanol, a secondary alcohol, is oxidized to cyclohexanone.

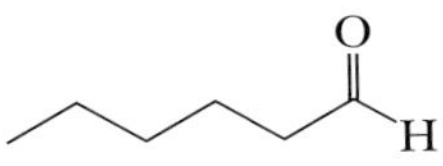

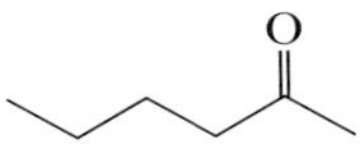

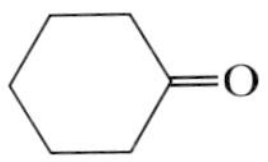

Hexanal   2-Hexanone   Cyclohexanone

Problem 10.11

Draw the product of treating each alcohol in Example 10.11 with chromic acid.

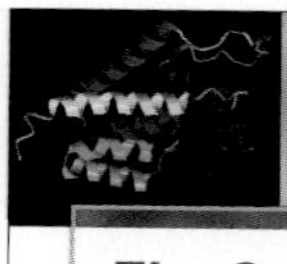

CONNECTIONS TO BIOLOGICAL CHEMISTRY

The Oxidation of Alcohols by NAD⁺

It should come as no surprise that biological systems do not use agents like potassium dichromate or the oxides of other transition metals for the oxidation of alcohols to aldehydes and ketones, or for the oxidation of aldehydes to carboxylic acids. What biological systems use instead is NAD^+, a compound derived from niacin, one of the B complex vitamins.

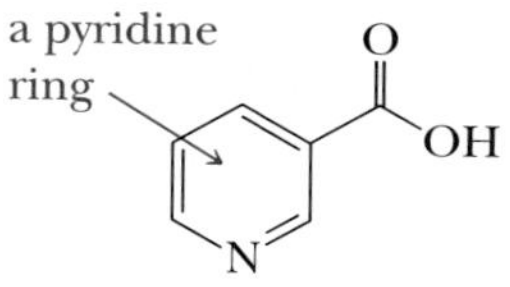

Nicotinic acid
(Niacin; Vitamin B6)

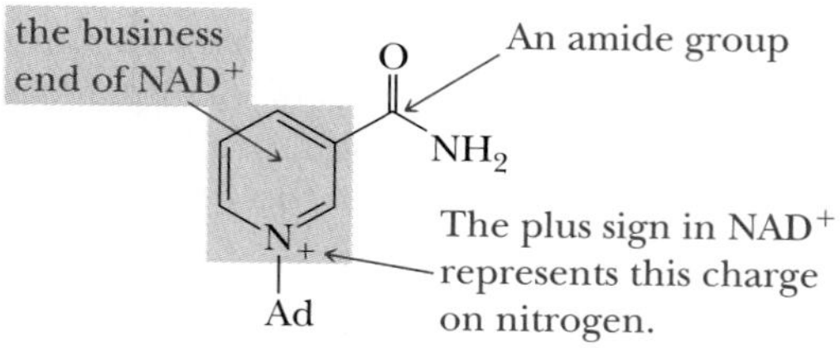

Nicotinamide adenine
dinucleotide (NAD^+)

In converting niacin to NAD^+, the body converts the carboxyl group to an amide group (the —OH of the carboxyl group is replaced by —NH_2). In addition, the nitrogen of the six-membered ring is bonded to the sugar ribose, which is in turn bonded to a molecule called adenosine diphosphate (ADP). We will study ribose and ADP in Chapters 25 and 28. For now, we abbreviate the combination of ribose and ADP as Ad. The function of the Ad portion of the molecule is to position NAD^+ on the surface of the enzyme in the proper orientation relative to the molecule it is to oxidize.

When NAD^+ functions as a biological oxidizing agent, it is reduced to NADH. In this transformation, NAD^+ gains one H and loses the positive charge on its nitrogen. The key concept here is that NAD^+ is a two-electron oxidizing agent, and in the process undergoes a two-electron reduction to NADH.

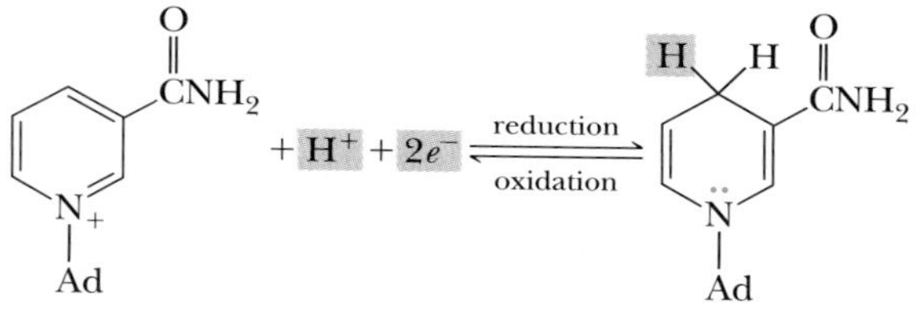

NAD^+
(oxidized form)

NADH
(reduced form)

NAD^+ serves as an oxidizing agent in a wide variety of enzyme-catalyzed reactions, two of which are shown here. The oxidation of ethanol to acetaldehyde is the first step in the reaction by which the liver detoxifies ethanol. The oxidation of lactate to pyruvate is one step in the process by which the body derives energy from the oxidation of carbohydrates. Lactate is the end-product of anaerobic (without oxygen) glycolysis.

$$CH_3CH_2OH + NAD^+ \xrightarrow{\text{alcohol dehydrogenase}} CH_3\overset{\overset{\displaystyle O}{\|}}{C}H + NADH + H^+$$

Ethanol

Ethanal
(Acetaldehyde)

$$CH_3\overset{\overset{\displaystyle OH}{|}}{C}HCOO^- + NAD^+ \xrightarrow{\text{lactate dehydrogenase}} CH_3\overset{\overset{\displaystyle O}{\|}}{C}COO^- + NADH + H^+$$

Lactate

Pyruvate

Following is a mechanism for the oxidation of an alcohol by alcohol dehydrogenase and NAD^+.

Mechanism *Oxidation of an Alcohol by NAD^+*

Arrow ①: A basic group, B^-, on the enzyme removes H^+ from the —OH group.

Arrow ②: The H—O sigma bond breaks as a C=O bond forms.

Arrow ③: Transfer of a hydride ion from the carbon bearing the —OH group to NAD^+ creates the new C—H bond in NADH. This is the oxidation-reduction step: The alcohol is oxidized, and NAD^+ is reduced.

Arrows ④–⑤: Electrons within the ring flow to the positively charged nitrogen.

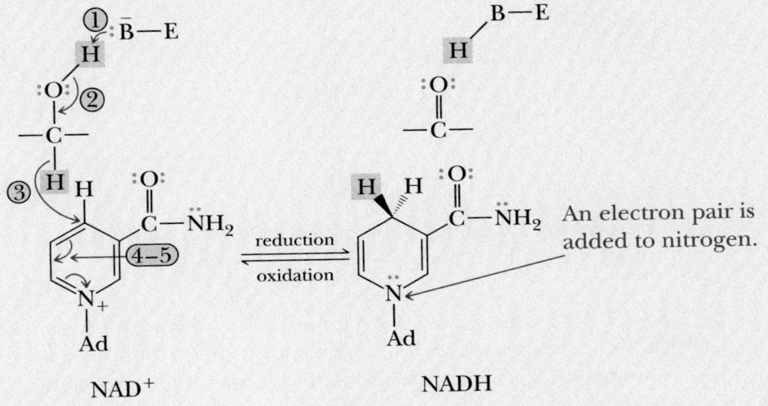

NAD⁺   NADH

An electron pair is added to nitrogen.

Because enzymes are chiral catalysts, NAD^+ oxidation takes place in a chiral environment, with the result that hydride transfer to NAD^+ is stereoselective; some enzymes catalyze the addition of hydride ion to the top face of NAD^+; others, to the bottom face. In the case of alcohol dehydrogenase, the hydride ion is transferred to the top face.

As we will see in Section 16.11, NADH can in turn reverse the process and transfer a hydride ion stereoselectively to the carbonyl group of an aldehyde or ketone, thus reducing these types of molecules to either primary or secondary alcohols respectively.

C. Periodic Acid Oxidation of Glycols

Periodic acid, H_5IO_6 (or alternatively $HIO_4 \cdot 2H_2O$), is a white crystalline solid, mp 122°C. Its major use in organic chemistry is for the cleavage of a glycol to two carbonyl groups. In the process, periodic acid is reduced to iodic acid.

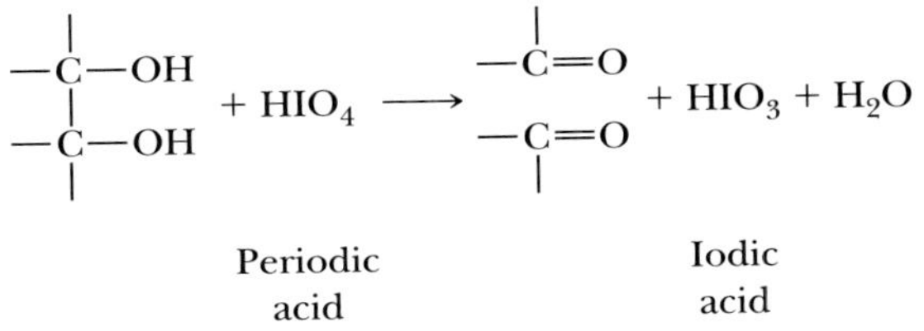

Periodic   Iodic
acid   acid

For example, periodic acid oxidizes *cis*-1,2-cyclohexanediol to hexanedial.

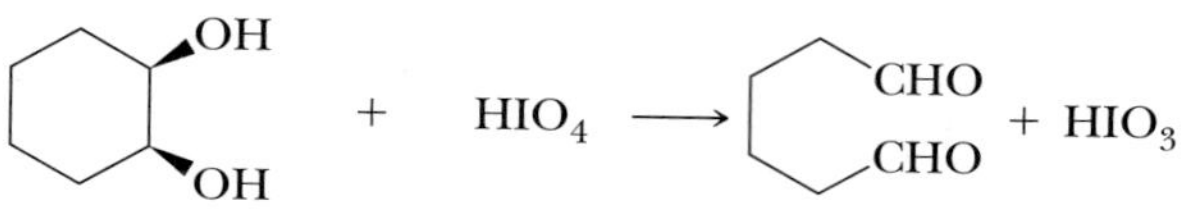

| *cis*-1,2-Cyclohexanediol | Periodic acid | Hexanedial | Iodic acid |

Mechanism Oxidation of a Glycol by Periodic Acid

Step 1: Reaction of the glycol with periodic acid gives a five-membered cyclic periodate.

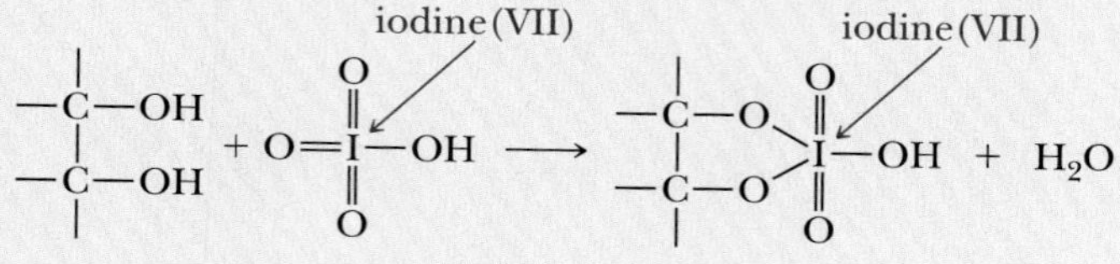

A cyclic periodate

There is no change in the oxidation state of either iodine or the glycol as a result of formation of the cyclic periodate.

Step 2: Redistribution of valence electrons within the cyclic periodate gives HIO_3 and two carbonyl groups. A result of this electron redistribution is an oxidation of the organic component and a reduction of the iodine-containing component.

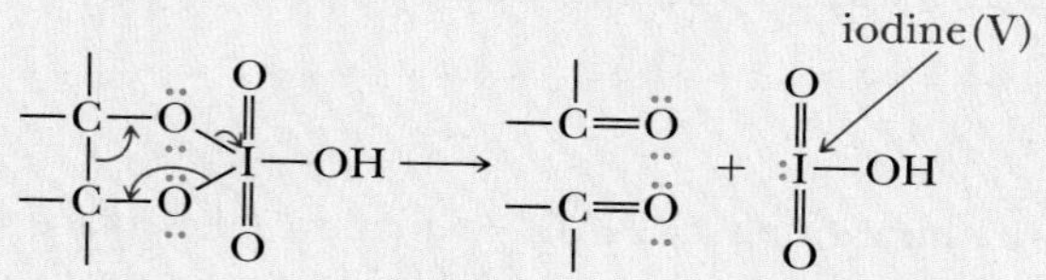

HIO_3
(Iodic acid)

This mechanism is consistent with the fact that HIO4 oxidations are restricted to glycols that can form a five-membered cyclic periodate. Any glycol that cannot form such a cyclic periodate is not oxidized by periodic acid. Following are structural formulas for two isomeric decalindiols. Only the cis glycol can form a cyclic periodate with periodic acid, and only the cis glycol is oxidized by this reagent.

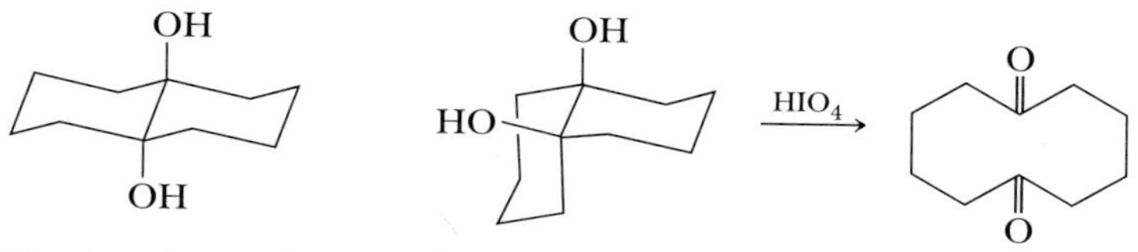

The *trans* isomer is unreactive toward periodic acid

The *cis* isomer forms a cyclic periodate and is cleaved

Example 10.12

What products are formed when each glycol is treated with HIO_4?

(a) [structure: cyclopentane with two OH groups, one on ring carbon and a CH$_2$OH]

(b) [structure: 2-methyl-2,3-butanediol showing OH, OH groups]

Solution

The bond between the carbons bearing the —OH groups is cleaved and each —OH group is converted to a carbonyl group.

(a) [structure: cyclopentanone ＝O + formaldehyde H–CHO]

(b) [structure: acetone + acetaldehyde H–CHO]

Problem 10.12

α-Hydroxyketones and α-hydroxyaldehydes are also oxidized by treatment with periodic acid.

[reaction scheme: α-hydroxyaldehyde with CHO and OH groups, reagents 1. H_2O, 2. HIO_4]

It is not the α-hydroxyketone or aldehyde, however, that undergoes reaction with periodic acid but the hydrate formed by addition of water to the carbonyl group of the α-hydroxyketone or aldehyde. Write a mechanism for the oxidation of this α-hydroxyaldehyde by HIO_4.

10.9 Thiols

A. Structure

The functional group of a **thiol** is an **—SH (sulfhydryl) group** bonded to an sp^3 hybridized carbon. Figure 10.4 shows a Lewis structure and a ball-and-stick model of methanethiol, CH_3SH, the simplest thiol. The C—S—H bond angle in methanethiol is 100.3°. By way of comparison, the H—S—H bond angle in H_2S is 93.3°. If a sulfur atom were bonded to two other atoms by fully hybridized sp^3 hybrid orbitals, bond angles about sulfur would be approximately 109.5°. If, instead, a sulfur atom were bonded to two other atoms by unhybridized $3p$ orbitals, bond angles would be approximately 90°. The fact that the C—S—H bond angle in methanethiol is 100.3° and the H—S—H bond angle in H_2S is 93.3° indicates that there is considerably more p-character (and less s-character) in the bonding orbitals of divalent sulfur than there is in those of divalent oxygen.

(a) H—C—S—H (with H above and below carbon, lone pairs on S)

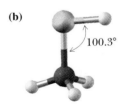

(b) 100.3°

Figure 10.4
Methanethiol, CH_3SH.
(a) Lewis structure and
(b) ball-and-stick model.

Thiol A compound containing an —SH (sulfhydryl) group bonded to an sp^3 hybridized carbon.

B. Nomenclature

In the older literature, thiols are often referred to as **mercaptans,** which literally means mercury capturing. They react with Hg^{2+} in aqueous solution to give sulfide salts as insoluble precipitates. Thiophenol, C_6H_5SH, for example, gives $(C_6H_5S)_2Hg$.

In the IUPAC system, thiols are named by selecting as the parent alkane the longest chain of carbon atoms that contains the —SH group. To show that the compound is a thiol, retain the final *-e* in the name of the parent alkane and add the suffix *-thiol.* The location of the —SH group takes precedence over alkyl groups and halogens in numbering the parent chain.

In the IUPAC system, —OH takes precedence over —SH in both numbering and naming. In compounds containing these two functional groups, an —SH group is indicated by the IUPAC prefix *sulfanyl-.* Alternatively, it may be indicated by the common-name prefix *mercapto-.*

Common names for simple thiols are derived by naming the alkyl group bonded to —SH and adding the word *mercaptan.*

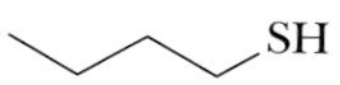

1-Butanethiol
(Butyl mercaptan)

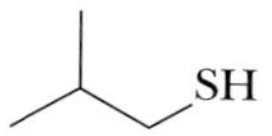

2-Methyl-1-propanethiol
(Isobutyl mercaptan)

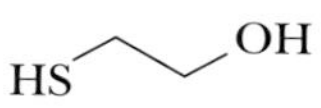

2-Sulfanylethanol
(2-Mercaptoethanol)

Example 10.13

Write names for these thiols.

(a) 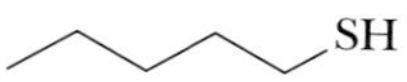   SH   (b) 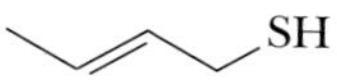   SH

Solution

(a) 1-Pentanethiol (pentyl mercaptan)
(b) (*E*)-2-Butene-1-thiol (*trans*-2-butene-1-thiol)

Problem 10.13

Write IUPAC names for these thiols.

(a) 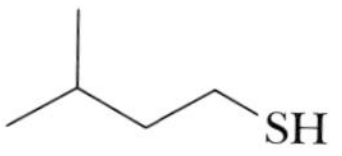   SH   (b) 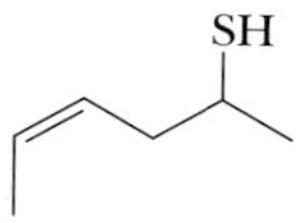   SH

C. Physical Properties

The most outstanding physical characteristic of low-molecular-weight thiols is their stench. The scent of skunks is primarily the result of the following two thiols.

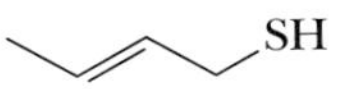

2-Butene-1-thiol 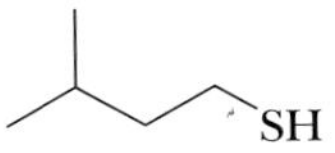   3-Methyl-1-butanethiol
                          (Isopentyl mercaptan)

A blend of low-molecular-weight thiols is added to natural gas as odorants. The most common of these is 2-methyl-2-propanethiol (*tert*-butyl mercaptan) because it is the most resistant to oxidation and has the greatest soil penetration. 2-Propanethiol is also used for this purpose, usually as a blend with *tert*-butyl mercaptan.

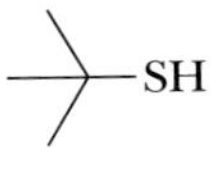

2-Methyl-2-propanethiol   2-Propanethiol
(*tert*-Butyl mercaptan)   (Isopropyl mercaptan)

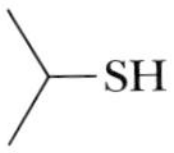

Because of the very low polarity of the S—H bond, thiols show little association by hydrogen bonding. Consequently, they have lower boiling points and are less soluble in water and other polar solvents than alcohols of comparable molecular weights. Table 10.3 gives names and boiling points for three low-molecular-weight thiols. Shown for comparison are boiling points for alcohols that contain the same number of carbon atoms.

In Section 10.2, we illustrated the importance of hydrogen bonding in alcohols by comparing the boiling points of ethanol (bp 78°C) and its constitutional isomer dimethyl ether (bp −24°C). By comparison, the boiling point of ethanethiol is 35°C, and that of its constitutional isomer dimethyl sulfide is 37°C.

$$CH_3CH_2SH \qquad CH_3SCH_3$$

Ethanethiol   Dimethyl sulfide
  bp 35°C   bp 37°C

The fact that the boiling points of these constitutional isomers are almost identical indicates that little or no association by hydrogen bonding occurs between thiol molecules.

D. Thiols in Biological Molecules

The thiol group is found in the amino acid L-cysteine. L-Cysteine is important because the thiol groups of pairs of cysteines are oxidized to disulfide bonds (Section 10.9G), which are a major factor in stabilizing the three-dimensional structure of protein

Table 10.3   Boiling Points of Thiols and Alcohols of the Same Number of Carbon Atoms

Thiol	bp (°C)	Alcohol	bp (°C)
Methanethiol	6	Methanol	65
Ethanethiol	35	Ethanol	78
1-Butanethiol	98	1-Butanol	117

molecules. The thiol group of cysteine functions as a nucleophile in certain enzyme mechanisms. In addition, it binds the metal in certain metal-containing enzymes.

$$HS\diagup\diagdown\underset{\overset{|}{NH_3^+}}{\diagup}COO^-$$

L-Cysteine

E. Preparation

The most common preparation of thiols, RSH, depends on the high nucleophilicity of the hydrosulfide ion, HS^- (Section 9.4B). Sodium hydrosulfide is prepared by bubbling H_2S through a solution of NaOH in water or aqueous ethanol. Reaction of HS^- with a haloalkane gives a thiol.

$$CH_3(CH_2)_8CH_2I + Na^+SH^- \xrightarrow[\text{ethanol}]{S_N2} CH_3(CH_2)_8CH_2SH + Na^+I^-$$

1-Iododecane　　　Sodium　　　　　1-Decanethiol
　　　　　　　hydrosulfide

In practice, the scope and limitations of this reaction are governed by the limitations of the S_N2 reaction and by competition between substitution and β-elimination (Section 9.9). The reaction is most useful for preparation of thiols from primary halides. Yields are lower from secondary halides because of the competing β-elimination (E2) reaction. With tertiary halides, β-elimination (E2) predominates, and the alkene formed by dehydrohalogenation is the major product.

In a commercial application of thiol formation by this nucleophilic substitution, the sodium salt of thioglycolic acid is prepared by the reaction of sodium hydrosulfide and sodium iodoacetate.

$$Na^+ SH^- + ICH_2\overset{\overset{\displaystyle O}{\|}}{C}O^- Na^+ \xrightarrow{S_N2} HSCH_2\overset{\overset{\displaystyle O}{\|}}{C}O^- Na^+ + Na^+ I^-$$

Sodium　　　　Sodium iodoacetate　　　Sodium mercaptoacetate
hydrosulfide　　　　　　　　　　　　(Sodium thioglycolate)

The sodium and ammonium salts of thioglycolic acid are used in cold waving of hair. These compounds work by breaking the disulfide bonds of hair proteins, which maintain the overall structure of hair (Section 27.6C). The disulfide bonds are subsequently reformed by oxidation (Section 10.9G) in a second step. The calcium salt of thioglycolic acid is used as a depilatory; that is, it is used to remove body hair.

$$(HSCH_2COO^-)_2Ca^{2+}$$

Calcium mercaptoacetate
(Calcium thioglycolate)

F. Acidity

Hydrogen sulfide is a stronger acid than water.

$$H_2O + H_2O \rightleftharpoons HO^- + H_3O^+ \qquad pK_a = 15.7$$

$$H_2S + H_2O \rightleftharpoons HS^- + H_3O^+ \qquad pK_a = 7.0$$

Similarly, thiols are stronger acids than alcohols. Compare, for example, the pK_a values of ethanol and ethanethiol in dilute aqueous solution.

$$CH_3CH_2OH + H_2O \rightleftharpoons CH_3CH_2O^- + H_3O^+ \qquad pK_a = 15.9$$

$$CH_3CH_2SH + H_2O \rightleftharpoons CH_3CH_2S^- + H_3O^+ \qquad pK_a = 8.5$$

The greater acidity of thiols compared to alcohols is a combination of two factors. First, S—H bonds are generally weaker than O—H bonds, which facilitates removal of the S—H proton by a base. Second, sulfur (a third-period element) is larger than oxygen (a second-period element), which means that the negative charge on an alkylsulfide ion (RS$^-$) is delocalized over a larger area and is, therefore, more stable than the negative charge on an alkoxide ion (RO$^-$).

Thiols are sufficiently strong acids that when dissolved in aqueous sodium hydroxide, they are converted completely to alkylsulfide salts.

$$CH_3CH_2SH + Na^+OH^- \longrightarrow CH_3CH_2S^-Na^+ + H_2O$$

$pK_a = 8.5$			$pK_a = 15.7$
(Stronger acid)	(Stronger base)	(Weaker base)	(Weaker acid)

To name salts of thiols, give the name of the cation first, followed by the name of the alkyl group to which is attached the suffix -*sulfide*. For example, the sodium salt derived from ethanethiol is named sodium ethylsulfide.

G. Oxidation

Many of the chemical properties of thiols stem from the fact that the sulfur atom of a thiol is oxidized easily to several higher oxidation states, the most common of which are shown in the following flow diagram.

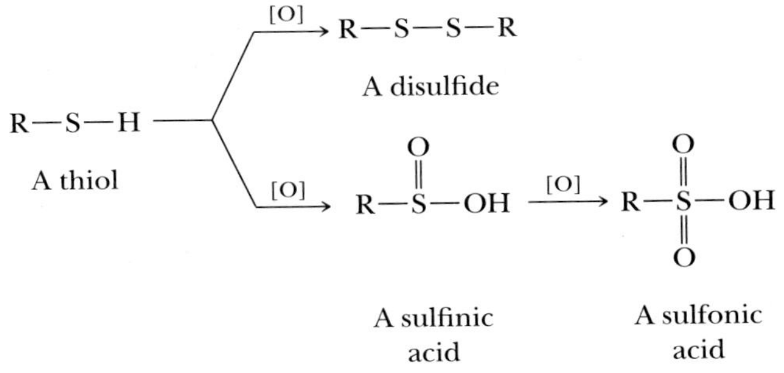

Each oxidation requires a specific oxidizing agent to avoid over-oxidation. There are other oxidation states, too, but they are not very stable. Note that the valence shell of sulfur contains 8 electrons in a thiol and a disulfide, 10 electrons in a sulfinic acid, and 12 electrons in a sulfonic acid (Section 1.2E).

The most common oxidation-reduction reaction of sulfur compounds in biological systems is interconversion between a thiol and a disulfide. The functional group of a disulfide is an —S—S— group.

$$2RSH + I_2 \longrightarrow RSSR + 2HI$$

A thiol   A disulfide

Thiols are also oxidized to disulfides by molecular oxygen

$$2RSH + \tfrac{1}{2}O_2 \longrightarrow RSSR + H_2O$$

A thiol   A disulfide

In fact, thiols are so susceptible to oxidation that they must be protected from contact with air during storage.

The disulfide bond is an important structural feature stabilizing the tertiary structure of many proteins (Section 27.6C).

Summary

The functional group of an alcohol (Section 10.1A) is an —**OH** (**hydroxyl**) group bonded to an sp^3 hybridized carbon. Alcohols are classified as 1°, 2°, or 3° depending on whether the —OH group is bonded to a primary, secondary, or tertiary carbon. IUPAC names of alcohols (Section 10.1B) are derived by changing the suffix of the parent alkane from -*e* to -*ol*. The chain is numbered from the direction that gives the carbon bearing the —OH the lower number. In compounds containing other functional groups of higher precedence, the presence of —OH is indicated by the prefix *hydroxy*-. Common names for alcohols are derived by naming the alkyl group bonded to —OH and adding the word *alcohol*.

Alcohols are polar compounds with oxygen bearing a partial negative charge and both the α-carbon and the hydroxyl hydrogen bearing partial positive charges (Section 10.2). Because of intermolecular association by **hydrogen bonding**, the boiling points of alcohols are higher than those of hydrocarbons of comparable molecular weight. Because of increased dispersion forces, the boiling points of alcohols increase with increasing molecular weight. Alcohols interact with water by hydrogen bonding and, therefore, are more soluble in water than hydrocarbons of comparable molecular weight.

According to the **principle of microscopic reversibility** (Section 10.6), the sequence of transition states and reactive intermediates (that is, the mechanism) for any reversible reaction must be the same, but in reverse order, for the reverse reaction as for the forward reaction.

A **thiol** (Section 10.9A) is a sulfur analog of an alcohol; it contains an —**SH** (**sulfhydryl**) group in place of an —OH group. Thiols are named in the same manner as alcohols, but the suffix -*e* of the parent alkane is retained and -*thiol* added (Section 10.9B). Common names for thiols are derived by naming the alkyl group bonded to —SH and adding the word *mercaptan*. In compounds containing other functional groups of higher precedence, the presence of —SH is indicated by the prefix *mercapto*-. Because the S—H bond is almost nonpolar, the physical properties of thiols are more like those of hydrocarbons of comparable molecular weight (Section 10.9C).

Key Reactions

ORGANIC
Chemistry·⚛·Now™

Click Flashcards to review the
Key Reactions of Alcohols

1. Acidity of Alcohols (Section 10.3)

In dilute aqueous solution, methanol and ethanol are comparable in acidity to water; secondary and tertiary alcohols are somewhat weaker acids.

$$CH_3OH + H_2O \rightleftharpoons CH_3O^- + H_3O^+ \qquad pK_a = 15.5$$

2. Reaction with Active Metals (Section 10.4)

Alcohols react with Li, Na, K, and other active metals to form metal alkoxides, which are nearly the same or somewhat stronger bases than the alkali metal hydroxides such as NaOH and KOH.

$$2CH_3OH + 2Na \longrightarrow 2CH_3O^-Na^+ + H_2$$

3. Reaction with HCl, HBr, and HI (Section 10.5A)

Primary alcohols react by an S_N2 mechanism.

$$\text{\textasciitilde\textasciitilde OH} + \text{HBr} \xrightarrow{\text{reflux}} \text{\textasciitilde\textasciitilde Br} + H_2O$$

Tertiary alcohols react by an S_N1 mechanism with formation of a carbocation intermediate.

$$\text{\textgreater\!-OH} + \text{HCl} \xrightarrow{25°C} \text{\textgreater\!-Cl} + H_2O$$

Secondary alcohols may react by an S_N2 or an S_N1 mechanism, depending on the alcohol and experimental conditions. Primary alcohols with extensive β-branching react by an S_N1 mechanism involving formation of a rearranged carbocation.

$$\text{\textgreater\!\textless OH} + \text{HBr} \longrightarrow \underset{\text{Br}}{\text{\textgreater\!\textless}} + H_2O$$

4. Reaction with PBr₃ (Section 10.5B)

Although some rearrangement may occur with this reagent, it is less likely than in the reaction of an alcohol with HBr.

$$\underset{\text{OH}}{3CH_3\overset{|}{C}HCH_3} + PBr_3 \longrightarrow \underset{\text{Br}}{3CH_3\overset{|}{C}HCH_3} + H_3PO_3$$

5. Reaction with SOCl₂ and SOBr₂ (Section 10.5C)

This is often the method of choice for converting a primary or secondary alcohol to an alkyl chloride or alkyl bromide.

$$\text{\textasciitilde\textasciitilde\textasciitilde OH} + SOCl_2 \xrightarrow{\text{pyridine}} \text{\textasciitilde\textasciitilde\textasciitilde Cl} + SO_2 + HCl$$

6. Acid-Catalyzed Dehydration (Section 10.6)

When isomeric alkenes are possible, the major product is generally the more substituted alkene (Zaitsev's rule). Rearrangements are common with secondary alcohols and also with primary alcohols with extensive β-branching.

$$\underset{\substack{\text{2-Butanol}}}{\underset{\text{OH}}{CH_3CH_2\overset{|}{C}HCH_3}} \xrightarrow[\text{heat}]{85\% \; H_3PO_4} \underset{\substack{\text{2-Butene}\\(80\%)}}{CH_3CH{=}CHCH_3} + \underset{\substack{\text{1-Butene}\\(20\%)}}{CH_3CH_2CH{=}CH_2} + H_2O$$

7. Pinacol Rearrangement (Section 10.7)

Dehydration of a glycol involves formation of a carbocation intermediate, rearrangement, and loss of H^+ to give an aldehyde or ketone.

$$\underset{}{\overset{\text{HO}\quad\text{OH}}{\text{\textgreater\!\!\textless}}} \xrightarrow{H_2SO_4} \overset{O}{\text{\textbackslash\!C\!/}} + H_2O$$

8. Oxidation of a Primary Alcohol to a Carboxylic Acid (Section 10.8A)

A primary alcohol is oxidized to a carboxylic acid by chromic acid.

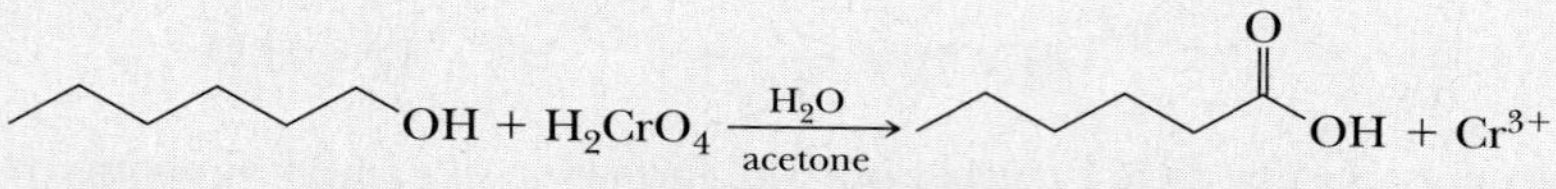

9. Oxidation of a Secondary Alcohol to a Ketone (Section 10.8A and 10.8B)

A secondary alcohol is oxidized to a ketone by chromic acid or by PCC.

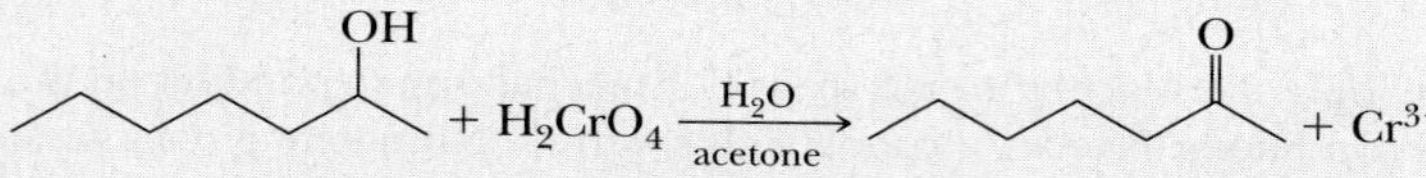

10. Oxidation of a Primary Alcohol to an Aldehyde (Section 10.8B)

The oxidation of a primary alcohol to an aldehyde is most conveniently carried out using pyridinium chlorochromate (PCC).

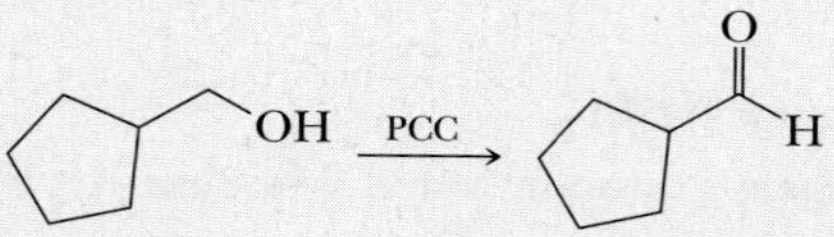

11. Oxidative Cleavage of a Glycol (Section 10.8C)

HIO_4 reacts with a glycol to form a five-membered cyclic periodate intermediate that undergoes carbon-carbon bond cleavage to form two carbonyl groups.

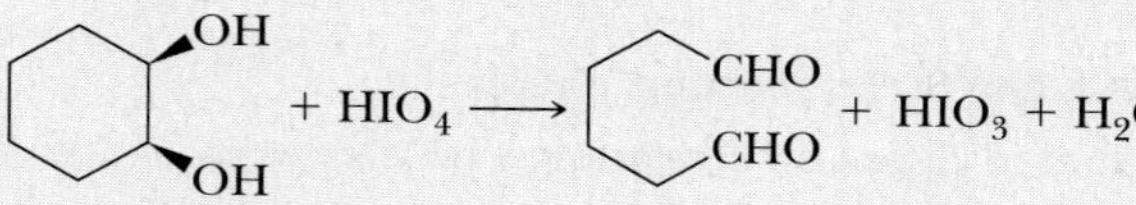

12. Acidity of Thiols (Section 10.9F)

Thiols are weak acids, pK_a 8–9, but considerably stronger than alcohols, pK_a 15.5–18.

$$CH_3CH_2SH + H_2O \rightleftharpoons CH_3CH_2S^- + H_3O^+ \qquad pK_a = 8.5$$

13. Oxidation of Thiols to Disulfides (Section 10.9G)

Oxidation by weak oxidizing agents such as O_2 and I_2 gives disulfides.

$$2RSH + \tfrac{1}{2}O_2 \longrightarrow RSSR + H_2O$$

Problems

ORGANIC Chemistry•Now™

Assess your understanding of this chapter's topics with additional quizzing and conceptual-based problems at **http://now.brookscole.com/bfi4**

Structure and Nomenclature

10.14 Which are secondary alcohols?

(a)

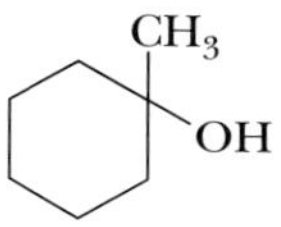

(b) $(CH_3)_3COH$

(c)

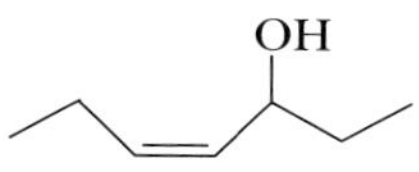

(d)

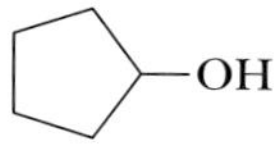

10.15 Name each compound.

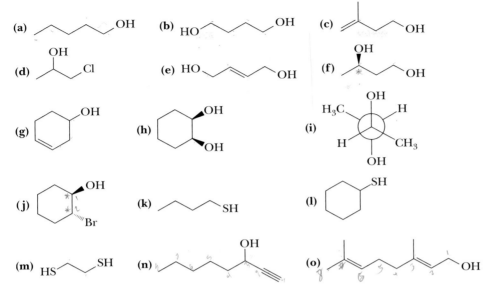

10.16 Write a structural formula for each compound.

(a) Isopropyl alcohol
(b) Propylene glycol
(c) 5-Methyl-2-hexanol
(d) 2-Methyl-2-propyl-1,3-propanediol
(e) 1-Chloro-2-hexanol
(f) *cis*-3-Isobutylcyclohexanol
(g) 2,2-Dimethyl-1-propanol
(h) 2-Mercaptoethanol
(i) Allyl alcohol
(j) *trans*-2-Vinylcyclohexanol
(k) (*Z*)-5-Methyl-2-hexen-1-ol
(l) 2-Propyn-1-ol
(m) 3-Chloro-1,2-propanediol
(n) *cis*-3-Pentene-1-ol

10.17 Name and draw structural formulas for the eight constitutional isomeric alcohols with molecular formula $C_5H_{12}O$. Classify each alcohol as primary, secondary, or tertiary. Which are chiral?

Physical Properties of Alcohols

10.18 Arrange these compounds in order of increasing boiling point (values in °C are −42, 78, 138, and 198)

(a) 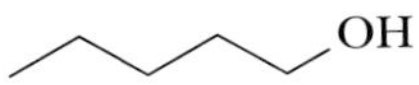    (b) 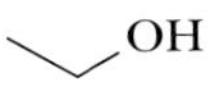   (c) 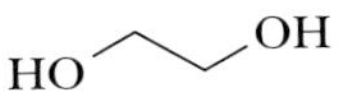 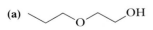   (d)

10.19 Arrange these compounds in order of increasing boiling point (values in °C are −42, −24, 78, and 118).

(a) CH_3CH_2OH (b) CH_3OCH_3 (c) $CH_3CH_2CH_3$ (d) CH_3COOH

10.20 Which compounds can participate in hydrogen bonding with water? State which compounds can act only as hydrogen-bond donors, which can act only as hydogen-bond acceptors, and which can act as a both hydrogen-bond donor and hydrogen-bond acceptors.

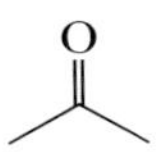

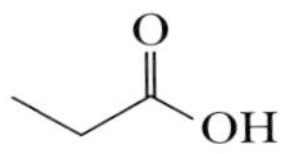

10.21 From each pair of compounds, select the one that is more soluble in water.

(a) CH_2Cl_2 or CH_3OH

(b)
$$CH_3\overset{\displaystyle O}{\overset{\|}{C}}CH_3 \quad\text{or}\quad CH_3\overset{\displaystyle CH_2}{\overset{\|}{C}}CH_3$$

(c) CH_3CH_2Cl or NaCl

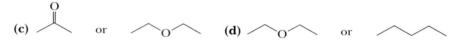

(d) (structure with SH) or (structure with OH)

(e) (structure with OH) or (structure with C=O)

10.22 Arrange the compounds in each set in order of decreasing solubility in water.

(a) Ethanol, butane, diethyl ether   (b) 1-Hexanol, 1,2-hexanediol, hexane

10.23 Each compound given in this problem is a common organic solvent. From each pair of compounds, select the solvent with the greater solubility in water.

(a) CH_2Cl_2 or CH_3CH_2OH (b) Et_2O or EtOH

(c) (acetone structure) or (ethoxy ether structure)   (d) (dimethoxy/ether structure) or (alkane structure)

10.24 The decalinols A and B can be equilibrated using aluminum isopropoxide in 2-propanol (isopropyl alcohol) containing a small amount of acetone.

$$\text{HO—(decalin) } \underset{\text{acetone}}{\overset{Al[OCH(CH_3)_2]_3}{\rightleftharpoons}} \text{ (decalin)—OH}$$

A   B

Assume a value of ΔG^0 (equatorial to axial) for cyclohexanol is 4.0 kJ (0.95 kcal)/mol, calculate the percent of each decalinol in the equilibrium mixture at 25°C.

Acid-Base Reactions of Alcohols

10.25 Complete the following acid-base reactions. Show all valence electrons on the interacting atoms, and show by the use of curved arrows the flow of electrons in each reaction.

(a) $CH_3CH_2OH + H_3O^+ \longrightarrow$

(b) $CH_3CH_2OCH_2CH_3 + HOSOH \longrightarrow$ (with O above and below S)

(c) (pentanol structure) OH + HI $\longrightarrow$

(d) (structure with C=O) OH + HOSOH $\longrightarrow$ (with O above and below S)

(e) (cyclohexyl)—OH + $BF_3 \longrightarrow$

(f) (alkene with + charge) + $H_2O \longrightarrow$

10.26 Select the stronger acid from each pair, and explain your reasoning. For each stronger acid, write a structural formula for its conjugate base.

(a) H_2O or H_2CO_3 (b) CH_3OH or CH_3COOH
(c) CH_3CH_2OH or $CH_3C\equiv CH$ (d) CH_3CH_2OH or CH_3CH_2SH

10.27 From each pair, select the stronger base. For each stronger base, write a structural formula of its conjugate acid.

(a) OH^- or CH_3O^- (each in H_2O) **(b)** $CH_3CH_2O^-$ or $CH_3C{\equiv}C^-$
(c) $CH_3CH_2S^-$ or $CH_3CH_2O^-$ **(d)** $CH_3CH_2O^-$ or NH_2^-

10.28 In each equilibrium, label the stronger acid, the stronger base, the weaker acid, and the weaker base. Also estimate the position of each equilibrium.

(a) $CH_3CH_2O^- + CH_3C{\equiv}CH \rightleftharpoons CH_3CH_2OH + CH_3C{\equiv}C^-$
(b) $CH_3CH_2O^- + HCl \rightleftharpoons CH_3CH_2OH + Cl^-$
(c) $CH_3COOH + CH_3CH_2O^- \rightleftharpoons CH_3COO^- + CH_3CH_2OH$

Reactions of Alcohols

10.29 Write equations for the reaction of 1-butanol with each reagent. Where you predict no reaction, write NR.

(a) Na metal **(b)** HBr, heat **(c)** HI, heat
(d) PBr_3 **(e)** $SOCl_2$, pyridine **(f)** $K_2Cr_2O_7$, H_2SO_4, H_2O, heat
(g) HIO_4 **(h)** PCC **(i)** CH_3SO_2Cl, pyridine

10.30 Write equations for the reaction of 2-butanol with each reagent listed in Problem 10.29. Where you predict no reaction, write NR.

10.31 Draw structural formulas for the major organic products of each reaction.

(a)
OH + H2CrO4

(b)
OH + SOCl2

(c)
OH + HCl

(d)
HO
OH + HBr
(excess)

(e)
OH
+ H2CrO4

(f)
OH
OH
+ HIO4

(g)
1. OsO4, H2O2
2. HIO4

(h)
OH + SOCl2

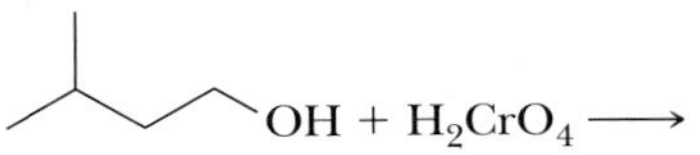

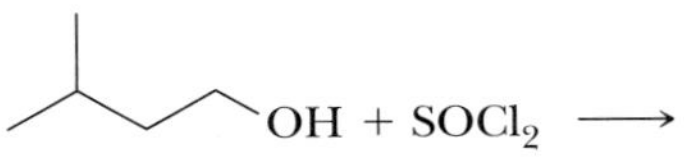

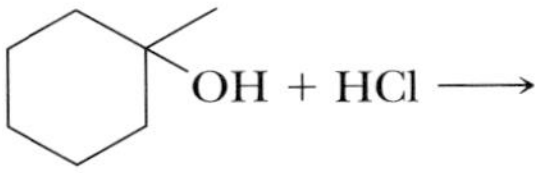

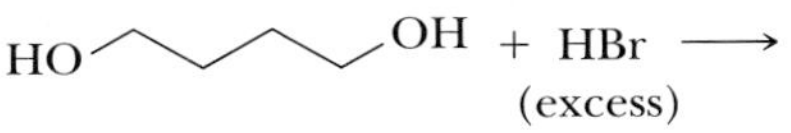

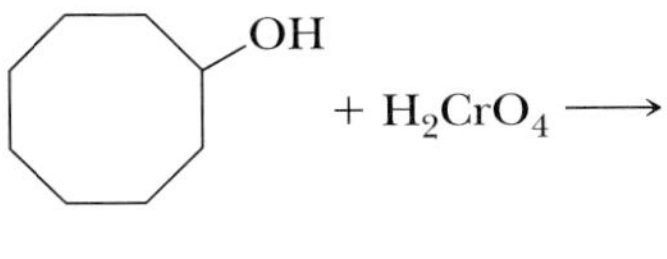

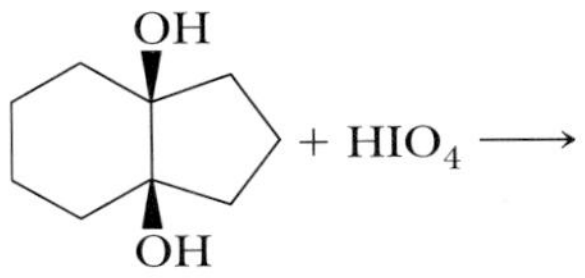

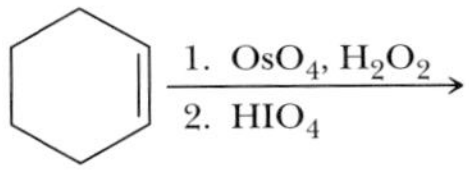

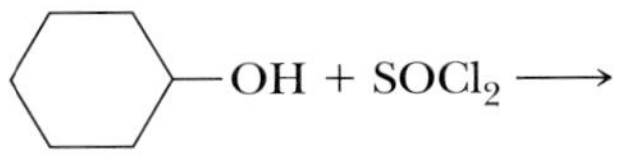

10.32 When (*R*)-2-butanol is left standing in aqueous acid, it slowly loses its optical activity. Account for this observation.

10.33 Two diastereomeric sets of enantiomers, A/B and C/D, exist for 3-bromo-2-butanol.

CH3
HO
H
H
Br
CH3
A

CH3
H
OH
Br
H
CH3
B

CH3
H
OH
H
Br
CH3
C

CH3
HO
H
Br
H
CH3
D

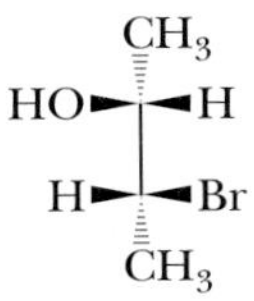

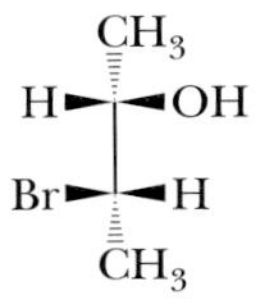

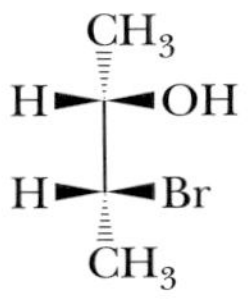

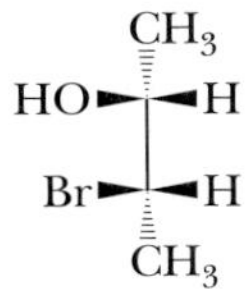

When enantiomer A or B is treated with HBr, only racemic 2,3-dibromobutane is formed; no meso isomer is formed. When enantiomer C or D is treated with HBr, only meso 2,3-dibromobutane is formed; no racemic 2,3-dibromobutane is formed. Account for these observations.

10.34 Acid-catalyzed dehydration of 3-methyl-2-butanol gives three alkenes: 2-methyl-2-butene, 3-methyl-1-butene, and 2-methyl-1-butene. Propose a mechanism to account for the formation of each product.

10.35 Show how you might bring about the following conversions. For any conversion involving more than one step, show each intermediate compound.

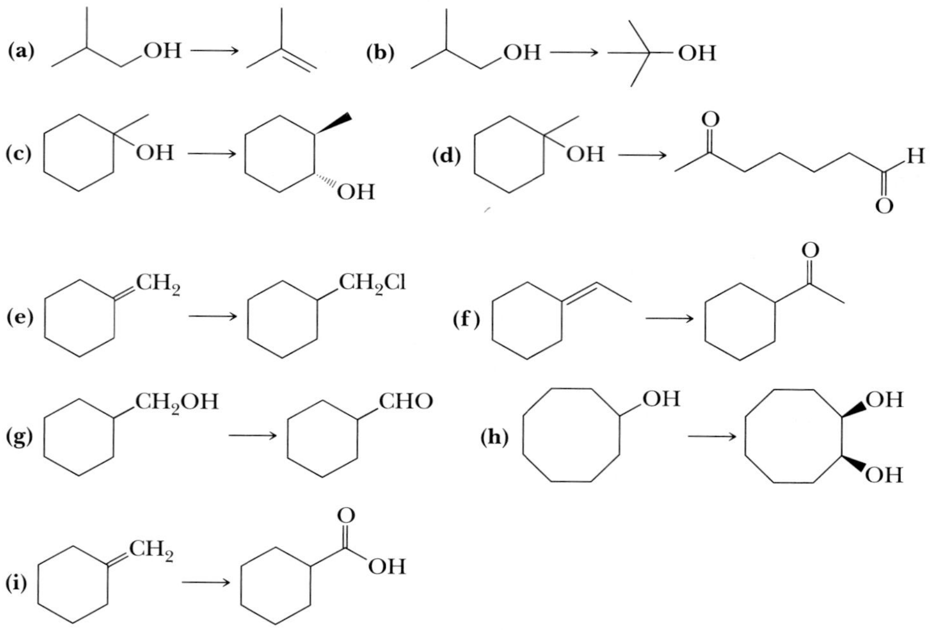

Pinacol Rearrangement

10.36 Propose a mechanism for the following pinacol rearrangement catalyzed by boron trifluoride etherate.

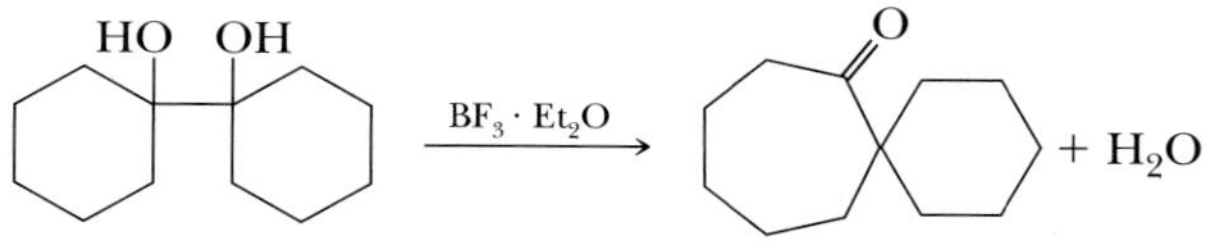

Synthesis

10.37 Alkenes can be hydrated to form alcohols by (1) hydroboration followed by oxidation with alkaline hydrogen peroxide and (2) acid-catalyzed hydration. Compare the product formed from each alkene by sequence (1) with those formed from (2).

(a) Propene   (b) *cis*-2-Butene   (c) *trans*-2-Butene
(d) Cyclopentene   (e) 1-Methylcyclohexene

10.38 Show how each alcohol or diol can be prepared from an alkene.

(a) 2-Pentanol   (b) 1-Pentanol   (c) 2-Methyl-2-pentanol
(d) 2-Methyl-2-butanol (e) 3-Pentanol   (f) 3-Ethyl-3-pentanol
(g) 1,2-Hexanediol

10.39 Dihydropyran is synthesized by treating tetrahydrofurfuryl alcohol with an arenesulfonic acid, $ArSO_3H$. Propose a mechanism for this conversion.

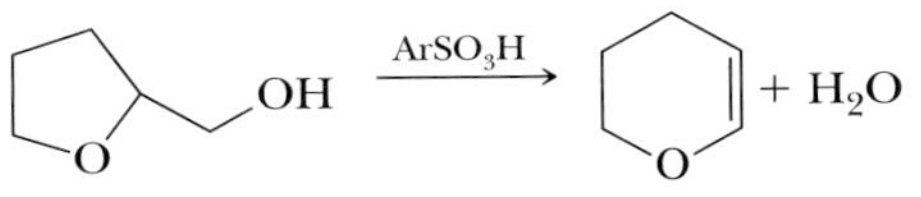

Tetrahydrofurfuryl
alcohol

Dihydropyran

10.40 Show how to convert propene to each of these compounds, using any inorganic reagents as necessary:

(a) Propane
(b) 1,2-Propanediol
(c) 1-Propanol
(d) 2-Propanol
(e) Propanal
(f) Propanone
(g) Propanoic acid
(h) 1-Bromo-2-propanol
(i) 3-Chloropropene
(j) 1,2,3-Trichloropropane
(k) 1-Chloropropane
(l) 2-Chloropropane
(m) 2-Propen-1-ol
(n) Propenal

10.41 (a) How many stereoisomers are possible for 4-methyl-1,2-cyclohexanediol?
(b) Which of the possible stereoisomers are formed by oxidation of (S)-4-methylcyclohexene with osmium tetroxide?
(c) Is the product formed in part (b) optically active or optically inactive?

10.42 Show how to bring about this conversion in good yield.

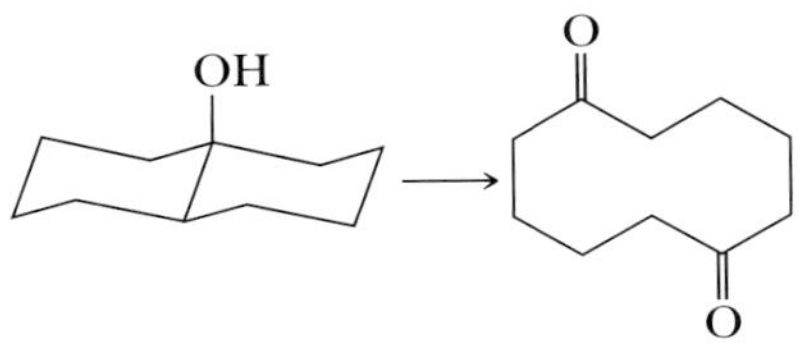

10.43 The tosylate of a primary alcohol normally undergoes an S_N2 reaction with hydroxide ion to give a primary alcohol. Reaction of this tosylate, however, gives a compound of molecular formula $C_7H_{12}O$.

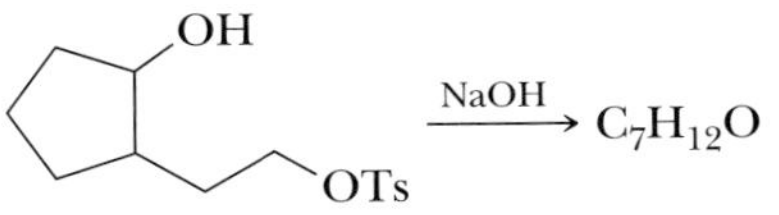

Propose a structural formula for this compound and a mechanism for its formation.

10.44 Chrysanthemic acid occurs as a mixture of esters in flowers of the chrysanthemum (pyrethrum) family. Reduction of chrysanthemic acid to its alcohol (Section 17.6A) followed by conversion of the alcohol to its tosylate gives chrysanthemyl tosylate. Solvolysis (Section 9.3) of the tosylate gives a mixture of artemesia and yomogi alcohols.

Chrysanthemic acid   Chrysanthemyl alcohol

$$\xrightarrow[\text{DMSO}]{H_2O}$$

Chrysanthemyl tosylate   Artemesia alcohol   Yomogi alcohol

Propose a mechanism for the formation of these alcohols from chrysanthemyl tosylate.

10.45 Show how to convert cyclohexene to each compound in good yield.

10.46 Hydroboration of the following bicycloalkene followed by oxidation in alkaline hydroperoxide is both stereoselective and regioselective. The product is a single alcohol in better than 95% yield.

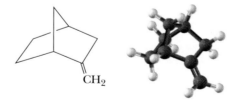

Propose a structural formula for this alcohol and account for the stereo- and regioselectivity of its formation. *Hint:* Examine a molecular model of this alkene and see if you can determine which face of the double bond is more accessible to hydroboration.

Looking Ahead

10.47 Compounds that contain an N—H group associate by hydrogen bonding.

(a) Do you expect this association to be stronger or weaker than that of compounds containing an O—H group?

(b) Based on your answer to part (a), which would you predict to have the higher boiling point, 1-butanol or 1-butanamine?

10.48 Ethanol (CH_3CH_2OH) and dimethyl ether (CH_3OCH_3) are constitutional isomers. (a) Predict which of the two has the higher boiling point. (b) Predict which of the two is the more soluble in water.

10.49 Following are structural formulas for phenol and cyclohexanol along with the acid dissociation constants for each.

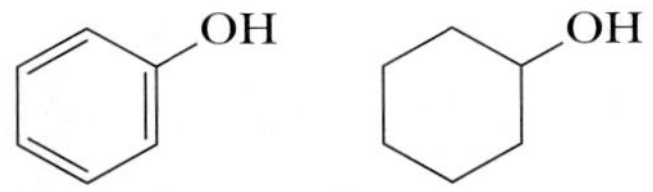

Phenol   Cyclohexanol
pK_a 9.96 pK_a 18

Propose an explanation for the fact that phenol is a considerably stronger acid than cyclohexanol.

11

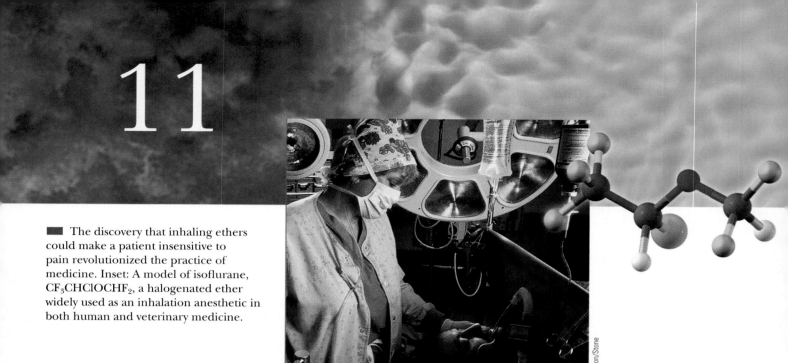

The discovery that inhaling ethers could make a patient insensitive to pain revolutionized the practice of medicine. Inset: A model of isoflurane, $CF_3CHClOCHF_2$, a halogenated ether widely used as an inhalation anesthetic in both human and veterinary medicine.

Allan Levenson/Stone

Ether A compound containing an oxygen atom bonded to two carbon atoms.

ORGANIC
Chemistry Now™

Look for this logo in the chapter and go to Organic ChemistryNow at **http://now.brookscole.com/bfi4** for tutorials, simulations, and problems.

Ethers, Sulfides, and Epoxides

In this chapter, we first discuss the structure, nomenclature, physical properties, and chemical properties of ethers and then compare their physical properties with those of isomeric alcohols. Next, we study the preparation and chemical properties of a group of cyclic ethers called epoxides. As we shall see, their most important reactions involve nucleophilic substitution. This chapter continues the discussion of S_N1 and S_N2 reaction mechanisms begun in Chapter 9 and continued into Chapter 10.

11.1 Structure of Ethers

The functional group of an **ether** is an atom of oxygen bonded to two carbon atoms. Figure 11.1 shows a Lewis structure and a ball-and-stick model of dimethyl ether, CH_3OCH_3, the simplest ether. In dimethyl ether, two sp^3 hybrid orbitals of oxygen form sigma bonds to the two carbon atoms. The other two sp^3 hybrid orbitals of oxygen each contain an unshared pair of electrons. The C—O—C bond angle in dimethyl ether is 110.3°, a value close to the tetrahedral angle of 109.5°.

In still other ethers, the ether oxygen is bonded to sp^2 hybridized carbons. In ethoxyethene (ethyl vinyl ether), for example, the ether oxygen is bonded to one sp^3 and one sp^2 hybridized carbon.

Ethoxyethene $CH_3CH_2—O—CH=CH_2$
(Ethyl vinyl ether)

11.2 Nomenclature of Ethers

In the IUPAC system, ethers are named by selecting the longest carbon chain as the parent alkane and naming the —OR group bonded to it as an **alkoxy** substituent. Common names are derived by listing the alkyl groups bonded to oxygen in alphabetical order and adding the word "ether." Following are the IUPAC names and, in parentheses, the common names for three low-molecular-weight ethers.

$CH_3CH_2OCH_2CH_3$

Ethoxyethane   (1*R*,2*R*)-2-Ethoxycyclohexanol   2-Methoxy-2-methylpropane
(Diethyl ether)   (*trans*-2-Ethoxycyclohexanol)   (*tert*-Butyl methyl ether)

Chemists almost invariably use common names for low-molecular-weight ethers. For example, although ethoxyethane is the IUPAC name for $CH_3CH_2OCH_2CH_3$, it is rarely called that; rather, it is called diethyl ether, ethyl ether, or even more commonly, simply ether. The abbreviation for *tert*-butyl methyl ether is MTBE, after the common name methyl *tert*-butyl ether (incorrectly alphabetized, as you will recognize).

Three other ethers deserve special mention. 2-Methoxyethanol and 2-ethoxyethanol, more commonly known as Methyl Cellosolve and Cellosolve, are good polar protic solvents in which to carry out organic reactions and are also used commercially in some paint strippers. *Di*ethylene *gly*col di*m*ethyl *e*ther, more commonly known by its acronym, diglyme, is a common solvent for hydroboration and $NaBH_4$ reductions.

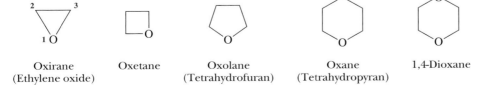

2-Methoxyethanol   2-Ethoxyethanol   Diethylene glycol dimethyl ether
(Methyl cellosolve)   (Cellosolve)   (Diglyme)

Cyclic ethers are given special names. The presence of an oxygen atom in a saturated ring is indicated by the prefix *ox-*, and ring sizes from three to six are indicated by the endings *-irane*, *-etane*, *-olane*, and *-ane*, respectively. Several of these smaller ring cyclic ethers are more often referred to by their common names, here shown in parentheses. Numbering of the atoms of the ring begins with the oxygen atom. These compounds and others in which there is a heteroatom (noncarbon atom) in the ring are called **heterocycles.**

Oxirane   Oxetane   Oxolane   Oxane   1,4-Dioxane
(Ethylene oxide)   (Tetrahydrofuran)   (Tetrahydropyran)

(a)

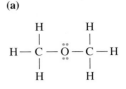

(b)

110.3°

Figure 11.1
Dimethyl ether, CH_3OCH_3.
(a) Lewis structure and
(b) ball-and-stick model.

Alkoxy group An —OR group, where R is an alkyl group.

ORGANIC
Chemistry•Now™
Click Tutorials to review the
Nomenclature of Ethers

Heterocycle A cyclic compound whose ring contains more than one kind of atom. Ethylene oxide, for example, is a heterocycle whose ring contains two carbon atoms and one oxygen atom.

Example 11.1

Write IUPAC and common names for these ethers.

(a) 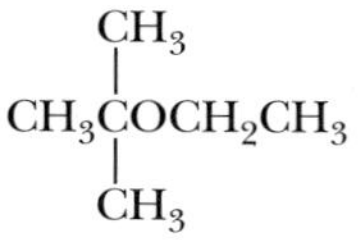 $CH_3\overset{\displaystyle CH_3}{\underset{\displaystyle CH_3}{\overset{|}{\underset{|}{C}}}}OCH_2CH_3$ (b) 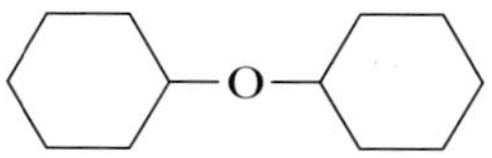 [cyclohexyl]—O—[cyclohexyl]   (c) $CH_2{=}CHOCH_3$

Solution

(a) 2-Ethoxy-2-methylpropane. Its common name is *tert*-butyl ethyl ether.
(b) Cyclohexoxycyclohexane. Its common name is dicyclohexyl ether.
(c) Methoxyethene. Its common name is methyl vinyl ether.

Problem 11.1

Write IUPAC and common names for these ethers.

(a) 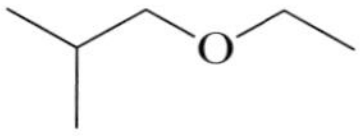 [structure]   (b) 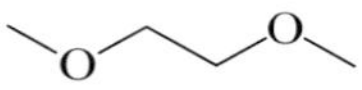 [structure]   (c) 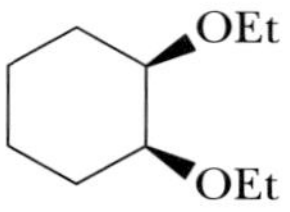 [cyclohexane with OEt and OEt substituents]

11.3 Physical Properties of Ethers

Ethers are polar molecules in which oxygen bears a partial negative charge and each attached carbon bears a partial positive charge (Figure 11.2). However, only weak dipole-dipole interactions exist between ether molecules in the liquid state. Consequently, boiling points of ethers are much lower than those of alcohols of comparable molecular weight (Table 11.1) and are close to those of hydrocarbons of comparable molecular weight (Table 2.5).

Table 11.1 Boiling Points and Solubilities in Water of Some Ethers and Alcohols of Comparable Molecular Weight

Structural Formula	Name	Molecular Weight	bp (°C)	Solubility in Water
CH_3CH_2OH	Ethanol	46	78	Infinite
CH_3OCH_3	Dimethyl ether	46	−24	7.8 g/100 g
$CH_3CH_2CH_2CH_2OH$	1-Butanol	74	117	7.4 g/100 g
$CH_3CH_2OCH_2CH_3$	Diethyl ether	74	35	8.0 g/100 g
$HOCH_2CH_2CH_2CH_2OH$	1,4-Butanediol	90	230	Infinite
$CH_3CH_2CH_2CH_2CH_2OH$	1-Pentanol	88	138	2.3 g/100 g
$CH_3OCH_2CH_2OCH_3$	Ethylene glycol dimethyl ether	90	84	Infinite
$CH_3CH_2CH_2CH_2OCH_3$	Butyl methyl ether	88	71	Slight

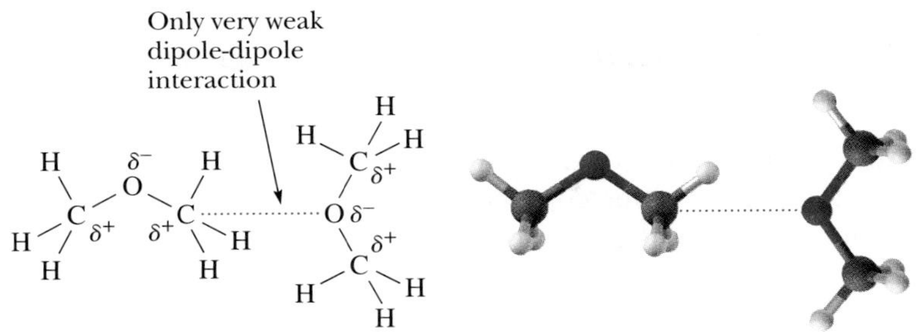

Only very weak
dipole-dipole
interaction

Figure 11.2
Although ethers are polar
compounds, there are only
weak dipole-dipole interactions
between their molecules in the
liquid state.

Because ethers cannot act as hydrogen bond donors, they are much less soluble
in water than alcohols. However, they can act as hydrogen bond acceptors (Figure
11.3), which makes them more water-soluble than hydrocarbons of comparable
molecular weight and shape (compare data in Tables 2.5 and 11.1).

Example 11.2

Arrange these compounds in order of increasing solubility in water.

Ethylene glycol   Diethyl ether   Hexane
dimethyl ether

Solution

Water is a polar solvent. Hexane, a nonpolar hydrocarbon, has the lowest solubility in
water. Both diethyl ether and ethylene glycol dimethyl ether are polar compounds
because of the presence of the polar C—O—C bond, and each interacts with water
as a hydrogen bond acceptor. Because ethylene glycol dimethyl ether has more sites
within its molecules for hydrogen bonding than diethyl ether, it is more soluble in
water than diethyl ether.

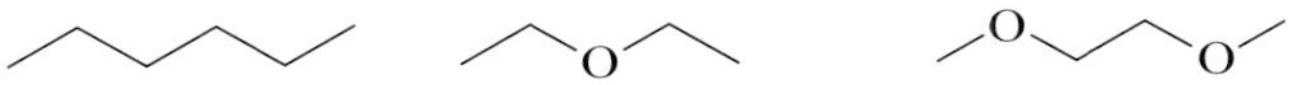

Insoluble   8g/100g water   Soluble in all proportions

Problem 11.2

Arrange these compounds in order of increasing boiling point.

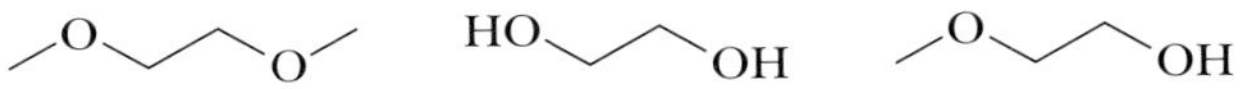

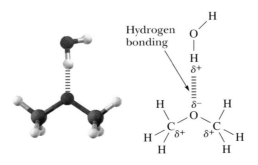

Hydrogen
bonding

Figure 11.3
Ethers are hydrogen bond
acceptors only. They are not
hydrogen bond donors.

11.4 Preparation of Ethers

A. Williamson Ether Synthesis

Williamson ether synthesis A general method for the synthesis of dialkyl ethers by an S_N2 reaction between an alkyl halide and an alkoxide ion.

The most common general method for the synthesis of ethers, the **Williamson ether synthesis,** involves nucleophilic displacement of a halide ion or other good leaving group by an alkoxide ion.

$$\underset{\substack{\text{Sodium}\\\text{isopropoxide}}}{\overset{\overset{\displaystyle CH_3}{|}}{CH_3CHO^-Na^+}} \ + \ \underset{\substack{\text{Iodomethane}\\\text{(Methyl iodide)}}}{CH_3I} \ \xrightarrow{\;S_N2\;} \ \underset{\substack{\text{2-Methoxypropane}\\\text{(Isopropyl methyl ether)}}}{\overset{\overset{\displaystyle CH_3}{|}}{CH_3CHOCH_3}} \ + \ Na^+I^-$$

In planning a Williamson ether synthesis, it is essential to use a combination of reactants that maximizes nucleophilic substitution and minimizes the competing β-elimination (E2, Section 9.9B). Yields of ether are highest when the halide to be displaced is on a methyl or a primary carbon. Yields are low in the displacement from secondary halides (because of competing β-elimination), and the Williamson ether synthesis fails altogether with tertiary halides (because β-elimination by an E2 mechanism is the exclusive reaction). For example, *tert*-butyl methyl ether can be prepared by the reaction of potassium *tert*-butoxide and bromomethane.

$$\underset{\substack{\text{Potassium}\\\textit{tert}\text{-butoxide}}}{\overset{\overset{\displaystyle CH_3}{|}}{\underset{\underset{\displaystyle CH_3}{|}}{CH_3CO^-K^+}}} \ + \ \underset{\substack{\text{Bromomethane}\\\text{(Methyl bromide)}}}{CH_3Br} \ \xrightarrow{\;S_N2\;} \ \underset{\substack{\text{2-Methoxy-2-methylpropane}\\(\textit{tert}\text{-Butyl methyl ether})}}{\overset{\overset{\displaystyle CH_3}{|}}{\underset{\underset{\displaystyle CH_3}{|}}{CH_3COCH_3}}} \ + \ K^+Br^-$$

With the alternative combination of sodium methoxide and 2-bromo-2-methylpropane, no ether is formed; 2-methylpropene, formed by dehydrohalogenation, is the only product.

$$\underset{\substack{\text{2-Bromo-2-}\\\text{methylpropane}}}{\overset{\overset{\displaystyle CH_3}{|}}{\underset{\underset{\displaystyle CH_3}{|}}{CH_3CBr}}} \ + \ \underset{\substack{\text{Sodium}\\\text{methoxide}}}{CH_3O^-Na^+} \ \xrightarrow{\;E2\;} \ \underset{\substack{\text{2-Methylpropene}}}{\overset{\overset{\displaystyle CH_3}{|}}{CH_3C=CH_2}} \ + \ CH_3OH + Na^+Br^-$$

Example 11.3

Show the combination of alcohol and haloalkane that can best be used to prepare each ether by the Williamson ether synthesis.

(a) **(b)**

Solution

(a) Treat 2-propanol with sodium metal to form sodium isopropoxide. Then treat this metal alkoxide with 1-bromobutane.

1-Bromobutane Sodium isopropoxide

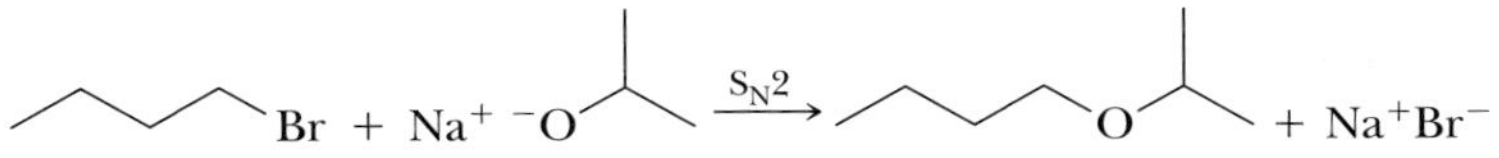

The alternative combination of sodium butoxide and 2-bromopropane gives considerably more elimination product.

(b) Treat (*S*)-2-butanol with sodium metal to form the sodium alkoxide. This reaction involves only the O—H bond and does not affect the chiral center. Then, treat this sodium alkoxide with a haloethane, for example ethyl iodide (EtI), to give the desired product.

(*S*)-2-Butanol (*S*)-2-Ethoxyethane

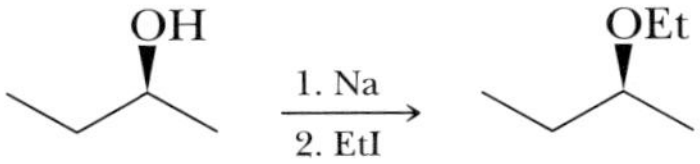

An alternative synthesis is to convert the (*R*)-2-butanol to its tosylate (Section 10.5D) followed by treatment with sodium ethoxide.

(*R*)-2-Butanol

Product of E2 (major product)

(*S*)-2-Ethoxybutane (product of S_N2 reaction; a minor product)

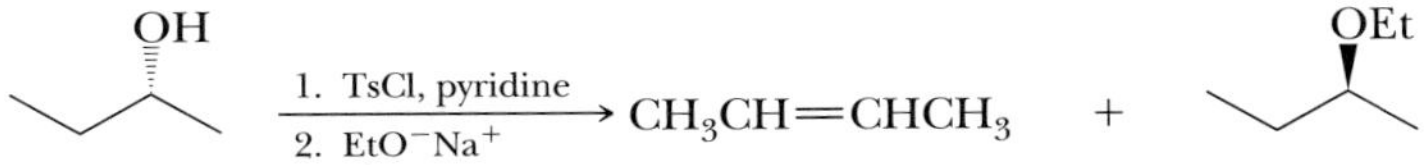

This synthesis, however, gives only a low yield of the desired product. Recall from Section 9.9B that when a 2° halide or tosylate is treated with a strong base/good nucleophile, E2 is the major reaction.

Problem 11.3

Show how you might use the Williamson ether synthesis to prepare each ether.

(a) **(b)**

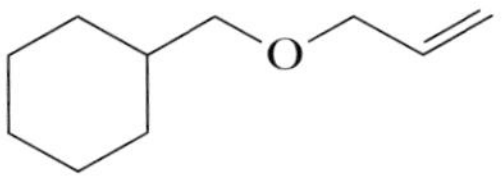

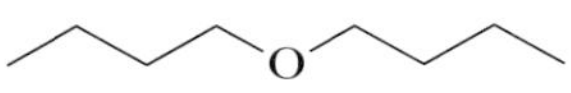

B. Acid-Catalyzed Dehydration of Alcohols

Diethyl ether and several other commercially available ethers are synthesized on an industrial scale by the acid-catalyzed dehydration of primary alcohols. Intermolecular dehydration of ethanol for example, gives diethyl ether.

$$2CH_3CH_2OH \xrightarrow[140°C]{H_2SO_4} CH_3CH_2OCH_2CH_3 + H_2O$$

Ethanol Diethyl ether

Mechanism   *Acid-Catalyzed Intermolecular Dehydration of a Primary Alcohol*

Step 1: Proton transfer from the acid catalyst to the hydroxyl group gives an oxonium ion, which converts OH^-, a poor leaving group, into H_2O, a better leaving group.

$$CH_3CH_2\!-\!\ddot{O}\!-\!H \;+\; H\!-\!\ddot{O}\!-\!\overset{\overset{\displaystyle O}{\|}}{\underset{\underset{\displaystyle O}{\|}}{S}}\!-\!O\!-\!H \;\underset{\text{reversible}}{\overset{\text{fast and}}{\rightleftharpoons}}\; CH_3CH_2\!-\!\overset{+}{\underset{\underset{\displaystyle H}{|}}{\ddot{O}}}\!-\!H \;+\; \overset{-}{:}\ddot{O}\!-\!\overset{\overset{\displaystyle O}{\|}}{\underset{\underset{\displaystyle O}{\|}}{S}}\!-\!O\!-\!H$$

An oxonium ion

Step 2: Nucleophilic displacement of H_2O by the OH group of a second alcohol molecule gives a new oxonium ion.

$$CH_3CH_2\!-\!\ddot{O}\!-\!H \;+\; CH_3CH_2\!-\!\overset{+}{\underset{\underset{\displaystyle H}{|}}{\ddot{O}}}\!-\!H \;\xrightarrow{\;S_N2\;}\; CH_3CH_2\!-\!\overset{+}{\underset{\underset{\displaystyle H}{|}}{\ddot{O}}}\!-\!CH_2CH_3 \;+\; \underset{\underset{\displaystyle H}{|}}{:\ddot{O}}\!-\!H$$

A new oxonium ion

Step 3: Proton transfer from the new oxonium ion to H_2O completes the reaction.

$$CH_3CH_2\!-\!\overset{+}{\underset{\underset{\displaystyle H}{|}}{\ddot{O}}}\!-\!CH_2CH_3 \;+\; \underset{\underset{\displaystyle H}{|}}{:\ddot{O}}\!-\!H \;\underset{\text{transfer}}{\overset{\text{proton}}{\rightleftharpoons}}\; CH_3CH_2\!-\!\ddot{O}\!-\!CH_2CH_3 \;+\; H\!-\!\overset{+}{\underset{\underset{\displaystyle H}{|}}{\ddot{O}}}\!-\!H$$

Note that the acid is a true catalyst in this reaction. One proton is used in Step 1 but another is generated in Step 3.

Yields of ethers from the acid-catalyzed intermolecular dehydration of alcohols are highest for symmetrical ethers formed from unbranched primary alcohols. Examples of symmetrical ethers formed in good yield by this method are dimethyl ether, diethyl ether, and dibutyl ether. From secondary alcohols, yields of ether are lower because of competition from acid-catalyzed dehydration (Section 10.6). In the case of tertiary alcohols, dehydration to an alkene is the only reaction.

Example 11.4

Explain why this reaction does not give a good yield of ethyl hexyl ether.

$$\text{(structure)}\;OH \;+\; HO\text{(structure)} \;\xrightarrow[\text{heat}]{H_2SO_4}\; \text{(structure)}O\text{(structure)} \;+\; H_2O$$

Solution

From this reaction, we expect a mixture of three ethers: diethyl ether, ethyl hexyl ether, and dihexyl ether.

Problem 11.4

Show how ethyl hexyl ether might be prepared by a Williamson ether synthesis.

C. Acid-Catalyzed Addition of Alcohols to Alkenes

Under suitable conditions, alcohols can be added to the carbon-carbon double bond of an alkene to give an ether. The usefulness of this method of ether synthesis is limited to the interaction of alkenes that form stable carbocations and methanol or primary alcohols. An example is the commercial synthesis of *tert*-butyl methyl ether. 2-Methylpropene and methanol are passed over an acid catalyst to give the ether.

$$\underset{\begin{array}{c}CH_3 \\ | \\ CH_3C=CH_2 \end{array}}{} + CH_3OH \xrightarrow[\text{catalyst}]{\text{acid}} \underset{\begin{array}{c}CH_3 \\ | \\ CH_3COCH_3 \\ | \\ CH_3 \end{array}}{}$$

2-Methoxy-2-methylpropane
(*tert*-Butyl methyl ether)

Mechanism *Acid-Catalyzed Addition of an Alcohol to an Alkene*

Step 1: Proton transfer from the acid catalyst to the alkene gives a carbocation intermediate.

$$\underset{\begin{array}{c}CH_3 \\ | \\ CH_3C=CH_2 \end{array}}{} + H-\overset{+}{\underset{|}{\overset{\cdot\cdot}{O}}}-CH_3 \rightleftharpoons \underset{\begin{array}{c}CH_3 \\ | \\ CH_3\overset{+}{C}CH_3 \end{array}}{} + \underset{\underset{H}{|}}{\overset{\cdot\cdot}{O}}-CH_3$$

Step 2: Reaction of the carbocation intermediate (an electrophile) with the alcohol (a nucleophile) gives an oxonium ion.

$$\underset{\begin{array}{c}CH_3 \\ | \\ CH_3\overset{+}{C}CH_3 \end{array}}{} + H\overset{\cdot\cdot}{O}CH_3 \rightleftharpoons \underset{\begin{array}{c}CH_3 \\ | \\ CH_3CCH_3 \\ | \\ \overset{+}{O} \\ H \quad CH_3 \end{array}}{}$$

Step 3: Proton transfer to solvent (in this case methanol) completes the reaction.

$$CH_3-\overset{\cdot\cdot}{\underset{\cdot\cdot}{O}}-H + \underset{\begin{array}{c}CH_3 \\ | \\ CH_3CCH_3 \\ | \\ \overset{+}{O} \\ H \quad CH_3 \end{array}}{} \rightleftharpoons CH_3-\overset{+}{\underset{|}{\overset{\cdot\cdot}{O}}}-H + \underset{\begin{array}{c}CH_3 \\ | \\ CH_3CCH_3 \\ | \\ \overset{\cdot\cdot}{O} \\ CH_3 \end{array}}{}$$

At one time MTBE was added to gasoline under a mandate from the Environmental Protection Agency to add "oxygenates," which make gasoline burn more smoothly (it raises the octane number) and lower exhaust emissions. As an octane-improving additive, MTBE is superior to ethanol (the additive in gasohol). A blend of 15% MTBE with gasoline improves octane rating by approximately 5 units. Unfortunately, because MTBE is much more soluble in water than gasoline, it has gotten into the water table in many places, in some cases because of leaky gas station storage tanks. It has been detected in lakes, reservoirs, and water supplies, in some cases at concentrations that exceed limits for both "taste and odor" and human health. Consequently, its use as a gasoline additive is being phased out.

11.5 Reactions of Ethers

Ethers resemble hydrocarbons in their resistance to chemical reaction. They do not react with oxidizing agents such as potassium dichromate or potassium permanganate. They are stable toward even very strong bases, and, except for tertiary alkyl ethers, they are not affected by most weak acids at moderate temperatures. Because of their good solvent properties and general inertness to chemical reaction, ethers are excellent solvents in which to carry out many organic reactions.

A. Acid-Catalyzed Cleavage by Concentrated HX

Cleavage of dialkyl ethers requires both a strong acid and a good nucleophile, hence the use of concentrated aqueous HI (57%) or HBr (48%). Dibutyl ether, for example, reacts with hot concentrated HBr to give two molecules of 1-bromobutane.

$$\text{\scriptsize (Dibutyl ether)} + 2HBr \xrightarrow{\text{heat}} 2 \text{\scriptsize } Br + H_2O$$

Dibutyl ether   1-Bromobutane

Concentrated HCl (38%) is far less effective in cleaving dialkyl ethers, primarily because Cl^- is a weaker nucleophile in water than either I^- or Br^-.

The mechanism of acid-catalyzed cleavage of dialkyl ethers depends on the nature of the carbons bonded to oxygen. If both carbons are primary, cleavage involves an S_N2 reaction in which a halide ion is the nucleophile. Otherwise cleavage is by an S_N1 reaction.

Mechanism *Acid-Catalyzed Cleavage of a Dialkyl Ether*

Step 1: Proton transfer from the acid catalyst to the oxygen atom of the ether gives an oxonium ion.

$$CH_3CH_2-\overset{\cdot\cdot}{\underset{\cdot\cdot}{O}}-CH_2CH_3 \;+\; H-\overset{+}{\underset{\underset{H}{|}}{O}}-H \;\underset{\text{reversible}}{\overset{\text{fast and}}{\rightleftharpoons}}\; CH_3CH_2-\overset{+}{\underset{\underset{H}{|}}{O}}-CH_2CH_3 \;+\; \overset{\cdot\cdot}{\underset{\underset{H}{|}}{O}}-H$$

An oxonium ion

Step 2: Nucleophilic displacement by halide ion on the primary carbon cleaves the C—O bond; the leaving group is CH_3CH_2OH, a weak base and a poor nucleophile.

$$\overset{\cdot\cdot}{\underset{\cdot\cdot}{Br}}{:}^- + CH_3CH_2-\overset{+}{\underset{\underset{H}{|}}{O}}-CH_2CH_3 \xrightarrow{S_N2} CH_3CH_2-\overset{\cdot\cdot}{\underset{\cdot\cdot}{Br}}{:} \;+\; {:}\overset{\cdot\cdot}{\underset{\underset{H}{|}}{O}}-CH_2CH_3$$

This cleavage produces one molecule of bromoalkane and one molecule of alcohol. In the presence of excess concentrated HBr, the alcohol is converted to a second molecule of bromoalkane by another S_N2 process (Section 9.3).

Tertiary, allylic, and benzylic ethers are particularly susceptible to cleavage by acid, often under quite mild conditions. Tertiary butyl ethers, for example, are cleaved by aqueous HCl at room temperature. Proton transfer from the acid to the oxygen atom of the ether produces an oxonium ion, which then cleaves to produce a particularly stable 3°, allylic, or benzylic carbocation. Reaction of the carbocation with Cl⁻ completes the reaction.

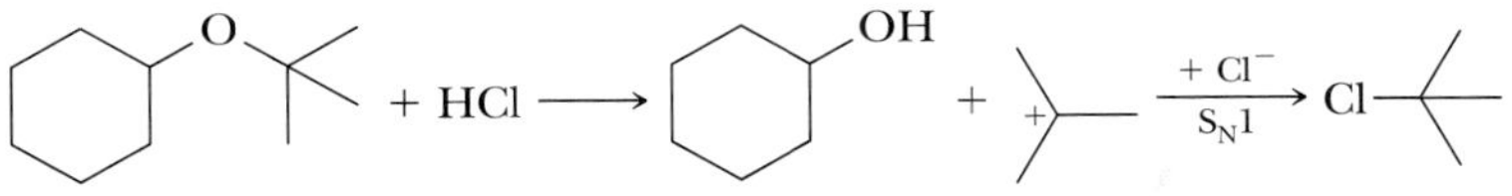

A *tert*-butyl ether   A 3° carbocation
                                              intermediate

Example 11.5

Account for the fact that treating most methyl ethers with concentrated HI gives CH_3I and ROH as the initial major products rather than CH_3OH and RI, as illustrated by the following reaction.

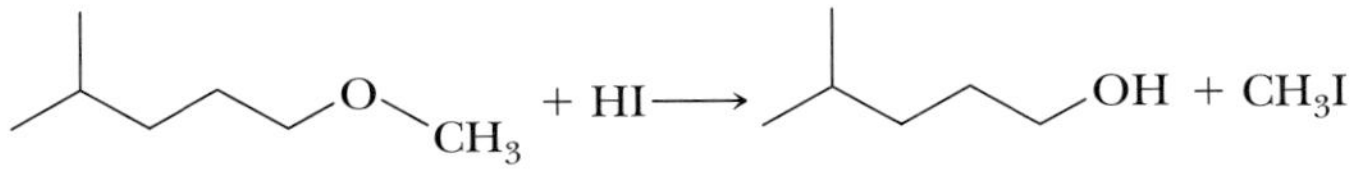

Solution

The first step is protonation of the ether oxygen to give an oxonium ion. Cleavage is by an S_N2 pathway on the less hindered methyl carbon.

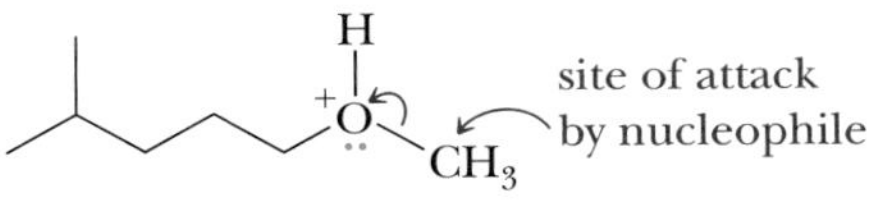

Problem 11.5

Account for the fact that treatment of *tert*-butyl methyl ether with a limited amount of concentrated HI gives methanol and *tert*-butyl iodide rather than methyl iodide and *tert*-butyl alcohol.

Example 11.6

Draw structural formulas for the major products of each reaction.

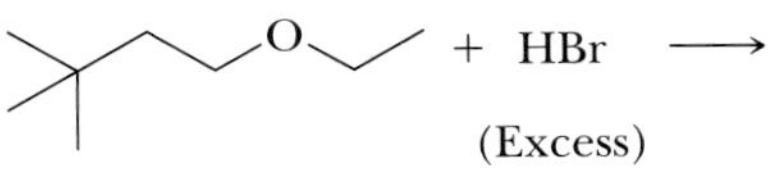

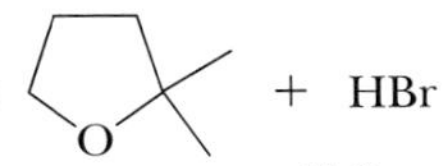

Solution

(a) Cleavage on either side of the ether oxygen by an S_N2 mechanism gives an alcohol and a bromoalkane. Reaction of the alcohol then gives a second molecule of alkyl bromide.

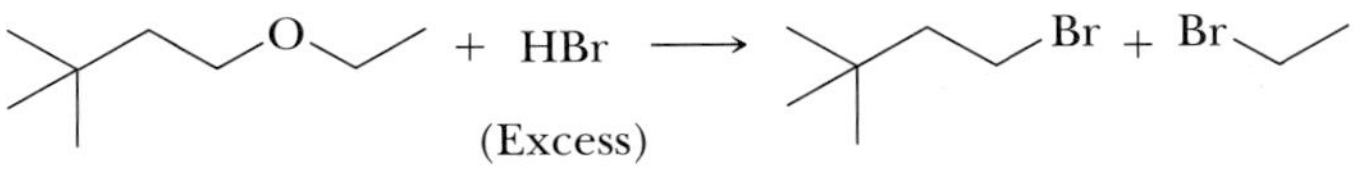

(b) Proton transfer to the ether oxygen followed by cleavage gives a 3° carbocation intermediate, which may then (1) react with bromide ion to give a bromoalcohol or (2) lose a proton to give an unsaturated alcohol.

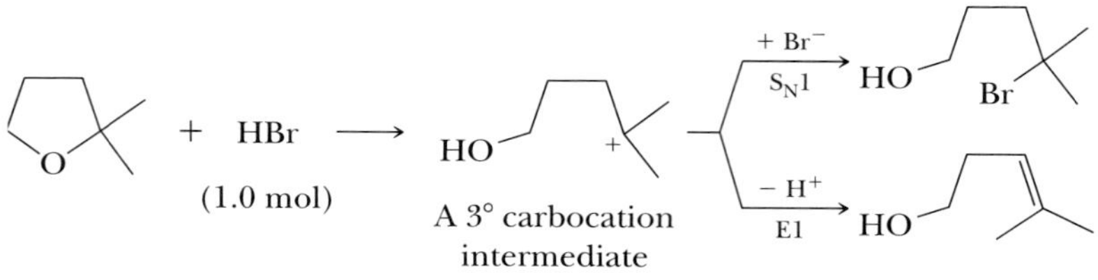

Problem 11.6

Draw structural formulas for the major products of each reaction.

(a) CH₃COCH₃ (with CH₃ groups) + HBr (Excess) ⟶   **(b)** 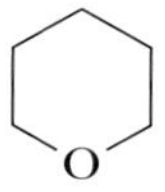 (pyran ring) + HBr (Excess) ⟶

B. Oxidation of Ethers: Formation of Hydroperoxides

Two hazards must be avoided when working with diethyl ether and other low-molecular-weight ethers. First, the commonly used ethers have low boiling points and are highly flammable, a dangerous combination. Consequently, open flames and electric appliances with sparking contacts must be avoided where ethers are being used (laboratory refrigerators and ovens are frequent causes of ignition). Because diethyl ether is so volatile (its boiling point is 35°C), it should be used in a fume hood to prevent the build-up of vapors and possible explosion. Second, anhydrous ethers react with molecular oxygen at a C—H bond adjacent to the ether oxygen to form **hydroperoxides,** which are dangerous because they are explosive.

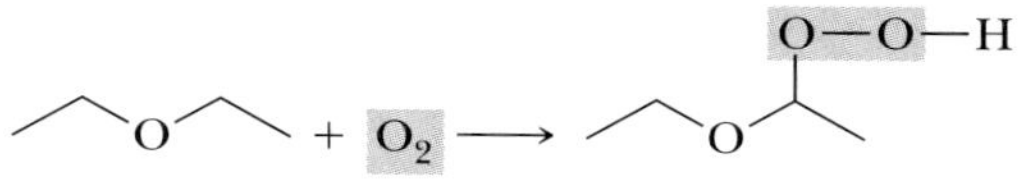

Hydroperoxide A compound containing an —OOH group.

Diethyl ether   A hydroperoxide

Hydroperoxidation proceeds by a radical chain mechanism (see "Chemical Connections: Antioxidants" in Chapter 8). Rates of hydroperoxide formation increase dramatically if the C—H bond adjacent to oxygen is secondary, as for example in diisopropyl ether, because of favored generation of a relatively stable 3° radical intermediate next to oxygen.

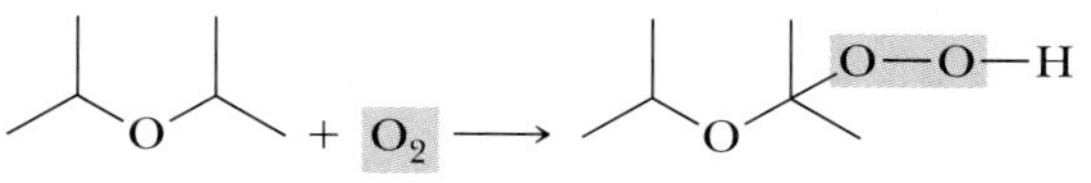

Diisopropyl ether   A hydroperoxide

This hydroperoxide precipitates from solution as a waxy solid and is particularly dangerous.

Hydroperoxides in ethers can be detected by shaking a small amount of the ether with an acidified 10% aqueous solution of potassium iodide, KI, or by using starch iodine paper with a drop of acetic acid. Peroxides oxidize iodide ion to iodine, I_2, which gives a yellow color to the solution. Hydroperoxides can be removed by treating them with a reducing agent. One effective procedure is to shake the hydroperoxide-contaminated ether with a solution of iron(II) sulfate in dilute aqueous sulfuric acid. You should never use ethers past their expiration date, and you should properly dispose of them before they expire.

11.6 Silyl Ethers as Protecting Groups

When dealing with organic compounds containing two or more functional groups, it is often necessary to protect one functional group (to prevent its reaction) while carrying out a reaction at another functional group. Suppose, for example, that you wish to convert 4-pentyn-1-ol to 4-heptyn-1-ol.

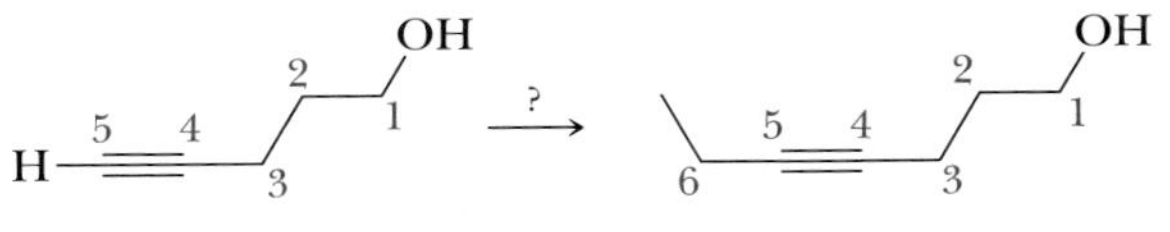

4-Pentyn-1-ol   4-Heptyn-1-ol

The new carbon-carbon bond can be formed by treating the acetylide anion (Section 7.5) of 4-pentyn-1-ol with bromoethane. 4-Pentyn-1-ol, however, contains two acidic hydrogens, one on the hydroxyl group (pK_a 16–18) and the other on the carbon-carbon triple bond (pK_a 25). Treatment of this compound with one equivalent of $NaNH_2$ forms the alkoxide anion (the —OH group is the stronger acid) rather than the acetylide anion.

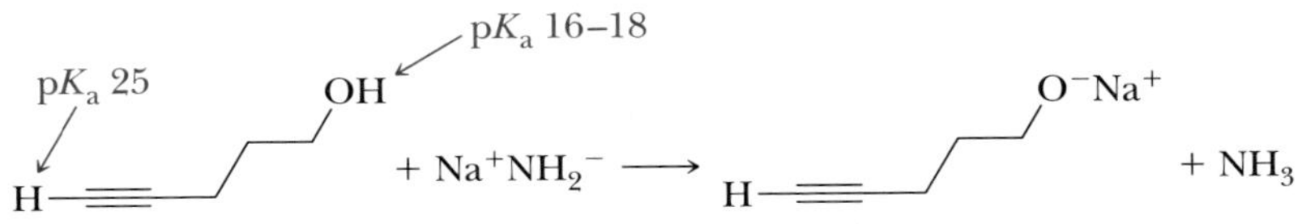

Alkoxide anion

To carry out the synthesis of 4-heptyne-1-ol, we must first protect the —OH group to prevent its reaction with sodium amide. A good protecting group is

- Easily added to the sensitive functional group.
- Resistant to the reagents used to transform the unprotected functional group or groups.
- Easily removed to regenerate the original functional group.

Chemists have devised protecting groups for most functional groups, and we will encounter several of them in this text. In this section, we concentrate on the most common type of hydroxyl-protecting group, namely silyl ethers.

Silicon is in Group 4A of the Periodic Table, immediately below carbon. Like carbon, silicon also forms tetravalent compounds such as the following.

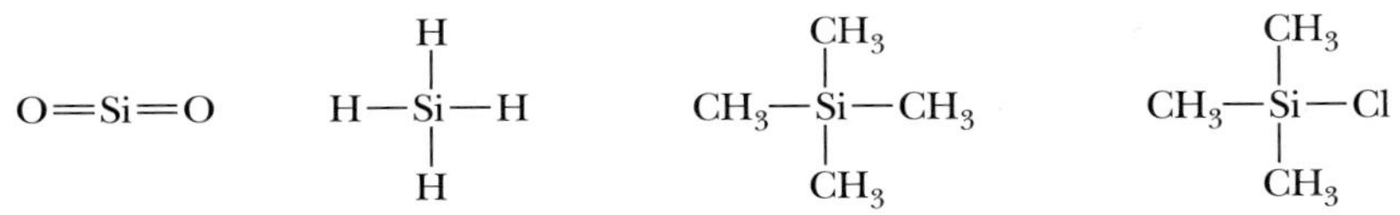

Silicon dioxide   Silane   Tetramethylsilane   Chlorotrimethylsilane

An —OH group can be converted to a silyl ether treating it with a trialkylsilyl chloride in the presence of an amine base. For example, treating an alcohol with chlorotrimethylsilane in the presence of a tertiary amine, such as triethylamine or pyridine, gives a trimethylsilyl ether.

$$
RCH_2OH + Cl-\underset{\underset{CH_3}{|}}{\overset{\overset{CH_3}{|}}{Si}}-CH_3 + Et_3N \longrightarrow RCH_2O-\underset{\underset{CH_3}{|}}{\overset{\overset{CH_3}{|}}{Si}}-CH_3 + Et_3NH^+Cl^-
$$

Chlorotri-   A trimethylsilyl   Triethylammonium
methylsilane   ether   chloride

The function of the tertiary amine is to catalyze the reaction by forming some of the more nucleophilic alkoxide ion and to neutralize the HCl formed during the reaction.

Replacement of one of the methyl groups of the trimethylsilyl group by *tert*-butyl gives the *tert*-butyldimethylsilyl (TBDMS) group, which is considerably more stable than the trimethylsilyl group. Other common silyl protecting groups are the triethylsilyl (TES) and triisopropylsilyl (TIPS) groups

Trimethylsilyl chloride (TMSCl)	Triethylsilyl chloride (TESCl)	*t*-Butyldimethylsilyl chloride (TBDMSCl)	Triisopropylsilysl chloride (TIPSCl)

Silyl ethers are unaffected by most oxidizing and reducing agents, and are stable to most nonaqueous acids and bases. The TBDMS group is stable in aqueous solution within the pH range 2 to 12, which makes it one of the most widely used —OH protecting groups. Silyl ether blocking groups are most easily removed by treatment with fluoride ion, generally in the form of tetrabutylammonium fluoride, $Bu_4N^+F^-$.

$$
RCH_2O-Si\overset{|}{\underset{|}{\big<}} + F^- \xrightarrow[\text{THF}]{Bu_4N^+F^-} RCH_2OH + F-Si\overset{|}{\underset{|}{\big<}}
$$

A TBDMS-protected
alcohol

This cleavage of the protecting group depends on the fact that a silicon-fluorine sigma bond is considerably stronger (582 kJ/mol) than a silicon-oxygen sigma bond (368 kJ/mol). In fact, the Si—F sigma bond is one of the strongest sigma bonds known. The large difference in bond strengths between Si—O and Si—F bonds drives the removal reaction to completion.

We can use a silyl ether in the following way to convert 4-pentyn-1-ol to 4-heptyn-1-ol. Treating 4-pentyn-1-ol with chlorotrimethylsilane in the presence of triethylamine gives the trimethylsilyl ether. Treatment of the terminal alkyne with sodium amide followed by bromoethane forms the new carbon-carbon bond. Subsequent

removal of the TMS protecting group with tetrabutylammonium fluoride gives the
desired 4-heptyne-1-ol.

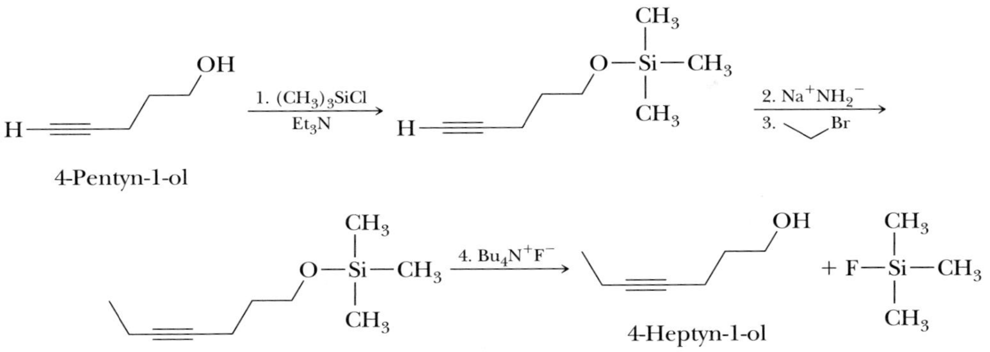

4-Pentyn-1-ol

4-Heptyn-1-ol

Example 11.7

Compare the polarity of the C—Cl bond in $(CH_3)_3C$—Cl with the polarity of the
Si—Cl bond in $(CH_3)_3Si$—Cl.

Solution

The difference in electronegativity between carbon and chlorine is $3.0 - 2.5 = 0.5$;
that between silicon and chlorine is $3.0 - 1.8 = 1.2$; a Si—Cl bond is more polar
than a C—Cl bond.

Problem 11.7

The trimethylsilyl protecting group is easily removed in aqueous solution containing
a trace of acid. Propose a mechanism for this reaction. (Note that a TBDMS protecting group is stable under these conditions because of the greater steric crowding
around silicon created by the *t*-butyl group.)

11.7 Epoxides: Structure and Nomenclature

Although **epoxides** are technically classed as ethers, we discuss them separately because of their exceptional chemical reactivity compared with other ethers. Simple
epoxides are named as derivatives of oxirane, the parent epoxide. Where the epoxide
is a part of another ring system, it is named using the prefix *epoxy-*.

Epoxide A cyclic ether in which
oxygen is one atom of a three-
membered ring.

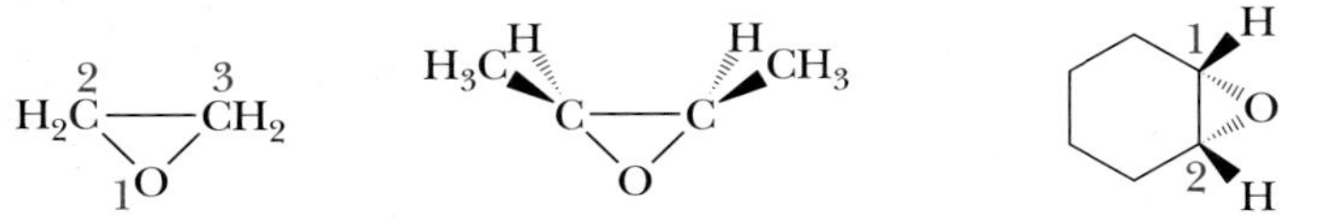

Oxirane
(Ethylene oxide)

cis-2,3-Dimethyloxirane
(*cis*-2-Butene oxide)

1,2-Epoxycyclohexane
(Cyclohexene oxide)

Common names of epoxides are derived by giving the name of the alkene from which the epoxide is formally derived followed by the word "oxide" an example is *cis*-2-butene oxide.

11.8 Synthesis of Epoxides

A. Ethylene Oxide

Ethylene oxide, one of the few epoxes synthesized on an industrial scale, is prepared by passing a mixture of ethylene and air (or oxygen) over a silver catalyst.

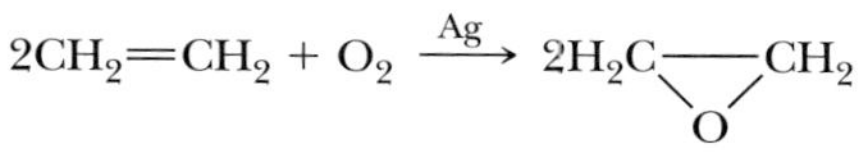

$$2CH_2{=}CH_2 + O_2 \xrightarrow{\text{Ag}} 2H_2C{-\!\!\!-}CH_2$$

Oxirane
(Ethylene oxide)

This method fails when applied to other low-molecular-weight alkenes.

B. Oxidation of Alkenes with Peroxycarboxylic Acids

The most common laboratory method for the synthesis of epoxides from alkenes is oxidation with a peroxycarboxylic acid (a peracid). Three of the most widely used peroxyacids are *meta*-chloroperoxybenzoic acid (MCPBA), the magnesium salt of monoperoxyphthalic acid (MMPP), and peroxyacetic acid.

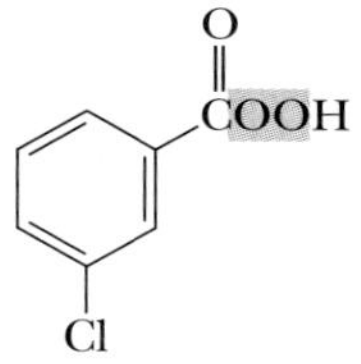

meta-Chloroperoxybenzoic acid
(MCPBA)

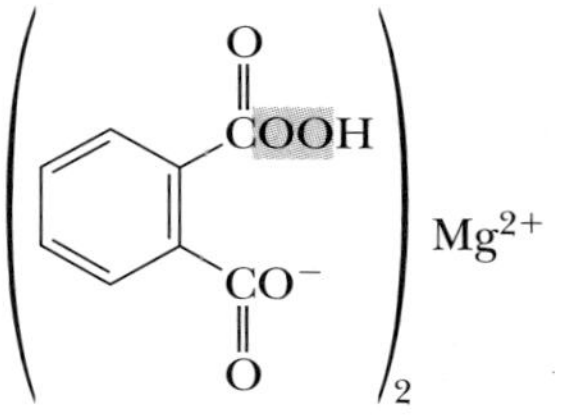

Magnesium monoperoxyphthalate
(MMPP)

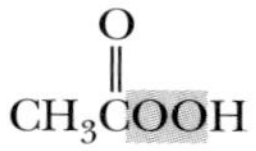

Peroxyacetic acid
(Peracetic acid)

Following is a balanced equation for the epoxidation of cyclohexene by a peroxycarboxylic acid, RCO_3H. In the process, the peroxycarboxylic acid is reduced to a carboxylic acid.

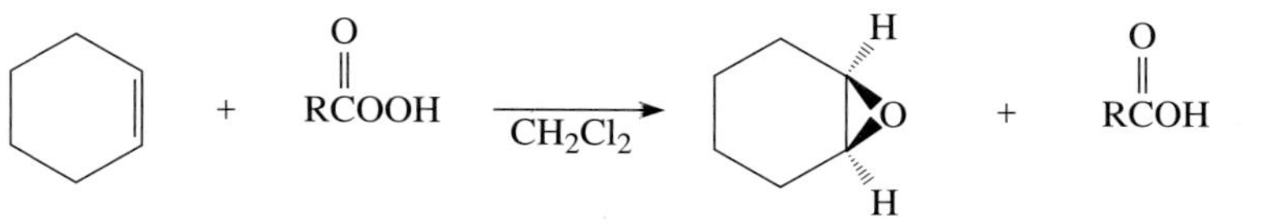

Cyclohexene

A peroxycarboxylic acid

1,2–Epoxycyclohexane
(Cyclohexene oxide)

A carboxylic acid

For an alkene that shows *cis-trans* isomerism, epoxidation is also stereospecific: The stereochemistry of the product depends on the stereochemistry of the starting alkene. Epoxidation of *cis*-2-butene, for example, yields only the meso compound *cis*-2,3-dimethyloxirane, and epoxidation of *trans*-2-butene yields only the enantiomers of *trans*-2,3-dimethyloxirane.

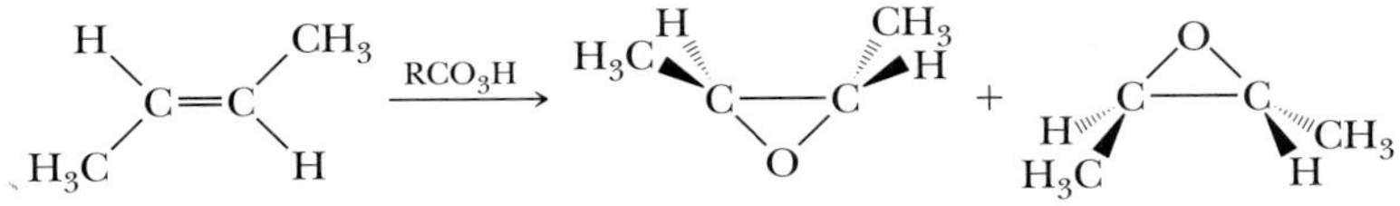

trans-2-Butene   *trans*-2,3-Dimethyloxirane
(A pair of enantiomers)

A mechanism for epoxidation by a peroxyacid must take into account the following facts. (1) The reaction takes place in nonpolar solvents, which means that the reaction cannot involve the formation of ions or any species with significant separation of unlike charges. (2) The reaction is stereospecific, with complete retention of the alkene configuration, which means that even though the pi bond of the carbon-carbon double bond is broken, at no time is there free rotation about the remaining sigma bond. Following is a mechanism consistent with these observations.

Mechanism   *Epoxidation of an Alkene by RCO₃H*

Arrow ①: Interaction of the pi electrons of the carbon-carbon double bond with the end oxygen atom of the peroxyacid and formation of a new C—O bond.

Arrows ② and ③: Redistribution of electron pairs within the peroxyacid.

Arrow ④: Formation of the second carbon-oxygen bond.

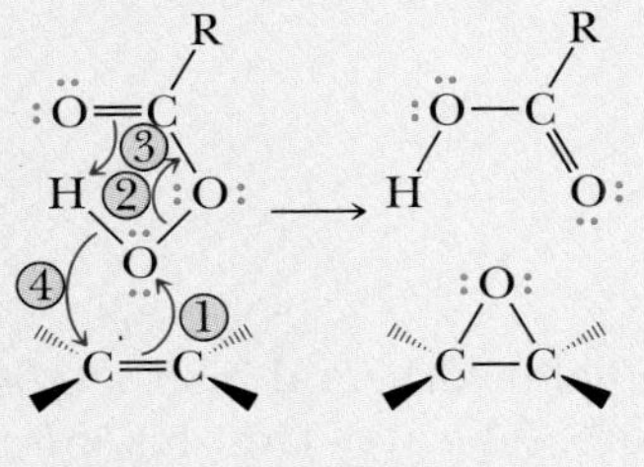

The numbering of the arrows in this mechanism does not imply an order in which covalent bonds are broken and made. Rather, they are meant as a guide to help you understand the mechanism. It is thought that the entire combination of bond-making and bond-breaking steps is concerted, or nearly so.

C. Internal Nucleophilic Substitution in Halohydrins

A second general method for the preparation of epoxides from alkenes involves (1) treating the alkene with chlorine or bromine in water to form a chlorohydrin or bromohydrin followed by (2) treating the halohydrin with a base to bring about intramolecular displacement of Cl⁻. By these steps, propene is first converted to 1-chloro-2-propanol and then to methyloxirane (propylene oxide).

$$CH_3CH{=}CH_2 \xrightarrow{Cl_2,\ H_2O} \underset{\underset{Cl}{|}}{CH_3CH}{-}CH_2 \xrightarrow{NaOH,\ H_2O} CH_3CH\overset{O}{-}CH_2$$

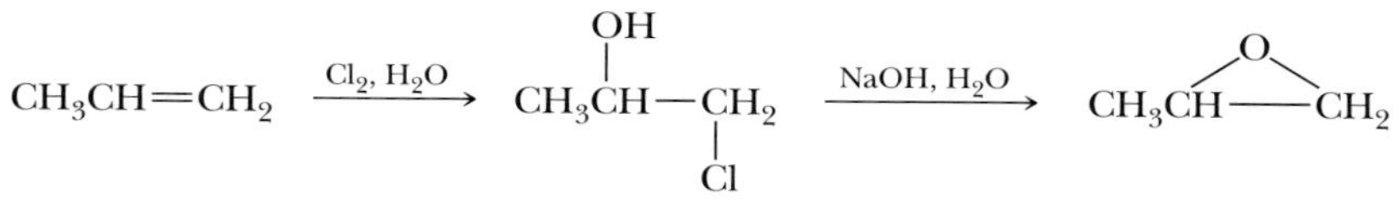

Propene 1-Chloro-2-propanol (a chlorohydrin) (racemic) Methyloxirane (Propylene oxide) (racemic)

We studied the reaction of alkenes with chlorine or bromine in water to form halohydrins in Section 6.3E and saw that it is both regioselective and stereoselective (for an alkene that shows *cis-trans* isomerism, it is also stereospecific). Conversion of a halohydrin to an epoxide with base is stereoselective as well and can be viewed as an internal S_N2 reaction. Hydroxide ion or other base abstracts a proton from the halohydrin hydroxyl group to form an alkoxide ion, a good nucleophile, which then displaces halogen on the adjacent carbon. As with all S_N2 reactions, attack of the nucleophile is from the backside of the C—X bond and causes inversion of configuration at the site of substitution.

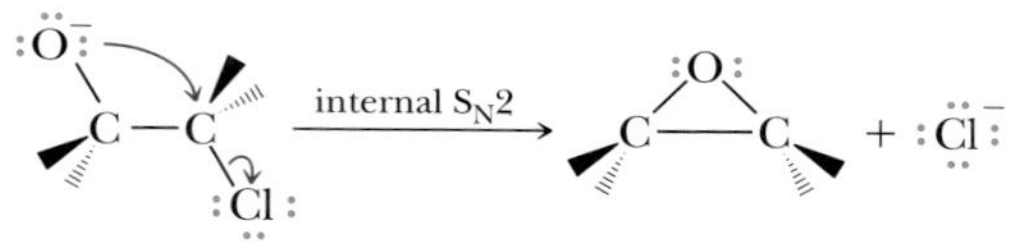

An epoxide

Note that this displacement of halide by the alkoxide ion can also be viewed as an intramolecular variation of the Williamson ether synthesis (Section 11.4A). In this case, the displacing alkoxide and leaving halide ions are on adjacent carbon atoms.

Example 11.8

Conversion of an alkene to a halohydrin and internal displacement of a halide ion by an alkoxide ion are both stereoselective. Use this information to demonstrate that the configuration of the alkene is preserved in the epoxide. As an illustration, show that reaction of *cis*-2-butene by this two-step sequence gives *cis*-2,3-dimethyloxirane (*cis*-2-butene oxide).

Solution

Addition of HOCl to an alkene occurs by stereoselective anti addition of —OH and —Cl to the double bond (Section 6.3E). The conformation of this product is also the conformation necessary for stereoselective backside displacement of the halide ion by alkoxide ion. Thus, a *cis* alkene gives a *cis* disubstituted oxirane, and the transformation from alkene to epoxide is stereospecific.

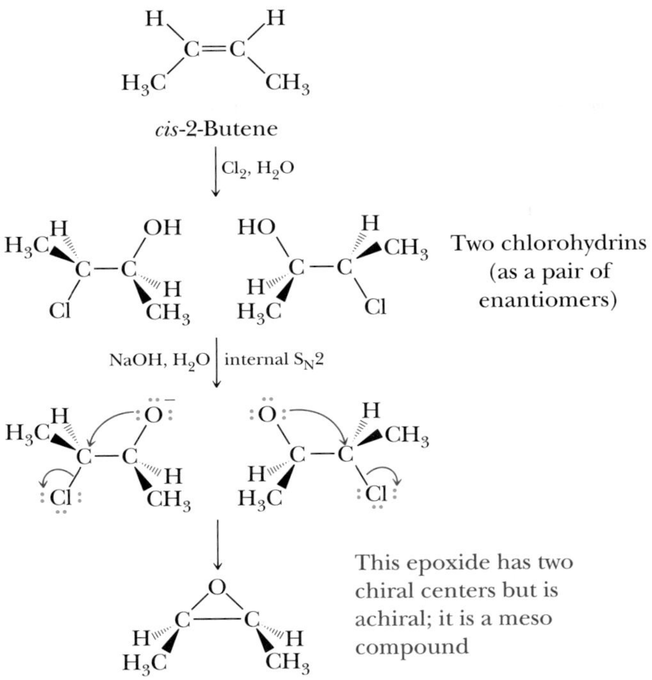

cis-2-Butene

Cl$_2$, H$_2$O

Two chlorohydrins (as a pair of enantiomers)

NaOH, H$_2$O | internal S$_N$2

This epoxide has two chiral centers but is achiral; it is a meso compound

cis-2,3-Dimethyloxirane
(achiral; a meso compound)

Problem 11.8

Consider the possibilities for stereoisomerism in the bromohydrin and epoxide formed from *trans*-2-butene.

(a) How many stereoisomers are possible for the bromohydrin? Which of the possible bromohydrin stereoisomers are formed by treating *trans*-2-butene with bromine in water?

(b) How many stereoisomers are possible for the epoxide? Which of the possible stereoisomers is/are formed in this two-step sequence?

D. Sharpless Asymmetric Epoxidation

One of the most useful organic reactions discovered in the last several decades is the titanium-catalyzed asymmetric epoxidation of primary allylic alcohols developed by Professor Barry Sharpless, then at Stanford University. The reagent consists of *tert*-butyl hydroperoxide, titanium tetraisopropoxide [Ti(O-iPr)$_4$], and diethyl tartrate. Recall from Section 3.4A that tartaric acid has two chiral centers and exists as three stereoisomers: a pair of enantiomers and a meso compound. The form of tartaric acid used in the Sharpless epoxidation is either pure (+)-diethyl tartrate or its enantiomer, (−)-diethyl tartrate. The *tert*-butyl hydroperoxide is the oxidizing agent and must be present in molar amounts. Titanium tetraisopropoxide and diethyl tartrate combine to make the active catalyst and are present in lesser amounts, generally 5 to 10 mole percent.

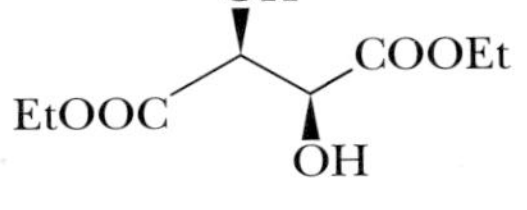

(2S,3S)-(−)-Diethyl tartrate

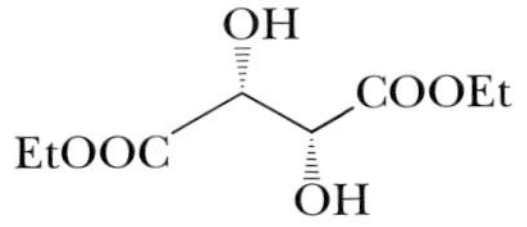

(2R,3R)-(+)-Diethyl tartrate

What is remarkable about the Sharpless epoxidation is that it is stereospecific based on the diethyl tartrate added; either enantiomer of an epoxide can be produced depending on which enantiomer of diethyl tartrate is used. If (−)-enantiomer is used, the product is enantiomer A. If (+)-enantiomer is used, the product is enantiomer B.

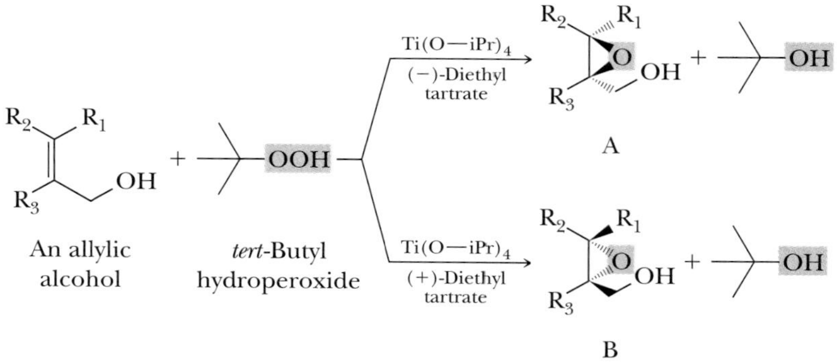

An allylic alcohol   *tert*-Butyl hydroperoxide

A

B

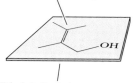

With (−)-diethyl tartrate, oxygen is delivered to the top face

With (+)-diethyl tartrate, oxygen is delivered to the bottom face

When predicting the stereochemistry of a Sharpless epoxidation product, you will find it helpful to draw the allylic alcohol in the same orientation each time, for example, as shown on the left.

When drawn in this manner, the (−)-tartrate catalyzes delivery of the epoxide oxygen from the top face of the alkene; the (+)-tartrate catalyzes its delivery from the bottom face.

The mechanism of this catalyzed epoxidation has been studied in detail and involves formation of a chiral complex in which the carbonyl oxygen of diethyl tartrate displaces one of the isopropoxide groups on titanium. When the R—OOH oxidizing agent is added, it displaces a second isopropoxide group. Finally, the oxygen of the allylic alcohol displaces a third isopropoxide group. Thus, although neither the alkene nor the ROOH oxidizing agent is chiral, both are now held in a fixed stereochemical relationship to the other in the chiral environment created by the diethyl tartrate-titanium complex. In this chiral environment, oxygen is now delivered to either the top face or the bottom face of the alkene, depending on which enantiomer of diethyl tartrate is present.

For their pioneering work in the field of enantioselective synthesis, Sharpless (along with William Knowles and Ryoji Noyori) received the 2001 Nobel Prize in chemistry.

Example 11.9

Draw the expected products of Sharpless epoxidation of each allylic alcohol using (+)-diethyl tartrate as the chiral catalyst.

(a) 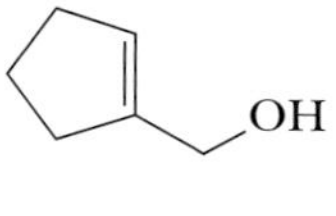   OH

(b) 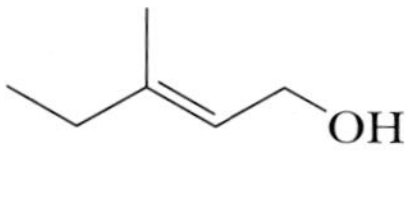   OH

Solution

(a) 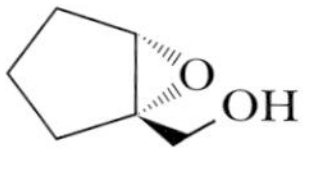   O OH

(b) In this solution, the carbon skeleton of the allylic alcohol is first reoriented to match the orientation in the template, the epoxidation is completed, and then the carbon skeleton is reoriented to match the original drawing.

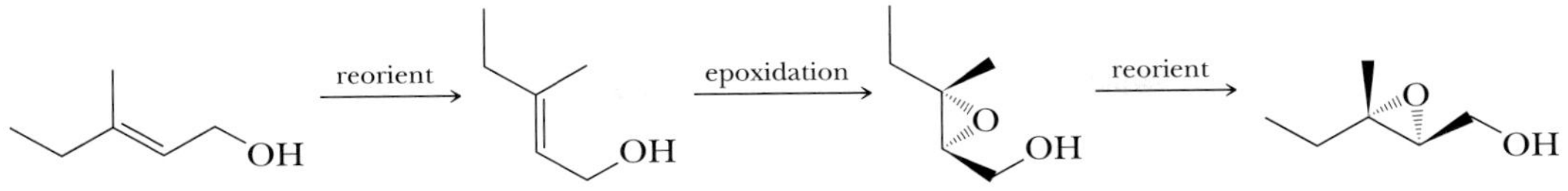

Problem 11.9

Draw the expected products of Sharpless epoxidation of each allylic alcohol using (+)-diethyl tartrate as the chiral catalyst.

(a)

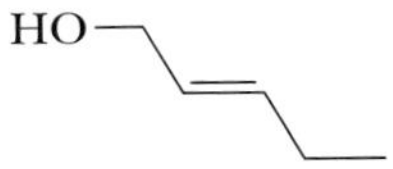

(b)

11.9 Reactions of Epoxides

Because of the strain associated with the three-membered ring, epoxides undergo a variety of ring-opening reactions, the characteristic feature of which is nucleophilic substitution at one of the carbons of the epoxide ring with the oxygen atom as the leaving group.

Characteristic
reaction of
epoxides:

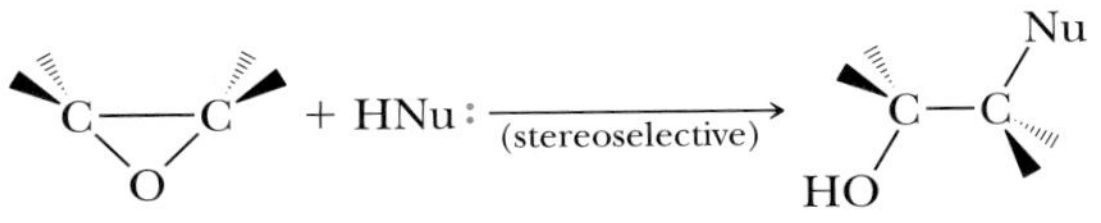

A. Acid-Catalyzed Ring Opening

In the presence of an acid catalyst, such as sulfuric acid, epoxides are hydrolyzed to glycols. As an example, acid-catalyzed hydrolysis of oxirane gives 1,2-ethanediol (ethylene glycol).

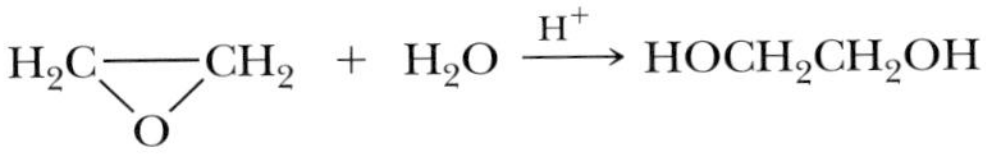

Oxirane
(Ethylene oxide)

1,2-Ethanediol
(Ethylene glycol)

Mechanism   *Acid-Catalyzed Hydrolysis of an Epoxide*

Step 1: Proton transfer from the acid catalyst to oxygen of the epoxide gives a bridged oxonium ion intermediate.

Step 2: Backside attack of H_2O on the protonated epoxide (a bridged oxonium ion) opens the three-membered ring.

Step 3: Proton transfer to solvent completes the formation of the glycol.

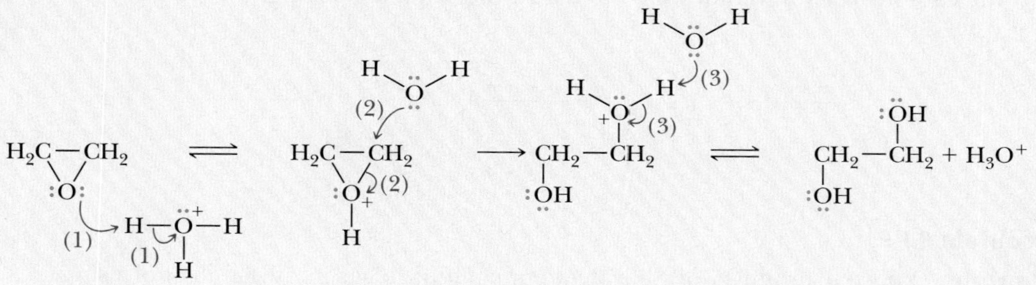

Attack of a nucleophile on a protonated epoxide shows an anti stereoselectivity typical of S_N2 reactions; the nucleophile attacks anti to the leaving hydroxyl group, and the —OH groups in the glycol thus formed are anti. As a result, hydrolysis of an epoxycycloalkane yields a *trans*-1,2-cycloalkanediol.

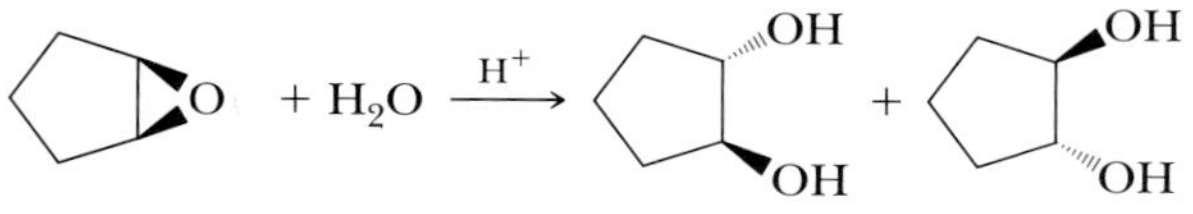

1,2-Epoxycyclopentane
(Cyclopentene oxide)
(achiral)

trans-1,2-Cyclopentanediol
(a racemic mixture)

Note the similarity in ring opening of this bridged oxonium ion intermediate and the bridged halonium ion intermediate in electrophilic addition of halogens or X_2/H_2O to an alkene (Sections 6.3D and 6.3E). In each case, the intermediate is a three-membered ring with a heteroatom bearing a positive charge, and attack of the nucleophile is anti to the leaving group.

Because there is some carbocation character developed in the transition state for an acid-catalyzed epoxide ring opening, attack of the nucleophile on unsymmetrical epoxides occurs preferentially at the carbon better able to bear a partial positive charge.

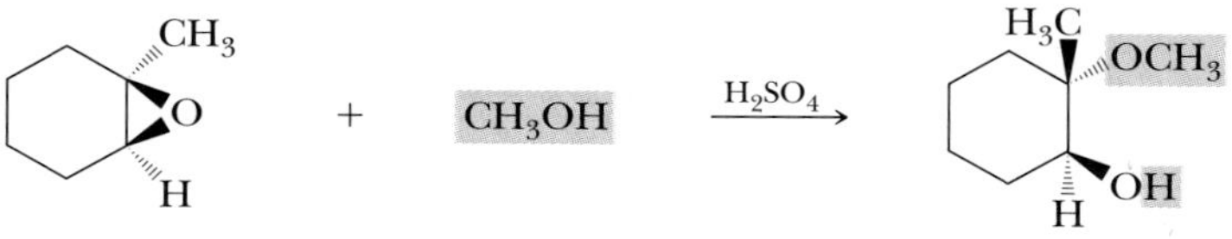

1-Methyl-1,2-epoxycyclohexane

2-Methoxy-2-methylcyclohexanol

The stereochemistry of acid-catalyzed ring openings is S_N2-like in that attack of the nucleophile is from the side opposite the bridged oxonium ion intermediate. The regiochemistry, however, is S_N1-like. Because of the partial carbocation character of the transition state, attack of the nucleophile on the oxonium ion intermediate occurs preferentially at the more substituted carbon. That is, attack occurs at the site better able to bear the partial positive charge that develops on carbon in the transition state in analogy to attack on a bridged bromonium ion.

At this point, let us compare the stereochemistry of the glycol formed by acid-catalyzed hydrolysis of an epoxide with that formed by oxidation of an alkene with osmium tetroxide (Section 6.5A). Each reaction sequence is stereoselective but gives a different stereoisomer. Acid-catalyzed hydrolysis of cyclopentene oxide gives *trans*-1,2-cyclopentanediol; osmium tetroxide oxidation of cyclopentene gives *cis*-1,2-cyclopentanediol. Thus, a cycloalkene can be converted to either a *cis* glycol or a *trans* glycol by the proper choice of reagents.

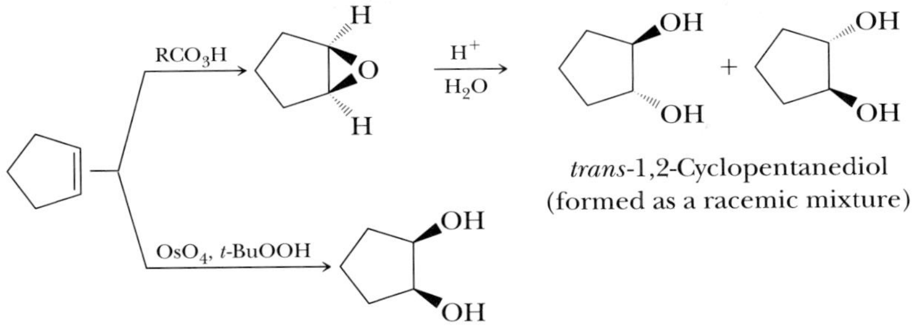

trans-1,2-Cyclopentanediol
(formed as a racemic mixture)

cis-1,2-Cyclopentanediol
(achiral)

B. Nucleophilic Ring Opening

Ethers are not normally susceptible to reaction with nucleophiles. Epoxides, however, are different. Because of the strain associated with a three-membered ring, epoxides undergo ring-opening reactions with a variety of nucleophiles. Good nucleophiles attack an epoxide ring by an S_N2 mechanism and show an S_N2-like regioselectivity; that is, the nucleophile attacks at the less hindered carbon. Following is an equation for the reaction of methyloxirane (propylene oxide) with sodium methoxide in methanol.

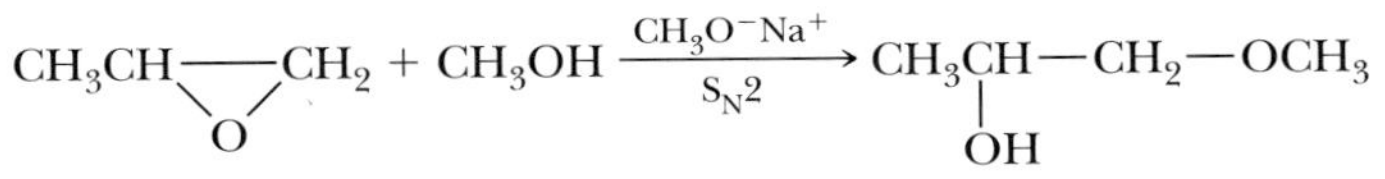

Mechanism *Nucleophilic Opening of an Epoxide Ring*

Step 1: Backside attack of the nucleophile on the less hindered carbon of the highly strained epoxide opens the ring and displaces O^-.

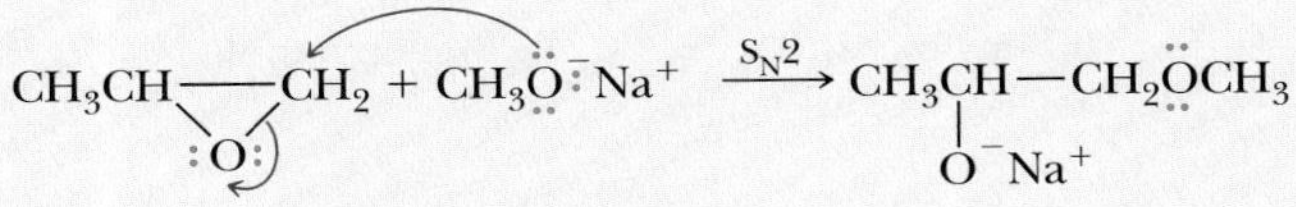

Step 2: Proton transfer completes the reaction.

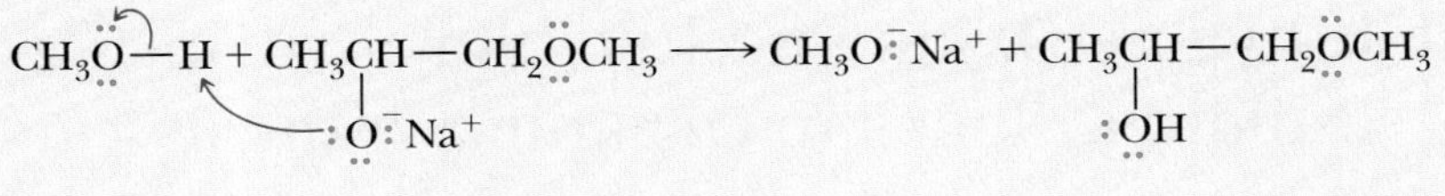

The nucleophilic ring opening of epoxides is also stereoselective; as expected of an S_N2 reaction, attack of the nucleophile is anti to the leaving group. An illustration is the reaction of cyclohexene oxide with sodium methoxide in methanol to give *trans*-2-methoxycyclohexanol.

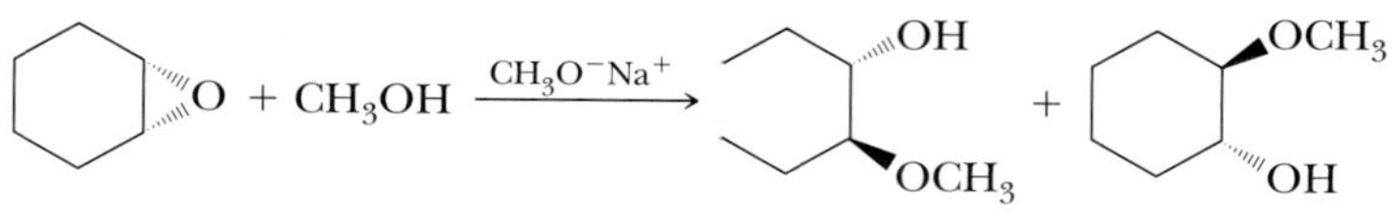

Cyclohexene oxide

trans-2-Methoxycyclohexanol
(a racemic mixture)

The value of epoxides lies in the number of nucleophiles that bring about ring opening and the combinations of functional groups that can be prepared from them. The most important of these ring-opening reactions are summarized in the following chart.

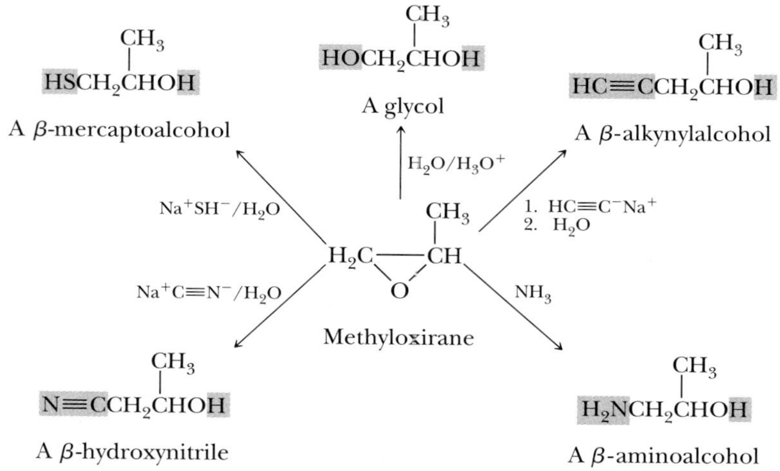

HSCH$_2$CHOH (CH$_3$)

A β-mercaptoalcohol

HOCH$_2$CHOH (CH$_3$)

A glycol

HC≡CCH$_2$CHOH (CH$_3$)

A β-alkynylalcohol

H$_2$C——CH (CH$_3$)
    \O/

Methyloxirane

N≡CCH$_2$CHOH (CH$_3$)

A β-hydroxynitrile

H$_2$NCH$_2$CHOH (CH$_3$)

A β-aminoalcohol

Finally, treatment with LiAlH$_4$ reduces an epoxide to an alcohol. Lithium aluminum hydride is similar to sodium borohydride, NaBH$_4$, in that it is a donor of hydride ion, H:$^-$, which is both a strong base and a good nucleophile. In the reduction of a substituted epoxide by LiAlH$_4$, attack of the hydride ion occurs preferentially at the less hindered carbon of the epoxide, an observation consistent with S$_N$2 reactivity.

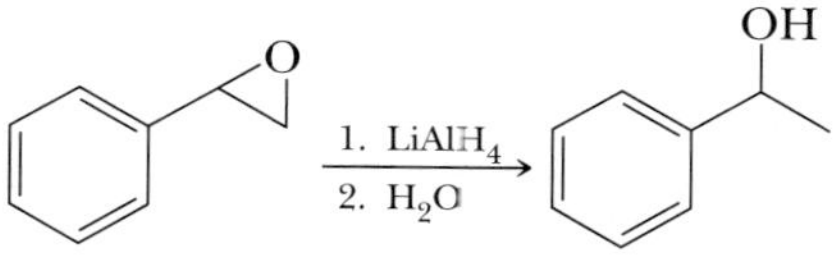

Phenyloxirane
(Styrene oxide)

1-Phenylethanol

11.10 Ethylene Oxide and Epichlorohydrin: Building Blocks in Organic Synthesis

Ethylene oxide is a valuable building block for organic synthesis because each carbon of its two-carbon skeleton has a functional group. Following is a flow chart illustrating some of the functional groups and types of molecules that can be generated from this building block. The key to recognizing a structural unit derived from ethylene oxide is the presence of an Nu—CH$_2$—CH$_2$—OH group. In cases where the —OH group is subsequently modified by replacement with another nucleophile, you will find the group Nu—CH$_2$—CH$_2$—Nu. The most widely used nucleophiles used in modification of the —OH group are ammonia, 1° amines, and 2° amines (Section 1.3B). We have seen all of these reactions before, but not in this form.

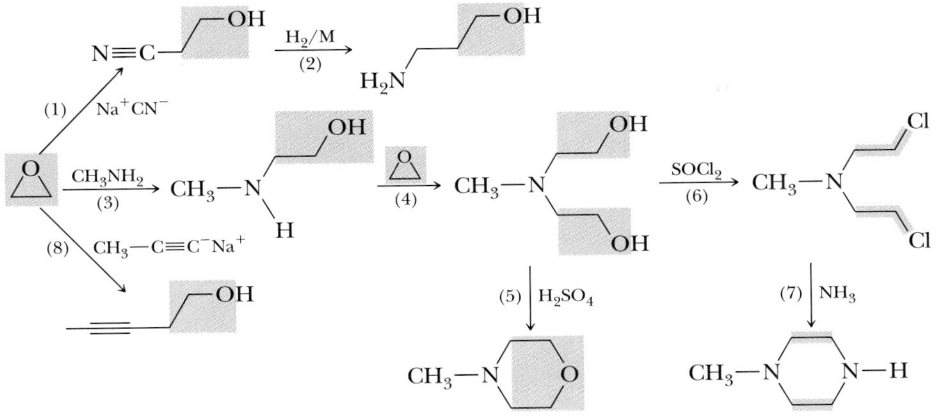

As you study this flow chart, notice the following points.

- Reactions (1) and (8) use carbon nucleophiles to open the three-membered ring and thus form new carbon-carbon bonds.
- Reaction (2) is a catalytic reduction of the carbon-nitrogen triple bond to a 1° amine. Just as a carbon-carbon triple bond can be reduced to a carbon-carbon single bond by hydrogen in the presence of a transition metal catalyst, a carbon-nitrogen triple bond can be similarly reduced.
- Reactions (3) and (4) are openings of the epoxide ring by nitrogen nucleophiles.
- Reaction (5) is an intramolecular acid-catalyzed dehydration of a 1,5-diol to give a cyclic ether.
- Reaction (7) involves two successive S$_N$2 reactions to form a nitrogen-containing ring.

An example of a compound, part of which is derived from the two-carbon skeleton of ethylene oxide, is the local anesthetic procaine.

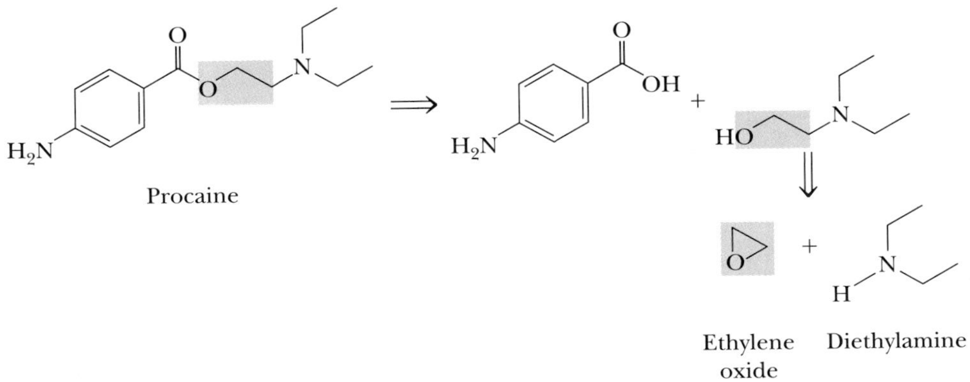

The hydrochloride salt of procaine is marketed under the trade name Novocaine. We will show how to complete the synthesis of procaine when we study the derivatives of carboxylic acids in Chapter 18.

The epoxide epichlorohydrin is also a valuable synthetic intermediate because each of its three carbons contains a reactive functional group.

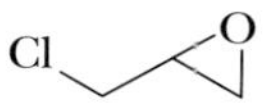

Cl⌁⟨O⟩

Epichlorohydrin

Epichlorohydrin is an oily liquid, bp 118°C. It is insoluble in water and nonpolar hydrocarbon solvents, but soluble in polar aprotic solvents such as diethyl ether and dichloromethane. Epichlorohydrin is synthesized industrially by the following series of three reactions.

Step 1: Allylic halogenation by a radical chain mechanism (Section 8.6A).

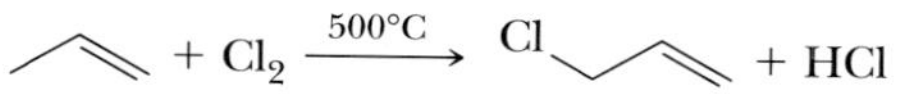

$$\text{Propene} + Cl_2 \xrightarrow{500°C} \text{Cl} \diagup\diagdown + HCl$$

Propene   3-Chloropropene
                (Allyl chloride)

Step 2: Treating the haloalkene with chlorine in water gives a chlorohydrin (Section 6.3E).

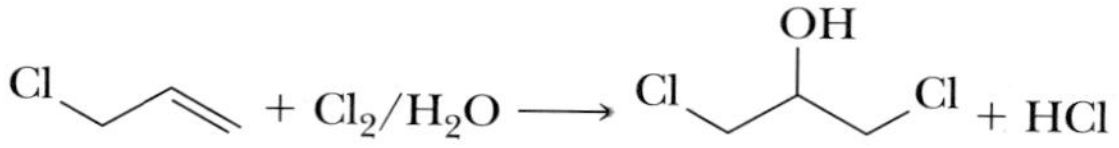

$$\text{Cl}\diagup\diagdown + Cl_2/H_2O \longrightarrow \text{Cl}\diagup\overset{OH}{\diagdown}\diagup\text{Cl} + HCl$$

Step 3: Treating the chlorohydrin with calcium hydroxide brings about an internal S_N2 reaction and gives epichlorohydrin.

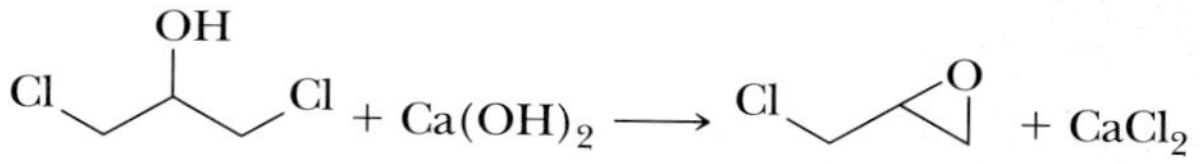

$$\text{Cl}\diagup\overset{OH}{\diagdown}\diagup\text{Cl} + Ca(OH)_2 \longrightarrow \text{Cl}\diagup\langle O\rangle + CaCl_2$$

3-Chloro-1,2-epoxypropane
(Epichlorohydrin)
(racemic)

The characteristic structural feature of a product derived from epichlorohydrin is a three-carbon unit with —OH on the middle carbon, and a carbon, nitrogen, oxygen, or sulfur nucleophile bonded to the two end carbons.

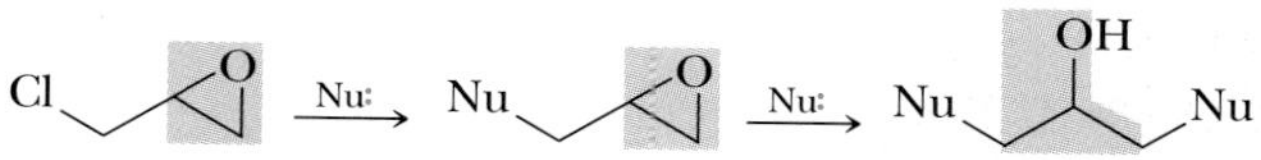

$$\text{Cl}\diagup\langle O\rangle \xrightarrow{Nu:} \text{Nu}\diagup\langle O\rangle \xrightarrow{Nu:} \text{Nu}\diagup\overset{OH}{\diagdown}\diagup\text{Nu}$$

Epichlorohydrin

An example of a compound that contains the three-carbon skeleton of epichlorohydrin is nadolol, a β-adrenergic blocker with vasodilating activity.

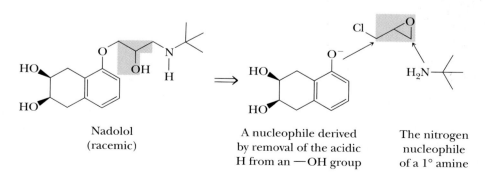

Nadolol
(racemic)

A nucleophile derived
by removal of the acidic
H from an —OH group

The nitrogen
nucleophile
of a 1° amine

Members of this class of compounds have received enormous clinical attention because of their effectiveness in treating hypertension (high blood pressure), migraine headaches, glaucoma, ischemic heart disease, and certain cardiac arrhythmias. Shown in this retrosynthetic analysis are the two nucleophiles used in the synthesis of nadolol. We will show how to complete the synthesis of nadolol when we study the chemistry of benzene and its derivatives in Chapter 21 and 22.

We are not concerned at this stage with how these nucleophiles are generated or, if there are two nucleophiles used, which nucleophile is added first or reacts at which site. Our concern with ethylene oxide and epichlorohydrin at this stage of the course is only that you recognize the structural features in a target molecule that might be derived from these building blocks. Call it pattern recognition if you will. Later, after we study the chemistry of other functional groups, we will discuss in detail the chemistry of how the target molecules are synthesized in the laboratory.

11.11 Crown Ethers

In the early 1960s, Charles Pedersen of DuPont discovered a family of cyclic polyethers derived from ethylene glycol and substituted ethylene glycols. Compounds of this structure are named **crown ethers** because one of their most stable conformations resembles the shape of a crown. These ethers are named by the system devised by Pedersen. The parent name *crown* is preceded by a number describing the size of the ring and followed by a number describing the number of oxygen atoms in the ring, as for example, 18-crown-6.

Crown ether   A cyclic polyether derived from ethylene glycol and substituted ethylene glycols. Crown ethers are excellent phase-transfer catalysts.

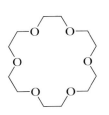

18-Crown-6
(a cyclic hexamer)

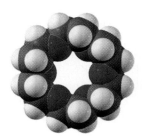

Space-filling model, viewed from above

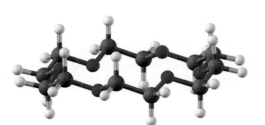

Ball-and-stick model, viewed through an edge

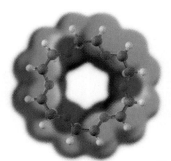

Electrostatic potential map showing the electron-rich interior and the nonpolar exterior

For his work, Pedersen shared the 1987 Nobel Prize for chemistry with Donald J. Cram of the United States and Jean-Marie Lehn of France.

The most significant structural feature of crown ethers is that the diameter of the cavity created by the repeating oxygen atoms of the ring is comparable to the diameter of alkali metal ions. The diameter of the cavity in 18-crown-6, for example, is approximately the diameter of a potassium ion. When a potassium ion is inserted into the cavity of 18-crown-6, the unshared electron pairs on the six oxygens of the crown ether are close enough to the potassium ion to provide very effective solvation for K^+:

Diameter of cavity created by
six oxygens is approximately
260 to 320 pm

Diameter of
K^+ is 266 pm

Ion	Diameter (pm)
Li^+	136
Na^+	194
K^+	266
Rb^+	294
Mg^{2+}	164
Ca^{2+}	286

A complex of K^+
and 18-crown-6

18-Crown-6 forms somewhat weaker complexes with rubidium ion (a somewhat larger ion) and with sodium ion (a somewhat smaller ion). It does not coordinate to any appreciable degree with lithium ion (a considerably smaller ion). 12-Crown-4, however, with its smaller cavity, does form a strong complex with lithium ion.

The cavity of a crown ether is a polar region, and the unshared pairs of electrons on the oxygen atoms lining the cavity provide effective solvation for alkali metal ions. The outer surface of the crown is nonpolar and hydrocarbon-like, and, thus, crown ethers and their alkali metal ion complexes dissolve readily in nonpolar organic solvents.

Crown ethers have proven to be particularly valuable for the same reasons as phase-transfer catalysts (Section 9.11), namely, their ability to cause inorganic salts to dissolve in nonpolar aprotic organic solvents such as methylene chloride, hexane, and benzene. Potassium permanganate, for example, does not dissolve in benzene. If 18-crown-6 is added to benzene, the solution takes on the purple color characteristic of permanganate ion. The crown-potassium ion complex is soluble in benzene and brings permanganate ion into solution with it. The resulting "purple benzene" is a valuable reagent for the oxidation of water-insoluble organic compounds.

Crown ethers have also proven valuable in nucleophilic displacement reactions. The cations of potassium salts, such as KF, KCN, or KN_3, are very tightly bound within the solvation cavity of 18-crown-6 molecules. The anions, however, are only weakly solvated, and because of the geometry of cation binding within the cavity of the crown, only loose ion pairing occurs between the anion and cation. Thus, in nonpolar aprotic solvents, these anions are without any appreciable solvent shell and are, therefore, highly reactive as nucleophiles. The nucleophilicity of F^-, CN^-, N_3^-, and other anions in nonpolar aprotic solvents containing an 18-crown-6 equals and often exceeds their nucleophilicity in polar aprotic solvents such as DMSO and acetonitrile.

11.12 Thioethers

A. Nomenclature

Sulfide The sulfur analog of an ether; a molecule containing a sulfur atom bonded to two carbon atoms.

To derive the IUPAC name of a thioether, select the longest carbon chain as the parent alkane and name the sulfur-containing substitutent as an *alkylsulfanyl* group. To derive a common name, list the groups bonded to sulfur and add the word **sulfide** to show the presence of the —S— group:

Ethylsulfanylethane
(Diethyl sulfide)

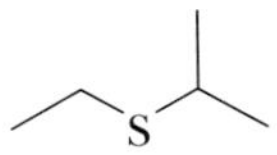

2-Ethylsulfanylpropane
(Ethyl isopropyl sulfide)

The functional group of a **disulfide** is an —S—S— group. IUPAC names of disulfides are derived by selecting the longest carbon chain as the parent alkane and indicating the disulfide-containing substitutent as an *alkyldisulfanyl* group. Common names of disulfides are derived by listing the names of the groups bonded to sulfur and adding the word *disulfide:*

Disulfide   A molecule containing an —S—S— group.

Ethyldisulfanylethane
(Diethyl disulfide)

B. Preparation of Sulfides

Symmetrical sulfides, RSR (also called symmetrical thioethers), are prepared by treating one mole of Na_2S (where S^{2-} is the nucleophile) with two moles of haloalkane.

$$2RX + Na_2S \longrightarrow RSR + 2NaX$$

A sulfide

This same reaction can also be used to prepare five- and six-membered cyclic sulfides. Treating a 1,4-dihaloalkane with Na_2S gives a five-membered cyclic sulfide; treating a 1,5-dihaloalkane with Na_2S gives a six-membered ring.

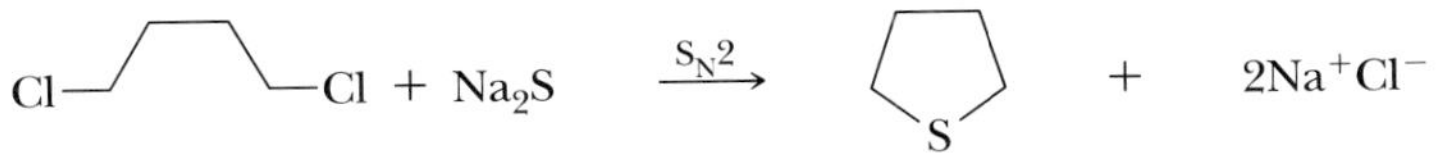

1,4-Dichlorobutane

Thiolane
(Tetrahydrothiophene)

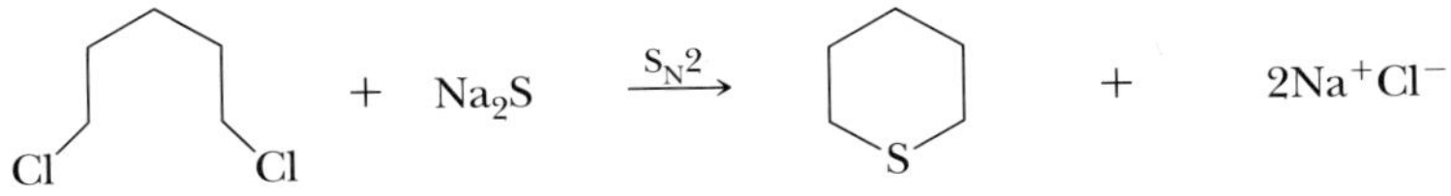

1,5-Dichloropentane

Thiane
(Tetrahydrothiopyran)

Unsymmetrical sulfides, RSR′, are prepared by converting a thiol to a sodium salt with either sodium hydroxide or sodium ethoxide and then allowing the salt to react with a haloalkane.

$$CH_3(CH_2)_8CH_2S^-Na^+ + CH_3I \xrightarrow{S_N2} CH_3(CH_2)_8CH_2SCH_3 + Na^+I^-$$

Sodium 1-decanethiolate

1-Methylsulfanyldecane
(Decyl methyl sulfide)

This method of thioether formation is the sulfur analog of the Williamson ether synthesis (Section 11.4A).

Note that all these reactions leading to sulfides (thioethers) are direct applications of nucleophilic substitution reactions (Chapter 9).

C. Oxidation of Sulfides

Many of the properties of sulfides stem from the fact that divalent sulfur is a reducing agent; it is easily oxidized to two higher oxidation states. Treatment of a sulfide with one mole of 30% aqueous hydrogen peroxide at room temperature gives a sulfoxide, as illustrated by oxidation of methyl phenyl sulfide to methyl phenyl sulfoxide. Several other oxidizing agents, including sodium periodate, $NaIO_4$, also bring about the same conversion. Treatment of a sulfoxide with $NaIO_4$ brings about its oxidation to a sulfone.

Methyl phenyl sulfide   Methyl phenyl sulfoxide   Methyl phenyl sulfone

Dimethyl sulfoxide (DMSO) is manufactured on an industrial scale by air oxidation of dimethyl sulfide in the presence of oxides of nitrogen.

Dimethyl sulfide   Dimethyl sulfoxide

Summary

An **ether** (Section 11.1) contains an atom of oxygen bonded to two carbon atoms. In the IUPAC name, the parent chain is named and the —OR group is named as an *alkoxy* substituent (Section 11.2). Common names are derived by naming the two groups bonded to oxygen followed by the word "ether." Heterocyclic ethers have an oxygen atom as one of the members of a ring. Ethers are weakly polar compounds (Section 11.3) and associate by weak dipole-dipole interactions and dispersion forces. The boiling points of ethers are close to those of hydrocarbons of comparable molecular weight but are much lower than those of the corresponding alcohols. Because ethers are hydrogen bond acceptors, they are more soluble in water than are hydrocarbons of comparable molecular weight.

Crown ethers (Section 11.11) are cyclic polyethers having 12 or more atoms in a ring. The cavity of a crown ether is a polar region, and the unshared pairs of electrons on the ether oxygens can solvate alkali metal ions. The cavity of 18-crown-6, for example, has approximately the same diameter as the potassium ion. The outer surface of a crown ether is nonpolar and hydrocarbon-like. Crown ethers are valuable for their ability to cause ionic compounds to dissolve in nonpolar organic solvents.

Thioethers (Section 11.12) are named as alkylsulfanyl alkanes. Common names for thioethers are derived by naming the two groups bonded to sulfur followed by the word "sulfide."

Key Reactions

1. Williamson Ether Synthesis (Section 11.4A)

The Williamson ether synthesis is a general method for the synthesis of dialkyl ethers by an S_N2 reaction between an alkyl halide and an alkoxide ion.

$$CH_3\underset{CH_3}{\overset{CH_3}{\underset{|}{\overset{|}{C}}}}O^-K^+ + CH_3Br \xrightarrow{S_N2} CH_3\underset{CH_3}{\overset{CH_3}{\underset{|}{\overset{|}{C}}}}OCH_3 + K^+Br^-$$

Yields are highest with methyl, 1° alkyl halides, and 1° allylic halides. They are considerably lower with 2° halides because of competition from E2 elimination. The Williamson ether synthesis reaction fails altogether with 3° halides.

ORGANIC Chemistry•Now™
Click Flashcards to review the **Key Reactions of Ethers, Sulfides, and Epoxides**

2. Acid-Catalyzed Dehydration of Alcohols (Section 11.4B)

Yields are highest for symmetrical ethers formed from unbranched primary alcohols.

$$2CH_3CH_2OH \xrightarrow[140°C]{H_2SO_4} CH_3CH_2OCH_2CH_3 + H_2O$$

3. Acid-Catalyzed Addition of Alcohols to Alkenes (Section 11.4C)

Proton transfer to the alkene generates a carbocation. Nucleophilic addition of an alcohol to the carbocation followed by proton transfer to the solvent gives the ether.

$$CH_3\overset{CH_3}{\overset{|}{C}}{=}CH_2 + CH_3OH \xrightarrow{\text{acid catalyst}} CH_3\underset{CH_3}{\overset{CH_3}{\underset{|}{\overset{|}{C}}}}OCH_3$$

4. Acid-Catalyzed Cleavage of Dialkyl Ethers (Section 11.5A)

Cleavage of ethers requires both a strong acid and a good nucleophile, hence the use of concentrated HBr and HI.

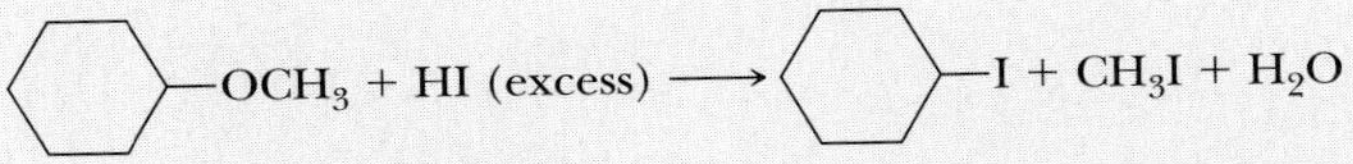

Cleavage of primary and secondary alkyl ethers is by an S_N2 pathway. Cleavage of tertiary alkyl ethers is by an S_N1 pathway.

5. Reaction of Alcohols with Chloro-*tert*-butyldimethylsilane (Section 11.6B)

The *tert*-butyldimethylsilyl (t-BuMe$_2$Si—) group is used to protect primary and secondary alcohols.

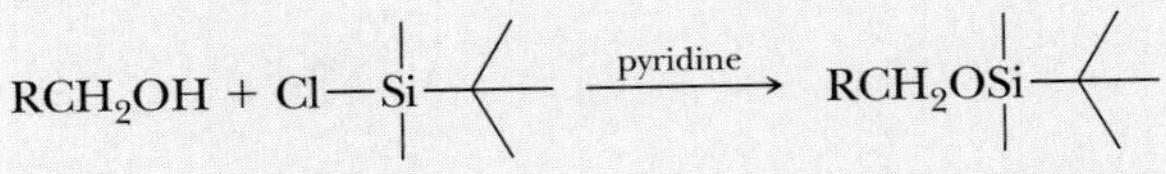

Chloro-*tert*-butyl dimethylsilane — A *tert*-butyldimethylsilyl ether

The protecting group is removed by treating the silyl ether with fluoride ion to regenerate the original alcohol.

6. Oxidation of Alkenes by Peroxycarboxylic acids (Section 11.8B)

Three commonly used peroxycarboxylic acid oxidizing agents are *meta*-chloroperoxybenzoic acid, the magnesium salt of monoperoxyphthalic acid, and peroxyacetic acid. Each reagent oxidizes an alkene to an epoxide.

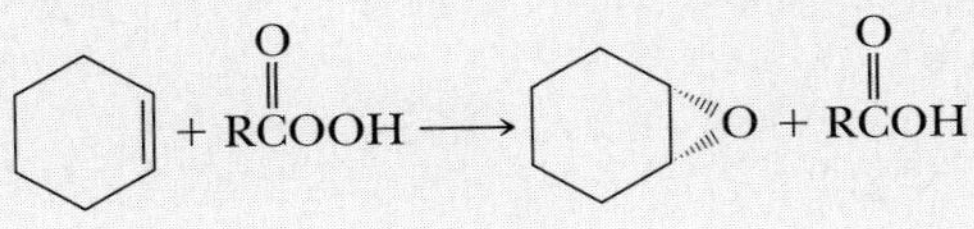

7. Synthesis of Epoxides from Halohydrins (Section 11.8C)

Formation of the halohydrin and the following intramolecular S_N2 reaction are both stereoselective (the configuration of the alkene is retained in the epoxide) and stereospecific (for alkenes that show *cis,trans* isomerism, the configuration of the epoxide depends on the configuration of the alkene).

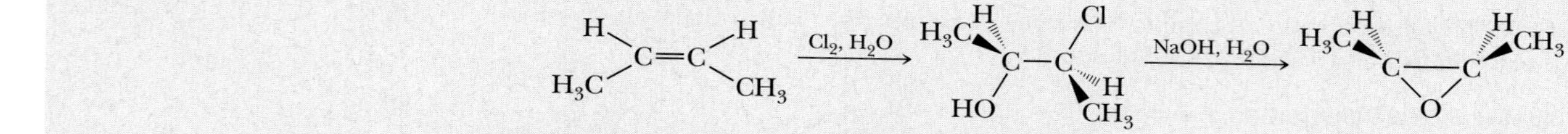

8. Sharpless Asymmetric Epoxidation (Section 11.8D)

Oxidation of the carbon-carbon double bond of a 1° allylic alcohol by *tert*-butyl hydroperoxide in the presence of a chiral catalyst consisting of either (+)- or (−)-diethyl tartrate and titanium tetraisopropoxide gives an enantiomerically pure epoxide. The enantiomer formed depends on which enantiomer of diethyl tartrate is used in the catalyst.

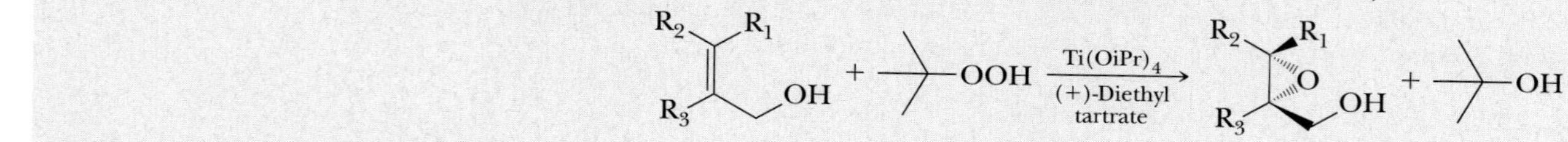

A 1° allylic   *tert*-Butyl
alcohol   hydroperoxide

9. Acid-Catalyzed Hydrolysis of Epoxides (Section 11.9A)

Hydrolysis of an epoxide derived from a cycloalkene gives a *trans* glycol.

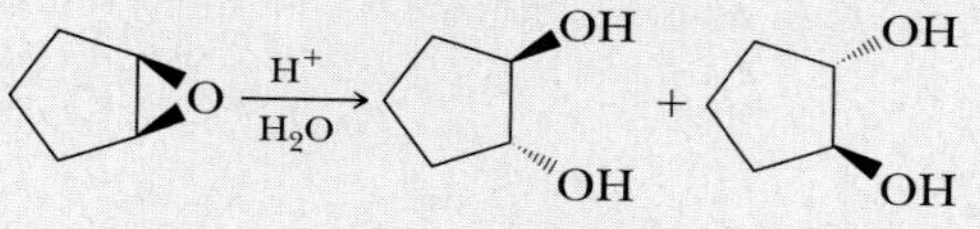

(racemic)

10. Nucleophilic Ring Opening of Epoxides (Section 11.9B)

Attack on the epoxide is regioselective with the nucleophile attacking the less substituted carbon of the epoxide.

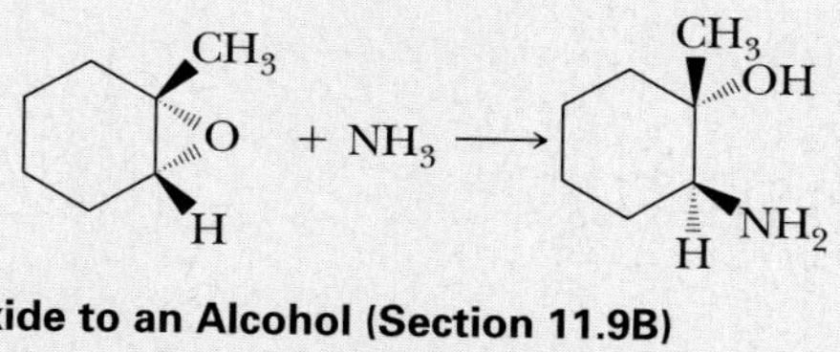

11. Reduction of an Epoxide to an Alcohol (Section 11.9B)

Regioselective hydride ion transfer from lithium aluminum hydride to the less hindered carbon of the epoxide gives an alcohol.

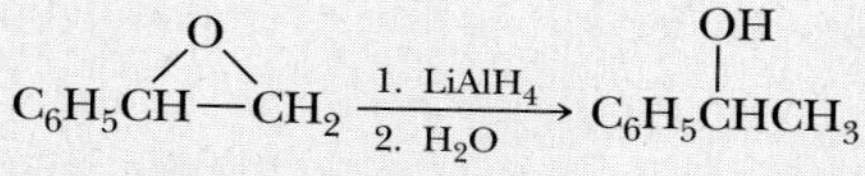

12. Oxidation of Sulfides (Section 11.12C)

Oxidation of a sulfide gives either a sulfoxide or a sulfone, depending on the oxidizing agent and experimental conditions. Air oxidation of dimethyl sulfide is a commercial route to dimethyl sulfoxide, a polar aprotic solvent.

$$CH_3-S-CH_3 + O_2 \xrightarrow[\text{nitrogen}]{\text{oxides of}} CH_3-\overset{\overset{\text{O}}{\|}}{S}-CH_3$$

Problems

Structure and Nomenclature

11.10 Write names for these compounds. Where possible, write both IUPAC names and common names.

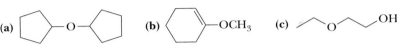

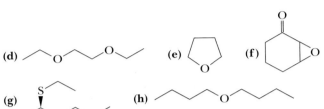

11.11 Draw structural formulas for these compounds.

(a) 2-(1-Methylethoxy)propane
(b) *trans*-2,3-Diethyloxirane
(c) *trans*-2-Ethoxycyclopentanol
(d) Ethenyloxyethene
(e) Cyclohexene oxide
(f) 3-Cyclopropyloxy-1-propene
(g) (*R*)-2-Methyloxirane
(h) 1,1-Dimethoxycyclohexane

Physical Properties

11.12 Each compound given in this problem is a common organic solvent. From each pair of compounds, select the solvent with the greater solubility in water.

(a) CH_2Cl_2 and EtOH   **(b)** Et_2O and EtOH

(c) 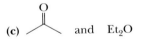 and Et_2O **(d)** Et_2O and

11.13 Account for the fact that tetrahydrofuran (THF) is very soluble in water, whereas the solubility of diethyl either in water is only 8 g/100 mL water.

11.14 Because of the Lewis base properties of ether oxygen atoms, crown ethers are excellent complexing agents for Na^+, K^+, and NH_4^+. What kind of molecule might serve as a complexing agent for Cl^- or Br^-?

Preparation of Ethers

11.15 Write equations to show a combination of reactants to prepare each ether. Which ethers can be prepared in good yield by a Williamson ether synthesis? If there are any that cannot be prepared by the Williamson method, explain why not.

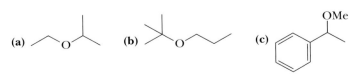

(d)   **(e)** OEt   **(f)**

11.16 Propose a mechanism for this reaction.

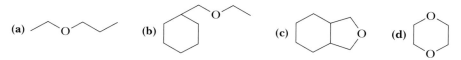

(racemic)

Reactions of Ethers

11.17 Draw structural formulas for the products formed when each compound is heated at reflux in concentrated HI.

(a)   **(b)**   **(c)**   **(d)**

11.18 Following is an equation for the reaction of diisopropyl ether and oxygen to form a hydroperoxide.

Diisopropyl ether   A hydroperoxide

Formation of an ether hydroperoxide is a radical chain reaction.

(a) Write a pair of chain propagation steps that accounts for the formation of this ether hydroperoxide. Assume that initiation is by a radical, R·.

(b) Account for the fact that hydroperoxidation of ethers is regioselective; that is, reaction occurs preferentially at a carbon adjacent to the ether oxygen.

Synthesis and Reactions of Epoxides

11.19 Triethanolamine, $(HOCH_2CH_2)_3N$, is a widely used biological buffer, with maximum buffering capacity at pH 7.8. Propose a synthesis of this compound from ethylene oxide and ammonia.

11.20 Ethylene oxide is the starting material for the synthesis Cellosolve, an important industrial solvent. Propose a mechanism for this reaction.

$$H_2C\text{---}CH_2 + CH_3CH_2OH \xrightarrow{H_2SO_4} CH_3CH_2OCH_2CH_2OH$$

Oxirane   2-Ethoxyethanol
(Ethylene oxide)   (Cellosolve)

11.21 Ethylene oxide is the starting material for the synthesis of 1,4-dioxane. Propose a mechanism for each step in this synthesis.

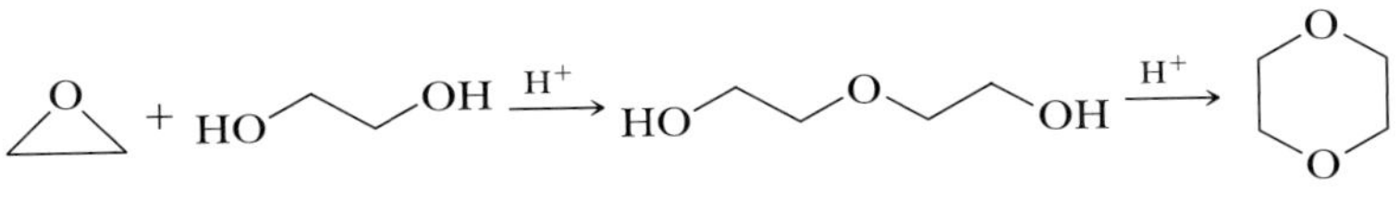

1,4-Dioxane

11.22 Propose a synthesis for 18-crown-6. If a base is used in your synthesis, does it make a difference whether it is lithium hydroxide or potassium hydroxide?

11.23 Predict the structural formula of the major product of the reaction of 2,2,3-trimethyloxirane with each set of reagents.

   (a) MeOH/MeO⁻Na⁺   (b) MeOH/H⁺   (c) Me₂NH

11.24 The following equation shows the reaction of *trans*-2,3-diphenyloxirane with hydrogen chloride in benzene to form 2-chloro-1,2-diphenylethanol.

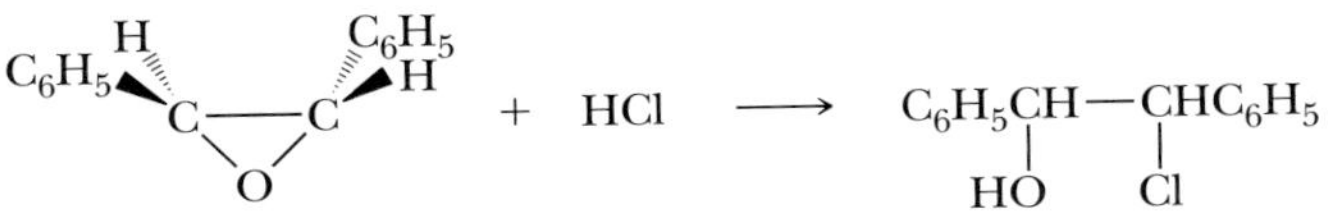

trans-2,3-Diphenyloxirane   2-Chloro-1,2-diphenylethanol

   (a) How many stereoisomers are possible for 2-chloro-1,2-diphenylethanol?
   (b) Given that opening of the epoxide ring in this reaction is stereoselective, predict which of the possible stereoisomers of 2-chloro-1,2-diphenylethanol is/are formed in the reaction.

11.25 Propose a mechanism to account for this rearrangement.

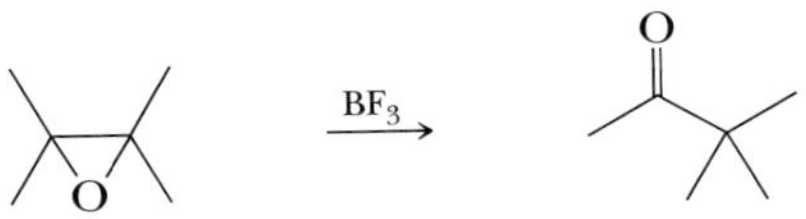

Tetramethyloxirane   3,3-Dimethyl-2-butanone

11.26 Acid-catalyzed hydrolysis of the following epoxide gives a *trans* diol.

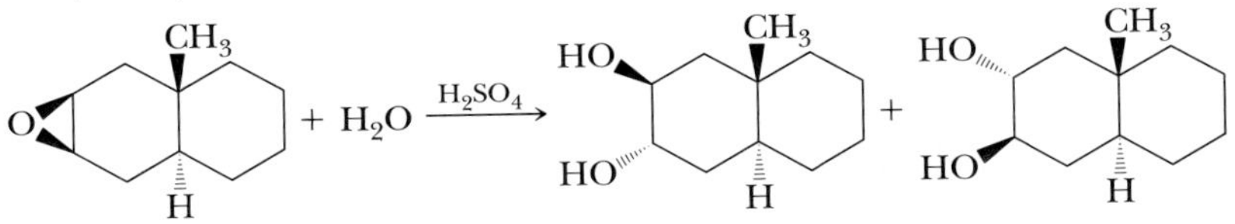

Only this glycol   This glycol is
is formed   not formed

Of the two possible *trans* diols, only one is formed. How do you account for this stereoselectivity?

11.27 Following are two reaction sequences for converting 1,2-diphenylethylene into 2,3-diphenyloxirane.

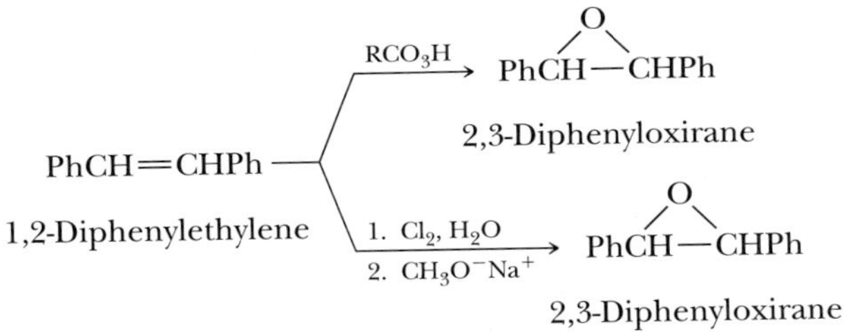

Suppose that the starting alkene is *trans*-1,2-diphenylethylene.

(a) What is the configuration of the oxirane formed in each sequence?
(b) Will the oxirane formed in either sequence rotate the plane of polarized light? Explain.

11.28 The following enantiomer of a chiral epoxide is an intermediate in the synthesis of the insect pheromone frontalin.

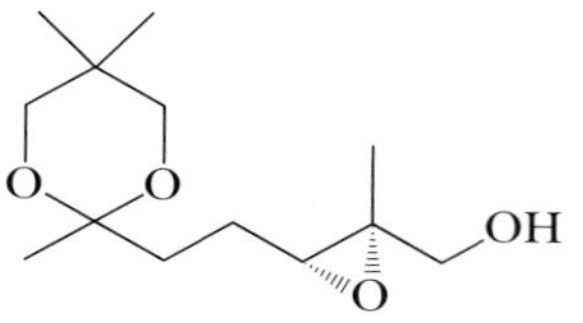

Show how this enantiomer can be prepared from an allylic alcohol, using the Sharpless epoxidation.

11.29 Human white cells produce an enzyme called myeloperoxidase. This enzyme catalyzes the reaction between hydrogen peroxide and chloride ion to produce hypochlorous acid, HOCl, which reacts as if it were Cl^+OH^-. When attacked by white cells, cholesterol gives a chlorohydrin as the major product.

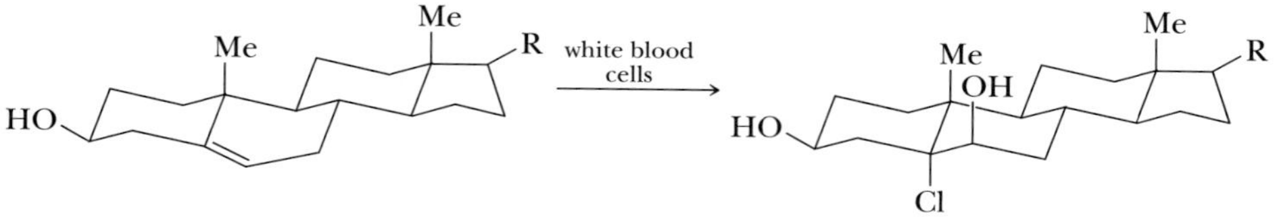

Cholesterol   Cholesterol chlorohydrin

(a) Propose a mechanism for this reaction. Account for both its regioselectivity and stereoselectivity.
(b) On standing or (much more rapidly) on treatment with base, the chlorohydrin is converted to an epoxide. Show the structure of the epoxide and a mechanism for its formation. This epoxide is believed to be involved in induction of certain cancers.

11.30 Propose a mechanism for the following acid-catalyzed rearrangement.

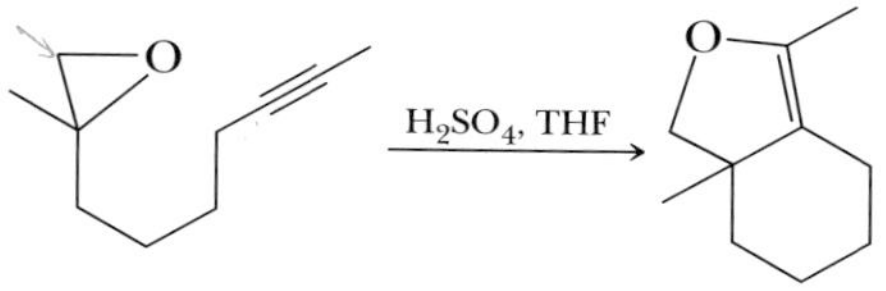

H_2SO_4, THF

Synthesis

11.31 Show reagents and experimental conditions to synthesize the following compounds from 1-propanol (any derivative of 1-propanol prepared in one part of this problem may be used for the synthesis of another part of the problem).

(a) Propanal	**(b)** Propanoic acid
(c) Propene	**(d)** 2-Propanol
(e) 2-Bromopropane	**(f)** 1-Chloropropane
(g) 1,2-Dibromopropane	**(h)** Propyne
(i) 2-Propanone	**(j)** 1-Chloro-2-propanol
(k) Methyloxirane	**(l)** Dipropyl ether
(m) Isopropyl propyl ether	**(n)** 1-Mercapto-2-propanol
(o) 1-Amino-2-propanol	**(p)** 1,2-Propanediol

11.32 Starting with *cis*-3-hexene, show how to prepare the following diols.

(a) Meso 3,4-hexanediol   **(b)** Racemic 3,4-hexanediol

11.33 Show reagents to convert cycloheptene to each of the following compounds.

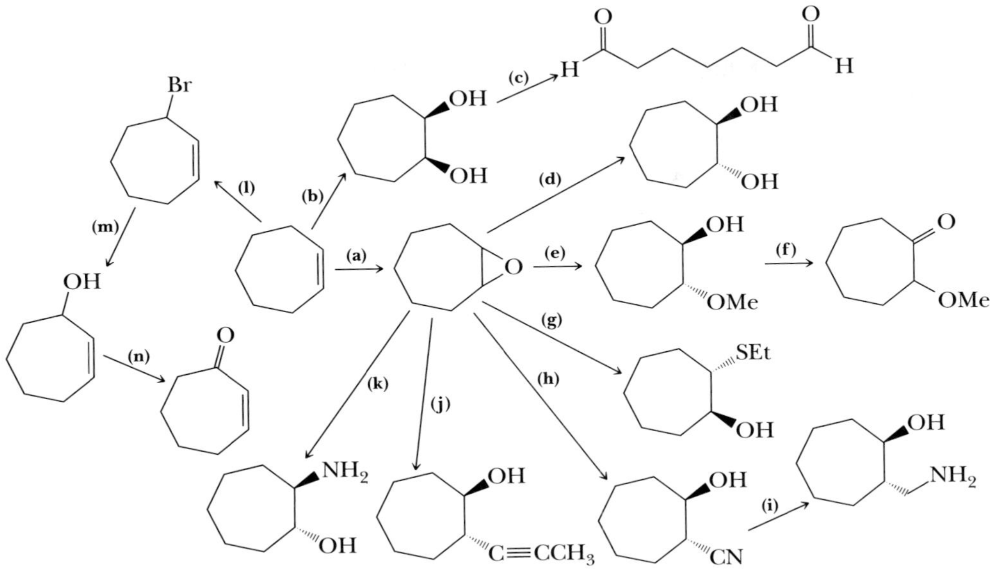

11.34 Show reagents to convert bromocyclopentane to each of the following compounds.

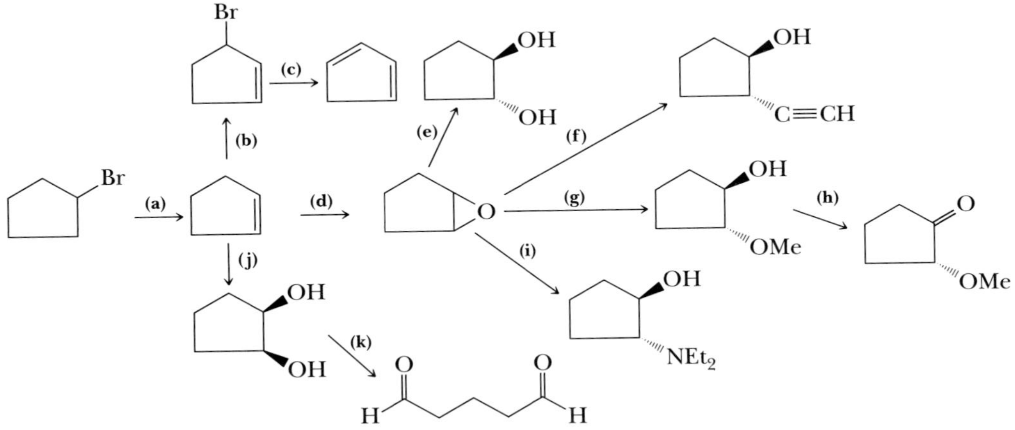

11.35 Given the following retrosynthetic analysis, show how to synthesize the target molecule from styrene and 1-chloro-3-methyl-2-butene.

Styrene   1-Chloro-3-methyl-2-butene

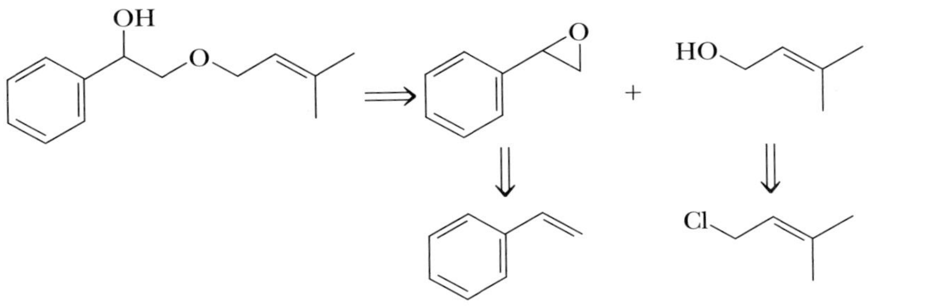

11.36 Starting with acetylene and ethylene oxide as the only sources of carbon atoms, show how to prepare these compounds.

(a) 3-Butyn-1-ol   **(b)** 3-Hexyn-1,6-diol   **(c)** 1,6-Hexanediol
(d) (Z)-3-Hexen-1,6-diol   **(e)** (E)-3-Hexen-1,6-diol   **(f)** Hexanedial

11.37 Following are the steps in the industrial synthesis of glycerin.

$$CH_2\!=\!CHCH_3 \xrightarrow[\text{heat}]{Cl_2} A\ (C_3H_5Cl) \xrightarrow{\text{NaOH, H}_2\text{O}} B\ (C_3H_6O) \xrightarrow{Cl_2,\ H_2O}$$

Propene

$$C\ (C_3H_7ClO_2) \xrightarrow[\text{heat}]{Ca(OH)_2} D\ (C_3H_6O_2) \xrightarrow{H_2O,\ HCl} HOCH_2\overset{\overset{\displaystyle OH}{|}}{C}HCH_2OH$$

1,2,3-Propanetriol
(glycerol, glycerin)

Provide structures for all intermediate compounds and describe the type of mechanism by which each is formed.

11.38 Gossyplure, the sex pheromone of the pink bollworm, is the acetic ester of 7,11-hexadeca-dien-1-ol. The active pheromone has the Z configuration at the C7-C8 double bond and is a mixture of E,Z isomers at the C11-C12 double bond. Shown here is the Z,E isomer.

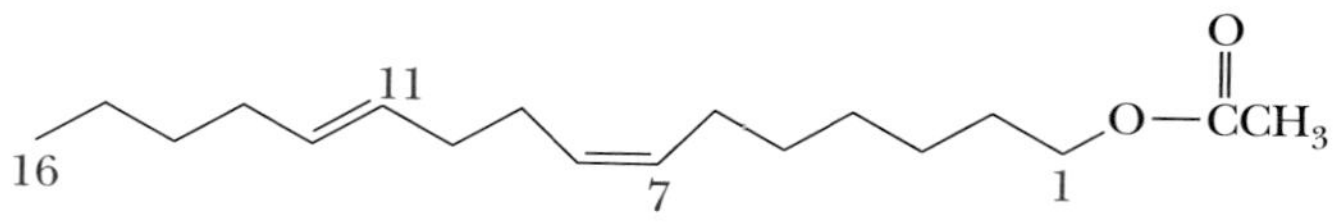

(7Z,11E)-7,11-hexadecadienyl acetate

Following is a retrosynthetic analysis for (7Z,11E)-7,11-hexadecadien-1-ol, which then led to a successful synthesis of gossyplure.

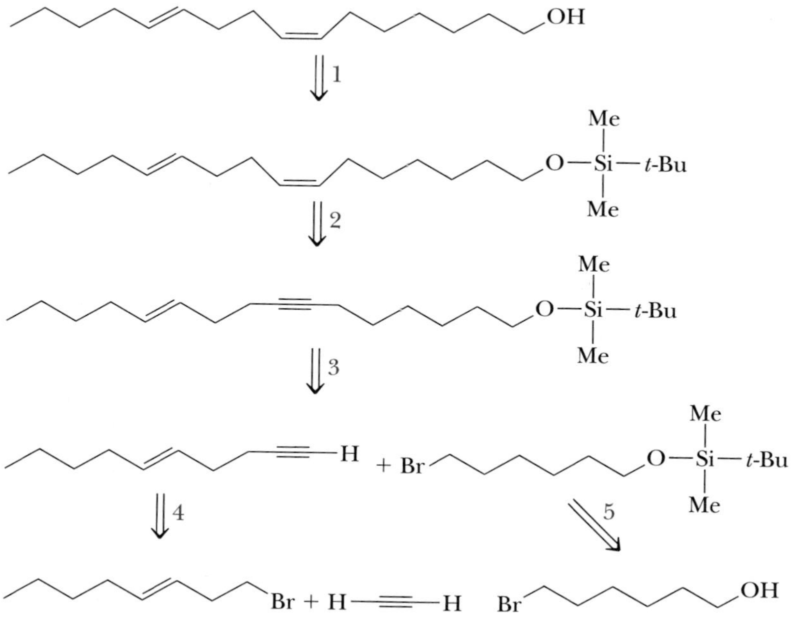

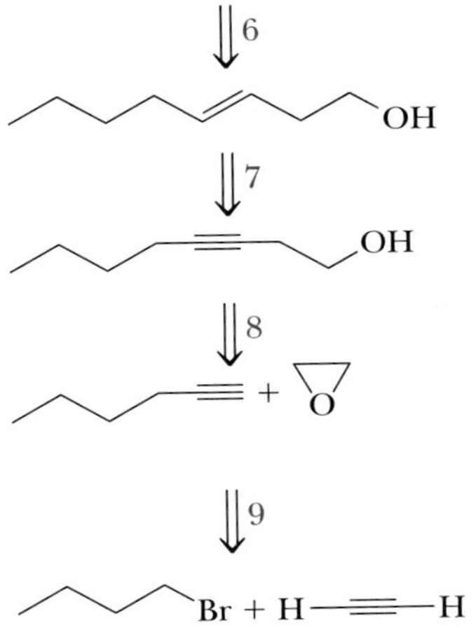

(a) Suggest reagents and experimental conditions for each step in this synthesis.
(b) Why is it necessary to protect the —OH group of 6-bromo-1-hexanol?
(c) How might you modify this synthesis to prepare the 7*Z*,11*Z* isomer of gossyplure?

11.39 Epichlorohydrin (Section 11.10) is a valuable synthetic intermediate because each of its three carbons contains a reactive group. Following is the first step in its synthesis from propene. Propose a mechanism for this step.

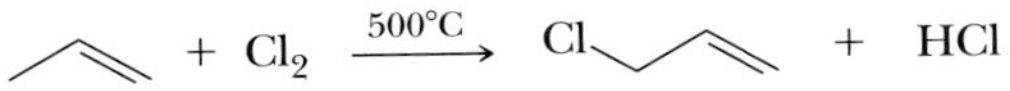

Propene

3-Chloropropene
(Allyl chloride)

11.40 Each of these drugs contains one or more building blocks derived from either ethylene oxide or epichlorohydrin.

(a)

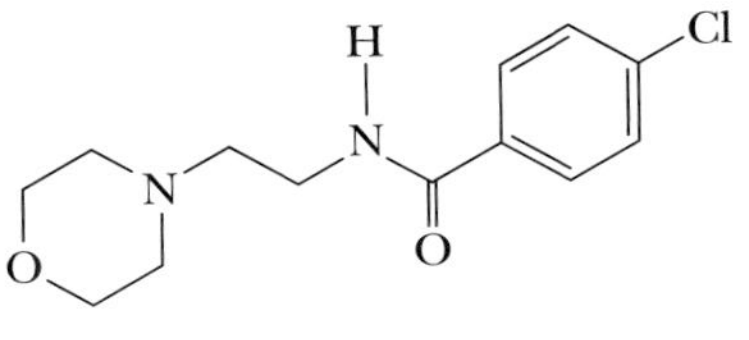

Moclobemide
(an antidepressant)

(b)

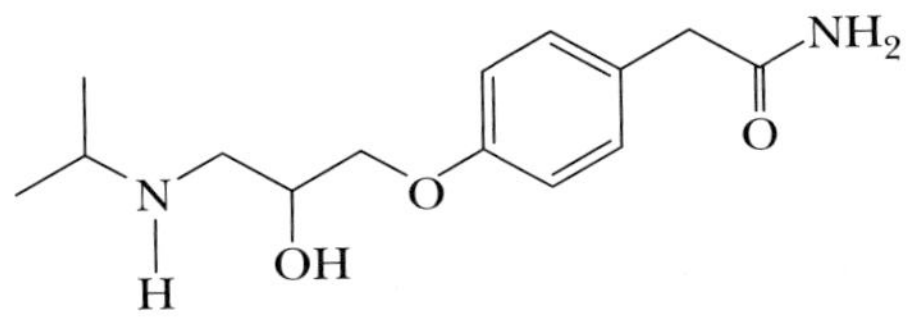

Atenolol
(an antihypertensive)

(c)

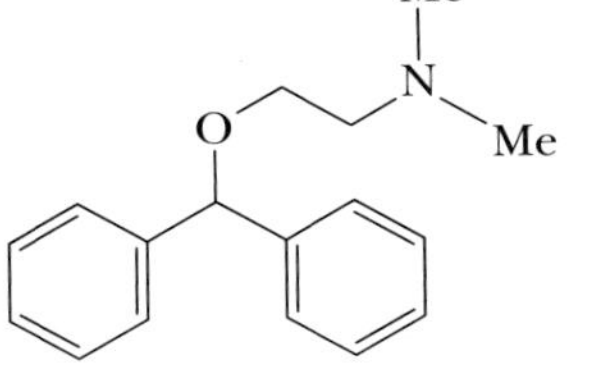

Diphenhydramine
(Benadryl, an antihistamine)

(d)

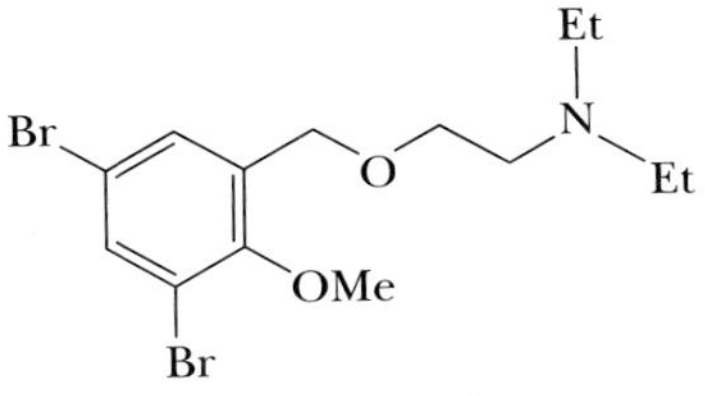

Spasmolytol
(an antispasmodic)

(e)

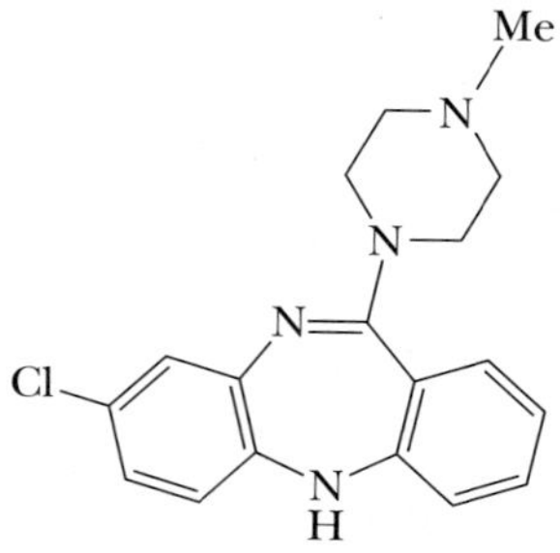

Clozapine
(an antischizophrenic)

(f)

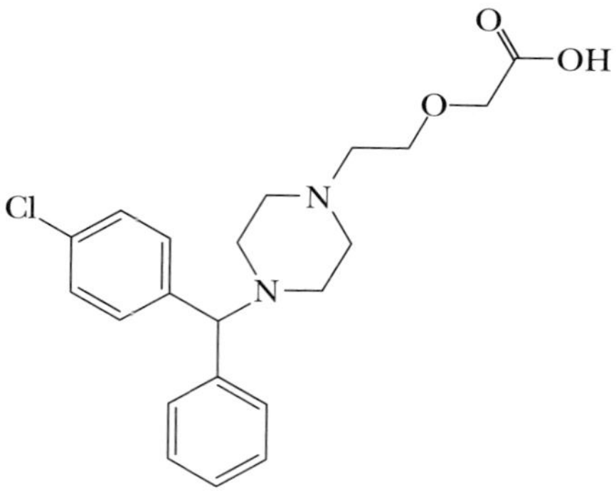

Cetirizine
(Zyrtec, an antihistamine)
(racemic)

Identify the part of each molecule that can be derived from one or the other of these building blocks and propose structural formulas for the nucleophile(s) that can be used along with either ethylene oxide or epichlorohydrin to synthesize each molecule. We will learn about the actual syntheses of each molecule in later chapters.

Looking Ahead

10.41 Aldehydes and ketones react with one molecule of an alcohol to form compounds called hemiacetals, in which there is one hydroxyl group and one ether-like group. Reaction of a hemiacetal with a second molecule of alcohol gives an acetal and a molecule of water. We study this reaction in Chapter 16.

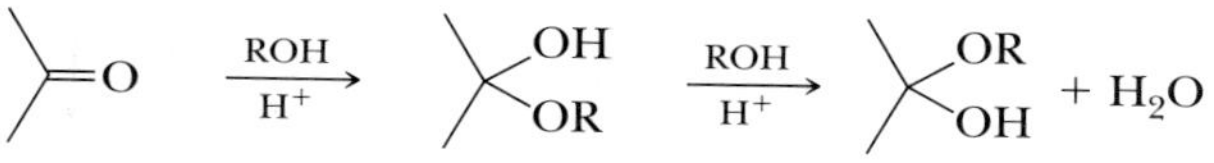

| The carbonyl group of an aldehyde or ketone | A hemiacetal (has an —OH and an —OR group to the same carbon) | An acetal (has two —OR groups to the same carbon) |

Draw structural formulas for the hemiacetal and acetal formed from these reagents. The stoichiometry of each reaction is given in the problem.

(a)

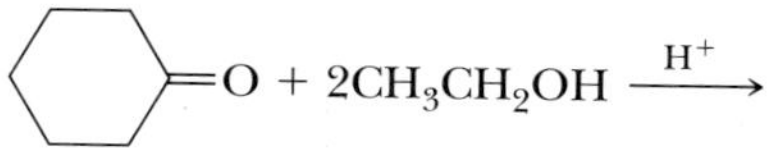

Cyclohexanone   Ethanol

(b)

Cyclohexanone   Ethylene glycol

(c)

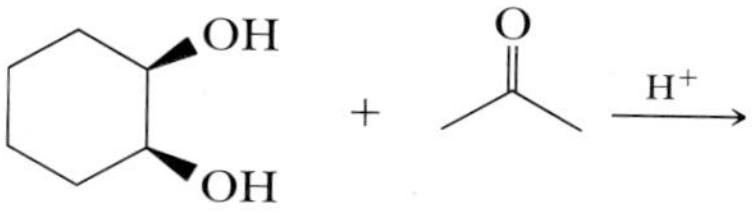

cis-1,2-Cyclohexanediol   Acetone

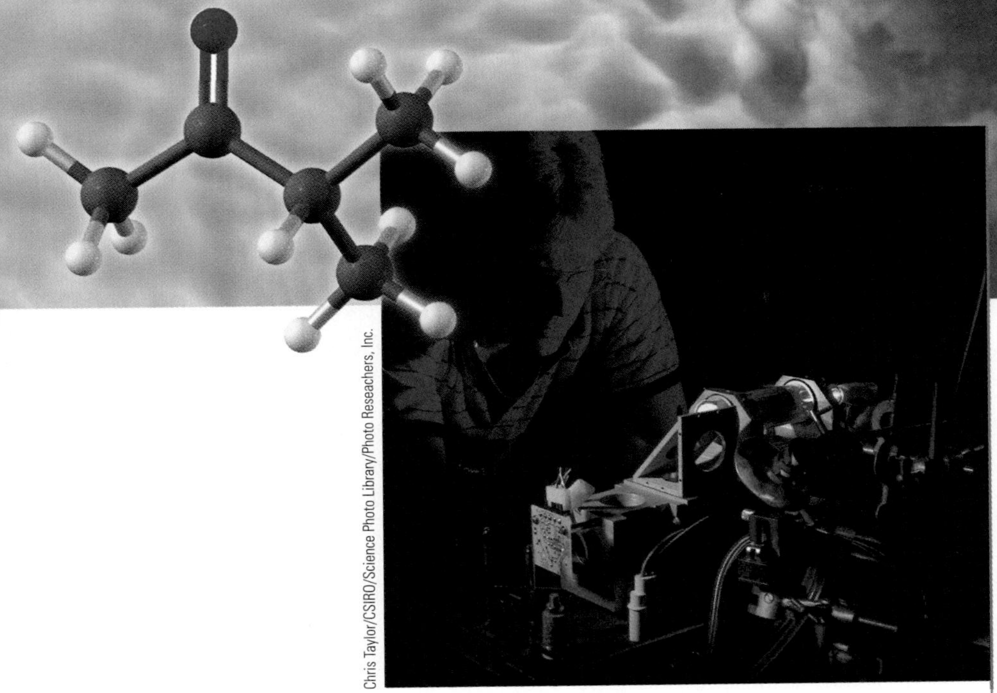

Chris Taylor/CSIRO/Science Photo Library/Photo Researchers, Inc.

12

■ A scientist working with a Fourier transform infrared spectrometer. Inset: A model of 3-methyl-2-butanone. For an IR spectrum of this compound, see Figure 12.2.

Outline

Infrared Spectroscopy

Determination of molecular structure is one of the central themes of organic chemistry. For this purpose, chemists today rely almost exclusively on instrumental methods, four of which we discuss in this text. We begin in this chapter with infrared (IR) spectroscopy. Then, in Chapters 13 and 14, we introduce nuclear magnetic resonance (NMR) spectroscopy and mass spectrometry (MS). A brief introduction to ultraviolet-visible spectroscopy is contained in Chapter 20, which deals with conjugated systems.

12.1 Electromagnetic Radiation

Gamma rays, x-rays, ultraviolet light, visible light, infrared radiation, microwaves, and radio waves are all types of **electromagnetic radiation.** Because electromagnetic radiation behaves as a wave traveling at the speed of light, it can be described in terms of its wavelength and its frequency.

Table 12.1 summarizes **wavelengths (λ), frequencies,** and energies of various regions of the electromagnetic spectrum. The wavelengths of visible light fall in the range 400–700 nm. Infrared radiation (which can be felt as heat but not seen) falls in the range 2–15 μm.

Frequency, the number of full cycles of a wave that pass a given point in a second, is given the symbol ν (Greek nu) and is reported in **hertz (Hz),** which has the units s^{-1}. Wavelength and frequency are inversely

Electromagnetic radiation Light and other forms of radiant energy.

Wavelength (λ) The distance between consecutive peaks on a wave.

Frequency The number of full cycles of a wave that pass a given point in a second; it is given the symbol ν (Greek nu) and reported in **hertz (Hz)**, which has the units s^{-1}.

Hertz (Hz) The unit in which frequency is measured: s^{-1} (read "per second").

ORGANIC Chemistry•Now™
Look for this logo in the chapter and go to Organic ChemistryNow at **http://now.brookscole.com/bfi4** for tutorials, simulations, and problems.

Table 12.1   Wavelengths, Frequencies, and Energies of Some Regions of the Electromagnetic Spectrum

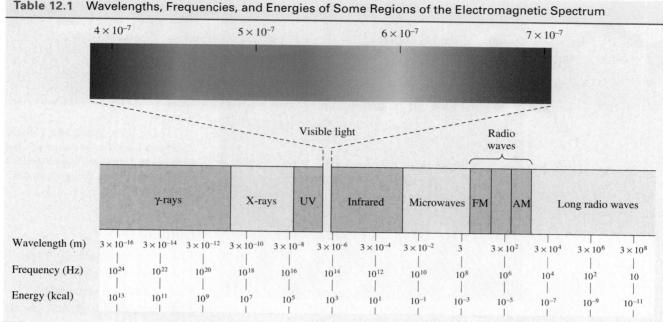

Wavelength (m)	3×10^{-16}	3×10^{-14}	3×10^{-12}	3×10^{-10}	3×10^{-8}	3×10^{-6}	3×10^{-4}	3×10^{-2}	3	3×10^{2}	3×10^{4}	3×10^{6}	3×10^{8}
Frequency (Hz)	10^{24}	10^{22}	10^{20}	10^{18}	10^{16}	10^{14}	10^{12}	10^{10}	10^{8}	10^{6}	10^{4}	10^{2}	10
Energy (kcal)	10^{13}	10^{11}	10^{9}	10^{7}	10^{5}	10^{3}	10^{1}	10^{-1}	10^{-3}	10^{-5}	10^{-7}	10^{-9}	10^{-11}

proportional, and one can be calculated from the other using the following relationship:

$$\lambda \nu = c$$

where c is the velocity of light, 3.00×10^{8} m/s. For example, consider the infrared radiation of wavelength 1.5×10^{-5} m (15 μm). The frequency of this radiation is 2.0×10^{13} Hz.

$$\nu = \frac{3 \times 10^{8}\,\text{m/s}}{1.5 \times 10^{-5}\,\text{m}} = 2.0 \times 10^{13}\,\text{Hz}$$

An alternative way to describe electromagnetic radiation is in terms of its properties as a stream of particles called **photons.** The energy in a mole of photons is related to the frequency of the radiation by the equations

$$E = h\nu = h\frac{c}{\lambda}$$

Table 12.2   Common Units Used to Express Wavelength (λ)

Unit	Relation to Meter
Meter (m)	—
Millimeter (mm)	$1\ \text{mm} = 10^{-3}\ \text{m}$
Micrometer (μm)	$1\ \mu\text{m} = 10^{-6}\ \text{m}$
Nanometer (nm)	$1\ \text{nm} = 10^{-9}\ \text{m}$
Angstrom (Å)	$1\ \text{Å} = 10^{-10}\ \text{m}$

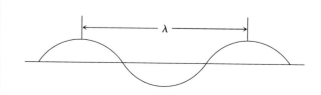

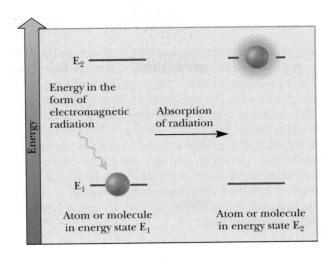

Figure 12.1
Absorption of energy in the form of electromagnetic radiation causes an atom or molecule in energy state E_1 to change to a higher energy state E_2.

where E is the energy in kJ (kcal)/mol and h is Planck's constant, 3.99×10^{-13} kJ (9.537×10^{-14} kcal) $\cdot$ s $\cdot$ mol^{-1}.

Wavelength is usually expressed in the SI base unit of meters. Other derived units commonly used to express wavelength are given in Table 12.2.

12.2 Molecular Spectroscopy

An atom or molecule can be made to undergo a transition from energy state E_1 to a higher energy state E_2 by irradiating the atom or molecule with electromagnetic radiation corresponding to the energy difference between states E_1 and E_2 as illustrated schematically in Figure 12.1. When the atom or molecule returns from state E_2 to state E_1, an equivalent amount of energy is emitted.

Molecular spectroscopy is the experimental process of measuring which frequencies of radiation are absorbed or emitted by a particular substance and then attempting to correlate patterns of energy absorption or emission with details of molecular structure. Table 12.3 summarizes the regions of the electromagnetic spectrum of most interest to us and the relationships of each to changes in atomic and molecular energy levels.

Molecular spectroscopy The study of which frequencies of radiation are absorbed or emitted by a particular substance and the correlation of these frequencies with details of molecular structure.

Table 12.3	Types of Energy Transitions Resulting from Absorption of Energy from Three Regions of the Electromagnetic Spectrum		
Region of Electromagnetic Spectrum	**Frequency (hertz)**	**Type of Spectroscopy**	**Absorption of Electromagnetic Radiation Results in Transitions Between**
Radio frequency	$3 \times 10^7 - 9 \times 10^8$	Nuclear magnetic resonance	Nuclear spin levels
Infrared	$1 \times 10^{13} - 1 \times 10^{14}$	Infrared	Vibrational energy levels
Ultraviolet-visible	$2.5 \times 10^{14} - 1.5 \times 10^{15}$	Ultraviolet-visible	Electronic energy levels

12.3 Infrared Spectroscopy

Infrared spectroscopy provides a direct method of detection of certain functional groups in a molecule. In this chapter, we first develop a basic understanding of the theory behind infrared spectroscopy, and then we concentrate on the interpretation of spectra and the information they can provide about details of molecular structure. Infrared spectroscopy depends on the absorption of infrared light in the wavelength range 2.5×10^{-6} to 2.5×10^{-5} m by vibrating bonds within molecules. Infrared spectroscopy is useful to organic chemists not only for the determination of molecular structure but also for many other applications. For example, infrared spectroscopy is used to identify certain illegal substances in forensic science, and a type of infrared frequency measurement is used as an important probe of atmospheric and mineral composition in space exploration programs such as the recent Mars rover missions.

A. The Vibrational Infrared Spectrum

Infrared (IR) Spectroscopy A spectroscopic technique in which a compound is irradiated with infrared radiation, absorption of which causes covalent bonds to change from a lower vibration state to a higher one. Infrared spectroscopy is particularly valuable for determining the kinds of functional groups present in a molecule.

Vibrational infrared region The portion of the infrared region that extends from 4000 to 400 cm^{-1}.

Wavenumbers $\overline{\nu}$ The frequency of electromagnetic radiation expressed as the number of waves per centimeter, with units cm^{-1} (read: reciprocal centimeters).

Organic molecules are flexible. As we discussed in Chapter 2, atoms and groups of atoms can rotate about single covalent bonds. In addition, covalent bonds can stretch and bend just as if their atoms were joined by flexible springs. **Infrared spectroscopy,** also called **IR spectroscopy,** probes stretching and bending vibrations of organic molecules.

The **vibrational infrared region,** which extends from 2.5×10^{-6} to 2.5×10^{-5} m in wavelength, is used for infrared spectroscopy. Radiation in this region is most commonly referred to by its frequency in **wavenumbers,** $\overline{\nu}$, the number of waves per centimeter, with units cm^{-1} (read: reciprocal centimeters). The frequency in wavenumbers is the reciprocal of the wavelength in centimeters, or the frequency (ν) in hertz divided by c, the speed of light.

$$\overline{\nu} = \frac{1}{\lambda \text{ (cm)}} = \frac{10^{-2} \text{ (m·cm}^{-1})}{\lambda \text{ (m)}} = \frac{\nu}{c}$$

When expressed in frequencies, the vibrational region of the infrared spectrum extends from 4000 to 400 cm^{-1}.

$$\overline{\nu} = \frac{10^{-2} \text{ m·cm}^{-1}}{2.5 \times 10^{-6} \text{ m}} = 4000 \text{ cm}^{-1} \qquad \overline{\nu} = \frac{10^{-2} \text{ m·cm}^{-1}}{2.5 \times 10^{-5} \text{ m}} = 400 \text{ cm}^{-1}$$

An advantage of using frequencies is that they are directly proportional to energy; the higher the frequency, the higher the energy of the radiation.

Figure 12.2 is an infrared spectrum of 3-methyl-2-butanone. The horizontal axis at the bottom of the chart paper is calibrated in frequency (wavenumbers, cm^{-1}); that at the top is calibrated in wavelength (micrometers, μm). The frequency scale is often divided into two or more regions. For all spectra reproduced in this text, it is divided into three linear regions: 4000–2200 cm^{-1}, 2200–1000 cm^{-1}, and 1000–400 cm^{-1}. The vertical axis measures transmittance (the fraction of light transmitted), with 100% at the top and 0% at the bottom. Thus, the baseline for an infrared spectrum (100% transmittance of radiation through the sample, 0% absorption) is at the top of the chart paper, and absorption of radiation corresponds to a trough or valley. Strange as it may seem, we commonly refer to infrared absorptions as peaks, even though they are conventionally displayed pointing downward.

The spectrum in Figure 12.2 was recorded using a neat sample, which means the pure liquid. A few drops of the liquid are placed between two sodium chloride discs

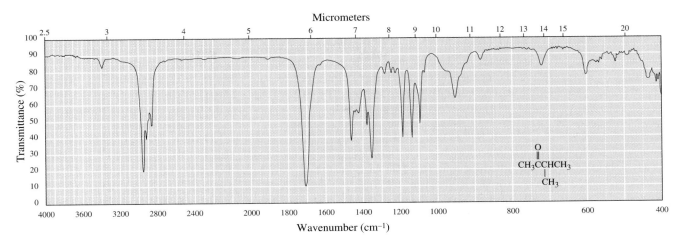

and spread to give a thin film through which infrared radiation is then passed. Liquid and solid samples may also be dissolved in carbon tetrachloride or another solvent with minimal infrared absorption and analyzed in a liquid sampling cell. Still another way to obtain the infrared spectrum of a solid is to mix it with potassium bromide and then compress the mixture into a thin transparent wafer, which is placed in the beam of the spectrophotometer. Both NaCl and KBr are transparent to infrared radiation and can be used as windows and optics for infrared spectroscopy. (These materials are very easily damaged by moisture.) Infrared spectra of gas samples are determined using specially constructed gas-handling cells.

Example 12.1

Some infrared spectrophotometers are calibrated to record spectra on an ordinate that is linear in wavelength (μm), whereas others record them on an ordinate that is linear in frequency (cm^{-1}). Carry out the following conversions (note the convenient formula for converting between μm and cm^{-1}: $\bar{\nu}\lambda = 10^4$).

(a) 7.05 μm to cm^{-1} **(b)** 3.35 μm to cm^{-1} **(c)** 3280 cm^{-1} to μm

Solution

(a) 1418 cm^{-1} **(b)** 2985 cm^{-1} **(c)** 3.05 μm

Problem 12.1

Which is higher in energy?

(a) Infrared radiation of 1715 cm^{-1} or of 2800 cm^{-1}?
(b) Radio-frequency radiation of 300 MHz or of 60 MHz?

B. Molecular Vibrations

Atoms joined by covalent bonds are not permanently fixed in one position but rather undergo continual vibrations relative to each other. The energies associated with these vibrations are quantized, which means that, within a molecule, only specific vibrational energy levels are allowed. The energies associated with transitions between vibrational energy levels in most covalent molecules correspond to frequencies in the infrared region, $4000-400$ cm^{-1}.

For a molecule to absorb this radiation, its vibration must cause *a periodic change in the bond dipole moment.* If two charges are connected by a spring, a change in distance between the charges corresponds to a change in dipole moment. In general, the greater the bond dipole, the greater the change in dipole moment caused by a vibration. Any vibration that leads to a substantial change in dipole moment is said to be **infrared active.** The greater the change is, the more intense the absorption will be. Covalent bonds whose vibration does not result in a change in bond dipole moment, for example, as a result of symmetry in the molecule, are said to be infrared inactive. The carbon-carbon double and triple bonds in symmetrically substituted alkenes and alkynes, for example 2,3-dimethyl-2-butene and 2-butyne, do not absorb infrared radiation because vibration does not result in a substantial bond dipole change.

Infrared active Any molecular vibration that leads to a substantial change in dipole moment so is observed in an IR spectrum.

$$H_3C \quad\quad CH_3$$
$$\diagdown\;\diagup$$
$$C{=}C \quad\quad\quad H_3C{-}C{\equiv}C{-}CH_3$$
$$\diagup\;\diagdown$$
$$H_3C \quad\quad CH_3$$

2,3-Dimethyl-2-butene   2-Butyne

For a nonlinear molecule containing n atoms, $3n - 6$ allowed fundamental vibrations exist. For a molecule as simple as ethanol, CH_3CH_2OH, there are 21 fundamental vibrations, and for hexanoic acid, $CH_3(CH_2)_4COOH$, there are 54. Thus, even for relatively simple molecules, a large number of vibrational energy levels exist, and the patterns of energy absorption for these and larger molecules are very complex.

The simplest vibrational motions in molecules giving rise to absorption of infrared radiation are stretching and bending motions. Figure 12.3 illustrates the fundamental stretching and bending vibrations for a methylene group.

A different technique called **Raman spectroscopy** is complementary to infrared spectroscopy in that infrared-inactive vibrations are seen in Raman spectra, while Raman-inactive vibrations are the ones that are infrared active.

Raman spectroscopy A vibrational molecular spectroscopy that is complementary to infrared (IR) spectroscopy in that infrared inactive vibrations are seen in Raman spectra.

C. Characteristic Absorption Patterns

Analysis of the modes of vibration for a molecule is very complex because all the atoms contribute to the vibrational modes. However, we can make useful generalizations about where absorptions due to particular vibrational modes will appear in an

Figure 12.3
Fundamental stretching and bending vibrations for a methylene group.

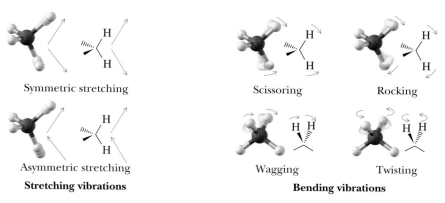

Symmetric stretching   Scissoring   Rocking

Asymmetric stretching   Wagging   Twisting

Stretching vibrations   **Bending vibrations**

infrared spectrum by considering each individual bond and ignoring other bonds in the molecule. As a simplifying assumption, let us consider two covalently bonded atoms as two vibrating masses connected by a spring. The total energy is proportional to the frequency of vibration. The frequency of a stretching vibration is given by the following equation, which is derived from Hooke's law for a vibrating spring.

$$\bar{\nu} = 4.12 \sqrt{\frac{K}{\mu}}$$

Here $\bar{\nu}$ is the frequency of the vibration in wavenumbers (cm^{-1}); K is the force constant of the bond, a measure of the bond's strength, in dynes per centimeter; and μ is the "reduced mass" of the two atoms, $(m_1 m_2)/(m_1 + m_2)$, where m is the mass of the atoms in grams.

Force constants for single, double, and triple bonds are approximately 5, 10, and 15×10^5 dynes/cm, respectively, thus approximately in the ratio $1 : 2 : 3$. Using the value for the force constant for a single bond, we calculate the frequency for the stretching vibration of a single bond between ^{12}C and ^1H as follows:

For ^{12}C—^1H stretching:

$$\text{Reduced mass} = 12 \times 1 /(12 + 1) = 0.923 \text{ g/atom}$$

and

$$\bar{\nu} = 4.12 \sqrt{\frac{K}{\mu}} = 4.12 \sqrt{\frac{5 \times 10^5}{0.923}} = 3032 \text{ cm}^{-1}$$

The experimentally determined value for the frequency of an alkyl C—H stretching vibration is approximately 3000 cm^{-1}. Given the simplifying assumptions made in this calculation and the fact that the value of the force constant for a single bond is an average value, the agreement between the calculated value and the experimental value is remarkably good. Although frequencies calculated in this manner can be close to the experimental values, they are generally not accurate enough for precise determination of molecular structure.

Hooke's law predicts that the *position* of the absorption of a stretching vibration in an IR spectrum depends both on the strength of the vibrating bond and on the masses of the atoms connected by the bond. The stronger the bond is and the lighter the atoms are, the higher the frequency of the stretching vibration will be. As we saw earlier, the *intensity* of an absorption depends primarily on the change in dipole of the vibrating bond.

Example 12.2

Calculate the stretching frequency in wavenumbers for a carbon-carbon double bond. Assume that each carbon is the most abundant isotope, namely ^{12}C.

Solution

Assume a force constant of 10×10^5 dynes per centimeter for C=C. The calculated frequency is 1682 cm^{-1}, a value close to the experimental value of 1650 cm^{-1}.

$$\bar{\nu} = 4.12 \sqrt{\frac{10 \times 10^5}{12 \times 12/(12 + 12)}} = 1682 \text{ cm}^{-1}$$

Problem 12.2

Without doing the calculation, which member of each pair do you expect to occur at the higher frequency?

(a) C=O or C=C stretching   (b) C=O or C—O stretching
(c) C≡C or C=O stretching   (d) C—H or C—Cl stretching

Detailed interpretation of most infrared spectra is difficult because of the complexity of vibrational modes. In addition to the fundamental vibrational modes we have described, other types of absorptions occur, resulting in so-called overtone and coupling peaks that are usually quite weak.

To one skilled in the interpretation of infrared spectra, the absorption patterns can yield an enormous amount of information about chemical structure. However, we have neither the time nor the need to develop this level of competence. The value of infrared spectra for us is that they can be used to determine the presence or absence of certain functional groups. A carbonyl group, for example, typically shows strong absorption at approximately $1630-1820$ cm^{-1}. The position of absorption for a particular carbonyl group depends on whether it is an aldehyde, ketone, carboxylic acid, or ester, or, if it is in a ring, it depends on the size of the ring. In this chapter, we discuss how structural variations, such as ring size or other factors, affect this value.

D. Correlation Tables

Data on absorption patterns of functional groups are collected in tables called **correlation tables.** Table 12.4 lists infrared absorptions for the types of bonds and functional groups we deal with most often. A cumulative correlation table can be found in Appendix 6. In these tables, the intensity of a particular absorption is often referred to as strong (s), medium (m), or weak (w). In general, bonds between C and O where the electronegativity difference is largest have the largest dipole moments and tend to give the strongest infrared absorptions.

In general, we pay most attention to the region from 3500 to 1500 cm^{-1} because the stretching and bending vibrations for most functional groups are found in this region. Vibrations in the region 1500 to 400 cm^{-1} are more complex and far more difficult to analyze. Because even slight variations in molecular structure and absorp-

ORGANIC
Chemistry Now™

Click Tutorials to review **Infrared Stretching Frequencies of Selected Functional Groups**

Table 12.4   Infrared Stretching Frequencies of Selected Functional Groups

Bond	Stretching Frequency (cm^{-1})	Intensity
O—H	3200–3650	Weak to strong (strongest when H-bonded)
N—H	3100–3550	Medium
C—H	2700–3300	Weak to medium
C=C	1600–1680	Weak to medium
C=O	1630–1820	Strong
C—O	1000–1250	Strong

tion patterns are most obvious in this region, it is often called the **fingerprint region.**
If two compounds have even slightly different structures, the differences in their infrared spectra are most clearly discernible in this region.

Fingerprint region Vibrations in the region 1500 to 400 cm^{-1} of IR spectra are complex and difficult to analyze but are characteristic for different molecules.

Example 12.3

What functional group is most likely present if a compound shows IR absorption at these frequencies?

(a) 1705 cm^{-1} **(b)** 2950 cm^{-1}

Solution

(a) A C=O group   **(b)** An aliphatic C—H group

Problem 12.3

A compound shows strong, very broad IR absorption in the region 3300–3600 cm^{-1} and strong, sharp absorption at 1715 cm^{-1}. What functional group accounts for both of these absorptions?

Example 12.4

Propanone and 2-propen-1-ol are constitutional isomers. Show how to distinguish between them by IR spectroscopy.

$$CH_3-\overset{\overset{\textstyle O}{\|}}{C}-CH_3 \qquad CH_2=CH-CH_2-OH$$

Propanone   2-Propen-1-ol
(Acetone)   (Allyl alcohol)

Solution

Only propanone shows strong absorption in the C=O stretching region, 1630–1820 cm^{-1}. Alternatively, only 2-propen-1-ol shows strong absorption in the O—H stretching region, 3200–3650 cm^{-1}.

Problem 12.4

Propanoic acid and methyl ethanoate are constitutional isomers. Show how to distinguish between them by IR spectroscopy.

$$CH_3CH_2\overset{\overset{\textstyle O}{\|}}{C}OH \qquad CH_3\overset{\overset{\textstyle O}{\|}}{C}OCH_3$$

Propanoic acid   Methyl ethanoate
                  (Methyl acetate)

12.4 Interpreting Infrared Spectra

A. Alkanes

Infrared spectra of alkanes are usually simple with few peaks, the most common of which are given in Table 12.5.

Table 12.5   Infrared Absorptions of Alkanes, Alkenes, and Alkynes and Arenes

Hydrocarbon	Vibration	Frequency (cm^{-1})	Intensity
Alkane			
C—H	Stretching	2850–3000	Medium
CH$_2$	Bending	1450–1475	Medium
CH$_3$	Bending	1375 and 1450	Weak to medium
C—C	(Not useful for interpretation—too many bands)		
Alkene			
C—H	Stretching	3000–3100	Weak to medium
C=C	Stretching	1600–1680	Weak to medium
Alkyne			
C—H	Stretching	3300	Medium to strong
C≡C	Stretching	2100–2250	Weak
Arene			
C—H	Stretching	3030	Weak to medium
C=C	Stretching	1450–1600	Medium
C—H	Bending	690–900	Strong

Figure 12.4 is an infrared spectrum of decane. The strong peak with multiple splitting between 2850 and 3000 cm^{-1} is characteristic of alkane C—H stretching; it is strong in this spectrum because there are so many C—H bonds and no other functional groups. The other prominent peaks correspond to methylene bending at 1465 cm^{-1} and methyl bending at 1380 cm^{-1}.

B. Alkenes

An easily recognized alkene absorption is the vinylic C—H stretching slightly to the left of 3000 cm^{-1} (i.e., higher frequency). Also characteristic of alkenes is C=C stretching at 1600–1680 cm^{-1}. This vibration, however, is often weak and difficult to observe: The more symmetrical the alkene is, the weaker the absorption will be. Both vinylic C—H stretching and C=C stretching can be seen in the infrared spectrum of

Figure 12.4

Infrared spectrum of decane (neat, salt plates).

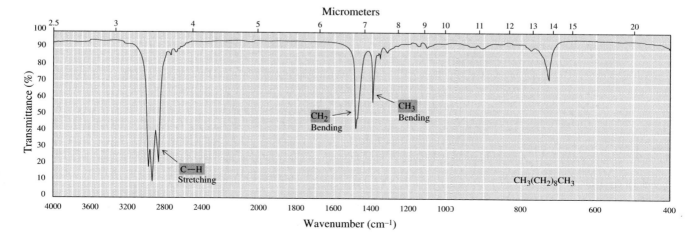

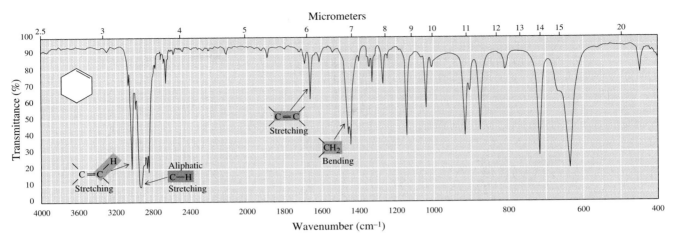

Figure 12.5

Infrared spectrum of cyclo-hexene (neat, salt plates).

cyclohexene (Figure 12.5). Also visible are the aliphatic C—H stretching near 2900 cm^{-1} and methylene bending near 1440 cm^{-1}.

C. Alkynes

Alkyne ≡C—H stretching occurs near 3300 cm^{-1}, at higher frequency than for either alkyl —C—H or vinylic =C—H stretching. This peak is usually sharp and strong. The (sp-1s) CH bond is unusually strong and therefore has a higher force constant than alkene (sp^2-1s) C—H bonds, which in turn absorb at higher frequency than the even weaker alkane (sp^3-1s) C—H bonds. Also observed in terminal alkynes is absorption near 2150 cm^{-1} owing to C≡C stretching. Both of these peaks can be seen in the infrared spectrum of 1-octyne (Figure 12.6). For internal alkynes, the C≡C stretching absorption is often very weak or completely absent (in symmetric alkynes) because stretching of this bond results in little or no change in the bond dipole moment (Section 12.3B).

D. Arenes (Benzene and Its Derivatives)

Aromatic rings show a medium to weak peak in the C—H stretching region at approximately 3030 cm^{-1} characteristic of (sp^2-1s) =C—H bonds. In addition, aromatic rings show strong absorption in the region 690–900 cm^{-1} as a result of

Figure 12.6

Infrared spectrum of 1-octyne.

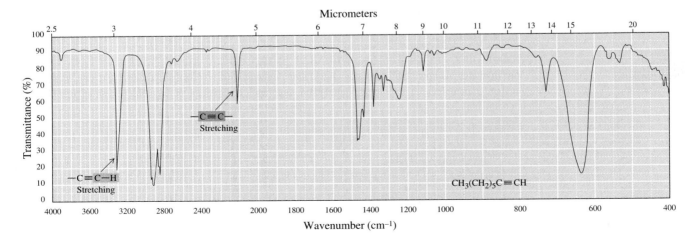

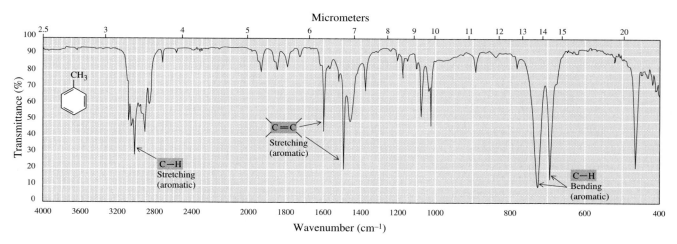

Figure 12.7
Infrared spectrum of toluene.

out-of-plane C—H bending. Finally, these compounds show several absorptions owing to C=C stretching between 1450 and 1600 cm^{-1}. Actually, these are complex vibrational modes of the entire ring. Some modes involve all atoms moving in and out (breathing), whereas others involve some atoms moving in and others moving out. The intensities of these peaks can vary depending on the symmetry of ring substitution patterns. Each of these characteristic absorption patterns can be seen in the infrared spectrum of toluene (Figure 12.7).

E. Alcohols

Both the position of the O—H stretching absorption and its intensity depend on the extent of hydrogen bonding. Under conditions where there is extensive hydrogen bonding between alcohol molecules (in pure alcohol or in concentrated solutions of the alcohol), the O—H stretching absorption occurs as a broad peak at 3200–3500 cm^{-1}. The variety of hydrogen-bonded states in different molecules leads to this broadening. The "free" O—H stretch near 3650 cm^{-1} is seen only in very dilute solution in non-hydrogen-bonding solvents. The C—O stretching absorption appears in the range 1000–1250 cm^{-1} (Table 12.6).

Shown in Figure 12.8 is an infrared spectrum of neat 1-hexanol. The hydrogen-bonded O—H stretching appears as a broad band of strong intensity centered at 3340 cm^{-1}. The C—O stretching appears at 1058 cm^{-1}, a value characteristic of primary alcohols.

F. Ethers

Ethers have an oxygen atom bonded to two carbon atoms. Either or both of the carbon atoms may be sp^3 hybridized, sp^2 hybridized, or sp hybridized. In the simplest ether, dimethyl ether, both carbons are sp^3 hybridized. In diphenyl ether, both

Table 12.6 Infrared Absorptions of Alcohols

Bond	Frequency (cm^{-1})	Intensity
O—H (free)	3600–3650	Weak
O—H (hydrogen bonded)	3200–3500	Medium, broad
C—O	1000–1250	Medium

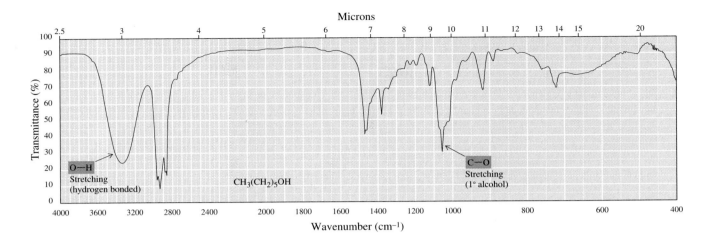

Figure 12.8

Infrared spectrum of 1-hexanol (neat, salt plates).

carbons are sp^2 hybridized, and in ethyl vinyl ether, one carbon is sp^3 hybridized and the other is sp^2 hybridized.

$$CH_3CH_2—O—CH_2CH_3$$

$$CH_3CH_2—O—CH=CH_2$$

Ethoxyethane
(Diethyl ether)

Diphenyl ether

Ethoxyethene
(Ethyl vinyl ether)

The C—O stretching absorptions of ethers are similar to those observed in alcohols. Dialkyl ethers typically show a single absorption in the region between 1000 and 1100 cm^{-1} as can be seen in the infrared spectrum of dibutyl ether (Figure 12.9).

Aromatic ethers (compounds in which the ether oxygen is bonded to one or more benzene rings) and vinyl ethers (compounds in which the ether oxygen is bonded to one or more sp^2 hybridized carbon of a C=C bond) typically show two C—O stretching vibrations, one at either end of the range for C—O stretching. Anisole (Figure 12.10), for example, shows C—O stretching vibrations at 1050 cm^{-1} (sp^3 C—O) and 1250 cm^{-1} (sp^2 C—O). Ethers in which one of the bonds is attached

Figure 12.9

Infrared spectrum of dibutyl ether.

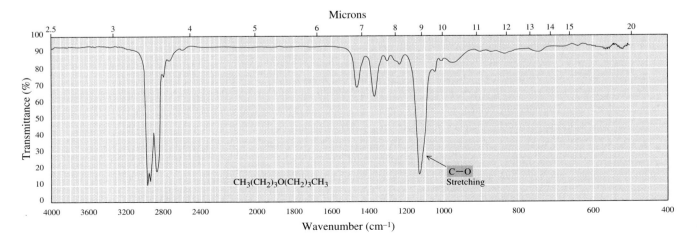

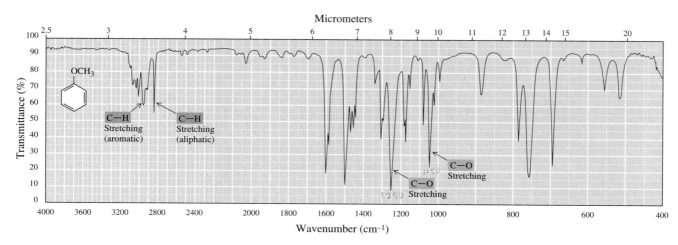

Figure 12.10

Infrared spectrum of anisole.

to an sp^2 hybridized carbon typically also have a band in the region between 1200 and 1250 cm^{-1}.

The presence or absence of O—H stretching at 3200 to 3500 cm^{-1} for hydrogen-bonded O—H can be used to distinguish between an ether and an isomeric alcohol. A C—O stretching absorption is also present in esters (see Section 12.4J). In this case, the presence or absence of C=O stretching can be used to distinguish between an ether and an ester.

G. Amines

The most important and readily observed infrared absorptions of primary and secondary amines are the result of N—H stretching vibrations and appear in the region 3300–3500 cm^{-1}. Like O—H bonds, N—H bonds become broader and shift to longer wavelength when they take part in hydrogen bonding. Primary amines have two bands in this region: one caused by symmetric stretching and the other by asymmetric stretching. The two N—H stretching absorptions characteristic of a primary amine can be seen in the IR spectrum of 1-butanamine (Figure 12.11). Secondary

Figure 12.11

Infrared spectrum of 1-butanamine, a primary amine.

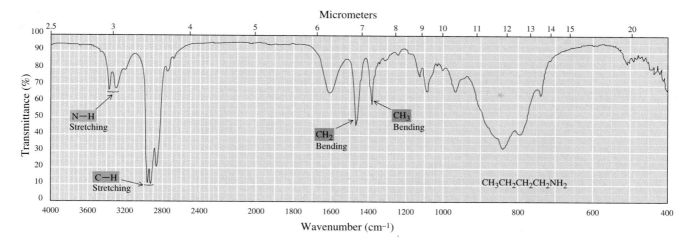

amines give only one absorption in this region. Tertiary amines have no N—H and, therefore, are transparent in this region of the infrared spectrum.

H. Aldehydes and Ketones

Aldehydes and ketones show characteristic strong infrared absorption between 1630 and 1820 cm^{-1} associated with the stretching vibration of the carbon-oxygen double bond. The stretching vibration for the carbonyl group of menthone occurs at 1705 cm^{-1} (Figure 12.12).

Because few other bond vibrations absorb energy between 1630 and 1820 cm^{-1}, absorption in this region of the spectrum is a reliable means for confirming the presence of a carbonyl group. Because several different functional groups contain a carbonyl group, it is often not possible to tell from absorption in this region alone whether the carbonyl-containing compound is an aldehyde, a ketone, a carboxylic acid (Section 12.4I), an ester (Section 12.4J), an amide (Section 12.4J), or an anhydride (Section 12.4J). However, other absorptions such as the C—O stretch in esters can help distinguish these groups. Aldehydes frequently have a weak but very distinctive absorption at 2720 cm^{-1} caused by the stretching of the C—H of the CHO group.

The position of the C=O stretching vibration is quite sensitive to the molecular environment of the carbonyl group, as illustrated by comparing these cycloalkanones. Cyclohexanone, which has very little angle strain shows absorption at 1715 cm^{-1}. As ring size decreases and angle strain increases, the C=O absorption shifts to a higher frequency as shown in the following series.

| 1715 cm^{-1} | 1745 cm^{-1} | 1780 cm^{-1} | 1850 cm^{-1} |

The presence of an adjacent carbon-carbon double bond or benzene ring in **conjugation** with the carbonyl group results in a shift of the C=O absorption to a lower frequency, as seen by comparing the carbonyl stretching frequencies of the following molecules. Conjugation occurs when the pi electrons of adjacent pi bonds

Conjugation A situation in which two multiple bonds are separated by a single bond. Alternatively, a series of overlapping *p* orbitals. 1,3-Butadiene, for example, is a conjugated diene, and 3-butene3-one is a conjugated enone.

Figure 12.12
Infrared spectrum of menthone.

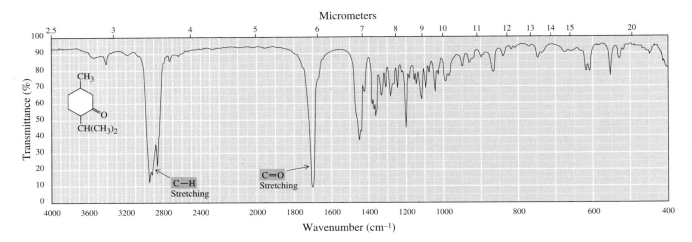

interact with each other and is a consequence of the special properties of extended pi systems (Chapter 20). Conjugation also moves the C=C stretch to the right (lower frequency) and increases its intensity.

1717 cm^{-1} $\qquad$ 1690 cm^{-1} $\qquad$ 1700 cm^{-1}

I. Carboxylic Acids

A carboxyl group gives rise to two characteristic absorptions in the infrared spectrum. One of these occurs in the region $1700-1725 \text{ cm}^{-1}$ and is associated with the stretching vibration of the carbonyl group. This range of absorption is essentially the same as that for the carbonyl group of aldehydes and ketones, but it is usually broader in the case of the carboxyl carbonyl because of intermolecular hydrogen bonding. The other infrared absorption characteristic of a carboxyl group is a peak between 2500 and 3300 cm^{-1} owing to the stretching vibration of the O—H group, which often overlaps the C—H stretching absorptions. The O—H absorption is generally very broad as a result of hydrogen bonding between molecules of the carboxylic acid. Both C=O and O—H stretches can be seen in the infrared spectrum of pentanoic acid in Figure 12.13.

J. Derivatives of Carboxylic Acids

In Chapter 18, you will learn about several important derivatives of carboxylic acids. The most important are amides, esters, anhydrides, and nitriles.

$\overset{\displaystyle O}{\overset{\|}{RCNH_2}}$	$\overset{\displaystyle O}{\overset{\|}{RCOR'}}$	$\overset{\displaystyle O\ \ O}{\overset{\| \ \ \|}{RCOCR'}}$	$RC{\equiv}N$
An amide	An ester	An acid anhydride	A nitrile

Figure 12.13

Infrared spectrum of pentanoic acid.

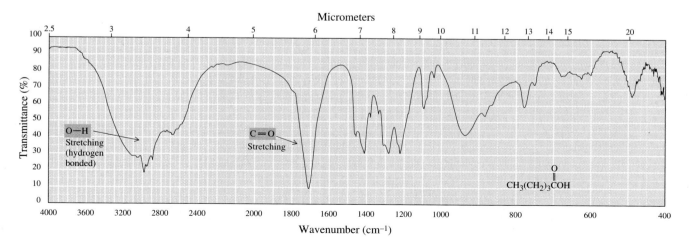

The most important infrared absorption of carboxylic acids and their derivatives is the result of the C=O stretching vibration. Infrared spectroscopic data for these derivatives and the other carbonyl-containing compounds are summarized in Table 12.7.

The carbonyl stretching of amides occurs at 1630–1680 cm^{-1}, at a lower frequency than for other carbonyl compounds. Primary and secondary amides show N—H stretching in the region 3200–3400 cm^{-1}; primary amides ($RCONH_2$) usually show two N—H absorptions, whereas secondary amides (RCONHR) show only a single N—H absorption (Figure 12.14).

Table 12.7   Infrared Absorptions of Molecules Containing Carbonyl Groups

Carbonyl Group		Vibration	Frequency (cm^{-1})	Intensity
O‖ RCR′	Ketones (Section 16.1)			
	C=O	Stretching	1630–1820	Strong
O‖ RCH	Aldehydes (Section 16.1)			
	C=O	Stretching	1630–1820	Strong
	C—H	Stretching	2720	Weak
O‖ RCOH	Carboxylic Acids (Section 17.1)			
	C=O	Stretching	1700–1725	Strong
	O—H	Stretching	2500–3300	Strong (broad)
O‖ RCNH₂	Amides (Section 18.1D)			
	C=O	Stretching	1630–1680	Strong
	N—H	Stretching	3200, 3400	Medium
	(1° amides have two N—H stretches)			
	(2° amides have one N—H stretch)			
O‖ RCOR′	Carboxylic Esters (Section 18.1C)			
	C=O	Stretching	1735–1800	Strong
	sp^2 C—O	Stretching	1200–1250	Strong
	sp^3 C—O	Stretching	1000–1100	Strong
O O‖ ‖ RCOCR	Acid Anhydrides (Section 18.1B)			
	C=O	Stretching	1740–1760 and 1800–1850	Strong
	C—O	Stretching	900–1300	Strong
RC≡N	Nitriles (Section 18.1E)			
	C≡N	Stretching	2200–2250	Medium

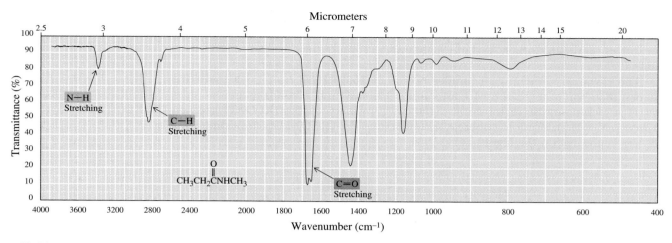

Figure 12.14

Infrared spectrum of
N-methylpropanamide (a
secondary amide).

The functional group of a carboxylic ester, most commonly referred to as simply an ester, is a carbonyl group bonded to an —OR group.

$$CH_3\overset{\overset{\displaystyle O}{\|}}{C}{-}OCH_2CH_3 \qquad CH_3(CH_2)_3\overset{\overset{\displaystyle O}{\|}}{C}{-}OCH(CH_3)_2$$

Ethyl ethanoate   Isopropyl pentanoate
(Ethyl acetate)

Note that one of the carbons of the C—O—C group is sp^2 hybridized and the other is sp^3 hybridized.

Esters display strong C=O stretching absorption in the region between 1735 and 1800 cm^{-1}. As in ketones, this band is shifted to higher frequency in smaller rings and to lower frequency by adjacent double bonds. In addition, esters also display strong C—O stretching absorptions in the region 1000–1100 cm^{-1} for the sp^3 C—O stretch and 1200–1250 cm^{-1} for the sp^2 C—O stretch (Figure 12.15). Ethers in which one of the carbons attached to oxygen is sp^2 hybridized also show this band.

Figure 12.15

Infrared spectrum of ethyl
butanoate.

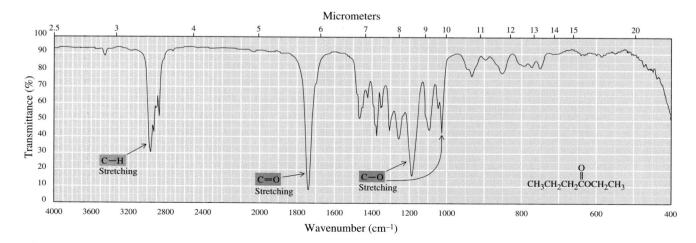

Anhydrides have two carbonyl stretching absorptions, one near 1760 cm^{-1} and the other near 1810 cm^{-1}. In addition, anhydrides display strong C—O stretching absorption in the region 900–1300 cm^{-1}. Nitriles can be distinguished by a strong C≡N stretching absorption at 2200–2250 cm^{-1}.

12.5 Solving Infrared Spectral Problems

The following steps may prove helpful as a systematic approach to solving IR problems.

Step 1: Check the region around 3000 cm^{-1}. Absorption in this region is caused by C—H stretching. Absorption is generally to the right of 3000 cm^{-1} for sp^3 C—H stretching of alkanes and to the left for the sp^2 C—H stretching of alkenes and aromatic rings.

Step 2: Is there a strong, broad band in the region of 3500 cm^{-1}? If yes, then the molecule contains an —OH group either of an alcohol or a carboxylic acid. If there is no absorption around 1700 cm^{-1} (caused by a carbonyl group), then the functional group is an —OH group of an alcohol. If there is a peak around 1700 cm^{-1}, then the functional group may be a carboxyl group. One or two peaks in the 3500 cm^{-1} region at somewhat lower frequency than for —OH may indicate a 2° or 1° amine, respectively.

Step 3: Is there a sharp peak in the region 1630–1820 cm^{-1}? If yes, there is a C=O group present. This peak will probably be the strongest peak in the spectrum. If no peak is present in this region, there is no C=O present. The type of carbonyl-containing functional group can often be determined by looking for the presence or absence of an aldehyde C—H stretch at 2720 cm^{-1}, a carboxyl O—H stretch around 3500 cm^{-1}, and a C—O stretch around 1000–1250 cm^{-1}.

Finally, here is a note of caution on interpreting infrared spectra. Even though it is possible to obtain a great deal of valuable information about a compound from its infrared spectrum, it is often very difficult to determine its structure based solely on this information, so other types of information should be sought.

Summary

The **vibrational infrared** spectrum (Section 12.3A) extends from 4000 to 400 cm^{-1}. To be **infrared active** (Section 12.3B), a bond must be polar, and its vibration must cause a substantial change in the dipole moment of a bond. There are $3n - 6$ allowed fundamental vibrations for a nonlinear molecule containing n atoms. The simplest vibrations that give rise to absorption of infrared radiation are stretching and bending vibrations. Stretching may be symmetrical or asymmetrical.

The frequency of vibration for an infrared-active bond can be derived from Hooke's law for the vibration of a simple harmonic oscillator such as a vibrating spring (Section 12.3C). Hooke's law predicts that the frequency of vibration increases when (1) the bond strength increases and (2) the reduced mass of the vibrating system decreases.

A correlation table (Section 12.3D) is a list of the absorption patterns of functional groups. The intensity of a peak is referred to as strong (s), medium (m), or weak (w). Bending and stretching vibrations for most functional groups appear in the region 3500–1500 cm^{-1}. The region 1500–400 cm^{-1} is called the **fingerprint region.**

12.5 Following are infrared spectra of methylenecyclopentane and 2,3-dimethyl-2-butene. Assign each compound its correct spectrum.

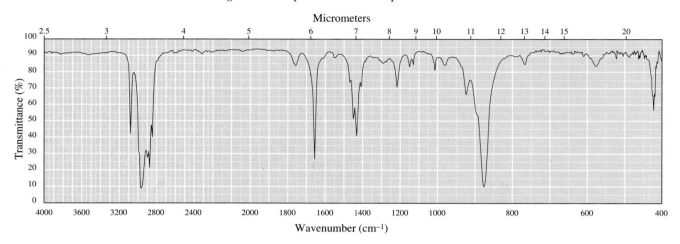

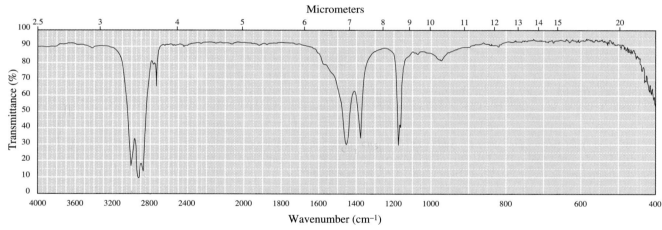

12.6 Following are infrared spectra of nonane and 1-hexanol. Assign each compound its correct spectrum.

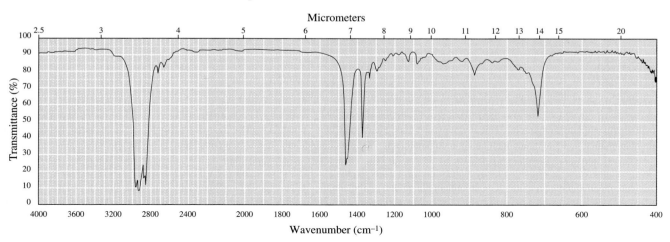

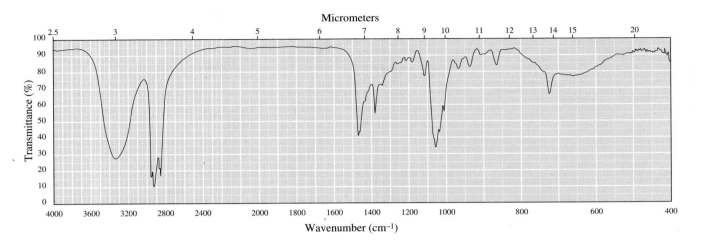

12.7 Following are infrared spectra of 2-methyl-1-butanol and *tert*-butyl methyl ether. Assign each compound its correct spectrum.

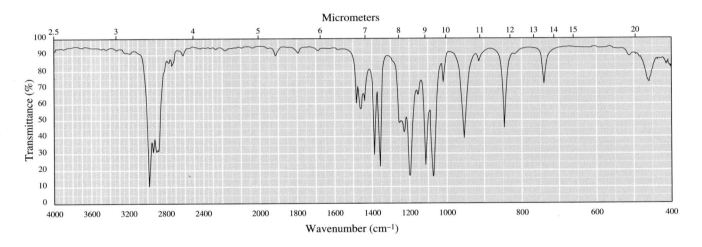

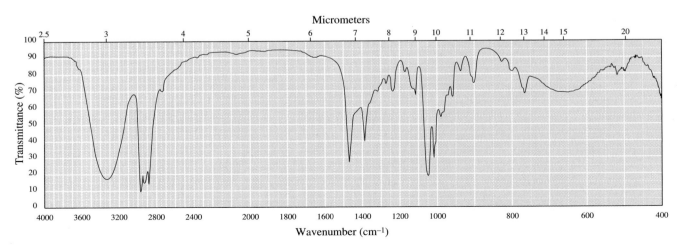

12.8 The IR C≡C stretching absorption in symmetrical alkynes are usually absent. Why is this so?

12.9 Explain the fact that the C—O stretch in ethers and esters occurs at 1000–1100 cm^{-1} when the C is sp^3 hybridized, but at 1250 cm^{-1} when it is sp^2 hybridized.

12.10 A compound has strong infrared absorptions at the following frequencies. Suggest likely functional groups that may be present.

(a) 1735, 1250, and 1100 cm^{-1}
(b) 1745 cm^{-1} but not 1000–1250 cm^{-1}
(c) 1710 and 2500–3400 (broad) cm^{-1}
(d) A single band at about 3300 cm^{-1}
(e) 3600 and 1050 cm^{-1}
(f) 1100 cm^{-1} but not 3300–3650 cm^{-1}

12.11 Show how IR spectroscopy can be used to distinguish between the compounds in each set.

(a) and

(b) and

(c) and

(d) and

(e) and

(f) and

(g) and

(h) and

(i) and

(j) and

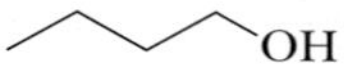

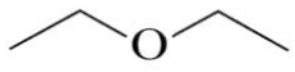

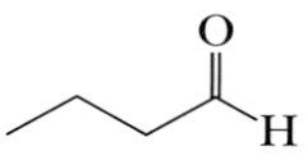

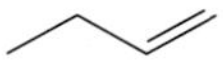

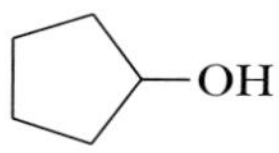

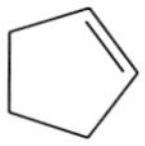

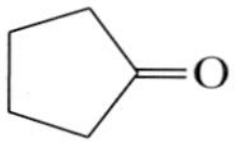

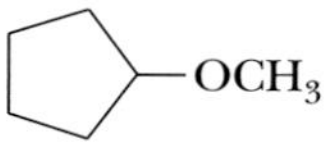

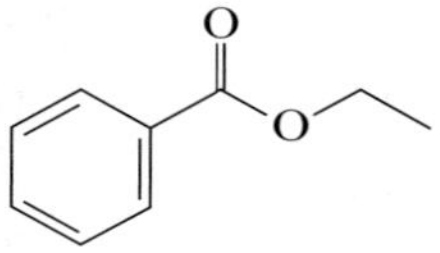

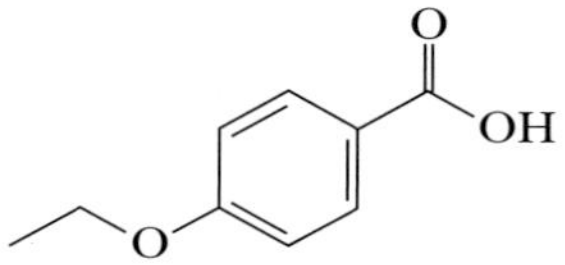

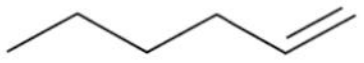

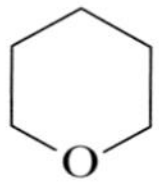

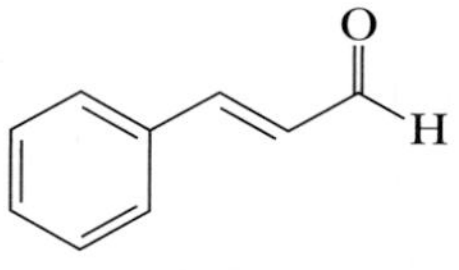

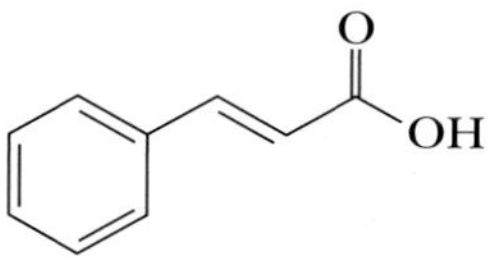

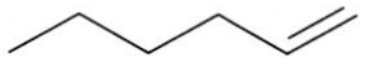

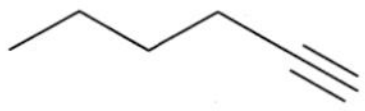

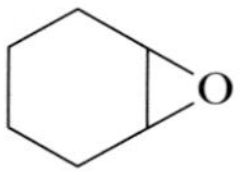

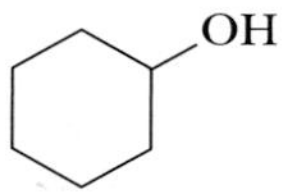

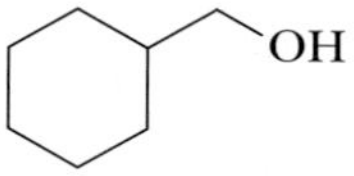

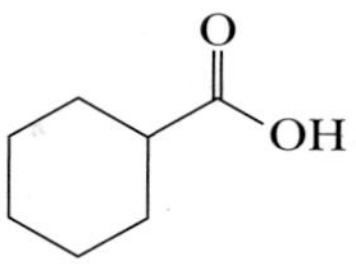

13

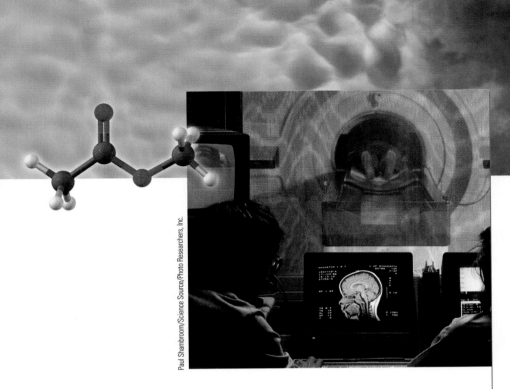

Paul Shambroom/Science Source/Photo Researchers, Inc.

Nuclear Magnetic Resonance Spectroscopy

I n this chapter, we concentrate on absorption of radio-frequency radiation by nuclei and the resulting transitions between energy levels, better known as **nuclear magnetic resonance (NMR) spectroscopy.** Felix Bloch and Edward Purcell, both of the United States, first detected the phenomenon of nuclear magnetic resonance in 1946. They shared the 1952 Nobel Prize for physics. Nuclear magnetic resonance (NMR) spectroscopy was developed in the late 1950s and, within a decade, became the single most important technique available to chemists for the determination of molecular structure. Nuclear magnetic resonance spectroscopy gives us information about the number and types of atoms in a molecule: for example, about hydrogens using ^{1}H-NMR spectroscopy, and about carbons using ^{13}C-NMR spectroscopy. It can also give us substantial information about the connectivity of the atoms and, in many cases, can allow determination of the structure of a molecule with no additional information.

■ Magnetic resonance imaging is a useful medical diagnostic tool. Inset: A model of methyl acetate. For a ^{1}H-NMR spectrum of methyl acetate, see Figure 13.5.

Outlines

ORGANIC
Chemistry⚛Now™

Look for this logo in the chapter and go to Organic ChemistryNow at **http://now.brookscole.com/bfi4** for tutorials, simulations, and problems.

Table 13.1 Spin Quantum Numbers and Allowed Nuclear Spin States for Selected Isotopes of Elements Common to Organic Compounds

Element	^1H	^2H	^{12}C	^{13}C	^{14}N	^{16}O	^{31}P	^{32}S
Nuclear spin quantum number (I)	$\frac{1}{2}$	1	0	$\frac{1}{2}$	1	0	$\frac{1}{2}$	0
Number of spin states	2	3	1	2	3	1	2	1

13.1 Nuclear Spin States

You are already familiar from general chemistry with the concepts that (1) an electron has a spin quantum number of $\frac{1}{2}$, with allowed values of $+\frac{1}{2}$ and $-\frac{1}{2}$, and that (2) a moving charge has an associated magnetic field. In effect, an electron behaves as if it is a tiny bar magnet and has a magnetic moment. The same effect holds for certain atomic nuclei.

Any atomic nucleus that has an odd mass number, an odd atomic number, or both, also has a spin and a resulting nuclear magnetic moment. The allowed nuclear spin states are determined by the spin quantum number, I, of the nucleus. A nucleus with spin quantum number I has $2I + 1$ spin states. Our focus in this chapter is on nuclei of ^1H and ^{13}C, isotopes of the two elements most common to organic compounds. Each has a nuclear spin quantum number of $\frac{1}{2}$ and therefore has $2(\frac{1}{2}) + 1 = 2$ allowed spin states. Quantum numbers and allowed nuclear spin states for these nuclei and those of other elements common to organic compounds are shown in Table 13.1. Note that ^{12}C, ^{16}O and ^{32}S each have a spin quantum number of zero and only one allowed nuclear spin state; these nuclei are inactive in NMR spectroscopy.

13.2 Orientation of Nuclear Spins in an Applied Magnetic Field

Within a sample containing ^1H and ^{13}C atoms, the orientations of the nuclear magnetic moments associated with their nuclear spins are completely random. When placed between the poles of a powerful magnet of field strength B_0, however, interac-

Figure 13.1

^1H and ^{13}C nuclei with spin $+\frac{1}{2}$ are aligned with the applied magnetic field, B_0, and are in the lower spin energy state; those with spin $-\frac{1}{2}$ are aligned against the applied magnetic field and are in the higher spin energy state.

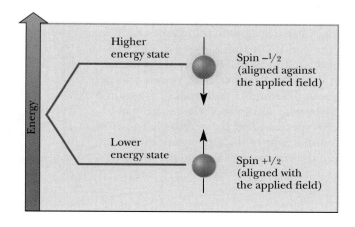

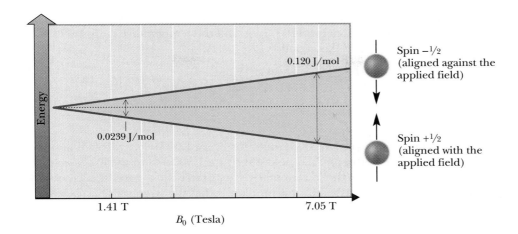

Figure 13.2
The energy difference between the allowed nuclear spin states increases linearly with applied field strength. Values shown here are for ^{1}H nuclei.

tions between the nuclear spins and the applied magnetic field are quantized, with the result that only certain orientations of nuclear magnetic moments are allowed. For ^{1}H and ^{13}C nuclei, only two orientations are allowed as illustrated in Figure 13.1. By convention, nuclei with spin $+\frac{1}{2}$ are aligned with the applied magnetic field and are in the lower energy state; nuclei with spin $-\frac{1}{2}$ are aligned against the applied magnetic field and are in the higher energy state.

The most important NMR physical concept from the point of view of molecular structure determination is that *the difference in energy between nuclear spin states for a given nucleus is proportional to the strength of the magnetic field experienced by that nucleus* (Figure 13.2). We will come back to this concept several times in this chapter. At an applied field strength of 7.05 T, which is readily available with present-day superconducting electromagnets, the difference in energy between nuclear spin states for ^{1}H is approximately 0.120 J (0.0286 cal)/mol, (corresponding to electromagnetic radiation of 300 MHz). At 7.05 T, the energy difference in nuclear spin states for ^{13}C nuclei is approximately 0.030 J (0.00715 cal)/mol (corresponding to radiation of 75 MHz). Advanced commercial instruments now operate at fields more than three times this value; the operating frequencies are proportional to the field. Sensitivities are more than proportionally higher!

To put these values for nuclear spin energy levels in perspective, energies for transitions between vibrational energy levels observed in infrared (IR) spectroscopy are 8 to 63 kJ (2 to 15 kcal)/mol. Those between electronic energy levels in ultraviolet-visible spectroscopy are 167 to 585 kJ (40 to 140 kcal)/mol. Nuclear transitions involve only small energies, on the order of a few hundredths of a calorie!

Note: the SI unit for magnetic field strength is the **tesla (T)**. A unit still in common use, however, is the gauss (G). Values of T and G are related by the equation 1 T = 10^4 G.

Example 13.1

Calculate the ratio of nuclei in the higher spin state to those in the lower spin state, N_h / N_l, for ^{1}H at 25°C in an applied field strength of 7.05 T.

Solution

Use the equation given in Section 2.6B for the relationship between the difference in energy states and equilibrium constant. In this problem, this relationship has the form

$$\Delta G^0 = -RT \ln \frac{N_h}{N_l}$$

The difference in energy between the higher and lower nuclear spin states in an applied field of 7.05 T is approximately 0.120 J/mol, and the temperature is 25 + 273 = 298 K. Substituting these values in this equation gives

$$\ln \frac{N_h}{N_l} = \frac{-\Delta G^0}{RT} = \frac{-0.120 \, \text{J} \cdot \text{mol}^{-1}}{8.314 \, \text{J} \cdot \text{K}^{-1} \cdot \text{mol}^{-1} \times 298 \, \text{K}} = -4.843 \times 10^{-5}$$

$$\frac{N_h}{N_l} = 0.9999516 = \frac{1.000000}{1.000048}$$

From this calculation, we determine that, for every 1,000,000 hydrogen atoms in the higher energy state in this applied field, there are 1,000,048 in the lower energy state. The excess population of the lower energy state under these conditions is only 48 per million! What is important about this number is that the strength of an NMR signal is proportional to the population difference. The greater the difference is, the stronger the signal will be.

Problem 13.1

Calculate the ratio of nuclei in the higher spin state to those in the lower spin state, N_h/N_l, for ^{13}C at 25°C in an applied field strength of 7.05 T. The difference in energy between the higher and lower nuclear spin states in this applied field is approximately 0.030 J (0.00715 cal)/mol.

13.3 Nuclear Magnetic "Resonance"

As we have seen, when nuclei with spin quantum number $\frac{1}{2}$ are placed in an applied magnetic field, a small majority of nuclear spins are aligned with the applied field in the lower energy state. When nuclei in the lower energy spin state are irradiated with a radio frequency of the appropriate energy, they absorb the energy, and nuclear spins flip from the lower energy state to the higher energy state, the only other allowed spin state.

To visualize the mechanism by which a spinning nucleus absorbs energy and the meaning of resonance in this context, think of the nucleus as if it were really spinning. When an applied field of strength B_0 is turned on, the nucleus becomes aligned with the applied field in an allowed spin energy state. The nucleus then begins to precess as shown in Figure 13.3(a) and traces out a cone-shaped surface in much the same manner as a spinning top or gyroscope traces out a cone-shaped surface as it precesses in the earth's gravitational field. We can express the rate of precession as a frequency in hertz.

If the precessing nucleus is irradiated with electromagnetic radiation at the precession frequency, then the two frequencies couple, energy is absorbed, and the nuclear spin "flips" from spin state $+\frac{1}{2}$ (with the applied field) to spin state $-\frac{1}{2}$ (against the applied field) as illustrated in Figure 13.3(b). For ^{1}H in an applied magnetic field of 7.05 T, the frequency of precession is approximately 300 MHz. For ^{13}C in the same field, it is approximately 75 MHz. **Resonance** in this context is the absorption of electromagnetic radiation by a precessing nucleus and the resulting flip of its nuclear spin from the lower energy state to the higher energy state. The spectrometer detects this absorption of electromagnetic radiation and records it as a **signal.** The process is quantized, so that only electromagnetic radiation of precisely the correct frequency causes a nuclear spin to flip. Electromagnetic radiation of a frequency that is too low or too high is not absorbed.

Resonance in NMR spectroscopy The absorption of electromagnetic radiation by a precessing nucleus and the resulting "flip" of its nuclear spin from the lower energy state to the higher energy state.

Signal A recording in an NMR spectrum of a nuclear magnetic resonance.

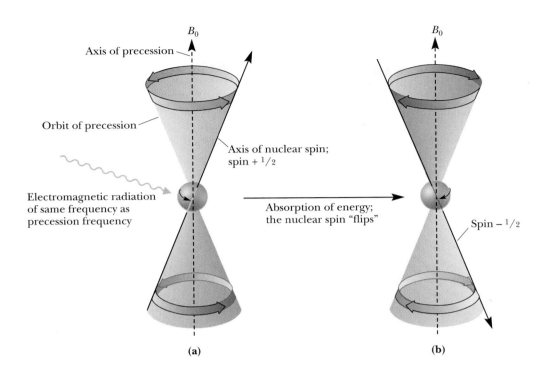

B_0

Axis of precession

Orbit of precession

Axis of nuclear spin;
spin $+ 1/2$

Electromagnetic radiation
of same frequency as
precession frequency

Absorption of energy;
the nuclear spin "flips"

B_0

Spin $- 1/2$

(a)   **(b)**

Figure 13.3
The origin of nuclear
magnetic "resonance."
(a) Precession of a spinning
nucleus in an applied
magnetic field. (b) Absorption
of electromagnetic radiation
occurs when the frequency
of radiation is equal to the
frequency of precession.

If we were dealing with ^{1}H nuclei isolated from all other atoms and electrons, any combination of applied field and electromagnetic radiation that produces a signal for one hydrogen nucleus would produce a signal for all hydrogen nuclei. In other words, hydrogens would be indistinguishable. Hydrogens in an organic molecule, however, are not isolated; they are surrounded by electron density.

A key physical principle for NMR is that circulating electrons induce a magnetic field. This is the principle behind electromagnets and electric motors. The direction of electron movement dictates the orientation of the induced magnetic field. Of equal importance to NMR, the converse is also true. An applied magnetic field induces electrons to circulate, and the orientation of the field dictates the direction of circulation. You will learn more details of these relationships in your physics classes. The important point for our purposes is that an applied magnetic field induces the electron density in a molecule to circulate. The spin states of underlying nuclei are, in turn, influenced to a small but measurable degree by the magnetic field created by the induced electron density circulation. The circulation of electron density in a molecule in an applied magnetic field is called a **diamagnetic current.**

It turns out that a molecule's sigma-bonding electron density is induced to circulate in a direction that creates a small magnetic field that directly *opposes* the applied magnetic field. As a result of the opposing magnetic fields, the nuclei within the circulating electron density experience a magnetic field that is slightly *smaller* than the actual applied field. In other words, nuclei underneath circulating sigma-bonding electron density are *shielded* to a small degree from the applied magnetic field. This nuclear **shielding** is called **diamagnetic shielding.** Although the diamagnetic shielding created by circulating electron density is generally orders of magnitude weaker than the applied fields used in NMR spectroscopy, it is nonetheless significant.

The degree of shielding depends on several factors, which we will take up later. Nevertheless, what we need to establish now is that as the shielding becomes greater,

Diamagnetic current in NMR
The circulation of electron
density in a molecule in an
applied magnetic field.

Shielding in NMR Also called
diagmagnetic shielding; the
term refers to the reduction in
magnetic field experienced by a
nucleus underneath electron
density induced to circulate when
the molecule is placed in a strong
magnetic field.

the net magnetic field present at a nucleus becomes smaller, so the energy of electromagnetic radiation required to bring that nucleus into resonance (i.e., "flip its spin") also decreases. Energy is proportional to electromagnetic radiation frequency, and resonance frequencies are plotted on an NMR spectrum. Putting all of these ideas together, we see that a nucleus that is more shielded will come into resonance at lower frequency than a nucleus that is less shielded. **Deshielding** is the term commonly used to express the concept of less shielding. The relationship of *greater* shielding leading to resonance at *lower* frequency is fundamental to understanding NMR spectra, but most students find it challenging to remember correctly. Make sure you have a clear understanding of this concept before going on.

Chemical shift (δ) The shift in parts per million of an NMR signal relative to the signal of TMS.

The difference in resonance frequencies caused by differing amounts of shielding is called **chemical shift.** The differences in resonance frequencies among the various hydrogen nuclei within a molecule attributable to shielding/deshielding are generally very small. The difference between the resonance frequencies of hydrogens in chloromethane compared with those in fluoromethane, for example, under an applied field of 7.05 T is only 360 Hz. Considering that the radio-frequency radiation used at this applied field is approximately 300 MHz, the difference in resonance frequencies between these two sets of hydrogens is only slightly greater than 1 **part per million** (1 **ppm**) compared with the irradiating frequency.

$$\frac{360 \text{ Hz}}{300 \times 10^6 \text{ Hz}} = \frac{1.2}{10^6} = 1.2 \text{ ppm}$$

It is customary to measure the resonance frequencies of individual nuclei relative to the resonance frequency of nuclei in a reference compound. The reference compound now universally accepted for ^{1}H-NMR and ^{13}C-NMR spectroscopy is tetramethylsilane (TMS), which is assigned a chemical shift of 0 ppm by convention.

$$\begin{array}{c} \text{CH}_3 \\ | \\ \text{H}_3\text{C}-\text{Si}-\text{CH}_3 \\ | \\ \text{CH}_3 \end{array}$$

Tetramethylsilane (TMS)

To standardize reporting of NMR data for both ^{1}H and ^{13}C spectra, the chemical shift (δ), in parts per million, is defined as the frequency shift from either the hydrogens or the carbons in TMS divided by the operating frequency of the spectrometer. Thus, by definition, chemical shift is independent of the operating frequency of the spectrometer. On the chart paper used to record NMR spectra, chemical shift values are shown in increasing order to the left of the TMS signal.

13.4 An NMR Spectrometer

The essential elements of an NMR spectrometer are a powerful magnet, a radio-frequency generator, a radio-frequency detector, and a sample tube (Figure 13.4).

The sample is dissolved in a solvent, most commonly carbon tetrachloride (CCl_4), deuterochloroform ($CDCl_3$), or deuterium oxide (D_2O), which have no protons and do not interfere in ^{1}H spectra. The sample cell is a small glass tube suspended in the magnetic field and set spinning on its long axis to ensure that all parts of the sample experience a homogeneous applied magnetic field. In the simplest form of the ^{1}H-NMR experiment, the absorption of electromagnetic radiation is measured as different ^{1}H nuclei are excited from their $+\frac{1}{2}$ spin states to their $-\frac{1}{2}$ spin

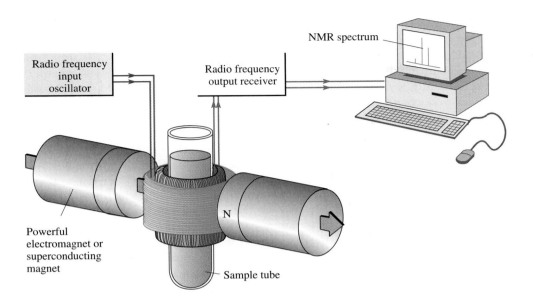

Figure 13.4
Schematic diagram of a nuclear magnetic resonance spectrometer.

states. The frequencies at which the absorptions occur are in the radio-frequency region of the electromagnetic spectrum. The observed absorption frequencies are plotted as peaks relative to the TMS standard on a ppm scale.

Modern Fourier transform NMR (FT-NMR) spectrometers can increase the power of the NMR technique significantly. An FT-NMR spectrometer operates in the following way. The magnetic field is held constant, and the sample is irradiated with a short pulse (approximately 10^{-5} s) of radio-frequency energy that flips the spins of all susceptible nuclei simultaneously. As each nucleus returns to its equilibrium state, it emits a sine wave at the frequency of its resonance. The intensity of the sine wave decays with time and falls to zero as nuclei return to their equilibrium state. A computer records this intensity-versus-time information and then uses a mathematical algorithm called a Fourier transform (FT) to convert it to intensity-versus-frequency information. An FT-NMR spectrum can be recorded in less than two seconds. A particular advantage of FT-NMR spectroscopy is that a large number of spectra (as many as

Figure 13.5
^{1}H-NMR spectrum of methyl acetate.

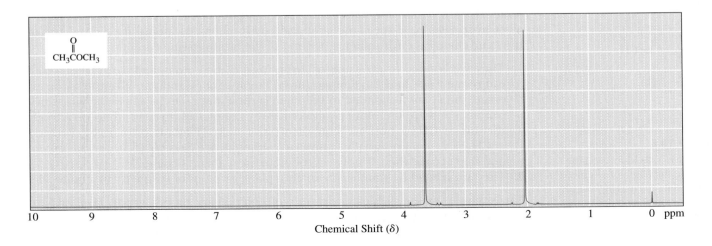

several thousand per sample) can be recorded and digitally summed to give a time-averaged spectrum. Instrumental electronic noise is random and partially cancels out when spectra are time-averaged, but sample signals accumulate and become much stronger relative to the electronic noise. The net result is that good NMR spectra can be obtained with very little sample. All NMR spectra shown in this text were recorded and displayed using FT techniques.

Figure 13.5 shows a 300-MHz ^{1}H-NMR spectrum of methyl acetate. The lower axis is δ, in parts per million. The small signal at δ 0 is caused by the hydrogens of the TMS reference, a small amount of which was added to the sample. The remainder of the spectrum consists of two signals: one for the three hydrogens on the methyl adjacent to oxygen and one for the three hydrogens on the methyl adjacent to the carbonyl group. It is not our purpose at the moment to determine which hydrogens give rise to which signal but only to recognize the form in which an NMR spectrum is recorded and the origin of the calibration marks.

Here is a note on terminology. If a signal is shifted toward the left on the chart paper (larger chemical shift), we say that it is shifted **downfield.** A downfield shift corresponds to decreased shielding around a nucleus, that is, deshielding. If a signal is shifted toward the right (smaller chemical shift), we say that it is shifted **upfield** and therefore corresponds to increased shielding around a nucleus.

Downfield A signal of an NMR spectrum that is shifted toward the left (larger chemical shift) on the chart paper.

Upfield A signal of an NMR spectrum that is shifted toward the right (smaller chemical shift) on the chart paper.

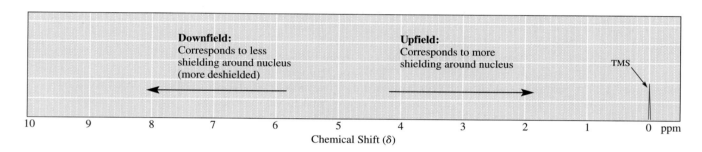

13.5 Equivalent Hydrogens

Equivalent hydrogens Hydrogens that have the same chemical environment.

Given the structural formula of a compound, how do we know how many signals to expect in its ^{1}H-NMR spectrum? The answer is that equivalent hydrogens give the same ^{1}H-NMR signal; nonequivalent hydrogens give different ^{1}H-NMR signals. **Equivalent hydrogens** have the same chemical environment. H atoms are equivalent (in the same chemical environment) if either of the two following conditions exist:

1. They are bonded to the same sp^3 hybridized carbon atom, and that carbon atom can rotate freely at room temperature. The rapid bond rotation means that, on average, the H atoms bonded to the same carbon atom see the same chemical environment and are therefore equivalent. For example, all three H atoms on a freely rotating —CH$_3$ group are equivalent, and both H atoms on a freely rotating —CH$_2$— group are usually equivalent, although see Section 13.10 for the exception that occurs when —CH$_2$— groups are near a chiral center.
2. They are related by symmetry, namely a plane or point of symmetry in a molecule.

For example, in the 2-chloropropane molecule, all six methyl group H atoms are equivalent. The methyl groups are freely rotating, and they are related by a plane of symmetry as shown (Figure 13.6). A convenient way to determine whether hydrogen

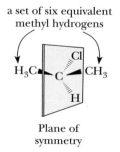

a set of six equivalent methyl hydrogens

Plane of symmetry

Figure 13.6

Structure of 2-chloropropane showing the plane of symmetry responsible for making the two methyl groups, and therefore the six methyl group H atoms, equivalent.

atoms are equivalent is to use the replacement test. In your mind, replace each of the hydrogen atoms in question with a test atom, let us say chlorine. If in each case the same molecule is produced through the replacement, then the original hydrogen atoms are equivalent.

Example 13.2

State the number of types of equivalent hydrogens in each compound and the number of hydrogens in each set.
(a) 2-Methylpropane   **(b)** 2-Methylbutane

Solution

(a) 2-Methylpropane contains two sets of equivalent hydrogens: a set of nine equivalent primary hydrogens and one tertiary hydrogen.
(b) 2-Methylbutane contains four sets of equivalent hydrogens. Nine primary hydrogens are in this molecule: one set of three and one set of six. To see that there are two sets, note that replacement by chlorine of any hydrogen in the set of three gives 1-chloro-3-methylbutane. Replacement by chlorine of any hydrogen in the set of six gives 1-chloro-2-methylbutane. In addition, the molecule contains a set of two equivalent secondary hydrogens and one tertiary hydrogen.

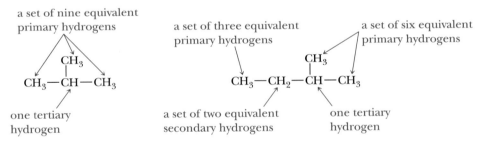

Problem 13.2

State the number of sets of equivalent hydrogens in each compound and the number of hydrogens in each set.
(a) 3-Methylpentane   **(b)** 2,2,4-Trimethylpentane

Here are four organic compounds, each of which has one set of equivalent hydrogens and gives one signal in its ^{1}H-NMR spectrum.

$\overset{\displaystyle O}{\overset{\displaystyle \|}{CH_3CCH_3}}$	$ClCH_2CH_2Cl$	⬠	$\begin{array}{c} H_3C \\ \\ H_3C \end{array}C{=}C\begin{array}{c} CH_3 \\ \\ CH_3 \end{array}$
Propanone (Acetone)	1,2-Dichloroethane	Cyclopentane	2,3-Dimethyl-2-butene

Molecules with two or more sets of equivalent hydrogens give rise to a different resonance signal for each set, as illustrated by these four compounds.

$$CH_3CHCl \quad (with\ Cl\ above)$$

1,1-Dichloroethane
(2 signals)

Cyclopentanone
(2 signals)

$$Cl, CH_3 / C=C / H, H$$

(Z)-1-Chloropropene
(3 signals)

Cyclohexene
(3 signals)

You should see immediately that valuable information about molecular structure can be obtained simply by counting the number of signals in the ^{1}H-NMR spectrum of a compound. Consider, for example, the two constitutional isomers of molecular formula $C_2H_4Cl_2$. The compound 1,2-dichloroethane has one set of equivalent hydrogens and one signal in its ^{1}H-NMR spectrum. Its constitutional isomer 1,1-dichloroethane has two sets of equivalent hydrogens and two signals in its ^{1}H-NMR spectrum. Thus, simply counting signals allows you to distinguish between these two compounds.

Example 13.3

Each compound gives only one signal in its ^{1}H-NMR spectrum. Propose a structural formula for each compound.
(a) C_2H_6O　　**(b)** $C_3H_6Cl_2$　　**(c)** C_6H_{12}　　**(d)** C_4H_6

Solution

(a) CH_3OCH_3　　**(b)** CH_3CCH_3 (with Cl above and Cl below)　　**(c)** (cyclohexane) or $H_3C, CH_3 / C=C / H_3C, CH_3$　　**(d)** $CH_3C{\equiv}CCH_3$

Problem 13.3

Each compound gives only one signal in its ^{1}H-NMR spectrum. Propose a structural formula for each compound.
(a) C_3H_6O　　**(b)** C_5H_{10}　　**(c)** C_5H_{12}　　**(d)** $C_4H_6Cl_4$

13.6 Signal Areas

We have just seen that the number of signals in a ^{1}H-NMR spectrum gives us information about the number of sets of equivalent hydrogens. The relative areas of these signals provide additional information. As a spectrum is being run, the instrument's computer numerically measures the area under each signal. In the spectra shown in this text, this information is displayed in the form of a line of integration superposed on the original spectrum. The vertical rise of the line of integration over each signal is proportional to the area under that signal, which, in turn, is proportional to the number of equivalent hydrogens giving rise to that signal. Figure 13.7 shows an integrated ^{1}H-NMR spectrum of *tert*-butyl acetate, $C_6H_{12}O_2$. The spectrum shows signals at δ 1.44 and 1.95. The integrated signal heights are 23 + 67, or 90 chart divisions, which correspond to 12 hydrogens. From these numbers, we calculate that (23/90) $\times$ 12 or 3 hydrogens are in one set, and (67/90) $\times$ 12 or 9 hydrogens are in the second set. An alternative way of indicating integration is a numerical readout given over each signal, and you may see this on many ^{1}H-NMR spectra you encounter.

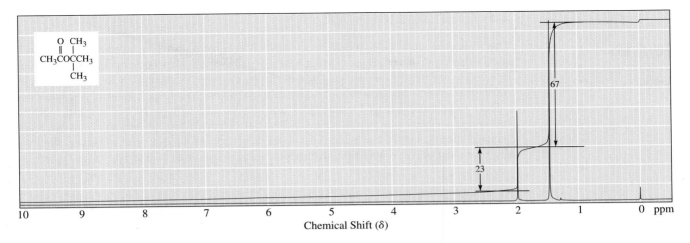

Figure 13.7
^{1}H-NMR spectrum of *tert*-butyl acetate showing the integration. The total vertical rise of 90 chart divisions corresponds to 12 hydrogens, 9 in one set and 3 in the other.

Example 13.4

Following is a ^{1}H-NMR spectrum for a compound of molecular formula $C_9H_{10}O_2$. From the integration, calculate the number of hydrogens giving rise to each signal.

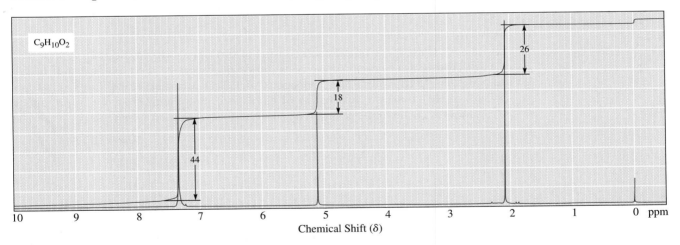

Solution

The total vertical rise in the line of integration is 88 chart divisions and corresponds to 10 hydrogens. From these numbers, we calculate that $44/88 \times 10$, or 5, of the hydrogens give rise to the signal at δ 7.34. By similar calculations, the signals at δ 5.08 and 2.06 correspond to two hydrogens and three hydrogens, respectively.

Problem 13.4

The line of integration of the two signals in the ^{1}H-NMR spectrum of a ketone of molecular formula $C_7H_{14}O$ rises 62 and 10 chart divisions, respectively. Calculate the number of hydrogens giving rise to each signal, and propose a structural formula for this ketone.

13.7 Chemical Shift

ORGANIC **Chemistry•Now™**
Click Tutorials to explore **Chemical Shift**

The chemical shift for a signal in a ^{1}H-NMR spectrum can give valuable information about the type of hydrogens giving rise to that signal. Hydrogens on methyl groups bonded to sp^3 hybridized carbons, for example, give signals near δ 0.8 to 1.0. Hydrogens on methyl groups bonded to a carbonyl carbon give signals near δ 2.1 to 2.3 (notice the signals near 2.0 ppm in Figures 13.5 and 13.7), and hydrogens on a methyl group bonded to oxygen give signals near δ 3.7 to 3.9 (Figure 13.5). Shown in Figure 13.8 are average chemical shifts for most of the types of hydrogens we deal with in this course. Notice that most of these values fall within a rather narrow range of 0 to 10 δ units (ppm).

Example 13.5

Following are structural formulas for two constitutional isomers of molecular formula $C_6H_{12}O_2$.

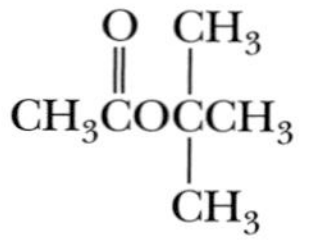

(1)

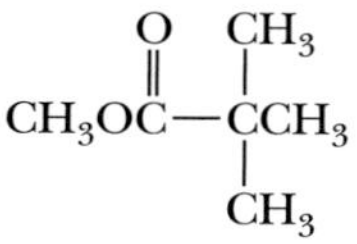

(2)

(a) Predict the number of signals in the ^{1}H-NMR spectrum of each isomer.
(b) Predict the ratio of areas of the signals in each spectrum.
(c) Show how you can distinguish between these isomers on the basis of chemical shift.

Solution

(a) The ^{1}H-NMR spectrum of each consists of two signals (b) in the ratio 9:3, or 3:1. (c) Distinguish between these constitutional isomers by the chemical shift of the single —CH_3 group. The hydrogens of CH_3O are deshielded (appear farther downfield) than the hydrogens of $CH_3C{=}O$. See Figure 13.8 for approximate values for each chemical shift. Experimentally determined values are

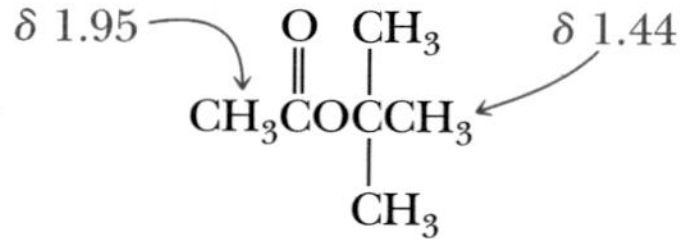

(1)

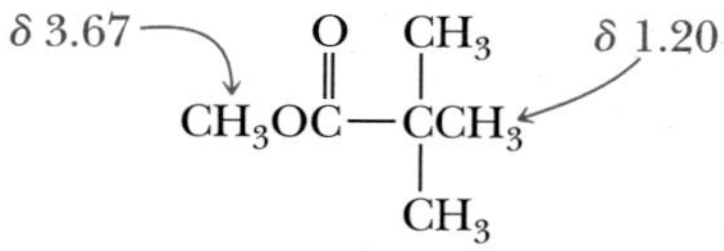

(2)

Please see Figure 13.7 for the spectrum of *tert*-butyl acetate (1).

Problem 13.5

Following are two constitutional isomers of molecular formula $C_4H_8O_2$.

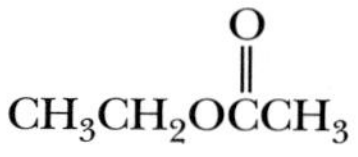

(1)

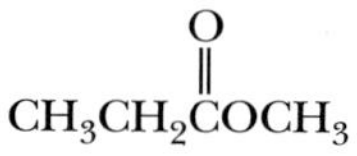

(2)

(a) Predict the number of signals in the ^{1}H-NMR spectrum of each isomer.
(b) Predict the ratio of areas of the signals in each spectrum.
(c) Show how you can distinguish between these isomers on the basis of chemical shift.

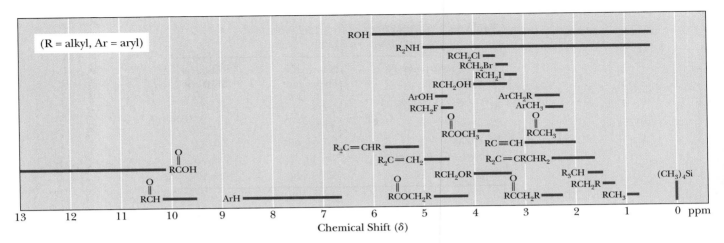

Figure 13.8
Average values of chemical shifts of representative types of hydrogens. These values are approximate. Other atoms in the molecules may cause signals to appear outside of these ranges.

The chemical shift of a particular type of hydrogen depends primarily on the extent of shielding it experiences. Shielding, in turn, depends on three factors: (1) the electronegativity of nearby atoms, (2) the hybridization of the adjacent atoms, and (3) the magnetic induction within an adjacent pi bond. Let us consider these factors one at a time.

A. Electronegativity of Nearby Atoms

As illustrated in Table 13.2 for the chemical shift of methyl hydrogens in the series CH_3—X, the greater the electronegativity of X is, the greater the chemical shift will be. The effect of an electronegative substituent falls off quickly with distance. The effect of an electronegative substituent two atoms away is only about 10% of that when it is on the adjacent atom. The effect of an electronegative substituent three atoms away is almost negligible.

Electronegativity and chemical shift are related in the following way. The presence of an electronegative atom or group reduces electron density on atoms bonded

Table 13.2 Dependence of Chemical Shift of CH_3X on the Electronegativity of X

CH_3—X	Electronegativity of X	Chemical Shift (δ) of Methyl Hydrogens
CH_3F	4.0	4.26
CH_3OH	3.5	3.47
CH_3Cl	3.1	3.05
CH_3Br	2.8	2.68
CH_3I	2.5	2.16
$(CH_3)_4C$	2.1	0.86
$(CH_3)_4Si$	1.8	0.00 (by definition)

Table 13.3	The Effect of Hybridization on Chemical Shift		
Type of Hydrogen (R = alkyl)		**Name of Hydrogen**	**Chemical Shift (δ)**
RCH_3, R_2CH_2, R_3CH		Alkyl	0.8–1.7
$R_2C{=}C(R)CHR_2$		Allylic	1.6–2.6
$RC{\equiv}CH$		Acetylenic	2.0–3.0
$R_2C{=}CHR$, $R_2C{=}CH_2$		Vinylic	4.6–5.7
RCHO		Aldehydic	9.5–10.1

to it and therefore their shielding. This effect deshields nearby atoms and causes them to resonate farther downfield, that is, with a larger chemical shift.

B. Hybridization of Adjacent Atoms

Hydrogens bonded to an sp^3 hybridized carbon typically absorb at δ 0.8 to 1.7. Vinylic hydrogens (those on a carbon of a carbon-carbon double bond) are considerably deshielded and resonate at δ 4.6 to 5.7 (Table 13.3). Part of the explanation for the greater deshielding of vinylic hydrogens compared with alkyl hydrogens lies in the hybridization of carbon. Because a sigma-bonding orbital of an sp^2 hybridized carbon has more s-character than a sigma-bonding orbital of an sp^3 hybridized carbon (33% compared with 25%), an sp^2 hybridized carbon atom is more electronegative. Vinylic hydrogens are deshielded by this electronegativity effect and resonate farther downfield relative to alkyl hydrogens. Similarly, acetylene and aldehyde hydrogens also appear farther downfield compared with alkyl hydrogens.

However, differences in chemical shifts of vinylic and acetylenic hydrogens cannot be accounted for on the basis of the hybridization of carbon alone. If the chemical shift of vinylic hydrogens (δ 4.6–5.7) were caused entirely by the hybridization of carbon, then the chemical shift of acetylenic hydrogens should be even greater than that of vinylic hydrogens. Yet the chemical shift of acetylenic hydrogens is only δ 2.0 to 3.0. It seems that either the chemical shift of acetylenic hydrogens is abnormally small or the chemical shift of vinylic hydrogens is abnormally large. In either case, another factor must be contributing to the magnitude of the chemical shift. Theoretical and experimental evidence suggest that the chemical shifts of hydrogens attached to pi-bonded carbons are influenced not only by the relative electronegativities of the sp^2- and sp-hybridized carbon atoms but also by magnetic induction from pi bonds.

C. Diamagnetic Effects from Pi Bonds

To understand the influence of pi bonds on the chemical shift of an acetylenic hydrogen, imagine that the carbon-carbon triple bond is oriented as shown in Figure 13.9 with respect to the applied field. Because of magnetic induction, the applied field induces a circulation of the pi electrons, which in turn produces an induced magnetic field. Given the geometry of an alkyne and the cylindrical nature of its pi electron cloud, the induced magnetic field is shielding in the vicinity of the acetylenic hydrogen. Therefore, lower frequency electromagnetic radiation is required to make an acetylenic hydrogen resonate; the local magnetic field induced in the pi bonds shifts the signal of an acetylenic hydrogen upfield to a smaller δ value.

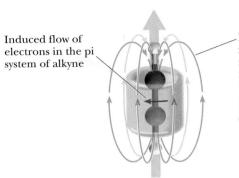

Induced flow of electrons in the pi system of alkyne

Induced local magnetic field of the pi electrons is against the applied field; it requires lower frequency radiation to bring an acetylenic hydrogen into resonance.

Applied field, B_0

Figure 13.9
A magnetic field induced in the pi bonds of a carbon-carbon triple bond shields an acetylenic hydrogen and shifts its signal upfield.

The effect of the induced circulation of pi electrons on a vinylic hydrogen (Figure 13.10) is opposite to that on an acetylenic hydrogen. The direction of the induced magnetic field in the pi bond of a carbon-carbon double bond is parallel to the applied field in the region of the vinylic hydrogens. The induced magnetic field deshields vinylic hydrogens and, thus, shifts their signal downfield to a larger δ value. The presence of the pi electrons in the carbonyl group has a similar effect on the chemical shift of the hydrogen of an aldehyde group.

The effects of the pi electrons in benzene are even more dramatic than in alkenes. All six hydrogens of benzene are equivalent, and its ^{1}H-NMR spectrum is a sharp singlet at δ 7.27. Hydrogens bonded to a substituted benzene ring appear in the region δ 6.5 to 8.5. Few other hydrogens absorb in this region; thus, aryl hydrogens are quite easily identifiable by their distinctive chemical shifts, as much as 2 ppm higher than comparably substituted alkenes.

That aryl hydrogens absorb even farther downfield than vinylic hydrogens is accounted for by the existence of a **ring current,** a special property of aromatic rings (Figure 13.11). When the plane of an aromatic ring tumbles in an applied magnetic field, the applied field causes the pi electrons to circulate around the ring, giving rise to the so-called ring current. This induced ring current has associated with it a magnetic field that opposes the applied field in the middle of the ring but reinforces the applied field on the outside of the ring. Thus, given the position of aromatic hydrogens relative to the induced ring current, they are deshielded and come into resonance at a larger chemical shift.

Ring current   An applied magnetic field causes the pi electrons of an aromatic ring to circulate, giving rise to the so-called ring current and an associated magnetic field that opposes the applied field in the middle of the ring but reinforces the applied field on the outside of the ring.

Figure 13.10
A magnetic field induced in the pi bond of a carbon-carbon double bond deshields vinylic hydrogens and shifts their signals downfield.

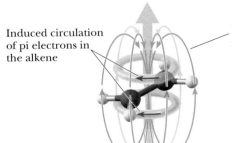

Induced circulation of pi electrons in the alkene

Induced local magnetic field of the pi electrons reinforces the applied field; it requires higher frequency radiation to bring a vinylic hydrogen into resonance.

Applied field, B_0

Figure 13.11

The magnetic field induced by circulation of pi electrons in an aromatic ring deshields the hydrogens of the aromatic ring and shifts their signal downfield.

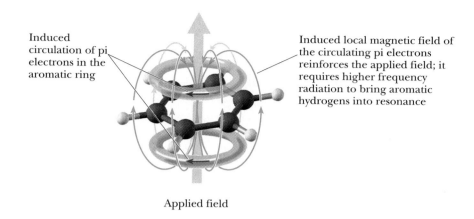

Induced circulation of pi electrons in the aromatic ring

Induced local magnetic field of the circulating pi electrons reinforces the applied field; it requires higher frequency radiation to bring aromatic hydrogens into resonance

Applied field

13.8 Signal Splitting and the ($n + 1$) Rule

We have now seen three kinds of information that can be derived from examination of a ^{1}H-NMR spectrum.

1. From the number of signals, we can determine the number of sets of equivalent hydrogens.
2. From integration of signal areas, we can determine the relative numbers of hydrogens giving rise to each signal.
3. From the chemical shift of each signal, we derive information about the types of hydrogens in each set.

A fourth kind of information can be derived from the splitting pattern of each signal. Consider, for example, the ^{1}H-NMR spectrum of 1,1-dichloroethane shown in Figure 13.12. This molecule contains two sets of equivalent hydrogens. According to what we have learned so far, we predict two signals with relative areas 3 : 1 corresponding to the three hydrogens of the —CH_3 group and the one hydrogen of the —$CHCl_2$ group, respectively. You see from the spectrum, however, that there are, in fact, six peaks. These peaks are named by how the signal is split: two peaks are a doublet, three peaks are a triplet, and so on. The grouping of two peaks at δ 2.0 is the

Figure 13.12

^{1}H-NMR spectrum of 1,1-dichloroethane.

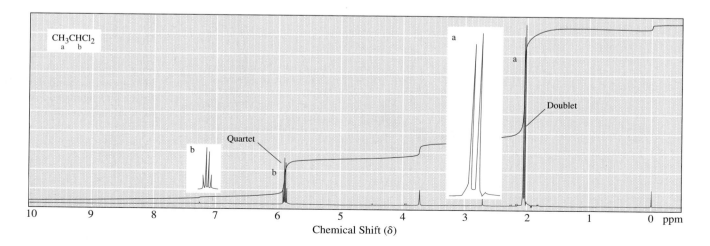

CH_3CHCl_2
a   b

a

a

Doublet

b

Quartet

b

| 10 | 9 | 8 | 7 | 6 | 5 | 4 | 3 | 2 | 1 | 0 ppm |

Chemical Shift (δ)

signal for the three hydrogens of the —CH_3 group, and the grouping of four peaks at δ 5.9 is the signal for the single hydrogen of the —$CHCl_2$ group. We say that the CH_3 resonance at δ 2.1 is split into a doublet and that the CH resonance at δ 5.9 is split into a quartet.

In many situations, the degree of **signal splitting** can be predicted on the basis of the **($n + 1$) rule.** According to this rule, if a hydrogen has n hydrogens nonequivalent to it but equivalent among themselves on the same or adjacent atom(s), its ^{1}H-NMR signal is split into ($n + 1$) peaks.

Let us apply the ($n + 1$) rule to the analysis of the spectrum of 1,1-dichloroethane. The three hydrogens of the —CH_3 group have one nonequivalent neighbor hydrogen ($n = 1$), and, therefore, their signal is split into a doublet. The single hydrogen of the —$CHCl_2$ group has a set of three nonequivalent neighbor hydrogens ($n = 3$), and its signal is split into a quartet.

($n + 1$) rule   If a hydrogen has n hydrogens nonequivalent to it but equivalent among themselves on the same or adjacent atom(s), its ^{1}H-NMR signal is split into ($n + 1$) peaks.

For these hydrogens, $n = 1$; their signal is split into ($1 + 1$) or two peaks—a **doublet.**

For this hydrogen, $n = 3$; its signal is split into ($3 + 1$) or four peaks—a **quartet.**

$$CH_3 - CH - Cl$$
$$|$$
$$Cl$$

Example 13.6

Predict the number of signals and the splitting pattern of each signal in the ^{1}H-NMR spectrum of each molecule.

$$\text{(a) } CH_3\overset{\overset{\displaystyle O}{\|}}{C}CH_2CH_3 \qquad \text{(b) } CH_3CH_2\overset{\overset{\displaystyle O}{\|}}{C}CH_2CH_3 \qquad \text{(c) } CH_3\overset{\overset{\displaystyle O}{\|}}{C}CH(CH_3)_2$$

Solution

The sets of equivalent hydrogens in each molecule are labeled a, b, and c. Molecule (a) has three sets of equivalent hydrogens; its ^{1}H-NMR spectrum shows a singlet, a quartet, and a triplet in the ratio 3 : 2 : 3. Molecule (b) has two sets of equivalent hydrogens; its ^{1}H-NMR spectrum shows a triplet and a quartet in the ratio 3 : 2. Molecule (c) has three sets of equivalent hydrogens; its ^{1}H-NMR spectrum shows a singlet, a septet, and a doublet in the ratio 3 : 1 : 6.

(a) singlet / quartet / triplet

$$\text{(a) } CH_3 - \overset{\overset{\displaystyle O}{\|}}{\underset{a}{C}} - \underset{b}{CH_2} - \underset{c}{CH_3}$$

(b) triplet / quartet

$$\text{(b) } \underset{a}{CH_3} - \underset{b}{CH_2} - \overset{\overset{\displaystyle O}{\|}}{C} - \underset{b}{CH_2} - \underset{a}{CH_3}$$

(c) singlet / septet / doublet

$$\text{(c) } \underset{a}{CH_3} - \overset{\overset{\displaystyle O}{\|}}{C} - \underset{b}{CH}\underset{c}{(CH_3)_2}$$

Problem 13.6

Following are pairs of constitutional isomers. Predict the number of signals in the ^{1}H-NMR spectrum of each isomer and the splitting pattern of each signal.

$$\text{(a) } CH_3OCH_2\overset{\overset{\displaystyle O}{\|}}{C}CH_3 \quad \text{and} \quad CH_3CH_2\overset{\overset{\displaystyle O}{\|}}{C}OCH_3 \qquad \text{(b) } CH_3\overset{\overset{\displaystyle Cl}{|}}{\underset{\underset{\displaystyle Cl}{|}}{C}}CH_3 \quad \text{and} \quad ClCH_2CH_2CH_2Cl$$

Spin-spin coupling An interaction in which nuclear spins of adjacent atoms influence each other and lead to the spitting of NMR signals.

Coupling constant (J) The separation on an NMR spectrum (in hertz) between adjacent peaks in a multiplet and a quantitative measure of the influence of the spin-spin coupling with adjacent nuclei.

13.9 The Origins of Signal Splitting

^{1}H-NMR signals are split into multiple peaks when molecules contain nonequivalent hydrogen atoms that are separated by no more than three bonds. The multiple peaks are the result of **spin-spin coupling** between ^{1}H nuclei, an interaction in which nuclear spins of adjacent atoms influence each other.

The nuclear spin and hence the chemical shift of the atom labeled H_a in Figure 13.13 is influenced by the adjacent atom H_b, whose nuclear spin might be aligned with or against an applied magnetic field in a ^{1}H-NMR spectrometer. Because of spin-spin coupling, alignment of the H_b nuclear spin *with* the applied magnetic field results in a slightly different chemical shift of the signal for H_a compared to the situation in which the H_b nuclear spin is aligned *against* the applied magnetic field. Across the population of molecules in a sample, there will be similar numbers of molecules having each spin alignment for H_b. Any single molecule gives rise to a single signal for H_a, but the spectrum of the entire sample shows both. The result is that the signal for the H_a atom appears in the spectrum as a **doublet.** In this hypothetical example, the signal for H_b will also be split into a similar doublet owing to H_a because the effect operates in both directions. A **coupling constant** (J) is the separation on an NMR spectrum between adjacent peaks in a multiplet and is a quantitative measure of the influence of the magnetic moments of adjacent hydrogen nuclei. The magnitude of a coupling constant is expressed in hertz (Hz); for protons in ^{1}H-NMR spectroscopy, it is generally in the range 0 to 18 Hz. The value of J depends only on fields caused by other nuclei within a molecule, and so it is independent of the applied field strength.

The coupling constant for two hydrogens on adjacent sp^3 hybridized carbon atoms is approximately 7 Hz. For a spectrometer operating at 300 MHz, a coupling constant of 10 Hz corresponds to only 0.023 ppm. Because peaks with this and comparable values of J are so narrowly spaced, splitting patterns from spectra taken at 300 MHz and higher are often very difficult to see by inspection of the spectra. It is, therefore, common practice to retrace certain signals in expanded form so that splitting patterns are easier to observe (Figure 13.14).

Given in Table 13.4 are approximate values for coupling constants for hydrogens on singly and doubly bonded carbon atoms.

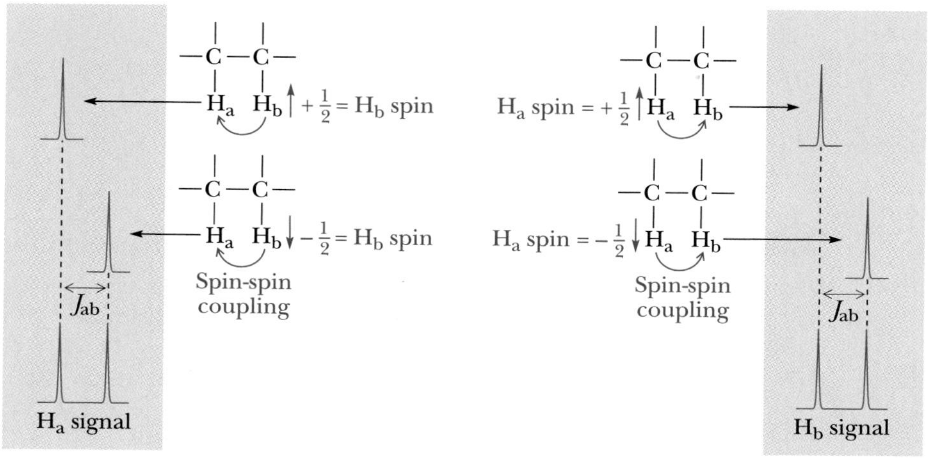

Figure 13.13
Illustration of spin-spin coupling that gives rise to signal splitting in ^{1}H-NMR spectra.

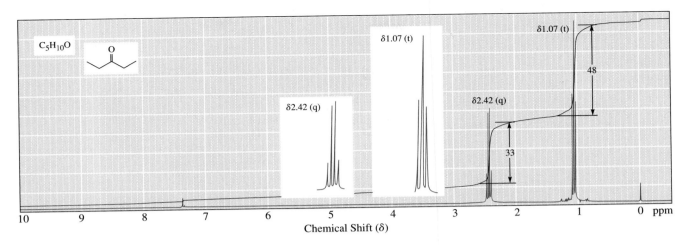

Figure 13.14
The quartet-triplet ^{1}H-NMR signals of 3-pentanone showing the original trace and a scale expansion to show the signal splitting pattern more clearly.

Table 13.4 **Approximate Values of J for Compounds Containing Alkyl and Alkenyl Groups**

6–8 Hz	8–14 Hz	0–5 Hz	0–5 Hz
11–18 Hz	5–10 Hz	0–5 Hz	8–11 Hz

A. Predicting Peak Intensities

As stated previously, in the general case, n equivalent H atoms will cause signal splitting into $n + 1$ peaks. The relative intensities of these peaks are predicted by analyzing all possible spin state combinations. There are $n + 1$ different spin state combinations of n spins aligning with or against an applied magnetic field. The probability of a molecule having a given set of spins is proportional to the number of possible spin alignments giving rise to that spin state. The arrows in Figure 13.15 are particularly helpful in understanding this very important concept, each arrow representing the spin alignment of a ^{1}H nucleus. If there is just one H_b nucleus to consider, there are only two possibilities (↑ or ↓), both of roughly equal probability, leading to a doublet with a 1:1 ratio of peaks for the signal of H_a (left). Two equivalent H_b nuclei can have three different possible combinations that occur in a 1:2:1 ratio (middle), while three equivalent H_b nuclei can have four possible combinations that occur in a 1:3:3:1 ratio (right).

Figure 13.15
The origins of signal splitting patterns. Each arrow represents an H_b nuclear spin orientation.

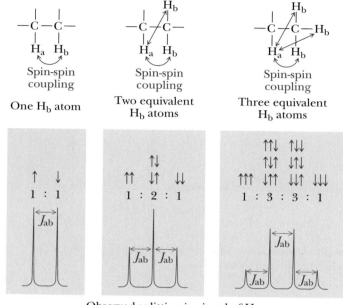

Observed splitting in signal of H_a

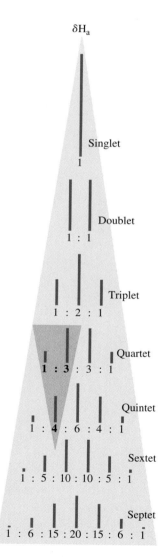

Figure 13.16
Pascal's triangle. As illustrated by the highlighted entries, each entry is the sum of the values immediately above it to the left and the right.

Alternatively, the ratio of peak areas in any multiplet can be derived from a mathematical mnemonic device called **Pascal's triangle** (Figure 13.16). Here is a note of caution in counting the number of peaks in a multiplet: If the signal of a particular hydrogen is of low intensity compared with others in the spectrum, it may not be possible to distinguish some of the smaller side peaks because of electronic noise in the baseline.

B. Physical Basis for the (*n* + 1) Rule

Coupling of nuclear spins is mediated through intervening bonds. The extent of coupling is related to a number of factors, including the number of bonds between the H atoms in question. H atoms with more than three bonds between them generally do not exhibit noticeable coupling, although longer range coupling can be seen in some cases. A common type of coupling involves the H atoms on two C atoms that are bonded to each other. These H atoms are three bonds apart, and this type of coupling is referred to as **vicinal coupling** as illustrated in Figure 13.17.

An important factor in vicinal coupling is the dihedral angle α between the C—H sigma bonds and whether or not it is fixed (Figure 13.18). Coupling is maximized when the angle α is 0° and 180°, and is minimized when α is 90°. Bonds that rotate rapidly at room temperature do not have a fixed angle between adjacent C—H bonds, so an average angle and an average coupling are observed. This latter concept is important for the interpretation of ^{1}H-NMR spectra for alkanes and other flexible molecules.

It should be noted here that all the nuclei of adjacent hydrogens couple. It is only when coupling is between nonequivalent hydrogens that signal splitting results; coupling between equivalent hydrogens, whether they are on the same or adjacent carbons, does not produce signal splitting.

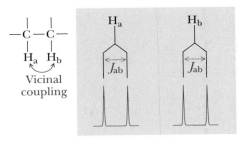

Figure 13.17
Vicinal coupling between two nonequivalent H atoms.

C. More Complex Splitting Patterns

So far, we have concentrated on spin-spin coupling with only one other nonequivalent set of H atoms. However, more complex situations often arise in which a set of H atoms is coupled to more than one set of nonequivalent H atoms in molecules that do not have rapid bond rotation. In these situations, the coupling from adjacent nonequivalent sets of H atoms *combines* to give more complex signal splitting patterns. Use of a tree diagram is helpful in understanding splitting in these cases. In a tree diagram, the different couplings are applied sequentially.

For example, the atom labeled H_b in Figure 13.19 is adjacent to nonequivalent atoms H_a and H_c on either side, so the resulting coupling will give rise to a so-called doublet of doublets, in other words, a signal with four peaks. Here, the signal for H_b is split into a doublet with coupling constant J_{ab} by H_a, and this doublet is split into a doublet of doublets with coupling constant J_{bc} by H_c. If there were no other H atoms in the molecule to be considered, then the signal for H_a would be a doublet with coupling constant J_{ab} and the signal for H_c would be a doublet with coupling constant J_{bc}. This analysis assumes that the H_a-H_b and H_b-H_c coupling constants, J_{ab} and J_{bc}, are different. If J_{ab} and J_{bc} are equal, the peaks overlap, a situation discussed in Section 13.9F.

If H_c is a set of two equivalent H atoms and H_a is still a single H atom, then the observed coupling would be a doublet of triplets, in other words, a signal with six peaks. Again we are assuming that $J_{ab} \neq J_{bc}$. The tree diagram in Figure 13.20 shows the complex pattern that results from this type of splitting. If there were no other H atoms in the molecule to be considered, then the signal for H_a would be a doublet with coupling constant J_{ab} and the signal for H_c would be a doublet with coupling constant J_{bc}. The peaks for H_a and H_b would each integrate to a relative value of one H atom, while the peaks for H_c would integrate to a relative value of 2 H atoms.

In the general case, a signal will be split into $(n + 1) \times (m + 1)$ peaks for an H atom that is coupled to a set of n H atoms with one coupling constant, and to a set of m H atoms with another coupling constant. Note that, in a tree diagram, you get the same splitting patterns no matter which order the two coupling constants are analyzed in.

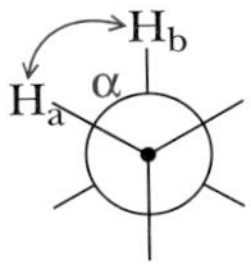

Figure 13.18
The strength of spin-spin coupling between adjacent H atoms is related to the dihedral angle between them. Coupling is a maximum when α is 0° or 180°, and is a minimum when α is 90°.

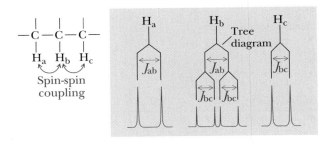

Figure 13.19
Coupling that arises when H_b is split by two different nonequivalent H atoms H_a and H_c. This analysis assumes no other coupling in the molecule and that $J_{ab} \neq J_{bc}$.

Figure 13.20

Complex coupling that arises when H_b is split by H_a and two equivalent atoms H_c. Again, this analysis assumes no other coupling in the molecule and that $J_{ab} \neq J_{bc}$.

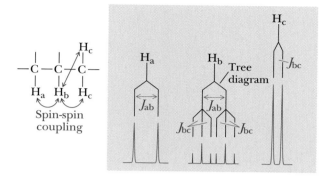

Figure 13.21

Geminal coupling that occurs when two H atoms on the same carbon atom are not equivalent. This is most common in unsymmetrical alkenes and cyclic molecules.

D. Bond Rotation

Because the angle between C—H bonds determines the extent of coupling in a molecule, bond rotation is a key parameter. In alkanes and other molecules with relatively free rotation about C—C sigma bonds, H atoms bonded to the same C atom in —CH_2— and —CH_3 groups are generally equivalent because of the rapid bond rotation. An exception is when a —CH_2— is adjacent to a chiral center, a situation that is discussed in Section 13.10. However, when there is restricted bond rotation, as in alkenes or cyclic structures, H atoms bonded to the same C atom may not be equivalent, especially if the molecule is not symmetrical. Nonequivalent ^1H nuclei on the *same* C atom will couple to each other and cause splitting. This is referred to as **geminal coupling** (Figure 13.21). Geminal coupling constants are generally small, on the order of 0 to 5 Hz.

Because of the restricted rotation about C=C bonds, the alkenyl (vinylic) H atoms of unsymmetrical alkenes are not equivalent; in other words, they are in unique chemical environments. For example, ethyl propenoate (ethyl acrylate) is an unsymmetrical terminal alkene; therefore, the three alkenyl H atoms are nonequivalent (Figure 13.22). As a result, they all couple with each other. In alkenes, *trans* coupling generally results in larger coupling constants (J_{trans} = 11–18 Hz) compared to *cis* coupling (J_{cis} = 5–10 Hz), with geminal coupling being by far the smallest (J_{gem} = 0–5 Hz). Unless a high-resolution spectrum is taken, the geminal coupling constant is so small that it is often difficult to see in terminal alkenes. In the spectrum of ethyl propenoate, the

Figure 13.22

300 MHz ^1H-NMR spectrum of ethyl propenoate.

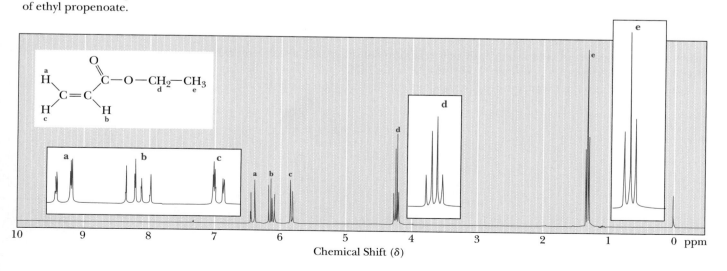

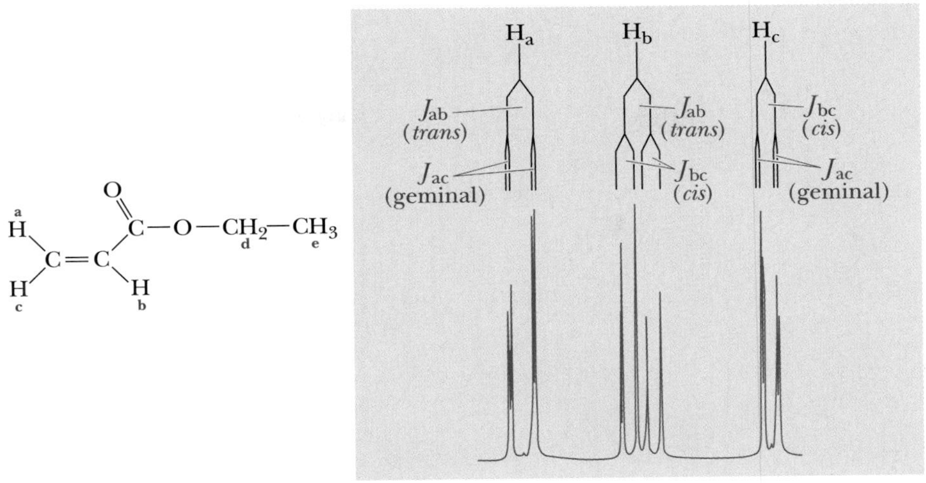

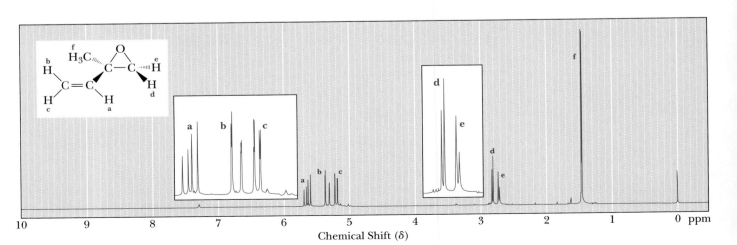

Figure 13.23
Tree diagrams for the complex coupling seen for the alkenyl H atoms in the ^{1}H-NMR spectrum of ethyl propenoate.

geminal coupling is only visible upon close inspection of the signals labeled a and c. You should be able to recognize the characteristic ethyl group pattern of a quartet integrating to two H atoms ($-CH_2-$, H_d) and a triplet integrating to three H atoms ($-CH_3$, H_e). Tree diagrams are provided in Figure 13.23 to help decipher patterns of the alkenyl signals.

Cyclic structures often exhibit restricted rotation about their C—C sigma bonds and can have constrained conformations. The result is that the two H atoms on $-CH_2-$ groups in cyclic molecules can be nonequivalent, leading to complex spin-spin coupling. Substituted epoxides such as 2-methyl-2-vinyloxirane provide a good example (Figure 13.24). The two H atoms on the three-membered epoxide ring are nonequivalent. H_d is *cis* to the vinyl group and *trans* to the methyl group, while H_e is the reverse. Because they are in different chemical environments, they are nonequivalent and exhibit geminal coupling (Figure 13.25). The geminal coupling constant is small but discernable in the spectrum because the signals for both H_d and H_e are doublets. Vinyl H atom H_a is split by both H_b (*trans* coupling) and H_c (*cis* coupling), giving rise to a doublet of doublets, or four peaks. H_b is split by H_a (*trans* coupling)

Figure 13.24
300 MHz ^{1}H-NMR spectrum of 2-methyl-2-vinyl oxirane. The two H atoms on the oxirane ring are nonequivalent, so they exhibit geminal coupling.

Chemical Shift (δ)

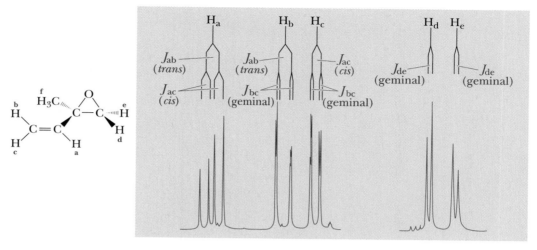

Figure 13.25
Tree diagrams that indicate the complex coupling seen in ¹H-NMR signals for the vinyl group and the oxirane ring H atoms of 2-methy-2-vinyloxirane.

along with H_c, the latter geminal coupling constant is so small that it is barely discernable at this resolution. H_c is split by H_a (*cis* coupling) as well as H_b (geminal coupling). The singlet near 1.5 ppm that integrates to three H atoms is the methyl group labeled f.

E. Coincidental Overlap

Here is a word of caution: Quite often, because peaks can overlap by coincidence, there are fewer *distinguishable* peaks in a signal than predicted. Coincidental peak overlap can occur in any molecule, but it is especially common with flexible alkyl chains. In addition, some coupling constants are so small that peak splitting is hard to see in a spectrum. Thus, the predicted number of peaks using the $(n + 1) \times (m + 1)$ rule should be considered the maximum that *might* be observed. Detailed analysis using extremely high resolution spectrometers is often required to distinguish all the peaks in a highly split signal. You should note also that the types of splitting patterns we have described are applicable only when the separation between coupled signals is much greater than the coupling constant. When this is not the case, spectra can become much more complex.

F. Complex Coupling in Flexible Molecules

Coupling in molecules having unrestricted bond rotation is often simplified to give only $m + n + 1$ peaks, not the expected $(n + 1) \times (m + 1)$. In other words, the number of peaks actually observed for a signal is the number of adjacent hydrogens + 1, no matter how many different sets of equivalent H atoms this represents. The explanation is that bond rotation averages the coupling constants throughout molecules with freely rotating bonds and tends to make them very similar in the 6- to 8-Hz range for H atoms on freely rotating sp^3 hybridized C atoms.

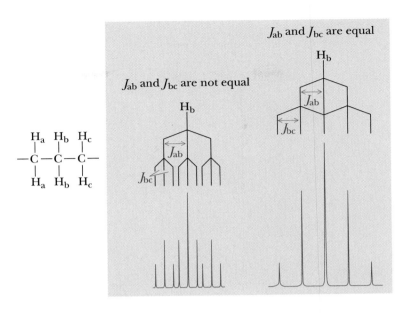

Figure 13.26
Simplification of signal splitting that occurs when coupling constants are the same.

Very similar or identical coupling constants simplify splitting patterns. For example, in the hypothetical unsymmetrical molecule depicted in Figure 13.26, the central H_b atoms are coupled to both H_a atoms as well as both H_c atoms. If $J_{ab} \neq J_{bc}$, this would lead to a triplet of triplets or nine peaks in the signal for H_b. However, if the coupling constants are identical so that $J_{ab} = J_{bc}$, the splitting pattern overlaps considerably to generate only five peaks in the signal for H_b. In the general case, simplification because of very similar or identical J values gives a number of peaks equal to the number of adjacent H atoms + 1, regardless of patterns of equivalence.

A good example of peak overlap occurs in the spectrum of 1-chloro-3-iodopropane (Figure 13.27). The signal for the H atoms of the central —CH_2— group (labeled c on the spectrum) is split by the H atoms on both of the other —CH_2— groups, raising the possibility of splitting into $3 \times 3 = 9$ peaks. However,

Figure 13.27
300-MHz ^{1}H-NMR spectrum of 3-chloro-1-iodopropane.

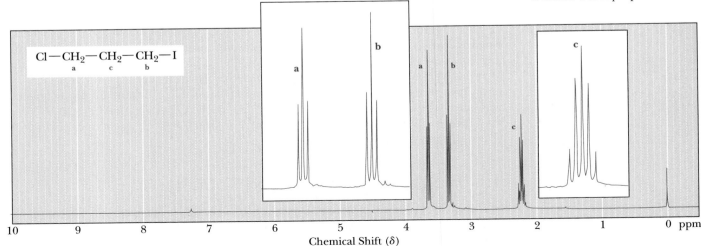

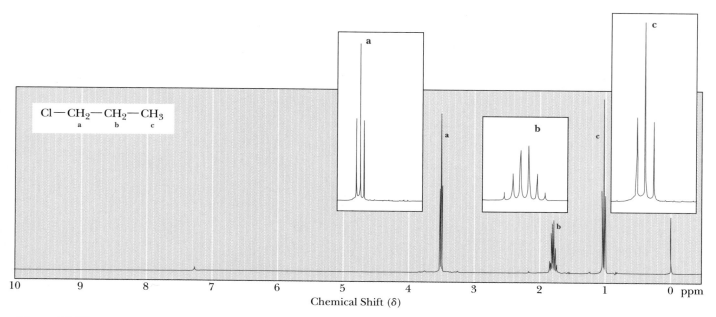

Figure 13.28
300-MHz ^{1}H-NMR spectrum of 1-chloropropane.

because the values for J_{ab} and J_{bc} are so similar, only $4 + 1 = 5$ peaks are distinguishable in the spectrum for the H_c signal as a result of peak overlap.

Another common example is the kind of splitting of the signal for the central $—CH_2—$ in a $—CH_2—CH_2—CH_3$ group, such as occurs in the molecule 1-chloropropane, $Cl—CH_2—CH_2—CH_3$ (Figure 13.28). A maximum of $3 \times 4 = 12$ peaks would be possible for the central $—CH_2—$ signal (labeled b in Figure 13.28), but because the coupling constants are very similar, only $5 + 1 = 6$ peaks are distinguishable.

G. Fast Exchange

Hydrogen atoms bonded to oxygen or nitrogen atoms can exchange with each other faster than the time it takes to acquire a ^{1}H-NMR spectrum. This process is greatly facilitated by even traces of acid or base in a sample. Important affected functional groups include carboxylic acids, alcohols, amines, and amides. Fast exchange has two important consequences. First, signals for exchanging H atoms are generally broad singlets that do not take part in splitting with other signals. Second, the signal will disappear altogether if D_2O or a deuterated alcohol is added to the sample because the H atoms will be replaced with D atoms, which are ^{1}H-NMR silent. This latter phenomenon can be used to identify signals from exchangeable H atoms by taking spectra with and without added D_2O. Note that these same exchangeable H atoms also can take part in hydrogen bonds, the presence of which can alter chemical shift in a concentration-dependent fashion.

13.10 Stereochemistry and Topicity

The discussion of the number of equivalent hydrogens given in the beginning of this chapter is slightly oversimplified because stereochemistry can affect chemical shift. Depending on the symmetry of the molecule, otherwise equivalent atoms may be **homotopic, enantiotopic,** or **diastereotopic.** The simplest way to visualize the topicity of a molecule (that is, which of the these classes it falls into) is by mentally substituting one of the atoms or groups of atoms by an isotope, and then deciding whether the resulting compound would be (a) the same or (b) different from its mirror image or whether (c) diastereomers are possible. Depending on the outcome of the test, the atoms or groups are homotopic, enantiotopic, or diastereotopic, respectively.

Consider the following molecules.

Dichloromethane
(achiral)

Achiral

Substitution does not create a chiral center; therefore hydrogens are homotopic.

Chlorofluoromethane
(achiral)

Chiral

Substitution produces a chiral center; therefore, hydrogens are enantiotopic. Both hydrogens are prochiral; one is pro-*R*-chiral, the other is pro-*S*-chiral.

If one hydrogen in dichloromethane is substituted with one deuterium, an achiral compound results. This molecule is identical to its mirror image, and the two hydrogens in dichloromethane are equivalent and homotopic. Homotopic groups have identical chemical shifts in all environments.

If one hydrogen of chlorofluoromethane is substituted with deuterium, the resulting molecule is chiral, and not identical to its mirror image. The two hydrogens in this compound are therefore enantiotopic. Enantiotopic hydrogens have identical chemical shifts *except in chiral environments.* In a chiral solvent, for example, the two hydrogens would have different chemical shifts. While the distinction between homotopic and enantiotopic compounds is of little practical consequence in NMR spectroscopy, the two hydrogens in chlorofluoromethane can be distinguished by enzymes, which also provide a chiral environment. The CH_2 hydrogens in this molecule are said to be **prochiral.**

The compound 2-butanol presents a more complex situation.

Pro-*R*   Pro-*S*
2-Butanol
(chiral)

Chiral

Substitution creates a chiral center that is diastereomeric; therefore, hydrogens are diastereotopic. Both hydrogens are prochiral; one is pro-*R*-chiral, the other is pro-*S*-chiral.

If a hydrogen on one of the methyl groups on carbon-3 of 3-methyl-2-butanol is substituted with a deuterium, a new chiral center is created. Because there is already one chiral center, diastereomers are now possible. Thus, the methyl groups on carbon-3 of 3-methyl-2-butanol are diastereotopic. Diastereotopic hydrogens have different chemical shifts under all conditions, which can lead to unexpected complexity in spectra of simple compounds. The ^{1}H-NMR spectrum of 3-methyl-2-butanol is shown in

Homotopic groups Atoms or groups on an atom that give an achiral molecule when one of the groups is replaced by another group. The hydrogens of the CH_2 group of propane, for example, are homotopic. Replacing either one of them with deuterium gives 2-deuteropropane, which is achiral. Homotopic groups have identical chemical shifts under all conditions.

Enantiotopic groups Atoms or groups on an atom that give a chiral center when one of the groups is replaced by another group. A pair of enantiomers results. The hydrogens of the CH_2 group of ethanol, for example, are enantiotopic. Replacing one of them by deuterium gives (*R*)-1-deuteroethanol; replacing the other gives (*S*)-1-deuteroethanol. Enantiotopic groups have identical chemical shifts in achiral environments but different chemical shifts in chiral environments.

Diastereotopic groups Atoms or groups on an atom that are bonded to an atom that is bonded to two nonidentical groups, one of which contains a chiral center. When one of the atoms or groups is replaced by another group, a new chiral center is created and a set of diastereomers results. The hydrogens of the CH_2 group of 2-butanol, for example, are diastereotopic. Diastereotopic groups have different chemical shifts under all conditions, although the differences are only seen for diastereotopic hydrogens very close to a chiral center.

CHEMICAL CONNECTIONS

Magnetic Resonance Imaging

The NMR phenomenon was discovered and explained by physicists in the 1950s, and, by the 1960s, it had become an invaluable analytical tool for chemists. By the early 1970s, scientists realized that imaging of parts of the body using NMR could be a valuable addition to diagnostic medicine. Because the term "nuclear magnetic resonance" sounds to many people as if the technique might involve radioactive material, health care personnel call the technique magnetic resonance imaging (MRI). MRI has become so important, that in 2003, the Nobel Prize for medicine or physiology was awarded to Paul Lauterbur and Peter Mansfield for their discoveries that lead to practical MRI.

The body contains several nuclei that, in principle, could be used for MRI. Of these, hydrogens, most of which come from the hydrogens of water, triglycerides (Section 26.1), and membrane phospholipids (Section 26.5) give the most useful signals. Phosphorus MRI is also used in diagnostic medicine.

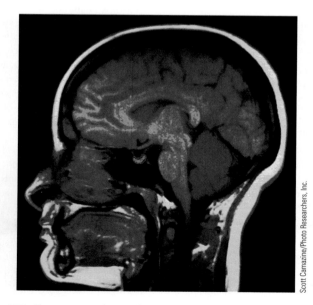

■ Computer-enhanced MRI scan of a normal human brain with pituitary gland highlighted.

Scott Camazine/Photo Researchers, Inc.

Recall that in NMR spectroscopy, energy in the form of radio-frequency radiation is absorbed by nuclei in the sample. Relaxation time is a characteristic time at which excited nuclei give up this energy and relax to their ground state.

In 1971, it was discovered that relaxation of water in certain cancerous tumors takes much longer than the relaxation of water in normal cells. Thus, if a relaxation image of the body could be obtained, it might be possible to identify tumors at an early stage. Subsequent work demonstrated that many tumors can be identified in this way. Another important application of MRI is in the examination of the brain and spinal cord. White and gray matter are easily distinguished by MRI, which is useful in the study of such diseases as multiple sclerosis. Magnetic resonance imaging and x-ray imaging are, in many cases, complementary. The hard, outer layer of bone is essentially invisible to MRI but shows up extremely well in x-ray images, whereas soft tissue is nearly transparent to x-rays but shows up in MRI.

The key to any medical imaging technique is to know which part of the body gives rise to which signal. In MRI, spatial information is encoded using magnetic field gradients. Recall that a linear relationship exists between the frequency at which a nucleus resonates and the intensity of the magnetic field. In ^{1}H-NMR spectroscopy, we use a homogeneous magnetic field, in which all equivalent hydrogens absorb at the same radio frequency and have the same chemical shift. In MRI, the patient is placed in a magnetic field gradient that can be varied from place to place. Nuclei in the weaker magnetic field gradient absorb at a lower frequency. Nuclei elsewhere in the stronger magnetic field absorb at a higher frequency. In a magnetic field gradient, a correlation exists between the absorption frequency of a nucleus and its position in space. A gradient along a single axis images a plane. Two mutually perpendicular gradients image a line segment, and three mutually perpendicular gradients image a point. In practice, more complicated procedures are used to obtain magnetic resonance images, but they are all based on the idea of magnetic field gradients.

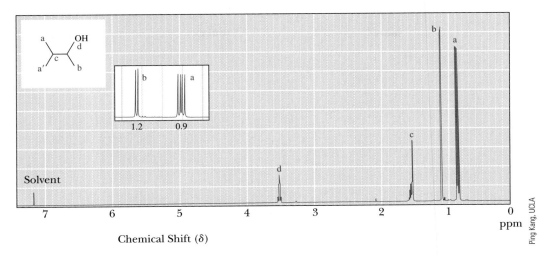

Figure 13.29
^{1}H-NMR spectrum of 3-methyl-2-butanol (500 MHz). The methyl groups on carbon-3 are diastereotopic and therefore nonequivalent. They appear as two doublets.

Figure 13.29. The methyl groups on carbon-3 are nonequivalent and give two doublets rather than one doublet of twice the intensity, which would be expected if they were equivalent.

Any molecule with a chiral center near two otherwise identical groups on a carbon with a third substituent has the potential for diastereotopicity. Of course, like any other nonequivalent groups, diastereotopic groups may have very similar or accidentally identical chemical shifts. Generally, the shift differences fall off rapidly with increasing distance from the chiral center.

Example 13.7

Indicate whether the highlighted hydrogens in the following compounds are homotopic, enantiotopic, or diastereotopic.

(a)   (b) CH_3CH_2Br (c)

Solution

(a) Diastereotopic (near a chiral center). These hydrogens will have different chemical shifts.
(b) Enantiotopic. These hydrogens will have the same chemical shift except in chiral environments.
(c) Enantiotopic [see part (b)].

Problem 13.7

Following is a ^{1}H-NMR spectrum of 2-butanol. Explain why the CH_2 protons appear as a complex multiplet rather than as a simple quintet.

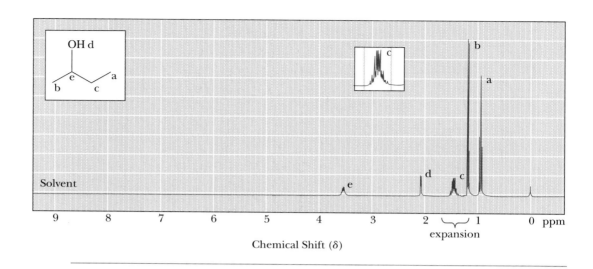

Chemical Shift (δ)

13.11 ^{13}C-NMR

The development of ^{13}C-NMR spectroscopy lagged behind ^{1}H-NMR spectroscopy primarily because of two problems. One is the particularly low natural abundance of ^{13}C (only 1.1%) and the resulting weak signal. The second problem is that the magnetic moment of ^{13}C is considerably smaller than that of ^{1}H, which causes the population of the higher and lower nuclear spin states to differ by much less than for ^{1}H. Taken in combination, these two factors mean that ^{13}C-NMR signals in natural samples (those not artificially enriched with carbon-13) are only about 10^{-4} times the strength of ^{1}H-NMR signals. Even though ^{1}H-NMR spectroscopy became a routine analytical tool in the mid-1960s, it was not until 20 years later with the development of FT-NMR techniques that ^{13}C-NMR spectroscopy became widely available as a routine analytical tool.

As with ^{1}H-NMR spectra, splitting patterns in ^{13}C-NMR spectra are also explained according to the $(n + 1)$ rule. Because, in natural abundance, only 1.1% of carbon atoms are ^{13}C, almost all ^{13}C atoms in a molecule have only magnetically inactive ^{12}C next to them; therefore, ^{13}C—^{13}C signal splitting is not normally observed. However, the signal from a ^{13}C nucleus is split by the hydrogens bonded to it. The signal for a ^{13}C atom with three attached hydrogens is split to a quartet, that for an atom of ^{13}C with two attached hydrogens is split to a triplet, and so on. The ^{13}C—H signal splitting provides important information about the number of hydrogen atoms bonded to carbon. The disadvantage of ^{13}C—H signal splitting is that coupling constants of between 100 and 250 Hz are common. Coupling constants of this magnitude correspond to 1.33 to 3.33 ppm at 75 MHz, which means that the overlap among signals can be significant and that splitting patterns are very often difficult to determine. In addition, there are smaller but significant couplings from hydrogens that are not directly bonded to the carbon, but are separated by two or three bonds. This extensive splitting causes the already weak signals of the ^{13}C to split into many smaller peaks that are easily lost in the noise. For this reason, the most common mode of operation of a ^{13}C-NMR spectrometer is a hydrogen-decoupled mode. (See Problem 13.27 for an interesting problem on the use of coupling constants to determine orbital hybridization.)

In the hydrogen-decoupled mode, the sample is irradiated with two different radio frequencies. The first radio frequency is used to excite the ^{13}C nuclei. The second

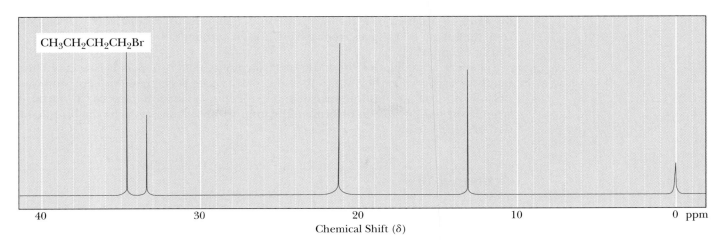

CH$_3$CH$_2$CH$_2$CH$_2$Br

40   30   20   10   0 ppm

Chemical Shift (δ)

Figure 13.30
Hydrogen-decoupled ^{13}C-NMR spectrum of 1-bromobutane.

is a broad spectrum of frequencies that causes all hydrogens in the molecule to undergo rapid transitions among their nuclear spin states. On the time scale of a ^{13}C-NMR spectrum, each hydrogen is in a time average of the two states, with the result that ^{1}H-^{13}C spin-spin interactions are not observed. The term for this process is spin-spin decoupling. In a hydrogen-decoupled spectrum, all ^{13}C signals appear as singlets. The hydrogen-decoupled ^{13}C-NMR spectrum of 1-bromobutane (Figure 13.30) consists of four singlets.

Figure 13.31 shows approximate chemical shifts for ^{13}C-NMR. Notice how much wider the range of chemical shifts is for ^{13}C-NMR than for ^{1}H-NMR. Most chemical shifts for ^{1}H-NMR fall within a rather narrow range of 0 to 10 ppm; however, those for ^{13}C-NMR cover 0 to 220 ppm. Because of this expanded scale, it is very unusual to find any two nonequivalent carbons in the same molecule with identical chemical shifts. Most commonly, each different type of carbon within a molecule has a distinct signal clearly resolved from all other signals.

Notice further that the chemical shift of carbonyl carbons is quite distinct from those of sp^3 hybridized carbons and of other types of sp^2 hybridized carbons. The

Figure 13.31
^{13}C-NMR chemical shifts of representative groups. These values are approximate. Other atoms in the molecules may cause signals to appear outside of these ranges.

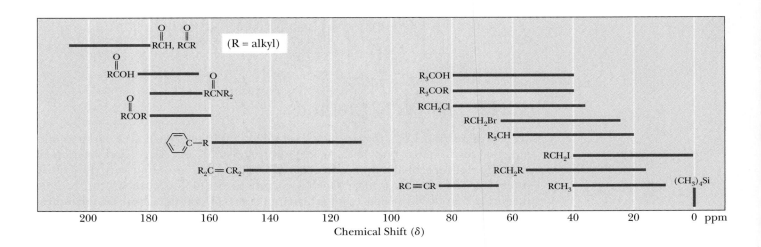

200   180   160   140   120   100   80   60   40   20   0 ppm

Chemical Shift (δ)

presence or absence of a carbonyl carbon is quite easy to recognize in a ^{13}C-NMR spectrum. Note that signals from sp^2 hybridized carbons fall in a distinctive range of 100 to 160 ppm.

A great advantage of ^{13}C-NMR spectroscopy is that it is possible to count the number of types of carbon atoms in a molecule. There is one caution here, however. Because of certain complications, including the long relaxation times of ^{13}C nuclei, it is generally not possible to determine the number of carbons of each type by integration of signal areas.

Example 13.8

Predict the number of signals in a proton-decoupled ^{13}C-NMR spectrum of each compound.

(a) $CH_3\overset{\overset{\displaystyle O}{\|}}{C}OCH_3$ (b) $CH_3CH_2CH_2\overset{\overset{\displaystyle O}{\|}}{C}CH_3$ (c) $CH_3CH_2\overset{\overset{\displaystyle O}{\|}}{C}CH_2CH_3$

Solution

Following are the number of signals in the proton-decoupled spectrum of each compound, along with the chemical shifts of each signal. The chemical shifts of the carbonyl carbons are quite distinctive (Figure 13.31) and in these examples occur at δ 171.37, 208.85, and 211.97.

(a) Methyl acetate: three signals (δ 171.37, 51.53, and 20.63)
(b) 2-Pentanone: five signals (δ 208.85, 45.68, 29.79, 17.35, and 13.68)
(c) 3-Pentanone: three signals (δ 211.97, 35.45, and 7.92)

Problem 13.8

Explain how to distinguish between the members of each pair of constitutional isomers based on the number of signals in the proton-decoupled ^{13}C-NMR spectrum of each member.

(a) and   (b) and

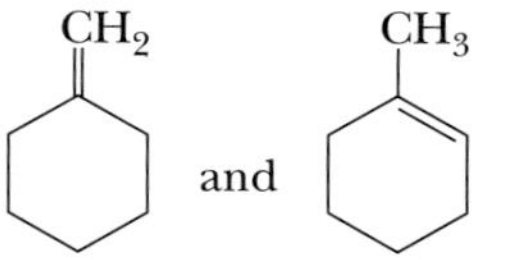

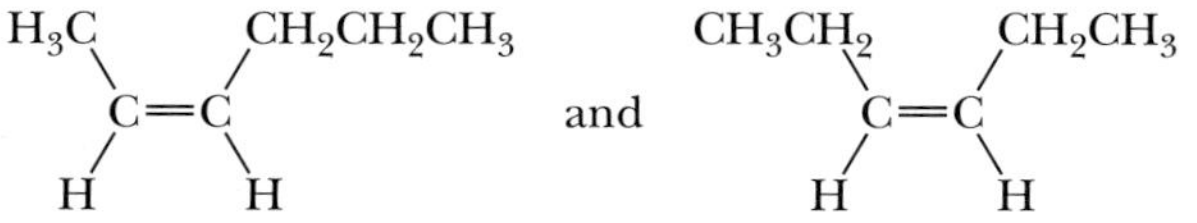

13.12 The DEPT Method

We saw in Section 13.11 that there is spin-spin coupling between a ^{13}C atom and its attached hydrogens, but, because coupling constants are large and overlap of peaks is considerable, proton-coupled ^{13}C-NMR spectra are often very difficult to interpret. For these reasons, ^{13}C-NMR spectra are commonly run in the proton-decoupled mode, in which case information on C/H ratios is lost. **Distortionless Enhancement by Polarization Transfer,** or **DEPT** as it is more commonly known, provides a way to reacquire this information with good signal strength. DEPT uses

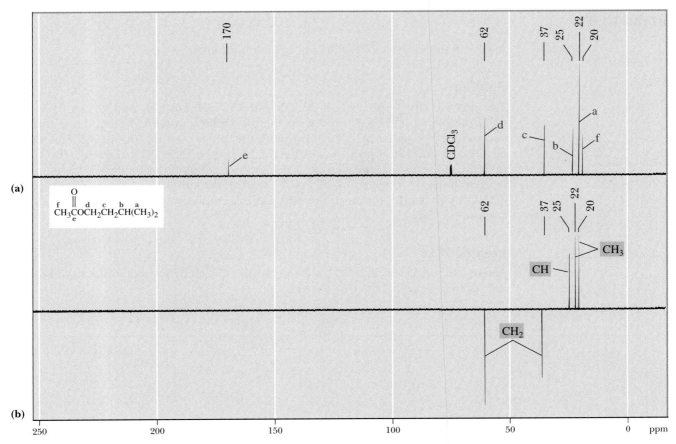

Figure 13.32
^{13}C-NMR spectra of isopentyl acetate. (a) Proton-decoupled spectrum and (b) DEPT spectrum.

complex sequences of pulses in both the ^{1}H and ^{13}C resonance ranges. (Pulse sequences are specific sequences of radio-frequency pulses with precisely controlled duration, intensity, and time between pulses and other important parameters.) As a result, the ^{13}C signals for CH$_3$, CH$_2$, and CH exhibit different "phases." Signals for CH$_3$ and CH carbons are recorded as positive signals, and those for CH$_2$ carbons are recorded as negative signals with one pulse sequence. By using a slightly different pulse sequence, CH$_3$ and CH signals can be distinguished. In the DEPT technique, a carbon with no attached hydrogens, such as a carbonyl or a quaternary carbon, gives no signal.

DEPT spectra may be displayed in several ways. In the most common variation, CH$_3$, CH$_2$, and CH signals are recorded on one spectrum. The first trace shows CH$_3$ and CH as positive signals, and the second trace shows CH$_2$ as negative signals. Usually, the CH carbons absorb at a lower field than the CH$_3$, so there is no ambiguity. Shown in Figure 13.32(a) is a proton-decoupled ^{13}C-NMR spectrum of isopentyl acetate showing six signals. Figure 13.32(b) is a DEPT spectrum showing color-coded signals corresponding to CH$_3$, CH$_2$, and CH groups. Note that the carbonyl carbon appears in the proton-decoupled spectrum but does not appear in the DEPT spectrum because it has no attached hydrogens.

Example 13.9

Assign all signals in the ^{13}C-NMR spectrum of isopentyl acetate.

Solution

The positive DEPT signals at δ 20, 22, and 25 represent CH_3 and CH groups. The taller methyl signal at δ 22 represents the two equivalent methyl groups (a), and the shorter methyl signal at δ 20 represents the single methyl group (f). The signal at δ 25, the only other positive DEPT signal, represents the CH group (b). The signal at δ 62 represents (d), the CH_2 group nearer to and more deshielded by the adjacent oxygen atom. The signal at δ 37 represents (c), the other CH_2 group. The signal at δ 170, which is not present in the DEPT spectrum, represents the carbon (e) of the carbonyl group.

Problem 13.9

Assign all signals in the ^{13}C-NMR spectrum of 4-methyl-2-pentanone. (a) Proton-decoupled spectrum, (b) DEPT spectrum.

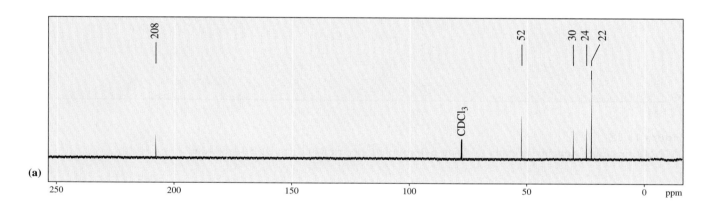

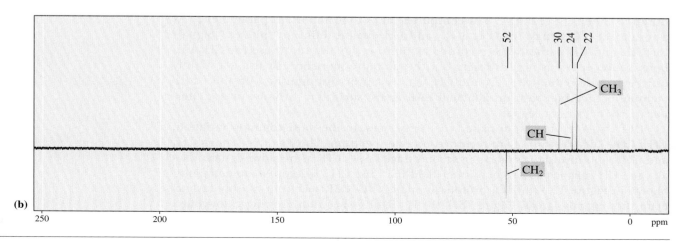

13.13 Interpretation of NMR Spectra

A. Alkanes

All hydrogens in alkanes are in very similar chemical environments; therefore, ^{1}H-NMR chemical shifts of alkane hydrogens fall within a narrow range of δ 0.8 to 1.7. Chemical shifts for alkane carbons in ^{13}C-NMR spectroscopy fall within the considerably wider range of δ 10 to 60.

B. Alkenes

The chemical shifts of vinylic hydrogens are larger than those of alkane hydrogens and typically fall in the range δ 4.6 to 5.7. Vinylic hydrogens are deshielded by the sp^2 hybridized carbons of the double bond (Section 13.9B) and the local magnetic field induced in the pi bond of alkenes (Section 13.9C). The splitting pattern observed in the ^{1}H-NMR spectrum of vinyl acetate (Figure 13.33) is typical of monosubstituted alkenes. The singlet at δ 2.12 represents the three hydrogens of the methyl group. The terminal vinylic hydrogens appear at δ 4.58 and δ 4.90. The internal vinylic hydrogen, which normally appears in the range δ 5.0 to 5.7, is shifted farther downfield to δ 7.30 as a result of deshielding by the adjacent electronegative oxygen atom of the ester.

As shown in Table 13.4, coupling constants are generally larger for *trans* vinylic hydrogens (11–18 Hz) than for *cis* vinylic hydrogens (5–10 Hz), and it is often possible to distinguish between *cis* and *trans* alkenes by an analysis of their coupling constants. It is also possible to distinguish between vicinal hydrogens and geminal hydrogens ($=CH_2$), the latter having small coupling constants, generally in the 0 to 5 Hz range.

The signal of each vinylic hydrogen in vinyl acetate is predicted to be a doublet of doublets. The signal for H_c, for example, is split to a doublet by coupling with H_a and further split to a doublet of doublets by coupling with H_b. For H_a and H_b, the geminal coupling is so small that it is not visible at this resolution, so their signals appear as doublets. Higher resolution would reveal the geminal coupling.

Figure 13.33
^{1}H-NMR spectrum of vinyl acetate.

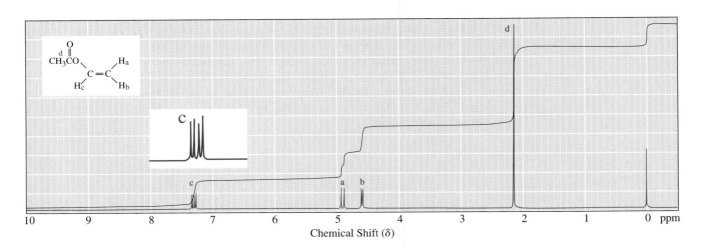

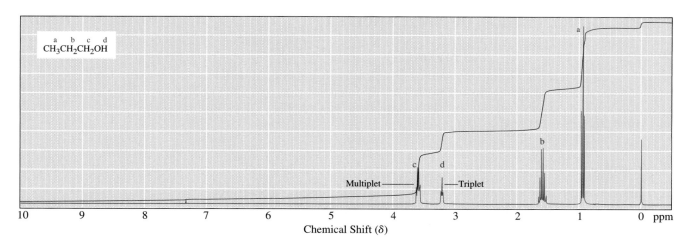

Figure 13.34
^{1}H-NMR spectrum of 1-propanol. The hydroxyl hydrogen appears at δ 3.18 as a narrowly spaced triplet. The signal of hydrogens on carbon 1 of 1-propanol appears as a quartet at δ 3.56 (split by the two CH_2 hydrogens and the one OH hydrogen).

The sp^2 hybridized carbons of alkenes give ^{13}C-NMR signals in the range δ 100 to 150 ppm (Figure 13.31), which is considerably downfield from sp^3 hybridized carbons.

C. Alcohols

The chemical shift of a hydroxyl hydrogen in an ^{1}H-NMR spectrum is variable and depends on the purity of the sample, the solvent, the concentration, and the temperature. It often appears in the range δ 3.0 to 4.0, but depending on experimental conditions, it may appear as far upfield as δ 0.5. Hydrogens on the carbon bearing the —OH group are deshielded by the electron-withdrawing inductive effect of the oxygen atom, and their absorptions also typically appear in the range δ 3.4 to 4.0. Shown in Figure 13.34 is the ^{1}H-NMR spectrum of 1-propanol.

Signal splitting between the hydrogen on O—H and its neighbors on the adjacent —CH_2— group is seen in the ^{1}H-NMR spectrum of 1-propanol. However, this splitting is rarely seen. The reason is that most samples of alcohol contain traces of acid, base, or other impurities that catalyze the transfer of the hydroxyl proton from the oxygen of one alcohol molecule to the oxygen of another alcohol molecule. This fast exchange decouples the hydroxyl proton from all other protons in the molecule (Section 13.9G). For this same reason, the hydroxyl proton does not usually split the signal of any α-hydrogens.

D. Ethers

The most distinctive feature of the ^{1}H-NMR spectra of ethers is the chemical shift of hydrogens on the carbons bonded to the ether oxygen. Resonance signals for this type of hydrogen fall in the range δ 3.3 to 4.0, which corresponds to a downfield shift of approximately 2.4 units compared with their normal position in alkanes. The chemical shifts of **H—C—O—** hydrogens in ethers are similar to those seen for comparable **H—C—OH** hydrogens of alcohols.

E. Aldehydes and Ketones

Aldehyde hydrogens typically appear between δ 9.5 and δ 10.1 in ^{1}H-NMR spectra. Because almost nothing else absorbs in this region, it is very useful for identification. Hydrogens on an α-carbon of an aldehyde or ketone appear around δ 2.2 to 2.6. The carbonyl carbons of aldehydes and ketones have characteristic positions in the ^{13}C-NMR between δ 180 and δ 215 (and can be distinguished from carboxylic acid derivatives, which absorb at higher field).

F. Carboxylic Acids and Esters

Hydrogens on the α-carbon to a carboxyl group in acids and esters appear in a ^{1}H-NMR spectrum in the range δ 2.0 to 2.6. The hydrogen of a carboxyl group gives a very distinctive signal in the range δ 10 to 13, downfield of most other types of hydrogens, even farther downfield than that of an aldehyde hydrogen (δ 9.5–10.1), and serves to distinguish carboxyl hydrogens from most other types of hydrogens. The ^{1}H-NMR signal for the carboxyl hydrogen of 2-methylpropanoic acid, for example, appears at δ 13.2 and is shown at the left in Figure 13.35.

The ^{13}C resonance of the carboxyl carbon in acids and esters appears in the range δ 165 to 185 and at a distinctly higher field than that in ketones. Hydrogens α to an ester oxygen are strongly deshielded and resonate between δ 3.7 and δ 4.7, lower than in alcohols and ethers.

Figure 13.35
^{1}H-NMR spectrum of 2-methylpropanoic acid (isobutyric acid).

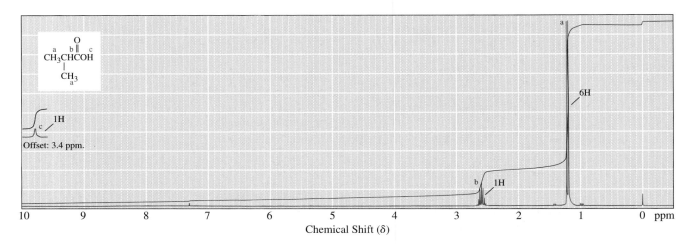

How To *Solve NMR Spectral Problems*

ORGANIC
Chemistry ⚛ Now™
Click Tutorials to explore
Index of Hydrogen Deficiency

One of the first steps in determining molecular structure is establishing the molecular formula. In the past, this task was most commonly done by elemental analysis, combustion analysis to determine percent composition, molecular weight determination, and so forth. More commonly today, molecular weight and molecular formula are determined by mass spectrometry (Chapter 14). In the examples that follow, we assume that the molecular formula of any unknown compound has already been determined, and we proceed from that point using spectral analysis to determine a structural formula.

A. Molecular Formula

Valuable information about the structural formula of an unknown compound can be obtained by inspecting its molecular formula, which gives its index of hydrogen deficiency. Refer to Chapter 5 to review this technique. A molecular formula can often be obtained from the mass spectrum (see Chapter 14).

B. From an ¹H-NMR Spectrum to a Structural Formula

The following steps may prove helpful as a systematic approach to solving spectral problems.

Step 1: Molecular formula and index of hydrogen deficiency. Examine the molecular formula, calculate the index of hydrogen deficiency, and deduce what information you can about the presence or absence of rings or pi bonds.

Step 2: Number of signals. Count the number of signals to determine the number of sets of equivalent hydrogens present in the compound.

Step 3: Integration. Use the integration and the molecular formula to determine the numbers of hydrogens present in each set.

Step 4: Pattern of chemical shifts. Compare signal chemical shifts with reference tables to determine which functional groups may be present. Keep in mind that these are broad ranges and that hydrogens of each type may be shifted either farther upfield or farther downfield, depending on the details of the molecular structure.

Types of Hydrogens	Descriptive Name	Chemical Shift (δ)
RCH_3 RCH_2R R_3CH	Alkyl hydrogens	0.8–1.7
$R_2C{=}CRCHR_2$	Allylic hydrogens	1.6–2.6
$\overset{\overset{\displaystyle O}{\|\|}}{RCH_2CR}$	Hydrogens on a sp^3 carbon adjacent to a carbonyl group	2.2–2.6
RCH_2OH RCH_2OR	Hydrogens on a carbon adjacent to an sp^3-hybridized oxygen	3.3–4.0
$R_2C{=}CH_2$ $R_2C{=}CHR$	Vinylic hydrogens	4.6–5.7
ArH	Aryl hydrogens	6.5–8.5
$\overset{\overset{\displaystyle O}{\|\|}}{RCH}$	Aldehyde hydrogens	9.5–10.1
$\overset{\overset{\displaystyle O}{\|\|}}{RCOH}$	Carboxyl hydrogens	10–13

Step 5: Signal splitting patterns. Examine splitting patterns for information about the number of nearest nonequivalent hydrogen neighbors.

Step 6: Structural formula. Construct possible molecules from the functional groups present, their relative signal integrations, and any other information you are given, especially the molecular formula and other spectra. Confirm the correct structure by making sure all the available information matches.

Spectral Problem 1: Molecular formula $C_5H_{10}O$

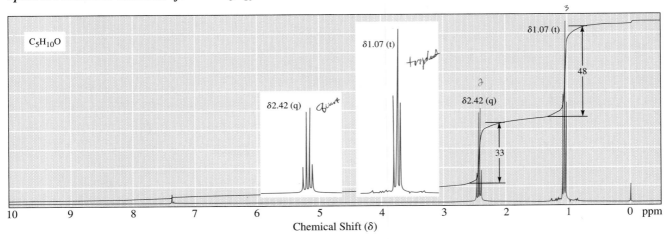

Analysis of Spectral Problem 1

Step 1: Molecular formula and index of hydrogen deficiency. The reference compound is C_5H_{12}; therefore, the index of hydrogen deficiency is 1 and the molecule contains either one ring or one pi bond.

Step 2: Number of signals. There are two signals (a triplet and a quartet) and, therefore, two sets of equivalent hydrogens.

Step 3: Integration. From the integration, the hydrogens in each set are in the ratio 3:2. Because there are 10 hydrogens, 6H must give rise to the signal at $\delta 1.07$, and 4H must give rise to the signal at $\delta 2.42$.

Step 4: Pattern of chemical shifts. The signal at $\delta 1.07$ is in the alkyl region and, based on its chemical shift, most probably indicates a methyl group. No signal occurs between $\delta 4.6$ and $\delta 5.7$; there are no vinylic hydrogens. If a carbon-carbon double bond is in the molecule, there are no hydrogens on it (that is, it is tetrasubstituted). The chemical shift of the four protons at $\delta 2.42$ is consistent with two CH_2 groups next to a carbonyl group.

Step 5: Signal splitting patterns. The methyl signal at $\delta 1.07$ is split into a triplet (t); it must have two neighbors, indicating $-CH_2CH_3$. The signal at $\delta 2.42$ is split into a quartet (q); it must have three neighbors. An ethyl group accounts for these two signals. No other signals occur in the spectrum; therefore, there are no other types of hydrogens in the molecule.

Step 6: Structural formula. Put this information together to arrive at the following structural formula. The chemical shift of the methylene group ($-CH_2-$) at $\delta 2.42$ is consistent with an alkyl group adjacent to a carbonyl group.

$$CH_3-CH_2-\overset{\overset{\displaystyle O}{\|}}{C}-CH_2-CH_3$$

3-Pentanone

Spectral Problem 2: Molecular formula $C_7H_{14}O$

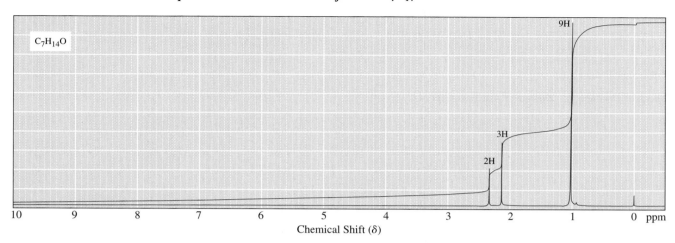

Analysis of Spectral Problem 2

Step 1: Molecular formula and index of hydrogen deficiency. The index of hydrogen deficiency is 1; the compound contains one ring or one pi bond.

Step 2: Number of signals. There are three signals and, therefore, three sets of equivalent hydrogens.

Step 3: Integration. Reading from right to left, there are 9, 3, and 2 hydrogens in these signals.

Step 4: Pattern of chemical shifts. The signal at δ 1.01 is characteristic of a methyl group adjacent to an sp^3 hybridized carbon. The signals at δ 2.11 and δ 2.32 are characteristic of alkyl groups adjacent to a carbonyl group.

Step 5: Signal splitting pattern. All signals are singlets (s). Therefore, none of the groups has hydrogens on neighboring carbons.

Step 6: Structural formula. The compound is 4,4-dimethyl-2-pentanone.

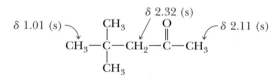

4,4-Dimethyl-2-pentanone

G. Amines

The chemical shifts of amine hydrogens, like those of hydroxyl hydrogens (Section 13.13C), vary between δ 0.5 and δ 5.0, depending on experimental conditions. As in alcohols, exchange is fast enough that spin-spin splitting between amine hydrogens and hydrogens on adjacent α-carbons is averaged. Thus, amine hydrogens generally appear as broad singlets. Coupling to ^{14}N (beyond the scope of this text) causes these signals to broaden. Hydrogens α to the amine nitrogen appear around δ 2.5 ppm, about 1 ppm higher than for hydrogens α to oxygen in ethers and alcohols.

Carbons bonded to nitrogen appear in the ^{13}C-NMR spectrum approximately 20 ppm lower than in alkanes of comparable structure, but about 20 ppm above carbons attached to oxygen in ethers or alcohols.

Here is a final word. We have barely scratched the surface of what NMR can do. Something called the nuclear Overhauser enhancement (NOE) can determine distances between atoms in molecules that are near each other in three-dimensional space even if there are more than three bonds separating them. In addition, using modern instruments, spectra that examine multiple parameters simultaneously can be produced to yield immense amounts of information about even very complicated molecules. Such spectra are plotted on more than one axis so they are referred to as multidimensional spectra. Multidimensional spectra are used to deduce structure and conformation of molecules ranging from small organic molecules to large biological macromolecules such as proteins, DNA, and RNA.

Summary

Absorption of electromagnetic energy leads to spectroscopic tools that are important for the determination of structures of organic molecules. These absorptions can be from electronic transitions (UV-visible spectra (Section 20.2), vibrations (IR spectra, Chapter 12) and nuclear magnetic spin transitions (NMR spectra at radio frequencies, Section 13.1).

Nuclei of ^{1}H and ^{13}C have a nuclear spin quantum number of $\frac{1}{2}$ and allowed nuclear spin states of $+\frac{1}{2}$ and $-\frac{1}{2}$ (Section 13.3). In the presence of an applied magnetic field, B_0, nuclei with spin $+\frac{1}{2}$ are aligned with the applied field and are in the lower energy state; nuclei with spin $-\frac{1}{2}$ are aligned against the applied field and are in the higher energy state (Section 13.4).

When placed in a powerful magnetic field (Section 13.2), ^{1}H and ^{13}C nuclei become aligned in an allowed spin state and precess about the applied field. **Resonance** is the absorption of electromagnetic radiation by a precessing nucleus, and the resulting flip of its nuclear spin from the lower energy spin state to the higher energy spin state (Section 13.3). An NMR spectrometer records such resonance as a **signal** and plots the irradiation frequencies where absorption takes place scaled by the strength of the applied magnetic field using the units of ppm (Section 13.4).

The experimental conditions required to cause nuclei to resonate are affected by the local chemical and magnetic environment. Electrons around a hydrogen or carbon create local magnetic fields that shield the nuclei of these atoms from the applied field (Section 13.3). Any factor that increases the exposure of nuclei to an applied field is said to **deshield** them and shifts their signal downfield to a larger δ value. Conversely, any factor that decreases the exposure of nuclei to an applied field is said to **shield** them and shifts their signal upfield to a smaller δ value.

Equivalent hydrogens within a molecule have identical chemical shifts (Section 13.5). The chemical shift of a particular set of equivalent hydrogens depends primarily on three factors: (1) nearby electronegative atoms have a deshielding effect; (2) the the greater the percent of s-character in a hybrid orbital the greater the deshielding effect of the atom to which the orbital belongs; and (3) induced local magnetic fields in pi bonds either add to or subtract from the applied field. The area of a ^{1}H-NMR signal is proportional to the number of equivalent hydrogens giving rise to that signal (Section 13.6).

The resonance signals in ^{1}H-NMR spectra are reported by how far they are shifted from the resonance signal of the 12 equivalent hydrogens in **tetramethylsilane (TMS).** The resonance signals in ^{13}C-NMR spectra are reported by how far they are shifted from the resonance signal of the four equivalent carbons in TMS. **Chemical shift, δ,** is defined as the frequency shift from TMS divided by the operating frequency of the spectrometer (Section 13.7).

According to the **($n + 1$) rule,** if a hydrogen has n hydrogens nonequivalent to it but equivalent among themselves on the same or adjacent atom(s), its ^{1}H-NMR signal will be split into ($n + 1$) peaks (Section 13.8). Splitting patterns are commonly referred to as singlets (s), doublets (d), triplets (t), quartets (q), quintets, and multiplets (m). The relative intensities of peaks in a multiplet can be predicted from an analysis of spin combinations for adjacent hydrogens (Table 13.4) or from the mnemonic device called **Pascal's triangle** (Figure 13.16). When a hydrogen is coupled to more than one set of adjacent hydrogen atoms, the couplings combine. In the general case, if a hydrogen is coupled to a set of n hydrogens on one side and a set of m hydrogens on the other, the signal will be split into a maximum of ($n + 1$)($m + 1$) peaks. In molecules that are rigid, for example alkenes or cyclic molecules, all the ($n + 1$)($m + 1$) peaks can often be seen. However, because coupling constants can be similar, especially in flexible molecules, this splitting can simplify to a number of observed peaks that is equal to the number of adjacent H atoms + 1, regardless of patterns of equivalence.

A **coupling constant (J)** is the distance between adjacent peaks in a multiplet and is reported in hertz (Section 13.11). The value of J depends only on internal fields within a molecule and is independent of the spectrometer field.

Groups of atoms in which substitution of one atom by an isotope creates an achiral molecule are called **homotopic** (Section 13.12). Those in which such substitution produces a chiral molecule are **enantiotopic.** Those molecules in which substitution produces diastereomers are called **diastereotopic.** Homotopic groups always have identical chemical shifts. Enantiotopic groups also do, except in a chiral environment. Diastereotopic groups, however, are nonequivalent in all environments.

Four important types of structural information can be obtained from a ^{1}H-NMR spectrum.

- From the number of signals, we can determine the number of sets of equivalent hydrogens.
- From the integration of signal areas, we can determine the relative numbers of hydrogens in each set.
- From the chemical shift of each signal, we can derive information about the chemical environment of the hydrogens in each set.
- From the splitting pattern of each signal, we can derive information about the number and chemical equivalency of hydrogens on the same and adjacent carbon atoms, in other words the connectivities between different groups on the molecule.

^{13}C-NMR spectra (Section 13.13) are commonly recorded in a hydrogen-decoupled instrumental mode. In this mode, all ^{13}C signals appear as singlets. The DEPT method can be used to identify CH_3, CH_2, and CH signals separately (Section 13.14).

Problems

ORGANIC
Chemistry.⚛.Now™

Assess your understanding of this chapter's topics with additional quizzing and conceptual-based problems at **http://now.brookscole.com/ bfi4**

13.10 Calculate the index of hydrogen deficiency of these compounds.

(a) Aspirin, $C_9H_8O_4$
(b) Ascorbic acid (vitamin C), $C_6H_8O_6$
(c) Pyridine, C_5H_5N
(d) Urea, CH_4N_2O
(e) Cholesterol, $C_{27}H_{46}O$
(f) Dopamine, $C_8H_{11}NO_2$

Interpretation of ^{1}H-NMR and ^{13}C-NMR Spectra

13.11 Complete the following table. Which nucleus requires the least energy to flip its spin at this applied field? Which nucleus requires the most energy?

Nucleus	Applied Field (tesla, T)	Radio Frequency (MHz)	Energy (J/mol)
^{1}H	7.05	300	_____
^{13}C	7.05	75.5	_____
^{19}F	7.05	282	_____

13.12 The natural abundance of ^{13}C is only 1.1%. Furthermore, its sensitivity in NMR spectroscopy (a measure of the energy difference between a spin aligned with or against an applied magnetic field) is only 1.6% that of ^{1}H. What are the relative signal intensities expected for the ^{1}H-NMR and ^{13}C-NMR spectra of the same sample of $Si(CH_3)_4$?

13.13 Following are structural formulas for three constitutional isomers of molecular formula $C_7H_{16}O$ and three sets of ^{13}C-NMR spectral data. Assign each constitutional isomer its correct spectral data.

(a) $CH_3CH_2CH_2CH_2CH_2CH_2CH_2OH$

(b)
$$\begin{array}{c} OH \\ | \\ CH_3CCH_2CH_2CH_2CH_3 \\ | \\ CH_3 \end{array}$$

(c)
$$\begin{array}{c} OH \\ | \\ CH_3CH_2CCH_2CH_3 \\ | \\ CH_2CH_3 \end{array}$$

Spectrum 1	Spectrum 2	Spectrum 3
74.66	70.97	62.93
30.54	43.74	32.79
7.73	29.21	31.86
	26.60	29.14
	23.27	25.75
	14.09	22.63
		14.08

13.14 Following are structural formulas for the *cis* isomers of 1,2-, 1,3-, and 1,4-dimethylcyclo-hexane and three sets of ^{13}C-NMR spectral data. Assign each constitutional isomer its correct spectral data.

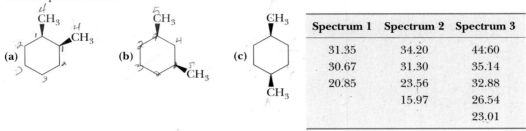

	Spectrum 1	Spectrum 2	Spectrum 3
	31.35	34.20	44.60
	30.67	31.30	35.14
	20.85	23.56	32.88
		15.97	26.54
			23.01

13.15 Following are structural formulas, dipole moments, and ^{1}H-NMR chemical shifts for acetonitrile, fluoromethane, and chloromethane.

$$CH_3C{\equiv}N \qquad CH_3F \qquad CH_3Cl$$

Acetonitrile	Fluoromethane	Chloromethane
3.92 D	1.85 D	1.87 D
δ 1.97	δ 4.26	δ 3.05

(a) How do you account for the fact that the dipole moments of fluoromethane and chloromethane are almost identical even though fluorine is considerably more electronegative than chlorine?

(b) How do you account for the fact that the dipole moment of acetonitrile is considerably greater than that of either fluoromethane or chloromethane?

(c) How do you account for the fact that the chemical shift of the methyl hydrogens in acetonitrile is considerably less than that for either fluoromethane or chloromethane?

13.16 Following are three compounds of molecular formula $C_4H_8O_2$, and three ^{1}H-NMR spectra. Assign each compound its correct spectrum and assign all signals to their corresponding hydrogens.

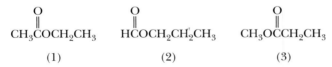

$$\underset{(1)}{CH_3COCH_2CH_3} \qquad \underset{(2)}{HCOCH_2CH_2CH_3} \qquad \underset{(3)}{CH_3OCCH_2CH_3}$$

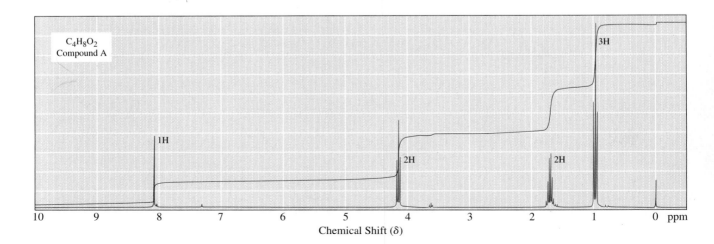

$C_4H_8O_2$
Compound A

1H

2H

2H

3H

10   9   8   7   6   5   4   3   2   1   0 ppm

Chemical Shift (δ)

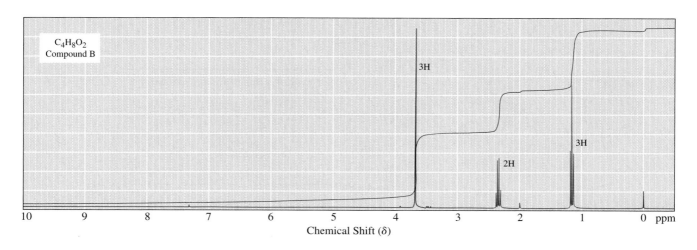

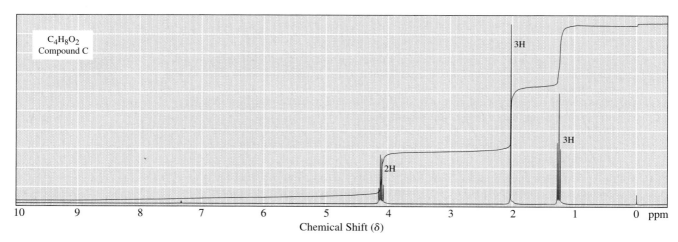

13.17 Following are ^{1}H-NMR spectra for compounds D, E, and F, each of molecular formula C_6H_{12}. Each readily decolorizes a solution of Br_2 in CCl_4. Propose structural formulas for compounds D, E, and F, and account for the observed patterns of signal splitting.

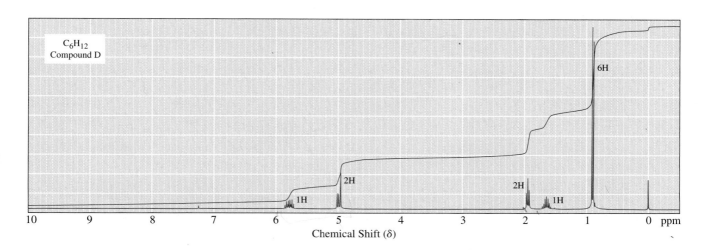

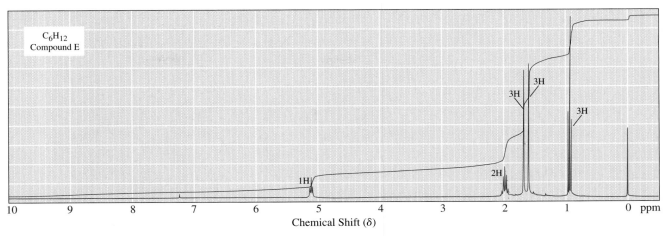

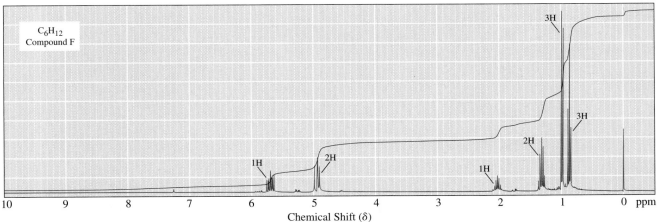

13.18 Following are ¹H-NMR spectra for compounds G, H, and I, each of molecular formula $C_5H_{12}O$. Each is a liquid at room temperature, is slightly soluble in water, and reacts with sodium metal with the evolution of a gas.

  (a) Propose structural formulas of compounds G, H, and I.
  (b) Explain why there are four lines between $\delta\, 0.86$ and 0.90 for compound G.
  (c) Explain why the 2H multiplets at 1.5 and 3.5 ppm for compound H are so complex.

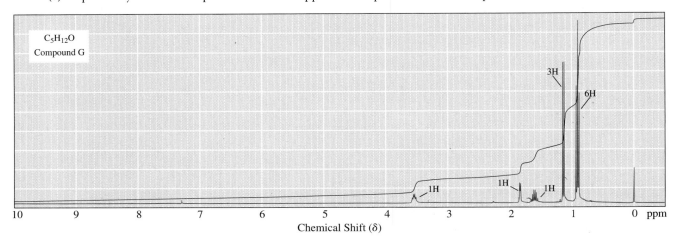

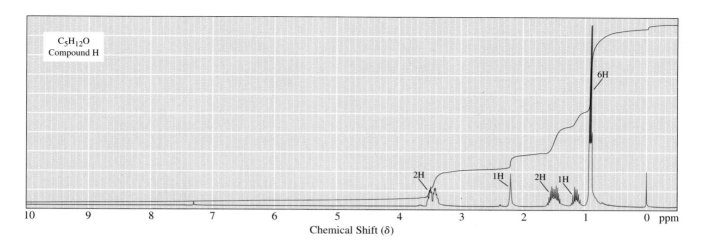

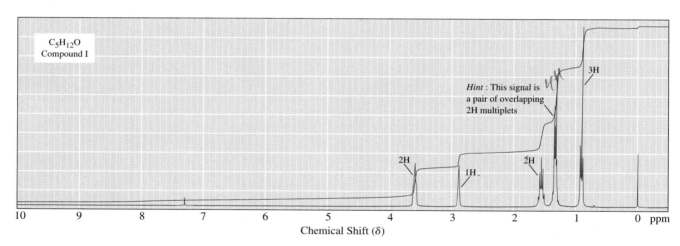

13.19 Propose a structural formula for compound J, molecular formula C_3H_6O, consistent with the following ^{1}H-NMR spectrum.

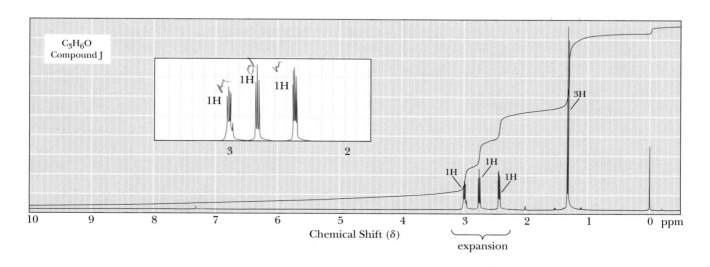

13.20 Compound K, molecular formula $C_6H_{14}O$, readily undergoes acid-catalyzed dehydration when warmed with phosphoric acid to give compound L, molecular formula C_6H_{12}, as the major organic product. The ^{1}H-NMR spectrum of compound K shows signals at δ 0.90 (t, 6H), 1.12 (s, 3H), 1.38 (s, 1H), and 1.48 (q, 4H). The ^{13}C-NMR spectrum of compound K shows signals at 72.98, 33.72, 25.85, and 8.16. Deduce the structural formulas of compounds K and L.

13.21 Compound M, molecular formula $C_5H_{10}O$, readily decolorizes Br_2 in CCl_4 and is converted by H_2/Ni into compound N, molecular formula $C_5H_{12}O$. Following is the ^{1}H-NMR spectrum of compound M. The ^{13}C-NMR spectrum of compound M shows signals at 146.12, 110.75, 71.05, and 29.38. Deduce the structural formulas of compounds M and N.

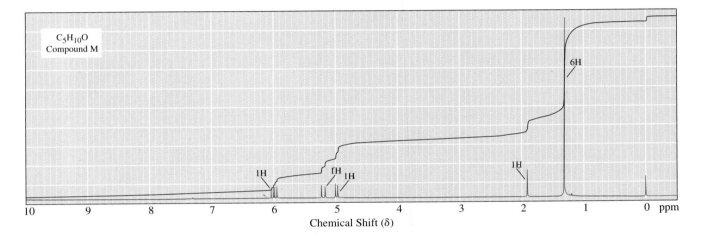

13.22 Following is the ^{1}H-NMR spectrum of compound O, molecular formula C_7H_{12}. Compound O reacts with bromine in carbon tetrachloride to give a compound of molecular formula $C_7H_{12}Br_2$. The ^{13}C-NMR spectrum of compound O shows signals at 150.12, 106.43, 35.44, 28.36, and 26.36. Deduce the structural formula of compound O.

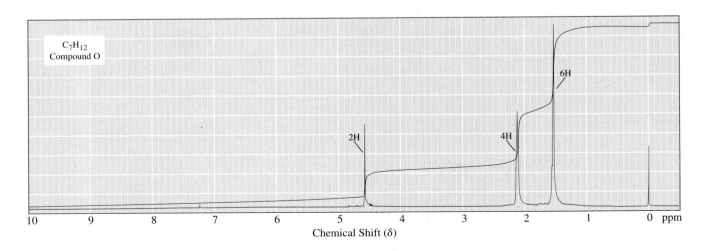

13.23 Treatment of compound P with BH_3 followed by $H_2O_2/NaOH$ gives compound Q. Following are ^{1}H-NMR spectra for compounds P and Q along with ^{13}C-NMR spectral data. From this information, deduce structural formulas for compounds P and Q.

$$C_7H_{12} \xrightarrow[\text{2. H}_2\text{O}_2,\text{ NaOH}]{\text{1. BH}_3} C_7H_{14}O$$

(P)   (Q)

	^{13}C-NMR	
	(P)	**(Q)**
	132.38	72.71
	32.12	37.59
	29.14	28.13
	27.45	22.68

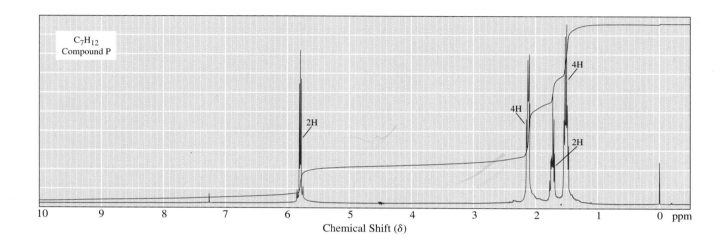

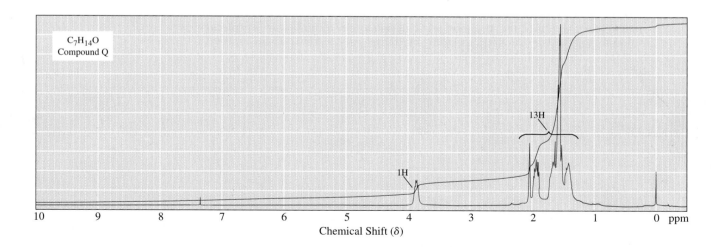

13.24 The ^{1}H-NMR of compound R, $C_6H_{14}O$, consists of two signals: δ 1.1 (doublet) and δ 3.6 (septet) in the ratio 6 : 1. Propose a structural formula for compound R consistent with this information.

13.25 Following are eight structural formulas along with the ^{13}C-NMR and DEPT-NMR spectral information. Given this information, assign each carbon in each compound its correct ^{13}C chemical shift.

(a) CH₃CH₂CH₂CHCH₃ (with Br on CH)

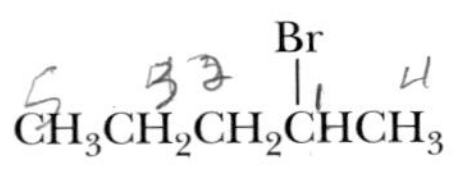

^{13}C	DEPT
51.55	CH
43.22	CH₂
26.46	CH₂
21.00	CH₃
13.40	CH₃

(b) CH₃CH₂C=CH₂ (with CH₃ on C)

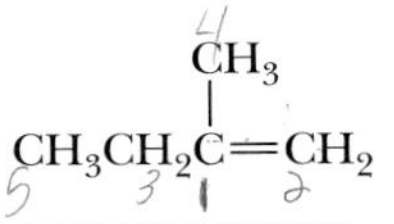

^{13}C	DEPT
147.70	—
108.33	CH₂
30.56	CH₂
22.47	CH₃
12.23	CH₃

(c) CH₂=CHCH₂CHCH₃ (with CH₃ on CH)

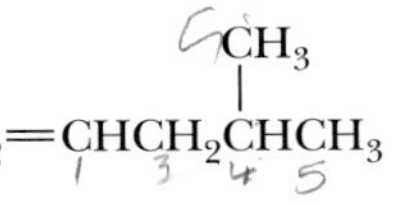

^{13}C	DEPT
137.81	CH
115.26	CH₂
43.35	CH₂
28.12	CH
22.26	CH₃

(d) CH₃CCH₂Br (with CH₃ above and CH₃ below on central C)

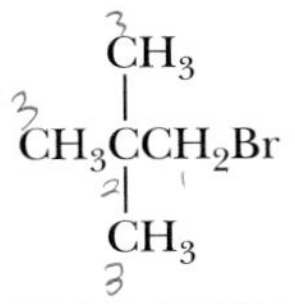

^{13}C	DEPT
49.02	CH₂
33.15	—
28.72	CH₃

(e) CH₃CH₂CCH₂CH₃ (ketone, O double bond)

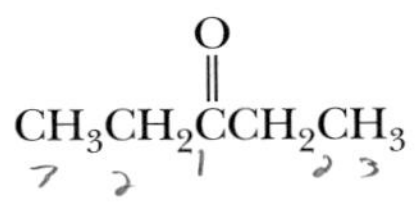

^{13}C	DEPT
207.8	—
35.1	CH₂
7.5	CH₂

(f) CH₃CH₂CCH₃ (ketone, O double bond)

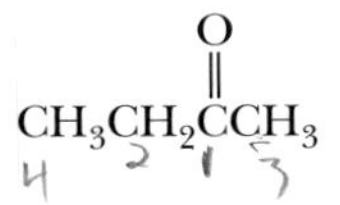

^{13}C	DEPT
208.7	—
37.6	CH₂
30.1	CH₃
9.2	CH₃

(g) CH₃CHCOCH₃ (ester, O double bond; with CH₃ on CH)

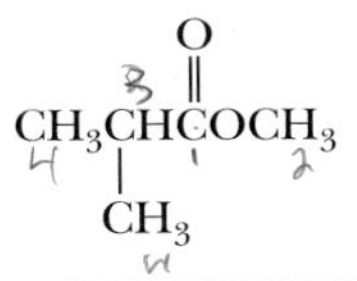

^{13}C	DEPT
177.48	—
51.50	CH₃
33.94	CH
19.01	CH₃

(h) CH₃COCH₂CH₂CHCH₃ (ester, O double bond; with CH₃ on CH)

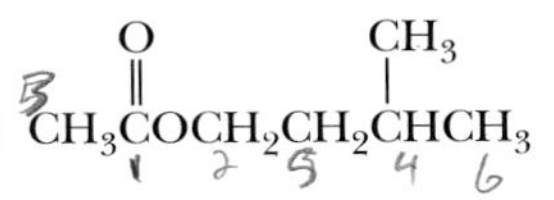

^{13}C	DEPT
171.17	—
63.12	CH₂
37.21	CH₃
25.05	CH
24.45	CH₃
21.02	CH₃

13.26 Write structural formulas for the following compounds.

(a) $C_2H_4Br_2$: δ 2.5 (d, 3H) and 5.9 (q, 1H)
(b) $C_4H_8Cl_2$: δ 1.60 (d, 3H), 2.15 (m, 2H), 3.72 (t, 2H), and 4.27 (m, 1H)
(c) $C_5H_8Br_4$: δ 3.6 (s, 8H)
(d) C_4H_8O: δ 1.0 (t, 3H), 2.1 (s, 3H), and 2.4 (quartet, 2H)
(e) $C_4H_8O_2$: δ 1.2 (t, 3H), 2.1 (s, 3H), and 4.1 (quartet, 2H); contains an ester
(f) $C_4H_8O_2$: δ 1.2 (t, 3H), 2.3 (quartet, 2H), and 3.6 (s, 3H); contains an ester
(g) C_4H_9Br: δ 1.1 (d, 6H), 1.9 (m, 1H), and 3.4 (d, 2H)
(h) $C_6H_{12}O_2$: δ 1.5 (s, 9H) and 2.0 (s, 3H)
(i) $C_7H_{14}O$: δ 0.9 (t, 6H), 1.6 (sextet, 4H), and 2.4 (t, 4H)
(j) $C_5H_{10}O_2$: δ 1.2 (d, 6H), 2.0 (s, 3H), and 5.0 (septet, 1H)
(k) $C_5H_{11}Br$: δ 1.1 (s, 9H) and 3.2 (s, 2H)
(l) $C_7H_{15}Cl$ δ 1.1 (s, 9H) and 1.6 (s, 6H)

13.27 The percent s-character of carbon participating in a C—H bond can be established by measuring the ^{13}C—^1H coupling constant and using the relationship

$$\text{Percent } s\text{-character} = 0.2 \, J(^{13}C - {}^1H)$$

The ^{13}C—^1H coupling constant observed for methane, for example, is 125 Hz, which gives 25% s-character, the value expected for an sp^3 hybridized carbon atom.

(a) Calculate the expected ^{13}C—^1H coupling constant in ethylene and acetylene.

(b) In cyclopropane, the ^{13}C—^1H coupling constant is 160 Hz. What is the hybridization of carbon in cyclopropane?

13.28 Ascaridole is a natural product that has been used to treat intestinal worms. Explain why the two methyls on the isopropyl group in ascaridole appear in its ^1H-NMR spectrum as four lines of equal intensity, with two sets of two each separated by 7 Hz.

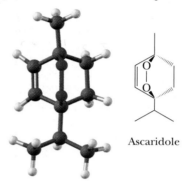

Ascaridole

13.29 The ^{13}C-NMR spectrum of 3-methyl-2-butanol shows signals at δ 17.88 (CH_3), 18.16 (CH_3), 20.01 (CH_3), 35.04 (carbon-3), and 72.75 (carbon-2). Account for the fact that each methyl group in this molecule gives a different signal.

13.30 Sketch the NMR spectrum you would expect from a partial molecule with the following parameters.

$$H_a = 1.0 \text{ ppm}$$
$$H_b = 3.0 \text{ ppm}$$
$$H_c = 6.0 \text{ ppm}$$
$$J_{ab} = 5.0 \text{ ppm}$$
$$J_{bc} = 8.0 \text{ ppm}$$
$$J_{ac} = 1.0 \text{ ppm}$$

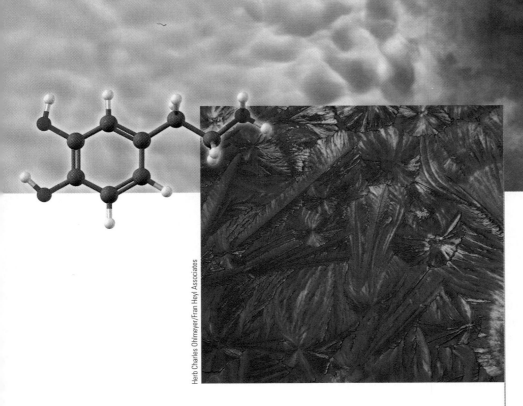

Crystals of dopamine viewed under polarized light. For a partial mass spectrum of dopamine, see Figure 14.2. Inset: A model of dopamine.

Herb Charles Ohlmeyer/Fran Heyl Associates

Outlines

Mass Spectrometry

Mass spectrometry is an analytical technique for measuring the mass-to-charge ratio (m/z) of ions, originally positive ions, now both positive and negative. The principles of mass spectrometry were first recognized in 1898. In 1911, J. J. Thomson recorded the first mass spectrum, that of neon, and discovered that this element can be separated into a more abundant isotope, ^{20}Ne, and a less abundant isotope, ^{22}Ne. Using improved instrumentation, F. W. Aston showed that most of the naturally occurring elements are mixtures of isotopes. It was found, for example, that approximately 75% of chlorine atoms in nature are ^{35}Cl, and 25% are ^{37}Cl. Mass spectrometry did not come into common use until the 1950s, when commercial instruments that offered high resolution, reliability, and relatively inexpensive maintenance became available. Today, mass spectrometry is our most valuable analytical tool for the determination of accurate molecular masses. Furthermore, extensive information about the molecular formula and structure of a compound can be obtained from analysis of its mass spectrum. Mass spectrometry is becoming increasingly important in biochemistry as well; sequencing of proteins using this technique alone allows protein structures to be determined on a virtually single-cell scale.

Mass spectrometry   An analytical technique for measuring the mass-to-charge ratio (m/z) of ions.

ORGANIC
Chemistry Now™

Look for this logo in the chapter and go to Organic ChemistryNow at **http://now.brookscole.com/bfi4** for tutorials, simulations, and problems.

14.1 A Mass Spectrometer

A mass spectrometer (Figure 14.1) is designed to do three things:

1. Convert neutral atoms or molecules into a beam of positive or negative ions.
2. Separate the ions on the basis of their mass-to-charge (m/z) ratio.
3. Measure the relative abundance of each type of ion.

From this information, we can determine both the molecular mass and the molecular formula of an unknown compound. In addition, we can obtain valuable clues about the molecular structure of the compound.

There are many types of mass spectrometers; we have space in this text to describe only the simplest. In the first-generation spectrometers, a vaporized sample in an evacuated ionization chamber is bombarded with high-energy electrons that cause electrons to be stripped from molecules of the sample, giving positively charged ions. Increasingly, radical anions (in which an extra electron has been added to a molecule) are studied; these are beyond the scope of this text. The resulting positive ions are accelerated by a series of negatively charged accelerator plates into an analyzing chamber inside a magnetic (electric in some spectrometers) field perpendicular to the direction of the ion beam. The magnetic field causes the ion beam to curve. The radius of curvature of each ion depends on the charge on the ion (z), its mass (m), the accelerating voltage, and the strength of the magnetic field. A mass spectrum is a plot of relative ion abundance versus m/z ratio.

Samples of gases and volatile liquids can be introduced directly into the ionization chamber. Because the interior of a mass spectrometer is kept at a high vacuum, volatile liquids and even some solids are vaporized. For less volatile liquids and solids, the sample may be placed on the tip of a heated probe that is then inserted directly into the ionization chamber. Another extremely useful method for introducing a sample into the ionization chamber is to link a gas chromatograph (GC) or liquid chromatograph (LC) directly to the mass spectrometer. Each fraction eluted from the chromatograph enters directly into the ionization chamber of the mass spectrometer.

Once in the ionization chamber, molecules of the sample are bombarded with a stream of high-energy electrons emitted from a hot filament and are then accelerated by

Figure 14.1

Schematic diagram of an electron ionization mass spectrometer (EI MS).

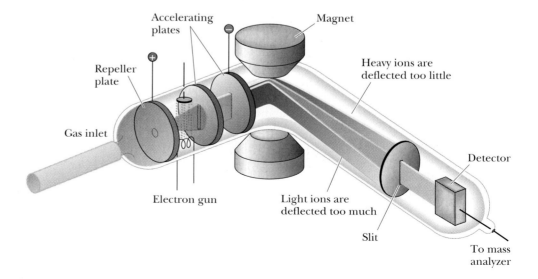

an electric field to energies of approximately 70 eV [1 eV = 96.49 kJ (23.05 kcal)/mol]. Collisions between molecules of the sample and these high-energy electrons result in loss of electrons from sample molecules to form positive ions. A **molecular ion, M⁺.**, is the species formed by removal of a single electron from a molecule. A molecular ion belongs to a class of ions called **radical cations.** When methane, for example, is bombarded with high-energy electrons, an electron is dislodged from a molecule to give a molecular ion of m/z 16.

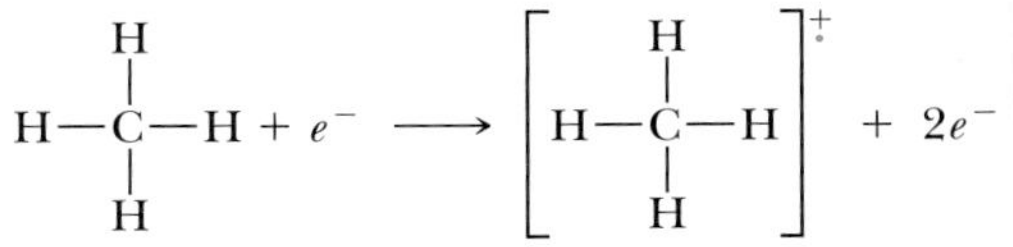

Molecular ion
(a radical cation)

Which electron is lost in forming the molecular ion is determined by the **ionization potential** of the atom or molecule. Ionization potentials for most organic molecules are between 8 and 15 eV. The potentials are at the lower end of this range both for nonbonding electrons of oxygen and nitrogen and for pi electrons in unsaturated compounds such as alkenes, alkynes, and aromatic hydrocarbons. Ionization potentials for sigma electrons, such as those of C—C, C—H, and C—O sigma bonds, are at the higher end of the range.

For our purposes, it doesn't matter which electron is lost because, in general, the radical cation is delocalized throughout the molecule. Therefore, we write the molecular formula of the parent molecule in brackets with a plus sign to show that it is a cation and with a dot to show that it has an odd number of electrons. See, for example, the molecular ion for ethyl isopropyl ether, shown on the left. At times, however, we will find it useful to depict the radical cation localized in a certain position to better understand its reactions, as in the formula on the right.

$$[CH_3CH_2OCH(CH_3)_2]^{+\cdot} \qquad CH_3CH_2\overset{\cdot\,+}{\underset{\cdot\cdot}{O}}CH(CH_3)_2$$

As we shall see in Section 14.2D, a molecular ion can undergo fragmentation to form a variety of smaller cations (which themselves may undergo further fragmentation), as well as radicals and smaller molecules. Only charged fragments are detected.

After molecular ions and their fragments form, a positively charged repeller plate directs the ions toward a series of negatively charged accelerator plates, producing a rapidly traveling ion beam. The ion beam is then focused by one or more slits and passes into a mass analyzer where it enters a magnetic field perpendicular to the direction of the ion beam. The magnetic field causes the ion beam to curve. Cations with larger values of m/z are deflected less than those with smaller m/z values. By varying either the accelerating voltage or the strength of the magnetic field, cations of the same m/z ratio can be focused on a detector, where the ion current is recorded. Modern detectors are capable of detecting single ions and of scanning a desired mass-to-charge region in a few tenths of a second or less.

A **mass spectrum** is a plot of the relative abundance of each cation versus mass-to-charge ratio. The peak resulting from the most abundant cation is called the **base peak** and is assigned an arbitrary intensity of 100. The relative abundances of all other cations in a mass spectrum are reported as percentages of the base peak. Figure 14.2 shows a partial mass spectrum of dopamine, a neurotransmitter in the brain's caudate nucleus, a center involved with coordination and integration of fine muscle movement. A deficiency of dopamine is an underlying biochemical defect in Parkinson's disease.

Molecular ion (M⁺.) The radical cation formed by removal of a single electron from a parent molecule in a mass spectrometer.

Radical cation A species formed when a neutral molecule loses one electron; it contains both an odd number of electrons and a positive charge.

Ionization potential (IP) The minimum energy required to remove an electron from an atom or molecule to a distance where there is no electrostatic interaction between the resulting ion and electron.

Mass spectrum A plot of the relative abundance of ions versus their mass-to-charge ratio.

Base peak The peak caused by the most abundant ion in a mass spectrum; the most intense peak. It is assigned an arbitrary intensity of 100.

Figure 14.2

A partial mass spectrum of dopamine showing all peaks with intensity equal to or greater than 0.5% of the base peak.

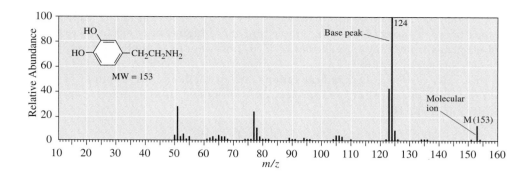

As can be seen in the following table, the number of peaks recorded depends on the sensitivity of the detector. If we record all peaks with intensity equal to or greater than 0.5% of the base peak, as in Figure 14.2, we find 45 peaks for dopamine. If we record all peaks with intensity equal to or greater than 0.05% of the base peak, we find 120 peaks.

Number of Peaks Recorded in a Mass Spectrum of Dopamine	
Peak Intensity Relative to Base Peak (%)	**Number of Peaks Recorded**
> 5	8
> 1	31
> 0.5	45
> 0.05	120

The technique we have described is called **electron ionization mass spectrometry (EI-MS).** This technique was the first developed and for a time was the one most widely used. It is limited, however, to relatively low-molecular-weight compounds that are vaporized easily in the evacuated ionization chamber. In recent years, a revolution in ionization techniques has extended the use of mass spectrometry to very high molecular-weight compounds and others that cannot be vaporized directly. Among the new techniques is fast-atom bombardment (FAB), which uses high-energy particles, such as xenon atoms accelerated to keV energies, to bombard a dispersion of a compound in a nonvolatile matrix, producing ions of the compound and expelling them into the gas phase. A second technique is matrix-assisted laser desorption ionization mass spectrometry (MALDI), which uses photons from an energetic laser for the same purpose. A third technique is chemical ionization (CI), which uses gas-phase acid-base reactions to produce ions. CI is particularly useful for identifying the molecular mass of a base (Brønsted-Lowry or Lewis) as its conjugate acid, MH^+. In addition, electrospray ionization mass spectrometry (ESI-MS) has become increasingly popular.

In ESI-MS, a solution of the analyte is introduced directly (for example, from a liquid chromatograph) through a charged capillary into a high-vacuum chamber. After entry into the vacuum chamber, the analyte exists in small charged droplets that rapidly evaporate, leaving the charge concentrated on the analyte molecule. The m/z ratios of the charged molecules are then determined. ESI is a relatively gentle method of generating charged molecular ions in a vacuum. ESI-MS is therefore particularly effective for the ionization of biological macromolecules, large molecules such as polysaccharides (Chapter 25), proteins (Chapter 27), and nucleotides (Chapter 28), allowing determination of their molecular mass and major fragments without the complications caused by the unavoidable overfragmentation seen with harsher ionization methods. MALDI is also a relatively gentle ionization technique that is now commonly used for the ionization of biological macromolecules.

14.2 Features of a Mass Spectrum

To understand a mass spectrum, we need to understand the relationships between mass spectra and resolution, the presence of isotopes, and the fragmentation of molecules and molecular ions in both the ionization chamber and the analyzing chamber.

A. Resolution

An important operating characteristic of a mass spectrometer is its **resolution,** that is, how well it separates ions of different mass. **Low-resolution mass spectrometry** refers to instruments capable of distinguishing among ions of different nominal mass; that is, ions that differ by one or more atomic mass units (amu). **High-resolution mass spectrometry** refers to instruments capable of distinguishing among ions that differ in precise mass by as little as 0.0001 amu.

To illustrate, compounds of molecular formulas C_3H_6O and C_3H_8O have nominal masses of 58 and 60 and can be resolved by low-resolution mass spectrometry. The compounds C_3H_8O and $C_2H_4O_2$, however, have the same nominal mass of 60 and cannot be distinguished by low-resolution mass spectrometry. If we calculate the precise mass of each compound using the data in Table 14.1, we see that they differ by 0.03642 amu and can be distinguished by high-resolution mass spectrometry. Observation of a molecular ion with a mass of 60.058 or 60.021 would establish the identities of C_3H_8O and $C_2H_4O_2$, respectively.

Resolution In mass spectrometry, a measure of how well a mass spectrometer separates ions of different mass.

Low-resolution mass spectrometry Instrumentation that is capable of separating only ions that differ in mass by 1 or more amu.

High-resolution mass spectrometry Instrumentation that is capable of separating ions that differ in mass by as little as 0.0001 amu.

Molecular Formula	Nominal Mass	Precise Mass
C_3H_8O	60	60.05754
$C_2H_4O_2$	60	60.02112

B. The Presence of Isotopes

In the mass spectrum of dopamine (Figure 14.2), the molecular ion appears at m/z 153. If you look more closely at this mass spectrum, you see a small peak at m/z 154, from an ion one amu heavier than the molecular ion of dopamine. This peak is actually the sum of four separate peaks, each of amu 154 and each corresponding to the presence in the ion of a single heavier isotope of H, C, N, or O in dopamine. Because this peak corresponds to an ion one amu heavier than the molecular ion, it is called an M + 1 peak. We are concerned in this section primarily with M + 1 and M + 2 peaks.

Table 14.1   Precise Masses and Natural Abundances of Isotopes Relative to 100 Atoms of the Most Abundant Isotope

Element	Atomic Weight	Isotope	Precise Mass (amu)	Relative Abundance
Hydrogen	1.0079	^{1}H	1.00783	100
		^{2}H	2.01410	0.016
Carbon	12.011	^{12}C	12.0000	100
		^{13}C	13.0034	1.11
Nitrogen	14.007	^{14}N	14.0031	100
		^{15}N	15.0001	0.38
Oxygen	15.999	^{16}O	15.9949	100
		^{17}O	16.9991	0.04
		^{18}O	17.9992	0.20
Sulfur	32.066	^{32}S	31.9721	100
		^{33}S	32.9715	0.78
		^{34}S	33.9679	4.40
Chlorine	35.453	^{35}Cl	34.9689	100
		^{37}Cl	36.9659	31.98
Bromine	79.904	^{79}Br	78.9183	100
		^{81}Br	80.9163	98.0

Virtually all the elements common to organic compounds, including H, C, N, O, S, Cl, and Br, are mixtures of isotopes. Exceptions are fluorine, phosphorus, and iodine, which occur in nature exclusively as ^{19}F, ^{31}P, and ^{127}I. Table 14.1 shows average atomic weights for the elements most common to organic compounds along with the masses and relative abundances in Nature of the stable isotopes of each. In this table, the relative abundances are tabulated according to the number of atoms of the heavier isotope per 100 atoms of the most abundant isotope. Naturally occurring carbon, for example, is 98.90% ^{12}C and 1.10% ^{13}C. Thus, there are 1.11 atoms of carbon-13 in nature for every 100 atoms of carbon-12.

$$1.10 \times \frac{100}{98.90} = 1.11 \text{ atoms } {}^{13}\text{C per 100 atoms } {}^{12}\text{C}$$

Example 14.1

Calculate the precise mass of each ion to five significant figures. Unless otherwise indicated, use the mass of the most abundant isotope of each element.

(a) $[CH_2Cl_2]^{+\cdot}$ (b) $[{}^{13}CH_2Cl_2]^{+\cdot}$ (c) $[CH_2Cl{}^{37}Cl]^{+\cdot}$

Solution

(a) $12.0000 + 2(1.00783) + 2(34.9689) = 83.953$ (b) 84.957 (c) 85.951

Problem 14.1

Calculate the nominal mass of each ion. Unless otherwise indicated, use the mass of the most abundant isotope of each element.

(a) $[CH_3Br]^{+\cdot}$ **(b)** $[CH_3{}^{81}Br]^{+\cdot}$ **(c)** $[{}^{13}CH_3Br]^{+\cdot}$

C. Relative Abundance of M, M + 2, and M + 1 Peaks

The most common elements giving rise to significant M + 2 peaks are chlorine, bromine, and oxygen. Chlorine in nature is 75.77% ^{35}Cl and 24.23% ^{37}Cl. Thus, a ratio of M to M + 2 peaks of approximately 3:1 indicates the presence of a single chlorine atom in the compound. Similarly, bromine in nature is 50.5% ^{79}Br and 49.5% ^{81}Br; a ratio of M to M + 2 of approximately 1:1 indicates the presence of a single bromine atom in the compound. The contribution of ^{18}O is only 0.2%, but it makes the major contribution to the M + 2 peak in compounds containing only C, H, N, and O. Sulfur is the only other element common to organic compounds that gives a significant M + 2 peak.

Let us use pentane, C_5H_{12}, to illustrate the relationship between M and M + 1 peaks. Pentane has a nominal mass of 72, and its molecular ion appears at m/z 72. In any sample of pentane, there is a probability that there will be a molecule in which one of the atoms of carbon is ^{13}C, the heavier isotope of carbon. This molecule has a nominal mass of 73, and its molecular ion will appear at m/z 73. Similarly, there is a probability that there will be a molecule in which one of the atoms of hydrogen is the heavier isotope of hydrogen, namely deuterium, ^2H. The probability of each of these isotope substitutions occurring is related to the natural abundance of each isotope in the following way.

$$\%(M + 1) = \Sigma \, (\% \text{ abundance of heavier isotope} \times \text{number of atoms in the formula})$$

Using this formula, we calculate that the relative intensity of the M + 1 peak for pentane is

$$(M + 1) = \Sigma \, (1.11 \times 5C + 0.016 \times 12H) = 5.55 + 0.19 = 5.74\% \text{ of molecular ion peak}$$

Notice that the M + 1 peak for pentane is almost entirely from ^{13}C. The same is true for other compounds containing only C and H. Because M + 1 peaks are relatively low in intensity compared to the molecular ion peak and often difficult to measure with any precision, they are not useful for accurate determinations of molecular formulas. M + 1 and M + 2 peaks, however, can be useful for getting a rough idea of the number of carbons, oxygens, sulfurs, and halogens. For example, the spectrum of chloroethane has peaks at m/z 64 and 66 (corresponding to $C_2H_5{}^{35}Cl$ and $C_2H_5{}^{37}Cl$, respectively), in a characteristic 3:1 ratio. Figures 14.3 and 14.4 illustrate this for a chloro and a bromo compound.

In contrast, 1-bromopropane has peaks at 122 and 124 for $C_3H_7{}^{79}Br$ and $C_3H_7{}^{81}Br$ in about a 1:1 ratio. This ratio of M and M + 2 peaks is characteristic of monobrominated compounds. The M and M + 2 peaks in chlorides and bromides in their distinctive ratios of 3:1 and 1:1, respectively, are very distinctive and allow almost immediate identification of monochloro and monobromo compounds.

Figure 14.3

Mass spectrum of chloroethane.

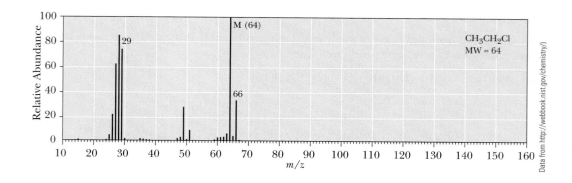

D. Fragmentation of Molecular Ions

To attain high efficiency of molecular ion formation and to give reproducible mass spectra, it is common to use electrons with energies of 70 eV [approximately 6750 kJ (1600 kcal)/mol] in EI. This energy is sufficient not only to dislodge one or more electrons from a molecule but also to cause extensive fragmentation because it is well in excess of bond dissociation enthalpies in organic molecules. These fragments may be unstable as well and, in turn, break apart into even smaller fragments.

The molecular ions for some compounds have a sufficiently long lifetime in the analyzing chamber that they are observed in the mass spectrum, sometimes as the base (most intense) peak. Molecular ions of other compounds have a shorter lifetime and are present in low abundance or not at all. As a result, the mass spectrum of a compound ionized with one of the harsher ionization methods such as EI or CI typically (but not always) consists of a peak for the molecular ion and series of peaks for fragment ions. The fragmentation pattern and relative abundances of ions are unique for each compound under a given set of ionizing conditions and are characteristic of that compound. Fragmentation patterns give us valuable information about molecular structure.

Figure 14.4

Mass spectrum of 1-bromopropane.

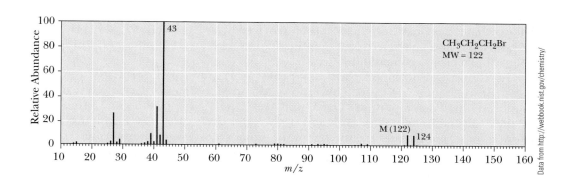

A great deal of the chemistry of ion fragmentation can be understood in terms of the formation and relative stabilities of carbocations in solution. Where fragmentation occurs forming new carbocations, the mode of fragmentation that gives the most stable carbocation is favored. Thus, the probability of fragmentation to form a new carbocation increases in the following familiar order.

$$CH_3^+ < 1° < 2° \cong 1° \text{ allylic/benzylic} < 3° \cong 2° \text{ allylic/benzylic} < 3° \text{ allylic/benzylic}$$

Increasing hydrocarbon carbocation stability ⟶

Molecular rearrangements are also characteristic of certain types of functional groups.

14.3 Interpreting Mass Spectra

ORGANIC
Chemistry-ꝉ-Now™

Click Tutorials to review the
Generation and Interpretation of Mass Spectra

Chemists often use mass spectra primarily for the determination of molecular mass and molecular formula. Very rarely do they attempt a full interpretation of a mass spectrum, which can be very time consuming, difficult, and dependent on the experimental details of the ionization. The mass spectrum of dopamine (Figure 14.2), for example, contains at least 45 peaks with intensity equal to or greater than 0.5% of the intensity of the base peak. We have neither the need nor the time to attempt to interpret this level of complexity. Rather, we concentrate in this section on the fragmentation mechanisms giving rise to major peaks.

As we now look at typical mass spectra of the classes of organic compounds we have seen so far, keep the following two points in mind. They provide valuable information about the molecular composition of an unknown compound.

1. The only elements giving rise to significant M + 2 peaks are ^{18}O (0.2%), ^{34}S (4.40%), ^{37}Cl (32%), and ^{81}Br (98%). If no large M + 2 peak is present, then these elements are absent.

2. Is the mass of the molecular ion odd or even? According to the **nitrogen rule,** if a compound has an odd number of nitrogen atoms, its molecular ion will appear at an odd m/z value. Conversely, if a compound has an even number of nitrogen atoms (including zero), its molecular ion will appear at an even m/z value. This rule is most helpful when there is an odd number of nitrogens. You may need additional experimental information to establish the presence of an even number of nitrogens.

Nitrogen rule   A rule stating that the molecular ion of a compound with an odd number of nitrogen atoms has at an odd m/z ratio; if there are zero or an even number of nitrogen atoms, the molecular ion has an even m/z ratio.

A. Alkanes

Two rules will help you interpret the mass spectra of alkanes. (1) Fragmentation tends to occur toward the middle of unbranched chains rather than at the ends. (2) The differences in energy among allylic, benzylic, tertiary, secondary, primary, and methyl carbocations in the gas phase are much greater than the differences among comparable radicals. Therefore, where alternative modes of fragmentation are possible, the more stable carbocation tends to form in preference to the more stable radical.

Unbranched alkanes fragment to form a series of cations differing by 14 amu (a CH_2 group), with each fragment formed by a one-bond cleavage having an odd mass

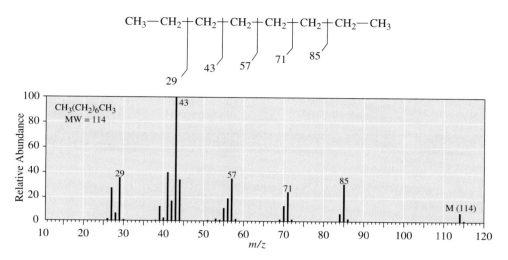

Figure 14.5

Mass spectrum of octane.

number. The mass spectrum of octane (Figure 14.5), for example, shows a peak for the molecular ion (m/z 114), as well as peaks for $C_6H_{13}^+$ (m/z 85), $C_5H_{11}^+$ (m/z 71), $C_4H_9^+$ (m/z 57), $C_3H_7^+$ (m/z 43), and $C_2H_5^+$ (m/z 29). These correspond to loss of ethyl, propyl, and butyl. Fragmentation of the CH_2—CH_3 bond is not observed; there is no peak corresponding to a methyl cation (m/z 15), nor is there one corresponding to a heptyl cation (loss of methyl, m/z 99). In mass spectrometry, fragmentations are shown by lines through the bond that is cleaved with an angled part toward the fragment that bears the charge.

Fragmentation of branched-chain alkanes leads preferentially to the formation of secondary and tertiary carbocations, and because these cations are more easily formed than methyl and primary carbocations, extensive fragmentation is likely. For this reason, the molecular ion of branched-chain hydrocarbons is often very weak or absent entirely from the spectrum. The molecular ion corresponding to m/z 114 is not observed, for example, in the mass spectrum of the highly-branched 2,2,4-trimethylpentane (Figure 14.6). The base peak for this hydrocarbon is at m/z 57, which corresponds to the *tert*-butyl cation ($C_4H_9^+$). Other prominent peaks are at m/z 43, (isopropyl cation) and m/z 41 (allyl cation (CH_2=$CHCH_2^+$)).

Figure 14.6

Mass spectrum of 2,2,4-trimethylpentane. The peak for the molecular ion is of such low intensity that it does not appear in this spectrum.

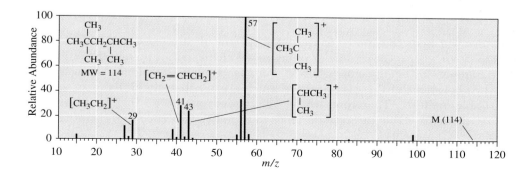

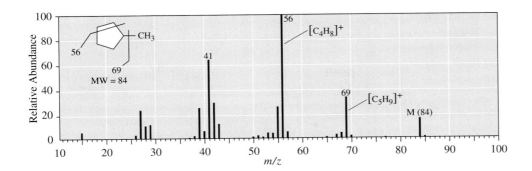

Figure 14.7
Mass spectrum of methylcy-clopentane.

Sometimes peaks that seem to defy the rules of chemical logic we have encountered so far occur in a mass spectrum. For example, the prominent peak at m/z 29 in the mass spectrum of 2,2,4-trimethylpentane (Figure 14.6) is consistent with the ethyl cation, $CH_3CH_2^+$. There is, however, no ethyl group in the parent molecule! This cation must be formed by some combination of fragmentation and rearrangement quite beyond anything that we have seen up to this point; such rearrangements are common at the high energies of electron-impact mass spectra. Exploration of this point is beyond the scope of this text.

The most common fragmentation patterns of cycloalkanes are loss of side chains and loss of ethylene, $CH_2=CH_2$. The peak at m/z 69 in the mass spectrum of methyl-cyclopentane (Figure 14.7) is the result of the loss of the one-carbon side chain to give the cyclopentyl cation, $C_5H_9^+$. The base peak at m/z 56 is caused by the loss of ethylene and corresponds to a cation of molecular formula $C_4H_8^+$. Note here that one-carbon cleavages of alkanes and cycloalkanes give fragments with odd mass numbers; two-bond cleavages give fragments with even mass numbers.

Example 14.2

The base peak at m/z 56 in the mass spectrum of methylcyclopentane corresponds to loss of ethylene to give a radical cation of molecular formula $C_4H_8^{+\cdot}$. Propose a structural formula for this radical cation and show how it might be formed.

Solution

Following is a structural formula for a molecular ion that might be formed in the ionizing chamber. In it, a single electron has been dislodged from a carbon-carbon single bond to give a 1° radical and a 2° carbocation. Rearrangement of bonding electrons in this radical cation gives ethylene and a new radical cation.

Molecular ion
(a radical cation, m/z 84)

Ethylene

A new radical cation
(a 2° carbocation, m/z 56)

Problem 14.2

Propose a structural formula for the cation of m/z 41 observed in the mass spectrum of methylcyclopentane.

Figure 14.8

Mass spectrum of 1-butene.

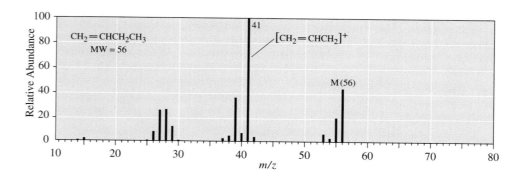

B. Alkenes

Alkenes characteristically show a strong molecular ion peak, most probably formed by removal of one pi electron from the double bond. Furthermore, they cleave readily to form resonance-stabilized allylic cations, such as the allyl cation seen at m/z 41 in the mass spectrum of 1-butene (Figure 14.8).

   Cyclohexenes undergo fragmentation to give a 1,3-diene and an alkene in a process that is the reverse of a Diels-Alder reaction (Section 24.6). The terpene limonene, a disubstituted cyclohexene, for example, fragments by a reverse Diels-Alder reaction to give two molecules of 2-methyl-1,3-butadiene (isoprene): one formed as a neutral diene and the other formed as a diene radical cation. Note here that the two-bond cleavage of this hydrocarbon gives fragments with even mass numbers.

| Limonene | A neutral diene | A radical cation |
| m/z 136 | m/z 68 | m/z 68 |

C. Alkynes

As with alkenes, alkynes show a strong peak for the molecular ion. Their fragmentation patterns are also similar to those of alkenes. One of the most prominent peaks in the mass spectrum of most alkynes is from the resonance-stabilized 3-propynyl (propargyl) cation (m/z 39) or a substituted propargyl cation.

$$HC\equiv C\!-\!CH_2{}^+ \longleftrightarrow H\overset{+}{C}\!=\!C\!=\!CH_2$$

3-Propynyl cation
(Propargyl cation)

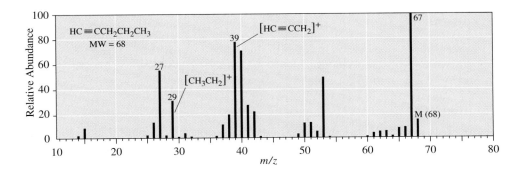

Figure 14.9
Mass spectrum of 1-pentyne.

Both the molecular ion, m/z 68, and the propargyl cation, m/z 39, are seen in the mass spectrum of 1-pentyne (Figure 14.9). Also seen is the ethyl cation, m/z 29.

D. Alcohols

The intensity of the molecular ion from primary and secondary alcohols is normally quite low, and there usually is no molecular ion detectable for tertiary alcohols. One of the most common fragmentation patterns for alcohols is loss of a molecule of water to give a peak corresponding to the molecular ion minus 18 (M − 18). Another common pattern is loss of an alkyl group from the carbon bearing the —OH group to form a resonance-stabilized oxonium ion and an alkyl radical. The oxonium ion is particularly stable because of delocalization of charge.

$$R'-\underset{\underset{R''}{|}}{\overset{\overset{R}{|}}{C}}-\overset{+}{\underset{\cdot\cdot}{O}}-H \longrightarrow R\cdot \ + \ R'-\underset{\underset{R''}{|}}{C}=\overset{+}{\underset{\cdot\cdot}{O}}-H \longleftrightarrow R'-\overset{+}{\underset{\underset{R''}{|}}{C}}-\overset{\cdot\cdot}{\underset{\cdot\cdot}{O}}-H$$

Molecular ion   A radical   A resonance-stabilized oxonium ion
(a radical cation)

Each of these patterns is found in the mass spectrum of 1-butanol (Figure 14.10). The molecular ion appears at m/z 74. The prominent peak at m/z 56 corresponds to loss of a molecule of water from the molecular ion (M − 18). The base peak at m/z 31 corresponds to cleavage of a propyl group (M − 43) from the carbon bearing the —OH group. The propyl cation is visible at m/z 43.

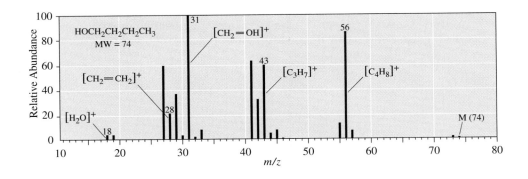

Figure 14.10
Mass spectrum of 1-butanol.

Example 14.3

A low-resolution mass spectrum of 2-methyl-2-butanol (MW 88) shows 16 peaks. The molecular ion is absent. Account for the formation of peaks at m/z 73, 70, 59, and 55, and propose a structural formula for each cation.

Solution

The peak at m/z 73 (M − 15) corresponds to loss of a methyl radical from the molecular ion. The peak at m/z 59 (M − 29) corresponds to loss of an ethyl radical. Loss of water as a neutral molecule from the molecular ion gives an alkene of m/z 70 (M − 18) as a radical cation. Loss of methyl from this radical cation gives an allylic carbocation of m/z 55.

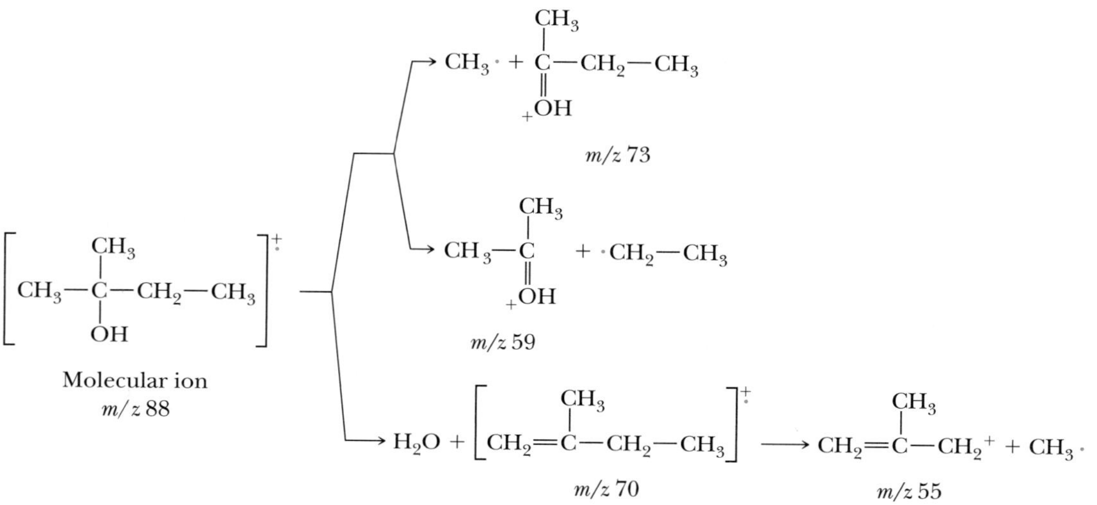

Problem 14.3

The low-resolution mass spectrum of 2-pentanol shows 15 peaks. Account for the formation of the peaks at m/z 73, 70, 55, 45, 43, and, 41.

E. Aldehydes and Ketones

A characteristic fragmentation pattern of aliphatic aldehydes and ketones is cleavage of one of the bonds to the carbonyl group (α-cleavage). α-Cleavage of 2-octanone, for example, gives carbonyl-containing ions of m/z 43 and 113. α-Cleavage of the aldehyde proton gives an M − 1 peak, which is often quite distinct and provides a useful way to distinguish between an aldehyde and a ketone.

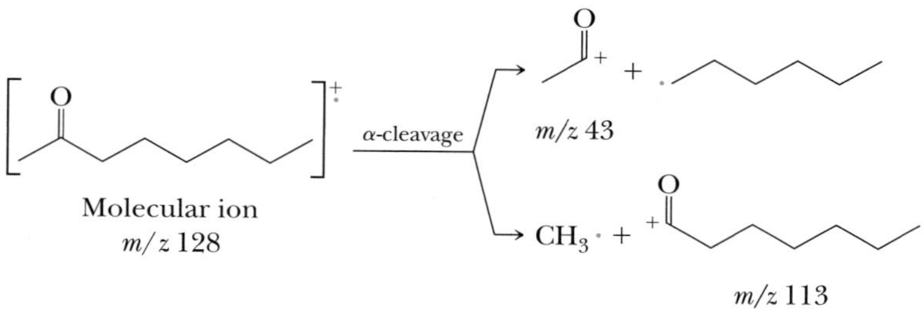

Aldehydes and ketones with a sufficiently long carbon chain show a fragmentation called a McLafferty rearrangement. In a **McLafferty rearrangement** of an aldehyde or ketone, the carbonyl oxygen abstracts a hydrogen five atoms away to give an alkene and a new radical cation. Because McLafferty rearrangements involve cleavage of two bonds, molecules of even m/z give fragments of even m/z.

Mechanism *McLafferty Rearrangement of a Ketone*

Reaction occurs through a six-membered ring transition state.

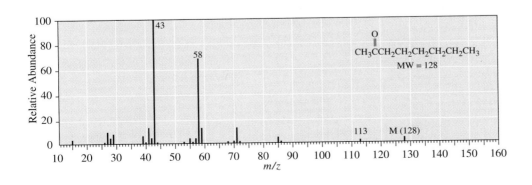

Molecular ion
m/z 86

m/z 58

McLafferty rearrangement of 2-octanone, for example, gives 1-pentene and a radical cation of m/z 58, which is the enol of acetone (Section 16.9).

Molecular ion
m/z 128

m/z 58

The results of both α-cleavage and McLafferty rearrangement can be seen in the mass spectrum of 2-octanone (Figure 14.11).

F. Carboxylic Acids, Esters, and Amides

The molecular ion peak from a carboxylic acid is generally observed, although it is often very weak. The most common fragmentation patterns are α-cleavage of the carboxyl group to give the ion $[COOH]^+$ of m/z 45 and McLafferty rearrangement. The base peak is very often the result of the McLafferty rearrangement product.

Figure 14.11

Mass spectrum of 2-octanone. Ions of m/z 43 and 113 result from α-cleavage. The ion at m/z 58 results from McLafferty rearrangement.

$$\left[\begin{array}{c} O \\ \parallel \\ \diagup\diagup\diagdown\diagdown C_{OH} \end{array} \right]^{+} \xrightarrow{\alpha\text{-cleavage}} \diagup\diagdown_{\cdot} + [O{=}C{-}O{-}H]^{+}$$

Molecular ion
m/z 88

m/z 45

Mechanism   *McLafferty Rearrangement of a Carboxylic Acid*

$$\left[\begin{array}{c} H \\ O \\ \diagdown C \diagdown \\ OH \end{array} \right]^{+} \xrightarrow[\text{rearrangement}]{\text{McLafferty}} \parallel + \left[\begin{array}{c} H_{\diagdown O} \\ \diagdown C \diagdown \\ OH \end{array} \right]^{+}$$

Molecular ion
m/z 88

m/z 60

Each of these patterns is seen in the mass spectrum of butanoic acid (Figure 14.12).

Esters and amides also generally show discernible molecular ion peaks. Like carboxylic acids, their most characteristic fragmentation patterns are α-cleavage and McLafferty rearrangement, both of which can be seen in the mass spectrum of methyl butanoate (Figure 14.13). Peaks at *m/z* 71 and 59 are the result of α-cleavage.

$$\left[\begin{array}{c} O \\ \parallel \\ \diagup\diagup\diagdown\diagdown C_{OCH_3} \end{array} \right]^{+} \xrightarrow{\alpha\text{-cleavage}}$$

Molecular ion
m/z 102

$$\diagup\diagup\diagdown\diagdown \stackrel{O}{C_+} + \cdot OCH_3$$
m/z 71

$$\diagup\diagdown_{\cdot} + \stackrel{O}{{}_+C_{OCH_3}}$$
m/z 59

The peak at *m/z* 74 is the result of McLafferty rearrangement.

$$\left[\begin{array}{c} H \\ O \\ \diagdown C \diagdown \\ OCH_3 \end{array} \right]^{+} \xrightarrow[\text{rearrangement}]{\text{McLafferty}} \parallel + \left[\begin{array}{c} H_{\diagdown O} \\ \diagdown C \diagdown \\ OCH_3 \end{array} \right]^{+}$$

Molecular ion
m/z 102

m/z 74

Figure 14.12

Mass spectrum of butanoic acid. Common fragmentation patterns of carboxylic acids are α-cleavage to give the ion [COOH]$^+$ of *m/z* 45 and McLafferty rearrangement.

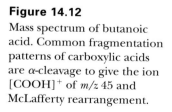

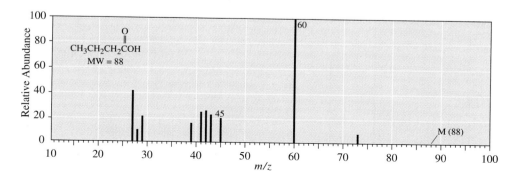

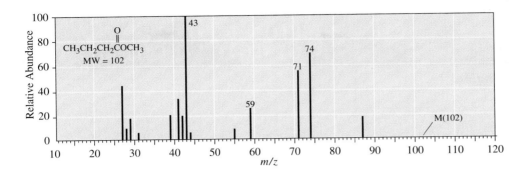

Figure 14.13
Mass spectrum of methyl butanoate. Characteristic fragmentation patterns of esters are α-cleavage and McLafferty rearrangement.

G. Aromatic Hydrocarbons

The mass spectra of most aromatic hydrocarbons show an intense molecular ion peak. The mass spectrum of toluene (Figure 14.14), for example, shows a large molecular ion peak at m/z 92.

The mass spectra of toluene and most other alkylbenzenes show a fragment ion of m/z 91. Although it might seem that the most likely structure for this ion is that of the benzyl cation, experimental evidence suggests a molecular rearrangement to form the more stable tropylium ion (Section 21.2E). In the tropylium ion, an aromatic cation, the positive charge is delocalized equally over all seven carbon atoms of the cycloheptatrienyl ring.

$$\left[\underset{}{\text{C}_6\text{H}_5-\text{CH}_3}\right]^{+\cdot} \xrightarrow{-\text{H}\cdot} \text{tropylium}$$

Toluene radical cation

Tropylium cation (m/z 91)

H. Amines

Of the compounds containing C, H, N, O, and the halogens, only those containing an odd number of nitrogen atoms have a molecular ion of odd m/z ratio. Thus, mass spectrometry can be a particularly valuable tool for identifying amines. The molecular ion for aliphatic amines, however, is often very weak. The most characteristic fragmentation of amines, and the one that often gives the base peak, is β-cleavage. Where alternative

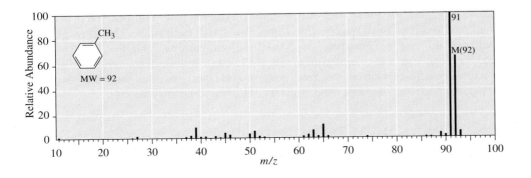

Figure 14.14
Mass spectrum of toluene. Prominent are the intense molecular ion peak at m/z 92 and the tropylium cation at m/z 91.

CONNECTIONS TO
BIOLOGICAL CHEMISTRY

Mass Spectrometry of Biological Macromolecules

As a result of enormous instrumental advances made during the last few years, mass spectrometry is becoming the method of choice for determining the structures of biological macromolecules, in particular, proteins and DNA. The Nobel Prize in chemistry in 2002 was awarded to John B. Fenn and Koichi Tanaka for the development of the electrospray ionization (ESI) and MALDI methods that have revolutionized structure determination of these molecules. The key advantage of using mass spectrometry for analysis of proteins and nucleic acids is that the entire process can be automated so that information about a large number of molecules can be obtained rapidly. A growing list of DNA sequences at the genome level is now available, and attention is increasingly being turned toward acquiring sequence information at the protein level with

high throughput. Mass spectrometry will certainly play a major role in this effort.

For example, it is now possible to obtain complete amino acid sequences on polypeptides (proteins, biological polyamides, see Chapter 27) of substantial length. The mass spectrometer cleaves polypeptides into fragments of varying length. Although there are many cleavage modes possible, the main cleavage is at peptide (amide) bonds. Both fragments (from the *N*-terminal and *C*-terminal part) can usually be identified. A mass analyzer determines the mass of each fragment. Because each amino acid has a slightly different mass (except for two isomeric amino acids leucine and isoleucine), the exact amino acid composition of the fragment can be determined. Powerful computers align overlapping fragments and determine the exact sequence.

$$H_2N-\underset{\underset{H}{|}}{\overset{\overset{R}{|}}{C}}-\overset{\overset{O}{\|}}{C}+N-\underset{\underset{H}{|}}{\overset{\overset{R}{|}}{C}}-\overset{\overset{O}{\|}}{C}+N-\underset{\underset{H}{|}}{\overset{\overset{R}{|}}{C}}-\overset{\overset{O}{\|}}{C}+N-\underset{\underset{H}{|}}{\overset{\overset{R}{|}}{C}}-\overset{\overset{O}{\|}}{C}+N-\underset{\underset{H}{|}}{\overset{\overset{R}{|}}{C}}-\overset{\overset{O}{\|}}{COH}$$

For larger proteins, the mass spectrometric method is usually preceded by enzymatic cleavage into fragments. However, in the most advanced systems, the individual fragments do not need to be separated. They are injected into a tandem mass spectrometer (called an MS-MS). The first segment of the instrument separates the individual fragments by mass, and the second segment fragments each peptide and sequences it separately. The MS-MS can also be used to separate, and then further

fragment, the primary ions obtained in the first stage to obtain sequence information directly. Enormous proteins can be sequenced on a picomole or lower concentration level. This method is particularly helpful because it can be used on mixtures (and ultimately, it is believed, on whole cell contents). Similar techniques are also now used for determining the sequence of DNA, the genetic material. Further discussion about applications to macromolecules may be found in Chapters 27 and 28.

possibilities for β-cleavage exist, it is generally the largest R group that is lost. In contrast to nitrogen-free molecules, single bond fragments from compounds that have one or an odd number of nitrogens give compounds with even mass. The most prominent peak in the mass spectrum of 3-methyl-1-butanamine (Figure 14.15) is from $[CH_2=NH_2]^+$, m/z 30, from β-cleavage. β-Cleavage is also characteristic of secondary and tertiary amines. Complex rearrangement and fragmentation processes give the m/z 30 peak as a major fragment even from secondary and tertiary amines.

$$CH_3-\underset{\underset{CH_3}{|}}{CH}-CH_2-CH_2-\overset{+}{N}H_2 \xrightarrow{\text{β-cleavage}} CH_3-\underset{\underset{CH_3}{|}}{CH}-CH_2\cdot + CH_2=\overset{+}{N}H_2$$

$$(m/z\ 30)$$

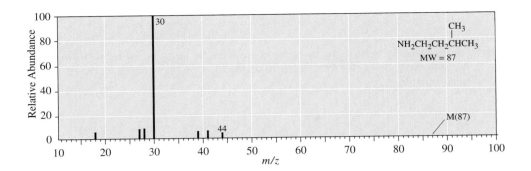

Figure 14.15
Mass spectrum of 3-methyl-1-butanamine (isopentylamine). The most characteristic fragmentation pattern of aliphatic amines is β-cleavage.

14.4. Mass Spectrometry in the Organic Synthesis Laboratory and Other Applications

Mass spectrometry, especially when interfaced with separation methods such as gas chromatography (GC-MS) and liquid chromatography (LC-MS), represents a powerful and rapid method of identifying compounds in reaction product mixtures. As a result, GC-MS and LC-MS are taking on increased importance as the primary method of routine molecule identification in the organic synthesis laboratory, both in industrial and academic settings.

Mass spectrometry is also becoming increasingly important for other practical applications. Some devices used to screen luggage in airports use mass spectrometry to identify traces of known explosives. Drug testing of athletes and advanced forensic science uses GC-MS to identify traces of pharmaceuticals or illicit drugs in blood samples. Looking toward the future, with the advent of powerful new-generation mass spectrometers that are increasingly small and inexpensive, there will be a dramatic increase in the use of mass spectrometry in the identification of molecules in many aspects of modern life, not just in the research laboratory.

Summary

A **mass spectrum** (Section 14.1) is a plot of relative ion abundance versus mass-to-charge (m/z) ratio. The **base peak** is the most intense peak in a mass spectrum. A **molecular ion, $M^{+\cdot}$,** is a radical cation derived from the parent molecule by loss of one electron. Numerous methods of ionization exist, including electron impact (EI), fast-atom bombardment (FAB), chemical ionization (CI), matrix-assisted laser desorption ionization (MALDI), and electrospray ionization (ESI).

Low-resolution mass spectrometry distinguishes between ions that differ in nominal mass, that is, ions that differ by 1 amu (Section 14.2A). **High-resolution mass spectrometry** distinguishes between ions that differ by as little as 0.0001 amu.

M + 1 and higher peaks in a mass spectrum are caused by heavier isotopes (Section 14.2B). The abundance of these higher mass-to-charge peaks relative to the molecular ion peak provides information about the elemental composition of the molecular ion (Section 14.2C). The presence of a single chlorine atom, for example, is indicated by M to M + 2 peaks in a ratio of 3:1. According to the **nitrogen rule,** if a compound has an odd number of nitrogen atoms, its molecular ion will have an odd m/z value (Section 14.3).

The mass spectrum of a compound typically consists of a peak for the molecular ion and a series of peaks for fragment ions (Section 14.2D). The fragmentation pattern and relative abundances of ions are unique for each compound and are characteristic of that compound. Fragments formed by cleavage of one bond have odd mass if they contain no nitrogen; those formed from the cleavage of two bonds have even mass. Many of the observed fragmentation patterns can be understood in terms of the relative stability of carbocations (Section 14.2D). Where alternative modes of fragmentation are possible, the more stable carbocation tends to be formed in preference to the more stable radical. Mass spectrometry can give significant information about structures of compounds. (Section 14.3) and is increasingly used in the forensic laboratory (Section 14.4). Mass spectral techniques can be used to determine the amino acid sequences of proteins and DNA.

Problems

14.4 Draw acceptable Lewis structures for the molecular ion (radical cation) formed from the following molecules when each is bombarded by high-energy electrons in a mass spectrometer.

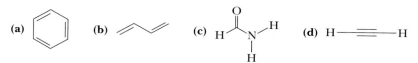

14.5 The molecular ion for compounds containing only C, H, and O always has an even mass-to-charge value. Why is this so? What can you say about the mass-to-charge ratio of ions that arise from fragmentation of one bond in the molecular ion? From fragmentation of two bonds in the molecular ion?

14.6 For which compounds containing a heteroatom (an atom other than carbon or hydrogen) does the molecular ion have an even-numbered mass and for which does it have an odd-numbered mass?

(a) A chloroalkane of molecular formula $C_nH_{2n+1}Cl$
(b) A bromoalkane of molecular formula $C_nH_{2n+1}Br$
(c) An alcohol of molecular formula $C_nH_{2n+1}OH$
(d) A primary amine of molecular formula $C_nH_{2n+1}NH_2$
(e) A thiol of molecular formula $C_nH_{2n+1}SH$

14.7 The so-called nitrogen rule states that if a compound has an odd number of nitrogen atoms, the value of m/z for its molecular ion will be an odd number. Why is this so?

14.8 Both $C_6H_{10}O$ and C_7H_{14} have the same nominal mass, namely 98. Show how these compounds can be distinguished by the m/z ratio of their molecular ions in high-resolution mass spectrometry.

14.9 Show how the compounds of molecular formula C_6H_9N and C_5H_5NO can be distinguished by the m/z ratio of their molecular ions in high-resolution mass spectrometry.

14.10 What rule would you expect for the m/z values of fragment ions resulting from the cleavage of one bond in a compound with an odd number of nitrogen atoms?

14.11 Determine the probability of the following in a natural sample of ethane.

(a) One carbon in an ethane molecule is ^{13}C.
(b) Both carbons in an ethane molecule are ^{13}C.
(c) Two hydrogens in an ethane molecule are replaced by deuterium atoms.

14.12 The molecular ions of both $C_5H_{10}S$ and $C_6H_{14}O$ appear at m/z 102 in low-resolution mass spectrometry. Show how determination of the correct molecular formula can be made from the appearance and relative intensity of the M + 2 peak of each compound.

14.13 In Section 14.3, we saw several examples of fragmentation of molecular ions to give resonance-stabilized cations. Make a list of these resonance-stabilized cations, and write important contributing structures of each. Estimate the relative importance of the contributing structures in each set.

14.14 Carboxylic acids often give a strong fragment ion at m/z (M − 17). What is the likely structure of this cation? Show by drawing contributing structures that it stabilized by resonance.

14.15 For primary amines with no branching on the carbon bearing the nitrogen, the base peak occurs at m/z 30. What cation does this peak represent? How is it formed? Show by drawing contributing structures that this cation is stabilized by resonance.

14.16 The base peak in the mass spectrum of propanone (acetone) occurs at m/z 43. What cation does this peak represent?

14.17 A characteristic peak in the mass spectrum of most aldehydes occurs at m/z 29. What cation does this peak represent? (No, it is not an ethyl cation, $CH_3CH_2^+$.)

14.18 Predict the relative intensities of the M and M + 2 peaks for the following.

    (a) CH_3CH_2Cl **(b)** CH_3CH_2Br **(c)** $BrCH_2CH_2Br$ **(d)** CH_3CH_2SH

14.19 The mass spectrum of compound A shows the molecular ion at m/z 85, an M + 1 peak at m/z 86 of approximately 6% abundance relative to M, and an M + 2 peak at m/z 87 of less than 0.1% abundance relative to M.

    (a) Propose a molecular formula for compound A.
    (b) Draw at least 10 possible structural formulas for this molecular formula.

14.20 The mass spectrum of compound B, a colorless liquid, shows these peaks in its mass spectrum. Determine the molecular formula of compound B, and propose a structural formula for it.

m/z	Relative Abundance
43	100 (base)
78	23.6 (M)
79	1.00
80	7.55
81	0.25

14.21 Write molecular formulas for the five possible molecular ions of m/z 88 containing the elements C, H, N, and O.

14.22 Write molecular formulas for the five possible molecular ions of m/z 100 containing only the elements C, H, N, and O.

14.23 The molecular ion in the mass spectrum of 2-methyl-1-pentene appears at m/z 84. Propose structural formulas for the prominent peaks at m/z 69, 55, 41, and 29.

14.24 Following is the mass spectrum of 1,2-dichloroethane.

    (a) Account for the appearance of an M + 2 peak with approximately two thirds the intensity of the molecular ion peak.
    (b) Predict the intensity of the M + 4 peak.
    (c) Propose structural formulas for the cations of m/z 64, 63, 62, 51, 49, 27, and 26.

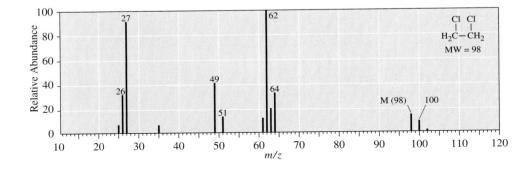

14.25 Following is the mass spectrum of 1-bromobutane.

    (a) Account for the appearance of the M + 2 peak of approximately 95% of the intensity of the molecular ion peak.

    (b) Propose structural formulas for the cations of m/z 57, 41, and 29.

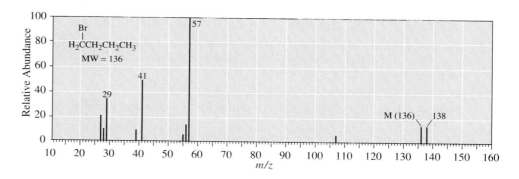

14.26 Following is the mass spectrum of bromocyclopentane. The molecular ion m/z 148 is of such low intensity that it does not appear in this spectrum. Assign structural formulas for the cations of m/z 69 and 41.

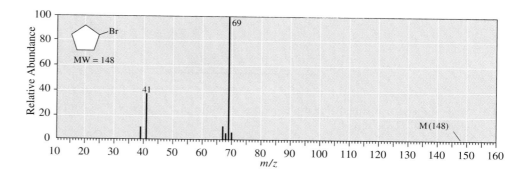

14.27 Following is the mass spectrum of an unknown compound. The two highest peaks are at m/z 120 and 122. Suggest a structure for this compound. (Data from http://webbook. nist.gov/chemistry/)

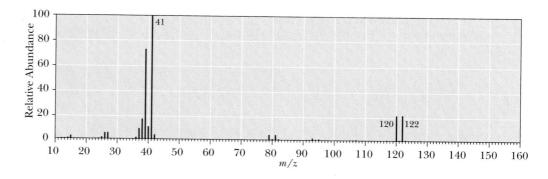

14.28 Following is the mass spectrum of 3-methyl-2-butanol. The molecular ion m/z 88 does not appear in this spectrum. Propose structural formulas for the cations of m/z 45, 43, and 41.

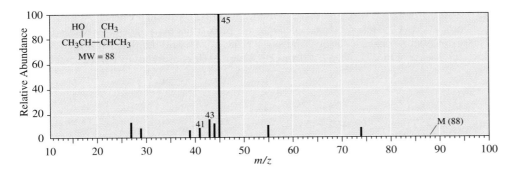

14.29 The following is the mass spectrum of compound C, C_3H_8O. Compound C is infinitely soluble in water, undergoes reaction with sodium metal with the evolution of a gas, and undergoes reaction with thionyl chloride to give a water-insoluble chloroalkane. Propose a structural formula for compound C, and write equations for each of its reactions.

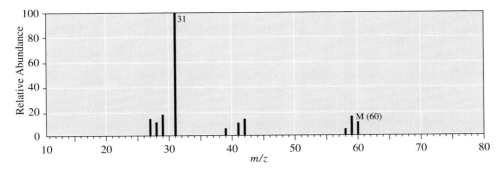

14.30 Following are mass spectra for the constitutional isomers 2-pentanol and 2-methyl-2-butanol. Assign each isomer its correct spectrum.

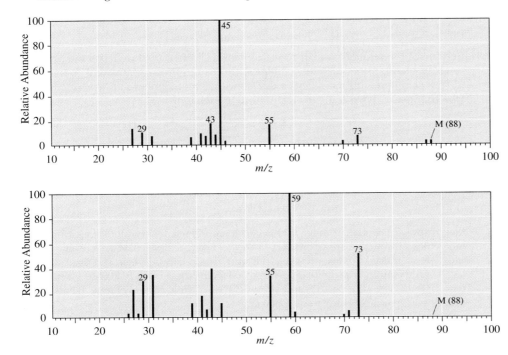

14.31 2-Methylpentanal and 4-methyl-2-pentanone are constitutional isomers of molecular formula $C_6H_{12}O$. Each shows a molecular ion peak in its mass spectrum at m/z 100. Spectrum A shows significant peaks at m/z 85, 58, 57, 43, and 42. Spectrum B shows significant peaks at m/z 71, 58, 57, 43, and 29. Assign each compound its correct spectrum.

2-Methylpentanal   4-Methyl-2-pentanone

14.32 Account for the presence of peaks at m/z 135 and 107 in the mass spectrum of 4-methoxybenzoic acid (*p*-anisic acid).

14.33 Account for the presence of the following peaks in the mass spectrum of hexanoic acid, $CH_3(CH_2)_4COOH$.

(a) m/z 60
(b) A series of peaks differing by 14 amu at m/z 45, 59, 73, and 87
(c) A series of peaks differing by 14 amu at m/z 29, 43, 57, and 71

14.34 All methyl esters of long-chain aliphatic acids (for example, methyl tetradecanoate, $C_{13}H_{27}COOCH_3$) show significant fragment ions at m/z 74, 59, and 31. What are the structures of these ions? How are they formed?

14.35 Propylbenzene, $C_6H_5CH_2CH_2CH_3$, and isopropyl benzene, $C_6H_5CH(CH_3)_2$, are constitutional isomers of molecular formula C_9H_{12}. One of these compounds shows prominent peaks in its mass spectrum at m/z 120 and 105. The other shows prominent peaks at m/z 120 and 91. Which compound has which spectrum?

14.36 Account for the formation of the base peaks in these mass spectra.

(a) Isobutylamine, m/z 44   (b) Diethylamine, m/z 58

14.37 Because of the sensitivity of mass spectrometry, it is often used to detect the presence of drugs in blood, urine, or other biological fluids. Tetrahydrocannabinol (nominal mass 314), a component of marijuana, exhibits two strong fragment ions at m/z 246 and 231 (the base peak). What is the likely structure of each ion?

14.38 Electrospray mass spectrometry is a recently developed technique for looking at large molecules with a mass spectrometer. In this technique, molecular ions, each associated with one or more H^+ ions, are prepared under mild conditions in the mass spectrometer. As an example, a protein (P) with a molecular mass of 11,812 gives clusters of the type $(P + 8H)^{8+}$, $(P + 7H)^{7+}$, and $(P + 6H)^{6+}$. At what mass-to-charge values do these three clusters appear in the mass spectrum?

© Clive Druett: Papilio/CORBIS

15

Organometallic Compounds

I n this chapter, we undertake our first discussion of a broad class of organic compounds called **organometallic compounds,** compounds that contain a carbon-metal bond. In recent years, there has been an enormous explosion in our understanding of their chemistry, particularly as stereospecific (and often enantioselective) reagents for synthetic chemistry. We have already seen one example in the Sharpless enantioselective epoxidation of alkenes (Section 11.8D).

This chapter cannot possibly cover the wealth of organometallic reagents and catalysts that have been developed for synthetic organic chemistry, particularly during the last few years. We focus, therefore, on transformations that are fundamental to synthetic chemistry. Organomagnesium, lithium, and copper reagents have been selected because of their historical importance and their continued use in modern organic synthesis, particularly in their reactions with carbonyl compounds that are the focus of the next several chapters. Several more recent reactions of organometallic compounds are introduced in the chapter on C—C bond formation and organic synthesis (Chapter 24).

Organometallic compound   A compound that contains a carbon-metal bond.

ORGANIC
Chemistry ⚛ Now™
Look for this logo in the chapter and go to Organic ChemistryNow at
http://now.brookscole.com/bfi4
for tutorials, simulations, and problems.

15.1 Organomagnesium and Organolithium Compounds

We begin with organomagnesium and organolithium compounds and concentrate on their formation and basicity. We discuss their use in organic synthesis in more detail in later chapters, particularly in Chapters 16 and 18.

RMgX

RLi

An organomagnesium compound (a Grignard reagent)

An organolithium compound

A. Formation and Structure

Organomagnesium compounds are among the most readily available, easily prepared, and easily handled organometallics. They are commonly named **Grignard reagents** after the French chemist Victor Grignard (1871–1935), who was awarded the 1912 Nobel Prize for chemistry for their discovery and application to organic synthesis.

Grignard reagents are typically prepared by the slow addition of an alkyl, aryl, or alkenyl (vinylic) halide to a stirred suspension of a slight excess of magnesium metal in an ether solvent, most commonly diethyl ether or tetrahydrofuran (THF). Organic iodides and bromides generally react very rapidly under these conditions, whereas most organic chlorides react more slowly. Bromides are the most common starting materials for preparation of Grignard reagents. It is common to use the higher-boiling THF (bp 67°C) to prepare Grignard reagents from the less reactive organic halides. Generally there is an induction period at the beginning of the reaction caused by the presence of traces of moisture and a thin oxide coating on the surface of the magnesium. When reaction starts, it is exothermic, and the remaining organic halide is added at a rate sufficient to maintain a gentle reflux of the ether.

Butylmagnesium bromide, for example, is prepared by treating 1-bromobutane in diethyl ether with magnesium metal. Aryl Grignard reagents, such as phenylmagnesium bromide, are prepared in the same manner. These reactions are referred to as oxidative additions, because they result in an increase in the formal oxidation state of magnesium by two, that is, from Mg(0) to Mg(II).

1-Bromobutane

Butylmagnesium bromide (an alkyl Grignard reagent)

Bromobenzene

Phenylmagnesium bromide (an aryl Grignard reagent)

Even though the equation for formation of Grignard reagents looks simple, the mechanism is considerably more complicated and involves radicals. We have no need in this course to discuss the mechanism for their formation.

Grignard reagents form on the surface of the metal and dissolve as coordination complexes solvated by ether. In this ether-soluble complex, magnesium acts as a Lewis acid, and the ether acts as a Lewis base.

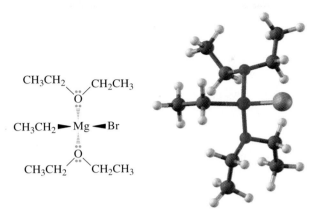

Ethylmagnesium bromide dietherate

Organolithium reagents are prepared by treating an alkyl, aryl, or alkenyl halide with two equivalents of lithium metal, as illustrated by the preparation of butyllithium. In this reaction, a solution of 1-chlorobutane in pentane is added to lithium wire at $-10°C$.

$$\text{Cl} + 2\text{Li} \xrightarrow{\text{pentane}} \text{Li} + \text{LiCl}$$

1-Chlorobutane   Butyllithium

Organolithium compounds are very reactive as nucleophiles in carbonyl addition reactions even at very low temperatures. They are also powerful and effective bases. For these reasons, they are now widely used in modern synthetic chemistry. However, they react rapidly with atmospheric oxygen and moisture and, therefore, must be used under an inert atmosphere of N_2 or Ar, which decreases their convenience.

The carbon-metal bonds in Grignard and organolithium reagents are best described as polar covalent, with carbon bearing a partial negative charge and the metal bearing a partial positive charge. Shown in Table 15.1 are electronegativity differences (Pauling scale, Table 1.5) between carbon and various metals. From this difference, we can estimate the percent ionic character of each carbon-metal bond.

Organolithium and organomagnesium bonds have the highest partial ionic character, whereas those of organocopper and organomercury compounds are lower. These compounds do not behave as salts. Organolithium reagents, for example, which have the highest percent partial ionic character, dissolve in nonpolar hydrocarbon solvents such as pentane because they self-assemble into well-ordered aggregates, $(\text{RLi})_x$, that present a nonpolar surface to the surrounding solvent.

An important aspect of the metals listed in Table 15.1 is that they all have electronegativities that are considerably lower than carbon. This means that the polarity of the carbon metal bond places a partial negative charge on carbon and a partial positive charge on the metal. The partial negative charge makes the carbon atom both basic and nucleophilic. When Grignard and organolithium reagents are prepared, the carbon atom bearing the halogen is transformed from an electrophilic center (partial positive charge) in the haloalkane, alkene, or arene to a nucleophilic center (partial negative charge) in the organometallic compound. In the structural

Table 15.1　Percent Ionic Character of Some C—M Bonds

δ− δ+ C—M Bond	Difference in Electronegativity	Percent Ionic Character*
C—Li	2.5 − 1.0 = 1.5	60
C—Mg	2.5 − 1.2 = 1.3	52
C—Al	2.5 − 1.5 = 1.0	40
C—Zn	2.5 − 1.6 = 0.9	36
C—Sn	2.5 − 1.8 = 0.7	28
C—Cu	2.5 − 1.9 = 0.6	24
C—Hg	2.5 − 1.9 = 0.6	24

*Percent ionic character $= \dfrac{E_C - E_M}{E_C} \times 100$

formula of butylmagnesium bromide on the right, the carbon-magnesium bond is shown as ionic to emphasize its nucleophilic character.

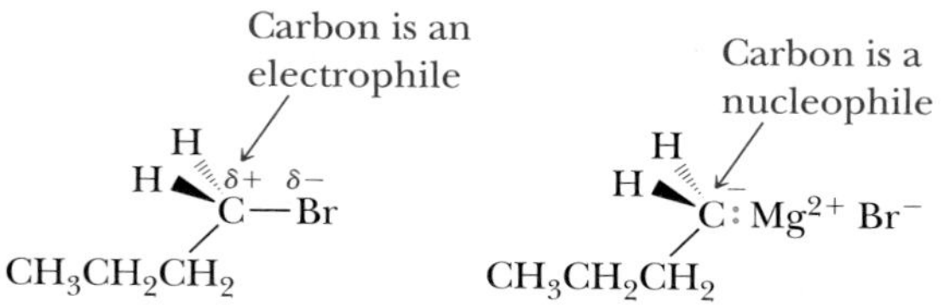

As nucleophiles, these compounds react with the electrophilic carbon atom of the carbonyl groups of aldehydes and ketones (Chapter 16) and of carboxylic esters and acid chlorides (Chapter 18). Herein lies the value of organomagnesium and organolithium reagents in synthetic organic chemistry—as carbon-centered nucleophiles, they enable the formation of new carbon-carbon bonds.

B. Reaction with Proton Acids

Both Grignard and organolithium compounds are very strong bases and react readily with any acid (proton donor) stronger than the alkane from which they are derived. Ethylmagnesium bromide, for example, reacts instantly with water, which donates a proton to give ethane and magnesium salts. This reaction is an example of a much stronger acid and a much stronger base reacting to give a weaker acid and a weaker base (Section 4.4). Ethane is evolved from the reaction mixture as a gas.

$$\overset{\delta-}{CH_3CH_2}-\overset{\delta+}{MgBr} + \overset{\delta+}{H}-\overset{\delta-}{OH} \longrightarrow CH_3CH_2-H + Mg^{2+} + OH^- + Br^- \qquad pK_{eq} = -35$$
$$K_{eq} = 10^{35}$$

	pK_a 15.7	pK_a 51	
Stronger base	Stronger acid	Weaker acid	Weaker base

Following are several classes of proton donors that react readily with Grignard and organolithium reagents. Because they react so readily with these types of compounds, Grignard and organolithium compounds cannot be prepared from any organohalogen that also contains one of these functional groups. Nor can they be prepared from any organohalogen compound that also contains a nitro or carbonyl group because they also react with these groups.

R_2NH	$RC{\equiv}CH$	ROH	HOH	$ArOH$	RSH	$RCOOH$
pK_a 38–40	pK_a 25	pK_a 16–18	pK_a 15.7	pK_a 9–10	pK_a 8–9	pK_a 4–5
1° and 2° amines	Terminal alkynes	Alcohols	Water	Phenols	Thiols	Carboxylic acids

Example 15.1

Write an equation for the acid-base reaction between ethylmagnesium iodide and an alcohol. Use curved arrows to show the flow of electrons in this reaction. In addition, show by using appropriate pK_a values that this reaction is an example of a stronger acid and stronger base reacting to give a weaker acid and weaker base.

Solution

The alcohol is the stronger acid, and the partially negatively charged ethyl group is the stronger base.

$$CH_3CH_2{-}MgI \;+\; H{-}\ddot{O}R \;\longrightarrow\; CH_3CH_2{-}H \;+\; MgI^+ \;{:}\ddot{O}R$$

A Grignard reagent (stronger base)	An alcohol pK_a 16–18 (stronger acid)	An alkane pK_a 51 (weaker acid)	A magnesium alkoxide (weaker base)

Problem 15.1

Explain how these Grignard reagents would react with molecules of their own kind to "self-destruct".

(a)

HO⁓⁓MgBr

(b)

(alkyne)⁓⁓⁓MgBr

C. Reaction with Oxiranes

As we saw in Section 11.9, the oxirane ring is so strained that it undergoes ring-opening reactions with a variety of nucleophiles. We can now add to the list of reactive nucleophiles Grignard and organolithium reagents. Butylmagnesium bromide, for example, reacts with oxirane (ethylene oxide) to give a magnesium alkoxide, which, on treatment with aqueous acid, gives 1-hexanol.

$$\sim\!\!\sim\!\!MgBr \;+\; \overset{:\ddot{O}:}{\triangle} \;\longrightarrow\; \sim\!\!\sim\!\!\sim\!:\ddot{O}:^- MgBr^+ \;\xrightarrow[H_2O]{HCl}\; \sim\!\!\sim\!\!\sim\!\ddot{O}H$$

Butylmagnesium bromide	Ethylene oxide	A magnesium alkoxide	1-Hexanol

As illustrated in this example, the product of treatment of a Grignard reagent with oxirane followed by protonation of the alkoxide is a primary alcohol with a carbon chain two carbons longer than the original chain. In reaction of a substituted oxirane, the major product corresponds to attack of the Grignard reagent on the less hindered carbon of the three-membered ring in an S_N2-like reaction. Treatment of racemic methyloxirane (propylene oxide) with phenylmagnesium bromide, for example, followed by workup in aqueous acid gives racemic 1-phenyl-2-propanol. The reaction does not work well if one or more of the oxirane carbons is quaternary.

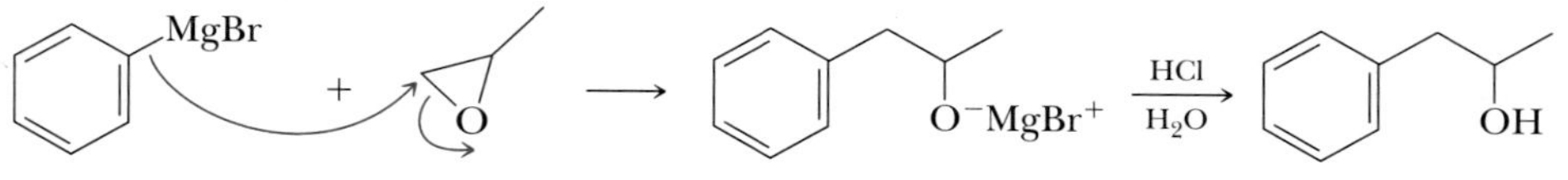

| Phenylmagnesium bromide | Methyloxirane (Propylene oxide) (racemic) | A magnesium alkoxide | 1-Phenyl-2-propanol (racemic) |

Example 15.2

Show how to prepare each alcohol from an organohalogen compound and an oxirane.

(a)

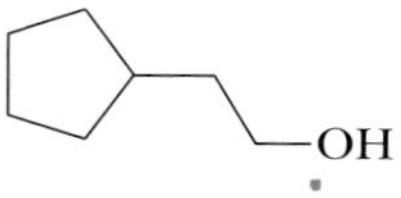

(b)

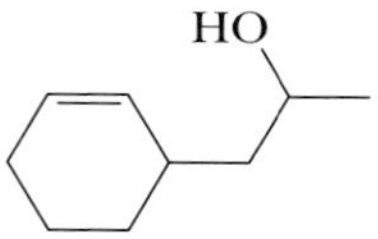

Solution

Shown is a retrosynthetic analysis for each compound followed by a synthesis.

(a)

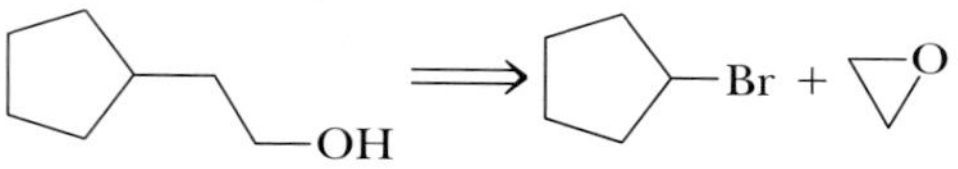

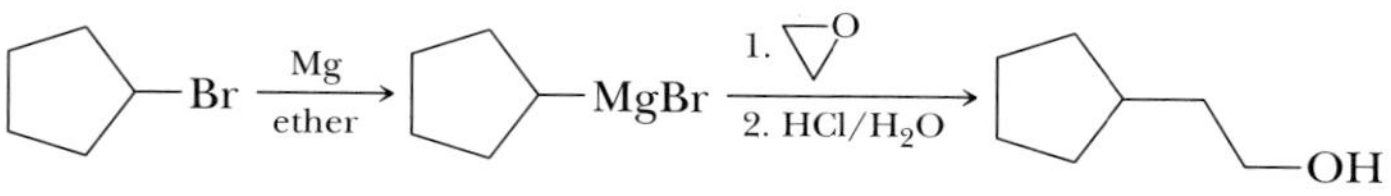

(b)

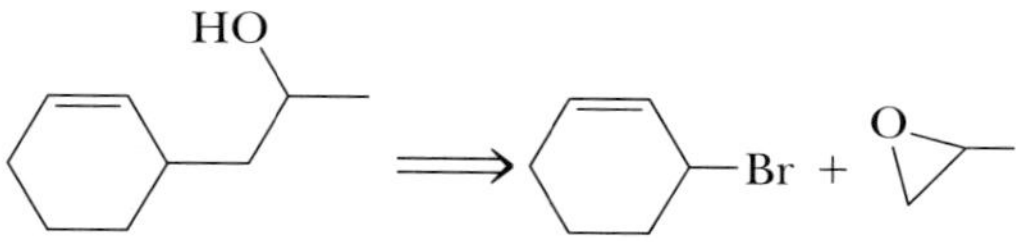

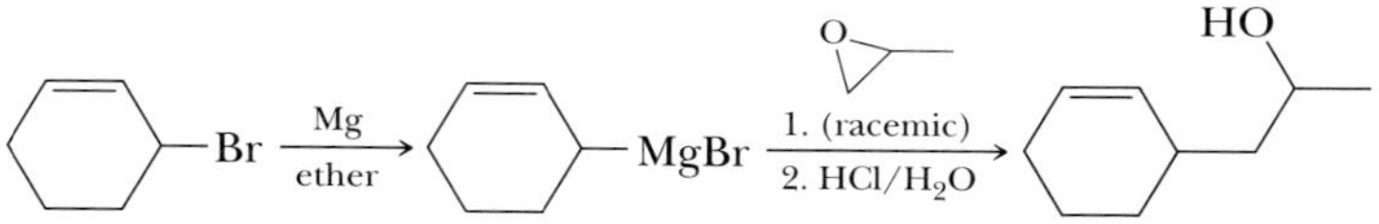

(a racemic mixture of two pairs of enantiomers)

Problem 15.2

Recalling the reactions of alcohols from Chapter 10, show how to synthesize each compound from an organohalogen compound and an oxirane, followed by a transformation of the resulting hydroxyl group to the desired oxygen-containing functional group.

(a)

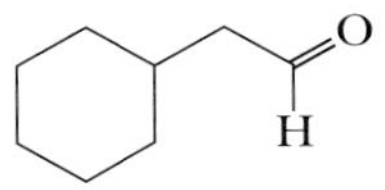

(b)

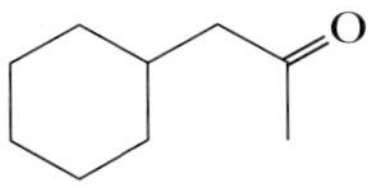

15.2 Lithium Diorganocopper (Gilman) Reagents

A. Formation and Structure

An important use of organolithium reagents (Section 15.1) is in the preparation of diorganocopper reagents, often called Gilman reagents after Henry Gilman (1893–1986) of Iowa State University who was the first to develop their chemistry. They are easily prepared by treatment of an alkyl, aryl, or alkenyllithium compound with copper(I) iodide, as illustrated by the preparation of lithium dibutylcopper from butyllithium.

$$2CH_3CH_2CH_2CH_2Li + CuI \xrightarrow[\text{or THF}]{\text{diethyl ether}} (CH_3CH_2CH_2CH_2)_2Cu^- \ Li^+ + LiI$$

| Butyllithium | Copper(I) iodide | Lithium dibutylcopper (a Gilman reagent) |

Gilman reagents consist of two organic groups associated with a copper(I) ion giving a negatively charged species, which is the source of the carbon nucleophile. Lithium ion is associated with this negatively charged species as the counter ion.

B. Coupling with Organohalogen Compounds

Gilman reagents are especially valuable for the formation of new carbon-carbon bonds by a coupling reaction with an alkyl chloride, bromide, or iodides (fluorides are unreactive under these conditions) as illustrated by the following preparation of 2-methyl-1-dodecene. Notice that only one of the Gilman-reagent alkyl groups is transferred in the reaction. Because Gilman reagents are ultimately prepared from halides, this leads to effective coupling of two halides.

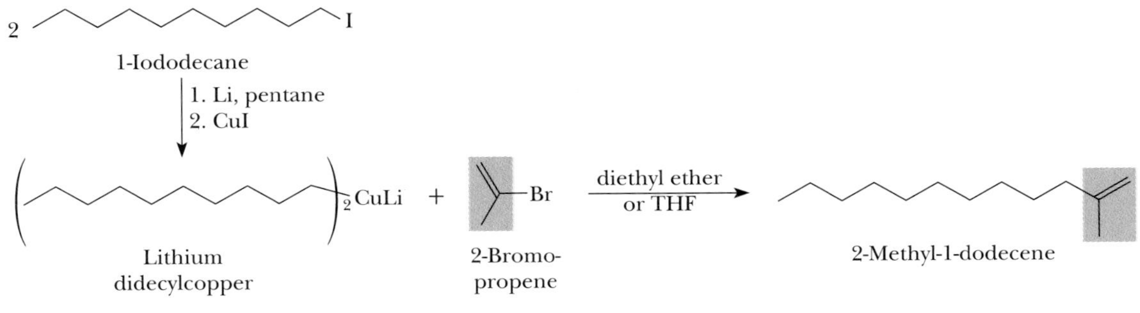

1-Iododecane

1. Li, pentane
2. CuI

Lithium didecylcopper

2-Bromo-propene

diethyl ether or THF

2-Methyl-1-dodecene

This example illustrates the coupling of an alkyl halide, a nucleophile, with a vinylic halide, an electrophile. Vinylic halides are normally quite unreactive toward nucleophilic displacement. Thus, the lithium diorganocopper reaction shown here is unique.

Gilman reagents giving the best yields of coupling products are those prepared from methyl, primary alkyl, allylic, vinylic, and aryl halides via the corresponding organolithium compounds. Yields are lower with secondary and tertiary alkyl halides.

Coupling with a vinylic halide is stereospecific; the configuration of the carbon-carbon double bond is preserved, as illustrated by the synthesis of *trans*-5-tridecene.

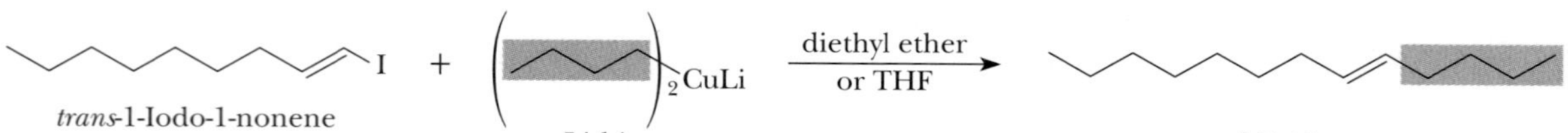

trans-1-Iodo-1-nonene   Lithium dibutylcopper   *trans*-5-Tridecene

A variation on the preparation of Gilman reagents is to use a Grignard reagent in the presence of a catalytic amount of Cu(I). Zoecon Corporation has developed a synthesis of 150-kg batches of the housefly sex attractant muscalure by treating (*Z*)-1-bromo-9-octadecene with pentylmagnesium bromide in the presence of catalytic amounts of Cu(I). The starting bromoalkene is easily prepared from the readily available (*Z*)-9-octadecenoic acid (oleic acid, Section 26.1). Yields of muscalure are nearly quantitative.

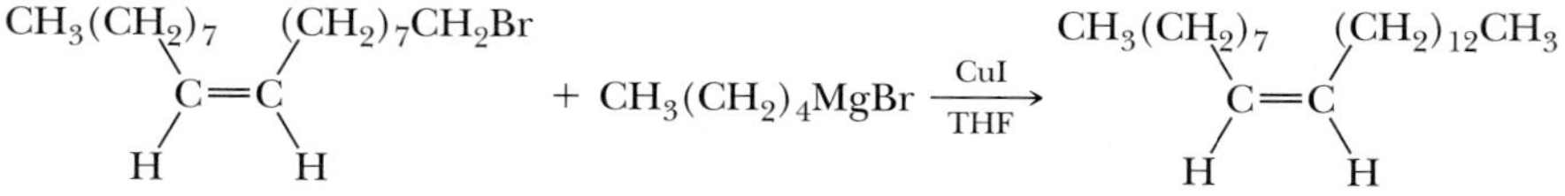

(*Z*)-1-Bromo-9-octadecene   (*Z*)-9-Tricosene (Muscalure)

The mechanism of these coupling reactions is not fully understood and is the subject of active investigation.

Example 15.3

Show how to bring about each conversion using a lithium diorganocopper reagent.
(a) 1-Bromocyclohexene to 1-methylcyclohexene
(b) 1-Bromo-2-methylpropane to 2,5-dimethylhexane using the bromoalkane as the only source of carbon

Solution

(a) Treat 1-bromocyclohexene with lithium dimethylcopper.

1-Bromocyclohexene   Lithium dimethylcopper   1-Methylcyclohexene

(b) Treat 1-bromo-2-methylpropane with lithium diisobutylcopper, itself prepared from 1-bromo-2-methylpropane.

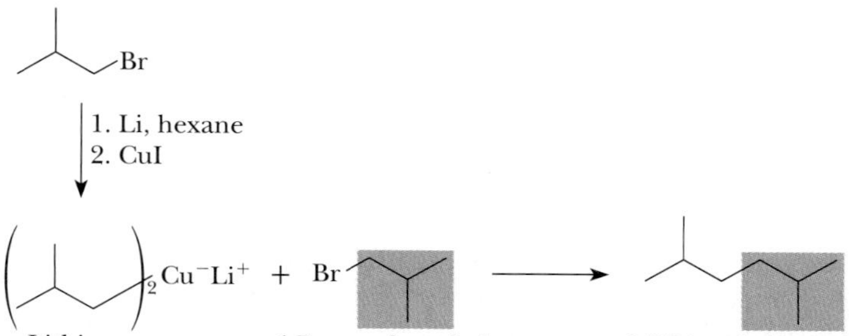

Lithium diisobutylcopper   1-Bromo-2-methyl-propane (Isobutyl bromide)   2,5-Dimethylhexane

Problem 15.3

Show how to bring about each conversion using a lithium diorganocopper reagent.

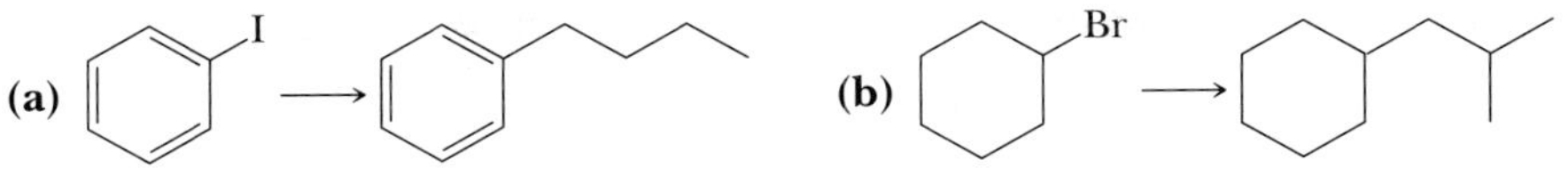

(a) ⟶

(b) ⟶

C. Reaction with Oxiranes

The reaction of epoxides with Gilman reagents is an important method for the formation of new carbon-carbon bonds. Like organolithium compounds and Grignard reagents, these compounds bring about regioselective ring opening of substituted epoxides at the less substituted carbon to give alcohols. Treatment of racemic styrene oxide with lithium divinylcopper, for example, followed by work-up in aqueous acid gives racemic 1-phenyl-3-buten-1-ol.

Styrene oxide
(racemic)

1-Phenyl-3-buten-1-ol
(racemic)

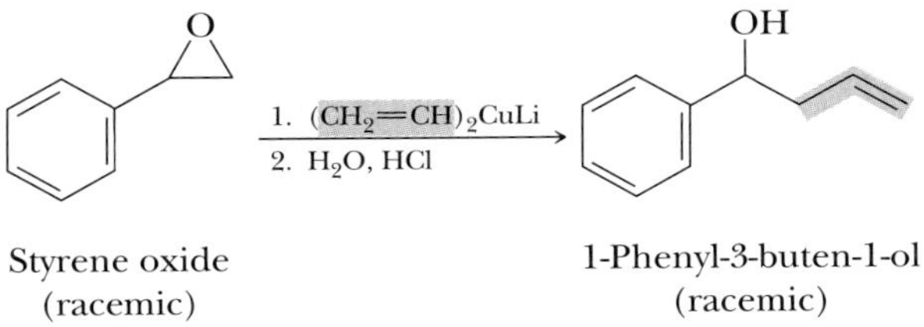

Example 15.4

Show two combinations of epoxide and Gilman reagent that can be used to prepare racemic 1-phenyl-3-hexanol.

Solution

The carbon bearing the hydroxyl group must have been one of the carbon atoms of the epoxide ring. The second carbon of the epoxide was either the one to the right of the carbon now bearing the —OH or the one to the left of it. In these solutions, the phenyl group is written C_6H_5—. Either route would be satisfactory.

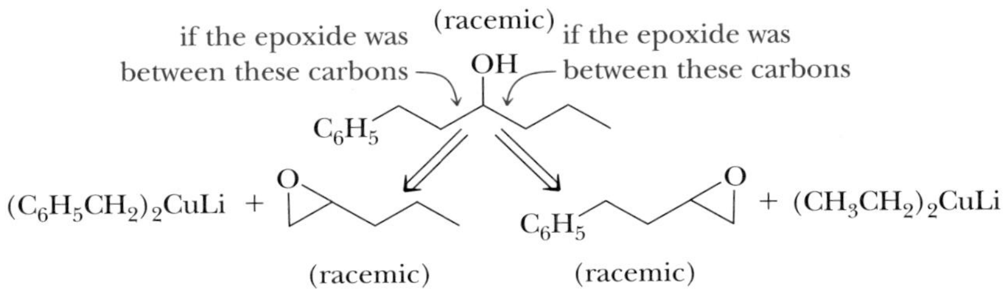

Problem 15.4

Show how to prepare each Gilman reagent in Example 15.4 from an appropriate alkyl halide and each epoxide from an appropriate alkene.

15.3 Carbenes and Carbenoids

Carbene   A neutral molecule that contains a carbon atom surrounded by only six valence electrons (R_2C:).

A **carbene**, R_2C:, is a neutral molecule in which a carbon atom is surrounded by only six valence electrons. Because they are electron-deficient, carbenes are highly reactive and behave as electrophiles. As we will see, one of their most important types of reactions is with alkenes (nucleophiles) to give cyclopropanes.

A. Methylene

Photolysis   Cleavage by light.

Thermolysis   Cleavage by heating.

The simplest carbene is methylene, CH_2, prepared by **photolysis** (cleavage by light) or **thermolysis** (cleavage on heating) of diazomethane, CH_2N_2, an explosive, toxic gas.

$$H_2\overset{+}{C}=\overset{..}{N}=\overset{..}{\underset{..}{N}}: \longleftrightarrow H_2\overset{..}{\underset{..}{C}}-\overset{+}{N}\equiv N: \xrightarrow{h\nu} H_2C: \quad + \quad :N\equiv N:$$

Methylene
(the simplest
carbene)

In the lowest electronic state of most carbenes, carbon is sp^2 hybridized with the unshared pair of electrons occupying the third sp^2 orbital. The unhybridized $2p$ orbital lies perpendicular to the plane created by the three sp^2 orbitals. Note that this orbital description of methylene is very much like that of a carbocation (Section 6.3A). In both species, carbon is sp^2 hybridized with a vacant $2p$ orbital. Methylene in this electronic state resembles a hybrid of a carbocation and a carbanion in that it has both a vacant p orbital and a lone pair.

Orbital structure
of CH_2 (methylene)

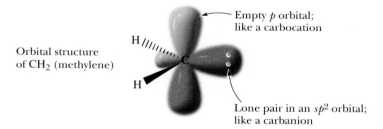

Empty p orbital;
like a carbocation

Lone pair in an sp^2 orbital;
like a carbanion

Methylene generated in this manner reacts with all C—H and C=C bonds and is so nonselective that it is of little synthetic use.

B. Dichlorocarbene

Although we often think of chlorine atoms as electron-withdrawing substituents, dichlorocarbene is much more stable and chemoselective than free methylene because resonance with the lone pairs on chlorine partially satisfies the electron deficiency on carbon.

$$:\overset{..}{\underset{..}{Cl}}-\overset{..}{\underset{\underset{Cl}{|}}{C}} \longleftrightarrow :\overset{..}{\underset{..}{Cl}}\overset{..}{\underset{\overset{+}{Cl}}{C^-}} \longleftrightarrow :\overset{+}{\underset{..}{Cl}}=\overset{..}{\underset{\underset{Cl}{\backslash}}{C^-}}$$

Dichlorocarbene can be prepared by treating chloroform with potassium *tert*-butoxide, removing the elements of HCl. The resulting carbene reacts cleanly with alkenes to give dichlorocyclopropanes. Addition of a dihalocarbene to an alkene shows syn stereoselectivity.

$$CHCl_3 \quad + \quad (CH_3)_3CO^-K^+ \quad \longrightarrow \quad Cl_2C: \quad + \quad (CH_3)_3COH + K^+Cl^-$$

| Trichloromethane (Chloroform) | Potassium *tert*-butoxide | Dichlorocarbene | *tert*-Butyl alcohol |

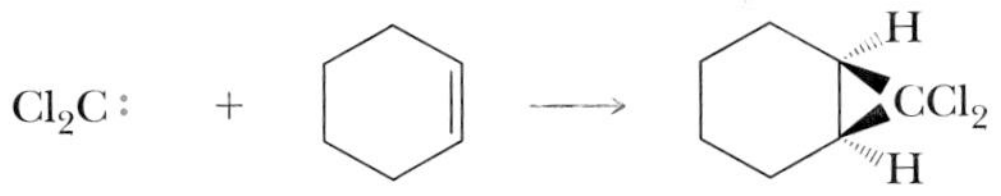

$$Cl_2C: \quad + \quad \bigcirc \quad \longrightarrow \quad \overset{H}{\underset{H}{\diagdown}}CCl_2$$

Dichlorocarbene   A dichlorocyclopropane

Reaction of a *cis* alkene with a dihalocarbene gives only a *cis* dihalocyclopropane as illustrated by the reaction of *cis*-3-hexene with dichlorocarbene. Similarly, reaction of a *trans* alkene gives only a *trans* dihalocyclopropane.

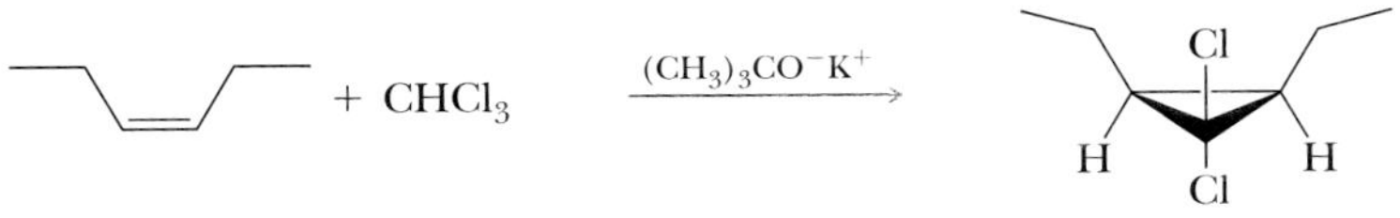

cis-3-Hexene

cis-1,1-Dichloro-2,3-diethylcyclopropane

Mechanism   *Formation of Dichlorocarbene and Its Reaction with Cyclohexene*

Taken together, Steps 1 and 2 result in α-elimination of H and Cl; that is, both atoms are eliminated from the same carbon. We have seen many examples of β-elimination, where hydrogen and a leaving group are eliminated from neighboring carbons. There are very few examples of α-elimination, and they are possible only where no β-hydrogen exists.

Step 1: Treatment of chloroform, which is somewhat acidic because of its three electron-withdrawing chlorine atoms, with potassium *tert*-butoxide gives the trichloromethide anion.

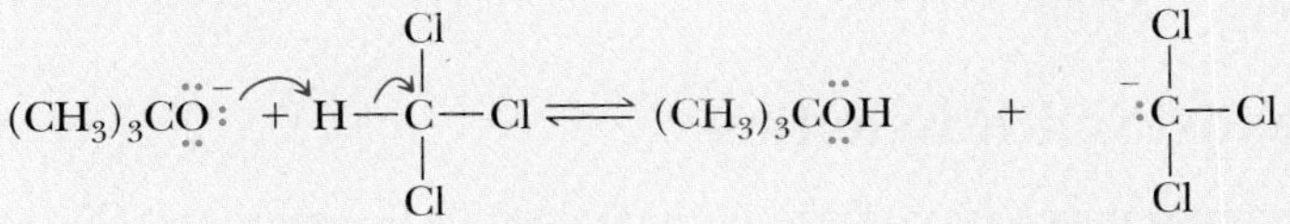

$$(CH_3)_3CO:^- + H-\overset{\overset{Cl}{|}}{\underset{\underset{Cl}{|}}{C}}-Cl \rightleftharpoons (CH_3)_3COH \quad + \quad :\overset{\overset{Cl}{|}}{\underset{\underset{Cl}{|}}{C}}-Cl$$

Trichloromethide anion

Step 2: Elimination of Cl⁻ from CCl₃⁻ gives dichlorocarbene.

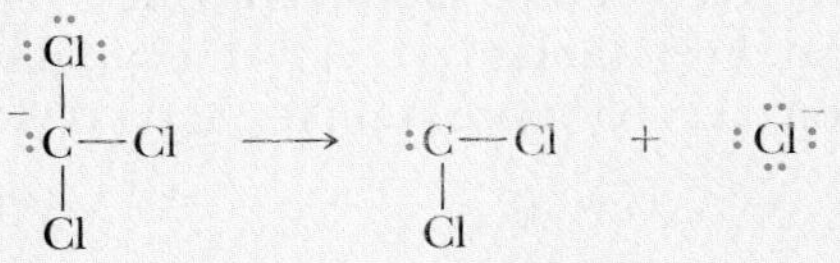

$$:\overset{\overset{:Cl:}{|}}{\underset{\underset{Cl}{|}}{C}}-Cl \quad \longrightarrow \quad :C-Cl \quad + \quad :Cl:^-$$

Dichlorocarbene

Step 3: Syn addition of dichlorocarbene to cyclohexene gives a dichlorocyclopropane.

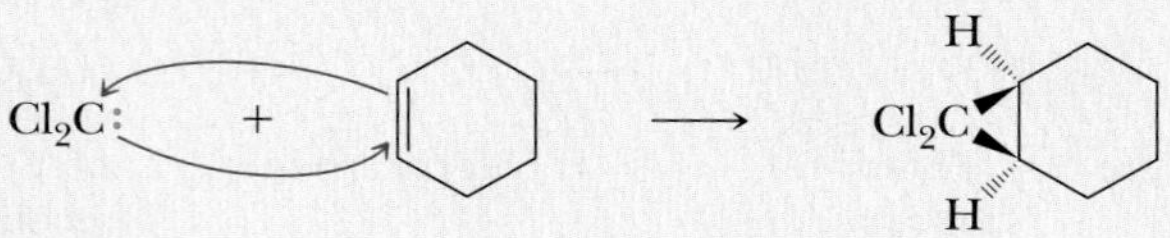

Dichlorocarbene   Cyclohexene   A dichlorocyclopropane
(an electrophile)   (a nucleophile)

Example 15.5

Predict the product from the following reaction.

$$CHBr_3 + (CH_3)_3CO^-K^+ \quad + \quad$$

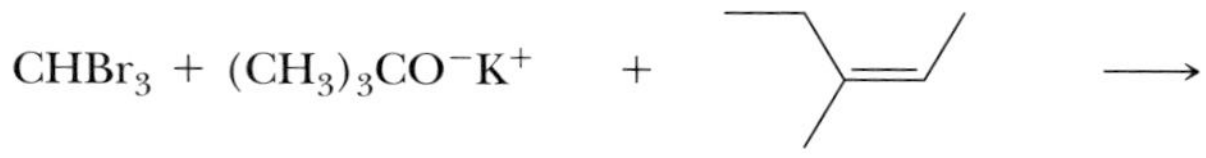

(Z)-3-Methyl-2-pentene

Solution

Bromoform gives dibromocarbene, which reacts stereospecifically with the alkene to give a dibromocyclopropane. The product is racemic.

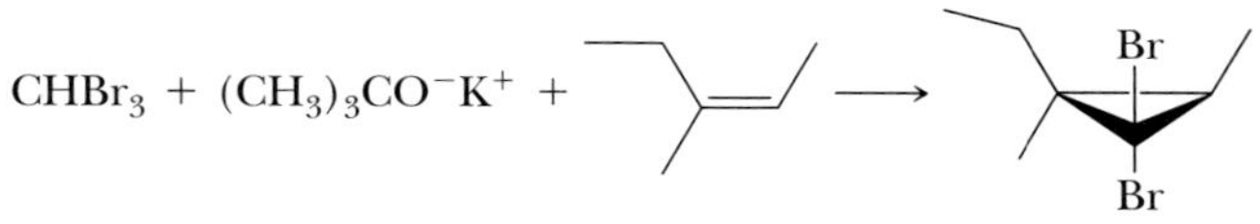

Problem 15.5

Predict the product of the following reaction.

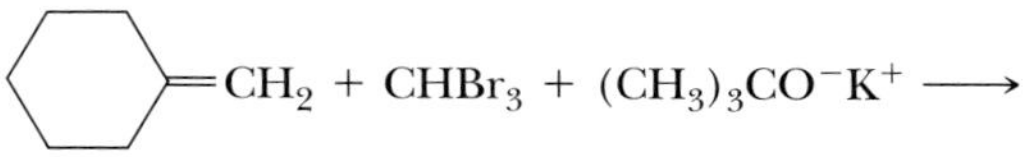

C. The Simmons-Smith Reaction

Although methylene prepared from diazomethane itself is not synthetically useful, addition of methylene to an alkene can be accomplished using a reaction first reported by the American chemists Howard Simmons and Ronald Smith. The Simmons-Smith reaction uses diiodomethane and zinc dust activated by a small amount of copper (a so-called "zinc-copper couple") to produce iodomethylzinc iodide, in a reaction reminiscent of a Grignard reaction. Even though we show the Simmons-Smith reagent here as ICH_2ZnI, its structure is considerably more complex and not fully understood.

$$CH_2I_2 \quad + \quad Zn(Cu) \quad \xrightarrow[\text{diethyl ether}]{} \quad ICH_2ZnI$$

Diiodomethane   Zinc-copper   Iodomethylzinc iodide
                        couple   (Simmons-Smith reagent)

This organozinc compound reacts with a wide variety of alkenes to give cyclopropanes.

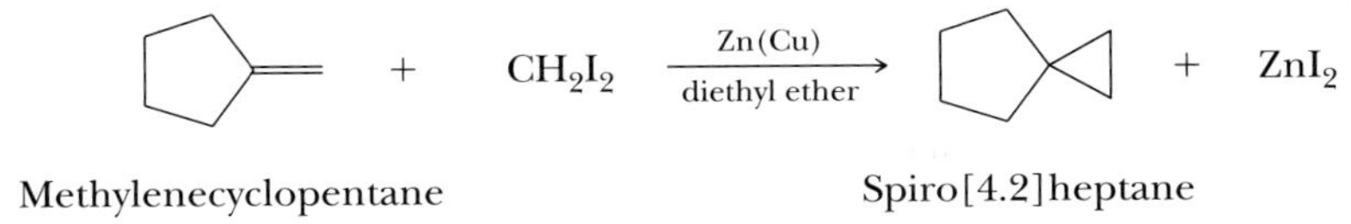

Methylenecyclopentane   Spiro[4.2]heptane

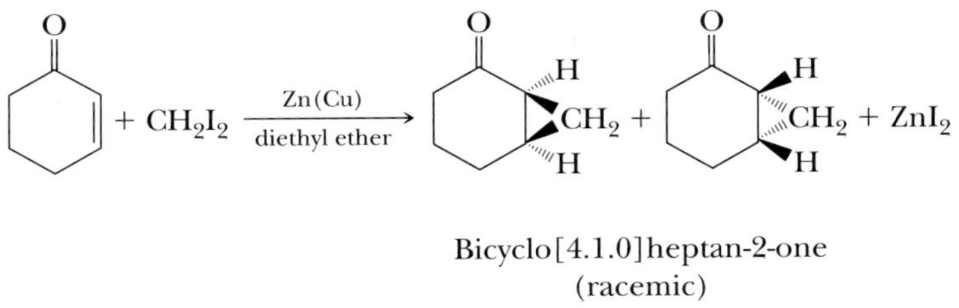

Bicyclo[4.1.0]heptan-2-one
(racemic)

Mechanism   *The Simmons-Smith Reaction with an Alkene*

Although an α-elimination from the Simmons-Smith reagent to give methylene would in principle be possible, the reagent is much more selective than free methylene. Instead, the organozinc compound reacts directly with the alkene by a concerted mechanism to give the cyclopropane-containing product. The Simmons-Smith reagent is an example of a **carbenoid,** a compound that delivers the elements of a carbene without actually producing a free carbene.

Carbenoid   A compound that delivers the elements of a carbene without actually producing a free carbene.

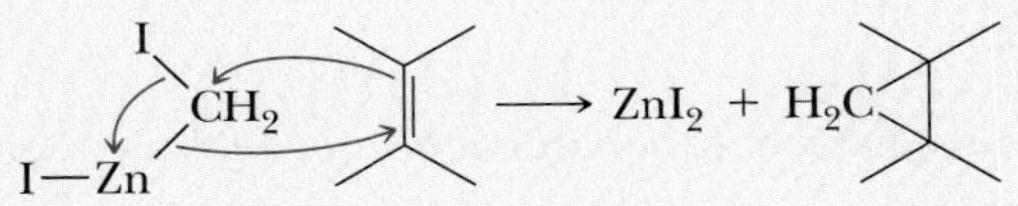

Example 15.6

Draw a structural formula for the product of treating each alkene with the Simmons-Smith reagent.

(a)   (b)

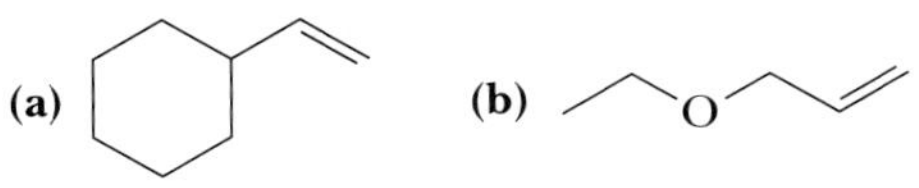

Solution

Reaction at each carbon-carbon double bond forms a cyclopropane ring.

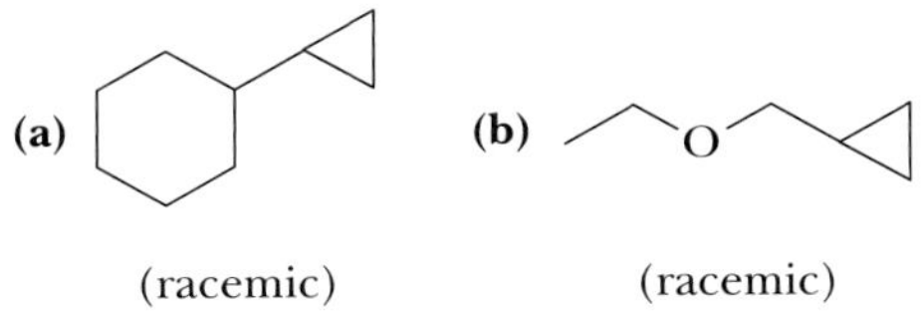

(a)   (b)

(racemic)   (racemic)

Problem 15.6

Show how the following compound could be prepared from any compound containing ten carbons or fewer.

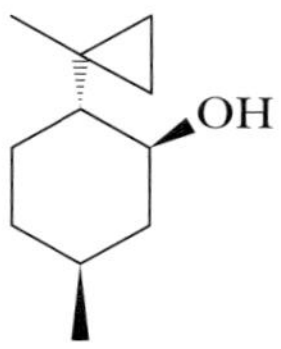

Summary

An **organometallic compound** is one that contains a carbon-metal bond (Introduction). The key feature of many of these reagents is that carbon of the carbon-metal bond carries a partial negative charge. The partial negative charge on carbon makes it basic and nucleophilic; the latter property can be exploited in organic synthesis in the construction of carbon-carbon bonds. Organomagnesium compounds are named **Grignard reagents** after their discoverer, Victor Grignard (Section 15.1). Grignard reagents and organolithium compounds (Section 15.1) react with a wide range of functional groups, including epoxides. Organolithium compounds also react with Cu(I) salts to give useful reagents called Gilman reagents (Section 15.2). These react with organic halides to form new carbon-carbon bonds. Divalent carbon species called **carbenes,** (Section 15.3) or their organometallic-complexed equivalents called **carbenoids** (Section 15.3C) are useful for preparation of cyclopropanes.

Key Reactions

ORGANIC Chemistry•Now™
Click Flashcards to review the **Key Reactions of Organometallic Compounds**

1. Formation of Organomagnesium (Grignard) and Organolithium Compounds (Section 15.1A)

Organomagnesium compounds are prepared by treating an alkyl, aryl, or alkenyl (vinylic) halide with magnesium in diethyl ether or THF. Organolithium compounds are prepared by treating an alkyl, aryl, or alkenyl halide with lithium in pentane or other hydrocarbon solvent.

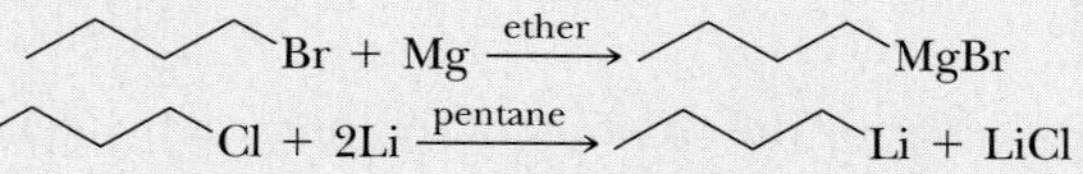

2. Reaction of RMgX and RLi with Proton Donors (Section 15.1B)

Both organomagnesium and organolithium compounds are strong bases and react with any proton donor stronger than the alkane from which the organolithium or magnesium compound is derived. Water or other proton donors must be completely excluded during their preparation and use.

$$CH_3CH_2MgBr + H_2O \longrightarrow CH_3CH_2{-}H + Mg(OH)Br$$

3. Reaction of a Grignard Reagent with an Epoxide (Section 15.1C)

Treatment of a Grignard reagent with an epoxide followed by hydrolysis of the magnesium alkoxide salt in aqueous acid gives an alcohol with its carbon chain extended by two carbon atoms.

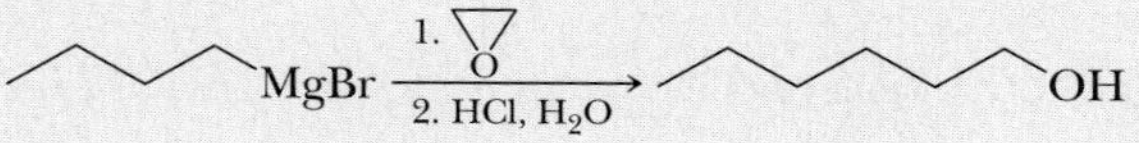

4. Formation of Gilman Reagents (Section 15.2A)

Lithium diorganocopper (Gilman) reagents are prepared by treating an organolithium compound with copper(I) iodide.

$$2CH_3CH_2CH_2CH_2Li + CuI \xrightarrow[\text{or THF}]{\text{ether}} (CH_3CH_2CH_2CH_2)_2Cu^- \; Li^+ + LiI$$

5. Treatment of a Gilman Reagent with an Alkyl, Aryl, or Alkenyl Halide (Section 15.2B)

Coupling of a Gilman reagent with an alkyl, alkenyl, or aryl halide results in formation of a new carbon-carbon bond.

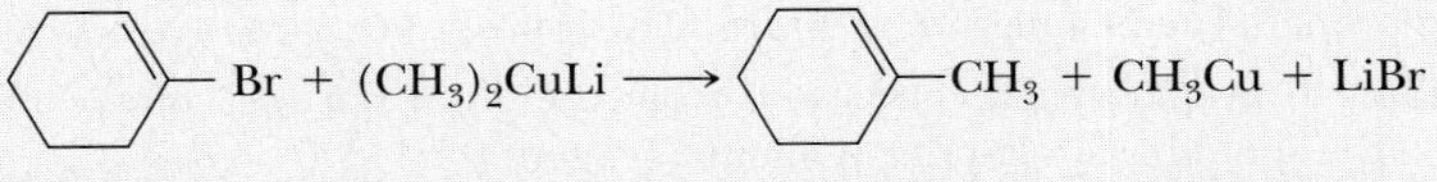

6. Reaction of Dichloro- or Dibromocarbene with an Alkene (Section 15.3B)

The dihalocarbene is generated by treatment of $CHCl_3$ or $CHBr_3$ with a strong base such as potassium *tert*-butoxide. Addition of the dihalocarbene to an alkene shows syn stereospecificity.

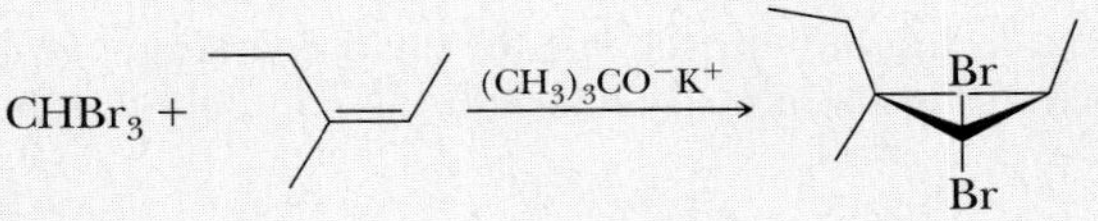

(racemic)

7. The Simmons-Smith Reaction (Section 15.3C)

Treatment of CH_2I_2 with a zinc-copper couple generates an organozinc compound, known as the Simmons-Smith reagent, which reacts with alkenes to give cyclopropanes.

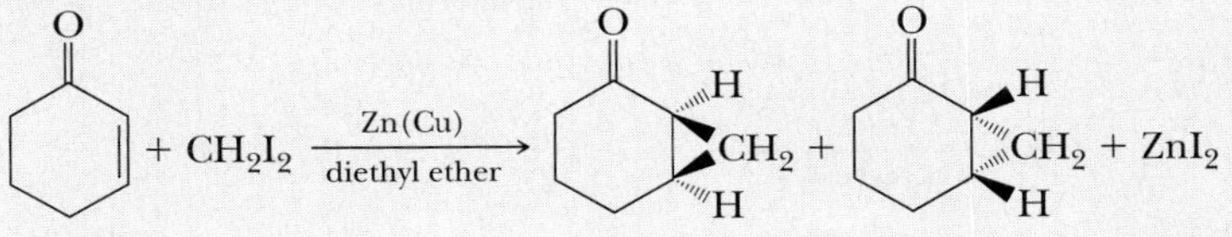

(racemic)

Problems

15.7   Complete these reactions involving lithium diorganocopper (Gilman) reagents.

(a)

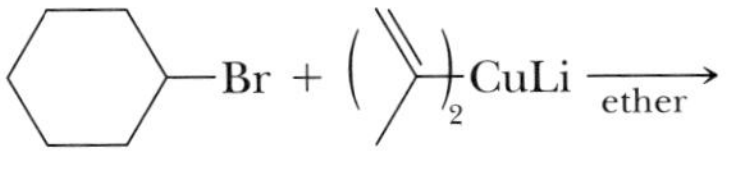

(b)

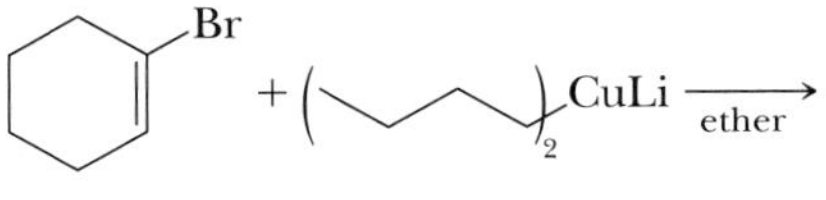

(c)

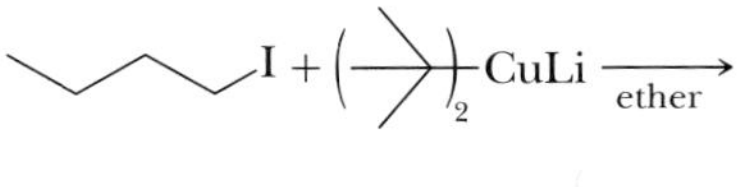

(d) 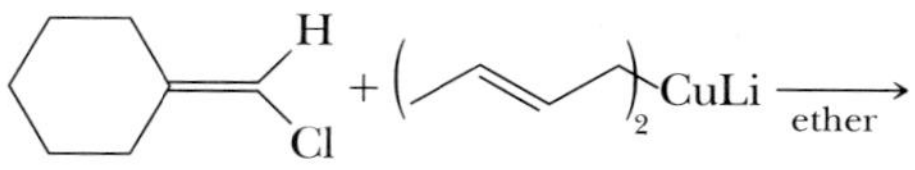 + ()$_2$CuLi $\xrightarrow{\text{ether}}$

15.8 Show how to convert 1-bromopentane to each of these compounds using a lithium diorganocopper (Gilman) reagent. Write an equation, showing structural formulas, for each synthesis.

(a) Nonane   (b) 3-Methyloctane   (c) 2,2-Dimethylheptane
(d) 1-Heptene   (e) 1-Octene

15.9 In Problem 15.8, you used a series of lithium diorganocopper (Gilman) reagents. Show how to prepare each Gilman reagent from an appropriate alkyl or vinylic halide.

15.10 Show how to prepare each compound from the given starting compound through the use of a lithium diorganocopper (Gilman) reagent.

(a) 4-Methylcyclopentene from 4-bromocyclopentene
(b) (Z)-2-Undecene from (Z)-1-bromopropene
(c) 1-Butylcyclohexene from 1-iodocyclohexene
(d) 1-Decene from 1-iodooctane
(e) 1,8-Nonadiene from 1,5-dibromopentane

15.11 The following is a retrosynthetic scheme for the preparation of *trans*-2-allylcyclohexanol. Show reagents to bring about the synthesis of this compound from cyclohexane.

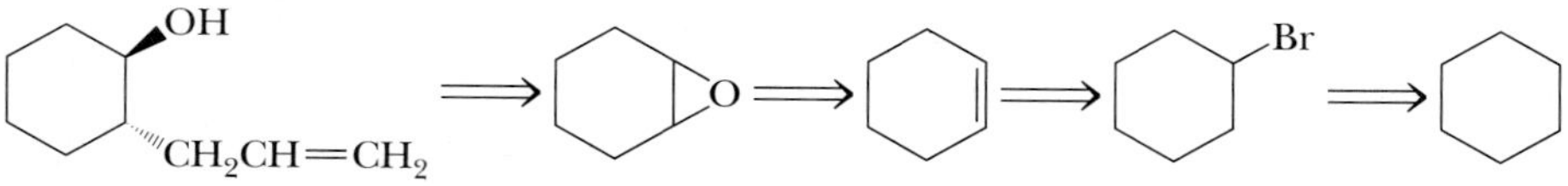

15.12 Complete these equations.

(a) $CH_3CH_2CH_2C{\equiv}CH + CH_3CH_2MgBr \xrightarrow{\text{diethyl ether}}$

(b) 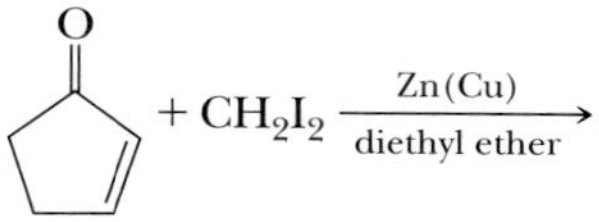 $+ CH_2I_2 \xrightarrow[\text{diethyl ether}]{\text{Zn(Cu)}}$

(c) 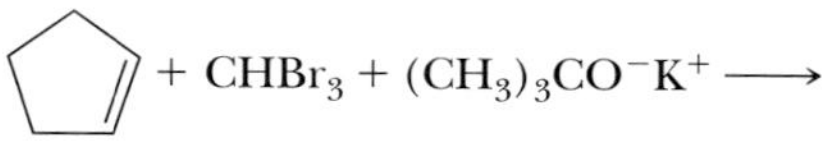 $+ CHBr_3 + (CH_3)_3CO^-K^+ \longrightarrow$

(d) 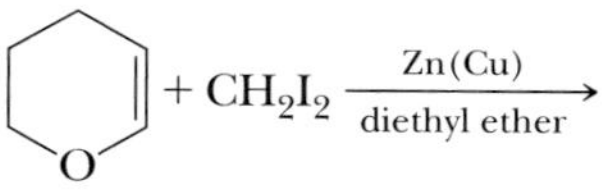 $+ CH_2I_2 \xrightarrow[\text{diethyl ether}]{\text{Zn(Cu)}}$

(e) 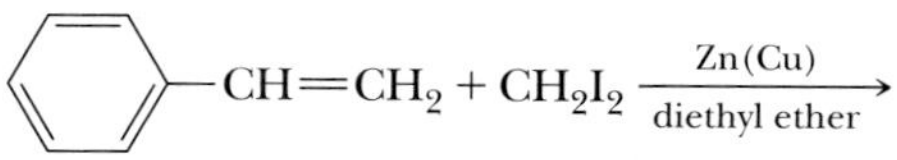 $-CH{=}CH_2 + CH_2I_2 \xrightarrow[\text{diethyl ether}]{\text{Zn(Cu)}}$

15.13 Reaction of the following cycloalkene with the Simmons-Smith reagent is stereospecific and gives only the isomer shown. Suggest a reason for this stereospecificity.

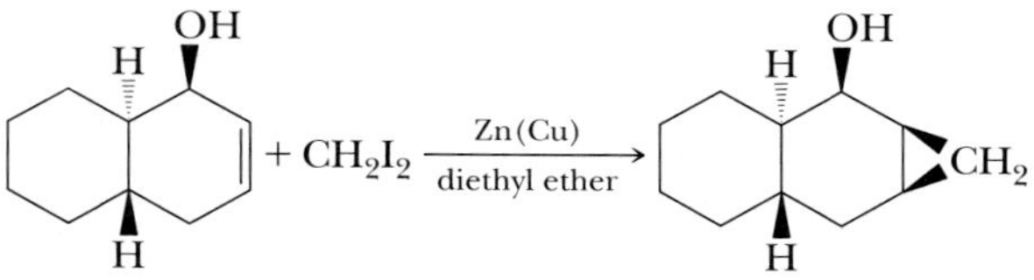

15.14 Show how the following compound can be prepared in good yield.

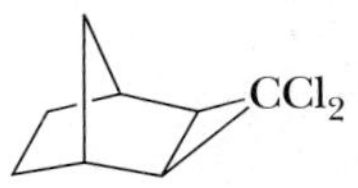

15.15 Show the product of the following reaction (do not concern yourself with which side of the ring is attacked).

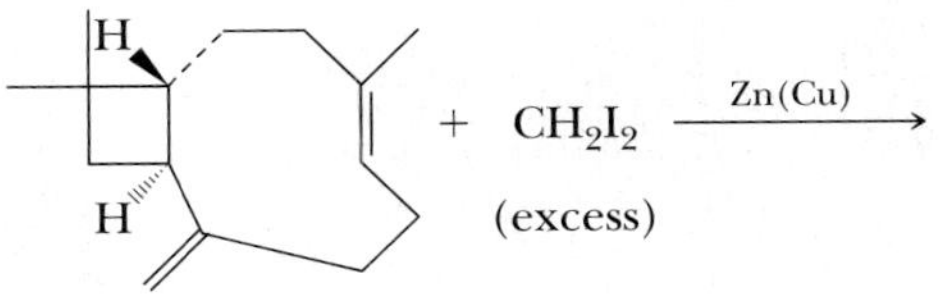

$$+ \ CH_2I_2 \ \xrightarrow{\ Zn(Cu)\ }$$

(excess)

Caryophyllene

15.16 Show how the spiro[2.2]pentane can be prepared in one step from organic compounds containing three carbons or less and any necessary inorganic reagents or solvents.

Looking Ahead

15.17 One of the most important uses for Grignard reagents is their addition to carbonyl compounds to give new carbon-carbon bonds (Section 16.5). In this reaction, the carbon of the organometallic compound acts as a nucleophile to add to the positive carbon of the carbonyl.

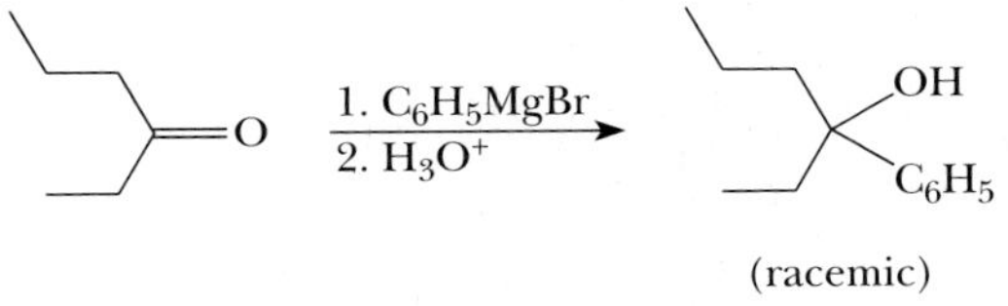

$$\xrightarrow[\text{2. } H_3O^+]{\text{1. } C_6H_5MgBr}$$

(racemic)

(a) Give a mechanism for the first step of the reaction.
(b) Explain the function of the acid in the second step.

15.18 Organolithium compounds react with carbonyl compounds in a way that is similar to that of Grignard reactions. Suggest a product of the following reaction.

$$\xrightarrow[\text{2. } H_3O^+]{\text{1. } CH_3Li} \ \ ?$$

C_6H_5

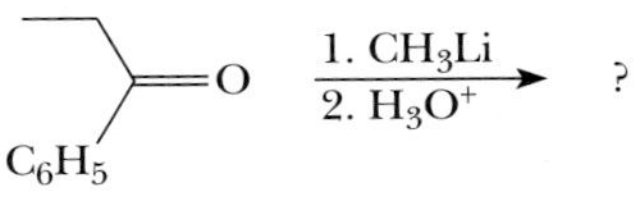

15.19 1-Bromobutane can be converted into either of the two products shown by a suitable choice of reagents. Give reagents and conditions for each reaction.

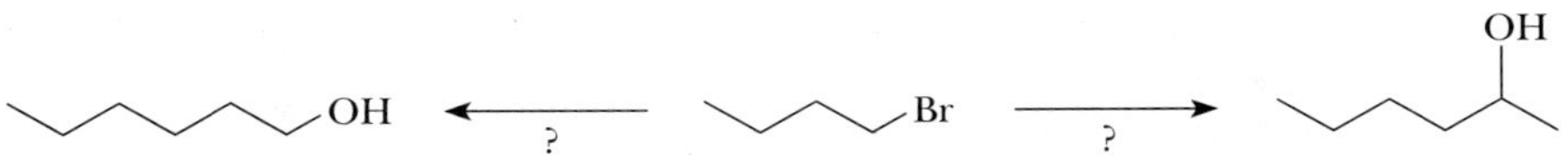

16

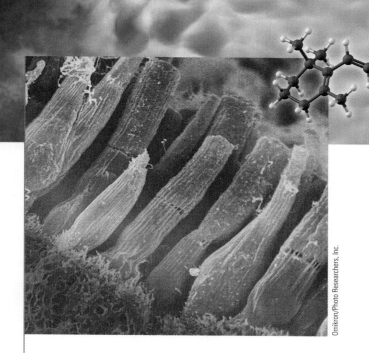

■ Rod cells in the human eye. Inset: A model of 11-*cis*-retinal, an oxidized form of vitamin A. For the reaction of 11-*cis*-retinal with opsin to form visual purple, see Section 16.8A.

Omikron/Photo Researchers, Inc.

Outline

ORGANIC
Chemistry · Now™

Look for this logo in the chapter and go to Organic ChemistryNow at **http://now.brookscole.com/bfi4** for tutorials, simulations, and problems.

Aldehydes and Ketones

In this and several of the following chapters, we study the physical and chemical properties of compounds containing the carbonyl group, C=O. Because the carbonyl group is the functional group of aldehydes, ketones, and carboxylic acids and their derivatives, it is one of the most important functional groups in organic chemistry. The chemical properties of this functional group are straightforward, and an understanding of its few characteristic reaction themes leads very quickly to an understanding of a wide variety of organic reactions.

16.1 Structure and Bonding

The functional group of an **aldehyde** is a carbonyl group bonded to a hydrogen atom and a carbon atom (Section 1.3C). In methanal (always called formaldehyde), the simplest aldehyde, the carbonyl group is bonded to two hydrogen atoms. In other aldehydes, it is bonded to one hydrogen atom and one carbon atom. Following are Lewis structures for formaldehyde and ethanal (which is always called acetaldehyde).

$$\overset{\displaystyle :O:}{\underset{\displaystyle HCH}{\|}}$$

Methanal
(Formaldehyde)

$$\overset{\displaystyle :O:}{\underset{\displaystyle CH_3CH}{\|}}$$

Ethanal
(Acetaldehyde)

$$\overset{\displaystyle :O:}{\underset{\displaystyle CH_3CCH_3}{\|}}$$

Propanone
(Acetone)

The functional group of a **ketone** is a carbonyl group bonded to two carbon atoms (Section 1.3C). The simplest ketone is propanone, which is always called acetone.

According to valence bond theory, the carbon-oxygen double bond consists of one sigma bond formed by the overlap of sp^2 hybrid orbitals of carbon and oxygen and one pi bond formed by the overlap of parallel $2p$ orbitals. The two nonbonding pairs of electrons on oxygen lie in the remaining sp^2 hybrid orbitals (Figure 1.19).

16.2 Nomenclature

A. IUPAC Nomenclature

IUPAC names for aldehydes and ketones follow the familiar pattern of selecting as the parent alkane the longest chain of carbon atoms that contains the functional group. We show the aldehyde group by changing the suffix -*e* of the parent alkane to -*al* (Section 2.5). Because the carbonyl group of an aldehyde can only appear at the end of a parent chain and because numbering must start with it as carbon-1, its position is unambiguous; there is no need to use a number to locate it.

For unsaturated aldehydes, the presence of a carbon-carbon double or triple bond is indicated by the infix -*en*- or -*yn*-. As with other molecules with both an infix and a suffix, the location of the group corresponding to the suffix determines the numbering pattern.

3-Methylbutanal

2-Propenal
(Acrolein)

(2*E*)-3,7-Dimethyl-2,6-octadienal
(Geranial)

For cyclic molecules in which —CHO is bonded directly to the ring, the molecule is named by adding the suffix -*carbaldehyde* to the name of the ring. The atom of the ring to which the aldehyde group is bonded is numbered 1.

Cyclopentane-
carbaldehyde

trans-4-Hydroxycyclo-
hexanecarbaldehyde

Among the aldehydes for which the IUPAC system retains common names are benzaldehyde and cinnamaldehyde, as well as formaldehyde and acetaldehyde. Note here the alternative ways of writing the phenyl group. In benzaldehyde, it is written as a line-angle formula; in cinnamaldehyde, it is written C_6H_5—.

Benzaldehyde   *trans*-3-Phenyl-2-propenal
(Cinnamaldehyde)

In the IUPAC system, ketones are named by selecting as the parent alkane the longest chain that contains the carbonyl group and then indicating its presence by changing the suffix from -*e* to -*one* (Section 2.5). The parent chain is numbered from the direction that gives the carbonyl carbon the smaller number. The IUPAC system retains the common names acetone, acetophenone, and benzophenone.

Propanone   Acetophenone   Benzophenone   1-Phenyl-1-pentanone
(Acetone)

Benzaldehyde is found in the kernels of bitter almonds. Cinnamaldehyde is found in Ceylon and Chinese cinnamon oils.

Example 16.1

Write IUPAC names for each compound. Specify the configuration of all chiral centers in (a) and (c).

(a)   (b)   (c)

Solution

(a) The parent chain is pentane. The name is (*S*)-3-ethyl-2-hydroxypentanal.
(b) Number the six-membered ring beginning with the carbonyl carbon. The IUPAC name is 3-methyl-2-cyclohexenone.
(c) The name (2*S*,5*R*)-2-isopropyl-5-methylcyclohexanone provides a complete description of the configuration of each stereocenter as well as the *trans* relationship between the isopropyl and methyl groups. The common name of this compound is menthone.

Problem 16.1

Write the IUPAC name for each compound. Specify the configuration in (c).

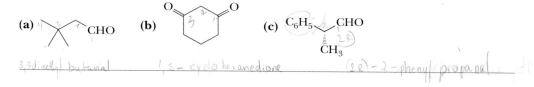

(a)   CHO   (b)   (c)   C₆H₅   CHO
                             CH₃

Example 16.2

Write structural formulas for all ketones with molecular formula $C_6H_{12}O$, and give each its IUPAC name. Which of these ketones are chiral?

Solution

Following are line-angle formulas and IUPAC names for the six ketones with this molecular formula. Only 3-methyl-2-pentanone is chiral; the R enantiomer is drawn here.

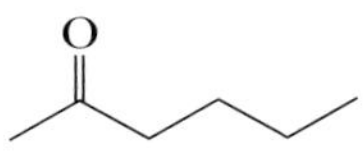

2-Hexanone 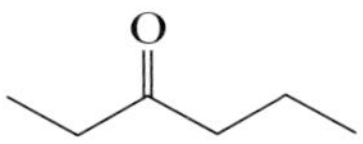   3-Hexanone 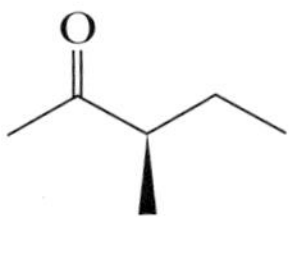   (R)-3-Methyl-2-pentanone

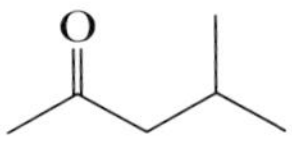

4-Methyl-2-pentanone 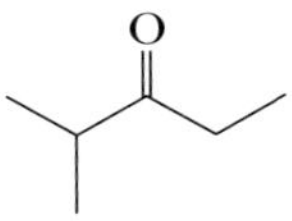   2-Methyl-3-pentanone 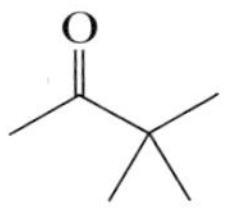   3,3-Dimethyl-2-butanone

Problem 16.2

Write structural formulas for all aldehydes with molecular formula $C_6H_{12}O$, and give each its IUPAC name. Which of these aldehydes are chiral?

B. IUPAC Names for More Complex Aldehydes and Ketones

In naming compounds that contain more than one functional group that might be indicated by a suffix, the IUPAC system has established an **order of precedence of functions.** Table 16.1 gives the order of precedence for the functional groups we have studied so far.

Order of precedence of functions
A ranking of functional groups in order of priority for the purposes of IUPAC nomenclature.

Table 16.1   Increasing Order of Precedence of Six Functional Groups

Functional Group	Suffix if Higher Priority	Prefix if Lower Priority	Example When the Functional Group Has Lower Priority	
Carboxyl	-oic acid	—		
Aldehyde	-al	oxo-	3-Oxopropanoic acid	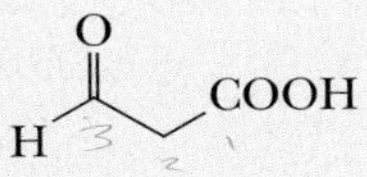
Ketone	-one	oxo-	3-Oxobutanoic acid	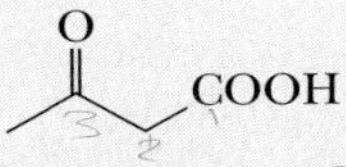
Alcohol	-ol	hydroxy-	4-Hydroxybutanoic acid	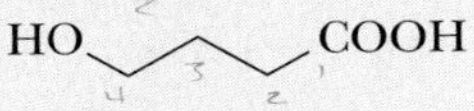
Amino	-amine	amino-	3-Aminobutanoic acid	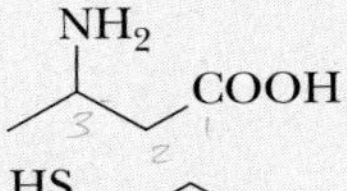
Sulfhydryl	-thiol	mercapto	2-Mercaptoethanol	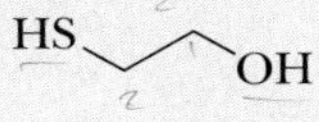

(Increasing precedence)

Example 16.3

Write the IUPAC name for each compound, being certain to specify configuration where appropriate.

(a) *[structure, with handwritten annotations: O O, numbers, "ketone", "aldehyde"]*

(b) HO—CHO / OH *[handwritten: "hydroxy", "(R,S)", "hydroxyl 1", "(2S)-2,3-dihydroxypropanal"]*

(c) *[structure with OH and O, handwritten: "one", "6 R", numbers, "(6R)-6-hydroxyheptanone"]*

Solution

(a) 3-Oxobutanal. An aldehyde has higher precedence than a ketone. The presence of the carbonyl group of the ketone is indicated by the prefix *oxo-* (Table 16.1).

(b) (*R*)-2,3-Dihydroxypropanal. Its common name is glyceraldehyde. Glyceraldehyde is the simplest carbohydrate (Section 25.1).

(c) (*R*)-6-Hydroxy-2-heptanone.

Problem 16.3

Write the IUPAC name for each compound.

(a) HO— —OH *[handwritten: "one", numbers]*   (b) *[cyclohexane ring with two O, handwritten "1","2", "cyclohexanedione"]*   (c) H₂N— —CHO *[handwritten: numbers, "4-aminobutanal"]*

C. Common Names

The common name for an aldehyde is derived from the common name of the corresponding carboxylic acid by dropping the word *acid* and changing the suffix -*ic* or -*oic* to -*aldehyde*. Because we have not yet studied common names for carboxylic acids, we are not in a position to discuss common names for aldehydes. We can illustrate how they are derived, however, by reference to a few common names with which you are familiar. The name formaldehyde is derived from formic acid; the name acetaldehyde is derived from acetic acid.

$$\begin{array}{cccc} O && O && O && O \\ \| && \| && \| && \| \\ HCH && HCOH && CH_3CH && CH_3COH \end{array}$$

Formaldehyde   Formic acid   Acetaldehyde   Acetic acid

Common names for ketones are derived by naming the two alkyl or aryl groups bonded to the carbonyl group as separate words, followed by the word *ketone*.

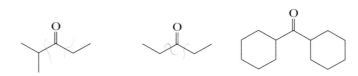

Ethyl isopropyl ketone   Diethyl ketone   Dicyclohexyl ketone

16.3 Physical Properties

Oxygen is more electronegative than carbon (3.5 compared with 2.5); therefore, a carbon-oxygen double bond is polar, with oxygen bearing a partial negative charge and carbon bearing a partial positive charge. In addition, the resonance structure shown on the right emphasizes the reactivity of the carbonyl oxygen as a Lewis base and the carbonyl carbon as a Lewis acid. The bond dipole moment of a carbonyl group is 2.3 D (Table 1.7).

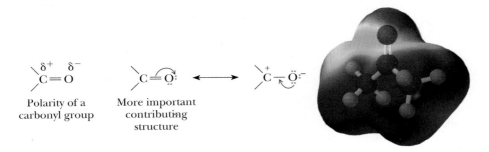

Polarity of a carbonyl group

More important contributing structure

■ An electrostatic potential map for acetone. Note the large negative charge (red) on the carbonyl oxygen and the positive charges (blue) on the three carbons.

Because of the polarity of the carbonyl group, aldehydes and ketones are polar compounds and interact in the pure liquid by dipole-dipole interactions; they have higher boiling points than nonpolar compounds of comparable molecular weight (Table 16.2).

Of the compounds listed in Table 16.2, pentane and diethyl ether have the lowest boiling points. Diethyl ether is a polar molecule, but because of steric hindrance, only weak dipole-dipole interactions exist between its molecules (Section 11.3). Both butanal and 2-butanone are polar molecules, and, because of the intermolecular attraction between their carbonyl groups, their boiling points are higher than those of pentane and diethyl ether. Alcohols (Section 10.2) and carboxylic acids (Section 17.3) are polar molecules that can associate by hydrogen bonding, the strongest intermolecular interaction. The boiling points of 1-butanol and propanoic acid are significantly higher than those of butanal and 2-butanone, compounds whose molecules cannot associate by hydrogen bonding.

The oxygen atoms of the carbonyl groups of aldehydes and ketones act as hydrogen bond acceptors with water; therefore, low-molecular-weight aldehydes and

Table 16.2 Boiling Points of Six Compounds of Comparable Molecular Weight

Name	Structural Formula	Molecular Weight (g/mol)	Boiling Point (°C)
Diethyl ether	$CH_3CH_2OCH_2CH_3$	74	34
Pentane	$CH_3CH_2CH_2CH_2CH_3$	72	36
Butanal	$CH_3CH_2CH_2CHO$	72	76
2-Butanone	$CH_3CH_2COCH_3$	72	80
1-Butanol	$CH_3CH_2CH_2CH_2OH$	74	117
Propanoic acid	CH_3CH_2COOH	74	141

Table 16.3 Physical Properties of Selected Aldehydes and Ketones

IUPAC Name	Common Name	Structural Formula	Boiling Point (°C)	Solubility (g/100 g water)
Methanal	Formaldehyde	HCHO	−21	Infinite
Ethanal	Acetaldehyde	CH_3CHO	20	Infinite
Propanal	Propionaldehyde	CH_3CH_2CHO	49	16
Butanal	Butyraldehyde	$CH_3CH_2CH_2CHO$	76	7
Hexanal	Caproaldehyde	$CH_3(CH_2)_4CHO$	129	Slight
Propanone	Acetone	CH_3COCH_3	56	Infinite
2-Butanone	Ethyl methyl ketone	$CH_3COCH_2CH_3$	80	26
3-Pentanone	Diethyl ketone	$CH_3CH_2COCH_2CH_3$	101	5

ketones are more soluble in water than are nonpolar compounds of comparable molecular weight. Listed in Table 16.3 are boiling points and solubilities in water for several low-molecular-weight aldehydes and ketones.

16.4 Reactions

One of the most common reaction themes of a carbonyl group is addition of a nucleophile to form a tetrahedral carbonyl addition compound. In the following general reaction, the nucleophilic reagent is written as Nu^- to emphasize the presence of its unshared pair of electrons. Notice how the pi bond of the carbonyl group breaks as the nucleophile attacks, changing the carbon atom hybridization state, all the while maintaining four bonds to carbon in the tetrahedral carbonyl addition compound.

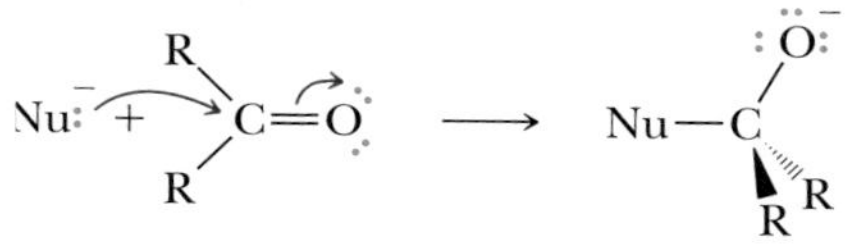

Tetrahedral carbonyl addition compound

A second common reaction theme of a carbonyl group is reaction with a proton or other Lewis acid to form a resonance-stabilized cation. Protonation increases the electron deficiency of the carbonyl carbon and makes it more reactive toward nucleophiles. This cation then reacts with nucleophiles to give a tetrahedral carbonyl addition compound.

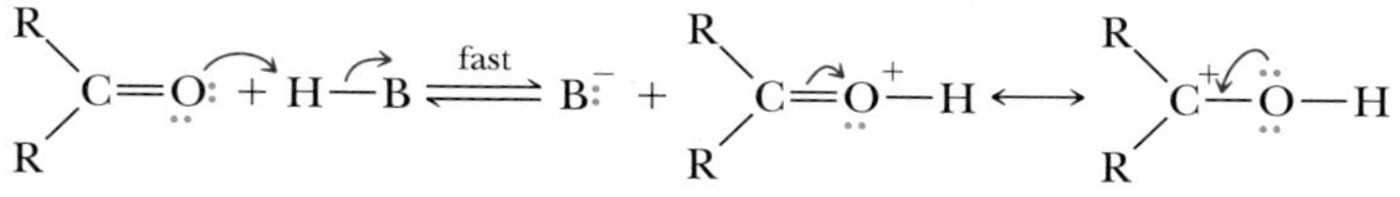

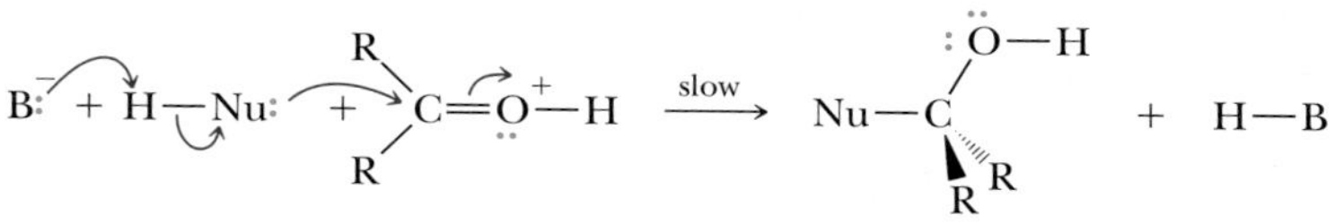

Tetrahedral carbonyl addition compound

Both of these reactivities can be predicted based on the large bond dipole of the carbonyl group. Using acetaldehyde as an example, Lewis acids such as protons are attracted to the partial negative charge (red color of the electrostatic potential map) of the carbonyl oxygen, while electron-rich nucleophiles are attracted to the partial positive charge (blue color of the electrostatic potential map) of the carbonyl carbon.

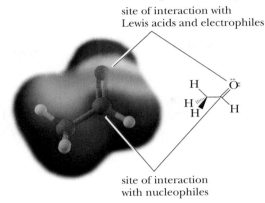

site of interaction with
Lewis acids and electrophiles

site of interaction
with nucleophiles

Often, the tetrahedral product of nucleophile addition to a carbonyl is a new chiral center. If none of the starting materials is chiral, then the two enantiomers will be created in equal amounts because the nucleophile will approach the carbonyl from either side with equal probability. As a result, the carbonyl addition product will consist of a racemic mixture.

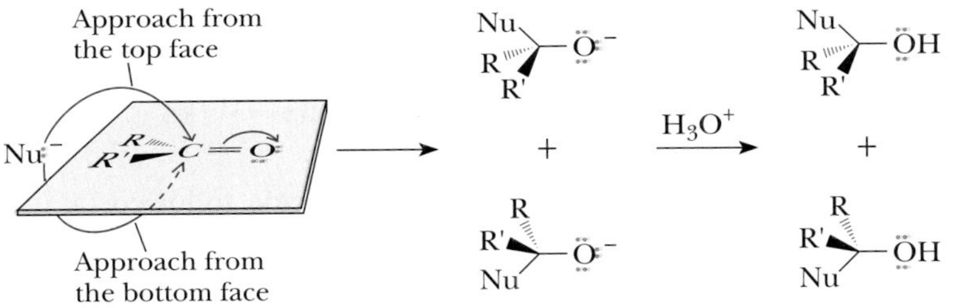

Approach from
the top face

Approach from
the bottom face

A new chiral center
is created

A racemic mixture

16.5 Addition of Carbon Nucleophiles

In this section, we examine reactions of aldehydes and ketones with the following types of carbon nucleophiles.

RMgX   RLi $RC{\equiv}C{:}^-$ $^-{:}C{\equiv}N{:}$

A Grignard   An organolithium   An anion of a   Cyanide ion
reagent   reagent   terminal alkyne

From the perspective of the organic chemist, addition of a carbon nucleophile is the most important type of carbonyl addition reaction because a new carbon-carbon bond is formed in the process.

A. Addition of Grignard Reagents

The special value of Grignard reagents (Section 15.1) is that they provide excellent ways to form new carbon-carbon bonds. Given the difference in electronegativity between carbon and magnesium ($2.5 - 1.2 = 1.3$), the carbon-magnesium bond of

Carbanion An anion in which carbon has an unshared pair of electrons and bears a negative charge.

a Grignard reagent is polar covalent with carbon bearing a partial negative charge and magnesium bearing a partial positive charge. In its reactions, a Grignard reagent behaves as a **carbanion.** A carbanion is a good nucleophile and adds to the carbonyl group of an aldehyde or ketone to form a tetrahedral carbonyl addition compound. The driving force for these reactions is the attraction of the partial negative charge on the carbon of the organometallic compound for the partial positive charge on the carbonyl carbon. The alkoxide ions formed in these reactions are strong bases (Section 10.4) and, when treated with an aqueous acid such as HCl or aqueous NH_4Cl during work-up, form alcohols. In the following examples, the magnesium oxygen bond is written $—O^-[MgBr]^+$ to emphasize its ionic character.

Caution: New chiral centers are often created in Grignard reactions with aldehydes or ketones. When neither the aldehyde/ketone nor the Grignard reagent is chiral but the product has a new chiral center, a racemic mixture is formed.

Addition to Formaldehyde Gives a Primary Alcohol

Treatment of a Grignard reagent with formaldehyde followed by hydrolysis in aqueous acid gives a primary alcohol.

$$CH_3CH_2—MgBr \ + \ H—\overset{\overset{\displaystyle :O:}{\|}}{C}—H \ \xrightarrow{\text{ether}} \ CH_3CH_2—CH_2 \ \xrightarrow[H_2O]{HCl} \ CH_3CH_2—CH_2 \ + \ Mg^{2+}$$

Formaldehyde · · · · · · A magnesium alkoxide · · · · · · 1-Propanol (a 1° alcohol)

Addition to an Aldehyde (Except Formaldehyde) Gives a Secondary Alcohol

Treatment of a Grignard reagent with any other aldehyde followed by hydrolysis in aqueous acid gives a 2° alcohol.

Acetaldehyde (an aldehyde) · · · · · · A magnesium alkoxide · · · · · · 1-Cyclohexylethanol (a 2° alcohol; racemic)

In this example, the product is chiral and is formed as a racemic mixture.

Addition to a Ketone Gives a Tertiary Alcohol

Treatment of a Grignard reagent with a ketone followed by hydrolysis in aqueous acid gives a tertiary alcohol.

Phenyl-magnesium bromide · · · · · Acetone (a ketone) · · · · · A magnesium alkoxide · · · · · 2-Phenyl-2-propanol (a 3° alcohol)

Example 16.4

Racemic 2-phenyl-2-butanol can be synthesized by three different combinations of a Grignard reagent and a ketone. Show each combination.

Solution

In each solution, curved arrows show formation of the new carbon-carbon bond and the alkoxide ion. The new carbon-carbon bond formed by each set of reagents is labeled in the final product.

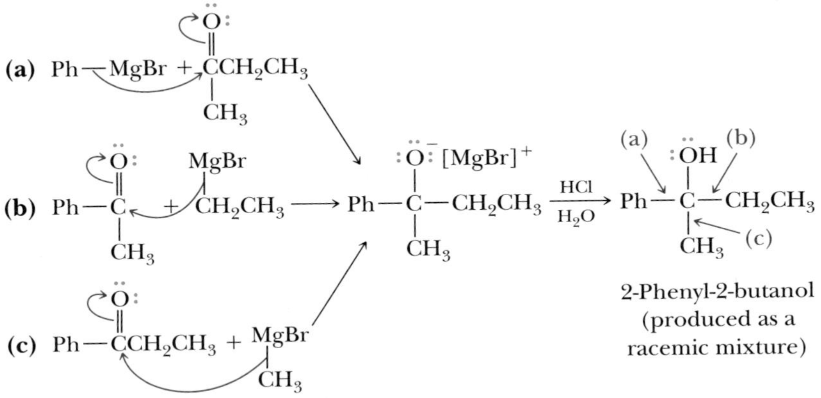

2-Phenyl-2-butanol
(produced as a
racemic mixture)

Problem 16.4

Show how these four products can be synthesized from the same Grignard reagent.

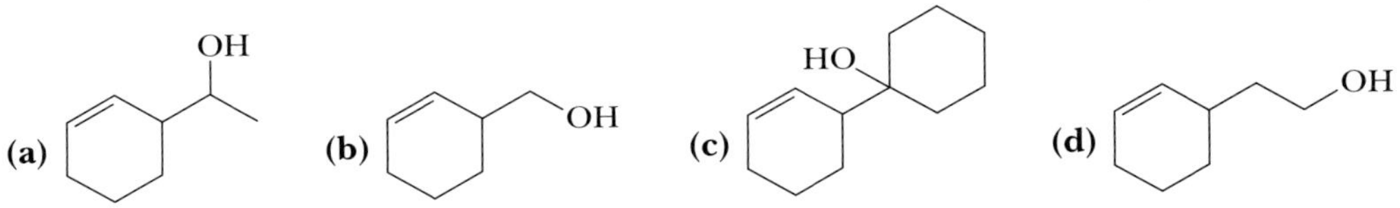

(a) (racemic)   (b) (racemic)   (c) (racemic)   (d) (racemic)

B. Addition of Organolithium Compounds

Organolithium compounds are generally more reactive in carbonyl addition reactions than organomagnesium compounds and typically give higher yields of products. They are more troublesome to use, however, because they must be prepared and used under an atmosphere of nitrogen or other inert gas. The following synthesis illustrates the use of an organolithium compound to form a sterically hindered tertiary alcohol.

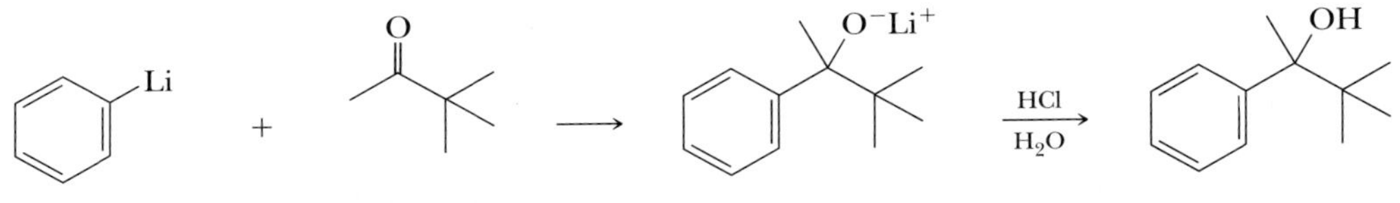

Phenyllithium   3,3-Dimethyl-2-butanone   A lithium
alkoxide (racemic)   3,3-Dimethyl-2-phenyl-2-butanol
(racemic)

C. Addition of Anions of Terminal Alkynes

The anion of a terminal alkyne is a nucleophile (Section 7.5) and adds to the carbonyl group of an aldehyde or ketone to form a tetrahedral carbonyl addition compound. In the following example, addition of sodium acetylide to cyclohexanone followed by hydrolysis in aqueous acid gives 1-ethynylcyclohexanol.

Sodium   Cyclohexanone   A sodium   1-Ethynylcyclohexanol
acetylide   alkoxide

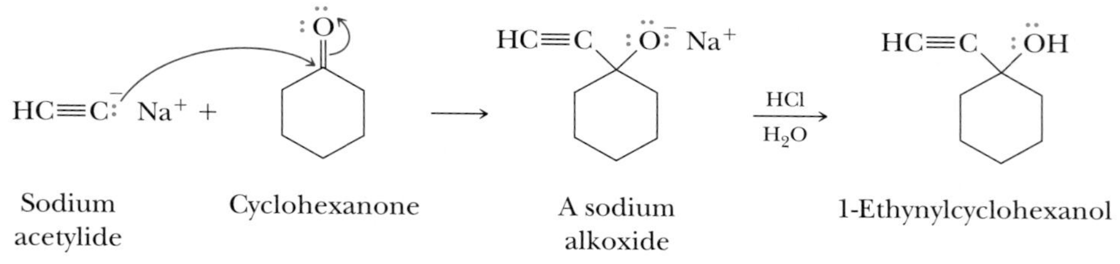

These addition compounds (alkynyl alcohols) contain both a hydroxyl group and a carbon-carbon triple bond, each of which can be further modified.

Acid-catalyzed hydration (Section 7.7B) of 1-ethynylcyclohexanol gives an α-hydroxyketone. Alternatively, hydroboration followed by oxidation with alkaline hydrogen peroxide (Section 7.7A) gives a β-hydroxyaldehyde.

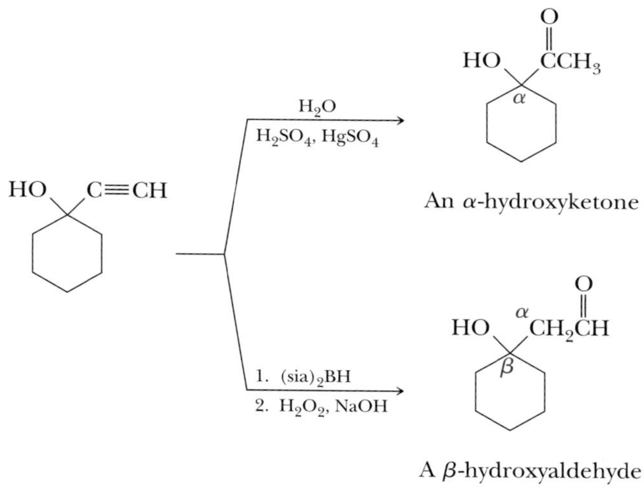

An α-hydroxyketone

A β-hydroxyaldehyde

This example illustrates two of the most valuable reactions of alkynes: (1) addition of the anion of a terminal alkyne to the carbonyl group of an aldehyde or ketone gives an alkynyl alcohol and (2) hydration of a terminal alkyne gives either an aldehyde or ketone depending on the alkyne and the method of hydration.

D. Addition of Hydrogen Cyanide

Cyanohydrin A molecule containing an —OH group and a —CN group bonded to the same carbon.

Hydrogen cyanide, HCN, adds to the carbonyl group of an aldehyde or ketone to form a tetrahedral carbonyl addition compound called a **cyanohydrin.** For example, HCN adds to acetaldehyde to form acetaldehyde cyanohydrin in 75% yield. We study the naming of compounds containing the nitrile group in Section 18.1E.

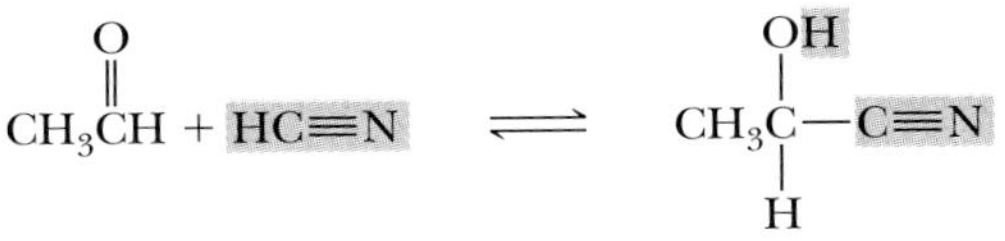

2-Hydroxypropanenitrile
(Acetaldehyde cyanohydrin;
produced as a racemic mixture)

Addition of hydrogen cyanide proceeds by way of cyanide ion. Because HCN is a weak acid, pK_a 9.31, the concentration of cyanide ion in aqueous HCN is too low for cyanohydrin formation to proceed at a reasonable rate. For this reason, cyanohydrin formation is generally carried out by dissolving NaCN or KCN in water and adjusting the pH of the solution to approximately 10.0, giving a solution in which HCN and CN^- are present in comparable concentrations.

Mechanism *Formation of a Cyanohydrin*

Step 1: Nucleophilic addition of cyanide ion to the carbonyl carbon gives a tetrahedral carbonyl addition compound.

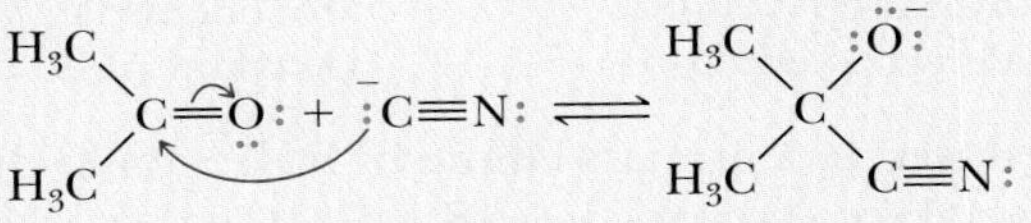

Step 2: Proton transfer from HCN gives the cyanohydrin and generates a new cyanide ion.

ORGANIC **Chemistry•Now**™ Click Mechanisms to view an animation of the **Formation of a Cyanohydrin**

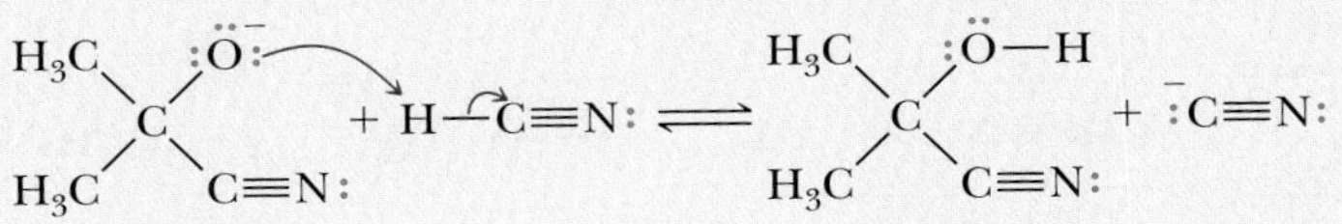

For aldehydes and most aliphatic ketones, the position of equilibrium favors cyanohydrin formation. For many aryl ketones (ketones in which the carbonyl carbon is bonded to a benzene ring) and sterically hindered aliphatic ketones, however, the position of equilibrium favors starting materials; cyanohydrin formation is not a useful reaction for these types of compounds. The following synthesis of ibuprofen, for example, failed because the cyanohydrin was formed only in low yield.

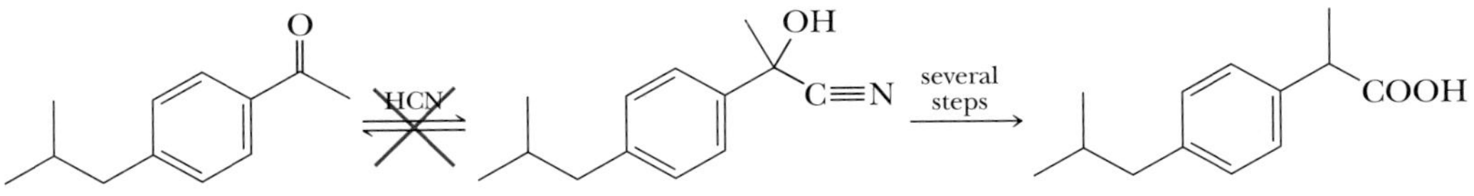

4-Isobutylacetophenone A cyanohydrin Ibuprofen

Benzaldehyde cyanohydrin (mandelonitrile) provides an interesting example of a chemical defense mechanism in the biological world. This substance is synthesized by millipedes (*Apheloria corrigata*) and stored in special glands. When a millipede is threatened, the cyanohydrin is released from the storage gland and undergoes enzyme-catalyzed reversal of cyanohydrin formation to produce HCN, which is then released to ward off predators. The quantity of HCN emitted by a single millipede is sufficient to kill a small mouse. Mandelonitrile is also found in bitter almond and peach pits. Its function there is unknown, as is how millipedes survive exposure to hydrogen cyanide.

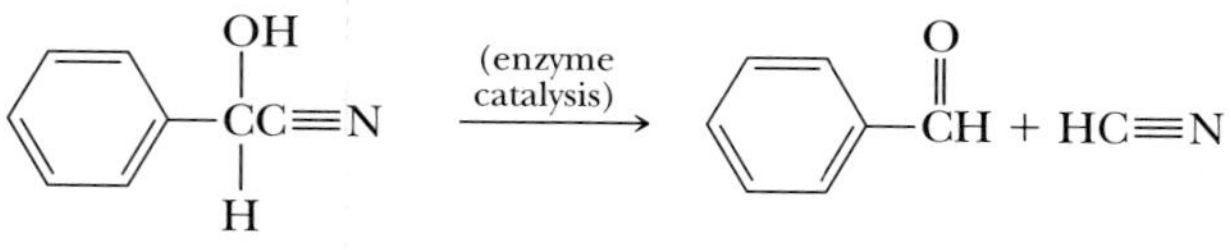

Benzaldehyde cyanohydrin
(Mandelonitrile)

Benzaldehyde

The value of cyanohydrins as synthetic intermediates lies in the new functional groups into which they can be converted. First, the secondary or tertiary alcohol group of the cyanohydrin may undergo acid-catalyzed dehydration to form an unsaturated nitrile. For example, acid-catalyzed dehydration of acetaldehyde cyanohydrin gives acrylonitrile, the monomer from which polyacrylonitrile (Orlon, Table 29.1) is made.

2-Hydroxypropanenitrile
(Acetaldehyde cyanohydrin)

Propenenitrile
(Acrylonitrile)

Second, a nitrile is reduced to a primary amine by hydrogen in the presence of nickel or other transition metal catalyst (Section 18.10C). Catalytic reduction of benzaldehyde cyanohydrin, for example, gives 2-amino-1-phenylethanol.

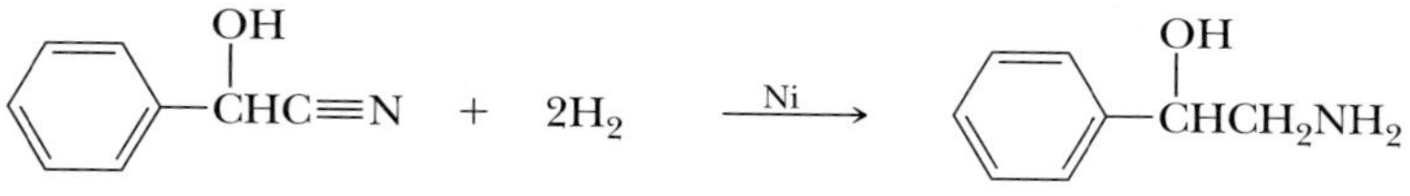

Benzaldehyde cyanohydrin

2-Amino-1-phenylethanol

As we shall see in Section 18.4E, hydrolysis of a nitrile in the presence of an acid catalyst gives a carboxylic acid. Thus, even though nitriles are little used themselves, they are valuable intermediates for the synthesis of other functional groups.

16.6 The Wittig Reaction

Ylide A neutral molecule with positive and negative charges on adjacent atoms.

In 1954, Georg Wittig reported a method for the synthesis of alkenes from aldehydes and ketones using compounds called phosphonium **ylides** (Chapter 1, Chemical Connections: "The Octet Rule"). For his pioneering study and development of this reaction into a major synthetic tool, Professor Wittig shared the 1979 Nobel Prize for chemistry. (The other recipient was Herbert C. Brown for his studies of hydroboration and the chemistry of organoboron compounds.) A Wittig synthesis is illustrated by the conversion of cyclohexanone to methylenecyclohexane. In this reaction, a C=O double bond is converted to a C=C double bond.

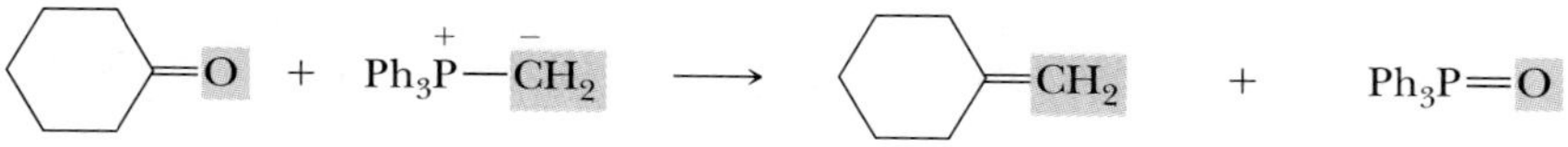

Cyclohexanone   A phosphonium
ylide

Methylenecyclohexane   Triphenylphosphine
oxide

We study the Wittig reaction in two stages: first, the formation and structure of phosphonium ylides and, second, the reaction of a phosphonium ylide with the carbonyl group of an aldehyde or ketone to give an alkene.

Phosphorus is the second element in Group 5A of the Periodic Table and, like nitrogen, has five electrons in its valence shell. Examples of trivalent phosphorus compounds are phosphine, PH_3, and triphenylphosphine, Ph_3P. Phosphine is a highly toxic, flammable gas. Triphenylphosphine is a colorless, odorless solid. Because phosphorus is below nitrogen in the Periodic Table, phosphines are weaker bases than amines and also good nucleophiles (Section 9.4B). Treatment of a phosphine with a primary or secondary alkyl halide gives a phosphonium salt by an S_N2 pathway.

$$Ph_3P \colon \quad + \quad CH_3 {-} I \colon \quad \xrightarrow{S_N2} \quad Ph_3\overset{+}{P}{-}CH_3 \colon \overset{..}{\underset{..}{I}} \colon {}^{-}$$

Triphenylphosphine   Methyltriphenylphosphonium iodide
                          (an alkyltriphenylphosphonium salt)

Because phosphines are also weak bases, treatment of a tertiary halide with a phosphine gives largely an alkene by an E2 pathway.

α-Hydrogen atoms on the alkyl group of an alkyltriphenylphosphonium ion are weakly acidic and can be removed by reaction with a very strong base, typically butyllithium (BuLi), sodium hydride (NaH), or sodium amide ($NaNH_2$).

$$CH_3CH_2CH_2CH_2 \colon Li^{+} \; + \; H{-}CH_2{-}\overset{+}{P}Ph_3 \; I^{-} \; \longrightarrow \; CH_3CH_2CH_2CH_3 \; + \; {}^{-}\colon CH_2{-}\overset{+}{P}Ph_3 \; + \; LiI$$

Butyllithium   An alkyltriphenyl-   Butane   A phosphonium
                    phosphonium iodide   ylide

The product of removal of a proton from an alkyltriphenylphosphonium ion is a phosphonium ylide. The important feature of a phosphonium ylide is that the deprotonated carbon atom bears considerable partial negative charge, making it a strong carbon-based nucleophile, analogous to species such as Grignard and organolithium reagents. Just like Grignard and organolithium reagents, the deprotonated carbon of phosphonium ylides readily reacts with the electrophilic carbon atom of aldehyde and ketone carbonyl groups.

Mechanism   *The Wittig Reaction*

Step 1: Reaction of a nucleophilic phosphonium ylide with the electrophilic carbonyl carbon of an aldehyde or ketone gives a dipolar intermediate called a **betaine** that then collapses to a four-membered oxaphosphetane ring.

$$
\begin{array}{ccc}
\colon \overset{\frown}{O} {=} CR_2 & \left[\begin{array}{c} \colon \overset{..}{\underset{..}{O}} {-} CR_2 \\ \overset{+}{Ph_3P} {-} CH_2 \end{array} \right] & \colon \overset{..}{\underset{..}{O}} {-} CR_2 \\
\overset{+}{Ph_3P} {-} \overset{..-}{CH_2} & & Ph_3P {-} CH_2 \\
& A\ betaine & An\ oxaphosphetane
\end{array}
$$

The name for this four-membered ring system is derived by combination of the following units: *oxa-* to show that it contains oxygen, *-phosph-* to show that it contains phosphorus, *-et-* to show that it is a four-membered ring, and *-ane* to show only carbon-carbon single bonds in the ring. Oxaphosphetanes can be isolated at low temperature.

Betaine   A neutral molecule with nonadjacent positive and negative charges. An example of a betaine is the intermediate formed by addition of a Wittig reagent to an aldehyde or ketone.

Step 2: Decomposition of the oxaphosphetane gives triphenylphosphine oxide and an alkene.

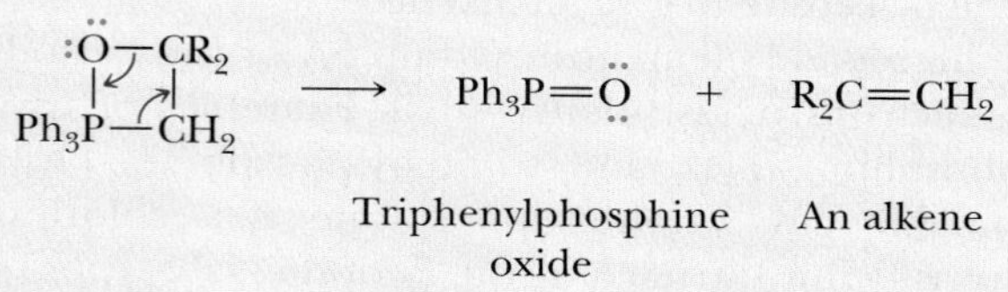

$$:\overset{..}{O} - CR_2 \quad \longrightarrow \quad Ph_3P{=}\overset{..}{\underset{..}{O}} \quad + \quad R_2C{=}CH_2$$
$$Ph_3P - CH_2$$

Triphenylphosphine   An alkene
oxide

The driving force for a Wittig reaction is the formation of the very strong phosphorus-oxygen bond in triphenylphosphine oxide.

The Wittig reaction is effective with a wide variety of aldehydes and ketones and with ylides derived from a wide variety of primary, secondary, and allylic halides as shown by the following examples.

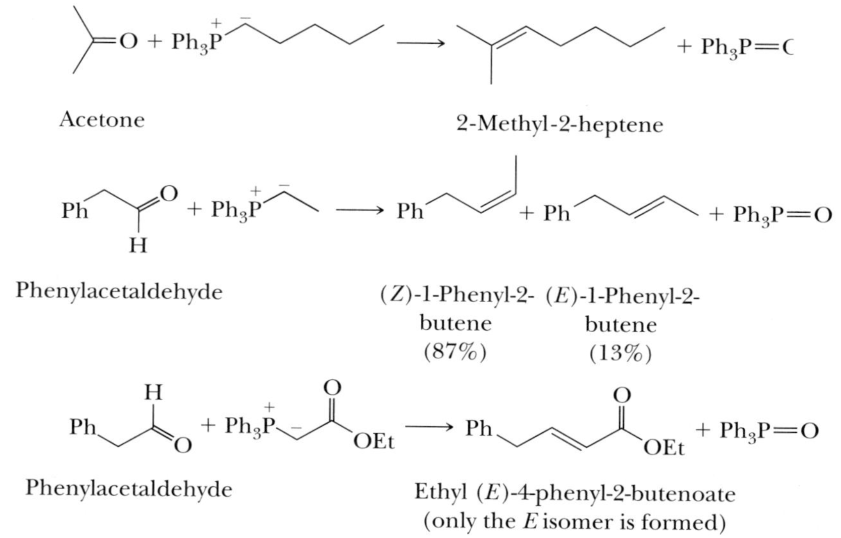

Acetone   2-Methyl-2-heptene

Phenylacetaldehyde   (Z)-1-Phenyl-2- (E)-1-Phenyl-2-
                                          butene   butene
                                          (87%)   (13%)

Phenylacetaldehyde   Ethyl (E)-4-phenyl-2-butenoate
                                    (only the E isomer is formed)

As illustrated by the second and third examples, some Wittig reactions are *Z* selective, while others are *E* selective. As a general rule, those Wittig reagents with anion-stabilizing substituents, such as a carbonyl group, adjacent to the negative charge are *E* selective. We refer to these ylides as being stabilized. We refer to ylides without an adjacent anion-stabilizing group as being unstabilized; unstabilized ylides are *Z* selective. We can write the following resonance contributing structures for a carbonyl-stabilized ylide.

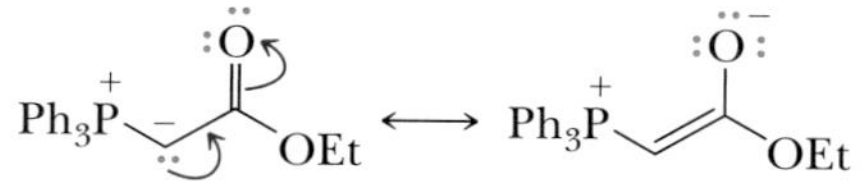

Resonance contributing structures for an
ylide stabilized by an adjacent carbonyl group

Stabilization of the ylide through resonance decreases its reactivity, allowing an equilibrium to be established during the product-determining step that favors the more stable *E* isomer.

Because the Wittig reaction is so useful for the preparation of alkenes, chemists have explored several variations of it. One of the most useful of these, known as the

Horner-Emmons-Wadsworth modification, uses a phosphonate ester derived from an α-haloester or ketone to generate the Wittig carbanion.

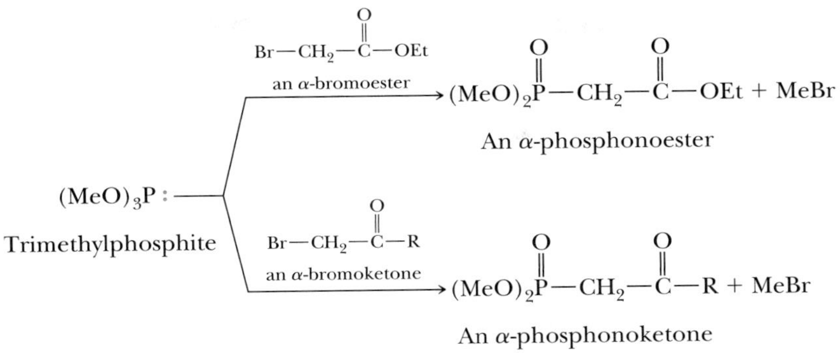

The α-phosphonoesters or ketones used in this variation of the Wittig reaction are formed by two successive S_N2 reactions. Trimethylphosphite is an excellent nucleophile and readily displaces bromine from an α-bromoester or α-bromoketone by an S_N2 reaction. Bromide ion then is the nucleophile in the second S_N2 reaction that generates the α-phosphonoester.

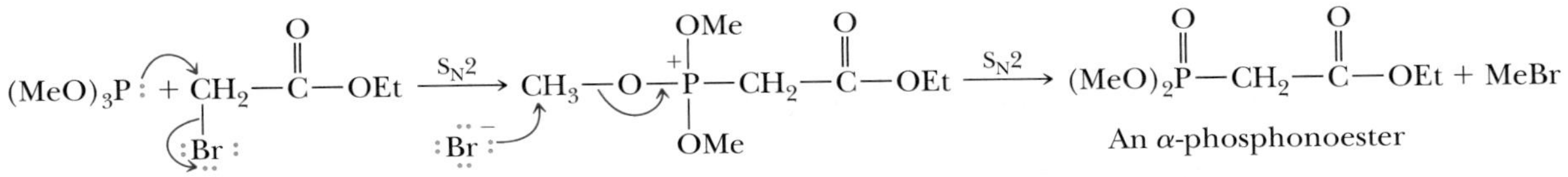

Treatment of a phosphonoester with a strong base followed by an aldehyde or ketone gives an alkene, in this case either an α,β-unsaturated ester, ketone, or aldehyde. A particular advantage of using a phosphonate-stabilized carbanion as the Wittig reagent is that the resulting alkene is either entirely or almost entirely the E isomer; that is, phosphonate-stabilized Wittig reagents are almost exclusively E selective. Another advantage of phosphonate-stabilized ylides is that the by-product, dimethylphosphate anion, is water-soluble and, therefore, easily separated from the desired organic product.

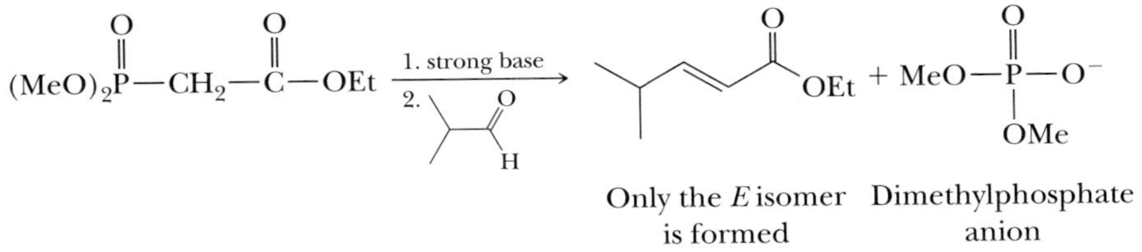

Example 16.5

Show how this alkene can be synthesized by a Wittig reaction.

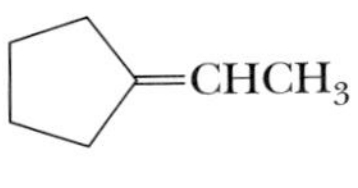

Solution

Starting materials are either cyclopentanone and the triphenylphosphonium ylide derived from bromoethane, or acetaldehyde and the triphenylphosphonium ylide derived from chlorocyclopentane. Either route is satisfactory.

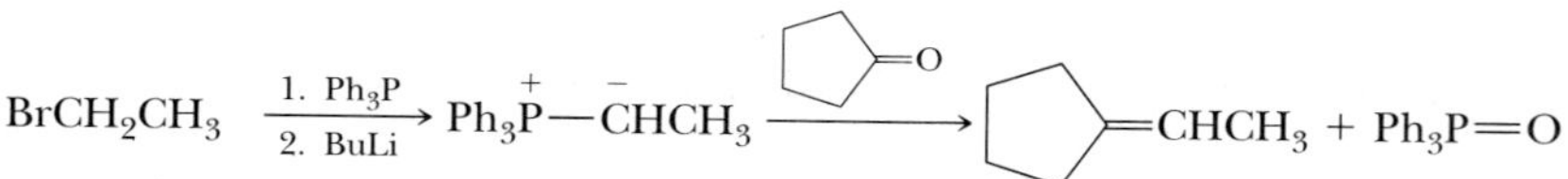

$$BrCH_2CH_3 \xrightarrow[\text{2. BuLi}]{\text{1. Ph}_3\text{P}} Ph_3\overset{+}{P}-\overset{-}{C}HCH_3 \longrightarrow \text{cyclopentylidene}=CHCH_3 + Ph_3P=O$$

Bromoethane

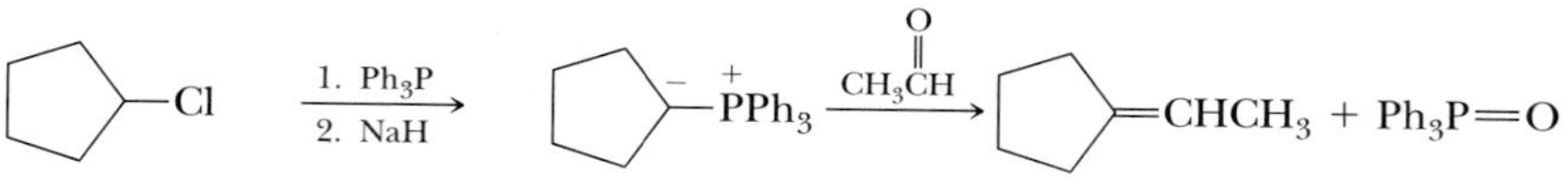

$$\text{cyclopentane}-Cl \xrightarrow[\text{2. NaH}]{\text{1. Ph}_3\text{P}} \text{cyclopentane}\overset{-}{}\overset{+}{P}Ph_3 \xrightarrow{CH_3\overset{O}{\overset{\|}{C}}H} \text{cyclopentylidene}=CHCH_3 + Ph_3P=O$$

Chlorocyclopentane

Problem 16.5

Show how each alkene can be synthesized by a Wittig reaction (there are two routes to each).

(a) 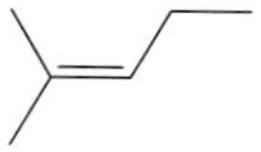 [structure] (b)  [structure]=CH₂

16.7 Addition of Oxygen Nucleophiles

A. Addition of Water: Formation of Carbonyl Hydrates

Addition of water (hydration) to a carbonyl group of an aldehyde or ketone forms a geminal, commonly abbreviated gem-diol.

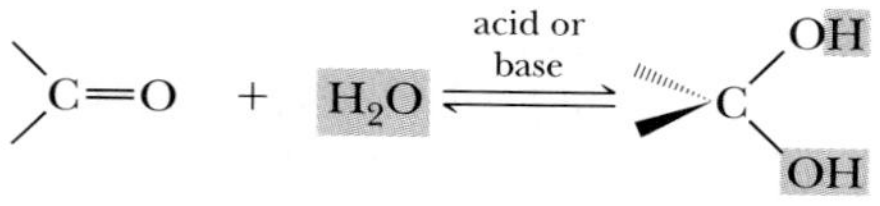

$$\begin{array}{c}\diagdown \\ \diagup\end{array}C=O \;+\; H_2O \underset{}{\overset{\text{acid or}}{\underset{\text{base}}{\rightleftharpoons}}} \begin{array}{c} OH \\ | \\ C \\ | \\ OH\end{array}$$

Carbonyl group
of an aldehyde
or ketone

A hydrate
(a gem-diol)

A gem-diol is commonly referred to as the hydrate of the corresponding aldehyde or ketone. These compounds are unstable and are rarely isolated. This reaction is catalyzed by acids and bases. The mechanism is identical to that for the addition of alcohols, which is discussed next.

Hydration of an aldehyde or ketone is readily reversible, and the diol can eliminate water to regenerate the aldehyde or ketone. In most cases, equilibrium strongly favors the carbonyl group. For a few simple aldehydes, however, the 1,1-diol is favored. For example, when formaldehyde is dissolved in water at 20°C, the position of equilibrium is such that it is more than 99% hydrated.

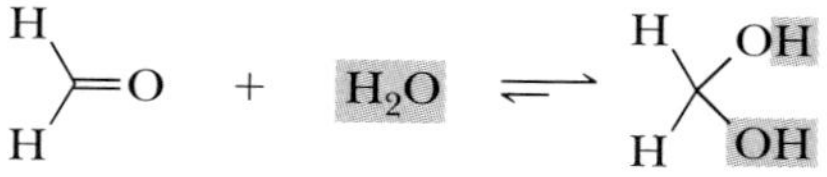

$$\begin{array}{c} H \\ \diagdown \\ \diagup \\ H\end{array}C=O \;+\; H_2O \;\rightleftharpoons\; \begin{array}{c} H \quad OH \\ \diagdown\diagup \\ \diagup\diagdown \\ H \quad OH\end{array}$$

Formaldehyde

Formaldehyde
hydrate
(>99%)

A 37% solution of formaldehyde in water, called formalin, is commonly used to pre-serve biological specimens.

In contrast, an aqueous solution of acetone consists of less than 0.1% of the diol at equilibrium.

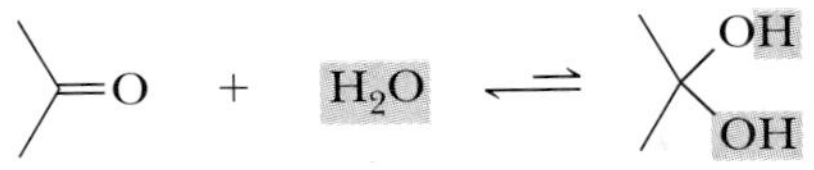

Acetone
(99.9%)

2,2-Propanediol
(0.1%)

B. Addition of Alcohols: Formation of Acetals

Alcohols add to aldehydes and ketones in the same manner as described for water. Addition of one molecule of alcohol to the carbonyl group of an aldehyde or ketone forms a **hemiacetal** (a half-acetal).

A hemiacetal

Hemiacetal A molecule containing an —OH and an —OR or —OAr group bonded to the same carbon.

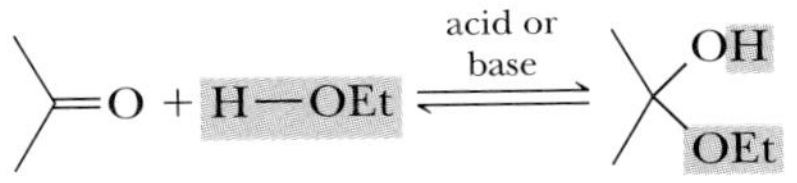

The functional group of a hemiacetal is a carbon bonded to an —OH group and an —OR group.

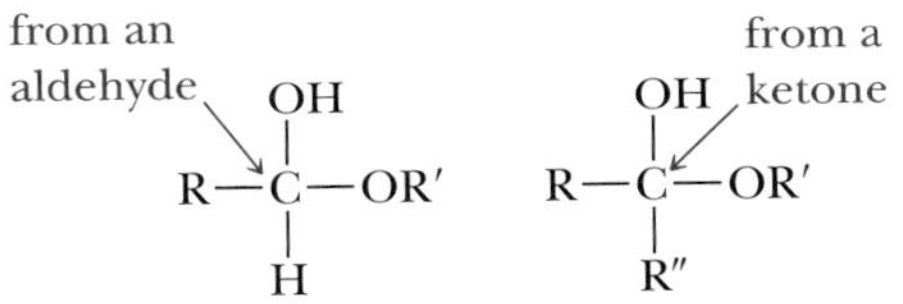

Hemiacetals

Hemiacetals are generally unstable and are only minor components of an equilib-rium mixture, except in one very important type of compound. When a hydroxyl group is part of the same molecule that contains the carbonyl group, and a five- or six-membered ring can form, the compound exists almost entirely in the cyclic hemi-acetal form. Recall that five- and six-membered rings have relatively little ring strain (Section 6.6B). In the following example, (*S*)-4-hydroxypentanal already has a chiral center, and a new chiral center is created upon formation of the hemiacetal.

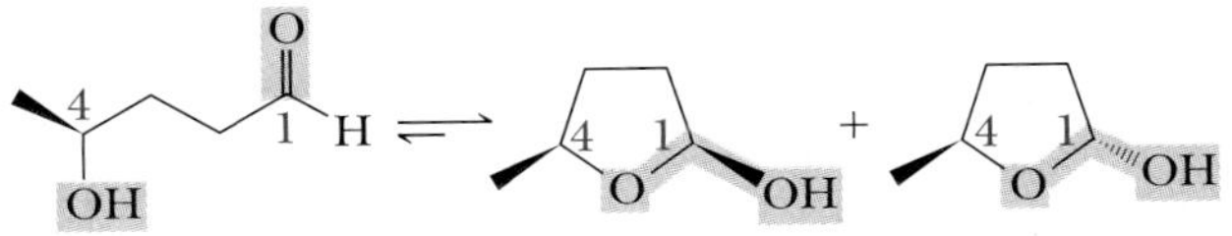

(*S*)-4-Hydroxypentanal

Cyclic hemiacetals
(major forms present at equilibrium)

Because of the original chiral center, the product hemiacetals in this example are diastereomers, not enantiomers, and are not necessarily produced in equal amounts.

Simple carbohydrates, all of which are polyhydroxyaldehydes or polyhydroxyke-tones, exist in solution predominantly as cyclic hemiacetals. Because carbohydrates have several hydroxyl groups, they could potentially form rings of different sizes.

ORGANIC
Chemistry⚛Now™
Click Mechanisms to view an animation of the **Formation of a Cyclic Hemiacetal**

Generally, only five- and six-membered cyclic hemiacetals (the most strain-free types of rings) are produced to an appreciable extent. The new chiral center created in carbohydrate cyclic hemiacetals can have either an R or an S configuration. The carbon atom at the new chiral center of a carbohydrate cyclic hemiacetal is given the special name of **anomeric carbon** and corresponds to the carbonyl carbon atom in the open chain form. The two different cyclic hemiacetals are called **anomers,** and the configuration of each designated as α or β depending on whether the hemiacetal —OH group is on the same side of the ring as the terminal —CH_2OH substituent (β anomer) or on the opposite side (α anomer). Note that owing to the presence of multiple chiral centers, the anomers are diastereomers, not enantiomers.

D-Glucose, the most important carbohydrate in mammalian metabolism, exists as a six-membered cyclic hemiacetal form, as both α and β anomers.

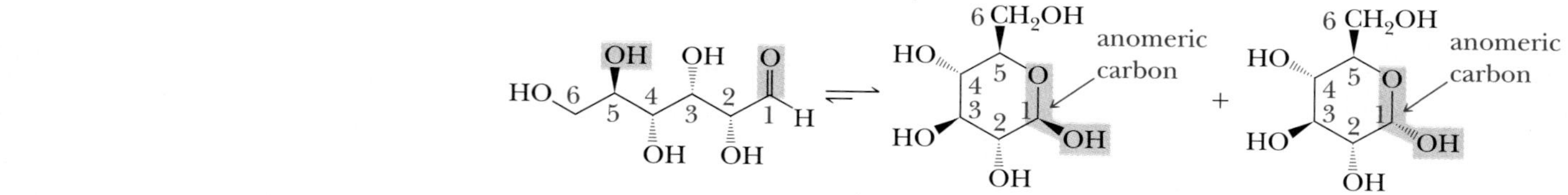

D-Glucose (open chain form)

β Anomer of D-glucose cyclic hemiacetal (predominates at equilibrium)

α Anomer of D-glucose cyclic hemiacetal

At equilibrium, the β anomer of D-glucose predominates, because the —OH group of the anomeric carbon is in the more stable equatorial position of the more stable chair conformation. In α-D-glucose, the —OH group on the anomeric carbon is axial.

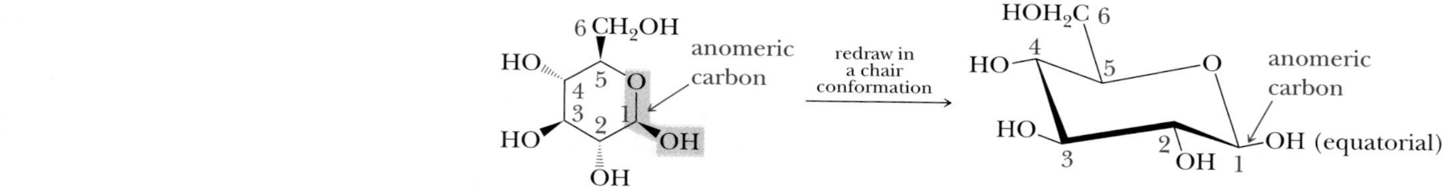

β Anomer of D-glucose cyclic hemiacetal

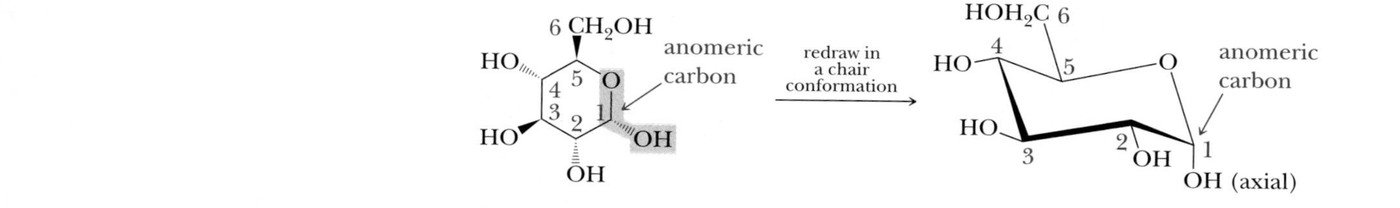

α Anomer of D-glucose cyclic hemiacetal

We discuss the chemistry of carbohydrate cyclic hemiacetals in more detail in Chapter 25.

Formation of hemiacetals is catalyzed by bases such as hydroxide or alkoxide. The function of the catalyst is to remove a proton from the alcohol, making it a better nucleophile.

Mechanism *Base-Catalyzed Formation of a Hemiacetal*

Step 1: Proton transfer from HOR to the base gives the alkoxide.

$$\text{B:}^{-} + \text{H}-\overset{..}{\underset{..}{\text{O}}}\text{R} \underset{\text{reversible}}{\overset{\text{fast and}}{\rightleftharpoons}} \text{B}-\text{H} + \text{:}\overset{..}{\underset{..}{\text{O}}}\text{R}$$

Step 2: Attack of RO⁻ on the carbonyl carbon gives a tetrahedral carbonyl addition compound.

$$\underset{\text{CH}_3}{\text{CH}_3}-\overset{\overset{..}{\underset{..}{\text{O}}}}{\underset{}{\text{C}}}-\text{CH}_3 + \text{:}\overset{..}{\underset{..}{\text{O}}}-\text{R} \rightleftharpoons \text{CH}_3-\overset{\overset{:\overset{..}{\text{O}}^{-}}{|}}{\underset{\underset{:\overset{..}{\text{O}}\text{R}}{|}}{\text{C}}}-\text{CH}_3$$

A tetrahedral carbonyl
addition compound

Step 3: Proton transfer from the alcohol to the O⁻ gives the hemiacetal and regenerates the base catalyst.

$$\text{CH}_3-\overset{\overset{:\overset{..}{\text{O}}:^{-}}{|}}{\underset{\underset{:\overset{..}{\text{O}}\text{R}}{|}}{\text{C}}}-\text{CH}_3 + \text{H}-\overset{..}{\underset{..}{\text{O}}}\text{R} \rightleftharpoons \text{CH}_3-\overset{\overset{:\overset{..}{\text{O}}\text{H}}{|}}{\underset{\underset{:\overset{..}{\text{O}}\text{R}}{|}}{\text{C}}}-\text{CH}_3 + \text{:}^{-}\overset{..}{\underset{..}{\text{O}}}\text{R}$$

Formation of hemiacetals can also be catalyzed by acid, most commonly sulfuric acid, *p*-toluenesulfonic acid, or hydrogen chloride.

Mechanism *Acid-Catalyzed Formation of a Hemiacetal*

Step 1: Proton transfer from HA, the acid catalyst, to the carbonyl oxygen gives the conjugate acid of the aldehyde or ketone.

$$\text{CH}_3-\overset{\overset{:\overset{..}{\text{O}}:}{\|}}{\underset{}{\text{C}}}-\text{CH}_3 + \text{H}-\text{A} \underset{\text{reversible}}{\overset{\text{fast and}}{\rightleftharpoons}} \text{CH}_3-\overset{\overset{+\overset{..}{\text{O}}\diagdown^{\text{H}}}{\|}}{\underset{}{\text{C}}}-\text{CH}_3 + \text{:}\text{A}^{-}$$

The function of the acid catalyst, here represented by H—A, is to protonate the carbonyl oxygen, thus rendering the carbonyl carbon more electrophilic and more susceptible to attack by the nucleophilic oxygen atom of the alcohol.

Step 2: Attack of ROH on the carbonyl carbon gives a tetrahedral carbonyl addition compound.

$$\text{CH}_3-\overset{\overset{+\overset{..}{\text{O}}\diagdown^{\text{H}}}{\|}}{\underset{}{\text{C}}}-\text{CH}_3 + \text{H}-\overset{..}{\underset{..}{\text{O}}}-\text{R} \rightleftharpoons \text{CH}_3-\overset{\overset{:\overset{..}{\text{O}}-\text{H}}{|}}{\underset{\underset{\text{H}\diagup\overset{+}{\underset{..}{\text{O}}}\diagdown\text{R}}{|}}{\text{C}}}-\text{CH}_3$$

A tetrahedral carbonyl
addition compound

Acetal A molecule containing
two —OR or —OAr groups
bonded to the same carbon.

Step 3: Proton transfer from the oxonium ion to A⁻ gives the hemiacetal and regenerates
the acid catalyst.

$$CH_3-\overset{\overset{\displaystyle :\ddot{O}H}{|}}{\underset{\underset{\displaystyle A\!:\,}{\overset{+}{\underset{H}{\overset{\displaystyle \ddot{O}}{|}}}\!\!-\!\!R}}{C}}-CH_3 \;\rightleftharpoons\; CH_3-\overset{\overset{\displaystyle :\ddot{O}H}{|}}{\underset{\underset{\displaystyle :\ddot{O}R}{|}}{C}}-CH_3 + H-A$$

Hemiacetals react further with alcohols to form **acetals** plus a molecule of water.

$$\underset{\text{A hemiacetal}}{\overset{\text{OH}}{\underset{\text{OEt}}{\diagup\!\!\!\times\!\!\!\diagdown}}} + H-OEt \;\overset{H^+}{\rightleftharpoons}\; \underset{\text{A diethyl acetal}}{\overset{\text{OEt}}{\underset{\text{OEt}}{\diagup\!\!\!\times\!\!\!\diagdown}}} + H_2O$$

The formation of acetals and its reverse is catalyzed by acids, not by bases, because
the OH group cannot be displaced by nucleophiles.

The functional group of an acetal is a carbon bonded to two —OR groups.

$$\underset{\underset{\displaystyle H}{|}}{\overset{\overset{\displaystyle \overset{\text{from an}}{\text{aldehyde}}\;OR'}{|}}{R-C}}-OR' \qquad \underset{\underset{\displaystyle R''}{|}}{\overset{\overset{\displaystyle OR'\;\overset{\text{from a}}{\text{ketone}}}{|}}{R-C}}-OR'$$

Acetals

The mechanism for the acid-catalyzed conversion of a hemiacetal to an acetal is
divided into four steps. As you study this mechanism, note that acid H—A is a true
catalyst in this reaction. It is used in Step 1, but another H—A is generated in Step 4.
The latter steps of this mechanism are very similar to those for hemiacetal formation.

Mechanism   *Acid-Catalyzed Formation of an Acetal*

Step 1: Proton transfer from the acid, H—A, to the hemiacetal OH group gives an oxonium ion.

$$\underset{\underset{\displaystyle H}{|}}{\overset{\overset{\displaystyle H\ddot{O}:}{|}}{R-C}}-\ddot{O}CH_3 + H-A \;\rightleftharpoons\; \underset{\underset{\displaystyle H}{|}}{\overset{\overset{\displaystyle H\overset{+}{\underset{\ddot{}}{O}}H}{|}}{R-C}}-\ddot{O}CH_3 + A\!:^-$$

An oxonium ion

Step 2: Loss of water gives a new, resonance-stabilized cation.

$$\underset{\underset{\displaystyle H}{|}}{\overset{\overset{\displaystyle H\overset{+}{\underset{\ddot{}}{O}}H}{|}}{R-C}}-\ddot{O}CH_3 \;\rightleftharpoons\; \underset{\underset{\displaystyle H}{|}}{\overset{\overset{}{}}{R-C}}\!\!=\!\!\overset{+}{\ddot{O}}CH_3 \;\longleftrightarrow\; \underset{\underset{\displaystyle H}{|}}{\overset{\overset{+}{}}{R-C}}-\ddot{O}CH_3 + H_2\ddot{O}:$$

A resonance-stabilized cation

Step 3: Reaction of the resonance-stabilized cation (an electrophile) with methanol (a nucleophile) gives the conjugate acid of the acetal.

$$CH_3 \overset{..}{\underset{..}{O}}H + R \overset{+}{\underset{H}{\overset{|}{C}}} \overset{..}{\underset{..}{O}}CH_3 \rightleftharpoons R \overset{\overset{H}{\underset{..}{\overset{+}{O}}} CH_3}{\underset{H}{\overset{|}{C}}} \overset{..}{\underset{..}{O}}CH_3$$

A protonated acetal

Step 4: Proton transfer from the protonated acetal to A⁻ gives the acetal and generates a new molecule of the acid catalyst.

$$A^{..-} + R \overset{\overset{H}{\underset{..}{\overset{+}{O}}} CH_3}{\underset{H}{\overset{|}{C}}} \overset{..}{\underset{..}{O}}CH_3 \overset{(4)}{\rightleftharpoons} R \overset{:\overset{..}{O}CH_3}{\underset{H}{\overset{|}{C}}} \overset{..}{\underset{..}{O}}CH_3 + H{-}A$$

An acetal

Formation of acetals is often carried out using the alcohol as the solvent and dissolving either dry HCl (hydrogen chloride gas) or *p*-toluenesulfonic acid in the alcohol. Because the alcohol is both a reactant and solvent, it is present in large molar excess, which forces the equilibrium to the right and favors acetal formation. Note that this reaction is completely reversible. Addition of excess water to an acetal causes hydrolysis to the ketone.

An excess of alcohol pushes the equilibrium toward formation of the acetal

Removal of water favors formation of the acetal

$$R \overset{O}{\overset{||}{-}} C {-} R + 2EtOH \overset{H^+}{\rightleftharpoons} R \overset{OEt}{\underset{R}{\overset{|}{C}}} {-} OEt + H_2O$$

A diethyl acetal

In another experimental technique to force the equilibrium to the right, water is removed from the reaction vessel as an **azeotrope** by distillation using a **Dean-Stark trap** (Figure 16.1).

In this method for preparing an acetal, the aldehyde or ketone, alcohol, acid catalyst, and benzene are brought to reflux. The component in this mixture with the lowest boiling point is an azeotrope, bp 69°C, consisting of 91% benzene and 9% water. This vapor is condensed and collected in a side trap where it separates into two layers. At room temperature, the composition of the upper, less dense layer is 99.94% benzene and 0.06% water. The composition of the lower, more dense layer is almost the reverse, 0.07% benzene and 99.93% water. As reflux continues, benzene from the top layer is returned to the refluxing mixture, and water is drawn off at the bottom through a stopcock. A Dean-Stark trap "pumps" water out of the reaction mixture, thus forcing the equilibrium to the right. This same apparatus is used in many other reactions where water needs to be removed, as for example in formation of enamines (Section 19.4A).

Azeotrope A liquid mixture of constant composition with a boiling point that is different from that of any of its components.

Figure 16.1

A Dean-Stark trap for removing water by azeotropic distillation with benzene. Toluene or xylene can be used if a higher reaction temperature is desired.

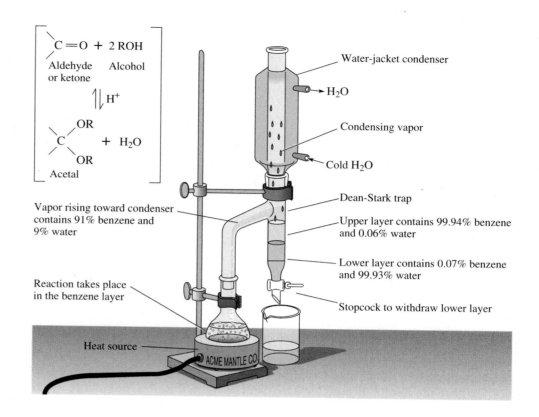

Example 16.6

Show the reaction of the carbonyl group of each aldehyde or ketone with one molecule of alcohol to give a hemiacetal and then with a second molecule of alcohol to give an acetal.

Note that in part (b), ethylene glycol is a diol, and one molecule of it provides both —OH groups.

Solution

Given are structural formulas of the hemiacetal and then the acetal.

(a) Hemiacetal → Acetal (b) Hemiacetal → Cyclic acetal

Problem 16.6

Hydrolysis of an acetal in aqueous acid gives an aldehyde or ketone and two molecules of alcohol or one molecule of a diol. Draw the structural formulas for the products of hydrolysis of the following acetals in aqueous acid.

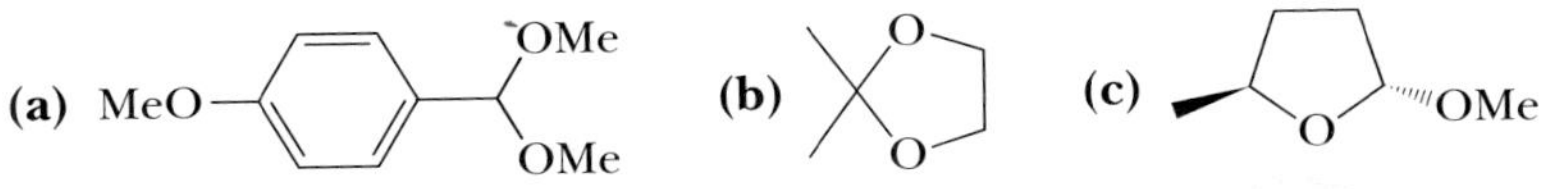

(a) MeO—⟨aromatic ring⟩—CH(OMe)₂

(b) ⟨structure: dimethyl dioxolane⟩

(c) ⟨structure: tetrahydrofuran with OMe⟩

Like ethers (Section 11.5), acetals are unreactive to bases, hydride-reducing agents such as LiAlH₄ and NaBH₄, Grignard and other organometallic reagents, oxidizing agents (except, of course, for those involving the use of aqueous acid), and catalytic reduction. This lack of reactivity is because acetals have no sp^2 hybridized electrophilic carbon atom to react with nucleophiles. Because of their lack of reactivity toward these reagents and ready hydrolysis in aqueous acid, acetals are often used to reversibly "protect" the carbonyl groups of aldehydes and ketones while reactions are carried out on other functional groups in the molecule.

C. Acetals as Carbonyl-Protecting Groups

The use of acetals as carbonyl-protecting groups is illustrated by the synthesis of 5-hydroxy-5-phenylpentanal from benzaldehyde and 4-bromobutanal.

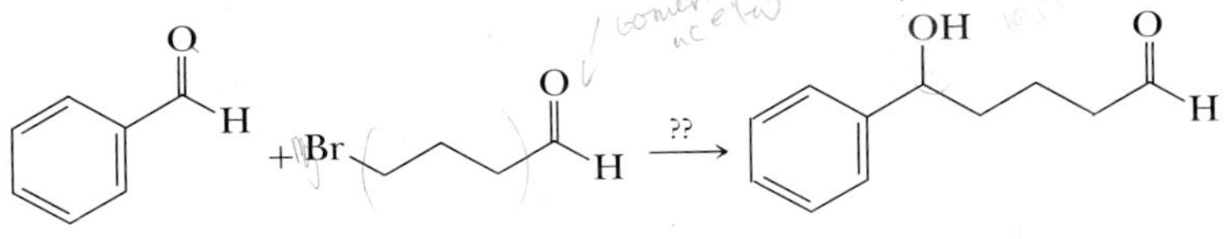

Benzaldehyde   4-Bromobutanal   5-Hydroxy-5-phenylpentanal

One obvious way to form a new carbon-carbon bond between these two molecules is to treat benzaldehyde with the Grignard reagent from 4-bromobutanal. A Grignard reagent formed from 4-bromobutanal, however, reacts immediately with the carbonyl group of another molecule of 4-bromobutanal, causing it to self-destruct during preparation. A way to avoid this problem is to protect its carbonyl group by conversion to an acetal. Cyclic acetals are often used because they are particularly easy to prepare.

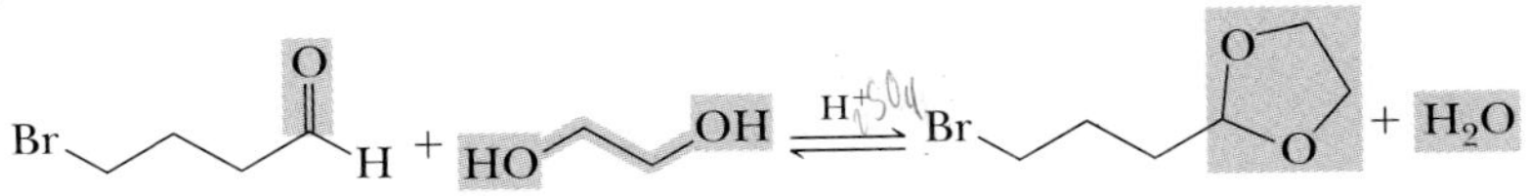

A cyclic acetal

Treatment of the protected bromoaldehyde with magnesium in diethyl ether followed by addition of benzaldehyde gives a chiral magnesium alkoxide as a racemic mixture.

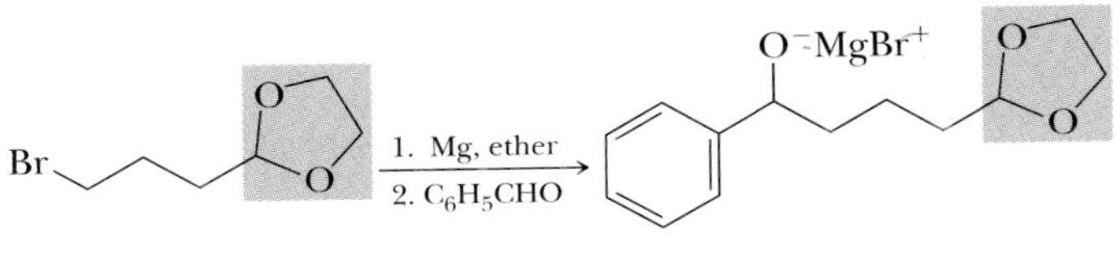

A cyclic acetal   A chiral magnesium alkoxide
                       (produced as a racemic mixture)

Treatment of the magnesium alkoxide with aqueous acid accomplishes two things. First, protonation of the alkoxide anion gives the desired 2° hydroxyl group; second, hydrolysis of the cyclic acetal regenerates the carbonyl group of the aldehyde.

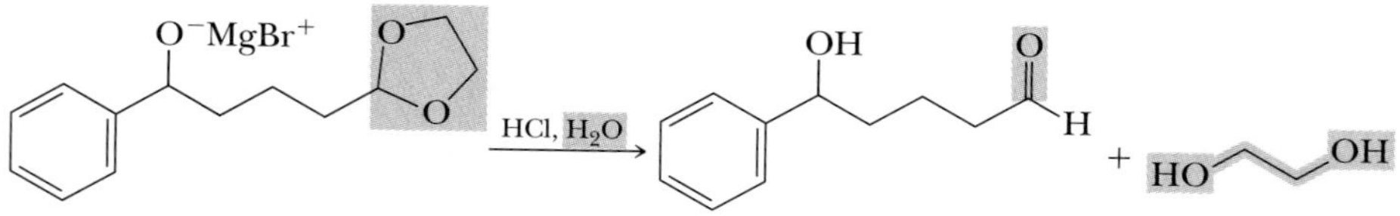

5-Hydroxy-5-phenylpentanal
(formed as a racemic mixture)

D. Tetrahydropyranyl Ethers: Protecting an Alcohol as an Acetal

We have just seen in Section 16.7C that an aldehyde or ketone can be protected by conversion to an acetal. A similar strategy can be used to protect a primary or secondary alcohol. Treatment of the alcohol with dihydropyran in the presence of an acid catalyst, commonly anhydrous HCl or a sulfonic acid, RSO_3H, converts the alcohol into a tetrahydropyranyl (THP) ether.

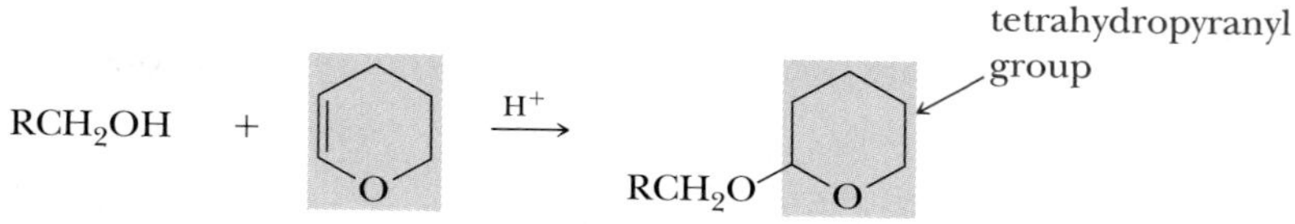

Dihydropyran   A tetrahydropyranyl ether

Because the THP group is an acetal, it is stable in neutral and basic solutions and to most oxidizing and reducing agents. It is removed easily by treatment with dilute aqueous acid to regenerate the original primary or secondary alcohol.

Example 16.7

Write a mechanism for the formation of a THP ether from a primary alcohol RCH_2OH catalyzed by a sulfonic acid $ArSO_3H$.

Solution

Step 1: Dihydropyran (a vinyl or enol ether) is weakly basic and is protonated to give a resonance-stabilized cation.

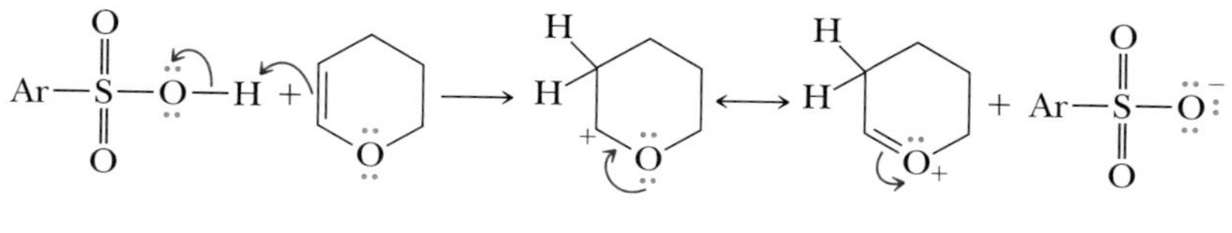

A resonance-stabilized cation

Step 2: Reaction of the resonance-stabilized cation (an electrophile) with the alcohol (a nucleophile) gives an oxonium ion.

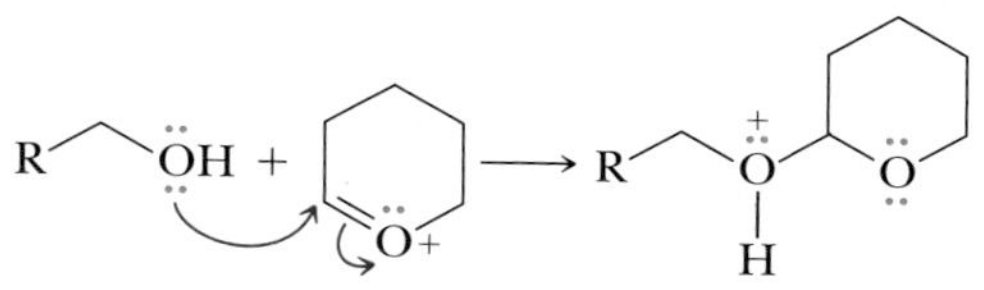

Step 3: Proton transfer from the oxonium ion to $ArSO_3^-$ completes the reaction.

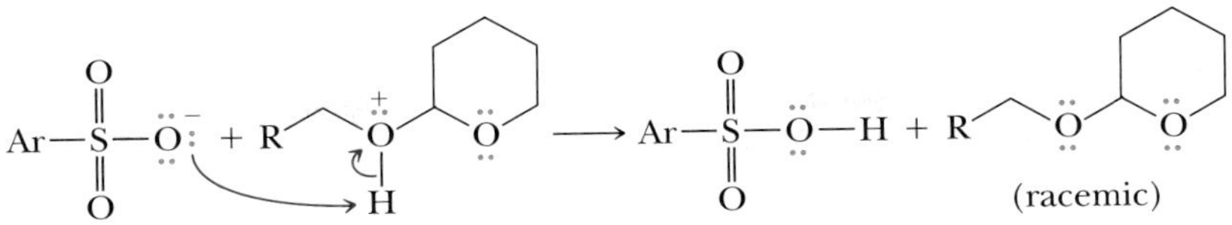

(racemic)

Problem 16.7

Write a mechanism for the acid-catalyzed hydrolysis of a THP ether to regenerate the original alcohol. Into what compound is the THP group converted?

16.8 Addition of Nitrogen Nucleophiles

A. Ammonia and Its Derivatives

Ammonia, primary aliphatic amines (RNH_2), and primary aromatic amines ($ArNH_2$) react with the carbonyl group of aldehydes and ketones in the presence of an acid catalyst to give an **imine** or, alternatively, a **Schiff base**.

Imine   A compound containing a carbon-nitrogen double bond, $R_2C=NR'$; also called a Schiff base.

Schiff base   An alternative name for an imine.

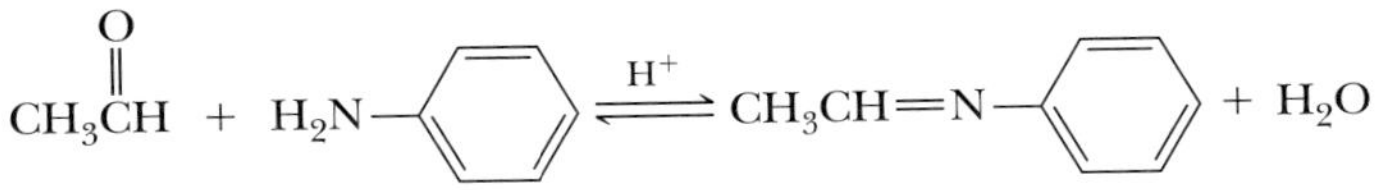

Acetaldehyde	Aniline	An imine (a Schiff base)

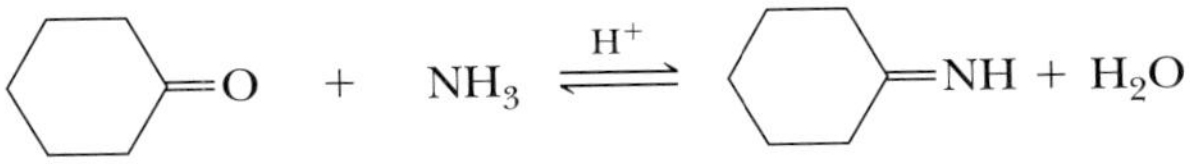

Cyclohexanone	Ammonia	An imine (a Schiff base)

Mechanism   *Formation of an Imine from an Aldehyde or Ketone*

Step 1: The nitrogen atom of ammonia or a primary amine, both good nucleophiles, adds to the carbonyl carbon to give a tetrahedral carbonyl addition compound.

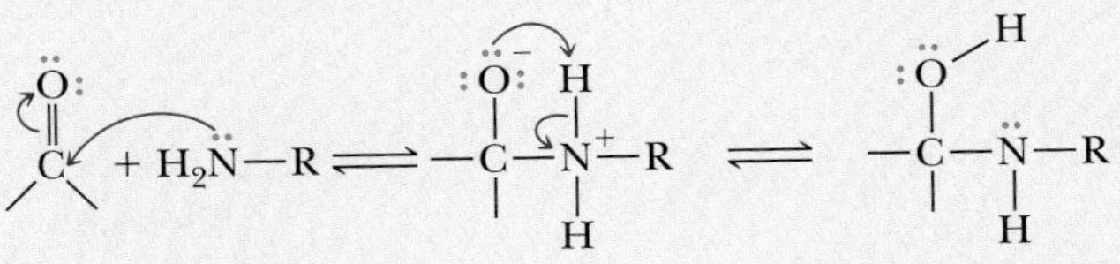

A tetrahedral carbonyl
addition compound

This reaction is analogous to the reaction of an aldehyde or ketone with water to form a hydrate, and the reaction with an alcohol to form a hemiacetal.

Step 2: Protonation of the OH group followed by loss of water and proton transfer from nitrogen to the solvent gives the imine.

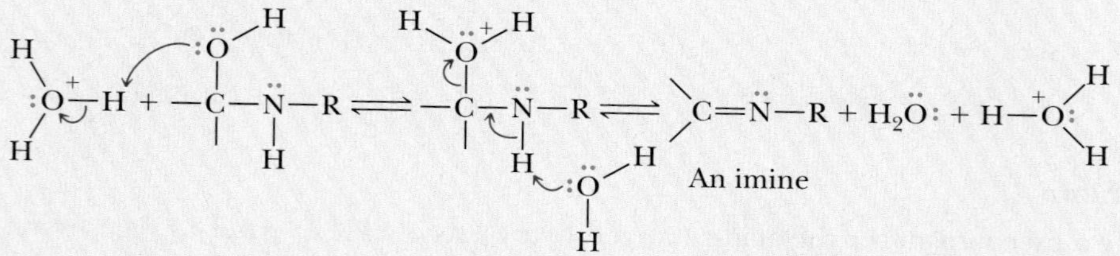

An imine

Acid-catalyzed dehydration of the addition compound in Step 2 is the slow, rate-determining step.

Reductive amination A method for preparing amines by treating an aldehyde or ketone with an amine in the presence of a reducing agent.

One of the chief values of imines is that the carbon-nitrogen double bond can be reduced by hydrogen in the presence of a nickel or other transition metal catalyst to a carbon-nitrogen single bond. By this two-step reaction, called **reductive amination,** a primary amine is converted to a secondary amine by way of an imine as illustrated by the conversion of cyclohexylamine to dicyclohexylamine.

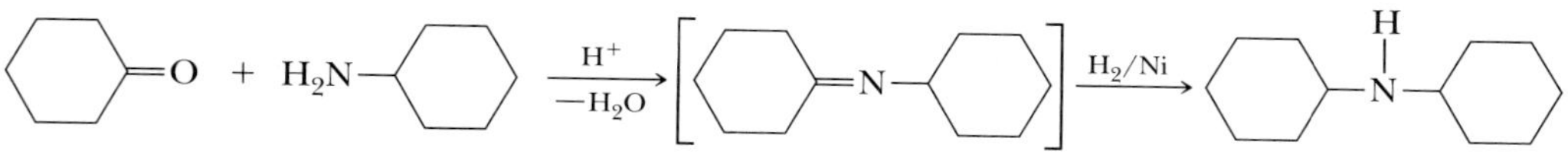

Cyclohexanone   Cyclohexylamine   (an imine)   Dicyclohexylamine

Imines are usually unstable unless the C=N group is part of an extended system of conjugation (e.g., rhodopsin) and are generally not isolated.

To give but one example of the importance of imines in biological systems, vitamin A aldehyde (retinal) is bound to the protein opsin in the human retina in the form of an imine. The primary amino group of opsin for this reaction is provided by the side chain of the amino acid lysine (Section 27.1). The imine is called rhodopsin or visual purple.

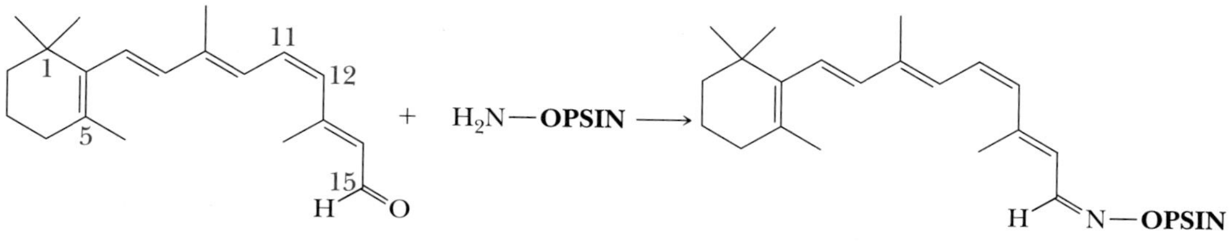

11-*cis*-Retinal   Rhodopsin
(Visual purple)

Absorption of photons by rhodopsin causes a *cis* to *trans* isomerization of the double bond at carbon 11, and the resulting change in molecular shape leads to creation of a nerve impulse that forms the basis of mammalian vision.

Example 16.8

Write a structural formula for the imine formed in each reaction.

(a)

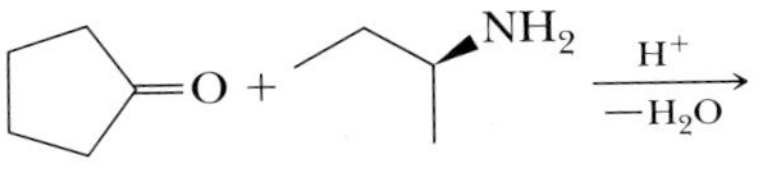

(b)

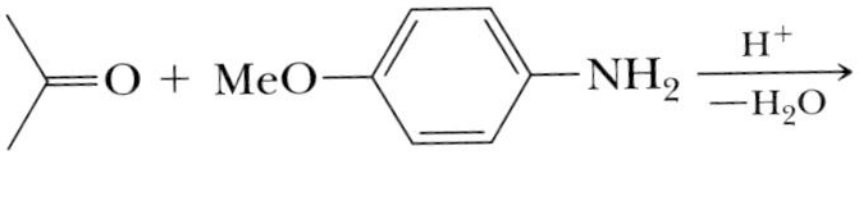

Solution

Given is a structural formula for each imine.

(a) 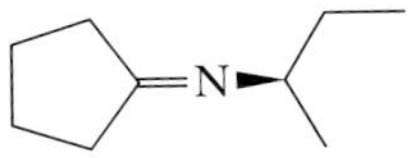   (b)

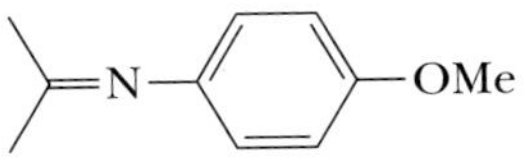

Problem 16.8

Acid-catalyzed hydrolysis of an imine gives an amine and an aldehyde or ketone. When one equivalent of acid is used, the amine is converted to an ammonium salt. Write structural formulas for the products of hydrolysis of the following imines using one equivalent of HCl.

(a) 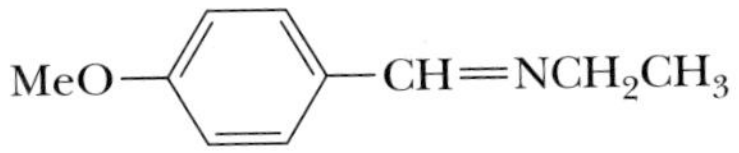 $MeO-\!\!\langle\ \rangle\!\!-CH=NCH_2CH_3$ (b)

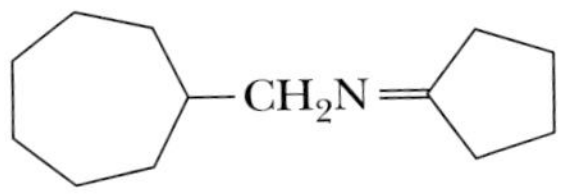

Secondary amines react with aldehydes and ketones to form enamines. The name **enamine** is derived from *-en-* to indicate the presence of a carbon-carbon double bond and *-amine* to indicate the presence of an amino group. An example is enamine formation between cyclohexanone and piperidine, a cyclic secondary amine. Water is removed by a Dean-Stark trap (Section 16.7B), which forces the equilibrium to the right.

Enamine An unsaturated compound derived by the reaction of an aldehyde or ketone and a secondary amine followed by loss of H_2O; $R_2C\!=\!CR\!-\!NR_2$.

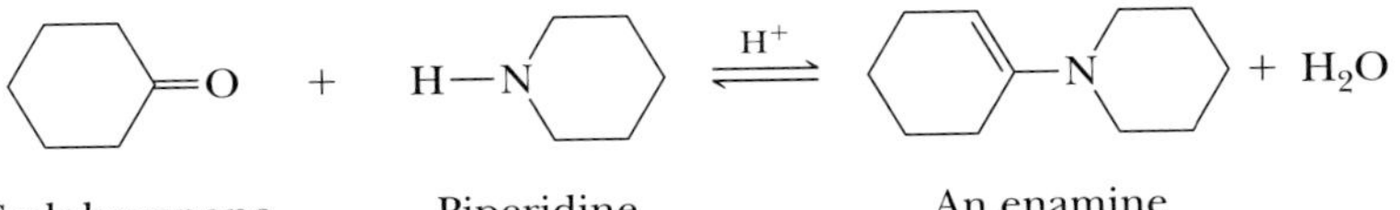

Cyclohexanone   Piperidine   An enamine
                  (a secondary amine)

The mechanism for formation of an enamine is very similar to that for the formation of an imine. In Step 1, nucleophilic addition of the secondary amine to the carbonyl carbon of the aldehyde or ketone followed by proton transfer from nitrogen to oxygen gives a tetrahedral carbonyl addition compound. Acid-catalyzed dehydration in Step 2 gives the enamine. It is at this stage that enamine formation differs from imine formation. The nitrogen has no proton to lose. Instead, a proton is lost from the α-carbon of the ketone or aldehyde portion of the molecule in an elimination reaction.

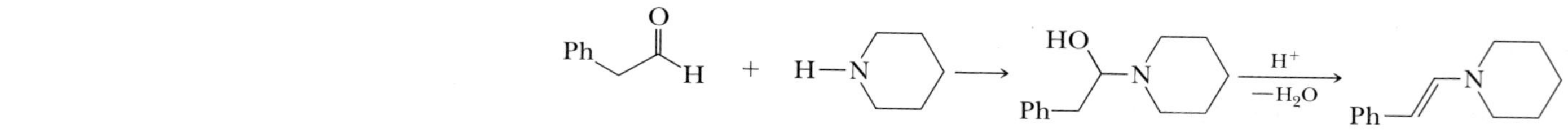

Phenylacetaldehyde   Piperidine   A tetrahedral carbonyl addition compound   An enamine (a mixture of E,Z isomers)

We will return to the chemistry of enamines and their use in synthesis in Section 19.5.

B. Hydrazine and Related Compounds

Aldehydes and ketones react with hydrazine to form compounds called hydrazones as illustrated by treating cyclopentanone with hydrazine.

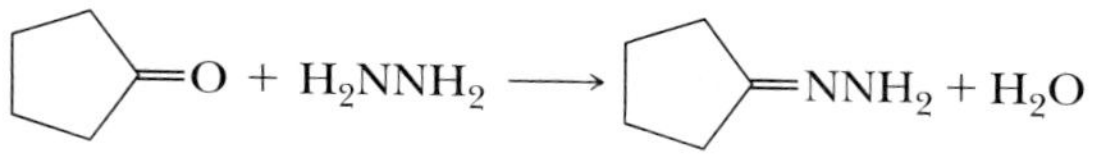

Hydrazine   A hydrazone

Table 16.4 lists several other derivatives of ammonia and hydrazine that react with aldehydes and ketones to give imines.

The chief value of the nitrogen nucleophiles listed in Table 16.4 is that most aldehydes and ketones react with them to give crystalline solids with sharp melting points. Historically, these derivatives often provided a convenient way to identify liquid aldehydes or ketones. Now, these compounds are more readily identified by IR and NMR spectroscopy.

Table 16.4 Derivatives of Ammonia and Hydrazine Used for Forming Imines

Reagent, H_2N—R	Name of Reagent	Name of Derivative Formed
H_2N—OH	Hydroxylamine	Oxime
H_2N—NH— (phenyl)	Phenylhydrazine	Phenylhydrazone
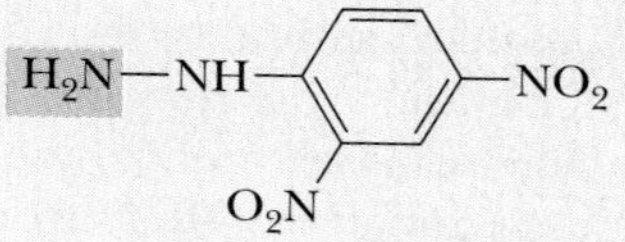 H_2N—NH— (2,4-dinitrophenyl, with NO_2 and O_2N)	2,4-Dinitrophenylhydrazine	2,4-Dinitrophenylhydrazone
H_2N—NHCNH$_2$ (with C=O)	Semicarbazide	Semicarbazone

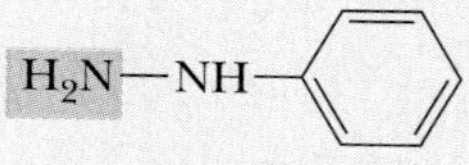

16.9 Keto-Enol Tautomerism

A. Acidity of α-Hydrogens

A carbon atom adjacent to a carbonyl group is called an **α-carbon,** and hydrogen atoms bonded to it are called **α-hydrogens.**

α-**Carbon**   A carbon atom adjacent to a carbonyl group.

α-**Hydrogen**   A hydrogen on a carbon alpha to a carbonyl group.

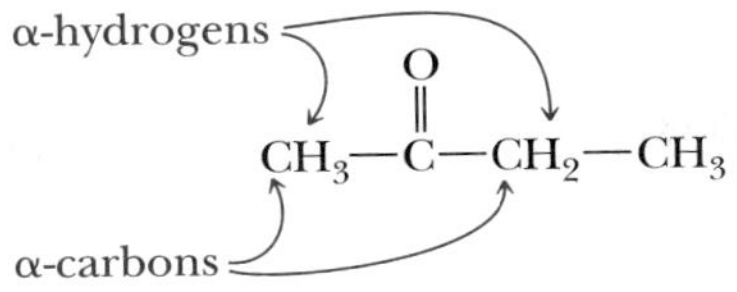

Because carbon and hydrogen have comparable electronegativities, a C—H bond normally has little polarity. In addition, carbon does not have a high electronegativity (compare it, for example, with oxygen, which has an electronegativity of 3.5), so that an anion based on carbon is relatively unstable. As a result, a hydrogen bonded to carbon usually shows very low acidity. The situation is different, however, for hydrogens alpha to a carbonyl group. α-Hydrogens are more acidic than acetylenic, vinylic, and alkane hydrogens but less acidic than —OH hydrogens of alcohols.

Type of Bond	pK_a
CH_3CH_2O—H	16
$CH_3\overset{\displaystyle O}{\overset{\displaystyle \|}{C}}CH_2$—H	20
$CH_3C{\equiv}C$—H	25
$CH_2{=}CH$—H	44
CH_3CH_2—H	51

The greater acidity of α-hydrogens arises because the negative charge on the resulting **enolate anion** is delocalized by resonance, thus stabilizing it relative to an alkane, alkene, or alkyne anion:

Enolate anion   An anion derived by loss of a hydrogen from a carbon alpha to a carbonyl group; the anion of an enol.

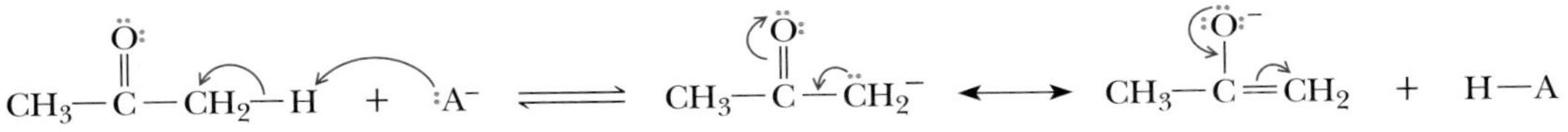

Resonance-stabilized enolate anion

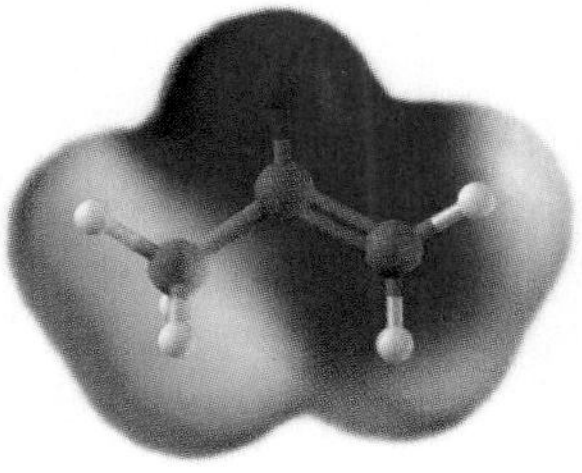

■ An electrostatic potential map of an enolate anion.

An enolate anion is also stabilized by the electron-withdrawing inductive effect of the electronegative oxygen. Recall that we used these same factors in Section 4.5 to explain the greater acidity of carboxylic acids compared with alcohols.

Example 16.9

Predict the position of the following equilibrium.

$$\text{(Acetophenone)} \quad \text{C}_6\text{H}_5-\overset{\text{O}}{\overset{\|}{\text{C}}}-\text{CH}_3 + \text{EtO}^-\text{Na}^+ \underset{\text{ethanol}}{\overset{}{\rightleftharpoons}} \text{C}_6\text{H}_5-\overset{\text{O}^-\text{Na}^+}{\overset{|}{\text{C}}}=\text{CH}_2 + \text{EtOH}$$

Acetophenone

Solution

The pK_a of ethanol is approximately 16 (Table 4.1). Assume that the pK_a of acetophenone is approximately equal to that of acetone, that is, about 20. Ethanol is the stronger acid; therefore, the equilibrium lies to the left.

$$\text{C}_6\text{H}_5-\overset{\text{O}}{\overset{\|}{\text{C}}}-\text{CH}_3 + \text{EtO}^-\text{Na}^+ \rightleftharpoons \text{C}_6\text{H}_5-\overset{\text{O}^-\text{Na}^+}{\overset{|}{\text{C}}}=\text{CH}_2 + \text{EtOH}$$

Weaker acid
pK_a 20

Stronger acid
pK_a 16

Problem 16.9

Predict the position of the following equilibrium.

$$\text{C}_6\text{H}_5-\overset{\text{O}}{\overset{\|}{\text{C}}}-\text{CH}_3 + \text{Na}^+\text{NH}_2^- \rightleftharpoons \text{C}_6\text{H}_5-\overset{\text{O}^-\text{Na}^+}{\overset{|}{\text{C}}}=\text{CH}_2 + \text{NH}_3$$

Acetophenone

When an enolate anion reacts with a proton donor, it may do so either on oxygen or on the α-carbon. Protonation of the enolate anion on the α-carbon gives the original molecule in what is called the **keto form.** Protonation on oxygen gives an **enol form.** Because the anion can be derived by loss of a proton from the enol form, it is called an enolate anion.

$$\text{A}^- + \text{CH}_3-\overset{\text{O}}{\overset{\|}{\text{C}}}-\text{CH}_3 \overset{\text{H}-\text{A}}{\longleftarrow} \left[\text{CH}_3-\overset{\text{O}}{\overset{\|}{\text{C}}}-\overset{..}{\text{C}}\text{H}_2 \longleftrightarrow \text{CH}_3-\overset{\text{O}^-}{\overset{|}{\text{C}}}=\text{CH}_2 \right] \overset{\text{H}-\text{A}}{\longrightarrow} \text{CH}_3-\overset{\text{OH}}{\overset{|}{\text{C}}}=\text{CH}_2 + \text{A}^-$$

Keto form
Resonance-stabilized enolate anion
Enol form

Enol formation can also be catalyzed by acid. The only difference between the base-catalyzed and acid-catalyzed reactions is the order of proton addition and elimination. In acid-catalyzed reactions, a proton is added first; in base-catalyzed reactions, a proton is removed first.

Mechanism *Acid-Catalyzed Equilibration of Keto and Enol Tautomers*

Step 1: Rapid and reversible proton transfer from the acid catalyst, H—A, to the carbonyl oxygen gives the conjugate acid of the ketone as a resonance-stabilized oxonium ion.

Keto form

The conjugate acid of the ketone

Step 2: Proton transfer from the α-carbon to the base, A⁻, gives the enol and generates a new molecule of the acid catalyst.

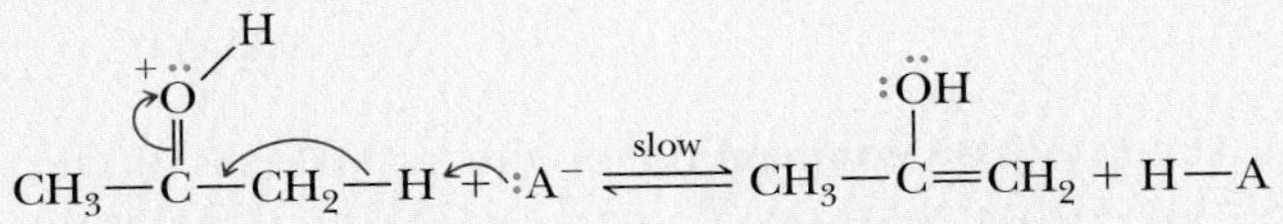

Enol form

B. The Position of Equilibrium in Keto-Enol Tautomerism

Aldehydes and ketones with at least one α-hydrogen are in equilibrium with their enol forms. We first encountered this type of equilibrium in our study of the hydroboration-oxidation and acid-catalyzed hydration of alkynes in Section 7.7. As we see in Table 16.5, the position of keto-enol equilibrium for simple aldehydes and ketones lies far on the side of the keto form, primarily because carbon-hydrogen single

Table 16.5 The Position of Keto-Enol Equilibrium for Some Simple Aldehydes and Ketones*

Keto Form	Enol Form	% Enol at Equilibrium
CH_3CH (O)	$CH_2=CH$ (OH)	6×10^{-5}
CH_3CCH_3 (O)	$CH_3C=CH_2$ (OH)	6×10^{-7}
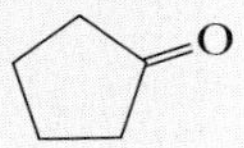 (cyclopentanone) —O	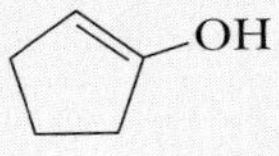 (cyclopentenol) —OH	1×10^{-6}
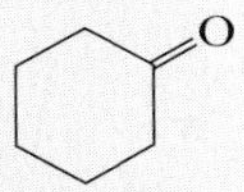 (cyclohexanone) —O	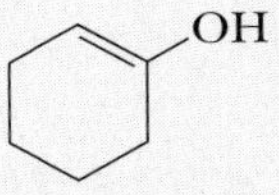 (cyclohexenol) —OH	4×10^{-5}

*Data from J. March, *Advanced Organic Chemistry*, 4th ed., Wiley Interscience, New York, 1992, p. 70.

bonds are about as strong as oxygen-hydrogen single bonds but a carbon-oxygen double bond is stronger than a carbon-carbon double bond.

For certain types of molecules, the enol form may be the major form and, in some cases, the only form present at equilibrium. For β-diketones, such as 1,3-cyclohexanedione and 2,4-pentanedione, where an α-carbon lies between two carbonyl groups, the position of equilibrium shifts in favor of the enol form.

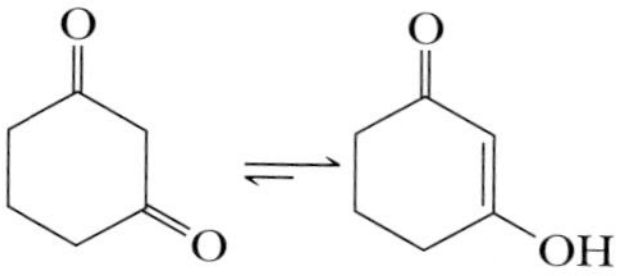

1,3-Cyclohexanedione

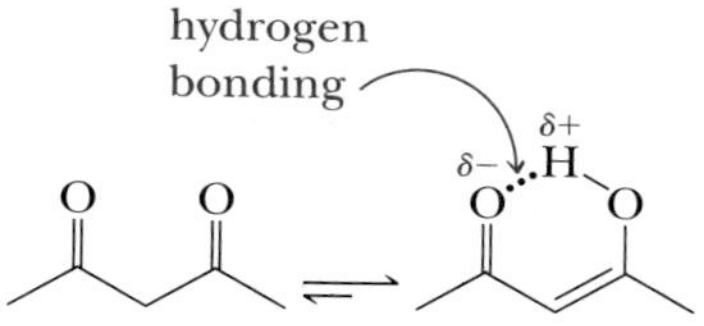

20%   80%

2,4-Pentanedione
(Acetylacetone)

Conjugation A situation that occurs when the electrons of adjacent pi bonds interact with each other; that is, when two double bonds are separated by one single bond.

These enols are stabilized by **conjugation** of the pi systems of the carbon-carbon double bond and the carbonyl group. The enol of 2,4-pentanedione, an open-chain β-diketone, is further stabilized by intramolecular hydrogen bonding.

Example 16.10

Write two enol forms for each compound. Which enol of each has the larger concentration at equilibrium?

(a) 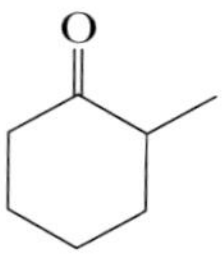   (b)

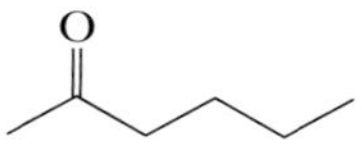

Solution

In each case, the major enol form has the more substituted (and more stable) double bond.

(a) 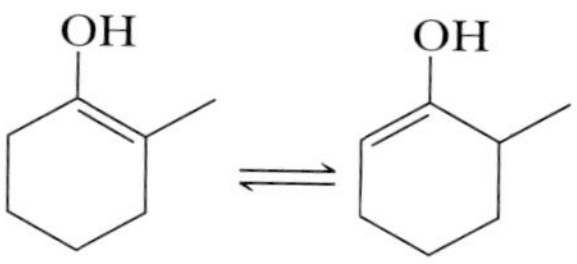   Major enol   (b) 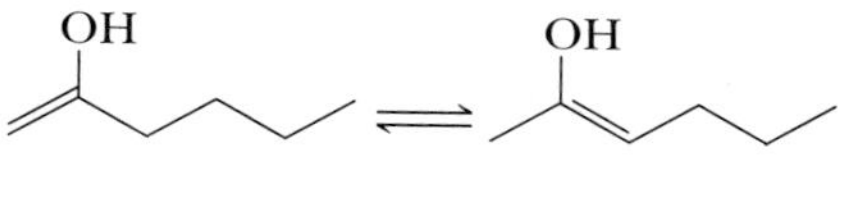   Major enol

Problem 16.10

Draw a structural formula for the keto form of each enol.

(a) 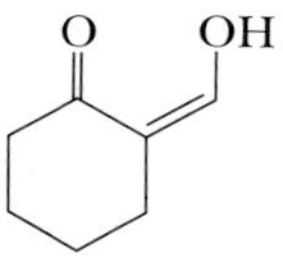   (b) 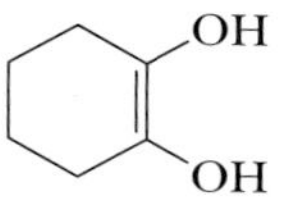   (c)

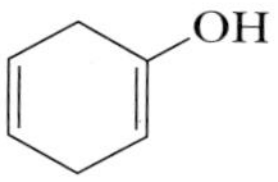

16.10 Oxidation

A. Oxidation of Aldehydes

Aldehydes are oxidized to carboxylic acids by a variety of common oxidizing agents, including chromic acid and molecular oxygen. In fact, aldehydes are one of the most easily oxidized of all functional groups. Oxidation by chromic acid is illustrated by the conversion of hexanal to hexanoic acid (for the mechanism of this oxidation, review Section 10.8A).

$$\text{Hexanal} \quad \xrightarrow{\text{H}_2\text{CrO}_4} \quad \text{Hexanoic acid}$$

Aldehydes are also oxidized to carboxylic acids by Ag(I) ion. One laboratory procedure is to shake a solution of the aldehyde in aqueous ethanol or tetrahydrofuran with a slurry of Ag_2O.

$$\text{Vanillin} + Ag_2O \xrightarrow[\text{NaOH}]{\text{THF, H}_2\text{O}} \xrightarrow[\text{H}_2\text{O}]{\text{HCl}} \text{Vanillic acid} + Ag$$

Tollens' reagent, another form of Ag(I), is prepared by dissolving silver nitrate in water, adding sodium hydroxide to precipitate Ag(I) as Ag_2O, and then adding aqueous ammonia to redissolve silver(I) as the silver-ammonia complex ion.

$$Ag^+NO_3^- + 2NH_3 \underset{}{\overset{\text{NH}_3,\,\text{H}_2\text{O}}{\rightleftharpoons}} Ag(NH_3)_2^+NO_3^-$$

When Tollens' reagent is added to an aldehyde, the aldehyde is oxidized to a carboxylic anion, and Ag(I) is reduced to metallic silver. If this reaction is carried out properly, silver precipitates as a smooth, mirror-like deposit, hence the name silver-mirror test. Ag(I) is rarely used at the present time for the oxidation of aldehydes because of the cost of silver and because other, more convenient methods exist for this oxidation. This reaction, however, is still used for silvering glassware, including mirrors. In this process, formaldehyde or glucose (Section 25.1) is generally used as the aldehyde to reduce Ag(I).

Aldehydes are also oxidized to carboxylic acids by molecular oxygen and by hydrogen peroxide.

$$2\ \text{Benzaldehyde} + O_2 \longrightarrow 2\ \text{Benzoic acid}$$

Reaction with oxygen is a radical chain reaction (Section 8.7). Molecular oxygen is the least expensive and most readily available of all oxidizing agents. On an industrial scale, air oxidation of organic compounds, including aldehydes, is very common. Air oxidation of aldehydes can also be a problem. Aldehydes that are liquid at room

Tollens' reagent A solution prepared by dissolving Ag_2O in aqueous ammonia; used for selective oxidation of an aldehyde to a carboxylic acid.

Charles D. Winters

■ A silvered mirror has been deposited in the inside of this flask by the reaction of an aldehyde with Tollens' reagent.

temperature are so sensitive to oxidation by molecular oxygen that they must be protected from contact with air during storage. Often this is done by sealing the aldehyde in a container under an atmosphere of nitrogen.

Example 16.11

Draw a structural formula for the product formed by treating each compound with Tollens' reagent followed by acidification with aqueous HCl.
(a) Pentanal   **(b)** Cyclopentanecarbaldehyde

Solution

The aldehyde group in each compound is oxidized to a carboxyl group.

Pentanoic acid   Cyclopentanecarboxylic acid

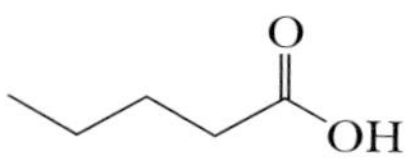

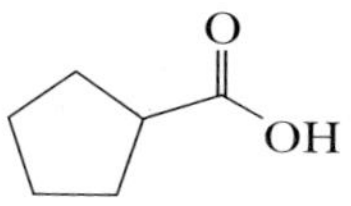

Problem 16.11

Complete the equations for these oxidations.
(a) Hexanal + H_2O_2 $\longrightarrow$ **(b)** 3-Phenylpropanal + Tollens' reagent $\longrightarrow$

B. Oxidation of Ketones

In contrast to aldehydes, ketones are oxidized only under rather special conditions. For example, they are not normally oxidized by chromic acid or potassium permanganate. In fact, chromic acid is used routinely to oxidize secondary alcohols to ketones in good yield (Section 10.8A).

Ketones undergo oxidative cleavage, via their enol form, when treated with potassium dichromate, potassium permanganate, and other strong oxidants at higher temperatures and higher concentrations of acid or base. The carbon-carbon double bond of the enol is cleaved to form two carboxyl or ketone groups, depending on the substitution pattern of the original ketone. An important industrial application of this reaction is oxidation by nitric acid of cyclohexanone to hexanedioic (adipic) acid, one of the two monomers required for the synthesis of the polymer nylon 66 (Section 29.5A).

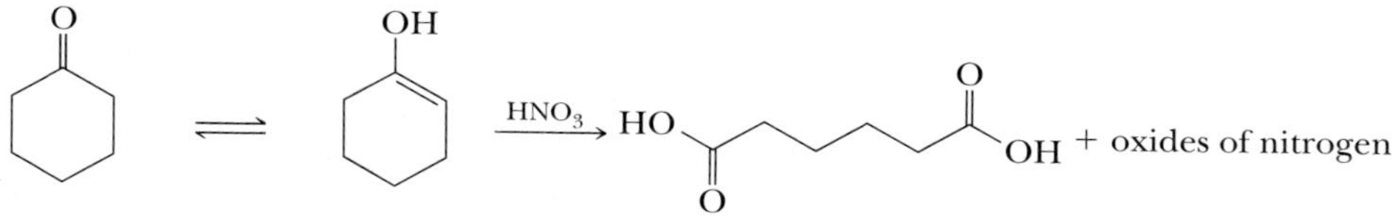

Cyclohexanone   Cyclohexanone   Hexanedioic acid
  (keto form)   (enol form)   (Adipic acid)

As the oxidation of cyclohexanone shows, this reaction is most useful for oxidation of symmetrical cycloalkanones, which yield a single product. Most other ketones give mixtures of products because the enol can form on either side of the carbonyl group.

16.11 Reduction

Aldehydes are reduced to primary alcohols, and ketones are reduced to secondary alcohols. In addition, both aldehyde and ketone carbonyl groups can be reduced to —CH$_2$— groups.

Aldehydes	Can Be Reduced to	Ketones	Can Be Reduced to
$\overset{O}{\overset{\|\|}{RCH}}$	RCH$_2$OH RCH$_3$	$\overset{O}{\overset{\|\|}{RCR'}}$	$\overset{OH}{\overset{\|}{RCHR'}}$ RCH$_2$R'

A. Metal Hydride Reductions

By far the most common laboratory reagents for reduction of the carbonyl group of an aldehyde or ketone to a hydroxyl group are sodium borohydride, lithium aluminum hydride (LAH), and their derivatives. These compounds behave as sources of **hydride ion,** a very strong nucleophile.

Na$^+$ H—B$^-$—H   Li$^+$ H—Al$^-$—H   H:$^-$

Sodium   Lithium aluminum   Hydride ion
borohydride   hydride (LAH)

Hydride ion   A hydrogen atom with two electrons in its valence shell; H:$^-$.

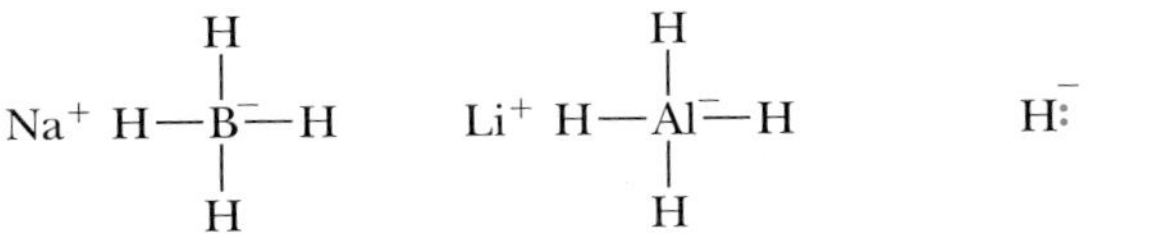

Lithium aluminum hydride is a very powerful reducing agent; it reduces not only the carbonyl groups of aldehydes and ketones rapidly but also those of carboxylic acids (Section 17.6A) and their functional derivatives (Section 18.10). Sodium borohydride is a less reactive and, therefore, much more selective reagent, reducing only aldehydes and ketones rapidly. Neither reagent reduces alkenes or alkynes to alkanes.

Reductions using sodium borohydride are most commonly carried out in aqueous methanol, in pure methanol, or in ethanol. The initial product of reduction is a tetraalkyl borate, which, on warming with water, is converted to an alcohol and sodium borate salts. One mole of sodium borohydride reduces four moles of aldehyde or ketone.

$$4R\overset{O}{\overset{\|\|}{C}}H + NaBH_4 \xrightarrow{\text{methanol}} (RCH_2O)_4B^-Na^+ \xrightarrow{H_2O} 4RCH_2OH + \text{borate salts}$$

A tetraalkyl borate

CONNECTIONS TO BIOLOGICAL CHEMISTRY

NADH: The Biological Equivalent of a Hydride Reducing Agent

For the reduction of aldehydes and ketones to alcohols, biological systems use NADH, a reagent whose results are equivalent to our laboratory hydride-reducing agents. As an example, the final step in alcoholic fermentation—the process by which yeast converts carbohydrates such as glucose to ethanol and carbon dioxide—is the enzyme-catalyzed reduction of acetaldehyde to ethanol.

$$CH_3-\overset{\overset{\displaystyle O}{\|}}{C}-H + NADH + H_3O^+ \xrightarrow[\text{dehydrogenase}]{\text{alcohol}} CH_3CH_2OH + NAD^+ + H_2O$$

Acetaldehyde   Ethanol

Alcoholic fermentation is the basis for the brewing of beers and the fermentation of grape sugar in wine making.

As another example, the end product of glycolysis is pyruvate and the reduced coenzyme NADH. In the absence of an adequate supply of oxygen (anaerobic metabolism) to reoxidize NADH to NAD^+ and thereby allow glycolysis to continue, cells use the reduction of pyruvate to lactate as a way to regenerate NAD^+:

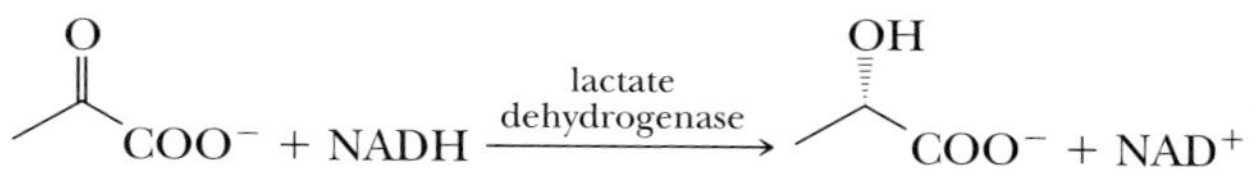

Anyone who exercises to the point of consuming all available oxygen knows the cramps and fatigue associated with the buildup of lactate in muscles. With rest and a renewed supply of oxygen, the concentration of lactate decreases rapidly and muscle cramps are relieved. Lactate production by anaerobic organisms during fermentation is responsible for the taste of sour milk and the characteristic taste and fragrance of sauerkraut (fermented cabbage).

In still other aldehyde and ketone reductions, biological systems use NADPH as a reducing agent. This molecule, which is a phosphate ester of NADH, functions in the same manner as NADH as a biological reducing agent.

$$\underset{\text{Pyruvate}}{\overset{\overset{\displaystyle O}{\|}}{\diagup}\!\!\diagdown_{COO^-}} + NADH \xrightarrow[\text{dehydrogenase}]{\text{lactate}} \underset{(S)\text{-Lactate}}{\overset{\displaystyle OH}{\diagup}\!\!\diagdown_{COO^-}} + NAD^+$$

Mechanism   *NADH Reduction of the Carbonyl Group of an Aldehyde or Ketone*

First the carbonyl-containing compound and NADH are positioned on the surface of the enzyme catalyst in a highly specific relationship to each other. Then follows a redistribution of valence electrons, one part of which is the transfer of a hydrogen atom with its pair of electrons (in effect a hydride ion) from NADH to the carbonyl compound.

Arrows 1 and 2: Electrons within the ring flow from nitrogen.

Arrow 3: Transfer of a hydride ion from the $-CH_2-$ of the six-membered ring to the carbonyl carbon creates the new C—H bond to the carbonyl carbon.

Arrow 4: The C=O pi bond breaks as the new C—H bond forms.

Arrow 5: An acidic group, —BH, on the surface of the enzyme transfers a proton to the newly formed alkoxide ion to complete formation of the hydroxyl group of the product.

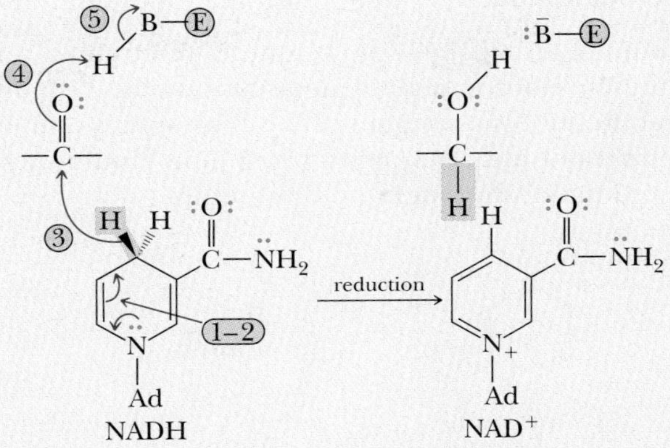

The enzyme-catalyzed reduction of pyruvate is completely stereoselective; in muscle tissue, only the *S* enantiomer of lactate is produced. This stereoselectivity arises because the reduction takes place in a chiral environment created by the enzyme. At the actual reduction step, both pyruvate and NADH are positioned precisely on the chiral surface of the enzyme with the result that the hydride ion from NADH can be delivered only to one face of pyruvate, in this case producing only the *S* enantiomer.

Unlike sodium borohydride, $LiAlH_4$ reacts violently with water, alcohols, and other protic solvents to liberate hydrogen gas and form metal hydroxides and alkoxides. Therefore, reductions of aldehydes and ketones using this reagent must be carried out in aprotic solvents, most commonly diethyl ether or tetrahydrofuran. The stoichiometry for $LiAlH_4$ reductions is the same as that for sodium borohydride reductions: one mole of $LiAlH_4$ per four moles of aldehyde or ketone. Because of the formation of gelatinous aluminum salts, aqueous acid or base work-up is usually used to dissolve these salts.

$$4RC(=O)R + LiAlH_4 \xrightarrow{\text{ether}} (R_2CHO)_4Al^-Li^+ \xrightarrow[\text{H}^+ \text{ or OH}^-]{H_2O} 4RCH(OH)R + \text{aluminum salts}$$

A tetraalkyl aluminate

Mechanism *Sodium Borohydride Reduction of an Aldehyde or Ketone*

The key step in the metal hydride reduction of an aldehyde or ketone is transfer of a hydride ion (a nucleophile) from the reducing agent to the carbonyl carbon (an electrophile) to form a tetrahedral carbonyl addition compound. In the reduction of an aldehyde or ketone to an alcohol, only the hydrogen atom bonded to carbon comes from the hydride-reducing agent; the hydrogen atom bonded to oxygen comes from water during hydrolysis of the metal alkoxide salt.

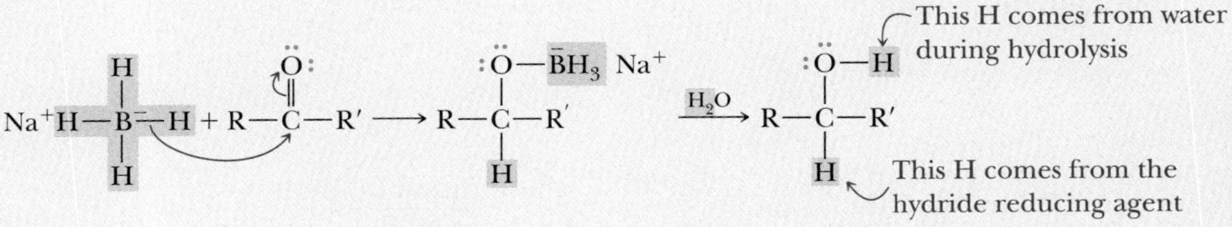

ORGANIC **Chemistry Now**™

Click Mechanisms to view an animation of the **Reduction of a Ketone by $NaBH_4$**

B. Catalytic Reduction

The carbonyl group of an aldehyde or ketone is reduced to a hydroxyl group by hydrogen in the presence of a transition metal catalyst, most commonly finely divided platinum or nickel. Reductions are generally carried out at temperatures from 25 to 100°C and at pressures of hydrogen from 1 to 5 atm. Under such conditions, cyclohexanone is reduced to cyclohexanol.

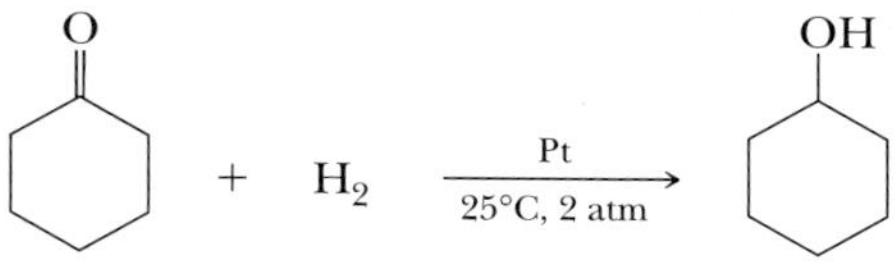

Cyclohexanone   Cyclohexanol

Catalytic reduction of aldehydes and ketones is simple to carry out, yields are generally very high, and isolation of the final product is very easy. A disadvantage is that some other functional groups are also reduced under these conditions, for example, carbon-carbon double and triple bonds.

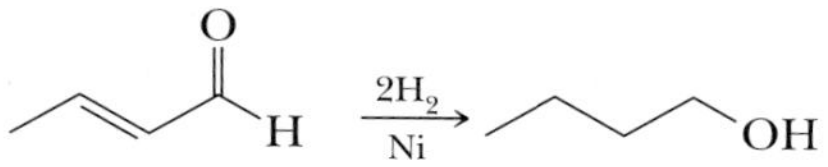

trans-2-Butenal
(Crotonaldehyde)   1-Butanol

It is generally easier to hydrogenate a carbon-carbon double bond than the carbon-oxygen double bond of an aldehyde or ketone. For this reason, it is often possible, by proper choice of metal catalyst and reaction conditions, to bring about selective catalytic reduction of a C=C bond in the presence of an aldehyde or ketone carbonyl group.

The following equations illustrate selective reduction of a carbonyl group in the presence of a carbon-carbon double bond and, alternatively, selective reduction of a carbon-carbon double bond in the presence of a carbonyl group using rhodium on powdered charcoal as a catalyst.

Selective reduction of a carbonyl group:

$$\text{RCH}=\text{CHCR}' \xrightarrow[\text{2. H}_2\text{O}]{\text{1. NaBH}_4} \text{RCH}=\text{CHCHR}'$$

Selective reduction of a carbon-carbon double bond:

$$\text{RCH}=\text{CHCR}' + \text{H}_2 \xrightarrow{\text{Rh}} \text{RCH}_2\text{CH}_2\text{CR}'$$

Example 16.12

Complete these reductions.

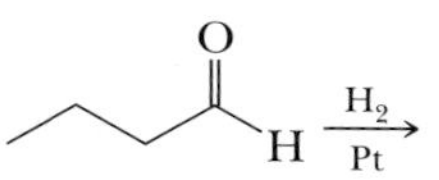

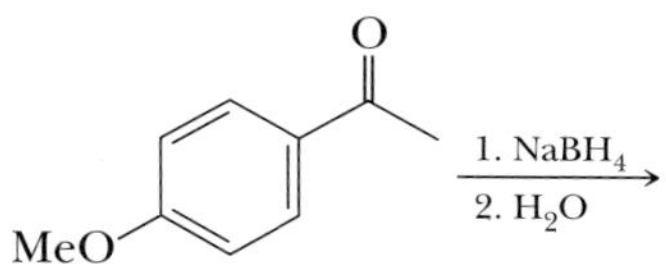

Solution

The carbonyl group of the aldehyde in (a) is reduced to a primary alcohol, and the carbonyl group of the ketone in (b) is reduced to a secondary alcohol.

(a) [structure: CH₂CH₂CH₂CH₂OH chain]   (b) [structure: 4-methoxyphenyl-CH(OH)CH₃]
MeO
(racemic)

Problem 16.12

What aldehyde or ketone gives these alcohols on reduction with NaBH₄?

(a) [cyclohexyl-CH₂OH structure]   (b) [phenyl-CH₂CH₂OH structure]   (c) [structure with two OH groups]

(mixture of
three stereoisomers)

C. Reduction of a Carbonyl Group to a Methylene Group

Several methods are available for reducing the carbonyl group of an aldehyde or ketone to a methylene group ($-CH_2-$). One of the first discovered was refluxing the aldehyde or ketone with amalgamated zinc (zinc with a surface layer of mercury) in concentrated HCl.

[structure: 2-hydroxyphenyl ketone] $\xrightarrow{\text{Zn(Hg), HCl}}$ [structure: 2-hydroxyphenyl alkyl]

This reaction is known as the **Clemmensen reduction** after the German chemist, E. Clemmensen, who developed it in 1912. The mechanism of Clemmensen reduction, although not well understood, involves electrons from the Zn to reduce the carbonyl group.

Because the Clemmensen reduction requires the use of concentrated HCl, it cannot be used to reduce a carbonyl group in a molecule that also contains acid-sensitive groups, such as a tertiary alcohol that might undergo dehydration, or an acetal that is hydrolyzed so that its resulting carbonyl group is also reduced.

The **Wolff-Kishner reduction,** discovered independently by N. Kishner in 1911 and L. Wolff in 1912 and reported within months of Clemmensen's discovery, is an alternative method for reduction of a carbonyl group to a methylene group. In this reduction, a mixture of the aldehyde or ketone, hydrazine, and concentrated potassium hydroxide is refluxed in a high-boiling solvent such as diethylene glycol (bp 245°C).

[structure: acetophenone] $+ H_2NNH_2 \xrightarrow[\substack{\text{diethylene glycol} \\ \text{(reflux)}}]{\text{KOH}}$ [structure: ethylbenzene] $+ N_2 + H_2O$

Hydrazine

Clemmensen reduction
Reduction of the C=O group of an aldehyde or ketone to a CH₂ group using Zn(Hg) and HCl.

Wolff-Kishner reduction
Reduction of the C=O group of an aldehyde or ketone to a CH₂ group using hydrazine and base.

More recently it has been found possible to bring about the same reaction in dimethyl sulfoxide (DMSO) with potassium *tert*-butoxide and hydrazine at room temperature. These conditions are referred to as the Huang Minlon modification.

Mechanism *Wolff-Kishner Reduction*

Step 1: Reaction of the carbonyl group of the aldehyde or ketone with hydrazine gives a hydrazone (Section 16.8B).

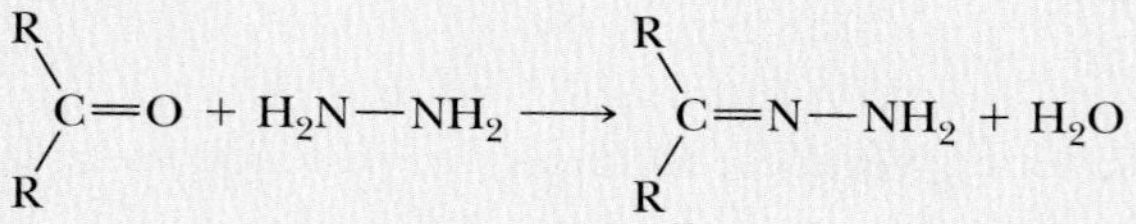

$$\underset{R}{\overset{R}{\diagdown}} C{=}O + H_2N{-}NH_2 \longrightarrow \underset{R}{\overset{R}{\diagdown}} C{=}N{-}NH_2 + H_2O$$

Hydrazine   A hydrazone

Step 2: Base-catalyzed tautomerism gives an isomer with an N=N double bond (compare keto-enol tautomerism, Section 16.9).

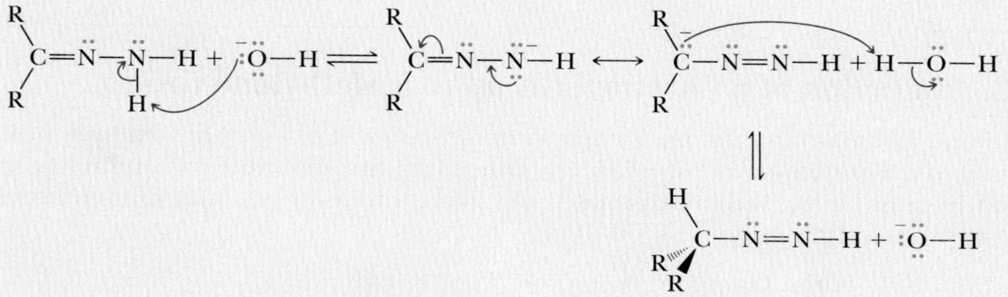

Step 3: Proton transfer to hydroxide ion followed by loss of N_2 gives a carbanion. Proton transfer from water to the carbanion gives the hydrocarbon and hydroxide ion.

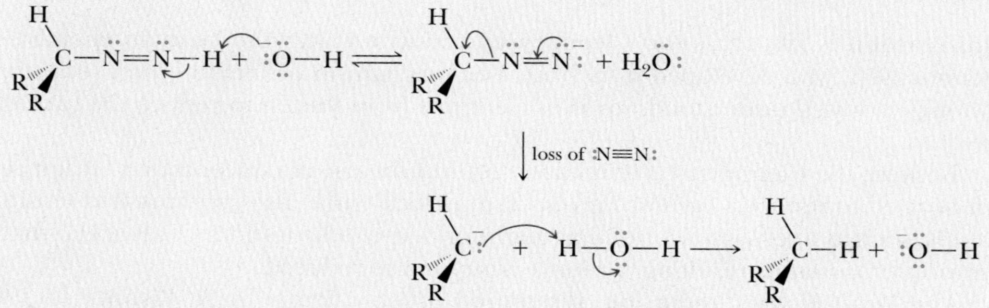

loss of :N≡N:

A carbanion

Each of the reductions has its special conditions, advantages, and disadvantages. The Clemmensen reduction cannot be used in the presence of groups sensitive to concentrated acid, the Wolff-Kishner reduction cannot be used in the presence of groups sensitive to concentrated base. However, the carbonyl group of almost any aldehyde or ketone can be reduced to a methylene group by one of these methods.

Example 16.13

Complete the following reactions.

(a)

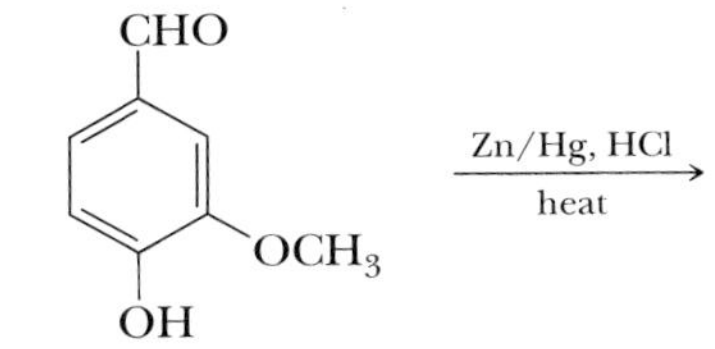

CHO

OCH₃

OH

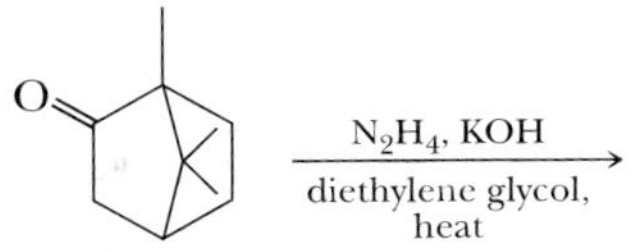

Vanillin
(from vanilla beans)

$\xrightarrow[\text{heat}]{\text{Zn/Hg, HCl}}$

(b)

O

Camphor

$\xrightarrow[\substack{\text{diethylene glycol,} \\ \text{heat}}]{N_2H_4,\ KOH}$

Solution

Reaction (a) is a Clemmensen reduction, and reaction (b) is a Wolff-Kishner reduction.

(a)

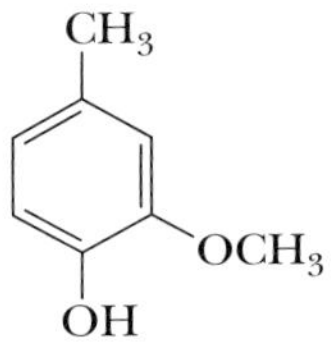

CH₃

OCH₃

OH

(b)

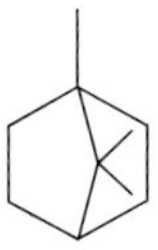

Problem 16.13

Complete the following reactions.

(a)

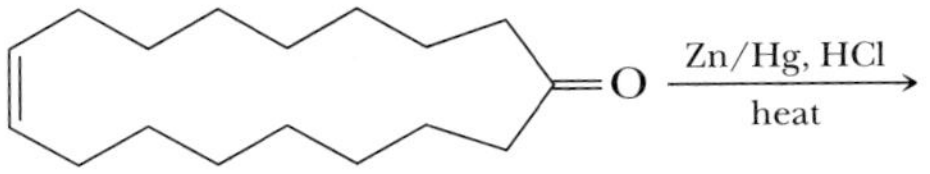

O

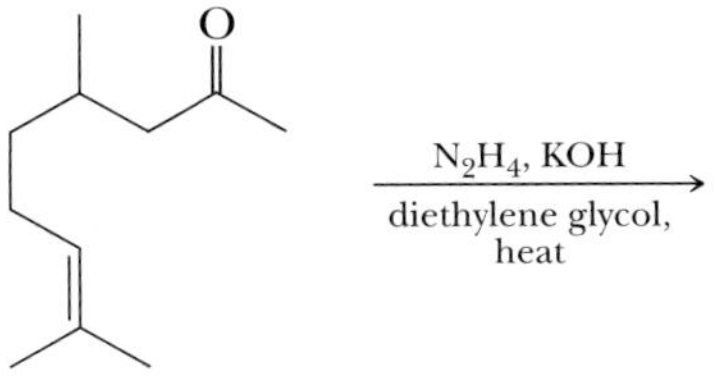

Civetone
(from the civet cat; used
in perfumery)

$\xrightarrow[\text{heat}]{\text{Zn/Hg, HCl}}$

(b)

O

Citronellal
(from citronella and
lemon grass oils)

$\xrightarrow[\substack{\text{diethylene glycol,} \\ \text{heat}}]{N_2H_4,\ KOH}$

16.12 Reactions at an α-Carbon

A. Racemization

When enantiomerically pure (either R or S) 3-phenyl-2-butanone is dissolved in ethanol, no change occurs in the optical activity of the solution over time. If, however, a trace of either acid (for example, aqueous or gaseous HCl) or base (for example, sodium ethoxide) is added, the optical activity of the solution begins to decrease gradually and eventually drops to zero. When 3-phenyl-2-butanone is isolated from this solution, it is found to be a racemic mixture. Furthermore, the rate of racemization is proportional to the concentration of acid or base. These observations can be explained by a rate-determining acid- or base-catalyzed formation of an achiral enol intermediate. Tautomerism of the achiral enol to the chiral keto form generates the R and S enantiomers with equal probability.

(*R*)-3-Phenyl-2-butanone   An achiral enol   (*S*)-3-Phenyl-2-butanone

Racemization by this mechanism occurs only at α-carbon stereocenters with at least one α-hydrogen.

B. Deuterium Exchange

When an aldehyde or ketone with one or more α-hydrogens is dissolved in an aqueous solution enriched with D_2O and also containing catalytic amounts of either D^+ or OD^-, exchange of α-hydrogens occurs at a rate that is proportional to the concentration of the acid or base catalyst. We account for incorporation of deuterium by proposing a rate-determining acid- or base-catalyzed enolization followed by incorporation of deuterium as the enol form converts to the keto form.

$$\underset{\text{Acetone}}{CH_3\overset{O}{\overset{\|}{C}}CH_3} + 6D_2O \underset{}{\overset{D^+ \text{ or } OD^-}{\rightleftharpoons}} \underset{\text{Acetone-}d_6}{CD_3\overset{O}{\overset{\|}{C}}CD_3} + 6HOD$$

Deuterium exchange has two values. First, by observing changes in hydrogen ratios before and after deuterium exchange, it is possible to determine the number of exchangeable α-hydrogens in a molecule. Second, exchange of α-hydrogens is a convenient way to introduce an isotopic label into molecules.

In naming compounds, the presence of deuterium is shown by the symbol "d," and the number of deuterium atoms is shown by a subscript following it. In addition to acetone-d_6, more than 225 deuterium-labeled compounds are available commercially, in isotopic enrichments of up to 99.8 atom % D. Among these are

$CDCl_3$	$CD_3\overset{O}{\overset{\|}{C}}OD$	$CH_3\overset{O}{\overset{\|}{C}}OD$	$NaBD_4$	CH_3CH_2OD
Chloroform-d	Acetic-d_3 acid-d	Acetic acid-d	Sodium borodeuteride	Ethanol-d

Deuterated solvents, such as $CDCl_3$, acetone-d_6, and benzene-d_6, are used as solvents in ^{1}H-NMR spectroscopy because they lack protons that might otherwise obscure the spectrum of the compound of interest.

C. α-Halogenation

Aldehydes and ketones with at least one α-hydrogen react at the α-carbon with bromine and chlorine to form α-haloaldehydes and α-haloketones as illustrated by bromination of acetophenone.

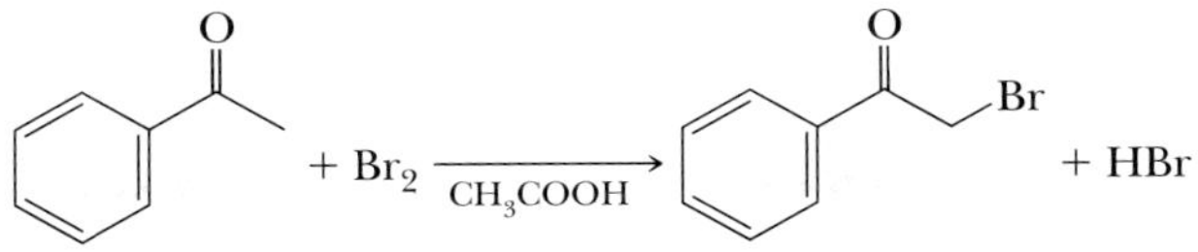

Acetophenone   α-Bromoacetophenone

Bromination or chlorination at an α-carbon is catalyzed by both acid and base. For acid-catalyzed halogenation, acid generated by the reaction catalyzes further reaction. The slow step of acid-catalyzed halogenation is formation of an enol. This is followed by rapid reaction of the double bond with halogen to give the α-haloketone.

Mechanism *Acid-Catalyzed α-Halogenation of a Ketone*

Step 1: Acid-catalyzed keto-enol tautomerism gives the enol.

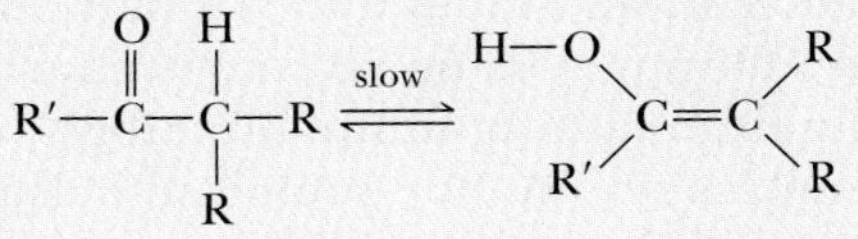

Step 2: Nucleophilic attack of the enol on the halogen molecule, X$_2$, gives the conjugate acid of an α-haloketone.

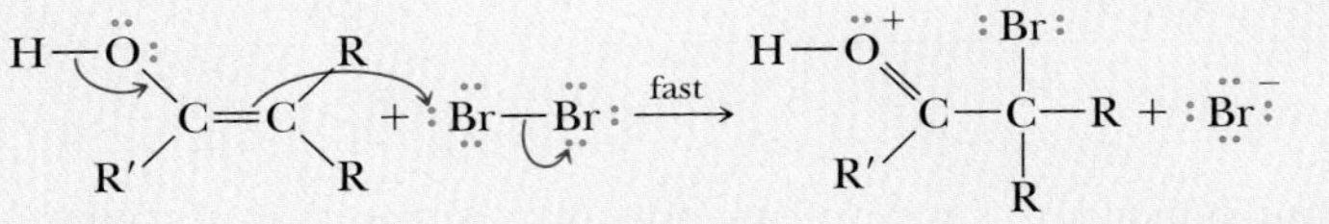

Step 3: Proton transfer to a base, in this case a molecule of acetic acid, gives the α-haloketone.

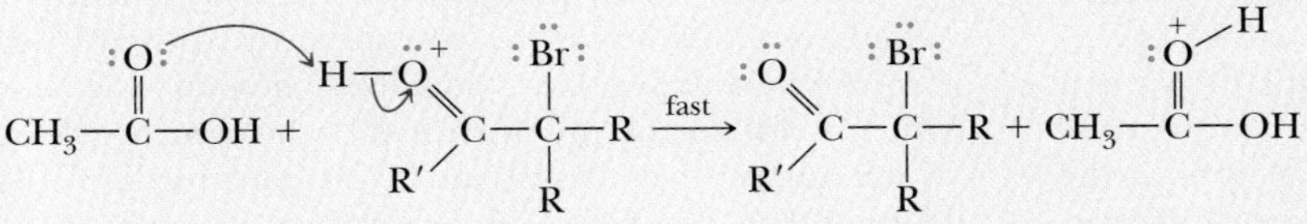

An α-bromoketone

The slow step in base-promoted α-halogenation is removal of an α-hydrogen by base to form an enolate anion, which then reacts with halogen by nucleophilic displacement to form the final product. This procedure for α-halogenation produces HX as a by-product, and, in order to keep the solution basic, it is necessary to add slightly more than one mole of base per mole of aldehyde or ketone. Because base is a reactant required in equimolar amounts, we say that this reaction is base-promoted rather than base-catalyzed.

Mechanism *Base-Promoted α-Halogenation of a Ketone*

Step 1: Proton transfer from the α-carbon to the base gives a resonance-stabilized enolate anion.

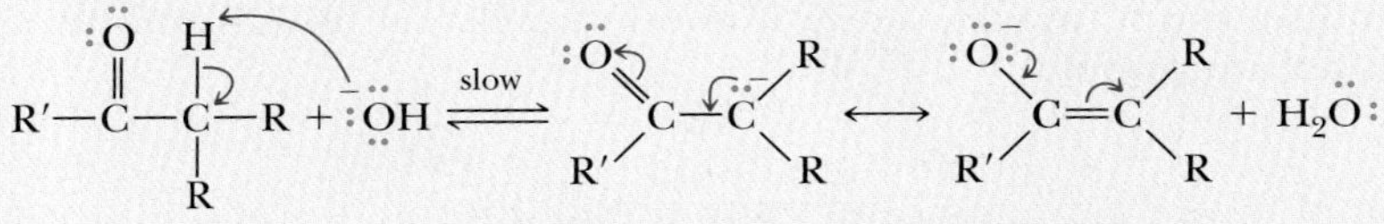

Resonance-stabilized enolate anion

ORGANIC
Chemistry Now™

Click Mechanisms to view an animation of the **Acid-Catalyzed α-Halogenation of a Ketone**

Step 2: Nucleophilic attack of the enolate anion on halogen gives an α-bromoketone.

A major difference exists between acid-catalyzed and base-promoted α-halogenation. In principle, both can lead to polyhalogenation. In practice, the rate of acid-catalyzed introduction of a second halogen is considerably less than the rate of the first halogenation because the electronegative α-halogen decreases the basicity of the carbonyl oxygen toward protonation. Thus, it is generally possible to stop acid-catalyzed halogenation at a single substitution. For base-promoted halogenation, each successive halogenation is more rapid than the previous one because introduction of an electronegative halogen atom on an α-carbon further increases the acidity of remaining α-hydrogens; thus, each successive α-hydrogen is removed more rapidly than the previous one. For this reason, base-promoted halogenation is generally not a useful synthetic reaction.

Summary

An **aldehyde** (Section 16.1) contains a carbonyl group bonded to a hydrogen atom and a carbon atom. A **ketone** contains a carbonyl group bonded to two carbons. An aldehyde is named by changing -*e* of the parent alkane to -*al* (Section 16.2). A ketone is named by changing -*e* of the parent alkane to -*one* and using a number to locate the carbonyl group. In naming compounds that contain more than one functional group, the IUPAC system has established an **order of precedence of functions** (Section 16.2B). If the carbonyl group of an aldehyde or ketone is lower in precedence than other functional groups in the molecule, it is indicated by the infix -*oxo*-.

Aldehydes and ketones are polar compounds (Section 16.3) and interact in the pure state by dipole-dipole interactions; they have higher boiling points and are more soluble in water than nonpolar compounds of comparable molecular weight.

One of the most common reaction themes of aldehydes and ketones is addition of a nucleophile to the carbonyl carbon to form a tetrahedral carbonyl addition compound (Section 16.4). Often, a new stereocenter is created by this reaction. When none of the starting materials is chiral, a racemic mixture is formed. Many of these reactions form new carbon-carbon bonds, making this a very important class of reactions in organic synthesis.

The carbon atom adjacent to a carbonyl group is called an **α-carbon** (Section 16.9A), and a hydrogen bonded to it is called an **α-hydrogen.** The pK_a of an α-hydrogen of an aldehyde or ketone is approximately 20, which makes it less acidic than alcohols but more acidic than terminal alkynes.

Key Reactions

1. Reaction with Grignard Reagents (Section 16.5A)

Treating formaldehyde with a Grignard reagent followed by hydrolysis gives a primary alcohol. Similar treatment of any other aldehyde gives a secondary alcohol. Treatment of a ketone gives a tertiary alcohol.

2. Reaction with Organolithium Reagents (Section 16.5B)

Reactions of aldehydes and ketones with organolithium reagents are similar to those with Grignard reagents.

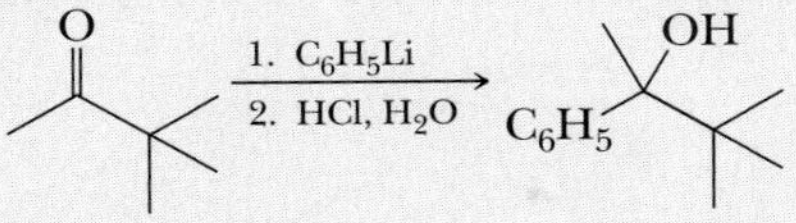

(produced as a
racemic mixture)

3. Reaction with Anions of Terminal Alkynes (Section 16.5C)

Treating an aldehyde or ketone with the alkali metal salt of a terminal alkyne followed by hydrolysis gives an α-alkynylalcohol.

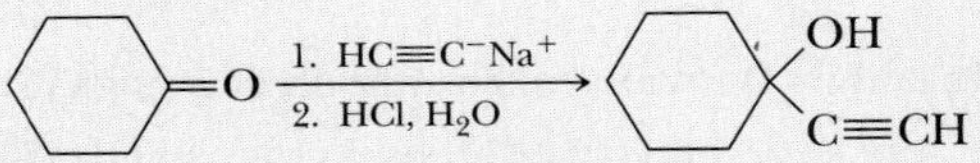

4. Reaction with HCN to Form Cyanohydrins (Section 16.5D)

For aldehydes and most sterically unhindered aliphatic ketones, equilibrium favors formation of the cyanohydrin. For aryl ketones, equilibrium favors starting materials, and little cyanohydrin is obtained.

$$\underset{\text{O}}{\overset{\text{O}}{\underset{\|}{\text{C}_6\text{H}_5\text{CH}}}} + \text{HC}\equiv\text{N} \xrightarrow{\text{NaCN}} \underset{\text{OH}}{\text{C}_6\text{H}_5\text{CHC}\equiv\text{N}}$$

(Formed as a
racemic mixture)

5. The Wittig Reaction (Section 16.6)

Treating an aldehyde or ketone with a triphenylphosphonium ylide gives an oxaphosphetane intermediate, which fragments to give triphenylphosphine oxide and an alkene.

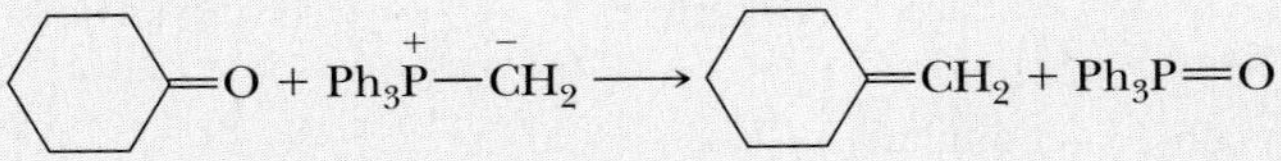

6. The Horner-Emmons-Wadsworth Modification of the Wittig Reaction (Section 16.6)

This modification of the original Wittig reaction uses a phosphonate ester derived from an α-haloester, aldehyde, or ketone to generate the Wittig carbanion and shows very high E selectivity.

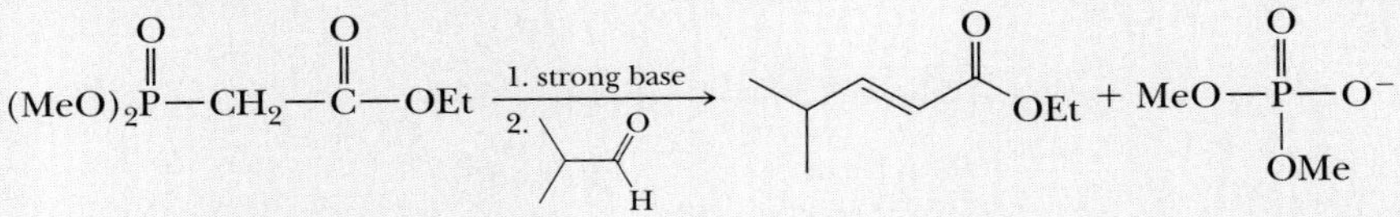

7. Hydration (Section 16.7A)

The degree of hydration is greater for aldehydes than for ketones.

$$\underset{\text{H}}{\overset{\text{O}}{\underset{\|}{\text{HCH}}}} + \text{H}_2\text{O} \rightleftharpoons \underset{\text{H}}{\overset{\text{OH}}{\underset{|}{\text{HCOH}}}}$$

(>99%)

8. Addition of Alcohols to Form Hemiacetals (Section 16.7B)

Hemiacetals are only minor components of an equilibrium mixture of aldehyde or ketone and alcohol, except where the —OH and the C=O are parts of the same molecule and a five- or six-membered ring can form. The reaction is catalyzed by acid or base.

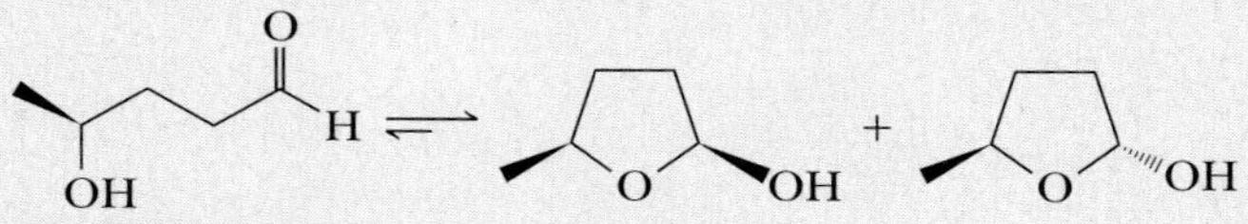

9. Addition of Alcohols to Form Acetals (Section 16.7B)

Formation of acetals is catalyzed by acid. Acetals are stable to water and aqueous base but are hydrolyzed in aqueous acid. Acetals are valuable as carbonyl-protecting groups.

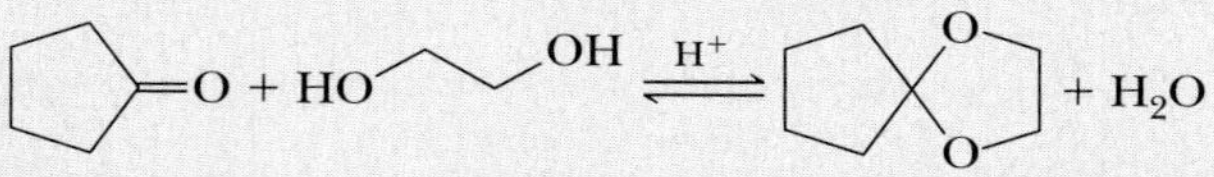

10. Addition of Ammonia and Its Derivatives: Formation of Imines (Section 16.8A)

Addition of ammonia or a primary amine to the carbonyl group of an aldehyde or ketone forms a tetrahedral carbonyl addition compound. Loss of water from this intermediate gives an imine.

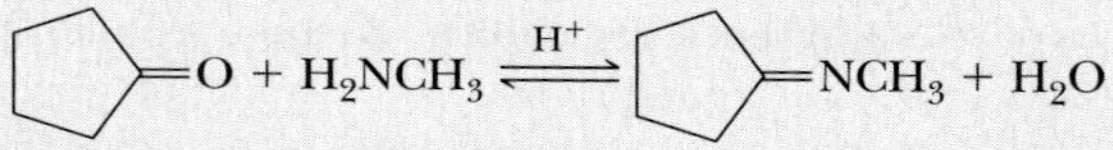

11. Addition of Secondary Amines: Formation of Enamines (Section 16.8A)

Addition of a secondary amine to the carbonyl group of an aldehyde or ketone forms a tetrahedral carbonyl addition intermediate. Acid-catalyzed dehydration of this intermediate gives an enamine.

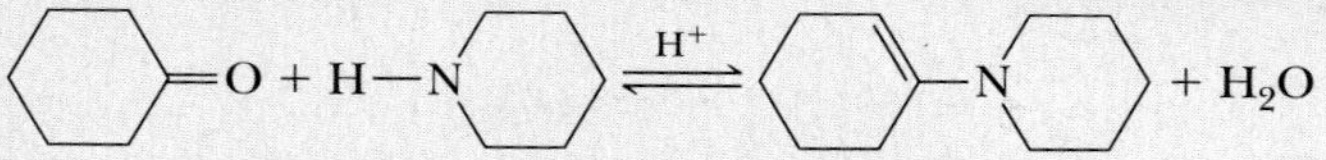

12. Addition of Hydrazine and Its Derivatives (Section 16.8B)

Treating an aldehyde or ketone with hydrazine gives a hydrazone.

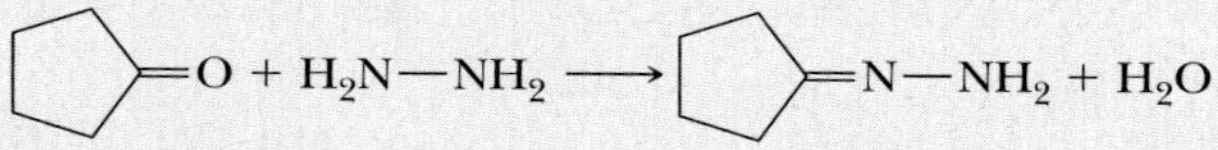

Derivatives of hydrazine react similarly.

13. Keto-Enol Tautomerism (Section 16.9B)

The keto form predominates at equilibrium, except for those aldehydes and ketones in which the enol is stabilized by resonance or hydrogen bonding.

$$CH_3CCH_3 \rightleftharpoons CH_3C=CH_2$$

Keto form   Enol form
(approx 99.9%)

14. Oxidation of an Aldehyde to a Carboxylic Acid (Section 16.10A)

The aldehyde group is among the most easily oxidized functional groups. Oxidizing agents include H_2CrO_4, $KMnO_4$, Ag_2O, Tollens' reagent, H_2O_2, and O_2.

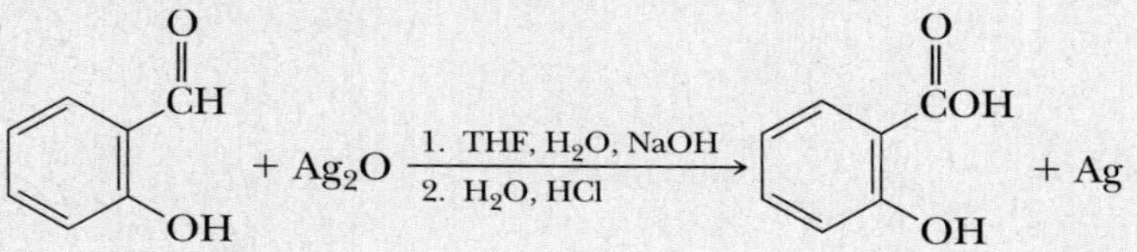

15. Metal Hydride Reduction (Section 16.11A)

Both $LiAlH_4$ and $NaBH_4$ are selective in that neither reduces isolated carbon-carbon double or triple bonds.

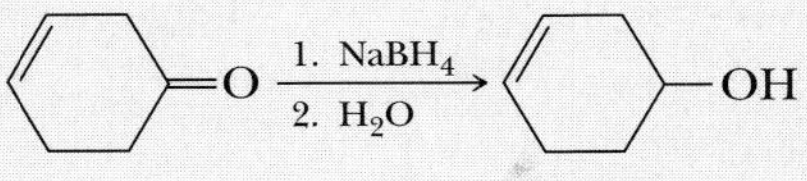

(Formed as a
racemic mixture)

16. Catalytic Reduction (Section 16.11B)

Catalytic reduction of the carbonyl group of an aldehyde or ketone to a hydroxyl group is simple to carry out and yields of the alcohols are high. A disadvantage of this method is that some other functional groups, including carbon-carbon double and triple bonds, may also be reduced.

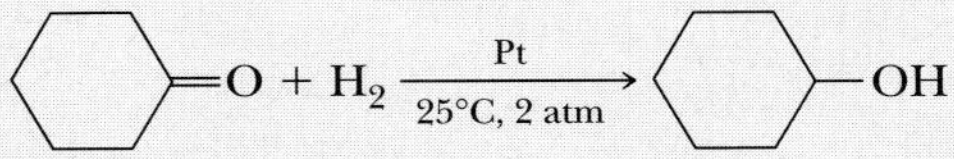

17. Clemmensen Reduction (Section 16.11C)

Reduction of the carbonyl group of an aldehyde or ketone using amalgamated zinc in the presence of concentrated hydrochloric acid gives a methylene group.

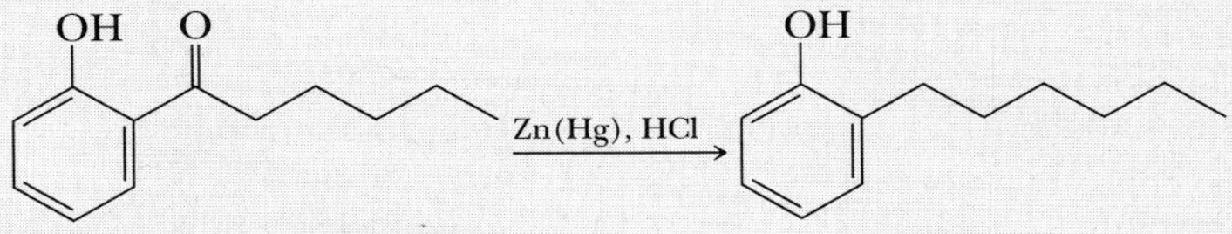

18. Wolff-Kishner Reduction (Section 16.11C)

Formation of a hydrazone followed by treatment with base, commonly KOH in diethylene glycol or potassium *tert*-butoxide in dimethyl sulfoxide, reduces the carbonyl group of an aldehyde or ketone to a methylene group.

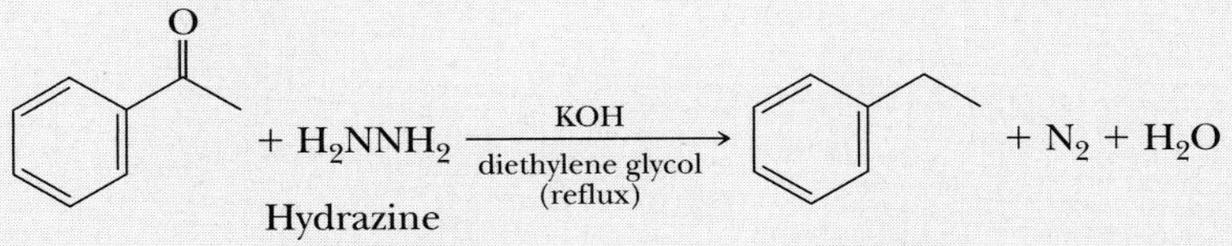

19. Deuterium Exchange at an α-Carbon (Section 16.12B)

Acid- or base-catalyzed deuterium exchange at an α-carbon involves formation of an enol or enolate anion intermediate:

$$CH_3CCH_3 + 6D_2O \underset{}{\overset{DCl}{\rightleftharpoons}} CD_3CCD_3 + 6HOD$$

20. Halogenation at an α-Carbon (Section 16.12C)

The rate-determining step in acid-catalyzed α-halogenation is the formation of an enol. In base-promoted α-halogenation, it is formation of an enolate anion.

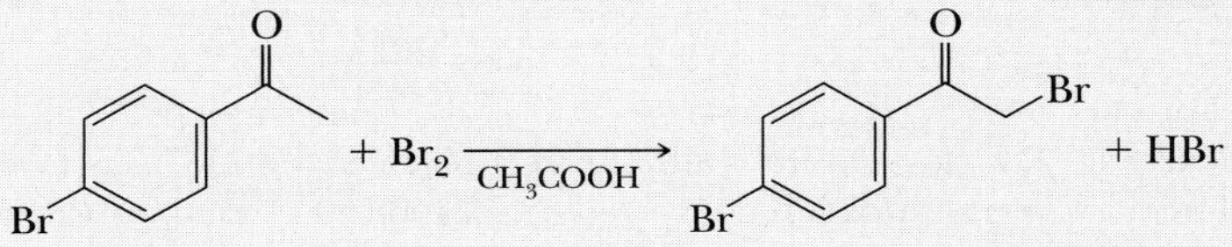

Problems

Structure and Nomenclature

16.14 Name each compound, showing stereochemistry where relevant.

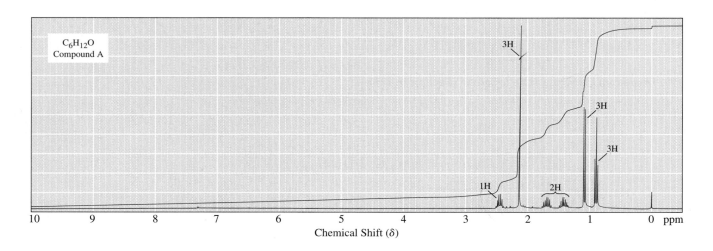

16.15 Draw a structural formula for each compound.

(a) 1-Chloro-2-propanone   (b) 3-Hydroxybutanal
(c) 4-Hydroxy-4-methyl-2-pentanone   (d) 3-Methyl-3-phenylbutanal
(e) 1,3-Cyclohexanedione   (f) 3-Methyl-3-buten-2-one
(g) 5-Oxohexanal   (h) 2,2-Dimethylcyclohexanecarbaldehyde
(i) 3-Oxobutanoic acid

16.16 The infrared spectrum of Compound A, $C_6H_{12}O$, shows a strong, sharp peak at 1724 cm^{-1}. From this information and its ^{1}H-NMR spectrum, deduce the structure of compound A.

16.17 Following are ^{1}H-NMR spectra for compounds B ($C_6H_{12}O_2$) and C ($C_6H_{10}O$). On warming in dilute acid, compound B is converted to compound C. Deduce the structural formulas for compounds B and C.

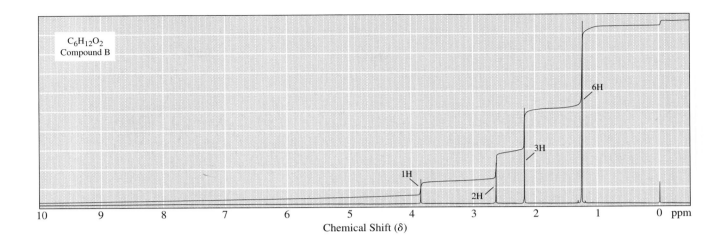

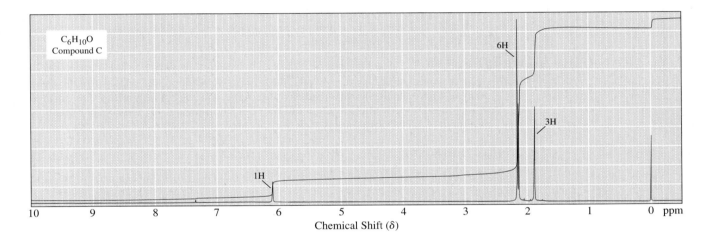

Addition of Carbon Nucleophiles

16.18 Draw structural formulas for the product formed by treating each compound with propylmagnesium bromide followed by aqueous HCl.

(a) CH_2O (b)  [epoxide structure]   (c) 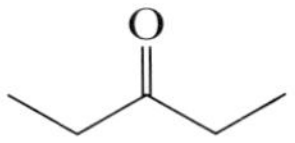 [ketone structure]   (d) 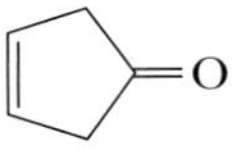 [cyclopentenone structure]

16.19 Suggest a synthesis for the following alcohols starting from an aldehyde or ketone and an appropriate Grignard reagent. Below each target molecule is the number of combinations of Grignard reagent and aldehyde or ketone that might be used.

(a) 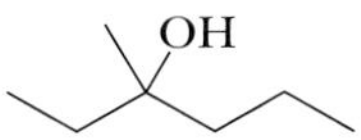 [structure with OH]

3 Combinations
(racemic)

(b) 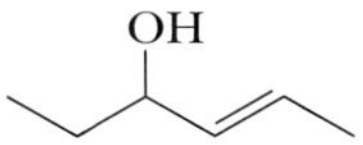 [structure with OH]

2 Combinations
(racemic)

(c) 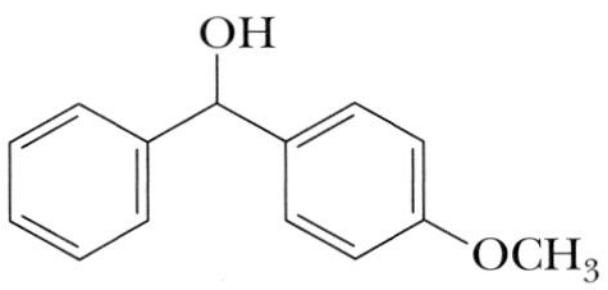 [diaryl carbinol structure with OH and OCH$_3$]

2 Combinations
(racemic)

16.20 Show how to synthesize the following alcohol using 1-bromopropane, propanal, and ethylene oxide as the only sources of carbon atoms.

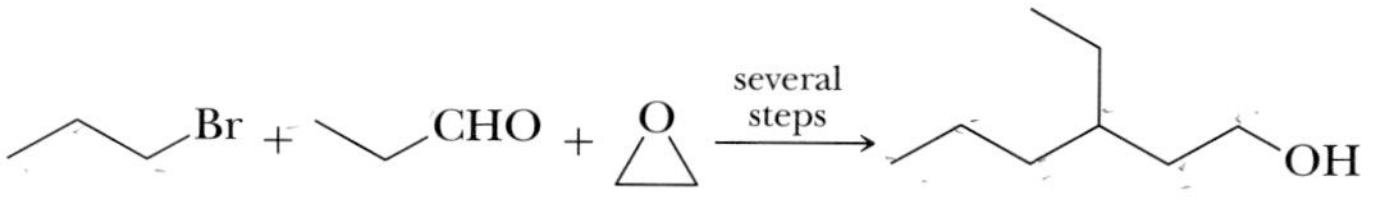

(racemic)

16.21 1-Phenyl-2-butanol is used in perfumery. Show how to synthesize this alcohol from bromobenzene, 1-butene, and any necessary inorganic reagents.

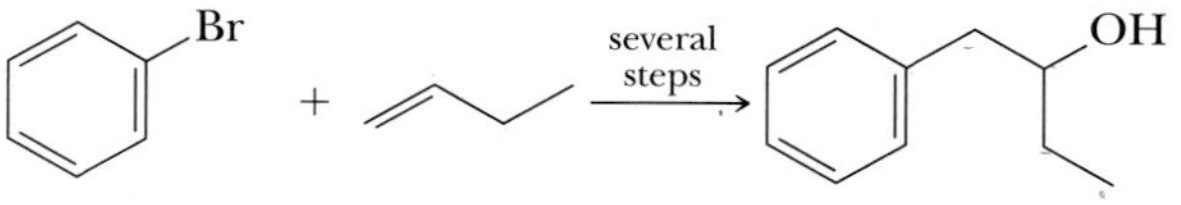

Bromobenzene   1-Butene   1-Phenyl-2-butanol
(racemic)

16.22 With organolithium and organomagnesium compounds, approach to the carbonyl carbon from the less hindered direction is generally preferred. Assuming this is the case, predict the structure of the major product formed by reaction of methylmagnesium bromide with 4-*tert*-butylcyclohexanone.

Wittig Reaction

16.23 Draw structural formulas for (1) the alkyltriphenylphosphonium salt formed by treatment of each haloalkane with triphenylphosphine, (2) the phosphonium ylide formed by treatment of each phosphonium salt with butyllithium, and (3) the alkene formed by treatment of each phosphonium ylide with acetone.

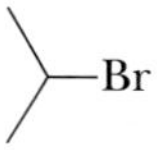

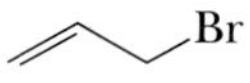

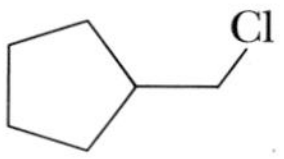

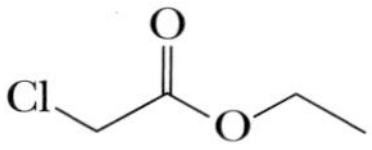

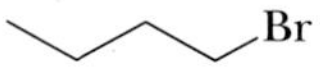

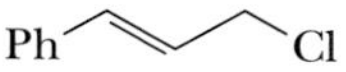

16.24 Show how to bring about the following conversions using a Wittig reaction.

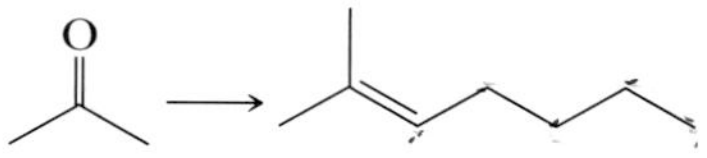

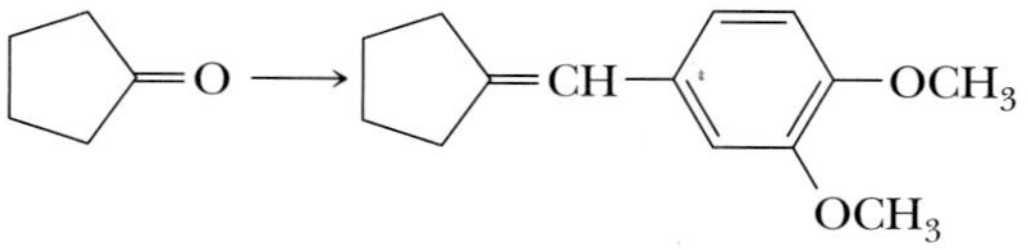

16.25 The Wittig reaction can be used for the synthesis of conjugated dienes, as for example, 1-phenyl-1,3-pentadiene.

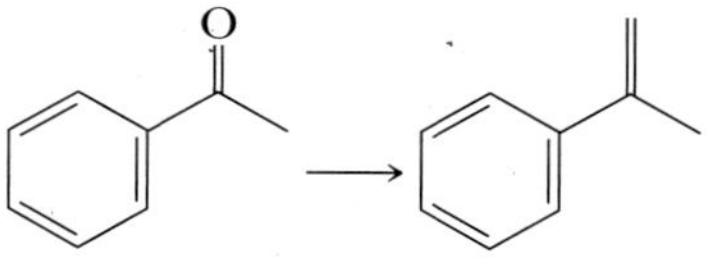

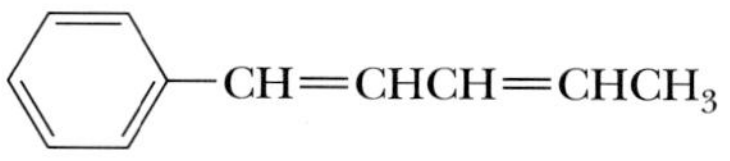

1-Phenyl-1,3-pentadiene

Propose two sets of reagents that might be combined in a Wittig reaction to give this conjugated diene.

16.26 Wittig reactions with the following α-chloroethers can be used for the synthesis of aldehydes and ketones.

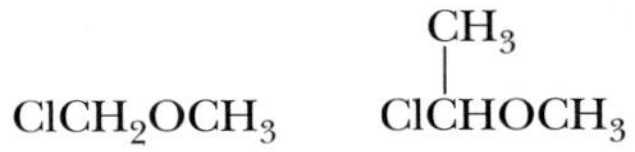

$$
\text{ClCH}_2\text{OCH}_3 \qquad \text{Cl}\overset{\overset{\displaystyle CH_3}{|}}{\text{CH}}\text{OCH}_3
$$

(A)   (B)

(a) Draw the structure of the triphenylphosphonium salt and Wittig reagent formed from each chloroether.
(b) Draw the structural formula of the product formed by treating each Wittig reagent with cyclopentanone. Note that the functional group is an enol ether or, alternatively, a vinyl ether.
(c) Draw the structural formula of the product formed on acid-catalyzed hydrolysis of each enol ether from part (b).

16.27 It is possible to generate sulfur ylides in a manner similar to that used to produce phosphonium ylides. For example, treating a sulfonium salt with a strong base gives the sulfur ylide.

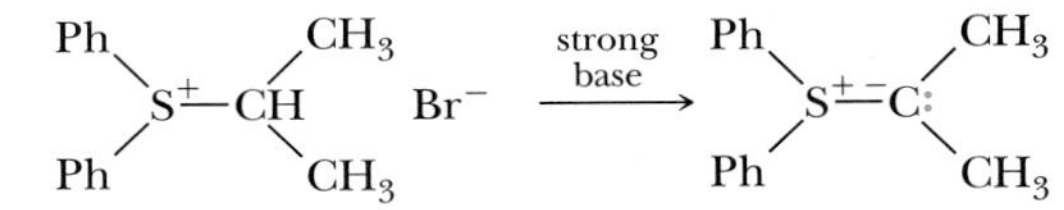

A sulfonium bromide salt   A sulfur ylide

Sulfur ylides react with ketones to give epoxides. Suggest a mechanism for this reaction.

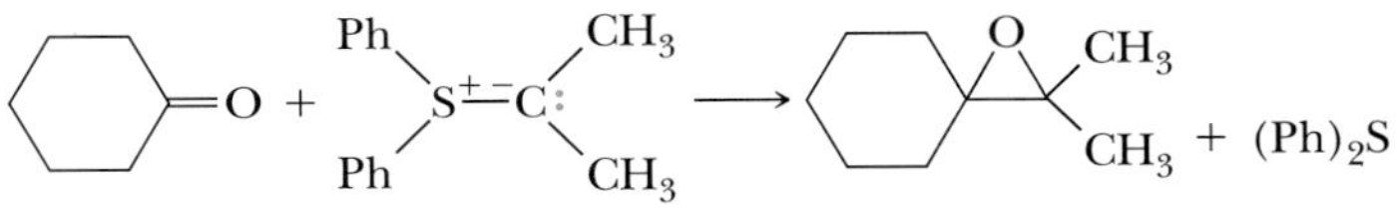

16.28 Propose a structural formula for Compound D and for the product, $C_9H_{14}O$, formed in this reaction sequence.

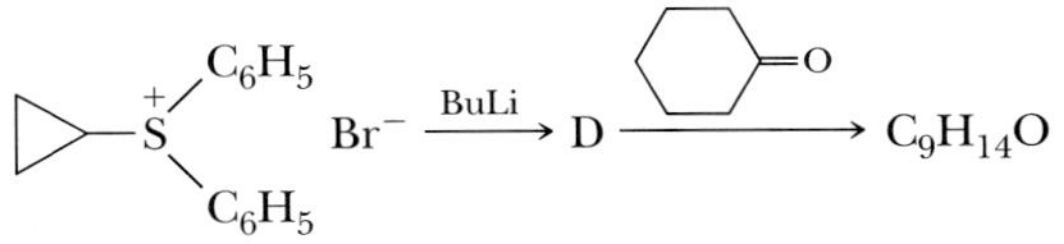

Addition of Oxygen Nucleophiles

16.29 5-Hydroxyhexanal forms a six-membered cyclic hemiacetal, which predominates at equilibrium in aqueous solution.

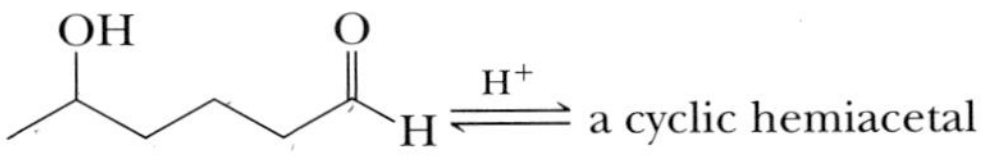

5-Hydroxyhexanal

(a) Draw a structural formula for this cyclic hemiacetal.
(b) How many stereoisomers are possible for 5-hydroxyhexanal?
(c) How many stereoisomers are possible for this cyclic hemiacetal?
(d) Draw alternative chair conformations for each stereoisomer and label groups axial or equatorial. Also predict which of the alternative chair conformations for each stereoisomer is the more stable.

16.30 Draw structural formulas for the hemiacetal and then the acetal formed from each pair of reactants in the presence of an acid catalyst.

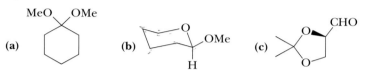

(a) 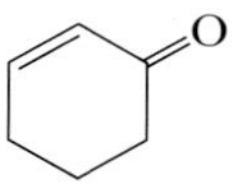 + EtOH   (b) 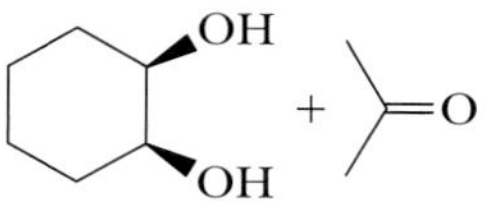   (c) 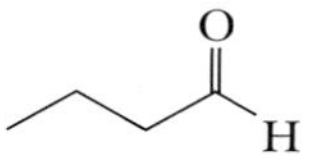 + MeOH

16.31 Draw structural formulas for the products of hydrolysis of the following acetals in aqueous HCl,

(a)   (b)   (c)

16.32 Propose a mechanism to account for the formation of a cyclic acetal from 4-hydroxypentanal and one equivalent of methanol. If the carbonyl oxygen of 4-hydroxypentanal is enriched with oxygen-18, do you predict that the oxygen label appears in the cyclic acetal or in the water?

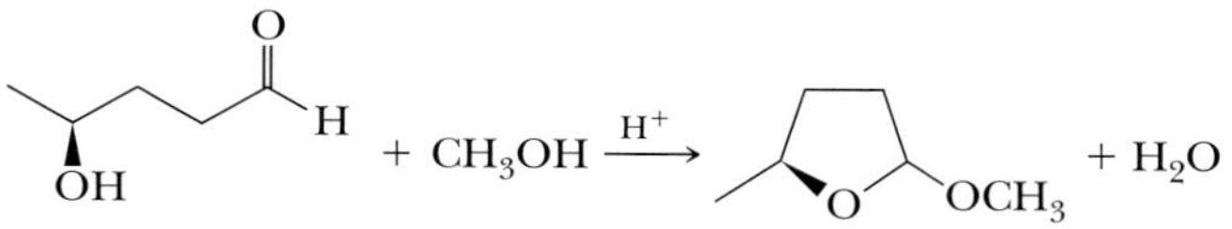

$$\text{4-Hydroxypentanal} + CH_3OH \xrightarrow{H^+} \text{A cyclic acetal (racemic)} + H_2O$$

4-Hydroxypentanal   A cyclic acetal
                                        (racemic)

16.33 Propose a mechanism for this acid-catalyzed hydrolysis.

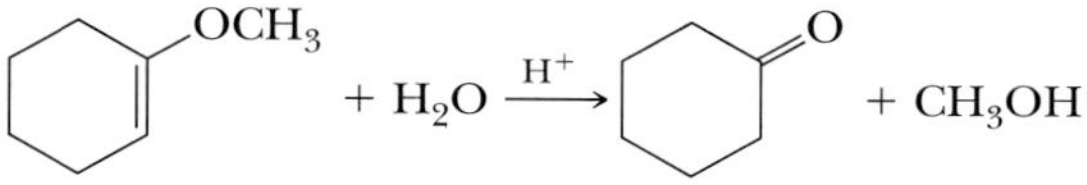

$$ + H_2O \xrightarrow{H^+} + CH_3OH$$

16.34 In Section 11.5 we saw that ethers, such as diethyl ether and tetrahydrofuran, are quite resistant to the action of dilute acids and require hot concentrated HI or HBr for cleavage. However, acetals in which two ether groups are linked to the same carbon undergo hydrolysis readily, even in dilute aqueous acid. How do you account for this marked difference in chemical reactivity toward dilute aqueous acid between ethers and acetals?

16.35 Show how to bring about the following conversion.

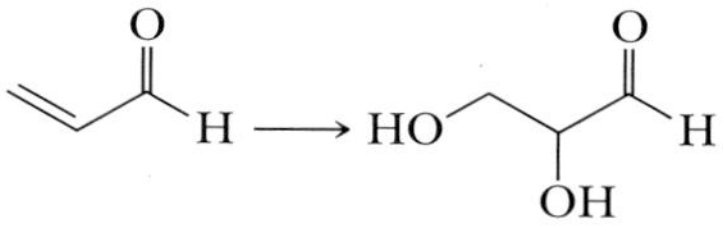

(racemic)

16.36 A primary or secondary alcohol can be protected by conversion to its tetrahydropyranyl ether. Why is formation of THP ethers by this reaction limited to primary and secondary alcohols?

16.37 Which of these molecules will cyclize to give the insect pheromone frontalin?

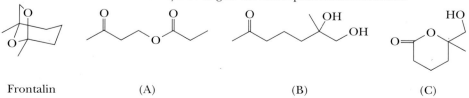

Frontalin   (A)   (B)   (C)

Addition of Nitrogen Nucleophiles

16.38 Draw a structural formula for the product of each acid-catalyzed reaction.

(a) Phenylacetaldehyde + hydrazine $\longrightarrow$
(b) Cyclopentanone + semicarbazide $\longrightarrow$
(c) Acetophenone + 2,4-dinitrophenylhydrazine $\longrightarrow$
(d) Benzaldehyde + hydroxylamine $\longrightarrow$

16.39 Following are structural formulas for amphetamine and methamphetamine.

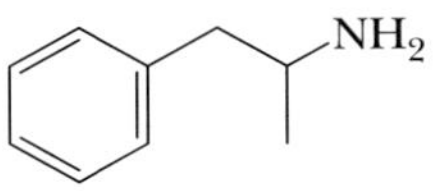

Amphetamine (racemic)

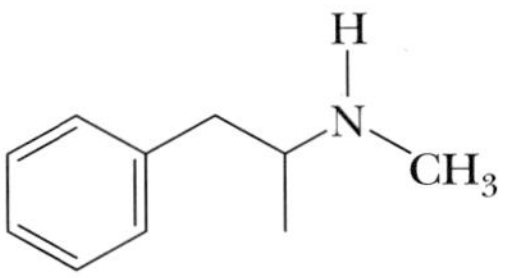

Methamphetamine (racemic)

The major central nervous system effects of amphetamine and amphetamine-like drugs are locomotor stimulation, euphoria and excitement, stereotyped behavior, and anorexia. Show how each drug can be synthesized by reductive amination of an appropriate aldehyde or ketone and amine. For the structural formulas of several more anorexics, see Problems 23.62 and 23.63.

16.40 Following is the final step in the synthesis of the antiviral drug Rimantidine (Problem 7.19).

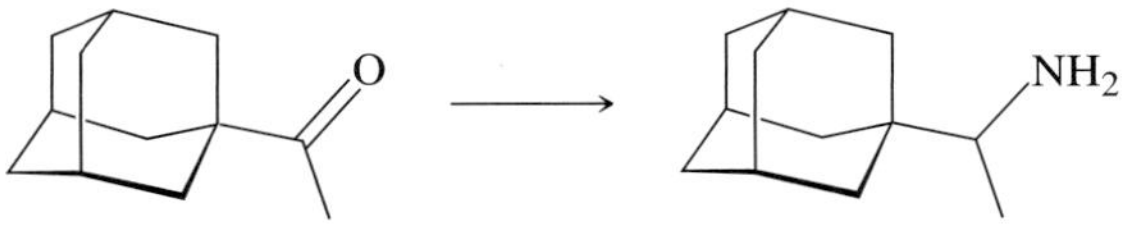

(a) Describe experimental conditions to bring this conversion.
(b) Is Rimantidine chiral? How many stereoisomers are possible for it?

Keto-Enol Tautomerism

16.41 The following molecule belongs to a class of compounds called enediols; each carbon of the double bond carries an —OH group. Draw structural formulas for the α-hydroxyketone and the α-hydroxyaldehyde with which this enediol is in equilibrium.

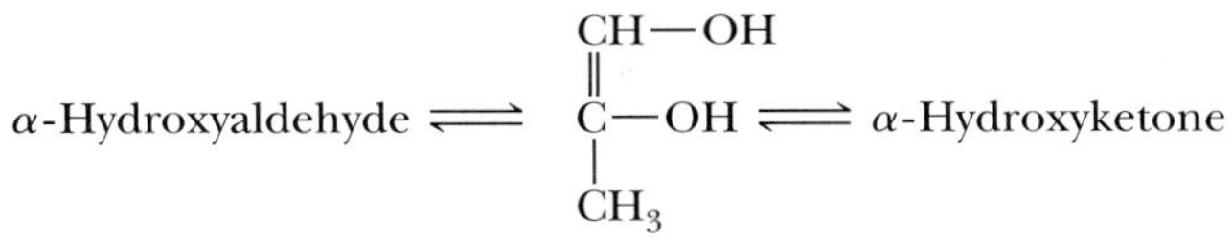

$$\alpha\text{-Hydroxyaldehyde} \rightleftharpoons \begin{matrix} CH-OH \\ \| \\ C-OH \\ | \\ CH_3 \end{matrix} \rightleftharpoons \alpha\text{-Hydroxyketone}$$

An enediol

16.42 When *cis*-2-decalone is dissolved in ether containing a trace of HCl, an equilibrium is established with *trans*-2-decalone. The latter ketone predominates in the equilibrium mixture.

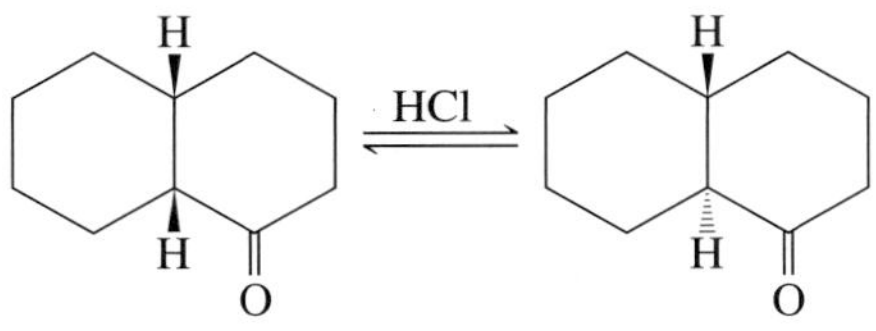

cis-2-Decalone   *trans*-2-Decalone

Propose a mechanism for this isomerization and account for the fact that the *trans* isomer predominates at equilibrium.

Oxidation/Reduction of Aldehydes and Ketones

16.43 Draw a structural formula for the product formed by treating butanal with each reagent.

(a) $LiAlH_4$ followed by H_2O (b) $NaBH_4$ in CH_3OH/H_2O
(c) H_2/Pt (d) $Ag(NH_3)_2^+$ in NH_3/H_2O
(e) H_2CrO_4, heat   (f) $HOCH_2CH_2OH$, HCl
(g) $Zn(Hg)/HCl$ (h) N_2H_4, KOH at 250°C
(i) $C_6H_5NH_2$ (j) $C_6H_5NHNH_2$

16.44 Draw a structural formula for the product of the reaction of acetophenone with each reagent given in Problem 16.43.

Reactions at an α-Carbon

16.45 The following bicyclic ketone has two α-carbons and three α-hydrogens. When this molecule is treated with D_2O in the presence of an acid catalyst, only two of the three α-hydrogens exchange with deuterium. The α-hydrogen at the bridgehead does not exchange.

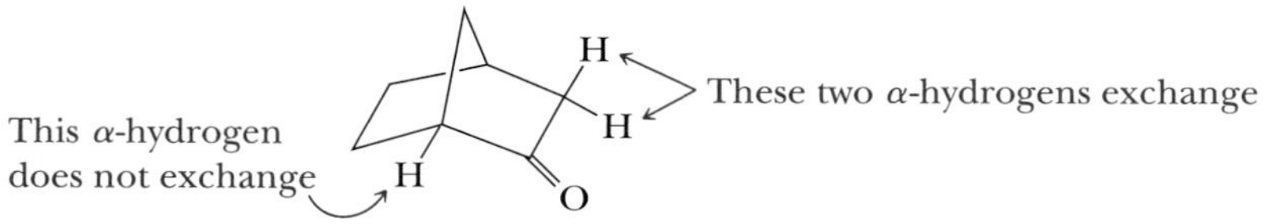

How do you account for the fact that two α-hydrogens do exchange but the third does not? You will find it helpful to build models of the enols by which exchange of α-hydrogens occurs.

16.46 Propose a mechanism for this reaction.

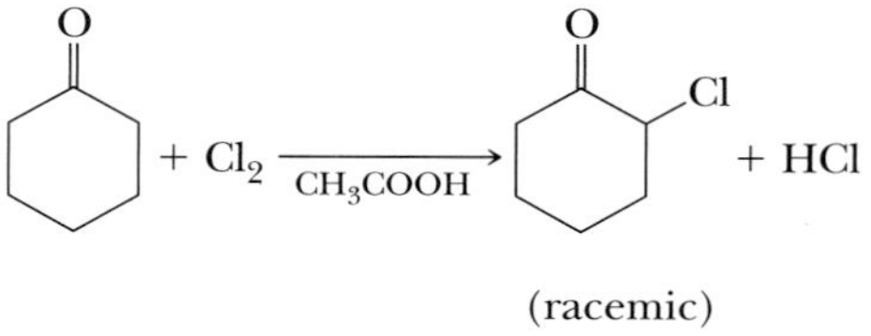

(racemic)

16.47 The base-promoted rearrangement of an α-haloketone to a carboxylic acid, known as the Favorskii rearrangement, is illustrated by the conversion of 2-chlorocyclohexanone to cyclopentanecarboxylic acid.

A proposed
intermediate

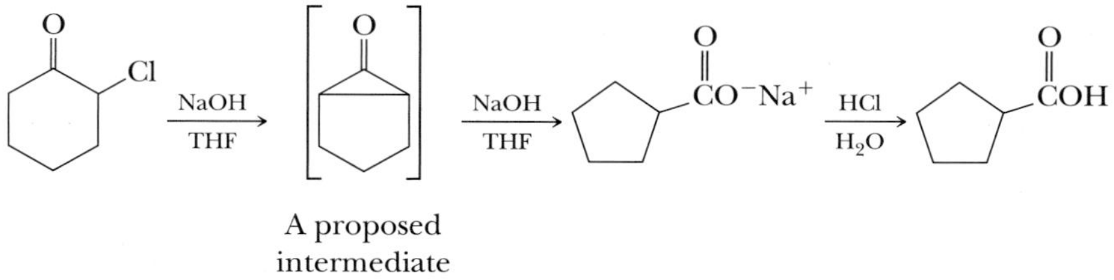

It is proposed that NaOH first converts the α-haloketone to the substituted cyclo-propanone shown in brackets, and then to the sodium salt of cyclopentanecarboxylic acid.

(a) Propose a mechanism for base-promoted conversion of 2-chlorocyclohexanone to the proposed intermediate.
(b) Propose a mechanism for base-promoted conversion of the proposed intermediate to sodium cyclopentanecarboxylate.

16.48 If the Favorskii rearrangement of 2-chlorocyclohexanone is carried out using sodium ethoxide in ethanol, the product is ethyl cyclopentanecarboxylate.

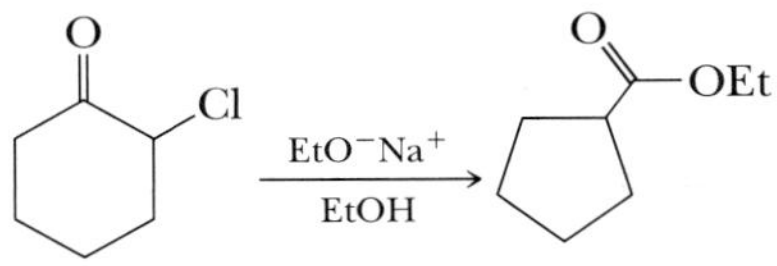

Propose a mechanism for this reaction.

16.49 (*R*)-Pulegone, readily available from pennyroyal oil, is an important enantiopure building block for organic syntheses. Propose a mechanism for each step in this transformation of pulegone.

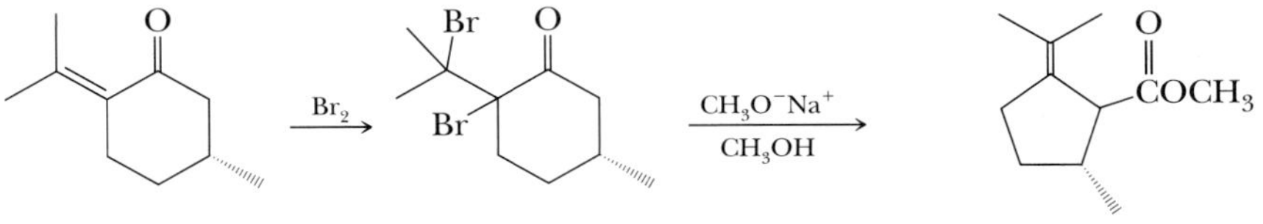

(*R*)-(+)-Pulegone

16.50 (*R*)-Pulegone is converted to (*R*)-citronellic acid by addition of HCl followed by treatment with NaOH.

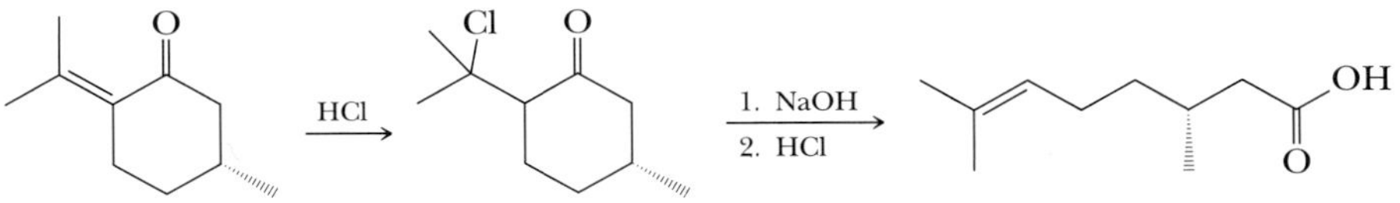

(*R*)-Pulegone   (*R*)-Citronellic acid

Propose a mechanism for each step in this transformation, and account for the regioselectivity of HCl addition.

Synthesis

16.51 Starting with cyclohexanone, show how to prepare these compounds. In addition to the given starting material, use any other organic or inorganic reagents as necessary.

(a) Cyclohexanol
(b) Cyclohexene
(c) *cis*-1,2-Cyclohexanediol
(d) 1-Methylcyclohexanol
(e) 1-Methylcyclohexene
(f) 1-Phenylcyclohexanol
(g) 1-Phenylcyclohexene
(h) Cyclohexene oxide
(i) *trans*-1,2-Cyclohexanediol

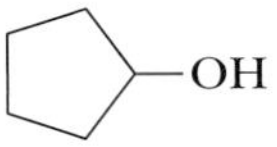

16.52 Show how to convert cyclopentanone to these compounds. In addition to cyclopentanone, use any other organic or inorganic reagents as necessary.

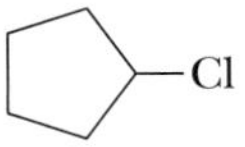

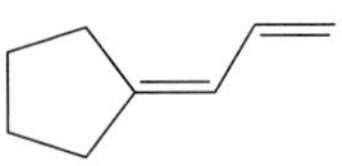

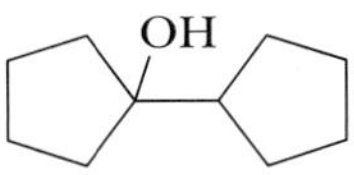

16.53 Disparlure is a sex attractant of the gypsy moth (*Porthetria dispar*). It has been synthesized in the laboratory from the following *Z* alkene.

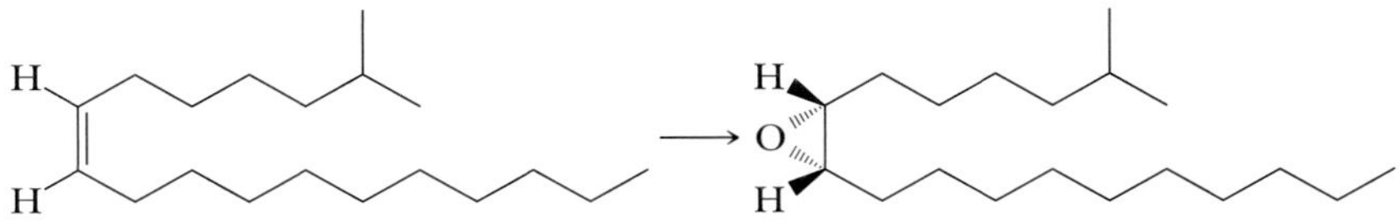

(*Z*)-2-Methyl-7-octadecene   Disparlure

(a) Propose two sets of reagents that might be combined in a Wittig reaction to give the indicated *Z* alkene.

(b) How might the *Z* alkene be converted to disparlure?

(c) How many stereoisomers are possible for disparlure? How many stereoisomers are formed in the sequence you chose?

16.54 Propose structural formulas for compounds A, B, and C in the following conversion. Show also how to prepare compound C by a Wittig reaction.

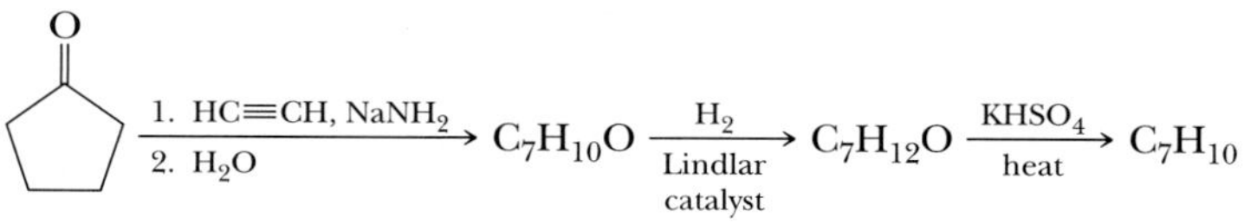

        (A)   (B)   (C)

16.55 Following is a retrosynthetic scheme for the synthesis of *cis*-3-penten-2-ol.

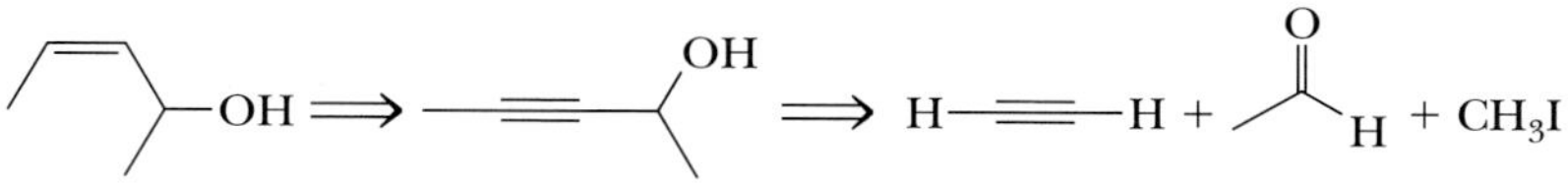

   (racemic)   (racemic)   Acetylene   Acetal-   Iodo-
                                           dehyde methane

Write a synthesis for this compound from acetylene, acetaldehyde, and iodomethane.

16.56 Following is the structural formula of Surfynol, a defoaming surfactant. Describe the synthesis of this compound from acetylene and a ketone. How many stereoisomers are possible for Surfynal?

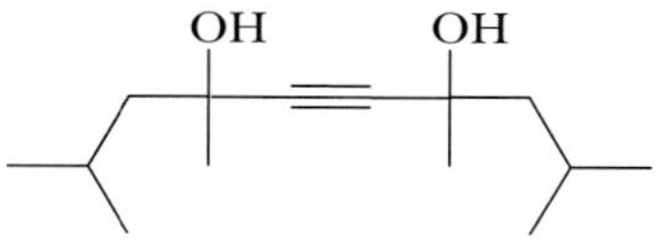

                       Surfynol

16.57 Propose a mechanism for this isomerization.

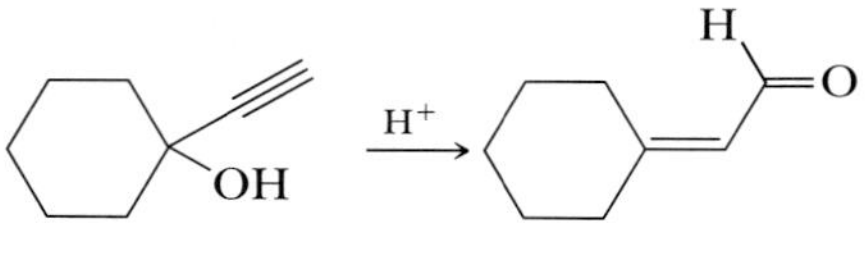

16.58 Propose a mechanism for this isomerization.

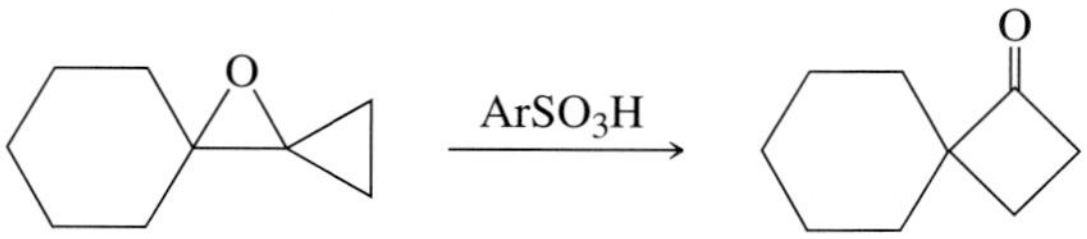

16.59 Starting with acetylene and 1-bromobutane as the only sources of carbon atoms, show how to synthesize the following.

(a) meso-5,6-Decanediol   (b) racemic 5,6-Decanediol
(c) 5-Decanone   (d) 5,6-Epoxydecane
(e) 5-Decanol   (f) Decane
(g) 6-Methyl-5-decanol   (h) 6-Methyl-5-decanone

16.60 Following are the final steps in one industrial synthesis of vitamin A acetate.

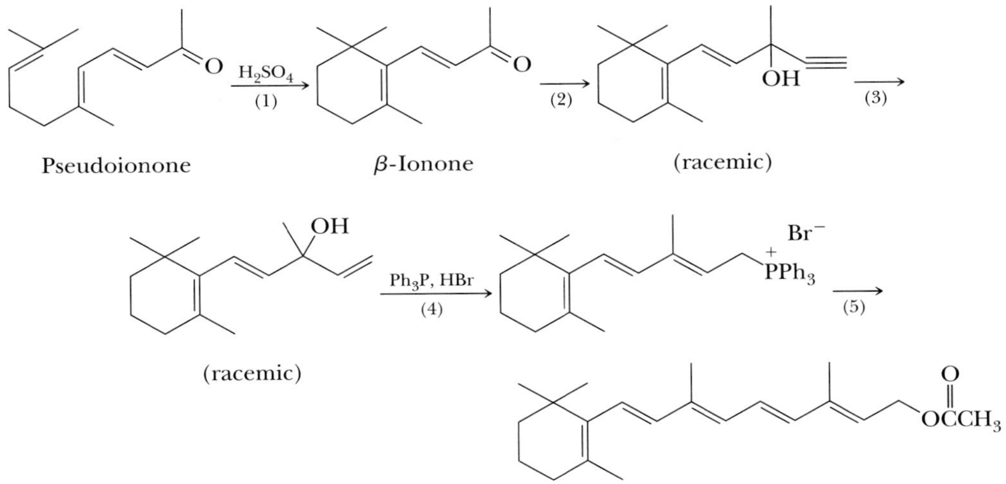

Pseudoionone   β-Ionone   (racemic)

(racemic)

Vitamin A acetate

(a) Propose a mechanism for the acid-catalyzed cyclization in Step 1.
(b) Propose reagents to bring about Step 2.
(c) Propose reagents to bring about Step 3.
(b) Propose a mechanism for formation of the phosphonium salt in Step 4.
(d) Show how Step 5 can be completed by a Wittig reaction.

16.61 Following is the structural formula of the principal sex pheromone of the Douglas fir tussock moth (*Orgyia pseudotsugata*), a severe defoliant of the fir trees of western North America.

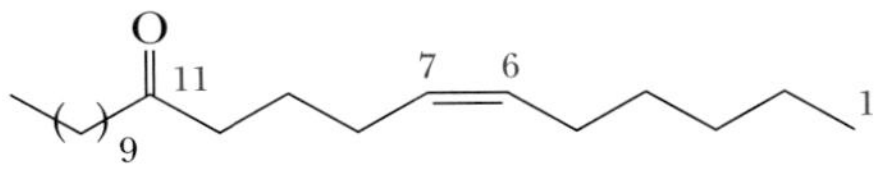

(Z)-6-Heneicosene-11-one

Several syntheses of this compound have been reported, starting materials for three of which are given here,

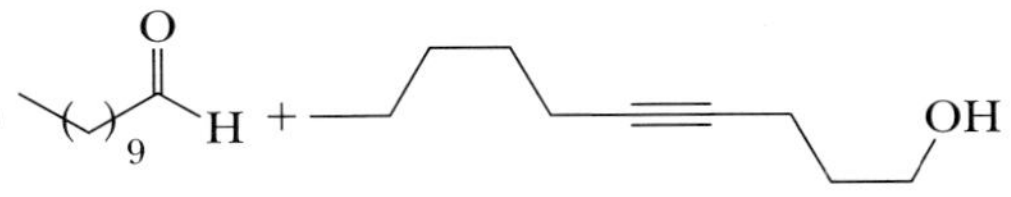

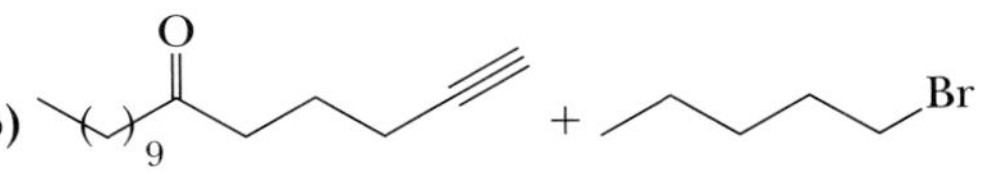

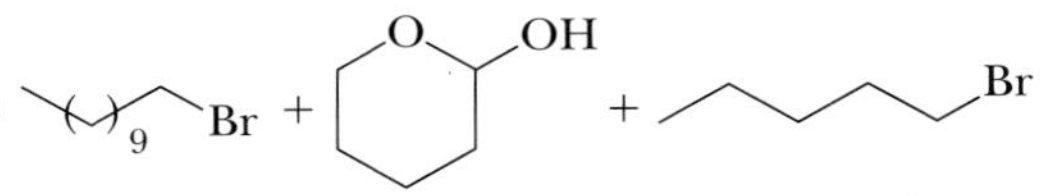

Show a series of steps by which each set of starting materials could be converted into the target molecule.

16.62 Both (*S*)-citronellal and isopulegol are naturally occurring terpenes (Section 5.4). When (*S*)-citronellal is treated with tin(IV) chloride (a Lewis acid) followed by neutralization with aqueous ammonium chloride, isopulegol is obtained in 85% yield.

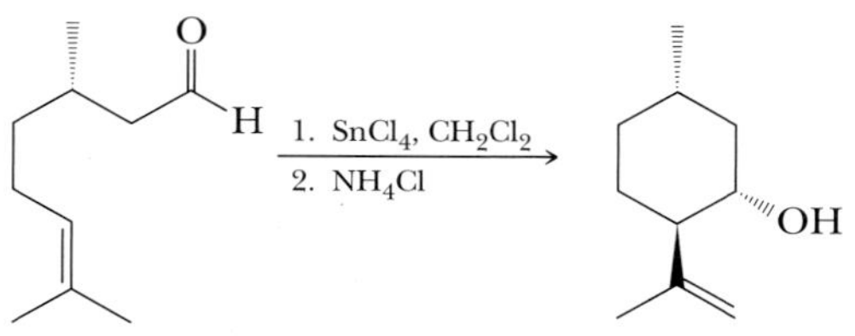

(S)-Citronellal
$(C_{10}H_{18}O)$

Isopulegol
$(C_{10}H_{18}O)$

(a) Show that both compounds are terpenes.
(b) Propose a mechanism for the conversion of (S)-citronellal to isopulegol.
(c) How many stereocenters are present in isopulegol? How many stereoisomers are possible for a molecule with this number of stereocenters?
(d) Isopulegol is formed as a single stereoisomer. Account for the fact that only a single stereoisomer is formed.

16.63 At some point during the synthesis of a target molecule, it may be necessary to protect an —OH group (that is, to prevent its reacting). In addition to the trimethylsilyl, *tert*-butyl-dimethylsilyl, and other trialkylsilyl groups described in Section 11.6, and the tetrahydropyranyl group described in Section 16.7D, the ethoxyethyl group may also be used as a protecting group.

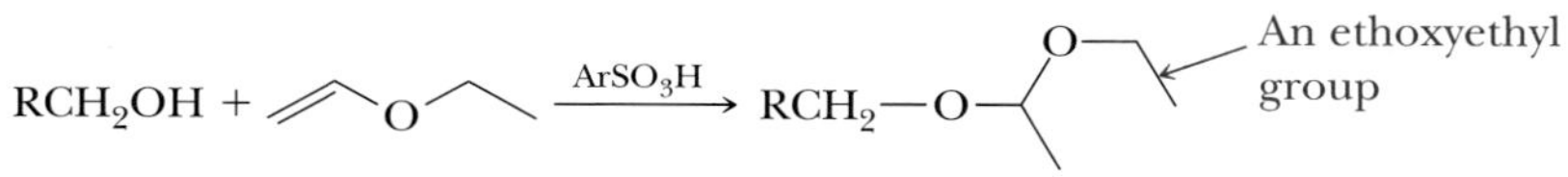

Ethyl vinyl ether

(a) Propose a mechanism for the acid-catalyzed formation of the ethoxyethyl protecting group.
(b) Suggest an experimental procedure whereby this protecting group can be removed to regenerate the unprotected alcohol.

16.64 Both 1,2-diols and 1,3-diols can be protected by treatment with 2-methoxypropene according to the following reaction.

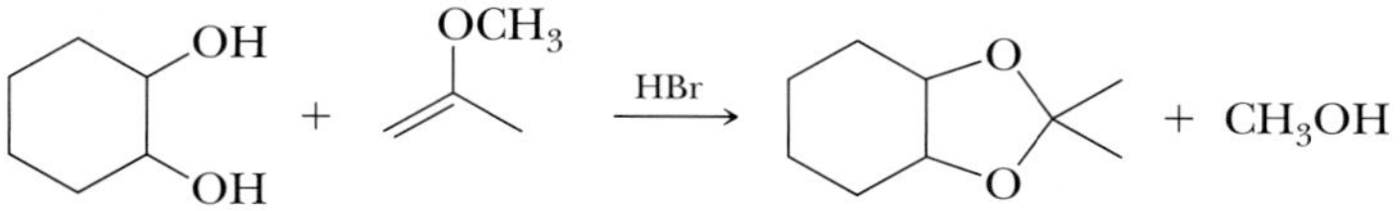

2-Methoxypropene   A protected 1,2-diol

(a) Propose a mechanism for the formation of this protected diol.
(b) Suggest an experimental procedure where this protecting group can be removed to regenerate the unprotected diol.

Looking Ahead

16.65 All rearrangements we have discussed so far have involved generation of an electron-deficient carbon followed by a 1,2-shift of an atom or group of atoms from an adjacent atom to the electron-deficient carbon. Rearrangements by a 1,2-shift can also occur following the generation of an electron-deficient oxygen. Propose a mechanism for the acid-catalyzed rearrangement of cumene hydroperoxide to phenol and acetone.

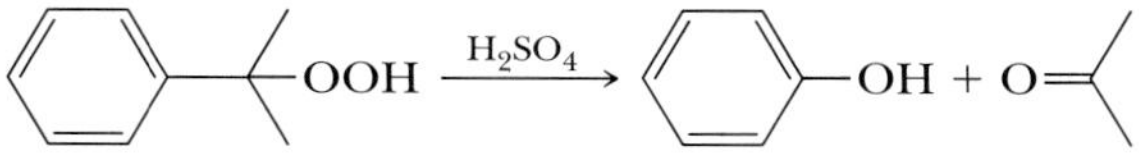

Cumene
hydroperoxide

Phenol   Acetone

16.66 In dilute aqueous base, (R)-glyceraldehyde is converted into an equilibrium mixture of (R,S)-glyceraldehyde and dihydroxyacetone. Propose a mechanism for this isomerization.

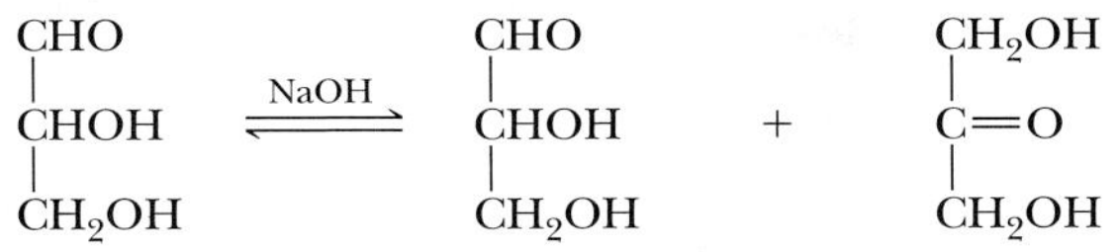

(R)-Glyceraldehyde   (R,S)-Glyceraldehyde   Dihydroxyacetone

16.67 Treatment of β-D-glucose with methanol in the presence of an acid catalyst converts it into a mixture of two compounds called methyl glucosides (Section 25.3A).

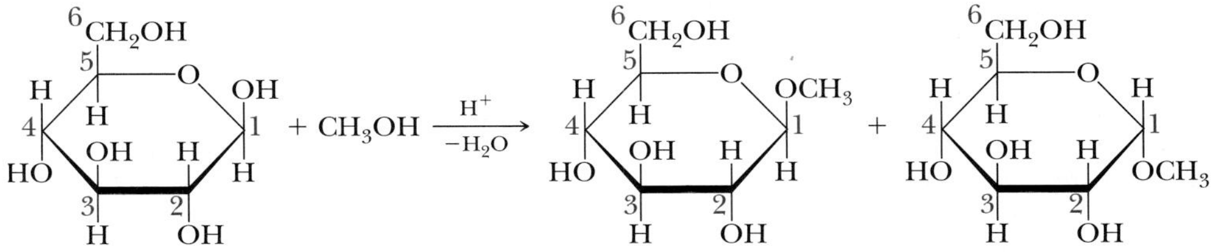

β-D-Glucose   Methyl β-D-glucoside   Methyl α-D-glucoside

In these representations, the six-membered rings are drawn as planar hexagons.

(a) Propose a mechanism for this conversion, and account for the fact that only the —OH on carbon 1 is transformed into an —OCH₃ group.
(b) Draw the more stable chair conformation for each product.
(c) Which of the two products has the chair conformation of greater stability? Explain.

16.68 Treating a Grignard reagent with carbon dioxide followed by aqueous HCl gives a carboxylic acid.

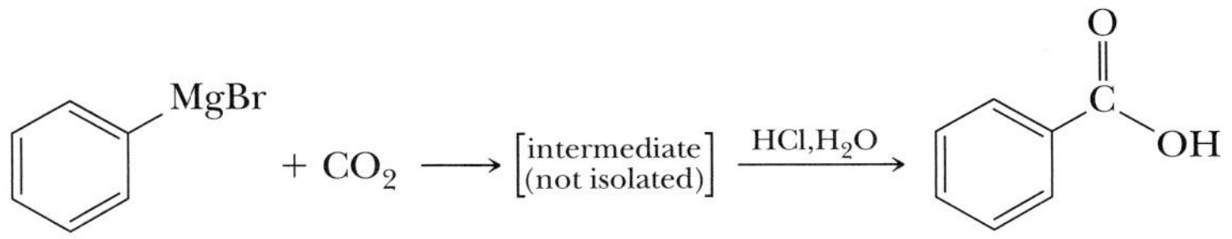

Propose a structural formula for the bracketed intermediate and a mechanism for its formation.

16.69 As we saw in Chapter 6, carbon-carbon double bonds are attacked by electrophiles but not by nucleophiles. An exception to this generalization is the reactivity of α,β-unsaturated aldehydes and ketones toward nucleophiles. Even though an isolated carbon-carbon double bond does not react with 2° amines such as dimethylamine, 3-buten-2-one reacts readily by regioselective addition.

$$Et_2NH + CH_2{=}CH{-}CH_2{-}CH_3 \longrightarrow \text{ no reaction}$$

Diethylamine   3-Buten-2-one
(Methyl vinyl ketone)

Account for the addition of nucleophiles to the carbon-carbon double bond of an α,β-unsaturated aldehyde or ketone and the regioselectivity of the addition.

16.70 Ribose, a carbohydrate with the formula shown, forms a cyclic hemiacetal, which, in principle, could contain either a four-membered, a five-membered, or a six-membered ring. When D-ribose is treated with methanol in the presence of an acid catalyst, two cyclic acetals, A and B, are formed, both with molecular formula $C_6H_{12}O_5$. These are separated, and each is treated with sodium periodate (Section 10.8C) followed by dilute aqueous acid. Both A and B yield the same three products in the same ratios.

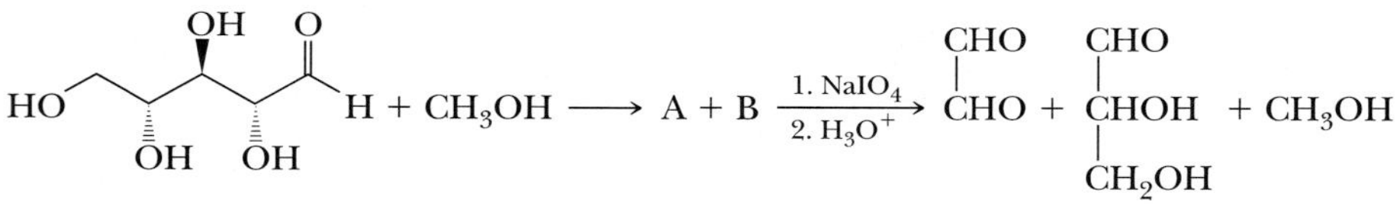

D-Ribose
$(C_5H_{10}O_5)$

Isomeric cydic
acetals with molecular
formula $C_6H_{12}O_5$

From this information, deduce whether the cyclic hemiacetal formed by D-ribose is four membered, five membered, or six membered.

16.71 The favorite nuclide used in positron emission tomography (PET scan) to follow glucose metabolism is fluorine-18, which decays by positron emission to oxygen-18 and has a half-life of 110 minutes. Fluorine-18 is administered in the form of fludeoxyglucose F-18; the product of this molecule's decay is glucose.

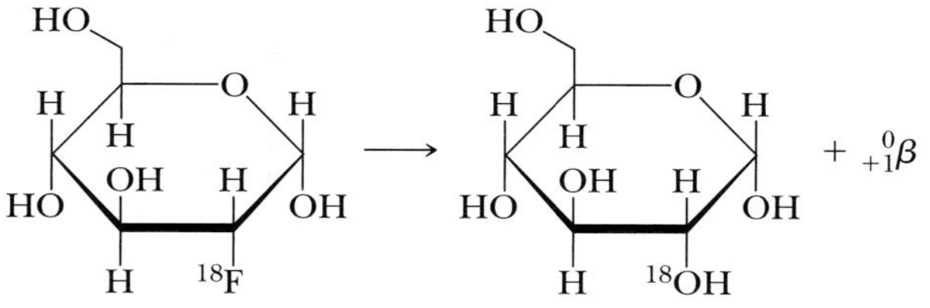

Fludeoxyglucose F-18

D-Glucose   A positron

Draw the alternative chair conformations for fludeoxyglucose F-18, and select the more stable of the two.

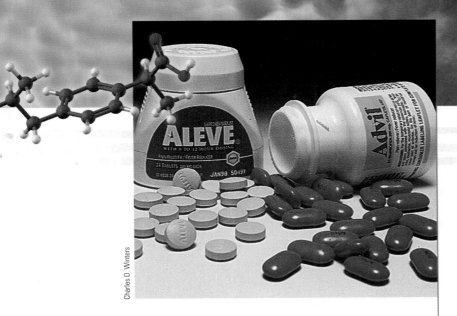

Charles D. Winters

17

Outline

Carboxylic Acids

The most important chemical property of carboxylic acids, another class of organic compounds containing the carbonyl group, is their acidity. Furthermore, carboxylic acids form numerous important derivatives, including esters, amides, anhydrides, and acid halides. In this chapter, we study carboxylic acids themselves; in Chapter 18, we study their derivatives.

17.1 Structure

The functional group of a carboxylic acid is the **carboxyl group** (Section 1.3D), so named because it is made up of a *carb*onyl group and a hydr*oxyl* group. Following is a Lewis structure of the carboxyl group as well as three alternative representations for it.

$$-C\overset{\displaystyle O}{\underset{\displaystyle O-H}{}} \qquad \overset{\displaystyle O}{\underset{\displaystyle O-H}{}} \qquad -COOH \qquad -CO_2H$$

Alternative representations of a carboxyl group

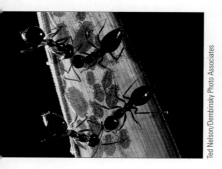

Ted Nelson/Dembinsky Photo Associates

■ Formic acid was first
obtained in 1670 from the
destructive distillation of ants,
whose Latin genus is *Formica*.
It is one of the components of
the venom injected by sting-
ing ants.

The general formula for an aliphatic carboxylic acid is RCOOH; the general formula
for an aromatic carboxylic acid is ArCOOH.

17.2 Nomenclature

A. IUPAC System

The IUPAC name of a carboxylic acid is derived from that of the longest carbon
chain that contains the carboxyl group by dropping the final -*e* from the name of the
parent alkane and adding the suffix -*oic* followed by the word "acid" (Section 2.5).
The chain is numbered beginning with the carbon of the carboxyl group. Because
the carboxyl carbon is understood to be carbon 1, there is no need to give it a number. The IUPAC system retains the common names formic acid and acetic acid, which
are always used to refer to these acids.

HCOOH   CH₃COOH

Methanoic acid   Ethanoic acid   3-Methylbutanoic acid
(Formic acid)   (Acetic acid)   (Isovaleric acid)

If the carboxylic acid contains a carbon-carbon double or triple bond, change the infix from -*an*- to -*en*- or -*yn*- as the case may be to indicate the presence of the multiple
bond and show the location of the multiple bond by a number.

Propenoic acid   *trans*-2-Butenoic acid   *trans*-3-Phenylpropenoic acid
(Acrylic acid)   (Crotonic acid)   (Cinnamic acid)

In the IUPAC system, a carboxyl group takes precedence over most other functional groups (Table 16.1), including hydroxyl groups, amino groups, and the carbonyl groups of aldehydes and ketones. As illustrated in the following examples, an
—OH group is indicated by the prefix *hydroxy*-; an —NH₂ group, by *amino*-; and the
C=O group of an aldehyde or ketone, by *oxo*-.

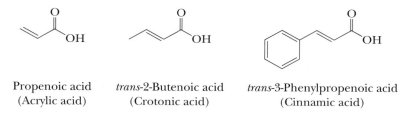

(*R*)-5-Hydroxyhexanoic acid   5-Oxohexanoic acid   4-Aminobutanoic acid

Dicarboxylic acids are named by adding the suffix -*dioic acid* to the name of the
carbon chain that contains both carboxyl groups. The numbers of the carboxyl carbons are not indicated because they can be only at the ends of the parent chain. Following are IUPAC and common names for several important aliphatic dicarboxylic
acids.

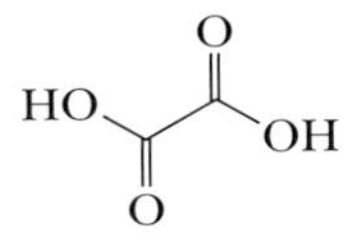

Ethanedioic acid
(Oxalic acid)

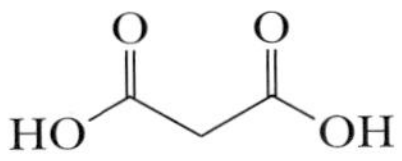

Propanedioic acid
(Malonic acid)

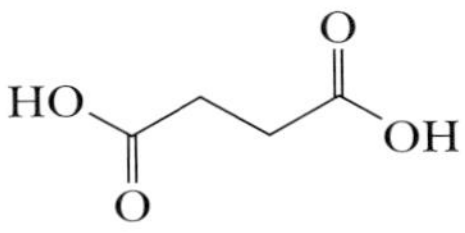

Butanedioic acid
(Succinic acid)

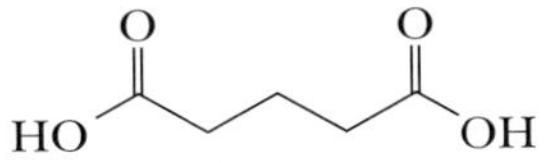

Pentanedioic acid
(Glutaric acid)

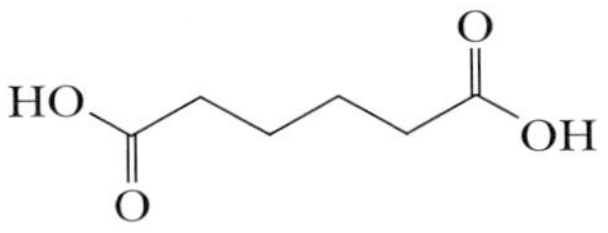

Hexanedioic acid
(Adipic acid)

The name *oxalic acid* is derived from one of its sources in the biological world, namely, plants of the genus *Oxalis*, one of which is rhubarb. Adipic acid is one of the two monomers required for the synthesis of the polymer nylon 66 (Section 29.5A). A mnemonic phrase for remembering the common names for the dicarboxylic acids oxalic through adipic is *Oh my, such good apples.*

A carboxylic acid containing a carboxyl group bonded to a cycloalkane ring is named by giving the name of the ring and adding the suffix -*carboxylic acid.* The atoms of the ring are numbered beginning with the carbon bearing the —COOH group.

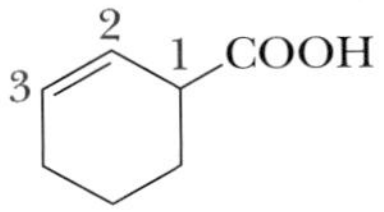

2-Cyclohexenecarboxylic
acid

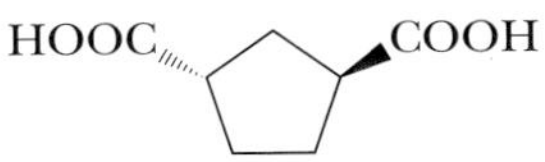

trans-1,3-Cyclopentanedicarboxylic
acid

The simplest aromatic carboxylic acid is benzoic acid. Derivatives are named by using numbers to show the location of substituents relative to the carboxyl group. Certain aromatic carboxylic acids have common names by which they are more usually known. For example, 2-hydroxybenzoic acid is more often called salicylic acid, a name derived from the fact that this aromatic carboxylic acid was first isolated from the bark of the willow, a tree of the genus *Salix.*

Benzoic
acid

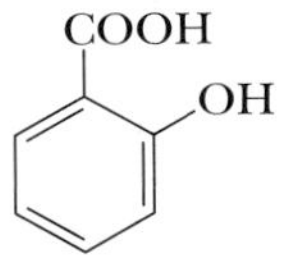

2-Hydroxybenzoic
acid
(Salicylic acid)

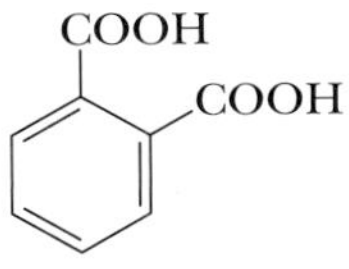

1,2-Benzenedicarboxylic
acid
(Phthalic acid)

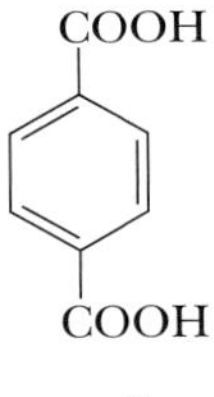

1,4-Benzenedicarboxylic
acid
(Terephthalic acid)

Aromatic dicarboxylic acids are named by adding the words *dicarboxylic acid* to "benzene," for example, 1,2-benzenedicarboxylic acid and 1,4-benzenedicarboxylic acid. Each is more usually known by its common name: phthalic acid and terephthalic acid, respectively. Terephthalic acid is one of the two organic components required for the synthesis of the textile fiber known as Dacron polyester, or Dacron (Section 29.5B).

Table 17.1 Several Aliphatic Carboxylic Acids—Their Common Names and Derivations

Structure	IUPAC Name	Common Name	Derivation
HCOOH	Methanoic acid	Formic acid	Latin: *formica*, ant
CH_3COOH	Ethanoic acid	Acetic acid	Latin: *acetum*, vinegar
CH_3CH_2COOH	Propanoic acid	Propionic acid	Greek: *propion*, first fat
$CH_3(CH_2)_2COOH$	Butanoic acid	Butyric acid	Latin: *butyrum*, butter
$CH_3(CH_2)_3COOH$	Pentanoic acid	Valeric acid	Latin: *valeriana*, a flowering plant
$CH_3(CH_2)_4COOH$	Hexanoic acid	Caproic acid	Latin: *caper*, goat
$CH_3(CH_2)_6COOH$	Octanoic acid	Caprylic acid	Latin: *caper*, goat
$CH_3(CH_2)_8COOH$	Decanoic acid	Capric acid	Latin: *caper*, goat
$CH_3(CH_2)_{10}COOH$	Dodecanoic acid	Lauric acid	Latin: *laurus*, laurel
$CH_3(CH_2)_{12}COOH$	Tetradecanoic acid	Myristic acid	Greek: *myristikos*, fragrant
$CH_3(CH_2)_{14}COOH$	Hexadecanoic acid	Palmitic acid	Latin: *palma*, palm tree
$CH_3(CH_2)_{16}COOH$	Octadecanoic acid	Stearic acid	Greek: *stear*, solid fat
$CH_3(CH_2)_{18}COOH$	Eicosanoic acid	Arachidic acid	Greek: *arachis*, peanut

B. Common Names

Aliphatic carboxylic acids, many of which were known long before the development of structural theory and IUPAC nomenclature, are named according to their source or for some characteristic property. Table 17.1 lists several of the unbranched aliphatic carboxylic acids found in the biological world along with the common name and Latin or Greek derivation of each. Those of 16, 18, and 20 carbon atoms are particularly abundant in fats and oils (Section 26.1) and in the phospholipid components of biological membranes (Section 26.5).

When common names are used, the Greek letters α, β, γ, δ, and so forth are often added as a prefix to locate substituents. The α-position in a carboxylic acid is the one next to the carboxyl group; an α-substituent in a common name is equivalent to a 2-substituent in an IUPAC name. GABA is an inhibitory neurotransmitter in the central nervous system of humans. Alanine is one of the 20 protein-derived amino acids.

4-Aminobutanoic acid
(γ-Aminobutyric acid, GABA)

(S)-2-Aminopropanoic acid
[(S)-α-Aminopropionic acid;
L-alanine]

In common names, the presence of a ketone carbonyl in a substituted carboxylic acid is indicated by the prefix *keto-*, illustrated by the common name β-ketobutyric acid. This substituted carboxylic acid is also named acetoacetic acid. In deriving this name, 3-oxobutanoic acid is regarded as a substituted acetic acid. In the common nomenclature, the substituent is named an **aceto group,** $CH_3CO—$.

Aceto group A $CH_3CO—$ group; also called an acetyl group.

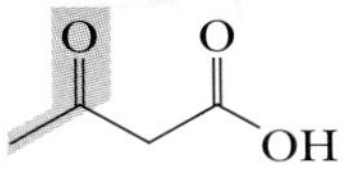

3-Oxobutanoic acid
(β-Ketobutyric acid;
acetoacetic acid)

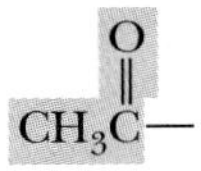

Acetyl group
(An aceto group)

Example 17.1

Write the IUPAC name for each carboxylic acid.

(a)

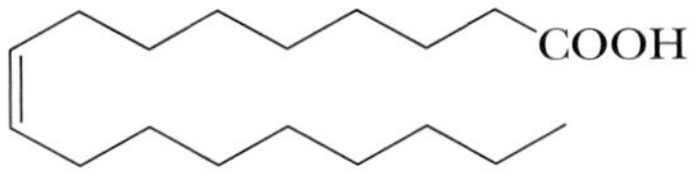

(b)

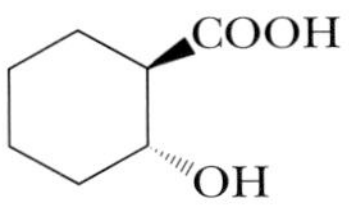

(racemic)

(c)

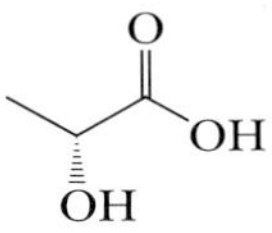

(d)

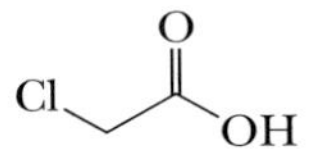

Solution

(a) (Z)-9-Octadecenoic acid (oleic acid)
(b) *trans*-2-Hydroxycyclohexanecarboxylic acid
(c) (R)-2-Hydroxypropanoic acid [(R)-lactic acid]
(d) Chloroacetic acid

Problem 17.1

Each of these carboxylic acids has a well-recognized common name. A derivative of glyceric acid is an intermediate in glycolysis. Maleic acid is an intermediate in the tricarboxylic acid (TCA) cycle. Mevalonic acid is an intermediate in the biosynthesis of steroids. Write the IUPAC name for each compound. Be certain to specify configuration.

(a) 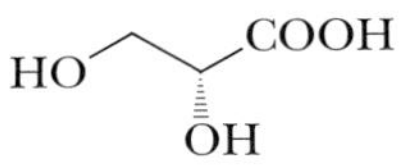 Glyceric acid

(b) 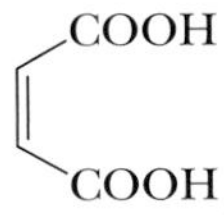 Maleic acid

(c) 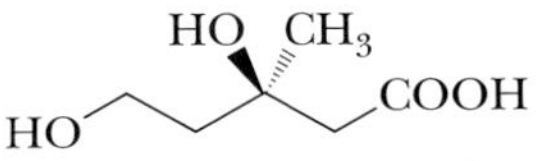 Mevalonic acid

---

17.3 Physical Properties

In the liquid and solid states, carboxylic acids are associated by hydrogen bonding into dimers, as shown here for acetic acid in the liquid state.

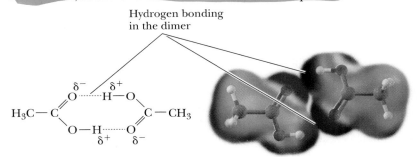

Hydrogen bonding
in the dimer

$$H_3C-C \overset{\delta^-}{\underset{O-H}{O}} \cdots \overset{\delta^+}{\underset{\delta^+}{H-O}} \cdots \overset{}{\underset{\delta^-}{O}} C-CH_3$$

Electrostatic potential map
of the acetic acid dimer

CHEMICAL CONNECTIONS

From Willow Bark to Aspirin and Beyond

The first drug developed for widespread use was aspirin, one of today's most common pain relievers. Americans alone consume approximately 80 billion tablets of aspirin a year! The story of the development of this modern pain reliever goes back more than 2000 years. In 400 BCE, the Greek physician Hippocrates recommended chewing bark of the willow tree to alleviate the pain of childbirth and to treat eye infections.

The active component of willow bark was found to be salicin, a compound composed of salicyl alcohol bonded to a unit of β-D-glucose (Section 25.1). Hydrolysis of salicin in aqueous acid gives salicyl alcohol, which can be oxidized to salicylic acid. Salicylic acid proved to be an even more effective reliever of pain, fever, and inflammation than salicin, without its extremely bitter taste. Unfortunately, patients quickly recognized salicylic acid's major side effect: It causes severe irritation of the mucous membrane lining of the stomach.

Salicin   Salicyl alcohol   Salicylic acid

In the search for less irritating but still effective derivatives of salicylic acid, chemists at the Bayer division of I. G. Farben in Germany in 1883 prepared acetylsalicylic acid and gave it the name aspirin.

Salicylic acid   Acetylsalicylic acid (Aspirin)

Aspirin proved to be less irritating to the stomach than salicylic acid and also more effective in relieving the pain and inflammation of rheumatoid arthritis. Aspirin, however, is still irritating to the stomach and frequent use can cause duodenal ulcers in susceptible persons.

In the 1960s, in a search for even more effective and less irritating analgesics and antiinflammatory drugs, chemists at the Boots Pure Drug Company in England synthesized a series of compounds related in structure to salicylic acid. Among them, they discovered an even more potent compound, which they named ibuprofen. Soon thereafter, Syntex Corporation in the United States developed naproxen, the active ingredient in Aleve. Each compound has one chiral center and can exist as a pair of enantiomers. For each drug, the physiologically active form is the *S* enantiomer.

(*S*)-Naproxen

(*S*)-Ibuprofen

In the 1960s, it was discovered that aspirin acts by inhibiting cyclooxygenase (COX), a key enzyme in the conversion of arachidonic acid to prostaglandins (Section 26.3). With this discovery, it became clear why only one enantiomer of ibuprofen and naproxen is active: only the *S* enantiomer of each has the correct handedness to bind to COX and inhibit its activity.

Recently it was recognized that there are actually two cyclooxygenases; one is more important for the inflammation pathway, and the other affects the stability of blood vessels. Aspirin and other nonsteroidal antiinflammatory drugs (NSAIDs) inhibit both, which is why they can cause gastrointestinal bleeding. New drugs (Celebrex, Vioxx) have been developed that inhibit only the inflammatory enzyme pathway and are remarkably effective for suppression of inflammation (for example, in arthritis) without the gastrointestinal side effects.

Table 17.2   Boiling Points and Solubilities in Water of Selected Carboxylic Acids, Alcohols, and Aldehydes of Comparable Molecular Weight

Structure	Name	Molecular Weight (g/mol)	Boiling Point (°C)	Solubility (g/100 g H_2O)
CH_3COOH	Acetic acid	60.1	118	Infinite
$CH_3CH_2CH_2OH$	1-Propanol	60.1	97	Infinite
CH_3CH_2CHO	Propanal	58.1	48	16.0
$CH_3(CH_2)_2COOH$	Butanoic acid	88.1	163	Infinite
$CH_3(CH_2)_3CH_2OH$	1-Pentanol	88.1	137	2.3
$CH_3(CH_2)_3CHO$	Pentanal	86.1	103	Slight
$CH_3(CH_2)_4COOH$	Hexanoic acid	116.2	205	1.0
$CH_3(CH_2)_5CH_2OH$	1-Heptanol	116.2	176	0.2
$CH_3(CH_2)_5CHO$	Heptanal	114.1	153	0.1

Carboxylic acids have significantly higher boiling points than other types of organic compounds of comparable molecular weight, such as alcohols, aldehydes, and ketones. For example, butanoic acid (Table 17.2) has a higher boiling point than either 1-pentanol or pentanal. The higher boiling points of carboxylic acids result from their polarity and from the fact that they form very strong intermolecular hydrogen bonds.

Carboxylic acids also interact with water molecules by hydrogen bonding through both the carbonyl and hydroxyl groups. Because of greater hydrogen-bonding interactions, carboxylic acids are more soluble in water than alcohols, ethers, aldehydes, and ketones of comparable molecular weight. The solubility of a carboxylic acid in water decreases as its molecular weight increases. We account for this trend in the following way. A carboxylic acid consists of two regions of distinctly different polarity: a polar **hydrophilic** carboxyl group and, except for formic acid, a nonpolar **hydrophobic** hydrocarbon chain. The hydrophilic carboxyl group increases water solubility; the hydrophobic hydrocarbon chain decreases water solubility.

Hydrophilic   From the Greek, meaning water loving.

Hydrophobic   From the Greek, meaning water fearing.

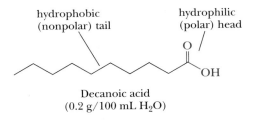

hydrophobic (nonpolar) tail

hydrophilic (polar) head

Decanoic acid (0.2 g/100 mL H_2O)

Electrostatic potential map

The first four aliphatic carboxylic acids (formic, acetic, propanoic, and butanoic acids) are infinitely soluble in water because the hydrophobic character of the hydrocarbon chain is more than counterbalanced by the hydrophilic character of the carboxyl group. As the size of the hydrocarbon chain increases relative to the size of the hydrophilic group, water solubility decreases. The solubility of hexanoic acid is 1.0 g/100 g H_2O, while that of decanoic acid is only 0.2 g/100 g H_2O.

One other physical property of carboxylic acids must be mentioned. The liquid carboxylic acids from propanoic acid to decanoic acid have extremely foul odors, about as bad as those of thiols, though different. Butanoic acid is found in stale perspiration, is a major component of "locker room odor," and provides the characteristic odor of regurgitated milk. Pentanoic acid smells even worse, and goats, which secrete C_6, C_8, and C_{10} acids, are not famous for their pleasant odors.

17.4 Acidity

A. Acid Ionization Constants

Carboxylic acids are weak acids. Values of K_a for most unsubstituted aliphatic and aromatic carboxylic acids fall within the range 10^{-4} to 10^{-5}. The value of K_a for acetic acid, for example, is 1.74×10^{-5}. Its pK_a is 4.76.

$$CH_3COOH + H_2O \rightleftharpoons CH_3COO^- + H_3O^+ \qquad K_a = \frac{[CH_3COO^-][H_3O^+]}{[CH_3COOH]} = 1.74 \times 10^{-5}$$

$$pK_a = 4.76$$

As we discussed in Sections 4.5C and 4.5D, the greater acidity of carboxylic acids (pK_a 4–5) compared with alcohols (pK_a 16–18) is because resonance stabilizes the carboxylate anion by delocalizing its negative charge, and because of the electron-withdrawing inductive effect of the carbonyl group. There is no comparable resonance or inductive stabilization of alkoxide ions.

We saw in Section 4.5D that substitution at the α-carbon of an atom or group of atoms of higher electronegativity than carbon further increases the acidity of carboxylic acids by the inductive effect. Compare, for example, the acidity of acetic acid (pK_a 4.76) and chloroacetic acid (pK_a 2.86). To see the effects of multiple halogen substitution, compare the values of pK_a for acetic acid with its mono-, di-, and trichloro derivatives. A single chlorine substituent increases acid strength by nearly 100. Trichloroacetic acid, the strongest of the three acids, is a stronger acid than H_3PO_4.

Formula:	CH_3COOH	$ClCH_2COOH$	$Cl_2CHCOOH$	Cl_3CCOOH
Name:	Acetic acid	Chloroacetic acid	Dichloroacetic acid	Trichloroacetic acid
pK_a:	4.76	2.86	1.48	0.70

Increasing acid strength →

We also see an example of the inductive effect in a comparison of the relative acidities of benzoic acid and acetic acid. Because of the stronger electron-withdrawing inductive effect of the sp^2 hybridized carbon of the benzene ring compared with the sp^3 hybridized carbon of the methyl group, benzoic acid is a stronger acid than acetic acid; its K_a is approximately four times that of acetic acid.

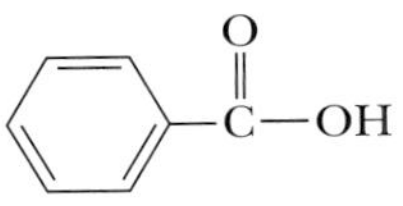

Benzoic acid
pK_a 4.19

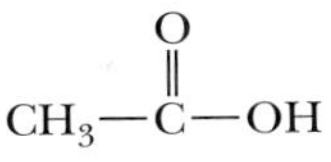

Acetic acid
pK_a 4.76

Example 17.2

Which is the stronger acid in each pair?

(a)

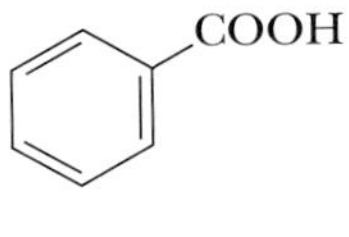

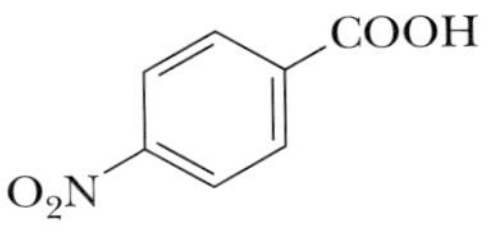

Benzoic acid   4-Nitrobenzoic acid

(b)

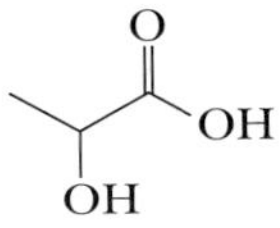

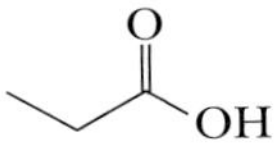

2-Hydroxypropanoic acid   Propanoic acid
(Lactic acid)

Solution

(a) 4-Nitrobenzoic acid (pK_a 3.42) is a considerably stronger acid than benzoic acid (pK_a 4.19) because of the electron-withdrawing inductive effect of the nitro group.

(b) 2-Hydroxypropanoic acid (pK_a 3.08) is a stronger acid than propanoic acid (pK_a 4.87) because of the electron-withdrawing inductive effect of the adjacent hydroxyl oxygen.

Problem 17.2

Which is the stronger acid in each pair?

(a) CH_3COOH or CH_3SO_3H

Acetic acid   Methanesulfonic
acid

(b)

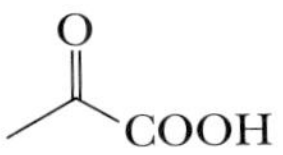

2-Oxopropanoic acid   Propanoic acid
(Pyruvic acid)

Here is one final point about carboxylic acids. When a carboxylic acid is dissolved in an aqueous solution, the form of the carboxylic acid present depends on the pH of the solution in which it is dissolved. Consider typical carboxylic acids, which have pK_a values in the range of 4.0 to 5.0. When the pH of the solution is equal to the pK_a of the carboxylic acid (that is, the pH of the solution is in the range 4.0 to 5.0), the acid and its anion (its conjugate base) are present in equal concentrations. If the pH of the solution is adjusted to 2.0 or lower by the addition of a strong acid, the carboxylic acid then is present in solution almost entirely as RCOOH. If, on the other hand, the pH of the solution is adjusted to 7.0 or higher, the carboxylic acid is present almost entirely as its anion. Thus, even in a neutral solution (pH 7.0), a carboxylic acid is present predominantly as its anion.

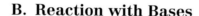

| predominant species when the pH of the solution is 2.0 or less | present in equal concentrations when the pH of the solution is equal to the pK_a of the acid | predominant species when the pH of the solution is 7.0 or greater |

A carboxylate anion has a negative charge, so in biological systems, molecules with substantial numbers of carboxylate anions have considerable negative charge. Because of the hydrophilic character of carboxylate anions, molecules with a large number of them are highly water-soluble.

Sodium benzoate and calcium propanoate are used as preservatives in baked goods.

B. Reaction with Bases

All carboxylic acids, whether soluble or insoluble in water, react with NaOH, KOH, and other strong bases to form water-soluble salts.

Benzoic acid (slightly soluble in water) + NaOH $\xrightarrow{H_2O}$ Sodium benzoate (60 g/100 mL water) + H_2O

Sodium benzoate, a fungal growth inhibitor, is often added to baked goods "to retard spoilage." Calcium propanoate is also used for the same purpose. Carboxylic acids also form water-soluble salts with ammonia and amines:

Benzoic acid (slightly soluble in water) + NH_3 $\xrightarrow{H_2O}$ Ammonium benzoate (20 g/100 mL water)

Carboxylic acids react with sodium bicarbonate and sodium carbonate to form water-soluble sodium salts and carbonic acid (a weaker acid). Carbonic acid, in turn, decomposes to give water and carbon dioxide, which evolves as a gas.

$$CH_3COOH + NaHCO_3 \longrightarrow CH_3COO^-Na^+ + H_2CO_3$$
$$H_2CO_3 \longrightarrow CO_2 + H_2O$$
$$\overline{CH_3COOH + NaHCO_3 \longrightarrow CH_3COO^-Na^+ + CO_2 + H_2O}$$

Salts of carboxylic acids are named in the same manner as the salts of inorganic acids; the cation is named first, and then the anion is named. The name of the anion is derived from the name of the carboxylic acid by dropping the suffix -ic acid and adding the suffix -ate.

Example 17.3

Complete each acid-base reaction and name the carboxylic salt formed.

(a) $\diagup\diagdown\diagup$ COOH + NaOH $\longrightarrow$

(b) $\underset{\text{OH}}{\diagup\diagdown}$ COOH + NaHCO$_3$ $\longrightarrow$

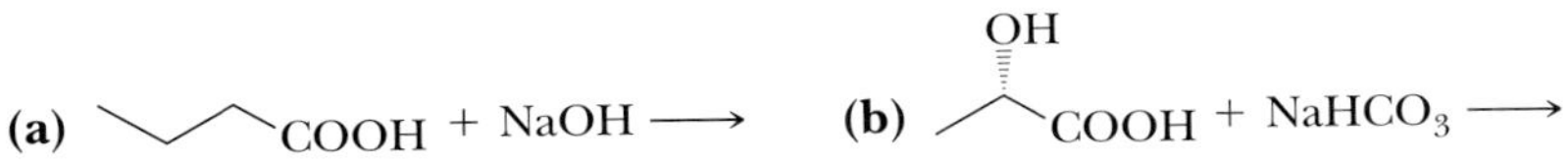

(S)-Lactic acid

Solution

Each carboxylic acid is converted to its sodium salt. In (b), carbonic acid is formed; it decomposes to carbon dioxide and water.

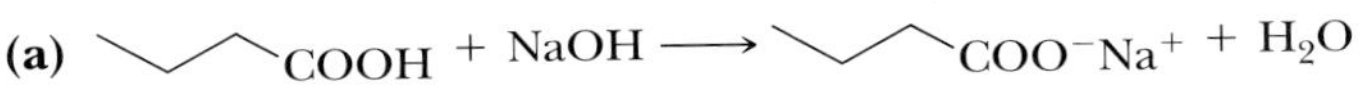

(a) $\diagup\diagdown\diagup$ COOH + NaOH $\longrightarrow$ $\diagup\diagdown\diagup$ COO$^-$Na$^+$ + H$_2$O

Butanoic acid   Sodium butanoate

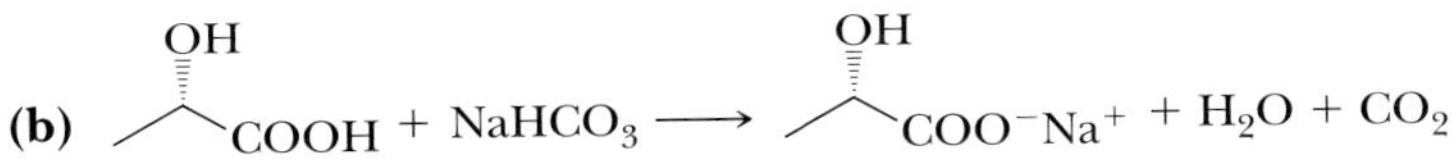

(b) $\underset{\text{OH}}{\diagup\diagdown}$ COOH + NaHCO$_3$ $\longrightarrow$ $\underset{\text{OH}}{\diagup\diagdown}$ COO$^-$Na$^+$ + H$_2$O + CO$_2$

(S)-Lactic acid   Sodium (S)-lactate

Problem 17.3

Write equations for the reaction of each acid in Example 17.3 with ammonia, and name the carboxylic salt formed.

A consequence of the water solubility of carboxylic acid salts is that water-insoluble carboxylic acids can be converted to water-soluble ammonium or alkali metal salts and extracted into aqueous solution. The salt, in turn, can be transformed back to the free carboxylic acid by addition of HCl, H$_2$SO$_4$, or other strong acid. These reactions allow an easy separation of carboxylic acids from water-insoluble nonacidic compounds.

Shown in Figure 17.1 is a flow chart for the separation of benzoic acid, a water-insoluble carboxylic acid, from benzyl alcohol, a nonacidic compound. First, the mixture of benzoic acid and benzyl alcohol is dissolved in diethyl ether. When the ether solution is shaken with aqueous NaOH or other strong base, benzoic acid is converted to its water-soluble sodium salt. Then the ether and aqueous phases are separated. The ether solution is distilled, yielding first diethyl ether (bp 35°C) and then benzyl alcohol (bp 205°C). The aqueous solution is acidified with HCl, and benzoic acid precipitates as a crystalline solid (mp 122°C) and is recovered by filtration.

Figure 17.1

Flow chart for separation of benzoic acid from benzyl alcohol.

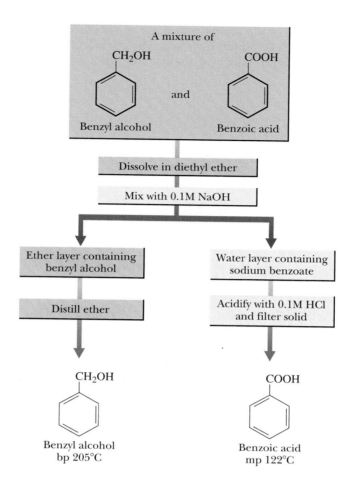

17.5 Preparation of Carboxylic Acids

We have already seen how carboxylic acids are prepared by oxidation of primary alcohols (Section 10.8) and aldehydes (Section 16.10A). We mention here two additional methods, a general one and an important industrial one.

A. Addition of Carbon Dioxide to a Grignard Reagent

Treating a Grignard reagent with carbon dioxide gives the magnesium salt of a carboxylic acid, which, on protonation with aqueous acid, gives a carboxylic acid.

Thus, carbonation of a Grignard reagent is a convenient way to convert an alkyl or aryl halide to a carboxylic acid.

B. Industrial Synthesis of Acetic Acid—Transition Metal Catalysis

Yearly production of acetic acid in the United States is approximately 10^7 kg, a volume that ranks it at the top of the list of organic chemicals manufactured by the U.S. chemical industry. The first industrial synthesis of acetic acid was commercialized in 1916 in Canada and Germany, using acetylene as a feedstock. The process involved two stages: (1) hydration of acetylene to acetaldehyde, followed by (2) oxidation of acetaldehyde to acetic acid by molecular oxygen, catalyzed by cobalt(III) acetate.

$$HC{\equiv}CH + H_2O \xrightarrow[\text{HgSO}_4]{\text{H}_2\text{SO}_4} \left[\overset{OH}{\underset{}{CH_2{=}CH}} \right] \longrightarrow CH_3\overset{O}{\overset{\|}{CH}} \xrightarrow[\text{Co}^{3+}]{\text{O}_2} CH_3\overset{O}{\overset{\|}{COH}}$$

Acetylene   Enol of   Acetaldehyde   Acetic acid
                           acetaldehyde

The technology of producing acetic acid from acetylene is simple, yields are high, and these factors made this procedure the major route to acetic acid for over 50 years. Acetylene was prepared by the reaction of calcium carbide with water. Calcium carbide, in turn, was prepared by heating calcium oxide (from limestone, $CaCO_3$) with coke (from coal) to between 2000 and 2500°C in an electric furnace.

$$CaO + 2C \xrightarrow{\text{2500°C}} CaC_2 \xrightarrow{\text{2H}_2\text{O}} HC{\equiv}CH + Ca(OH)_2$$

Calcium   Calcium   Acetylene
oxide   carbide

This preparation of calcium carbide requires enormous amounts of energy so, as the cost of energy rose, acetylene ceased to be an economical feedstock from which to manufacture acetic acid.

As an alternative feedstock, chemists and chemical engineers turned to ethylene, already available in huge quantities from the refining of natural gas and petroleum. The process of producing acetic acid from ethylene depends on the fact, known since 1894, that, in the presence of catalytic amounts of Pd^{2+} and Cu^{2+} salts, ethylene is oxidized by molecular oxygen to acetaldehyde.

$$2CH_2{=}CH_2 + O_2 \xrightarrow[\text{(Wacker process)}]{\text{Pd}^{2+}, \text{ Cu}^{2+}} 2CH_3\overset{O}{\overset{\|}{CH}}$$

The first chemical plant to use ethylene oxidation for the manufacture of acetaldehyde was built in Germany by Wacker-Chemie in 1959, and the process itself became known as the **Wacker process.**

In another approach to the synthesis of acetic acid, chemists turned to a route based on carbon monoxide, a readily available raw material. The carbonylation of methanol is exothermic. The challenge was to find a catalytic system that would bring about this reaction.

$$CH_3OH + \boxed{CO} \longrightarrow CH_3\overset{O}{\overset{\|}{COH}} \qquad \Delta H^0 = -138 \text{ kJ}(33 \text{ kcal})/\text{mol}$$

In 1973, the Monsanto Company in the United States developed a process for the carbonylation of methanol in the presence of small amounts of soluble rhodium(III) salts, HI, and H_2O. In this process, methanol and CO react continuously in the liquid phase at between 150 and 160°C under a pressure of 30 atm. In the following mechanism, Steps 1 and 2 represent the preparation of species for the catalytic cycle. Steps 3 and 4 are the catalytic cycle; in them, methanol and carbon monoxide are converted to acetic acid.

Mechanism *Rhodium-Catalyzed Carbonylation of Methanol*

Step 1: Reaction of methanol with HI gives methyl iodide.

$$CH_3OH + HI \longrightarrow CH_3I + H_2O$$

Step 2: Methyl iodide and the rhodium-carbonyl complex undergo oxidative addition to form a complex containing a methyl-rhodium bond.

$$CH_3I + Rh(CO)I_2 \longrightarrow [CH_3-Rh(CO)I_3]^-$$

A methyl-rhodium
carbonyl complex

Step 3: Carbon monoxide is inserted into the methyl-rhodium bond to give an acyl-rhodium bond.

$$[CH_3-Rh(CO)I_3]^- + CO \longrightarrow [CH_3\overset{\overset{\displaystyle O}{\|}}{C}-Rh(CO)I_3]^-$$

An acyl-rhodium
carbonyl complex

Step 4: Methanolysis of the acyl-rhodium bond gives acetic acid and regenerates the methyl-rhodium carbonyl complex.

$$[CH_3\overset{\overset{\displaystyle O}{\|}}{C}-Rh(CO)I_3]^- + CH_3OH \longrightarrow CH_3\overset{\overset{\displaystyle O}{\|}}{C}OH + [CH_3-Rh(CO)I_3]^-$$

The **Monsanto process** is based on methanol, which is readily available by catalytic reduction of carbon monoxide. Carbon monoxide and hydrogen, a mixture called **synthesis gas** (Section 2.10C), are in turn available from the reaction of water with methane, coal, and various petroleum products. Synthesis gas is now the major source of both methanol and acetic acid, and it is likely that it will be a major feedstock for the production of other organics in the decades ahead.

17.6 Reduction

The carboxyl group is one of the organic functional groups most resistant to reduction. It is not affected by catalytic hydrogenation under conditions that easily reduce aldehydes and ketones to alcohols and that reduce alkenes and alkynes to alkanes. The most common reagent for the reduction of carboxylic acids to primary alcohols is the very powerful reducing agent lithium aluminum hydride (Section 16.11A).

A. Lithium Aluminum Hydride

Lithium aluminum hydride, $LiAlH_4$ (LAH), reduces a carboxylic acid to a primary alcohol in excellent yield, although heating is required. LAH is usually dissolved in diethyl ether or tetrahydrofuran (THF). When carboxylic acids react with $LiAlH_4$, the initial product is a tetraalkoxy aluminate ion, which is then treated with water to give the primary alcohol and lithium and aluminum hydroxides. These hydroxides are insoluble in diethyl ether and THF and are removed by filtration. Evaporation of the solvent then yields the primary alcohol.

3-Cyclopentene-
carboxylic acid   4-Hydroxymethylcyclopentene

Alkenes are generally not affected by metal hydride reducing agents. These reagents function as hydride ion donors; that is, as nucleophiles, and alkenes are not attacked by nucleophiles.

In the reduction of a carboxyl group by lithium aluminum hydride, the first hydride ion reacts with the carboxyl hydrogen to give H_2. The resulting carboxylate anion reacts with a hydride ion at the carbonyl carbon atom, most likely with the assistance of Li^+ and Al complexes acting as Lewis acids. The details of this step are not certain, and heating is usually required. Following hydride reaction, departure of an oxygen atom as an aluminum oxide species produces an intermediate aldehyde. The aldehyde immediately reacts under the reaction conditions (Section 16.11A) to give a tetraalkoxy aluminate ion that hydrolyzes in an aqueous workup to yield the final product alcohol. Following are balanced equations for treatment of a carboxylic acid with $LiAlH_4$ to form a tetraalkoxy aluminate ion, followed by its hydrolysis in water.

In the reduction of a carboxyl group, two hydrogens from $LiAlH_4$ are delivered to the carbonyl group. The hydrogen on the hydroxyl group of the product is provided by water or by aqueous acid during work-up. The mechanism of lithium aluminum hydride reduction of carboxyl derivatives is presented in Section 18.10.

B. Selective Reduction of Other Functional Groups

Because carboxyl groups are not affected by the conditions of catalytic hydrogenation, which normally reduce aldehydes, ketones, alkenes, and alkynes, it is possible to selectively reduce these functional groups to alcohols or alkanes in the presence of carboxyl groups.

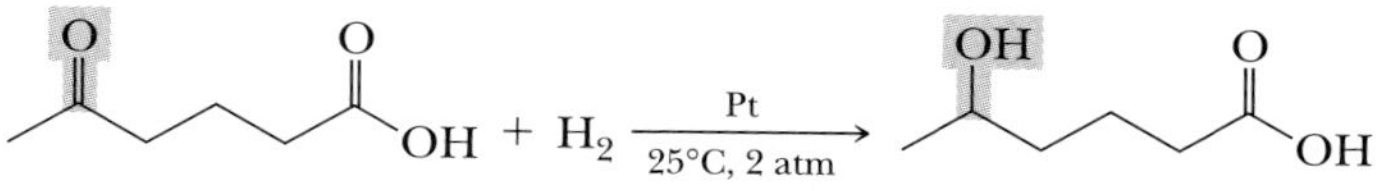

5-Oxohexanoic acid   5-Hydroxyhexanoic acid
(racemic)

We saw in Section 16.11A that both $LiAlH_4$ and $NaBH_4$ reduce aldehydes and ketones to alcohols. Only $LiAlH_4$, however, reduces carboxyl groups. Thus, it is possible to reduce an aldehyde or ketone carbonyl group selectively in the presence of a carboxyl group by using the less reactive $NaBH_4$ as the reducing agent. An example is the selective reduction of the following ketoacid to a hydroxyacid.

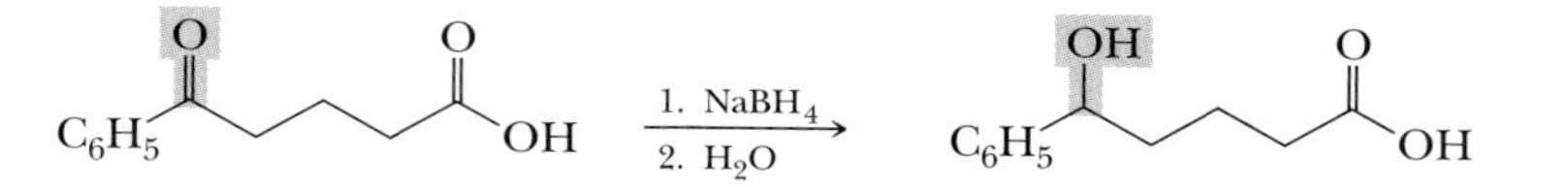

5-Oxo-5-phenylpentanoic acid   5-Hydroxy-5-phenylpentanoic acid
(racemic)

17.7 Esterification

A. Fischer Esterification

Fischer esterification The process of forming an ester by refluxing a carboxylic acid and an alcohol in the presence of an acid catalyst, commonly H_2SO_4, $ArSO_3H$, or HCl.

Esters can be prepared by treating a carboxylic acid with an alcohol in the presence of an acid catalyst, most commonly sulfuric acid, $ArSO_3H$, or gaseous HCl. Conversion of a carboxylic acid and alcohol to an ester is given the special name **Fischer esterification** after the German chemist, Emil Fischer (1852–1919), whose name is also firmly established in the chemistry of carbohydrates. An example of Fischer esterification is the treatment of acetic acid with ethanol in the presence of concentrated sulfuric acid, which gives ethyl acetate and water.

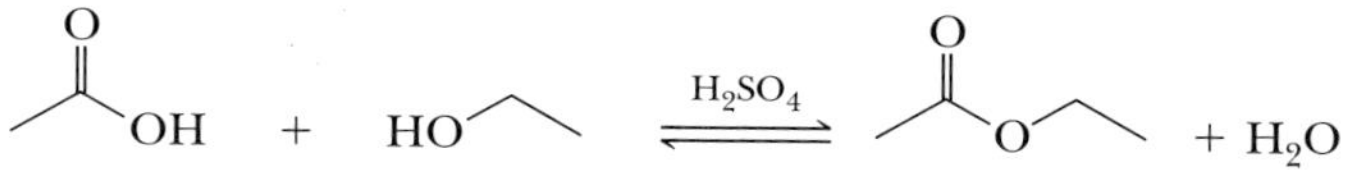

Ethanoic acid   Ethanol   Ethyl ethanoate
(Acetic acid) (Ethyl alcohol)   (Ethyl acetate)

We study the structure, nomenclature, and reactions of esters in detail in Chapter 18. In this chapter, we discuss only their preparation from carboxylic acids.

Acid-catalyzed esterification is reversible, and generally the quantities of both carboxylic acid and alcohol remaining at equilibrium are appreciable. If, for example, 60.1 g (1.00 mol) of acetic acid and 60.1 g (1.00 mol) of 1-propanol are heated under reflux in the presence of a few drops of concentrated sulfuric acid until equilibrium is reached, the reaction mixture contains approximately 0.67 mol each of propyl acetate and water, and 0.33 mol each of acetic acid and 1-propanol. Thus, at equilibrium, about 67% of the carboxylic acid and alcohol are converted to the desired ester.

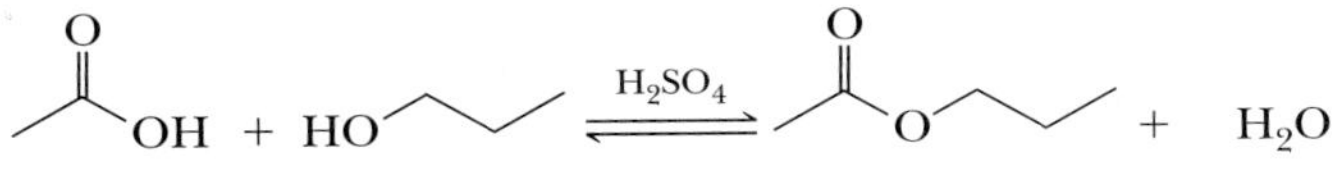

	Acetic acid	1-Propanol	Propyl acetate	
Initial:	1.00 mol	1.00 mol	0.00 mol	0.00 mol
Equilibrium:	0.33 mol	0.33 mol	0.67 mol	0.67 mol

By control of reaction conditions, it is possible to use Fischer esterification to prepare esters in high yields. If the alcohol is inexpensive compared with the carboxylic acid, a large excess of it can be used to drive the equilibrium to the right and achieve a high conversion of carboxylic acid to its ester. Alternatively, water can be removed by azeotropic distillation and a Dean-Stark trap (Figure 16.1).

Example 17.4

Complete the equation for each Fischer esterification reaction.

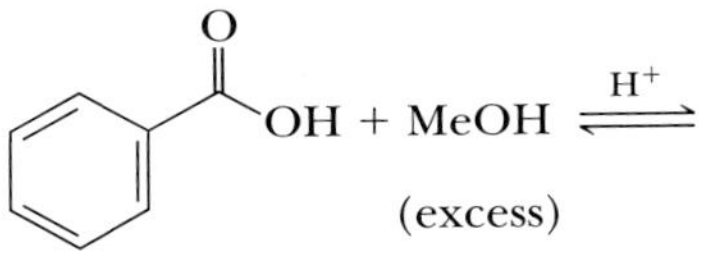

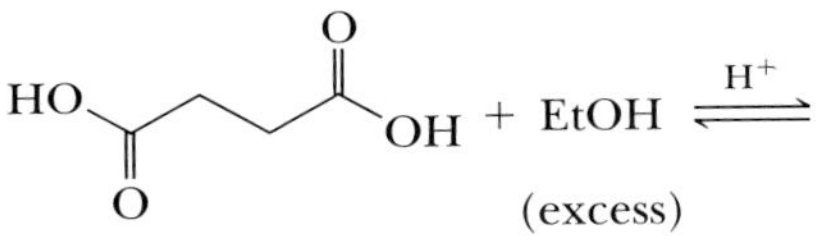

Solution

Following is the structural formula for the ester produced in each reaction.

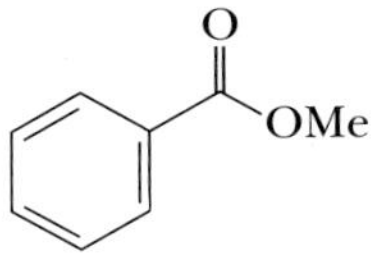

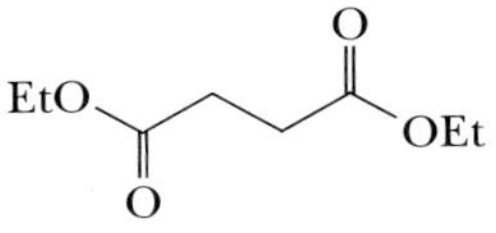

Problem 17.4

Complete the equation for each Fischer esterification.

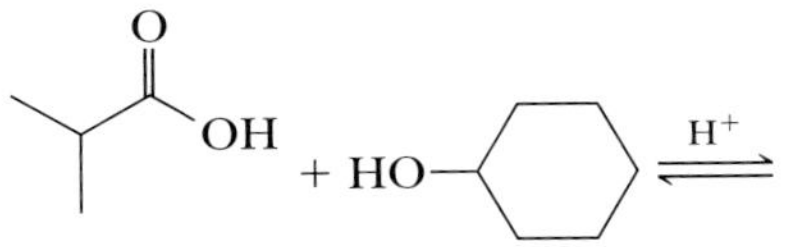

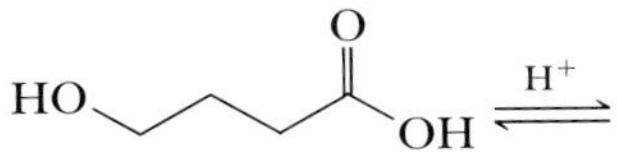

B. Mechanism of Fischer Esterification

It is important that you understand the mechanism of Fischer esterification thoroughly because it is a model for many other reactions of carboxylic acids presented in this chapter as well as for the reactions of functional derivatives of carboxylic acids presented in Chapter 18.

CHEMICAL CONNECTIONS

The Pyrethrins: Natural Insecticides of Plant Origin

Pyrethrin is a natural insecticide obtained from the powdered flower heads of several species of *Chrysanthe-*

Pyrethrin I

mum, particularly *C. cinerariaefolium.* The active substances in pyrethrum, principally pyrethrins I and II, are contact poisons for insects and cold-blooded vertebrates. Because their concentrations in the pyrethrum powder used in chrysanthemum-based insecticides are nontoxic to plants and higher animals, pyrethrum powder has found wide use in household and livestock sprays as well as in dusts for edible plants. Pyrethrins I and II are esters of chrysanthemic acid.

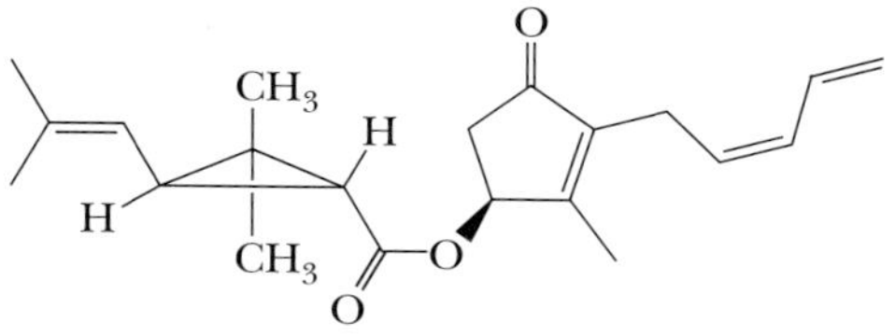

Although pyrethrum powders are effective insecticides, the active substances in them are destroyed rapidly in the environment. In an effort to develop synthetic compounds as effective as these natural insecticides but with greater biostability, chemists prepared a series of esters related in structure to chrysanthemic acid. Among the synthetic pyrethrenoids now in common use in household and agricultural products are the following.

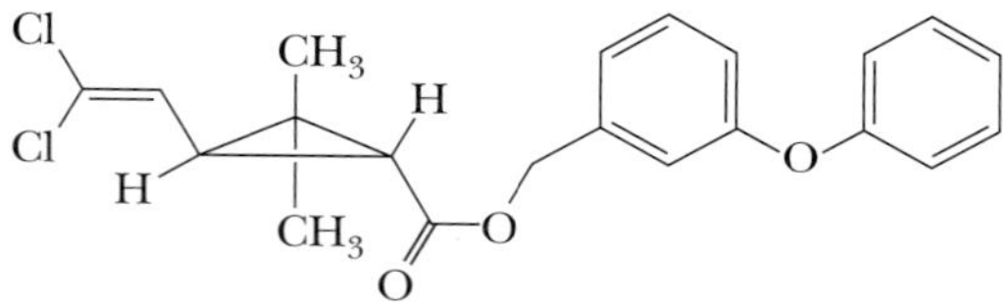

Permethrin

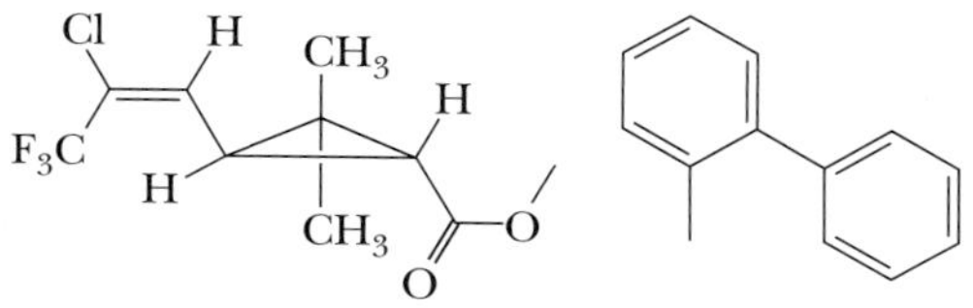

Bifenthrin

As you study this mechanism, note the following points.

- In the examples of Fischer esterification, the catalyst is shown as a proton acid (for example, H_2SO_4 or $ArSO_3H$). The actual catalyst that initiates the reaction, however, is the conjugate acid of the alcohol, ROH_2^+, used in the esterification.
- Steps 1–5 of Fischer esterification are closely analogous to the acid-catalyzed reaction of aldehydes and ketones with alcohols to form hemiacetals (Section 16.7B).
- Steps 6–9 are closely analogous to acid-catalyzed dehydration of alcohols, except that in Fischer esterification, H^+ is lost from oxygen; in acid-catalyzed dehydration of an alcohol, H^+ is lost from carbon. Loss of H^+ from oxygen is much easier than from carbon.
- The result of Fischer esterification is nucleophilic substitution, but the mechanism is an addition-elimination sequence and is quite different from either the S_N2 or S_N1 substitution mechanisms discussed in Chapter 9.

Mechanism *Fischer Esterification*

ORGANIC
Chemistry·-Now™

Click Mechanisms to view an animation of **Fischer Esterification**

① Proton transfer from the acid catalyst to the carbonyl oxygen increases the electrophilicity of the carbonyl carbon . . .

② which is then attacked by the nucleophilic oxygen atom of the alcohol . . .

③ to form an oxonium ion.

④ Proton transfer from the oxonium ion to a second molecule of alcohol . . .

⑤ gives a tetrahedral carbonyl addition intermediate (TCAI).

⑥ Proton transfer to one of the —OH groups of the TCAI . . .

⑦ gives a new oxonium ion.

⑧ Loss of water from this oxonium ion . . .

⑨ gives the ester and water.

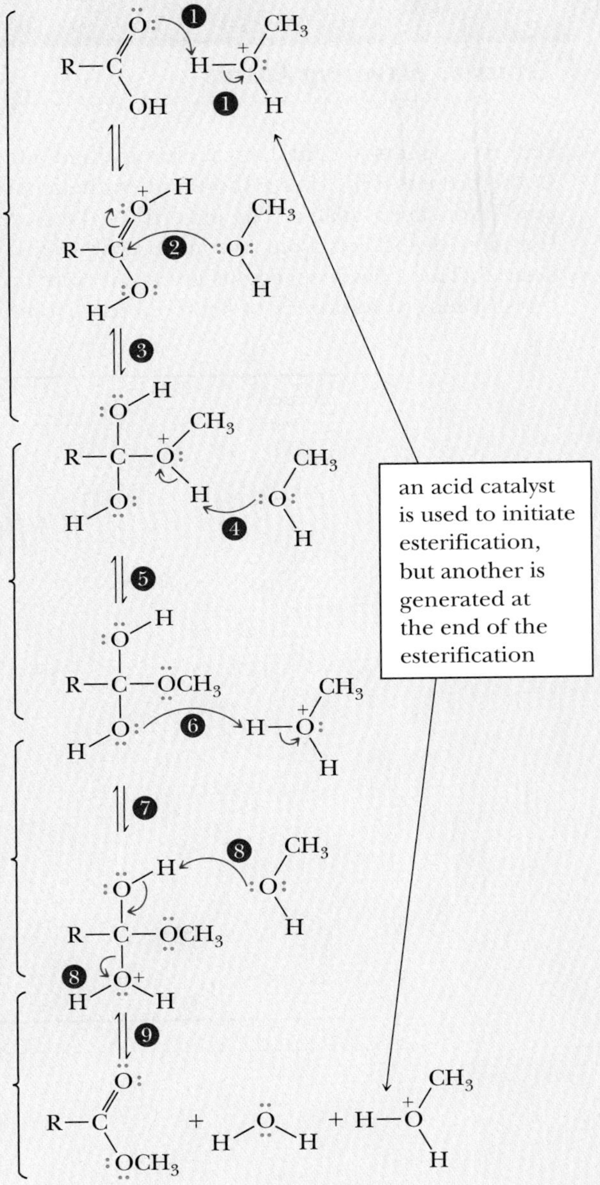

an acid catalyst is used to initiate esterification, but another is generated at the end of the esterification

C. Formation of Methyl Esters Using Diazomethane

Treating a carboxylic acid with diazomethane, usually in ether solution, converts the carboxylic acid under mild conditions and in very high yield to its methyl ester.

$$\underset{\text{Diazomethane}}{\overset{\displaystyle O}{\underset{\|}{\text{RCOH}}}} \; + \; CH_2N_2 \; \xrightarrow{\text{ether}} \; \underset{\text{A methyl ester}}{\overset{\displaystyle O}{\underset{\|}{\text{RCOCH}_3}}} \; + \; N_2$$

CHEMICAL CONNECTIONS

Esters as Flavoring Agents

Flavoring agents are the largest class of food additives. At the present time, over a thousand synthetic and natural flavors are available. The majority of these are concentrates or extracts from the material whose flavor is desired. These flavoring agents are often complex mixtures of tens to hundreds of compounds. A number of flavoring agents, many of them esters, however, are synthesized industrially. Many of these synthetic flavoring agents are major components of the natural flavors, and adding only one or a few of them is sufficient to make ice cream, soft drinks, or candy taste naturally flavored.

Structure	Name	Flavor
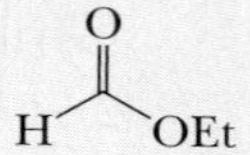	Ethyl formate	Rum
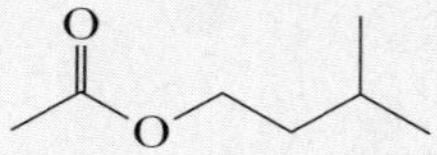	(3-Methyl)butyl acetate (Isopentyl acetate)	Banana
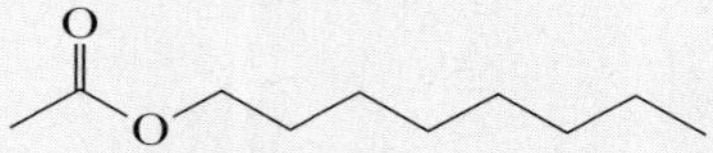	Octyl acetate	Orange
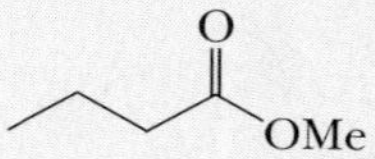	Methyl butanoate	Apple
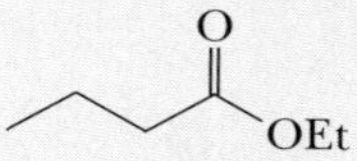	Ethyl butanoate	Pineapple
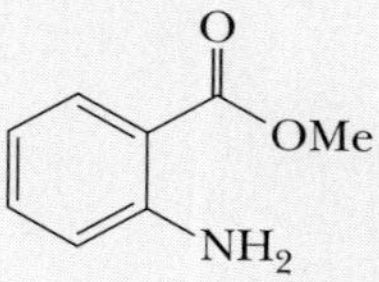	Methyl 2-aminobenzoate (Methyl anthranilate)	Grape

Diazomethane, a potentially explosive, toxic yellow gas, is best represented as a hybrid of two resonance contributing structures.

$$H-\overset{\underset{|}{H}}{\overset{..}{C}}-N\equiv N: \longleftrightarrow H-\overset{\underset{|}{H}}{C}=N=\overset{..}{N}:$$

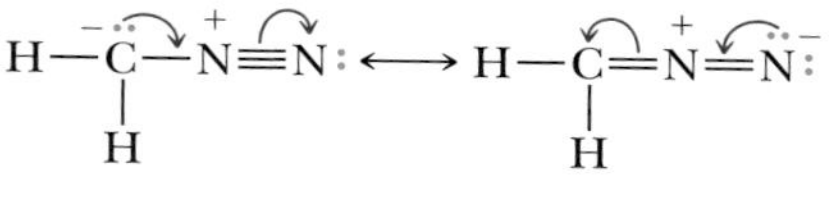

Diazomethane
(a resonance hybrid of two
important contributing structures)

Mechanism   *Formation of a Methyl Ester Using Diazomethane*

The reaction of a carboxylic acid with diazomethane occurs in two steps.

Step 1: Proton transfer from the carboxyl group to diazomethane gives a carboxylate anion and methyldiazonium cation.

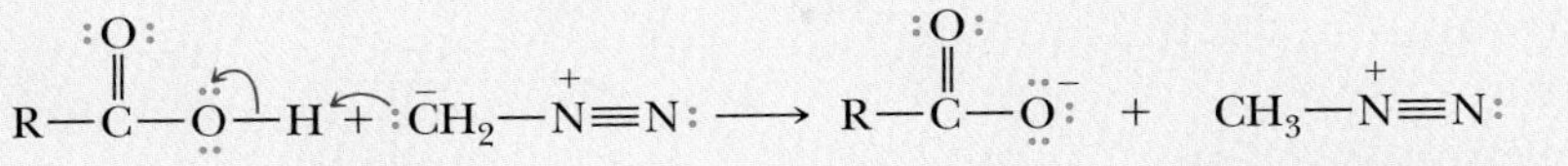

$$R-\overset{\overset{\displaystyle :O:}{\|}}{C}-\overset{..}{\underset{..}{O}}-H + :\overset{-}{C}H_2-\overset{+}{N}\equiv N: \longrightarrow R-\overset{\overset{\displaystyle :O:}{\|}}{C}-\overset{..}{\underset{..}{O}}:^- + CH_3-\overset{+}{N}\equiv N:$$

A carboxylate   Methyldiazonium
anion   cation

Step 2: Nucleophilic displacement of N_2, an extraordinarily good leaving group, gives the methyl ester.

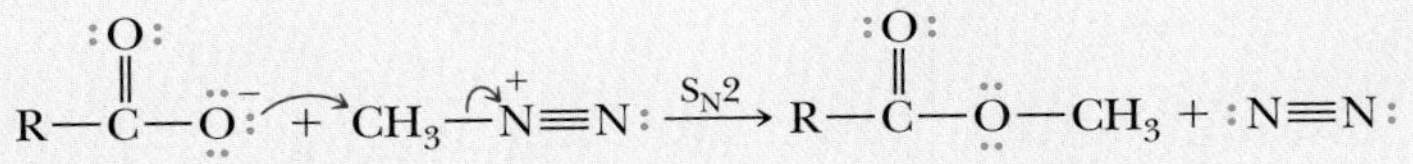

$$R-\overset{\overset{\displaystyle :O:}{\|}}{C}-\overset{..}{\underset{..}{O}}:^- + CH_3-\overset{+}{N}\equiv N: \xrightarrow{S_N2} R-\overset{\overset{\displaystyle :O:}{\|}}{C}-\overset{..}{\underset{..}{O}}-CH_3 + :N\equiv N:$$

Because of the hazards associated with the use of diazomethane, it is used only where other means of preparation of methyl esters are too harsh, and, even then, it is used only in small quantities.

ORGANIC
Chemistry Now™
Click Mechanisms to view an animation of the **Formation of a Methyl Ester with Diazomethane**

17.8 Conversion to Acid Chlorides

The functional group of an acid halide is a carbonyl group bonded to a halogen atom. Among the acid halides, acid chlorides are the most frequently used in the laboratory and in industrial organic chemistry.

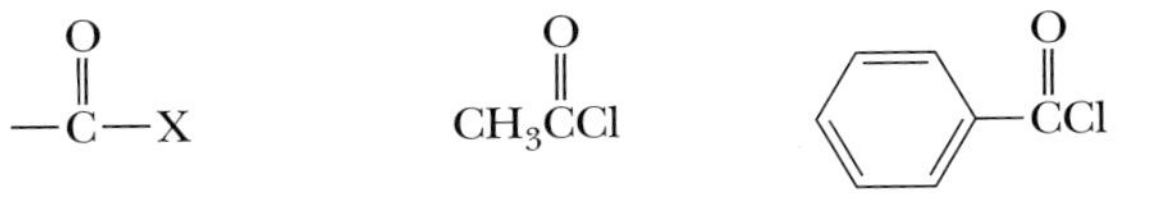

Functional group   Acetyl chloride   Benzoyl chloride
of an acid halide

We study the nomenclature, structure, and characteristic reactions of acid halides in Chapter 18. Here, we are concerned only with their synthesis from carboxylic acids.

Acid chlorides are most often prepared by treating a carboxylic acid with thionyl chloride, the same reagent used to convert an alcohol to a chloroalkane (Section 10.5C).

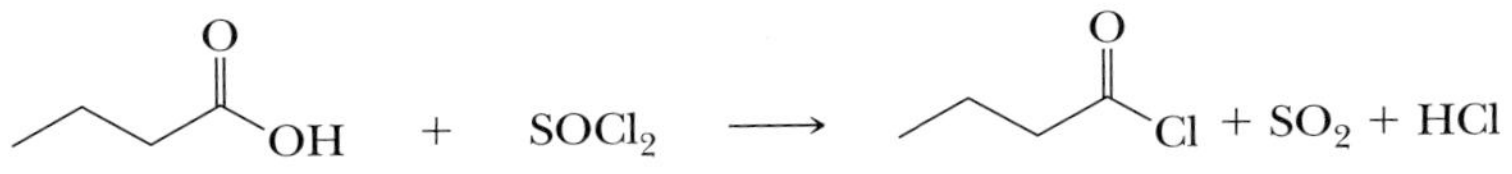

Butanoic acid   Thionyl chloride   Butanoyl chloride

The mechanism for the reaction of thionyl chloride with a carboxylic acid to form an acid chloride is very similar to that presented in Section 10.5C for the conversion of an alcohol to a chloroalkane.

Mechanism *Reaction of a Carboxylic Acid with Thionyl Chloride*

Step 1: Reaction of the carboxyl group with thionyl chloride gives an acyl chlorosulfite. The effect of this step is to convert OH⁻, a poor leaving group, into a chlorosulfite group, a good leaving group.

$$R-\overset{O}{\overset{\|}{C}}-O-H + Cl-\overset{O}{\overset{\|}{S}}-Cl \longrightarrow R-\overset{O}{\overset{\|}{C}}-\underset{\smile}{O}-\overset{O}{\overset{\|}{S}}-Cl + H-Cl$$

A chlorosulfite
group

Step 2: Attack of chloride ion on the carbonyl carbon of the chlorosulfite gives a tetrahedral carbonyl addition intermediate. The collapse of this intermediate gives the acid chloride, sulfur dioxide, and chloride ion.

$$R-\overset{:O:}{\overset{\|}{C}}-\overset{:O:}{\overset{\|}{O}}-S-\overset{..}{\underset{..}{Cl}}: + :\overset{..}{\underset{..}{Cl}}:^{-} \longrightarrow \left[R-\overset{:O:^{-}}{\underset{:Cl:}{\overset{|}{C}}}-\overset{:O:}{\overset{|}{O}}-S-\overset{..}{\underset{..}{Cl}}: \right] \longrightarrow R-\overset{:O:}{\overset{\|}{C}}-\overset{..}{\underset{..}{Cl}}: + SO_2 + :\overset{..}{\underset{..}{Cl}}:^{-}$$

A tetrahedral carbonyl
addition intermediate

Note that, in the second step of this mechanism, chloride ion adds to the carbonyl carbon to form a tetrahedral carbonyl addition intermediate, which then collapses by loss of the chlorosulfite group to give the acid chloride, as we saw in Section 17.2. Addition to a carbonyl carbon to form a tetrahedral carbonyl addition intermediate followed by elimination of a leaving group is a theme common to a great many reactions of carboxylic acids and their derivatives, and we shall see much more of this theme in Chapters 18 and 19.

Example 17.5

Complete the equation for each reaction.

(a) (pentanoic acid) $\overset{O}{\overset{\|}{C}}$OH + SOCl₂ ⟶   **(b)** (acrylic acid) $\overset{O}{\overset{\|}{C}}$OH + SOCl₂ ⟶

Solution

Following are the products of each reaction.

(a) (pentanoyl chloride) $\overset{O}{\overset{\|}{C}}$Cl + SO₂ + HCl   **(b)** (acryloyl chloride) $\overset{O}{\overset{\|}{C}}$Cl + SO₂ + HCl

Problem 17.5

Complete the equation for each reaction.

(a) (o-methoxybenzoic acid: COOH and OCH₃ groups on benzene ring) + SOCl₂ ⟶   **(b)** (cyclohexanol) OH + SOCl₂ ⟶

17.9 Decarboxylation

A. β-Ketoacids

Decarboxylation is the loss of CO_2 from the carboxyl group of a molecule. Almost any carboxylic acid, heated to a very high temperature, undergoes thermal decarboxylation.

$$RCOH \xrightarrow[\text{heat}]{\text{decarboxylation}} RH + CO_2$$

Decarboxylation Loss of CO_2 from a carboxyl group.

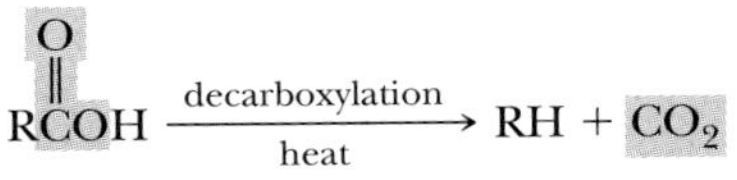

Most carboxylic acids, however, are quite resistant to moderate heat and melt or even boil without decarboxylation. Exceptions are carboxylic acids that have a carbonyl group β to the carboxyl group. This type of carboxylic acid undergoes decarboxylation quite readily on mild heating. For example, warming 3-oxobutanoic acid brings about its decarboxylation to give acetone and carbon dioxide.

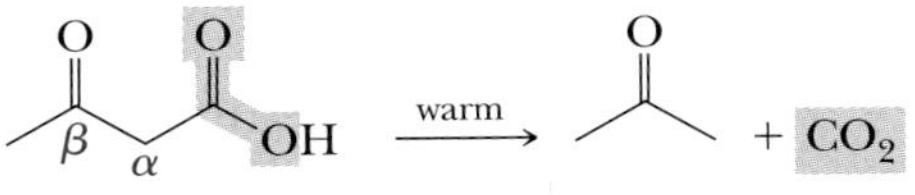

3-Oxobutanoic acid
(Acetoacetic acid)

Acetone

Decarboxylation on moderate heating is a unique property of 3-oxocarboxylic acids (β-ketoacids) and is not observed with other classes of ketoacids.

Mechanism *Decarboxylation of a β-Ketocarboxylic Acid*

Step 1: Redistribution of six electrons in a cyclic six-membered transition state gives carbon dioxide and an enol.

Step 2: Keto-enol tautomerism (Section 16.9B) of the enol gives the more stable keto form of the product.

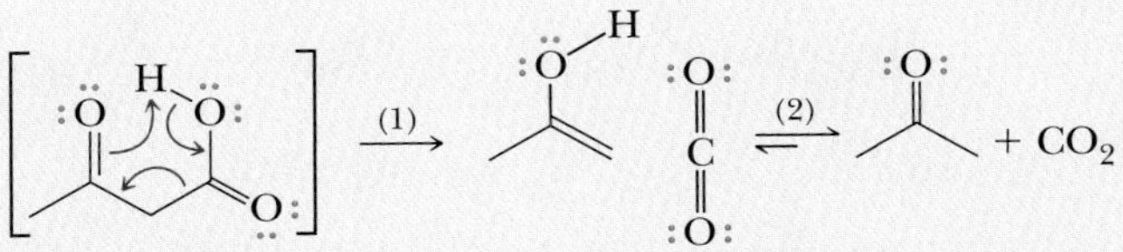

(A cyclic six-membered transition state)

Enol of a ketone

A ketone

ORGANIC
Chemistry Now™
Click Mechanisms to view an animation of the **Decarboxylation of a β-Ketocarboxylic Acid**

A hydrogen bond between the carboxyl hydrogen atom and the β-carbonyl oxygen promotes the reaction by favoring a conformation on the path to the six-membered ring transition state. Through this conformational stabilization, the molecules have a much higher probability of undergoing reaction, and the reaction occurs rapidly at moderate temperatures.

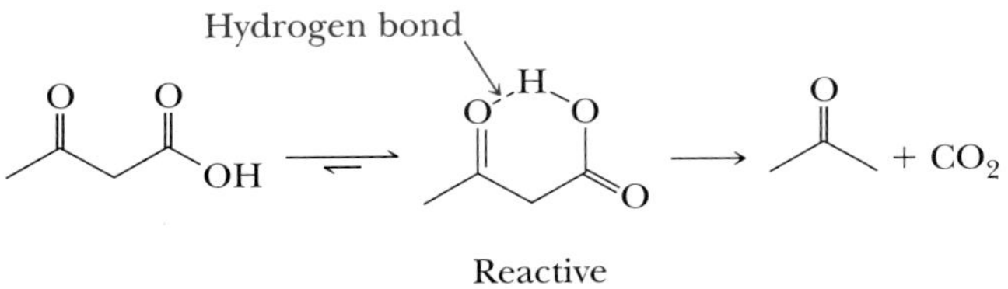

Hydrogen bond

Reactive conformation

CONNECTIONS TO BIOLOGICAL CHEMISTRY

Ketone Bodies and Diabetes Mellitus

3-Oxobutanoic acid (acetoacetic acid) and its reduction product, 3-hydroxybutanoic acid, are synthesized in the liver from acetyl-CoA, a product of the metabolism of fatty acids and certain amino acids. 3-Hydroxybutanoic acid and 3-oxobutanoic acid are known collectively as ketone bodies.

The concentration of ketone bodies in the blood of healthy, well-fed humans is approximately 0.01 mM/L. However, in persons suffering from starvation or diabetes mellitus, the concentration of ketone bodies may increase to as much as 500 times normal. Under these conditions, the concentration of acetoacetic acid increases to the point where it undergoes spontaneous decarboxylation to form acetone and carbon dioxide. Acetone is not metabolized by humans and is excreted through the kidneys and the lungs. The odor of acetone is responsible for the characteristic "sweet smell" of the breath of severely diabetic patients.

3-Oxobutanoic acid
(Acetoacetic acid)

3-Hydroxybutanoic acid
(β-Hydroxybutyric acid)

An important example of decarboxylation of a β-ketoacid in the biological world occurs during the oxidation of foodstuffs in the tricarboxylic acid (TCA) cycle. One of the intermediates in this cycle is oxalosuccinic acid, which undergoes spontaneous decarboxylation to produce α-ketoglutaric acid. Only one of the three carboxyl groups of oxalosuccinic acid has a carbonyl group in the β-position to it, and this carboxyl group is lost as CO_2.

Only this carboxyl
has a C=O
β to it.

Oxalosuccinic acid

α-Ketoglutaric acid

B. Malonic Acid and Substituted Malonic Acids

The presence of a ketone or aldehyde carbonyl group beta to the carboxyl group is sufficient to facilitate decarboxylation. In the more general reaction, decarboxylation is facilitated by the presence of any carbonyl group at the β-position, including that of a carboxyl group or ester. Malonic acid and substituted malonic acids, for example, undergo thermal decarboxylation, as illustrated by the decarboxylation of malonic acid when it is heated slightly above its melting point of 135–137°C.

$$HOCCH_2COH \xrightarrow{140°-150°C} CH_3COH + CO_2$$

Propanedioic acid
(Malonic acid)

The mechanism of decarboxylation of malonic acids is very similar to what we have just seen for the decarboxylation of β-ketoacids. In each case, formation of a cyclic, six-membered transition state involving rearrangement of three electron pairs gives the enol form of a carboxylic acid, which is in turn tautomerized to the carboxylic acid.

Mechanism *Decarboxylation of a β-Dicarboxylic Acid*

Step 1: Redistribution of six electrons in a cyclic six-membered transition state gives carbon dioxide and the enol form of a carboxyl group.

Step 2: Keto-enol tautomerism (Section 16.9B) of the enol gives the more stable keto form of the carboxyl group.

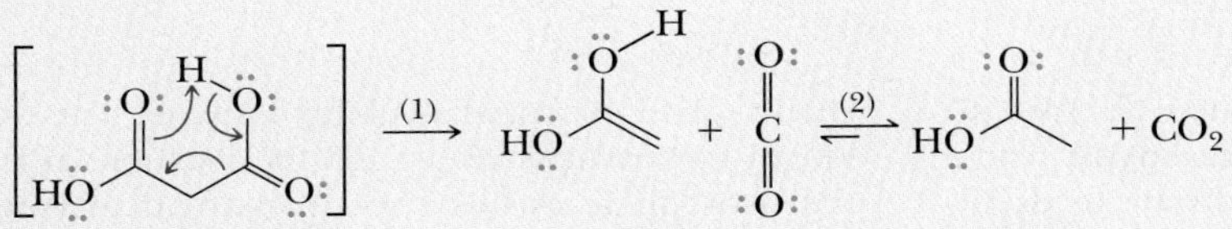

A cyclic six-membered transition state — Enol of a carboxylic acid — A carboxylic acid

ORGANIC Chemistry Now™
Click Mechanisms to view an animation of the **Decarboxylation of a β-Dicarboxylic Acid**

Example 17.6

Each of these carboxylic acids undergoes thermal decarboxylation.

(a) 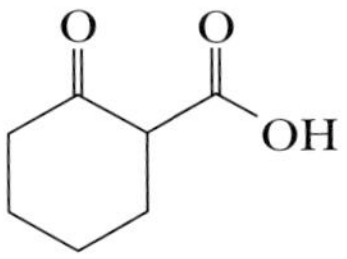

(b) 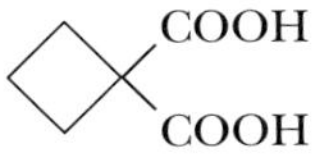

(c) 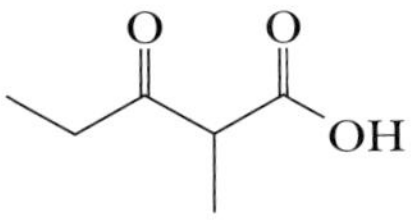

Draw a structural formula for the enol intermediate and final product formed in each reaction.

Solution

Following is a structural formula for the enol intermediate and the final product of each decarboxylation.

(a) 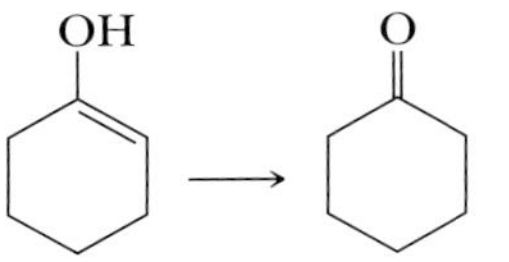

(b) 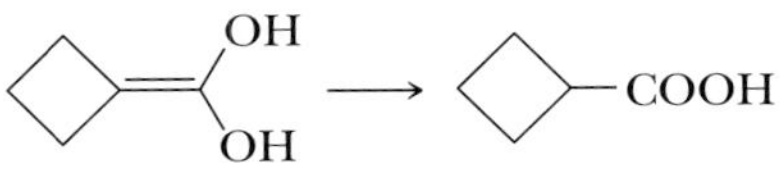

(c) 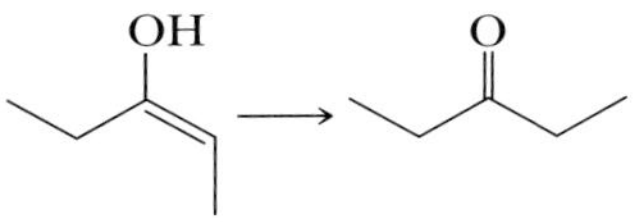

Problem 17.6

Account for the observation that the following β-ketoacid can be heated for extended periods at temperatures above its melting point without noticeable decarboxylation.

Summary

The functional group of a carboxylic acid (Section 17.1) is the **carboxyl group,** —COOH. IUPAC names of carboxylic acids (Section 17.2A) are derived from the parent alkane by dropping the suffix -*e* and adding -*oic acid*. Dicarboxylic acids are named by adding -*dioic acids*.

Carboxylic acids are polar compounds (Section 17.3) and, in the liquid and solid states, are associated by hydrogen bonding into dimers. Carboxylic acids have higher boiling points and are more soluble in water than alcohols, aldehydes, ketones, and ethers of comparable molecular weight. A carboxylic acid consists of two regions of distinctly different polarity; a polar **hydrophilic** carboxyl group, which increases solubility in water, and a nonpolar **hydrophobic** hydrocarbon

chain, which decreases solubility in water. The low-molecular-weight carboxylic acids are infinitely soluble in water because the hydrophilic carboxyl group more than counterbalances the hydrophobic hydrocarbon chain. As the size of the carbon chain increases, however, the hydrophobic group becomes dominant, and solubility in water decreases.

Values of pK_a for aliphatic carboxylic acids are in the range 4.0–5.0 (Section 17.4A). The greater acidity of carboxylic acids compared with alcohols is explained by resonance stabilization of a carboxylate anion relative to an alkoxide ion and the electron-withdrawing inductive effect of the carbonyl group. Electron-withdrawing substituents near the carboxyl group increase its acidity.

Key Reactions

1. Acidity of Carboxylic Acids (Section 17.4A)

Values of pK_a for most unsubstituted aliphatic and aromatic carboxylic acids are within the range pK_a 4–5.

$$CH_3\overset{O}{\overset{\|}{C}}OH + H_2O \rightleftharpoons CH_3\overset{O}{\overset{\|}{C}}O^- + H_3O^+ \qquad K_a = 1.74 \times 10^{-5}$$

The presence of electron-withdrawing groups near the carboxyl group decreases its pK_a (increases its acidity).

2. Reaction of Carboxylic Acids with Bases (Section 17.4B)

Carboxylic acids form water-soluble salts with alkali metal hydroxides, carbonates, and bicarbonates, as well as with ammonia and aliphatic and aromatic amines.

3. Carbonation of a Grignard reagent (Section 17.5A)

Adding CO_2 to a Grignard reagent followed by acidification provides a useful route to carboxylic acids.

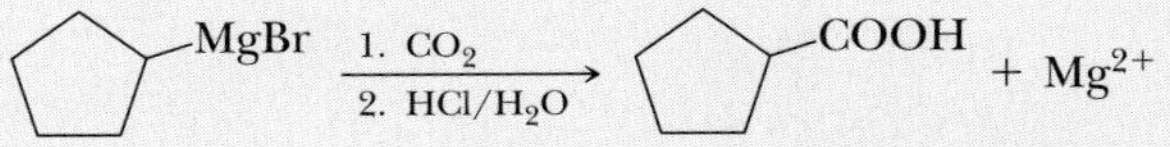

Cyclopentane-
carboxylic acid

4. Industrial Preparation of Acetic Acid by the Carbonylation of Methanol (Section 17.5B)

$$CO + 2H_2 \xrightarrow{\text{catalyst}} CH_3OH \xrightarrow[\substack{\text{rhodium(III)}\\ \text{HI, H}_2\text{O}}]{\text{CO}} CH_3\overset{\displaystyle O}{\overset{\|}{C}}OH$$

5. Reduction by Lithium Aluminum Hydride (Section 17.6A)

Lithium aluminum hydride reduces a carboxyl group to a primary alcohol.

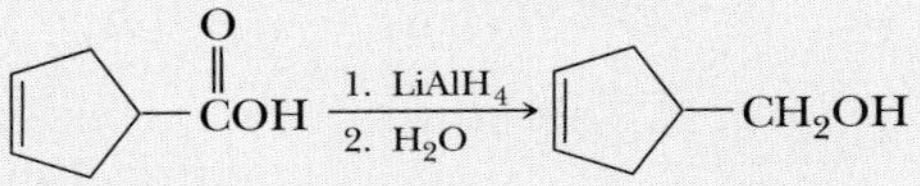

6. Fischer Esterification (Section 17.7A)

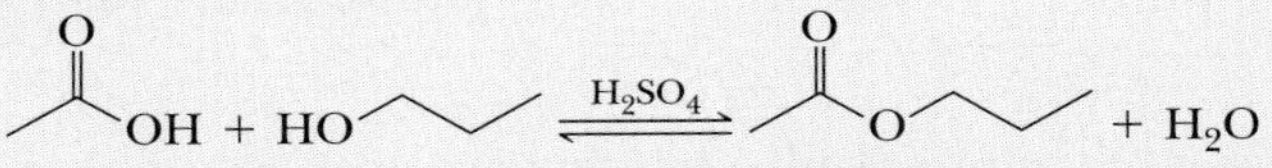

Fischer esterification is reversible. To achieve high yields of ester, it is necessary to force the equilibrium to the right. One way to accomplish this is to use an excess of alcohol; another is to remove water by azeotropic distillation using a Dean-Stark trap.

7. Reaction with Diazomethane (Section 17.7C)

Diazomethane is used to form methyl esters from carboxylic acids.

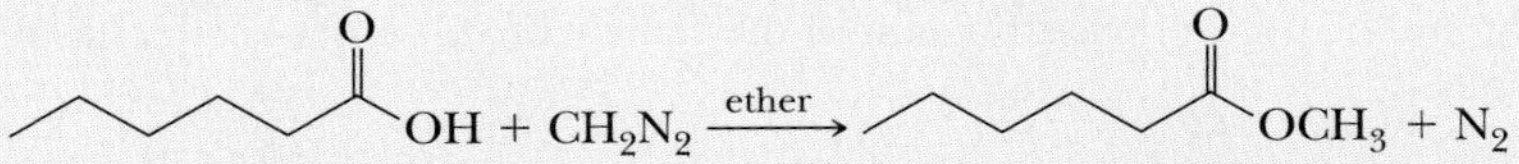

Because diazomethane is explosive and poisonous, it is used only when other means of preparation of methyl esters are not suitable.

8. Conversion to Acid Halides (Section 17.8)

Acid chlorides, the most common and widely used of the acid halides, are prepared by treating a carboxylic acid with thionyl chloride.

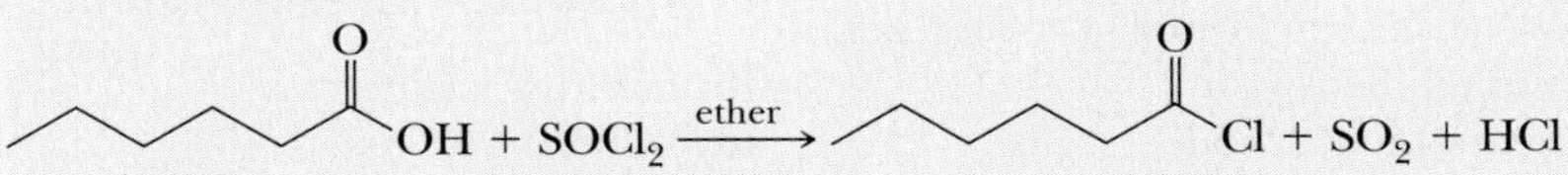

9. Decarboxylation of β-Ketoacids (Section 17.9A)

The mechanism of decarboxylation involves redistribution of bonding electrons in a cyclic, six-membered transition state.

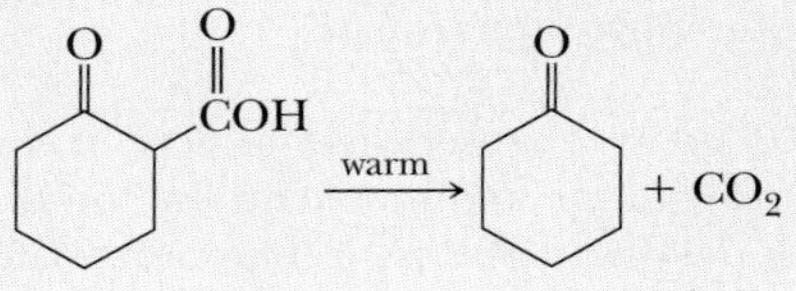

10. Decarboxylation of β-Dicarboxylic Acids (Section 17.9B)

The mechanism of decarboxylation of a β-dicarboxylic acid is similar to that for decarboxylation of a β-ketoacid.

$$\text{HOCCH}_2\text{COH} \xrightarrow{\text{heat}} \text{CH}_3\text{COH} + \text{CO}_2$$

Problems

Structure and Nomenclature

17.7   Write the IUPAC name of each compound, showing stereochemistry where relevant:

(a) 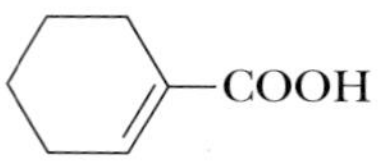 [structure] —COOH   (b) 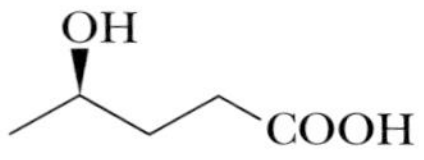 [structure with OH] COOH   (c) 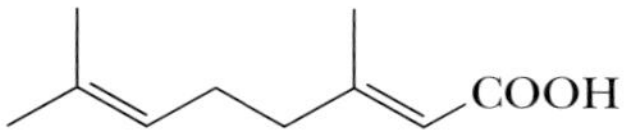 [structure] COOH

(d)  [structure] COOH   (e) 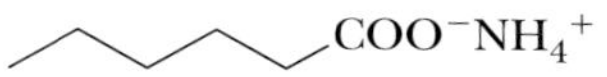 [structure] COO⁻NH₄⁺   (f) 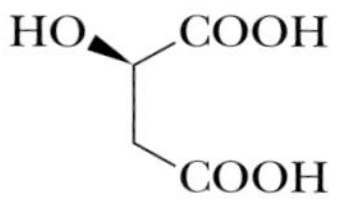 [structure with HO, COOH] COOH

17.8   Draw a structural formula for each compound.
    (a) Phenylacetic acid   **(b)** 4-Aminobutanoic acid
    (c) 3-Chloro-4-phenylbutanoic acid   **(d)** Propenoic acid (acrylic acid)
    (e) (Z)-3-Hexenedioic acid   **(f)** 2-Pentynoic acid
    (g) Potassium phenylacetate   **(h)** Sodium oxalate
    (i) 2-Oxocyclohexanecarboxylic acid   **(j)** 2,2-Dimethylpropanoic acid

17.9   Megatomoic acid, the sex attractant of the female black carpet beetle, has the following structure.

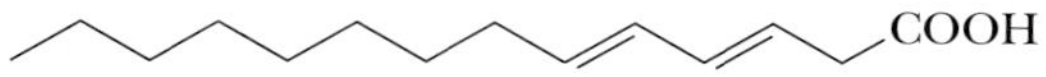

[structure] COOH

    (a) What is its IUPAC name?
    (b) State the number of stereoisomers possible for this compound.

17.10   Draw a structural formula for each salt.
    (a) Sodium benzoate   **(b)** Lithium acetate
    (c) Ammonium acetate   **(d)** Disodium adipate
    (e) Sodium salicylate   **(f)** Calcium butanoate

17.11   The monopotassium salt of oxalic acid is present in certain leafy vegetables, including rhubarb. Both oxalic acid and its salts are poisonous in high concentrations. Draw the structural formula of monopotassium oxalate.

17.12   Potassium sorbate is added as a preservative to certain foods to prevent bacteria and molds from causing food spoilage and to extend the foods' shelf life. The IUPAC name of potassium sorbate is potassium (2E,4E)-2,4-hexadienoate. Draw a structural formula for potassium sorbate.

17.13 Zinc 10-undecenoate, the zinc salt of 10-undecenoic acid, is used to treat certain fungal infections, particularly *Tinea pedis* (athlete's foot). Draw a structural formula for this zinc salt.

17.14 On a cyclohexane ring, an axial carboxyl group has a conformational energy of 5.9 kJ (1.4 kcal)/mol relative to an equatorial carboxyl group. Consider the equilibrium for the alternative chair conformations of *trans*-1,4-cyclohexanedicarboxylic acid. Draw the less stable chair conformation on the left of the equilibrium arrows and the more stable chair on the right. Calculate ΔG^0 for the equilibrium as written, and calculate the ratio of the more stable chair to the less stable chair at 25°C.

Physical Properties

17.15 Arrange the compounds in each set in order of increasing boiling point.

(a) $CH_3(CH_2)_5COOH$ $CH_3(CH_2)_6CHO$ $CH_3(CH_2)_6CH_2OH$

(b)

17.16 Acetic acid has a boiling point of 118°C, whereas its methyl ester has a boiling point of 57°C. Account for the fact that the boiling point of acetic acid is higher than that of its methyl ester, even though acetic acid has a lower molecular weight.

17.17 Given here are ^{1}H-NMR and ^{13}C-NMR spectral data for nine compounds. Each compound shows strong absorption between 1720 and 1700 cm^{-1}, and strong, broad absorption over the region 2500–3300 cm^{-1}. Propose a structural formula for each compound. Refer to Appendices 4, 5, and 6 for spectral correlation tables.

(a) $C_5H_{10}O_2$

^{1}H-NMR	^{13}C-NMR
0.94 (t, 3H)	180.71
1.39 (m, 2H)	33.89
1.62 (m, 2H)	26.76
2.35 (t, 2H)	22.21
12.0 (s, 1H)	13.69

(b) $C_6H_{12}O_2$

^{1}H-NMR	^{13}C-NMR
1.08 (s, 9H)	179.29
2.23 (s, 2H)	47.82
12.1 (s, 1H)	30.62
	29.57

(c) $C_5H_8O_4$

^{1}H-NMR	^{13}C-NMR
0.93 (t, 3H)	170.94
1.80 (m, 2H)	53.28
3.10 (t, 1H)	21.90
12.7 (s, 2H)	11.81

(d) $C_5H_8O_4$

^{1}H-NMR	^{13}C-NMR
1.29 (s, 6H)	174.01
12.8 (s, 2H)	48.77
	22.56

(e) $C_4H_6O_2$

^{1}H-NMR	^{13}C-NMR
1.91 (d, 3H)	172.26
5.86 (d, 1H)	147.53
7.10 (m, 1H)	122.24
12.4 (s, 1H)	18.11

(f) $C_3H_4Cl_2O_2$

^{1}H-NMR	^{13}C-NMR
2.34 (s, 3H)	171.82
11.3 (s, 1H)	79.36
	34.02

(g) $C_5H_8Cl_2O_2$

^{1}H-NMR	^{13}C-NMR
1.42 (s, 6H)	180.15
6.10 (s, 1H)	77.78
12.4 (s, 1H)	51.88
	20.71

(h) $C_5H_9BrO_2$

^{1}H-NMR	^{13}C-NMR
0.97 (t, 3H)	176.36
1.50 (m, 2H)	45.08
2.05 (m, 2H)	36.49
4.25 (t, 1H)	20.48
12.1 (s, 1H)	13.24

(i) $C_4H_8O_3$

^{1}H-NMR	^{13}C-NMR
2.62 (t, 2H)	177.33
3.38 (s, 3H)	67.55
3.68 (s, 2H)	58.72
11.5 (s, 1H)	34.75

Preparation of Carboxylic Acids

17.18 Complete each reaction.

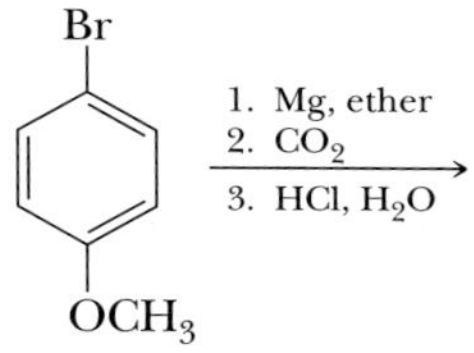

17.19 Show how to bring about each conversion in good yield.

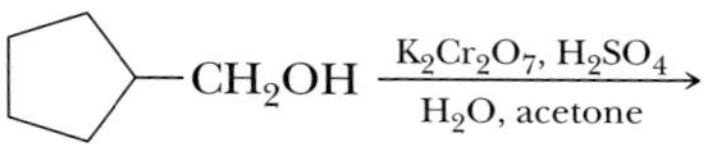

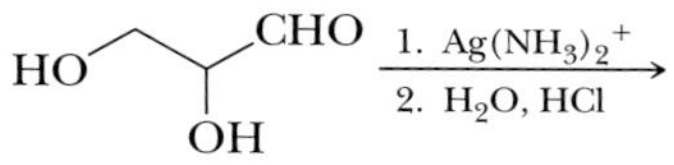

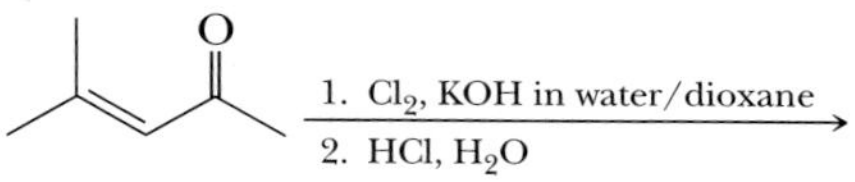

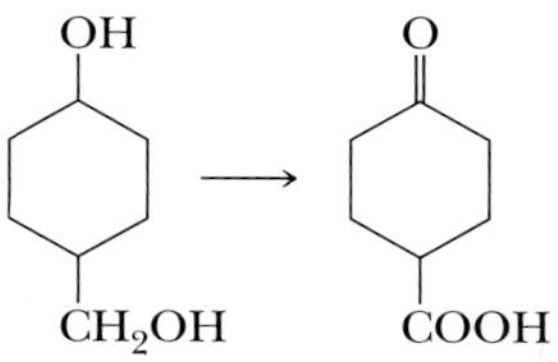

(c) C_6H_5 ⌒ OH ⟶ C_6H_5 ⌒ COOH

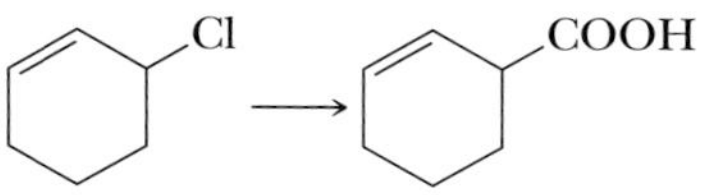

17.20 Show how to prepare pentanoic acid from each compound.

(a) 1-Pentanol   **(b)** Pentanal   **(c)** 1-Pentene   **(d)** 1-Butanol
(e) 1-Bromopropane   **(f)** 1-Hexene

17.21 Draw the structural formula of a compound of the given molecular formula that, on oxidation by potassium dichromate in aqueous sulfuric acid, gives the carboxylic acid or dicarboxylic acid shown.

(a) $C_6H_{14}O \xrightarrow{\text{oxidation}}$ [structure: pentanoic acid] OH **(b)** $C_6H_{12}O \xrightarrow{\text{oxidation}}$ [structure: acid] OH

(c) $C_6H_{14}O_2 \xrightarrow{\text{oxidation}}$ HO—[structure: adipic acid diketone]—OH

17.22 Show the reagents and experimental conditions necessary to bring about each conversion in good yield.

(a) [cyclopentanol] OH ⟶ [cyclopentane] COOH

(b) $\underset{\underset{CH_3}{|}}{\overset{\overset{CH_3}{|}}{CH_3COH}} \longrightarrow \underset{\underset{CH_3}{|}}{\overset{\overset{CH_3}{|}}{CH_3CCOOH}}$

(c) $\underset{\underset{CH_3}{|}}{\overset{\overset{CH_3}{|}}{CH_3COH}} \longrightarrow \overset{\overset{CH_3}{|}}{CH_3CHCOOH}$

(d) $\underset{\underset{CH_3}{|}}{\overset{\overset{CH_3}{|}}{CH_3COH}} \longrightarrow \overset{\overset{CH_3}{|}}{CH_3CHCH_2COOH}$

(e) $CH_3CH{=}CHCH_3 \longrightarrow CH_3CH{=}CHCH_2COOH$

17.23 Succinic acid can be synthesized by the following series of reactions from acetylene. Show the reagents and experiential conditions necessary to carry out this synthesis.

$H{-}{\equiv}{-}H \longrightarrow$ HO—[2-butyne-1,4-diol]—OH $\longrightarrow$ HO—[1,4-butanediol]—OH $\longrightarrow$ HO—[butanedioic acid]—OH

| Acetylene | 2-Butyne-1,4-diol | 1,4-Butanediol | Butanedioic acid (Succinic acid) |

17.24 The reaction of an α-diketone with concentrated sodium or potassium hydroxide to give the salt of an α-hydroxyacid is given the general name benzil-benzilic acid rearrangement. It is illustrated by the conversion of benzil to sodium benzilate and then to benzilic acid.

$$Ph{-}\overset{\overset{O}{\|}}{C}{-}\overset{\overset{O}{\|}}{C}{-}Ph + NaOH \xrightarrow{H_2O} \underset{\underset{Ph}{|}}{Ph{-}\overset{\overset{HO}{|}}{C}{-}\overset{\overset{O}{\|}}{C}{-}O^-Na^+} \xrightarrow[H_2O]{HCl} \underset{\underset{Ph}{|}}{Ph{-}\overset{\overset{HO}{|}}{C}{-}\overset{\overset{O}{\|}}{C}{-}OH}$$

| Benzil | Sodium benzilate | Benzilic acid |
| (an α-diketone) | | |

Propose a mechanism for this rearrangement.

Acidity of Carboxylic Acids

17.25 Select the stronger acid in each set.
  (a) Phenol (pK_a 9.95) and benzoic acid (pK_a 4.19)
  (b) Lactic acid (K_a 8.4×10^{-4}) and ascorbic acid (K_a 7.9×10^{-5})

17.26 In each set, assign the acid its appropriate pK_a.

(a)  [structure: benzene with COOH] and [structure: benzene with SO₃H] (pK_a 4.19 and 0.70)

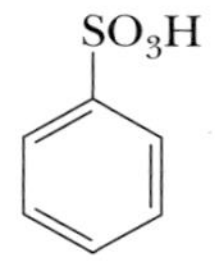

(b) 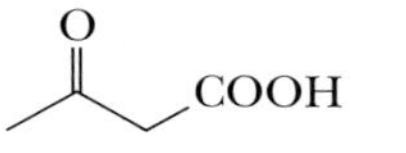 [structure with O and COOH] and [structure with O and COOH] (pK_a 3.58 and 2.49)

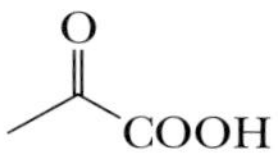

(c) CH_3CH_2COOH and $N\equiv CCH_2COOH$ (pK_a 4.78 and 2.45)

17.27 Low-molecular-weight dicarboxylic acids normally exhibit two different pK_a values. Ionization of the first carboxyl group is easier than the second. This effect diminishes with molecular size, and, for adipic acid and longer chain dicarboxylic acids, the two acid ionization constants differ by about one pK unit.

Dicarboxylic Acid	Structural Formula	pK_{a1}	pK_{a2}
Oxalic	$HOOCCOOH$	1.23	4.19
Malonic	$HOOCCH_2COOH$	2.83	5.69
Succinic	$HOOC(CH_2)_2COOH$	4.16	5.61
Glutaric	$HOOC(CH_2)_3COOH$	4.31	5.41
Adipic	$HOOC(CH_2)_4COOH$	4.43	5.41

Why do the two pK_a values differ more for the shorter chain dicarboxylic acids than for the longer chain dicarboxylic acids?

17.28 Complete the following acid-base reactions.

(a)  [benzene]–$CH_2COOH + NaOH \longrightarrow$ (b) $CH_3CH=CHCH_2COOH + NaHCO_3 \longrightarrow$

(c) 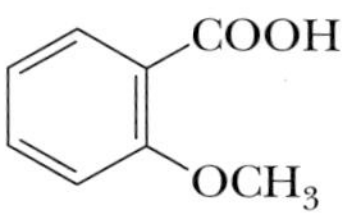 [benzene with COOH and OCH₃] $+ NaHCO_3 \longrightarrow$ (d) $CH_3\overset{OH}{CH}COOH + H_2NCH_2CH_2OH \longrightarrow$

(e) $CH_3CH=CHCH_2COO^- Na^+ + HCl \longrightarrow$
(f) $CH_3CH_2CH_2CH_2Li + CH_3COOH \longrightarrow$
(g) $CH_3CH_2CH_2CH_2MgBr + CH_3CH_2OH \longrightarrow$

17.29 The normal pH range for blood plasma is 7.35–7.45. Under these conditions, would you expect the carboxyl group of lactic acid (pK_a 3.08) to exist primarily as a carboxyl group or as a carboxylic anion? Explain.

17.30 The K_{a1} of ascorbic acid is 7.94×10^{-5}. Would you expect ascorbic acid dissolved in blood plasma (pH 7.35–7.45) to exist primarily as ascorbic acid or as ascorbate anion? Explain.

17.31 Excess ascorbic acid is excreted in the urine, the pH of which is normally in the range 4.8–8.4. What form of ascorbic acid would you expect to be present in urine of pH 8.4, free ascorbic acid or ascorbate anion? Explain.

Reactions of Carboxylic Acids

17.32 Give the expected organic product when phenylacetic acid, PhCH₂COOH, is treated with each reagent.

(a) SOCl₂   (b) NaHCO₃, H₂O
(c) NaOH, H₂O   (d) CH₃MgBr (one equivalent)
(e) LiAlH₄ followed by H₂O   (f) CH₂N₂
(g) CH₃OH + H₂SO₄ (catalyst)

17.33 Show how to convert *trans*-3-phenyl-2-propenoic acid (cinnamic acid) to each compound.

(a) 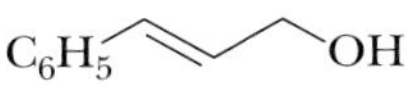 C₆H₅⌁⌁OH   (b) 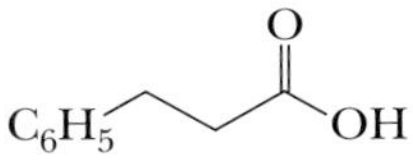 C₆H₅⌁⌁⌁OH   (c) 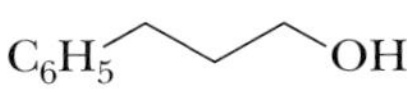 C₆H₅⌁⌁OH

17.34 Show how to convert 3-oxobutanoic acid (acetoacetic acid) to each compound.

(a) 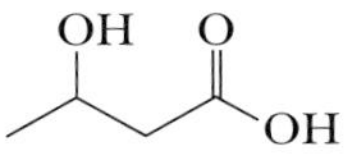 [structure] OH O / OH
(racemic)
(b) 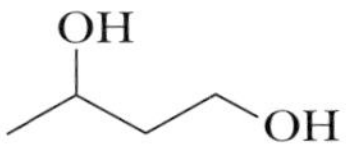 [structure] OH / OH
(racemic)
(c) 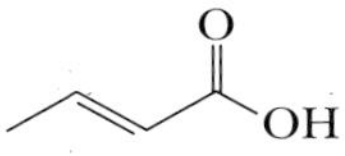 [structure] O / OH

17.35 Complete these examples of Fischer esterification. Assume that the alcohol is present in excess.

(a) 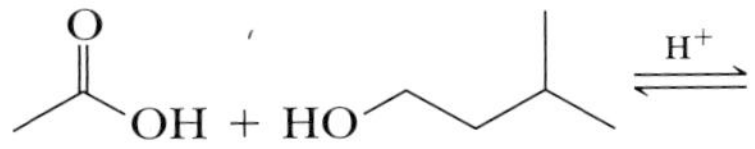 [structure] O / OH + HO⌁⌁ ⇌ (with H⁺)

(b) 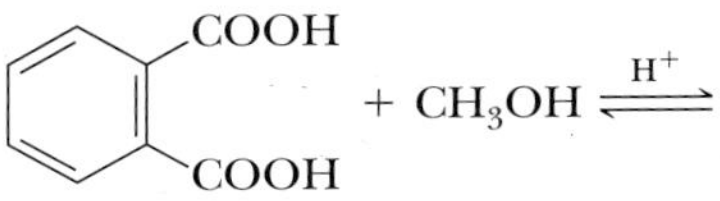 [structure] COOH / COOH + CH₃OH ⇌ (with H⁺)

(c) 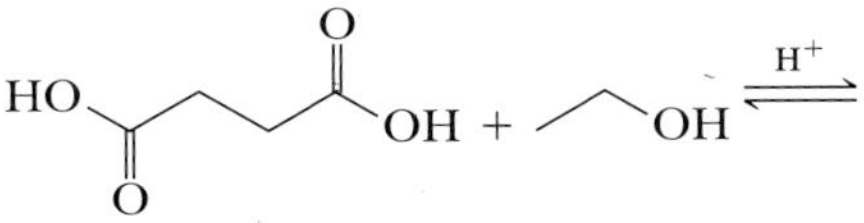 [structure] HO / OH + ⌁OH ⇌ (with H⁺)

17.36 Benzocaine, a topical anesthetic, is prepared by treatment of 4-aminobenzoic acid with ethanol in the presence of an acid catalyst followed by neutralization. Draw a structural formula for benzocaine.

17.37 Name the carboxylic acid and alcohol from which each ester is derived.

(a) 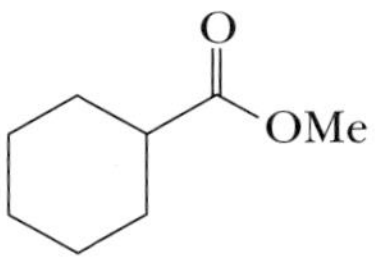 [structure] O / OMe

(b) 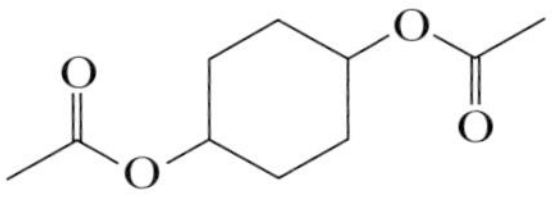 [structure] O / O / O

(c) 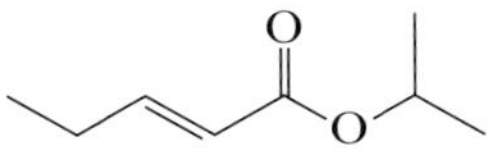 [structure] O / O

(d) 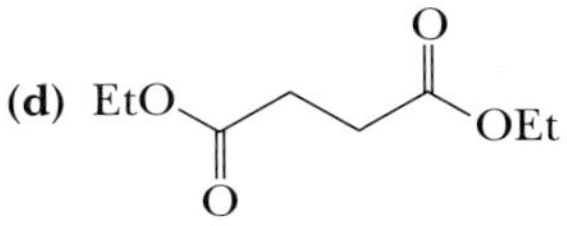 EtO [structure] O / O / OEt

17.38 When 4-hydroxybutanoic acid is treated with an acid catalyst, it forms a lactone (a cyclic ester). Draw the structural formula of this lactone, and propose a mechanism for its formation.

17.39 Fischer esterification cannot be used to prepare *tert*-butyl esters. Instead, carboxylic acids are treated with 2-methylpropene in the presence of an acid catalyst to generate them.

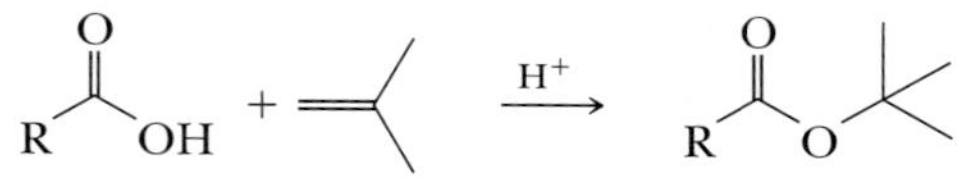

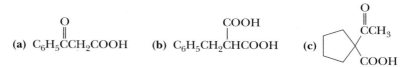

2-Methylpropene   A *tert*-butyl ester
(Isobutylene)

(a) Why does the Fischer esterification fail for the synthesis of *tert*-butyl esters?

(b) Propose a mechanism for the 2-methylpropene method.

17.40 Draw the product formed on thermal decarboxylation of each compound.

(a) $C_6H_5\overset{O}{\overset{\|}{C}}CH_2COOH$ (b) $C_6H_5CH_2\overset{COOH}{\overset{|}{C}H}COOH$ (c)

17.41 When heated, carboxylic salts in which there is a good leaving group on the carbon beta to the carboxylate group undergo decarboxylation/elimination to give an alkene.

(a)

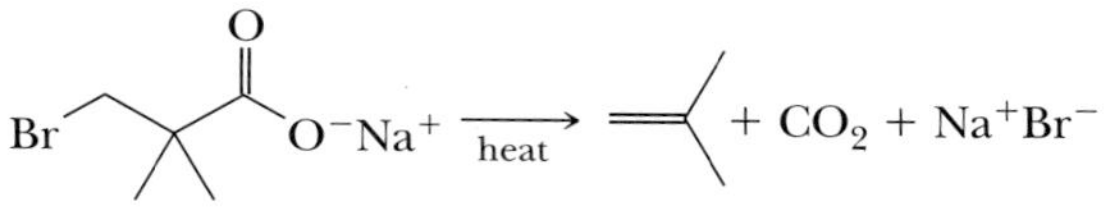

(b)

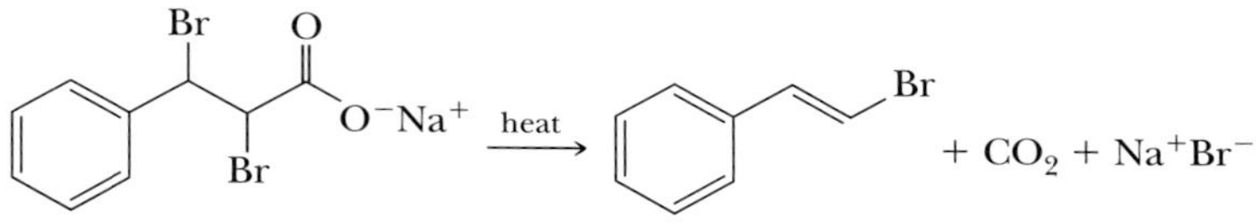

Propose a mechanism for this type of decarboxylation/elimination. Compare the mechanism of these decarboxylations with the mechanism for decarboxylation of β-ketoacids; in what way(s) are the mechanisms similar?

17.42 Show how cyclohexanecarboxylic acid could be synthesized from cyclohexane in good yield.

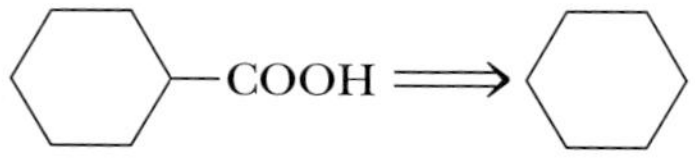

Looking Ahead

17.43 In Section 17.7B, we suggested that the mechanism of Fischer esterification of carboxylic acids is a model for the reactions of functional derivatives of carboxylic acids. One of these reactions is that of an acid chloride with water (Section 18.4A).

$$R-\overset{O}{\overset{\|}{C}}-Cl + H_2O \longrightarrow R-\overset{O}{\overset{\|}{C}}-OH + HCl$$

Suggest a mechanism for this reaction.

17.44 We have studied Fischer esterification, in which a carboxylic acid is reacted with an alcohol in the presence of an acid catalyst to form an ester. Suppose that you start instead with a dicarboxylic acid such as terephthalic acid and a diol such as ethylene glycol. Show how Fischer esterification in this case can lead to a macromolecule with a molecular weight several thousands of times that of the starting materials.

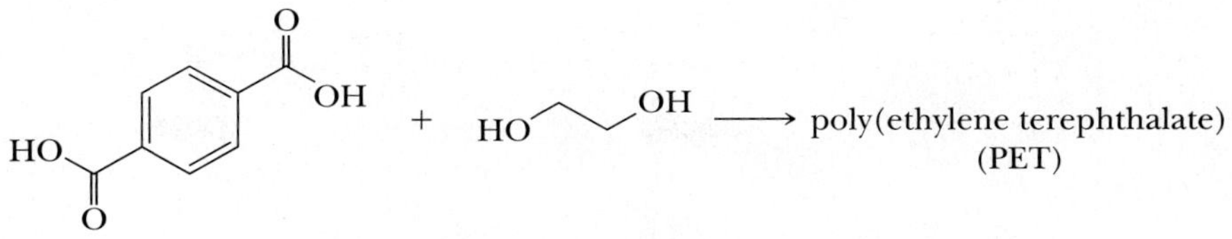

1,4-Benzenedicarboxylic acid	1,2-Ethanediol
(Terephthalic acid)	(Ethylene glycol)

As we shall see in Section 29.5B, the material produced in this reaction is a high-molecular-weight polymer, which can be fabricated into Mylar films, and into the textile fiber known as Dacron polyester.

18

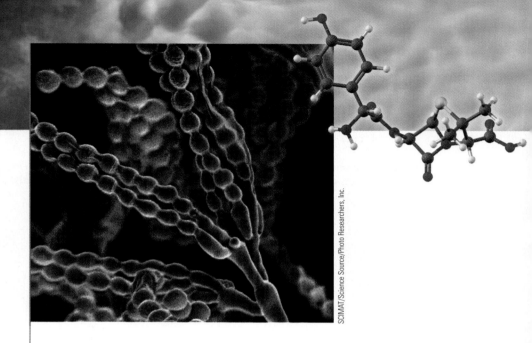

Colored scanning electron micrograph of *Penicillium s.* fungus. The stalklike objects are condiophores to which are attached numerous round condia. The condia are the fruiting bodies of the fungus. Inset: A model of amoxicillin. See Chemical Connections: "The Penicillins and Cephalosporins: *β*-Lactam Antibiotics."

SCIMAT/Science Source/Photo Researchers, Inc.

Functional Derivatives of Carboxylic Acids

In this chapter, we study five classes of organic compounds, each related to the carboxyl group: acid halides, acid anhydrides, esters, amides, and nitriles.

Under the general formula of each functional group is an illustration to show you how the group is formally related to a carboxylic acid. Formal loss of —OH from a carboxyl group and H— from H—Cl, for example, gives an acid chloride. Similarly, loss of —OH from a carboxyl group and H— from ammonia gives an amide.

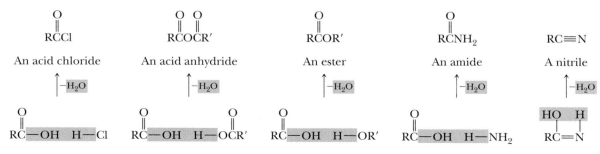

18.1 Structure and Nomenclature

A. Acid Halides

ORGANIC
Chemistry-ᐧ-Now™

Look for this logo in the chapter and go to Organic ChemistryNow at **http://now.brookscole.com/ bfi4** for tutorials, simulations, and problems.

The functional group of an **acid halide** (acyl halide) is an **acyl group (RCO—)** bonded to a halogen atom. Acid chlorides are the most common acid halides.

Acyl group An RCO— or ArCO— group.

An acyl group 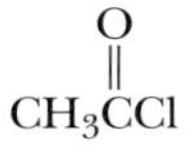 | Ethanoyl chloride (Acetyl chloride) 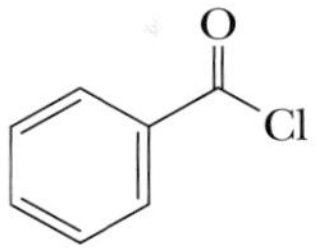 | Benzoyl chloride 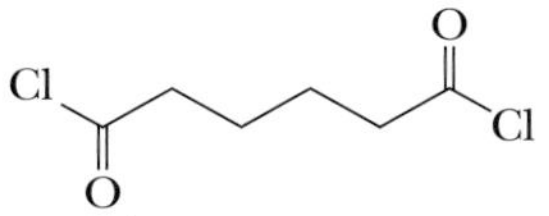 | Hexanedioyl chloride (Adipoyl chloride)

Acid halides are named by changing the suffix *-ic acid* in the name of the parent carboxylic acid to *-yl halide*.

Similarly, replacement of —OH in a sulfonic acid by chlorine gives a derivative called a **sulfonyl chloride.** Following are structural formulas for two sulfonic acids and the acid chloride derived from each.

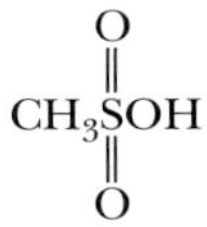

Methanesulfonic acid 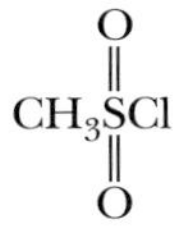 | Methanesulfonyl chloride (MsCl) 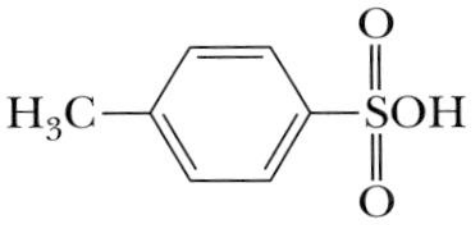 | *p*-Toluenesulfonic acid 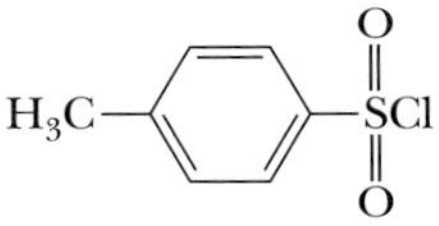 | *p*-Toluenesulfonyl chloride (Tosyl chloride, TsCl)

B. Acid Anhydrides

Carboxylic Anhydrides

The functional group of a **carboxylic anhydride** is two acyl groups bonded to an oxygen atom. These compounds are called acid anhydrides because they are formally derived from two carboxylic acids by the loss of water. An anhydride may be symmetrical (two identical acyl groups), or it may be mixed (two different acyl groups). Anhydrides are named by replacing the word *acid* in the name of the parent carboxylic acid with the word *anhydride.*

Acetic anhydride 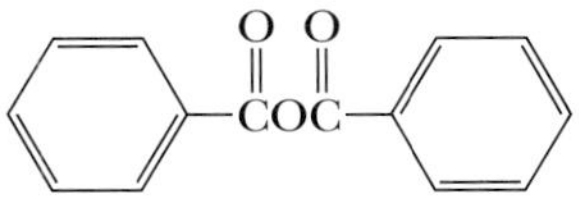 | Benzoic anhydride

Cyclic anhydrides are named from the dicarboxylic acids from which they are derived. Here are the cyclic anhydrides derived from succinic acid, maleic acid, and phthalic acid.

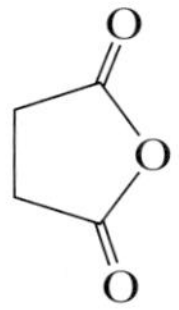

Succinic anhydride 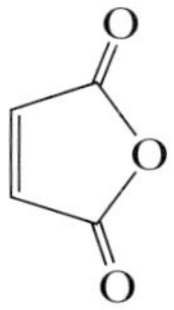 | Maleic anhydride 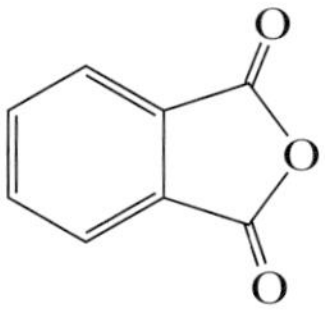 | Phthalic anhydride

Phosphoric Anhydrides

Because of the special importance of anhydrides of phosphoric acid in biological chemistry, we include them here to show their similarity with the anhydrides of carboxylic acids. The functional group of a **phosphoric anhydride** is two phosphoryl groups bonded to an oxygen atom. Here are structural formulas for two anhydrides of phosphoric acid and the ions derived by ionization of each acidic hydrogen.

Diphosphoric acid
(Pyrophosphoric acid)

Diphosphate ion
(Pyrophosphate ion)

Triphosphoric acid

Triphosphate ion

C. Esters

Esters of Carboxylic Acids

The functional group of a **carboxylic ester** is an acyl group bonded to —OR or —OAr. Both IUPAC and common names of esters are derived from the names of the parent carboxylic acids. The alkyl or aryl group bonded to oxygen is named first, followed by the name of the acid in which the suffix -*ic acid* is replaced by the suffix -*ate*.

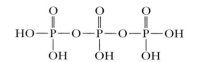

$$CH_3\overset{\displaystyle O}{\overset{\|}{C}}OCH_2CH_3$$

Ethyl ethanoate
(Ethyl acetate)

Diethyl butanedioate
(Diethyl malonate)

Lactones: Cyclic Esters

Lactone A cyclic ester.

Cyclic esters are called **lactones.** The IUPAC system has developed a set of rules for naming these compounds. Nonetheless, the simplest lactones are still named by dropping the suffix -*ic acid* or -*oic acid* from the name of the parent carboxylic acid and adding the suffix -*olactone*. The location of the oxygen atom in the ring is indicated by a number if the IUPAC name of the acid is used, or by a Greek letter α, β, γ, δ, ε, and so forth, if the common name of the acid is used.

Click Tutorials to review the
Nomenclature of Esters

(*S*)-3-Butanolactone
(β-Butyrolactone)

4-Butanolactone
(γ-Butyrolactone)

6-Hexanolactone
(ε-Caprolactone)

Esters of Phosphoric Acid

Phosphoric acid has three —OH groups and forms mono-, di-, and triesters, which are named by giving the name(s) of the alkyl or aryl group(s) bonded to oxygen followed by the word *phosphate,* as for example dimethyl phosphate. In more complex phosphoric esters, it is common to name the organic molecule and then indicate the presence of the phosphoric ester using either the word *phosphate* or the prefix *phospho-*. On the right are two phosphoric esters, each of special importance in the biological world.

CHEMICAL CONNECTIONS

From Cocaine to Procaine and Beyond

Cocaine is an alkaloid present in the leaves of the South American coca plant *Erythroxylon coca*. It was first isolated in 1880, and soon thereafter its property as a local anesthetic was discovered. Cocaine was introduced into medicine and dentistry in 1884 by two young Viennese physicians, Sigmund Freud and Karl Koller. Unfortunately, the use of cocaine can create a dependence, as Freud himself observed when he used it to wean a colleague from morphine and thereby produced one of the first documented cases of cocaine addiction.

Cocaine

Cocaine reduces fatigue, permits greater physical endurance, and gives a feeling of tremendous confidence and power. In some of the Sherlock Holmes stories, the great detective injects himself with a 7% solution of cocaine to overcome boredom.

After determining cocaine's structure, chemists could ask, "How is the structure of cocaine related to its anesthetic effects? Can the anesthetic effects be separated from the habituation effect?" If these questions could be answered, it might be possible to prepare synthetic drugs with the structural features essential for the anesthetic activity but without those giving rise to the undesirable effects. Chemists focused on three structural features of cocaine: its benzoic ester, its basic nitrogen atom, and something of its carbon skeleton. This search resulted in 1905 in the synthesis of procaine, which almost immediately replaced cocaine in dentistry and surgery. Lidocaine was introduced in 1948 and today is one of the most widely used local anesthetics. More recently, other members of the "caine" family of local anesthetics have been introduced, for example etidocaine. All of these local anesthetics are administered as their water-soluble hydrochloride salts.

| Procaine (Novocain) | Lidocaine (Xylocaine) | Etidocaine (Duranest; racemic) |

Thus, seizing on clues provided by nature, chemists have been able to synthesize drugs far more suitable for a specific function than anything known to be produced by nature itself.

Dimethyl phosphate   Glyceraldehyde 3-phosphate   Pyridoxal phosphate   Phosphoenolpyruvate

Glyceraldehyde 3-phosphate is an intermediate in glycolysis, the metabolic pathway by which glucose is converted to pyruvate. Pyridoxal phosphate is one of the metabolically active forms of vitamin B_6. Each of these esters is shown as it is ionized at pH 7.4,

From Moldy Clover to a Blood Thinner

In 1933, a disgruntled farmer delivered a pail of unclotted blood to the laboratory of Dr. Karl Link at the University of Wisconsin and tales of cows bleeding to death from minor cuts. Over the next couple of years, Link and his collaborators discovered that when cows are fed moldy clover, their blood clotting is inhibited, and they bleed to death from minor cuts and scratches. From the moldy clover they isolated the anticoagulant dicoumarol, a substance that delays or prevents blood clotting. Dicoumarol exerts its anticoagulation effect by interfering with vitamin K activity. Within a few years after its discovery, dicoumarol became widely used to treat victims of heart attack and others at risk for developing blood clots.

Dicoumarol is a derivative of coumarin, a lactone that gives sweet clover its pleasant smell. Coumarin, which does not interfere with blood clotting, is converted to dicoumarol as sweet clover becomes moldy.

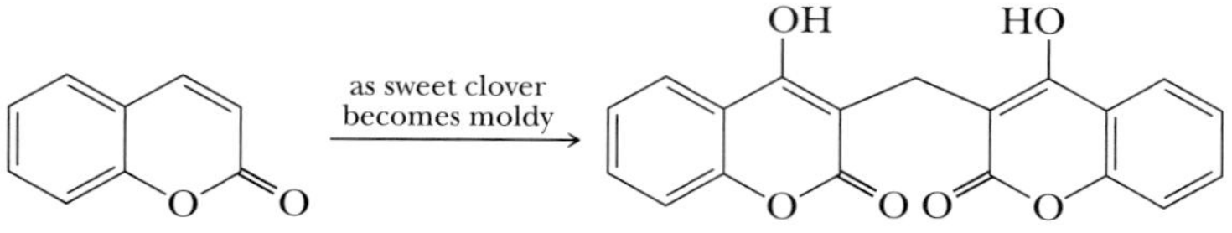

Coumarin
(from sweet clover)

Dicoumarol
(an anticoagulant)

In a search for even more potent anticoagulants, Link developed warfarin (named for the Wisconsin Alumni Research Foundation), now used primarily as a rat poison. When rats consume it, their blood fails to clot, and they bleed to death. Warfarin is also used as a blood anticoagulant in humans. The *S* enantiomer shown here is more active than the *R* enantiomer. The commercial product is sold as a racemic mixture. The synthesis of racemic warfarin is described in Problem 19.56.

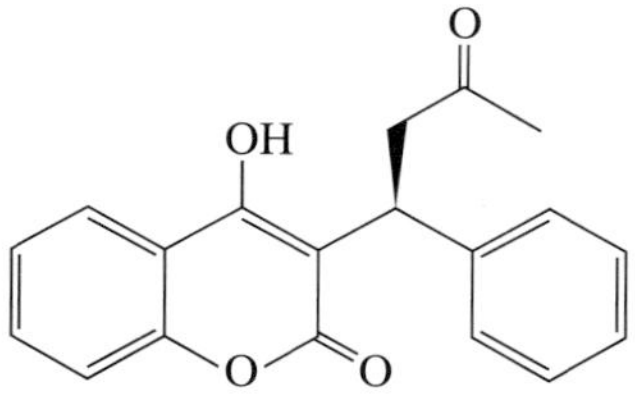

(*S*)-Warfarin
(A synthetic anticoagulant)

the pH of blood plasma; the two hydroxyl groups of these phosphoryl groups are ionized giving each a charge of -2.

D. Amides and Imides

The functional group of an **amide** is an acyl group bonded to a nitrogen atom. Amides are named by dropping the suffix *-oic acid* from the IUPAC name of the parent acid, or *-ic acid* from its common name, and adding *-amide*. If the nitrogen atom of an amide is bonded to an alkyl or aryl group, the group is named, and its location on nitrogen is indicated by *N-*. Two alkyl or aryl groups on nitrogen are indicated by *N,N*-di-. *N,N*-Dimethylformamide (DMF) is a widely used polar aprotic solvent (Section 9.2).

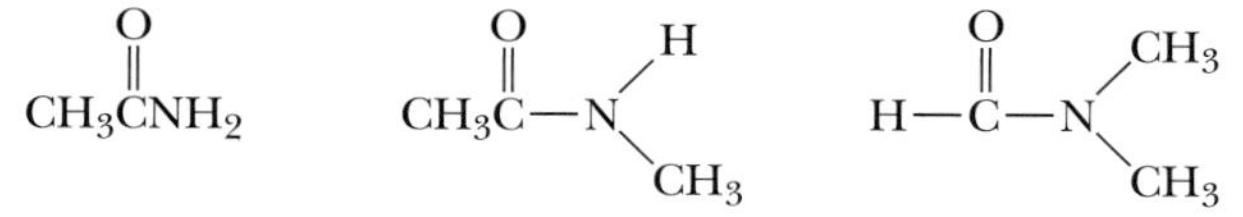

Acetamide
(a 1° amide)

N-Methylacetamide
(a 2° amide)

N,N-Dimethyl-
formamide (DMF)
(a 3° amide)

Cyclic amides are given the special name **lactam.** Their names are derived in a manner similar to those of lactones, with the difference that the suffix *-lactone* is replaced by *-lactam*.

Lactam   A cyclic amide.

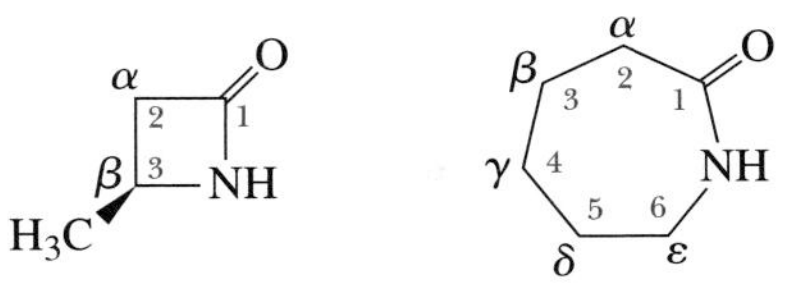

(*S*)-3-Butanolactam   6-Hexanolactam
(β-Butyrolactam)   (ε-Caprolactam)

The functional group of an **imide** is two acyl groups bonded to nitrogen. Both succinimide and phthalimide are cyclic imides.

Imide   A functional group in which two acyl groups, RCO— or ArCO—, are bonded to a nitrogen atom.

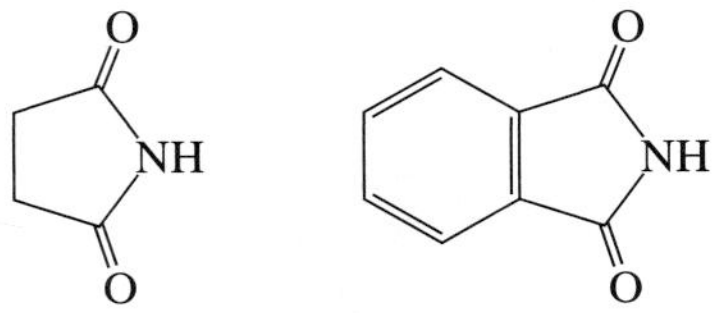

Succinimide   Phthalimide

ORGANIC
Chemistry··Now™
Click Tutorials to review the
Nomenclature of Amides

Example 18.1

Write the IUPAC name for each compound.

(a)

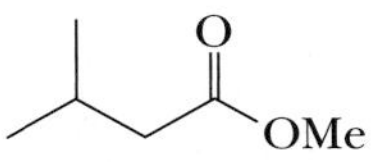

(b)

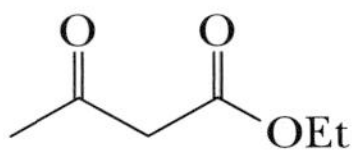

(c)

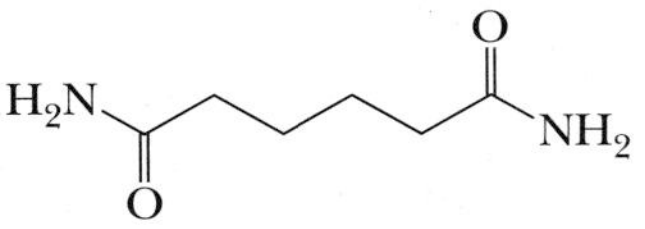

(d)

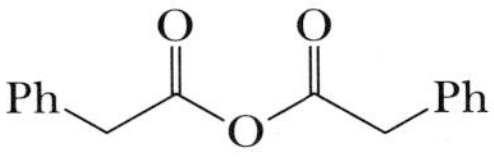

Solution

Given first is the IUPAC name and then, in parentheses, the common name.
(a) Methyl 3-methylbutanoate (methyl isovalerate, from isovaleric acid)
(b) Ethyl 3-oxobutanoate (ethyl β-ketobutyrate, from β-ketobutyric acid); also named ethyl acetoacetate (Section 19.5)
(c) Hexanediamide (adipamide, from adipic acid)
(d) Phenylethanoic anhydride (phenylacetic anhydride, from phenylacetic acid)

Problem 18.1

Draw a structural formula for each compound.
(a) *N*-Cyclohexylacetamide   (b) 1-Methylpropyl methanoate
(c) Cyclobutyl butanoate   (d) *N*-(1-Methylheptyl)succinimide
(e) Diethyl adipate   (f) 2-Aminopropanamide

The Penicillins and Cephalosporins: β-Lactam Antibiotics

The penicillins were discovered in 1928 by the Scottish bacteriologist Sir Alexander Fleming. As a result of the brilliant experimental work of Sir Howard Florey, an Australian pathologist, and Ernst Chain, a German chemist who fled Nazi Germany, penicillin G was introduced into the practice of medicine in 1943. For their pioneering work in developing one of the most effective antibiotics of all time, Fleming, Florey, and Chain were awarded the 1945 Nobel Prize in medicine or physiology.

The mold from which Fleming discovered penicillin was *Penicillium notatum,* a strain that gives a relatively low yield of penicillin. It was replaced in commercial production of the antibiotic by *P. chrysogenum,* a strain cultured from a mold found growing on a grapefruit in a market in Peoria, Illinois.

The structural feature common to all penicillins is a β-lactam ring fused to a five-membered thiazolidine ring.

The penicillins
differ in the
group bonded to
the acyl carbon

Amoxicillin
(a β-lactam antibiotic

The penicillins owe their antibacterial activity to a common mechanism that inhibits the biosynthesis of a vital part of bacterial cell walls.

Soon after the penicillins were introduced into medical practice, penicillin-resistant strains of bacteria began to appear and have since proliferated. One approach to combating resistant strains is to synthesize newer, more effective penicillins. Among those developed are ampicillin, methicillin, and amoxicillin. Another approach is

to search for newer, more effective β-lactam antibiotics. At the present time, the most effective of these are the cephalosporins, the first of which was isolated from the fungus *Cephalosporium acremonium.*

The cephalosporins differ in the
group bonded to the acyl carbon and
the side chain of the thiazine ring

Cephalexin
(Keflex)

The cephalosporin antibiotics have an even broader spectrum of antibacterial activity than the penicillins and are effective against many penicillin-resistant bacterial strains. However, resistance to the cephalosporins is now also widespread.

A common mechanism of resistance in bacteria involves their production of a specific enzyme, called a β-lactamase, that catalyzes the hydrolysis of the β-lactam ring, which is common to all penicillins and cephalosporins. Several compounds have been found that inhibit this enzyme, and now drugs based on these compounds can be taken in combination with penicillins and cephalosporins to restore their effectiveness when resistance is known to be a problem. The commonly prescribed formulation called Augmentin is a combination of a β-lactamase inhibitor and a penicillin. It is used as a second line of defense against childhood ear infections when resistance is suspected. Most children know it as the white liquid with the banana taste.

CONNECTIONS TO BIOLOGICAL CHEMISTRY

The Unique Structure of Amide Bonds

Amides have structural characteristics that are unique among carboxylic acid derivatives. Had you been asked in Chapter 1 to describe the geometry of an amide bond, you probably would have predicted bond angles of 120° about the carbonyl carbon and 109.5° about a tetrahedral amide nitrogen. You would have been correct about bond angles to the carbonyl carbon but wrong about the bond angles to the amide nitrogen. In the late 1930s, Linus Pauling discovered that the bond angles about the nitrogen atom of an amide bond in proteins are close to 120°; the amide nitrogen is trigonal planar and sp^2 hybridized rather than tetrahedral and sp^3 hybridized. We now know that amides are best represented as a hybrid of three resonance contributing structures.

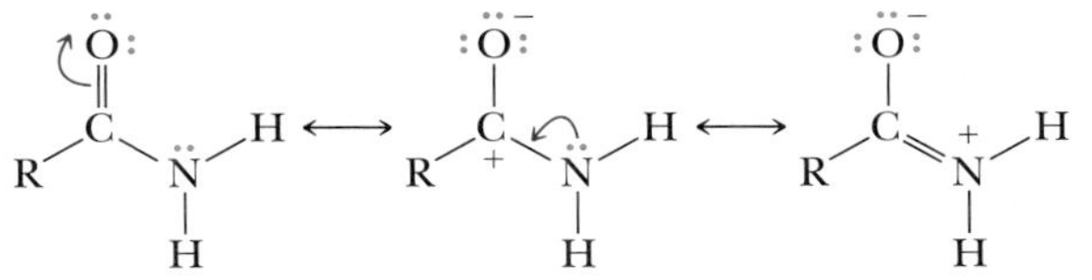

This contributing structure places a double bond between C and N

The fact that the six atoms of an amide bond are planar with bond angles of 120° means that the resonance structure on the right makes a significant contribution to the hybrid, and that the hybrid looks very much like this third structure. Inclusion of the third contributing structure explains why the amide nitrogen is sp^2 hybridized and therefore trigonal planar. Also, the presence of a partial double bond (pi bond) in the resonance hybrid indicates the presence of a restricted bond rotation about the C—N bond. The measured C—N bond rotation barrier in amides is approximately 63–84 kJ (15–20 kcal)/mol, large enough so that, at room temperature, rotation about the C—N bond is restricted. In addition, because the lone pair on nitrogen is delocalized into the pi bond, it is not as available for interacting with protons and other Lewis acids. Thus, amide nitrogens are not basic. In fact, in acid solution, amides are protonated on the carbonyl oxygen atom, rather than on the nitrogen (review Example 4.2). Finally, delocalization of the nitrogen lone pair reduces the electrophilic character (partial positive charge) on the carbonyl carbon, thus reducing the susceptibility of amides to nucleophilic attack.

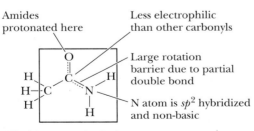

Amides protonated here

Less electrophilic than other carbonyls

Large rotation barrier due to partial double bond

N atom is sp^2 hybridized and non-basic

All of the atoms in the box are in the same plane

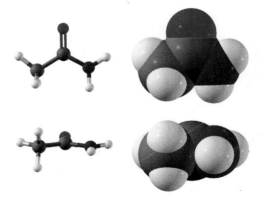

The amide —NH group is a good hydrogen bond donor, while the amide carbonyl is a good hydrogen bond acceptor, allowing both primary and secondary amides to form strong hydrogen bonds.

As we will see in Chapter 27, the ability of amides to participate in both intermolecular and intramolecular hydrogen bonding is an important factor in determining the three-dimensional structure of polypeptides and proteins.

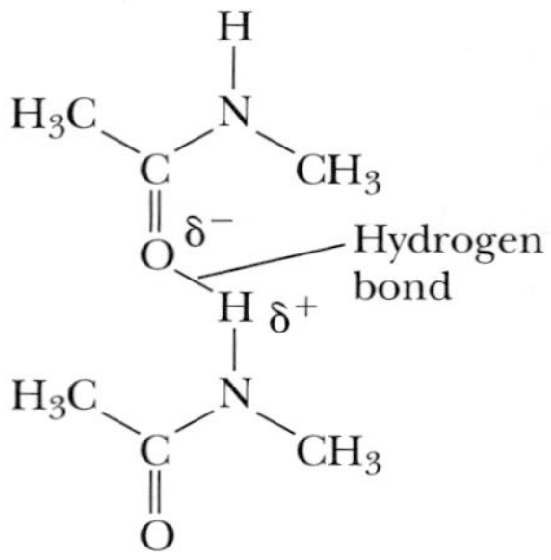

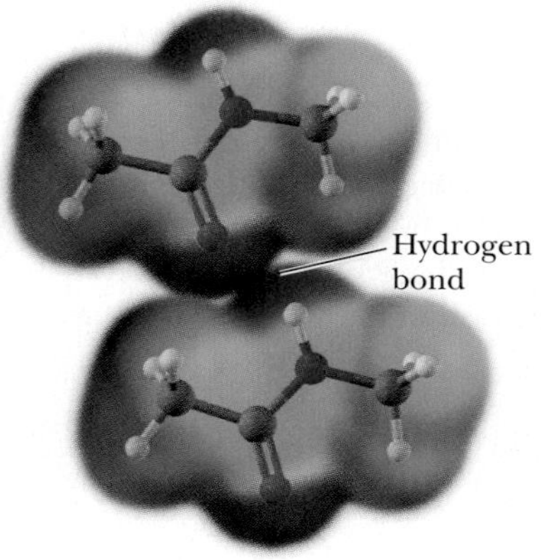

E. Nitriles

The functional group of a **nitrile** is a cyano (C≡N) group bonded to a carbon atom. IUPAC names follow the pattern alkanenitrile: for example, ethanenitrile. Common names are derived by dropping the suffix *-ic* or *-oic acid* from the name of the parent carboxylic acid and adding the suffix *-onitrile*.

$$CH_3C \equiv N$$

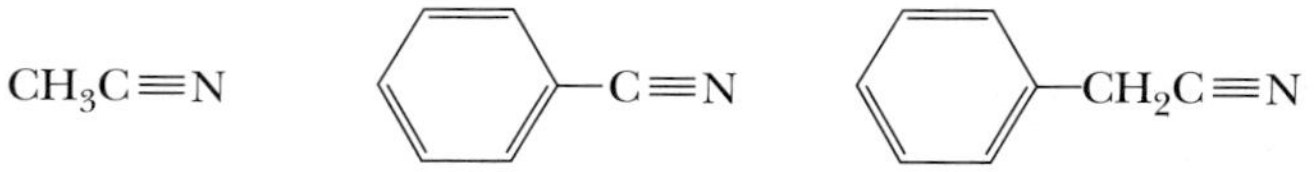

Ethanenitrile
(Acetonitrile)

Benzonitrile

Phenylethanenitrile
(Phenylacetonitrile)

18.2 Acidity of Amides, Imides, and Sulfonamides

Following are structural formulas of a primary amide, a sulfonamide, and two cyclic imides, along with pK_a values for each.

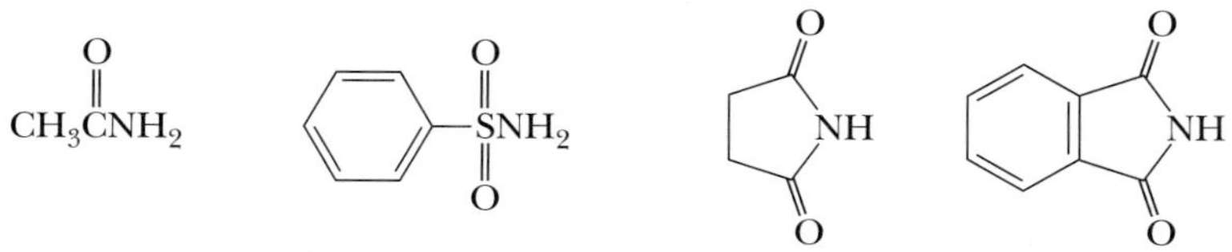

Acetamide
pK_a 15–17

Benzenesulfonamide
pK_a 10

Succinimide
pK_a 9.7

Phthalimide
pK_a 8.3

Values of pK_a for amides of carboxylic acids are in the range of 15–17, which means that they are comparable in acidity to alcohols. Amides show no evidence of acidity in aqueous solution; that is, water-insoluble amides do not react with aqueous solutions of NaOH or other alkali metal hydroxides to form water-soluble salts.

Imides (pK_a 8–10) are considerably more acidic than amides and readily dissolve in 5% aqueous NaOH by forming water-soluble salts. We account for the acidity of imides in the same manner as for the acidity of carboxylic acids (Section 17.4), namely the imide anion is stabilized by delocalization of its negative charge. The more important contributing structures for the anion formed by ionization of an imide delocalize the negative charge on nitrogen and the two carbonyl oxygens.

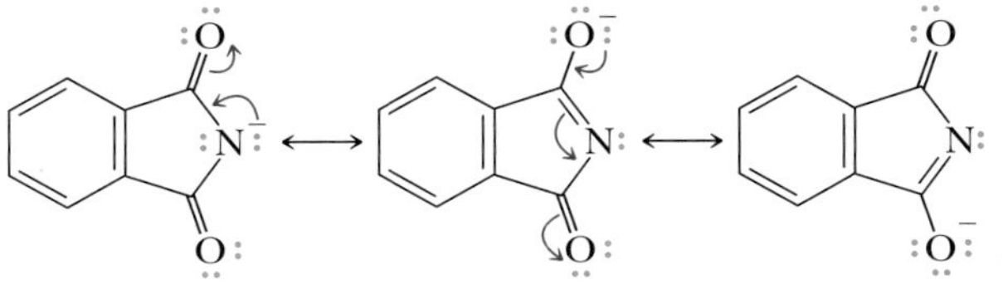

A resonance-stabilized anion

Sulfonamides derived from ammonia and primary amines are also sufficiently acidic to dissolve in aqueous solutions of NaOH or other alkali metal hydroxides by forming water-soluble salts. The pK_a of benzenesulfonamide is approximately 10. We account for the acidity of sulfonamides in the same manner as for imides, namely the resonance stabilization of the resulting anion.

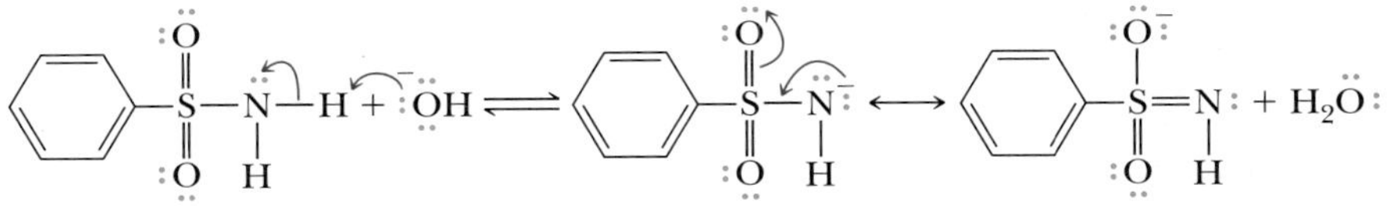

Benzenesulfonamide   A resonance-stabilized anion

Example 18.2

Phthalimide is insoluble in water. Will phthalimide dissolve in aqueous NaOH?

Solution

Phthalimide is the stronger acid, and NaOH is the stronger base. The position of equilibrium, therefore, lies to the right. Phthalimide dissolves in aqueous NaOH by forming a water-soluble sodium salt.

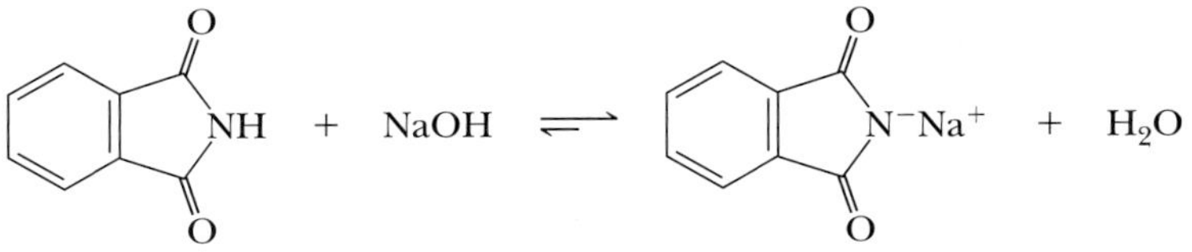

$$pK_{eq} = -7.4$$
$$K_{eq} = 2.5 \times 10^7$$

pK_a 8.3 pK_a 15.7
(stronger acid)   (stronger base)   (weaker base)   (weaker acid)

Problem 18.2

Will phthalimide dissolve in aqueous sodium bicarbonate?

The noncaloric artificial sweetener, saccharin, is an imide. The imide hydrogen of saccharin is sufficiently acidic that it reacts with sodium hydroxide and aqueous ammonia to form water-soluble salts. The ammonium salt is used to make liquid sweeteners. Saccharin is used in solid form as the Ca^{2+} salt.

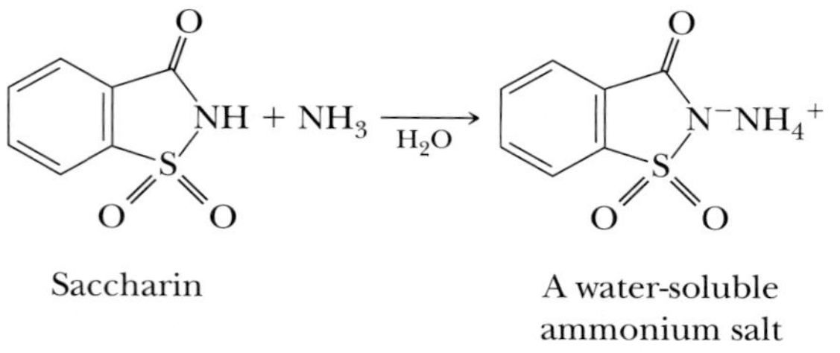

Saccharin   A water-soluble
                             ammonium salt

Saccharin is approximately 500 times sweeter than sugar and, at one time, was the most important noncaloric sweetener used in foods. At the present time, the most widely used noncaloric artificial sweetener is aspartame (Nutrasweet). For the structure of aspartame, see Problem 27.49.

18.3 Characteristic Reactions

In this and subsequent sections, we examine the interconversions of various carboxylic acid derivatives. All these reactions begin with formation of a tetrahedral carbonyl addition intermediate. The first step of this reaction is exactly analogous to the addition of alcohols to aldehydes and ketones (Section 16.7B). This reaction can be carried out under basic conditions, in which a negative or reactive neutral nucleophile adds directly to the carbonyl carbon. The tetrahedral carbonyl addition intermediate formed then adds a proton from a proton donor, HA. The result of this reaction is nucleophilic addition.

Nucleophilic
addition
(basic conditions):

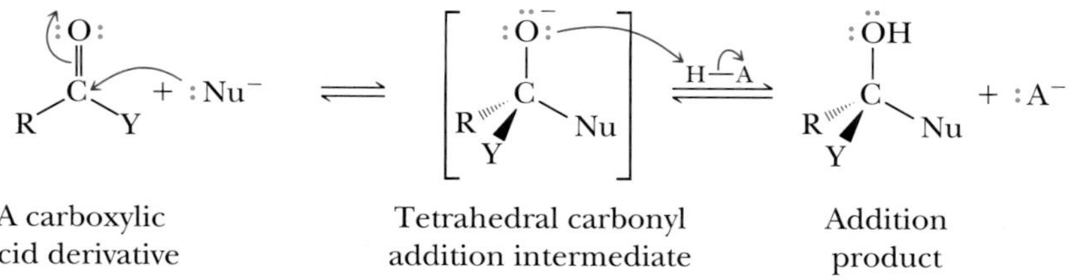

A carboxylic   Tetrahedral carbonyl   Addition
acid derivative   addition intermediate   product

As with aldehydes and ketones, this reaction can also be catalyzed by acid, in which case protonation of the carbonyl oxygen precedes the attack of the nucleophile.

Nucleophilic
addition
(acid conditions):

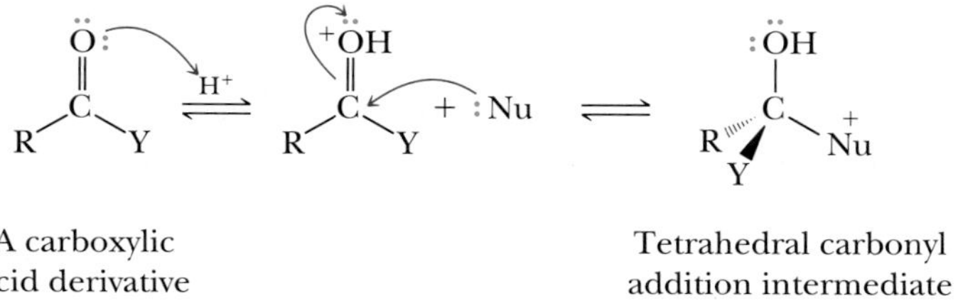

A carboxylic   Tetrahedral carbonyl
acid derivative   addition intermediate

For functional derivatives of carboxylic acids, the fate of the tetrahedral carbonyl addition intermediate is quite different from that of aldehydes and ketones; the intermediate collapses to expel the leaving group Lv and regenerate the carbonyl group. The result of this addition-elimination sequence is **nucleophilic acyl substitution.**

Nucleophilic acyl substitution:

$$\text{R}-\overset{\overset{\displaystyle ::O:}{\|}}{\underset{\text{Lv}}{C}} + :\text{Nu}^- \longrightarrow \left[\underset{\underset{\displaystyle \text{Lv}}{R}}{\overset{\overset{\displaystyle :\ddot{O}:^-}{|}}{C}} \text{Nu} \right] \longrightarrow \text{R}-\overset{\overset{\displaystyle ::O:}{\|}}{\underset{\text{Nu}}{C}} + :\text{Lv}^-$$

Tetrahedral carbonyl Substitution
addition intermediate product

Nucleophilic acyl substitution A reaction in which a nucleophile bonded to the carbon of an acyl group is replaced by another nucleophile.

The major difference between these two types of carbonyl addition reactions is that aldehydes and ketones do not have a group that can leave as a relatively stable anion. They undergo only nucleophilic acyl addition. The four carboxylic acid derivatives we study in this chapter have a leaving group, Y, that can leave as a relatively stable anion or as a neutral species.

In this general reaction, we show the nucleophile and the leaving group as anions. This need not be the case. Neutral molecules, such as water, alcohols, ammonia, and amines, may also serve as nucleophiles and leaving groups in this reaction, mainly when it is carried out under acid-catalyzed conditions. We show the leaving groups here as anions, however, to illustrate an important point about leaving groups: The weaker the base (the more stable the anion), the better the leaving group (Section 9.4F).

$$\text{R}_2\text{N}^- \qquad \text{RO}^- \qquad \overset{\overset{\displaystyle O}{\|}}{\text{RCO}^-} \qquad \text{X}^-$$

Increasing leaving ability →

← **Increasing basicity**

The weakest base in the series, and the best leaving group, is halide ion; acid halides are the most reactive toward nucleophilic acyl substitution. The strongest base, and the poorest leaving group, is amide ion; amides are the least reactive toward nucleophilic acyl substitution. Acid halides and acid anhydrides are so reactive that they are not found in nature. Esters and amides, however, are universally present.

The degree of partial positive charge on the carbonyl carbon increases in order from amide to ester to acid anhydride to acid chloride. This increase is shown clearly by the increasing amount of blue on the carbonyl carbon in the electrostatic potential maps.

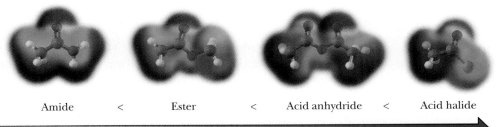

Amide < Ester < Acid anhydride < Acid halide

Increasing reactivity toward nucleophilic acyl substitution →

Given this trend, the carbonyl carbon of an acid chloride is most susceptible of the four functional groups to attack by a nucleophile; the carbonyl carbon of the amide is the least susceptible to nucleophilic attack.

Taken together, the combined effects of leaving group ability and susceptibility to nucleophilic attack reinforce each other. Acid chlorides are the most susceptible to nucleophilic attack, and chloride is the best leaving group, making it clear that acid chlorides have the highest reactivity of the four toward nucleophilic substitution reactions. Amides are overall the least reactive.

$$
\begin{array}{cccc}
\overset{\text{O}}{\overset{\|}{\text{RCNH}_2}} & \overset{\text{O}}{\overset{\|}{\text{RCOR}'}} & \overset{\text{O}\quad\text{O}}{\overset{\|\quad\|}{\text{RCOCR}'}} & \overset{\text{O}}{\overset{\|}{\text{RCX}}} \\
\text{Amide} & \text{Ester} & \text{Anhydride} & \text{Acid halide}
\end{array}
$$

> **Increasing reactivity toward nucleophilic acyl substitution** ⟶

In addition, many reactions of the less reactive carboxyl derivatives occur by acid catalysis. In these reactions, the carbonyl group is protonated in the first step, which increases its electrophilicity and facilitates nucleophilic attack. In addition, the leaving group is protonated by the acid in a later step to decrease its basicity and make it a better leaving group. We will see detailed mechanisms for many examples of both acid- and base-catalyzed reactions in this chapter. We saw the acid-catalyzed reversible reaction of acids with alcohols in Section 17.7A.

18.4 Reaction with Water: Hydrolysis

A. Acid Chlorides

Low-molecular-weight acid chlorides react very rapidly with water to form carboxylic acids and HCl.

$$
\overset{\text{O}}{\overset{\|}{\text{CH}_3\text{CCl}}} + \text{H}_2\text{O} \longrightarrow \overset{\text{O}}{\overset{\|}{\text{CH}_3\text{COH}}} + \text{HCl}
$$

Acetyl chloride

Higher molecular-weight acid halides are less soluble and, consequently, react less rapidly with water. Because the mechanisms for hydrolysis of acid chlorides and anhydrides are identical, we show the mechanism only for acid anhydrides (Section 18.4B).

B. Acid Anhydrides

Anhydrides are generally less reactive than acid chlorides. However the lower molecular-weight anhydrides also react readily with water to form two molecules of carboxylic acid.

$$
\overset{\text{O}\quad\text{O}}{\overset{\|\quad\|}{\text{CH}_3\text{COCCH}_3}} + \text{H}_2\text{O} \longrightarrow \overset{\text{O}}{\overset{\|}{\text{CH}_3\text{COH}}} + \overset{\text{O}}{\overset{\|}{\text{HOCCH}_3}}
$$

Acetic anhydride

Mechanism   *Hydrolysis of an Acid Anhydride*

The following mechanism is divided into two stages: first, nucleophilic addition of water forming a tetrahedral carbonyl addition intermediate, and second, collapse of this intermediate by elimination of acetate ion, a moderate base, and a good leaving group. This type of addition/elimination mechanism is characteristic of nucleophilic acyl substitution reactions.

Step 1: Protonation of a carbonyl oxygen (not shown) followed by addition of H_2O to one of the carbonyl groups gives a tetrahedral carbonyl addition intermediate.

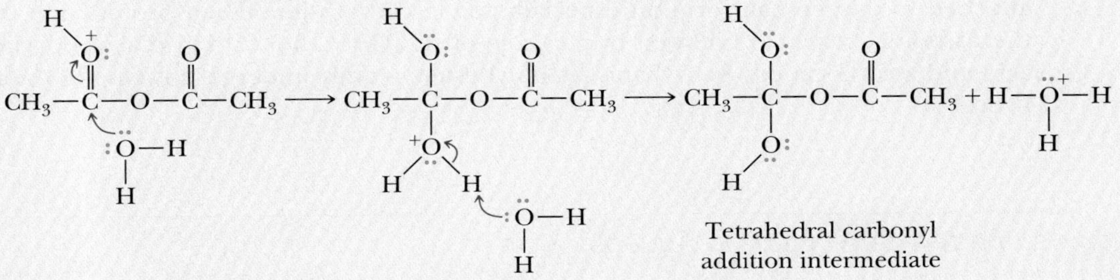

Tetrahedral carbonyl
addition intermediate

This reaction is acid catalyzed but will occur without added acid. After a few molecules of anhydride have reacted with water, molecules of the organic acid are produced and then catalyze further reaction.

Step 2: Protonation of the leaving group followed by collapse of the tetrahedral carbonyl addition intermediate gives two molecules of carboxylic acid and generates another acid catalyst.

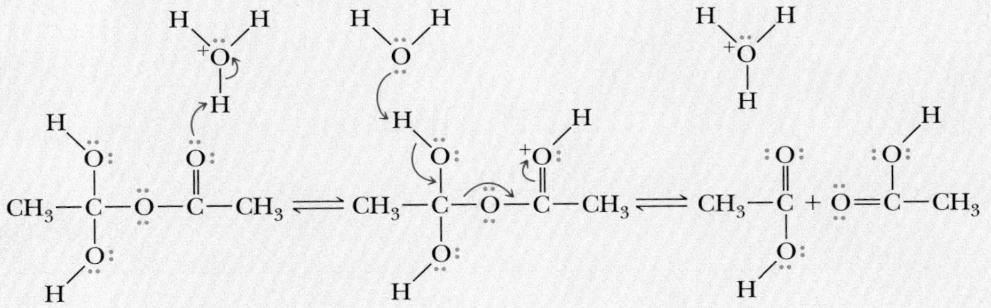

ORGANIC
Chemistry ⚛ Now™
Click Mechanisms to view
an animation of the **Acid
Anhydride Hydrolysis**

C. Esters

Esters are hydrolyzed only very slowly, even in boiling water. Hydrolysis becomes considerably more rapid, however, when they are refluxed in aqueous acid or base. We discussed acid-catalyzed (Fischer) esterification in Section 17.7A and pointed out that it is an equilibrium reaction. Hydrolysis of esters in aqueous acid is also an equilibrium reaction and proceeds by the same nucleophilic addition/elimination mechanism as esterification, except in reverse. The role of the acid catalyst is to protonate the carbonyl oxygen. In doing so, it increases the electrophilic character of the carbonyl carbon toward attack by water to form a tetrahedral carbonyl addition intermediate. Collapse of this intermediate gives the carboxylic acid and an alcohol. In this reaction, acid is a catalyst; it is consumed in the first step, but another is generated at the end of the reaction.

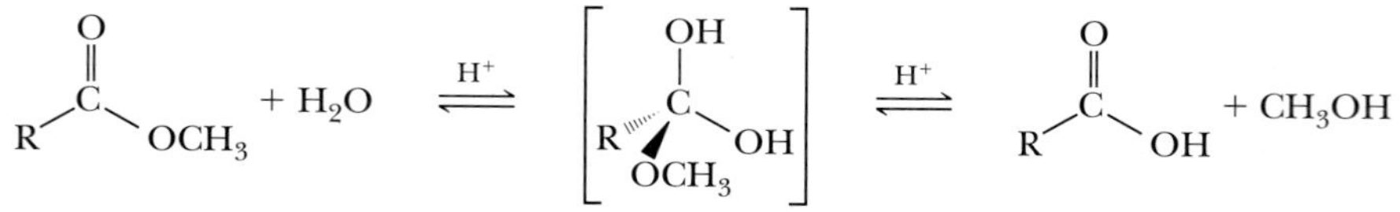

$$R-C(=O)-OCH_3 + H_2O \underset{H^+}{\overset{H^+}{\rightleftharpoons}} \left[R-C(OH)_2-OCH_3 \right] \underset{H^+}{\overset{H^+}{\rightleftharpoons}} R-C(=O)-OH + CH_3OH$$

Tetrahedral carbonyl
addition intermediate

Although formation of a tetrahedral carbonyl addition intermediate is the most common mechanism for the hydrolysis of esters in aqueous acid, alternative mechanistic pathways are followed in special cases. Such a case occurs when the alkyl group attached to oxygen can form an especially stable carbocation. Then protonation of the carbonyl oxygen is followed by cleavage of the O—C bond to give a carboxylic acid and a carbocation. Benzyl and *tert*-butyl esters readily undergo this type of ester hydrolysis.

Mechanism *Hydrolysis of a tert-Butyl Ester in Aqueous Acid*

Step 1: Proton transfer to the carbonyl oxygen gives a cation, which is the conjugate acid of the ester.

Step 2: Rearrangement of electron pairs in the cation intermediate gives the carboxylic acid and *tert*-butyl cation. This cation then either reacts with water to give *tert*-butyl alcohol (S_N1) or transfers a proton to water to give 2-methylpropene (E1).

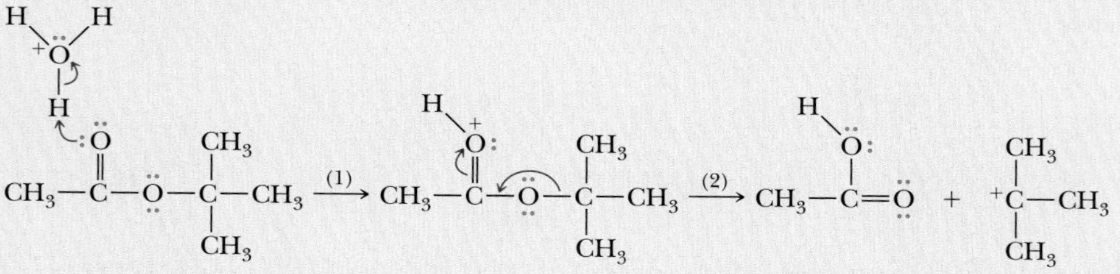

tert-Butyl cation

Hydrolysis of esters may also be carried out using hot aqueous base, such as aqueous NaOH.

$$RC(=O)OCH_3 + NaOH \xrightarrow{H_2O} RC(=O)O^-Na^+ + CH_3OH$$

Saponification Hydrolysis of an ester in aqueous NaOH or KOH to an alcohol and the sodium or potassium salt of a carboxylic acid.

Hydrolysis of esters in aqueous base is often called **saponification**, a reference to the use of this reaction in the manufacture of soaps (Section 26.2A). Although the carbonyl carbon of an ester is not strongly electrophilic, hydroxide ion is a good nucleophile and adds to the carbonyl carbon to form a tetrahedral carbonyl addition intermediate, which in turn collapses to give a carboxylic acid and an alkoxide ion. The carboxylic acid reacts with the alkoxide ion or other base present to form a carboxylate anion. Thus, each mole of ester hydrolyzed requires one mole of base.

Mechanism *Hydrolysis of an Ester in Aqueous Base (Saponification)*

Step 1: Addition of hydroxide ion to the carbonyl carbon of the ester gives a tetrahedral carbonyl addition intermediate.

Step 2: Collapse of this intermediate gives a carboxylic acid and an alkoxide ion.

Step 3: Proton transfer between the carboxyl group and the alkoxide ion gives the carboxylate anion. This strongly exothermic acid-base reaction drives the whole reaction to completion.

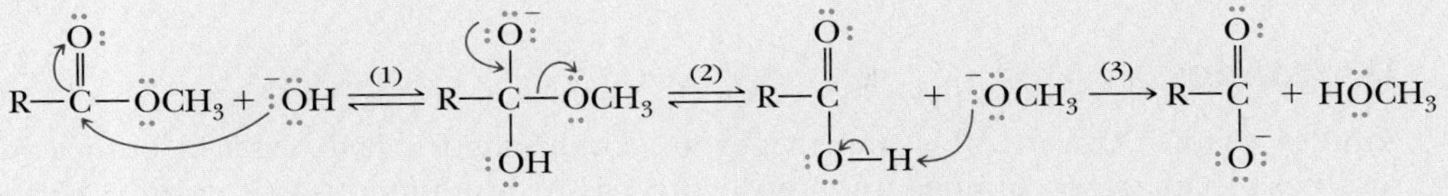

There are two major differences between hydrolysis of esters in aqueous acid and aqueous base.

1. For hydrolysis of an ester in aqueous acid, acid is required in only catalytic amounts. For hydrolysis in aqueous base, base in required in stoichiometric amounts because it is a reactant, not a catalyst.
2. Hydrolysis of an ester in aqueous acid is reversible, but hydrolysis in aqueous base is irreversible because a carboxylate anion (weakly electrophilic, if at all) is not attacked by ROH (a weak nucleophile).

Other acid derivatives react with base in an identical manner to esters.

ORGANIC **Chemistry Now**™
Click Mechanisms to view an animation of the **Ester Hydrolysis in Aqueous Base**

Example 18.3

Complete and balance equations for the hydrolysis of each ester in aqueous sodium hydroxide. Show all products as they are ionized under these conditions.

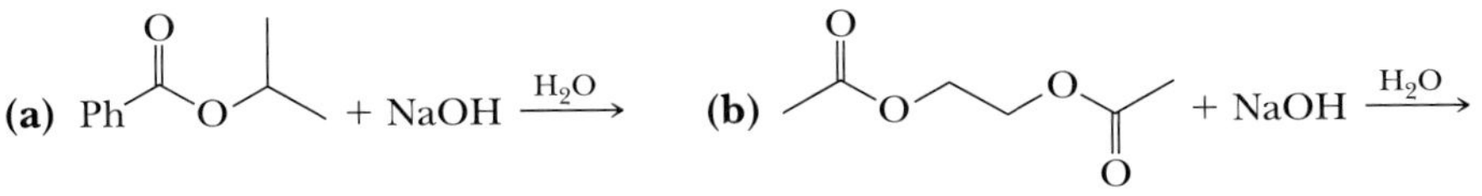

Solution

The products of hydrolysis of (a) are benzoic acid and 2-propanol. In aqueous NaOH, benzoic acid is converted to its sodium salt. Therefore, one mole of NaOH is required for hydrolysis of one mole of this ester. Compound (b) is a diester of ethylene glycol. Two moles of NaOH are required for its hydrolysis.

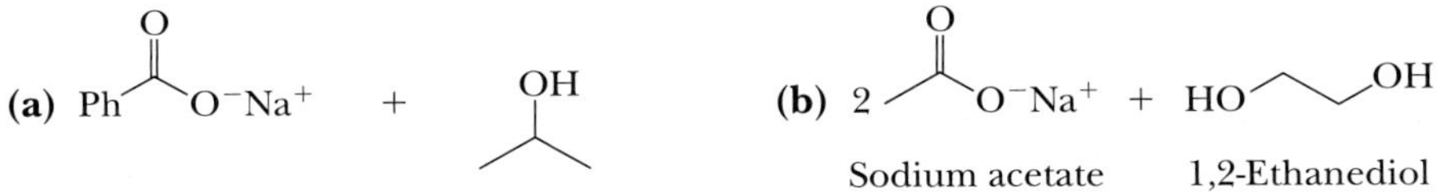

Sodium benzoate, 2-Propanol (Isopropyl alcohol), Sodium acetate, 1,2-Ethanediol (Ethylene glycol)

Problem 18.3

Complete and balance equations for the hydrolysis of each ester in aqueous solution; show each product as it is ionized under the indicated experimental conditions.

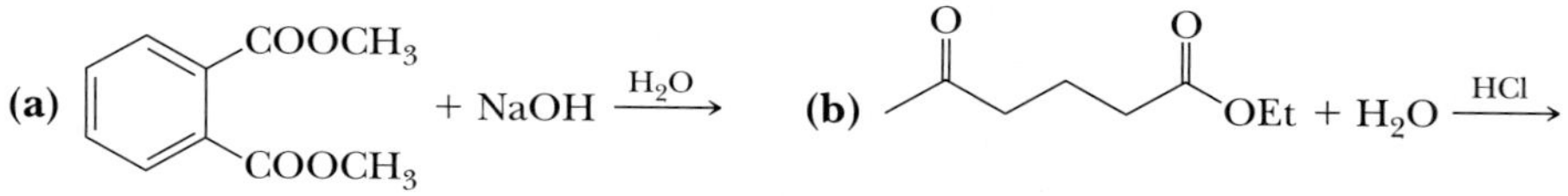

(a) [structure: benzene ring with two COOCH$_3$ groups] + NaOH $\xrightarrow{H_2O}$

(b) [structure: diketone/ester chain] OEt + H$_2$O $\xrightarrow{HCl}$

D. Amides

Amides require considerably more vigorous conditions for hydrolysis in both acid and base than esters. Amides undergo hydrolysis in hot aqueous acid to give a carboxylic acid and an ammonium ion. Hydrolysis is driven to completion by the acid-base reaction between ammonia or the amine and acid to form an ammonium salt. One mole of acid is required per mole of amide.

[structure of (R)-2-Phenylbutanamide] + H$_2$O + HCl $\xrightarrow[\text{heat}]{H_2O}$ [structure of (R)-2-Phenylbutanoic acid] + NH$_4{}^+$Cl$^-$

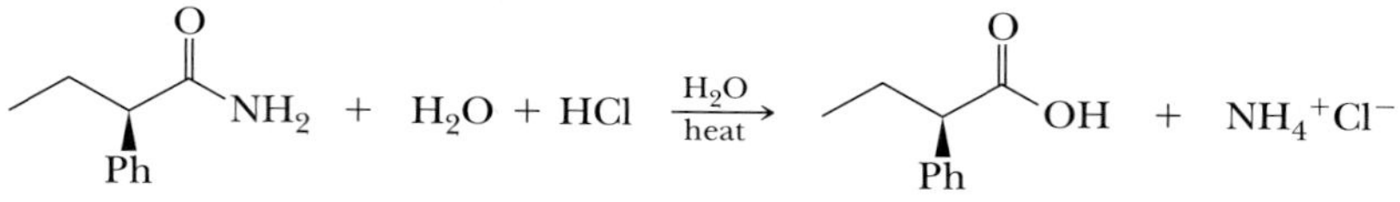

(R)-2-Phenylbutanamide   (R)-2-Phenylbutanoic acid

In aqueous base, the products of amide hydrolysis are a carboxylate salt and ammonia or an amine. Hydrolysis in aqueous base is driven to completion by the acid-base reaction between the resulting carboxylic acid and base to form a salt. One mole of base is required per mole of amide.

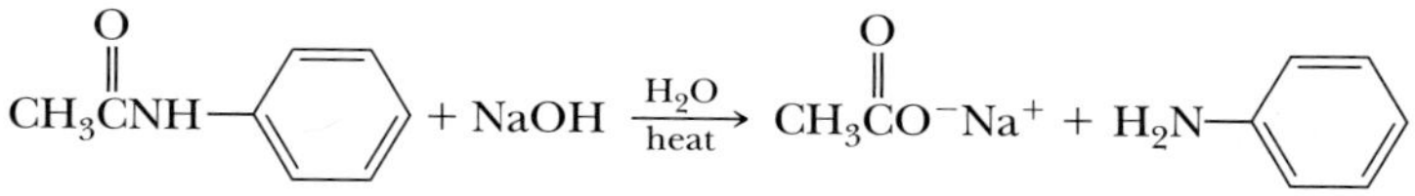

[structure] CH$_3$CNH—[benzene ring] + NaOH $\xrightarrow[\text{heat}]{H_2O}$ CH$_3$CO$^-$Na$^+$ + H$_2$N—[benzene ring]

N-Phenylethanamide   Sodium acetate   Aniline
(N-Phenylacetamide,
Acetanilide)

The steps in the mechanism for the hydrolysis of amides in aqueous acid are similar to those for the hydrolysis of esters in aqueous acid.

Mechanism *Hydrolysis of an Amide in Aqueous Acid*

Step 1: Protonation of the carbonyl oxygen gives a resonance-stabilized cation intermediate.

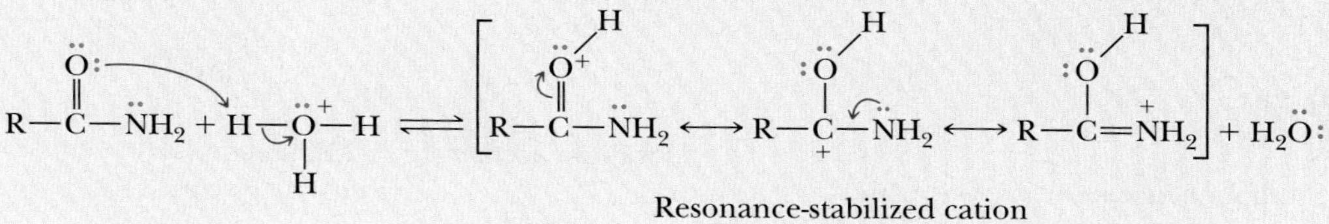

[mechanism structures]

R—C—NH$_2$ + H—O$^+$—H $\rightleftharpoons$ [R—C—NH$_2$ $\longleftrightarrow$ R—C—NH$_2$ $\longleftrightarrow$ R—C=NH$_2$] + H$_2$O:

Resonance-stabilized cation

The role of the proton in this step is to protonate the carbonyl oxygen to increase the electrophilic character of the carbonyl carbon.

Step 2: Addition of water to the carbonyl carbon of the cation intermediate followed by proton transfer gives a tetrahedral carbonyl addition intermediate.

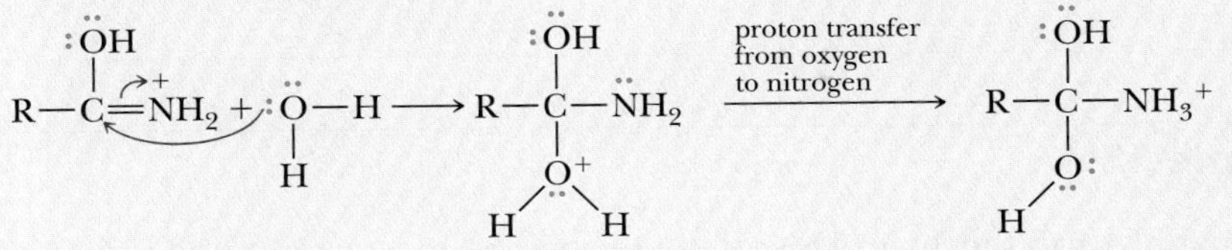

Tetrahedral carbonyl
addition intermediate

Step 3: Collapse of the tetrahedral carbonyl addition intermediate coupled with proton transfer gives a carboxylic acid and ammonium ion.

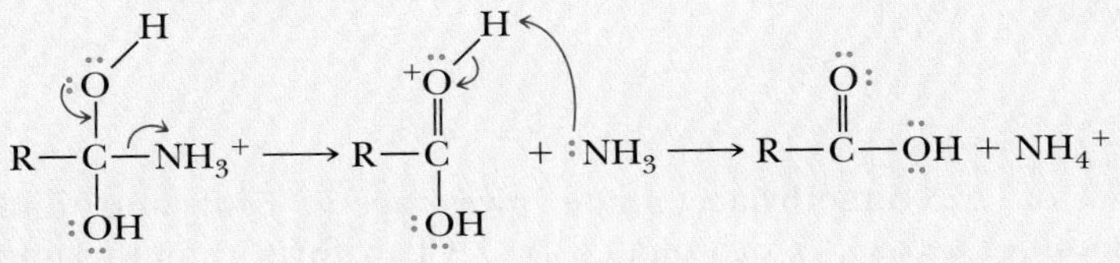

ORGANIC
Chemistry⋅ᵢ⋅Now™
Click Mechanisms to view
an animation of the **Acid
Hydrolysis of an Amide**

Note that the leaving group in this step is a neutral amine (a weaker base), a far better leaving group than an amide ion (a much stronger base).

The mechanism for the hydrolysis of amides in aqueous base is similar to that for the hydrolysis of esters in aqueous base.

Mechanism   *Hydrolysis of an Amide in Aqueous Base*

Step 1: Addition of hydroxide ion to the carbonyl carbon gives a tetrahedral carbonyl addition intermediate as with esters.

Step 2: Collapse of the tetrahedral carbonyl addition intermediate gives a carboxylic acid and ammonia.

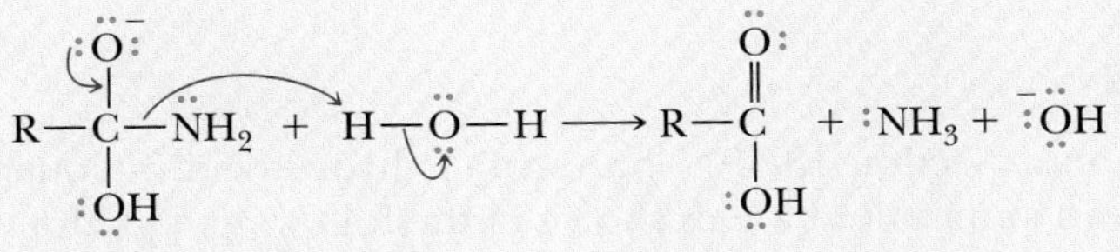

Tetrahedral carbonyl
addition intermediate

Loss of nitrogen and proton transfer from water to nitrogen are concerted so that the leaving group is not amide ion, NH_2^-, a stronger base, and poorer leaving group, but rather ammonia, NH_3, a weaker base, and better leaving group.

Step 3: Proton transfer to form the carboxylate anion and water. Hydrolysis is driven to completion by this acid-base reaction.

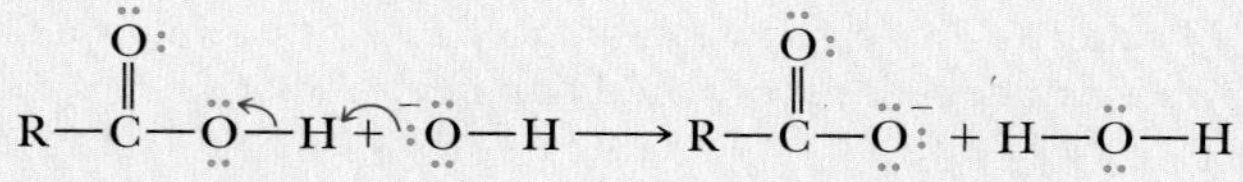

$$R-\overset{\overset{\displaystyle \ddot{O}:}{\|}}{C}-\overset{..}{\underset{..}{O}}\overset{\frown}{-}H\overset{+}{}:\overset{-}{\underset{..}{O}}-H \longrightarrow R-\overset{\overset{\displaystyle \ddot{O}:}{\|}}{C}-\overset{..}{\underset{..}{O}}:^{-} + H-\overset{..}{\underset{..}{O}}-H$$

Example 18.4

Write equations for the hydrolysis of these amides in concentrated aqueous HCl. Show all products as they exist in aqueous HCl, and the number of moles of HCl required for hydrolysis of each amide.

(a) $CH_3\overset{\overset{\displaystyle O}{\|}}{C}N(CH_3)_2$ (b)

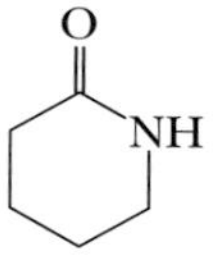

Solution

(a) Hydrolysis of *N,N*-dimethylacetamide gives acetic acid and dimethylamine. Dimethylamine, a base, is protonated by HCl to form dimethylammonium ion and is shown in the balanced equation as dimethylammonium chloride. One mole of HCl is required per mole of amide.

$$CH_3\overset{\overset{\displaystyle O}{\|}}{C}N(CH_3)_2 + H_2O + HCl \xrightarrow{heat} CH_3\overset{\overset{\displaystyle O}{\|}}{C}OH + (CH_3)_2NH_2{}^+Cl^-$$

(b) Hydrolysis of this δ-lactam gives the protonated form of 5-aminopentanoic acid. One mole of HCl is required per mole of amide.

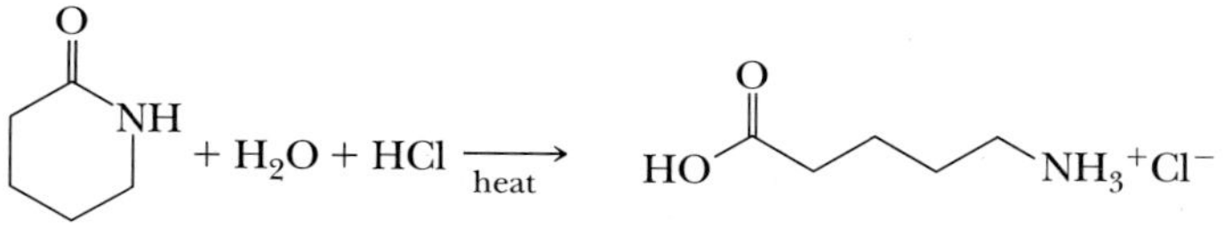

Problem 18.4

Complete equations for the hydrolysis of the amides in Example 18.4 in concentrated aqueous NaOH. Show all products as they exist in aqueous NaOH and the number of moles of NaOH required for hydrolysis of each amide.

E. Nitriles

The cyano group (Section 18.5E) of a nitrile is hydrolyzed in aqueous acid to a carboxyl group and ammonium ion as shown in the following equation.

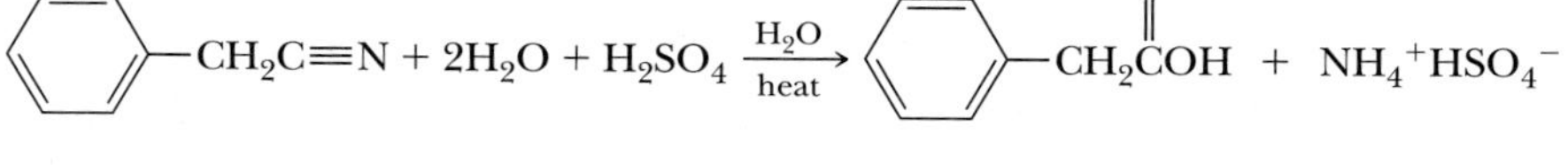

Phenylacetonitrile   Phenylacetic acid   Ammonium hydrogen sulfate

In hydrolysis of a cyano group in aqueous acid, protonation of the nitrogen atom gives a cation that reacts with water to give an imidic acid (the enol of an amide). Keto-enol tautomerism of the imidic acid gives an amide. The amide is then hydrolyzed, as already described, to a carboxylic acid and an ammonium ion.

$$R-C\equiv N + H_2O \xrightarrow{\;H^+\;} \underset{\substack{\text{An imidic acid}\\ \text{(enol of an amide)}}}{R-\overset{\overset{\displaystyle OH}{|}}{C}=NH} \;\rightleftharpoons\; \underset{\text{An amide}}{R-\overset{\overset{\displaystyle O}{\|}}{C}-NH_2}$$

The reaction conditions required for acid-catalyzed hydrolysis of a cyano group are typically more vigorous than those required for hydrolysis of an amide, and in the presence of excess water, a cyano group is hydrolyzed first to an amide and then to a carboxylic acid. It is possible to stop at the amide by using sulfuric acid as a catalyst and one mole of water per mole of nitrile. Selective hydrolysis of a nitrile to an amide, however, is not a good method for the preparation of amides. They are better prepared from acid chlorides, acid anhydrides, or esters.

Hydrolysis of a cyano group in aqueous base gives a carboxylate anion and ammonia. The reaction is driven to completion by the acid-base reaction between the carboxylic acid and base to form a carboxylate anion. Acidification of the reaction mixture during workup converts the carboxylate anion to the carboxylic acid.

$$\underset{\text{Undecanenitrile}}{CH_3(CH_2)_9C\equiv N} + H_2O + NaOH \xrightarrow[\text{heat}]{H_2O} \underset{\text{Sodium undecanoate}}{CH_3(CH_2)_9\overset{\overset{\displaystyle O}{\|}}{C}O^-Na^+} + NH_3$$

$$\Big\downarrow\, HCl \mid H_2O$$

$$\underset{\text{Undecanoic acid}}{CH_3(CH_2)_9\overset{\overset{\displaystyle O}{\|}}{C}OH} + NaCl + NH_4Cl$$

Mechanism   *Hydrolysis of a Cyano Group to an Amide in Aqueous Base*

Hydrolysis of a cyano group in aqueous base involves initial formation of the anion of an imidic acid, which, after proton transfer from water, undergoes keto-enol tautomerism to give an amide. The amide is then hydrolyzed by aqueous base, as we have seen earlier, to the carboxylate anion and ammonia.

Step 1: Addition of hydroxide ion to the carbon of the cyano group followed by proton transfer from water gives an imidic acid.

$$HO^- + \underset{\text{A nitrile}}{R-C\equiv N:} \longrightarrow R-\overset{\overset{\displaystyle :\ddot{O}H}{|}}{C}=N:^- \xrightarrow{\;H-\ddot{O}-H\;} \underset{\text{An imidic acid}}{R-\overset{\overset{\displaystyle :\ddot{O}H}{|}}{C}=N-H} + :\ddot{O}H^-$$

Step 2: Tautomerism of the imidic acid gives the amide.

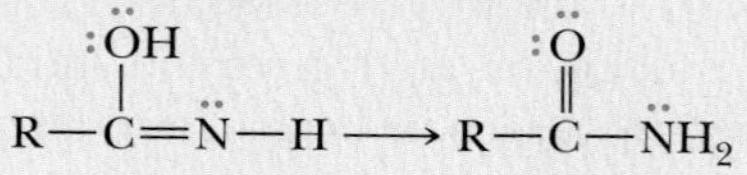

$$R-\overset{:\ddot{O}H}{C}=\ddot{N}-H \longrightarrow R-\overset{:\ddot{O}}{\underset{}{C}}-\ddot{N}H_2$$

An imidic acid   An amide

The acid-catalyzed reaction proceeds similarly; the only difference is in the order of proton transfers.

Hydrolysis of nitriles is a valuable route to the synthesis of carboxylic acids from primary or secondary alkyl halides. In this route, one carbon in the form of a cyano group (Table 8.1) is added to a carbon chain and then converted to a carboxyl group.

$$CH_3(CH_2)_8CH_2Cl \xrightarrow[\substack{ethanol, \\ water}]{KCN} CH_3(CH_2)_9C\equiv N \xrightarrow[heat]{H_2SO_4,\ H_2O} CH_3(CH_2)_9\overset{O}{\overset{\|}{C}}OH$$

1-Chlorodecane   Undecanenitrile   Undecanoic acid

Hydrolysis of cyanohydrins, which are obtained by the addition of HCN to an aldehyde or ketone (Section 16.5D), provides a valuable route to α-hydroxycarboxylic acids, as illustrated by the synthesis of mandelic acid.

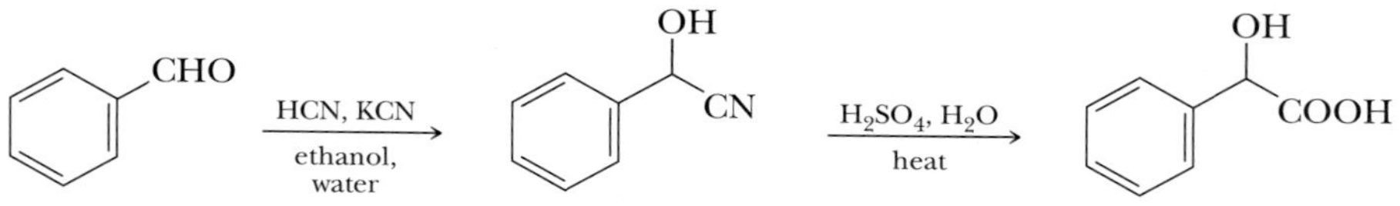

Benzaldehyde   Benzaldehyde cyanohydrin   2-Hydroxyphenylacetic acid
                               (Mandelonitrile)   (Mandelic acid)
                                 (racemic)   (racemic)

Example 18.5

Show how to bring about the following conversions using as one step the hydrolysis of a cyano group.

(a)

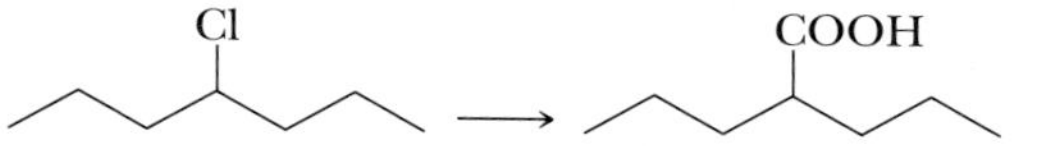

4-Chloroheptane   2-Propylpentanoic acid
                          (Valproic acid)

(b)

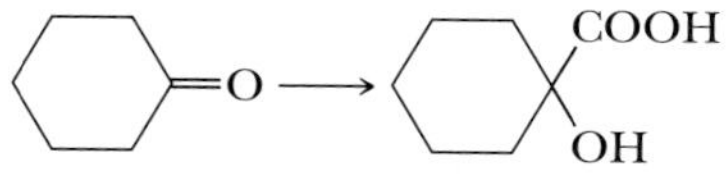

COOH / OH

Solution

(a) Treatment of 4-chloroheptane with KCN in aqueous ethanol gives a nitrile. Hydrolysis of the cyano group in aqueous sulfuric acid gives the product.

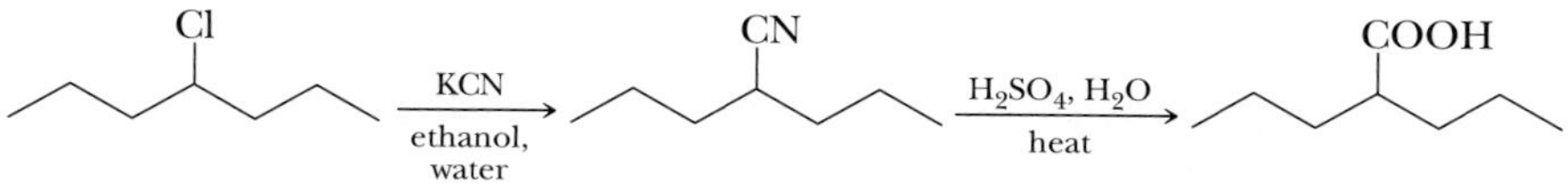

This synthesis can also be accomplished by conversion of the chloroalkane to a Grignard reagent followed by carbonation and hydrolysis in aqueous acid.

(b) Treatment of cyclohexanone with HCN/KCN in aqueous ethanol gives a cyanohydrin. Hydrolysis of the cyano group in concentrated sulfuric acid gives the carboxyl group of the product.

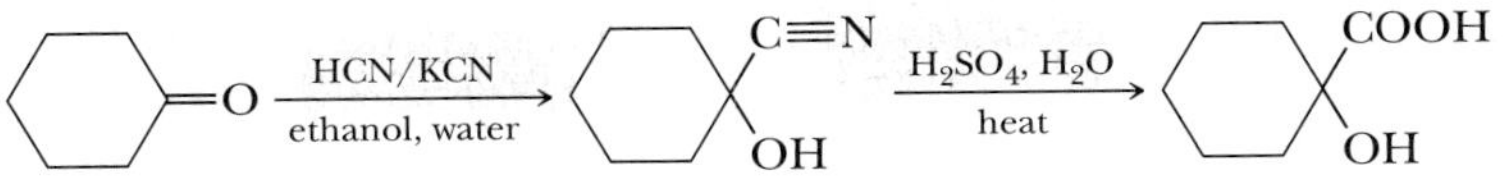

Problem 18.5

Synthesis of nitriles by nucleophilic displacement of halide from an alkyl halide is practical only with primary and secondary alkyl halides. It fails with tertiary alkyl halides. Why? What is the major product of the following reaction?

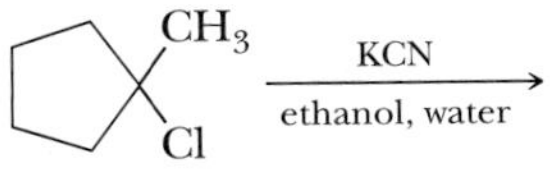

18.5 Reaction with Alcohols

A. Acid Halides

An acid halide reacts with an alcohol to give an ester.

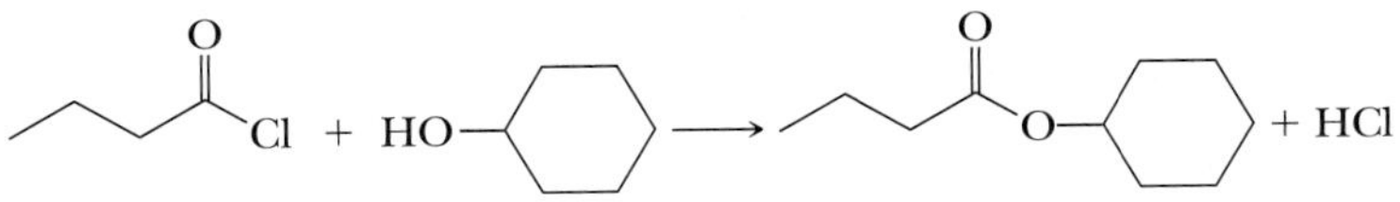

Butanoyl chloride   Cyclohexanol   Cyclohexyl butanoate

Because acid halides are so reactive toward even weak nucleophiles such as alcohols, no catalyst is necessary for these reactions.

In cases in which the alcohol or resulting ester is sensitive to acid, the reaction can be carried out in the presence of a tertiary amine to neutralize the HCl as it is formed. The tertiary amine also catalyzes the reaction by deprotonating the alcohol to form an alkoxide ion, a better nucleophile. The amines most commonly used for this purpose are pyridine and triethylamine.

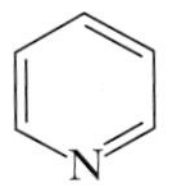

Pyridine   Triethylamine

When used for this purpose, each amine is converted to its hydrochloride salt. Pyridine, for example, is converted to pyridinium chloride, as illustrated by its use in the synthesis of isoamyl benzoate.

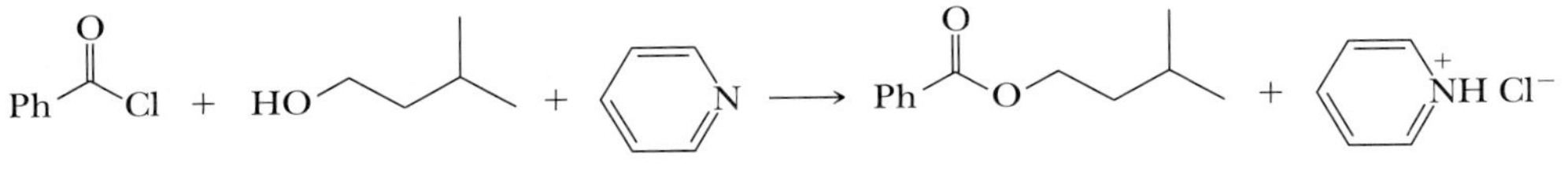

| Benzoyl chloride | 3-Methyl-1-butanol (Isoamyl alcohol) | Pyridine | 3-Methylbutyl benzoate (Isoamyl benzoate) | Pyridinium chloride |

Sulfonic acid esters are prepared by the reaction of an alkane- or arenesulfonyl chloride with an alcohol or phenol. Two of the most common sulfonyl chlorides are *p*-toluenesulfonyl chloride, abbreviated TsCl, and methanesulfonyl chloride, abbreviated MsCl (Section 18.1A).

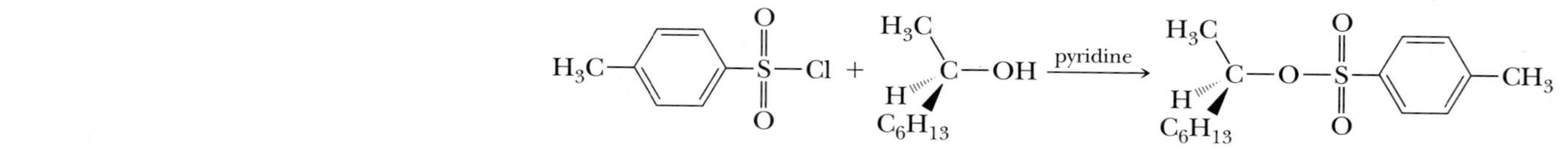

p-Toluenesulfonyl chloride (Tosyl chloride; TsCl) | (*R*)-2-Octanol | (*R*)-2-Octyl *p*-toluenesulfonate [(*R*)-2-Octyl tosylate]

As discussed in Section 9.4F, a special value of *p*-toluenesulfonic (tosylate) and methanesulfonic (mesylate) esters is that, in forming them, an —OH is converted from a poor leaving group (hydroxide ion) in nucleophilic displacement to an excellent leaving group, the *p*-toluenesulfonate (tosylate) or methanesulfonate (mesylate) anions.

B. Acid Anhydrides

Acid anhydrides react with alcohols to give one mole of ester and one mole of a carboxylic acid.

$$CH_3\overset{O}{\overset{\|}{C}}O\overset{O}{\overset{\|}{C}}CH_3 + HOCH_2CH_3 \longrightarrow CH_3\overset{O}{\overset{\|}{C}}OCH_2CH_3 + CH_3\overset{O}{\overset{\|}{C}}OH$$

Acetic anhydride | Ethanol | Ethyl acetate | Acetic acid

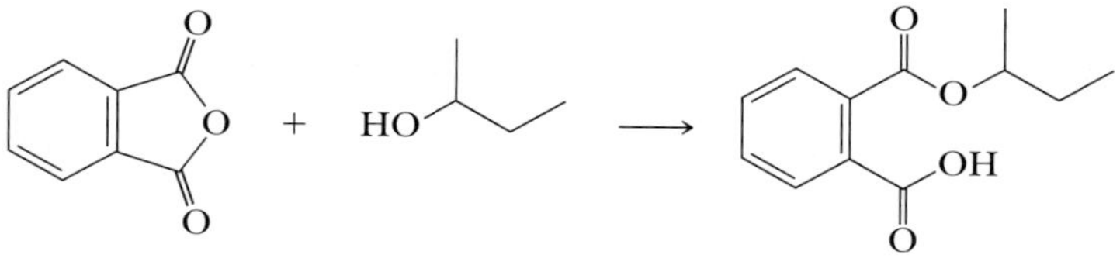

Phthalic anhydride | 2-Butanol (*sec*-Butyl alcohol) (racemic) | 1-Methylpropyl hydrogen phthalate (*sec*-Butyl hydrogen phthalate) (racemic)

Thus, the reaction of an alcohol with an anhydride is a useful method for the synthesis of esters. This reaction is catalyzed by acids and by tertiary amines.

Aspirin is synthesized on an industrial scale by the reaction of acetic anhydride and salicylic acid.

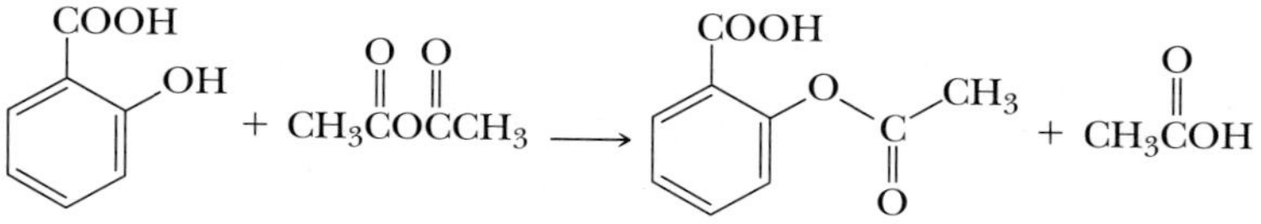

2-Hydroxybenzoic acid (Salicylic acid) | Acetic anhydride | Acetylsalicylic acid (Aspirin) | Acetic acid

C. Esters

Transesterification Exchange of the —OR or —OAr group of an ester for another —OR or —OAr group.

Esters react with alcohols in an acid-catalyzed reaction called **transesterification.** For example, it is possible to convert methyl acrylate to butyl acrylate by heating the methyl ester with 1-butanol in the presence of an acid catalyst.

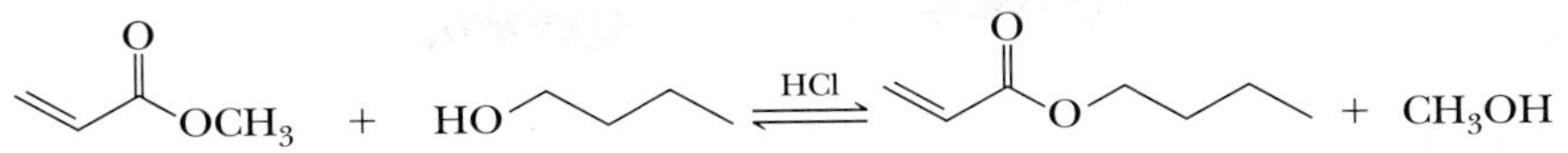

Methyl propenoate
(Methyl acrylate)
(bp 81°C)

1-Butanol
(bp 117°C)

Butyl propenoate
(Butyl acrylate)
(bp 147°C)

Methanol
(bp 65°C)

The acids most commonly used for transesterification are HCl as a gas bubbled into the reaction medium and *p*-toluenesulfonic acid.

Transesterification is an equilibrium reaction and can be driven in either direction by control of experimental conditions. For example, in the reaction of methyl acrylate with 1-butanol, transesterification is carried out at a temperature slightly above the boiling point of methanol (the lowest boiling component in the mixture). Methanol distills from the reaction mixture, thus shifting the position of equilibrium in favor of butyl acrylate. Conversely, reaction of butyl acrylate with a large excess of methanol shifts the equilibrium to favor formation of methyl acrylate.

Example 18.6

Complete the following transesterification reactions (the stoichiometry of each is given in the problem).

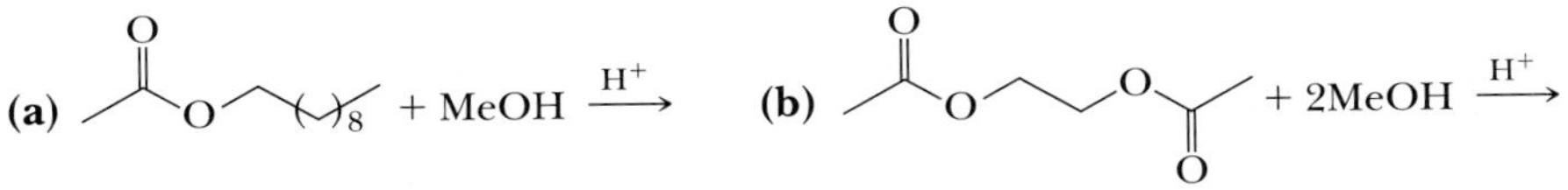

Solution

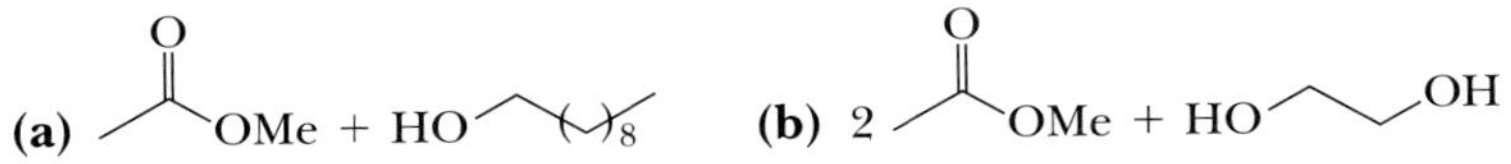

Problem 18.6

Complete the following transesterification reaction (the stoichiometry is given in the equation).

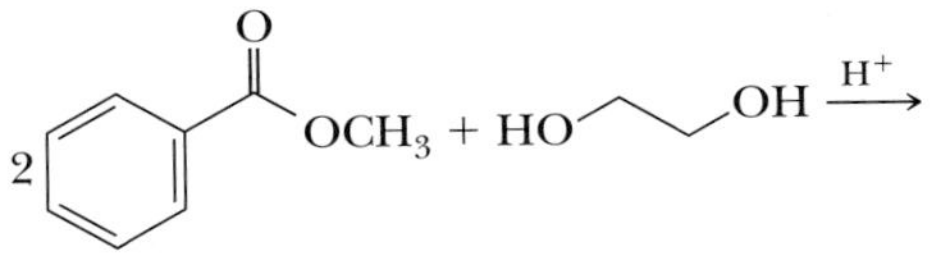

D. Amides

Amides, the least reactive of the functional derivatives of carboxylic acids, do not react with alcohols. Thus, the reaction of an amide with an alcohol cannot be used to prepare an ester.

18.6 Reactions with Ammonia and Amines

A. Acid Halides

Acid halides react readily with ammonia and 1° and 2° amines to form amides. For complete conversion of an acid halide to an amide, two equivalents of ammonia or amine are used, one to form the amide and one to neutralize the hydrogen halide formed.

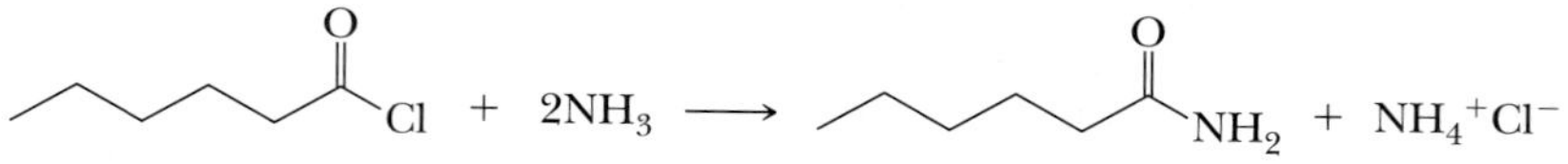

| Hexanoyl chloride | Ammonia | | Hexanamide | Ammonium chloride |

Mechanism   *Reaction of Acetyl Chloride and Ammonia*

Step 1: Nucleophilic addition of ammonia to the carbonyl carbon followed by a proton transfer gives a tetrahedral carbonyl addition intermediate.

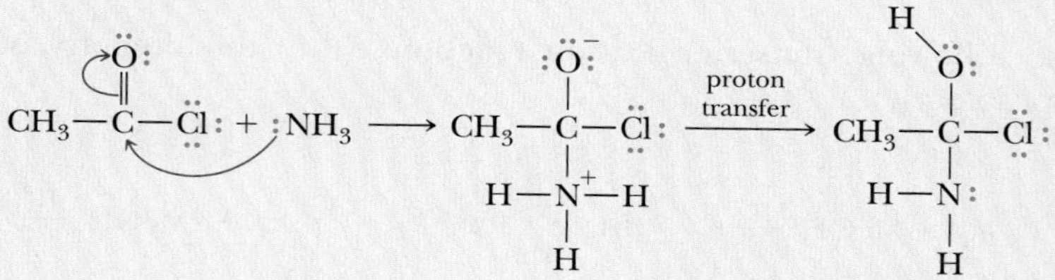

Tetrahedral carbonyl addition intermediate

Step 2: Collapse of this intermediate gives chloride ion, an amide, and an ammonium ion.

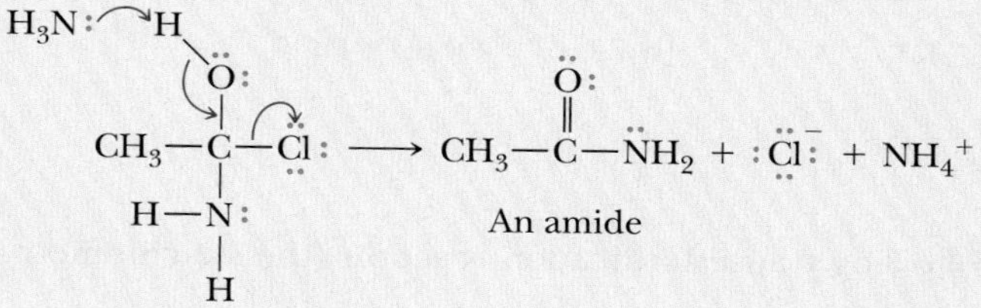

An amide

B. Acid Anhydrides

Acid anhydrides react with ammonia and 1° and 2° amines to form amides. As with acid halides, two moles of amine are required; one mole to form the amide and one mole to neutralize the carboxylic acid byproduct.

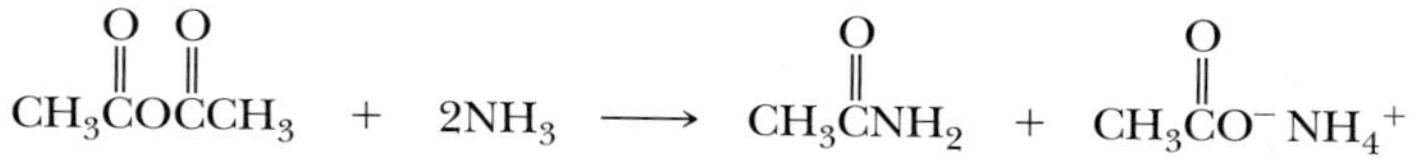

| Acetic anhydride | Ammonia | | Ethanamide (Acetamide) | Ammonium acetate |

Alternatively, if the amine used to make the amide is expensive, a tertiary amine such as triethylamine may be used to neutralize the carboxylic acid.

C. Esters

Esters react with ammonia and with 1° and 2° amines to form amides.

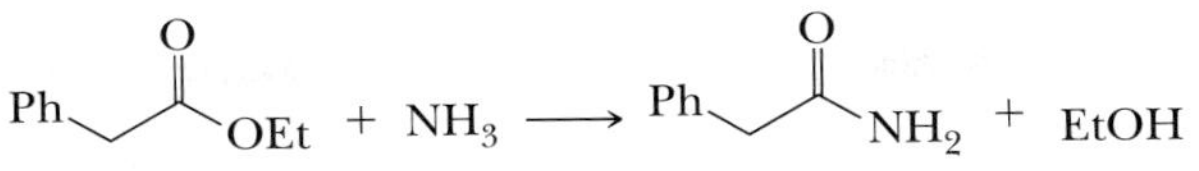

$$Ph\!\!\diagdown\!\!\overset{O}{\underset{OEt}{||}} + NH_3 \longrightarrow Ph\!\!\diagdown\!\!\overset{O}{\underset{NH_2}{||}} + EtOH$$

Ethyl phenylacetate   Phenylacetamide   Ethanol

Because an alkoxide anion is a poor leaving group compared with either a halide or a carboxylate ion, esters are less reactive toward ammonia, 1° amines, and 2° amines than are acid halides or acid anhydrides. The reaction often requires heating or high concentrations of amine.

D. Amides

Amides do not react with ammonia or primary or secondary amines.

Example 18.7

Complete the following reactions (the stoichiometry of each reaction is given in the equation).

(a) 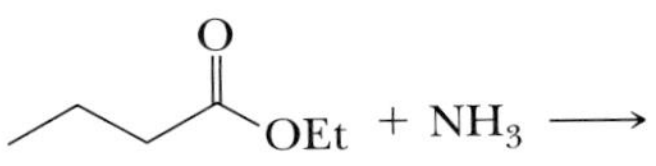 $\overset{O}{\underset{OEt}{||}}$ + NH$_3$ $\longrightarrow$ **(b)** EtO$\overset{O}{\underset{}{||}}$OEt + 2NH$_3$ $\longrightarrow$

Ethyl butanoate   Diethyl carbonate

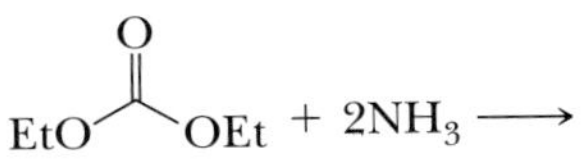

Solution

(a) 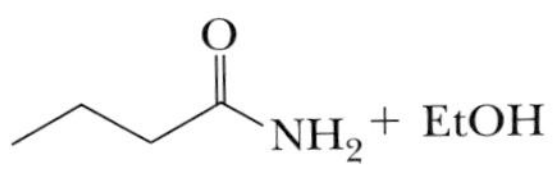 $\overset{O}{\underset{NH_2}{||}}$ + EtOH   **(b)** 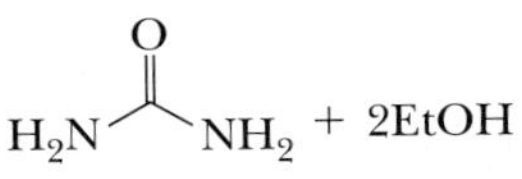 H$_2$N$\overset{O}{\underset{}{||}}NH_2$ + 2EtOH

Butanamide   Urea

Problem 18.7

Complete and balance equations for the following reactions (the stoichiometry of each reaction is given in the equation).

(a) 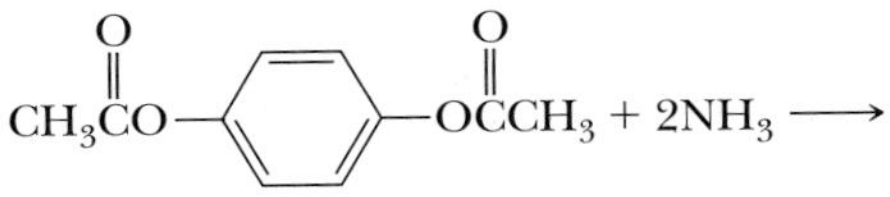 CH$_3$CO—⟨benzene ring⟩—OCCH$_3$ + 2NH$_3$ $\longrightarrow$ **(b)** 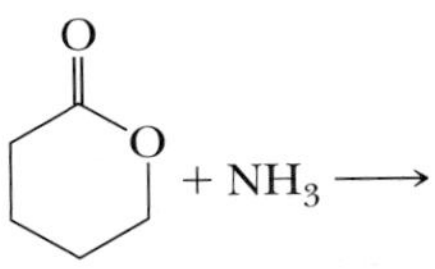 ⟨lactone ring⟩ + NH$_3$ $\longrightarrow$

18.7 Reaction of Acid Chlorides with Salts of Carboxylic Acids

Acid chlorides react with salts of carboxylic acids to give anhydrides. Most commonly used are the sodium or potassium salts.

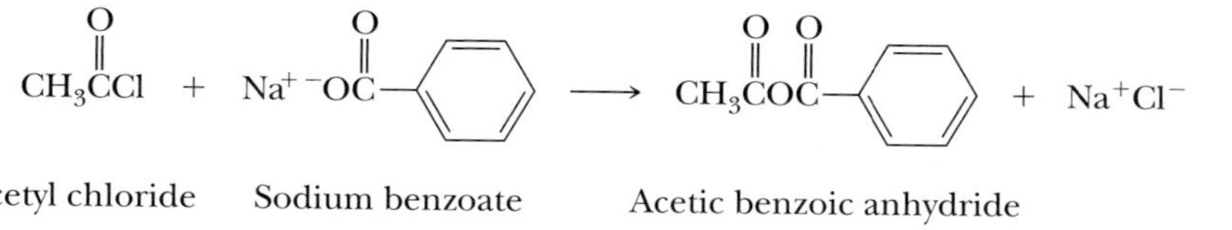

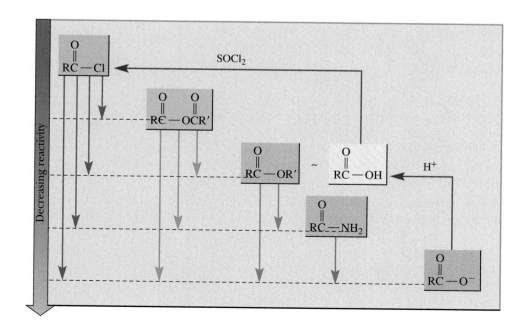

| Acetyl chloride | Sodium benzoate | Acetic benzoic anhydride |

Reaction of an acid halide with a carboxylate anion of a carboxylic acid is a particularly useful method for synthesis of mixed anhydrides.

18.8 Interconversion of Functional Derivatives

We have seen throughout the past several sections that acid chlorides are the most reactive toward nucleophilic acyl substitution and that carboxylate anions are the least reactive. Another useful way to think about the reactions of the functional derivatives of carboxylic acids is summarized in Figure 18.1.

Any functional group lower in this figure can be prepared from any functional group above it by treatment with an appropriate oxygen or nitrogen nucleophile. An acid chloride, for example, can be converted to an acid anhydride, an ester, an amide, or a carboxylic acid. Acid anhydrides, esters, and amides, however, do not react with chloride ion to give acid chlorides.

ORGANIC
Chemistry⚛Now™
Active Figure 18.1
Reactivities of carboxyl derivatives toward nucleophilic acyl substitution. A more reactive derivative may be converted to a less reactive derivative by treatment with an appropriate reagent. Treatment of a carboxylic acid with thionyl chloride (the acid chloride of sulfurous acid) converts it to the more reactive acid chloride. Carboxylic acids are about as reactive as esters under acidic conditions, but they are converted to unreactive carboxylates under basic conditions. **See a simulation based on this figure and take a short quiz on the concepts.**

18.9 Reactions with Organometallic Compounds

A. Grignard Reagents

Treating a formic ester with two moles of a Grignard reagent followed by hydrolysis of the magnesium alkoxide salt in aqueous acid gives a secondary alcohol.

$$\underset{\substack{\text{An ester of}\\\text{formic acid}}}{\text{HCOCH}_3} + 2\text{RMgX} \longrightarrow \underset{\substack{\text{magnesium}\\\text{alkoxide}\\\text{salt}}}{} \xrightarrow{\text{H}_2\text{O, HCl}} \underset{\substack{\text{A 2°}\\\text{alcohol}}}{\text{HC}-\text{R}} + \text{CH}_3\text{OH}$$

Treating an ester other than a formate with a Grignard reagent gives a tertiary alcohol in which two of the groups bonded to the carbon bearing the —OH group are the same.

$$\underset{\substack{\text{An ester of any acid}\\\text{other than formic acid}}}{\text{CH}_3\text{COCH}_3} + 2\text{RMgX} \longrightarrow \underset{\substack{\text{magnesium}\\\text{alkoxide}\\\text{salt}}}{} \xrightarrow{\text{H}_2\text{O, HCl}} \underset{\substack{\text{A 3°}\\\text{alcohol}}}{\text{CH}_3\text{C}-\text{R}} + \text{CH}_3\text{OH}$$

Reaction of an ester with a Grignard reagent involves formation of two successive tetrahedral carbonyl addition intermediates.

Mechanism *Reaction of an Ester with a Grignard Reagent*

Steps 1 & 2: The reaction begins with addition of one mole of Grignard reagent to the carbonyl carbon to form a tetrahedral carbonyl addition intermediate. Because an alkoxide ion is a moderately good leaving group from a tetrahedral carbonyl addition intermediate, this intermediate collapses in Step 2 to give a new carbonyl-containing compound and a magnesium alkoxide salt.

$$\underset{①}{\text{CH}_3-\overset{\overset{①\curvearrowright\ddot{\text{O}}:}{\|}}{\text{C}}-\ddot{\text{O}}\text{CH}_3} + \text{R}-\text{MgX} \longrightarrow \underset{\substack{\text{A magnesium salt}\\\text{(a tetrahedral carbonyl}\\\text{addition intermediate)}}}{\text{CH}_3-\overset{\overset{②\ddot{\text{O}}^{\bar{}}\,[\text{MgX}]^+}{|}}{\underset{\underset{②}{\overset{|}{\text{R}}}}{\text{C}}}-\ddot{\text{O}}\text{CH}_3} \longrightarrow \underset{\underset{\text{A ketone}}{}}{\text{CH}_3-\overset{\overset{\ddot{\text{O}}:}{\|}}{\underset{\underset{\text{R}}{|}}{\text{C}}}} + \text{CH}_3\ddot{\text{O}}^{\bar{}}\,[\text{MgX}]^+$$

Steps 3 & 4: This new carbonyl-containing compound then reacts in Step 3 with a second mole of Grignard reagent to form a second tetrahedral carbonyl addition compound, which, after hydrolysis in aqueous acid (Step 4), gives a tertiary alcohol (or a secondary alcohol if the starting ester was a formate).

$$\underset{\underset{\text{A ketone}}{}}{\text{CH}_3-\overset{\overset{③\curvearrowright\ddot{\text{O}}:}{\|}}{\underset{\underset{\text{R}}{|}}{\text{C}}} + \overset{③}{\text{R}}-\text{MgX}} \longrightarrow \underset{\underset{\text{Magnesium salt}}{}}{\text{CH}_3-\overset{\overset{:\ddot{\text{O}}^{\bar{}}\,[\text{MgX}]^+}{|}}{\underset{\underset{\text{R}}{|}}{\text{C}}}-\text{R}} \xrightarrow[④]{\text{H}_2\text{O, HCl}} \underset{\underset{\text{A 3° alcohol}}{}}{\text{CH}_3-\overset{\overset{:\ddot{\text{O}}\text{H}}{|}}{\underset{\underset{\text{R}}{|}}{\text{C}}}-\text{R}}$$

ORGANIC
Chemistry Now™

Click Mechanisms to view an animation of the **Reaction of a Grignard Reagent with an Ester**

It is important to realize that it is not possible to use RMgX and an ester to prepare a ketone; the intermediate ketone is more reactive than the ester and reacts immediately with the Grignard reagent to give a tertiary alcohol.

Example 18.8

Complete each Grignard reaction.

(a) 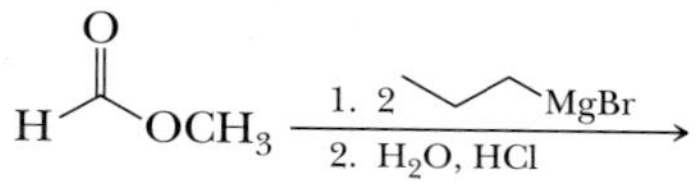 (structure) H—C(=O)—OCH₃ $\xrightarrow[\text{2. H}_2\text{O, HCl}]{\text{1. 2 }\;\;\;\;\text{MgBr}}$

(b) 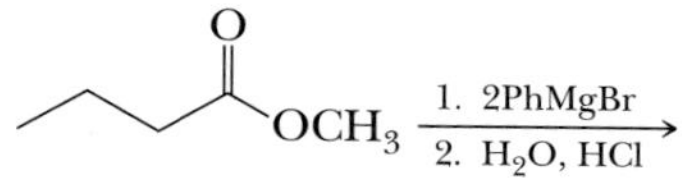 (structure) —C(=O)—OCH₃ $\xrightarrow[\text{2. H}_2\text{O, HCl}]{\text{1. 2PhMgBr}}$

Solution

Sequence (a) gives a secondary alcohol, and sequence (b) gives a tertiary alcohol.

(a) 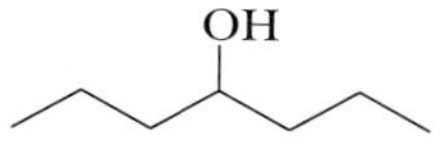 (structure with OH secondary alcohol)

(b) 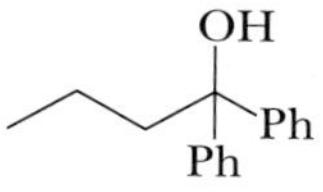 (structure with OH, Ph, Ph tertiary alcohol)

Problem 18.8

Show how to prepare each alcohol by treating an ester with a Grignard reagent.

(a) 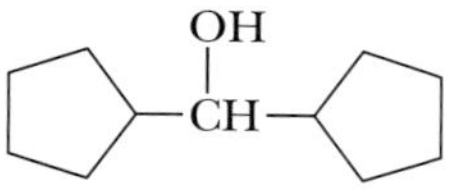 (cyclopentyl)—CH(OH)—(cyclopentyl)

(b) 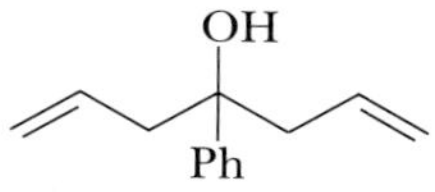 (structure with OH and Ph)

B. Organolithium Compounds

Organolithium compounds are even more powerful nucleophiles than Grignard reagents and react with esters to give the same types of secondary and tertiary alcohols as shown for Grignard reagents, often in higher yields.

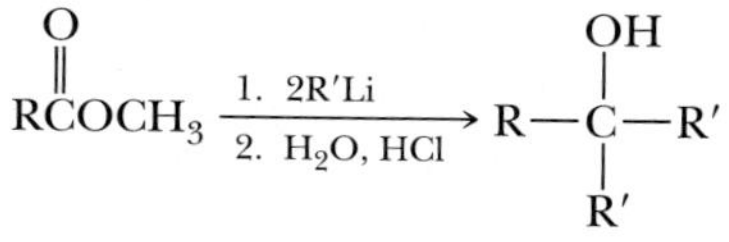

$$\text{RCOCH}_3 \xrightarrow[\text{2. H}_2\text{O, HCl}]{\text{1. 2R'Li}} \text{R}-\overset{\overset{\displaystyle \text{OH}}{|}}{\underset{\underset{\displaystyle \text{R}'}{|}}{\text{C}}}-\text{R}'$$

C. Lithium Diorganocuprates

Acid chlorides react readily with lithium diorganocopper (Gilman) reagents to give ketones, as illustrated by the conversion of pentanoyl chloride to 2-hexanone. The reaction is carried out at $-78°C$ in either diethyl ether or tetrahydrofuran. Following hydrolysis in aqueous acid, the ketone is isolated in good yield.

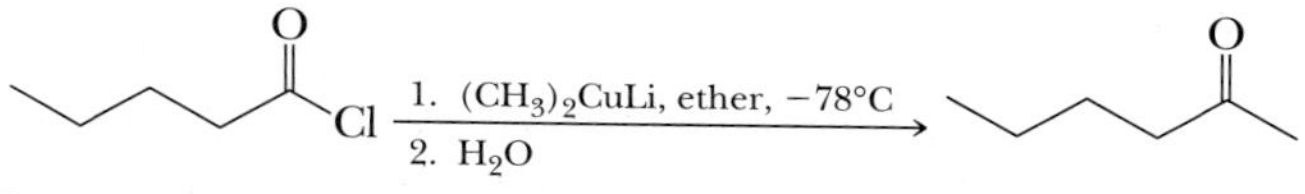

(structure) —C(=O)—Cl $\xrightarrow[\text{2. H}_2\text{O}]{\text{1. (CH}_3)_2\text{CuLi, ether, }-78°\text{C}}$ (structure) 2-Hexanone

Pentanoyl chloride   2-Hexanone

Notice that, under these conditions, the ketone does not react further. This contrasts with the reaction of an ester with a Grignard reagent or organolithium compound, where the intermediate ketone reacts with a second mole of the organometallic compound to give an alcohol. The reason for this difference in reactivity is that the tetrahedral carbonyl addition intermediate in a diorganocuprate reaction is stable at $-78°C$; it survives until the workup causes it to decompose to the ketone, at which point the Gilman reagent has been destroyed.

R_2CuLi reagents react readily only with the very reactive acid chlorides; they do not react with aldehydes, ketones, esters, amides, acid anhydrides, or nitriles. The following compound contains both an acid chloride and an ester group. When treated with lithium dimethylcopper, only the acid chloride reacts.

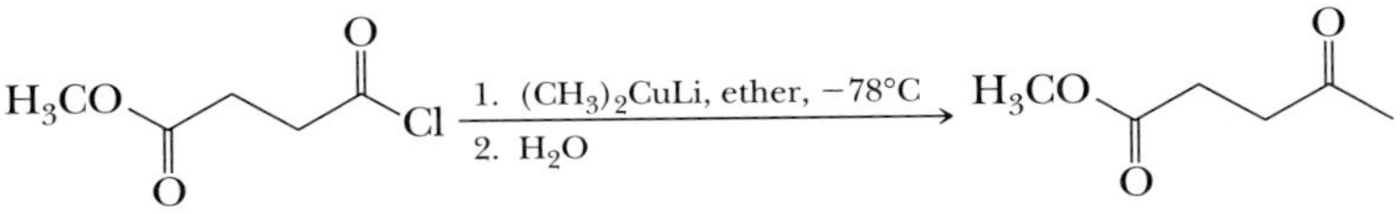

Example 18.9

Show how to bring about each conversion in good yield.

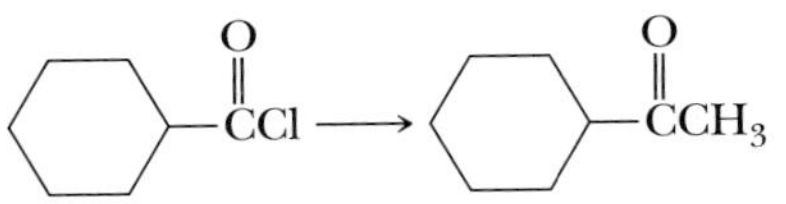

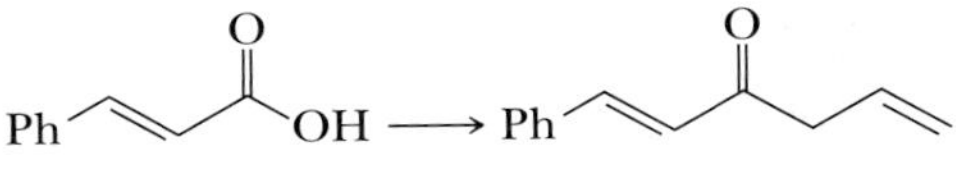

Solution

(a) Treat the acid chloride with lithium dimethylcopper followed by H_2O.

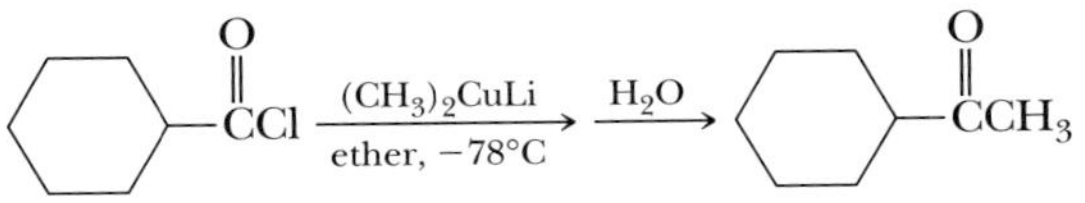

(b) Treat the carboxylic acid with thionyl chloride to form the acid chloride, followed by treatment with lithium diallylcopper and then aqueous acid.

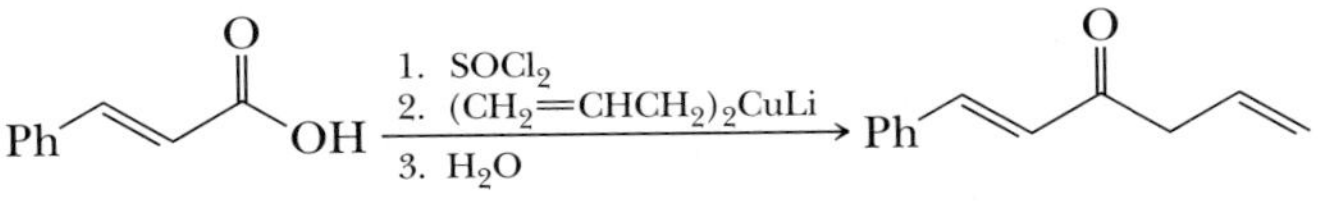

Problem 18.9

Show how to bring about each conversion in good yield.

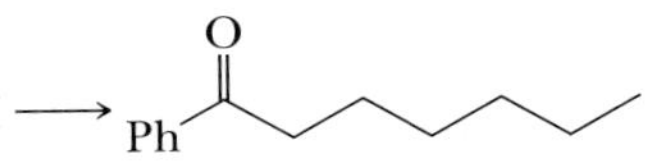

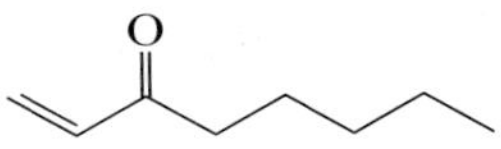

18.10 Reduction

Most reductions of carbonyl compounds, including aldehydes and ketones, are now accomplished by transfer of hydride ions from boron or aluminum hydrides. We have already seen the use of sodium borohydride to reduce the carbonyl group of aldehydes and ketones to hydroxyl groups (Section 16.11A) and the use of lithium aluminum hydride to reduce not only aldehyde and ketone carbonyl groups but also carboxyl groups to hydroxyl groups (Section 17.6A).

A. Esters

Lithium aluminum hydride reduces an ester to two alcohols; the alcohol derived from the acyl group is primary and is usually the objective of the reduction.

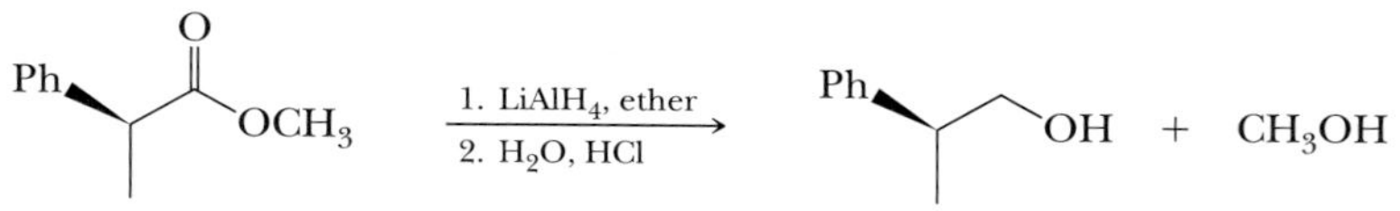

Methyl(*S*)-2-phenylpropanoate (*S*)-2-Phenyl-1-propanol Methanol

Mechanism *Reduction of an Ester by Lithium Aluminum Hydride*

As you study this mechanism, note that Steps 1 and 3 are closely analogous to the reaction of Grignard reagents with an ester, with the exception that a hydride ion rather than a carbanion is being donated to the carbonyl carbon.

Steps 1 & 2: Nucleophilic addition of hydride ion to the carbonyl carbon gives a tetrahedral carbonyl addition intermediate. The hydride ion is not free but is donated by the AlH_4^- ion. Collapse of this intermediate by loss of alkoxide ion in Step 2 gives a new carbonyl-containing compound.

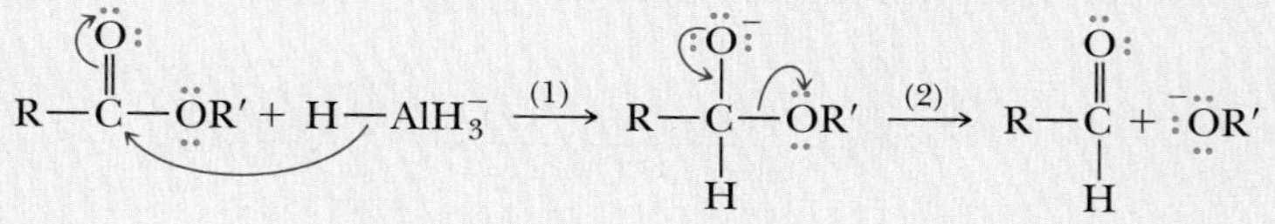

A tetrahedral carbonyl addition intermediate

Step 3 & 4: Nucleophilic addition of a second hydride ion to the newly formed carbonyl group gives an alkoxide ion. Treatment of this alkoxide ion with water in Step 4 gives the alcohol product.

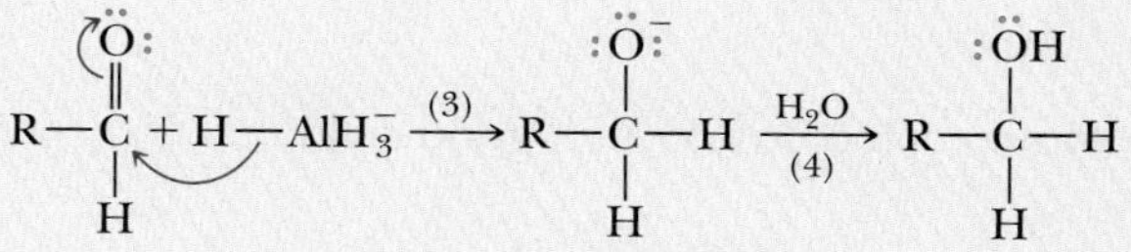

A 1° alcohol

ORGANIC Chemistry•Now™
Click Mechanisms to view an animation of the **Reduction of an Ester by $LiAlH_4$**

Sodium borohydride is not normally used to reduce esters because the reaction is very slow. Because of this lower reactivity of sodium borohydride toward esters, it is possible to reduce the carbonyl group of an aldehyde or ketone to a hydroxyl group with this reagent without reducing an ester or carboxyl group in the same molecule.

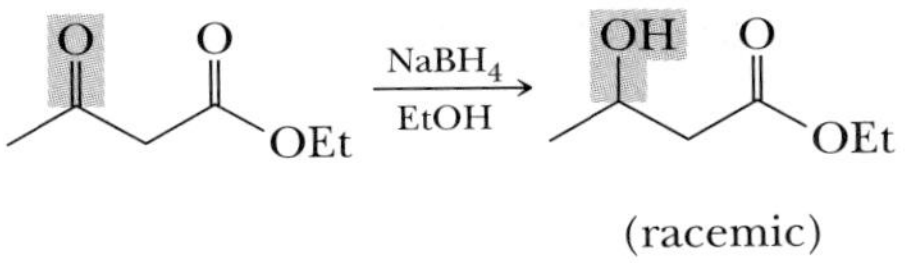

(racemic)

   Reduction of an ester to a primary alcohol can be viewed as two successive hydride ion transfers, as shown in the mechanism we have just presented. Chemists wondered if it might be possible to modify the structure of the reducing agent so as to reduce an ester to an aldehyde and no further. A useful modified hydride-reducing agent developed for this purpose is diisobutylaluminum hydride (DIBALH).

$[(CH_3)_2CHCH_2]_2AlH$

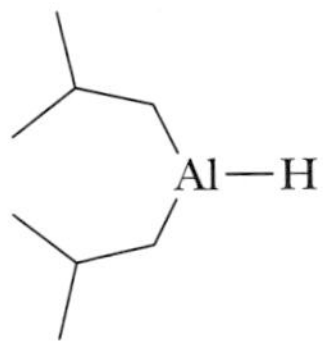

Diisobutylaluminum hydride (DIBALH)

DIBALH reductions are typically carried out in toluene or hexane at $-78°C$ (dry ice/acetone temperature) followed by warming to room temperature and addition of aqueous acid to hydrolyze the aluminum salts and liberate the aldehyde. Reduction of esters using DIBALH has become a valuable method for the synthesis of aldehydes, as illustrated by the synthesis of hexanal.

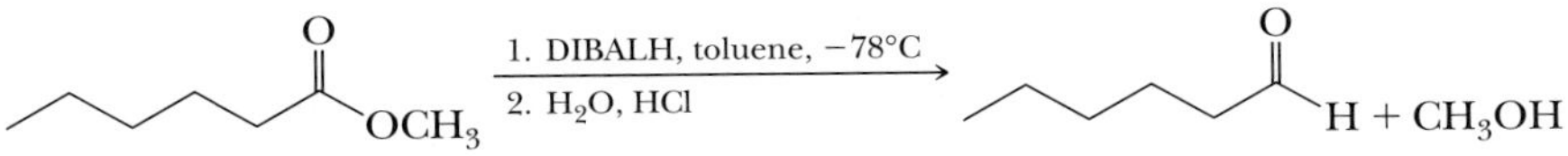

Methyl hexanoate   Hexanal

   If a DIBALH reduction of an ester is carried out at room temperature, the ester is reduced to a primary alcohol. At low temperature, the tetrahedral carbonyl addition intermediate does not eliminate alkoxide ion, and the more reactive aldehyde is not formed until after workup, when the hydride ion has been destroyed. Thus, temperature control is critical for the selective reduction of an ester to an aldehyde.

B. Amides

Lithium aluminum hydride reduction of amides can be used to prepare 1°, 2°, or 3° amines, depending on the degree of substitution of the amide.

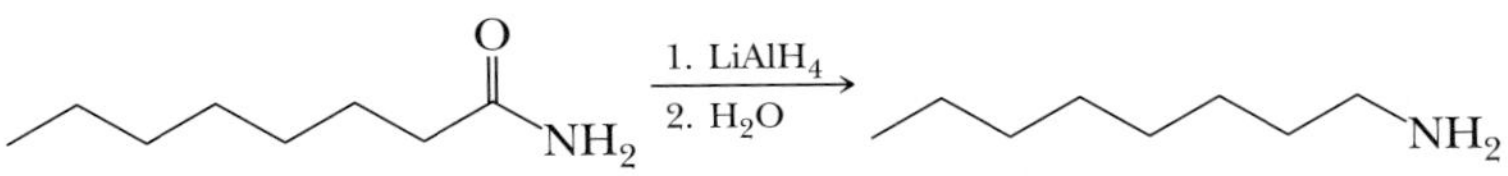

Octanamide   1-Octanamine

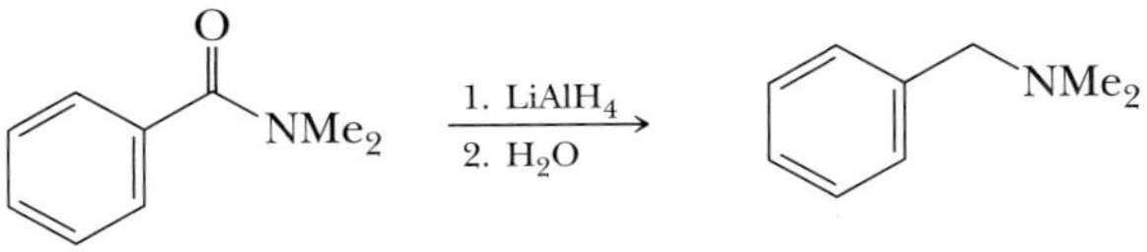

N,N-Dimethylbenzamide   *N,N*-Dimethylbenzylamine

The mechanism for the reduction of an amide to an amine is shown here divided into four steps.

Mechanism *Reduction of an Amide by Lithium Aluminum Hydride*

Steps 1 & 2: Transfer of a hydride ion from AlH_4^- to the carbonyl carbon followed by a Lewis acid-base reaction between $—O^-$ (a Lewis base) and AlH_3 (a Lewis acid) forms an oxygen-aluminum bond.

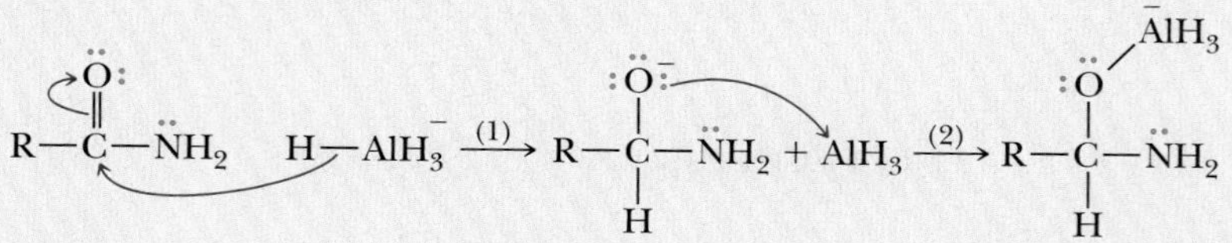

Step 3 & 4: Rearrangement of electron pairs ejects H_3AlO^{2-} and generates an iminium ion. Because aluminum hydroxides are somewhat acidic, H_3AlO^{2-} is a reasonably good leaving group. In the final step of the reduction, the iminium ion adds a second hydride ion to complete the reduction.

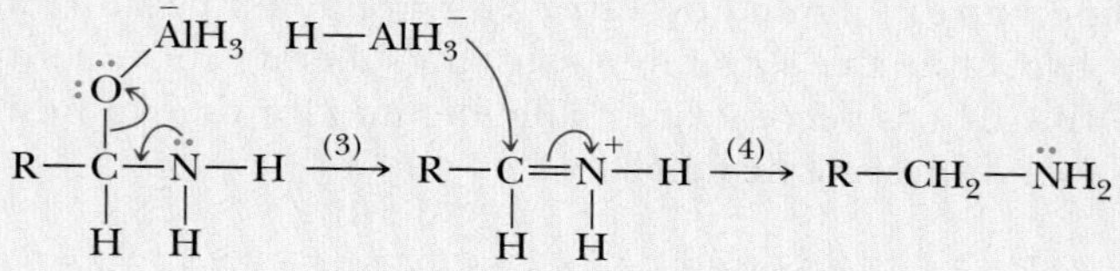

An iminium ion   A 1° amine

Example 18.10

Show how to bring about each conversion.

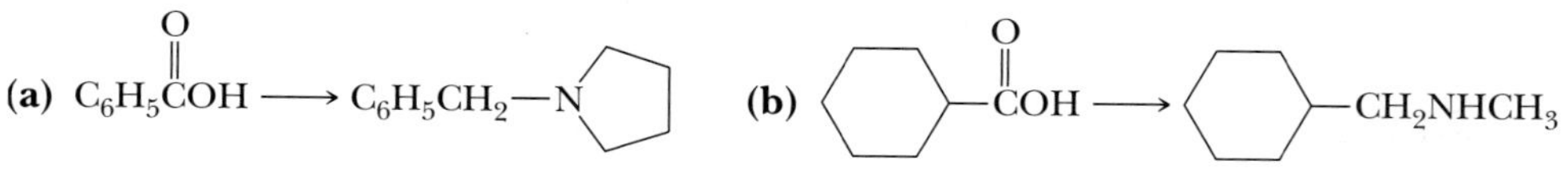

Solution

The key in each part is to convert the carboxylic acid to an amide and then to reduce the amide with $LiAlH_4$. The amide can be prepared by treating the carboxylic acid with $SOCl_2$ to give the acid chloride (Section 17.8) and then treating the acid chloride with an amine (Section 18.6A). Alternatively, the carboxylic acid can be converted to an ethyl ester by Fischer esterification (Section 17.7A), and the ester can then be treated with an amine to give the amide. Solution (a) uses the acid chloride route, and solution (b) uses the ester route.

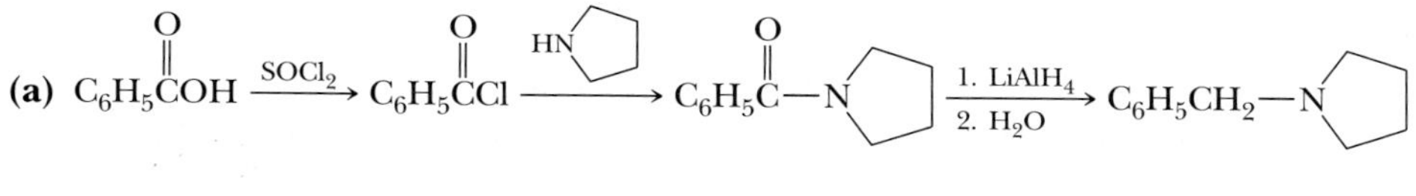

(b)

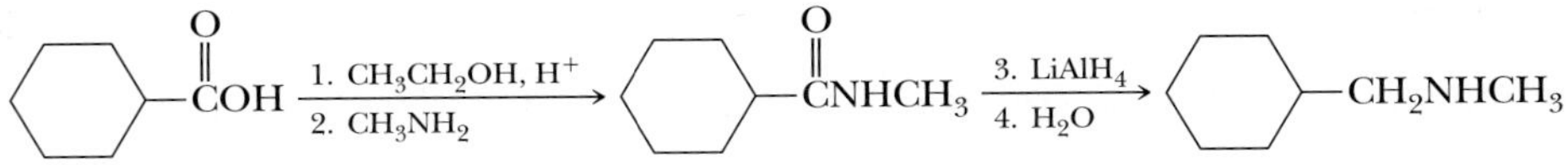

Problem 18.10

Show how to convert hexanoic acid to each amine.

(a)

(b)

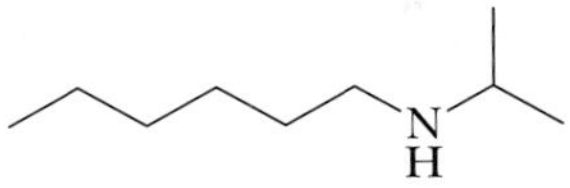

Example 18.11

Show how to convert phenylacetic acid to each compound.

(a)

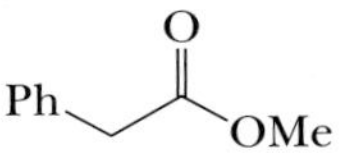

(b)

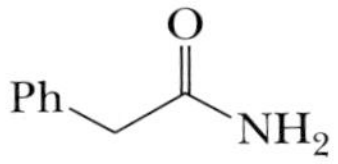

(c)

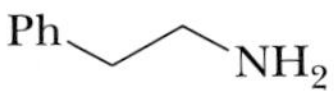

(d)

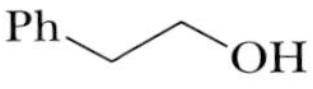

Solution

Prepare methyl ester (a) by Fischer esterification (Section 17.7A) of phenylacetic acid with methanol. Then treat this ester with ammonia to prepare amide (b). Alternatively, treat phenylacetic acid with thionyl chloride (Section 17.8) to give an acid chloride. Then treat this acid chloride with ammonia to give amide (b). Reduction of the amide (b) by LiAlH$_4$ gives the primary amine (c). Similar reduction of either phenylacetic acid or ester (a) gives alcohol (d).

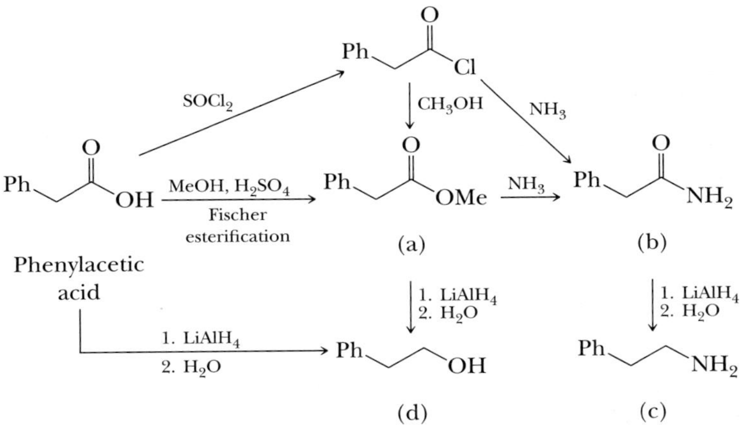

Problem 18.11

Show how to convert (R)-2-phenylpropanoic acid to each compound.

(a)

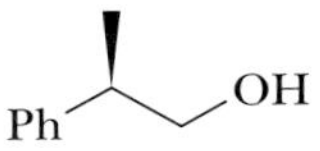

(R)-2-Phenyl-1-propanol

(b)

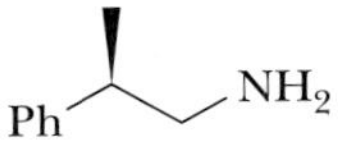

(R)-2-Phenyl-1-propanamine

C. Nitriles

Lithium aluminum hydride reduces the cyano group of a nitrile to a primary amino group.

$$CH_3CH{=}CH(CH_2)_4C{\equiv}N \xrightarrow[\text{2. } H_2O]{\text{1. LiAlH}_4} CH_3CH{=}CH(CH_2)_4CH_2NH_2$$

6-Octenenitrile   6-Octen-1-amine

Reduction of cyano groups is useful for the preparation of primary amines only.

Summary

The functional group of an **acid halide** (Section 18.1A) is an acyl group bonded to a halogen atom. The most common and widely used of these are the acid chlorides. The functional group of a **carboxylic anhydride** (Section 18.1B) is two acyl groups bonded to an oxygen. The functional group of a **carboxylic ester** (Section 18.1C) is an acyl group bonded to —OR or —OAr. A cyclic ester is given the name **lactone.** Phosphoric acid has three —OH groups and can form mono-, di-, and triesters. The functional group of an **amide** (Section 18.1D) is an acyl group bonded to a nitrogen. A cyclic amide is given the name **lactam.** The functional group of an **imide** is two acyl groups bonded to a nitrogen. The functional group of a **nitrile** (Section 18.1E) is a cyano (—C≡N) group bonded to a carbon atom.

Values of pK_a for amides of carboxylic acids are 15–17, which means that they are comparable in acidity to alcohols (Section 18.2). Values of pK_a for imides are 8–10, which means that they dissolve in aqueous NaOH to form water-soluble sodium salts. Sulfonamides derived from ammonia and primary amines have pK_a values of approximately 10 and dissolve in aqueous NaOH to form water-soluble sodium salts.

A common reaction theme of functional derivatives of carboxylic acids is nucleophilic addition to the carbonyl carbon to form a tetrahedral carbonyl addition intermediate, which then collapses to regenerate the carbonyl group. The result is **nucleophilic acyl substitution** (Section 18.3). Listed in order of increasing reactivity toward nucleophilic acyl substitution, these functional derivatives are

$$\underset{RCNH_2}{\overset{O}{\underset{\|}{}}} \quad \underset{RCOR'}{\overset{O}{\underset{\|}{}}} \quad \underset{RCOCR'}{\overset{O \quad O}{\underset{\| \quad \|}{}}} \quad \underset{RCCl}{\overset{O}{\underset{\|}{}}}$$

Increasing chemical reactivity →

Any more reactive functional derivative can be converted to any less reactive functional derivative by reaction with an appropriate oxygen or nitrogen nucleophile (Section 18.8).

Key Reactions

1. Acidity of Imides (Section 18.2)

Imides (pK_a 8–10) dissolve in aqueous NaOH by forming water-soluble sodium salts.

Insoluble in water   A water-soluble
                                       sodium salt

Imides are stronger acids than amides because the imide anion is stabilized by delocalization of the negative charge onto the two carbonyl oxygens.

2. Acidity of Sulfonamides (Section 18.2)

Sulfonamides (pK_a 9–10) dissolve in aqueous NaOH by forming water-soluble salts. The stability of the sulfonamide anion is the result of resonance stabilization of the anion.

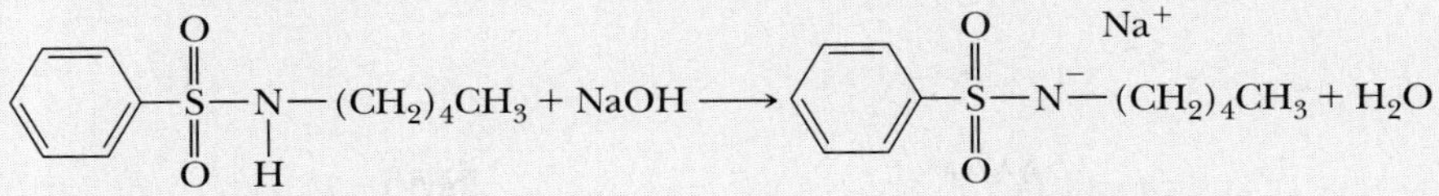

Insoluble in water   A water-soluble salt

3. Hydrolysis of an Acid Chloride (Section 18.4A)

Low-molecular-weight acid chlorides react vigorously with water.

$$CH_3CCl + H_2O \longrightarrow CH_3COH + HCl$$

Higher molecular-weight acid chlorides react less rapidly.

4. Hydrolysis of an Acid Anhydride (Section 18.4B)

Low-molecular-weight acid anhydrides react readily with water:

$$CH_3COCCH_3 + H_2O \longrightarrow CH_3COH + HOCCH_3$$

Higher molecular-weight acid anhydrides react less rapidly.

5. Hydrolysis of an Ester (Section 18.4C)

Esters are hydrolyzed only in the presence of acid or base. Acid is a catalyst. Base is required in an equimolar amount.

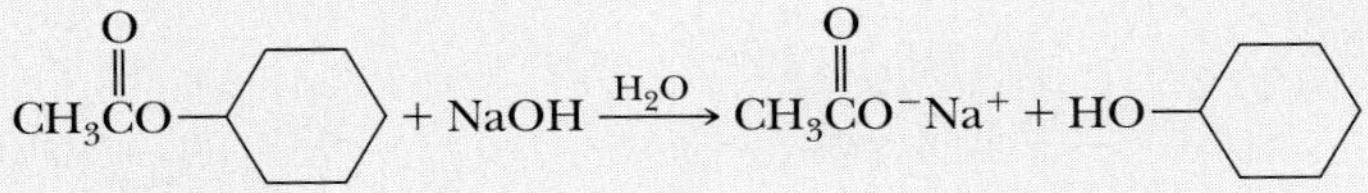

6. Hydrolysis of an Amide (Section 18.4D)

Either acid or base is required in an amount equivalent to that of the amide.

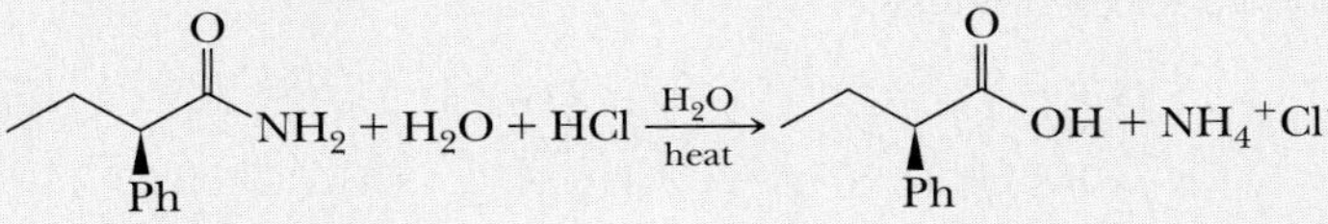

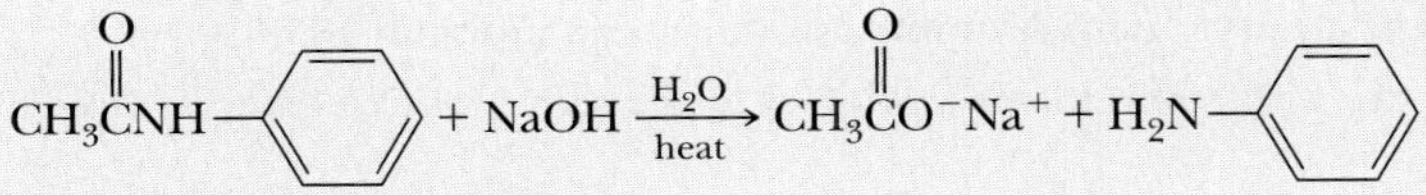

7. Hydrolysis of a Nitrile (Section 18.4E)

Either acid or base is required in an amount equivalent to that of the nitrile.

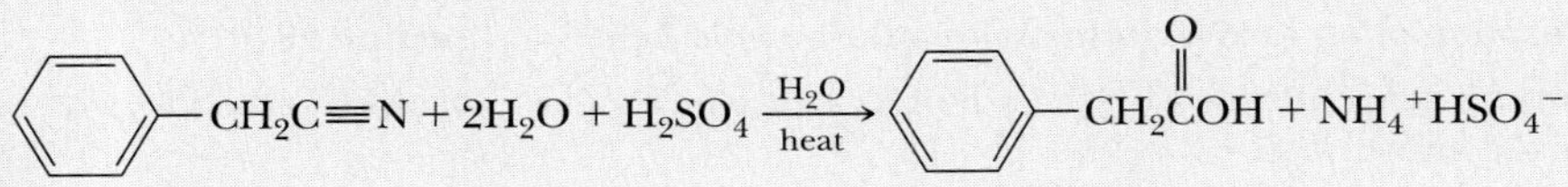

$$CH_3(CH_2)_9C{\equiv}N + H_2O + NaOH \xrightarrow[\text{heat}]{H_2O} CH_3(CH_2)_9CO^-Na^+ + NH_3$$

8. Reaction of an Acid Chloride with an Alcohol (Section 18.5A)

Treating an acid chloride with an alcohol gives an ester plus HCl.

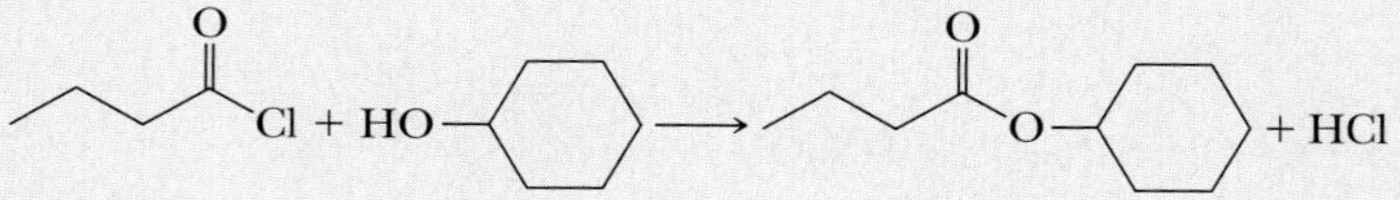

Preparation of an acid-sensitive ester is carried out using an equimolar amount of triethyl-amine or pyridine to neutralize the HCl.

9. Reaction of an Acid Anhydride with an Alcohol (Section 18.5B)

Treating an acid anhydride with an alcohol gives one mole of ester and one mole of carboxylic acid.

$$CH_3COCCH_3 + HOEt \longrightarrow CH_3COEt + CH_3COH$$

10. Reaction of an Ester with an Alcohol: Transesterification (Section 18.5C)

Transesterification requires an acid catalyst and an excess of alcohol to drive the reaction to completion.

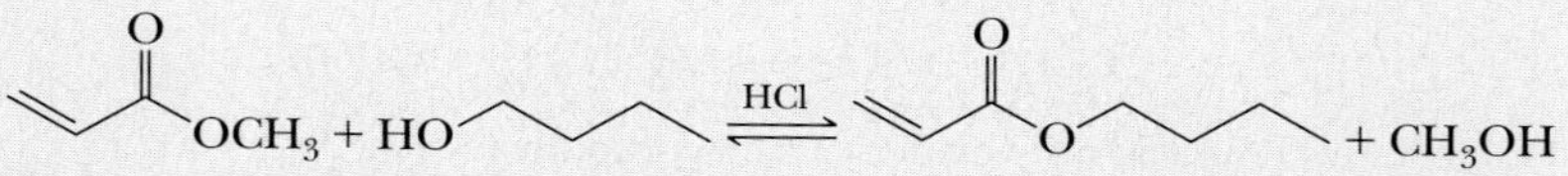

11. Reaction of an Acid Chloride with Ammonia or an Amine (Section 18.6A)

Reaction requires two moles of ammonia or amine, one to form the amide and one to neutralize the HCl byproduct.

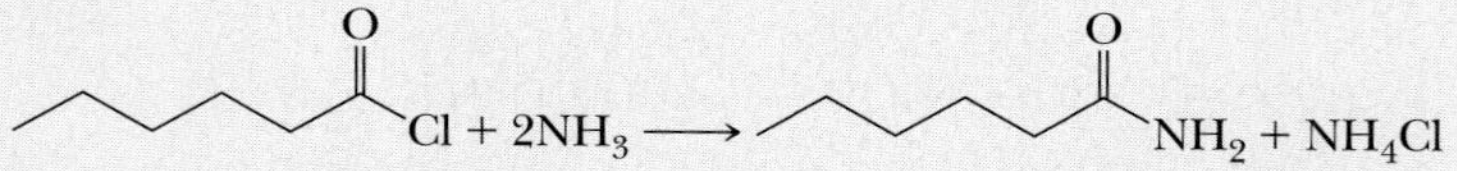

12. Reaction of an Acid Anhydride with Ammonia or an Amine (Section 18.6B)

Reaction requires two moles of ammonia or amine, one to form the amide and one to neutralize the carboxylic acid byproduct.

$$CH_3COCCH_3 + 2NH_3 \longrightarrow CH_3CNH_2 + CH_3CO^-NH_4^+$$

13. Reaction of an Ester with Ammonia or an Amine (Section 18.6C)

Treating an ester with ammonia or a primary or secondary amine gives an amide.

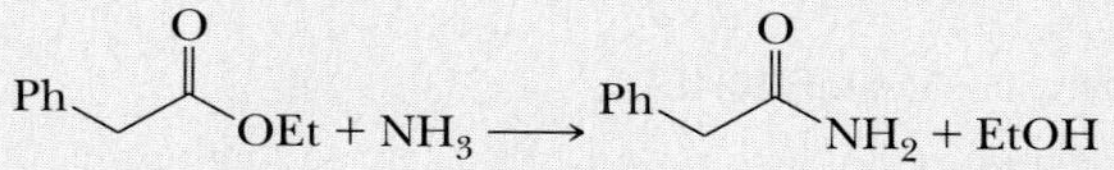

14. Reaction of an Acid Chloride with a Carboxylic Acid Salt (Section 18.7)

Treating an acid chloride with the salt of a carboxylic acid is a valuable method for synthesizing mixed anhydrides.

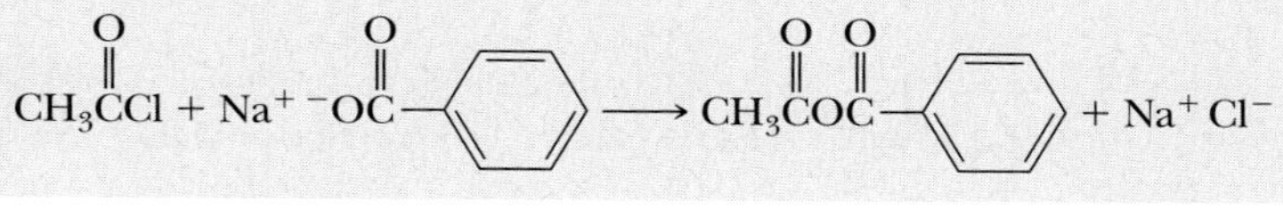

15. Reaction of an Ester with a Grignard Reagent (Section 18.9A)

Treating a formic ester with a Grignard reagent followed by hydrolysis gives a secondary alcohol. Treating of any other ester with a Grignard reagent gives a tertiary alcohol.

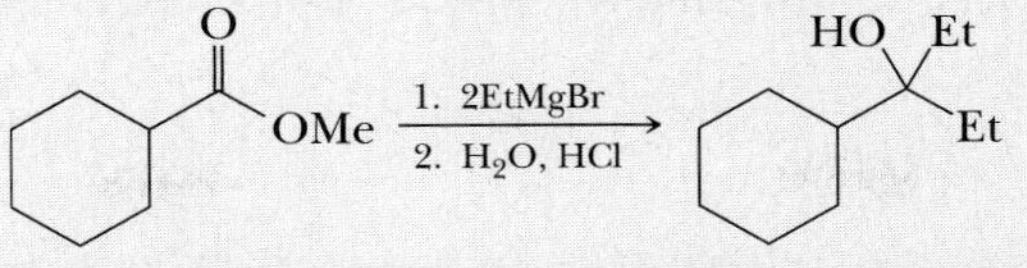

16. Reaction of an Acid Chloride with a Lithium Diorganocuprate (Section 18.9C)

Acid chlorides react readily with lithium diorganocuprates at $-78°C$ to give ketones.

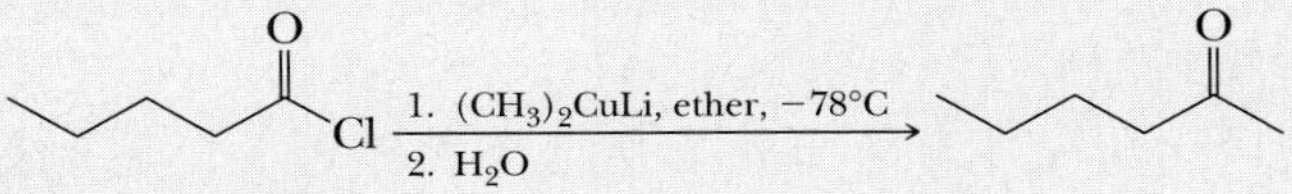

17. Reduction of an Ester (Section 18.10A)

Reduction of an ester by lithium aluminum hydride gives two alcohols:

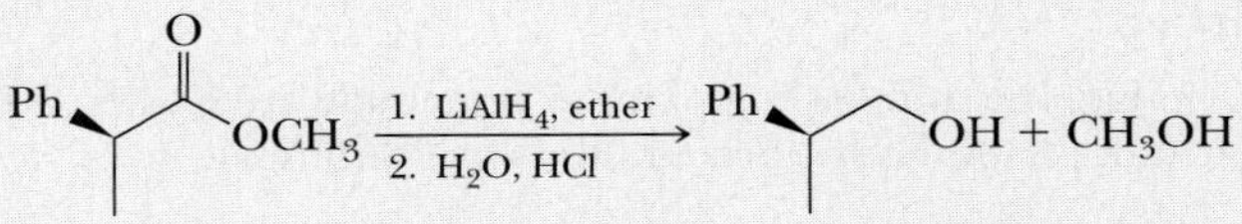

Reduction by diisobutylaluminum hydride (DIBALH) at low temperature gives an aldehyde and an alcohol.

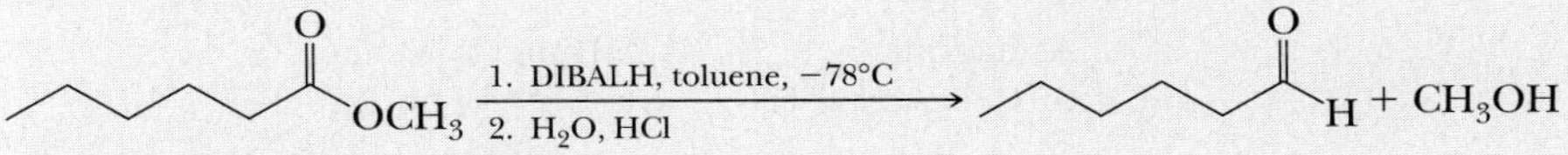

18. Reduction of an Amide (Section 18.10B)

Reduction of an amide by lithium aluminum hydride gives an amine.

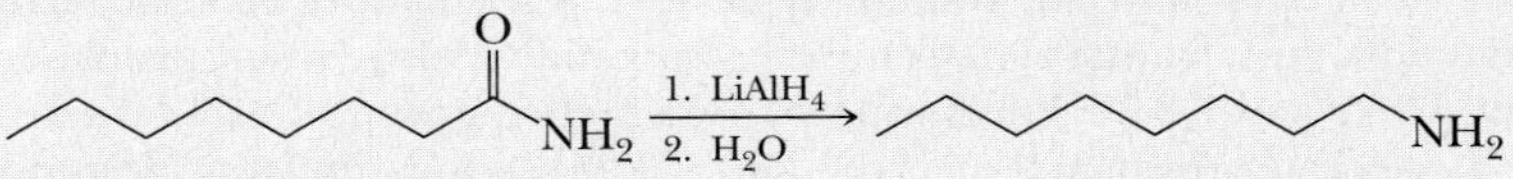

19. Reduction of a Nitrile (Section 18.10C)

Reduction of a cyano group by lithium aluminum hydride gives a primary amino group.

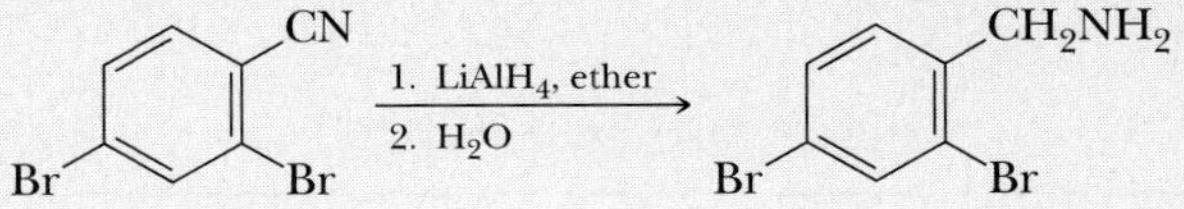

Problems

Structure and Nomenclature

18.12 Draw a structural formulas for each compound.

- **(a)** Dimethyl carbonate
- **(b)** Benzonitrile
- **(c)** Isopropyl 3-methylhexanoate
- **(d)** Diethyl oxalate
- **(e)** Ethyl (*Z*)-2-pentenoate
- **(f)** Butanoic anhydride
- **(g)** Dodecanamide
- **(h)** Ethyl 3-hydroxybutanoate
- **(i)** Octanoyl chloride
- **(j)** Diethyl *cis*-1,2-cyclohexanedicarboxylate
- **(k)** Methanesulfonyl chloride
- **(l)** *p*-Toluenesulfonyl chloride

18.13 Write the IUPAC name for each compound.

(a) Ph$\overset{O}{\overset{\|}{C}}O\overset{O}{\overset{\|}{C}}$Ph   (b) Ph—$\overset{\overset{O}{\|}}{\underset{\underset{O}{\|}}{S}}$—NH$_2$ (c) [structure with O and NHCH$_3$]

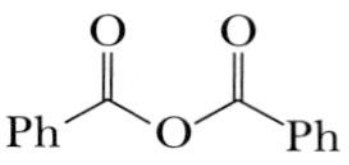

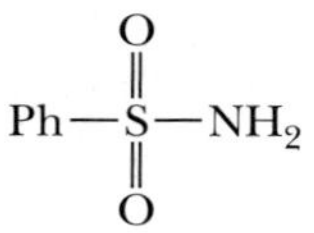

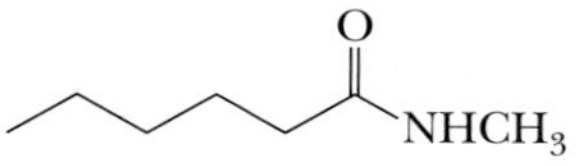

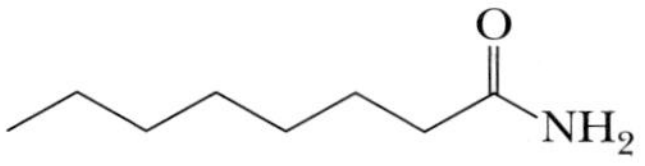

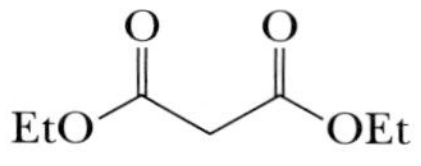

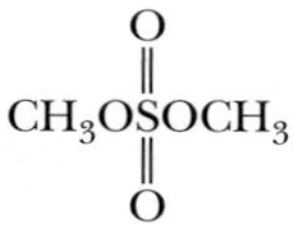

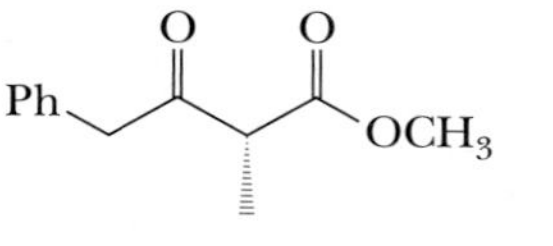

(d) [structure with O and NH$_2$]   (e) EtO$\overset{O}{\overset{\|}{C}}$$\overset{O}{\overset{\|}{C}}$OEt   (f) CH$_3O\overset{\overset{O}{\|}}{\underset{\underset{O}{\|}}{S}}OCH_3$

(g) Ph[structure]OCH$_3$ (h) Cl[structure]Cl   (i) CH$_3$(CH$_2$)$_5$CN

Physical Properties

18.14 Both the melting point and boiling point of acetamide are higher than those of its *N,N*-dimethyl derivative. How do you account for these differences?

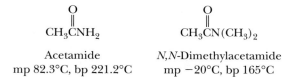

$$CH_3\overset{O}{\overset{\|}{C}}NH_2 \qquad\qquad CH_3\overset{O}{\overset{\|}{C}}N(CH_3)_2$$

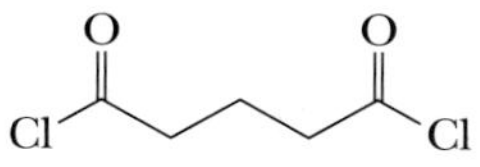

Acetamide   N,N-Dimethylacetamide
mp 82.3°C, bp 221.2°C   mp −20°C, bp 165°C

Spectroscopy

18.15 Each hydrogen of a primary amide typically has a separate ^{1}H-NMR resonance, as illustrated by the separate signals for the two amide hydrogens of propanamide, which fall at δ 6.22 and δ 6.58. Furthermore, each methyl group of *N,N*-dimethylformamide has a separate resonance (δ3.88 and δ3.98). How do you account for these observations?

18.16 Propose a structural formula for compound A, C$_7$H$_{14}$O$_2$, consistent with its ^{1}H-NMR and IR spectra.

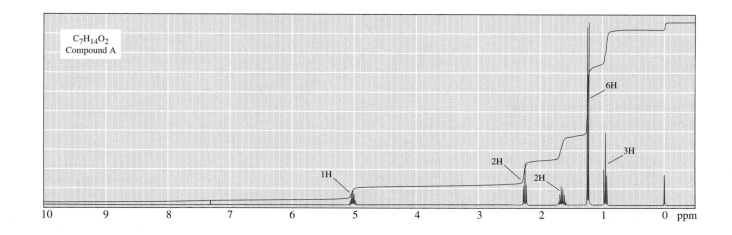

C$_7$H$_{14}$O$_2$
Compound A

6H
3H
2H
2H
1H

10   9   8   7   6   5   4   3   2   1   0 ppm

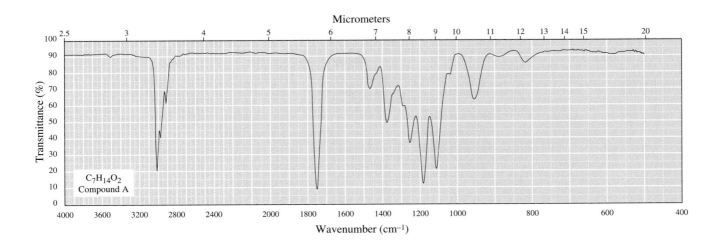

18.17 Propose a structural formula for compound B, $C_6H_{13}NO$, consistent with its ^{1}H-NMR and IR spectra.

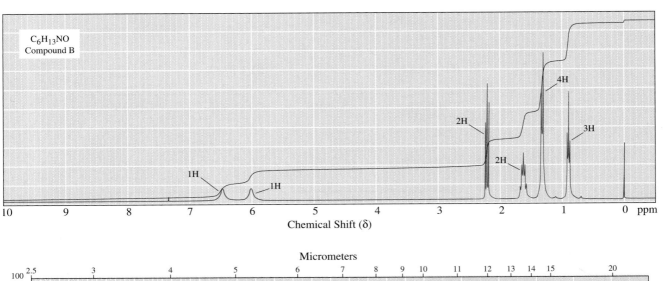

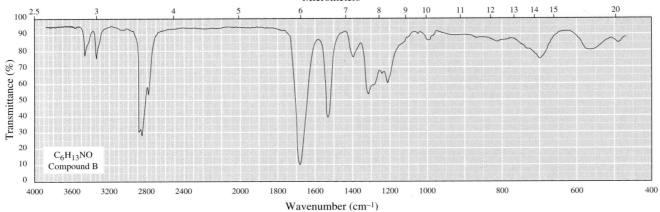

18.18 Propose a structural formula for each compound consistent with its ^{1}H-NMR and ^{13}C-NMR spectra.

(a) $C_5H_{10}O_2$

(b) $C_7H_{14}O_2$

^{1}H-NMR	^{13}C-NMR
0.96 (d, 6H)	161.11
1.96 (m, 1H)	70.01
3.95 (d, 2H)	27.71
8.08 (s, 1H)	19.00

^{1}H-NMR	^{13}C-NMR
0.92 (d, 6H)	171.15
1.52 (m, 2H)	63.12
1.70 (m, 1H)	37.31
2.09 (s, 3H)	25.05
4.10 (t, 2H)	22.45
	21.06

(c) $C_6H_{12}O_2$

(d) $C_7H_{12}O_4$

^{1}H-NMR	^{13}C-NMR
1.18 (d, 6H)	177.16
1.26 (t, 3H)	60.17
2.51 (m, 1H)	34.04
4.13 (q, 2H)	19.01
	14.25

^{1}H-NMR	^{13}C-NMR
1.28 (t, 6H)	166.52
3.36 (s, 2H)	61.43
4.21 (q, 4H)	41.69
	14.07

(e) $C_4H_7ClO_2$

(f) $C_4H_6O_2$

^{1}H-NMR	^{13}C-NMR
1.68 (d, 3H)	170.51
3.80 (s, 3H)	52.92
4.42 (q, 1H)	52.32
	21.52

^{1}H-NMR	^{13}C-NMR
2.29 (m, 2H)	177.81
2.50 (t, 2H)	68.58
4.36 (t, 2H)	27.79
	22.17

Reactions

18.19 Draw a structural formula for the principal product formed when benzoyl chloride is treated with each reagent.

(a) 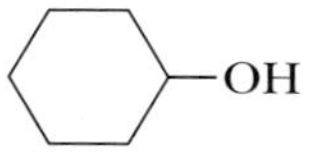 $\langle\rangle$—OH

(b) 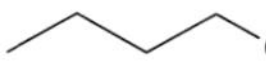 $\diagup\diagdown\diagup$OH, pyridine

(c) 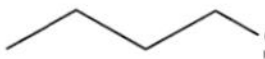 $\diagup\diagdown\diagup$SH, pyridine

(d) 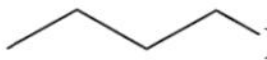 $\diagup\diagdown\diagup$NH$_2$ (two equivalents)

(e) 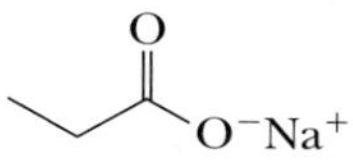 image of acetate structure with O$^-$Na$^+$

(f) $(CH_3)_2CuLi$, then H_3O^+

(g) CH_3O—$\langle\rangle$—NH$_2$, pyridine

(h) C_6H_5MgBr (two equivalents), then H_3O^+

18.20 Draw a structural formula of the principal product formed when ethyl benzoate is treated with each reagent.

(a) H_2O, NaOH, heat   (b) H_2O, H_2SO_4, heat
(c) $CH_3CH_2CH_2CH_2NH_2$ (d) DIBALH ($-78°C$), then H_2O
(e) $LiAlH_4$, then H_2O (f) C_6H_5MgBr (two equivalents), then HCl/H_2O

18.21 The mechanism for hydrolysis of an ester in aqueous acid involves formation of a tetra-hedral carbonyl addition intermediate. Evidence in support of this mechanism comes from an experiment designed by Myron Bender. He first prepared ethyl benzoate en-riched with oxygen-18 in the carbonyl oxygen and then carried out acid-catalyzed hy-drolysis of the ester in water containing no enrichment in oxygen-18. If he stopped the experiment after only partial hydrolysis and isolated the remaining ester, the recovered ethyl benzoate had lost a portion of its enrichment in oxygen-18. In other words, some exchange had occurred between oxygen-18 of the ester and oxygen-16 of water. Show how this observation bears on the formation of a tetrahedral carbonyl addition interme-diate during acid-catalyzed ester hydrolysis.

18.22 Predict the distribution of oxygen-18 in the products obtained from hydrolysis of ethyl benzoate labeled in the ethoxy oxygen under the following conditions.

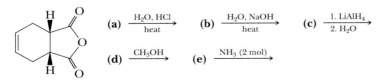

(a) In aqueous NaOH   (b) In aqueous HCl
(c) What distribution would you predict if the reaction were done with the *tert*-butyl ester in HCl?

18.23 Draw a structural formula for the principal product formed when benzamide is treated with each reagent.

(a) H_2O, HCl, heat   (b) NaOH, H_2O, heat   (c) $LiAlH_4$, then H_2O

18.24 Draw a structural formula of the principal product formed when benzonitrile is treated with each reagent.

(a) H_2O (one equivalent), H_2SO_4, heat   (b) H_2O (excess), H_2SO_4, heat
(c) NaOH, H_2O, heat   (d) $LiAlH_4$, then H_2O

18.25 Show the product expected when the following unsaturated δ-ketoester is treated with each reagent.

(a) $\dfrac{H_2 \text{ (1 mol)}}{Pd, EtOH}$ (b) $\dfrac{NaBH_4}{CH_3OH}$ (c) $\dfrac{1.\ LiAlH_4,\ THF}{2.\ H_2O}$ (d) $\dfrac{1.\ DIBALH,\ -78°}{2.\ H_2O}$

18.26 The reagent diisobutylaluminum hydride (DIBALH) reduces esters to aldehydes. When nitriles are treated with DIBALH, followed by mild acid hydrolysis, the product is also an aldehyde. Propose a mechanism for this reduction.

18.27 Show the product of treating this anhydride with each reagent.

(a) $\dfrac{H_2O, HCl}{heat}$ (b) $\dfrac{H_2O, NaOH}{heat}$ (c) $\dfrac{1.\ LiAlH_4}{2.\ H_2O}$

(d) $\dfrac{CH_3OH}{}$ (e) $\dfrac{NH_3 \text{ (2 mol)}}{}$

18.28 The analgesic acetaminophen is synthesized by treating 4-aminophenol with one equiva-lent of acetic anhydride. Draw a structural formula for acetaminophen.

18.29 Treating choline with acetic anhydride gives acetylcholine, a neurotransmitter. Write an equation for the formation of acetylcholine.

$$(CH_3)_3\overset{+}{N}CH_2CH_2OH$$

Choline

18.30 Nicotinic acid, more commonly named niacin, is one of the B vitamins. Show how nicotinic acid can be converted to (a) ethyl nicotinate and then to (b) nicotinamide.

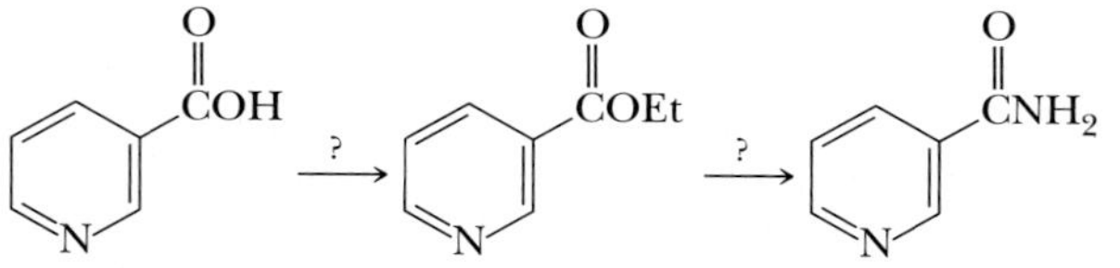

Nicotinic acid   Ethyl nicotinate   Nicotinamide
(Niacin)

18.31 Complete each reaction.

(a)

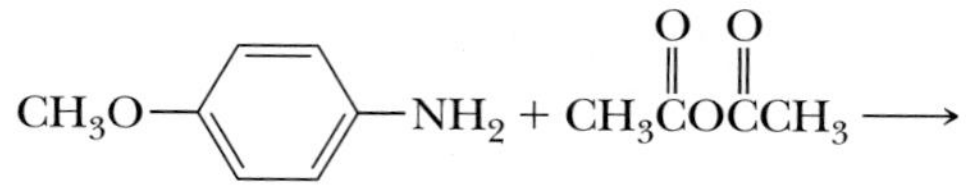

(b)

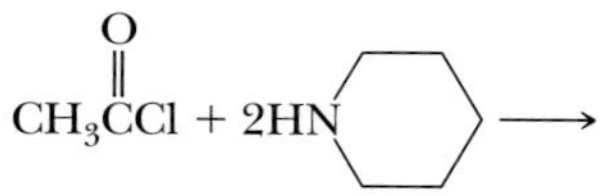

(c)

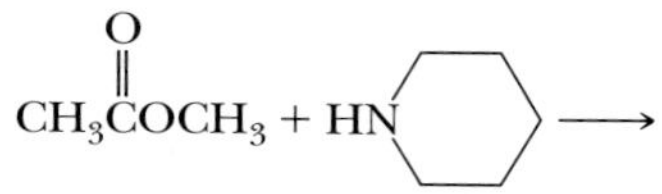

(d)

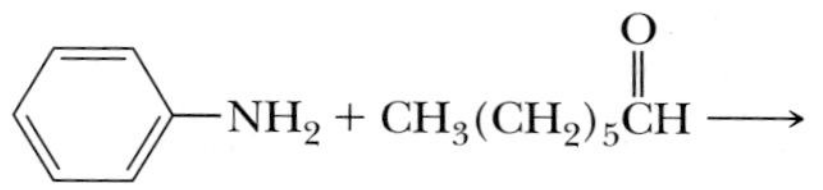

18.32 Show the product of treating γ-butyrolactone with each reagent.

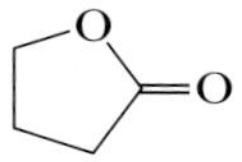

(a) NH₃ ⟶

(b) $\dfrac{\text{1. LiAlH}_4}{\text{2. H}_2\text{O}}$

(c) $\dfrac{\text{1. 2PhMgBr, ether}}{\text{2. H}_2\text{O, HCl}}$

(d) NaOH $\dfrac{\text{H}_2\text{O}}{\text{heat}}$

(e) $\dfrac{\text{1. 2CH}_3\text{Li, ether}}{\text{2. H}_2\text{O, HCl}}$

(f) $\dfrac{\text{1. DIBALH, ether, }-78°\text{C}}{\text{2. H}_2\text{O, HCl}}$

18.33 Show the product of treating the following γ-lactam with each reagent.

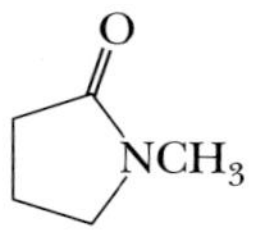

(a) $\dfrac{\text{H}_2\text{O, HCl}}{\text{heat}}$

(b) $\dfrac{\text{H}_2\text{O, NaOH}}{\text{heat}}$

(c) $\dfrac{\text{1. LiAlH}_4}{\text{2. H}_2\text{O}}$

18.34 Draw structural formulas for the products of complete hydrolysis of meprobamate, phenobarbital, and pentobarbital in hot aqueous acid.

(a)

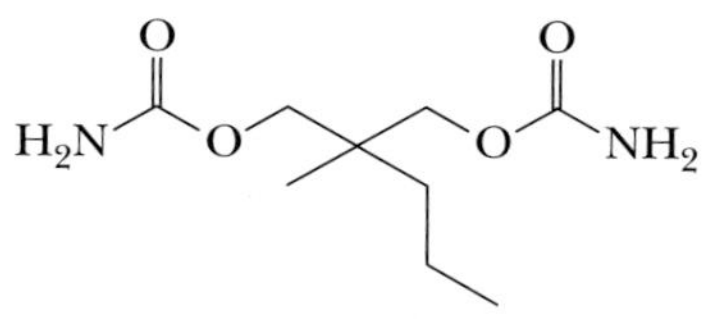

(b)

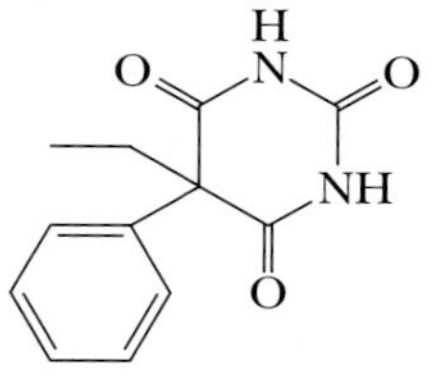

(c)

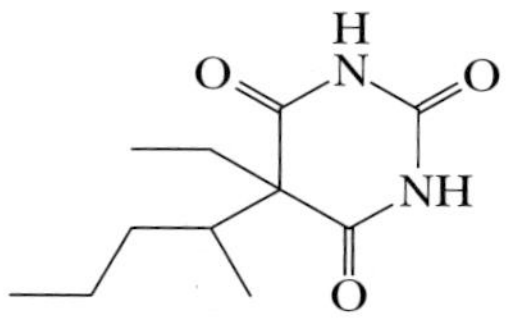

Meprobamate   Phenobarbital   Pentobarbital
(racemic)   (racemic)   (racemic)

Meprobamate is a tranquilizer prescribed under 58 different trade names, including Equanil and Miltown. Phenobarbital is a long-acting sedative, hypnotic, and anticonvulsant.

Luminal is one of over a dozen names under which it is prescribed. Pentobarbital is a short-acting sedative, hypnotic, and anticonvulsant. Nembutal is one of several trade names under which it is prescribed.

Synthesis

18.35 *N,N*-Diethyl-*m*-toluamide (DEET) is the active ingredient in several common insect repellents. Propose a synthesis for DEET from 3-methylbenzoic acid.

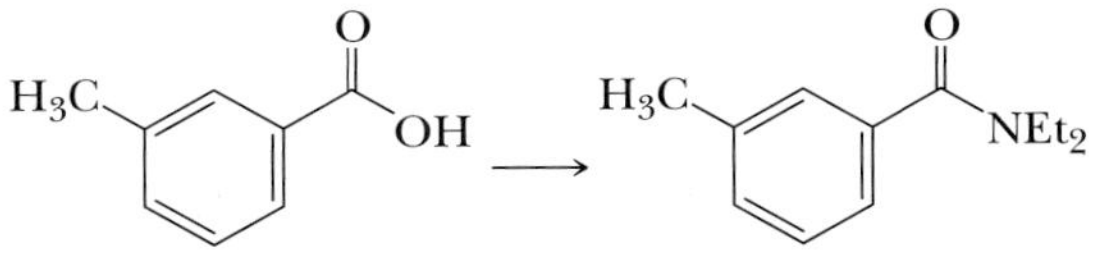

3-Methylbenzoic acid   *N,N*-Diethyl-*m*-toluamide
(*m*-Toluic acid)   (DEET)

18.36 Isoniazid, a drug used to treat tuberculosis, is prepared from pyridine-4-carboxylic acid. How might this synthesis be carried out?

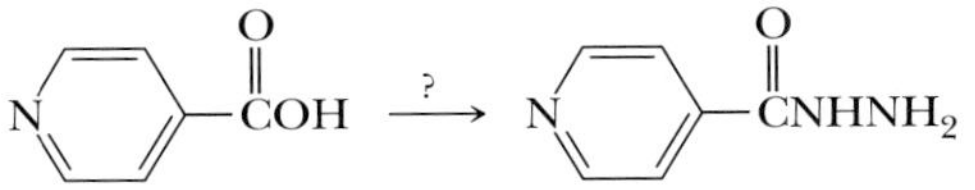

Pyridine-4-   Pyridine-4-
carboxylic acid   carboxylic acid hydrazide
                              (Isoniazid)

18.37 Show how to convert phenylacetylene to allyl phenylacetate.

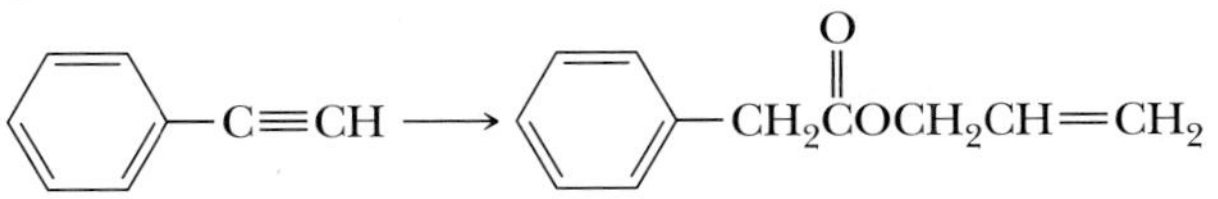

Phenylacetylene   Allyl phenylacetate

18.38 A step in a synthesis of PGE$_1$ (prostaglandin E$_1$, alprostadil) is the reaction of a trisubstituted cyclohexene with bromine to form a bromolactone. Propose a mechanism for formation of this bromolactone and account for the observed stereochemistry of each substituent on the cyclohexane ring.

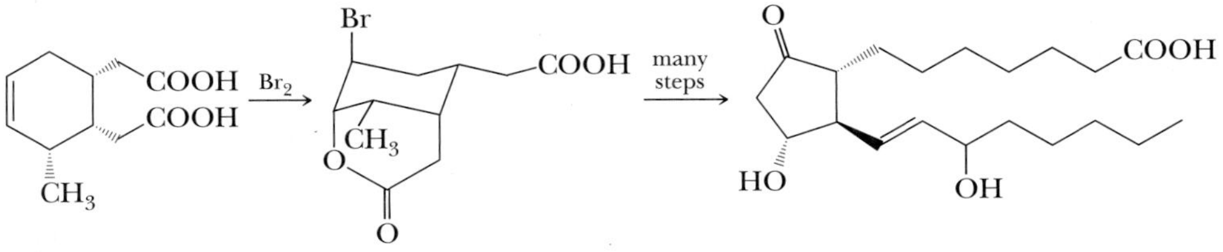

A bromolactone   PGE$_1$ (alprostadil)

Alprostadil is used as a temporary therapy for infants born with congenital heart defects that restrict pulmonary blood flow. It brings about dilation of the ductus arteriosus, which in turn increases blood flow in the lungs and blood oxygenation.

18.39 Barbiturates are prepared by treating a derivative of diethyl malonate with urea in the presence of sodium ethoxide as a catalyst. Following is an equation for the preparation of barbital, a long-duration hypnotic and sedative, from diethyl diethylmalonate and urea.

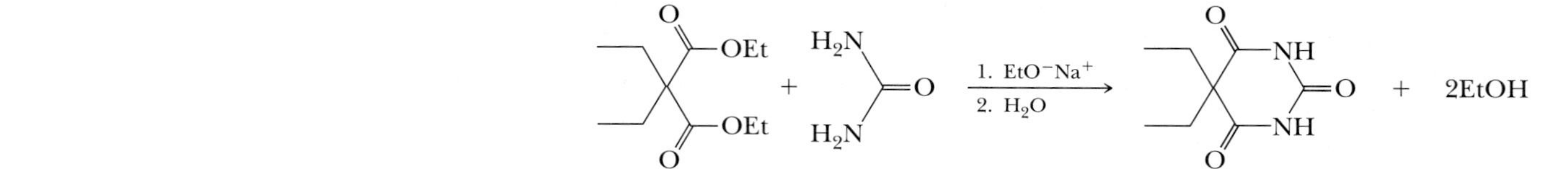

Diethyl   Urea   5,5-Diethylbarbituric acid
diethylmalonate   (Barbital)

Barbital is prescribed under one of a dozen or more trade names.

(a) Propose a mechanism for this reaction.
(b) The pK_a of barbital is 7.4. Which is the most acidic hydrogen in this molecule? How do you account for its acidity?

18.40 The following compound is one of a group of β-chloroamines, many of which have anti-tumor activity. Describe a synthesis of this compound from anthranilic acid and ethylene oxide.

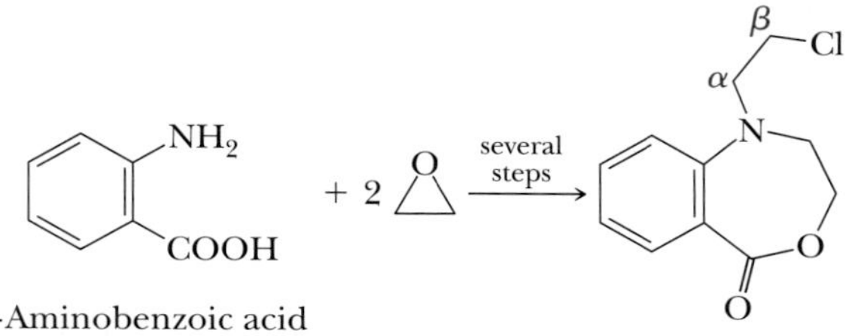

2-Aminobenzoic acid
(Anthranilic acid)

A β-chloramine

18.41 Show how to synthesize 5-nonanone from 1-bromobutane as the only organic starting material.

18.42 Procaine (its hydrochloride is marketed as Novocain) was one of the first local anesthetics for infiltration and regional anesthesia. See Chemical Connections: "From Cocaine to Procaine and Beyond." According to the following retrosynthetic scheme, procaine can be synthesized from 4-aminobenzoic acid, ethylene oxide, and diethylamine as sources of carbon atoms.

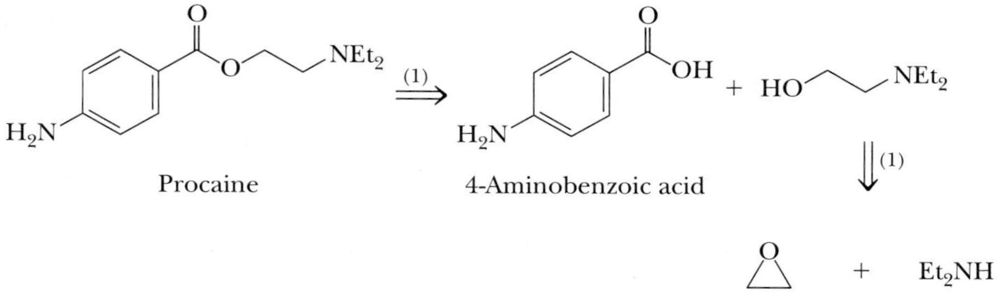

Procaine   4-Aminobenzoic acid

Ethylene oxide   Diethylamine

Provide reagents and experimental conditions to carry out the synthesis of procaine from these three compounds.

18.43 The following sequence of steps converts (*R*)-2-octanol to (*S*)-2-octanol.

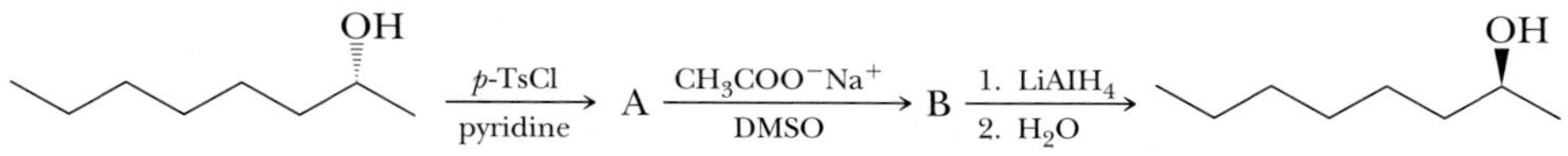

(*R*)-2-Octanol   (*S*)-2-Octanol

Propose structural formulas for intermediates A and B, specify the configuration of each, and account for the inversion of configuration in this sequence.

18.44 Reaction of a primary or secondary amine with diethyl carbonate under controlled conditions gives a carbamic ester.

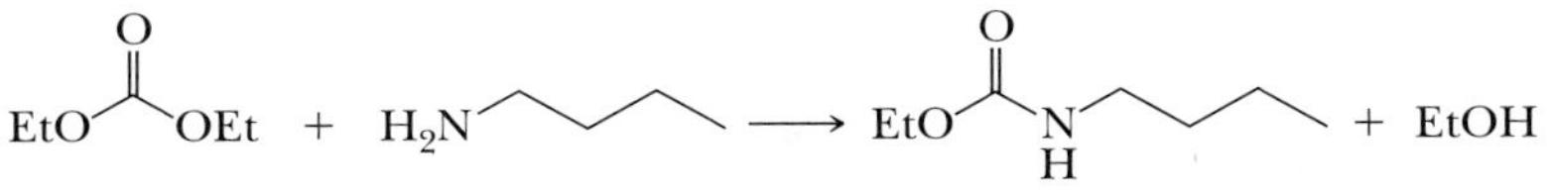

| Diethyl carbonate | Butylamine | Ethyl N-butylcarbamate |

Propose a mechanism for this reaction.

18.45 Several sulfonylureas, a class of compounds containing $RSO_2NHCONHR$, are useful drugs as orally active replacements for injected insulin in patients with adult-onset diabetes. These drugs decrease blood glucose concentrations by stimulating β cells of the pancreas to release insulin and by increasing the sensitivity of insulin receptors in peripheral tissues to insulin stimulation.

Tolbutamide is synthesized by the reaction of the sodium salt of p-toluenesulfonamide and ethyl N-butylcarbamate (see Problem 18.44 for the synthesis of this carbamic ester). Propose a mechanism for this step.

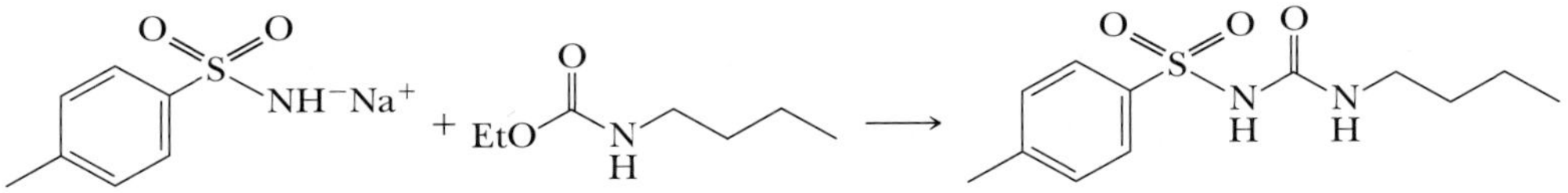

| Sodium salt of p-toluenesulfonamide | A carbamic ester | Tolbutamide (Oramide, Orinase) |

18.46 Following are structural formulas for two more widely used sulfonylurea hypoglycemic agents.

| (a) Tolazamide (Tolamide, Tolinase) | (b) Gliclazide (Diamicron) |

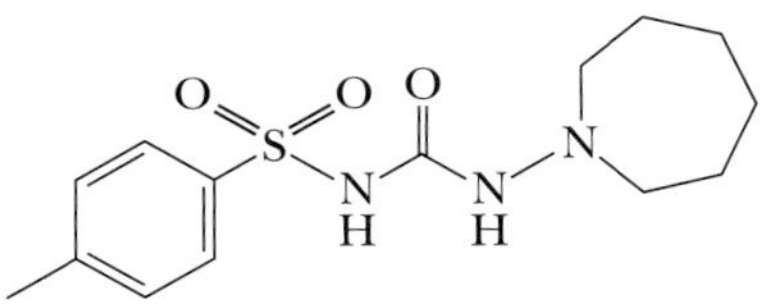

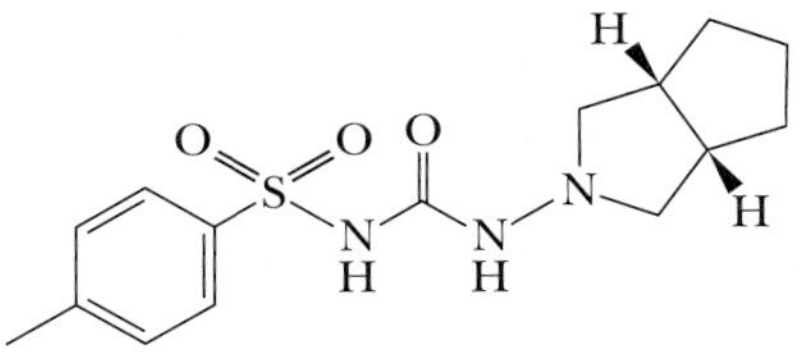

Show how each might be synthesized by converting an appropriate amine to a carbamic ester and then treating the carbamate with the sodium salt of a substituted benzenesulfonamide.

18.47 Amantadine is effective in preventing infections caused by the influenza A virus and in treating established illnesses. It is thought to block a late stage in the assembly of the virus. Amantadine is synthesized by treating 1-bromoadamantane with acetonitrile in sulfuric acid to give N-adamantylacetamide, which is then converted to amantadine.

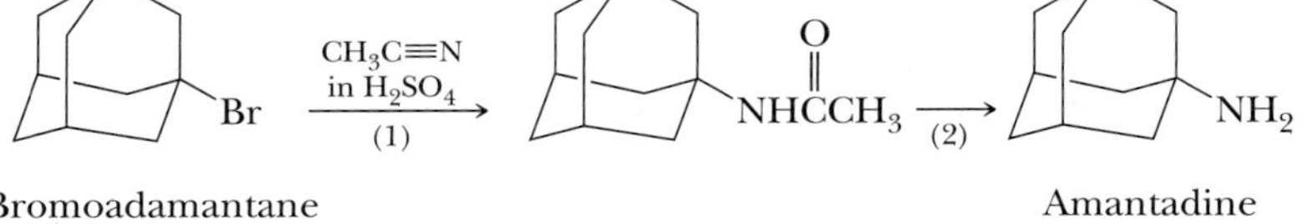

1-Bromoadamantane   Amantadine

(a) Propose a mechanism for the transformation in Step 1.
(b) Describe experimental conditions to bring about Step 2.

18.48 In a series of seven steps, (*S*)-malic acid is converted to the bromoepoxide shown on the right in 50% overall yield. This synthesis is enantioselective—of the stereoisomers possible for the bromoepoxide, only one is formed.

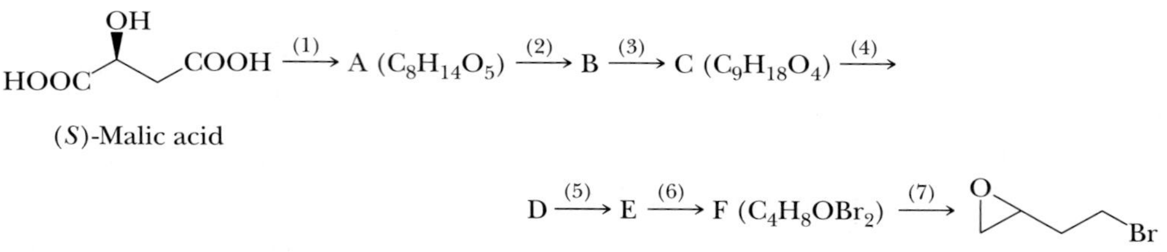

(*S*)-Malic acid

A bromoepoxide

Steps/reagents: 1. CH_3CH_2OH, H^+ 3. $LiAlH_4$, then H_2O 6. H_2O, CH_3COOH

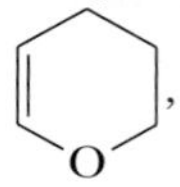

2. , H^+ 4. TsCl, pyridine   7. KOH

5. NaBr, DMSO

In thinking about the chemistry of these steps, you will want to review the use of dihydropyran as an —OH protecting group (Section 16.7D) and the use the *p*-toluenesulfonyl chloride to convert the —OH, a poor leaving group, into a tosylate, a good leaving group (Section 10.5D).

(a) Propose structural formulas for intermediates A through F, and specify the configuration at each chiral center.

(b) What is the configuration of the chiral center in the bromoepoxide? How do you account for the stereoselectivity of this seven-step conversion?

18.49 Following is a retrosynthetic analysis for the synthesis of the herbicide (*S*)-Metolachlor from 2-ethyl-6-methylaniline, chloroacetic acid, acetone, and methanol.

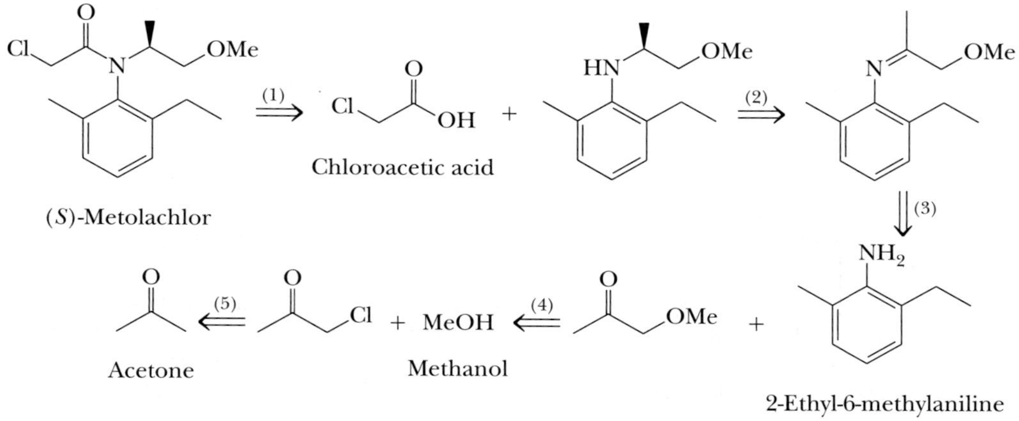

(*S*)-Metolachlor

Chloroacetic acid

Acetone   Methanol

2-Ethyl-6-methylaniline

Show reagents and experimental conditions for the synthesis of Metolachlor from these four organic starting materials. Your synthesis will most likely give a racemic mixture. The chiral catalyst used by Novartis for reduction in Step 2 gives 80% enantiomeric excess of the *S* enantiomer.

18.50 Following is a retrosynthetic analysis for the anthelmintic (against worms) diethylcarbamazine.

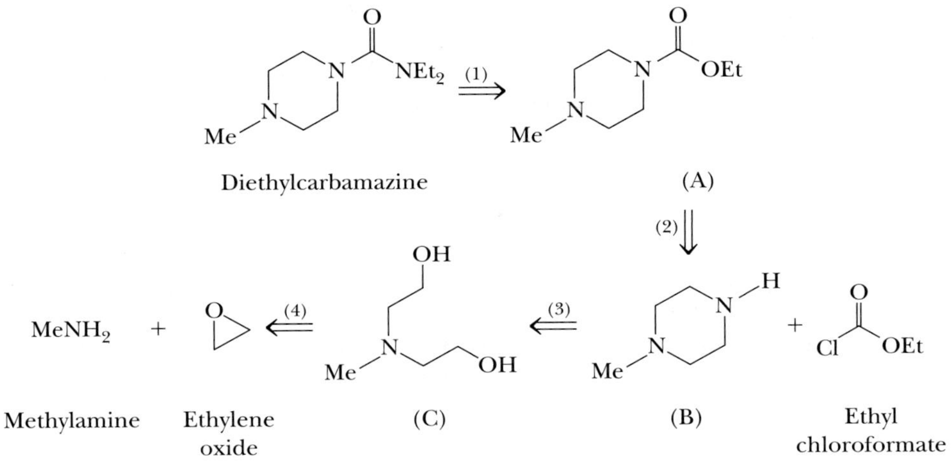

Diethylcarbamazine (A)

MeNH$_2$ + Ethylene oxide → (C) → (B) + Ethyl chloroformate

Methylamine   Ethylene oxide   (C)   (B)   Ethyl chloroformate

Diethylcarbamazine is used chiefly against nematodes, small cylindrical or slender threadlike worms such as the common roundworm, which are parasitic in animals and plants. Given this retrosynthetic analysis, propose a synthesis of diethylcarbamazine from the four named starting materials.

18.51 Given this retrosynthetic analysis, propose a synthesis for the antidepressant moclobemide.

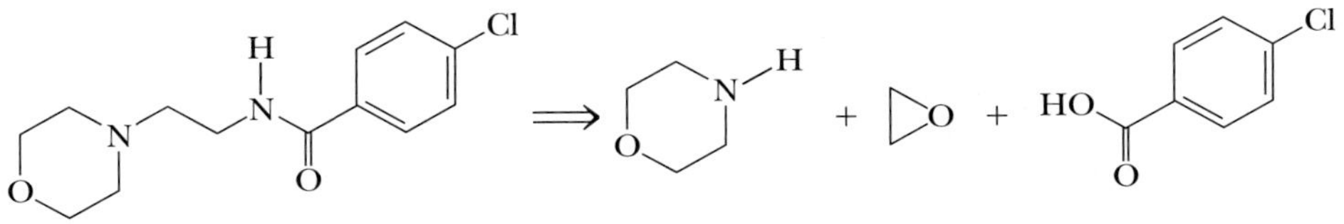

Moclobemide   Morpholine   Ethylene oxide p-Chlorobenzoic acid

18.52 Propose a synthesis for diphenhydramine starting from benzene, benzoic acid, and 2-(N,N-dimethylamino)ethanol.

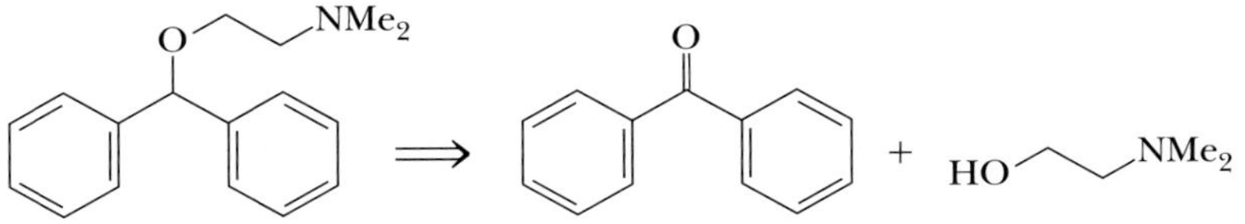

Diphenhydramine   Benzophenone   2-(N,N-Dimethylamino)-ethanol

The hydrochloride salt of diphenhydramine, best known by its trade name of Benadryl, is an antihistamine.

18.53 Propose a synthesis of the topical anesthetic cyclomethycaine from 4-hydroxybenzoic acid and 2-methylpiperidine and any other necessary reagents.

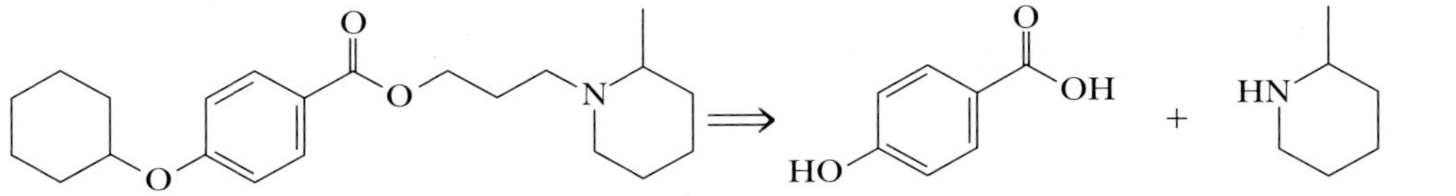

Cyclomethycaine   4-Hydroxybenzoic acid   2-Methylpiperidine

18.54 Following is an outline of a synthesis of bombykol, the sex attractant of the male silk-worm moth. Of the four stereoisomers possible for this conjugated diene, the 10-*trans*-12-*cis* isomer show here is over 10^6 times more potent as a sex attractant than any of the other three possible stereoisomers.

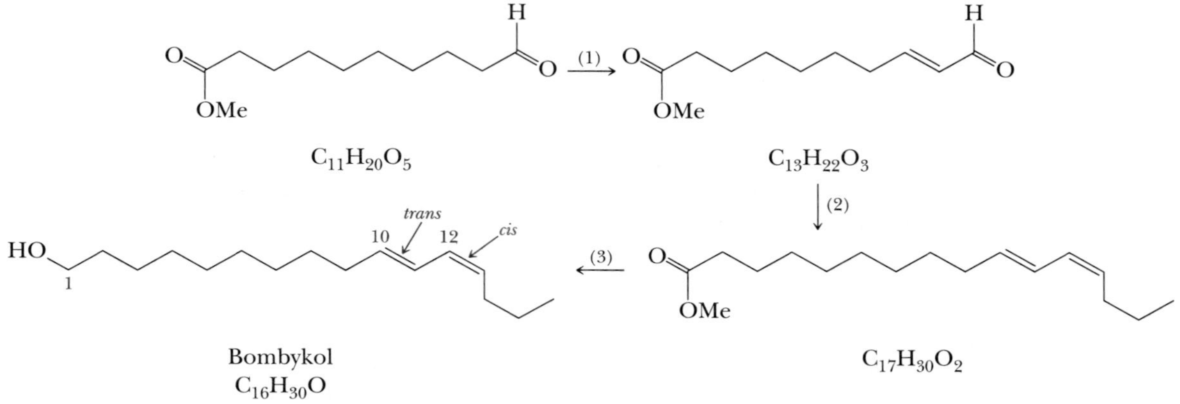

$C_{11}H_{20}O_5$

$C_{13}H_{22}O_3$

Bombykol
$C_{16}H_{30}O$

$C_{17}H_{30}O_2$

Show how this synthesis might be accomplished, and explain how your proposed synthesis is stereoselective for the 10-*trans*-12-*cis* isomer.

18.55 In Problem 7.28, we saw this step in Johnson's synthesis of the steroid hormone progesterone.

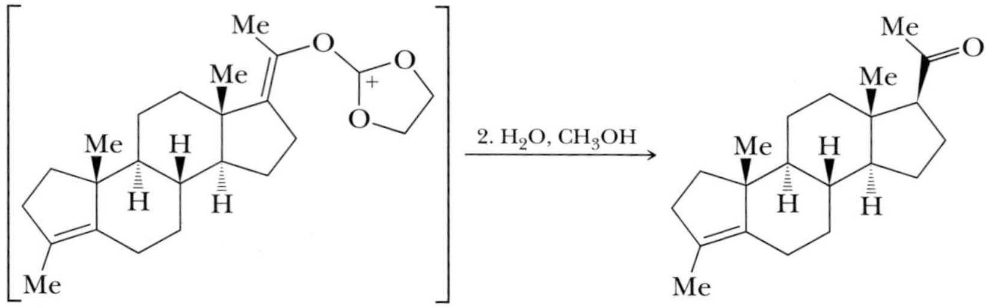

B

C

Propose a mechanism for this step in the synthesis.

Looking Ahead

18.56 We have seen two methods for converting a carboxylic acid and an amine into an amide. Suppose that you start instead with a dicarboxylic acid such as hexanedioic acid and a diamine such as 1,6-hexanediamine. Show how amide formation in this case can lead to a polymer (a macromolecule of molecular weight several thousands of times that of the starting materials).

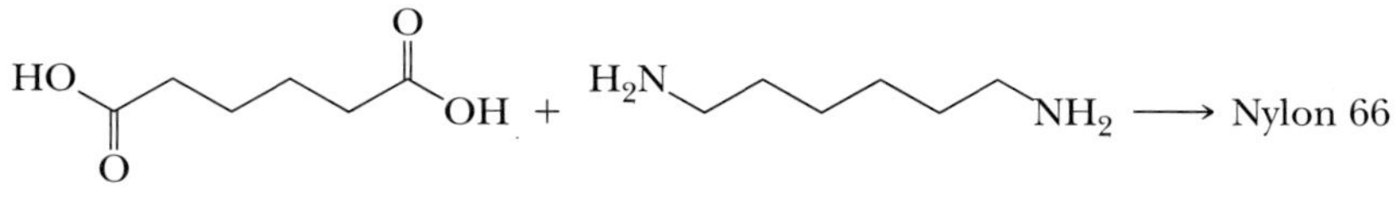

Hexanedioic acid
(Adipic acid)

1,6-Hexanediamine
(Hexamethylenediamine)

As we shall see in Section 29.5A, the material produced in this reaction is the high-molecular-weight polymer nylon-66, so named because it is synthesized from two 6-carbon starting materials.

18.57 Using the same reasoning as in Problem 18.56, show how amide formation between this combination of dicarboxylic acid and diamine will also lead to a polymer, in this case Kevlar.

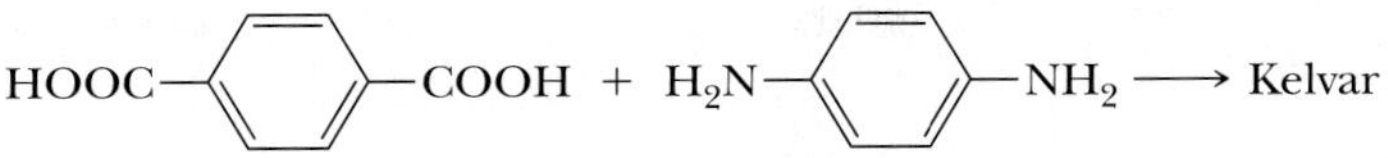

$$HOOC-\langle \rangle-COOH \ + \ H_2N-\langle \rangle-NH_2 \longrightarrow \text{Kelvar}$$

1,4-Benezenedicarboxylic acid   1,6-Benezenediamine
(Terephthalic acid)

18.58 A urethane is a molecule in which a carbonyl group is part of an ester and an amide (it is an amide in one direction, an ester in the other direction). Propose a mechanism for the reaction of an isocyanate with an alcohol to form a urethane.

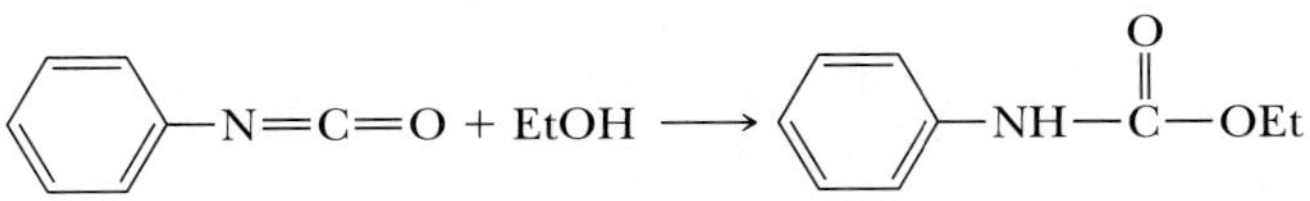

$$\langle \rangle-N{=}C{=}O + EtOH \longrightarrow \langle \rangle-NH-\overset{\overset{\displaystyle O}{\|}}{C}-OEt$$

Phenylisocyanate   Ethanol   A urethane

18.59 Suppose that you start with a diisocyanate and a diol. Show how their reaction can lead to a polymer called a polyurethane (Section 29.5D).

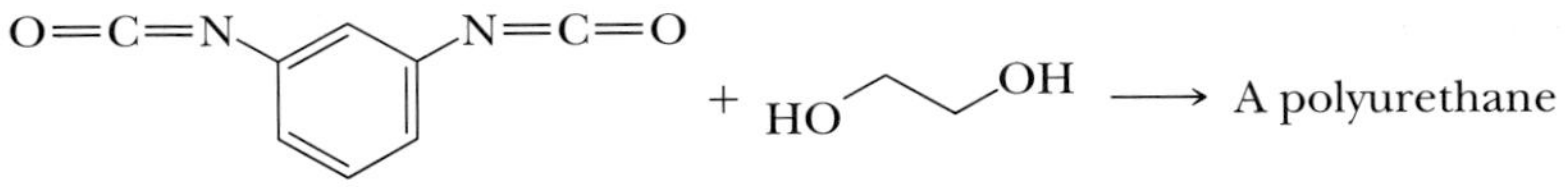

$$O{=}C{=}N \diagdown \overset{N{=}C{=}O}{\langle \rangle} \quad + \ HO \diagup\diagdown OH \longrightarrow \text{A polyurethane}$$

A diisocyanate   Ethylene glycol

19

(Herb Ohymeyer/Fran Heyl Associates)

■ Crystals of tamoxifen (Problems 19.41, 21.61, and 21.62) viewed under polarizing light. Inset: A model of tamoxifen.

Outline

Enolate Anions and Enamines

In this chapter, we continue the chemistry of carbonyl compounds. In Chapters 16 through 18, we concentrated on the carbonyl group itself and on nucleophilic additions to the carbonyl carbon to form tetrahedral carbonyl addition intermediates and on products derived from collapse of this intermediate. In this chapter, we expand on the chemistry of carbonyl-containing compounds and consider the consequences of the acidity of α-hydrogens and the formation of enolate anions.

19.1 Formation and Reactions of Enolate Anions: An Overview

Following is the resonance-stabilized **enolate anion** (Section 16.9A) formed by treating acetaldehyde with base. This anion is best represented as a hybrid of two contributing structures. Of these, the structure with the negative charge on the more electronegative oxygen atom makes the greater contribution to the hybrid. Note that although the majority of the negative charge is on the carbonyl oxygen, there is still a significant partial negative charge on the alpha carbon.

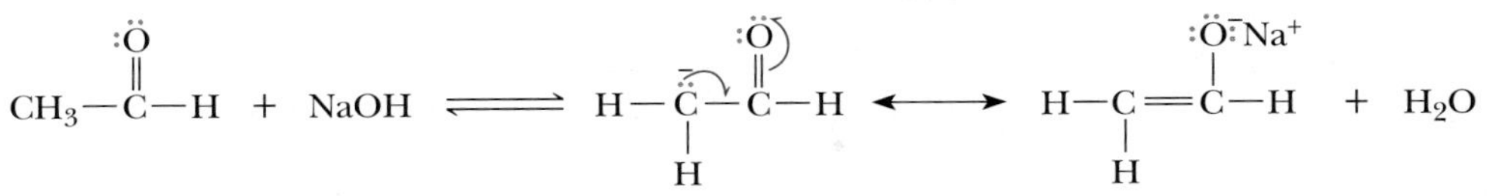

The majority of the negative
charge in the hybrid is on oxygen

Enolate anions are important synthetic reagents because they react at carbon to create new carbon-carbon bonds in two types of reactions. First, they can function as nucleophiles in S_N2 reactions as shown in this general reaction.

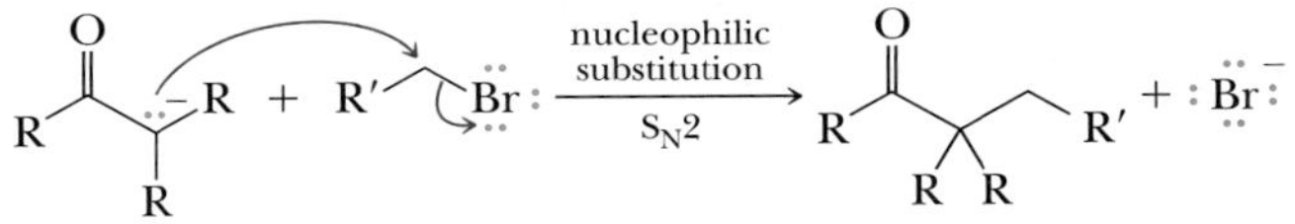

An enolate   A 1° haloalkane
anion   or sulfonate

Second, they function as nucleophiles in carbonyl addition reactions. Here, we show nucleophilic addition of an enolate anion to the carbonyl carbon of a ketone.

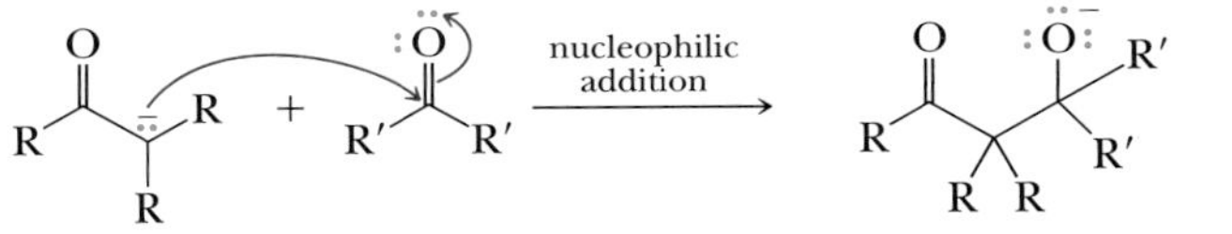

An enolate anion   A ketone   A tetrahedral carbonyl
                                    addition intermediate

Enolate anions also add in this manner to the carbonyl groups of aldehydes and esters.

As shown by the charge distribution on the electrostatic potential map, the majority of negative charge of an enolate anion is on the carbonyl oxygen. If reaction were to occur at the carbonyl oxygen, the product would be a vinyl ether, whereas reaction at the α-carbon leads to alkylation.

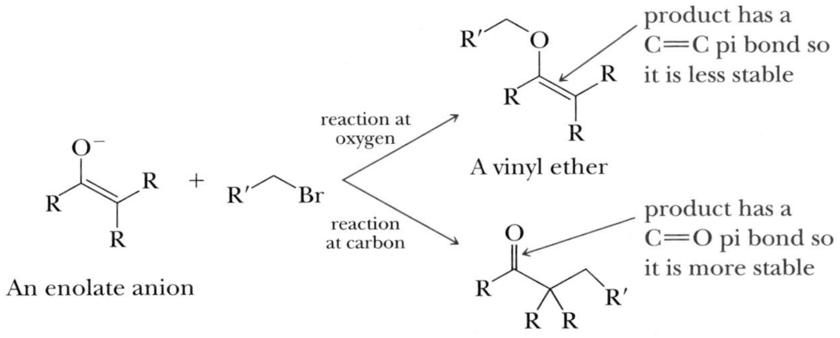

product has a
C=C pi bond so
it is less stable

reaction at
oxygen

A vinyl ether

An enolate anion

reaction
at carbon

product has a
C=O pi bond so
it is more stable

Despite this charge distribution, enolate anions react primarily at carbon for two reasons. First, there is always a counterion such as the Li^+ or Na^+ ion associated with the enolate anion. These counterions are more tightly associated with the oxygen atom than the alpha carbon. As a result, the counterion to some degree blocks the approaching electrophile, thus reducing the likelihood of a productive collision with the oxygen. In fact, enolates are thought to exist in solution as larger aggregates containing several counterions associated with several enolate oxygen atoms, effectively amplifying this effect.

The second reason enolates react at carbon is based on product thermodynamics. We have already seen that, other factors being equal, reactions at equilibrium will favor products with stronger bonds. If an enolate anion were to react at the alpha carbon, the product would contain a C=O pi bond. If it were to react at the carbonyl oxygen, the product would contain a C=C pi bond. In general, C=O bonds are stronger than C=C bonds. (Recall, for example, the relative percentages of keto and enol forms present at equilibrium for simple aldehydes and ketones, Section 16.9B.) Thus, enolate anions react primarily at the alpha carbon to form new carbon-carbon bonds.

19.2 Aldol Reaction

Unquestionably, the most important reaction of enolate anions derived from aldehydes and ketones is their nucleophilic addition to the carbonyl group of another molecule of the same or different compound, as illustrated by the following reactions.

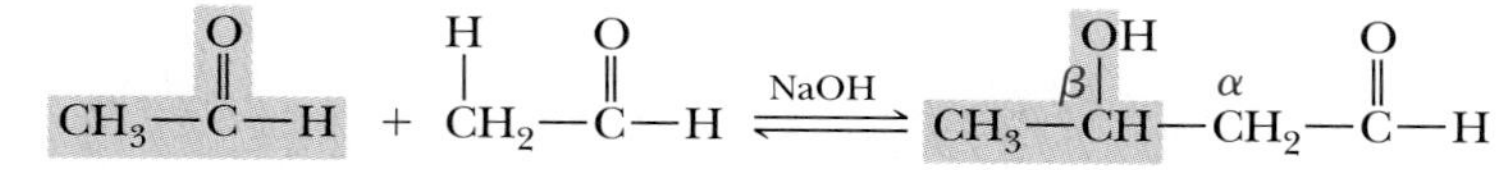

$$
\underset{\substack{\text{Ethanal} \\ \text{(Acetaldehyde)}}}{CH_3-\overset{\overset{\text{O}}{\|}}{C}-H} + \underset{\substack{\text{Ethanal} \\ \text{(Acetaldehyde)}}}{\overset{\overset{\text{H}}{|}}{CH_2}-\overset{\overset{\text{O}}{\|}}{C}-H} \xrightarrow{\text{NaOH}} \underset{\substack{\text{3-Hydroxybutanal} \\ \text{(a } \beta\text{-hydroxyaldehyde;} \\ \text{formed as a racemic} \\ \text{mixture)}}}{CH_3-\overset{\overset{\text{OH}}{\overset{\beta|}{}}}{CH}-\overset{\alpha}{CH_2}-\overset{\overset{\text{O}}{\|}}{C}-H}
$$

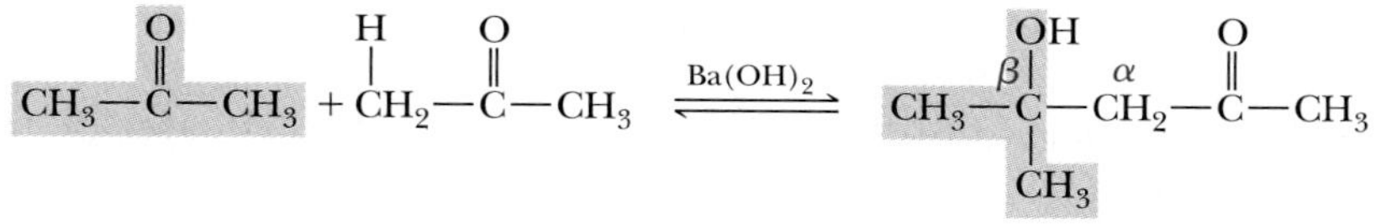

$$
\underset{\substack{\text{Propanone} \\ \text{(Acetone)}}}{CH_3-\overset{\overset{\text{O}}{\|}}{C}-CH_3} + \underset{\substack{\text{Propanone} \\ \text{(Acetone)}}}{\overset{\overset{\text{H}}{|}}{CH_2}-\overset{\overset{\text{O}}{\|}}{C}-CH_3} \xrightarrow{\text{Ba(OH)}_2} \underset{\substack{\text{4-Hydroxy-4-methyl-2-pentanone} \\ \text{(a } \beta\text{-hydroxyketone)}}}{CH_3-\underset{\underset{\text{CH}_3}{|}}{\overset{\overset{\text{OH}}{\overset{\beta|}{}}}{C}}-\overset{\alpha}{CH_2}-\overset{\overset{\text{O}}{\|}}{C}-CH_3}
$$

Although such reactions may be catalyzed by either acid or base, base catalysis is more common.

The common name of the product derived from the reaction of acetaldehyde in base is aldol because it is both an *ald*ehyde and an alcoh*ol*. Aldol is also the generic name given to any product formed in this type of reaction. The product of an **aldol reaction** is a β-hydroxyaldehyde or a β-hydroxyketone.

The key step in a base-catalyzed aldol reaction is nucleophilic addition of the enolate anion of one carbonyl-containing molecule to the carbonyl group of another to form a tetrahedral carbonyl addition intermediate. This mechanism is illustrated by the aldol reaction between two molecules of acetaldehyde. Notice in this three-step mechanism that OH^- is a catalyst; an OH^- is used in Step 1, but another OH^- is generated in Step 3.

Mechanism *Base-Catalyzed Aldol Reaction*

Step 1: Removal of an α-hydrogen by base gives a resonance-stabilized enolate anion.

$$H-\ddot{\overset{..}{O}}{:}^- + H-CH_2-\overset{\overset{\displaystyle \ddot{O}:}{\|}}{C}-H \rightleftharpoons H-\overset{..}{\underset{..}{O}}-H + \left[{}^-{:}CH_2-\overset{\overset{\displaystyle \ddot{O}:}{\|}}{C}-H \longleftrightarrow CH_2{=}\overset{\overset{\displaystyle {:}\ddot{O}{:}^-}{|}}{C}-H \right]$$

pK_a 20 pK_a 15.7   An enolate anion
(weaker acid)   (stronger acid)

Given the relative acidities of the two acids in this equilibrium, the position of this equilibrium lies considerably to the left.

Step 2: Nucleophilic addition of the enolate anion to the carbonyl carbon of another aldehyde (or ketone) gives a tetrahedral carbonyl addition intermediate.

$$CH_3-\overset{\overset{\displaystyle \ddot{O}:}{\|}}{C}-H + {}^-{:}CH_2-\overset{\overset{\displaystyle {:}O{:}}{\|}}{C}-H \rightleftharpoons CH_3-\overset{\overset{\displaystyle {:}\ddot{O}{:}^-}{|}}{C}H-CH_2-\overset{\overset{\displaystyle {:}O{:}}{\|}}{C}-H$$

A tetrahedral carbonyl
addition intermediate

Step 3: Reaction of the tetrahedral carbonyl addition intermediate with a proton donor gives the aldol product as a racemic mixture and generates a new base catalyst.

$$CH_3-\overset{\overset{\displaystyle {:}\ddot{O}{:}^-}{|}}{C}H-CH_2-\overset{\overset{\displaystyle {:}O{:}}{\|}}{C}-H + H-\ddot{\overset{..}{O}}H \rightleftharpoons CH_3-\overset{\overset{\displaystyle {:}OH}{|}}{C}H-CH_2-\overset{\overset{\displaystyle {:}O{:}}{\|}}{C}-H + {}^-{:}\ddot{O}H$$

(Formed as a
racemic mixture)

Mechanism *Acid-Catalyzed Aldol Reaction*

The mechanism of an acid-catalyzed aldol reaction involves three steps, the first two of which are preparation of the aldehyde or ketone for formation of the new carbon-carbon bond. The key step is attack of the enol of one molecule on the protonated carbonyl group of a second molecule.

Step 1: Keto and enol forms of one molecule of the aldehyde or ketone undergo acid-catalyzed equilibration (Section 16.9B).

$$CH_3-\overset{\overset{\displaystyle O}{\|}}{C}-H \underset{}{\overset{HA}{\rightleftharpoons}} CH_2{=}\overset{\overset{\displaystyle OH}{|}}{C}-H$$

Step 2: Proton transfer from the acid, HA, to the carbonyl oxygen of a second molecule of aldehyde or ketone gives an oxonium ion.

$$CH_3-\overset{\overset{\displaystyle {:}\ddot{O}{:}}{\|}}{C}-H + H-A \longrightarrow CH_3-\overset{\overset{\displaystyle \overset{+}{O}{:}-H}{\|}}{C}-H + {:}A^-$$

Step 3: Attack by the enol of one molecule on the protonated carbonyl group of another molecule forms the new carbon-carbon bond. This attack is followed by proton transfer to A^- to regenerate the acid catalyst.

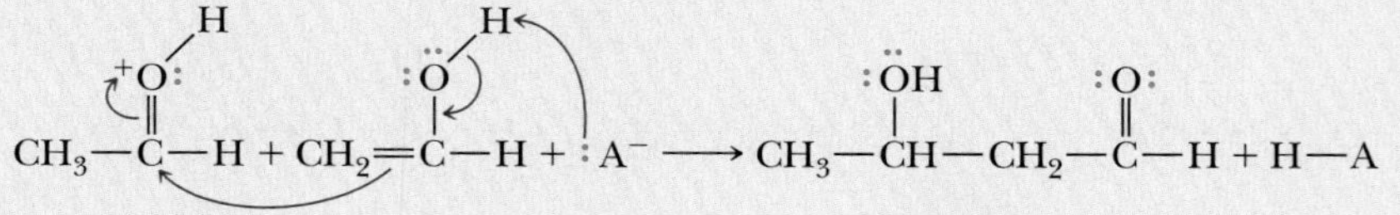

$$CH_3-\overset{+\overset{\cdot\cdot}{O}\cdot H}{\underset{\|}{C}}-H + CH_2{=}\overset{\cdot\cdot\overset{H}{O}\cdot}{\underset{\|}{C}}-H + :A^- \longrightarrow CH_3-\overset{:\overset{\cdot\cdot}{O}H}{\underset{|}{CH}}-CH_2-\overset{:O:}{\underset{\|}{C}}-H + H-A$$

(Formed as a racemic mixture)

You might compare the mechanisms of the acid- and base-catalyzed aldol reactions. Under base catalysis, the carbon-carbon bond-forming step involves attack of an enolate anion (a nucleophile) on the uncharged carbonyl carbon (an electrophile) of a second molecule of the aldehyde or ketone. Under acid catalysis, it involves attack of the enol (a nucleophile) of one molecule on the protonated carbonyl group (an electrophile) of the second molecule.

It is quite common to create chiral products during aldol reactions, as well as in the other enolate reactions we discuss in this chapter. The products will be formed as racemic mixtures unless one of the reactants is chiral and present as a single enantiomer. In cases when two chiral centers are created in the reaction, four stereoisomers are produced as two 1:1 mixtures of enantiomers. A great deal of work has gone into learning how to carry out aldol and other enolate reactions that give predominantly a single enantiomer product.

Example 19.1

Draw the product of the base-catalyzed aldol reaction of each compound.
(a) Butanal   **(b)** Cyclohexanone

Solution

The aldol product is formed by nucleophilic addition of the α-carbon of one molecule to the carbonyl carbon of another.

(a)

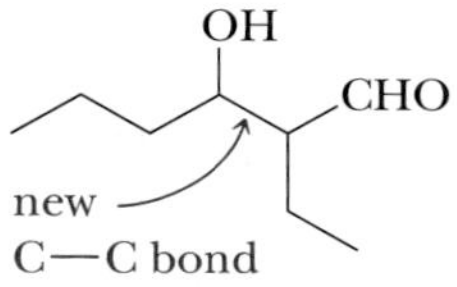

new
C—C bond

(four stereoisomers;
formed as two
racemic mixtures)

(b)

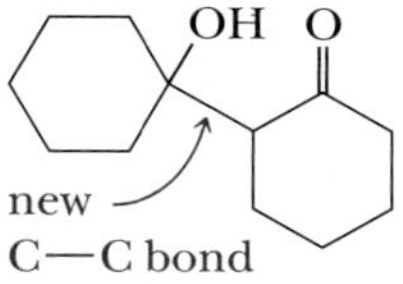

new
C—C bond

(four stereoisomers;
formed as two
racemic mixtures)

Problem 19.1

Draw the product of the base-catalyzed aldol reaction of each compound.
(a) Phenylacetaldehyde   **(b)** Cyclopentanone

β-Hydroxyaldehydes and β-hydroxyketones are very easily dehydrated, and often the conditions necessary to bring about an aldol reaction are sufficient to cause dehydration. Dehydration is particularly facile in the case of acid-catalyzed aldol reactions. Alternatively, dehydration can be brought about by warming the aldol product in dilute acid. The major product from dehydration of an aldol is one in which the carbon-carbon double bond is conjugated with the carbonyl group; that is, the product is an α,β-unsaturated aldehyde or ketone.

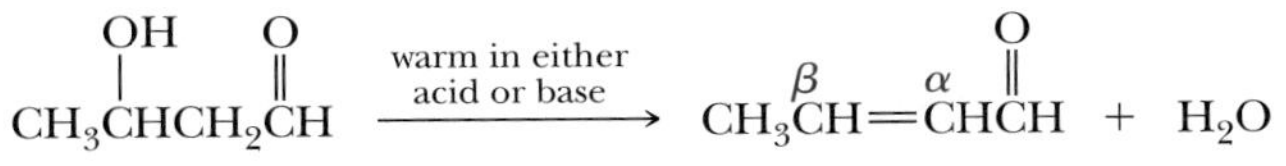

$$\underset{\substack{\text{OH} \quad\quad \text{O}}}{CH_3CHCH_2CH} \xrightarrow[\text{acid or base}]{\text{warm in either}} \underset{\beta \quad\quad \alpha}{CH_3CH=CHCH} + H_2O$$

An α,β-unsaturated
aldehyde

As we will see in Chapter 20, conjugated systems are more stable in general than unconjugated ones; therefore, the product of an aldol dehydration is usually the α,β-unsaturated one.

Aldol reactions are readily reversible, especially when catalyzed by base, and there is generally little aldol product present at equilibrium, especially for aldol reactions of ketones. Equilibrium constants for dehydration, however, are generally large so that, if reaction conditions are sufficiently vigorous to bring about dehydration, good yields of product can be obtained.

Mechanism   *Acid-Catalyzed Dehydration of an Aldol Product*

Step 1: Proton transfer from the acid catalyst, here shown as H_3O^+, to the enol form of the aldol product gives an oxonium ion.

Step 2: Concerted loss of water from the oxonium ion gives the conjugate acid of the final product.

Step 3: Proton transfer from the conjugate acid of the final product to solvent completes the reaction and generates a new H_3O^+.

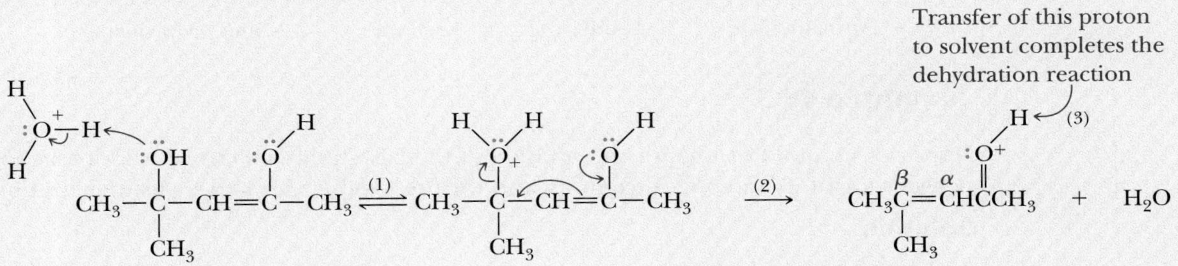

Transfer of this proton
to solvent completes the
dehydration reaction

The enol of the ketone

Protonated α,β-unsaturated ketone

Example 19.2

Draw the product of dehydration of each aldol product in Example 19.1.

Solution

Loss of H_2O from aldol product (a) gives an α,β-unsaturated aldehyde; loss of H_2O from (b) gives an α,β-unsaturated ketone.

(a) 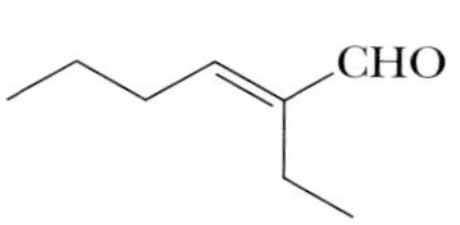   **(b)**

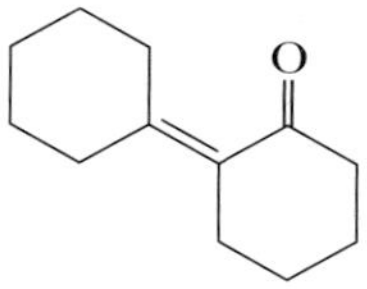

Problem 19.2
Draw the product of dehydration of each aldol product in Problem 19.1.

The reactants in the key step of an aldol reaction are an enolate anion and an enolate anion acceptor. In self-reactions, both roles are played by one kind of molecule. **Crossed aldol reactions** are also possible, as for example the crossed aldol reaction between acetone and formaldehyde. Formaldehyde cannot provide an enolate anion because it has no α-hydrogen, but it can function as a particularly good enolate anion acceptor because its carbonyl group is unhindered. Acetone forms an enolate anion, but its carbonyl group, which is bonded to two alkyl groups, is less reactive than that of formaldehyde. Consequently, the crossed aldol reaction between acetone and formaldehyde gives 4-hydroxy-2-butanone.

4-Hydroxy-2-butanone

As this example illustrates, for a crossed aldol reaction to be successful, one of the two reactants should have no α-hydrogen so that an enolate anion does not form. It also helps if the compound with no α-hydrogen has the more reactive carbonyl, for example, an aldehyde. If these requirements are not met, a complex mixture of products results. Following are examples of aldehydes that have no α-hydrogens and can be used in crossed aldol reactions.

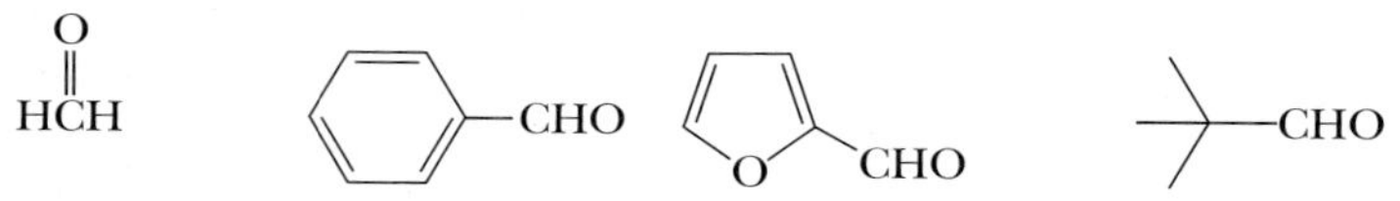

Formaldehyde   Benzaldehyde   Furfural   2,2-Dimethylpropanal

Example 19.3

Draw a structural formula for the product of the base-catalyzed crossed aldol reaction between furfural and cyclohexanone and for the product formed by its dehydration.

Solution

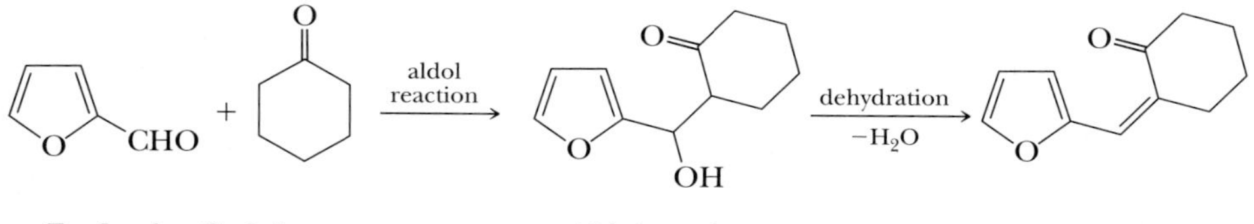

Furfural   Cyclohexanone   Aldol product
(racemic)

Problem 19.3
Draw the product of the base-catalyzed crossed aldol reaction between benzaldehyde and 3-pentanone and the product formed by its dehydration.

Nitro groups can be introduced into aliphatic compounds by way of an aldol reaction between the anion of a nitroalkane and an aldehyde or a ketone. The α-hydrogens of nitroalkanes are sufficiently acidic that they are removed by bases such as aqueous NaOH and KOH. The pK_a of nitromethane, for example, is 10.2. The acidity of the α-hydrogen of a nitroalkane is caused by the stabilization of the resulting anion by delocalization of its negative charge into the nitro group.

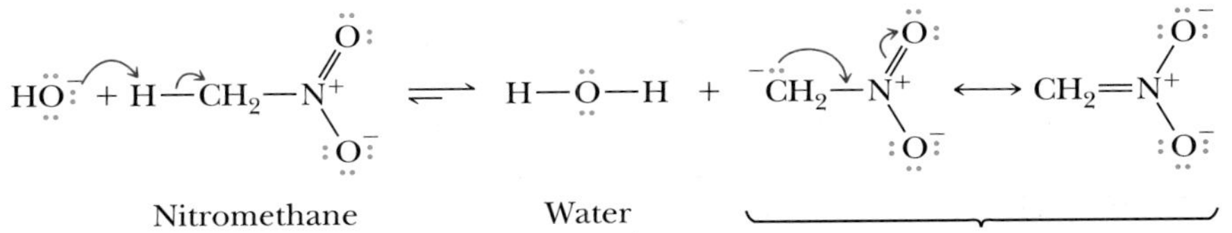

Nitromethane	Water	
pK_a 10.2	pK_a 15.7	
(stronger acid)	(weaker acid)	Resonance-stabilized anion

Following is an aldol reaction between nitromethane and cyclohexanone. Reduction of the nitro group in the aldol product thus formed is a convenient synthetic route to β-aminoalcohols.

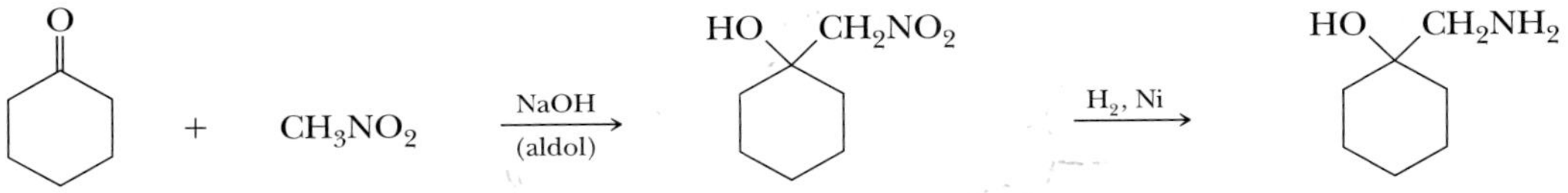

Cyclohexanone   Nitromethane   1-(Nitromethyl)cyclohexanol   1-(Aminomethyl)cyclohexanol

When both the enolate anion and the carbonyl group to which it adds are in the same molecule, aldol reaction results in the formation of a ring. This type of **intramolecular aldol reaction** is particularly useful for the formation of five- and six-membered rings. Because they are the most stable rings, five- and six-membered rings form much more readily than four-, seven-, or larger membered rings. Intramolecular aldol reaction of 2,7-octanedione via enolate anion α_3, for example, gives a five-membered ring. Intramolecular aldol reaction of this same compound via enolate anion α_1 would give a seven-membered ring. In the case of 2,7-octadienone, the five-membered ring forms in preference to the seven-membered ring.

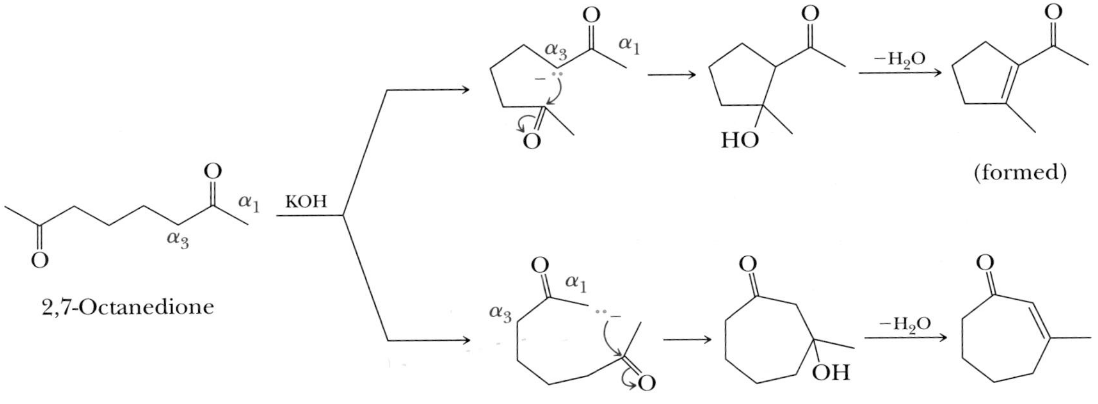

In general, smaller rings form faster than larger rings because the reacting groups are closer together. However, the formation of three- and four-membered rings is disfavored because of the strain in these rings.

Following is another example in which either a four-membered ring (via an enolate anion at α_3) or a six-membered ring (via an enolate anion at α_1) could be formed. Because of the greater stability of six-membered rings compared to four-membered rings, the six-membered ring is formed exclusively in this intramolecular aldol reaction.

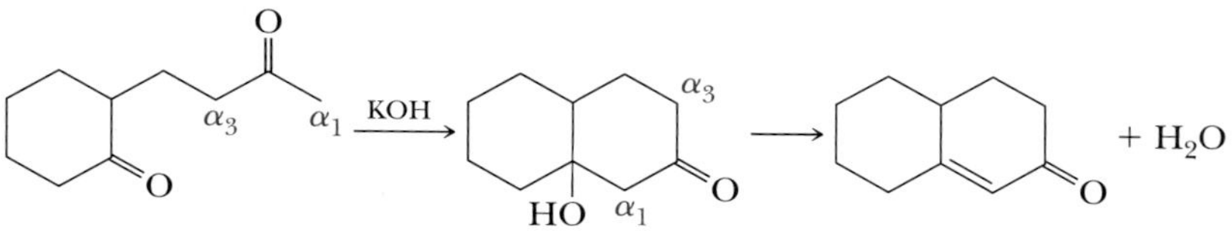

(racemic)   (four stereoisomers as   (racemic)
two racemic mixtures)

19.3 Claisen and Dieckmann Condensations

A. Claisen Condensation

In Chapter 18, we described reactions of esters, all of which take place at the carbonyl carbon and involve nucleophilic acyl substitution by an addition/elimination. In this section, we examine a second type of reaction characteristic of esters, namely one that involves both formation of an enolate anion and its participation in nucleophilic acyl substitution. One of the first of these reactions discovered is the **Claisen condensation,** named after the German chemist Ludwig Claisen (1851–1930). A Claisen condensation is illustrated by the reaction of two molecules of ethyl acetate in the presence of sodium ethoxide followed by acidification to give ethyl acetoacetate.

Ethyl ethanoate   Ethyl 3-oxobutanoate   Ethanol
(Ethyl acetate)   (Ethyl acetoacetate)

As this example illustrates, the product of a Claisen condensation is a **β-ketoester.**

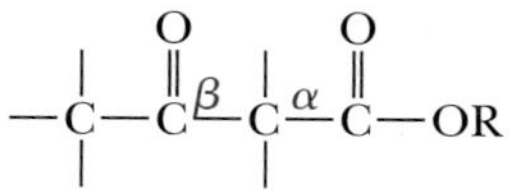

A β-ketoester

Like aldol reactions, Claisen condensations require a base. Aqueous bases, such as NaOH, however, cannot be used in Claisen condensations because they would bring about the hydrolysis of the ester. Rather, the bases most commonly used in Claisen condensations are nonaqueous bases, such as sodium ethoxide in ethanol and sodium methoxide in methanol.

Claisen condensation of two molecules of ethyl propanoate gives the following β-ketoester as a racemic mixture.

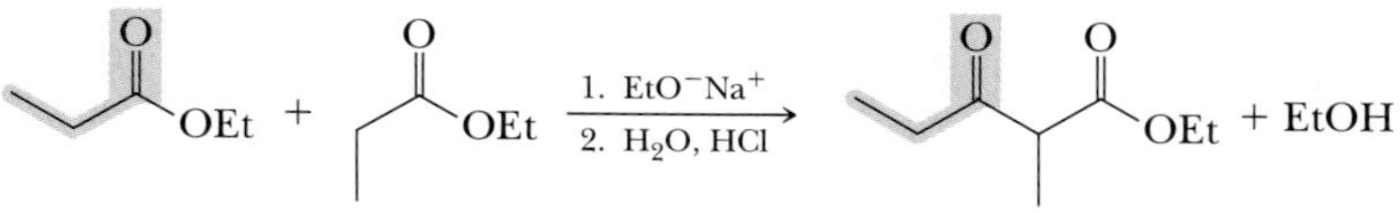

Ethyl   Ethyl   Ethyl 2-methyl-3-oxopentanoate
propanoate   propanoate   (formed as a racemic mixture)

The first steps of a Claisen condensation bear a close resemblance to the first steps of an aldol reaction (Section 19.1). The carbon-carbon bond-forming step in each reaction involves nucleophilic addition of an enolate anion to the carbonyl group of another molecule.

Mechanism   *Claisen Condensation*

Step 1: Removal of an α-hydrogen by base gives a resonance-stabilized enolate anion.

$$EtO^{-} \; + \; H-CH_2-COEt \;\rightleftharpoons\; EtO-H \; + \; {}^{-}CH_2-COEt \longleftrightarrow CH_2=COEt$$

pK$_a$ 22   pK$_a$ 15.9   Resonance-stabilized enolate anion
(weaker acid)   (stronger acid)   (stronger base)

Because the α-hydrogen of an ester is the weaker acid and ethoxide ion is the weaker base, the position of equilibrium for this step lies very much toward the left; the concentration of enolate anion is very low compared with that of ethoxide ion and ester. Thus, there is an excess of ester to react with the small amount of enolate anion that forms.

Step 2: Attack of the enolate anion of one ester on the carbonyl carbon of another ester gives a tetrahedral carbonyl addition intermediate.

$$CH_3-\overset{O}{\underset{}{C}}-OEt + {}^{-}CH_2-COEt \rightleftharpoons \left[CH_3-\overset{O^{-}}{\underset{OEt}{C}}-CH_2-\overset{O}{\underset{}{C}}-OEt \right]$$

A tetrahedral carbonyl
addition intermediate

Unlike similar intermediates in aldol reactions, this intermediate (a hemiacetal anion) has an ethoxy leaving group.

Step 3: Collapse of the tetrahedral carbonyl addition intermediate and ejection of ethoxide ion gives a β-ketoester.

$$CH_3-\overset{O^{-}}{\underset{OEt}{C}}-CH_2-\overset{O}{\underset{}{C}}-OEt \rightleftharpoons CH_3-\overset{O}{\underset{}{C}}-CH_2-\overset{O}{\underset{}{C}}-OEt + EtO^{-}$$

Step 4: The position of equilibrium for Steps 1–3 lies very much on the side of starting materials. The overall condensation, however, is driven to completion by the acid-base reaction between the β-ketoester (the stronger acid) and ethoxide ion (the stronger base) to give ethanol (the weaker acid) and the anion of the β-ketoester (the weaker base).

$$EtO^{-} + CH_3-\overset{O}{\underset{}{C}}-CH-\overset{O}{\underset{H}{C}}-OEt \rightleftharpoons CH_3-\overset{O}{\underset{}{C}}-\overset{-}{C}H-\overset{O}{\underset{}{C}}-OEt + EtOH$$

pK$_a$ 10.7   (weaker base)   pK$_a$ 15.9
(stronger acid)   (weaker acid)

Overall, the Claisen condensation involves consumption of a stronger base, in this case ethoxide, and creation of a weaker base, the resonance-stabilized enolate anion of the β-ketoester. One molecule of the original base is consumed for every two molecules of ester that react. This is in contrast to an aldol reaction, in which base is catalytic (not consumed).

Step 5: Acidification of the enolate anion gives the β-ketoester.

$$CH_3-\overset{O}{\overset{\|}{C}}-\overset{..}{\overset{-}{CH}}-\overset{O}{\overset{\|}{C}}OEt +H_3O^+ \xrightarrow{HCl,\ H_2O} CH_3-\overset{O}{\overset{\|}{C}}-CH_2-\overset{O}{\overset{\|}{C}}OEt + H_2O$$

From an analysis of this mechanism, we see that the structural feature required for a successful Claisen condensation is an ester with two α-hydrogens, one to form the initial enolate anion and the second to form the enolate anion of the resulting β-ketoester.

Example 19.4

Show the product of the Claisen condensation of ethyl butanoate in the presence of sodium ethoxide followed by acidification with aqueous HCl.

Solution

The new bond formed in a Claisen condensation is between the carbonyl group of one ester molecule and the α-carbon of another.

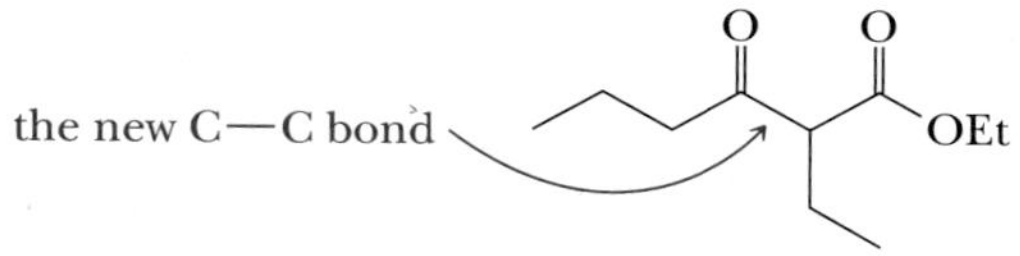

the new C—C bond

Ethyl 2-ethyl-3-oxohexanoate
(formed as a racemic mixture)

Problem 19.4

Show the product of Claisen condensation of ethyl 3-methylbutanoate in the presence of sodium ethoxide followed by acidification with aqueous HCl.

B. Dieckmann Condensation

An intramolecular Claisen condensation of a dicarboxylic ester to give a five- or six-membered ring is given the special name of **Dieckmann condensation.** In the presence of one equivalent of sodium ethoxide, for example, diethyl hexanedioate (diethyl adipate) undergoes an intramolecular condensation to form a five-membered ring.

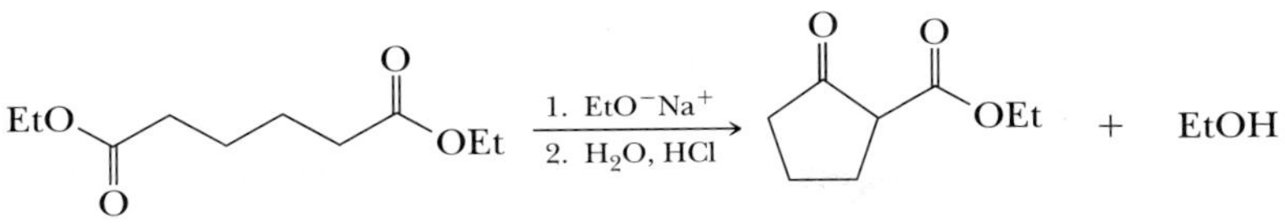

Diethyl hexanedioate
(Diethyl adipate)

Ethyl 2-oxocyclo-
pentanecarboxylate
(formed as a racemic mixture)

The mechanism of a Dieckmann condensation is identical to the mechanism we have described for the Claisen condensation. An anion formed at the α-carbon of one ester group adds to the carbonyl of the other ester group to form a tetrahedral carbonyl addition intermediate. This intermediate ejects ethoxide ion to regenerate the carbonyl group. Cyclization is followed by formation of the conjugate base of the β-ketoester, just as in the Claisen condensation. The β-ketoester is isolated after acidification with aqueous acid.

C. Crossed Claisen Condensations

In a **crossed Claisen condensation** between two different esters, each with two α-hydrogens, a mixture of four β-ketoesters is possible; therefore, crossed Claisen condensations of this type are not synthetically useful. Such condensations are useful, however, if appreciable differences in reactivity exist between the two esters, as for example when one of the esters has no α-hydrogens and can function only as an enolate anion acceptor. Following are four examples of esters without α-hydrogens.

Ethyl formate    Diethyl carbonate 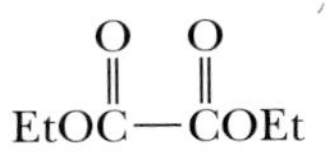   Diethyl ethanedioate   Ethyl benzoate
(Diethyl oxalate)

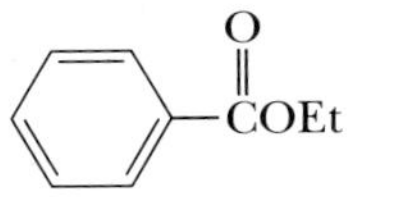

Crossed Claisen condensations of this type are usually carried out by using the ester with no α-hydrogens in excess. In the following illustration, methyl benzoate is used in excess.

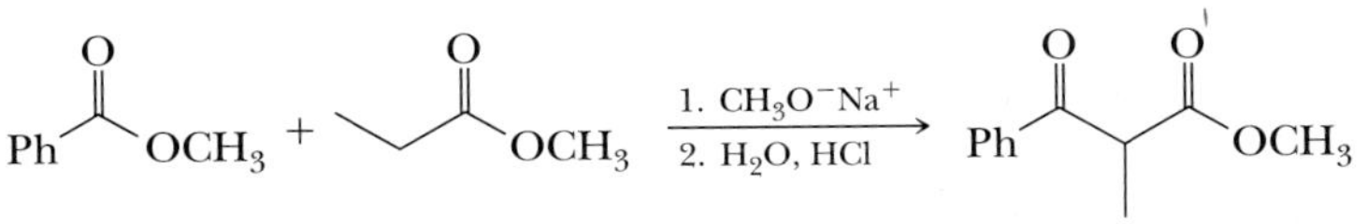

Methyl
benzoate

Methyl
propanoate

Methyl 2-methyl-3-oxo-
3-phenylpropanoate
(racemic)

Example 19.5

Complete the equation for this crossed Claisen condensation.

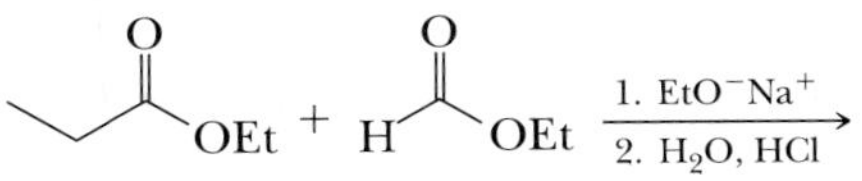

Solution

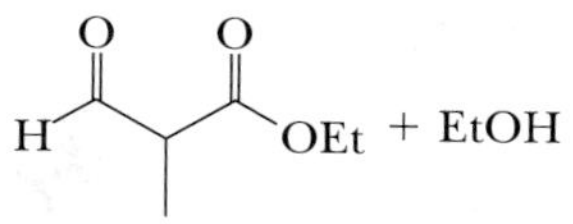

(racemic)

Problem 19.5

Complete the equation for this crossed Claisen condensation.

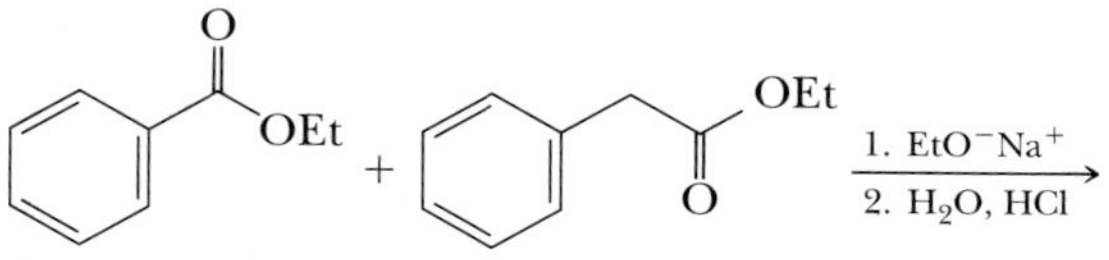

(excess)

D. Hydrolysis and Decarboxylation of β-Ketoesters

Recall from Section 18.4C that hydrolysis of an ester in aqueous sodium hydroxide (saponification) followed by acidification of the reaction mixture with aqueous HCl converts an ester to a carboxylic acid. Recall also from Section 17.9 that β-ketoacids and β-dicarboxylic acids (substituted malonic acids) readily undergo decarboxylation (lose CO_2) when heated. Both the Claisen and Dieckmann condensations yield esters of β-ketoacids. The following equations illustrate the results of a Claisen condensation followed by hydrolysis of the ester, acidification, and decarboxylation.

Claisen condensation:

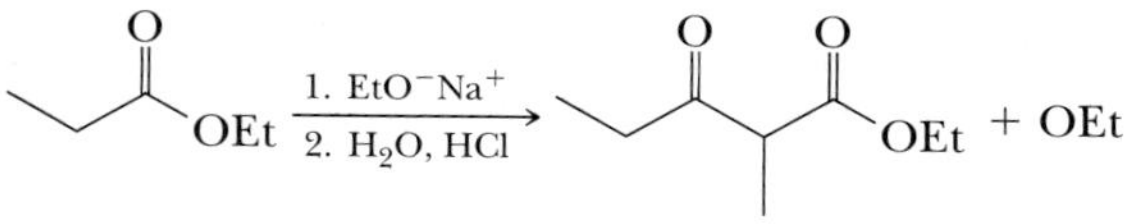

(formed as a
racemic mixture)

Saponification followed by acidification:

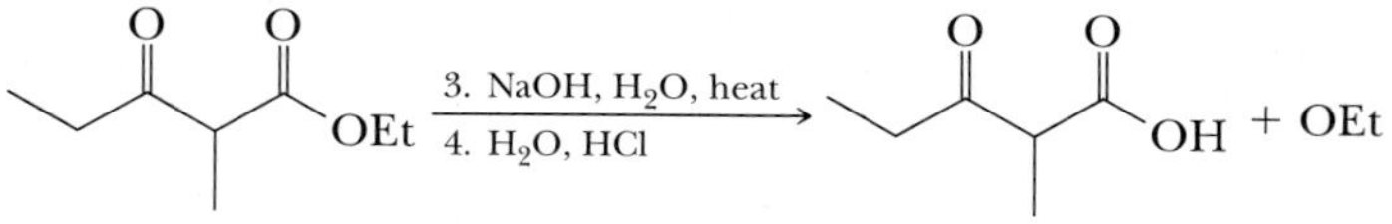

(racemic)

Decarboxylation:

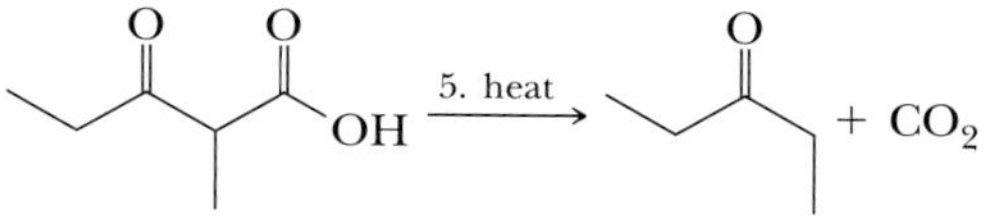

The result of this five-step sequence is reaction between two molecules of ester (one furnishing a carbonyl group and the other furnishing an enolate anion) to give a ketone and carbon dioxide. In the general reaction, both ester molecules are the same, and the product is a symmetrical ketone.

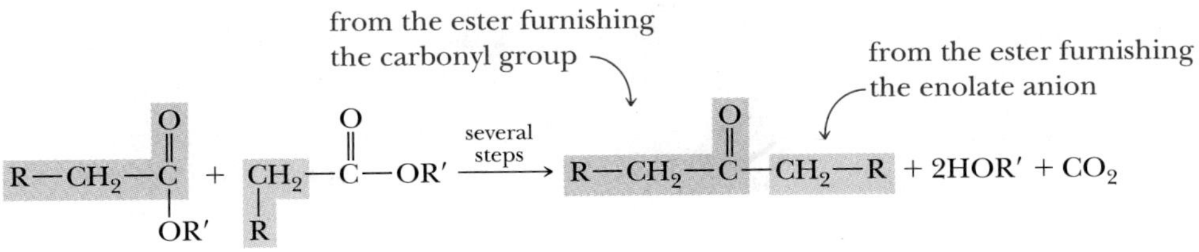

from the ester furnishing the carbonyl group

from the ester furnishing the enolate anion

$$R-CH_2-\overset{\overset{\displaystyle O}{\|}}{\underset{\underset{\displaystyle OR'}{|}}{C}} + \underset{\underset{\displaystyle R}{|}}{CH_2}-\overset{\overset{\displaystyle O}{\|}}{C}-OR' \xrightarrow{\text{several steps}} R-CH_2-\overset{\overset{\displaystyle O}{\|}}{C}-CH_2-R + 2HOR' + CO_2$$

The same sequence of reactions starting with a crossed Claisen condensation gives an unsymmetrical ketone.

Example 19.6

Each set of compounds undergoes a Claisen or Dieckmann condensation followed by acidification, saponification, acidification, and thermal decarboxylation.

(a) PhCOEt + CH₃COEt

(b)

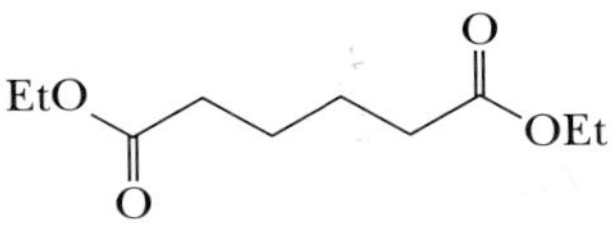

Draw a structural formula of the product isolated after completion of this reaction sequence.

Solution

Steps 1 and 2 bring about a crossed Claisen or Dieckmann condensation to give a β-ketoester. Steps 3 and 4 bring about hydrolysis of the β-ketoester to give a β-ketoacid, and Step 5 brings about decarboxylation to give a ketone.

(a) $\xrightarrow{(1,2)}$ PhCCH₂COEt $\xrightarrow{(3,4)}$ PhCCH₂COH $\xrightarrow{(5)}$ PhCCH₃

(b) $\xrightarrow{(1,2)}$
(racemic)   (racemic)

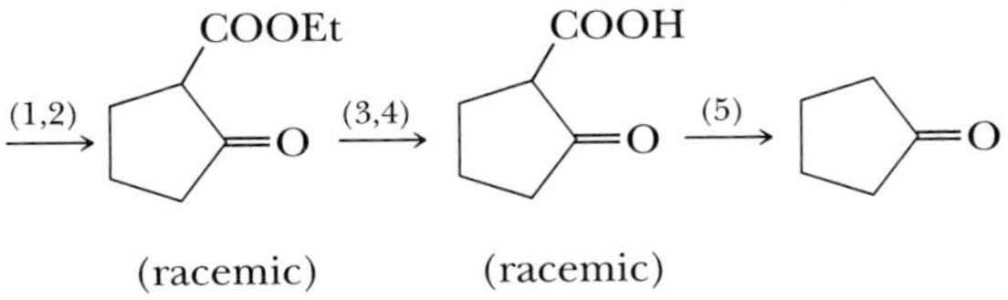

Problem 19.6

Show how to convert benzoic acid to 3-methyl-1-phenyl-1-butanone (isobutyl phenyl ketone) by the following synthetic strategies, each of which uses a different type of reaction to form the new carbon-carbon bond to the carbonyl group of benzoic acid.

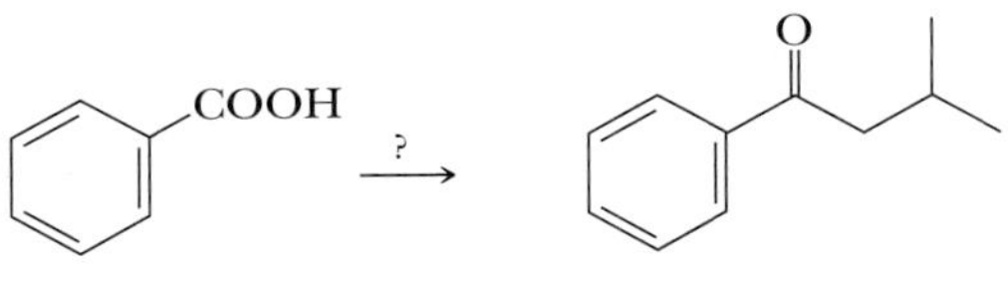

Benzoic acid   3-Methyl-1-phenyl-1-butanone

(a) A lithium diorganocopper (Gilman) reagent   (b) A Claisen condensation

CHEMICAL CONNECTIONS

Drugs that Lower Plasma Levels of Cholesterol

Coronary artery disease is the leading cause of death in the United States and other Western countries, where about one half of all deaths can be attributed to atherosclerosis. Atherosclerosis results from the buildup of fatty deposits called plaque on the inner walls of arteries. A major component of plaque is cholesterol derived from low-density-lipoproteins (LDL), which circulate in blood plasma. Because more than one half of total body cholesterol in humans is synthesized in the liver from acetyl-CoA, intensive efforts have been directed toward finding ways to inhibit this synthesis. The rate-determining step in cholesterol biosynthesis is reduction of (S)-3-hydroxy-3-methylglutaryl CoA to (R)-mevalonic acid. This reduction is catalyzed by the enzyme HMG-CoA reductase and requires two moles of NADPH per mole of HMG-CoA.

Beginning in the early 1970s, researchers at the Sankyo Company in Tokyo screened more than 8000 strains of microorganisms and in 1976 announced the isolation of mevastatin, a potent inhibitor of HMG-CoA reductase, from culture broths of the fungus *Penicillium citrinum*. The same compound was isolated by researchers at Beecham Pharmaceuticals in England from cultures of *Penicillium brevicompactum*. Soon thereafter, a second, more active compound called lovastatin was isolated at the Sankyo Company from the fungus *Monascus ruber*, and at Merck Sharpe & Dohme from *Aspergillus terreus*. Both mold metabolites are extremely effective in lowering plasma concentrations of LDL. The active form of each is the 5-hydroxycarboxylic acid formed by hydrolysis of the δ-lactone.

(S)-3-Hydroxy-3-methyl glutaryl-CoA (HMG-CoA) → A hemithioacetal intermediate formed by the first NADPH reduction → → (R)-Mevalonate

19.4 Claisen and Aldol Condensations in the Biological World

Carbonyl condensations are among the most widely used reactions in the biological world for the assembly of new carbon-carbon bonds. One source of carbon atoms for the synthesis of biomolecules is **acetyl-CoA,** a thioester of acetic acid and the thiol group of coenzyme A (Problem 25.41). In this section, we examine the series of reactions by which the carbon skeleton of acetic acid is converted to isopentenyl pyrophosphate, a key intermediate in the synthesis of terpenes, cholesterol, steroid hormones, and bile acids. Note that, in the discussion that follows, we will not be concerned with the mechanism by which each of these enzyme-catalyzed reactions occurs. Rather, our concern is in recognizing the types of reactions that take place.

Soon thereafter, Merck developed a synthesis for simvastatin (Zocor), which came onto the market in the late 1980s and is still used worldwide for the control of plasma cholesterol levels. These drugs and several synthetic modifications now available inhibit HMG-CoA reductase by forming an enzyme-inhibitor complex that prevents further catalytic action of the enzyme. It is reasoned that the 3,5-dihydroxycarboxylic acid part of each drug binds tightly to the enzyme because it mimics the hemithioacetal intermediate formed by the first reduction of HMG-CoA.

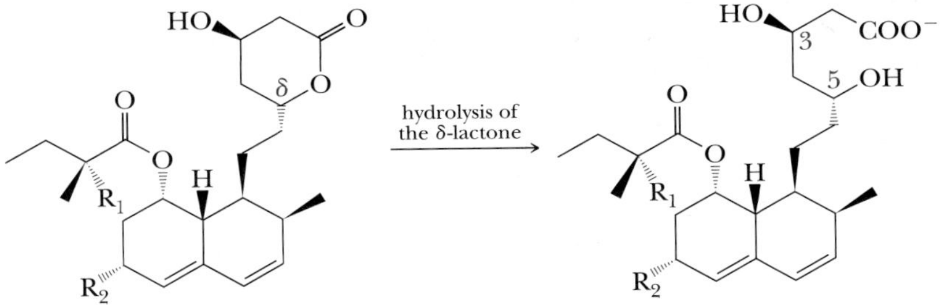

R$_1$ = R$_2$ = H, mevastatin
R$_1$ = H, R$_2$ = CH$_3$, lovastatin (Mevacor)
R$_1$ = R$_2$ = CH$_3$, simvastatin (Zocor)

The active form of each drug

Systematic studies have shown the importance of each part of the drug for effectiveness. It has been found, for example, that the carboxylate anion (—COO$^-$) is essential, as are both the 3—OH and 5—OH groups. Insertion of a bridging unit other than —CH$_2$—CH$_2$— between carbon 5 and the two fused six-membered rings reduces potency, as does almost any modification of the six-membered rings and their pattern of substitution.

In a Claisen condensation catalyzed by the enzyme thiolase, acetyl-CoA is converted to its enolate anion, which then adds to the carbonyl group of a second molecule of acetyl-CoA to give a tetrahedral carbonyl addition intermediate. Collapse of this intermediate by elimination of coenzyme A anion (CoAS$^-$) gives acetoacetyl-CoA. Subsequent proton transfer to coenzyme A anion gives coenzyme A. The mechanism of this reaction is identical to that of the Claisen condensation (Section 19.3A).

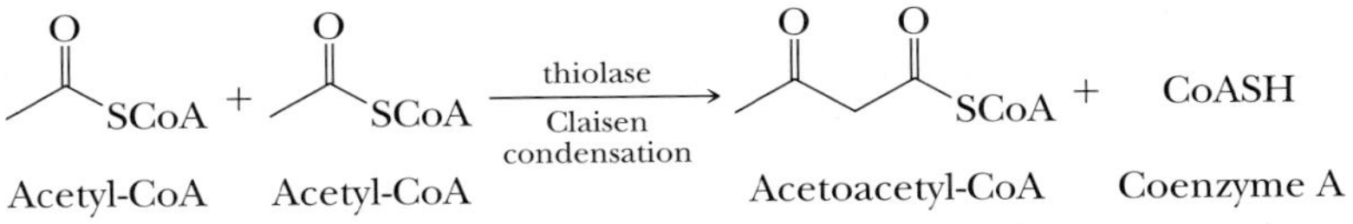

Acetyl-CoA   Acetyl-CoA   Acetoacetyl-CoA   Coenzyme A

Enzyme-catalyzed aldol reaction with a third molecule of acetyl-CoA on the ketone carbonyl of acetoacetyl-CoA gives (S)-3-hydroxy-3-methylglutaryl-CoA.

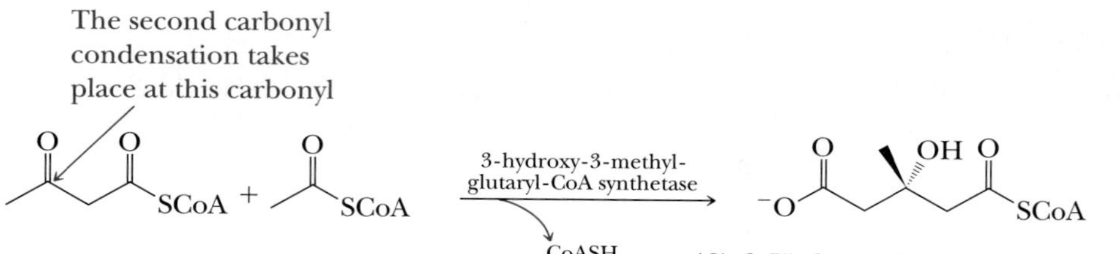

The second carbonyl condensation takes place at this carbonyl

(S)-3-Hydroxy-3-methylglutaryl-CoA

Note three features of this second carbonyl-condensation reaction.

- The reaction is completely enantioselective; only the S enantiomer is formed. Condensation takes place in a chiral environment created by the enzyme, 3-hydroxy-3-methylglutaryl-CoA synthetase, which induces the formation of one enantiomer of the product to the exclusion of the other.
- Hydrolysis of the thioester group of acetyl-CoA is coupled with the aldol reaction.
- The carboxyl group is shown as it is ionized at pH 7.4, the approximate pH of blood plasma and many cellular fluids.

Enzyme-catalyzed reduction by NADPH (a phosphorylated form of NADH) of the thioester group of 3-hydroxy-3-methylglutaryl-CoA to a primary alcohol gives mevalonic acid, here shown as its anion.

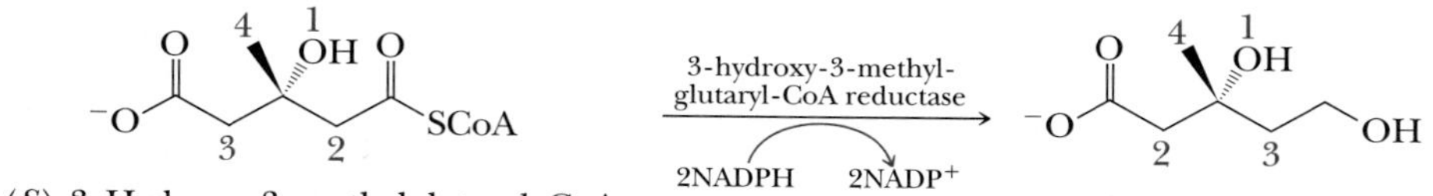

(S)-3-Hydroxy-3-methylglutaryl-CoA (R)-Mevalonate

In this reduction, note that a change occurs in the designation of configuration from S to R, not because of any change in configuration at the chiral center, but rather because priorities 2 and 3 become reversed as a result of the reduction.

Enzyme-catalyzed transfer of a phosphate group from adenosine triphosphate (ATP) to the 3-hydroxyl group of mevalonate gives a phosphoric ester at carbon 3. Enzyme-catalyzed transfer of a pyrophosphate group from a second molecule of ATP gives a pyrophosphoric ester at carbon 5. Enzyme-catalyzed β-elimination from this molecule results in loss of CO_2 and PO_4^{3-}, both good leaving groups.

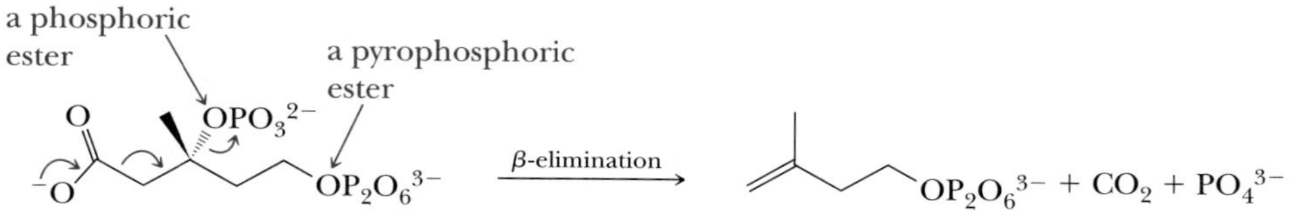

a phosphoric ester

a pyrophosphoric ester

(R)-3-Phospho-5-pyrophosphomevalonate   Isopentenyl pyrophosphate

Isopentenyl pyrophosphate has the carbon skeleton of isoprene, the unit into which terpenes can be divided (Section 5.4). This compound is, in fact, a key intermediate in the synthesis of terpenes, as well as of cholesterol and steroid hormones. We shall return to the chemistry of isopentenyl pyrophosphate in Section 26.4B and discuss its conversion to cholesterol and terpenes.

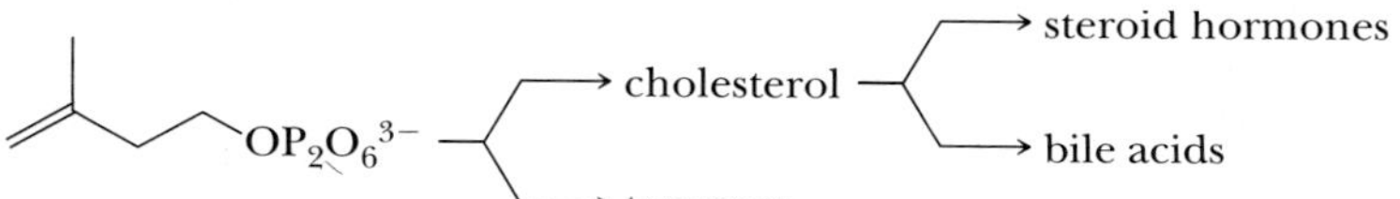

Isopentenyl pyrophosphate

19.5 Enamines

Enamines are formed by the reaction of a secondary amine with an aldehyde or ketone (Section 16.8A). The secondary amines most commonly used for this purpose are pyrrolidine and morpholine.

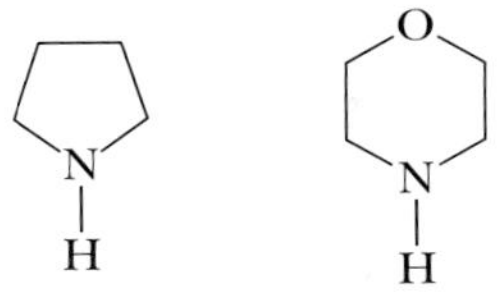

Pyrrolidine   Morpholine

Example 19.7

Draw structural formulas for the aminoalcohol and enamine formed in the following reactions.

(a) 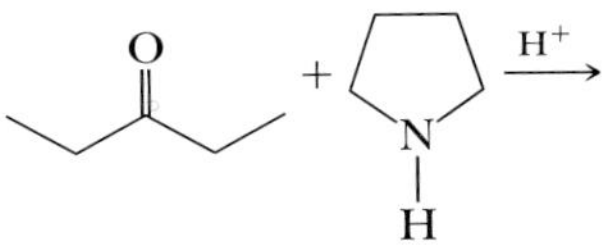 ![diagram]   (b) 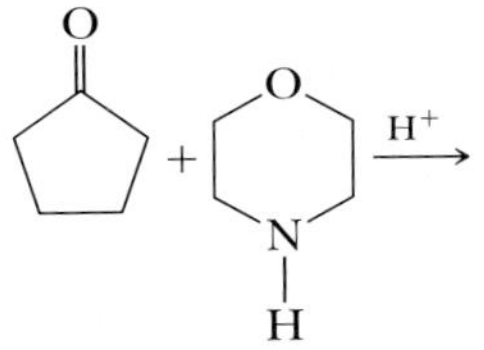 ![diagram]

Solution

(a) 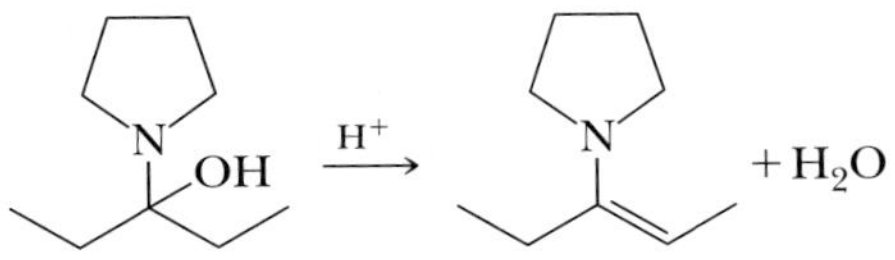 An aminoalcohol   An enamine   +H₂O

(b) 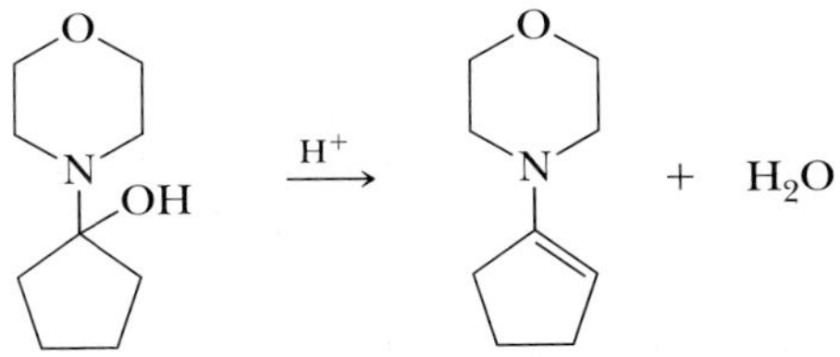 An aminoalcohol   An enamine   + H₂O

Problem 19.7

Following are structural formulas for two enamines.

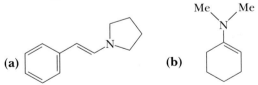

(a)   (b)

Draw structural formulas for the secondary amine and carbonyl compound from which each enamine is derived.

The particular value of enamines in synthetic organic chemistry is the fact that the β-carbon of an enamine is a nucleophile by virtue of the conjugation of the carbon-carbon double bond with the electron pair on nitrogen. Enamines resemble enols and enolate anions in their reactions.

The use of enamines as synthetic intermediates for the alkylation and acylation at the α-carbon of aldehydes and ketones was pioneered by Gilbert Stork of Columbia University. This use of enamines is called the Stork enamine reaction.

A. Alkylation of Enamines

Enamines readily undergo S_N2 reactions with methyl and primary alkyl halides, α-haloketones, and α-haloesters. Enamines are superior to enolate anions for these reactions because they are less basic and consequently give higher ratios of substitution to elimination products. In addition, they also give more alkylation on carbon than do enolate anions.

Mechanism *Alkylation of an Enamine*

Step 1: Treatment of the enamine with one equivalent of an alkylating agent gives an iminium halide.

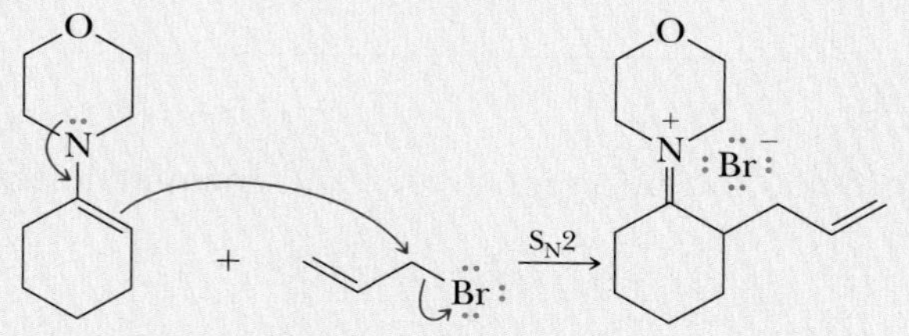

The morpholine enamine of cyclohexanone	3-Bromopropene (Allyl bromide)	An iminium bromide (formed as a racemic mixture)

Step 2: Hydrolysis of the iminium salt gives the alkylated ketone and regenerates morpholine as its hydrocyloride salt.

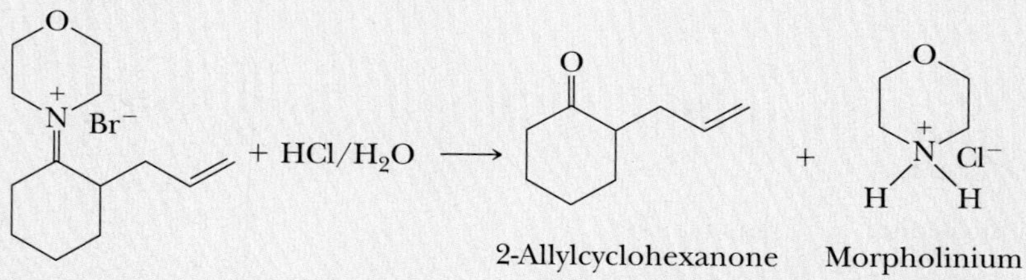

(racemic)   + HCl/H₂O ⟶   2-Allylcyclohexanone (formed as a racemic mixture)   +   Morpholinium chloride

---

Example 19.8

Show how to use an enamine to bring about this synthesis.

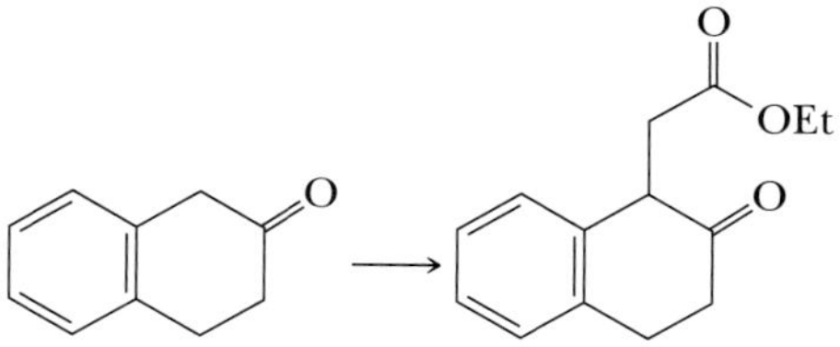

(as a racemic mixture)

Solution

Prepare an enamine by treating the ketone with either morpholine or pyrrolidine. The intermediate aminoalcohol can undergo dehydration in two directions. The direction shown here is favored because of the stabilization gained by conjugation of the carbon-carbon double bond of the enamine with the aromatic ring. Treatment of the enamine with ethyl 2-chloroacetate followed by hydrolysis of the iminium chloride in aqueous hydrochloric acid gives the product.

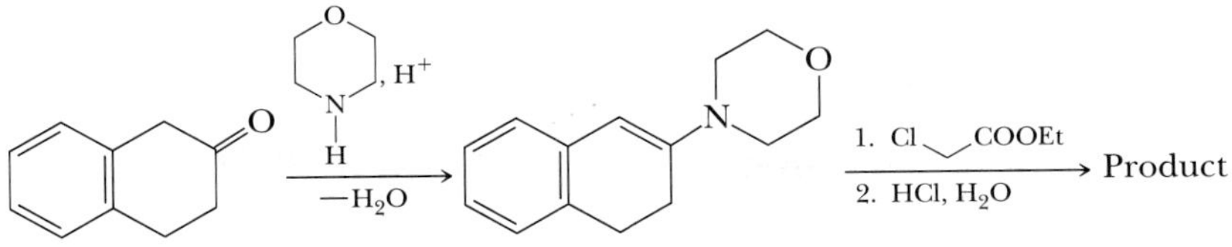

Problem 19.8

Write a mechanism for the hydrolysis of the following iminium chloride in aqueous HCl.

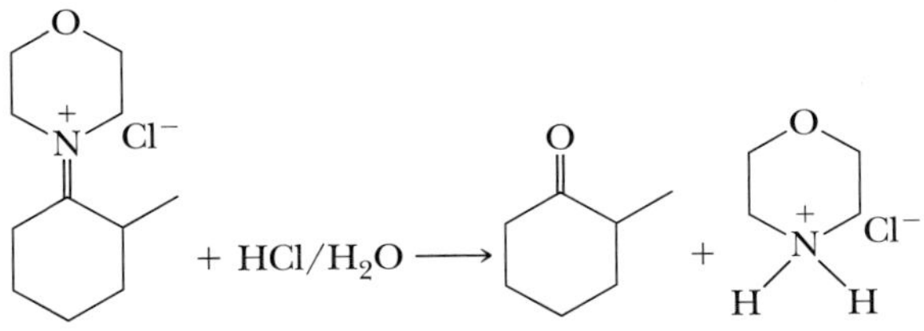

(racemic)   (racemic)

B. Acylation of Enamines

Enamines undergo acylation when treated with acid chlorides and acid anhydrides. The reaction is a nucleophilic acyl substitution as illustrated by the conversion of cyclohexanone, via its pyrrolidine enamine, to 2-acetylcyclohexanone.

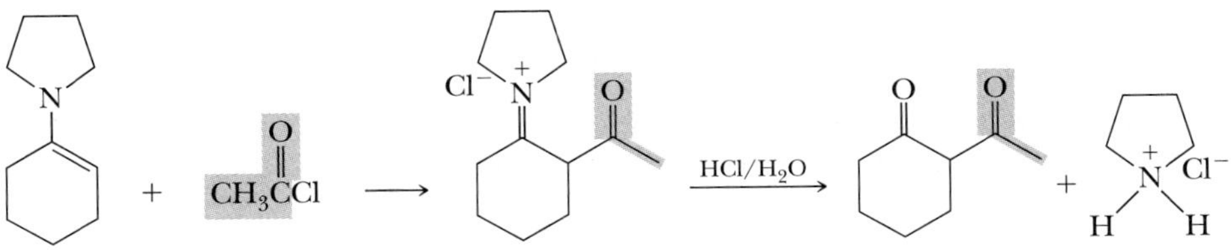

| The pyrrolidine enamine of cyclohexanone | Acetyl chloride | An iminium chloride (racemic) | 2-Acetylcyclo-hexanone (racemic) | |

Thus, we can attach an acyl group to the α-carbon of an aldehyde or ketone using its enamine as an intermediate. The process of introducing an acyl group onto an organic molecule is called **acylation.**

Acylation The process of introducing an acyl group, RCO— or ArCO—, onto an organic molecule.

Example 19.9

Show how to use an enamine to bring about this synthesis.

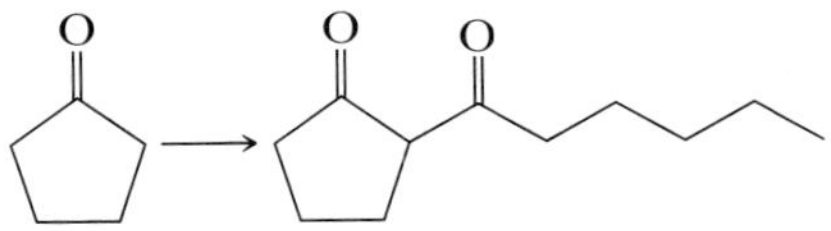

(as a racemic mixture)

Solution

Treating cyclopentanone with pyrrolidine gives an enamine. Treating the enamine with hexanoyl chloride followed by hydrolysis in aqueous HCl gives the desired β-diketone.

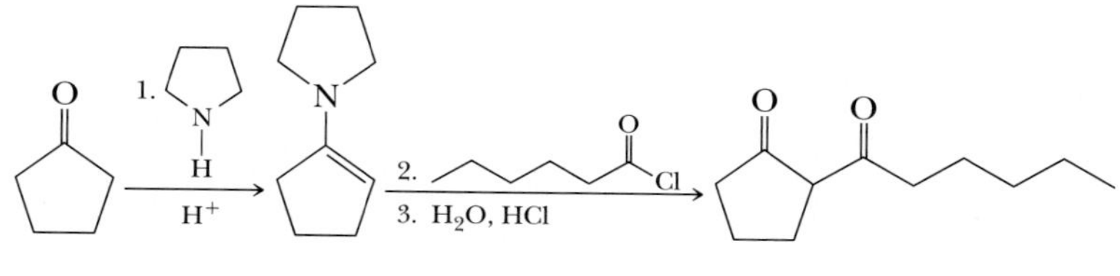

(formed as a racemic mixture)

Problem 19.9

Show how to use alkylation or acylation of an enamine to convert acetophenone to the following compounds.

(a) 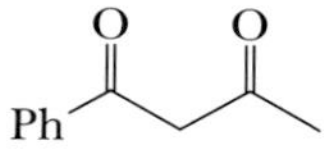 Ph

(b) 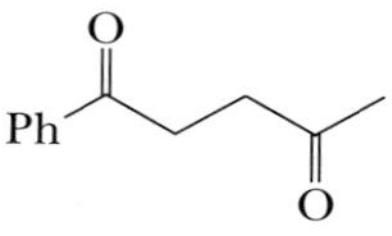 Ph

(c) 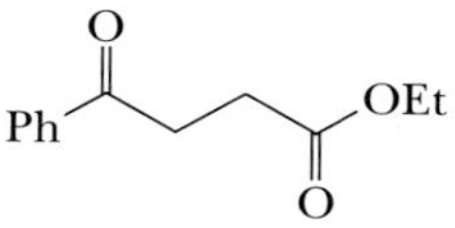 Ph   OEt

---

19.6 Acetoacetic Ester Synthesis

Acetoacetic ester and other β-ketoesters are such versatile starting materials for the formation of new carbon-carbon bonds because of

1. The acidity of α-hydrogens (pK_a 10–11) between the two carbonyl groups
2. The nucleophilicity of the enolate anion resulting from loss of an α-hydrogen
3. The ability of the product to undergo decarboxylation after hydrolysis of the ester

The **acetoacetic ester synthesis** is useful for the preparation of monosubstituted and disubstituted acetones of the following types.

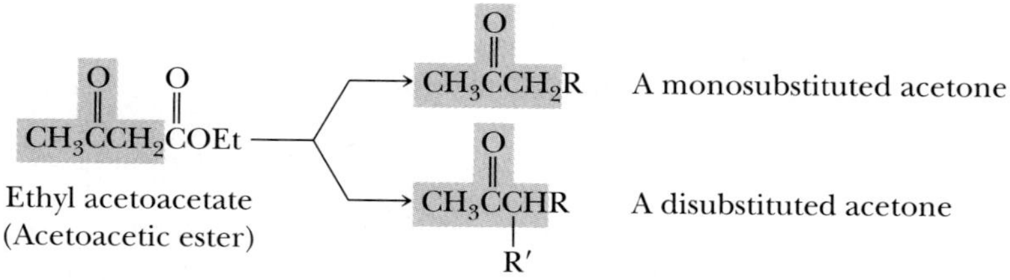

$$CH_3CCH_2COEt$$

Ethyl acetoacetate
(Acetoacetic ester)

$$CH_3CCH_2R \quad \text{A monosubstituted acetone}$$

$$CH_3CCHR \quad \text{A disubstituted acetone}$$
R′

We have already seen the chemistry of the individual steps in this synthesis, but we have not put them together in this particular sequence. Let us illustrate the acetoacetic ester synthesis by choosing 5-hexen-2-one as a target molecule. The three carbons shown in color are provided by ethyl acetoacetate. The remaining three carbons represent the —R group of a substituted acetone.

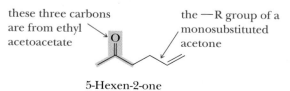

these three carbons are from ethyl acetoacetate

the —R group of a monosubstituted acetone

5-Hexen-2-one

1. The methylene hydrogens of ethyl acetoacetate (pK_a 10.7) are more acidic than those of ethanol (pK_a 15.9); therefore, ethyl acetoacetate is converted completely to its anion by sodium ethoxide or other alkali metal alkoxides.

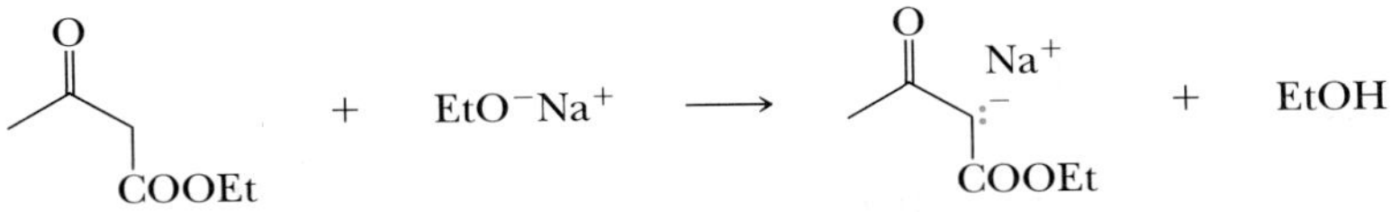

Ethyl acetoacetate	Sodium ethoxide	Sodium salt of	Ethanol
pK_a 10.7	(stronger base)	ethyl acetoacetate	pK_a 15.9
(stronger acid)		(weaker base)	(weaker acid)

2. The enolate anion of ethyl acetoacetate is a nucleophile and reacts by an S_N2 pathway with methyl and primary alkyl halides, α-haloketones, and α-haloesters. Secondary halides give lower yields, and tertiary halides undergo E2 elimination exclusively. In the following example, the anion of ethyl acetoacetate is alkylated with allyl bromide.

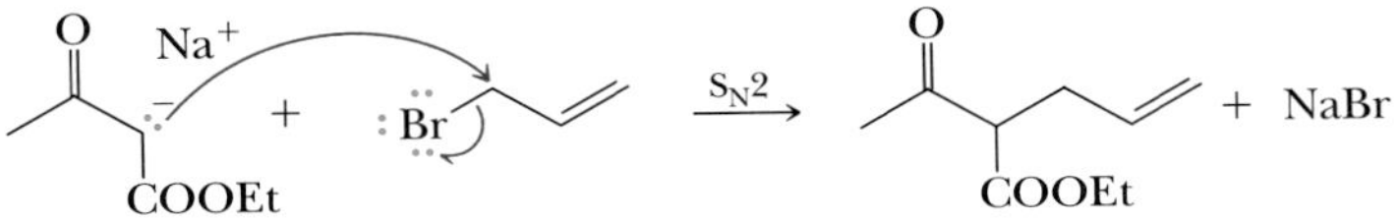

3-Bromopropene   (formed as a
(Allyl bromide)   racemic mixture)

3, 4. Hydrolysis of the alkylated acetoacetic ester in aqueous NaOH followed by acidification with aqueous HCl (Section 18.4C) gives a β-ketoacid.

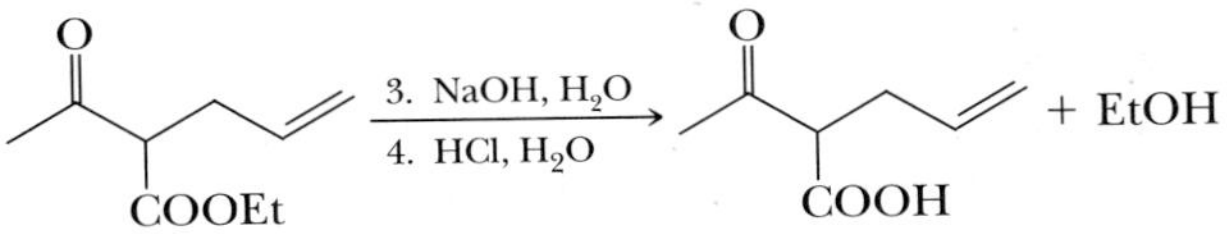

(a racemic mixture)

5. Heating the β-ketoacid brings about decarboxylation (Section 17.9A) to give 5-hexen-2-one.

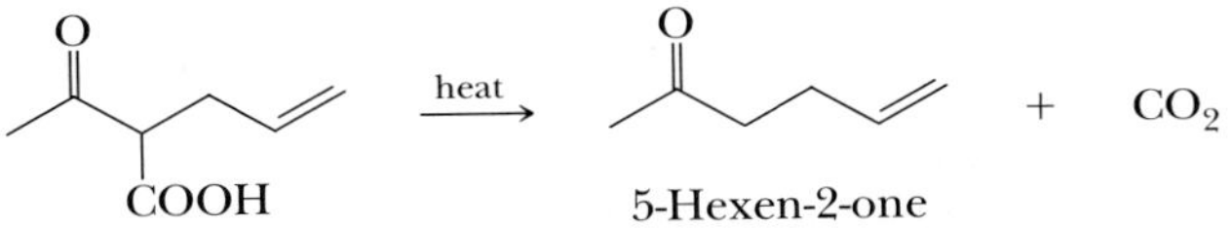

5-Hexen-2-one
(a monosubstituted acetone)

A disubstituted acetone can be prepared by interrupting this sequence after Step 2, treating the monosubstituted acetoacetic ester with a second equivalent of base, carrying out a second alkylation, and then proceeding with Steps 3–5.

1'. Treatment with a second equivalent of base gives a second enolate anion.

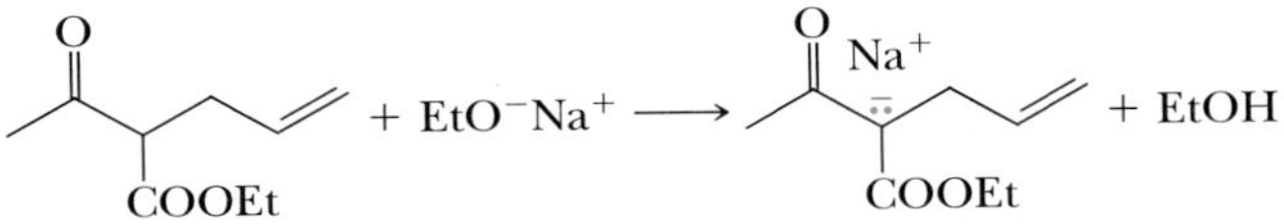

(a racemic mixture)

2′. Treatment of this enolate anion with an alkyl halide completes second alkylation.

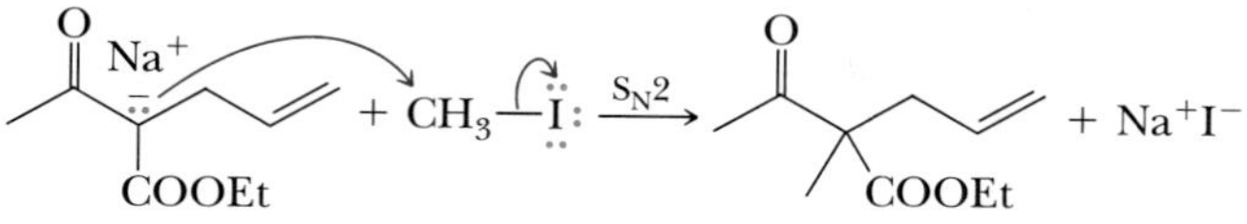

(formed as a
racemic mixture)

3, 4, 5. Hydrolysis of the ester in aqueous base followed by acidification and heating
gives the ketone.

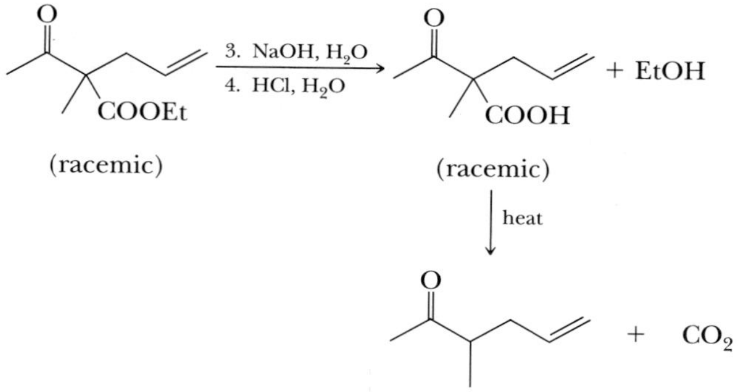

3-Methyl-5-hexen-2-one
(a disubstituted acetone; racemic)

Example 19.10

Show how the acetoacetic ester synthesis can be used to prepare this ketone.

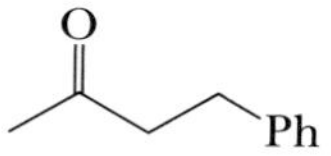

4-Phenyl-2-butanone

Solution

Determine which three carbons of the product originate from ethyl acetoacetate;
then establish the location on the carbon chain of the —COOH lost in decarboxylation, and finally verify the bond formed in the alkylation step. On the basis of this
analysis, determine that the starting materials are ethyl acetoacetate and a benzyl
halide.

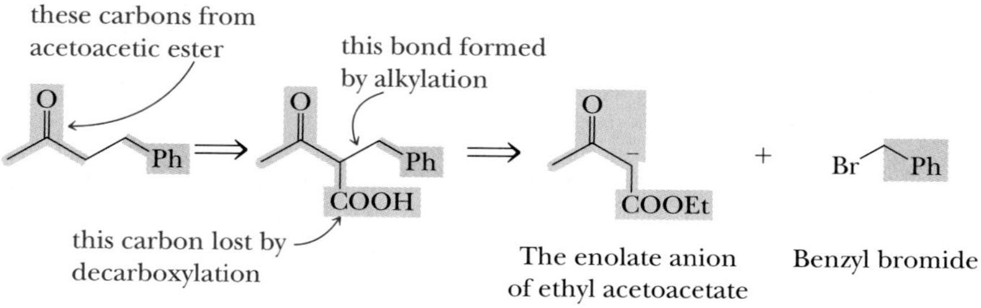

The enolate anion   Benzyl bromide
of ethyl acetoacetate

Now combine these reagents in the following way to prepare the desired ketone.

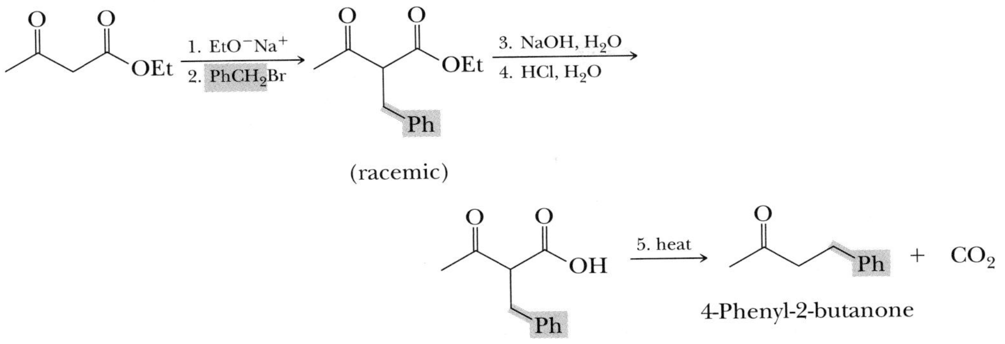

(racemic)

4-Phenyl-2-butanone

(racemic)

Problem 19.10

Show how the acetoacetic ester synthesis can be used to prepare these compounds.

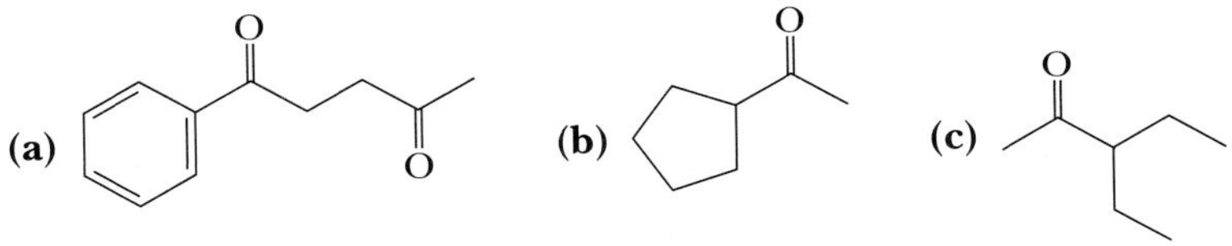

(a)   (b)   (c)

We have described what is commonly known as the acetoacetic ester synthesis and have illustrated the use of ethyl acetoacetate as the starting reagent. This same synthetic strategy is applicable to any β-ketoester, as for example those that are available by the Claisen (Section 19.3A) and Dieckmann (Section 19.3B) condensations. Following are structural formulas for two β-ketoesters that can be made to undergo (1) formation of an enolate anion, (2) alkylation or acylation, (3) hydrolysis followed by (4) acidification, and finally (5) decarboxylation just as we have shown for ethyl acetoacetate.

Ethyl 2-oxocyclopentanecarboxylate
(racemic)

Ethyl 2-methyl-3-oxopentanoate
(racemic)

Example 19.11

Show how to convert racemic ethyl 2-oxocyclopentanecarboxylate to racemic 2-allyl-cyclopentanone.

Solution

Treat this β-ketoester with one equivalent of sodium ethoxide to form an anion followed by alkylation of the anion with one equivalent of an allyl halide. Subsequent hydrolysis of the ester in aqueous base followed by acidification and thermal decarboxylation gives the desired product.

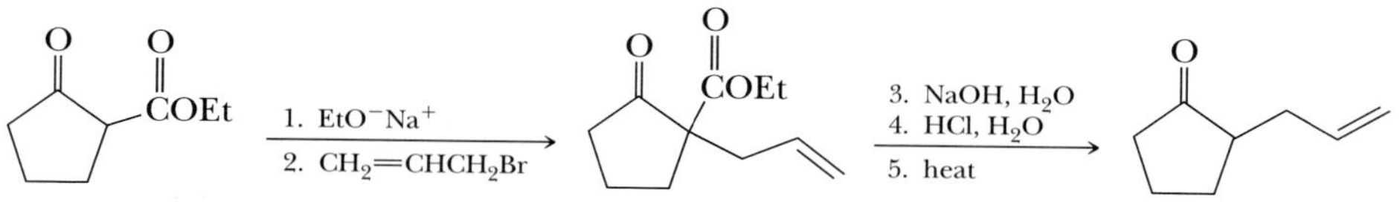

Ethyl 2-oxocyclo-
pentanecarboxylate
(racemic)

2-Allylcyclopentanone
(racemic)

Problem 19.11

Show how to convert racemic ethyl 2-oxocyclopentanecarboxylate to this compound.

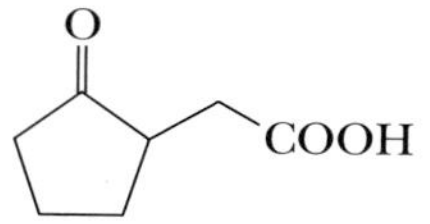

(racemic)

19.7 Malonic Ester Synthesis

The factors that make malonic esters and other β-diesters such versatile starting materials for formation of new carbon-carbon bonds are the same as those we have already seen for the acetoacetic ester synthesis, namely

1. The acidity of α-hydrogens (pK_a 13–14) between the two carbonyl groups
2. The nucleophilicity of the enolate anion resulting from loss of such an α-hydrogen
3. The ability of the product to undergo decarboxylation after hydrolysis of the ester

The **malonic ester synthesis** is useful for the preparation of monosubstituted and disubstituted acetic acids of the following types.

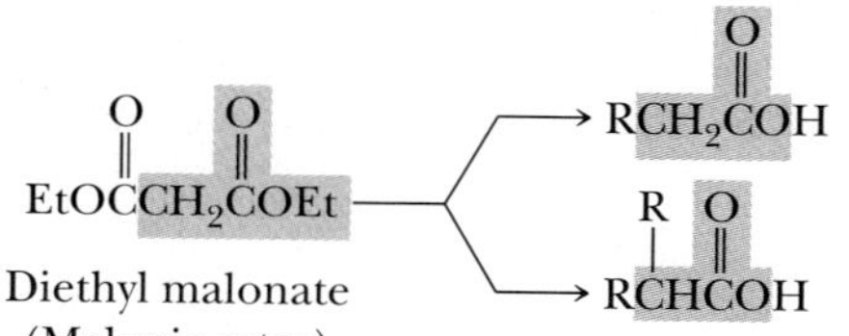

A monosubstituted
acetic acid

A disubstituted
acetic acid

Diethyl malonate
(Malonic ester)

As with the acetoacetic ester synthesis, we have already encountered all the important chemistry of the malonic ester synthesis, although not in this particular pattern. Let us illustrate this synthesis by choosing 5-methoxypentanoic acid as a target molecule. The two carbons shown in color are provided by diethyl malonate. The remaining three carbons and the methoxy group represent the —R group of a monosubstituted acetic acid.

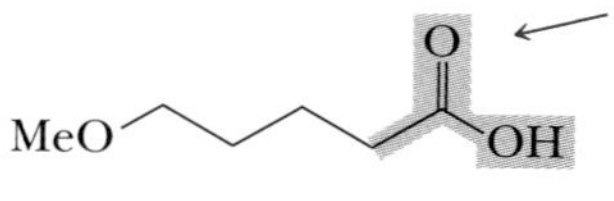

two carbons
from diethyl malonate

5-Methoxypentanoic acid

1. The α-hydrogens of diethyl malonate (pK_a 13.3) are more acidic than ethanol (pK_a 15.9); therefore, diethyl malonate is converted completely to its anion by sodium ethoxide or other alkali metal alkoxide.

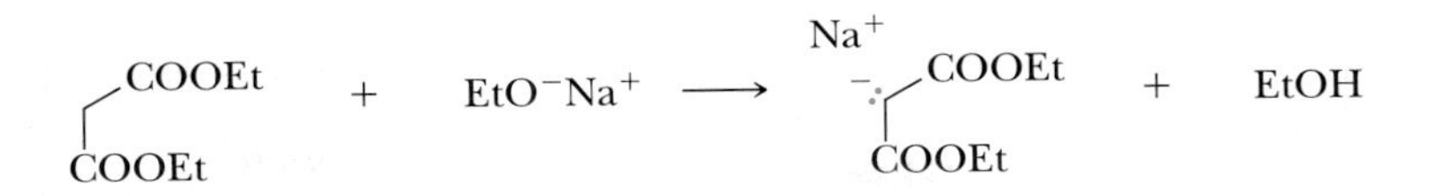

Diethyl malonate	Sodium ethoxide	Sodium salt of	Ethanol
pK_a 13.3	(stronger base)	diethyl malonate	pK_a 15.9
(stronger acid)			(weaker acid)

2. The enolate anion of diethyl malonate is a nucleophile and reacts by an S_N2 pathway with methyl and primary alkyl halides, α-haloketones, and α-haloesters. In the following example, the anion of diethyl malonate is alkylated with 1-bromo-3-methoxypropane.

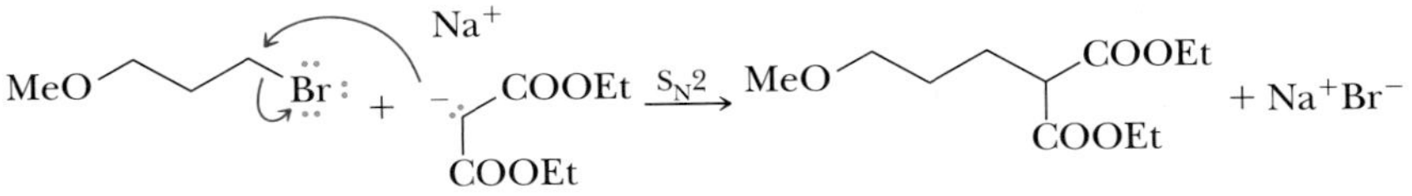

3, 4. Hydrolysis of the alkylated malonic ester in aqueous NaOH followed by acidification with aqueous HCl gives a β-dicarboxylic acid.

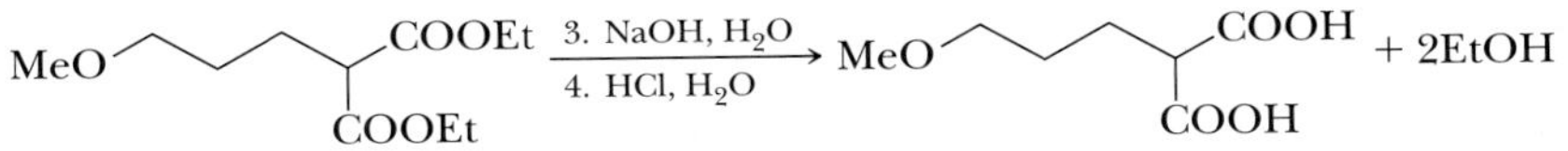

5. Heating the β-dicarboxylic acid slightly above its melting point brings about decarboxylation and gives 5-methoxypentanoic acid.

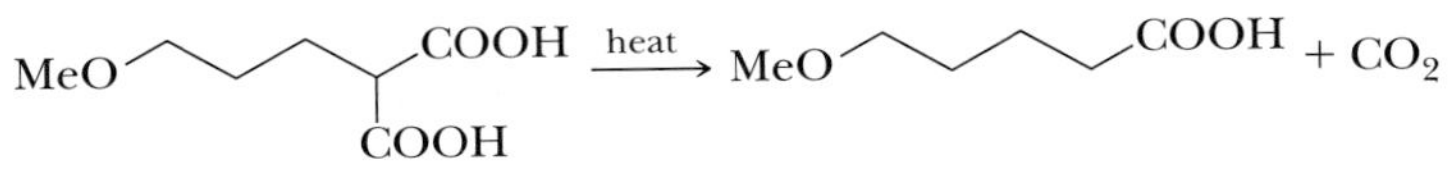

5-Methoxypentanoic acid

A disubstituted acetic acid can be prepared by interrupting the previous sequence after Step 2, treating the monosubstituted diethyl malonate with a second equivalent of base, carrying out a second alkylation, and then proceeding with Steps 3–5.

Example 19.12

Show how the malonic ester synthesis can be used to prepare 3-phenylpropanoic acid.

Solution

Determine which two carbons of the product originate from diethyl malonate, the location on the carbon chain of the —COOH lost in decarboxylation, and finally the bond formed in the alkylation step. On the basis of this analysis, determine that the starting materials are diethyl malonate and a benzyl halide.

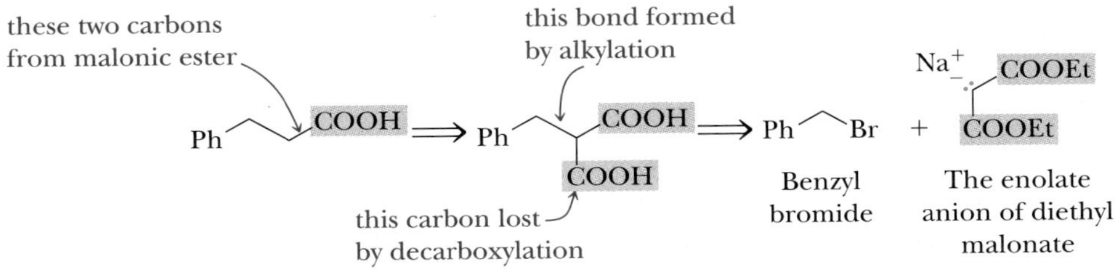

Ibuprofen: The Evolution of an Industrial Synthesis

A major consideration in any industrial synthesis is atom efficiency; it is most efficient to use only reagents whose atoms appear in the final product. An example of the evolution of syntheses with greater and greater atom efficiency is the industrial synthesis of ibuprofen.

Synthesis I

One of the first industrial syntheses of ibuprofen used the following sequence to introduce a methyl group on

the carboxyl side chain of 4-isobutylphenylacetic acid. Fischer esterification of 4-isobutylphenylacetic acid followed by treatment of the ethyl ester with diethyl carbonate in the presence of sodium ethoxide in a crossed Claisen condensation (Section 19.3C) gives the anion of a substituted malonic ester B. Alkylation of this anion (Section 19.7) followed by hydrolysis gives a disubstituted malonic acid D. Decarboxylation (Section 17.9B) of D gives ibuprofen.

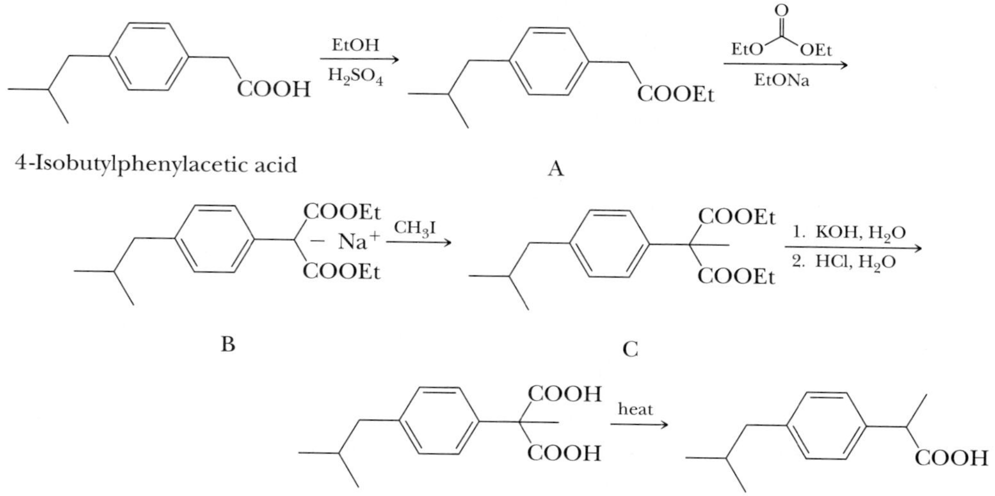

4-Isobutylphenylacetic acid

A

B

C

D

Ibuprofen
(formed as a
racemic mixture)

Even though this synthesis gives ibuprofen in quite good yield, it wastes carbons; only 13 of the 18 carbons in intermediate C appear in ibuprofen.

Synthesis II

An alternative route with greater atom efficiency also starts with 4-isobutylacetophenone. Treating this ketone

Now combine these reagents in the following way to get the desired product.

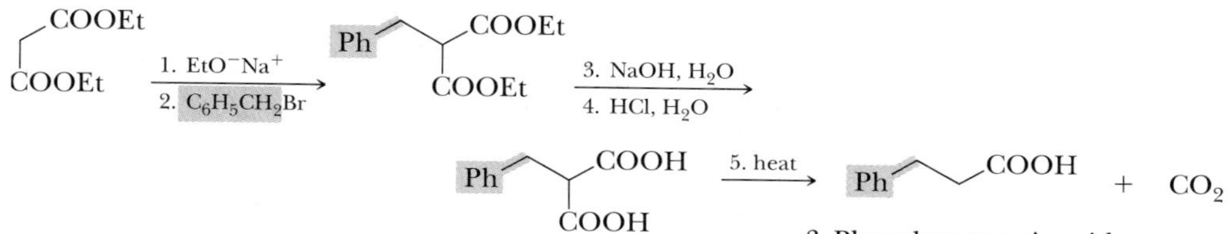

3-Phenylpropanoic acid

double bond for nucleophilic attack in a Michael reaction is the presence of the adjacent carbonyl group. One important resonance structure of the α,β-unsaturated carbonyl compounds puts positive charge at the end (in this case, the β-carbon) of the double bond, making it resemble a carbonyl group in its reactivity. Thus nucleophiles can add to this type of double bond, which we call "activated" for this reason.

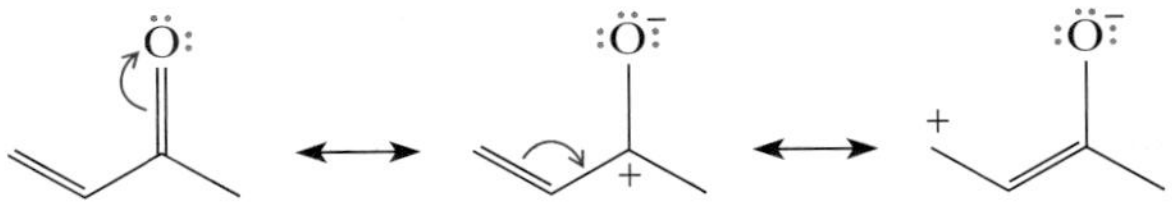

Although the major fraction of the partial positive charge (blue) of an α,β-unsaturated aldehyde or ketone is on the carbonyl carbon, there is nevertheless a significant partial positive charge on the beta carbon.

Note that aldol, Claisen, and Dieckmann condensations all give primary products with oxygens in a 1,3 relationship. The Michael reaction with enolate anions gives products with oxygens in a 1,5 relationship. These relationships are a consequence of the polarization of the reagents. In aldol, Claisen, and Dieckmann condensations, the carbonyl carbon is positive, and the α-position is negative.

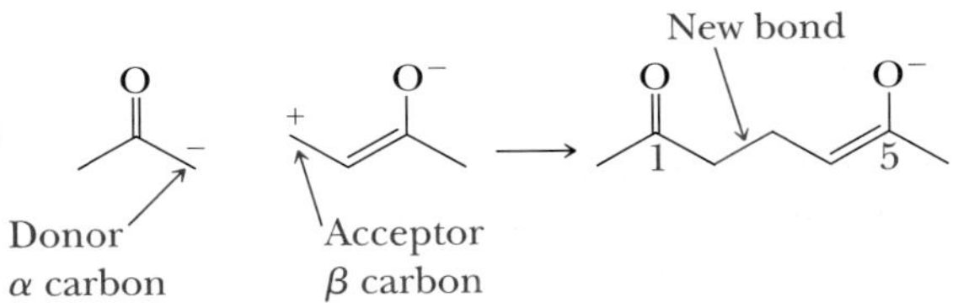

In a Michael reaction, the positive polarization of the carbonyl carbon is transmitted two carbons farther by the double bond.

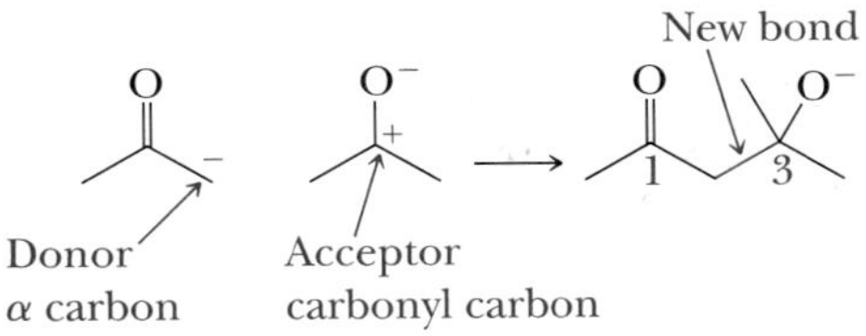

The Michael reaction takes place with a wide variety of α,β-unsaturated carbonyl compounds as well as with α,β-unsaturated nitriles and nitro compounds. The most commonly used types of nucleophiles in Michael reactions are summarized in Table 19.1. The bases most commonly used to generate the nucleophile are metal alkoxides, pyridine, and piperidine. It is important to realize that other nucleophiles can undergo similar additions to the beta carbon of unsaturated carbonyl compounds, for example, amines, alcohols, and water.

We can write the following general mechanism for a Michael reaction. Note that in Step 3 the base, B⁻, is regenerated, in accord with the experimental observation that a Michael reaction requires only a catalytic amount of base rather than a molar equivalent.

Table 19.1 Combinations of Reagents for Effective Michael Reactions

These Types of α,β-Unsaturated Compounds Are Nucleophile Acceptors in Michael Reactions		These Types of Compounds Provide Effective Nucleophiles for Michael Reactions	
$CH_2=CHCH$ (O)	Aldehyde	$CH_3CCH_2CCH_3$ (O, O)	β-Diketone
$CH_2=CHCCH_3$ (O)	Ketone	CH_3CCH_2COEt (O, O)	β-Ketoester
$CH_2=CHCOEt$ (O)	Ester	CH_3CCH_2CN (O)	β-Ketonitrile
$CH_2=CHCNH_2$ (O)	Amide	$EtOCCH_2COEt$ (O, O)	β-Diester
		(pyrrolidine ring, N)	
$CH_2=CHC≡N$	Nitrile	 $CH_3C=CH_2$	Enamine
$CH_2=CHNO_2$	Nitro compound	NH_3, RNH_2, R_2NH	Amine

Mechanism *Michael Reaction—Conjugate Addition of Enolate Anions*

Step 1: Treating H—Nu with base gives the nucleophile, Nu:⁻.

$$Nu{-}H + :B^- \;\rightleftharpoons\; Nu:^- + H{-}B$$

Base

Step 2: Nucleophilic addition of Nu:⁻ to the β-carbon of the conjugated system gives a resonance-stabilized enolate anion.

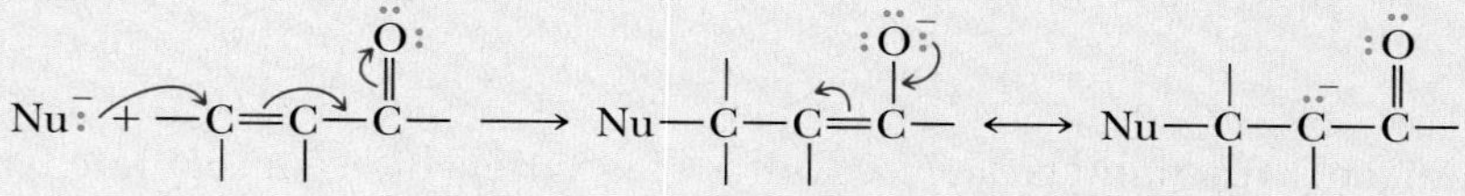

$$Nu:^- + -C=C-C- \longrightarrow Nu-C-C=C- \longleftrightarrow Nu-C-C-C-$$

Resonance-stabilized enolate anion

Step 3: Proton transfer from H—B gives the enol.

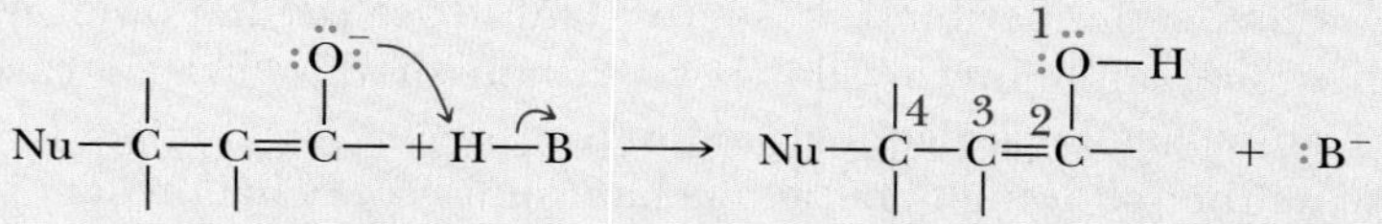

$$Nu-C-C=C- + H{-}B \longrightarrow Nu-C-C=C- + :B^-$$

An enol
(a product of 1,4-addition)

The enol formed in Step 3 corresponds to 1,4-addition to the conjugated system of the α,β-unsaturated carbonyl compound. It is because this intermediate is formed that the Michael reaction is classified as a 1,4- or conjugate addition.

Step 4: Tautomerization (Section 16.9B) of the less stable enol form gives the more stable keto form.

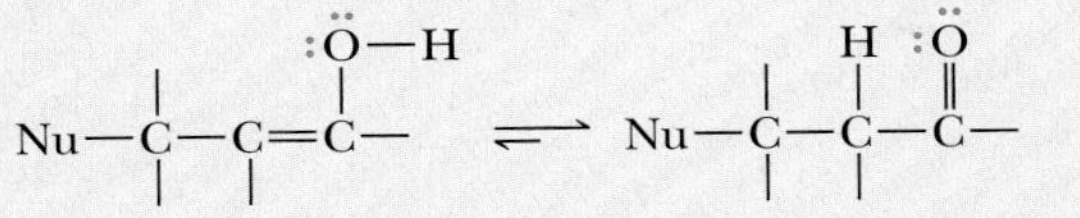

Less stable enol form   More stable keto form

ORGANIC
Chemistry Now™
Click Mechanisms to view an animation of the **Michael Reaction**

Example 19.13

Draw a structural formula for the product formed by treating each set of reactants with sodium ethoxide in ethanol under conditions of the Michael reaction.

Solution

(a) 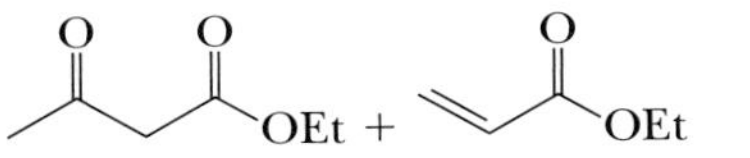   (b) 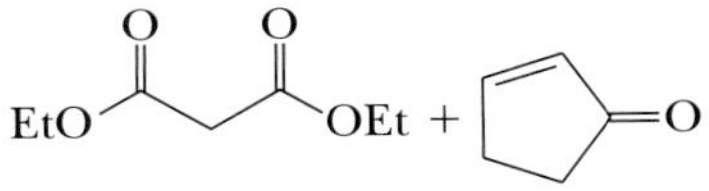 EtOOC

(formed as a 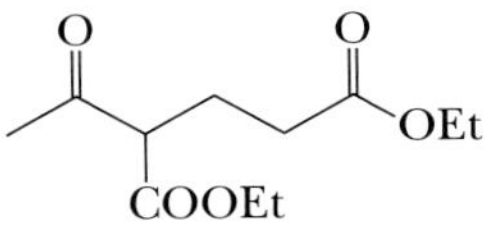   (formed as a
racemic mixture)   racemic mixture)

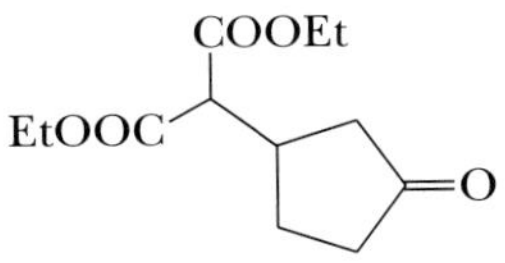

Problem 19.13

Show the product formed from each Michael product in the solution to Example 19.13 after (1) hydrolysis in aqueous NaOH, (2) acidification, and (3) thermal decarboxylation of each β-ketoacid or β-dicarboxylic acid. These reactions illustrate the usefulness of the Michael reaction for the synthesis of 1,5-dicarbonyl compounds.

Example 19.14

Show how the series of reactions in Example 19.13 and Problem 19.13 (Michael reaction, hydrolysis, acidification, and thermal decarboxylation) can be used to prepare 2,6-heptanedione.

Solution

As shown in the following retrosynthetic analysis, this molecule can be constructed from the carbon skeletons of ethyl acetoacetate and methyl vinyl ketone.

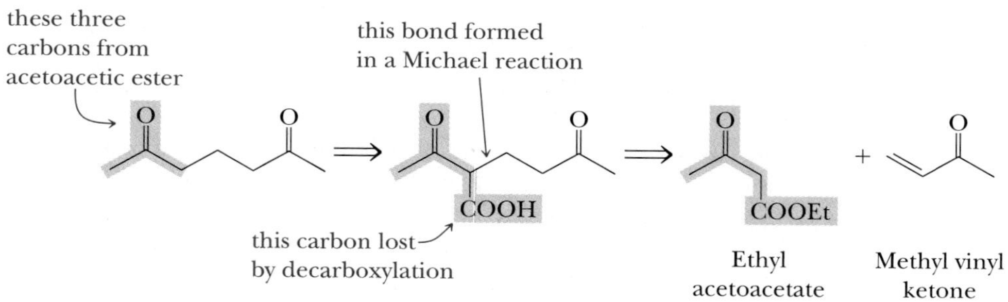
these three
carbons from
acetoacetic ester
this bond formed
in a Michael reaction
O
O
O
O
O
O
COOH
COOEt
this carbon lost
by decarboxylation

Ethyl acetoacetate

Methyl vinyl ketone

Following are the steps in their conversion to 2,6-heptanedione.

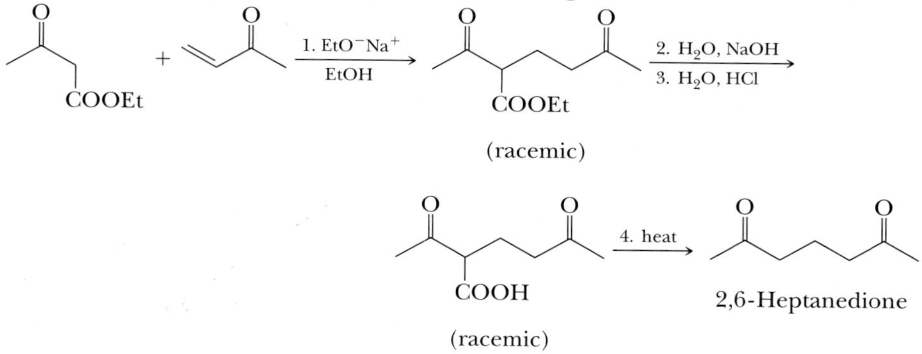
O
O
COOEt
1. EtO⁻Na⁺
EtOH
COOEt
(racemic)
2. H₂O, NaOH
3. H₂O, HCl
COOH
(racemic)
4. heat

2,6-Heptanedione

Problem 19.14

Show how the sequence of Michael reaction, hydrolysis, acidification, and thermal decarboxylation can be used to prepare pentanedioic acid (glutaric acid).

As noted in Table 19.1, enamines also participate in Michael reactions as illustrated by the addition of the enamine of cyclohexanone to acrylonitrile, $CH_2{=}CHCN$.

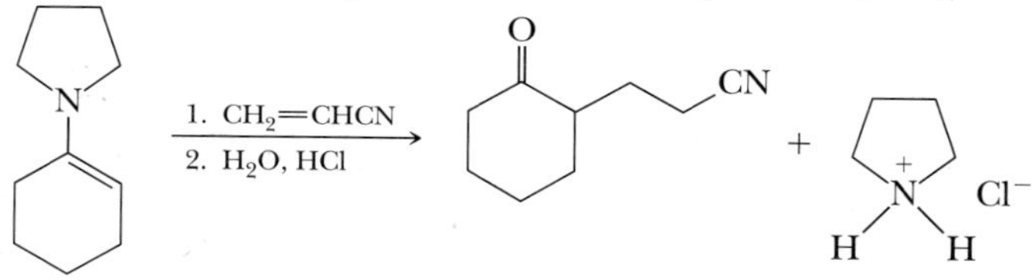
N
1. CH₂=CHCN
2. H₂O, HCl
O
CN
N
Cl⁻
H
H

Pyrrolidine enamine of cyclohexanone

(formed as a racemic mixture)

A final word about addition of nucleophiles to α,β-unsaturated carbonyl compounds. The Michael reaction is an example of 1,4-addition (conjugate addition) to an α,β-unsaturated carbonyl compound. In general, resonance-stabilized enolate anions and enamines are weak bases, react slowly, and give 1,4-addition products. Organolithium and organomagnesium compounds, on the other hand, are strong bases, react rapidly, and give primarily 1,2-addition products; that is, they give products formed by addition to the carbonyl carbon.

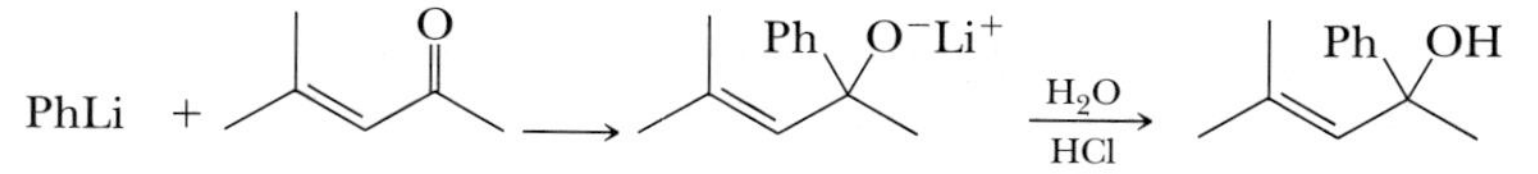
PhLi
O
Ph
O⁻Li⁺
H₂O
HCl
Ph
OH

Phenyllithium

4-Methyl-3-penten-2-one

4-Methyl-2-phenyl-3-penten-2-ol (racemic)

Why do the nucleophiles listed in Table 19.1 react with conjugated carbonyl compounds by 1,4-addition rather than 1,2-addition? The answer has to do with **kinetic control** versus **thermodynamic control** of product formation. It has been shown that 1,2-addition of nucleophiles to the carbonyl carbon of α,β-unsaturated carbonyl compounds is faster than conjugate addition. If formation of the 1,2-addition product is irreversible, then that is the product observed. If, however, formation of the 1,2-addition product is reversible, then an equilibrium is established between the more rapidly formed but less stable 1,2-addition product and the more slowly formed but more stable 1,4-addition product. As mentioned at the beginning of the chapter, a carbon-oxygen double bond is stronger than a carbon-carbon double bond. Thus, under conditions of thermodynamic (equilibrium) control, the more stable 1,4-Michael addition product is formed.

Kinetic control   Experimental conditions under which the composition of the product mixture is determined by the relative rates of formation of each product.

Thermodynamic control   Experimental conditions that permit the establishment of equilibrium between two or more products of a reaction. The composition of the product mixture is determined by the relative stabilities of the products.

1,2-Addition
(less stable product)

1,4-Addition
(more stable product)

Michael reaction with an α,β-unsaturated ketone followed by an intramolecular aldol reaction has proven to be a valuable method for the synthesis of 2-cyclohexenones. An especially important example of a Michael-aldol sequence is the **Robinson annulation,** in which treatment of a cyclic ketone, β-ketoester, or β-diketone with an α,β-unsaturated ketone in the presence of a base catalyst forms a cyclohexenone ring fused to the original ring. When the following racemic β-ketoester, for example, is treated with methyl vinyl ketone in the presence of sodium ethoxide in ethanol, the Michael adduct forms and then, in the presence of sodium ethoxide, undergoes a base-catalyzed intramolecular aldol reaction followed by dehydration to give a racemic substituted cyclohexenone.

Ethyl 2-oxocyclo-
hexanecarboxylate
(racemic)

3-Buten-2-one
(Methyl vinyl
ketone)

(racemic)

(four stereoisomers as
two racemic mixtures)

(racemic)

Example 19.15

Draw structural formulas for the lettered compounds in the following synthetic sequence.

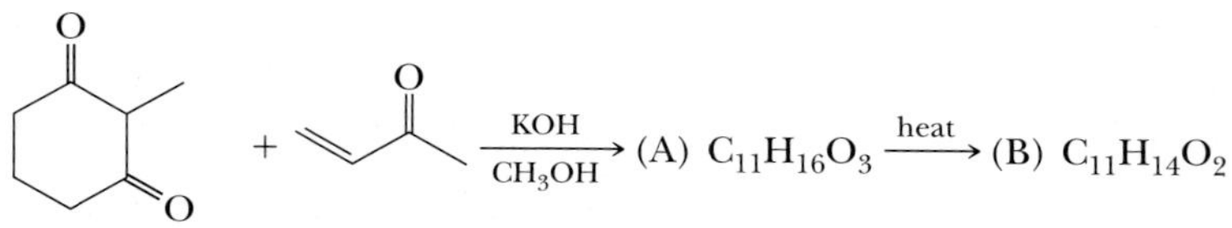

Solution

The product is the result of Michael addition to an α,β-unsaturated ketone followed by base-catalyzed aldol reaction and dehydration.

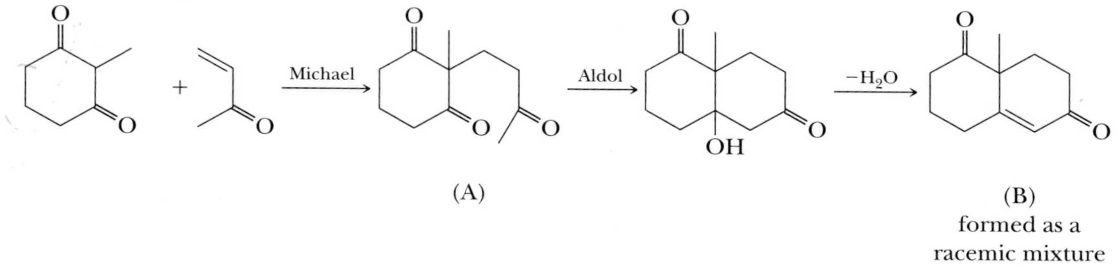

(A)

(B)
formed as a
racemic mixture

Problem 19.15

Show how to bring about the following conversion.

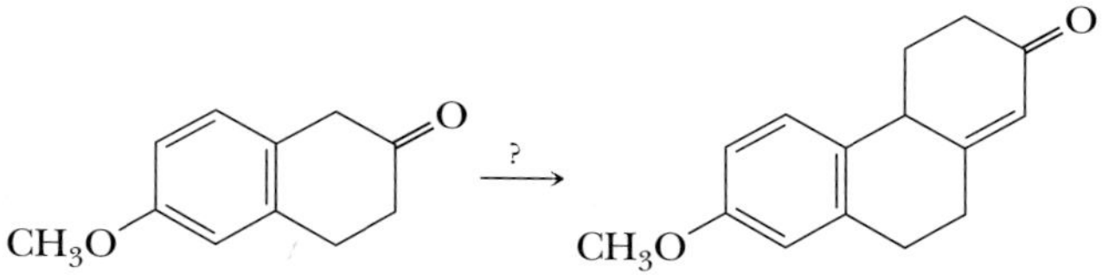

B. Conjugate Addition of Lithium Diorganocopper Reagents

Lithium diorganocopper reagents undergo 1,4-addition to α,β-unsaturated aldehydes and ketones in a reaction that is closely related to the Michael reaction. Yields are highest with primary alkyl, allylic, vinyl, and aryl organocopper reagents.

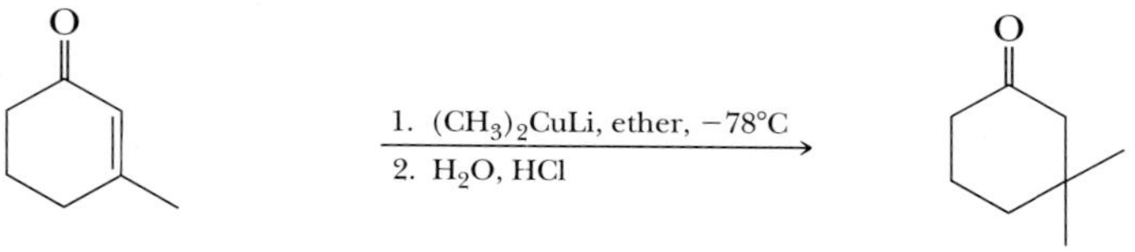

3-Methyl-2-cyclohexenone

3,3-Dimethylcyclohexanone

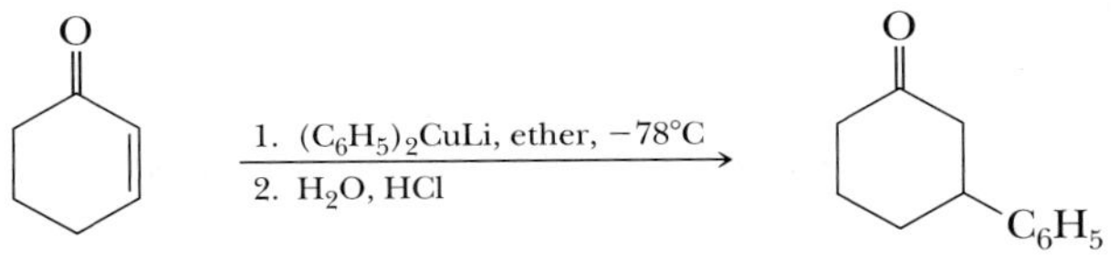

2-Cyclohexenone   3-Phenylcyclohexanone
                               (formed as a
                               racemic mixture)

Lithium diorganocopper reagents are unique among organometallic compounds in that they give almost exclusively 1,4-addition, which makes them very valuable reagents in synthetic organic chemistry. The mechanism of conjugate addition of lithium diorganocopper reagents is not fully understood.

Example 19.16

Propose two syntheses of 4-octanone, each involving conjugate addition of a lithium diorganocopper reagent.

Solution

A lithium diorganocopper reagent adds to the beta carbon of an α,β-unsaturated aldehyde or ketone. Therefore, locate each carbon beta to the carbonyl group in this target molecule and disconnect at those points.

Synthesis 1:

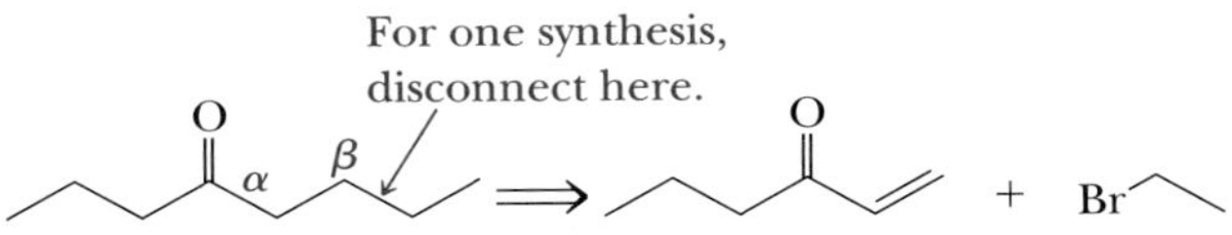

4-Octanone   1-Hexen-3-one   Bromoethane

For this synthesis, add lithium diethylcopper to 1-hexen-3-one.

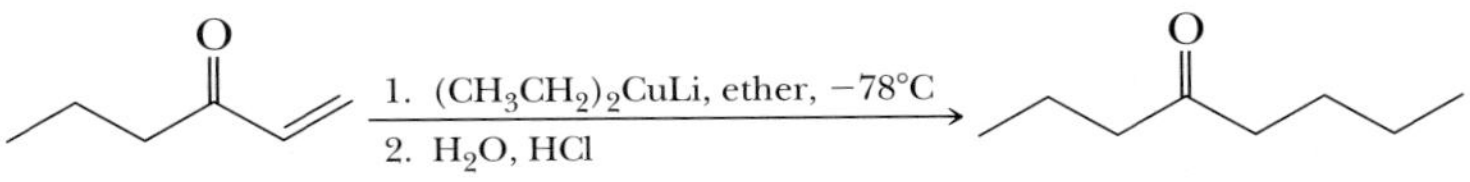

1-Hexen-3-one   4-Octanone

Synthesis 2:

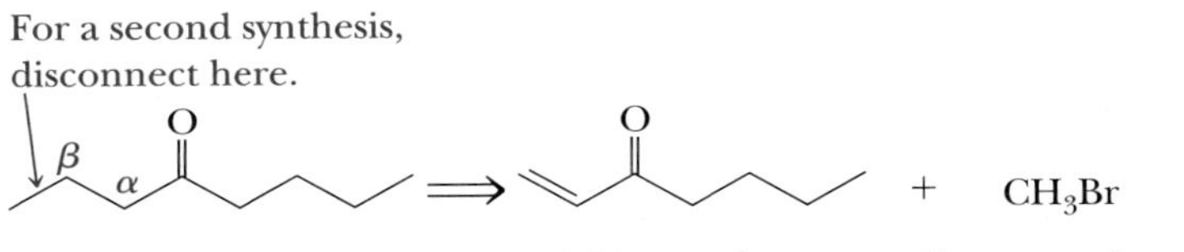

4-Octanone   1-Hepten-3-one   Bromomethane

For this synthesis, add lithium dimethylcopper to 1-hepten-3-one.

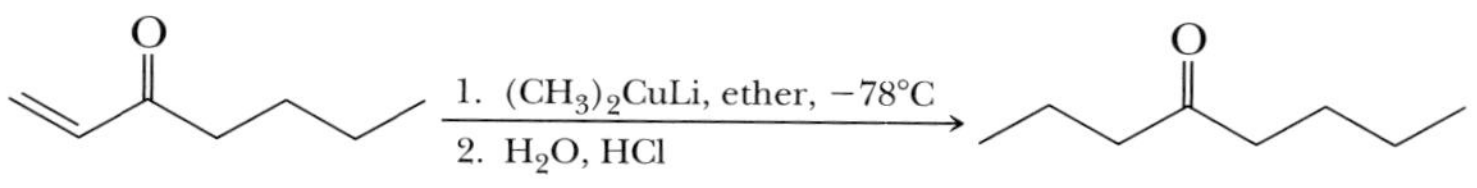
O
1. $(CH_3)_2CuLi$, ether, −78°C
2. H_2O, HCl
O
1-Hepten-3-one
4-Octanone

Problem 19.16

Propose two syntheses of 4-phenyl-2-pentanone, each involving conjugate addition of a lithium diorganocopper reagent.

Summary

An **enolate anion** is an anion formed by removing an α-hydrogen from a carbonyl-containing compound (Section 19.1). Aldehydes, ketones, and esters can be converted to their enolate anions by treatment with a strong base such as NaOEt in EtOH.

Acetyl-CoA (Section 19.4) is the source of the carbon atoms for the synthesis of terpenes, cholesterol, steroid hormones, and fatty acids. Key intermediates in the synthesis of these biomolecules are mevalonic acid and isopentenyl pyrophosphate.

Key Reactions

ORGANIC Chemistry•Now™

Click Flashcards to review the **Key Reactions of Enolate Anions**

1. Aldol Reaction (Section 19.2)

The aldol reaction involves nucleophilic addition of the enolate anion of one aldehyde or ketone to the carbonyl group of another aldehyde or ketone. The product of an aldol reaction is a β-hydroxyaldehyde or a β-hydroxyketone.

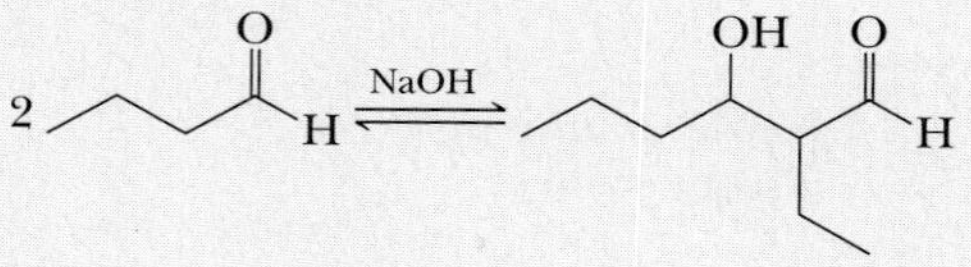
2
O
H
NaOH
OH
O
H

(formed as a racemic mixture of four stereoisomers)

2. Dehydration of the Product of an Aldol Reaction (Section 19.2)

Dehydration of the β-hydroxyaldehyde or ketone from an aldol reaction occurs very readily under acidic or basic conditions and gives an α,β-unsaturated aldehyde or ketone.

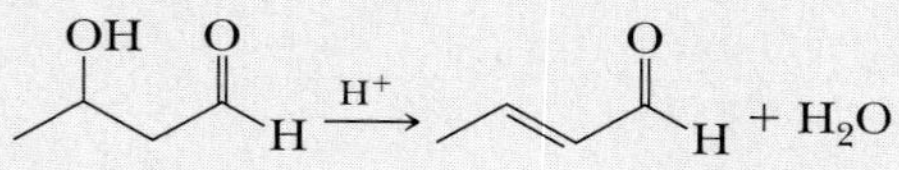
OH
O
H
H^+
O
H
+ H_2O

3. Claisen Condensation (Section 19.3A)

The product of a Claisen condensation is a β-ketoester. Condensation occurs by nucleophilic acyl substitution in which the attacking nucleophile is the enolate anion of an ester.

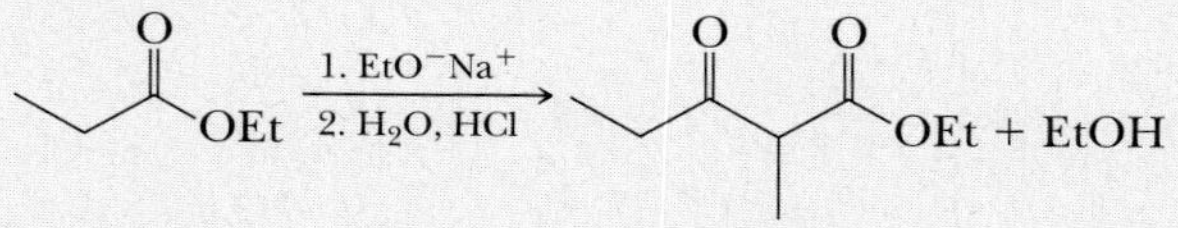
O
OEt
1. EtO^-Na^+
2. H_2O, HCl
O
O
OEt + EtOH

(formed as a racemic mixture)

4. Dieckmann Condensation (Section 19.3B)

An intramolecular Claisen condensation is called a Dieckmann condensation.

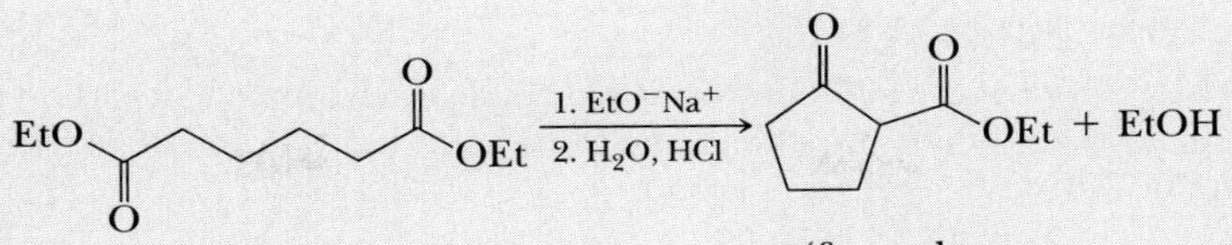
EtO
OEt
1. EtO^-Na^+
2. H_2O, HCl
OEt + EtOH

(formed as a racemic mixture)

5. Alkylation of an Enamine Followed by Hydrolysis (Section 19.5A)

Enamines are reactive nucleophiles with methyl and primary alkyl halides, α-haloketones, and α-haloesters.

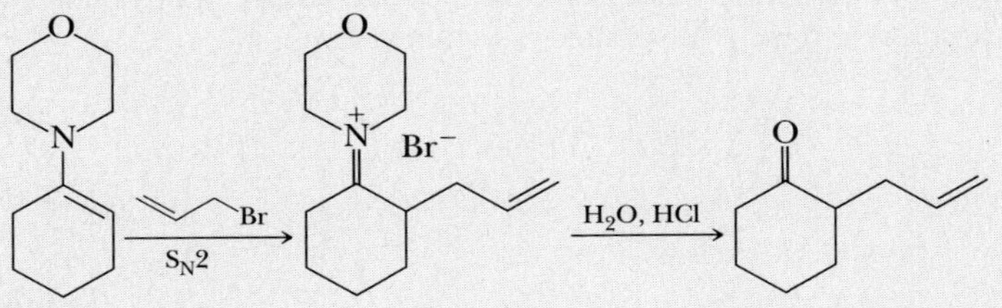
Br
S_N2
Br^-
H_2O, HCl

(formed as a racemic mixture)

6. Acylation of an Enamine Followed by Hydrolysis (Section 19.5B)

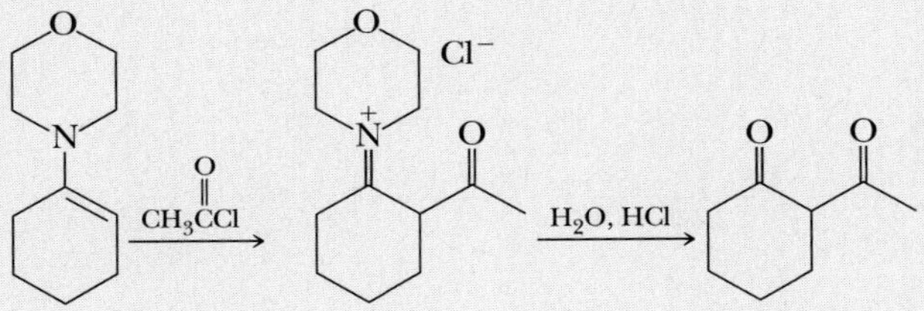
CH_3CCl
Cl^-
H_2O, HCl

(formed as a racemic mixture)

7. Acetoacetic Ester Synthesis (Section 19.6)

This sequence is useful for the synthesis of monosubstituted and disubstituted acetones.

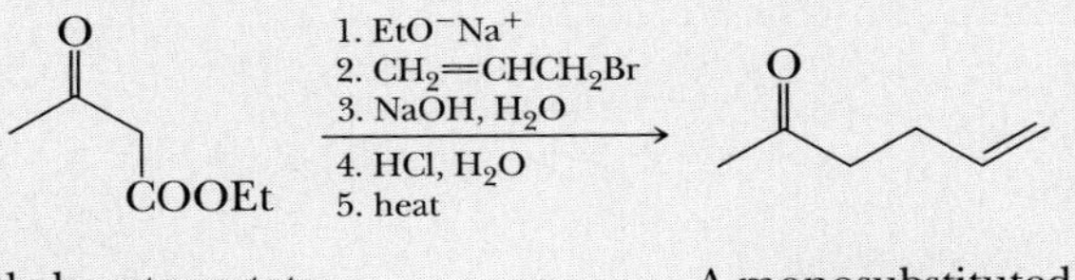
COOEt
1. EtO^-Na^+
2. $CH_2=CHCH_2Br$
3. NaOH, H_2O
4. HCl, H_2O
5. heat

Ethyl acetoacetate — A monosubstituted acetone

8. Malonic Ester Synthesis (Section 19.7)

This sequence is useful for the synthesis of monosubstituted and disubstituted acetic acids.

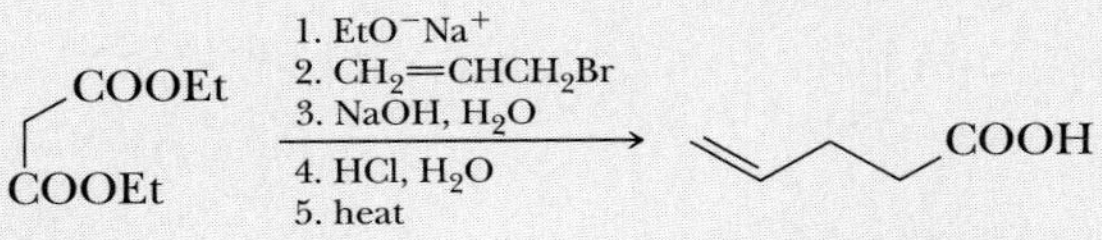
COOEt
COOEt
1. EtO^-Na^+
2. $CH_2=CHCH_2Br$
3. NaOH, H_2O
4. HCl, H_2O
5. heat
COOH

Diethyl malonate — A monosubstituted acetic acid

9. Michael Reaction (Section 19.8A)

A Michael reaction involves the addition of a weakly basic nucleophile to a carbon-carbon double bond made electrophilic by conjugation with the carbonyl group of an aldehyde, ketone, or ester or with a nitro or cyano group.

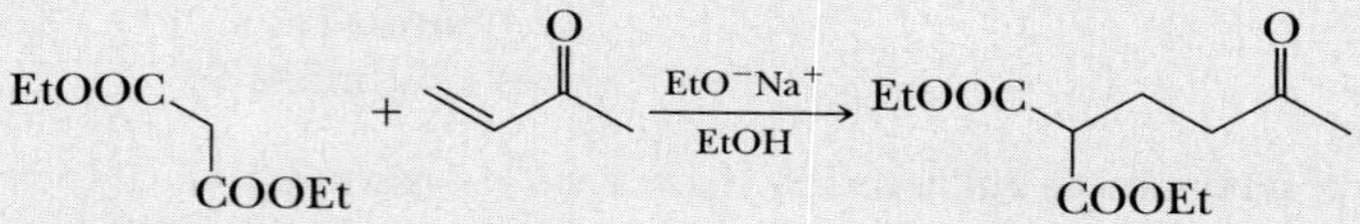

10. Robinson Annulation (Section 19.8A)

A Robinson annulation comprises a Michael reaction followed by an intramolecular aldol reaction and dehydration to form a substituted 2-cyclohexenone.

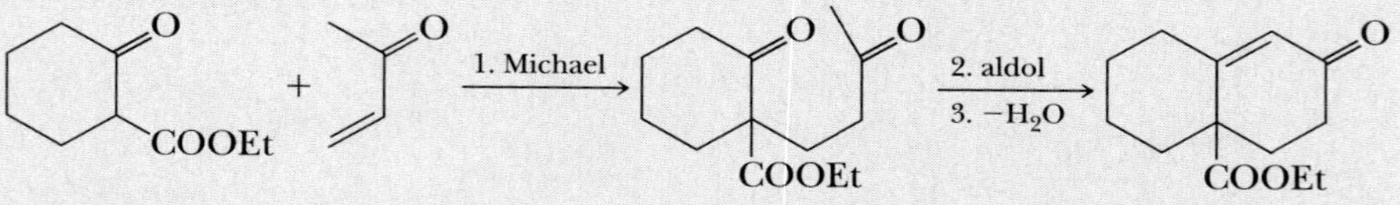

(formed as a racemic mixture)

11. Conjugate Addition of Lithium Diorganocopper Reagents (Section 19.8B)

In a reaction closely related to the Michael reaction, lithium diorganocopper reagents undergo conjugate addition to the electrophilic double bond of α,β-unsaturated aldehydes and ketones.

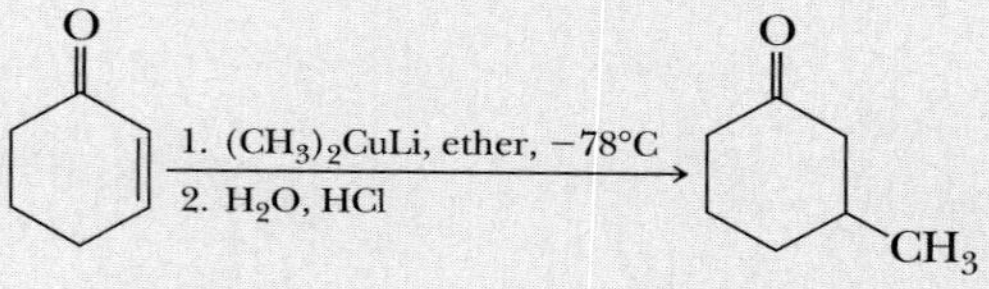

1. $(CH_3)_2CuLi$, ether, $-78°C$
2. H_2O, HCl

(formed as a racemic mixture)

Problems

The Aldol Reaction

19.17 Draw a structural formula for the product of the aldol reaction of each compound and for the α,β-unsaturated aldehyde or ketone formed from dehydration of each aldol product.

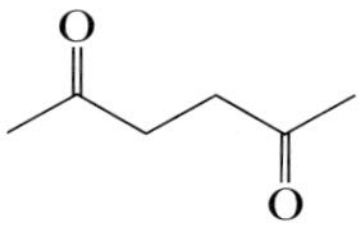

19.18 Draw a structural formula for the product of each crossed aldol reaction and for the compound formed by dehydration of each aldol product.

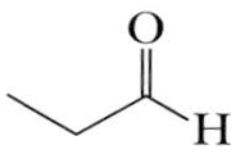

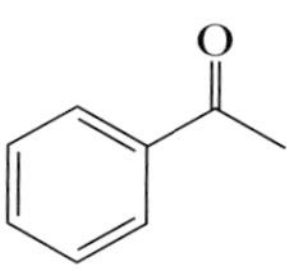

(a) 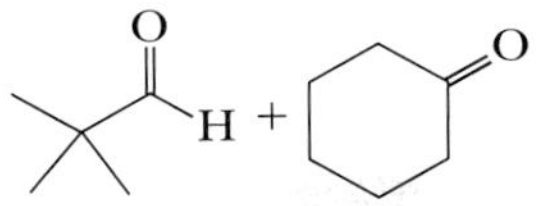   **(b)**

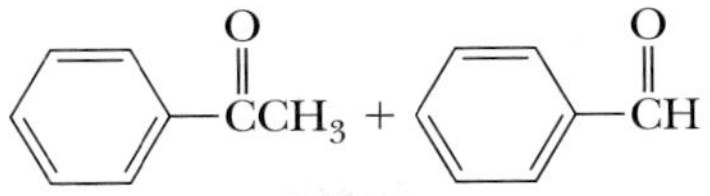

(c) 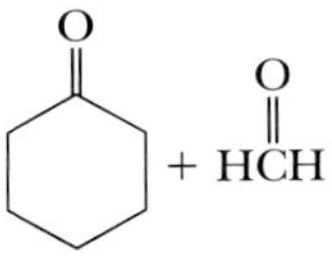   **(d)** PhCHO +   CHO

19.19 When a 1:1 mixture of acetone and 2-butanone is treated with base, six aldol products are possible. Draw a structural formula for each product.

19.20 Show how to prepare each α,β-unsaturated ketone by an aldol reaction followed by dehydration of the aldol product.

(a) 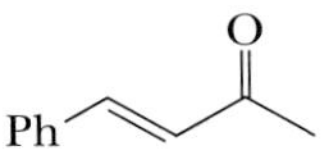 Ph   **(b)**

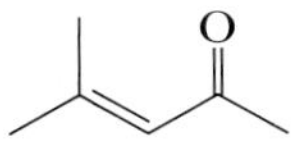

19.21 Show how to prepare each α,β-unsaturated aldehyde by an aldol reaction followed by dehydration of the aldol product.

(a) 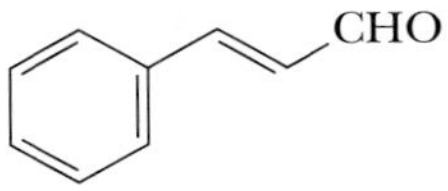   **(b)**

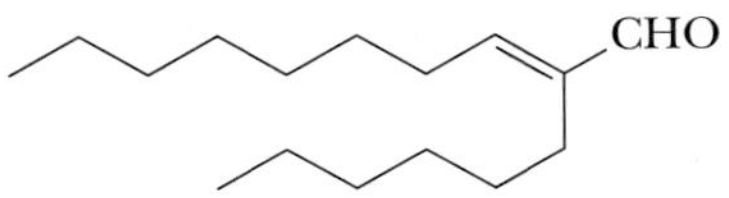

19.22 When treated with base, the following compound undergoes an intramolecular aldol reaction to give a product containing a ring (yield 78%).

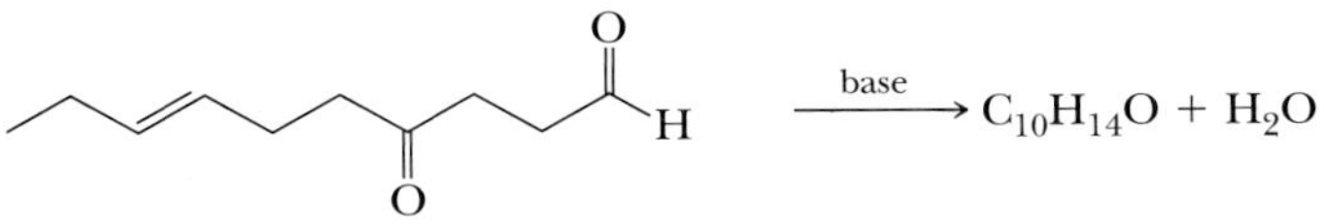

$\xrightarrow{\text{base}}$ C$_{10}$H$_{14}$O + H$_2$O

Propose a structural formula for this product.

19.23 Cyclohexene can be converted to 1-cyclopentenecarbaldehyde by the following series of reactions.

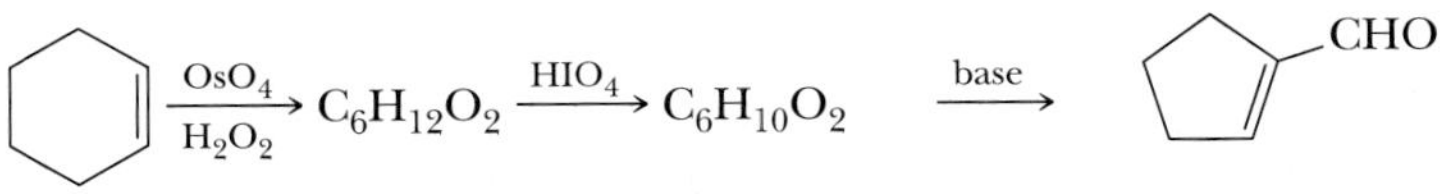

$\xrightarrow[\text{H}_2\text{O}_2]{\text{OsO}_4}$ C$_6$H$_{12}$O$_2$ $\xrightarrow{\text{HIO}_4}$ C$_6$H$_{10}$O$_2$ $\xrightarrow{\text{base}}$ CHO

1-Cyclopentenecarbaldehyde

Propose a structural formula for each intermediate compound.

19.24 Propose a structural formula for each lettered compound.

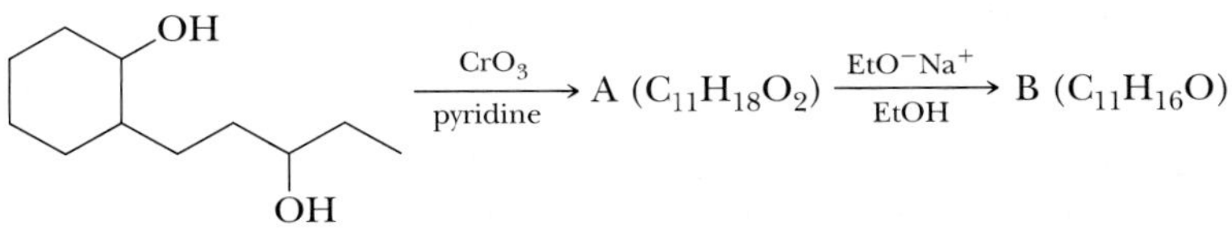

$\xrightarrow[\text{pyridine}]{\text{CrO}_3}$ A (C$_{11}$H$_{18}$O$_2$) $\xrightarrow[\text{EtOH}]{\text{EtO}^-\text{Na}^+}$ B (C$_{11}$H$_{16}$O)

19.25 How might you bring about the following conversion?

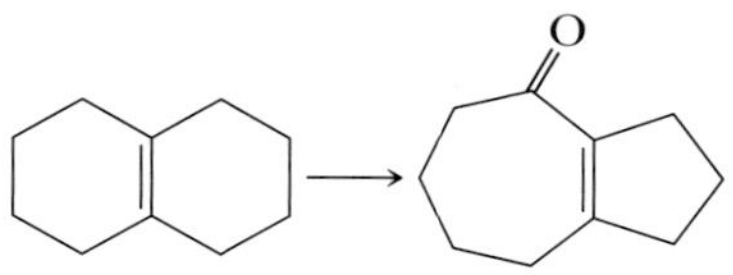

19.26 Pulegone, $C_{10}H_{16}O$, a compound from oil of pennyroyal, has a pleasant odor midway between peppermint and camphor. Treatment of pulegone with steam produces acetone and 3-methylcyclohexanone.

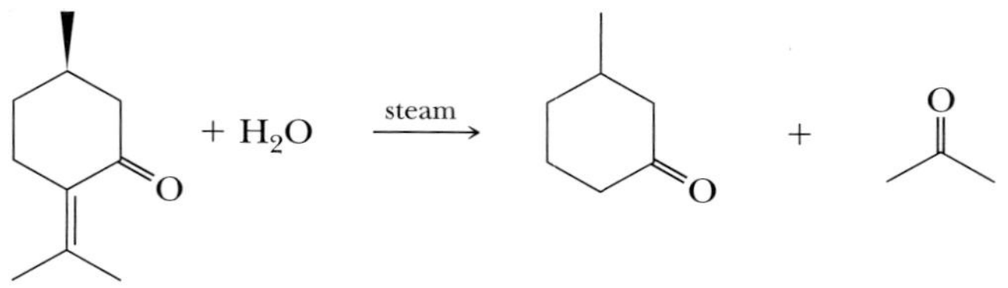

      Pulegone   3-Methylcyclohexanone   Acetone

  (a) Natural pulegone has the configuration shown. Assign an *R* or *S* configuration to its chiral center.

  (b) Propose a mechanism for the steam hydrolysis of pulegone to the compounds shown.

  (c) In what way does this steam hydrolysis affect the configuration of the chiral center in pulegone? Assign an *R* or *S* configuration to the 3-methylcyclohexanone formed in this reaction.

19.27 Propose a mechanism for this acid-catalyzed aldol reaction and the dehydration of the resulting aldol product.

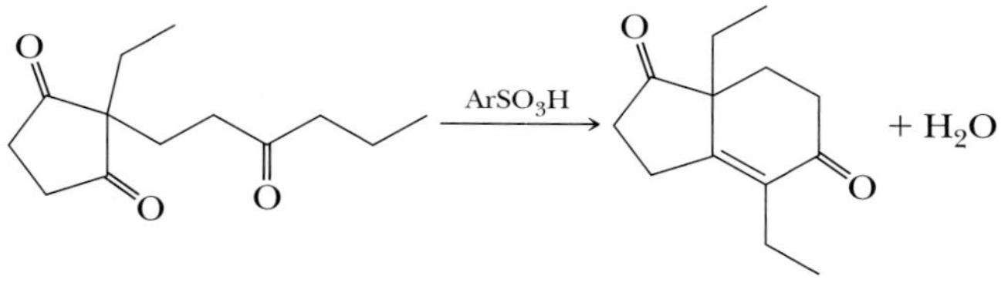

(racemic)

The Claisen Condensation

19.28 Show the product of Claisen condensation of each ester.

  (a) Ethyl phenylacetate in the presence of sodium ethoxide

  (b) Methyl hexanoate in the presence of sodium methoxide

19.29 When a 1:1 mixture of ethyl propanoate and ethyl butanoate is treated with sodium ethoxide, four Claisen condensation products are possible. Draw a structural formula for each product.

19.30 Draw structural formulas for the β-ketoesters formed by Claisen condensation of ethyl propanoate with each ester.

(a) EtOC—COEt   **(b)** PhCOEt   **(c)** HCOEt

19.31 Draw a structural formula for the product of saponification, acidification, and decarboxylation of each β-ketoester formed in Problem 19.30.

19.32 The Claisen condensation can be used as one step in the synthesis of ketones, as illustrated by this reaction sequence.

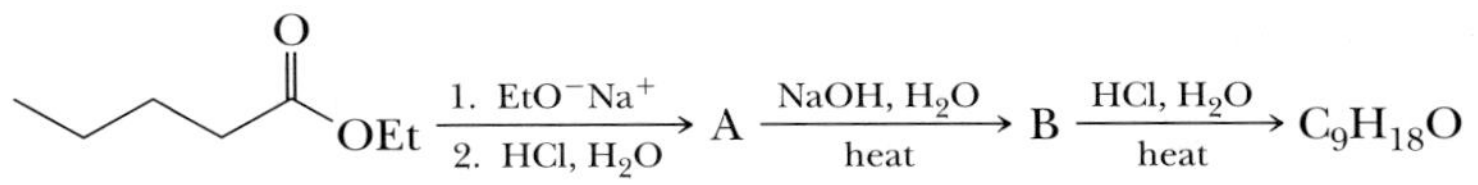

Propose structural formulas for compounds A and B and the ketone formed in this sequence.

19.33 Propose a synthesis for each ketone, using as one step in the sequence a Claisen condensation and the reaction sequence illustrated in Problem 19.32.

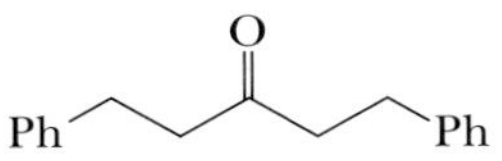

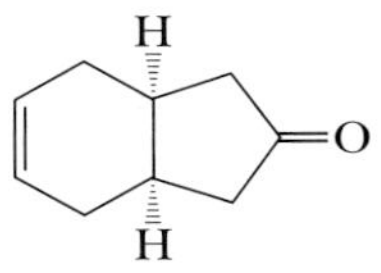

19.34 Propose a mechanism for the following conversion.

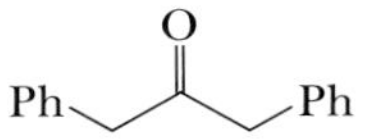

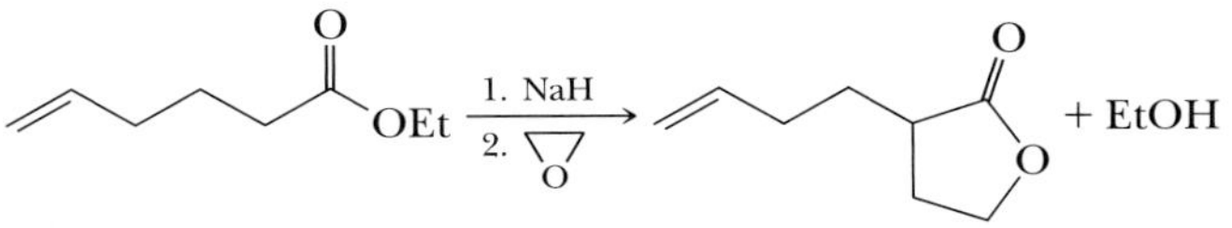

(racemic)

19.35 Claisen condensation between diethyl phthalate and ethyl acetate followed by saponification, acidification, and decarboxylation forms a diketone, $C_9H_6O_2$.

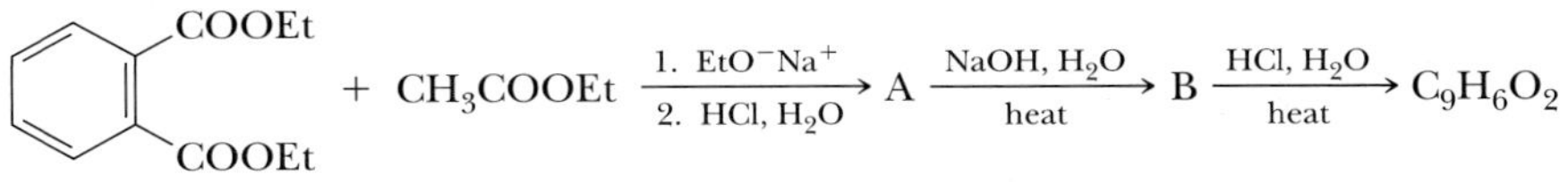

Diethyl phthalate   Ethyl acetate

Propose structural formulas for compounds A, B, and the diketone.

19.36 In 1887, the Russian chemist Sergei Reformatsky at the University of Kiev discovered that treatment of an α-haloester with zinc metal in the presence of an aldehyde or ketone followed by hydrolysis in aqueous acid results in formation of a β-hydroxyester. This reaction is similar to a Grignard reaction in that a key intermediate is an organometallic compound, in this case a zinc salt of an ester enolate anion. Grignard reagents, however, are so reactive that they undergo self-condensation with the ester.

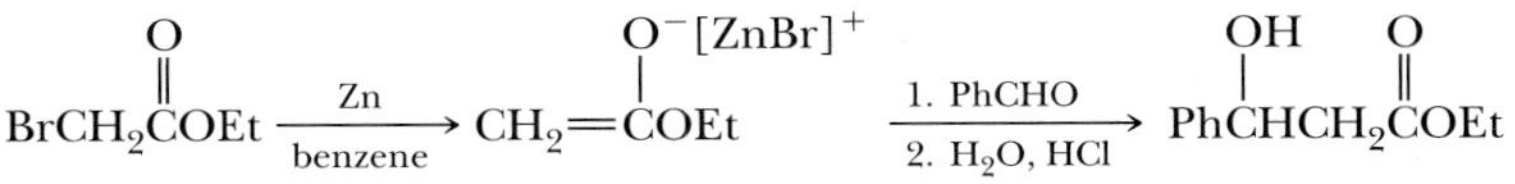

Zinc salt of an
enolate anion

A β-hydroxyester
(racemic)

Show how a Reformatsky reaction can be used to synthesize these compounds from an aldehyde or ketone and an α-haloester.

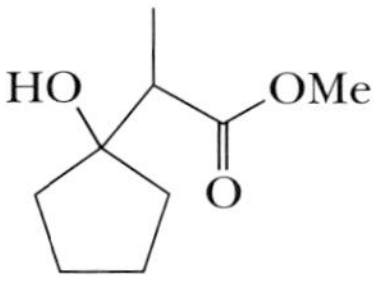

(four stereoisomers as
two racemic mixtures)

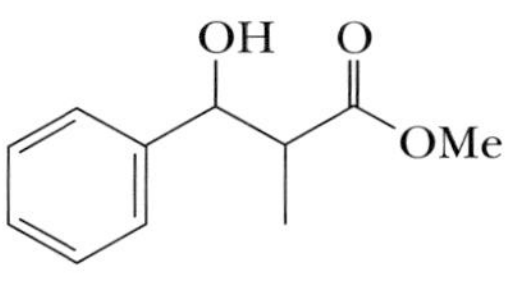

(four stereoisomers as
two racemic mixtures)

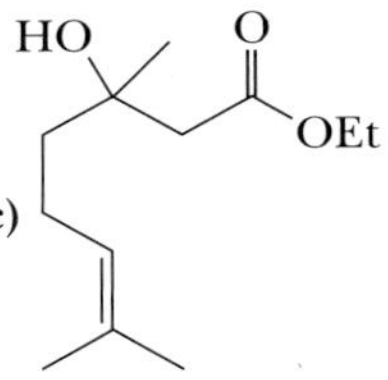

(racemic)

19.37 Many types of carbonyl condensation reactions have acquired specialized names, after the 19th century organic chemists who first studied them. Propose mechanisms for the following named condensations.

(a) Perkin condensation: Condensation of an aromatic aldehyde with an acid anhydride

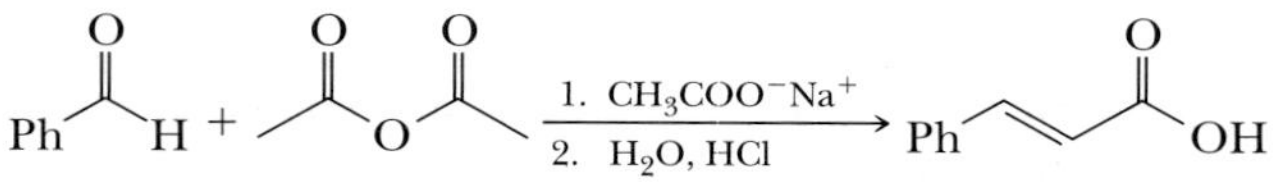

Cinnamic acid

(b) Darzens condensation: Condensation of an α-haloester with a ketone or an aromatic aldehyde

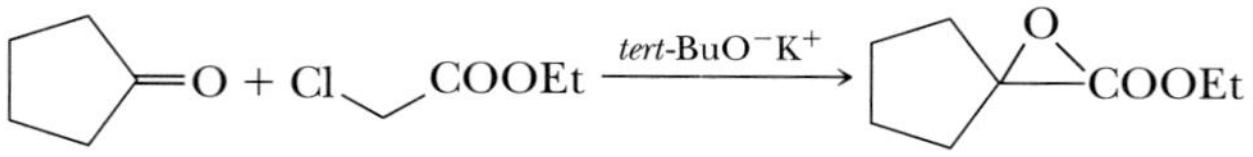

Enamines

19.38 When 2-methylcyclohexanone is treated with pyrrolidine, two isomeric enamines are formed.

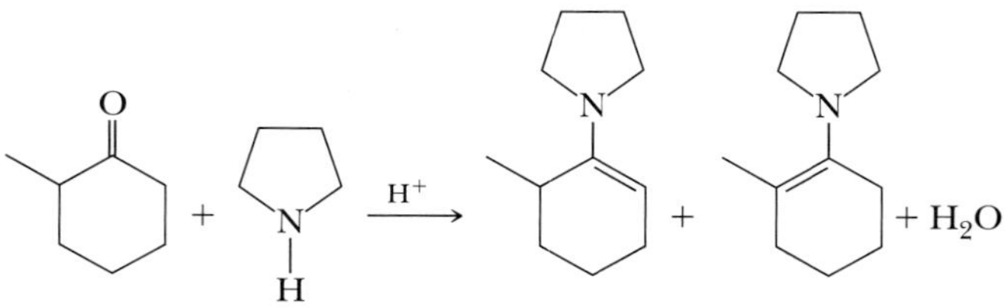

A (85%)   B (15%)

Why is enamine A with the less substituted double bond the thermodynamically favored product? (You will find it helpful to examine the models of these two enamines.)

19.39 Enamines normally react with methyl iodide to give two products: one arising from alkylation at nitrogen and the second arising from alkylation at carbon. For example,

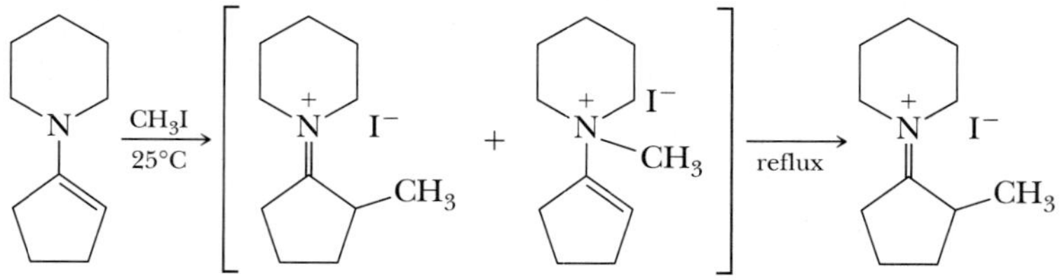

Product of   Product of   (racemic)
C-alkylation   N-alkylation

Heating the mixture of C-alkylation and N-alkylation products gives only the product from C-alkylation. Propose a mechanism for this isomerization.

19.40 Propose a mechanism for the following conversion.

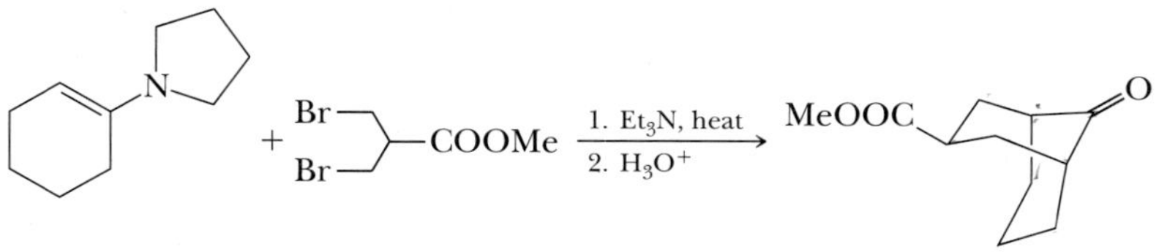

19.41 The following intermediate was needed for the synthesis of tamoxifen, a widely used anti-estrogen drug for treating estrogen-dependent cancers such as breast and ovarian cancer.

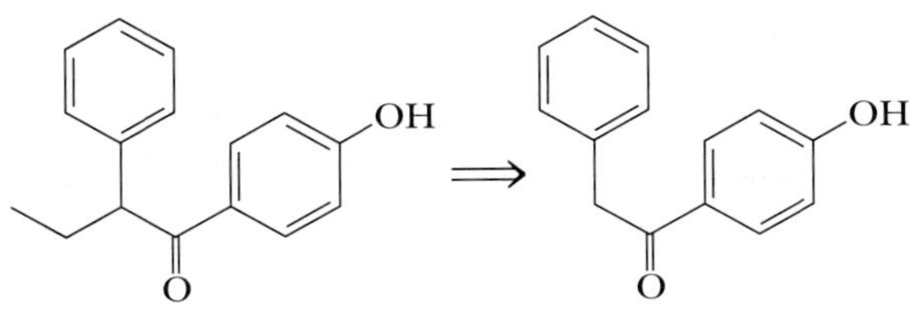

Needed for the   (A)
synthesis of tamoxifen

Propose a synthesis for this intermediate from compound A.

19.42 Propose a mechanism for the following reaction.

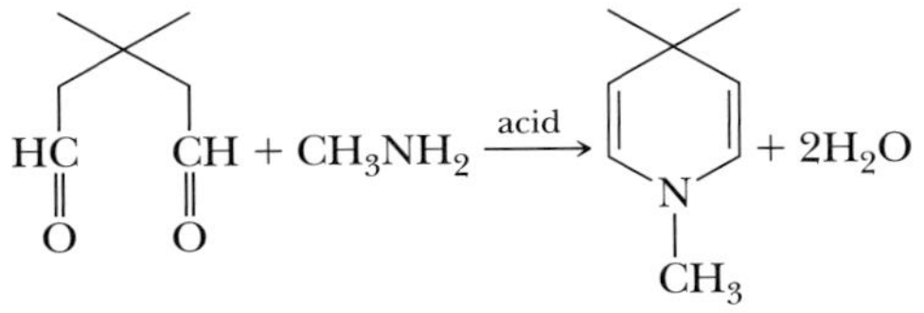

Acetoacetic Ester and Malonic Ester Syntheses

19.43 Propose syntheses of the following derivatives of diethyl malonate, each of which is a starting material for synthesis of a barbiturate.

(a) 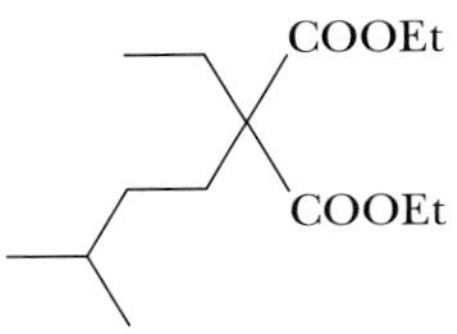   (b)

Needed for the   Needed for the
synthesis of amobarbital   synthesis of secobarbital

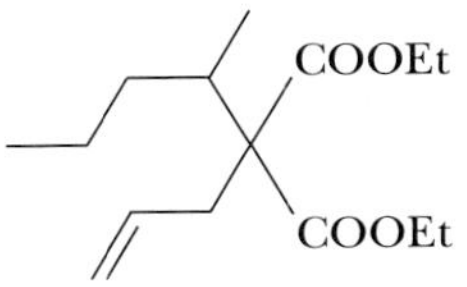

19.44 2-Propylpentanoic acid (valproic acid) is an effective drug for treatment of several types of epilepsy, particularly absence seizures, which are generalized epileptic seizures characterized by brief and abrupt loss of consciousness. Propose a synthesis of valproic acid starting with diethyl malonate.

19.45 Show how to synthesize the following compounds using either the malonic ester synthesis or the acetoacetic ester synthesis.

(a) 4-Phenyl-2-butanone   (b) 2-Methylhexanoic acid
(c) 3-Ethyl-2-pentanone   (d) 2-Propyl-1,3-propanediol
(e) 4-Oxopentanoic acid   (f) 3-Benzyl-5-hexene-2-one
(g) Cyclopropanecarboxylic acid (h) Cyclobutyl methyl ketone

19.46 Propose a mechanism for formation of 2-carbethoxy-4-butanolactone and then 4-butanolactone (γ-butyrolactone) in the following sequence of reactions.

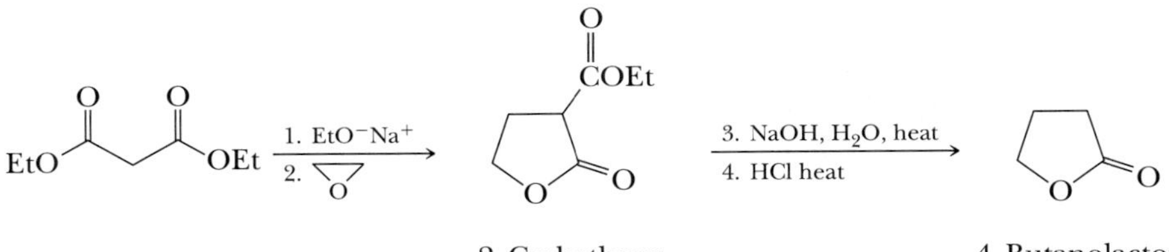

2-Carbethoxy-   4-Butanolactone
4-butanolactone   (γ-Butyrolactone)

19.47 Show how the scheme for formation of 4-butanolactone in Problem 19.46 can be used to synthesize lactones (a) and (b), each of which has a peach odor and is used in perfumery. As sources of carbon atoms for these syntheses, use diethyl malonate, ethylene oxide, 1-bromoheptane, and 1-nonene.

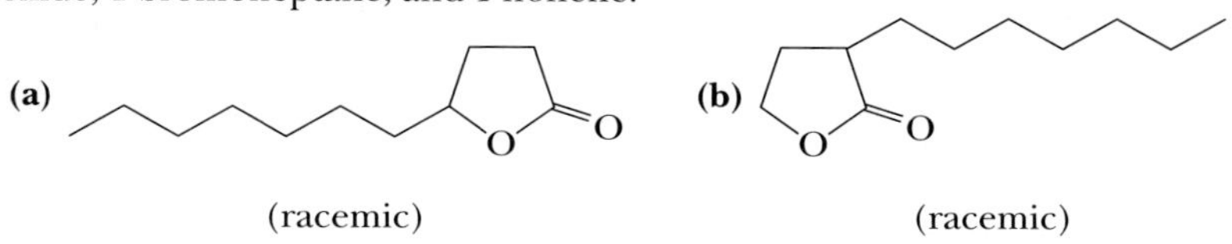

(a)   **(b)**

(racemic)   (racemic)

Michael Reactions

19.48 The following synthetic route is used to prepare an intermediate in the total synthesis of the anticholinergic drug benzilonium bromide.

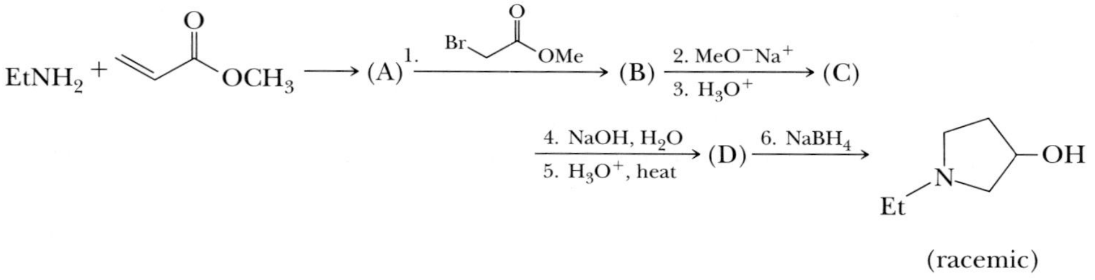

$$EtNH_2 + \overset{O}{\underset{}{\overset{}{}}}OCH_3 \longrightarrow (A) \xrightarrow[\text{Br}]{1.} \overset{O}{\underset{}{}}OMe \; (B) \xrightarrow[3. \; H_3O^+]{2. \; MeO^-Na^+} (C)$$

$$\xrightarrow[\text{5. } H_3O^+, \text{ heat}]{\text{4. NaOH, } H_2O} (D) \xrightarrow{\text{6. NaBH}_4} $$

Et
(racemic)

Propose structural formulas for intermediates A, B, C, and D.

19.49 Propose a mechanism for formation of the bracketed intermediate, and for the bicyclic ketone formed in the following reaction sequence.

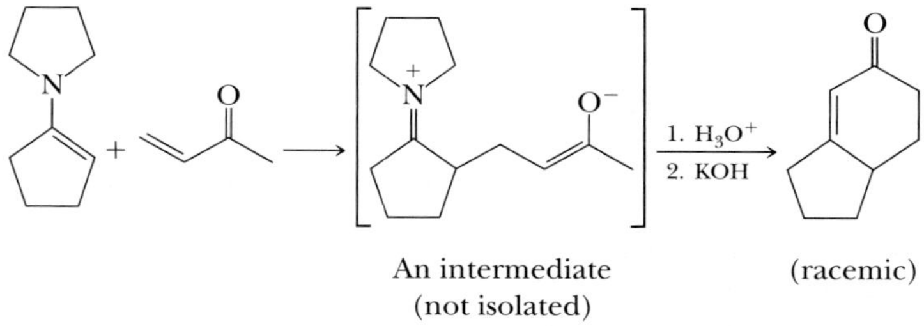

$$\xrightarrow[\text{2. KOH}]{1. \; H_3O^+}$$

An intermediate   (racemic)
(not isolated)

Synthesis

19.50 Show experimental conditions by which to carry out the following synthesis.

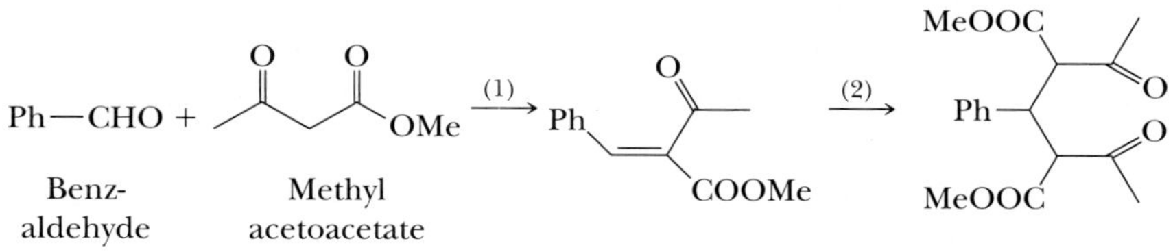

$$Ph-CHO + \overset{O\;\;\;\;O}{\underset{}{}}OMe \xrightarrow{(1)} Ph \xrightarrow{(2)} $$

Benz-   Methyl   COOMe   MeOOC
aldehyde   acetoacetate

19.51 Nifedipine (Procardia, Adalat) belongs to a class of drugs called calcium channel blockers and is effective in the treatment of various types of angina, including that induced by exercise. Show how nifedipine can be synthesized from 2-nitrobenzaldehyde, methyl acetoacetate, and ammonia. (*Hint*: Review the chemistry of your answers to Problems 19.42 and 19.50 and then combine that chemistry to solve this problem.)

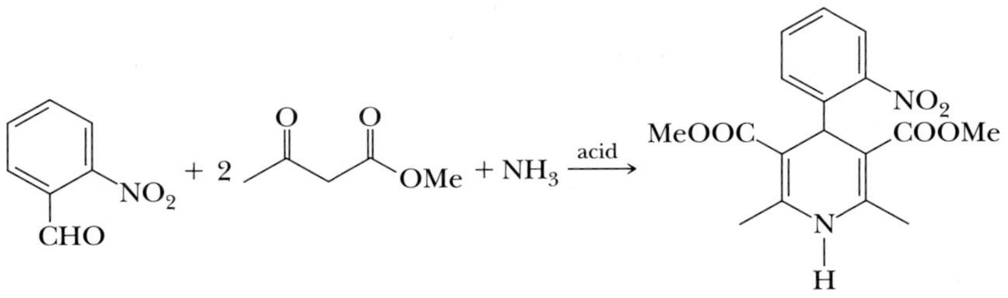

Nifedipine

19.52 The compound 3,5,5-trimethyl-2-cyclohexenone can be synthesized using acetone and ethyl acetoacetate as sources of carbon atoms. New carbon-carbon bonds in this synthesis are formed by a combination of aldol reactions and Michael reactions. Show reagents and conditions by which this synthesis might be accomplished.

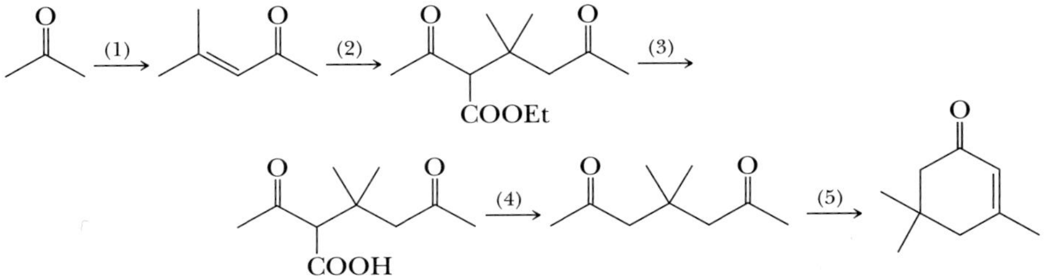

3,5,5-Trimethyl-
2-cyclohexenone

19.53 The following β-diketone can be synthesized from cyclopentanone and an acid chloride using an enamine reaction.

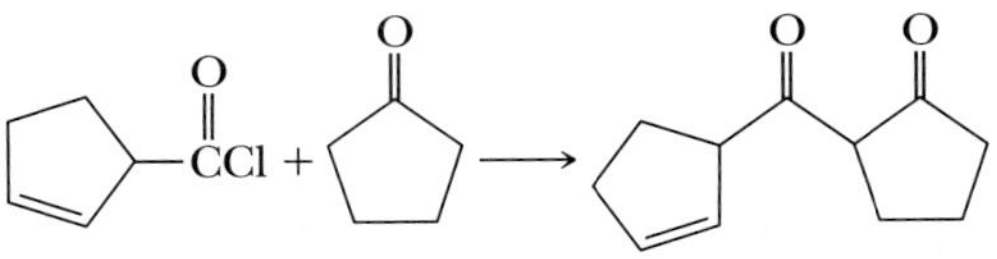

(racemic)

(a) Propose a synthesis of the starting acid chloride from cyclopentene.
(b) Show the steps in the synthesis of the β-diketone using a morpholine enamine.

19.54 Oxanamide is a mild sedative belonging to a class of molecules called oxanamides. As seen in this retrosynthetic scheme, the source of carbon atoms for the synthesis of oxanamide is butanal.

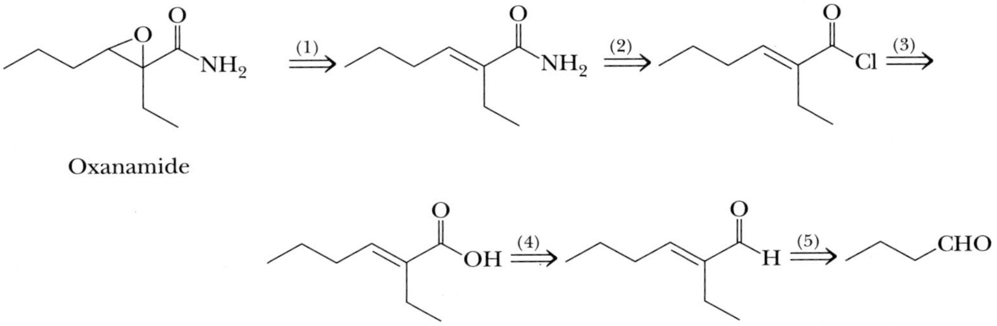

Oxanamide

Butanal

(a) Show reagents and experimental conditions by which oxanamide can be synthesized from butanal.

(b) How many chiral centers are there in oxanamide? How many stereoisomers are possible for this compound?

19.55 The widely used anticoagulant warfarin (see Chemical Connections: "From Moldy Clover to a Blood Thinner" in Chapter 18) is synthesized from 4-hydroxycoumarin, benzaldehyde, and acetone as shown in this retrosynthesis. Show how warfarin is synthesized from these reagents.

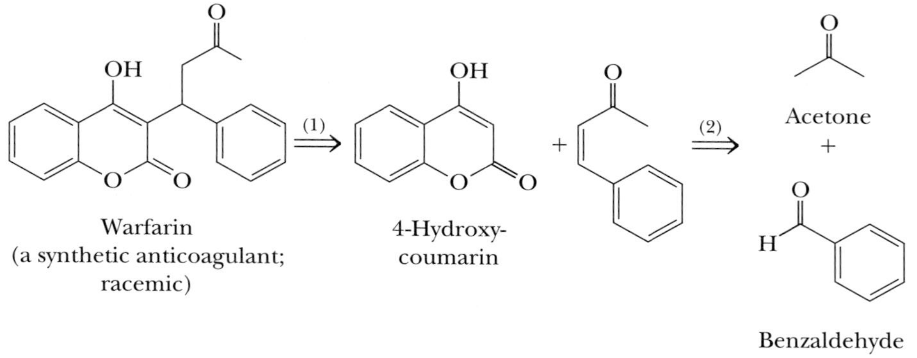

Warfarin
(a synthetic anticoagulant; racemic)

4-Hydroxy-coumarin

Acetone
+

Benzaldehyde

19.56 Following is a retrosynthetic analysis for an intermediate in the industrial synthesis of vitamin A.

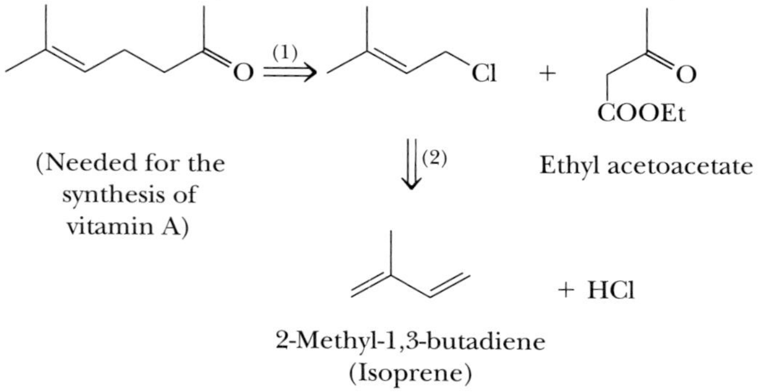

(Needed for the synthesis of vitamin A)

Ethyl acetoacetate

2-Methyl-1,3-butadiene
(Isoprene)

+ HCl

(a) Addition of one mole of HCl to isoprene gives 4-chloro-2-methyl-2-butene as the major product. Propose a mechanism for this addition and account its regioselectivity.

(b) Propose a synthesis of the vitamin A precursor from this allylic chloride and ethyl acetoacetate.

19.57 Following are the steps in one of the several published synthesis of frontalin, a pheromone of the western pine beetle.

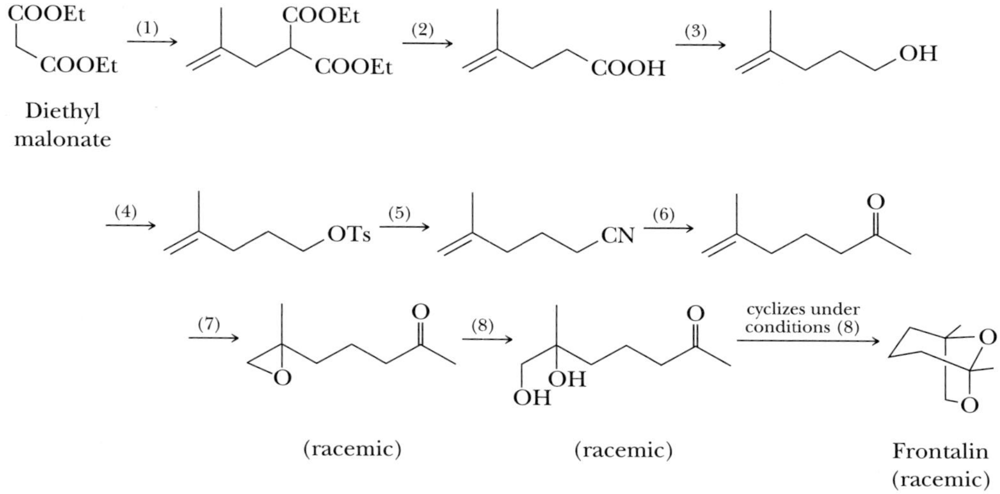

Diethyl malonate

(racemic)

(racemic)

Frontalin
(racemic)

(a) Propose reagents for Steps 1–8.

(b) Propose a mechanism for the cyclization of the ketodiol from Step 8 to frontalin.

19.58 2-Ethyl-1-hexanol was needed for the synthesis of the sunscreen octyl *p*-methylcinnamate (See Chapter 23, Chemical Connections: "Sunscreens and Sunblocks"). Show how this alcohol could be synthesized (a) by an aldol condensation of butanal and (b) by a malonic ester synthesis starting with diethyl malonate.

19.59 Gabapentin, an anticonvulsant used in the treatment of epilepsy, is structurally related to the neurotransmitter 4-aminobutanoic acid (GABA).

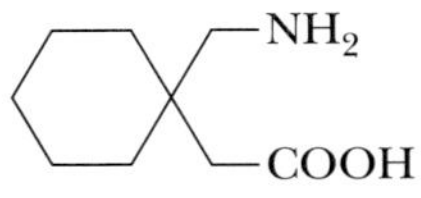

Gabapentin

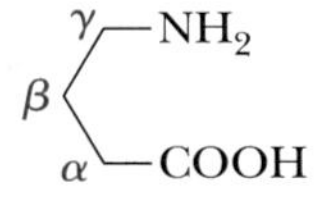

4-Aminobutanoic acid
(γ-Aminobutyric acid, GABA)

Gabapentin was designed specifically to be more lipophilic than GABA and, therefore, more likely to cross the blood-brain barrier, the lipid-like protective membrane that surrounds the capillary system in the brain and prevents hydrophilic (water-loving) compounds from entering the brain by passive diffusion. Given the following retrosynthetic analysis, propose a synthesis for gabapentin.

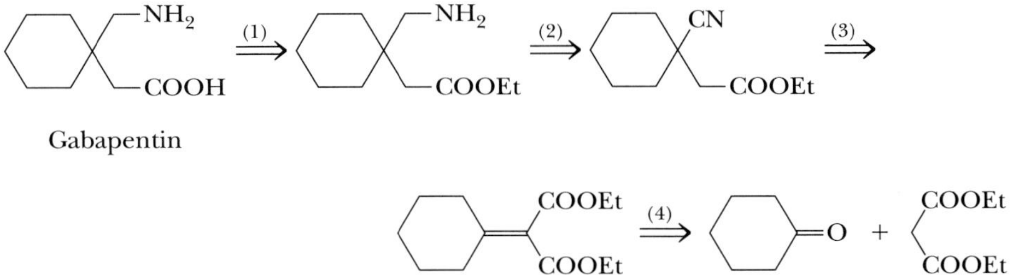

Gabapentin

Cyclohexanone   Diethyl
malonate

19.60 The following three derivatives of succinimide are anticonvulsants and have found use in the treatment of epilepsy, particularly petit mal seizures.

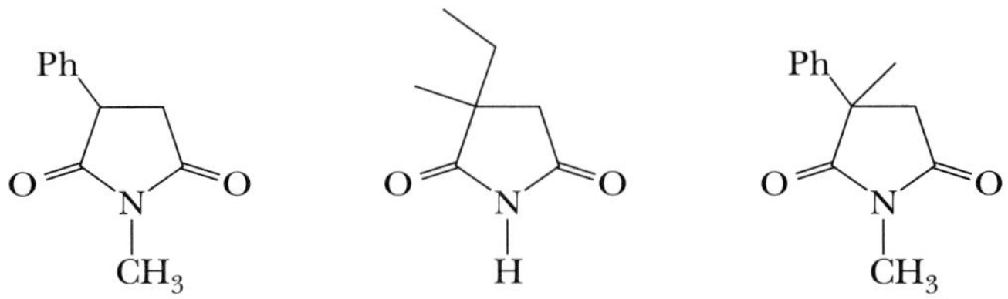

Methsuximide
(racemic)

Ethosuximide
(racemic)

Phensuximide
(racemic)

Following is a synthesis of phensuximide.

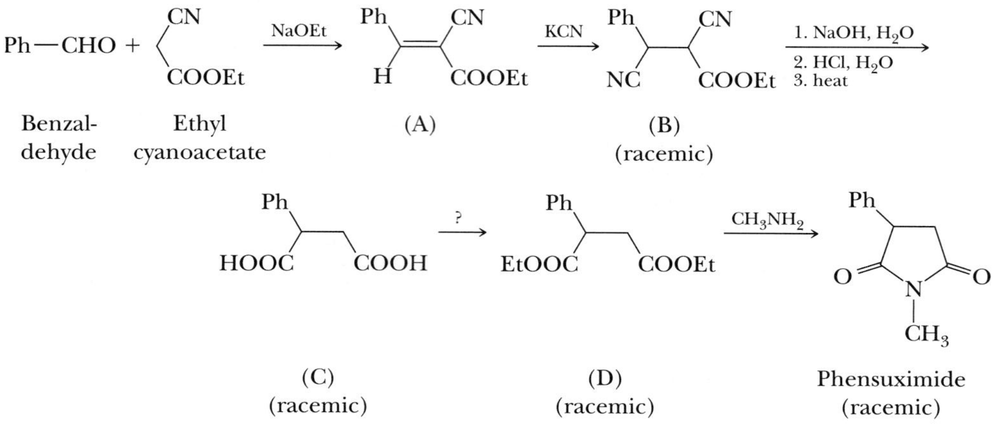

(a) Propose a mechanism for the formation of (A).

(b) What (person's) name is given to this type of reaction involved in the conversion of (A) to (B)?

(c) Describe the chemistry involved in the conversion of (B) to (C). You need not present detailed mechanisms. Rather, state what is accomplished by treating (B) first with NaOH and then with HCl followed by heating.

(d) Propose experimental conditions for the conversion of (C) to (D).

(e) Propose a mechanism for the conversion of (D) to phensuximide.

(f) Show how this same synthetic strategy can be used to prepare ethosuximide and methsuximide.

(g) Of these three anticonvulsants, one is considerably more acidic than the other two. Which is the most acidic compound? Estimate its pK_a and account for its acidity. How does its acidity compare with that of phenol? With that of acetic acid?

19.61 The analgesic meperidine (Demerol) was developed in the search for analgesics without the addictive effects of morphine. As shown in these structural formulas, it represents a simplification of morphine's structure.

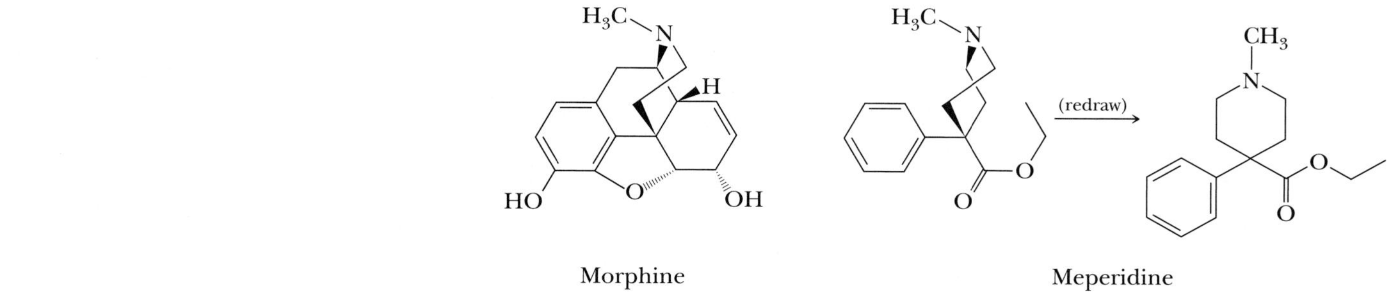

Morphine   Meperidine

Meperidine is prepared by treating phenylacetonitrile with one mole of *bis* (*N*-2-chloroethyl)methylamine (a nitrogen mustard) in the presence of two moles of sodium hydride to give (A). Refluxing (A) with concentrated sodium hydroxide followed by neutralization of the reaction mixture with dilute HCl gives (B). Treating (B) with ethanol in the presence of one equivalent of HCl gives meperidine as its hydrochloride salt.

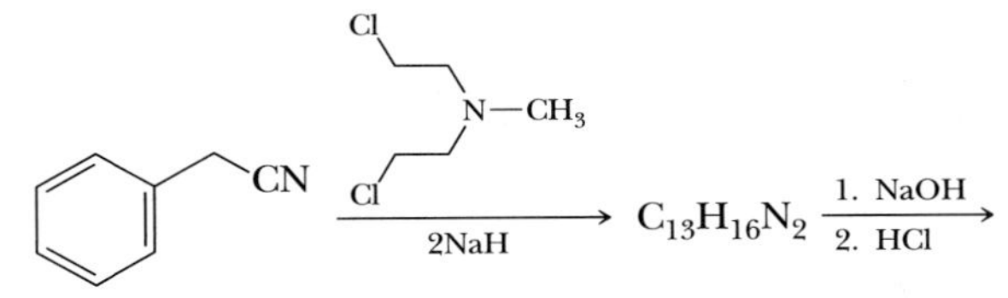

Phenylacetonitrile

(A)

$$C_{13}H_{17}NO_2 \xrightarrow{\text{EtOH, HCl}} C_{15}H_{21}NO_2 \cdot HCl$$

(B)   Meperidine hydrochloride

(a) Propose structural formulas for (A) and (B).

(b) Propose a mechanism for the formation of (A).

19.62 Verapamil (Effexor), a coronary artery vasodilator, is used in the treatment of angina caused by insufficient blood flow to cardiac muscle. Even though its effect on coronary

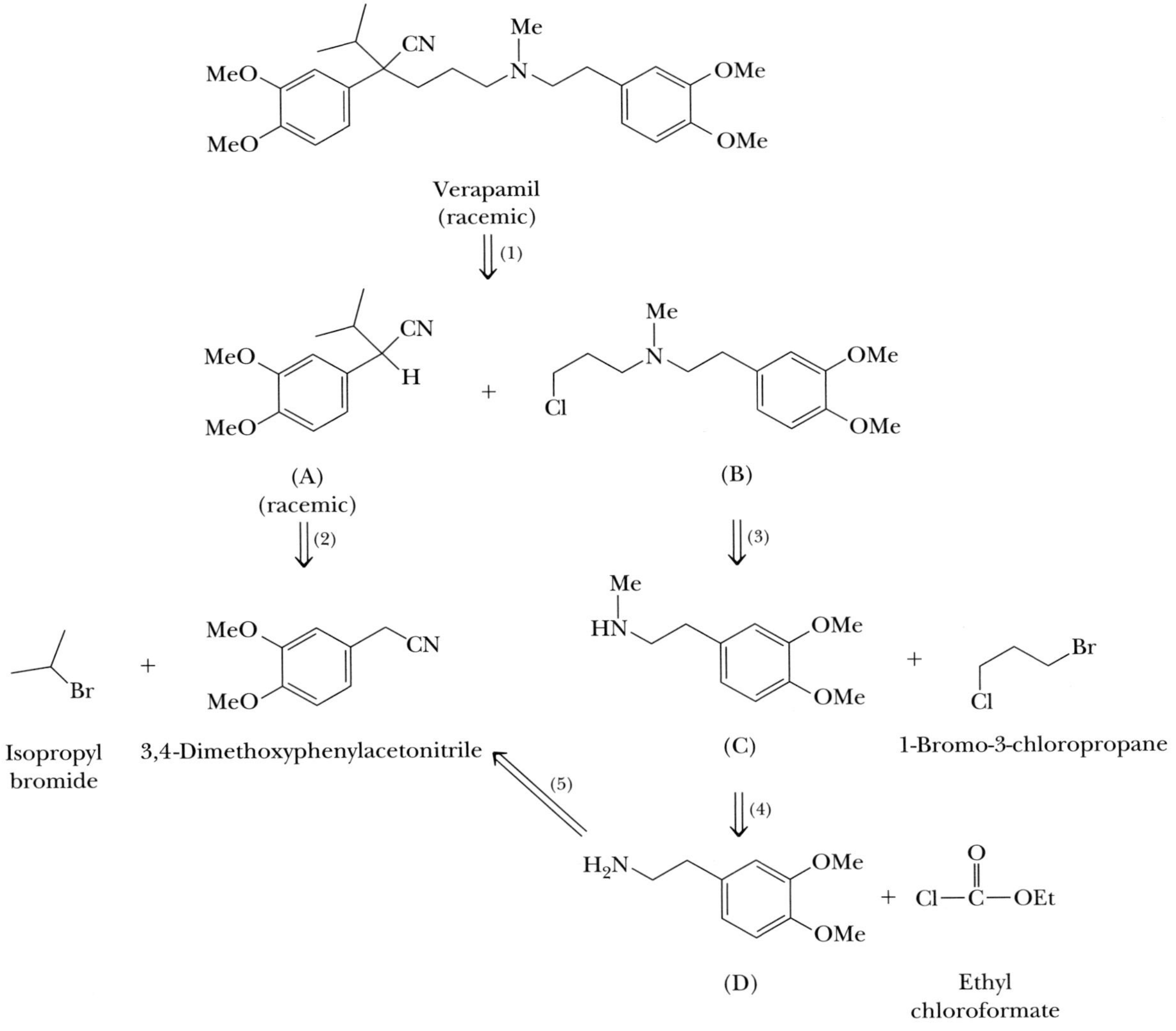

Verapamil
(racemic)

(A)
(racemic)

(B)

Isopropyl   3,4-Dimethoxyphenylacetonitrile   (C)   1-Bromo-3-chloropropane
bromide

(D)   Ethyl
chloroformate

vasculature tone was recognized over 30 years ago, it has only been more recently that its role as a calcium channel blocker has become understood. On the previous page is a retrosynthetic analysis leading to a convergent synthesis; it is convergent because (A) and (B) are made separately and then combined (i.e., the route converges) to give the final product. Convergent syntheses are generally much more efficient than those in which the skeleton is built up stepwise.

(a) Given this retrosynthetic analysis, propose a synthesis for verapamil from the four named starting materials.

(b) It requires two steps to convert (D) to (C). The first is treatment of (D) with ethyl chloroformate. What is the product of this first step? What reagent can be used to convert this product to (C)?

(c) How do you account for the regioselectivity of the nucleophilic displacement involved in converting (C) to (B)?

19.63 Based on this retrosynthetic analysis, propose a synthesis of the anticoagulant (inhibits blood clotting) diphenadione.

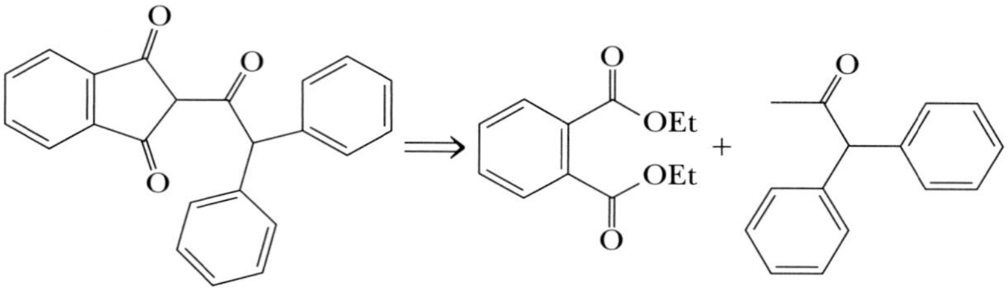

Diphenadione   Diethyl phthalate

Because of its anticoagulant activity for blood, this compound is used as a rodenticide. For the story of the discovery of the anticoagulant dicoumarin, see Chemical Connections: "From Moldy Clover to a Blood Thinner" in Chapter 18.

19.64 Following are two possible retrosynthetic analyses for the anticholinergic drug cycrimine. Fill in the details of each potential synthesis.

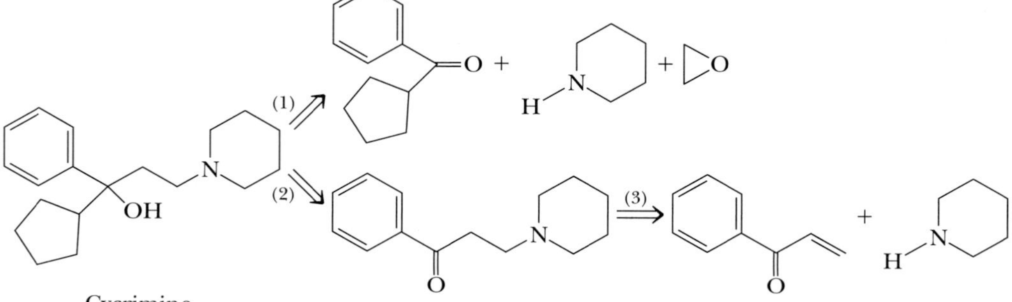

Cycrimine
(racemic)

19.65 Show how the tranquilizer valnoctamide can be synthesized using diethyl malonate as the source of the carboxamide group.

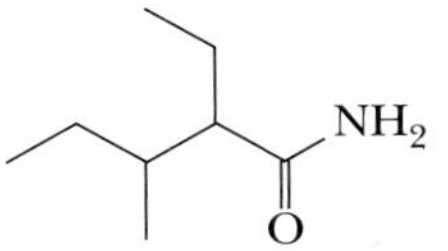

2-Ethyl-3-methylpentanamide
(Valnoctamide; racemic)

19.66 In Problem 7.28 we saw this two-step sequence in Johnson's synthesis of the steroid hormone progesterone. Propose a structural formula for the intermediate formed in Step 3, and a mechanism for its conversion in Step 4 to progesterone.

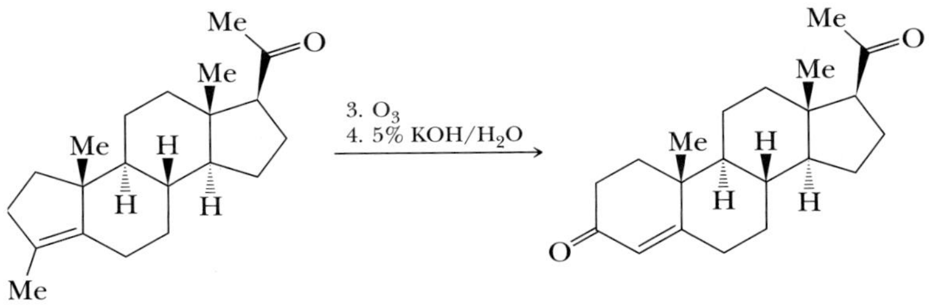

3. O₃
4. 5% KOH/H₂O

C   Progesterone

19.67 Monensin, a polyether antibiotic, was isolated from a strain of *Streptomyces cinamonensis* in 1967, and its structure was determined shortly thereafter.

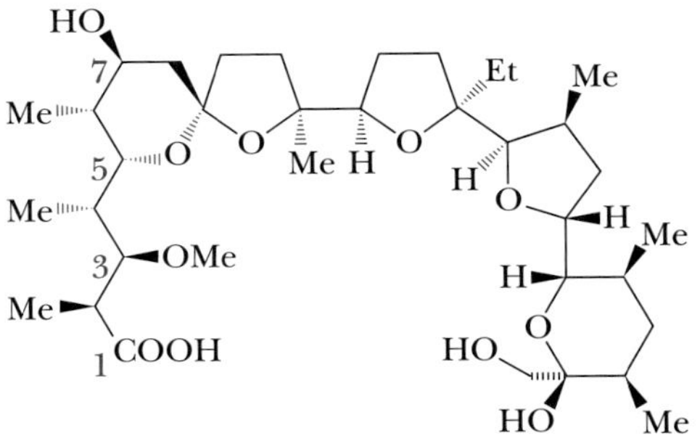

Monensin

This molecule exhibits a broad-spectrum anticoccidial activity and, since its introduction in 1971, has been used to treat coccidial infections in poultry and as an additive in cattle feed. Monensin owes its antibiotic activity to the fact that it can complex cations and transport them through lipid membranes. Its core of oxygen atoms functions as a set of Lewis bases and coordinates with cations (Lewis acids). When complexed in this manner, the outer surface of the monensin-cation complex is hydrophobic, which facilitates transport of the complex across a lipid interface (in this regard it functions like the crown ethers described in Section 11.11).

In the synthesis of monensin, Y. Kishi chose to create the molecule in sections and then join them to create the target molecule. Following is an outline of the steps by which he created the seven-carbon-chain building block on the left side of the molecule.

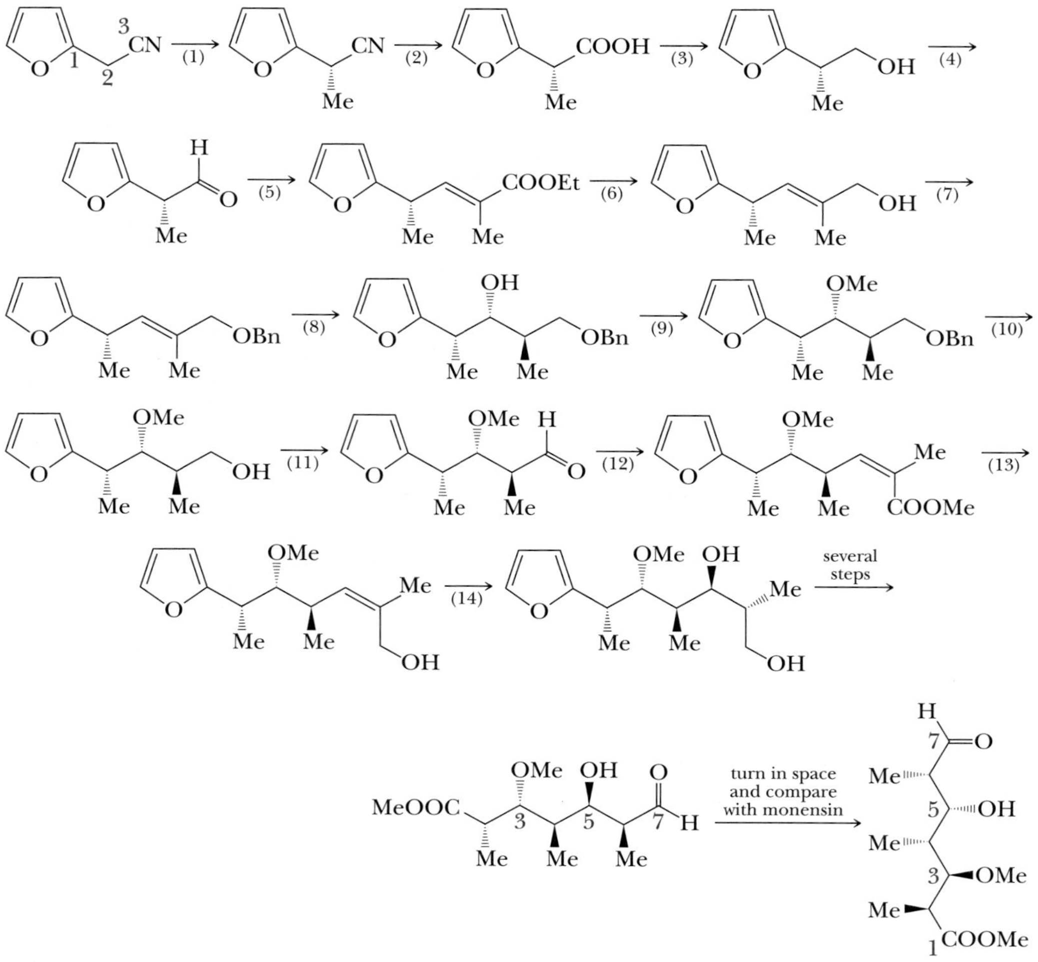

Propose a reagent or reagents for Steps 1–14. Note that this fragment contains five chiral centers. We are not asking you to predict or to rationalize the stereochemistry of each step, but rather only to propose a reagent or type of reagent to bring about each step.

© Kevin Schafer/CORBIS

Turmeric flower. Curcumin, an orange-yellow powder isolated from the spice tumeric and responsible for much of the color of curry, has recently been found to retard tumor growth. See Chemical Connections "Curry and Cancer" on page 796. Inset: a model of curcumin.

Outline

20.1 Stability of Conjugated Dienes

20.2 Electrophilic Addition to Conjugated Dienes

20.3 UV-Visible Spectroscopy

Conjugated Systems

In Chapters 5 and 6, we discussed the structure and characteristic reactions of alkenes. We limited this discussion to molecules containing isolated double bonds. In this chapter, we extend our study of molecules with pi bonds to include molecules that contain two or more adjacent double bonds. Such compounds are called **conjugated.** The important feature of conjugated systems is that all the adjacent $2p$ orbitals combine with each other. The result is that the pi electrons are not localized between just two carbon atoms. Instead, they are best thought of as being *delocalized* throughout the entire conjugated pi orbital systems.

Conjugated A conjugated diene or carbonyl is one in which the double bonds are separated by one single bond.

Cumulated A cumulated diene is one in which two double bonds share an *sp*-hybridized carbon.

ORGANIC
Chemistry⚛Now™
Look for this logo in the chapter and go to Organic ChemistryNow at
http://now.brookscole.com/bfi4
for tutorials, simulations, and problems.

20.1 Stability of Conjugated Dienes

Dienes are compounds that contain two carbon-carbon double bonds. Dienes can be divided into three groups: unconjugated, conjugated, and cumulated. An **unconjugated diene** is one in which the double bonds are separated by two or more single bonds. A **conjugated diene** is one in which the double bonds are separated by one single bond. A **cumulated diene** is one in which two double bonds share an *sp*-hybridized carbon. Because of the geometry of this carbon, the *p* orbitals of the two double bonds do not overlap in a cumulated diene and are not conjugated.

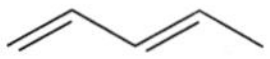

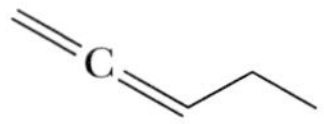

1,4-Pentadiene
(an unconjugated diene)

1,3-Pentadiene
(a conjugated diene)

1,2-Pentadiene
(a cumulated diene)

Example 20.1

Which of these molecules contain conjugated double bonds?

(a) 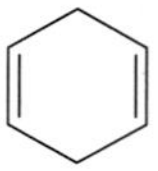   (b)    (c)    (d)

Solution

Compounds (b) and (c) contain conjugated double bonds. The double bonds in compounds (a) and (d) are unconjugated.

Problem 20.1

Which of these terpenes (Section 5.4) contains conjugated double bonds?

(a) 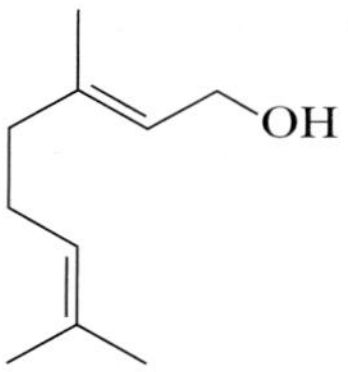   (b) 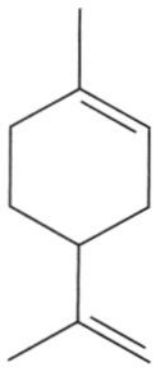   (c)

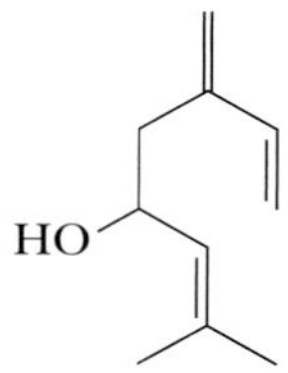

Geraniol   Limonene   An aggregating
pheromone of
bark beetles

Given in Table 20.1 are heats of hydrogenation for several alkenes and conjugated dienes. By using these data, we can compare the relative stabilities of conjugated and unconjugated dienes.

Table 20.1 Heats of Hydrogenation of Several Alkenes and Conjugated Dienes

Name	Structural Formula	ΔH^0 kJ (kcal)/mol
1-Butene		−127 (−30.3)
1-Pentene		−126 (−30.1)
cis-2-Butene	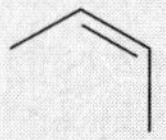	−120 (−28.6)
trans-2-Butene		−115 (−27.6)
1,3-Butadiene		−237 (−56.5)
trans-1,3-Pentadiene	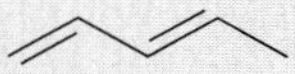	−226 (−54.1)
1,4-Pentadiene		−254 (−60.8)

The simplest conjugated diene is 1,3-butadiene, but because this molecule has only four carbon atoms, it has no unconjugated constitutional isomer. However, we can estimate the effect of conjugation of two double bonds in this molecule in the following way. The heat of hydrogenation of 1-butene is -127 kJ (-30.3 kcal)/mol. A molecule of 1,3-butadiene has two terminal double bonds, each with the same degree of substitution as the one double bond in 1-butene; therefore, we might predict that the heat of hydrogenation of 1,3-butadiene should be $2(-127$ kJ/mol) or -254 kJ (-60.6 kcal)/mol. However, the observed heat of hydrogenation of 1,3-butadiene is -237 kJ (-56.5 kcal)/mol, a value 17 kJ (4.1 kcal)/mol less than estimated.

$$2 \text{ } + 2H_2 \xrightarrow{\text{catalyst}} 2 \text{ } \quad \Delta H^0 = 2(-127 \text{ kJ/mol}) = -254 \text{ kJ/mol}$$

$$\text{ } + 2H_2 \xrightarrow{\text{catalyst}} \text{ } \quad \Delta H^0 = -237 \text{ kJ/mol}$$

Both reactions are exothermic and give the same product; the more stable compound (lower in enthalpy) releases less heat upon hydrogenation. The conclusion is that conjugation of two double bonds in 1,3-butadiene gives an extra stability to the molecule of approximately 17 kJ (4.1 kcal)/mol. These energy relationships are displayed graphically in Figure 20.1.

Calculations of this type for other conjugated and unconjugated dienes give similar results: Conjugated dienes are more stable than isomeric unconjugated dienes by approximately 14.5 to 17 kJ (3.5 to 4.1 kcal)/mol. The effects of conjugation on stability are even more general. Compounds containing conjugated double bonds, not just those in dienes, are more stable than isomeric compounds containing unconjugated double bonds. For example, 2-cyclohexenone is more stable than its isomer 3-cyclohexenone.

2-Cyclohexenone (more stable) 3-Cyclohexenone (less stable)

Figure 20.1

Conjugation of double bonds in butadiene gives the molecule an additional stability of approximately 17 kJ (4.1 kcal)/mol.

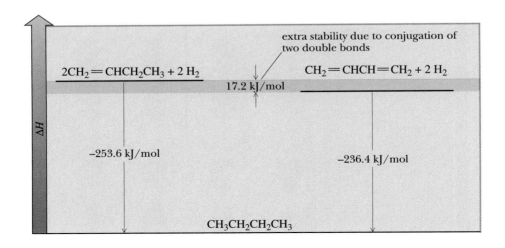

The additional stability of conjugated dienes relative to unconjugated dienes arises from delocalization of electron density in the conjugated diene. In two unconjugated double bonds, each pair of pi electrons is localized between two carbons. In a conjugated diene, however, the four pi electrons are delocalized over the set of four parallel $2p$ orbitals. As we have seen many times before, delocalization leads to increased stability.

According to the molecular orbital model, the conjugated system of a diene is described as a set of four pi molecular orbitals arising from combination of four $2p$ atomic orbitals. The key idea here is that in conjugated systems, the adjacent $2p$ orbitals overlap in space, even between the p orbitals on C2 and C3 in butadiene. As a result, they all combine to produce pi molecular orbitals that cover all the atoms of the conjugated system, in this case the four carbon atoms. These MOs have zero, one, two, and three nodes, respectively, as illustrated in Figure 20.2. In the ground state, all four pi electrons lie in pi-bonding MOs. Because the lowest two MOs are at lower energies than that of an isolated pi bond, the net heat given off by filling these orbitals is more than would be the case for two isolated

Figure 20.2

Structure of 1,3-butadiene— molecular orbital model. Combination of four parallel $2p$ atomic orbitals gives two pi-bonding MOs and two pi-antibonding MOs. In the ground state, each pi-bonding MO is filled with two spin-paired electrons. The pi-antibonding MOs are unoccupied.

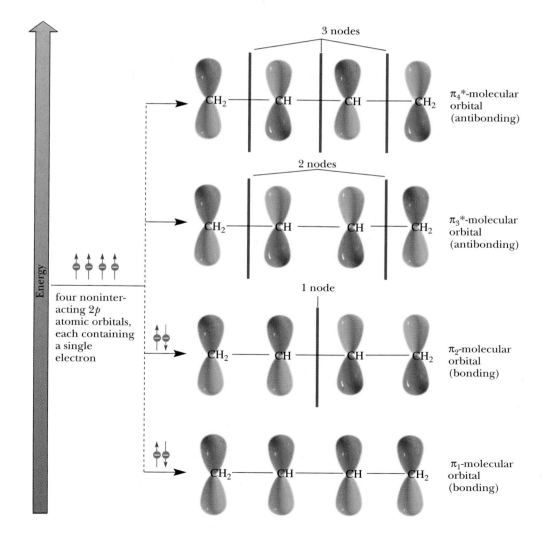

pi bonds. Note that the electrons in these filled MOs are delocalized over the entire pi orbital system. This pi electron delocalization is the hallmark of conjugated systems and can be used to explain the spectroscopy and reactivity of conjugated molecules.

Example 20.2

Using data from Table 20.1, estimate the extra stability that results from the conjugation of double bonds in *trans*-1,3-pentadiene.

Solution

Compare the sum of heats of hydrogenation of 1-pentene and *trans*-2-butene with the heat of hydrogenation of *trans*-1,3-pentadiene. Conjugation of double bonds in *trans*-1,3-pentadiene imparts an added stability of approximately 15 kJ (3.6 kcal)/mol.

Problem 20.2

Estimate the stabilization gained as a result of conjugation when 1,4-pentadiene is converted to *trans*-1,3-pentadiene. Note that the answer is not as simple as comparing the heats of hydrogenation of 1,4-pentadiene and *trans*-1,3-pentadiene. Although the double bonds are moved from unconjugated to conjugated, the degree of substitution of one of the double bonds is also changed, in this case from a monosubstituted double bond to a *trans* disubstituted double bond. To answer this question, you must separate the effect that is the result of conjugation from that caused by a change in the degree of substitution.

20.2 Electrophilic Addition to Conjugated Dienes

Conjugated dienes undergo two-step electrophilic addition reactions just like simple alkenes (Section 6.3). However, certain features are unique to the reactions of conjugated dienes.

A. 1,2-Addition and 1,4-Addition

Addition of one mole of HBr to 1,3-butadiene at $-78°C$ gives a mixture of two constitutional isomers, 3-bromo-1-butene and 1-bromo-2-butene.

$$CH_2=CH-CH=CH_2 + HBr \xrightarrow{-78°C} CH_2=CH-\underset{\underset{H}{|}}{\overset{\overset{Br}{|}}{C}}H-CH_2 + CH_2-CH=CH-\underset{\underset{H}{|}}{CH_2}$$

1,3-Butadiene   3-Bromo-1-butene   1-Bromo-2-butene

(90%)   (10%)

(1,2-addition)   (1,4-addition)

The designations "1,2-" and "1,4-" used here to describe additions to conjugated dienes and other systems do not refer to IUPAC nomenclature. Rather, they refer to the four-atom system of two conjugated double bonds and indicate that addition takes place at either carbons 1 and 2 or carbons 1 and 4 of the four-atom system. For example, the Michael reaction (Section 19.7A) is referred to as a 1,4-addition. The bromobutenes formed by addition of one mole of HBr to butadiene can in turn undergo addition of a second mole of HBr to give a mixture of dibromobutanes. Our concern at this point is only with the products of the first reaction.

Addition of one mole of Br_2 at −15°C also gives a mixture of 1,2-addition and 1,4-addition products.

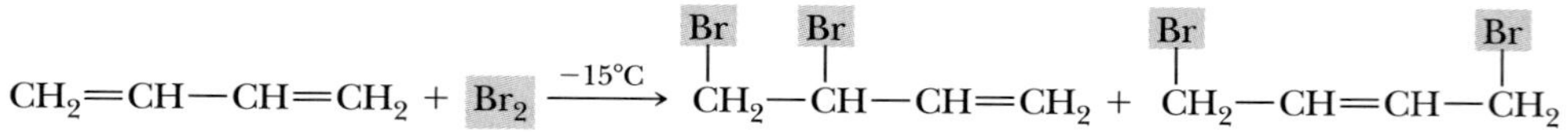

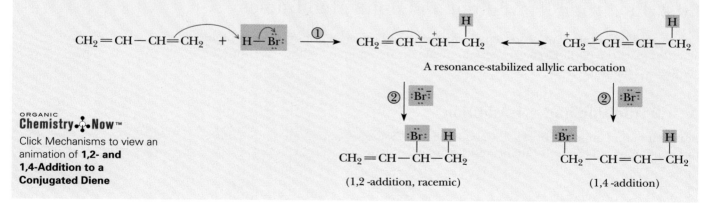

Br
Br
Br
Br
$CH_2{=}CH{-}CH{=}CH_2 + Br_2 \xrightarrow{-15°C} CH_2{-}CH{-}CH{=}CH_2 + CH_2{-}CH{=}CH{-}CH_2$

1,3-Butadiene — 3,4-Dibromo-1-butene (54%, racemic) (1,2-addition) — 1,4-Dibromo-2-butene (46%) (1,4-addition)

We can account for the formation of isomeric products in the addition of HBr and Br_2 by the following mechanism.

Mechanism *1,2- and 1,4-Addition to a Conjugated Diene*

Step 1: Electrophilic addition is initiated by reaction of a terminal carbon of one of the double bonds with HBr to give an allylic carbocation intermediate (Section 9.4E), which can best be represented as a resonance hybrid of two contributing structures. Formation of this very stable cation is the rate-determining step.

Step 2: Reaction of bromide at one of the carbons bearing partial positive charge gives the 1,2-addition product; reaction at the other gives the 1,4-addition product.

A resonance-stabilized allylic carbocation
(1,2 -addition, racemic)
(1,4 -addition)

ORGANIC **Chemistry•Now™**

Click Mechanisms to view an animation of **1,2- and 1,4-Addition to a Conjugated Diene**

Example 20.3

Addition of one mole of HBr to 2,4-hexadiene gives a mixture of 4-bromo-2-hexene and 2-bromo-3-hexene. No 5-bromo-2-hexene is formed. Account for the formation of the first two bromoalkenes and for the fact that the third bromoalkene is not formed.

Solution

2,4-Hexadiene is a conjugated diene, and you can expect products from both 1,2-addition and 1,4-addition. Reaction of the diene with HBr at C2 of the diene in Step 1, the rate-determining step, gives a resonance-stabilized 2° allylic carbocation intermediate. Reaction of this intermediate in Step 2 at one of the carbons bearing a partial positive charge gives 4-bromo-2-hexene, a 1,2-addition product; reaction at the other gives 2-bromo-3-hexene, a 1,4-addition product.

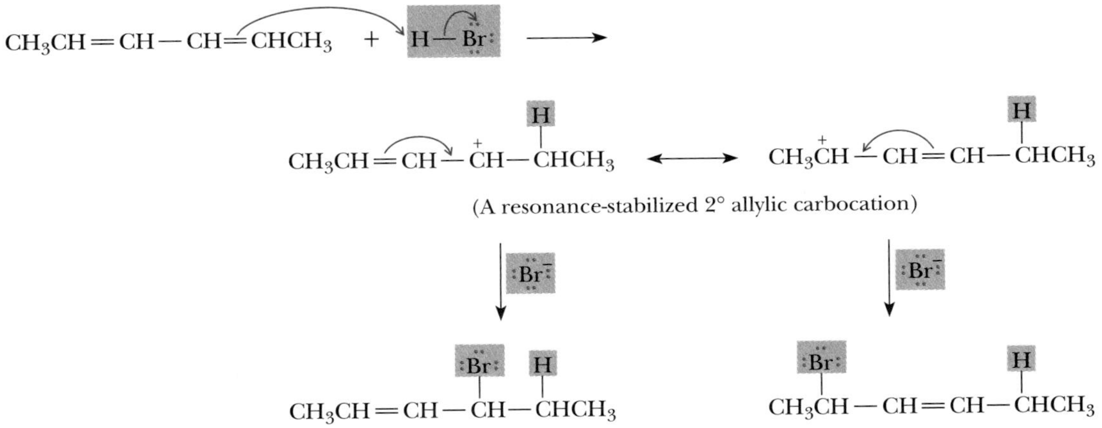

$$CH_3CH=CH-CH=CHCH_3 \quad + \quad H-Br: \longrightarrow$$

$$CH_3CH=CH-\overset{+}{C}H-CHCH_3 \quad \longleftrightarrow \quad CH_3\overset{+}{C}H-CH=CH-CHCH_3$$

(A resonance-stabilized 2° allylic carbocation)

| :Br⁻ | :Br⁻ |

$$CH_3CH=CH-\overset{Br}{\underset{|}{C}}H-CHCH_3 \qquad CH_3\overset{Br}{\underset{|}{C}}H-CH=CH-CHCH_3$$

4-Bromo-2-hexene   2-Bromo-3-hexene
(1,2-addition, racemic)   (1,4-addition, racemic)

Formation of 5-bromo-2-hexene requires reaction of the diene with HBr to give a secondary, nonallylic carbocation by protonation at C3. The activation energy for formation of this less stable 2° carbocation is considerably greater than that for formation of the resonance-stabilized 2° allylic carbocation; therefore, formation of this carbocation and the resulting 5-bromo-2-hexene does not compete effectively with formation of the observed products.

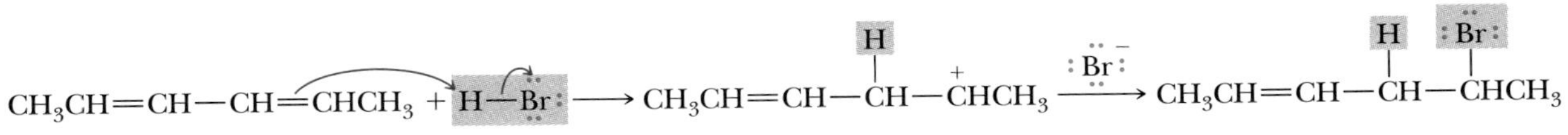

$$CH_3CH=CH-CH=CHCH_3 + H-Br: \longrightarrow CH_3CH=CH-CH-\overset{+}{C}HCH_3 \overset{:Br:^-}{\longrightarrow} CH_3CH=CH-\overset{H}{\underset{|}{C}H}-\overset{:Br:}{\underset{|}{C}HCH_3}$$

(2° carbocation, but not allylic)   5-Bromo-2-hexene
(not formed)

Problem 20.3

Predict the product(s) formed by addition of one mole of Br_2 to 2,4-hexadiene.

B. Kinetic Versus Thermodynamic Control of Electrophilic Addition

We saw in the previous section that electrophilic addition to conjugated dienes gives a mixture of 1,2-addition and 1,4-addition products. Following are some additional experimental observations about the products of electrophilic additions to 1,3-butadiene.

1. For addition of HBr at $-78°C$ and addition of Br_2 at $-15°C$, the 1,2-addition products predominate over the 1,4-addition product. Generally at lower temperatures, the 1,2-addition products predominate over 1,4-addition products.
2. For addition of HBr and Br_2 at higher temperatures (generally 40–60°C), the 1,4-addition products predominate.
3. If the products of low-temperature addition are allowed to remain in solution and then are warmed to a higher temperature, the composition of the product changes over time and becomes identical to that obtained when the reaction is carried out at higher temperature. The same result can be accomplished at the higher temperature in a far shorter time by adding a Lewis acid catalyst, such as $FeCl_3$ or $ZnCl_2$, to the mixture of low-temperature addition products. Thus, under these higher temperature conditions, an equilibrium is established between 1,2- and 1,4-addition products in which 1,4-addition products predominate.
4. If either the pure 1,2- or pure 1,4-addition product is dissolved in an inert solvent at the higher temperature and a Lewis acid catalyst is added, an equilibrium mixture of 1,2- and 1,4-addition products form. The same equilibrium mixture is obtained regardless of which isomer is used as the starting material.

Chemists interpret these experimental results using the twin concepts of kinetic control and equilibrium control of reactions (see Section 19.7A).

To review briefly, for reactions under **kinetic (rate) control,** the distribution of products is determined by the relative rates of formation of each. We see the operation of kinetic control in the following way. At lower temperatures, the reaction is essentially irreversible, and no equilibrium is established between 1,2- and 1,4-addition products. The 1,2-addition product predominates under these conditions because the rate of 1,2-addition is greater than that of 1,4-addition.

ORGANIC Chemistry Now™
Click Simulations to explore **Kinetic Versus Thermodynamic Control of Chemical Reactions**

For reactions under **thermodynamic (equilibrium) control,** the distribution of products is determined by the relative stability of each. We see the operation of thermodynamic control in the following way. At higher temperatures, the reaction is reversible, and an equilibrium is established between 1,2- and 1,4-addition products. The percentage of each product present at equilibrium is in direct relation to the relative thermodynamic stability of that product. The fact that the 1,4-addition product predominates at equilibrium means that it is thermodynamically more stable than the 1,2-addition product.

Relationships between kinetic and thermodynamic control for electrophilic addition of HBr to 1,3-butadiene are illustrated graphically in Figure 20.3. The structure shown in the Gibbs free energy well in the center of Figure 20.3 is the resonance-stabilized allylic cation intermediate formed by proton transfer from HBr to C1 of 1,3-butadiene. The dashed lines in this intermediate show the partial double bond character between C2 and C3 and between C3 and C4 in the resonance hybrid. To the left of this intermediate is the activation energy for its reaction with bromide ion to form the less stable 1,2-addition product; to the right is the activation energy for its reaction with bromide ion to form the more stable 1,4-addition product. As shown in Figure 20.3, the activation energy for 1,2-addition is less than that for 1,4-addition, and the 1,2-addition product is therefore favored under kinetic control. The 1,4-addition product is the more stable and is favored when the reaction is under thermodynamic control.

To complete our discussion of electrophilic addition to conjugated dienes and of kinetic versus thermodynamic control, we need to ask the following questions.

1. Why is the 1,2-addition product (the less stable product) formed more rapidly at lower temperatures? First, we need to look at the resonance-stabilized allylic carbocation intermediate and determine which Lewis structure makes the greater

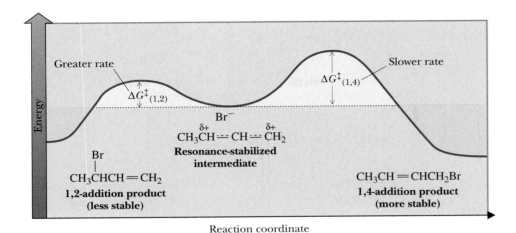

Figure 20.3
Kinetic versus thermodynamic control. A plot of Gibbs free energy versus reaction coordinate for Step 2 in the electrophilic addition of HBr to 1,3-butadiene. The resonance-stabilized allylic carbocation intermediate reacts with bromide ion by way of the transition state on the left to give the 1,2-addition product. It reacts with bromide ion by way of the alternative transition state on the right to give the 1,4-addition product.

contribution to the hybrid. We must consider the degree of substitution of both the positive carbon and the carbon-carbon double bond in each contributing structure.

$$CH_2{=}CH{-}\overset{+}{C}H{-}CH_3 \quad \longleftrightarrow \quad \overset{+}{C}H_2{-}CH{=}CH{-}CH_3$$

Less substituted double bond   **More substituted double bond**
Secondary carbocation   Primary carbocation

A secondary carbocation is more stable than a primary carbocation. If the degree of substitution of the carbon bearing the positive charge were the more important factor, the Lewis structure on the left would make the greater contribution to the hybrid. However, a more substituted double bond is more stable than a less substituted double bond (Section 6.6B). If the degree of substitution of the carbon-carbon double bond were the more important factor, the Lewis structure on the right would make the greater contribution to the hybrid.

We know from other experimental evidence that the location of the positive charge in the allylic carbocation is more important than the location of the double bond. Therefore, in the hybrid, the greater fraction of positive charge is on the secondary carbon. Reaction with bromide ion occurs more rapidly at this carbon, giving 1,2-addition, simply because it has a greater density of positive charge. The electrostatic potential map shows that the positive charge (blue) is more intense on the secondary carbon.

2. Is the 1,2-addition product also formed more rapidly at higher temperatures, even though the 1,4-addition product predominates under these conditions? The answer is yes. The factors affecting the structure of a resonance-stabilized allylic carbocation intermediate and the reaction of this intermediate with a nucleophile are not greatly affected by changes in temperature.

3. Why is the 1,4-addition product the thermodynamically more stable product? The answer to this question has to do with the relative degree of substitution of double bonds. In general, the greater the degree of substitution of a carbon-carbon double bond, the greater the stability of the compound or ion containing it. Following are pairs of 1,2- and 1,4-addition products. In each case, the more stable alkene is the 1,4-addition product.

■ Electrostatic potential map of the allylic carbocation formed by protonating 1,3-butadiene.

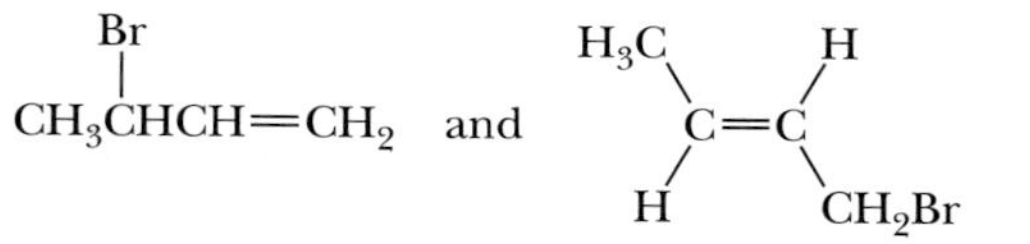

3-Bromo-1-butene
(less stable alkene)

(E)-1-Bromo-2-butene
(more stable alkene)

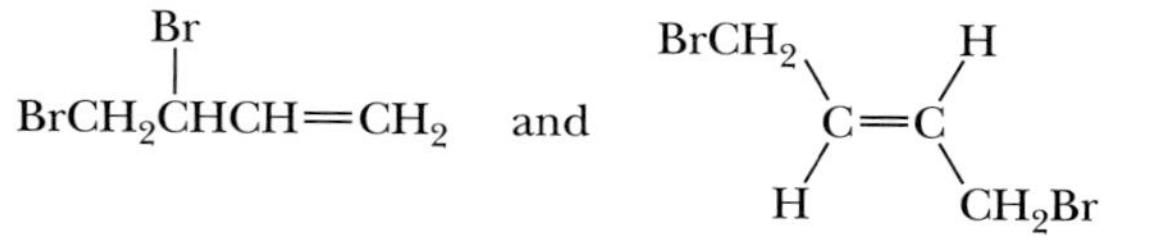

3,4-Dibromo-1-butene
(less stable alkene)

(E)-1,4-Dibromo-2-butene
(more stable alkene)

However, there are cases where the 1,2-addition product is more stable and would be the product of thermodynamic control. For example, addition of bromine to 1,4-dimethyl-1,3-cyclohexadiene under conditions of thermodynamic control gives 3,4-dibromo-1,4-dimethylcyclohexene because its trisubstituted double bond is more stable than the disubstituted double bond of the 1,4-addition product.

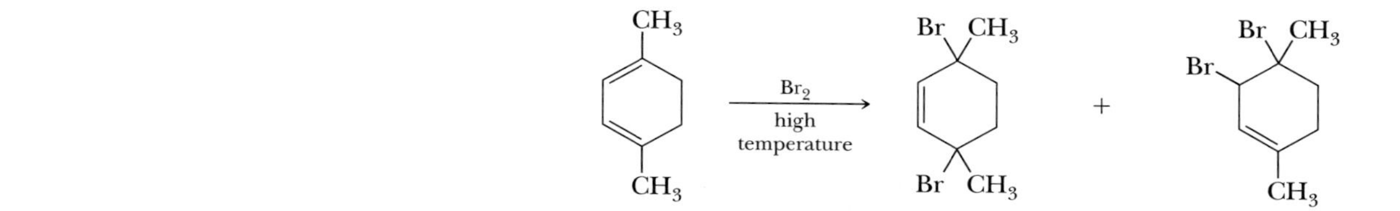

1,4-Addition product
(less stable)

1,2-Addition product
(more stable)

4. What is the mechanism by which the thermodynamically less stable product is converted to the thermodynamically more stable product at higher temperatures? To answer this question, we must look at the relationships between kinetic energy, potential energy, and activation energy. On collision, a part of the kinetic energy (the energy of motion) is transformed into potential energy. If the increase in potential energy is equal to or greater than the activation energy for reaction, then reaction may occur. At the higher temperatures for electrophilic addition of HBr and Br_2 to conjugated dienes, collisions are sufficiently energetic that ionization of the 1,2-addition product occurs to re-form the resonance-stabilized allylic carbocation intermediate. It then reacts with bromide ion again to form the thermodynamically more stable 1,4-addition product. At lower temperatures, however, the increase in potential energy on collision is not sufficient to overcome the potential energy barrier to bring about this ionization.

5. Is it a general rule that, where two or more products are formed from a common intermediate, the thermodynamically less stable product is formed at a greater rate? The answer is no. Whether the thermodynamically more or less stable product is formed at a greater rate from a common intermediate depends very much on the particular reaction and the reaction conditions.

20.3 UV-Visible Spectroscopy

An important property of conjugated systems is that they absorb energy in the ultraviolet-visible region of the spectrum as a result of electronic transitions (Table 12.1). In this section, we study the information this absorption gives us about the conjugation of carbon-carbon and carbon-oxygen double bonds and their substitution.

A. Introduction

The region of the electromagnetic spectrum covered by most ultraviolet spectrophotometers is from 200 to 400 nm, a region commonly referred to as the **near ultraviolet.** Wavelengths shorter than 200 nm require special instrumentation and are not used routinely. The region covered by most visible spectrophotometers runs from 400 nm (violet) to 700 nm (red), with extensions into the (near) IR region to 800 or 1000 nm available on many instruments.

Example 20.4

Calculate the energy of radiation at either end of the near-ultraviolet spectrum, that is, at 200 nm and 400 nm (review Section 12.1).

Solution

Use the relationship $E = hc/\lambda$. Be certain to express the dimension of length in consistent units.

$$E = \frac{hc}{\lambda} = 3.99 \times 10^{-13} \frac{\text{kJ} \times \text{s}}{\text{mol}} \times 3.00 \times 10^{8} \frac{\text{m}}{\text{s}} \times \frac{1}{200 \times 10^{-9}\,\text{m}} = 598 \text{ kJ (143 kcal)/mol}$$

By a similar calculation, the energy of radiation of wavelength 400 nm is found to be 299 kJ (71.5 kcal)/mol.

Problem 20.4

Wavelengths in ultraviolet-visible spectroscopy are commonly expressed in nanometers; wavelengths in infrared spectroscopy are sometimes expressed in micrometers. Carry out the following conversions.
(a) 2.5 μm to nanometers   (b) 200 nm to micrometers

Wavelengths and corresponding energies for near-ultraviolet and visible radiation are summarized in Table 20.2.

Table 20.2 Wavelengths and Energies of Near Ultraviolet and Visible Radiation

Region of Spectrum	Wavelength (nm)	Energy	
		kJ/mol	kcal/mol
Near Ultraviolet	200–400	299–598	71.5–143
Visible	400–700	171–299	40.9–71.5

Ultraviolet and visible spectral data are recorded as plots of **absorbance** (**A**) on the vertical axis versus wavelength on the horizontal axis.

$$\text{Absorbance } (A) = \log \frac{I_0}{I}$$

where I_0 is the intensity of radiation incident on the sample, and I is the intensity of the radiation transmitted through the sample. The quantity $(I/I_0) \times 100$ is called **percent transmittance;** many spectrophotometers read in this scale.

Polyenes

Typically, UV-visible spectra consist of a small number of broad absorption bands, sometimes just one. Figure 20.4 is an ultraviolet absorption spectrum of 2,5-dimethyl-2,4-hexadiene. Absorption of ultraviolet radiation by this conjugated diene begins at wavelengths below 200 nm and continues to almost 270 nm, with maximum absorption at 242 nm. This spectrum is reported as a single absorption peak using the notation λ_{max} 242 nm.

The extent of absorption of ultraviolet-visible radiation is proportional to the number of molecules capable of undergoing the observed electronic transition; therefore, ultraviolet-visible spectroscopy can be used for quantitative analysis of samples. The relationship between absorbance, concentration, and length of the sample cell (cuvette), is known as the **Beer-Lambert law.** The proportionality constant in this equation is given the name **molar absorptivity** (ε) or extinction coefficient.

Molar absorptivity (ε) The absorbance of a 1 M solution of a compound.

Absorbance (**A**) A quantitative measure of the extent to which a compound absorbs radiation of a particular wavelength. $A = \log (I_0/I)$ where I_0 is the incident radiation and I is the transmitted radiation.

$$\text{Beer-Lambert Law:} \quad A = \varepsilon c l$$

where A is the **absorbance** (unitless), ε is the molar absorptivity (in moles per liter per centimeter, $M^{-1}cm^{-1}$), c is the concentration of solute (in moles per liter, M), and l is the length of the sample cell, or cuvette (in centimeters, cm).

The molar absorptivity is a characteristic property of a compound and is not affected by its concentration or the length of the light path. Values range from zero to 10^6 $M^{-1}cm^{-1}$. Values above 10^4 $M^{-1}cm^{-1}$ correspond to high-intensity absorptions; values below 10^4 $M^{-1}cm^{-1}$, to low-intensity absorptions. The molar absorptivity of 2,5-dimethyl-2,4-hexadiene, for example, is 13,100 $M^{-1}cm^{-1}$, a high-intensity absorption.

Figure 20.4

Ultraviolet spectrum of 2,5-dimethyl-2,4-hexadiene (in methanol).

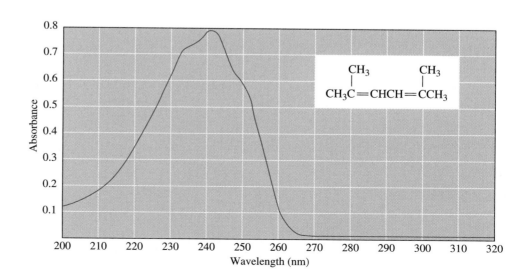

Example 20.5

The molar absorptivity of 2,5-dimethyl-2,4-hexadiene in methanol is 13,100 $M^{-1}cm^{-1}$. What concentration of this diene in methanol is required to give an absorbance of 1.6? Assume a light path of 1.00 cm. Calculate concentration in these units.

(a) Moles per liter   (b) Milligrams per milliliter

Solution

Solve the Beer-Lambert equation for concentration and substitute appropriate values for length, absorbance, and molar absorptivity.

(a) $c = \dfrac{A}{l \times \varepsilon} = \dfrac{1.6}{1.00 \text{ cm} \times 13{,}100 \text{ L} \cdot \text{mol}^{-1} \cdot \text{cm}^{-1}} = 1.22 \times 10^{-4} \text{ mol/L}$

(b) The molecular weight of 2,5-dimethyl-2,4-hexadiene is 110 g/mol. The concentration of the sample in milligrams per milliliter is

$$1.22 \times 10^{-4} \frac{\text{mol}}{\text{L}} \times \frac{110 \text{ g}}{\text{mol}} \times \frac{1 \text{ L}}{1000 \text{ mL}} \times \frac{1000 \text{ mg}}{\text{g}} = 1.34 \times 10^{-2} \text{ mg/mL}$$

Problem 20.5

The visible spectrum of β-carotene ($C_{40}H_{56}$, MW 536.89, the orange pigment in carrots) dissolved in hexane shows intense absorption maxima at 463 nm and 494 nm, both in the blue-green region. Because light of these wavelengths is absorbed by β-carotene, we perceive the color of this compound as that of the complement to blue-green, namely, red-orange.

β-Carotene
λ_{max} 463 nm (log ε 5.10); 494 nm (log ε 4.77)

Calculate the concentration in milligrams per milliliter of β-carotene that gives an absorbance of 1.8 at 463 nm.

Carbonyls

Simple aldehydes and ketones show only weak absorption in the ultraviolet region of the spectrum owing to an n to π^* electronic transition of the carbonyl group. If, however, the carbonyl group is conjugated with one or more carbon-carbon double bonds, intense absorption ($\varepsilon = 8{,}000 - 20{,}000$ $M^{-1}cm^{-1}$) occurs as a result of a π to π^* transition; as with polyenes, the position of absorption is shifted to longer wavelengths, and the molar absorptivity, ε, of the absorption maximum increases sharply. For the α,β-unsaturated ketone 3-penten-2-one, for example, λ_{max} is 224 nm (log ε 4.10).

2-Pentanone
λ_{max} 180 nm (ε 900)

3-Penten-2-one
λ_{max} 224 nm (ε 12,590)

Acetophenone
λ_{max} 246 nm (ε 9,800)

CHEMICAL CONNECTIONS

Curry and Cancer

Curcumin is a natural dye from the root of *Curcuma longa* L. In pure form, it is an orange-yellow crystalline powder that is isolated from the spice turmeric, one of the major ingredients of curry. Its color is a result of the highly conjugated system in curcumin (it is probable that the molecule is actually enolized as shown). It has been known for some time that curcumin retards the growth of new cancers by inhibiting the formation of blood vessels that are necessary for the cancers to grow (angiogenesis). Recently, Korean biochemists have shown that curcumin acts by inhibiting an enzyme that is important to angiogenesis. So curry may be good for you!

Curcumin
(Natural Yellow-3, Turmeric yellow)

The greater the extent of conjugation of unsaturated systems with the carbonyl group, the more the absorption maximum is shifted toward the visible region of the spectrum.

Like the carbonyl groups of simple aldehydes and ketones, the carboxyl group shows only weak absorption in the ultraviolet spectrum unless it is conjugated with a carbon-carbon double bond or an aromatic ring.

B. The Origin of Transitions Between Electronic Energy Levels

ORGANIC
Chemistry Now™

Click Simulations to explore the relationship between **Molecular Orbital Energies and UV-Visible Spectroscopy**

Absorption of radiation in the ultraviolet-visible spectrum results in promotion of electrons from a lower energy, occupied MO to a higher energy, unoccupied MO. The energy of this radiation is generally insufficient to affect electrons in the much lower energy, sigma-bonding molecular orbitals. It is, however, sufficient to cause an electron in a pi orbital to be promoted to an antibonding pi* orbital (called a $\pi \rightarrow \pi^*$ transition), especially an electron in a conjugated pi system. Following are three examples of conjugated systems.

$$CH_2=CH-CH=CH_2 \qquad CH_2=CH-\overset{\overset{\displaystyle O}{\|}}{C}-CH_3$$

1,3-Butadiene   3-Buten-2-one   Benzaldehyde

As an example of a $\pi \rightarrow \pi^*$ transition, consider ethylene. The double bond in ethylene consists of one sigma bond formed by combination of sp^2 orbitals and one pi bond formed by combination of $2p$ orbitals. The relative energies of the pi-

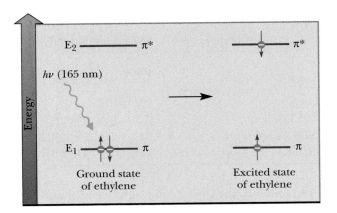

Figure 20.5

A $\pi \rightarrow \pi^*$ transition in excitation of ethylene. Absorption of ultraviolet radiation causes a transition of an electron from a pi-bonding MO in the ground state to a pi-antibonding MO in the excited state. There is no change in electron spin.

bonding and pi-antibonding molecular orbitals are shown schematically in Figure 20.5. The $\pi \rightarrow \pi^*$ transitions for simple, unconjugated alkenes occur below 200 nm (at 165 nm for ethylene). Because these transitions occur at extremely short wavelengths, they are not observed in conventional ultraviolet spectroscopy and, therefore, are not useful to us for determining molecular structure.

For 1,3-butadiene, the difference in energy between the highest occupied pi molecular orbital and the lowest unoccupied pi-antibonding molecular orbital is less than it is for ethylene with the result that a $\pi \rightarrow \pi^*$ transition for 1,3-butadiene (Figure 20.6) takes less energy (occurs at longer wavelength) than that for ethylene. This transition for 1,3-butadiene occurs at 217 nm.

Electronic excitations in molecules are accompanied by changes in vibrational or rotational energy levels. The energy levels for these excitations are considerably smaller than the energy differences between electronic excitations. These transitions are superposed on the electronic excitations, which results in a large number of absorption peaks so closely spaced that the spectrophotometer cannot resolve them. For this reason, UV-visible absorption peaks usually are much broader than IR absorption peaks.

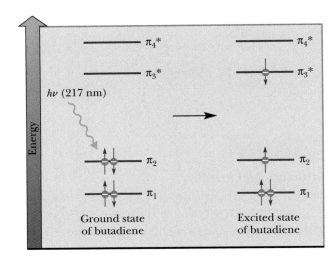

Figure 20.6

Electronic excitation of 1,3-butadiene; a $\pi \rightarrow \pi^*$ transition.

Table 20.3   Wavelengths and Energies Required for $\pi \rightarrow \pi^*$ Transitions of Ethylene and Three Conjugated Polyenes

Name	Structural Formula	λ_{max} (nm)	Energy [kJ (kcal)/mol]
Ethylene	$CH_2{=}CH_2$	165	724 (173)
1,3-Butadiene	$CH_2{=}CHCH{=}CH_2$	217	552 (132)
(3E)-1,3,5-Hexatriene	$CH_2{=}CHCH{=}CHCH{=}CH_2$	268	448 (107)
(3E,5E)-1,3,5,7-Octatetraene	$CH_2{=}CH(CH{=}CH)_2CH{=}CH_2$	290	385 (92)

In general, the greater the number of double bonds in conjugation, the longer the wavelength of ultraviolet radiation absorbed. Shown in Table 20.3 are wavelengths and energies required for $\pi \rightarrow \pi^*$ transitions in several conjugated alkenes.

Summary

Compounds containing conjugated double bonds are more stable than isomeric compounds containing unconjugated double bonds (Section 20.1). The extra stability of two conjugated double bonds arises because the overlap of four parallel $2p$ orbitals results in delocalization of electron density over the entire pi framework.

Conjugated dienes undergo electrophilic addition to give 1,2-addition **and 1,4-addition** products (Section 20.2A). The intermediate in electrophilic addition to a conjugated diene is a resonance-stabilized **allylic carbocation.**

For a reaction under **kinetic (rate) control** (Section 20.2B), the distribution of products is determined by the relative rates of formation of each. For a reaction under **thermodynamic (equilibrium) control,** the distribution of products is determined by the relative stabilities of each.

Conjugated systems absorb energy in the UV to visible range. In general, the longer the conjugated system is, the longer the wavelength of the absorbed radiation will be.

Key Reactions

ORGANIC
Chemistry-⚡-Now™

Click Flashcards to review the **Key Reactions of Conjugated Systems**

1. Electrophilic Addition to Conjugated Dienes (Section 20.2)
The ratio of 1,2- to 1,4-addition products depends on whether the reaction is under kinetic control or thermodynamic control.

$$CH_2{=}CHCH{=}CH_2 + HBr \longrightarrow \underset{Br}{\overset{|}{CH_3CHCH}}{=}CH_2 + CH_3CH{=}CHCH_2Br$$

Products at $-78°C$ (kinetic control):	90%	10%
Products at $40°C$ (thermodynamic control):	15%	85%

Problems

ORGANIC
Chemistry-⚡-Now™

Assess your understanding of this chapter's topics with additional quizzing and conceptual-based problems at **http://now.brookscole.com/bfi4**

Structure and Stability

20.6   If an electron is added to 1,3-butadiene, into which molecular orbital does it go? If an electron is removed from 1,3-butadiene, from which molecular orbital is it taken?

20.7   Draw all important contributing structures for the following allylic carbocations; then rank the structures in order of relative contributions to each resonance hybrid.

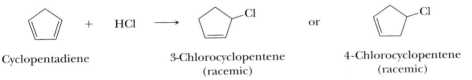

(a)   **(b)** $CH_2{=}CHCH{=}CHCH_2{}^+$ **(c)** $CH_3\overset{CH_3}{\underset{+}{C}}CH{=}CH_2$

Electrophilic Addition to Conjugated Dienes

20.8 Predict the structure of the major product formed by 1,2-addition of HCl to 2-methyl-1,3-butadiene (isoprene).

20.9 Predict the major product formed by 1,4-addition of HCl to isoprene.

20.10 Predict the structure of the major 1,2-addition product formed by reaction of one mole of Br_2 with isoprene. Also predict the structure of the major 1,4-addition product formed under these conditions.

20.11 Which of the two molecules shown do you expect to be the major product formed by 1,2-addition of HCl to cyclopentadiene? Explain.

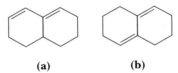

Cyclopentadiene   3-Chlorocyclopentene   4-Chlorocyclopentene
                              (racemic)   (racemic)

20.12 Predict the major product formed by 1,4-addition of HCl to cyclopentadiene.

20.13 Draw structural formulas for the two constitutional isomers of molecular formula $C_5H_6Br_2$ formed by adding one mole of Br_2 to cyclopentadiene.

20.14 What are the expected kinetic and thermodynamic products from addition of one mole of Br_2 to the following dienes?

(a)   **(b)**

Ultraviolet-Visible Spectra

20.15 Show how to distinguish between 1,3-cyclohexadiene and 1,4-cyclohexadiene by ultraviolet spectroscopy.

1,3-Cyclohexadiene   1,4-Cyclohexadiene

20.16 Pyridine exhibits a UV transition of the type $n \rightarrow \pi^*$ at 270 nm. In this transition, one of the unshared electrons on nitrogen is promoted from a nonbonding MO to a pi*-antibonding MO. What is the effect on this UV peak if pyridine is protonated?

Pyridine   Pyridinium ion

20.17 The weight of proteins or nucleic acids in solution is commonly determined by UV spectroscopy using the Beer-Lambert law. For example, the ε of double-stranded DNA at 260 nm is 6670 $M^{-1}cm^{-1}$. The formula weight of the repeating unit in DNA (650 Daltons) can be used as the molecular weight. What is the weight of DNA in 2.0 mL of aqueous buffer if the absorbance, measured in a 1-cm cuvette, is 0.75?

20.18 A sample of adenosine triphosphate (ATP) (MW 507, $\varepsilon = 14{,}700$ $M^{-1}cm^{-1}$ at 257 nm) is dissolved in 5.0 mL of buffer. A 250-μL aliquot is removed and placed in a 1-cm cuvette with sufficient buffer to give a total volume of 2.0 mL. The absorbance of the sample at 257 nm is 1.15. Calculate the weight of ATP in the original 5.0-mL sample.

20.19 The following equilibrium was discussed in Section 20.1.

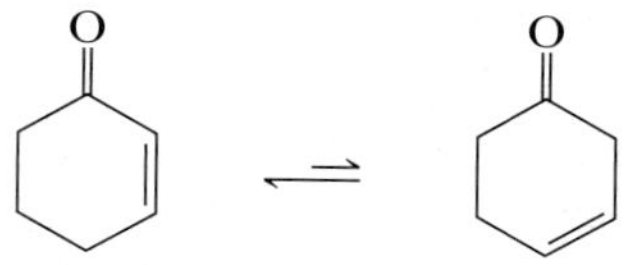

2-Cyclohexenone   3-Cyclohexenone
(more stable)   (less stable)

(a) Give a mechanism for this reaction under either acidic or basic conditions.

(b) Explain the position of the equilibrium.

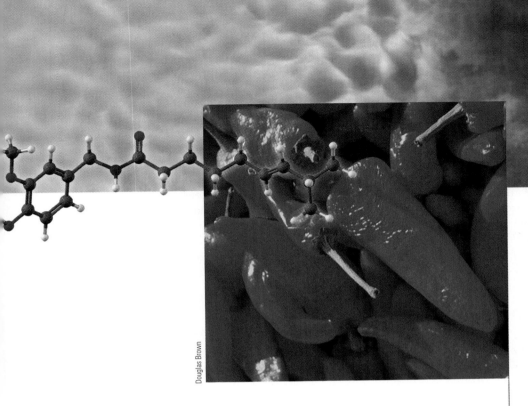

Douglas Brown

Benzene and the Concept of Aromaticity

Benzene, a colorless compound with a melting point of 6°C and a boiling point of 80°C, was first isolated by Michael Faraday in 1825 from the oily residue that collected in the illuminating gas lines of London. Benzene's molecular formula, C_6H_6, suggests a high degree of unsaturation. Compared with an alkane of molecular formula C_6H_{14}, its index of hydrogen deficiency is four, which can be met by an appropriate combination of rings, double bonds, and triple bonds. For example, a compound of molecular formula C_6H_6 might have four double bonds, or three double bonds and one ring, or two double bonds and two rings, or one triple bond and two rings, and so on.

Considering benzene's high degree of unsaturation, it might be expected to show many of the reactions characteristic of alkenes and alkynes. Yet, benzene is remarkably unreactive! It does not undergo the addition, oxidation, and reduction reactions characteristic of alkenes and alkynes. For example, benzene does not react with bromine, hydrogen bromide, or other reagents that usually add to

ORGANIC
Chemistry⋅⚛⋅Now™

Look for this logo in the chapter and go to Organic ChemistryNow at
http://now.brookscole.com/bfi4
for tutorials, simulations, and problems.

carbon-carbon double and triple bonds. It is not oxidized by chromic acid under conditions that readily oxidize alkenes and alkynes. When benzene reacts, it does so by substitution, in which a hydrogen atom is replaced by another atom or group of atoms.

Aromatic compound A term used initially to classify benzene and its derivatives. More accurately, it is used to classify any compound that meets the Hückel criteria for aromaticity (Section 21.2A).

As noted in Chapter 5, the term *aromatic* was originally used to classify benzene and its derivatives because many of them have distinctive odors. The term *aromatic,* as it is now used, refers instead to the fact that these compounds are highly unsaturated and unexpectedly stable toward reagents that attack alkenes and alkynes. The term *arene* is used to describe aromatic hydrocarbons, by analogy with alkane, alkene, and alkyne. Benzene is the parent arene. Just as a group derived by removal of an H from an alkane is called an alkyl group and given the symbol R—, a group derived by removal of an H from an arene is called an **aryl group** and given the symbol **Ar—**.

21.1 The Structure of Benzene

Let us put ourselves in the mid-19th century and examine the evidence on which chemists attempted to build a model for the structure of benzene. First, because the molecular formula of benzene is C_6H_6, it seemed clear that the molecule must be highly unsaturated. Yet, benzene does not show the chemical properties of alkenes, the only unsaturated hydrocarbons known at that time. Benzene does undergo chemical reactions, but its characteristic reaction is substitution rather than addition. When benzene is treated with bromine in the presence of ferric chloride, for example, only one compound of molecular formula C_6H_5Br is formed.

$$C_6H_6 + Br_2 \xrightarrow{FeCl_3} C_6H_5Br + HBr$$

Benzene   Bromobenzene

Chemists concluded, therefore, that all six hydrogens of benzene must be equivalent. When bromobenzene is treated with bromine in the presence of ferric chloride as a catalyst, three isomeric dibromobenzenes are formed.

$$C_6H_5Br + Br_2 \xrightarrow{FeCl_3} C_6H_4Br_2 + HBr$$

Bromobenzene   Three isomeric
                               dibromobenzenes

For chemists in the mid-19th century, the problem was to incorporate these observations, along with the accepted tetravalence of carbon, into a structural formula for benzene. Before we examine these proposals, we should note that the problem of the structure of benzene and other aromatic hydrocarbons has occupied the efforts of chemists for over a century. Only since the 1930s has a general understanding of this problem been realized.

A. Kekulé's Model of Benzene

The first structure for benzene was proposed by August Kekulé in 1865 and consisted of a six-membered ring with one hydrogen bonded to each carbon. Although Kekulé's original structural formula provided for the equivalency of the C—H and C—C bonds, it was inadequate because all the carbon atoms were trivalent. To maintain the tetravalence of carbon, Kekulé proposed in 1872 that the ring contains three double bonds that shift back and forth so rapidly that the two forms cannot be separated. Kekulé regarded this interconversion as an equilibrium, each structure in which has become known as a **Kekulé structure.**

Kekulé structures for benzene

Now, more than 135 years after the time of Kekulé, we are apt to misunderstand what scientists in his time knew and did not know. For example, it is a given to us that covalent bonds consist of one or more pairs of shared electrons. We must remember, however, that it was not until 1897 that J. J. Thomson, professor of physics at the Cavendish Laboratory of Cambridge University, discovered the electron. Thomson was awarded the 1906 Nobel Prize in physics. That the electron played any role in chemical bonding did not become clear for another 30 years. Thus, at the time Kekulé made his proposal for the structure of benzene, the existence of electrons and their role in chemical bonding was completely unknown.

Kekulé's proposal accounted nicely for the fact that bromination of benzene gives only one bromobenzene, and that bromination of bromobenzene gives three isomeric dibromobenzenes.

Bromobenzene   Three isomeric dibromobenzenes

Although his proposal was consistent with many experimental observations, it did not totally solve the problem and was contested for years. The major objection was that it did not account for the unusual chemical behavior of benzene. If benzene contains three double bonds, Kekulé's critics argued, why doesn't it show reactions typical of alkenes? Why, for example, doesn't benzene add three moles of bromine to form 1,2,3,4,5,6-hexabromocyclohexane? We now understand the surprising unreactivity of benzene on the basis of two complimentary descriptions, the molecular orbital model and the resonance model.

B. The Molecular Orbital Model of Benzene

The concepts of hybridization of atomic orbitals and the theory of resonance, developed by Linus Pauling in the 1930s, provided the first adequate description of the structure of benzene. The carbon skeleton of benzene forms a regular hexagon with C—C—C and H—C—C bond angles of 120°. For this type of bonding, carbon uses sp^2 hybrid orbitals. Each carbon forms sigma bonds to two adjacent carbons by overlap of sp^2-sp^2 hybrid orbitals, and one sigma bond to hydrogen by overlap of sp^2-$1s$ orbitals. As determined experimentally, all carbon-carbon bonds are 139 pm in length, a value almost midway between the length of a single bond

Figure 21.1

The molecular orbital representation of the pi bonding in benzene.

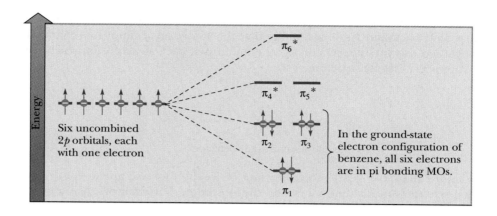

between sp^3 hybridized carbons (154 pm) and a double bond between sp^2 hybridized carbons (133 pm).

Each carbon also has a single unhybridized $2p$ orbital that is perpendicular to the plane of the ring and contains one electron. According to molecular orbital theory, the combination of these six parallel $2p$ atomic orbitals gives a set of six pi MOs, three pi-bonding MOs, and three pi-antibonding MOs. Figure 21.1 shows these six molecular orbitals and their relative energies. Note that π_2 and π_3 MOs are degenerate (they have the same energy) bonding orbitals. Similarly, π^*_4 and π^*_5 are a degenerate pair of pi-antibonding MOs.

In the ground-state electron configuration of benzene, the six electrons of the pi system occupy the three bonding MOs (Figure 21.1). According to molecular orbital calculations, the great stability of benzene results from the fact that these three bonding MOs are much lower in energy when compared with the six uncombined $2p$ atomic orbitals.

It is common to represent the pi system of benzene as one torus (a donut-shaped region) above the plane of the ring and a second torus below it, as shown in Figure 21.2.

ORGANIC
Chemistry ⚛ Now™

Click Simulations to explore the relationship between **Structure and Molecular Orbitals in Aromatic Systems**

Figure 21.2

The pi system of benzene. (a) The carbon-hydrogen framework. The six $2p$ orbitals, each with one electron, are shown uncombined. (b) Overlap of parallel $2p$ orbitals forms a continuous pi cloud, shown by one torus above the plane of the ring and a second torus below the plane of the ring.

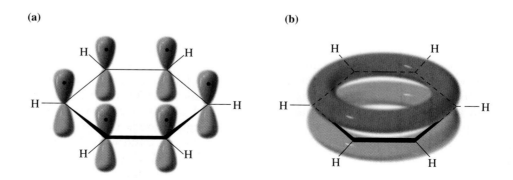

Although this picture is useful in thinking about the electron density of the pi system, you must use it with caution because it represents only the lowest-lying pi-bonding molecular orbital.

C. The Resonance Model of Benzene

One of the postulates of resonance theory is that, when a molecule or ion can be represented by two or more contributing structures, it is not adequately represented by any single contributing structure. We represent benzene as a hybrid of two equivalent contributing structures, often referred to as Kekulé structures.

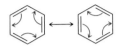

Benzene as a hybrid of two equivalent
contributing structures

Each Kekulé structure makes an equal contribution to the hybrid and thus the C—C bonds are neither single nor double bonds but something intermediate. We recognize that neither of these contributing structures exists (they are merely alternative ways to pair $2p$ orbitals with no reason to prefer one or the other) and that the actual structure is a superposition of both. Nevertheless, chemists continue to use a single contributing structure to represent this molecule because it is as close as we can come to an accurate structure within the limitations of classical valence bond structures and the tetravalence of carbon.

Resonance energy The difference in energy between a resonance hybrid and the most stable of its hypothetical contributing structures in which electrons are localized on particular atoms and in particular bonds.

One way to estimate the **resonance energy** of benzene is to compare the heats of hydrogenation of cyclohexene and benzene. Cyclohexene is readily reduced to cyclohexane by hydrogen in the presence of a transition metal catalyst (Section 6.6A).

$$\text{Cyclohexene} + H_2 \xrightarrow[\text{1–2 atm}]{\text{Ni}} \text{Cyclohexane} \qquad \Delta H^0 = -119.7 \text{ kJ } (-28.6 \text{ kcal})/\text{mol}$$

Benzene is reduced very slowly under these conditions to cyclohexane. It is reduced more rapidly when heated and under a pressure of several hundred atmospheres of hydrogen.

$$\text{Benzene} + 3H_2 \xrightarrow[\text{200–300 atm}]{\text{Ni}} \text{Cyclohexane} \qquad \Delta H^0 = -208 \text{ kJ } (-49.8 \text{ kcal})/\text{mol}$$

Catalytic hydrogenation of an alkene is an exothermic reaction (Section 6.6B). The heat of hydrogenation per double bond varies somewhat with the degree of substitution of the particular alkene; for cyclohexene, $\Delta H^0 = -119.7$ kJ (-28.6 kcal)/mol. If we consider benzene to be 1,3,5-cyclohexatriene, a hypothetical unsaturated compound with alternating single and double bonds, we calculate that $\Delta H^0 = 3(-119.7$ kJ/mol) $= -359$ kJ (-85.8 kcal)/mol. The ΔH^0 for reduction of benzene to cyclohexane is -208 kJ (-49.8 kcal)/mol, considerably less than that calculated for 1,3,5-cyclohexatriene. The difference between these values, 151 kJ (36.0 kcal)/mol, is the **resonance energy of benzene.** Note that the product of both reductions is cyclohexane

Figure 21.3

The resonance energy of benzene as determined by comparison of the heats of hydrogenation of cyclohexene, benzene, and the hypothetical compound 1,3,5-cyclohexatriene.

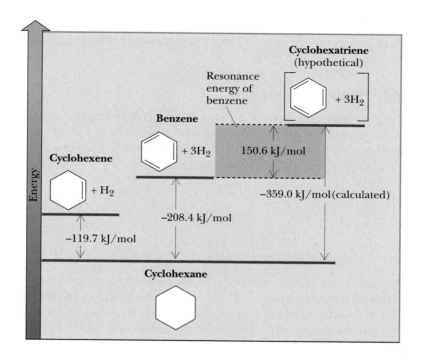

and that both reductions are exothermic. Therefore, the lower heat of hydrogenation for benzene confirms that it is more stable than 1,3,5-cyclohexatriene. These experimental results are shown graphically in Figure 21.3.

Several other experimental determinations of the resonance energy of benzene have been performed using different model compounds, and although these determinations differ somewhat in their results, they all agree that the resonance stabilization of benzene is large. Following are resonance energies for several other aromatic hydrocarbons.

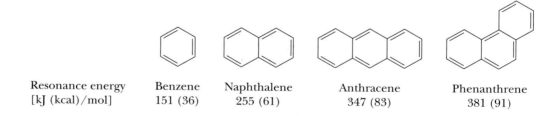

| Resonance energy [kJ (kcal)/mol] | Benzene 151 (36) | Naphthalene 255 (61) | Anthracene 347 (83) | Phenanthrene 381 (91) |

21.2 The Concept of Aromaticity

The molecular orbital and resonance theories are powerful tools with which chemists can understand the unusual stability of benzene and its derivatives. According to resonance theory, benzene is best represented as a hybrid of two equivalent contributing structures. By analogy, cyclobutadiene and cyclooctatetraene can also be represented as hybrids of two equivalent contributing structures. Is either of these compounds aromatic?

Cyclobutadiene as a
hybrid of two equivalent
contributing structures

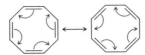

Cyclooctatetraene as a hybrid of
two equivalent contributing structures

The answer for both compounds is no. Repeated attempts to isolate cyclobutadiene have all failed. It was not until 1965 that it was finally synthesized, and, even then, it could only be detected if trapped at 4 K ($-269°C$). Cyclobutadiene is a highly unstable compound and does not show any of the chemical and physical properties we associate with aromatic compounds. Cyclooctatetraene has chemical properties typical of alkenes. It reacts readily with halogens and halogen acids, as well as with mild oxidizing and reducing agents.

We are then faced with the broad question: "What are the fundamental principles underlying aromatic character?" In other words, what are the structural characteristics of unsaturated compounds that have a large resonance energy and do not undergo reactions typical of alkenes but rather undergo substitution reactions?

A. The Hückel Criteria for Aromaticity

The underlying criteria for aromaticity were recognized in the early 1930s by Erich Hückel, a German chemical physicist. He carried out MO energy calculations for monocyclic, planar molecules in which each atom of the ring has one $2p$ orbital available for forming sets of molecular orbitals. His calculations demonstrated that monocyclic, planar molecules with a closed loop of 2, 6, 10, 14, 18, ... pi electrons in a fully conjugated system should be aromatic. These numbers are generalized in the **($4n + 2$) pi electron rule,** where n is a positive integer (0, 1, 2, 3, 4, ...). Conversely, monocyclic, planar molecules with $4n$ pi electrons (4, 8, 12, 16, 20, ...) are especially unstable and are said to be antiaromatic. We will have more to say about antiaromaticity shortly. Hückel's **criteria for aromaticity** are summarized as follows. To be aromatic, a compound must:

1. Be cyclic.
2. Have one p orbital on each atom of the ring.
3. Be planar or nearly planar so that there is continuous or nearly continuous overlap of all p orbitals of the ring.
4. Have a closed loop of ($4n + 2$) pi electrons in the cyclic arrangement of p orbitals.

To appreciate the reasons for aromaticity and antiaromaticity, we must examine MO energy diagrams for the molecules and ions we will consider in this and the following section. The relative energies of the pi MOs for planar, monocyclic, fully conjugated systems can be constructed quite easily using the **Frost circle,** or inscribed polygon method. To construct such a diagram, draw a circle and then inscribe in it a polygon of the same number of sides as the ring in question. Inscribe the polygon in such a way that one of its vertices is at the bottom of the circle. The relative energies of the MOs in the ring are then given by the points where the vertices touch the circle. Those MOs below the horizontal line through the center of the circle are bonding MOs. Those on the horizontal line are nonbonding MOs, and those above the line are antibonding MOs.

Hückel criteria for aromaticity
To be aromatic, a monocyclic compound must have one p orbital on each atom of the ring, be planar or nearly so, and have ($4n + 2$) pi electrons in the cyclic arrangement of p orbitals.

Frost circle A graphic method for determining the relative energies of pi MOs for planar, fully conjugated, monocyclic compounds.

Figure 21.4

Frost circles showing the number and relative energies of the pi MOs for planar, fully conjugated four-, five-, and six-membered rings.

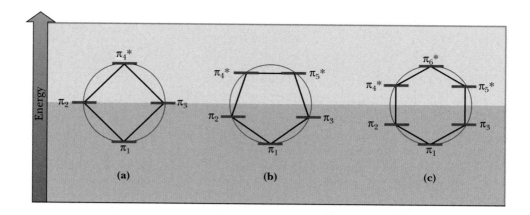

Figure 21.4 shows Frost circles describing the MOs of monocyclic, planar, and fully conjugated four-, five-, and six-membered rings. This apparently magical method works because it reproduces geometrically the mathematical solutions to the wave equation.

Example 21.1

Construct a Frost circle for a planar seven-membered ring with one $2p$ orbital on each atom of the ring, and show the relative energies of its seven pi molecular orbitals. Which are bonding MOs, which are antibonding, and which are nonbonding?

Solution

Of the seven pi molecular orbitals, three are bonding, and four are antibonding.

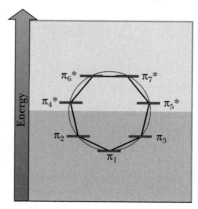

Problem 21.1

Construct a Frost circle for a planar eight-membered ring with one $2p$ orbital on each atom of the ring, and show the relative energies of its eight pi molecular orbitals. Which are bonding MOs, which are antibonding, and which are nonbonding?

B. Aromatic Hydrocarbons

Cyclobutadiene, benzene, and cyclooctatetraene are the first members of a family of molecules called annulenes. An **annulene** is a cyclic hydrocarbon with a continuous alternation of single and double bonds. The name of an annulene is derived by showing the number of atoms in the ring in brackets followed by the word *annulene*. Named as annulenes, cyclobutadiene, benzene, and cyclooctatetraene are [4]annulene, [6]annulene, and [8]annulene, respectively. These compounds, however, are rarely named as annulenes.

Annulene A cyclic hydrocarbon with a continuous alternation of single and double bonds.

Beginning in the 1960s, Franz Sondheimer and his colleagues, first at the Weizmann Institute in Israel and later at the University of London, synthesized a number of larger annulenes, primarily to test the validity of Hückel's criteria for aromaticity. They found, for example, that both [14]annulene and [18]annulene are aromatic, as predicted by Hückel. [18]Annulene has a resonance energy of approximately 418 kJ (100 kcal)/mol. Notice that, for these annulenes to achieve planarity, several of the carbon-carbon double bonds in each must have the *trans* configuration.

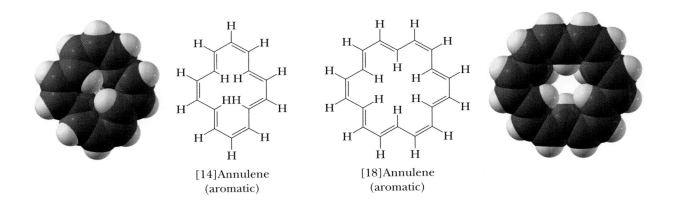

[14]Annulene
(aromatic)

[18]Annulene
(aromatic)

In these larger annulenes, there are two sets of equivalent hydrogens: those that point outward from the ring and those that point inward to the center of the ring. The fact is that these two sets of equivalent hydrogens have quite different ^{1}H-NMR chemical shifts.

The protons on benzene and other arenes are deshielded and appear far downfield (usually around 7–8 ppm) because of the induced ring current that occurs in aromatic molecules (Section 13.7C). The effect of induced ring current is characteristic not only of benzene and its derivatives but also of all compounds that meet the Hückel criteria for aromaticity. This concept of a circulating ring current and of an induced magnetic field predicts that hydrogen atoms on the outside of the ring should come into resonance with a downfield shift. It also predicts that a hydrogen atom in the inside of the ring should come into resonance farther upfield. Of course, no hydrogens are on the inside of the benzene ring, but with larger aromatic annulenes, as for example [18]annulene, there are both "inside" hydrogens and "outside" hydrogens. The degree of the upfield chemical shift of the inside hydrogens of [18]annulene is remarkable. They come into resonance at δ −3.00; that is, at 3.00 δ units upfield (to the right) of the TMS standard.

Local induced magnetic field of circulating pi electrons opposes the applied field inside the ring. The six hydrogens inside the ring come into resonance at a lower frequency; at δ –3.00.

Local induced magnetic field of circulating pi electrons reinforces the applied field outside the ring. The twelve hydrogens outside the ring come into resonance at a higher frequency; at δ 9.3.

[18]Annulene
(aromatic)

Example 21.2

Which hydrogens have a larger chemical shift, the six hydrogens of benzene or the eight hydrogens of cyclooctatetraene? Explain.

Solution

Benzene is an aromatic compound; its six equivalent hydrogens appear as a sharp singlet at δ 7.27. Cyclooctatetraene does not meet the Hückel criteria for aromaticity because it has $4n$ pi electrons and is nonplanar. Therefore, the eight equivalent hydrogens of the cyclooctatetraene ring appear as a singlet at δ 5.8 in the region of vinylic hydrogens (δ 4.6–δ 5.7).

Problem 21.2

Which compound gives a signal in the ¹H—NMR spectrum with a larger chemical shift, furan or cyclopentadiene? Explain.

According to Hückel's criteria, [10]annulene should be aromatic; it is cyclic, has one $2p$ orbital on each carbon of the ring, and has $4(2) + 2 = 10$ electrons in its pi system. It has been found, however, that this molecule shows reactions typical of alkenes and is, therefore, classified as nonaromatic. The reason for its lack of aromaticity lies in the fact that the ten-membered ring is too small to accommodate the two hydrogens that point inward toward the center of the ring. Nonbonded interaction between these two hydrogens forces the ring into a nonplanar conformation in which the overlap of all ten $2p$ orbitals is no longer continuous. Therefore, because [10]annulene is not planar, it is not aromatic.

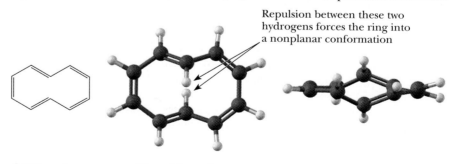

Repulsion between these two hydrogens forces the ring into a nonplanar conformation

[10]Annulene   Viewed from above   Viewed through an edge

What is remarkable is that if the two hydrogen atoms facing inward toward the center of the ring in [10]annulene are replaced by a CH_2 group, the ring is now able to assume a conformation close enough to planar that it becomes aromatic.

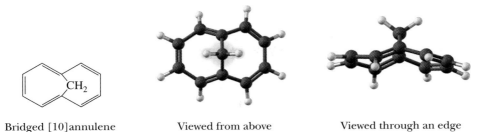

Bridged [10]annulene   Viewed from above   Viewed through an edge

C. Antiaromatic Hydrocarbons

According to the Hückel criteria, monocyclic, planar molecules with $4n$ pi electrons (4, 8, 12, 16, 20, . . .) are especially unstable and are said to be antiaromatic. By these criteria, cyclobutadiene with 4 pi electrons is antiaromatic. Using the Frost circle energy diagram from Figure 21.4, we can construct a molecular orbital energy diagram for cyclobutadiene (Figure 21.5).

In the ground-state electron configuration of cyclobutadiene, two pi electrons fill the π_1-bonding MO. The third and fourth pi electrons are unpaired and lie in the π_2- and π_3-nonbonding MOs. The existence of these two unpaired electrons in planar cyclobutadiene makes this molecule highly unstable and reactive compared to butadiene, a noncyclic molecule containing two conjugated double bonds. It has been found that cyclobutadiene is not planar, but slightly puckered with two shorter bonds and two longer bonds, which makes the two degenerate orbitals no longer equivalent; nevertheless, it retains some apparent diradical character.

Cyclooctatetraene shows reactions typical of alkenes and is classified as nonaromatic. X-ray studies show clearly that the most stable conformation of the molecule is a nonplanar "tub" conformation with two distinct types of carbon-carbon bonds: four longer carbon-carbon single bonds and four shorter carbon-carbon double bonds. The four single bonds are equal in length to the single bonds between sp^2 hybridized carbons (approximately 146 pm), and the four double bonds are equal in length to

Antiaromatic compound A monocyclic compound that is planar or nearly so, has one p orbital on each atom of the ring, and has $4n$ pi electrons in the cyclic arrangement of overlapping p orbitals, where n is an integer. Antiaromatic compounds are especially unstable.

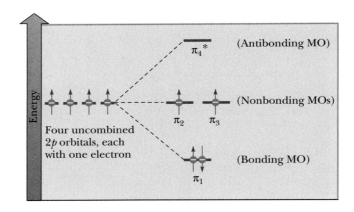

Figure 21.5
Molecular orbital energy diagram for cyclobutadiene. In the ground state, two electrons are in the low-lying π_1-bonding MO. The remaining two electrons are unpaired and occupy the degenerate π_2- and π_3-nonbonding MOs.

double bonds in alkenes (approximately 133 pm). In the tub conformation, the overlap of 2p orbitals on carbons forming double bonds is excellent, but almost no overlap occurs between 2p orbitals at the ends of carbon-carbon single bonds because these 2p orbitals are not parallel. Thus the pi system in cyclooctatetraene is not conjugated despite having continuous sp^2 hybridized carbon atoms.

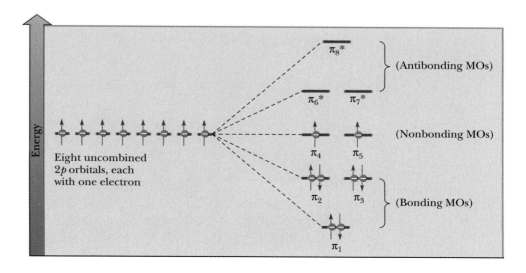

1,3,5,7-Cyclooctatetraene
(tub conformation with alternating
single and double bonds)

Viewed from above

Viewed through
an edge

To appreciate why planar cyclooctatetraene would be classified as antiaromatic, we need to examine the MO energy diagram for an eight-membered ring containing eight pi electrons in a cyclic, fully conjugated ring. You constructed a Frost circle for this ring in answer to Problem 21.1. Note that the most stable conformation of cyclooctatetraene is not planar, but if it were planar, the Frost circle you constructed would be its MO energy diagram. The molecular orbital energy diagram for planar cyclooctatetraene is shown in Figure 21.6. In the ground state, six pi electrons fill the three low-lying π_1-, π_2-, and π_3-bonding MOs. The remaining two pi electrons are unpaired and lie in the degenerate π_4- and π_5-nonbonding MOs. Because of these two unpaired electrons, planar cyclooctatetraene, if it existed, would be classified as antiaromatic. Cyclooctatetraene, however, is large enough to pucker into a nonplanar conformation and become nonaromatic.

If [16]annulene were planar, it too would be antiaromatic. The size of the ring, however, is large enough that it can pucker into a nonplanar conformation in which

Figure 21.6
Molecular orbital energy diagram for a planar conformation of cyclooctatetraene. Three pairs of electrons fill the three low-lying pi-bonding molecular orbitals. Two electrons are unpaired in degenerate pi-nonbonding molecular orbitals.

CHEMICAL CONNECTIONS

Capsaicin, for Those Who Like It Hot

Capsaicin, the pungent principle from the fruit of various species of peppers (from the genus *Capsicum* and the family Solanaceae), was isolated in 1876, and its structure was determined in 1919.

Capsaicin
(from various types of peppers)

The inflammatory properties of capsaicin are well known; as little as one drop in 5 L of water can be detected by the human tongue. We all know of the burning sensation in the mouth and sudden tearing in the eyes caused by a good dose of hot chili peppers. Capsaicin-containing extracts from these flaming foods are also used in sprays to ward off dogs or other animals that might nip at your heels while you are running or cycling.

Ironically, capsaicin is able to cause pain and relieve it as well. Currently, two capsaicin-containing creams, Mioton and Zostrix, are prescribed to treat the burning pain associated with postherpetic neuralgia, a complication of shingles. They are also prescribed for diabetics to relieve persistent foot and leg pain.

The mechanism by which capsaicin relieves these pains is not fully understood. The suggestion has been made that after application, the nerve endings in the area responsible for the transmission of pain remain temporarily numb. Capsaicin remains bound to specific receptor sites on these pain-transmitting neurons, blocking them from further action. Eventually, capsaicin is removed from these receptor sites, but in the meantime, its presence provides needed relief from pain.

Chuck Pefley/Stone/Getty Images

■ Red chili peppers being dried.

B. Acidity of Phenols

Phenols and alcohols both contain a hydroxyl group. Phenols, however, are grouped as a separate class of compounds because their chemical properties are quite different from those of alcohols. One of the most important of these differences is that phenols are significantly more acidic than alcohols. The acid ionization constant of phenol is 10^6 times larger than that of ethanol.

$$C_6H_5{-}OH + H_2O \rightleftharpoons C_6H_5{-}O^- + H_3O^+ \qquad K_a = 1.1 \times 10^{-10} \qquad pK_a = 9.95$$

$$CH_3CH_2OH + H_2O \rightleftharpoons CH_3CH_2O^- + H_3O^+ \qquad K_a = 1.3 \times 10^{-16} \qquad pK_a = 15.9$$

Table 21.1   Relative Acidities of 0.1 M Solutions of Ethanol, Phenol, and HCl

Acid Ionization Equation	[H$^+$]	pH
$CH_3CH_2OH + H_2O \rightleftharpoons CH_3CH_2O^- + H_3O^+$	1×10^{-7}	7.0
$C_6H_5OH + H_2O \rightleftharpoons C_6H_5O^- + H_3O^+$	3.3×10^{-6}	5.4
$HCl + H_2O \rightleftharpoons Cl^- + H_3O^+$	0.1	1.0

Another way to compare the relative acid strengths of ethanol and phenol is to look at the hydrogen ion concentration and pH of a 0.1 M aqueous solution of each (Table 21.1). For comparison, the hydrogen ion concentration and pH of 0.1 M HCl are also included.

In aqueous solution, alcohols are neutral substances, and the hydrogen ion concentration of 0.1 M ethanol is the same as that of pure water. A 0.1 M solution of phenol is slightly acidic and has a pH of 5.4. By contrast, 0.1 M HCl, a strong acid (completely ionized in aqueous solution), has a pH of 1.0.

The greater acidity of phenol is a result of the greater stability of the phenoxide ion compared with an alkoxide ion. The negative charge on the phenoxide ion is delocalized by resonance. The two contributing structures on the left place the negative charge on oxygen. The three contributing structures on the right place it on the ortho and para positions of the ring. Taken together, these contributing structures delocalize the negative charge of the phenoxide ion over four atoms. There is no possibility for delocalization of charge in an alkoxide ion.

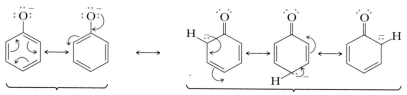

These two Kekulé structures   These three contributing structures delocalize
are equivalent   the negative charge onto carbon atoms of the ring

Note that although the charge-delocalization resonance model gives us a way of understanding why phenol is a stronger acid than ethanol, it does not provide us with any quantitative means of predicting just how much stronger an acid it might be. To find out how much stronger one acid is compared with another, we must determine their pK_a values experimentally and compare them.

Ring substituents, particularly halogens and nitro groups, have marked effects on the acidities of phenols by a combination of inductive and resonance effects. Both m-cresol and p-cresol are weaker acids than phenol itself; m-chlorophenol and p-chlorophenol are stronger acids than phenol.

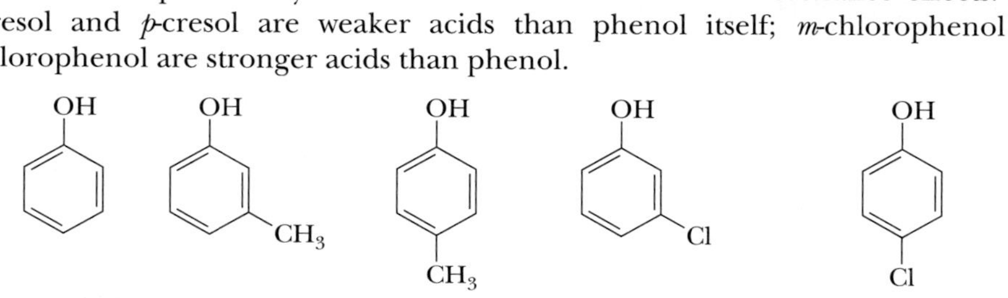

Phenol	m-Cresol	p-Cresol	m-Chlorophenol	p-Chlorophenol
pK_a 9.95	pK_a 10.01	pK_a 10.17	pK_a 8.85	pK_a 9.18

The acid-weakening effect of alkyl-substituted phenols can be understood in the following way. The sp^2-hybridized carbon of an aromatic ring is more electronegative than the sp^3-hybridized atom of an alkyl substituent. Alkyl substituents are electron releasing toward the aromatic ring. Because they are electron releasing, they destabilize phenoxide ion-contributing structures and in effect reduce the acidity of substituted phenols.

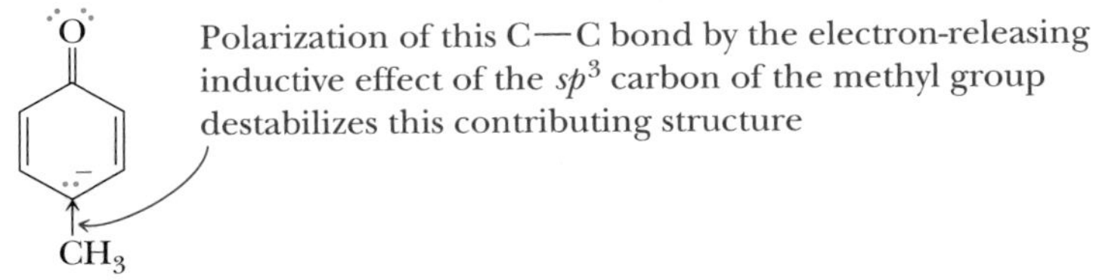

Polarization of this C—C bond by the electron-releasing inductive effect of the sp^3 carbon of the methyl group destabilizes this contributing structure

The inductive effect of the halogens is opposite to that of alkyl substituents. Because the halogens are more electronegative than carbon, they withdraw electron density from the aromatic ring and stabilize the halophenoxide ion compared to phenoxide ion itself. Fluorine, the most electronegative halogen, has the greatest acid-strengthening effect in halophenols; the effect is less for chlorophenols and still less for bromophenols.

We find the operation of both the inductive and resonance effects in the nitrophenols.

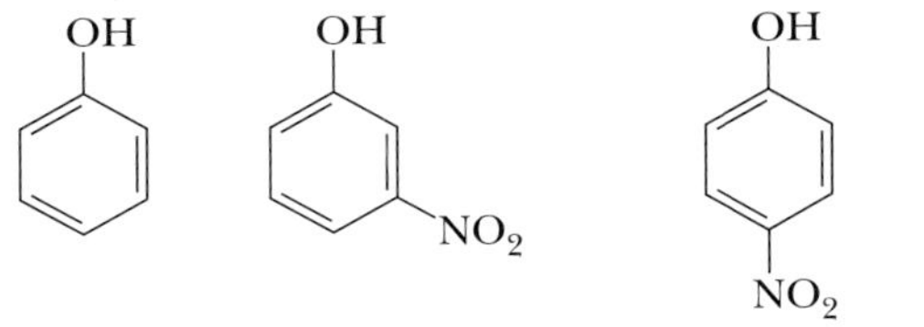

Phenol
pK_a 9.95

m-Nitrophenol
pK_a 8.28

p-Nitrophenol
pK_a 7.15

Both m-nitrophenol and p-nitrophenol are stronger acids than phenol. The acid-strengthening effect of the nitro group is greater in the para position, even though it is farther away from the —OH group. Part of the acid-strengthening property of the nitro group is the result of its electron-withdrawing inductive effect. In addition, nitro substitution in the ortho or para positions increases acidity because the negative charge of the phenoxide ion is delocalized onto an oxygen of the nitro group as shown in the contributing structure on the right.

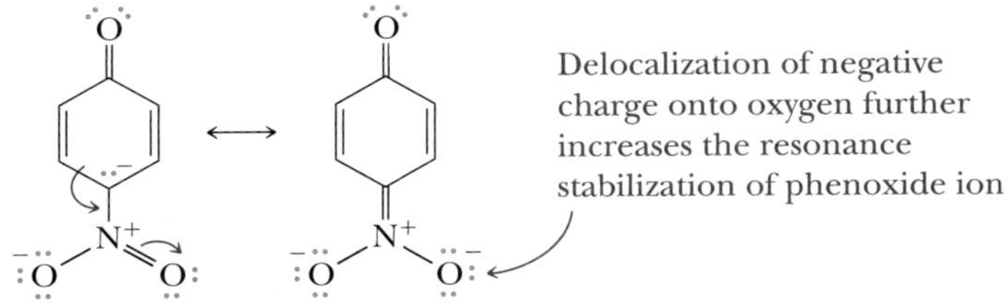

Delocalization of negative charge onto oxygen further increases the resonance stabilization of phenoxide ion

The combined inductive and resonance acid-strengthening effects of the nitro group are such that 2,4,6-trinitrophenol (picric acid) is a stronger acid than phosphoric acid or the hydrogen sulfate ion.

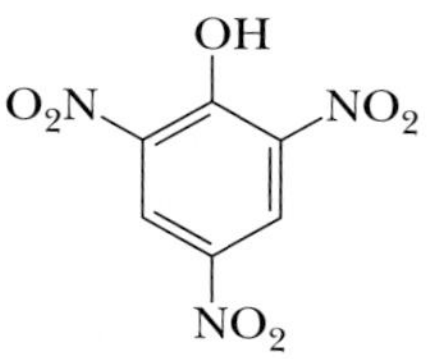

2,4,6-Trinitrophenol (Picric acid) pK_a 0.38	Phosphoric acid pK_a 2.1	Hydrogen sulfate ion pK_a 1.92

Example 21.6

Arrange these compounds in order of increasing acidity: 2,4-dinitrophenol, phenol, and benzyl alcohol.

Solution

Benzyl alcohol, a primary alcohol, has a pK_a of approximately 16–18 (Section 10.3). The pK_a of phenol is 9.95. Nitro groups are electron withdrawing and increase the acidity of the phenolic —OH group. In order of increasing acidity, they are

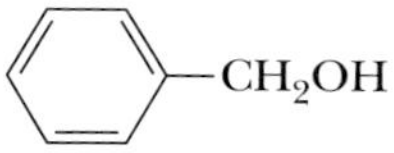

Benzyl alcohol pK_a 16–18	Phenol pK_a 9.95	2,4-Dinitrophenol pK_a 3.96

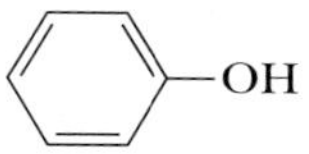

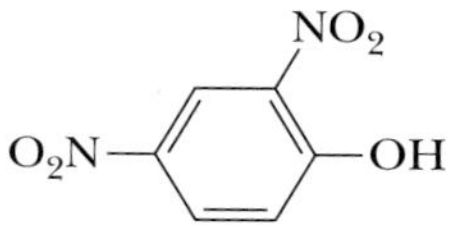

Problem 21.6

Arrange these compounds in order of increasing acidity: 2,4-dichlorophenol, phenol, cyclohexanol.

C. Acid-Base Reactions of Phenols

Phenols are weak acids and react with strong bases, such as NaOH, to form water-soluble salts.

Phenol pK_a 9.95 (stronger acid)	Sodium hydroxide	Sodium phenoxide	Water pK_a 15.7 (weaker acid)

Most phenols do not react with weaker bases such as sodium bicarbonate; they do not dissolve in aqueous sodium bicarbonate. Here you would do well to review Section 4.4.

Carbonic acid is a stronger acid than phenol, and consequently the equilibrium for the reaction of phenol and bicarbonate ion lies far to the left.

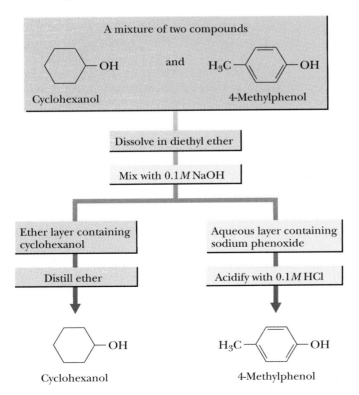

Phenols do, however, form water-soluble salts with sodium carbonate, a stronger base than sodium bicarbonate.

The fact that phenols are weakly acidic whereas alcohols are neutral provides a very convenient way to separate water-insoluble phenols from water-insoluble alcohols. Suppose that we want to separate 4-methylphenol (*p*-cresol) from cyclohexanol. Each is only slightly soluble in water; therefore, they cannot be separated on the basis of their water solubility. They can be separated, however, on the basis of their differences in acidity. First, the mixture of the two is dissolved in diethyl ether or some other water-immiscible solvent. Next, the ether solution is placed in a separatory funnel and shaken with dilute aqueous NaOH. Under these conditions, 4-methylphenol reacts with NaOH and is converted to a water-soluble phenoxide salt. The upper layer in the separatory funnel is now diethyl ether (density 0.74 g/cm³) containing only dissolved cyclohexanol. The lower aqueous layer contains the dissolved phenoxide salt. The layers are separated, and removal of the ether (bp 35°C) by distillation leaves pure cyclohexanol (bp 161°C). Acidification of the aqueous phase with 0.1 *M* HCl or other strong acid converts the phenoxide salt to 4-methylphenol, which is more soluble in ether than in water and can be extracted with ether and recovered in pure form. These experimental steps are summarized in the following flow chart.

D. Preparation of Alkyl-Aryl Ethers

Alkyl-aryl ethers can be prepared from a phenoxide salt and an alkyl halide (the Williamson synthesis, Section 11.4A). They cannot be prepared from an aryl halide and alkoxide salt, however, because aryl halides are quite unreactive under the conditions of Williamson synthesis; they do not undergo nucleophilic displacement by either S_N1 or S_N2 mechanisms.

Alkyl-aryl ethers are often synthesized by phase-transfer catalysis (Section 9.10). Both the alkyl halide and phenol are dissolved in dichloromethane; then, the solution is mixed with an aqueous solution of sodium hydroxide, and a phase-transfer catalyst such as tetrabutylammonium bromide, $Bu_4N^+Br^-$, is added. Phenol, a poor nucleophile, reacts with sodium hydroxide in the aqueous phase to form the phenoxide ion, a good nucleophile. The phase-transfer catalyst transports phenoxide ion to the dichloromethane phase where it reacts with the alkyl halide to form an ether. The following example illustrates the Williamson synthesis of alkyl-aryl ethers.

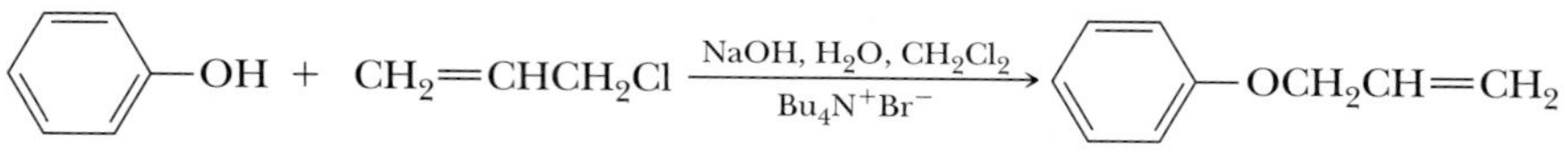

Phenol   3-Chloropropene
(Allyl chloride)

Phenyl 2-propenyl ether
(Allyl phenyl ether)

The synthesis of anisole illustrates the use of dimethyl sulfate as a methylating agent.

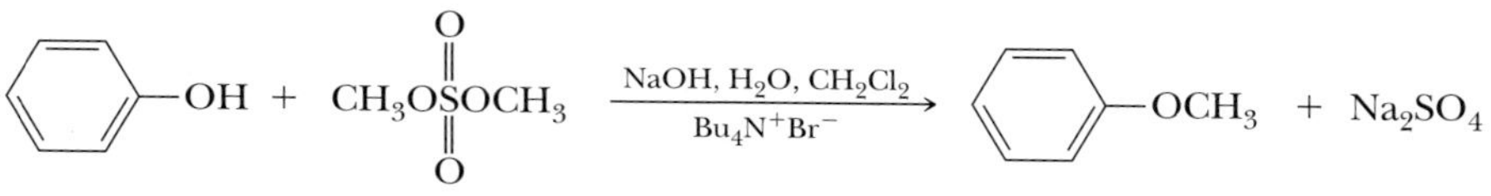

Phenol   Dimethyl sulfate

Methyl phenyl ether
(Anisole)

An alkyl-aryl ether, ArOR, is cleaved by hydrohalic acids, HX, to form an alkyl halide and a phenol.

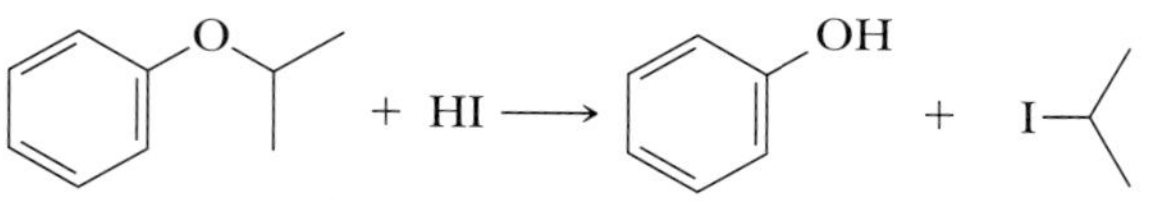

2-Phenoxypropane
(Isopropyl phenyl ether)

Phenol   2-Iodopropane
(Isopropyl iodide)

This cleavage illustrates the fact that nucleophilic substitution is not likely to occur at an aromatic carbon, and that phenols, unlike alcohols, are not converted to aryl halides by treatment with concentrated HCl, HBr, or HI.

E. Kolbe Carboxylation: Synthesis of Salicylic Acid

Phenoxide ions react with carbon dioxide to give a carboxylic acid salt as shown by the industrial synthesis of salicylic acid, the starting material for the production of aspirin (Section 18.5B). Phenol is dissolved in aqueous NaOH, and this solution is then saturated with CO_2 under pressure to give sodium salicylate.

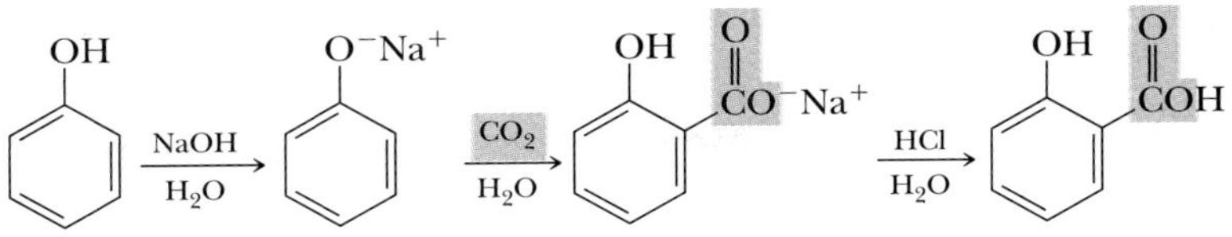

Phenol   Sodium   Sodium salicylate   Salicylic acid
         phenoxide

This process is referred to as high-pressure carboxylation of sodium phenoxide. Upon acidification of the alkaline solution, salicylic acid is isolated as a solid, mp 157–159°C.

The importance of salicylic acid in industrial organic chemistry is demonstrated by the fact that over 6×10^6 kg of aspirin are synthesized in the United States each year.

Mechanism   *Kolbe Carboxylation of Phenol*

The phenoxide ion reacts like an enolate anion; it is a strong nucleophile. In Step 1, nucleophilic attack of the phenoxide anion on a carbonyl group of carbon dioxide gives a substituted cyclohexadienone intermediate. Keto-enol tautomerism of this intermediate in Step 2 gives salicylate anion. Note that the enol in this case, owing to its aromatic character, is the more stable of the two tautomers.

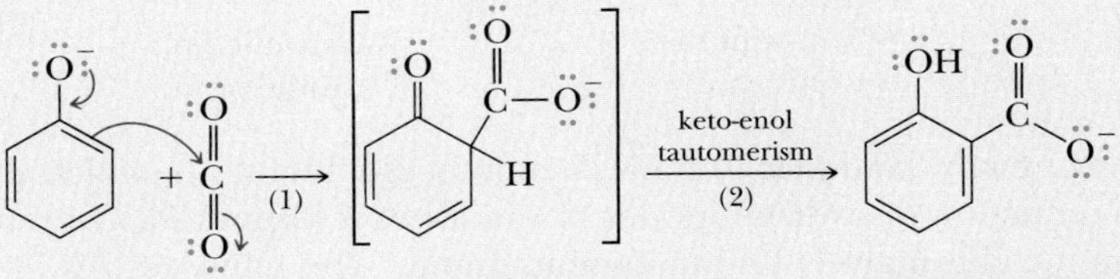

Sodium   A cyclohexadienone   Salicylate anion
phenoxide   intermediate

ORGANIC
Chemistry•¦•Now™
Click Mechanisms to view an animation of the **Kolbe Carboxylation of Phenol**

F. Oxidation to Quinones

Because of the presence of the electron-donating —OH group on the ring, phenols are susceptible to oxidation by a variety of strong oxidizing agents. For example, oxidation of phenol itself by potassium dichromate gives 1,4-benzoquinone (*p*-quinone).

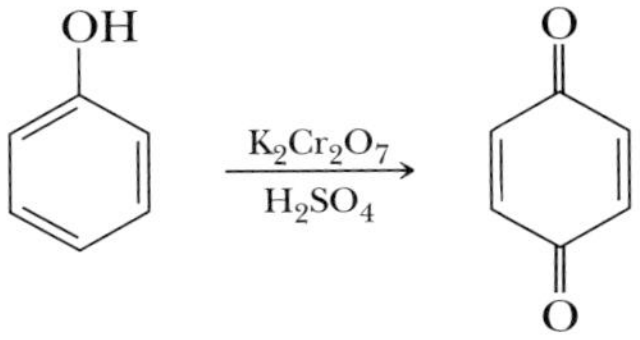

Phenol   1,4-Benzoquinone
             (*p*-Quinone)

By definition, a quinone is a cyclohexadienedione. Those with carbonyl groups ortho to each other are called *o*-quinones; those with carbonyl groups para to each other are called *p*-quinones.

Quinones can also be obtained by oxidation of 1,2-benzenediol (catechol) or 1,4-benzenediol (hydroquinone).

1,2-Benzenediol (Catechol) → $\xrightarrow[H_2SO_4]{K_2Cr_2O_7}$ → 1,2-Benzoquinone (o-Quinone)

1,4-Benzenediol (Hydroquinone) → $\xrightarrow[H_2SO_4]{K_2Cr_2O_7}$ → 1,4-Benzoquinone (p-Quinone)

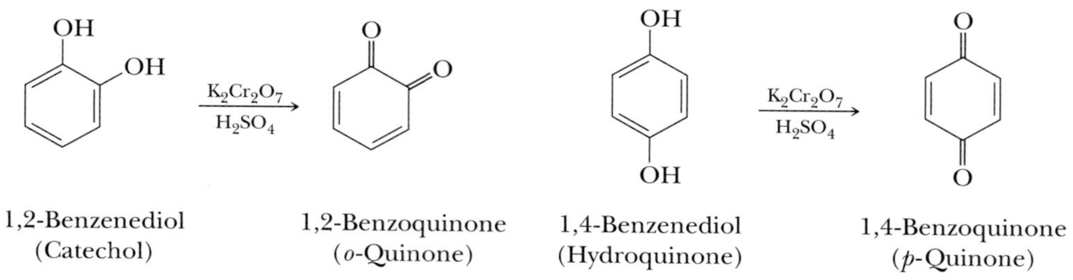

Perhaps the most important chemical property of quinones is that they are readily reduced to benzenediols. For example, *p*-quinone is readily reduced to hydroquinone by sodium dithionite in neutral or alkaline solution. There are other ways to carry out this reduction. The point is that it can be done very easily, as can the corresponding oxidation of a hydroquinone.

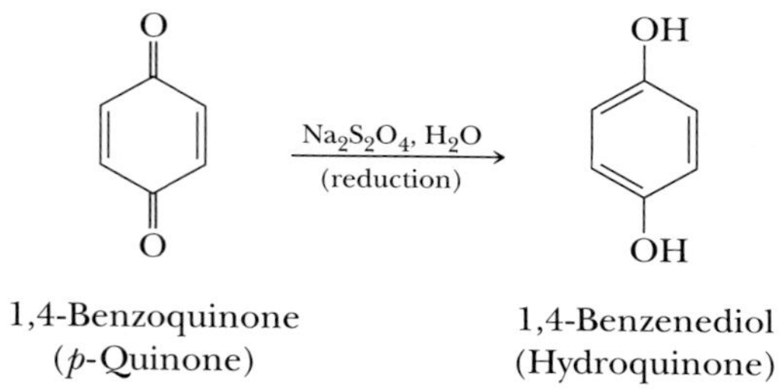

1,4-Benzoquinone (*p*-Quinone) → $\xrightarrow[\text{(reduction)}]{Na_2S_2O_4, H_2O}$ → 1,4-Benzenediol (Hydroquinone)

There are many examples in both chemistry and biology in which the reversible oxidation/reduction of hydroquinones or quinones is important. One such example is coenzyme Q, alternatively known as ubiquinone. The name of this important biomolecule is derived from the Latin *ubique* (everywhere) + quinone.

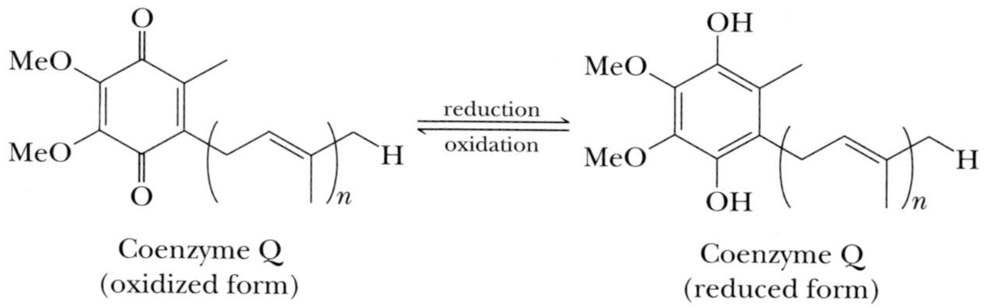

Coenzyme Q (oxidized form) $\underset{\text{oxidation}}{\overset{\text{reduction}}{\rightleftarrows}}$ Coenzyme Q (reduced form)

Coenzyme Q, a carrier of electrons in the respiratory chain, contains a long hydrocarbon chain of between 6 and 10 isoprene units that serves to anchor it firmly in the nonpolar environment of the mitochondrial inner membrane. The oxidized form of coenzyme Q is a two-electron oxidizing agent. In subsequent steps of the respiratory chain, the reduced form of coenzyme Q transfers these two electrons to another link until they are eventually delivered to a molecule of oxygen, which is in turn reduced to water.

Another quinone important in biological systems is vitamin K_2. This compound was discovered in 1935 as a result of a study of newly hatched chicks with a fatal disease in which their blood was slow to clot. It was later discovered that the delayed clotting time of blood was caused by a deficiency of prothrombin. We now know that a prothrombin deficiency is, in turn, caused by a deficiency in vitamin K_2, which is essential to the synthesis of prothrombin in the liver. The natural form of vitamin K_2 has a chain of five to eight isoprene units bonded to a 1,4-naphthoquinone ring. The following structure shows seven isoprene units in the side chain.

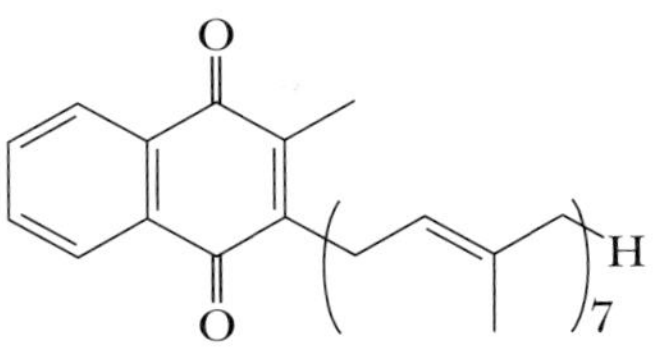

Vitamin K_2

The natural vitamins of the K family have for the most part been replaced by synthetic preparations in food supplements. Menadione, one such synthetic material with vitamin K activity, has only hydrogen in place of the long alkyl side chain. Menadione is prepared by chromic acid oxidation of 2-methylnaphthalene under mild conditions.

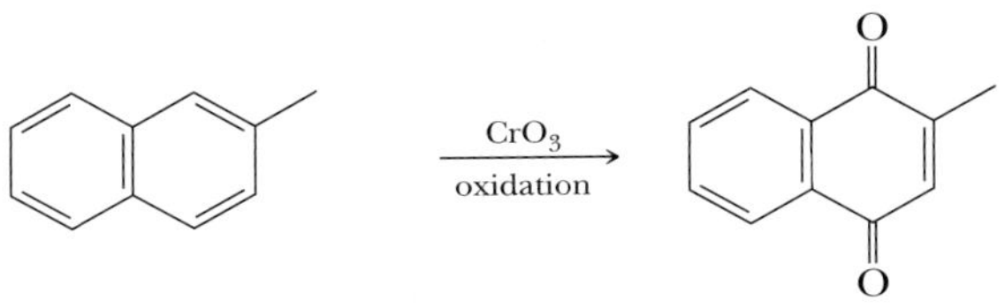

2-Methylnaphthalene   2-Methyl-1,4-naphthoquinone
                                    (Menadione)

A commercial process that uses a quinone is black-and-white photography. Black-and-white film is coated with an emulsion containing silver bromide or silver iodide crystals, which become activated by exposure to light. The activated silver ions are reduced in the developing stage to metallic silver by hydroquinone, which at the same time is oxidized to quinone. Following is an equation showing the relationship between these species.

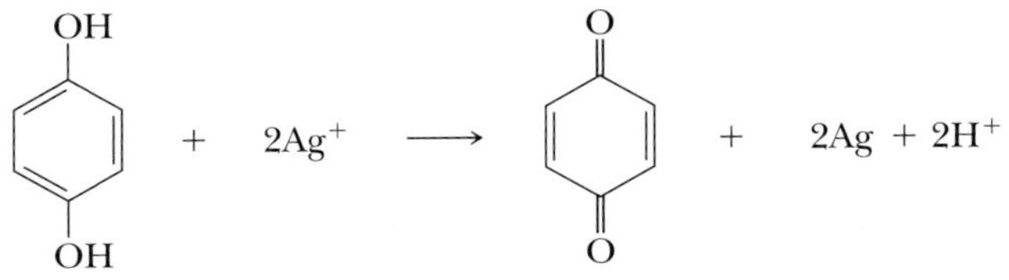

1,4-Benzenediol   1,4-Benzoquinone
(Hydroquinone)   (*p*-Quinone)

All silver halide not activated by light and then reduced by interaction with hydroquinone is removed in the fixing process, and the result is a black image (a negative) left by deposited metallic silver where the film has been struck by light. Other compounds are now used to reduce "light-activated" silver bromide, but the result is the same—a deposit of metallic silver in response to exposure of film to light.

21.5 Reactions at a Benzylic Position

In this section, we study two reactions of substituted aromatic hydrocarbons that occur preferentially at the **benzylic position.**

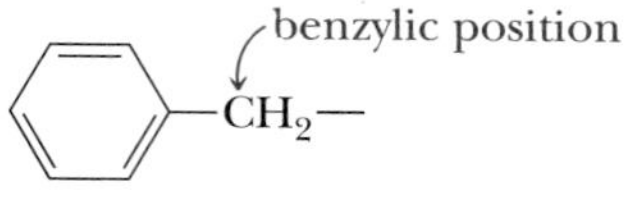

Benzyl group

Benzylic position An sp^3 hybridized carbon bonded to a benzene ring.

Reactions involving alkyl side chains of aromatic compounds occur preferentially at the benzylic position for two reasons. First, the benzene ring is especially resistant to reaction with many of the reagents that normally attack alkanes. Second, benzylic cations and benzylic radicals are easily formed because of resonance stabilization of these intermediates. A benzylic cation or radical is a hybrid of five contributing structures: two Kekulé structures and three that delocalize the positive charge (or the lone electron) onto carbons of the aromatic ring. Following are contributing structures for a benzylic cation. Similar contributing structures can be written for a benzylic radical and anion. Benzylic contributing structures are closely analogous to allylic structures in stabilizing cations, radicals, and anions.

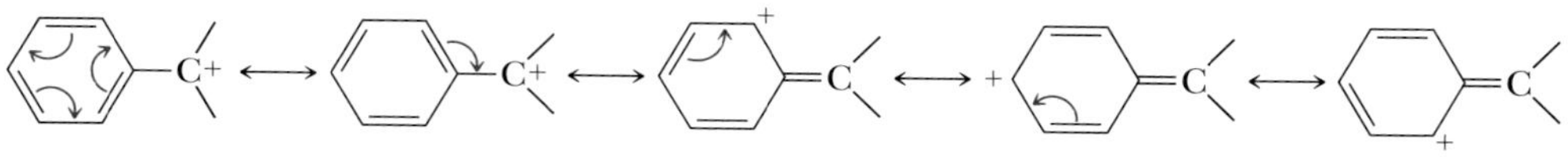

The benzyl cation as a hybrid of five contributing structures

A. Oxidation

Benzene is unaffected by strong oxidizing agents, such as H_2CrO_4 and $KMnO_4$. However, when toluene is treated with these oxidizing agents under quite vigorous conditions, the side-chain methyl group is oxidized to a carboxyl group to give benzoic acid.

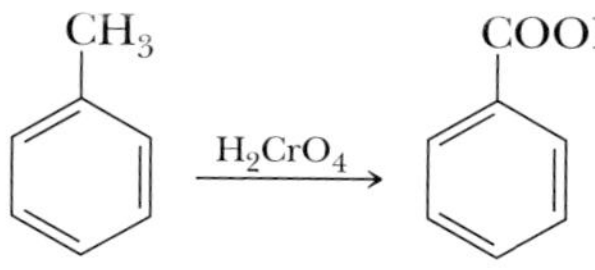

Toluene   Benzoic acid

Halogen and nitro substituents on an aromatic ring are unaffected by these oxidations. 2-Chloro-4-nitrotoluene, for example, is oxidized to 2-chloro-4-nitrobenzoic acid.

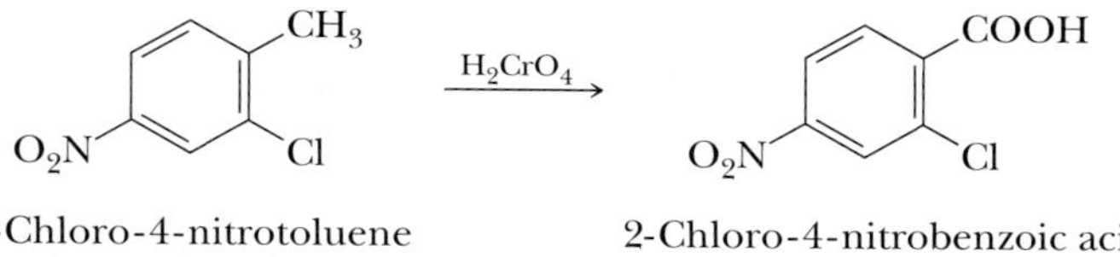

2-Chloro-4-nitrotoluene   2-Chloro-4-nitrobenzoic acid

Ethyl and isopropyl side chains are also oxidized to carboxyl groups. The side chain of *tert*-butylbenzene, however, is not oxidized.

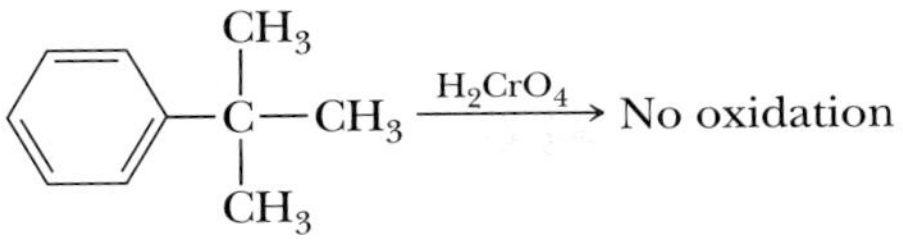

tert-Butylbenzene

From these observations, we conclude that if a benzylic hydrogen exists, then the benzylic carbon is oxidized to a carboxyl group, and all other carbons of the side chain are removed as CO_2. If no benzylic hydrogen exists, as in the case of *tert*-butylbenzene, no oxidation of the side chain occurs.

If more than one alkyl side chain exists, each is oxidized to —COOH. Oxidation of *m*-xylene gives 1,3-benzenedicarboxylic acid, more commonly named isophthalic acid.

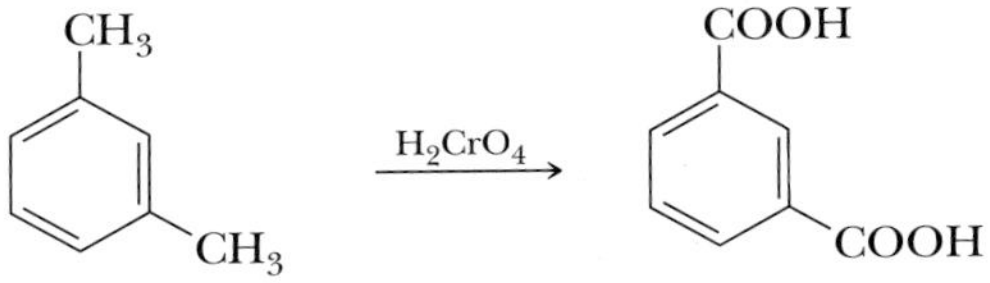

m-Xylene   1,3-Benzenedicarboxylic acid
                                    (Isophthalic acid)

Example 21.7

Draw structural formulas for the product of vigorous oxidation of 1,4-dimethylbenzene (*p*-xylene) by H_2CrO_4.

Solution

Both alkyl groups are oxidized to —COOH groups. The product is terephthalic acid, one of two monomers required for the synthesis of Dacron polyester and Mylar (Section 29.5B).

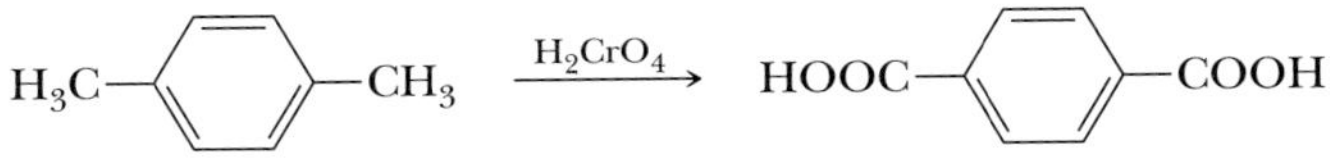

1,4-Dimethylbenzene   1,4-Benzenedicarboxylic acid
    (*p*-Xylene)   (Terephthalic acid)

Problem 21.7

Predict the products resulting from vigorous oxidation of each compound by H_2CrO_4.

(a) 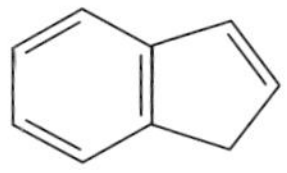   (b)

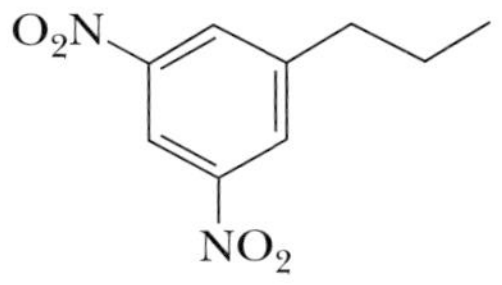

Studying these side-chain oxidations and formulating mechanisms for them is difficult. Available evidence, however, supports the formation of unstable intermediates that are either benzyl radicals or benzylic carbocations.

Naphthalene is oxidized to phthalic acid by molecular oxygen in the presence of a vanadium(V) oxide (vanadium pentoxide) catalyst.

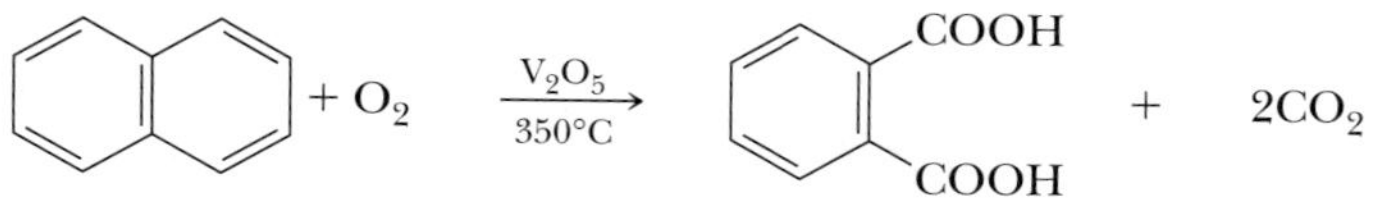

$$\text{Naphthalene} + O_2 \xrightarrow[350°C]{V_2O_5} \text{1,2-Benzenedicarboxylic acid} + 2CO_2$$

Naphthalene   1,2-Benzenedicarboxylic acid
                           (Phthalic acid)

This conversion, which is the basis for an industrial synthesis of this aromatic dicarboxylic acid, illustrates the ease of oxidation of condensed benzene rings compared with benzene itself.

B. Halogenation

Reaction of toluene with chlorine in the presence of heat or light results in formation of chloromethylbenzene and HCl.

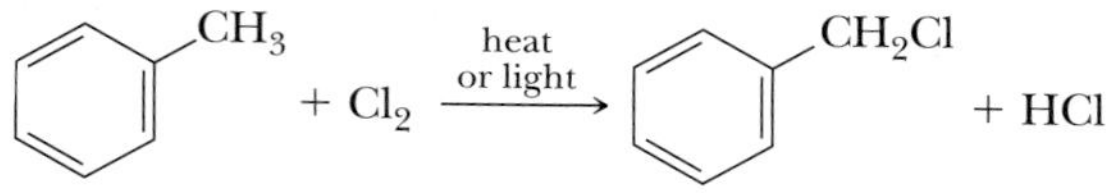

$$\text{Toluene} + Cl_2 \xrightarrow{\text{heat or light}} \text{Chloromethylbenzene} + HCl$$

Toluene   Chloromethylbenzene
                       (Benzyl chloride)

Bromination is easily accomplished by using *N*-bromosuccinimide (NBS) in the presence of a peroxide catalyst.

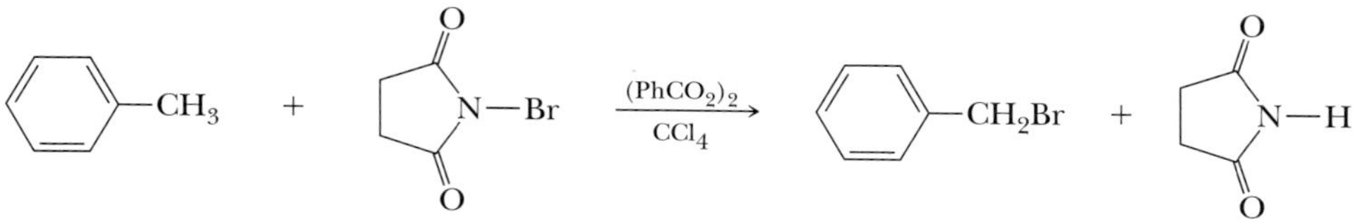

Toluene   *N*-Bromosuccinimide   Bromomethylbenzene   Succinimide
               (NBS)   (Benzyl bromide)

Halogenation of a larger alkyl side chain is highly regioselective, as illustrated by the halogenation of ethylbenzene. When treated with NBS, the only monobromo organic product formed is 1-bromo-1-phenylethane. This regioselectivity is dictated by the resonance stabilization of the benzylic radical intermediate. The mechanism of radical bromination at a benzylic position is identical to that for allylic bromination (Section 8.6A).

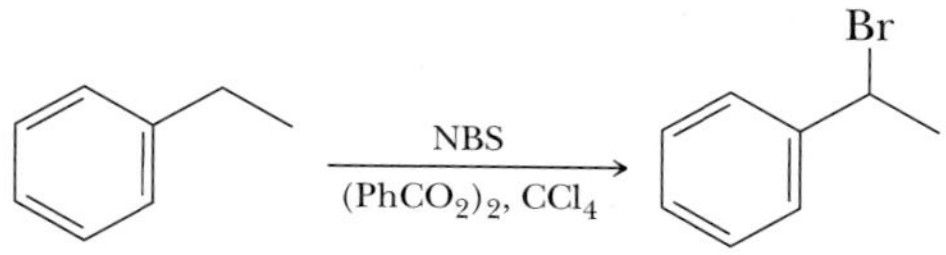

$$\text{Ethylbenzene} \xrightarrow[(PhCO_2)_2, CCl_4]{NBS} \text{1-Bromo-1-phenylethane}$$

Ethylbenzene   1-Bromo-1-phenylethane
                           (racemic)

When ethylbenzene is treated with chlorine under radical reaction conditions, two products are formed in the ratio of 9 : 1.

Ethylbenzene + Cl₂ → heat or light → 1-Chloro-1-phenylethane (90%) (racemic) + 1-Chloro-2-phenylethane (10%) + HCl

Thus, chlorination of alkyl side chains is also regioselective but not to the same high degree as bromination. Recall that we observed this same pattern in the regioselectivities of bromination and chlorination of alkanes (Section 8.4A).

Combining the information on product distribution for bromination and chlorination of hydrocarbons, we conclude that the order of stability of radicals is

$$\text{methyl} < 1° < 2° < 3° < \text{allylic} \cong \text{benzylic}$$

Increasing radical stability →

This order reflects the C—H bond dissociation enthalpies (BDE) for formation of these radicals (Appendix 3).

C. Hydrogenolysis of Benzyl Ethers

Among ethers, benzylic ethers are unique in that they are cleaved under the conditions of catalytic hydrogenation as illustrated by the **hydrogenolysis** of benzyl hexyl ether. In this illustration, the benzyl group is converted to toluene, and the alkyl group is converted to an alcohol.

Hydrogenolysis Cleavage of a single bond by H_2, most commonly accomplished by treating a compound with H_2 in the presence of a transition metal catalyst.

This bond is cleaved.

Benzyl hexyl ether + H₂ → Pd/C → 1-Hexanol + Toluene

Hydrogenolysis is the cleavage of a single bond by H_2. In the hydrogenolysis of a benzylic ether, the single bond between the benzylic carbon and its attached oxygen is cleaved and replaced by a carbon-hydrogen bond.

Benzyl ethers are formed by treatment of an alcohol or phenol with benzyl chloride in the presence of a base such as triethylamine or pyridine. The particular value of benzylic ethers is that they can serve as protecting groups for the —OH groups of alcohols and phenols. Suppose, for example, we want to treat 2-allylphenol with diborane followed by hydrogen peroxide to bring about non-Markovnikov hydration of the carbon-carbon double bond. This scheme will not give the desired result because the phenolic —OH group is sufficiently acidic to react with BH_3 and destroy it. The desired product can be prepared, however, by protection of the phenolic —OH group as the benzylic ether, hydroboration/oxidation of the carbon-carbon double bond, and hydrogenolysis of the benzylic ether.

2-(2-Propenyl)phenol
(2-Allylphenol)

2-(3-Hydroxypropyl)phenol

Summary

Benzene and its alkyl derivatives are classified as **aromatic hydrocarbons,** or **arenes.** The structure of benzene, proposed by August Kekulé in 1865, represented benzene as two rapidly interconverting Kekulé structures (Section 21.1A). The concepts of hybridization of atomic orbitals and the theory of resonance (Section 21.1B), developed by Linus Pauling in the 1930s, provided the first adequate description of the structure of benzene. According to the **molecular orbital model,** the six $2p$ atomic orbitals of the sp^2 hybridized ring carbon atoms combine to give three pi-bonding MOs and three pi-antibonding MOs. In the ground state, the six pi electrons of benzene lie in the three pi-bonding MOs. The **resonance energy** of benzene (Section 21.1C), as calculated from experimental values for heats of hydrogenation of benzene and cyclohexene, is approximately 151 kJ (36 kcal)/mol.

According to the **Hückel criteria for aromaticity** (Section 21.2A), a monocyclic compound is aromatic if it (1) has one p orbital on each atom of the ring, (2) is planar so that overlap of all p orbitals of the ring is continuous or nearly continuous, and (3) has $(4n + 2)$ pi electrons in the cyclic, overlapping arrangement of p orbitals. An **annulene** (Section 21.2B) is a cyclic hydrocarbon with an alternation of single and double bonds. Many have been synthesized to test the validity of Hückel's criteria for aromaticity. It has been found, for example, that [14]annulene and [18]annulene are aromatic as predicted. **Antiaromatic compounds** (Section 21.2C) have only $4n$ pi electrons in a monocyclic, planar system of continuously overlapping p orbitals.

A **heterocyclic aromatic compound** (Section 21.2D) contains one or more atoms other than carbon in an aromatic ring. Particularly abundant in the biological world are derivatives of the heterocyclic aromatic amines pyridine, pyrimidine, imidazole, and pyrrole.

The **cyclopropenyl cation,** the **cyclopentadienyl anion,** and the **cycloheptatrienyl cation** (Section 21.2E) each meet the Hückel criteria for aromaticity and are particularly stable hydrocarbon ions.

Aromatic compounds are named by the IUPAC system. The common names toluene, xylene, cumene, styrene, phenol, aniline, and benzoic acid (Section 21.3) are retained. The C_6H_5— group is named **phenyl,** and the $C_6H_5CH_2$— group is named **benzyl.** Two substituents on a benzene ring may be located by numbering the atoms of the ring or by using the locators **ortho (o), meta (m),** and **para (p). Polynuclear aromatic hydrocarbons** (Section 21.3C) contain two or more fused benzene rings. Particularly abundant are naphthalene, anthracene, phenanthrene, and their derivatives.

The functional group of a **phenol** (Section 21.4A) is an —OH group bonded to a benzene ring. Phenol and its derivatives are weak acids, pK_a approximately 10. The greater acidity of phenols substituted with electron-withdrawing groups, for example NO_2, is accounted for by a combination of **inductive** and **resonance effects.**

Reactions of aromatic compounds containing alkyl side chains occur preferentially at the benzylic carbon (Section 21.5). Benzylic cations and radicals are especially stable because of delocalization of their positive charge or unpaired electron, respectively, onto the ortho and para positions of the aromatic ring.

Key Reactions

ORGANIC
Chemistry Now™

Click Flashcards to review the
Key Reactions of Phenols

1. Acidity of Phenols (Section 21.4B)

Phenols are weak acids, pK_a approximately 10. Ring substituents may increase or decrease acidity by a combination of resonance and inductive effects.

$$K_a = 1.1 \times 10^{-10} \qquad pK_a = 9.95$$

2. Reaction of Phenols with Strong Bases (Section 21.4C)

Water-insoluble phenols react quantitatively with strong bases to form water-soluble salts.

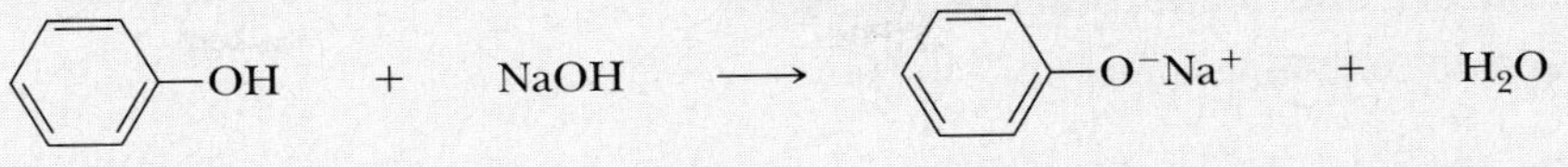

Phenol	Sodium	Sodium	Water
pK_a 9.95	hydroxide	phenoxide	pK_a 15.7
(stronger acid)			(weaker acid)

3. Kolbe Synthesis: Carboxylation of Phenols (Section 21.4E)

Nucleophilic addition of a phenoxide ion to carbon dioxide gives a substituted cyclohexa-dienone, which then undergoes keto-enol tautomerism to regenerate the aromatic ring.

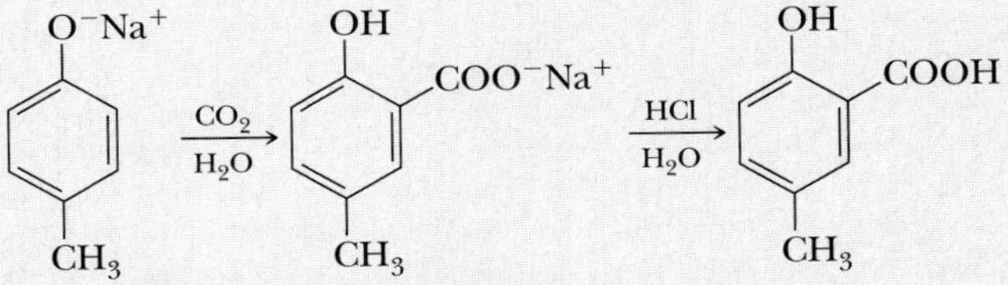

4. Oxidation of Phenols to Quinones (Section 21.4F)

Oxidation by H_2CrO_4 gives 1,2-quinones (*o*-quinones) or 1,4-quinones (*p*-quinones), depending on the structure of the particular phenol.

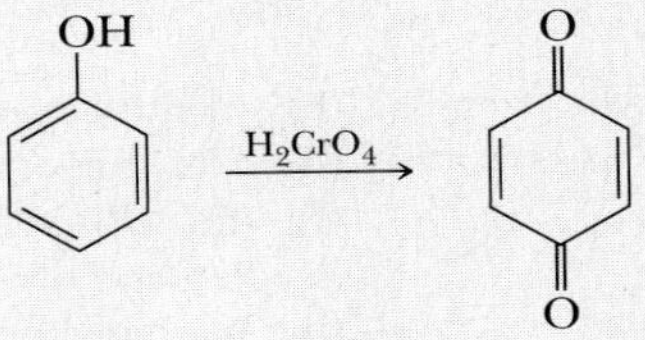

Phenol   1,4-Benzoquinone
                  (*p*-Quinone)

5. Oxidation at a Benzylic Position (Section 21.5A)

A benzylic carbon bonded to at least one hydrogen is oxidized to a carboxyl group.

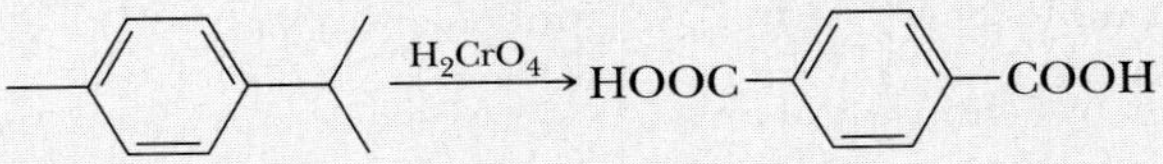

6. Halogenation at a Benzylic Position (Section 21.5B)

Halogenation is regioselective for a benzylic position and occurs by a radical chain mechanism. Bromination shows a higher regioselectivity for a benzylic position than chlorination.

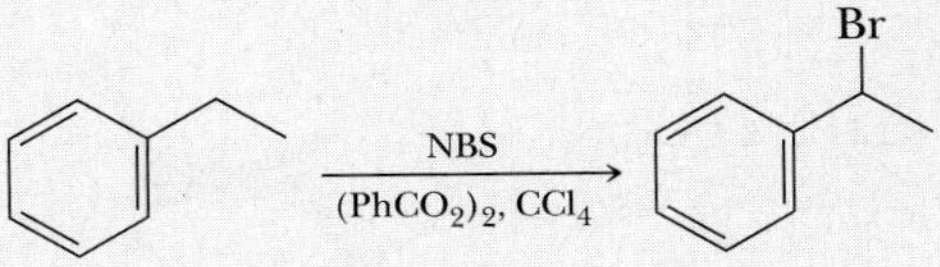

(racemic)

7. Hydrogenolysis of Benzylic Ethers (Section 21.5C)

Benzylic ethers are cleaved under the conditions of catalytic hydrogenation.

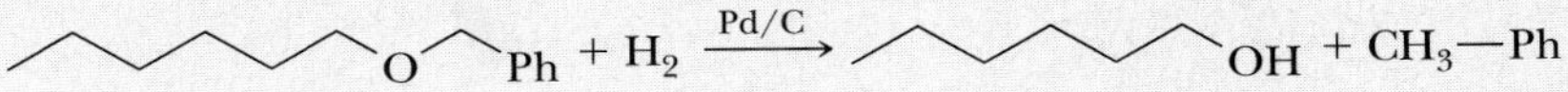

Problems

Nomenclature and Structural Formulas

21.8 Name the following compounds and ions.

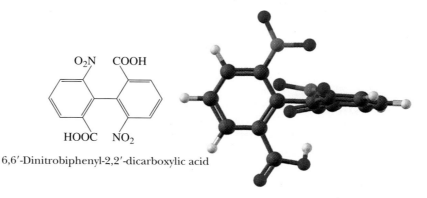

21.9 Draw a structural formula for each compound.

(a) 1-Bromo-2-chloro-4-ethylbenzene (b) *m*-Nitrocumene
(c) 4-Chloro-1,2-dimethylbenzene (d) 3,5-Dinitrotoluene
(e) 2,4,6-Trinitrotoluene (f) (2S, 4R)-4-Phenyl-2-pentanol
(g) *p*-Cresol (h) Pentachlorophenol
(i) 1-Phenylcyclopropanol (j) Triphenylmethane
(k) Phenylethylene (styrene) (l) Benzyl bromide
(m) 1-Phenyl-1-butyne (n) (*E*)-3-Phenyl-2-propen-1-ol

21.10 Draw a structural formula for each compound.

(a) 1-Nitronaphthalene (b) 1,6-Dichloronaphthalene
(c) 9-Bromoanthracene (d) 2-Methylphenanthrene

21.11 Molecules of 6,6′-dinitrobiphenyl-2,2′-dicarboxylic acid have no tetrahedral chiral center, and yet they can be resolved to a pair of enantiomers. Account for this chirality.

6,6′-Dinitrobiphenyl-2,2′-dicarboxylic acid

Resonance in Aromatic Compounds

21.12 Following each name is the number of Kekulé structures that can be drawn for it. Draw these Kekulé structures, and show, using curved arrows, how the first contributing structure for each molecule is converted to the second and so forth.

(a) Naphthalene (3)   **(b)** Phenanthrene (5)

21.13 Each molecule in this problem can be drawn as a hybrid of five contributing structures: two Kekulé structures and three that involve creation and separation of unlike charges. Draw these five contributing structures for each molecule.

(a) Chlorobenzene   **(b)** Phenol   **(c)** Nitrobenzene

21.14 Following are structural formulas for furan and pyridine.

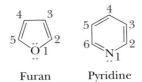

Furan   Pyridine

(a) Draw four contributing structures for furan that place a positive charge on oxygen and a negative charge first on carbon 3 of the ring and then on each other carbon of the ring.
(b) Draw three contributing structures for pyridine that place a negative charge on nitrogen and a positive charge first on carbon 2, then on carbon 4, and finally carbon 6.

The Concept of Aromaticity

21.15 State the number of *p* orbital electrons in each molecule or ion.

21.16 Which of the molecules and ions given in Problem 21.15 are aromatic according to the Hückel criteria? Which, if planar, would be antiaromatic?

21.17 Construct MO energy diagrams for the cyclopropenyl cation, radical, and anion. Which of these species is aromatic according to the Hückel criteria?

21.18 Naphthalene and azulene are constitutional isomers of molecular formula $C_{10}H_8$.

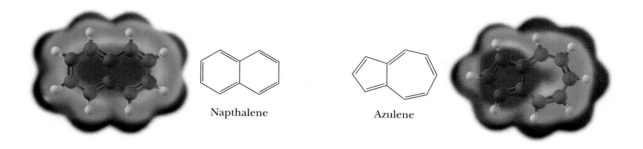

Napthalene   Azulene

Naphthalene is a colorless solid with a dipole moment of zero. Azulene is a solid with an intense blue color and a dipole moment of 1.0 D. Account for the difference in dipole moments of these constitutional isomers.

Spectroscopy

21.19 Compound A (C_9H_{12}) shows prominent peaks in its mass spectrum at m/z 120 and 105. Compound B (also C_9H_{12}) shows prominent peaks at m/z 120 and 91. On vigorous oxidation with chromic acid, both compounds give benzoic acid. From this information, deduce the structural formulas of compounds A and B.

21.20 Compound C shows a molecular ion at m/z 148 and other prominent peaks at m/z 105 and 77. Following are its infrared and ^{1}H-NMR spectra.

(a) Deduce the structural formula of compound C.
(b) Account for the appearance of peaks in its mass spectrum at m/z 105 and 77.

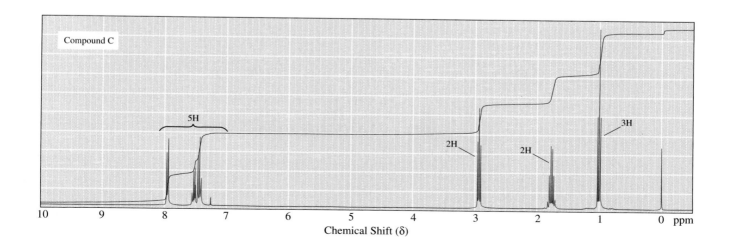

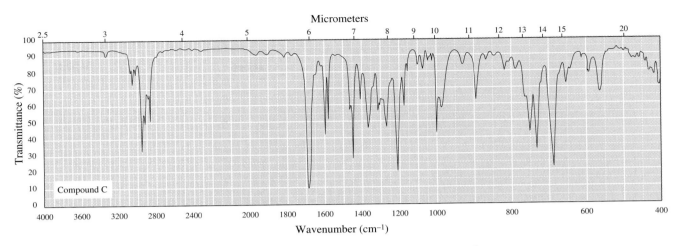

Compound C

21.21 Following are IR and ¹H-NMR spectra of compound D. The mass spectrum of compound D shows a molecular ion peak at *m/z* 136, a base peak at *m/z* 107, and other prominent peaks at *m/z* 118 and 59.

(a) Propose a structural formula for compound D.
(b) Propose structural formulas for ions in the mass spectrum at *m/z* 118, 107, and 59.

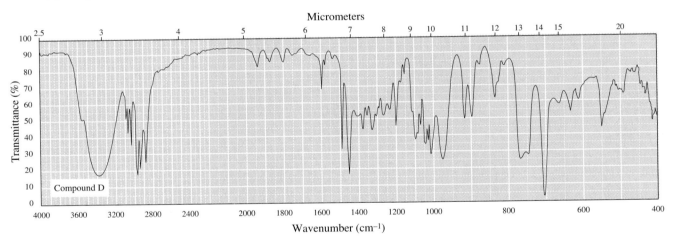

Compound D

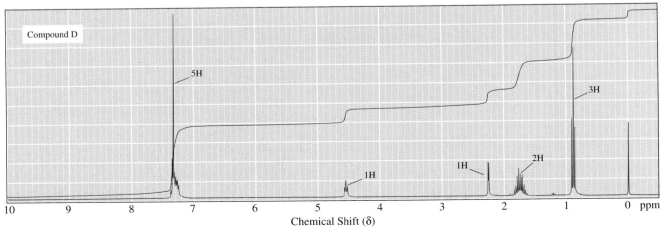

Compound D

21.22 Compound E ($C_8H_{10}O_2$) is a neutral solid. Its mass spectrum shows a molecular ion at m/z 138 and prominent peaks at M-1 and M-17. Following are IR and ^{1}H-NMR spectra of compound E. Deduce the structure of compound E.

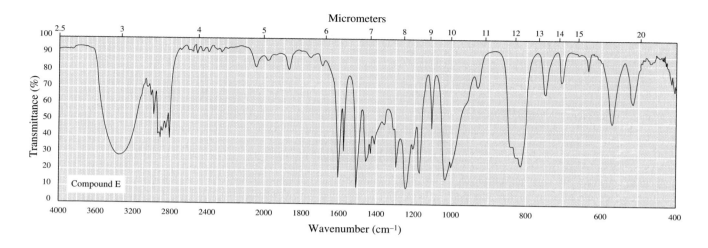

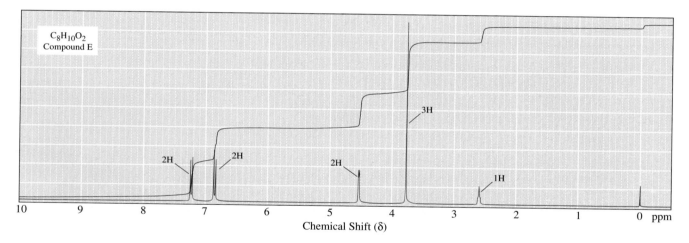

21.23 Following are ^{1}H-NMR and ^{13}C-NMR spectral data for compound F ($C_{12}H_{16}O$). From this information, deduce the structure of compound F.

^{1}H-NMR	^{13}C-NMR	
0.83 (d, 6H)	207.82	50.88
2.11 (m, 1H)	134.24	50.57
2.30 (d, 2H)	129.36	24.43
3.64 (s, 2H)	128.60	22.48
7.2–7.4 (m, 5H)	126.86	

21.24 Following are ^{1}H-NMR and ^{13}C-NMR spectral data for compound G ($C_{10}H_{10}O$). From this information, deduce the structure of compound G.

^{1}H-NMR	^{13}C-NMR	
2.50 (t, 2H)	210.19	126.82
3.05 (t, 2H)	136.64	126.75
3.58 (s, 2H)	133.25	45.02
7.1–7.3 (m, 4H)	128.14	38.11
	127.75	28.34

21.25 Compound H ($C_8H_6O_3$) gives a precipitate when treated with hydroxylamine in aqueous ethanol, and a silver mirror when treated with Tollens' solution. Following is its ^{1}H-NMR spectrum. Deduce the structure of compound H.

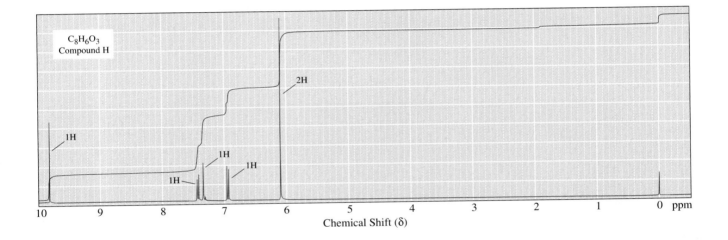

21.26 Compound I ($C_{11}H_{14}O_2$) is insoluble in water, aqueous acid, and aqueous NaHCO$_3$ but dissolves readily in 10% Na$_2$CO$_3$ and 10% NaOH. When these alkaline solutions are acidified with 10% HCl, compound I is recovered unchanged. Given this information and its ^{1}H-NMR spectrum, deduce the structure of compound I.

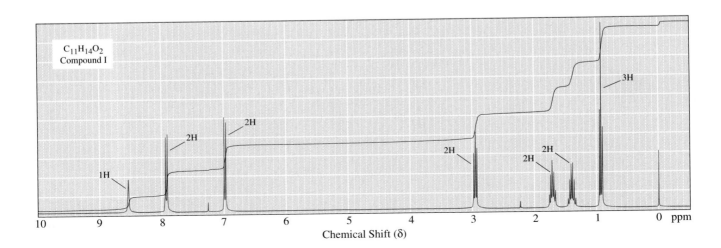

21.27 Propose a structural formula for compound J ($C_{11}H_{14}O_3$) consistent with its ^{1}H-NMR and infrared spectra.

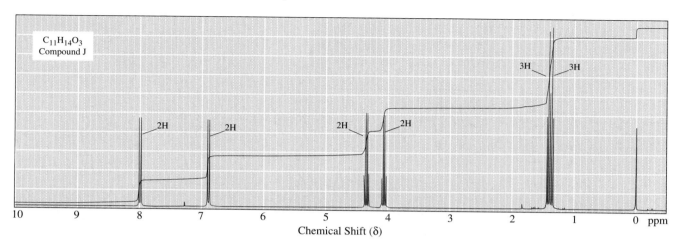

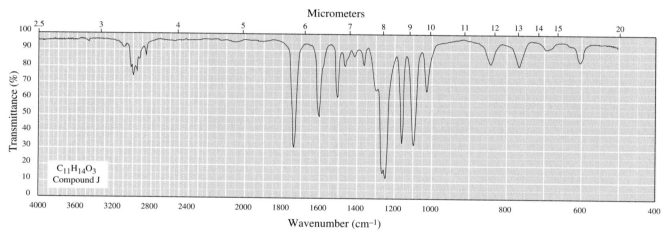

21.28 Propose a structural formula for the analgesic phenacetin, molecular formula $C_{10}H_{13}NO_2$, based on its ^{1}H-NMR spectrum.

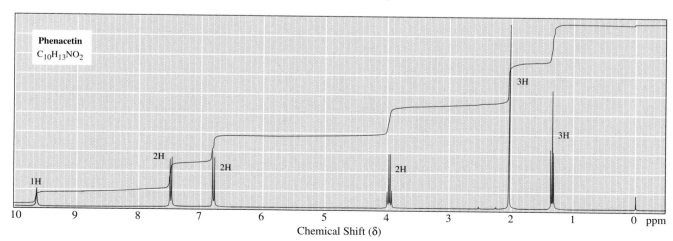

21.29 Compound K, $C_{10}H_{12}O_2$, is insoluble in water, 10% NaOH, and 10% HCl. Given this information and the following ^{1}H-NMR and ^{13}C-NMR spectral information, deduce the structural formula of Compound K.

^{1}H-NMR	^{13}C-NMR	
2.10 (s, 3H)	206.51	114.17
3.61 (s, 2H)	158.67	55.21
3.77 (s, 3H)	130.33	50.07
6.86 (d, 2H)	126.31	29.03
7.12 (d, 2H)		

21.30 Propose a structural formula for each compound given these NMR data.

(a) $C_9H_9BrO_2$ (b) C_8H_9NO (c) $C_9H_9NO_3$

^{1}H-NMR	^{13}C-NMR
1.39 (t, 3H)	165.73
4.38 (q, 2H)	131.56
7.57 (d, 2H)	131.01
7.90 (d, 2H)	129.84
	127.81
	61.18
	14.18

^{1}H-NMR	^{13}C-NMR
2.06 (s, 3H)	168.14
7.01 (t, 1H)	139.24
7.30 (m, 2H)	128.51
7.59 (d, 2H)	122.83
9.90 (s, 1H)	118.90
	23.93

^{1}H-NMR	^{13}C-NMR
2.10 (s, 3H)	168.74
7.72 (d, 2H)	166.85
7.91 (d, 2H)	143.23
10.3 (s, 1H)	130.28
12.7 (s, 1H)	124.80
	118.09
	24.09

21.31 Given here are ^{1}H-NMR and ^{13}C-NMR spectral data for two compounds. Each shows strong, sharp absorption between 1700 and 1720 cm^{-1}, and strong, broad absorption over the region 2500–3000 cm^{-1}. Propose a structural formula for each compound.

(a) $C_{10}H_{12}O_3$ (b) $C_{10}H_{10}O_2$

^{1}H-NMR	^{13}C-NMR
2.49 (t, 2H)	173.89
2.80 (t, 2H)	157.57
3.72 (s, 3H)	132.62
6.78 (d, 2H)	128.99
7.11 (d, 2H)	113.55
12.4 (s, 1H)	54.84
	35.75
	29.20

^{1}H-NMR	^{13}C-NMR
2.34 (s, 3H)	167.82
6.38 (d, 1H)	143.82
7.18 (d, 1H)	139.96
7.44 (d, 2H)	131.45
7.56 (d, 2H)	129.37
12.0 (s, 1H)	127.83
	111.89
	21.13

Acidity of Phenols

21.32 Account for the fact that p-nitrophenol (K_a 7.0×10^{-8}) is a stronger acid than phenol (K_a 1.1×10^{-10}).

21.33 Account for the fact that water-insoluble carboxylic acids (pK_a 4–5) dissolve in 10% aqueous sodium bicarbonate (pH 8.5) with the evolution of a gas but that water-insoluble phenols (pK_a 9.5–10.5) do not dissolve in 10% sodium bicarbonate.

21.34 Match each compound with its appropriate pK_a value.

   (a) 4-Nitrobenzoic acid, benzoic acid, 4-chlorobenzoic acid
     $pK_a = 4.19$, 3.98, and 3.41

   (b) Benzoic acid, cyclohexanol, phenol
     $pK_a = 18.0$, 9.95, and 4.19

   (c) 4-Nitrobenzoic acid, 4-nitrophenol, 4-nitrophenylacetic acid
     $pK_a = 7.15$, 3.85, and 3.41

21.35 Arrange the molecules and ions in each set in order of increasing acidity (from least acidic to most acidic).

   (a) 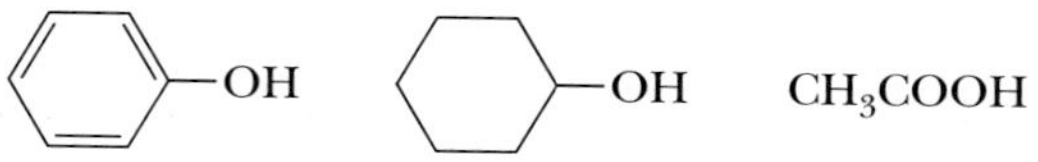 ⬡—OH　⬡—OH　CH_3COOH

   (b)  ⬡—OH　HCO_3^-　H_2O

   (c) 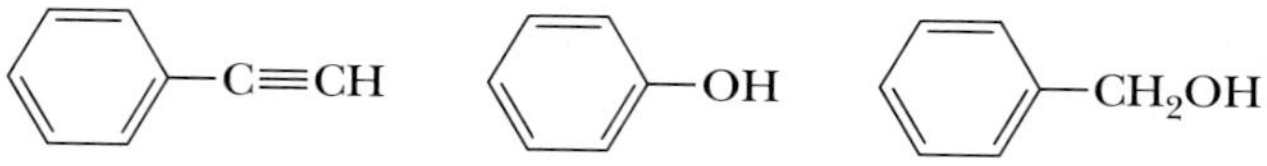 ⬡—C≡CH　⬡—OH　⬡—CH_2OH

21.36 Explain the trends in the acidity of phenol and the monofluoro derivatives of phenol.

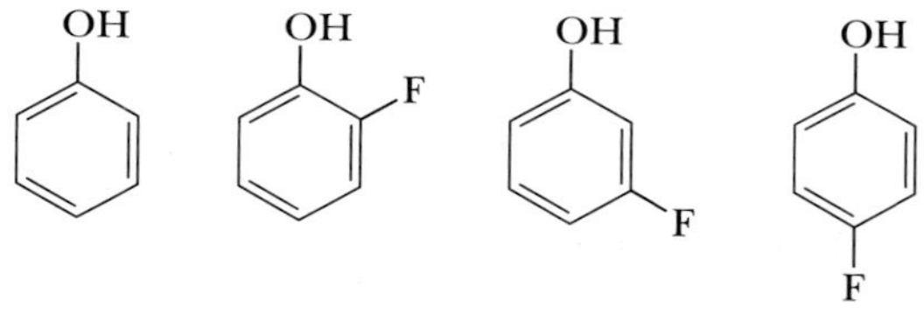

   pK_a 10.0　　pK_a 8.81　　pK_a 9.28　　pK_a 9.81

21.37 Suppose you wish to determine the inductive effects of a series of functional groups, for example Cl, Br, CN, COOH, and C_6H_5. Is it best to use a series of ortho-, meta-, or para-substituted phenols? Explain.

21.38 From each pair, select the stronger base.

   (a) 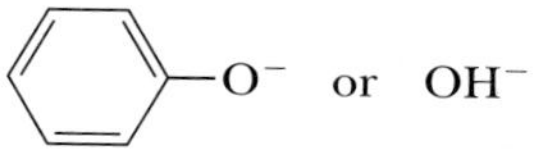 ⬡—O^-　or　OH^-　　**(b)** 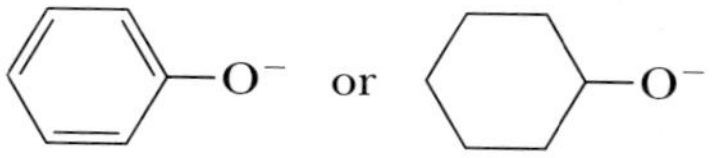 ⬡—O^-　or　⬡—O^-

   (c) 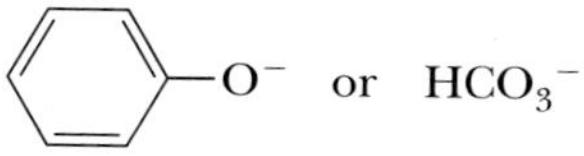 ⬡—O^-　or　HCO_3^-　　**(d)** 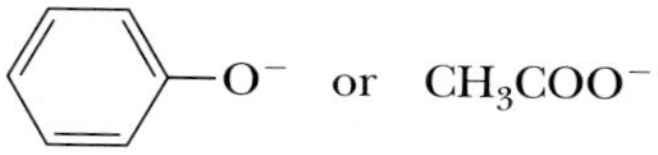 ⬡—O^-　or　CH_3COO^-

21.39 Describe a chemical procedure to separate a mixture of benzyl alcohol and *o*-cresol and to recover each in pure form.

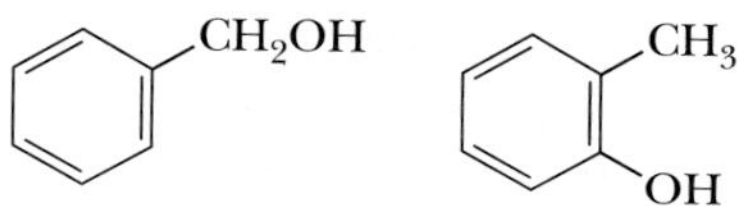

     ⬡—CH_2OH　　⬡(CH_3)(OH)

     Benzyl alcohol　　*o*-Cresol

21.40 The compound 2-hydroxypyridine, a derivative of pyridine, is in equilibrium with 2-pyridone. 2-Hydroxypyridine is aromatic. Does 2-pyridone have comparable aromatic character? Explain.

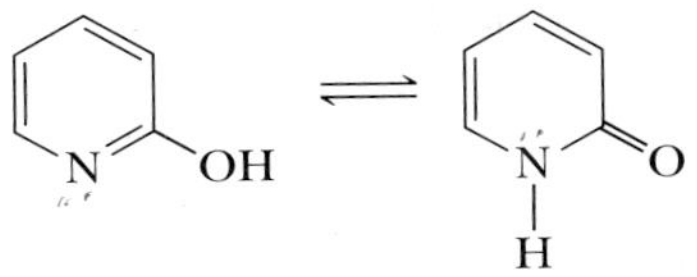

2-Hydroxypyridine   2-Pyridone

Reactions at the Benzylic Position

21.41 Write a balanced equation for the oxidation of *p*-xylene to 1,4-benzenedicarboxylic acid (terephthalic acid) using potassium dichromate in aqueous sulfuric acid. How many milligrams of H_2CrO_4 are required to oxidize 250 mg of *p*-xylene to terephthalic acid?

21.42 Each of the following reactions occurs by a radical chain mechanism.

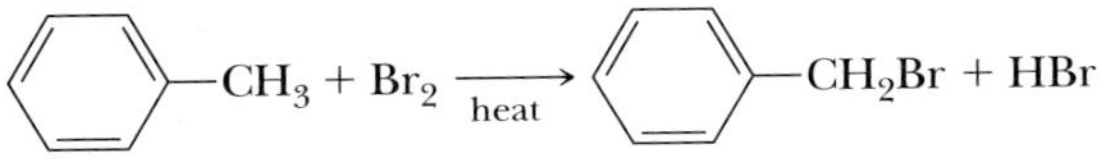

Toluene   Benzyl bromide

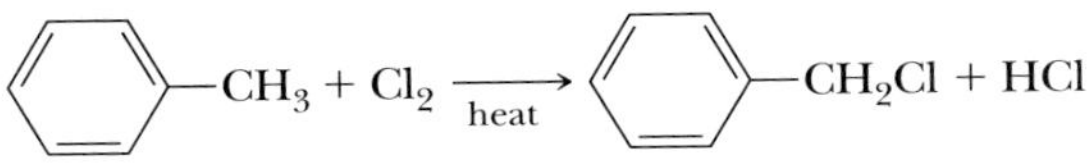

Toluene   Benzyl chloride

(a) Calculate the heat of reaction, ΔH^0, in kilojoules per mole for each reaction. (Consult Appendix 3 for bond dissociation enthalpies.)
(b) Write a pair of chain propagation steps for each mechanism, and show that the net result of each pair is the observed reaction.
(c) Calculate ΔH^0 for each chain propagation step, and show that the sum for each pair of steps is identical with the ΔH^0 value calculated in part (a).

21.43 Following is an equation for iodination of toluene.

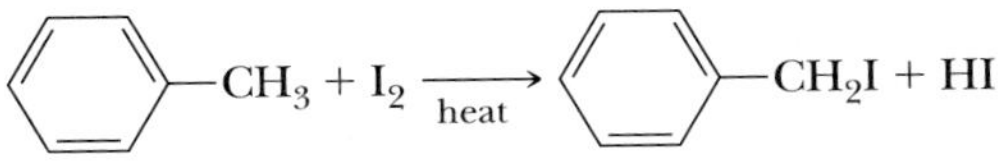

Toluene   Benzyl iodide

This reaction does not take place. All that happens under experimental conditions for the formation of radicals is initiation to form iodine radicals, I·, followed by termination to reform I_2. How do you account for these observations?

21.44 Although most alkanes react with chlorine by a radical chain mechanism when reaction is initiated by light or heat, benzene fails to react under the same conditions. Benzene cannot be converted to chlorobenzene by treatment with chlorine in the presence of light or heat.

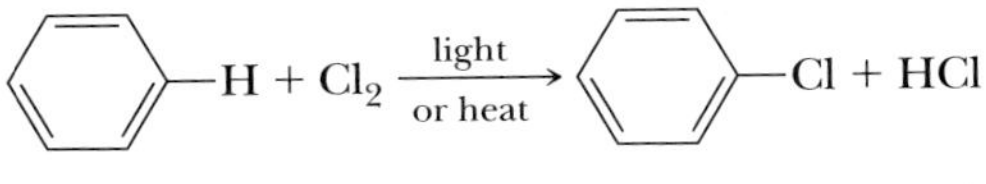

(a) Explain why benzene fails to react under these conditions. (Consult Appendix 3 for relevant bond dissociation enthalpies.)
(b) Explain why the bond dissociation enthalpy of a C—H bond in benzene is significantly greater than that in alkanes.

21.45 Following is an equation for hydroperoxidation of cumene.

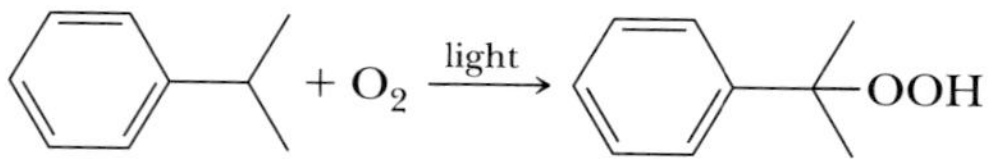

Cumene   Cumene hydroperoxide

Propose a radical chain mechanism for this reaction. Assume that initiation is by an unspecified radical, R·.

21.46 Para-substituted benzyl halides undergo reaction with methanol by an S_N1 mechanism to give a benzyl ether. Account for the following order of reactivity under these conditions.

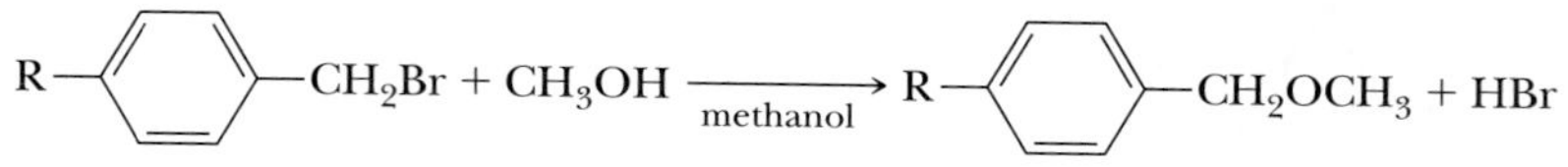

Rate of S_N1 reaction: R = $CH_3O > CH_3 > H > NO_2$

21.47 When warmed in dilute sulfuric acid, 1-phenyl-1,2-propanediol undergoes dehydration and rearrangement to give 2-phenylpropanal.

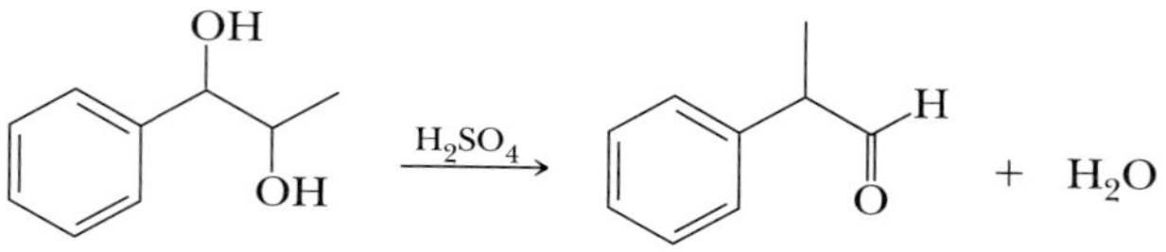

1-Phenyl-1,2-propanediol (racemic)   2-Phenylpropanal (racemic)

(a) Propose a mechanism for this example of a pinacol rearrangement (Section 10.7).
(b) Account for the fact that 2-phenylpropanal is formed rather than its constitutional isomer, 1-phenyl-1-propanone.

21.48 In the chemical synthesis of DNA and RNA, hydroxyl groups are normally converted to triphenylmethyl (trityl) ethers to protect the hydroxyl group from reaction with other reagents.

$$RCH_2OH + Ph_3CCl \xrightarrow{\text{tertiary amine}} RCH_2OCPh_3 + HCl$$

Ph$_3$CCl: Triphenylmethyl chloride (Trityl chloride)

RCH$_2$OCPh$_3$: A triphenylmethyl ether (A trityl ether)

HCl: neutralized by the tertiary amine

Triphenylmethyl ethers are stable to aqueous base but are rapidly cleaved in aqueous acid.

$$RCH_2OCPh_3 + H_2O \xrightarrow{H^+} RCH_2OH + Ph_3COH$$

(a) Why are triphenylmethyl ethers so readily hydrolyzed by aqueous acid?
(b) How might the structure of the triphenylmethyl group be modified to increase or decrease its acid sensitivity?

Synthesis

21.49 Using ethylbenzene as the only aromatic starting material, show how to synthesize the following compounds. In addition to ethylbenzene, use any other necessary organic or inorganic chemicals. Any compound already synthesized in one part of this problem may then be used to make any other compound in the problem.

(a) 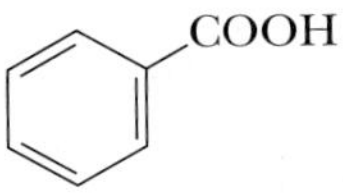 [benzoic acid structure, COOH]

(b) 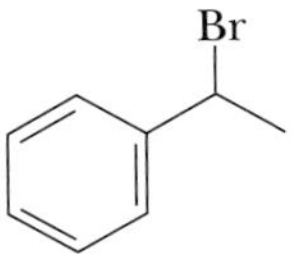 [1-bromoethylbenzene structure, Br] (racemic)

(c) 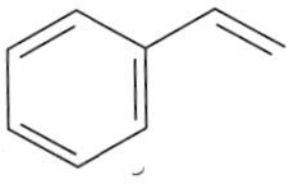 [styrene structure]

(d) 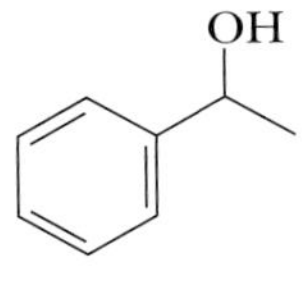 [1-phenylethanol structure, OH] (racemic)

(e) 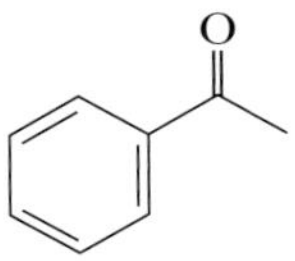 [acetophenone structure, O]

(f) 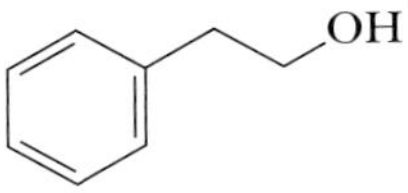 [2-phenylethanol structure, OH]

(g) 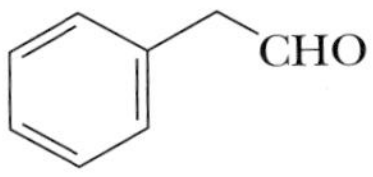 [phenylacetaldehyde structure, CHO]

(h) 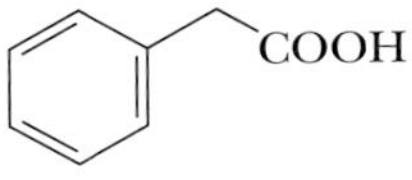 [phenylacetic acid structure, COOH]

(i) 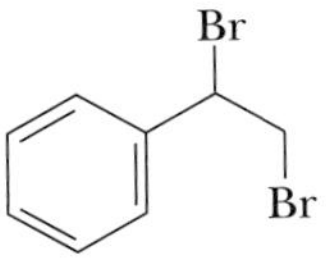 [1,2-dibromo structure, Br, Br] (racemic)

(j) 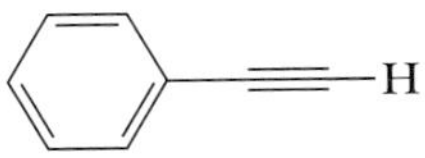 [phenylacetylene structure, ═H]

(k) 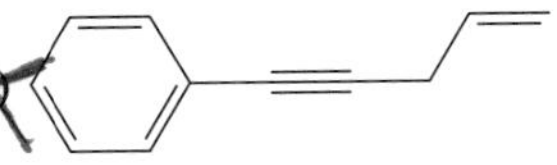 [enyne structure]

(l) 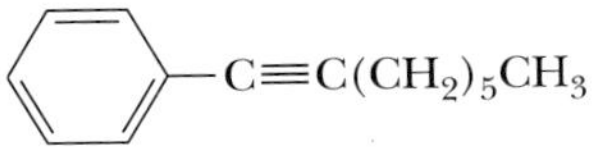 [structure] $-C\equiv C(CH_2)_5CH_3$

(m) 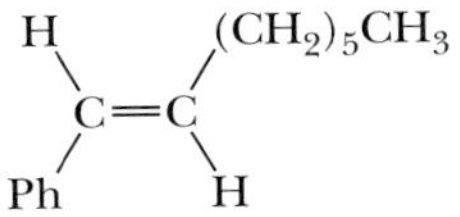 $\begin{array}{cc} H & (CH_2)_5CH_3 \\ \diagdown C = C \diagup \\ Ph & H \end{array}$

(n) 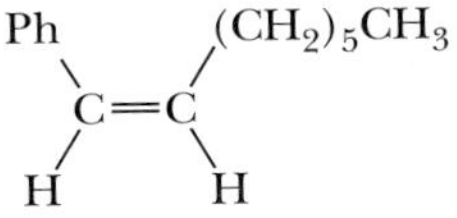 $\begin{array}{cc} Ph & (CH_2)_5CH_3 \\ \diagdown C = C \diagup \\ H & H \end{array}$

21.50 Show how to convert 1-phenylpropane into the following compounds. In addition to this starting material, use any necessary inorganic reagents. Any compound synthesized in one part of this problem may be used to make any other compound in the problem.

(a) 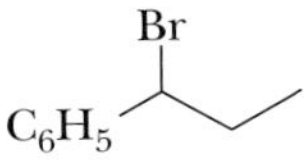 C_6H_5 [structure with Br] (racemic)

(b) 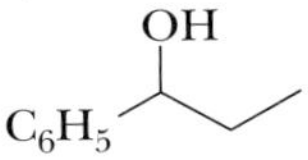 C_6H_5 [structure with OH] (racemic)

(c) 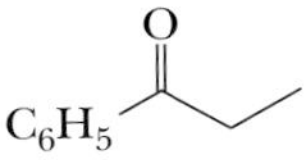 C_6H_5 [structure with O]

(d) 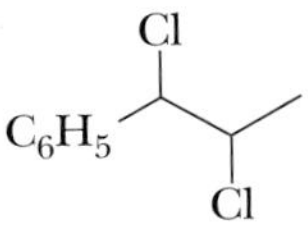 C_6H_5 [structure with Cl, Cl] (racemic)

(e) 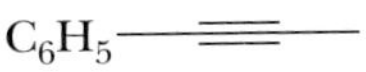 C_6H_5 [alkyne structure]

(f) 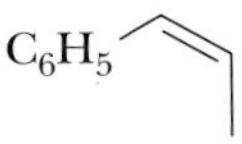 C_6H_5 [cis-alkene structure]

(g) 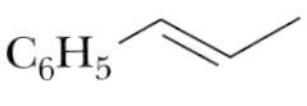 C_6H_5 [trans-alkene structure]

(h) 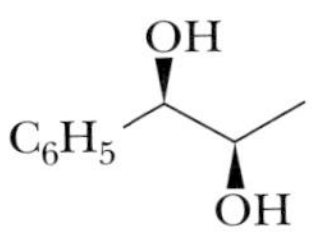 C_6H_5 [structure with OH, OH] (racemic)

(i) 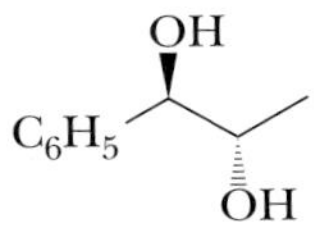 C_6H_5 [structure with OH, OH] (racemic)

21.51 Carbinoxamine is a histamine antagonist, specifically an H_1-antagonist. The maleic acid salt of the levorotatory isomer is sold as the prescription drug Rotoxamine.

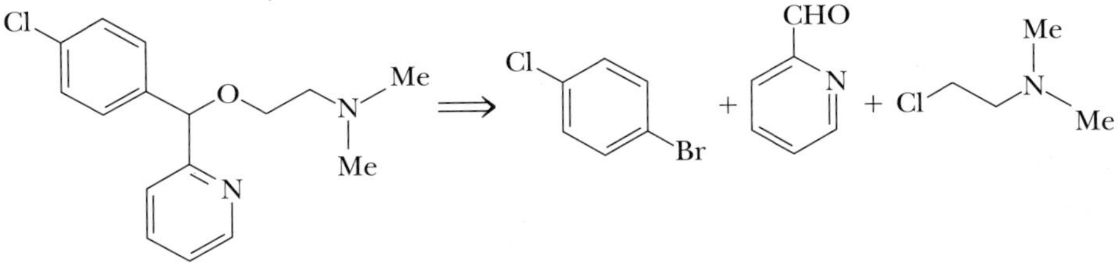

[Carbinoxamine structure with Cl-phenyl, pyridine, O-CH2CH2-N(Me)Me] $\Rightarrow$ [Cl-phenyl-CH2Br] + [pyridine-CHO] + [Cl-CH2CH2-N(Me)Me]

Carbinoxamine

(a) Propose a synthesis of carbinoxamine. (*Note:* Aryl bromides form Grignard reagents much more readily than do aryl chlorides.)

(b) Is carbinoxamine chiral? If so how many stereoisomers are possible? Which of the possible stereoisomers are formed in this synthesis?

21.52 Cromolyn sodium, developed in the 1960s, has been used to prevent allergic reactions primarily affecting the lungs, as for example exercise-induced emphysema. It is thought to block the release of histamine, which prevents the sequence of events leading to swelling, itching, and constriction of bronchial tubes. Cromolyn sodium is synthesized in the following series of steps. Treatment of one mole of epichlorohydrin (Section 11.10) with two moles of 2,6-dihydroxyacetophenone in the presence of base gives I. Treatment of I with two moles of diethyl oxalate in the presence of sodium ethoxide gives a diester II. Saponification of the diester with aqueous NaOH gives cromolyn sodium.

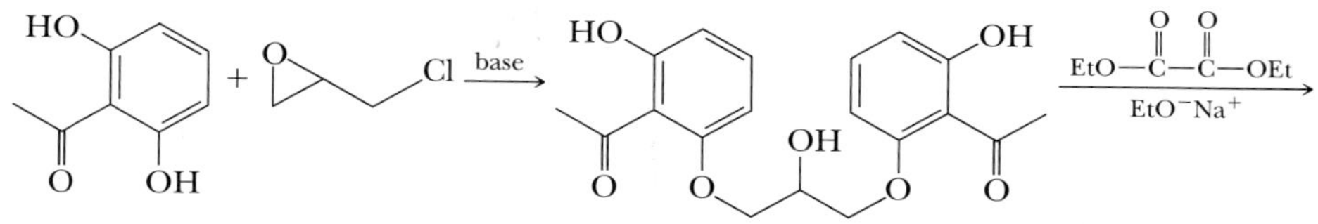

2,6-Dihydroxy-   Epichloro-   I
acetophenone   hydrin

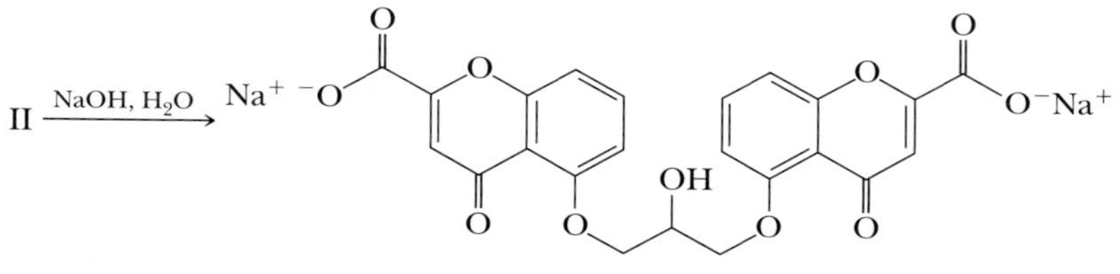

Cromolyn sodium

(a) Propose a mechanism for the formation of compound I.

(b) Propose a structural formula for compound II and a mechanism for its formation.

(c) Is cromolyn sodium chiral? If so, which of the possible stereoisomers are formed in this synthesis?

21.53 The following stereospecific synthesis is part of the scheme used by E. J. Corey of Harvard University in the synthesis of erythronolide B, the precursor of the erythromycin antibiotics. In this remarkably simple set of reactions, the relative configurations of five chiral centers are established.

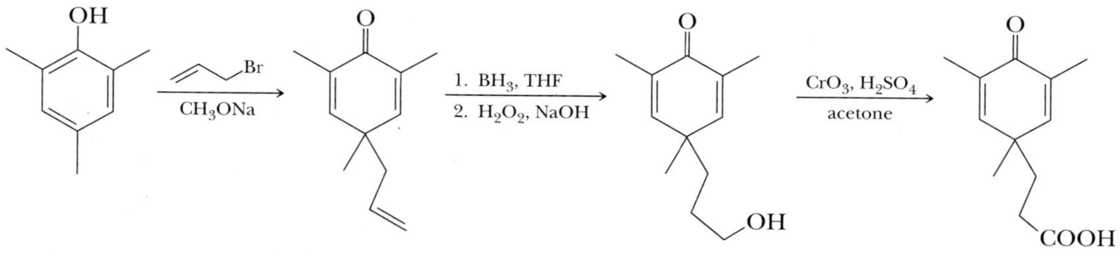

2,4,6-Trimethyl-   (A)   (B)   (C)
phenol

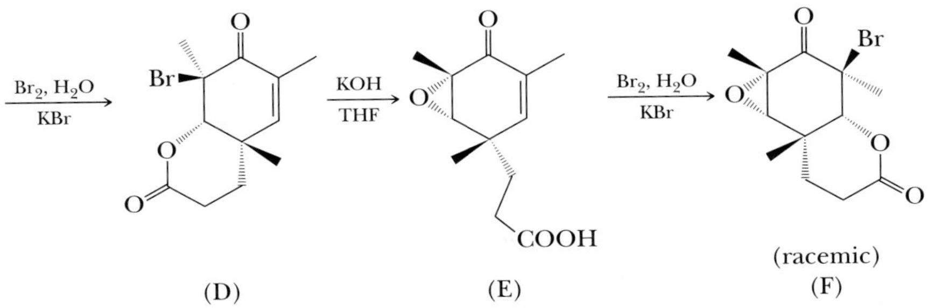

(D) (E) (F)
(racemic)

(a) Propose a mechanism for the conversion of 2,4,6-trimethylphenol to compound A.

(b) Account for the stereoselectivity and regioselectivity of the three steps in the conversion of compound C to compound F.

(c) Is compound F produced in this synthesis as a single enantiomer or as a racemic mixture? Explain.

21.54 Following is an outline of one of the first syntheses of the antidepressant fluoxetine (Prozac).

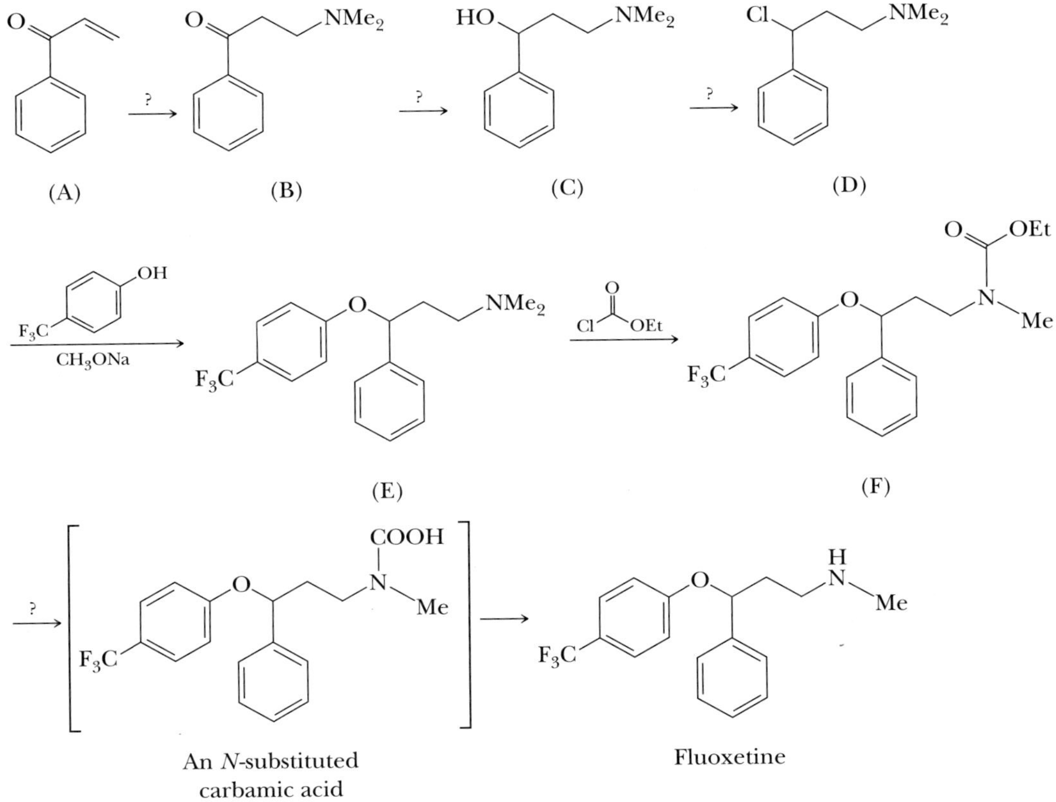

(A) (B) (C) (D)

(E) (F)

An *N*-substituted
carbamic acid

Fluoxetine

(a) Propose a reagent for the conversion of (A) to (B).

(b) Propose a reagent for the conversion of (B) to (C).

(c) Propose a reagent for the conversion of (C) to (D).

(d) Propose a mechanism for the conversion of (E) to (F). The reagent used in this synthesis is ethyl chloroformate. The other product of this conversion is chloromethane, CH_3Cl. Your mechanism should show how the CH_3Cl is formed.

(e) Propose a reagent or reagents to bring about the conversion of (F) to fluoxetine. Note that the bracketed intermediate formed in this step is an *N*-substituted carbamic acid. Such compounds are unstable and break down to carbon dioxide and an amine.

(f) Is fluoxetine chiral? If so, which of the possible stereoisomers are formed in this synthesis?

21.55 Following is a synthesis for the antiarrhythmic drug bidisomide. The symbol Bn is an abbreviation for the benzyl group, $C_6H_5CH_2$—.

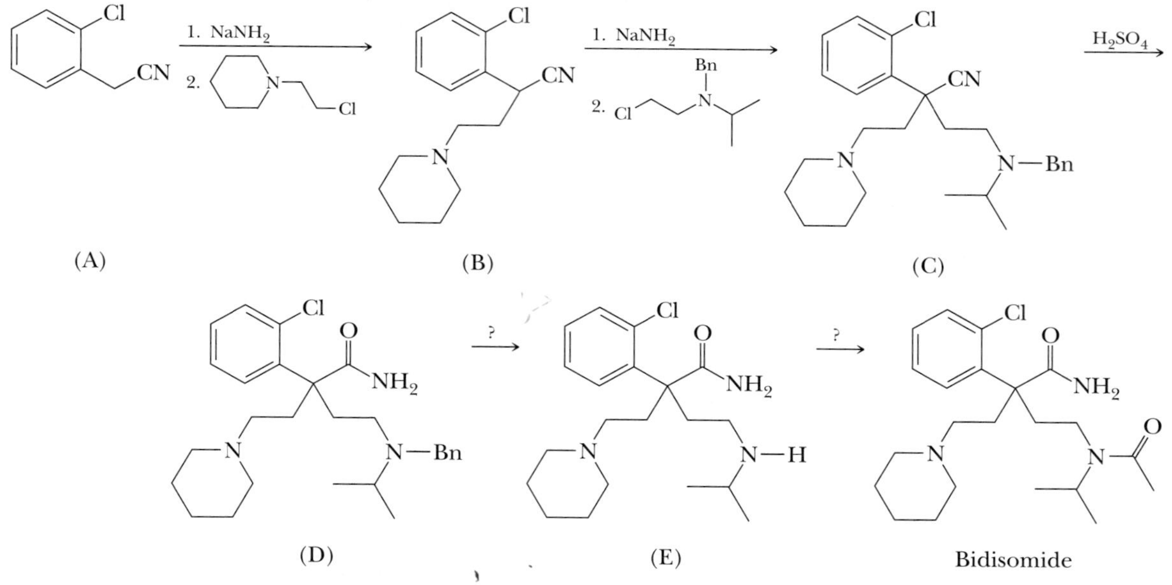

(A)   (B)   (C)

(D)   (E)   Bidisomide

(a) Propose mechanisms for the conversion of (A) to (B) and of (B) to (C). What is the function of sodium amide in each reaction?

(b) Why is it necessary to incorporate the benzyl group on the chloroamine used to convert (B) to (C)?

(c) Propose a reagent or reagents for the removal of the benzyl group in the conversion of (D) to (E).

(d) Propose a reagent for the conversion of (E) to bidisomide.

(e) Is bidisomide chiral? If so, which of the possible stereoisomers are formed in this synthesis?

21.56 A finding that opened a route to β-blockers was the discovery that β-blocking activity is retained if an oxygen atom is interposed between the aromatic ring and the side chain. To see this difference, compare the structures of labetalol (Problem 22.55) and propranolol. Thus, alkylation of phenoxide ions can be used as a way to introduce this side chain. The first of this new class of drugs was propranolol.

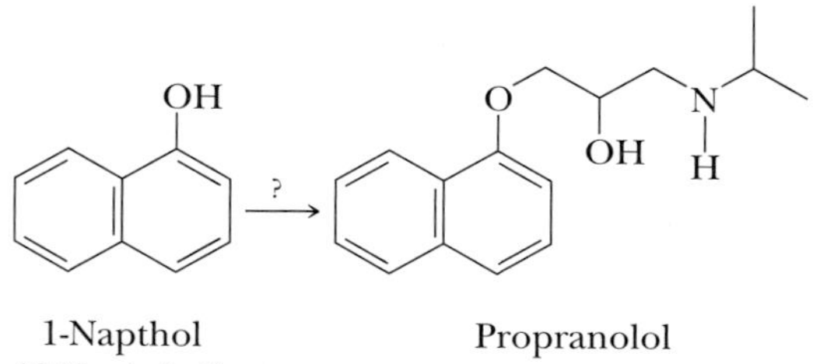

1-Napthol
(β-Naphthol)

Propranolol

(a) Show how propanolol can be synthesized from 1-naphthol, epichlorohydrin (Section 11.10), and isopropylamine.

(b) Is propranolol chiral? If so, which of the possible stereoisomers are formed in this synthesis?

21.57 Side effects of propranolol (Problem 21.56) include disturbances of the central nervous system (CNS) such as fatigue, sleep disturbances (including insomnia and nightmares), and depression. Pharmaceutical companies wondered if this drug could be redesigned to eliminate or at least reduce these side effects. Propranolol, it was reasoned, enters the CNS by passive diffusion because of the lipidlike character of its naphthalene ring. The challenge, then, was to design a more hydrophilic drug that does not cross the blood-brain barrier but still retains a β-adrenergic antagonist property. A product of this research is atenolol, a potent β-adrenergic blocker that is hydrophilic enough that it crosses the blood-brain barrier to only a very limited extent. Atenolol is now one of the most widely used β-blockers.

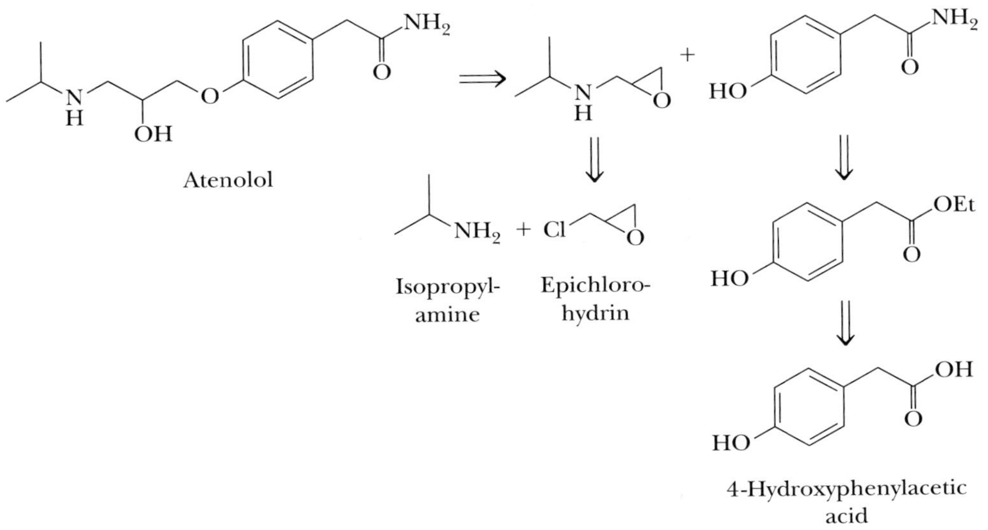

Atenolol

Isopropyl-   Epichloro-
amine   hydrin

4-Hydroxyphenylacetic
acid

(a) Given this retrosynthetic analysis, propose a synthesis for atenolol from the three named starting materials.
(b) Note that the amide functional group is best made by amination of the ester. Why was this route chosen rather than conversion of the carboxylic acid to its acid chloride and then treatment of the acid chloride with ammonia?
(c) Is atenolol chiral? If so, which of the possible stereoisomers are formed in this synthesis?

21.58 In certain clinical situations, there is need for an injectable β-blocker with a short biological half-life. The clue to development of such a drug was taken from the structure of atenolol, whose corresponding carboxylic acid (the product of hydrolysis of its amide) has no β-blocking activity. Substitution of an ester for the amide group and lengthening the carbon side chain by one methylene group resulted in esmolol. Its ester group is hydrolyzed quite rapidly to a carboxyl group by serum esterases under physiological conditions. This hydrolysis product has no β-blocking activity. Propose a synthesis for esmolol from 4-hydroxycinnamic acid, epichlorohydrin, and isopropylamine.

Esmolol

4-Hydroxycinnamic   Isopropyl-   Epichloro-
acid   amine   hydrin

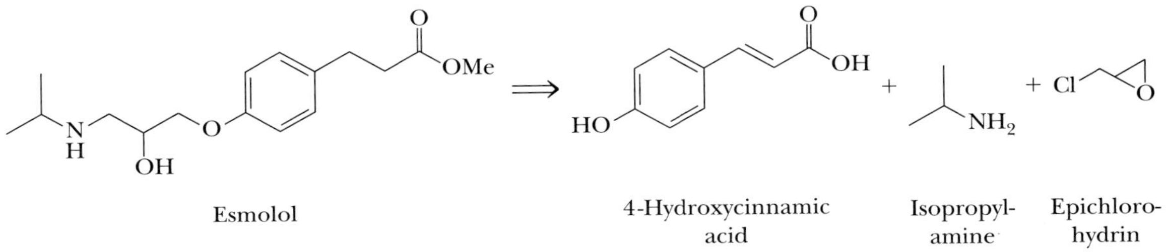

(a) Propose a synthesis for esmolol from 4-hydroxycinnamic acid, epichlorohydrin, and isopropylamine.
(b) Is esmolol chiral? If so, which of the possible stereoisomers are formed in this synthesis?

21.59 Following is an outline of a synthesis of the bronchodilator carbuterol, a beta-2 adrenergic blocker with high selectivity for airway smooth muscle receptors.

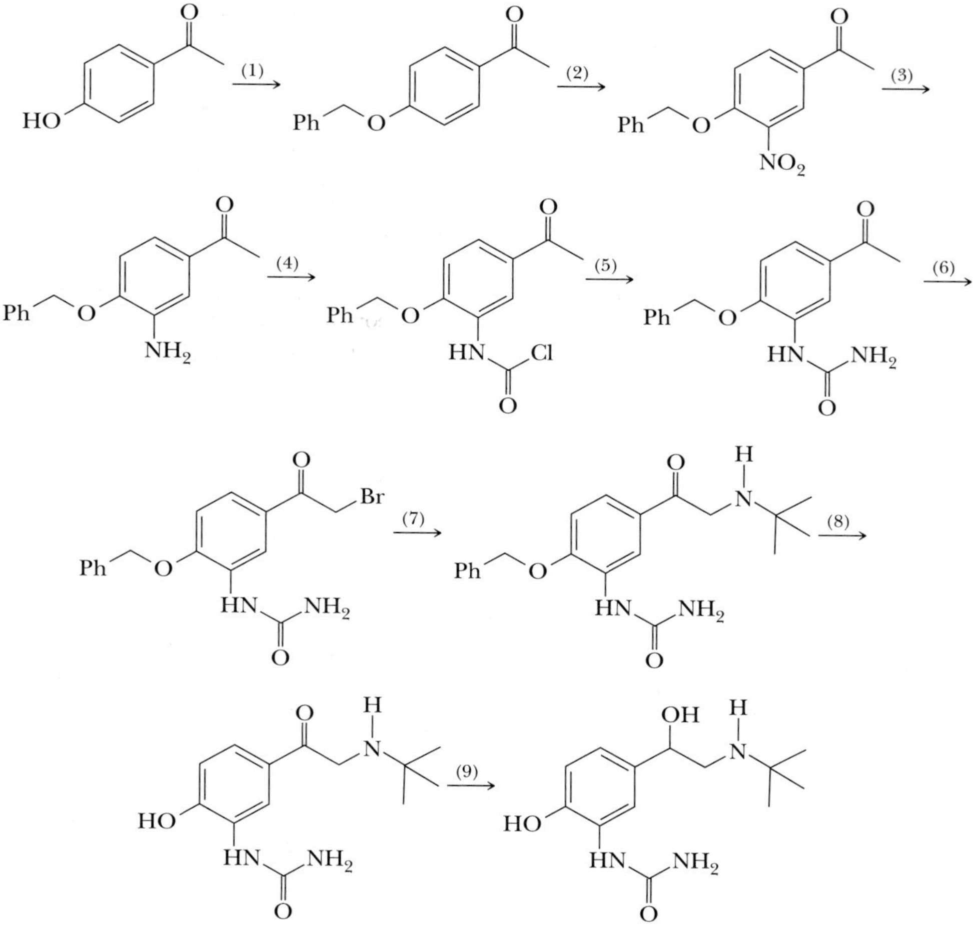

Carbuterol

(a) Propose reagents to bring about each step.
(b) Why is it necessary to add the benzyl group, PhCH$_2$—, as a blocking group in Step 1?
(c) Suggest a structural relationship between carbuterol and ephedrine.
(d) Is carbuterol chiral? If so, which of the possible stereoisomers are formed in this synthesis?

21.60 Following is a synthesis for albuterol (Proventil), currently one of the most widely used inhalation bronchodilators.

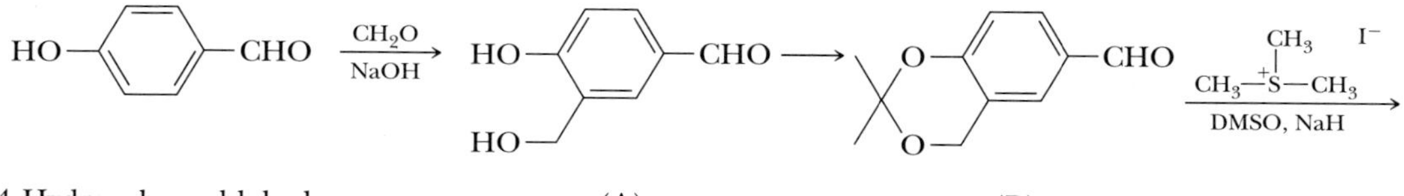

4-Hydroxybenzaldehyde   (A)   (B)

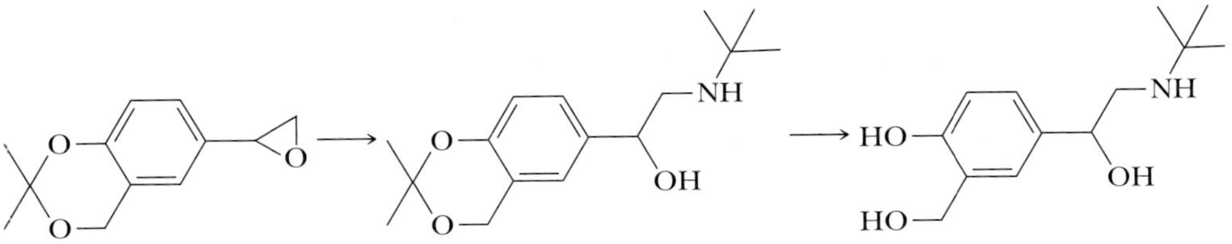

(C)   (D)   Albuterol

(a) Propose a mechanism for conversion of 4-hydroxybenzaldehyde to (A).
(b) Propose reagents and experimental conditions for conversion of (A) to (B).
(c) Propose a mechanism for the conversion of (B) to (C). *Hint:* Think of trimethylsulfonium iodide as producing a sulfur equivalent of a Wittig reagent.
(d) Propose reagents and experimental conditions for the conversion of (C) to (D).
(e) Propose reagents and experimental conditions for the conversion of (D) to albuterol.
(f) Is albuterol chiral? If so, which of the possible stereoisomers are formed in this synthesis?

21.61 Estrogens are female sex hormones, the most potent of which is β-estradiol.

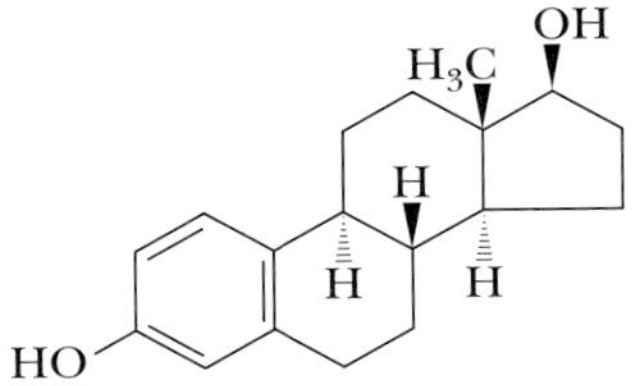

β-Estradiol

In recent years, there have been intense efforts to design and synthesize molecules that will bind to estrogen receptors. One target of this research has been nonsteroidal estrogen antagonists, compounds that interact with estrogen receptors and block the effects of both endogenous and exogenous estrogens. A feature common to one type of nonsteroidal estrogen antagonist is the presence of a 1,2-diphenylethylene with one of the benzene rings bearing a dialkylaminoethoxyl substituent. The first nonsteroidal estrogen antagonist of this type to achieve clinical importance was tamoxifen, now an important drug in the treatment of breast cancer. Tamoxifen has the Z configuration as shown here.

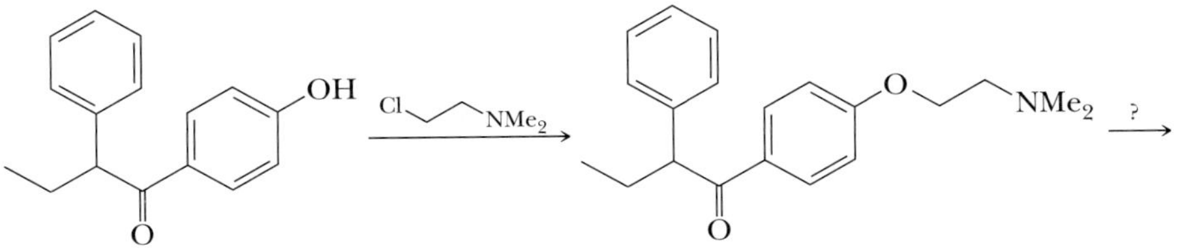

(A)   (B)

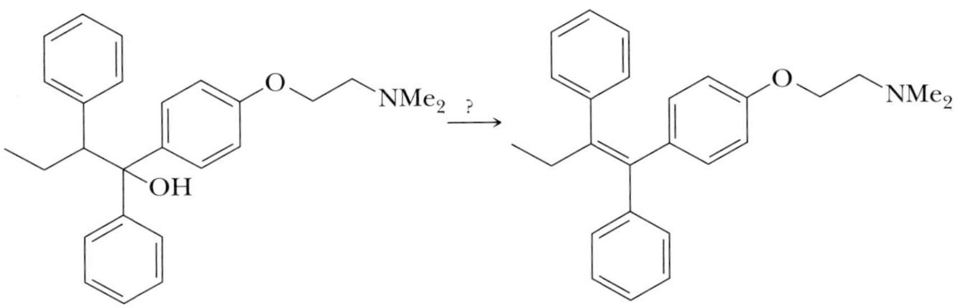

(C)   Tamoxifen

Propose reagents for the conversion of (A) to tamoxifen. *Note:* The final step in this synthesis gives a mixture of *E* and *Z* isomers.

21.62 Following is a synthesis for toremifene, a nonsteroidal estrogen antagonist whose structure is closely related to that of tamoxifen.

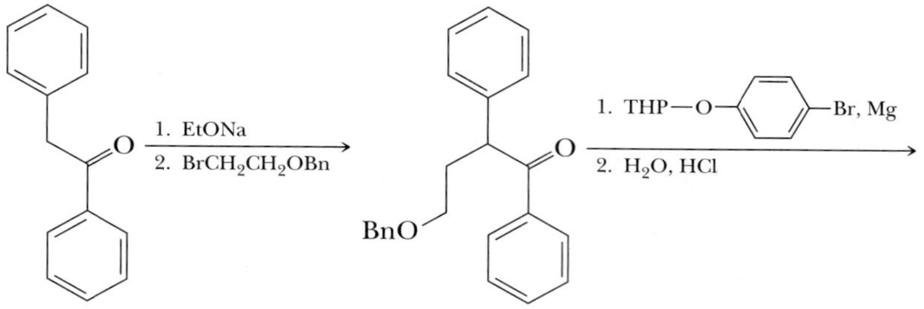

(A)   (B)

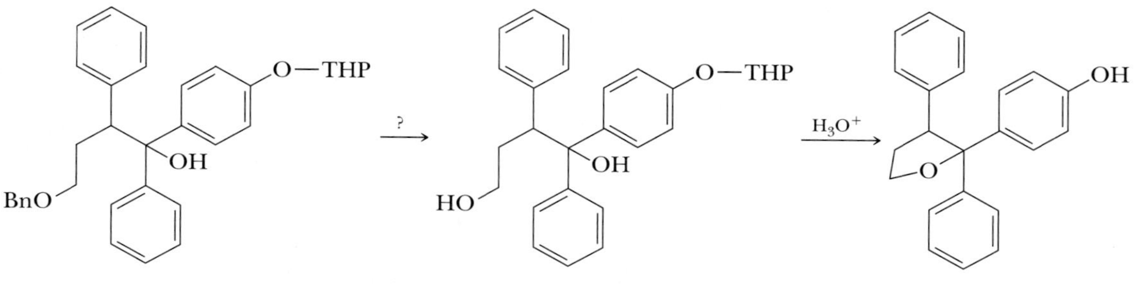

(C)   (D)   (E)

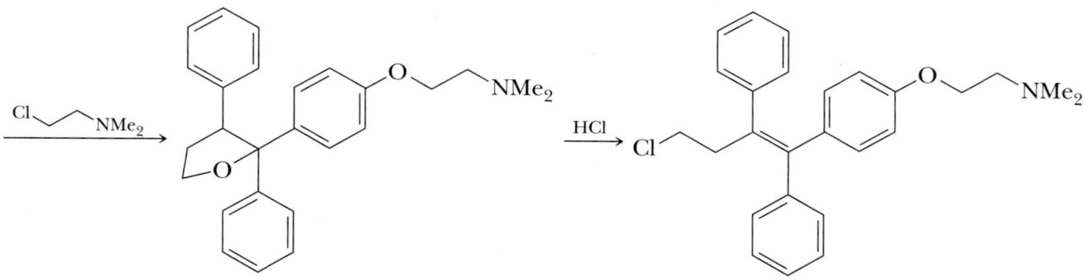

(F)   Toremifene

(a) This synthesis makes use of two blocking groups, the benzyl (Bn) group and the tetrahydropyranyl (THP) group. Draw a structural formula of each group, and describe the experimental conditions under which it is attached and removed.

(b) Discuss the chemical logic behind the use of each blocking group in this synthesis.

(c) Propose a mechanism for the conversion of (D) to (E).

(d) Propose a mechanism for the conversion of (F) to toremifene.

(e) Is toremifene chiral? If so, which of the possible stereoisomers are formed in this synthesis?

Charles D. Winters

■ 2,6-Di-*tert*-4-methylphenol, alternatively known as butylated hydroxytoluene or BHT (see Problem 22.23) is often used as an antioxidant to retard spoilage. Inset: A model of BHT.

Outlines

Reactions of Benzene and Its Derivatives

By far the most characteristic reaction of aromatic compounds is substitution at a ring carbon. In this reaction, one of the ring hydrogens is replaced by another atom or group of atoms. Some groups that can be introduced directly on the ring are the halogens, the nitro (—NO_2) group, the sulfonic acid (—SO_3H) group, alkyl (—R) groups, and acyl (RCO—) groups. Each of these substitution reactions is represented in the following equations.

Halogenation:

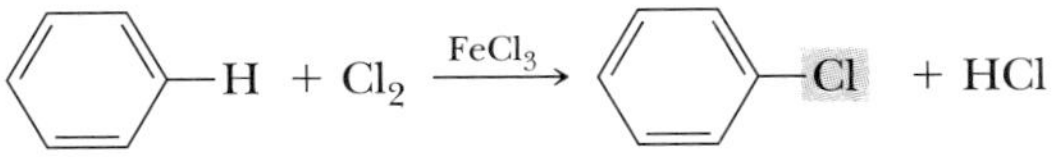

$$\text{C}_6\text{H}_5\text{—H} + \text{Cl}_2 \xrightarrow{\text{FeCl}_3} \text{C}_6\text{H}_5\text{—Cl} + \text{HCl}$$

Chlorobenzene

Nitration:

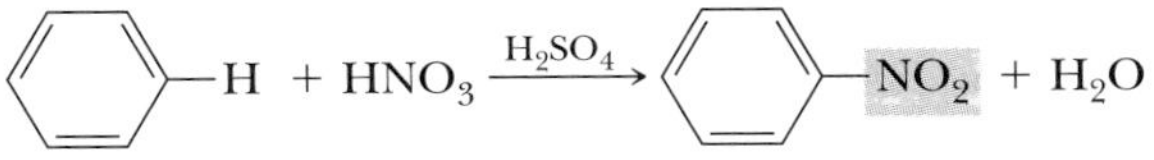

$$\text{C}_6\text{H}_5\text{—H} + \text{HNO}_3 \xrightarrow{\text{H}_2\text{SO}_4} \text{C}_6\text{H}_5\text{—NO}_2 + \text{H}_2\text{O}$$

Nitrobenzene

ORGANIC
Chemistry Now™

Look for this logo in the chapter and go to Organic ChemistryNow at **http://now.brookscole.com/bfi4** for tutorials, simulations, and problems.

Sulfonation:

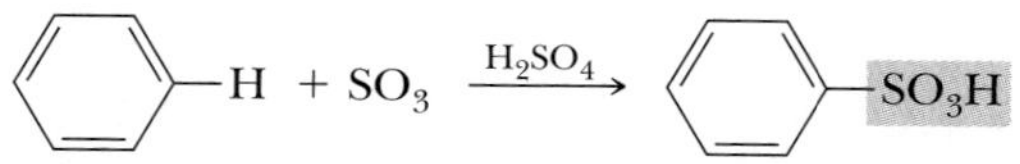

Benzenesulfonic acid

Alkylation:

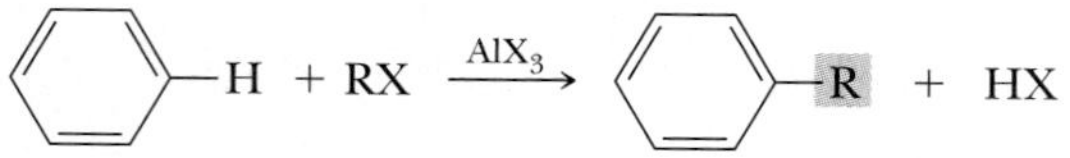

An alkylbenzene

Acylation:

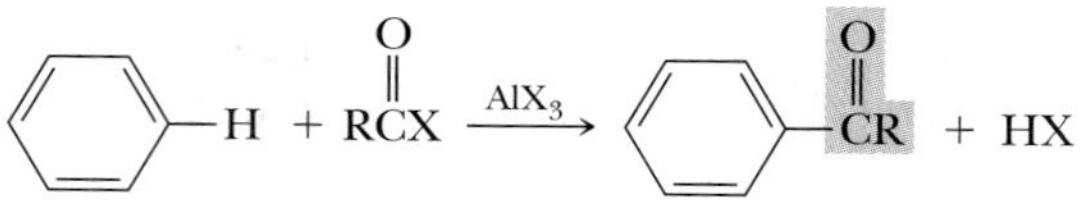

An acylbenzene

We will take these reactions one at a time and examine their common mechanistic theme.

22.1 Electrophilic Aromatic Substitution

Electrophilic aromatic substitution A reaction in which there is substitution of an electrophile, E^+, for a hydrogen on an aromatic ring.

ORGANIC Chemistry Now™
Click Tutorials for an overview of **Electrophilic Aromatic Substitution**

In an **electrophilic aromatic substitution,** a hydrogen atom of an aromatic ring is replaced by an electrophile, E^+.

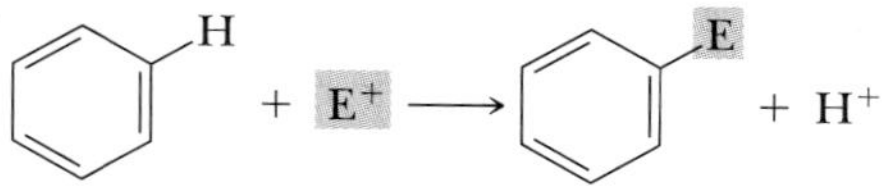

We study several common types of electrophiles, how each is generated, and the mechanism by which it replaces hydrogen on an aromatic ring.

A. Chlorination and Bromination

Chlorine alone does not react with benzene, in contrast to its instantaneous addition to cyclohexene. However, in the presence of a Lewis acid catalyst, such as ferric chloride or aluminum chloride (Section 4.6), benzene reacts with chlorine to give chlorobenzene and HCl. As shown in the following mechanism, this reaction involves a series of Lewis acid/base reactions.

Mechanism *Electrophilic Aromatic Substitution—Chlorination*

Step 1: Reaction between chlorine and the Lewis acid catalyst gives a molecular complex with a positive charge on chlorine and a negative charge on iron. Redistribution of electrons in this complex generates a **chloronium ion,** Cl^+, a very strong electrophile, as part of an ion pair.

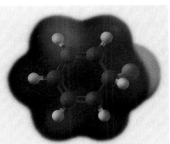

Cation Intermediate

Chlorine (a Lewis base)   Ferric chloride (a Lewis acid)   A molecular complex with a positive charge on chlorine and a negative charge on iron   An ion pair containing the chloronium ion

Step 2: Attack of the chloronium ion (a strong electrophile) on the pi system (a weak nucleophile) of the aromatic ring gives a resonance-stabilized cation intermediate, here represented as a hybrid of three contributing structures. Notice that the positive charge is located primarily at the ortho and para positions of the resonance-stabilized cation intermediate. This distribution of positive charge is clearly visible in the electrostatic potential surface of the cation intermediate, as the blue color at the ortho and para positions.

slow, rate determining

Step 3: Proton transfer from the cation intermediate to $FeCl_4^-$ forms HCl, regenerates the Lewis acid catalyst, and gives chlorobenzene.

fast

Cation intermediate   Chlorobenzene

ORGANIC
Chemistry⚬❀⚬**Now**™
Click Mechanisms to view an animation of the **Electrophilic Aromatic Substitution with Cl**

Treating benzene with bromine in the presence of ferric chloride or aluminum chloride gives bromobenzene and HBr. The mechanism of this reaction is the same as that for the chlorination of benzene.

We can write the following general two-step mechanism for electrophilic aromatic substitution. The first and rate-determining step is attack of the strong electrophile, E^+, on the weakly nucleophilic pi electrons of the aromatic ring to give a resonance-stabilized cation intermediate. The second and faster step, loss of H^+ from the cation intermediate, regenerates aromaticity in the ring and gives the product.

Step 1:

slow, rate determining

Electrophile   Resonance-stabilized cation intermediate

Step 2:

fast

The major difference between addition of halogen to an alkene and halogen substitution on an aromatic ring centers on the fate of the cationic intermediate formed in the first step of each reaction. Recall from Section 6.3D that addition of chlorine or bromine to an alkene is a two-step process, the first and slower step of

Figure 22.1
Energy diagram for the reaction of benzene with bromine. Formation of the addition product results in loss of the resonance stabilization of the aromatic ring. Formation of a substitution product regenerates the resonance-stabilized aromatic ring.

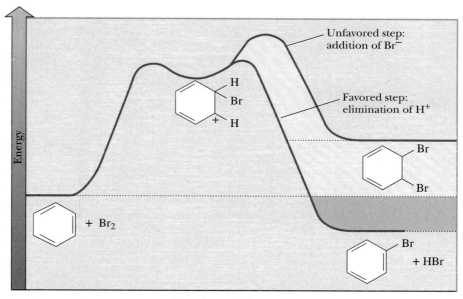

which is formation of a bridged halonium ion intermediate. This cationic intermediate then reacts with chloride or bromide ion to complete the addition. With aromatic compounds, the cationic intermediate instead loses H^+ to regenerate the aromatic ring and regain the large resonance stabilization. No such resonance stabilization is regained in the case of an alkene. The energy diagram in Figure 22.1 shows both addition and substitution reactions of benzene. Addition causes loss of the aromatic resonance energy and is disfavored except under extreme circumstances.

B. Nitration and Sulfonation

The sequence of steps for nitration and sulfonation of benzene is similar to that for chlorination and bromination. For nitration, the **nitronium ion,** NO_2^+, a very strong electrophile, is generated by the reaction between nitric acid and sulfuric acid.

Mechanism   *Formation of the Nitronium Ion*

Step 1: Proton transfer from sulfuric acid to the OH group of nitric acid gives the conjugate acid of nitric acid.

$$HSO_3-\overset{..}{\underset{..}{O}}-H + H-\overset{..}{\underset{..}{O}}-N^+ \rightleftharpoons HSO_4^- + \overset{H}{\underset{H}{\diagdown}}\overset{+}{O}-N^+$$

Sulfuric acid   Nitric acid   Conjugate acid
of nitric acid

Step 2: Loss of water from this conjugate acid gives the nitronium ion, a very strong electrophile.

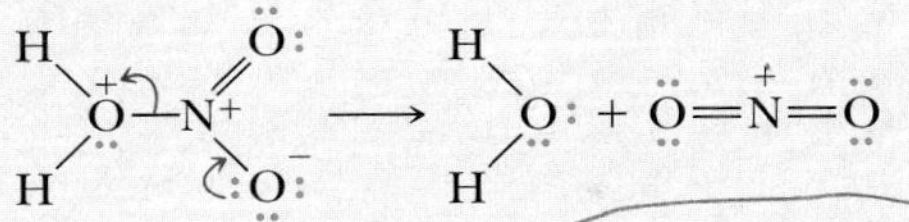

The nitronium ion

ORGANIC
Chemistry·ᐧ·Now™
Click Mechanisms to view an
animation of **Aromatic Nitration**

Example 22.1

Write a stepwise mechanism for the nitration of benzene.

Solution

The nitronium ion (a strong electrophile) attacks the benzene ring (a weak nucleophile) in Step 1 to give a resonance-stabilized cation intermediate. Proton transfer from this intermediate to either H_2O or HSO_4^- in Step 2 regenerates the aromatic ring and gives nitrobenzene.

Step 1:

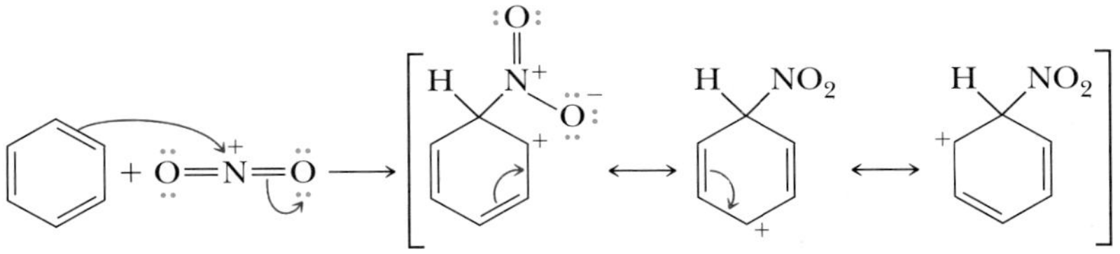

Resonance-stabilized cation intermediate

Step 2:

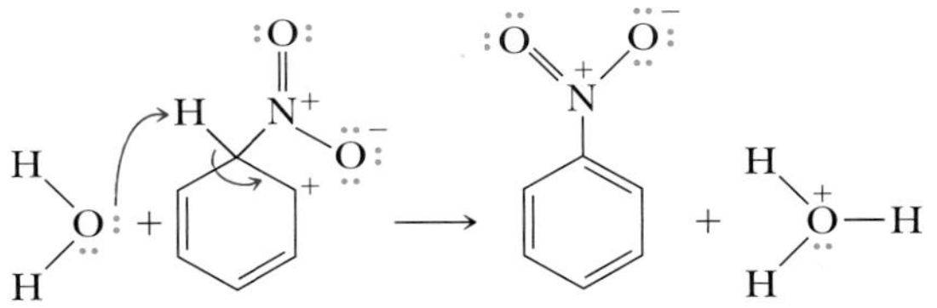

An important feature of nitration is that the resulting nitro group can be reduced to a primary amino group, $-NH_2$, by hydrogenation in the presence of a transition metal catalyst such as nickel, palladium, or platinum under fairly mild conditions.

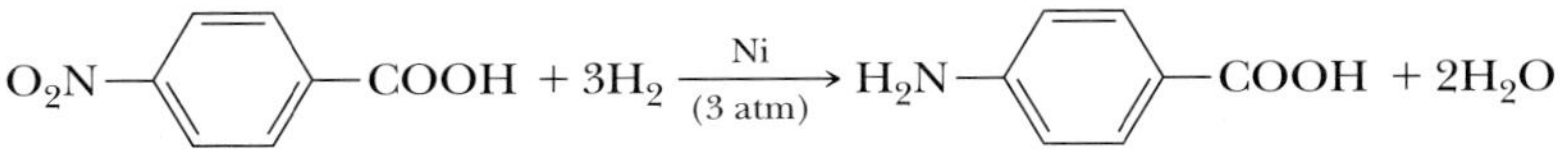

4-Nitrobenzoic acid   4-Aminobenzoic acid

As illustrated by this example, neither a $-COOH$ nor an aromatic ring is reduced under these conditions. Catalytic reduction of a nitro group has the potential disadvantage, however, that other susceptible groups such as carbon-carbon double bonds and aldehyde and ketone carbonyl groups may also be reduced.

Alternatively, a nitro group can be reduced to a primary amino group by a metal in aqueous acid. The most commonly used metal-reducing agents are iron, zinc, and tin in dilute HCl. The reductant is electrons from the metal. When reduced with a metal and hydrochloric acid, the amine is obtained as a salt, which is then treated with strong base to liberate the free amine.

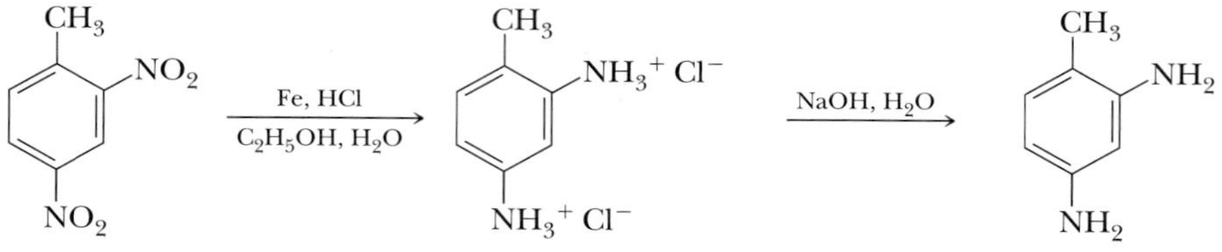

2,4-Dinitrotoluene

4-Methyl-1,3-benzenediamine
(2,4-Diaminotoluene)

The reduction of a nitro group to an amino group is important because the amino group cannot be substituted directly onto an aromatic ring; it must be done by this indirect method.

Sulfonation of benzene is carried out using concentrated sulfuric acid containing dissolved sulfur trioxide (fuming sulfuric acid). The electrophile is either SO_3 or HSO_3^+, depending on experimental conditions.

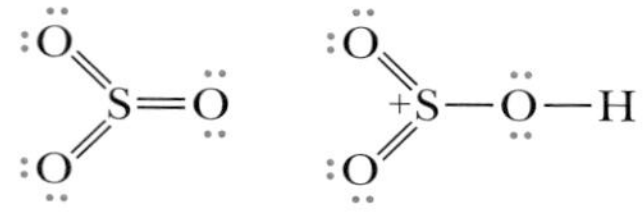

In the following equation, the sulfonating agent is shown as sulfur trioxide.

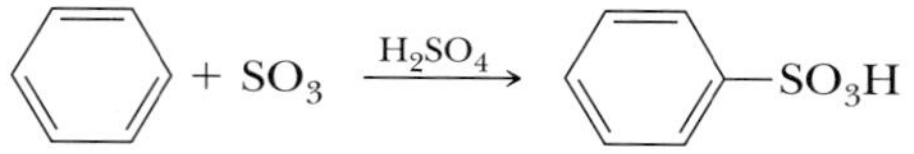

Benzene   Benzenesulfonic acid

Problem 22.1

Write the stepwise mechanism for sulfonation of benzene by hot, concentrated sulfuric acid. In this reaction, the electrophile is SO_3 formed as shown in the following equation.

$$H_2SO_4 \rightleftharpoons SO_3 + H_2O$$

C. Friedel-Crafts Alkylation and Acylation

Alkylation of aromatic hydrocarbons was discovered in 1877 by the French chemist Charles Friedel and a visiting American chemist, James Crafts. They discovered that mixing benzene, an alkyl halide, and $AlCl_3$ results in the formation of an alkylbenzene and HX. **Friedel-Crafts alkylation,** which is among the most important methods for forming new carbon-carbon bonds to aromatic rings, is illustrated here by the reaction of benzene with 2-chloropropane in the presence of aluminum chloride.

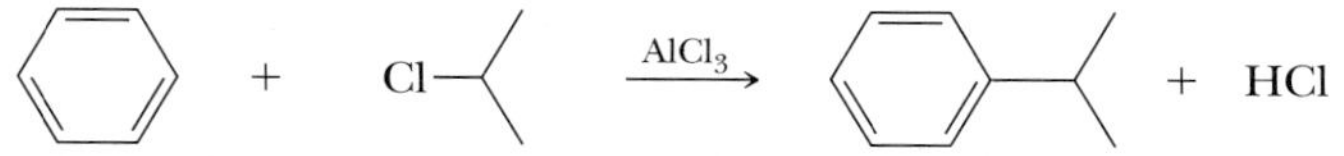

Friedel-Crafts reaction An electrophilic aromatic substitution in which a hydrogen of an aromatic ring is replaced by an alkyl or acyl group.

Benzene   2-Chloropropane   Cumene
(Isopropyl chloride)   (Isopropylbenzene)

The mechanism for Friedel-Crafts alkylation, like that for halogenation, nitration, and sulfonation, involves the attack of the aromatic ring by a strong electrophile, in this case a carbocation formed by reaction between the alkyl halide and the Lewis acid catalyst.

Mechanism *Friedel-Crafts Alkylation*

Step 1: The alkyl halide (a Lewis base) and aluminum chloride (a Lewis acid) form a complex in which aluminum has a negative charge and the halogen of the alkyl halide has a positive charge. The alkyl group can also be written as a carbocation. It is unlikely, however, that a free carbocation is actually formed, especially in the case of the relatively unstable primary and secondary carbocations. Nonetheless, we often represent the reactive intermediate as a carbocation to simplify the notation of the mechanism.

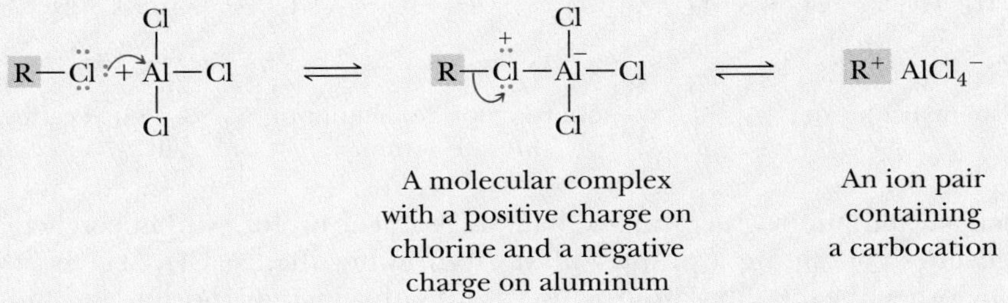

A molecular complex
with a positive charge on
chlorine and a negative
charge on aluminum

An ion pair
containing
a carbocation

Step 2: Reaction of the carbocation (a strong electrophile) with the pi electrons (a weak nucleophile) of the aromatic ring gives a resonance-stabilized cation intermediate.

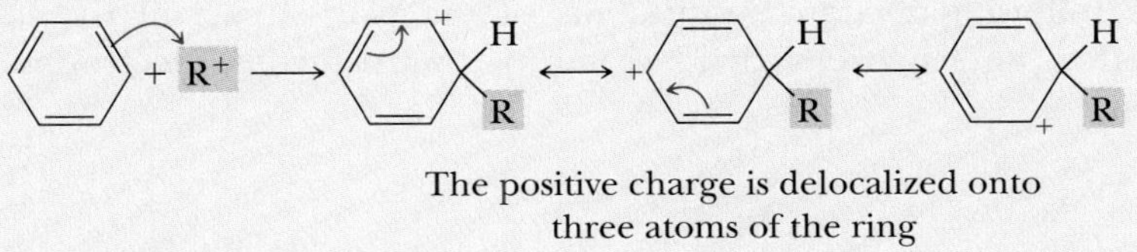

The positive charge is delocalized onto
three atoms of the ring

Step 3: Proton transfer regenerates the ring aromaticity.

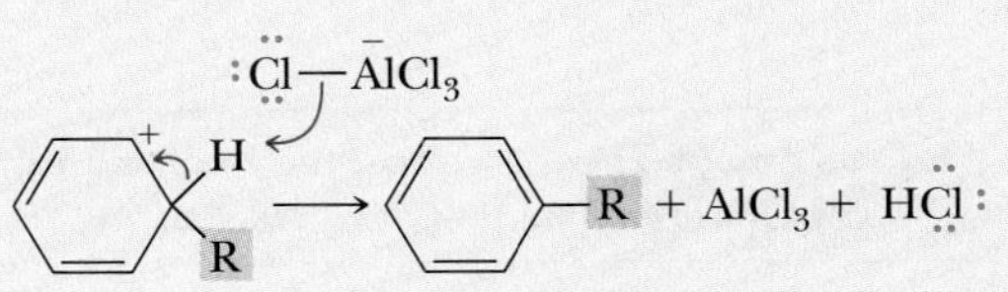

Halogen atoms on sp^2-hybridized carbons (vinylic and aryl halides) do not react to produce electrophiles under conditions of the Friedel-Crafts alkylation because of the high activation energy required to form these carbocations.

There are three major limitations on the Friedel-Crafts alkylation. The first is the possibility for rearrangement of the alkyl group, which occurs in the following way. Friedel-Crafts alkylation involves the generation of a carbocation, and, as we have already seen in Section 6.3C, carbocations may rearrange to a more stable carbocation.

Carbocation rearrangements are common in Friedel-Crafts alkylations. For example, reaction of benzene with 1-chloro-2-methylpropane (isobutyl chloride) gives only 2-methyl-2-phenylpropane (*tert*-butylbenzene).

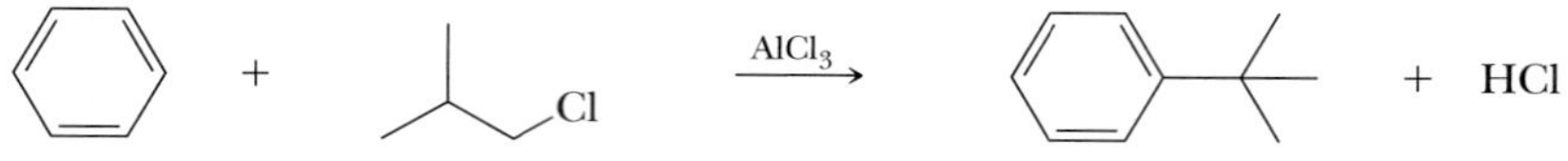

1-Chloro-2-methylpropane
(Isobutyl chloride)

2-Methyl-2-phenylpropane
(*tert*-Butylbenzene)

In this case, the isobutyl chloride/AlCl₃ complex rearranges directly to the *tert*-butyl cation/AlCl₄⁻ ion pair, which is the electrophile in this example of Friedel-Crafts alkylation.

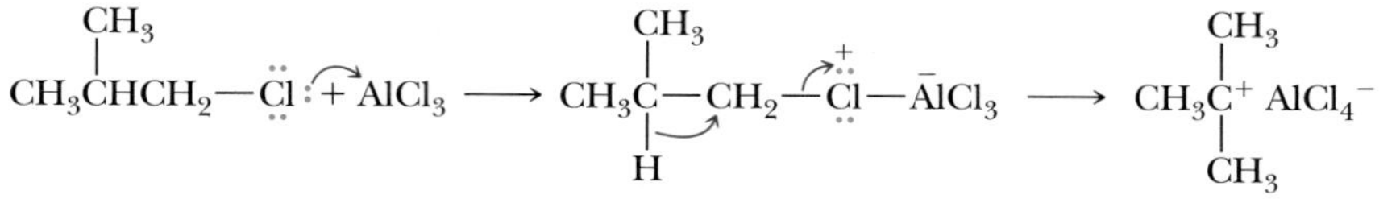

Isobutyl chloride

Isobutyl chloride/aluminum
chloride complex

tert-Butyl cation/
AlCl₄⁻ ion pair

In practice, alkylation with primary halides is not a useful synthetic reaction, and alkylbenzenes containing a primary alkyl group other than —CH₂CH₃ must be prepared by other means. Alkylation is useful for introducing isopropyl, *tert*-butyl, and other alkyl groups, the cations of which tend not to rearrange.

A second limitation on Friedel-Crafts alkylation is that it fails altogether on benzene rings bearing one or more strongly electron-withdrawing groups. As we shall see in the following section, substituents of the type shown in the table have a dramatic effect on a benzene ring's reactivity toward further electrophilic aromatic substitution.

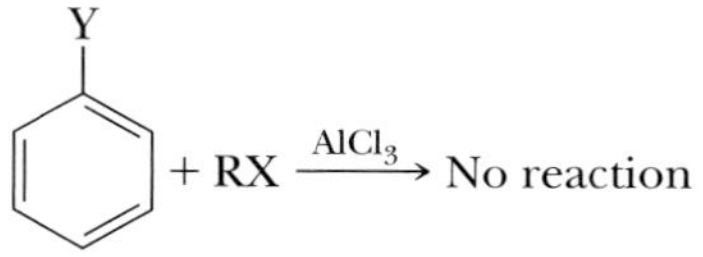

When Y Equals Any of These Groups, the Benzene Ring Does Not Undergo Friedel-Crafts Alkylation				
$\overset{O}{\overset{\|}{-CH}}$	$\overset{O}{\overset{\|}{-CR}}$	$\overset{O}{\overset{\|}{-COH}}$	$\overset{O}{\overset{\|}{-COR}}$	$\overset{O}{\overset{\|}{-CNH_2}}$
—SO₃H	—C≡N	—NO₂	—NR₃⁺	
—CF₃	—CCl₃			

The third limitation on Friedel-Crafts alkylation is that it is hard to stop the reaction at monoalkylation because the alkylated product is more reactive than benzene itself. We will discuss reactivity in detail in Section 22.2, but, in general, alkylated benzenes are more reactive than unsubstituted compounds. This limitation can be overcome if it is feasible to use a large excess of benzene, often as both the solvent and as the reactant.

Friedel and Crafts also discovered that treating an aromatic hydrocarbon with an acyl halide (Section 18.1A) in the presence of aluminum chloride gives a ketone. An RCO— group is known as an acyl group; hence, reaction of an aromatic hydrocarbon with an acyl halide is known as **Friedel-Crafts acylation,** as illustrated by the reaction of benzene and acetyl chloride in the presence of aluminum chloride to form acetophenone.

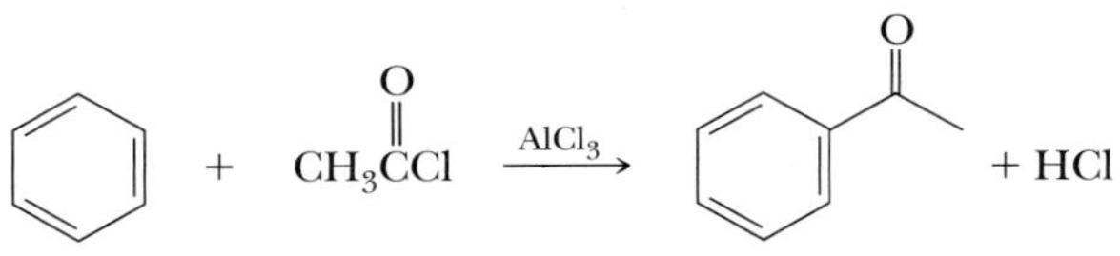

| Benzene | Acetyl chloride (an acyl halide) | | Acetophenone (a ketone) |

The fact that an acylbenzene is less reactive than the starting material (unreactive in most cases) overcomes the third limitation of the alkylation.

The following example of electrophilic aromatic substitution involves intramolecular acylation to form a six-membered ring.

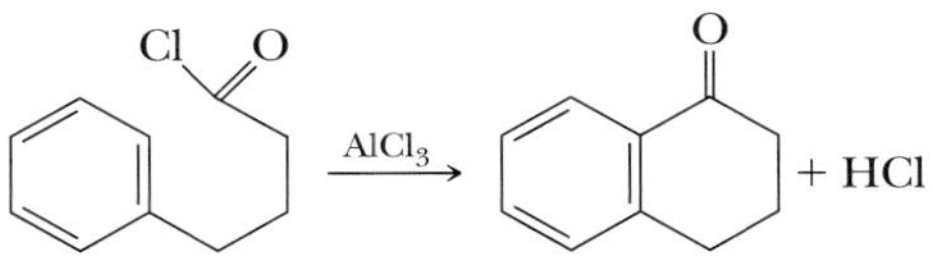

| 4-Phenylbutanoyl chloride | | α-Tetralone |

Mechanism   *Friedel-Crafts Acylation—Generation of an Acylium Ion*

Step 1: Friedel-Crafts acylation begins with the donation of a pair of electrons from the halogen of the acyl halide to aluminum chloride to form a molecular complex similar to what we drew for Friedel-Crafts alkylations. In this complex, halogen has a positive formal charge and aluminum has a negative formal charge.

Step 2: Redistribution of electrons of the carbon-chlorine bond gives an ion pair containing an **acylium ion.**

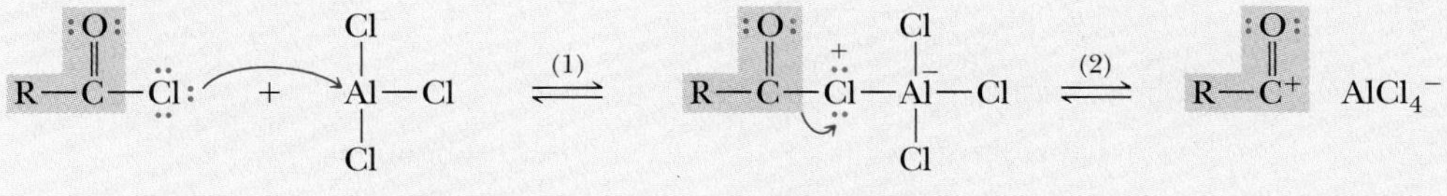

| An acyl chloride (a Lewis base) | Aluminum chloride (a Lewis acid) | A molecular complex with a positive charge on chlorine and a negative charge on aluminum | An ion pair containing an acylium ion |

Acylium ion   A resonance-stabilized cation with the structure $[RC{=}O]^+$ or $[ArC{=}O]^+$. The positive charge is delocalized over both the carbonyl carbon and the carbonyl oxygen.

Of the two major contributing structures that can be drawn for an acylium ion, the one with complete valence shells for both carbon and oxygen makes the greater contribution to the hybrid.

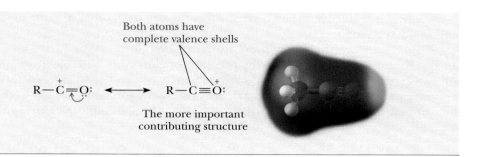

ORGANIC
Chemistry·ᵢ·Now™
Click Mechanisms to view an
animation of the **Friedel-Crafts Acylation**

Both atoms have
complete valence shells

The more important
contributing structure

Friedel-Crafts acylation is free of a second major limitation on Friedel-Crafts alkylations: Acyl cations do not undergo rearrangement. Thus, the carbon skeleton of an acyl halide is transferred unchanged to the aromatic ring.

Example 22.2

Write a structural formula for the product from Friedel-Crafts alkylation or acylation of benzene for each compound.

(a)　$C_6H_5CH_2Cl$ 　(b)　$C_6H_5\overset{O}{\overset{\|}{C}}Cl$

　　Benzyl chloride　　　Benzoyl chloride

Solution

(a) Benzyl chloride in the presence of a Lewis acid catalyst gives the benzyl cation (an electrophile), which then attacks benzene (a weak nucleophile) followed by proton transfer to give diphenylmethane. In this example, the benzyl cation, although primary, cannot rearrange.

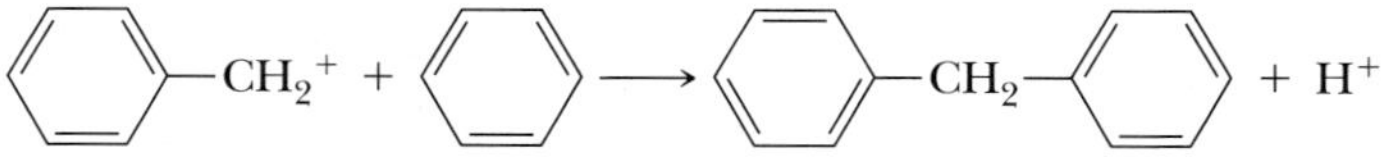

$$\text{Benzyl cation} + \text{benzene} \longrightarrow \text{Diphenylmethane} + H^+$$

　　　Benzyl cation　　　　　　Diphenylmethane

(b) Treating benzoyl chloride with aluminum chloride gives an acylium ion (an electrophile). Reaction of this cation with the pi electrons of the aromatic ring (a weak nucleophile) followed by proton transfer gives benzophenone.

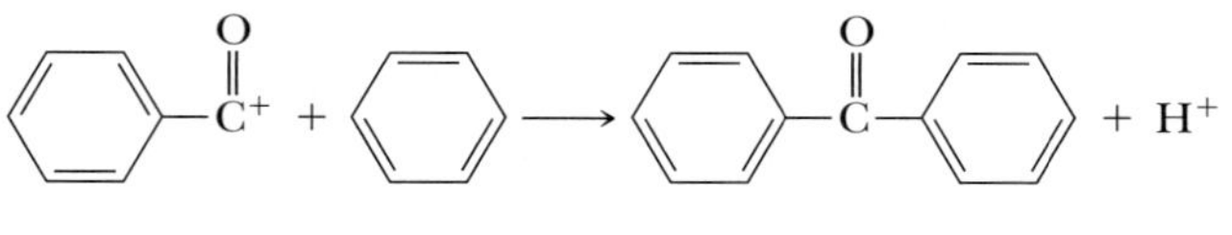

$$\text{Benzoyl cation} + \text{benzene} \longrightarrow \text{Benzophenone} + H^+$$

　　　Benzoyl cation　　　　　　Benzophenone

Problem 22.2

Write a structural formula for the product from Friedel-Crafts alkylation or acylation of benzene with each compound.

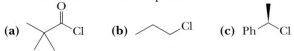

(a)　　　(b)　　　(c) Ph

A special value of Friedel-Crafts acylations in synthesis is for the preparation of unrearranged alkylbenzenes, as illustrated by the preparation of isobutylbenzene.

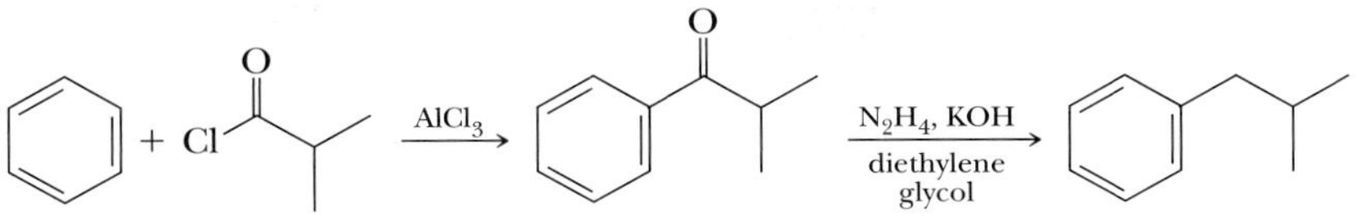

2-Methyl-
propanoyl
chloride

2-Methyl-1-
phenyl-1-propanone

Isobutylbenzene

Treating benzene with 2-methylpropanoyl chloride in the presence of aluminum chloride gives 2-methyl-1-phenyl-1-propanone. Wolff-Kishner or Clemmensen reduction of the carbonyl group to a methylene group (Section 16.11C) gives isobutylbenzene.

D. Other Electrophilic Aromatic Alkylations

After the discovery that Friedel-Crafts alkylations and acylations involve cationic electrophiles, it was realized that the same reactions can be accomplished by other combinations of reagents and catalysts. We study two of these reactions: generation of carbocations from alkenes and from alcohols.

As we saw in Section 6.3, treatment of an alkene with a strong acid, most commonly HX, H_2SO_4, H_3PO_4, or HF/BF_3, generates a carbocation. Cumene, an intermediate in the industrial synthesis of both acetone and phenol (Problem 16.65), is synthesized industrially by treating benzene with propene in the presence of phosphoric acid as a catalyst.

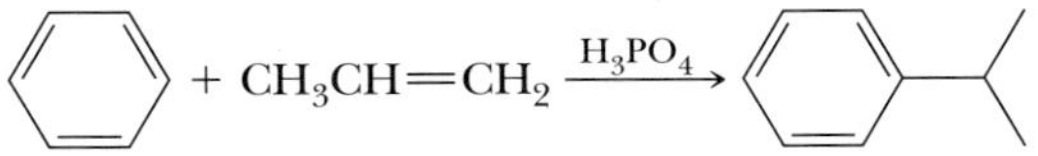

Benzene   Propene   Cumene

Alkylation with an alkene can also be carried out with a Lewis acid catalyst. Treatment of benzene with cyclohexene in the presence of aluminum chloride gives phenylcyclohexane.

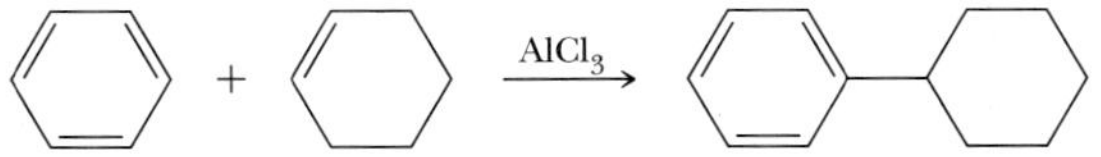

Benzene   Cyclohexene   Phenylcyclohexane

Carbocations can also be generated by treatment of an alcohol with H_2SO_4, H_3PO_4, or HF (Section 10.5).

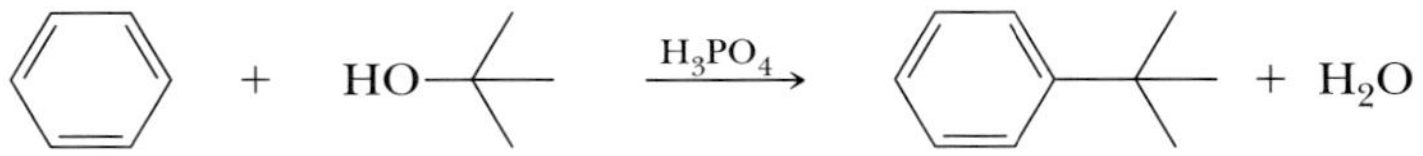

Benzene   2-Methyl-2-propanol
(*tert*-Butyl alcohol)

2-Methyl-2-
phenylpropane
(*tert*-Butylbenzene)

Example 22.3

Write a mechanism for the formation of isopropylbenzene (cumene) from benzene and propene in the presence of phosphoric acid.

Solution

Step 1: Proton transfer from phosphoric acid to propene gives the isopropyl cation.

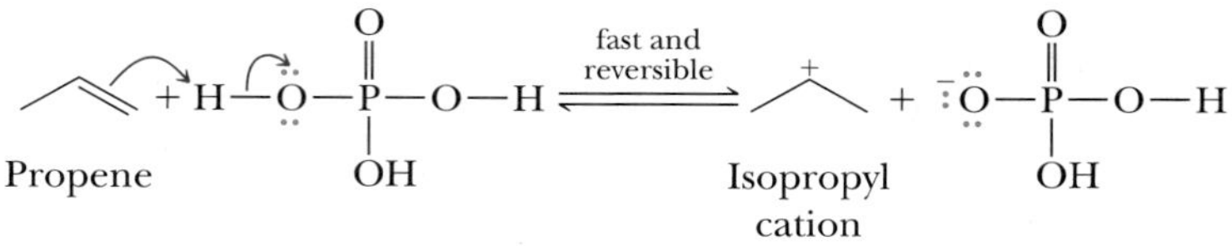

Step 2: Reaction of the isopropyl cation with the pi electrons of the benzene ring gives a resonance-stabilized carbocation intermediate.

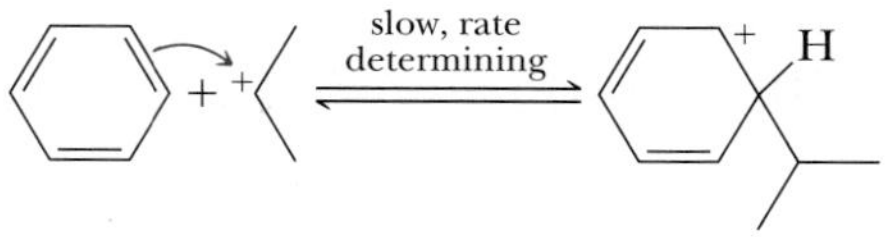

Step 3: Proton transfer to dihydrogen phosphate ion gives cumene.

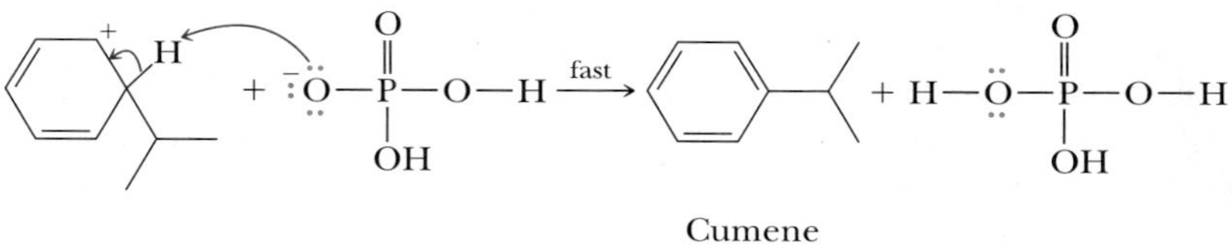

Problem 22.3

Write a mechanism for the formation of *tert*-butylbenzene from benzene and *tert*-butyl alcohol in the presence of phosphoric acid.

22.2 Disubstitution and Polysubstitution

A. Effects of a Substituent Group on Further Substitution

In electrophilic aromatic substitution of a monosubstituted benzene, three products are possible: The new group may become oriented ortho, meta, or para to the existing group. Table 22.1 shows the orientation on nitration of a series of monosubstituted benzenes.

Based on the information in Table 22.1 and other studies like it, we can make the following generalizations about the manner in which existing groups influence further substitution reactions.

1. *Substituents affect the orientation of new groups.* Certain substituents (for example, —OCH$_3$ and —Cl) direct an incoming group preferentially to the ortho and para positions; other substituents (for example, —NO$_2$ and —COOH) direct it preferentially to the meta position. In other words, substituents on a benzene ring can be classified as **ortho-para directing** or as **meta directing.**

Table 22.1 Orientation on Nitration of Monosubstituted Benzenes

Substituent	ortho	meta	para	ortho + para	meta
—OCH$_3$	44	—	55	99	trace
—CH$_3$	58	4	38	96	4
—Cl	70	—	30	100	trace
—Br	37	1	62	99	1
—COOH	18	80	2	20	80
—CN	19	80	1	20	80
—NO$_2$	6.4	93.2	0.3	6.7	93.2

2. *Substituents affect the rate of further substitution.* Certain substituents cause the rate of a second substitution to be greater than that for benzene itself, whereas other substituents cause the rate of a second substitution to be lower than that for benzene. In other words, groups on a benzene ring can be classified as **activating** or **deactivating** toward further substitution.

These directing and activating-deactivating effects can be seen by comparing the products and rates of nitration of anisole and nitration of benzoic acid. Nitration of anisole proceeds at a rate considerably greater than that for benzene (the methoxy group is activating), and the product is a mixture of *o*-nitroanisole and *p*-nitroanisole (the methoxy group is ortho-para directing).

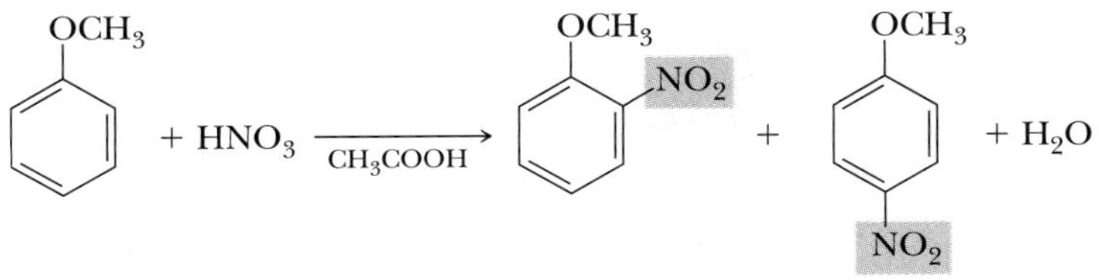

Activating group Any substituent on a benzene ring that causes the rate of electrophilic aromatic substitution to be greater than that for benzene.

Deactivating group Any substituent on a benzene ring that causes the rate of electrophilic aromatic substitution to be lower than that for benzene.

Anisole   *o*-Nitroanisole (44%)   *p*-Nitroanisole (55%)

Quite another situation is seen in the nitration of benzoic acid. First, the reaction requires the more reactive fuming nitric acid and a higher temperature than for benzene. Because nitration of benzoic acid proceeds much more slowly than nitration of benzene itself, we say that a carboxyl group is strongly deactivating. Second, the product formed consists of approximately 80% of the meta isomer and 20% of the ortho and para isomers combined; thus we say that the carboxyl group is meta directing.

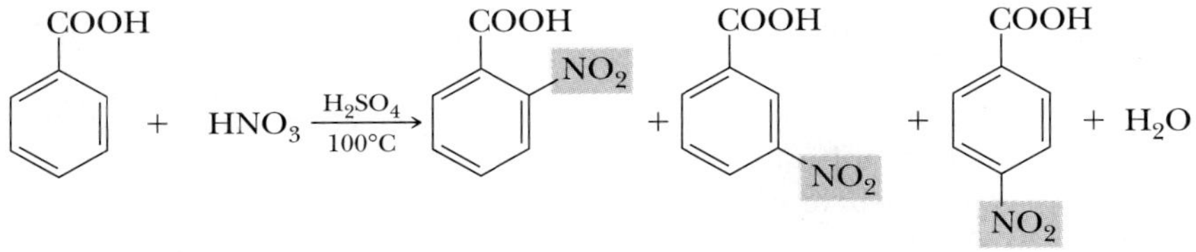

Benzoic acid   *o*-Nitro-benzoic acid (18%)   *m*-Nitro-benzoic acid (80%)   *p*-Nitro-benzoic acid (2%)

Table 22.2   Directing Effects of Substituents on Further Substitution

Ortho-Para Directing	Strongly activating	$-\ddot{N}H_2$	$-\ddot{N}HR$	$-\ddot{N}R_2$	$-\ddot{O}H$	$-\ddot{O}R$	
	Moderately activating	$-\overset{\displaystyle O}{\overset{\|}{\ddot{N}HCR}}$	$-\overset{\displaystyle O}{\overset{\|}{\ddot{N}HCAr}}$	$-\overset{\displaystyle O}{\overset{\|}{\ddot{O}CR}}$	$-\overset{\displaystyle O}{\overset{\|}{\ddot{O}CAr}}$		
	Weakly activating	$-R$	⬡				
	Weakly deactivating	$-\ddot{F}:$	$-\ddot{C}l:$	$-\ddot{B}r:$	$-\ddot{I}:$		
Meta Directing	Moderately deactivating	$-\overset{\displaystyle O}{\overset{\|}{CH}}$	$-\overset{\displaystyle O}{\overset{\|}{CR}}$	$-\overset{\displaystyle O}{\overset{\|}{COH}}$	$-\overset{\displaystyle O}{\overset{\|}{COR}}$	$-\overset{\displaystyle O}{\overset{\|}{CNH_2}}$	$-\overset{\displaystyle O}{\underset{\underset{\displaystyle O}{\|}}{\overset{\|}{S OH}}}$ $-C\equiv N$
	Strongly deactivating	$-NO_2$	$-NH_3{}^+$	$-CF_3$	$-CCl_3$		

Relative importance in directing further substitution ⬆

Listed in Table 22.2 are the directing and activating-deactivating effects for the major functional groups with which we are concerned in this text. If we compare these ortho-para and meta directors for structural similarities and differences, we can make the following generalizations.

1. Alkyl groups, phenyl groups, and substituents in which the atom bonded to the ring has an unshared pair of electrons are ortho-para directing. All other substituents are meta directing.
2. All ortho-para directing groups are activating. The exception to this generalization is the halogens, which are weakly deactivating.

The fact that alkyl groups are weakly activating is why it is difficult to stop Friedel-Crafts alkylations at monoalkylation. When a first alkyl group is introduced onto an aromatic ring, the ring is activated toward further alkylation and, unless reaction conditions are very carefully controlled, a mixture of di-, tri-, and polyalkylation products is formed. Friedel-Crafts acylations, on the other hand, never go beyond monoacylation because an acyl group is deactivating toward further substitution.

We can illustrate the usefulness of these generalizations by considering the synthesis of two different disubstituted derivatives of benzene. Suppose we wish to prepare *m*-bromonitrobenzene from benzene. This conversion can be done in two steps: nitration and bromination. If the steps are carried out in just that order, the major product is indeed *m*-bromonitrobenzene. The nitro group is a meta director and, therefore, directs bromination to a meta position.

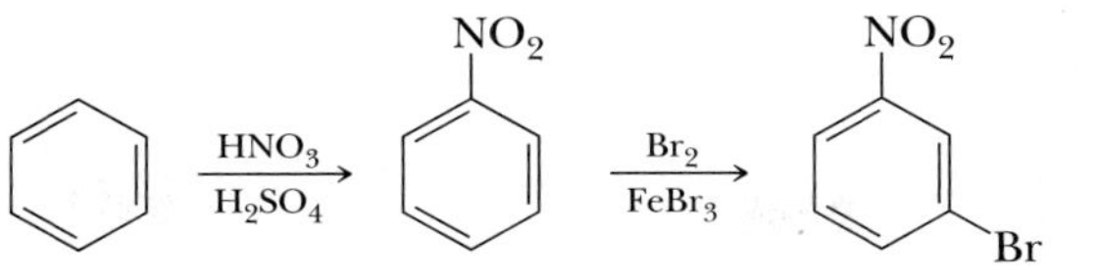

Nitrobenzene   *m*-Bromonitro-
                                 benzene

If, however, we reverse the order of the steps and first form bromobenzene, we now have an ortho-para directing group on the ring, and nitration takes place preferentially at the ortho and para positions.

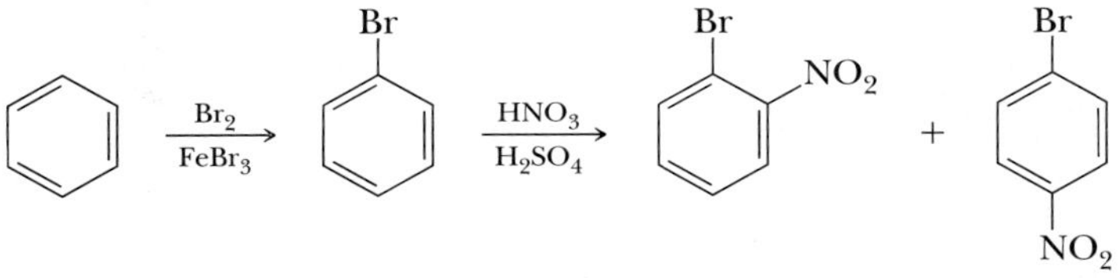

Bromobenzene   *o*-Bromonitro-   *p*-Bromonitro-
                       benzene   benzene

As another example of the importance of the order in electrophilic aromatic substitutions, consider the conversion of toluene to *p*-nitrobenzoic acid. The nitro group can be introduced with a nitrating mixture of nitric and sulfuric acids. The carboxyl group can be produced by oxidation of the methyl group of toluene (Section 21.5A). Nitration of toluene yields a product with the two substituents in the desired para relationship. Nitration of benzoic acid, on the other hand, yields a product with the substituents meta to each other.

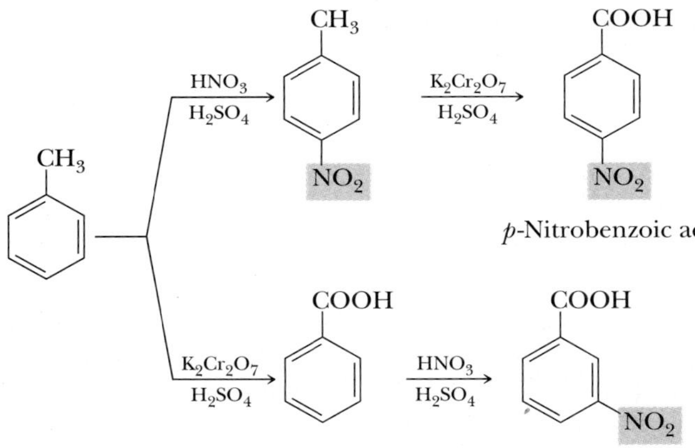

p-Nitrobenzoic acid

m-Nitrobenzoic acid

Again, we see that the order in which the reactions are performed is critical.

Note that, in this last example, we showed nitration of toluene producing only the para isomer. Because methyl is an ortho-para directing group, both the ortho and para isomers are formed (Table 22.1). In problems of this type in which you are asked to prepare the para isomer, assume that both ortho and para isomers are formed but that there are physical methods by which they can be separated and the desired isomer obtained.

Example 22.4

Complete these electrophilic aromatic substitution reactions. Where you predict meta substitution, show only the meta product. Where you predict ortho-para substitution, show both products.

(a) [structure: benzene with Br] + H_2SO_4 $\xrightarrow{heat}$ (b) [structure: benzene with SO_3H] + HNO_3 $\xrightarrow{H_2SO_4}$

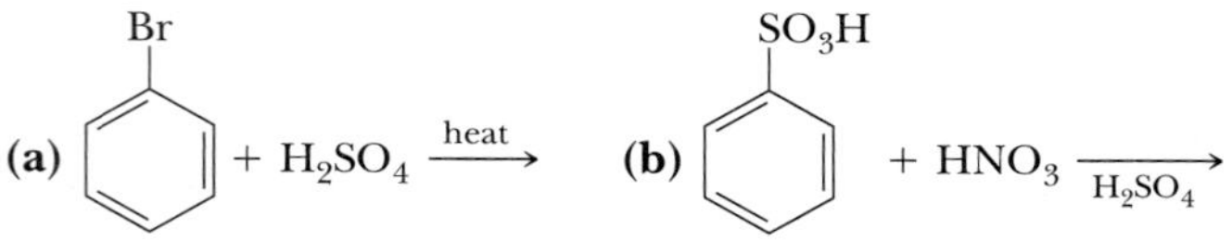

Solution

Bromine in (a) is ortho-para directing and weakly deactivating. The sulfonic acid group in (b) is meta directing and moderately deactivating.

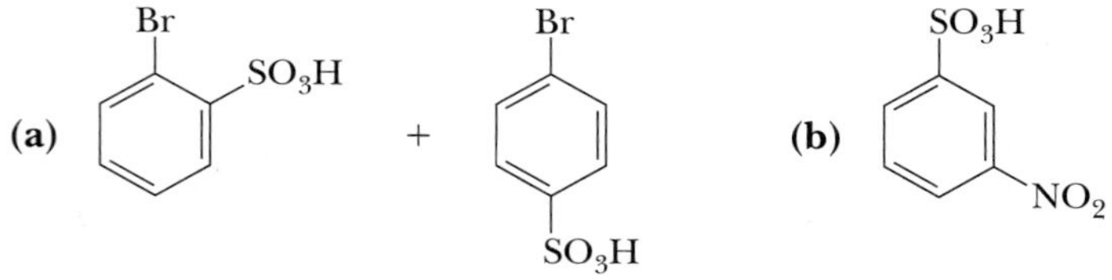

(a) [structure: benzene with Br and SO_3H ortho]   +   [structure: benzene with Br and SO_3H para]   (b) [structure: benzene with SO_3H and NO_2 meta]

o-Bromobenzene-   p-Bromobenzene-   m-Nitrobenzene-
sulfonic acid   sulfonic acid   sulfonic acid

Problem 22.4

Draw structural formulas for the product of nitration of each compound. Where you predict ortho-para substitution, show both products.

(a) [structure: benzene with $-\overset{O}{\overset{\|}{C}}OCH_3$]   (b) [structure: benzene with $-O\overset{O}{\overset{\|}{C}}CH_3$]

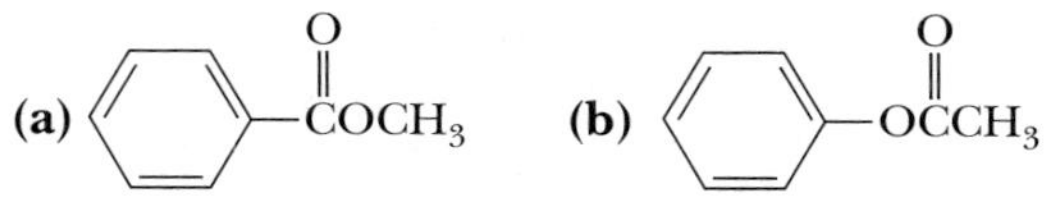

B. Theory of Directing Effects

As we have just seen, a group on a benzene ring exerts a major effect on the pattern of further substitution. We can account for these patterns by starting with the general mechanism first presented in Section 22.1 for electrophilic aromatic substitution and carrying it a step further to consider how groups already present on the ring affect the energetics of further substitution. In this regard, we need to consider both resonance and inductive effects and the relative importance of each.

Nitration of Anisole

The rate of electrophilic aromatic substitution is limited by the slowest step in the mechanism. For the nitration of anisole, and for almost every other substitution we consider, the slow and rate-determining step is attack of the electrophile on the aromatic ring. The rate of this step depends on the stability of the transition state for this step. The more stable the transition state, the faster the rate-determining step and, thus, the overall reaction.

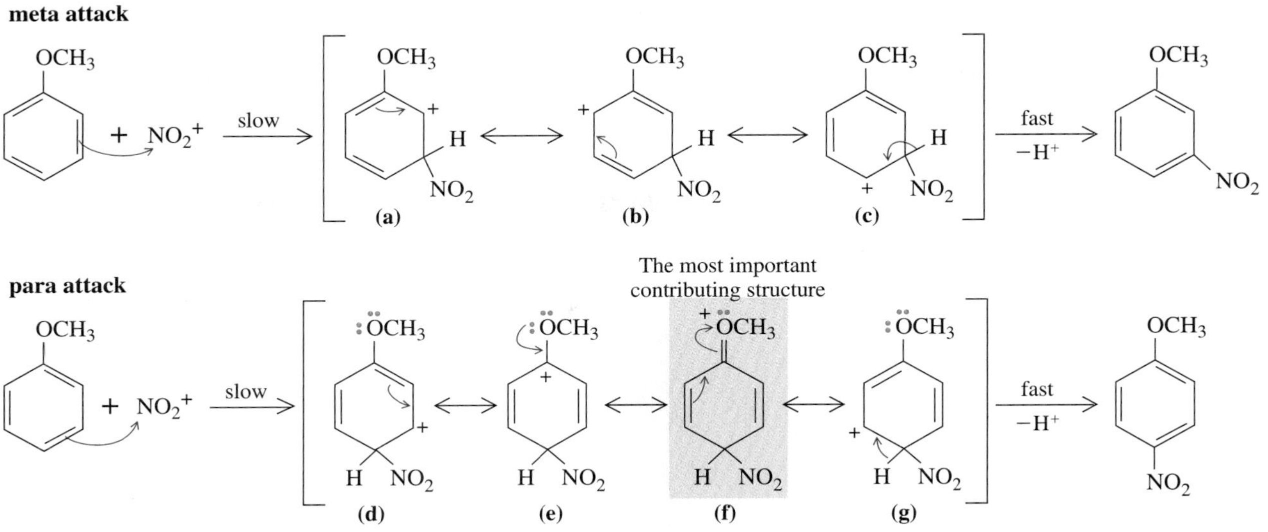

Figure 22.2
Nitration of anisole. Electrophilic attack meta and para to the methoxy group.

Shown in Figure 22.2 is the cation intermediate formed by attack of the nitronium ion meta to the methoxy group. Also shown in the figure is the cation intermediate formed by attack para to the methoxy group. Note that, in terms of electronic effects, structural formulas for the cation formed by attack ortho to the methoxy group are essentially the same as those for para attack, so, for convenience, we deal only with para attack. The cation intermediate formed by meta attack is a hybrid of contributing structures (a), (b), and (c). The cation intermediate formed by para attack is a hybrid of contributing structures (d), (e), (f), and (g). For each orientation, we can draw three contributing structures that place the positive charge on carbon atoms of the benzene ring. These three structures are the only ones that can be drawn for meta attack. However, for para attack (and for ortho attack as well), a fourth contributing structure, (f), can be drawn that involves an unshared pair of electrons on the oxygen atom of the methoxy group and places a positive charge on this oxygen. Structure (f) contributes more than structures (d), (e), or (g) because, in it, all atoms have complete octets. Because the cation formed by ortho or para attack on anisole has a greater degree of charge delocalization, and therefore a lower activation energy for its formation, nitration of anisole occurs faster in the ortho and para positions.

Nitration of Benzoic Acid

Shown in Figure 22.3 are resonance-stabilized cation intermediates formed by attack of the nitronium ion meta to the carboxyl group and then para to it. Each cation in Figure 22.3 is a hybrid of three contributing structures; no additional ones can be drawn. Now we need to compare the relative resonance stabilization of each hybrid. If we draw a Lewis structure for the carboxyl group showing the partial positive

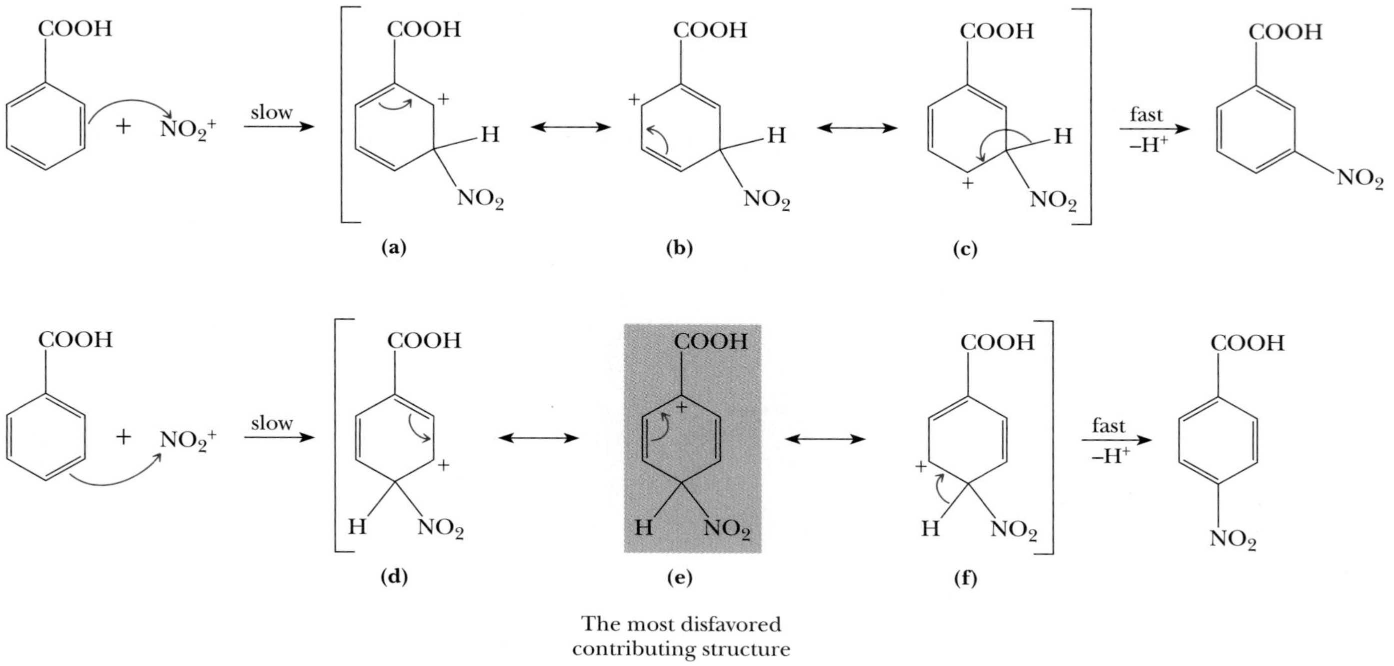

(a)   **(b)**   **(c)**

(d)   **(e)**   **(f)**

The most disfavored
contributing structure

Figure 22.3
Nitration of benzoic acid. Electrophilic attack meta and para to the carboxyl group.

charge on the carboxyl carbon, we see that contributing structure (e) in Figure 22.3 places positive charges on adjacent atoms.

(e)

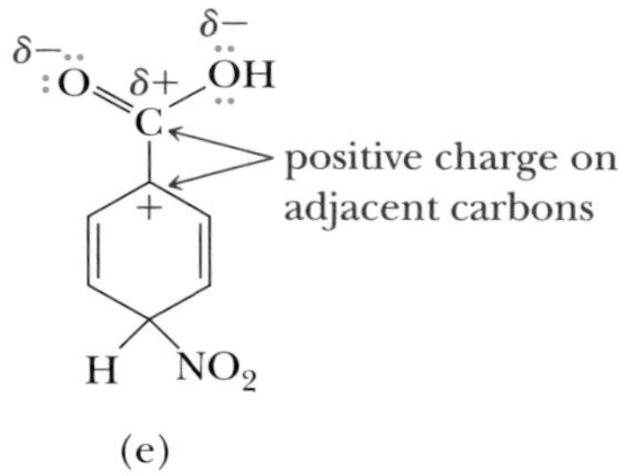

Because of the electrostatic repulsion thus generated, this structure is very unstable and makes only a negligible contribution to the hybrid.

None of the contributing structures for meta attack places positive charges on adjacent atoms. As a consequence, resonance stabilization of the cation for meta attack is greater than that for para (or ortho) attack. Stated alternatively, the activation energy for meta attack is less than that for para attack.

Comparison of the entries in Table 22.1 shows that almost all the ortho-para directing groups have an unshared pair of electrons on the atom bonded to the aromatic ring. Thus, the directing effect of these groups is primarily attributable to the ability of the atom bonded to the ring to further delocalize the positive charge of the cation intermediate formed when electrophilic attack occurs at the ortho or para positions. Recall that, all things being equal, delocalization of a charge stabilizes a charged species.

To account for the fact that alkyl groups are also ortho-para directing, we need to consider their inductive effect on stability of the cation intermediate. In the case of alkyl groups, there is an inductive polarization of electrons from the alkyl substituent toward the cationic ring of the intermediate. This polarization amounts to a further delocalization of the positive charge, thereby stabilizing the cationic intermediate. The alkyl groups are activating because, compared to benzene alone, the cationic intermediates are lower in energy with alkyl substituents at ortho or para positions. Recall that we used the electron-releasing inductive effect (that is, hyperconjugation) of alkyl groups in Section 6.3A to account for the relative stabilities of methyl, primary, secondary, and tertiary carbocations as well.

The inductive polarization of electrons from alkyl groups is most effective at delocalizing the positive charge of the cation intermediate when the alkyl group is bonded directly to the ring atom carrying significant positive charge. In other words, when the alkyl group is ortho or para to the location of the incoming electrophile.

A part of the positive charge on the cation intermediate is delocalized onto methyl.

C. Theory of Activating-Deactivating Effects

We account for the activating-deactivating effects of substituent groups by much the same combination of resonance and inductive effects.

1. Any resonance effect, such as that of —NH$_2$, —OH, and —OR, that delocalizes the positive charge of the cation intermediate lowers the activation energy for its formation and has an activating effect toward further electrophilic aromatic substitution.
2. Any resonance or inductive effect, such as that of —NO$_2$, —C≡N, —C=O, —SO$_2$—, and —SO$_3$H, that decreases electron density on the ring deactivates the ring to further substitution.
3. Any inductive effect, such as that of —CH$_3$ or another alkyl group, that releases electron density toward the cationic intermediate activates the ring toward further substitution.
4. Any inductive effect, such as that of a halogen, —NR$_3^+$, —CCl$_3$, and—CF$_3$, that decreases electron density on the ring deactivates the ring to further substitution.

The halogens represent an interesting combination of the resonance and inductive effects, the two operating in opposite directions. Recall from Table 22.2 that halogens are ortho-para directing but, unlike other ortho-para directors listed in the table, they are weakly deactivating. These observations can be accounted for in the following way.

1. The inductive effect of halogens. The halogens are relatively electronegative and have an electron-withdrawing inductive effect. Aryl halides, therefore, react more slowly in electrophilic aromatic substitution than benzene.
2. The resonance effect of halogens. When a halogen-substituted aromatic ring is attacked by an electrophile to form a cation intermediate, a halogen ortho or para to the site of electrophilic attack can help to stabilize the cation intermediate by delocalization of the positive charge.

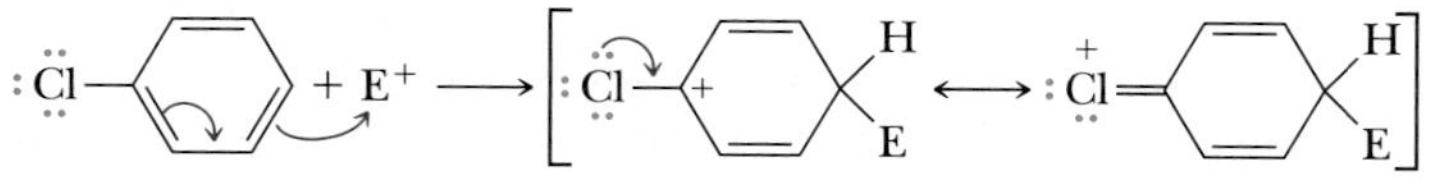

Thus, the inductive and resonance effects of the halogens are counter to each other, but the former is somewhat stronger than the latter. The net effect of this opposition is that the halogens are weakly deactivating but ortho-para directing.

Example 22.5

Predict the major product of each electrophilic aromatic substitution.

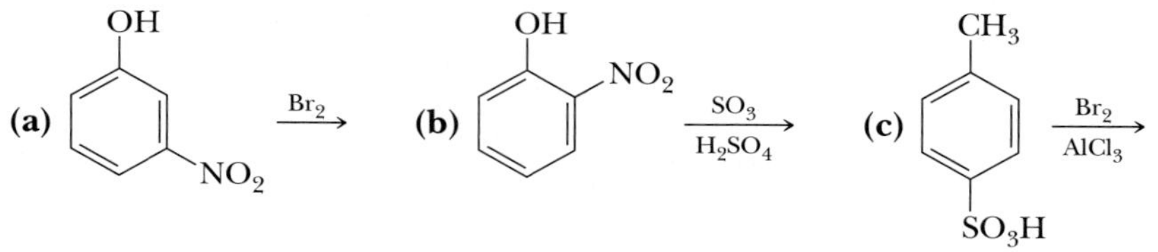

Solution

The key to predicting orientation of electrophilic aromatic substitution on each molecule is that ortho-para directing groups activate the ring toward further substitution, whereas meta directing groups deactivate. Therefore, where there is competition between ortho-para and meta directing groups, ortho-para directing groups win out. In these examples, the major product is that resulting from substitution ortho or para to the activating group. For (a) the next substitution is directed ortho/para to the strongly activating —OH group; the isomer with bromine between the —OH and —NO$_2$ groups is a very minor product because of steric hindrance. For (b) the incoming group is directed ortho and para to the strongly activating —OH group. For (c) the next substitution is directed ortho to the weakly activating —CH$_3$ group.

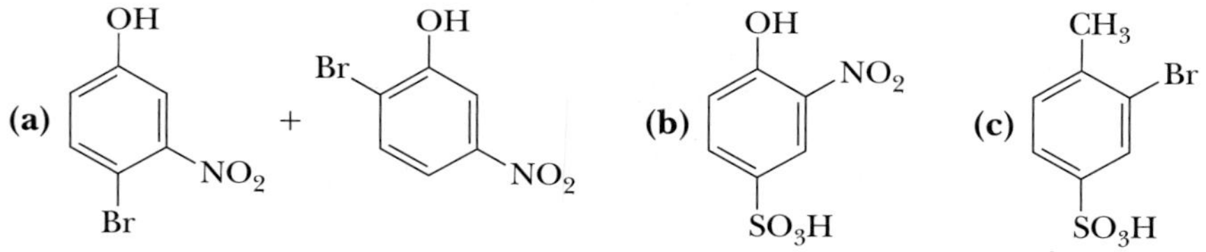

Problem 22.5

Predict the major product(s) of each electrophilic aromatic substitution.

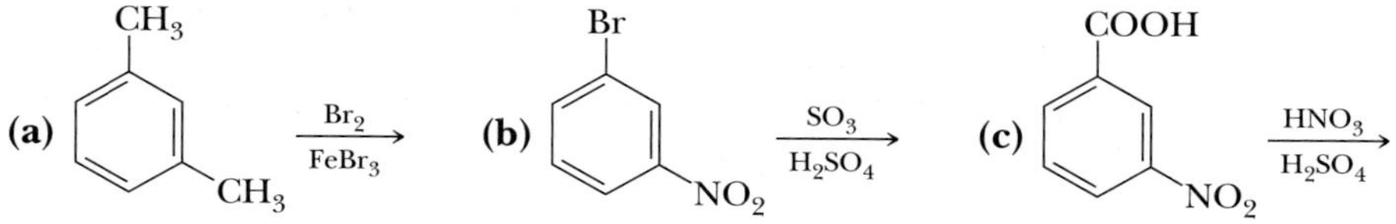

22.3 Nucleophilic Aromatic Substitution

One of the important chemical characteristics of aryl halides is that they undergo relatively few reactions involving the carbon-halogen bond. Aryl halides, for example, do not undergo substitution by either of the S$_N$1 or S$_N$2 pathways that are characteristic of nucleophilic aliphatic substitutions. They do, however, undergo **nucleophilic aromatic substitution** under certain conditions but by mechanisms quite different from those for nucleophilic aliphatic substitutions. Nucleophilic aromatic substitution reactions are far less common than electrophilic aromatic substitution reactions and have only limited usefulness in the synthesis of organic compounds. We study these reactions not only for their synthetic usefulness but also for the additional insights they give us into the unique chemical properties of aromatic compounds.

> **Nucleophilic aromatic substitution** A reaction in which a nucleophile, most commonly a halogen, on an aromatic ring is replaced by another nucleophile.

A. Nucleophilic Substitution by Way of a Benzyne Intermediate

An apparent exception to the generalization about the lack of reactivity of aryl halides to nucleophilic substitution is an industrial process for the synthesis of phenol from chlorobenzene. When heated at 300°C under high pressure with aqueous NaOH, chlorobenzene is converted to sodium phenoxide. Neutralization of this salt with aqueous acid gives phenol.

Chlorobenzene + 2NaOH $\xrightarrow[\text{pressure, 300°C}]{H_2O}$ Sodium phenoxide + NaCl + H$_2$O

In later technological developments, the discovery was made that chlorobenzene can be hydrolyzed to phenol by steam under pressure at 500°C. Each of these reactions appears to involve nucleophilic substitution of —OH for —Cl on the benzene ring. However, this reaction is not as simple as it might seem, as is illustrated by the reaction of substituted halobenzenes with NaOH. For example, o-chlorotoluene under these conditions gives a mixture of 2-methylphenol (o-cresol) and 3-methylphenol (m-cresol).

$\xrightarrow[\text{2. HCl, H}_2\text{O}]{\text{1. NaOH, heat, pressure}}$

2-Methylphenol + 3-Methylphenol
(o-Cresol) (m-Cresol)

The same type of reaction can be brought about by the use of sodium amide in liquid ammonia. Under these conditions, for example, p-chlorotoluene gives a mixture of 4-methylaniline (p-toluidine) and 3-methylaniline (m-toluidine) in approximately equal amounts.

+ NaNH$_2$ $\xrightarrow[(-33°C)]{NH_3(l)}$ + NaCl

4-Methylaniline 3-Methylaniline
(p-Toluidine) (m-Toluidine)

Benzyne intermediate A reactive intermediate formed by β-elimination from adjacent carbon atoms of a benzene ring and having a triple bond in the benzene ring. The second pi bond of the benzyne triple bond is formed by weak overlap of coplanar sp^2 orbitals on adjacent carbons.

The difference in this reaction compared with other substitution reactions we have dealt with so far is that the entering group appears not only at the position occupied by the leaving group but also at a position adjacent to it.

To account for these experimental observations, it has been proposed that an elimination of HX occurs to form a **benzyne intermediate** that then undergoes nucleophilic addition to the triple bond to give the products observed.

Mechanism   *Nucleophilic Aromatic Substitution via a Benzyne Intermediate*

Step 1: Dehydrohalogenation of the benzene ring gives a benzyne intermediate.

$$CH_3\text{-ring} + :NH_2^- \longrightarrow CH_3\text{-ring} + :NH_3 + :\ddot{C}\ddot{l}:^-$$

A benzyne intermediate

Step 2: Nucleophilic addition of amide ion to a carbon of the benzyne triple bond gives a carbanion intermediate. Addition to either carbon of the "triple" is possible.

$$CH_3\text{-ring} + :NH_2^- \longrightarrow CH_3\text{-ring}\text{-}NH_2$$

A carbanion intermediate

Step 3: Proton transfer from ammonia to the carbanion intermediate gives one of the observed substitution products and generates a new amide ion.

$$CH_3\text{-ring-}NH_2 + H-NH_2 \longrightarrow CH_3\text{-ring-}NH_2 + :NH_2^-$$

3-Methylaniline
(*m*-Toluidine)

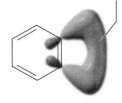

pi bonding, but of reduced strength because of poor orbital overlap

A benzyne intermediate

The bonding in a benzyne intermediate and also the reason for its extremely reactive nature can be pictured in the following way. According to molecular orbital theory, the benzene ring retains its planarity, pi bonding, and aromatic character. The adjacent sp^2 orbitals formerly bonding to a halogen and a hydrogen now overlap to form the second pi bond of the benzyne triple bond. The problem is that the atomic orbitals forming this pi bond are not parallel as in acetylene and unstrained alkynes but, rather, lie at an angle of 120° to the bond axis connecting them. Consequently, the overlap between these orbitals is reduced. Reduced overlap, in turn, means a weaker and more reactive pi bond. Therefore, the second pi bond of the benzyne intermediate undergoes addition very readily to form two new and stronger sigma bonds.

B. Nucleophilic Substitution by Addition-Elimination

Aromatic halides are normally quite inert to the types of nucleophiles that readily displace halide ions from alkyl halides. However, when an aromatic compound contains strong electron-withdrawing nitro groups ortho or para (or both) to the halogen, nucleophilic aromatic substitution occurs quite readily. For example, when 1-chloro-2,4-dinitrobenzene is heated at reflux in aqueous sodium carbonate followed by treatment with aqueous acid, it is converted in nearly quantitative yield to 2,4-dinitrophenol.

1-Chloro-2,4-
dinitrobenzene

Sodium 2,4-dinitro-
phenoxide

2,4-Dinitrophenol

One application of this reaction is the synthesis of 2,4-dinitrophenylhydrazine, a common reagent used to prepare derivatives of aldehydes and ketones (Section 16.8B).

1-Chloro-2,4-
dinitrobenzene

Hydrazine

2,4-Dinitro-
phenylhydrazine

This type of nucleophilic aromatic substitution for halogen has been studied extensively, and it has been determined that reaction occurs in two steps: nucleophilic addition followed by elimination. For the majority of reactions of this type, addition of the nucleophile in Step 1 is the slow, rate-determining step. Elimination of halide ion in Step 2 gives the product. This reaction thus resembles reactions of carboxylic acid derivatives in that it proceeds by an addition-elimination mechanism rather than by direct substitution.

Mechanism   *Nucleophilic Aromatic Substitution by Addition-Elimination*

Step 1: The nucleophile adds to the aromatic ring at the carbon bearing the halogen. This addition places a negative charge on the ring, which is stabilized by resonance interaction with the nitro or other strong electron-withdrawing groups in the ortho or para positions to the halogen. Such intermediates are named **Meisenheimer complexes** after the German chemist who first characterized them. Note that nitro groups on both ortho and para positions participate in delocalization of the negative charge in the complex.

ORGANIC
Chemistry Now™

Click Mechanisms to view an animation of **Nucleophilic Aromatic Substitution**

Step 2: Elimination of halide ion regenerates the aromatic ring and gives the observed product.

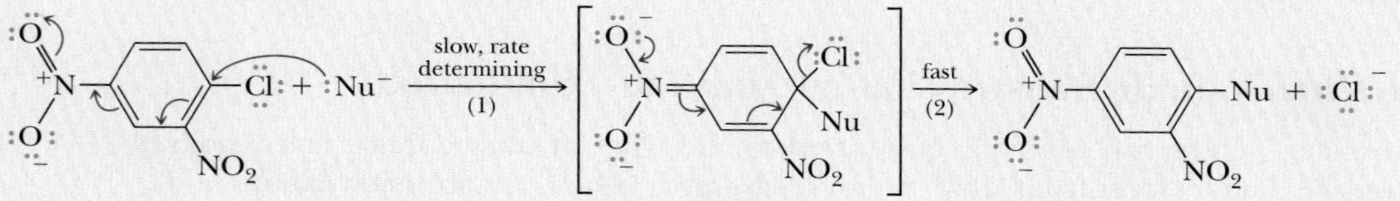

A Meisenheimer complex

Example 22.6

What is the state of hybridization of each ring carbon atom in the Meisenheimer complex just shown?

Solution

The carbon atom bonded to both the leaving group and entering nucleophile (—Cl and —Nu in the structure shown) is sp^3 hybridized. The other five carbons of the ring are sp^2 hybridized.

Problem 22.6

In S_N2 reactions of alkyl halides, the order of reactivity is RI > RBr > RCl > RF. Alkyl iodides are considerably more reactive than alkyl fluorides, often by factors as great as 10^6. All 1-halo-2,4-dinitrobenzenes, however, react at approximately the same rate in nucleophilic aromatic substitutions. Account for this difference in relative reactivities.

Summary

A characteristic reaction of aromatic compounds is **electrophilic aromatic substitution** (Section 22.1). This reaction begins with attack on the aromatic ring of an electrophile to give a resonance-stabilized cation intermediate. Loss of a proton from this intermediate regenerates the aromatic ring and completes the substitution.

Substituents on an aromatic ring influence both the site of further substitution and its rate (Section 22.2A). Substituent groups that direct an incoming group preferentially to the ortho and para positions are known as **ortho-para directors.** Those that direct an incoming group preferentially to the meta position are known as **meta directors.** Groups that cause the rate of further substitution to be faster than that for benzene are said to be **activating**; those that cause the rate of further substitution to be slower than that for benzene are said to

be **deactivating.** A mechanistic rationale for directing effects is based on a consideration of the degree of resonance stabilization of the possible cation intermediates formed on reaction of the aromatic ring and the electrophile (Section 22.2B). Groups able to stabilize a cation intermediate, either by their inductive effect or resonance effect, are activators and ortho-para directors. Groups that destabilize a cation intermediate, either by their inductive effect or resonance effect, are deactivators and meta directors (Sections 22.2B and 22.2C). The halogens constitute an exception in that they are ortho-para directors even though they are weak deactivators.

Aromatic halogen compounds undergo nucleophilic substitution reactions by two mechanisms. One involves a benzyne intermediate (Section 22.3A) and the other involves an addition-elimination sequence (Section 22.3B).

Key Reactions

1. Halogenation (Section 22.1A)

The electrophile is a halonium ion formed as an ion pair by interaction of chlorine or bromine with a Lewis acid.

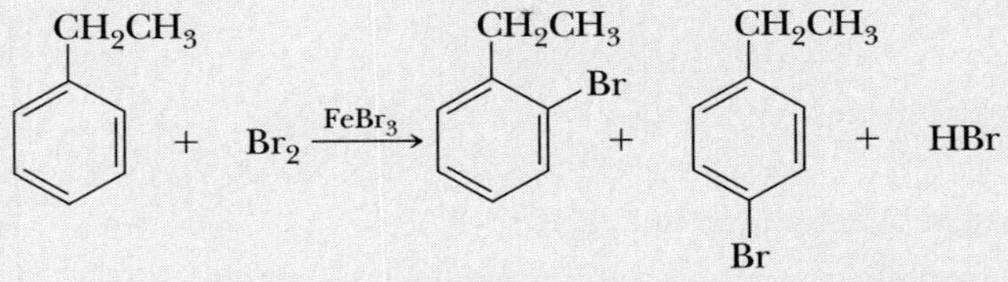

Halogenation of an aromatic ring substituted by strongly activating groups (such as —OH, —OR, and —NH$_2$) does not require a Lewis acid catalyst.

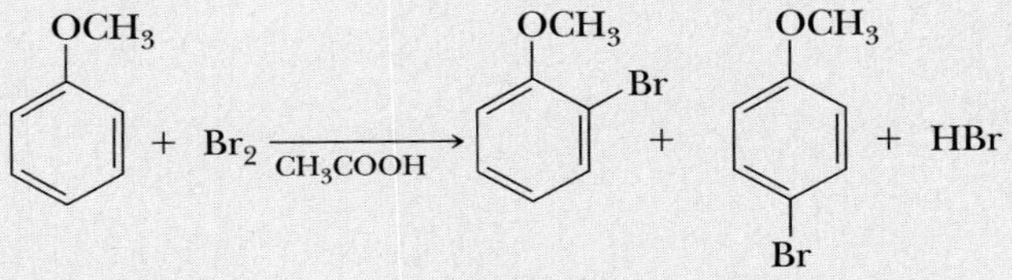

2. Nitration (Section 22.1B)

The attacking electrophile is the nitronium ion, NO$_2{}^+$, formed by interaction of nitric acid and sulfuric acid.

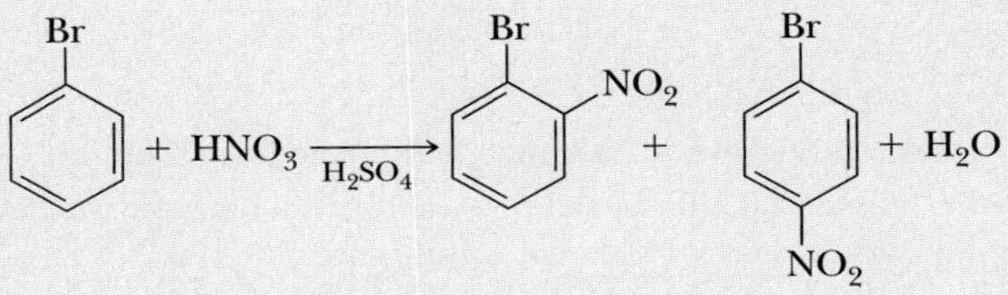

3. Sulfonation (Section 22.1B)

The attacking electrophile is either sulfur trioxide, SO$_3$, or HSO$_3{}^+$ depending on experimental conditions.

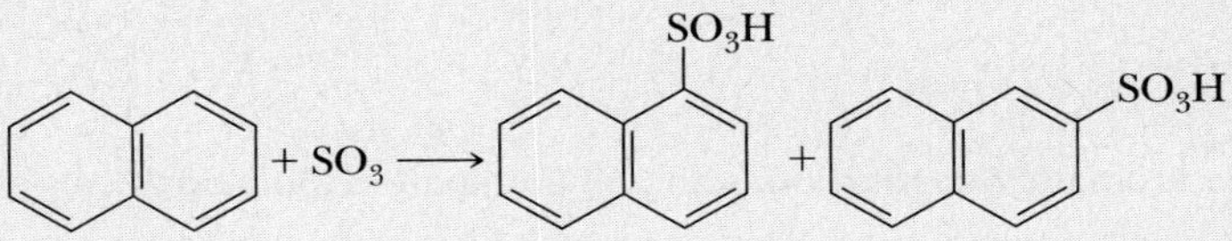

4. Friedel-Crafts Alkylation (Section 22.1C)

The attacking electrophile is a carbocation formed as an ion pair by interaction of an alkyl halide with a Lewis acid. Rearrangements from a less stable carbocation to a more stable carbocation are common.

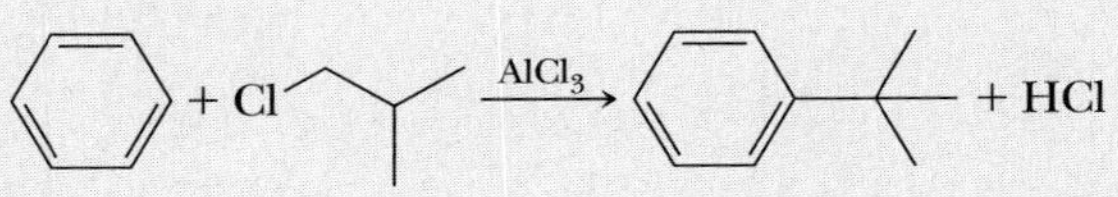

Friedel-Crafts alkylation fails with compounds much less reactive than benzene.

5. Friedel-Crafts Acylation (Section 22.1C)

The attacking electrophile is an acyl cation (an acylium ion) formed as an ion pair by interaction of an acyl halide with a Lewis acid.

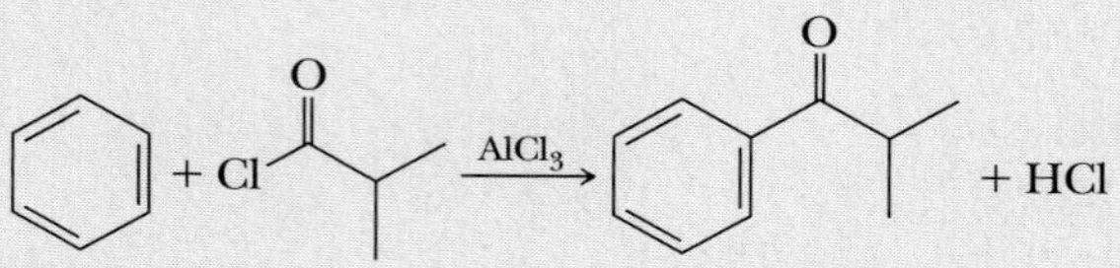

6. Alkylation Using an Alkene (Section 22.1D)

The attacking electrophile is a carbocation formed by interaction of the alkene with a Brønsted or Lewis acid.

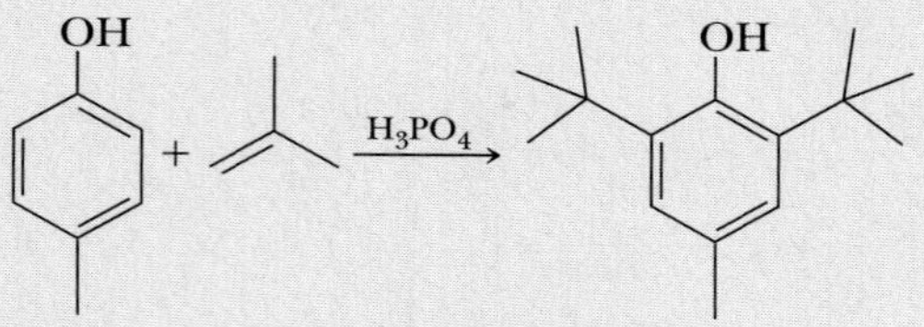

7. Alkylation Using an Alcohol (Section 22.1D)

The attacking electrophile is a carbocation formed by treatment of the alcohol with a Brønsted or Lewis acid.

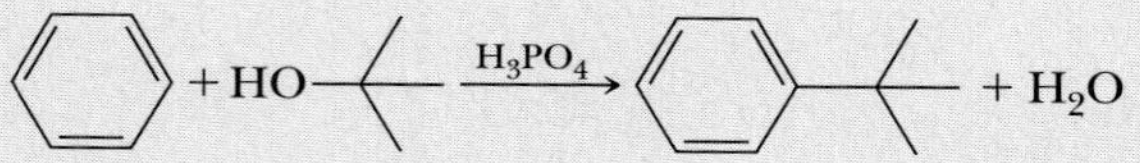

8. Nucleophilic Aromatic Substitution: A Benzyne Intermediate (Section 22.3A)

Elimination of HX from an aryl halide by strong base forms a benzyne intermediate, which undergoes nucleophilic addition to give the substitution product(s).

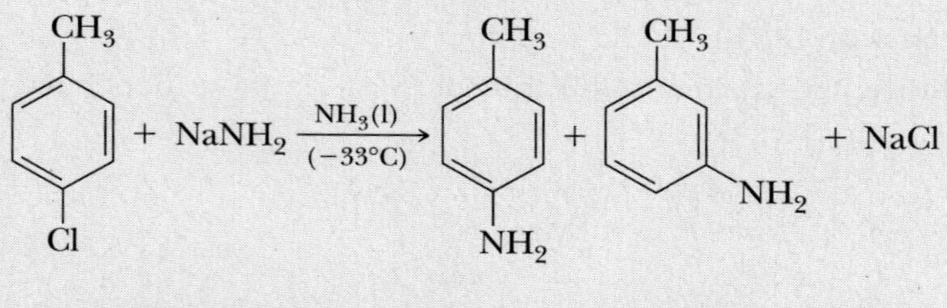

9. Nucleophilic Aromatic Substitution: Addition-Elimination (Section 22.3B)

Addition of the nucleophile to the carbon bearing the leaving group forms a tetrahedral intermediate from which halide ion is ejected to regenerate the aromatic ring. This type of aromatic substitution is made possible by strong electron-withdrawing groups, most commonly nitro groups, located ortho and para to the halogen.

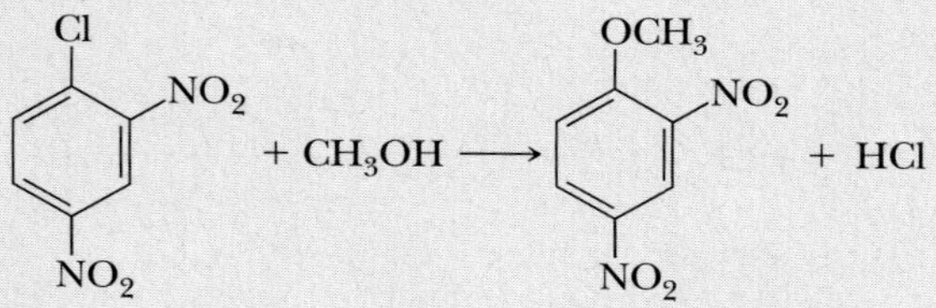

Problems

Electrophilic Aromatic Substitution: Monosubstitution

22.7 Write a stepwise mechanism for each of the following reactions. Use curved arrows to show the flow of electrons in each step.

(a) + Cl$_2$ $\xrightarrow{\text{FeCl}_3}$ + HCl

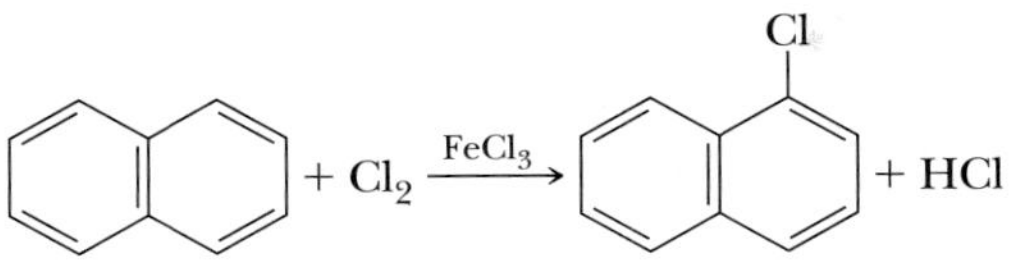

(b) + Cl $\xrightarrow{\text{AlCl}_3}$ + HCl

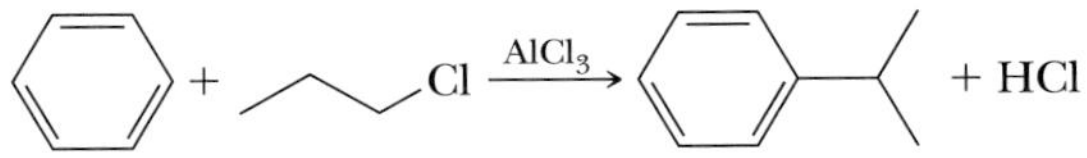

(c) + Cl $\xrightarrow{\text{SnCl}_4}$ + HCl

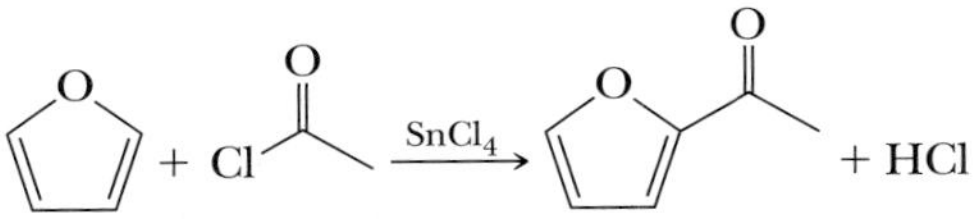

(d) + CH$_2$Cl$_2$ $\xrightarrow{\text{AlCl}_3}$ + 2HCl

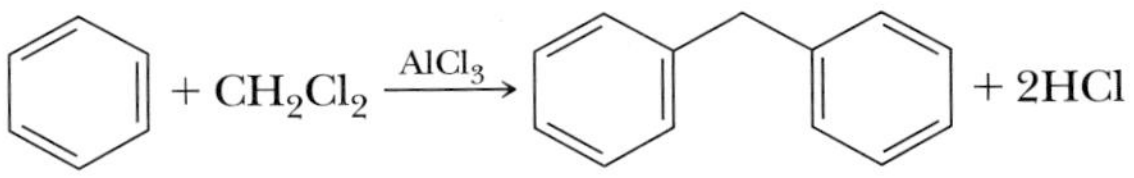

22.8 Pyridine undergoes electrophilic aromatic substitution preferentially at the 3 position as illustrated by the synthesis of 3-nitropyridine.

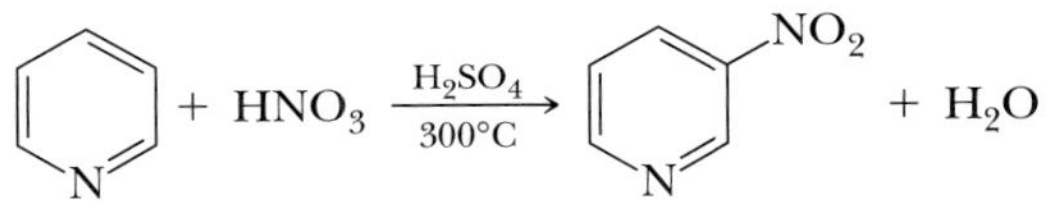

+ HNO$_3$ $\xrightarrow[\text{300°C}]{\text{H}_2\text{SO}_4}$ + H$_2$O

  Pyridine   3-Nitropyridine

Under these acidic conditions, the species undergoing nitration is not pyridine but its conjugate acid. Write resonance contributing structures for the intermediate formed by attack of NO$_2^+$ at the 2, 3, and 4 positions of the conjugate acid of pyridine. From examination of these intermediates, offer an explanation for preferential nitration at the 3 position.

22.9 Pyrrole undergoes electrophilic aromatic substitution preferentially at the 2 position as illustrated by the synthesis of 2-nitropyrrole.

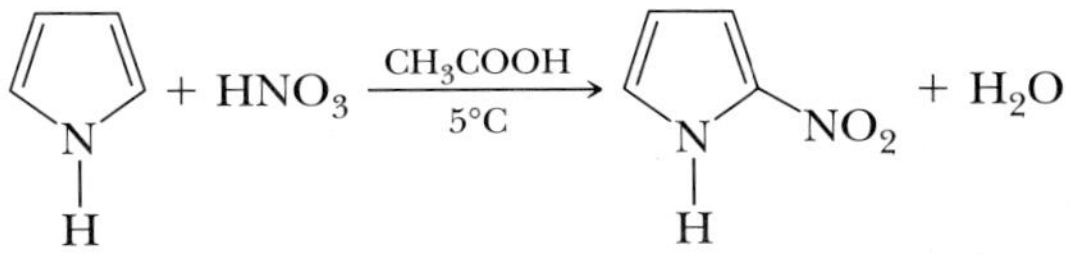

+ HNO$_3$ $\xrightarrow[\text{5°C}]{\text{CH}_3\text{COOH}}$ + H$_2$O

  Pyrrole   2-Nitropyrrole

Write resonance contributing structures for the intermediate formed by attack of NO$_2^+$ at the 2 and 3 positions of pyrrole. From examination of these intermediates, offer an explanation for preferential nitration at the 2 position.

22.10 Addition of *m*-xylene to the strongly acidic solvent HF/SbF$_5$ at −45°C gives a new species, which shows ^{1}H-NMR resonances at δ 2.88 (3H), 3.00 (3H), 4.67 (2H), 7.93 (1H), 7.83 (1H), and 8.68 (1H). Assign a structure to the species giving this spectrum.

ORGANIC
Chemistry Now™
Assess your understanding of this chapter's topics with additional quizzing and conceptual-based problems at
http://now.brookscole.com/bfi4

22.11 Addition of *tert*-butylbenzene to the strongly acidic solvent HF/SbF_5 followed by aqueous work-up gives benzene. Propose a mechanism for this dealkylation reaction. What is the other product of the reaction?

22.12 What product do you predict from the reaction of SCl_2 with benzene in the presence of $AlCl_3$? What product results if diphenyl ether is treated with SCl_2 and $AlCl_3$?

22.13 Other groups besides H^+ can act as leaving groups in electrophilic aromatic substitution. One of the best is the trimethylsilyl group, Me_3Si—. For example, treatment of $Me_3SiC_6H_5$ with CF_3COOD rapidly forms DC_6H_5. What are the properties of a silicon-carbon bond that allows you to predict this kind of reactivity?

Disubstitution and Polysubstitution

22.14 The following groups are ortho-para directors.

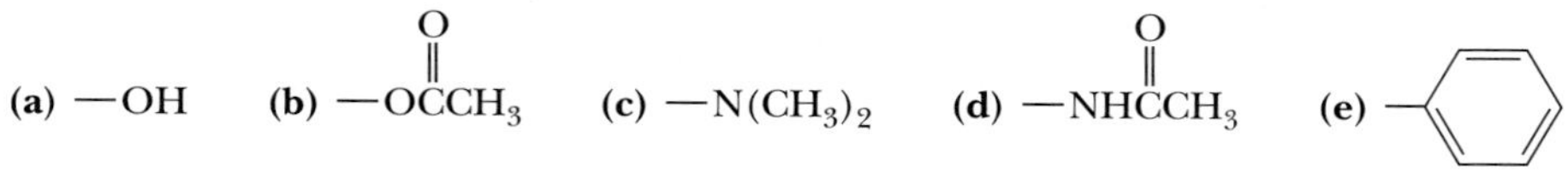

$$\text{(a)} \ -OH \quad \text{(b)} \ -O\overset{\overset{\displaystyle O}{\|}}{C}CH_3 \quad \text{(c)} \ -N(CH_3)_2 \quad \text{(d)} \ -NH\overset{\overset{\displaystyle O}{\|}}{C}CH_3 \quad \text{(e)}$$

Draw a contributing structure for the resonance-stabilized cation formed during electrophilic aromatic substitution that shows the role of each group in stabilizing the intermediate by further delocalizing its positive charge.

22.15 Predict the major product or products from treatment of each compound with HNO_3/H_2SO_4.

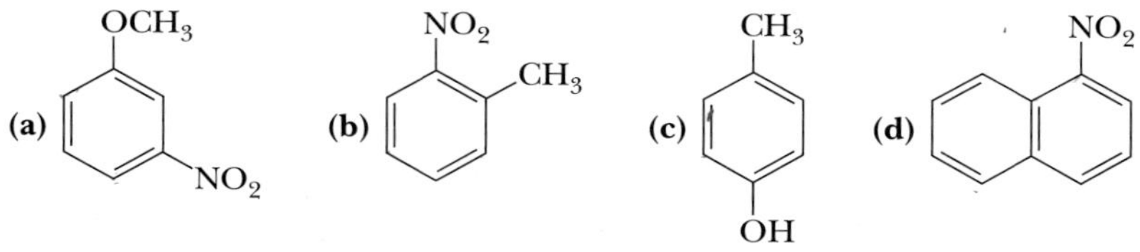

22.16 How do you account for the fact that *N*-phenylacetamide (acetanilide) is less reactive toward electrophilic aromatic substitution than aniline?

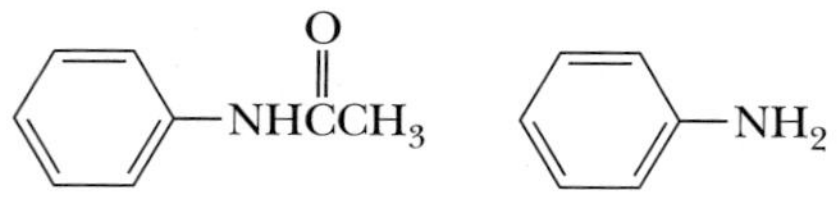

N-Phenylacetamide
(Acetanilide)

Aniline

22.17 Propose an explanation for the fact that the trifluoromethyl group is almost exclusively meta directing.

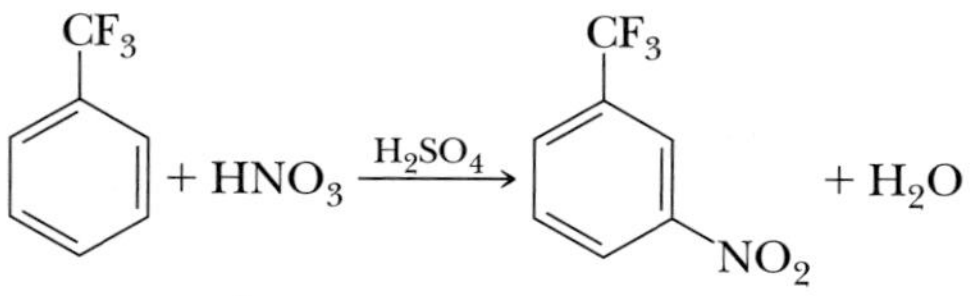

22.18 Suggest a reason why the nitroso group, $-N{=}O$, is ortho-para directing whereas the nitro group, $-NO_2$, is meta directing.

22.19 Arrange the compounds in each set in order of decreasing reactivity (fastest to slowest) toward electrophilic aromatic substitution.

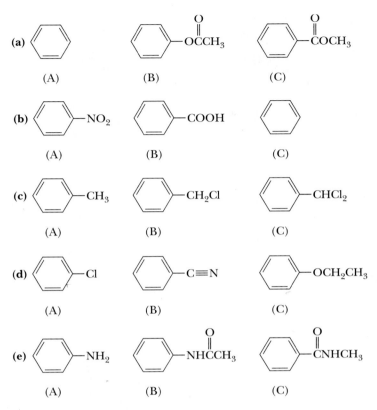

22.20 For each compound, indicate which group on the ring is the more strongly activating and then draw a structural formula of the major product formed by nitration of the compound.

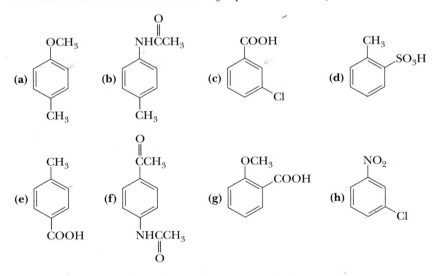

22.21 The following molecules each contain two aromatic rings.

Which ring in each undergoes electrophilic aromatic substitution more readily? Draw the major product formed on nitration.

22.22 Reaction of phenol with acetone in the presence of an acid catalyst gives a compound known as bisphenol A, which is used in the production of epoxy and polycarbonate resins (Section 29.5). Propose a mechanism for the formation of bisphenol A.

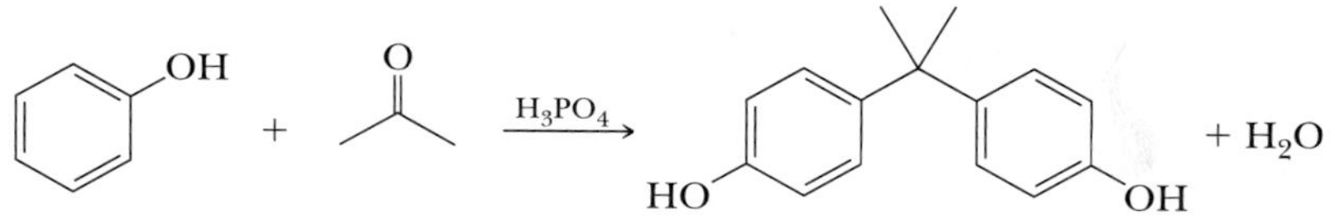

Phenol   Acetone   Bisphenol A

22.23 2,6-Di-*tert*-butyl-4-methylphenol, alternatively known as butylated hydroxytoluene (BHT), is used as an antioxidant in foods to "retard spoilage" (Section 8.7). BHT is synthesized industrially from 4-methylphenol by reaction with 2-methylpropene in the presence of phosphoric acid. Propose a mechanism for this reaction.

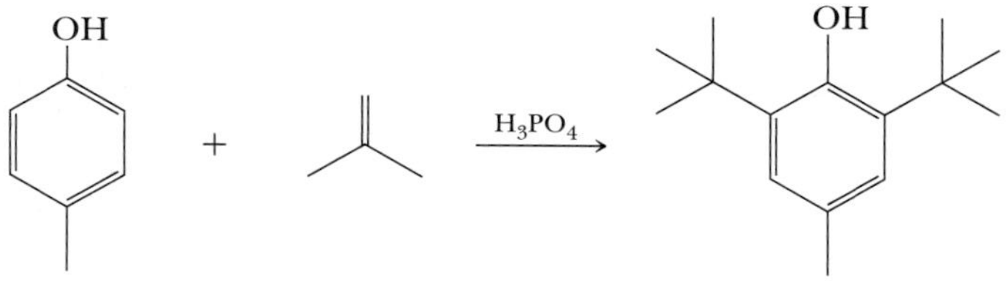

4-Methylphenol   2-Methylpropene   2,6-Di-*tert*-butyl-4-methylphenol
(*p*-Cresol)   "Butylated hydroxytoluene"
                                              (BHT)

22.24 The insecticide DDT is prepared by the following route. Suggest a mechanism for this reaction. The abbreviation DDT is derived from the common name **d**ichloro**d**iphenyl-**t**richloroethane.

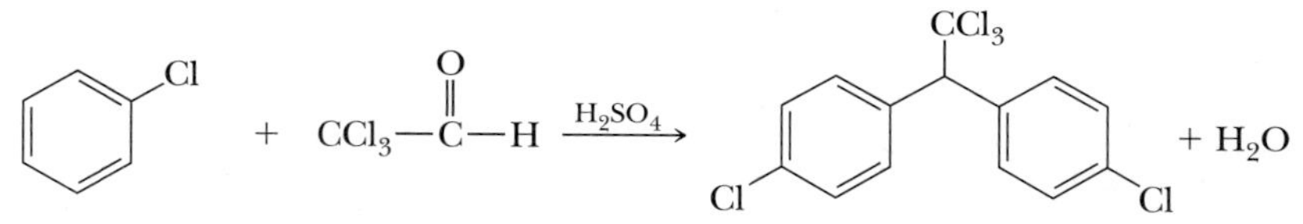

Chlorobenzene   Trichloro-   DDT
                   acetaldehyde

22.25 Treatment of salicylaldehyde (2-hydroxybenzaldehyde) with bromine in glacial acetic acid at 0°C gives a compound with molecular formula $C_7H_4Br_2O_2$, which is used as a topical fungicide and antibacterial agent. Propose a structural formula for this compound.

22.26 Propose a synthesis for 3,5-dibromo-2-hydroxybenzoic acid (3,5-dibromosalicylic acid) from phenol.

22.27 Treatment of benzene with succinic anhydride in the presence of polyphosphoric acid gives the following γ-ketoacid. Propose a mechanism for this reaction.

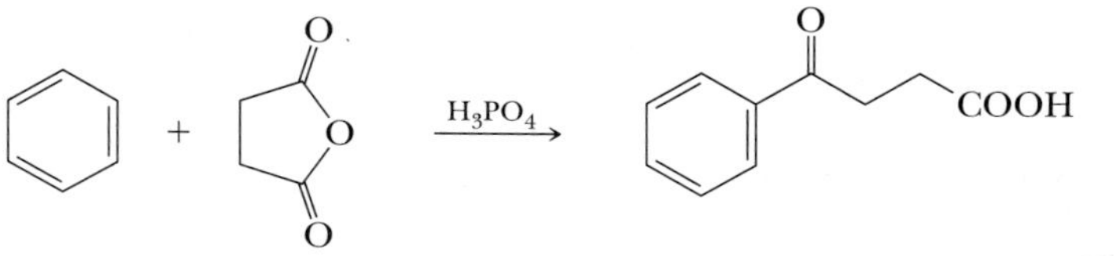

Succinic anhydride   4-Oxo-4-phenylbutanoic acid

Nucleophilic Aromatic Substitution

22.28 Following are the final steps in the synthesis of trifluralin B, a pre-emergent herbicide.

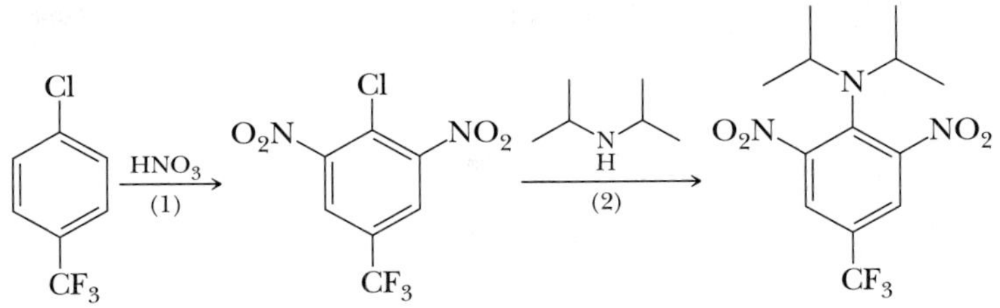

Trifluralin B

(a) Account for the orientation of nitration in Step 1.

(b) Propose a mechanism for the substitution reaction in Step 2.

22.29 A problem in dyeing fabrics is the degree of fastness of the dye to the fabric. Many of the early dyes were surface dyes; that is, they did not bond to the fabric with the result that they tended to wash off after repeated laundering. Indigo, for example, which gives the blue color to blue jeans, is a surface dye. Color fastness can be obtained by bonding a dye to the fabric. The first such dyes were the so-called reactive dyes, developed in the 1930s for covalent bonding dyes containing —NH_2 groups to cotton, wool, and silk fabrics. In the first stage of the first-developed method for reactive dyeing, the dye is treated with cyanuryl chloride, which links the two through the amino group of the dye. The remaining chlorines are then displaced by the —OH groups of cotton (cellulose) or the —NH_2 groups of wool or silk (both proteins).

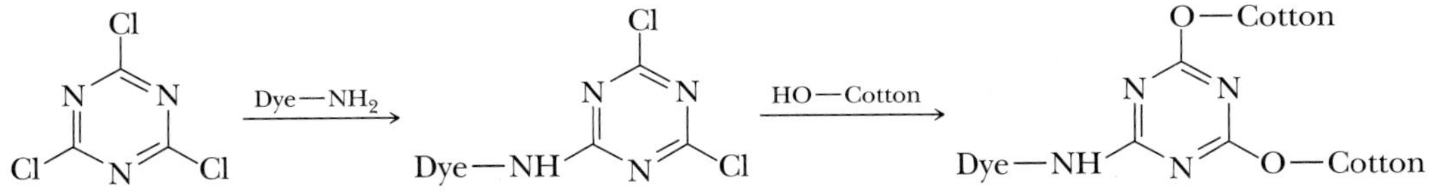

Cyanuryl chloride   A reactive dye   Dye covalently bonded to cotton

Propose a mechanism for the displacement of a chlorine from cyanuryl chloride by (a) the NH_2 group of a dye and (b) by an —OH group of cotton.

Syntheses

22.30 Show how to convert toluene to these compounds.

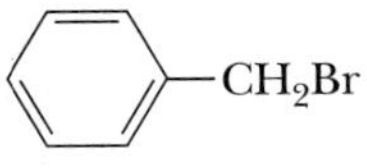

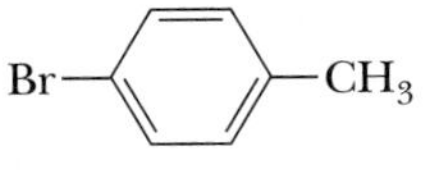

22.31 Show how to prepare each compound from 1-phenyl-1-propanone.

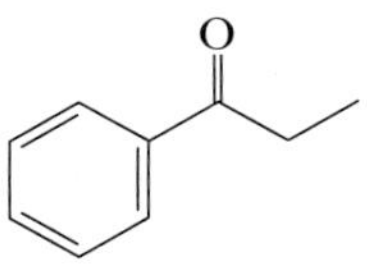

1-Phenyl-1-propanone   (racemic)

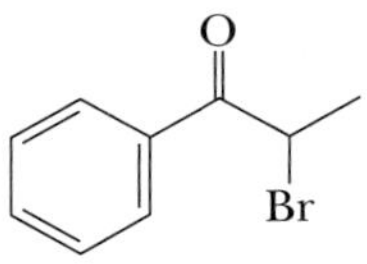

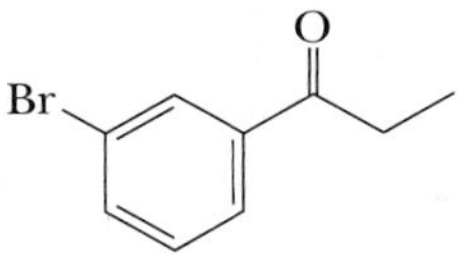

22.32 Show how to convert toluene to (a) 2,4-dinitrobenzoic acid and (b) 3,5-dinitrobenzoic acid.

22.33 Show reagents and conditions to bring about the following conversions.

(a)

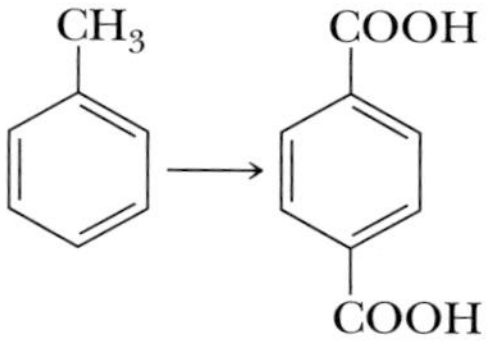

(b)

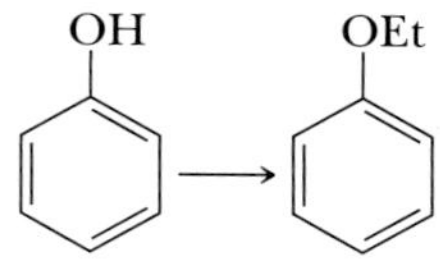

(c)

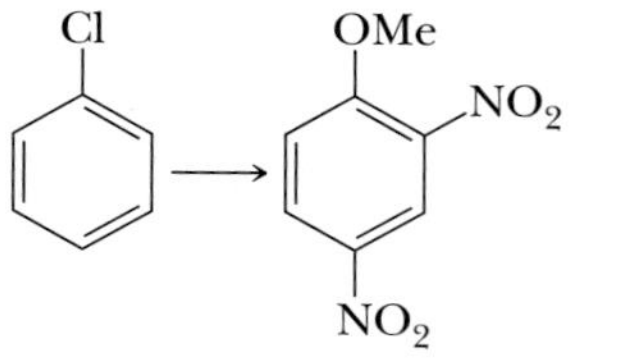

(d)

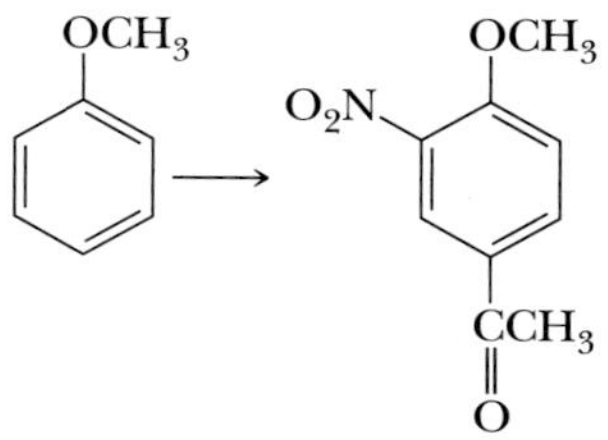

(e)

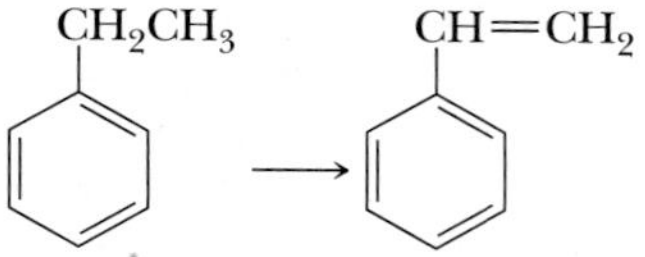

22.34 Propose a synthesis of triphenylmethane from benzene, as the only source of aromatic rings, and any other necessary reagents.

22.35 Propose a synthesis for each compound from benzene.

(a)

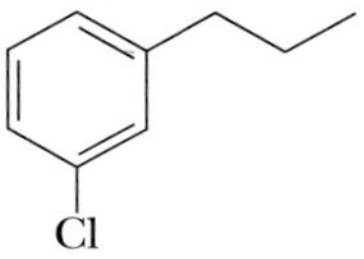

(b)

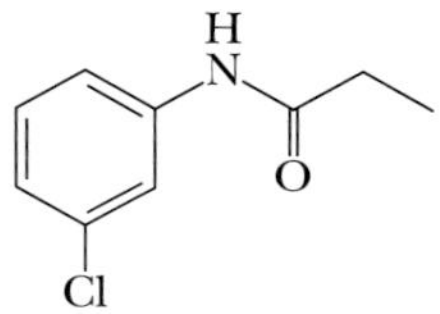

22.36 The first widely used herbicide for the control of weeds was 2,4-dichlorophenoxyacetic acid (2,4-D). Show how this compound might be synthesized from phenol and chloroacetic acid by way of the given chlorinated phenol intermediate.

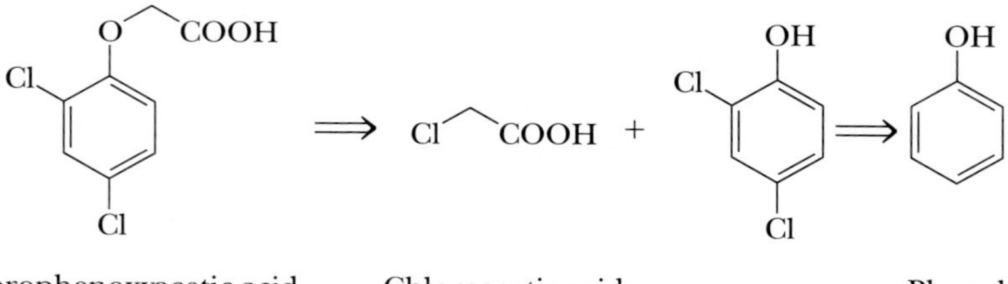

2,4-Dichlorophenoxyacetic acid
(2,4-D)

Chloroacetic acid

Phenol

22.37 Phenol is the starting material for the synthesis of 2,3,4,5,6-pentachlorophenol, known alternatively as pentachlorophenol or more simply as penta. At one time, penta was widely used as a wood preservative for decks, siding, and outdoor wood furniture. Draw the structural formula for pentachlorophenol, and describe its synthesis from phenol.

22.38 Starting with benzene, toluene, or phenol as the only sources of aromatic rings, show how to synthesize the following. Assume in all syntheses that mixtures of ortho-para products can be separated into the desired isomer.

(a) 1-Bromo-3-nitrobenzene
(b) 1-Bromo-4-nitrobenzene
(c) 2,4,6-Trinitrotoluene (TNT)
(d) *m*-Chlorobenzoic acid

(e) *p*-Chlorobenzoic acid (f) *p*-Dichlorobenzene
(g) *m*-Nitrobenzenesulfonic acid

22.39 3,5-Dibromo-4-hydroxybenzenesulfonic acid is used as a disinfectant. Propose a synthesis of this compound from phenol.

22.40 Propose a synthesis for 3,5-dichloro-2-methoxybenzoic acid starting from phenol.

22.41 The following compound used in perfumery has a violet-like scent. Propose a synthesis of this compound from benzene.

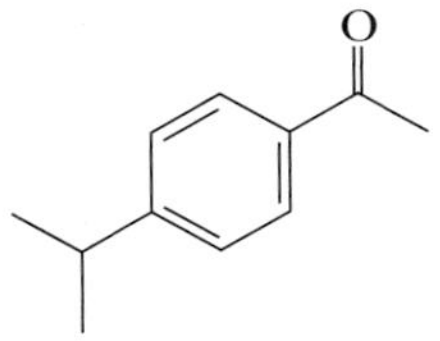

4-Isopropylacetophenone

22.42 Cancer of the prostate is the second leading cause of cancer deaths among American males, exceeded only by lung cancer. One treatment of prostate cancer is based on the fact that testosterone and androsterone (both androgens) enhance the proliferation of prostate tumors. The drug flutamide (an antiandrogen) reduces the level of androgens in target tissues and is currently used to prevent and treat prostate cancer.

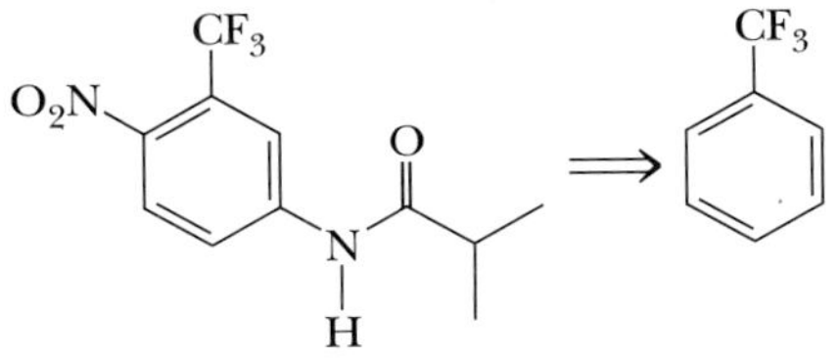

Flutamide   (Trifluoromethyl)
(Eulexin)   benzene

Propose a synthesis of flutamide from (trifluoromethyl)toluene.

22.43 The compound 4-isobutylacetophenone is needed for the synthesis of ibuprofen (See Chemical Connections: "Ibuprofen: The Evolution of an Industrial Synthesis" in Chapter 19). Propose a synthesis of 4-isobutylacetophenone from benzene and any other necessary reagents.

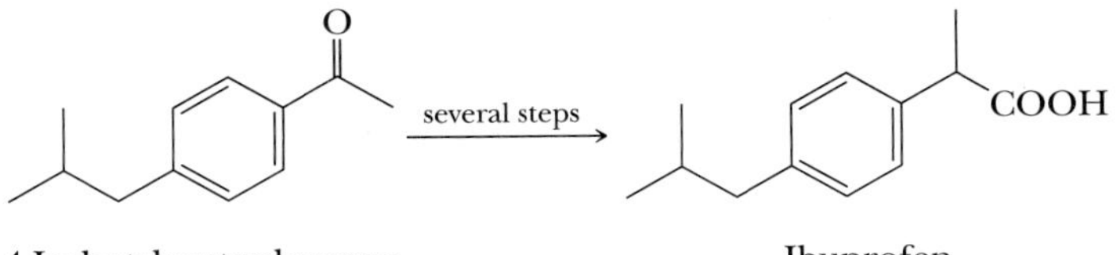

4-Isobutylacetophenone   Ibuprofen
(racemic)

22.44 Following is the structural formula of musk ambrette, a synthetic musk, essential in perfumes to enhance and retain odor. Propose a synthesis of this compound from *m*-cresol (3-methylphenol).

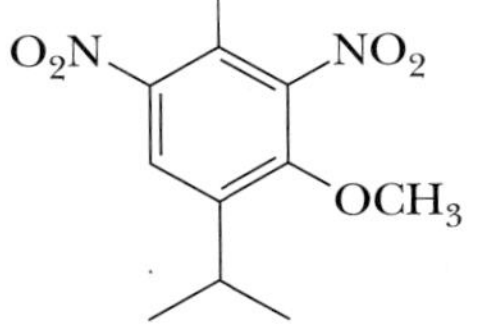

22.45 Propose a synthesis of this compound starting from toluene and phenol.

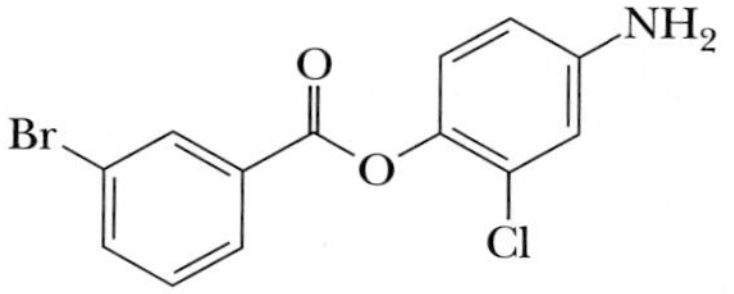

22.46 When certain aromatic compounds are treated with formaldehyde, CH_2O, and HCl, the CH_2Cl group is introduced onto the ring. This reaction is known as chloromethylation.

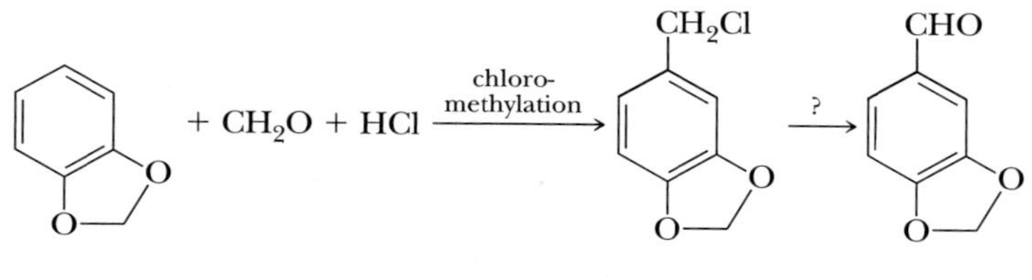

Piperonal

(a) Propose a mechanism for this example of chloromethylation.
(b) The product of this chloromethylation can be converted to piperonal, which is used in perfumery and in artificial cherry and vanilla flavors. How might the CH_2Cl group of the chloromethylation product be converted to a CHO group?

22.47 Following is a retrosynthetic analysis for the acaracide (killing mites and ticks) and fungicide dinocap. Given this analysis, propose a synthesis for dinocap from phenol and 1-octene.

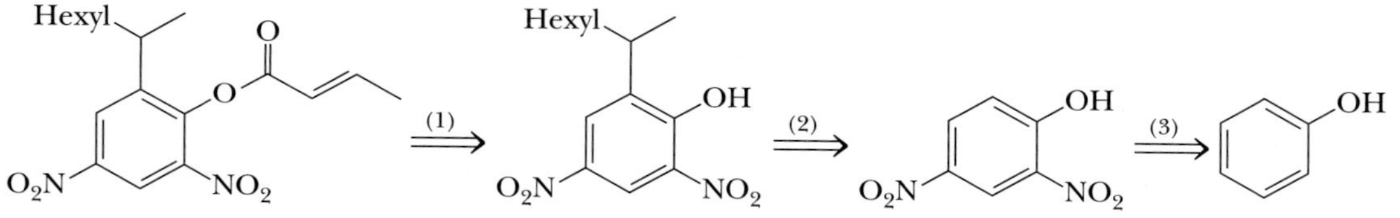

Dinocap

(a) Given this analysis, propose a synthesis for dinocap.
(b) Is dinocap chiral? If so, which of the possible stereoisomers are formed in this synthesis?

22.48 Following is the structure of miconazole, the active antifungal agent in a number of over-the-counter preparations, including Monistat, which are used to treat vaginal yeast infections. One of the compounds needed for the synthesis of miconazole is the trichloro derivative of toluene shown on its right.

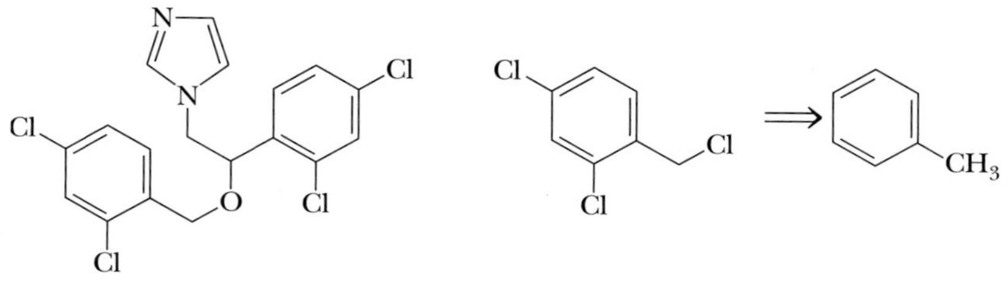

Miconazole   2,4-Dichloro-1-   Toluene
                            chloromethylbenzene

(a) Show how this derivative can be synthesized from toluene.
(b) How many stereoisomers are possible for miconazole?

22.49 Bupropion, the hydrochloride of which was first marketed in 1985 by Burroughs-Wellcome, now GlaxoWellcome, is an antidepressant sold under the trade name Wellbutrin. During clinical trials, it was discovered that smokers, after one to two weeks on the drug, reported that their craving for tobacco lessened. Further clinical trials confirmed this finding, and the drug was marketed in 1997 under the trade name Zyban as an aid in smoking cessation.

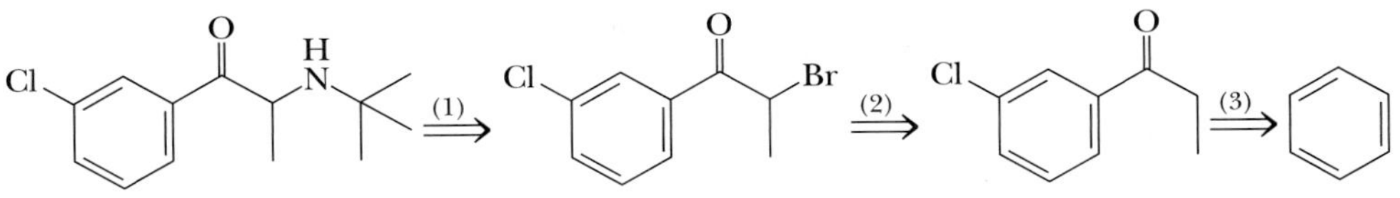

Bupropion

(a) Given this retrosynthetic analysis, propose a synthesis for bupropion.
(b) Is buprion chiral? If so, how many of the possible stereoisomers are formed in this synthesis?

22.50 Diazepam, better known as Valium, is a central nervous system (CNS) sedative/hypnotic. As a sedative, it diminishes activity and excitement and thereby has a calming effect. In 1976, based on the number of new and refilled prescriptions processed, diazepam was the most prescribed drug in the United States.

Following is a retrosynthetic analysis for a synthesis of diazepam. Note that the formation of compound B involves a Friedel-Crafts acylation. In this reaction it is necessary to protect the 2° amine by prior treatment with acetic anhydride. The acetyl-protecting group is then removed by treatment with aqueous NaOH followed by careful acidification with HCl.

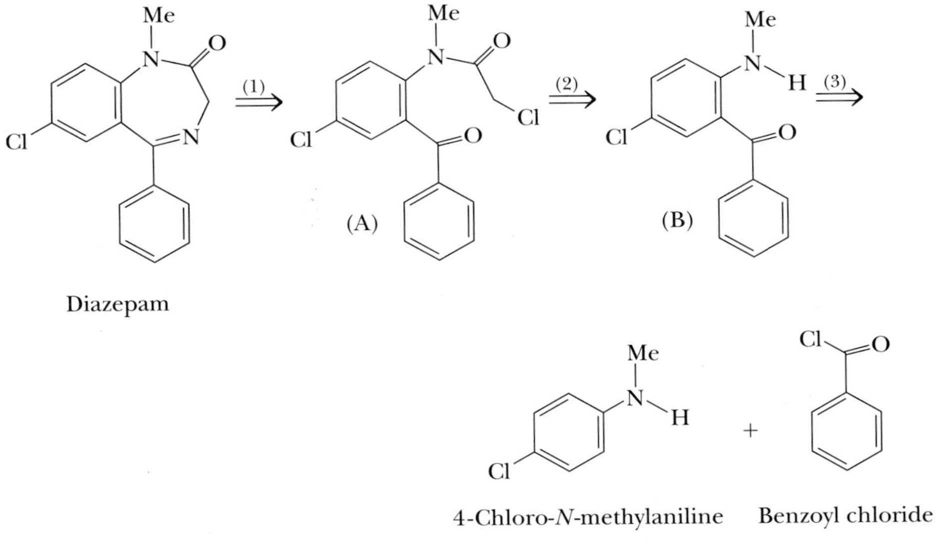

Diazepam   (A)   (B)

4-Chloro-*N*-methylaniline   Benzoyl chloride

(a) Given this retrosynthetic analysis, propose a synthesis for diazepam.
(b) Is diazepam chiral? If so, how many of the possible stereoisomers are formed in this synthesis?

22.51 The antidepressant amitriptyline inhibits the re-uptake of norepinephrine and serotonin from the synaptic cleft. Because the re-uptake of these neurotransmitters is inhibited, their effects are potentiated; they remain available to interact with serotonin and norepinephrine receptor sites longer and continue to cause excitation of serotonin- and norepinephrine-mediated neural pathways. Following is a synthesis for amitriptyline.

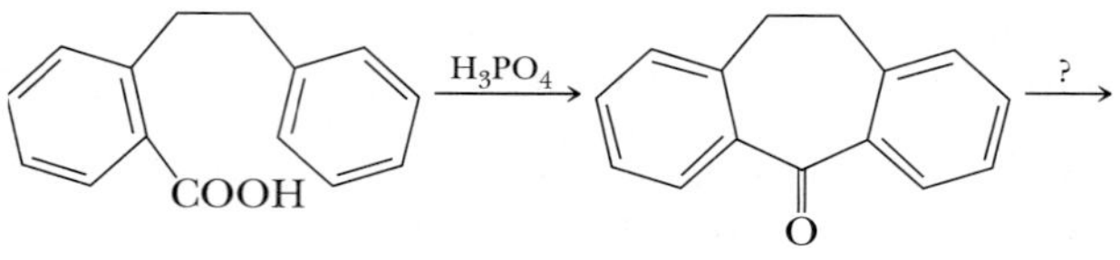

(A)   (B)

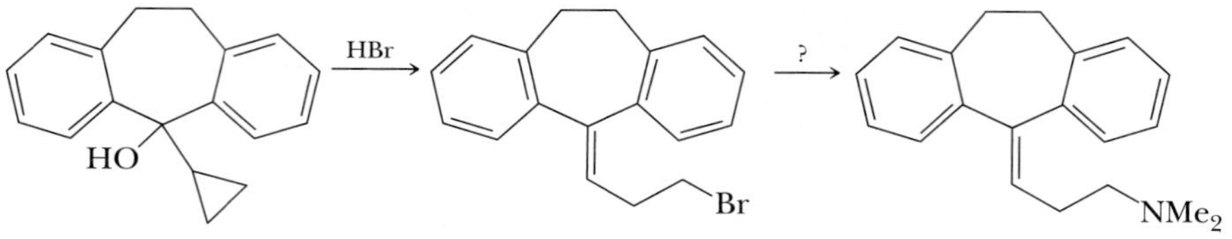

(C)   (D)   Amitriptyline

(a) Propose a mechanism for the conversion of (A) to (B).

(b) Propose reagents for the conversion of (B) to (C).

(c) Propose a mechanism for the conversion of (C) to (D). *Note:* It is not acceptable to propose a primary carbocation as an intermediate.

(d) Propose a reagent for the conversion of (D) to amitriptyline.

(e) Is amitriptyline chiral? If so how many of the possible stereoisomers are formed in this synthesis?

22.52 Show how the antidepressant venlafaxine (Effexor) can be synthesized from these readily available starting materials. Is venlafaxine chiral? If so, how many of the possible stereoisomers are formed in this synthesis?

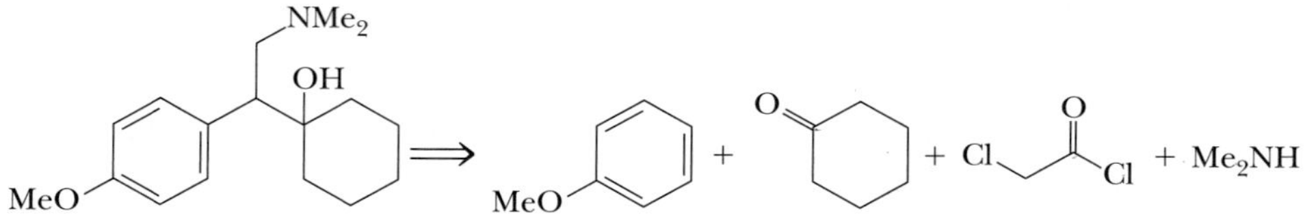

Venlafaxine   Anisole   Cyclohex-anone   Chloroacetyl chloride   Dimethyl-amine

22.53 One potential synthesis of the antiinflammatory and analgesic drug nabumetone is chloromethylation (Problem 22.46) of 2-methoxynaphthalene followed by an acetoacetic ester synthesis (Section 19.6).

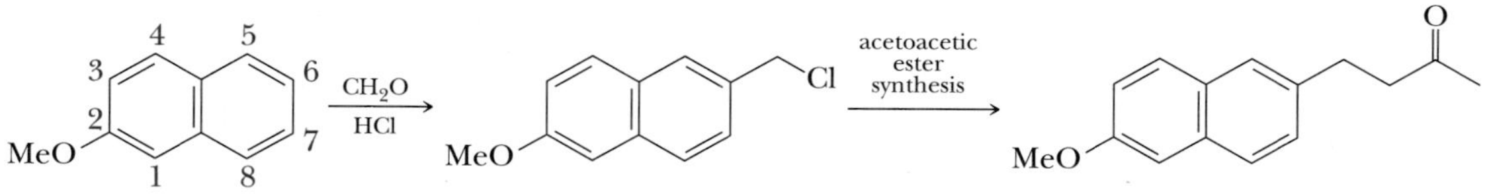

2-Methoxynaphthalene   Nabumetone

(a) Account for the regioselectivity of chloromethylation at carbon 6 rather than at carbons 5 or 7.

(b) Show steps in the acetoacetic ester synthesis by which the synthesis of nabumetone is completed.

22.54 The analgesic, soporific, and euphoriant properties of the dried juice obtained from unripe seed pods of the opium poppy *Papaver somniferum* have been known for centuries. By the beginning of the 19th century, the active principle, morphine, had been isolated and its structure determined. Even though morphine is one of modern medicine's most effective painkillers, it has two serious disadvantages. First, it is addictive. Second, it depresses the respiratory control center of the central nervous system. Large doses of morphine (or heroin which is *N*-acetylmorphine) can lead to death by respiratory failure.

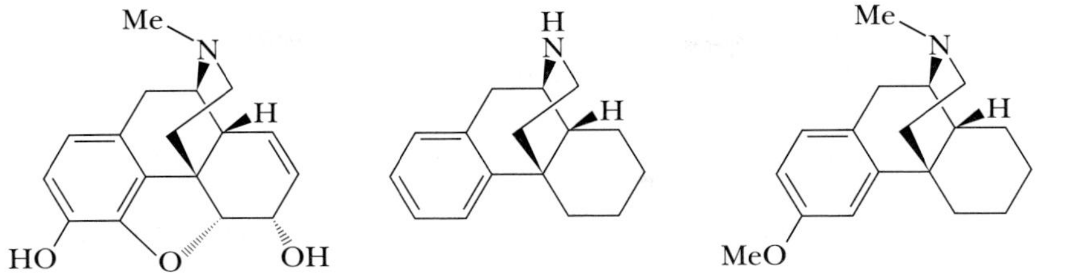

Morphine   Morphinan $(+/-)$ = Racemethorphan
                                      $(+)$ = Dextromethorphan
                                      $(-)$ = Levomethorphan

For these reasons, chemists have sought to produce painkillers related in structure to morphine, but without these serious disadvantages. One strategy has been to modify the carbon-nitrogen skeleton of morphine in the hope of producing medications equally effective but with reduced side effects. One target of this synthetic effort was morphinan, the bare morphine skeleton. Among the compounds thus synthesized, racemethorphan (the racemic mixture) and levomethorphan (the levorotatory enantiomer) proved to be very potent analgesics. Interestingly, the dextrorotatory enantiomer, dextromethorphan, has no analgesic activity. It does, however, show approximately the same antitussive (cough-suppressing) activity as morphine and is, therefore, used extensively in cough remedies.

Following is a synthesis of racemethorphan.

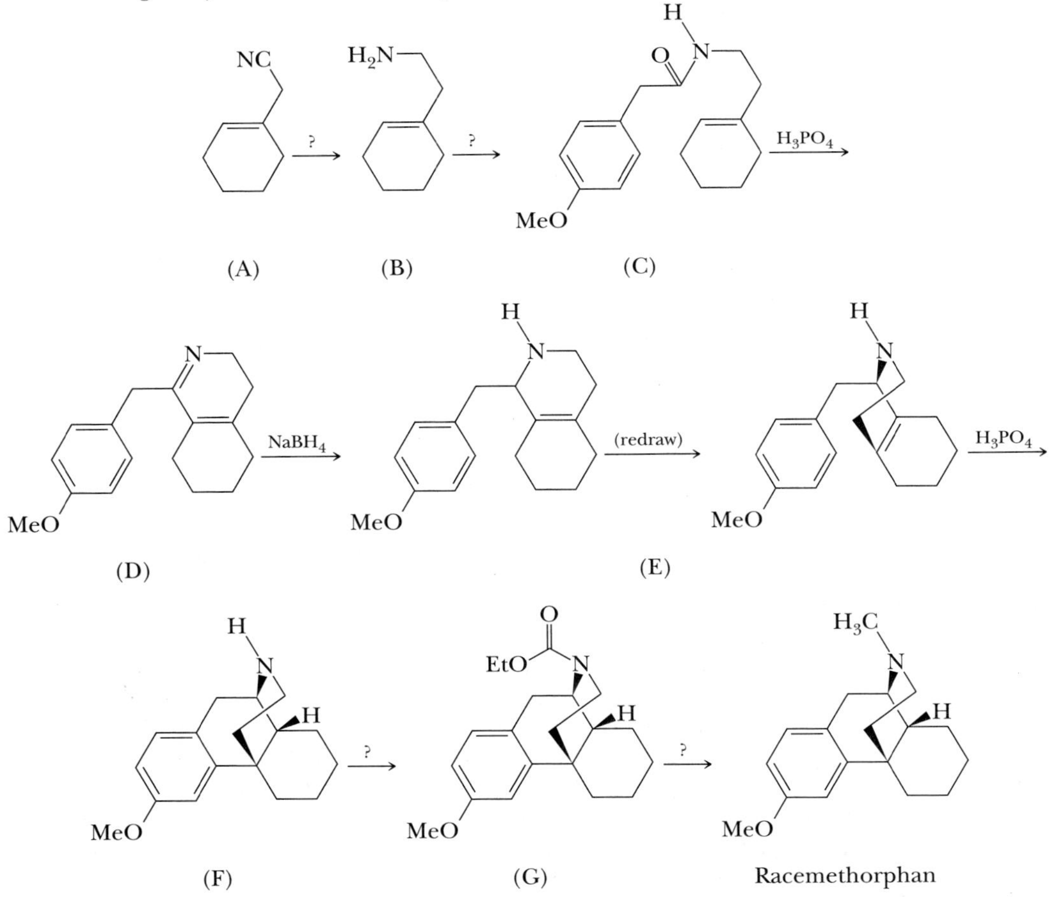

(A)   (B)   (C)

(D)   (E)

(F)   (G)   Racemethorphan

(a) Propose a reagent for the conversion of (A) to (B).
(b) Propose a reagent for the conversion of (B) to (C).
(c) Propose a mechanism for the conversion of (C) to (D).
(d) Propose a mechanism for the conversion of (E) to (F).
(e) Propose a reagent for the conversion of (F) to (G).
(f) Propose a reagent for the conversion of (G) to racemethorphan.

22.55 Following is the structural formula of the antihypertensive drug labetalol, a nonspecific β-adrenergic blocker with vasodilating activity. Members of this class have received enormous clinical attention because of their effectiveness in treating hypertension (high blood pressure), migraine headaches, glaucoma, ischemic heart disease, and certain cardiac arrhythmias. This retrosynthetic analysis involves disconnects to the α-haloketone (B) and the amine (C). Each is in turn derived from a simpler, readily available precursor.

Labetalol

(A)

(B)   +   (C)   (D)

(E)

Salicylic acid

(F)

PhCH₂Cl

Benzyl chloride

(a) Given this retrosynthetic analysis, propose a synthesis for labetalol from salicylic acid and benzyl chloride. *Note:* The conversion of salicylic acid to (E) involves a Friedel-Crafts acylation in which the phenolic —OH must be protected by treatment with acetic anhydride to prevent the acylation of the —OH group. The protecting group is later removed by treatment with KOH followed by acidification.
(b) Labetalol has two chiral centers and, as produced in this synthesis, is a racemic mixture of the four possible stereoisomers. The active stereoisomer is dilevalol, which has the *R,R* configuration at its chiral centers. Draw a structural formula of dilevalol showing the configuration of each chiral center.

22.56 Propose a synthesis for the antihistamine *p*-methyldiphenhydramine, given this retrosynthetic analysis. Is *p*-methyldiphenhydramine chiral? If so, how many of the possible stereoisomers are formed in this synthesis?

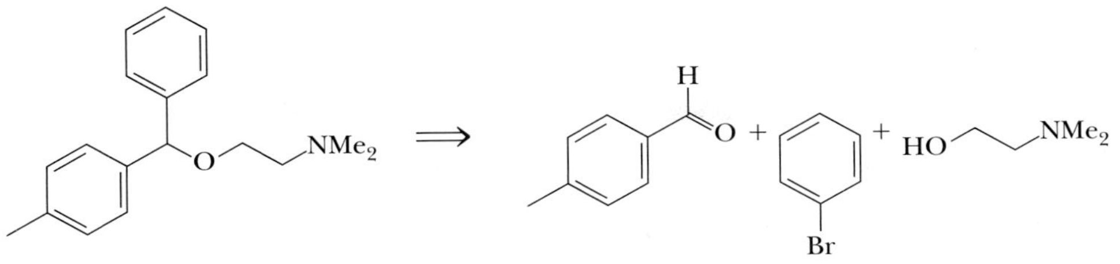

p-Methyldiphenhydramine

22.57 Meclizine is an antiemetic (it helps prevent or at least lessen the throwing up associated with motion sickness, including seasickness). Among the names of its over-the-counter preparations are Bonine, Sea-Legs, Antivert, and Navicalm.

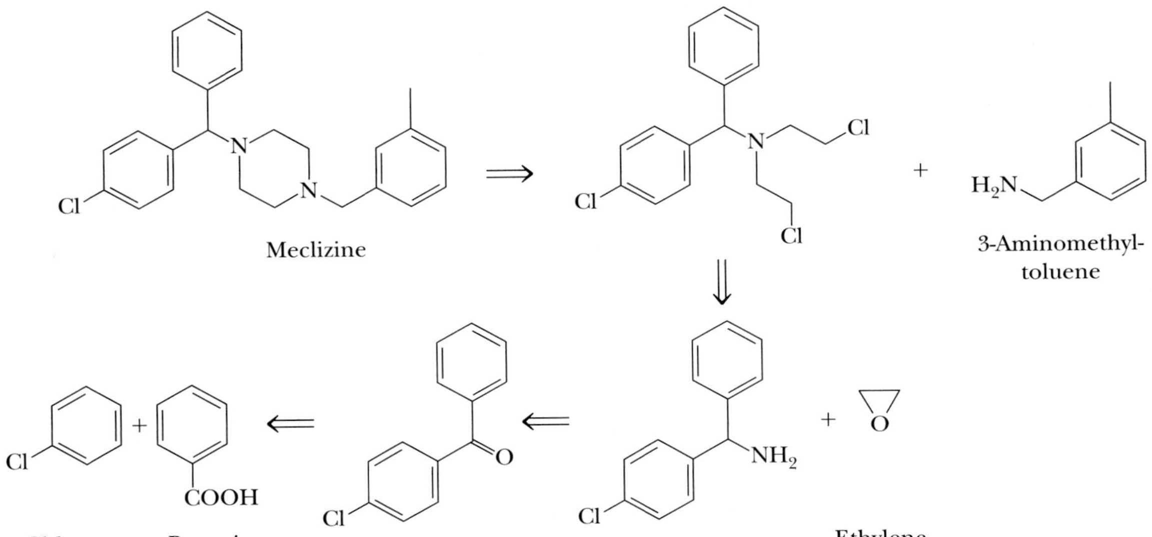

Meclizine   3-Aminomethyl-
toluene

Chloro-   Benzoic   Ethylene
benzene   acid   oxide

(a) Given this retrosynthetic analysis, show how meclizine can be synthesized from the four named organic starting materials.

(b) Is meclizine chiral? If so, how many of the possible stereoisomers are formed in this synthesis?

22.58 Spasmolytol, as its name suggests, is an antispasmodic. Given this retrosynthetic analysis, propose a synthesis for spasmolytol from salicylic acid, ethylene oxide, and diethylamine.

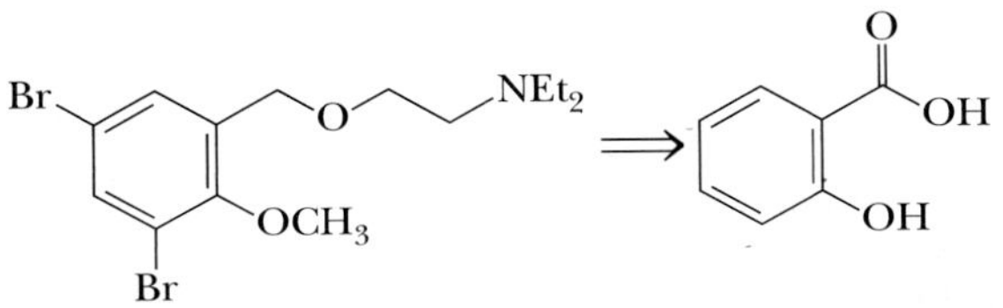

Spasmolytol   Salicyclic acid

22.59 Among the first antipsychotic drugs for the treatment of schizophrenia was haloperidol (Haldol), a competitive inhibitor of dopamine receptor sites in the central nervous system.

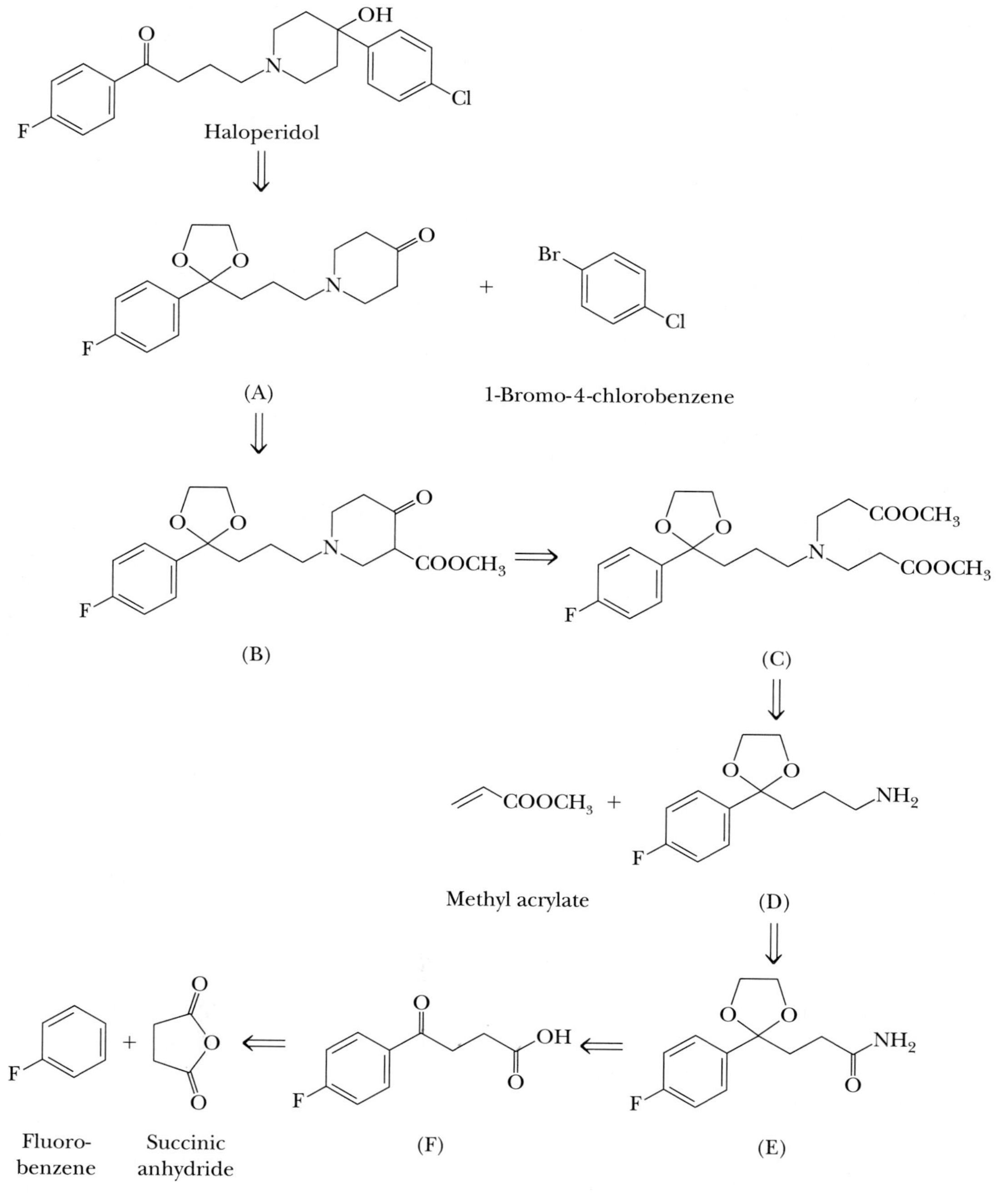

(a) Given the this retrosynthetic analysis, propose a synthesis for haloperidol.

(b) Is haloperidol chiral? If so, how many of the possible stereoisomers are formed in this synthesis?

22.60 A newer generation of antipsychotics, among them clozapine, are now used to treat the symptoms of schizophrenia. These drugs are more effective than earlier drugs in improving patient response in the areas of social withdrawal, apathy, memory, comprehension, and judgment. They also produce fewer side effects such as seizures and tardive dyskinesia (involuntary body movements). In the following synthesis of clozapine, Step 1 is an Ulmann coupling, a type of nucleophilic aromatic substitution that uses a copper catalyst.

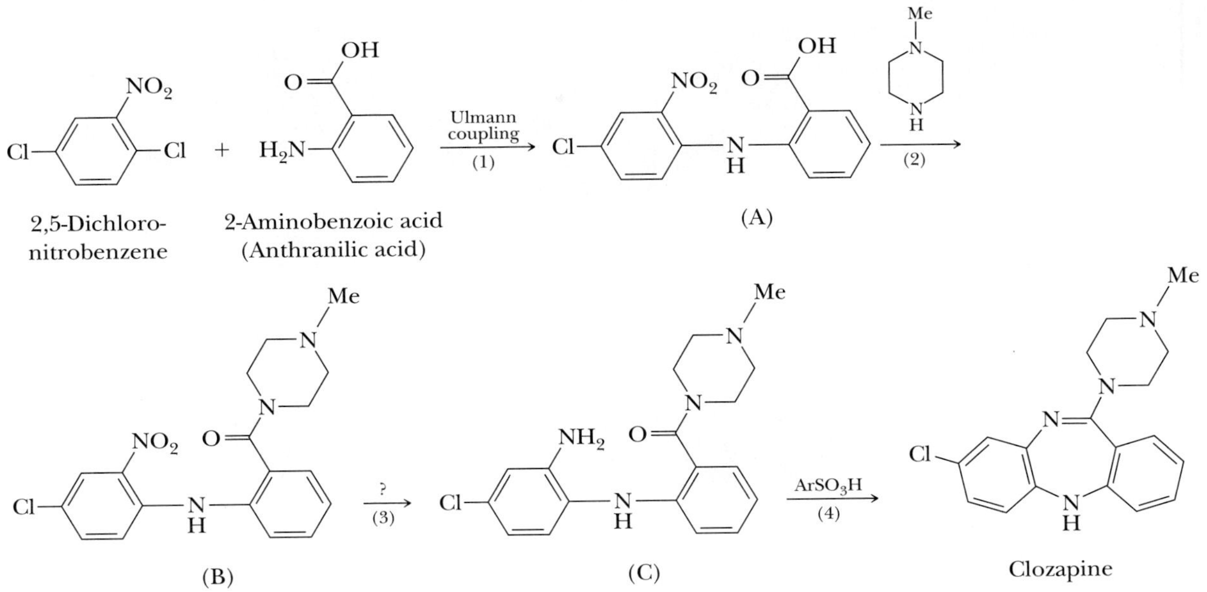

2,5-Dichloro-nitrobenzene

2-Aminobenzoic acid (Anthranilic acid)

(A)

(B)

(C)

Clozapine

(a) Show how you might bring about formation of the amide in Step 2.
(b) Propose a reagent for Step 3.
(c) Propose a mechanism for Step 4.
(d) Is clozapine chiral? If so, how many of the possible stereoisomers are formed in this synthesis?

22.61 Proparacaine is one of a class of -caine local anesthetics.

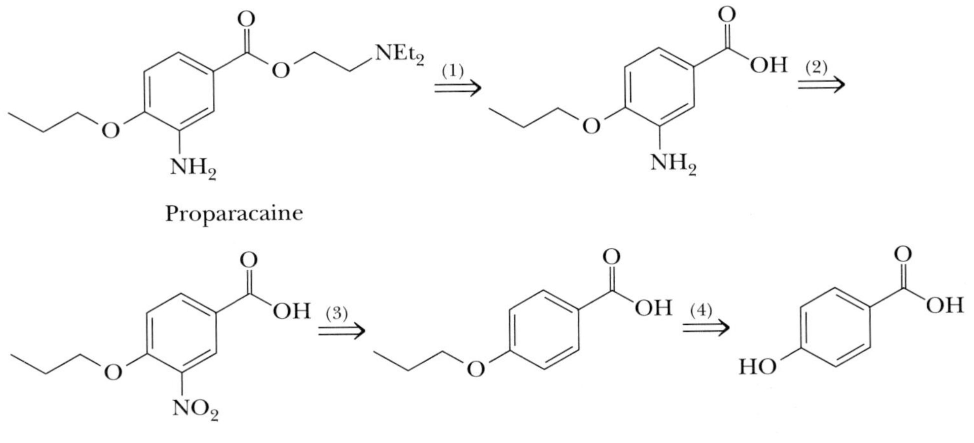

Proparacaine

(a) Given this retrosynthetic analysis, propose a synthesis of proparacaine from 4-hydroxybenzoic acid.
(d) Is proparacaine chiral? If so, how many of the possible stereoisomers are formed in this synthesis?

23

Frank Orel/Stone/Getty

■ Morphine, a potent painkiller isolated from the ripe seed heads of the opium poppy, has been a lead drug for chemists in search of potent but less addicting synthetic painkillers. See Problem 23.21. Insert: A model of morphine.

Outlines

ORGANIC
Chemistry⋅⚛⋅Now™

Look for this logo in the chapter and go to Organic ChemistryNow at **http://now.brookscole.com/bfi4** for tutorials, simulations, and problems.

Aliphatic amine An amine in which nitrogen is bonded only to alkyl groups.

Amines

Carbon, hydrogen, and oxygen are the three most common elements in organic compounds. Because of the wide distribution of amines in the biological world, nitrogen is the fourth most common element in organic compounds. The most important chemical properties of amines are their basicity and their nucleophilicity.

23.1 Structure and Classification

Amines are derivatives of ammonia in which one or more hydrogens are replaced by alkyl or aryl groups. Amines are classified as primary, secondary, or tertiary, depending on the number of carbon atoms bonded directly to nitrogen (Section 1.3B).

$$CH_3-\overset{..}{N}H_2 \qquad CH_3-\overset{..}{N}H \qquad CH_3-\overset{CH_3}{\underset{CH_3}{\overset{|}{N}}:|}$$

Methylamine
(a 1° amine)

Dimethylamine
(a 2° amine)

Trimethylamine
(a 3° amine)

Amines are further divided into aliphatic and aromatic amines. In an **aliphatic amine,** all carbons bonded to nitrogen are derived from alkyl groups; in an **aromatic amine,** one or more of the groups bonded to nitrogen are aryl groups.

Aniline	*N*-Methylaniline	Benzyldimethylamine
(a 1° aromatic amine)	(a 2° aromatic amine)	(a 3° aliphatic amine)

An amine in which the nitrogen atom is part of a ring is classified as a **heterocyclic amine.** When the nitrogen is part of an aromatic ring (Section 21.2D), the amine is classified as a **heterocyclic aromatic amine.** Following are structural formulas for two heterocyclic aliphatic amines and two heterocyclic aromatic amines.

Pyrrolidine	Piperidine	Pyrrole	Pyridine
(heterocyclic aliphatic amines)		(heterocyclic aromatic amines)	

Aromatic amine An amine in which nitrogen is bonded to one or more aryl groups.

Heterocyclic amine An amine in which nitrogen is one of the atoms of a ring.

Heterocyclic aromatic amine An amine in which nitrogen is one of the atoms of an aromatic ring.

Example 23.1

Alkaloids are basic nitrogen-containing compounds of plant origin, many of which are physiologically active when administered to humans. Ingestion of coniine, isolated from water hemlock, can cause weakness, labored respiration, paralysis, and eventually death. Coniine is the toxic substance in the "poison hemlock" used in the death of Socrates. In small doses, nicotine is an addictive stimulant. In larger doses, it causes depression, nausea, and vomiting. In still larger doses, it is a deadly poison. Solutions of nicotine in water are used as insecticides. Cocaine is a central nervous system stimulant obtained from the leaves of the coca plant.

Alkaloid A basic nitrogen-containing compound of plant origin, many of which are physiologically active when administered to humans.

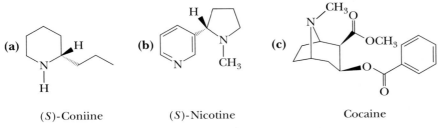

(a)	(b)	(c)
(*S*)-Coniine	(*S*)-Nicotine	Cocaine

Classify each amino group in these alkaloids according to type (primary, secondary, tertiary, aliphatic, aromatic, heterocyclic).

Solution

(a) A secondary heterocyclic aliphatic amine
(b) A tertiary heterocyclic aliphatic amine and a heterocyclic aromatic amine
(c) A tertiary heterocyclic aliphatic amine

Problem 23.1

Identify all carbon chiral centers in coniine, nicotine, and cocaine.

23.2 Nomenclature

A. Systematic Names

Systematic names for aliphatic amines are derived just as they are for alcohols. The suffix -*e* of the parent alkane is dropped and is replaced by -*amine*.

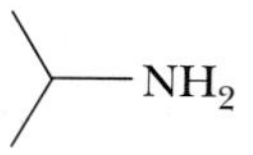

2-Propanamine 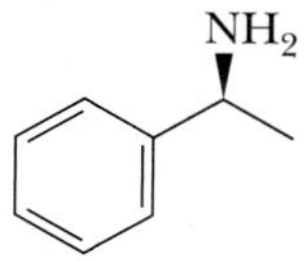   (*S*)-1-Phenylethanamine 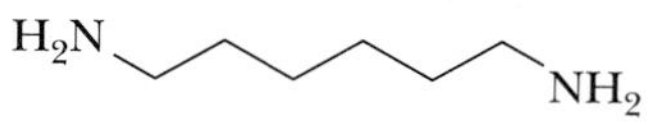   1,6-Hexanediamine

---

Example 23.2

Write systematic names for these amines.

(a) $CH_3(CH_2)_5NH_2$ **(b)** 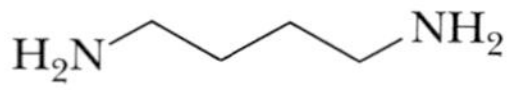   **(c)**

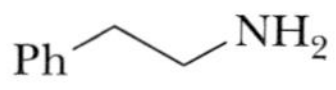

Solution

(a) 1-Hexanamine   **(b)** 1,4-Butanediamine   **(c)** 2-Phenylethanamine

Problem 23.2

Write structural formulas for these amines.
(a) 2-Methyl-1-propanamine   **(b)** Cyclohexanamine   **(c)** (*R*)-2-Butanamine

---

IUPAC nomenclature retains the common name aniline for $C_6H_5NH_2$, the simplest aromatic amine. Its simple derivatives are named using the prefixes *o*-, *m*-, and *p*-, or numbers to locate substituents. Several derivatives of aniline have common names that are still widely used. Among these are toluidine for a methyl-substituted aniline and anisidine for a methoxyl-substituted aniline.

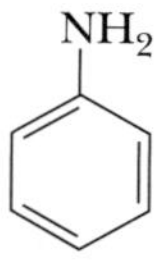

Aniline    4-Nitroaniline   4-Methylaniline   3-Methoxyaniline
         (*p*-Nitroaniline)   (*p*-Toluidine)   (*m*-Anisidine)

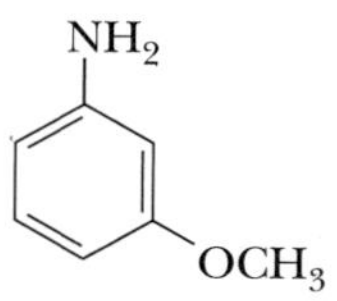

Secondary and tertiary amines are commonly named as *N*-substituted primary amines. For unsymmetrical amines, the largest group is taken as the parent amine; then the smaller group(s) attached to nitrogen are named, and their location is indicated by the prefix *N* (indicating that they are bonded to nitrogen).

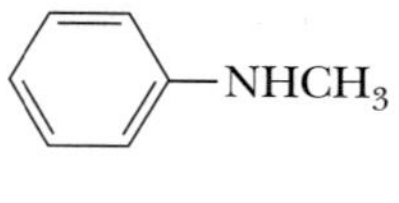

N-Methylaniline 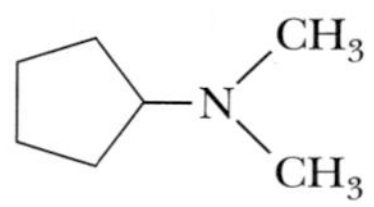   *N,N*-Dimethyl-
                     cyclopentanamine

Following are names and structural formulas for four heterocyclic aromatic amines, the common names of which have been retained in the IUPAC system.

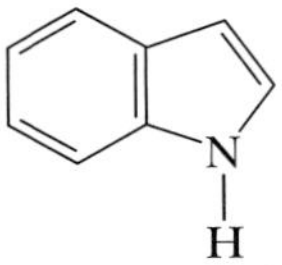

Indole 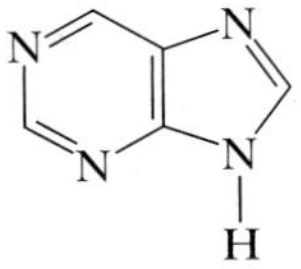 Purine 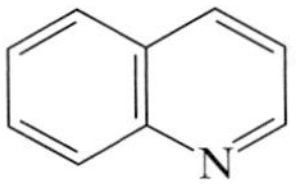 Quinoline 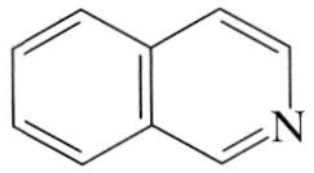 Isoquinoline

Among the various functional groups discussed in this text, the —NH_2 group is one of the lowest in precedence (Table 16.1). The following compounds each contain a functional group of higher precedence than the amino group, and, accordingly, the amino group is indicated by the prefix *amino-*.

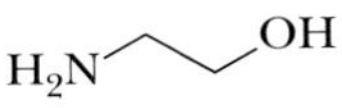

2-Aminoethanol 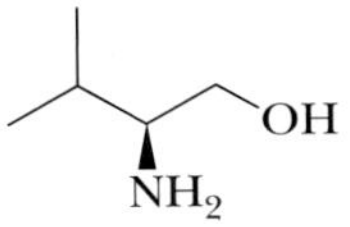 (*S*)-2-Amino-3-methyl-1-butanol 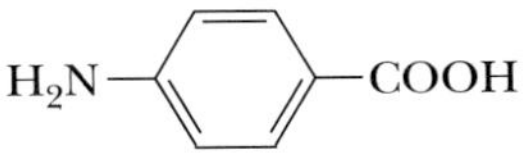 4-Aminobenzoic acid

B. Common Names

Common names for most aliphatic amines are derived by listing the alkyl groups bonded to nitrogen in alphabetical order in one word ending in the suffix *-amine*, that is, they are named as alkylamines.

CH_3NH_2
Methylamine

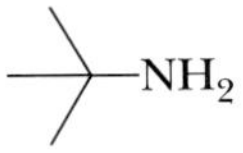

—NH_2
tert-Butylamine

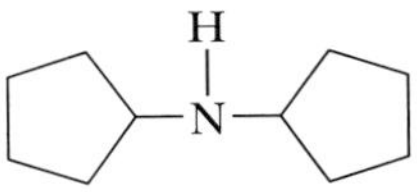

Dicyclopentylamine

Et_3N
Triethylamine

Example 23.3

Write structural formulas for these amines.
(a) Isopropylamine **(b)** Cyclohexylmethylamine **(c)** Benzylamine

Solution

(a) 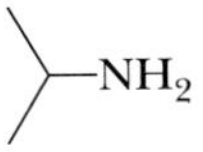 —NH_2 **(b)** 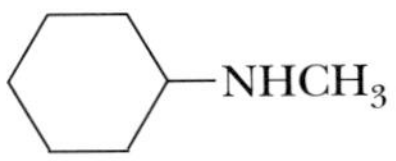 —$NHCH_3$ **(c)** 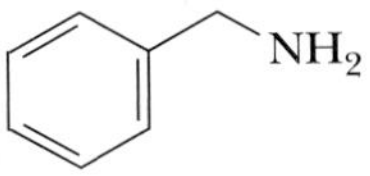 NH_2

Problem 23.3

Write structural formulas for these amines.
(a) Isobutylamine **(b)** Triphenylamine **(c)** Diisopropylamine

---

When four atoms or groups of atoms are bonded to a nitrogen atom, the compound is named as a salt of the corresponding amine. The ending *-amine* (or *-aniline*, *pyridine,* and so on) is replaced by *-ammonium* (or *anilinium, pyridinium,* and so on), and the name of the anion is added.

$$Et_3NH^+ \; Cl^-$$

 NH CH_3COO^-

| Triethylammonium chloride | Pyridinium acetate |

Example 23.4

Write the IUPAC name and, where possible, a common name for each compound.

(a) $(C_6H_5)_2NH$ **(b)** 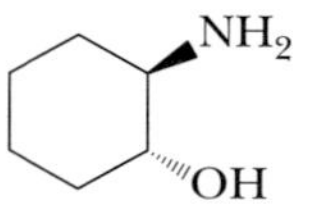 [structure with NH$_2$ and OH on cyclohexane] (racemic) **(c)** 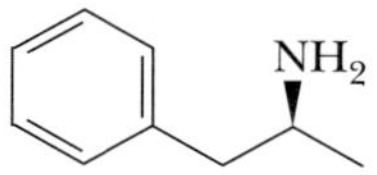 [structure with NH$_2$]

Solution

(a) Diphenylamine **(b)** *trans*-2-Aminocyclohexanol
(c) Its systematic name is (*S*)-1-phenyl-2-propanamine. Its common name is amphetamine. The dextrorotatory isomer of amphetamine (shown here) is a central nervous system stimulant and is manufactured and sold under several trade names. The salt with sulfuric acid is marketed as Dexedrine sulfate.

Problem 23.4

Write the IUPAC and, where possible, a common name for each compound.

(a) [structure with OH, O, NH$_2$] **(b)** H_2N [structure with O, OH] **(c)** [structure with NH$_2$]

An ion containing a nitrogen atom bonded to any combination of four alkyl or aryl groups is classified as a **quaternary (4°) ammonium ion,** Compounds containing such ions have properties characteristic of salts. Cetylpyridinium chloride is used as a topical antiseptic and disinfectant.

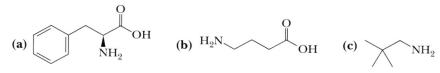

| Tetramethylammonium chloride | Tetradecylpyridinium chloride (Cetylpyridinium chloride) | Benzyltrimethylammonium hydroxide |

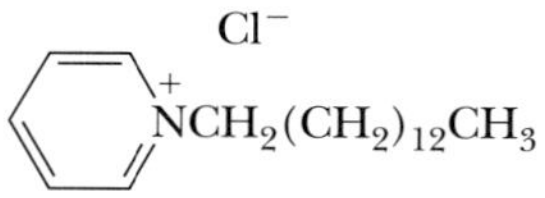

23.3 Chirality of Amines and Quaternary Ammonium Ions

The geometry of a nitrogen atom bonded to three other atoms or groups of atoms is trigonal pyramidal (Section 1.4). The sp^3-hybridized nitrogen atom is at the apex of the pyramid, and the three groups bonded to it extend downward to form the

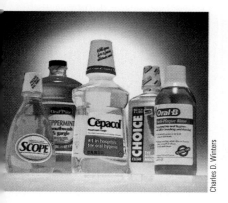

■ Several over-the-counter mouthwashes contain an *N*-alkylpyridinium chloride as an antibacterial agent.

Charles D. Winters

Quaternary (4°) ammonium ion
An ion in which nitrogen is bonded to four carbons and bears a positive charge.

ORGANIC
Chemistry⋅⋅Now™

Click Tutorials to review the
Nomenclature of Amines

triangular base of the pyramid. If we consider the unshared pair of electrons on nitrogen as a fourth group, then the arrangement of "groups" around nitrogen is approximately tetrahedral. Because of this geometry, an amine with three different groups bonded to nitrogen is chiral and can exist as a pair of enantiomers, as illustrated by the nonsuperposable mirror images of ethylmethylamine. In assigning configuration to these enantiomers, the group of lowest priority on nitrogen is the unshared pair of electrons.

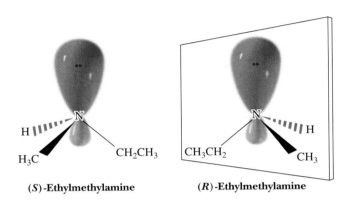

(S)-Ethylmethylamine   **(R)-Ethylmethylamine**

In principle, a chiral amine can be resolved; that is, it can be separated into a pair of enantiomers. Except for special cases, however, the enantiomers cannot be resolved because they undergo rapid interconversion by a process known as pyramidal inversion. **Pyramidal inversion** is the rapid oscillation of a nitrogen atom from one side of the plane of the three atoms bonded to it to the other side of that plane.

To visualize this process, imagine the sp^3-hybridized nitrogen atom lying above the plane of the three atoms to which it is bonded. In the transition state for pyramidal inversion, the nitrogen atom and the three groups to which it is bonded become coplanar, and the molecule becomes achiral. In this planar transition state, nitrogen is sp^2 hybridized, and its lone pair of electrons lies in its unhybridized $2p$ orbital. Nitrogen then completes the inversion, becomes sp^3 hybridized again, and now lies below the plane of the three atoms to which it is bonded.

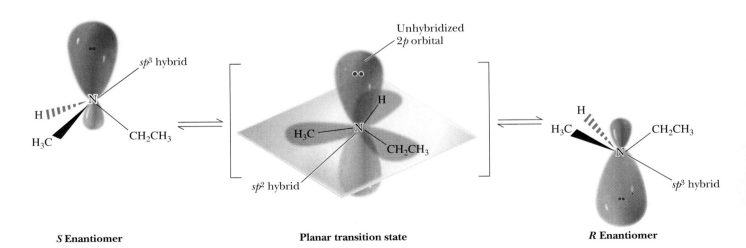

S **Enantiomer**   **Planar transition state**   *R* **Enantiomer**

As a result of pyramidal inversion, a chiral amine quite literally turns itself inside out, like an umbrella in a strong wind, and in the process becomes a racemic mixture. The activation energy for pyramidal inversion of simple amines is about 25 kJ (6 kcal)/mol. For ammonia at room temperature, the rate of nitrogen inversion is approximately 2×10^{11} s^{-1}. For simple amines, the rate is less rapid but nonetheless sufficient to make resolution impossible.

Pyramidal inversion is not possible for quaternary ammonium ions, and their salts can be resolved.

S Enantiomer   *R* Enantiomer

Phosphorus, in the same family as nitrogen, forms trivalent compounds called phosphines, which also have trigonal pyramidal geometry. The activation energy for pyramidal inversion of trivalent phosphorus compounds is considerably greater than it is for trivalent compounds of nitrogen, with the result that a number of chiral phosphines have been resolved.

23.4 Physical Properties

Amines are polar compounds, and both primary and secondary amines form intermolecular hydrogen bonds (Figure 23.1).

An N----H—N hydrogen bond is weaker than an O----H—O hydrogen bond because the difference in electronegativity between nitrogen and hydrogen $(3.0 - 2.1 = 0.9)$ is less than that between oxygen and hydrogen $(3.5 - 2.1 = 1.4)$. The effect of intermolecular hydrogen bonding can be illustrated by comparing the boiling points of methylamine and methanol. Both are polar molecules and interact in the pure liquid by hydrogen bonding. Because hydrogen bonding is stronger in methanol than in methylamine, methanol has the higher boiling point.

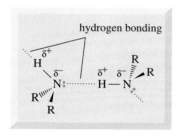

Figure 23.1
Intermolecular association by hydrogen bonding in primary and secondary amines. Nitrogen is approximately tetrahedral in shape with the axis of the hydrogen bond along the fourth position of the tetrahedron.

	CH$_3$CH$_3$	**CH$_3$NH$_2$**	**CH$_3$OH**
MW (g/mol)	30.1	31.1	32.0
bp (°C)	−88.6	−6.3	65.0

All classes of amines form hydrogen bonds with water and are more soluble in water than hydrocarbons of comparable molecular weight. Most low-molecular-weight amines are completely soluble in water (Table 23.1). Higher molecular-weight amines are only moderately soluble or insoluble.

CHEMICAL CONNECTIONS

The Poison Dart Frogs of South America

The Noanamá and Embrá peoples of the jungles of western Colombia have used poison blow darts for centuries, perhaps millennia. The poisons are obtained from the skin secretions of several brightly colored frogs of the genus *Phyllobates* (*neará* and *kokoi* in the language of the native peoples). A single frog contains enough poison for up to 20 darts. For the most poisonous species (*Phyllobates terribilis*), just rubbing a dart over the frog's back suffices to charge the dart with poison.

Scientists at the National Institutes of Health in the United States became interested in studying these poisons when it was discovered that they act on cellular ion channels, which would make them useful tools in basic research on mechanisms of ion transport. A field station was, therefore, established in western Colombia to collect the relatively common poison dart frogs. From 5000 frogs, 11 mg of two toxins, given the names batrachotoxin and batrachotoxinin A, was isolated. These names are derived from *batrachos*, the Greek word for frog. A combination of NMR spectroscopy, mass spectrometry, and single-crystal x-ray diffraction was used to determine the structures of these compounds.

Juan M. Renjifo/Animals Animals

■ Poison dart frog, *Phyllobates terribilis*.

Batrachotoxin and batrachotoxinin A are among the most lethal poisons ever discovered. It is estimated that as little as 200 μg of batrachotoxin is sufficient to induce irreversible cardiac arrest in a human being. It has been determined that they act by causing voltage-

Batrachotoxin

Batrachotoxinin A

gated Na^+ channels in nerve and muscle cells to be blocked in the open position, which leads to a huge influx of Na^+ ions into the affected cell.

The batrachotoxin story illustrates several common themes in drug discovery. First, information about the kinds of biologically active compounds and

their sources is often obtained from the native peoples of a region. Second, tropical rain forests are a rich source of structurally complex, biologically active substances. Third, the entire ecosystem, not just the plants, is a potential source of fascinating organic molecules.

Table 23.1   Physical Properties of Selected Amines

Name	Structural Formula	mp (°C)	bp (°C)	Solubility in Water
Ammonia	NH_3	−78	−33	Very soluble
Primary Amines				
Methylamine	CH_3NH_2	−95	−6	Very soluble
Ethylamine	$CH_3CH_2NH_2$	−81	17	Very soluble
Propylamine	$CH_3CH_2CH_2NH_2$	−83	48	Very soluble
Isopropylamine	$(CH_3)_2CHNH_2$	−95	32	Very soluble
Butylamine	$CH_3(CH_2)_3NH_2$	−49	78	Very soluble
Benzylamine	$C_6H_5CH_2NH_2$	—	185	Very soluble
Cyclohexylamine	$C_6H_{11}NH_2$	−17	135	Slightly soluble
Secondary Amines				
Dimethylamine	$(CH_3)_2NH$	−93	7	Very soluble
Diethylamine	$(CH_3CH_2)_2NH$	−48	56	Very soluble
Tertiary Amines				
Trimethylamine	$(CH_3)_3N$	−117	3	Very soluble
Triethylamine	$(CH_3CH_2)_3N$	−114	89	Slightly soluble
Aromatic Amines				
Aniline	$C_6H_5NH_2$	−6	184	Slightly soluble
Heterocyclic Aromatic Amines				
Pyridine	C_5H_5N	−42	116	Very soluble

23.5 Basicity

Like ammonia, all amines are weak bases, and aqueous solutions of amines are basic. The following acid-base reaction between an amine and water is written using curved arrows to emphasize that, in these proton-transfer reactions, the unshared pair of electrons on nitrogen forms a new covalent bond with hydrogen and displaces hydroxide ion.

$$CH_3{-}\overset{\overset{\displaystyle H}{|}}{\underset{\underset{\displaystyle H}{|}}{N}}{:} \;+\; H{-}\overset{\cdot\cdot}{\underset{\cdot\cdot}{O}}{-}H \;\rightleftharpoons\; CH_3{-}\overset{\overset{\displaystyle H}{|}}{\underset{\underset{\displaystyle H}{|}}{N^{+}}}{-}H \quad {:}\overset{\cdot\cdot}{\underset{\cdot\cdot}{O}}{-}H^{-}$$

Methylamine   Methylammonium hydroxide

It is common to discuss the basicity of amines by reference to the acid ionization constant of the corresponding conjugate acid, as illustrated for the ionization of the methylammonium ion.

$$CH_3NH_3{}^+ + H_2O \rightleftharpoons CH_3NH_2 + H_3O^+$$

$$K_a = \frac{[CH_3NH_2][H_3O^+]}{[CH_3NH_3^+]} = 2.29 \times 10^{-11} \qquad pK_a = 10.64$$

Table 23.2 Acid Strengths, pK_a, of the Conjugate Acids of Selected Amines

Amine	Structure	pK_a of Conjugate Acid
Ammonia	NH_3	9.26
Primary Amines		
Methylamine	CH_3NH_2	10.64
Ethylamine	$CH_3CH_2NH_2$	10.81
Cyclohexylamine	$C_6H_{11}NH_2$	10.66
Secondary Amines		
Dimethylamine	$(CH_3)_2NH$	10.73
Diethylamine	$(CH_3CH_2)_2NH$	10.98
Tertiary Amines		
Trimethylamine	$(CH_3)_3N$	9.81
Triethylamine	$(CH_3CH_2)_3N$	10.75
Aromatic Amines		
Aniline	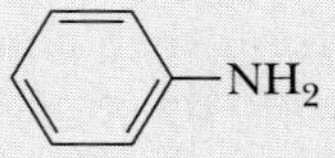	4.63
4-Methylaniline	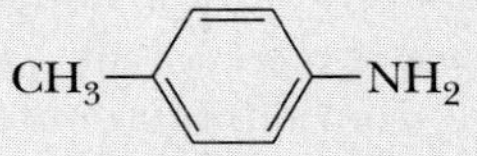	5.08
4-Chloroaniline	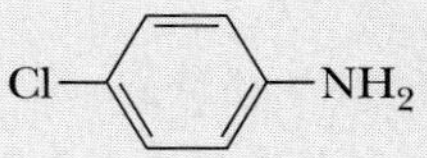	4.15
4-Nitroaniline	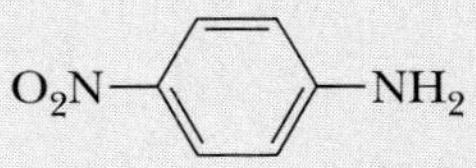	1.0
Heterocyclic Aromatic Amines		
Pyridine	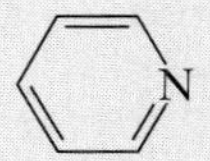	5.25
Imidazole	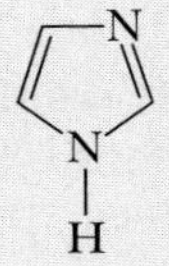	6.95

Table 23.2 gives values of pK_a for the conjugate acids of selected amines. Keep in mind that the weaker the conjugate acid (the larger its pK_a), the greater the basicity of the amine.

Example 23.5

Predict the position of equilibrium for this acid-base reaction.

$$CH_3NH_2 + CH_3COOH \rightleftharpoons CH_3NH_3^+ + CH_3COO^-$$

Solution

Use the approach we developed in Section 4.4 to predict the position of equilibrium in acid-base reactions. Equilibrium favors reaction of the stronger acid with the stronger base to give the weaker acid and weaker base.

$$CH_3NH_2 + CH_3COOH \rightleftharpoons CH_3NH_3^+ + CH_3COO^- \qquad pK_{eq} = -5.88$$
$$pK_a\ 4.76 \qquad pK_a\ 10.64 \qquad\qquad K_{eq} = 7.6 \times 10^5$$
$$\text{(stronger} \qquad \text{(weaker}$$
$$\text{acid)} \qquad\quad \text{acid)}$$

Problem 23.5

Predict the position of equilibrium for this acid-base reaction.

$$CH_3NH_3^+ + H_2O \rightleftharpoons CH_3NH_2 + H_3O^+$$

A. Aliphatic Amines

All aliphatic amines have about the same base strength, pK_a of the conjugate acid 10.0–11.0, and are slightly stronger bases than ammonia. The increase in basicity compared with ammonia can be attributed to the greater stability of an alkylammonium ion, as for example $RCH_2NH_3^+$ compared with the ammonium ion, NH_4^+. This greater stability arises from the electron-releasing effect of alkyl groups and the resulting partial delocalization of the positive charge from nitrogen onto carbon in the alkylammonium ion.

Positive charge is partially
delocalized onto the alkyl group.

$$R\!-\!CH_2 \overset{\delta+}{\longrightarrow} \overset{\overset{\textstyle H}{|}}{\underset{\underset{\textstyle H}{|}}{N}}\!\!\overset{\delta+}{-}\!H$$

Recall that we invoked a similar argument in Section 6.3A to account for the effect of alkyl groups in stabilizing carbocations.

B. Aromatic Amines

Aromatic amines are considerably weaker bases than aliphatic amines. Compare, for example, values of pK_a for aniline and cyclohexylamine. The ionization constant for the conjugate acid of aniline is larger (the smaller the value of pK_a, the weaker the base) than that for cyclohexylamine by a factor of 10^6.

$$\text{Cyclohexyl}\!-\!NH_2 + H_2O \rightleftharpoons \text{Cyclohexyl}\!-\!NH_3^+\ OH^- \qquad pK_a = 10.66;\ K_a = 2.19 \times 10^{-11}$$

Cyclohexylamine   Cyclohexylammonium hydroxide

$$\text{Aryl}\!-\!NH_2 + H_2O \rightleftharpoons \text{Aryl}\!-\!NH_3^+\ OH^- \qquad pK_a = 4.63;\ K_a = 2.34 \times 10^{-5}$$

Aniline   Anilinium hydroxide

Aromatic amines are less basic than aliphatic amines because of a combination of two factors. First is the resonance stabilization of the free base form of aromatic amines.

| Interaction of the electron pair on nitrogen with the pi system of the aromatic ring | No resonance is possible with alkylamines. |

For aniline and other arylamines, the resonance stabilization is the result of the interaction of the unshared pair on nitrogen with the pi system of the aromatic ring. The resonance energy of benzene is approximately 151 kJ (36 kcal)/mol. For aniline, it is 163 kJ (39 kcal)/mol. Because of this resonance interaction, the electron pair on nitrogen is less available for reaction with acid. No such resonance stabilization is possible for alkylamines. Therefore, the electron pair on the nitrogen of an alkylamine is more available for reaction with an acid; alkylamines are stronger bases than arylamines.

The second factor contributing to the decreased basicity of aromatic amines is the electron-withdrawing inductive effect of the sp^2-hybridized carbons of the aromatic ring compared with the sp^3-hybridized carbons of aliphatic amines. The unshared pair of electrons on nitrogen in an aromatic amine is pulled toward the ring and, therefore, is less available for protonation to form the conjugate acid of the amine. These factors are the same two that operate to make phenoxide ion less basic than alkoxide ions (Section 21.4B).

Electron-releasing groups (for example, methyl, ethyl, and other alkyl groups) increase the basicity of aromatic amines, whereas electron-withdrawing groups (for example, nitro, and carbonyl groups) decrease their basicity. The decrease in basicity on halogen substitution is the result of the electron-withdrawing inductive effect of the electronegative halogen. The decrease in basicity on nitro substitution is caused by a combination of inductive and resonance effects as can be seen by comparing the pK_a values for the conjugate acids of 3-nitroaniline and 4-nitroaniline.

| 3-Nitroaniline pK_a 2.47 | 4-Nitroaniline pK_a 1.0 |

The basicity-decreasing effect of nitro substitution in the 3-position is almost entirely the result of its inductive effect, whereas that of nitro substitution in the 4-position is attributable to both inductive and resonance effects. In the case of para substitution (as well as ortho substitution), delocalization of the lone pair on the amino nitrogen involves not only the carbons of the aromatic ring but also oxygen atoms of the nitro group.

delocalization of the nitrogen lone pair onto the oxygen atoms of the nitro group

CONNECTIONS TO BIOLOGICAL CHEMISTRY

The Planarity of —NH₂ Groups on Aromatic Rings

An important aspect of the molecular structure of bio-molecules is that amino groups bonded to an aromatic

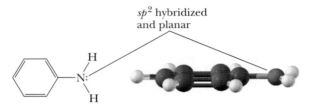

sp² hybridized and planar

Aniline   Aniline (viewed through an edge)

ring are *sp²* hybridized and planar. As described for aniline in the preceding section, this structure is the result of resonance delocalization of the amino group lone pair into the aromatic ring. Resonance delocalization requires the nitrogen to be *sp²* hybridized with the lone pair in a *2p* orbital to allow overlap with the aromatic pi system.

The planarity of amino groups on aromatic rings has a profound influence on the properties and folding of nucleic acids. Three of the four common aromatic heterocyclic amine bases of nucleic acid bases have planar amino groups (Figure 1). Not only does the *sp²* hybridization of the amino group allow for a relatively flat overall base structure (perfect for stacking), but the geometry of the planar amino group is ideal for making specific, highly directional hydrogen bonds with the complementary base.

Figure 1

The structures of the T—A and C—G base pairs showing the locations of planar —NH₂ groups bonded to the aromatic bases as well as the specific patterns of hydrogen bonds responsible for recognition between complementary strands of DNA.

The DNA duplex (Chapter 28) stores genetic information, based largely on patterns of specific hydrogen bonds. Adenine (A)-thymine (T) base pairs have two hydrogen bonds between them; guanine (G)-cytosine (C) base pairs have three hydrogen bonds (Figure 1). Moreover, the pattern of hydrogen bond donors and acceptors is different for the two base pairs. These two distinct patterns of hydrogen bonding make possible the recognition of DNA strands of the complementary sequence. In addition, the hydrogen bonds help hold two complementary DNA strands together in the famil-

iar double helix (Figure 2). If the amino groups on the DNA bases were not sp^2 hybridized and planar, the DNA base pairs would not be flat enough to stack in the interior of the assembled helix.

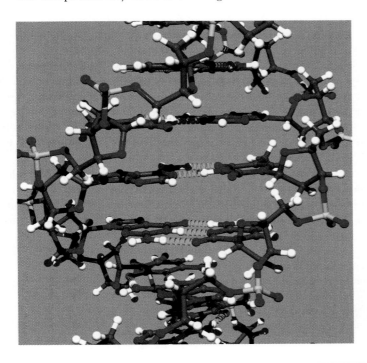

Figure 2
A section of a DNA double helix illustrating the planarity of the bases, including the planarity of the —NH$_2$ groups. Base-pair hydrogen bonds are shown as green broken lines.

C. Heterocyclic Aromatic Amines

Heterocyclic aromatic amines are weaker bases than heterocyclic aliphatic amines. Compare, for example, the pK_a values for the conjugate acids of piperidine, pyridine, and imidazole.

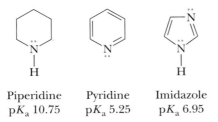

Piperidine   Pyridine   Imidazole
pK_a 10.75   pK_a 5.25   pK_a 6.95

We discussed the structure and bonding in pyridine and imidazole in Section 21.2D. In accounting for the relative basicities of these and other heterocyclic aromatic amines, it is important to determine first if the unshared pair of electrons on nitrogen is or is not a part of the $(4n + 2)$ pi electrons giving rise to aromaticity. In the case of pyridine, the unshared pair of electrons is not a part of the aromatic sextet. Rather, it lies in an sp^2 hybrid orbital in the plane of the ring and perpendicular to the six $2p$ orbitals containing the aromatic sextet.

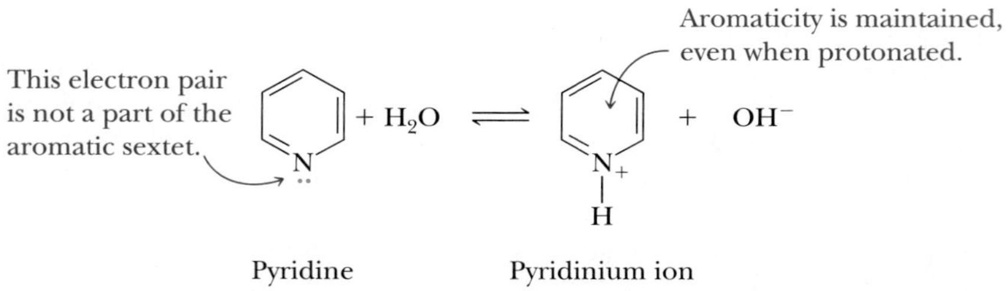

Aromaticity is maintained, even when protonated.

This electron pair is not a part of the aromatic sextet.

Pyridine + H_2O $\rightleftharpoons$ Pyridinium ion + OH^-

Proton transfer from water or other acid to pyridine does not involve the electrons of the aromatic sextet. Why, then, is pyridine a considerably weaker base than aliphatic amines? The answer is that the unshared pair of electrons on the pyridine nitrogen lies in a relatively electronegative sp^2 hybrid orbital, whereas in aliphatic amines, the unshared pair lies in an sp^3 hybrid orbital. This effect decreases markedly the basicity of the electron pair on an sp^2-hybridized nitrogen compared with that on an sp^3 hybridized nitrogen.

There are two nitrogen atoms in imidazole, each with an unshared pair of electrons. One unshared pair lies in a $2p$ orbital and is an integral part of the $(4n + 2)$ pi electrons of the aromatic system. The other unshared pair lies in an sp^2 hybrid orbital and is not a part of the aromatic sextet; this pair of electrons functions as the proton acceptor.

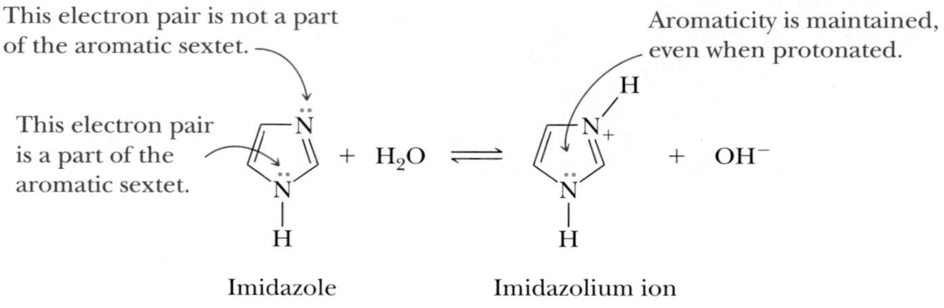

This electron pair is not a part of the aromatic sextet.

This electron pair is a part of the aromatic sextet.

Aromaticity is maintained, even when protonated.

Imidazole + H_2O $\rightleftharpoons$ Imidazolium ion + OH^-

As is the case with pyridine, the unshared pair of electrons functioning as the proton acceptor in imidazole lies in an sp^2 hybrid orbital and has markedly decreased basicity compared with an unshared pair of electrons in an sp^3 hybrid orbital. The positive charge on the imidazolium ion is delocalized on both nitrogen atoms of the ring; therefore, imidazole is a stronger base than pyridine.

Example 23.6

Select the stronger base in each pair of amines.

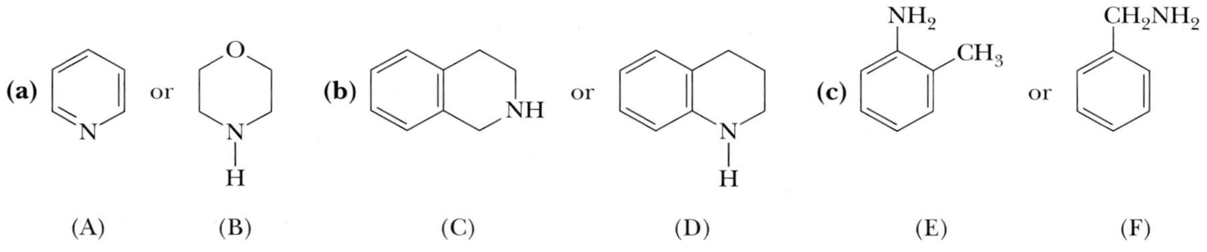

(a) (A) or (B)   (b) (C) or (D)   (c) (E) or (F)

Solution

(a) Morpholine (B) is the stronger base (conjugate acid pK_a 8.2). It has a basicity comparable to that of secondary aliphatic amines. Pyridine (A), a heterocyclic aromatic amine (pK_a 5.25), is considerably less basic than aliphatic amines.

(b) Tetrahydroisoquinoline (C) has a basicity comparable to that of secondary aliphatic amines (pK_a ~10.8) and is the stronger base. Tetrahydroquinoline (D) has a basicity comparable to that of *N*-substituted anilines (pK_a ~4.4) and is the weaker base.

(c) Benzylamine (F) is the stronger base (pK_a 9.6). Its basicity is comparable to that of other aliphatic amines. The basicity of *o*-toluidine (E), an aromatic amine, is comparable to that of aniline (pK_a 4.6).

Problem 23.6

Select the stronger acid from each pair of compounds.

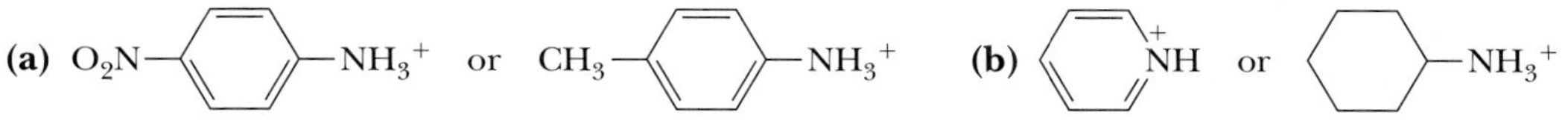

(A)   (B)   (C)   (D)

D. Guanidine

Guanidine, pK_a 13.6, is the strongest base among neutral organic compounds; it is almost as basic as hydroxide ion. Alternatively, its conjugate acid is a weaker acid than almost any other protonated amine.

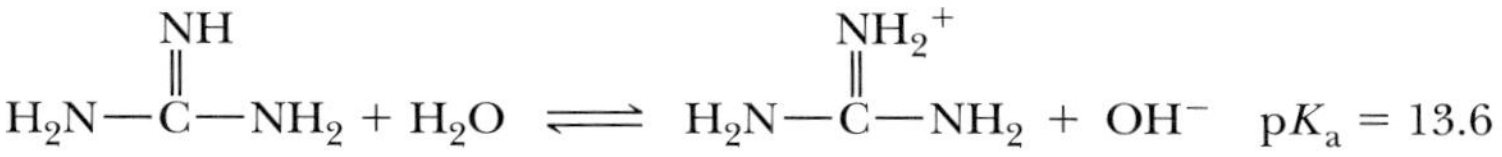

Guanidine   Guanidinium ion

The remarkable basicity of guanidine is attributed to the fact that the positive charge on the guanidinium ion is delocalized equally over the three nitrogen atoms as shown by these three equivalent contributing structures.

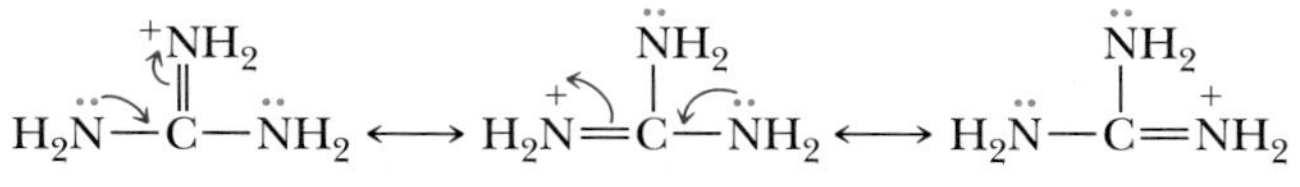

Three equivalent contributing structures

23.6 Reactions with Acids

Amines, whether soluble or insoluble in water, react quantitatively with strong acids to form water-soluble salts as illustrated by the reaction of norepinephrine (noradrenaline) with aqueous HCl to form a hydrochloride salt.

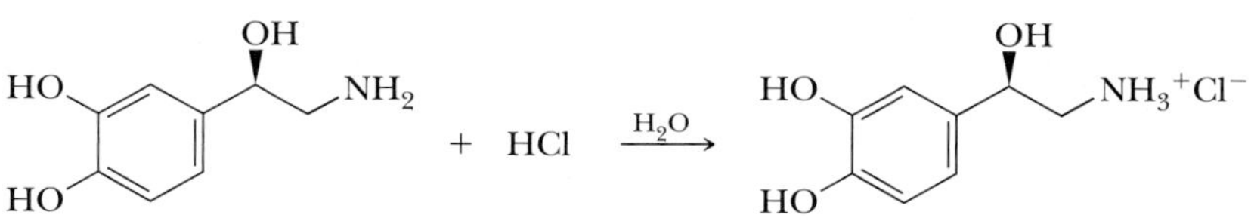

(*R*)-Norepinephrine   (*R*)-Norepinephrine hydrochloride
(only slightly soluble in water)   (a water-soluble salt)

Norepinephrine, secreted by the medulla of the adrenal gland, is a neurotransmitter. The suggestion has been made that it acts in those areas of the brain that mediate emotional behavior.

Example 23.7

Complete each acid-base reaction and name the salt formed.
(a) $Et_2NH + HCl \longrightarrow$ (b) $PhCH_2NH_2 + CH_3COOH \longrightarrow$

Solution

(a) $Et_2NH_2{}^+Cl^-$
Diethylammonium
chloride

(b) $PhCH_2NH_3{}^+ CH_3COO^-$
Benzylammonium
acetate

Problem 23.7

Complete each acid-base reaction and name the salt formed.

(a) $Et_3N + HCl \longrightarrow$ (b)  NH + $CH_3COOH \longrightarrow$

Example 23.8

Following are two structural formulas for (S)-serine, one of the building blocks of proteins (Chapter 27).

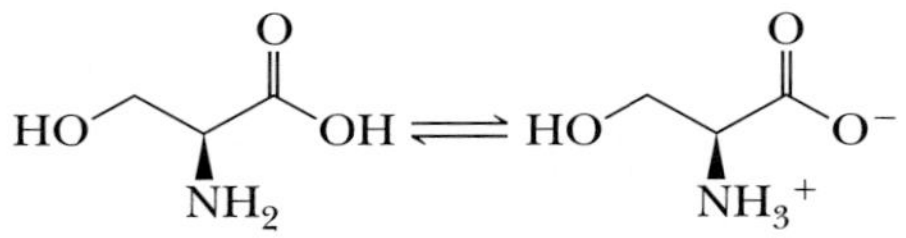

(A)   (B)

Is (S)-serine better represented by structural formula A or B?

Solution

Structural formula A contains both an amino group (a base) and a carboxyl group (an acid). Proton transfer from the stronger acid (—COOH) to the stronger base (—NH₂) gives an internal salt; therefore, B is the better representation for (S)-serine. Within the field of amino acid chemistry, the internal salt represented by B is called a **zwitterion** (Section 27.2).

Problem 23.8

Following are structural formulas for propanoic acid and the conjugate acids of iso-propylamine and alanine, along with pK_a values for each functional group.

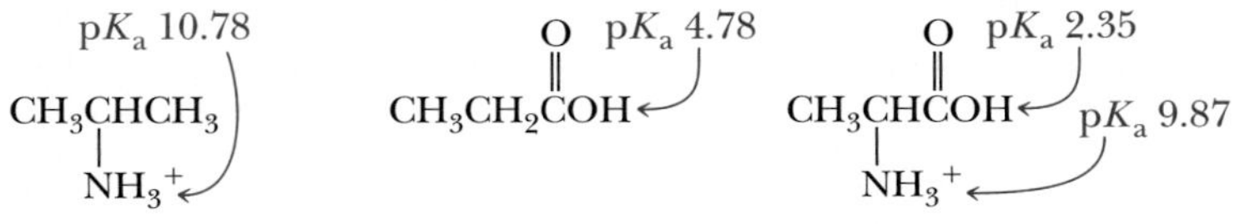

Conjugate acid of   Propanoic acid   Conjugate acid of
isopropylamine   alanine

(a) How do you account for the fact that the —NH$_3^+$ group of the conjugate acid of alanine is a stronger acid than the —NH$_3^+$ group of the conjugate acid of isopropylamine?

(b) How do you account for the fact that the —COOH group of the conjugate acid of alanine is a stronger acid than the —COOH group of propanoic acid?

The basicity of amines and the solubility in water of amine salts can be used to separate water-insoluble amines from water-insoluble, nonbasic compounds. Shown in Figure 23.2 is a flow chart for the separation of aniline from acetanilide, a neutral compound.

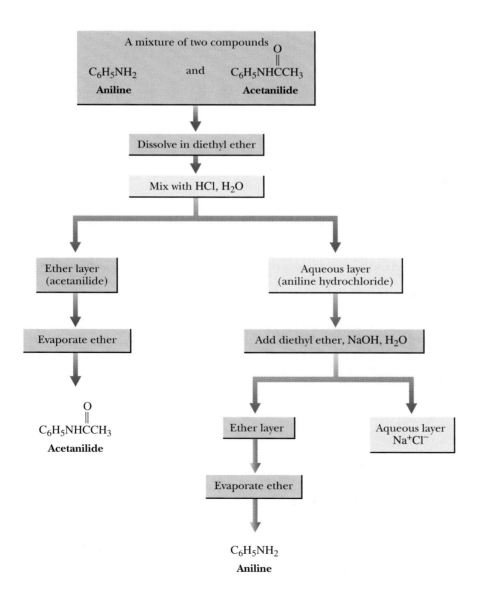

Figure 23.2
Separation and purification of an amine and a neutral compound.

Example 23.9

Here is a flow chart for the separation of a mixture of a primary aliphatic amine (RNH$_2$, pK_a 10.8), a carboxylic acid (RCOOH, pK_a 5), and a phenol (ArOH, pK_a 10). Assume that each is insoluble in water but soluble in diethyl ether. The mixture is separated into fractions A, B, and C. Which fraction contains the amine, which the carboxylic acid, and which the phenol?

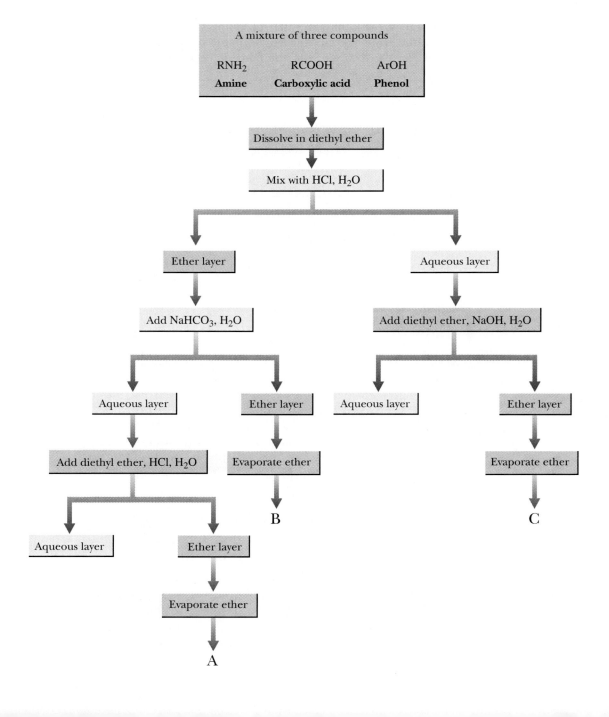

Solution

Fraction C contains RNH_2, fraction B contains ArOH, and fraction A contains RCOOH.

Problem 23.9

In what way(s) might the results of the separation and purification procedure outlined in Example 23.9 be different if the following conditions exist?
(a) Aqueous NaOH is used in place of aqueous $NaHCO_3$.
(b) The starting mixture contains an aromatic amine, $ArNH_2$, rather than an aliphatic amine, RNH_2.

23.7 Preparation

The synthesis of amines is primarily a problem of how to form a carbon-nitrogen bond and, if the newly formed nitrogen-containing compound is not already an amine, how to convert it into an amine. We have already seen the following methods for the preparation of amines.

1. Nucleophilic ring opening of epoxides by ammonia and amines (Section 11.9B)
2. Addition of nitrogen nucleophiles to the carbonyl group of aldehydes and ketones to form imines (Section 16.8)
3. Reduction of imines to amines (Section 16.8)
4. Reduction of amides by $LiAlH_4$ (Section 18.10B)
5. Reduction of nitriles to primary amines (Section 18.10C)
6. Nitration of arenes followed by reduction of the nitro group to a primary amine (Section 22.1B)

In this chapter, we present two additional methods for the preparation of amines.

A. Alkylation of Ammonia and Amines

Surely one of the most direct synthetic routes to an amine would seem to be treating an alkyl halide with ammonia or an amine. Reaction between these two compounds by an S_N2 mechanism gives an alkylammonium salt, as illustrated by treatment of bromomethane, MeBr, with ammonia to give methylammonium bromide.

$$MeBr + NH_3 \xrightarrow{S_N2} MeNH_3{}^+ Br^-$$

Methylammonium
bromide

Unfortunately, reaction does not stop at this stage but continues to give a complex mixture of products as shown in the following equation.

$$MeBr + NH_3 \longrightarrow MeNH_3{}^+Br^- + Me_2NH_2{}^+Br^- + Me_3NH^+Br^- + Me_4N^+Br^-$$

This mixture is formed in the following way. Proton transfer between ammonia and methylammonium ion gives ammonium ion and methylamine, also a good nucleophile, which then undergoes reaction with bromomethane to give dimethylammonium bromide. A second proton transfer reaction converts the dimethylammonium ion to dimethylamine, yet another good nucleophile, which also participates in nucleophilic substitution, and so on.

$$\text{MeNH}_3{}^+\text{Br}^- + \text{NH}_3 \underset{\xrightarrow{\text{proton transfer}}}{\rightleftharpoons} \text{MeNH}_2 + \text{NH}_4{}^+\text{Br}^-$$

Methylammonium   Methylamine
bromide

$$\text{MeBr} + \text{MeNH}_2 \xrightarrow{\text{S}_\text{N}2} \text{Me}_2\text{NH}_2{}^+\text{Br}^-$$

Dimethylammonium bromide

$$\text{Me}_2\text{NH}_2{}^+\text{Br}^- + \text{NH}_3 \underset{\xrightarrow{\text{proton transfer}}}{\rightleftharpoons} \text{Me}_2\text{NH} + \text{NH}_4{}^+ \text{Br}^-$$

Dimethylamine

The final product from such a series of nucleophilic substitution and proton transfer reactions is a tetraalkylammonium halide. The relative proportions of the various alkylation products depend on the ratio of alkyl halide to ammonia in the reaction mixture. Whatever the starting mixture, however, the product is almost invariably a mixture of alkylated products. For this reason, alkylation of ammonia or amines is not a generally useful laboratory method for the preparation of more complex amines. However, primary amines are easily prepared because ammonia is inexpensive and can be used in large excess. Other amines can also be prepared in this way if the nucleophilic amine is inexpensive enough to be used in large excess.

B. Alkylation of Azide Ion

As we have just seen, alkylation of ammonia or amines is generally not a useful method for the preparation of amines. One strategy for eliminating the problem of overalkylation is to use a form of nitrogen that can function as a nucleophile, but that is no longer an effective nucleophile after it has formed a new carbon-nitrogen bond. One such nucleophilic form of nitrogen is the azide ion, $\text{N}_3{}^-$. Alkyl azides are easily prepared from sodium or potassium azide and a primary or secondary alkyl halide by an $\text{S}_\text{N}2$ reaction. Azides are, in turn, reduced to primary amines by a variety of reducing agents, including lithium aluminum hydride.

$$\text{N}_3{}^- \qquad {}^-:\!\ddot{\text{N}}\!=\!\overset{+}{\text{N}}\!=\!\ddot{\text{N}}\!:^- \qquad\qquad \text{RN}_3 \qquad \text{R}-\ddot{\text{N}}\!=\!\overset{+}{\text{N}}\!=\!\ddot{\text{N}}\!:^-$$

Azide ion (a good nucleophile)   An alkyl azide

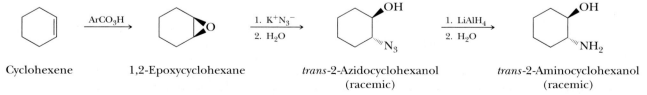

Benzyl chloride   Benzyl azide   Benzylamine

The azide ion can also be used for stereoselective ring opening of epoxides. Reduction of the resulting β-azidoalcohol gives a β-aminoalcohol as illustrated by the conversion of cyclohexene to *trans*-2-aminocyclohexanol.

Cyclohexene   1,2-Epoxycyclohexane   *trans*-2-Azidocyclohexanol   *trans*-2-Aminocyclohexanol
                                                  (racemic)   (racemic)

Oxidation of cyclohexene by a peroxyacid (Section 11.8B) gives an epoxide. Stereoselective nucleophilic attack by azide ion anti to the leaving oxygen of the epoxide ring

(Section 11.9B) followed by reduction of the azide with lithium aluminum hydride gives racemic *trans*-2-aminocyclohexanol.

Example 23.10

Show how to convert 4-methoxybenzyl chloride to each amine.

(a)

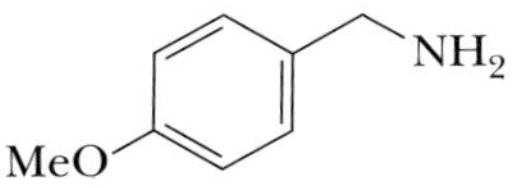

(b)

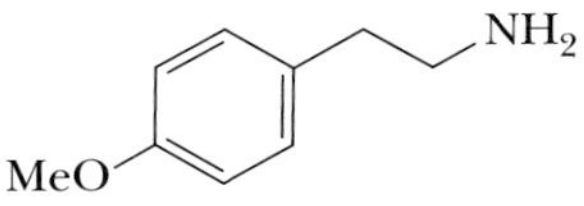

Solution

(a) Two methods might be used: (1) alkylation of NH_3 using a large molar excess of NH_3 to reduce the extent of overalkylation or (2) nucleophilic displacement of chloride using azide ion (from NaN_3) followed by $LiAlH_4$ reduction of the azide. Of these methods, nucleophilic displacement by azide is the more convenient on a laboratory scale.

(b) Nucleophilic displacement of chloride by cyanide ion is followed by reduction of the cyano group with lithium aluminum hydride.

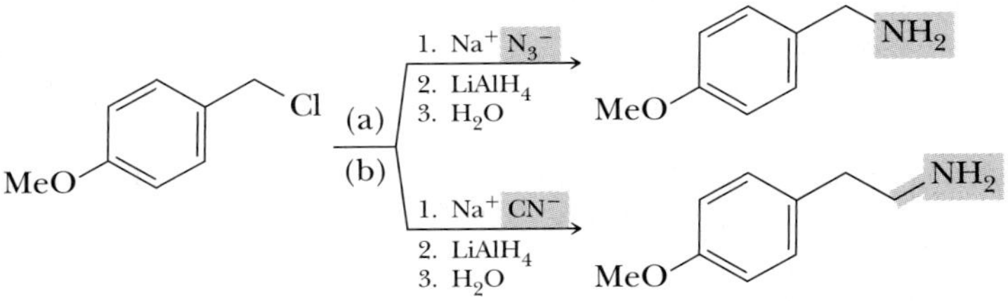

Problem 23.10

Show how to bring about each conversion in good yield. In addition to the given starting material, use any other reagents as necessary.

(a)

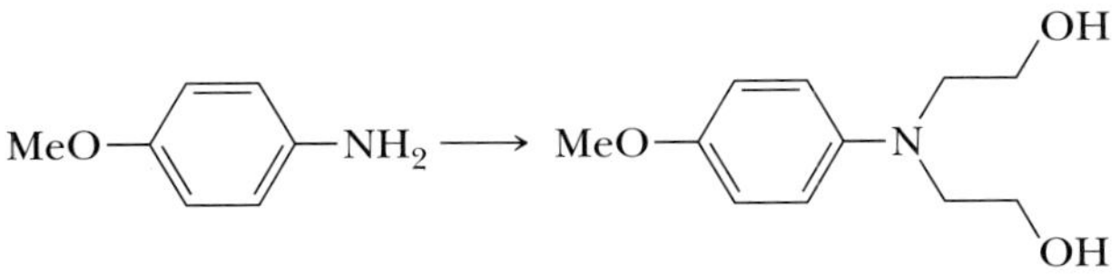

(b)

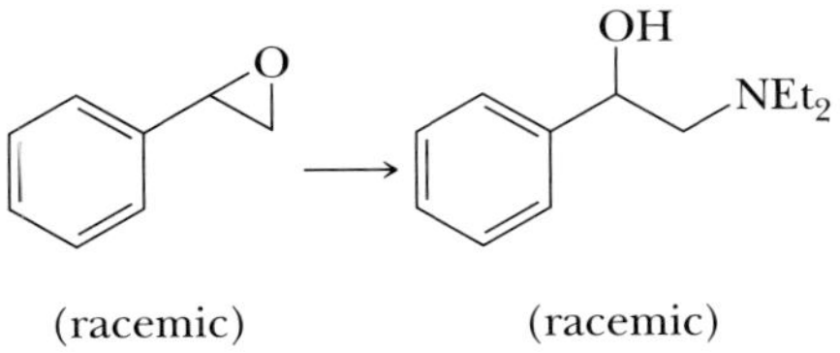

(racemic)   (racemic)

23.8 Reaction with Nitrous Acid

Nitrous acid, HNO_2, is an unstable compound that is prepared by adding sulfuric or hydrochloric acid to an aqueous solution of sodium nitrite, $NaNO_2$. Nitrous acid is a weak oxygen acid and ionizes according to the following equation.

$$HNO_2 + H_2O \rightleftharpoons H_3O^+ + NO_2^- \qquad pK_a = 3.37$$

Nitrous acid undergoes reaction with amines in different ways, depending on whether the amine is primary, secondary, or tertiary and whether it is aliphatic or aromatic. These reactions are all related by the facts that nitrous acid (1) participates in proton-transfer reactions and (2) is a source of the nitrosyl cation, a weak electrophile.

Mechanism *Formation of the Nitrosyl Cation*

Step 1: Protonation of the OH group of nitrous acid gives an oxonium ion.

Step 2: Loss of water gives the nitrosyl cation, here represented as a hybrid of two contributing structures.

$$H{-}\ddot{O}{-}\ddot{N}{=}\ddot{O}{:} + H^+ \overset{(1)}{\rightleftharpoons} H{-}\overset{+}{\underset{H}{\ddot{O}}}{-}\ddot{N}{=}\ddot{O}{:} \overset{(2)}{\rightleftharpoons} H{-}\underset{H}{\ddot{O}}{:} \ + \ ^+{:}N{=}\ddot{O}{:} \longleftrightarrow {:}N{\equiv}\ddot{O}{:}^+$$

The nitrosyl cation

A. Tertiary Aliphatic Amines

When treated with nitrous acid, tertiary aliphatic amines, whether water-soluble or water-insoluble, are protonated to form water-soluble salts. No further reaction occurs beyond salt formation. This reaction is of no practical use.

B. Tertiary Aromatic Amines

Tertiary aromatic amines are bases and can also form salts with nitrous acid. An alternative pathway, however, is open to tertiary aromatic amines, namely, electrophilic aromatic substitution. The nitrosyl cation, a very weak electrophile, reacts only with aromatic rings containing strongly activating, ortho-para directing groups, such as the hydroxyl and dialkylamino groups. When treated with nitrous acid, these compounds undergo nitrosation, predominantly in the para position to give blue or green aromatic nitroso compounds.

$$Me_2N{-}\langle\bigcirc\rangle \xrightarrow[\text{2. NaOH, H}_2\text{O}]{\text{1. NaNO}_2\text{, HCl, 0–5°C}} Me_2N{-}\langle\bigcirc\rangle{-}N{=}O$$

N,N-Dimethylaniline   *N,N*-Dimethyl-4-nitrosoaniline

C. Secondary Aliphatic and Aromatic Amines

Secondary amines, whether aliphatic or aromatic, undergo reaction with nitrous acid to give *N*-nitrosoamines as illustrated by the reaction of piperidine with nitrous acid.

$$\langle\bigcirc\rangle N{-}H + HNO_2 \longrightarrow \langle\bigcirc\rangle N{-}N{=}O \ + \ H_2O$$

Piperidine   *N*-Nitrosopiperidine

Mechanism *Reaction of a 2° Amine with the Nitrosyl Cation to Give an N-Nitrosamine*

Step 1: Reaction of the 2° amine (a nucleophile) with the nitrosyl cation (an electrophile) gives an *N*-nitrosammonium ion.

Step 2: Proton transfer to solvent gives the *N*-nitrosamine.

ORGANIC
Chemistry Now™
Click Mechanisms to view an animation of the **Nitrosyl Cation Reaction with an Amine**

N-Nitrosamines are of little synthetic or commercial value. They have received considerable attention in recent years, however, because many of them are potent carcinogens. Following are structural formulas of two *N*-nitrosamines, each of which is a known carcinogen.

N-Nitrosodimethylamine
(found in cigarette smoke
and when bacon "preserved"
with sodium nitrite is fried)

N-Nitrosopyrrolidine
(formed when bacon "preserved"
with sodium nitrite is fried)

Common practice within the food industry has been to add sodium nitrite to processed meats to "retard spoilage," that is, to inhibit the growth of *Clostridium botulinum,* the bacterium responsible for botulism poisoning. Although this practice was well grounded before the days of adequate refrigeration, it is of questionable value today. Sodium nitrite is also added to prevent red meats from turning brown. If you buy some nice red hamburger in a food market and find it is gray or brown inside, you can be sure that the outside has been treated with sodium nitrite. Controversy over the use of sodium nitrite has been generated by the demonstration that nitrite ion in the presence of acid converts secondary amines to *N*-nitrosamines and that many *N*-nitrosamines are powerful carcinogens. This demonstration led in turn to pressure by consumer groups to force the Food and Drug Administration (FDA) to ban the use of nitrite additives in foods. The strength of the argument to ban nitrites was weakened with the finding that enzymes in our mouths and intestinal tracts have the ability to catalyze the reduction of nitrate to nitrite. Nitrate ion is normally found in a wide variety of foods and in drinking water. To date, there is no evidence that nitrite as a food additive poses any risk not already present through our existing dietary habits. The FDA has established the current permissible level of sodium nitrite in processed meats as 50 to 125 ppm (that is, 50–125 μg nitrite per gram of cured meat).

D. Primary Aliphatic Amines

Treatment of a primary aliphatic amine with nitrous acid results in the loss of nitrogen, N_2, and the formation of substitution, elimination, and rearrangement products as illustrated by the treatment of butylamine with nitrous acid.

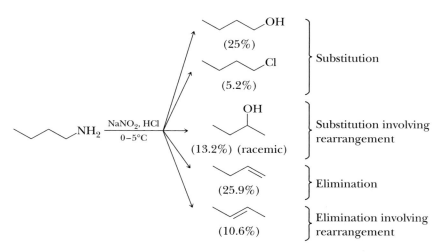

Diazonium ion An ArN_2^+ or RN_2^+ ion.

The mechanism by which this mixture of products is formed involves formation of a **diazonium ion.** The conversion of a primary amine to a diazonium ion is called **diazotization.**

Mechanism   *Reaction of a 1° Amine with Nitrous Acid*

ORGANIC
Chemistry Now™

Click Mechanisms to view an animation of the **Reaction of a Primary Amine with Nitrous Acid**

Step 1: Reaction of a 1° amine with the nitrosonium ion from nitrous acid gives an *N*-nitrosamine, which undergoes keto-enol tautomerism (Section 16.9B) to give a diazotic acid, so named because it has two (*di-*) nitrogen (*-azot-*) atoms within its structure.

$$R—\overset{\cdot\cdot}{N}H_2 \; + \; \overset{+}{:}N{=}\overset{\cdot\cdot}{O}{:} \longrightarrow R—\overset{\overset{\displaystyle H}{|}}{N}—\overset{\cdot\cdot}{N}{=}\overset{\cdot\cdot}{O}{:} \overset{\text{keto-enol}}{\underset{\text{tautomerism}}{\rightleftharpoons}} R—\overset{\cdot\cdot}{N}{=}\overset{\cdot\cdot}{N}—\overset{\cdot\cdot}{O}—H$$

A 1° aliphatic amine   An *N*-nitrosamine   A diazotic acid

Step 2: Protonation of the diazotic acid followed by loss of H_2O gives a diazonium ion.

$$R—\overset{\cdot\cdot}{N}{=}\overset{\cdot\cdot}{N}—\overset{\cdot\cdot}{O}—H \underset{}{\overset{H^+}{\rightleftharpoons}} R—\overset{\cdot\cdot}{N}{=}\overset{\cdot\cdot}{N}—\overset{+}{\overset{|}{O}}—H \overset{-H_2O}{\longrightarrow} R—\overset{+}{N}{\equiv}N: \longrightarrow R^+ \; + \; :N{\equiv}N:$$

A diazotic acid   A diazonium ion   A carbo-
                                                                         cation

---

Aliphatic diazonium ions are unstable, even at 0°C, and immediately lose nitrogen to give carbocations and nitrogen gas. The driving force for this reaction is the fact that N_2 is one of the best leaving groups because it is an extraordinarily weak base. It is removed from the reaction mixture as a gas as it is formed. The carbocation now has open to it the three reactions in the repertoire of aliphatic carbocations: (1) loss of a proton to give an alkene, (2) reaction with a nucleophile to give a substitution product, and (3) rearrangement to a more stable carbocation and then reaction further by (1) or (2).

Because treatment of a primary aliphatic amine with HNO_2 gives a mixture of products, it is generally not a useful reaction. An exception is the Tiffeneau-

Demjanov reaction in which a cyclic β-aminoalcohol is treated with nitrous acid to give a ring-expanded ketone, with evolution of nitrogen.

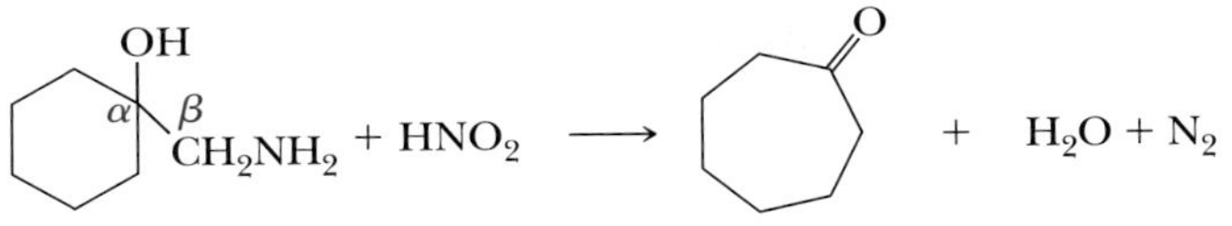

A β-aminoalcohol   Cycloheptanone

We account for this molecular rearrangement as shown in the following mechanism.

Mechanism *The Tiffeneau-Demjanov Reaction*

Step 1: Reaction of the 1° amine with nitrous acid gives a diazonium ion.

Step 2: Simultaneous loss of N_2 and rearrangement by a 1,2-shift gives the conjugate acid of the final product as a resonance-stabilized cation.

Step 3: Proton transfer from this cation to solvent completes the reaction.

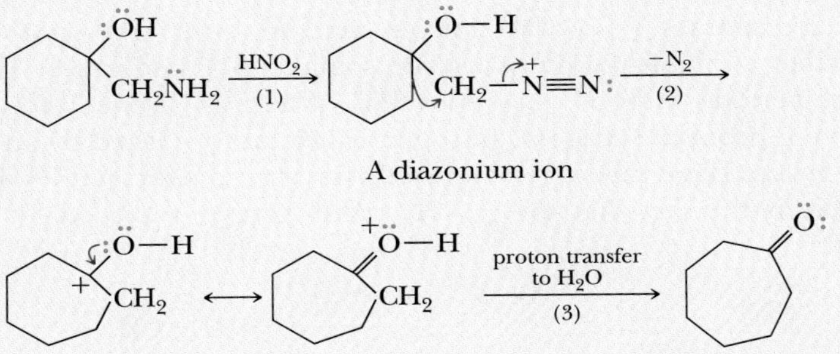

A diazonium ion

A resonance-stabilized cation   Cycloheptanone

The driving force for this molecular rearrangement is precisely what we already saw for other cation rearrangements: transformation of a less stable cation into a more stable cation. This reaction is analogous to the pinacol rearrangement (Section 10.7) with the leaving group being N_2 rather than H_2O.

Example 23.11

The following sequence of reactions gives cyclooctanone. Propose a structural formula for compound A and a mechanism for its conversion to cyclooctanone.

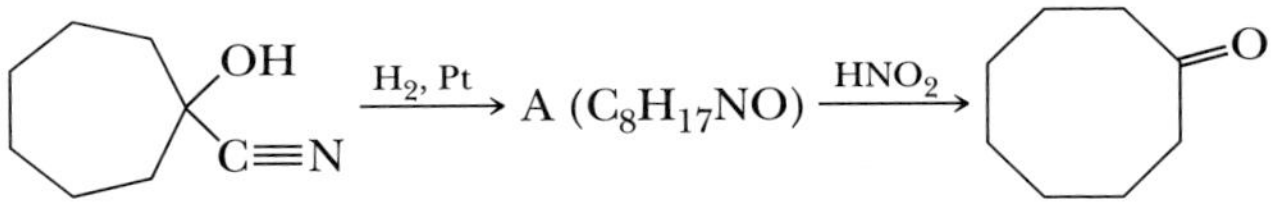

Solution

Catalytic hydrogenation using hydrogen over a platinum catalyst reduces the carbon-nitrogen triple bond to a single bond (Section 18.10C) and gives a β-aminoalcohol.

Treatment of the β-aminoalcohol with nitrous acid results in loss of N_2 and expansion of the seven-membered ring to an eight-membered cyclic ketone.

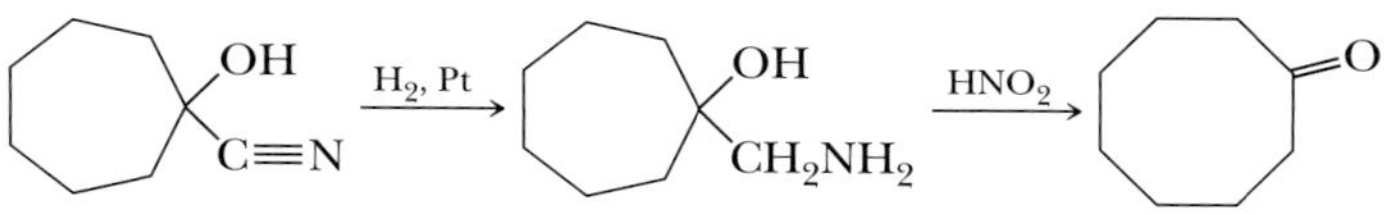

Problem 23.11

How might you bring about this conversion?

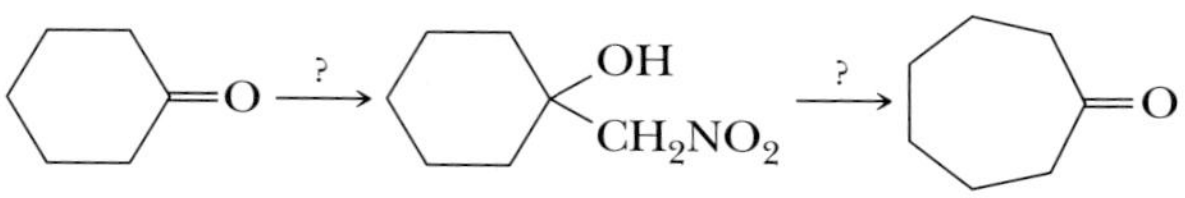

E. Primary Aromatic Amines

Primary aromatic amines react with nitrous acid to form arenediazonium salts, which unlike their aliphatic counterparts, are stable at 0°C and can be kept in solution for short periods without decomposition. When an arenediazonium salt is treated with an appropriate reagent, nitrogen is lost and replaced by another atom or functional group. What makes reactions of primary aromatic amines with nitrous acid so valuable is the fact that the —NH_2 group can be replaced by the groups shown.

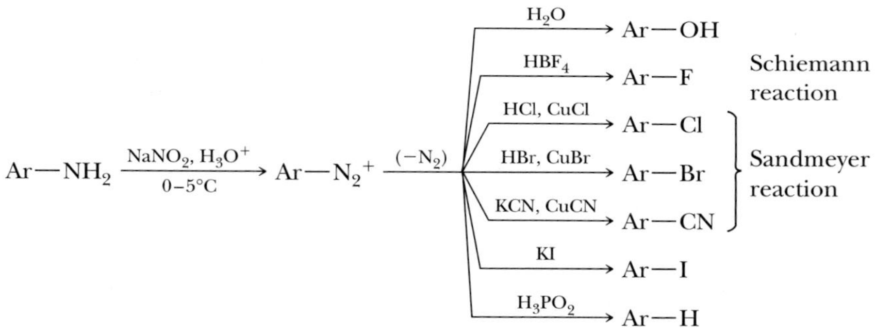

Aromatic amines can be converted to phenols by first forming the arenediazonium salt in aqueous sulfuric acid and then heating the solution. In this manner, 2-bromo-4-methylaniline is converted to 2-bromo-4-methylphenol.

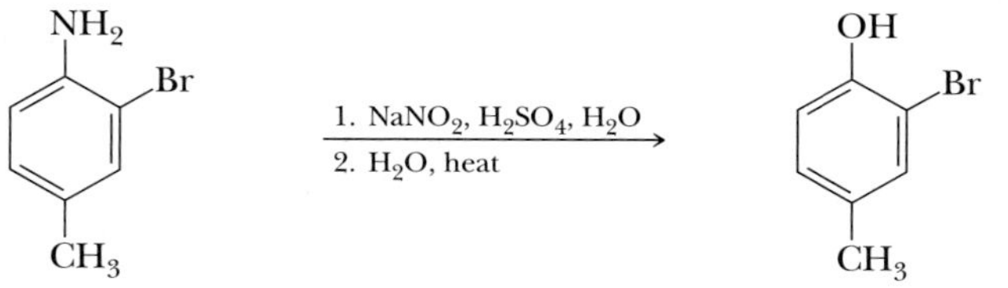

2-Bromo-4-methylaniline   2-Bromo-4-methylphenol

The intermediate in the decomposition of an arenediazonium ion in water is an aryl cation, which then undergoes reaction with water to form the phenol.

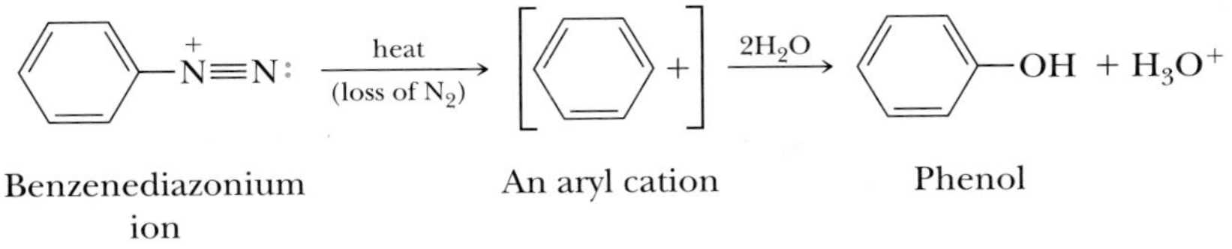

Benzenediazonium   An aryl cation   Phenol
ion

Note that the aryl cation is so unstable that it can be formed only with N_2 as the leaving group. This reaction of arenediazonium salts represents the main laboratory preparation of phenols.

Example 23.12

What reagents and experimental conditions will bring about each step in the conversion of toluene to 4-hydroxybenzoic acid?

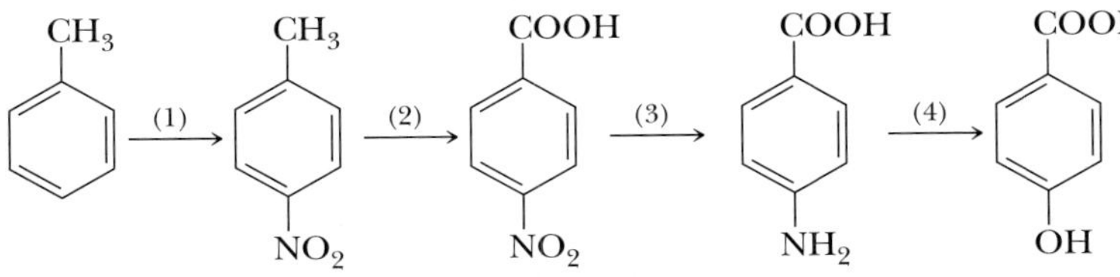

Solution

Step 1: Nitration of the aromatic ring (Section 22.1B) using HNO_3 in H_2SO_4 followed by separation of the ortho and para isomers gives 4-nitrotoluene.

Step 2: Oxidation at the benzylic carbon (Section 21.5A) using $K_2Cr_2O_7$ in H_2SO_4 gives 4-nitrobenzoic acid.

Step 3: Catalytic reduction of the nitro group (Section 22.1B) to an amino group using H_2 in the presence of Ni or other transition metal catalyst gives 4-aminobenzoic acid. Alternatively, reduction of the nitro group to a primary amine can be brought about using Zn, Sn, or Fe in aqueous HCl followed by aqueous NaOH.

Step 4: Reaction of the aromatic amine with $NaNO_2$ in aqueous H_2SO_4 followed by heating gives 4-hydroxybenzoic acid.

Problem 23.12

Show how to convert toluene to 3-hydroxybenzoic acid using the same set of reactions as in Example 23.12, but changing the order in which two or more of the steps are carried out.

The **Schiemann reaction** is the most common method for the introduction of fluorine onto an aromatic ring. It is carried out by treatment of a primary aromatic amine with sodium nitrite in aqueous HCl followed by addition of HBF_4 or $NaBF_4$. The diazonium fluoroborate salt precipitates and is collected and dried. Heating the dry salt brings about its decomposition to an aryl fluoride, nitrogen, and boron trifluoride. The Schiemann reaction is also thought to involve an aryl cation intermediate.

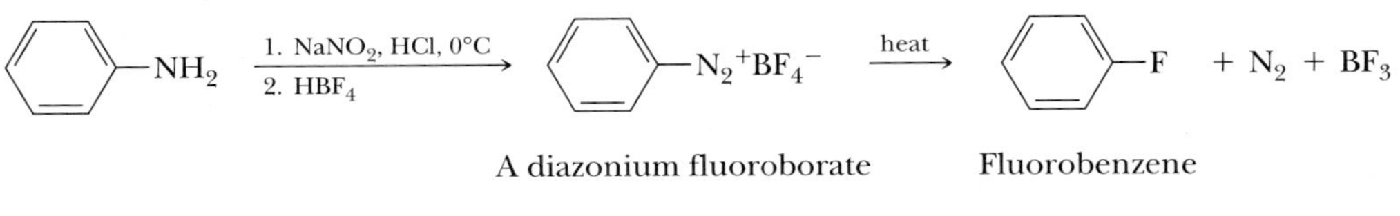

A diazonium fluoroborate   Fluorobenzene

Treatment of a primary aromatic amine with nitrous acid followed by heating with HCl/CuCl, HBr/CuBr, or KCN/CuCN results in replacement of the diazonium group by —Cl, —Br, or —CN, respectively, and is known as the **Sandmeyer reaction.** The Sandmeyer reaction fails when attempted with CuI or CuF.

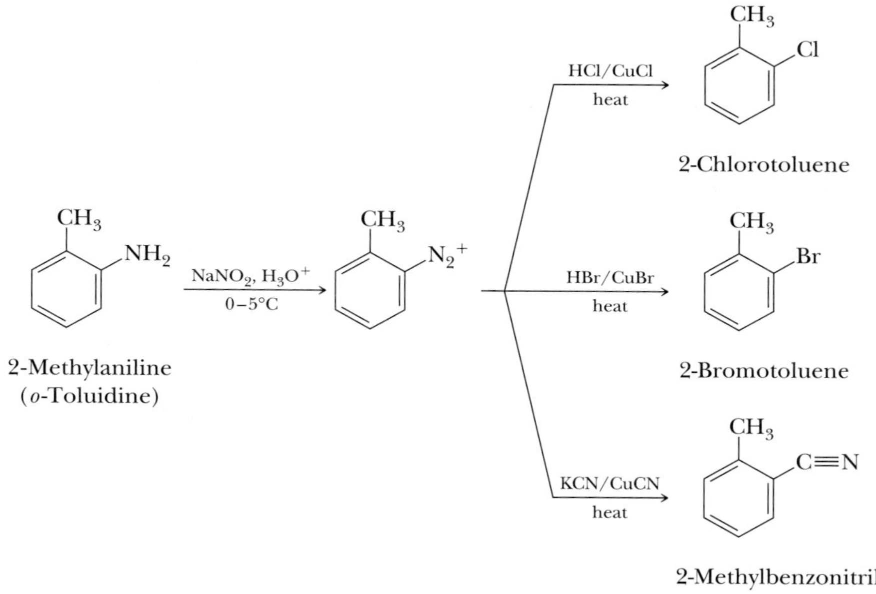

2-Methylaniline
(*o*-Toluidine)

2-Chlorotoluene

2-Bromotoluene

2-Methylbenzonitrile

Treating an arenediazonium ion with iodide ion, generally from potassium iodide, is the best and most convenient method for introducing iodine onto an aromatic ring.

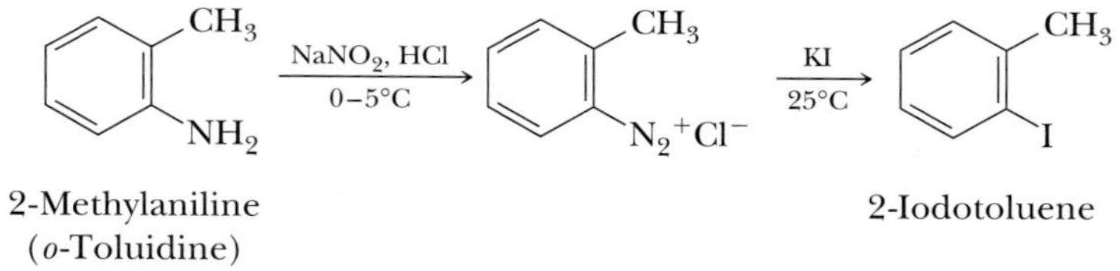

2-Methylaniline
(*o*-Toluidine)

2-Iodotoluene

Treating an arenediazonium ion with hypophosphorous acid, H_3PO_2, results in reduction of the diazonium group and its replacement by —H as illustrated by the conversion of aniline to 1,3,5-trichlorobenzene. Recall that —NH$_2$ is a powerful activating group (Section 22.2A). Treating aniline with chlorine requires no catalyst and gives 2,4,6-trichloroaniline. To complete the conversion, the —NH$_2$ group is removed by treatment with nitrous acid followed by hypophosphorous acid to give 1,3,5-trichlorobenzene.

Aniline   2,4,6-Trichloroaniline   2,4,6-Trichlorobenzene-
diazonium chloride   1,3,5-Trichlorobenzene

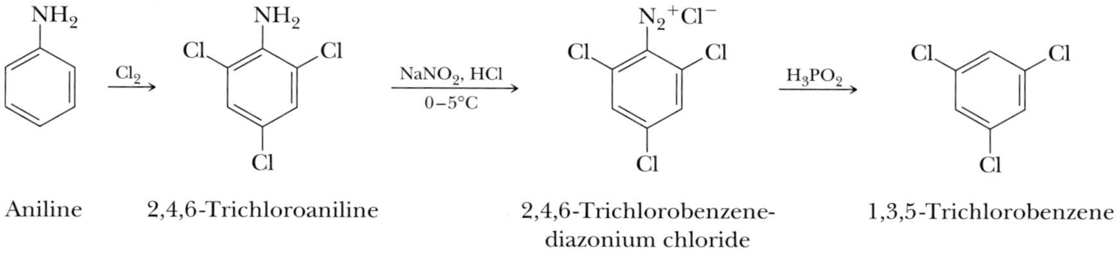

Example 23.13

Show reagents and conditions to convert toluene to 3-bromo-4-methylphenol.

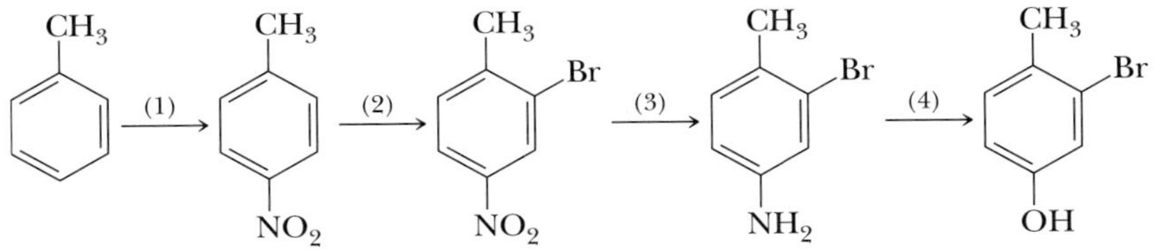

Solution

Step 1: HNO_3 in H_2SO_4. Methyl is ortho-para directing and slightly activating.

Step 2: Treat 4-nitrotoluene with bromine in the presence of $FeCl_3$.

Step 3: Reduce the nitro group either using H_2/Ni or using Sn, Zn, or Fe in aqueous HCl followed by aqueous NaOH.

Step 4: Diazotize the amine with $NaNO_2$ in aqueous sulfuric acid followed by warming of the solution replaces $-N_2^+$ by $-OH$.

Problem 23.13

Starting with 3-nitroaniline, show how to prepare the following compounds.
(a) 3-Nitrophenol   (b) 3-Bromoaniline   (c) 1,3-Dihydroxybenzene (resorcinol)
(d) 3-Fluoroaniline   (e) 3-Fluorophenol   (f) 3-Hydroxybenzonitrile

23.9 Hofmann Elimination

When a quaternary ammonium halide is treated with moist silver oxide (a slurry of Ag_2O in H_2O), silver halide precipitates, leaving a solution of a quaternary ammonium hydroxide.

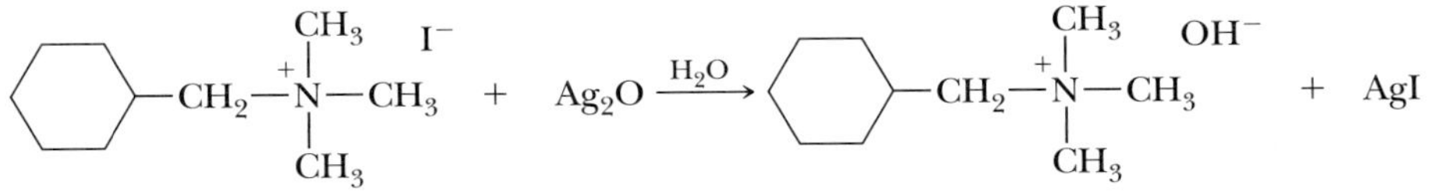

(Cyclohexylmethyl)trimethyl-   Silver   (Cyclohexylmethyl)trimethyl-
ammonium iodide   oxide   ammonium hydroxide

In the mid-19th century, Augustus Hofmann discovered that when a quaternary ammonium hydroxide is heated, it decomposes to an alkene, a tertiary amine, and water. Thermal decomposition of a quaternary ammonium hydroxide to an alkene is known as **Hofmann elimination.**

(Cyclohexylmethyl)trimethyl-   Methylenecyclohexane   Trimethylamine
ammonium hydroxide

Hofmann elimination When treated with a strong base, a quaternary ammonium halide undergoes β-elimination by an E2 mechanism to give the less substituted alkene as the major product.

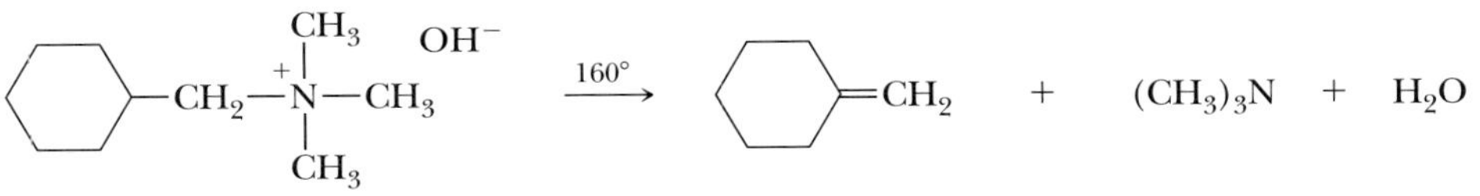

The Hofmann elimination has most of the characteristics of an E2 reaction (Section 9.8). First, Hofmann eliminations are concerted, meaning that bond-breaking and bond-forming steps occur simultaneously, or nearly so. Second, Hofmann eliminations are stereoselective for anti elimination, meaning that —H and the leaving group must be anti to each other. The following mechanism illustrates the concerted nature of bond forming and bond breaking, and the anti arrangement of —H and the trialkylamino group.

Mechanism   *The Hofmann Elimination*

Removal of a β-hydrogen by base, collapse of the electron pair of the C—H bond to become the pi bond of the alkene, and loss of the trialkylamino group occur simultaneously. The reaction shows anti stereoselectivity.

When we studied E2 reactions of alkyl halides in Section 9.8, we saw that a β-hydrogen must be anti to the leaving group. If only one β-hydrogen meets this requirement, then the double bond is formed in that direction. If, however, two β-hydrogens meet this requirement, then elimination follows Zaitsev's rule: Elimination occurs preferentially to form the more substituted double bond.

$$CH_3CH_2CHCH_3 \xrightarrow[\text{E2}]{CH_3CH_2O^-Na^+} CH_3CH{=}CHCH_3 + CH_3CH_2CH{=}CH_2$$
(with Br on the second carbon)

(75%)   (25%)

Thermal decomposition of quaternary ammonium hydroxides is different because elimination occurs preferentially to form the least substituted double bond. Thermal decomposition of *sec*-butyltrimethylammonium hydroxide, for example, gives 1-butene as the major product.

$$CH_3CH_2CHCH_3 \xrightarrow[\text{heat}]{\text{E2}} CH_3CH{=}CHCH_3 + CH_3CH_2CH{=}CH_2 + (CH_3)_3N + H_2O$$
(with $\overset{+}{N}(CH_3)_3$ group and HO^-)

(5%)   (95%)

Elimination reactions that give the less substituted alkene as the major product are said to follow the **Hofmann rule.**

Hofmann rule Any β-elimination that occurs preferentially to give the less substituted alkene as the major product.

Example 23.14

Draw a structural formula of the major alkene formed in each β-elimination.

(a) (structure with NH₂) 1. CH₃I (excess), K₂CO₃ 2. Ag₂O, H₂O 3. heat →

(racemic)

(b) (structure with I) CH₃O⁻Na⁺ →

(racemic)

Solution

Thermal decomposition of the quaternary ammonium hydroxide in (a) follows Hofmann elimination and gives 1-octene as the major product. E2 elimination from an alkyl iodide in (b) by sodium methoxide follows Zaitsev's rule and gives *trans*-2-octene as the major product.

(a) $\diagup\!\diagdown\!\diagup\!\diagdown\!\diagup\!\diagdown$ + $(CH_3)_3N + H_2O$ (b) $\diagup\!\diagdown\!\diagup\!\diagdown\!\diagup\!\diagdown$ + CH_3OH

       1-Octene   *trans*-2-Octene

Problem 23.14

The procedure of methylation of amines and thermal decomposition of quaternary ammonium hydroxides was first reported by Hofmann in 1851, but its value as a means of structure determination was not appreciated until 1881 when he published a report of its use to determine the structure of piperidine. Following are the results obtained by Hofmann.

$$C_5H_{11}N \xrightarrow[\substack{1.\ CH_3I\ (excess),\ K_2CO_3 \\ 2.\ Ag_2O,\ H_2O \\ 3.\ heat}]{} C_7H_{15}N \xrightarrow[\substack{4.\ CH_3I\ (excess),\ K_2CO_3 \\ 5.\ Ag_2O,\ H_2O \\ 6.\ heat}]{} CH_2{=}CHCH_2CH{=}CH_2$$

   Piperidine   (A)   1,4-Pentadiene

(a) Show that these results are consistent with the structure of piperidine (Section 23.1).
(b) Propose two additional structural formulas (excluding stereoisomers) for $C_5H_{11}N$ that are also consistent with the results obtained by Hofmann.

In summary, both Hofmann and Zaitsev eliminations are always preferentially anti. If only one β-hydrogen is anti to the leaving group, then that one will be removed. If more than one β-hydrogen is anti, then there will be competition between Hofmann and Zaitsev elimination.

1. Eliminations involving a negatively charged leaving group, for example Cl^-, Br^-, I^-, and OTs^-, almost always follow Zaitsev's rule, unless a bulky base is used.
2. Eliminations involving a neutral leaving group, for example $N(CH_3)_3$ and $S(CH_3)_2$, almost always follow Hofmann's rule.
3. The bulkier the base, the greater the percentage of Hofmann product; compare, for example, $(CH_3)_3CO^-K^+$, which gives mostly Hofmann elimination, with $CH_3O^-Na^+$, which gives mostly Zaitsev elimination.

One of the likeliest explanations for the formation of the less stable carbon-carbon double bond is that Hofmann elimination is governed largely by steric factors, namely the bulk of the $-NR_3^+$ group. The hydroxide ion preferentially approaches and removes the least hindered α-hydrogen and gives the least substituted alkene as product. For the same reason, bulky bases, such as $(CH_3)_3CO^-K^+$, also give Hofmann elimination from alkyl halides.

23.10 Cope Elimination

Treatment of a tertiary amine with hydrogen peroxide results in oxidation of the amine to an amine oxide.

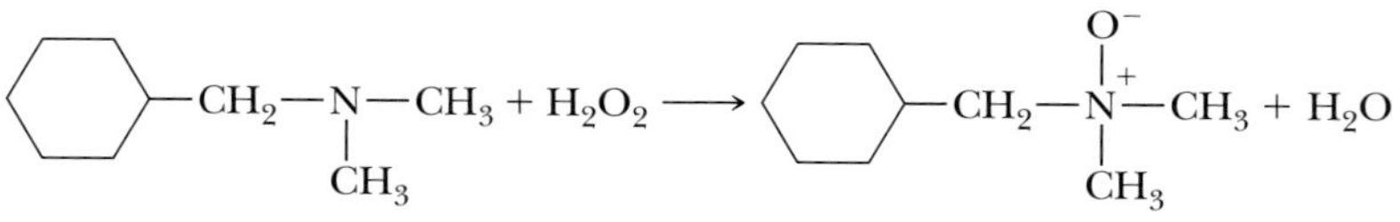

A 3° amine   An amine oxide

When an amine oxide with at least one β-hydrogen is heated, it undergoes thermal decomposition to form an alkene and an N,N-dialkylhydroxylamine. Thermal decomposition of an amine oxide to an alkene is known as a **Cope elimination** after its discoverer, Arthur C. Cope, of the Massachusetts Institute of Technology.

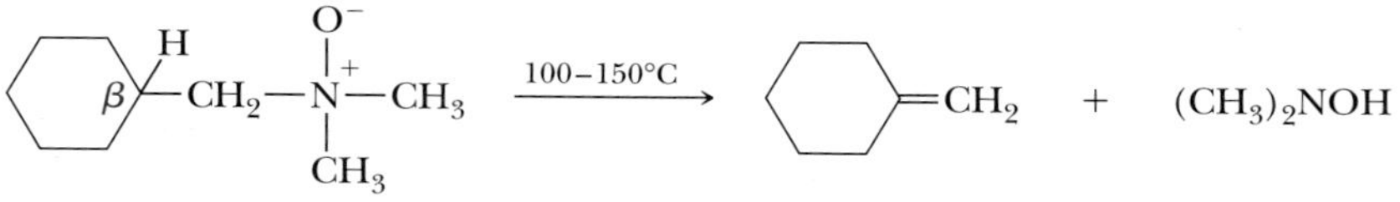

Methylenecyclohexane N,N-Dimethyl-
hydroxylamine

All experimental evidence indicates that the Cope elimination is syn stereoselective and concerted.

Mechanism   *The Cope Elimination*

The transition state involves a planar or nearly planar arrangement of the five participating atoms and a cyclic flow of three pairs of electrons. Elimination shows syn stereoselectivity.

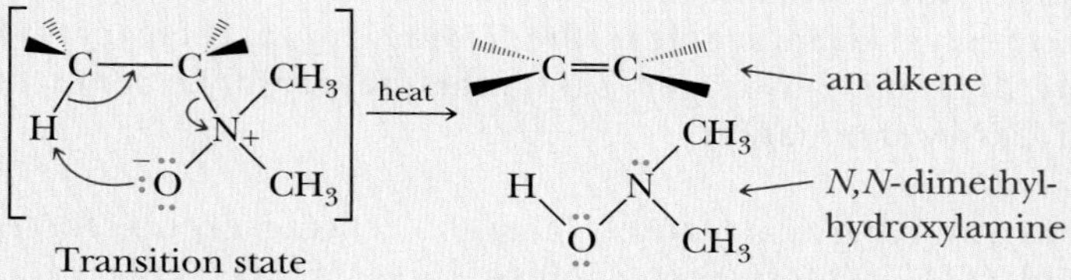

Transition state

If two or more syn β-hydrogens can be removed in a Cope elimination, there is little preference for one over another except when the double bond is conjugated with an aromatic ring. Therefore, as a method of preparation of alkenes, Cope eliminations are best used where only one alkene is possible.

Example 23.15

When 2-dimethylamino-3-phenylbutane is treated with hydrogen peroxide and then made to undergo a Cope elimination, the major alkene formed is 2-phenyl-2-butene.

$$CH_3CHCHCH_3 \xrightarrow[\text{2. heat}]{\text{1. } H_2O_2} CH_3C=CHCH_3 + (CH_3)_2NOH$$

with $N(CH_3)_2$ and Ph substituents

2-Dimethylamino-
3-phenylbutane

2-Phenyl-2-butene

(a) How many stereoisomers are possible for 2-dimethylamino-3-phenylbutane?
(b) How many stereoisomers are possible for 2-phenyl-2-butene?
(c) Suppose that the starting amine is the $2R,3S$ isomer. What is the configuration of the product?

Solution

(a) There are two chiral centers in the starting amine. Four stereoisomers are possible: two pair of enantiomers.
(b) There is one carbon-carbon double bond about which stereoisomerism is possible. Two stereoisomers are possible: one E,Z pair.
(c) Following is a stereodrawing of the $2R,3S$ stereoisomer showing a syn conformation of the dimethylamino group and the β-hydrogen. Cope elimination on this stereoisomer gives (E)-2-phenyl-2-butene.

$$\xrightarrow[\text{2. heat}]{\text{1. } H_2O_2}$$

2R,3S isomer

(E)-2-Phenyl-2-butene

Problem 23.15

In Example 23.15, you considered the product of Cope elimination from the $2R,3S$ stereoisomer of 2-dimethylamino-3-phenylbutane. What is the product of a Cope elimination from the following stereoisomers? What is the product of a Hofmann elimination from each stereoisomer?

(a) $2S,3R$ stereoisomer   (b) $2S,3S$ stereoisomer

Summary

Amines are classified as 1°, 2°, or 3°, depending on the number of carbon atoms bonded to nitrogen (Section 23.1). In an **aliphatic amine,** all carbon atoms bonded to nitrogen are derived from alkyl groups. In an **aromatic amine,** one or more of the groups bonded to nitrogen are aromatic rings. A **heterocyclic amine** is one in which the nitrogen atom is part of a ring. A **heterocyclic aromatic amine** is one in which the nitrogen atom is part of an aromatic ring.

In systematic nomenclature, aliphatic amines are named *alkanamines* (Section 23.2A). A cation in which a nitrogen is bonded to four alkyl or aryl groups is named as a quaternary ammonium ion. In common nomenclature (Section 23.2B), aliphatic amines are named *alkylamines;* the alkyl groups are listed in alphabetical order in one word ending in the suffix -*amine.*

Amines in which nitrogen is bonded to three different groups are chiral (Section 23.3) and, in principle, can be resolved. In practice, however, they undergo rapid **pyramidal inversion** with the result that a chiral amine is converted into a racemic mixture. Pyramidal inversion is not possible for quaternary ammonium ions, and chiral quaternary ammonium ions can be resolved. Chiral phosphines invert slowly and have been prepared and resolved.

Amines are polar compounds, and primary and secondary amines associate by intermolecular hydrogen bonding (Section 23.4). An N····H—N hydrogen bond is weaker than an O····H—O hydrogen bond and, therefore, amines have lower boiling points than alcohols of comparable molecular weight and structure. All classes of amines form hydrogen bonds with water and are more soluble in water than hydrocarbons of comparable molecular weight.

Amines are weak bases, and aqueous solutions of amines are basic (Section 23.5). It is common to discuss the acid-base properties of amines by reference to the pK_a for their corresponding conjugate acids.

Most aliphatic amines have comparable basicity, and all are slightly stronger bases than ammonia (Section 23.5A). For representative aliphatic amines, values of pK_a are in the range 10 to 11; aliphatic amines are fully protonated in aqueous solution at pH 7.0. Aromatic amines are considerably weaker bases than aliphatic amines. For representative aromatic amines, val-ues of pK_a are in the range 3 to 5; aromatic amines are not protonated in aqueous solution at pH 7.0. Unprotonated aromatic amines have resonance stabilization attributable to interaction of the unshared pair of electrons on nitrogen with the pi system of the aromatic ring. This interaction decreases the availability of the electron pair for protonation.

Heterocyclic aromatic amines are considerably weaker bases than aliphatic amines because the unshared pair of electrons on nitrogen giving rise to basicity lies in an sp^2 hybrid orbital (Section 23.5C). The electron-withdrawing inductive effect of an sp^2-hybridized nitrogen compared with an sp^3-hybridized nitrogen is responsible for the decreased basicity of heterocyclic aromatic amines.

All amines, whether soluble or insoluble in water, react quantitatively with strong acids to form water-soluble salts (Section 23.6). The basicity of amines and the solubility of amine salts in water can be used to separate water-insoluble amines from water-insoluble nonbasic compounds.

Key Reactions

1. Basicity of Aliphatic Amines (Section 23.5A)

Aliphatic amines are slightly stronger bases than ammonia owing to the electron-releasing effect of alkyl groups and partial delocalization of positive charge in the alkylammonium ion.

$$CH_3NH_2 + H_2O \rightleftharpoons CH_3NH_3^+ + OH^- \qquad pK_a = 10.64$$

2. Basicity of Aromatic Amines (Section 23.5B)

Aromatic amines are considerably weaker bases than aliphatic amines. Resonance stabilization by interaction of the unshared electron pair on nitrogen with the pi system decreases its availability for protonation.

$$pK_a = 4.63$$

3. Basicity of Heterocyclic Aromatic Amines (Section 23.5C)

Heterocyclic aromatic amines are considerably weaker bases than aliphatic amines.

$$pK_a = 6.95$$

4. Reaction of Amines with Strong Acids (Section 23.6)

All amines react quantitatively with strong acids to form water-soluble salts.

Insoluble in water   A water-soluble salt

5. Alkylation of Ammonia and Amines (Section 23.7A)

This method is seldom used for preparation of pure amines because of overalkylation and the difficulty of separating products.

$$\text{—CH}_2\text{NH}_2 + \text{CH}_3\text{I} \longrightarrow \text{—CH}_2\overset{\overset{\text{H}}{|}}{\underset{\underset{\text{H}}{|}}{\overset{+}{\text{N}}}}\text{CH}_3 \quad \text{I}^-$$

6. Alkylation of Azide Ion Followed by Reduction (Section 23.7B)

Azides are prepared by treatment of a primary or secondary alkyl halide, or an epoxide with NaN_3, and are reduced to primary amines by a variety of reducing agents including lithium aluminum hydride.

$$\begin{array}{ccc} & \xrightarrow[\text{2. H}_2\text{O}]{\text{1. K}^+\text{N}_3^-} & \xrightarrow[\text{2. H}_2\text{O}]{\text{1. LiAlH}_4} \\ & \text{(racemic)} & \text{(racemic)} \end{array}$$

7. Nitrosation of Tertiary Aromatic Amines (Section 23.8B)

The nitrosyl cation is a very weak electrophile and participates in electrophilic aromatic substitution only with highly activated aromatic rings.

$$(\text{CH}_3)_2\text{N}\text{—}\underset{}{\bigcirc}\xrightarrow[\text{2. NaOH, H}_2\text{O}]{\text{1. NaNO}_2\text{, HCl, 0–5°C}} (\text{CH}_3)_2\text{N}\text{—}\underset{}{\bigcirc}\text{—N}=\text{O}$$

8. Formation of N-Nitrosamines from Secondary Amines (Section 23.8C)

Reaction of the nitrosyl cation (an electrophile) with a 2° amine (a nucleophile) gives an *N*-nitrosamine.

$$\underset{}{\bigcirc}\text{N—H} + \text{HNO}_2 \longrightarrow \underset{}{\bigcirc}\text{N—N}=\text{O} + \text{H}_2\text{O}$$

9. Reaction of Primary Aliphatic Amines with Nitrous Acid (Section 23.8D)

Treating a primary aliphatic amine with nitrous acid gives an unstable diazonium salt that loses N_2 to give a carbocation. The carbocation may (1) lose a proton to give an alkene, (2) react with a nucleophile, or (3) rearrange, followed by (1) or (2).

$$\text{RCH}_2\text{NH}_2 \xrightarrow[\text{0–5°C}]{\text{NaNO}_2\text{, HCl}} [\text{RCH}_2\text{N}_2^+] \longrightarrow \text{RCH}_2^+ + \text{N}_2$$

10. Reaction of Cyclic β-Aminoalcohols with Nitrous Acid (Section 23.8D)

Treating a cyclic β-aminoalcohol with nitrous acid leads to rearrangement and a ring-expanded ketone.

$$\underset{\text{CH}_2\text{NH}_2}{\overset{\text{OH}}{\bigcirc}} + \text{HNO}_2 \longrightarrow \underset{}{\bigcirc}\overset{\text{O}}{} + \text{H}_2\text{O} + \text{N}_2$$

11. Formation of Arenediazonium Salts (Diazotization) (Section 23.8E)

Arenediazonium salts are stable in aqueous solution at 0°C for short periods.

$$\underset{}{\bigcirc}\text{—NH}_2 + \text{HNO}_2 \xrightarrow[\text{0°C}]{\text{HCl}} \underset{}{\bigcirc}\text{—N}_2^+\text{Cl}^- + \text{H}_2\text{O}$$

12. Conversion of a Primary Arylamine to a Phenol (Section 23.8E)

Formation of an arenediazonium salt followed by loss of nitrogen gives an aryl cation intermediate, which then reacts with water to give a phenol.

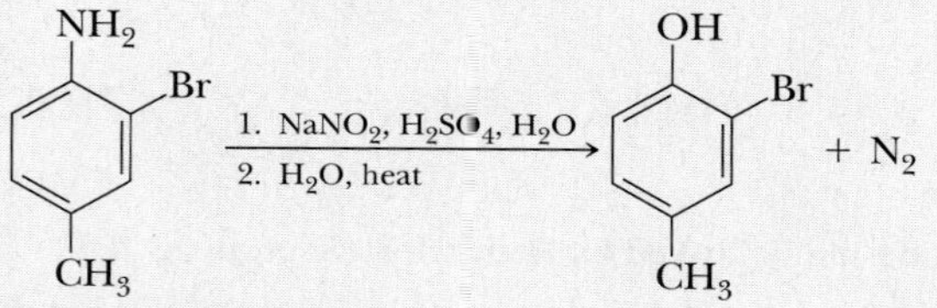

13. Schiemann Reaction (Section 23.8E)

Heating an arenediazonium fluoroborate is the most common synthetic method for introduction of fluorine onto an aromatic ring.

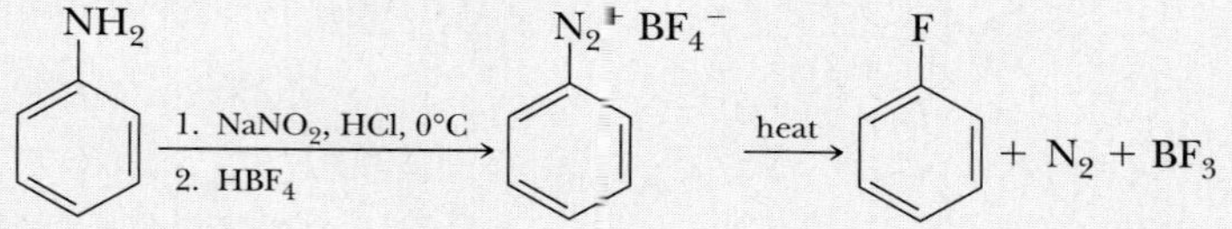

14. Sandmeyer Reaction (Section 23.8E)

Treatment of an arenediazonium salt with CuCl, CuBr, or CuCN results in replacement of the diazonium group by —Cl, —Br, or —CN, respectively.

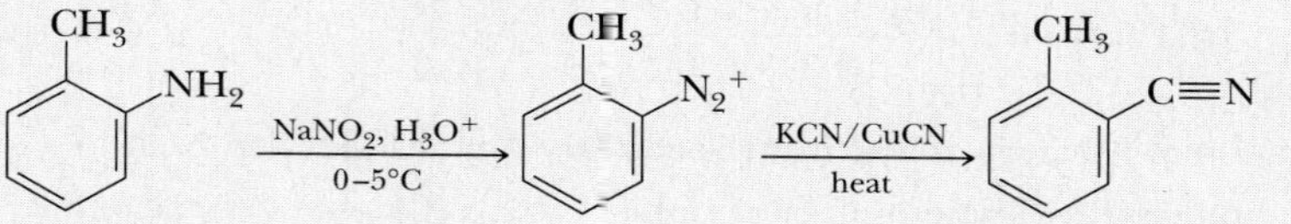

15. Reaction of an Arenediazonium Salt with KI (Section 23.8E)

Treatment of an arenediazonium salt with KI is the most convenient method for introducing iodine onto an aromatic ring.

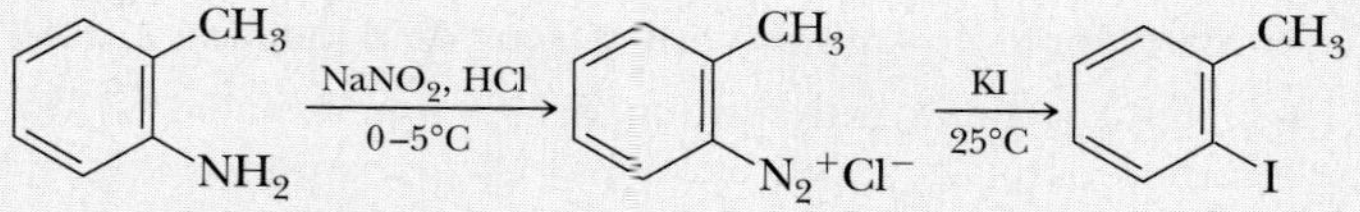

16. Reduction of an Arenediazonium Salt with Hypophosphorous Acid (Section 23.8E)

An —NO_2 or —NH_2 group can be used to control orientation of further substitution and then removed after it has served its purpose.

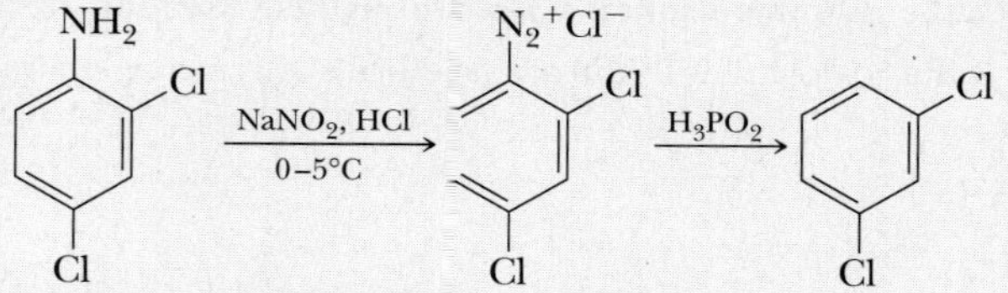

17. Hofmann Elimination (Section 23.9)

Anti stereoselective elimination of quaternary ammonium hydroxides occurs preferentially to form the least substituted carbon-carbon double bond (Hofmann's rule).

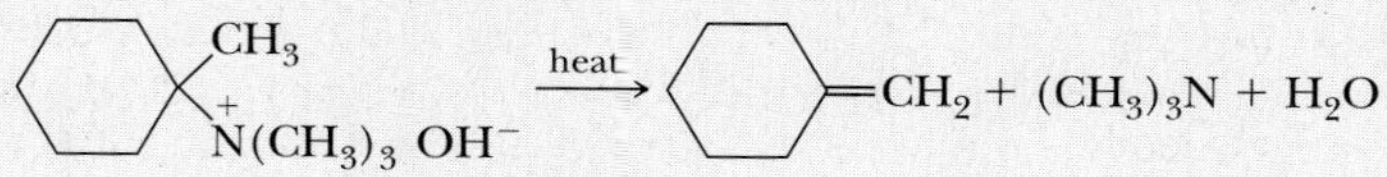

18. Cope Elimination: Pyrolysis of a Tertiary Amine Oxide (Section 23.10)

Elimination is syn stereoselective and involves a cyclic flow of six electrons in a planar transition state.

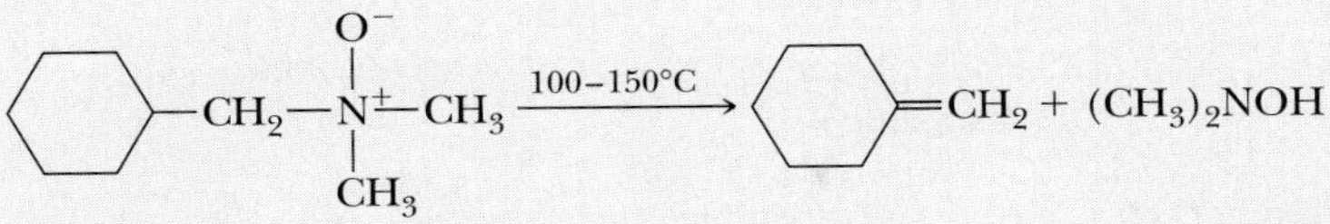

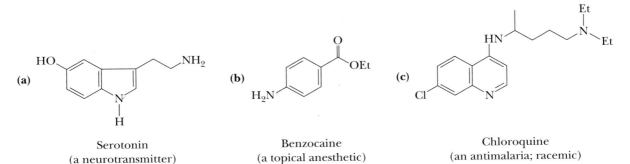

Problems

Structure and Nomenclature

ORGANIC **Chemistry Now™**
Assess your understanding of this chapter's topics with additional quizzing and conceptual-based problems at **http://org.brookscole.com/bfi4**

23.16 Draw a structural formula for each amine and amine derivative.

- **(a)** *N,N*-Dimethylaniline
- **(b)** Triethylamine
- **(c)** *tert*-Butylamine
- **(d)** 1,4-Benzenediamine
- **(e)** 4-Aminobutanoic acid
- **(f)** (*R*)-2-Butanamine
- **(g)** Benzylamine
- **(h)** *trans*-2-Aminocyclohexanol
- **(i)** 1-Phenyl-2-propanamine (amphetamine)
- **(j)** Lithium diisopropylamide (LDA)
- **(k)** Benzyltrimethylammonium hydroxide (Triton B)

23.17 Give an acceptable name for these compounds.

(a)

(b)

(c)

(d)

(e)

(f)

(g)

(h)

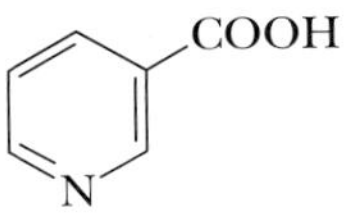

23.18 Classify each amine as primary, secondary, or tertiary and as aliphatic or aromatic.

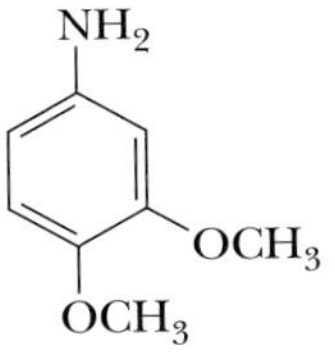

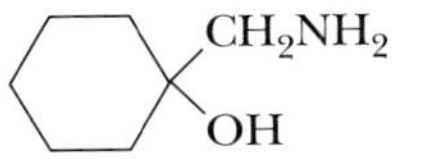

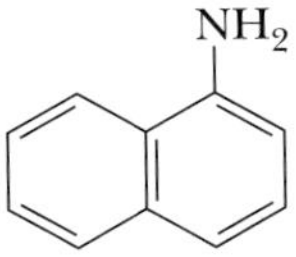

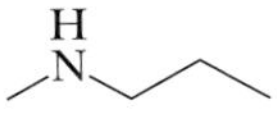

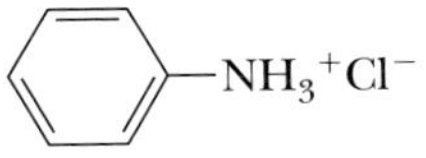

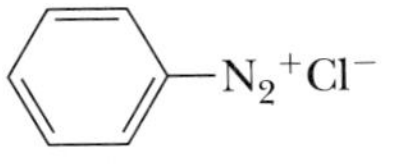

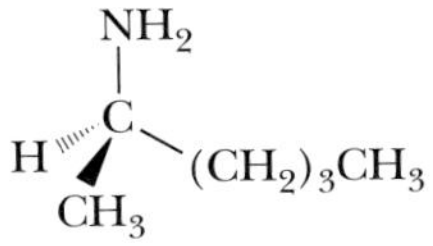

(a)

Serotonin
(a neurotransmitter)

(b)

Benzocaine
(a topical anesthetic)

(c)

Chloroquine
(an antimalaria; racemic)

23.19 Epinephrine is a hormone secreted by the adrenal medulla. Among its actions, it is a bronchodilator. Albuterol, sold under several trade names, including Proventil and Salbumol, is one of the most effective and widely prescribed antiasthma drugs. The *R* enantiomer of albuterol is 68 times more effective in the treatment of asthma than the *S* enantiomer.

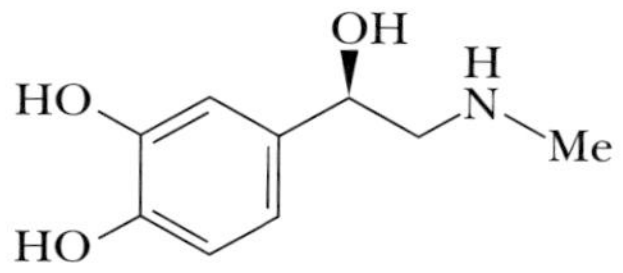

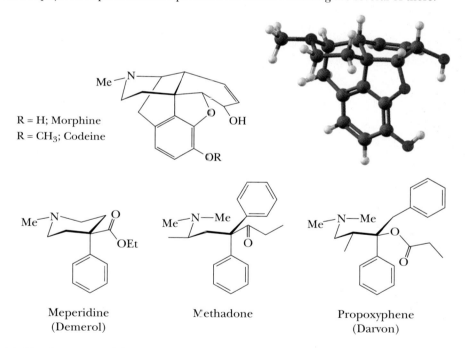

(R)-Epinephrine
(Adrenaline)

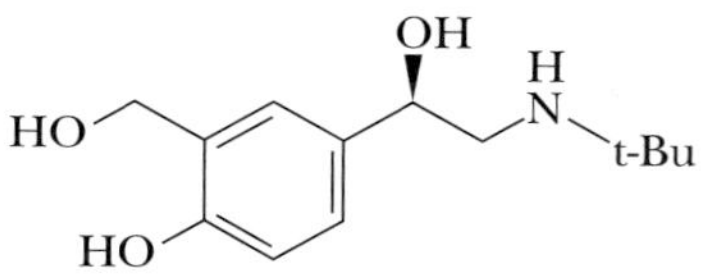

(R)-Albuterol

(a) Classify each as a primary, secondary, or tertiary amine.
(b) Compare the similarities and differences between their structural formulas.

23.20 Draw the structural formula for a compound with the given molecular formula.

(a) A 2° arylamine, C_7H_9N

(b) A 3° arylamine, $C_8H_{11}N$

(c) A 1° aliphatic amine, C_7H_9N

(d) A chiral 1° amine, $C_4H_{11}N$

(e) A 3° heterocyclic amine, $C_6H_{11}N$

(f) A trisubstituted 1° arylamine, $C_9H_{13}N$

(g) A chiral quaternary ammonium salt, $C_6H_{16}NCl$

23.21 Morphine and its O-methylated derivative codeine are among the most effective painkillers known. However, they possess two serious drawbacks: They are addictive, and repeated use induces a tolerance to the drug. Many morphine analogs have been prepared in an effort to find drugs that are equally effective as painkillers but that have less risk of physical dependence and potential for abuse. Following are several of these.

R = H; Morphine
R = CH₃; Codeine

Meperidine
(Demerol)

Methadone

Propoxyphene
(Darvon)

(a) List the structural features common to each of these molecules.
(b) The Beckett-Casey rules are a set of empirical rules to predict the structure of molecules that bind to morphine receptors and act as analgesics. According to these rules, to provide an effective morphine-like analgesia, a molecule must have (1) an aromatic ring attached to (2) a quaternary carbon and (3) a nitrogen at a distance equal to two carbon-carbon single bond lengths from the quaternary center. Show that these structural requirements are present in the molecules given in this problem.

23.22 Following is a structural formula of desosamine, a sugar component of several macrolide antibiotics, including the erythromycins. The configuration shown is that of the natural or D isomer. Erythromycin is produced by a strain of *streptomyces erythreus* found in a soil sample from the Philippine Archipelago.

Desosamine

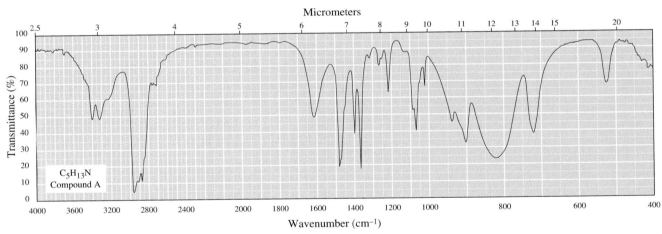

(a) Name all functional groups in desosamine.

(b) How many chiral centers are present in desosamine? How many stereoisomers are possible for it? How many pairs of enantiomers are possible for it?

(c) Draw the alternative chair conformations for desosamine. In each, label which groups are equatorial and which are axial.

(d) Which of the alternative chair conformations for desosamine is the more stable? Explain.

Spectroscopy

23.23 Account for the formation of the base peaks in these mass spectra.

   (a) Isobutylmethylamine, m/z 30   (b) Diethylamine, m/z 58

23.24 Propose a structural formula for compound A, $C_5H_{13}N$, given its IR and ^{1}H-NMR spectra.

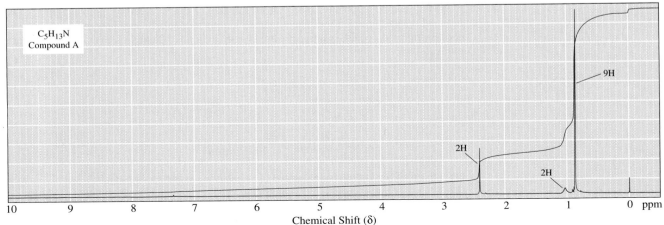

Basicity of Amines

23.25 Select the stronger base from each pair of compounds.

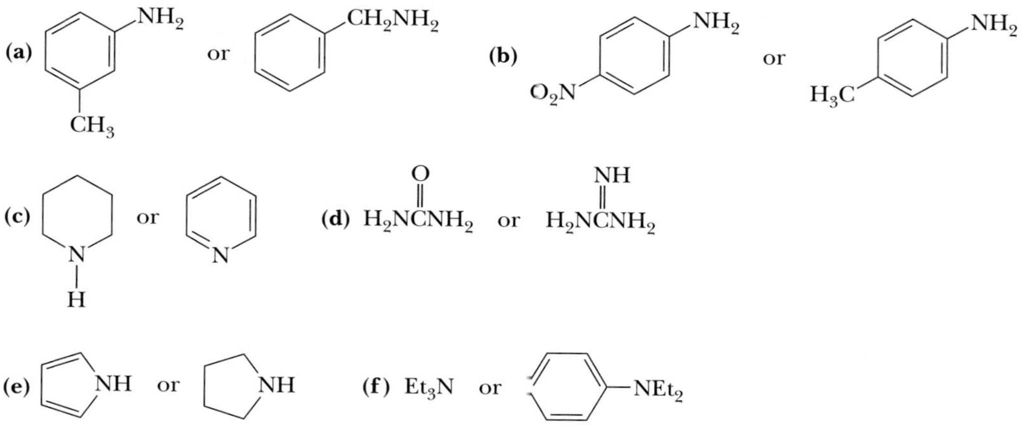

(a) [structure: aryl-NH₂ with CH₃] or [structure: benzyl CH₂NH₂] (b) [structure: O₂N-aryl-NH₂] or [structure: H₃C-aryl-NH₂]

(c) [piperidine, N–H] or [pyridine] (d) $H_2N\overset{O}{\overset{\|}{C}}NH_2$ or $H_2N\overset{NH}{\overset{\|}{C}}NH_2$

(e) [pyrrole] NH or [pyrrolidine] NH (f) Et_3N or [aryl]—NEt_2

23.26 The pK_a of the conjugate acid of morpholine is 8.33.

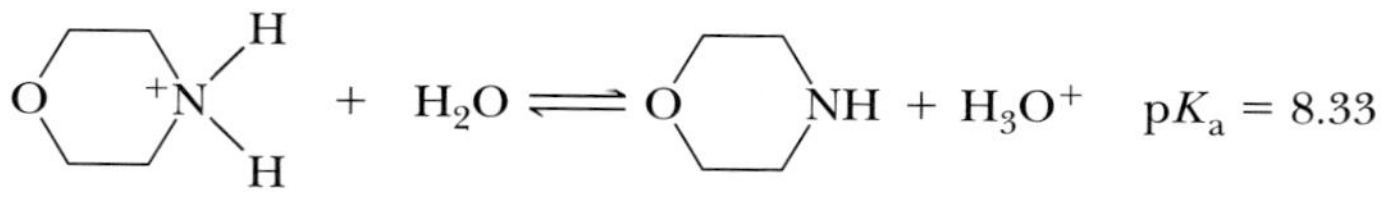

[structure: Morpholinium ion] $+ H_2O \rightleftharpoons$ [structure: Morpholine] $NH + H_3O^+$ p$K_a = 8.33$

Morpholinium ion   Morpholine

(a) Calculate the ratio of morpholine to morpholinium ion in aqueous solution at pH 7.0.
(b) At what pH are the concentrations of morpholine and morpholinium ion equal?

23.27 Which of the two nitrogens in pyridoxamine (a form of vitamin B₆) is the stronger base? Explain your reasoning.

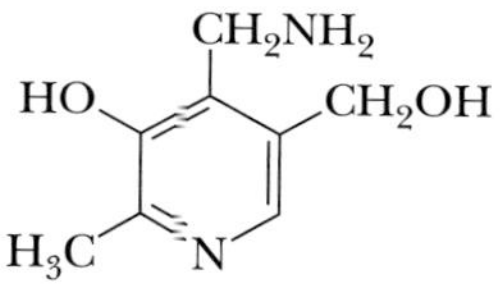

[structure: Pyridoxamine with CH₂NH₂, CH₂OH, HO, H₃C groups]

Pyridoxamine
(Vitamin B₆)

23.28 Epibatidine, a colorless oil isolated from the skin of the Ecuadorian poison frog *Epipedobates tricolor* has several times the analgesic potency of morphine. It is the first chlorine-containing, nonopioid (nonmorphine-like in structure) analgesic ever isolated from a natural source.

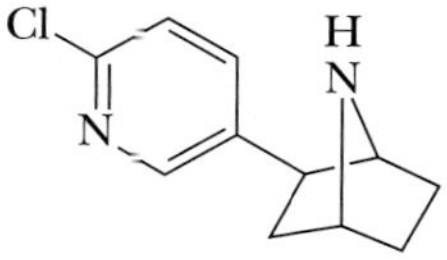

[structure: Epibatidine]

Epibatidine

(a) Which of the two nitrogen atoms of epibatidine is the more basic?
(b) Mark all chiral centers in this molecule.

23.29 Aniline (conjugate acid pK_a 4.63) is a considerably stronger base than diphenylamine (pK_a 0.79). Account for these marked differences.

23.30 Complete the following acid-base reactions and predict the direction of equilibrium (to the right or to the left) for each. Justify your prediction by citing values of pK_a for the stronger and weaker acid in each equilibrium. For values of acid ionization constants, consult Table 23.2 (Acid Strengths, pK_a, of the Conjugate Acids of Selected Amines), and Appendix 2 (Acid Ionization Constants for the Major Classes of Organic Acids). Where no ionization constants are given, make the best estimate from the information given in the reference tables and sections.

(a) CH_3COOH + [pyridine] $\rightleftharpoons$

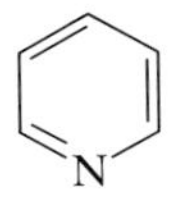

Acetic acid   Pyridine

(b) 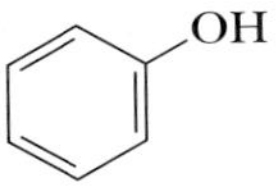 [phenol] OH + Et_3N $\rightleftharpoons$

Phenol   Triethylamine

(c) $PhC{\equiv}CH$ + NH_3 $\rightleftharpoons$

Phenylacetylene   Ammonia

(d) $PhC{\equiv}CH$ + $iPr_2N^-Li^+$ $\rightleftharpoons$

Phenylacetylene   Lithium diisopropylamide (LDA)

(e) $PhCO^-Na^+$ + $Et_3NH^+Cl^-$ $\rightleftharpoons$

Sodium benzoate   Triethylammonium chloride

(f) 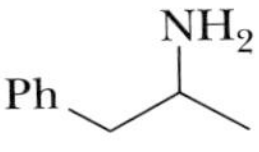 [Ph–CH(NH$_2$)CH$_3$] + [CH$_3$CH(OH)COOH] $\rightleftharpoons$

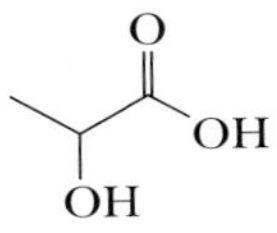

1-Phenyl-2-propanamine (Amphetamine)   2-Hydroxypropanoic acid (Lactic acid)

(g) 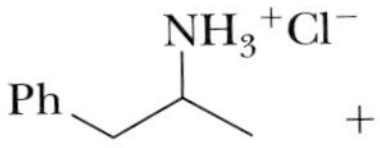 [Ph–CH(NH$_3^+$Cl$^-$)CH$_3$] + $NaHCO_3$ $\rightleftharpoons$

Amphetamine hydrochloride   Sodium bicarbonate

(h) 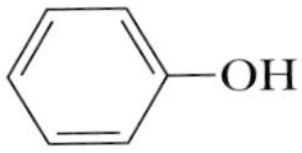 [phenyl]–OH + $Me_4N^+OH^-$ $\rightleftharpoons$

Phenol   Tetramethylammonium hydroxide

23.31 Quinuclidine and triethylamine are both tertiary amines. Quinuclidine, however, is a considerably stronger base than triethylamine. Stated alternatively, the conjugate acid of quinuclidine is a considerably weaker acid than the conjugate acid of triethylamine.

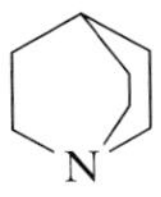

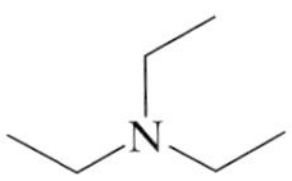

Quinuclidine   Triethylamine
pK_a 10.6 pK_a 8.6

Propose an explanation for these differences in acidity/basicity.

23.32 Suppose that you have a mixture of these three compounds. Devise a chemical procedure based on their relative acidity or basicity to separate and isolate each in pure form.

H_3C—⟨⟩—NO_2 H_3C—⟨⟩—NH_2 H_3C—⟨⟩—OH

4-Nitrotoluene    4-Methylaniline   4-Methylphenol
                        (*p*-Toluidine)   (*p*-Cresol)

Preparation of Amines

23.33 Propose a synthesis of 1-hexanamine from the following.

(a) A bromoalkane of six carbon atoms
(b) A bromoalkane of five carbon atoms

23.34 Show how to convert each starting material into benzylamine in good yield.

(a) 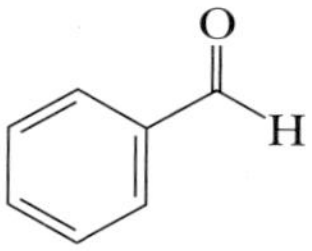 benzaldehyde   (b) 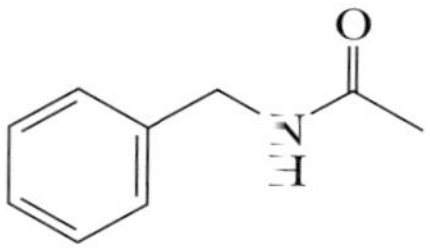 N-benzylacetamide   (c) 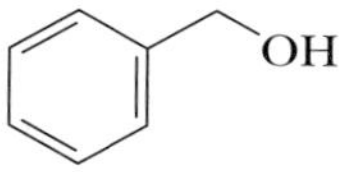 benzyl alcohol

(d) 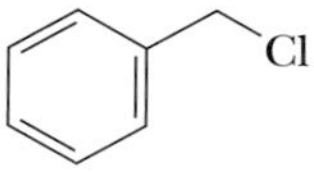 benzyl chloride   (e) 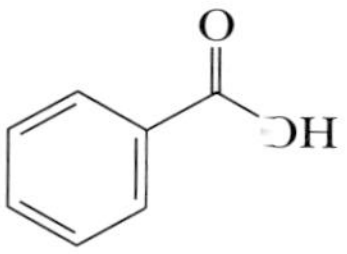 benzoic acid   (f) 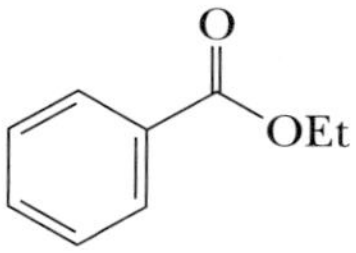 ethyl benzoate

Reactions of Amines

23.35 Treating trimethylamine with 2-chloroethyl acetate gives acetylcholine as its chloride. Acetylcholine is a neurotransmitter.

$$Me_3N + CH_3\overset{\overset{\displaystyle O}{\|}}{C}OCH_2CH_2Cl \longrightarrow C_7H_{16}ClNO_2$$

Acetylcholine chloride

Propose a structural formula for this quaternary ammonium salt and a mechanism for its formation.

23.36 *N*-Nitrosamines by themselves are not significant carcinogens. However, they are activated in the liver by a class of iron-containing enzymes (members of the cytochrome P-450 family). Activation involves the oxidation of a C—H bond next to the amine nitrogen to a C—OH group.

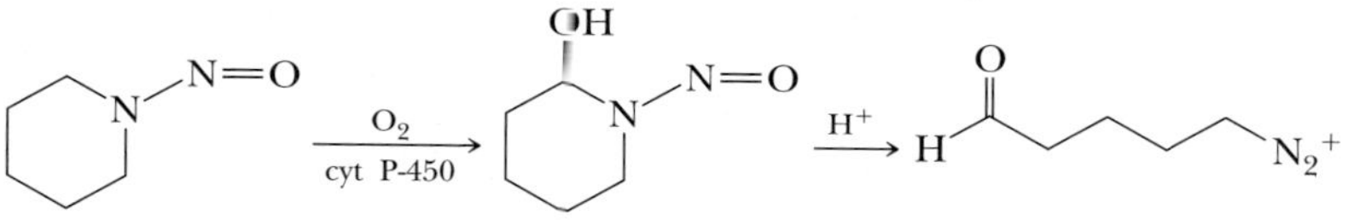

N-Nitroso-piperidine $\xrightarrow[\text{cyt P-450}]{O_2}$ 2-Hydroxy-N-nitrosopiperidine $\xrightarrow{H^+}$ An alkyl diazonium ion (a carcinogen)

Show how this hydroxylation product can be transformed into an alkyl diazonium ion, an active alkylating agent and therefore a carcinogen, in the presence of an acid catalyst.

23.37 Marked similarities exist between the mechanism of nitrous acid deamination of β-aminoalcohols and the pinacol rearrangement. Following are examples of each.

Nitrous acid
deamination of
a β-aminoalcohol:

$$\text{—NH}_2 \xrightarrow{\text{NaNO}_2,\ \text{HCl}} \text{=O} + \text{N}_2 + \text{H}_2\text{O}$$

Pinacol
rearrangement:

$$\text{—OH} \xrightarrow{\text{H}_2\text{SO}_4} \text{—CHO} + \text{H}_2\text{O}$$

(a) Analyze the mechanism of each rearrangement and list their similarities.
(b) Why does the first reaction, but not the second, give ring expansion?
(c) Suggest a β-aminoalcohol that would give cyclohexanecarbaldehyde as a product?

23.38 (S)-Glutamic acid is one of the 20 amino acid building blocks of polypeptides and proteins (Chapter 27). Propose a mechanism for the following conversion.

$$\xrightarrow[\text{0–5°C}]{\text{NaNO}_2,\ \text{HCl}}$$

(S)-Glutamic acid   (The S enantiomer)

23.39 The following sequence of methylation and Hofmann elimination was used in the determination of the structure of this bicyclic amine. Compound B is a mixture of two isomers.

$$\xrightarrow[\substack{\text{1. CH}_3\text{I} \\ \text{2. Ag}_2\text{O, H}_2\text{O} \\ \text{3. heat}}]{} \text{C}_{10}\text{H}_{19}\text{N} \xrightarrow[\substack{\text{4. CH}_3\text{I} \\ \text{5. Ag}_2\text{O, H}_2\text{O} \\ \text{6. heat}}]{} \text{C}_8\text{H}_{12}$$

($C_9H_{17}N$)   (A)   (B)

(a) Propose structural formulas for compounds A and B.
(b) Suppose you were given the structural formula of compound B but only the molecular formulas for compound A and the starting bicyclic amine. Given this information, is it possible, working backward, to arrive at an unambiguous structural formula for compound A? For the bicyclic amine?

23.40 Propose a structural formula for the compound, $C_{10}H_{16}$, and account for its formation.

$$\xrightarrow{\text{1. CH}_3\text{I, 2 moles}} \xrightarrow{\text{2. H}_2\text{O}_2} \xrightarrow{\text{3. heat}} \text{C}_{10}\text{H}_{16}$$

23.41 An amine of unknown structure contains one nitrogen and nine carbon atoms. The ^{13}C-NMR spectrum shows only five signals, all between 20 and 60 ppm. Three cycles of Hofmann elimination sequence [(1) CH$_3$I; (2) Ag$_2$O, H$_3$O; (3) heat] give trimethylamine and 1,4,8-nonatriene. Propose a structural formula for the amine.

23.42 The pyrolysis of acetic esters to give an alkene and acetic acid is thought to involve a planar transition state and cyclic redistribution of $(4n + 2)$ electrons. Propose a mechanism for pyrolysis of the following ester.

$$\xrightarrow{\text{500°C}}$$

Butyl acetate   1-Butene   Acetic acid

Synthesis

23.43 Propose steps for the following conversions using a reaction of a diazonium salt in at least one step of each conversion.

(a) Toluene to 4-methylphenol (*p*-cresol)   (b) Nitrobenzene to 3-bromophenol
(c) Toluene to *p*-cyanobenzoic acid   (d) Phenol to *p*-iodoanisole
(e) Acetanilide to *p*-aminobenzylamine   (f) Toluene to 4-fluorobenzoic acid
(g) 3-Methylaniline (*m*-toluidine) to 2,4,6-tribromobenzoic acid

23.44 Show how to bring about each step in this synthesis of the herbicide propranil.

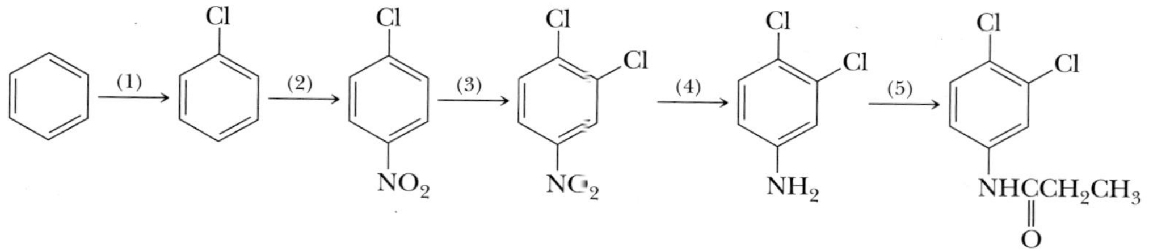

Propranil

23.45 Show how to bring about each step in the following synthesis.

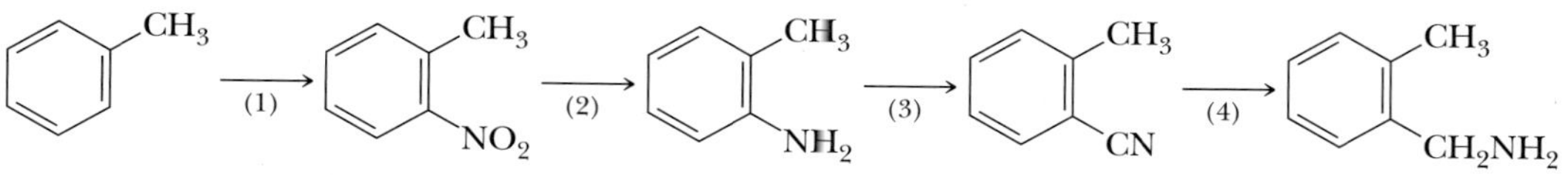

23.46 Show how to bring about this synthesis.

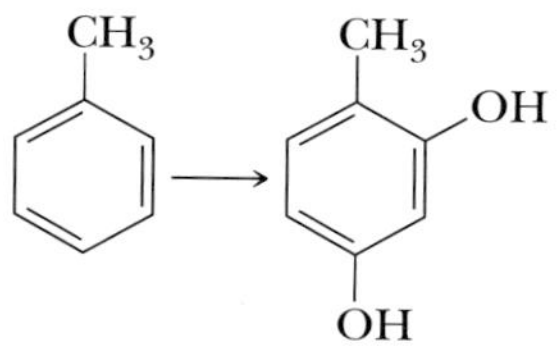

23.47 Show how to bring about each step in the following synthesis.

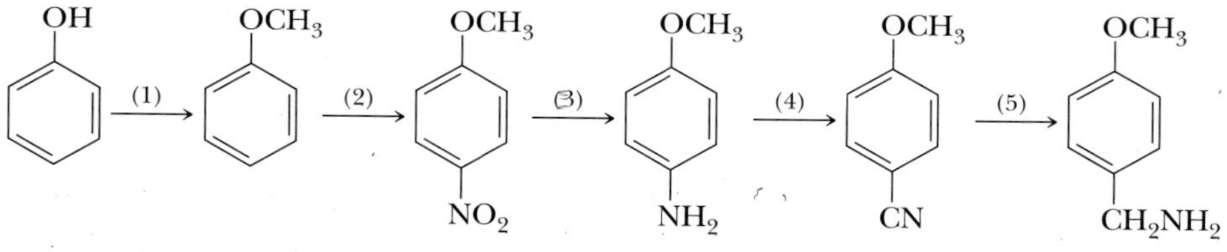

23.48 Methylparaben is used as a preservative in foods, beverages, and cosmetics. Provide a synthesis of this compound from toluene.

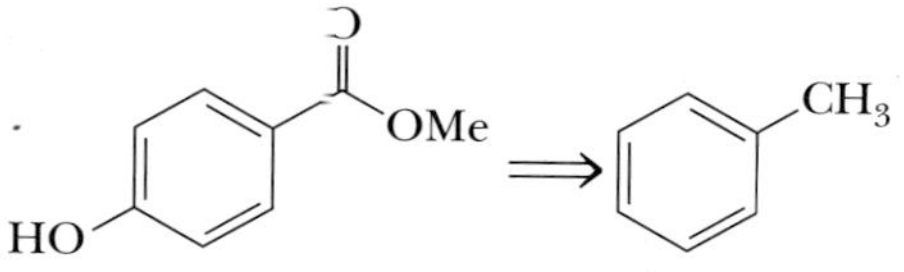

Methyl *p*-hydroxybenzoate
(Methylparaben)

23.49 Given the following retrosynthetic analysis, show how to synthesize the following tertiary amine as a racemic mixture from benzene and any necessary reagents.

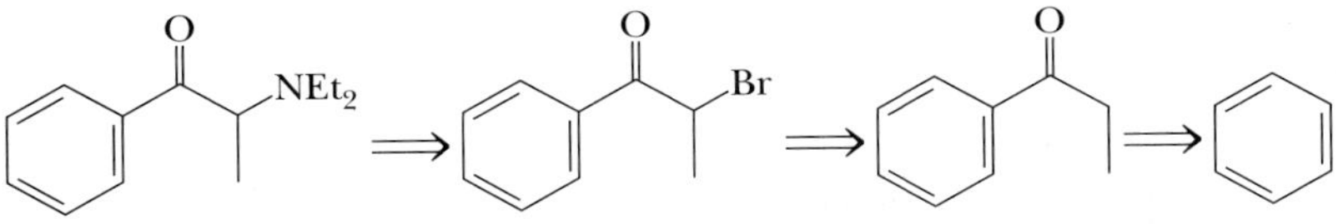

23.50 N-Substituted morpholines are a building block in many drugs. Show how to synthesize N-methylmorpholine given this retrosynthetic analysis.

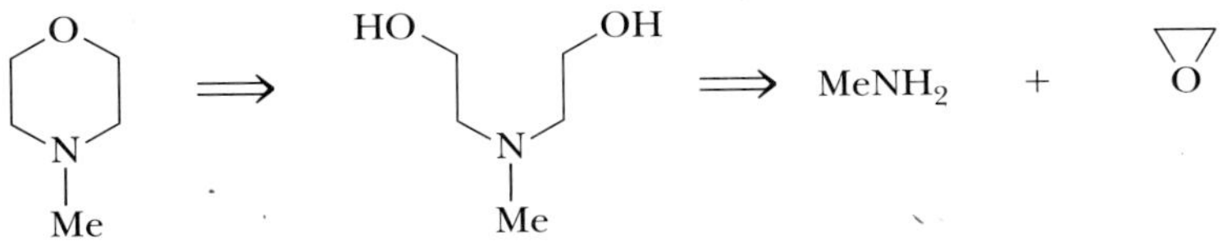

N-Methylmorpholine   Methylamine   Ethylene oxide

23.51 Propose a synthesis for the systemic agricultural fungicide tridemorph from dodecanoic acid (lauric acid) and propene. How many stereoisomers are possible for tridemorph?

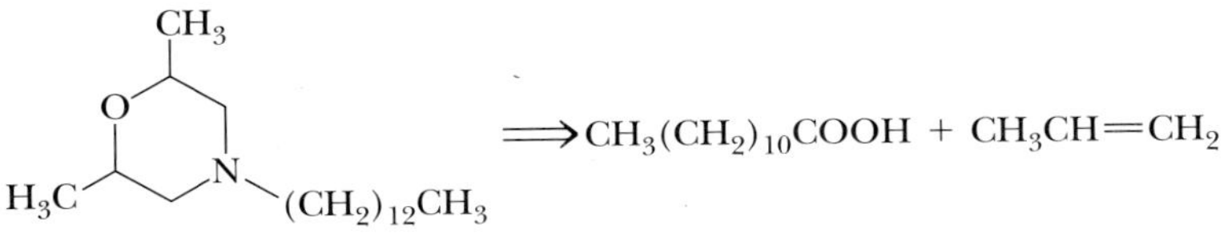

$$\Longrightarrow CH_3(CH_2)_{10}COOH + CH_3CH=CH_2$$

Tridemorph   Dodecanoic acid   Propene
                    (Lauric acid)

23.52 The Ritter reaction is especially valuable for the synthesis of 3° alkanamines. In fact, there are few alternative routes to them. This reaction is illustrated by the first step in the following sequence. In the second step, the Ritter product is hydrolyzed to the amine.

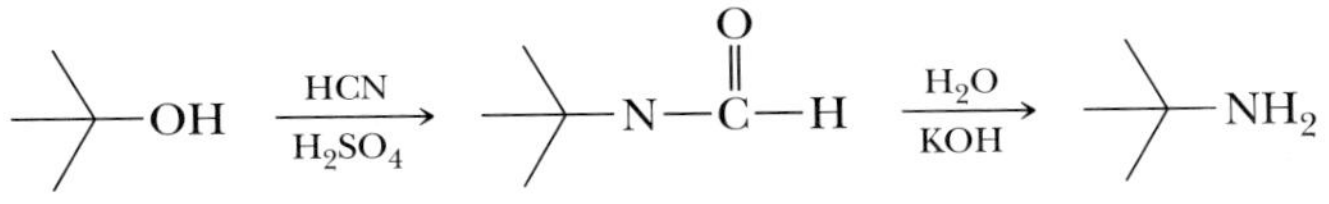

Ritter product

(a) Propose a mechanism for the Ritter reaction.
(b) What is the product of a Ritter reaction using acetonitrile, CH_3CN, instead of HCN followed by reduction of the Ritter product with lithium aluminum hydride?

23.53 Several diamines are building blocks for the synthesis of pharmaceuticals and agrochemicals. Show how both 1,3-propanediamine and 1,4-butanediamine can be prepared from acrylonitrile.

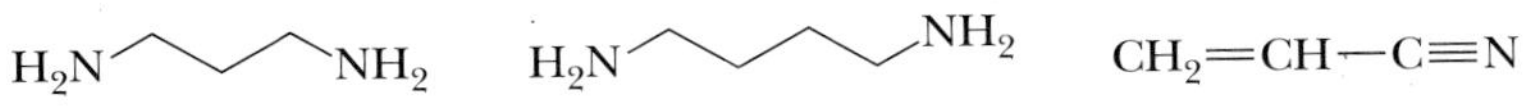

$$CH_2=CH-C\equiv N$$

1,3-Propanediamine   1,4-Butanediamine   Acrylonitrile

23.54 Given the following retrosynthetic analysis, show how the intravenous anesthetic 2,6-diisopropylphenol (Propofol) can be synthesized from phenol.

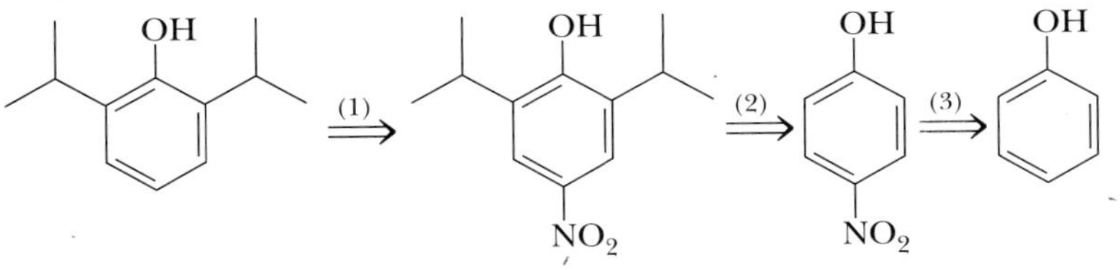

2,6-Diisopropylphenol   Phenol
(Propofol)

23.55 Following is a retrosynthetic analysis for propoxyphene, the hydrochloride salt of which is Darvon. The naphthalenesulfonic acid salt of propoxyphene is Darvon-N. The configuration of the carbon in Darvon bearing the hydroxyl group is S, and the configuration of the other stereocenter is R. Its enantiomer has no analgesic properties, but it is used as a cough suppressant.

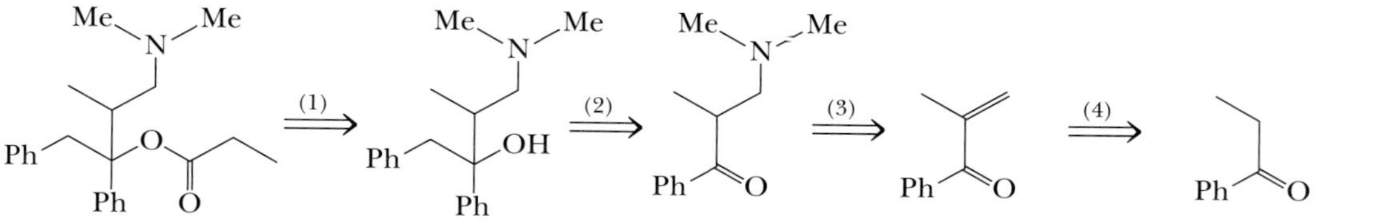

Propoxyphene

1-Phenyl-1-propanone
(Propiophenone)

(a) Propose a synthesis for propoxyphene from 1-phenyl-1-propanone and any other necessary reagents.

(b) Is propoxyphene chiral? If so, which of the possible stereoisomers are formed in this synthesis?

23.56 Following is a retrosynthetic analysis for ibutilide, a drug used to treat cardiac arrhythmia. In this scheme, Hept is an abbreviation for the 1-heptyl group.

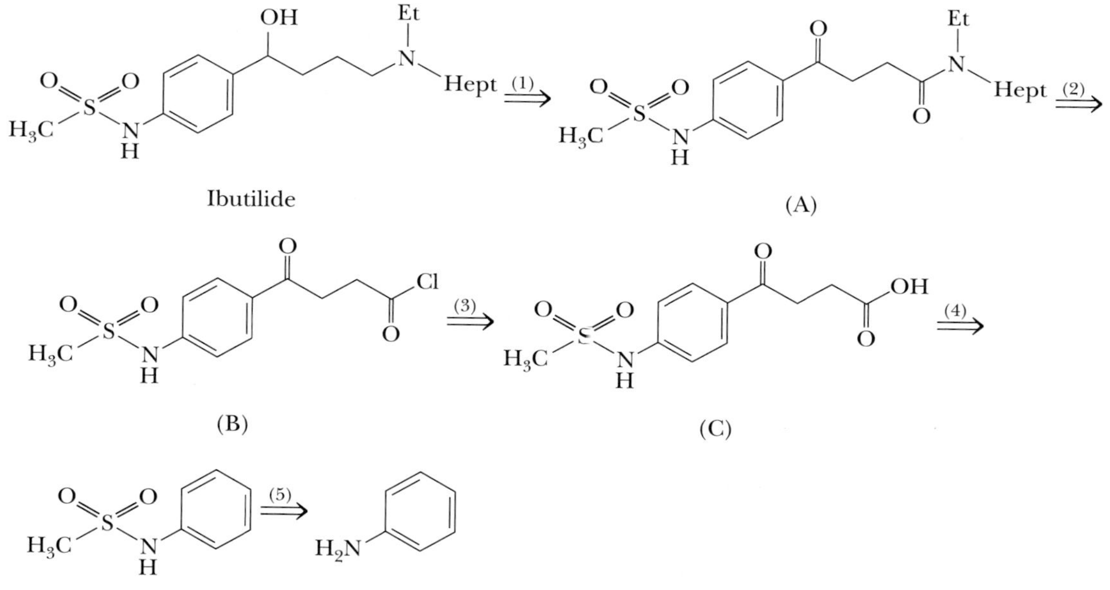

Ibutilide

(A)

(B)

(C)

(D)

Aniline

(a) Propose a synthesis for ibutilide starting with aniline, methanesulfonyl chloride, succinic anhydride, and N-ethyl-1-heptanamine.

(b) Is isobutilide chiral? If so, which of the possible stereoisomers are formed in this synthesis?

23.57 Propose a synthesis for the antihistamine histapyrrodine.

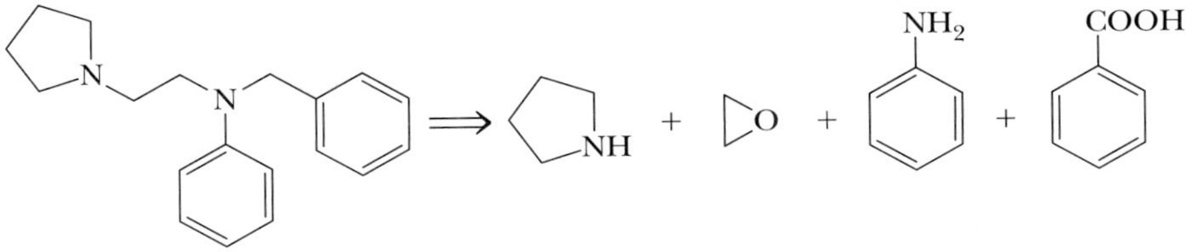

Histapyrrodine   Pyrrolidine   Ethylene   Aniline   Benzoic
                                      oxide   acid

23.58 Following is a retrosynthesis for the coronary vasodilator ganglefene.

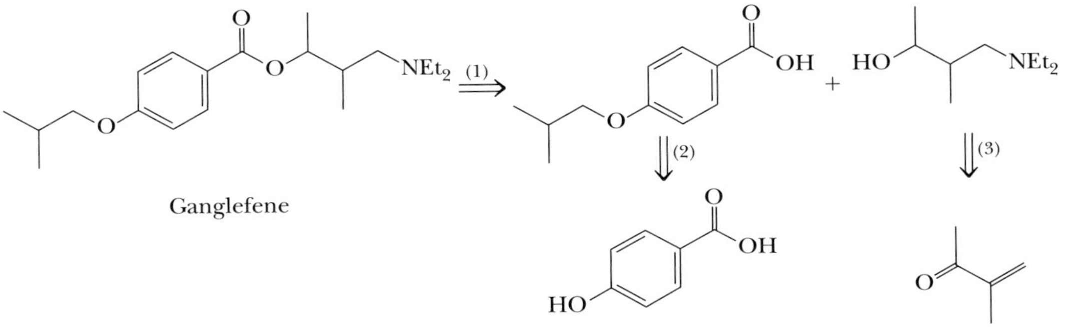

Ganglefene

4-Hydroxybenzoic acid   3-Methyl-3-buten-2-one

(a) Propose a synthesis for ganglefene from 4-hydroxybenzoic acid and 3-methyl-3-buten-2-one.

(b) Is ganglefene chiral? If so, which of the possible stereoisomers are formed in this synthesis?

23.59 Moxisylyte, an α-adrenergic blocker, is used as a peripheral vasodilator. Propose a synthesis for this compound from thymol, which occurs in the volatile oils of members of the thyme family. Thymol is made industrially from *m*-cresol.

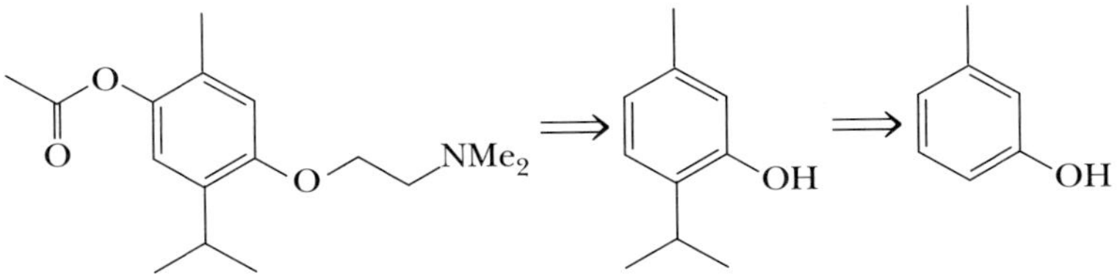

Moxisylyte   Thymol   *m*-Cresol

23.60 Propose a synthesis of the local anesthetic ambucaine from 4-nitrosalicylic acid, ethylene oxide, diethylamine, and 1-bromobutane.

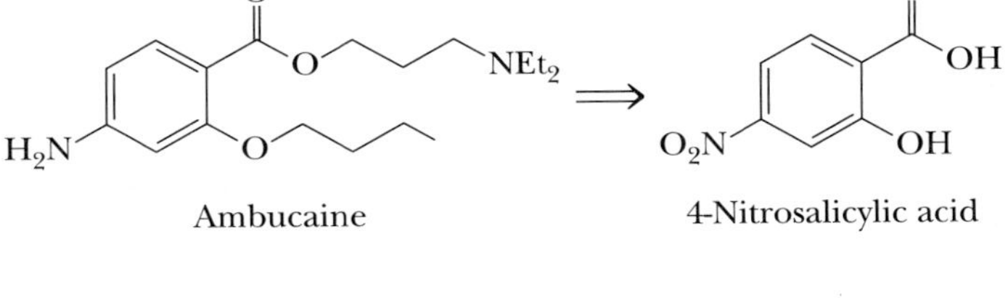

Ambucaine   4-Nitrosalicylic acid

23.61 Given this retrosynthetic analysis, propose a synthesis for the local anesthetic hexylcaine.

Hexylcaine   Benzoic acid   Propene   Cyclohexylamine
(racemic)

23.62 Following is an outline for a synthesis of the anorexic (appetite suppressant) fen-fluramine. This compound was one of the two ingredients in Phen-Fen, a weight-loss preparation now banned because of its potential to cause irreversible heart valve damage.

Fenfluramine

(a) Propose reagents and conditions for Step 1. Account for the fact that the CF_3 group is meta directing.
(b) Propose reagents and experimental conditions for Steps 2 and 3.
(c) An alternative procedure for preparing the amine of Step 3 is reductive amination of the corresponding ketone. What is reductive amination? Why might this two-step route for formation of the amine be preferred over the one-step reductive amination?
(d) Propose reagents for Steps 4 and 5.
(e) Is fenfluramine chiral? If so, which of the possible stereoisomers are formed in this synthesis?

23.63 Following is a series of anorexics (appetite suppressants). As you study their structures, you will surely be struck by the sets of characteristic structural features.

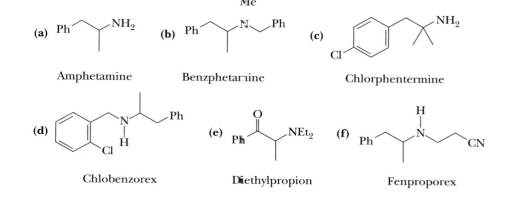

(a) Amphetamine   (b) Benzphetamine   (c) Chlorphentermine

(d) Chlobenzorex   (e) Diethylpropion   (f) Fenproporex

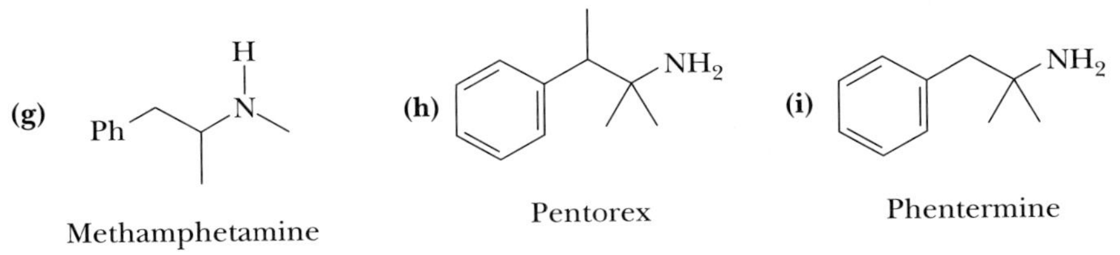

(g) Methamphetamine

(h) Pentorex

(i) Phentermine

(a) Knowing what you do about the synthesis of amines, including the Ritter reaction (Problem 23.52), suggest a synthesis for each compound.

(b) Which of these compounds are chiral?

23.64 The drug sildefanil, sold under the trade name Viagra, is a potent inhibitor of phosphodiesterase V (PDE V), an enzyme found in high levels in the corpus carvenosum of the penis. Inhibitors of this enzyme enhance vascular smooth muscle relaxation and are used for treatment of male impotence. Following is an outline for a synthesis of sildefanil.

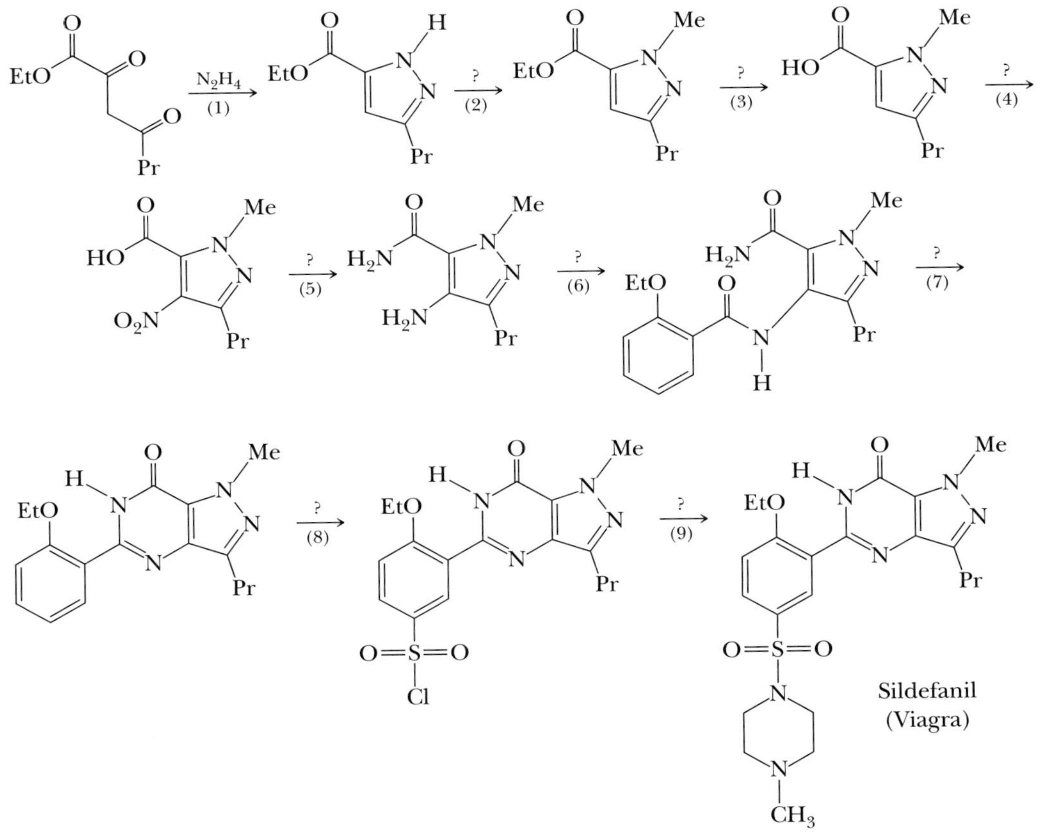

(a) Propose a mechanism for Step 1.

(b) The five-membered nitrogen-containing ring formed in Step 1 is named pyrazole. Show that, according to the Hückel criteria for aromaticity, pyrrazole can be classified as an aromatic compound.

(c) Propose a reagent or reagents for Steps 2–7 and 9.

(d) Show how the reagent for Step 6 can be prepared from salicylic acid (2-hydroxybenzoic acid). Salicylic acid, the starting material for the synthesis of aspirin and a number of other pharmaceuticals, is readily available by the Kolbe carboxylation of phenol (Section 21.4E).

(e) Chlorosulfonic acid, ClSO$_3$H, the reagent used in Step 8 is not described in the text. Given what you have studied about other types of electrophilic aromatic substitutions (Section 22.1), propose a mechanism for the reaction in Step 8.

(f) Propose a structural formula for the reagent used in Step 9 and show how it can be prepared from methylamine and ethylene oxide.

(g) Is sildefanil chiral? If so, which of the possible stereoisomers are formed in this synthesis?

23.65 Radiopaque imaging agents are substances administered either orally or intravenously that absorb x-rays more strongly than body material. One of the best known of these is barium sulfate, the key ingredient in the so-called barium cocktail for imaging of the gastrointestinal tract. Among other x-ray contrast media are the so-called triiodoaromatics. You can get some idea of the imaging for which they are used from the following selection of trade names: Angiografin, Gastrografin, Cardiografin, Cholegrafin, Renografin, and Urografin. Following is a synthesis for diatrizoic acid from benzoic acid.

Diatrizoic acid

(a) Provide reagents and experimental conditions for Steps (1), (2), (3), and (5).

(b) Iodine monochloride, ICl, a black crystalline solid with a mp of 27.2°C and a bp of 97°C, is prepared by mixing equimolar amounts of I$_2$ and Cl$_2$. Propose a mechanism for the iodination of 3-aminobenzoic acid by this reagent.

23.66 Show how the synthetic scheme developed in Problem 23.65 can be modified to synthesize this triiodobenzoic acid x-ray contrast agent.

Iodipamide

23.67 A diuretic is a compound that causes increased urination and thereby reduces fluid volume in the body. An important use of diuretics in clinical medicine is in the reduction of the fluid build up, particularly in the lungs, that is associated with congestive heart failure. It is also used as an antihypertensive; that is, to reduce blood pressure. Furosemide, an exceptionally potent diuretic, is prescribed under 30 or more trade names, the best known of which is Lasix. The synthesis of furosemide begins with treatment of 2,4-dichlorobenzoic acid with chlorosulfonic acid in a reaction called chlorosulfonation. The product of this reaction is then treated with ammonia, followed by heating with furfurylamine.

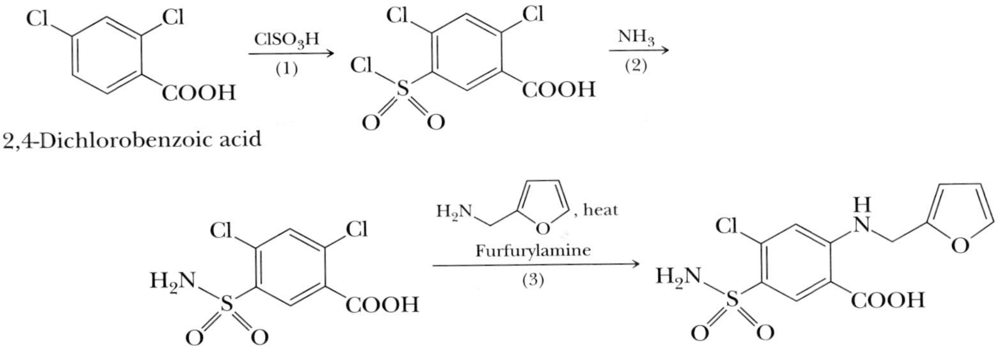

2,4-Dichlorobenzoic acid

Furosemide

(a) Propose a synthesis of 2,4-dichlorobenzoic acid from toluene.
(b) Propose a mechanism for the chlorosulfonation reaction in Step (1).
(c) Propose a mechanism for Step (3).
(d) Is furosamide chiral? If so, which of the possible stereoisomers are formed in this synthesis?

23.68 Among the newer generation diuretics is bumetanide, prescribed under several trade names, including Bumex and Fordiuran. Following is an outline of a synthesis of this drug.

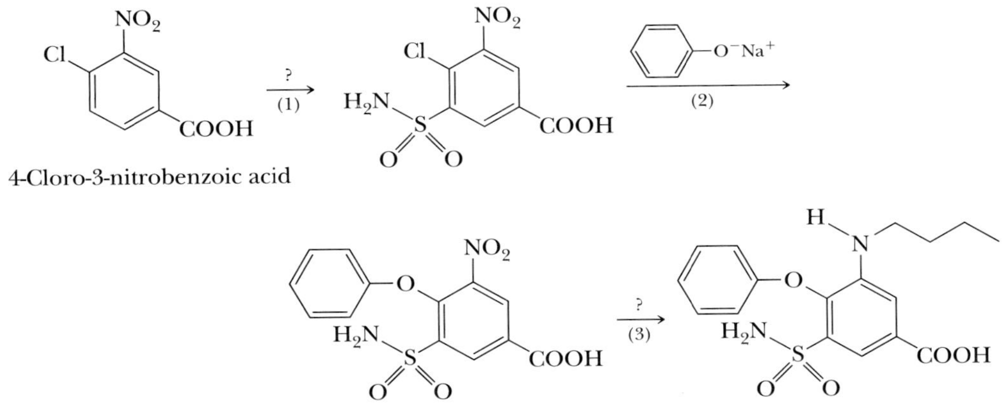

4-Cloro-3-nitrobenzoic acid

Bumetanide

(a) Propose a synthesis of 4-chloro-3-nitrobenzoic acid from toluene.
(b) Propose reagents for Step (1). (*Hint:* It requires more than one reagent.)
(c) Propose a mechanism for reaction (2).
(d) Propose reagents for Step (3). (*Hint:* It too requires more than one reagent.)
(e) Is bumetanide chiral? If so, which of the possible stereoisomers are formed in this synthesis?

23.69 Of the early antihistamines, most had a side effect of mild sedation; they made one sleepy. More recently, there has been introduced a new generation of nonsedating antihistamines known as histamine H_1 receptor antagonists. One of the most widely prescribed of these is fexofenadine (Allegra). This compound is nonsedating because the polarity of its carboxylic anion prevents it from crossing the blood-brain barrier. Following is a retrosynthetic analysis for the synthesis of fexofenadine.

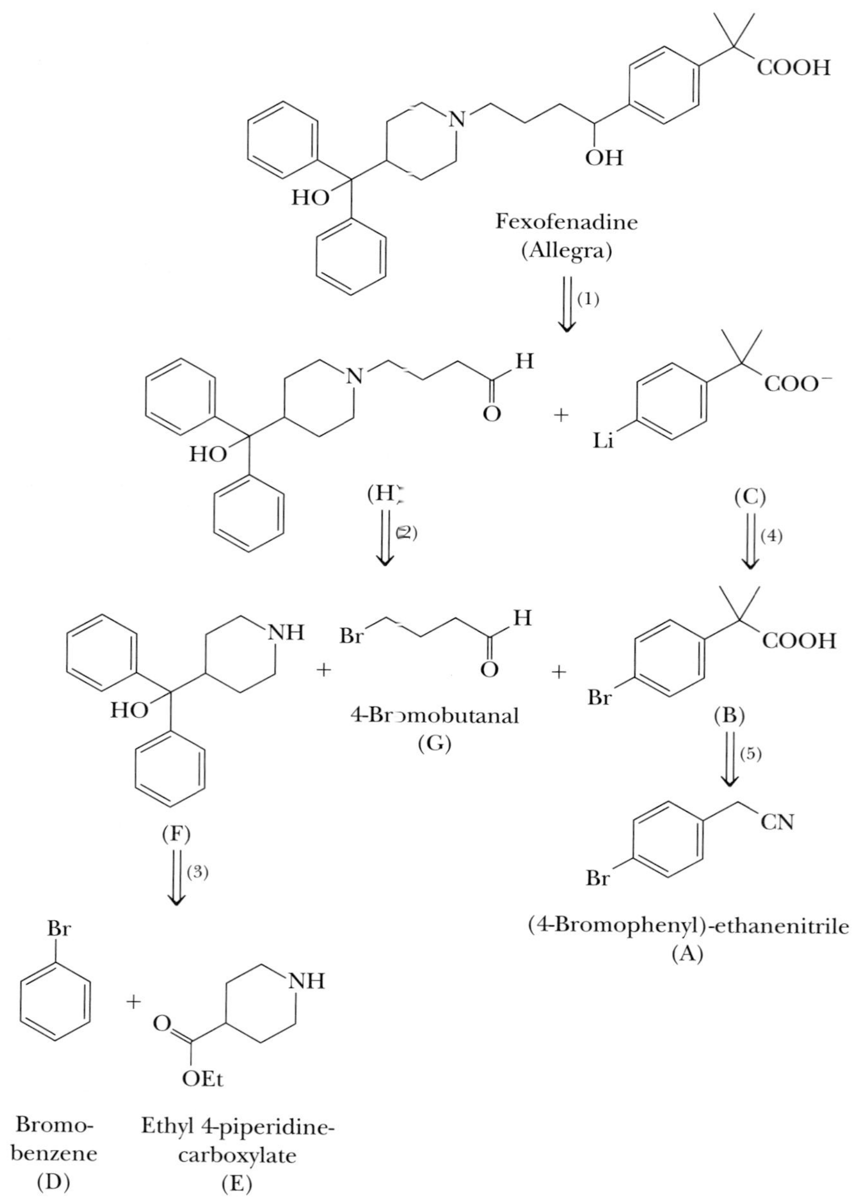

COOH
N
OH
HO
Fexofenadine
(Allegra)
(1)
N
H
O
+
COO⁻
Li
HO
(H)
(2)
(C)
(4)
NH
Br
H
+
O
+
COOH
HO
4-Bromobutanal
(G)
Br
(B)
(5)
CN
(F)
(3)
Br
(4-Bromophenyl)-ethanenitrile
(A)
Br
NH
+
O
OEt
Bromo-
benzene
(D)
Ethyl 4-piperidine-
carboxylate
(E)

(*Note:* The organolithium reagent C cannot be made directly from B because the presence of the carboxyl group in B would lead to intermolecular destruction of the reagent by an acid-base reaction. In practice, B is first converted to its sodium salt by treatment with sodium hydride, NaH, and then the organolithium reagent is prepared.)

(a) Given this retrosynthetic analysis, propose a synthesis for fexofenadine from the four named starting materials.

(b) Is fexofenadine chiral? If so, which of the possible stereoisomers are formed in this synthesis?

23.70 Sotalol is a β-adrenergic blocker used to treat certain types of cardiac arrhythmias. Its hydrochloride sale is marketed under several trade names, including Betapace. Following is a retrosynthetic analysis.

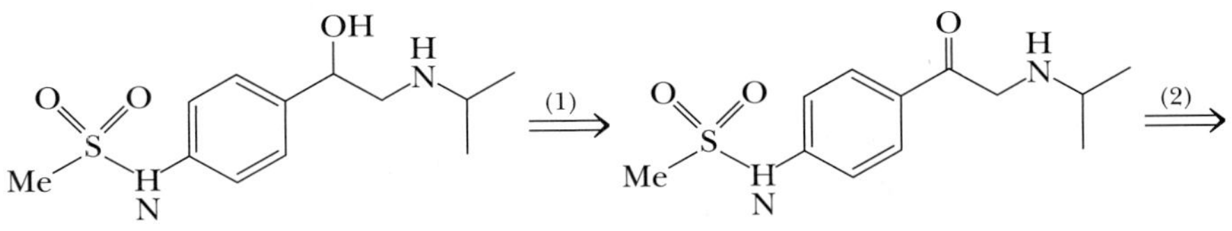

Sotalol

(A)

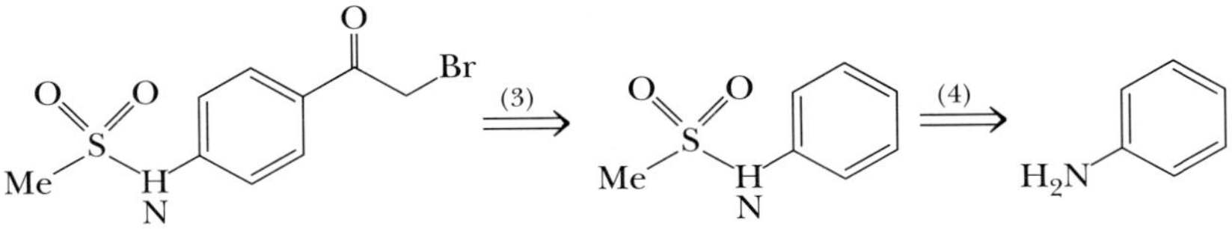

(B)

(C)

Aniline

(a) Propose a synthesis for sotalol from aniline.

(b) Is sotalol chiral? If so, which if the possible stereoisomers are formed in this synthesis?

24

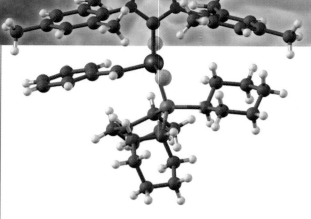

■ A ruthenium-containing organometallic catalyst for alkene metathesis reactions. See Section 24.5 for the structure of this catalyst and a discussion of this reaction.

ORGANIC
Chemistry••Now™

Look for this logo in the chapter and go to Organic ChemistryNow at **http://now.brookscole.com/bfi4** for tutorials, simulations, and problems.

Carbon-Carbon Bond Formation and Synthesis

Organic chemists have learned over the course of the last one hundred twenty or so years to synthesize amazingly complex molecules. In recent years, particularly, they have focused on compounds of medicinal interest, and a large number of current pharmaceuticals are synthesized from simpler compounds. Many of these pharmaceuticals are natural products or their analogs, and others are either simpler analogs or unrelated compounds that have been found to be active against certain organisms or diseased cells, specific cellular receptors, or specific enzyme targets. A key development that has allowed synthesis of these compounds has been the discovery of many novel methods of carbon-carbon bond formation. It is now possible, using a combination of new and classical methods to carry out synthesis of molecules with sensitive functionality and amazingly complex carbon skeletons from simple and inexpensive starting materials, often with excellent stereo- and regiocontrol. In this chapter, we make a dramatic leap from the more classical organic reactions covered in previous chapters of this book to survey several particularly useful methods of carbon-carbon bond formation, some of which represent very recent developments. We have room for only a few representative examples out of the wealth of carbon-carbon bond forming reactions that are now available to the modern synthetic chemist. Finally, a number of problems based on modern organic syntheses are given to illustrate the use of these reactions and their combination with other reactions.

24.1 Carbon-Carbon Bond-Forming Reactions from Earlier Chapters

Let us list the methods of carbon-carbon bond formation that you have already studied as a review. All these reactions should be available to you for synthetic problems.

Nucleophilic displacement of a leaving group by a carbon nucleophile

- Gilman (organocuprate) reagents (Section 15.2B). The leaving group may be the oxygen of an epoxide with Grignard reagents, organolithium reagents (Section 15.1C), and Gilman reagents (Section 15.2C).
- Alkyne (Sections 7.5 and 9.1) and cyanide (Sections 9.1 and 9.5) anions. The leaving group can be the oxygen of an epoxide (Section 11.9B).
- Enolate anion alkylations (acetoacetic ester and malonic ester synthesis; Sections 19.5 and 19.6).
- Enamine alkylations (Section 19.4A).

Nucleophilic addition to a carbonyl or a carboxyl group

- Grignard reagents (Sections 16.5A and 18.8A), organolithium reagents (Sections 16.5B and 18.8B), and Gilman reagents (Section 18.8C).
- Alkyne (Section 16.5C) and cyanide (Section 16.5D) anions.
- Aldol reactions (Section 19.1).
- Claisen (Section 19.2A) and Dieckmann (Section 19.2B) condensations.
- Enamine acylations (Section 19.4B).
- Wittig reaction (Section 16.6 for C=C double bonds).

Conjugate addition to α,β-unsaturated carbonyl compounds

- Michael reaction (Section 19.7A)

Carbene/carbenoid additions (Section 15.3)

Aromatic substitution

- Friedel-Crafts alkylation and acylation of aromatics (Section 22.1C)
- Reaction of cyanide with aromatic diazonium compounds (Section 23.8E)

This list already seems to include many different reactions, but you will find that the new reactions in this chapter add greatly to the power and synthetic functionality and are of somewhat different character.

24.2 Organometallic Compounds

We introduced organometallic compounds in Chapter 15. In the next few sections, we discuss several reactions of transition metals that are particularly useful for preparation of new carbon-carbon bonds.

Oxidative addition   Addition of a reagent to a metal center causing it to add two substituents and to increase its oxidation state by two.

Reductive elimination   Elimination of two substituents at a metal center, causing the oxidation state of the metal to decrease by two.

Ligand   A Lewis base bonded to a metal atom in a coordination compound. It may bond strongly or weakly.

Two extremely important reactions of metals and metal compounds are **oxidative addition** and its complement, **reductive elimination.** In oxidative addition, a reagent adds to a metal, causing its coordination to increase by two; reductive elimination is the reverse. These reactions are called oxidative or reductive because the formal charge of the metal changes by two during the reaction. Oxidative addition can occur with a metal coordinated with one or more **ligands** (**L**$_n$, where n is the number); it can also occur with a free metal, M(0). Alkyl halides, hydrogen, halogens, and many other types of compounds can take part in these reactions. The reactivity of different substrates depends greatly on the metal.

$$ML_n + X_2 \; \underset{\substack{\text{reductive} \\ \text{elimination}}}{\overset{\substack{\text{oxidative} \\ \text{addition}}}{\rightleftarrows}} \; \begin{matrix} X \\ \diagdown \\ \diagup \\ X \end{matrix} ML_n$$

These sections focus on transformations that illustrate the power of modern organometallic reagents to carry out reactions that cannot be accomplished in other ways. The interface between organic and inorganic chemistry and the development of new organometallic reagents and catalysts is one of the most exciting areas of chemical research and development today.

24.3 Organopalladium Reagents— The Heck Reaction

A. The Nature of the Reaction

In the early 1970s, Richard Heck, at the Hercules Company and later at the University of Delaware, discovered a palladium-catalyzed reaction in which the carbon group of a haloalkene or haloarene is substituted for a hydrogen on the carbon-carbon double bond (a vinylic hydrogen) of an alkene. This reaction, now known as the **Heck reaction,** is particularly valuable in synthetic organic chemistry because it is the only general method yet discovered for this type of substitution.

| Haloalkene or haloarene | Alkene | Base | Substituted alkene | Conjugate acid of the base |

Substitution for a vinylic hydrogen by the Heck reaction is highly regioselective; formation of the new carbon-carbon bond most commonly occurs at the less substituted carbon of the double bond. In addition, where an E or Z configuration is possible at the new carbon-carbon double bond of the product, the Heck reaction is highly stereoselective, often giving almost exclusively the E configuration of the product.

Bromobenzene   Methyl 2-propenoate (Methyl acrylate)   Methyl (*E*)-3-phenyl-2-propenoate (Methyl cinnamate)

In addition, the Heck reaction is completely stereospecific with regard to the haloalkene; the configuration of the double bond in the haloalkene is preserved.

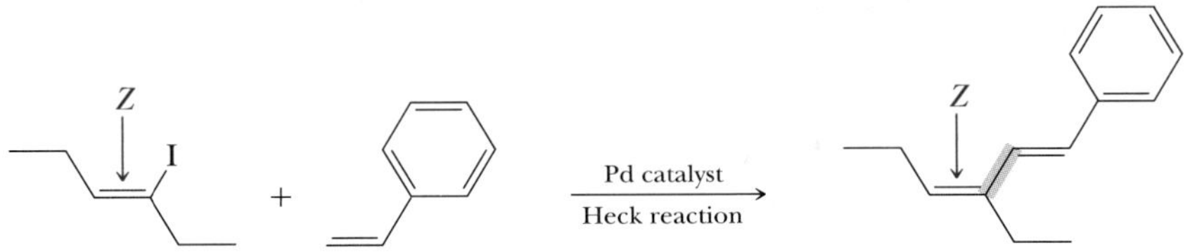

(Z)-3-Iodo-3-hexene   Phenylethene   (1E,3Z)-3-Ethyl-1-phenyl-1,3-hexadiene
                        (Styrene)

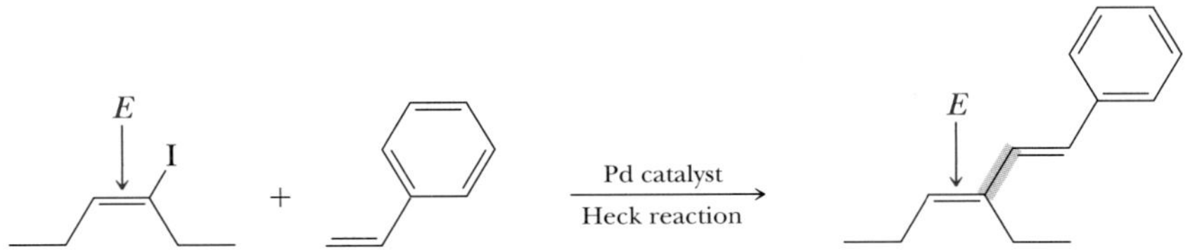

(E)-3-Iodo-3-hexene   Phenylethene   (1E,3E)-3-Ethyl-1-phenyl-1,3-hexadiene
                        (Styrene)

Preparation of the Catalyst

The form of the palladium catalyst most commonly added to the reaction medium is palladium(II) acetate, $Pd(OAc)_2$. This and other Pd(II) compounds are better termed precatalysts because the catalytically active form of the metal is a complex of Pd(0) formed *in situ* by reduction of Pd(II) to Pd(0).

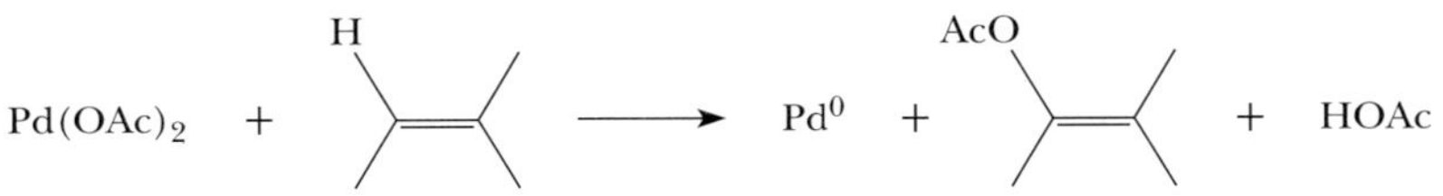

Palladium(II)   An alkene   Oxidized
   acetate   (reductant)   alkene

Reaction of Pd(0) with good ligands, L, gives the actual Heck catalyst, PdL_2. Without the ligand, Pd(0) is insoluble.

$$Pd^0 \quad + \quad 2\,L \quad \longrightarrow \quad PdL_2$$

Ligand   The Heck
                  catalyst

The Organic Halogen Compound

The most common halides used in Heck reactions are aryl, heterocyclic, benzylic, and vinylic iodides and bromides, with iodides being generally most reactive. The reactivity of substrates with leaving groups on sp^2 carbons contrasts with nucleophilic substitution reactions, where such substrates are essentially unreactive. Alkyl halides in which there is an acidic beta hydrogen are rarely used because of the ease with which they undergo β-elimination under conditions of the Heck reaction to form alkenes. Triflates (trifluoromethanesulfonates, CF_3SO_2O-), which are easily prepared by treating an alcohol with trifluoromethanesulfonyl chloride, are also excellent substrates.

$$
\underset{\substack{\text{Trifluoromethanesulfonyl chloride}}}{\text{CF}_3\overset{\overset{\displaystyle O}{\|}}{\underset{\underset{\displaystyle O}{\|}}{S}}\text{—Cl}} \quad + \quad \underset{\text{Alcohol}}{\text{HO—R}} \quad \longrightarrow \quad \underset{\substack{\text{A trifluoromethanesulfonate}\\\text{(a triflate)}}}{\text{CF}_3\overset{\overset{\displaystyle O}{\|}}{\underset{\underset{\displaystyle O}{\|}}{S}}\text{—OR}} \quad + \quad \text{HCl}
$$

The halide or triflate (RX) reacts with PdL_2 by oxidative addition to give a square planar Pd(II) species, which is the reaction intermediate.

$$
\underset{\substack{\text{The Heck}\\\text{catalyst}}}{PdL_2} \quad + \quad RX \quad \longrightarrow \quad \overset{\displaystyle R}{\underset{\displaystyle L}{\overset{\displaystyle |}{\underset{\displaystyle |}{L\text{—}Pd\text{—}X}}}}
$$

A particular advantage of the Heck reaction is the wide range of functional groups, including alcohols, ethers, aldehydes, ketones, and esters, that may be present elsewhere in the organic halogen compound or alkene without reacting themselves or affecting the Heck reaction.

The Alkene

The reactivity of the alkene is a function of steric crowding about the carbon-carbon double bond. Ethylene and monosubstituted alkenes are the most reactive; the greater the degree of substitution on the double bond is, the slower the reaction and the lower the yield of product will be. These steric effects also control the regiochemistry of the addition, with the alkyl group adding to the less hindered carbon of the alkene.

The Base

Commonly used bases are tertiary amines such as triethylamine, Et_3N, sodium or potassium acetate, and sodium hydrogen carbonate.

The Solvent

Polar aprotic solvents (Section 9.2) such as N,N-dimethylformamide (DMF), acetonitrile, and dimethyl sulfoxide (DMSO) are commonly used. It is also possible to carry out some Heck reactions in aqueous methanol. The polar solvents are needed to dissolve the $Pd(OAc)_2$ at the beginning of the reaction.

The Ligands Coordinating with Pd(0)

Among the most common ligands, L, used for coordination of the Pd(0) is triphenylphosphine, $(C_6H_5)_3P$. Many other ligands can be used as well, including chiral ones such as BINAP (Section 6.7) that can lead to a significant excess of a single enantiomer in the case of chiral products.

B. Mechanism of the Reaction

The mechanism of the Heck reaction is divided into two stages: formation of the Heck catalyst and the catalytic cycle. As you study the catalytic cycle, note in particular that both Steps 2 and 4 are syn stereoselective; the reaction will not proceed if these syn relationships cannot be obtained. Step 2 involves syn addition of R and PdL_2X to the double bond. Step 4 involves syn elimination of H and the Pd(II) species to generate a new double bond. These syn additions and eliminations contrast with most of the addition and elimination reactions we have seen, which prefer the anti geometry. Additions of boron hydrides (Section 6.4, hydroboration) and osmium tetroxide (Section 6.5A) or ozone (Section 6.5B) to alkenes are some examples of syn additions that you have already seen.

Mechanism   *The Heck Reaction*

Stage 1: Formation of the Heck Catalyst, PdL$_2$

A two-electron reduction of Pd(II) to Pd(0) accompanied by its complex formation with two molecules of a ligand, L, gives the Heck catalyst, PdL$_2$. A common reducing agent is triethylamine or, as in the following example, the alkene itself. Because the catalyst is present only in small amounts, an insignificant amount of the alkene is lost to this reaction. In the reaction shown here, L is triphenylphosphine, $(C_6H_5)_3P$. As mentioned previously, this is actually a two-step reaction: reduction of the palladium, followed by reaction of the palladium with the ligand. We show the two steps combined here for simplicity.

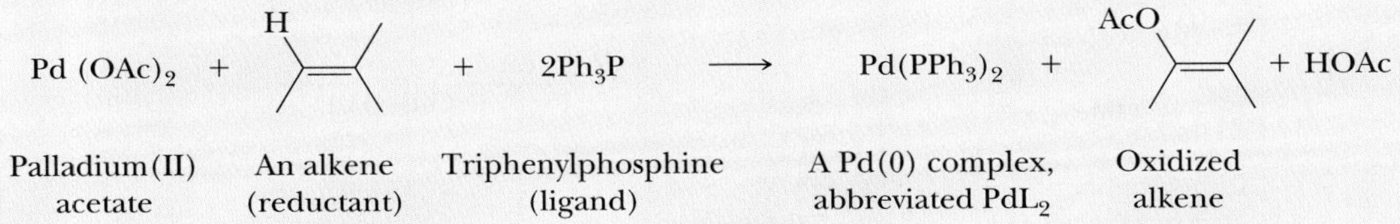

| Palladium(II) acetate | An alkene (reductant) | Triphenylphosphine (ligand) | A Pd(0) complex, abbreviated PdL$_2$ | Oxidized alkene |

Stage 2: The Catalytic Cycle

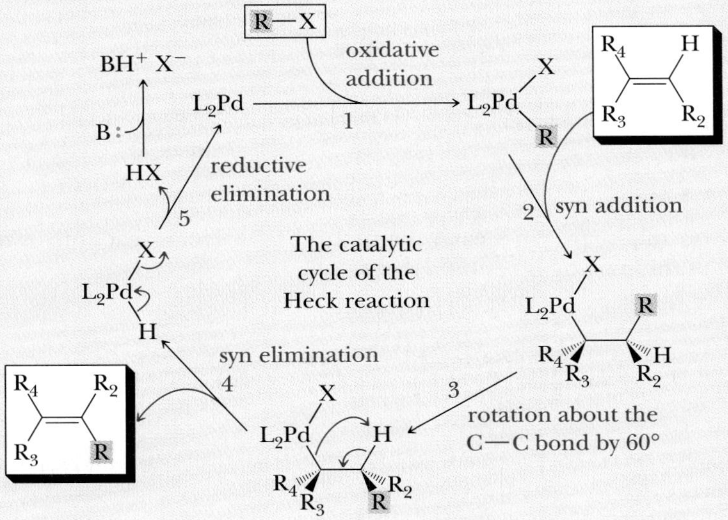

The catalytic cycle of the Heck reaction involves five steps. In Step 1, oxidative addition of the haloalkene or haloarene, RX, to PdL$_2$ gives a tetracoordinated Pd(0) complex containing both R and X groups bonded to Pd. Syn addition of the R and PdL$_2$X of this complex to the alkene gives an intermediate in which Pd is bonded to the more substituted carbon for steric reasons. (Because of the long Pd-C bond, the palladium is sterically less demanding than the organic group, so ends up on the more hindered carbon.) This intermediate must undergo internal rotation about the central carbon-carbon single bond in Step 3 to place H and PdL$_2$X syn to each other. Syn elimination of H and PdL_2X in Step 4 gives the new alkene and $HPdL_2X$. Reductive elimination in Step 5 releases the acid HX and regenerates the PdL_2 catalyst. HX is then neutralized by the added base.

In this cycle, the alkene, organohalogen compound, and base are required in equimolar amounts; the Pd(0) species is required in only a catalytic amount. Note also the inversion of the configuration (R_2 and R_3 are originally *cis* to each other but in the product are *trans*). This inversion is a consequence of the consecutive syn addition and elimination steps. The complete mechanism for this reaction has additional intermediates (involving pi complexes of the alkene with the palladium), but those shown here are the important ones for understanding the reaction and its stereochemistry.

Example 24.1

Complete these Heck reactions.

(a)

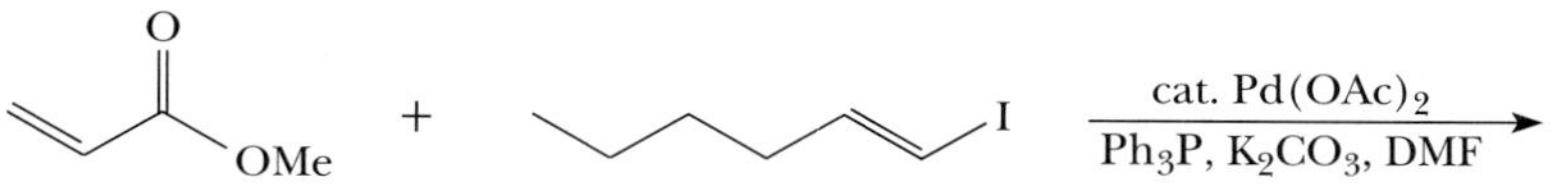

(b)

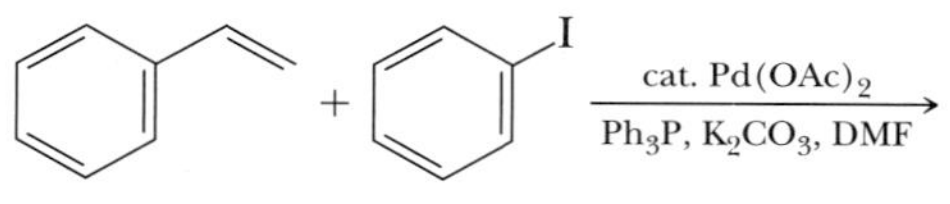

Solution

In (a), 1-iodohexene has the *E* configuration, and this double bond retains its configuration in the product. Furthermore, the carbon-carbon double bond adjacent to the ester in the product now has the possibility for *cis,trans* isomerism. The Heck reaction is highly stereoselective, and this double bond has the more stable *E* configuration as well. In (b), the major product is (*E*)-1,2-diphenylethene.

(a)   Methyl (2*E*,4*E*)-2,4-nonadienoate

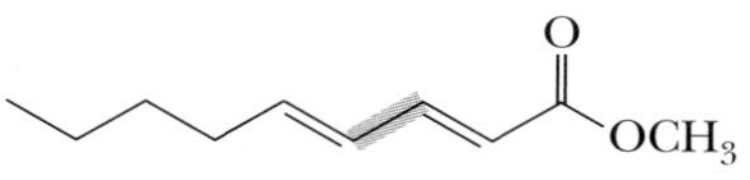

(b)   (*E*)-1,2-Diphenylethene
(*trans*-Stilbene)

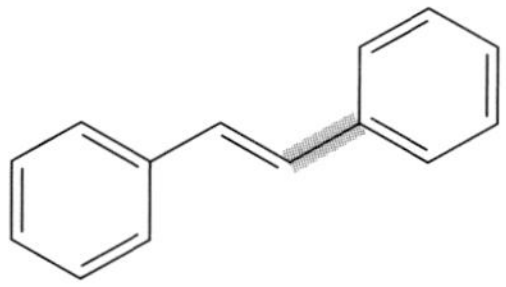

Problem 24.1

Show how you might prepare each compound by a Heck reaction using methyl 2-propenoate as the starting alkene.

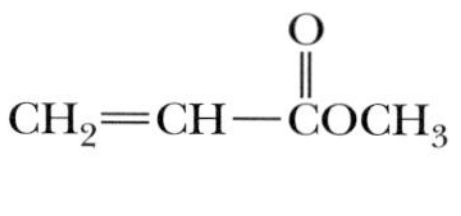

$CH_2{=}CH{-}\overset{\overset{\displaystyle O}{\|}}{C}OCH_3$

Methyl 2-propenoate
(Methyl acrylate)

(a) (the *E* isomer)

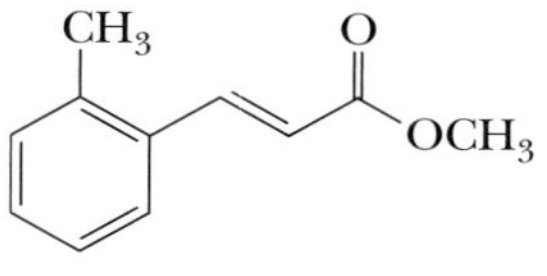

(b) (the 2*E*,4*Z* isomer)

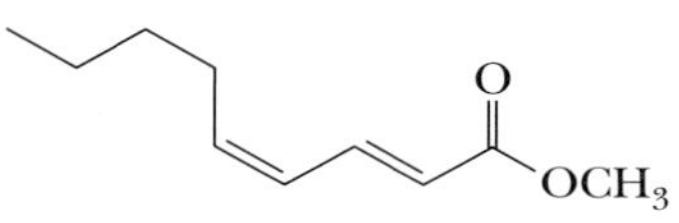

The usual pattern in a Heck reaction of acyclic alkenes is replacement of one of the hydrogens on the double bond by an organo group. If the organopalladium group attacks the double bond so that the R group in the original RX is bonded to a carbon that lacks a hydrogen, or if the only syn hydrogen is on a neighboring carbon, the double bond shifts away from the original position. Note that the product of the following reaction contains a chiral center, but, because it is formed from achiral reagents in an achiral environment, it is formed as a racemic mixture.

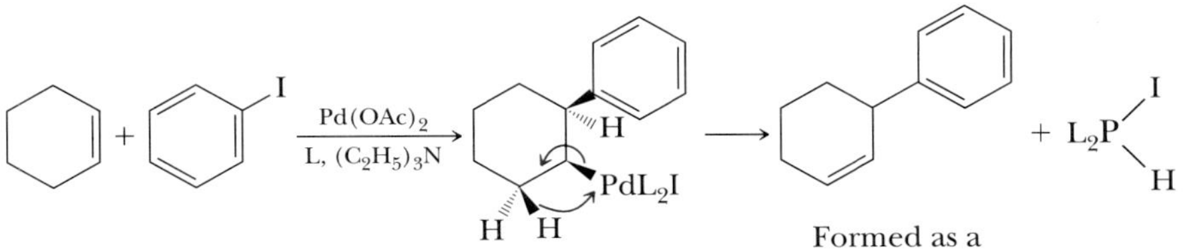

Formed as a
racemic mixture

As mentioned earlier, a particularly valuable feature of the Heck reaction is that, when used with a chiral ligand, it can give chiral products in significant enantiomeric excess (ee). In the following, the chirality is provided by the chiral ligand (*R*)-BINAP (Section 6.7C).

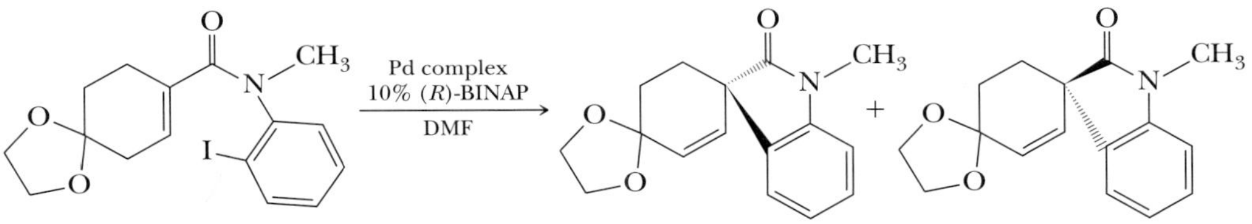

This enantiomer
is formed in 71% ee

For this reaction to yield a chiral product, the hydrogen eliminated cannot be on the carbon on which the aryl substituent ends up because, if this were the case, the substituent would be attached to a double bond, and the product would be achiral.

Because of the chiral ligand, the activation energy for the transition state in the syn addition to the alkene (Step 2 of the catalytic cycle) is different depending on which side of the alkene the metal complex approaches (the two transition states are diastereomers). This difference in activation energy means that approach to one side of the alkene is favored and results in an excess of one enantiomer of the product. Note that this reaction is not a normal Heck reaction in that it forms a carbon-carbon bond to the more substituted carbon, and the double bond shifts. Attack at the other carbon, because of the requirement for syn elimination, cannot lead to a normal Heck product. The reason the attack takes place at the more substituted carbon in this case is beyond the scope of this course.

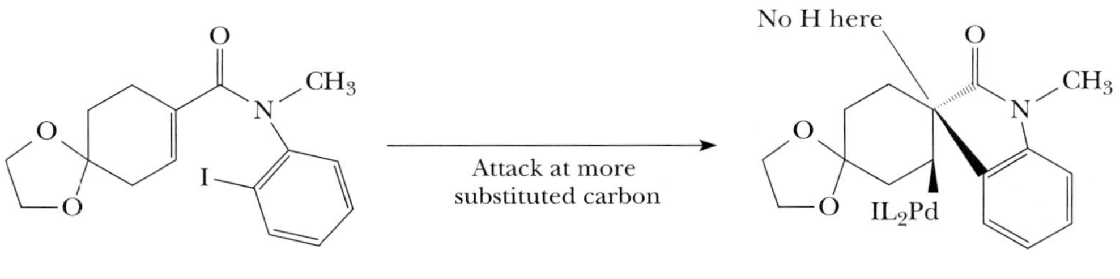

No H here

Attack at more
substituted carbon

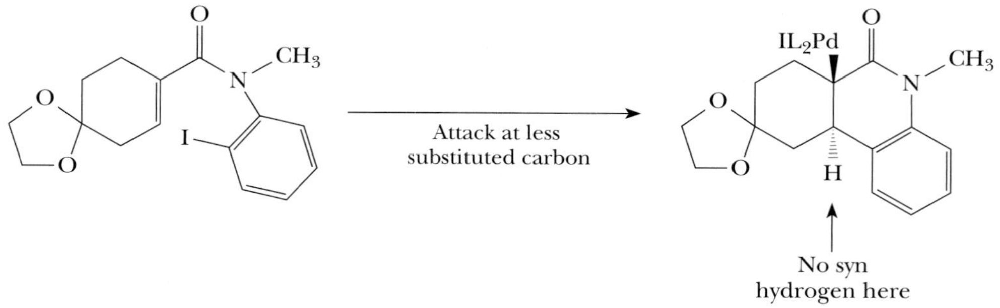

Attack at less substituted carbon

No syn hydrogen here

Example 24.2

Heck reaction of bromobenzene and (*E*)-3-hexene gives a mixture of (*Z*)-3-phenyl-3-hexene and (*E*)-4-phenyl-2-hexene in roughly equal amounts. Account for the formation of these two products.

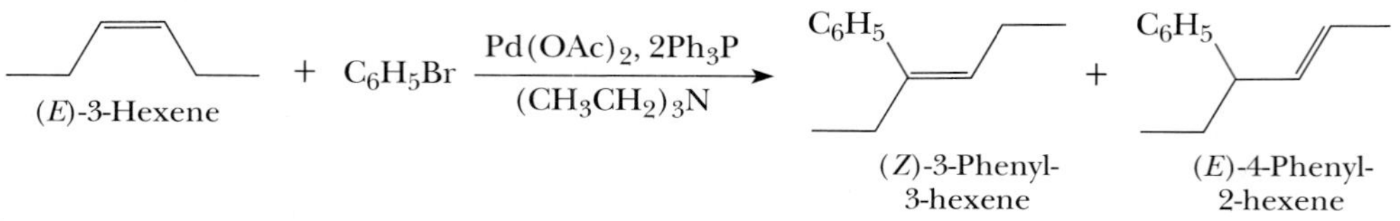

(*E*)-3-Hexene

$\xrightarrow[\text{(CH}_3\text{CH}_2)_3\text{N}]{\text{Pd(OAc)}_2, 2\text{Ph}_3\text{P}}$

(*Z*)-3-Phenyl-3-hexene

(*E*)-4-Phenyl-2-hexene

Solution

Syn addition gives the product shown. After rotation, syn elimination of the H on the original double bond gives (*Z*)-3-phenyl-3-hexene; syn elimination on the neighboring carbon (in its most stable conformation) gives (*E*)-4-phenyl-2-hexene.

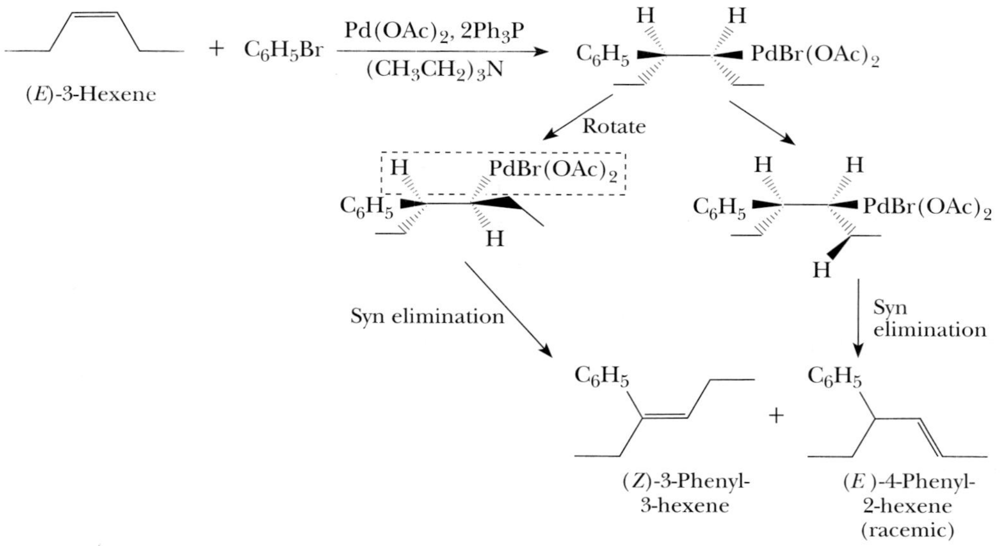

(*E*)-3-Hexene

Rotate

Syn elimination

Syn elimination

(*Z*)-3-Phenyl-3-hexene

(*E*)-4-Phenyl-2-hexene (racemic)

Problem 24.2

Give reagents and conditions for the following reaction.

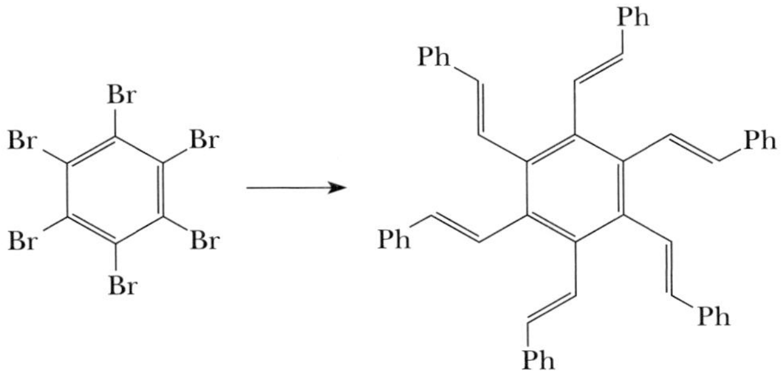

24.4 Organopalladium Reagents— The Suzuki Coupling

A. Characteristics of the Reaction

A second and very versatile method of forming C—C bonds using palladium catalysis is called the **Suzuki coupling,** developed by Professor Akira Suzuki of Hokkaido University. The Suzuki coupling uses a boron compound ($R'-BY_2$) and an alkenyl, aryl, or alkynyl halide or triflate (RX) as the carbon sources, with a palladium salt as the catalyst. Bromides and iodides are the most commonly used halides; chlorides are less reactive. Alkyl halides can sometimes be used but are subject to elimination. A base is also required. The boron compound can be a borane (R'_3B), a borate ester ($R'B(OR)_2$), or a boric acid ($R'B(OH)_2$), where R' is alkyl, alkenyl, or aryl. The general reaction is shown in the following scheme, where X is halide or triflate and Y is alkyl, alkoxyl, or OH. A list of the types of components that can be used is given in Table 24.1. This reaction is one of the principal methods now used to prepare biaryls.

$$RX + R'\text{-}BY_2 \xrightarrow[\text{Base}]{PdL_4} R\text{-}R' + XBY_2$$

Boranes are easily prepared from alkenes or alkynes by hydroboration (Section 6.4); borates are made from aryl or alkyl lithium compounds and trimethyl borate, among other routes.

Table 24.1 Suzuki Coupling Components Where One of the Organoboron Compounds Couples with One of the Coupling Reagents Shown

Organoboron Compounds	Coupling reagents X = halide or triflate
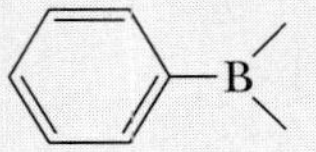	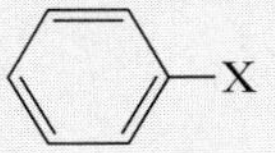
RCH=CH—B⟨	RCH=CH—X
Alkyl—B⟨	RC≡CH—X
	Alkyl-X (Difficult)

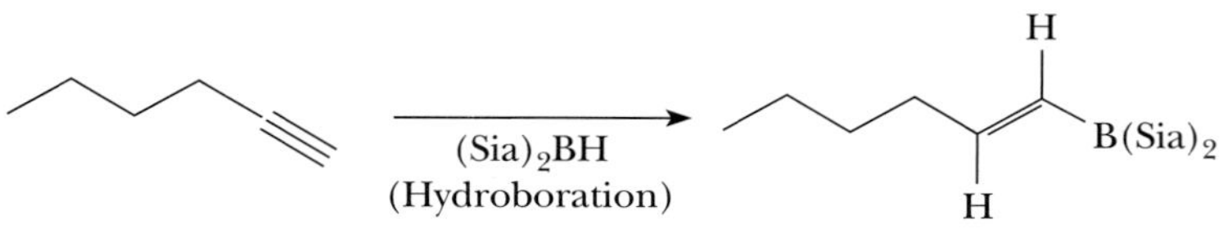

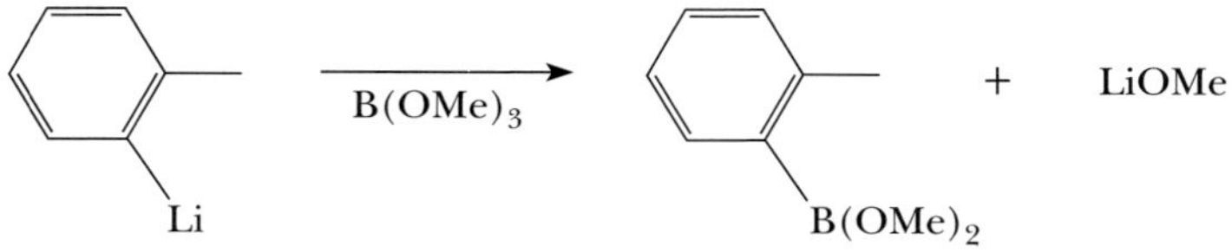

Following are three examples of the reaction that show its versatility.

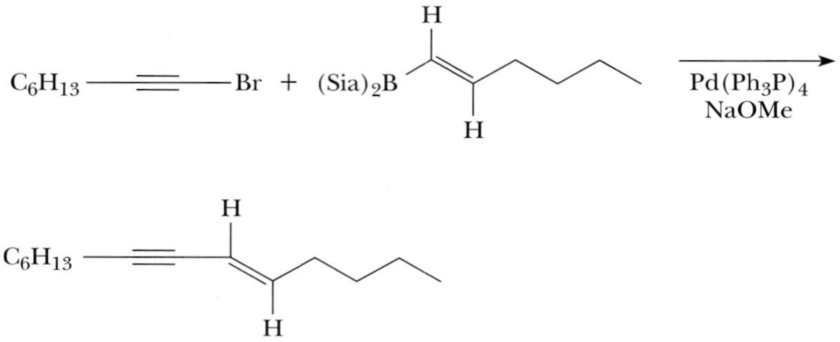

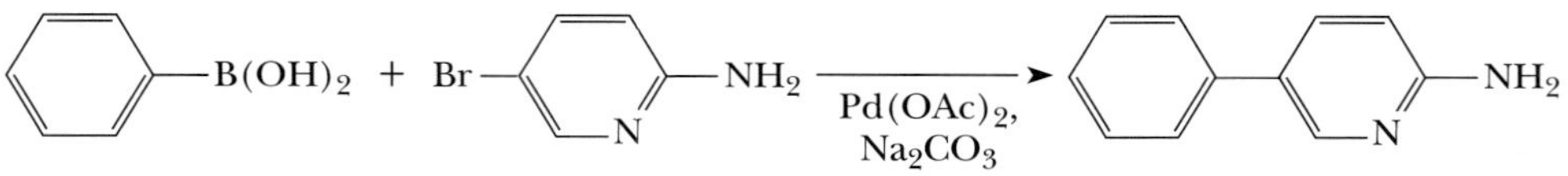

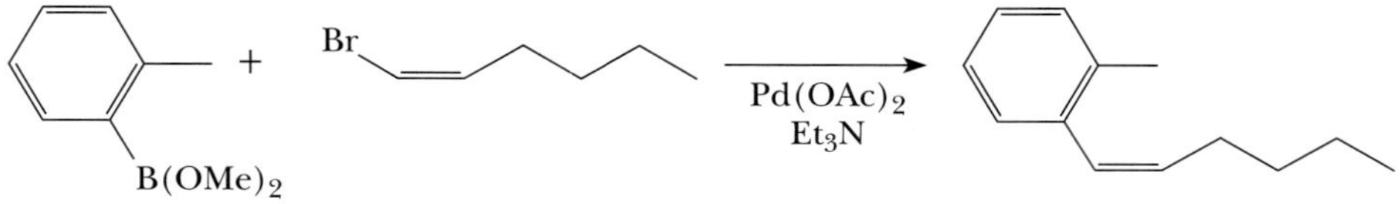

B. Mechanism

The mechanism of the reaction involves a **transmetallation,** in which the substituent on the borane replaces the ligand on the palladium, followed by a reductive elimination of the palladium to form the new C—C bond. The base may serve as a new, labile ligand for the palladium, or when the reagent is a borane, base may activate it by coordination.

Ligand exchange:

$$R—X \xrightarrow{\text{PdL}_n} R—Pd—X \xrightarrow{\text{RO}^-} R—Pd—OR$$

Borane activation:

$$R_3B \xrightarrow{\text{RO}^-} R_3B^-—OR$$

Reaction:

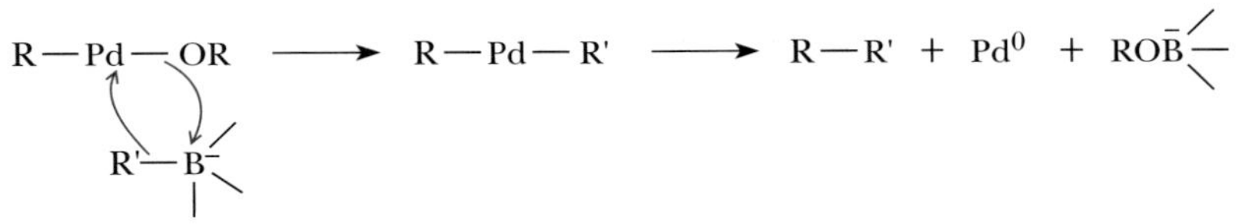

Example 24.3

Show how the following penicillin analog can be prepared from the indicated starting material and any other necessary compounds.

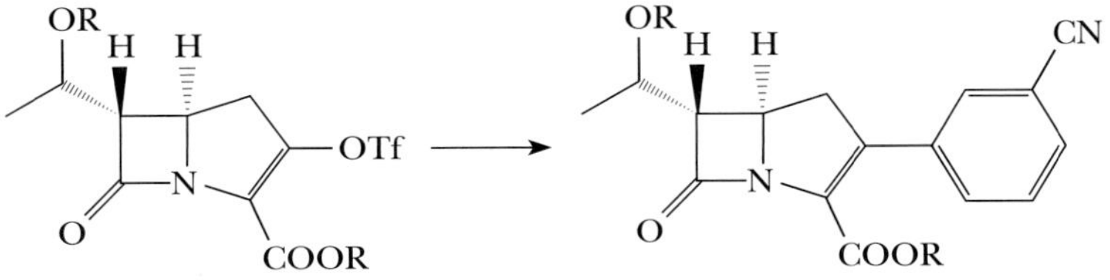

Solution

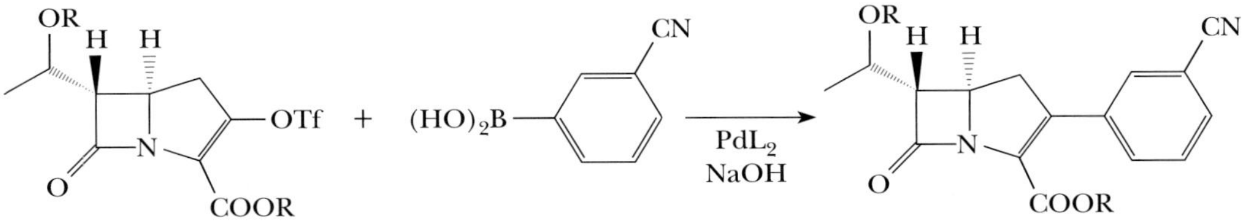

Problem 24.3

Show how the following compound can be prepared from starting materials containing eight carbons or less.

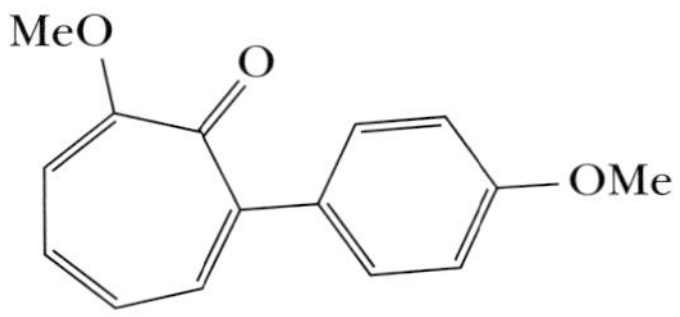

24.5 Alkene Metathesis

Recently, a novel catalytic reaction leading to alkene metathesis has been developed. Robert Grubbs of the California Institute of Technology and Richard Schrock of the Massachusetts Institute of Technology made major contributions to this chemistry. Together their work has provided a remarkably easy and general way to generate carbon-carbon double bonds, even in complex molecules. In an **alkene metathesis** reaction, two alkenes interchange the carbons attached to their double bonds.

Alkene metathesis   In an alkene metathesis reaction, two alkenes interchange the carbons attached to their double bonds.

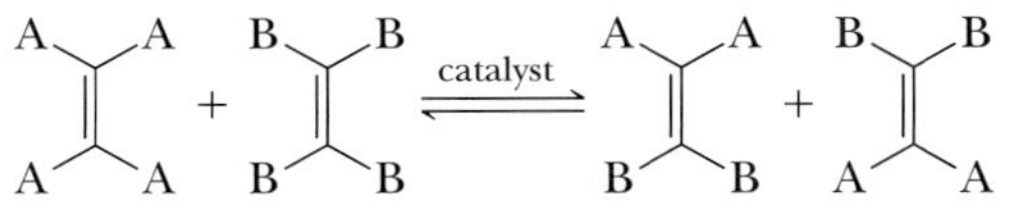

A. Stable Nucleophilic Carbenes

We discussed carbenes and carbenoids (derivatives of divalent carbon) in Section 15.3, where we saw that these compounds provide one of the best routes to three-membered rings, making two C—C bonds in the process. Certain carbenes with strongly electron-donating substituents are particularly stable. Their stability can be enhanced further by adding sterically bulky substituents that hinder self-reactions. For example, the following cyclic carbene is stable enough to isolate. In this case, the large 2,4,6-trimethylphenyl substituents protect the carbene from attack by nucleophiles or oxygen. Rather than being electron-deficient like most carbenes, these compounds are nucleophiles because of the strong electron donation by the nitrogens. Because of their nucleophilicity, they are excellent ligands (resembling phosphines) for certain transition metals.

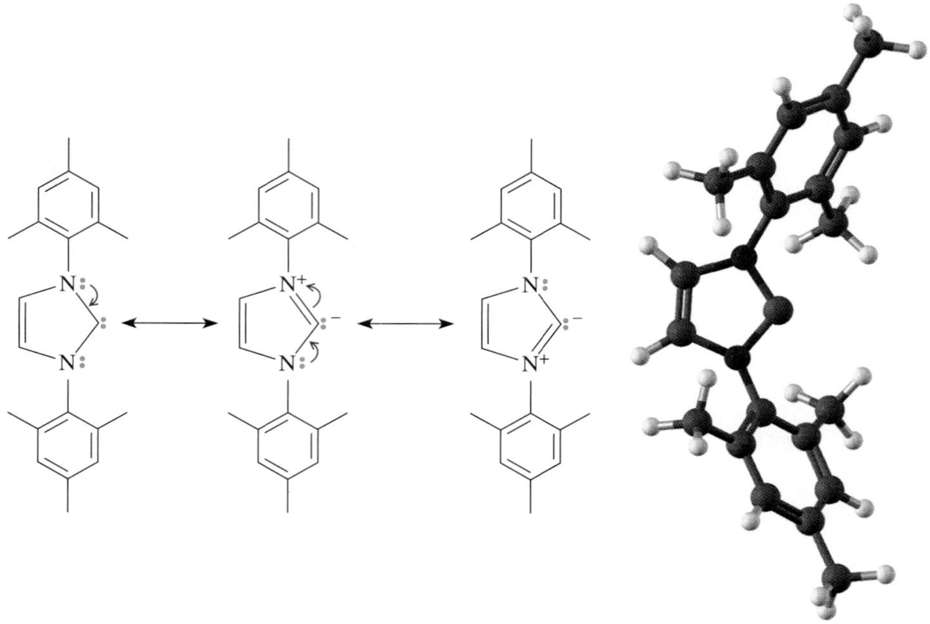

B. Ring-Closing Alkene Metathesis Using Nucleophilic Carbene Catalysts

These stable carbenes (and others that are less stable) provide ligands for certain metals that are catalysts for the alkene metathesis reaction. As we saw at the beginning of this section, this reaction is an equilibrium. However, it can be an effective means of forming new carbon-carbon double bonds if the equilibrium can be driven in the desired direction. For example, if the reaction involves two 2,2-disubstituted alkenes of the type $R_2C=CH_2$, one of the products is ethylene. Loss of gaseous ethylene drives the reaction to the right, giving a single alkene as product.

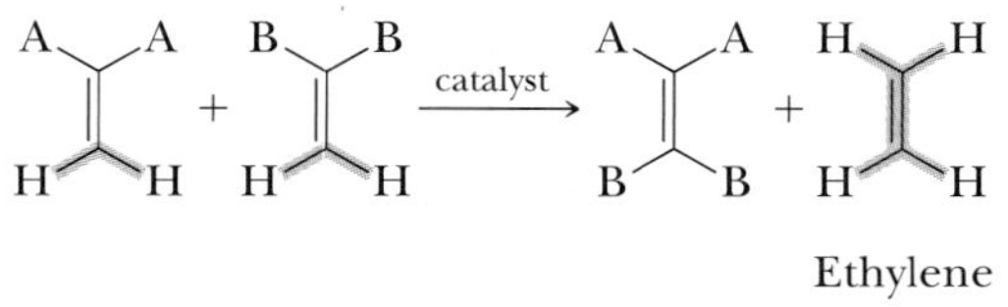

Ethylene

A particularly useful variant of this reaction uses a starting material in which both alkenes are in the same molecule. In this case, the product is a cycloalkene, and the reaction is called ring-closing alkene metathesis. Ring sizes up to 26 and higher have been prepared by ring-closing alkene metathesis. This reaction is amazingly general and synthetically useful.

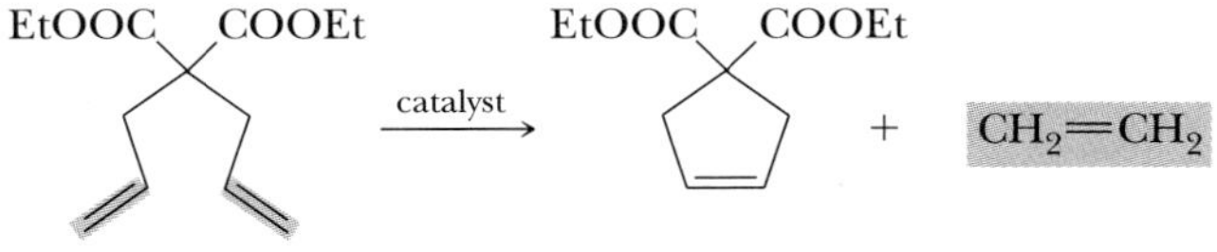

Example 24.4

Show how the following compound can be prepared from an acyclic diene.

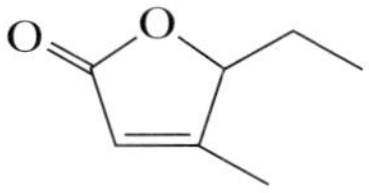

Solution

Ring-closing alkene metathesis gives the product in one step.

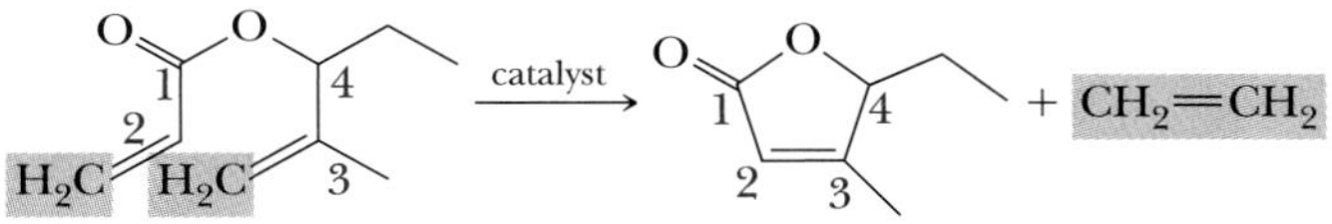

Problem 24.4

Show the product of the following reaction.

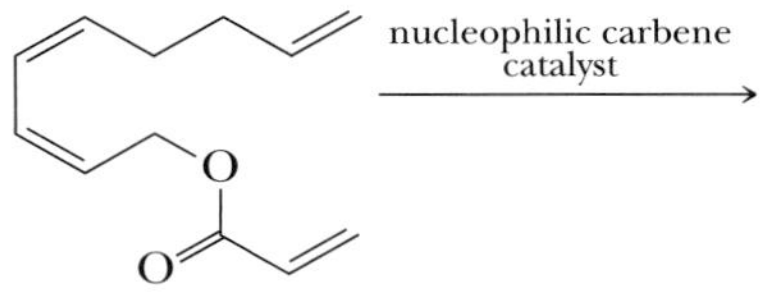

A particularly useful alkene metathesis catalyst consists of ruthenium complexes with a nucleophilic carbene and another carbenoid ligand, $C_6H_5CH=$. For a model of the catalyst shown here, see the opening page of this chapter.

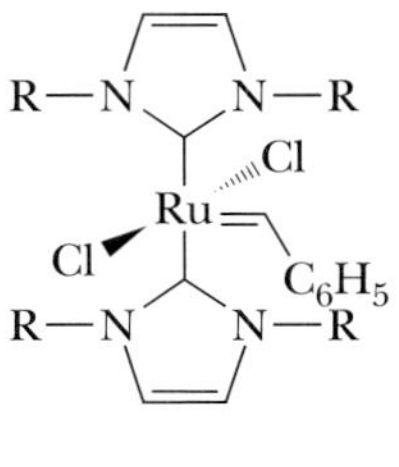

C. Mechanism of the Metathesis Reaction

Like the Heck reaction, the mechanism of the alkene metathesis reaction also involves a catalytic cycle. A key step involves addition of the metallocarbenoid to the alkene to give a four-membered metallacycle. This metallacycle is unstable and can either revert to starting material or eliminate an alkene in the opposite direction to give a new alkene. Addition is not regioselective; consequently, all possible combinations of R_1 and R_2 result. In this scheme, the catalyst is R_1CH=[M], where [M] is the metal with its ligands.

cis or trans   A metallacycle

In this section, we have concentrated on the use of transition-metal nucleophilic-carbene catalysts to bring about ring-closing alkene metathesis reactions. These same types of compounds can also be used to catalyze a remarkable reaction called ring-opening alkene metathesis polymerization (ROMP). A special value of ROMP is that it can be used to prepare highly unsaturated polymers. For a discussion of the ROMP techniques, see Section 29.6E.

24.6 The Diels-Alder Reaction

In 1928 Otto Diels and Kurt Alder in Germany discovered a unique reaction of conjugated dienes: They undergo cycloaddition reactions with certain types of carbon-carbon double and triple bonds. For their discovery and subsequent studies of this reaction, Diels and Alder were jointly awarded the 1950 Nobel Prize for chemistry.

The compound with the double or triple bond that reacts with the diene in a Diels-Alder reaction is given the special name of **dienophile** (diene-loving), and the product of a Diels-Alder reaction is given the special name of **Diels-Alder adduct.** The designation **cycloaddition** refers to the fact that two reactants add together to give a cyclic product. Following are two examples of Diels-Alder reactions: one with a compound containing a carbon-carbon double bond, and the other containing a carbon-carbon triple bond.

Dienophile A compound containing a double bond (consisting of one or two C, N, or O atoms) that can react with a conjugated diene to give a Diels-Alder adduct.

Diels-Alder adduct A cyclohexene resulting from the cycloaddition reaction of a diene and a dienophile.

Cycloaddition reaction A reaction in which two reactants add together in a single step to form a cyclic product. The best known of these is the Diels-Alder reaction.

1,3-Butadiene   3-Buten-2-one
(a diene)   (a dienophile)

4-Cyclohexenyl methyl ketone
(a Diels-Alder adduct)
(racemic mixture)

1,3-Butadiene   Diethyl 2-butynedioate
(a diene)   (a dienophile)

Diethyl 1,4-cyclohexadiene-
1,2-dicarboxylate
(a Diels-Alder adduct)

Note that the four carbon atoms of the diene and two carbon atoms of the dienophile combine to form a six-membered ring. Note further that there are two more sigma bonds and two fewer pi bonds in the product than in the reactants. This exchange of two (weaker) pi bonds for two (stronger) sigma bonds is a major driving force in Diels-Alder reactions.

We can write a Diels-Alder reaction in the following way, showing only the carbon skeletons of the diene and dienophile. In this representation, curved arrows are used to show that two new sigma bonds are formed, three pi bonds are broken, and one new pi bond is formed. It must be emphasized here that in this particular case, the curved arrows in this diagram are not meant to show a mechanism. Rather they are intended to show which bonds are broken, which new bonds are formed, and how many electrons are involved (six in this case).

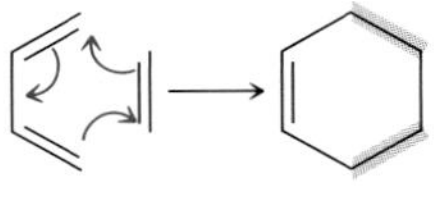

The special values of the reaction discovered by Diels and Alder are that (1) it is one of the simplest reactions that can be used to form six-membered rings, (2) it is one of few reactions that can be used to form two new carbon-carbon bonds at the same time, and, as we will see later in this section, (3) it is completely stereospecific and quite regioselective. For these reasons, the Diels-Alder reaction has proved to be enormously valuable in synthetic organic chemistry.

Example 24.5

Draw a structural formula for the Diels-Alder adduct formed by reaction of each diene and dienophile pair.
(a) 1,3-Butadiene and propenal   **(b)** 2,3-Dimethyl-1,3-butadiene and 3-buten-2-one

Solution

First draw the diene and dienophile so that each molecule is properly aligned to form a six-membered ring. Then complete the reaction to form the six-membered ring Diels-Alder adduct.

(a)

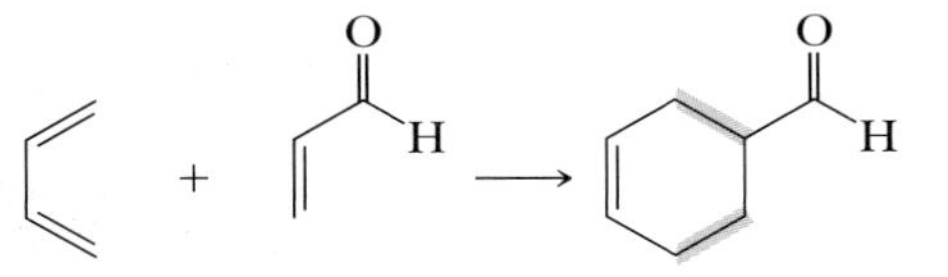

1,3-Butadiene   Propenal   (racemic mixture)
(Acrolein)

(b)

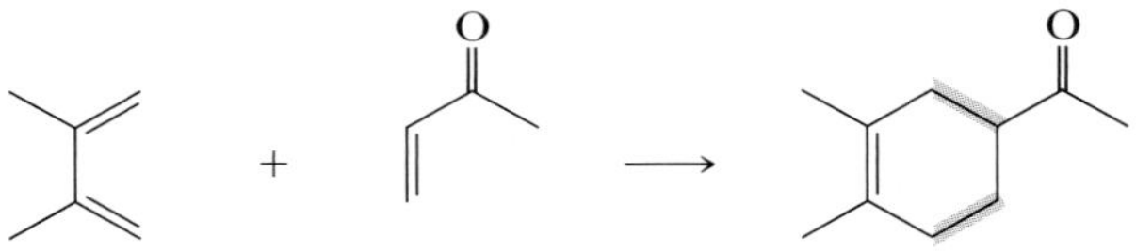

2,3-Dimethyl-   3-Buten-2-one   (racemic mixture)
1,3-butadiene   (Methyl vinyl ketone)

Problem 24.5

What combination of diene and dienophile undergoes Diels-Alder reaction to give each adduct?

(a)

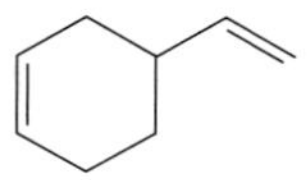

(racemic)

(b)

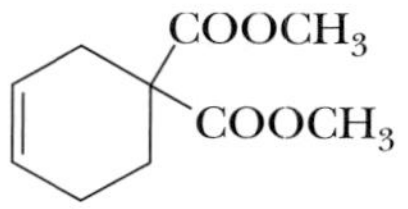

COOCH$_3$

COOCH$_3$

(c)

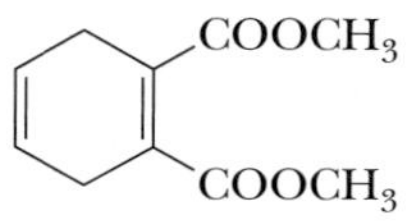

COOCH$_3$

COOCH$_3$

Now let us look more closely at the scope and limitations, stereochemistry, and mechanism of Diels-Alder reactions.

A. The Diene Must Be Able to Assume an s-*Cis* Conformation

We can illustrate the significance of conformation of the diene by reference to 1,3-butadiene. For maximum stability of a conjugated diene, overlap of the four unhybridized 2p orbitals making up the pi system must be complete, a condition that occurs only when all four carbon atoms of the diene lie in the same plane. If the carbon skeleton of 1,3-butadiene is planar, the six atoms bonded to the skeleton of the diene are also contained in the same plane. Because of conjugation, bond rotation is actually somewhat restricted around the central single bond: If the atoms are not coplanar, conjugation is imperfect or broken completely. There are two planar conformations of 1,3-butadiene, called the **s-*trans* conformation** and the **s-*cis* conformation** where the designation **s** refers to the carbon-carbon single bond of the diene. Of these, the s-*trans* conformation is slightly lower in energy and, therefore, slightly more stable.

Although s-*trans*-1,3-butadiene is the more stable conformation, s-*cis*-1,3-butadiene is the reactive conformation in Diels-Alder reactions. In the s-*cis* conformation, carbon atoms 1 and 4 of the conjugated system are close enough to react with the carbon-carbon double or triple bond of the dienophile and to form a six-membered ring. In the s-*trans* conformation, they are too far apart for this to happen.

The energy barrier for interconversion of the s-*trans* and s-*cis* conformations for 1,3-butadiene is low, approximately 11.7 kJ (2.8 kcal)/mol; consequently, 1,3-butadiene can still be a reactive diene in Diels-Alder reactions.

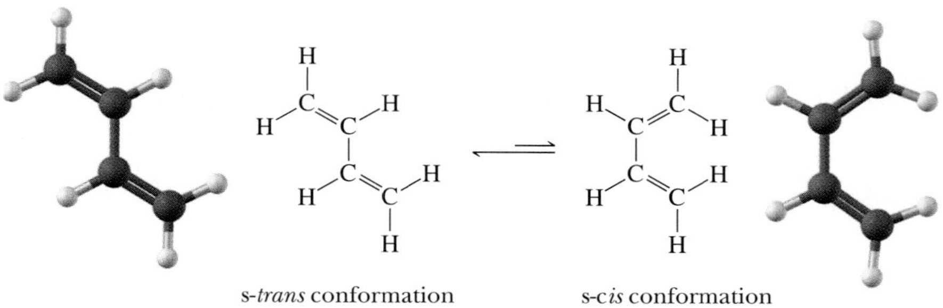

s-*trans* conformation
(Lower in energy)

s-*cis* conformation
(Higher in energy)

(2Z,4Z)-2,4-Hexadiene is unreactive in Diels-Alder reactions because it is prevented by steric hindrance from assuming the required s-*cis* conformation.

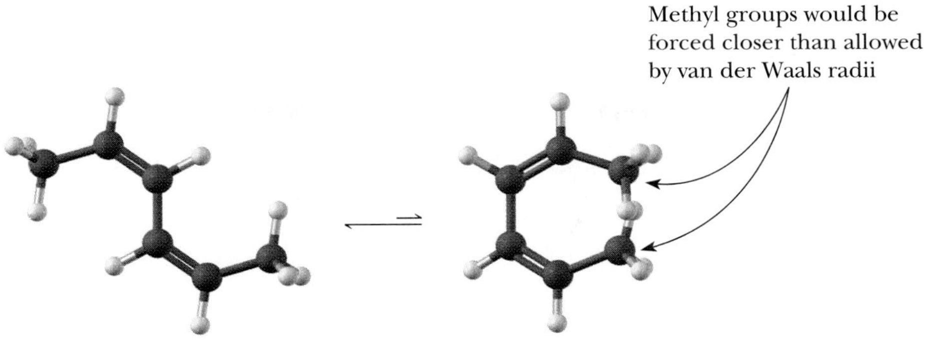

Methyl groups would be
forced closer than allowed
by van der Waals radii

s-*trans* conformation
(Lower in energy)

s-*cis* conformation
(Higher in energy)

(2Z,4Z)-2,4-Hexadiene

Example 24.6

Which molecules can function as dienes in Diels-Alder reactions?

(a) [structure with CH₂] (b) [structure] (c) [structure with two CH₂]

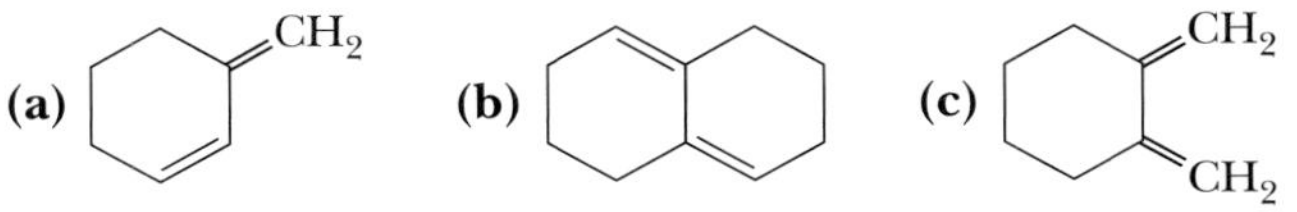

Solution

The dienes in both (a) and (b) are fixed in the s-*trans* conformation and, therefore, are not capable of participation in Diels-Alder reactions. The diene in (c) is fixed in the s-*cis* conformation and, therefore, has the proper orientation to participate in Diels-Alder reactions.

Problem 24.6

Which molecules can function as dienes in Diels-Alder reactions?

(a) [structure] (b) [structure] (c) [structure]

B. The Effect of Substituents on Rate

The simplest example of a Diels-Alder reaction is that between 1,3-butadiene and ethylene, both gases at room temperature. Although this reaction does occur, it is very slow and takes place only if the reactants are heated at 200°C under pressure.

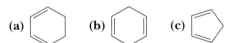

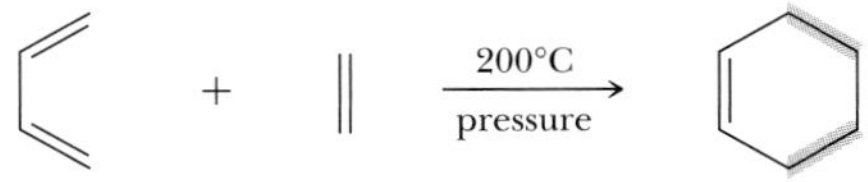

$$\text{[diene]} + \text{[ethylene]} \xrightarrow[\text{pressure}]{200°C} \text{[cyclohexene]}$$

1,3-Butadiene   Ethylene   Cyclohexene

Diels-Alder reactions are facilitated by a combination of electron-withdrawing substituents on one of the reactants and electron-releasing substituents on the other.

Most commonly, the dienophile is electron-deficient and the diene is electron-rich. For example, placing a carbonyl group (electron-withdrawing because of the partial positive charge on its carbon) on the dienophile facilitates the reaction. To illustrate, 1,3-butadiene and 3-buten-2-one form a Diels-Alder adduct when heated at 140°C.

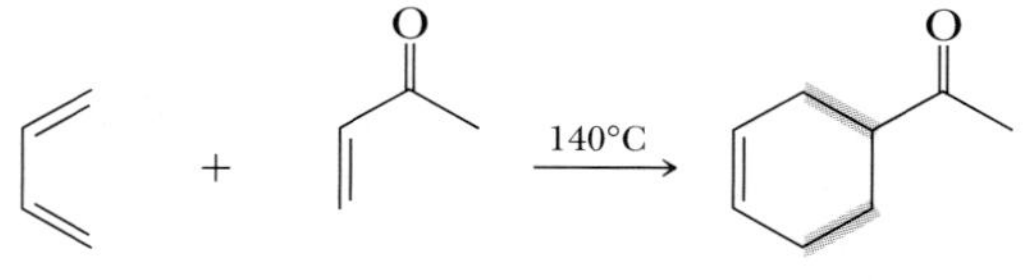

1,3-Butadiene   3-Buten-2-one   (racemic)

Placing electron-releasing methyl groups on the diene further facilitates reaction; 2,3-dimethyl-1,3-butadiene and 3-buten-2-one form a Diels-Alder adduct at 30°C.

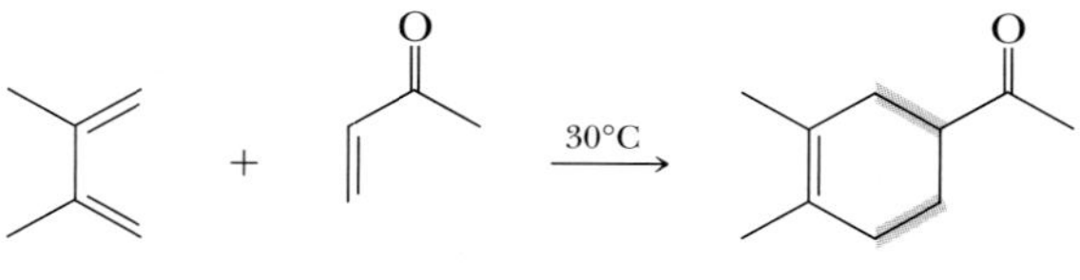

2,3-Dimethyl-   3-Buten-2-one   (racemic)
1,3-butadiene

Several of the electron-releasing and electron-withdrawing groups most commonly encountered in Diels-Alder reactions are given in Table 24.2. Note that the ester group can be either electron-donating or -withdrawing depending on whether the oxygen or the carbonyl is attached to the double bond.

C. Diels-Alder Reactions Can Be Used to Form Bicyclic Systems

Conjugated cyclic dienes, in which the double bonds are of necessity held in an s-*cis* conformation, are highly reactive in Diels-Alder reactions. Two particularly useful dienes for this purpose are cyclopentadiene and 1,3-cyclohexadiene. In fact, cyclopentadiene is reactive both as a diene and as a dienophile, and, on standing at room temperature, it forms a Diels-Alder self-adduct known by the common name

Table 24.2   Electron-Releasing and Electron-Withdrawing Groups

Electron-Releasing Groups	Electron-Withdrawing Groups
$-CH_3$	$\overset{O}{\overset{\|}{-CH}}$ (aldehyde)
$-CH_2CH_3$	
$-CH(CH_3)_2$	$\overset{O}{\overset{\|}{-CR}}$ (ketone)
$-C(CH_3)_3$	$\overset{O}{\overset{\|}{-COH}}$ (carboxyl)
$-R$ (other alkyl groups)	
$-OR$ (ether)	$\overset{O}{\overset{\|}{-COR}}$ (ester)
	$-NO_2$ (nitro)
$\overset{O}{\overset{\|}{-OCR}}$ (ester)	$-C\equiv N$ (cyano)

dicyclopentadiene. When dicyclopentadiene is heated to 170°C, a reverse Diels-Alder reaction takes place, and cyclopentadiene is reformed.

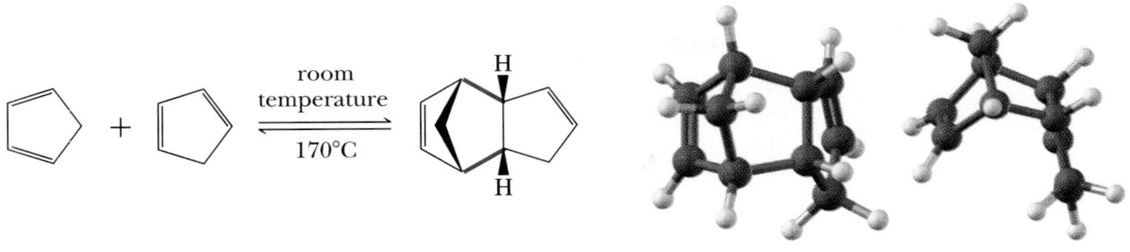

| Diene | Dienophile | Dicyclopentadiene (endo form) | From top | From side |

The terms "endo" and "exo" are used for bicyclic Diels-Alder adducts to describe the orientation of substituents of the dienophile in relation to the two-carbon diene-derived bridge. *Exo* (Greek, outside) substituents are on the opposite side from the diene-derived bridge; *endo* (Greek, within) substituents are on the same side.

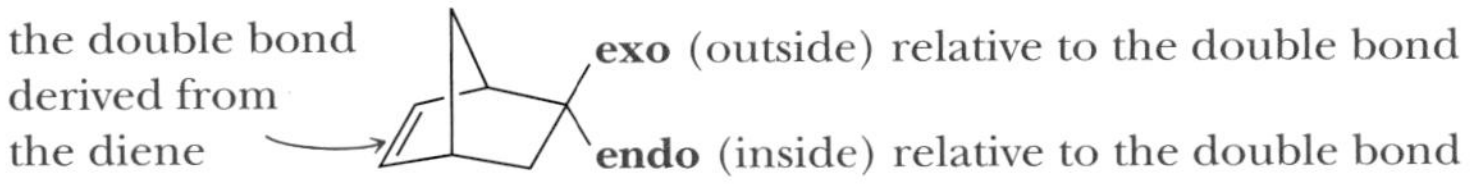

the double bond derived from the diene

exo (outside) relative to the double bond

endo (inside) relative to the double bond

For Diels-Alder reactions under kinetic control, the endo orientation of the dienophile is favored. Treatment of cyclopentadiene with methyl propenoate (methyl acrylate) gives the endo adduct almost exclusively. The exo adduct is not formed. Diels-Alder reactions are not always so stereoselective.

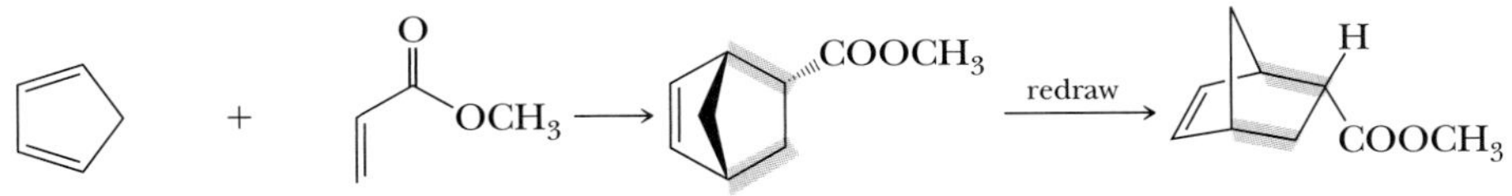

| Cyclopentadiene | Methyl propenoate | Methyl bicyclo[2.2.1]hept-5-en-endo-2-carboxylate (racemic) |

D. The Configuration of the Dienophile Is Retained

The reaction is completely stereospecific at the dienophile. If the dienophile is a *cis* isomer, then the substituents *cis* to each other in the dienophile are *cis* in the Diels-Alder adduct. Conversely, if the dienophile is a *trans* isomer, substituents that are *trans* in the dienophile are *trans* in the adduct.

Dimethyl *cis*-2-butenedioate (a *cis* dienophile)

Dimethyl *cis*-4-cyclohexene 1,2-dicarboxylate

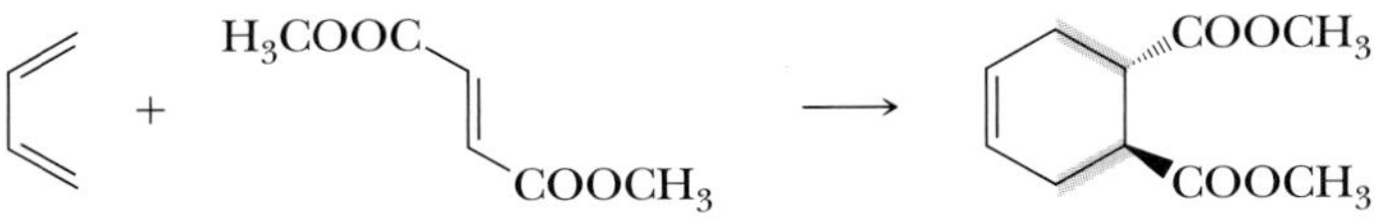

Dimethyl *trans*-2-butenedioate (a *trans* dienophile)

Dimethyl *trans*-4-cyclohexene 1,2-dicarboxylate (racemic)

E. The Configuration at the Diene Is Retained

The reaction is also completely stereospecific at the diene. Groups on the 1 and 4 positions of the diene retain their relative orientation.

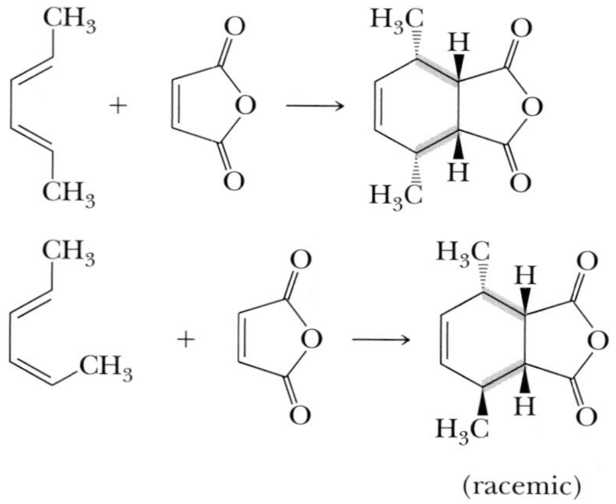

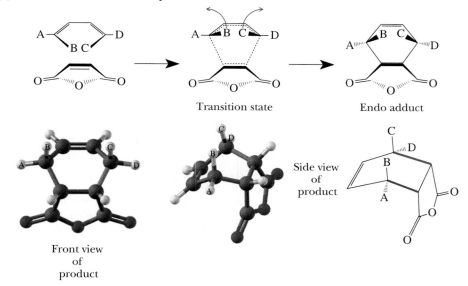

(racemic)

A picture of the transition state will help clarify the reason for this. Bonds being formed in the transition state are shown as dashed red lines; bonds being broken are shown as dashed blue lines. The groups that are inside on the diene end up on the opposite side from the dienophile.

Transition state

Endo adduct

Side view of product

Front view of product

F. Mechanism—The Diels-Alder Reaction Is a Pericyclic Reaction

As chemists probed for details of the Diels-Alder reaction, they found no evidence for participation of either ionic or radical intermediates. Thus, the Diels-Alder reaction is unlike any reaction we have studied thus far. To account for the stereoselectivity and the lack of evidence for either ionic or radical intermediates, chemists have proposed that the reaction takes place in a single step during which there is a cyclic redistribution of electrons. During this cyclic redistribution, bond forming and bond breaking are concerted (simultaneous). Such reactions that take place in a single

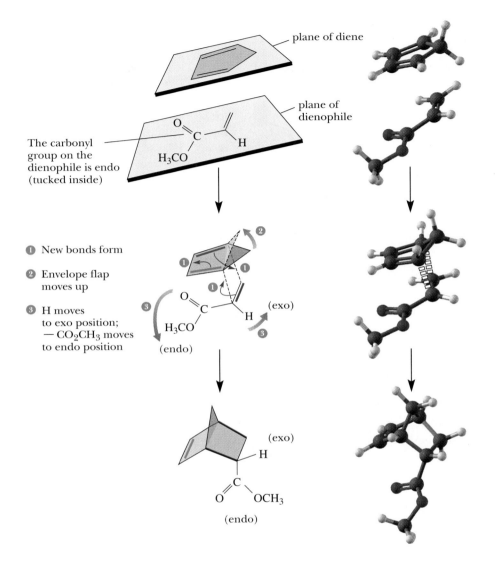

Figure 24.1
Mechanism of the Diels-Alder reaction. The diene and dienophile approach each other in parallel planes, one above the other, with the substituents on the dienophile *endo* to the diene. There is overlap of the pi orbitals of each molecule and syn addition of each molecule to the other. As (1) new sigma bonds form in the transition state, (2) the —CH$_2$— on the diene rotates upward, (3) the hydrogen atom of the dienophile becomes exo and the ester group becomes endo.

step, without intermediates, and involve a cyclic redistribution of bonding electrons are called **pericyclic reactions.** We can envision a Diels-Alder reaction taking place as shown in Figure 24.1.

Pericyclic reaction   A reaction that takes place in a single step, without intermediates, and involves a cyclic redistribution of bonding electrons.

Example 24.7

Complete the following Diels-Alder reaction, showing the stereochemistry of the product.

$$\square\!\!\!\!\square + \begin{array}{c} \text{COOCH}_3 \\ \| \\ \text{COOCH}_3 \end{array} \longrightarrow$$

Solution

Reaction of cyclopentadiene with this dienophile forms a disubstituted bicyclo[2.2.1]hept-2-ene. The two ester groups are *cis* in the dienophile, and, given the stereoselectivity of the Diels-Alder reaction, they are *cis* and endo in the product.

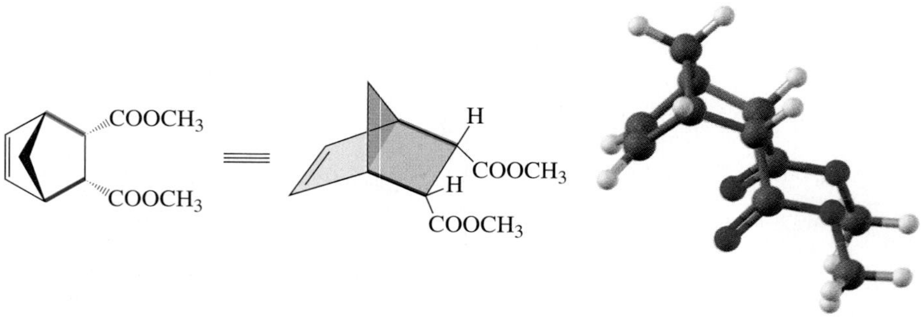

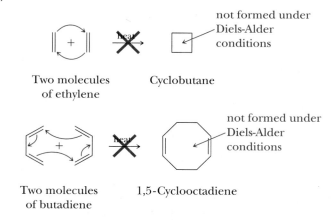

Problem 24.7

What diene and dienophile might you use to prepare the following racemic Diels-Alder adduct?

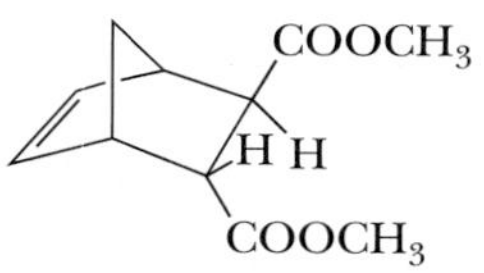

G. A Word of Caution About Electron Pushing

We developed a mechanism of the Diels-Alder reaction and used curved arrows to show the flow of electrons that takes place in the process of bond breaking and bond forming. Diels-Alder reactions involve a four-carbon diene and a two-carbon dienophile and are termed [4 + 2] cycloadditions. We can write similar electron-pushing mechanisms for the dimerization of ethylene by a [2 + 2] cycloaddition to form cyclobutane, and for the dimerization of butadiene by a [4 + 4] cycloaddition to form 1,5-cyclooctadiene.

not formed under
Diels-Alder
conditions

Two molecules
of ethylene

Cyclobutane

not formed under
Diels-Alder
conditions

Two molecules
of butadiene

1,5-Cyclooctadiene

Although [2 + 2] and [4 + 4] cycloadditions bear a formal relationship to the Diels-Alder reaction, neither, in fact, takes place under the thermal conditions required for Diels-Alder reactions. These cycloadditions do occur, but only under different (and usually much more vigorous) experimental conditions and by quite different mechanisms. There is more to some of these reactions than the use of curved arrows might suggest.

24.7 Pericyclic Reactions and Transition State Aromaticity

The Diels-Alder reaction is a concerted reaction involving a redistribution of six electrons in a cyclic transition state. The key point is that there are six electrons and the transition state is cyclic. We can place the transition state for this reaction in a larger context of transition state aromaticity. Recall the Hückel criteria for aromaticity (Section 21.2A): the presence of $(4n + 2)$ pi electrons in a ring that is planar and fully conjugated. Just as aromaticity imparts a stability to certain types of molecules and ions, the presence of $(4n + 2)$ electrons imparts a stability to certain types of transition states. Reactions involving 2, 6, 10, 14, ... electrons in a cyclic transition state have especially low activation energies and take place particularly readily. In contrast, the reactions in Section 24.6G that involve $4n$ electrons do not take place readily. Just as the Hückel theory of aromaticity gives us a clearer understanding of the stability of certain types of molecules and ions, it also gives a clearer understanding of certain types of reactions and their transition states. Transition states involving a cyclic redistribution of $4n$ electrons, on the other hand, are antiaromatic and have especially high activation energies. R. B. Woodward (Harvard), Roald Hoffmann (then at Harvard, now at Cornell), Kenichi Fukui (Kyoto University), and Howard Zimmerman (University of Wisconsin) provided the key insights into this specificity; Hoffmann and Fukui were awarded the Nobel Prize for this work in 1981 (after the death of Woodward). For this reason, the dimerization of ethylene to give cyclobutane (a four-electron transition state) and of 1,3-butadiene to give 1,5-cyclooctadiene (an eight-electron transition state) do not occur.

We have seen six examples of reactions that proceed by cyclic, six-electron transition states:

1. The decarboxylation of β-ketoacids and β-dicarboxylic acids (Section 17.9)
2. The Cope elimination of amine N-oxides (Section 23.10)
3. The Diels-Alder reaction (Section 24.6)
4. Addition of osmium tetroxide to alkenes (Section 6.5A)
5. Addition of ozone to alkenes (and the resulting fragmentation) (Section 6.5B)
6. Pyrolysis of esters (Problem 23.42)

We now look at two more examples of reactions that proceed by aromatic transition states.

A. The Claisen Rearrangement

The Claisen rearrangement transforms allyl phenyl ethers to o-allylphenols. Heating allyl phenyl ether, for example, the simplest member of this class of compounds, at 200–250°C results in a Claisen rearrangement to form o-allyphenol. In this rearrangement, an allyl group migrates from a phenolic oxygen to a carbon atom ortho to it. Carbon-14 labeling, here shown in color, has demonstrated that, during a Claisen rearrangement, carbon 3 of the allyl group becomes bonded to the ring carbon ortho to the phenolic oxygen.

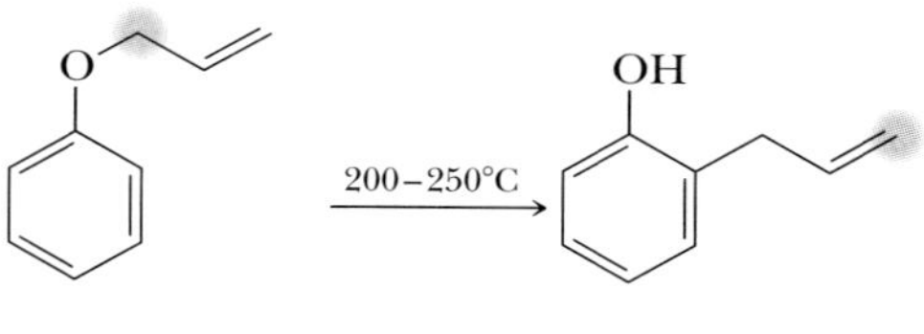

Allyl phenyl ether 2-Allylphenol

The mechanism of a Claisen rearrangement involves a concerted redistribution of six electrons in a cyclic transition state. The product of this rearrangement is a substituted cyclohexadienone, which undergoes keto-enol tautomerism to reform the aromatic ring. A new carbon-carbon bond is formed in the process.

Mechanism *The Claisen Rearrangement*

Step 1: Redistribution of six electrons in a cyclic transition state gives a cyclohexadienone intermediate. Dashed red lines indicate bonds being formed in the transition state, and dashed blue lines indicate bonds being broken.

Step 2: Keto-enol tautomerization restores the aromatic character of the ring.

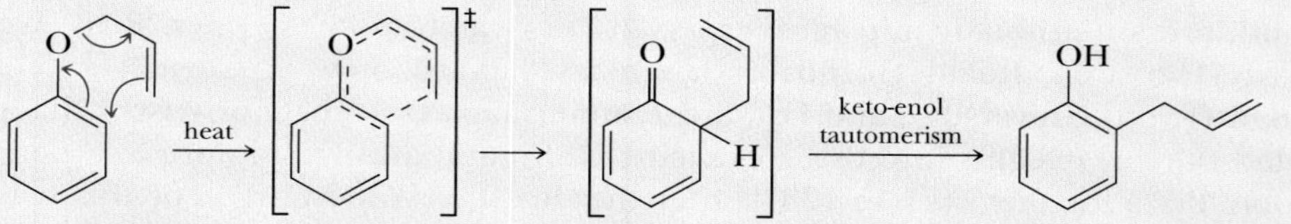

Allyl phenyl ether | Transition state | A cyclohexadienone intermediate | 2-Allylphenol

ORGANIC **Chemistry•Now™**
Click Mechanisms to view an animation of the **Claisen Rearrangement**

Thus, we see that the transition state for the Claisen rearrangement bears a close resemblance to that for the Diels-Alder reaction. Both involve a concerted redistribution of six electrons in a cyclic transition state.

Example 24.8

Predict the product of Claisen rearrangement of *trans*-2-butenyl phenyl ether.

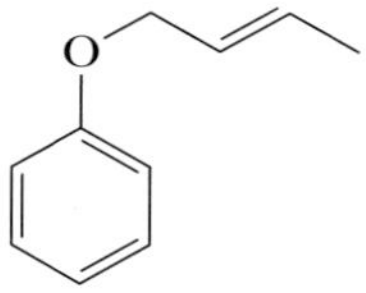

trans-2-Butenyl phenyl ether

Solution

In the six-membered transition state for this rearrangement, carbon 3 of the allyl group becomes bonded to the ortho position of the ring.

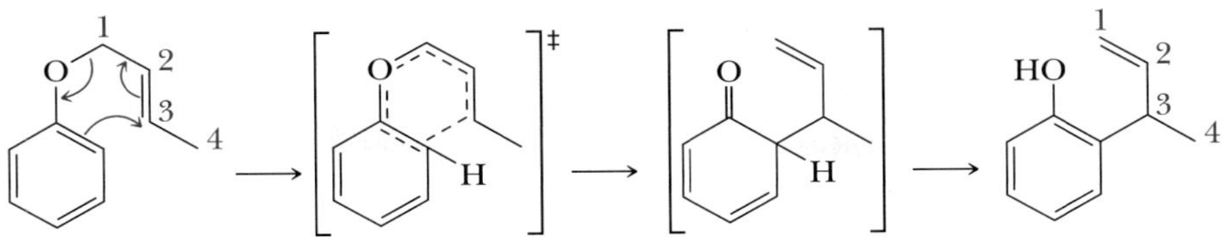

Transition state — A cyclohexadienone intermediate — (racemic)

Problem 24.8

Show how to synthesize allyl phenyl ether and 2-butenyl phenyl ether from phenol and appropriate alkenyl halides.

B. The Cope Rearrangement

The Cope rearrangement of 1,5-dienes also takes place via a cyclic six-electron transition state. In this example, the product is an equilibrium mixture of isomeric dienes. The favored product is the diene on the right, which contains the more highly substituted double bonds.

Mechanism   *The Cope Rearrangement*

Redistribution of six electrons in a cyclic transition state converts a 1,5-diene to an isomeric 1,5-diene.

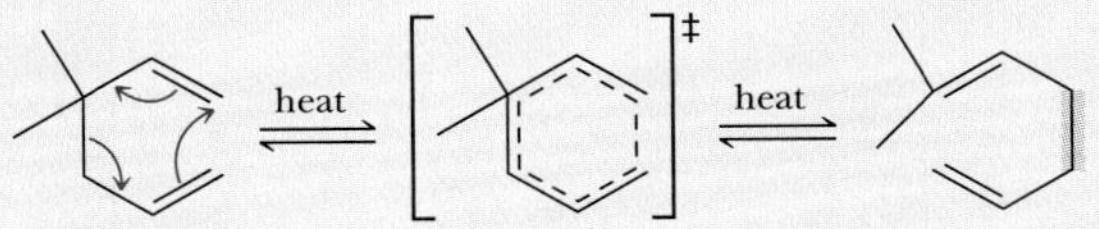

3,3-Dimethyl-1,5-hexadiene — Transition state — 6-Methyl-1,5-heptadiene

Example 24.9

Propose a mechanism for the following Cope rearrangement.

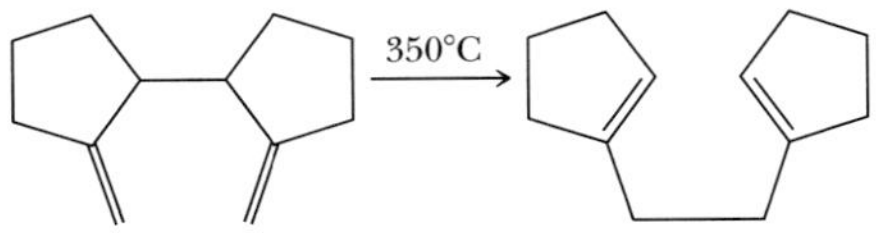

Solution

Redistribution of six electrons in a cyclic transition state gives the observed product.

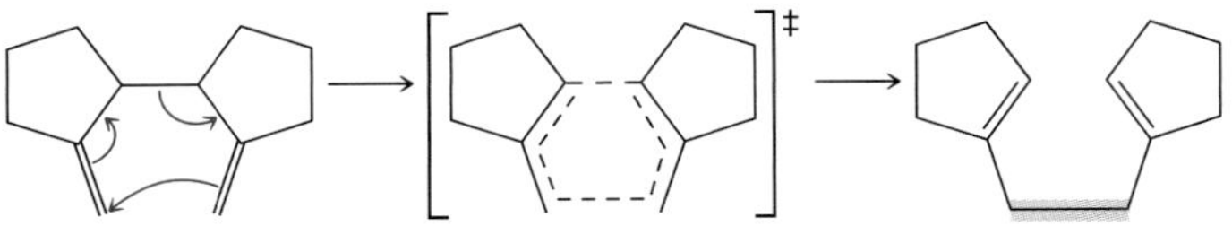

Transition state

CHEMICAL CONNECTIONS

Singlet Oxygen

Molecular oxygen differs from most other stable molecules by having two unpaired electrons. This comes about because, during the "Aufbau" of molecular oxygen, there are two highest occupied molecular orbitals (π^*) of the same energy and only two electrons to put in them; thus by Hund's rule, the lowest energy ("ground") state of oxygen has one of these electrons in each orbital and with the same spin. The unpaired electrons cause oxygen to be paramagnetic (attracted by a magnetic field). Most molecules are diamagnetic (unaffected by a magnetic field).

The ground state of oxygen is often called a diradical, but, in fact, oxygen does not really behave as a diradical (it is much less reactive than most radicals). Like protons, electrons have a magnetic moment; two electrons can take on three orientations in a magnetic field, and the ground state of oxygen is thus called a "triplet" state and given the symbol $^3O_2$. Oxygen is one of an extremely small number of compounds that has a triplet ground state.

However, the triplet state is not the only electronic configuration oxygen can have, only the lowest. There are also two low-energy excited electronic states of oxygen in which the electrons are paired. Only the lower of these, with an energy of 92 kJ (22 kcal)/mol above the ground state, has an appreciable lifetime. Chemists refer to this species as singlet oxygen, $^1O_2$. Its lifetime varies with solvent from about 4 μs to

almost 0.1 s, and it has a very different reactivity from $^3O_2$.

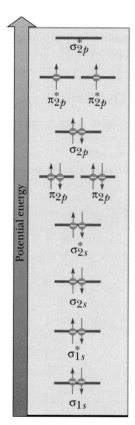

■ Molecular orbital energy diagram for the O_2 molecule.

Problem 24.9

Propose a mechanism for the following Cope rearrangement.

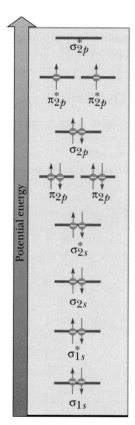

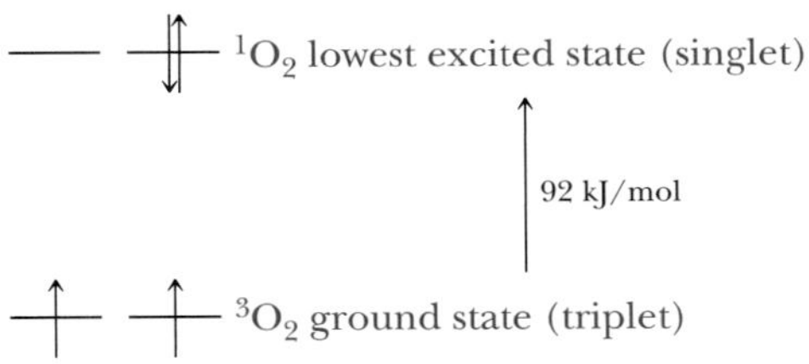

$^1O_2$ lowest excited state (singlet)

92 kJ/mol

$^3O_2$ ground state (triplet)

Singlet oxygen reacts as a dienophile in Diels-Alder reactions. For example, it reacts with cyclohexadiene to give the Diels-Alder adduct peroxide. This is a general and synthetically useful reaction.

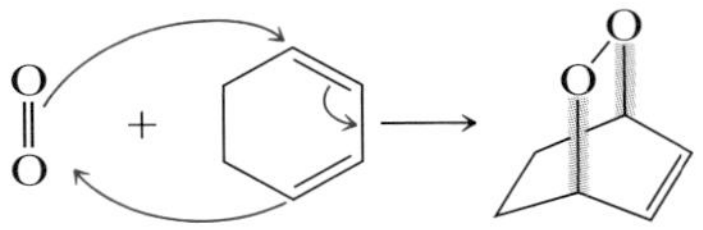

Singlet
oxygen ($^1O_2$)

Singlet oxygen is most often produced photochemically using a dye or other compound called a **photosensitizer** (Sens). Photosensitizers absorb a photon of light ($h\nu$), often in the visible region of the spectrum, and produce an electronically excited state (Sens*). This state can transfer its energy to $^3O_2$, producing $^1O_2$. This process can be very efficient. The resulting singlet oxygen can then react with dienes and other types of reactive acceptors. 1,4-Dimethylnaphthalene, for example, reacts with singlet oxygen by a Diels-

Alder reaction to form an endoperoxide. This endoperoxide can be used to store singlet oxygen and then to release it on warming in a reverse Diels-Alder reaction.

$$Sens \xrightarrow{h\nu} Sens^*$$

$$Sens^* + {}^3O_2 \longrightarrow Sens + {}^1O_2$$

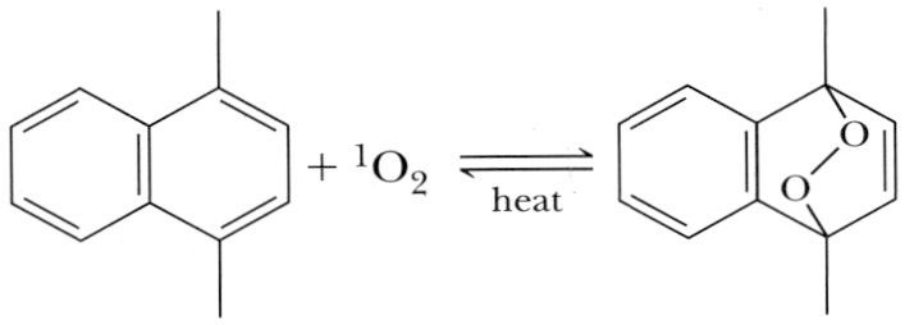

1,4-Dimethyl-
naphthalene

An endoperoxide

Singlet oxygen can also react with many biological materials such as unsaturated lipids. This reaction, called **photosensitized oxidation,** can cause damage to or death of an organism. The biological damage caused by photosensitizers, light, and oxygen can be used to kill tumor cells selectively. This **photodynamic therapy** has recently been approved by the Food and Drug Administration, especially for esophageal and certain other types of lung cancer. It is also under investigation for bladder, larynx, and other types of cancers.

24.8 The Synthesis of Enantiomerically Pure Target Molecules

As we have mentioned repeatedly throughout the text, the synthesis of chiral products from achiral starting materials in an achiral environment invariably leads to a racemic mixture of products. Nature achieves the synthesis of single enantiomers by using enzymes, which create a chiral environment in which reaction takes place. Enzymes, in fact, show such high enantiomeric and diastereomeric selectivity that the result of an enzyme-catalyzed reaction is invariably only a single one of all possible stereoisomers. Chemists have developed chiral catalysts that produce chiral products. However, these catalysts are often far less stereoselective than nature's enzyme catalysts, although great progress has been made in this field in recent years. How then do chemists achieve the synthesis of single enantiomers uncontaminated by their mirror images?

One strategy they use is resolution (Section 3.8A) to separate enantiomers and recover each in pure form. The most common methods for resolution depend on (1) the different physical properties of diastereomeric salts, (2) the use of enzymes as resolving agents, and (3) chromatography on a chiral substrate. While resolution is effective in preparing pure enantiomers, half of all product prepared to the point of resolution, namely the unwanted enantiomer, is lost in the process. Thus, this strategy for the preparation of single enantiomers wastes starting materials and reagents.

We illustrate an alternative strategy, namely **asymmetric induction,** by E. J. Corey's preparation of a key intermediate in his synthesis of prostaglandins. In asymmetric induction, the reactive functional group of an achiral molecule is placed in a chiral environment by reacting it with a **chiral auxiliary.** The strategy is that the chiral auxiliary then exerts a control over the stereoselectivity of the desired reaction. The chiral auxiliary chosen by Corey was 8-phenylmenthol. This molecule has three chiral centers and can exist as a mixture of $2^3 = 8$ possible stereoisomers. It was prepared in enantiomerically pure form from naturally occurring, enantiomerically pure menthol.

Menthol
(enantiomerically pure)

8-Phenylmenthol
(a chiral auxillary)

The initial step in Corey's prostaglandin synthesis was a Diels-Alder reaction between a substituted cyclopentadiene and the double bond of an acrylate ester. By binding the achiral acrylate reactant to enantiomerically pure 8-phenylmenthol, Corey thus placed the carbon-carbon double bond of the dienophile in a chiral environment. The result was that the diene approached the carbon-carbon double bond of the acrylate preferentially from one direction.

Achiral   Enantiomerically pure   97%   3%

A remarkable feature of this reaction is that it creates three chiral centers. Two of the chiral centers, namely those at the two ring junctions, are established by the

Diels-Alder reaction. The third, namely the endo position of the ester group, is also established by the Diels-Alder reaction. Without the chiral auxiliary 8-phenylmenthyl group, two of the eight possible stereoisomers would be produced, namely the pair of enantiomers shown. Although both enantiomers of the bicyclic products were formed in Corey's scheme, they were formed in the ratio of 93:7, and the desired enantiomer could be separated in pure form. In subsequent steps, the 8-phenylmenthyl ester was hydrolyzed, and the pure enantiomer was converted to the so-called Corey lactone (Problem 24.47) and then to enantiomerically pure prostaglandin $F_{2\alpha}$.

A third strategy for the preparation of single stereoisomers is illustrated by Gilbert Stork's synthesis of prostaglandin $F_{2\alpha}$. Stork began his synthesis with the naturally occurring, enantiomerically pure sugar D-erythrose (Section 25.1). This four-carbon sugar has the R configuration at each of its chiral centers.

D-Erythrose

With these two chiral centers thus established, he then used a series of well-understood reactions to synthesize his target molecule in enantiomerically pure form.

Summary

The Heck reaction (Section 24.3) is unique in that it allows substitution of an alkyl, aryl, or alkenyl group for a hydrogen on a carbon-carbon double bond and is not affected by most other functional groups in the organohalogen compound. The Suzuki reaction (Section 24.4) is used to couple aryl and alkenyl boron compounds with alkyl, aryl and alkenyl halides and triflates in the presence of a palladium complex catalyst and a base. The alkene metathesis reaction (Section 24.5) is unique in that it can be used for ring-closing reactions or polymerizations (Section 24.5C).

The Diels-Alder reaction (Section 24.6) is a **cycloaddition** between a conjugated diene and a dienophile to give a six-membered ring. Dienophiles contain either a double or triple bond. Diels-Alder reactions are facilitated by electron-withdrawing substituents on one of the reactants (either the diene or dienophile) and electron-releasing substituents on the other reactant. The mechanism is described as a **pericyclic reaction;** that is, it takes place in a single step, without intermediates, and involves redistribution of bonding electrons in a cyclic transition state. The Diels-Alder reaction is stereospecific: (1) The configuration of the dienophile is retained; if substituents on the dienophile are *cis* (or *trans*), they remain *cis* (or *trans*) in the product; (2) the configuration of the diene is retained; and (3) formation of the **endo** adduct is favored. The Diels-Alder reaction is just one of a number of pericyclic reactions that take place in a single step, usually with a cyclic mechanism involving six electrons.

Enantiomerically pure products can be obtained by the wasteful process of resolution, or more efficiently by use of a chiral auxiliary or use of chiral starting materials from natural sources.

Key Reactions

1. The Heck Reaction (Section 24.3)

In a palladium(0)-catalyzed reaction, the carbon group of a haloalkene (a vinylic halide) or haloarene is substituted for a hydrogen on a carbon-carbon double bond (a vinylic hydrogen) of an alkene. Reaction generally proceeds with a high degree of both stereoselectivity and regioselectivity.

ORGANIC
Chemistry⋅⦁⋅Now ™
Click Flashcards to review the
Key Reactions of Conjugated Systems

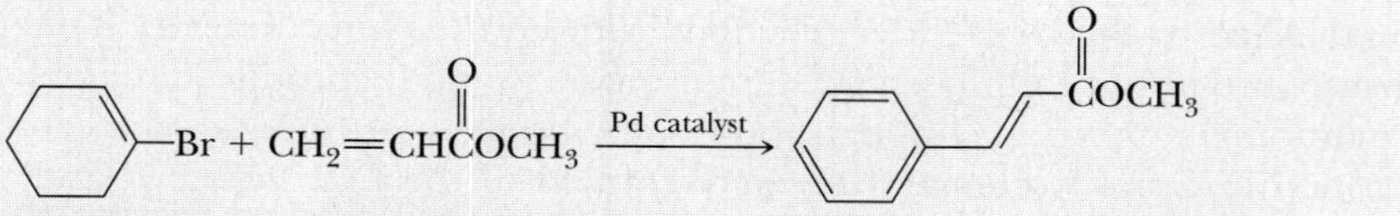

2. The Suzuki Coupling (Section 24.4)

The Suzuki reaction is a palladium-catalyzed reaction of an organoboron compound with an organic halide or triflate.

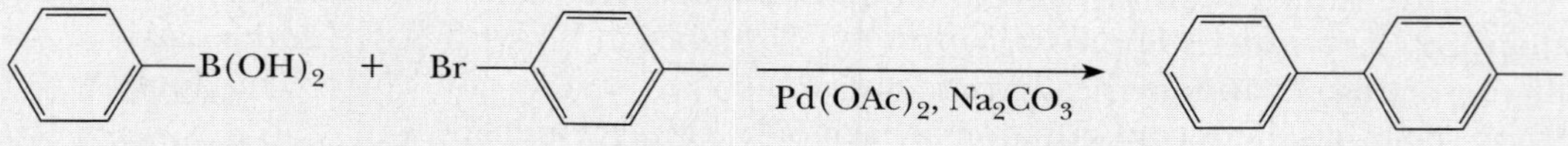

3. Alkene Metathesis (Section 24.5)

The alkene metathesis reaction is an organometallic-catalyzed reaction in which two alkenes exchange carbons of their double bonds. In a ring-closing alkene methathesis reaction, both alkenes are in the same molecule, and the product is a cycloalkene. Catalysts with Ru and Mo are often used; a nucleophilic carbene complex of Ru is particularly useful.

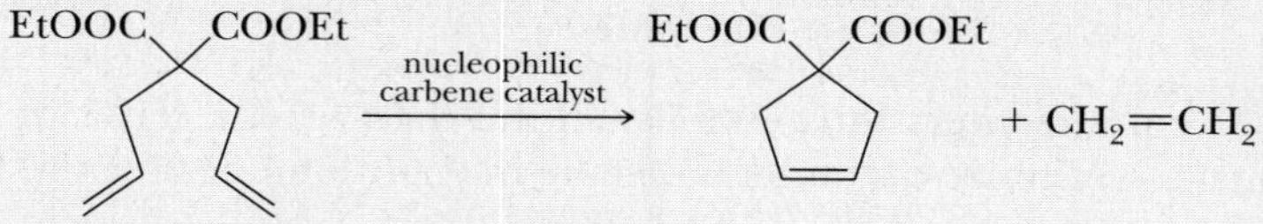

4. The Diels-Alder Reaction: A Pericyclic Reaction (Section 24.6)

A Diels-Alder reaction takes place in a single step, without intermediates, and involves a redistribution of six pi electrons in a cyclic transition state. The configuration of the diene and dienophile is preserved. Formation of the endo adduct is favored.

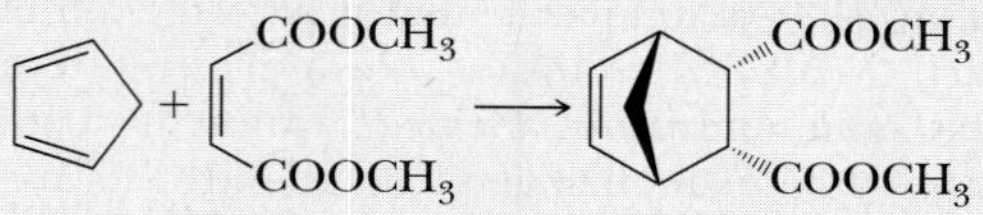

5. The Claisen Rearrangement: A Pericyclic Reaction (Section 24.7A)

The Claisen rearrangement transforms an allyl phenyl ether to an ortho-substituted phenol.

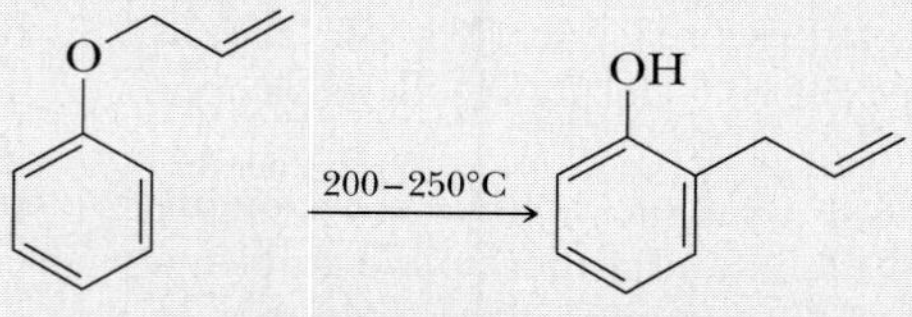

6. The Cope Rearrangement: A Pericyclic Reaction (Section 24.7B)

The Cope rearrangement converts a 1,5-diene to give an isomeric 1,5-diene.

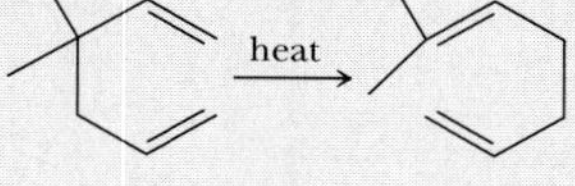

Problems

The Heck Reaction

24.10 As has been demonstrated in the text, when the starting alkene has CH_2 as its terminal group, the Heck reaction is highly stereoselective for formation of the E isomer. Here, the benzene ring is abbreviated C_6H_5—. Show how the mechanism proposed in the text allows you to account for this stereoselectivity.

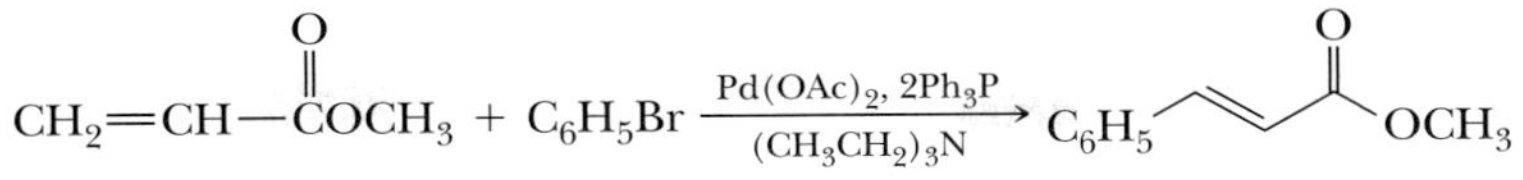

$$CH_2{=}CH{-}\overset{O}{\overset{\|}{C}}OCH_3 + C_6H_5Br \xrightarrow[\text{(CH}_3\text{CH}_2)_3\text{N}]{\text{Pd(OAc)}_2, \text{ 2Ph}_3\text{P}} C_6H_5 \diagup\!\!\diagdown\!\!\diagup \overset{O}{\overset{\|}{C}}OCH_3$$

24.11 The following reaction involves two sequential Heck reactions. Draw structural formulas for each organopalladium intermediate formed in the sequence and show how the final product is formed. Note from the molecular formula given under each structural formula that this conversion corresponds to a loss of H and I from the starting material. Acetonitrile, CH_3CN, is the solvent.

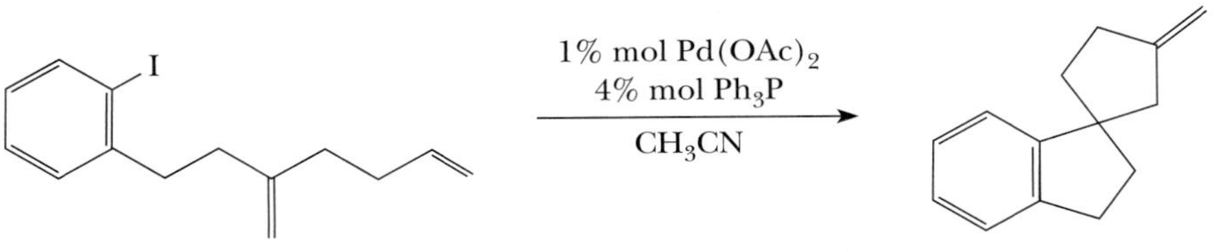

$C_{14}H_{17}I$ → $\xrightarrow[\text{CH}_3\text{CN}]{\begin{array}{c}1\%\ \text{mol Pd(OAc)}_2\\4\%\ \text{mol Ph}_3\text{P}\end{array}}$ → $C_{14}H_{16}$

24.12 Complete these Heck reactions.

(a) 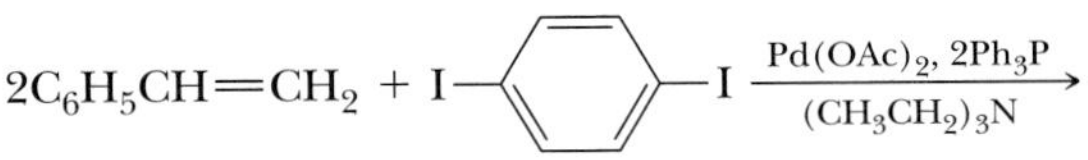 $2C_6H_5CH{=}CH_2 + $ I—⬡—I $\xrightarrow[\text{(CH}_3\text{CH}_2)_3\text{N}]{\text{Pd(OAc)}_2, \text{ 2Ph}_3\text{P}}$

(b) 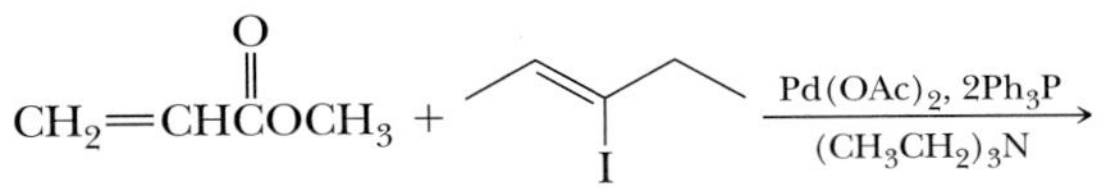 $CH_2{=}CH\overset{O}{\overset{\|}{C}}OCH_3 + $ $\xrightarrow[\text{(CH}_3\text{CH}_2)_3\text{N}]{\text{Pd(OAc)}_2, \text{ 2Ph}_3\text{P}}$

24.13 Treatment of cyclohexene with iodobenzene under the conditions of the Heck reaction might be expected to give 1-phenylcyclohexene. The exclusive product, however, is 3-phenylcyclohexene. Account for the formation of this product.

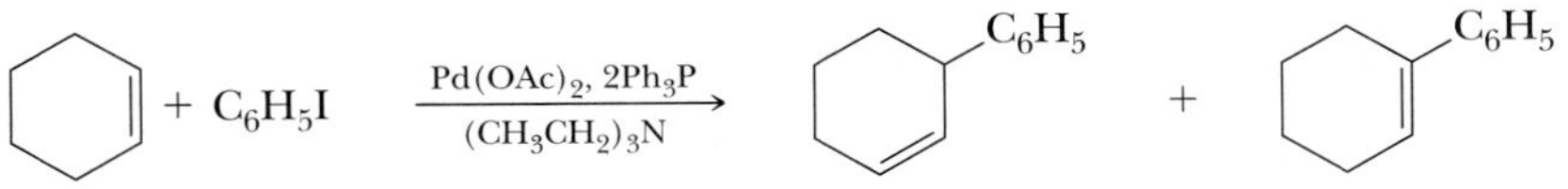

⬡ + C_6H_5I $\xrightarrow[\text{(CH}_3\text{CH}_2)_3\text{N}]{\text{Pd(OAc)}_2, \text{ 2Ph}_3\text{P}}$

3-Phenylcyclohexene   1-Phenylcyclohexene
(racemic)   (not formed)

24.14 Account for the formation of the product and for the *cis* stereochemistry of its ring junction. (The function of silver carbonate is to enhance the rate of reaction.)

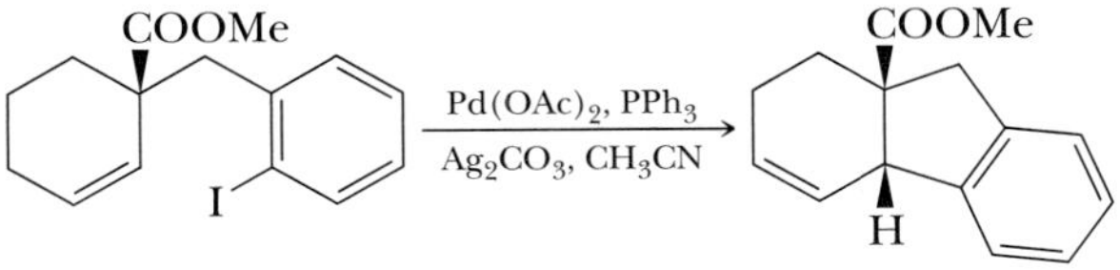

$\xrightarrow[\text{Ag}_2\text{CO}_3, \text{ CH}_3\text{CN}]{\text{Pd(OAc)}_2, \text{ PPh}_3}$

86%

24.15 Account for the formation of the following product, including the *cis* stereochemistry at the ring junction.

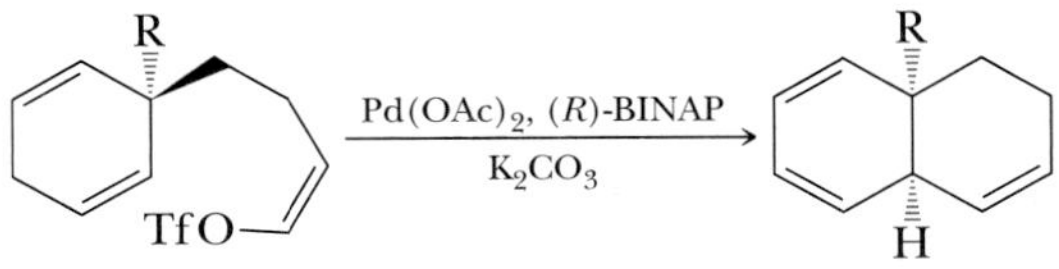

$\xrightarrow[\text{K}_2\text{CO}_3]{\text{Pd(OAc)}_2, \text{ }(R)\text{-BINAP}}$

24.16 The aryl diene undergoes sequential Heck reactions to give a product with molecular formula $C_{15}H_{18}$. Propose a structural formula for this product.

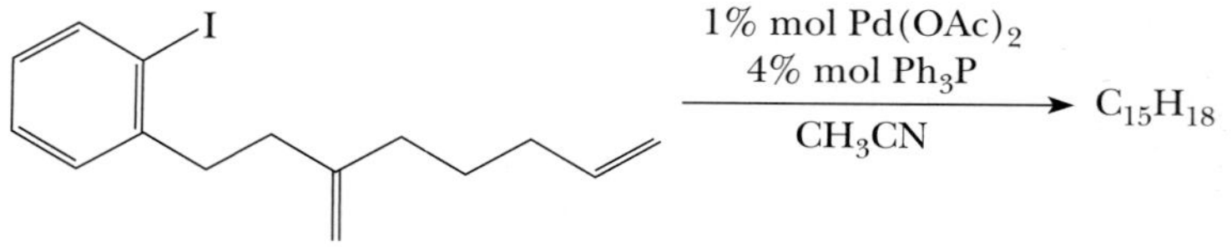

$$\xrightarrow[\text{CH}_3\text{CN}]{\substack{1\% \text{ mol Pd(OAc)}_2 \\ 4\% \text{ mol Ph}_3\text{P}}} C_{15}H_{18}$$

$(C_{15}H_{19}I)$

24.17 Heck reactions take place with alkynes as well as alkenes. The following conversion involves an intramolecular Heck reaction followed by an intermolecular Heck. Propose structural formulas for the palladium-containing intermediates involved in this reaction.

$$+ \quad \text{COOMe} \quad \xrightarrow{\text{Heck reaction}}$$

Me

Me

MeOOC

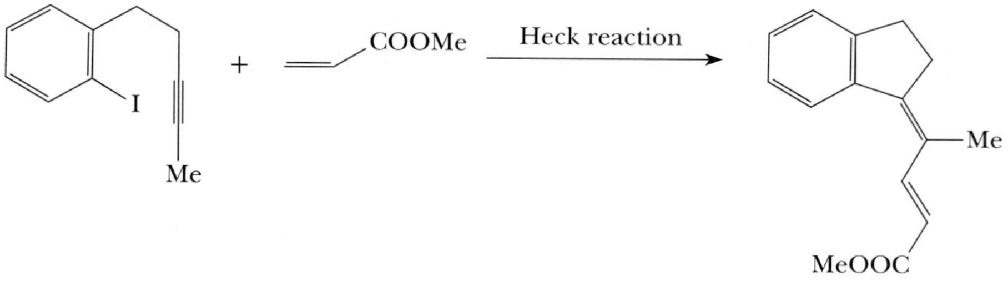

24.18 The following conversion involves sequential Heck reactions. Propose structural formulas for the palladium-containing intermediates involved in this reaction.

EtOOC

I

EtOOC

SiMe$_3$

$$\xrightarrow{\substack{\text{Heck} \\ \text{reaction}}}$$

EtOOC

EtOOC

SiMe$_3$

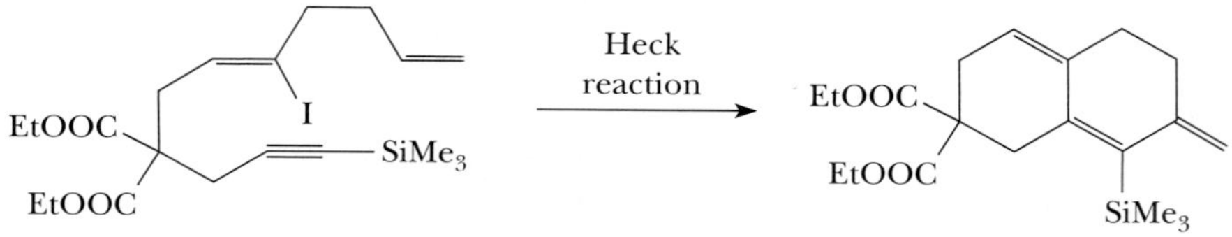

24.19 The following transformation involves a series of four consecutive Heck reactions and the formation of the four-ring steroid nucleus (Section 26.4) as a racemic mixture. Propose structural formulas for the palladium-containing intermediates involved in this reaction.

EtOOC

EtOOC

I

$$\xrightarrow{\text{Heck reaction}}$$

EtOOC

EtOOC

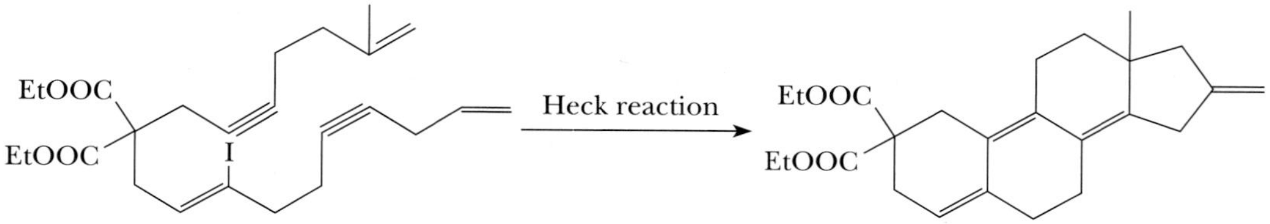

24.20 Suggest reagents and the other fragment that could be used to carry out the indicated conversion.

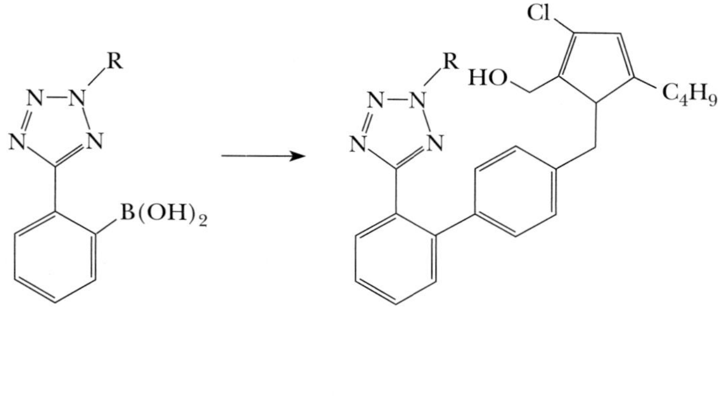

R

N—N

N   N

B(OH)$_2$

$$\longrightarrow$$

R   HO   Cl

N—N   C$_4$H$_9$

N   N

24.21 Show how the following compound could be prepared by a Suzuki reaction.

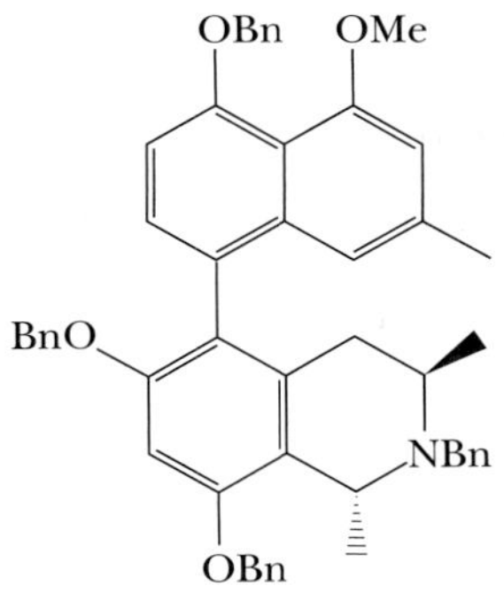

24.22 Show the sequence of Heck reactions by which the following conversion takes place. Note from the molecular formula given under each structural formula that this conversion corresponds to a loss of H and I from the starting material.

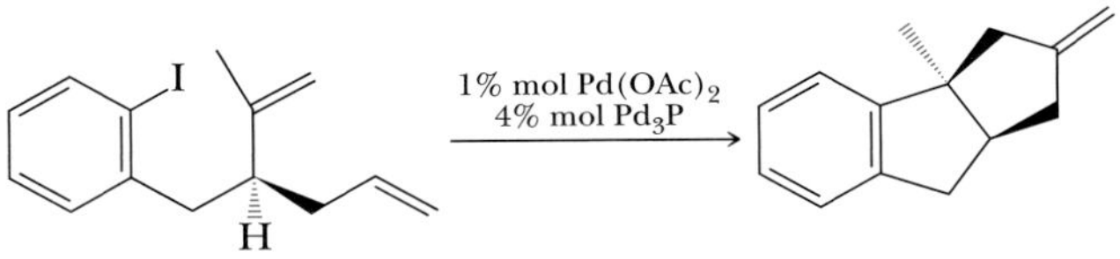

$(C_{14}H_{17}I)$ $(C_{14}H_{16})$

24.23 The cyclic ester (lactone) Exaltolide has a musk-like fragrance and is used as a fixative in perfumery. Show how this compound could be synthesized from the indicated starting material. Give the structure of R.

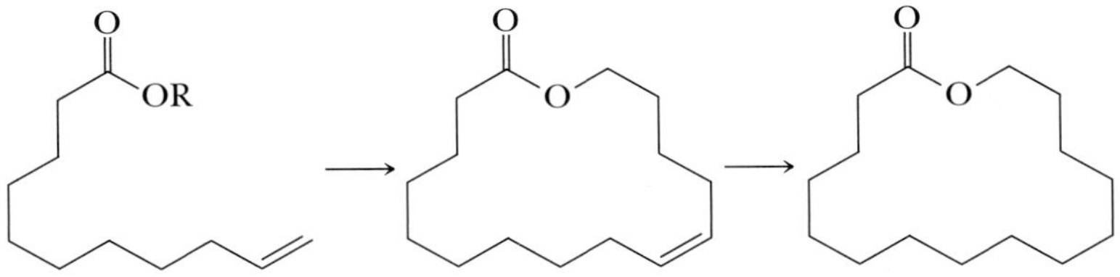

Exaltolide

24.24 Predict the product of each alkene metathesis reaction using a Ru-nucleophilic carbene catalyst.

(a) 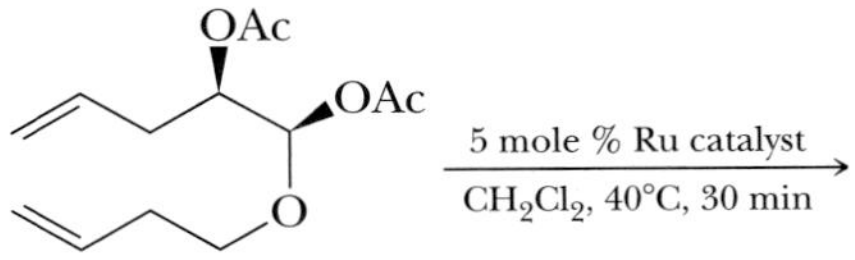   5 mole % Ru catalyst
       CH_2Cl_2, 40°C, 30 min

(b) 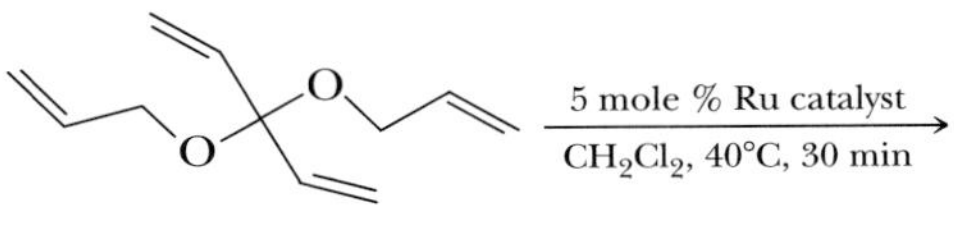   5 mole % Ru catalyst
       CH_2Cl_2, 40°C, 30 min

Diels-Alder Reaction

24.25 Draw structural formulas for the products of reaction of cyclopentadiene with each dienophile.

(a) CH_2=CHCl   (b) CH_2=CHCOCH$_3$ (with C=O)   (c) HC≡CH   (d) CH_3OCC≡$CCOCH_3$ (with two C=O)

24.26 Propose structural formulas for compounds A and B and specify the configuration of compound B.

$$\text{(cyclopentadiene)} + CH_2=CH_2 \xrightarrow{200°C} C_7H_{10} \xrightarrow[\text{2. }(CH_3)_2S]{\text{1. }O_3} C_7H_{10}O_2$$

(A)   (B)

24.27 Under certain conditions, 1,3-butadiene can function both as a diene and a dienophile. Draw a structural formula for the Diels-Alder adduct formed by reaction of 1,3-butadiene with itself.

24.28 1,3-Butadiene is a gas at room temperature and requires a gas-handling apparatus to use in a Diels-Alder reaction. Butadiene sulfone is a convenient substitute for gaseous 1,3-butadiene. This sulfone is a solid at room temperature (mp 66°C) and, when heated above its boiling point of 110°C, decomposes by a reverse Diels-Alder reaction to give s-cis-1,3-butadiene and sulfur dioxide. Draw Lewis structures for butadiene sulfone and SO_2; then, show by curved arrows the path of this reaction, which resembles a reverse Diels-Alder reaction.

$$\text{(butadiene sulfone)} \xrightarrow{140°C} \text{(1,3-butadiene)} + SO_2$$

Butadiene sulfone   1,3-Butadiene   Sulfur dioxide

24.29 The following triene undergoes an intramolecular Diels-Alder reaction to give the product shown. Show how the carbon skeleton of the triene must be coiled to give this product, and show by curved arrows the redistribution of electron pairs that takes place to give the product.

$$\text{(triene)} \xrightarrow{160°C} \text{(bicyclic product)}$$

24.30 The following triene undergoes an intramolecular Diels-Alder reaction to give a bicyclic product. Propose a structural formula for the product. Account for the observation that the Diels-Alder reaction given in this problem takes place under milder conditions (at lower temperature) than the analogous Diels-Alder reaction shown in Problem 24.29.

$$\text{(triene)} \xrightarrow{0°C} \text{Diels-Alder adduct}$$

24.31 The following compound undergoes an intramolecular Diels-Alder reaction to give a bicyclic product. Propose a structural formula for the product.

$$\text{(compound)} \xrightarrow{heat} \text{An intramolecular Diels-Alder adduct}$$

24.32 Draw a structure formula for the product of this Diels-Alder reaction, including the stereochemistry of the product.

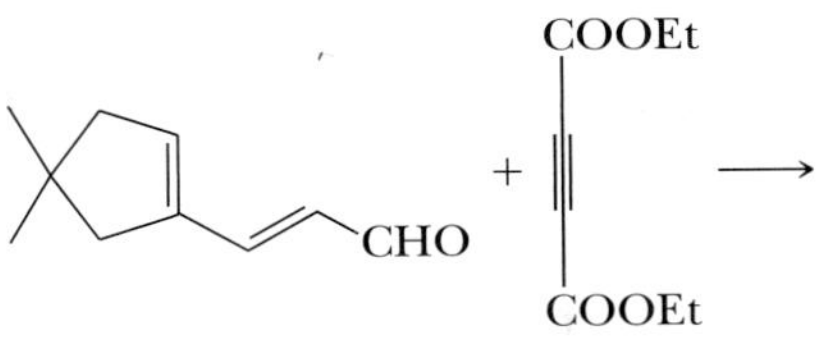

24.33 Following is a retrosynthetic analysis for the dicarboxylic acid shown on the left.

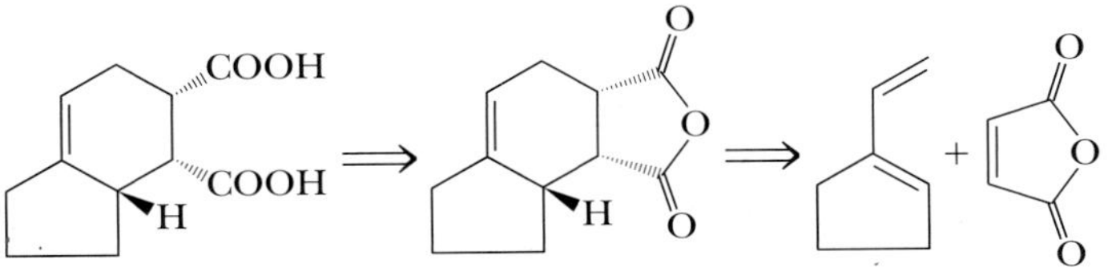

(a) Propose a synthesis of the diene from cyclopentanone and acetylene.
(b) Rationalize the stereochemistry of the target dicarboxylic acid.

24.34 One of the published syntheses of warburganal begins with the following Diels-Alder reaction. Propose a structure for compound A.

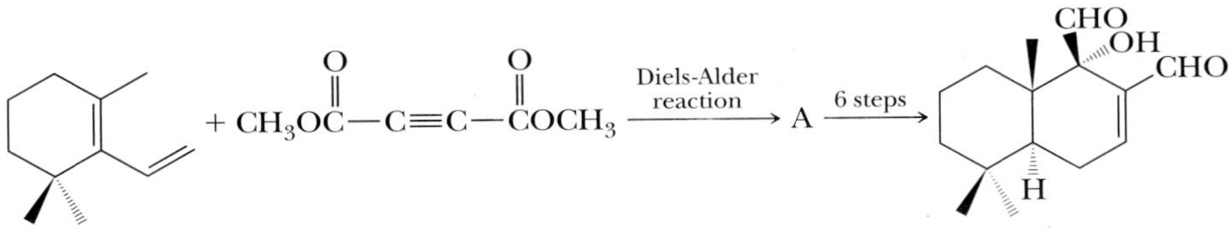

Warburganal
(racemic)

24.35 The Diels-Alder reaction is not limited to making six-membered rings with only carbon atoms. Predict the products of the following reactions that produce rings with atoms other than carbon in them.

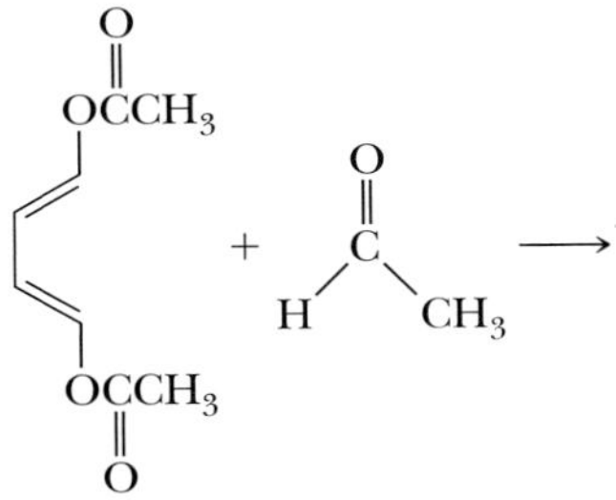

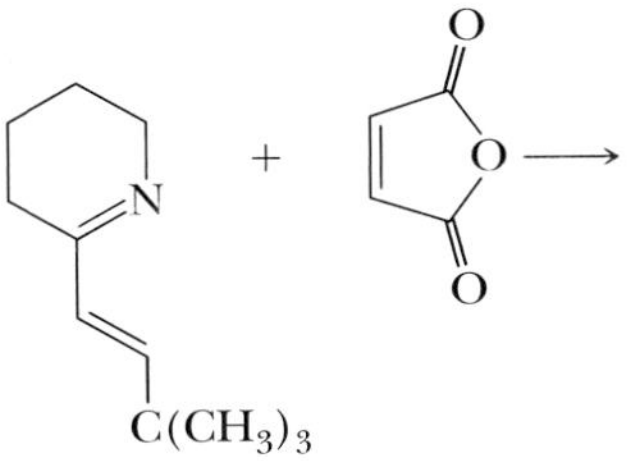

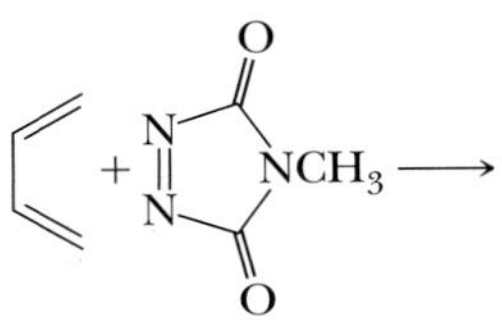

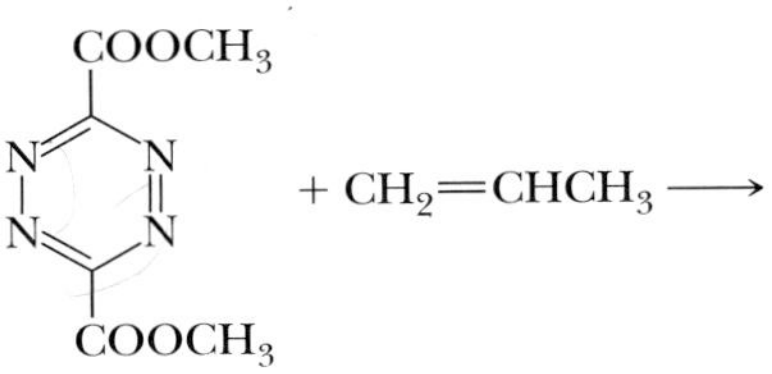

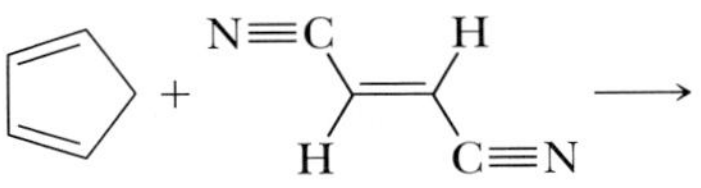

24.36 The first step in a synthesis of dodecahedrane involves a Diels-Alder reaction between the cyclopentadiene derivative (1) and dimethyl acetylenedicarboxylate (2). Show how these two molecules react to form the dodecahedrane synthetic intermediate (3).

$$+ \text{ CH}_3\text{OOCC} \equiv \text{CCOOCH}_3 \longrightarrow$$

Cyclopentadienyl-
cyclopentadiene
(1)

Dimethyl acetylene-
dicarboxylate
(2)

COOCH$_3$
COOCH$_3$
(3)

24.37 Bicyclo[2.2.1]-2,5-heptadiene can be prepared in two steps from cyclopentadiene and vinyl chloride. Provide a mechanism for each step.

$$+ \text{ CH}_2 = \text{CHCl} \xrightarrow{\text{heat}}$$

H
Cl

$$\xrightarrow[\text{C}_2\text{H}_5\text{OH}]{\text{C}_2\text{H}_5\text{ONa}}$$

Bicyclo[2.2.1]-2,5-heptadiene

24.38 Treatment of anthranilic acid with nitrous acid gives an intermediate, A, that contains a diazonium ion and a carboxylate group. When this intermediate is heated in the presence of furan, a bicyclic compound is formed. Propose a structural formula for compound A and a mechanism for the formation of the bicyclic product.

COOH

$$\xrightarrow{\text{NaNO}_2, \text{ HCl}} \text{[A]} \longrightarrow$$

NH$_2$

Anthranilic
acid

O

$+ \text{ CO}_2 + \text{N}_2$

24.39 All attempts to synthesize cyclopentadienone yield only a Diels-Alder adduct. Cycloheptatrienone, however, has been prepared by several methods and is stable. *Hint:* Consider important resonance contributing structures.

2   ⟶ a Diels-Alder adduct

Cyclopentadienone

Cycloheptatrienone

(a) Draw a structural formula for the Diels-Alder adduct formed by cyclopentadienone.

(b) How do you account for the marked difference in stability of these two ketones?

24.40 Following is a retrosynthetic scheme for the synthesis of the tricyclic diene on the left. Show how to accomplish this synthesis from 2-bromopropane, cyclopentadiene, and 2-cyclohexenone.

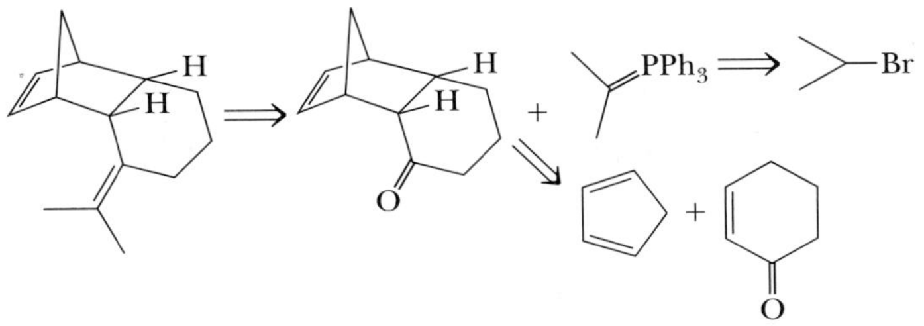

24.41 Claisen rearrangement of an allyl phenyl ether with substituent groups in both ortho positions leads to the formation of a para-substituted product. Propose a mechanism for the following rearrangement.

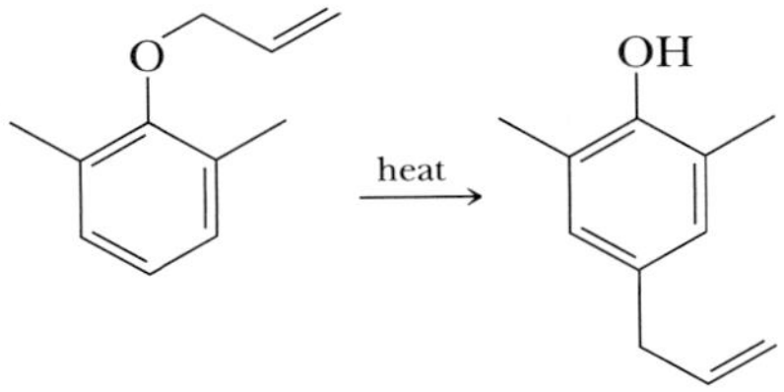

24.42 Following are three examples of Cope rearrangements of 1,5-dienes. Show that each product can be formed in a single step by a mechanism involving redistribution of six electrons in a cyclic transition state.

(a)

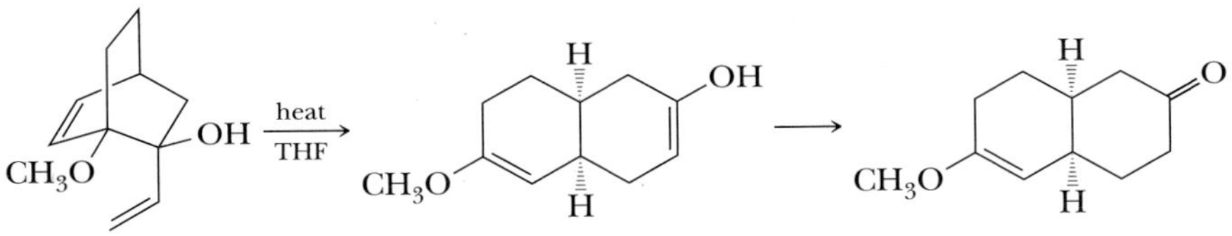

(b)

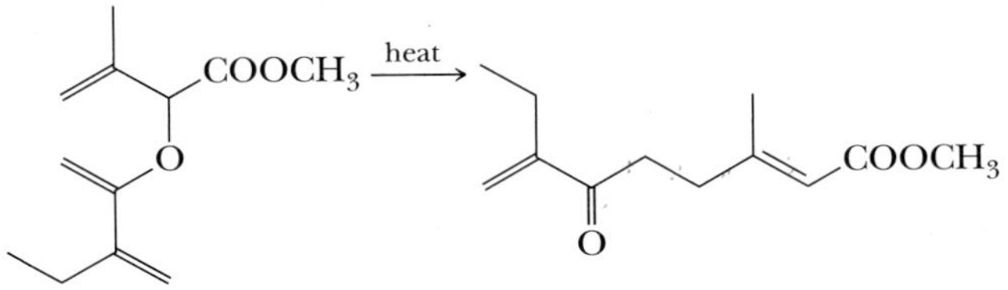

(c)

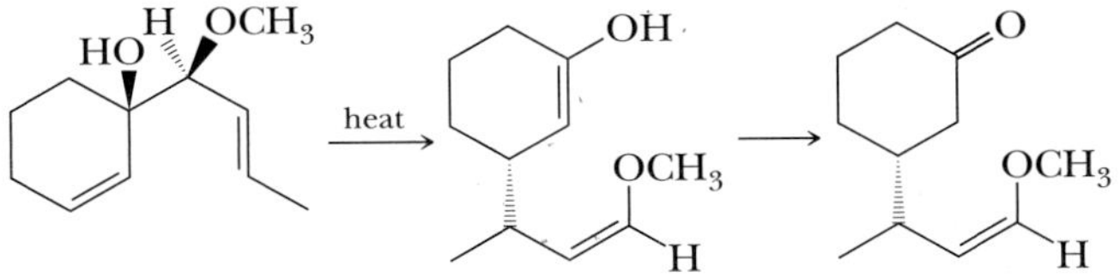

24.43 The following transformation is an example of the Carroll reaction, named after the English chemist M. F. Carroll, who first reported it. Propose a mechanism for this reaction.

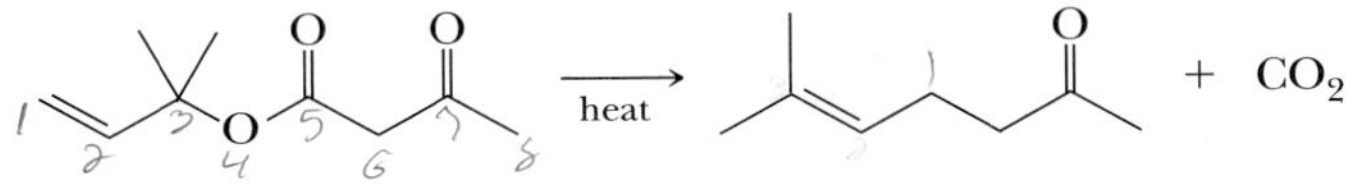

6-Methyl-5-hepten-2-one

24.44 Show the product of the following reaction. Include stereochemistry.

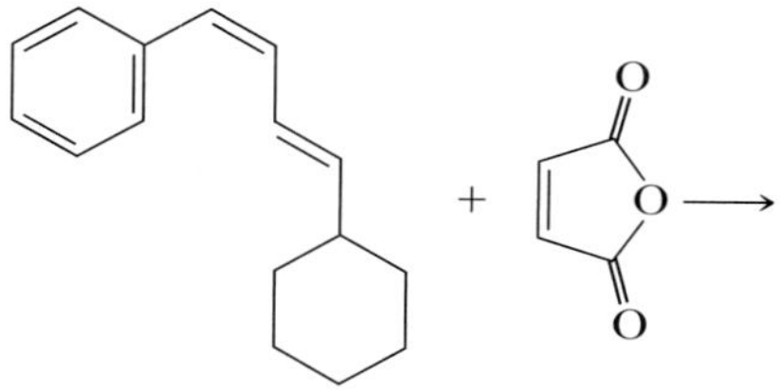

24.45 Following is a synthesis for the antifungal agent tolciclate.

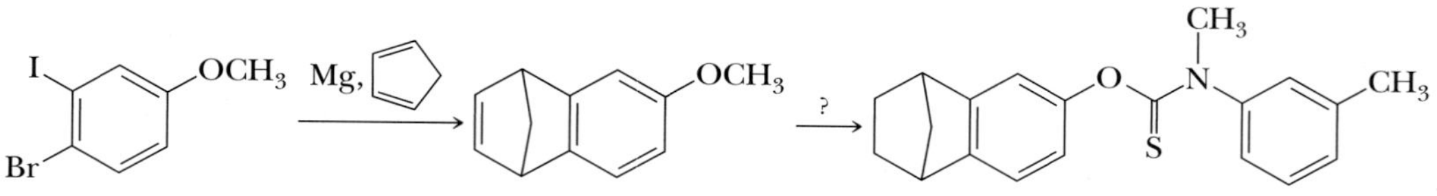

4-Bromo-3-
iodoanisole

(A)

Tolciclate

(a) Propose a mechanism for formation of (A).

(b) Show how (A) can be converted to tolciclate. Use 3-methyl-*N*-methylaniline as the source of the amine nitrogen and thiophosgene, $Cl_2C=S$, as the source of the $C=S$ group.

24.46 Ascaridole is a natural product from Oil of Chenopodium (from *chenopodium ambrosioides*, also called American wormseed, Mexican tea, epazote [from Nahuatl words for skunk and sweat!] or ambrosia; the herb is used in seasoning in Yucatán cuisine) that has been used to treat intestinal worms. After World War II, the German population was near starvation, and intestinal worms were a major problem. G. O. Schenck (then in Heidelberg, later Göttingen, and then Mühlheim) devised a remarkable industrial-scale synthesis of this compound in a bombed-out lot in Heidelberg. Using chlorophyll extracted from spinach, he used large carboys with a methanol solution of α-terpinene (isolated from natural oils such as cardamom) with air bubbling and sunlight to produce large amounts of this compound.

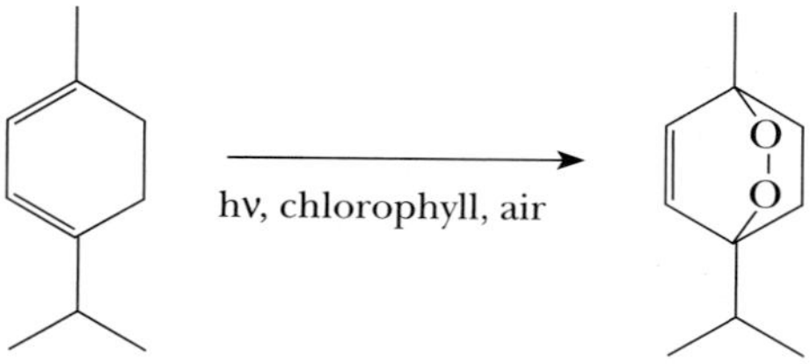

hv, chlorophyll, air

α-Terpinene

Ascaridole

Suggest a mechanism for this reaction.

24.47 The following transformation can be accomplished by two reactions we have studied in this chapter. Name the type of reaction used in each step.

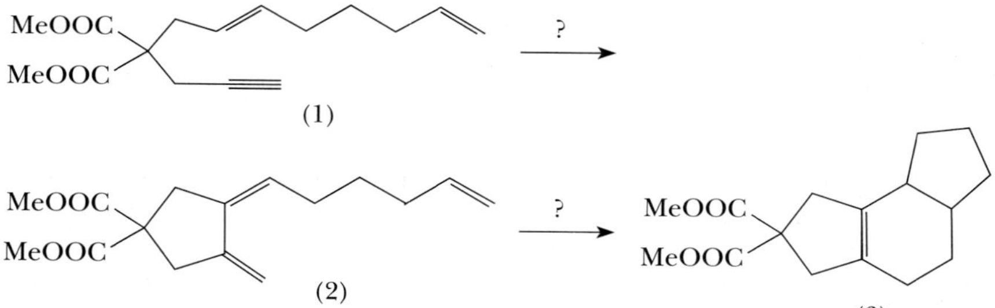

MeOOC
MeOOC
(1)

?

MeOOC
MeOOC
(2)

?

MeOOC
MeOOC
(3)

Synthesis

The following problems are based on relatively recent total syntheses of important natural products. Many such syntheses are outlined in compendia of synthetic reactions. Particularly valuable in preparing these problems were *Classics in Total Synthesis*, K. C. Nicolaou and E. J. Sorensen, Wiley-VCH Weinheim, New York, Basel, Cambridge, Tokyo, 1996; *Classics in Total Synthesis II*, K. C. Nicolaou and S. A. Snyder, Wiley-VCH Verlag GMBH, Weinheim (2003).

24.48 Following is an outline of the stereospecific synthesis of the "Corey lactone." Professor E. J. Corey (Harvard University) describes it in this way. "The first general synthetic route to all the known prostaglandins was developed by way of bicyclo[2.2.1]heptene intermediates. The design was guided by the requirements that the route be versatile enough to allow the synthesis of many analogs and also allow early resolution. This synthesis has been used on a large scale and in laboratories throughout the world; it has been applied to the production of countless prostaglandin analogs." Corey was awarded the 1990 Nobel Prize for chemistry for the development of retrosynthetic analysis for synthetic production of complex molecules. See E. J. Corey and Xue-Min Cheng, *The Logic of Chemical Synthesis*, John Wiley & Sons, New York, 1989, p. 255. For the structure of the prostaglandins, see Section 26.3. *Note:* The wavy lines in compound C indicate that the stereochemistry of —Cl and —CN groups was not determined. (The conversion of (D) to (E) involves an oxidation of the ketone group to a lactone by the Baeyer-Villiger reaction, which we have not studied in this text.)

(a) What is the function of sodium hydride, NaH, in the first step? What is the pK_a of cyclopentadiene? How do you account for its remarkable acidity?

(b) By what type of reaction is (B) converted to (C)?

(c) What is the function of the carbon dioxide added to the reaction mixture in Step 2 of the conversion of (E) to (F)? *Hint:* What happens when carbon dioxide is dissolved in water? Why not just use HCl?

(d) The tributyltin hydride, $(Bu)_3SnH$, used in the conversion of (H) to (I) reacts via a radical chain reaction; the first step involves a reaction with an radical initiator to form $(Bu)_3Sn\cdot$. Suggest a mechanism for the rest of the reaction.

(e) The Corey lactone contains four chiral centers with the relative configurations shown. In what step or steps in this synthesis is the configuration of each chiral center determined? Propose a mechanism to account for the observed stereospecificity of the relevant steps.

(f) Compound (F) was resolved using (+)-ephedrine. Following is the structure of (−)-ephedrine, the naturally occurring stereoisomer. What is meant by "resolution" and what is the rationale for using a chiral, enantiomerically pure amine for the resolution of (F)?

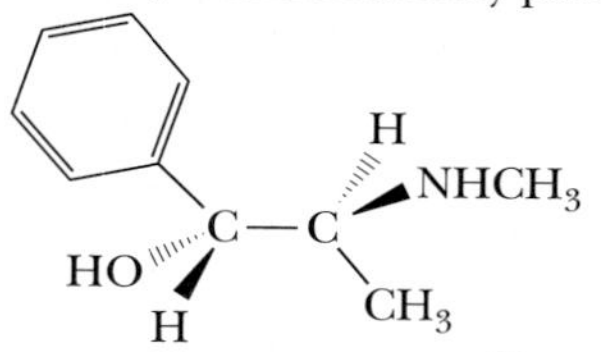

Ephedrine $[\alpha]_D^{21}$ −41

(g) You have not studied the Baeyer-Villiger reaction (D to E). The mechanism involves nucleophilic reaction of the peroxyacid with the carbonyl, followed by a rearrangement much like that involved in the hydroboration reaction (Section 6.4). Write a mechanism for this reaction.

Note: By resolving at this stage, one half of the material is discarded. A more efficient route would be to have an earlier resolution; in fact, Corey solved this problem in a very elegant way by using an enantioselective Diels-Alder with the alkene in the form of an acrylate ester of enantiomerically pure 8-phenylmenthol. Asymmetric induction gave a product with a diastereoselectivity of 97:3. So rather than resolving, he was able to get the correct stereoisomer directly!

24.49 Chapman's (O. L. Chapman, then at Iowa State, later UCLA) classic total synthesis of (±)-Carpanone is so remarkably simple that it is used as an undergraduate laboratory preparation. It is modeled on a possible biosynthetic route for this lignan-derived natural product. Phenol oxidations figure prominently in many such biosyntheses of natural products. In one step, this reaction creates no less than five contiguous chiral centers, all in the correct relative configuration.

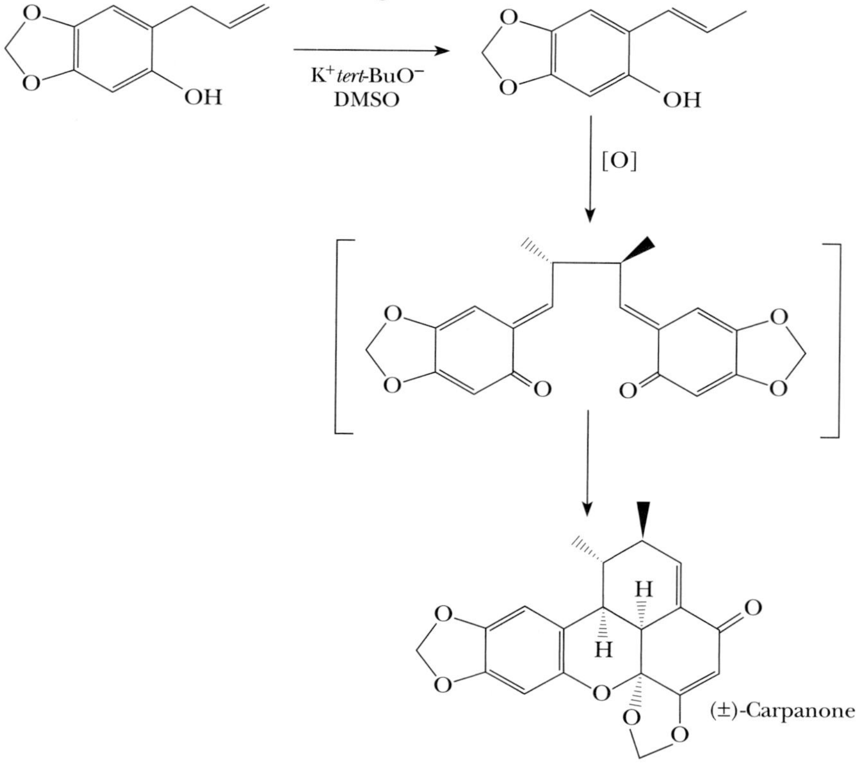

(±)-Carpanone

(a) Give a mechanism for the first step of the reaction and explain why it goes in the direction it does.

(b) The oxidation step uses a palladium salt. Suggest a mechanism for this coupling, which you have not encountered. (*Hint:* Do not concern yourself with the role of the metal except as an acceptor of electrons.)

(c) The third step is spontaneous. Give a mechanism for this reaction and show how it accounts for the stereochemistry of the final product.

(d) Would you expect the product to be racemic or a single enantiomer?

24.50 Gilvocarcin M is isolated from *Streptomyces* strains and has strong antitumor activity.

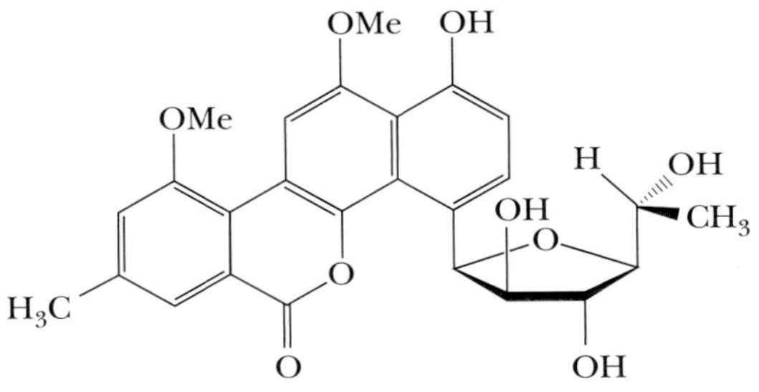

Gilvocarcin M

Suzuki and coworkers were able to carry out the total synthesis of naturally occurring (−)-gilvocarcin M. Their synthesis included the following steps. (The wavy line means stereochemistry is unspecified or a mixture.) The stereochemistry of the product appears to be counterintuitive (apparent attack from the more hindered side). The reason is that the reaction involves initial O-alkylation followed by a rearrangement that need not concern us.

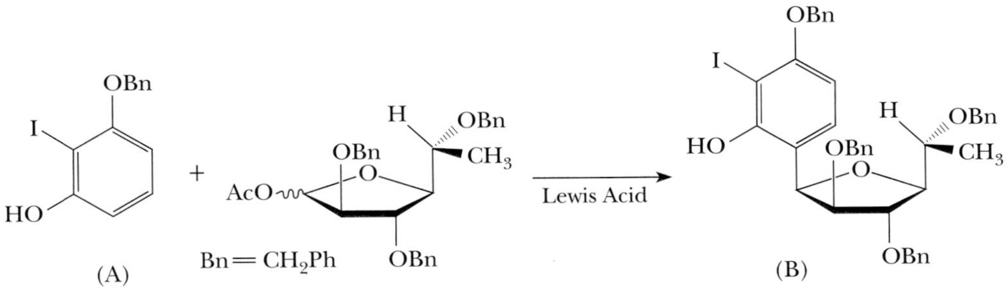

(a) This reaction gives both high regioselectivity and stereoselectivity. What other products might have been expected?

The next step involves triflation and treatment with butyl lithium.

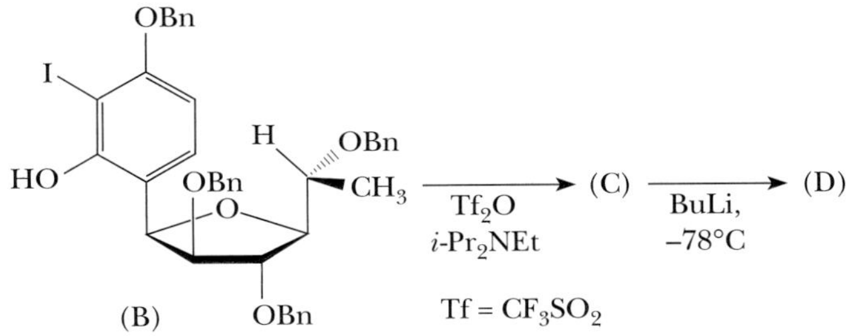

(b) Give a structure for C.

(c) Give a structure for D. This reaction requires that you know that lithium reagents can interchange with aryl halides:

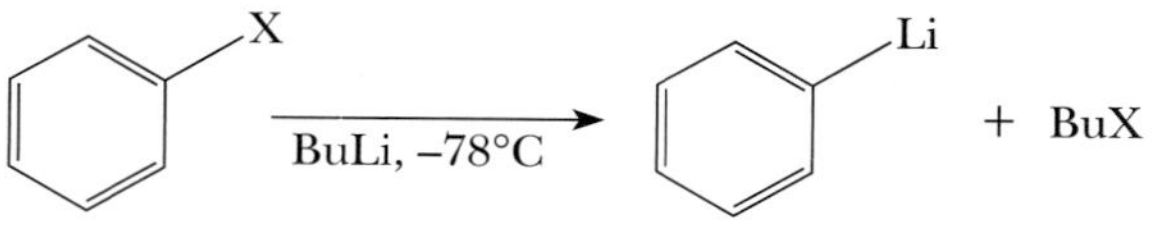

Recall that OTf is an excellent leaving group. You may wish to review Section 24.3A. The reaction yielding D is carried out in the presence of 2-methoxyfuran. D decomposes under the conditions to a compound E that instantly reacts with the furan to give F.

(d) Give a structure for E and the mechanism of D to E.

(e) Give a mechanism for E to F.

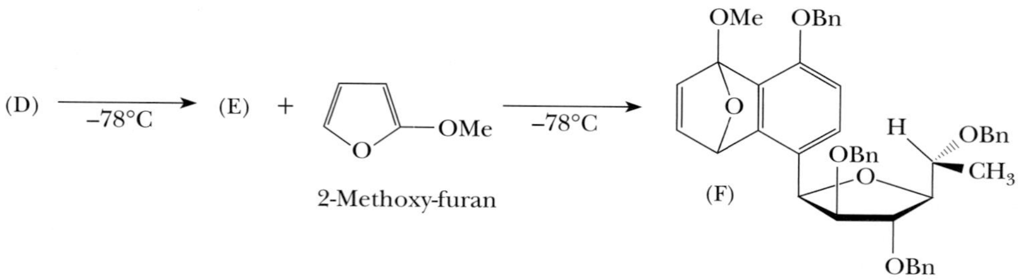

2-Methoxy-furan

F is unstable and undergoes ring opening on workup to give G.

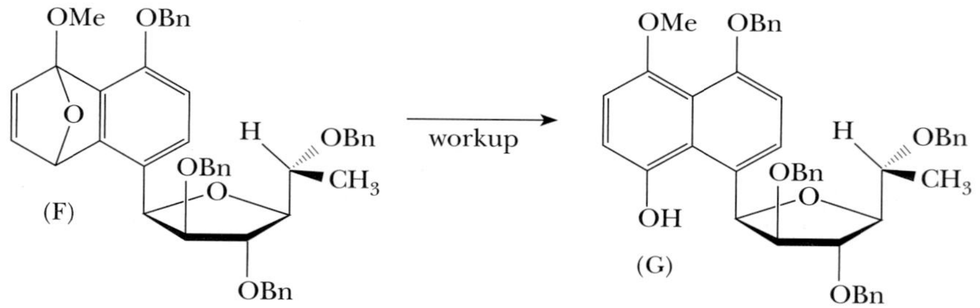

(f) Give a mechanism for F to G.

The next step involves conversion of G to H.

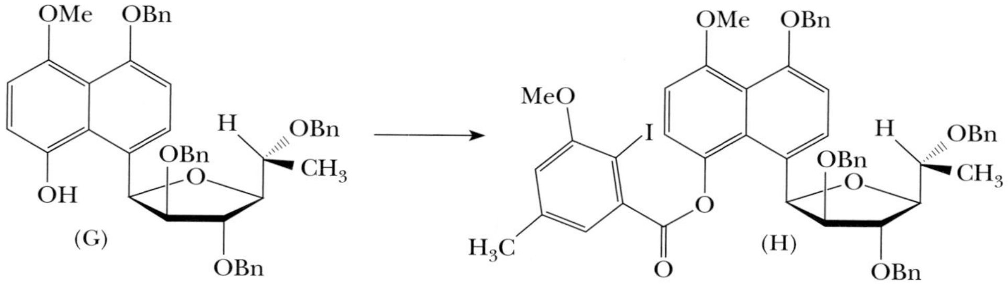

(g) Give reagents and conditions required for G to H.

Formation of the final tetracyclic ring involves conversion of H to I.

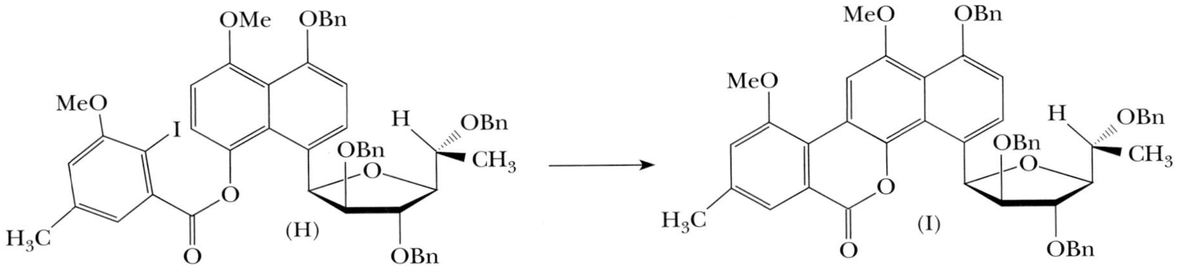

(h) Give reagents and conditions required for H to I.

I is then is converted to (−)-gilvocarcin M, the natural enantiomer.

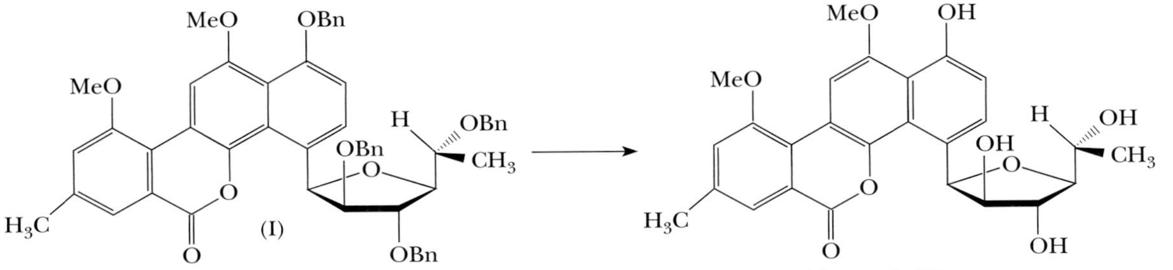

Gilvocarcin M

(i) What reagents could be used for this reaction?

(j) Comment on the probable source of the chiral centers in this synthesis. Note that the chirality was not created in any of the reaction steps. You can find a possible readily available and inexpensive source (see Chapter 25, Carbohydrates).

(k) Given reactions that are later in the sequence, why is it necessary to protect some of the OH groups as the benzyl ether? What side reactions would occur without this protection? Starting with OH groups, how would you add these protecting groups?

24.51 Vancomycin is an important antibiotic. It is isolated from the bacterium *Streptomyces orientalis* and functions by inhibiting bacterial mucopeptide synthesis. It is a last line of defense against the resistant Staph organisms that are now common in hospitals.

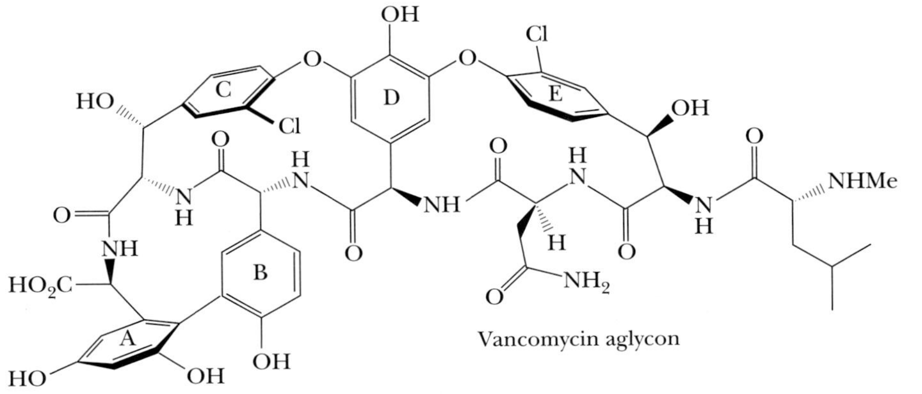

Vancomycin aglycon

In 1999, Professor Dale Boger (The Scripps Research Institute) reported a synthesis of vancomycin **aglycon** (aglycon = lacking a sugar) involving the following steps, among others. Compound I was prepared from simple starting materials by a series of steps involving forming amide bonds.

Aglycon Lacking a sugar.

(a) Suggest reasonable precursors and show how the bonds could be formed (the actual reagents used have not been introduced, but they work in a similar way to those you know).

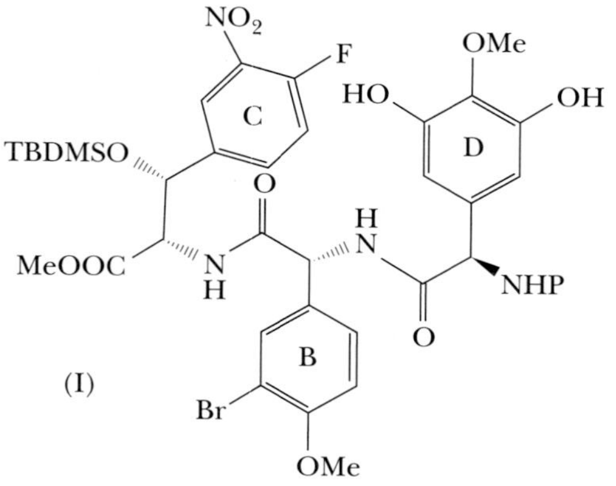

(I)

P = Protecting Group

I was then converted into II.

(b) Give reagents for this reaction and suggest the mechanism.

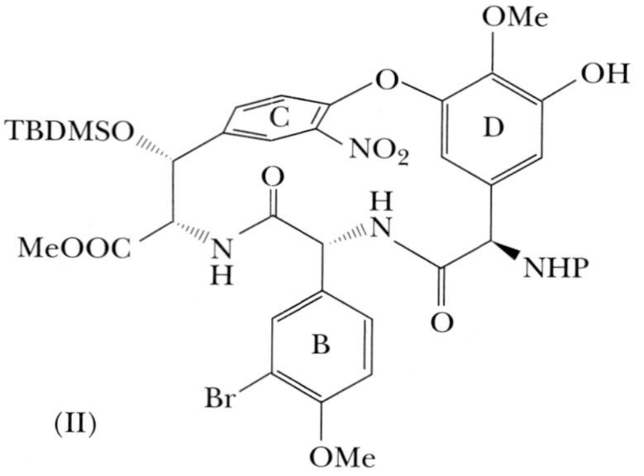

(II)

P = Protecting Group

One of the interesting features of this synthesis is that Ring C in compound II (and subsequent compounds in this synthesis) has extremely hindered rotation. As a result, compound II exists as two atropisomers (Section 3.2) that are interconverted only at 140°C.

(c) Show these two isomers.

II was then converted to III.

(d) Suggest reagents to accomplish this transformation.

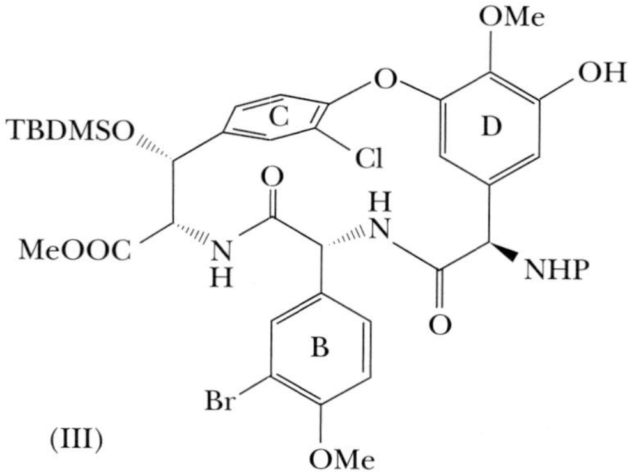

(III)

P = Protecting Group

III was then converted to IV.

(e) Suggest reagents and the ring A fragment that could be used for this reaction.

Closure of an amide link between the amine on ring A (after removal of the protecting group) and the carbomethoxy group above it led to a precursor of vancomycin.

(f) Show the ring closure reaction of the deprotected free amino group and its mechanism.

Another interesting feature of this synthesis is that ring A and B also form atropisomers. These can be converted into a 3:1 mixture of the desired and undesired atropisomers on heating at 120°C.

(g) Draw these atropisomers and show that only one can be converted to vancomycin. The synthesis of the aglycon was completed by functional manipulation and addition of ring E by chemistry similar to that detailed earlier. Yet another set of atropisomers (this time of ring E) was formed! However, this one was more easily equilibrated than the others; model studies had shown that the activation barrier for this set of atropisomers should be lower that that of the others.

24.52 E. J. Corey's 1964 total synthesis of α-caryophyllene (essence of cloves) solves a number of problems of construction of unusual-sized rings.

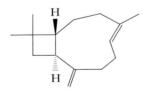

α-Caryophyllene

The first step uses an efficient photochemical [2 + 2] reaction. The desired stereochemistry and regiochemistry had been predicted based on model reactions.

(a) [2 + 2] Reactions are quite common in photochemical reactions. Would this reaction be predicted to occur in the ground state?

The next steps follow. Basic alumina is a chromatography support that will often act as a base catalyst.

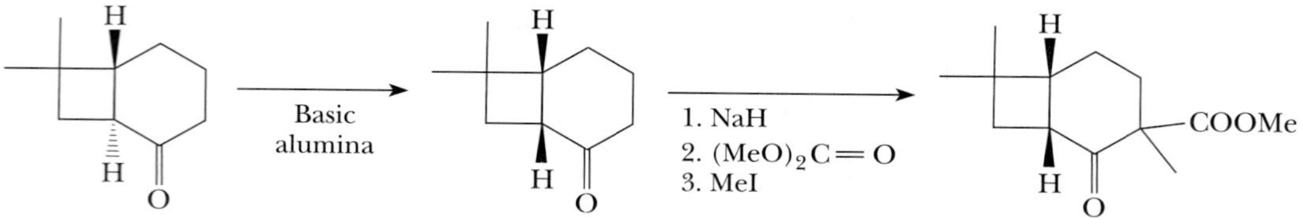

(b) What is the mechanism of the first step?
(c) What is the mechanism of the second step?
(d) Look at later steps in the synthesis. Does the stereochemistry of the added carbomethoxy group matter?

The next steps are shown here.

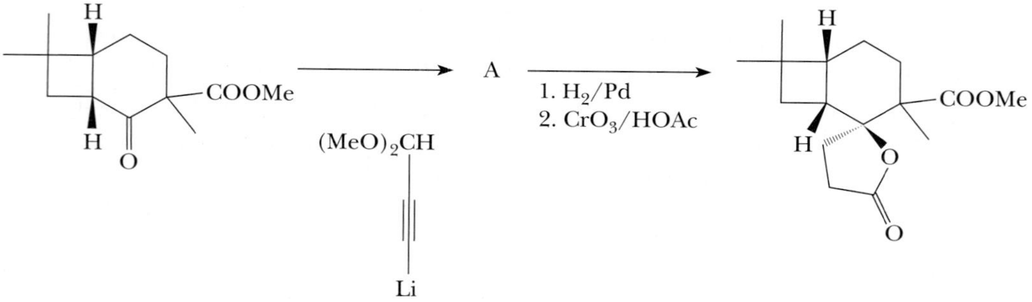

(e) What is the structure of compound A?
(f) Give a mechanism for the formation of the cyclized product.

Here are the next steps.

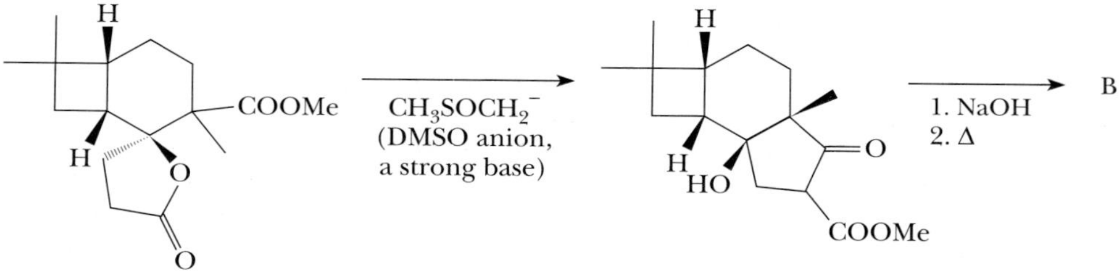

(g) Give a mechanism for the first step. (*Hint:* Attack on the lactone carbonyl may be the first step.)
(h) Give a structure for product B.

The following two steps are next.

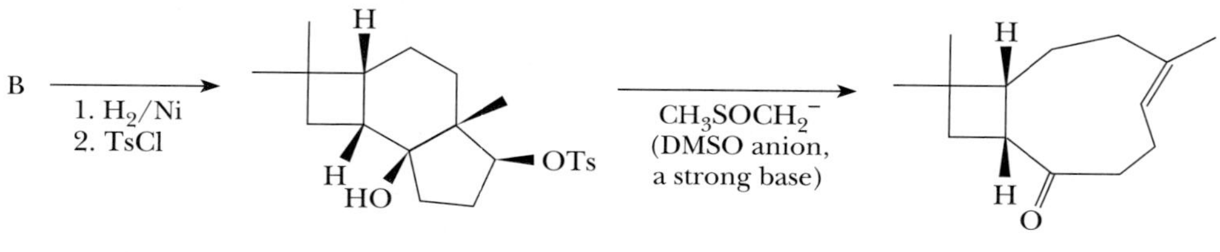

(i) Show the reactions of B.

(j) Write a mechanism for the ring-opening reaction. *Hint:* Note the presence of an acidic proton and a good leaving group in the molecule.

The synthesis was completed by the following steps.

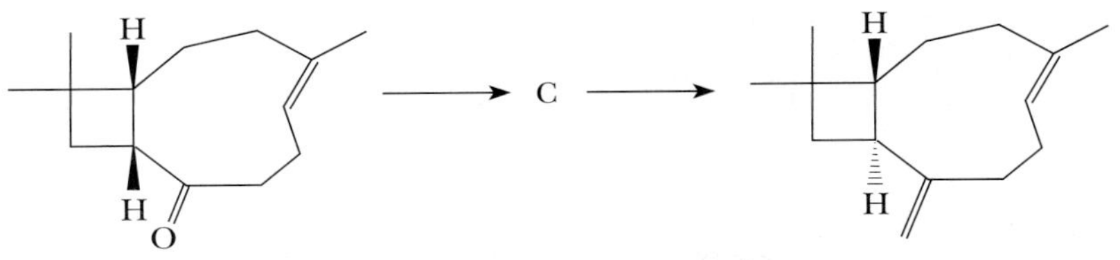

(k) What is C?

(l) What reagents would you use for these transformations?

24.53 Over the past several decades, chemists have developed a number of synthetic methodologies for the synthesis of steroid hormones. One of these, developed by Lutz Tietze at the Institut für Organische Chemie der Georg-August-Universität, Göttingen, Germany, used a double Heck reaction to create ring B of the steroid nucleus. As shown in the following retrosynthetic analysis, a key intermediate in his synthesis is Compound 1. Two Heck reaction disconnects of this intermediate give compounds (2) and (3). Compound (2) contains the aromatic ring that becomes ring A of estrone. Compound (3) contains the fused five- and six-membered rings that become rings C and D of estrone.

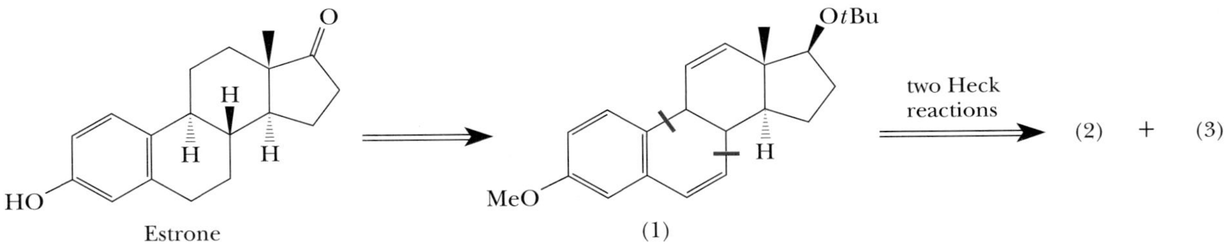

(a) Name the types of functional groups in estrone.

(b) How many chiral centers are present in estrone?

(c) Propose structural formulas for Compounds (2) and (3).

(d) Show how your proposals for Compounds (2) and (3) can be converted to Compound (1). *Note:* In the course of developing this synthesis, Tietze discovered that vinylic bromides and iodides are more reactive in Heck reactions than aryl bromides and iodides.

(e) In the course of the double Heck reactions, two new chiral centers are created. Assume that in Compound (3), the precursor to rings C and D of estrone, the fusion of rings C and D is *trans* and that the angular methyl group is above the plane of the ring. Given this stereochemistry, predict the stereochemistry of Compound (1) formed by the double Heck reaction.

(f) To convert (1) to estrone, the *tert*-butyl ether on ring D must be converted to a ketone. How might this transformation be accomplished?

25

Christi Carter/Grant Heilman, Inc.

Foxglove (*Digitalis purpurea*), an ornamental flowering plant, is the source of digitoxin and digitalis, medicines used in cardiology to reduce pulse rate, regularize heart rhythm, and strengthen heart beat. Inset: Digitoxose, a monosaccharide obtained on hydrolysis of digitoxin. See Problem 25.15.

Outline

Carbohydrates

C arbohydrates are the most abundant organic compounds in the plant world. They act as storehouses of chemical energy (glucose, starch, glycogen); are components of supportive structures in plants (cellulose), crustacean shells (chitin), and connective tissues in animals (acidic polysaccharides); and are essential components of nucleic acids (D-ribose and 2-deoxy-D-ribose). Carbohydrates make up about three fourths of the dry weight of plants. Animals (including humans) get their carbohydrates by eating plants, but they do not store much of what they consume. Less than 1% of the body weight of animals is made up of carbohydrates.

The name *carbohydrate* means hydrate of carbon and derives from the formula $C_n(H_2O)_m$. Following are two examples of carbohydrates with molecular formulas that can be written alternatively as hydrates of carbon.

Glucose (blood sugar): $C_6H_{12}O_6$, or alternatively $C_6(H_2O)_6$
Sucrose (table sugar): $C_{12}H_{22}O_{11}$, or alternatively $C_{12}(H_2O)_{11}$

Not all carbohydrates, however, have this general formula. Some contain too few oxygen atoms to fit this formula, and some others contain too many oxygens. Some also contain nitrogen. The term carbohydrate has become so firmly rooted in chemical nomenclature that, although not completely accurate, it persists as the name for this class of compounds.

At the molecular level, most **carbohydrates** are polyhydroxyaldehydes, polyhydroxyketones, or compounds that yield either of these after hydrolysis. Therefore, the chemistry of carbohydrates is essentially the chemistry of hydroxyl groups and carbonyl groups, and of the acetal bonds formed between these two functional groups.

The fact that carbohydrates have only two types of functional groups, however, belies the complexity of their chemistry. All but the simplest carbohydrates contain multiple chiral centers. For example, glucose, the most abundant carbohydrate in the biological world, contains one aldehyde group, one primary and four secondary hydroxyl groups, and four chiral centers. Working with molecules of this complexity presents enormous challenges to organic chemists and biochemists alike.

Carbohydrate A polyhydroxyaldehyde, a polyhydroxyketone, or a substance that gives these compounds on hydrolysis.

25.1 Monosaccharides

A. Structure and Nomenclature

Monosaccharides have the general formula $C_nH_{2n}O_n$ with one of the carbons being the carbonyl group of either an aldehyde or a ketone. The most common monosaccharides have three to eight carbon atoms. The suffix *-ose* indicates that a molecule is a carbohydrate, and the prefixes *tri-*, *tetr-*, and *pent-*, and so forth indicate the number of carbon atoms in the chain. Monosaccharides containing an aldehyde group are classified as **aldoses;** those containing a ketone group are classified as **ketoses.**

ORGANIC
Chemistry Now™
Click Molecular Models to examine **Models of Carbohydrates**

Monosaccharide A carbohydrate that cannot be hydrolyzed to a simpler carbohydrate.

Aldose A monosaccharide containing an aldehyde group.

Ketose A monosaccharide containing a ketone group.

Monosaccharides Classified by Number of Carbon Atoms	
Name	**Formula**
Triose	$C_3H_6O_3$
Tetrose	$C_4H_8O_4$
Pentose	$C_5H_{10}O_5$
Hexose	$C_6H_{12}O_6$
Heptose	$C_7H_{14}O_7$
Octose	$C_8H_{16}O_8$

There are only two trioses: the aldotriose glyceraldehyde and the ketotriose dihydroxyacetone.

```
CHO              CH₂OH
|                |
CHOH             C=O
|                |
CH₂OH            CH₂OH

Glyceraldehyde   Dihydroxyacetone
(an aldotriose)  (a ketotriose)
```

George Semple

■ 1,3-Dihydroxypropanone, more commonly known as dihydroxyacetone, is the active ingredient in artificial tanning agents such as Man-Tan and Magic Tan.

Often the designations *aldo-* and *keto-* are omitted, and these molecules are referred to simply as trioses, tetroses, and the like.

Glyceraldehyde is a common name; the IUPAC name for this monosaccharide is 2,3-dihydroxypropanal. Similarly, dihydroxyacetone is a common name; its IUPAC name is 1,3-dihydroxypropanone. The common names for these and other monosaccharides, however, are so firmly rooted in the literature of organic chemistry and biochemistry that they are used almost exclusively to refer to these compounds. Therefore, throughout our discussions of the chemistry and biochemistry of carbohydrates, we use the names most common in the literature of chemistry and biochemistry.

B. Fischer Projection Formulas

Glyceraldehyde contains a chiral center and therefore exists as a pair of enantiomers.

CHO
H►C◄OH
CH₂OH

(R)-Glyceraldehyde

CHO
HO►C◄H
CH₂OH

(S)-Glyceraldehyde

Fischer projection   A two-dimensional representation for showing the configuration of chiral centers; horizontal lines represent bonds projecting forward, and vertical lines represent bonds projecting to the rear.

Chemists commonly use two-dimensional representations called **Fischer projections** (Section 3.4C) to show the configuration of carbohydrates. Following is an illustration of how a three-dimensional representation is converted to a Fischer projection.

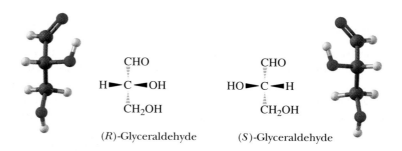

CHO
H►C◄OH
CH₂OH

(R)-Glyceraldehyde
(three-dimensional
representation)

convert to a
Fischer projection
⟶

CHO
H——OH
CH₂OH

(R)-Glyceraldehyde
(Fischer projection)

The horizontal segments of a Fischer projection represent bonds directed toward you, and the vertical segments represent bonds directed away from you. The only atom in the plane of the paper is the chiral center.

C. D- and L-Monosaccharides

Even though the *R,S* system is widely accepted today as a standard for designating configuration, the configuration of carbohydrates as well as those of amino acids and many other compounds in biochemistry is commonly designated by the D,L system proposed by Emil Fischer in 1891. At that time, it was known that one enantiomer of glyceraldehyde has a specific rotation of $+13.5$; the other has a specific rotation of -13.5. Fischer proposed that these enantiomers be designated D and L (for dextro- and levorotatory) but he had no experimental way to determine which enantiomer has which specific rotation. Fischer, therefore, did the only possible thing—he made an arbitrary assignment. He assigned the dextrorotatory enantiomer an arbitrary configuration and named it D-glyceraldehyde. He named its enantiomer L-glyceraldehyde.

$$
\begin{array}{cc}
\text{CHO} & \text{CHO} \\
\text{H}\!-\!\!-\!\!\text{OH} & \text{HO}\!-\!\!-\!\!\text{H} \\
\text{CH}_2\text{OH} & \text{CH}_2\text{OH} \\
\text{D-Glyceraldehyde} & \text{L-Glyceraldehyde} \\
[\alpha]_D^{25} = +13.5 & [\alpha]_D^{25} = -13.5
\end{array}
$$

Fischer could have been wrong, but by a stroke of good fortune he was correct, as proven in 1952 by a special application of X-ray crystallography.

D- and L-glyceraldehyde serve as reference points for the assignment of relative configuration to all other aldoses and ketoses. The reference point is the chiral center farthest from the carbonyl group. Because this chiral center is always the next to the last carbon on the chain, it is called the **penultimate carbon.** A **D-monosaccharide** has the same configuration at its penultimate carbon as D-glyceraldehyde (its —OH is on the right when written as a Fischer projection); an **L-monosaccharide** has the same configuration at its penultimate carbon as L-glyceraldehyde (its —OH is on the left).

Table 25.1 shows names and Fischer projections for all D-aldotetroses, pentoses, and hexoses. Each name consists of three parts. The letter D specifies the configuration of the penultimate carbon. Prefixes such as *rib-*, *arabin-*, and *gluc-* specify the configuration of all other chiral centers in the monosaccharide. The suffix *-ose* shows that the compound is a carbohydrate.

The three most abundant hexoses in the biological world are D-glucose, D-galactose, and D-fructose. The first two are D-aldohexoses; the third is a D-2-ketohexose. Glucose, by far the most common hexose, is also known as dextrose because it is dextrorotatory. Other names for this monosaccharide are grape sugar and blood sugar. Human blood normally contains 65–110 mg of glucose/100 mL of blood. Glucose is synthesized by chlorophyll-containing plants using sunlight as a source of energy. In the process called photosynthesis, plants convert carbon dioxide from the air and water from the soil to glucose and oxygen.

$$
6\text{CO}_2 \;+\; 6\text{H}_2\text{O} \;+\; \text{energy} \;\xrightarrow[\text{chlorophyll}]{\text{sunlight}}\; \text{C}_6\text{H}_{12}\text{O}_6 \;+\; 6\text{O}_2
$$

Carbon dioxide   Water   Glucose   Oxygen

D-Fructose is found combined with glucose in the disaccharide sucrose (table sugar, Section 25.4A). D-Galactose is obtained with glucose in the disaccharide lactose (milk sugar, Section 25.4B).

D-Ribose and 2-deoxy-D-ribose, the most abundant pentoses in the biological world, are essential building blocks of nucleic acids; D-ribose in ribonucleic acids (RNA) and 2-deoxy-D-ribose in deoxyribonucleic acids (DNA).

D-Monosaccharide A monosaccharide that, when written as a Fischer projection, has the —OH on its penultimate carbon to the right.

L-Monosaccharide A monosaccharide that, when written as a Fischer projection, has the —OH on its penultimate carbon to the left.

Table 25.1   Configurational Relationships Among the Isomeric D-Aldotetroses, D-Aldopentoses, and D-Aldohexoses

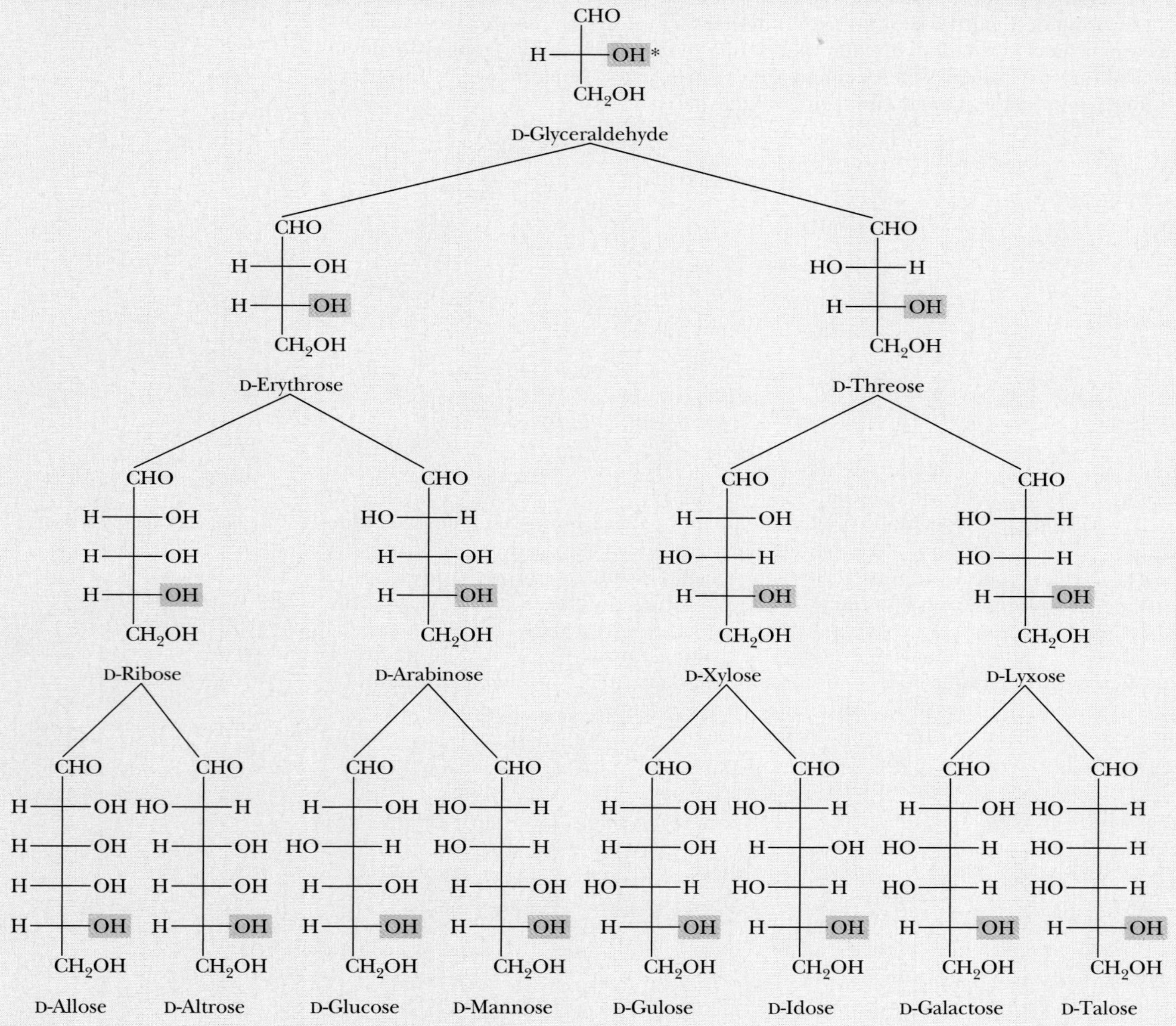

CHO
H——OH *
CH₂OH

D-Glyceraldehyde

CHO
H——OH
H——OH
CH₂OH

D-Erythrose

CHO
HO——H
H——OH
CH₂OH

D-Threose

CHO
H——OH
H——OH
H——OH
CH₂OH

D-Ribose

CHO
HO——H
H——OH
H——OH
CH₂OH

D-Arabinose

CHO
H——OH
HO——H
H——OH
CH₂OH

D-Xylose

CHO
HO——H
HO——H
H——OH
CH₂OH

D-Lyxose

CHO
H——OH
H——OH
H——OH
H——OH
CH₂OH

D-Allose

CHO
HO——H
H——OH
H——OH
H——OH
CH₂OH

D-Altrose

CHO
H——OH
HO——H
H——OH
H——OH
CH₂OH

D-Glucose

CHO
HO——H
HO——H
H——OH
H——OH
CH₂OH

D-Mannose

CHO
H——OH
H——OH
HO——H
H——OH
CH₂OH

D-Gulose

CHO
HO——H
H——OH
HO——H
H——OH
CH₂OH

D-Idose

CHO
H——OH
HO——H
HO——H
H——OH
CH₂OH

D-Galactose

CHO
HO——H
HO——H
HO——H
H——OH
CH₂OH

D-Talose

*The configuration of the reference —OH on the penultimate carbon is shown in color.

Example 25.1

Draw Fischer projections for the four aldotetroses. Which are D-monosaccharides, which are L-monosaccharides, and which are enantiomers? Refer to Table 25.1, and write the name of each aldotetrose.

Solution

Following are Fischer projections for the four aldotetroses. The letters D- and L- refer to the configuration of the penultimate carbon, which, in the case of aldotetroses, is carbon 3. In the Fischer projection of a D-aldotetrose, the —OH on carbon 3 is on the right, and, in an L-aldotetrose, it is on the left. The erythroses are each diastereomers of each of the threoses.

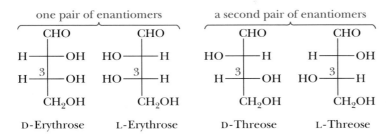

Problem 25.1

Draw Fischer projections for all 2-ketopentoses. Which are D-2-ketopentoses, which are L-2-ketopentoses, and which are enantiomers?

D. Amino Sugars

Amino sugars contain an —NH_2 group in place of an —OH group. Only three amino sugars are common in nature: D-glucosamine, D-mannosamine, and D-galactosamine.

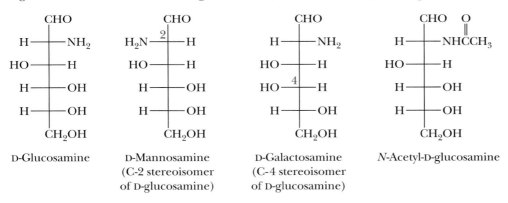

N-Acetyl-D-glucosamine, a derivative of D-glucosamine, is a component of many polysaccharides, including chitin, the hard shell-like exoskeleton of lobsters, crabs, shrimp, and other shellfish. Many other amino sugars are components of naturally occurring antibiotics.

E. Physical Properties

Monosaccharides are colorless, crystalline solids, although they often crystallize with difficulty. Because hydrogen bonding is possible between their polar —OH groups and water, all monosaccharides are very soluble in water. They are only slightly soluble in ethanol and are insoluble in nonpolar solvents such as diethyl ether, chloroform, and benzene.

25.2 The Cyclic Structure of Monosaccharides

We saw in Section 16.7B that aldehydes and ketones react with alcohols to form hemiacetals. We also saw that cyclic hemiacetals form very readily when hydroxyl and carbonyl groups are part of the same molecule and their interaction can form a five- or six-membered ring. For example, 4-hydroxypentanal forms a five-membered cyclic hemiacetal.

4-Hydroxypentanal   A cyclic hemiacetal

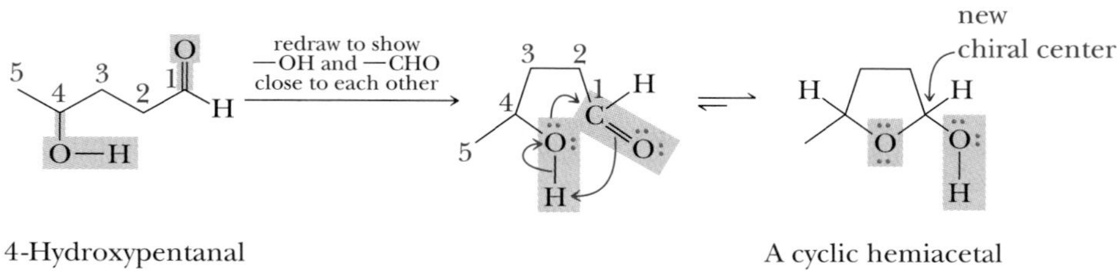

Note that 4-hydroxypentanal contains one chiral center and that a second chiral center is generated at carbon 1 as a result of hemiacetal formation.

Monosaccharides have hydroxyl and carbonyl groups in the same molecule. As a result, they too exist almost exclusively as five- and six-membered cyclic hemiacetals.

A. Haworth Projections

A common way of representing the cyclic structure of monosaccharides is the **Haworth projection,** named after the English chemist Sir Walter N. Haworth (1937 Nobel Prize for chemistry). In a Haworth projection, a five- or six-membered cyclic hemiacetal is represented as a planar pentagon or hexagon, as the case may be, lying perpendicular to the plane of the paper. Groups bonded to the carbons of the ring then lie either above or below the plane of the ring. The new chiral center created in forming the cyclic structure is called an **anomeric carbon.** Stereoisomers that differ in configuration only at the anomeric carbon are called **anomers.** The anomeric carbon of an aldose is carbon 1; that of the most common ketoses is carbon 2.

Haworth projections are most commonly written with the anomeric carbon to the right and the hemiacetal oxygen to the back (Figure 25.1). In the terminology of carbohydrate chemistry, the designation β means that the —OH on the anomeric carbon of the cyclic hemiacetal is on the same side of the ring as the terminal —CH$_2$OH. Conversely, the designation α means that the —OH on the anomeric

ORGANIC
Chemistry·⚛·Now™

Click Tutorials to review **Haworth Projections**

Haworth projection   A way to view furanose and pyranose forms of monosaccharides. The ring is drawn flat and most commonly viewed through its edge with the anomeric carbon on the right and the oxygen atom of the ring in the rear.

Anomeric carbon   The hemiacetal or acetal carbon of the cyclic form of a carbohydrate.

Anomers   Carbohydrates that differ in configuration only at their anomeric carbons.

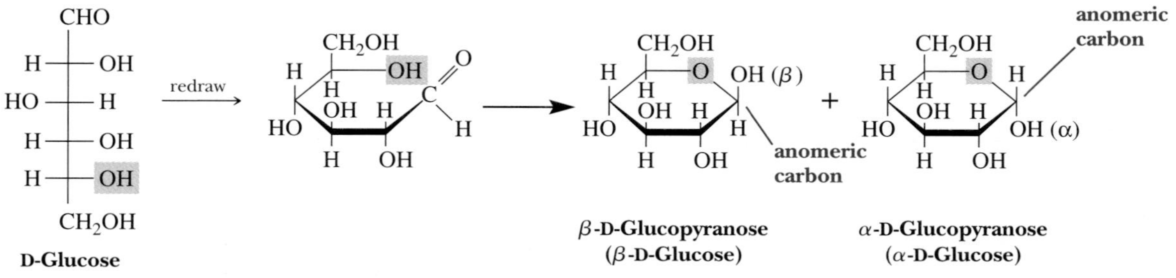

Figure 25.1
Haworth projections for α-D-glucopyranose and β-D-glucopyranose.

carbon of the cyclic hemiacetal is on the side of the ring opposite from the terminal —CH₂OH.

A six-membered hemiacetal ring is indicated by the infix *-pyran-*, and a five-membered hemiacetal ring is indicated by the infix *-furan-*, The terms **furanose** and **pyranose** are used because monosaccharide five- and six-membered rings correspond to the heterocyclic compounds furan and pyran.

Furanose A five-membered cyclic form of a monosaccharide.

Pyranose A six-membered cyclic form of a monosaccharide.

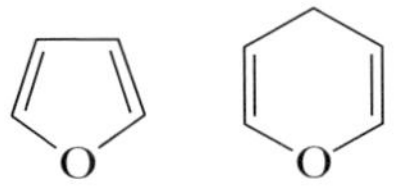

Furan   Pyran

Because the α and β forms of glucose are six-membered cyclic hemiacetals, they are named α-D-glucopyranose and β-D-glucopyranose. These infixes are not always used in monosaccharide names, however. Thus, the glucopyranoses, for example, are often named simply α-D-glucose and β-D-glucose.

You would do well to remember the configuration of groups on the Haworth projections of α-D-glucopyranose and β-D-glucopyranose as reference structures. Knowing how the open-chain configuration of any other aldohexoses differs from that of D-glucose, you can then construct its Haworth projection by reference to the Haworth projection of D-glucose.

Example 25.2

Draw Haworth projections for the α and β anomers of D-galactopyranose.

Solution

One way to arrive at these projections is to use the α and β forms of D-glucopyranose as reference and to remember (or discover by looking at Table 25.1) that D-galactose differs from D-glucose only in the configuration at carbon 4. Thus, begin with the Haworth projections shown in Figure 25.1 and then invert the configuration at carbon 4.

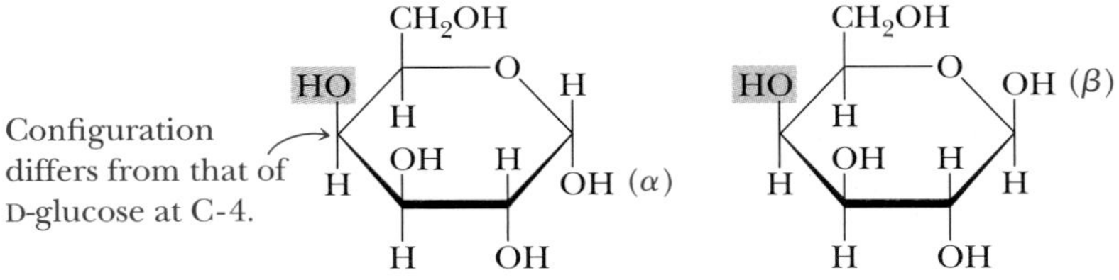

α-D-Galactopyranose
(α-D-Galactose)

β-D-Galactopyranose
(β-D-Galactose)

Problem 25.2

Mannose exists in aqueous solution as a mixture of α-D-mannopyranose and β-D-mannopyranose. Draw Haworth projections for these molecules.

Aldopentoses also form cyclic hemiacetals. The most prevalent forms of D-ribose and other pentoses in the biological world are furanoses. Following are Haworth projections for α-D-ribofuranose (α-D-ribose) and β-2-deoxy-D-ribofuranose (β-2-deoxy-D-ribose). The prefix 2-deoxy indicates the absence of oxygen at carbon 2.

CHEMICAL CONNECTIONS

L-*Ascorbic Acid (Vitamin C)*

The structure of L-ascorbic acid (vitamin C) resembles that of a monosaccharide. In fact, this vitamin is synthesized both biochemically by plants and some animals and commercially from D-glucose. Humans do not have the enzymes required for this synthesis and, therefore, we must obtain it in the food we eat or as a vitamin supplement. Approximately 66 million kilograms of vitamin C are synthesized every year in the United States.

L-Ascorbic acid is very easily oxidized to L-dehydroascorbic acid, a diketone. Both L-ascorbic acid and L-dehydroascorbic acid are physiologically active and are found together in most body fluids.

L-Ascorbic acid
(Vitamin C)

L-Dehydroascorbic
acid

Ascorbic acid is one of the most important antioxidants (the H in the enolic OH is weakly bonded and easily abstracted by radicals). One of the most important roles it plays may be to replenish the lipid-soluble antioxidant α-tocopherol by transferring a hydrogen atom to the tocopherol radical, formed by reaction with radicals in the autoxidation process (see Section 8.7).

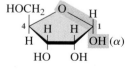

α-D-**Ribofuranose**
(α-D-**Ribose**)

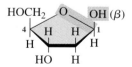

β-**2-Deoxy-D-ribofuranose**
(β-**2-Deoxy-D-ribose**)

Units of D-ribose and 2-deoxy-D-ribose in nucleic acids and most other biological molecules are found almost exclusively in the β-configuration.

Other monosaccharides also form five-membered cyclic hemiacetals. Following are the five-membered cyclic hemiacetals of fructose.

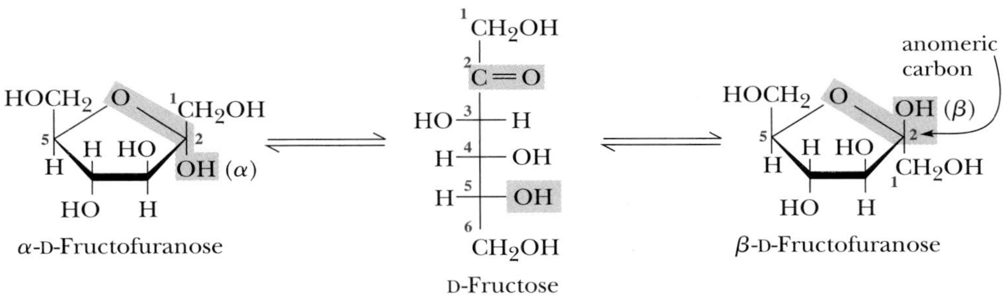

α-D-Fructofuranose   D-Fructose   β-D-Fructofuranose

The β-D-fructofuranose form is found in the disaccharide sucrose (Section 25.4A).

B. Conformation Representations

A five-membered ring is so close to being planar that Haworth projections are adequate representations of furanoses. For pyranoses, however, the six-membered ring is more accurately represented as a chair conformation. Following are structural formulas for α-D-glucopyranose and β-D-glucopyranose drawn as chair conformations. Also shown is the open chain or free aldehyde form with which the cyclic hemiacetal forms are in equilibrium in aqueous solution.

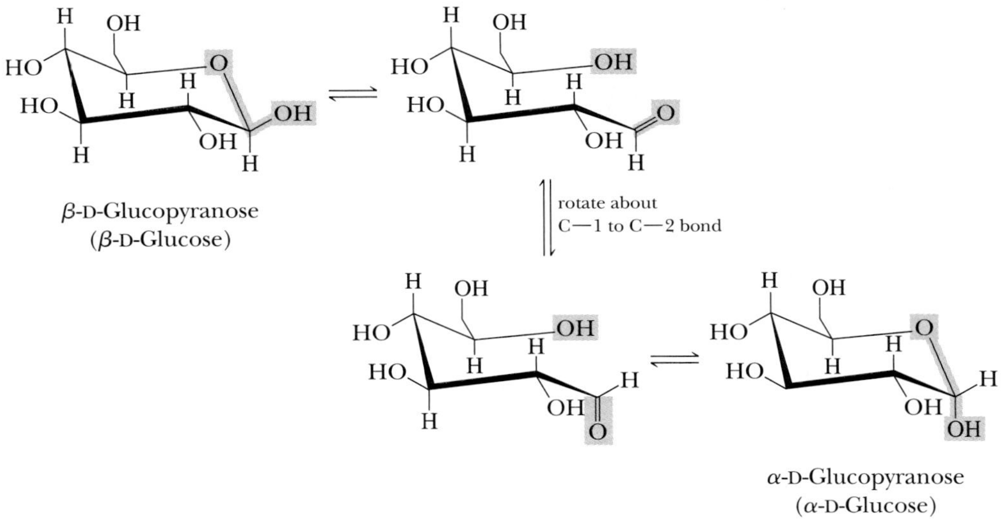

β-D-Glucopyranose
(β-D-Glucose)

rotate about
C—1 to C—2 bond

α-D-Glucopyranose
(α-D-Glucose)

Notice that each group, including the anomeric —OH, on the chair conformation of β-D-glucopyranose is equatorial. Notice also that the —OH group on the anomeric carbon is axial in α-D-glucopyranose. Because of the equatorial orientation of the —OH on its anomeric carbon, β-D-glucopyranose is more stable and predominates in aqueous solution.

    At this point, you should compare the relative orientations of groups on the D-glucopyranose ring in the Haworth projection and the chair conformation. The orientations of groups on carbons 1 through 5 of β-D-glucopyranose, for example, are up, down, up, down, and up in both representations.

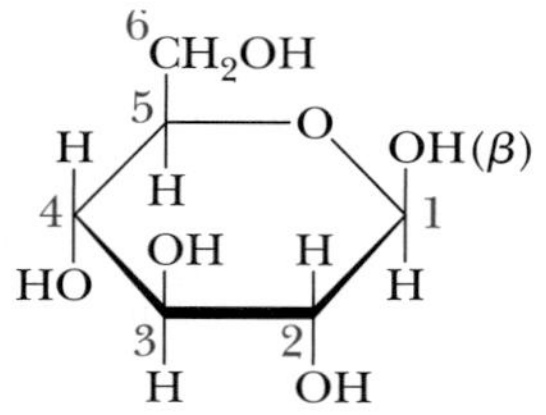

β-D-Glucopyranose
(Haworth projection)

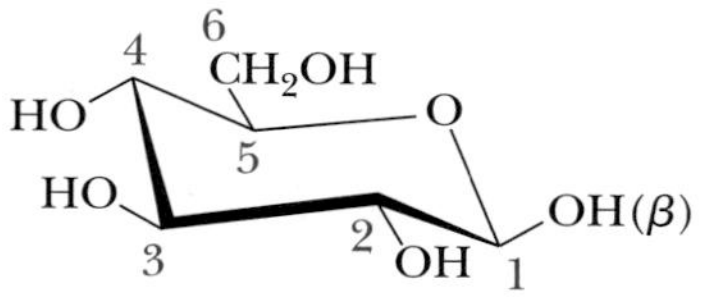

β-D-Glucopyranose
(chair conformation)

Example 25.3

Draw chair conformations for α-D-galactopyranose and β-D-galactopyranose. Label the anomeric carbon in each.

Solution

D-Galactose differs in configuration from D-glucose only at carbon 4. Therefore, draw the α and β forms of D-glucopyranose and then interchange the positions of the —OH and —H groups on carbon 4.

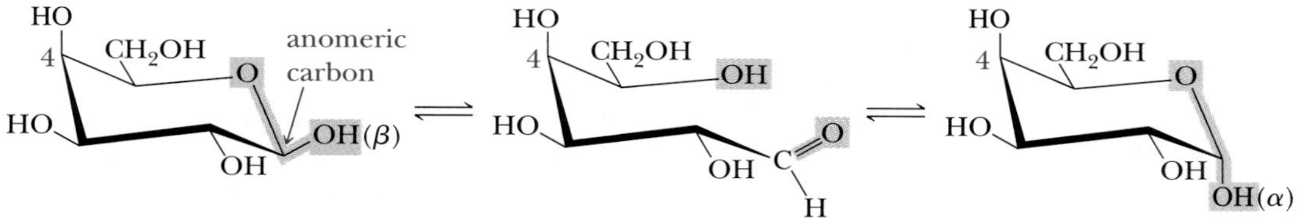

β-D-Galactopyranose
(β-D-Galactose)
$[\alpha]_D = +52.8$

D-Galactose

α-D-Galactopyranose
(α-D-Galactose)
$[\alpha]_D = +150.7$

Problem 25.3

Draw chair conformations for α-D-mannopyranose and β-D-mannopyranose. Label the anomeric carbon in each.

C. Mutarotation

Mutarotation The change in specific rotation that occurs when an α or β hemiacetal form of a carbohydrate in aqueous solution is converted to an equilibrium mixture of the two forms.

Mutarotation is the change in specific rotation that accompanies the interconversion of α- and β-anomers in aqueous solution. As an example, a solution prepared by dissolving crystalline α-D-glucopyranose in water shows an initial rotation of +112, which gradually decreases to an equilibrium value of +52.7 as α-D-glucopyranose reaches an equilibrium with β-D-glucopyranose. A solution of β-D-glucopyranose also undergoes mutarotation, during which the specific rotation changes from an initial value of +18.7 to the same equilibrium value of +52.7. The equilibrium mixture consists of 64% β-D-glucopyranose and 36% α-D-glucopyranose. It contains only traces (0.003%) of the open-chain form. Mutarotation is common to all carbohydrates that exist in hemiacetal forms.

25.3 Reactions of Monosaccharides

A. Formation of Glycosides (Acetals)

We saw in Section 16.7B that treatment of an aldehyde or ketone with one molecule of alcohol gives a hemiacetal, and that treatment of the hemiacetal with a molecule of alcohol gives an acetal. Treatment of monosaccharides, all of which exist almost exclusively in a cyclic hemiacetal form, also gives acetals, as illustrated by the reaction of β-D-glucopyranose with methanol.

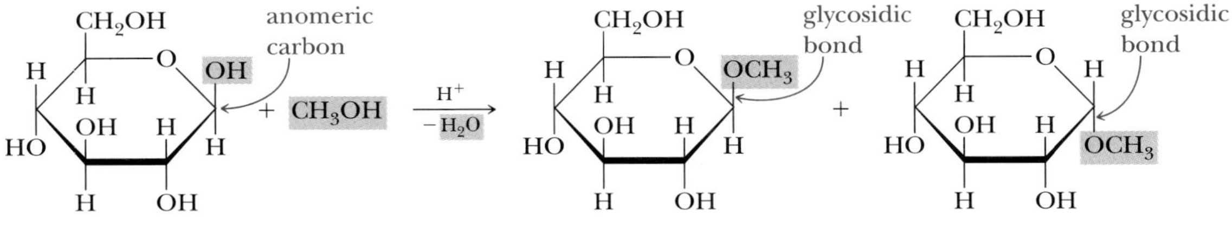

β-D-Glucopyranose
(β-D-Glucose)

Methyl β-D-glucopyranoside
(Methyl β-D-glucoside)

Methyl α-D-glucopyranoside
(Methyl α-D-glucoside)

A cyclic acetal derived from a monosaccharide is called a **glycoside,** and the bond from the anomeric carbon to the —OR group is called a **glycosidic bond.** Mutarotation is not possible in a glycoside because an acetal is no longer in equilibrium with the open-chain carbonyl-containing compound. Glycosides are stable in water and aqueous base, but like other acetals (Section 16.7), they are hydrolyzed in aqueous acid to an alcohol and a monosaccharide.

Glycosides are named by listing the alkyl or aryl group bonded to oxygen followed by the name of the carbohydrate in which the ending -e is replaced by -ide. For example, the glycosides derived from β-D-glucopyranose are named β-D-glucopyranosides; those derived from β-D-ribofuranose are named β-D-ribofuranosides.

Glycoside A carbohydrate in which the —OH on its anomeric carbon is replaced by —OR.

Glycosidic bond The bond from the anomeric carbon of a glycoside to an —OR group.

Example 25.4

Draw a structural formula for methyl β-D-ribofuranoside (methyl β-D-riboside). Label the anomeric carbon and the glycosidic bond.

Solution

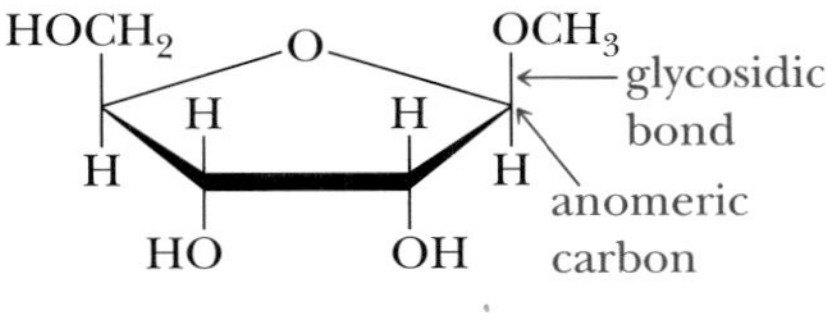

Problem 25.4

Draw a Haworth projection and a chair conformation for methyl α-D-mannopyranoside (methyl α-D-mannoside). Label the anomeric carbon and the glycosidic bond.

Just as the anomeric carbon of a cyclic hemiacetal undergoes reaction with the —OH group of an alcohol to form a glycoside, it also undergoes reaction with the

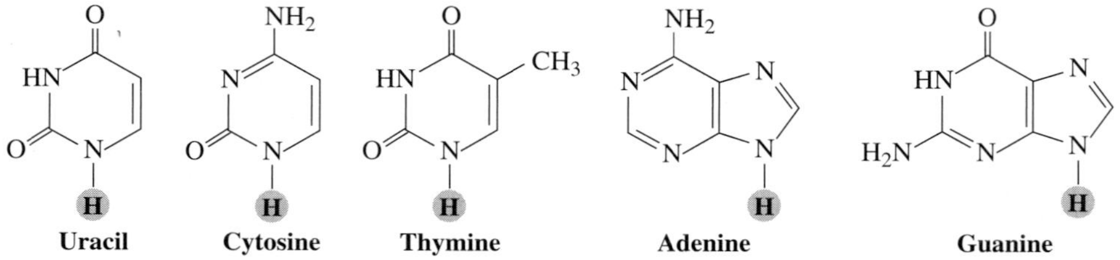

Figure 25.2
Structural formulas of the five most important pyrimidine and purine bases found in DNA and RNA. The hydrogen atom shown in color is lost in forming an *N*-glycoside.

Uracil   Cytosine   Thymine   Adenine   Guanine

N—H group of an amine to form an *N*-glycoside. Especially important in the biological world are the *N*-glycosides formed between D-ribose and 2-deoxy-D-ribose, each as a furanose, and the heterocyclic aromatic amines uracil, cytosine, thymine, adenine, and guanine (Figure 25.2). *N*-Glycosides of these pyrimidine and purine bases are structural units of nucleic acids (Chapter 28).

Example 25.5

Draw a structural formula for cytidine, the *β-N*-glycoside formed between D-ribofuranose and cytosine.

Solution

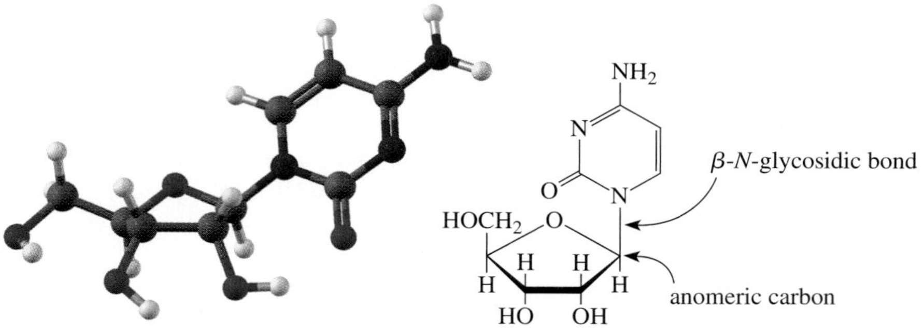

Problem 25.5

Draw a structural formula for the *β-N*-glycoside formed between 2-deoxy-D-ribofuranose and adenine.

B. Reduction to Alditols

Alditol The product formed when the C=O group of a monosaccharide is reduced to a CHOH group.

The carbonyl group of a monosaccharide can be reduced to a hydroxyl group by a variety of reducing agents, including sodium borohydride and hydrogen in the presence of a transition metal catalyst. The reduction products are known as **alditols.** Reduction of D-glucose gives D-glucitol, more commonly known as D-sorbitol. Note that D-glucose is shown here in the open-chain form. Only a small amount of this form is present in solution, but, as it is reduced, the equilibrium between cyclic hemiacetal forms and the open-chain form shifts to replace it.

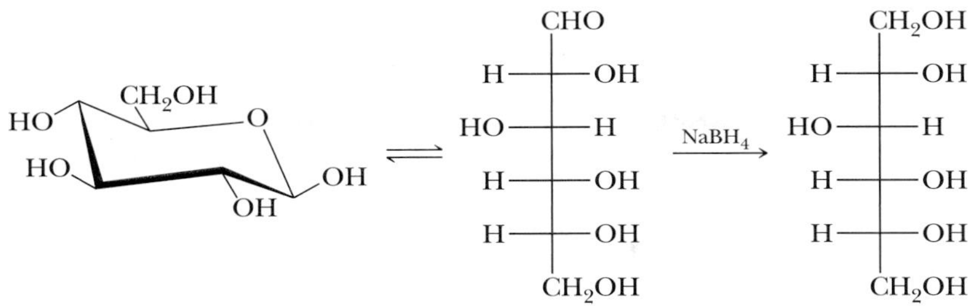

CHO   CH₂OH

β-D-Glucopyranose   D-Glucose   D-Glucitol (D-Sorbitol)

Sorbitol is found in the plant world in many berries and in cherries, plums, pears, apples, seaweed, and algae. It is about 60% as sweet as sucrose (table sugar) and is used in the manufacture of candies and as a sugar substitute for diabetics. D-Sorbitol is an important food additive, usually added to prevent dehydration of foods and other materials on exposure to air because it binds water strongly.

Other alditols common in the biological world are erythritol, D-mannitol, and xylitol. Xylitol is used as a sweetening agent in "sugarless" gum, candy, and sweet cereals.

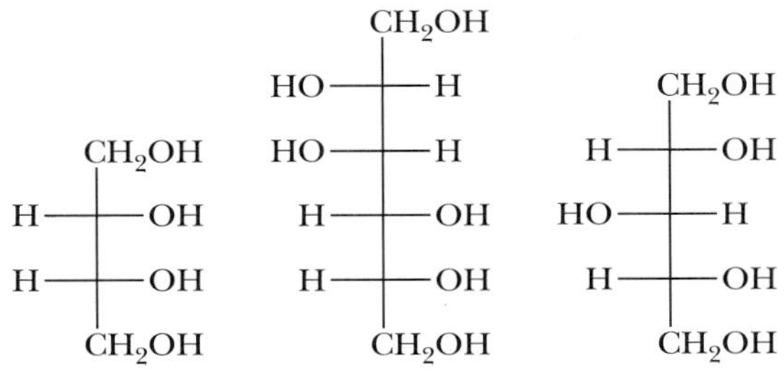

Erythritol   D-Mannitol   Xylitol

C. Oxidation to Aldonic Acids: Reducing Sugars

As we saw in Section 16.10A, aldehydes (RCHO) are oxidized to carboxylic acids (RCOOH) by several oxidizing agents, including oxygen, O_2. Similarly, the aldehyde group of an aldose can be oxidized, under basic conditions, to a carboxylate group. Oxidizing agents for this purpose include bromine in aqueous calcium carbonate (Br_2, $CaCO_3$, H_2O) and Tollens' solution [$Ag(NH_3)_2^+$]. Under these conditions, the cyclic form of an aldose is in equilibrium with the open-chain form, which is then oxidized by the mild oxidizing agent. D-Glucose, for example, is oxidized to D-gluconate (the anion of D-gluconic acid).

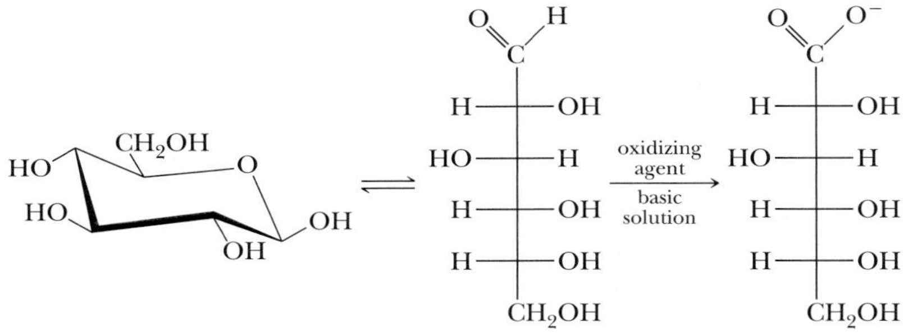

β-D-Glucopyranose (β-D-Glucose)   D-Glucose   D-Gluconate

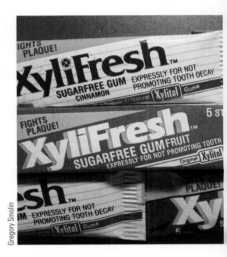

■ Many "sugar-free" products contain sugar alcohols, such as D-sorbitol and xylitol.

Gregory Smolin

Aldonic acid The product formed when the —CHO group of an aldose is oxidized to a —COOH group.

Reducing sugar A carbohydrate that reacts with an oxidizing agent to form an aldonic acid. In this reaction, the carbohydrate reduces the oxidizing agent.

Any carbohydrate that reacts with an oxidizing agent to form an **aldonic acid** is classified as a **reducing sugar** (it reduces the oxidizing agent).

Surprisingly, 2-ketoses are also reducing sugars. Carbon 1 (a CH_2OH group) of a 2-ketose is not oxidized directly. Rather, under the basic conditions of this oxidation, a 2-ketose is in equilibrium with an aldose by way of an enediol intermediate. The aldose is then oxidized by the mild oxidizing agent.

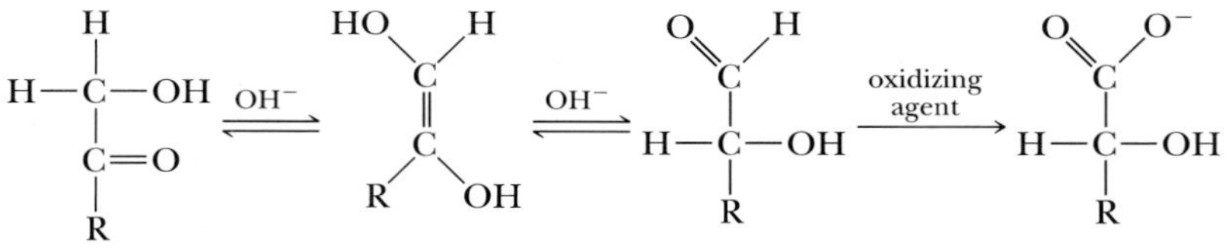

A 2-ketose   An enediol   An aldose   An aldonate

D. Oxidation to Uronic Acids

Enzyme-catalyzed oxidation of the primary hydroxyl group at carbon 6 of a hexose yields a uronic acid. Enzyme-catalyzed oxidation of D-glucose, for example, yields D-glucuronic acid, shown here in both its open-chain and cyclic hemiacetal forms.

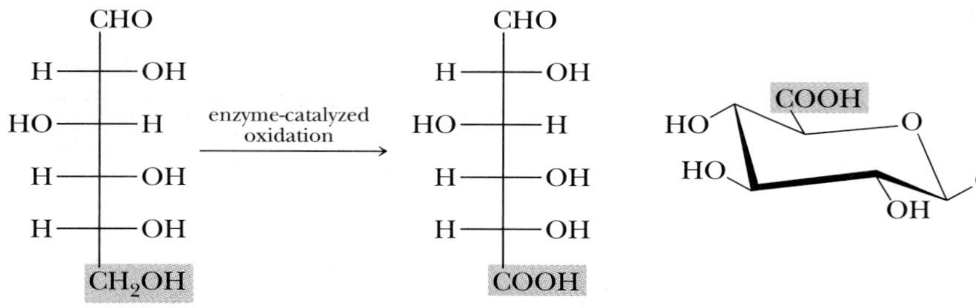

D-Glucose   Fischer projection   Chair conformation

D-Glucuronic acid
(a uronic acid)

D-Glucuronic acid is widely distributed in both the plant and animal world. In humans, it is an important component of the acidic polysaccharides of connective tissues (Section 25.6). It is also used by the body to detoxify foreign hydroxyl-containing compounds, such as phenols and alcohols. In the liver, these compounds are converted to glycosides of glucuronic acid (glucuronides) and excreted in the urine. The intravenous anesthetic propofol, for example, is converted to the following glucuronide and excreted in the urine.

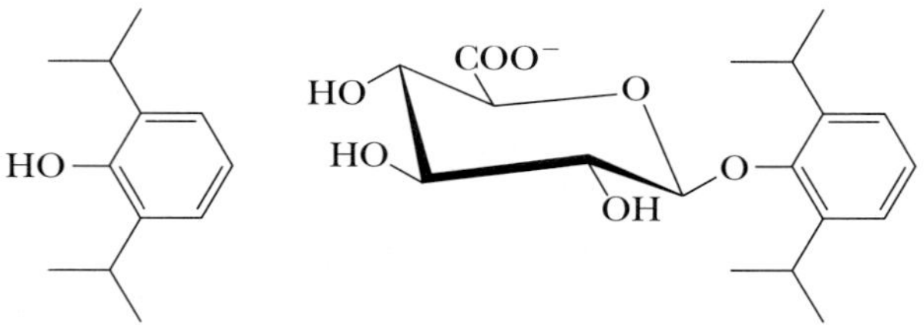

Propofol   A urine-soluble glucuronide

E. Oxidation by Periodic Acid

Oxidation by periodic acid has proven useful in structure determinations of carbohydrates, particularly in determining the size of glycoside rings. Recall from Section 10.8C that periodic acid cleaves the carbon-carbon bond of a glycol in a reaction that proceeds through a cyclic periodic ester. In this reaction, iodine(VII) of periodic acid is reduced to iodine(V) of iodic acid.

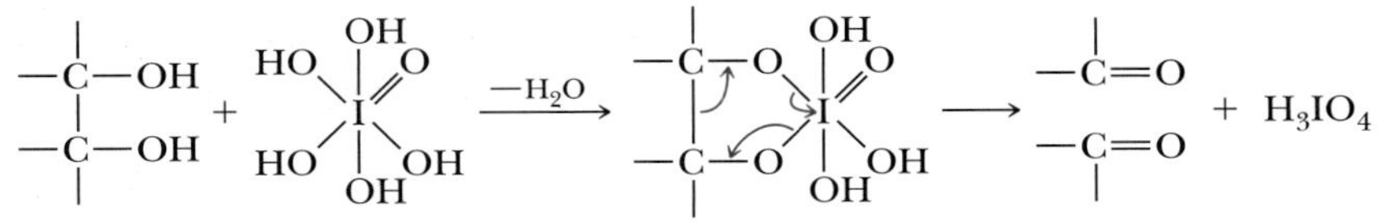

A glycol Periodic acid A cyclic periodic ester Iodic acid

Periodic acid also cleaves carbon-carbon bonds of α-hydroxyketones and α-hydroxyaldehydes by a similar mechanism. Following are abbreviated structural formulas for these functional groups and the products of their oxidative cleavage by periodic acid. As a way to help you understand how each set of products is formed, each carbonyl in a starting material is shown as a hydrated intermediate that is then oxidized. In this way, each oxidation can be viewed as analogous to oxidation of a glycol.

α-Hydroxyketone Hydrated intermediate

α-Hydroxyketone Hydrated intermediate

α-Hydroxyaldehyde Hydrated intermediate

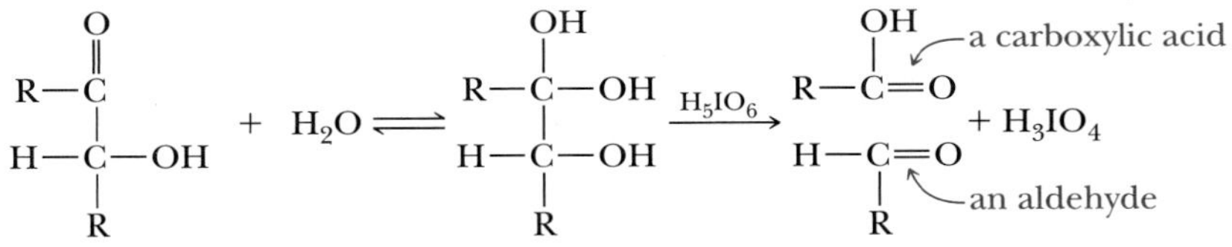

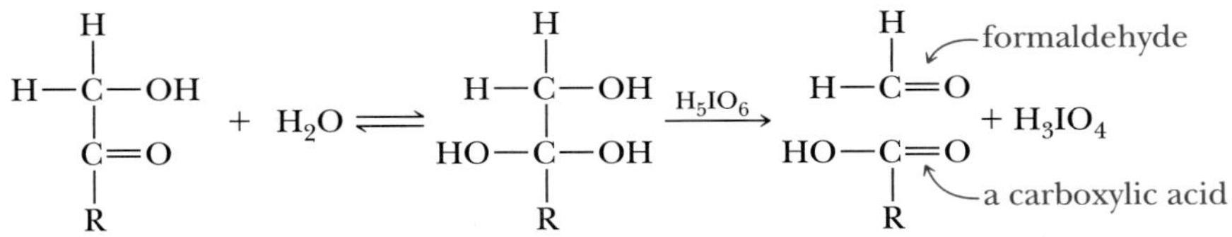

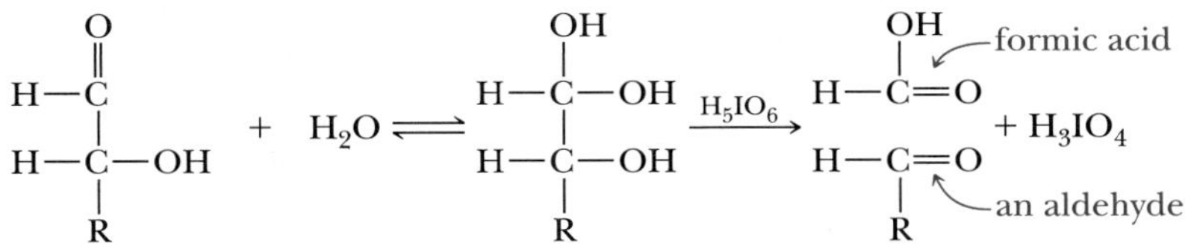

As an example of the usefulness of this reaction in carbohydrate chemistry, oxidation of methyl β-D-glucoside consumes two moles of periodic acid and produces one mole of formic acid. This stoichiometry and the formation of formic acid are possible only if —OH groups are on three adjacent carbon atoms.

CHEMICAL CONNECTIONS

Testing for Glucose

The analytical procedure most often performed in a clinical chemistry laboratory is the determination of glucose in blood, urine, or other biological fluids. This is true because of the high incidence of diabetes mellitus. Approximately 15 million known diabetics live in the United States, and it is estimated that another one million are undiagnosed.

Diabetes mellitus is characterized by insufficient blood levels of the hormone insulin. If the blood concentration of insulin is too low, muscle and liver cells do not absorb glucose from the blood, which, in turn, leads to increased levels of blood glucose (hyper-

glycemia), impaired metabolism of fats and proteins, ketosis, and possible diabetic coma. A rapid test for blood glucose levels is critical for early diagnosis and effective management of this disease. In addition to being rapid, a test must also be specific for D-glucose; it must give a positive test for glucose but not react with any other substance normally present in biological fluids.

Blood glucose levels are now measured by an enzyme-based procedure using the enzyme glucose oxidase. This enzyme catalyzes the oxidation of β-D-glucose to D-gluconic acid.

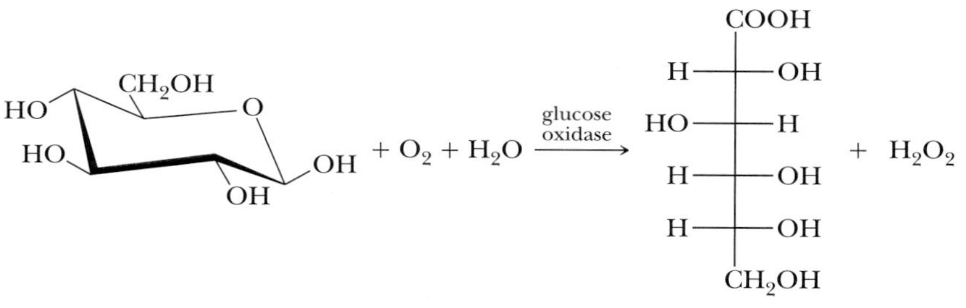

β-D-Glucopyranose
(β-D-Glucose)

D-Gluconic acid   Hydrogen peroxide

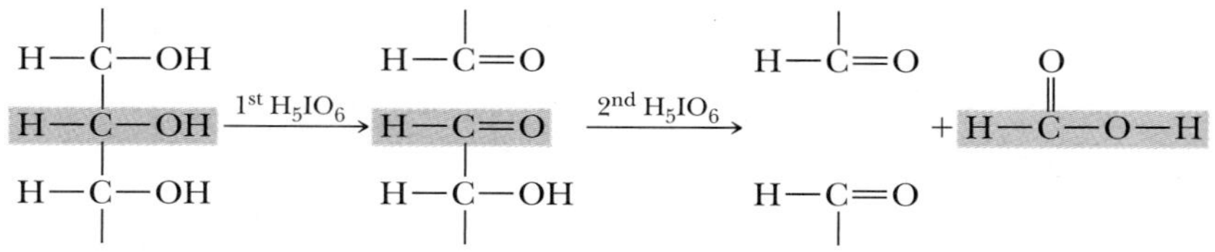

This is evidence that methyl β-D-glucoside is indeed a pyranoside.

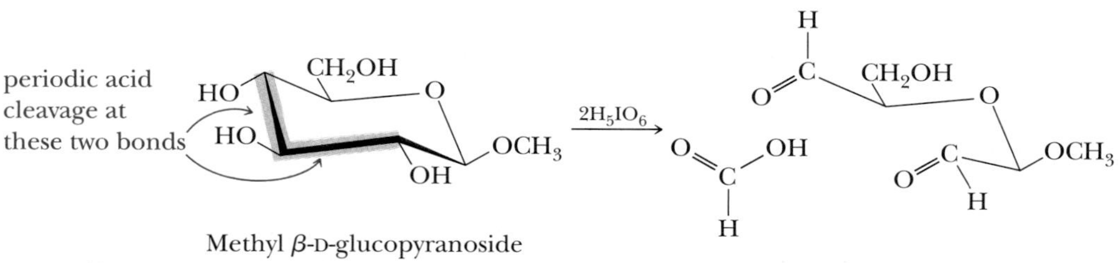

Methyl β-D-glucopyranoside

Methyl β-D-fructoside consumes only one mole of periodic acid and produces neither formaldehyde nor formic acid. Thus, oxidizable groups exist on adjacent carbons

Glucose oxidase is specific for β-D-glucose. Therefore, complete oxidation of any sample containing both β-D-glucose and α-D-glucose requires conversion of the α form to the β form. Fortunately, this interconversion

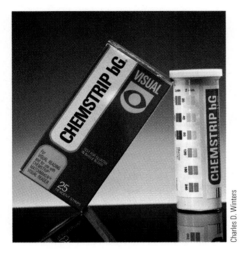

■ A test kit for the presence of glucose in urine.

Charles D. Winters

is rapid and complete in the short time required for the test.

Molecular oxygen, O_2, is the oxidizing agent in this reaction and is reduced to hydrogen peroxide, H_2O_2. In one procedure, hydrogen peroxide formed in the glucose oxidase-catalyzed reaction is used to oxidize colorless *o*-toluidine to a colored product in a reaction catalyzed by the enzyme peroxidase. The concentration of the colored oxidation product is determined spectrophotometrically and is proportional to the concentration of glucose in the test solution.

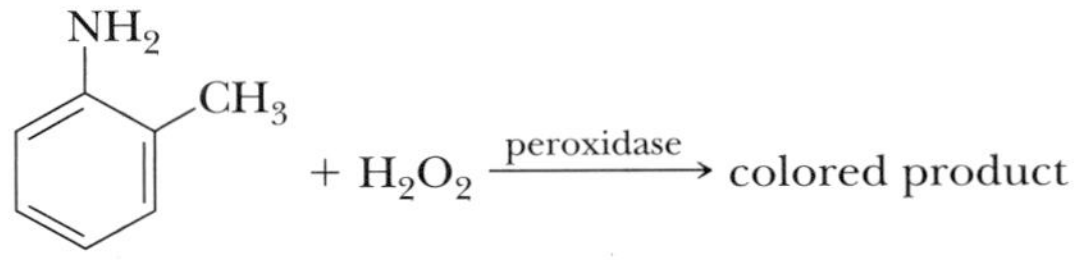

$$\text{2-Methylaniline } (o\text{-Toluidine}) + H_2O_2 \xrightarrow{\text{peroxidase}} \text{colored product}$$

2-Methylaniline
(*o*-Toluidine)

Several commercially available test kits use the glucose oxidase reaction for qualitative determination of glucose in urine.

only at one site in the molecule. The fructoside, therefore, must be a five-membered ring (a fructofuranoside).

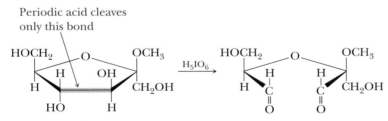

Periodic acid cleaves only this bond

H_5IO_6

Methyl β-D-fructofuranoside

25.4 Disaccharides and Oligosaccharides

Most carbohydrates in nature contain more than one monosaccharide unit. Those that contain two units are called **disaccharides,** those that contain three units are called **trisaccharides,** and so forth. The general term **oligosaccharide** is often

Disaccharide A carbohydrate containing two monosaccharide units joined by a glycosidic bond.

Oligosaccharide A carbohydrate containing four to ten monosaccharide units, each joined to the next by a glycosidic bond.

used for carbohydrates that contain from four to ten monosaccharide units. Carbohydrates containing larger numbers of monosaccharide units are called **polysaccharides.**

In a disaccharide, two monosaccharide units are joined together by a glycosidic bond between the anomeric carbon of one unit and an —OH of the other. Three important disaccharides are sucrose, lactose, and maltose.

Polysaccharide A carbohydrate containing a large number of monosaccharide units, each joined to the next by one or more glycosidic bonds.

A. Sucrose

Sucrose (table sugar) is the most abundant disaccharide in the biological world. It is obtained principally from the juice of sugar cane and sugar beets. In sucrose, carbon 1 of α-D-glucopyranose is joined to carbon 2 of β-D-fructofuranose by an α-1,2-glycosidic bond.

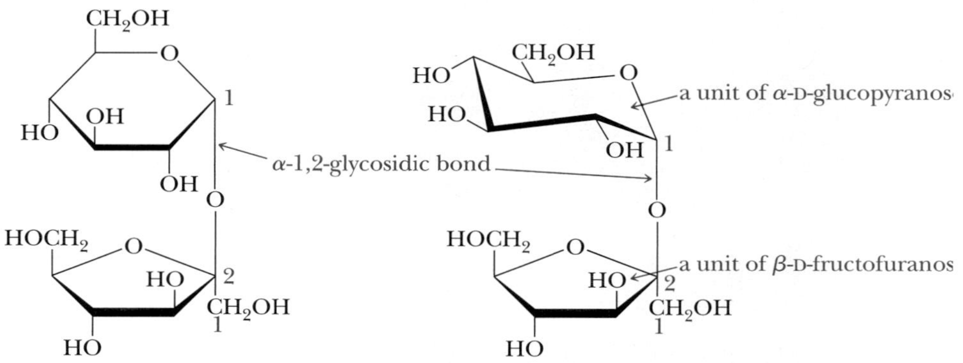

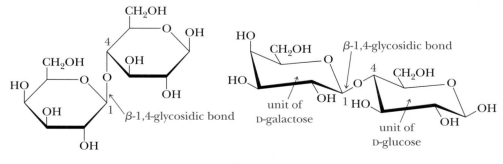

Sucrose

Note that glucose is a six-membered (pyranose) ring, whereas fructose is a five-membered (furanose) ring. Because the anomeric carbons of both the glucopyranose and fructofuranose units are involved in formation of the glycosidic bond, sucrose is a nonreducing sugar.

B. Lactose

Lactose is the principal sugar present in milk. It makes up about 5–8% of human milk and 4–6% of cow's milk. It consists of D-galactopyranose bonded by a β-1,4-glycosidic bond to carbon 4 of D-glucopyranose. Lactose is a reducing sugar.

■ These products help individuals with lactose intolerance meet their calcium needs.

Charles D. Winters

Lactose

C. Maltose

Maltose derives its name from its presence in malt, the juice from sprouted barley and other cereal grains (from which beer is brewed). Maltose consists of two molecules of D-glucopyranose joined by an α-1,4-glycosidic bond between carbon 1 (the anomeric carbon) of one unit and carbon 4 of the other unit. Following are representations for β-maltose, so named because the —OH on the anomeric carbon of the glucose unit on the right is beta.

Maltose

Maltose is a reducing sugar because the hemiacetal group on the right unit of D-glucopyranose is in equilibrium with the free aldehyde and can be oxidized to a carboxylic acid.

D. Relative Sweetness of Some Carbohydrate and Artificial Sweeteners

Although all monosaccharides are sweet to the taste, some are sweeter than others (Table 25.2). D-Fructose tastes the sweetest, even sweeter than sucrose (table sugar, Section 25.4A). The sweet taste of honey is attributable largely to D-fructose and D-glucose. Lactose (Section 25.4B) has almost no sweetness. It occurs in many milk products and is sometimes added to foods as a filler. Some people lack an enzyme that allows them to tolerate lactose well; they should avoid these foods.

Table 25.2 Relative Sweetness of Some Carbohydrates and Artificial Sweetening Agents*

Carbohydrate	Sweetness Relative to Sucrose	Artificial Sweetener	Sweetness Relative to Sucrose
Fructose	1.74	Saccharin	450
Invert sugar	1.25	Acesulfame-K	200
Sucrose (table sugar)	1.00	Aspartame	160
Honey	0.97		
Glucose	0.74		
Maltose	0.33		
Galactose	0.32		
Lactose (milk sugar)	0.16		

*We have no mechanical way to measure sweetness. Such testing is done by having a group of people taste solutions of varying sweetness and ranking them in order by taste.

CHEMICAL CONNECTIONS

A, B, AB, and O Blood Group Substances

Membranes of animal plasma cells have large numbers of relatively small carbohydrates bound to them. In fact, the outsides of most plasma cell membranes are literally "sugar-coated." These membrane-bound carbohydrates are part of the mechanism by which cell types recognize each other and, in effect, act as biochemical markers.

Typically, these membrane-bound carbohydrates contain from 4 to 17 monosaccharide units consisting primarily of relatively few monosaccharides, including D-galactose, D-mannose, L-fucose, N-acetyl-D-glucosamine, and N-acetyl-D-galactosamine. L-Fucose is a 6-deoxyaldohexose.

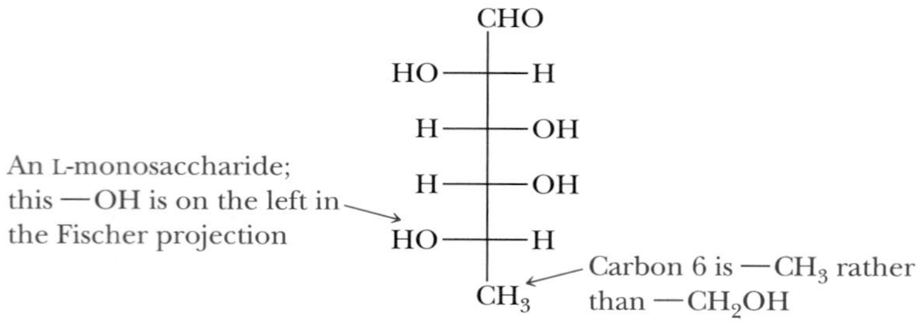

An L-monosaccharide; this —OH is on the left in the Fischer projection →

CHO
HO——H
H——OH
H——OH
HO——H
CH₃

Carbon 6 is —CH₃ rather than —CH₂OH

L-Fucose

Among the first discovered and best understood of these membrane-bound carbohydrates are those of the ABO blood group system, discovered in 1900 by Karl Landsteiner (1868–1943). Whether an individual has type A, B, AB, or O blood is genetically determined and depends on the type of trisaccha-

ride or tetrasaccharide bound to the surface of the person's red blood cells. The monosaccharides of each blood group and the type of glycosidic bond joining them are shown in the figure. The configurations of the glycosidic bonds are shown in parentheses.

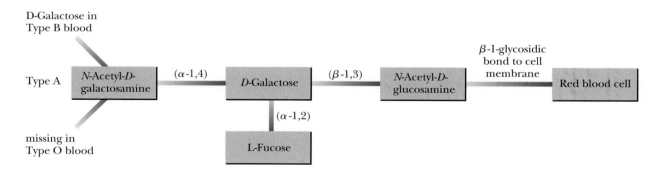

Example 25.6

Draw a chair conformation for the β anomer of a disaccharide in which two units of D-glucopyranose are joined by an α-1,6-glycosidic bond.

Solution

First draw a chair conformation of α-D-glucopyranose. Then connect the anomeric carbon of this monosaccharide to carbon 6 of a second D-glucopyranose unit by an α-glycosidic bond. The resulting molecule is either α or β depending on the orientation of the —OH group on the reducing end of the disaccharide. The disaccharide shown here is β.

Breads, grains, and pasta are sources of starches.

Problem 25.6

Draw a chair conformation for the α form of a disaccharide in which two units of D-glucopyranose are joined by a β-1,3-glycosidic bond.

25.5 Polysaccharides

Polysaccharides consist of large numbers of monosaccharide units bonded together by glycosidic bonds. Three important polysaccharides, all made up of glucose units, are starch, glycogen, and cellulose.

A. Starch: Amylose and Amylopectin

Starch is used for energy storage in plants. It is found in all plant seeds and tubers and is the form in which glucose is stored for later use. Starch can be separated into two principal polysaccharides: amylose and amylopectin. Although the starch from each plant is unique, most starches contain 20–25% amylose and 75–80% amylopectin.

Figure 25.3
Amylopectin is a branched polymer of approximately 10,000 D-glucose units joined by α-1,4-glycosidic bonds. Branches consist of 24–30 D-glucose units started by α-1, 6-glycosidic bonds.

High-Fructose Corn Syrup

If you read the labels of soft drinks and other artificially sweetened food products, you will find that many of them contain high-fructose corn syrup. Looking at Table 25.3, you will see one reason for its use: Fructose is more than 70% sweeter than sucrose (table sugar).

The production of high-fructose corn syrup begins with the partial hydrolysis of corn starch catalyzed by the enzyme α-amylase. This enzyme catalyzes the

hydrolysis of α-glucosidic bonds and breaks corn starch into small polysaccharides called dextrins. The enzyme glucoamylase then catalyzes the hydrolysis of the dextrins to D-glucose. Finally, enzyme-catalyzed isomerization of D-glucose gives D-fructose. Several billion pounds of high-fructose corn syrup are produced each year in this way for use by the food processing industry.

$$\text{Corn starch} \xrightarrow{\alpha\text{-amylase}} \text{Dextrins} \xrightarrow{\text{Glucoamylase}} \text{D-Glucose} \xrightarrow[\text{isomerase}]{\text{glucose}} \text{D-Fructose}$$

| (a polysaccharide of | Oligosaccharides of |
| α-D–glucose units) | 6–10 α-D-glucose units |

Complete hydrolysis of both amylose and amylopectin yields only D-glucose. Amylose is composed of unbranched chains of up to 4000 D-glucose units joined by α-1,4-glycosidic bonds. Amylopectin contains chains up to 10,000 D-glucose units also joined by α-1,4-glycosidic bonds. In addition, there is considerable branching from this linear network. At branch points, new chains of 24 to 30 units are started by α-1,6-glycosidic bonds (Figure 25.3).

B. Glycogen

Glycogen is the energy-reserve carbohydrate for animals. Like amylopectin, glycogen is a branched polysaccharide of approximately 10^6 glucose units joined by α-1,4- and α-1,6-glycosidic bonds. The total amount of glycogen in the body of a well-nourished adult human is about 350 g, divided almost equally between liver and muscle.

C. Cellulose

Cellulose, the most widely distributed plant skeletal polysaccharide, constitutes almost half of the cell wall material of wood. Cotton is almost pure cellulose. Cellulose is a linear polysaccharide of D-glucose units joined by β-1,4-glycosidic bonds (Figure 25.4). It has an average molecular weight of 400,000 g/mol, corresponding

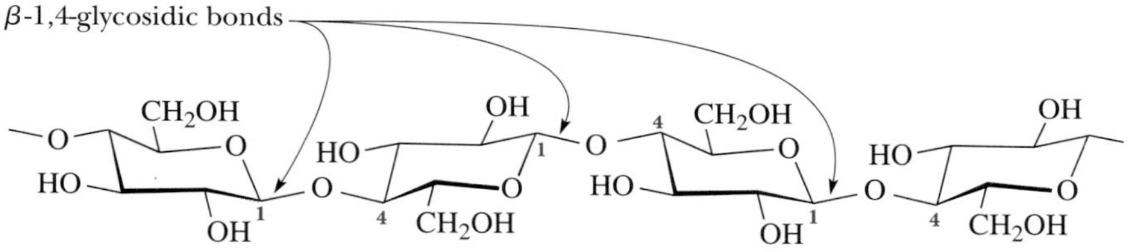

Figure 25.4
Cellulose is a linear polysaccharide of up to 2200 units of D-glucose joined by β-1, 4-glycosidic bonds.

to approximately 2200 glucose units per molecule. Cellulose molecules act very much like stiff rods, a feature that enables them to align themselves side by side into well-organized water-insoluble fibers in which the OH groups form numerous intermolecular hydrogen bonds. This arrangement of parallel chains in bundles gives cellulose fibers their high mechanical strength. It is also the reason cellulose is insoluble in water. When a piece of cellulose-containing material is placed in water, there are not enough water molecules on the surface of the fiber to pull individual cellulose molecules away from the strongly hydrogen-bonded fiber.

Humans and other animals cannot use cellulose as food because our digestive systems do not contain β-glucosidases, enzymes that catalyze hydrolysis of β-glucosidic bonds. Instead, we have only α-glucosidases; hence, the polysaccharides we use as sources of glucose are starch and glycogen. On the other hand, many bacteria and microorganisms do contain β-glucosidases and so can digest cellulose. Termites are fortunate (much to our regret) to have such bacteria in their guts and can use wood as their principal food. Ruminants (cud-chewing animals) and horses can also digest grasses and hay because β-glucosidase-containing microorganisms are present in their alimentary systems.

D. Textile Fibers from Cellulose

Cotton is almost pure cellulose. Both rayon and acetate rayon are made from chemically modified cellulose and were the first commercially important synthetic textile fibers. In the production of rayon, cellulose-containing materials are treated with carbon disulfide, CS_2, in aqueous sodium hydroxide. In this reaction, some of the —OH groups on a cellulose fiber are converted to the sodium salt of a xanthate ester, which causes the fibers to dissolve in alkali as a viscous colloidal dispersion.

an —OH group in
a cellulose fiber

Cellulose—OH $\xrightarrow{\text{NaOH}}$ Cellulose—O:⁻ Na⁺ $\xrightarrow{:S=C=S:}$ Cellulose—O—S—S:⁻ Na⁺

Cellulose
(insoluble in water)

Sodium salt of a xanthate ester
(a viscous colloidal suspension)

The solution of cellulose xanthate is separated from the alkali insoluble parts of wood and then forced through a spinneret, a metal disc with many tiny holes, into dilute sulfuric acid to hydrolyze the xanthate ester groups and precipitate regenerated cellulose. Regenerated cellulose extruded as a filament is called viscose rayon thread.

In the industrial synthesis of acetate rayon, cellulose is treated with acetic anhydride.

A glucose unit in
a cellulose fiber

Acetic
anhydride

A fully acetylated glucose unit

Acetylated cellulose is then dissolved in a suitable solvent, precipitated, and drawn into fibers known as acetate rayon. Today, acetate rayon fibers rank fourth in production in the United States, surpassed only by Dacron polyester, nylon, and rayon.

25.6 Acidic Polysaccharides

Acidic polysaccharides—a group of polysaccharides that contain carboxyl groups and/or sulfuric ester groups—play important roles in the structure and function of connective tissues. There is no single general type of connective tissue. Rather, there are a large number of highly specialized forms, such as cartilage, bone, synovial fluid, skin, tendons, blood vessels, intervertebral disks, and cornea. Most connective tissues are made up of collagen, a structural protein, in combination with a variety of acidic polysaccharides that interact with collagen to form tight or loose networks.

A. Hyaluronic Acid

Hyaluronic acid is the simplest acidic polysaccharide present in connective tissue. It has a molecular weight of between 10^5 and 10^7 g/mol and contains from 3000 to 100,000 repeating units, depending on the organ in which it occurs. It is most abundant in embryonic tissues and in specialized connective tissues such as synovial fluid, the lubricant of joints in the body, and the vitreous humor of the eye, where it provides a clear, elastic gel that maintains the retina in its proper position.

The repeating disaccharide unit in hyaluronic acid is D-glucuronic acid linked by a β-1,3-glycosidic bond to N-acetyl-D-glucosamine.

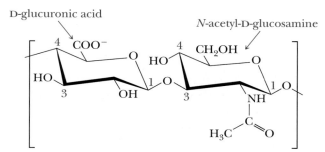

The repeating unit of hyaluronic acid

B. Heparin

Heparin is a heterogeneous mixture of variably sulfonated polysaccharide chains, ranging in molecular weight from 6000 to 30,000 g/mol. This acidic polysaccharide is synthesized and stored in mast cells of various tissues, particularly the liver, lungs, and gut. Heparin has many biological functions, the best known and understood of which is its anticoagulant activity. It binds strongly to antithrombin III, a plasma protein involved in terminating the clotting process.

The repeating monosaccharide units of heparin are N-acetyl-D-glucosamine, D-glucuronic acid, D-glucosamine, and L-ioduronic acid bonded by a combination of

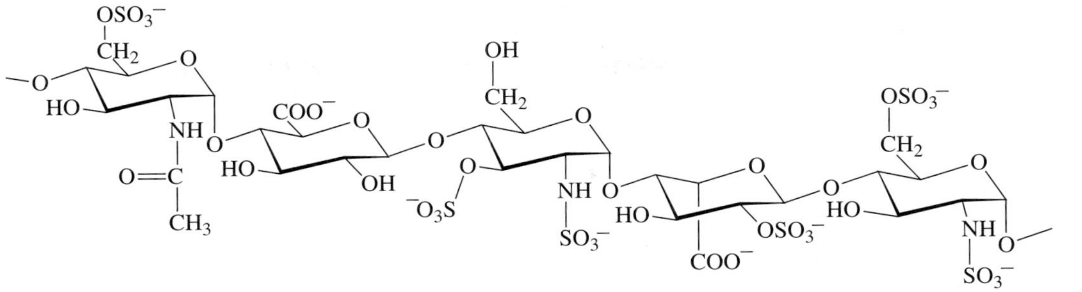

Figure 25.5
A pentasaccharide unit of heparin.

α-1,4- and β-1,4-glycosidic bonds. Figure 25.5 shows a pentasaccharide unit of heparin that binds to and inhibits the enzymatic activity of antithrombin III.

Summary

Monosaccharides (Section 25.1A) are polyhydroxyaldehydes or polyhydroxyketones. They have the general formula $C_nH_{2n}O_n$, where n varies from 3 to 8. Their names contain the suffix -*ose*. The prefixes *tri-*, *tetr-*, *pent-*, and so on show the number of carbon atoms in the chain. The prefix *aldo-* shows an aldehyde, and the prefix *keto-* shows a ketone. In a **Fischer projection** (Section 25.1B) of a monosaccharide, the carbon chain is written vertically with the most highly oxidized carbon toward the top. Horizontal lines show groups projecting above the plane of the page; vertical lines show groups projecting behind the plane of the page.

The **penultimate carbon** is the next to last on the carbon chain of a Fischer projection of a monosaccharide (Section 25.1C). A monosaccharide that has the same configuration at its penultimate carbon as D-glyceraldehyde is called a **D-monosaccharide;** one that has the same configuration at its penultimate carbon as L-glyceraldehyde is called an **L-monosaccharide.**

Monosaccharides exist primarily as cyclic hemiacetals (Section 25.2). The new chiral center resulting from hemiacetal formation is referred to as the **anomeric carbon,** and the stereoisomers thus formed are called **anomers.** A six-membered cyclic hemiacetal is called a **pyranose;** a five-membered cyclic hemiacetal is called a **furanose.** The symbol β indicates that the —OH on the anomeric carbon is on the same side of the ring as the terminal —CH₂OH. The symbol α indicates that —OH on the anomeric carbon is on the opposite side from the terminal —CH₂OH. Furanoses and pyranoses can be drawn as **Haworth projections** (Section 25.2A). Pyranoses can also be shown as chair conformations (Section 25.2B). **Mutarotation** (Section 25.2C) is the change in specific rotation that accompanies formation of an equilibrium mixture of α- and β-anomers in aqueous solution.

A **glycoside** (Section 25.3A) is an acetal derived from a monosaccharide. The name of the glycoside is composed of the name of the alkyl or aryl group bonded to the acetal oxygen atom followed by the name of the monosaccharide in which the terminal -*e* has been replaced by -*ide*.

An **alditol** (Section 25.3B) is a polyhydroxy compound formed by reduction of the carbonyl group of a monosaccharide to an hydroxyl group. Reduction of D-glucose, for example, gives D-glucitol. An **aldonic acid** (Section 25.3C) is a carboxylic acid formed by oxidation of the aldehyde group of an aldose. Oxidation of D-glucose, for example, gives D-gluconic acid. **Reducing sugars** (Section 25.3C) are oxidized by mild oxidizing agents to aldonic acids.

A **disaccharide** (Section 25.4) contains two monosaccharide units joined by a glycosidic bond. Terms applied to carbohydrates containing larger numbers of monosaccharides are trisaccharide, tetrasaccharide, oligosaccharide, and polysaccharide. Sucrose is a disaccharide containing D-glucose joined to D-fructose by a 1,2-glycosidic bond. Lactose is a disaccharide consisting of D-galactose joined to D-glucose by a β-1,4-glycosidic bond. Maltose is a disaccharide of two molecules of D-glucose joined by an α-1,4-glycosidic bond.

Starch (Section 25.5A) can be separated into two fractions given the names amylose and amylopectin. Amylose is a linear polymer of up to 4000 units of D-glucopyranose joined by α-1,4-glycosidic bonds. Amylopectin is a highly branched polymer of D-glucopyranose joined by α-1,4-glycosidic bonds and, at branch points, by α-1,6-glycosidic bonds. Glycogen (Section 25.5B), the reserve carbohydrate of animals, is a highly branched polymer of D-glucopyranose joined by α-1,4-glycosidic bonds and, at branch points, by α-1,6-glycosidic bonds. Cellulose (Section 25.5C), the skeletal polysaccharide of plants, is a linear polymer of D-glucopyranose joined by β-1,4-glycosidic bonds. Rayon (Section 25.5D) is made from chemically modified and regenerated cellulose. Acetate rayon is made by acetylation of cellulose.

The carboxyl and sulfate groups of acidic polysaccharides (Section 25.6) are ionized to $-COO^-$ and $-SO_3^-$ at the pH of body fluids, which gives these polysaccharides net negative charges.

Key Reactions

1. Formation of Cyclic Hemiacetals (Section 25.2A)

A monosaccharide existing as a five-membered ring is a furanose; one existing as a six-membered ring is a pyranose. A pyranose is most commonly drawn as either a Haworth projection or a chair conformation.

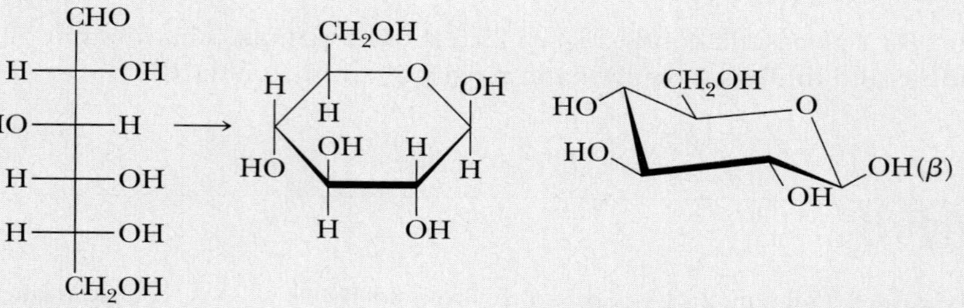

D-Glucose $\longrightarrow$ β-D-Glucopyranose (β-D-Glucose)

2. Mutarotation (Section 25.2C)

Anomeric forms of a monosaccharide are in equilibrium in aqueous solution. Mutarotation is the change in specific rotation that accompanies this equilibration.

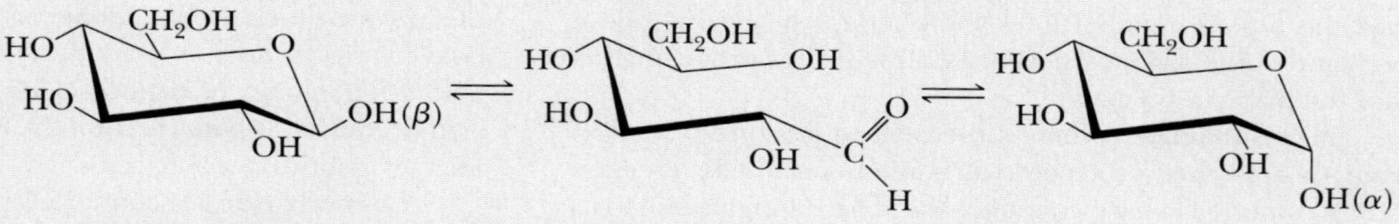

β-D-Glucopyranose $[\alpha]_D^{25}$ +18.7 $\rightleftharpoons$ Open-chain form $\rightleftharpoons$ α-D-Glucopyranose $[\alpha]_D^{25}$ +112

3. Formation of Glycosides (Section 25.3A)

Treatment of a monosaccharide with an alcohol in the presence of an acid catalyst forms a cyclic acetal called a glycoside. The bond to the new $-OR$ group is called a glycosidic bond.

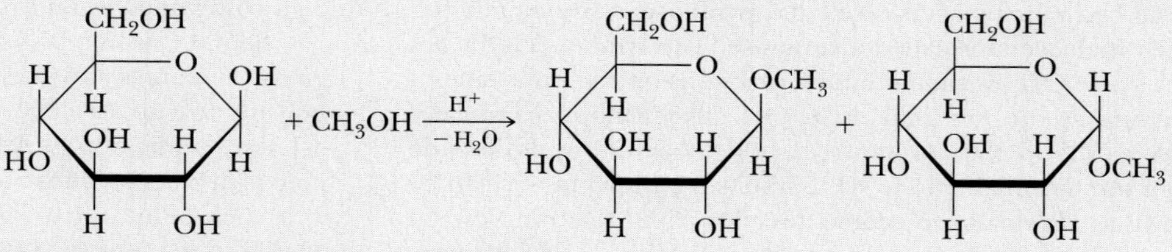

$+ CH_3OH \xrightarrow[-H_2O]{H^+}$

4. Formation of N-Glycosides (Section 25.3A)

N-Glycosides formed between a monosaccharide and a heterocyclic aromatic amine are especially important in the biological world.

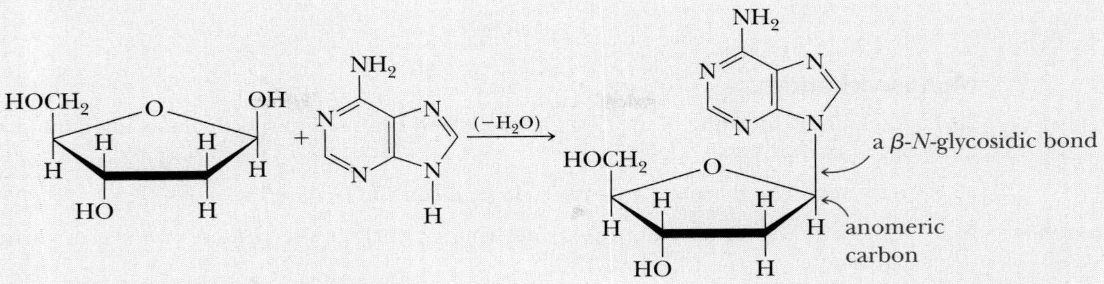

a β-N-glycosidic bond

anomeric carbon

5. Reduction to Alditols (Section 25.3B)

Reduction of the carbonyl group of an aldose or ketose to a hydroxyl group yields a polyhydroxy compound called an alditol.

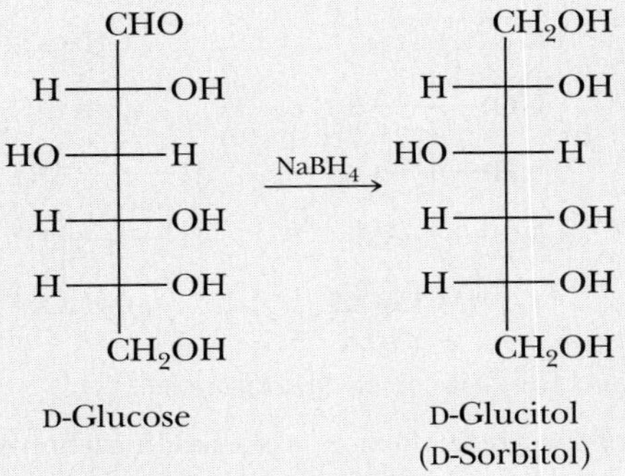

D-Glucose

D-Glucitol
(D-Sorbitol)

6. Oxidation to an Aldonic Acid (Section 25.3C)

Oxidation of the aldehyde group of an aldose to a carboxyl group by a mild oxidizing agent gives a polyhydroxycarboxylic acid called an aldonic acid.

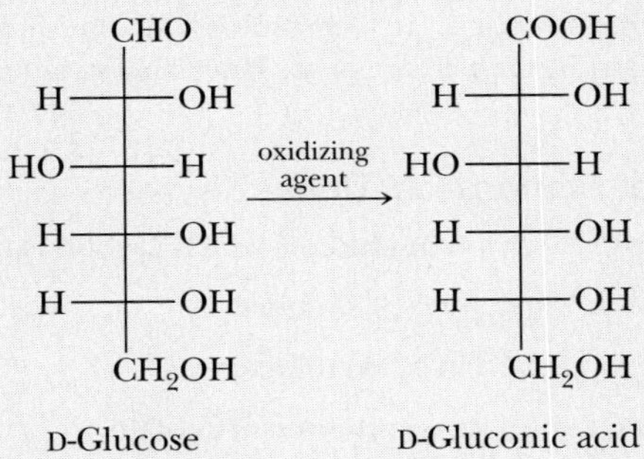

D-Glucose

D-Gluconic acid

7. Oxidation by Periodic Acid (Section 25.3E)

Periodic acid oxidizes and cleaves carbon-carbon bonds of glycol, α-hydroxyketone, and α-hydroxyaldehyde groups.

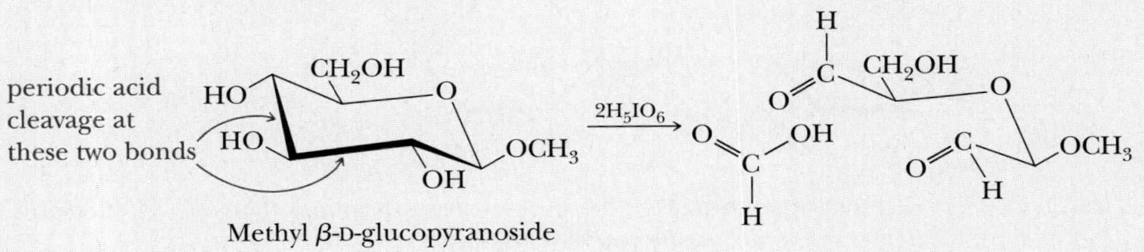

periodic acid cleavage at these two bonds

Methyl β-D-glucopyranoside

Problems

Monosaccharides

25.7 Explain the meaning of the designations D and L as used to specify the configuration of monosaccharides.

25.8 How many chiral centers are present in D-glucose? In D-ribose?

25.9 Which carbon of an aldopentose determines whether the pentose has a D or L configuration?

25.10 How many aldooctoses are possible? How many D-aldooctoses are possible?

25.11 Which compounds are D-monosaccharides? Which are L-monosaccharides?

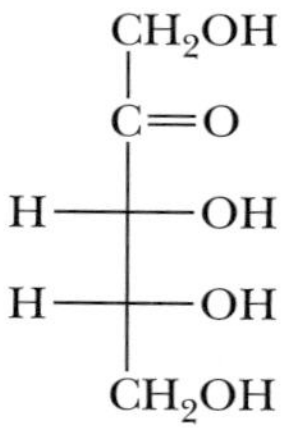

25.12 Write Fischer projections for L-ribose and L-arabinose.

25.13 What is the meaning of the prefix *deoxy-* as it is used in carbohydrate chemistry?

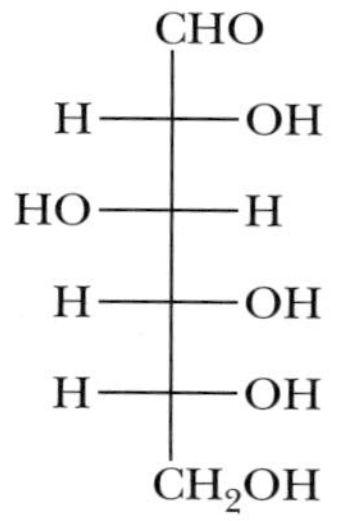

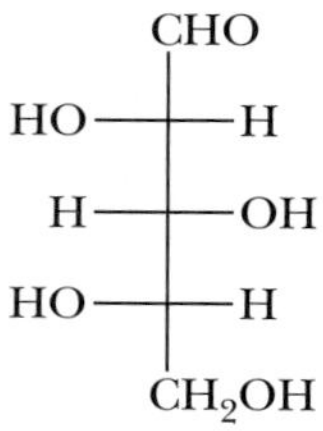

25.14 Give L-fucose (Chemical Connections: "A, B, AB, and O Blood Group Substances") a name incorporating the prefix *deoxy-* that shows its relationship to galactose.

25.15 2,6-Dideoxy-D-altrose, known alternatively as D-digitoxose, is a monosaccharide obtained on hydrolysis of digitoxin, a natural product extracted from foxglove (*Digitalis purpurea*). Digitoxin is used in cardiology to reduce pulse rate, regularize heart rhythm, and strengthen heart beat. Draw the structural formula of 2,6-dideoxy-D-altrose.

The Cyclic Structure of Monosaccharides

25.16 Define the term *anomeric carbon*. In glucose, which carbon is the anomeric carbon?

25.17 Define the terms (a) *pyranose* and (b) *furanose.*

25.18 Which is the anomeric carbon in a 2-ketohexose?

25.19 Are α-D-glucose and β-D-glucose enantiomers? Explain.

25.20 Convert each Haworth projection to an open-chain form and then to a Fischer projection. Name the monosaccharide you have drawn.

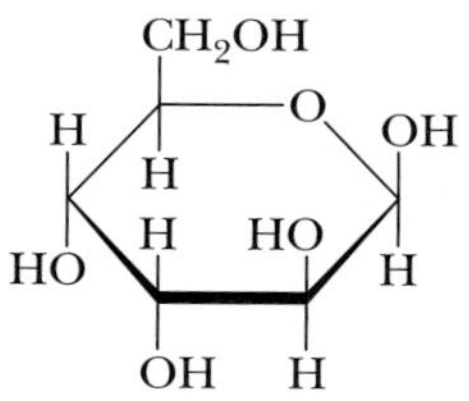

25.21 Convert each chair conformation to an open-chain form and then to a Fischer projection. Name the monosaccharide you have drawn.

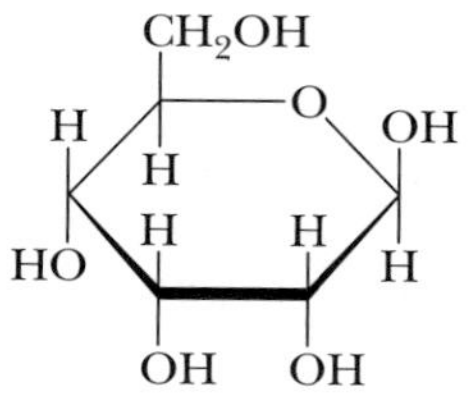

HO CH₂OH
(a)   O
HO
HO
OH

(b) HO CH₂OH
     O
     OH
HO   OH

25.22 Explain the phenomenon of mutarotation with reference to carbohydrates. By what means is it detected?

25.23 The specific rotation of α-D-glucose is +112.2.

   (a) What is the specific rotation of α-L-glucose?

   (b) When α-D-glucose is dissolved in water, the specific rotation of the solution changes from +112.2 to +52.7. Does the specific rotation of α-L-glucose also change when it is dissolved in water? If so, to what value does it change?

Reactions of Monosaccharides

25.24 Draw Fischer projections for the product(s) formed by reaction of D-galactose with the following. In addition, state whether each product is optically active or inactive.

   (a) NaBH₄ in H₂O   **(b)** H₂/Pt   **(c)** HNO₃, warm
   (d) Br₂/H₂O/CaCO₃   **(e)** H₅IO₆   **(f)** C₆H₅NH₂

25.25 Repeat Problem 25.24 using D-ribose.

25.26 An important technique for establishing relative configurations among isomeric aldoses and ketoses is to convert both terminal carbon atoms to the same functional group. This can be done either by selective oxidation or reduction. As a specific example, nitric acid oxidation of D-erythrose gives meso-tartaric acid (Section 3.4B). Similar oxidation of D-threose gives (2S,3S)-tartaric acid. Given this information and the fact that D-erythrose and D-threose are diastereomers, draw Fischer projections for D-erythrose and D-threose. Check your answers against Table 25.1.

25.27 There are four D-aldopentoses (Table 25.1). If each is reduced with NaBH₄, which yield optically active alditols? Which yield optically inactive alditols?

25.28 Name the two alditols formed by NaBH₄ reduction of D-fructose.

25.29 One pathway for the metabolism of D-glucose 6-phosphate is its enzyme-catalyzed conversion to D-fructose 6-phosphate. Show that this transformation can be accomplished as two enzyme-catalyzed keto-enol tautomerisms.

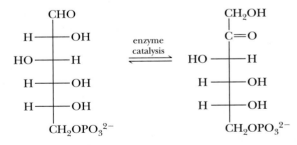

D-Glucose 6-phosphate   D-Fructose 6-phosphate

25.30 L-Fucose, one of several monosaccharides commonly found in the surface polysaccharides of animal cells, is synthesized biochemically from D-mannose in the following eight steps.

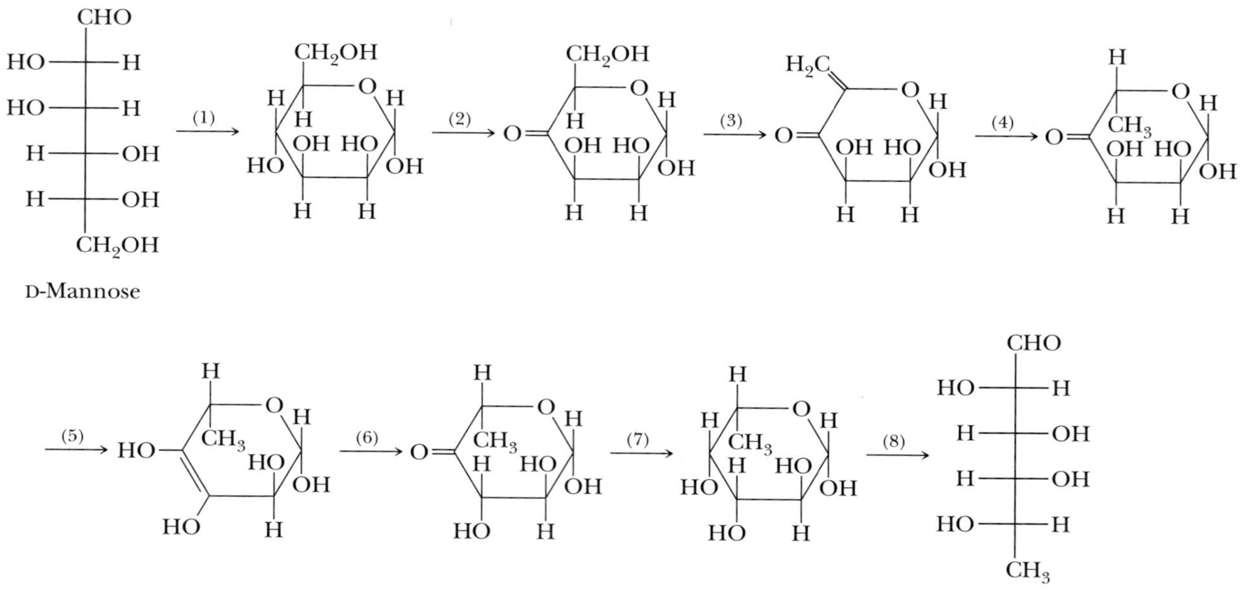

D-Mannose

L-Fucose

(a) Describe the type of reaction (that is, oxidation, reduction, hydration, dehydration, and so on) involved in each step.

(b) Explain why this monosaccharide derived from D-mannose now belongs to the L series.

25.31 What is the difference in meaning between the terms glycosidic bond and glucosidic bond?

25.32 Treatment of methyl β-D-glucopyranoside with benzaldehyde forms a six-membered cyclic acetal. Draw the most stable conformation of this acetal. Identify each new chiral center in the acetal.

25.33 Vanillin (4-hydroxy-3-methoxybenzaldehyde), the principal component of vanilla, occurs in vanilla beans and other natural sources as a β-D-glucopyranoside. Draw a structural formula for this glycoside, showing the D-glucose unit as a chair conformation.

25.34 Hot water extracts of ground willow and poplar bark are an effective pain reliever. Unfortunately, the liquid is so bitter that most persons refuse it. The pain reliever in these infusions is salicin, a β-glycoside of D-glucopyranose and the phenolic —OH group of 2-(hydroxymethyl)phenol. Draw a structural formula for salicin, showing the glucose ring as a chair conformation.

25.35 Draw structural formulas for the products formed by hydrolysis at pH 7.4 (the pH of blood plasma) of all ester, thioester, amide, anhydride, and glycoside groups in acetyl coenzyme A. Name as many of the products as you can.

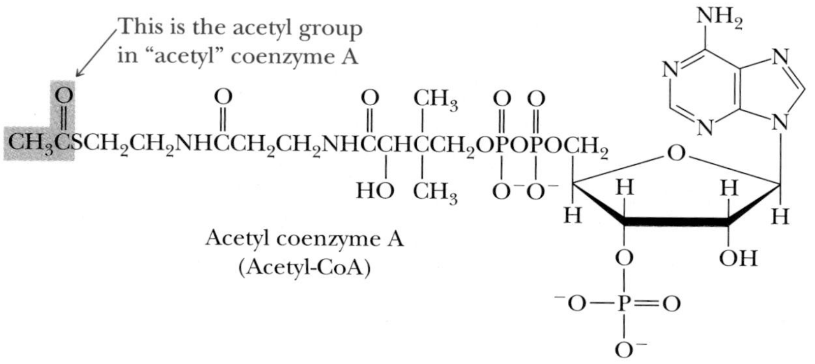

This is the acetyl group in "acetyl" coenzyme A

Acetyl coenzyme A
(Acetyl-CoA)

Disaccharides and Oligosaccharides

25.36 In making candy or sugar syrups, sucrose is boiled in water with a little acid, such as lemon juice. Why does the product mixture taste sweeter than the starting sucrose solution?

25.37 Trehalose is found in young mushrooms and is the chief carbohydrate in the blood of certain insects. Trehalose is a disaccharide consisting of two D-monosaccharide units, each joined to the other by an α-1,1-glycosidic bond.

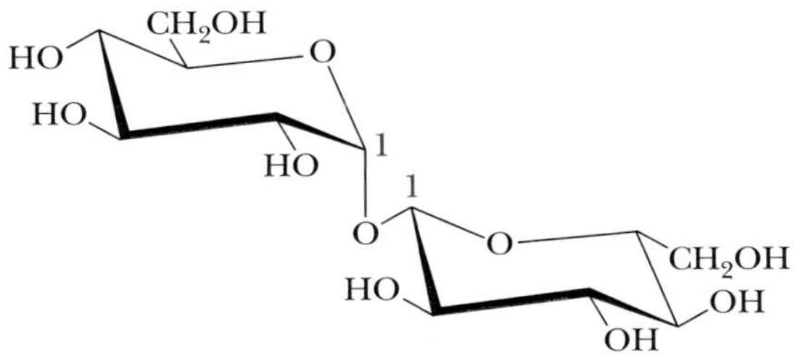

Trehalose

(a) Is trehalose a reducing sugar?
(b) Does trehalose undergo mutarotation?
(c) Name the two monosaccharide units of which trehalose is composed.

25.38 The trisaccharide raffinose occurs principally in cottonseed meal.

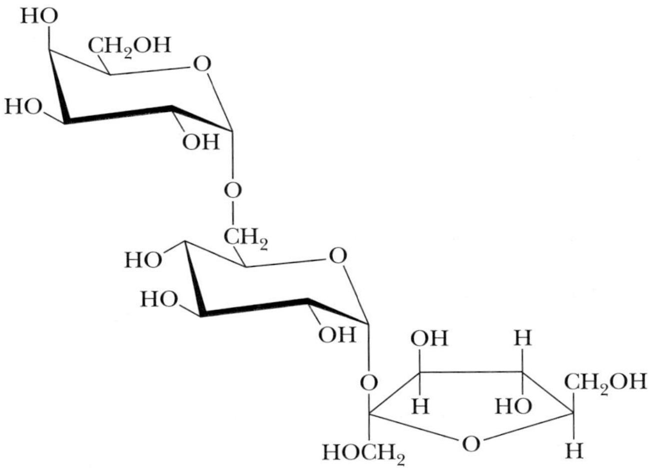

Raffinose

(a) Name the three monosaccharide units in raffinose.
(b) Describe each glycosidic bond in this trisaccharide.
(c) Is raffinose a reducing sugar?
(d) With how many moles of periodic acid will raffinose react?

25.39 Amygdalin is a toxic component in the pits of bitter almonds, peaches, and apricots.

(a) Name the two monosaccharide units in amygdalin and describe the glycoside bond by which they are joined.
(b) Account for the fact that on hydrolysis of amygdalin in warm aqueous acid liberates benzaldehyde and HCN.

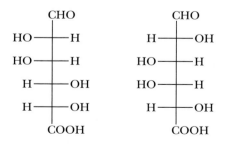

Amygdalin

Polysaccharides

25.40 What is the difference in structure between oligo- and polysaccharides?

25.41 Why is cellulose insoluble in water?

25.42 Consider *N*-acetyl-D-glucosamine (Section 25.1D).

    (a) Draw a chair conformation for the α- and β-pyranose forms of this monosaccharide.
    (b) Draw a chair conformation for the disaccharide formed by joining two units of the pyranose form of *N*-acetyl-D-glucosamine by a β-1,4-glycosidic bond. If you drew this correctly, you have the structural formula for the repeating dimer of chitin, the structural polysaccharide component of the shell of lobster and other crustaceans.

25.43 Propose structural formulas for the following polysaccharides.

    (a) Alginic acid, isolated from seaweed, is used as a thickening agent in ice cream and other foods. Alginic acid is a polymer of D-mannuronic acid in the pyranose form joined by β-1,4-glycosidic bonds.
    (b) Pectic acid is the main component of pectin, which is responsible for the formation of jellies from fruits and berries. Pectic acid is a polymer of D-galacturonic acid in the pyranose form joined by α-1,4-glycosidic bonds.

```
        CHO                       CHO
   HO ──┼── H              H ──┼── OH
   HO ──┼── H             HO ──┼── H
    H ──┼── OH            HO ──┼── H
    H ──┼── OH             H ──┼── OH
       COOH                     COOH
```

    D-Mannuronic acid   D-Galacturonic acid

25.44 Digitalis is a preparation made from the dried seeds and leaves of the purple foxglove, *Digitalis purpurea*, a plant native to southern and central Europe and cultivated in the United States. The preparation is a mixture of several active components, including digitalin. Digitalis is used in medicine to increase the force of myocardial contraction and as a conduction depressant to decrease heart rate (the heart pumps more forcefully but less often).

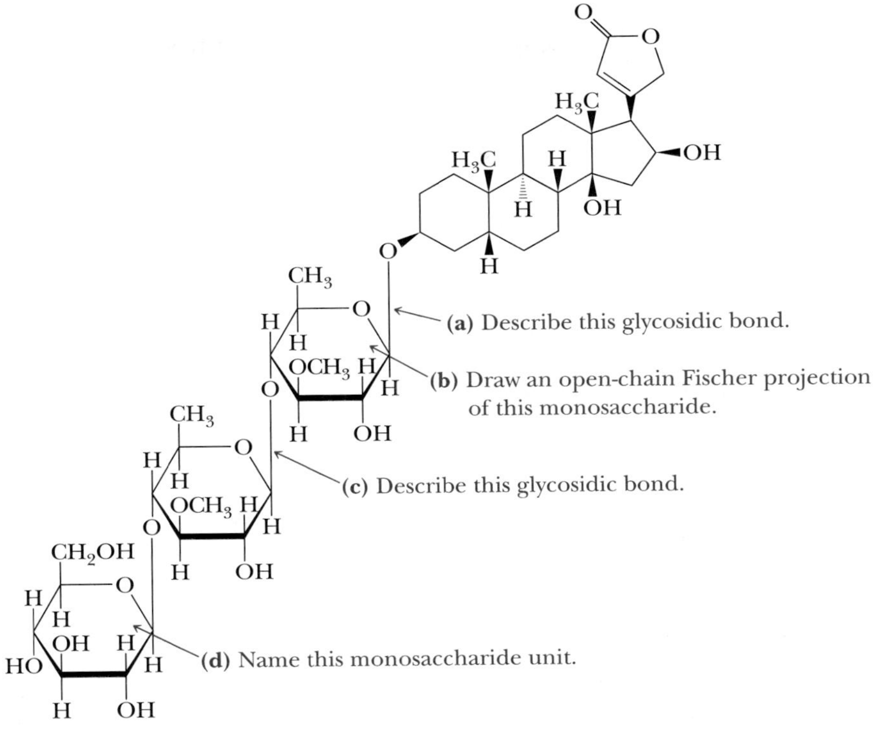

(a) Describe this glycosidic bond.

(b) Draw an open-chain Fischer projection of this monosaccharide.

(c) Describe this glycosidic bond.

(d) Name this monosaccharide unit.

Digitalin

25.45 Following is the structural formula of ganglioside GM$_2$, a macromolecular glycolipid (meaning that it contains lipid and monosaccharide units joined by glycosidic bonds).

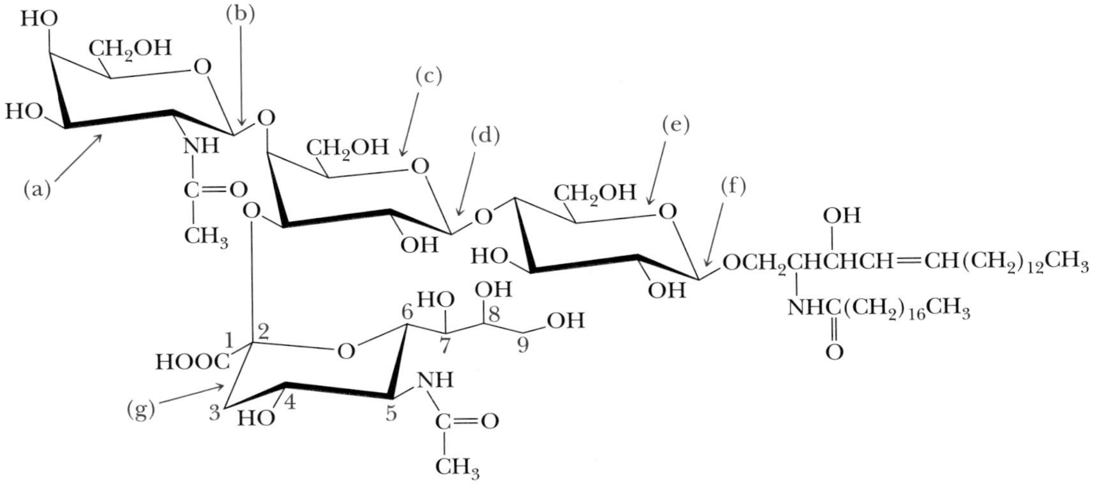

Ganglioside GM$_2$ or Tay-Sachs ganglioside

In normal cells, this and other gangliosides are synthesized continuously and degraded by lysosomes, which are cell organelles containing digestive enzymes. If pathways for the degradation of gangliosides are inhibited, the gangliosides accumulate in the central nervous system causing all sorts of life-threatening consequences. In inherited diseases of ganglioside metabolism, death usually occurs at an early age. Diseases of ganglioside metabolism include Gaucher's disease, Niemann-Pick disease, and Tay-Sachs disease. Tay-Sachs disease is a hereditary defect that is transmitted as an autosomal recessive gene. The concentration of ganglioside GM_2 is abnormally high in this disease because the enzyme responsible for catalyzing the hydrolysis of glycosidic bond (b) is absent.

(a) Name this monosaccharide unit.
(b) Describe this glycosidic bond (α or β, and between which carbons of each unit).
(c) Name this monosaccharide unit.
(d) Describe this glycosidic bond.
(e) Name this monosaccharide unit.
(f) Describe this glycosidic bond.
(g) This unit is *N*-acetylneuraminic acid, the most abundant member of a family of amino sugars containing nine or more carbons and distributed widely throughout the animal kingdom. Draw the open-chain form of this amino sugar. Do not be concerned with the configuration of the five chiral centers in the open-chain form.

25.46 Hyaluronic acid acts as a lubricant in the synovial fluid of joints. In rheumatoid arthritis, inflammation breaks hyaluronic acid down to smaller molecules. Under these conditions, what happens to the lubricating power of the synovial fluid?

25.47 The anticlotting property of heparin is partly the result of the negative charges it carries.

(a) Identify the functional groups that provide the negative charges.
(b) Which type of heparin is a better anticoagulant, one with a high or a low degree of polymerization?

25.48 Keratin sulfate is an important component of the cornea of the eye. Following is the repeating unit of this acidic polysaccharide.

(a) From what monosaccharides or derivatives of monosaccharides is keratin sulfate made?
(b) Describe the glycosidic bond in this repeating disaccharide unit.
(c) What is the net charge on this repeating disaccharide unit at pH 7.0?

25.49 Following is a chair conformation for the repeating disaccharide unit in chondroitin 6-sulfate. This biopolymer acts as the flexible connecting matrix between the tough protein filaments in cartilage. It is available as a dietary supplement, often combined with D-glucosamine sulfate. Some believe this combination can strengthen and improve joint flexibility.

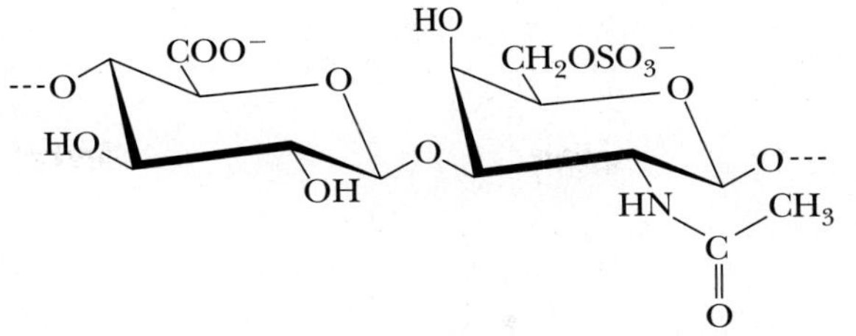

(a) From what two monosaccharide units is the repeating disaccharide unit of chondroitin 6-sulfate derived?

(b) Describe the glycosidic bond between the two units.

26

Sea lions are marine mammals that require a heavy layer of fat in order to survive in cold waters. Inset: Linoleic acid, a major component of unsaturated triglycerides and phospholipids (Sections 26.1 and 26.5).

Outline

Lipid A biomolecule isolated from plant or animal sources by extraction with nonpolar organic solvents, such as diethyl ether and hexane.

ORGANIC
Chemistry ⚛ Now™

Look for this logo in the chapter and go to Organic ChemistryNow at **http://now.brookscole.com/bfi4** for tutorials, simulations, and problems.

Triglyceride (triacylglycerol) An ester of glycerol with three fatty acids.

Lipids

Lipids are a heterogeneous group of naturally occurring organic compounds (many related to fats and oils), classified together on the basis of their common solubility properties. Lipids are insoluble in water but soluble in nonpolar aprotic organic solvents, including diethyl ether, dichloromethane, and acetone.

Lipids are divided into two main groups. First are those lipids that contain both a relatively large nonpolar hydrophobic region, most commonly aliphatic in nature, and a polar hydrophilic region. Found among this group are fatty acids, triglycerides, phospholipids, prostaglandins, and the fat-soluble vitamins. Second are those lipids that contain the tetracyclic ring system called the steroid nucleus, including cholesterol, steroid hormones, and bile acids. In this chapter, we describe the structures and biological functions of each group of lipids.

26.1 Triglycerides

Animal fats and vegetable oils, the most abundant naturally occurring lipids, are triesters of glycerol and long-chain carboxylic acids. Fats and oils are also referred to as **triglycerides** or **triacylglycerols.** Hydrolysis of a triglyceride in aqueous base followed by acidification gives glycerol and three fatty acids.

$$\begin{matrix} & \overset{O}{\underset{\|}{}} \\ O & CH_2OCR \\ \overset{\|}{R'COCH} & \overset{O}{\underset{\|}{}} \\ & CH_2OCR'' \end{matrix} \quad \xrightarrow[\text{2. HCl, H}_2\text{O}]{\text{1. NaOH, H}_2\text{O}} \quad \begin{matrix} CH_2OH \\ HOCH \\ CH_2OH \end{matrix} \quad + \quad \begin{matrix} RCOOH \\ R'COOH \\ R''COOH \end{matrix}$$

A triglyceride   1,2,3-Propanetriol   Fatty acids
                              (Glycerol, glycerin)

A. Fatty Acids

More than 500 different **fatty acids** have been isolated from various cells and tissues. Given in Table 26.1 are common names and structural formulas for the most abundant of these. The number of carbons in a fatty acid and the number of carbon-carbon double bonds in its hydrocarbon chain are shown by two numbers separated by a colon. In this notation, linoleic acid, for example, is designated as an 18:2 fatty acid; its 18-carbon chain contains two carbon-carbon double bonds. Following are several characteristics of the most abundant fatty acids in higher plants and animals.

Fatty acid A long, unbranched-chain carboxylic acid, most commonly of 12 to 20 carbons, derived from the hydrolysis of animal fats, vegetable oils, or the phospholipids of biological membranes.

1. Nearly all fatty acids have an even number of carbon atoms, most between 12 and 20, in an unbranched chain.
2. The three most abundant fatty acids in nature are palmitic acid (16:0), stearic acid (18:0), and oleic acid (18:1).
3. In most unsaturated fatty acids, the *cis* isomer predominates; the *trans* isomer is rare.
4. Unsaturated fatty acids have lower melting points than their saturated counterparts. The greater the degree of unsaturation, the lower the melting point (see

Table 26.1 The Most Abundant Fatty Acids in Animal Fats, Vegetable Oils, and Biological Membranes

Carbon Atoms/ Double Bonds*	Structure	Common Name	Melting Point (°C)
Saturated Fatty Acids			
12:0	$CH_3(CH_2)_{10}COOH$	Lauric acid	44
14:0	$CH_3(CH_2)_{12}COOH$	Myristic acid	58
16:0	$CH_3(CH_2)_{14}COOH$	Palmitic acid	63
18:0	$CH_3(CH_2)_{16}COOH$	Stearic acid	70
20:0	$CH_3(CH_2)_{18}COOH$	Arachidic acid	77
Unsaturated Fatty Acids			
16:1	$CH_3(CH_2)_5CH{=}CH(CH_2)_7COOH$	Palmitoleic acid	1
18:1	$CH_3(CH_2)_7CH{=}CH(CH_2)_7COOH$	Oleic acid	16
18:2	$CH_3(CH_2)_4(CH{=}CHCH_2)_2(CH_2)_6COOH$	Linoleic acid	−5
18:3	$CH_3CH_2(CH{=}CHCH_2)_3(CH_2)_6COOH$	Linolenic acid	−11
20:4	$CH_3(CH_2)_4(CH{=}CHCH_2)_4(CH_2)_2COOH$	Arachidonic acid	−49

*The first number is the number of carbons in the fatty acid; the second is the number of carbon-carbon double bonds in its hydrocarbon chain.

Polyunsaturated fatty acid A fatty acid with two or more carbon-carbon double bonds in its hydrocarbon chain.

Connections to Biological Chemistry: "The Importance of *cis* Double Bonds in Fats versus Oils" in Section 5.4). Compare, for example, the melting points of linoleic acid, a **polyunsaturated fatty acid,** and stearic acid, a saturated fatty acid.

Charles D. Winters

◼ Among the components of beeswax is triacontyl palmitate, $CH_3(CH_2)_{14}COO(CH_2)_{29}CH_3$, an ester of palmitic acid.

Example 26.1

Draw a structural formula of a triglyceride derived from one molecule each of palmitic acid, oleic acid, and stearic acid, the three most abundant fatty acids in the biological world.

Solution

In this structure, palmitic acid is esterified at carbon 1 of glycerol; oleic acid, at carbon 2; and stearic acid, at carbon 3.

oleate (18:1)

palmitate (16:0)

$$CH_3(CH_2)_7CH=CH(CH_2)_7COCH$$

$$CH_2OC(CH_2)_{14}CH_3$$

stearate (18:0)

$$CH_2OC(CH_2)_{16}CH_3$$

A triglyceride

Problem 26.1

(a) How many constitutional isomers are possible for a triglyceride containing one molecule each of palmitic acid, oleic acid, and stearic acid?

(b) Which of these constitutional isomers are chiral?

B. Physical Properties

Oil When used in the context of fats and oils, a mixture of triglycerides that is liquid at room temperature.

Fat A mixture of triglycerides that is semisolid or solid at room temperature.

The physical properties of a triglyceride depend on its fatty acid components. In general, the melting point of a triglyceride increases as the number of carbons in its hydrocarbon chains increases and as the number of carbon-carbon double bonds decreases. Triglycerides rich in oleic acid, linoleic acid, and other unsaturated fatty acids are generally liquids at room temperature and are called **oils,** as for example corn oil and olive oil. Olive oil, which contains mainly the monounsaturated oleic acid, solidifies in the refrigerator, whereas the more unsaturated corn oil will not. Triglycerides rich in palmitic, stearic, and other saturated fatty acids are generally semisolids or solids at room temperature and are called **fats,** as for example human fat and butter fat. Fats of land animals typically contain approximately 40–50% saturated fatty acids by weight (Table 26.2). Most plant oils, on the other hand, contain 20% or less saturated fatty acids and 80% or more unsaturated fatty acids. The notable exception to this generalization about plant oils are the **tropical oils** (as for example coconut and palm oils), which are considerably richer in low-molecular-weight saturated fatty acids.

The lower melting points of triglycerides rich in unsaturated fatty acids are related to differences in three-dimensional shape between the hydrocarbon chains of their unsaturated and saturated fatty acid components. Shown in Figure 26.1 is a space-filling model of tristrearin, a saturated triglyceride. In this model, the hydrocarbon

Table 26.2   Grams of Fatty Acid per 100 g of Triglyceride of Several Fats and Oils*

| Fat or Oil | Saturated Fatty Acids | | | Unsaturated Fatty Acids | |
	Lauric (12:0)	Palmitic (16:0)	Stearic (18:0)	Oleic (18:1)	Linoleic (18:2)
Human fat	—	24.0	8.4	46.9	10.2
Beef fat	—	27.4	14.1	49.6	2.5
Butter fat	2.5	29.0	9.2	26.7	3.6
Coconut oil	45.4	10.5	2.3	7.5	trace
Corn oil	—	10.2	3.0	49.6	34.3
Olive oil	—	6.9	2.3	84.4	4.6
Palm oil	—	40.1	5.5	42.7	10.3
Peanut oil	—	8.3	3.1	56.0	26.0
Soybean oil	0.2	9.8	2.4	28.9	50.7

*Only the most abundant fatty acids are given; other fatty acids are present in lesser amounts.

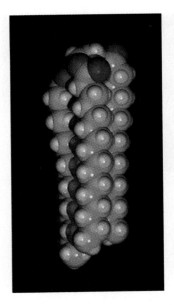

Figure 26.1
Tristearin, a saturated triglyceride.

Polyunsaturated triglyceride   A triglyceride having several carbon-carbon double bonds in the hydrocarbon chains of its three fatty acids.

chains lie parallel to each other, giving the molecule an ordered, compact shape. Because of this compact three-dimensional shape and the resulting strength of the dispersion forces between hydrocarbon chains of adjacent molecules, triglycerides rich in saturated fatty acids have melting points above room temperature.

The three-dimensional shape of an unsaturated fatty acid is quite different from that of a saturated fatty acid. Recall from Section 26.1A that unsaturated fatty acids of higher organisms are predominantly of the *cis* configuration; *trans* configurations are rare. Figure 26.2 shows a space-filling model of a **polyunsaturated triglyceride** derived from one molecule each of stearic acid, oleic acid, and linoleic acid. Each double bond in this polyunsaturated triglyceride has the *cis* configuration.

Polyunsaturated triglycerides have a less ordered structure and do not pack together as closely or as compactly as saturated triglycerides. Intramolecular and intermolecular dispersion forces are weaker, with the result that polyunsaturated triglycerides have lower melting points than their saturated counterparts.

C. Reduction of Fatty Acid Chains

For a variety of reasons, in part convenience and in part dietary preference, conversion of oils to fats has become a major industry. The process is called **hardening** of oils and involves catalytic reduction (Section 6.6A) of some or all carbon-carbon double bonds. In practice, the degree of hardening is carefully controlled to produce fats of a desired consistency. The resulting fats are sold for kitchen use (Crisco, Spry, and others). Margarine and other butter substitutes are produced by partial hydrogenation of polyunsaturated oils derived from corn, cottonseed, peanut, and soybean oils. To the hardened oils are added β-carotene (to give the final product a yellow color and make it look like butter), salt, and about 15% milk by volume to form the final emulsion. Vitamins A and D are also often added. Because the product at this stage is tasteless, acetoin and diacetyl are often added. These two compounds mimic the characteristic flavor of butter.

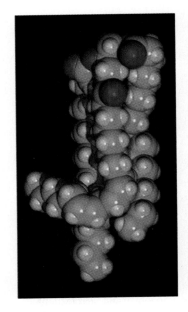

Figure 26.2
A polyunsaturated triglyceride.

Charles D. Winters

■ Liquid vegetable oils contain mostly unsaturated fatty acids.

Soap A sodium or potassium salt of a fatty acid.

3-Hydroxy-2-butanone
(Acetoin)

2,3-Butanedione
(Diacetyl)

26.2 Soaps and Detergents

A. Structure and Preparation of Soaps

Natural **soaps** are prepared most commonly from a blend of tallow and coconut oils. In the preparation of tallow, the solid fats of cattle are melted with steam, and the tallow layer formed on the top is removed. The preparation of soaps begins by boiling these triglycerides with sodium hydroxide. The reaction that takes place is called **saponification** (Latin: *saponem*, soap). At the molecular level, saponification corresponds to base-promoted hydrolysis of the ester groups in triglycerides (Section 18.4C). The resulting soaps contain mainly the sodium salts of palmitic, stearic, and oleic acids from tallow and the sodium salts of lauric and myristic acids from coconut oil.

A triglyceride + 3NaOH $\xrightarrow{\text{saponification}}$ 1,2,3-Propanetriol (Glycerol; glycerin) + Sodium soaps

After hydrolysis is complete, sodium chloride is added to precipitate the soap as thick curds. The water layer is then drawn off, and glycerol is recovered by vacuum distillation. The crude soap contains sodium chloride, sodium hydroxide, and other impurities. These are removed by boiling the curd in water and reprecipitating with more sodium chloride. After several purifications, the soap can be used without further processing as an inexpensive industrial soap. Other treatments transform the crude soap into pH-controlled cosmetic soaps, medicated soaps, and the like.

B. How Soaps Clean

Soap owes its remarkable cleansing properties to its ability to act as an emulsifying agent. Because the long hydrocarbon chains of natural soaps are insoluble in water, they tend to cluster in such a way as to minimize their contact with surrounding water molecules. The polar carboxylate groups, on the other hand, tend to remain in contact with the surrounding water molecules. Thus, in water, soap molecules spontaneously cluster into **micelles** (Figure 26.3).

Micelle A spherical arrangement of organic molecules in water solution clustered so that their hydrophobic parts are buried inside the sphere and their hydrophilic parts are on the surface of the sphere and in contact with water.

Most of the things we commonly think of as dirt (such as grease, oil, and fat stains) are nonpolar and insoluble in water. When soap and this type of dirt are mixed together, as in a washing machine, the nonpolar hydrocarbon inner parts of the soap micelles "dissolve" the nonpolar dirt molecules. In effect, new soap micelles are formed, this time with nonpolar dirt molecules in the center (Figure 26.4). In this way, nonpolar organic grease, oil, and fat are "dissolved" and washed away in the polar wash water.

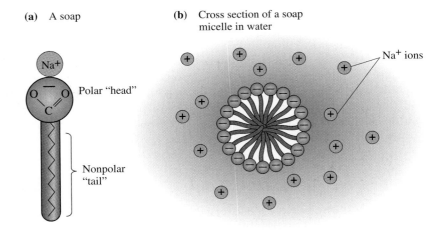

(a) A soap

Na$^+$

Polar "head"

Nonpolar "tail"

(b) Cross section of a soap micelle in water

Na$^+$ ions

Figure 26.3
Soap micelles. Nonpolar (hydrophobic) hydrocarbon chains are clustered in the interior of the micelle, and polar (hydrophilic) carboxylate groups are on the surface of the micelle. Soap micelles repel each other because of their negative surface charges.

Soaps are not without their disadvantages. Foremost among these is the fact that they form insoluble salts when used in water containing Ca(II), Mg(II), or Fe(III) ions (hard water).

$$2CH_3(CH_2)_{14}COO^- Na^+ \quad + \quad Ca^{2+} \longrightarrow [CH_3(CH_2)_{14}COO^-]_2Ca^{2+} \quad + \quad 2Na^+$$

A sodium soap
(soluble in water as micelles)

Calcium salt of a fatty acid
(insoluble in water)

These calcium, magnesium, and iron salts of fatty acids create problems, including rings around the bathtub, films that spoil the luster of hair, and grayness and roughness that build up on textiles after repeated washings.

C. Synthetic Detergents

After the cleansing action of soaps was understood, synthetic detergents could be designed. Molecules of a good detergent must have a long hydrocarbon chain, preferably 12 to 20 carbon atoms long, and a polar group at one end of the molecule that does not form insoluble salts with the Ca(II), Mg(II), or Fe(III) ions present in hard water. Chemists recognized that these essential characteristics of a soap could be produced in a molecule containing a sulfate or sulfonate group instead of a carboxylate group. Calcium, magnesium, and iron salts of monoalkylsulfuric and sulfonic acids are much more soluble in water than comparable salts of fatty acids.

The most widely used synthetic detergents are the linear alkylbenzenesulfonates (LAS). One of the most common of these is sodium 4-dodecylbenzenesulfonate. To prepare this type of detergent, a linear alkylbenzene is treated with sulfuric acid (Section 22.1B) to form an alkylbenzenesulfonic acid. The sulfonic acid is then neutralized with NaOH; the product is mixed with builders and spray-dried to give a smooth-flowing powder. The most common builder is sodium silicate.

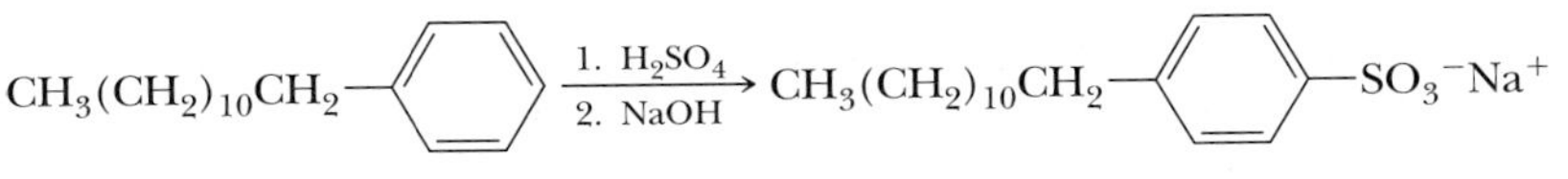

Soap micelle with "dissolved" grease

Grease

Soap

Figure 26.4
A soap micelle with a "dissolved" oil or grease droplet.

$$CH_3(CH_2)_{10}CH_2 - \underset{\text{Dodecylbenzene}}{\bigcirc} \xrightarrow[\text{2. NaOH}]{\text{1. H}_2\text{SO}_4} CH_3(CH_2)_{10}CH_2 - \underset{\substack{\text{Sodium 4-dodecylbenzenesulfonate} \\ \text{(an anionic detergent)}}}{\bigcirc} - SO_3^- Na^+$$

Alkylbenzenesulfonate detergents were introduced in the late 1950s, and today they command close to 90% of the market once held by natural soaps.

Among the most common additives to detergent preparations are foam stabilizers, bleaches, and optical brighteners. A foam stabilizer frequently added to liquid soaps but not laundry detergents (for obvious reasons: think of a top-loading washing machine with foam spewing out the lid!) is the amide prepared from dodecanoic acid (lauric acid) and 2-aminoethanol (ethanolamine). The most common bleach is sodium perborate tetrahydrate, which decomposes at temperatures above 50°C to give hydrogen peroxide, the actual bleaching agent.

$$CH_3(CH_2)_{10}\overset{\overset{\displaystyle O}{\|}}{C}NHCH_2CH_2OH \qquad O{=}B{-}O{-}O^-Na^+\cdot 4H_2O$$

N-(2-Hydroxyethyl)dodecanamide
(a foam stabilizer)

Sodium perborate tetrahydrate
(a bleach)

Also added to laundry detergents are optical brighteners, known also as optical bleaches. They are absorbed into fabrics and, after absorbing ambient light, fluoresce with a blue color, offsetting the yellow color caused by fabric aging. Quite literally, these optical brighteners produce a "whiter-than-white" appearance. You most certainly have observed the effects of optical brighteners if you have seen the glow of "white" shirts or blouses when exposed to black light (UV radiation).

26.3 Prostaglandins

Prostaglandin A member of the family of compounds having the 20-carbon skeleton of prostanoic acid.

The **prostaglandins** are a family of compounds all having the 20-carbon skeleton of prostanoic acid.

Prostanoic acid

The story of the discovery and structure determination of these remarkable compounds began in 1930 when gynecologists Raphael Kurzrok and Charles Lieb reported that human seminal fluid stimulates contraction of isolated uterine muscle. A few years later in Sweden, Ulf von Euler confirmed this report and noted that human seminal fluid also produces contraction of intestinal smooth muscle and lowers blood pressure when injected into the bloodstream. Von Euler proposed the name *prostaglandin* for the mysterious substance(s) responsible for these diverse effects because it was believed at the time that they were synthesized in the prostate gland. Although we now know that prostaglandin production is by no means limited to the prostate gland, the name nevertheless has stuck.

Prostaglandins are not stored as such in target tissues. Rather, they are synthesized in response to specific physiological triggers. Starting materials for the biosynthesis of prostaglandins are polyunsaturated fatty acids of 20 carbons, stored until needed as membrane phospholipid esters. In response to a physiological trigger, the ester is hydrolyzed, the fatty acid is released, and the synthesis of prostaglandins is

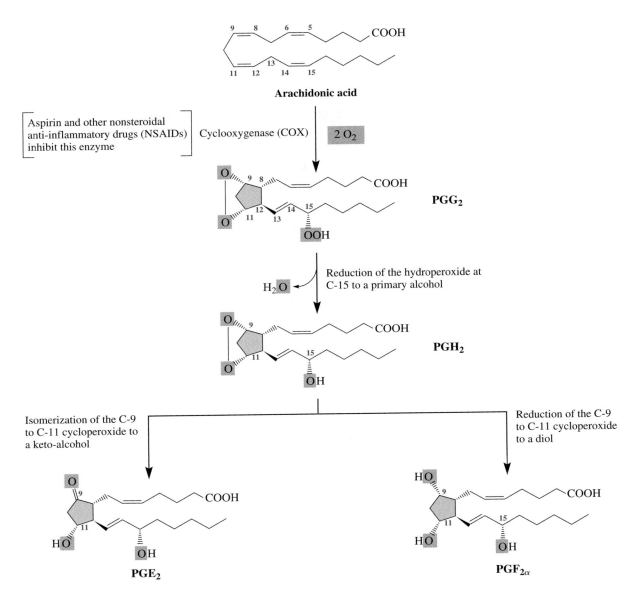

Figure 26.5

Key intermediates in the conversion of arachidonic acid to PGE$_2$ and PGF$_{2\alpha}$. PG stands for prostaglandin. The letters E, F, G, and H are different types of prostaglandins.

initiated. Figure 26.5 outlines the steps in the synthesis of several prostaglandins from arachidonic acid. A key step in this biosynthesis is the enzyme-catalyzed reaction of arachidonic acid with two molecules of O$_2$ to form prostaglandin G$_2$ (PGG$_2$). The anti-inflammatory and anticlotting effects of aspirin and other nonsteroidal anti-inflammatory drugs (NSAIDs) results from their ability to inhibit the enzyme that catalyzes this step.

Research on the involvement of prostaglandins in reproductive physiology and the inflammatory process has produced several clinically useful prostaglandin derivatives. The observations that PGF$_{2\alpha}$ stimulates contractions of uterine smooth muscle led to a synthetic derivative that is used as a therapeutic abortifacient. A problem with the use of the natural prostaglandins for this purpose is that they are very rapidly degraded within the body. In the search for less rapidly degraded prostaglandins, a number of analogs have been prepared, one of the most effective of which is carboprost. This synthetic prostaglandin is 10 to 20 times more potent than the natural $PGF_{2\alpha}$ and is only slowly degraded in the body. The comparison of these two prostaglandins illustrates how a simple change in the structure of a drug can make a significant change in its effectiveness.

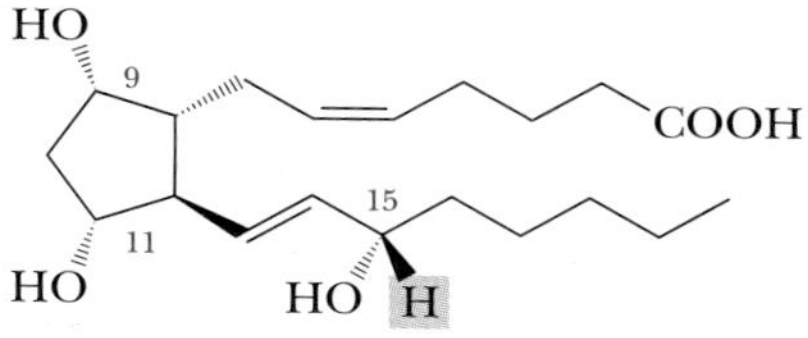

$PGF_{2\alpha}$

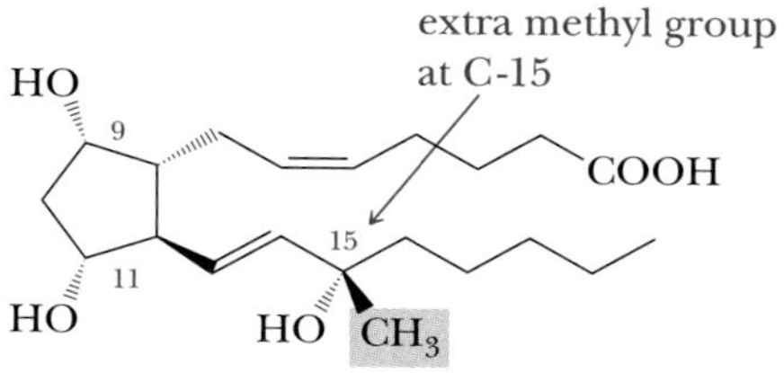

Carboprost
(15S)-15-Methyl-$PGF_{2\alpha}$

The PGEs along with several other PGs suppress gastric ulceration and appear to heal gastric ulcers. The PGE_1 analog, misoprostol, is currently used primarily for prevention of ulceration associated with aspirin-like NSAIDs (partly caused by their inhibition of clotting).

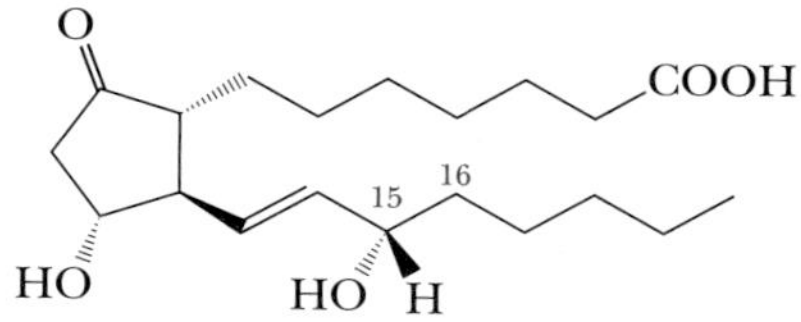

PGE_1

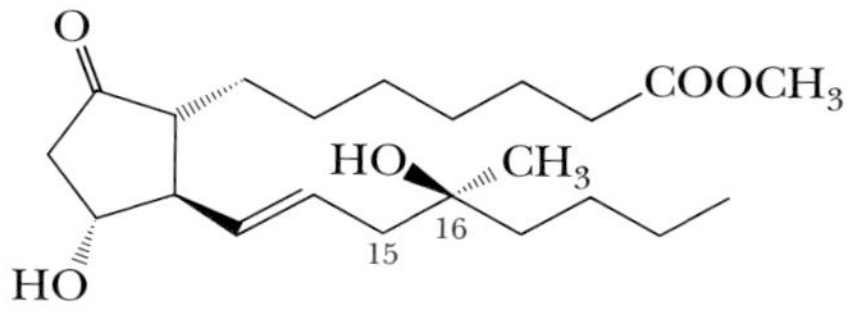

Misoprostol

Prostaglandins are members of an even larger family of compounds called **eicosanoids.** Eicosanoids contain 20 carbons and are derived from fatty acids. They include not only the prostaglandins but also the leukotrienes, thromboxanes, and prostacyclins. The eicosanoids are extremely widespread, and members of this family of compounds have been isolated from almost every tissue and body fluid.

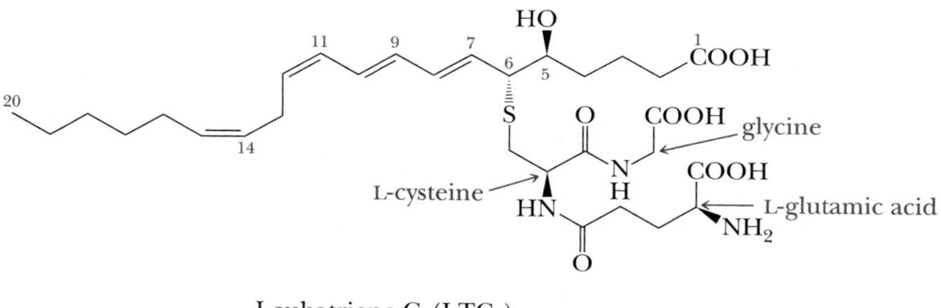

Leukotriene C_4(LTC_4)
(a smooth muscle constrictor)

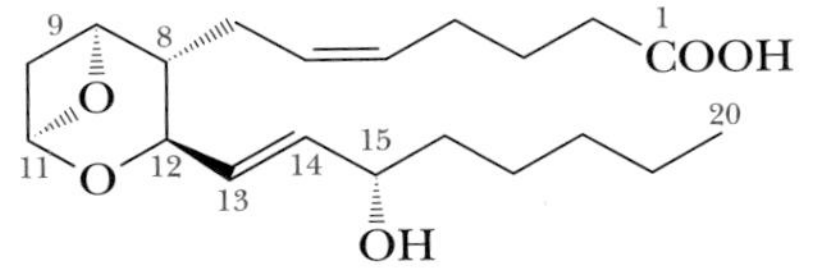

Thromboxane A$_2$
(a potent vasoconstrictor)

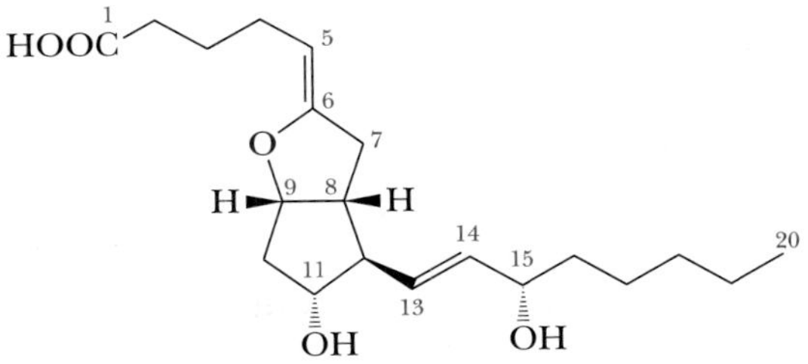

Prostacyclin
(a platelet aggregation inhibitor)

Leukotrienes are derived from arachidonic acid and are found primarily in leukocytes (white blood cells). Leukotriene C$_4$ (LTC$_4$), a typical member of this family, has three conjugated double bonds (hence the suffix *-triene*) and contains the amino acids L-cysteine, glycine, and L-glutamic acid (Chapter 27). An important physiological action of LTC$_4$ is constriction of smooth muscles, especially those of the lungs. The synthesis and release of LTC$_4$ is prompted by allergic reactions. Drugs that inhibit the synthesis of LTC$_4$ show promise for the treatment of the allergic reactions associated with asthma. Thromboxane A$_2$ is a very potent vasoconstrictor; its release triggers the irreversible phase of platelet aggregation and constriction of injured blood vessels. It is thought that aspirin and aspirin-like drugs act as mild anticoagulants because they inhibit cyclooxygenase, the enzyme that initiates the synthesis of thromboxane A$_2$.

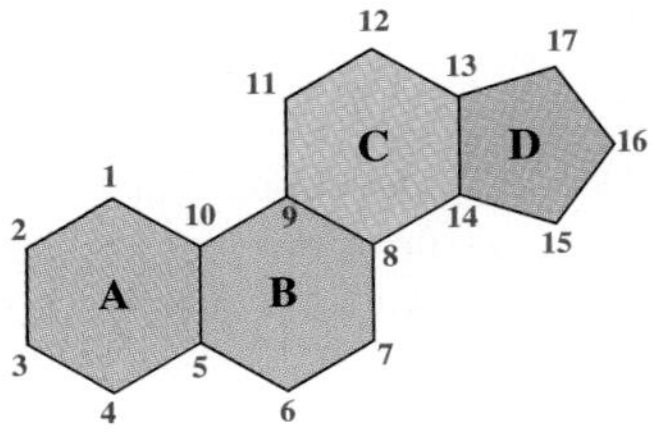

26.4 Steroids

Steroids are a group of plant and animal lipids that have the tetracyclic ring system shown in Figure 26.6. The features common to the tetracyclic ring system of most naturally occurring steroids are illustrated in Figure 26.7.

1. The fusion of rings is *trans,* and each atom or group at a ring junction is axial. Compare, for example, the orientations of —H at carbon 5 and —CH$_3$ at carbon 10.
2. The pattern of atoms or groups along the points of ring fusion (carbons 5 to 10 to 9 to 8 to 14 to 13) is nearly always *trans*-anti-*trans*-anti-*trans.*
3. Because of the *trans*-anti-*trans*-anti-*trans* arrangement of atoms or groups along the points of ring fusion, the tetracyclic steroid ring system is nearly flat and quite rigid.
4. Many steroids have axial methyl groups at carbons 10 and 13 of the tetracyclic ring system.

Figure 26.6
The tetracyclic ring system characteristic of steroids.

Steroid A plant or animal lipid having the characteristic tetracyclic ring structure of the steroid nucleus, namely three six-membered rings and one five-membered ring.

Methyl groups at C–10 and C–13 are axial and above the plane of the rings

C/D *trans*
B/C *trans*
A/B *trans*

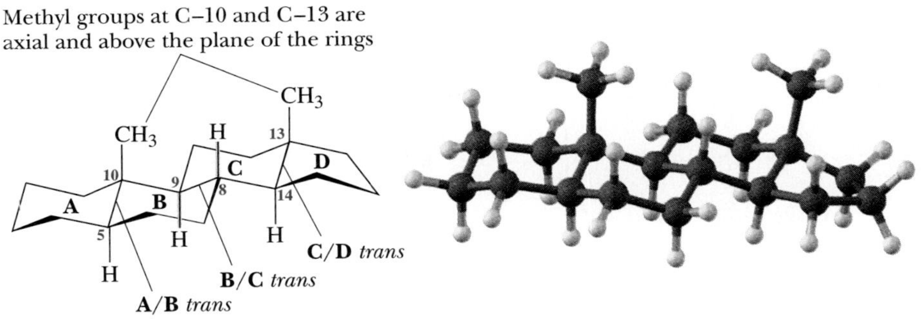

Figure 26.7
Features common to the tetracyclic ring system of many steroids.

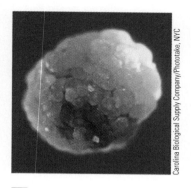

■■■ Human gallstones are almost pure cholesterol; this gallstone is about 0.5 cm in diameter.

Low-density lipoprotein (LDL)
Plasma particles, density 1.02–1.06 g/mL, consisting of approximately 26% proteins, 50% cholesterol, 21% phospholipids, and 4% triglycerides.

High-density lipoprotein (HDL)
Plasma particles, density 1.06–1.21 g/mL, consisting of approximately 33% proteins, 30% cholesterol, 29% phospholipids, and 8% triglycerides.

Estrogen   A steroid hormone, such as estrone and estradiol, that mediates the development of sexual characteristics in females.

A. Structure of the Major Classes of Steroids

Cholesterol

Cholesterol is a white, water-insoluble, waxy solid found in blood plasma and in all animal tissues. This substance is an integral part of human metabolism in two ways: (1) It is an essential component of biological membranes. The body of a healthy adult contains approximately 140 g of cholesterol, about 120 g of which is present in membranes. Membranes of the central and peripheral nervous systems, for example, contain about 10% cholesterol by weight. (2) It is the compound from which sex hormones, adrenocorticoid hormones, bile acids, and vitamin D are synthesized. Thus, cholesterol is, in a sense, the parent steroid.

Cholesterol has eight chiral centers, and a molecule with this number of chiral centers can exist as 2^8, or 256, stereoisomers (128 pairs of enantiomers). Only one of these stereoisomers is known to exist in nature: the stereoisomer with the configuration shown in Figure 26.8.

Cholesterol is insoluble in blood plasma but can be transported as a plasma-soluble complex formed with proteins called lipoproteins. **Low-density lipoproteins (LDL,** often called "bad cholesterol") transport cholesterol from the site of its synthesis in the liver to the various tissues and cells of the body where it is to be used. Primarily, the cholesterol associated with LDLs builds up in atherosclerotic deposits in blood vessels. **High-density lipoproteins (HDL,** often called "good cholesterol") transport excess and unused cholesterol from cells back to the liver for its degradation to bile acids and eventual excretion in the feces. It is thought that HDLs retard or reduce atherosclerotic deposits.

Steroid Hormones

Given in Table 26.3 are representations of each major class of steroid hormones, along with the principal functions of each.

The two most important female sex hormones, or **estrogens,** are estrone and estradiol. In addition, progesterone, another type of steroid hormone, is essential for preparing the uterus for implantation of a fertilized egg. After the role of progesterone in inhibiting ovulation was understood, its potential as a possible

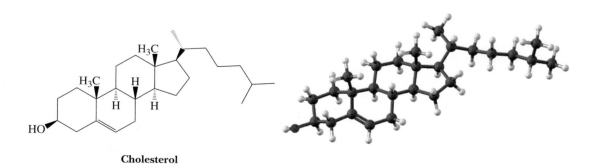

Cholesterol

Figure 26.8
Cholesterol is found in blood plasma and in all animal tissues.

Table 26.3 Selected Steroid Hormones

Structure	Source and Major Effects
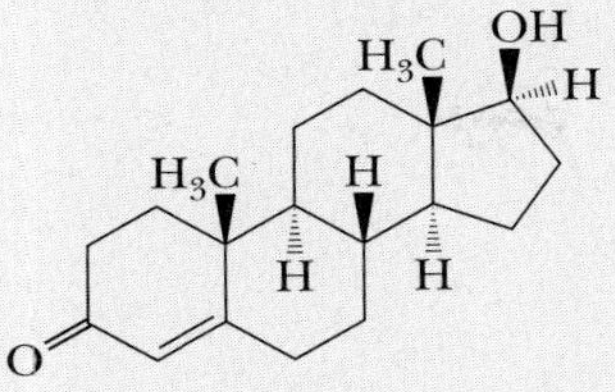 Testosterone 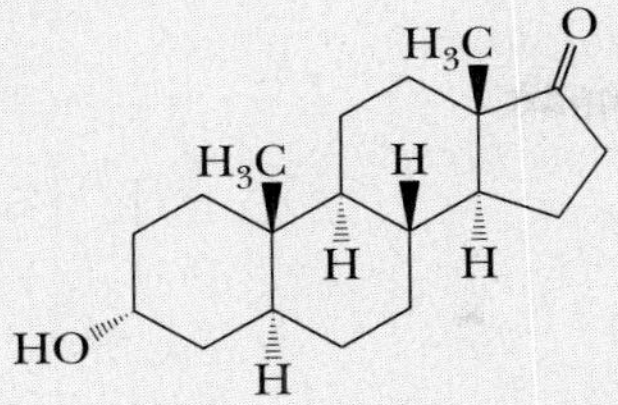 Androsterone	Androgens (male sex hormones)— synthesized in the testes; responsible for development of male secondary sex characteristics
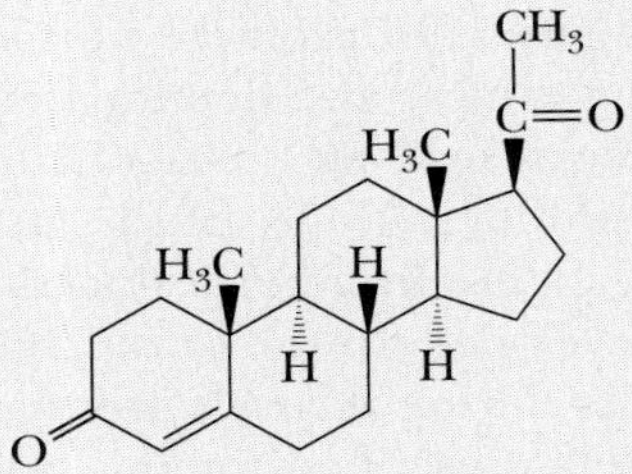 Progesterone 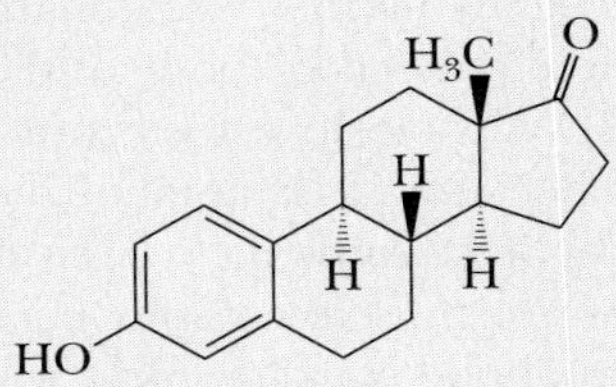 Estrone	Estrogens (female sex hormones)— synthesized in the ovaries; responsible for development of female secondary sex characteristics and control of the menstrual cycle
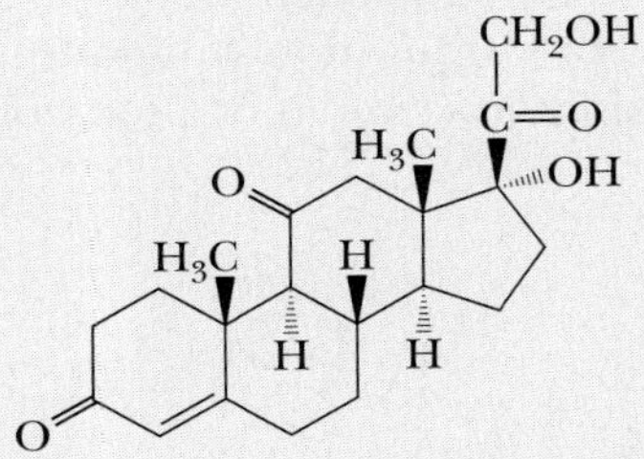 Cortisone 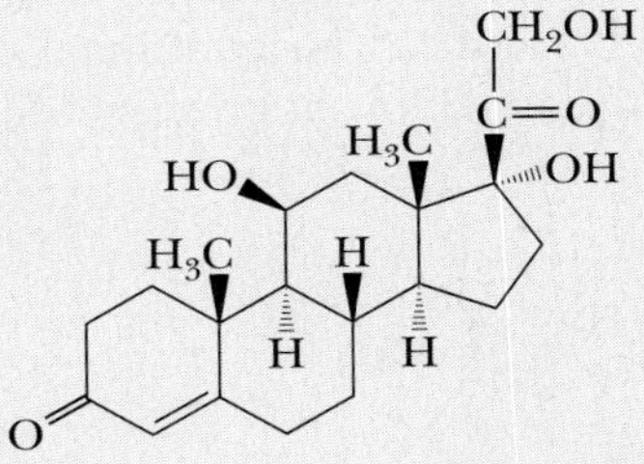 Cortisol	Glucocorticoid hormones— synthesized in the adrenal cortex; regulate metabolism of carbohydrates, decrease inflammation, and are involved in the reaction to stress
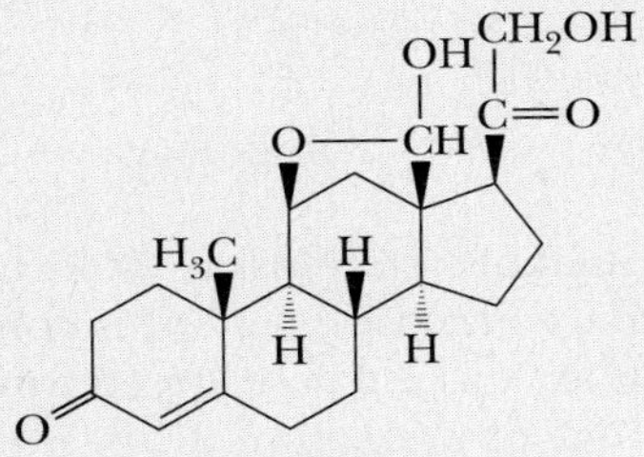 Aldosterone	A mineralocorticoid hormone— synthesized in the adrenal cortex; regulates blood pressure and volume by stimulating the kidneys to absorb Na^+, Cl^-, and HCO_3^-

contraceptive was realized. Progesterone itself is relatively ineffective when taken orally. As a result of a massive research program in both industrial and academic laboratories, many synthetic progesterone-mimicking steroids became available in the 1960s. When taken regularly, these drugs prevent ovulation yet allow women to maintain a normal menstrual cycle. Some of the most effective of these preparations contain a progesterone analog, such as norethindrone, combined with a smaller amount of an estrogen-like material to help prevent irregular menstrual flow during prolonged use of contraceptive pills.

"Nor" refers to the absence of a methyl group here. The methyl group is present in ethindrone.

Norethindrone
(a synthetic progesterone analog)

Androgen A steroid hormone, such as testosterone, that mediates the development of sexual characteristics of males.

The chief function of testosterone and other **androgens** is to promote normal growth of male reproductive organs (primary sex characteristics) and development of the characteristic deep voice, pattern of body and facial hair, and musculature (secondary sex characteristics). Although testosterone produces these effects, it is not active when taken orally because it is metabolized in the liver to an inactive steroid. A number of oral **anabolic steroids** have been developed for use in rehabilitation medicine, particularly when muscle atrophy occurs during recovery from an injury. Among the synthetic anabolic steroids most widely prescribed for this purpose are methandrostenolone and stanozolol. The structural formula of methandrostenolone differs from that of testosterone by introduction of (1) a methyl group at carbon 17, and (2) an additional carbon-carbon double bond between carbons 1 and 2. In stanozolol, ring A is modified by attachment of a pyrazole ring.

Anabolic steroid A steroid hormone, such as testosterone, that promotes tissue and muscle growth and development.

Methandrostenolone   Stanozolol

Among certain athletes, the misuse of anabolic steroids to build muscle mass and strength, particularly for sports that require explosive action, is common. The risks associated with abuse of anabolic steroids for this purpose are enormous: heightened aggressiveness, sterility, impotence, and risk of premature death from complications of diabetes, coronary artery disease, and liver cancer.

Bile Acids

Bile acid A cholesterol-derived detergent molecule, such as cholic acid, which is secreted by the gallbladder into the intestine to assist in the absorption of dietary lipids.

Shown in Figure 26.9 is a structural formula for cholic acid, a constituent of human bile. The molecule is shown as an anion, as it is ionized in bile and intestinal fluids. **Bile acids,** or more properly, bile salts, are synthesized in the liver, stored in the gallbladder, and secreted into the intestine, where their function is to emulsify dietary fats and thereby aid in their absorption and digestion. Furthermore, bile salts are the end products of the metabolism of cholesterol

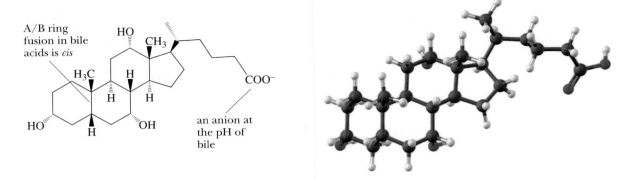

Figure 26.9
Cholic acid, an important constituent of human bile. Each six-membered ring is in a
chair conformation.

and, thus, are a principal pathway for the elimination of this substance from
the body. A characteristic structural feature of bile salts is a *cis* fusion of rings A
and B.

B. Biosynthesis of Cholesterol

The biosynthesis of cholesterol illustrates a point we first made in our introduction to the structure of terpenes (Section 5.4). In building large molecules,
one of the common patterns in the biological world is to begin with one or
more smaller subunits, join them by an iterative process, and then chemically
modify the completed carbon skeleton by oxidation, reduction, cross-linking,
addition, elimination, or related processes to give a biomolecule with a unique
identity.

The building block from which all carbon atoms of steroids are derived is the
two-carbon acetyl group of acetyl-CoA (Problem 25.35). The American biochemist,
Konrad Bloch, who shared the 1964 Nobel Prize for medicine or physiology
with German biochemist Feodor Lynen for their discoveries concerning the
biosynthesis of cholesterol and fatty acids, showed that 15 of the 27 carbon
atoms of cholesterol are derived from the methyl group of acetyl-CoA. The remaining 12 carbon atoms are derived from the carbonyl group of acetyl-CoA
(Figure 26.10).

A remarkable feature of this synthetic pathway is that the biosynthesis of cholesterol from acetyl-CoA is completely stereoselective; it is synthesized as only one of 256
possible stereoisomers. We cannot duplicate this exquisite degree of stereoselectivity
in the laboratory. Cholesterol is, in turn, the key intermediate in the synthesis of most
other steroids.

cholesterol
- bile acids (e.g., cholic acid)
- sex hormones (e.g., testosterone and estrone)
- mineralocorticoid hormones (e.g., aldosterone)
- glucocorticoid hormones (e.g., cortisone)

Drugs that Lower Plasma Levels of Cholesterol

Coronary artery disease is the leading cause of death in the United States and other Western countries, where about one half of all deaths can be attributed to atherosclerosis. Atherosclerosis results from the buildup and autoxidation of fatty deposits called plaque on the inner walls of arteries. A major component of plaque is cholesterol derived from low-density lipoproteins (LDL), which circulate in blood plasma. Because more than one half of total body cholesterol in humans is synthesized in the liver from acetyl CoA, intensive efforts have been directed toward finding ways to inhibit this synthesis. The rate-determining step in cholesterol biosynthesis is reduction of 3-hydroxy-3-methylglutaryl CoA (HMG-CoA) to mevalonic acid. This four-electron reduction is catalyzed by the enzyme HMG-CoA reductase and requires two moles of NADPH per mole of HMG-CoA.

Beginning in the early 1970s, researchers at the Sankyo Company in Tokyo screened more than 8000 strains of microorganisms and in 1976 announced the isolation of mevastatin, a potent inhibitor of HMG-CoA reductase, from culture broths of the fungus *Penicillium citrinum*. The same compound was isolated by researchers at Beecham Pharmaceuticals in England from cultures of *Penicillium brevicompactum*. Soon thereafter, a second, more active compound called lovastatin was isolated at the Sankyo Company from the fungus *Monascus ruber*, and at Merck Sharpe & Dohme from *Aspergillus terreus*. Both mold metabolites are extremely effective in lowering plasma concentrations of LDL. The active form of each is the 5-hydroxycarboxylic acid formed by hydrolysis of the δ-lactone.

$R_1 = R_2 = H$, mevastatin
$R_1 = H$, $R_2 = CH_3$, lovastatin (Mevacor)
$R_1 = R_2 = CH_3$, simvastatin (Zocor)

The active form of each drug

Soon thereafter, Merck developed a synthesis for simvastatin (Zocor). It came onto the market in the late 1980s and is still used worldwide for the control of plasma cholesterol levels.

It is thought that these drugs and several synthetic modifications now available inhibit HMG-CoA reductase by forming an enzyme-inhibitor complex that prevents further catalytic action of the enzyme. It is reasoned that the 3,5-dihydroxycarboxylic acid part of each drug binds tightly to the enzyme because it mimics the hemithioacetal intermediate formed after the first two-electron reduction of HMG-CoA.

3-Hydroxy-3-methyl-glutaryl-CoA (HMG-CoA)

A hemithioacetal intermediate formed by the first two-electron reduction

Mevalonate

Systematic studies have shown the importance of each part of the drug for effectiveness. It has been found, for example, that the carboxylate anion is essential, and both the 3-OH and 5-OH groups must be free (not masked as ethers). Insertion of a bridging unit other than —CH$_2$—CH$_2$— between carbon 5 and the bicyclo[4.4.0] ring system reduces potency as does almost any modification of the bicyclic ring system and its pattern of substitution.

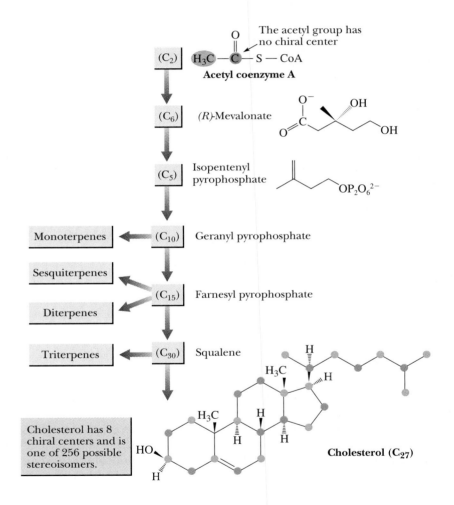

Figure 26.10

Several key intermediates in the synthesis of cholesterol from acetyl groups of acetyl CoA. Eighteen moles of acetyl CoA are required for the synthesis of one mole of cholesterol.

26.5 Phospholipids

A. Structure

Phospholipids, or phosphoacylglycerols as they are more properly named, are the second most abundant group of naturally occurring lipids. They are found almost exclusively in plant and animal membranes, which typically consist of about 40–50%

Phospholipid A lipid containing glycerol esterified with two molecules of fatty acid and one molecule of phosphoric acid.

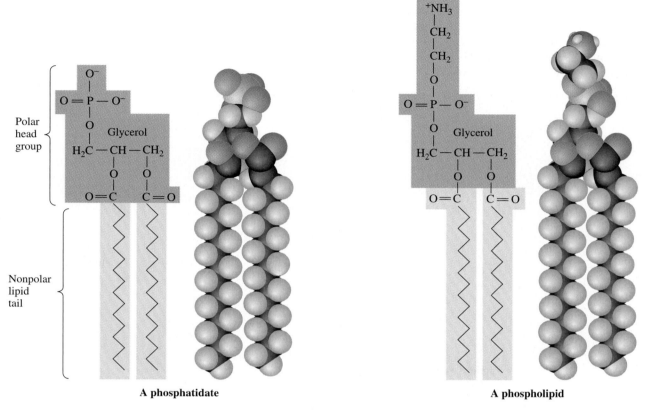

Polar head group

Nonpolar lipid tail

A phosphatidate

A phospholipid

Figure 26.11

In a phosphatidic acid, glycerol is esterified with two molecules of fatty acid and one molecule of phosphoric acid. Further esterification of the phosphoric acid group with a low-molecular-weight alcohol gives a phospholipid.

Figure 26.12
Space-filling model of a lecithin.

phospholipids and 50–60% proteins. The most abundant phospholipids are derived from a phosphatidic acid (Figure 26.11).

The fatty acids most common in phosphatidic acids are palmitic and stearic acids (both fully saturated) and oleic acid (one double bond in the hydrocarbon chain). Further esterification of a phosphatidic acid with a low-molecular-weight alcohol gives a phospholipid. Several of the most common alcohols forming phospholipids are given in Table 26.4. All functional groups in this table and in Figure 26.11 are shown as they are ionized at pH 7.4, the approximate pH of blood plasma and of many other biological fluids. Under these conditions, each phosphate group bears a negative charge and each amino group bears a positive charge.

B. Lipid Bilayers

Figure 26.12 shows a space-filling model of a lecithin (a phosphatidylcholine). It and other phospholipids are elongated, almost rodlike molecules, with the nonpolar (hydrophobic) hydrocarbon chains lying roughly parallel to one another and the polar (hydrophilic) phosphoric ester group pointing in the opposite direction.

Table 26.4 Low-Molecular-Weight Alcohols Most Common to Phospholipids

Alcohols Found in Phospholipids		
Structural Formula	**Name**	**Name of Phospholipid**
$HOCH_2CH_2NH_2$	Ethanolamine	Phosphatidylethanolamine (Cephalin)
$HOCH_2CH_2\overset{+}{N}(CH_3)_3$	Choline	Phosphatidylcholine (Lecithin)
$HOCH_2CHCOO^-$ | NH_3^+	Serine	Phosphatidylserine
(inositol ring structure)	Inositol	Phosphatidylinositol

All of these products contain lecithin.

When placed in aqueous solution, phospholipids spontaneously form a **lipid bilayer** (Figure 26.13) in which polar head groups lie on the surface, giving the bilayer an ionic coating. Nonpolar hydrocarbon chains of fatty acids lie buried within the bilayer. This self-assembly of phospholipids into a bilayer is a spontaneous process, driven by two types of noncovalent forces: (1) hydrophobic effects, which result when nonpolar hydrocarbon chains cluster together and exclude water molecules, and (2) electrostatic interactions, which result when polar head groups interact with water and other polar molecules in the aqueous environment.

Lipid bilayer A back-to-back arrangement of phospholipid monolayers, often forming a closed vesicle or membrane.

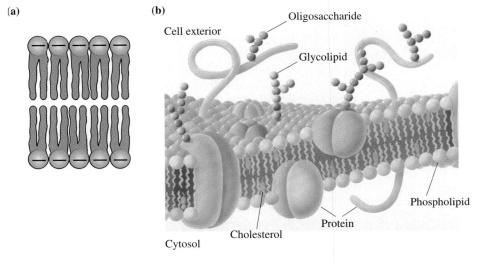

Figure 26.13
Fluid-mosaic model of a biological membrane, showing the lipid bilayer and membrane proteins oriented on the inner and outer surfaces of the membrane and penetrating the entire thickness of the membrane.

CHEMICAL CONNECTIONS

Snake Venom Phospholipases

The venoms of certain snakes contain enzymes called phospholipases. These enzymes catalyze the hydrolysis of carboxylic ester bonds of phospholipids. The venom of the eastern diamondback rattlesnake (*Crotalus adamanteus*) and the Indian cobra (*Naja naja*) both contain phospholipase PLA_2, which catalyzes the hydrolysis of esters at carbon 2 of phospholipids. The breakdown product of this hydrolysis, a lysolecithin, acts as a detergent and dissolves the membranes of red blood cells causing them to rupture. Indian cobras kill several thousand people each year.

A phospholipid

A lysolecithin

■ Milking an Indian cobra for its venom.

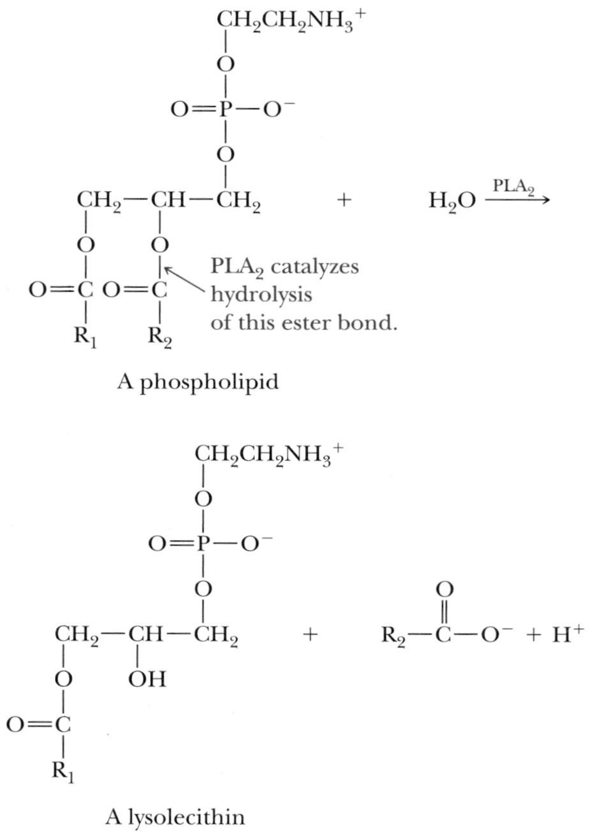

Dan McCoy/Rainbow

Recall from Section 26.2B that formation of soap micelles is driven by these same noncovalent forces; the polar (hydrophilic) carboxylate groups of soap molecules lie on the surface of the micelle and associate with water molecules, and the nonpolar (hydrophobic) hydrocarbon chains cluster within the micelle and thus are removed from contact with water.

The arrangement of hydrocarbon chains in the interior of a phospholipid bilayer varies from rigid to fluid, depending on the degree of unsaturation of the hydrocarbon chains themselves. Saturated hydrocarbon chains tend to lie parallel and closely packed, leading to a rigidity of the bilayer. Unsaturated hydrocarbon chains, on the other hand, have one or more *cis* double bonds, which cause "kinks" in the chains. As a result, they do not pack as closely and with as great an order as saturated chains.

The disordered packing of unsaturated hydrocarbon chains leads to fluidity of the bilayer.

Biological membranes are made of lipid bilayers. The most satisfactory current model for the arrangement of phospholipids, proteins, and cholesterol in plant and animal membranes is the **fluid-mosaic model** proposed in 1972 by S. J. Singer and G. Nicolson (Figure 26.13). The term "mosaic" signifies that the various components in the membrane coexist side by side, as discrete units, rather than combining to form new molecules or ions. "Fluid" signifies that the same sort of fluidity exists in membranes that we have already seen for lipid bilayers. Furthermore, the protein components of membranes "float" in the bilayer and can move laterally along the plane of the membrane.

Fluid-mosaic model A biological membrane that consists of a phospholipid bilayer with proteins, carbohydrates, and other lipids on the surface and embedded in the surface of the bilayer.

26.6 Fat-Soluble Vitamins

Vitamins are divided into two broad classes on the basis of solubility, those that are fat-soluble (and hence classed as lipids) and those that are water-soluble. The fat-soluble vitamins include A, D, E, and K.

A. Vitamin A

Vitamin A, or retinol, occurs only in the animal world, where the best sources are cod-liver oil and other fish-liver oils, animal liver, and dairy products. Vitamin A in the form of a precursor, or provitamin, is found in the plant world in a group of tetraterpene (C_{40}) pigments called carotenes. The most common of these is β-carotene, abundant in carrots but also found in some other vegetables, particularly yellow and green ones. β-Carotene has activity as an antioxidant; one of its functions in green plants is to quench singlet oxygen, which can be produced as a byproduct of photosynthesis. β-Carotene has no vitamin A activity; however, after ingestion, it is cleaved at the central carbon-carbon double bond followed by reduction of the newly formed aldehyde to give retinol (vitamin A).

Cleavage of this
C=C gives vitamin A.

β-Carotene

enzyme-catalyzed cleavage
and reduction in the liver

CH$_2$OH

Retinol (Vitamin A)

Probably the best understood role of vitamin A is its participation in the visual cycle in rod cells. In a series of enzyme-catalyzed reactions (Figure 26.14), retinol

Figure 26.14
The primary chemical reaction of vision in rod cells is absorption of light by rhodopsin followed by isomerization of a carbon-carbon double bond from a *cis* configuration to a *trans* configuration.

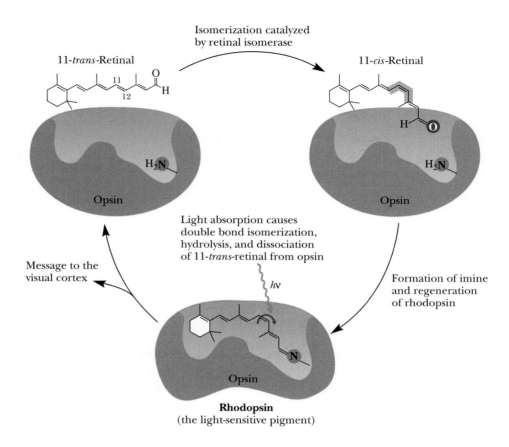

Rhodopsin
(the light-sensitive pigment)

undergoes a two-electron oxidation to all-*trans*-retinal, isomerization about the carbon 11 to 12 double bond to give 11-*cis*-retinal, and formation of an imine (Section 16.8) with the —NH$_2$ from a lysine unit of the protein, opsin. The product of these reactions is rhodopsin, a highly conjugated pigment that shows intense absorption in the blue-green region of the visual spectrum.

The primary event in vision is absorption of light by rhodopsin in rod cells of the retina of the eye to produce an electronically excited molecule. Within several picoseconds (1 ps = 10^{-12} s), the excess electronic energy is converted to vibrational and rotational energy, and the 11-*cis* double bond is isomerized to the more stable 11-*trans* double bond. This isomerization triggers a conformational change in opsin that causes firing of neurons in the optic nerve and produces a visual image. Coupled with this light-induced change is hydrolysis of rhodopsin to give 11-*trans*-retinal and free opsin. At this point, the visual pigment is bleached and in a refractory period. Rhodopsin is regenerated by a series of enzyme-catalyzed reactions that converts 11-*trans*-retinal to 11-*cis*-retinal and then to rhodopsin. The visual cycle is shown in abbreviated form in Figure 26.14.

B. Vitamin D

Vitamin D is the name for a group of structurally related compounds that play a major role in the regulation of calcium and phosphorus metabolism. A deficiency of vitamin D in childhood is associated with rickets, a mineral-metabolism disease that leads to bone defects that form bowlegs, knock-knees, and enlarged joints. Vitamin D_3, the most abundant form of the vitamin in the circulatory system, is produced in the skin of mammals by the action of ultraviolet radiation on 7-dehydrocholesterol (cholesterol with a double bond between carbons 7 and 8). In the liver, vitamin D_3 undergoes an enzyme-catalyzed, two-electron oxidation at carbon 25 of the side chain to form 25-hydroxyvitamin D_3; the oxidizing agent is molecular oxygen, O_2. 25-Hydroxyvitamin D_3 undergoes further oxidation in the kidneys, also by O_2, to form 1,25-dihydroxyvitamin D_3, the hormonally active form of the vitamin.

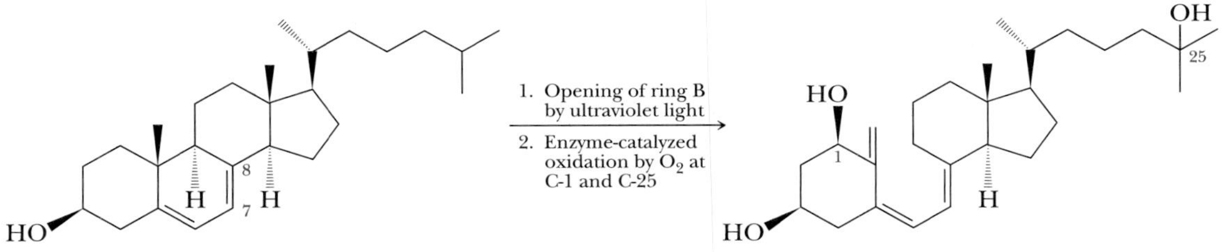

1. Opening of ring B by ultraviolet light
2. Enzyme-catalyzed oxidation by O_2 at C-1 and C-25

7-Dehydrocholesterol

1,25-Dihydroxyvitamin D_3

C. Vitamin E

Vitamin E was first recognized in 1922 as a dietary factor essential for normal reproduction in rats, hence its name tocopherol from the Greek: *tocos*, birth, and *pherein*, to bring about. Vitamin E is a group of compounds of similar structure, the most active of which is α-tocopherol. This vitamin occurs in fish oil, in other oils such as cottonseed and peanut oil, and in leafy green vegetables. The richest source of vitamin E is wheat germ oil.

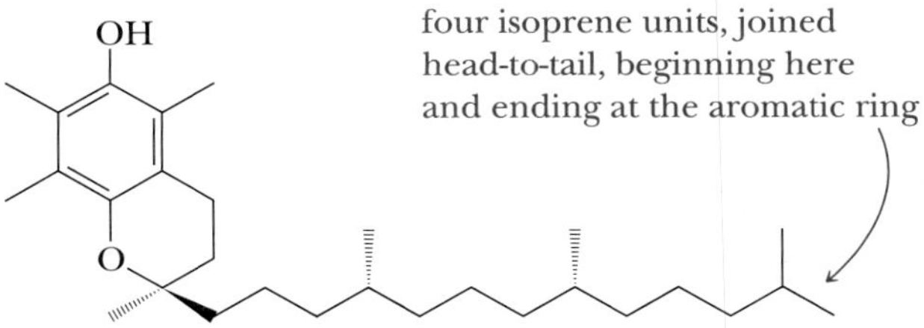

four isoprene units, joined head-to-tail, beginning here and ending at the aromatic ring

Vitamin E (α-Tocopherol)

In the body, vitamin E functions as an antioxidant; it traps peroxy radicals of the type HOO· and ROO· formed as a result of enzyme-catalyzed oxidation by molecular oxygen of the unsaturated hydrocarbon chains in membrane phospholipids (see Chemical Connection: "Antioxidants" in Section 8.7). There is speculation that peroxy radicals play a role in the aging process and that vitamin E and other antioxidants may retard that process. Vitamin E is also necessary for the proper development and function of the membranes of red blood cells.

D. Vitamin K

The name of this vitamin comes from the German word *Koagulation*, signifying its important role in the blood-clotting process. A vitamin K deficiency results in slowed blood clotting.

CHEMICAL CONNECTIONS

Vitamin K, Blood Clotting, and Basicity

Vitamin K is a fat-soluble vitamin, which must be obtained from the diet. A vitamin K deficiency results in slowed blood clotting, which can be a serious threat to a wounded animal or human. In the process of blood

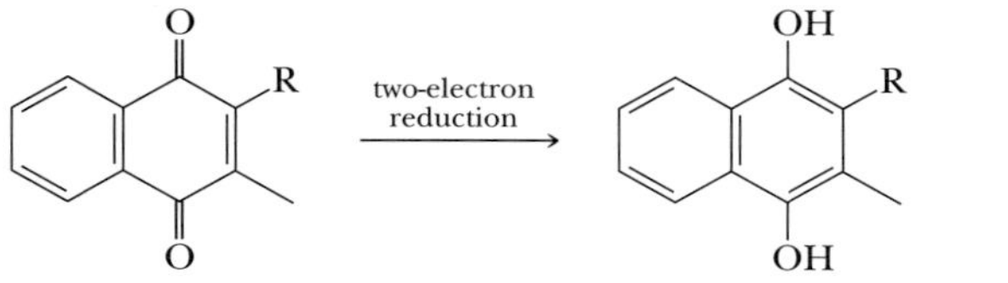

Vitamin K_1 quinone　　　　Vitamin K_1 hydroquinone

clotting, the natural vitamin, a quinone, is converted to its active hydroquinone form by reduction.

In the presence of vitamin K, O_2, CO_2, and an enzyme called microsomal carboxylase, side chains of glutamate units in prothrombin, a protein essential for blood clotting, are modified by addition of carboxyl groups (from CO_2) to form γ-carboxyglutamate units. Note that this reaction is the reverse of the decarboxylation of β-dicarboxylic (malonic) acids seen in Section 17.9B. The two carboxyl groups of the chemically modified glutamate now form a tight bidentate ("two teeth") complex with Ca^{2+} during the blood-clotting process. Although there is more to be understood about blood clotting, it is at least clear that if prothrombin is not carboxylated, it does not bind calcium, and blood does not clot.

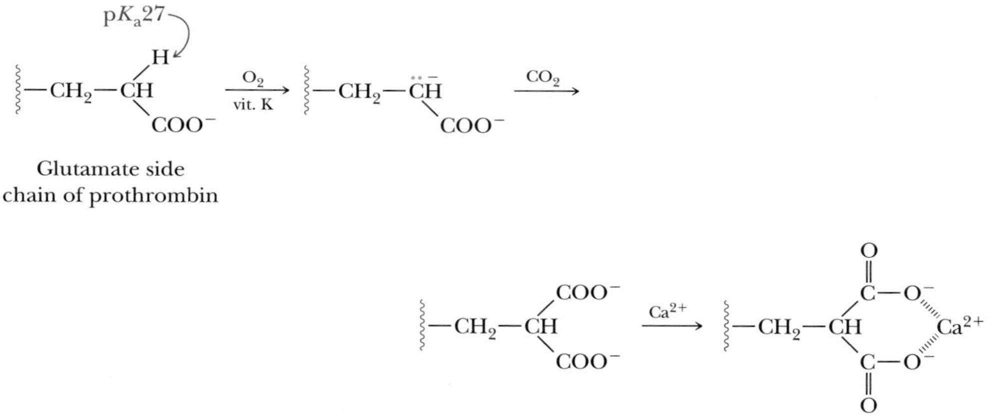

Glutamate side
chain of prothrombin

Carboxylated glutamate
side chain

Carboxylated glutamate side
chain binding calcium ion

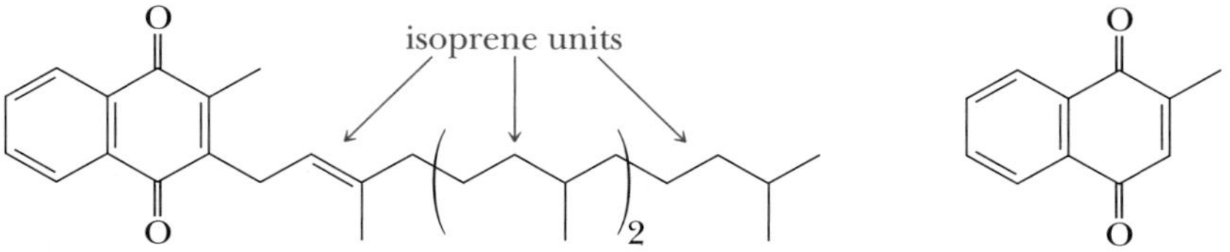

Vitamin K_1

Menadione
(a synthetic vitamin K analog)

　　Natural vitamins of the K family have, for the most part, been replaced in vitamin supplements by synthetic preparations. Menadione, one such synthetic material with vitamin-K activity, has only hydrogen in the place of the alkyl chain.

These facts have been known for many years. Until quite recently, however, the role of vitamin K_1 in this process remained a mystery. The problem was this. The anion of vitamin K_1 hydroquinone is a weak base (pK_a of approximately 9) derived from a phenol. To remove a proton from a glutamate side chain requires a very strong base derived from a conjugate acid of pK_a approximately 27. How can molecular oxygen increase the base strength of vitamin K_1 hydroquinone by 18 orders of magnitude?

Recently Paul Dowd discovered by that the vitamin K_1 hydroquinone anion reacts with oxygen to give the peroxide anion intermediate 1, which is converted to compound 2. Notice that compound 2 contains a weak O—O bond in a highly strained four-membered ring. Compound 2 then rearranges to vitamin K_1 base, a strong, sterically hindered alkoxide base.

Vitamin K_1
hydroquinone anion
(a weak base)

1

2

Vitamin K_1 base
(a very strong base)

The weak O—O bond is, in this way, replaced by a stronger C—O bond in vitamin K_1 base. The extra stability provides the driving force for turning a weak phenoxide base into a strong alkoxide base, which is able to remove a proton from glutamate side chains. This makes possible the addition of CO_2 to form γ-carboxyglutamate side chains, which bind calcium ions during the clotting cascade. Thus, it is now understood why O_2, CO_2, and vitamin K_1 are essential for this phase of blood clotting. Synthetic vitamin K analogs are as effective in this process as naturally occurring vitamin K_1

Summary

Lipids are a heterogeneous class of compounds grouped together on the basis of their solubility properties; they are insoluble in water and soluble in diethyl ether, acetone, and dichloromethane. Carbohydrates, amino acids, and proteins are largely insoluble in these organic solvents.

Triglycerides (**triacylglycerols**), the most abundant lipids, are triesters of glycerol and fatty acids (Section 26.1). **Fatty acids** (Section 26.1A) are long-chain carboxylic acids derived from the hydrolysis of fats, oils, and the phospholipids of biological membranes. The melting point of a triglyceride increases as (1) the length of its hydrocarbon chains increases and (2) its degree of saturation increases. Triglycerides rich in saturated fatty acids are generally solids at room temperature; those rich in unsaturated fatty acids are generally oils at room temperature (Section 26.1B).

Soaps are sodium or potassium salts of fatty acids (Section 26.2A). In water, soaps form **micelles,** which "dissolve" nonpolar organic grease and oil. Natural soaps precipitate as water-insoluble salts with Mg^{2+}, Ca^{2+}, and Fe^{3+} ions in hard water. The most common and most widely used synthetic detergents (Section 26.2C) are linear alkylbenzenesulfonates.

Prostaglandins are a group of compounds having the 20-carbon skeleton of prostanoic acid (Section 26.3). They are synthesized, in response to physiological triggers, from phospholipid-bound arachidonic acid and other 20-carbon fatty acids.

Steroids are a group of plant and animal lipids that have a characteristic tetracyclic structure of three six-membered rings and one five-membered ring (Section 26.4). **Cholesterol** is an integral part of animal membranes, and it is the compound from which human sex hormones, adrenocorticoid hormones, bile acids, and vitamin D are biosynthesized. **Low-density lipoproteins** (**LDLs**) transport cholesterol from the site of its synthesis in the liver to tissues and cells where it is to be used. **High-density lipoproteins** (**HDLs**) transport cholesterol from cells back to the liver for its degradation to bile acids and eventual excretion in the feces.

Oral contraceptives contain a synthetic progestin, for example norethindrone, which prevents ovulation yet allows women to maintain an otherwise normal menstrual cycle. A variety of synthetic **anabolic steroids** are available for use in rehabilitation medicine where muscle tissue has weakened or deteriorated as a result of an injury. **Bile acids** differ from most other steroids in that they have a *cis* configuration at the junction of rings A and B.

The carbon skeleton of cholesterol and those of all biomolecules derived from it originate with the acetyl group (a C_2 unit) of acetyl-CoA (Section 26.4B).

Phospholipids (Section 26.5A), the second most abundant group of naturally occurring lipids, are derived from phosphatidic acids, compounds containing glycerol esterified with two molecules of fatty acid and a molecule of phosphoric acid. Further esterification of the phosphoric acid part with a low-molecular-weight alcohol, most commonly ethanolamine, choline, serine, or inositol, gives a phospholipid. When placed in aqueous solution, phospholipids spontaneously form **lipid bilayers** (Section 26.5B).

According to the **fluid-mosaic model** (Section 26.5B), membrane phospholipids form lipid bilayers with membrane proteins associated with the bilayer as both peripheral and integral proteins.

Vitamin A (Section 26.6A) occurs only in the animal world. The carotenes of the plant world are tetraterpenes (C_{40}) and are cleaved, after ingestion, into vitamin A. The best-understood role of vitamin A is its participation in the visual cycle.

Vitamin D (Section 26.6B) is synthesized in the skin of mammals by the action of ultraviolet radiation on 7-dehydrocholesterol. This vitamin plays a major role in the regulation of calcium and phosphorus metabolism. **Vitamin E** (Section 26.6C) is a group of compounds of similar structure, the most active of which is α-tocopherol. In the body, vitamin E functions as an antioxidant. **Vitamin K** (Section 26.6D) is required for the clotting of blood.

Problems

Fatty Acids and Triglycerides

26.2 Define the term *hydrophobic*.

26.3 Identify the hydrophobic and hydrophilic region(s) of a triglyceride.

26.4 Explain why the melting points of unsaturated fatty acids are lower than those of saturated fatty acids.

26.5 Which would you expect to have the higher melting point, glyceryl trioleate or glyceryl trilinoleate?

26.6 Draw a structural formula for methyl linoleate. Be certain to show the correct configuration of groups about each carbon-carbon double bond.

26.7 Explain why coconut oil is a liquid triglyceride, even though most of its fatty acid components are saturated.

26.8 It is common now to see "contains no tropical oils" on cooking oil labels, meaning that the oil contains no palm or coconut oil. What is the difference between the composition of tropical oils and that of vegetable oils, such as corn oil, soybean oil, and peanut oil?

26.9 What is meant by the term *hardening* as applied to vegetable oils?

26.10 How many moles of H_2 are used in the catalytic hydrogenation of one mole of a triglyceride derived from glycerol and equal portions of stearic acid, linoleic acid, and arachidonic acid?

26.11 Characterize the structural features necessary to make a good synthetic detergent.

26.12 Following are structural formulas for a cationic detergent and a neutral detergent. Account for the detergent properties of each.

$$CH_3(CH_2)_6CH_2\overset{+}{\underset{CH_2C_6H_5}{\overset{|}{\underset{|}{N}}}}CH_3 \quad Cl^-$$

Benzyldimethyloctylammonium chloride
(a cationic detergent)

$$HOCH_2\overset{HOCH_2}{\underset{HOCH_2}{\overset{|}{\underset{|}{C}}}}CH_2O\overset{O}{\overset{||}{C}}(CH_2)_{14}CH_3$$

Pentaerythrityl palmitate
(a neutral detergent)

26.13 Identify some of the detergents used in shampoos and dish-washing liquids. Are they primarily anionic, neutral, or cationic detergents?

26.14 Show how to convert palmitic acid (hexadecanoic acid) into the following.

(a) Ethyl palmitate   (b) Palmitoyl chloride
(c) 1-Hexadecanol (cetyl alcohol)   (d) 1-Hexadecanamine
(e) *N,N*-Dimethylhexadecanamide

26.15 Palmitic acid (hexadecanoic acid) is the source of the hexadecyl (cetyl) group in the following compounds. Each is a mild surface-acting germicide and fungicide and is used as a topical antiseptic and disinfectant. They are examples of quaternary ammonium detergents, commonly called "quats."

Cetylpyridinium chloride   Benzylcetyldimethylammonium chloride

(a) Cetylpyridinium chloride is prepared by treating pyridine with 1-chlorohexadecane (cetyl chloride). Show how to convert palmitic acid to cetyl chloride.
(b) Benzylcetyldimethylammonium chloride is prepared by treating benzyl chloride with *N,N*-dimethyl-1-hexadecanamine. Show how this tertiary amine can be prepared from palmitic acid.

26.16 Lipases are enzymes that catalyze the hydrolysis of esters, especially esters of glycerol. Because enzymes are chiral catalysts, they catalyze the hydrolysis of only one enantiomer of a racemic mixture. For example, porcine pancreatic lipase catalyzes the hydrolysis of only one enantiomer of the following racemic epoxyester. Calculate the number of grams of epoxyalcohol that can be obtained from 100 g of racemic epoxyester by this method.

A racemic mixture   This enantiomer   This epoxyalcohol
                       is recovered   is obtained in
                       unhydrolyzed   pure form

Prostaglandins

26.17 Examine the structure of PGF$_{2\alpha}$. Identify all chiral centers and all double bonds about which *cis,trans* isomerism is possible, and state the number of stereoisomers possible for a molecule of this structure.

26.18 Following is the structure of unoprostone, a compound patterned after the natural prostaglandins (Section 26.3). Rescula, the isopropyl ester of unoprostone, is an antiglaucoma drug used to treat ocular hypertension. Compare the structural formula of this synthetic prostaglandin with that of $PGF_{2\alpha}$.

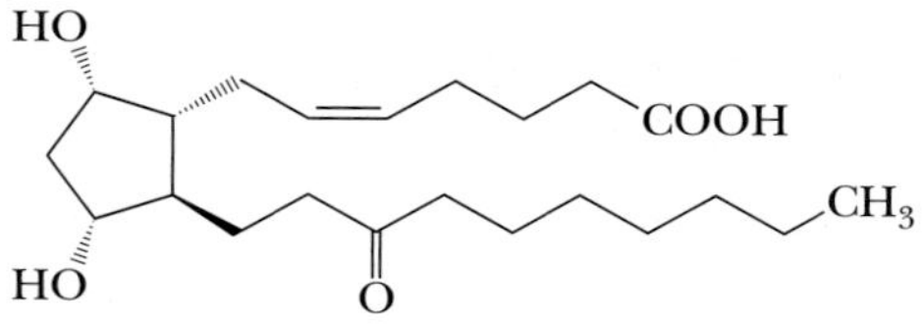

Unoprostone
(antiglaucoma)

26.19 Doxaprost, an orally active bronchodilator patterned after the natural prostaglandins (Section 26.3), is synthesized in the following series of reactions starting with ethyl 2-oxocyclopentanecarboxylate. Except for the Nef reaction in Step 8, we have seen examples of all other types of reactions involved in this synthesis.

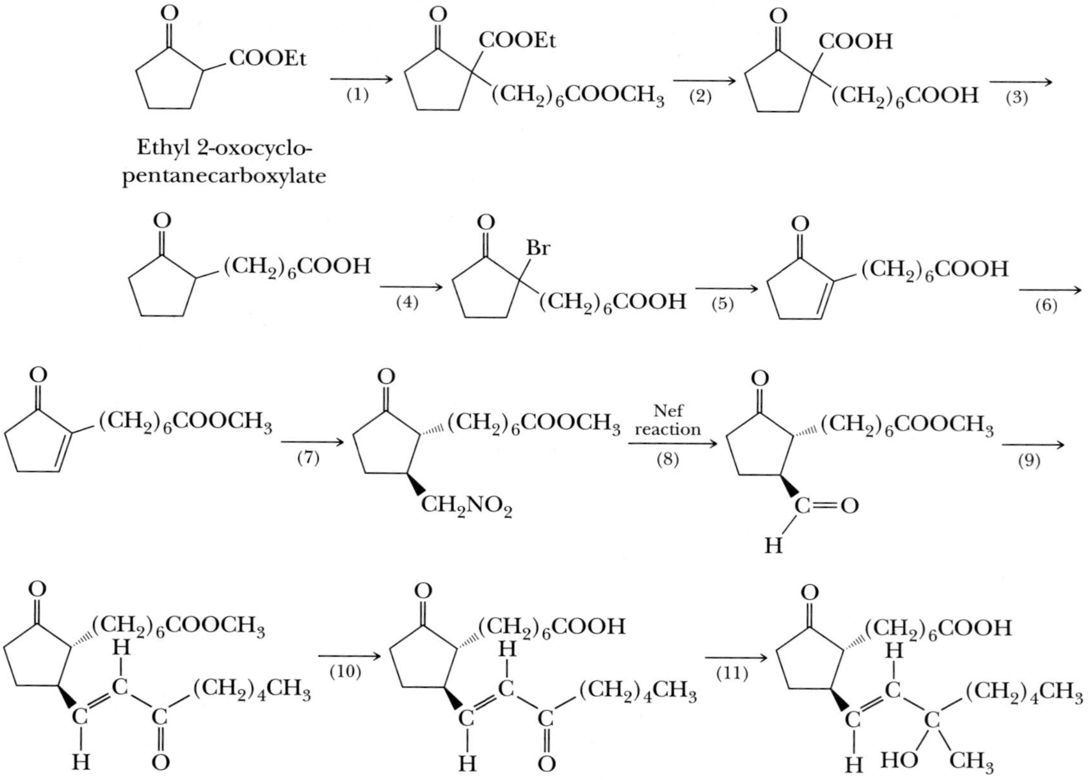

Ethyl 2-oxocyclo-
pentanecarboxylate

Doxaprost
(an orally active
bronchodilator)

(a) Propose a set of experimental conditions to bring about the alkylation in Step 1. Account for the regioselectivity of the alkylation, that is, that it takes place on the carbon between the two carbonyl groups rather than on the other side of the ketone carbonyl.

(b) Propose experimental conditions to bring about Steps 2 and 3.

(c) Propose experimental conditions for bromination of the ring in Step 4 and dehydrobromination in Step 5.

(d) Write equations to show that Step 6 can be brought about using either methanol or diazomethane (CH_2N_2) as a source of the —CH_3 in the methyl ester.

(e) Describe experimental conditions to bring about Step 7 and account for the fact that the *trans* isomer is formed in this step.

(f) Step 9 is done by a Wittig reaction. Suggest a structural formula for a Wittig reagent that gives the product shown.

(g) Name the type of reaction involved in Step 10.

(h) Step 11 can best be described as a Grignard reaction with methylmagnesium bromide under very carefully controlled conditions. In addition to the observed reaction, what other Grignard reactions might take place in Step 11?

(i) Assuming that the two side chains on the cyclopentanone ring are *trans*, how many stereoisomers are possible from this synthetic sequence?

Steroids

26.20 Draw the structural formula for the product formed by treatment of cholesterol with H_2/Pd; with Br_2.

26.21 Both low-density lipoproteins (LDL) and high-density lipoproteins (HDL) consist of a core of triacylglycerols and cholesterol esters surrounded by a single phospholipid layer. Draw the structural formula of cholesteryl linoleate, one of the cholesterol esters found in this core.

26.22 Examine the structural formulas of testosterone (a male sex hormone) and progesterone (a female sex hormone). What are the similarities in structure between the two? What are the differences?

26.23 Examine the model of cholic acid (Problem 2.63) and account for the ability of this and other bile salts to emulsify fats and oils and thus aid in their digestion.

26.24 Following is a structural formula for cortisol (hydrocortisone). Draw a stereorepresentation of this molecule showing the conformations of the five- and six-membered rings.

Cortisol
(Hydrocortisone)

26.25 Much of our understanding of conformational analysis has arisen from studies on the reactions of rigid steroid nuclei. For example, the concept of *trans*-diaxial ring opening of epoxycyclohexanes was proposed to explain the stereoselective reactions seen with steroidal epoxides. Predict the product when each of the following steroidal epoxides is treated with $LiAlH_4$.

(c) 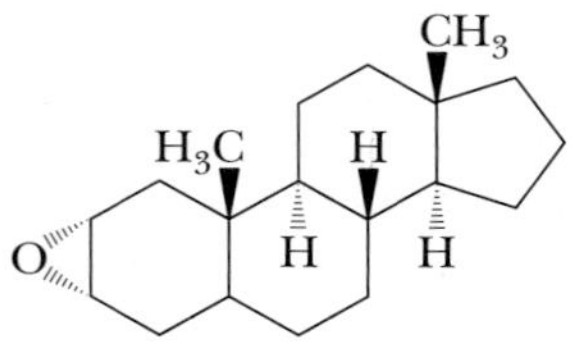   (d)

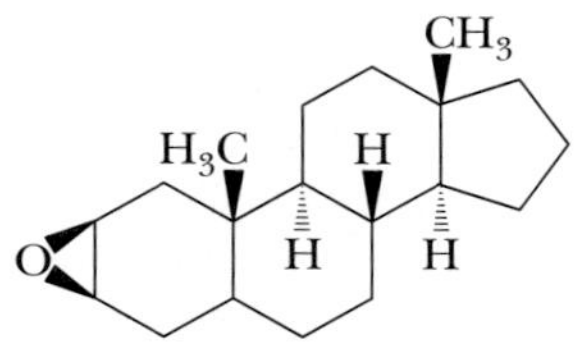

Phospholipids

26.26 Draw the structural formula of a lecithin containing one molecule each of palmitic acid and linoleic acid.

26.27 Identify the hydrophobic and hydrophilic region(s) of a phospholipid.

26.28 The hydrophobic effect is one of the most important noncovalent forces directing the self-assembly of biomolecules in aqueous solution. The hydrophobic effect arises from tendencies of biomolecules (1) to arrange polar groups so that they interact with the aqueous environment by hydrogen bonding and (2) to arrange nonpolar groups so that they are shielded from the aqueous environment. Show how the hydrophobic effect is involved in directing the following.

   (a) The formation of micelles by soaps and detergents
   (b) The formation of lipid bilayers by phospholipids

26.29 How does the presence of unsaturated fatty acids contribute to the fluidity of biological membranes?

26.30 Lecithins can act as emulsifying agents. The lecithin of egg yolk, for example, is used to make mayonnaise. Identify the hydrophobic part(s) and the hydrophilic part(s) of a lecithin. Which parts interact with the oils used in making mayonnaise? Which parts interact with the water?

Fat-Soluble Vitamins

26.31 Examine the structural formula of vitamin A, and state the number of *cis,trans* isomers possible for this molecule.

26.32 The form of vitamin A present in many food supplements is vitamin A palmitate. Draw the structural formula of this molecule.

26.33 Examine the structural formulas of vitamin A, 1,25-dihydroxy-D_3, vitamin E, and vitamin K_1 (Section 26.6). Do you expect them to be more soluble in water or in dichloromethane? Do you expect them to be soluble in blood plasma?

Amino Acids and Proteins

We begin this chapter with a study of amino acids, compounds whose chemistry is built on amines (Chapter 23) and carboxylic acids (Chapter 17). We concentrate in particular on the acid-base properties of amino acids because these properties are so important in determining many of the properties of proteins, including the catalytic functions of enzymes. With this understanding of the chemistry of amino acids, we then examine the structure of proteins themselves.

27.1 Amino Acids

A. Structure

An **amino acid** is a compound that contains both a carboxyl group and an amino group. Although many types of amino acids are known, the **α-amino acids** are the most significant in the biological world because they are the monomers from which proteins are constructed. We have already introduced amino acids (see Connections to Biological Chemistry: "Amino Acids" in Chapter 3). A general structural formula of an α-amino acid is shown in Figure 27.1.

■ Spider silk is a fibrous protein that exhibits unmatched strength and toughness. Insert: Alanine and Glycine are major components of the fibrous protein of spider silk. See Chemical Connections: "Spider Silk."

ORGANIC
Chemistry ⚛ Now™
Look for this logo in the chapter and go to Organic ChemistryNow at **http://now.brookscole.com/bfi4** for tutorials, simulations, and problems.

Amino acid   A compound that contains both an amino group and a carboxyl group.

α-Amino acid   An amino acid in which the amino group is on the carbon adjacent to the carboxyl group.

$$\underset{\substack{|\\ NH_2}}{RCHCOH} \qquad \underset{\substack{|\\ NH_3^+}}{RCHCO^-}$$
$$\overset{\displaystyle O}{\|} \qquad\qquad \overset{\displaystyle O}{\|}$$

     (a)   (b)

Figure 27.1

An α-amino acid. (a) Un-ionized form and (b) internal salt (zwitterion) form.

Figure 27.2

The enantiomers of alanine. The vast majority of α-amino acids in the biological world have the L configuration at the α-carbon.

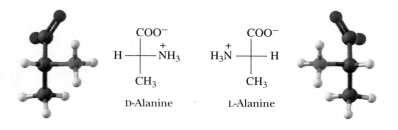

D-Alanine   L-Alanine

Zwitterion An internal salt of an amino acid; the carboxylate anion is negatively charged, and the ammonium group is positively charged.

Although Figure 27.1(a) is a common way of writing structural formulas for amino acids, it is not accurate because it shows an acid (—COOH) and a base (—NH₂) within the same molecule. These acidic and basic groups react with each other to form a dipolar ion or internal salt [Figure 27.1(b)]. The internal salt of an amino acid is given the special name **zwitterion**. Note that a zwitterion has no net charge; it contains one positive charge and one negative charge.

Because they exist as zwitterions, amino acids have many of the properties associated with salts. They are crystalline solids with high melting points and are fairly soluble in water but insoluble in nonpolar organic solvents such as ether and hydrocarbon solvents.

B. Chirality

To review, with the exception of glycine, H_2NCH_2COOH, all protein-derived amino acids have at least one chiral center and, therefore, are chiral. Figure 27.2 shows Fischer projection formulas for the enantiomers of alanine. The vast majority of carbohydrates in the biological world are of the D-series, whereas the vast majority of α-amino acids in the biological world are of the L-series.

The alternative R,S convention is also used to specify the configurations of amino acids. According to this convention, L-alanine is designated (S)-alanine. Because D- and L- are used more commonly to describe the configuration of amino acids, we use this convention throughout the remainder of the chapter.

C. Protein-Derived Amino Acids

Table 27.1 gives common names, structural formulas, and standard three-letter and one-letter abbreviations for the 20 common L-amino acids found in proteins. The amino acids in this table are divided into four categories: those with nonpolar side chains, polar but un-ionized side chains, acidic side chains, and basic side chains. The following structural features of these amino acids should be noted.

1. All 20 of these protein-derived amino acids are α-amino acids, meaning that the amino group is located on the carbon alpha to the carboxyl group.
2. For 19 of the 20 amino acids, the α-amino group is primary. Proline is different; its α-amino group is secondary.
3. With the exception of glycine, the α-carbon of each amino acid is a chiral center. Although not shown in this table, all 19 chiral amino acids have the same relative configuration at the α-carbon. In the D,L convention, all are L-amino acids. According to the R,S convention, amino acid α-carbons, with the exception of cysteine, have the S configuration. Because of priority rules, the presence of the sulfhydryl group on the side chain of L-cysteine gives the chiral center the R configuration.

Table 27.1   The 20 Common Amino Acids Found in Proteins

Nonpolar Side Chains

Alanine
(Ala, A)

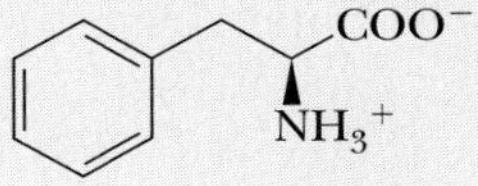

Glycine
(Gly, G)

Isoleucine
(Ile, I)

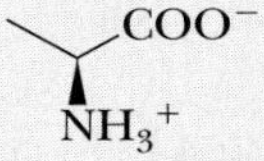

Leucine
(Leu, L)

Methionine
(Met, M)

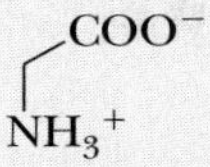

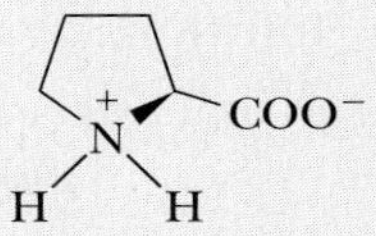

Phenylalanine
(Phe, F)

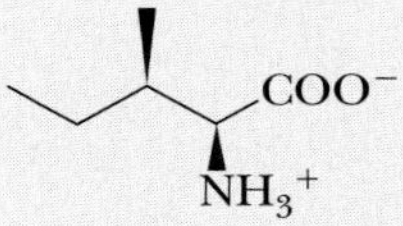

Proline
(Pro, P)

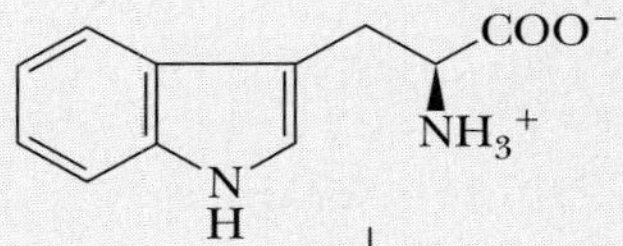

Tryptophan
(Trp, W)

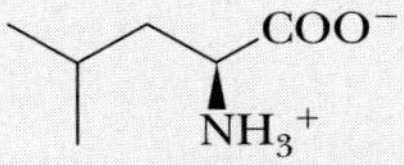

Valine
(Val, V)

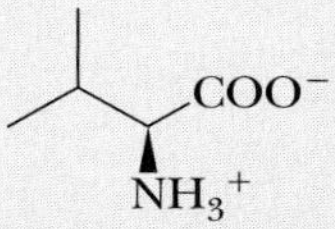

Polar Side Chains

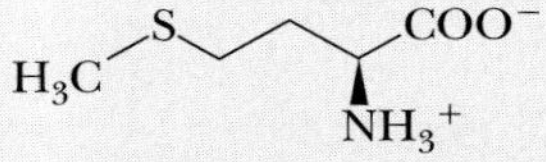

Asparagine
(Asn, N)

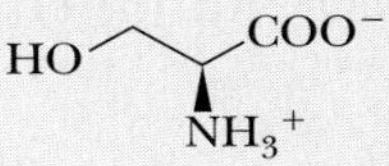

Glutamine
(Gln, Q)

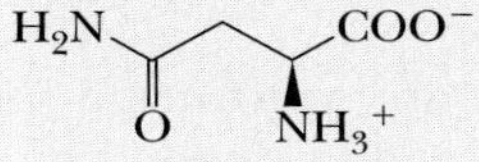

Serine
(Ser, S)

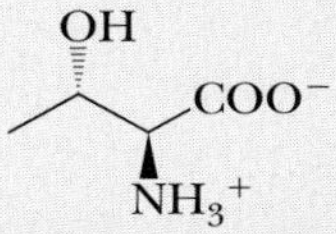

Threonine
(Thr, T)

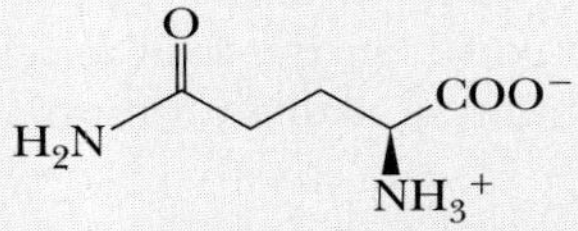

Acidic Side Chains   Basic Side Chains

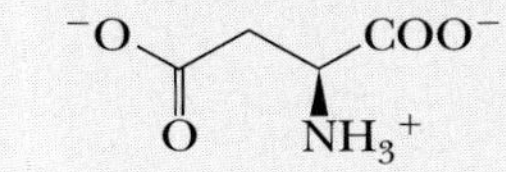

Aspartic acid
(Asp, D)

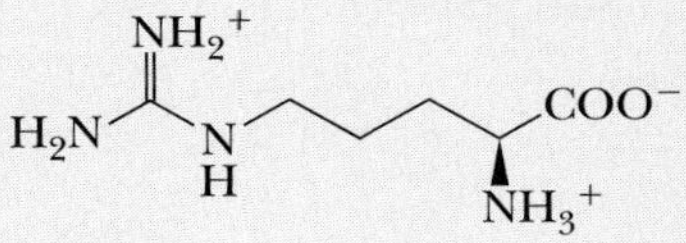

Glutamic acid
(Glu, E)

Cysteine
(Cys, C)

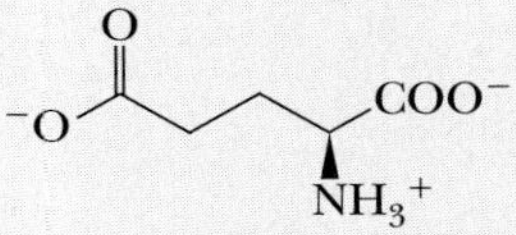

Tyrosine
(Tyr, Y)

Arginine
(Arg, R)

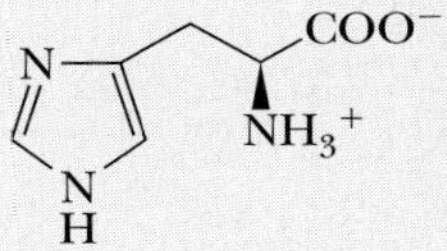

Histidine
(His, H)

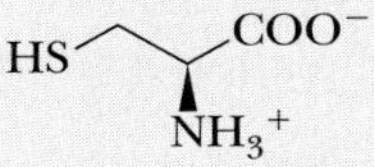

Lysine
(Lys, K)

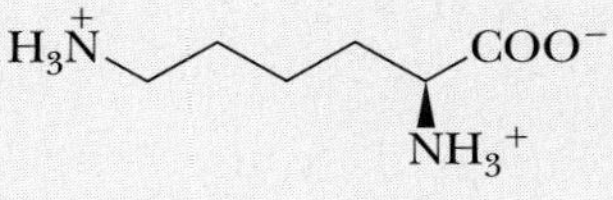

* Each ionizable group is shown in the form present in highest concentration at pH 7.0.

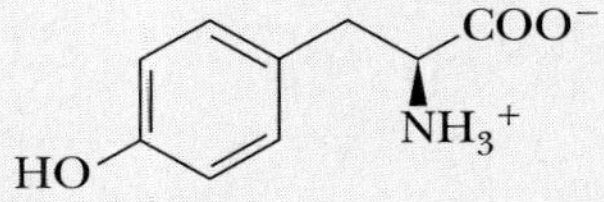

4. Isoleucine and threonine contain a second chiral center. Four stereoisomers are possible for each amino acid, but only one is found in proteins.

5. The sulfhydryl group of cysteine, the imidazole group of histidine, and the phenolic hydroxyl of tyrosine are partially ionized at pH 7.0, but the ionic form is not the major form present at this pH.

Example 27.1

Of the 20 protein-derived amino acids shown in Table 27.1, how many contain: (a) an aromatic ring, (b) a side-chain hydroxyl group, (c) a phenolic —OH group, and (d) sulfur?

Solution

(a) Phenylalanine, tryptophan, tyrosine, and histidine contain aromatic rings.
(b) Serine and threonine contain side-chain hydroxyl groups.
(c) Tyrosine contains a phenolic —OH group.
(d) Methionine and cysteine contain sulfur.

Problem 27.1

Of the 20 protein-derived amino acids shown in Table 27.1, which contain (a) no chiral center, (b) two chiral centers?

D. Some Other Common L-Amino Acids

Although the vast majority of plant and animal proteins are constructed from just these 20 α-amino acids, many other amino acids are also found in nature. Ornithine and citrulline, for example, are found predominantly in the liver and are an integral part of the urea cycle, the metabolic pathway that converts ammonia to urea.

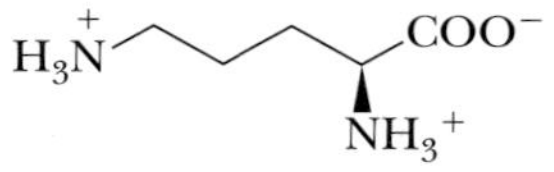

L-Ornithine

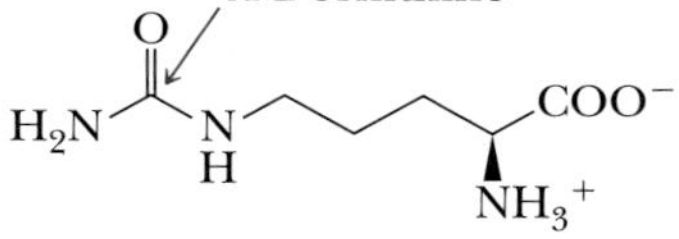

L-Citrulline

Thyroxine and triiodothyronine, two of several hormones derived from the amino acid tyrosine, are found in thyroid tissue. Their principal function is to stimulate metabolism in other cells and tissues.

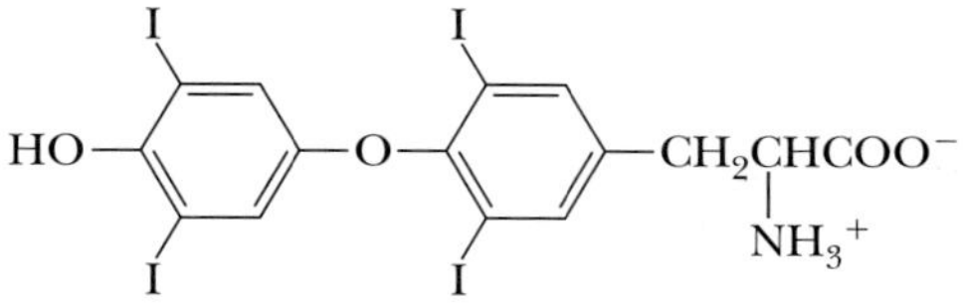

L-Thyroxine, T_4

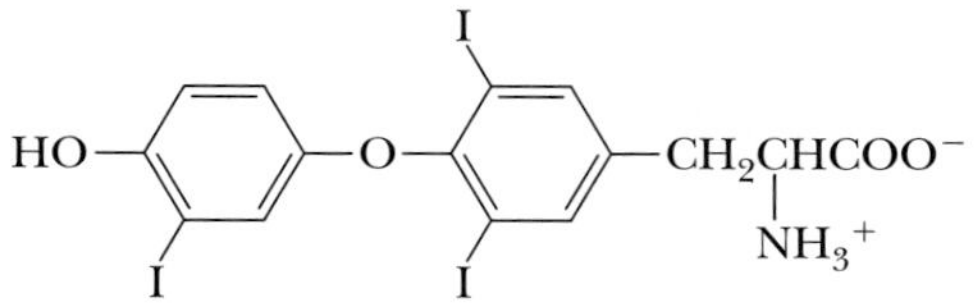

L-Triiodothyronine, T_3

4-Aminobutanoic acid is found in high concentration (0.8 mM) in the brain but in no significant amounts in any other mammalian tissue. It is synthesized in neural tissue by decarboxylation of the α-carboxyl group of glutamic acid and is a neurotransmitter in the central nervous system of invertebrates and in humans as well.

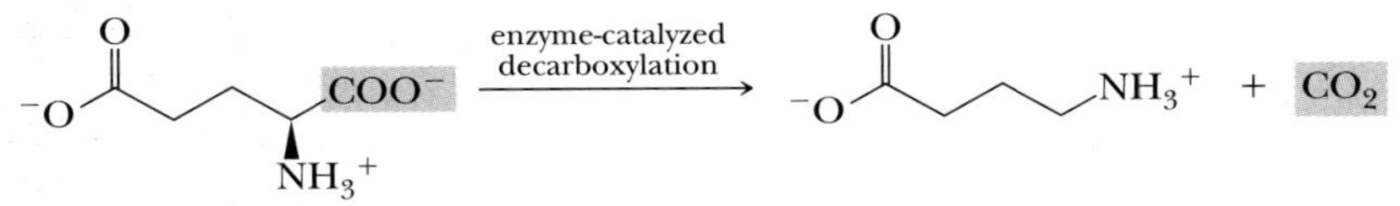

Glutamic acid

4-Aminobutanoic acid
(γ-Aminobutyric acid, GABA)

Only L-amino acids are found in proteins, and only rarely are D-amino acids a part of the metabolism of higher organisms. Several D-amino acids, however, along with their L-enantiomers, are found in lower forms of life. D-Alanine and D-glutamic acid, for example, are structural components of the cell walls of certain bacteria. Several D-amino acids are also found in peptide antibiotics.

27.2 Acid-Base Properties of Amino Acids

A. Acidic and Basic Groups of Amino Acids

Among the most important chemical properties of amino acids are their acid-base properties; all are weak polyprotic acids because of their —COOH and —NH$^+$ groups. Given in Table 27.2 are pK_a values for each ionizable group of the 20 protein-derived amino acids.

Table 27.2   pKa Values for Ionizable Groups of Amino Acids

Amino Acid	pK_a of α-COOH	pK_a of α-NH$_3^+$	pK_a of Side Chain	Isoelectric Point (pI)
Alanine	2.35	9.87	—	6.11
Arginine	2.01	9.04	12.48	10.76
Asparagine	2.02	8.80	—	5.41
Aspartic acid	2.10	9.82	3.86	2.98
Cysteine	2.05	10.25	8.00	5.02
Glutamic acid	2.10	9.47	4.07	3.08
Glutamine	2.17	9.13	—	5.65
Glycine	2.35	9.78	—	6.06
Histidine	1.77	9.18	6.10	7.64
Isoleucine	2.32	9.76	—	6.04
Leucine	2.33	9.74	—	6.04
Lysine	2.18	8.95	10.53	9.74
Methionine	2.28	9.21	—	5.74
Phenylalanine	2.58	9.24	—	5.91
Proline	2.00	10.60	—	6.30
Serine	2.21	9.15	—	5.68
Threonine	2.09	9.10	—	5.60
Tryptophan	2.38	9.39	—	5.88
Tyrosine	2.20	9.11	10.07	5.63
Valine	2.29	9.72	—	6.00

Acidity of α-Carboxyl Groups

The average value of pK_a for an α-carboxyl group of a protonated amino acid is 2.19. Thus, the α-carboxyl group is a considerably stronger acid than acetic acid (pK_a 4.76) and other low-molecular-weight aliphatic carboxylic acids. This greater acidity is accounted for by the electron-withdrawing inductive effect of the adjacent $-NH_3^+$ group. Recall that we used similar reasoning in Section 17.4A to account for the relative acidities of acetic acid and its mono-, di-, and trichloroderivatives.

The ammonium group has an electron-withdrawing inductive effect.

$$\text{RCHCOOH} + H_2O \rightleftharpoons \text{RCHCOO}^- + H_3O^+ \qquad pK_a = 2.19$$
$$\overset{|}{\underset{NH_3^+}{}} \qquad\qquad \overset{|}{\underset{NH_3^+}{}}$$

Acidity of Side-Chain Carboxyl Groups

Owing to the electron-withdrawing inductive effect of the α-NH_3^+ group, the side-chain carboxyl groups of protonated aspartic and glutamic acids are also stronger acids than acetic acid (pK_a 4.76). Notice that this acid-strengthening inductive effect decreases with increasing distance of the $-COOH$ from the $-NH_3^+$. Compare the acidities of the α-COOH of alanine (pK_a 2.35), the β-COOH of aspartic acid (pK_a 3.86), and the γ-COOH of glutamic acid (pK_a 4.07).

Acidity of α-Ammonium Groups

The average value of pK_a for an α-ammonium group, $-NH_3^+$, is 9.47 compared with an average value of 10.60 for primary aliphatic ammonium ions (Section 23.5A). Thus, the α-ammonium group of an amino acid is a slightly stronger acid than a primary aliphatic ammonium ion. Conversely, an α-amino group is a slightly weaker base than a primary aliphatic amine.

$$\text{RCHCOO}^- + H_2O \rightleftharpoons \text{RCHCOO}^- + H_3O^+ \qquad pK_a = 9.47$$
$$\overset{|}{\underset{NH_3^+}{}} \qquad\qquad \overset{|}{\underset{NH_2}{}}$$

$$CH_3CHCH_3 + H_2O \rightleftharpoons CH_3CHCH_3 + H_3O^+ \qquad pK_a = 10.60$$
$$\overset{|}{\underset{NH_3^+}{}} \qquad\qquad \overset{|}{\underset{NH_2}{}}$$

Basicity of the Guanidine Group of Arginine

The side-chain guanidine group of arginine is a considerably stronger base than an aliphatic amine. As we saw in Section 23.5D, guanidine (pK_a 13.6) is the strongest base of any neutral organic compound. The remarkable basicity of the guanidine group of arginine is attributed to the large resonance stabilization of the protonated form relative to the neutral form.

The guanidinium ion side chain of arginine is a hybrid of three contributing structures

No resonance stabilization without charge separation

$pK_a = 12.48$

Basicity of the Imidazole Group of Histidine

Because the imidazole group on the side chain of histidine contains six π electrons in a planar, fully conjugated ring, imidazole is classified as a heterocyclic aromatic amine (Section 21.2D). The unshared pair of electrons on one nitrogen is a part of the aromatic sextet, whereas that on the other nitrogen is not. The pair of electrons that is not part of the aromatic sextet is responsible for the basic properties of the imidazole ring. Protonation of this nitrogen produces a resonance-stabilized cation.

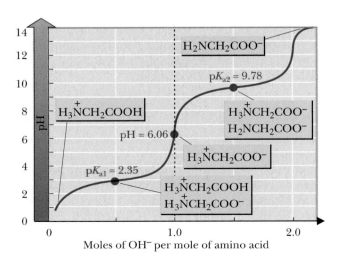

Resonance-stabilized imidazolium cation

B. Titration of Amino Acids

Values of pK_a for the ionizable groups of amino acids are most commonly obtained by acid-base titration and by measuring the pH of the solution as a function of added base (or added acid, depending on how the titration is done). To illustrate this experimental procedure, consider a solution containing 1.00 mol of glycine to which has been added enough strong acid that both the amino and carboxyl groups are fully protonated. Next, this solution is titrated with 1.00 M NaOH; the volume of base added and the pH of the resulting solution are recorded and then plotted as shown in Figure 27.3.

The most acidic group and the one to react first with added sodium hydroxide is the carboxyl group. When exactly 0.50 mol of NaOH has been added, the carboxyl group is half neutralized. At this point, the concentration of the zwitterion equals that of the positively charged ion, and the pH of 2.35 equals the pK_a of the carboxyl group (pK_{a1}).

Figure 27.3
Titration of glycine with sodium hydroxide.

$$\text{At pH} = pK_{a1} \quad [\overset{+}{H_3N}CH_2COOH] = [\overset{+}{H_3N}CH_2COO^-]$$

Positive ion   Zwitterion

The end point of the first part of the titration is reached when 1.00 mol of NaOH has been added. At this point, the predominant species present is the zwitterion, and the observed pH of the solution is 6.06.

The next section of the curve represents titration of the —NH_3^+ group. When another 0.50 mole of NaOH has been added (bringing the total to 1.50 mol), half of the —NH_3^+ groups are neutralized and converted to —NH_2. At this point, the concentrations of the zwitterion and negatively charged ion are equal, and the observed pH is 9.78, the pK_a of the amino group of glycine (pK_{a2}).

$$\text{At pH} = pK_{a2} \quad [\overset{+}{H_3N}CH_2COO^-] = [H_2NCH_2COO^-]$$

Zwitterion   Negative ion

The second end point of the titration is reached when a total of 2.00 mol of NaOH have been added and glycine is converted entirely to an anion.

C. Isoelectric Point

Isoelectric point (pI) The pH at which an amino acid, polypeptide, or protein has no net charge.

Titration curves such as that for glycine permit us to determine pK_a values for the ionizable groups of an amino acid. They also permit us to determine another important property: isoelectric point. **Isoelectric point, pI,** for an amino acid is the pH at which the majority of molecules in solution have a net charge of zero (they are zwitterions). By examining the titration curve, you can see that the isoelectric point for glycine falls half way between the pK_a values for the carboxyl and amino groups.

$$pI = \tfrac{1}{2}(pK_a \ \alpha\text{-COOH} + pK_a \ \alpha\text{-NH}_3^+)$$

$$= \tfrac{1}{2}(2.35 + 9.78) = 6.06$$

At pH 6.06, the predominant form of glycine molecules is the dipolar ion; furthermore, at this pH, the concentration of positively charged glycine molecules equals the concentration of negatively charged glycine molecules.

Given a value for the isoelectric point of an amino acid, it is possible to estimate the charge on that amino acid at any pH. For example, the charge on tyrosine at pH 5.63, its isoelectric point, is zero. A small fraction of tyrosine molecules are positively charged at pH 5.00 (0.63 unit less than its pI), and virtually all are positively charged at pH 3.63 (2.00 units less than its pI). As another example, the net charge on lysine is zero at pH 9.74. At pH values smaller than 9.74, an increasing fraction of lysine molecules are positively charged.

D. Electrophoresis

Electrophoresis The process of separating compounds on the basis of their electric charge.

Electrophoresis, a process of separating compounds on the basis of their electric charges, is used to separate and identify mixtures of amino acids and proteins. Electrophoretic separations can be carried out using paper, starch, polyacrylamide and agarose gels, and cellulose acetate as solid supports. In paper electrophoresis, a paper strip saturated with an aqueous buffer of predetermined pH serves as a bridge between two electrode vessels (Figure 27.4). Next, a sample of amino acids is applied as a spot on the paper strip. When an electrical potential is then applied to the electrode vessels, amino acids migrate toward the electrode carrying the charge opposite

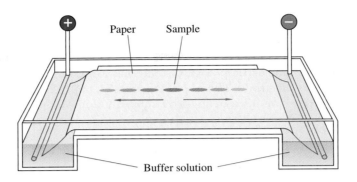

Figure 27.4
An apparatus for electrophoresis of a mixture of amino acids. Those with a negative charge move toward the positive electrode; those with a positive charge move toward the negative electrode; those with no charge remain at the origin.

to their own. Molecules having a high charge density move more rapidly than those with a lower charge density. Any molecule already at its isoelectric point remains at the origin. After separation is complete, the strip is dried and sprayed with a dye to make the separated components visible.

A dye commonly used to detect amino acids is ninhydrin (1,2,3-indanetrione monohydrate). Ninhydrin reacts with α-amino acids to produce an aldehyde, carbon dioxide, and a purple-colored anion. This reaction is used very commonly in both qualitative and quantitative analysis of amino acids.

$$
\underset{\substack{\text{An } \alpha\text{-amino}\\ \text{acid}}}{\text{RCHCO}^-\ \underset{\text{NH}_3^+}{|}} + 2\ \underset{\text{Ninhydrin}}{\text{(structure)}} \longrightarrow \underset{\text{Purple-colored anion}}{\text{(structure)}} + \text{RCH} + \text{CO}_2 + \text{H}_3\text{O}^+
$$

Nineteen of the 20 protein-derived α-amino acids have primary amino groups and give the same purple-colored ninhydrin-derived anion. Proline, a secondary amine, gives a different, orange-colored compound.

Example 27.2

The isoelectric point of tyrosine is 5.63. Toward which electrode does tyrosine migrate on paper electrophoresis at pH 7.0?

Solution

On paper electrophoresis at pH 7.0 (more basic than its isoelectric point), tyrosine has a net negative charge and migrates toward the positive electrode.

Problem 27.2

The isoelectric point of histidine is 7.64. Toward which electrode does histidine migrate on paper electrophoresis at pH 7.0?

Example 27.3

Electrophoresis of a mixture of lysine, histidine, and cysteine is carried out at pH 7.64. Describe the behavior of each amino acid under these conditions.

Solution

The isoelectric point of histidine is 7.64. At this pH, histidine has a net charge of zero and does not move from the origin. The pI of cysteine is 5.02; at pH 7.64 (more basic than its isoelectric point), cysteine has a net negative charge and moves toward the positive electrode. The pI of lysine is 9.74; at pH 7.64 (more acidic than its isoelectric point), lysine has a net positive charge and moves toward the negative electrode.

Problem 27.3

Describe the behavior of a mixture of glutamic acid, arginine, and valine on paper electrophoresis at pH 6.0.

27.3 Polypeptides and Proteins

Peptide bond The special name given to the amide bond formed between the α-amino group of one amino acid and the α-carboxyl group of another amino acid.

Dipeptide A molecule containing two amino acid units joined by a peptide bond.

Tripeptide A molecule containing three amino acid units, each joined to the next by a peptide bond.

Polypeptide A macromolecule containing many amino acid units, each joined to the next by a peptide bond.

In 1902, Emil Fischer proposed that proteins are long chains of amino acids joined together by amide bonds between the α-carboxyl group of one amino acid and the α-amino group of another. For these amide bonds, Fischer proposed the special name **peptide bond.** Figure 27.5 shows the peptide bond formed between serine and alanine in the dipeptide serylalanine.

Peptide is the name given to a short polymer of amino acids. Peptides are classified by the number of amino acid units in the chain. A molecule containing 2 amino acids joined by an amide bond is called a **dipeptide.** Those containing 3 to 10 amino acids are called **tripeptides, tetrapeptides, pentapeptides,** and so on. Molecules containing more than 10 but fewer than 20 amino acids are called **oligopeptides.** Those containing several dozen or more amino acids are called **polypeptides.** **Proteins** are biological macromolecules of molecular weight 5000 or greater, consisting of one or more polypeptide chains. The distinctions in this terminology are not precise.

By convention, polypeptides are written from the left, beginning with the amino acid having the free $—NH_3^+$ group and proceeding to the right toward the amino

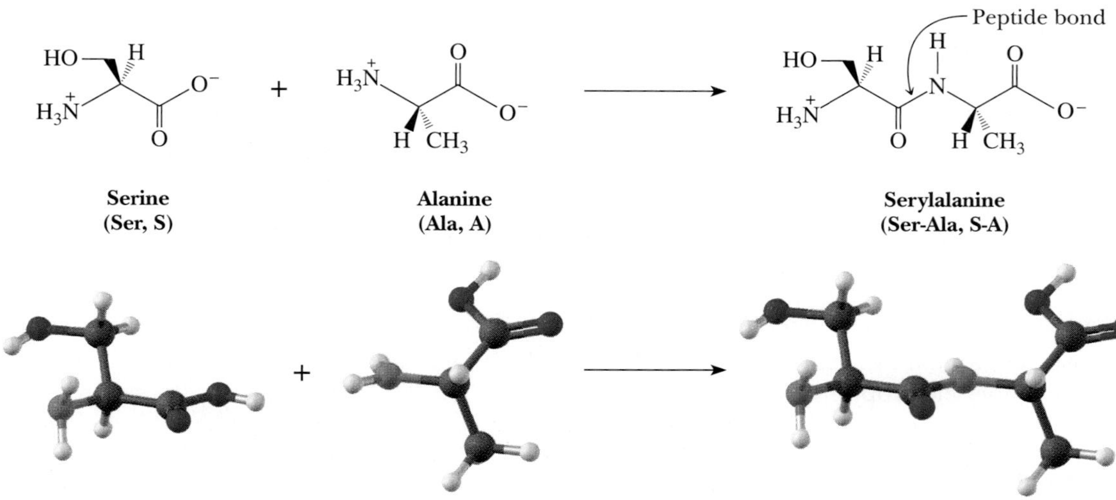

Serine
(Ser, S)

Alanine
(Ala, A)

Serylalanine
(Ser-Ala, S-A)

Figure 27.5
The peptide bond in serylalanine.

acid with the free —COO⁻ group. The amino acid with the free —NH₃⁺ group is called the **N-terminal amino acid** and that with the free —COO⁻ group is called the **C-terminal amino acid.** Notice the repeating pattern in the peptide chain of N—α-carbon—carbonyl, etc.

N-Terminal amino acid The amino acid at the end of a polypeptide chain having the free —NH₂ group.

C-Terminal amino acid The amino acid at the end of a polypeptide chain having the free —COOH group.

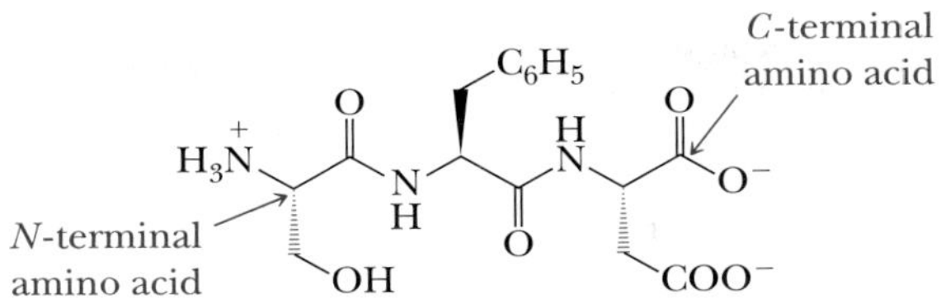

Ser-Phe-Asp

Example 27.4

Draw a structural formula for Cys-Arg-Met-Asn. Label the N-terminal amino acid and the C-terminal amino acid. What is the net charge on this tetrapeptide at pH 6.0?

Solution

The backbone of this tetrapeptide is a repeating sequence of nitrogen—α-carbon—carbonyl. The net charge on this tetrapeptide at pH 6.0 is +1.

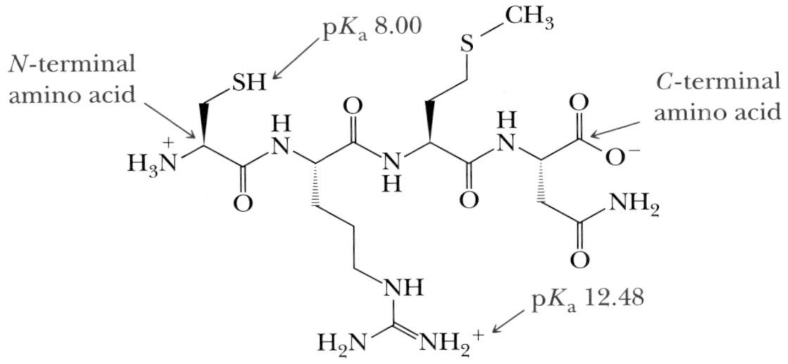

Problem 27.4

Draw a structural formula for Lys-Phe-Ala. Label the N-terminal amino acid and the C-terminal amino acid. What is the net charge on this tripeptide at pH 6.0?

27.4 Primary Structure of Polypeptides and Proteins

The **primary (1°) structure** of a polypeptide or protein refers to the sequence of amino acids in its polypeptide chain. In this sense, primary structure is a complete description of all covalent bonding in a polypeptide or protein.

In 1953, Frederick Sanger of Cambridge University, England, reported the primary structure of the two polypeptide chains of the hormone insulin. Not only was

Primary structure of proteins The sequence of amino acids in the polypeptide chain, read from the N-terminal amino acid to the C-terminal amino acid.

this a remarkable achievement in analytical chemistry, but it also clearly established that the molecules of a given protein all have the same amino acid composition and the same amino acid sequence. Today, the amino acid sequences of over 20,000 different proteins are known.

A. Amino Acid Analysis

The first step for determining the primary structure of a polypeptide is hydrolysis and quantitative analysis of its amino acid composition. Recall from Section 18.4D that amide bonds are very resistant to hydrolysis. Typically, samples of protein are hydrolyzed in 6 M HCl in sealed glass vials at 110°C for 24 to 72 hours. This hydrolysis can be done in a microwave oven in a shorter time. After the polypeptide is hydrolyzed, the resulting mixture of amino acids is analyzed by ion-exchange chromatography.

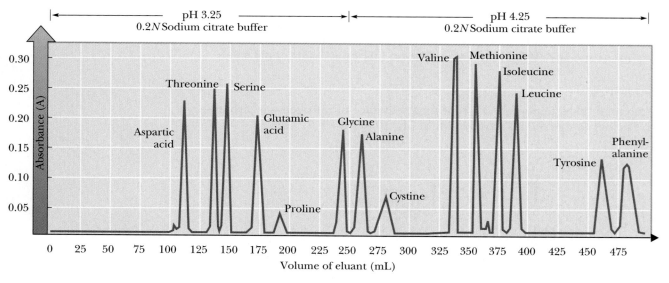

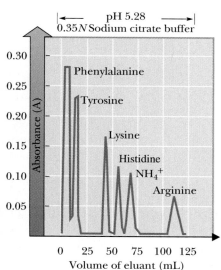

Figure 27.6

Analysis of a mixture of amino acids by ion-exchange chromatography using Amberlite IR-120, a sulfonated polystyrene resin. The resin contains phenyl-$SO_3^-Na^+$ groups. The amino acid mixture is applied to the column at low pH (3.25) under which conditions the acidic amino acids (Asp, Glu) are weakly bound to the resin, and the basic amino acids (Lys, His, Arg) are tightly bound. Sodium citrate buffers at two different concentrations, and three different values of pH are used to elute the amino acids from the column. Cysteine is determined as cystine, Cys-S-S-Cys, the disulfide of cysteine.

Amino acids are detected as they emerge from the column by reaction with ninhydrin (Section 27.2D) followed by absorption spectroscopy. Current procedures for hydrolysis of polypeptides and analysis of amino acid mixtures have been refined to the point where it is possible to obtain amino acid composition from as little as 50 nanomoles (50×10^{-9} mole) of polypeptide. Figure 27.6 shows the analysis of a polypeptide hydrolysate by ion-exchange chromatography. Note that during hydrolysis, the side-chain amide groups of asparagine and glutamine are hydrolyzed, and these amino acids are detected as aspartic acid and glutamic acid. For each glutamine or asparagine hydrolyzed, an equivalent amount of ammonium chloride is formed.

B. Sequence Analysis

After the amino acid composition of a polypeptide has been determined, the next step is to determine the order in which the amino acids are joined in the polypeptide chain. The most common sequencing strategy is to cleave the polypeptide at specific peptide bonds (using, for example, cyanogen bromide or certain proteolytic enzymes), determine the sequence of each fragment (using, for example, the Edman degradation), and then match overlapping fragments to arrive at the sequence of the polypeptide.

Cyanogen Bromide

Cyanogen bromide (BrCN) is specific for cleavage of peptide bonds formed by the carboxyl group of methionine (Figure 27.7). The products of this cleavage are a substituted γ-lactone (Section 18.1C) derived from the *N*-terminal portion of the polypeptide and a second fragment containing the *C*-terminal portion of the polypeptide.

   A three-step mechanism can be written for this reaction. The strategy for cyanogen bromide cleavage depends on chemical manipulation of the leaving ability of the sulfur atom of methionine. Because CH_3S^- is the anion of a weak acid, it is a very poor leaving group, just as OH^- is a poor leaving group (Section 9.4F). Yet, just as the oxygen atom of an alcohol can be transformed into a better leaving group by converting it into an oxonium ion (by protonation), so too can the sulfur atom of methionine be transformed into a better leaving group by converting it into a sulfonium ion.

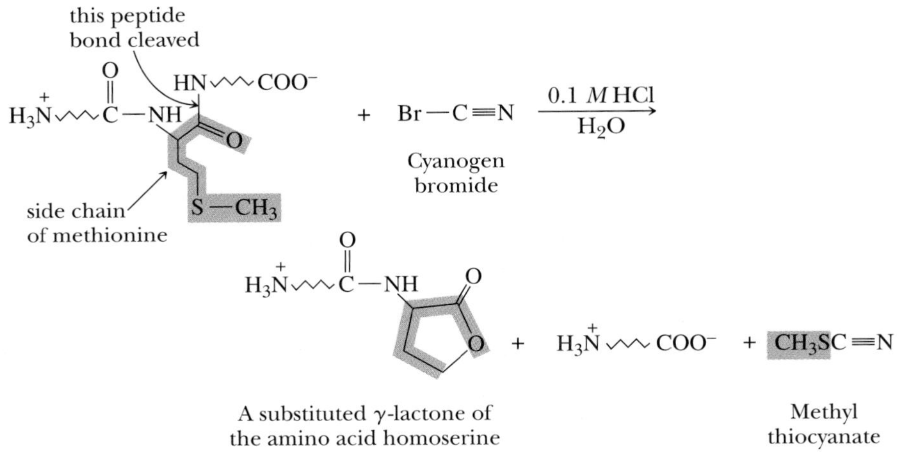

Figure 27.7

Cleavage by cyanogen bromide, BrCN, of a peptide bond formed by the carboxyl group of methionine.

Mechanism *Cleavage of a Peptide Bond at Methionine by Cyanogen Bromide*

Step 1: Reaction is initiated by nucleophilic attack of the divalent sulfur atom of methionine on the carbon of cyanogen bromide displacing bromide ion. The product of this nucleophilic displacement is a sulfonium ion.

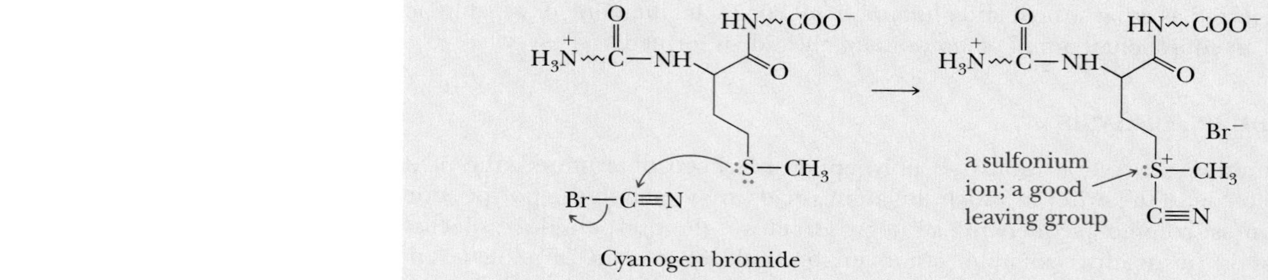

Cyanogen bromide

Step 2: An internal S_N2 reaction in which the oxygen of the methionine carbonyl group attacks the γ-carbon and displaces methyl thiocyanate gives a five-membered ring. Note that the oxygen of a carbonyl group is at best a weak nucleophile. This displacement is facilitated, however, because the sulfonium ion is a very good leaving group and because of the ease with which a five-membered ring is formed.

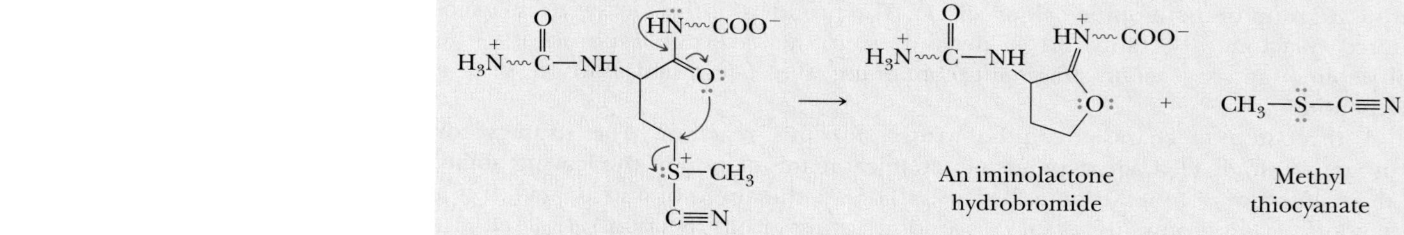

An iminolactone hydrobromide

Methyl thiocyanate

Step 3: Hydrolysis of the imino group gives a γ-lactone derived from the N-terminal end of the original polypeptide.

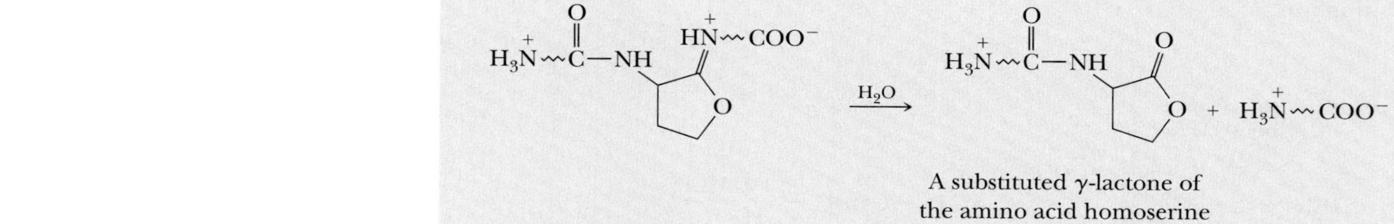

A substituted γ-lactone of the amino acid homoserine

Enzyme-Catalyzed Hydrolysis of Peptide Bonds

A group of proteolytic enzymes, among them trypsin and chymotrypsin, can be used to catalyze the hydrolysis of specific peptide bonds. Trypsin catalyzes the hydrolysis of peptide bonds formed by the carboxyl groups of arginine and lysine; chymotrypsin catalyzes the hydrolysis of peptide bonds formed by the carboxyl groups of phenylalanine, tyrosine, and tryptophan (Table 27.3).

Table 27.3	Cleavage of Specific Peptide Bonds Catalyzed by Trypsin and Chymotrypsin	
Enzyme	**Catalyzes Hydrolysis of Peptide Bond Formed by Carboxyl Group of**
Trypsin	Arginine, lysine
Chymotrypsin	Phenylalanine, tyrosine, tryptophan

Example 27.5

Which of these dipeptides are hydrolyzed by trypsin? By chymotrypsin?
(a) Arg-Glu-Ser   **(b)** Phe-Gly-Lys

Solution

(a) Trypsin catalyzes the hydrolysis of peptide bonds formed by the carboxyl groups of lysine and arginine. Therefore, the peptide bond between arginine and glutamic acid is hydrolyzed in the presence of trypsin.

$$\text{Arg-Glu-Ser} + \text{H}_2\text{O} \xrightarrow{\text{trypsin}} \text{Arg} + \text{Glu-Ser}$$

Chymotrypsin catalyzes the hydrolysis of peptide bonds formed by the carboxyl groups of phenylalanine, tyrosine, and tryptophan. Because none of these three aromatic amino acids is present, tripeptide (a) is not affected by chymotrypsin.

(b) Tripeptide (b) is not affected by trypsin. Although lysine is present, its carboxyl group is at the *C*-terminal end and is not involved in peptide bond formation. Tripeptide (b) is hydrolyzed in the presence of chymotrypsin.

$$\text{Phe-Gly-Lys} + \text{H}_2\text{O} \xrightarrow{\text{chymotrypsin}} \text{Phe} + \text{Gly-Lys}$$

Problem 27.5

Which of these tripeptides are hydrolyzed by trypsin? By chymotrypsin?
(a) Tyr-Gln-Val   **(b)** Thr-Phe-Ser   **(c)** Thr-Ser-Phe

Edman Degradation

Of the various chemical methods developed for determining the amino acid sequence of a polypeptide, the one most widely used today is the **Edman degradation,** introduced in 1950 by Pehr Edman of the University of Lund, Sweden. In this procedure, a polypeptide is treated with phenyl isothiocyanate, $\text{C}_6\text{H}_5\text{N}{=}\text{C}{=}\text{S}$, and then with acid. The effect of Edman degradation is to remove the *N*-terminal amino acid selectively as a substituted phenylthiohydantoin (Figure 27.8), which is then separated and identified.

The key feature of the Edman degradation is successive $\text{C}{=}\text{N}$, $\text{C}{=}\text{O}$, and $\text{C}{=}\text{O}$ addition reactions.

Edman degradation A method for selectively cleaving and identifying the *N*-terminal amino acid of a polypeptide chain.

Figure 27.8

Edman degradation. Treatment of a polypeptide with phenyl isothiocyanate followed by acid selectively cleaves the N-terminal amino acid as a substituted phenylthiohydantoin.

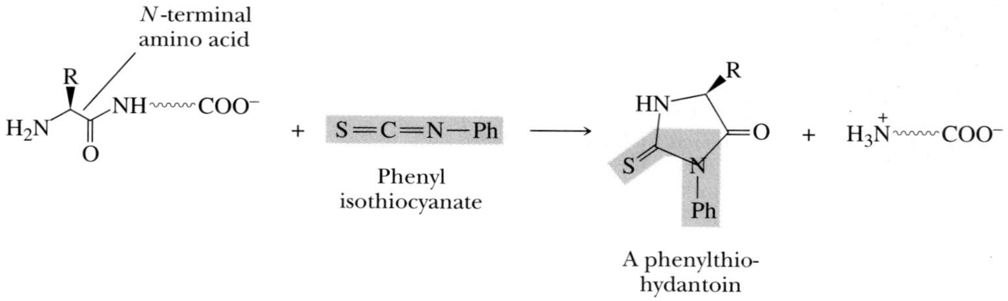

Mechanism   *Edman Degradation—Cleavage of an N-Terminal Amino Acid*

Step 1: Nucleophilic addition of the *N*-terminal amino group to the C=N bond of phenyl isothiocyanate gives a derivative of *N*-phenylthiourea.

A derivative of
N-phenylthiourea

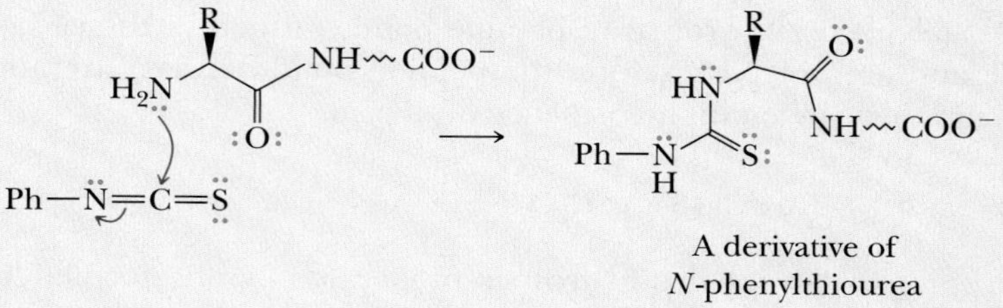

Step 2: Heating the derivatized polypeptide in HCl at 100°C results in nucleophilic addition of sulfur to the carbonyl of the adjacent amide group to give a tetrahedral carbonyl addition intermediate, which collapses to give a thiazolinone ring derived from the *N*-terminal amino acid.

A thiazolinone

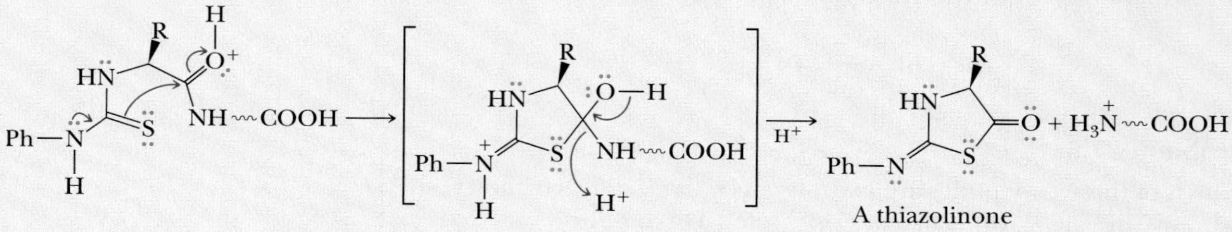

Step 3: The thiazolinone ring undergoes isomerization by ring opening followed by reclosure to give a more stable phenylthiohydantoin, which is separated and identified by chromatography.

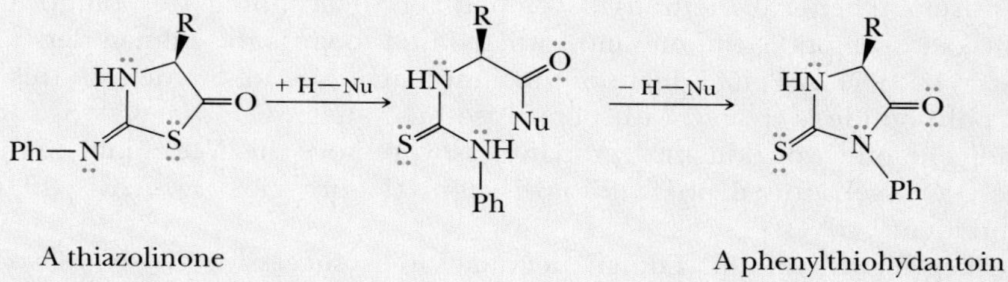

A thiazolinone   A phenylthiohydantoin

The special value of the Edman degradation is that it cleaves the *N*-terminal amino acid from a polypeptide without affecting any other bonds in the chain. Furthermore, Edman degradation can be repeated on the shortened polypeptide, causing the next amino acid in the sequence to be cleaved and identified. In practice, it is now possible to sequence as many as the first 20 to 30 amino acids in a polypeptide by this method using only a few milligrams of material.

Most polypeptides in nature are longer than 20 to 30 amino acids, the practical limit to the number of amino acids that can be sequenced by repetitive Edman degradation. The special value of cleavage with cyanogen bromide, trypsin, and chymotrypsin is that a long polypeptide chain can be cleaved at specific peptide bonds into smaller polypeptide fragments, and each fragment can then be sequenced separately. The entire procedure, from the chemical cleavage step to the separation of the product can be automated. Samples of a pure protein can be added to a machine and the first several *N*-terminal amino acids determined automatically.

Example 27.6

Deduce the amino acid sequence of a pentapeptide from the following experimental results. Note that under the column labeled "Amino Acid Composition," the amino acids are listed in alphabetical order. In no way does this listing give any information about primary structure.

Experimental Procedure	Amino Acid Composition
Pentapeptide	Arg, Glu, His, Phe, Ser
Edman degradation	Glu
Hydrolysis catalyzed by chymotrypsin	
Fragment A	Glu, His, Phe
Fragment B	Arg, Ser
Hydrolysis catalyzed by trypsin	
Fragment C	Arg, Glu, His, Phe
Fragment D	Ser

Solution

Edman degradation cleaves Glu from the pentapeptide; therefore, glutamic acid must be the *N*-terminal amino acid.

Glu-(Arg, His, Phe, Ser)

Fragment A from chymotrypsin-catalyzed hydrolysis contains Phe. Because of the specificity of chymotrypsin, Phe must be the *C*-terminal amino acid of fragment A. Fragment A also contains Glu, which we already know is the *N*-terminal amino acid. From these observations, conclude that the first three amino acids in the chain must be Glu-His-Phe and then write the following partial sequence:

Glu-His-Phe-(Arg, Ser)

The fact that trypsin cleaves the pentapeptide means that Arg must be within the pentapeptide chain; it cannot be the *C*-terminal amino acid. Therefore, the complete sequence must be

Glu-His-Phe-Arg-Ser

Problem 27.6

Deduce the amino acid sequence of an undecapeptide (11 amino acids) from the experimental results shown in the table.

Experimental Procedure	Amino Acid Composition
Undecapeptide	Ala, Arg, Glu, Lys$_2$, Met, Phe, Ser, Thr, Trp, Val
Edman degradation	Ala
Trypsin-catalyzed hydrolysis	
Fragment E	Ala, Glu, Arg
Fragment F	Thr, Phe, Lys
Fragment G	Lys
Fragment H	Met, Ser, Trp, Val
Chymotrypsin-catalyzed hydrolysis	
Fragment I	Ala, Arg, Glu, Phe, Thr
Fragment J	Lys$_2$, Met, Ser, Trp, Val
Treatment with cyanogen bromide	
Fragment K	Ala, Arg, Glu, Lys$_2$, Met, Phe, Thr, Val
Fragment L	Trp, Ser

Sequencing by Mass Spectrometry

As mentioned in Chapter 14, mass spectrometry is increasingly being used for the direct sequencing of proteins in very small quantities. This topic was fully addressed in Connections to Biological Chemistry: "Mass Spectra of Biological Macromolecules" in Section 14.3 and will not be further developed here.

Sequencing Proteins from the Coding Nucleotide Sequence

As it has become easier to determine nucleotide sequences (see Section 28.5), it is now often easier to sequence the nucleotide that codes for a protein than to sequence the protein itself. In some cases, this has lead to discovery of new proteins of unknown function. Often comparison of the revealed protein sequences with those of known proteins from simpler organisms discloses sequence homologies that suggest the function of the new proteins. Comparisons with yeast, whose genome was one of the first to be sequenced, and for which most of the proteins coded for have known functions, have been particularly fruitful.

27.5 Synthesis of Polypeptides

A. The Problems

The problem in peptide synthesis is to join the carboxyl group of amino acid 1 (aa$_1$) by an amide (peptide) bond to the amino group of amino acid 2 (aa$_2$).

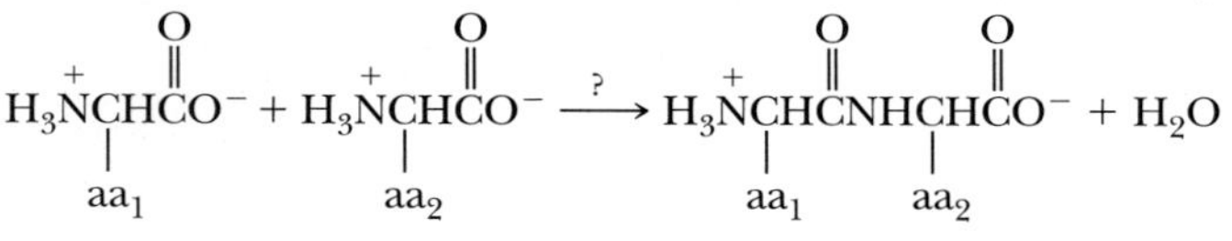

$$\overset{+}{H_3}NCHCO^- + \overset{+}{H_3}NCHCO^- \xrightarrow{\ ?\ } \overset{+}{H_3}NCHCNHCHCO^- + H_2O$$

$$\underset{aa_1}{\qquad} \underset{aa_2}{\qquad} \underset{aa_1}{\qquad} \underset{aa_2}{\qquad}$$

B. The Strategy

A rational strategy for the synthesis of peptide bonds and polypeptides requires three steps.

1. Protect the α-amino group of amino acid aa₁ to reduce its nucleophilicity so that it does not participate in nucleophilic addition to the carboxyl group of either aa₁ or aa₂.
2. Protect the α-carboxyl group of amino acid aa₂ so that it is not susceptible to nucleophilic attack by the α-amino group of another molecule of aa₂.
3. Activate the α-carboxyl group of amino acid aa₁ so that it is susceptible to nucleophilic attack by the α-amino group of aa₂.

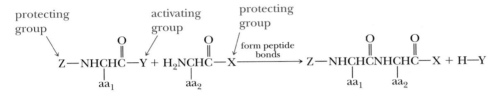

After dipeptide aa₁—aa₂ has been formed, the protecting group Z can be removed, and chain growth can be continued from the *N*-terminal end of the dipeptide. Alternatively, the protecting group X can be removed and chain growth can be continued from the *C*-terminal end. The range of protecting groups and activating groups is large, and experimental conditions have been found to attach and remove them as desired.

C. Amino-Protecting Groups

The most common strategy for protecting amino groups and reducing their nucleophilicity is to react them with carbonyl derivatives. The reagents most commonly used for this purpose are benzyloxycarbonyl chloride and di-*tert*-butyl dicarbonate. In the terminology adopted by the IUPAC, the benzyloxycarbonyl group is given the symbol Z, and the *tert*-butoxycarbonyl group is given the symbol BOC—.

O‖ PhCH₂OCCl	O‖ PhCH₂OC—	O O‖ ‖ (CH₃)₃COCOCOC(CH₃)₃	O‖ (CH₃)₃COC—
Benzyloxycarbonyl chloride	Benzyloxycarbonyl (Z—) group	Di-*tert*-butyl dicarbonate	*tert*-Butoxycarbonyl (BOC—) group

Treatment of an amino group with either of these reagents forms a new functional group called a carbamate. A carbamate is an ester of carbamic acid; that is, it is an ester of the monoamide of carbonic acid.

$$
\underset{\substack{\text{Benzyloxycarbonyl}\\\text{chloride}\\\text{(Z—Cl)}}}{\text{PhCH}_2\text{OCCl}} + \underset{\substack{\text{Alanine}}}{\overset{+}{\text{H}_3}\text{NCHCO}^-} \xrightarrow[\text{2. HCl, H}_2\text{O}]{\text{1. NaOH}} \underset{\substack{\textit{N}\text{-Benzyloxycarbonylalanine}\\\text{(Z—Ala)}\\\text{A carbamate}}}{\text{PhCH}_2\text{OCNHCHCOH}}
$$

The special advantage of the carbamate group is that it is stable to dilute base but can be removed by treatment with HBr in acetic acid.

$$PhCH_2OCNH\text{—peptide} \xrightarrow[CH_3COOH]{HBr} PhCH_2Br + CO_2 + H_3\overset{+}{N}\text{—peptide}$$

A Z-protected peptide   Benzyl bromide   Unprotected peptide

A study of the mechanism for removal of this protecting group has shown that the reaction is first order in $[H^+]$ and involves formation of a carbocation and a carbamic acid. A carbamic acid spontaneously loses carbon dioxide to form the free amine. The carbocation reacts with an available nucleophile such as halide ion to form an alkyl halide.

Mechanism   *Acid-Catalyzed Removal of a Benzyloxycarbonyl Protecting Group*

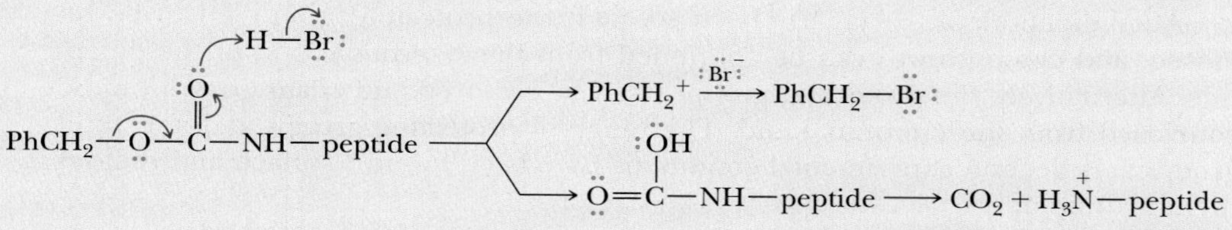

A carbamic acid

Note that because acid-catalyzed removal of this protecting group is carried out in nonaqueous media, there is no danger of simultaneous acid-catalyzed hydrolysis of peptide (amide) bonds within the newly synthesized polypeptide. This relationship exists because water is required for hydrolysis of a peptide bond.

The benzyloxycarbonyl group can also be removed by treatment with H_2 in the presence of a transition metal catalyst (hydrogenolysis, Section 21.5C). In hydrogenolysis of a Z-protecting group, one product is toluene. The other is a carbamic acid, which undergoes spontaneous decarboxylation to give carbon dioxide and the unprotected peptide.

$$PhCH_2OCNH\text{—peptide} + H_2 \xrightarrow{Pd} PhCH_3 + CO_2 + H_2N\text{—peptide}$$

A Z-protected peptide   Toluene   Unprotected peptide

D. Carboxyl-Protecting Groups

Carboxyl groups are most often protected by conversion to methyl, ethyl, or benzyl esters. Methyl and ethyl esters are prepared by Fischer esterification (Section 17.7A)

and are removed by hydrolysis in aqueous base (Section 18.4C) under mild conditions. Benzyl esters are conveniently removed by hydrogenolysis with H_2 over a palladium or platinum catalyst (Section 21.5C). Benzyl groups can also be removed by treatment with HBr in acetic acid.

E. Peptide-Bond Forming Reactions

The reagent most commonly used to bring about peptide bond formation is 1,3-dicyclohexylcarbodiimide (DCC). This reagent is the anhydride of a disubstituted urea, and, when treated with water, it is converted to N,N'-dicyclohexylurea (DCU).

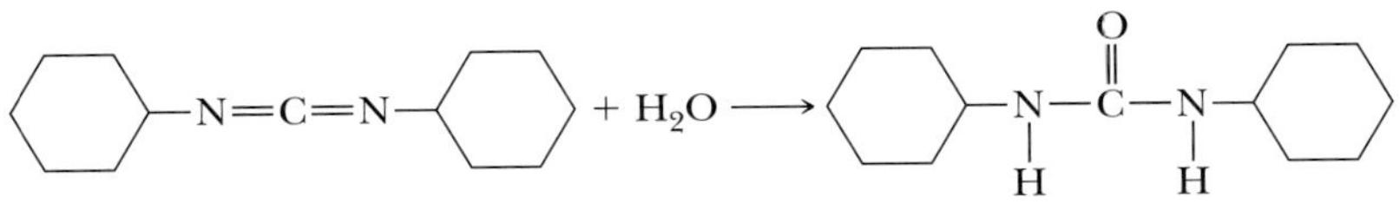

1,3-Dicyclohexylcarbodiimide
(DCC)

N,N'-Dicyclohexylurea
(DCU)

When an amino-protected aa_1 and a carboxyl-protected aa_2 are treated with DCC, this reagent acts as a dehydrating agent; it removes —OH from the carboxyl group and —H from the amino group to form an amide bond. More specifically, DCC activates the α-carboxyl group of aa_1 toward nucleophilic acyl substitution by converting its —OH group into a better leaving group.

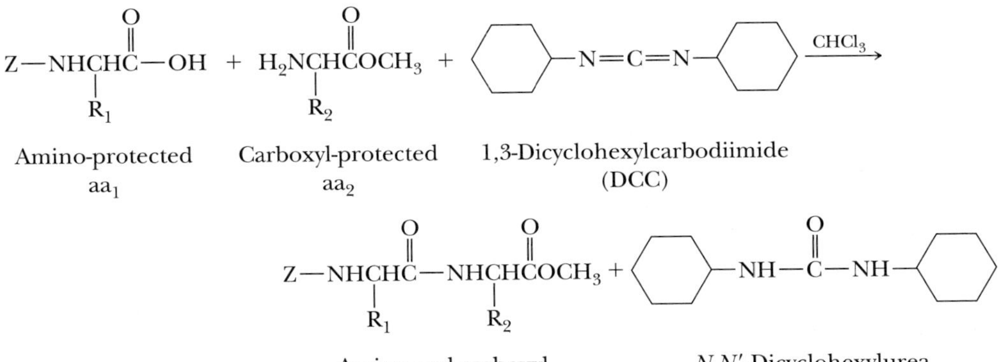

Amino-protected
aa_1

Carboxyl-protected
aa_2

1,3-Dicyclohexylcarbodiimide
(DCC)

Amino and carboxyl
protected dipeptide

N,N'-Dicyclohexylurea
(DCU)

An abbreviated mechanism for this intermolecular dehydration is shown in Figure 27.9. An acid-base reaction in Step 1 between the carboxyl group of aa_1 and a nitrogen of DCC followed in Step 2 by addition of the carboxylate anion to the C=N double bond results in electrophilic addition to a C=N double bond. The O-acylisourea formed is the nitrogen analog of a mixed anhydride. Nucleophilic addition of the amino group of aa_2 to the carbonyl group of the O-acylisourea in Step 3 generates a tetrahedral carbonyl addition intermediate that collapses in Step 4 to give a dipeptide and DCU.

Mechanism *Dicyclohexylcarbodiimide and Formation of a Peptide Bond*

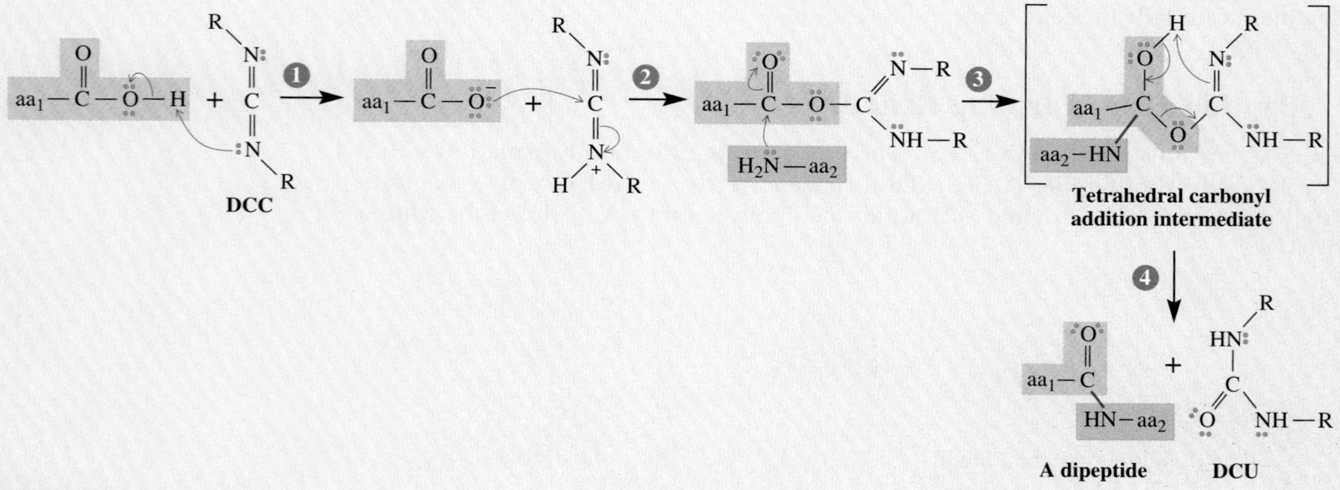

Figure 27.9

The role of 1,3-dicyclohexylcarbodiimide (DCC) in the formation of a peptide bond between an amino-protected amino acid (aa$_1$) and a carboxyl-protected amino acid (aa$_2$).

F. Solid-Phase Synthesis

A major problem associated with polypeptide synthesis is purification of intermediates after each protection, activation, coupling, and deprotection step. If unreacted starting materials are not removed after each step, the final product is contaminated by polypeptides missing one or more amino acids. The required purification steps are not only laborious and time consuming, but they also inevitably result in some loss of the desired product. These losses become especially severe in the synthesis of larger polypeptides.

A major advance in polypeptide synthesis came in 1962 when R. Bruce Merrifield of Rockefeller University described a solid-phase synthesis (alternatively called polymer-supported synthesis) of the tetrapeptide, Leu-Ala-Gly-Ala, by a technique that now bears his name. Merrifield was awarded the 1984 Nobel Prize for chemistry for his work in developing the solid-phase method for peptide synthesis.

The solid support used by Merrifield was a type of polystyrene in which about 5% of the phenyl groups carry a chloromethyl (—CH$_2$Cl) group in their para positions (Figure 27.10). These chloromethyl groups, like all benzylic halides, are particularly reactive in nucleophilic substitution reactions.

In the Merrifield method, the *C*-terminal amino acid is joined as a benzyl ester to the solid polymer support, and then the polypeptide chain is extended one amino acid at a time from the *N*-terminal end. The advantage of polypeptide synthesis on a solid support is that the polymer beads with the peptide chains anchored on them are completely insoluble in the solvents used in the synthesis. Furthermore, excess reagents (for example, DCC) and byproducts (for example, DCU) are removed after each step simply by washing the polymer beads. When synthesis is completed, the

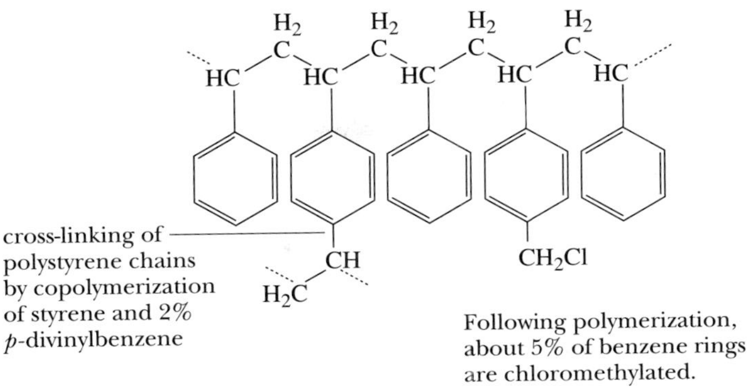
cross-linking of polystyrene chains by copolymerization of styrene and 2% p-divinylbenzene
Following polymerization, about 5% of benzene rings are chloromethylated.

Figure 27.10
The support used for the Merrifield solid-phase synthesis is a chloromethylated polystyrene resin.

polypeptide is released from the polymer beads by cleavage of the benzyl ester. The steps in solid-phase synthesis of a polypeptide are summarized in Figure 27.11.

Thanks to automation, the synthesis of polypeptides is now a routine procedure in chemical research. It is common for researchers to order several peptides at a time for use in fields as diverse as medicine, biology, material science, and biomedical engineering.

A dramatic illustration of the power of the solid-phase method was the synthesis of the enzyme ribonuclease by Merrifield in 1969. The synthesis involved 369 chemical reactions and 11,931 operations, all of which were performed by an automated machine and without any intermediate isolation stages. Each of the 124 amino acids was added as an *N-tert*-butoxycarbonyl derivative and coupled using DCC. Cleavage from the resin and removal of all protective groups gave a mixture that was purified by ion-exchange chromatography. The specific activity of the synthetic enzyme was 13–24% of that of the natural enzyme. The fact that the specific activity of the synthetic enzyme was lower than that of the natural enzyme was probably attributable to the presence of polypeptide byproducts closely related to but not identical to the natural enzyme. Synthesizing ribonuclease (124 amino acids) requires forming 123 peptide bonds. If each peptide bond is formed in 99% yield, the yield of homogeneous polypeptide is $0.99^{123} = 29\%$. If each peptide bond is formed in 98% yield, the yield is 8%. Thus, even with yields as high as 99% in each peptide bond-forming step, a large portion of the synthetic polypeptides have one or more sequence defects. Many of these, nonetheless, may be fully or partially active.

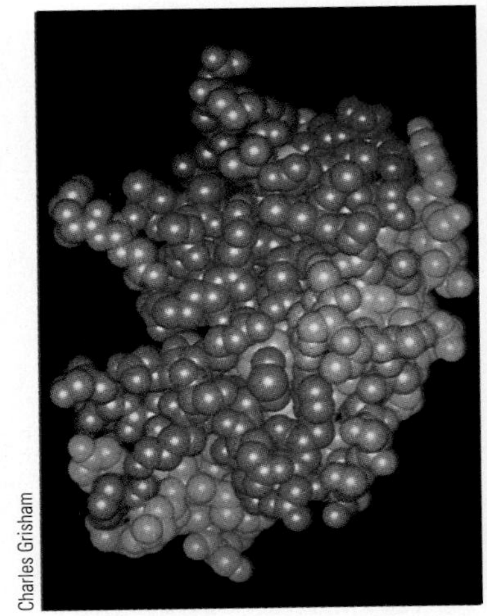
Charles Grisham

A model of the protein ribonuclease A. The purple segments are regions of α-helix and the yellow segments are regions of β-pleated sheet, both of which are described in Section 27.6. Other colors represent loop regions.

27.6 Three-Dimensional Shapes of Polypeptides and Proteins

A. Geometry of a Peptide Bond

In the late 1930s, Linus Pauling began a series of studies to determine the geometry of a peptide bond. One of his first and most important discoveries was that a peptide bond itself is planar. As shown in Figure 27.12, the four atoms of a peptide bond and the two α-carbons joined to it all lie in the same plane.

Had you been asked in Chapter 1 to describe the geometry of a peptide bond, you probably would have predicted bond angles of 120° about the carbonyl carbon and 109.5° about the amide nitrogen. However, as fully discussed in Connections to Biological Chemistry: "The Unique Structure of Amide Bonds" in Chapter 18, both

Figure 27.11

Steps in the Merrifield
solid-phase polypeptide
synthesis.

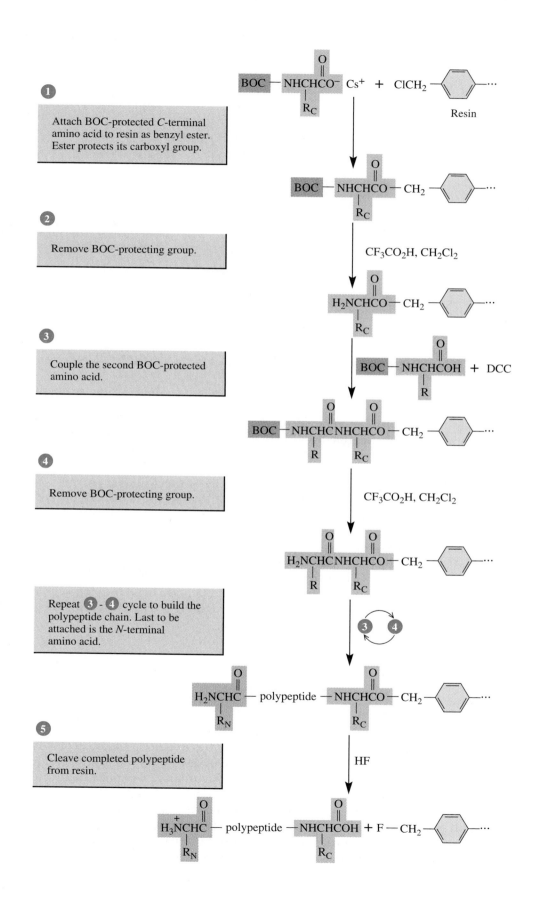

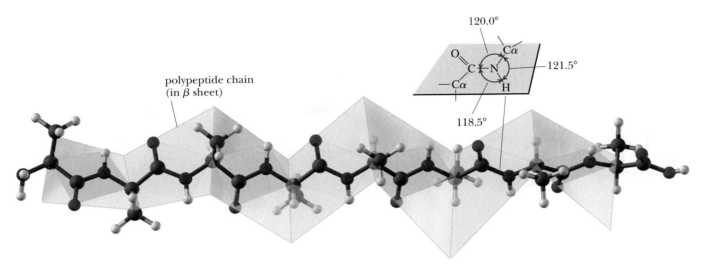

Figure 27.12
Planarity of a peptide bond. Bond angles about the carbonyl carbon and the amide nitrogen
are approximately 120°.

atoms are actually planar with approximately 120° bond angles about each because of
resonance of the nitrogen lone pair with the carbonyl. Two configurations are possible for the atoms of a planar peptide bond. In one, the two α-carbons are *cis* to each
other; in the other, they are *trans* to each other. The *trans* configuration is more
favorable because the α-carbons with the bulky groups bonded to them are farther
from each other than they are in the *cis* configuration. Almost all peptide bonds in
naturally occurring proteins studied to date have the *trans* configuration. Proline is
found *cis* most of the time, and there are some well-known examples of other *cis*
peptide bonds as well.

Secondary structure of proteins
The ordered arrangements
(conformations) of amino acids in
localized regions of a polypeptide
or protein.

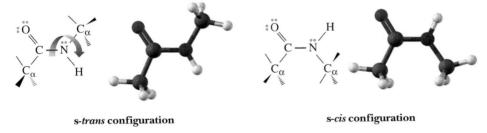

s-*trans* configuration   **s-*cis* configuration**

B. Secondary Structure

Secondary (2°) structure refers to ordered arrangements (conformations) of amino
acids in localized regions of a polypeptide or protein molecule. The first studies of
polypeptide conformations were carried out by Linus Pauling and Robert Corey beginning in 1939. They assumed that in conformations of greatest stability, all atoms in
a peptide bond lie in the same plane, and there is hydrogen bonding between the
N—H of one peptide bond and the C=O of another, as shown in Figure 27.13.
    On the basis of model building, Pauling proposed that two types of secondary
structure should be particularly stable: the α-helix and the antiparallel β-pleated
sheet. X-ray crystallography has validated this prediction completely.

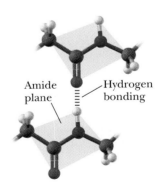

Amide   Hydrogen
plane   bonding

Figure 27.13
Hydrogen bonding between
amide groups.

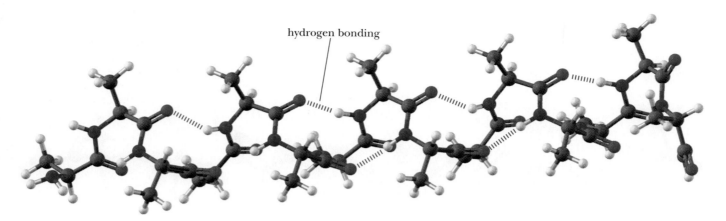

hydrogen bonding

Figure 27.14
An α-helix. The peptide chain is repeating units of L-alanine.

The α-Helix

α-Helix A type of secondary structure in which a section of polypeptide chain coils into a spiral, most commonly a right-handed spiral.

In an **α-helix** pattern shown in Figure 27.14, a polypeptide chain is coiled in a spiral. As you study this section of α-helix, note the following.

1. The helix is coiled in a clockwise, or right-handed, manner. Right-handed means that if you turn the helix clockwise, it twists away from you. In this sense, a right-handed helix is analogous to the right-handed thread of a common wood or machine screw.
2. There are 3.6 amino acids per turn of the helix.
3. Each peptide bond is *trans* and planar.
4. The N—H group of each peptide bond points roughly downward, parallel to the axis of the helix, and the C=O of each peptide bond points roughly upward, also parallel to the axis of the helix.
5. The carbonyl group of each peptide bond is hydrogen-bonded to the N—H group of the peptide bond four amino acid units away from it. Hydrogen bonds are shown as dashed lines.
6. All R— groups point outward from the helix.

Almost immediately after Pauling proposed the α-helix conformation, other researchers proved the presence of α-helix conformations in keratin, the protein of hair and wool. It soon became obvious that the α-helix is one of the fundamental folding patterns of polypeptide chains.

The β-Pleated Sheet

β-Pleated sheet A type of secondary structure in which sections of polypeptide chains are aligned parallel or antiparallel to one another.

An antiparallel **β-pleated sheet** consists of extended polypeptide chains with neighboring chains running in opposite (antiparallel) directions. In a parallel β-pleated sheet, the polypeptide chains run in the same direction. Unlike the α-helix arrangement, N—H and C=O groups lie in the plane of the sheet and are roughly perpendicular to the long axis of the sheet. The C=O group of each peptide bond is hydrogen-bonded to the N—H group of a peptide bond of a neighboring chain (Figure 27.15).

As you study this section of β-pleated sheet, note the following.

1. The three polypeptide chains lie adjacent to each other and run in opposite (antiparallel) directions.
2. Each peptide bond is planar, and the α-carbons are *trans* to each other.

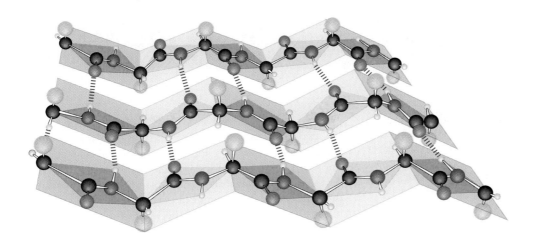

Figure 27.15
β-Pleated sheet conformation with three polypeptide chains running in opposite (antiparallel) directions. Hydrogen bonding between chains is indicated by dashed lines.

3. The C=O and N—H groups of peptide bonds from adjacent chains point at each other and are in the same plane so that hydrogen bonding is possible between adjacent polypeptide chains.
4. The R-groups on any one chain alternate, first above and then below the plane of the sheet, and so on.

The β-pleated sheet conformation is stabilized by hydrogen bonding between N—H groups of one chain and C=O groups of an adjacent chain. By comparison, the α-helix is stabilized by hydrogen bonding between N—H and C=O groups within the same polypeptide chain.

C. Tertiary Structure

Tertiary (3°) structure refers to the overall folding pattern and arrangement in space of all atoms in a single polypeptide chain. No sharp dividing line exists between secondary and tertiary structures. Secondary structure refers to the spatial arrangement of amino acids close to one another on a polypeptide chain, whereas tertiary structure refers to the three-dimensional arrangement of all atoms of a polypeptide chain. Among the most important factors in maintaining 3° structure are disulfide bonds, hydrophobic interactions, hydrogen bonding, and salt linkages.

Disulfide bonds (Section 10.9G) play an important role in maintaining tertiary structure. Disulfide bonds are formed between side chains of two cysteine units by oxidation of their thiol groups (—SH) to form a disulfide bond. Treatment of a disulfide bond with a reducing agent regenerates the thiol groups.

Tertiary structure of proteins
The three-dimensional arrangement in space of all atoms in a single polypeptide chain.

Spider Silk

Spider silk has some remarkable properties. Research is currently concentrated on the strong dragline silk that forms the spokes of a web of the Golden Orb Weaver (*Nephila clavipes*). This silk has three times the impact strength of Kevlar and is 30% more flexible than nylon. The commercial application of spider silk is not a novel concept. Eighteenth-century French entrepreneur Bon de Saint-Hilaire attempted to mass-produce silk in his high-density spider farms but failed because of cannibalism among his territorial arachnid workers. In contrast, native New Guineans continue to successfully collect and utilize spider silk for a wide range of applications including bags and fishing nets. Today, the only way to obtain large amounts of silk is to extract it from the abdomens of immobilized spiders, but scientific advances make the mass production and industrial application of spider silk increasingly possible.

Biologically produced dragline silk is a combination of two liquid proteins, Spidroin 1 and 2, which become oriented and solidify as they travel through a complex duct system in the spider's abdomen. These proteins are composed largely of alanine and glycine, the two smallest amino acids. Although glycine comprises almost 42% of each protein, the short, 5 to 10 peptide chains of alanine, which account for 25% of each protein's composition, are more important for the properties. Nuclear magnetic resonance (NMR) techniques have vastly improved the level of understanding of spider silk's structure, which was originally determined by x-ray crystallography. NMR data of spidroins containing deuterium-tagged alanine have shown that all alanines are configured into β-pleated sheets. Furthermore, the NMR data suggest that 40% of the alanine β-sheets are highly structured while the other 60% are less oriented, forming fingers that reach out from each individual strand. These fingers are believed to join the oriented alanine β-sheets and the glycine-rich, amorphous "background" sectors of the polypeptide.

Currently, genetically modified *Escherichia coli* is used to mass-produce Spidroin 1 and 2. However, DNA redundancy initially caused synthesis problems when the spider genes were transposed into the bacteria. The *E. coli* did not transcribe some of the codons in the same way that spider cells would, forcing scientists to modify the DNA. When the proteins could be synthesized, it was necessary to develop a system to mimic the natural production of spider silk while preventing the silk from contacting the air and subsequently hardening. After the two proteins are separated from the *E. coli,* they are drawn together into methanol through separate needles. Another approach is to dissolve the silk in formic acid or to add codons for hydrophilic amino acids, in this case histidine and arginine, to keep the artificial silk pliable. The industrial and practical applications of spider silk will not be fully known until it can be abiotically synthesized in large quantities.

■ Golden orb weaver.

Tom Bean/Stone/Getty

Based on a Chem 30H honors paper by Paul Celestre, UCLA.

Figure 27.16 shows the amino acid sequence of human insulin. This protein consists of two polypeptide chains: an A chain of 21 amino acids and a B chain of 30 amino acids. The A chain is bonded to the B chain by two interchain disulfide bonds. An intrachain disulfide bond also connects the cysteine units at positions 6 and 11 of the A chain.

Figure 27.16
Human insulin. The A chain of 21 amino acids and B chain of 30 amino acids are connected by interchain disulfide bonds between A7 and B7 and between A20 and B19. In addition, a single intrachain disulfide bond occurs between A6 and A11.

As an example of 2° and 3° structure, let us look at the three-dimensional structure of myoglobin—a protein found in skeletal muscle and particularly abundant in diving mammals, such as seals, whales, and porpoises. Myoglobin and its structural relative, hemoglobin, are the oxygen storage and transport molecules of vertebrates. Hemoglobin binds molecular oxygen in the lungs and transports it to myoglobin in muscles. Myoglobin stores molecular oxygen until it is required for metabolic oxidation.

Myoglobin consists of a single polypeptide chain of 153 amino acids. Myoglobin also contains a single heme unit. Heme consists of one Fe^{2+} ion coordinated in a square planar array with the four nitrogen atoms of a molecule of porphyrin (Figure 27.17).

Determination of the three-dimensional structure of myoglobin represented a milestone in the study of molecular architecture. For their contribution to this research, John C. Kendrew and Max F. Perutz, both of Britain, shared the 1962 Nobel Prize for chemistry. The secondary and tertiary structures of myoglobin are shown in Figure 27.18. The single polypeptide chain is folded into a complex, almost boxlike shape.

Following are important structural features of the three-dimensional shape of myoglobin.

1. The backbone consists of eight relatively straight sections of α-helix, each separated by a bend in the polypeptide chain. The longest section of α-helix has 24 amino acids, the shortest has 7. Some 75% of the amino acids are found in these eight regions of α-helix.

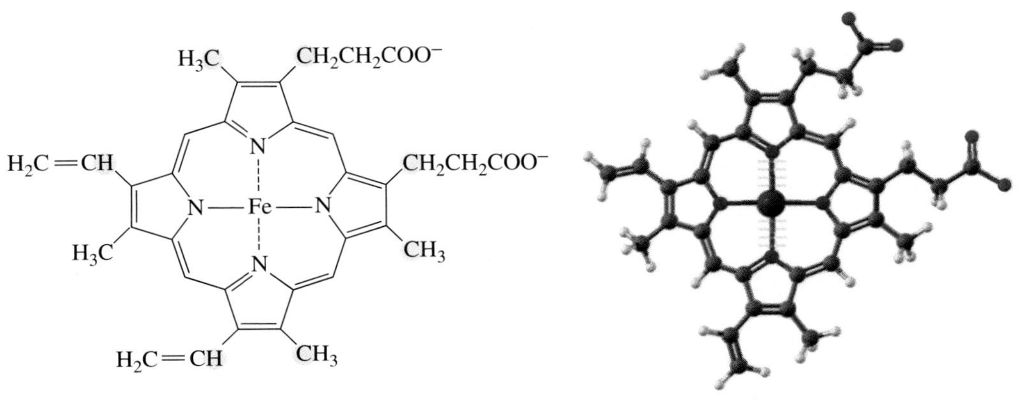

Figure 27.17
The structure of heme, found in myoglobin and hemoglobin.

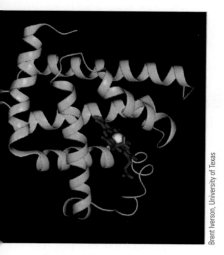

Brent Iverson, University of Texas

Figure 27.18
Ribbon model of myoglobin.
The polypeptide chain is
shown in yellow, the heme
ligand in red, and the Fe atom
as a white sphere.

2. Hydrophobic side chains of phenylalanine, alanine, valine, leucine, isoleucine, and methionine are clustered in the interior of the molecule where they are shielded from contact with water. **Hydrophobic interactions** are a major factor in directing the folding of the polypeptide chain of myoglobin into this compact, three-dimensional shape.

3. The outer surface of myoglobin is coated with hydrophilic side chains, such as those of lysine, arginine, serine, glutamic acid, histidine, and glutamine, which interact with the aqueous environment by **hydrogen bonding.** The only polar side chains that point to the interior of the myoglobin molecule are those of two histidine units, which point inward toward the heme group.

4. Oppositely charged amino acid side chains close to each other in the three-dimensional structure interact by electrostatic attractions called **salt linkages.** An example of a salt linkage is the attraction of the side chains of lysine ($-NH_3^+$) and glutamic acid ($-COO^-$).

The tertiary structures of hundreds of proteins have also been determined. It is clear that proteins contain α-helix and β-pleated sheet structures, but that wide variations exist in the relative amounts of each. Lysozyme, with 129 amino acids in a single polypeptide chain, has only 25% of its amino acids in α-helix regions. Cytochrome, with 104 amino acids in a single polypeptide chain, has no α-helix structure but does contain several regions of β-pleated sheet. Yet, whatever the proportions of α-helix, β-pleated sheet, or other periodic structure, most nonpolar side chains of water-soluble proteins are directed toward the interior of the molecule, whereas polar side chains are on the surface of the molecule and in contact with the aqueous environment. Note that this arrangement of polar and nonpolar groups in water-soluble proteins very much resembles the arrangement of polar and nonpolar groups of soap molecules in micelles (Figure 26.3). It also resembles the arrangement of phospholipids in lipid bilayers (Figure 26.13).

Example 27.7

With which of the following amino acid side chains can the side chain of threonine form hydrogen bonds?

(a) Valine   (b) Asparagine   (c) Phenylalanine
(d) Histidine   (e) Tyrosine   (f) Alanine

Solution

The side chain of threonine contains a hydroxyl group that can participate in hydrogen bonding in two ways: Its oxygen has a partial negative charge and can function as a hydrogen bond acceptor, and its hydrogen has a partial positive charge and can function as a hydrogen bond donor. Therefore, the side chain of threonine can form hydrogen bonds with the side chains of tyrosine, asparagine, and histidine.

Problem 27.7

At pH 7.4, with what amino acid side chains can the side chain of lysine form salt linkages?

D. Quaternary Structure

Quaternary structure The arrangement of polypeptide monomers into a noncovalently bonded aggregate.

Most proteins of molecular weight greater than 50,000 consist of two or more noncovalently linked polypeptide chains. The arrangement of protein monomers into an aggregation is known as **quaternary (4°) structure.** A good example is hemoglobin, a protein that consists of four separate polypeptide chains: two α-chains of 141 amino acids each and two β-chains of 146 amino acids each. The quaternary structure of hemoglobin is shown in Figure 27.19.

A major factor stabilizing the aggregation of protein subunits is the **hydrophobic effect.** When separate polypeptide chains fold into compact three-dimensional shapes to expose polar side chains to the aqueous environment and shield nonpolar side chains from water, hydrophobic "patches" still appear on the surface, in contact with water. These patches can be shielded from water if two or more monomers assemble so that their hydrophobic patches are in contact. The numbers of subunits of several proteins of known quaternary structure are shown in Table 27.4. Other important factors include correctly located complementary hydrogen bonding and charged sites on different subunits. The formation of aggregates of well-defined structure based on specific structural units on the subunits is being explored in the new field of molecular recognition.

Hydrophobic effect The tendency of nonpolar groups to cluster to shield themselves from contact with an aqueous environment.

Table 27.4 Quaternary Structure of Selected Proteins	
Protein	**Number of Subunits**
Alcohol dehydrogenase	2
Aldolase	4
Hemoglobin	4
Lactate dehydrogenase	4
Insulin	6
Glutamine synthetase	12
Tobacco mosaic virus protein disc	17

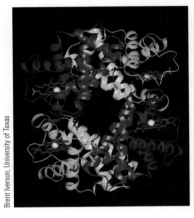

Brent Iverson, University of Texas

Figure 27.19
Ribbon model of hemoglobin. The α-chains are shown in purple, the β-chains in yellow, the heme ligands in red, and the Fe atoms as white spheres.

Summary

Amino acids are compounds that contain both an amino group and a carboxyl group (Section 27.1A). A **zwitterion** is an internal salt of an amino acid. With the exception of glycine, all protein-derived amino acids are chiral (Section 27.1B). In the D,L convention, all are L-amino acids. In the R,S convention, 18 are (S)-amino acids. Although cysteine has the same absolute configuration, it is an (R)-amino acid because of the manner in which priorities are assigned about the tetrahedral chiral center. Isoleucine and threonine contain a second chiral center. The 20 protein-derived amino acids are commonly divided into four categories (Section 27.1C): nine with nonpolar side chains, four with polar but un-ionized side chains, four with acidic side chains, and three with basic side chains.

The **isoelectric point, pI,** of an amino acid, polypeptide, or protein is the pH at which it has no net charge (Section 27.2C). **Electrophoresis** is the process of separating compounds on the basis of their electric charge (Section 27.2D).

Compounds having a high charge density move more rapidly than those with a lower charge density. Any amino acid or protein in a solution with a pH that equals the pI of the compound remains at the origin.

A **peptide bond** is the special name given to the amide bond formed between α-amino acids (Section 27.3). A **polypeptide** is a biological macromolecule containing many amino acids, each joined to the next by a peptide bond. By convention, the sequence of amino acids in a polypeptide is written beginning with the **N-terminal amino acid** toward the **C-terminal amino acid. Primary (1°) structure** of a polypeptide is the sequence of amino acids in the polypeptide chain (Section 27.4).

In solid-phase synthesis (Section 27.5F), or polymer-supported synthesis of polypeptides, the C-terminal amino acid is joined to a chloromethylated polystyrene resin as a benzyl ester. The polypeptide chain is then extended one amino acid at a time from the N-terminal end. When synthesis is completed, the polypeptide chain is released from the solid support by cleavage of the benzyl ester.

A peptide bond is planar (Section 27.6A), that is, the four atoms of the amide and the two α-carbons of a peptide bond lie in the same plane. Bond angles about the amide nitrogen and the amide carbonyl carbon are 120°. **Secondary (2°) structure** (Section 26.7B) refers to the ordered arrangement (conformations) of amino acids in localized regions of a polypeptide or protein. Two types of secondary structure are the α-helix and the β-pleated sheet. **Tertiary (3°) structure** (Section 27.6C) refers to the overall folding pattern and arrangement in space of all atoms in a single polypeptide chain. **Quaternary (4°) structure** (Section 27.6D) is the arrangement of polypeptide monomers into a noncovalently bonded aggregate.

Key Reactions

1. Acidity of an α-Carboxyl Group (Section 27.2A)

An α-COOH (pK_a approximately 2.19) of a protonated amino acid is a considerably stronger acid than acetic acid (pK_a 4.76) or other low-molecular-weight aliphatic carboxylic acid owing to the electron-withdrawing inductive effect of the α-NH_3^+ group.

$$\underset{\underset{NH_3^+}{|}}{RCHCOOH} + H_2O \rightleftharpoons \underset{\underset{NH_3^+}{|}}{RCHCOO^-} + H_3O^+ \qquad pK_a = 2.19$$

2. Acidity of an α-Ammonium Group (Section 27.2A)

An α-NH_3^+ group (pK_a approximately 9.47) is a slightly stronger acid than a primary aliphatic ammonium ion (pK_a approximately 10.76).

$$\underset{\underset{NH_3^+}{|}}{RCHCOO^-} + H_2O \rightleftharpoons \underset{\underset{NH_2}{|}}{RCHCOO^-} + H_3O^+ \qquad pK_a = 9.47$$

3. Reaction of an α-Amino Acid with Ninhydrin (Section 27.2D)

Treatment of an α-amino acid with ninhydrin gives a purple-colored solution. Treatment of proline with ninhydrin gives an orange-colored solution.

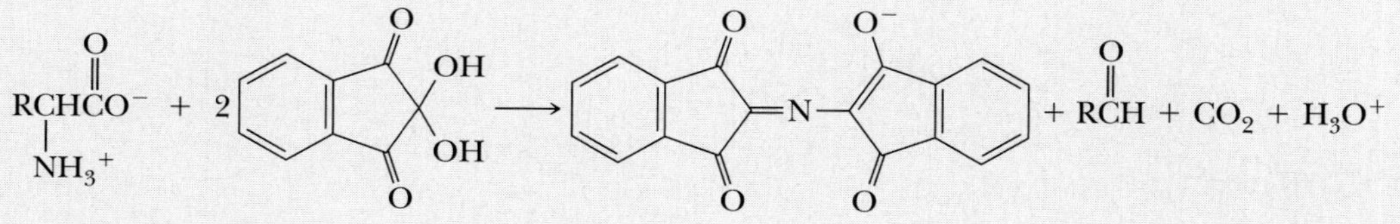

An α-amino   Ninhydrin   Purple-colored anion
acid

4. Cleavage of a Peptide Bond by Cyanogen Bromide (Section 27.4B)

Cleavage is regioselective for a peptide bond formed by the carboxyl group of methionine.

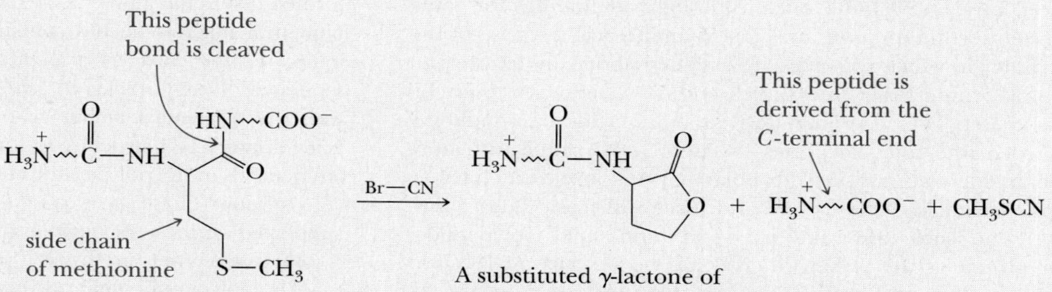

A substituted γ-lactone of
the amino acid homoserine

5. Edman Degradation (Section 27.4B)

Treatment with phenyl isothiocyanate followed by acid removes the *N*-terminal amino acid as a substituted phenylthiohydantoin, which is then separated and identified.

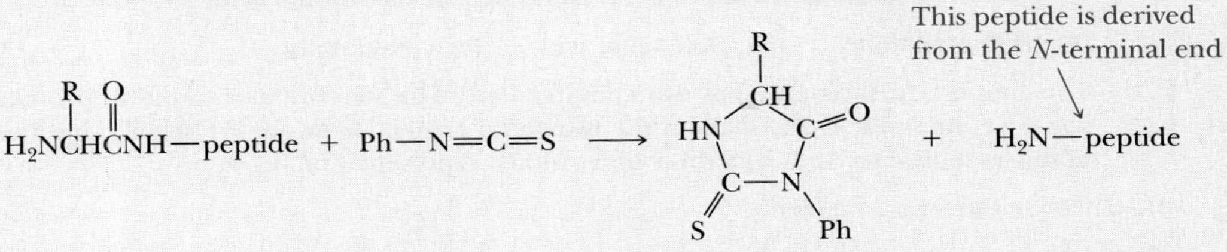

This peptide is derived from the *N*-terminal end

$$
\text{H}_2\text{NCHCNH}\text{—peptide} \; + \; \text{Ph}\text{—N}{=}\text{C}{=}\text{S} \; \longrightarrow \; \text{(A phenylthiohydantoin)} \; + \; \text{H}_2\text{N—peptide}
$$

Phenyl
isothiocyanate

A phenylthiohydantoin

6. The Benzyloxycarbonyl (*Z*—) Protecting Group (Section 27.5C)

The benzyloxycarbonyl protecting group is prepared by treatment of an unprotected α-NH$_2$ group with benzyloxycarbonyl chloride. It is removed by treatment with HBr in acetic acid or by hydrogenolysis.

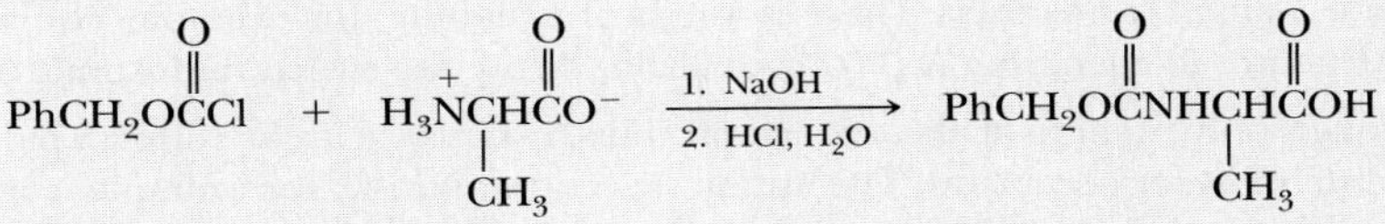

$$
\text{PhCH}_2\text{OCCl} \; + \; \overset{+}{\text{H}_3}\text{NCHCO}^- \; \xrightarrow[\text{2. HCl, H}_2\text{O}]{\text{1. NaOH}} \; \text{PhCH}_2\text{OCNHCHCOH}
$$

Benzyloxycarbonyl
chloride
(Z—Cl)

Alanine

N-Benzyloxycarbonylalanine
(Z—Ala)

7. Peptide Bond Formation Using 1,3-Dicyclohexylcarbodiimide (Section 27.5E)

This substituted carbodiimide is a dehydrating agent and is converted to a disubstituted urea. The reaction is efficient and yields are generally very high.

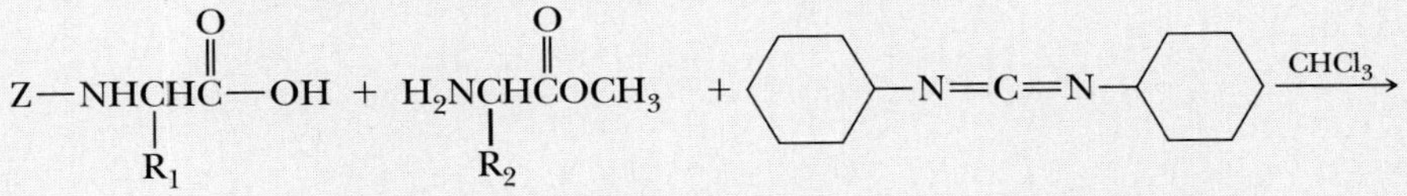

$$
\text{Z—NHCHC—OH} \; + \; \text{H}_2\text{NCHCOCH}_3 \; + \; \text{(cyclohexyl)—N}{=}\text{C}{=}\text{N—(cyclohexyl)} \; \xrightarrow{\text{CHCl}_3}
$$

Amino protected
aa$_1$

Carboxyl protected
aa$_2$

1,3-Dicyclohexylcarbo-
diimide (DCC)

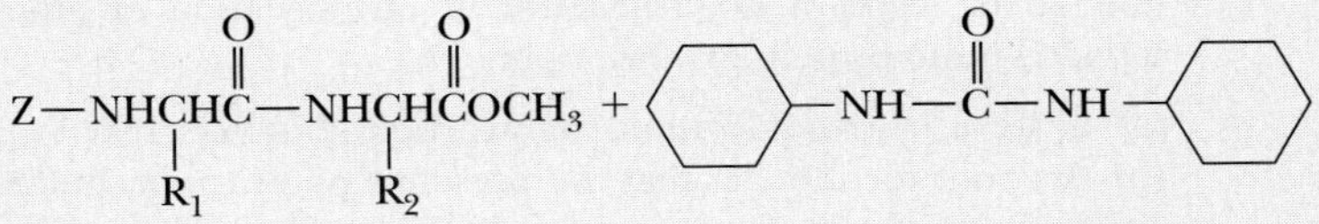

$$
\text{Z—NHCHC—NHCHCOCH}_3 \; + \; \text{(cyclohexyl)—NH—C—NH—(cyclohexyl)}
$$

Amino and carboxyl
protected dipeptide

N,N′-Dicyclohexylurea
(DCU)

Problems

Amino Acids

27.8   What amino acid does each abbreviation stand for?

(a) Phe   (b) Ser   (c) Asp   (d) Gln
(e) His   (f) Gly   (g) Tyr

ORGANIC
Chemistry Now™

Assess your understanding of this chapter's topics with additional quizzing and conceptual-based problems at
http://org.brookscole.com/bfi4

27.9   The configuration of the chiral center in α-amino acids is most commonly specified using the D,L convention. It can also be identified using the R,S convention (Section 3.3). Does the chiral center in L-serine have the R or the S configuration?

27.10   Assign an R or S configuration to the chiral center in each amino acid.

    (a) L-Phenylalanine   **(b)** L-Glutamic acid   **(c)** L-Methionine

27.11   The amino acid threonine has two chiral centers. The stereoisomer found in proteins has the configuration 2S,3R about the two chiral centers. Draw (a) a Fischer projection of this stereoisomer and (b) a three-dimensional representation.

27.12   Define the term *zwitterion*.

27.13   Draw zwitterion forms of these amino acids.

    (a) Valine   **(b)** Phenylalanine   **(c)** Glutamine

27.14   Why are Glu and Asp often referred to as acidic amino acids?

27.15   Why is Arg often referred to as a basic amino acid? Which two other amino acids are also basic amino acids?

27.16   What is the meaning of the alpha as it is used in α-amino acid?

27.17   Several β-amino acids exist. There is a unit of β-alanine, for example, contained within the structure of coenzyme A (Problem 25.35). Write the structural formula of β-alanine.

27.18   Although only L-amino acids occur in proteins, D-amino acids are often a part of the metabolism of lower organisms. The antibiotic actinomycin D, for example, contains a unit of D-valine, and the antibiotic bacitracin A contains units of D-asparagine and D-glutamic acid. Draw Fischer projections and three-dimensional representations for these three D-amino acids.

27.19   Histamine is synthesized from one of the 20 protein-derived amino acids. Suggest which amino acid is its biochemical precursor and the type of organic reaction(s) involved in its biosynthesis (for example, oxidation, reduction, decarboxylation, nucleophilic substitution).

Histamine

27.20   As discussed in Chemical Connections: "Vitamin K, Blood Clotting, and Basicity" in Section 26.6D, vitamin K participates in carboxylation of glutamic acid residues of the blood-clotting protein prothrombin.

    (a) Write a structural formula for γ-carboxyglutamic acid.
    (b) Account for the fact that the presence of γ-carboxyglutamic acid escaped detection for many years; on routine amino acid analyses, only glutamic acid was detected.

27.21   Both norepinephrine and epinephrine are synthesized from the same protein-derived amino acid. From which amino acid are they synthesized, and what types of reactions are involved in their biosynthesis?

    (a) Norepinephrine   **(b)** Epinephrine (Adrenaline)

27.22 From which amino acid are serotonin and melatonin synthesized, and what types of reactions are involved in their biosynthesis?

(a) Serotonin

(b) Melatonin

Acid-Base Behavior of Amino Acids

27.23 Draw a structural formula for the form of each amino acid most prevalent at pH 1.0.

(a) Threonine   (b) Arginine   (c) Methionine   (d) Tyrosine

27.24 Draw a structural formula for the form of each amino most prevalent at pH 10.0.

(a) Leucine   (b) Valine   (c) Proline   (d) Aspartic acid

27.25 Write the zwitterion form of alanine and show its reaction with the following.

(a) 1 mol NaOH   (b) 1 mol HCl

27.26 Write the form of lysine most prevalent at pH 1.0 and then show its reaction with the following. Consult Table 27.2 for pK_a values of the ionizable groups in lysine.

(a) 1 mol NaOH   (b) 2 mol NaOH   (c) 3 mol NaOH

27.27 Write the form of aspartic acid most prevalent at pH 1.0 and then show its reaction with the following. Consult Table 27.2 for pK_a values of the ionizable groups in aspartic acid.

(a) 1 mol NaOH   (b) 2 mol NaOH   (c) 3 mol NaOH

27.28 Given pK_a values for ionizable groups from Table 27.2, sketch curves for the titration of (a) glutamic acid with NaOH and (b) histidine with NaOH.

27.29 Draw a structural formula for the product formed when alanine is treated with the following reagents.

(a) Aqueous NaOH   (b) Aqueous HCl
(b) CH_3CH_2OH, H_2SO_4 (b) $(CH_3CO)_2O$, CH_3COONa

27.30 For lysine and arginine, the isoelectric point, pI, occurs at a pH where the net charge on the nitrogen-containing groups is +1 and balances the charge of −1 on the α-carboxyl group. Calculate pI for these amino acids.

27.31 For aspartic and glutamic acids, the isoelectric point occurs at a pH where the net charge on the two carboxyl groups is −1 and balances the charge of +1 on the α-amino group. Calculate pI for these amino acids.

27.32 Account for the fact that the isoelectric point of glutamine (pI 5.65) is higher than the isoelectric point of glutamic acid (pI 3.08).

27.33 Enzyme-catalyzed decarboxylation of glutamic acid gives 4-aminobutanoic acid (Section 27.1D). Estimate the pI of 4-aminobutanoic acid.

27.34 Guanidine and the guanidino group present in arginine are two of the strongest organic bases known. Account for their basicity.

27.35 At pH 7.4, the pH of blood plasma, do the majority of protein-derived amino acids bear a net negative charge or a net positive charge? Explain.

27.36 Do the following compounds migrate to the cathode or to the anode on electrophoresis at the specified pH?

(a) Histidine at pH 6.8

(b) Lysine at pH 6.8

(c) Glutamic acid at pH 4.0

(d) Glutamine at pH 4.0

(e) Glu-Ile-Val at pH 6.0

(f) Lys-Gln-Tyr at pH 6.0

27.37 At what pH would you carry out an electrophoresis to separate the amino acids in each mixture?

(a) Ala, His, Lys   (b) Glu, Gln, Asp   (c) Lys, Leu, Tyr

27.38 Examine the amino acid sequence of human insulin (Figure 27.16), and list each Asp, Glu, His, Lys, and Arg in this molecule. Do you expect human insulin to have an isoelectric point nearer that of the acidic amino acids (pI 2.0–3.0), the neutral amino acids (pI 5.5–6.5), or the basic amino acids (pI 9.5–11.0)?

27.39 A chemically modified guanidino group is present in cimetidine (Tagamet), a widely prescribed drug for the control of gastric acidity and peptic ulcers. Cimetidine reduces gastric acid secretion by inhibiting the interaction of histamine with gastric H_2 receptors. In the development of this drug, a cyano group was added to the substituted guanidino group to alter its basicity. Do you expect this modified guanidino group to be more basic or less basic than the guanidino group of arginine? Explain.

$$N-CN$$
$$\parallel$$
$$H_3C \quad\quad CH_2SCH_2CH_2NHCNHCH_3$$
$$HN \diagdown N$$

Cimetidine
(Tagamet)

27.40 Draw a structural formula for the product formed when alanine is treated with the following reagents.

(a) C₆H₅—CCl, $(CH_3CH_2)_3N$ (b) (indane-1,3-dione-2,2-diol)

(c) C₆H₅—CH₂OCCl, NaOH   (d) $(CH_3)_3COCOCOC(CH_3)_3$, NaOH

(e) Product(c) + L-Alanine ethyl ester + DCC

(f) Product(d) + L-Alanine ethyl ester + DCC

Primary Structure of Polypeptides and Proteins

27.41 If a protein contains four different SH groups, how many different disulfide bonds are possible if only a single disulfide bond is formed? How many different disulfides are possible if two disulfide bonds are formed?

27.42 How many different tetrapeptides can be made under the following conditions?

(a) The tetrapeptide contains one unit each of Asp, Glu, Pro, and Phe.

(b) All 20 amino acids can be used, but each only once.

27.43 A decapeptide has the following amino acid composition.

$$Ala_2, \text{ Arg, Cys, Glu, Gly, Leu, Lys, Phe, Val}$$

Partial hydrolysis yields the following tripeptides.

Cys-Glu-Leu + Gly-Arg-Cys + Leu-Ala-Ala + Lys-Val-Phe + Val-Phe-Gly

One round of Edman degradation yields a lysine phenylthiohydantoin. From this information, deduce the primary structure of this decapeptide.

27.44 Following is the primary structure of glucagon, a polypeptide hormone of 29 amino acids. Glucagon is produced in the α-cells of the pancreas and helps maintain blood glucose levels in a normal concentration range.

```
1           5               10              15              20              25          29
His-Ser-Glu-Gly-Thr-Phe-Thr-Ser-Asp-Tyr-Ser-Lys-Tyr-Leu-Asp-Ser-Arg-Arg-Ala-Gln-Asp-Phe-Val-Gln-Trp-Leu-Met-Asn-Thr
```

Which peptide bonds are hydrolyzed when this polypeptide is treated with each reagent?

(a) Phenyl isothiocyanate   **(b)** Chymotrypsin   **(c)** Trypsin   **(d)** Br—CN

27.45 A tetradecapeptide (14 amino acid residues) gives the following peptide fragments on partial hydrolysis. From this information, deduce the primary structure of this polypeptide. Fragments are grouped according to size.

Pentapeptide Fragments	Tetrapeptide Fragments
Phe-Val-Asn-Gln-His	Gln-His-Leu-Cys
His-Leu-Cys-Gly-Ser	His-Leu-Val-Glu
Gly-Ser-His-Leu-Val	Leu-Val-Glu-Ala

27.46 Draw a structural formula of these tripeptides. Mark each peptide bond, the *N*-terminal amino acid, and the *C*-terminal amino acid.

(a) Phe-Val-Asn   **(b)** Leu-Val-Gln

27.47 Estimate the pI of each tripeptide on Problem 27.46.

27.48 Glutathione (G-SH), one of the most common tripeptides in animals, plants, and bacteria, is a scavenger of oxidizing agents. In reacting with oxidizing agents, glutathione is converted to G-S-S-G.

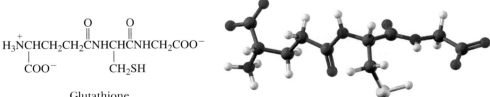

Glutathione

(a) Name the amino acids in this tripeptide.
(b) What is unusual about the peptide bond formed by the *N*-terminal amino acid?
(c) Write a balanced half-reaction for the reaction of two molecules of glutathione to form a disulfide bond. Is glutathione a biological oxidizing agent or a biological reducing agent?
(d) Write a balanced equation for reaction of glutathione with molecular oxygen, O_2, to form G-S-S-G and H_2O. Is molecular oxygen oxidized or reduced in this process?

27.49 Following are a structural formula and a ball-and-stick model for the artificial sweetener aspartame. Each amino acid has the L configuration.

Aspartame is present in many artificially sweetened foods and beverages.

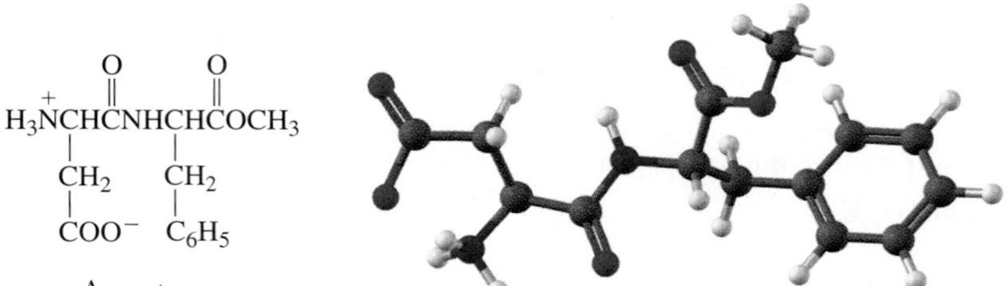

Charles D. Winters

$$\overset{+}{H_3}NCHCNHCHCOCH_3$$

with the two carbonyl O groups, CH_2 and CH_2 substituents, COO^- and C_6H_5

Aspartame

(a) Name the two amino acids in this molecule.
(b) Estimate the isoelectric point of aspartame.
(c) Draw structural formulas for the products of hydrolysis of aspartame in 1 M HCl.

27.50 2,4-Dinitrofluorobenzene, very often known as Sanger's reagent after the English chemist Frederick Sanger who popularized its use, reacts selectively with the *N*-terminal amino group of a polypeptide chain. Sanger was awarded the 1958 Nobel Prize for chemistry for his work in determining the primary structure of bovine insulin. One of the few persons to be awarded two Nobel Prizes, he also shared the 1980 award in chemistry with American chemists, Paul Berg and Walter Gilbert, for the development of chemical and biological analyses of DNAs.

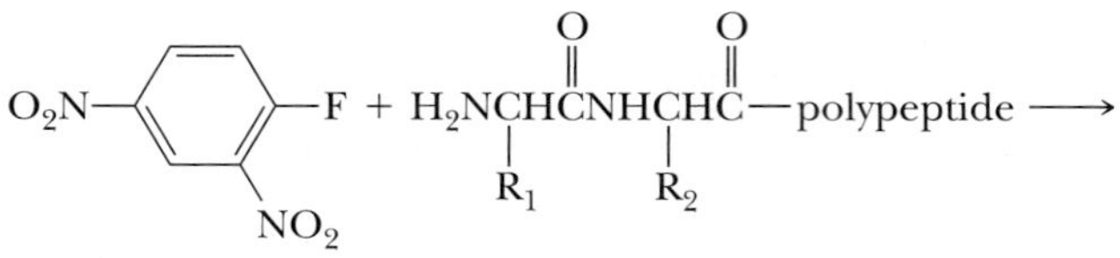

$$O_2N-\overbrace{}-F \ + \ H_2NCHCNHCHC-\text{polypeptide} \longrightarrow$$
with R_1, R_2 substituents, NO_2

polypeptide chain in which the
N-terminal amino acid is labeled
with a 2,4-dinitrophenyl group

2,4-Dinitro-
fluorobenzene

(*N*-Terminal end of
a polypeptide chain)

Following reaction with 2,4-dinitrofluorobenzene, all amide bonds of the polypeptide chain are hydrolyzed, and the amino acid labeled with a 2,4-dinitrophenyl group is separated by either paper or column chromatography and identified.

(a) Write a structural formula for the product formed by treatment of the *N*-terminal amino group with Sanger's reagent and propose a mechanism for its formation.
(b) When bovine insulin is treated with Sanger's reagent followed by hydrolysis of all peptide bonds, two labeled amino acids are detected: glycine and phenylalanine. What conclusions can be drawn from this information about the primary structure of bovine insulin?
(c) Compare and contrast the structural information that can be obtained from use of Sanger's reagent with that from use of the Edman degradation.

Synthesis of Polypeptides

27.51 In a variation of the Merrifield solid-phase peptide synthesis, the amino group is protected by a fluorenylmethoxycarbonyl (FMOC) group. This protecting group is removed by treatment with a weak base such as the secondary amine, piperidine. Write a balanced equation and propose a mechanism for this deprotection.

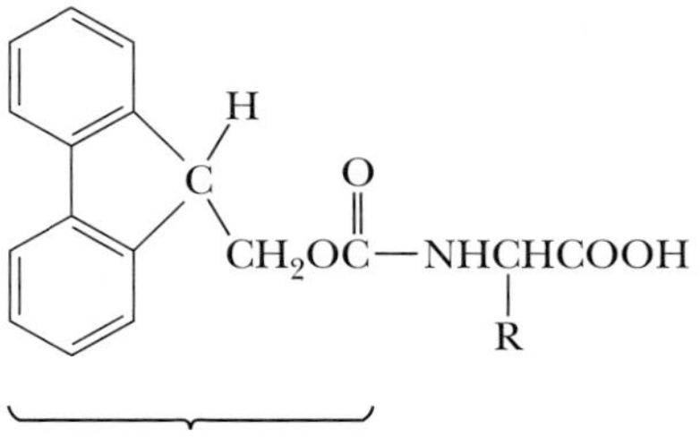

Structure of the fluorenylmethoxycarbonyl (FMOC) group, with fluorene ring system bearing $-CH_2OC(=O)-NHCHCOOH$ with R substituent.

Fluorenylmethoxy-
carbonyl (FMOC) group

27.52 The BOC-protecting group may be added by treatment of an amino acid with di-*tert*-butyl dicarbonate as shown in the following reaction sequence. Propose a mechanism to account for formation of these products.

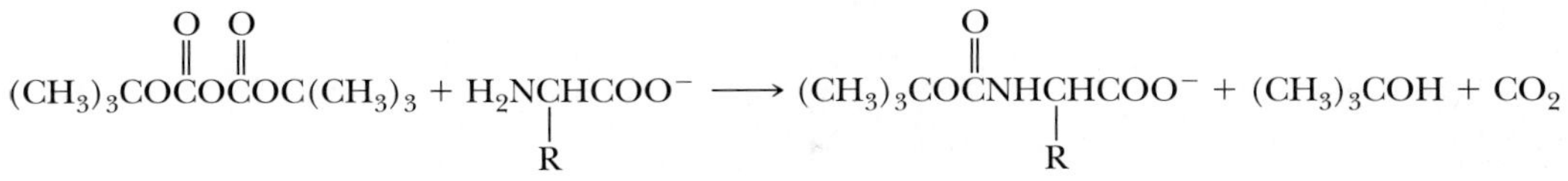

$$(CH_3)_3COCOCOC(CH_3)_3 + H_2NCHCOO^- \longrightarrow (CH_3)_3COCNHCHCOO^- + (CH_3)_3COH + CO_2$$
$$\underset{R}{\mid} \qquad\qquad\qquad\qquad \underset{R}{\mid}$$

Di-*tert*-butyl dicarbonate   BOC-amino acid

27.53 The side-chain carboxyl groups of aspartic acid and glutamic acid are often protected as benzyl esters.

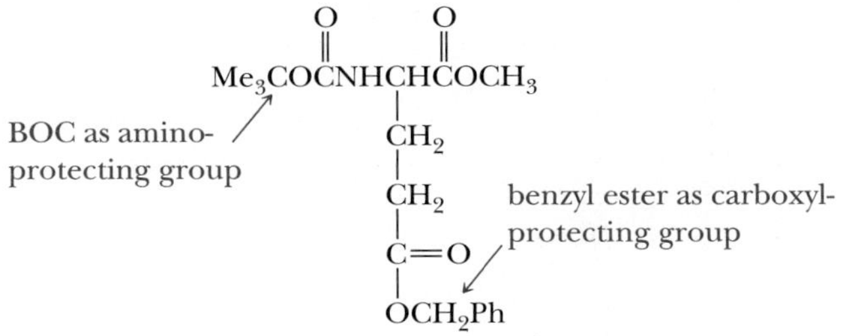

$$Me_3COCNHCHCOCH_3$$

BOC as amino-
protecting group
$$CH_2$$
$$CH_2 \quad \text{benzyl ester as carboxyl-}$$
$$\underset{}{\mid} \quad \text{protecting group}$$
$$C=O$$
$$\mid$$
$$OCH_2Ph$$

  (a) Show how to convert the side-chain carboxyl group to a benzyl ester using benzyl chloride as a source of the benzyl group.
  (b) How do you deprotect the side-chain carboxyl under mild conditions without removing the BOC-protecting group at the same time?

Three-Dimensional Shapes of Polypeptides and Proteins

27.54 Examine the α-helix conformation. Are amino acid side chains arranged all inside the helix, all outside the helix, or randomly?

27.55 Distinguish between intermolecular and intramolecular hydrogen bonding between the backbone groups on polypeptide chains. In what type of secondary structure do you find intermolecular hydrogen bonds? In what type do you find intramolecular hydrogen bonding?

27.56 Many plasma proteins found in an aqueous environment are globular in shape. Which amino acid side chains would you expect to find on the surface of a globular protein and in contact with the aqueous environment? Which would you expect to find inside, shielded from the aqueous environment? Explain.

  (a) Leu   **(b)** Arg   **(c)** Ser   **(d)** Lys   **(e)** Phe

27.57 Denaturation of a protein is a physical change, the most readily observable result of which is loss of biological activity. Denaturation stems from changes in secondary, tertiary, and quaternary structure through disruption of noncovalent interactions including hydrogen bonding and hydrophobic interactions. Three common denaturing agents are sodium dodecyl sulfate (SDS), urea, and heat. What kinds of noncovalent interactions might each reagent disrupt?

28

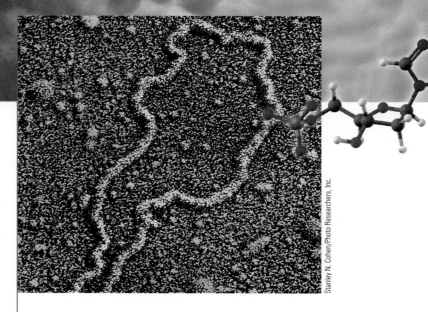

Stanley N. Cohen/Photo Researchers, Inc.

■ False-colored transmission electron micrograph of a plasmid DNA. If the cell wall of a bacterium such as *Escherichia coli* is partially digested and the cell is then osmotically shocked by dilution with water, its contents are extruded to the exterior. Shown here is the bacterial chromosome surrounding the cell. Inset: 2′-Deoxyadenosine 5′-monophosphate (dAMP), a building block of DNA (Section 28.2).

Outline

Nucleic acid   A biopolymer containing three types of monomer units: heterocyclic aromatic amine bases derived from purine and pyrimidine, the monosaccharides D-ribose or 2-deoxy-D-ribose, and phosphoric acid.

Nucleic Acids

The organization, maintenance, and regulation of cellular function require a tremendous amount of information, all of which must be processed each time a cell is replicated. With very few exceptions, genetic information is stored and transmitted from one generation to the next in the form of deoxyribonucleic acids (DNA). Genes, the hereditary units of chromosomes, are long stretches of double-stranded DNA. If the DNA in a human chromosome in a single cell were uncoiled, it would be approximately 1.8 meters long!

Genetic information is expressed in two stages: transcription from DNA to ribonucleic acids (RNA) and then translation for the synthesis of proteins.

$$\text{DNA} \xrightarrow{\text{transcription}} \text{RNA} \xrightarrow{\text{translation}} \text{proteins}$$

Thus, DNA is the repository of genetic information in cells, whereas RNA serves in the transcription and translation of this information, which is then expressed through the synthesis of proteins.

In this chapter, we examine the structure of nucleosides and nucleotides and the manner in which these monomers are covalently bonded to form **nucleic acids.** Then, we examine the manner in which genetic information is encoded on molecules of DNA, the function of the three types of ribonucleic acids, and finally how the primary structure of a DNA molecule is determined.

Using chemistry that we will not discuss in the text, DNA can be synthesized on a solid phase analogous to the method described for the synthesis of oligopeptides [Section 27.5(f)]. In fact, both DNA and RNA can now be synthesized with high yield and the entire process has been automated. It is commonplace for molecular biologists to place an order for a specific DNA or RNA on the Internet and have the sample arrive one or two days later. Such efficiency, made possible by automation and high-yielding synthetic reactions, has revolutionized molecular biology and biotechnology.

28.1   Nucleosides and Nucleotides

Controlled hydrolysis of nucleic acids yields three components: heterocyclic aromatic amine bases, the monosaccharides D-ribose or 2-deoxy-D-ribose (Section 25.1A), and phosphate ions. The five heterocyclic aromatic amine bases most common to nucleic acids are shown in Figure 28.1. Uracil, cytosine, and thymine are referred to as pyrimidine bases after the name of the parent base; adenine and guanine are referred to as purine bases.

A **nucleoside** is a compound containing D-ribose or 2-deoxy-D-ribose bonded to a heterocyclic aromatic amine base by a *β-N*-glycosidic bond (Section 25.3A). The monosaccharide component of DNA is 2-deoxy-D-ribose, whereas that of RNA is D-ribose. The glycosidic bond is between C-1$'$ (the anomeric carbon) of ribose or 2-deoxyribose and N-1 of a pyrimidine base or N-9 of a purine base. Figure 28.2 shows a structural formula for uridine, a nucleoside derived from ribose and uracil.

A **nucleotide** is a nucleoside in which a molecule of phosphoric acid is esterified with a free hydroxyl of the monosaccharide, most commonly either the 3$'$-hydroxyl or the 5$'$-hydroxyl. A nucleotide is named by giving the name of the parent nucleoside followed by the word "monophosphate." The position of the phosphoric ester is specified by the number of the carbon to which it is bonded. Figure 28.3 shows a structural formula of 5$'$-adenosine monophosphate, AMP. Monophosphoric esters are diprotic acids with pK_a values of approximately 1 and 6. Therefore, at pH 7, the

Nucleoside   A building block of nucleic acids, consisting of D-ribose or 2-deoxy-D-ribose bonded to a heterocyclic aromatic amine base by a β-*N*-glycosidic bond.

Nucleotide   A nucleoside in which a molecule of phosphoric acid is esterified with an —OH of the monosaccharide, most commonly either the 3$'$-OH or the 5$'$-OH.

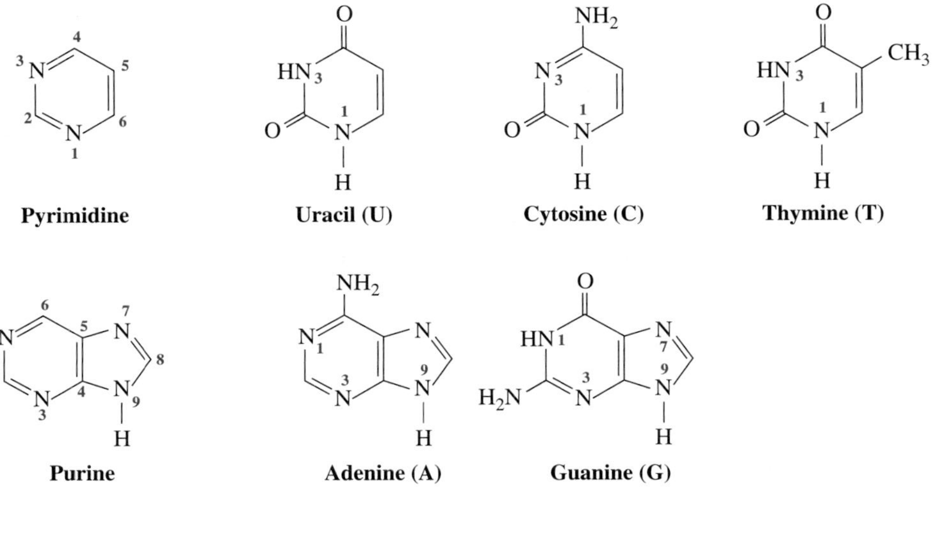

Figure 28.1

Names and one-letter abbreviations for the heterocyclic aromatic amine bases most common to DNA and RNA. Bases are numbered according to the patterns of the parent compounds, pyrimidine and purine.

Pyrimidine   **Uracil (U)**   **Cytosine (C)**   **Thymine (T)**

Purine   **Adenine (A)**   **Guanine (G)**

Figure 28.2
Uridine, a nucleoside. Atom numbers on the monosaccharide rings are primed to distinguish them from atom numbers on the heterocyclic aromatic amine bases.

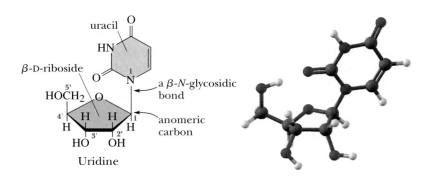

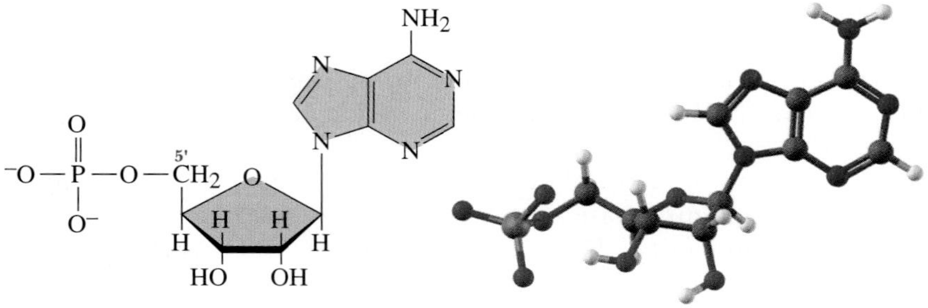

Uridine

Figure 28.3
Adenosine 5'-monophosphate, AMP. The phosphate group is fully ionized at pH 7.0 giving this nucleotide a charge of -2.

two hydrogens of a phosphoric monoester are fully ionized giving a nucleotide a charge of -2.

Nucleoside monophosphates can be further phosphorylated to form nucleoside diphosphates and nucleoside triphosphates. Shown in Figure 28.4 is a structural formula for adenosine 5'-triphosphate, ATP.

Nucleoside diphosphates and triphosphates are also polyprotic acids and are extensively ionized at pH 7.0. pK_a values of the first three ionization steps for adenosine triphosphate are less than 5.0. The value of pK_{a4} is approximately 7.0. Therefore, at pH 7.0, approximately 50% of adenosine triphosphate is present as ATP^{4-}, and 50% is present as ATP^{3-}.

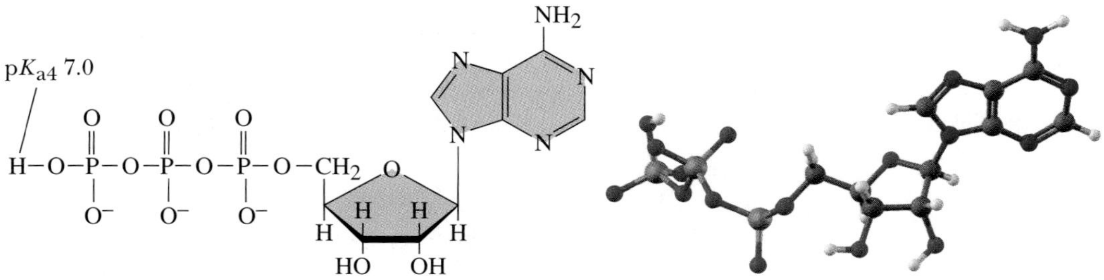

pK_{a4} 7.0

Figure 28.4
Adenosine triphosphate, ATP.

Example 28.1

Draw a structural formula for each nucleotide.
(a) 2′-Deoxycytidine 5′-diphosphate   **(b)** 2′-Deoxyguanosine 3′-monophosphate

Solution

(a) Cytosine is joined by a β-N-glycosidic bond between N-1 of cytosine and C-1 of the cyclic hemiacetal form of 2-deoxy-D-ribose. The 5′-hydroxyl of the pentose is bonded to a phosphate group by an ester bond, and this phosphate is, in turn, bonded to a second phosphate group by an anhydride bond.

(b) Guanine is joined by a β-N-glycosidic bond between N-9 of guanine and C-1 of the cyclic hemiacetal form of 2-deoxy-D-ribose. The 3′-hydroxyl group of the pentose is joined to a phosphate group by an ester bond.

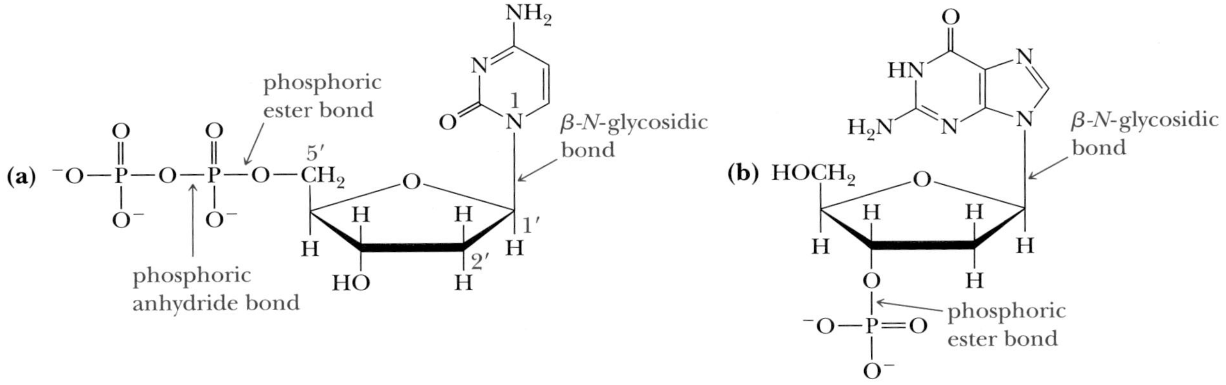

Problem 28.1

Draw a structural formula for each nucleotide.
(a) 2′-Deoxythymidine 5′-monophosphate
(b) 2′-Deoxythymidine 3′-monophosphate

28.2 The Structure of DNA

In Chapter 27 we saw that the four levels of structural complexity in polypeptides and proteins are primary, secondary, tertiary, and quaternary structures. There are three levels of structural complexity in nucleic acids. Although these levels are somewhat comparable to those in polypeptides and proteins, they also differ in significant ways.

A. Primary Structure—The Covalent Backbone

Deoxyribonucleic acids consist of a backbone of alternating units of deoxyribose and phosphate in which the 3′-hydroxyl of one deoxyribose unit is joined by a phosphodiester bond to the 5′-hydroxyl of another deoxyribose unit (Figure 28.5).

Figure 28.5
A tetranucleotide section of a
single-stranded DNA.

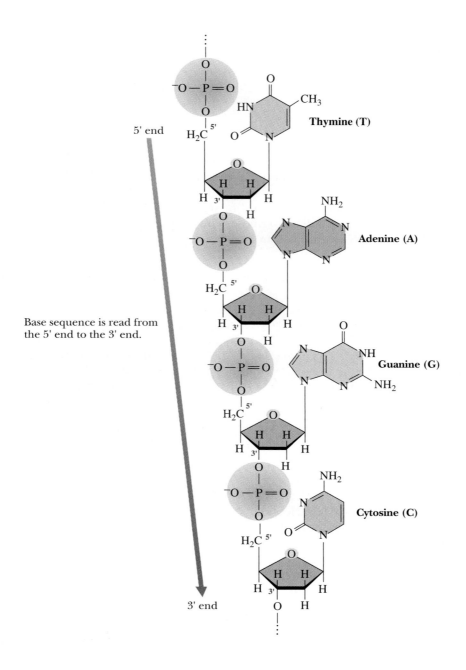

Base sequence is read from
the 5' end to the 3' end.

Primary structure of nucleic acids
The sequence of bases along the
pentose-phosphodiester backbone
of a DNA or RNA molecule read
from the 5' end to the 3' end.

This pentose-phosphodiester backbone is constant throughout an entire DNA molecule. A heterocyclic aromatic amine base—adenine, guanine, thymine, or cytosine—is bonded to each deoxyribose unit by a β-N-glycosidic bond. **Primary structure** of DNA refers to the order of heterocyclic bases along the pentose-phosphodiester backbone. The sequence of bases is usually written from the 5' end to the 3' end.

The Search for Antiviral Drugs

The search for antiviral drugs has been more difficult than the search for antibacterial drugs primarily because viral replication depends on the metabolic processes of the invaded cell. Thus, antiviral drugs are also likely to cause harm to the cells that harbor the virus. The challenge in developing antiviral drugs has been to understand the biochemistry of viruses and to develop drugs that target processes specific to them. Compared with the large number of antibacterial drugs that are available, there are only a handful of antiviral drugs, and they have nowhere near the effectiveness that antibiotics have on bacterial infections.

Acyclovir is one of the first of a new family of drugs for the treatment of infectious diseases caused by DNA viruses called herpesvirus. Herpes infections in humans are of two kinds: herpes simplex type 1, which gives rise to mouth and eye sores, and herpes simplex type 2, which gives rise to serious genital infections. Acyclovir is highly effective against herpesvirus-caused genital infections. The drug is activated *in vivo* by conversion of the primary —OH (which corresponds to the 5′-OH of a riboside or a deoxyriboside) to a triphosphate. Because

of its close resemblance to deoxyguanosine triphosphate, an essential precursor for DNA synthesis, acyclovir triphosphate is taken up by viral DNA polymerase to form an enzyme-substrate complex on which no 3′-OH exists for replication to continue. Thus, the enzyme-substrate complex is no longer active (it is a dead-end complex), viral replication is disrupted, and the virus is destroyed.

Perhaps the best known of the new viral antimetabolites is zidovudine (azidothymidine, AZT), an analog of deoxythymidine in which the 3′-OH has been replaced by an azido group, N_3. AZT is effective against HIV-1, a retrovirus that is the causative agent of AIDS. It is converted *in vivo* by cellular enzymes to the 5′-triphosphate, recognized as deoxythymidine 5′-triphosphate by viral RNA-dependent DNA polymerase (reverse transcriptase), and added to a growing DNA chain. There it stops chain elongation because no 3′-OH exists on which to add the next deoxynucleotide. AZT owes its effectiveness to the fact that it binds more strongly to viral reverse transcriptase than it does to human DNA polymerase.

Acyclovir
(drawn to show its structural
relationship to 2-deoxyguanosine)

Zidovudine
(Azidothymidine; AZT)

Example 28.2

Draw a structural formula for the DNA dinucleotide TG that is phosphorylated at the 5′ end only.

Solution

Problem 28.2

Draw a structural formula for the section of DNA that contains the base sequence CTG and is phosphorylated at the 3′ end only.

B. Secondary Structure—The Double Helix

By the early 1950s, it was clear that DNA molecules consist of chains of alternating units of deoxyribose and phosphate joined by 3′,5′-phosphodiester bonds with a base attached to each deoxyribose unit by a β-*N*-glycosidic bond. In 1953, the American

■ Watson and Crick with their model of DNA.

biologist James D. Watson and the British physicist Francis H. C. Crick proposed a double-helix model for the **secondary structure** for DNA. Watson, Crick, and Maurice Wilkins shared the 1962 Nobel Prize for physiology or medicine for "their discoveries concerning the molecular structure of nucleic acids, and its significance for information transfer in living material." Although Rosalind Franklin also played an important part in this research, she did not share in the Nobel Prize because of her death in 1958.

The **Watson-Crick model** was based on molecular modeling and two lines of experimental observations: chemical analyses of DNA base compositions and mathematical analyses of x-ray diffraction patterns of crystals of DNA.

Base Composition

At one time, it was thought that the four principal bases occur in the same ratios and perhaps repeat in a regular pattern along the pentose-phosphodiester backbone of DNA for all species. However, more precise determinations of base composition by Erwin Chargaff revealed that bases do not occur in the same ratios (Table 28.1).

Researchers drew the following conclusions from this and related data. To within experimental error,

1. The mole-percent base composition in any organism is the same in all cells of the organism and is characteristic of the organism.
2. The mole-percents of adenine (a purine base) and thymine (a pyrimidine base) are equal. The mole-percents of guanine (a purine base) and cytosine (a pyrimidine base) are also equal.
3. The mole-percents of purine bases (A + G) and pyrimidine bases (C + T) are equal.

Analyses of X-Ray Diffraction Patterns

Additional information about the structure of DNA emerged when x-ray diffraction photographs taken by Rosalind Franklin and Maurice Wilkins were analyzed. These diffraction patterns revealed that, even though the base composition of DNA isolated from different organisms varies, DNA molecules themselves are remarkably uniform in thickness. They are long and fairly straight, with an outside diameter of approximately 2000 pm, and not more than a dozen atoms thick. Furthermore, the crystallographic pattern repeats every 3400 pm. Herein lay one of the chief problems to be solved. How could the molecular dimensions of DNA be so regular even though the relative percentages of the various bases differ so widely? With this accumulated information, the stage was set for the development of a hypothesis about DNA structure.

Secondary structure of nucleic acids The ordered arrangement of nucleic acid strands.

Watson-Crick model A double-helix model for the secondary structure of a DNA molecule.

Rosalind Franklin (1920–1958). In 1951 she joined the Biophysical Laboratory at King's College, London, where she used x-ray diffraction methods to study DNA. She is credited with discoveries that established the density of DNA and its helical conformation. Her work was important to the model of DNA developed by Watson and Crick. She died in 1958 at age 37, and because the Nobel Prize is never awarded posthumously, she did not share in the 1962 Nobel Prize for physiology or medicine with Watson, Crick, and Wilkins. Although her relation with Watson and Crick was initially strained, Watson said, "we later came to appreciate . . . the struggles the intelligent woman faces to be accepted by the scientific world which often regards women as mere diversions from serious thinking."

Table 28.1	Comparison in Base Composition, in Mole-Percent, of DNA from Several Organisms						
	Purines		**Pyrimidines**				**Purines/**
Organism	**A**	**G**	**C**	**T**	**A/T**	**G/C**	**Pyrimidines**
Human	30.4	19.9	19.9	30.1	1.01	1.00	1.01
Sheep	29.3	21.4	21.0	28.3	1.04	1.02	1.03
Yeast	31.7	18.3	17.4	32.6	0.97	1.05	1.00
E. coli	26.0	24.9	25.2	23.9	1.09	0.99	1.04

The Watson–Crick Double Helix

The heart of the Watson-Crick model is the postulate that a molecule of DNA is a complementary **double helix.** It consists of two antiparallel polynucleotide strands coiled in a right-handed manner about the same axis to form a double helix. As illustrated in the ribbon models in Figure 28.6, chirality is associated with a double helix; left-handed and right-handed double helices are related by reflection just as enantiomers are related by reflection.

To account for the observed base ratios and uniform thickness of DNA, Watson and Crick postulated that purine and pyrimidine bases project inward toward the axis of the helix and are always paired in a very specific manner. According to scale models, the dimensions of an adenine-thymine base pair are almost identical to the dimensions of a guanine-cytosine base pair, and the length of each pair is consistent with the core thickness of a DNA strand (Figure 28.7). Thus, if the purine base in one strand is adenine, then its complement in the antiparallel strand must be thymine. Similarly, if the purine in one strand is guanine, its complement in the antiparallel strand must be cytosine. The "fits" between the TA pair and between the CG pair are remarkable and represent another important example of molecular recognition and supramolecular complex formation. The resulting hydrogen bonding holds the two strands together very strongly.

A significant feature of Watson and Crick's model is that no other base-pairing is consistent with the observed thickness of a DNA molecule. A pair of pyrimidine bases is too small to account for the observed thickness, whereas a pair of purine bases is too large. Thus, according to the Watson-Crick model, the repeating units in a double-stranded DNA molecule are not single bases of differing dimensions but rather base pairs of almost identical dimensions.

To account for the periodicity observed from x-ray data, Watson and Crick postulated that base pairs are stacked one on top of the other with a distance of 340 pm between base pairs and with ten base pairs in one complete turn of the helix. There is one complete turn of the helix every 3400 pm. Shown in Figure 28.8 is a ribbon model of double-stranded **B-DNA,** the predominant form of DNA in dilute aqueous solution and thought to be the most common form in nature.

In the double helix, the bases in each base pair are not directly opposite one another across the diameter of the helix but rather are slightly displaced. This displacement and the relative orientation of the glycosidic bonds linking each base to the sugar-phosphate backbone leads to two differently sized grooves, a major groove and a minor groove (Figure 28.8). Each groove runs along the length of the cylindrical column of the double helix. The major groove is approximately 2200 pm wide; the minor groove is approximately 1200 pm wide.

Double helix A type of secondary structure of DNA molecules in which two antiparallel polynucleotide strands are coiled in a right-handed manner about the same axis.

Figure 28.6
A DNA double helix has a chirality associated with the helix. Right-handed and left-handed double helices of otherwise identical DNA chains are nonsuperposable mirror images.

Figure 28.7
Base-pairing between adenine and thymine (A-T) and between guanine and cytosine (G-C). An A-T base pair is held by two hydrogen bonds, whereas a G-C base pair is held by three hydrogen bonds.

Thymine ⚌ Adenine

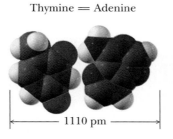

|← 1110 pm →|

Cytosine ⚌ Guanine

|← 1080 pm →|

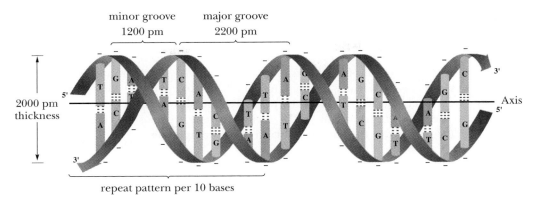

minor groove
1200 pm

major groove
2200 pm

2000 pm
thickness

Axis

repeat pattern per 10 bases

Figure 28.8
Ribbon model of double-stranded B-DNA. Each ribbon shows the pentose-phosphodiester backbone of a single-stranded DNA molecule. The strands are antiparallel; one strand runs to the left from the 5′ end to the 3′ end, the other runs to the right from the 5′ end to the 3′ end. Hydrogen bonds are shown by three dotted lines between each G-C base pair and two dotted lines between each A-T base pair.

Figure 28.9 shows more detail of an idealized B-DNA double helix. The major and minor grooves are clearly recognizable in this model.

Other forms of secondary structure are known that differ in the distance between stacked base pairs and in the number of base pairs per turn of the helix. One of the most common of these, **A-DNA,** also a right-handed helix, is thicker than B-DNA and has a repeat distance of only 2900 pm. There are ten base pairs per turn of the helix with a spacing of 290 pm between base pairs.

Example 28.3

One strand of a DNA molecule contains the base sequence 5′-ACTTGCCA-3′. Write its complementary base sequence.

Solution

Remember that the base sequence is always written from the 5′ end of the strand to the 3′ end, that A pairs with T, and that G pairs with C. In double-stranded DNA, the two strands run in opposite (antiparallel) directions so that the 5′ end of one strand is associated with the 3′ end of the other strand.

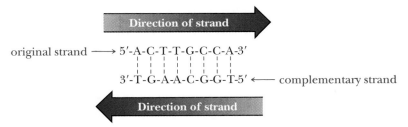

original strand ⟶ 5′-A-C-T-T-G-C-C-A-3′

3′-T-G-A-A-C-G-G-T-5′ ⟵ complementary strand

Written from the 5′ end, the complementary strand is 5′-TGGCAAGT-3′.

Problem 28.3
Write the complementary DNA base sequence for 5′-CCGTACGA-3′.

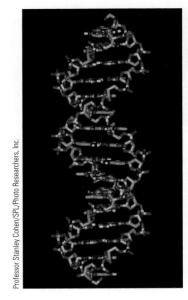

Figure 28.9
An idealized model of B-DNA.

C. Tertiary Structure—Supercoiled DNA

The length of a DNA molecule is enormously greater than its diameter, and the extended molecule is quite flexible. A DNA molecule is said to be relaxed if it has no twists other than those imposed by its secondary structure. Said another way, relaxed DNA does not have a clearly defined tertiary structure. We consider two types of **tertiary structure,** one type induced by perturbations in circular DNA and a second type introduced by coordination of DNA with nuclear proteins called histones. Tertiary structure, whatever the type, is referred to as **supercoiling.**

Supercoiling of Circular DNA

Circular DNA is a type of double-stranded DNA in which the two ends of each strand are joined by phosphodiester bonds [Figure 28.10(a)]. This type of DNA, the most prominent form in bacteria and viruses, is also referred to as circular duplex (because it is double-stranded) DNA. One strand of circular DNA may be opened, partially unwound, and then rejoined. The unwound section introduces a strain into the molecule because the nonhelical gap is less stable than hydrogen-bonded, base-paired helical sections. The strain can be localized in the nonhelical gap. Alternatively, it may be spread uniformly over the entire circular DNA by introduction of **superhelical twists,** one twist for each turn of a helix unwound. The circular DNA shown in Figure 28.10(b) has been unwound by four complete turns of the helix. The strain introduced by this unwinding is spread uniformly over the entire molecule by introduction of four superhelical twists [Figure 28.10(c)]. Interconversion of relaxed and supercoiled DNA is catalyzed by groups of enzymes called topoisomerases and gyrases.

Supercoiling of Linear DNA

Supercoiling of linear DNA in plants and animals takes another form and is driven by interaction between negatively charged DNA molecules and a group of positively charged proteins called **histones.** Histones are particularly rich in lysine and arginine, and at the pH of most body fluids, they have an abundance of positively charged sites along their length. The complex between negatively charged DNA and positively charged histones is called **chromatin.** Histones associate to form core particles about which double-stranded DNA then wraps. Further coiling of DNA produces the chromatin found in cell nuclei.

Tertiary structure of nucleic acids
The three-dimensional arrangement of all atoms of a nucleic acid, commonly referred to as supercoiling.

Circular DNA   A type of double-stranded DNA in which the 5′ and 3′ ends of each strand are joined by phosphodiester groups.

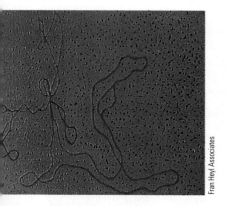

Fran Heyl Associates

■ Supercoiled DNA from a mitochondrion.

Histone   A protein, particularly rich in the basic amino acids lysine and arginine, that is found associated with DNA molecules.

Figure 28.10
Relaxed and supercoiled DNA.
(a) Circular DNA is relaxed.
(b) One strand is broken, unwound by four turns, and the ends then rejoined. The strain of unwinding is localized in the nonhelical gap.
(c) Supercoiling by four twists distributes the strain of unwinding uniformly over the entire molecule of circular DNA.

(a) Relaxed: circular (duplex) DNA

(b) Slightly strained: circular (duplex) DNA with four twists of the helix unwound

(c) Strained: supercoiled circular DNA

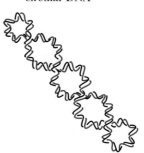

CHEMICAL CONNECTIONS

The Fountain of Youth

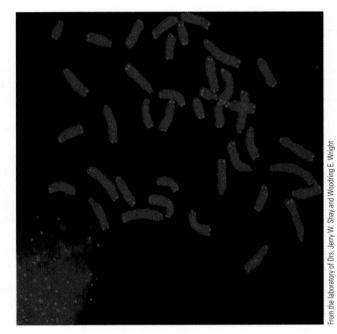

From the laboratory of Drs. Jerry W. Shay and Wooding E. Wright

■ Telomers are the repeating DNA strings (TTAGGC in vertebrates) that cap chromosomes.

In 1997 a sheep named Dolly became the first mammal to be born through cloning. By the time she was three years old, however, her genetic material was aging at the rate of the six-year-old sheep from which she was cloned. It turned out that Dolly had shortened telomeres, and—as a consequence—her genetic structure was considerably "older" than Dolly herself.

Telomeres are the physical ends of eukaryotic chromosomes, and they surrender a bit of themselves during each cell division because the DNA-replicating machinery normally fails to copy the DNA at a chromosome's tips. This process limits the total number of divisions a cell can undergo because the telomeres are ultimately consumed. In egg and cancer cells, among others, an enzyme called telomerase rebuilds the telomere cap after each division, making such cells effectively immortal. Consequently, telomerase may be the fountain of youth because it is responsible for the extension of telomeres in most species. If telomerase activity is limited, as it is in normal cells, the telomeres will shorten, and cells will age like Dolly's. Telomeres are the repeating DNA strings (TTAGGG in vertebrates) that cap chromosomes.

Researchers are trying to alter the integrity of telomeres by controlling the expression and location of telomerase. Preserving telomere caps may rejuvenate aging organisms by enabling cells to survive additional divisions. Conversely, doing away with telomerase activity in tumors may cause cancer cells to cease their endless replications and die.

Based on a Chem30H honors paper by James Stinebaugh, UCLA.

28.3 Ribonucleic Acids

Ribonucleic acids are similar to deoxyribonucleic acids in that they, too, consist of long, unbranched chains of nucleotides joined by phosphodiester groups between the 3′-hydroxyl of one pentose and the 5′-hydroxyl of the next. There are, however, three major differences in structure between RNA and DNA.

1. The pentose unit in RNA is β-D-ribose rather than β-2-deoxy-D-ribose.
2. The pyrimidine bases in RNA are uracil and cytosine rather than thymine and cytosine (Figure 28.1).
3. RNA is single-stranded rather than double-stranded.

Table 28.2 Types of RNA Found in Cells of *E. coli*

Type	Molecular-Weight Range (g/mol)	Number of Nucleotides	Percentage of Cell RNA
mRNA	25,000–1,000,000	75–3000	2
tRNA	23,000–30,000	73–94	16
rRNA	35,000–1,100,000	120–2904	82

Cells contain up to eight times as much RNA as DNA; in contrast to DNA, RNA occurs in different forms and in multiple copies of each form. RNA molecules are classified, according to their structure and function, into three major types: ribosomal RNA, transfer RNA, and messenger RNA. The molecular weight, number of nucleotides, and percent cellular abundance of these types in cells of *Escherichia coli* are summarized in Table 28.2.

A. Ribosomal RNA

Ribosomal RNA (rRNA) A ribonucleic acid found in ribosomes, the sites of protein synthesis.

The bulk of **ribosomal RNA (rRNA)** is found in the cytoplasm in subcellular particles called ribosomes, which contain about 60% RNA and 40% protein. Ribosomes are the sites in cells at which protein synthesis takes place.

B. Transfer RNA

Transfer RNA (tRNA) A ribonucleic acid that carries a specific amino acid to the site of protein synthesis on ribosomes.

Transfer RNA (tRNA) molecules have the lowest molecular weight of all nucleic acids. They consist of 73–94 nucleotides in a single chain. The function of tRNA is to carry amino acids to the sites of protein synthesis on the ribosomes. Each amino acid has at least one tRNA dedicated specifically to this purpose. Several amino acids have more than one. In the transfer process, the amino acid is joined to its specific tRNA by an ester bond between the α-carboxyl group of the amino acid and the 3′ hydroxyl group of the ribose unit at the 3′ end of the tRNA.

C. Messenger RNA

Messenger RNA (mRNA) A ribonucleic acid that carries coded genetic information from DNA to the ribosomes for the synthesis of proteins.

Messenger RNAs (mRNA) are present in cells in relatively small amounts and are very short-lived. They are single-stranded, and their synthesis is directed by information encoded on DNA molecules. Double-stranded DNA is unwound, and a complementary strand of mRNA is synthesized along one strand of the DNA template, beginning from the 3′ end. The synthesis of mRNA from a DNA template is called transcription because genetic information contained in a sequence of bases of DNA is transcribed into a complementary sequence of bases on mRNA. The name "messenger" is derived from the function of this type of RNA, which is to carry coded genetic information from DNA to the ribosomes for the synthesis of proteins.

Example 28.4

Following is a base sequence from a portion of DNA. Write the sequence of bases of the mRNA synthesized using this section of DNA as a template.

3'-A-G-C-C-A-T-G-T-G-A-C-C-5'

Solution

RNA synthesis begins at the 3' end of the DNA template and proceeds toward the 5' end. The complementary mRNA strand is formed using the bases C, G, A, and U. Uracil (U) is the complement of adenine (A) on the DNA template.

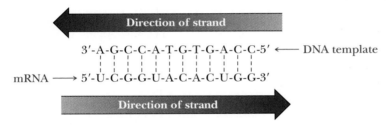

Reading from the 5' end, the sequence of mRNA is 5'-UCGGUACACUGG-3'.

Problem 28.4

Here is a portion of the nucleotide sequence in phenylalanine tRNA.

3'-ACCACCUGCUCAGGCCUU-5'

Write the nucleotide sequence of its DNA complement.

28.4 The Genetic Code

A. Triplet Nature of the Code

It was clear by the early 1950s that the sequence of bases in DNA molecules constitutes the store of genetic information and directs the synthesis of messenger RNA, which, in turn, directs the synthesis of proteins. However, the statement that "the sequence of bases in DNA directs the synthesis of proteins" presents the following problem. How can a molecule containing only four variable units (adenine, cytosine, guanine, and thymine) direct the synthesis of molecules containing up to 20 variable units (the protein-derived amino acids)? How can an alphabet of only four letters code for the order of letters in the 20-letter alphabet that occurs in proteins?

An obvious answer is that there is not one base but rather a combination of bases coding for each amino acid. If the code consists of nucleotide pairs, there are $4^2 = 16$ combinations; this code is more extensive, but it is still not extensive enough to code for 20 amino acids. If the code consists of nucleotides in groups of three, there are $4^3 = 64$ combinations, which is more than enough to code for the primary structure of

Codon   A triplet of nucleotides on mRNA that directs incorporation of a specific amino acid into a polypeptide sequence.

a protein. This appears to be a very simple solution to a system that must have taken eons to evolve. Yet proof now exists, from comparison of gene (nucleic acid) and protein (amino acid) sequences, that nature does indeed use this simple three-letter or triplet code to store genetic information. A triplet of nucleotides is called a **codon.**

B. Deciphering the Genetic Code

The next question is, which of the 64 triplets code for which amino acid? In 1961 Marshall Nirenberg provided a simple experimental approach to the problem, based on the observation that synthetic polynucleotides direct polypeptide synthesis in much the same manner as do natural mRNAs. Nirenberg incubated ribosomes, amino acids, tRNAs, and appropriate protein-synthesizing enzymes. With only these components, there was no polypeptide synthesis. However, when he added synthetic polyuridylic acid (poly U), a polypeptide of high molecular weight was synthesized. What was more important, the synthetic polypeptide contained only phenylalanine. With this discovery, the first element of the genetic code was deciphered: the triplet UUU codes for phenylalanine.

Similar experiments were carried out with different synthetic polyribonucleotides. It was found, for example, that polyadenylic acid (poly A) leads to the synthesis of polylysine, and that polycytidylic acid (poly C) leads to the synthesis of polyproline. By 1964, all 64 codons had been deciphered (Table 28.3).

Table 28.3   The Genetic Code—mRNA Codons and the Amino Acid Each Codon Directs

First Position (5'-end)	Second Position				Third Position (3'-end)
	U	**C**	**A**	**G**	
U	UUU Phe	UCU Ser	UAU Tyr	UGU Cys	U
	UUC Phe	UCC Ser	UAC Tyr	UGC Cys	C
	UUA Leu	UCA Ser	UAA Stop	UGA Stop	A
	UUG Leu	UCG Ser	UAG Stop	UGG Trp	G
C	CUU Leu	CCU Pro	CAU His	CGU Arg	U
	CUC Leu	CCC Pro	CAC His	CGC Arg	C
	CUA Leu	CCA Pro	CAA Gln	CGA Arg	A
	CUG Leu	CCG Pro	CAG Gln	CGG Arg	G
A	AUU Ile	ACU Thr	AAU Asn	AGU Ser	U
	AUC Ile	ACC Thr	AAC Asn	AGC Ser	C
	AUA Ile	ACA Thr	AAA Lys	AGA Arg	A
	AUG* Met	ACG Thr	AAG Lys	AGG Arg	G
G	GUU Val	GCU Ala	GAU Asp	GGU Gly	U
	GUC Val	GCC Ala	GAC Asp	GGC Gly	C
	GUA Val	GCA Ala	GAA Glu	GGA Gly	A
	GUG Val	GCG Ala	GAG Glu	GGG Gly	G

*AUG also serves as the principal initiation codon.

C. Properties of the Genetic Code

Several features of the genetic code are evident from a study of Table 28.3.

1. Only 61 triplets code for amino acids. The remaining three (UAA, UAG, and UGA) are signals for chain termination; they signal to the protein-synthesizing machinery of the cell that the primary sequence of the protein is complete. The three chain termination triplets are indicated in Table 28.3 by "Stop."
2. The code is degenerate, which means that several amino acids are coded for by more than one triplet. Only methionine and tryptophan are coded for by just one triplet. Leucine, serine, and arginine are coded for by six triplets, and the remaining amino acids are coded for by two, three, or four triplets.
3. For the 15 amino acids coded for by two, three, or four triplets, only the third letter of the code varies. For example, glycine is coded for by the triplets GGA, GGG, GGC, and GGU.
4. There is no ambiguity in the code, meaning that each triplet codes for only one amino acid.

   Finally, we must ask one last question about the genetic code. Is the code universal, that is, is it the same for all organisms? Every bit of experimental evidence available today from the study of viruses, bacteria, and higher animals, including humans, indicates that the code is universal. Furthermore, the fact that it is the same for all these organisms means that it has been the same over billions of years of evolution.

Example 28.5

During transcription, a portion of mRNA is synthesized with the following base sequence.

$$5'\text{-AUG-GUA-CCA-CAU-UUG-UGA-}3'$$

(a) Write the nucleotide sequence of the DNA from which this portion of mRNA was synthesized.
(b) Write the primary structure of the polypeptide coded for by this section of mRNA.

Solution

(a) During transcription, mRNA is synthesized from a DNA strand, beginning from the $3'$ end of the DNA template. The DNA strand must be the complement of the newly synthesized mRNA strand.

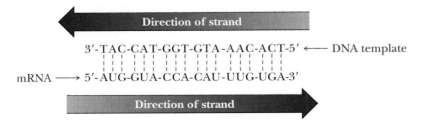

Note that the codon UGA codes for termination of the growing polypeptide chain; therefore, the sequence given in this problem codes for a pentapeptide only.

(b) The sequence of amino acids is shown in the following mRNA strand.

5'-AUG-GUA-CCA-CAU-UUG-UGA-3'
met—val—pro—his—leu—stop

Problem 28.5

The following section of DNA codes for oxytocin, a polypeptide hormone.

3'ACG-ATA-TAA-GTT-TTA-ACG-GGA-GAA-CCA-ACT-5'

(a) Write the base sequence of the mRNA synthesized from this section of DNA.
(b) Given the sequence of bases in part (a), write the primary structure of oxytocin.

28.5   Sequencing Nucleic Acids

As recently as 1975, the task of determining the primary structure of a nucleic acid was thought to be far more difficult than determining the primary structure of a protein. Nucleic acids, it was reasoned, contain only four different units, whereas proteins contain 20 different units. With only four different units, there are fewer specific sites for selective cleavage, distinctive sequences are more difficult to recognize, and there is greater chance of ambiguity in the assignment of sequence. Two breakthroughs reversed this situation. First was the development of a type of electrophoresis called polyacrylamide gel electrophoresis, a technique so sensitive that it is possible to separate nucleic acid fragments that differ from one another in only a single nucleotide. The second breakthrough was the discovery of a class of enzymes called restriction endonucleases, isolated chiefly from bacteria.

A. Restriction Endonucleases

Restriction endonuclease An enzyme that catalyzes hydrolysis of a particular phosphodiester bond within a DNA strand.

A **restriction endonuclease** recognizes a set pattern of four to eight nucleotides and cleaves a DNA strand by hydrolysis of the linking phosphodiester bonds at any site that contains that particular sequence. Close to 1000 restriction endonucleases have been isolated and their specificities characterized; each cleaves at specific sites, often with unique specificity. *E. coli,* for example, has a restriction endonuclease, EcoRI, that recognizes the hexanucleotide sequence, GAATTC, and cleaves it between G and A.

cleavage here

5'---G-A-A-T-T-C---3' $\xrightarrow{\text{EcoRI}}$ 5'---G + 5'-A-A-T-T-C---3'

Note that the action of restriction endonucleases is analogous to the action of trypsin (Section 27.4B), which catalyzes hydrolysis of amide bonds formed by the carboxyl groups of Lys and Arg, and of chymotrypsin, which catalyzes cleavage of amide bonds formed by the carboxyl groups of Phe, Tyr, and Trp.

Example 28.6

The following is a section of the gene coding for bovine rhodopsin along with several restriction endonucleases, their recognition sequences, and their hydrolysis sites. Which endonucleases will catalyze cleavage of this section of DNA?

5′-GCCGTCTACAACCCGGTCATCTACTATCATGATCAACAAGCAGTTCCGGAACT-3′

Enzyme	Recognition Sequence	Enzyme	Recognition Sequence
AluI	AG↓CT	HpaII	C↓CGG
BalI	TGG↓CCA	MboI	↓GATC
FnuDII	CG↓CG	NotI	GC↓GGCCGC
HeaIII	GG↓CC	SacI	GAGCT↓C

Solution

Only restriction endonucleases HpaII and MboI catalyze cleavage of this polynucleotide: HpaII at two sites and MboI at one site.

5′-GCCGTCTACAACCCGGTCATCTACTATCATGATCAACAAGCAGTTCCGGAACT-3′

Problem 28.6

The following is another section of the bovine rhodopsin gene. Which of the endonucleases given in Example 28.6 will catalyze cleavage of this section?

5′-ACGTCGGGTCGTCGTCCTCTCGCGGTGGTGAGTCTTCCGGCTCTTCT-3′

B. Methods for Sequencing Nucleic Acids

Any sequencing of DNA begins with site-specific cleavage of double-stranded DNA by one or more restriction endonucleases into smaller fragments called **restriction fragments.** Each restriction fragment is then sequenced separately, overlapping base sequences are identified, and the entire sequence of bases is then deduced.

Two methods for sequencing restriction fragments have been developed. The first of these, developed by Allan Maxam and Walter Gilbert and known as the **Maxam-Gilbert method,** depends on base-specific chemical cleavage. The second method, developed by Frederick Sanger and known as the **chain termination** or **dideoxy method,** depends on interruption of DNA-polymerase catalyzed synthesis. Sanger and Gilbert shared the 1980 Nobel Prize for biochemistry for their "development of chemical and biochemical analysis of DNA structure." Sanger's dideoxy method is currently more widely used, and it is on this method that we concentrate.

Sanger dideoxy method A method developed by Frederick Sanger for sequencing DNA molecules.

C. DNA Replication *in Vitro*

To appreciate the rationale for the dideoxy method, we must first understand certain aspects of the biochemistry of DNA replication. During replication, the sequence of nucleotides in one strand is copied as a complementary strand to form the second strand of a double-stranded DNA molecule. Synthesis of the complementary strand is catalyzed by the enzyme DNA polymerase. DNA polymerase will also carry out this synthesis *in vitro* using single-stranded DNA as a template, provided that both the four deoxynucleotide triphosphate (dNTP) monomers and a primer are present. A

Figure 28.11
DNA polymerase catalyzes the synthesis *in vitro* using single-stranded DNA as a template provided that both the four dNTP monomers and a primer are present. The primer provides a short stretch of double-stranded DNA by base-pairing with its complement on the single stranded DNA.

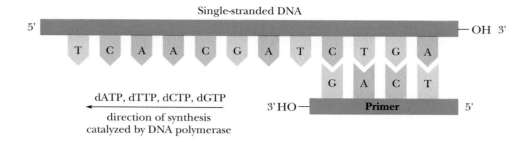

primer is an oligonucleotide capable of forming a short section of double-stranded DNA (dsDNA) by base-pairing with its complement on a single-stranded DNA (ssDNA). Because a new DNA strand grows from its 5′ to 3′ end, the primer must have a free 3′-OH group to which the first nucleotide of the growing chain is added (Figure 28. 11).

D. The Chain Termination or Dideoxy Method

The key to the chain termination method is the addition to the synthesizing medium of a 2′,3′-dideoxynucleoside triphosphate (ddNTP). Because a ddNTP has no —OH group at the 3′ position, it cannot serve as an acceptor for the next nucleotide to be added to the growing polynucleotide chain. Thus, chain synthesis is terminated at any point where a ddNTP becomes incorporated, hence the designation chain termination method.

A 2′,3′-dideoxynucleoside triphosphate
(ddNTP)

In the chain termination method, a single-stranded DNA of unknown sequence is mixed with primer and divided into four separate reaction mixtures. To each reaction mixture are added all four dNTPs, one of which is labeled in the 5′ phosphoryl group with ^{32}P so that the newly synthesized fragments can be visualized by autoradiography.

$$^{32}_{15}P \longrightarrow \: ^{32}_{16}S + \text{Beta particle} + \text{Gamma rays}$$

Also added to each reaction mixture are DNA polymerase and one of the four ddNTPs. The ratio of dNTPs to ddNTP in each reaction mixture is adjusted so that incorporation of a ddNTP takes place infrequently. In each reaction mixture, DNA synthesis takes place; however, in a given population of molecules, synthesis is interrupted at every possible site (Figure 28.12).

When gel electrophoresis of each reaction mixture is completed, a piece of x-ray film is placed over the gel, and gamma rays released by radioactive decay of ^{32}P darken the film and create a pattern on it that is an image of the resolved oligonucleotides. The base sequence of the complement to the original single-stranded template is then read directly from bottom to top of the developed film.

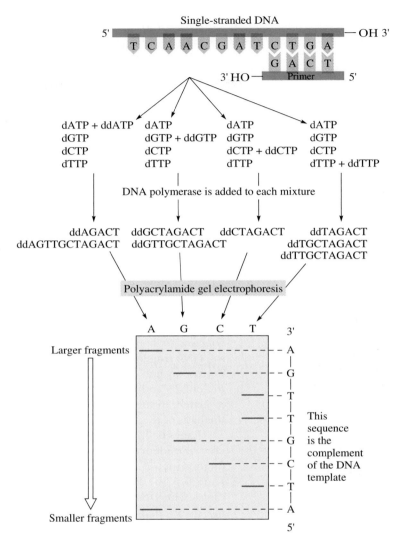

Figure 28.12
The chain termination or dideoxy method of DNA sequencing. The primer-DNA template is divided into four separate reaction mixtures. To each mixture is added the four dNTPs, DNA polymerase, and one of the four ddNTPs. Synthesis is interrupted at every possible site. The mixtures of oligonucleotides are separated by polyacrylamide gel electrophoresis. The base sequence of the DNA complement is read from the bottom to the top (from the 5′ end to the 3′ end) of the developed gel.

If the complement of the DNA template is 5′ A–T–C–G–T–T–G–A–3′
Then the original DNA template must be 5′–T–C–A–A–C–G–A–T–3′

A variation on this method is to use a single reaction mixture with each of the four ddNTPs labeled with a different fluorescent indicator. Each label is then detected by its characteristic spectrum. Automated DNA sequencing machines using this variation are capable of sequencing up to 10,000 base pairs per day.

E. Sequencing the Human Genome

As nearly everyone knows, the sequencing of the human genome was announced in the spring of 2000 by two competing groups, the Human Genome Project, a loosely linked consortium of publicly funded groups, and a private company called Celera Genomics. Actually, this milestone didn't represent a complete sequence, but a so-called rough draft, comprising about 85% of the entire genome. The methodology used is based on a refinement of the techniques described earlier using massively

CHEMICAL
CONNECTIONS

DNA Fingerprinting

Each human being has a genetic makeup consisting of approximately 3 billion base pairs of nucleotides, and, except for identical twins, the base sequence of DNA in one individual is different from that of every other individual. As a result, each person has a unique DNA fingerprint. To determine a DNA fingerprint, a sample of DNA from a trace of blood, skin, or other tissue is treated with a set of restriction endonucleases, and the

9 8 7 6 5 4 3 2 1

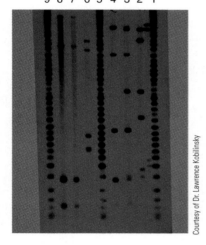

Courtesy of Dr. Lawrence Kobilinsky

■ A DNA "fingerprint."

5′ end of each restriction fragment is labeled with phosphorus-32. The resulting ^{32}P-labeled restriction fragments are then separated by polyacrylamide gel electrophoresis and visualized by placing a photographic plate over the developed gel.

In the DNA fingerprint patterns shown in the figure, lanes 1, 5, and 9 represent internal standards or control lanes. They contain the DNA fingerprint pattern of a standard virus treated with a standard set of restriction endonucleases. Lanes 2, 3, and 4 were used in a paternity suit. The DNA fingerprint of the mother in lane 4 contains five bands, which match with five of the six bands in the DNA fingerprint of the child in lane 3. The DNA fingerprint of the alleged father in lane 2 contains six bands, three of which match with bands in the DNA fingerprint of the child. Because the child inherits only half of its genes from the father, only half of the child's and father's DNA fingerprints are expected to match. In this instance, the paternity suit was won on the basis of the DNA fingerprint matching.

Lanes 6, 7, and 8 contain DNA fingerprint patterns used as evidence in a rape case. Lanes 7 and 8 are DNA fingerprints of semen obtained from the rape victim. Lane 6 is the DNA fingerprint pattern of the alleged rapist. The DNA fingerprint patterns of the semen do not match that of the alleged rapist and excluded the suspect from the case.

parallel separations of fragments by electrophoresis in capillary tubes. The Celera approach used some 300 of the fastest sequencing machines in parallel, each operating on many parallel DNA fragments. Supercomputers were used to assemble and compare millions of overlapping sequences. Figure 28.13 shows the apparatus used by the Celera group for this milestone achievement.

It is now much easier to obtain nucleic acid sequences than protein sequences. Sequencing a gene immediately gives the investigator the sequence of the protein produced by that gene because the code is known. Because many simpler organisms have been sequenced and the functions of many of the coded proteins are known, it is possible to take the sequence of a protein of unknown function and determine sequence homologies with the vast number of proteins of known function in a repository known as the Protein Data Bank, and thereby make an educated guess about the function of a protein the new gene codes for. This technology is causing a revolution in chemistry, biology, and medicine.

Figure 28.13
Photo of sequencing equipment.

© David Parker/Photo Researchers, Inc.

This achievement represents the beginning of a new era of molecular medicine, in which specific genetic deficiencies leading to inherited diseases will be understood on a molecular basis, and new therapies targeted at shutting down undesired genes or turning on desired ones will be developed. In addition, different individuals may respond differently to different drug therapies. In the future, medicines may be specifically tailored to the genetic makeup of individual patients.

Summary

Nucleic acids are composed of three types of monomer units: heterocyclic aromatic amine bases derived from purine and pyrimidine, the monosaccharides D-ribose or 2-deoxy-D-ribose, and phosphate ions (Section 28.1). A **nucleoside** is a compound containing D-ribose or 2-deoxy-D-ribose bonded to a heterocyclic aromatic amine base by a β-N-glycosidic bond. A **nucleotide** is a nucleoside in which a molecule of phosphoric acid is esterified with an —OH of the monosaccharide, most commonly either the 3'-OH or the 5'-OH. Nucleoside mono-, di-, and triphosphates are strong polyprotic acids and are extensively ionized at pH 7.0. At pH 7.0, adenosine triphosphate, for example, is a 50:50 mixture of ATP^{3-} and ATP^{4-}.

The **primary structure** of **deoxyribonucleic acids** (**DNA**) consists of units of 2-deoxyribose bonded by 3',5'-phosphodiester bonds (Section 28.2A). A heterocyclic aromatic amine base is attached to each deoxyribose unit by a β-N-glycosidic bond. The sequence of bases is read from the 5' end of the polynucleotide strand to the 3' end.

The heart of the **Watson-Crick model** is the postulate that a molecule of DNA consists of two antiparallel polynucleotide strands coiled in a right-handed manner about the same axis to form a **double helix** (Section 28.2B). Purine and pyrimidine bases point inward toward the axis of the helix and are always paired G-C and A-T. In **B-DNA**, base pairs are stacked one on top of another with a spacing of 340 pm and ten base pairs per

3400-pm helical repeat. In **A-DNA**, bases are stacked with a spacing of 290 pm between base pairs and 10 base pairs per 2900-pm helical repeat.

The **tertiary structure** of DNA is commonly referred to as **supercoiling** (Section 28.2C). **Circular DNA** is a type of double-stranded DNA in which the ends of each strand are joined by phosphodiester groups. Opening of one strand followed by partial unwinding and rejoining the ends introduces strain in the nonhelical gap. This strain can be spread over the entire molecule of circular DNA by introduction of **superhelical twists**. **Histones** are particularly rich in lysine and arginine and, therefore, have an abundance of positive charges. The association of DNA and histones produces a structure called **chromatin.**

There are two important differences between the primary structure of ribonucleic acids (RNA) and DNA (Section 28.3). (1) The monosaccharide unit in RNA is D-ribose. (2) Both RNA and DNA contain the purine bases adenine (A) and guanine (G), and the pyrimidine base cytosine (C). As the fourth base, however, RNA contains uracil (U), whereas DNA contains thymine (T).

The genetic code (Section 28.4) consists of nucleosides in groups of three; that is, it is a triplet code. Only 61 triplets code for amino acids; the remaining three code for termination of polypeptide synthesis.

Restriction endonucleases recognize a set pattern of four to eight nucleotides and cleave a DNA strand by hydrolysis of the linking phosphodiester bonds at any site that contains that particular sequence (Section 28.5A). In the **chain termination** or **dideoxy method** of DNA sequencing developed by Frederick Sanger (Section 28.5D), a primer-DNA template is divided into four separate reaction mixtures. To each is added the four dNTPs, one of which is labeled with ^{32}P. Also added are DNA polymerase and one of the four ddNTPs. Synthesis is interrupted at every possible site. The mixtures of newly synthesized oligonucleotides are separated by polyacrylamide gel electrophoresis and visualized by autoradiography. The base sequence of the DNA complement to the original DNA template is read from the bottom to the top (from the 5' end to the 3' end) of the developed photographic plate.

Problems

Nucleosides and Nucleotides

28.7 A pioneer in designing and synthesizing antimetabolites that could destroy cancer cells was George Hitchings at Burroughs Wellcome Company. In 1942 he initiated a program to discover DNA antimetabolites, and in 1948 he and Gertrude Elion synthesized 6-mercaptopurine, a successful drug for treating acute leukemia. Another DNA antimetabolite synthesized by Hutchings and Elion, was 6-thioguanine. Hitchings and Elion along with Sir James W. Black won the 1988 Nobel Prize for physiology or medicine for their discoveries of "important principles of drug treatment." In each drug, the oxygen at carbon 6 of the parent molecule is replaced by divalent sulfur. Draw structural formulas for the enethiol (the sulfur equivalent of an enol) forms of 6-mercaptopurine and 6-thioguanine.

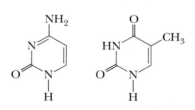

6-Mercaptopurine   6-Thioguanine

28.8 Following are structural formulas for cytosine and thymine. Draw two additional tautomeric forms for cytosine and three additional tautomeric forms for thymine.

Cytosine (C)   Thymine (T)

28.9 Draw a structural formula for a nucleoside composed of the following.

  (a) α-D-Ribose and adenine   **(b)** β-2-Deoxy-D-ribose and cytosine

28.10 Nucleosides are stable in water and in dilute base. In dilute acid, however, the glycosidic bond of a nucleoside undergoes hydrolysis to give a pentose and a heterocyclic aromatic amine base. Propose a mechanism for this acid-catalyzed hydrolysis.

28.11 Explain the difference in structure between a nucleoside and a nucleotide.

28.12 Draw a structural formula for each nucleotide, and estimate its net charge at pH 7.4, the pH of blood plasma.

(a) 2′-Deoxyadenosine 5′-triphosphate (dATP)
(b) Guanosine 3′-monophosphate (GMP)
(c) 2′-Deoxyguanosine 5′-diphosphate (dGDP)

28.13 Cyclic-AMP, first isolated in 1959, is involved in many diverse biological processes as a regulator of metabolic and physiological activity. In it, a single phosphate group is esterified with both the 3′ and 5′ hydroxyls of adenosine. Draw a structural formula of cyclic-AMP.

The Structure of DNA

28.14 Why are deoxyribonucleic acids called acids? What are the acidic groups in their structure?

28.15 Human DNA contains approximately 30.4% A. Estimate the percentages of G, C, and T and compare them with the values presented in Table 28.1.

28.16 Draw a structural formula of the DNA tetranucleotide 5′-A-G-C-T-3′. Estimate the net charge on this tetranucleotide at pH 7.0. What is the complementary tetranucleotide to this sequence?

28.17 List the postulates of the Watson-Crick model of DNA secondary structure.

28.18 The Watson-Crick model is based on certain experimental observations of base composition and molecular dimensions. Describe these observations and show how the Watson-Crick model accounts for each.

28.19 If you read J. D. Watson's account of the discovery of the structure of DNA, *The Double Helix*, you will find that for a time in their model-building studies, he and Crick were using alternative (and incorrect, at least in terms of their final model of the double helix) tautomeric structures for some of the heterocyclic bases.

(a) Write at least one alternative tautomeric structure for adenine.
(b) Would this structure still base-pair with thymine, or would it now base-pair more efficiently with a different base? If so, identify that base?

28.20 Compare the α-helix of proteins and the double helix of DNA in these ways.

(a) The units that repeat in the backbone of the polymer chain.
(b) The projection in space of substituents along the backbone (the R groups in the case of amino acids; purine and pyrimidine bases in the case of double-stranded DNA) relative to the axis of the helix.

28.21 Discuss the role of the hydrophobic interactions in stabilizing the following.

(a) Double-stranded DNA   (b) Lipid bilayers
(c) Soap micelles

28.22 Name the type of covalent bond(s) joining monomers in these biopolymers.

(a) Polysaccharides   (b) Polypeptides
(c) Nucleic acids

28.23 In terms of hydrogen bonding, which is more stable, an A-T base pair or a G-C base pair?

28.24 At elevated temperatures, nucleic acids become denatured, that is they unwind into single-stranded DNA. Account for the observation that the higher the G-C content of a nucleic acid, the higher the temperature required for its thermal denaturation.

28.25 Write the DNA complement for 5′-ACCGTTAAT-3′. Be certain to label which is the 5′ end and which is the 3′ end of the complement strand.

28.26 Write the DNA complement for 5′-TCAACGAT-3′.

Ribonucleic Acids

28.27 Compare the degree of hydrogen bonding in the base pair A-T found in DNA with that in the base pair A-U found in RNA.

28.28 Compare DNA and RNA is these ways.

(a) Monosaccharide units (b) Principal purine and pyrimidine bases
(c) Primary structure (d) Location in the cell
(e) Function in the cell

28.29 What type of RNA has the shortest lifetime in cells?

28.30 Write the mRNA complement for 5′-ACCGTTAAT-3′. Be certain to label which is the 5' end and which is the 3′ end of the mRNA strand.

28.31 Write the mRNA complement for 5′-TCAACGAT-3′.

The Genetic Code

28.32 What does it mean to say that the genetic code is degenerate?

28.33 Write the mRNA codons for the following.

(a) Valine (b) Histidine (c) Glycine

28.34 Aspartic acid and glutamic acid have carboxyl groups on their side chains and are called acidic amino acids. Compare the codons for these two amino acids.

28.35 Compare the structural formulas of the aromatic amino acids phenylalanine and tyrosine. Compare also the codons for these two amino acids.

28.36 Glycine, alanine, and valine are classified as nonpolar amino acids. Compare their codons. What similarities do you find? What differences do you find?

28.37 Codons in the set CUU, CUC, CUA, and CUG all code for the amino acid leucine. In this set, the first and second bases are identical, and the identity of the third base is irrelevant. For what other sets of codons is the third base also irrelevant, and for what amino acid(s) does each set code?

28.38 Compare the codons with a pyrimidine, either U or C, as the second base. Do the majority of the amino acids specified by these codons have hydrophobic or hydrophilic side chains?

28.39 Compare the codons with a purine, either A or G, as the second base. Do the majority of the amino acids specified by these codons have hydrophilic or hydrophobic side chains?

28.40 What polypeptide is coded for by this mRNA sequence?

5′-GCU-GAA-GUC-GAG-GUG-UGG-3′

28.41 The alpha chain of human hemoglobin has 141 amino acids in a single polypeptide chain. Calculate the minimum number of bases on DNA necessary to code for the alpha chain. Include in your calculation the bases necessary for specifying termination of polypeptide synthesis.

28.42 In HbS, the human hemoglobin found in individuals with sickle-cell anemia, glutamic acid at position 6 in the beta chain is replaced by valine.

(a) List the two codons for glutamic acid and the four codons for valine.
(b) Show that one of the glutamic acid codons can be converted to a valine codon by a single substitution mutation, that is by changing one letter in one codon.

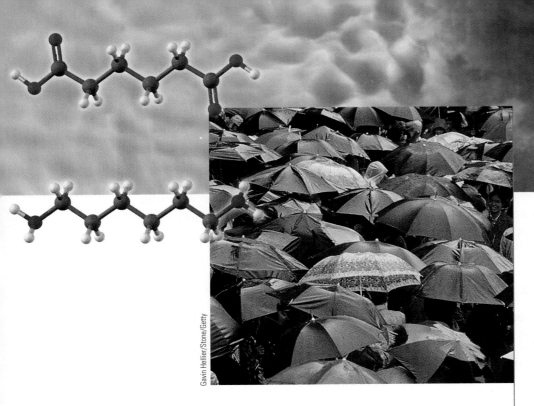

Gavin Hellier/Stone/Getty

■ Sea of umbrellas on a rainy day in Shanghai, China. Inset: Models of adipic acid and hexamethylenediamine, the two monomers of nylon 66.

Organic Polymer Chemistry

The technological advancement of any society is inextricably tied to the materials available to it. Indeed, historians have used the emergence of new materials as a way of establishing a timeline to mark the development of human civilization. As part of the search to discover new materials, scientists have made increasing use of organic chemistry for the preparation of synthetic polymers. The versatility afforded by these polymers allows for the creation and fabrication of materials with ranges of properties unattainable using such materials as wood, metals, and ceramics. Deceptively simple changes in the chemical structure of a given polymer, for example, can change its mechanical properties from those of a sandwich bag to those of a bulletproof vest. Furthermore, structural changes can introduce properties never before imagined in organic polymers. For example, using well-defined organic reactions, one type of polymer can be made into an insulator (for example, the rubber sheath that surrounds electrical cords), or, if treated differently, it can be made into an electrical conductor with a conductivity nearly equal to that of metallic copper!

The years since the 1930s have seen extensive research and development in polymer chemistry, and an almost explosive growth in plastics, coatings, and rubber technology has created a worldwide multibillion-

ORGANIC
Chemistry🞂Now™
Look for this logo in the chapter and go to Organic ChemistryNow at
http://now.brookscole.com/bfi4
for tutorials, simulations, and problems.

dollar industry. A few basic characteristics account for this phenomenal growth. First, the raw materials for synthetic polymers are derived mainly from petroleum. With the development of petroleum-refining processes, raw materials for the synthesis of polymers became generally cheap and plentiful. Second, within broad limits, scientists have learned how to tailor polymers to the requirements of the end use. Third, many consumer products can be fabricated more cheaply from synthetic polymers than from competing materials, such as wood, ceramics, and metals. For example, polymer technology created the water-based (latex) paints that have revolutionized the coatings industry; plastic films and foams have done the same for the packaging industry. The list could go on and on as we think of the manufactured items that are everywhere around us in our daily lives.

29.1 The Architecture of Polymers

Polymer From the Greek, *poly + meros*, many parts. Any long-chain molecule synthesized by linking together many single parts called monomers.

Monomer From the Greek, *mono + meros*, single part. The simplest nonredundant unit from which a polymer is synthesized.

Plastic A polymer that can be molded when hot and retains its shape when cooled.

Thermoplastic A polymer that can be melted and molded into a shape that is retained when it is cooled.

Thermoset plastic A polymer that can be molded when it is first prepared, but once cooled, hardens irreversibly and cannot be remelted.

Polymers (Greek: *poly + meros*, many parts) are long-chain molecules synthesized by linking **monomers** (Greek: *mono + meros*, single part) through chemical reactions. The molecular weights of polymers are generally high compared with those of common organic compounds and typically range from 10,000 g/mol to more than 1,000,000 g/mol. The architectures of these macromolecules can also be quite diverse. Types of polymer architecture include linear and branched chains as well as those with comb, ladder, and star structures (Figure 29.1). Additional structural variations can be achieved by introducing covalent cross links between individual polymer chains.

Linear and branched polymers are often soluble in solvents, such as chloroform, benzene, toluene, DMSO, and THF. In addition, many linear and branched polymers can be melted to form highly viscous liquids. In polymer chemistry, the term **plastic** refers to any polymer that can be molded when hot and retains its shape when cooled. **Thermoplastics** are polymers that can be melted and become sufficiently fluid that they can be molded into shapes that are retained when they are cooled. **Thermosetting plastics** or thermosets can be molded when they are first prepared, but once they cool, they harden irreversibly and cannot be remelted. Because of these very different physical characteristics, thermoplastics and thermosets must be processed differently and are used in very different applications.

The single most important property of polymers at the molecular level is the size and shape of their chains. A good example of the importance of size is a comparison of paraffin wax, a natural polymer, and polyethylene, a synthetic polymer. These two distinct materials have identical repeat units, namely —CH_2—, but differ greatly in chain size. Paraffin wax has between 25 and 50 carbon atoms per chain, whereas polyethylene has between 1000 and 3000 carbon atoms per chain. Paraffin wax, such as in birthday candles, is soft and brittle but polyethylene, such as in plastic beverage

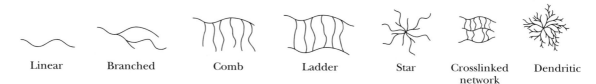

| Linear | Branched | Comb | Ladder | Star | Crosslinked network | Dendritic |

Figure 29.1
Various polymer architectures.

bottles, is strong, flexible, and tough. These vastly different properties arise directly from the difference in size and molecular architecture of the individual polymer chains.

29.2 Polymer Notation and Nomenclature

We show the structure of a polymer by placing parentheses around the **repeat unit,** which is the smallest molecular fragment that contains all the nonredundant structural features of the chain. Thus, the structure of an entire polymer chain can be reproduced by repeating the enclosed structure in both directions. A subscript n, called the **average degree of polymerization,** is placed outside the parentheses to indicate that this unit is repeated n times.

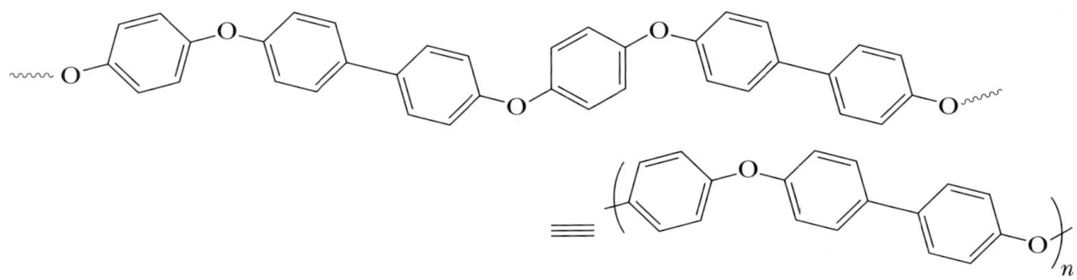

The polymers formed from symmetric monomer units, such as polyethylene, $+CH_2CH_2 \rightarrow_{\overline{n}}$, and polytetrafluoroethylene, $+CF_2CF_2 \rightarrow_{\overline{n}}$, are an exception to this notation. Although the simplest repeat units are the $-CH_2-$ and $-CF_2-$ groups, respectively, we show two methylene groups and two difluoromethylene groups because they originate from ethylene $(CH_2{=}CH_2)$ and tetrafluoroethylene $(CF_2{=}CF_2)$, the monomer units from which these polymers are derived.

The most common method of naming a polymer is to attach the prefix *poly-* to the name of the monomer from which the polymer is derived, as for example polyethylene and polystyrene. In the case of a more complex monomer, or where the name of the monomer is more than one word, as for example the monomer vinyl chloride, parentheses are used to enclose the name of the monomer.

| Polystyrene | 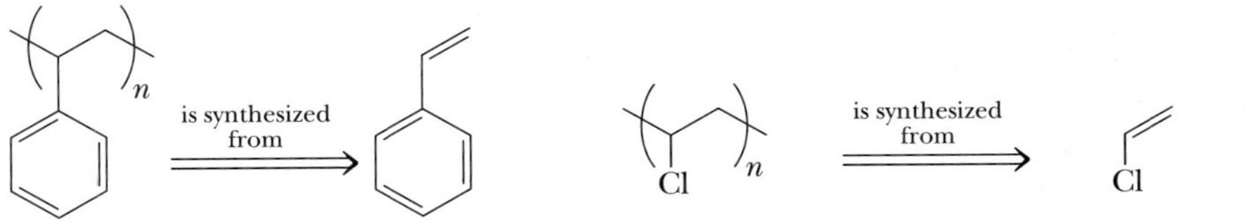 is synthesized from | Styrene | Poly(vinyl chloride) (PVC) | is synthesized from | Vinyl chloride |

Example 29.1

Given the following structure, determine the polymer's repeat unit, redraw the structure using the simplified parenthetical notation, and name the polymer.

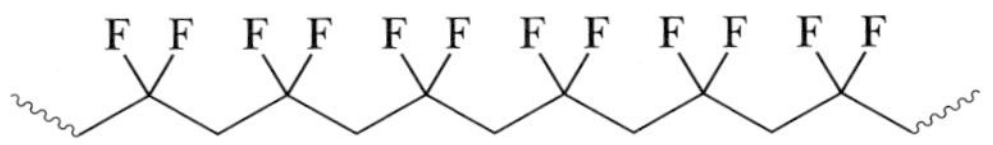

Solution

The repeat unit is —CH₂CF₂— and the polymer is written $+CH_2CF_2\rightarrow_n$. The repeat unit is derived from 1,1-difluoroethylene, and the polymer is named poly(1,1-difluoroethylene). This polymer is used in microphone diaphragms.

Problem 29.1

Given the following structure, determine the polymer's repeat unit, redraw the structure using the simplified parenthetical notation, and name the polymer.

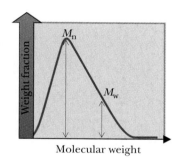

Figure 29.2
The distribution of molecular weights in a given polymer sample.

29.3 Molecular Weights of Polymers

All synthetic polymers and most naturally occurring polymers are mixtures of individual polymer molecules of variable molecular weights. When defining molecular weights in polymer chemistry, the two most common definitions are the number average and weight average molecular weights. The **number average molecular weight, M_n,** is calculated by counting the number of polymer chains of a particular molecular weight, multiplying each number by the molecular weight of its chain, summing these values, and dividing by the total number of polymer chains. The **weight average molecular weight, M_w,** is calculated by recording the total weight of each chain of a particular length, summing these weights, and dividing by the total weight of the sample. Because the larger chains in a sample weigh more than the smaller chains, the weight average molecular weight is skewed to higher values, and M_w is always greater than M_n (Figure 29.2).

Both M_n and M_w are useful values, and their ratio, M_w/M_n, called the **polydispersity index,** provides a measure of the breadth of the molecular-weight distribution. When the M_w/M_n ratio is equal to one, all the polymer molecules in a sample are the same length, and the polymer is said to be **monodisperse.** No synthetic polymers are ever monodisperse unless the individual molecules are carefully fractionated using time-consuming, rigorous separation techniques based on molecular size. On the other hand, natural polymers, such as polypeptides and DNA, that are formed using biological processes are monodisperse polymers.

29.4 Polymer Morphology—Crystalline Versus Amorphous Materials

Crystalline domain An ordered crystalline region in the solid state of a polymer. Also called crystallites.

Amorphous domain A disordered, noncrystalline region in the solid state of a polymer.

Polymers, like small organic molecules, tend to crystallize upon precipitation or as they are cooled from a melt. Acting to inhibit this tendency are their very large molecules, which tend to slow diffusion, and their sometimes complicated or irregular structures, which prevent efficient packing of the chains. The result is that polymers in the solid state tend to be composed of both ordered **crystalline domains** (crystallites) and disordered **amorphous domains.** The relative amounts of crystalline and amorphous domains differ from polymer to polymer and often depend upon the manner in which the material is processed.

High degrees of crystallinity are most often found in polymers with regular, compact structures and strong intermolecular forces, such as hydrogen bonding and dipolar interactions. The temperature at which crystallites melt corresponds to the **melt transition (T_m)** of the polymer. As the degree of crystallinity of a polymer increases, its T_m increases, and it becomes more opaque owing to scattering of the light by the crystalline domains. There is also a corresponding increase in strength and stiffness with increase in crystallinity. For example, poly(6-amino-hexanoic acid) has a $T_m = 223°C$. At and well above room temperature, this polymer is a hard durable material that does not undergo any appreciable change in properties even on a very hot summer afternoon. Its uses range from textile fibers to shoe heels.

Amorphous domains are characterized by the absence of long-range order. Highly amorphous polymers are sometimes referred to as glassy polymers. Because they lack crystalline domains that scatter light, amorphous polymers are transparent. In addition, they are typically weak polymers, in terms both of their greater flexibility and smaller mechanical strength. On being heated, amorphous polymers are transformed from a hard glass to a soft, flexible, rubbery state. The temperature at which this transition occurs is called the **glass transition temperature (T_g).** Amorphous polystyrene, for example, has a $T_g = 100°C$. At room temperature, it is a rigid solid used for drinking cups, foamed packaging materials, disposable medical wares, tape reels, and so forth. If it is placed in boiling water, it becomes soft and rubbery.

This relationship between mechanical properties and the degree of crystallinity can be illustrated by poly(ethylene terephthalate) (PET).

Poly(ethylene terephthalate) (PET)

PET can be made with a percent of crystalline domains ranging from 0% to about 55%. Completely amorphous PET is formed by cooling the melt quickly. By prolonging the cooling time, more molecular diffusion occurs, and crystallites form as the chains become more ordered. The differences in mechanical properties between these forms of PET are substantial. PET with a low degree of crystallinity is used for plastic beverage bottles, whereas fibers drawn from highly crystalline PET are used for textile fibers and tire cords.

Rubber materials must have low T_g values to behave as **elastomers (elastic polymers).** If the temperature drops below its T_g value, then the material is converted to a rigid glassy solid, and all elastomeric properties are lost. A poor understanding of this behavior of polymers contributed to the *Challenger* spacecraft disaster in 1985. The elastomeric O-rings used to seal the solid booster rockets had a T_g value around 0°C. When the temperature dropped to an unanticipated low on the morning of the *Challenger* launch, the O-ring seals dropped below their T_g value and consequently changed from elastomers to rigid glasses, losing any sealing capabilities. The rest is tragic history. The physicist Richard Feynman sorted this out publicly in a famous televised hearing in which he put a *Challenger*-type O-ring in ice water and showed that its elasticity was lost!

Melt transition, T_m The temperature at which crystalline regions of a polymer melt.

Glass transition temperature, T_g The temperature at which a polymer undergoes the transition from a hard glass to a rubbery state.

Elastomer A material that, when stretched or otherwise distorted, returns to its original shape when the distorting force is released.

29.5 Step-Growth Polymerizations

Step-growth polymerization A polymerization in which chain growth occurs in a stepwise manner between difunctional monomers as, for example, between adipic acid and hexamethylenediamine to form nylon 66. Also called condensation polymerization.

Condensation polymerization A polymerization in which chain growth occurs in a stepwise manner between difunctional monomers. Also called step-growth polymerization.

Polymerizations in which chain growth occurs in a stepwise manner are called **step-growth** or **condensation polymerizations.** Step-growth polymers are formed by reaction between difunctional molecules, with each new bond created in a separate step. During polymerization, monomers react with monomers to form dimers, dimers react with dimers to form tetramers, tetramers react with monomers to form pentamers, and so on. This stepwise construction of polymer chains has important consequences for both their molecular weights and molecular-weight distributions. Probability tells us that the most abundant species tend to co-condense. Thus, at the early stages of polymerization, small chains are most likely to react with monomers or other small chains to generate many low-molecular-weight oligomers rather than a small number of high-molecular-weight polymers. This tendency persists until most monomer units are used up. As a result, high-molecular-weight polymers are not produced until very late in the reaction, typically past 99% conversion of monomers to higher molecular-weight chains. Only at this point is there the probability of larger chains reacting with one another to form high-molecular-weight polymer molecules. This restriction points to an important distinction between small-molecule organic reactions and step-growth polymerizations. Although a reaction that typically yields 85% of the desired product is considered "good" in organic synthesis, the same reaction is essentially useless for step-growth polymerizations because high-molecular-weight polymers are rarely formed at such low conversions.

There are two common types of step-growth processes: (1) reaction between A-A and B-B type monomers to give $+$(A-A-B-B$)_n$ polymers and (2) the self-condensation of A-B monomers to give $+$(A-B$)_n$ polymers. In each case, an A functional group reacts exclusively with a B functional group, and a B functional group reacts exclusively with an A functional group. New covalent bonds in step-growth polymerizations are generally formed by polar reactions, as for example nucleophilic acyl substitution. In this section, we discuss five types of step-growth polymers: polyamides, polyesters, polycarbonates, polyurethanes, and epoxy resins.

A. Polyamides

Polyamide A polymer in which each monomer unit is joined to the next by an amide bond, as, for example, nylon 66.

In the years following World War I, a number of chemists recognized the need for developing a basic knowledge of polymer chemistry. One of the most creative of these was Wallace M. Carothers. In the early 1930s, Carothers and his associates at E. I. DuPont de Nemours & Company began fundamental research into the reactions of aliphatic dicarboxylic acids and diols. From adipic acid and ethylene glycol, they obtained a polyester of high molecular weight that could be drawn into fibers.

Hexanedioic acid   1,2-Ethanediol   Poly(ethylene adipate)
(Adipic acid)   (Ethylene glycol)

These first polyester fibers had melt transitions (T_m) too low for use as textile fibers, and they were not investigated further. Carothers then turned his attention to the reactions of dicarboxylic acids and diamines to form polyamides and, in 1934, synthesized nylon 66, the first purely synthetic fiber. Nylon 66 is so named because it is synthesized from two different monomers, each containing six carbon atoms.

In the synthesis of nylon 66, hexanedioic acid (adipic acid) and 1,6-hexanedi-amine (hexamethylenediamine) are dissolved in aqueous ethanol where they react to form a one-to-one salt called nylon salt. Nylon salt is then heated in an autoclave to 250°C where the internal pressure rises to about 15 atm. Under these conditions, —COO⁻ groups from adipic acid and —NH₃⁺ groups from hexamethylenediamine react with loss of H_2O to form amide groups.

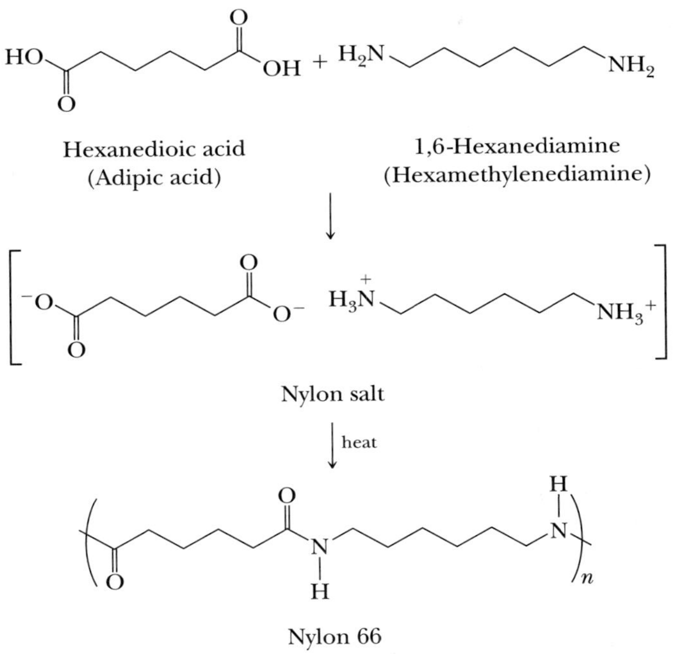

Hexanedioic acid
(Adipic acid)

1,6-Hexanediamine
(Hexamethylenediamine)

Nylon salt

heat

Nylon 66

Nylon 66 formed under these conditions has a T_m of 250–260°C and has a molecular-weight range of 10,000 to 20,000 g/mol.

In the first stage of fiber production, crude nylon 66 is melted, spun into fibers, and cooled. Next, the melt-spun fibers are **cold-drawn** (drawn at room temperature) to about four times their original length to increase their degree of crystallinity. As the fibers are drawn, individual polymer molecules become oriented in the direction of the fiber axis, and hydrogen bonds form between carbonyl oxygens of one chain and amide hydrogens of another chain (Figure 29.3).

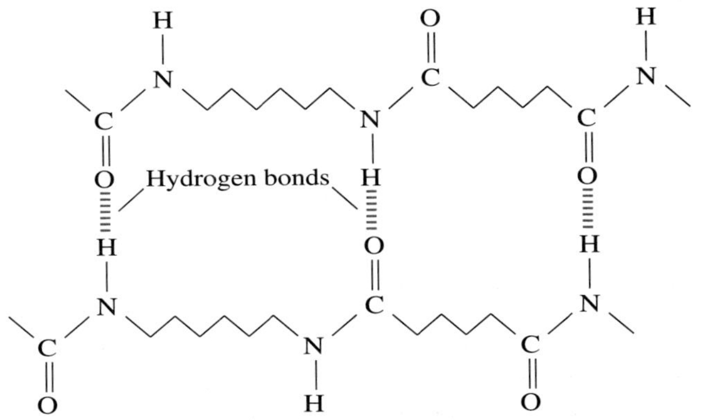

Figure 29.3

The structure of cold-drawn nylon 66. Hydrogen bonds between adjacent polymer chains provide additional tensile strength and stiffness to the fibers.

The effects of orientation of polyamide molecules on the physical properties of the fiber are dramatic; both tensile strength and stiffness are increased markedly. Cold-drawing is an important step in the production of most synthetic fibers.

The current raw material base for the production of nylon 66 is benzene, which is derived almost entirely from catalytic cracking and reforming of petroleum. Catalytic reduction of benzene to cyclohexane followed by catalyzed air oxidation gives a mixture of cyclohexanol and cyclohexanone. Oxidation of this mixture by nitric acid gives adipic acid.

Benzene $\xrightarrow[\text{catalyst}]{3H_2}$ Cyclohexane $\xrightarrow[\text{catalyst}]{O_2}$ [Cyclohexanol + Cyclohexanone] $\xrightarrow{HNO_3}$ Hexanedioic acid (Adipic acid)

Adipic acid, in turn, is a starting material for the synthesis of hexamethylenediamine. Treatment of adipic acid with ammonia gives an ammonium salt, which, when heated, gives adipamide. Catalytic reduction of adipamide gives hexamethylenediamine. Thus, carbon sources for the production of nylon 66 are derived entirely from petroleum, which unfortunately is not a renewable resource.

Ammonium hexanedioate (Ammonium adipate) $\xrightarrow{\text{heat}}$ Hexanediamide (Adipamide) $\xrightarrow[\text{catalyst}]{4H_2}$

1,6-Hexanediamine (Hexamethylenediamine)

The nylons are a family of polymers, the members of which have subtly different properties that suit them to one use or another. The two most widely used members of this family are nylon 66 and nylon 6. Nylon 6 is so named because it is synthesized from caprolactam, a six-carbon monomer. In the synthesis of nylon 6, caprolactam is partially hydrolyzed to 6-aminohexanoic acid and then heated to 250°C to bring about polymerization. Nylon 6 is fabricated into fibers, brush bristles, rope, high-impact moldings, and tire cords.

Caprolactam $\xrightarrow[\text{2. heat}]{\text{1. partial hydrolysis}}$ Nylon 6

Charles D. Winters

■ Bulletproof vests have a thick layer of Kevlar.

Based on extensive research into relationships between molecular structure and bulk physical properties, scientists at DuPont reasoned that a polyamide containing aromatic rings would be stiffer and stronger than either nylon 66 or nylon 6. In early

1960, DuPont introduced Kevlar, a polyaromatic amide (**aramid**) fiber synthesized from terephthalic acid and *p*-phenylenediamine.

Aramid A polyaromatic amide; a polymer in which the monomer units are an aromatic diamine and an aromatic dicarboxylic acid.

$$n\text{HOC}\text{---}\text{COH} + n\text{H}_2\text{N}\text{---}\text{NH}_2 \longrightarrow \left(\text{C}\text{---}\text{CNH}\text{---}\text{NH}\right)_n$$

1,4-Benzenedicarboxylic acid
(Terephthalic acid)

1,4-Benzenediamine
(*p*-Phenylenediamine)

Kevlar

One of the remarkable features of Kevlar is its light weight compared with other materials of similar strength. For example, a 3-in. cable woven of Kevlar has a strength equal to that of a similarly woven 3-in. steel cable. Whereas the steel cable weighs about 20 lb/ft, the Kevlar cable weighs only 4 lb/ft. Kevlar now finds use in such articles as anchor cables for offshore drilling rigs and reinforcement fibers for automobile tires. Kevlar is also woven into a fabric that is so tough that it can be used for bulletproof vests, jackets, and raincoats.

B. Polyesters

Recall that, in the early 1930s, Carothers and his associates had concluded that polyester fibers from aliphatic dicarboxylic acids and ethylene glycol were not suitable for textile use because their melting points are too low. Winfield and Dickson at the Calico Printers Association in England further investigated polyesters in the 1940s and reasoned that a greater resistance to rotation in the polymer backbone would stiffen the polymer, raise its melting point, and thereby lead to a more acceptable polyester fiber. To create stiffness in the polymer chain, they used 1,4-benzenedicarboxylic acid (terephthalic acid). Polymerization of this aromatic dicarboxylic acid with ethylene glycol gives poly(ethylene terephthalate), abbreviated PET (also PETE).

Polyester A polymer in which each monomer unit is joined to the next by an ester bond, as, for example, poly(ethylene terephthalate).

$$\text{HO}\text{---}\text{OH} + \text{HO}\diagdown\diagup\text{OH} \xrightarrow{\text{heat}} \left(\text{---}\text{O}\diagdown\diagup\text{O}\right)_n + 2n\text{H}_2\text{O}$$

1,4-Benzenedicarboxylic acid
(Terephthalic acid)

1,2-Ethanediol
(Ethylene glycol)

Poly(ethylene terephthalate)
(Dacron, Mylar)

The crude polyester can be melted, extruded, and then cold-drawn to form the textile fiber Dacron polyester, outstanding features of which are its stiffness (about four times that of nylon 66), very high strength, and remarkable resistance to creasing and wrinkling. Because the early Dacron polyester fibers were harsh to the touch owing to their stiffness, they were usually blended with cotton or wool to make acceptable textile fibers. Newly developed fabrication techniques now produce less harsh Dacron polyester textile fibers. PET is also fabricated into Mylar films and recyclable plastic beverage containers.

Ethylene glycol for the synthesis of PET is obtained by air oxidation of ethylene to ethylene oxide (Section 11.8A) followed by hydrolysis to the glycol (Section 11.9A). Ethylene is, in turn, derived entirely from cracking either petroleum or ethane derived from natural gas (Section 2.10A). Terephthalic acid is obtained by oxidation of *p*-xylene, an aromatic hydrocarbon obtained along with benzene and toluene from catalytic cracking and reforming of naphtha and other petroleum fractions (Section 2.10B).

Charles D. Winters

■ Mylar can be made into extremely strong films. Because the film has very tiny pores, it is used for balloons that can be inflated with helium; the helium atoms diffuse only slowly through the pores of the film.

$$CH_2{=}CH_2 \xrightarrow[\text{catalyst}]{O_2} \underset{\text{Oxirane}}{CH_2{-}CH_2} \xrightarrow{H^+, H_2O} \underset{\text{1,2-Ethanediol}}{HOCH_2CH_2OH}$$

Ethylene   Oxirane   1,2-Ethanediol
                           (Ethylene oxide)   (Ethylene glycol)

$$H_3C{-}\langle\;\rangle{-}CH_3 \xrightarrow[\text{catalyst}]{O_2} HOC{-}\langle\;\rangle{-}COH$$

p-Xylene   Terephthalic acid

C. Polycarbonates

Polycarbonates, the most familiar of which is Lexan, are a class of commercially important engineering polyesters. In the production of Lexan, an aqueous solution of the disodium salt of bisphenol A (Problem 22.22) is brought into contact with a solution of phosgene dissolved in dichloromethane. The two solutions are immiscible, and no reaction occurs until tetrabutylammonium chloride or other phase-transfer catalyst (Section 9.10) is added. The tetrabutylammonium cation carries the bisphenol A anion into the dichloromethane phase where it reacts smoothly with phosgene to form the polymer. The tetrabutylammonium ion then carries chloride ion back to the aqueous phase.

$$Na^+{}^-O{-}\langle\;\rangle{-}\underset{CH_3}{\overset{CH_3}{C}}{-}\langle\;\rangle{-}O^-Na^+ + Cl{-}\overset{O}{\overset{\|}{C}}{-}Cl \longrightarrow \left(\langle\;\rangle{-}\underset{CH_3}{\overset{CH_3}{C}}{-}\langle\;\rangle{-}O{-}\overset{O}{\overset{\|}{C}}{-}O\right)_n + 2n\,NaCl$$

Disodium salt   Phosgene   Lexan
of bisphenol A   (a polycarbonate)

Lexan is a tough, transparent polymer with high-impact and tensile strengths, and it retains its properties over a wide temperature range. It has found significant use in sporting equipment, such as bicycle, football, motorcycle, and snowmobile helmets as well as hockey and baseball catchers' face masks. In addition it is used to make light, impact-resistant housings for household appliances and automobile and aircraft equipment and to manufacture safety glass and unbreakable windows.

■ A polycarbonate hockey mask.

D. Polyurethanes

A urethane, or carbamate, is an ester of carbamic acid, H_2NCOOH. Carbamates are most commonly prepared by treatment of an isocyanate with an alcohol.

$$RN{=}C{=}O + R'OH \longrightarrow RNH\overset{O}{\overset{\|}{C}}OR'$$

An isocyanate   A carbamate

Polyurethanes consist of flexible polyester or polyether units (blocks) alternating with rigid urethane units (blocks). The rigid urethane blocks are derived from a diisocyanate, commonly a mixture of 2,4- and 2,6-toluene diisocyanate. The more flexible blocks are derived from low-molecular-weight (MW 1000–4000) polyesters or polyethers with —OH groups at each end of the polymer chain. Polyurethane fibers are fairly soft and elastic and have found use as Spandex and Lycra, the "stretch" fabrics used in bathing suits, leotards, and undergarments.

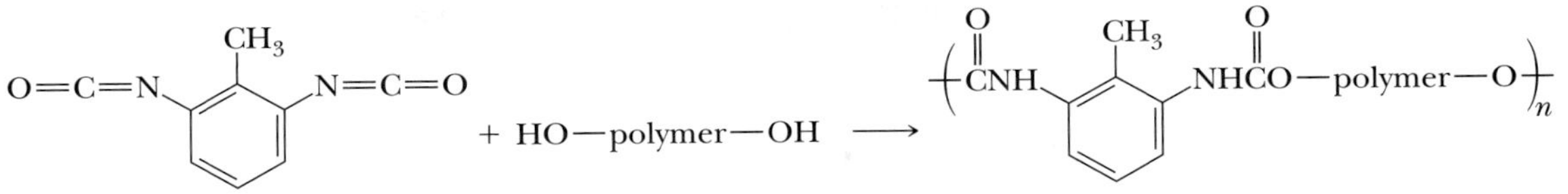

2,6-Toluene diisocyanate   Low-molecular-weight   A polyurethane
                                  polyester or polyether

Polyurethane foams for upholstery and insulating materials are made by adding small amounts of water during polymerization. Water reacts with isocyanate groups to produce gaseous carbon dioxide, which then acts as the foaming agent.

An isocyanate   A carbamic acid
                       (unstable)

E. Epoxy Resins

Epoxy resins are materials prepared by a polymerization in which one monomer contains at least two epoxy groups. Within this range, there are a large number of polymeric materials possible, and epoxy resins are produced in forms ranging from low-viscosity liquids to high-melting solids. The most widely used epoxide monomer is the diepoxide prepared by treatment of one mole of bisphenol A (Problem 22.22) with two moles of epichlorohydrin (Section 11.10). To prepare the following epoxy resin, the diepoxide monomer is treated with 1,2-ethanediamine (ethylene diamine). This component is usually called the "catalyst" in the two-component formulations that can be bought in any hardware store; it is also the component with the pungent smell. It is not a catalyst but a reagent.

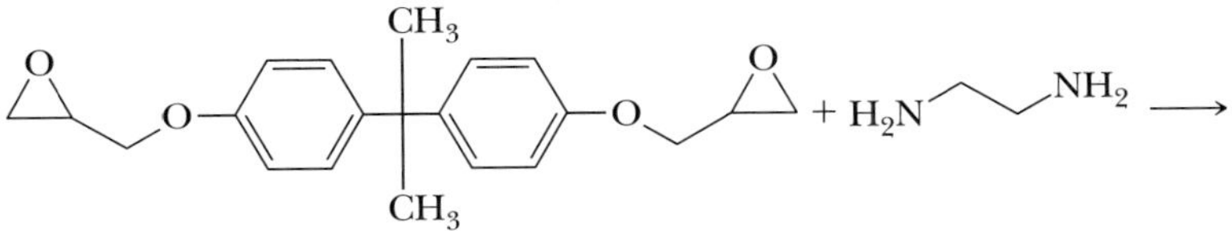

A diepoxide   A diamine

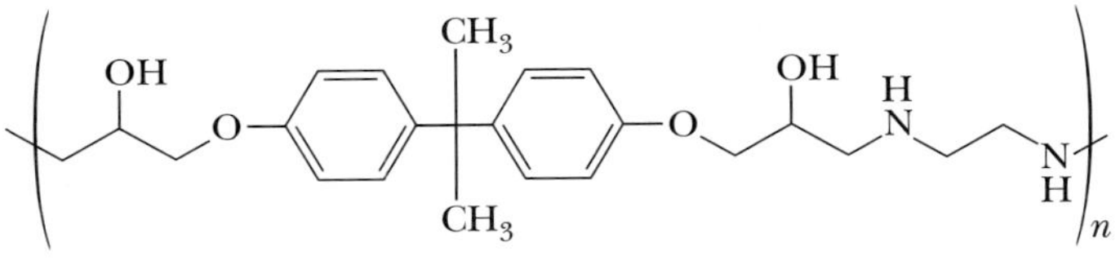

An epoxy resin

Epoxy resins are widely used as adhesives and insulating surface coatings. They have good electrical insulating properties, which leads to their use for encapsulating electrical components ranging from integrated circuit boards to switch coils and insulators for power transmission systems. They are also used as composites with other materials, such as glass fiber, paper, metal foils, and other synthetic fibers to create structural components for jet aircraft, rocket motor casings, and so on.

Example 29.2

Write a mechanism for the acid-catalyzed polymerization of 1,4-benzenediisocyanate and ethylene glycol. To simplify your mechanism, consider only the reaction of one —NCO group with one —OH group.

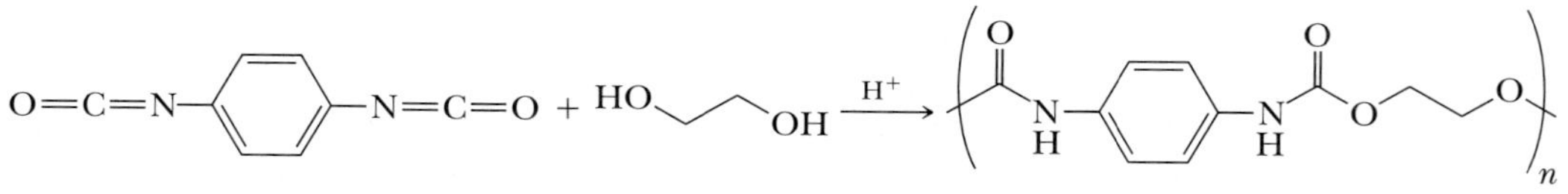

1,4-Benzenediisocyanate   Ethylene glycol   A polyurethane

Solution

A mechanism is shown in three steps. Proton transfer in Step 1 from the acid, HA, to nitrogen followed by addition of ROH to the carbonyl carbon in Step 2 gives an oxonium ion. Proton transfer in Step 3 from the oxonium ion to A^- gives the carbamate ester.

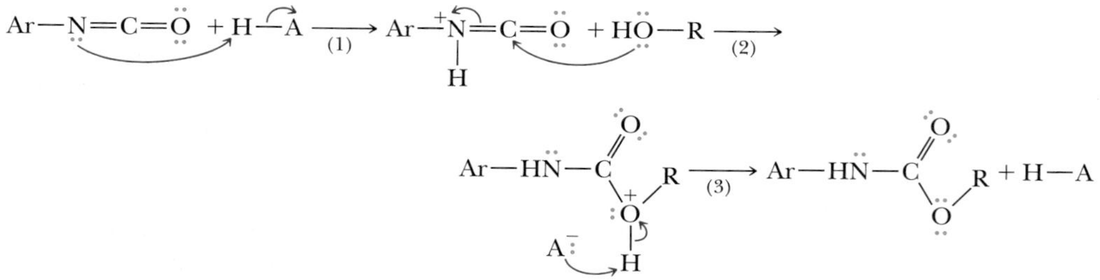

Problem 29.2

Write the repeating unit of the polymer formed from the following reaction and propose a mechanism for its formation.

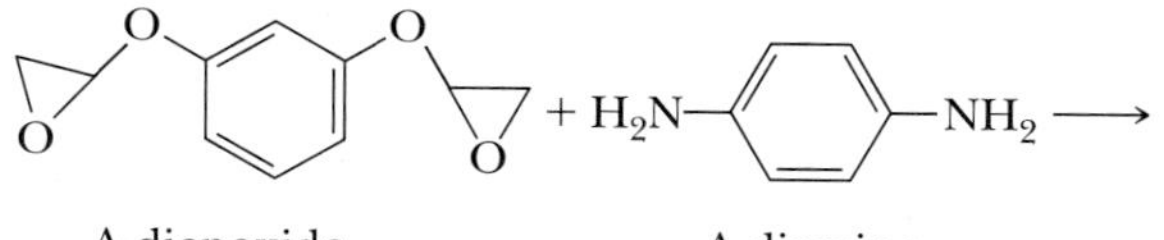

A diepoxide   A diamine

F. Thermosetting Polymers

Thermosetting polymers are composed of long chains that are cross linked by covalent bonds. In effect, a thermosetting polymer is one giant molecule. The first thermosetting polymer was produced by Leo Baekeland (1863–1944) in 1907 by reacting phenol with formaldehyde to form the following three-dimensional structure. The product, known as Bakelite, is a good electrical insulator.

Stitches That Dissolve

Medical science has advanced very rapidly in the last few decades. Some procedures considered routine today, such as organ transplantation and the use of lasers in surgery, were unimaginable 60 years ago. As the technological capabilities of medicine have grown, the demand for synthetic materials that can be used inside the body has increased as well. Polymers have many of the characteristics of an ideal biomaterial: They are lightweight and strong, are inert or biodegradable depending on their chemical structure, and have physical properties (softness, rigidity, elasticity) that are easily tailored to match those of natural tissues. Carbon-carbon backbone polymers are degradation resistant and are used widely in permanent organ and tissue replacements.

Whereas most medical uses of polymeric materials require biostability, applications that use the biodegradable nature of some macromolecules have been developed. An example is the use of poly(glycolic acid) and glycolic acid/lactic acid copolymers as absorbable sutures, which go under the trade name Lactomer.

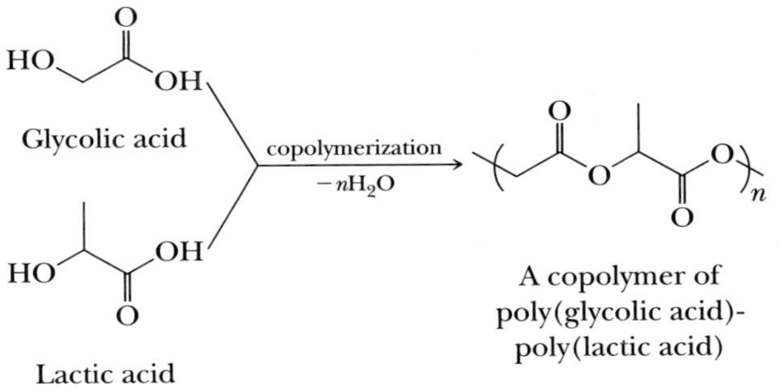

Glycolic acid

Lactic acid

copolymerization
$-n\mathrm{H_2O}$

A copolymer of
poly(glycolic acid)-
poly(lactic acid)

Traditional suture materials such as catgut must be removed by a health care specialist after they have served their purpose. Stitches of Lactomer, however, are hydrolyzed slowly over a period of approximately two weeks, and by the time the torn tissues have fully healed, the stitches have fully degraded, and no suture removal is necessary. Glycolic and lactic acids formed during hydrolysis of the stitches are metabolized and excreted by existing biochemical pathways.

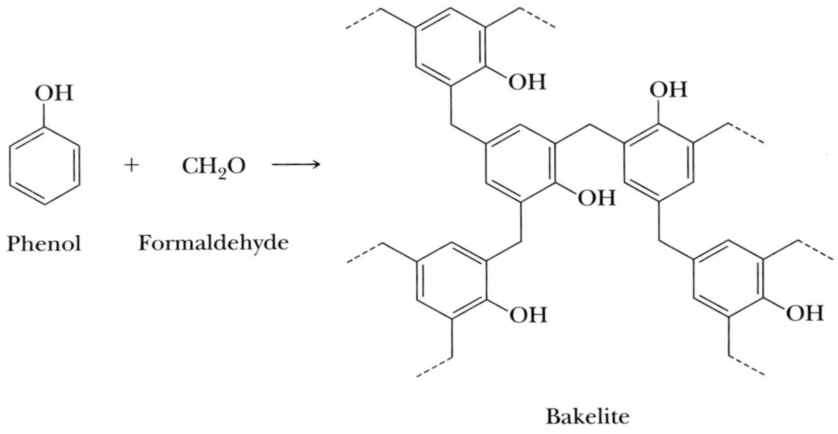

Phenol   Formaldehyde $+$ $\mathrm{CH_2O}$ $\longrightarrow$

Bakelite

In the preparation of a thermoset, one of the monomers must be trifunctional. In the case of Bakelite, the trifunctional monomer is phenol. Alkyd thermosets are polyesters of an organic diacid, HOOC—R—COOH, and a trialcohol such as glycerol. Urea-formaldehyde thermosets are polyamides in which one molecule of urea, H_2N—CO—NH_2, can condense with up to four molecules of formaldehyde.

Manufacture of thermosets begins with a fluid mixture of the two monomers. The fluid is first shaped and then polymerized, either by heating or by being mixed with an initiator. The product of the polymerization is a network of covalently bonded atoms that is a solid, even at high temperatures. When heated to high temperatures, thermoset polymers char and decompose, but they do not melt.

29.6 Chain-Growth Polymerizations

From the perspective of the chemical industry, the single most important reaction of alkenes is **chain-growth polymerization,** a type of polymerization in which monomer units are joined together without loss of atoms. An example is the formation of polyethylene from ethylene.

Chain-growth polymerization A polymerization that involves sequential addition reactions, either to unsaturated monomers or to monomers possessing other reactive functional groups.

$$n CH_2{=}CH_2 \xrightarrow{\text{catalyst}} {+}CH_2CH_2{+}_{\overline{n}}$$

Ethylene Polyethylene

The mechanism of this type of polymerization differs greatly from the mechanism of step-growth polymerizations. In the latter, all monomers plus the polymer endgroups possess equally reactive functional groups, allowing for all possible combinations of reactions to occur, including monomer with monomer, dimer with dimer and so forth. In contrast, chain-growth polymerizations involve endgroups possessing reactive intermediates that react with a monomer only. The reactive intermediates used in chain-growth polymerizations include radicals, carbanions, carbocations, and organometallic complexes.

The number of monomers that undergo chain-growth polymerizations is large and includes such compounds as alkenes, alkynes, allenes, isocyanates, and cyclic compounds, such as lactones, lactams, ethers, and epoxides. We concentrate on the chain-growth polymerizations of ethylene and substituted ethylenes and show how these compounds can be polymerized by radical, cation, anion, and organometallic-mediated mechanisms.

An alkene

Table 29.1 lists several important polymers derived from ethylene and substituted ethylenes along with their common names and most important uses.

A. Radical Chain-Growth Polymerizations

Among the initiators used for radical chain-growth polymerizations are diacyl peroxides, such as dibenzoyl peroxide, which decompose as shown upon heating. In the first step, homolytic cleavage of the weak O—O peroxide bond yields two

■ The low thermal conductivity of polystyrene makes it a good insulating material.

Table 29.1 Polymers Derived from Ethylene and Substituted Ethylenes

Monomer Formula	Common Name	Polymer Name(s) and Common Uses
$CH_2{=}CH_2$	Ethylene	Polyethylene, Polythene; break-resistant containers and packaging materials
$CH_2{=}CHCH_3$	Propylene	Polypropylene, Herculon; textile and carpet fibers
$CH_2{=}CHCl$	Vinyl chloride	Poly(vinyl chloride), PVC; construction tubing
$CH_2{=}CCl_2$	1,1-Dichloroethylene	Poly(1,1-dichloroethylene), SaranWrap; food packaging
$CH_2{=}CHCN$	Acrylonitrile	Polyacrylonitrile, Orlon; acrylics and acrylates
$CF_2{=}CF_2$	Tetrafluoroethylene	Poly(tetrafluoroethylene), PTFE; Teflon, nonstick coatings
$CH_2{=}CHC_6H_5$	Styrene	Polystyrene, Styrofoam; insulating materials
$CH_2{=}CHCOOCH_2CH_3$	Ethyl acrylate	Poly(ethyl acrylate); latex paints
$CH_2{=}\underset{\underset{CH_3}{\mid}}{C}COOCH_3$	Methyl methacrylate	Poly(methyl methacrylate), Lucite, Plexiglas; glass substitutes

acyloxy radicals. Each acyloxy radical then decomposes to form an aryl radical and CO_2.

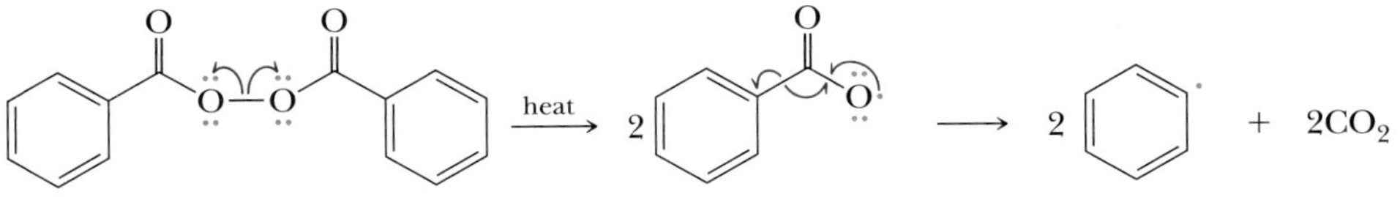

Dibenzoyl peroxide   A benzoyloxy radical   A phenyl radical

Another common class of initiators used in radical polymerizations are azo compounds, such as azoisobutyronitrile (AIBN), which decompose upon heating or by the absorption of UV light to produce alkyl radicals and nitrogen gas.

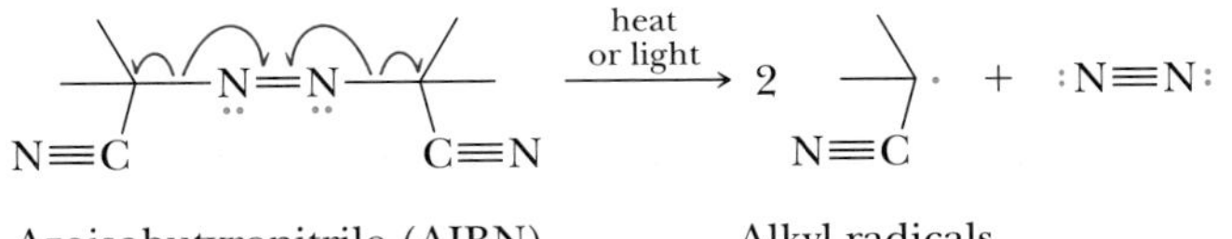

Azoisobutyronitrile (AIBN)   Alkyl radicals

The chain initiation, propagation, and termination steps for radical polymerization of a substituted ethylene monomer are shown for the monomer $RCH{=}CH_2$. Dissociation of the initiator produces a radical that reacts with the double bond of a monomer. Once initiated, the chains continue to propagate through successive additions of monomers.

Mechanism   *Radical Polymerization of a Substituted Ethylene*

Step 1: Initiation: Radicals form from nonradical compounds.

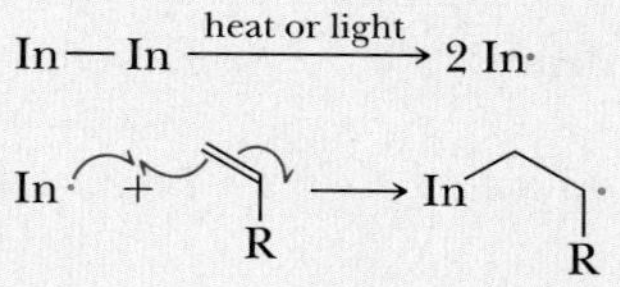

$$In-In \xrightarrow{\text{heat or light}} 2\ In\cdot$$

Step 2: Propagation: Reaction of a radical and a molecule gives a new radical.

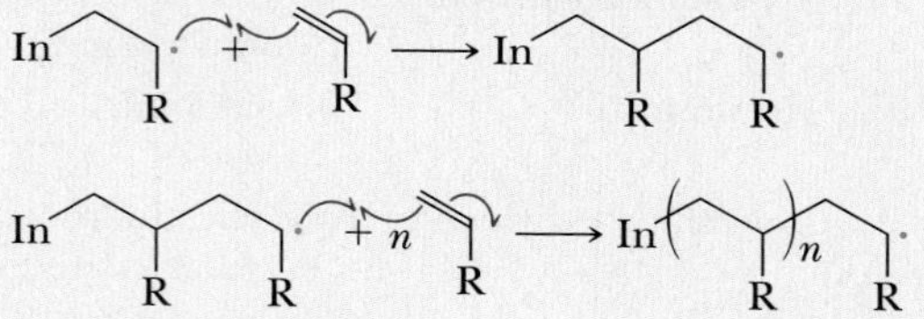

Step 3: Chain termination: Radicals are destroyed.

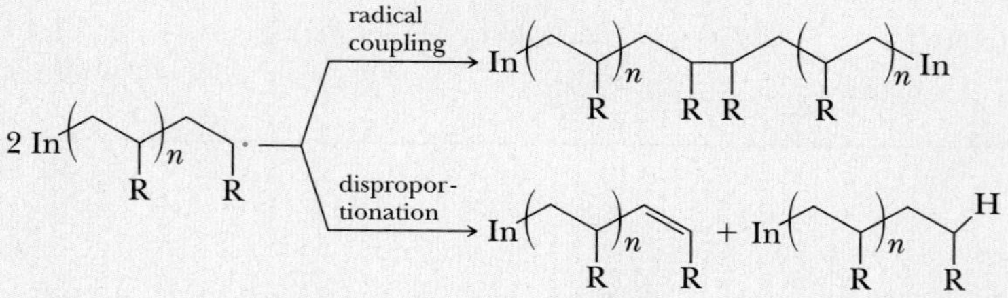

In radical reactions, the chain termination involves combination of radicals to produce a nonradical molecule or molecules. One common termination step is **radical coupling** to form a new carbon-carbon bond linking two growing polymer chains. This type of termination step is a diffusion-controlled process that occurs without an activation energy barrier. Another common termination process is **disproportionation,** which involves the abstraction of a hydrogen atom from the beta position to the propagating radical of one chain by the radical endgroup of another chain. This process results in two dead chains, one terminated in an alkyl group and the other in an alkenyl group.

Radical reactions with double bonds almost always give the more stable (more substituted) radical. Because additions are biased in this fashion, the polymerizations of vinyl monomers tend to yield polymers with head-to-tail linkages. Vinyl polymers made by radical processes generally have no more than 1–2% head-to-head linkages.

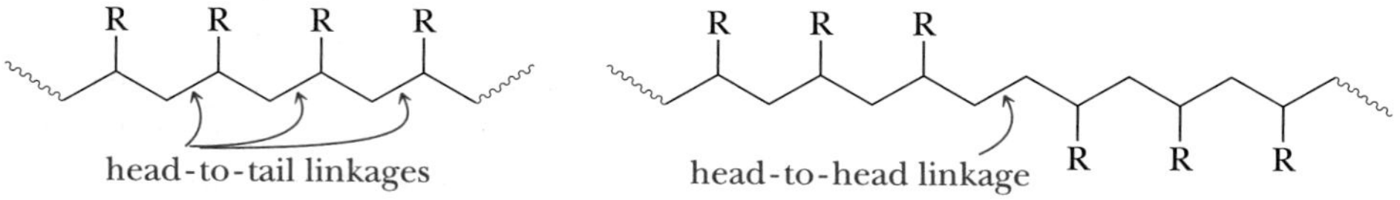

Disproportionation   A termination process that involves the abstraction of a hydrogen atom from the beta position of the propagating radical of one chain by the radical endgroup of another chain.

head-to-tail linkages   head-to-head linkage

Because organic radicals are highly reactive species, it is not surprising that radical polymerizations are often complicated by unwanted side reactions. A frequently observed side reaction is hydrogen abstraction by the radical endgroup from a grow-ing polymer chain, a solvent molecule, or another monomer. These side reactions are called **chain-transfer reactions** because the activity of the endgroup is "trans-ferred" from one chain to another.

Chain transfer is illustrated by radical polymerization of ethylene. Polyethyl-ene formed by radical polymerization exhibits a number of butyl branches on the polymer main chain. These four-carbon branches are generated in a "back-biting" chain-transfer reaction in which the radical endgroup abstracts a hydrogen from the fourth carbon back (the fifth carbon in the chain). Abstraction of this hydrogen is particularly facile because the transition state associated with the process can adopt a conformation like that of a chair cyclohexane. Continued polymerization of monomer from this new radical center leads to branches four carbons long.

Chain-transfer reaction   The transfer of reactivity of an end-group from one chain to another during a polymerization.

A six-membered transition
state leading to
1,5-hydrogen abstraction

As a result of these various abstraction reactions, polymers synthesized by radi-cal processes can have highly branched structures. The number of butyl branches depends on the relative stability of the propagating-radical endgroup and varies depending on the polymer. Polyethylene chains propagate through highly reactive primary radicals, which tend to be susceptible to 1,5-hydrogen abstraction reac-tions; these polymers typically have 15 to 30 branches per 500 monomer units. In contrast, polystyrene chains propagate through substituted benzyl radicals, which are stabilized by delocalization of the unpaired electron into the aromatic ring. These stabilized radicals are less likely to undergo hydrogen abstraction reactions. Polystyrene typically exhibits only one branch per 4000 to 10,000 monomer units.

The first commercial process for ethylene polymerization used peroxide cata-lysts at temperatures of 500°C and pressures of 1000 atm and produced a soft, tough polymer known as low-density polyethylene (LDPE). At the molecular level, chains of LDPE are highly branched owing to chain-transfer reactions. Because this extensive chain branching prevents polyethylene chains from packing efficiently, LDPE is largely amorphous and transparent, with only a small amount of crystallites of a size too small to scatter light. LDPE has a density between 0.91 and 0.94 g/cm^3 and a melt transition temperature (T_m) of about 108°C. Because its T_m is only slightly above 100°C, it cannot be used for products that will be exposed to boil-ing water.

Approximately 65% of all low-density polyethylene is used for the manufacture of films. Fabrication of LDPE films is done by a blow-molding technique illustrated in Figure 29.4. A tube of molten LDPE along with a jet of compressed air is forced through an opening and blown into a giant, thin-walled bubble. The film is then cooled and taken up onto a roller. This double-walled film can be slit down the side to give LDPE film or it can be sealed at points along its length to make LDPE bags. LDPE film is inexpensive, which makes it ideal for trash bags and for packaging for such consumer items as baked goods, vegetables, and other produce.

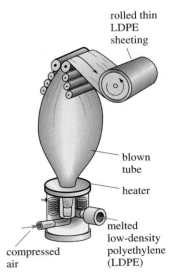

Figure 29.4
Fabrication of a LDPE film.

CHEMICAL CONNECTIONS

Organic Polymers That Conduct Electricity

The influence of chemical structure on the properties of an organic compound is clearly seen in the electrical conducting properties of certain organic polymers. Most organic polymers are insulators. For example, polytetrafluoroethylene with the repeating unit —CF_2CF_2— or poly(vinyl chloride) with the repeating unit —CH_2CHCl— have conductivities of 10^{-18} S/cm. On the other end of the scale, the conductivity of copper is almost 10^6 S/cm.

Can organic polymers approach the conductivity of copper? Research carried out over the last 20 years shows that the answer is yes. When acetylene is passed through a solution containing certain transition metal catalysts, it can be polymerized to a shiny film of polyacetylene.

Polyacetylene

By itself, polyacetylene is not a conductor. However, by a process called doping, which involves introducing small amounts of electron-donating or electron-accepting compounds, it is possible to produce a polyacetylene that shows a conductivity of 1.5×10^5 S/cm.

The purpose of the doping agent is either to remove electrons from the pi system (*p*-doping) or add electrons to the pi system (*n*-doping). A *p*-doped polyacetylene can be represented as a conjugated polyalkene chain containing positively charged carbons at several points along the chain.

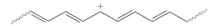

A *p*-doped polyacetylene

We can think of the positive charge as a defect that can move to the left or to the right along the polymer chain, thus giving rise to conductivity.

In crude polyacetylene, the polymer chains are jumbled, pointing in all directions. However, by stretching the film, the chains can be made to line up in a more ordered fashion. The conductivity of doped and oriented polyacetylene chains is greater along the direction of the chain than it is perpendicular to the chain. This result suggests that it is much easier for electrons to travel along a chain than to hop from one chain to the next.

Applications for conducting organic polymers are beginning to be developed. A rechargeable battery with electrodes of *p*-doped and *n*-doped polyacetylene already has been produced. Given the atomic weight of carbon, organic polymer batteries should be lighter than nickel-cadmium or lead-acid batteries. Weight is an important consideration if battery powered electric cars are ever to be made practical. In addition, many metals used in today's batteries (mercury, nickel, lead) are toxic. If research leads to practical organic batteries, waste disposal problems could be considerably lessened.

B. Ziegler-Natta Chain-Growth Polymerizations

An alternative method for polymerization of alkenes, which does not involve radicals, was developed by Karl Ziegler of Germany and Giulio Natta of Italy in the 1950s. For their pioneering work, they were awarded the 1963 Nobel Prize in chemistry. The early Ziegler-Natta catalysts were highly active, heterogeneous catalysts composed of a $MgCl_2$ support, a Group 4B transition metal halide, such as $TiCl_4$, and an alkylaluminum compound, such as $Al(CH_2CH_3)_2Cl$. These catalysts bring about polymerization of ethylene and propylene at 1 to 4 atm and at temperatures as low as 60°C. Polymerizations under these conditions do not involve radicals.

$$CH_2 = CH_2 \xrightarrow[\text{MgCl}_2]{\text{TiCl}_4/\text{Al(CH}_2\text{CH}_3)_2\text{Cl}} \left(\text{\large{\vphantom{X}}} \right)_n$$

Ethylene   Polyethylene

The active catalyst in a Ziegler-Natta polymerization is thought to be an alkyltitanium compound, which is formed by alkylation of the titanium halide by $Al(CH_2CH_3)_2Cl$ on the surface of a $MgCl_2/TiCl_4$ particle. Once formed, this species repeatedly inserts ethylene into the titanium-carbon bond to yield polyethylene.

Mechanism   *Ziegler-Natta Catalysis of Ethylene Polymerization*

Step 1: A titanium-ethyl bond forms.

$MgCl_2/TiCl_4$ particle

$$-\text{Ti}-\text{Cl} \ + \ \text{Cl}-\text{Al} \longrightarrow -\text{Ti} \ + \ \text{Cl}-\text{Al} \overset{\text{Cl}}{<}$$

Diethylaluminum chloride

Step 2: Ethylene inserts into the titanium-carbon bond.

$$-\text{Ti} \ + \ CH_2 = CH_2 \longrightarrow -\text{Ti}$$

Over 2.5×10^{11} kg of polyethylene are produced worldwide every year using optimized Ziegler-Natta catalysts, and large-scale reactors can yield up to 1.25×10^5 kg of polyethylene per hour. Production of polymer at this scale is partly attributable to the mild conditions required for a Ziegler-Natta polymerization and the fact that the polymer obtained has substantially different physical and mechanical properties from that obtained by radical polymerization. Polyethylene from Ziegler-Natta systems, termed high-density polyethylene (HDPE), has a higher density (0.96 g/cm³) and T_m (133°C) than low-density polyethylene, is three to ten times stronger, and is opaque rather than transparent. This added strength and opacity is the result of a much lower degree of chain branching and the resulting higher degree of crystallinity of HDPE compared with LDPE.

Approximately 45% of all HDPE used in the United States is blow molded. In blow molding, a short length of HDPE tubing is placed in an open die

The Stock Market/Corbis

■ Polyethylene films are produced by extruding the molten plastic through a ring-like gap and inflating the film into a balloon.

Figure 29.5
Blow molding of a HDPE
container.

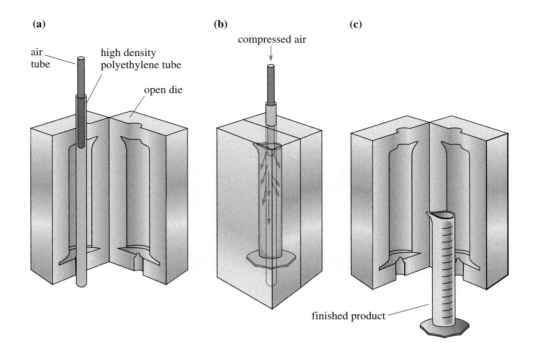

(a)   **(b)**   **(c)**

[Figure 29.5(a)], and the die is closed, sealing the bottom of the tube. Compressed air is then forced into the hot polyethylene/die assembly, and the tubing is literally blown up to take the shape of the mold [Figure 29.5(b)]. After cooling, the die is opened [Figure 29.5(c)], and there is the container!

Even greater improvements in properties of HDPE can be realized through special processing techniques. In the melt state, HDPE chains adopt random coiled conformations similar to those of cooked spaghetti. Engineers have developed special extrusion techniques that force the individual polymer chains of HDPE to uncoil and adopt an extended linear conformation. These extended chains then align with one another to form highly crystalline materials. HDPE processed in this fashion is stiffer than steel and has approximately four times the tensile strength of steel! Because the density of polyethylene ($\approx 1.0\,\text{g/cm}^3$) is considerably less than that of steel ($8.0\,\text{g/cm}^3$), these comparisons of strength and stiffness are even more favorable if they are made on a weight basis.

In recent years, there have been several important advances made in catalysts used in Ziegler-Natta type polymerizations. One of the most important has been the discovery of soluble complexes that catalyze the polymerization of ethylene and propylene at extraordinary rates. Because these new homogeneous catalysts are substantially different in structure from the early Ziegler-Natta systems, these polymerizations are referred to as **coordination polymerizations.** Catalysts for coordination polymerizations are frequently formed by allowing bis-(cyclopentadienyl) dimethylzirconium, [$Cp_2Zr(CH_3)_2$], to react with methaluminoxane (MAO). MAO is a complex mixture of methylaluminum oxide oligomers, [$—(CH_3)AlO—)n$], formed by allowing trimethylaluminum to react with small amounts of water. It is thought that MAO activates the zirconium by abstracting a methyl anion to form a zirconium cation that is the active polymerization catalyst.

Mechanism *Homogeneous Catalysis for Ziegler-Natta Coordination Polymerization*

Step 1: The zirconium catalyst is activated.

Bis(cyclopentadienyl)-dimethylzirconium + MAO ⟶ A zirconium cation (the active form of the catalyst) +

Step 2: Ethylene inserts into the zirconium-carbon bond.

$$Zr-CH_3 + H_2C=CH_2 \longrightarrow Zr-CH_2CH_2CH_3$$

Some of these coordination-polymerization catalysts polymerize up to 20,000 ethylene monomer units per second, a rate reached only by enzyme-catalyzed biological reactions. Another important characteristic of these catalysts is that they show high reactivity toward 1-alkenes, allowing the formation of copolymers, such as that of ethylene and 1-hexene.

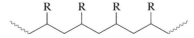

Ethylene   1-Hexene   Ethylene-(1-hexene) copolymer

Copolymers of this type with these moderate length branches (C_4, C_6, and so on) are called linear low-density polyethylene, or LLDPE. These are useful materials because they have many of the properties of LDPE made from radical reactions but are formed at the substantially milder conditions associated with Ziegler-Natta polymerizations.

C. Stereochemistry and Polymers

Thus far, we have written the formula of a substituted ethylene polymer in the following manner and have not been concerned with the configuration of each chiral center along the chain.

Nevertheless, the relative configurations of these chiral centers are important in determining the properties of a polymer. Polymers with identical configurations at all

Figure 29.6
Relative configurations of chiral centers in polymers with different tacticities.

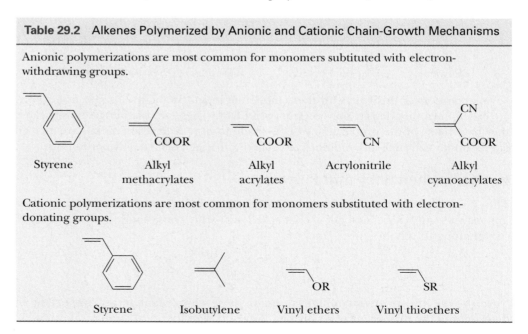

Isotactic polymer
(identical configurations)

Syndiotactic polymer
(alternating configurations)

Atactic polymer
(random configurations)

Isotactic polymer A polymer with identical configurations (either all *R* or all *S*) at all chiral centers along its chain, as, for example, isotactic polypropylene.

Syndiotactic polymer A polymer with alternating *R* and *S* configurations at the chiral centers along its chain, as, for example, syndiotactic polypropylene.

Atactic polymer A polymer with completely random configurations at the chiral centers along its chain, as, for example, atactic polypropylene.

chiral centers along the chain are called **isotactic** polymers. Those with alternating configurations are called **syndiotactic** polymers, and those with completely random configurations are called **atactic** polymers (Figure 29.6).

In general, the more stereoregular the chiral centers are, that is, the more highly isotactic or highly syndiotactic the polymer is, the more crystalline it is. A random placement of the substituents, such as in atactic materials, results in a polymer that cannot pack well and is usually highly amorphous. Atactic polystyrene, for example, is an amorphous glass, whereas isotactic polystyrene is a crystalline fiber-forming polymer with a high melt transition. The control over the relative configuration, or tacticity, along a polymer backbone is, therefore, an area of considerable interest in modern polymer synthesis.

D. Ionic Chain-Growth Polymerizations

Chain-growth polymers can also be synthesized using reactions that rely on either anionic or cationic species in the propagation steps. The choice of ionic procedure depends greatly on the electronic nature of the monomers to be polymerized. Vinyl monomers with electron-withdrawing groups, which stabilize carbanions, are used in anionic polymerizations, whereas vinyl monomers with electron-donating groups, which stabilize cations, are used in cationic polymerizations (Table 29.2).

Table 29.2 Alkenes Polymerized by Anionic and Cationic Chain-Growth Mechanisms

Anionic polymerizations are most common for monomers subtituted with electron-withdrawing groups.

Styrene	Alkyl methacrylates	Alkyl acrylates	Acrylonitrile	Alkyl cyanoacrylates
	COOR	COOR	CN	CN / COOR

Cationic polymerizations are most common for monomers substituted with electron-donating groups.

Styrene	Isobutylene	Vinyl ethers	Vinyl thioethers
		OR	SR

Styrene is conspicuous among the monomers given in Table 29.2 because it can be polymerized using either anionic or cationic techniques as well as radical techniques. This characteristic particular to styrene is attributable to the fact that the phenyl group can stabilize cationic, anionic, and radical benzylic intermediates.

Resonance stabilization of a
benzylic cation endgroup

Resonance stabilization of a
benzylic anion endgroup

Anionic Polymerizations

Anionic polymerizations can be initiated by addition of a nucleophile to an activated alkene. The most common nucleophiles used for this purpose are metal alkyls, such as methyl- or *sec*-butyllithium. The newly formed carbanion then acts as a nucleophile and adds to another monomer unit, and the propagation continues.

Mechanism *Initiation of Anionic Polymerization of Alkenes*

Step 1: Polymerization is initiated by addition of a nucleophile, shown here as a carbanion derived from an organolithium compound, to an activated carbon-carbon double bond to give a carbanion.

Step 2: This carbanion adds to the activated double bond of a second alkene molecule to give a dimer.

Step 3: Chain growth continues to give a polymer

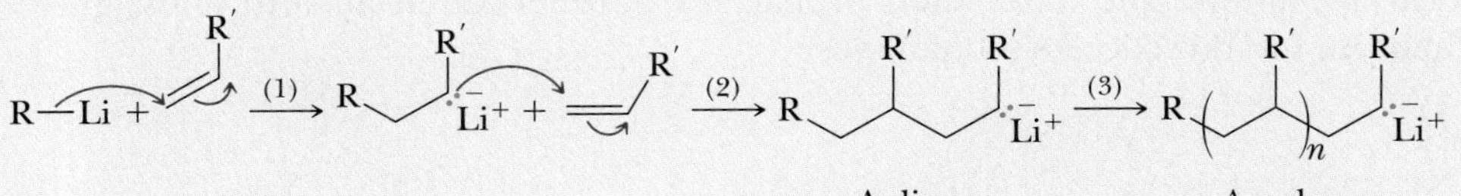

A dimer

A polymer

An alternative method for the initiation of anionic polymerizations involves a one-electron reduction of the monomer by lithium or sodium to form a radical anion. The radical anion thus formed is either further reduced to form a dianion, or dimerizes to form a dimer dianion.

Mechanism *Initiation of Anionic Polymerization of Butadiene*

Step 1: A one-electron reduction of the diene by lithium metal gives a radical anion.

Step 2: One-electron reduction of this radical anion gives a monomer dianion.

Step 3: Alternatively, radical coupling gives a dimer dianion.

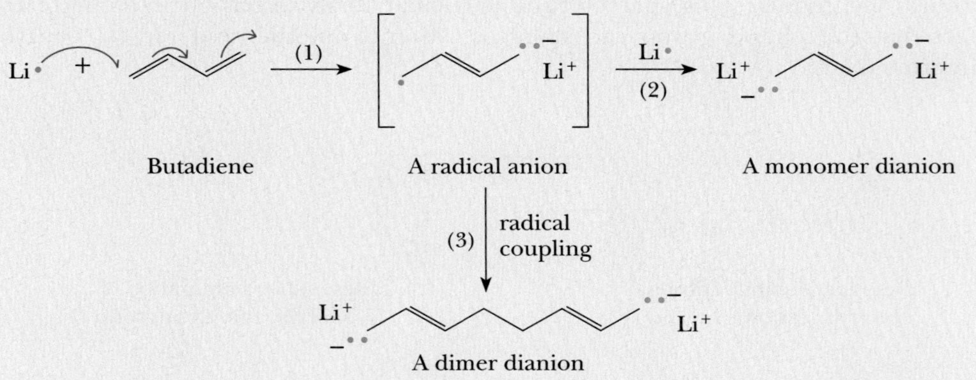

| Butadiene | A radical anion | A monomer dianion |

A dimer dianion

In either case, a single initiator can now propagate chains from both ends, by virtue of its two active endgroup carbanions. These reactions are heterogeneous and involve transfer of the electron from the surface of the metal. To improve the efficiency of this process, soluble reducing agents such as sodium naphthalide are used. Sodium undergoes electron-transfer reactions with extended aromatic compounds, such as naphthalene, to form soluble radical anions.

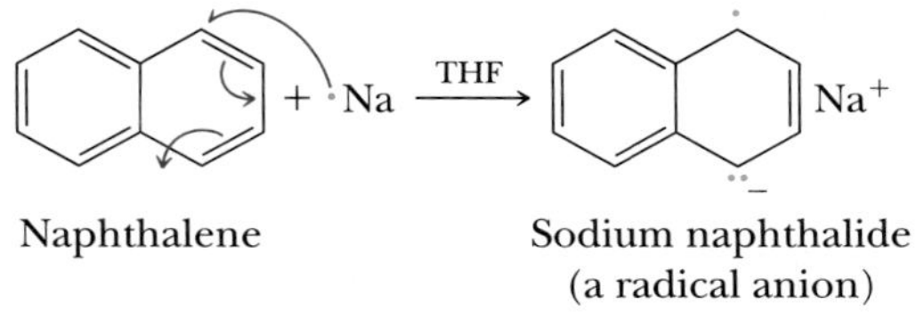

Naphthalene   Sodium naphthalide
                               (a radical anion)

The naphthalide radical anion is a powerful reducing agent. For example, styrene undergoes a one-electron reduction to form the styryl radical anion, which couples to form a dianion. The latter then propagates polymerization at both ends, growing chains in two directions simultaneously.

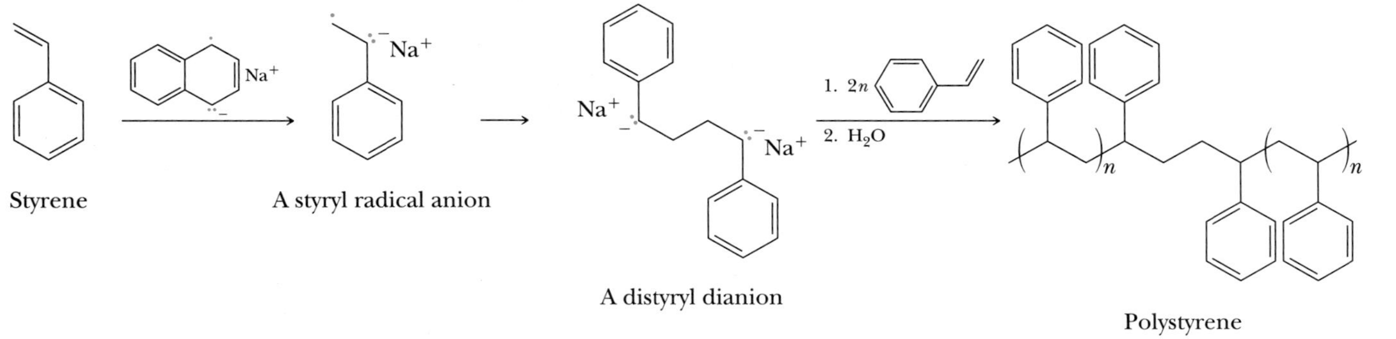

Styrene   A styryl radical anion   A distyryl dianion   Polystyrene

The propagation characteristics of anionic polymerizations are similar to those of radical polymerizations, but with the important difference that many of the chain-transfer and chain-termination reactions that plague radical processes are absent. Furthermore, because the propagating chain ends carry the same charge, bimolecular coupling and disproportionation reactions are also averted. An interesting set of circumstances arises when chain-transfer and chain-termination steps are no longer significant. Under these conditions, polymer chains are initiated and continue to grow until either all the monomer is consumed or some external agent is added to terminate the chains. Polymerizations of this type are called **living polymerizations** because they will restart if more monomer is added after it is initially consumed.

The absence of chain-transfer and chain-termination steps in living polymerizations has far-reaching consequences. One of the most visible of these is in the area of molecular-weight control. The molecular weight of a polymer originating from living polymerizations is determined directly by the monomer-to-initiator ratio. It is, therefore, relatively easy to obtain polymers of a well-defined size simply by controlling the stoichiometry of the reagents. In contrast, the average sizes of polymer chains formed from nonliving, chain-growth processes (radical, Ziegler-Natta, and so on) vary from system to system and are determined by the ratio of the rate of propagation to the rate of termination. In most cases, precise control over the molecular weight of the product obtained in nonliving systems is not possible because it is very difficult to change one of the rates involved without affecting the other.

After consumption of the monomer under living, anionic conditions, electrophilic terminating agents can be added to functionalize the chain ends. Examples of terminating reagents include CO_2 and ethylene oxide, which, after protonation, form carboxylic acid and alcohol-terminated chains, respectively.

In a similar fashion, polymer chains with functional groups at both ends, called **telechelic polymers,** can be prepared by addition of these same reagents (CO_2, ethylene oxide, and so on) to solutions of chains with two active ends initiated by sodium naphthalide.

Example 29.3

Show how to prepare polybutadiene that is terminated at both ends with carboxylate groups.

Solution

Form a growing chain with two active endgroups by treatment of butadiene with two moles of lithium metal to form a dianion followed by addition of monomer units and formation of a living polymer. Cap the active endgroups with a carboxylate group by treatment of the living polymer with carbon dioxide.

Living polymer A polymer chain that continues to grow without chain-termination steps until either all of the monomer is consumed or some external agent is added to terminate the chain. The polymer chains will continue to grow if more monomer is added.

Telechelic polymer A polymer in which its growing chains are terminated by formation of new functional groups at both ends of its chains. These new functional groups are introduced by adding reagents, such as CO_2 or ethylene oxide, to the growing chains.

The Chemistry of Superglue

Used in everything from model planes to passenger planes, Superglue is one of the best-known modern glues. Indeed, anyone who has unwittingly glued his or her fingers together while using the product can acknowledge its remarkable (and insidious) adhesive properties. The curing process that facilitates these properties is a chain polymerization reaction. The ingredient that gives Superglue its adhesive ability is methyl cyanoacrylate. This compound is just one member of a larger family of cyanoacrylates with the following general structure.

$$H_2C=C\begin{array}{c} C\equiv N \\ \\ C=O \\ | \\ O \\ | \\ R \end{array}$$

Contrary to popular understanding, Superglue does not "air dry." In fact, cyanoacrylates cure (convert from liquid to solid) in the presence of weak nucleophiles such as water. Under normal conditions, a thin layer of water is present on almost all surfaces. This accounts for many unintended adhesions involving appendages and/or expensive tools! The curing process involves an anionic chain polymerization reaction, which occurs as follows.

In the chain-initiating step, a weak nucleophile donates an electron pair to a cyanoacrylate monomer. The CH_2 group is highly electropositive as a result of the electron withdrawal by the cyano and ester groups, which makes this Michael reaction very easy. The addition produces a new anion, which is also a weak nucleophile and adds to another monomer, ultimately creating the powerfully adhesive polymer chains of cured Superglue.

Superglue has a rich history. It begins during World War II, when cyanoacrylates were discovered in an attempt to produce optically clear gun sights. Tests with Superglue's main ingredient failed for obvious reasons: The compound stuck to all instruments. Approximately ten years later, researchers at Eastman Kodak rediscovered cyanoacrylates, and Kodak first marketed them in 1958. It is said that researchers discovered Superglue's properties when they attempted to take the refractive index of the monomer. A refractometer, the instrument used for this purpose, has two prisms that come together on the liquid whose refractive index is to be determined. The prisms of the instrument were permanently glued together, and a new product was born!

$$H_2\ddot{O}: + H_2C=C\begin{array}{c} C\equiv N \\ \\ C=O \\ | \\ O \\ | \\ R \end{array} \longrightarrow H_2\overset{+}{O}-CH_2-\overset{C\equiv N}{\underset{CO_2R}{\overset{|}{\underset{|}{C}}:}} + H_2C=C\begin{array}{c} C\equiv N \\ \\ C=O \\ | \\ O \\ | \\ R \end{array} \xrightarrow{-H^+}$$

$$H\ddot{O}-CH_2-\overset{C\equiv N}{\underset{CO_2R}{\overset{|}{\underset{|}{C}}}}-CH_2-\overset{C\equiv N}{\underset{COOR}{\overset{|}{\underset{|}{C}}:}} \xrightarrow[+H^+]{etc.} H\ddot{O}-\left(CH_2-\overset{C\equiv N}{\underset{COOR}{\overset{|}{\underset{|}{C}}}}\right)_n H$$

Superglue has been improved since that time with stabilizers and other supplementary ingredients that reinforce its adhesive strength. In recent years, that same adhesive strength has captured attention in new fields. Medical-grade Superglues such as 2-octylcyanoacrylate and 2-butylcyanoacrylate are now commonly used as sutures in laceration repair. They have also proven effective in skin, bone, and cartilage grafts. Dentists use cyanoacrylates in dental cements and fillings, and paleontologists use them to reorganize fragile fossil specimens. Superglue has many fascinating applications, all facilitated by the anionic chain polymerization reaction.

Based on a Chem30H honors paper by Carrie Brubaker, UCLA.

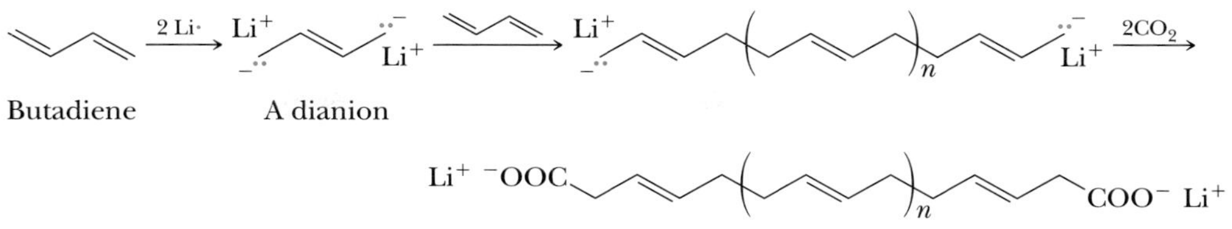

Butadiene   A dianion

Problem 29.3

Show how to prepare polybutadiene that is terminated at both ends with primary alcohol groups.

Cationic Polymerizations

Only alkenes with electron-donating substituents, such as alkyl, aryl, ether, thioether, and amino groups, undergo useful cationic polymerizations. The two most common methods of generating cationic initiators are (1) the reaction of a strong protic acid with an organic monomer and (2) the abstraction of a halide from the organic initiator by a Lewis acid. Cationic chain-growth polymerizations are generally effective only for monomers yielding relatively stable carbocations, that is monomers that form either 3° carbocations or cations stabilized by electron-donating groups, such as ether, thioether, or amino groups.

Initiation by protonation of an alkene requires the use of a strong acid with a nonnucleophilic anion to avoid 1,2-addition across the alkene double bond. Suitable acids with nonnucleophilic anions include HF/AsF_5 and HF/BF_3. In the following general equation, initiation is by proton transfer from $H^+BF_4^-$ to the alkene to form a tertiary carbocation, which then continues the cationic chain growth polymerization.

Mechanism   *Initiation of Cationic Polymerization of an Alkene by HF·BF₃*

Step 1: Proton transfer from the HF · BF₃ complex to the alkene gives a carbocation.

Step 2: Propagation continues the polymerization.

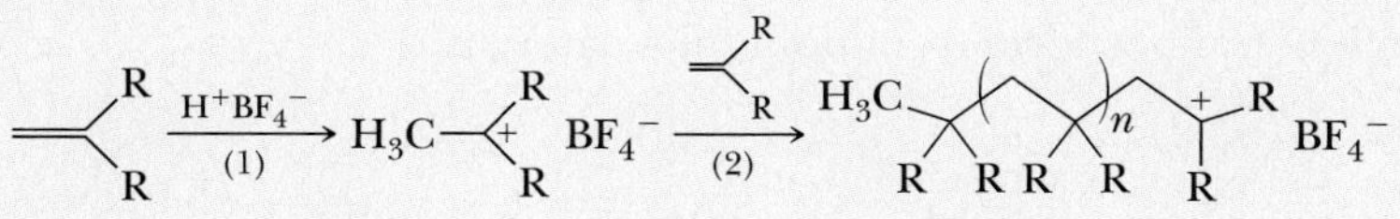

The second common method for generating carbocations involves the reaction between an alkyl halide and a Lewis acid, such as BF_3, $SnCl_4$, $AlCl_3$, $Al(CH_3)_2Cl$, and $ZnCl_2$. When a trace of water is present, the mechanism of initiation using some Lewis acids is thought to involve protonation of the alkene.

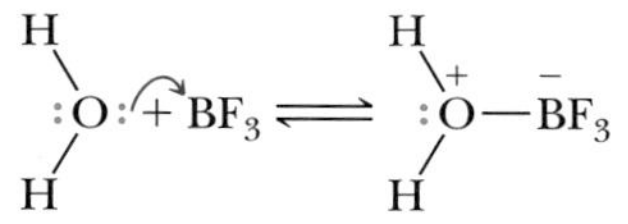

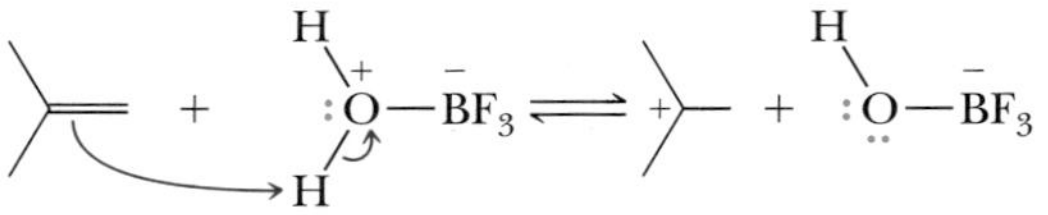

2-Methylpropene
(Isobutylene)

In the absence of water, the Lewis acid removes a halide ion from the alkyl halide to form the initiating carbocation.

Mechanism	*Initiation of Cationic Polymerization of an Alkene by a Lewis Acid*

Step 1: Reaction of the chloroalkane (a Lewis base) with tin(IV) chloride (a Lewis acid) gives a carbocation from which polymerization then proceeds.

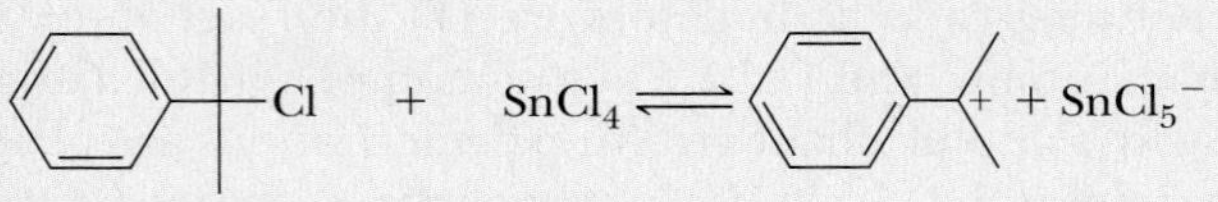

2-Chloro-2-phenylpropane

The polymerization of alkenes then propagates by the electrophilic attack of the carbocation on the double bond of the alkene monomer. The regiochemistry of the addition is determined by the formation of the more stable (the more highly substituted) carbocation.

Example 29.4

Write a mechanism for the polymerization of 2-methylpropene (isobutylene) initiated by treatment of 2-chloro-2-phenylpropane with $SnCl_4$. Label the initiation, propagation, and termination steps.

Solution

Chain initiation: Cations form from nonionic materials.

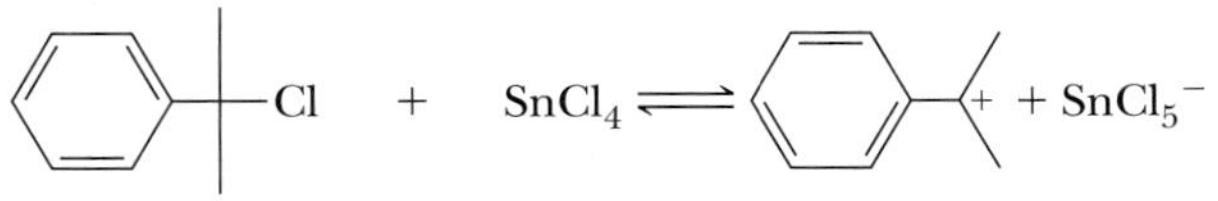

2-Chloro-2-phenylpropane

Chain propagation: A cation and a molecule react to give a new cation.

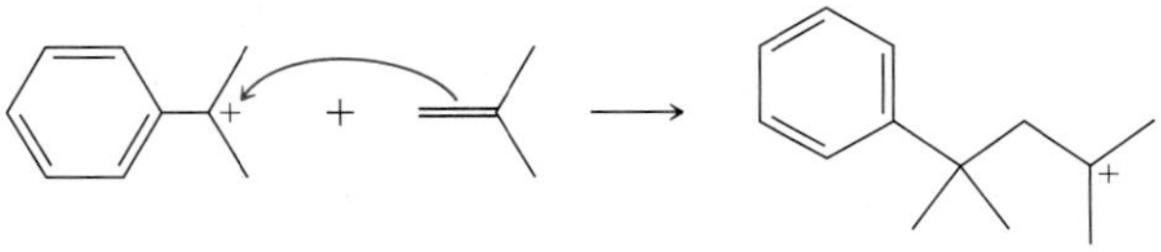

2-Methylpropene

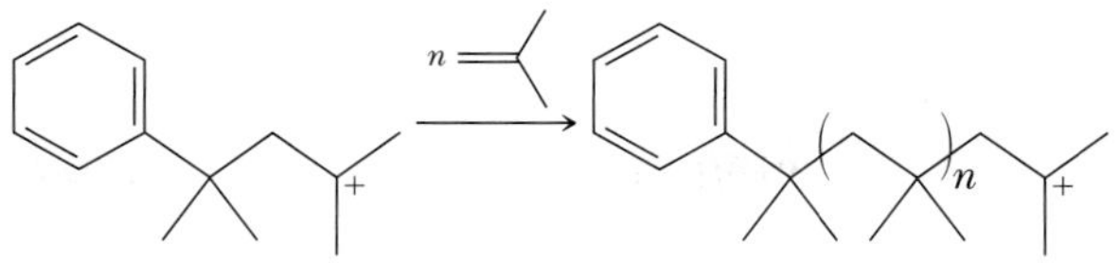

Chain termination: Destruction of cations.

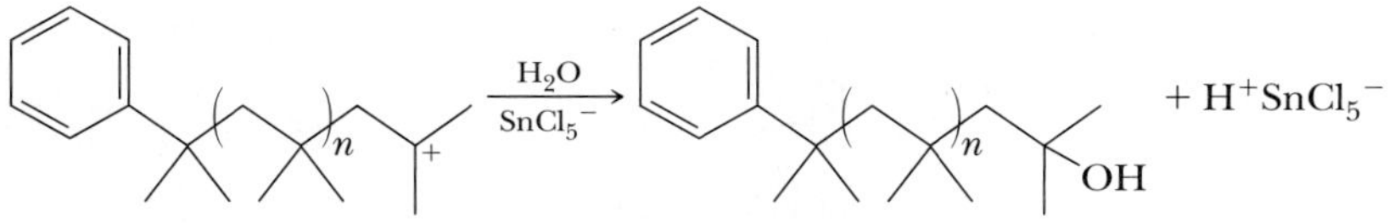

Problem 29.4

Write a mechanism for the polymerization of methyl vinyl ether initiated by 2-chloro-2-phenylpropane and $SnCl_4$. Label the initiation, propagation, and termination steps.

E. Ring-Opening Metathesis Polymerizations

During early investigations into the polymerizations of cycloalkenes by transition metal catalysts such as those used in Ziegler-Natta polymerizations, polymers of unexpected structures that contained the same number of double bonds as originally present in the monomers were formed. This process is illustrated by the polymerization of bicyclo[2.2.1]-2-heptene (norbornene).

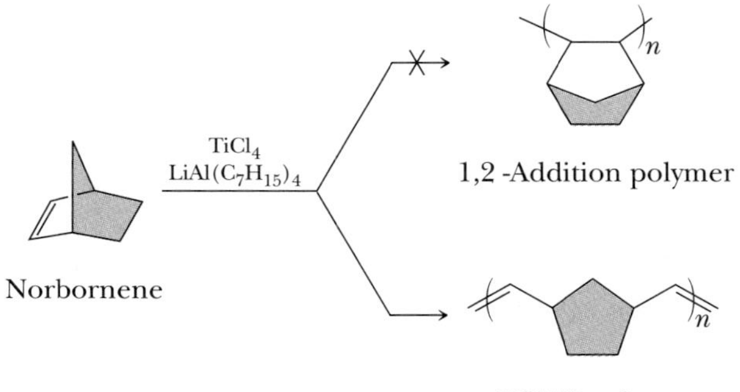

If reaction had proceeded in the same manner as Ziegler-Natta polymerization of ethylene and substituted ethylenes (Section 29.6B), a 1,2-addition polymer would have been formed. What is formed, however, is an unsaturated polymer in which the number of double bonds in the polymer is the same as that in the monomers polymerized. This process is called **ring-opening metathesis polymerization,** or ROMP, after the olefin metathesis involving reaction of acyclic alkenes and nucleophilic carbenes catalysts described in Section 24.5.

The fact that ROMP polymers are unsaturated requires that this polymerization proceed by a mechanism substantially different from that involved in polymerization of ethylene and substituted ethylenes by the same catalyst mixtures. Following lengthy and detailed studies, chemists discovered that ROMP involves the same metallacyclobutane species as in ring-closing alkene metathesis reactions (Section 24.5B). The intermediate metallacyclobutane derivative undergoes a ring-opening reaction

CHEMICAL CONNECTIONS

Recycling of Plastics

Polymers, in the form of plastics, are materials upon which our society is incredibly dependent. Durable and lightweight, plastics are probably the most versatile synthetic materials in existence; in fact, their current production in the United States exceeds that of steel. Plastics have come under criticism, however, for their role in the trash crisis. They comprise 21% of the volume and 8% of the weight of solid wastes, most of which is derived from disposable packaging and wrapping. Of the 2.5×10^7 kg of thermoplastic materials produced in 1993 in America, less than 2% was recycled.

Why aren't more plastics being recycled? The durability and chemical inertness of most plastics make them ideally suited for reuse. The answer to this question has more to do with economics and consumer habits than with technological obstacles. Because curbside pickup and centralized drop-off stations for recyclables are just now becoming common, the amount of used material available for reprocessing has traditionally been small. This limitation, combined with the need for an additional sorting and separation step, rendered the use of recycled plastics in manufacturing expensive compared with virgin materials. Until recently, consumers perceived products made from "used" materials as being inferior to new ones, so the market for recycled products has not been large. However, the increase in environmental concerns over the last few years has resulted in a greater demand for recycled products. As manufacturers adapt to satisfy this new market, plastic recycling will eventually catch up with the recycling of other materials, such as glass and aluminum.

Six types of plastics are commonly used for packaging applications. In 1988 manufacturers adopted recycling code numbers developed by the Society of the Plastics Industry. Because the plastics recycling industry still is not fully developed, only polyethylene terephthalate (PET) and high-density polyethylene (HDPE) are currently being recycled in large quantities, although outlets for the other plastics are being developed. Low-density polyethylene (LDPE), which accounts for about

Charles D. Winters

■ These students are wearing jackets made from recycled PET soda bottles.

40% of plastic trash, has been slow in finding acceptance with recyclers. Facilities for the reprocessing of poly(vinyl chloride) (PVC), polypropylene (PP), and polystyrene (PS) exist, but are still rare.

The process for the recycling of most plastics is simple, with separation of the desired plastics from other contaminants the most labor-intensive step. PET soft drink bottles, for example, usually have a paper label, adhesive, and an aluminum cap that must be removed before the PET can be reused. The recycling process begins with hand or machine sorting, after which the bottles are shredded into small chips. An air cyclone removes paper and other lightweight materials, and any remaining labels and adhesives are eliminated with a detergent wash. The PET chips are then dried, and aluminum, the final contaminant, is removed electrostatically. The PET produced by this method is 99.9% free of contaminants and sells for about half the price of the

Recycling Code	Polymer	Common Uses	Uses of Recycled Polymer
1 PET	Poly(ethylene terephthalate)	Soft drink bottles, household chemical bottles, films, textile fibers	Soft drink bottles, household chemical bottles, films, textile fibers
2 HDPE	High-density polyethylene	Milk and water jugs, grocery bags, bottles	Bottles, molded containers
3 PVC	Poly(vinyl chloride)	Shampoo bottles, pipes, shower curtains, vinyl siding, wire insulation, floor tiles, credit cards	Plastic floor mats
4 LDPE	Low-density polyethylene	Shrink wrap, trash and grocery bags, sandwich bags, squeeze bottles	Trash bags and grocery bags
5 PP	Polypropylene	Plastic lids, clothing fibers, bottle caps, toys, diaper linings	Mixed plastic components
6 PS	Polystyrene	Styrofoam cups, egg cartons, disposable utensils, packaging materials, appliances	Molded items such as cafeteria trays, rulers, Frisbees, trash cans, videocasettes
7	All other plastics and mixed plastics	Various	Plastic lumber, playground equipment, road reflectors

virgin material. Unfortunately, plastics with similar densities cannot be separated with this technology, nor can plastics composed of several polymers be broken down into pure components. However, recycled mixed plastics can be molded into plastic lumber that is strong, durable, and graffiti-resistant.

An alternative to this process, which uses only physical methods of purification, is chemical recycling. Eastman Kodak salvages large amounts of its PET film scrap by a transesterification reaction. The scrap is treated with methanol in the presence of an acid catalyst to give ethylene glycol and dimethyl terephthalate.

These monomers are purified by distillation or recrystallization and used as feed stocks for the production of more PET film.

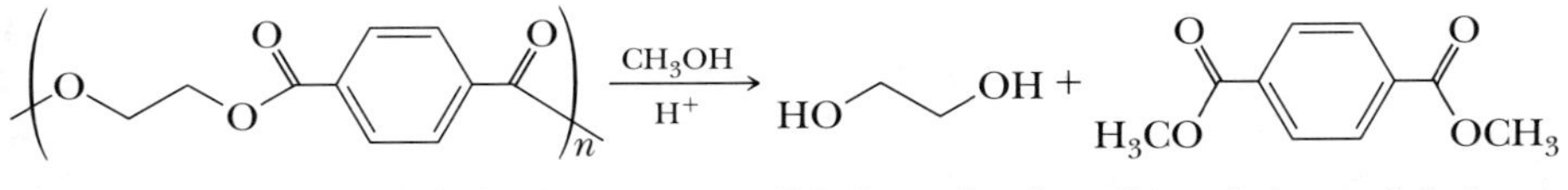

Poly(ethylene terephthalate)
(PET)   Ethylene glycol   Dimethyl terephthalate

to give a new substituted carbene. Repetition of these steps leads to the formation of the unsaturated polymer as illustrated here by ROMP of cyclopentene.

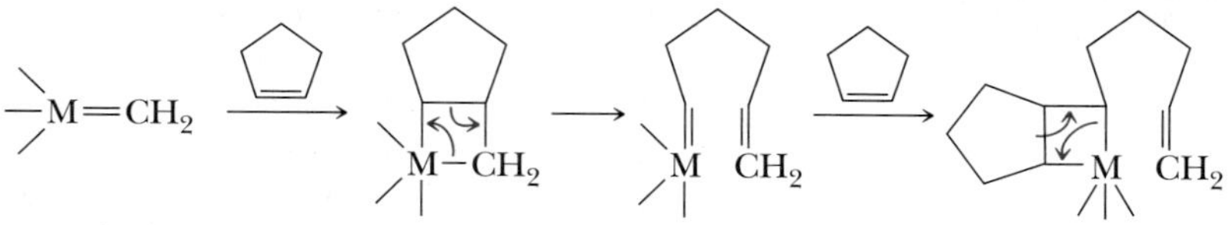

A metal carbene

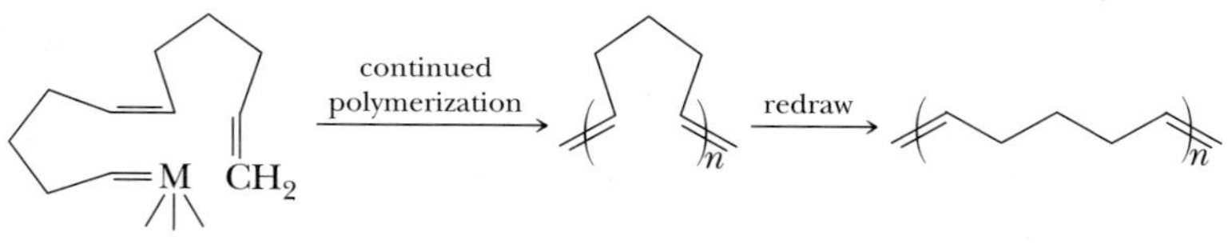

ROMP polymer from cyclopentene

All steps in ROMP are reversible, and the reaction is driven in the forward direction by the release of ring strain that accompanies the opening of the ring. The reactivity of the following cycloalkenes toward ROMP decreases in this order.

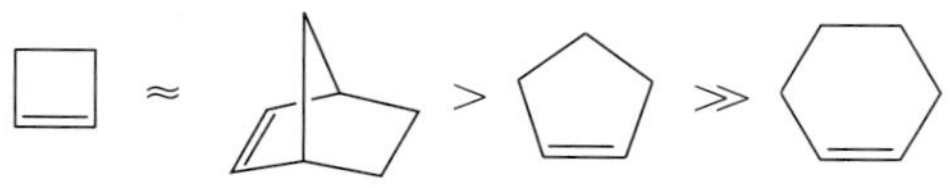

Ring strain   125(29.8)   113(27)   24.7(5.9)   5.9(1.4)
[kJ (kcal)/mol]

ROMP reactions are unique in that all the unsaturation present in the monomers is conserved in the polymeric product. This feature makes ROMP techniques especially attractive for the preparation of highly unsaturated, fully conjugated materials. One example is the direct preparation of polyacetylene by the ROMP technique through one of the double bonds of cyclooctatetraene. For further discussion of polyacetylene, see Chemical Connections: "Organic Polymers That Conduct Electricity" in this chapter.

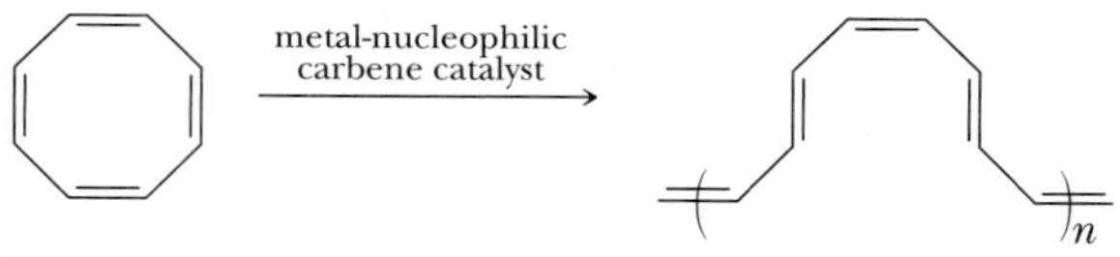

Cyclooctatetraene   Polyacetylene

An important polymer in electrooptical applications is poly(phenylene vinylene) (PPV), which has alternating phenyl and vinyl groups. One of the routes to this polymer starts with a substituted bicyclo[2.2.2]octadiene, which is polymerized using ROMP techniques to form a soluble, processable polymer. Heating the processed polymer results in elimination of two equivalents of acetic acid, which aromatizes the six-membered ring and completes the conjugation.

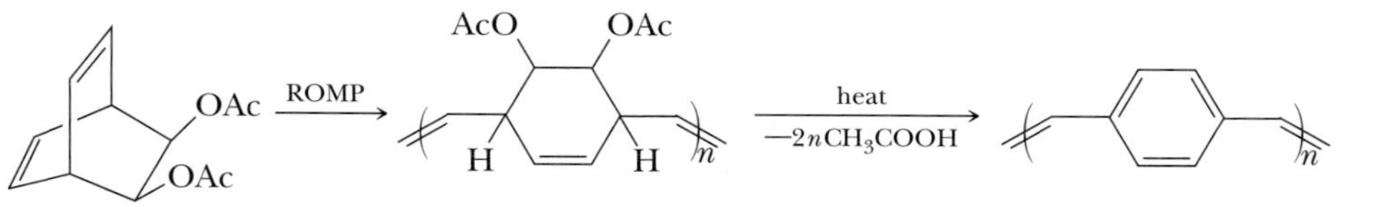

Poly(phenylene vinylene)

Summary

Polymerization is the process of joining together many small **monomers** into large, high-molecular-weight **polymers** (Section 29.1). The properties of polymeric materials depend on the structure of the repeat unit, molecular weight (Section 29.3), chain architecture, the presence or absence of crystalline phases (Section 29.4), tacticity (Section 29.6C), interchain order and packing, and the materials' morphology. **Step-growth polymerizations** involve the stepwise reaction of difunctional monomers (Section 29.5). Further structural variations, such as cross links and branches, can be introduced into the resulting polymer by the addition of multifunctional monomers to the reaction mixture. The formation of high-molecular-weight polymers from step-growth processes requires the use of reactions that proceed with very high yields. Important commercial polymers synthesized through step-growth processes include polyamides, polyesters, polycarbonates, polyurethanes, and epoxy resins.

Chain-growth polymerization proceeds by the sequential addition of monomer units to an active chain end group (Section 29.6). Important mechanisms for chain-growth polymerizations include radical, anionic, cationic, and transition metal-mediated processes. Chain-growth polymerizations involve initiation, propagation, and termination steps. **Chain-transfer steps** terminate one chain but simultaneously initiate the growth of another. Chain polymerizations that proceed without chain-transfer or chain-termination steps are called **living polymerizations.**

Among the transition metal-mediated polymerizations, the Ziegler-Natta polymerizations of ethylene and propylene are the most significant (Section 29.6B). These reactions proceed with high specificity to yield polymers that are stereoregular and highly linear. This regularity leads to highly crystalline polymers. When the chains are elongated and oriented through special processing procedures, a polymer with strength and stiffness greater than those of steel can be obtained. **Ring-opening metathesis polymerization (ROMP)** of strained cycloalkenes and bicycloalkenes is catalyzed by transition-metal nucleophilic-carbene complexes (Section 29.6E). The products of ROMP are unusual in that all the unsaturation present in the monomer is conserved in the resulting polymer. Hence, ROMP reactions are ideal for the formation of unsaturated or fully conjugated materials.

Key Reactions

1. Step-Growth Polymerization of a Dicarboxylic Acid and a Diamine Gives a Polyamide (Section 29.5A)

2. Step-Growth Polymerization of a Dicarboxylic Acid and a Diol Gives a Polyester (Section 29.5B)

3. Step-Growth Polymerization of a Diacyl Chloride and a Diol Gives a Polycarbonate (Section 29.5C)

4. Step-Growth Polymerization of a Diisocyanate and a Diol Gives a Polyurethane (Section 29.5D)

5. Step-Growth Polymerization of a Diepoxide and a Diamine Gives an Epoxy Resin (Section 29.5E)

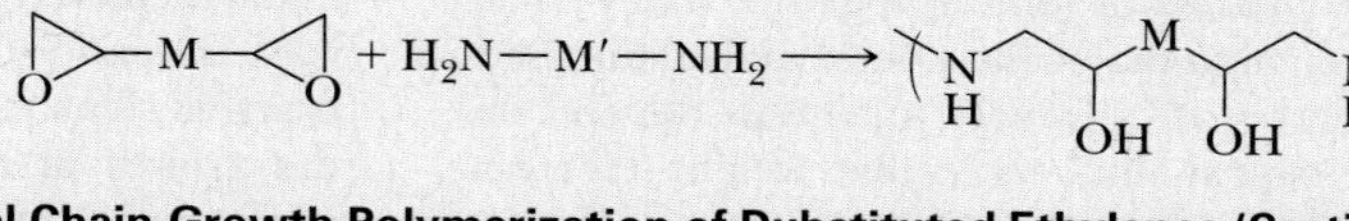

6. Radical Chain-Growth Polymerization of Dubstituted Ethylenes (Section 29.6A)

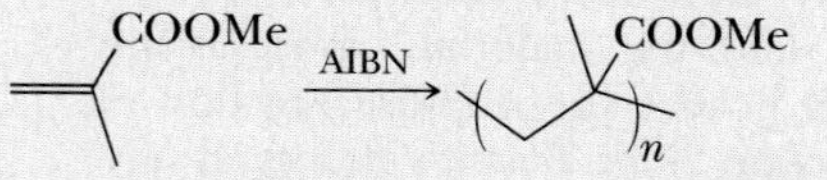

7. Titanium-Mediated (Ziegler-Natta) Chain-Growth Polymerization of Ethylene and Substituted Ethylenes (Section 29.6B)

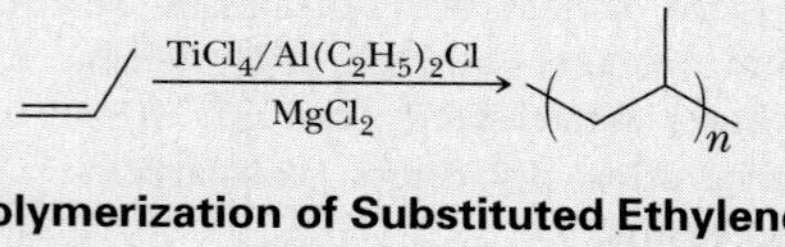

8. Anionic Chain-Growth Polymerization of Substituted Ethylenes (Section 29.6D)

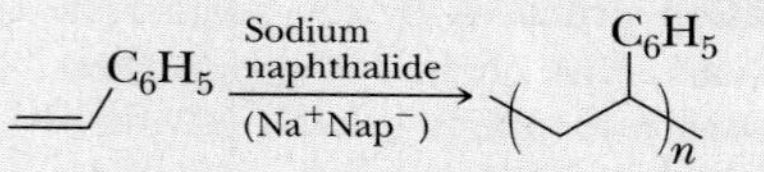

9. Cationic Chain-Growth Polymerization of Substituted Ethylenes (Section 29.6D)

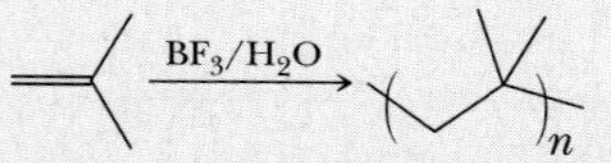

10. Ring-Opening Metathesis Polymerization (ROMP) (Section 29.6E)

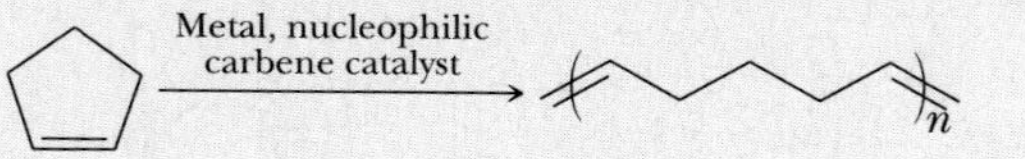

Problems

ORGANIC **Chemistry Now**™

Assess your understanding of this chapter's topics with additional quizzing and conceptual-based problems at **http://now.brookscole.com/bfi4**

Structure and Nomenclature

29.5 Name the following polymers.

(a)

(b)

(c)

(d)

(e)

(f)

(g)

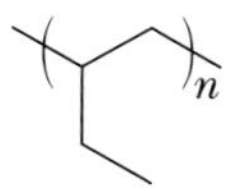

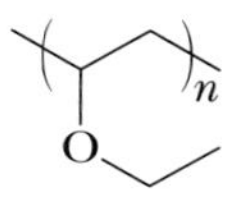

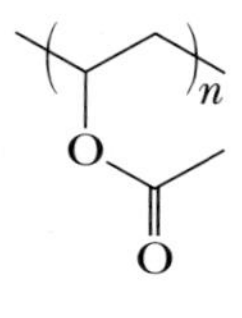

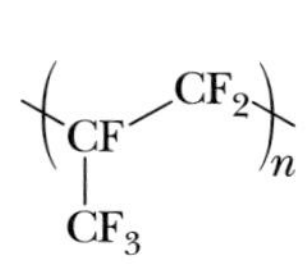

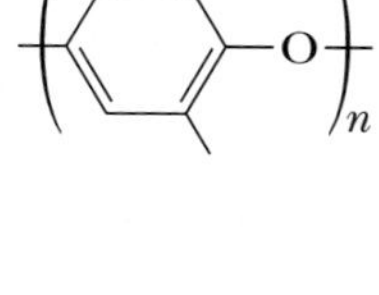

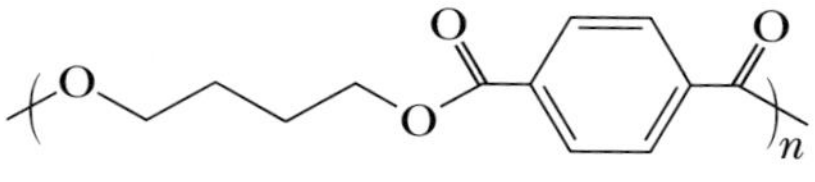

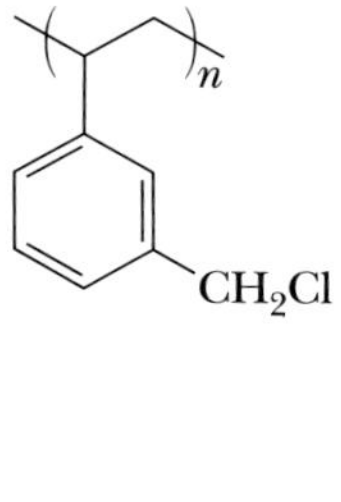

(h)

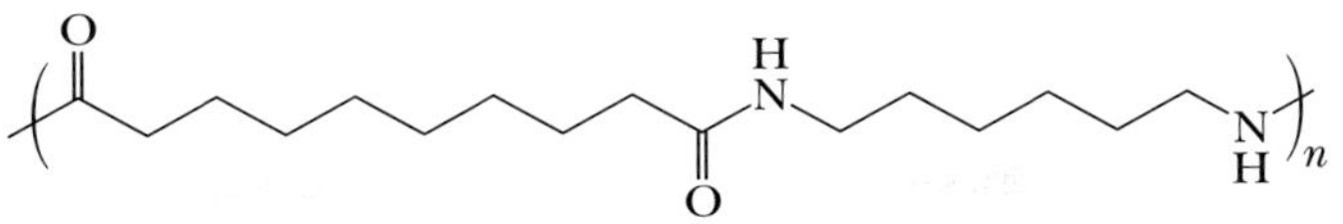

29.6 Draw the structure(s) of the monomer(s) used to make each polymer in Problem 29.5.

Step-Growth Polymerizations

29.7 Draw a structure of the polymer formed in the following reactions.

(a) 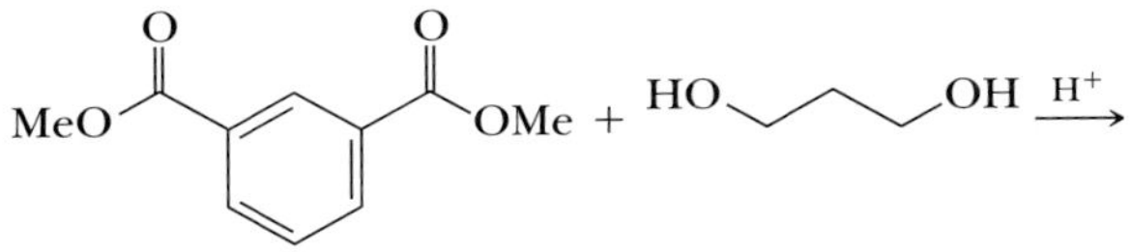 MeO—(benzene ring with two C=O groups)—OMe + HO~~~OH $\xrightarrow{H^+}$

(b) MeO—(benzene ring with two C=O groups)—OMe + HO~~~(with OH)~~~OH $\xrightarrow{H^+}$

(c) (β-lactone ring with O) $\xrightarrow{CF_3SO_3H}$ (d) 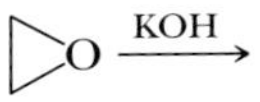 (epoxide ring with O) $\xrightarrow{KOH}$

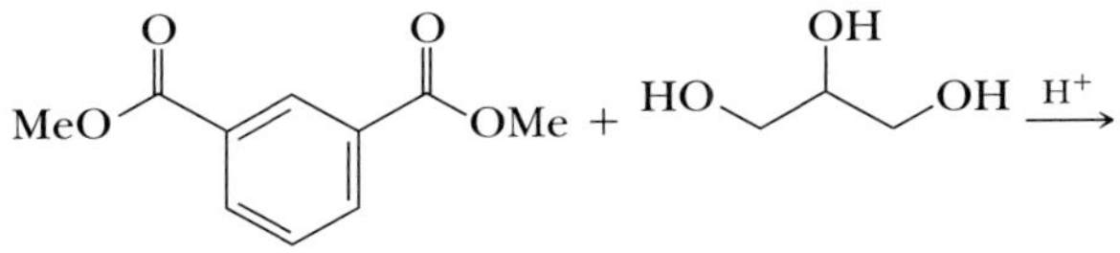

29.8 At one time, a raw material for the production of hexamethylenediamine was the pentose-based polysaccharides of agricultural wastes, such as oat hulls. Treatment of these wastes with sulfuric acid or hydrochloric acid gives furfural. Decarbonylation of furfural over a zinc-chromium-molybdenum catalyst gives furan. Propose reagents and experimental conditions for the conversion of furan to hexamethylenediamine.

oat hulls, corn cobs, sugar cane stalks, etc. $\xrightarrow[H_2O]{H_2SO_4}$ (furan ring with CHO) $\xrightarrow[\text{catalyst}]{Zn-Cr-Mo}$ (furan ring) $\xrightarrow{(1)}$ (THF ring) $\xrightarrow{(2)}$

Furfural   Furan   Tetrahydro-
                                        furan (THF)

$Cl(CH_2)_4Cl \xrightarrow{(3)} N{\equiv}C(CH_2)_4C{\equiv}N \xrightarrow{(4)} H_2N(CH_2)_6NH_2$

1,4-Dichloro-   Hexanedinitrile   1,6-Hexanediamine
butane   (Adiponitrile)   (Hexamethylenediamine)

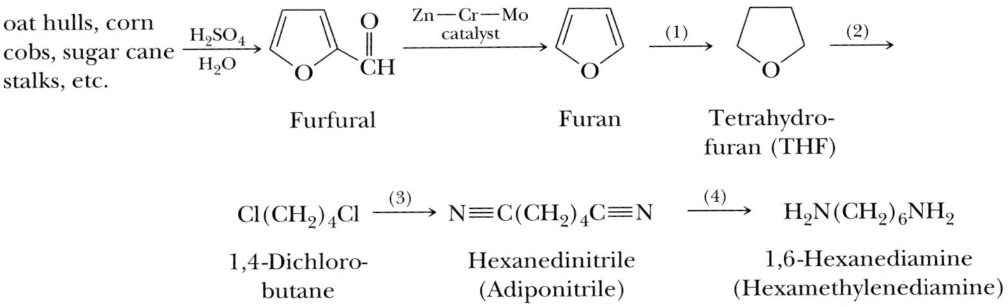

29.9 Another raw material for the production of hexamethylenediamine is butadiene derived from thermal and catalytic cracking of petroleum. Propose reagents and experimental conditions for the conversion of butadiene to hexamethylenediamine.

$CH_2{=}CHCH{=}CH_2 \xrightarrow{(1)} ClCH_2CH{=}CHCH_2Cl \xrightarrow{(2)}$

Butadiene   1,4-Dichloro-2-butene

$N{\equiv}CCH_2CH{=}CHCH_2C{\equiv}N \xrightarrow{(3)} H_2N(CH_2)_6NH_2$

3-Hexenedinitrile   1,6-Hexanediamine
                                   (Hexamethylenediamine)

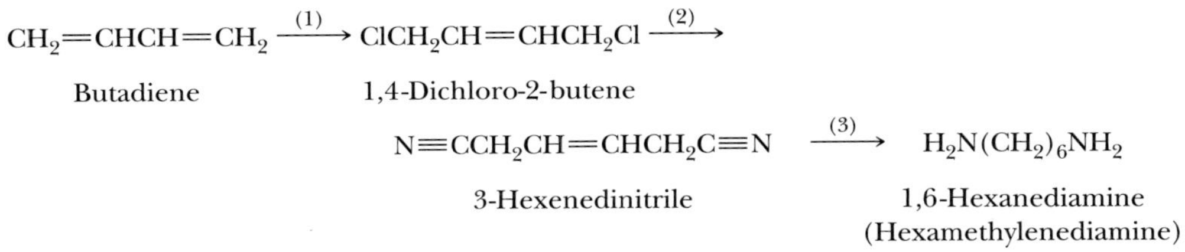

29.10 Propose reagents and experimental conditions for the conversion of butadiene to adipic acid.

1,3-Butadiene   Hexanedioic acid
                        (Adipic acid)

29.11 Polymerization of 2-chloro-1,3-butadiene under Ziegler-Natta conditions gives a synthetic elastomer called neoprene. All carbon-carbon double bonds in the polymer chain have the *trans* configuration. Draw the repeat unit in neoprene.

29.12 Poly(ethylene terephthalate) (PET) can be prepared by this reaction. Propose a mechanism for the step-growth reaction in this polymerization.

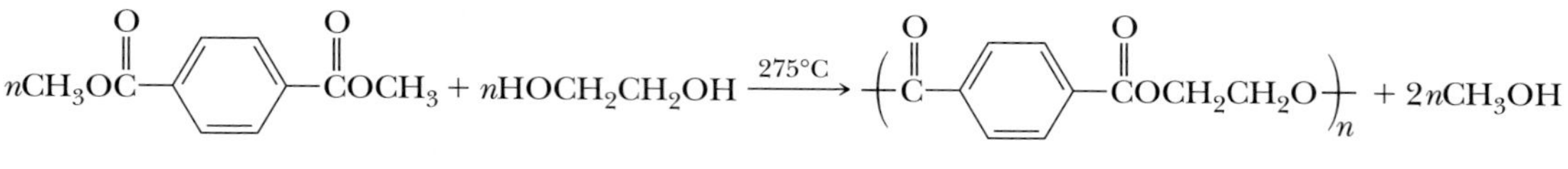

Dimethyl terephthalate   Ethylene glycol   Poly(ethylene terephthalate)   Methanol

29.13 Identify the monomers required for the synthesis of these step-growth polymers.

(a) 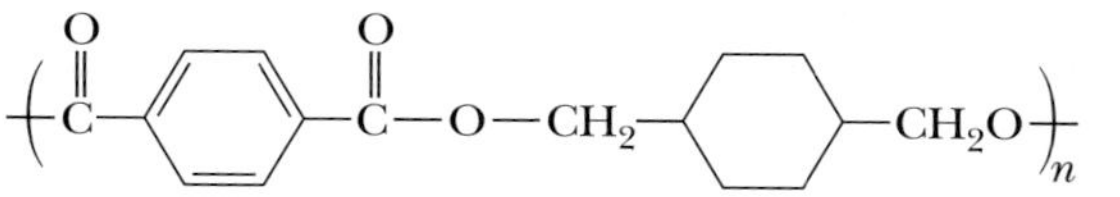   **(b)**

Kodel   Quiana
(a polyester)   (a polyamide)

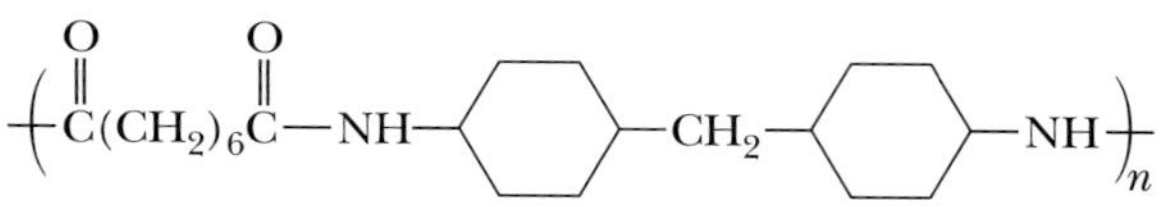

29.14 Nomex, another aromatic polyamide (compare aramid), is prepared by polymerization of 1,3-benzenediamine and the diacid chloride of 1,3-benzenedicarboxylic acid. The physical properties of the polymer make it suitable for high-strength, high-temperature applications such as parachute cords and jet aircraft tires. Draw a structural formula for the repeating unit of Nomex.

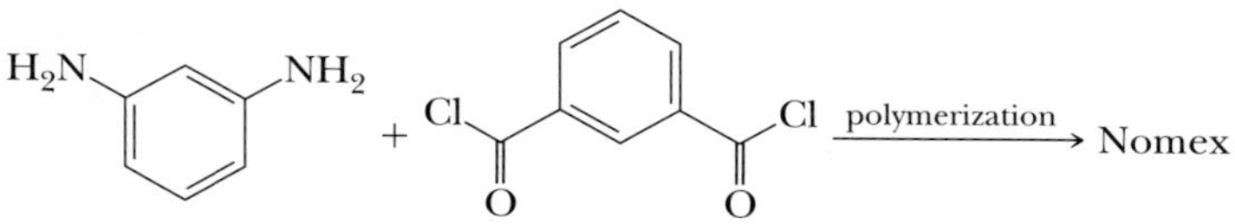

1,3-Benzenediamine   1,3-Benzene-
                            dicarbonyl chloride

29.15 Caprolactam, the monomer from which nylon 6 is synthesized, is prepared from cyclohexanone in two steps. In Step 1, cyclohexanone is treated with hydroxylamine to form cyclohexanone oxime. Treatment of the oxime with concentrated sulfuric acid in Step 2 gives caprolactam by a reaction called a Beckmann rearrangement. Propose a mechanism for the conversion of cyclohexanone oxime to caprolactam.

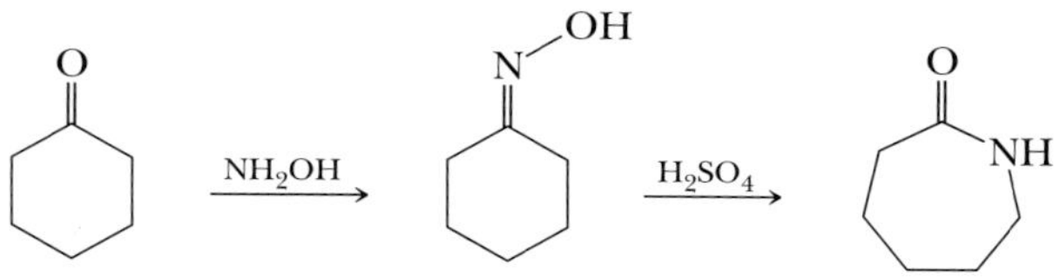

Cyclohexanone   Cyclohexanone   Caprolactam
                          oxime

29.16 Nylon 6,10 is prepared by polymerization of a diamine and a diacid chloride. Draw a structural formula for each reactant and for the repeat unit in this polymer.

29.17 Polycarbonates (Section 29.5C) are also formed by using a nucleophilic aromatic substitution route (Section 22.3B) involving aromatic difluoro monomers and carbonate ion. Propose a mechanism for this reaction.

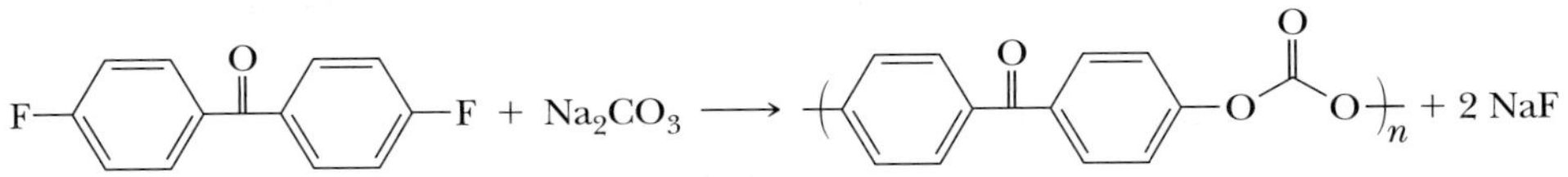

An aromatic   Sodium   A polycarbonate
difluoride   carbonate

29.18 Propose a mechanism for the formation of this polyphenylurea. To simplify your presentation of the mechanism, consider the reaction of one —NCO group with one —NH$_2$ group.

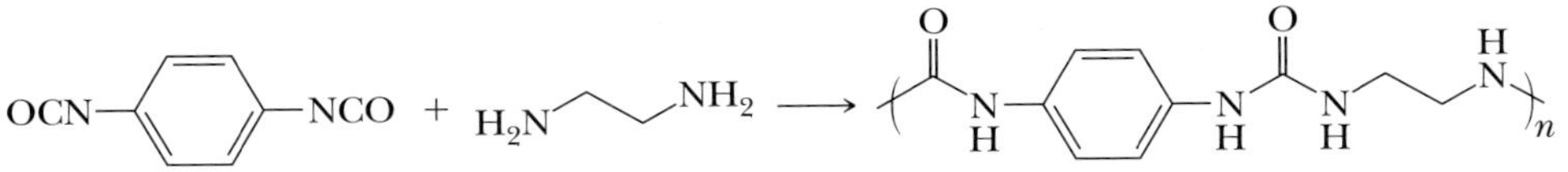

1,4-Benzenediisocyanate   1,2-Ethanediamine   Poly(ethylene phenylurea)

29.19 When equal molar amounts of phthalic anhydride and 1,2,3-propanetriol are heated, they form an amorphous polyester. Under these conditions, polymerization is regioselective for the primary hydroxyl groups of the triol.

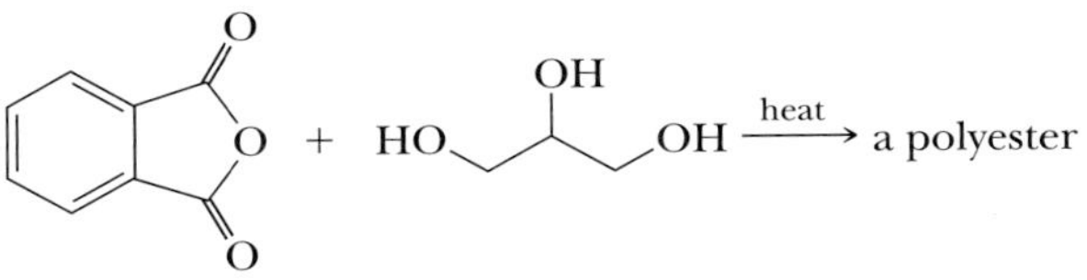

Phthalic anhydride   1,2,3-Propanetriol
(Glycerol)

(a) Draw a structural formula for the repeat unit of this polyester.

(b) Account for the regioselective reaction with the primary hydroxyl groups only.

29.20 The polyester from Problem 29.19 can be mixed with additional phthalic anhydride (0.5 mole of phthalic anhydride for each mole of 1,2,3-propanetriol in the original polyester) to form a liquid resin. When this resin is heated, it forms a hard, insoluble, thermosetting polyester called glyptal.

(a) Propose a structure for the repeat unit in glyptal.

(b) Account for the fact that glyptal is a thermosetting plastic.

29.21 Propose a mechanism for the formation of the following polymer.

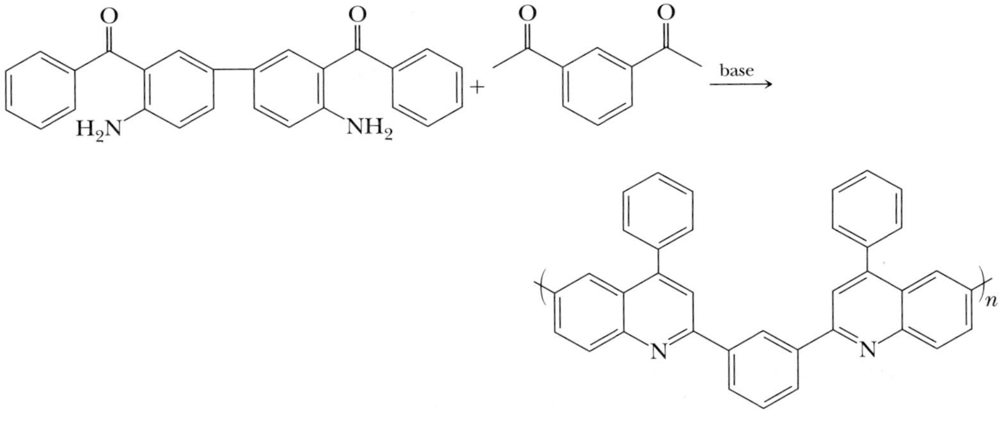

29.22 Draw a structural formula of the polymer resulting from base-catalyzed polymerization of each compound. Would you expect the polymers to be optically active? (*S*)-(+)-lactide is the dilactone formed from two molecules of (*S*)-(+)-lactic acid.

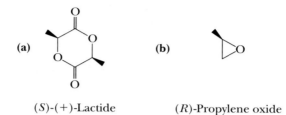

(a)   (b)

(*S*)-(+)-Lactide   (*R*)-Propylene oxide

29.23 Poly(3-hydroxybutanoic acid), a biodegradable polyester, is an insoluble, opaque material that is difficult to process into shapes. In contrast, the copolymer of 3-hydroxybutanoic acid and 3-hydroxyoctanoic acid is a transparent polymer that shows good solubility in a number of organic solvents. Explain the difference in properties between these two polymers in terms of their structure.

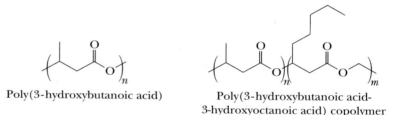

Poly(3-hydroxybutanoic acid)   Poly(3-hydroxybutanoic acid-
                                        3-hydroxyoctanoic acid) copolymer

Chain-Growth Polymerizations

29.24 How might you determine experimentally if a particular polymerization is propagating by a step-growth or a chain-growth mechanism?

29.25 Draw a structural formula for the polymer formed in the following reactions.

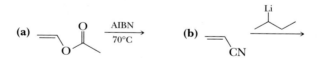

29.26 Select the monomer in each pair that is more reactive toward cationic polymerization.

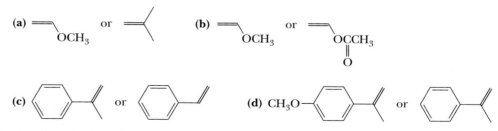

29.27 Polymerization of vinyl acetate gives poly(vinyl acetate). Hydrolysis of this polymer in aqueous sodium hydroxide gives the useful water-soluble polymer poly(vinyl alcohol). Draw the repeat units of both poly(vinyl acetate) and poly(vinyl alcohol).

29.28 Benzoquinone can be used to inhibit radical polymerizations. This compound reacts with a radical intermediate, R·, to form a less reactive radical that does not participate in chain propagation steps and, thus, breaks the chain.

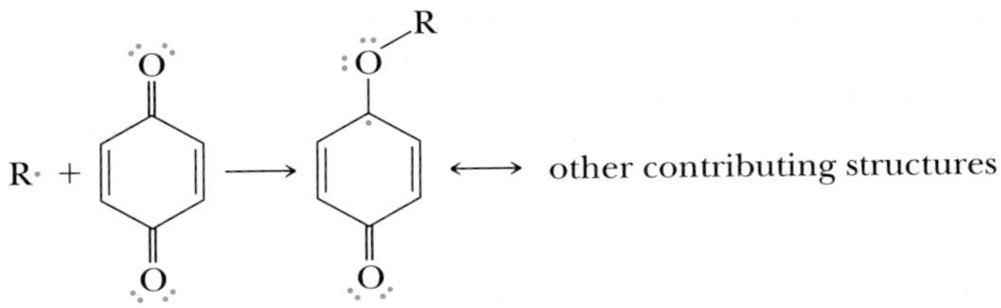

Draw a series of contributing structures for this less reactive radical and account for its stability.

29.29 Following is the structural formula of a section of polypropylene derived from three units of propylene monomer.

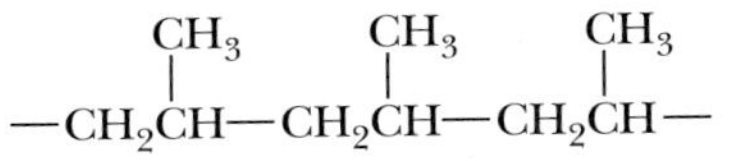

Polypropylene

Draw structural formulas for comparable sections of the following.

(a) Poly(vinyl chloride)   **(b)** Polytetrafluoroethylene
(c) Poly(methyl methacrylate)   **(d)** Poly(1,1-dichloroethylene)

29.30 Low-density polyethylene (LDPE) has a higher degree of chain branching than high-density polyethylene (HDPE). Explain the relationship between chain branching and density.

29.31 We saw how intramolecular chain transfer in radical polymerization of ethylene creates a four-carbon branch on a polyethylene chain. What branch is created by a comparable intramolecular chain transfer during radical polymerization of styrene?

29.32 Compare the densities of low-density polyethylene (LDPE) and high-density polyethylene (HDPE) with the densities of the liquid alkanes listed in Table 2.5. How might you account for the differences between them?

29.33 Natural rubber is the all-*cis* polymer of 2-methyl-1,3-butadiene (isoprene).

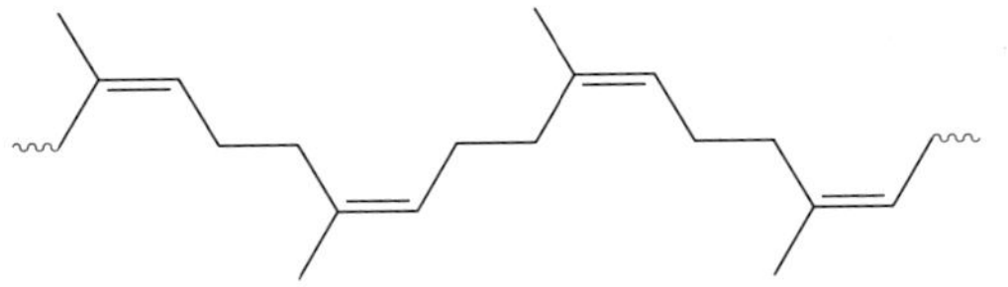

Poly(2-methyl-1,3-butadiene)
(Polyisoprene)

(a) Draw a structural formula for the repeat unit of natural rubber.
(b) Draw a structural formula of the product of oxidation of natural rubber by ozone followed by a workup in the presence of $(CH_3)_2S$. Name each functional group present in this product.
(c) The smog prevalent in many major metropolitan areas contains oxidizing agents, including ozone. Account for the fact that this type of smog attacks natural rubber (automobile tires and the like) but does not attack polyethylene or polyvinyl chloride.
(d) Account for the fact that natural rubber is an elastomer but the synthetic all-*trans* isomer is not.

29.34 Radical polymerization of styrene gives a linear polymer. Radical polymerization of a mixture of styrene and 1,4-divinylbenzene gives a cross-linked network polymer of the type shown in Figure 29.1. Show by drawing structural formulas how incorporation of a few percent 1,4-divinylbenzene in the polymerization mixture gives a cross-linked polymer.

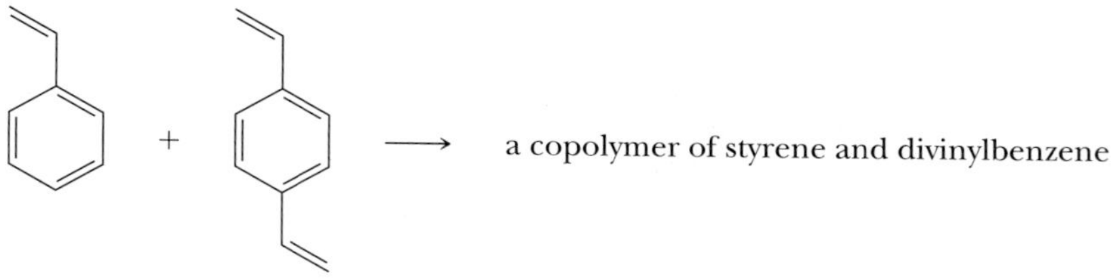

+ ⟶ a copolymer of styrene and divinylbenzene

Styrene   1,4-Divinylbenzene

29.35 One common type of cation exchange resin is prepared by polymerization of a mixture containing styrene and 1,4-divinylbenzene (Problem 29.34). The polymer is then treated with concentrated sulfuric acid to sulfonate a majority of the aromatic rings in the polymer.

(a) Show the product of sulfonation of each benzene ring.
(b) Explain how this sulfonated polymer can act as a cation exchange resin.

29.36 The most widely used synthetic rubber is a copolymer of styrene and butadiene called SB rubber. Ratios of butadiene to styrene used in polymerization vary depending on the end use of the polymer. The ratio used most commonly in the preparation of SB rubber for use in automobile tires is 1 mole styrene to 3 moles butadiene. Draw a structural formula of a section of the polymer formed from this ratio of reactants. Assume that all carbon-carbon double bonds in the polymer chain are in the *cis* configuration.

29.37 From what two monomer units is the following polymer made?

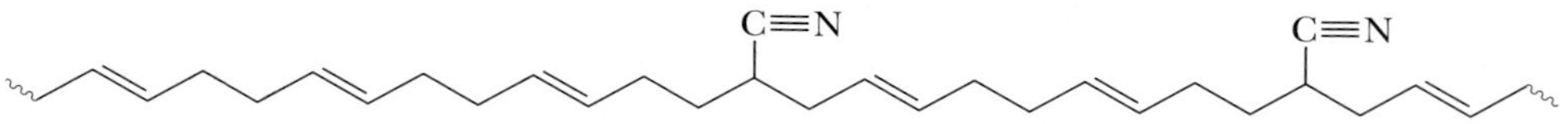

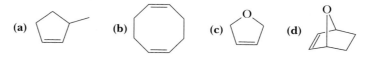

29.38 Draw the structure of the polymer formed from ring-opening metathesis polymerization (ROMP) of each monomer.

(a)   (b)   (c)   (d)

Thermodynamics and the Equilibrium Constant

For the equilibrium $A \rightleftharpoons B$ $\qquad K_{eq} = \dfrac{[B]}{[A]}$

$$\Delta G^0 = -RT \ln K_{eq}$$

R = molar gas constant = 8.3145 J (1.987 cal) $\cdot$ mol^{-1} $\cdot$ K^{-1}

T = in kelvin (K)

$$\%B = \dfrac{B}{A + B} \times 100$$

K_{eq}	ΔG^0 kJ/mol	ΔG^0 kcal/mol	$\ln K_{eq}$	$\log K_{eq}$	% B in Mixture
1	0.00	0.00	0.00	0.00	50.00
2	−1.72	−0.41	0.69	0.30	66.67
5	−3.97	−0.95	1.61	0.70	83.33
10	−5.69	−1.36	2.30	1.00	90.91
20	−7.41	−1.77	3.00	1.30	95.24
100	−11.4	−2.73	4.61	2.00	99.01
1,000	−17.1	−4.09	6.91	3.00	99.90
10,000	−22.8	−5.46	9.21	4.00	99.99

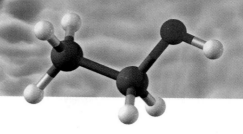

Major Classes of Organic Acids

Class and Example	Typical pK_a	Class and Example	Typical pK_a
Sulfonic acid	0–1	β-Ketoester	11
Carboxylic acid	3–5	Water $HO-H$	15.7
Arylammonium ion	4–5	Alcohol CH_3CH_2O-H	15–19
		Amide	15–19
Imide	8–9	Cyclopentadiene	16
Thiol CH_3CH_2S-H	8–12	α-Hydrogen of an aldehyde or ketone	18–20
Phenol	9–10	α-Hydrogen of an ester	23–25
Ammonium ion NH_3-H^+	9.24	Alkyne $HC{\equiv}C-H$	25
β-Diketone	10	Ammonia NH_2-H	38
Nitroalkane $H-CH_2NO_2$	10	Amine $[(CH_3)_2CH]_2N-H$	40
Alkylammonium ion $(CH_3CH_2)_3\overset{+}{N}-H$	10–12	Alkene $CH_2{=}CH-H$	44
		Alkane CH_3CH_2-H	51

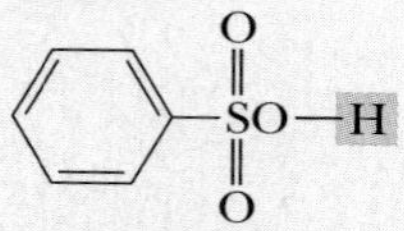

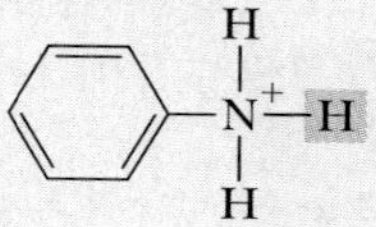

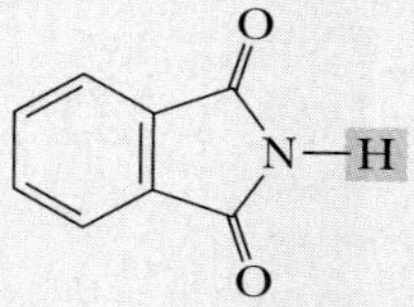

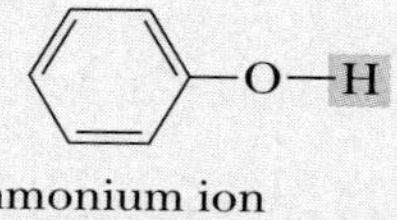

Sulfonic acid

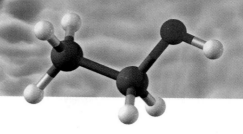

Carboxylic acid: CH_3CO-H

Arylammonium ion

Imide

Thiol: CH_3CH_2S-H

Phenol

Ammonium ion: NH_3-H^+

β-Diketone: $CH_3-\overset{O}{\overset{\|}{C}}-\overset{H}{\overset{\|}{CH}}-\overset{O}{\overset{\|}{C}}CH_3$

Nitroalkane: $H-CH_2NO_2$

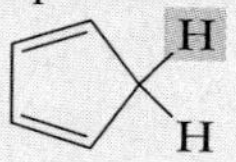

Alkylammonium ion: $(CH_3CH_2)_3\overset{+}{N}-H$

β-Ketoester: $CH_3-\overset{O}{\overset{\|}{C}}-\overset{H}{\overset{\|}{CH}}-\overset{O}{\overset{\|}{C}}OCH_2CH_3$

Water: $HO-H$

Alcohol: CH_3CH_2O-H

Amide: $CH_3\overset{O}{\overset{\|}{C}}N-H$

Cyclopentadiene

α-Hydrogen of an aldehyde or ketone: $CH_3\overset{O}{\overset{\|}{C}}CH_2-H$

α-Hydrogen of an ester: $CH_3CH_2O\overset{O}{\overset{\|}{C}}CH_2-H$

Alkyne: $HC{\equiv}C-H$

Ammonia: NH_2-H

Amine: $[(CH_3)_2CH]_2N-H$

Alkene: $CH_2{=}CH-H$

Alkane: CH_3CH_2-H

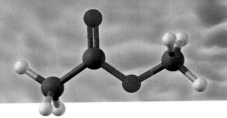

Bond Dissociation Enthalpies

Bond dissociation enthalpy (BDE) is defined as the amount of energy required to break a bond homolytically into two radicals in the gas phase at 25°C.

$$A\overset{\frown}{}B \longrightarrow A\cdot + B\cdot \qquad \Delta H^0[\text{kJ (kcal)/mol}]$$

Bond	ΔH^0	Bond	ΔH^0	Bond	ΔH^0
H—H bonds		**C—C multiple bonds**		**C—Br bonds**	
H—H	435 (104)	$CH_2{=}CH_2$	727 (174)	CH_3—Br	301 (72)
D—D	444 (106)	$HC{\equiv}CH$	966 (231)	C_2H_5—Br	301 (72)
				$(CH_3)_2CH$—Br	309 (74)
X—X bonds		**C—H bonds**		$(CH_3)_3C$—Br	305 (73)
F—F	159 (38)	CH_3—H	439 (105)	$CH_2{=}CHCH_2$—Br	247 (59)
Cl—Cl	247 (59)	C_2H_5—H	422 (101)	C_6H_5—Br	351 (84)
Br—Br	192 (46)	$(CH_3)_2CH$—H	414 (99)	$C_6H_5CH_2$—Br	263 (63)
I—I	151 (36)	$(CH_3)_3C$—H	405 (97)		
		$CH_2{=}CH$—H	464 (111)	**C—I bonds**	
H—X bonds		$CH_2{=}CHCH_2$—H	372 (89)	CH_3—I	242 (58)
H—F	568 (136)	C_6H_5—H	472 (113)	C_2H_5—I	238 (57)
H—Cl	431 (103)	$C_6H_5CH_2$—H	376 (90)	$(CH_3)_2CH$—I	238 (57)
H—Br	368 (88)	$HC{\equiv}C$—H	556 (133)	$(CH_3)_3C$—I	234 (56)
H—I	297 (71)			$CH_2{=}CHCH_2$—I	192 (46)
		C—F bonds		C_6H_5—I	280 (67)
		CH_3—F	481 (115)	$C_6H_5CH_2$—I	213 (51)
O—H bonds		C_2H_5—F	472 (113)		
HO—H	497 (119)	$(CH_3)_2CH$—F	464 (111)	**C—N single bonds**	
CH_3O—H	439 (105)	C_6H_5—F	531 (127)	CH_3—NH_2	355 (85)
C_6H_5O—H	376 (90)			C_6H_5—NH_2	435 (104)
		C—Cl bonds			
O—O bonds		CH_3—Cl	351 (84)	**C—O single bonds**	
HO—OH	213 (51)	C_2H_5—Cl	355 (85)	CH_3—OH	385 (92)
CH_3O—OCH_3	159 (38)	$(CH_3)_2CH$—Cl	355 (85)	C_6H_5—OH	468 (112)
$(CH_3)_3CO$—$OC(CH_3)_3$	159 (38)	$(CH_3)_3C$—Cl	355 (85)		
		$CH_2{=}CHCH_2$—Cl	288 (69)		
C—C single bonds		C_6H_5—Cl	405 (97)		
CH_3—CH_3	376 (90)	$C_6H_5CH_2$—Cl	309 (74)		
C_2H_5—CH_3	372 (89)				
$CH_2{=}CH$—CH_3	422 (101)				
$CH_2{=}CHCH_2$—CH_3	322 (77)				
C_6H_5—CH_3	435 (104)				
$C_6H_5CH_2$—CH_3	326 (78)				

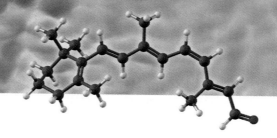

Characteristic ^{1}H-NMR Chemical Shifts

Type of Hydrogen (R = alkyl, Ar = aryl)	Chemical Shift (δ)*	Type of Hydrogen (R = alkyl, Ar = aryl)	Chemical Shift (δ)*
$(CH_3)_4Si$	0 (by definition)	RCH_2OH	3.4–4.0
R_2NH	0.5–5.0	RCH_2Br	3.4–3.6
ROH	0.5–6.0	RCH_2Cl	3.6–3.8
RCH_3	0.8–1.0	$\overset{\text{O}}{\overset{\|}{R C}}OCH_3$	3.7–3.9
RCH_2R	1.2–1.4		
R_3CH	1.4–1.7	$\overset{\text{O}}{\overset{\|}{R C}}OCH_2R$	4.1–4.7
$R_2C{=}CRCHR_2$	1.6–2.6		
$RC{\equiv}CH$	2.0–3.0	RCH_2F	4.4–4.5
		$ArOH$	4.5–4.7
$\overset{\text{O}}{\overset{\|}{R C}}CH_3$	2.1–2.3	$R_2C{=}CH_2$	4.6–5.0
		$R_2C{=}CHR$	5.0–5.7
$\overset{\text{O}}{\overset{\|}{R C}}CH_2R$	2.2–2.6	ArH	6.5–8.5
$ArCH_3$	2.2–2.5	$\overset{\text{O}}{\overset{\|}{R C}}H$	9.5–10.1
$ArCH_2R$	2.3–2.8		
RCH_2I	3.1–3.3	$\overset{\text{O}}{\overset{\|}{R C}}OH$	10–13
RCH_2OR	3.3–4.0		

*Values are relative to tetramethylsilane. Other atoms within the molecule may cause the signal to appear outside these ranges.

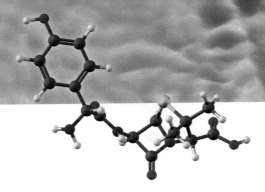

Characteristic ^{13}C-NMR Chemical Shifts

Type of Carbon	Chemical Shift (δ)	Type of Carbon	Chemical Shift (δ)
$(CH_3)_4Si$	0 (by definition)	C—R (aromatic)	110–160
RCH_2I	0–40		
RCH_3	10–40	$RCOR$ (O)	160–180
RCH_2R	15–55		
R_3CH	20–60	$RCNR_2$ (O)	165–180
RCH_2Br	25–65		
RCH_2Cl	35–80	$RCOH$ (O)	165–185
R_3COH	40–80		
R_3COR	40–80	RCH, RCR (O)	180–215
$RC{\equiv}CR$	65–85		
$R_2C{=}CR_2$	100–150		

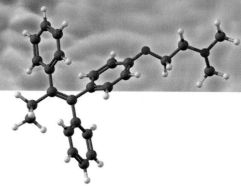

Characteristic Infrared Absorption Frequencies

Bonding		Frequency (cm^{-1})	Intensity*	Type of Vibration (Stretching unless noted)
C—H	alkane	2850–3000	m	
	—CH$_3$	1375 and 1450	w–m	bending
	—CH$_2$—	1450–1475	m	bending
	alkene	3000–3100	w–m	
		650–1000	s	out-of-plane bending
	alkyne	3300	m-s	
	arene	3030	w-m	
		690–900	s	out-of-plane bending
	aldehyde	2720	w	
C=C	alkene	1600–1680	w–m	
	arene	1450 and 1600	m	
C≡C	alkyne	2100–2250	w	
C—O	alcohol, ether, ester, carboxylic acid	1000–1100 (sp^3 C—O)	s	
		1200–1250 (sp^2 C—O)	s	
	anhydride	900–1300	s	
C=O	amide	1630–1680	s	
	carboxylic acid	1700–1725	s	
	ketone	1630–1820	s	
	aldehyde	1630–1820	s	
	ester	1735–1800	s	
	anhydride	1740–1760 and 1800–1850	s	
	acid chloride	1800	s	
O—H	alcohol, phenol			
	free	3600–3650	w	
	hydrogen bonded	3200–3500	m	
	carboxylic acid	2500–3300	s	
N—H	amine and amide	3100–3550	m–s	
C≡N	nitrile	2200–2250	m	

*m = medium, s = strong, w = weak

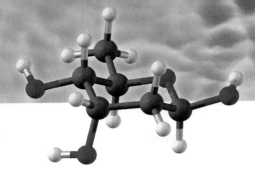

Electrostatic Potential Maps

The term *electronic structure* refers to the distribution of electron density in a molecule. According to the laws of quantum mechanics, electrons have no definite locations. Instead, they collectively produce a negatively charged region around a nucleus, measured by electron density in units of $e/\text{Å}^3$ (electrons per cubic angstrom). For an atom, density is high near the nucleus and vanishingly small far from the nucleus. Electrostatic potential maps are now easily computed for small molecules using desktop computers and various software packages. Electrostatic potential maps (elpots) in this text were produced using MacSpartan (Wavefunction, Inc.).

Electrostatic potential (elpot) maps provide a way to visualize the distribution of electron density in a molecule. Electrostatic potential is defined as the potential energy that a positively charged particle would experience in a molecule's presence. The electrostatic potential is made up of two parts.

1. The repulsive component (positive potential, repulsion) exerted by the positively charged nuclei
2. The attractive component (negative potential, attraction) exerted by the negatively charged electron cloud

Thus, electrostatic potential contains information about the entire electron distribution.

Electrostatic potential maps are color-coded. By convention, the most negative potential is red, and the most positive potential is blue. Intermediate potentials are coded accordingly (orange-yellow-green). While any surface might be chosen to display an electrostatic potential map, the most common is $0.002 \ e/\text{Å}^3$. Nearly all of a molecule's electron density lies within this surface, which corresponds almost exactly to how closely another molecule can approach without running into severe steric repulsive forces; that is, the surface corresponds almost exactly to the van der Waals surface of the molecule.

An electrostatic potential map for ethylene shows areas of high electron density to which electrophiles will be attracted (red) over the pi orbitals. There are four blue patches, one over each hydrogen; these regions are relatively electron-poor.

The methyl carbocation, CH_3^+, provides an even more dramatic visualization of an electrostatic potential. The entire ion is blue in color, corresponding to the net positive charge. The central atom is the deepest blue, corresponding to the location of the largest fraction of the positive charge.

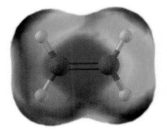

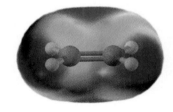

■ An electrotatic potential map of ethylene (top and side views).

■ An electrostatic potential map of the methyl cation (CH_3^+).

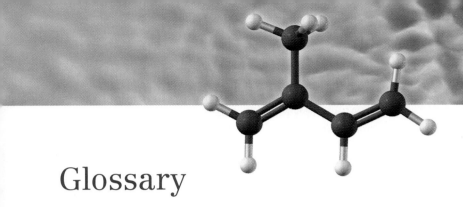

Glossary

Absolute configuration   (Section 3.3) Which of the two possible isomers an enantiomer is (i.e., whether it is the right- or left-handed isomer).

Absorbance (A)   (Section 20.3A) A quantitative measure of the extent to which a compound absorbs radiation of a particular wavelength. $A = \log (I_0/I)$ where I_0 is the incident radiation and I is the transmitted radiation.

Acetal   (Section 16.7B) A molecule containing two —OR or —OAr groups bonded to the same carbon.

Aceto group   (Section 17.2B) A CH_3CO— group; also called an acetyl group.

Achiral   (Section 3.2) An object that lacks chirality; an object that has no handedness.

Activating group   (Section 22.2A) Any substituent on a benzene ring that causes the rate of electrophilic aromatic substitution to be greater than that for benzene.

Activation energy   (Section 6.2A) The difference in Gibbs free energy between reactants and a transition state.

Acylation   (Section 19.5B) The process of introducing an acyl group, RCO— or ArCO— onto an organic molecule.

Acyl group   (Section 18.1A) An RCO— or ArCO— group.

Acylium ion   (Section 22.1C) A resonance-stabilized cation with the structure $[RC{=}O]^+$ or $[ArC{=}O]^+$. The positive charge is delocalized over both the carbonyl carbon and the carbonyl oxygen.

Addition reaction   (Section 6.1) A reaction in which two atoms or ions react with a double bond, forming a compound with the two new groups bonded to the carbons of the original double bond.

Aglycon   (Synthesis Problems, Chapter 24) Lacking a sugar.

Alcohol   (Section 1.3A) A compound containing an —OH (hydroxyl) group bonded to a carbon atom.

Aldehyde   (Section 1.3C) A compound containing a —CHO group.

Alditol   (Section 25.3B) The product formed when the $C{=}O$ group of a monosaccharide is reduced to a CHOH group.

Aldonic acid   (Section 25.3C) The product formed when the —CHO group of an aldose is oxidized to a —COOH group.

Aldose   (Section 25.1A) A monosaccharide containing an aldehyde group.

Aliphatic amine   (Section 23.1) An amine in which nitrogen is bonded only to alkyl groups.

Alkaloid   (Section 23.1) A basic nitrogen-containing compound of plant origin, many of which are physiologically active when administered to humans.

Alkene metathesis   (Section 24.5) A reaction in which two alkenes interchange the carbons attached to their double bonds.

Alkyl group   (Section 2.3A) A group derived by removing a hydrogen from an alkane; given the symbol R—.

Alkyne   (Section 7.1) An unsaturated hydrocarbon that contains one or more carbon-carbon triple bonds.

Alkylation reaction   (Section 7.5A) Any reaction in which a new carbon-carbon bond to an alkyl group is formed.

Allene   (Section 7.5B) The compound $CH_2{=}C{=}CH_2$. Any compound that contains adjacent carbon-carbon double bonds; that is, any molecule that contains a $C{=}C{=}C$ functional group.

Allyl   (Section 5.3B) a —$CH_2CH{=}CH_2$ group.

Allylic   (Section 9.4B) Next to a carbon-carbon double bond.

Allylic carbocation   (Section 9.4E) A carbocation in which an allylic carbon bears the positive charge.

Allylic carbon   (Section 8.6) A carbon adjacent to a carbon-carbon double bond.

Allylic substitution   (Section 8.6) Any reaction in which an atom or group of atoms is substituted for another atom or group of atoms at an allylic carbon.

Amino acid (Section 27.1A) A compound that contains both an amino group and a carboxyl group.

α-Amino acid (Section 27.1A) An amino acid in which the amino group is on the carbon adjacent to the carboxyl group.

Amino group (Section 1.3B) A compound containing an sp^3-hybridized nitrogen atom bonded to one, two, or three carbon atoms.

Amorphous domain (Section 29.4) A disordered, noncrystalline region in the solid state of a polymer.

Anabolic steroid (Section 26.4A) A steroid hormone, such as testosterone, which promotes tissue and muscle growth and development.

Androgen (Section 26.4A) A steroid hormone, such as testosterone, which mediates the development of sexual characteristics of males.

Angle strain (Section 2.6A) The strain that arises when a bond angle is either compressed or expanded compared to its optimal value.

Anion (Section 1.2A) An atom or group of atoms bearing a negative charge.

Annulene (Section 21.2B) A cyclic hydrocarbon with a continuous alternation of single and double bonds.

Anomeric carbon (Section 25.2A) The hemiacetal or acetal carbon of the cyclic form of a carbohydrate.

Anomers (Section 25.2A) Carbohydrates that differ in configuration only at their anomeric carbons.

Antiaromatic compound (Section 21.2C) A monocyclic compound that is planar or nearly so, has one p orbital on each atom of the ring, and has $4n$ pi electrons in the cyclic arrangement of overlapping p orbitals, where n is an integer. Antiaromatic compounds are especially unstable.

Antibonding molecular orbital (Section 1.8A) A molecular orbital in which electrons have a higher energy than they would in isolated atomic orbitals.

Anti conformation (Section 2.6A) A conformation about a single bond in which two groups on adjacent carbons lie at a dihedral angle of 180°.

Anti stereoselectivity (Section 6.3D) The addition of atoms or groups of atoms to opposite faces of a carbon-carbon double bond.

Aprotic acid (Section 4.6) An acid that is not a proton donor; an acid that is an electron pair acceptor in a Lewis acid-base reaction.

Aprotic solvent (Section 9.2) A solvent that cannot serve as a hydrogen-bond donor; nowhere in the molecule is a hydrogen bonded to an atom of high electronegativity. Common aprotic solvents are dichloromethane, diethyl ether, and dimethyl sulfoxide.

Aramid (Section 29.5A) A polyaromatic amide; a polymer in which the monomer units are an aromatic diamine and an aromatic dicarboxylic acid.

Arene (Introduction, Chapter 5) A term used to classify benzene and its derivatives.

Aromatic amine (Section 23.1) An amine in which nitrogen is bonded to one or more aryl groups.

Aromatic compound (Introduction, Chapter 21) A term used initially to classify benzene and its derivatives. More accurately, it is used to classify any compound that meets the Hückel criteria for aromaticity (Section 21.2A).

Aryl group (Ar—) (Introduction, Chapter 5) A group derived from an arene by removal of an H.

Atactic polymer (Section 29.6C) A polymer with completely random configurations at the chiral centers along its chain, as, for example, atactic polypropylene.

Atropisomers (Section 3.2) Enantiomers that lack a chiral center and differ because of hindered rotation.

Aufbau principle (Section 1.1A) Orbitals fill in order of increasing energy, from lowest to highest.

Autoxidation (Section 8.7) Air oxidation of materials such as unsaturated fatty acids.

Axial bond (Section 2.6B) A bond to a chair conformation of cyclohexane that extends from the ring parallel to the imaginary axis through the center of the ring; a bond that lies roughly perpendicular to the equator of the ring.

Azeotrope (Section 16.7B) A liquid mixture of constant composition with a boiling point that is different from that of any of its components.

Base peak (Section 14.1) The peak caused by the most abundant ion in a mass spectrum; the most intense peak. It is assigned an arbitrary intensity of 100.

Basicity (Section 9.4B) An equilibrium property measured by the position of equilibrium in an acid-base reaction, as for example the acid-base reaction between ammonia and water.

Benzyl group ($C_6H_5CH_2$—) (Section 21.3A) The group derived from toluene by removing a hydrogen from its methyl group.

Benzylic position (Section 21.5) An sp^3-hybridized carbon bonded to a benzene ring.

Benzyne intermediate (Section 22.3A) A reactive intermediate formed by β-elimination from adjacent carbon atoms of a benzene ring and having a triple bond in the benzene ring. The second pi bond of the benzyne triple bond is formed by the weak overlap of coplanar sp^2 orbitals on adjacent carbons.

Betaine (Section 16.6) A neutral molecule with nonadjacent positive and negative charges. An example of a betaine is the intermediate formed by addition of a Wittig reagent to an aldehyde or ketone.

Bicycloalkane (Section 2.7B) An alkane containing two rings that share two carbons.

Bile acid (Section 26.4A) A cholesterol-derived detergent molecule, such as cholic acid, which is secreted by the gallbladder into the intestine to assist in the absorption of dietary lipids.

Bimolecular reaction (Section 9.3) A reaction in which two species are involved in the rate-determining step.

Boat conformation (Section 2.6B) A nonplanar conformation of a cyclohexane ring in which carbons 1 and 4 of the ring are bent toward each other.

Bond dipole moment (μ) (Section 1.2B) A measure of the polarity of a covalent bond. The product of the charge on either atom of a polar bond times the distance between the atoms.

Bonding electrons (Section 1.2C) Valence electrons involved in forming a covalent bond (i.e., shared electrons).

Bonding molecular orbital (Section 1.8A) A molecular orbital in which electrons have a lower energy than they would in isolated atomic orbitals.

Bond length (Section 1.2B) The distance between atoms in a covalent bond in picometers (pm; 1 pm = 10^{-12} m) or Å (1Å = 10^{-10} m).

Brønsted-Lowry acid (Section 4.2) A proton donor.

Brønsted-Lowry base (Section 4.2) A proton acceptor.

Carbanion (Section 16.5A) An anion in which carbon has an unshared pair of electrons and bears a negative charge.

Carbene (Section 15.4) A neutral molecule that contains a carbon atom surrounded by only six valence electrons ($R_2C:$).

Carbenoid (Section 15.4C) A compound that delivers the elements of a carbene without actually producing a free carbene.

Carbocation (Section 6.3A) A species in which a carbon atom has only six electrons in its valence shell and bears a positive charge.

Carbohydrate (Section 25.1A) A polyhydroxyaldehyde, a polyhydroxyketone, or a substance that gives these compounds on hydrolysis.

α-Carbon (Section 16.9A) A carbon atom adjacent to a carbonyl group.

Carbonyl group (Section 1.3C) A C=O group.

Carboxyl group (Section 1.3D) A —COOH group.

Carboxylic acid (Section 1.3D) A compound containing a carboxyl, —COOH, group.

Cation (Section 1.2A) An atom or group of atoms bearing a positive charge.

Center of symmetry (Section 3.2) A point so situated that identical components of an object are located on opposite sides and equidistant from that point along any axis passing through it.

Chain-growth polymerization (Section 29.6) A polymerization that involves sequential addition reactions, either to unsaturated monomers or to monomers possessing other reactive functional groups.

Chain initiation (Section 8.5B) A step in a chain reaction characterized by the formation of reactive intermediates (radicals, anions, or cations) from nonradical or noncharged molecules.

Chain length (Section 8.5B) The number of times the cycle of chain propagation steps repeats in a chain reaction.

Chain propagation (Section 8.5B) A step in a chain reaction characterized by the reaction of a reactive intermediate and a molecule to give a new reactive intermediate.

Chain termination (Section 8.5B) A step in a chain reaction that involves destruction of reactive intermediates.

Chain-transfer reaction (Section 29.6A) The transfer of reactivity of an endgroup from one chain to another during a polymerization.

Chair conformation (Section 2.6B) The most stable nonplanar conformation of a cyclohexane ring; all bond angles are approximately 109.5°, and all bonds on adjacent carbons are staggered.

Chemical shift (δ) (Section 13.3) The shift in parts per million of an NMR signal relative to the signal of TMS.

Chiral (Section 3.2) From the Greek, *cheir*, hand; an object that is not superposable on its mirror image; an object that has handedness.

Chiral center (Section 3.2) A tetrahedral atom, most commonly carbon, that is bonded to four different groups; also called a chirality center.

Chlorofluorocarbons (CFCs, Freons) (Section 8.3B) Compounds of one or two carbons, chlorine, and fluorine, formerly used as refrigerants.

Chromatography (Section 3.8C) A separation method involving passing a vapor or solution mixture through a column packed with a material with different affinities for different components of the mixture.

Circular DNA (Section 28.2C) A type of double-stranded DNA in which the 5′ and 3′ ends of each strand are joined by phosphodiester groups.

Cis (Section 2.7A) A prefix meaning on the same side.

Cis,trans isomers (Sections 2.7 and 5.1C) Isomers that have the same order of attachment of their atoms but a different arrangement of their atoms in space owing to the presence of either a ring or a carbon-carbon double bond.

Clemmensen reduction (Section 16.11C) Reduction of the C=O group of an aldehyde or ketone to a CH_2 group using $Zn(Hg)$ and HCl.

Codon (Section 28.4A) A triplet of nucleotides on mRNA that directs incorporation of a specific amino acid into a polypeptide sequence.

Condensation polymerization (Section 29.5) A polymerization in which chain growth occurs in a stepwise manner between difunctional monomers. Also called step-growth polymerization.

Configuration (Section 2.7A) Refers to the arrangement of atoms about a stereocenter.

Configurational isomers (Section 3.1) Isomers that differ by the configuration of substituents on an atom.

Conformation (Section 2.6A) Any three-dimensional arrangement of atoms in a molecule that results by rotation about a single bond.

Conjugate acid (Section 4.2A) The species formed when a base accepts a proton from an acid.

Conjugate addition (Section 19.8) Addition of a nucleophile to the β-carbon of an α,β-unsaturated carbonyl compound. (Section 20.2A) Addition to carbons 1 and 4 of a conjugated diene.

Conjugate base (Section 4.2A) The species formed when an acid transfers a proton to a base.

Conjugated (Section 20.1) A conjugated diene or carbonyl is one in which the double bonds are separated by one single bond.

Conjugation (Section 12.4H) A situation in which two multiple bonds are separated by a single bond. Alternatively, a series of overlapping p orbitals. 1,3-butadiene, for example, is a conjugated diene, and 3-butene-2-one is a conjugated enone.

Constitutional isomers (Section 2.2) Compounds with the same molecular formula but a different connectivity of their atoms.

Contributing structures (Section 1.6A) Representations of a molecule or ion that differ only in the distribution of valence electrons.

Correlation tables (Section 12.3D) Tables of data on absorption patterns of functional groups.

Coupling constant (J) (Section 13.9) The separation on an NMR spectrum (in hertz) between adjacent peaks in a multiplet and a quantitative measure of the influence of the spin-spin coupling from adjacent nuclei.

Covalent bond (Section 1.2A) A chemical bond formed between two atoms by sharing one or more pairs of electrons.

Crystalline domain (Section 29.4) An ordered crystalline region in the solid state of a polymer. Also called a crystallite.

Cumulated (Section 20.1) A diene in which two double bonds share an sp-hybridized carbon.

Curved arrow (Section 1.6A) A symbol used to show the redistribution of valence electrons in resonance contributing structures or reactions, symbolizing movement of two electrons.

Cyanohydrin (Section 16.5D) A molecule containing an —OH group and a —CN group bonded to the same carbon.

Cycloaddition reaction (Section 24.6) A reaction in which two reactants add together in a single step to form a cyclic product. The best known of these is the Diels-Alder reaction.

Cycloalkane (Section 2.4) A saturated hydrocarbon that contains carbons joined to form a ring.

Deactivating group (Section 22.2A) Any substituent on a benzene ring that causes the rate of electrophilic aromatic substitution to be lower than that for benzene.

Decarboxylation (Section 17.9A) Loss of CO_2 from a carboxyl group.

Dehydration (Section 10.6) Elimination of water.

Dehydrohalogenation (Section 9.6) Removal of —H and —X from adjacent carbons; a type of β-elimination.

DEPT-NMR (Section 13.12) Distortionless enhancement by polarization transfer. An NMR technique for distinguishing among ^{13}C signals for CH_3, CH_2, CH, and quaternary carbons in ^{13}C-NMR spectra.

Deshielding (Section 13.3) The term used to express the concept of less shielding in NMR.

Dextrorotatory (Section 3.7B) Refers to a substance that rotates the plane of polarized light to the right.

Diamagnetic current in NMR (Section 13.3) The circulation of electron density in a molecule in an applied magnetic field.

Diastereomers (Section 3.4A) Stereoisomers that are not mirror images of each other; refers to relationships among two or more objects.

Diastereotopic groups (Section 13.10) Atoms or groups on an atom that are bonded to an atom that is bonded to two nonidentical groups, one of which contains a chiral center. When one of the atoms or groups is replaced by another group, a new chiral center is created and a set of diastereomers results. The hydrogens of the CH_2 group of 2-butanol, for example, are diastereotopic. Diastereotopic groups have different chemical shifts under all conditions.

Diaxial interactions (Section 2.6B) Refers to the steric strain arising from interaction between an axial substituent and an axial hydrogen (or other group) on the same side of a chair conformation of a cyclohexane ring.

Diazonium ion (Section 23.8D) An ArN_2^+ or RN_2^+ ion.

Dielectric constant (Section 9.2) A measure of a solvent's ability to insulate opposite charges from one another.

Diels-Alder adduct (Section 24.6) A cyclohexene resulting from the cycloaddition reaction of a diene and a dienophile.

Dienophile (Section 24.6) A compound containing a double bond (consisting of one or two C, N, or O atoms) that can react with a conjugated diene to give a Diels-Alder adduct.

Dihedral angle (Section 2.6A) The angle created by two intersecting planes.

Diol (Section 10.1B) A compound containing two hydroxyl groups.

Dipeptide (Section 27.3) A molecule containing two amino acid units joined by a peptide bond.

Dipole-dipole interaction (Section 10.2) The attraction between the positive end of one dipole and the negative end of another.

Disaccharide (Section 25.4) A carbohydrate containing two monosaccharide units joined by a glycosidic bond.

Dispersion forces (Section 2.7B) Very weak intermolecular forces of attraction resulting from the interaction between temporary induced dipoles.

Disproportionation (Section 29.6A) A termination process that involves the abstraction of a hydrogen atom from the beta position of the propagating radical of one chain by the radical endgroup of another chain.

Double-headed arrow (Section 1.5A) A symbol used to show that structures on either side of it are resonance contributing structures.

Double helix (Section 28.2B) A type of secondary structure of DNA molecules in which two antiparallel polynucleotide strands are coiled in a right-handed manner about the same axis.

Downfield (Section 13.4) A signal of an NMR spectrum that is shifted toward the left (larger chemical shift) on the chart paper.

_E_ (Section 5.3C) From the German, _entgegen,_ opposite. Specifies that groups of higher priority on the carbons of a double bond are on opposite sides.

E1 (Section 9.7A) A unimolecular β-elimination reaction.

E2 (Section 9.7B) A bimolecular β-elimination reaction.

Eclipsed conformation (Section 2.6A) A conformation about a carbon-carbon single bond in which the atoms or groups on one carbon are as close as possible to the atoms or groups on an adjacent carbon.

Edman degradation (Section 27.4B) A method for selectively cleaving and identifying the _N_-terminal amino acid of a polypeptide chain.

Elastomer (Section 29.4) A material that, when stretched or otherwise distorted, returns to its original shape when the distorting force is released.

Electromagnetic radiation (Section 12.1) Light and other forms of radiant energy.

Electronegativity (Section 1.2B) A measure of the force of an atom's attraction for electrons.

Electrophile (Section 6.3A) From the Greek meaning electron loving. Any species that can accept a pair of electrons to form a new covalent bond; alternatively, a Lewis acid.

Electrophilic aromatic substitution (Section 22.1) A reaction in which there is substitution of an electrophile, E^+, for a hydrogen on an aromatic ring.

Electrophoresis (Section 27.2D) The process of separating compounds on the basis of their electric charge.

β-Elimination (Introduction, Chapter 9) A reaction in which a molecule, such as HCl, HBr, HI, or HOH, is split out or eliminated from adjacent carbons.

Enamine (Section 16.8A) An unsaturated compound derived by the reaction of an aldehyde or ketone and a secondary amine followed by loss of H_2O; $R_2C=CR-NR_2$.

Enantiomeric excess (ee) (Section 3.7D) The difference between the percentage of two enantiomers in a mixture.

Enantiomers (Section 3.2) Stereoisomers that are nonsuperposable mirror images of each other; refers to a relationship between pairs of objects.

Enantioselective reaction (Section 6.7B) A reaction that produces one enantiomer in preference to the other.

Enantiotopic groups (Section 13.10) Atoms or groups on an atom that give a chiral center when one of the groups is replaced by another group. A pair of enantiomers results. The hydrogens of the CH_2 group of ethanol, for example, are enantiotopic. Replacing one of them by deuterium gives (R)-1-deuteroethanol; replacing the other gives (S)-1-deuteroethanol. Enantiotopic groups have identical chemical shifts in achiral environments but different chemical shifts in chiral environments.

Endergonic reaction (Section 6.2) A reaction in which the Gibbs free energy of the products is higher than that of the reactants. The position of equilibrium for an endergonic reaction favors starting materials.

Endothermic reaction (Section 6.2A) A reaction in which the enthalpy of the products is higher than the enthalpy of the reactants; a reaction in which heat is absorbed.

Energy diagram (Section 6.2A) A graph showing the changes in energy that occur during a chemical reaction; energy is plotted on the vertical axis, and reaction progress is plotted on the horizontal axis.

Enol (Section 7.7A) A compound containing a hydroxyl group bonded to a doubly bonded carbon atom.

Enolate anion (Section 16.9A) An anion derived by loss of a hydrogen from a carbon alpha to a carbonyl group; the anion of an enol.

Enthalpy change, ΔH^0 (Section 6.2A) The difference in total bond energy between reactants and products; a measure of bond making (exothermic) and bond breaking (endothermic).

Equatorial bond (Section 2.6B) A bond to a chair conformation of cyclohexane that extends from the ring roughly perpendicular to the imaginary axis through the center of the ring; a bond that lies roughly along the equator of a cyclohexane ring.

Equivalent hydrogens (Section 13.5) Hydrogens that have the same chemical environment.

Ester (Section 1.3E) A derivative of a carboxylic acid in which H of the carboxyl group is replaced by a carbon.

Estrogen (Section 26.4A) A steroid hormone, such as estrone and estradiol, that mediates the development of sexual characteristics in females.

Exergonic reaction (Section 6.2) A reaction in which the Gibbs free energy of the products is lower than that of the reactants. The position of equilibrium for an exergonic reaction favors products.

Exothermic reaction (Section 6.2A) A reaction in which the enthalpy of the products is lower than that of the reactants; a reaction in which heat is released.

E,Z system (Section 5.3C) A system to specify the configuration of groups about a carbon-carbon double bond.

Fat (Section 26.1B) A mixture of triglycerides that is semisolid or solid at room temperature.

Fatty acid (Section 26.1A) A long, unbranched-chain carboxylic acid, most commonly of 12 to 20 carbons, derived from the hydrolysis of animal fats, vegetable oils, or the phospholipids of biological membranes.

Fingerprint region (Section 12.3D) Vibrations in the region 1500 to 400 cm^{-1} of an IR spectrum.

Fischer esterification (Section 17.7A) The process of forming an ester by refluxing a carboxylic acid and an alcohol in the presence of an acid catalyst, commonly H_2SO_4, $ArSO_3H$, or HCl.

Fischer projection (Section 3.5C and Section 25.1B) A two-dimensional representation for showing the configuration of chiral centers; horizontal lines represent bonds projecting forward, and vertical lines represent bonds projecting to the rear.

Fishhook arrow (Section 8.5A) A barbed curved arrow used to show the change in position of a single electron.

Fluid-mosaic model (Section 26.5B) A biological membrane that consists of a phospholipid bilayer with proteins, carbohydrates, and other lipids on the surface and embedded in the bilayer.

Formal charge (Section 1.2D) The charge on an atom in a polyatomic ion or molecule.

Fourier transform NMR (FT-NMR) (Section 13.4) The modern NMR method that is based on a constant magnetic field, a short pulse of electromagnetic radiation, and a mathematical Fourier transform to produce the spectrum.

Frequency (ν) (Section 12.1) The number of full cycles of a wave that pass a given point in a second, and reported in hertz (Hz), which has the units s^{-1}.

Friedel-Crafts reaction (Section 22.1C) An electrophilic aromatic substitution in which a hydrogen of an aromatic ring is replaced by an alkyl or acyl group.

Frost circle (Section 21.2A) A graphic method for determining the relative energies of pi MOs for planar, fully conjugated, monocyclic compounds.

Functional group   (Section 1.3) An atom or group of atoms within a molecule that shows a characteristic set of physical and chemical properties.

Furanose   (Section 25.2A) A five-membered cyclic form of a monosaccharide.

Gauche conformation   (Section 2.6A) A conformation about a single bond of an alkane in which two groups on adjacent carbons lie at a dihedral angle of 60°.

Geminal coupling   (Section 13.9D) Spin-spin coupling that occurs between nonequivalent H atoms bonded to the same C atom. The H atoms are generally nonequivalent owing to restricted bond rotation in the molecule.

Gibbs free energy change (ΔG)   (Section 6.2A) A thermodynamic function relating enthalpy, entropy, and temperature, given by the equation $\Delta G = \Delta H - T\Delta S$. If $\Delta G > 0$, the position of equilibrium for the reaction favors the starting material(s). If $\Delta G < 0$, the position of equilibrium favors the product(s).

Glass transition temperature (T_g)   (Section 29.4) The temperature at which a polymer undergoes the transition from a hard glass to a rubbery state.

Glycol   (Section 6.5B) A compound with hydroxyl (—OH) groups on adjacent carbons.

Glycoside   (Section 25.3A) A carbohydrate in which the —OH on its anomeric carbon is replaced by —OR.

Glycosidic bond   (Section 25.3A) The bond from the anomeric carbon of a glycoside to an —OR group.

Ground-state electron configuration   (Section 1.1A) The lowest-energy electron configuration for an atom or molecule.

Haloalkane (alkyl halide)   (Section 8.1) A compound containing a halogen atom covalently bonded to an sp^3-hybridized carbon atom. Given the symbol R—X.

Haloalkene (vinylic halide)   (Section 8.1) A compound containing a halogen atom bonded to one of the carbons of a carbon-carbon double bond.

Haloarene (aryl halide)   (Section 8.1) A compound containing a halogen atom bonded to a benzene ring. Given the symbol Ar—X.

Haloform   (Section 8.2B) A compound of the type CHX_3 where X is a halogen.

Halohydrin   (Section 6.3E) A compound containing a halogen atom and a hydroxyl group on adjacent carbons; those containing Br and OH are bromohydrins, and those containing Cl and OH are chlorohydrins.

Hammond's postulate   (Section 8.5D) The structure of the transition state for an exothermic step looks more like the reactants of that step than the products. Conversely, the structure of the transition state for an endothermic step looks more like the products of that step than the reactants.

Haworth projection   (Section 25.2A) A way to view furanose and pyranose forms of monosaccharides. The ring is drawn flat and most commonly viewed through its edge with the anomeric carbon on the right and the oxygen atom of the ring in the rear.

Heat of combustion (ΔH^0)   (Section 2.9A) The heat released when one mole of a substance in its standard state (gas, liquid, solid) is oxidized completely to carbon dioxide and water.

Heat of reaction (ΔH^0)   (Section 6.2A) The difference in enthalpy between reactants and products. If the enthalpy of products is lower than that of the reactants, heat is released and the reaction is exothermic. If the enthalpy of the products is higher than that of the reactants, energy is absorbed, and the reaction is endothermic.

α-Helix   (Section 27.6B) A type of secondary structure in which a section of polypeptide chain coils into a spiral, most commonly a right-handed spiral.

Hemiacetal   (Section 16.7B) A molecule containing an —OH and an —OR or —OAr group bonded to the same carbon.

Hertz (Hz)   (Section 12.1) The unit is which frequency is measured: s^{-1} (read "per second").

Heterocyclic amine   (Section 23.1) An amine in which nitrogen is one of the atoms of a ring.

Heterocyclic aromatic amine   (Section 23.1) An amine in which nitrogen is one of the atoms of an aromatic ring.

Heterolytic cleavage   (Section 8.5A) Cleavage of a bond so that one fragment retains both electrons and the other retains none.

High-density lipoprotein (HDL)   (Section 26.4A) Plasma particles, density 1.06–1.21 g/mL, consisting of approximately 33% proteins, 30% cholesterol, 29% phospholipids, and 8% triglycerides.

High-resolution mass spectrometry   (Section 14.2A) Instrumentation that is capable of separating ions that differ in mass by as little as 0.0001 amu.

Histone   (Section 28.2C) A protein, particularly rich in the basic amino acids lysine and arginine, that is found associated with DNA molecules.

Hofmann elimination   (Section 23.9) When treated with a strong base, a quaternary ammonium halide undergoes β-elimination by an E2 mechanism to give the less-substituted alkene as the major product.

Hofmann rule   (Section 23.9) Any β-elimination that occurs preferentially to give the less substituted alkene as the major product.

Homolytic cleavage   (Section 8.5A) Cleavage of a bond so that each fragment retains one electron; formation of radicals.

Homotopic groups   (Section 13.10) Atoms or groups on an atom that give an achiral molecule when one of the groups is replaced by another group. The hydrogens of the CH_2 group of propane, for example, are homotopic. Replacing either one of them with deuterium gives 2-deuteropropane, which is achiral. Homotopic groups have identical chemical shifts under all conditions.

Hückel criteria for aromaticity   (Section 21.2A) To be aromatic, a monocyclic compound must have one p orbital on each atom of the ring, be planar or nearly so, and have $(4n + 2)$ pi electrons in the cyclic arrangement of p orbitals.

Hund's rule   (Section 1.1A) When orbitals of equivalent energy are available but there are not enough electrons to fill all of them completely, one electron is put in each before a second electron is added to any.

Hybridization   (Section 1.8B) The combination of atomic orbitals of different types.

Hybrid orbital   (Section 1.8B) An orbital formed by the combination of two or more atomic orbitals.

Hydration   (Section 6.3B) The addition of water.

Hydride ion   (Section 16.11A) A hydrogen atom with two electrons in its valence shell; $H:^-$.

Hydroboration-oxidation   (Section 6.4) A method for converting an alkene to an alcohol. The alkene is treated with borane (BH_3) to give a trialkylborane, which is then oxidized with alkaline hydrogen peroxide to give the alcohol.

α-Hydrogen   (Section 16.9A) A hydrogen on a carbon alpha to a carbonyl group.

Hydrogen bonding   (Section 10.2) The attractive interaction between a hydrogen atom bonded to an atom of high electronegativity (most commonly O or N) and a lone pair of electrons on another atom of high electronegativity (again, most commonly O or N).

Hydrogenolysis   (Section 21.5C) Cleavage of a single bond by H_2, most commonly accomplished by treating a compound with H_2 in the presence of a transition metal catalyst.

Hydrophilic   (Section 9.10) From the Greek, meaning water loving.

Hydrophobic   (Section 9.10) From the Greek, meaning water fearing.

Hydrophobic effect   (Section 27.6D) The tendency of nonpolar groups to cluster so as to shield themselves from contact with an aqueous environment.

Hydroxyl group   (Section 1.3A) An —OH group.

Hyperconjugation   (Section 6.3A) Interaction of electrons in a sigma-bonding orbital with the vacant $2p$ orbital of an adjacent positively charged carbon.

Imide   (Section 18.1D) A functional group in which two acyl groups, RCO— or ArCO— are bonded to a nitrogen atom.

Imine   (Section 16.8A) A compound containing a carbon-nitrogen double bond, $R_2C{=}NR'$; also called a Schiff base.

Index of hydrogen deficiency   (Section 5.1C) The sum of the number of rings and pi bonds in a molecule.

Inductive effect   (Section 4.5D, 6.3A) The polarization of the electron density of a covalent bond caused by the electronegativity of a nearby atom.

Infrared active   (Section 12.3B) Any molecular vibration that leads to a substantial change in dipole moment and is observed in an IR spectrum.

Infrared (IR) spectroscopy   (Introduction, Chapter 12) A spectroscopic technique in which a compound is irradiated with infrared radiation, absorption of which causes covalent bonds to change from a lower vibration state to a higher one. Infrared spectroscopy is particularly valuable for determining the kinds of functional groups present in a molecule.

Ionization potential (IP)   (Section 14.1) The minimum energy required to remove an electron from an atom or molecule to a distance where there is no electrostatic interaction between the resulting ion and electron.

Isoelectric point (pI)   (Section 27.2C) The pH at which an amino acid, polypeptide, or protein has no net charge.

Isomers   (Section 1.2C) Different compounds with the same molecular formula.

Isotactic polymer   (Section 29.6C) A polymer with identical configurations (either all R or all S) at all chiral centers along its chain, as, for example, isotactic polypropylene.

Keto-enol tautomerism   (Section 7.7A) A type of isomerism involving keto (from ketone) and enol tautomers.

Ketone   (Section 1.3C) A compound containing a carbonyl group bonded to two carbons.

Ketose (Section 25.1A) A monosaccharide containing a ketone group.

Kinetic control (Section 19.8A) Experimental conditions under which the composition of the product mixture is determined by the relative rates of formation of each product.

Lactam (Section 18.1D) A cyclic amide.

Lactone (Section 18.1C) A cyclic ester.

Leaving group (Introduction, Chapter 9) The group that is displaced in a substitution reaction or the Lewis base that is lost in an elimination reaction.

Levorotatory (Section 3.7B) Refers to a substance that rotates the plane of polarized light to the left.

Lewis acid (Section 4.6) Any molecule or ion that can form a new covalent bond by accepting a pair of electrons.

Lewis base (Section 4.6) Any molecule or ion that can form a new covalent bond by donating a pair of electrons.

Lewis dot structure (Section 1.1B) The symbol of an element surrounded by a number of dots equal to the number of electrons in the valence shell of the atom.

Ligand A Lewis base bonded to a metal atom in a coordination compound. It may bond strongly or weakly.

Lindlar catalyst (Section 7.8A) Finely powdered palladium metal deposited on solid calcium carbonate that has been specially modified with lead salts. Its particular use is as a catalyst for the reduction of an alkyne to a *cis* alkene.

Line-angle formula (Section 2.1) An abbreviated way to draw structural formulas in which vertices and line endings represent carbons.

Lipid (Introduction, Chapter 26) A biomolecule isolated from plant or animal sources by extraction with nonpolar organic solvents, such as diethyl ether and hexane.

Lipid bilayer (Section 26.5B) A back-to-back arrangement of phospholipid monolayers, often forming a closed vesicle or membrane.

Living polymer (Section 29.6D) A polymer chain that continues to grow without chain-termination steps until either all of the monomer is consumed or some external agent is added to terminate the chain. The polymer chains will continue to grow if more monomer is added.

Low-density lipoprotein (LDL) (Section 26.4A) Plasma particles, density 1.02–1.06 g/mL, consisting of approximately 26% proteins, 50% cholesterol, 21% phospholipids, and 4% triglycerides.

Low-resolution mass spectrometry (Section 14.2A) Instrumentation that is capable of separating only ions that differ in mass by 1 or more amu.

Markovnikov's rule (Section 6.3A) In the addition of HX, H_2O, or ROH to an alkene, hydrogen adds to the carbon of the double bond having the greater number of hydrogens.

Mass spectrometry (Introduction Chapter 14) An analytical technique for measuring the mass-to-charge ratio (m/z) of ions.

Mass spectrum (Section 14.1) A plot of the relative abundance of ions versus their mass-to-charge ratio.

Melt transition (T_m) (Section 29.4) The temperature at which crystalline regions of a polymer melt.

Mercaptan (Section 10.9B) A common name for a thiol; that is, any compound that contains an —SH (sulfhydryl) group.

Meso compound (Section 3.4B) An achiral compound possessing two or more chiral centers that also has chiral isomers.

Messenger RNA (mRNA) (Section 28.3C) A ribonucleic acid that carries coded genetic information from DNA to the ribosomes for the synthesis of proteins.

Meta (m) (Section 21.3B) Refers to groups occupying 1,3-positions on a benzene ring.

Methylene (Section 5.3B) a CH_2 group.

Micelle (Section 26.2B) A spherical arrangement of organic molecules in water solution clustered so that their hydrophobic parts are buried inside the sphere and their hydrophilic parts are on the surface of the sphere and in contact with water.

Molar absorptivity (ε) (Section 20.3A) The absorbance of a 1 M solution of a compound.

Molecular dipole moment (μ) (Section 1.5) The vector sum of individual bond dipoles.

Molecular ion (M^+) (Section 14.1) The cation formed by removal of a single electron from a parent molecule in a mass spectrometer.

Molecular orbital theory (Section 1.8A) A theory of chemical bonding in which electrons in molecules occupy molecular orbitals that extend over the entire molecule and are formed by the combination of the atomic orbitals that make up the molecule.

Molecular spectroscopy (Section 12.2) The study of which frequencies of radiation are absorbed or emitted by a particular substance and the correlation of these frequencies with details of molecular structure.

Monomer (Section 29.1) From the Greek, *mono + meros*, single part. The simplest nonredundant unit from which a polymer is synthesized.

Monosaccharide (Section 25.1A) A carbohydrate that cannot be hydrolyzed to a simpler carbohydrate.

D-Monosaccharide (Section 25.1C) A monosaccharide that, when written as a Fischer projection, has the —OH on its penultimate carbon to the right.

L-Monosaccharide (Section 25.1C) A monosaccharide that, when written as a Fischer projection, has the —OH on its penultimate carbon to the left.

Mutarotation (Section 25.2C) The change in specific rotation that occurs when an α or β hemiacetal form of a carbohydrate in aqueous solution is converted to an equilibrium mixture of the two forms.

(n + 1) rule (Section 13.8) If a hydrogen has n hydrogens nonequivalent to it but equivalent among themselves on the same or adjacent atom(s), its ^{1}H-NMR signal is split into $(n + 1)$ peaks.

Newman projection (Section 2.6A) A way to view a molecule by looking along a carbon-carbon single bond.

Nitrile (Section 18.1E) A compound containing a —C≡N (cyano) group bonded to a carbon atom.

Nitrogen rule (Section 14.3) A rule stating that the molecular ion of a compound with an odd number of nitrogen atoms has at an odd m/z ratio; if zero or an even number of nitrogen atoms, the molecular ion has an even m/z ratio.

Node (Section 1.7A) A point in space where the value of a wave function is zero.

Nonbonding electrons (Section 1.2C) Valence electrons not involved in forming covalent bonds. Also called unshared pairs or lone pairs.

Nonpolar covalent bond (Section 1.2B) A covalent bond between atoms whose difference in electronegativity is less than approximately 0.5.

Nuclear magnetic resonance (NMR) spectroscopy (Introduction, Chapter 13) A spectroscopic technique that gives information about the number and types of atoms in a molecule, for example, hydrogens (^{1}H-NMR) and carbons (^{13}C-NMR).

Nucleic acid (Introduction, Chapter 28) A biopolymer containing three types of monomer units: heterocyclic aromatic amine bases derived from purine and pyrimidine, the monosaccharides D-ribose or 2-deoxy-D-ribose, and phosphoric acid.

Nucleophile (Section 6.3A) From the Greek meaning nucleus loving. Any species that can donate a pair of electrons to form a new covalent bond; alternatively, a Lewis base.

Nucleophilic acyl substitution (Section 18.3) A reaction in which a nucleophile bonded to the carbon of an acyl group is replaced by another nucleophile.

Nucleophilic aromatic substitution (Section 22.3) A reaction in which a nucleophile, most commonly a halogen, on an aromatic ring is replaced by another nucleophile.

Nucleophilicity (Section 9.4B) A kinetic property measured by the rate at which a nucleophile causes nucleophilic substitution on a reference compound under a standardized set of experimental conditions.

Nucleophilic substitution (Introduction, Chapter 9) Any reaction in which one nucleophile is substituted for another at a tetravalent carbon atom.

Nucleoside (Section 28.1) A building block of nucleic acids, consisting of D-ribose or 2-deoxy-D-ribose bonded to a heterocyclic aromatic amine base by a β-N-glycosidic bond.

Nucleotide (Section 28.1) A nucleoside in which a molecule of phosphoric acid is esterified with an —OH of the monosaccharide, most commonly either the 3'—OH or the 5'—OH.

Octet rule (Section 1.1B) Group 1A–7A elements react to achieve an outer shell of eight valence electrons.

Oil (Section 26.1B) When used in the context of fats and oils, a mixture of triglycerides that is liquid at room temperature.

Oligosaccharide (Section 25.4) A carbohydrate containing four to ten monosaccharide units, each joined to the next by a glycosidic bond.

Optically active (Section 3.7A) Refers to a compound that rotates the plane of polarized light.

Optical purity (Section 3.7D) The specific rotation of a mixture of enantiomers divided by the specific rotation of the enantiomerically pure substance (expressed as a percent). Optical purity is numerically equal to enantiomeric excess, but experimentally determined.

Orbital (Section 1.1) A region of space that can hold two electrons.

Order of precedence of functions (Section 16.2B) A ranking of functional groups in order of priority for the purposes of IUPAC nomenclature.

Organic synthesis (Section 7.9) A series of reactions by which a set of organic starting materials is converted to a more complicated structure.

Organometallic compound (Introduction, Chapter 15) A compound that contains a carbon-metal bond.

Ortho (*o*) (Section 21.3B) Refers to groups occupying l,2-positions on a benzene ring.

Oxidation (Section 6.5A) The loss of electrons. Alternatively, either the loss of hydrogens, the gain of oxygens, or both.

Oxidative addition (Section 24.2) Addition of a reagent to a metal center causing it to add two substituents and to increase its oxidation state by two.

Oxonium ion (Section 6.3B) An ion in which oxygen bears a positive charge.

Oxymercuration-reduction (Section 6.3F) A method for converting an alkene to an alcohol. The alkene is treated with mercury(II) acetate followed by reduction with sodium borohydride.

Para (*p*) (Section 21.3B) Refers to groups occupying l,4-positions on a benzene ring.

Part per million (ppm) (Section 13.3) Units used on NMR spectra to record chemical shift relative to the TMS standard.

Pauli exclusion principle (Section 1.1A) No more than two electrons may be present in an orbital. If two electrons are present, their spins must be paired.

Peptide bond (Section 27.3) The special name given to the amide bond formed between the α-amino group of one amino acid and the α-carboxyl group of another amino acid.

Pericyclic reaction (Section 24.6F) A reaction that takes place in a single step, without intermediates, and involves a cyclic redistribution of bonding electrons.

Phase-transfer catalyst (Section 9.10) A substance that transfers ions from an aqueous phase into an organic phase and vice versa.

Phenol (Section 21.4A) A compound that contains an —OH bonded to a benzene ring; a benzenol.

Phenyl group (Introduction, Chapter 5) A group derived by removing an H from benzene; abbreviated C_6H_5— or Ph—.

Phospholipid (Section 26.5A) A lipid containing glycerol esterified with two molecules of fatty acid and one molecule of phosphoric acid.

Photodynamic therapy (Section 24.7B) Biological damage caused by photosensitizers, light, and oxygen, used to kill tumor and other cells.

Photolysis (Section 15.4A) Cleavage by light.

Photosensitizer (Section 24.7B) A compound that absorbs light and transfers the energy to another molecule.

Photons (Section 12.1) An alternative way to describe electromagnetic radiation as a stream of particles.

Pi (π) bond (Section 1.8D) A covalent bond formed by the overlap of parallel *p* orbitals.

Pi (π) molecular orbital (Section 1.8A) A molecular orbital formed by overlapping parallel *p* orbitals on adjacent atoms; its electron density lies above and below the line connecting the atoms.

Plane of symmetry (Section 3.2) An imaginary plane passing through an object dividing it so that one half is the mirror image of the other half.

Plane-polarized light (Section 3.7A) Light oscillating in only parallel planes.

Plastic (Section 29.1) A polymer that can be molded when hot and retains its shape when cooled.

β-Pleated sheet (Section 27.6B) A type of secondary structure in which sections of polypeptide chains are aligned parallel or antiparallel to one another.

Polar covalent bond (Section 1.2B) A covalent bond between atoms whose difference in electronegativity is between approximately 0.5 and 1.9.

Polarimeter (Section 3.7B) An instrument for measuring the ability of a compound to rotate the plane of polarized light.

Polarizability (Section 8.3B) A measure of the ease of distortion of the distribution of electron density about an atom or group in response to interaction with other molecules or ions. Fluorine, which has a high electronegativity and holds its electrons tightly, has a very low polarizability. Iodine, which has a lower electronegativity and holds its electrons less tightly, has a very high polarizability.

Polyamide (Section 29.5A) A polymer in which each monomer unit is joined to the next by an amide bond, as, for example, nylon 66.

Polycarbonate (Section 29.5C) A polyester in which the carboxyl groups are derived from carbonic acid.

Polyester (Section 29.5B) A polymer in which each monomer unit is joined to the next by an ester bond, as, for example, poly(ethylene terephthalate).

Polymer (Section 29.1) From the Greek, *poly* + *meros*, many parts. Any long-chain molecule synthesized by linking together many single parts called monomers.

Polynuclear aromatic hydrocarbon (PAH) (Section 21.3C) A hydrocarbon containing two or more fused benzene rings.

Polypeptide (Section 27.3) A macromolecule containing many amino acid units, each joined to the next by a peptide bond.

Polysaccharide (Section 25.5) A carbohydrate containing a large number of monosaccharide units, each joined to the next by one or more glycosidic bonds.

Polyunsaturated fatty acid (Section 26.1A) A fatty acid with two or more carbon-carbon double bonds in its hydrocarbon chain.

Polyunsaturated triglyceride (Section 26.1B) A triglyceride having several carbon-carbon double bonds in the hydrocarbon chains of its three fatty acids.

Polyurethane (Section 29.5D) A polymer containing the —$NHCO_2$— group as a repeating unit.

Primary (1°) amine (Section 1.3B) An amine in which nitrogen is bonded to one carbon and two hydrogens.

Primary structure of nucleic acids (Section 28.2A) The sequence of bases along the pentose-phosphodiester backbone of a DNA or RNA molecule read from the 5′ end to the 3′ end.

Primary structure of proteins (Section 27.4) The sequence of amino acids in the polypeptide chain, read from the N-terminal amino acid to the C-terminal amino acid.

Principle of microscopic reversibility (Section 10.6) This principle states that the sequence of transition states and reactive intermediates in the mechanism of any reversible reaction must be the same, but in reverse order, for the reverse reaction as for the forward reaction.

Prochiral hydrogens (Section 13.10) Refers to two hydrogens bonded to a carbon atom. When one or the other is replaced by a different atom, the carbon becomes a chiral center. The hydrogens of the CH_2 group of ethanol, for example, are prochiral. Replacing one of them by deuterium gives (R)-1-deuteroethanol; replacing the other gives (S)-1-deuteroethanol.

Pro-R-hydrogen (Section 13.10) Replacing this hydrogen by deuterium gives a chiral center with an R configuration.

Pro-S-hydrogen (Section 13.10) Replacing this hydrogen by deuterium gives a chiral center with an S configuration.

Prostaglandin (Section 26.3) A member of the family of compounds having the 20-carbon skeleton of prostanoic acid.

Protic acid (Section 4.6) An acid that is a proton donor in an acid-base reaction.

Protic solvent (Section 9.2) A solvent that is a hydrogen-bond donor; the most common protic solvents contain —OH groups. Common protic solvents are water, low-

molecular-weight alcohols, and low-molecular weight carboxylic acids.

Pyranose (Section 25.2A) A six-membered cyclic form of a monosaccharide.

Quantum mechanics (Section 1.7A) The branch of science that studies the interaction of matter and radiation.

Quaternary (4°) ammonium ion (Section 23.2B) An ion in which nitrogen is bonded to four carbons and bears a positive charge.

Quaternary structure (Section 27.6D) The arrangement of polypeptide monomers into a noncovalently bonded aggregate.

R (Section 3.3) From the Latin, *rectus*, straight, correct; used in the R,S convention to show that the order of priority of groups on a chiral center is clockwise.

R,S System (Section 3.3) A set of rules for specifying absolute configuration about a chiral center; also called the Cahn-Ingold-Prelog system.

Racemic mixture (Section 3.7C) A mixture of equal amounts of two enantiomers.

Radical (Section 8.3D) Any chemical species that contains one or more unpaired electrons.

Radical cation (Section 14.1) A species formed when a neutral molecule loses one electron; it contains both an odd number of electrons and a positive charge.

Radical inhibitor (Section 8.7) A compound such as a phenol that selectively reacts with radicals to remove them from a chain reaction and terminate the chain.

Raman spectroscopy (Section 12.3B) A vibrational molecular spectroscopy that is complementary to infrared (IR) spectroscopy

Reaction coordinate (Section 6.2A) A measure of the change in the positions of atoms during a reaction; plotted on the horizontal axis in a reaction energy diagram.

Reaction intermediate (Section 6.2A) A species, formed between two successive reaction steps, that lies in an energy minimum between the two transition states.

Reaction mechanism (Section 6.2) A step-by-step description of how a chemical reaction occurs.

Rearrangement (Section 6.3C) A change in connectivity of the atoms in a product compared with the connectivity of the same atoms in the starting material.

Reducing sugar (Section 25.3C) A carbohydrate that reacts with an oxidizing agent to form an aldonic acid. In this reaction, the carbohydrate reduces the oxidizing agent.

Reduction (Section 6.5A) The gain of electrons. Alternatively, either the gain of hydrogen, loss of oxygen, or both.

Reductive amination (Section 16.8A) A method for preparing amines by treating an aldehyde or ketone with an amine in the presence of a reducing agent.

Reductive elimination (Section 24.2) Elimination of two substituents at a metal center, causing the oxidation state of the metal to decrease by two.

Resolution (Sections 3.8 and 14.2A) Separation of a racemic mixture into its enantiomers; in mass spectrometry, a measure of how well a mass spectrometer separates ions of different mass.

Resonance (Section 1.6) A theory that many molecules are best described as a hybrid of several Lewis structures.

Resonance energy (Section 21.1C) The difference in energy between a resonance hybrid and the most stable of its hypothetical contributing structures in which electrons are localized on particular atoms and in particular bonds.

Resonance hybrid (Section 1.6A) A molecule, ion, or radical described as a composite of a number of contributing structures.

Resonance in NMR spectroscopy (Section 13.3) The absorption of electromagnetic radiation by a precessing nucleus and the resulting "flip" of its nuclear spin from the lower energy state to the higher energy state.

Restriction endonuclease (Section 28.5A) An enzyme that catalyzes the hydrolysis of a particular phosphodiester bond within a DNA strand.

Retrosynthesis (Section 7.9) A process of reasoning backwards from a target molecule to a suitable set of starting materials.

Ribosomal RNA (rRNA) (Section 28.3A) A ribonucleic acid found in ribosomes, the sites of protein synthesis.

Ring current (Section 13.7) An applied magnetic field causes the pi electrons of an aromatic ring to circulate, giving rise to the so-called ring current and an associated magnetic field that opposes the applied field in the middle of the ring but reinforces the applied field on the outside of the ring.

S (Section 3.3) From the Latin, *sinister*, left; used in the *R,S* convention to show that the order of priority of groups on a chiral center is counterclockwise.

Sanger dideoxy method (Section 28.5B) A method developed by Frederick Sanger for sequencing DNA molecules.

Saponification (Section 18.4C) Hydrolysis of an ester in aqueous NaOH or KOH to an alcohol and the sodium or potassium salt of a carboxylic acid.

Schiff base (Section 16.8A) An alternative name for an imine.

Secondary (2°) amine (Section 1.3B) An amine in which nitrogen is bonded to two carbons and one hydrogen.

Secondary structure of nucleic acids (Section 28.2B) The ordered arrangement of nucleic acid strands.

Secondary structure of proteins (Section 27.6A) The ordered arrangements (conformations) of amino acids in localized regions of a polypeptide or protein.

Shell (Section 1.1) A region of space around a nucleus that can be occupied by electrons, corresponding to a principal quantum number.

Shielding in NMR (Section 13.3) Also called diamagnetic shielding; the term refers to the reduction in magnetic field experienced by a nucleus underneath electron density induced to circulate when the molecule is placed in a strong magnetic field.

Sigma (σ) molecular orbital (Section 1.8A) A molecular orbital in which electron density is concentrated between two nuclei, along the axis joining them, and is cylindrically symmetrical.

Signal (Section 13.3) A recording in an NMR spectrum of a nuclear magnetic resonance.

Signal splitting in NMR (Section 13.8) Spin-spin coupling with adjacent nuclei splits NMR signals depending on the extent of coupling and the number of adjacent equivalent nuclei.

S_N1 reaction (Section 9.3) A unimolecular nucleophilic substitution reaction.

S_N2 reaction (Section 9.3) A bimolecular nucleophilic substitution reaction.

Soap (Section 26.2A) A sodium or potassium salt of a fatty acid.

Solvolysis (Section 9.3) A nucleophilic substitution in which the solvent is also the nucleophile.

sp Hybrid orbital (Section 1.8E) A hybrid atomic orbital formed by the combination of one s atomic orbital and one p atomic orbital.

sp^2 Hybrid orbital (Section 1.8D) A hybrid atomic orbital formed by the combination of one s atomic orbital and two p atomic orbitals.

sp^3 Hybrid orbital (Section 1.8C) A hybrid atomic orbital formed by the combination of one s atomic orbital and three p atomic orbitals.

Specific rotation (Section 3.7B) Observed rotation of the plane of polarized light when a sample is placed in a tube 1.0 dm in length and at a concentration of 1 g/100 mL for a solution, or at a concentration of g/mL (density) for a pure liquid.

Spin-spin coupling   (Section 13.9) An interaction in which nuclear spins of adjacent atoms influence each other and lead to the spitting of NMR signals.

Staggered conformation   (Section 2.6A) A conformation about a carbon-carbon single bond in which the atoms or groups on one carbon are as far apart as possible from atoms or groups on an adjacent carbon.

Step-growth polymerization   (Section 29.5) A polymerization in which chain growth occurs in a stepwise manner between difunctional monomers as, for example, between adipic acid and hexamethylenediamine to form nylon 66. Also called condensation polymerization.

Stereocenter   (Sections 2.7A and 3.2) An atom about which exchange of two groups produces a stereoisomer. A chiral center is one type of stereocenter.

Stereochemistry   (Section 3.1) The study of three-dimensional arrangements of atoms in molecules.

Stereoisomers   (Sections 2.7 and 3.1) Isomers that have the same molecular formula and the same connectivity of their atoms but a different orientation of their atoms in space.

Stereoselective reaction   (Section 6.3D) A reaction in which one stereoisomer is formed in preference to all others. A stereoselective reaction may be enantioselective or diastereoselective, as the case may be.

Stereospecific reaction   (Section 6.7A) A special type of stereoselective reaction in which the stereochemistry of the product is dependent on the stereochemistry of the starting material.

Steric hindrance   (Section 9.4D) The ability of groups, because of their size, to hinder access to a reaction site within a molecule.

Steric strain   (Section 2.6A) The strain that arises when nonbonded atoms separated by four or more bonds are forced closer to each other than their atomic (contact) radii would allow. Steric strain is also called nonbonded interaction strain, or van der Waals strain.

Steroid   (Section 26.4A) A plant or animal lipid having the characteristic tetracyclic ring structure of the steroid nucleus, namely three six-membered rings and one five-membered ring.

Substitution   (Section 8.4) A reaction in which an atom or group of atoms in a compound is replaced by another atom or group of atoms.

Syndiotactic polymer   (Section 29.6C) A polymer with alternating R and S configurations at the chiral centers along its chain, as, for example, syndiotactic polypropylene.

Syn stereoselective   (Section 6.4) The addition of atoms or groups of atoms to the same face of a carbon-carbon double bond.

Tautomers   (Section 7.7A) Constitutional isomers in equilibrium with each other that differ in the location of a hydrogen atom and a double bond relative to a heteroatom, most commonly O, N, or S.

Telechelic polymer   (Section 29.6D) A polymer in which its growing chains are terminated by formation of new functional groups at both ends of its chains. These new functional groups are introduced by adding reagents, such as CO_2 or ethylene oxide, to the growing chains.

***C*-Terminal amino acid**   (Section 27.3) The amino acid at the end of a polypeptide chain having the free —COOH group.

***N*-Terminal amino acid**   (Section 27.3) The amino acid at the end of a polypeptide chain having the free —NH_2 group.

Terpene   (Section 5.4) A compound whose carbon skeleton can be divided into two or more units identical with the carbon skeleton of isoprene.

Tertiary (3°) amine   (Section 1.3B) An amine in which nitrogen is bonded to three carbons.

Tertiary structure of nucleic acids   (Section 28.2C) The three-dimensional arrangement of all atoms of a nucleic acid, commonly referred to as supercoiling.

Tertiary structure of proteins   (Section 27.6C) The three-dimensional arrangement in space of all atoms in a single polypeptide chain.

Tesla (T)   (Section 13.2) The SI unit for magnetic field strength.

Thermodynamic control   (Section 19.8A) Experimental conditions that permit the establishment of equilibrium between two or more products of a reaction. The composition of the product mixture is determined by the relative stabilities of the products.

Thermolysis   (Section 15.4A) Cleavage by heating.

Thermoplastic   (Section 29.1) A polymer that can be melted and molded into a shape that is retained when it is cooled.

Thermoset plastic   (Section 29.1) A polymer that can be molded when it is first prepared, but once cooled, hardens irreversibly and cannot be remelted.

Thiol (Section 10.9A) A compound containing an —SH (sulfhydryl) group bonded to an sp^3-hybridized carbon.

Tollens' reagent (Section 16.10A) A solution prepared by dissolving Ag_2O in aqueous ammonia; used for selective oxidation of an aldehyde to a carboxylic acid.

Torsional strain (Section 2.6A) Strain that arises when nonbonded atoms separated by three bonds are forced from a staggered conformation to an eclipsed conformation. Torsional strain is also called eclipsed-interaction strain.

Trans (Section 2.7A) A prefix meaning across from.

Transesterification (Section 18.6C) Exchange of the —OR or —OAr group of an ester for another —OR or —OAr group.

Transfer RNA (tRNA) (Section 28.3B) A ribonucleic acid that carries a specific amino acid to the site of protein synthesis on ribosomes.

Transition state (Section 6.2A) An unstable species of maximum energy formed during the course of a reaction; a maximum on an energy diagram.

Triglyceride (triacylglycerol) (Section 26.1) An ester of glycerol with three fatty acids.

Triol (Section 10.1B) A compound containing three hydroxyl groups.

Tripeptide (Section 27.3) A molecule containing three amino acid units, each joined to the next by a peptide bond.

Twist-boat conformation (Section 2.6B) A nonplanar conformation of a cyclohexane ring that is twisted from and slightly more stable than a boat conformation.

Unimolecular reaction (Section 9.3) A reaction in which only one species is involved in the rate-determining step.

Unsaturated hydrocarbon (Introduction, Chapter 5) A hydrocarbon containing one or more carbon-carbon double or triple bonds. The three classes of unsaturated hydrocarbons are alkenes, alkynes, and arenes.

Upfield (Section 13.4) A signal of an NMR spectrum that is shifted toward the right (smaller chemical shift) on the chart paper.

Valence electrons (Section 1.1B) Electrons in the valence (outermost occupied) shell of an atom.

Valence shell (Section 1.1B) The outermost electron shell of an atom.

van der Waals forces (Section 8.3B) A group of intermolecular attractive forces including dipole-dipole, dipole-induced dipole, and induced dipole-induced dipole (dispersion) forces.

van der Waals radius (Section 8.3B) The minimum distance of approach to an atom that does not cause nonbonded interaction strain.

Vibrational infrared region (Section 12.3A) The portion of the infrared region that extends from 4000 to 400 cm^{-1}.

Vicinal coupling (Section 13.9B) A common type of spin-spin coupling involving the H atoms on two C atoms that are bonded to each other.

Vinyl group (Section 5.3B) a —CH=CH_2 group.

Vinylic carbocation (Section 7.6B) A carbocation in which the positive charge is on one of the carbons of a carbon-carbon double bond.

VSEPR (Section 1.4) A method for predicting bond angles based on the idea that electron pairs repel each other and keep as far apart as possible.

Watson-Crick model (Section 28.2B) A double-helix model for the secondary structure of a DNA molecule.

Wave function (Section 1.7A) A solution to a set of equations that defines the energy of an electron in an atom and the region of space it may occupy.

Wavelength (λ) (Section 12.1) The distance between consecutive peaks on a wave.

Wavenumbers ($\bar{\nu}$) (Section 12.3A) The frequency of electromagnetic radiation expressed as the number of waves per centimeter, with units cm^{-1} (read: reciprocal centimeters).

Wolff-Kishner reduction (Section 16.11C) Reduction of the C=O group of an aldehyde or ketone to a CH_2 group using hydrazine and a base.

Ylide (Section 16.6) A neutral molecule with positive and negative charges on adjacent atoms.

Z (Section 5.3C) From the German, *zusammen*, together. Specifies that groups of higher priority on the carbons of a double bond are on the same side.

Zaitsev's rule (Section 9.8) A rule stating that the major product of a β-elimination reaction is the most stable alkene; that is, it is the alkene with the greatest number of substituents on the carbon-carbon double bond.

Zwitterion (Section 27.1A) An internal salt of an amino acid; the carboxylate is negatively charged, and the ammonium group is positively charged.

Index

Periodic Table of the Elements

Group number,
U.S. system ——— 1A
IUPAC system ——— (1)

2A
(2)

Period number

KEY

79	Atomic number
Au	Symbol
Gold	Name
196.9665	Atomic mass

An element

☐ Metals
☐ Semimetals
☐ Nonmetals

Numbers in parentheses are mass numbers of radioactive isotopes.

1A (1)	2A (2)	3B (3)	4B (4)	5B (5)	6B (6)	7B (7)	8B (8)	8B (9)	8B (10)	1B (11)	2B (12)	3A (13)	4A (14)	5A (15)	6A (16)	7A (17)	8A (18)
1 **H** Hydrogen 1.0079																	2 **He** Helium 4.0026
3 **Li** Lithium 6.941	4 **Be** Beryllium 9.0122											5 **B** Boron 10.811	6 **C** Carbon 12.011	7 **N** Nitrogen 14.0067	8 **O** Oxygen 15.9994	9 **F** Fluorine 18.9984	10 **Ne** Neon 20.1797
11 **Na** Sodium 22.9898	12 **Mg** Magnesium 24.3050											13 **Al** Aluminum 26.9815	14 **Si** Silicon 28.0855	15 **P** Phosphorus 30.9738	16 **S** Sulfur 32.066	17 **Cl** Chlorine 35.4527	18 **Ar** Argon 39.948
19 **K** Potassium 39.0983	20 **Ca** Calcium 40.078	21 **Sc** Scandium 44.9559	22 **Ti** Titanium 47.88	23 **V** Vanadium 50.9415	24 **Cr** Chromium 51.9961	25 **Mn** Manganese 54.9380	26 **Fe** Iron 55.847	27 **Co** Cobalt 58.9332	28 **Ni** Nickel 58.693	29 **Cu** Copper 63.546	30 **Zn** Zinc 65.39	31 **Ga** Gallium 69.723	32 **Ge** Germanium 72.61	33 **As** Arsenic 74.9216	34 **Se** Selenium 78.96	35 **Br** Bromine 79.904	36 **Kr** Krypton 83.80
37 **Rb** Rubidium 85.4678	38 **Sr** Strontium 87.62	39 **Y** Yttrium 88.9059	40 **Zr** Zirconium 91.224	41 **Nb** Niobium 92.9064	42 **Mo** Molybdenum 95.94	43 **Tc** Technetium (98)	44 **Ru** Ruthenium 101.07	45 **Rh** Rhodium 102.9055	46 **Pd** Palladium 106.42	47 **Ag** Silver 107.8682	48 **Cd** Cadmium 112.411	49 **In** Indium 114.82	50 **Sn** Tin 118.710	51 **Sb** Antimony 121.757	52 **Te** Tellurium 127.60	53 **I** Iodine 126.9045	54 **Xe** Xenon 131.29
55 **Cs** Cesium 132.9054	56 **Ba** Barium 137.327	57 **La** Lanthanum 138.9055	72 **Hf** Hafnium 178.49	73 **Ta** Tantalum 180.9479	74 **W** Tungsten 183.85	75 **Re** Rhenium 186.207	76 **Os** Osmium 190.2	77 **Ir** Iridium 192.22	78 **Pt** Platinum 195.08	79 **Au** Gold 196.9665	80 **Hg** Mercury 200.59	81 **Tl** Thallium 204.3833	82 **Pb** Lead 207.2	83 **Bi** Bismuth 208.9804	84 **Po** Polonium (209)	85 **At** Astatine (210)	86 **Rn** Radon (222)
87 **Fr** Francium (223)	88 **Ra** Radium 227.0278	89 **Ac** Actinium (227)	104 **Rf** Rutherfordium (261)	105 **Db** Dubnium (262)	106 **Sg** Seaborgium (263)	107 **Bh** Bohrium (262)	108 **Hs** Hassium (265)	109 **Mt** Meitnerium (266)	110 **Ds** Darmstadtium (271)	111 **Rg** roentgenium (277)	112 — (277)	—	114 — (285)	—	116 — (289)		

Lanthanides 6

58 **Ce** Cerium 140.115	59 **Pr** Praseodymium 140.9076	60 **Nd** Neodymium 144.24	61 **Pm** Promethium (145)	62 **Sm** Samarium 150.36	63 **Eu** Europium 151.965	64 **Gd** Gadolinium 157.25	65 **Tb** Terbium 158.9253	66 **Dy** Dysprosium 162.50	67 **Ho** Holmium 164.9303	68 **Er** Erbium 167.26	69 **Tm** Thulium 168.9342	70 **Yb** Ytterbium 173.04	71 **Lu** Lutetium 174.967

Actinides 7

90 **Th** Thorium 232.0381	91 **Pa** Protactinium 231.0359	92 **U** Uranium 238.0289	93 **Np** Neptunium (237)	94 **Pu** Plutonium (244)	95 **Am** Americium (243)	96 **Cm** Curium (247)	97 **Bk** Berkelium (247)	98 **Cf** Californium (251)	99 **Es** Einsteinium (252)	100 **Fm** Fermium (257)	101 **Md** Mendelevium (258)	102 **No** Nobelium (259)	103 **Lr** Lawrencium (260)